CAMBRIDGE LIBRARY COLLECTION

Books of enduring scholarly value

Botany and Horticulture

Until the nineteenth century, the investigation of natural phenomena, plants and animals was considered either the preserve of elite scholars or a pastime for the leisured upper classes. As increasing academic rigour and systematisation was brought to the study of 'natural history', its subdisciplines were adopted into university curricula, and learned societies (such as the Royal Horticultural Society, founded in 1804) were established to support research in these areas. A related development was strong enthusiasm for exotic garden plants, which resulted in plant collecting expeditions to every corner of the globe, sometimes with tragic consequences. This series includes accounts of some of those expeditions, detailed reference works on the flora of different regions, and practical advice for amateur and professional gardeners.

Sweet's Hortus Britannicus

The first career of Robert Sweet (1783–1835) was as a gardener in private employment and as a nurseryman. He turned in 1826 to botanical writing, having already published *Hortus suburbanus Londinensis* (1818), and the first of the five-volume *Geraniaceae* (1820–30). The first edition of this work was published in 1826, and this revised second edition in 1830. Sweet uses Jussieu's 'natural' system of classification, but concedes that 'we still consider the addition of the Linnean Classes and Orders, of great use, as they are so readily attained by the young Botanist'. He provides nine two-column closely packed pages of source works in which images of the plants cited in this unillustrated work can be found, and which also testify to the breadth of his own research in producing a reference work which is comprehensive as a record of plants then growing and flowering in British gardens.

Cambridge University Press has long been a pioneer in the reissuing of out-of-print titles from its own backlist, producing digital reprints of books that are still sought after by scholars and students but could not be reprinted economically using traditional technology. The Cambridge Library Collection extends this activity to a wider range of books which are still of importance to researchers and professionals, either for the source material they contain, or as landmarks in the history of their academic discipline.

Drawing from the world-renowned collections in the Cambridge University Library and other partner libraries, and guided by the advice of experts in each subject area, Cambridge University Press is using state-of-the-art scanning machines in its own Printing House to capture the content of each book selected for inclusion. The files are processed to give a consistently clear, crisp image, and the books finished to the high quality standard for which the Press is recognised around the world. The latest print-on-demand technology ensures that the books will remain available indefinitely, and that orders for single or multiple copies can quickly be supplied.

The Cambridge Library Collection brings back to life books of enduring scholarly value (including out-of-copyright works originally issued by other publishers) across a wide range of disciplines in the humanities and social sciences and in science and technology.

Sweet's
Hortus Britannicus

Or, a Catalogue of Plants, Indigenous, or Cultivated in the Gardens of Great Britain, Arranged According to their Natural Orders

Robert Sweet

Cambridge University Press

CAMBRIDGE
UNIVERSITY PRESS

University Printing House, Cambridge, CB2 8BS, United Kingdom

Cambridge University Press is part of the University of Cambridge.
It furthers the University's mission by disseminating knowledge in the pursuit of education, learning and research at the highest international levels of excellence.

www.cambridge.org
Information on this title: www.cambridge.org/9781108079204

This edition first published 1830
This digitally printed version 2015

ISBN 978-1-108-07920-4 Paperback

SWEET'S HORTUS BRITANNICUS;

OR,

A CATALOGUE OF PLANTS,

INDIGENOUS, OR CULTIVATED IN THE GARDENS

OF

GREAT BRITAIN;

ARRANGED ACCORDING TO THEIR

NATURAL ORDERS,

WITH

REFERENCES TO THE LINNEAN CLASS AND ORDER TO WHICH EACH GENUS BELONGS;

THE WHOLE BROUGHT UP TO THE PRESENT TIME,

AND CONTAINS

Above 34,000 Plants;

By far the greatest Number ever before published in any Garden Catalogue of this or any other Country; with their Generic and Specific Names, Habitats, when introduced to this Country, Times of Flowering, Colours of the Flowers, Accentuations; and numerous other Arrangements, for the first Time introduced, with all the most useful Synonyms.

SECOND EDITION.

By ROBERT SWEET, F.L.S.

Author of Hortus Suburbanus Londinensis; Botanical Cultivator; Geraniaceæ; Cistineæ; British Flower Garden; Flora Australasica; Florist's Guide; the British Warblers, &c.

LONDON:

JAMES RIDGWAY, 169, PICCADILLY.

1830.

TILLING, PRINTER, GROSVENOR ROW, CHELSEA.

The rapid sale of the last edition, at once shows the preference given to the superiority of the Natural Arrangement, over that of the Linnean, for a Garden Catalogue, as it brings together all the plants that are nearest related to each other, and which have a general similarity by which any person, who already has a tolerable knowledge of plants, may give a good guess at first sight to what family, or Order, any plant or specimen they possess may belong to, or is related; this gives a pleasing idea, that is not to be met with by the Linnean Botanist, as he only looks at the Stamens and Styles to see what Class and Order it belongs to, without ever thinking to what else it is allied; but we still consider the addition of the Linnean Classes and Orders, of great use, as they are so readily attained by the young Botanist, though he should always keep the natural arrangement in view.

The pains we have taken with the present edition may be readily supposed, by the length of time it has taken us to get through it; and had it not been for the kindness and liberality of A. B. Lambert, Esq. who always admitted us to his Library with pleasure, we should certainly not have been able to have made it so complete; he is also so kind as to give us information of any new plants or importations that he hears or knows of, for which we now return him our sincere thanks.

To Mr. David Don, Librarian of the Linnean Society, we feel much obliged for his readiness in procuring us any books or specimens we wish to examine, and also inform us of any new publications that we may not have heard of.

The Linnean collection of specimens being now in the Linnean Society's collection, has enabled us to clear up many doubts and mistakes, by examining Linnaeus's original specimens, several of which we intend setting to rights by publishing figures of them in our British Flower Garden, many of the very common plants having been confused with others, or were considered in a doubtful state.

To Robert Barclay, Esq. of Bury Hill, we also feel greatly obliged for his kindness in sending us specimens of the numerous new plants that are continually flowering in his delightful garden, where rarities are without end; and we must not forget his gardener, Mr. D. Cameron, to whose merit may be attached the highly cultivated state of the plants, and to whom we are indebted for the names of many that have not yet flowered, and that we should not have otherwise known were in the country.

We have also to return our thanks to A. H. Haworth, Esq. for his liberality in presenting us with all his papers as soon as published, of the new and rare succulent plants and bulbs, published in Taylor's Philosophical Magazine.

To the Honourable and Reverend William Herbert we are indebted for the arrangement of the genus Crinum, and for the names of the hybrid species of Amaryllideæ raised by him, and those that he possesses, which have been raised in other collections, and the references by whom, or at what place they were raised, which is added in the column of Native, and the two parents from which they originated is given, where known: the whole of which are marked with W. H.; he has also been so kind as to furnish us with a list of many other new plants, that have flowered with him, and that had not before been known in our collections.

We also return our thanks to Mr. G. Charlwood, for his readiness in lending us any Botanical publications that we did not ourselves possess, his collection consisting of many rare and valuable botanical works.

To Mr. William Anderson, the worthy Curator of the Apothecaries' Company's Garden, at Chelsea, we are also much indebted, for his kindness in allowing us to copy his catalogues of newly introduced plants, and for his observations on their habit and culture; and to numerous other Gardeners and Nurserymen we return our best thanks, for their kindness and attention to forward our researches.

We flatter ourselves that our Subscribers will be well satisfied with the present edition, as it contains all the new arrangements and improvements up to the present time, that no other Catalogue, however new, can boast of; we have also been careful what we have inserted, not to put in thousands of plants that were never in the country, nor likely ever to be, a plan sometimes resorted to, to swell the size of the book, and to show a great number of names, but we believe there would be a difficulty to find the plants, should they be sought for. As for its being entirely free from errors, we will not assert, as we have no doubt but some may be found, but none, we believe, of much consequence. We now conclude, with our sincerest thanks to our numerous Subscribers, and hope we shall continue to merit their favour and approbation, as we shall always do our utmost endeavours to give them satisfaction.

October, 1830.

REFERENCE TO THE BOOKS,

IN WHICH THE FIGURES REFERRED TO, ARE PUBLISHED.

Abb. ins. The Natural History of the rarer Lepidopterous Insects of Georgia, collected from the observations of J. Abbott, by J. E. Smith, 1797. 4to.

Act. Stock. Kongl. Vetenskaps Academiens Handlingar. Stockholm, 1739. 8vo.

A.DC. Decandolle (Alphonso.) Monographie des Campanulees. Paris. 1830.

All. auct. Allioni (Carolus.) Auctuarium ad floram Pedemontanam. 4to. Taurini. 1789.

All. m. taur. Car. Allionii Miscellanea Philosophico-mathematica Societatis privatæ Taurinensis.

All. ped.. Car. Allionii Flora Pedemontana.

Alp. Ægypt. Alpinus (Prosper.) De Plantis Ægypti liber. 1 vol. 4to.

Alp. ex. Alpinus (Prosper.) De Plantis Exoticis, libri duo. 4to. 1629.

Amm. ruth. Amman (Johan.) Stirpium rariorum in imperio Rutheno sponte provenientium Icones et Descriptiones. 4to. Petropoli. 1739.

Amoen. ac. Caroli Linnæi Amoenitates academicæ. 8vo.

And.bot.rep. Andrews (Henry.) The Botanist's repository for new and rare plants. 4to. 1797. et seq.

And. geran. Andrews (Henry.) Geraniums, or a monagraph of the genus Geranium. 1805. seq. 4to.

And. heaths. Andrews (Henry.) Coloured engravings of Heaths, with botanical descriptions. fol. and 8vo. 1802, 1809, et seq.

Ann. bot. Annals of Botany, by C. Konig and J. Sims. 1805. 8vo.

Ann. mus. Annales du Museum d'Histoire naturelle, par les Professeurs de cet etablissement. 1802. seq. 4to.

Ap. to B. Field's trav. Appendix to Baron Field's Travels. 4to.

Ard. mem. Memorie di osservazioni, e di sperienze sopra la coltura, e gli usi di varie piante, da P. Arduino. Padova. 1766. 4to.

Ard. sag. pad. Arduino (Pietro.) Saggi di Padova.

Ard. spec. Arduino (Pietro.) Animadversionum botanicarum Specimen. 4to. Patavii. 1759.

Asso. arrag. De Asso (Ignatius.) Synopsis Stirpium indigenarum Arragoniæ. 4to. Marsiliæ. 1779.

Aub. Thou. obs. Aubert Du Petit Thouars Plantes des iles de l'Afrique australe. 4to. 1804.

Aub. gui. Histoire des plantes de la Guiane Francoise, par M. Fusée Aublet. 1775. 4to.

Balb. act. taur., *Balb. misc. taur.* } Miscellanea Botanica. 4to. ex Actis Academiæ Taurinensis exserta. 1804.

Balb. h. taur. Horti Academici Taurinensis stirpium minus cognitarum aut forte novarum icones et descriptiones. Taurini. 1811.

Banks ic. Kæmp. Banks (Josephus.) icones selectæ Plantarum quas in Japonia collegit et delineavit E. Kæmpfer. fol. 1791.

Banks. rel. Hous. Reliquia Houstounianæ. 4to. 1781.

Barr. ic., *Barr. rar.* } Plantæ per Galliam, Hispaniam, et Italiam, observatæ, Iconibus æneis exhibitæ a Jac. Barreliero. 1714. fol.

Bart. med. bot. Barton (Benj. Smith.) Collections for an essay towards a materia medica of the United States. 1 fasc. in 8vo. Philadelphia. 1798.

Bartr. it., *Bartr. trav.* } Bartram (William.) Travels through North and South Carolina, Georgia, and Florida. 8vo. Philadelphia. 1791.

Bauer. n. hol. Bauer (Ferdinand.) Illustrationes floræ novæ Hollandiæ. fol. 1813.

Bauh. hist. Bauhin (Johannes.) Historia Plantarum universalis. fol. 1651.

Bauh. prodr. Caspar Bauhini προδρομος Theatri Botanici. Basileæ. 1671. 4to.

Beau. fl. ow. Palisot de Beauvois flore des royaumes d'Oware et de Benin. fol. 1805.

Bell. act. taur. Bellardi (Ludovico.) Memoires de l'Academie Royale des Sciences de Turin. 4to. 1782.

Berg. act. holm. Bergius (Pehr. Jonas.) Kongl Vetanskaps Academiens Handlingar. 8vo. 1739.

Berg. cap. Petr. Jon. Bergii descriptiones Plantarum ex Capite bonæ spei. 1769. 8vo.

Berg. phyt. Bergeret Phytonomatotechnie universelle. fol.

Berr. in As. res. Berry in Asiatic Researches, or Transactions of the Society instituted in Bengal. 4to. 1788.

Besl. eyst. Besler (Basilius.) Hortus Eystettensis. fol. 1612.

Bieb. cent. Bieberstein (Marshall von.) Centuriæ plantarum rariorum Rossiæ meridionalis. Charkoviæ.

Big. med. bot. Bigelow (Jac.) American Medical Botany. 4to. 1817.

Biv. Bern. st. rar. Bivona Bernardi. Stirpium rariorum minusque cognitarum in Sicilia provenientium Descriptiones. 4to. 1813.

Biv. cent. pl. Bivona Bernardi (Antonin.) Sicularum Plantarum Centuria prima. 8vo. 1806.

Blackw. Blackwell (Elizabeth.) A curious Herbal, containing 500 Cuts of the useful Plants. fol. 1737.

Blum. fl. jav. Blume (C. L.) Flora Java, &c. fol. 1829.

Boc. Mus. Boccone (Paolo.) Museo di Piante rare della Sicilia, Malta, Corsica, Italia, Piemonte e Germania. 4to. 1697.

Boc. sic. rar. Boccone (Paolo.) Icones et Descriptiones rariorum Plantarum Siciliæ, Melitæ, Galiæ et Italiæ. 4to. 1674.

Boer. lugd. Herm. Boerhaave Index alter Plantarum quæ in horto Academiæ Lugduno Batavæ aluntur. 4to. 1720.

Boiss. fl. eur. De Boissieu (C. V.) Flore d'Europe. 8vo. 1805.

Bolt. fil. Bolton (James.) Filices Britanniæ. 4to. 1785.

Bonp. mon. Bonpland (Aimé.) Monographie des Melastomes et autres Plantes de cet ordre. fol. 1809.

Bonp. nav. Bonpland (Aimé.) Description des Plantes rares cultivés a Malmaison et a Navarre. fol. 1813.

Bory. it. / *Bory. voy.* } Bory de St. Vincent (J. B. M.) A Voyage en Espagne a travers les royaumes de Galice, Leon, etc. 8vo.

Bot. mag. Curtis's Botanical Magazine, continued by J. Sims, J. Bellenden Ker, and W. J. Hooker. 8vo.

Bot. reg. Edwards (Sydenham.) The Botanical Register. descr. by J. Bellenden Ker. 8vo. 1815. continued by J. Lindley.

Bot. rep. Andrews (Henry.) The Botanist's Repository for new and rare Plants. 4to. 1797.

Braam ic. chin. Braam's Icones Chinensis.

Bradl. Suc. Bradley (Richard.) Historia Plantarum succulentarum. 4to. 1716.

Braun. Salz. De Braune (Franz. Ant.) Salzburgische flora. 8vo. 1791.

Breyn. cent. Breynius (Jacobus.) Exoticarum Plantarum centuria. fol. 1678.

Breyn. ic. Breynii (J.) Icones rariorum et exoticarum Plantarum. 4to. 1739.

Brong. rham. Brongniart (Adolph.) Dissertatione des Rhamnees. 4to. Paris.

Brot. phyt. Brotero (Felix Avellar.) Phytographia Lusitaniæ selectior. fol. 1801.

Brown. gen. rem. Brown (Robert.) General Remarks, Geographical and Systematical, on the Botany of Terra Australis. 4to. 1814.

Br. Jam. / *Brown. Jam.* } Browne (Patrick.) The Civil and Natural History of Jamaica. fol. 1756.

Bruces. tr. Bruce (James.) Travels to discover the source of the Nile. 4to. 1790.

Buchoz. ic. col. Buchoz (Pierre Joseph.) Collection des fleures qui se cultivent tant dans les Jardins de la chine que dans ceux d'Europe. fol.

Bull. herb. Bulliard. Herbier de la France. fol. 1780.

Burm. afr. / *Burm. dec.* } Burmann (Johannes.) Rariorum Africanarum Plantarum. 4to. 1738.

Burm. Ind. Nic. Laur. Burmanni Flora Indica. 4to. 1768.

Burm. zeyl. Burmann (Johannes.) Thesaurus Zeylanicus. 4to. 1737.

Buxb. cent. Buxbaum (Joh. Christ.) Plantarum minus cognitarum Centuriæ quinque. 4to. 1728.

Camer. hort. Camerarius (Alexander.) Hortus Medicus et Philosophicus. 4to. 1588.

Cates. car. Catesby (Mark.) The Natural History of Carolina & Florida. fol. 1741.

Cav. diss. Cavanilles (Ant. Jos.) Monadelphiæ classis Dissertationes decem. 4to. 1785.

Cav. ic. Cavanilles (Ant. Jos.) Icones et Descriptiones Plantarum quæ aut sponte in Hispania crescunt aut in hortis hospitantur. fol. 1791.

Chabr. sci. Chabrey (Dominicus.) Omnium Stirpium Sciagraphia. fol. 1677.

Chois. p. hyp. / *Chois. m. h.* } Choisy (J. D.) Prodromus d' une monographie de la famille des Hypericinees. 4to. 1821.

Chois. selag. Choisy (J. D.) Memoire sur la famille des Selaginees. 4to. Geneva. 1823.

Clus. hist. Clusius seu l'Ecluse (Charles de.) Rariorum Plantarum Historia. fol. 1601.

Col. ecphr. Columna (Fabius.) Minus cognitarum stirpium Ecphrasis. 4to. 1616.

Col. phyt. Columna (Fabius.) Phytobasanos. 4to. 1592.

Col. h. rip. Colla Hortus Ripulensis. 4to.

Com. hort. Commelyn (Casp.) Horti Medici Amstelodamensis rariorum Plantarum descriptio et icones. fol. 1703.

Comm. petrop. Commelyn (Casp.) Commentarii Academiæ Scientarium imperialis Petropolitanæ. 4to. 1728.

Comm. prael. Commelyn (Casp.) Praeludia Botanica ad publicas Plantarum demonstrationes. 4to. 1703.

Comm. rar. Commelyn (Casp.) Horti Medici Amstelodamensis Plantæ rariores et exoticæ. 4to. 1706.

Cook. et. Cook (J.) Voyages, 2 vols. 4to. 1777.

Corn. can. Cornuti (Jacob.) Canadensium Plantarum aliarumque nondum editarum historia. 4to. 1635.

Crantz. Aust. Crantz (Hen. Joh. Nepom.) Stirpium austriacarum. 8vo. 1762.

Cup. pan. Cupani (Franciscus.) Catalogus plantarum Sicularum noviter detectarum. Panormi, 1652.

Curt. lond. Curtis (William.) A catalogue of the plants growing wild in the environs of London. fol. 1774. et seq.

Cyril. ic. fasc. / *Cyr. neap.* } Cyrilli (Dom.) Plantarum rariorum regni Neapolitani fasciculus. 1778. fol.

Dalech. lug. Dalechamps (Jacques.) Historia generalis plantarum. fol. 1586.

Dard. trait. ren. D'ardenne (Jean Paul Rome.) Traite des Renoncules. 8vo. 1746.

DC. an. mus. De Candolle (Augustin Pyramus.) Annales du Museum d'histoire naturelle. 4to. 1802.

DC. astr. De Candolle (Augustin Pyramus.) Astragalogia. 4to. et fol. 1802.

DC. dis. cact. De Candolle (Aug. Pyr.) Revue de la famille des Cactees. 4to. Paris. 1829.

DC. h. mons. De Candolle (Augustin Pyramus.) Catalogus plantarum horti botanici Monspeliensis addito observationum circa species novas aut non satis cognitas fasciculo. 8vo. 1813.

DC. ic. gall. rar. De Candolle (Augustin Pyramus.) Icones plantarum Galliæ rariorum. 4to. 1808.

DC. leg. mem. De Candolle (Augustin Pyramus.) Memoires sur la famille des Legumineusis. 4to. 1826.

DC. mem. mus. De Candolle (Augustin Pyramus.) Memoires du Museum d'histoire naturelle. 4to. 1815.

DC. mem. umb. De Candolle (Aug. Pyr.) Memoire sur la famille des Ombelliferes. 4to. Paris. 1829.

DC. plant. grass. De Candolle (Augustin Pyramus.) Plantarum historia succulentarum. Fasc. 28. 1799.

Debr. flor. nov. De Bry (Joannes Theodor.) Florilegium novum. fol. 1612.

Delil. ægyp. / *Del. fl. æg.* } Delile (Alire Rafeneau.) Memoires botaniques extraits de la Description de l'Egypte. fol. 1813.

De la Roch. dis. De La Roche (Daniel.) Specimens botanicum inaugurale sistens descriptiones plantarum aliquot novarum. 4to. 1766.

Deless. ic. sel. Delessert (Benj.) Icones selectæ plantarum in system. univers. descriptarum. 4to. 1820.

Descrip. de Egyp. Description de l'Egypt on Recueil des observations et des recherches faites pendant l'expedition de l'armée Francaise. 4to. et fol. 1810.

Desf. act. par. Desfontaines (Rene Louiche.) Mémoires de l'Académie Royale des Sciences. 4to. 1666.

Desf. atl. Desfontaines (Rene Louiche.) Flora atlantica. 4to. 1798.

Desv. j. bot. Desvaux. Journal de Botanique. 8vo. 1808.

D. et H. Dreves et Hayne. Choix de Plantes de l'Europe. fasc. 4. 1802.

Dill. elt. Dillenius (Joh. Jac.) Hortus elthamensis. fol. 1732.

Dill. musc. Dillenius (Joh. Jac.) Historia muscorum. 4to. 1741.

Dodart mem. Dodart (Denys.) Mémoires pour servir a l'Histoire des Plantes. fol. 1676.

Dod. pemp. Dodart (Denys.) Stirpium historiæ pemptades. fol. 1583.

Du Fresn. Val. Dufresne (Pierre.) Histoire naturelle et medicale de la famille des Valerianees. 4to. 1811.

Duh. arb. Duhamel du Monceau (Henry Louis.) Traite des arbres et arbustes qui cultivent en France en pleine terre. 4to. 1755.

Duham. ed. nov. Duhamel du Monceau (Henry Louis.) Traites des arbres et arbustes qui se cultivent en pleine terre en France. fol. 1801.

Dun. mon. Dunal (Michel Felix.) Monographie de la Famille des Anonacees. 4to. 1817.

Dunal. solan. Dunal (Michel Felix.) Histoire naturelle medicale et economique des Solanum et des genres qui ont ete confondus avec eux. 4to. 1813.

Ehret pict. Ehret (Georg. Dion.) Planta et Papiliones rariores. fol. 1748.

Eng. bot. English Botany. By J. E. Smith; the Figures by J. Sowerby. 8vo.

Exot. bot. Exotic Botany. By J. E. Smith; the Figures by J. Sowerby. 8vo.

Ferr. hesp. Ferrari (Joh. Bapt.) Hesperides, sive de malorum aureorum cultura et usu. fol. 1646.

Feuil. obs. / *Feuill. per.* } Feuillee (Louis.) Journal des observationes Physiques, Mathématiques et Botaniques faites dans l'Amerique meridionale. 4to. 1814.

Fl. dan. Icones plantarum sponte nascentium in regnis Daniæ et Norvegiæ, etc. fol. 1761.

Fl. lap. Linnæus s. von Linne (Carolus.) Flora Laponica. 8vo. 1737.

Flor. græ. Flora Græca J. E. Smith. fol. 1552.

Flor. mex. ic. ined. Sesse et Mocinno. Flora Mexicana Icones ineditæ.

Fl. per. Ruiz (Hippol.) et Pavon (Jos.) Flora Peruviana et Chilensis. 8vo. 1798.

Flugg. an. mus. Flugge (Johan.) Annales du Museum d'Histoire naturelle. 4to.

Fraser mon. Fraser (John.) A short history of the Agrostis cornucopiæ. fol. 1784.

Forsk. egyp. / *Forsk. dis.* } Forskaol (Petrus.) Flora Ægyptiaco arabica. 4to. 1775.

Forst. gen. Forster (George.) Characteres generum Plantarum quas in itinere ad insulas maris australis, etc. 4to. 1776.

Fourc. j. bot. Fourcroy (Ant. Franc.) La Médécine eclairee par les Sciences physiques, ou Journal des Deconvertes, etc. 8vo. 1791.

Froel. gent. Froelich (Jos. Aloys.) De Gentiana libellus. 8vo. 1796.

Fuchs. hist. Fuchs (Leonhard.) De historia Stirpium commentarii insignes. fol. 1542.

Gaert. carp. Gaertner (Car. Frid.) Supplementum Carpologiæ. 4to. 1805.

Gaert. fr. / *Gaert. sem.* } Gaertner (Josephus.) De fructibus et seminibus plantarum. 4to. 1788.

Garid. aix. / *Garid. prov.* } Garidel (Pierre Joseph.) Histoire des Plantes qui naissent en Province et principalement aux Environs d'Aix. fol. 1715.

Gay dis. las. Gay dissertatio Lasiopetalæ.

Gerard. gal. / *Ger. prov.* } Gerard (John.) Flora Gallo-provincialis. 8vo. 1761.

Ger. emac. Gerard (John.) The Herbal, enlarged by Th. Johnson. fol. 1633.

Ger. herb. Gerard (John.) The Herbal, or general history of Plants. fol. 1797.

Gilib. h. pl. Gilibert (Joh. Em.) Histoire des plantes d'Europe, ou elemens de botanique pratique. 3 vol. in 8vo. Lyon. 1806.

Giorn. pis. Giornale Pisano. 8vo. 1801.

Giseck. ic. Giseke (Paul Dieter.) et Schulze (J. Dom.) Icones plantarum. 1777.

Gmel. fl. bad. Gmelin (Carol. Christ.) Flora Badensis-Alsatica. 8vo. 1808.
Gmel. sib. Gmelin (Joh. Georg.) Flora sibirica. 4to. 1747.
Gou. fl. mons. Gouan (Anton.) Flora Monspeliaca. 8vo. 1765.
Gouan. ill. Gouan (Anton.) Illustrationes botanicæ. fol. 1773.
Hacq. pl. alp. car. Hacquet (Balth.) Plantæ alpinæ Carniolicæ. 4to. 1782.
Hall. all. Haller (Albert.) Allii genus naturale constitutum. 4to. 1745.
Hall. helv. / *Hall. hist.* } Haller (Albert.) Historia Stirpium indigenarum Helvetiæ. fol. 1768.
Hayn. term. bot. Hayne (Fried. Gottl.) et Willdenow (Carl. Ludv.) Termini botanici iconibus illustrati. 4to. 1799.
Hayne in ust. an. Hayne (Fried. Gottl.) in Usteri Annalen der Botanik. 8vo. 1791.
H. et B. n. gen. Humboldt, Bonpland et Kunth Nova Plantarum genera et species, etc. 4to. 1815.
H. et B. pl. aeq. Von Humboldt (Alexand. Fr. Henr.) et Bonpland (Amie) Plantes equinoxiales. fol. 1808.
Herbert. append. Herbert (William.) A Botanical arrangement of bulbous roots. 8vo. 1821.
Herm. lugd. Hermann (Paulus.) Floræ Lugduno-Batavæ flores. 8vo. 1690.
Herm. parad. Hermann (Paulus.) Paradisi Batavi prodromus. 12mo. 1689.
Hills. veg. syst. Hill (John.) Outlines of a system of vegetable generation. 8vo. 1758.
Hoff. c. gott. Hoffmann (Georg. Franz.) Programma Hortus Goettingensis. fol. 1793.
Hoff. deut. fl. Hoffmann (Georg. Franz.) Deutschlands flora, ou Flore de l'Allemagne. 12mo. 1791.
Hoff. phyt. Hoffmann (Georg. Franz.) Phytographische Blætter. 8vo.
Hoffm. salic. Hoffmann (Georg. Franz.) Historia salicum iconibus illustro. fol. 1785.
Hoff. umb. Hoffmann (Georg. Franz.) Plantarum umbelliferarum genera. 8vo. 1816.
Hoff. et Lk. fl. lus. / *Hoff. f. port.* } Hoffmansegg et Link. Flore portugaise. fol. 1806.
Hook. ex. flor. Hooker (W. J.) Exotic flora. 8vo.
H. et G. fil. Hooker (W. J.) and Greville (W. K.) Icones Filicum. fol. London. 1827. et seq.
Hop. in. sturm. fl. fas. Hoppe in Sturm Deutschland flora. 4to. 1798.
Horn. fl. dan. Hornemann (J. W.) Icones Plantarum sponte nascentium in regnis Daniæ et Norvegiæ, etc. fol. 1761.
Hort. Kew. Aiton (Willam.) Hortus Kewensis, editio prima. 8vo. 1789.
Hort. trans. Transactions of the Horticultural Society of London. 4to.
Host. gram. Host (Nicol. Thomas.) Icones et descriptiones Graminum austriacorum. fol. 1801.
Hout. pflanz. Houttuyn (Mart.) Des ritters von Linne Pflanzen system nach der anleitung des Houttuynschen werks. 8vo. 1777.
Hout. syst. Houttuyn (Mart.) Des ritters von Linné Pflanzen system nach der anleitung des Houttuynschen werks. 14 vol. 8vo. 1777—1788.
Icon. Kæmpf. / *Kæmpf. Icon.* } Kæmpfer (Engelbert.) Icones selectæ plantarum. fol. 1791.
Icon. Hort. Kew. Delineations of exotic Plants cultivated in the Royal Garden at Kew. Published by W. T. Aiton.
Isn. act. par. Isnard. Histoire de l'Academie Royale des Sciences, avec les Mémoires de Mathematique et de Physique.
Jacq. amer. Jacquin (Nic. Jos.) selectarum Stirpium Americanarum Historia. fol. 1763.
Jacq. am. pict. Jacquin (N. J.) Stirpium Americanarum Historia.
Jacq. aust. Von Jacquin (Nicol. Jos.) Floræ Austriacæ Icones. fol. 1773.
Jacq. coll. Von Jacquin (Nicol. Jos.) Collectanea ad Botanicam, etc. Spectantia. 4to. 1786.
Jacq. f. ecl. Von Jacquin (Jos. Franc.) Eclogæ Botanicæ. fol. 1811.
Jacq. frag. Von Jacquin (Nicol. Jos.) Fragmenta Botanica. fol. 1809.
Jacq. h. vind. Von Jacquin (Nicol. Jos.) Hortus Botanicus vindobonensis. fol. 1770.
Jacq. ic. Von Jacquin (Nicol. Jos.) Icones Plantarum rariorum. fol. 1781.
Jacq. misc. Von Jacquin (Nicol. Jos.) Miscellanea Austriaca ad Botanicam, etc. Spectantia. 4to. 1778.
Jacq. obs. Von Jacquin (Nicol. Jos.) Observationes botanicæ. fol. 1764.
Jacq. ox. Von Jacquin (Nicol. Jos.) Oxalis monographia iconibus illustrata. 4to. 1792.
Jacq. schoenb. Von Jacquin (Nicol. Jos.) Plantarum rariorum horti cæsari Schœnbrunensis. fol. 1797.
Jacq. stap. Von Jacquin (Nicol. Jos.) Stapeliæ cultæ. fol. 1806.
J. mem. ac. sci. par. Jussieu, Mémoires de la Société d'Histoire Naturelle de Paris. 4to. 1799.
Journ. sc. Ker (J. B.) Journal of Science and the Arts. 8vo.
Jung. ic. rar. Junghans (Phil. Casp.) Icones Plantarum ad vitam impressæ. fol. 1787.
Juss. an. Jussieu, Annales du Museum d'Histoire Naturelle. 4to. 1802.
Juss. rut. Jussieu (Adrien.) Mémoires sur les Rutacées. Paris. 1825. 4to.
Kæmp. am. Kæmpfer (Engelbert.) Amænitates exoticæ. 4to. 1712.
Kalm. it. Kalm (Pehr.) Resatil norra America. 8vo. 1753.
Ker's review. Ker (J. B.) Review of the Genera Amaryllis, Crinum, and Pancratium, in the Journal of Science and the Arts. 8vo.
Ker's strelitz. Ker (H. B.) Coloured Figures of the Genus Strelitzia. fol.
Knight's prot. Knight (Joseph.) Horticultural Essays on the natural order of Proteeæ. 4to. 1809.
Knor. del. Knorr (Georg. Wolfgang.) Deliciæ naturæ selectæ oder auserlesene. fol. 1766.
Knorr. thes. Knorr (Georg. Wolfgang.) The-

saurus rei herbariæ hortensisque universalis. fol. 1770.

Kon. an. bot. Annals of Botany, by C. Konig and J. Sims. 2 vol. 8vo. I. 1805. II. 1806.

Krock. siles. Krocker (Ant. Joh.) Flora Silesiaca renovata. 2 vol. 8vo.

Kth. mim. Kunth (Car. Sig.) Mimoses et autres plantes legumineuses du nouveau continent. fol. 1819.

Lab. nov. hol. Labillardiere (Jac. Jul.) Novæ Hollandiæ plantarum specimen. fol. 1804.

Lab. ic. pl. Syr. / *Labil. Syr.* } Labillardiere (Jac. Jul.) Icones plantarum Syriæ rariorum. 4to. 1791.

Lab. voy. Labillardiere (Jac. Jul.) Relation du Voyage a la recherche de La Peyrouse. 4to. 1798.

Lamb. cinch. Lambert (Aylm. Bourk.) Description of the Cinchona. 4to. 1797.

Lamb. pin. Lambert (Aylm. Bourk.) A Description of the genus Pinus. fol. 1803.

Lam. ill. Monet De La Mark (Jean Bapt.) Illustration des Genres. 1791.

Lap. pyr. ic. ined. Picot De La Peyrouse (Philippe.) Figures de la Flore des Pyrenees. fol. 1795.

Laroch. eryng. De La Roche (Francois.) Eryngiorum nec non generis novi alepideæ historia. fol. 1808.

Ledeb. fl. alt. Ledebour. Flora Altaica. fol. 1829.

Leers. fl. herb. Leers (Joh. Dan.) Flora Herbornensis. 8vo. 1789.

Lehm. nicot. Lehman (Joh. Georg. Chr.) Generis Nicotianarum historia. 4to. 1818.

L. Herit. ger. L'Heritier (Charles Louis.) Geraniologia sive Erodii, etc. historia. fol. 1787.

L. Her. sert. ang. L'Heritier (Charles Louis.) Sertum anglicum seu Plantæ rariores, etc. fol. 1788.

L. Her. stirp. L'Heritier (Charles Louis.) Stirpes novæ aut minus cognitæ. fol. 1784.

Lind. als. Von Lindern (Fr. Balth.) Hortus Alsaticus Plantas in Alsatia, nobili designans. 12mo. 1747.

Lindl. coll. Lindley (John.) Collectanea Botanica. fol.

Lind. digit. Lindley (John.) Digitalium Monographia. fol.

Lindl. ros. Lindley (John.) Rosarum Monographia, or a Botanical History of Roses. 8vo. 1820.

Lin. amœn. ac. Linnæus s. Von Linne (Carolus.) Amœnitates academicæ seu diss. antehæ seorsim editæ. 8vo. 1749.

Lin. fl. lap. Linnæus (Carolus.) Flora Lapponica. 1 vol. 8vo. 1737.

Linn. hort. clif. Linnæus (Carolus.) Hortus Cliffortianus. 1 vol. fol. 1737.

Lin. f. dec. Linnæus Filius (Carolus.) Plantarum rariorum horti Upsaliensis decas 1. 1762. fol. 1763.

Linn. fil. fasc. Linnæus Filius (Carolus.) Plantarum rariorum horti Upsaliensis. fasc. 1. fol. 1767.

Linn. sup. Linnæus Filius (Carolus.) Supplementum Plantarum. 1 vol. 8vo. 1781.

Linn. trans. Transactions of the Linnean Society of London. vol. 1. to 16. 4to.

Lk. et Ott. abb. Link et Otto. Abbildungen und Beschreibungen seltener Pflanzen im Berl. Garten. Berlin. 1821 et seq.

Lk. et Ott. dis. cac. Link et Otto. Cacti et Echinocacti. icones.

Link. et Ot. h. ber. Link et Otto. Hortus Berolinensis. fol.

Lobel. adv. De Lobel seu Lobelius (Mathias.) Adversariorum altera pars. 1 vol. fol. 1605.

Lob. ic. Lobelius (Mathias.) Stirpium Icones. 1 vol. 4to. 1591.

Lodd. bot. cab. Botanical Cabinet, by Messrs. Loddiges'. 4to.

Loef. it. rar. / *Loef. hisp.* } Loefling (Petr.) Iter Hispanicum, eller Resa til spanska, etc. 1 vol. 8vo. 1758.

Loes. prus. Loeselius (Joh.) Flora Prussica. 1 vol. 4to. 1703.

Loisel. gal. Loiseleur-Deslonchamps (J. L. A.) Flora Gallica. 2 vol. 12mo. I. 1806, II. 1807.

Magn. b. mons. Magnol (Petrus.) Botanicon Monspeliense. 1 vol. 12mo. 1686.

Marcg. bras. Marcgravius (Georg.) Historia rerum naturalium Brasiliæ. 1 vol. fol. 1648.

Mart. cent. / *Mart. dec.* } Martyn (John.) Historia Plantarum rariorum centuriæ 1 decas 1-5. fol. 1728.

Mart. fl. rust. Martyn (Thomas.) Flora rustica. 4 vol. 8vo. 1792-4.

Mart. n. pl. bras. Martius (C. F. P.) Nova Genera et Species Plantarum quas in itinere, par Brasiliam. ann. 1817-1820. 4to.

Mass. Stap. Masson (Francis.) Stapeliæ novæ fasc. fol. 1796.

Math. com. Mathiolus (P. Andr.) Commentarii in libros Dioscoridis de Medica Materia.

Meerb. ic. Meerburg (Nicol.) Plantarum selectarum Icones pictæ. fol. 1798.

Mentz. pug. Mentzel (Christ.) Pugillus Plantarum rariorum. fol. 1682.

Merian. surin. Merian (Maria Sibylla.) De metamorphosibus insectorum Surinamensium. 1 vol. fol. 1726.

Mich. amer. Michaux (André.) Flora Boreali Americana. 2 vol. 8vo. 1803.

Mich. arb. Michaux fils (André Franc.) Histoire des arbres forestiers de l'Amérique septentrionale. 3 vol. 4to. 1810-1813.

Mich. querc. Michaux (André.) Histoire des chênes de l'Amérique septentrionale. 1 vol. fol. 1801.

Mich. gen. Micheli (Petr. Anto.) Nova Plantarum genera. 1 vol. fol. 1729.

Mill. ic. Figures of Plants described in the Gardener's Dictionary, by Philip Miller. 2 vol. fol. 1760.

J. Mill. ic. Seven coloured Plates of Plants, by John Miller.

J. F. Mill. ic. Sixty coloured Plates of Plants and Animals, by Joh. Fred. Miller. fol.

Miss Laur. ros. / *M. L. R.* } A Collection of Roses from Nature, by Miss Lawrance (*now Mrs. Kearse*). 90 plates.

Moret. fl. vicent. Moretti (Gius.) Notizia

supra diverse piante da aggiungiarsi alla flora Vicentina. 4to. 1813.

Moris. hist. / *Moris. s.* / *Moris. ox.* Morison (Robert J.) Plantarum historia universalis Oxoniensis. 2 vol. fol. 1680.

Moris. umb. Morison (Robert.) Plantarum umbelliferarum distributio nova. 1 vol. fol. 1672.

Munt. phyt. cur. Munting (Abrah.) Phytographia curiosa. 1 vol. fol. 1702, 1713, 1727.

Murr. com. / *Mur. c. got.* Murray (Joh. Andr.) Commentarii Societatis regiæ Scientiarum Gœttingensis. 4to.

Nectoux voy. Nectoux (H.) Voyage dans la haute Egypte, avec des observations sur les diverses espèces de sené. fol. 1808.

N. et Mart. ac. b. Nees et Martius, in Mémoires de l'Académie royale des Sciences de Berlin. 4to. ab ann. 1770 ad 1816.

Nees. hor. ber. Nees von Esembeck (G. G.) Horæ physicæ Berolinensis collectæ ex symbolis virorum doctorum Link, Rudolphi, etc. 1 vol. fol. 1820.

Niss. act. par. Nissole in Histoire de l'Académie Royale des Sciences, avec les Mémoires de Mathématique et de Physique. 4to.

Nois. jard. fr. Noisette (Louis.) Le Jardin fruitier. 2 fasc. in 4to. Paris. 1813.

Nov. act. pet. Nova Acta Academiæ Scientiarum Imperialis Petropolitanæ.

Nov. com. pet. Novi Commentarii Academiæ Scientiarum Imperialis Petropolitanæ.

Oed. fl. dan. Von Oeder (Georg. Christ.) Icones plantarum sponte nascentium in regnis Daniæ et Norvegiæ, etc. fol.

Ort. dec. De Ortega (Cas Gomez.) Novarum aut rariorum Plantarum Horti Reg. Botan. Matrit. Descriptionum Decades. 4to.

Osb. it. Osbeck (Pehr.) Dagbock œf'er en ostendyck resa. 1 vol. 8vo.

Palisot d'owar. / *Pal. fl. d'ow.* Palisot de Beauvois. Flore des royaumes d'Oware et de Benin. fol. 1805 et seq.

Pall. act. pet. Pallas (Peter Simon.) In Acta Academiæ scientiarum imperialis Petropolitanæ. 4to.

Pall. it. ed. gal. Voyage dans l'Empire Russe. 8 vol. 8vo.

Pall. it. ed. germ. Pallas (P. S.) Reise durch verschiedene provinzen des Russischen Reichs. 3 vol. 4to. 1771—1776.

Pall. ross. Pallas (P. S.) Flora Rossica. fol. 2 vol. 1784, 1788.

Par. lond. / *Sal. par. lond.* Salisbury (Rich. Anth.) Paradisus Londinensis. 2 vol. 4to.

Park. par. Parkinson (John.) Paradisi in sole paradisus terrestris, or a garden of flowers. 1 vol. fol. 1629.

Park. theat. Parkinson (John.) Theatrum botanicum. 1 vol. fol. 1640.

Patters. iter. A Narrative of four journeys into the country of the Hottentots and Caffraria. By Will. Patterson. 4to. 1789.

Pet. eng. herb. / *Pet. h. brit.* Petiver (James.) Herbarii Britannici Catalogus. fol.

Pet. gaz. Petiver (James.) Gazophylacium naturæ et artis. fol. 1702, 1704.

Pet. Th. veg. afr. Aubert du Petit Thouars. Plantes des iles de l'Afrique australe. 3 fasc. in 4to. Paris. 1804—6.

Pio viol. Pio (Joh. Bapt.) De Violâ specimen. 4to. 1813.

Pis. bras. Piso (Guil.) Historia naturalis Brasiliæ. 1 vol. fol. 1648.

Piso. lib. Pisonis (Guil.) De Indiæ utriusque re naturali et medica libri 14. fol. 1658.

Plenck. ic. Plenck (Jos. Jac.) Icones plantarum medicinalium. fasc. fol. 1803 et seq.

Pl. rar. hung. / *W. et K. hung.* Waldstein (Franc.) et Kitaibel (Paul.) Descriptiones et Icones plantarum rariorum Hungariæ. 3 vol. fol.

Pluk. alm. Plukenett (Leonard.) Almagestum botanicum sive phyt. onomasticon. 1 vol. 4to. 1796.

Pluk. am. Plukenett (L.) Amaltheum botanicum. 1 vol. 4to. 1705.

Pluk. mant. Plukenett (L.) Almagesti botanici Mantissa. 1 vol. 4to. 1700.

Pluk. phyt. Plukenett (L.) Phytographia sive Stirpium illustriorum, etc. 4 vol. 4to. 1691.

Plum. amer. Plumier (Charles.) Description des Plantes de l'Amerique. 1 vol. fol. 1693 et 1713.

Plum. ed. B. / *Pl. ed. Burm.* Plantæ Americanæ à C. Plumier detectæ et à J. Burmanno editæ. 5 fasc. fol. 1755.

Plum. fil. Plumier (C.) Traité des fougères de l'Amerique. 1 vol. fol. 1705.

Plum. gen. Plumier (C.) Nova plantarum Americanarum genera. 1 vol. 4to. 1703.

Plum. ic. Plumier (C.) Plantarum Americanarum fasc. 10, continentes plantes quas olim C. Plumierius detexit et depinxit, Edidit Jo. Burmannus.

Pohl. pl. bras. Pohl Plantæ Brasiliæ. 4to. 1830.

Pollich Palat. Pollich (Joh. Adam.) Historia plantarum in Palatinatu spontè nascentium. 3 vol. 8vo. 1776.

Poit. et T. fl. Poiteau et Turpin. Flore Parisienne. 6 livr. fol. 1808.

Pursh amer. Flora Americæ Septentrionalis; or a systematic arrangement and description of the plants of North America. By Frederick Pursh. 2 vol. 8vo.

Quer. hisp. Query Martiner (Joseph.) Flora Espanola. 4 vol. 4to. 1762—1764.

Raii syn. / *Ray. syn.* Ray (Joh.) Synopsis Stirpium Britannicarum. 1 vol. 8vo.

Rauw. it. Rauwolf (Leonh.) Etlicher Schoner Ausslandischer Kreuter, deren in Seiner Rayss, in die Morgenlander Gethon, Gedacht wirt, lebendige Contrafactur.

Red. lil. / *Redoute lil.* Redoute (P. J.) Les Liliacées. 8 vol. fol. 1802—1816.

Redout. ros. Redoute (P. J.) Les Roses. 4to. et fol. 1817 et seq.

Reich. ac. Reichenbach (Ludov.) Monographia generis Aconiti et Delphinii. Lipsiæ. 1820. fol.

Reich. ic. Reichenbach (Ludov.) Icones et descriptiones plantarum. 4to. Lipsiæ. 1822 et seq.

Reliq. Houst. Reliquiæ Houstounianæ, in Bibliotheca Jos. Banks asservatis. 4to. 1781.

Renealm. spec. Pauli Renealmi Specimen Historiæ Plantarum. 1 vol. 4to. 1611.

Retz. obs. Retzius (And. Joh.) Observationes Botanicæ. 6 fasc. fol. 1774—1791.

Rheed. mal. } *Rhe.hort.mal.* } Van Rheede (Henricus.) Hortus Indicus Malabaricus. 12 vol. fol.

Rivin. monop. Rivinus (Aug. Quirinus.) Ordo Plantarum flore irregulari monopetalo. 1 vol. fol. 1690.

Riv. tetr. Rivinus (A. Q.) Ordo Plantarum flore irregulari tetrapetalo. 1 vol. fol. 1691.

Riv. pent. Rivinus (A. Q.) Ordo Plantarum flore irregulari pentapetalo. 1 vol. fol. 1699.

Rob. ic. Three hundred and nineteen plates of plants; engraved by Nic. Robert. A. Bosse, and Lud. de Chatillon. fol.

Rod. Scub. cer. Scubler (Gust.) Diss. sistens descriptiones cerealium cultorum. Tubing. 1818. 4to.

Rœm. arch. Rœmer (Joh. Jac.) Archiv für de Botanik. 3 vol. 4to. 1796, 1799, 1805.

Rœm. script. Rœmer (J. J.) Scriptores de Plantis Hispanicis, Lusitanicis, et Brasiliensis. 1 vol. 8vo. 1796.

Rœm. et Ust. mag. Rœmer (Joh. Jac.) et Usteri (Paul.) Magazin fur die Botanik. 4to. 1789—1790. Zurich.

Rosc. scitam. Roscoe's Figures of the order of Scitamineæ. fol.

Rosc. fl. ill. sea. Roscoe (Mrs. M.) Floral Illustrations of the Seasons. Liverpool. Royal 4to. 1829 et seq.

Roth. abh. Roth (Alb. Wilh.) Botanische abhandlungen und beobachtungen. 4to. 1787.

Rottb. gram. } *Rottb. desc. rar.* } Descriptiones et Icones Plantarum rariorum, etc. 1 vol. fol. 1773.

Roxb. corom. Roxburgh (William.) Plants of the coast of Coromandel, published under the direction of Sir Joseph Banks. fol.

Rudg. gui. Rudge (Edward.) Plantarum Guianeæ rariorum Icones et desc. fol. 1805.

Ruiz. et P. prod. Ruiz (Hippol.) et Pavon (Jos.) Floræ Peruvianeæ et Chilensis prodromus. fol.

Rumph. amb. Rumphius (Georg. Everh.) Herbarium Amboinense. 6 vol. fol. 1750-1755.

Russ. alep. Russell (Alex.) Natural History of Aleppo. 1 vol. 4to. 1756. ed. 2. 1794.

Sabb. hort. } *Sabb. h. rom.* } Sabbati (Liberatus.) Hortus Romanus. fol. vol. 3. 8vo. 1772, 1774, 1805.

Sal. stirp. rar. Salisbury (R. A.) Icones stirpium rariorum. 1 fasc. fol. 1791.

San. viag. Santi (Georg.) Viaggi al mont Amiata. 1 vol. 8vo. 1795. II. 1798. III. 1806.

Savi fl. pis. Savi (Gaetano.) Flora Pisana. 2 vol. 8vo. 1798.

Scheuch. gram. Scheuchzer (Johan.) Agrostographia, sive Graminum, Juncorum, Cyperorum, iisque affinium historia.

Scheuch. it. Scheuchzer (Joh. Jac.) Itinera Alpina. 4to. ed. 1. 1702—1708. Ed. 2; 1 vol. 4to. 1723.

Schk. caric. Schkuhr (Christ.) Histoire des Carex ou Laiches. 1 vol. 8vo. 1802.

Schkuhr filic. Schkuhr (Christ.) Abbildungen der Farrnkräuter. 4to.

Schk. hand. Schkuhr (C.) Botanisches handbuch. 3 vol. 8vo.

Schm. arb. } *Schmid. arb.* } Schmidt (Franz.) Œsterreichs allegemeine Baumzucht 2 Band. fol. 1792—1794.

Schm. ic. Schmiedel (Cas. Christ.) Icones Plantarum. 1 vol. fol. 1762.

Schousb. maroc. Schousboe (P. K. A.) Iagttagelser over voextriget i Marocco. 1 vol. 4to. 1800.

Schrad. d. hal. Schrader (Henr. Adolph.) De Halophytis Pallasii. 4to. 1810.

Schrad. gen. nov. Schrader (H.A.) Nova Plantarum genera. 1 fasc. fol. 1797.

Schrad. germ. Schrader (H.A.) Flora Germanica. 8vo. 1806.

Schr. s. han. Schrader (H.A.) Sertum Hannoveranum. 4 fasc. fol. 1795, 1796.

Schrad. verb. Schrader (H.A.) Monographia generis verbasci. 4to. 1813.

Schrad. ver. Schrader (H.A.) Commentatis de Veronicis spicatis. 8vo. 1803.

Schrank h. mon. } *Schrank pl. rar.* } Schrank (Fr. von Paula.) Plantæ rariores horti Monacensis. 1817, 1818, 1819.

Schreb. a. ups. Von Schreber (Joh. Christ. Dan.) Acta litteraria et scientarum Upsaliæ aut ab academia Upsaliensi publicata. 2 vol. 1720, 1816.

Sch. gram. Schreber (J. C. D.) Beschreibung der Grœser. fol. 1769—1779.

Scop. carn. Scopoli (Joh. Ant.) Flora Carniolica. 8vo. 1760. ed. 2. 2 vol. 8vo. 1772.

Scop. del. } *Scop. ins.* } Scopoli (J.A.) Deliciæ floræ Insubricæ. 3 vol. fol. 1786, 1788.

Seb. mus. } *Seb. thes.* } Seba (Alb.) Locupletissimi Rerum naturalium Thesauri Descriptio. 4 vol. fol. 1734—1765.

Second. d. chen. De Secondat (M.) Mémoires sur l'Histoire naturelle du Chéne. 1785. fol.

Segu. ver. Seguier (Joh. Franc.) Plantæ Veronenses. 3 vol. 8vo. 1745.

Ser. m. helv. Seringe (N. C.) Musée Helvétique d'histoire naturelle. 4to. 1818, 1820.

Shaw afric. } *Shaw barb.* } Shaw (Thomas.) Travels in Barbary and the Levant. 1 vol. fol. 1738.——Supplement. 1746.

Sinc. gr. wob. Sinclair (George.) Hortus gramineus Woburnensis. 1 vol. 8vo. 1825.

Slo. hist. } *Sloan. j.* } Sloane (Hans.) A Voyage to the Islands Madeira, Barbadoes, Nieves, St. Christopher's, and Jamaica; with the Natural History of the last of those Islands. 2 vol. fol. 1707.

Sm. exot. bot. Smith (J. Edward.) Exotic Botany. 8vo. 1804—1808.

Sm. ic. ined. Smith (J. E.) Plantarum Icones hactenus ineditæ. fasc. 1-3. 1789, 1791. fol.

Smith ic. pict. Smith (J. E.) Icones pictæ Plantarum rariorum. 3 fasc. fol. 1790—1793.

Sm. nov. hol. Smith (J. E.) A specimen of the Botany of New Holland. 1 vol. 4to. 1793.

Smith spic. Smith (J. E.) Spicilegium Botanicum. 2 fasc. fol. 1791, 1792.

Smith. tracts. Smith (J. E.) Tracts relating to Natural History. 1 vol. 8vo. 1798.

Sole's mints. Sole (William.) Menthæ Britannicæ; being a new botanical arrangement of all the British Mints hitherto discovered. 1 vol. fol. 1798.

Sonner. it. Sonnerat (P.) Voyage aux Indes orientales et à la Chine. 2 vol. 4to.

Sonner. n. guin. Voyage à la nouvelle Guinée. 1 vol. 4to. 1776.

Sparm. a. st. Sparrman (Andr.) Kongl. Vetenskaps Academiens Handlingar. Stockholm. 1739. seq. 8vo.

Spr. anl. Sprengel (Kurt.) Anleitung zur Kentmiss der Gewachse. 2 theil. 8vo. 1817, 1818.

Spreng. crypt. Sprengel (K.) Introduction to the study of Cryptogamous Plants, translated from the German. 8vo. 1813.

Spr. fl. hall. Sprengel (K.) Floræ halensis Tentamen novum. 1 vol. 8vo. 1806.

Spreng. umb. Sprengel (K.) Species umbelliferarum minùs cognitarum illustratæ. 1 vol. 4to. 1818.

Sternb. saxif. Sternberg (Graf. Casp.) Revisio Saxifragarum Iconibus illustrata. 1 vol. fol. 1810.

Stœr. lib. de. stram. Stœrck (Ant.) Libellus de Stramonio, Hyosciamo. Aconito. 8vo. 1763.

Sturm. deut. Sturm (Jacob.) Deutschland flora. 4to.

Swartz ic. Swartz (Olof.) Icones Plantarum Indiæ occidentalis. 1 fasc. fol. 1794.

Swart. obs. Swartz (O.) Observationes botanicæ. 1 vol. 8vo. 1791.

Sw. syn. fil. Swartz (O.) Synopsis filicum earum genera et species compl. 1 vol. 8vo. 1806.

Swt. br. fl. gar. Sweet (Robert.) The British Flower Garden, published in monthly numbers, containing full and accurate coloured figures and descriptions, with the best mode of cultivation, of the most choice and curious plants, that may be cultivated in the open air of Great Britain. 3 vol. royal 8vo.

Swt. br. fl. gar. s. 2. Sweet (R.) British Flower Garden, series 2, number 1 to 17, et seq.

Swt. cist. Sweet (R.) Cistineæ. The natural Order of Cistus, or Rock-Rose, complete in one volume, containing 112 beautifully coloured figures of the different species and varieties of this handsome family of plants.

Swt. flor. aust. Sweet (R.) Flora Australasica, or a selection of the most beautiful and interesting plants, natives of New Holland, and the South Sea Islands, containing 56 beautifully coloured figures from the living plants. 1 vol. royal 8vo. London, 1827-1828.

Swt. flor. guid. Sweet (R.) The Florist's Guide, or Cultivator's Directory, containing elegantly coloured figures of the choicest flowers cultivated by Florists, including Tulips, Hyacinths, Ranunculus', Carnations, Pinks, Auriculas, Polyanthus's, Georgina's, and Roses. 1827 et seq. royal 8vo. London.

Sweet ger. Sweet (R.) Geraniaceæ, or the natural Order of Geraniums, published in monthly parts, with accurate coloured figures and descriptions of the different kinds, and their mode of culture. royal 8vo. Five volumes complete.

Swert. flor. Swertius (Eman.) Florilegium tractans de variis floribus. 1 vol. fol. 1612.

Tab. ic. Tabernæmontanus (Jac. Theod.) Eicones Plantarum cur N. Bassæo. 4to. 1590.

Tabern. kraut. Tabernæmontanus (Jac.Theod.) Kræuterbuch. fol. 1588.

Ten. flor. med. Tenore (Michel.) Saggio sulle qualita medicinali delle plante Napolitanæ. 8vo. Napoli. 1808.

Tenor. fl. nap. Tenore (Michel.) Flora Napolitana. fol. 1811.

Ten. mem. croc. Tenore (M.) Memoires sui Crochi. Naples.

Thunb. eric. Thunberg (Car. Petr.) de Erica. 4to.

Thunb. ir. Thunberg (Car. Petr.) Dissertatis de Iris. 4to. 1782.

Thunb. jap. Thunberg (Car. Petr.) Flora Japonica. 8vo. 1784.

Thunb. voy. Thunberg (Car. Petr.) Voyage au Japon, etc. 8vo. 1796.

Till. pis. Tilli (Mich. Ang.) Catalogus horti Pisani. fol. 1723.

Tourn. voy. Pitton De Tournefort (Joseph.) Relation d'un voyage du Levant. 4to. 1717.

Trat. arch. Trattinick (Leop.) Archiv der Gewæchskunde. 4to. 1811.

Trattin. fl. aust. Trattinick (Leop.) Flora Austriaca sicca. fol. 1792.

Tratt. tab. Trattinick (Leop.) Observationes botanicæ Tabularium rei herbariæ illustrates. 4to. 1811.

Tratt. thes. bot. Trattinick (Leop.) Thesaurus Botanicus. fol. 1819.

Trev. delph. Treviranus (Lud. Christ.) De Delphinio et Aquilegia observationes. 4to. 1817.

Trew. ehret. Trew (Christ. Jac.) Plantæ selectæ ab Ehret pictæ. fol. 1750.

Trew. pl. rar. Trew (Christ. Jac.) Plantæ rariores, ed. J. C. Keller. fol. 1763.

Trin. ic. gram. Trinius (Car. Bern.) Icones Graminiæ.

Triumf. obs. Triumfetti (Joh. Bapt.) Observationes de ortu et vegetatione Plantarum. 4to. 1635.

Ust. N. ann. Usteri (Paul.) Neue annalen der Botanik. 8vo. 1794.

Vahl. ecl. Vahl (Martinus.) Eclogæ Americanæ. fol. 1796.

Vahl. ic. rar. Vahl (Martinus.) Icones Plantarum in eclogis descriptarum. fol. 1798.

Vahl. symb. Vahl (Martinus.) Symbolæ botanicæ. fol. 1790.

Vail. par. Vaillant (Sebastien.) Botanicon Parisiense. fol. 1727.

Vent. cels. Ventenat (Etienne Pierre.) Description des plantes nouvelles ou peu connues du jardin de J. M. Cels. fol. 1800.
Vent. choix. Ventenat (Etienne Pierre.) Choix des plantes. fol. 1803.
Vent. malm. Ventenat (Etienne Pierre.) Jardin de la Malmaison. fol. 1803.
Vill. c. stras. Villars (D.) Catalogue methodique du jardin de Strasbourg. 8vo. 1807.
Vill. dauph. / *Vill. delph.* } Villars (D.) Histoire des Plantes du Dauphine. 8vo. 1786.
Viv. frag. / *Vivian. fr. fl. it.* } Viviani (Dom.) Floræ Italicæ fragmenta. 4to. 1808.
Volck. norib. Volkamer (Jo. Georg.) Flora Noribergensis. 4to. 1700.
Wahl. helv. Wahlenberg (Georg.) De vegetatione et clymate Helvetiæ septentrionalis specimen. 8vo. 1813.
Wahl. lap. Wahlenberg (Georg.) Flora Lapponica. 8vo. 1812.
Wal. as. res. Wallich in Asiatic Researches, or Transactions of the Society instituted in Bengal. 4to. 1788.
Wall. pl. as. / *—— ind. pl.* } Wallich (Nathanael.) Plantæ asiaticæ rariores, ou description et figures d'un choix de plantes inédites de l'Inde orientale. fol. 1829 et seq.
Walth. hort. Walther (Aug. Frid.) Designatio plantarum horti ejus. 8vo. 1735.
Wang. amer. Von Wangenheim (Fried Ad. Jul.) Anpflanzung North Amerikanischer Holzarten.
Wats. dend. brit. Watson (W.) Dendrologia Britannica. 2 vol. 8vo.
Weig. obs. Weigel (Christ. Ehret.) Observationes Botanicæ. 4to. 1772.
Wei. et N. rub. Weihe (Aug.) et Nees (C. G.) Rubi germanici.
Weinm. phyt. Weinmann (Joh. Guil.) Phytanthoza Iconographica. fol. 1737.
Wend. ac. Wendland (H. Lud.) Commentatis de Acaciis aphyllis. 4to. 1820.
Wendl. col. Wendland (H. Lud.) Collectio Plantarum tum exoticarum quam indigenarum. 4to. 1805.
Wendl. er. Wendland (Joh. Christ.) Ericarum Icones et descriptiones. 4to. 1798.
Wend. h. herr. Wendland (Joh. Christ.) Hortus Herrenhusanus. fol. 1798.
Wend. s. han. Wendland. (J. C.) Sertum Hannoveranum. 4 fasc. fol.
Willd. arb. Willdenow (Car. Lud.) Berlinische Baumzucht. 8vo. 1797.
W. ber. mag. Willdenow (Car. Lud.) Berlinischer magazin oder gesammlete schriften, etc. 4to. 1765.
Willd. hort. ber. Willdenow (Car. Lud.) Hortus Berolinensis. fol. 1806.
Woodv. med. bot. Woodville (William.) Medical Botany. 3 vol. 4to. 1790.
Zanon. hist. Zanoni (Jacobus.) Rariorum stirpium historia. Ed. C. Monti. 1 vol. fol. 1742.
Zorn. ic. Zorn (Joh.) Icones Plantarum medicinalium. 8vo. 1779.

ABBREVIATIONS.

A. B. Annals of Botany.
A. DC. Alphonso De Candolle. Monographie des Campanulees.
Afz. Afzelius, a Swedish Professor.
A. G. Andrews's Geraniums.
A. H. Andrews's Heaths.
A. R. Andrews's Botanist's Repository.
All. Allíoni, an Italian Botanist.
Aub. Aublet, a French Traveller in Guiana.
B. Mr. Robert Brown, a celebrated English Botanist.
B. A. Barton's Flora Boreali Americana. 2 vol. 4to.
Bal. Balbis, a French Botanist.
Bau. Baumgarten.
B. C. Loddiges's Botanical Cabinet.
B. D. Bartling. Diosmeæ descriptæ et illustrate. 1824. 8vo.
Be. Bertoloni, an Italian Botanist.
Bes. Besser, a Russian Botanist.
BF. Bancroft.
B. F. G. Sweet's British Flower Garden.
Bhi. Bernhardi, a German Botanist.
B. M. Curtis's Botanical Magazine.
B. O. Beauvois. Flora d'Oware et de Benin.
Boj. Bojer, Professor of Botany in the Isle of France.
B. P. Brown's Prodromus Floræ novæ Hollandiæ.
B. R. Botanical Register.
Bro. Brogniart's Rhamneæ.
B. T. Burchell's Travels in Africa.
C. (Cav.) Cavanille's, a Spanish Botanist.
C. C. Colvill's Catalogue.
C. I. Cavanille's Icones.
C. L. T. Colebrooke, in Linnean Transactions.
C. S. Choisy. Memoire sur la Famille des Selaginees.
Cyr. Cyrilli, an Italian Botanist.
D. David Douglas, on the Plants of the West Coast of America.
DC. Decandolle (Augustin Pyramus.)
D. D. David Don, Librarian of the Linnean Society, Soho Square.
Desf. Desfontaines, a French Botanist.
Desv. (Dv.) Desvaux, a French Botanist.
DF. Dufresne's Valerianees.
D. L. (Lht.) David Lockhart, Botanical Gardener at Trinidad.
Dod. Dodonaeus, a Dutch Botanist.
D. P. David Don's Prodromus Floræ Nepalensis.
E. Ehrenberg, a Traveller in Egypt, &c.
E. B. Smith's English Botany.
E. C. Elliott's Flora of Carolina.
E. F. Smith's English Flora.
Ex. B. Smith's Exotic Botany.
F. Fischer, Curator of the Botanic Garden at St. Petersburgh.
F. D. Flora Danica.
F. E. Forskaol Flora Ægyptiaco-Arabica.
F. G. Flora Græca.
F. I. Flora Indica.
G. Gærtner, a celebrated German Carpologist.
G. B. Gray's Flora Britannica.
G. D. George Don, an English Botanist.
Gm. Gmelin, a Russian Botanist.
Gr. Graham, Regius Professor of Botany at Edinburgh.
G. R. Georgi, Geographische, etc. Bechreibung des Russichen Reichs.
H. Hornemann, a Danish Botanist.
H. A. Hooker's Flora Boreali Americana.
H. B. Hortus Bengalensis.
H. E. Hortus Epsomensis.
H. E. F. Hooker's Exotic Flora.
H. F. Hooker and Greville's Ferns.
H. G. Hoare's Geraniaceæ.
H. K. Hortus Kewensis.
H. M. Hooker's Miscellany.
H. M. N. Haworth's Miscellanea Naturalia.
H. P. Idem, in Taylor's Philosophical Magazine.
H. R. Idem, Saxifrages and Revisions of Succulent Plants.
H. S. Idem, Synopsis Plantarum Succulentarum.
H. S. S. Idem, Supplementum Plantarum Succulentarum.
H. S. L. Hortus Suburbanus Londinensis.
H. T. Transactions of the Horticultural Society.
H. U. Hoffman's Umbelliferæ.
J. Jussieu.
Jac. Jacquin.
J. E. (J. F.) Jacquin (Filius) Eclogæ Botanicæ.
J. I. Jacquin's Icones.
J. S. Jacquin's Stapeliæ.
K. F. Kaulfuss. Filices.
K. S. Kunth's Synopsis Plantarum.
L. Linnæus.
L. A. Lehman's Asperifoliæ.
L. A. Ledebour's Flora Altaica.
Lam. Lamarck.
L. C. Loddiges's Catalogue.
L. col. Lindley's Collectanea Botanica.
Led. Ledebour, a Russian Botanist.
L.en. Link's Enumeratio Plantarum.
LG. et Lag. Lagasca, a Spanish Botanist.
L'H. L'Heritier.
L. O. Link et Otto, Abbildungen et Beschreibungen seltener Pflanzen in Berlin Garten.
L. P. Lehman's Potentillæ.
L. R. Lindley, Rosarum Monographia.
L. T. Transactions of the Linnean Society.
M. Michaux. Flora Boreali Americana.
M. B. Marschall de Bieberstein, Flora Taurico-Caucasica.
M. C. Mackay's Catalogue.
Med. Medicus, a German Botanist of the last Century.
Mh. Mœnch, a German Botanist.

Mil. Miller, an English Gardener and Botanist.
M. N. Martius. Flora of Brazil.
M. P. Martius. Genera et Species Palmarum.
N. Nuttall's Genera of North American Plants.
N. A. Nee's Asters.
N. L. F. News of Literature and Fashion.
N. P. Nestler's Potentillæ.
Ot. Otto, Curator of the Botanic Garden at Berlin.
P. B. Pohl's Flora Brasiliana.
Ph. Pursh. Flora Americana Septentrionalis.
P. L. Salisbury in Paradisus Londinensis.
Poir. Poiret, a French Botanist.
Pr. Presl, a Bohemian Botanist.
P. S. Persoon's Synopsis Plantarum.
R. (Rox.) Roxburgh.
R. A. Reichenbach. Monographie Generis Aconiti et Delphinii.
Raf. Rafinesque Schmalz.
R. B. Roth's neue Beitrage zur Botanik.
R. C. Roxburgh's Plants of the Coast of Coromandel.
R. O. Richard's Orchideæ.
R. P. Ruiz et Pavon. Flora Peruviana.
R. S. Rœmer et Schultes Systema Vegetabilium.
R. U. (Rch.) Reichenbach Ubersicht der gattung Aconitum.
S. Schulte's Systema Vegetabilium.
S. C. Sweet's Cistineæ.
Sch. (Schl.) Schleicher, a Swiss Collector.
Sck. Schrank, a Bavarian Botanist.
Scop. Scopoli, an Italian Botanist.
Scr. Schrader, a German Botanist.
Sct. Schott, a Traveller in Brazil.
S. F. A. Sweet's Flora Australasica.
S. F. G. Sweet's Florist's Guide.
S. G. Sweet's Geraniaceæ.
Sie. Sieber, a Botanical Collector.
S. M. Sole's Mints.
S. O. Seidl et Opiz.
S. P. Salisbury's Prodromus.
S. S. Sprengel's Systema Vegetabilium.
S. T. Savi. Observationes in varias Trifoliorum Species.
Sw. Swartz, a Swedish Botanist.
T. C. Tenore. Catalogus Plantarum Horti Regii Neapolitani.
T. N. Idem, Flora Neapolitana.
T. P. Idem, Floræ Neapolitanæ Prodromi.
T. S. Idem Synopsis Floræ Neapolitanæ.
Th. Thunberg, a Swedish Botanical Traveller.
Tv. Treviranus, a German Botanist.
V. Ventenat, a French Botanist.
V. M. Ventenat's Jardin de Malmaison.
Viv. Viviani, an Italian Botanist.
W. Willdenow. Species Plantarum.
W. A. Idem, Historia Amaranthorum.
W.en. Idem, Enumeratio Plantarum Horti Bot. Berolinensis.
Wall. Wallich, Curator of the Botanic Garden at Calcutta.
W. C. Wendland. Collectio Plantarum.
W. D. Watson's Dendrologia Britannica.
W. H. William Herbert. Observations on the Amaryllideæ, &c.
W. K. Waldstein and Kitaibel. Descriptiones et icones plantarum rariorum Hungariæ.

1. 2. Flowering the first and second month of the year.
3. 6. Flowering from the third to the sixth month, and so on according to the different figures.
1. 12. Flowering all the year round, or at various seasons of the year, not having any settled period.
—— Denotes the same as the line above it.
.... Uncertain, or not properly known.
H. 5. Hardy large tree.
H. ♄. Hardy Shrub, or small tree.
H. ♄̶. Hardy Evergreen Shrub.
F. ♄. Frame Shrub, or requiring the protection of a Frame or Mats in severe frosty weather, but to be exposed when the weather is mild.
G. ♄. Greenhouse Shrub, requiring the protection of a Greenhouse in Winter.
S. ♄. Shrubs requiring to be kept in the Stove or Hothouse in Winter, but may be removed to the Greenhouse in Summer.
D. S. ♄. Dry Stove Shrub, a plant requiring very little water.
D. G. ♄. Dry Greenhouse Shrub, or a plant that requires but little water in Winter; many of which will succeed well in a good frame or pit, with the covering of a mat or two on the lights in severe frost, but to be continually exposed to the air in mild weather.
♄̶. Suffruticose, a dwarf Shrub, or a soft wooded Shrub.
♃. Herbaceous, perennial, dying down in Winter, and shooting up afresh the following Spring.
♄.◡. Climbing Shrub, a Shrub that requires support, either with a trellis, stick, or line, or any other convenient support.
♃.◡. Climbing, perennial, herbaceous plant, requiring the same sort of support as the last.
w. ♄. Water Shrub, or one that prefers growing in water, or in moist situations.
w. ♃. Perennial herbaceous plant, requiring to be grown in water, or in a moist situation.
♂. Biennial, seldom surviving more than two years.
⊙. Annual, requiring to be sown yearly.
b. Bulbous-rooted.
H. Hardy, surviving through the Winter without protection.
F. Frame, not quite hardy, but requiring the protection of a frame or mats in severe weather.
G. Greenhouse, requiring the protection of a Greenhouse in Winter, but to be well supplied with air in mild weather.
I. Intermediate between Stove and Greenhouse.
S. Stove, those require the protection of a Stove or Hothouse in Winter; the temperature should never be allowed to get below 60 degrees of Fahrenheit's Thermometer.
C. B. S. Cape of Good Hope.
N.S. W. New South Wales.
N. Holl. New Holland.
V. Diem. Isl. Van Dieman's Island.

Classis 1.

DICOTYLEDONEÆ seu EXOGENÆ.

* *Subclassis* 1. *THALAMIFLORÆ.*

ORDO I.

RANUNCULACEÆ. *DC. syst. nat.* 127.

Tribus 1. *CLEMATIDEÆ.*

Syst. Name.	*Color.*	*Eng. Name.*	*Native.*	*Introd.*	*Flower.*	*Habit.*	*Ref. to Figures.*
CLEMATIS. W.		**Virgin's Bower.** Polyandria Polygynia. L.					
1 erécta. DC.	(*wh.*)	upright.	Austria.	1597.	6. 8.	H. ♃.	Jacq. aust. 3. t. 291.
γ hispànica.	(*wh.*)	*Spanish.*					
2 flámmula. DC.	(*wh.*)	sweet-scented.	S. Europe.	1596.	7. 10.	H. ♄.⌣.	Lob. ic. 627. f. 1.
α rotundifòlia.	(*wh.*)	*round-leaved.*	———	———	———	H. ♄.⌣.	Ten. fl. neap. v. 1. t. 48.
β vulgàris.	(*wh.*)	*oblong-leaved.*	———	———	———	H. ♄.⌣.	
γ marítima.	(*wh.*)	*narrow-leaved.*	———	———	———	H. ♄.⌣.	Zanic. ist. 69. t. 129.
3 Massoniàna. DC.	(*wh.*)	Masson's.	C. B. S.	....	....	G. ♄.⌣.	
4 orientàlis. DC.	(*ye.*)	oriental.	Levant.	1731.	7. 10.	H. ♄.⌣.	Dill. elt. t. 119. f. 145.
5 glàuca. W.	(*ye.*)	glaucous.	Siberia.	....	7. 8.	H. ♄.⌣.	Willd. arb. 65. t. 4. f. 1.
6 paniculàta. DC.	(*wh.*)	panicled.	Japan.	1789.	6. 8.	H. ♄.⌣.	Houtt. pflanz. 7. f. 2.
7 chinénsis. DC.	(*wh.*)	Chinese.	China.	1820.	....	H. ♄.⌣.	Retz. obs. 2. n. 53. t. 2.
8 terniflòra. DC.	(*wh.*)	three-flowered.	———	1826.	....	H. ♄.⌣.	
9 Vitálba. DC.	(*wh.*)	Traveller's Joy.	England.	....	7. 9.	H. ♄.⌣.	Eng. bot. 612.
10 virginiàna. DC.	(*wh.*)	Virginian.	N. America.	1767.	6. 8.	H. ♄.⌣.	Wats. dend. brit. t. 74.
11 brasiliàna. DC.	(*wh.*)	Brazilian.	Brazil.	1820.	....	S. ♄.⌣.	Deless. ic. sel. 1. t. 1.
12 dioíca. DC.	(*wh.*)	Jamaica.	Jamaica.	1733.	5. 6.	S. ♄.⌣.	Slo. hist. 1. t. 128. f. 1.
13 americàna. DC.	(*wh.*)	West-Indian.	W. Indies.	....	....	S. ♄.⌣.	
14 coriàcea. DC.	(*wh.*)	leathery-ld.	N. S. W.	1821.	....	G. ♄.⌣.	
15 stenosèpala. DC.	(*wh.*)	narrow-sepaled.	———	1826.	5. 8.	G. ♄.⌣.	
16 aristàta. DC.	(*wh.*)	awned-anther'd.	———	1812.	5. 8.	G. ♄.⌣.	Bot. reg. 238.
17 hedysarifòlia. DC.	(*wh.*)	Hedysarum-ld.	Nepaul.	1819.	7. 2.	G. ♄.⌣.	Bot. reg. 599.
18 biternàta. DC.	(*wh.*)	biternate.	Japan.	1825.	7. 9.	H. ♄.⌣.	
19 triternàta. DC.	(*wh.*)	triternate-ld.	....	1806.	....	H. ♄.⌣.	
20 domínica. DC.	(*wh.*)	St. Domingo.	St. Domingo.	....	....	S. ♄.⌣.	Jac. am. p. 2. t. 261. f. 244.
21 brachiàta. DC.	(*wh.*)	armed.	C. B. S.	1804.	10. 12.	G. ♄.⌣.	Bot. reg. 97.
22 grandiflòra. DC.	(*wh.*)	great-flowered.	Sierra Leon.	1822.	....	S. ♄.⌣.	
23 dahùrica. DC.	(*pu.*)	Dahurian.	Dahuria.	1822.	6. 8.	H. ♄.⌣.	
24 zanzibárica.	(*wh.*)	Zanzibar.	Africa.	1826.	....	S. ♄.⌣.	
25 angustifòlia. DC.	(*wh.*)	narrow-leaved.	Siberia.	1787.	5. 9.	H. ♃.	Jacq. ic. 1. t. 104.
26 diversifòlia. DC.	(*pu.*)	various-leaved.	....	....	9. 10.	H. ♄.⌣.	Deless. ic. sel. 1. t. 4.
27 Viórna. DC.	(*pu.*)	leathery-flow'd.	N. America.	1730.	6. 9.	H. ♄.⌣.	Jacq. ecl. 1. t. 32.
28 cylíndrica. DC.	(*bl.*)	long-flowered.	———	1802.	7. 9.	H. ♄.⌣.	Bot. mag. 1160.
Viórna. A. R. 71. *nec aliorum.*							
29 Símsii.	(*pu.*)	Sims's.	———	1812.	6. 9.	H. ♄.⌣.	Bot. mag. 1816.
cordàta. B. M. *nec aliorum.*							
30 reticulàta. DC.	(*pu.*)	netted.	———	———	———	H. ♄.⌣.	Wats. dend. brit. t. 72.
31 japónica. DC.	(*pu.*)	Japan.	Japan.	1826.	....	H. ♄.⌣.	
32 integrifòlia. DC.	(*bl.*)	entire-leaved.	Hungary.	1596.	6. 8.	H. ♃.	Bot. mag. 65.
β ? angustifòlia.	(*bl.*)	*narrow entire ld.*	Siberia.		———	H. ♃.	
γ ? elongàta. DC.	(*bl.*)	*long-leaved.*	———	1823.	———	H. ♃.	Tratt. tab. 4. t. 178.

33 ochroleùca. DC. (st.) silky. N. America. 1767. 6. 7. H. ♃. Bot. mag. 1175.
34 gentianoídes. DC. (wh.) gentian-like. V. Diem. Isl. 1825. .... G. ♃. Deless. ic. sel. 1. t. 5.
35 flòrida. DC. (wh.) large-flowered. Japan. 1776. 4. 9. H. ♄.◡. Bot. mag. 834.
β flòre plèno (wh.) *double-flowered.* ——— —— —— H. ♄.
36 Viticélla. DC. (pu.) smooth-seeded. Spain. 1569. 6. 9. H. ♄.◡. Bot. mag. 565.
α purpùrea. (pu.) *purple-flowered.*
β cærùlea. (bl.) *blue-flowered.*
γ pulchélla. P.S. (bl.) *double-flowered.* ——— —— —— H. ♄.◡. Chabr. sci. p. 117. f. 5.
37 críspa. B.M. (pu.) curled-flower'd. N. America. 1726. 7. 9. H. ♄.◡. Bot. mag. 1892.
38 parviflòra. DC. (wh.) small-flowered. .... 1822. 6. 8. H. ♄.◡. Dc. pl. rar. gen. t. 12.
campaniflòra. Lodd. bot. cab. 987. *non* DC.
39 cirrhòsa. DC. (pu.) tendriled. S. Europe. 1596. 3. 4. H. ♄.◡. Flor. græc. t. 517.
40 pedicellàta. (st.) pedicled. Majorca. .... —— H. ♄.◡. Bot. mag. 1070.
cirrhòsa. B.M. nec aliorum. *cirrhòsa β. pedicellàta.* DC.
41 semitrilòba. DC. (st.) three-lobed. Spain. .... 9. 11. H. ♄.◡.
42 calycìna. H.K. (st.) evergreen. Minorca. 1783. 4. 6. H. ♄.◡. Bot. mag. 959.
baleàrica. DC.

ATRAG'ENE. W. ATRAG'ENE. Polyandria Polygynia. L.
1 austrìaca. B.M. (bl.) Austrian. Austria. 1792. 5. 7. H. ♄.◡. Bot. rep. 180.
alpìna. S.S. *Clématis alpìna.* DC.
2 sibírica. B.M. (wh.) Siberian. Siberia. 1753. 3. 5. H. ♄.◡. Bot. mag. 1951.
Clématis sibírica. DC.
3 ochoténsis. S.S. (wh.) linear-petaled. ——— 1822. —— H. ♄.◡.
Clématis ochoténsis. DC.
4 americàna. B.M. (pu.) American. N. America. 1797. 5. 6. H. ♄.◡. Bot. mag. 887.
Clématis verticillàris. DC.
5 occidentàlis. S.S. shining-leaved. .... 1824. .... H. ♄.◡.

NARAV'ELIA. DC. NARAV'ELIA. Polyandria Polygynia. L.
zeylánica. DC. .... Ceylon. Ceylon. 1796. .... S. ♄.◡.
β Roxbúrghii. DC. (ye.) *Roxburgh's.* Coromandel. —— .... S. ♄.◡. Roxb. cor. 2. t. 188.

Tribus II. *ANEMONEÆ.*

THALI'CTRUM. DC. MEADOW-RUE. Polyandria Polygynia. L.
1 contórtum. DC. (wh.) crook-seeded. Siberia. 1796. 6. 7. H. ♃.
2 aquilegifòlium. DC. .. Columbine-l[d]. Europe. 1731. 5. 7. H. ♃.
α álbum. (wh.) *white-stamened.* ——— —— —— H. ♃. Jacq. aust. t. 318.
β atropurpùreum. (pu.) *purple.* ——— —— —— H. ♃. Bot. mag. 1818.
γ formòsum. B.M. (pu.) *Sabine's.* ——— —— —— H. ♃. Bot. mag. 2025.
3 clavàtum. DC. (wh.) club-stamened. N. America. 1812. 6. 8. H. ♃.
4 mexicànum. DC. (gr.) Mexican. Mexico. 1826. 7. 9. H. ♃. Hern. mex. 236. ic.
5 corynéllum. DC. (wh.) purple-stemm'd. N. America. .... —— H. ♃.
6 polygàmum. S.S. (ye.) polygamous. ——— .... 5. 7. H. ♃.
7 revolùtum. DC. (ye.) revolute ——— 1806. 6. 7. H. ♃.
pubéscens. Ph. *non.* DC.
8 dioícum. DC. (ye.) diœcious. ——— 1759. —— H. ♃. Deles. ic. sel. 1. t. 8.
9 caroliniànum. DC. (ye.) Carolina. ——— .... —— H. ♃.
10 purpuráscens. DC. (pu.) dwarf purple. ——— 1699. —— H. ♃.
11 petaloídeum. DC. (wh.) Dahurian. Dahuria. 1799. —— H. ♃. Deles. ic. sel. 1. t. 9.
12 alpìnum. DC. (gr.) Alpine. Britain. .... 5. 7. H. ♃. Eng. bot. 262.
13 fœ'tidum. DC. (wh.) fœtid. Europe. 1640. —— H. ♃. W. et K. hung. 2. t. 174.
14 pubéscens. DC. (ye.) pubescent. ——— —— —— H. ♃.
15 acutilòbum. DC. (ye.) acute-lobed. Siberia. .... —— H. ♃. Deles. ic. sel. 1. t. 10.
16 divérgens. S.S. (gr.) divergent. ——— 1820. 6. 7. H. ♃.
17 Schweíggeri. S.S. (ye.) Schweigger's. .... 1826. —— H. ♃.
18 calàbricum. S.S. (ye.) Calabrian. Sicily. 1802. 6. 8. H. ♃. Moris. 1. s. 9. t. 20. f. 16.
19 squarròsum. DC. (ye.) squarrose Siberia. 1806. —— H. ♃.
20 sibíricum. DC. (ye.) Siberian. ——— 1775. —— H. ♃. Gært. fruct. 1. t. 74.
21 mìnus. DC. (ye.) lesser. Britain. .... 6. 7. H. ♃. Eng. bot. t. 11.
22 collìnum. DC. (ye.) hill. Germany. 1823. —— H. ♃.
23 saxátile. DC. (ye.) rock. Europe. .... 5. 7. H. ♃.
24 ambíguum. S.S. (ye.) ambiguous. Switzerland. 1820. —— H. ♃.
25 elàtum. DC. (ye.) tall. Hungary. 1794. 6. 8. H. ♃. Jacq. vind. 3. t. 95.
26 màjus. DC. (ye.) greater. England. .... —— H. ♃. Eng. bot. 611.
27 nùtans. DC. (ye.) nodding. Italy. .... —— H. ♃.
28 mèdium. DC. (ye.) middle. Hungary. 1789. —— H. ♃. Jacq. vind. 3. t. 96.

29 concínnum. DC.	(*wh.*)	neat.	....	....	6.	H. ♃.	
30 flexuòsum. Jac.	(*ye.*)	flexuose.	....	....	6. 8.	H. ♃.	
31 trig'ynum. DC.	(*ye.*)	three-styled.	Dahuria.	1823.	——	H. ♃.	
32 glaucéscens. DC.	(*pu.*)	loose-flowered.	Russia.	——	6.	H. ♃.	
33 galio'ides. DC.	(*ye.*)	Galium-leaved.	Europe.	1816.	5. 7.	H. ♃.	Deles. ic. sel. 1. t. 11.
34 angustifòlium. DC.	(*ye.*)	narrow-leaved.	Germany.	1739.	6. 7.	H. ♃.	Jacq. vind. 3. t. 43.
35 divaricàtum. S.S.	(*ye.*)	spreading.	....	——	——	H. ♃.	
36 cynapiifòlium. F.	(*ye.*)	Fool's-parsleyl'd	Siberia.	1821.	——	H. ♃	
37 oligospérmum. F.	(*ye.*)	few-seeded.	————	——	——	H. ♃.	
38 diffùsum. Sch.	(*ye.*)	diffuse.	....	——	——	H. ♃.	
39 lùcidum. DC.	(*ye.*)	shining.	Spain.	1739.	5. 7.	H. ♃.	Pluk. alm. t. 65. f. 5.
40 rosmarinifòlium. S.S.	(*ye.*)	Rosemary-l'd	Italy.	1826.	6. 7.	H. ♃.	
41 nìgricans. DC.	(*ye.*)	black.	Austria.	1798.	5. 7.	H. ♃.	Jacq. aust. 5. t. 421.
42 flàvum. DC.	(*ye.*)	common.	Britain.	....	——	H. ♃.	Eng. bot. 367.
43 vaginàtum. Desf.	(*ye.*)	sheathed.	Europe.	....	——	H. ♃.	
44 símplex. DC.	(*ye.*)	simple-stalked.	Sweden.	1778.	5. 6.	H. ♃.	Flor. dan. t. 244.
45 cinèreum. DC.	(*ye.*)	ash-coloured.	....	....	——	H. ♃.	
46 glàucum. DC.	(*ye.*)	glaucous	S. Europe.	1798.	6. 7.	H. ♃.	
47 rugòsum. DC.	(*ye.*)	rugged.	N. America.	1774.	7.	H. ♃.	
48 ranunculìnum. DC	(*ye.*)	Ranunculus-l'd.	————	1806.	6. 7.	H. ♃.	
49 tuberòsum. DC.	(*wh.*)	tuberous-rooted.	Spain.	1713.	6.	H. ♃.	Mill. ic. t. 265. f. 2.
50 anemonoídes. DC.	(*wh.*)	Anemone-like.	N. America.	1768.	4. 5.	H. ♃.	Bot. mag. 866.
Anemòne thalictròides. W.							
β flòre plèno.		*double-flowered.*	————	1822.	——	H. ♃.	
PULSATI'LLA. W.en.		PASQUE-FLOWER. Polyandria Polygynia.					
1 vernàlis. S.S.	(*pu.*)	spring.	N. Europe.	1752.	4.	H. ♃.	Swt. br. fl. gar. 205.
2 Hallèri. S.S.	(*bl.*)	Haller's.	Switzerland.	1816.	4. 5.	H. ♃.	All. ped. t. 80. f. 2.
3 cérnua. S.S.	(*bl.*)	drooping.	Japan.	1806.	4. 6.	H. ♃.	
4 pàtens. S.S.	(*va.*)	spreading.	Siberia.	1752.	6. 7.	H. ♃.	Breyn. ic. t. 61.
β ochrolèuca.	(*st.*)	*straw-coloured.*	————	——	5. 7.	H. ♃.	Bot. mag. 1994.
5 vulgàris. S.S.	(*bl.*)	common.	England.	....	——	H. ♃.	Eng. bot. 51.
β rùbra. DC.	(*re.*)	red-flowered.	Europe.	....	——	H. ♃.	Lob. ic. 282. f. 1.
6 intermèdia.	(*bl.*)	long-petal'd.	Switzerland.	1820.	——	H. ♃.	
Anemòne intermèdia. Schult. *longipétala.* Schleich.							
7 praténsis. S.S.	(*pu.*)	meadow.	Germany.	1731.	5.	H. ♃.	Flor. dan. t. 611.
8 obsolèta.	(*bl.*)	pale-flowered.	————	——	——	H. ♃.	Bot. mag. 1863.
9 albàna. S.S.	(*wh.*)	white-flowered.	Caucasus.	1827.	——	H. ♃.	
10 dahùrica. DC.	(*fl.*)	Dahurian.	Dahuria.	1823.	....	H. ♃.	
11 alpìna. S.S.	(*wh.*)	Alpine.	Europe.	1658.	7.	H. ♃.	Bot. mag. 2007.
12 micrántha.	(*wh.*)	small-flowered.	————	1827.	——	H. ♃.	Jacq. aust. t. 85.
Anemòne alpìna. Jacq. *alpìna β micrántha.* DC.							
13 apiifòlia.	(*ye.*)	Parsley-leaved.	————	....	——	H. ♃.	Jacq. misc. 2. t. 4.
Anemòne alpìna δ flavéscens. DC.							
14 sulphùrea.	(*st.*)	sulphur-coloured.	————	....	——	H. ♃.	Cam. epit. 393. ic.
Anemòne sulphùrea. L. *apiifòlia.* W.							
ANEM'ONE. W.en.		ANEM'ONE. Polyandria Polygynia. L.					
1 capénsis. DC.	(*pu.*)	Cape.	C.B.S.	1795.	3. 7.	G. ♄.	Bot. mag. 716.
Atragène capénsis. W.							
2 coronària. DC.	(*va.*)	Poppy.	Levant.	1596.	1. 12.	H. ♃.	Bot. mag. 841.
β flòre-plèno.	(*va.*)	*double flowered.*					
3 pavonìna. DC.	(*re.*)	acute-petaled.	————	....	4. 5.	H. ♃.	
α coccìnea.	(*sc.*)	*scarlet.*	————	....	——	H. ♃.	
β lilacìna.	(*li.*)	*lilac.*	————	....	——	H. ♃.	
γ flòre-plèno.	(*sc.*)	*double scarlet.*	————	....	——	H. ♃.	Clus. hist. 1. p. 261. ic.
4 stellàta. DC.	(*pu.*)	star.	Italy.	1597.	——	H. ♃.	Chabr. sciagr. 461. f. 3.
horténsis. B.M. 123.							
α versícolor.	(*v.*)	*various-coloured.*	————	——	——	H. ♃.	Bot. mag. 123.
β purpùrea.	(*pu.*)	*purple.*	————	——	——	H. ♃.	Swt. br. fl. gar. 112.
5 palmàta. DC.	(*ye.*)	palmated.	Portugal.	1597.	5. 6.	H. ♃.	Bot. reg. 200.
β flòre-plèno.	(*ye.*)	*double-flowered.*	————	——	——	H. ♃.	Chabr. sciagr. 461. f. 2.
6 parviflòra. DC.	(*wh.*)	small-flowered.	Labrador.	1823.	3. 5.	H. ♃.	Juss. ann. mus. 3. t. 21. f. 1
cuneifòlia. Ph.							
7 caroliniàna. DC.	(*pu.*)	slender.	Carolina.	——	5.	H. ♃.	
tenélla. Ph.							
8 apennìna. DC.	(*bl.*)	blue mountain.	England.	....	3. 4.	H. ♃.	Eng. bot. 1062.
9 cœrùlea. DC	(*bl.*)	blue.	Siberia.	1827.	——	H. ♃.	Deless. ic. sel 1. t. 14.
10 uralénsis. DC.	(*bl.*)	Uralian.	————	——	——	H. ♃.	
11 baldénsis. DC.	(*wh.*)	Strawberry-like.	Switzerland.	1792.	5.	H. ♃.	Jacq. ic. 1. t. 103.
fragífera. Jacq.							

12 nemoròsa. DC.	(*wh.*)	wood.	Britain.	....	3. 5.	H. ♃.	Eng. bot. 355.
β *flòre-plèno.*	(*wh.*)	*double-flowered.*	——	....	——	H. ♃.	Debry.fl.nov.t.29 et 56.
γ *quinquefòlia.* DC.	(*w.*)	*five-leaved.*	N. America.	1812.	——	H. ♃.	
13 Fischeriàna. DC.	(*wh.*)	Fischer's.	Siberia.	1827.	4. 5.	H. ♃.	
14 lancifòlia. DC.	(*wh.*)	lance-leaved.	N. America.	1823.	5. 7.	H. ♃.	
15 trifòlia. DC.	(*wh.*)	three-leaved.	France.	1597.	4. 5.	H. ♃.	Lob. ic. 281. f. 1.
16 ranunculoídes. DC.	(*ye.*)	yellow wood.	England.	....	3. 4.	H. ♃.	Eng. bot. 1484.
17 refléxa. DC.	(*wh.*)	reflex-flowered.	Siberia.	1824.	4. 5.	H. ♃.	Deless. ic. sel. 1. t. 15.
18 sylvéstris. DC.	(*wh.*)	Snowdrop.	Europe.	1596.	4. 5.	H. ♃.	Bot. mag. 54.
19 álba. DC.	(*wh.*)	five-sepaled.	Dahuria.	1818.	4. 5.	H. ♃.	—— 2167.
ochoténsis. F.							
20 virginiàna. DC.	(*wh.*)	Virginian.	N. America.	1722.	5. 6.	H. ♃.	Herm. par. t. 18.
21 pensylvánica. DC.	(*wh.*)	Pensylvanian.	——	1766.	——	H. ♃.	
22 dichótoma. DC.	(*wh.*)	forked.	——	1768.	——	H. ♃.	Lin. f. dec. 29. t. 15.
23 narcissiflòra. DC.	(*wh.*)	Narcissus-fl'd.	Europe.	1773.	5.	H. ♃.	Bot. mag. 1120.
24 umbellàta. DC.	(*wh.*)	umbel flower'd.	Levant.	1824.	——	H. ♃.	Deless. ic. sel. 1. t. 18.
25 sibírica. DC.	(*wh.*)	Siberian.	Siberia.	1804.	6.	H. ♃.	
HEPA'TICA. DC.		HEPA'TICA. Polyandria Polygynia. L.					
1 acutilòba. DC.	(*bl.*)	acute-lobed.	N. America.	1826.	3. 4.	H. ♃.	
2 americàna. DC.	(*bl.*)	American.	——	1819.	3. 4.	H. ♃.	Bot. reg 387.
3 trilòba. DC.	(*va.*)	three-lobed.	Europe.	1573.	2. 4.	H. ♃.	Eng. bot. t. 51.
α *cœrùlea.*	(*bl.*)	*blue.*					
β *cœrùleo-plèna.*	(*bl.*)	*double-blue.*					
γ *rùbra.*	(*re.*)	*red.*	——	——	——	——	Bot. mag. 10.
δ *rùbro-plèna.*	(*re.*)	*double-red.*					
ε *álba.*	(*wh.*)	*red-anthered white.*					
ζ *nívea.*	(*wh.*)	*snowy-white.*					
HYDRA'STIS. DC.		YELLOW-ROOT. Polyandria Polygynia. L.					
canadénsis. DC.	(*wh.*)	Canadian.	Canada.	1759	5. 6.	H. ♃.	Pict. hort. par. 37. t. 17.
KNOWLT'ONIA. DC.		KNOWLT'ONIA. Polyandria Polygynia. L.					
1 rígida. DC.	(*gr.*)	thick-leaved.	C. B. S.	1780.	3. 5.	G. ♃.	Vent. malm. p. 22. t. 22.
2 vesicatòria. DC.	(*gr.*)	blistering.	——	1691.	2. 4.	G. ♃.	Bot. mag. 775.
3 grácilis. DC.	(*gr.*)	slender.	——	1820.	——	G. ♃.	Deles. ic. sel. 1. t. 19.
4 hirsùta. DC.	(*gr.*)	hairy-leaved.	——	——	——	G. ♃.	Burm. afr. t. 51.
5 daucifòlia. DC.	(*gr.*)	Carrot-leaved.	——	1822.	....	G. ♃.	
AD'ONIS. DC.		AD'ONIS. Polyandria Polygynia. L.					
1 autumnàlis. DC.	(*sc.*)	Pheasant's-eye.	Britain.	....	5. 10.	H. ⊙.	Eng. bot. 308.
2 flàva. DC.	(*ye.*)	yellow.	France.	....	6. 7.	H. ⊙.	Tab. ic. 790. f. 1.
3 micrántha. DC.	(*ye.*)	small-flowered.	S. France.	1826.	5. 6.	H. ⊙.	
4 citrìna. DC.	(*st.*)	Citron-colour'd.	——	1822.	5. 6.	H. ⊙.	
5 flámmea. DC.	(*fl.*)	flame-coloured.	Austria.	1800.	6. 7.	H. ⊙.	Jac. aust. 4. t. 355.
6 æstivàlis. DC.	(*sc.*)	tall.	S. Europe.	1629.	——	H. ⊙.	Weinm. phyt. t. 27.
miniàta. Jac. aust. t. 354.							
7 vernàlis. DC.	(*ye.*)	vernal.	Europe.	1629.	3. 4.	H. ♃.	Bot. mag. 134.
8 apennìna. W. EN.	(*ye.*)	Apennine.	——	....	——	H. ♃.	Mentz. pug. t. 3. f. 1.
vernàlis β Mentzèlii. DC.							
9 pyrenàica. DC.	(*ye.*)	Pyrenean.	Pyrenees.	1820.	6. 7.	H. ♃.	Deless. ic. sel. 1. t. 21.
10 sibírica. S.S.	(*ye.*)	Siberian.	Siberia.	1827.	4. 5.	H. ♃.	
11 davùrica. S.S.	(*ye.*)	Davurian.	——	——	——	H. ♃.	Reich ic. t. 321.
12 distòrta. T.P.	(*ye.*)	distorted.	Naples.	——	——	H. ♃.	Ten. fl. med. p. 448. t. 1.

Tribus III. *RANUNCULEÆ.*

MYOS'URUS. DC.		MOUSE-TAIL. Pentandria Polygynia. L.					
mínimus. DC.	(*gr.*)	small.	Britain.	....	4. 5.	H. ⊙.	Eng. bot. 435.
CERATOCE'PHALUS.		CERATOCE'PHALUS. Pentandria Polygynia.					
1 falcàtus. DC.	(*ye.*)	sickle-seeded.	S. Europe.	1739.	5.	H. ⊙.	Jac. aust. t. 48.
2 orthocèras. DC.	(*ye.*)	straight-seeded.	Siberia.	1823.	——	H. ⊙.	Deles. ic. sel. 1. t. 23.
RANU'NCULUS. DC.		CROW-FOOT. Polyandria Polygynia. L.					
1 hederàceus. DC.	(*wh.*)	Ivy-leaved.	Britain.	....	5. 8.	H.*w* ♃.	Eng. bot. 2003.
2 tripartìtus. DC.	(*wh.*)	three-parted.	——	....	——	H.*w.* ♃.	DC. ic. g. rar. 1. t. 49.
β *obtusiflòrus.* DC.	(*w.*)	*blunt-flowered.*	——	....	——	H.*w.* ♃.	Pet. eng. herb. t. 39. f. 1.
3 aquátilis. DC.	(*wh.*)	water.	——	....	4. 8.	H.*w.* ♃.	Eng. bot. 101.
α *heterophy'llus.* DC.	(*w.*)	*various-leaved.*	——	....	——	H.*w.* ♃.	Tab. ic. 54. f. 2.
β *peltàtus.* DC.	(*wh.*)	*peltate-leaved.*	——	....	——	H.*w.* ♃.	Lob. ic. 2. p. 35. f. 2.

4	pantothríx. DC.	(wh.)	divided-leaved.	Britain.	....	4. 8.	H.w.♃.	Lob. ic 791. f. 1.
	α capillàceus. DC.	(wh.)	*fine-leaved.*	——	....	——	H.w.♃.	Pet. eng. herb. t. 39. f. 4.
	β cæspitòsus. DC.	(wh.)	*tufted.*	——	....	——	H.w ♃.	—— t. 39. f. 3.
	γ peucedanifòlius. DC.	(wh.)	*Fennel-ld.*	——	....	——	H w.♃.	—— t. 39. f. 2.
	fluviátilis. W.							
5	rutæfòlius. DC.	(wh.)	Rue-leaved.	Austria.	1759.	5. 7.	H. ♃.	Jac. coll. 1. t. 6 et 7.
6	glaciàlis. DC.	(wh.)	two-flowered.	Lapland.	1775.	6. 8.	H. ♃.	Flor. dan. t. 19.
7	Seguíeri. DC.	(wh.)	Seguier's.	Switzerland.	1820.	——	H. ♃.	Vill. dauph. 4. t. 49.
8	alpéstris. DC.	(wh.)	Alpine.	Scotland.	....	——	H. ♃.	Eng. bot. 2390.
9	crenàtus. DC.	(wh.)	crenate-leaved.	Hungary.	1822.	——	H. ♃.	W. et K. hung. 1. t. 10.
10	aconitifòlius. DC.	(wh.)	Aconite-leaved.	Europe.	1596.	5. 6.	H. ♃.	J. Ger. hist. 951. f. 4.
	α hùmilis. DC.	(wh.)	*dwarf.*	——	....	——	H. ♃.	Lob. adv. 300. f. 2.
	β crassicàulis. DC.	(w.)	*thick-stalked.*	——	....	——	H. ♃.	
11	platanifòlius. W.	(wh.)	Plane-tree-ld.	Germany.	1769.	6. 7.	H. ♃.	Flor. dan. t. 111.
	β flòre-plèno.	(wh.)	*double-flowered.*	——	——	——	H. ♃.	Bot. mag. 204.
	aconitifòlius. B.M. 204.							
12	lácerus. DC.	(wh.)	torn-leaved.	Piedmont.	....	——	H. ♃,	Bell. act. taur. 5. t. 8.
13	pállidus. DC.	(st.)	pale-yellow.	Hybrid.	....	——	H. ♃.	Schrank. hort. mon. t. 57.
14	pyrenæ̀us. DC.	(wh.)	Pyrenean.	Pyrenees.	....	4. 6.	H. ♃.	Deles. ic. sel. 1. t. 27. f. B.
	β bupleurifòlius. DC.	(w.)	*one-flowered.*	——	....	——	H. ♃.	Jac. misc. 1. t. 18. f. 1.
	γ plantagíneus. DC.	(wh.)	*many-flowered.*	——	....	——	H. ♃.	All ped. n. 1445. t. 76 f. 1.
15	angustifòlius. DC.	(wh.)	narrow-leaved.	——	1822.	——	H. ♃.	Deles. ic. sel. 1. t. 27. f. C.
16	amplexicàulis. DC.	(w.)	Plantain-leav'd.	——	1633.	4. 5.	H. ♃.	Bot. mag. 266.
17	parnassifòlius. DC.	(w.)	Parnassia-leavd.	——	1769.	6. 7.	H. ♃.	—— 386.
18	gramíneus. DC.	(ye.)	grass-leaved.	Wales.	....	4. 6.	H. ♃.	Eng. bot. 2306.
	β flòre-plèno.	(ye.)	*double-flowered.*	....		——	H. ♃.	Lob. ic. 671. f. 2.
19	bupleuròides. DC.	(ye.)	Hare's-ear-ld.	Portugal.	1826.	——	H. ♃.	
20	Língua. DC.	(ye.)	great spear-wort.	Britain.	....	6. 8.	H.w.♃.	Eng. bot. 100.
21	Flámmula. DC.	(ye.)	less. spear-wort.	——	....	6. 9.	H.w.♃.	—— 387.
	β serràta. DC.	(ye.)	*saw-leaved.*	——	....	——	H.w.♃.	Lob. ic. 670. f. 2.
	γ ovàta. DC.	(ye.)	*oval-leaved.*	——	....	——	H.w.♃.	
22	réptans. DC.	(ye.)	least spear-wort.	——	....	——	H.w.♃.	Flor. dan. t. 108.
23	filifórmis. M.	(ye.)	filiform.	N. America.	1816.	——	H.w.♃.	
24	uliginòsus. DC.	(ye.)	marsh.	Teneriffe.	1826.	6. 7.	H.w.⊙.	
25	bonariénsis. DC.	(ye.)	Buenos Ayres.	S. America.	1819.	——	H.w.⊙.	Deles. ic. sel. 1. t. 29.
26	polyphy̆llus. DC.	(ye.)	many-leaved.	Hungary.	——	5. 6.	H.w.⊙.	W. et K. pl. hung. 1. t. 45.
27	ophioglossifòlius.	(ye.)	fistulous-stalk'd.	S. Europe.	1822.	6.	H.w.⊙.	Vill. dauph. 4. t. 49.
28	nodiflórus. DC.	(ye.)	knot-flowered.	France.	1714.	5. 7.	H.w.⊙.	
	β dentàtus. DC.	(ye.)	*toothed.*	Hungary.	——	——	H.w.⊙.	Bot. mag. 2171.
29	salsuginòsus. DC.	(ye.)	Russian.	Russia.	1821.	——	H. ♃.	Jac. vind. 3. t. 31.
	ruthénicus. Jac. vind. 3. t. 31.							
30	Cymbalàriæ. Ph.	(ye.)	Ivy-like.	N. America.	1823.	——	H.w.♃.	
31	bullàtus. DC.	(ye.)	blistered.	S. Europe.	1640.	5. 8.	H. ♃.	
	α parviflòrus. DC.	(ye.)	*small-flowered.*	——	——	——	H. ♃.	Clus. hist. 1. f. 2.
	β grandiflòrus. DC.	(ye.)	*large-flowered.*	——	——	——	H. ♃.	—— 1. f. 1.
	γ flòre-plèno.	(ye.)	*double-flowered.*	——	——	——	H. ♃.	Corn. can. 95. ic.
32	chærophy̆llos. DC.	(ye.)	Chervil-leaved.	——	....	4. 6.	H. ♃.	Barr. ic. t. 581.
	α vulgàris. DC.	(ye.)	*common.*	——	....	——	H. ♃.	
	β gregàrius DC.	(ye.)	*villous.*	——	....	——	H. ♃.	
33	adscéndens. DC.	(ye.)	ascending.	Portugal.	1826.	5. 6.	H. ♃.	
34	millefoliàtus. DC.	(ye.)	numerous-ld.	Sicily.	1824.	4. 7.	H. ♃.	Flor. græc t. 521.
35	fumariæfòlius. DC.	(ye.)	Fumitory-leav'd.	....		——	H. ♃.	Desf. ic. pict. h. par. t. 74.
36	oxyspérmus. DC.	(ye.)	awned-seeded.	Tauria.	1822.	5. 6.	H. ♃.	
37	pedàtus. DC.	(ye.)	pedate.	Hungary.	1806.	5. 6.	H. ♃.	Bot. mag. 2229.
38	ill'yricus. DC.	(ye.)	Illryian.	Europe.	1596.	——	H. ♃.	Jacq. aust. t. 222.
39	monspelìacus. DC.	(ye.)	Montpelier.	Montpelier.	1816.	——	H. ♃.	Swt. br. fl. gar. 94.
	α angustilòbus. DC.	(y.)	*narrow-lobed.*	——	....	——	H. ♃.	Besl. eyst. 1. t. 13. f. 1.
	β cuneàtus. DC.	(ye.)	*wedge-lobed.*	——	....	——	H. ♃.	DC. ic. gall. rar. t. 50.
	γ rotundifòlius. DC.	(y.)	*round-leaved.*	——	....	——	H. ♃.	
40	asiàticus. DC.	(va.)	garden.	Levant.	1596.	——	H. ♃.	Dard. trait. ren. 1. t. 6.
	A vulgàris. DC.	(va.)	*common.*	——	——	——	H. ♃.	Mill. ic. t. 216.
	α pictus. S.F.G.	(r.w.)	*Quillafila.*	....		——	H. ♃.	Swt. flor. guid. 61.
	β incomparábilis.	(r.w.)	*Rose incomparable*	....		——	H. ♃.	—— 54.
	γ Victoriànus.	(bh.)	*Princess Alexandrina Victoria.*	....		——	H. ♃.	—— 12.
	δ míxtus.	(r.y.)	*Melange des beautes*	......		——	H. ♃.	—— 28.
	ε Juliusiànus.	(p.y.)	*Julius.*	....		——	H. ♃.	—— 68.
	ζ Galitzìni.	(p.y.)	*Prince Galitzin.*	....		——	H. ♃.	—— 6.
	η Douglásii.	(p.y.)	*Gadwin Douglas*	....		——	H. ♃.	—— 34.
	θ áureus.	(ye.)	*Beroth.*	....		——	H. ♃.	—— 37.
	ι Búrnsii.	(p.w.)	*Burns the Poet.*	....		——	H. ♃.	—— 18.

	κ àter.	(*d.p.*)	*Vereatre.*	....	....	5. 6.	H. ♃.	Swt. flor. guid. 43.
	λ hæmánthus.	(*cr.*)	*Xanthus.*	....	....	——	H. ♃.	—— 78.
	μ Agrícoli.	(*s.p.*)	*Agricola.*	....	....	——	H. ♃.	—— 73.
	ν Clarenceànus.	(*s.p.*)	*Duke of Clarence.*	....	....	——	H. ♃.	—— 83.
	ξ Carlosiànus.	(*s.p.*)	*Carlos.*	....	....	——	H. ♃.	—— 90.
	B *sanguìneus.* DC	(*va.*)	*Turban.*	Levant.	1596.	——	H. ♃.	Lob. ic. 672. f. 2.
	C *tenuilòbus.* DC.	(*va.*)	*slender-lobed.*	——	——	——	H. ♃.	Debr. flor. nov. t. 31.
	α atérrimus.	(*da.*)	*Viola le vrai noir.*	....	....	——	H. ♃.	Swt. flor. guid 60.
	β Nomiási.	(*pu.*)	*Nomias.*	....	....	——	H. ♃.	—— 24.
	γ variegàtus.	(*w.r.*)	*Oeillet parfait.*	....	....	——	H. ♃.	—— 40.
	δ càrus.	(*w.r.*)	*Cara.*	....	....	——	H. ♃.	—— 86.
41	crèticus. DC.	(*ye.*)	Cretan.	Candia.	1659.	4. 5.	H. ♃.	Clus. hist. 1. p. 239. f. 1.
42	cortusæfòlius. DC.	(*ye.*)	Cortusa-leaved.	Canaries.	1826.	4. 6.	H. ♃.	Deless. ic. sel. 1. t. 36.
43	Thòra. DC.	(*ye.*)	kidney-leaved.	AlpsEurope.	1710.	5. 6.	H. ♃.	Jacq. aust. 5. t. 442.
44	scutàtus. DC.	(*ye.*)	shield-leaved.	Hungary.	1822.	4. 5.	H. ♃.	Pl. rar. hung. 2. t. 187.
45	brevifòlius. DC.	(*ye.*)	short-leaved.	Naples.	1826.	——	H. ♃.	
46	hy'bridus. DC.	(*ye.*)	hybrid.	Austria.	1822.	6. 7.	H. ♃.	Sturm. deutsch. fl. ic.
47	auricomus. DC.	(*ye.*)	Goldilocks.	Britain.	....	3. 5.	H. ♃.	Eng. bot. 624.
48	cassùbicus. DC.	(*ye.*)	Cassubian.	Siberia.	1794.	6. 7.	H. ♃.	Bot. mag. 2267.
49	abortìvus. DC.	(*ye.*)	three-flowered.	N. America.	1713.	5. 8.	H. ♃.	
50	sceleràtus. DC.	(*ye.*)	Celery-leaved.	Britain.	....	——	H.*w.*⊙.	Eng. bot. 681.
51	lappònicus. DC.	(*ye.*)	Lapland.	Lapland.	1827.	4. 6.	H. ♃.	Sm. fl. lapp. ed. 2. t. 3 f. 5.
52	hyperbòreus. DC.	(*ye.*)	northern.	Siberia.	1822.	6. 8.	H. ♃.	Flor. dan. t. 31.
53	frígidus. DC.	(*ye.*)	frigid.	——	1826.	——	H. ♃.	Lax. n. c. ac. pet. v. 18. t. 8.
54	nivàlis. DC.	(*ye.*)	snowy.	Lapland.	1775.	——	H. ♃.	Fl. lap. ed. 1. t. 3. f. 2.
55	Sabìni. S.S.	(*ye.*)	Sabine's.	N. America.	1827.	——	H. ♃.	
56	árcticus. S.S.	(*ye.*)	arctic.	——	——	——	H. ♃.	
57	Púrshii. S.S.	(*ye.*)	Pursh's.	——	——	——	H. ♃.	
58	affìnis. S.S.	(*ye.*)	likened.	——	——	——	H. ♃.	
59	montànus. DC	(*ye.*)	mountain.	Austria.	——	——	H. ♃.	Jacq. aust. t. 325 et 6.
60	Villársii. DC.	(*ye.*)	Villars's.	France.	....	5. 7.	H. ♃.	Crantz. aust. 2. t. 4. f. 2?
61	Gouàni. DC.	(*ye.*)	Gouan's.	Pyrenees.	1818.	——	H. ♃.	Gouan ill. t. 17 f. 1. 2.
62	àcris. DC.	(*ye.*)	upright.	Britain.	....	6. 7.	H. ♃.	Eng. bot. 652.
	β flòre plèno.	(*ye.*)	*double-flowered.*	——	....	——	H. ♃.	Bot. mag. 215.
	γ sylváticus. DC.	(*ye.*)	*wood*	——	....	——	H. ♃.	
	δ multífidus. DC.	(*ye.*)	*multifid.*	——	....	——	H. ♃.	
63	Brùtius. DC.	(*ye.*)	Neapolitan.	Naples.	1828.	——	H. ♃.	Ten. fl. nap. 1. t. 50.
64	Stevèni. DC.	(*ye.*)	Steven's.	Poland.	1825.	——	H. ♃.	
65	caucàsicus. DC.	(*ye.*)	Caucasian.	Caucasus.	——	6. 8.	H. ♃.	
66	rùfulus. DC.	(*ye.*)	rufous-haired.	Portugal.	——	——	H. ♃.	
67	polyánthemos. DC.	(*ye.*)	many-flowered.	N. Europe.	1596.	5. 6.	H. ♃.	Cran. aus. 2. t. 4. f. 1. s.
68	nemoròsus. DC.	(*ye.*)	wood.	Switzerland.	1818.	5. 7.	H. ♃.	
69	Breyniánus. S.S.	(*ye.*)	Breynius's.	——	——	——	H. ♃.	
	áureus. Schl. *villòsus.* St. Am. fl. ag. 227. bon. t. 5.							
70	lanuginòsus. DC.	(*ye.*)	woolly-leaved.	S. Europe.	1683.	6. 7.	H. ♃.	Flor. dan. t. 397.
71	obtusifòlius. DC.	(*ye.*)	blunt-leaved.	Spain.	1825.	——	H. ♃.	
72	tuberòsus. DC.	(*ye.*)	tuberous.	Pyrenees.	1824.	5. 7.	H. ♃.	
73	napellifòlius. DC.	(*ye.*)	Napellus-leav'd.	Greece.	1825.	——	H. ♃.	
74	platyspérmus. DC.	(*ye.*)	flat-seeded.	Siberia.	1827.	——	H. ♃.	
75	disséctus. DC.	(*ye.*)	cut-leaved.	——	1823.	——	H. ♃.	
76	rèpens. DC.	(*ye.*)	creeping.	Britain.	....	5. 9.	H. ♃.	Eng. bot. 516.
	β flòre-plèno.	(*ye.*)	*double-flowered.*	——	....	——	H. ♃.	Tab. ic. 53. f. 1.
	γ glábratus. DC.	(*ye.*)	*shining-leaved.*	——	....	——	H.*w.*♃.	
77	lappàceus. DC.	(*ye.*)	New Holland.	N. S. W.	1822.	——	G. ♃.	
78	plebeius. DC.	(*ye.*)	plebeian.	——	1823.	5. 7.	G. ♃.	
79	hìrtus. DC.	(*ye.*)	hairy.	N. Zealand.	——	——	H. ♃.	
80	pensylvànicus. DC.	(*ye.*)	Pensylvanian.	N. America.	1785.	6. 7.	H. ♃.	Jacq. ic. 1. t. 105.
	canadénsis. Jac.							
81	fasciculàris. DC.	(*ye.*)	fascicled.	——	1823.	4. 5	H. ♃.	
82	marylándicus. DC.	(*ye.*)	Maryland.	——	1811.	5. 7	H. ♃.	
83	caroliniànus. DC.	(*ye.*)	Carolina.	——	1826.	5. 7.	H. ♃.	
84	tomentòsus. DC.	(*ye.*)	tomentose.	——	1820.	——	H. ♃.	
85	bulbòsus. DC.	(*ye.*)	bulbous.	Britain.	....	4. 6	H. ⊙	Eng. bot. 515.
	β flòre-plèno.	(*ye.*)	*double-flowered.*	——	....	——	H. ♃.	Lob. ic. 666. f. 2.
86	bracteàtus. Sch.	(*ye.*)	bracted.	Switzerland	1817.	——	H. ♃.	
87	Philonòtis. DC.	(*ye.*)	pale hairy.	Britain.	....	6. 10.	H. ⊙.	Curt. lond. 2. t. 40.
	α hirsùtus. E.B.	(*ye.*)	*hairy upright.*	——	....	——	H. ⊙.	Eng. bot. t. 1504.
	β intermèdius. DC.	(*ye.*)	*intermediate.*	France.	....	——	H. ⊙.	
	γ párvulus. DC.	(*ye.*)	*little upright.*	England.	....	——	H. ⊙.	Col. ecphr. t. 316.
88	tuberculàtus. DC.	(*ye.*)	tubercled.	....	1821.	5. 6.	H. ♃.	

89 Hornemánni. DC.	(*ye.*)	Hornemann's.	N. America.	1820.	5. 7.	H. ♃.	
tuberòsus. Horn. non. Lapeyr.							
90 arvénsis. DC.	(*ye.*)	corn.	Britain.	....	——	H. ⊙.	Eng. bot. 135.
91 muricàtus. DC.	(*ye.*)	prickly seeded.	S. Europe.	1683.	7. 8.	H. ⊙.	Lam. ill. t. 498.
92 echinàtus. V.	(*ye.*)	American.	N. America.	....	6. 7.	H. ⊙.	Vent. cels. t. 73.
93 parviflòrus. DC.	(*ye.*)	small-flowered.	England.	....	5. 6.	H. ⊙.	Eng. bot. 120.
94 trilòbus. DC.	(*ye.*)	three-lobed.	S. Europe	1824.	5. 7.	H. ⊙.	Desf. atl. 1. t. 113.
95 sessiliflòrus. DC.	(*ye.*)	sessile-flowered.	N. S. W.	1822.	——	H. ⊙.	
96 laciniàtus. DC.	(*ye.*)	jagged.	Transylvania.	1826.	6. 7.	H. ♃.	
97 crassicàulis. H.	(*ye.*)	thick-stemmed.	....	——	——	H. ♃.	
FIC'ARIA. DC.		PILEWORT.	Polyandria Polygynia. L.				
ranunculòides. DC.	(*ye.*)	vernal.	Britain.	....	3. 5.	H. ♃.	Eng. bot. 584.
vérna. P.S. *Ranúnculus Ficària.* W.							
β flòre-plèno.	(*ye.*)	*double-flowered.*	——	....	——	H. ♃.	

Tribus IV. *HELLEBOREÆ.*

CA'LTHA. DC.		MARSH-MARYGOLD.	Polyandria Polygynia L.				
1 palústris. DC.	(*ye.*)	common.	Britain.	....	3. 5.	H. *w.* ♃.	Eng. bot. 506.
β flòre-plèno.	(*ye.*)	*double-flowered.*	——	....	——	H. *w.* ♃.	Tab. ic. 751. f. 1.
2 mìnor. Mill.	(*ye.*)	small.	——	....	——	H. *w.* ♃.	—— 750. f. 2.
3 radìcans. DC.	(*ye.*)	creeping.	Scotland.	....	——	H. *w.* ♃.	Linn. trans. 8. t. 17.
4 integérrima. DC.	(*ye.*)	entire-leaved.	N. America.	1827.	5. 7.	H. *w.* ♃.	
5 parnassifòlia. DC.	(*ye.*)	Parnassia-leav[d].	——	1825.	——	H. *w.* ♃.	
ficariòides. Ph.							
6 árctica. S.S.	(*ye.*)	northern.	——	1827.	——	H. *w.* ♃.	
7 flabellifòlia. DC.	(*ye.*)	fan-leaved.	....	1818.	7. 8.	H. *w.* ♃.	Pur. fl. am. sept. 2. t. 17.
8 nátans. DC.	(*wh.*)	floating.	Siberia.	1823.	4. 5.	H. *w.* ♃.	Gmel. sib. 4. t. 82.
TRO'LLIUS. DC.		GLOBE-FLOWER.	Polyandria Polygynia. L.				
1 europæ'us. DC.	(*ye.*)	European.	Britain.	....	5. 6.	H. ♃.	Eng. bot. t. 28.
α altíssimus. DC.	(*ye.*)	*tall.*	——	....	——	H. ♃.	
β hùmilis. DC.	(*ye.*)	*dwarf.*	——	....	——	H. ♃.	
2 napellifòlius. DC.	(*ye.*)	intermediate.	....	....	——	H. ♃.	
3 caucàsicus. DC.	(*ye.*)	Caucasean.	Caucasus.	....	——	H. ♃.	
4 pátulus. DC.	(*ye.*)	spreading.	Siberia.	....	5. 7.	H. ♃.	Deless. ic. sel. 1. t. 44.
5 Ledeboùrii. S.S.	(*ye.*)	Ledebour's.	——	1827.	——	H. ♃.	Reich. ic. t. 272.
6 asiàticus. DC.	(*ye.*)	Asiatic.	——	1759.	——	H. ♃.	Bot. mag. 235.
7 americànus. DC.	(*ye.*)	American.	N. America.	1805.	5. 7.	H. ♃.	—— 1988.
láxus. Ph. *Gaissènia vérna.* Raf.							
ERA'NTHIS. DC.		WINTER-ACONITE.	Polyandria Polygynia. L.				
1 hyemàlis. DC.	(*ye.*)	common.	Europe.	1596.	1. 3.	H. ♃.	Bot. mag. 3.
2 sibírica. DC.	(*ye.*)	Siberian.	Siberia.	1824.	2. 4.	H. ♃.	
HELLE'BORUS. DC.		HELLEBORE.	Polyandria Polygynia. L.				
1 nìger. DC.	(*wh.*)	Christmas-Rose.	Europe.	1596.	1. 3.	H. ♃.	
α latifòlius.	(*wh.*)	*broad-leaved.*	——	——	——	H. ♃.	Bot. mag. 8.
β angustifòlius.	(*wh.*)	*narrow-leaved.*	——	——	——	H. ♃.	
2 purpuráscens.	(*pu.*)	purplish.	Hungary.	1817.	3. 4.	H. ♃.	Pl. rar. hung. 2. t. 101.
3 Boccòni. T.P.	(*gr.*)	Boccone's.	Sicily.	1826.	——	H. ♃.	Bocc. mus. 2. t. 11. f. 2.
4 odòrus. DC.	(*gr.*)	sweet-scented.	Hungary.	——	——	H. ♃.	
5 víridis. DC.	(*gr.*)	green-flowered.	Britain.	....	4. 6.	H. ♃.	Eng. bot. 200.
6 atrorùbens. DC.	(*pu.*)	dark-red.	Hungary.	1822.	1. 3.	H. ♃.	Pl. rar. hung. 3. t. 271.
7 dumetórum. DC.	(*gr.*)	bush.	——	1817.	3. 4.	H. ♃.	Swt. br. fl. gar. 109.
8 fœ'tidus. DC.	(*gr.*)	Bear's-foot.	England.	....	2. 4.	H. ♃.	Eng. bot. 613.
9 lívidus. DC.	(*br.*)	three-leaved.	Corsica.	1710.	1 5.	H. ♃.	Bot. mag. 72.
β integrilòbus. DC.	(*br.*)	*entire-lobed.*	——	——	——	H. ♃.	
CO'PTIS. DC.		CO'PTIS.	Polyandria Polygynia. L.				
trifòlia. DC.	(*wh.*)	three-leaved.	N. America.	1782.	6. 7.	H. ♃.	Lodd. bot. cab. t. 173.
ISOP'YRUM. DC.		ISOP'YRUM.	Polyandria Polygynia. L.				
1 thalictròides. DC.	(*wh.*)	Meadow-Rue[ld].	Italy.	1759.	3. 4.	H. ♃.	Jac. aust. 2. t. 105.
2 fumariòides. DC.	(*wh.*)	Fumitory-l[d].	Siberia.	1741.	6. 7.	H. ⊙.	Schkuhr hand. 2. t. 153.
3 grandiflòrum. F.	(*wh.*)	large-flowered.	Altay.	1804.	——	H. ♃.	

GARIDE'LLA. DC.		GARIDE'LLA. Decandria Trigynia. L.					
Nigellástrum. DC.	(*wh.*)	Nigella-leaved.	S. Europe.	1736.	6. 7.	H. ⊙	Bot. mag. 1266.
NIGE'LLA. DC.		FENNEL-FLOWER. Polyandria Pentagynia. L					
1 corniculàta. DC.	(*st.*)	horned.	....	1822.	6. 9.	H. ⊙.	
2 orientàlis. DC.	(*ye.*)	yellow.	Syria.	1699.	——	H. ⊙.	Bot. mag. 1264.
3 hispánica. DC.	(*bl.*)	Spanish.	Spain.	1629.	——	H. ⊙.	———— 1265.
4 arvénsis. DC.	(*wh.*)	field.	Europe.	1683	——	H. ⊙.	Bull. herb. t. 126.
5 satìva. DC.	(*wh.*)	small.	Egypt.	1548.	——	H. ⊙.	Flor. græc. t. 511.
β *crètica.* DC.	(*wh.*)	*Cretan.*	Crete.	——	——	H. ⊙.	Math. com. 581. f. IV.
γ *citrìna.* DC.	(*wh.*)	*yellow-seeded.*	....	——	——	H. ⊙.	Lob. ic. 741. f. 1.
6 damascèna. DC.	(*wh.*)	common.	S. Europe.	1570.	——	H. ⊙.	Bot. mag. 22.
β *flòre-plèno.*	(*wh.*)	*double-flowered.*	————	——	——	H. ⊙.	Clus. hist. 2. p. 208. f. 1.
7 coarctàta. DC.	(*wh.*)	dwarf.	....	1793.	——	H. ⊙.	
AQUIL'EGIA. DC.		COLUMBINE. Polyandria Pentagynia. L.					
1 vulgàris. DC.	(*va.*)	common.	Britain.	....	5. 7.	H. ♃.	Eng. bot. 297.
β *corniculàta.* DC.	(*va.*)	*downward-spur'd*		....	——	H. ♃	Barr. ic. 614. 615. 616. &c.
γ *invérsa.* DC.	(*va.*)	*upward-spurred.*		....	——	H. ♃.	———— 613.
δ *stellàta.* DC.	(*va.*)	*starry.*	....	....	——	H. ♃.	———— 619. 620. &c.
ε *degéner.* DC.	(*gr.*)	*green-flowered.*	....	....	——	H. ♃.	———— 608.
2 viscòsa. DC.	(*bl.*)	clammy.	S. Europe.	1752.	5. 6.	H. ♃.	Gouan. ill. t. 19.
3 sibírica. DC.	(*bl.*)	Siberian.	Siberia.	....	——	H. ♃.	Deles. ic. sel. 1. t. 47.
4 glandulòsa. F.	(*bl.*)	glandulous.	Altay.	1818.	——	H. ♃.	
α *díscolor.* DC.	(*b.w.*)	*two-coloured.*	————	——	——	H. ♃.	
β *cóncolor.* DC.	(*bl.*)	*one-coloured.*	————	——	——	H. ♃.	
5 alpìna. DC.	(*bl.*)	Alpine.	Switzerland.	1731.	——	H. ♃.	Swt. br. fl. gar. 3. t. 218.
6 pyrenaìca. DC.	(*bl.*)	Pyrenean.	Pyrenees.	1821.	——	H. ♃.	Trev. delph. 23. t. 2.
7 canadénsis. DC.	(*re.*)	Canadian.	N. America.	1640.	4. 5.	H. ♃.	Bot. mag. 246.
8 formòsa. F.	(*re.*)	handsome.	Kamtschatka.	1824.	——	H ♃.	
9 viridiflòra. DC.	(*gr.*)	green-flowered.	Siberia.	1780.	5. 6.	H. ♃.	Jac. ic. 1. t. 102.
10 dahùrica. DC.	(*pu.*)	Dahurian.	Dahuria.	1821.	——	H. ♃.	Deless. ic. sel. 1. t. 49.
11 Fischeriàna. DC.	(*pu.*)	Fischer's.	————	——	——	H. ♃.	
12 atropurpùrea. DC.	(*pu.*)	dark-purple.	Siberia.	....	5. 6.	H. ♃.	Botan. regist. 922.
13 hy'brida. DC.	(*p.y.*)	hybrid.	....	....	——	H. ♃.	Bot. mag. 1221.
DELPHI'NIUM. DC.		LARKSPUR. Polyandria Trigynia. L.					
1 Ajácis. DC.	(*va.*)	Rocket.	Tauria.	1573.	6. 9.	H. ⊙.	Spreng. geh. 1. t. 24. f. 1. 5
β *flòre-plèno.*	(*va.*)	*double-flowered.*	————	——	——	H. ⊙.	Clus. hist. 2. f. 2.
2 Consólida. DC.	(*va.*)	field.	England.	....	——	H. ⊙.	Eng. bot. 1839.
β *flòre-plèno.*	(*va.*)	*double-flowered.*	————	....	——	H. ⊙.	Besl. eyst. 2. t. 14. f. 1.
3 pubéscens. DC.	(*bl.*)	pubescent	S. France.	....	——	H. ⊙.	Flor. græc. t. 504.
4 Aconìti. DC.	(*pu.*)	Aconite-like.	Levant.	1801.	——	H. ⊙.	Vahl symb. 1. 40. t. 13.
5 ambíguum. DC.	(*bl.*)	doubtful.	Barbary.	1759.	7. 9.	H. ⊙.	
6 júnceum. DC.	(*bl.*)	Rush-like.	Italy.	1629.	——	H. ⊙.	Flor. græc. t. 506.
peregrìnum. Lin. *non* Lam.							
7 grácile. DC.	(*re.*)	slender.	Spain.	1826.	——	H. ⊙.	
8 cardiopètalum. DC.	(*bl.*)	heart-petaled.	Pyrenees.	1822.	——	H. ⊙.	Math. com. 557. ic.
peregrìnum. Lam. *non* Lin.							
9 vírgatum. DC.	(*bl.*)	twiggy.	Syria.	1824.	6. 9.	H. ⊙.	Deless. ic. sel. 1. t. 55.
10 grandiflòrum. DC.	(*bl.*)	great-flowered.	Siberia.	1741.	6. 9.	H. ♃.	Bot. mag. 1686.
β *chinénse.* F.	(*bl.*)	*Chinese.*	China.	1816.	——	H. ♃.	Bot. reg. 472.
γ *álbum.*	(*wh.*)	*white-flowered.*	————	——	——	H. ♃.	
δ *flòre-plèno.*	(*bl.*)	*double-flowered.*	————	——	——	H. ♃.	
11 cheilánthum. DC.	(*bl.*)	hairy-leaved.	Dahuria.	1818.	——	H. ♃.	Bot. reg. 473.
12 viréscens. N.	(*wh.*)	greenish.	N. America.	1827.	——	H. ♃.	
13 puníceum. DC.	(*pu.*)	dark-purple.	Tartary.	1785.	7. 9.	H. ♃.	
14 albiflòrum. DC.	(*wh.*)	white-flowered.	Armenia.	1823.	——	H. ♃.	Deless. ic. sel. 1. t. 58.
15 ochroleùcum. DC.	(*st.*)	sulphur-colour^{d}.	Iberia.	1817.	——	H. ♃.	
16 hy'bridum. DC.	(*bl.*)	hairy.	Siberia.	1794.	6. 9.	H. ♃.	
17 físsum. P. S.	(*bl.*)	cloven.	Hungary.	1816.	——	H. ♃.	Pl. rar. hung. 1. t. 81.
18 velutìnum. DC.	(*bl.*)	velvetty.	Italy.	1824.	——	H. ♃.	
19 pentágynum. DC.	(*bl.*)	five-styled.	Portugal.	1820.	——	H. ♃.	Pic. hort. par. t. 30.
20 Menziesii. DC.	(*bl.*)	Menzies'.	N. America.	1826.	6. 8.	H. ♃.	Bot. reg. 1192.
21 élegans. DC.	(*bl.*)	elegant.	————	1741.	——	H. ♃.	
β *multipléx.*	(*bl.*)	*common-double.*	————	——	——	H. ♃.	
22 tricórne. DC.	(*bl.*)	three-horned.	————	1806.	4. 6.	H. ♃.	Lodd. bot. cab. 306.
23 amœ'num. DC	(*bl.*)	pretty.	Siberia.	....	——	H. ♃	Gmel. sib. 4. t. 77.
24 azùreum. DC.	(*bl.*)	azure.	Carolina.	1805	5. 7.	H. ♃.	Deless. ic. sel. 1. t. 60.
25 exaltàtum. DC.	(*bl.*)	tall.	N. America.	1758.	7. 8.	H. ♃.	Mill. ic. t. 250. f. 2?
26 urceolàtum. DC.	(*bl.*)	hollow-leaved.	————	1806.	6. 7.	H. ♃.	Jac. ic. 1. t. 101.

27	revolùtum. DC.	(*bl.*)	revolute.	....	.....	7. 9.	H. ♃.	
28	mesoleùcum. DC.	(*bl.*)	white-lipped.	....	1824.	——	H. ♃.	Swt. br. fl. gar. ic.
29	palmatífidum. DC.	(*bl.*)	palmate.	Siberia.	.....	6. 9.	H. ♃.	
	α híspidum. DC.	(*bl.*)	*hispid.*	——	....	——	H. ♃.	Gmel. sib. 4. t. 79?
	β glábellum. DC.	(*bl.*)	*smooth.*	——	....	——	H. ♃.	—— 4. t. 75?
30	intermèdium. DC.	(*bl.*)	intermediate.	Europe.	1710	7. 8.	H. ♃.	
	α pilosíssimum. DC.	(*bl.*)	*villous.*	——	....	——	H. ♃.	Gmel. sib. 4. t. 80?
	β alpìnum. DC.	(*bl.*)	*Alpine.*	——	1819.	——	H. ♃.	Pl. rar. hung. 3. t. 246.
	γ leptostáchyum. DC.	(*bl.*)	*slender-spiked.*	——	....	——	H. ♃.	
	δ ranunculifòlium. DC.		*Ranunculus-lᴊ.*	——	....	——	H. ♃.	
	ε láxum. DC.	(*bl.*)	*loose-flowered.*	——	....	——	H. ♃.	Mill. ic. t. 119?
31	cuneàtum. DC.	(*bl.*)	wedge-leaved.	Siberia.	1815.	——	H. ♃.	Bot. reg. 327.
32	villòsum. DC.	(*bl.*)	villous.	....	1819.	——	H. ♃.	
33	dyctiocárpum. DC.	(*bl.*)	netted-capsuled.	Siberia.	——	——	H. ♃.	
34	laxiflòrum. DC.	(*bl.*)	loose-flowered.	——	——	6. 8.	H. ♃.	
35	montànum. DC.	(*bl.*)	mountain.	Pyrenees.	——	——	H. ♃.	
	β bracteòsum. DC.	(*bl.*)	*large-bracted.*	Switzerland.	——	——	H. ♃.	
36	dasycárpum. DC.	(*bl.*)	thick-capsuled.	Caucasus.	1824.	——	H. ♃.	
37	speciòsum. DC.	(*bl.*)	large-flowered.	——	1817.	8. 10.	H. ♃.	Deless. ic. sel. 1. t. 62.
38	spùrium. F.	(*bl.*)	spurious.	Siberia.	1817.	6. 9.	H. ♃.	
39	ucrànicum. Bes.	(*bl.*)	Ucranian.	——	1818.	——	H. ♃.	
40	pseudoperegrìnum. F.	(*bl.*)	Fischer's.	——	——	——	H. ♃.	
41	crassicàule. Gm.	(*bl.*)	thick-stemmed.	....	1822.	——	H. ♃.	
42	pállidum. F.	(*pa.*)	pale-flowered.	Siberia.	1818.	——	H. ♃.	
43	flexuòsum. DC.	(*bl.*)	flexuose.	Caucasus.	1817.	7. 9.	H. ♃.	Trev. delph. t. 1. et 2.
44	díscolor. F.	(*v.*)	two-coloured.	Siberia.	1819.	6. 8.	H. ♃.	
45	tríste. DC.	(*da.*)	dark-flowered.	——	1822.	——	H. ♃.	
46	Requiénii. DC.	(*pa.*)	Requien's.	S. Europe.	1819.	6. 7.	F. ♂.	Deles. ic. sel. 1 t 63.
47	píctum. DC.	(*pa.*)	painted.	——	1816.	——	H. ♂.	Swt. br. fl. gar. 2. 123.
48	Stasphysàgria. DC.	(*bl.*)	Stavesacre.	——	1596.	5. 6.	F. ♂.	Flor. græc. t. 508.
	ACON'ITUM. DC.		MONKSHOOD.	Polyandria Trigynia.				
1	A'nthora. L.	(*st.*)	wholesome.	Pyrenees.	1596.	6. 8.	H. ♃.	Rch. ac. p. 61. t. 1. f. A.
	β grandiflòrum. R.U.	(*st.*)	*large-flower'd.*	——	....	——	H. ♃.	—— p. 63. t. 1. f. B.
	γ glabriflòrum. R.U.	(*st.*)	smooth-flowd.	——	....	——	H. ♃.	
2	eulòphum. R.U.	(*st.*)	eatable.	Switzerland.	1824.	——	H. ♃.	Rch. ac. p. 69. t. 5.
	anthoróideum. DC. non R.U.							
3	nemoròsum. R.U.	(*ye.*)	wood.	Caucasus.	1822.	——	H. ♃.	—— p. 71. t. 6.
4	Dec ndòllii. R.U.	(*ye.*)	Decandolle's.	Pyrenees.	——	——	H. ♃.	—— p. 67. t. 3.
5	Jacquìni. R.U.	(*ye.*)	Jacquin's.	Austria.	....	——	H. ♃.	—— p. 65. t. 2.
	A'nthora. Jacq. aust. t. 382.							
6	Pallàsii. R.U.	(*st.*)	Pallas's.	Siberia.	1824.	——	H. ♃.	—— p. 72. t. 6. f. A. B.
7	inclinàtum.	(*st.*)	inclined.	——	....	——	H. ♃.	Ser. m. helv. 1. t. 15. f. 2.
	anthoroìdeum. Rch. ac. 68. t. 4. non DC.							
8	Kœlleànum. R.U.	(*bl.*)	Koelle's.	Germany.	....	5. 7.	H. ♃.	Rch. ac. t. 11. f. 1.
9	ta'uricum. R.U.	(*bl.*)	Taurian.	Tauria.	1752.	6. 8.	H. ♃.	—— t. 12. f. 2—3.
10	fírmum. R.U.	(*bl.*)	firm.	Germany.	1822.	——	H. ♃.	—— t. 14. f. 2.
11	acùtum. R.U.	(*bl.*)	acute.	——	——	——	H. ♃.	—— t. 14. f. 1.
12	læ'tum. R.U.	(*bl.*)	great-flowered.	Sweden.	1823.	——	H. ♃.	—— t. 13. f. 2.
13	Clùsii. R.U.	(*bl.*)	Clusius's.	Austria.	....	6. 7.	H. ♃.	Mor. his. 3. s. 12. t. 3. f. 17.
14	amœ'num. R.U.	(*bl.*)	delightful.	Sweden.	1824.	6. 8.	H. ♃.	Rch. ac. 93. t. 14. f. 1.
15	Hóppii. R.U.	(*bl.*)	Hoppe's.	Carinthia.	——	——	H. ♃.	
16	oligocárpum. R.U.	(*bl.*)	two-podded.	Styria.	1823.	——	H. ♃.	
17	callibòtrys. R.U.	(*bl.*)	long-spiked.	Moravia.	1824.	6. 7.	H. ♃.	Rch. ac. 98. t. 16. f. 1.
18	tenuifòlium. R.U.	(*bl.*)	slender-leaved.	Styria.	——	——	H. ♃.	
19	angustifòlium. R.U.	(*bl.*)	narrow-leaved.	Siberia.	——	——	H. ♃.	Rch. ac. 95. t. 15. f. 2.
20	eustàchyum. R.U.	(*bl.*)	loose-spiked.	M. Baldo.	——	——	H. ♃.	—— t. 15. f. 3.
21	Napéllus. R.U.	(*bl.*)	common.	Europe.	....	5. 7.	H. ♃.	Lob. ic. 679. f. 1.
22	compáctum. R.U.	(*bl.*)	compact.	Switzerland.	1823.	——	H. ♃.	
	β rubéllum. R.U.	(*re.*)	*red-flowered.*	——	——	——	H. ♃.	
23	Hallèri. R.U.	(*bl.*)	Haller's.	——	1824.	6. 7.	H. ♃.	Chabr. sci. p. 531. f. 2.
	β hùmile. R.U.	(*bl.*)	*dwarf.*	——	——	——	H. ♃.	
	γ ramòsum. R.U.	(*bl.*)	*branching.*	——	1822.	——	H. ♃.	
	δ bícolor. R.U.	(*w.b.*)	*two-coloured.*	——	——	——	H. ♃.	
24	virgàtum. R.U.	(*bl.*)	slender-branch'd.	——	1824.	——	H. ♃.	
25	venústum. R.U.	(*bl.*)	beautiful.	——	1823.	6. 8.	H. ♃.	
26	laxìflòrum. R.U.	(*bl.*)	loose-flowered.	——	1824.	——	H. ♃.	
27	Fúnckii. R.U.	(*bl.*)	Funck's.	Germany	——	——	H. ♃.	
28	Bráunii. R.U.	(*bl.*)	Braune's	——	——	6. 7.	H. ♃.	
29	trichocárpum. R.U.	(*bl.*)	three-podded.	——	——	——	H. ♃.	

30	pubéscens. R.U.	(*bl.*)	pubescent.	Switzerland.	....	5. 7.	H. ♃.	
31	Mielichóferi. R.U.	(*bl.*)	Mielichofer's.	Germany.	1823.	6. 8.	H. ♃.	
32	plicàtum. R.U.	(*bl.*)	plaited.	Sweden.	1824.	——	H. ♃.	
33	Kœhlèri. R.U.	(*bl.*)	Koehler's	——	1823.	——	H. ♃.	
34	elàtum. R.U.	(*bl.*)	tall.	Germany.	....	——	H. ♃.	
35	álbidum. R.U.	(*wh.*)	whitish.	....	1824.	——	H. ♃.	
36	neubergénse. R.U.	(*bl.*)	Styrian.	Styria.	....	——	H. ♃.	Clus. hist. 2. p. 96. f. 1.
	Napéllus. Jacq. aust. 4. t. 381.							
37	Meyèri. R.U.	(*bl.*)	Meyer's.	Germany.	1824.	——	H. ♃.	
38	Bernhardiànum.	(*bl.*)	Bernhardi's.	....	1823.	——	H. ♃.	
39	stríctum. R.U.	(*bl.*)	straight.	Siberia.	....	——	H. ♃.	Rch. ac. t. 17. f. 1.
	spicàtum. Donn. h. cant. ed. 10. p. 209.							
40	Sprengèlii. R.U.	(*bl.*)	Sprengel's.	....	1823.	6. 8.	H. ♃.	
41	Willdenòwii. R.U.	(*bl.*)	Willdenow's.	Carniola.	——	——	H. ♃.	
42	éminens. R.U.	(*bl.*)	eminent.	....	1822.	——	H. ♃.	
43	Schleichèri. R.U.	(*bl.*)	Schleicher's.	Switzerland.	....	5. 6.	H. ♃.	
	β *comòsum.* R.U.	(*bl.*)	*tufted.*	——	....	——	H. ♃.	Mor. his. 3.s.12.t.3.f.19.
44	microstáchyum. R.U.	(*b.*)	small-spiked.	Hungary.	1822.	6. 8.	H. ♃.	
45	formòsum. R.U.	(*bl.*)	handsome.	Germany.	1824.	——	H. ♃.	Rch. ac. t. 18. f. 2.
46	rígidum. R.U.	(*bl.*)	rigid.	Sweden.	——	——	H. ♃.	
47	commutàtum. R.U.	(*bl.*)	changed.	Switzerland.	——	——	H. ♃.	Rch. ac. t. 18. f. 3.
48	láxum. R.A.	(*bl.*)	loose.	Germany.	1823.	6. 8.	H. ♃.	Rch. ac. 97. t. 15. f. 4.
49	autumnàle. R.A.	(*bl.*)	autumnal.	——	——	——	H. ♃.	—— t. 17. f. 2.
50	amplifiórum. R.U.	(*bl.*)	large-flowered.	Austria.	1823.	——	H. ♃.	
51	máximum. DC.	(*bl.*)	large.	Kamtschatka.	——	7. 8.	H. ♃.	Ser. mus. 1. t. 15. f. 31. 32.
52	biflòrum. DC.	(*va.*)	two-flowered.	Altay.	1826.	6. 8.	H. ♂.	—— 1. t. 15. f. 45. 46.
	α *lutèo-cærùleum*	(*y.b.*)	*blue and yellow.*	——	——	——	H. ♂.	
	β *cærùleum.*	(*bl.*)	*blue-flowered.*	——	——	——	H. ♂.	Rch. ac. 74. t. 7. f. 1. 2.
53	fèrox. DC.	(*pu.*)	poison-rooted.	Nepaul.	1820.	——	H. ♃.	Ser. mus. 1. t. 15. f. 43. 44.
	viròsum. D.P.							
54	prodùctum. R.U.	(*bl.*)	long-lipped.	Siberia.	1824.	7. 8.	H. ♃.	Rch. ac. 75. t. 7. f 3.
55	semigaleàtum. R.U.	(*bl.*)	half-helmeted.	Kamtschatka.	——	——	H. ♃.	—— 77. t. 8.
56	uncinàtum. DC.	(*pu.*)	American.	N. America.	1768.	——	H. ♃.	
	β *Michauxiànum.*	(*pu.*)	Michaux's	——	——	——	H. ♃.	Bot. mag. 1119.
57	villòsum. R.U.	(*bl.*)	villous.	Siberia.	1825.	——	H. ♃.	
58	fláccidum. R.U.	(*bl.*)	flaccid.	——	....	——	H. ♃	Deles. ic. sel. 1. t. 65.
	ciliàre β *polytrìchum.* DC.							
59	volùbile. R.U.	(*bl.*)	twining.	——	1799.	——	H. ♃.	
60	tortuòsum. DC.	(*bl.*)	twisting.	....	1812.	——	H. ♃.	Ser. mus. 1. t. 15. f. 28. 29.
61	erióstemum. DC.	(*bl.*)	woolly-stamen'd.	Siberia.	1824.	6. 8.	H. ♃.	
62	recognìtum. R.U.	(*bl.*)	recognised.	....	——	——	H. ♃.	
63	ambíguum. R.U.	(*bl.*)	ambiguous.	Siberia.	1826.	——	H. ♃.	
64	tóxicum. R.U.	(*bl.*)	poisonous.	Germany.	——	——	H. ♃.	
65	refléxum. R.U.	(*bl.*)	reflexed.	Carniola.	1826.	6. 8.	H. ♃.	
66	cérnuum. R.U.	(*bl.*)	drooping-flow'd.	Germany.	1824.	6. 7.	H. ♃.	Chabr. sci. 531. f. 6.
67	acuminàtum. R.U.	(*bl.*)	acuminate.	M. Baldo.	1824.	——	H. ♃.	
68	parviflòrum. R.U.	(*bl.*)	small-flowered.	Switzerland.	——	——	H. ♃.	
69	hebégynum. DC.	(*bl.*)	downy-podded.	——	——	——	H. ♃.	
70	paniculàtum. R.U.	(*bl.*)	panicled.	——	....	5. 7.	H. ♃.	Chabr. sci. 531. f. 3.
71	laciniòsum. Sch.	(*bl.*)	jagged.	——	1822.	——	H. ♃.	
72	mólle. R.U.	(*bl.*)	soft.	——	1824.	——	H. ♃.	Ser. m. 1. t. 15. f. 35. 36.
73	gibbòsum. DC.	(*bl.*)	gibbous.	Caucasus.	1823.	——	H. ♃.	—— 1. t. 15. f. 14. 15.
74	hamàtum. R.U.	(*bl.*)	hooked.	Sweden.	1804.	6. 8.	H. ♃.	
75	palmatìfidum. R.U.	(*bl.*)	palmatifid.	M. Carpathian.	1819.	——	H. ♃.	
76	pyramidàle. R.U.	(*bl.*)	pyramidal.	Bohemia.	1596.	——	H. ♃.	
	α *densiflòrum.* R.U.	(*bl.*)	*close-flowered.*	——	——	——	H. ♃.	
	β *elongàtum.* R.U.	(*bl.*)	*elongated.*	——	....	——	H. ♃.	
	γ *bícolor.* R.U.	(*w.b.*)	*two-coloured.*	——	——	——	H. ♃.	
77	Stœrkiànum. R.U.	(*bl.*)	Stœrk's.	S. Europe.	1824.	——	H. ♃.	Stor. lib. ac. 69. ic.
	β *laxiflòrum.* R.U.	(*bl.*)	*loose-flowered.*	....	——	——	H. ♃.	Blackw. ic. 561.
	γ *bícolor.* R.U.	(*w.b.*)	*variegated.*	....	——	——	H. ♃.	
78	versícolor. R.U.	(*va.*)	various-colored.	Switzerland.	——	——	H. ♃.	
	β *bícolor.* R.U.	(*w.b.*)	*white and blue.*	——	....	——	H. ♃.	Lodd. bot. cab. 794.
79	decòrum. R.U.	(*bl.*)	neat.	Pyrenees.	1824.	——	H. ♃.	
80	exaltàtum. R.U.	(*bl.*)	lofty.	....	——	——	H. ♃.	Ser. m. 1. t. 15. f. 37. 38.
81	macránthum. R.U.	(*bl.*)	large-blossom'd.	Germany.	1826.	——	H. ♃.	
82	Cámmarum. R.U.	(*pu.*)	purple.	Europe.	1752.	——	H. ♃.	Jacq. aust. 5. t. 424.
83	illinìtum. R.U.	(*bl.*)	besmeared.	Switzerland.	1822.	——	H. ♃.	
84	Breiteriànum. R.U.	(*bl.*)	Breiter's.	....	1821.	——	H. ♃.	

85	Ottoniànum. R.U.	(*bl.*)	Otto's.	....	1823.	6. 8.	H. ♃.	
86	speciòsum. R.U.	(*bl.*)	specious.	....	1804.	7. 8.	H. ♃.	Ser. m. 1. t. 15. f. 26. 27.
87	bulbíferum. R.U.	(*bl.*)	bulb-bearing.	....	1824.	6. 8.	H. ♃.	
88	leucánthum. R.U.	(*w.*)	white-flowered.	....	....	——	H. ♃.	
89	lasiocárpum. R.U.	(*bl.*)	woolly-podded.	Hungary.	1822.	——	H. ♃.	
90	grácile. R.U.	(*bl.*)	slender.	Germany.	1824.	6. 7.	H. ♃.	
91	rhyncánthum. R.U.		beaked.	Hungary.	1804.	6. 8.	H. ♃.	
	β *bícolor*. R.U.	(*w.b.*)	*two-coloured.*	....	——	——	H. ♃.	
92	nasùtum. R.U.	(*bl.*)	large-nosed.	Caucasus.	1823.	——	H. ♃.	
93	rostràtum. R.U.	(*pu.*)	beaked.	Switzerland.	....	——	H. ♃.	Lodd. bot. cab. 203.
94	variegàtum. DC.	(*p.w.*)	variegated.	S. Europe.	1597.	——	H. ♃.	Clus. hist. 2. p. 98. f. 1.
95	míxtum. R.U.	(*p.w.*)	mixed.	....	....	——	H. ♃.	Rivin. pentap. ic.
96	álbum. R.U.	(*wh.*)	white.	Levant.	1752.	7. 8.	H. ♃.	
97	japónicum. DC.	(*va.*)	Japan.	Japan.	1790.	6. 9.	H. ♃.	Ser. m. 1. t. 15. f. 22. 23.
	α *cárneum.*	(*fl.*)	*flesh-coloured.*	———	——	——	H. ♃.	
	β *cærùleum.*	(*bl.*)	*blue-flowered.*	———	——	——	H. ♃.	
98	boreàle. R.U	(*st.*)	Siberian.	Siberia.	....	6. 8.	H. ♃.	
99	barbàtum. DC.	(*st.*)	bearded.	———	1807.	——	H. ♃.	Swt. br. fl. gar. v. 2. 164.
100	híspidum. DC.	(*st.*)	hispid.	———	1826.	——	H. ♃.	Gmel. sib. 4. t. 81.
	Gmelìni. R.U.							
101	squarròsum. DC.	(*st.*)	squarrose.	———	1823.	——	H. ♃.	
102	pállidum. R.U.	(*st.*)	pale-flowered.	....	1823.	——	H. ♃.	
103	zoóctonum. R.U.	(*st.*)	beast-bane.	Germany.	1824.	——	H. ♃.	
104	melóctonum. R.U.	(*st.*)	Badger-bane.	Piedmont.	1823.	——	H. ♃.	Chabr. sci. p. 531. f. 1.
105	pyrenaìcum. DC.	(*st.*)	Pyrenean.	Pyrenees.	1739.	——	H. ♃.	Cam. epit. 831. ic.
106	galéctonum. R.U.	(*st.*)	Weasel-bane.	Hungary.	1823.	——	H. ♃.	
107	moldávicum. R.U.	(*pu.*)	Moldavian.	Bohemia.	——	——	H. ♃.	
108	ochroleùcum. R.U.	(*st.*)	pale-white.	Caucasus.	1794.	——	H. ♃.	
109	ægòphonum. R.U.	(*st.*)	Goat-poison.	....	1822.	——	H. ♃.	
110	perniciòsum. R.U.	(*st.*)	pernicious.	France.	——	——	H. ♃.	Bois. fl. eur. 2. t. 369.
111	myóctonum. R.U.	(*st.*)	Mouse-bane.	Thuringia.	——	——	H. ♃.	
112	strictíssimum. R.U.	(*st.*)	straight.	Baruth.	....	——	H. ♃.	
113	altíssimum. R.U.	(*st.*)	tallest.	Germany.	....	——	H. ♃.	Tabern. ic. 583. f. 1.
114	réctum. R.U.	(*st.*)	upright.	Italy.	....	——	H. ♃.	Barrel. ic. 599.
115	vulpària. R.U.	(*st.*)	Fox-bane.	Germany.	1824.	——	H. ♃.	
	β *latifòlium*. R.U.	(*st.*)	*broad-leaved.*	Bavaria.	——	——	H. ♃.	
116	tragóctonum. R.U.	(*st.*)	Goat-bane.	Carinthia.	——	——	H. ♃.	
117	lupícida. R.U.	(*st.*)	Wolf-bane.	Switzerland.	1824.	——	H. ♃.	
118	arctóphonum. R.U.	(*st.*)	Bear-bane.	Styria.	1826.	——	H. ♃.	
119	lagóctonum. R.U.	(*st.*)	Hare-bane.	Bavaria.	——	——	H. ♃.	
120	Phthòra. R.U.	(*st.*)	destructive.	Hungary.	——	——	H. ♃.	
121	australe. R.U.	(*pu.*)	Carpathian.	M. Carpathian.	1815.	——	H. ♃.	Bot. mag. 2196.
	septentrionàle β *carpáthicum.* B.M.							
122	rubicùndum. F.	(*pu.*)	rubicund.	Russia.	1822.	——	H. ♃.	
123	cynóctonum. R.U.	(*st.*)	Dog-bane.	Germany.	1824.	——	H. ♃.	Blackw. ic. 563.
124	thely'phonum. R.U.	(*st.*)	Elwert's.	Switzerland.	——	——	H. ♃.	
125	therióphonum. R.U.	(*st.*)	park-bane.	Italy.	....	6. 7.	H. ♃.	Barrel. ic. 600.
126	lycóctonum. R.U.	(*st.*)	great-yellow.	Germany.	1596.	7. 8.	H. ♃.	Clus. hist. 5. p. 94. ic.
127	lupària. R.U.	(*st.*)	Wolf-wort.	———	....	——	H. ♃.	Trag. stirp. 247. ic.
128	septentrionàle. R.U.	(*bl.*)	northern.	N. Europe.	1800.	——	H. ♃.	Flor. dan. t. 123.

Tribus V. *PÆONIACEÆ.* *DC. prodr.* 1. 64.

	CIMICI'FUGA. L.		BUG-WORT. Polyandria Pentagynia. L.					
1	fœ'tida. W.	(*wh.*)	stinking.	Siberia.	1777.	6. 7.	H. ♃.	Linn. amœn. ac. 8. t. 4.
2	podocárpa.	(*wh.*)	American.	Carolina.	1822.	8. 9.	H. ♃.	
	americàna. M. *Act'æa podocárpa.* DC.							
3	palmàta. Ph.	(*wh.*)	palmated.	———	1812.	7. 8.	H. ♃.	Bot. mag. 1630.
4	cordifòlia. Ph.	(*wh.*)	heart-leaved.	———	——	6. 7.	H. ♃.	——— 2069.
	MACR'OTYS. Raf.		SNAKEROOT. Polyandria Monogynia. L.					
	racemòsa.	(*wh.*)	black-rooted.	N. America.	1732.	6. 7.	H. ♃.	Dill. elth. 79. t. 67. f. 78.
	Act'æa racemòsa. W.		*Cimicífuga serpentària.* Ph.					
	ACT'ÆA. Ph.		BANE-BERRIES. Polyandria Monogynia. L.					
1	spicàta. DC.	(*wh.*)	common.	Britain.	....	4 6.	H. ♃.	Eng. bot. 918.
2	rùbra. W.en.	(*wh.*)	red-berried.	N. America.	....	——	H. ♃.	
	brachypètala. β *rùbra.* DC.							

3 álba. Mill.	(*wh.*)	white-berried.	N. America.	....	4. 6.	H. ♃.	Corn. canad. t. 77.
brachypètala. a álba. DC.							
ZANTHORH'IZA. DC.		YELLOW-ROOT. Pentandria Polygynia. L.					
apiifòlia. DC.	(*pu.*)	Parsleyleaved.	N. America.	1766.	2. 4.	H. ♄.	Bot. mag. 1736.
PÆ'ONIA. DC.		PÆONY. Polyandria Digynia. L.					
1 Moután. H.K.		Chinese-tree.	China.	1789.	4. 6.	H. ♄.	
α papaveràcea. L.T.	(*w.*)	*Poppy-flowered.*	———	1806.	——	H. ♄.	Bot. rep. 463.
β Bánksii. L.T.	(*bl.*)	*common.*	———	1789.	——	H. ♄.	——— 448.
γ ròsea. L.T.	(*ro.*)	*rose-coloured.*	———	1794.	——	H. ♄.	——— 373.
2 èdulis. P.L.	(*va.*)	eatable-rooted.	....	....	5. 6.	H. ♃.	
albiflòra. L.T.							
α véstalis. L.T.	(*wh.*)	*virgin.*	Siberia.	1784.	——	H. ♃.	Andr. rep. 64.
β cándida. L.T.	(*wh.*)	*pale-flowered.*	———	....	——	H. ♃.	
γ tatárica. L.T.	(*bh.*)	*Tartarian.*	Tartary.	....	——	H. ♃.	Bot. reg. 42.
δ sibírica. L.T.	(*wh.*)	*Siberian.*	Siberia.	....	——	H. ♃.	
ε rubéscens. L.T.	(*bh.*)	*pale-red.*	....	....	——	H. ♃.	
ζ uniflòra. L.T.	(*wh.*)	*one-flowered.*	....	....	——	H. ♃.	Bot. mag. 1756.
η Whitlèji. L.T.	(*wh.*)	*double-white.*	China.	1808.	6. 7.	H. ♃.	Bot. reg. 630.
θ Hùmei. L.T.	(*re.*)	*double-crimson.*	———	1810.	——	H. ♃.	Bot. mag. 1768.
ι fràgrans. L.T.	(*re.*)	*Rose-scented.*	———	1805.	——	H. ♃.	Bot. reg. 485.
3 anómala. L.T.	(*re.*)	jagged-leaved.	Siberia.	1788.	5. 6.	H. ♃.	Bot. mag. 1754.
laciniata. Pall. ross. 2. t. 85.							
4 hy'brida. DC.	(*re.*)	intermediate.	Caucasus.	1822.	....	H. ♃.	
5 tenuifòlia. L.T.	(*cr.*)	slender-leaved.	Siberia.	1765.	5. 6.	H. ♃.	Bot. mag. 926.
6 officinàlis. L T.	(*va.*)	officinal.	Europe.	1548.	5. 6.	H. ♃.	
α Sabìni. L.T.	(*sc.*)	*Sabine's.*	———	....	——	H. ♃.	Bot. mag. 1784.
β ròsea. L.T.	(*ro.*)	*rose-coloured.*	———	....	——	H. ♃.	
γ blánda. L.T.	(*bh.*)	*blush-flowered.*	———	....	——	H. ♃.	
δ rubra. L.T.	(*cr.*)	*double-red.*	———	....	——	H. ♃.	Lob. ic. 684.
ε carnéscens. L.T.	(*fl.*)	*flesh-coloured.*	———	....	——	H. ♃.	
ζ álbicans. L.T.	(*bh.*)	*double-blush.*	———	....	——	H. ♃.	Park. par. f. 4.
7 corállina. L.T.	(*cr.*)	English.	England.	....	——	H. ♃.	Eng. bot. 1513.
8 daùrica. L.T.	(*fl.*)	Daurian.	Dauria.	1790.	——	H. ♃.	Bot. mag. 1441.
triternàta. DC. prodr.							
9 lobàta. DC.	(*sc.*)	lobed-leaved.	Portugal.	1821.	——	H. ♃.	Swt. br. fl. gar. 70.
10 Rússi. DC.	(*cr.*)	crimson.	Sicily.	....	——	H. ♃.	——— v. 2. t. 122.
11 pùbens. B M.	(*re.*)	pubescent.	....	1821.	——	H. ♃.	Bot. mag. 2264.
12 hùmilis. L.T.	(*re.*)	dwarf.	Spain.	1633.	5.	H. ♃.	——— 1422.
13 decòra. L.T.	(*re.*)	comely.	Turkey.	....	——	H. ♃.	
α Pallàsii. L.T.	(*re.*)	*Pallas's.*	....	....	——	H. ♃.	
β elàtior. L.T.	(*re.*)	*tall.*	....	....	——	H. ♃.	
14 arietìna. L.T.	(*re.*)	rams-horned.	....	....	——	H. ♃.	
15 crètica. B.R.	(*bh.*)	Cretan.	Crete.	....	——	H. ♃.	Bot. reg. 819.
16 peregrìna. L.T.	(*re.*)	Turkish.	Levant.	1588.	5. 6.	H. ♃.	
α byzantìna. L.T.	(*re.*)	*Levant.*	———	....	——	H. ♃.	Bot. mag. 1050.
β compácta. L.T.	(*re.*)	*compact.*	France ?	....	——	H. ♃.	
γ Grevíllii. L.T.	(*re.*)	*Greville's.*	....	....	——	H. ♃.	
17 paradóxa. L.T.	(*re.*)	paradoxical.	Levant.	....	——	H. ♃.	
α simpliciflòra.	(*re.*)	*single-flowered.*	———	....	——	H. ♃.	
β fimbriàta. L.T.	(*cr.*)	*double-flowered.*	———	....	——	H. ♃.	Swt. br. fl. gar. 19.
18 móllis. L.T.	(*re.*)	soft-leaved.	Siberia ?	....	——	H. ♃.	Bot. reg. 474.
19 villòsa. B.F.G.	(*wh.*)	villous.	....	1822.	——	H. ♃.	Swt. br. fl. gar. 113.
sessiliflòra. B.M. 2648.							

ORDO II.

DILLENIACEÆ. *DC. syst. nat.* 1. *p.* 395.

Tribus I. *DELIMACEÆ.* DC.

TETRA'CERA. DC.		TETRA'CERA. Polygamia Diœcia.					
1 volùbilis. DC.	(*ye.*)	climbing.	Barbadoes.	....	....	S. ♄.⌣.	Pluk. alm. 48. t. 146. f. 1.
2 jamaicénsis. DC.	(*ye.*)	Jamaica.	Jamaica.	1826.	....	S. ♄.⌣.	
3 obovàta. DC.	(*ye.*)	obovate-leaved.	Senegal	1822.	....	S. ♄.⌣.	
4 alnifòlia. DC.	(*ye.*)	Alder-leaved.	Sierra Leone.	——	....	S. ♄.⌣.	
5 potatòria. Afz.	(*ye.*)	water-bearing.	———	——	....	S. ♄.⌣.	
DOLIOCA'RPUS. DC.		DOLIOCA'RPUS. Polyandria Monogynia.					
1 Calínea. DC.	(*wh.*)	climbing.	Guiana.	1822.	....	S. ♄.⌣.	Aub. Guian. 1. t. 221.
Tetrácera calínea. W. *Calínea scándens.* Aub.							

DEL'IMA. DC. DEL'IMA. Polyandria Monogynia.
sarmentòsa. DC. sarmentose. Ceylon. 1822. .... S. ♄.◡. Lam. ill. t. 475.
Tetrácera sarmentòsa. w.
CURATE'LLA. DC. CURATE'LLA. Polyandria Digynia.
1 americàna. DC. (*wh.*) American. Guiana. .... S. ♄.◡. Aub. guian. 1. t. 232.
2 ? alàta. V. .... wing-petioled. ——— .... S. ♄.◡. Venten. choix. t. 49.

Tribus II. *DILLENEÆ.* DC.

PLEURA'NDRA. DC. PLEURA'NDRA. Polyandria Monogynia.
1 bracteàta. DC. (*ye.*) elegant. N. S. W. 1822. 4. 8. G. ♄. Deles. ic. sel. 1. t. 78.
2 nítida. DC. (*ye.*) shining. ——— ——— ——— G. ♄.
3 cneòrum, DC. (*ye.*) blunt-leaved. ——— 1824. 5. 7. G. ♄.
4 scàbra. DC. (*ye.*) rough-leaved. ——— ——— ——— G. ♄.
5 aciculàris. DC. (*ye.*) needle-leaved. ——— ——— 4. 7. G. ♄. Labill. nov. holl. 2. t. 144.
6 stricta. DC. (*ye.*) upright. ——— 1822. ——— G. ♄.
7 calycìna. DC. (*ye.*) linear-leaved. ——— ——— ——— G. ♄.
CANDO'LLEA. CANDO'LLEA. Polyadelphia Polyandria.
cuneifórmis. DC. (*ye.*) wedge-leaved. N. Holland. 1823. 5. 8. G. ♄. Bot. mag. 2711.
HIBBE'RTIA. DC. HIBBE'RTIA. Polyandria Mono-Pentagynia.
1 grossulariæfòlia. DC. (*y.*) Gooseberry-ld. N. Holland. 1803. 3. 8. G. ♄.◡. Bot. mag. 1218.
crenàta. A. R. *t.* 472.
2 volùbilis. DC. (*ye.*) twining. N. S. W. 1796. 5. 10. G. ♄.◡. Andr. rep. t. 126.
3 dentàta. DC. (*ye.*) tooth-leaved. N. Holland. 1814. 1. 8. G. ♄.◡. Bot. reg. 282.
4 salígna DC. (*ye.*) Willow-leaved. N. S. W. 1824. 5. 8. G. ♄.
5 virgàta DC. (*ye.*) twiggy. ——— 1822. .... G. ♄.
6 fasciculàta. DC. (*ye.*) bushy. ——— 1826. 5. 8. G. ♄.
7 lineàris. DC. (*ye.*) linear-leaved. ——— ——— .... G. ♄.
8 monógyna. DC. (*ye.*) one-styled. ——— ——— 5. 8. G. ♄.
9 pedunculàta. DC. (*ye.*) long-pedicled. ——— ——— 4. 9. G. ♄. Bot. reg. 1001.
corifòlia. Bot. mag. 2672.
WO'RMIA. DC. WO'RMIA. Polyandria Pentagynia.
dentàta. DC. (*ye.*) tooth-leaved. Ceylon. 1821. .... S. ♄. Thunb. linn. tr. 1. t. 20.
Dillènia dentàta. w.
COLBE'RTIA. DC. COLBE'RTIA. Polyandria Pentagynia.
1 coromandeliàna. DC. (*y.*) Coromandel. E. Indies. 1803. .... S. ♄. Roxb. cor. 1. t. 20.
Dillènia pentágynia. Roxb.
2 scabrélla. D. P. ...... rough. Nepaul. 1826. .... S. ♄.
DILL'ENIA. DC. DILL'ENIA. Polyandria Monogynia. L.
1 speciòsa. DC. (*wh.*) large-flowered. E. Indies. 1800. .... S. ♄. Exot. bot. t. 2. 3.
2 intègra. DC. .... entire-leaved. Ceylon. 1826. .... S. ♄. Thunb. linn. tr. 1. t. 18.

ORDO III.

WINTEREÆ. *Brown in DC. syst.* 1. *add. p.* 548.

ILL'ICIUM. DC. ANISEED-TREE. Polyandria Polygynia. L.
1 floridànum. DC. (*cr.*) red-flowered. Florida. 1766. 4. 6. F. ♄. Bot. mag. 439.
2 anisàtum. DC. (*ye.*) Chinese. China. .... ——— F. ♄. Lam. ill. t. 493. f. 2.
3 parviflòrum. DC. (*ye.*) small-flowered. Florida. 1790. ——— F. ♄. Vent. cels. t. 22.
DR'IMYS. DC. DR'IMYS. Polyandria Tetra-Octogynia.
Wintèri. DC. .... Winter-bark. Magellan. 1825. .... G. ♄. Forst. gen. p. 84. t. 42.
Wintèra aromàtica. w. *Winteràna aromàtica.* Sol. med. obs. 5. p. 46. t. 1.
TASMA'NNIA. DC. TASMA'NNIA. Diœcia Polyandria.
dipétala. B. .... two-petaled. N. Holland. 1824. 5. 8. G. ♄.

ORDO IV.

MAGNOLIACEÆ. *DC. syst. nat. v.* 1. *p.* 439.

MICH'ELIA. DC. MICH'ELIA. Polyandria Polygynia. L.
Champàca. DC. (*ye.*) sweet-scented. E. Indies. 1779. .... S. ♄. Lam. ill. t. 493.

MAGN'OLIA. DC.		MAGN'OLIA. Polyandria Polygynia. L.					
1 grandiflòra. DC.	(wh.)	Laurel-leaved.	Carolina.	1734.	6. 10.	H. ħ.	
a rotundifòlia.	(wh.)	*round-leaved.*	———	....	——	H. ħ.	
β obovàta. H.K.	(wh.)	*obovate-leaved.*	———	....	——	H. ħ.	
γ ellíptica. H.K.	(wh.)	*elliptic-leaved.*	———	....	——	H. ħ.	Mill. ic. v. 2. t. 172.
δ ferrugìnea. B.M.	(w.)	*ferruginous.*	———	....	——	H. ħ.	Tr. Eh. p. 8. t. 33-35.
ε lanceolàta. H.K.	(wh.)	*Exmouth.*	———	....	——	H. ħ.	Bot. mag. 1952.
2 glaùca. DC.	(wh.)	glaucous-leaved.	N. America.	1688.	6. 9.	H. ♄.	Mich. arb. t. 2.
3 longifòlia.	(wh.)	narrow-leaved.	———	....	——	H. ħ.	
glaùca β longifòlia. Ph.							
4 Thompsoniàna.	(wh.)	hybrid.	Hybrid.	1818.	——	H. ♄.	
5 umbrélla. DC.	(wh.)	Umbrella.	N. America.	1752.	5. 6.	H. ♄.	Mich. arb. t. 5.
tripètala. W.							
6 acuminàta. DC.	(pa.)	bluish-flowered.	———	1736.	5. 7.	H. ♄.	Bot. mag. 2427.
β Candòlii.	(st.)	*De Candolle's.*	———	....	——	H. ♄.	Sav.bi.ital.1819.n.47.ic.
7 auriculàta. DC.	(wh.)	ear-leaved.	———	1786.	4. 5.	H. ♄.	Bot. mag. 1206.
8 pyramidàta. DC.	(wh.)	small ear-leav^d.	———	1811.	——	H. ♄.	Bot. reg. 407.
9 macrophy'lla. DC.	(wh.)	large-leaved.	———	1800.	6. 7.	H. ♄.	Bot. mag. 2189.
10 cordàta. DC.	(st.)	heart-leaved.	———	1801.	——	H. ♄.	Bot. reg. 325.
11 conspícua. P.L.	(wh.)	Youlan.	China.	1789.	2. 5.	H. ♄.	Bot. mag. 1621.
Yulan. DC. Bonpl. nav. t. 20.							
12 Soulangeàna.	(p.w.)	Soulange's.	Hybrid.	1826.	——	H. ♄.	Swt. br. fl. gar. 260.
13 Kòbus. DC.	(p.w.)	slender.	China.	1804.	3. 4.	H. ♄.	Par. lond. 87.
tomentòsa. W. *grácilis.* P.L.							
14 obovàta. DC.	(p.w.)	obovate-leaved.	———	1790.	4. 6.	H. ♄.	Bot. mag. 390.
purpùrea. B.M. *discolor.* V.							
15 fuscàta. DC.	(br.)	brown-stalked.	———	1789.	——	G. ħ.	——— 1008.
16 annonæfòlia. P.L.	(br.)	small-flowered.	———	1804.	——	G. ħ.	Par. lond. 5.
17 pùmila. DC.	(wh.)	dwarf.	———	1786.	1. 12.	G. ħ.	Bot. mag. 977.
LIRIODE'NDRON. W.		TULIP-TREE. Polyandria Polygynia. L.					
tulipífera. W.	(ye.)	common.	N. America.	1663.	6. 7.	H. 5.	Bot. mag. 275.
a acutilòba. M.	(ye.)	*acute-lobed.*	———	——	——	H. 5.	
β obtusilòba. M.	(ye.)	*blunt-lobed.*	———	....	——	H. 5.	

ORDO V.

ANONACEÆ. *DC. syst. nat.* 1. *p.* 463.

AN'ONA. DC.		CUSTARD-APPLE. Polyandria Polygynia. L.					
1 muricàta. DC.	(st.)	Soursop.	W. Indies.	1656.	6. 8.	S. ♄.	Jacq. obs. 1. t. 5.
2 purpùrea. DC.	(pu.)	purple.	Mexico.	1826.	....	S. ♄.	Dunal. mon. p. 64. t. 2.
3 Humbóldtii. DC.	(ye.)	Humboldt's.	Cumana.	——	....	S. ♄.	——— p. 64. t. 3.
4 laurifòlia. DC.	(wh.)	laurel-leaved.	W. Indies.	1824.	6. 8.	S. ♄.	Catesb. car. 2. t. 67.
5 palústris. DC.	(ye.)	shining-leaved.	———	1731.	——	S. ♄.	Sloan. hist. 2. t. 228. f. 1.
6 longifòlia. DC.	(pu.)	long-leaved.	Guiana.	1820.	——	S. ♄.	Aub. gui. 1. t. 248.
7 punctàta. DC.	(st.)	spotted-fruited.	Cayenne.	1822.	——	S. ♄.	——— 1. t. 247.
8 paludòsa. DC.	(gr.)	marsh.	Guiana.	1803.	6. 8.	S. ♄.	——— 1. t. 246.
9 squamòsa. DC.	(st.)	Swe tsop.	S. America.	1731.	——	S. ♄.	Jacq. obs. 1. t. 6. f. 1.
10 cinèrea. DC.	(st.)	cinereous.	St. Thomas.	1823.	....	S. ♄.	
11 Cherimòlia. DC.	(wh.)	Cherimolli.	S. America.	1739.	7. 8.	S. ♄.	Bot. mag. 2011.
tripétala. W. B.M.							
12 reticulàta. DC.	(st.)	netted.	———	1690.	5. 7.	S. ♄.	Catesb. car. 2. t. 86.
13 mucòsa. DC.	(st.)	narrow-leaved.	Guiana.	....	....	S. ♄.	Rumph. amb. 1. t. 45.
14 glábra. DC.	(st.)	smooth-fruited.	Carolina.	1774.	7. 8.	S. ♄.	Catesb. car. 2. t. 64.
15 grandiflòra. DC.	(wh.)	great-flowered.	Mauritius.	1825.	....	S. ♄.	Dunal. mon. p. 75. t. 6.
16 amplexicaulis. DC.	(pu.)	stem-clasping.	———	1824.	....	S. ♄.	——— p. 76. t. 7.
17 asiàtica. DC.	....	Ceylon.	Ceylon.	....	....	S. ♄.	
18 senegalénsis. DC.	(st.)	Senegal.	Senegal.	1822.	....	S. ♄.	Deles. ic. sel. 1. t. 86.
MONOD'ORA. DC.		MONOD'ORA. Polyandria Monogynia.					
myrística. DC.	(wh.)	Nutmeg.	Jamaica.	....	....	S. ♄.	Gært. fr. 2. t. 125. f. 1.
ASI'MINA. DC.		ASI'MINA. Polyandria Polygynia.					
1 parviflòra. DC.	(pu.)	small-flowered.	N. America.	1806.	4. 5.	H. ♄.	Dun. mon. t. 9.
2 trilòba. DC.	(pu.)	trifid-fruited.	———	1736.	——	H. ♄.	Duh. arb. ed. 2. v. 2. t. 25.
Annòna trilòba. W.		*Porcèlia trilòba.* Ph.					
3 pygm'æa. DC.	(wh.)	dwarf.	———	1812.	——	H. ♄.	Dun. mon. t. 10.
Orchidocárpum pygm'æum. M.							
4 grandiflòra. DC.	(wh.)	large-flowered.	Florida.	1820.	——	H. ♄.	Dun. mon. t. 11.
Porcèlia grandiflòra. P.S.							

UV'ARIA. DC.		Uv'aria. Polyandria Polygynia. L.					
1 zeylànica. DC.	(*pu.*)	Ceylon.	E. Indies.	1794.	....	S. ♄.	Lam. ill. t. 495. f. 2.
2 Gærtnèri. DC.		Gærtner's.	———	....	....	S. ♄.	——— t. 495. f. 3.
3 lùtea. DC.	(*st.*)	yellow-fruited.	———	1822.	....	S. ♄.	Roxb. cor. 1. t. 36.
4 tomentòsa. DC.	(*br.*)	woolly.	———	——	....	S. ♄.	——— 1. t. 35.
5 velutina. DC.	(*br.*)	velvetty.	E. Indies.	1825.	....	S. ♄.	
6 lucida. Boj.		shining.	E. Africa.	——	....	S. ♄.	
UN'ONA. DC.		Un'ona. Polyandria Polygynia. L.					
1 Nàrum. DC.	(*re.*)	Rheede's.	E. Indies.	....	....	S. ♄.◡.	Rheed. mal. 2. t. 9.
2 fuscàta. DC.	(*br.*)	brown-flower'd.	Guiana.	1820.	....	S. ♄.	
3 acuminàta. DC.	(*br.*)	taper-pointed.	———	——	....	S. ♄.	
4 odoràta. DC.	(*st.*)	sweet-scented.	Java.	1804.	5. 7.	S. ♄.	Lam. ill. t. 495. f. 1.
5 longifòlia. DC.	(*ye.*)	long-leaved.	E. Indies.	1820.	....	S. ♄.	Sonn. voy. 4. t. 131.
6 aromática. DC.	(*pu.*)	aromatic.	Guiana.	——	....	S. ♄.	Aub. gui. 2. t. 243.
7 æthiòpica. DC.	(*wh.*)	Ethiopian.	Sierra Leone.	1822.	....	S. ♄.	Math. com. 1. p. 434. ic.
8 oxypétala. DC.	(*br.*)	sharp-petaled.	———	——	....	S. ♄.	
ARTAB'OTRYS. Br.		Artab'otrys. Polyandria Polygynia. L.					
odoratíssimus. B.R.	(*b.*)	sweet-scented.	China.	1758.	6. 7.	S. ♄.◡.	Bot. reg. 423.
Annòna hexapètala. W. *Uvària odoratíssima.* H.B.							
XYL'OPIA. DC.		Bitter-wood. Polyandria Polygynia. L.					
1 muricàta. DC.	(*br.*)	rough-fruited.	W. Indies.	1793.	....	S. ♄.	Brown. jam. 250. t. 5. f. 2.
2 frutéscens. DC.	(*br.*)	shrubby.	Guiana.	1823.	....	S. ♄.	Aub. gui. 1. t. 292.
3 glábra. DC.	(*br.*)	smooth-fruited.	W. Indies.	....	....	S. ♄.	Dun. mon. t. 19.
GUATT'ERIA. DC.		Guatt'eria. Polyandria Polygynia. L.					
1 cerasóides. DC.	(*br.*)	Cherry-fruited.	E. Indies.	1820.	....	S. ♄.	Roxb. cor. 1. t. 33.
2 suberòsa. DC.	(*wh.*)	cork-barked.	———	——	....	S. ♄.	——— 1. t. 34.
3 rùfa. DC.	(*br.*)	rufous.	India.	——	4. 8.	S. ♄.	Bot. reg. 836.
4 virgàta. DC.	(*wh.*)	Lance-wood.	Jamaica.	1793.	——	S. ♄.	Dun. mon. t. 31.
Uvària lanceolàta. W.							
5 laurifòlia. DC.	(*wh.*)	Laurel-leaved.	———	1820.	....	S. ♄.	——— t. 32.
Uvària laurifòlia. Sw.							

ORDO VI.

MENISPERMACEÆ. *DC. prodr.* 1. *p.* 95.

Tribus II. *MENISPERMEÆ.* DC.

CO'CCULUS. DC.		Co'cculus. Diœcia Hexandria.					
1 rotundifòlius. DC.	(*gr.*)	round-leaved.	....	1820.	....	S. ♄.◡.	
2 cordifòlius. DC.	(*gr.*)	heart-leaved.	E. Indies.	1822.	....	S. ♄.◡.	Rheed. mal. 7. t. 21.
3 suberòsus. DC.	(*gr.*)	cork-barked.	———	1790.	....	S. ♄.◡.	Gært. fr. 1. t. 70. f. 7.
4 Plukenètii. DC.	(*gr.*)	Plukenet's.	———	——	....	S. ♄.◡.	Pluk. man. 52. t. 345. f. 2.
5 incànus. L.T.	(*gr.*)	hoary.	———	1820.	....	S. ♄.◡.	
6 tomentòsus. L.T.	(*gr.*)	woolly-leaved.	———	——	....	S. ♄.◡.	
7 críspus. L.T.	(*gr.*)	curled.	———	1822.	....	S. ♄.◡.	Rumph. amb. 5 t. 44. f. 1.
8 palmàtus. DC.	(*gr.*)	palmated.	———	....	....	S. ♄.◡.	Berr. in As. res. 10. t. 5.
9 orbiculàtus. DC.	(*gr.*)	orbicular-leaved.	———	1790.	....	S. ♄.◡.	Rheed. mal. 11. t. 62.
10 laurifòlius. DC.	(*gr.*)	Laurel-leaved.	———	1820.	....	S. ♄.◡.	
COSCI'NIUM. C.L.T.		Cosci'nium. Diœcia Hexandria.					
fenestràtum. L.T.	(*gr.*)	knotted.	Ceylon.	1820.	....	S. ♄.◡.	Gært. fr. 1. t. 46. f. 5.
TILIAC'ORA. C.L.T.		Tiliac'ora. Diœcia Hexandria.					
racemòsa. L.T.	(*gr.*)	racemed.	E. Indies.	....	....	S. ♄.◡.	
WENDLA'NDIA. W.		Wendla'ndia. Heptandria Polygynia. W.					
populifòlia. W.	(*wh.*)	Poplar-leaved.	Carolina.	1759.	6. 7.	H. ♄.◡.	Wend. h. herr. 3. t. 16.
Menispérmum Carolìnum. L. *Cócculus Carolìnus.* DC.							
CISSA'MPELOS. DC.		Cissa'mpelos. Diœcia Monadelphia. L.					
1 Paréira. DC.	(*gr.*)	velvet-leaf.	S. America.	1733.	7. 8.	S. ♄.◡.	Lam. ill. t. 830.
2 microcárpa. DC.	(*gr.*)	small-fruited.	W. Indies.	1823.	....	S. ♄.◡.	
3 hirsùta. DC.	(*gr.*)	hairy.	Nepaul.	1820.	....	G. ♄.◡.	
4 Mauritiàna. DC.	(*gr.*)	hispid-branch'd.	Mauritius.	——	....	S. ♄.◡.	
5 Caapèba. DC.	(*gr.*)	nerved-leaved.	S. America.	1733.	7. 8.	S. ♄.◡.	Plum. ic. 67. f. 2.
6 ? capénsis. DC.	(*gr.*)	Cape.	C. B. S.	1775.	....	G. ♄.◡.	
MENISPE'RMUM. DC.		Moon-seed. Diœcia Dodecandria. L.					
1 canadénse. DC.	(*st.*)	Canadian.	N. America.	1691.	6. 7.	H. ♄.◡.	Bot. mag. 1910.

2 daùricum. DC.	(*st.*)	Daurian.	Dahuria.	1823.	....	H. ♄.◡.	Deles. ic. sel. 1. t. 100.
3 smilacìnum. DC.	(*st.*)	Smilax-leaved.	Carolina.	1776.	....	G. ♄.◡.	Jacq. ic. 3. t. 629.
Cissámpelos smilacìna. W.							
4 Lyòni. Ph.	(*st.*)	Lyon's.	N. America.	1812.	6. 7.	H. ♃ ◡.	
AB'UTA. DC.		AB'UTA. *Menispermi sp.* W.					
rufèscens. DC.	(*pu.*)	brown-leaved.	Guiana.	1822.	....	S. ♄.◡.	Aub. gui. 1. t. 250.

Tribus III. *SCHIZANDREÆ. DC. prodr.* 1. *p.* 104.

SCHIZA'NDRA. DC.		SCHIZA'NDRA. Monœcia Pentandria.					
coccìnea. DC.	(*sc.*)	scarlet-flower'd.	Carolina.	1806.	6. 7.	G. ♄.◡.	Bot mag. 1413.

ORDO VII.

BERBERIDEÆ. *DC. syst. nat. v.* 2. *p.* 1.

BE'RBERIS. W.		BARBERRY. Hexandria Monogynia. L.					
1 vulgàris. DC.	(*ye.*)	common.	England.	....	4. 5.	H. ♄.	Eng. bot. 49.
α rùbra.	(*ye.*)	*red-fruited.*	————	....	——	H. ♄.	
β lùtea. DC.	(*ye.*)	*yellow-fruited.*	....	....	——	H. ♄.	
γ violàcea. DC.	(*ye.*)	*violet-fruited.*	....	....	——	H. ♄.	
δ purpùrea. DC.	(*ye.*)	*purple-fruited.*	....	....	——	H. ♄.	
ε nìgra. DC.	(*ye.*)	*black-fruited.*	Levant.	....	——	H. ♄.	
ζ álba. DC.	(*ye.*)	*white-fruited.*	....	....	——	H. ♄.	
ϑ aspérma. DC.	(*ye.*)	*stoneless.*	....	....	——	H ♄.	
2 ibérica.	(*ye.*)	Iberian.	Iberia.	1790.	——	H. ♄.	
vulgàris β ibérica. DC.							
3 emarginàta. DC.	(*ye.*)	emarginate.	Siberia.	1820.	——	H. ♄.	
4 canadénsis. DC.	(*ye.*)	Canadian.	Canada.	1759.	4. 6.	H. ♄.	
5 sinénsis. DC.	(*ye.*)	China.	China.	1800.	——	H. ♄.	Wats. dend. brit. t. 26.
6 aristàta DC.	(*ye.*)	awned.	Nepaul.	1820.	6. 8.	H. ♄.	Bot. mag. 2549.
Chìtria. Bot. reg. 729.							
7 crética. DC.	(*ye.*)	Cretan.	Crete.	1759.	5. 6.	H. ♄.	Flor. græc. t. 342.
8 ilicifòlia. DC.	(*ye.*)	Holly-leaved.	Terra del Fuego.	1791.	7. 8.	H. ♄.	
9 asiática. DC.	(*ye.*)	Asiatic.	Nepaul.	1820.	6. 8.	H. ♄.	Deles. ic. sel. 2. t. 1.
10 Wallichiàna. DC.	(*ye.*)	Wallich's.	————	——	6. 7.	H. ♄.	
11 buxifòlia. DC.	(*ye.*)	Box-leaved.	Magellan.	1827.	....	G. ♄.	Lam. ill. t. 253. f. 3.
12 empetrifòlia. DC.	(*ye.*)	Empetrum-l[d].	————	——	....	G. ♄.	
13 heterophy'lla. DC.	(*ye.*)	various-leaved.	————	....	5. 6.	H. ♄.	Hook. exot. fl. t. 14.
14 sibírica. DC.	(*ye.*)	Siberian.	Siberia.	1790.	6. 7.	H. ♄.	Bot. reg. 487.
15 inérmis. DC.	(*ye.*)	spineless.	Magellan.	1827.	....	G. ♄.	
16 pinnata. B.R.	(*ye.*)	pinnate-leaved.	America.	1820.	3. 5.	G. ♄.	Bot. reg. 702.
Mahònia fasciculàris. DC.							
17 rèpens. B.R.	(*ye.*)	creeping-rooted.	N. America.	1822.	4. 6.	H. ♄.	Bot. reg. 1176.
18 nervòsa. P.	(*ye.*)	nerved-leaved.	————	1826.	....	H. ♄.	Pursh. fl. am. sept. 1. t. 5.
NAND'INA. DC.		NAND'INA. Hexandria Monogynia.					
doméstica. DC.	(*wh.*)	garden.	China.	1804.	6. 7.	G. ♄.	Bot. mag. 1109.
LEO'NTICE. DC.		LEO'NTICE. Hexandria Monogynia. L.					
1 Chrysógonum. DC.	(*ye.*)	Oak-leaved.	Levant.	1740.	3. 6.	H. ♃.	Barrel. ic. 1113.
2 Leontopétalum. DC.	(*y.*)	Lion's-leaf.	————	1597.	4. 5.	H. ♃.	Lam. ill. t. 254. f. 1.
3 vesicària. DC.	(*ye.*)	bladder-capsul'd.	Siberia.	1822.	——	H. ♃.	Pal. a. p. 1779. 2. t. 9. f. 4.
4 altaìca. DC.	(*ye.*)	Altay.	————	——	——	H. ♃.	—— p. 257. t. 8. f. 1. 2. 3.
5 thalictroìdes. DC.	(*ye.*)	Columbine-l[d].	N. America.	1755.	5.	H. ♃.	Mich. amer. 1. t. 21.
Caulophy'llum thalictròides. M.							
EPIM'EDIUM. DC.		BARRENWORT. Tetrandria Monogynia. L.					
alpìnum. DC.	(*pu.*)	Alpine.	England.	....	4. 5.	H. ♃.	Eng. bot. 438.
DIPHYLL'EIA. DC.		DIPHYLL'EIA. Hexandria Monogynia. L.					
cymòsa. DC.	(*wh.*)	blue-berried.	N. America.	1812.	5. 6.	H. ♃.	Bot. mag. 1666.

ORDO VIII.

PODOPHYLLEÆ. *DC. syst. nat.* 2. *p.* 31.

Tribus I. *PODOPHYLLEÆ.*

PODOPHY'LLUM. DC.		DUCK'S-FOOT. Polyandria Monogynia. L.					
peltàtum. DC.	(*wh.*)	May-apple.	N. America.	1664.	5.	H. ♃.	Bot. mag. 1819.
JEFFERS'ONIA. DC.		JEFFERS'ONIA. Octandria Monogynia.					
diphy'lla. DC.	(*wh.*)	two-leaved.	N. America.	1792.	5.	H. ♃.	Bot. mag. 1513.
Podophy'llum diphy'llum. W.							

Tribus II. *HYDROPELTIDEÆ.*

CABO'MBA. DC.		CABO'MBA. Hexandria Digynia. W.					
aquática. DC.	(*ye.*)	water.	Carolina.	1823.	....	S.*w.*♃.	Aubl. gui. 1. t. 124.
Néctris peltàta. Ph.							
HYDROPE'LTIS. DC.		HYDROPE'LTIS. Polyandria Polygynia.					
purpùrea. DC.	(*pu.*)	purple.	N. America.	1798.	7. 8.	F.*w.*♃.	Bot. mag. 1147.
Brasènia peltàta. Ph.							

ORDO IX.

NYMPHÆACEÆ. *DC. syst. nat.* 2. *p.* 34.

Tribus I. *NELUMBONEÆ.*

NELU'MBIUM. DC.		SACRED-BEAN. Polyandria Polygynia. L.					
1 speciòsum. DC.	(*bh.*)	Indian.	India.	1787.	6. 8.	S.*w.*♃.	Bot. mag. 903.
Cyámus Nelúmbo. Sm.		Exot. bot. 1. t. 31. 32.					
2 Támara.	(*ro.*)	Rheede's.	———	....	——	S.*w.*♃.	Rheed. mal. XI. t. 30.
3 cáspicum. F.	(*ro.*)	Fischer's.	Volga.	1821.	....	S.*w.*♃.	
4 lùteum. DC.	(*ye.*)	yellow.	N. America.	1810.	....	F.*w.*♃.	Poit.a.m. 13. t. 29. f. 42.
5 jamaicénse. DC.	....	Jamaica.	W. Indies.	1823.	....	S.*w.*♃.	

Tribus II. *NYMPHÆEÆ.*

EURY'ALE. DC.		EURY'ALE. Polyandria Monogynia. L.					
fèrox. DC.	(*re.*)	prickly.	India.	1809.	7. 9.	S.*w.*♃.	Bot. mag. 1447.
Anneslèa spinòsa. And. bot. rep. 618.							
NYMPHÆ'A. DC.		WATER-LILY. Polyandria Monogynia. L.					
1 scutifòlia. DC.	(*bl.*)	Cape blue.	C. B. S.	1792.	6. 9.	S.*w.*♃.	Bot. mag. 552.
cærùlea. Bot. rep. 197.							
2 cœrùlea. DC.	(*bl.*)	Egyptian blue.	Egypt.	1812.	——	S.*w.*♃.	Vent. malm. t. 6.
stellata. var. Bot. mag. 2058.							
3 cyànea. H.B.	(*bl.*)	Indian-blue.	E. Indies.	1809.	——	S.*w.*♃.	
Cahlàra. Donn. cant. ed. 7.							
4 stellàta. DC.	(*bl.*)	star-flowered.	———	1803.	——	S.*w.*♃.	Andr. rep. 330.
5 edulis. DC.	(*wh.*)	eatable.	———	....	——	S.*w.*♃.	
esculénta. H.B.							
6 rùbra. DC.	(*re.*)	red-flowered.	———	1803.	7. 8.	S.*w.*♃.	Bot. mag. 1280.
7 ròsea.	(*ro.*)	rose-coloured.	———	——	——	S.*w.*♃.	——— 1364.
rùbra. β ròsea. DC.							
8 pubéscens. DC.	(*wh.*)	Indian Lotus.	———	——	5. 8.	S.*w.*♃.	Rheed. h. mal. XI. t. 26.
9 Lòtus. DC.	(*wh.*)	Egyptian.	Egypt.	1802.	6. 9.	S.*w.*♃.	Del. fl. egyp. ill. t. 60. f. 1.
10 thérmalis. DC.	(*wh.*)	Hungarian.	Hungary.	——	——	S.*w.*♃.	Bot. mag. 797.
Lòtus. And. bot. rep. 391.							

11 ámpla. DC.	(*wh.*)	West-Indian.	W. Indies.	....	6. 9.	S.*w*.♃.	Plum. mss. 123. t. 4.
12 versícolor. DC.	(*bh.*)	changeable	E. Indies.	1807.	——	S.*w*.♃.	Bot. mag. 1189.
13 renifórmis. DC.	(*wh.*)	kidney-leaved.	Carolina.	1823.	7.	G.*w*.♃.	Deles. ic. sel. 2. t. 5.
14 álba. DC.	(*wh.*)	white.	Britain.	....	6. 7.	H.*w*.♃.	Eng. bot. 160.
15 odoràta. DC.	(*wh.*)	sweet-scented.	N. America.	1786.	7.	H.*w*.♃	Bot. mag. 819.
16 mìnor. DC.	(*wh.*)	smaller	——	1812.	——	H.*w*.♃.	—— 1652.
odoràta. β mìnor. B.M.							
17 nítida. DC.	(*wh.*)	cup-flowered.	Siberia.	1809.	7. 8.	H.*w*.♃.	Bot. mag. 1359.
18 pygmæ'a. DC.	(*wh.*)	pigmy.	China.	1805	5. 9.	H.*w*.♃.	—— 1525.
Castàlia pygmæ'a. Sal.		par. lond. 68.					
19 blánda. DC.	(*wh.*)	handsome.	W. Indies.	1827.	....	S.*w*.♃.	
N'UPHAR. DC.		N'UPHAR.	Polyandria Monogynia. L.				
1 lùtea. DC.	(*ye.*)	common.	Britain.	....	6. 7.	H.*w*.♃.	Eng. bot. 159.
Nymph'æa lùtea. W.							
2 pùmila. DC.	(*ye.*)	small-yellow.	Scotland	....	——	H.*w*.♃.	—— 2292.
mínima. E.B.							
3 Kalmiàna. DC.	(*ye.*)	Canadian.	N. America.	1807.	7. 8.	H.*w*.♃.	Bot. mag. 1243.
4 ádvena. DC.	(*ye.*)	stripe-flowered.	——	1772.	——	H.*w*.♃.	Bot. mag. 684.
Nymph'æa ádvena. W.							

ORDO X.

SARRACENIÆ.

SARRAC'ENIA. W.		SIDE-SADDLE-FLOWER.	Polyandria Monogynia. L.				
1 purpùrea. W.	(*pu.*)	purple.	N. America.	1640.	6. 7.	H. ♃.	Bot. mag. 849.
2 rùbra. W.	(*re.*)	red-flowered.	——	1786.	——	H. ♃.	Hook. ex. flor. 13.
psittácina. M ?							
3 variolaris. W.	(*st.*)	hook-leaved.	——	1803.	——	H. ♃.	Bot. mag. 1710.
adúnca. Ex. bot. t. 53. *mìnor* W.							
4 flàva. W.	(*ye.*)	yellow.	——	1752.	——	H. ♃.	—— 780.

ORDO XI.

PAPAVERACEÆ. *DC. syst. nat.* 2. *p.* 67.

PAP'AVER. DC.		POPPY.	Polyandria Monogynia.				
1 nudicáule. DC.	(*ye.*)	naked-stalked.	Siberia.	1730.	6. 8.	H. ♃.	Bot. mag. 1633.
β glábratum. DC.	(*ye.*)	*smooth-stalked.*	——	....	——	H. ♃.	
γ radicàtum. DC.	(*ye.*)	*small hairy.*	Norway.	....	——	H. ♃.	
2 rùbro-aurantìacum. F.		red orange.	Dahuria.	1822.	——	H. ♃.	Bot. mag. t. 2344.
3 microcárpum. DC.	(*br.*)	small-capsuled.	Kamtschatka.	——	——	H. ♃.	
4 pyrenáicum DC.	(*ye.*)	Pyrenean.	Pyrenees.	....	——	H. ♃.	Seg. ver. 1. t. 4. f. 4.
aurantìacum. Lois.							
α lùteum. DC.	(*ye.*)	*yellow-flowered.*	——	....	——	H. ♃.	Barr. ic. t. 764.
β puníceum. DC.	(*re.*)	*red-flowered.*	——	....	——	H. ♃.	
5 alpìnum. DC.	(*wh.*)	Alpine.	Austria.	1759.	——	H. ♃.	Swt. br. fl. gar. 217.
6 hy'bridum. DC.	(*re.*)	mongrel.	England.	....	6. 7.	H. ⊙.	Eng. bot. 43.
7 Argemòne. DC.	(*re.*)	rough.	Britain.	....	——	H. ⊙.	—— 643.
β marítimum. With.		*one-flowered.*	——	....	——	H. ⊙.	
8 dùbium. DC.	(*re.*)	smooth.	——	....	——	H. ⊙.	Eng. bot. 644.
9 Rh'œas. DC.	(*re.*)	common-corn.	——	....	——	H. ⊙.	—— 645.
α maculàtum.	(*re.*)	*spot-petaled.*	——	....	——	H. ⊙.	Curt. lond. t. 32.
β coccíneum.	(*sc.*)	*scarlet-flowered.*	——	....	——	H. ⊙.	Tab. ic. f. 1.
γ cárneum.	(*fl.*)	*flesh-coloured.*	——	....	——	H. ⊙.	Barr. obs. ic. t. 763.
δ álbum.	(*wh.*)	*white-flowered.*	——	....	——	H. ⊙.	Wein. phyt. t. 788. f. d.
ε álbo-marginàtum.		*white-edged.*	——	....	——	H. ⊙.	—— t. 790. f. e.
ζ coccíneo-marginàtum.		*scarlet-edged.*	——	....	——	H. ⊙.	—— t. 790. f. b.

10 lævigàtum. DC.	(*re.*)	slender.	Caucasus.	1823.	6. 7.	H. ⊙.	
11 arenàrium. DC.	(*re.*)	sand.	Siberia.	1825.	6. 8.	H. ⊙.	
12 Roubìæi. DC.	(*re.*)	Roubieu's.	Montpelier.	1823.	——	H. ⊙.	Vig. diss. 39. n. 4. t. 1. f. 1.
13 floribúndum. DC.	(*re.*)	many-flowered.	Levant.	1815.	6. 7.	H. ♂.	Bot. reg. 134.
14 hórridum. DC.	(*co.*)	prickly-stalk'd.	N. S. W.	1825.	6. 8.	H. ⊙.	Swt. br. fl. gar. 173.
15 orientàle. DC.	(*re.*)	oriental.	————	1714.	5. 6.	H. ♃.	Bot. mag. 57.
16 bracteàtum. DC.	(*re.*)	bracted.	Caucasus.	1821.	5. 8.	H. ♃.	Lind. coll. t. 23.
17 setígerum. DC.	(*wh.*)	bristle-pointed.	Italy.	1824.	——	H. ⊙.	Swt. br. fl. gar. 172.
18 somníferum. DC.	(*va.*)	garden.	England.	....	7. 8.	H. ⊙.	Lam. ill. t. 451.
* *nigrum.* DC.	(*va.*)	*black-seeded.*	....	....	——	H. ⊙.	Lob. ic. t. 274. f. 1.
β *flore-pleno.*	(*re.*)	*double-red.*	....	....	——	H. ⊙.	Swert. flor. 2. t. 22. f. 1.
γ *fimbriatum.*	(*va.*)	*fringed-flowered.*	....	....	——	H. ⊙.	Wein. phyt. t. 795. f. c.
** *album.* DC.	(*wh.*)	*white-seeded.*	England.	....	——	H. ⊙.	Eng. bot. 2145.
β *flore-pleno.*	(*wh.*)	*double-white.*	————	....	——	H. ⊙.	Swert. flor. 2. t. 22. f. 2.
γ *variegatum.*	(*v.*)	*variegated.*	————	....	——	H. ⊙.	
19 caucásicum. DC.	(*re.*)	Caucasean.	Caucasus.	1813.	6. 7.	H. ⊙.	Bot. mag. 1675.
20 armenìacum. DC.	(*re.*)	Armenian.	Armenia.	1815.	——	H. ⊙.	
Argemòne armenìaca. W.							
ARGEM'ONE. DC.		ARGEMÒNE.	Polyandria Monogynia.				
1 mexicàna. DC.	(*ye.*)	Mexican.	Mexico.	1592.	7. 8.	H. ⊙.	Bot. mag. 243.
2 ochroleùca. B.F.G.	(*st.*)	straw-coloured.	————	1827.	7. 10.	H. ⊙.	Swt. br. fl. gar. 242.
3 albiflòra. B.M.	(*wh.*)	white-flowered.	Louisiana.	1820.	——	H. ♃.	Bot. mag. 2342.
4 grandiflòra. B.F.G.	(*w.*)	large-flowered.	Mexico.	1827.	7. 10.	H. ♃.	Swt. br. fl. gar. 226.
MECONO'PSIS. DC.		MECONO'PSIS.	Polyandria Monogynia.				
cámbrica. DC.	(*st.*)	Welsh.	Wales.	....	5. 8.	H. ♃.	Eng. bot. 66.
Papàver cámbricum. W.							
SANGUIN'ARIA. DC.		PUCCOON.	Polyandria Monogynia.				
canadénsis. DC.	(*wh.*)	Bloodwort.	N. America.	1680.	3. 4.	H. ♃.	Bot. mag. 162.
MACLEA'YA. B.		MACLEA'YA.	Polyandria Monogynia.				
cordàta. B.	(*bh.*)	heart-leaved.	China.	1795.	5. 8.	H. ♃.	Bot. mag. 1905.
Boccònia cordàta. DC.							
BOCC'ONIA. DC.		BOCC'ONIA.	Dodecandria Monogynia.				
1 frutéscens. DC.	(*gr.*)	Tree Celandine.	W. Indies.	1739.	1. 4.	S. ♄.	Lodd. bot. cab. t. 83.
2 integrifòlia. DC.	(*gr.*)	entire-leaved.	Peru.	1822.	....	S. ♄.	H. et B. nov. pl. 1. t. 35.
ESCHSCHO'LTZIA. D.		ESCHSCHO'LTZIA.	Polyandria Monogynia.				
califórnica. DC.	(*ye.*)	Menzies'.	California.	1826.	7. 10.	H. ♃.	Swt. br. fl. gar. 265.
HUNNEMA'NNIA. B.F.G.		HUNNEMA'NNIA.	Polyandria Monogynia.				
fumariæfòlia. B.F.G.	(*y.*)	Fumitory-leav'd	Mexico.	1827.	7. 10.	H. ♄?	Swt. br. fl. gar. 276.
RŒM'ERIA. DC.		RŒM'ERIA.	Polyandria Monogynia.				
1 hy'brida. DC.	(*re.*)	violet-flowered.	England.	....	5. 7.	H. ⊙.	Eng. bot. 201.
Gláucium violàceum. H.K.							
2 refrácta. DC.	(*re.*)	pendulous-fruit.	Tauria.	1823.	——	H. ⊙.	Deles. ic. sel. 2. t. 8.
GLA'UCIUM. DC.		HORN-POPPY.	Polyandria Monogynia.				
1 flàvum. DC.	(*ye.*)	yellow.	Britain.	....	6. 10.	H. ♂.	Eng. bot. t. 8.
lùteum. H.K. *Chelidònium gláucium.* L.							
2 fúlvum. DC.	(*or.*)	orange.	S. Europe.	1802.	8. 9.	H. ♃.	Swt. br. fl. gar. 35.
3 trícolor. S.S.	(*v.*)	three-coloured.	Poland.	1828.	6. 9.	H. ⊙.	
4 corniculàtum. DC.	(*re.*)	various-colour'd	Europe.	....	6. 7.	H. ⊙.	
α *phæníceum.* DC.	(*re.*)	*red-flowered.*	England.	....	——	H. ⊙.	Eng. bot. 1433.
β *flaviflòrum.* DC.	(*ye.*)	*yellow-flowered.*	Tauria.	1823.	——	H. ⊙.	
CHELID'ONIUM. DC.		CELANDINE.	Polyandria Monogynia. L.				
1 màjus. DC.	(*ye.*)	common.	Britain.	....	4. 10.	H. ♃.	Eng. bot. 1581.
2 grandiflòrum. DC.	(*ye.*)	large-flowered.	Dahuria.	1823.	——	H. ♃.	
3 laciniàtum. DC.	(*ye.*)	jagged.	S. Europe.	....	——	H. ♃.	Mill. ic. 1. t. 92. f. 2.
HYPE'COUM. DC.		HYPE'COUM.	Tetrandria Digynia. L.				
1 procúmbens. DC	(*ye.*)	procumbent.	S. Europe.	1596.	6. 7.	H. ⊙.	Swt. br. fl. gar. 3. t. 217.
2 péndulum. DC.	(*ye.*)	pendulous.	S. France.	1640.	——	H. ⊙.	Mill. ic. t. 250. f. 1.
3 eréctum. DC.	(*ye.*)	erect.	Siberia.	1759.	5. 6.	H. ♂.	Amm. ruth. 58. t. 9.

ORDO XII.

FUMARIACEÆ. *DC. syst. nat.* 2. *p.* 105.

DICLʹYTRA. DC.		DICLʹYTRA.	Diadelphia Hexandria. L.				
1 cucullària. DC.	(*wh.*)	naked stalked.	N. America.	1731.	6. 7.	H. ♃.	Bot. mag. 1127.
Fumària cucullària. B.M.							
β *divaricàta.* DC.	(*wh.*)	*spreading-spur'd.*	N. America.	——	——	H. ♃.	
2 bracteòsa. DC.	(*wh.*)	leafy-bracted.	———	1823.	——	H. ♃.	
3 formòsa. DC.	(*re.*)	blush.	———	1796.	——	H. ♃.	Bot. mag. 1335.
4 exímia. DC.	(*re.*)	choice.	———	1812.	——	H. ♃.	Bot. reg. 51.
Fumària exímia. B.R.							
5 tenuifòlia. DC.	(*bh.*)	slender-leav'd.	Kamtschatka.	1824.	5. 7.	H. ♃	Deles. ic. sel. 2. t. 9. f. B.
6 canadénsis. DC.	(*pu.*)	Canadian.	Canada.	1823.	——	H. ♃.	
7 lachenalíæflòra. DC.	(*p.*)	Lachenalia-fld.	Siberia.	1824.	....	H. ♃.	Rudol. m. sc. pet. 1. t. 19.
ADLʹUMIA. DC.		ADLʹUMIA.	Diadelphia Hexandria. L.				
cirrhòsa. DC.	(*bh.*)	tendrilled.	N. America.	1778.	6. 9.	H. ♂.	Vent. choix. t. 19.
Fumària fungòsa. W.							
CYSTICAʹPNOS. DC.		CYSTICAʹPNOS.	Diadelphia Hexandria. L.				
africàna. DC.	(*bh.*)	African.	C. B. S.	1696.	6. 8.	H ⊙.	Wein. phyt. t. 521. f. c.
Fumària vesicària. W.							
CORYʹDALIS. DC.		CORYʹDALIS.	Diadelphia Hexandria. L.				
1 pauciflòra. DC.	(*pu.*)	few-flowered.	Siberia.	1823.	....	H. ♃.	
2 Marschalliàna. DC.	(*st.*)	Marschall's.	Tauria.	——	4. 5.	H. ♃.	
3 tuberòsa. DC.	(*pu.*)	hollow-rooted.	Europe.	1596.	2. 4.	H. ♃.	Bot. mag. 232.
β *albiflòra.*	(*wh.*)	*white-flowered.*	———		——	H. ♃.	——— 2340.
Fumària càva. albiflòra. B.M.							
4 fabàcea. DC.	(*pu.*)	Bean-leaved.	Germany.	1815.	——	H. ♃.	Horn. fl. dan. t. 1394.
5 caucásica. DC.	(*pu.*)	Caucasean.	Caucasus.	1823.	——	H. ♃.	
6 bulbòsa. DC.	(*pu.*)	solid-rooted.	Britain.	....	——	H. ♃.	Eng. bot. 1471.
Fumària sólida. E. B. *Coryʹdalis Hallèri.* W. en.							
7 angustifòlia. DC.	(*pu.*)	narrow-leaved.	Iberia.	1823.	——	H. ♃.	
8 bracteàta. DC.	(*st.*)	large-bracted.	Siberia.	1822.	——	H. ♃.	
9 nòbilis. DC.	(*st.*)	great-flowered.	———	1783.	5.	H. ♃.	Bot. reg. 395.
10 Geblèri. F.	(*bh.*)	Altay.	Altay.	1827.	5. 6.	H. ♃?	
11 pæoniæfòlia. DC.	(*pu.*)	Pæony-leaved.	Siberia.	1823.	——	H. ♃.	Gmel. sib. 4. t. 34.
12 glàuca. DC.	(*bh.*)	glaucous.	N. America.	1683.	7. 8.	H. ⊙.	Bot. mag. 179.
Fumària sempervìrens. W.							
13 strícta. DC.	(*pu.*)	upright.	Siberia.	1825.	5. 7.	H. ♃.	
14 sibírica. DC.	(*ye.*)	Siberian.	———	——	——	H. ♃.	Amm. ruth. n. 173. t. 20.
15 impàtiens. DC.	(*ye.*)	impatient.	Dahuria.	1823.	——	H. ⊙.	
16 àurea. DC.	(*ye.*)	golden.	N. America.	1812.	5. 7.	H. ♂.	Bot. reg. 66.
Fumària áurea. B.R.							
17 lùtea. P.S.	(*ye.*)	yellow.	England.	....	4. 10.	H. ♃.	Eng. bot. 588.
18 capnoídes. P.S.	(*wh.*)	white-flowered.	S. Europe.	1596.	5. 10.	H. ♂.	Pluk. alm. t. 90. f. 2.
19 acàulis. DC.	(*st.*)	stalkless.	Germany.	1825.	7. 9.	H. ⊙.	Jacq. ic. 3. t. 554.
20 uraléusis. DC.	(*ye.*)	Fischer's.	Siberia.	1823.	....	H. ♃.	
21 breviflòra. DC.	(*st.*)	short-flowered.	Kamtschatka.	——	5. 10.	H. ⊙.	
22 claviculàta. DC.	(*wh.*)	climbing.	Britain.	....	——	H. ⊙.	Eng. bot. 103.
SARCOCAʹPNOS. DC.		SARCOCAʹPNOS.	Diadelphia Hexandria. L.				
enneaphyʹlla. DC.	(*bh.*)	nine-leaved.	Spain.	1714.	5. 7.	H. ♃.	Barr. ic. t. 42.
Fumària enneaphyʹlla. W.							
FUMʹARIA. DC.		FUMITORY.	Diadelphia Hexandria. L.				
1 spicàta. DC.	(*re.*)	narrow-leaved.	S. Europe.	1714.	4. 6.	H. ⊙.	Lob. ic. t. 757. f. 2.
2 capreolàta. DC.	(*wh.*)	climbing.	Europe.	....	5. 9.	H. ⊙.	DC. ic. rar. 1. t. 34.
β? *Burchéllii.* DC.	(*wh.*)	*Burchell's.*	C. B. S.	1812.	——	H. ⊙.	
3 mèdia. DC.	(*pu.*)	intermediate.	Britain.	....	——	H. ⊙.	Eng. bot. 943.
capreolàta. E.B.							
4 prehénsibilis. K.	(*pu.*)	short-leaved.	Hungary.	1821.	6. 9.	H. ⊙.	
5 alexandrìna. E.	....	Egyptian.	Egypt.	1826.	——	H. ⊙.	
6 officinàlis. DC.	(*re.*)	common.	———	....	——	H. ⊙.	Eng. bot. 589.
β *grandiflòra* DC.	(*re.*)	*large-flowered.*	C. B. S.	1818.	——	H. ⊙.	
7 micrantha. DC.	(*wh.*)	small-blossom'd.	Spain.	1824.	4. 8.	H. ⊙.	
8 parviflòra. DC.	(*wh.*)	small-flowered.	England.	....	8. 9.	H. ⊙.	Eng. bot. 590.
9 Vaillántii. DC.	(*pu.*)	Vaillant's.	France.	....	5. 6.	H. ⊙.	Vail. bot. par. t. 10. f. 6.
10 densiflòra. DC.	(*pu.*)	close-flowered.	Montpelier.	1823.	——	H. ⊙.	
β *albìda.* DC.	(*wh.*)	*white-flowered.*	———	——	——	H. ⊙.	

ORDO XIII.

CRUCIFERÆ. *DC. syst. nat.* 2. *p.* 139.

Subordo I. PLEURORHIZEÆ. DC.

Tribus I. *ARABIDEÆ* seu Pleurorhizeæ Siliquosæ. DC.

MATHIOLA. DC.		Stock. Tetradynamia Siliquosa. L.					
1 simplicicáulis.	(*sc.*)	Brompton.	....	....	5. 8.	H. ♂.	
α coccìnea.	(*sc.*)	*scarlet.*	....	....	——	H. ♂.	Besl. eyst. 2. t. 3. f. 3.
β álba.	(*wh.*)	*white.*	....	....	——	H. ♂.	——— 2. t. 3. f. 2.
2 incàna. DC.	(*va.*)	queen.	England.	....	5. 11.	H. ♄.	Eng. bot. 1935.
α coccìnea.	(*sc.*)	*scarlet.*	———	....	——	H. ♄.	
β purpùrea.	(*pu.*)	*purple.*	———	....	——	H. ♄.	Wein. phyt. t. 642. f. 9.
γ álba.	(*wh.*)	*white.*	———	....	——	H. ♄.	Besl. eyst. 2. t. 3. f. 2.
δ multipléx.	(*va.*)	*double-flowered.*	———	....	——	H. ♄.	——— 2. t. 3. f. 1.
3 ánnua. DC.	(*va*)	ten-week.	S. Europe.	1731.	5 11.	H. ⊙.	
α coccìnea.	(*sc.*)	*scarlet.*	———	——	——	H. ⊙.	
β purpùrea.	(*pu.*)	*purple.*	———	——	——	H. ⊙	J. Bauh. hist. 2. f. 1.
δ rùbra.	(*re.*)	*red.*	———	——	——	H. ⊙.	
ε ròsea.	(*ro.*)	*rose-coloured.*	———	——	——	H. ⊙.	
ζ cárnea.	(*fl.*)	*flesh-coloured.*	———	——	——	H. ⊙.	
η fuscàta.	(*br.*)	*brown.*	———	——	——	H. ⊙.	
ϑ fúlva.	(*br.*)	*tawny.*	———	——	——	H. ⊙.	
ι lilacìna.	(*li.*)	*lilac.*	———	——	——	H. ⊙.	
κ álba.	(*wh.*)	*white.*	———	——	——	H. ⊙.	Dalech. lug. 802. f. 1.
λ variegàta.	(*v.*)	*variegated.*	———	——	——	H. ⊙.	
μ flòre plèno.	(*va.*)	*double-flowered.*	———	——	——	H. ⊙.	
4 glábra. DC.	(*va.*)	Wall-flower-l^d.		....	——	H. ♄.	Wein. phyt. t. 642. f. 2.
α álba.	(*wh.*)	*white-flowered.*	....	....	——	H. ♄.	Moris. ox. s. 3. t. 8. f. 2.
β purpùrea.	(*pu.*)	*purple.*	....	....	——	H. ♄.	
γ coccìnea.	(*sc.*)	*scarlet.*	....	....	——	H. ♄.	
δ flòre plèno.	(*va.*)	*double-flowered.*	....	....	——	H. ♄.	
5 græca. DC.	(*wh.*)	smooth-leaved.	Greece.	....	——	H. ⊙.	
6 fenestràlis. DC.	(*re.*)	window.	Crete.	1759.	7. 8	F. ♄.	Jacq. vind. 2. t. 179.
7 sinuàta. DC.	(*pu.*)	great sea.	England.	....	5. 8.	H. ♂.	Eng. bot. 462.
8 acàulis. DC.	(*pu.*)	stalkless.	Egypt.	1823.	——	H. ⊙.	
9 torulòsa. DC.	(*pu.*)	twisted-podded.	C. B. S.	1816.	——	G. ♄.	
10 tatárica. DC.	(*br.*)	Tartarian.	Tartary.	1826.	——	H. ♃.	Pall. it. 1. app. 117. t. O.
11 odoratíssima. DC.	(*br.*)	sweet-scented.	Tauria.	1797.	6. 7.	G. ♄.	Bot. mag. 1711.
β tanaicénsis. DC.	(*br.*)	*fragrant.*	Tanaim.	1822.	——	G. ♄.	
12 vària. DC.	(*br.*)	entire-leaved.	Greece.	....	5. 7.	G. ♄.	Fl. græc. t. 636.
13 trístis. DC.	(*br.*)	various-leaved.	S. Europe.	1768.	——	G. ♄.	Barr. ic. t. 803.
14 coronopifòlia. DC.	(*pu.*)	Buckshorn-l^d.	———	1825.	——	H. ♃.	Flor. græc. t. 637.
15 lívida. DC.	(*br.*)	livid.	Egypt.	1828.	——	H. ⊙.	Deles. ic. sel. 2. t. 12.
16 longipétala. DC.	(*br.*)	long-petaled.	Bagdad.	1819.	——	H. ⊙.	Vent. cels. t. 93.
17 tricuspidàta. DC.	(*pu.*)	three-forked.	Barbary.	1739.	7. 10.	H. ⊙.	Swt. br. fl. gar. 46.
18 parviflòra. DC.	(*pu.*)	small-flowered.	Morocco.	1799.	——	H. ⊙.	
CHEIRA'NTHUS. DC		Wall-flower. Tetradynamia Siliquosa. L.					
1 Chéiri. DC.	(*ye.*)	garden.	S. Europe.	1573.	4. 7.	H. ♄.	Schkuhr hand. 2. t. 184.
α vulgàris.	(*ye.*)	*common.*	———	——	——	H. ♄.	Besl. eyst. 2. t. 5. f. 3.
β flòre plèno.	(*ye.*)	*double-yellow.*	———	——	——	H. ♄.	——— 2. t. 4. f. 2. 3.
γ grandiflòrus.	(*ye.*)	*large-flowered.*	———	——	——	H. ♄.	——— 2. t. 5. f. 2.
δ máximus.	(*ye.*)	*largest.*	———	——	——	H. ♄.	——— 2. t. 5. f. 1.
ε serràtus.	(*ye.*)	*saw-leaved.*	———	——	——	H. ♄.	
ζ pátulus.	(*ye.*)	*double-spreading.*	———	——	——	H. ♄.	
ι ferrugíneus.	(*p.y.*)	*rusty-flowered.*	———	——	——	H. ♄.	
κ várius.	(*p.y*)	*various-coloured.*	———	——	——	H. ♄.	
λ sanguíneus.	(*p.y.*)	*bloody.*	———	——	——	H. ♄.	
μ hæmánthus.	(*p.y.*)	*double-bloody.*	———	——	——	H. ♄.	
ν flavéscens.	(*ye.*)	*pale-yellow.*	———	——	——	H. ♄.	
ξ thyrsoídeus.	(*ye.*)	*bunch-flowered.*	———	——	——	H. ♄.	

2 fruticulòsus. E.B.	(*ye.*)	common.	England.	....	4. 7.	H. ♄.	Eng. bot. 1934.
3 alpìnus. DC.	(*ye.*)	Alpine.	Norway.	1823.	——	H. ♃.	Wahl. fl. lap. t. 12. f. 1.
4 ochroleúcus. DC.	(*st.*)	pale yellow.	Switzerland.	1820.	6. 8.	H. ♃.	
decúmbens. W. en.							
5 tenuifòlius. DC.	(*ye.*)	fine-leaved.	Madeira.	1777.	5. 6.	F. ♄.	
6 mutábilis. DC.	(*li.*)	changeable.	———	——	3. 5.	F. ♄.	Bot. mag. 195.
7 longifólius. V.	(*li.*)	long-leaved.	———	1815.	9. 12.	F. ♄.	Vent. malm. t. 83.
8 scopárius. DC.	(*va.*)	rock.	Teneriffe.	——	3. 8.	F. ♄.	
α purpuráscens. DC.	(*p.*)	*purplish.*	———	——	——	F. ♄.	
β æruginòsus. DC.	(*br.*)	*rusty.*	———	——	——	F. ♄.	
γ cham'æleo. DC.	(*p.y.*)	*Chameleon.*	———	——	——	F. ♄.	Bot. reg. 219.
Cheíri, var. *Cham'æleon.* B.R.							
9 semperflòrens. DC.	(*w.*)	ever-blowing.	Barbary.	——	1. 12.	F. ♄.	
10 frutéscens. P.S.	(*wh.*)	frutescent.	———	——	3. 7.	F. ♄.	
11 linifòlius. DC.	(*pu.*)	flax-leaved.	Spain.	——	3. 8.	F. ♄.	
12 arboréscens. Sie.	(*pu.*)	arborescent.	Levant.	1827.	4. 8.	G. h.	
NASTU'RTIUM. DC.		NASTU'RTIUM.	Tetradynamia Siliquosa. L.				
officinále. DC.	(*wh.*)	Water Cress.	Britain.	....	5. 7.	H.*w.* ♃.	Eng. bot. 855.
Sisy'mbrium Nastúrtium. W.							
β præcòcius. DC.	(*wh.*)	*early.*	———	——	——	H.*w.* ♃.	Pet. h. brit. t. 47. f. 3.
2 sylvéstre. DC.	(*ye.*)	creeping.	———	....	6. 9.	H.*w.* ♃.	Eng. bot. 2324.
3 ancéps. DC.	(*ye.*)	flat-edged.	N. Europe.	1826.	——	H.*w.* ♃.	Flor. dan. t. 984.
4 palústre. DC.	(*ye.*)	marsh.	Britain.	....	——	H.*w.* ⊙.	Eng. bot. 1747.
Sisy'mbrium terréstre. E.B.							
5 sagittàtum. DC.	(*ye.*)	arrow-leaved.	Siberia.	1780.	5. 6.	H. ♃.	Jacq. ic. 1. t. 122.
6 lippizénse. DC.	(*ye.*)	Carinthian.	Carinthia.	1823.	——	H. ♃.	——— 3. t. 505.
7 pyrenàicum. DC.	(*ye.*)	Pyrenean.	Pyrenees.	1775.	——	H. ♃.	All. fl. ped. t. 18. f. 1.
8 amphíbium. DC.	(*ye.*)	amphibious.	Britain.	....	6. 8.	H.*w.* ♃.	Eng. bot. 1840.
α indivìsum. DC.	(*ye.*)	*brd. wat.-radish.*	———	....	——	H.*w.* ♃.	Pet. h. brit. t. 49. f 8.
β variifòlium. DC.	(*ye.*)	*jag. wat.-radish.*	———	....	——	H.*w.* ♃.	——— t. 49. f. 10.
9 benghalénse. DC.	(*wh.*)	Bengal.	Bengal.	....	6. 10.	H. ⊙.	
10 clandestìnum. DC.	(*wh.*)	small-flowered.	Brazil.	1821.	——	H. ⊙.	
11 índicum. DC.	(*gr.*)	Indian.	E. Indies.	——	——	H. ⊙.	
12 atrovìrens. H.	(*ye.*)	dark green.	China.	1826.	——	H. ⊙.	
LEPTOCA'RPÆA. DC.		LEPTOCA'RPÆA.	Tetradynamia Siliquosa. L.				
Lœsélii. DC.	(*ye.*)	Loeselius's.	Germany.	1683.	5. 8.	H. ⊙.	Jacq. aust. t. 324.
Sisy'mbrium Lœsélii. W. *Turrìtis Lœsélii.* H. K.							
NOTO'CERAS. DC.		NOTO'CERAS.	Tetradynamia Siliquosa. L.				
1 canariénse. DC.	(*ye.*)	Canary.	Canaries.	1779.	8. 9.	H. ⊙.	Jacq. fil. eclog. t. 111.
Ery'simum bicórne. W.							
2 hispánicum. DC.	(*ye.*)	Spanish.	Spain.	1822.	3. 10.	H. ⊙.	Deles. ic. sel. 2. t. 17.
3 quadricorne. DC.	(*ye.*)	four-horned.	Siberia.	1824.	5. 10.	H. ⊙.	——— 2. t. 16.
BARBAR'EA. DC.		WINTER-CRESS.	Tetradynamia Siliquosa. L.				
1 vulgàris. DC	(*ye.*)	common.	Britain.	....	5. 8.	H. ♃.	Eng. bot. 443.
Ery'simum Barbarèa. W.							
2 pr'æcox, DC.	(*ye.*)	early.	England.	....	4. 10.	H. ♃.	——— 1129.
3 táurica. DC.	(*ye.*)	Taurian.	Tauria.	1825.	7. 9.	H. ♃.	
4 ibèrica. DC.	(*ye.*)	Iberian.	Iberia.	1816.	5. 8.	H. ♃.	
Cheiránthus ibèricus. W. en.							
5 plantagínea. DC.		Plaintain-leav'd.	Levant.	1823.	——	H. ♃.	Deles. ic. sel. 2. t. 19.
STEVE'NIA. DC.		STEVE'NIA.	Tetradynamia Siliquosa. L.				
1 alyssóides. DC.	(*wh.*)	Alyssum-like.	Siberia..	1823.	6. 8.	H. ⊙.	Deles. ic. sel. 2. t. 20.
2 cheiranthóides. DC.	(*pu.*)	Wall-flow.-like.	———	——	——	H. ♃.	——— 2. t. 21.
BRA'YA. DC.		BRA'YA.	Tetradynamia Siliquosa. DC.				
1 alpìna. DC.	(*pu.*)	Alpine.	Carinthia.	1821.	4. 5.	H. ♃.	Deles. ic. sel. 2. t. 22.
2 glábella. DC.	(*pu.*)	smooth.	Arctic R.	1827.	——	H. ♃.	
TURRI'TIS. DC.		TOWER-MUSTARD.	Tetradynamia Siliquosa. L.				
1 glábra. DC.	(*wh.*)	long-podded.	England.	....	5. 6.	H. ⊙.	Eng. bot. 777.
β ramòsa. DC.	(*wh.*)	branching.	....	....	——	H. ⊙.	Clus. hist. 2. p. 126. f. 2.
2 salsuginòsa. DC.	(*wh.*)	shorter-podded.	Siberia.	1822.	——	H. ⊙.	Pal. it. 2. app. n. 114. t. V.
A'RABIS. DC.		WALL-CRESS.	Tetradynamia Siliquosa. L.				
1 vérna. DC.	(*pu.*)	vernal.	S. Europe.	1710.	4. 5.	H. ⊙.	Flor. græc. t. 641.
Hésperis vérna. W.							
2 roseà. DC.	(*ro.*)	rose-coloured.	Calabria.	1827.	——	H. ♂.	Deles. ic. sel. 2. t. 23.
3 alpìna. DC.	(*wh.*)	Alpine.	Pyrenees.	1596.	3. 5.	H. ♃.	Bot. mag. 226.
β Clusiàna. DC.	(*wh.*)	*Clusius's.*	———	——	——	H. ♃.	Clus. hist. 2. p. 125. f. 2.
γ nàna. DC.	(*wh.*)	*dwarf.*	Hungary.	....	——	H. ♃.	
4 álbida. DC.	(*wh.*)	early-flowered.	Tauria.	1798.	1. 5.	H. ♃.	Jacq. f. ecl. 1. t. 71.
caucàsica. B.M. 2046.							

5	undulàta. L. en.	(*wh.*)	wave-leaved.	S. Europe.	1823.	5. 8.	H. ⊙.	
6	thyrsòidea. DC.	(*wh.*)	bunch-flowered.	Levant.	1826.	——	H. ♃.	Flor. græc. t. 642.
7	longifòlia. DC.	(*wh.*)	long-leaved.	Persia.	1820.	——	H. ♃.	Deles. ic. sel. 2. t. 25.
8	móllis. DC.	(*wh.*)	soft.	Caucasus.	1825.	5. 7.	H. ♂.	
9	toxophy'lla. DC.	(*wh.*)	bow-leaved.	Volga.	1828.	——	H. ♂.	
10	auriculàta. DC.	(*wh.*)	ear-leaved.	S. Europe.	1805.	5. 6.	H. ⊙.	
	áspera. All. ped. t. 2. f. 2.							
11	saxátilis. DC.	(*wh.*)	rock.	————	——	——	H. ♂.	Vill. dauph. 3. t. 37.
12	crispàta. DC.	(*wh.*)	curled-leaved.	Carniola.	1816.	3. 4.	H. ♃.	
13	sagittàta. DC.	(*wh.*)	arrow-leaved.	Europe.	....	5. 7.	H. ♂.	
14	hirsùta. DC.	(*wh.*)	hairy.	Britain.	....	——	H. ♂.	Eng. bot. 587.
15	stenopètala. W. en.	(*w.*)	narrow-petal'd.	....	1822.	6. 8.	H. ⊙.	
16	curtisíliqua. DC.	(*wh.*)	short-podded.	Scania.	1825.	5. 7.	H. ♂.	
17	Alliònii. DC.	(*wh.*)	Allioni's.	Italy.	....	5. 7.	H. ♃.	
18	muràlis. DC.	(*wh.*)	dwarf.	Switzerland.	1824.	5. 6.	H. ♃.	
19	stricta. DC.	(*wh.*)	Bristol.	England.	....	——	H. ♃.	Eng. bot. 614.
20	ciliàta. DC.	(*wh.*)	ciliated.	Ireland.	....	6. 7.	H. ♂.	———— 1746.
	Turrítis álpina. E.B.							
21	incàna. DC.	(*wh.*)	hoary.	Switzerland.	1816.	5. 6.	H. ♂.	
22	Thaliàna. DC.	(*wh.*)	common.	Britain.	....	4. 6.	H. ⊙.	Eng. bot. 901.
23	serpyllifòlia. DC.	(*wh.*)	Thyme-leaved.	Pyrenees.	1824.	6. 7.	H. ♂.	
24	procúrrens. DC.	(*wh.*)	procurrent.	Hungary.	——	5. 6.	H. ♃.	W. et. Kit. hun. 2. t. 141.
25	præ'cox. DC.	(*wh.*)	early-flowering.	————	1822.	4. 6.	H. ♃.	
26	Schivereckiàna. DC.	(*w.*)	Schivereck's.	Austria.	1827.	——	H. ♃.	
27	petr'æa. DC.	(*wh.*)	short-podded.	Europe.	....	5. 7.	H. ♃.	
	α *Crantziàna.* DC.	(*w.*)	*Crantz's.*	Germany.	....	——	H. ♃.	Crantz. aust. t. 3. f. 2.
	β *híspida.* DC.	(*wh.*)	*hispid.*	Wales.	....	——	H. ♃.	Flor. dan. t. 386.
	γ *hastulàta.* DC.	(*wh.*)	*hastulate.*	Britain.	....	——	H. ♃.	Eng. bot. 409.
28	ambígua. DC.	(*wh.*)	ambiguous.	Siberia.	1824.	5. 7.	H. ♃.	
29	lyràta. DC.	(*wh.*)	lyre-leaved.	N. America.	1823.	5. 6.	H. ♃.	
30	arenòsa. DC.	(*ro.*)	purple.	Europe.	1798.	6. 7.	H. ♃.	Schranck. m. 3. t. 256.
31	Hallèri. DC.	(*wh.*)	Haller's.	————	1820.	5.	H. ♂.	W. et Kit. hun. 2. t. 120.
32	stolonífera. DC.	(*wh.*)	stoloniferous.	Carniola.	1824.	——	H. ♃.	Scop. carn. ed. 2. t. 39.
33	ovirènsis. DC.	(*ro.*)	reddish.	Carinthia.	——	5. 6.	H. ♃.	Jacq. ic. 1. t. 125.
34	cebennénsis. DC.	(*re.*)	pale-violet.	S. France.	1824.	5. 7.	H. ♂.	Deles. ic. sel. 2. t. 26.
35	lasiolòba. DC.	(*wh.*)	woolly-podded.	Mexico.	——	——	H. ♂.	
36	Turríta. DC.	(*wh.*)	tower.	England.	....	4. 5.	H. ♂.	Eng. bot. 178.
37	péndula. DC.	(*wh.*)	pendulous.	Siberia.	1752.	5. 6.	H. ⊙.	Jacq. vind. 3. t. 34.
38	Patriniàna. DC.	(*wh.*)	Patrin's.	————	1826.	——	H. ♂.	Deles. ic. sel 2. t. 27.
39	oxyòta. DC.	(*wh.*)	sharp-eared.	————	——	——	H. ♂.	
40	lùcida. DC.	(*wh.*)	shining-leaved.	Hungary.	1790.	6. 7.	H. ♃.	
41	canadénsis. DC.	(*wh.*)	sickle-podded.	N. America.	1768.	5. 7.	H. ♃.	Deles ic. sel. 2. t. 28.
42	pùmila. DC.	(*wh.*)	nodding.	Switzerland.	1658.	4. 7.	H. ♃.	Jacq. fl. aust. t. 281.
	nùtans. B.M. 2219. *ciliaris.* W. en.							
43	bellidifòlia. DC.	(*wh.*)	Daisy-leaved.	Pyrenees.	1773.	5. 6.	H. ♃.	———— 3. t. 280.
44	cœrùlea. DC.	(*bl.*)	blue.	————	1793.	6. 7.	H. ♃.	Sturm. deut. fl. ic.
45	collìna. DC.	(*wh.*)	hill.	Naples.	1823.	——	H. ♃.	
46	dasycárpa. DC.	(*wh.*)	thick-podded.	Poland.	1826.	5. 7.	H. ♃.	

MACROP'ODIUM. DC. MACROP'ODIUM. Tetradynamia Siliquosa. L.

	nivàle. DC.	(*wh.*)	Siberian.	Siberia.	1796.	6. 9.	H. ♃.	Pall. it. 2. ap. n. 113. t. U.
	Cardámine nivàlis. Pall.							

CARDA'MINE. DC. LADY'S SMOCK. Tetradynamia Siliquosa. L.

1	rhomboídea. DC.	(*wh.*)	rhomboid.	N. America.	1825.	5. 7.	H. *w.* ♃.	Pluk. amal. t. 435. f. 6.
2	rotundifòlia. DC.	(*wh.*)	round-leaved.	————	1823.	5. 6.	H. *w.* ♃.	
3	asarifòlia. DC.	(*wh.*)	kidney-leaved.	Italy.	1710.	6. 7.	H. *w.* ♃.	Bot. mag. 1735.
4	bellidifòlia. DC.	(*wh.*)	Daisy-leaved.	Scotland.	....	4. 6.	H. ♃.	
	α *petiolàris.* DC.	(*wh.*)	*long-petioled.*	Lapland.	....	——	H. ♃.	Œd. fl. Dan. t. 20.
	β *alpìna.* DC.	(*wh.*)	*Alpine.*	Scotland.	....	——	H. ♃.	Eng. bot. 2355.
5	resedifòlia. DC.	(*wh.*)	Rocket-leaved.	Germany.	1658.	7.	H. ⊙.	Sturm. fl. germ. ic.
6	microphy'lla. DC.	(*wh.*)	small-leaved.	Siberia.	1828.	5. 7.	H. ♃.	
7	africàna. DC.	(*wh.*)	African.	C. B. S.	1691.	5. 6.	G. ♃.	Pluk. alm. t. 101. f. 5.
8	trifòlia. DC.	(*wh.*)	three-leaved.	Europe.	1629.	3. 6.	H. *w.* ♃.	Bot. mag. 452.
9	amàra. DC.	(*wh.*)	bitter.	Britain.	....	4. 5.	H. *w.* ♃.	Eng. bot. 1000.
10	prorèpens. DC.	(*wh.*)	creeping.	Siberia.	1825.	4. 6.	H. *w.* ♃.	
11	praténsis. DC.	(*li.*)	Cuckoo-flower.	Britain.	....	——	H. *w.* ♃.	Eng. bot. 776.
	β *flòre plèno.*	(*li.*)	*double-flowered.*	————	....	——	H. *w.* ♃.	Clus. hist. 2. p. 129. f. 1.
12	dentàta. DC.	(*wh.*)	toothed.	Poland.	1823.	——	H. *w.* ♃.	
13	hirsùta. DC.	(*wh.*)	hairy.	Britain.	....	1. 12.	H. ⊙.	Eng. bot. 492.
14	sylvática. DC.	(*wh.*)	wood.	Europe.	....	5. 6.	H. ⊙.	Flor. dan. t. 735.

15 umbròsa. DC. (*wh.*) smooth. Europe. .... 4. 6. H. ⊙.
16 parviflòra. DC. (*wh.*) small-flowered. ——— 1822. 4. 5. H. ⊙. Gmel. fl. sib. 3. t. 64.
17 impàtiens. DC. (*wh.*) impatient. Britain. .... 4. 6. H. ⊙. Eng. bot. 80.
18 latifòlia. DC. (*pu.*) broad-leaved. Spain. 1710. 6. 8. H.*w.*♃. Herm. par. p. 203. t. 69.
19 macrophy'lla. DC. (*li.*) large-leaved. Siberia. 1824. —— H. ♃.
20 chelidònia. DC. (*pu.*) Celandine-l^d^. Italy. 1739. 3. 5. H. ♃. W. et Kit. hung. 2. t. 140.
21 thalictròides. DC. (*wh.*) Thalictrum-l^d^. Switzerland. 1824. 6. 7. H. ♂. All. ped. n. 951. t. 57. f. 1.
22 glaùca. DC. (*wh.*) glaucous. Calabria. 1824. 5. 7. H. ♃. Deles. ic. sel. 2. t. 31.
23 occúlta. DC. (*gr.*) hidden-flower'd. China. —— 5. 9. H. ⊙.

PTERONE'URUM. DC. PTERONE'URUM. Tetradynamia Siliquosa. L.
1 carnòsum. DC. (*wh.*) fleshy. Dalmatia. 1824. 6. 8. H. ♃. W. et Kit. hung. 2. t. 129.
2 græ'cum. DC. (*wh.*) Grecian. S. Europe. 1710. —— H. ⊙. Flor. græc. t. 631.
Cardámine græ'ca. W.

DENT'ARIA. DC. TOOTH-WORT. Tetradynamia Siliquosa. L.
1 polyphy'lla. DC. (*st.*) many-leaved. Hungary. 1817. 5. 6. H. ♃. W. et Kit. hung 2. t. 160.
2 enneaphy'lla. DC. (*wh.*) nine-leaved. Austria. 1656. —— H. ♃. Jacq. fl. aust. t 316.
3 glandulòsa. DC. (*pu.*) glandular. Hungary. 1815. —— H. ♃. W. et Kit. hung. 3. t. 272.
4 laciniàta. DC. (*wh.*) laciniate. N. America. 1823. 4. 6. H. ♃.
5 diphy'lla. DC. (*wh.*) two-leaved. ——— 1810. 5. 6. H. ♃. Bot. mag. 1465.
6 trifòlia. DC. (*wh.*) three-leaved. Hungary. 1824. —— H. ♃. W. et K. hung. 2. t. 139.
7 digitàta. DC. (*w.p.*) fingered. Switzerland. 1656. —— H. ♃. Bot. mag. 2202.
pentaphy'lla. B.M. *Cardamine pentaphy'lla.* H. K. ed. 2.
8 pinnàta. DC. (*wh.*) seven-leaved. ——— 1683. —— H. ♃. Boiss. fl. eur. t. 449.
9 quinquefòlia. DC. (*wh.*) five-leaved. Caucasus. 1823. 4. 5. H. ♃. Deles. ic. sel. 2. t. 33.
10 bulbífera. DC. (*bh.*) bulbiferous. England. .... —— H. ♃. Eng. bot. 309.
Cardámine bulbífera. H.K. ed. 2.
11 tenuifòlia. DC. (*pu.*) slender-leaved. Siberia. 1825. 5. 6. H. ♃. Gmel. sib. 3. n. 41. t. 65.
trífida. Lam. ill. t. 562. f. 2.

PA'RRYA. B. PA'RRYA. Tetradynamia Siliquosa.
1 árctica. B. (*pu.*) arctic. Melvill Isl. 1827. 5. 6. H. ♃.
2 arabidiflòra. (*pu.*) Arabis-flower'd. Siberia. 1798. —— H. ♃. Linn. am. ac. 2. t. L. f. 20.
Arabis grandiflòra. L. *Neurolòma arabidiflòrum.* DC.

PLATYPE'TALUM. B. PLATYPE'TALUM. Tetradynamia Siliquosa.
purpuráscens. B. (*pu.*) purplish. Melvill Isl. 1827. 5. 6. H. ♃.

Tribus II. *ALYSSINEÆ* seu Pleurorhizeæ Latiseptæ. DC.

LUN'ARIA. DC. HONESTY. Tetradynamia Siliculosa. L.
1 redivìva. DC. (*li.*) perennial. Germany. 1596. 4. 6. H. ♃. Hill veg. syst. t. 38. f. 1.
2 biénnis. DC. (*w.p.*) biennial. ——— 1570. —— H. ♂. ——— t. 38. f. 2.
ánnua. W.
α violàcea. (*pu.*) *violet-flowered.* ——— —— —— H. ♂.
β albiflòra. (*wh.*) *white-flowered.* ——— —— —— H. ♂.

RIC'OTIA. DC. RIC'OTIA. Tetradynamia Siliculosa. L.
Lunària. DC. Syrian. Syria. 1757. 6. 7. H. ⊙. Lam. ill t. 561.
ægyptìaca. Bot. Reg. 49.

FARS'ETIA. DC. FARS'ETIA. Tetradynamia Siliculosa. L.
1 ægyptìaca. DC. (*br.*) Stock-like. Levant. 1788. 6. 7. G. ♄. Desf. atl. 2. t. 160.
cheiranthòides. H.K. *Cheiránthus Farsètia.* L.

FIB'IGIA. Med. FIB'IGIA. Tetradynamia Siliculosa.
1 suffruticòsa. (*pu.*) suffrutescent. Persia. 1820. 4. 5. G. ♄. Vent. cels. t. 19.
Lunària suffruticòsa. Vent. *Farsètia suffruticòsa.* DC.
2 lunarioídes. (*ye.*) oriental. Greece. 1731. 6. 7. H. ♃. Tourn. ic ed. ger. 1 t. 30.
3 clypeàta. Med. (*ye.*) buckler-podded. S. Europe. 1596. —— H. ♂. Schkuhr. han. 2. t 181.
Aly'ssum clypeàtum. L. *Farsètia clypeàta.* DC.

BERTER'OA. DC. BERTER'OA. Tetradynamia Siliculosa. L.
1 incàna. DC. (*wh.*) hoary. Europe. 1640. 7. 9. H. ♂. Horn. fl. dan. 1461.
Aly'ssum incànum. W.
2 mutábilis. DC. (*bh.*) changeable. Levant. 1802. 7. 8. H. ♃. Vent. cels. t. 85.
3 oblíqua. DC. (*wh.*) oblique-podded. Naples. 1824. —— H. ♄. Flor. græc. t. 623.

AUBRI'ETIA. DC. AUBRI'ETIA. Tetradynamia Siliculosa. L.
1 deltoídea. DC. (*pu.*) spreading. Levant. 1710. 3. 5. H. ♃. Bot. mag. 126.
Farsètia deltoídea. H.K. *Aly'ssum deltoídeum.* L.
2 purpurea. DC. (*pu.*) tufted. ——— 1821. —— H. ♃. Swt. br. fl. gar. 207.

VESIC'ARIA. VESIC'ARIA. Tetradynamia Siliculosa. L.
1 utriculàta. DC. (*ye.*) smooth. Levant. 1739. 4 6. H. ♃. Bot. mag. 130.
Aly'ssum utriculàtum. B.M.
2 reticulàta. DC. (*ye.*) veined-podded. ——— .... 6. H. ♃. Tour. voy 2. p. 252. ic.

3 Ludoviciàna. DC. (*ye.*) Missouri. Missouri. 1825. 5. 6. H. ♃.
4 sinuàta. DC. sinuate-leaved. Spain. 1596. 4. 6. H. ♂. Schkuhr. han. 2. t. 181.
5 crética. DC. Cretan. Crete. 1739. 5. 8. H. ♃. Alp. ex. p.117 et 118. ic.
Aly'ssum créticum. W.
6 árctica. DC. (*ye.*) arctic. Greenland. 1826. 5. 6. H. ♃. Horn. fl. dan. t. 1520.
7 arenòsa. DC. (*ye.*) sand. Arctic Reg. —— 6. 8. H. ♃.
SCHIVERE'CKIA. DC. SCHIVERE'CKIA. Tetradynamia Siliculosa. L.
podòlica. DC. (*wh.*) canescent. Podolia. 1821. 5. 7. H. ♃. Swt. br. fl. gar. t. 77.
KO'NIGA. B. KO'NIGA. Tetradynamia Siliculosa.
1 marítima. DC. (*wh.*) sweet-scented. S. Europe. .... 5. 10. H. ♃. Eng. bot. 1729.
Aly'ssum marítimum. E. B. *halimifólium.* B.M. 101.
2 canariénsis. (*wh.*) Canary. Canaries. 1812. —— G. ♄.
Aly'ssum marítimum. β canariénse. DC.
β variegàta. (*wh.*) *striped-leaved.* ———— .... —— G. ♄.
ADYS'ETON. Scop. ADYS'ETON. Tetradynamia Siliculosa.
1 saxátile. (*ye.*) rock. Russia. 1710. 4. 5. H. ♄. Bot. mag. 159.
Aly'ssum saxátile. D. *Aurínia saxátilis.* Dv.
2 gemonénse. (*ye.*) canescent. Germany. —— —— H. ♄. Jacq. ic. 3. t. 503.
3 orientàle. (*ye.*) oriental. Crete. .... —— H. ♄. Flor. græc. t. 625.
4 spathulàtum. (*ye.*) spathulate. Siberia. 1821. 4. 6. H. ♄. Deles. ic. sel. 2. t. 37.
5 argénteum. (*ye.*) silvery. Switzerland. —— 5. 6. H. ♄. All. ped. t. 54. f. 3.
6 Bertolònii. (*ye.*) Bertoloni's. Italy. 1825. —— H. ♄.
7 muràle. (*ye.*) wall. Europe. 1801. 6. 8. H ♄. W. et Kit. hung. 1. t. 6.
8 obtusifòlium. (*ye.*) blunt-leaved. Siberia. 1812. —— H. ♄. Deles. ic. sel. 2. t. 38.
9 serpyllifòlium. (*ye.*) Thyme-leaved. S. Europe. 1823. 5. 7. H. ♄.
10 rèpens. (*ye.*) creeping. Transylvania. 1827. —— H. ♃.
11 vernàle. (*ye.*) vernal. .... 1823. 4. 5. H. ♃. Schrank hort. mon. t. 96.
12 tortuòsum. (*ye.*) twisted. Hungary. 1804. 6. 7. H. ♃. W. et Kit. hung. 1. t. 91.
13 alpéstre. (*ye.*) Italian. Italy. 1777. 6. 10 H. ♄. Flor. græc. t. 624.
14 nebrodènse. (*ye.*) Sicilian. Sicily. 1827. 5. 6. H. ♄.
15 Marschalliànum. (*ye.*) Marschall's. Caucasus. —— —— H. ♄.
16 montànum. SC. (*ye.*) mountain. Germany. 1713. 5. 8. H. ♃. Bot. mag. 419.
Aly'ssum montànum. B.M.
17 cuneifòlium. (*ye.*) wedge-leaved. Naples. 1825. 5. 8. H. ♃.
18 diffùsum. (*ye.*) spreading. ———— —— H. ♃.
19 Wulfeniànum. (*ye.*) Wulfen's. Carinthia. 1821. —— H. ♃.
20 Fischeriànum. (*ye.*) Fischer's. Siberia. 1828. —— H. ♃.
Aly'ssum Fischeriànum. DC.
ALY'SSUM. B. MADWORT. Tetradynamia Siliculosa. L.
1 umbellàtum. DC. (*ye.*) umbelled. Tauria. 1828. 5. 8. H. ⊙.
2 rostràtum. DC. (*ye.*) beaked. ———— 1824. 6. 9. H. ⊙. Stev. m. ac. P. 3. t. 15. f. 1.
3 micropètalum. DC. (*ye.*) small-petaled. Iberia. —— 5. 8. H. ⊙. Deles. ic. sel. 2. t. 39.
4 hirsùtum. DC. (*ye.*) hairy. ———— —— —— H. ⊙. ———— 2. t. 40.
5 campéstre. DC. (*ye.*) field. Europe. 1768. 6. 8. H. ⊙. Flor. græc. t. 626.
6 calycìnum. DC. (*ye.*) persistent-cal'd. ———— 1740. —— H. ⊙. Jacq. aust. t. 338.
7 mínimum. DC. (*ye.*) small. ———— 1791. —— H. ⊙. Tratt. thes. bot. t. 35.
ANODO'NTEA. DC. ANODO'NTEA. Tetradynamia Siliculosa.
1 edéntula. (*ye.*) toothless. Hungary. 1821. 5. 6. H. ♂. W. et K. hung. 1. t. 92.
Aly'ssum edéntulum. DC. *Adysèton edéntulum.* L. enum.
2 dasycárpa. (*ye.*) thick-podded. Siberia. 1827. 6. 10. H. ⊙. Trev. ber. mag. 1816. t. 2.
3 rupéstris. (*wh.*) silvery rock. Italy. 1824. 5. 10. H. ♃. Tenor. fl. nap. t. 60.
4 halimifòlia. (*wh.*) Halimus-leaved. ———— —— —— H. ♄. All. ped. n. 900. t. 54. f. 1.
Aly'ssum halimifòlium. DC.
5 spinòsa. (*wh.*) thorny. S. Europe. 1683. 6. 8. H. ♄. Barrel. ic. t. 808.
6 macrocárpa. (*wh.*) large-podded. ———— 1825. —— H. ♄. Deles. ic. sel. 2. t. 41.
7 canéscens. (*wh.*) hoary. Siberia. 1828. 4. 9. H. ♄.
MENI'OCUS. DC. MENI'OCUS. Tetradynamia Siliculosa. L.
línifolius. DC. (*ye.*) Flax-leaved. Levant. 1822. 5. 6. H. ⊙. Deles. ic. sel. 2. t. 42.
Aly'ssum línifolium. W.
CLYPE'OLA. DC. TREACLE-MUSTARD. Tetradynamia Siliculosa. L.
1 Jonthláspi. DC. (*ye.*) common. S. Europe. 1710. 5. 7. H. ⊙. Desv. j. bot. 3. t. 25. f. 7.
2 erióphora. DC. (*wh.*) woolly-podded. Spain. 1820. —— H. ⊙. ———— 3. t. 25. f. 10.
PELT'ARIA. DC. PELT'ARIA. Tetradynamia Siliculosa. L.
alliàcea. DC. (*wh.*) garlic-scented. Austria. 1601. 5. 7. H. ♃. Jacq. aust. 2. t. 123.
PETROCA'LLIS. DC. PETROCA'LLIS. Tetradynamia Siliculosa. L.
pyrenàica. DC. (*ro.*) Pyrenean. Pyrenees. 1759. 5. 6. H. ♃. Lodd. bot. cab. 635.
Dràba pyrenàica. B.M. 713.
DR'ABA. DC. WHITLOW-GRASS. Tetradynamia Siliculosa. L.
1 aizóides. DC. (*ye.*) sen-green. Wales. .... 2. 4. H. ♃. Eng. bot. 1271.

2	brachystèmon. DC.	(*ye.*)	short-stamened.	....	1731.	2. 4.	H. ♃.	Bot. mag. 170.
	aizòides. B.M. *ciliàris.* H.K.							
3	aizòon. DC.	(*ye.*)	hairy-podded.	Hungary.	1821.	——	H. ♃.	
4	cuspidàta. DC.	(*ye.*)	sharp-pointed.	Tauria.	——	4. 5.	H. ♃.	
5	bryóides. DC.	(*ye.*)	Bryum-like.	Caucasus.	——	3. 4.	H. ♃.	
6	bruniæfòlia. DC.	(*ye.*)	Brunia-leaved.	————	1826.	5. 6.	H. ♃.	
7	ericæfòlia. DC.	(*ye.*)	Heath-leaved.	————	1823.	——	H. ♃.	
8	pilòsa. DC.	(*ye.*)	pilose.	Siberia.	1825.	——	H. ♃.	
9	alpìna. DC.	(*ye.*)	Alpine.	Norway.	1823.	4. 5.	H. ♃.	Flor. dan. t. 56.
10	glaciàlis. DC.	(*ye.*)	starry-haired.	Siberia.	1825.	5. 6.	H. ♃.	
11	rèpens. DC.	(*ye.*)	creeping.	Caucasus.	1823.	——	H. ♃.	
12	Gmelíni. DC.	(*ye.*)	Gmelin's.	Siberia.	——	——	H. ♃.	Gmel. sib. 3. t. 56. f. 2.
13	tridentàta. DC.	(*ye.*)	three-toothed.	Caucasus.	1828.	——	H. ♃.	
14	muricélla. DC.	(*wh.*)	tufted.	Norway.	——		H. ♃.	Wahl. fl. lap. t. 11. f. 2.
15	incómpta. DC.	(*ye.*)	straggling.	Caucasus.	——	6.	H. ♃.	Deles. ic. sel. 2. t. 44.
16	oblongàta. DC.	(*wh.*)	oblong-leaved.	Baffin's-bay.	1820.	5. 6.	H. ♃.	
17	corymbòsa. DC.	(*wh.*)	corymbose.	————	——	——	H. ♃.	
18	hírta. DC.	(*wh.*)	hairy.	Lapland.	....	5. 7.	H. ♃.	Wahl. f. lap. t. 11. f. 3.
19	rupéstris. DC.	(*wh.*)	rock.	Scotland.	....	——	H. ♃.	Eng. bot. 1338.
	hírta. E.B. non Linn.							
20	nivàlis. DC.	(*wh.*)	snowy.	Norway.	1824.	5. 6.	H. ♃.	Flor. dan. t. 142.
	stellàta. F.D. nec aliorum.							
21	lappònica. DC.	(*wh.*)	Lapland.	Lapland.	——	——	H. ♃.	Wahl. fl. lap. t. 11. f. 3.
22	helvètica. DC.	(*wh.*)	Swiss.	Switzerland.	1824.	——	H. ♃.	
23	fladnizénsis. DC.	(*wh.*)	fringe-leaved.	————	——	6. 8.	H. ♃.	Jacq. misc. 1. t. 17. f. 1.
24	tomentòsa. DC.	(*wh.*)	woolly-leaved.	————	——	5. 7.	H. ♃.	Wahl. helv. p. 123. t. 3.
25	stellàta. DC.	(*wh.*)	starry.	————	1820.	——	H. ♃.	Jacq. obs. t. 4. f. 3.
	hírta. Jacq. aust. t. 432.							
26	contórta. DC.	(*wh.*)	twisted-podded.	Britain.	....	5. 6.	H. ♂.	Eng. bot. t. 388.
	incàna. E.B.							
27	confùsa. DC.	(*wh.*)	confused.	N. Europe.	....	——	H. ♂.	Flor. dan. t. 130.
	incàna. F.D.							
28	arabìsans. DC.	(*wh.*)	Arabis-like.	N. America.	1827.	——	H. ♂.	
29	daùrica. DC.	(*wh.*)	Daurian.	Dahuria.	1824.	——	H. ♂.	
30	áurea. DC.	(*ye.*)	golden.	Greenland.	——	——	H. ♂.	Flor. dan. t. 1460.
31	lùtea. DC.	(*ye.*)	yellow.	Caucasus.	1825.	5. 8.	H. ☉.	
32	grácilis. GR.	(*ye.*)	slender.	N. America.	1827.	——	H. ☉.	
33	nemoràlis. DC.	(*ye.*)	wood.	Europe, &c.	1759.	——	H. ☉.	Houtt. syst. 4. t. 60. f. 1.
34	muràlis. DC.	(*wh.*)	wall.	England.	....	4. 5.	H. ☉.	Eng. bot. 912.
	ERO'PHILA. DC.		ERO'PHILA.	Tetradynamia Siliculosa. L.				
1	americàna. DC.	(*wh.*)	American.	N. America.	1823.	3. 4.	H. ☉.	
2	vulgàris. DC.	(*wh.*)	common.	Britain.	....	——	H. ☉.	Flor. dan. t. 983.
	Dràba vérna. Eng. bot. 586.							
3	pr'æcox. DC.	(*wh.*)	round-podded.	Caucasus.	1824.	2. 4.	H. ☉.	
	KE'RNERA. Med.		KE'RNERA.	Tetradynamia Siliculosa.				
1	saxátilis.	(*wh.*)	rock.	Europe.	1775.	5. 7.	H. ♃.	Jacq. aust. 2. t. 128.
	myagròides. Med. *Cochleària saxátilis.* DC. *Myàgrum saxátile.* L.							
	β incìsa.	(*wh.*)	*cut-leaved.*	————	——	——	H. ♃.	Mori. ox. 2. s. 3. t. 17. f. ult.
2	auriculàta.	(*wh.*)	eared.	Pyrenees.	1824.	——	H. ♃.	Berg. phyt. 3. p. 140. ic.
	Cochleària auriculàta. DC.							
	ARMOR'ACIA. Plin.		HORSE-RADISH.	Tetradynamia Siliculosa.				
1	rusticàna. Bau.	(*wh.*)	common.	England.	....	5. 6.	H ♃.	Eng. bot. 2223.
	Cochleària Armoràcia. E.B.							
2	macrocárpa. Bau.	(*wh.*)	large-fruited.	Hungary.	1806.	6. 8.	H. ♃.	W. et K. hung. 2. t. 184.
	COCHLE'ARIA. DC.		SCURVY-GRASS.	Tetradynamia Siliculosa. L.				
1	glastifòlia. DC.	(*wh.*)	Woad-leaved.	S. Europe.	1648.	5. 7.	H. ♂.	Schuhr. hand. 2. t. 181.
2	ánglica. DC.	(*wh.*)	English.	England.	....	5.	H. ☉.	Eng. bot. 552.
3	officinàlis. DC.	(*wh.*)	common.	Britain.	....	4. 5.	H. ☉.	———— 551.
	β mìnor. DC.		*lesser.*	————	....	——	H. ☉.	
	γ rotundifòlia. DC.		*round-leaved.*	————	....	——	H. ☉.	
	grœnlándica. With. brit. 573. non. Lin.							
4	pyrenàica. DC.	(*wh.*)	Pyrenean.	Pyrenees.	....	6. 8.	H. ♂.	Deles. ic. sel. 2. t. 48.
5	grœnlándica. DC.	(*wh.*)	Greenland.	Scotland.	....	5. 6.	H. ♃.	Eng. bot. 2403.
6	dánica. DC.	(*wh.*)	Danish.	Britain.	....	——	H. ☉.	———— 696.
7	alpìna.	(*wh.*)	Alpine.	Ireland.	....	5. 6.	H. ♃.	
8	fenestràta. DC.	(*wh.*)	fenestrate.	Baffin's Bay.	1820.	4. 5.	H. ♃.	
9	grandiflòra. DC.	(*wh.*)	large-flowered.	Siberia.	1826.	5. 6.	H. ♃.	
10	integrifòlia. DC.	(*wh.*)	entire-leaved.	Altay.	1827.	——	H. ♃.	
11	acáulis. DC.	(*li.*)	stalkless.	Portugal.	1823.	3. 4.	H. ♃.	Jacq. ecl. t. 132.

Tribus III. *THLASPIDEÆ* seu Pleurorhizeæ Angustiseptæ. DC.

THLA'SPI. DC.		BASTARD-CRESS. Tetradynamia Siliculosa. L.					
1 latifòlium. DC.	(*wh.*)	broad-leaved.	Caucasus.	1828.	4. 5.	H. ♃.	Deles. ic. sel. 2. t. 51.
2 ceratocárpon. DC.	(*wh.*)	horn-podded.	Siberia.	1779.	7.	H. ⊙.	Scop. insub. 1. t. 4.
3 arvénse. DC.	(*wh.*)	Penny-cress.	Britain.	....	6. 7.	H. ⊙.	Eng. bot. 1659.
4 collìnum. DC.	(*wh.*)	hill.	Iberia.	1824.	4. 9.	H. ⊙.	
5 alliàceum. DC.	(*wh.*)	Garlic-scented.	S. Europe.	1714.	5. 7.	H. ⊙.	Jacq. ic. 1. t. 121.
6 stríctum.	(*wh.*)	upright.	Persia.	1827.	4. 9.	H. ⊙.	
7 perfoliàtum. DC.	(*wh.*)	perfoliate.	England.	....	4. 7.	H. ⊙.	Eng. bot. 2354.
8 montànum. DC.	(*wh.*)	mountain.	Europe.	....	5. 6.	H. ♃.	Jacq. aust. 3. t. 237.
9 alpéstre. DC.	(*wh.*)	Alpine.	England.	....	5. 7.	H. ♃.	Eng. bot. t. 81.
10 cochlearifórme. DC.	(*w.*)	Scurvy-grass.	Dahuria.	1827.	6. 7.	H. ♃.	
CA'PSELLA. DC.		SHEPHERDS-PURSE. Tetradynamia Siliculosa. L.					
1 Bùrsa-pastòris. DC.	(*w.*)	common.	Britain.	....	1. 12.	H. ⊙.	Eng. bot. 1485.
Thláspi Búrsa-pastòris. L.							
β *mìnor.* DC.	(*wh.*)	*small.*	———	....	——	H. ⊙.	Tabern. ic. 199.
γ *integrifòlia.* DC.	(*w.*)	*entire-leaved.*	———	....	——	H. ⊙.	Moris. ox. 2. t. 20. f. 1.
δ *coronopifòlia.* DC.	(*w.*)	*Buck's-horn-ld.*	———	....	——	H. ⊙.	
2 procúmbens. B.	(*wh.*)	procumbent.	Europe.	1823.	3. 5	H. ⊙.	Hill. veg. sys. 11. t. 42. f. 1.
Lepídium procúmbens. L. *Hutchínsia procúmbens.* DC.							
HUTCHI'NSIA. DC.		HUTCHI'NSIA. Tetradynamia Siliculosa. L.					
1 rotundifòlia. DC.	(*bh.*)	round-leaved.	Switzerland.	1759.	5. 7.	H. ♃.	Scop. carn. n. 805. t. 37.
Ibèris rotundifòlia. L.							
2 cepeæfòlia. DC.	(*pu.*)	toothed-leaved.	Carinthia.	1824.	4. 5.	H. ♃.	Jacq. misc. 2. p. 28. t. 1.
3 stylòsa. DC.	(*ro.*)	long-styled.	Naples.	1826.	3. 5.	H. ♃.	Bot. mag. 2772.
4 hastulàta. DC.	(*bh.*)	halbert-leaved.	Siberia.	1827.	——	H. ♃.	Gmel. sib. 3. t. 5.
5 calycìna. DC.	(*wh.*)	persistent-cal'd.	———	——	4. 5.	H. ♃.	
6 alpìna. DC.	(*wh.*)	Alpine.	Pyrenees.	1775.	4. 6.	H. ♃.	Jacq. aust. 2. t. 137.
Lepídium alpìnum. L.							
7 petr'æa. DC.		rock.	England.	....	3. 5.	H. ♂.	Eng. bot. 111.
Lepídium petr'æum. L.							
TEESD'ALIA. DC.		TEESD'ALIA. Tetradynamia Siliculosa. L.					
1 Ibèris. DC.		Shepherd's-cress.	England.	....	5. 7.	H. ⊙.	Eng. bot. 327
nudicáulis. H.K. *Ibèris nudicáulis.* L. *Guepínia Ibèris.* DC.							
2 Lepídium. DC.	(*wh.*)	equal-petal'd.	S. Europe.	....	5. 8.	H. ⊙.	Magn. ic. n. car. 247.
Lepídium nudicàule. L. *Teesdàlia regulàris.* L.T. *Guepínia Lepídium.* DV.							
IB'ERIS. DC.		CANDY-TUFT. Tetradynamia Siliculosa. L.					
1 semperflòrens. DC.	(*w.*)	broad-leaved.	Sicily.	1679.	1. 12.	G. ♄.	Moris. ox. 2. t. 25. f. 5.
2 gibraltárica. DC.	(*bh.*)	Gibraltar.	Gibraltar.	1732.	5. 6.	G. ♄.	Bot. mag. 124.
3 saxátilis. DC.	(*wh.*)	rock.	S. Europe.	1739.	4. 6.	H. ♄.	Gou. fl. mons. 177. t. 1.
4 corifòlia.	(*wh.*)	Coris-leaved.	———	——	——	H. ♄.	Clus. hist. 2. p. 132. ic.
saxátilis β corifòlia.		Bot. mag. 1642.					
5 pubéscens. DC.	(*wh.*)	pubescent.	....	1821.	——	H. ♄.	
6 subvelutìna. DC.	(*wh.*)	velvetty.	Spain.	1824.	——	H ♄.	
7 sempervìrens. DC.	(*wh.*)	evergreen.	Crete.	1731.	——	H. ♄.	Flor. græc. t. 620.
8 Garrexiàna. DC.	(*wh.*)	Allioni's.	Pyrenees.	....	6. 8.	H. ♄.	All. ped. n. 920. t. 40. f. 3.
9 conférta. DC.	(*wh.*)	crowded-leaved.	Spain.	1824.	——	H. ♄.	
10 amára. DC.	(*wh.*)	bitter.	England.	....	——	H. ⊙.	Eng. bot. 52.
11 pinnàta. DC.	(*wh.*)	wing-leaved.	S. Europe.	1596.	——	H. ⊙.	Lob. ic. t. 217. f. 2.
12 odoràta. DC.	(*wh.*)	sweet-scented.	Crete.	1806.	——	H. ⊙.	Swt. br. fl. gar. 50.
13 Lagascàna. DC.	(*wh.*)	Lagasca's.	Spain.	1824.	——	H. ⊙.	
14 umbellàta. DC.	(*pu.*)	purple.	S. Europe.	1596.	——	H. ⊙.	Bot. mag. 106.
15 linifòlia. DC.	(*pu.*)	Flax-leaved.	———	1759.	——	H. ♂.	Garid. aix. t. 105.
16 ciliàta. DC.	(*wh.*)	fringed.	———	....	6. 7.	H. ♂.	
17 táurica. DC.	(*wh.*)	Taurian.	Tauria.	1802.	5. 7.	H. ♂.	
18 violàcea. DC.	(*pu.*)	blunt-leaved.	....	1782.	6. 7.	H. ⊙.	
19 nàna. DC.	(*pu.*)	dwarf.	S. Europe.	1822.	5. 7.	H. ♂.	All. auct. p. 15. t. 2. f. 1.
20 spathulàta. DC.	(*pu.*)	spathulate.	Pyrenees.	1826.	6. 8.	H. ⊙.	Berg. phyt. ic.
21 Tenoreàna. DC.	(*wh.*)	Tenore's.	Naples.	——	5. 7.	H. ♄.	Swt. br. fl. gar. 88.
22 Pruìti. DC.	(*wh.*)	Sicilian.	Sicily.	1828.	——	H. ♄.	
23 contrácta. DC.	(*wh.*)	contracted.	Spain.	1824.	4. 6.	H. ♄.	
BISCUTE'LLA. DC.		BUCKLER-MUSTARD. Tetradynamia Siliculosa. L.					
1 auriculàta. DC.	(*ye.*)	ear-podded.	S. Europe.	1683.	6. 7.	H. ⊙.	DC. diss. n. 1. t. 1. f. 2.
2 erigerifòlia. DC.	(*ye.*)	Erigeron-leav'd.	Spain.	1824.	——	H. ⊙.	Deles. ic. sel. 2. t. 55.
3 híspida. DC.	(*ye.*)	hispid.	S. France.	1822.	——	H. ⊙.	Bot. mag. 2444.
4 cichoriifòlia. DC.	(*ye.*)	Succory-leaved.	Pyrenees.	——	——	H. ♃.	DC. diss. n. 4. t. 2.

5 lyràta. DC.	(*ye.*)	lyre-leaved.	Spain.	1799.	6. 7.	H. ⊙.	Bocc. sic. 45. t. 23.
6 raphanifòlia. DC.	(*ye.*)	Radish-leaved.	Sicily.	1824.	——	H. ⊙.	
7 marítima. DC.	(*ye.*)	sea.	Naples.	——	——	H. ⊙.	Tenor. fl. nap. t. 61.
8 ciliàta. DC.	(*ye.*)	fringe-podded.	Italy.	1790.	——	H. ⊙.	DC. ic. gal. rar. 1. t. 39.
coronopifòlia. W. non. Lin.							
9 depréssa. DC.	(*ye.*)	dwarf.	Egypt.	1823.	——	H. ⊙.	
10 microcárpa. DC.	(*ye.*)	small-podded.	Gibraltar.	——	5. 6.	H. ⊙.	
11 eriocárpa. DC.	(*ye.*)	woolly-podded.	Spain.	1826.	5. 7.	H. ⊙.	DC. diss. n. 12. t. 9. f. 2.
12 Colúmnæ. DC.	(*ye.*)	Columna's.	Apulia.	1824.	6. 7.	H. ⊙.	Col. ecphr. 1. t. 284. f. 1.
13 ápula. DC.	(*ye.*)	spear-leaved.	Italy.	1710.	——	H. ⊙.	Fl. græc. t. 629.
14 leiocárpa. DC.	(*ye.*)	smooth-capsul'd.	Levant.	1824.	——	H. ⊙.	Gært. fruct. 2. t. 141.
15 obovàta. DC.	(*ye.*)	obovate-leaved.	....	....	——	H. ⊙.	
16 montàna. DC.	(*ye.*)	mountain.	Spain.	1824.	3. 4.	H. ⊙.	Cavan. ic. 2. t. 177.
17 lævigàta. DC.	(*ye.*)	smooth-podded.	Europe.	1777.	——	H. ♃.	DC. ic. rar. 11. t. 38.
β *alpéstris.* W.en.	(*ye.*)	*Hungarian.*	Hungary.	1816.	——	H. ♃.	W.etKit.hung. 3. t. 228.
18 lùcida. DC.	(*ye.*)	glossy.	Italy.	1827.	4. 6.	H. ♃.	DC. diss. n. 20. t. 7.
19 coronopifòlia. DC.	(*ye.*)	Buck's-horn-l[d].	S. Europe.	1790.	——	H. ♃.	DC. diss. n. 22. t. 18.
20 ambígua. DC.	(*ye.*)	ambiguous.	————	....	——	H. ♃.	———— n. 23. t. 11. f. 1.
21 saxátilis. DC.	(*ye.*)	rock.	————	1824.	——	H. ♃.	———— n. 24. t. 10.
22 sempervìrens. DC.	(*ye.*)	downy-leaved.	Spain.	1784.	——	F. ♄.	Barr. ic. t. 841.

MEGACA'RPÆA. DC. MEGACA'RPÆA. Tetradynamia Siliculosa. L.

laciniàta. DC.	(*ye.*)	jagged-leaved.	Volga.	1821.	6. 7.	H. ♃.	DC. diss. bisc. n. 5. t. 3.
Biscutélla megalocárpa. F.							

Tribus IV. *EUCLIDIEÆ* seu Pleurorhizeæ Nucamentaceæ. DC.

EUCLI'DIUM. DC. EUCLI'DIUM. Tetradynamia Siliculosa. L.

1 syrìacum. DC.	(*wh.*)	Syrian.	Levant.	1778.	7. 8.	H. ⊙.	Jacq. aust. 1. t. 6.
Bùnias syrìaca. W.		*Anastática syrìaca.* Jacq.					
2 tatáricum. DC.	(*wh.*)	Tartarian.	Tartary.	1823.	4. 5.	H. ⊙.	Pall. it. 3. ap. t. U. f. 2.
Bùnias tatárica. W.							

OCHTH'ODIUM. DC. OCHTH'ODIUM. Tetradynamia Siliculosa. L.

ægyptìacum. DC.	(*ye.*)	Egyptian.	Egypt.	1787.	8.	H. ⊙.	Jacq vind. t. 145.
Rapístrum ægyptìacum. H.K.		*Bùnias ægyptìaca.* L.					

Tribus V. *ANASTATICEÆ* seu Pleurorhizeæ Septulatæ. DC.

ANASTA'TICA. DC. ROSE OF JERICHO. Tetradynamia Siliculosa. L.

Hierochúntina. DC.	(*w.*)	common.	Levant.	1597.	5. 8.	H. ⊙.	Jacq. vind. 1. t. 58.

Tribus VI. *CAKILINEÆ* seu Pleurorhizeæ Lomentaceæ. DC.

CAK'ILE. DC. CAK'ILE. Tetradynamia Siliculosa. L.

1 marítima. DC.	(*pu.*)	Sea Rocket.	Britain.	....	6. 9.	H. ⊙.	Eng. bot. 231.
β *sinuatifòlia.* DC.	(*p.*)	*sinuate-leaved.*	————	....	——	H. ⊙.	Horn. fl. dan. t. 1583.
2 americàna. DC.	(*li.*)	American.	America.	....	——	H. ⊙.	Tussac fl. ant. 1. f. 17.

CHORISP'ORA. DC. CHORISP'ORA. Tetradynamia Siliquosa. L.

1 tenélla. DC.	(*pu.*)	purple.	Tauria.	1780.	6. 7.	H. ⊙.	Pall. it. 3. ap. t. L. f. 3.
2 stricta. DC.	(*pu.*)	upright.	Siberia.	1825.	5. 7.	H. ⊙.	
3 sibírica. DC.	(*ye.*)	Siberian.	Siberia.	1822.	——	H ⊙.	Murr. com. gœtt. t. 11.
4 ibérica. DC.	(*ye.*)	strong-scented.	Iberia.	——	——	H. ⊙.	Deles. ic. sel. 2. t. 582.

Subordo II. NOTORHIZEÆ. *DC. syst.* 2. *p.* 438.

Tribus VII. *SISYMBREÆ* seu Notorhizeæ Siliquosæ. DC.

MALC'OMIA. DC. MALC'OMIA. Tetradynamia Siliquosa. L.

1 africàna. DC.	(*pu.*)	African.	Africa.	1747.	6. 7.	H. ⊙.	Buxb. cent. 4. t. 44.
2 taraxacifòlia. DC.	(*pu.*)	Dandelion-l[d].	....	1795.	——	H. ⊙.	
3 láxa. DC.	(*pu.*)	loose-branched.	Siberia.	1822.	——	H. ⊙.	

4 Chìa. DC.	(*pu.*)	dwf.-branching.	Chio.	1732.	5. 6.	H. ⊙.	Swt. br. fl. gar. 40.
5 marítima. DC.	(*p.w.*)	dwarf annual.	S. Europe.	1713.	——	H. ⊙.	Bot. mag. 166.
Cheiránthus marítimus. B.M.							
6 arenària. DC.	(*pu.*)	sand.	Barbary.	1804.	——	H. ⊙.	Desf. fl. atl. 2. t. 162.
Hésperis arenària. Desf.							
7 parviflòra. DC.	(*pu.*)	small-flowered.	S. Europe.	1822.	6. 8.	H. ⊙.	DC. ic. gall. rar. t. 35.
8 lyràta. DC.	(*pu.*)	rigid.	Levant.	1824.	5. 8.	H. ⊙.	Flor. græc. t. 635.
9 littòrea. DC.	(*li.*)	sea.	S. Europe.	1683.	6 11.	H. ♂.	Swt. br. fl. gar. 54.
10 pátula. DC.	(*pu.*)	spreading.	Spain.	1825.	5. 7.	H. ♃.	
11 lácera. DC.	(*pu.*)	cut-leaved.	S. Europe.	——	——	H. ⊙.	Desf. ann. mus. 11. t. 34.
Hésperis pinnatífida. Desf.							
HE'SPERIS. DC.		ROCKET. Tetradynamia Siliquosa. L.					
1 trístis. DC.	(*br.*)	night-smelling.	Europe.	1622.	4. 6.	H. ♂.	Bot. mag. 730.
β *màjor.*	(*st.*)	*larger.*	————	1827.	4. 7.	H. ♂.	
2 fràgrans. F.	(*pu.*)	evening-scent[d].	Siberia?	1821.	——	H. ♂.	Swt. br. fl. gar. 61.
3 laciniàta. DC.	(*pu.*)	laciniated.	Italy.	1820.	5. 6.	H. ♂.	All. ped. n. 985. t. 82. f. 1.
4 villòsa. DC.	(*pu.*)	villous.	Italy.	1827.	——	H. ♂.	
5 grandiflòra. B.M.	(*pu.*)	large-flowered.	....	1818.	5. 8.	H. ♂.	Bot. mag. 2683.
6 elàta. H.	(*pu.*)	tall.	....	1825.	——	H. ♂.	
7 runcinàta. DC.	(*bh.*)	runcinate.	Hungary.	1804.	5. 7.	H. ♂.	W. et Kit. hung. 2. t. 200.
8 bituminosa. W. en.	(*bh.*)	clammy.	....	1816.	——	H. ♂.	
9 matronàlis. L.	(*va.*)	common.	Europe.	1597.	5. 8.	H. ♃.	Lob. ic. t. 323. f. 2.
α *albiflòra.* DC.	(*wh.*)	*single-white.*	————	——	——	H. ♃.	Tabern. kraut. 692. f. 1.
β *álbo-plèna.*	(*wh.*)	*double-white.*	————	——	——	H. ♃.	
γ *purpùrea.* DC.	(*pu.*)	*single-purple.*	————	——	——	H. ♃.	Tabern. kraut. 692 f. 2.
δ *purpùreo-plèna.*	(*pu.*)	*double-purple.*	————	——	——	H. ♃.	Wein. phyt. t. 572. f. c.
ε *variegàta.* DC.	(*v.*)	*dble.-variegated.*	————	——	——	H. ♃.	Munt. phyt. cur. t. 186.
ζ *foliiflòra.* DC.	(*gr.*)	*green-flowered.*	————	——	——	H. ♃.	Wein. phyt. t. 572. f. a.
10 inodòra. L.	(*pu.*)	scentless.	Britain.	....	——	H. ♃.	Eng. bot. 731.
11 sibírica. L.	(*pu.*)	Siberian.	Siberia.	....	——	H. ♃.	
12 heterophy'lla. DC.	(*pu.*)	various-leaved.	Naples.	1825.	5. 7.	H. ♃.	
13 áprica. DC.	(*pu.*)	sun.	Siberia.	1823	5. 6.	H. ♃.	Deles. ic. sel. 2. t. 62.
14 ramosíssima. DC.	(*pu.*)	branched.	Barbary.	1827.	——	H. ⊙.	Desf. atl. 2. t. 161.
ANDREO'SKIA. DC.		ANDREO'SKIA. Tetradynamia Siliquosa. L.					
1 integrifòlia. DC.	(*wh.*)	entire-leaved.	Siberia.	1824.	6. 7.	H. ⊙.	
Sisy'mbrium integrifòlium. W. *Hésperis glándulòsa.* P.S.							
2 pectinàta DC.	(*wh.*)	pectinated.	Dahuria.	1827.	5. 6.	H. ♂.	
SISY'MBRIUM. DC.		SISY'MBRIUM. Tetradynamia Siliquosa. L.					
1 officinàle. DC.	(*ye.*)	Hedge-mustard.	Britain.	....	5. 7.	H. ⊙.	Eng. bot. 725.
Ery'simum officinàle. E.B.							
2 strictíssimum DC.	(*ye.*)	spear-leaved.	Europe.	1658.	6. 8.	H. ♃.	Jacq. aust. 2. t. 194.
Cheirínia strictíssima. L. en.							
3 jùnceum. DC.	(*ye.*)	rushy.	Hungary.	1804.	7.	H. ♃.	W. et Kit. hung. 3. t. 234.
Ery'simum júnceum. W. *Cheirínia júncea.* L. en.							
4 hispánicum. DC.	(*ye.*)	Spanish.	Spain.	1820.	5. 6.	H. ♂.	Jacq. ic. rar. 1. t. 124.
5 pùmilum. DC.	(*ye.*)	dwarf.	Caucasus.	——	4. 5.	H. ⊙.	
6 obtusángulum. DC.	(*ye.*)	blunt-angled.	Switzerland.	....	5. 8.	H. ⊙.	Moris. ox. 2. s. 3. t. 5. f. 10.
7 acutángulum. DC.	(*ye.*)	acute-angled.	Pyrenees.	1791.	6. 7.	H. ♂.	Jacq. vind. 3. t. 97.
sinapóides. H.K. *Sinápis pyrenàica.* Jacq.							
8 taraxacifòlium. DC.	(*y.*)	Dandelion-l[d].	France.	1823.	5. 6.	H. ♂.	DC. ic. rar. p. 11. t. 37.
9 contórtum. W. en.	(*ye.*)	twisted.	Spain.	1824.	——	H. ♂.	
10 affíne. W. en.	(*ye.*)	related.	....	1823.	5. 8.	H. ♂.	
11 austrìacum. DC.	(*ye.*)	Austrian.	Austria.	1799.	6. 8.	H. ⊙.	Jacq. aust. 3. t. 262.
12 eckartsbergénse. W.	(*y.*)	bending-podd'd.	————	——	——	H. ♂.	
13 Tillìeri. W.	(*ye.*)	Tillier's.	Switzerland.	1824.	6. 8.	H. ⊙.	
14 I'rio. DC.	(*ye.*)	London Rocket.	England.	....	5. 8.	H. ⊙.	Eng. bot. 1631.
15 glábrum. W. en.	(*ye.*)	smooth.	....	1823.	——	H. ⊙.	
16 gallicum. W. en.	(*ye.*)	French.	France.	——	——	H. ⊙.	
17 nítidum. DC.	(*ye.*)	glossy.	....	....	6. 8.	H. ⊙.	
18 subhastàtum. DC.	(*ye.*)	halbert-leaved.	Archipelago.	1822.	4. 5.	H. ⊙.	
19 Colùmnæ. W.	(*ye.*)	Columna's.	Italy.	1796.	6. 7.	H. ⊙.	Jacq. aust. t. 323.
20 altíssimum. L.	(*ye.*)	tall.	Siberia.	1759.	8.	H. ⊙.	Walth. hort. t. 122.
21 orientàle. L.	(*ye.*)	oriental.	Levant.	1739.	7. 8.	H. ⊙.	
22 pannònicum. DC.	(*st.*)	Hungarian.	Hungary.	1787.	——	H. ⊙.	Jacq. ic. rar. 1. t. 123.
23 lyràtum. DC.	(*ye.*)	lyrate.	C. B. S.	....	5. 8.	G. ♃.	Deles. ic. sel. 2. t. 64.
24 ásperum. DC.	(*ye.*)	rough-podded.	S. Europe.	1778.	5. 6.	H. ⊙.	J. Bauh. hist. 2. f. 3.
25 fugáx. DC.	(*ye.*)	fugacious.	Spain.	1820.	7. 8.	H. ♃.	
26 Sophìa. DC.	(*ye.*)	Flix-weed.	Britain.	....	7. 8.	H. ⊙.	Eng. bot. 963.

27 pérsicum. DC. (ye.) Persian. Persia. 1820. 7. 8. H. ⊙.
28 Richardsòni. (ye.) Richardson's. N. America. 1828. 6. 7. H. ⊙.
canescens. Richardson. *non* Nutt.
29 brachycarpum. DC. short-podded. ——— —— —— H. ⊙.
30 tripinnàtum. DC. (ye.) tripinnate. C. B. S. 1817. —— H. ⊙.
31 millefòlium. DC. (ye.) Milfoil-leaved. Canaries. 1779. 5. 9. G. ♄. Jacq. ic. rar. 1. t. 127.
32 tanacetifòlium. DC. (ye.) Tansy-leaved. Italy. 1731 6. 7. H. ♃. Mor. his. 2. s. 3. t. 6. f. 19.
33 supìnum. DC. (wh.) trailing. S. Europe. 1778. —— H. ⊙. Isn. act. Par. 1724. t. 18.
Ery'simum supìnum. L. en.
34 hirsùtum. DC. (ye.) hairy-podded. ——— 1822. —— H. ⊙.
Ery'simum hirsùtum. L. en.
35 polyceràtium. DC. (ye.) sinuate-leaved. ——— 1633. 6. 8 H. ⊙. Jacq. vind. t. 79.
Ery'simum polyceràtium. L. en.
36 rígidum. DC. (wh.) rigid. Tauria. 1824. 6. 8. H. ⊙. Pall. it. 3. ap. t. Mm. f. 1.
37 bursifòlium. DC. (wh.) various-leaved. ——— 1732. —— H. ⊙. Dill. elt. t. 148. f. 177.
38 pinnatífidum. DC. (wh.) pinnatifid. ——— 1822. 5. 8. H. ♃. All. ped. t. 57. f. 3.
39 erysimóides. DC. (wh.) Erysimum-like. Spain. 1827. 10. 12. H. ♃. Desf. alt. 2. t. 158.
40 contortuplicàtum. (bh.) twisted-podded. Cumana. —— 6. 8. H. ⊙. Deles. ic. sel. 2. t. 63. B.
Hésperis contortuplicàtus. Bieb. *Cheiránthus contortuplicàtus.* W.

ALLI'ARIA. DC. ALLI'ARIA. Tetradynamia Siliquosa. L.
1 officinàlis. DC. (wh.) garlick-scented. Britain. .... 5. 6. H. ♃. Eng. bot. 796.
Ery'simum Allìaria. E. B. *Hésperis Allìaria.* Lam.
2 brachycárpa. DC. (wh.) short-podded. Iberia. 1827. 5. 7. H. ♃.

ERY'SIMUM. DC. ERY'SIMUM. Tetradynamia Siliquosa. L.
1 siliculòsum. DC. (ye.) siliculose. Tauria. 1824. 5. 6. H. ♂. Deles. ic. sel. 2. t. 65.
2 sículum. DC. (ye.) Sicilian. Sicily. 1824. 6. 7. H. ♂.
3 sessiliflòrum. DC. (ye.) sessile-flower'd. Siberia. 1794. —— H. ♃. L. Her. strirp. 1. t. 44.
Cheiránthus quadrángulus. W. *Cheirínia sessiliflòra.* L. en. *Syrènia Lamárckii.* Andrz.
4 angustifòlium. DC. narrow-leaved. Hungary. 1800. 7. 8. H. ♂. W. et Kit. hung. 1. t. 98.
Cheirínia angustifòlia. L. en. *Syrènia Ehrárti.* Andrz.
5 cuspidàtum. DC. (ye.) sharp-pointed. Caucasus. 1800. 5. 6. H. ♂. Buxb. cent. 2. t. 33. f. 1.
Syrènia Biebersteínii. Andrz. *Cheiránthus cuspidàtus.* W.
6 leptosty'lum. DC. (ye.) slender-styled. Caucasus. 1825. —— H. ♂.
7 exaltàtum. DC. (ye.) tall. Poland. —— H. ♂.
8 suffruticòsum. DC. (st.) suffruticose. .... 1822. 4. 5. H. ♄.
9 stríctum. DC. (ye.) upright. Austria. 1795. 6. 8. H. ♂. Jacq. aust. 1. t. 73.
odoràtum. W. *hieracifòlium.* Jacq. *Cheirínia strícta.* L. en.
10 virgàtum. DC. (ye.) twiggy. Portugal. 1807. 6. 7. H. ♂. Fl. Dan. t. 923?
11 longisiliquòsum. DC. (y.) long-podded. Switzerland. 1823. —— H. ♂. DC. ic. gal. rar. t. 36.
12 hieracifòlium. DC. (ye.) Hawkweed-l[d]. Europe. —— 5. 6. H. ♂. J. Bauh. his. 2. t. 2.
13 Marschalliànum. (ye.) Marschall's. Poland. 1827. —— H. ♂.
14 pátulum. H. (ye.) spreading. .... 1824. —— H. ♂.
15 pùmilum. H. (ye.) dwarf. Switzerland. —— H. ♂.
16 áureum. DC. (ye.) golden. Caucasus. 1824. —— H. ♂. Deles. ic. sel. 2. t. 66.
17 ibéricum. DC. (ye.) Iberian. Iberia. 1803. 5. 8. H. ♂. Bot. mag. 835.
Cheiránthus armenìacus. B. M. *Cheiránthus ibéricus.* Bieb.
18 cheiranthóides. DC. (y.) treacle. Britain. .... 6. 8. H. ⊙. Eng. bot. 942.
19 repándum. DC. (ye.) small-flowered. S. Europe. 1772. 5. 6. H. ⊙. Jacq. aust. 1. t. 22.
20 helvéticum. DC. (ye.) Swiss. Switzerland. 1793. —— H. ♂. Jacq. vind. t. 9.
Cheiránthus helvéticus. Jacq. *Cheirínia helvética.* L. en.
21 canéscens. DC. (ye.) canescent. S. Europe. 1731. 5. 7. H. ♂. Barr. ic. t. 884.
diffusum. W. *Cheiránthus alpìnus.* Jacq. aust. 1. t. 75.
22 Andrzejowskiànum. (y.) Andrzejowski's. Tauria. 1828. —— H. ♂.
23 collìnum. DC. (ye.) hill. Caucasus. —— —— H. ♂.
24 leptophy'llum. DC. (ye.) slender-leaved. Iberia. 1827. —— H. ♂.
25 versícolor. DC. (va.) various-color'd. Persia. 1823. 5. 7. H. ♂.
Cheiránthus versícolor. M. B.
26 lanceolàtum. DC. (ye.) spear-leaved. ——— 1597. —— H. ♂. Jacq. aust. t. 74.
Cheiránthus erysimoídes. Jacq.
27 fírmum. (ye.) hoary-podded. Switzerland. 1824. —— H. ♂.
Cheirínia firma. L. en. *Cheiránthus fírmus.* W. en.
28 rhæ'ticum. DC. (ye.) Rhætian. Rhætia. —— —— H. ♂.
29 dùbium. DC. (ye.) doubtful. .... —— —— H. ♂.
30 longifòlium. DC. (ye.) long-leaved. Algiers. 1822. —— H. ♃.
31 grácile. DC. (ye.) slender. Caucasus. 1824. 6. 8. H. ♂.
32 ásperum. DC. (ye.) rough. Missouri. —— —— H. ♂.
33 strigòsum. DC. (ye.) strigose. Siberia. 1828. —— H. ♂.
34 Redòwskii. DC. (ye.) Redowski's. ——— 1825. —— H. ♂.

35 altíssimum. DC.	(*ye.*)	tallest.	Limbourg.	1822.	6. 7	H. ♂.	
Cheirínia altíssima. L.en.							
36 sabulosum. F.	(*ye.*)	gravel.	Siberia.	1823.	——	H. ♂.	
37 bìcolor. DC.	(*v.*)	two-coloured.	Switzerland.	1827.	——	H. ♂.	
CONRI'NGIA. L. en.		CONRI'NGIA.	Tetradynamia Siliquosa. L.				
1 alpìna. L. en.	(*wh.*)	Alpine.	Europe.	1793.	5. 6.	H. ♃.	Vill. dauph. 3. t. 36.
Brássica alpìna. L. *Ery'simum alpìnum.* DC.							
2 perfòliata. L. en.	(*wh.*)	perfoliate.	England.	....	——	H. ⊙.	Eng. bot. 1804.
Brássica orientàlis. E.B. *Ery'simum perfoliàtum.* DC.							
3 austrìaca.	(*st.*)	Austrian.	Austria.	....	——	H. ⊙.	Jacq. aust. t. 283.
Brássica austrìaca. Jacq. *Ery'simum austrìacum.* DC.							
STANLEYA. DC.		STANLEYA.	Tetradynamia Siliquosa.				
pinnatifida. DC.		pinnatifid.	Missouri.	1812.	5. 6.	H. ♃.	
Cleome pinnata. Ph.							

Tribus VIII. *CAMELINEÆ* seu Notorhizeæ Latiseptæ. DC.

CAMEL'INA. DC.		CAMEL'INA.	Tetradynamia Siliculosa. L.				
1 satìva. DC.	(*ye.*)	Gold of pleasure.	Britain.	....	5. 7.	H. ⊙.	Eng. bot. 1254.
Aly'ssum satìvum. E.B. *Myàgrum satìvum.* L.							
α pilòsa. DC.	(*ye.*)	*hairy.*	————	....	——	H. ⊙.	Moris.ox.2.s.3.t.21.f.2.
β glábrata. DC.	(*ye.*)	*smooth.*	....	....	——	H. ⊙.	Lind. als. 94. t. 1.
2 dentàta. DC.	(*ye.*)	tooth-leaved.	Europe.	1806.	6. 7.	H. ⊙.	J. Bauh.hist.2.p.893.ic.
3 pinnatìfida. H.	(*ye.*)	pinnatifid.	Spain.	——	——	H. ⊙.	
4 microcárpa. DC.	(*ye.*)	small-podded.	Poland.	1827.	——	H. ⊙.	Deles. ic. sel. 2. t. 69.
5 barbareæfò.ia.DC.	(*ye.*)	Barbarea-ld.	Siberia.	1824.	5. 7.	H. ♃.	———— 2. t. 70.
6 austrìaca. DC.	(*ye.*)	Austrian.	Austria.	1795.	6. 7.	H. ⊙.	Jacq. aust. 2. t. 111.
NE'SLIA. DC.		NE'SLIA.	Tetradynamia Siliculosa.				
paniculàta. DC.	(*ye.*)	panicled.	Europe.	1683.	6. 8.	H. ⊙.	Flor. dan. t. 204.
Rapístrum paniculàtum. Gært. *Myàgrum paniculàtum.* L.							

Tribus IX. *LEPIDINEÆ* seu Notorhizeæ Angustiseptæ. DC.

SENEBIE'RA. DC.		SENEBIE'RA.	Tetradynamia Siliculosa. L.				
1 pinnatífida. DC.	(*wh.*)	pinnatifid.	England.	....	7. 8.	H. ⊙.	Eng. bot. 248.
Corónopus dídyma. E.B. *Lepídium dídymum.* L.							
β incìsa. W. en.	(*wh.*)	*cut-leaved.*	America.	1817.	——	H. ⊙.	
2 Corónopus. DC.	(*wh.*)	common.	Britain.	....	6. 10.	H. ⊙.	Eng. bot. 1660.
Corónopus Ruéllii. E.B. *Cochleària Corónopus.* L.							
LEPI'DIUM. DC.		PEPPERWORT.	Tetradynamia Siliculosa. L.				
1 Drába DC.	(*wh.*)	Whitlow-grass.	Europe.	1596.	5. 6.	H. ♃.	Jacq. fl. aust. 4. t. 315.
Cochleària Drába. Jacq.							
2 chalepénse. DC.	(*wh.*)	Aleppo.	Aleppo.	1798.	5. 7.	H. ⊙.	Moris. ox. 2. s. 3. t. 25.
3 coronopifòlium.DC.	(*w.*)	Buckshorn-ld.	Volga.	1823.	——	H. ♃.	
4 satìvum. DC.	(*wh.*)	common Cress.	Persia.	1548.	5. 10.	H. ⊙.	Fl. græc. t. 616.
β críspum. DC.	(*wh.*)	*curled Cress.*	————	....	——	H. ⊙.	Chabr. sci. 289. f. 2.
γ latifòlium. DC.	(*wh.*)	*broad-leaved.*	————	....	——	H. ⊙.	Moris.ox.2.s.3. t.19.f.2.
5 campéstre. DC.	(*wh.*)	hoary field.	Britain.	....	6. 7.	H. ⊙.	Eng. bot. 1385.
Thláspi campéstre. E.B.							
6 hírtum. DC.	(*wh.*)	hairy.	————	....	——	H. ♂.	———— 1803.
Thláspi hírtum. E.B.							
7 spinòsum. DC.	(*wh.*)	prickly.	Levant.	1787.	4. 9.	H. ⊙.	Ard. spec. 2. t. 16.
8 virgínicum. DC.	(*wh.*)	Virginian.	America.	1713.	6. 7.	H. ⊙.	Schkuhr. hand. 2. t. 180.
9 subulàtum. DC.	(*wh.*)	awl-leaved.	Spain.	1739.	7. 8.	F. ♄.	Asso syn. arr. t. 6. f. 3.
10 ruderàle. DC.	(*wh.*)	narrow-leaved.	Britain.	....	6. 7.	H. ⊙.	Eng. bot. 1595.
11 incìsum. DC.	(*wh.*)	cut-leaved.	Tauria.	1824.	5. 6.	H. ⊙.	
12 vesicàrium. DC.	(*wh.*)	bladdered.	Iberia.	1821.	4. 8.	H. ⊙.	Hil.veg.syst.11.t.41.f.3.
13 angulòsum. DC.	(*wh.*)	angular-stalk'd.	Hungary.	1825.	——	H. ⊙.	
14 perfoliàtum. DC.	(*wh.*)	v..rious-leaved.	Austria.	1640.	4. 6.	H. ⊙.	Jacq. aust. 4. t. 346.
15 Cardámines. DC.	(*wh.*)	Spanish Cress.	Spain.	1789.	6. 7.	F. ♂.	Ard. spec. 1. t. 19.
16 divaricàtum. DC.	(*wh.*)	close-spiked.	C. B. S.	1774.	5. 8.	G. ♄.	
17 bonariénse. DC.	(*wh.*)	Buenos Ayres.	S. America.	1732.	5. 6.	H. ⊙.	Dill. elt. t. 286. f. 370.
18 pubéscens. DC.	(*wh.*)	pubescent.	Para.	1820.	5. 8.	H. ⊙.	
19 cuneifòlium. DC.	(*wh.*)	wedge-leaved.	N. S. W.	1820.	6. 8.	H. ⊙.	

20 piscídium. DC.	(*wh.*)	South Sea.	Society Isles.	1779.	9.	H. ⊙.	Mont. n. a. n. c. 6. t. 5. a.
21 desertòrum. Led.	(*wh.*)	desert.	Siberia.	1827.	——	H. ⊙.	
22 decúmbens. DC.	(*wh.*)	decumbent.	....	1820.	5. 8.	H. ⊙.	
23 oleràceum. DC.	(*wh.*)	eatable.	NewZealand.	....	——	H. ⊙.	
24 lyràtum. DC.	(*wh.*)	lyre-leaved.	Levant.	1759.	6. 7.	H. ⊙	Tourn. it. 2. t. 389.
25 latifòlium. DC.	(*wh.*)	broad-leaved.	Britain.	....	——	H. ♃.	Eng. bot. 182.
26 affìne. DC.	(*wh.*)	saw-leaved.	Siberia.	1823.	7. 8.	H. ♃.	
27 crassifòlium. DC.	(*wh.*)	thick-leaved.	Hungary.	——	5. 6.	H. ♃.	W. et Kit. hung. 1. t. 4.
28 suffruticòsum. DC.	(*w.*)	shrubby.	Spain.	1790.	7. 8.	F. ♄.	Cav. ic. 2. t. 161. f. 2.
graminifòlium. Cav.							
29 lineàre. DC.	(*wh.*)	linear-leaved.	Spain.	....	——	F. ♄.	
30 Ibèris. DC.	(*wh.*)	grass-leaved.	S. Europe.	1683.	6. 9.	H. ♃.	Moris. his. 2. s. 3. t. 21. f. 1.
graminifòlia. H.K.							
31 capénse. DC.	(*wh.*)	Cape.	C. B. S.	1818.	6. 7.	G. ♃.	
BIVON'ÆA. DC.		BIVON'ÆA.	Tetradynamia Siliculosa. L.				
lùtea. DC.	(*ye.*)	yellow.	Sicily.	1823.	4. 5.	H. ⊙.	Cup. pan. sic. 2. t. 256.
Thláspi lùteum. Biv.							
ÆTHION'EMA. DC.		ÆTHION'EMA.	Tetradynamia Siliculosa. L.				
1 saxàtile. DC.	(*w.p.*)	rock.	S. Europe.	1759.	6. 7.	H. ⊙.	Jacq. aust 3. t. 236.
2 grácile. DC.	(*wh.*)	slender.	Carniola.	1822.	——	H. ♄.	Boc. mus. p. 79. t. 70.
3 Buxbáumii. DC.	(*pu.*)	Buxbaum's.	Iberia.	1820.	——	H. ⊙.	Buxb. cent. 1. t. 5. f. 1.
4 monospérmum DC.	(*p.*)	one-seeded.	Spain.	1778.	7. 8.	H. ♂.	

Tribus X. *ISATIDEÆ* seu Notorhizeæ Nucamentaceæ. DC.

TAUSCH'ERIA. DC.		TAUSCH'ERIA.	Tetradynamia Siliculosa. L.				
1 lasiocárpa. DC.	(*wh.*)	villous-podded.	Siberia.	1824.	6. 7.	H. ⊙.	
2 gymnocárpa. DC.	(*wh.*)	smooth-podded.	——————	——	5. 6.	H. ⊙.	
IS'ATIS. DC.		WOAD.	Tetradynamia Siliculosa. L.				
1 armèna. DC.	(*ye.*)	Armenian.	Armenia.	1824.	6. 7.	H. ⊙.	Desv. j. 3. t. 25. f. 6.
2 latisilíqua. DC.	(*ye.*)	broad-podded.	Iberia.	1827.	5. 7.	H. ♂.	
3 lusitánica. DC.	(*ye.*)	Portugal.	Portugal.	1739.	——	H. ⊙.	
4 alpìna. DC.	(*ye.*)	Alpine.	Italy.	1800.	——	H. ♃.	All. ped. t. 86. f. 2.
5 pr'æcox. DC.	(*ye.*)	early-flowering.	Hungary.	1820.	4. 6.	H. ♂.	Trat. arch. 2. t. 68.
6 littoràlis. DC.	(*ye.*)	sea-side.	Tauria.	——	6. 7.	H. ⊙.	Deles. ic. sel. 2. t. 78.
7 hebecárpa. DC.	(*ye.*)	pubescent-pod[d].	Tauria.	1827.	——	H. ⊙.	—————— 2. t. 79.
8 dasycárpa. Led.	(*ye.*)	thick-podded.	Siberia.	——	5. 7.	H. ♂.	
9 tinctòria. DC.	(*ye.*)	common.	England.	....	5. 7.	H. ♂.	Eng. bot. 97.
β *satìva.* DC.	(*ye.*)	*smooth-leaved.*	....	....	——	H. ♂.	Blackw. herb. t. 246.
γ *hirsùta.* DC.	(*ye.*)	*hairy-leaved.*	....	....	——	H. ♂.	
δ *microcárpa.* DC.	(*ye.*)	*small-podded.*	....	....	——	H. ♂.	
10 campéstris. DC.	(*ye.*)	field.	Podolia.	1823.	——	H. ⊙.	
11 bannática. DC.	(*ye.*)	Bannatic.	Hungary.	——	——	H. ⊙.	
12 táurica. DC.	(*ye.*)	Taurian.	Tauria.	1827.	——	H. ♂.	
13 oblongàta. DC.	(*ye.*)	oblong-podded.	Siberia.	——	——	H. ♂.	
14 orientàlis. DC.	(*ye.*)	oriental.	Levant.	....	——	H. ⊙.	
15 canéscens. DC.	(*ye.*)	canescent.	S. Europe.	1825.	——	H. ♂.	Buxb. cent. 1. p. 4. t. 5.
16 aléppica. DC.	(*ye.*)	Aleppo.	Levant.	1739.	6. 7.	H. ⊙.	Scop. d. ins. 2. t. 16.
17 variegàta. Led.	(*ye.*)	variegated.	Siberia.	1827.	——	H. ♂.	
MY'AGRUM. DC.		MY'AGRUM.	Tetradynamia Siliculosa. L.				
perfoliàtum. DC.	(*ye.*)	perfoliate.	Europe.	1648.	6. 7.	H. ⊙.	Schkuhr. han. 2. t. 178.
SOBOLE'WSKIA. DC.		SOBOLE'WSKIA.	Tetradynamia Siliculosa. L.				
lithóphila. DC.	(*wh.*)	large-fruited.	Iberia.	1823.	5. 8.	H. ♂.	Deles. ic. sel. 2. t. 80.

Tribus XI. *ANCHONIEÆ* seu Notorhizeæ Lomentaceæ. DC.

GOLDBA'CHIA. DC.		GOLDBA'CHIA.	Tetradynamia Siliquosa.				
1 lævigàta. DC.	(*li.*)	smooth.	Astracan.	1823.	5. 6.	H. ⊙.	Deles. ic. sel. 2. t. 81.
2 torulòsa. DC.	(*li.*)	twisted-podded	Levant.	....	——	H. ⊙.	
3 tetragòna. Led.	(*li.*)	four-sided.	——————	1825.	——	H. ⊙.	
STERI'GMA. DC.		STERI'GMA.	Tetradynamia Siliquosa.				
1 tomentòsum. DC.	(*ye.*)	tomentose.	Caspia.	1823.	4. 5.	H. ♂.	Pall. it. 2. ap. t. K. f. 2.
Cheiránthus tomentòsus. W.							
2 torulòsum. DC.	(*ye.*)	twisted-podded.	Iberia.	——	——	H. ♂.	Buxb. cent. 5. t. 52. f. 2.

Subordo III. ORTHOPLOCEÆ. *DC. syst.* 2. *p.* 581.

Tribus XII. *BRASSICEÆ* seu Orthoploceæ Siliquosæ. DC.

BRA'SSICA. DC. CABBAGE. Tetradynamia Siliquosa. L.

1 olerácea. DC.	(*va.*)	common.	England.	....	4. 6.	H.	♂.	
A. SYLVE'STRIS. DC.	(*st.*)	WILD.	———	....	——	H.	♂.	Eng. bot. t. 637.
B. ACE'PHALA. DC.	(*st.*)	CAVALIER.	....	....	——	H.	♂.	
α ramòsa. DC.	(*st.*)	*thousand-headed*	....	....	——	H.	♂.	
β vulgàris. DC.	(*st.*)	*upright.*	....	....	——	H.	♂.	Moris. ox. s. 3. t. 1. f. 6.
* *víridis.* DC.	(*st.*)	*green.*	....	....	——	H.	♂.	Lob. ic. 243. f. 1.
** *purpuráscens.* DC.	(*st.*)	*purple.*	....	....	——	H.	♂.	Chabr. sciag. 270. f. 6.
γ quercifòlia. DC.	(*st.*)	*oak-leaved.*	....	....	——	H.	♂.	——— 270. f. 1.
δ sabéllica. DC.	(*st.*)	BORECOLE.	....	....	——	H.	♂.	——— 271. f. 1.
* *pinnàta.* DC.	(*st.*)	*winged-leaved.*	....	....	——	H.	♂.	Lob. ic. 246. f. 2.
** *purpuráscens.* DC.	(*st.*)	*purple.*	....	....	——	H	♂.	
*** *versícolor.* DC.	(*st.*)	*variegated.*	....	....	——	H.	♂.	
ε palmifòlia. DC.	(*st.*)	*palm-leaved.*	....	....	——	H.	♂.	
ζ costàta. DC.	(*wh.*)	*thick-ribbed.*	....	....	——	H.	♂.	
* *nepenthifórmis.* DC.	(*wh.*)	*tubular-ld.*	....	....	——	H.	♂.	Hort. trans. 5. p. 13. t. 1.
C. BULL'ATA. DC.	(*st.*)	SAVOY.	....	....	——	H.	♂.	Weinm. phyt. t. 261.
α vulgàris. DC.	(*st.*)	*common.*	....	....	——	H.	♂.	Lob. ic. t. 244. f. 1.
* *pr'æcox.* DC.	(*st.*)	*early.*	....	....	——	H.	♂.	
** *hùmilis.* DC.	(*st.*)	*dwarf.*	....	....	——	H.	♂.	
*** *turionénsis.* DC.	(*st.*)	*Tourraine.*	....	....	——	H.	♂.	
**** *auràta.* DC.	(*st.*)	*yellow.*	....	....	——	H.	♂.	
β oblónga. DC.	(*st.*)	*winter.*	....	....	——	H.	♂.	Lob. ic. t. 244. f. 2.
γ màjor. DC.	(*st.*)	*large.*	....	....	——	H.	♂.	
δ gemmífera. DC.	(*st.*)	*Brussel's-sprouts*	....	....	——	H.	♂.	
D. CAPIT'ATA. DC.	(*st.*)	HEADED.	....	....	——	H.	♂.	Lob. ic. 243. f. 2.
α depréssa. DC.	(*st.*)	*drum-head.*	....	....	——	H.	♂.	
β sphæ'rica. DC.	(*st.*)	*common.*	....	....	——	H.	♂.	
* *álba.* DC.	(*st.*)	*white.*	....	....	——	H.	♂.	
** *rùbra.* DC.	(*st.*)	*red.*	....	....	——	H.	♂.	Dod. pempt. 621. f. 2.
γ obovàta. DC.	(*st.*)	*Sugarloaf.*	....	....	——	H.	♂.	
δ ellíptica. DC.	(*st.*)	*early york.*	....	....	——	H.	♂.	
ε subacùta.	(*st.*)	*Battersea.*	....	....	——	H.	♂.	
ζ cónica. DC.	(*st.*)	*heart.*	....	....	——	H.	♂.	
E. CA'ULO-R'APA. DC.		TURNEP-CABBAGE.	....	....	——	H.	♂.	Dalech. lugd. 522. f. 3.
α commùnis. DC.	(*st.*)	*flat-leaved.*	....	....	4. 6.	H.	♂.	Lob. ic. 246. f. 1.
* *álba.* DC.	(*st.*)	*white-stemmed.*	....	....	——	H.	♂.	
** *purpuráscens.* DC.	(*st.*)	*violet-stemd.*	....	....	——	H.	♂.	Weinm. phyt. t. 264.
β? críspa. DC.	(*st.*)	*curl-leaved.*	....	....	——	H.	♂.	
F. BOTR'YTIS. DC.	(*st.*)	BROCCOLI.	....	....	——	H.	♂.	Lob. ic. 245. f. 1.
α cauliflòra. DC.	(*st.*)	*Cauliflower.*	....	....	——	H.	♂.	Weinm. phyt. t. 256.
β asparagóides. DC.	(*st.*)	*Broccoli.*	....	....	——	H.	♂.	Dalech. lugd. 522. f. 1.
* *commùnis.* DC.	(*st.*)	*white.*	....	....	——	H.	♂.	Chabr. sciagr. 269. f. 3.
** *violàcea.* DC.	(*w.s.*)	*violet.*	....	....	——	H.	♂.	
*** *víridis.*	(*w.s.*)	*green.*	....	....	——	H.	♂.	
**** *ramòsa.*	(*w.s.*)	*branching.*	....	....	——	H.	♂.	
***** *autumnàlis.*	(*st.*)	*Cape.*	....	....	——	H.	♂.	
2 campéstris. DC.	(*ye.*)	field.	England.	....	6.	H.	♂.	Eng. bot. 2224.
A. OLEI'FERA. DC.	(*ye.*)	COLE-SEED.	———	....	——	H.	♂.	Dod. pempt. 623. f. 1.
* *autumnàlis.* DC.	(*ye.*)	*autumnal.*	———	....	——	H.	♂.	
** *pr'æcox.* DC.	(*ye.*)	*early.*	———	....	——	H.	♂.	
B. PA'BULARIA. DC.	(*ye.*)	SHEEP'S.	———	....	——	H.	♂.	
C. N'APO-BRA'SSICA.		NAVEW.	———	....	——	H.	♂.	
α commùnis. DC.	(*ye.*)	*common.*	———	....	——	H.	♂.	
* *álba.* DC.	(*ye.*)	*white-rooted.*	———	....	——	H.	♂.	
** *purpuráscens.* DC.	(*ye.*)	*purple-rooted*	....	....	——	H.	♂.	
β Rutabàga. DC.	(*st.*)	*Swedish Turnep*		....	——	H.	♂.	
3 Ràpa. DC.	(*ye.*)	Turnep.	Britain.	....	4. 6.	H.	♂.	Eng. bot. 2176.
A. DEPRE'SSA. DC.		FLAT-ROOTED.	———	....	——	H.	♂.	Mart. fl. rust. t. 49. 50.
* *álba.* DC.	(*ye.*)	*white-round.*	———	....	——	H.	♂.	
** *flavéscens.* DC.	(*ye.*)	*yellow-round.*	———	....	——	H.	♂.	
*** *nígricans.* DC.	(*ye.*)	*dark-round.*	———	....	——	H.	♂.	
**** *punícea.* DC.	(*ye.*)	*red-round.*	———	....	——	H.	♂.	
***** *víridis.* DC.	(*ye.*)	*green-round.*	———	....	——	H.	♂.	
****** *pr'æcox.*	(*ye.*)	*early Dutch.*	———	....	——	H.	♂.	

B. OBLO'NGA. DC.	(*ye.*)	OBLONG-ROOTED.	Britain.	....	4. 6.	H. ♂.	Lob. ic. t. 197. f. 2.
C. OLEI'FERA. DC.	(*y.*)	OIL-BEARING.	Dauphiny.	....	——	H. ♂.	—— t. 298. f. 1.
4 Nàpus. DC.	(*ye.*)	Rape.	Britain.	....	5.	H. ♂.	Eng. bot. 2146.
A. OLEI'FERA. DC.	(*ye.*)	wild.	——	....	——	H. ♂.	Lob. ic. t. 200. f. 2.
B. ESCULE'NTA. DC.	(*ye.*)	esculent.	——	....	——	H. ♂.	—— 200. f. 1.
α *álba.* DC.	(*ye.*)	*white-rooted.*	——	....	——	H. ♂.	Blackw. herb. t. 410.
β *flàva.* DC.	(*ye.*)	*yellow-rooted.*	——	....	——	H. ♂.	
γ *nígricans.* DC.	(*ye.*)	*black-rooted.*	——	....	——	H. ♂.	
5 pr'æcox. DC.	(*st.*)	early.	S. Europe.	....	5. 6.	H. ⊙.	
6 chinénsis. DC.	(*ye.*)	Chinese.	China.	1770.	7.	H. ♂.	
7 violàcea. DC.	(*li.*)	violet-flowered.	——	....	——	H. ♂.	
8 incàna. DC.	(*st.*)	hoary.	Naples.	1824.	4. 5.	H. ♄.	Cup. panph. 1. t. 132.
9 baleàrica. DC.	(*st.*)	oak-leaved.	Minorca.	——	5. 7.	H. ♄.	Deles. ic. sel. 2. t. 86.
10 hy'brida. T.C.	(*st.*)	hybrid.	Naples.	——	——	H. ♂.	
11 Gravìnæ. DC.	(*ye.*)	pubescent.	Aprutia.	——	5. 6.	H. ♃.	Ten. fl. nap. t. 62.
12 pinnatífida. DC.	(*st.*)	pinnatifid.	Spain.	——	9. 10.	H. ♃.	Desf. fl. atl. 2. t. 165.
13 hùmilis. DC.	(*ye.*)	dwarf.	S. France.	——	4. 5.	H. ♃.	
14 repánda. DC.	(*ye.*)	wavy.	S. Europe.	1818.	6. 8.	H. ♃.	Vill. dauph. 3. t. 39.
15 Richérii. DC.	(*ye.*)	alpine.	——	1825.	5. 8.	H. ♃.	—— 3. t. 36.
16 monénsis. DC.	(*st.*)	Isle of Man.	Britain.	....	——	H. ⊙.	Eng. bot. 962.
17 Erucástrum. DC.	(*ye.*)	runcinate.	S. Europe.	1790.	——	H. ⊙.	Bull. herb. t. 331.
18 cheiránthos. DC.	(*ye.*)	Stock-like.	——	....	——	H. ♃.	Vill. dauph. 3. t. 36.
19 cheiranthiflòra. DC.	(*y.*)	Wall-flowered.	——	1820.	——	H. ⊙.	Willd. hort. ber. t. 19.
20 Tournefórtii. DC.	(*st.*)	Tournefort's.	Spain.	——	6. 7.	H. ⊙.	Gou. ill. p. 44. t. 20. f. A.
21 lævigàta. DC.	(*wh.*)	smooth.	——	1825.	——	H. ♂.	
22 valentìna. DC.	(*wh.*)	Spanish.	——	——	4. 5.	H. ⊙.	Barr. ic. t. 195. f. 1.
23 fruticulòsa. DC.	(*st.*)	frutescent.	S. Europe.	1820.	2. 4	H. ♄.	Cyr. pl. rar. 2. t. 1.
24 elongàta. DC.	(*ye.*)	stalk-leaved.	Hungary.	1801.	5. 10.	H. ♂.	W. et K. hung. 1. t. 28.
25 sabulària. DC.	(*ye.*)	gravel.	Portugal.	1818.	——	H. ⊙.	Brot. phyt. p. 97. t. 43.
26 odoràta. Sck.	(*ye.*)	sweet-scented.	....	1826.	——	H. ⊙.	
SIN'APIS. DC.		MUSTARD.	Tetradynamia Siliquosa. L.				
1 nìgra. DC.	(*ye.*)	common.	Britain.	....	4. 8.	H. ⊙.	Eng. bot. 969.
β *torulòsa.* DC.	(*ye.*)	*twisted-podded.*	——	....	——	H. ⊙	
γ *túrgida.* DC.	(*ye.*)	*turgid-podded.*	——	....	——	H. ⊙.	
δ *villòsa.* DC.	(*ye.*)	*villous-podded.*	——	....	——	H. ⊙.	
2 geniculàta. DC.	(*ye.*)	jointed.	Barbary.	1824.	——	H. ⊙.	
3 lævigàta. DC.	(*ye.*)	smooth.	Spain.	1769.	6. 7.	H. ♂.	
4 auriculàta. DC.	(*ye.*)	eared.	....	....	——	H. ⊙.	
5 integrifòlia. DC.	(*ye.*)	entire-leaved.	E. Indies.	1804.	7. 8.	H. ⊙.	W. hort. ber. t. 14.
6 júncea. DC.	(*ye.*)	fine-leaved.	China.	1710.	6. 7.	H. ⊙.	Jacq. h. vind. t. 171.
7 chinénsis. DC.	(*ye.*)	Chinese.	——	1782.	——	H. ⊙.	Ard. spec. 1. t. 10.
8 brassicàta. DC.	(*ye.*)	Cabbage-leav'd.	——	1801.	——	H. ⊙.	
9 cérnua. DC.	(*wh.*)	drooping.	China.	1824.	7. 9.	H. ⊙.	
10 pubéscens. DC.	(*ye.*)	downy.	Sicily.	1789.	——	H. ♃.	Ard. spec. 1. t. 9.
11 arvénsis. DC.	(*ye.*)	Charlock.	Britain.	....	4. 8.	H. ⊙.	Eng. bot. 1748.
β *incìsa.*	(*ye.*)	*cut-leaved.*	Europe.	....	——	H. ⊙.	Lob. ic. 198. f. 2.
12 orientàlis. DC.	(*ye.*)	oriental.	Levant.	1778.	6. 7.	H. ⊙.	Schk. hand. 1. t. 186.
13 táurica. DC.	(*ye.*)	Taurian.	Tauria.	1823.	4. 8.	H. ⊙.	
14 subbipinnatífida. DC.	(*ye.*)	subbipinnatifid.	Spain.	——	6. 7.	H. ⊙.	
15 Kàber. DC.	(*ye.*)	Persian.	Persia.	——	——	H. ⊙.	
16 Alliònii. DC.	(*ye.*)	Allioni's.	Egypt.	1789.	——	H. ⊙.	Jacq. h. vind. 2. t. 168.
17 túrgida. DC.	(*ye.*)	turgid-podded.	——	....	——	H. ⊙.	
18 incàna. DC.	(*ye.*)	hoary-jointed.	S. Europe.	1771.	7.	H. ♂.	Jacq. h. vind. 2. t. 169.
19 heterophy'lla. DC.	(*ye.*)	various-leaved.	Spain.	1820.	5. 6.	H. ♂.	
20 álba. DC.	(*ye.*)	white-seeded.	Britain.	....	4. 10.	H. ⊙.	Eng. bot. 1677.
21 híspida. DC.	(*ye.*)	hispid.	Spain.	1804.	6. 7.	H. ⊙.	
22 dissécta. DC.	(*ye.*)	cut-leaved.	——	——	4. 6.	H. ⊙.	
23 hastàta. DC.	(*ye.*)	hastate-leaved.	N. Holland.	1824.	7. 9.	H. ⊙.	
24 foliòsa DC.	(*ye.*)	leafy.	Levant.	1804.	5. 6.	H. ⊙.	
25 ápula. DC.	(*ye.*)	Apulian.	Italy.	1824.	6. 8.	H. ⊙.	
26 ramòsa. H.B.	(*ye.*)	branching.	E. Indies.	1804.	7. 9.	H. ⊙.	
27 virgàta. H.	(*ye.*)	twiggy.	....	1827.	6. 8.	H. ⊙.	
28 subulària.	(*ye.*)	awl-shaped.	....	——	——	H. ⊙.	
29 frutéscens. DC.	(*st.*)	frutescent.	Madeira.	1777.	12. 6.	G. ♄.	
MORICA'NDIA. DC.		MORICA'NDIA.	Tetradynamia Siliquosa. L.				
arvénsis. DC.	(*pu.*)	field.	S. Europe.	1739.	6.	H. ♃.	Swt. br. fl. gar. 278.
Brássica arvénsis. L.		*Túrritis arvénsis.* H.K.					
DIPLOTA'XIS. DC.		DIPLOTA'XIS.	Tetradynamia Siliquosa. L.				
1 crassifòlia. DC.	(*ye.*)	thick-leaved.	Sicily.	1824.	5. 8.	H. ♃.	

2 erucóides. DC.	(*ye.*)	dwarf.	S. Europe.	1736.	6. 7.	H. ⊙.	Jacq. vind. 2. t. 170.
Sinàpis erucóides. L.							
3 vírgata. DC.	(*ye.*)	slender.	Spain.	1820.	——	H. ⊙.	
4 cathólica. DC.	(*ye.*)	Spanish.	———	——	4. 5.	H. ⊙.	
Sisy'mbrium cathólicum. L.							
5 tenuifòlia. DC.	(*ye.*)	Wall rocket.	England.	....	7. 10.	H. ⊙.	Eng. bot. 525.
Sisy'mbrium tenuifòlium. E.B. *Sinàpis tenuifòlia.* H.K.							
6 muràlis. DC.	(*ye.*)	sand.	England.	....	7. 9.	H. ⊙.	Eng. bot. 1090.
Sisy'mbrium muràle. L. *Sinàpis muràlis.* H.K.							
7 vimínea. DC.	(*ye.*)	twiggy.	S. Europe.	....	5. 6.	H. ⊙.	Mor. his. ox. 2. t. 5. f. 8.
8 saxátilis. DC.	(*ye.*)	rock.	———	....	——	H. ♃.	
9 ramosíssima. DC.	(*ye.*)	branching.	....	1822.	——	H. ♃.	
ER'UCA. DC.		ER'UCA. Tetradynamia Siliquosa. L.					
1 satìva. DC.	(*wh.*)	stripe-flowered.	S. Europe.	1573.	7.	H. ⊙.	Fl. græc. t. 646. et 647.
Brássica Erùca. L. *Euzòmum satìvum.* L. en.							
2 híspida. DC.	(*wh.*)	hispid.	Naples.	1824.	——	H. ⊙.	
3 vesicària. DC.	(*wh.*)	bladdered.	Spain.	1820.	——	H. ⊙.	Asso. syn. arrag. 88. t. 4.
β *flaviflòra.* DC.	(*ye.*)	*yellow.*	———	——	——	H. ⊙.	

Tribus XIII. *VELLEÆ* seu Orthoploceæ Latiseptæ. DC.

VE'LLA. DC.		CRESS-ROCKET. Tetradynamia Siliculosa. L.					
pseudocy'tisus. DC.	(*y.*)	shrubby.	Spain.	1759.	4. 5.	H. ♄.	Bot. reg. 293.
CARRICHT'ERA. DC.		CARRICHT'ERA. Tetradynamia Siliculosa. L.					
Véllæ. DC.	(*st.*)	annual.	England.	....	6. 7.	H. ⊙.	Eng. bot. 1441.
Vélla ánnua. E.B.							
SUCCO'WIA. DC.		SUCCO'WIA. Tetradynamia Siliculosa. L.					
baleàrica. DC.	(*ye.*)	Minorca.	Minorca.	1781.	6. 7.	H. ⊙.	Jacq. vind. 2. t. 144.
Bùnias baleàrica. L.							

Tribus XIV. *PSYCHINEÆ* seu Orthoploceæ Angustiseptæ. DC.

PSYCH'INE. DC.		PSYCH'INE. Tetradynamia Siliculosa. L.					
stylòsa. DC.	(*wh.*)	striped-flower'd.	Barbary.	1822.	4. 5.	H. ⊙.	Desf. fl. atl. 2. t. 148.

Tribus XV. *ZILLEÆ* seu Orthoploceæ Nucamentaceæ. DC.

ZI'LLA. DC.		ZI'LLA. Tetradynamia Siliculosa. L.					
myagroídes. DC.	(*li.*)	spiny.	Egypt.	....	4. 5.	H. ♄.	Vent. malm. t. 16.
Bùnias spinòsa. Vent.							
CALEP'INA. DC.		CALEP'INA. Tetradynamia Siliculosa. L.					
Corvìni. DC.	(*wh.*)	annual.	S. Europe.	1820.	4. 6.	H. ⊙.	W. et Kit. hung. 2. t. 107.
Bùnias cochlearioìdes. W.							

Tribus XVI. *RAPHANEÆ* seu Orthoploceæ Lomentaceæ. DC.

CRA'MBE. DC.		COLEWORT. Tetradynamia Siliculosa. L.					
1 marítima. DC.	(*wh.*)	Sea Kale.	Britain.	....	5. 6.	H. ♃.	Eng. bot. 924.
2 pinnatífida. DC.	(*wh.*)	smooth-winged.	Siberia.	1759.	6. 7.	H. ♃.	Jacq. ic. 1. t. 128.
3 orientàlis. DC.	(*wh.*)	oriental.	Levant.	1752.	6. 7.	H. ♃.	
4 tatárica. DC.	(*wh.*)	Tartarian.	Hungary.	1789.	——	H. ♃.	Jacq. ic. 1. t. 129.
5 áspera. DC.	(*wh.*)	rough.	Tauria.	1823.	5. 6.	H. ♃.	Deles. ic. sel. 2. t. 91.
6 júncea. DC.	(*wh.*)	rushy.	Iberia.	1828.	——	H. ♃.	
7 cordifòlia. DC.	(*wh.*)	heart-leaved.	Caucasus.	——	6. 7.	H. ♃.	
8 hispánica. DC.	(*wh.*)	Spanish.	Spain.	1683.	——	H. ⊙.	Jacq. obs. 2. t. 41.
9 filifórmis. DC.	(*wh.*)	Patagonian.	Patagonia.	1796.	7. 8.	H. ⊙.	Jacq. ic. 3. t. 504.
10 fruticòsa. DC.	(*wh.*)	Madeira.	Madeira.	1777.	5. 11.	G. ♄.	
11 strigòsa. DC.	(*wh.*)	strigose shrubb.	Canaries.	1779.	5. 6.	G. ♄.	L'. her. stirp. 1. t. 72.

RAPI'STRUM. DC.		RAPI'STRUM. Tetradynamia Siliculosa. L.					
1 perénne. DC.	(ye.)	perennial.	Europe.	1789.	7.	H. ♃.	Jacq. aust. 5. t. 414.
Myàgrum perénne. L.		*Cakìle perénnis.* H.K.					
2 rugòsum. DC.	(ye.)	wrinkled.	———	1739.	6. 7.	H. ⊙.	All. ped. 1. t. 78.
Myàgrum rugòsum. L.		*Cakile rugosa.* H.K.					
3 orientàle. DC.	(ye.)	oriental.	Levant.	1795.	——	H. ⊙.	Flor. græc. t. 612.
ENARTHROCA'RPUS		ENARTHROCA'RPUS. Tetradynamia Siliquosa.					
1 lyrátus. DC.	(ye.)	yellow-strip'd.	Egypt.	1826.	6. 8.	H. ⊙.	Delil.fl.æg.105.t.36.f.1.
2 pterocárpus. DC.	....	wing-podded.	———	1828.	....	H. ⊙.	Deles. ic. sel. 2. t. 93.
RA'PHANUS. DC.		RADISH. Tetradynamia Siliquosa. L.					
1 satìvus. Mil.	(li.)	common.	China.	1548.	5. 6.	H. ⊙.	Fuchs. hist. 659. ic.
α pr'æcox.	(li.)	*early Salmon.*	———	——	——	H. ⊙.	
β macrorhìzus.	(li.)	*long Salmon.*	———	——	——	H. ⊙.	
γ purpùreus.	(li.)	*purple-rooted.*	———	——	——	H. ⊙.	
2 rotùndus. Mil.	(li.)	Turnep-rooted.	———	——	——	H. ⊙.	Lob. ic. t. 201. f. 1.
3 chinénsis. Mil.	(li.)	oil-seeded.	———	——	——	H. ⊙.	
4 nìger. DC.	(li.)	black Spanish.	———	——	——	H. ⊙.	Lob. ic. 202. f. 1.
α vulgàris. DC.	(li.)	*long-rooted.*	———	——	——	H. ⊙.	Wein. phyt. t. 860. f. c.
β rotúndus. DC.	(li.)	*round-rooted.*	———	——	——	H. ⊙.	——— t. 860. f. b.
γ grìseus. DC.	(li.)	*grey.*	———	——	——	H. ⊙.	
5 orbiculàris. Mil.	(li.)	white Spanish.	———	——	——	H. ⊙.	
6 caudàtus. DC.	(pu.)	long-podded.	Java.	1815.	5. 8.	H. ⊙.	Lin. f. dec. 1. t. 10.
7 Raphanístrum. DC.	(va.)	wild.	Britain.	....	4. 8.	H. ⊙.	Eng. bot. 856.
α álbum.	(wh.)	*white-flowered.*	———	....	——	H. ⊙.	Wein. phyt. t. 862. f. a.
β purpuráscens.	(pu.)	*purple-flowered.*	———	....	——	H. ⊙.	——— t. 862. f. b.
ζ flàvum.	(ye.)	*yellow-flowered.*	———	....	——	H. ⊙.	J. Bauh. h. 2. p. 844. ic.
8 Landra. DC.	(ye.)	Italian.	Italy.	1827.	4. 6.	H. ♂.	Deles. ic. sel. 2. t. 94.
9 marítimus. DC.	(ye.)	sea-side.	Britain.	....	5. 7.	H. ♂.	Eng. bot. 1643.

Subordo IV. SPIROLOBEÆ. *DC. syst.* 2. *p.* 670.

Tribus XVII. *BUNIADEÆ* seu Spirolobeæ Nucamentaceæ. DC.

B'UNIAS. DC.		B'UNIAS. Tetradynamia Siliculosa. L.					
1 Erucàgo. DC.	(ye.)	prickly-podded.	S. Europe.	1640.	6. 7.	H. ⊙.	Jacq. aust. 4. t. 340.
2 áspera. DC.	(ye.)	rough.	Portugal.	1823.	6. 7.	H. ⊙.	
3 orientàlis. DC.	(ye.)	oriental.	Levant.	1795.	5. 7.	H. ♃.	Schk. han. 2. t. 189.

Tribus XVIII. *ERUCARIEÆ* seu Spirolobeæ Lomentaceæ. DC.

ERUC'ARIA. DC.		ERUC'ARIA. Tetradynamia Siliquosa. L.					
1 aléppica. DC.	(pu.)	Aleppo.	Levant.	1680.	7. 8.	H. ⊙.	Vent. h. cels. t. 64.
Cordylocárpus lævigàtus. W.							
2 crassifòlia. DC.	(pu.)	thick-leaved.	Egypt.	1828.	7. 9.	H. ⊙.	Delil. pl. bot. t. 34. f. 1.

Subordo V. DIPLECOLOBEÆ. *DC. syst.* 2. *p.* 676.

Tribus XIX. *HELIOPHILEÆ* seu Diplecolobeæ Siliquosæ. L.

HELIO'PHILA. DC.		HELIO'PHILA. Tetradynamia Siliquosa. L.					
1 filifórmis. DC.	(pu.)	awl-podded.	C. B. S.	1786.	7. 9.	H. ⊙.	Lam. ill. t. 563. f. 3. ?.
2 dissécta. DC.	(bl.)	cut-leaved.	———	1822.	6. 8.	H. ⊙.	
3 tenuisíliqua. DC.	(bh.)	slender-podded.	———	——	——	H. ⊙.	Deles. ic. sel. 2. t. 96.
4 amplexicáulis. DC.	(bh.)	stem-clasping.	———	1774.	6. 9.	H. ⊙.	Jacq. frag. t. 64. f. 2.
5 rivàlis. DC.	(wh.)	rivulet.	———	1819.	——	H. ⊙.	
6 péndula. DC.	(wh.)	pendulous	———	1792.	——	H. ⊙.	Lam. ill. t. 562. f. 2.
pinnàta. Vent. malm. t. 113. non. L.							

7 trífida. DC.	(*pu.*)	trifid-leaved.	C. B. S.	1819.	6. 9.	H. ⊙.	
pinnàta. L. non. Vent.							
8 pusílla. DC.	(*wh.*)	small.	———	1824.	——	H. ⊙.	Pluk. mant. t. 432. f. 2.
9 lepidióides. L.en	(*wh.*)	Cress-like.	———	——	——	H. ⊙.	
10 diffùsa. DC.	(*wh.*)	spreading.	———	1820.	——	H. ⊙.	
11 Peltària. DC.	(*wh.*)	small-white.	———	——	——	H. ⊙.	
12 pilòsa. DC.	(*bl.*)	hairy.	———	1768.	——	H. ⊙.	
α integrifòlia. DC.	(*bl.*)	*entire-leaved.*	———	——	——	H. ⊙.	Jacq. ic. 3. t. 506.
β incìsa. DC.	(*bl.*)	*cut-leaved.*	———	——	——	H. ⊙.	Bot. mag. 496.
arabiòides. B.M.							
13 strícta. B.M.	(*bl.*)	upright.	———	1824.	——	H. ⊙.	Bot. mag. 2526.
14 digitàta. DC.	(*bl.*)	fingered.	———	1819.	——	H. ⊙.	Bot. reg. 838.
15 coronopifòlia. DC.	(*bl.*)	Buck's-horn-ld.	———	1778.	6. 7.	H. ⊙.	Herm. lugd. t. 367.
16 pectinàta. DC.	(*wh.*)	pectinated.	———	1819.	——	H. ⊙.	
17 fœniculàcea. DC.	(*pu.*)	Fennel-leaved.	———	1774.	6. 9.	H. ⊙.	
18 crithmifòlia. DC.	(*bh.*)	Samphire-ld.	———	1816.	——	H. ⊙.	Deles. ic. sel. 2. t. 97.
19 platysíliqua. DC.	(*bh.*)	flat-podded.	———	1774.	7. 8.	G. ♄.	
20 linearifòlia. DC.	(*bl.*)	linear-leaved.	———	1819.	——	G. ♄.	
21 scopària. DC.	(*bh.*)	Broom-like.	———	1802.	——	G. ♄.	Deles. ic. sel. 2. t. 98.
22 incàna. DC.	(*pu.*)	hoary.	———	1774.	5. 8.	G. ♄.	
23 cleomóides. DC.	(*wh.*)	Cleome-like.	———	1802.	1. 12.	G. ♄.	Deles. ic. sel. 2. t. 99.
Cheiránthus stríctus. W.							

Tribus XX. *SUBULARIEÆ* seu Diplocolobeæ Latiseptæ. DC.

SUBUL'ARIA. DC. AWLWORT. Tetradynamia Siliculosa. L.

aquática. DC.	(*wh.*)	water.	Britain.	....	7.	H.*w.*⊙.	Eng. bot. 732.

Subordo VI. SCHIZOPETALEÆ.

SCHIZOPE'TALUM. B.M. SCHIZOPE'TALUM. Tetradynamia Siliquosa. L.

Walkèri. B.M.	(*wh.*)	Walker's.	Chili.	1822.	7. 9.	H. ⊙.	Brown bot. reg. 752.

ORDO XIV.

CAPPARIDEÆ. *DC. prodr.* 1. *p.* 237.

Tribus I. *CLEOMEÆ.* DC.

PERI'TOMA. DC PERI'TOMA. Monadelphia Hexandria.

serrulàtum. DC.	(*pu.*)	saw-leaved.	Missouri.	1823.	6. 8.	H. ⊙.	
Cleòme serrulàta. Ph.							

GYNANDRO'PSIS. DC. GYNANDRO'PSIS. Monadelphia Hexandria.

1 sessilifòlia. DC.	(*wh.*)	sessile-leaved.	W. Indies.	1820.	6. 8.	H. ⊙.	
2 triphy'lla. DC.	(*wh.*)	three-leaved.	———	——	——	H. ⊙.	Herm. lugd. 565. ic
Cleòme triphy'lla. L.?							
3 pentaphy'lla. DC.	(*wh.*)	five-leaved.	India.	1640.	——	H. ⊙.	Jacq. vind. 1. t. 21.
Cleòme pentaphy'lla. B.M. 1681.							
4 candelàbrum.	(*wh.*)	chandelier.	W. Indies.	1824.	7. 9.	H. ⊙.	Bot. mag. 2656.
Cleòme candelàbrum. B.M.							
5 muricàta. Scr.	(*wh.*)	bristly.	....	1828.	——	S. ⊙.	
6 heterotrìcha. DC.	(*wh.*)	various-haired.	C. B. S.	1822.	——	H. ⊙.	

CLE'OME. DC. CLE'OME. Tetradynamia Siliquosa. L.

1 gigántea. DC.	(*gr.*)	gigantic.	S. America.	1774.	6. 8.	S. ♄.	Jacq. obs. 4. t. 76.
2 heptaphy'lla. DC.	(*wh.*)	seven-leaved.	W. Indies.	1820.	——	H. ⊙.	
3 spinòsa. DC.	(*wh.*)	white-flowered.	———	1731.	5. 10.	S. ♄.	Marc. bras. t. 34.
4 púngens. DC.	(*bh.*)	pink-flowered.	S. America.	1812.	7. 8.	S. ⊙.	W. hort. ber. 1. t. 18.
spinòsa. B.M. 1640. nec aliorum.							
5 pubescens. B.M.	(*wh.*)	pubescent.	....	1815.	7. 8.	H. ⊙.	Bot. mag. 1857.

6 ròsea. DC.	(*ro.*)	rose-coloured.	Brazil.	1825.	5. 8.	S. ⊙.	Bot. reg. 960.
7 monophy'lla. DC.	(*ye.*)	simple-leaved.	E. Indies.	1759.	——	H. ⊙.	Rheed. mal. 9. t. 34.
β *zeylánica.* DC.	(*ye.*)	*Ceylon.*	————	——	——	H. ⊙.	Burm. zeyl. t. 100. f. 2.
8 procúmbens. DC.	(*wh.*)	procumbent.	W. Indies.	1798.	——	S. ♃.	Jacq. amer. t. 120.
9 violàcea. DC.	(*vi.*)	violet-coloured.	Portugal.	1776.	6. 7.	H. ⊙.	Schk. han. 2. t. 189. f. 1.
10 Dilleniàna. DC.	(*ye.*)	Dillenius's.	Levant.	1732.	——	H. ⊙.	Dill. elt. t. 266. f. 345.
ornithopodioìdes. W.							
11 ibérica. DC.	(*ye.*)	Iberian.	Iberia.	1820.	——	H. ⊙.	
12 virgàta. DC.	(*ye.*)	twiggy.	Persia.	....	——	H. ⊙.	Buxb. cent. 1. t. 9. f. 2.
13 arábica. DC.	(*ye.*)	Arabian.	Arabia.	1794.	——	H. ⊙.	Lin. f. dec. t. 8.
14 flàva. DC.	(*ye.*)	yellow.	N. Holland.	1825.	7. 10.	H. ⊙.	
15 polygàma. DC.	(*wh.*)	polygamous.	W. Indies.	1822.	6. 7.	H. ⊙.	Slo. jam. t. 124. f. 1.
16 aculeata. DC.	(*wh.*)	prickly.	S. America.	——	——	H. ⊙.	
17 Houstòni. DC.	(*fl.*)	Houstoun's.	W. Indies.	1730.	——	S. ⊙.	Mart. dec. t. 45.
18 diffùsa. DC.	(*wh.*)	spreading.	Brazil.	1823.	——	H. ⊙.	
19 micrántha.	(*wh.*)	small-flowered.	....	1824.	6. 10.	S. ♄.	
POLANI'SIA. DC.		POLANI'SIA.	Dodecandria Monogynia. L.				
1 Chelidònii. DC.	(*ro.*)	Celandine-fld.	E. Indies.	1790.	7. 10.	S. ⊙.	
Cleòme Chelidònii. L.							
2 gravèolens. DC.	(*pu.*)	strong-scented.	N. America.	....	——	H. ⊙.	Corn. can. 131. ic.
3 viscòsa. DC.	(*ye.*)	viscid.	Ceylon.	1730.	——	S. ⊙.	Rheed. mal. 9. t. 23.
4 dodecándra. DC.	(*pu.*)	three-leaved.	————	1795.	——	S. ⊙.	Burm. zeyl. t. 100. f. 1.
Cleòme dodecándra. L.							
5 uniglandulòsa. DC.	(*w.*)	one-glanded.	S. America.	1823.	——	H. ⊙.	Cav. ic. 4. t. 306.

Tribus II. *CAPPAREÆ.* DC.

CRAT'ÆVA. DC.		CRAT'ÆVA.	Dodecandria Monogynia. L.				
1 gynándra. DC.	(*wh.*)	thin-leaved.	Jamaica.	1789.	....	S. ♄.	Pluk. alm. t. 147. f. 6.
2 Tàpia. DC.	(*wh.*)	Garlick Pear.	S. America.	1752.	....	S. ♄.	Plum. gen. t. 21.
3 Roxbúrghii. B.	(*wh.*)	Roxburgh's.	E. Indies.	1822.	....	S. ♄.	
Cápparis trifoliàta. H.B.							
RITCHI'EA. B.		RITCHI'EA.	Dodecandria Monogynia.				
fràgrans.	(*wh.*)	fragrant.	Sierra Leon.	1795.	6. 7.	S. ♄. ‿.	Bot. mag. 596.
Crat'æva fragans. B.M. *capparoides.* A.R. t. 176.							
NIEB'UHRIA. DC.		NIEB'UHRIA.	Polyandria Monogynia. L.				
1 cáfra. DC.	(*wh.*)	Caffrarian.	C. B. S.	1820.	....	G. ♄.	
2 madagascariénsis. DC.	(*w.*)	Madagascar.	Madagascar.	1822.	....	S. ♄.	
3 oblongifòlia. DC.	(*wh.*)	oblong-leaved.	E. Indies.	1823.	....	S. ♄. ‿.	
Cápparis heteroclìta. H.B.							
BO'SCIA. DC.		BO'SCIA.	Dodecandria Monogynia. L.				
senegalénsis. DC.	(*wh.*)	Senegal.	Senegal.	1822.	....	S. ♄.	Lam. ill. t. 395. f. 2.
CA'PPARIS. DC.		CAPER-TREE.	Polyandria Monogynia. L.				
1 mariàna. DC.	(*wh.*)	heart-leaved.	Timor.	1822.	....	S. ♄.	Jacq. Schœnb. t. 109.
2 rupéstris. DC.	(*wh.*)	rock.	Crete.	....	7. 10.	F. ♄. ‿.	Flor. græc. t. 487.
spinòsa. B.M. 291. nec aliorum.							
3 spinòsa. DC.	(*wh.*)	common.	S. Europe.	1596.	5. 8.	G. ♄. ‿.	Flor. græc. t. 486.
4 ovàta. W.	(*wh.*)	oval-leaved.	Mauritania.	....	——	G. ♄. ‿.	Boc. sic. t. 42. f. 3.
Fontanèsii. DC.							
5 herbàcea. DC.	(*wh.*)	herbaceous.	Tauria.	1819.	——	H. ♃.	
6 ægy'ptia. DC.	(*wh.*)	Egyptian.	Egypt.	1822.	....	G. ♄.	Delil. fl. eg. t. 31. f. 3.
7 aphy'lla. DC.	(*wh.*)	leafless.	E. Indies.	——	....	S. ♄.	
8 zeylànica. DC.	(*wh.*)	Ceylon.	Ceylon.	——	....	S. ♄. ‿.	
9 acuminàta. DC.	(*wh.*)	acuminate.	E. Indies.	1821.	....	S. ♄. ‿.	Braam ic. chin. t. 29.?
10 frondòsa. DC.	(*g.p.*)	large-leaved.	Carthagena.	1800.	....	S. ♄.	Jacq. amer. t. 104.
11 triflòra. DC.	(*wh.*)	three-flowered.	S. America.	....	....	S. ♄.	
12 cynophallóphora. DC.	(*wh.*)	Bay-leavd.	W. Indies.	1752.	....	S. ♄.	Jacq. amer. t. 98.
13 eustachiàna. DC.	(*pu.*)	purple-flower'd.	————	1822.	....	S. ♄.	—— ed. pict. t. 146.
14 salígna. DC.	....	Willow-leaved.	S. Cruz.	1807.	....	S. ♄.	
15 lineàris. DC.	(*wh.*)	linear-leaved.	W. Indies.	1793.	....	S. ♄.	Jacq. amer. t. 102.
16 verrucòsa. DC.	(*wh.*)	wart-podded.	Carthagena.	1823.	....	S. ♄.	———— t. 99.
17 pulchérrima. DC.	(*st.*)	handsome.	————	...	....	S. ♄.	Jacq. amer. t. 106.
arboréscens. Mil.							
18 amygdalina. DC.	(*wh.*)	Almond-leaved.	S. America.	1819.	....	G. ♄.	Plum. gen. 40. t. 16.

19 odoratíssima. DC.	(*pu.*)	sweet-scented.	Caracas.	1814.	....	S. ♄.	Jacq. schœn. t. 110.
20 tenuisíliqua. DC.	(*wh.*)	slender-podded.	Carthagena.	1822.	....	S. ♄.	Jacq. amer. t. 105.
21 ferrugínea. DC.	(*wh.*)	ferruginous.	Jamaica.	....	....	S. ♄.	——— t. 100.
octándra. Jacq.							
22 Bréynia. DC.	(*wh.*)	Oleaster-leav'd.	W. Indies.	1752.	....	G. ♄.◡.	Jacq. amer. t. 103.?
23 jamaicénsis. DC.	(*wh.*)	Jamaica.	———	1793.	....	S. ♄.◡.	——— t. 101.
24 torulòsa. DC.	(*wh.*)	twisting.	———	1819.	....	S. ♄.◡.	
STEPHA'NIA. DC.		STEPHA'NIA.	Hexandria Monogynia.				
cleomóides. DC.	(*o.y.*)	Cleome-like.	Caracas.	1823.	....	S. ♄.	Jacq. schœn. t. 111.
Cápparis paradóxa. Jacq. *Sterίphoma cleomóides.* s.s.							
MORIS'ONIA. DC.		MORIS'ONIA.	Monadelphia Polyandria. L.				
americàna. DC.	(*wh.*)	American.	Caribees.	1824.	....	S. ♄.	Jacq. amer. t. 97.

ORDO XV.

RESEDACEÆ. *DC. theor. ed.* 1. *p.* 214.

RE'SEDA. W.		RE'SEDA.	Dodecandria Trigynia. L.				
1 Lutèola. W.	(*st.*)	Dyers weed.	Britain.	....	6. 7.	H. ☉.	Eng. bot. 320.
2 críspata. L. en.	(*st.*)	curled-leaved.	S. Europe.	1820.	——	H. ☉.	
3 viréscens. s.s.	(*st.*)	greenish.	Spain.	——	——	H. ☉.	
4 canéscens. s.s.	(*wh.*)	hoary.	Egypt.	1597.	5. 7.	H. ♃.	Clus. hist. 1. t. 295.
5 gláuca. s.s.	(*wh.*)	glaucous.	S. Europe.	1700.	——	F. ♃.	Pluk. alm. t. 107. f. 2.
6 scopària. s.s.	(*wh.*)	Broom-like.	Teneriffe.	1815.	8. 9.	G. ♄.	
7 dipétala. s.s.	(*wh.*)	two-petaled.	C. B. S.	1774.	8.	G. ♂.	
8 linifòlia. s.s.	(*wh.*)	flax-leaved.	S. Europe.	1820.	7. 8.	F. ♂.	
9 subulàta. Del.	(*wh.*)	awl-leaved.	Egypt.	1827.	6. 8.	H. ♂.	
10 chinénsis. s.s.	(*wh.*)	Chinese.	China.	1824.	——	H. ♂.	
11 sesamóides. s.s.	(*wh.*)	spear-leaved.	France.	1787.	——	H. ☉.	All. ped. 2. t. 88. f. 3.
12 mediterrànea. s.s.	(*w.*)	Mediterranean.	Palestine.	1791.	6. 9.	H. ☉.	Jacq. ic. 3. t. 475.
13 odoràta. s.s.	(*st.*)	Mignonette.	Egypt.	1752.	6. 10.	H. ☉.	Bot. mag. 29.
β *frutéscens.* B.R.	(*st.*)	*tree mignonette*........		——	——	G. ♄.	Bot. reg. 227.
14 Phyteùma. s.s.	(*wh.*)	trifid.	S. Europe.	——	6. 9.	H. ☉.	Jacq. aust. 2. t. 132.
15 incìsa. T.C.	(*wh.*)	cut-leaved.	Italy.	1825.	——	H. ☉.	
16 ramosíssima. s.s.	(*wh.*)	branching.	Spain.	1816.	——	H. ♃.	
17 saxátilis. s.s.	(*wh.*)	rock.	———	——	——	H. ♃.	
stricta. P.S.							
18 pruinòsa. s.s.	(*wh.*)	papillose-ld.	Egypt.	1823.	——	H. ♃.	
19 lùtea. s.s.	(*st.*)	Base-rocket.	Britain.	....	7. 8.	H. ♂.	Eng. bot. 321.
20 undàta. s.s.	(*wh.*)	wave-leaved.	Spain.	1739.	6. 8.	H. ♃.	Barrel. rar. t. 587.
21 álba. s.s.	(*wh.*)	upright-white.	S. Europe.	1596.	5. 10.	H. ♂.	Lobel. ic. 222.
22 fruticulòsa. s.s.	(*wh.*)	frutescent.	Spain.	1794.	9.	G. ♄.	Jacq. ic. 3. t. 474.
23 bipinnàta. s.s.	(*wh.*)	bipinnate.	———	1816.	6. 8.	G. ♄.	
24 myriophy'lla. T.C.	(*st.*)	numerous-ld.	Italy.	1825.	——	H. ♃.	

ORDO XVI.

DATISCEÆ. *Brown in app. to Denham and Clap. trav. p.* 25.

DATI'SCA. L.		DATI'SCA.	Diœcia Dodecandria. L.				
1 cannabìna. L.	(*ye.*)	Hemp-like.	Levant.	1640.	7. 9.	H. ♃.	Schkuhr han. 3. t. 336.
2 hírta. W.	(*ye.*)	hairy.	Pensylvania.	1826.	9. 10.	H. ♃.	

ORDO XVII.

FLACOURTIANEÆ. *DC. prodr.* 1. *p.* 255.

Tribus I. *PATRISIEÆ.* DC.

RYAN'ÆA. DC.		RYAN'ÆA. Polyandria Monogynia.					
speciòsa. DC.	(*re.*)	beautiful.	Trinidad.	1822.	....	S. ♄.	Vahl. ecl. 1. p. 51. t. 9.
Patrísia pyrífera. P.S.							

Tribus II. *FLACOURTIEÆ.* DC.

FLACOU'RTIA. DC.		FLACOU'RTIA. Diœcia Polyandria. L.					
1 Ramóntchi. DC.	(*wh.*)	shining-leaved.	Madagascar.	1775.	6. 7.	S. ♄.	L. Her. strip. 59. t. 30.
2 sápida. DC.	(*wh.*)	esculent.	E. Indies.	1800.	....	S. ♄.	Roxb. cor. 1. t. 69.
3 inérmis. DC.	(*wh.*)	unarmed.	Moluccas.	1814.	10. 2.	S. ♄.	Bot. reg. ined.
4 sepiària. DC.	(*wh.*)	obovate-leaved.	Coromandel.	1820.	....	S. ♄.	Roxb. cor. 1. t. 68.
5 cataphrácta. DC.	(*wh.*)	many-spined.	E. Indies.	1804.	....	S. ♄.	
6 rotundifòlia. H.B.	(*wh.*)	round-leaved.	———	1823.	....	S. ♄.	
7 flavéscens. DC.	(*st.*)	yellow-flower'd.	Guinea.	1780.	....	S. ♄.	
8 rhamnóides. DC.	(*wh.*)	Rhamnus like.	C. B. S.	1819.	....	G. ♄.	

Tribus III. *KIGGELARIEÆ.* DC.

KIGGEL'ARIA. DC.		KIGGEL'ARIA. Diœcia Decandria. L.					
1 africàna. DC.	(*wh.*)	African.	C. B. S.	1683.	5. 6.	G. ♄.	Lam. ill. t. 821.
2 integrifòlia. DC.	(*wh.*)	entire-leaved.	———	1819.	———	G. ♄.	Jacq. ic. rar. t. 628.
MELI'CYTUS. DC.		MELI'CYTUS. Diœcia Pentandria.					
ramiflòrus. DC.	....	branch-flow'rd.	N. Zealand.	1822.	....	G. ♄.	Lam. ill. t. 812. f. 1.

ORDO XVIII.

BIXINEÆ. *DC. prodr.* 1. *p.* 259.

BI'XA. DC.		ANOTTA. Polyandria Monogynia. L.					
1 purpùrea.	(*pu.*)	purple.	E. Indies.	1820.	....	S. ♄.	
2 Orellàna. DC.	(*ro.*)	heart-leaved.	S. America.	1690.	5. 8.	S. ♄.	Bot. mag. 1456.
3 Urucuràna. W.	(*ro.*)	white-leaved.	Brazil.	1823.	....	S. ♄.	
LÆ'TIA. DC.		LÆ'TIA. Polyandria Monogynia. L.					
Thámnia. DC.	......	shining-leaved.	Jamaica	....	....	S. ♄.	Browne jam. t. 25. f. 2.
PRO'CKIA. DC.		PRO'CKIA. Polyandria Monogynia. L.					
1 Crùcis. DC.	(*gr.*)	oval-leaved.	Santa Cruz.	....	....	S. ♄.	Vahl symb. 3. t. 64.
2 serràta. DC.	(*gr.*)	saw-leaved.	Montserrat.	1822.	....	S. ♄.	
3 theæfórmis. DC.	(*gr.*)	Tea-leaved.	Bourbon.	———	....	S. ♄.	Bory voy. 2. t. 24.
Lùdia heterophy'lla. Bory. non. Lam.							
L'UDIA. DC.		L'UDIA. Polyandria Monogynia. L.					
1 heterophy'lla. DC.	(*gr.*)	various-leaved.	Mauritius.	1822.	....	S. ♄.	Lam. ill. t. 466. f. 1. 2.
2 sessiliflòra. DC.	(*bl.*)	sessile-flower'd.	———	———	....	S. ♄.	Jacq. schœnb. 1. t. 112.
tuberculàta. Jacq.							

ORDO XIX.

CISTINEÆ. *DC. prodr.* 1. *p.* 263.

CI'STUS. DC.		ROCK-ROSE. Polyandria Monogynia. L.					
1 candidíssimus. DC.	(*ro.*)	Canary Island.	Canaries.	1817.	6. 9.	F ♄.	Swt. Cist. t. 3.
2 vaginàtus. DC.	(*ro.*)	oblong-leaved.	Teneriffe.	1779.	4. 8.	F. ♄.	——— t. 9.

3 álbidus. DC. (ro.) white-leaved. S. Europe. 1640. 5. 9. H. ♄. Swt. cist. t. 31.
4 incànus. DC. (li.) hoary-leaved. —— 1596. 5. 8. F. ♄. —— t. 44.
5 canéscens. S.C. (pu.) narrow hoary. —— .... —— F. ♄. —— t. 45.
6 villòsus. DC. (pu.) villous. —— 1640. —— F. ♄. —— t. 35.
7 rotundifòlius. S.C. (pu.) round-leaved. —— .... —— F. ♄. —— t. 75.
8 undulàtus. DC. (pu.) wave-leaved. —— .... —— H. ♄. —— t. 63.
9 heterophy'llus. DC. (p.) various-leaved. Algiers. .... —— F. ♄. —— t. 6.
10 críspus. DC. (pu.) curled-leaved. S. Europe. 1656. —— F. ♄. —— t. 22.
11 crèticus. DC. (pu.) Cretan. Levant. 1731. —— F. ♄. Flor. græc. t. 495.
12 purpureus. DC. (pu.) purple. —— .... 6. 8. F. ♄. Swt. cist. t. 17.
13 cymòsus. DC. (li.) cyme-flowered. —— .... —— F. ♄. —— t. 90.
incànus. Flor. græc. t. 494. *nec aliorum.*
14 parviflòrus. DC. (li.) small-flowered. —— 1821. —— F. ♄. Swt. cist. t. 14.
15 Cupaniànus. S.S. (wh.) heart-leaved. Sicily. .... 6.10. H. ♄. —— t. 70.
16 asperifòlius. S.C. (wh.) rough-leaved. .... —— H. ♄. —— t. 87.
17 acutifòlius. S.C. (wh.) acute-leaved. S. Europe. .... 5. 8. H. ♄. —— t. 78.
18 salvifòlius. DC. (wh.) Sage leaved. Europe. 1548. 5. 9. H. ♄. —— t. 54.
19 obtusifòlius. S.C. (wh.) blunt-leaved. Levant. .... —— F. ♄. —— t. 42.
20 oblongifòlius. S.C. (wh.) oblong-leaved. .... 6. 10. H. ♄. —— t. 67.
21 corbariénsis. DC. (wh.) mountain. S. France. .... 5. 6. H. ♄. —— t. 8.
22 platysépalus. S.C. (wh.) broad-sepaled. Levant. .... 6. 8. F. ♄. —— t. 47.
23 florentinus. DC. (wh.) Florentine. Italy. .... 6. 7. H. ♄. —— t. 59.
24 monspeliénsis. DC. (w.) Montpelier. S. Europe. 1656. —— F. ♄. —— t. 27.
25 Lèdon. DC. (wh.) many-fl[d]. gum. S. France. 1730. 6. 8. H. ♄. Duham. arb. 1. t. 66.
26 hirsùtus. DC. (wh.) hairy. Spain. 1656. —— H. ♄. Swt. cist. t. 19.
27 psilosépalus. S.C. (wh.) smooth-calyxed. .... —— H. ♄. —— t. 33
28 láxus. DC. (wh.) waved-leaved. Spain. 1656. —— H. ♄. —— t. 12.
29 populifòlius. S.C. (wh.) Poplar-leaved. —— —— 5. 6. H. ♄. —— t. 23.
30 latifòlius. S.C. (wh.) broad-leaved. Barbary. .... —— F. ♄. —— t. 15.
31 laurifòlius. DC. (wh.) Laurel-leaved. S. Europe. 1731. 6. 8. H. ♄. —— t. 52.
32 c'yprius. DC. (wh.) common-gum. Cyprus. .... —— H. ♄. —— t. 39.
ladaníferus. Bot. mag. 112. *nec. aliorum.* C. *undulatus.* L. en?
33 ladaníferus. DC. (wh.) flat-leaved gum. Spain. 1629. —— F. ♄.
α albiflòrus. DC. (wh.) *white-flowered.* —— —— —— F. ♄. Swt. cist. t. 84.
β maculàtus. DC. (w.p.) *spot-flowered.* —— —— —— F. ♄. —— t. 1.
34 Clùsii. DC. (wh.) Clusius's. Spain. .... —— H. ♄. —— t. 32.

HELIA'NTHEMUM. DC. SUN-ROSE. Polyandria Monogynia. L.

1 Libanòtis. DC. (wh.) Rosemary-like. Spain. 1752. 6. F. ♄. Barr. ic. 294.
2 umbellàtum. DC. (wh.) umbel-flowered. S. Europe. 1731. 6. 8. F. ♄. Swt. cist. t. 5.
3 ocymóides. DC. (ye.) Basil-like. —— .... —— F. ♄. —— t. 13.
Cístus sampsucifòlius. Cav. ic. 1. t. 96.
4 alyssóides. DC. (ye.) Alyssum-like. —— .... —— F. ♄. Vent. choix. t. 20.
5 rugòsum. DC. (ye.) rugose. Portugal. .... —— F. ♄. Swt. cist. t. 65.
6 scabròsum. DC. (ye.) rough. Italy. 1775. —— F. ♄. —— t. 81.
7 algarvénse. DC. (ye.) Algarvian. Portugal. 1800. 7. 8. F. ♄. —— t. 40.
Cístus algarvénsis. B.M. 627.
8 formòsum. DC. (ye.) beautiful. —— 1780. 5. 7. F. ♄. Swt. cist. t. 50.
Cístus formòsus. B.M. 264.
9 atriplicifòlium. DC. (y.) Orach-leaved. Spain. 1656. 6. 7. F. ♄. Barr. ic. t. 292.
10 microphy'llum. S.C. (y.) small-leaved. S. Europe. .... 6.10. F. ♄. Swt. cist. t. 96.
11 lasiánthum. DC. (ye.) downy. Portugal. 1822. 7. 9. F. ♄.
12 involucràtum. DC. (ye.) short-pedicled. S. Europe. 1826. 6. 8. F. ♄.
13 cándidum. S.C. (ye.) white-leaved. Spain. 1822. —— F. ♄. Swt. cist. t. 25.
14 cheiranthóides. DC. (y.) Stock-like. —— 1800. 7. 8. F. ♄. —— t. 107.
15 halimifòlium. S.C. (ye.) Sea-Purslane-l[d] —— 1656. —— F. ♄. —— t. 4.
16 rosmarinifòlium. DC. (y.) Rosemary-l[d]. N. America. 1823. —— H. ♃.
17 glomeràtum. DC. (ye.) cluster-flower'd. New Spain. —— 6. 9. F. ♄. Swt. cist. t.
18 ramuliflòrum. DC. (ye.) branch-flowers Carolina. 1823. —— H. ♃.
19 canadénse. DC. (ye.) Canadian. Canada. 1799. —— H. ♃. Swt. cist. t. 21.
20 caroliniànum. DC. (ye.) Carolina. Carolina. 1823. —— F. ♃. —— t. 99.
21 brasiliénse. DC. (st.) Brazilian. Brazil. —— 6. 8. F. ♄. —— t. 43.
22 polygalæfòlium. S.C. (ye.) Milkwort-leav[d]. —— —— —— F. ♄. —— t. 11.
23 lignòsum. S.C. (ye.) woody-stemm'd. S. Europe. 1810. —— F. ♄. —— t. 46.
24 Tuberària. DC. (ye.) Plantain-leav'd. —— 1752. —— F. ♃. —— t. 18.
25 plantagíneum. DC. (ye.) Plantain like. —— 1823. —— H. ⊙.
26 guttàtum. DC. (ye.) spotted-flower[d]. England. .... 6. 8. H. ⊙. Eng. bot. 544.
27 eriocáulon. DC. (ye.) woolly-stalked. Spain. 1823. —— H. ⊙. Swt. cist. t. 30.
28 punctàtum. DC. (ye.) punctated. France. 1816. —— H. ⊙. —— t. 61.
29 lunulàtum. DC. (ye.) lunulate-mark[d]. Piedmont. .... —— H. ♄. All. auct. p. 30. t. 2. f. 3.

30	villòsum. DC.	(st.)	villous.	Spain.	1823.	——	H. ⊙.	
31	nilóticum. DC.	(st.)	Nile.	Egypt.	....	——	H. ⊙.	
32	ledifòlium. DC.	(st.)	Ledum-leaved.	England.	....	——	H. ⊙.	Swt. cist. t. 41.
33	intermèdium. DC.	(ye.)	intermediate.	Spain.	1759.	——	H. ⊙.	Cav. ic. t. 144.
	salicifòlium. Cav.							
34	denticulàtum. DC.	(ye.)	tooth-leaved.	S. France.	1828.	——	H. ⊙.	
35	salicifòlium. DC.	(ye.)	Willow-leaved.	——	——	——	H. ⊙.	Swt. cist. t. 71.
36	sanguíneum. DC.	(ye.)	bloody-stalked.	——	1823.	——	H. ⊙.	
	retrofráctum. P.S.							
37	ægyptìacum. DC.	(ye.)	Egyptian.	Egypt.	1764.	——	H. ⊙.	Jacq. obs. 3. t. 68.
38	Líppii. DC.	(ye.)	small-petaled.	——	1820.	——	F. ♄.	
39	ellípticum. DC.	(st.)	elliptic-leaved.	Levant.	1827.	——	F. ♄.	Swt. cist. t. 108.
40	confértum. DC.	(ye.)	close-flowered.	Teneriffe.	....	6. 7.	F. ♄.	
41	canariénse. DC.	(ye.)	Canary.	Canaries.	1790.	——	F. ♄.	Jacq. ic. 1. t. 97.
42	ericóides. DC.	(ye.)	Heath-leaved.	S. Europe.	....	6. 8.	F. ♄.	Cav. ic. 2. t. 172.
43	Fumàna. DC.	(ye.)	Heath-like.	——	1752.	——	F. ♄.	Swt. cist. t. 16.
44	procúmbens. DC.	(ye.)	procumbent.	——	....	——	F. ♄.	—— t. 68.
45	arábicum. DC.	(sc.)	Arabian.	Levant.	1826.	——	F. ♄.	—— t. 97.
46	læ'vipes. DC.	(ye.)	cluster-leaved.	——	1690.	——	G. ♄.	—— t. 24.
47	víride. DC.	(ye.)	green-leaved.	Sicily.	1828.	——	F. ♄.	
48	juniperìnum. DC.	(ye.)	Juniper-leaved.	——	....	——	G. ♄.	Barr. ic. t. 443.
49	Barreliéri. DC.	(ye.)	Barrelier's.	Italy.	1822.	——	F. ♄.	—— t. 416.
50	thymifòlium. DC.	(ye.)	Thyme-leaved.	Spain.	1658.	——	F. ♄.	Swt. cist. t. 102.
51	glutinòsum. DC.	(ye.)	clammy.	——	1790.	——	F. ♄.	—— t. 83.
52	mólle. DC.	(ye.)	soft.	——	....	——	F. ♄.	Cav. ic. 3. t. 262. f. 2.
53	origanifòlium. DC.	(ye.)	Marjoram-ld.	——	1795.	——	F. ♄.	—— 3. t. 262. f. 1.
54	dichótomum. DC.	(ye.)	forked.	——	....	——	F. ♄.	—— 3. t. 263. f. 1.
55	œlándicum. DC.	(ye.)	smooth-leaved.	Europe.	1816.	——	H. ♄.	Swt. cist. t. 85.
56	alpéstre. DC.	(ye.)	alpine.	——	....	——	H. ♄.	—— t. 2.
57	penicillàtum. DC.	(ye.)	pencilled.	S. Europe.	....	——	H. ♄.	
58	obovàtum. DC.	(ye.)	obovate-leaved.	Spain.	1826.	——	F. ♄.	
59	itálicum. DC.	(ye.)	Italian.	Italy.	1779.	7. 9.	H. ♄.	Barr. rar. t. 366.
60	vineàle. DC.	(ye.)	slender-trailing.	Europe.	1823.	——	H. ♄.	Swt. cist. t. 77.
61	cànum. DC.	(ye.)	hoary.	——	1772.	6. 7.	F. ♄.	—— t. 56.
62	marifòlium. DC.	(ye.)	Marum-leaved	——	....	——	F. ♄.	Barr. rar. t. 441.
63	pulchéllum. S.C.	(ye.)	neat.	S. Europe.	....	——	H. ♄.	Swt. cist. t. 71.
64	rotundifòlium. DC.	(ye.)	round-leaved.	Spain.	1826.	——	F. ♄.	Cav. ic. 2. t. 142.
	Cístus nummulàrius. Cav. *non* L.							
65	crassifòlium. DC.	(ye.)	thick-leaved.	——	——	——	F. ♄.	
66	cinéreum. DC.	(ye.)	gray.	Spain.	....	——	F. ♄.	Cav. ic. 2. t. 141.
67	squammàtum. DC.	(ye.)	scaly.	——	1815.	——	F. ♄.	—— 2. t. 139.
68	lavandulæfòlium. DC.	(ye.)	Lavender-ld.	Levant.	1739.	——	F. ♄.	Barr. ic. t. 288.
	β *syrìacum*. DC.	(ye.)	*Syrian*.	Syria.	——	——	F. ♄.	Jacq. ic. 1. t. 96.
69	stœchadifòlium. DC.	(y.)	woolly-leaved.	Portugal.	1819.	——	F. ♄.	
70	cròceum. DC.	(ye.)	Saffron-colourd.	Spain.	——	——	F. ♄.	Swt. cist. t. 53.
71	nudicáule. DC.	(ye.)	naked-stalked.	——	1826.	——	F. ♄.	
72	gláucum. DC.	(ye.)	glaucous-leaved.	——	1815.	6. 8.	F. ♄.	Cav. ic. 3. t. 261.
73	leptophy'llum. DC.	(ye.)	narrow-leaved.	——	....	7. 8.	F. ♄.	Swt. cist. t. 20.
74	serpyllifòlium. DC.	(ye.)	Serpyllum-ld.	England.	....	5. 9.	H. ♄.	—— t. 60.
75	vulgáre. DC.	(ye.)	common.	Britain.	....	——	H. ♄.	—— t. 34.
	β *flòre plèno*.	(ye.)	*double-flowered*.	——	....	——	H. ♄.	—— t. 64.
76	surrejànum. DC.	(ye.)	dotted-leaved.	England.	....	7. 10.	H. ♄.	—— t. 28.
	β *tomentòsum*.	(ye.)	*woolly-leaved*.	——	....	——	H. ♄.	—— t. 34. f. 3.
77	ovàtum. DC.	(ye.)	ovate-leaved.	Geneva.	....	6. 8.	H. ♄.	Viv. frag. 1. t. 8. f. 2.
78	lùcidum. H.	(ye.)	glossy-leaved.	....	1827.	——	H. ♄.	
79	táuricum. S.C.	(p.y.)	Taurian.	Tauria.	1824.	——	H. ♄.	Swt. cist. t. 105.
80	Andersòni. S.C.	(ye.)	Anderson's.	Hybrid.	1827.	5. 10.	H. ♄.	—— t. 89.
81	grandiflòrum. DC.	(ye.)	large-flowered.	Pyrenees.	1800.	——	H. ♄.	—— t. 69.
82	sampsucifòlium.	(p.y.)	rough dwarf.	Montpelier.	....	——	H. ♄.	Bot. mag. 1803.
	Cístus sampsucifòlius. B.M. *non* Cav.							
83	nummulàrium. DC.	(ye.)	Money-wort.	——	1752.	——	H. ♄.	Swt. cist. t. 80.
84	angustifòlium. DC.	(ye.)	distinct-petaled.	....	1800.	——	H. ♄.	Jacq. vind. 3. t. 53.
85	hírtum. DC.	(ye.)	bristly-calyxed.	Spain.	1759.	——	F. ♄.	Swt. cist. t. 109.
86	barbàtum. P.S.	(ye.)	bearded.	S. Europe.	1820.	——	H. ♄.	—— t. 73.
87	eriosépalon. S.C.	(p.y.)	woolly-sepaled.	——	....	——	H. ♄.	—— t. 76.
88	rotàtum.	....	rotate.	....	1828.	....	F. ♄.	
89	racemòsum. DC.	(wh.)	long-racemed.	S. Europe.	....	——	F. ♄.	Swt. cist. t. 82.
90	pilòsum. DC.	(wh.)	hairy.	——	1731.	——	F. ♄.	—— t. 49.
91	lineàre. DC.	(wh.)	linear-leaved.	——	1824.	6. 8.	F. ♄.	—— t. 48.

92 virgàtum. DC. (*p.r.*) slender. Barbary. —— —— F. ♄. Swt. cist. t. 79.
93 apennìnum. DC. (*wh.*) Apennine. Europe. —— —— H. ♄. ——— t. 62.
94 híspidum. DC. (*wh.*) hispid. S. France. —— —— H. ♄.
95 pulverulèntum. DC. (*w.*) powdered. ——— .... —— H. ♄. Swt. cist. t. 29.
96 canéscens. S.C. (*cr.*) canescent. .... —— H. ♄. ——— t. 51.
97 rhodánthum. DC. (*ro.*) rose-flowered. Spain. .... —— H. ♄. ——— t. 7.
98 diversifòlium. S.C. (*re.*) various-leaved. .... 5. 9. F. ♄. ——— t. 95.
β multiplèx. S.C. (*re.*) *full-flowered.* .... —— F. ♄. ——— t. 98.
99 cùpreum. S.C. (*co.*) copper-colour'd. Hybrid. .... —— F. ♄. ——— t. 66.
100 versícolor. S.C. (*p.r.*) various-colour'd. ——— .... —— H. ♄. ——— t. 26.
101 venústum. S.C. (*sc.*) charming. .... —— H. ♄. ——— t. 10.
β flòre plèno. (*sc.*) *double-flowered.* .... —— H. ♄.
102 variegàtum. S.C. (*r.w.*) striped-flower'd. Hybrid. .... —— F. ♄. Swt. cist. t. 38.
103 lanceolàtum. S.C. (*w.*) spear-leaved. .... —— H. ♄. ——— t. 100.
104 polifòlium. DC. (*wh.*) white-mountain. England. .... 5. 7. H. ♄. ——— t. 88.
105 confùsum. S.C. (*wh.*) confused. Europe. .... —— H. ♄. ——— t. 91.
106 macránthon. S.C. (*w.*) great-flowered. ——— .... —— H. ♄. ——— t. 103
β multiplèx. S.C. (*wh.*) *full-flowered.* ——— .... —— H. ♄. ——— t. 104.
107 mutábile. DC. (*p.r.*) changeable. Spain. 1795. 6. 7. H. ♄. ——— t. 106.
108 ròseum. DC. (*ro.*) rosy. S. Europe. .... —— H ♄. ——— t. 55.
β multiplèx. SC. (*ro.*) *double-rose.* ——— .... —— H. ♄. ——— t. 86
109 Milléri. S.C. (*sa.*) Miller's. .... —— H. ♄. ——— t. 101.
110 hyssopifòlium. DC. (*v.*) Hyssop-leaved. Italy. .... —— F. ♄.
α crocàtum. S.C. (*sa.*) *saffron-colour'd.* ——— .... —— F. ♄. Swt. cist. t. 92.
β cùpreum. S.C. (*co.*) *copper-coloured.* ——— .... —— F. ♄. ——— t. 58.
γ multiplèx. S.C. (*co.*) *full-flowered.* .... —— F. ♄. ——— t. 72.
111 fœ'tidum. DC. (*wh.*) Bryony-scented. .... —— H. ♄. Jacq. ic. 1. t. 98.
112 sulphùreum. DC. (*st.*) sulphur-color'd. Spain. 1815. —— H. ♄. Swt. cist. t. 37.
113 stramíneum. S.C. (*st.*) straw-coloured. .... —— H. ♄. ——— t. 93.
β multiplèx. S.C. (*st.*) *full-flowered.* .... —— H. ♄. ——— t. 94.

HUDS'ONIA. DC. HUDS'ONIA. Dodecandria Monogynia. L.
1 ericóides. DC. (*ye.*) Heath-like. N. America. 1805. 5. 7. F. ♄. Swt. cist. t. 36.
2 tomentòsa. N. (*ye.*) tomentose. ——— 1826. 6. 8. F. ♄. ——— t. 57.

LECH'EA. DC. LECH'EA. Tri-Dodecandria Trigynia.
1 villòsa. DC. (*ye.*) villous. Canada. 1780. 7. 8. H. ♃. Lam. ill. t. 52. f. 2.
màjor. Ph. *non* Lin. *mìnor* L. *non* Ph.
2 mìnor. DC. (*ye.*) lesser. ——— 1802. —— H. ♃. Lam. ill. t. 52. f. 1.
3 thymifòlia. DC. (*ye.*) Thyme-leaved. N. America. 1823. —— H. ♃.

ORDO XX.

VIOLARIEÆ. *DC. prodr.* 1. *p.* 287.

Tribus I. *VIOLEÆ.* DC.

CALY'PTRION. DC. CALY'PTRION. Pentandria Monogynia.
Aublétii. DC. (*wh.*) Aublet's. S. America. 1820. .... S. ♄.◡. Aub. gui. 2. t. 319.
Vìola Hybánthus. Aub. *Jonídium Aublétii.* R. S.

NOISE'TTIA. DC. NOISE'TTIA. Pentandria Monogynia.
1 longifòlia. DC. (*wh.*) long-leaved. Cayenne. 1822. .... S. ♄. Nov. g. am. t. 409. b. f. 2.
Jonídium longifòlium. R.S.
2 pyrifòlia. Mart. (*wh.*) Pear-leaved. Brazil. 1822. .... S. ♄.◡. Mart. n. pl. bras. ic.

V'IOLA. DC. VIOLET. Pentandria Monogynia.
1 pedàta. DC. (*bl.*) cut-leaved. N. America. 1759. 5. 8. H. ♃. Swt. br. fl. gar. t. 69.
2 flabellifòlia. B.C. (*b.p.*) fan-leaved. ——— 1822. —— H. ♃. Lod. bot. cab. 777.
pedàta β atropurpùrea. DC. *pedàta* var. *bícolor.* Ph.
3 digitàta. Ph. (*bl.*) finger-leaved. Virginia. 1824. —— H. ♃.
4 palmàta. DC. (*bl.*) palmated. ——— 1752. 5. 6. H. ♃. Bot. mag. 535.
5 asarifolia. DC. (*bl.*) Asarum-leaved. ——— 1823. —— H. ♃.
6 papilionàcea. DC. (*bl.*) variegated. ——— 1800. —— H. ♃.
7 oblíqua. Ph. (*bl.*) oblique-flower'd. ——— 1762. —— H. ♃.
8 cucullàta. Ph. (*bl.*) hollow-leaved. ——— 1772. 5. 7. H. ♃. Bot. mag. 1795.
9 pinnàta. R.S. (*pu.*) winged-leaved. Europe. 1752. 5. 6. H. ♃. All. m. taur. 3. t. 5. f. 2.
10 multífida. R.S. (*pu.*) multifid Siberia. —— —— H. ♃. Gmel. sib. 4. t. 48. f 4.

11 variegàta. DC.	(*pu.*)	variegated-l[d].	Dahuria.	1818.	5. 6.	H. ♃.	
12 primulæfòlia. DC.	(*bl.*)	Primrose-l[d].	N. America.	1783.	4. 6.	H. ♃.	
13 lanceolàta. DC.	(*wh.*)	spear-leaved.	———	1759.	6. 7.	F. ♃.	Swt. br. fl. gar. 2. t.174.
14 Patrínii. DC.	(*bl.*)	Siberian.	Siberia.	....	——	H. ♃.	
15 cæspitòsa. D.P.	(*pu.*)	tufted.	Nepaul.	1825.	5. 8.	F. ♃.	
Patrínii γ napaulénsis. DC. *primulæfolia.* Buch. non L.							
16 palmàris. D.P.	(*ye.*)	long-petioled.	———	——	——	F. ♃.	
17 ovàta. DC.	(*bl.*)	oval-leaved.	N. America.	1822.	——	H. ♃.	
18 sagittàta. DC.	(*b.w.*)	arrow-leaved.	———	1775.	——	H. ♃.	
19 dentàta. Ph.	(*bl.*)	large-toothed.	Pensylvania.	1828.	——	H. ♃.	
20 betonicæfòlia. DC.	(*pu.*)	Betony-leaved.	N. S. W.	1823.	——	F. ♃.	
21 Fischèri.	(*pu.*)	Fischer's.	Siberia.	1823.	5. 7.	H ♃.	
suávis. F. *non*, M.B.							
22 palústris. DC.	(*pa.*)	marsh.	Britain.	....	5. 6.	H. ♃.	Eng. bot. 444.
23 blánda. DC.	(*wh.*)	white-flowered.	N. America.	1802.	5. 7.	H. ♃.	W. hort. ber. 1. t. 24.
24 rotundifòlia. DC.	(*ye.*)	round-leaved.	———	1800.	——	H. ♃.	
25 soròria. W.	(*pu.*)	white-rooted.	———	1802.	4. 6.	H. ♃.	W. hort. ber. 1. t. 72.
26 áspera. DC.	(*wh.*)	rough.	Nepaul.	1825.	5. 8.	F. ♃.	
27 hírta. E.B.	(*bl.*)	hairy.	England.	....	——	H. ♃.	Eng. bot. 894.
28 ambígua. R.S.	(*pu.*)	ambiguous.	Danube.	1822.	——	H. ♃.	W. et K. hung. t. 190.
29 campéstris. M.B.	(*pu.*)	field.	Tauria.	1824.	——	H. ♃.	
30 hirsùta. R.S.	(*bl.*)	hairy-stalked.	Russia.	1824.	1. 12.	H. ♃.	
31 collìna. R.S.	(*li.*)	hill.	Volhynia.	——	5. 8.	H. ♃.	
32 cordàta. S.S.	(*bl.*)	heart-leaved.	N. America.	——	5. 7.	H. ♃.	
33 suávis. M.B.	(*p.b.*)	fragrant.	Siberia.	1823.	3. 5.	H. ♃.	
34 odoràta. DC.	(*vi.*)	sweet-scented.	Britain.	....	3. 10.	H. ♃.	Eng. bot. t. 619.
α violàcea.	(*vi.*)	*violet-coloured.*	———	....	——	H. ♃.	
β rosea.	(*ro.*)	*rose-coloured.*	———	....	——	H. ♃.	
γ alba.	(*wh.*)	*white-flowered.*	———	....	——	H. ♃.	
δ cærùlea.	(*bl.*)	*blue-flowered.*	———	....	——	H. ♃.	
ε purpùreo-plèna.	(*pu.*)	*double-purple.*	———	....	——	H. ♃.	
ζ álbo-plèna.	(*wh.*)	*double-white.*	———	....	——	H. ♃.	
η cærùleo-plèna.	(*bl.*)	*double-blue.*	———	....	——	H. ♃.	
θ pállido-plèna.	(*p.b.*)	*Neapolitan.*	Naples.	....	——	H. ♃.	
35 uliginòsa. DC	(*pu.*)	fen.	Carniola.	1822.	4. 5.	H. ♃.	
36 mirábilis. DC.	(*w.l.*)	broad-leaved.	Europe.	1732	6. 8.	H. ♃.	Jacq. fl. aust. t. 19.
37 striàta. DC.	(*st.*)	streaked.	N. America.	1772.	——	H. ♃.	
38 ochroleùca. DC.	(*st.*)	sulphur-color'd.	———	....	——	H. ♃.	
39 canìna. DC.	(*bl.*)	dog's.	Britain.	....	4. 6	H. ♃.	Eng. bot. 620.
40 sylvéstris. R.S.	(*p.b.*)	wood.	Hungary.	1823.	——	H. ♃.	
41 Riviniàna. S.S.	(*bl.*)	Rivini's.	Europe.	....	4. 6.	H. ♃.	
42 ericetòrum. L.en.	(*bl.*)	heath.	Germany.	1826.	——	H. ♃.	
43 neglécta. R.S.	(*p.b.*)	neglected.	Bohemia.	1824.	——	H. ♃.	
44 débilis. DC.	(*p.b.*)	weak.	N. America.	1820.	——	H. ♃.	
45 arenària. DC.	(*bl.*)	sand.	Europe.	——	——	H. ♃.	Pio viol. p. 20. t. 1. f. 2.
Alliònii. R.S. *nummularifòlia.* Schl.							
46 gláuca. M.B.	(*pu.*)	glaucous.	Russia.	1820.	——	H. ♃.	
47 pùmila. DC.	(*p.b.*)	small.	Europe.	——	——	H. ♃.	Vill. c. stras. t. 5.
48 flavicórnis. E.F.	(*bl.*)	yellow-spurred.	Britain.	....	5. 7.	H. ♃.	Raii synops. t. 24. f. 1.
49 montàna. S.S.	(*bl.*)	mountain.	Europe.	1683.	5. 6.	H. ♃.	Bot. mag. 1595.
50 stricta. H.	(*p.b.*)	upright.	———	1822.	——	H. ♃.	
Hornemanniàna. R.S.							
51 persicifòlia. S S.	(*bl.*)	Peach-leaved.	S. Europe.	1818.	——	H. ♃.	Schkuhr hand. 3. t. 269.
52 láctea. R.S.	(*pa.*)	cream-coloured.	England.	....	——	H. ♃.	Eng. bot. 445.
53 Rúppii. S.S.	(*pa.*)	Piedmont.	Piedmont.	1820.	——	H. ♃.	Lodd. bot. cab. 686.
54 Broussonetiàna. R.S.		Broussonet's.	Teneriffe.	1825.	——	G. ♄.	
55 arboréscens. DC.	(*pa.*)	shrubby.	S. Europe.	1779.	4. 5.	G. ♄.	Barr. ic. 568.
56 epípsila. S.S.	(*p.b.*)	creeping.	Siberia.	1821.	4. 6.	H. ♃.	
57 fennìca. F.	(*bl.*)	Siberian.	———	1827.	——	H. ♃.	
58 Selkírkii. DC.	......	Selkirk's.	Montreal.	1823.	——	H. ♃.	
59 py′gmæa. DC.	(*p.b.*)	pigmy.	Andes.	1822.	7. 11.	H. ♃.	
60 biflòra. DC.	(*ye.*)	two-flowered.	Europe.	1752.	——	H. ♃.	Flor. dan. 46.
61 Nuttállii. DC.	(*ye.*)	Nuttal's.	N. America.	1812.	5. 6.	H. ♃.	
62 hastàta. DC.	(*ye.*)	hastate.	———	1823.	5. 7.	H. ♃.	
63 albiflòra. L. en.	(*wh.*)	white-flowered.	———	....	——	H. ♃.	
64 canadénsis. DC.	(*w.r.*)	Canadian.	———	1783.	——	H. ♃.	
65 pubéscens. DC.	(*ye.*)	downy.	———	1772.	6. 7.	H. ♃.	Swt. br. fl. gar. 223.
66 eriocárpa. DC.	(*ye.*)	woolly-capsul'd.	———	1820.	——	H. ♃.	——— 102.
67 uniflòra. DC.	(*ye.*)	Siberian.	Siberia.	1774.	——	H. ♃.	Gmel. sib. 4. t. 48. f. 5.

68 chrysántha. Scr. (*ye.*) golden-flowered......... 1823. 5. 8. H. ♃.
69 nummularifòlia. DC. (*b.*) money-wort. Piedmont. —— —— H. ♃. All.ped.n.1640.t.9.f.4.
70 alpìna. DC. (*pu.*) alpine. Austria. 1826. —— H. ♃. Jacq. obs. 1. t. 11.
71 cornùta. DC. (*bl.*) horned. Pyrenees. 1776. 5. 6. H. ♃. Bot. mag. 791.
72 cenísia. DC. (*bl.*) Mount-Cenis. Switzerland. 1759. 6. 7. H. ♃. All. ped. 2. t. 22. f. 6.
73 valdéria. R.S. (*pu.*) fringed-leaved. ——— 1805. 5. 6. H. ♃. ——— 2. t. 24. f. 3.
74 calcaràta. R.S. (*bl.*) spurred. Europe. 1752. 3. 6. H. ♃.
75 Villarsiàna. R.S. (*bl.*) Villars's. S. France. .... —— H. ♃. Vill. cat. arg. t. 5.
76 Zóysii. R.S. (*ye.*) crenated. Europe. .... 7. 9. H. ♃. Jacq. coll. 4. t. 11. f. 1.
77 amœ'na. DC. (*pu.*) purple. Scotland. .... 5. 8. H. ♃.
78 altáica. DC. (*ye.*) Tartarian. Altay. 1805. 3. 6. H. ♃. Bot. reg. 54.
grandiflòra. L. *non* DC.
β *purpùrea.* DC. (*pu.*) *purple-flowered.* Altay. .... 3. 6. H. ♃. Swt. br. fl. gar. ic.
79 prostràta. R.S. (*st.*) prostrate. Teneriffe. 1824. —— H. ⊙.
80 sudética. W.en. (*ye.*) large-flowered. Europe. .... 5. 8. H. ♃.
grandiflòra. DC. fl. fr. 5. p. 620.
81 lùtea. E.B. (*ye.*) large yellow. Britain. .... 5. 8. H. ♃. Eng. bot. 721.
82 saxátilis. R.S. (*st.*) rock. Bohemia. 1824. —— H. ♃. Swt. br. fl. gar. t.
83 rothomagénsis. DC. (*b.*) Rouen. France. 1783. —— H. ♃. Bot. mag. 1498.
84 trícolor. R.S. (*b.y.*) Heart's-ease. Britain. .... 4. 11. H. ♂. Œd. fl. dan. t. 623.
85 bannática. R.S. (*st.*) intermediate. Hungary. .... 5. 8. H. ⊙.
86 arvénsis. R.S. (*st.*) corn. Britain. .... —— H. ⊙.
87 declinàta. R.S. (*bl.*) declined. Hungary. 1820. —— H. ♂. W. et K. hung. 3. t. 223.
88 Kitaibeliàna. R.S. (*st.*) Kitaibel's. Switzerland. 1824. 4. 6. H. ⊙.

ERP'ETION B.F.G. SPURLESS VIOLET. Pentandria Monogynia.
1 hederàceum. (*b.w.*) Ivy-leaved. V. Diem. Isl. 1825. 5. 9. F. ♃. Lab. nov. holl. 1. t. 91.
Víola hederàcea. Lab.
2 renifórme. B.F.G. (*b.w.*) kidney-leaved. N. S. W. 1823. —— F. ♃. Swt. br. fl. gar. 170.
Víola renifórmis. B.

S'OLEA. DC. S'OLEA. Pentandria Monogynia.
cóncolor. DC. (*gr.*) green-flowered. N. America. 1788. 6. 7. H. ♃. Linn. trans. 6. t. 28.

POMB'ALIA. DC. POMB'ALIA. Pentandria Monogynia.
'Itubu. DC. (*wh.*) white Ipecacuanha. Brazil. 1822. 7. 8. S. ♃. H. et. B. 5. t. 496. f. 2.
β *calceolària.* (*wh.*) *villous-stemmed.* ——— —— —— S. ♃. Bot. mag. 2453
Jonídium Ipecacuánha. β. B.M. 2453.

JONI'DIUM. DC. JONI'DIUM. Pentandria Monogynia.
1 stríctum. DC. (*wh.*) upright. St. Domingo. 1820. .... S. ♃.
2 polygalæfòlium. DC. (*st.*) Milkwort-l^d. S. America. 1797. 4. 8. G. ♃. Vent. malm. 27.
3 verbenàceum. DC. (*p.b.*) Vervain-like. ——— 1822. 6. 8. H. ⊙. H. et B. V. 379. t. 497.

Tribus II. *ALSODINEÆ.* DC.

ALSODE'IA. DC. ALSODE'IA. Pentandria Monogynia.
1 arbòrea. DC. (*wh.*) tree. Madagascar. 1823. .... S. ♄.
2 latifòlia. DC. (*wh.*) broad-leaved. ——— —— .... S. ♄. Pet. Th. veg. af. t. 18. f. 2.

CERA'NTHERA. DC. CERA'NTHERA. Pentandria Monogynia.
dentàta. DC. (*wh.*) toothed. Sierra Leone. 1822. .... S. ♄. Beauv. fl. d'ow. 2. t. 65.

Tribus III. *SAUVAGEÆ.* DC.

SAUVAG'ESIA. DC. SAUVAG'ESIA. Pentandria Monogynia.
1 erécta. DC. (*wh.*) upright. S. America. 1823. 7. 9. S. ⊙. Jacq. am. pict. t. 77.
2 nùtans. DC. (*wh.*) nodding. Madagascar. —— —— S. ⊙.

ORDO XXI.

DROSERACEÆ. *DC. prodr.* 1. *p.* 317.

DR'OSERA. DC. SUNDEW. Pentandria Pentagynia. L.
1 acáulis. DC. (*wh.*) stalkless. C. B. S. 1821. 7. 8. G.*w.* ♃.
2 pauciflòra. DC. (*wh.*) few-flowered. ——— —— —— G.*w.* ♃.

3 cuneifòlia. DC.	(*w.r.*)	wedge-leaved.	C. B. S.	1824.	6. 8.	G.*w.*♃.	
α alba.	(*wh.*)	*white-flowered.*	———	——	——	G.*w.*♃.	
β rubra.	(*re.*)	*red-flowered.*	———	——	——	G.*w.*♃.	
4 rotundifòlia. DC.	(*wh.*)	round-leaved.	Britain.	....	7. 8.	H.*w.*♃.	Eng. bot. 867.
5 longifòlia. L. *intermèdia.* DC.	(*wh.*)	intermediate.	England.	....	——	H.*w.*♃.	——— 868.
6 americàna. W. en.	(*wh.*)	American.	N. America.	1824.	——	H.*w.*♃.	
7 ánglica. DC. *longifòlia.* D. et H. 3. t. 75. A.	(*wh.*)	great.	———	....	——	H.*w.*♃.	Eng. bot. 869.
8 lineàris. DC.	(*wh.*)	linear-leaved.	N. America.	1822.	5. 6.	H.*w.*♃.	
9 filifórmis. Ph.	(*pu.*)	thready-leaved.	———	1811.	——	H.*w.*♃.	
ALDROVA'NDA. DC.		ALDROVA'NDA.	Pentandria Pentagynia. L.				
vesiculòsa. DC.	(*wh.*)	whorl-leaved.	S. Europe.	1826.	6. 8.	H.*w.*⊙.	Lam. ill. t. 220.
B'YBLIS. DC.		B'YBLIS.	Pentandria Monogynia.				
liniflòra. DC.	(*bl.*)	blue-flowered.	N. S. W.	1803.	5. 6.	G.*w.*♃.?	Salisb. par. 95.
DION'ÆA. DC.		DION'ÆA.	Decandria Monogynia. L.				
Muscípula. DC.	(*wh.*)	Venus Flytrap.	Carolina.	1768.	7. 8.	G. ♃.	Bot. mag. 785.
PARNA'SSIA. DC.		PARNA'SSIA.	Pentandria Tetragynia. L.				
1 palústris. DC.	(*wh.*)	marsh.	Britain.	....	7. 8.	H.*w.*♃.	Eng. bot. 82.
2 parviflòra. DC.	(*wh.*)	small-flowered.	N. America.	....	5. 6.	H.*w.*♃.	
3 caroliniàna. DC.	(*wh.*)	Carolina.	Carolina.	1802.	5. 8.	H.*w.*♃.	Bot. mag. 1459.
4 asarifòlia. DC.	(*wh.*)	Asarum-leaved.	N. America.	1812.	7. 8.	H.*w.*♃.	Vent. malm. t. 39.

ORDO XXII.

POLYGALEÆ. *DC. prodr.* 1. *p.* 231.

POLY'GALA. DC.		MILKWORT.	Diadelphia Octandria. L.				
1 oppositifòlia. DC.	(*pu.*)	opposite-leav'd.	C. B. S.	1790.	1. 12.	G. ♄.	Bot. reg. 636.
2 cordifòlia. DC.	(*pu.*)	heart-leaved.	———	1791.	——	G. ♄.	
3 tetragòna. DC. *attenuàta.* Lod. bot. cab. 1000. *oppositifolia β màjor.* B.R. 1146.	(*pu.*)	square-stalked.	———	1820.	——	G. ♄.	
4 nummulària. DC.	(*pu.*)	Moneywort-ld.	C. B. S.	1812.	——	G. ♄.	
5 latifòlia. B.R. *cordifòlia.* B.M. *non* DC.	(*pu.*)	broad-leaved.	———	1820.	——	G. ♄.	Bot. reg. 645.
6 borboniæfòlia. DC. *oppositifòlia.* B.M. *non* DC.	(*p.*)	Borbonia-leav'd.	———	1790.	——	G. ♄.	
7 grandiflòra. B.C.	(*pu.*)	large-flowered.	———	1825.	4. 9.	G. ♄.	Lodd. bot. cab. 1227.
8 myrtifòlia. DC.	(*pu.*)	Myrtle-leaved.	———	1707.	5. 8.	G. ♄.	Bot. reg. 669.
9 liguläris. DC.	(*pu.*)	ligulate-leaved.	———	1820.	——	G. ♄.	——— 637.
10 intermèdia. DC.	(*pu.*)	intermediate.	———	....	5. 10.	G. ♄.	
11 bracteolàta. DC.	(*pu.*)	spear-leaved.	———	1713.	——	G. ħ.	Bot. mag. 345.
12 Linkiàna. *tenuifòlia.* L. en. *non* W.	(*pa.*)	slender-leaved.	———	1820.	——	G. ħ.	
13 umbellàta.	(*pu.*)	hispid-stalked.	———	——	——	G. ħ.	Burm. afr. t. 73. f. 5.
14 Burmánni. DC.	(*pu.*)	Burman's.	———	....	——	G. ♄.	——— t. 73. f. 4.
15 simpléx. DC.	(*pu.*)	simple-stalked.	———	1820.	——	G. ♄.	
16 affìnis. DC.	(*pu.*)	blunt-leaved.	———	1825.	6. 9.	G. ħ.	
17 speciòsa. DC.	(*pu.*)	showy.	———	1814.	——	G. ♄.	Bot. reg. 150.
18 teretifòlia. DC.	(*pu.*)	columnar-ld.	———	1791.	5. 8.	G. ♄.	Andr. rep. 370.
19 Garcìni. DC.	(*pu.*)	slender-branch'd	———	....	5. 10.	G. ♄.	Burm. afr. t. 73. f. 3.
20 màjor. DC.	(*pu.*)	large Austrian.	Austria.	1739.	7. 8.	H. ♃.	Jacq. aust. 5. t. 413.
21 vulgàris. DC.	(*va.*)	common.	Britain.	....	5. 7.	H. ♃.	Eng. bot. 76.
α vèra. DC.	(*r.b.*)	*true.*	———	....	——	H. ♃.	Flor. dan. t. 516.
β pubéscens. DC.	(*r.b.*)	*pubescent.*	———	....	——	H. ♃.	
γ elàta. DC.	(*r.b.*)	*tall.*	———	....	——	H. ♃.	Schk. handb. t. 194.
δ Verviàna. DC.	(*wh.*)	*white-flowered.*	———	....	——	H. ♃.	
ε acutifòlia. DC.	(*bl.*)	*blue-flowered.*	———	....	——	H. ♃.	
ζ angustifòlia. DC.	(*ro.*)	*rose-coloured.*	———	....	——	H. ♃.	
η grandiflòra. DC.	(*re.*)	*red-flowered.*	———	....	——	H. ♃.	
22 amàra. DC.	(*bl.*)	bitter.	Europe.	1775.	——	H. ♃.	Bot. mag. 2437.
23 monspelìaca. DC.	(*fl.*)	Montpelier.	Montpelier.	1823.	6. 9.	H. ⊙.	DC. ic. rar. 1. t. 9.
24 incarnàta. DC.	(*fl.*)	flesh-coloured.	N. America.	1812.	6. 7.	H. ⊙.	
25 cruciàta. DC.	(*ro.*)	four-leaved.	———	1739.	——	H. ⊙.	
26 brevifòlia. DC.	(*re.*)	short-leaved.	———	1824.	——	H. ⊙.	
27 fastigiàta. DC.	(*re.*)	close-branched.	———	——	——	H. ⊙.	

28 lùtea. DC.	(*ye.*) golden.	N. America.	1739.	6. 7.	H. ⊙.	Pluk. am. t. 438. f. 6.
29 nàna. DC.	(*gr.*) dwarf.	———	1815.	7. 8.	H. ⊙.	
30 purpùrea. DC.	(*ro.*) purple.	———	1739.	7. 9.	H. ⊙.	
sanguínea. Mich.						
31 sanguínea. DC.	(*cr.*) bloody.	———	——	——	H. ⊙.	
32 ambígua. DC.	(*pu.*) ambiguous.	———	1824.	6. 7.	H. ⊙.	
33 verticillàta. DC.	(*w.r.*) whorl-leaved.	———	1739.	——	H. ⊙.	Pluk. am. t. 438. f. 4.
34 paniculàta. DC.	(*pu.*) panicled.	S. America.	1822.	7. 10.	S. ⊙.	Bot. reg. 761.
35 graminifòlia. DC.	(*st.*) grass-leaved.	N. America	——	6. 7.	F. ♃.	
36 álba. N.	(*wh.*) white-flowered.	Louisiana.	1827.	6. 7.	H. ♃.	
37 rubélla. DC.	(*re.*) reddish.	N. America.	1828.	——	H. ♃.	
38 Sénega. DC.	(*bh.*) Rattlesnake-rt.	———	1739.	——	H. ♃.	Bot. mag. 1051.
39 paucifòlia. DC.	(*pu.*) naked-stalked.	———	1791.	5. 6.	H. ♃.	——— 2852.
40 commutàta.	(*pu.*) frutescent.	———	——	——	H. ♄.	
purpùrea. H.K. nec aliorum.						
41 Chamæbúxus. DC.	(*w.*) Box-leaved.	Europe.	1658.	——	H. ♄.	Bot. mag. 316.
SALOM'ONIA. DC.	SALOM'ONIA.	Monadelphia Tetrandria.				
cantoniénsis. DC.	(*wh.*) Canton.	China.	1824.	6. 8.	H. ⊙.	
COMESPE'RMA. DC.	COMESPE'RMA.	Monadelphia Octandria.				
1 ericìna. DC.	(*pu.*) Heath-like.	N. S. W.	1822.	5. 8.	G. ♄.	
2 virgàta. DC.	(*pu.*) twiggy.	———	1826.	——	G. ♄.	Lab. nov. holl. 2. t. 159.
3 coridifòlia. F.T.	(*pu.*) Coris-leaved.	———	1822.	——	G. ♄.	
MURA'LTIA. DC.	MURA'LTIA.	Diadelphia Octandria. L.				
1 Heistèria. DC.	(*pu.*) Furze-leaved.	C. B. S.	1787.	1. 12.	G. ħ.	Bot. mag. 340.
Poly'gala Heistèria. B.M.						
2 alopecuróides. DC.	(*p.*) Fox-tail.	———	1800.	5. 8.	G. ħ.	Bot. mag. 1006.
Poly'gala alopecuróides. B.M.						
3 squarròsa. DC.	(*li.*) squarrose.	———	1821.	——	G. ħ.	
4 stipulàcea. DC.	(*pu.*) stipuled.	———	1801.	4. 9.	G. ħ.	Bot. mag. 1715.
5 juniperifòlia. DC.	(*re.*) Juniper-leaved.	———	....	——	G. ♄.	
6 diffùsa. DC.	(*li.*) diffuse.	———	1800.	1. 12.	G. ♄.	
7 vírgata. DC.	(*pu.*) twiggy.	———	——	——	G. ♄.	
8 linophy'lla. DC.	(*pu.*) Flax leaved.	———	1812.	5. 8.	G. ♄.	
9 macrocèras. DC.	(*pu.*) long-horned.	———	——	——	G. ♄.	
10 míxta. DC.	(*li.*) Heath-leaved.	———	1791.	1. 12.	G. ħ.	Bot. mag. 1714.
Poly'gala míxta. B.M.						
α rùbra.	(*li.*) *pale red.*	———	——	——	G. ħ.	
β álba.	(*wh.*) *white-flowered.*	———	——	——	G. ħ.	
11 ciliàris. DC.	(*li.*) ciliated.	———	1822.	5. 7.	G. ħ.	
12 hùmilis. DC.	(*pu.*) humble.	———	1817.	5. 8.	G. ♄.	Lodd. bot. cab. 420.
Poly'gala hùmilis. B.C.						
13 pubéscens. DC.	(*li.*) pubescent.	———	1823.	——	G. ħ.	
MU'NDIA. DC.	MU'NDIA.	Diadelphia Octandria. L.				
spinòsa. DC.	(*li.*) spiny.	C. B. S.	1780.	1. 5.	G. ħ.	
α latifòlia. DC.	(*li.*) *broad-leaved.*	———	——	——	G. ħ.	
β angustifòlia. DC.	(*li.*) *narrow-leaved.*	———	——	——	G. ħ.	
SECURID'ACA. DC.	SECURID'ACA.	Diadelphia Octandria. L.				
1 virgàta. DC.	(*v.*) slender.	Jamaica.	1822.	....	S ħ.‿.	Plum.ed.Bur.t.248.f.1
2 volubilis. DC	(*re.*) climbing.	W. Indies.	1739.	....	S. ħ.‿.	Jacq.amer.t.183.f.38.
3 erécta. DC.	(*pu.*) upright.	———	1824	....	S. ħ.	———t.183.f.39.
4 paniculàta. DC.	(*st.*) panicled.	Cayenne.	1820.	....	S. ħ.	Lam. ill. t. 599. f. 2.

ORDO XXIII.

TREMANDREÆ. *DC. prodr.* 1. *p.* 343.

TETRATH'ECA. DC.	TETRATH'ECA.	Octandria Monogynia.				
1 júncea. DC.	(*w.p.*) rushy.	N. S. W.	1803.	7. 8.	G. ♄.	Sm. nov. hol. 1. t 2.
2 glandulòsa. DC.	(*re.*) glandular.	V. Diemen.	1822.	——	G. ♄.	Lab. nov. hol. 1. t. 123.
3 pilòsa. DC.	(*re.*) hairy.	———	——	——	G. ♄.	——— 1. t. 122.
. *glandulòsa.* Ex. bot. 21.						
4 ericifòlia. DC.	(*r.w.*) Heath-leaved.	N. S. W.	——	——	G. ♄.	Sm. exot. bot. t. 20.
5 rubiæóides. F.T.	(*ro.*) Rubia-like.	———	1823.	6. 8.	G. ♄.	

ORDO XXIV.

PITTOSPOREÆ. *DC. prodr.* 1. *p.* 345.

BILLARDIE'RA. DC.		APPLE-BERRY.	Pentandria Monogynia. L.				
1 scàndens. DC.	(*st.*)	downy-fruited.	N. S. W.	1790.	6. 8.	G. ♄.◡.	Swt. fl. aust. t. 54.
2 mutábilis. DC.	(*p.s.*)	changeable.	———	1795.	5. 9.	G. ♄.◡.	Bot. mag. 1313.
3 longiflòra. DC.	(*st.*)	blue-fruited.	V. Diemen.	1810.	——	G. ♄.◡.	——— 1507.
4 fusifórmis. DC.	(*bl.*)	long-fruited.	———	1822.	——	G. ♄.◡.	Lab. nov. holl. t. 90.
5 angustifòlia. DC.	(*st.*)	narrow-leaved.	N. S. W.	——	——	G. ♄.◡.	
PITTO'SPORUM. DC.		PITTO'SPORUM.	Pentandria Monogynia. L.				
1 coriàceum. DC.	(*wh.*)	thick-leaved.	Madeira.	1787.	5. 6.	G. ♄.	Andr. rep. 151.
2 viridiflòrum. DC.	(*gr.*)	green-flowered.	C. B. S.	1806.	——	G. ♄.	Bot. mag. 1684.
3 Tobìra. DC.	(*wh.*)	glossy-leaved.	China.	1804.	3. 8.	G. ♄.	——— 1396.
4 undulàtum. DC.	(*wh.*)	wave-leaved.	N. S. W.	1789.	2. 6.	G. ♄.	Bot. reg. 16.
5 revolùtum. DC.	(*st.*)	revolute-leaved.	———	1795.	6. 7.	G. ♄.	——— 186.
6 tomentòsum. DC.	(*st.*)	woolly-leaved.	N. Holland.	1825.	6. 7.	G. ♄.	Swt. fl. aust. t. 33.
7 fúlvum. DC.	(*st.*)	tawny.	N. S. W.	1824.	1. 5.	G. ♄.	——— t. 25.
8 tenuifòlium. G.	(*st.*)	Willow-leaved.	———	1818.	——	G. ♄.	Gært. fr. 1. t. 59.
9 ferrugíneum. DC.	(*st.*)	rusty-leaved.	Guinea.	1787.	2. 5.	S. ♄.	Bot. mag. 2075.
BURS'ARIA. DC.		BURS'ARIA.	Pentandria Monogynia. L.				
spinòsa. DC.	(*wh.*)	thorny.	N. S. W.	1793.	8. 12.	G. ♄.	Bot.. mag. 1767.
Itea spinosa. And. rep. t. 314.							
SEN'ACIA. DC.		SEN'ACIA.	Pentandria Monogynia. L.				
1 undulàta. DC.	(*wh.*)	wave-leaved.	Mauritius.	1785.	....	S. ♄.	
2 nepalénsis. DC.	(*wh.*)	flat-leaved.	Nepaul.	1826.	....	G. ♄.	

ORDO XXV.

FRANKENIACEÆ. *DC. prodr.* 1. *p.* 349.

FRANK'ENIA. DC.		SEA HEATH.	Hexandria Monogynia. L.				
1 pulverulénta. DC.	(*re.*)	powdery.	England.	....	7.	H. ♃.	Eng. bot. 2222.
2 nodiflòra. DC.	(*re.*)	knotted-flow'r'd.	C. B. S.	1818.	6. 8.	G. ♃.	Lam. ill. t. 262. f. 4.
3 Nóthria. DC.	(*re.*)	Cape.	———	1816.	——	G. ♃.	Berg. cap. 171. t. 1. f. 2.
Nóthria rèpens. Berg.							
4 læ'vis. DC.	(*re.*)	smooth.	England.	....	7. 8.	H. ♃.	Eng. bot. 205.
5 intermèdia. DC.	(*re.*)	sea-side.	S. Europe.	1820.	6. 8.	H. ♃.	Mich. gen. t. 22. f. 2.
hirsùta. var. *calábrica.* L.							
6 hìspida. DC.	(*re.*)	hairy.	Siberia.	1789.	——	H. ♃.	Flor. græc. t. 313.
hirsùta. var. *crética.* L.							
7 ericifòlia. DC.	(*re.*)	Heath-leaved.	Canaries.	1815.	——	G. ♄.	
8 pauciflòra. DC.	(*ro.*)	few-flowered.	N. Holland.	1825.	8. 10.	G. ♄.	

ORDO XXVI.

CARYOPHYLLEÆ. *DC. prodr.* 1. *p.* 351.

Tribus I. *SILENEÆ.* DC.

GYPSO'PHILA. DC.		GYPSO'PHILA.	Decandria Digynia. L.				
1 Struthium. DC.	(*wh.*)	fleshy-leaved.	Spain.	1729.	7. 8.	H. ♃.	Barr. ic. 64. t. 119.
2 fastigiàta. DC.	(*wh.*)	angular-leaved.	Europe.	1759.	6. 7.	H. ♃.	Gmel. sib. 4. t. 61. f. 1.
3 dichótoma. DC.	(*wh.*)	forked.	Galicia.	1826.	6. 8.	H. ♃.	
4 arenària. DC.	(*wh.*)	sand.	Hungary.	1801.	7. 8.	H. ♃.	W. et. K. hung. t. 41.
5 viscòsa. DC.	(*bh.*)	clammy.	Levant.	1773.	6. 8.	H. ⊙.	Mur. c. got. 1783. t. 3.
6 altíssima. DC.	(*wh.*)	upright.	Siberia.	1759.	7. 8.	H. ♃.	Gmel. sib. 4. t. 60.
7 latifòlia. F.	(*wh.*)	broad-leaved.	———	1824.	6. 9.	H. ♃.	

8	perfoliàta. DC.	(*wh.*)	perfoliate.	Spain.	1732.	6. 9.	H. ♃.	Dill. elt. t. 276.
	β tomentòsa. DC.	(*wh.*)	*woolly.*	Tauria.	——	——	H. ♃.	Barr. ic. t. 1002.
9	scorzonerifòlia. DC.	(*w.*)	Scorzonera-ld.	————	1823.	——	H. ♃.	
10	acutifòlia. DC.	(*wh.*)	sharp-leaved.	Caucasus.	1821.	——	H. ♃.	
11	paniculàta. DC.	(*wh.*)	panicled.	Siberia.	1759.	6. 7.	H. ♃.	Jacq. aust. 5. t. app. 1
12	adscéndens. DC.	(*wh.*)	ascending.	Levant.	1819.	——	H. ♃.	Jacq. h. vind. 2. t. 138.
13	élegans. DC.	(*bh.*)	elegant.	Tauria.	1820.	6. 8.	H. ⊙.	Swt. br. fl. gar.
14	Stevèni. DC.	(*wh.*)	Steven's.	Iberia.	1817.	——	H. ♃.	
15	tenuifòlia. DC.	(*ro.*)	slender-leaved.	Caucasus.	1824.	——	H. ♃.	
16	répens. DC.	(*wh.*)	creeping.	Pyrenees.	1774.	7. 9.	H. ♃.	Bot. mag. 1448.
17	dùbia. DC	(*wh.*)	doubtful.	....	1815.	5. 9.	H. ♃.	
18	prostràta. DC.	(*bh.*)	trailing.	Siberia.	1759.	7. 9.	H. ♃.	Bot. mag. 1281.
19	muràlis. DC.	(*li.*)	wall.	Europe.	1739.	6. 10.	H. ⊙.	Flor. græc. t. 381.
20	serotìna. W. en.	(*wh.*)	late-flowering.	Magdeburg.	1816.	——	H. ⊙.	
21	glomeràta. DC.	(*wh.*)	round-headed.	Tauria.	1804.	6. 7.	H. ♃.	
22	saxífraga. DC.	(*bh.*)	small.	Germany.	1774.	7. 8.	H. ♃.	Exot. bot. 2. t. 90.
23	rígida. L.	(*li.*)	rigid.	France.	1769.	6. 8.	H. ♃.	Flor. græc. t. 382.
	DIA'NTHUS. DC.		PINK. Decandria Digynia. L.					
1	prolífer. DC.	(*ro.*)	proliferous.	England.	....	7. 8.	H. ⊙.	Eng. bot. 956.
2	diminùtus. W.	(*ro.*)	small-flowered.	S. Europe.	1771.	——	H. ⊙.	
3	Armèria. DC.	(*re.*)	Deptford.	England.	....	7. 9.	H. ⊙.	Eng. bot. 317.
4	pseùdo-armèria.	(*re.*)	canescent.	Tauria.	1820.	——	H. ♃.	Bot. mag. 2288.
5	armerioídes. DC.	(*re.*)	Thrift-like.	New Jersey.	——	——	H. ⊙.	
6	barbàtus. DC.	(*va.*)	Sweet-William.	Germany.	1573.	6. 7.	H. ♃.	Bot. mag. 207.
	β flòre plèno.	(*va.*)	*double.*	————	——	——	H. ♃.	
7	latifòlius. DC.	(*cr.*)	broad-leaved.	....	....	5. 11.	H. ♃.	Swt. br. fl. gar. t. 2.
8	aggregàtus. DC.	(*re.*)	clustered.	....	1820.	6. 8.	H. ♃.	
9	japònicus. DC.	(*re.*)	Japanese.	China.	1804.	6. 10.	F. ♃.	Thunb. jap. t. 23.
10	cephalòtes. DC.	(*re.*)	close-headed.	....	....	6. 9.	H. ♃.	
11	capitàtus. DC.	(*re.*)	headed.	Tauria.	1794.	6. 7.	H. ♃.	
12	polymórphus. DC.	(*re.*)	variable.	————	1820.	6. 8.	H. ♃.	
13	diutìnus. L. en.	(*re.*)	pedicled.	Hungary.	——	——	H. ♃.	Swt. br. fl. gar.
14	Balbísii. DC.	(*re.*)	Balbis's.	Genoa.	1827.	7. 10.	H. ♃.	Swt. br. fl. gar.
	glaucophy'llus. H.							
15	gigánteus. DC.	(*re.*)	gigantic.	Bulgaria.	——	6. 10.	H. ♃.	Swt. br. fl. gar. 288.
16	ferrugíneus. DC.	(*st.*)	rusty.	Pyrenees.	1756.	7. 9.	H. ♃.	Barr. ic. t. 497.
17	Carthusianòrum. DC.	(*re.*)	Carthusian.	Europe.	1573.	——	H. ♃.	Flor. græc. t. 392.
18	atrorùbens. All.	(*re.*)	dark-red.	Italy.	1802.	——	H. ♃.	Jacq. ic. 3. t. 467.
19	nànus.	(*cr.*)	dwarf.	Switzerland.	1820.	——	H. ♃.	
	Carthusianòrum γ nanus. DC.							
20	arbòreus. DC.	(*re.*)	arborescent.	Crete.	——	——	F. ♄.	Flor. græc. t. 406.
21	juniperìnus. DC.	(*re.*)	Juniper-leaved.	Greece.	1825.	——	F. ♄.	
22	fruticòsus. DC.	(*re.*)	frutescent.	Crete.	1815.	6. 9.	F. ♄.	Flor. græc. t. 407.
23	suffruticòsus. DC.	(*re.*)	shrubby.	....	1804.	6. 7.	F. ♄.	
24	rupícola. DC.	(*ro.*)	Sicilian.	Sicily.	1825.	6. 8.	H. ♃.	Ten. fl. napol. t. 39.
	Bisigniàni. T.							
25	caroliniànus. DC.	(*re.*)	Carolina.	N. America.	1811.	6. 9.	F. ♃.	
26	caryophylloídes. DC.	(*re.*)	Clove-like.	....	1827.	——	H. ♃.	
27	ásper. W. en.	(*re.*)	rough-stalked.	Switzerland.	1802.	7. 8.	H. ♃.	
28	collìnus. W. en.	(*re.*)	hill.	Hungary.	1800.	7. 9.	H. ♃.	Parad. lond. 62.
29	umbellàtus. DC.	(*re.*)	umbelled.	....	1825.	——	H. ♃.	
30	campéstris. DC.	(*re.*)	field.	Tauria.	1815.	7. 8.	H. ♃.	Bot. mag. 1876.
31	alpéstris. DC.	(*re.*)	pasture.	————	1820.	——	H. ♃.	Balb. act. taur. 7. t. 1.
32	nítidus. DC.	(*re.*)	glossy.	Hungary.	1821.	6. 8.	H. ♃.	W. et K. hung. 2. t. 191.
33	hírtus. DC.	(*re.*)	hairy.	Dauphiny.	——	——	H. ♃.	Vill. delph. 3. t. 46.
34	guttàtus. DC.	(*pu.*)	spotted.	Caucasus.	1816.	7. 9.	H. ♃.	
35	versícolor. DC.	(*s.r.*)	various-colour'd	Russia.	1821.	——	H. ♃.	
36	praténsis. DC.	(*st.*)	meadow.	Tauria.	——	6. 8.	H. ♃.	
	ochroleùcus. L. en. *non* P. S. *tataricus.* F.							
37	chinénsis. DC.	(*va.*)	China.	China.	1713.	7. 9.	H. ♂.	Bot. mag. 28.
	β flòre-plèno.	(*va.*)	*double-flowered.*	————	....	——	H. ♂.	
38	montànus. DC.	(*pu.*)	mountain.	Caucasus.	1803.	6. 9.	H. ♃.	Bot. mag. t. 1162.
	díscolor. B.M.							
39	ochroleùcus. DC.	(*st.*)	straw-coloured.	Levant.	1821.	——	H. ♃.	
40	Caryophy'llus. DC.	(*va.*)	Clove.	England.	....	6. 8.	H. ♃.	Eng. bot. 214.
	β multiplèx,	(*cr.*)	*double Clove.*	....	....	——	H. ♃.	
	γ rùbro-cínctus.	(*wh.*)	*Hogg's Beauty of Middlesex Picotee.*			——	H. ♃.	Swt. flor. guid. 70.
	δ inimitábilis.	(*w.c.*)	*Hird's Inimitable Picotee.*		....	——	H. ♃.	———— t. 80.
	ε striátulus.	(*y.r.*)	*Rosalie de Rohan Picotee.*		....	——	H. ♃.	———— t. 25.

ζ *purpùreo-cínctus.* (*y.p.*) *Erasmus Picotee.* .... 6. 8. H. ♃. Swt. flor. guid. 59.
η *Clintòniæ.* (*r.w.*) *Hufton's Dutchess of Newcastle Carnation.* H. ♃. —— t. 14.
θ *triúmphans.* (*p.w.*) *Pardoe's Ace of Trumps Carnation.* —— H. ♃. —— t. 36.
ι *cærùleo-purpùreus.* (*p.w.*) *Cordon Bleu Carnation.* .... —— H. ♃. —— t. 64.
κ *Franklìni.* (*r.p.w.*) *Franklin's Queen of Hearts Carnation.* H. ♃. —— t. 20.
λ *puéllulus.* (*c.r.w.*) *Pucelle de Gand Carnation.* .... —— H. ♃. —— t. 67.
μ *Leopoldi.* (*c.p.w.*) *Ives's Prince Leopold Carnation.* .. —— H. ♃. —— t. 31.
ν *iricolor.* (*c.p.w.*) *Cartwright's Rainbow Carnation*... —— H. ♃. —— t. 9.
ξ *Bacchus.* (*s.p.w.*) *Davey's Bacchus Carnation.* .... —— H. ♃. —— t. 44.
ο *Tarrara.* (*s.p.w.*) *Hall's Tarrara Carnation.* .... —— H. ♃. —— t. 21.
π *pulchella.* (*s.p.w.*) *Franklin's Tartar Carnation.* .... —— H. ♃. Bot. mag. 39.
ρ *fruticòsus.* (*va.*) *tree-Carnation.* .... —— H. ♄.
ς *imbricatus.* (*d.r.*) *Wheat-ear.* .... —— H. ♃. Bot. mag. 1622.
41 sylvéstris. (*re.*) short-scaled. Jurassa. .... 6. 7. H. ♃. Jacq. ic. rar. t. 82.
virgíneus. B.M. 1740. *non* L.
42 longicáulis. DC. (*re.*) long-stalked. Naples. 1824. —— H. ♃.
43 monadélphus. DC. (*wh.*) procumbent. Levant. .... —— H. ♃. Vent. h. cels. t. 39.
procúmbens. P.S.
44 suavéolens. DC. (*sp.*) sweet-scented. .... 1823. —— H. ♃.
45 Poiretiànus. DC. (*pu.*) Poiret's. .... —— H. ♃.
β *flòre-plèno.* (*pu.*) *double-flowering*. .... —— H. ♃.
46 Liboschitziànus. DC. Armenian. Armenia. 1823. 6. 8. H. ♃.
petr'æus. M.B. *non* DC.
47 sylváticus. DC. (*cr.*) wood. Ratisbon. 1815. —— H. ♃.
48 arbúscula. B.R. (*re.*) China tree. China. 1824. 7. 10. H. ♄. Bot. reg. 1086.
49 pomeridiànus. DC. (*st.*) afternoon. Palestine. 1804. —— H. ♃. Parad. lond. 57.
50 leptopétalus. DC. (*wh.*) narrow-petaled. Caucasus. 1814. —— H. ♃. Bot. mag. 1739.
51 púngens. DC. (*wh.*) pungent. Spain. 1781. 8. 10. H. ♃.
52 bícolor. DC. (*w.p.*) two-coloured. Tauria. 1818. 6. 8. H. ♃.
53 furcàtus. DC. (*fl.*) forked. Piedmont. 1820. —— H. ♃. Balb. act taur. 7. f. 2.
54 virgíneus. DC. (*re.*) virgin. Montpelier. 1732. —— H. ♃. Dill. elt. 401. t. 385?
55 deltóides. DC. (*re.*) maiden. Britain. .... 6. 10 H. ♃. Eng. bot. 61.
56 gláucus. L. (*wh.*) glaucous. —— .... —— H. ♃.
57 crenàtus. B.R. (*wh.*) long-cupped. C. B. S. 1817. 8. G. ♃. Bot. reg. 256.
58 rígidus. DC. (*pa.*) rigid. Tauria. 1802. 6. 10. H. ♃.
59 clavàtus. DC. (*fl.*) clubbed. .... 1824. 6. 8. H. ♃.
60 suávis. DC. (*fl.*) sweet. .... 1820. 6. 7. H. ♃.
61 rèpens. DC. (*re.*) creeping. Siberia. 1825. —— H. ♃.
62 cæ'sius. DC. (*fl.*) tufted. Britain. .... —— H. ♃. Eng. bot. 62.
63 alpinus. DC. (*re.*) Alpine. Austria. 1759. —— H. ♃. Bot. mag. 1205.
64 glaciàlis. DC. (*re.*) acute-leaved. France. 1820. —— H. ♃.
65 élegans. DC. (*re.*) elegant. Levant. 1825. —— F. ♃.
66 multipunctàtus. DC. (*sp.*) spotted. —— —— 6. 9. F. ♃.
67 Hornemánni. DC. (*fl.*) Horneman's. Italy. 1822. 6. 7. H. ♃.
68 Sternbérgii. DC. (*re.*) Sternberg's. .... 1825. —— F. ♃.
69 petr'æus. DC. (*wh.*) rock. Hungary. 1804. 7. 8. H. ♃. W. et. K. hung. 3. t. 222.
β *Símsii.* (*wh.*) *Sims's.* —— —— —— H. ♃. Bot. mag. 1204.
70 ibéricus. DC. (*pu.*) Iberian. Iberia. 1820. 6. 8. H. ♃.
Willdenòwii. L. en.
71 ruthènicus. DC. (*re.*) Russian. Russia. —— —— H. ♃.
72 dentòsus. F. (*re.*) toothed. Siberia. 1826. —— H. ♃.
73 Buchtorménsis. F. (*re.*) Buchtormian. Russia. —— —— H. ♃.
74 gállicus. DC. (*wh.*) French. France. .... —— H. ♃. DC. ic. gall. rar. t. 41.
75 álbens. DC. (*wh.*) white Cape. C. B. S. 1787. 8. G. ♃.
76 serrulàtus. DC. (*ro.*) serrulate. Tunis. 1826. —— H. ♃.
77 pulchéllus. P.S. (*re.*) pretty. .... —— H. ♃.
78 dùbius. H. (*w.p.*) doubtful. .... 5. 6. H. ♃.
pr'æcox. W. en.
79 plumàrius. DC. (*wh.*) common. Europe. 1629. 6. 8. H. ♃.
β *heròicus.* (*w.p.*) *Styles's Hero Pink*....... .... —— H. ♃. Swt. flor. guid. 17.
γ *Cheeseànus.* (*w.p.*) *Cheese's Miss Cheese*...... .... —— H. ♃. —— t. 13.
δ *Houseànus.* (*w.p.*) *House's Woodland Beauty* .... —— H. ♃. —— t. 47.
ε *Uxbridgeànus.* (*w.r.*) *Dry's Earl of Uxbridge* .. .... —— H. ♃. —— t. 55.
ζ *Aclándæ.* (*w.p.*) *Knight's Lady Acland*.... .... —— H. ♃. —— t. 40.
η *íncl'ytus.* (*w.p.*) *Smith's Champion* .... —— H. ♃. —— t. 27.
θ *Barrátti.* (*w.p.*) *Barratt's Conqueror*...... .... —— H. ♃. —— t. 32.
ι *regàlis.* (*w.p.*) *Davey's Roi de pourpre* .. .... —— H. ♃. —— t. 33.
80 adulterìnus. (*f.p.*) hybrid. Hybrid. .... —— H. ♃.
β *Juliéti.* (*f.p.*) *Davey's Juliet.* .... —— H. ♃. Swt. flor. guid. 8.

81 Penry'næ. (*pu.*) Lady Penryn's. Hybrid. .... 6. 8. H. ♃.
β purpùreus. (*pu.*) *Davey's Lady Penryn.*—— .... —— H. ♃. Swt. flor. guid. 41.
82 horténsis. W.en. (*wh.*) garden. Europe. 1805. 6. 8. H. ♃.
83 caucàseus. DC. (*re.*) Caucasean. Caucasus. 1803. 6. 9. H. ♃. Bot. mag. 795.
84 saxátilis. DC. (*wh.*) rock. France. .... —— H. ♃.
85 squarròsus. DC. (*wh.*) squarrose. Tauria. 1823. 6. 8. H. ♃. Bieb. cent. 1. t. 33.
86 Mussìni. DC. (*wh.*) Mussin's. Caucasus. 1825. —— H. ♃.
87 fràgrans. B.M. (*wh.*) fragrant. ———— 1804. 6. 7. H. ♃. Bot. mag. 2067.
88 punctàtus. DC. (*sp.*) spot-flowered. .... 1825. —— H. ♃. Lodd. bot. cab. 896.
89 serotìnus. DC. (*wh.*) late-flowering. Hungary. 1804. 7. 8. H. ♃. W. et. K. hung. 2. t. 172.
90 arenàrius. DC. (*wh.*) sand. N. Europe. .... 8. H. ♃. Bot. mag. 2038.
91 prostràtus. DC. (*re.*) prostrate. C. B. S. .... 8. 9. G. ♃. Jacq. schœnb. 3. t. 271.
92 fimbriàtus. DC. (*re.*) fringed. Iberia. 1802. 6. 8. H. ♄. Bot. mag. 1069.
93 monspessulànus. DC. (*re.*) Montpelier. Pyrenees. 1764. 7. 8. H. ♃.
94 erubéscens. Tr. (*bh.*) blush-flowered. .... 1825. —— H. ♃.
95 superbus. DC. (*re.*) superb. Europe. 1596. 7. 9. H. ♃. Bot. mag. 297.
96 Fischèri. DC. (*re.*) Fischer's. Russia. 1826. 6. 8. H. ♃. Swt. br. fl. gar. 245.

VACC'ARIA. Dod. VACC'ARIA. Decandria Digynia.

1 sessilifòlia. (*re.*) sessile-leaved. Europe. 1596. 7. 8. H. ⊙. Flor. græc. t. 380.
Saponària Vaccària. B.M. 2290. *Gypsóphila Vaccària.* F.G.
2 perfoliàta. (*re.*) perfoliate. E. Indies. 1820. —— H. ⊙.
Saponària perfoliàta. DC.

SAPON'ARIA. DC. SOAPWORT. Decandria Digynia. L.

1 officinàlis. DC. (*bh.*) common. England. .... 7. 10. H. ♃. Eng. bot. t. 1060.
β flòre plèno. (*bh.*) *double-flowered.* .... —— H. ♃.
2 ocymoídes. DC. (*ro.*) Basil-leaved. Europe. 1768. 5. 7. H. ♃. Bot. mag. 154.
3 glutinòsa. DC. (*re.*) small-flowered. Tauria. 1823. —— H. ♂. —— —— 2855.
4 pórrigens. DC. (*bh.*) hairy. Levant. 1680. 7. 8. H. ⊙. Jacq. vind. 2. t. 109.
5 orientàlis. DC. (*re.*) small annual. ———— 1732. 6. 8. H. ⊙. Dill. elt. t. 167. f. 204.
6 depréssa. DC. (*ro.*) depressed. Sicily. 1828. —— H. ♃. Biv. Bon. t. 163. f. 1.
7 cæspitòsa. DC. (*ro.*) tufted. Pyrenees. 1827. —— H. ♃.
8 Smíthii. DC. (*re.*) Smith's. Levant. 1828. —— F. ♃. Flor. græc. t. 389.
cæspitòsa. F.G. non DC.
9 lùtea. DC. (*ye.*) yellow. Switzerland. 1804. —— H. ♃. All. ped. t. 23. f. 1.
10 bellidifòlia. DC. (*ye.*) Daisy-leaved. Italy. .... —— H. ♃. Smith spic. t. 5.

CUC'UBALUS. DC. CAMPION. Decandria Trigynia. L.

baccíferus. DC. (*gr.*) berry-bearing. England. .... 6. 7. H. ♃. Eng. bot. 1577.

SIL'ENE. DC. CATCHFLY. Decandria Trigynia. L.

1 acáulis. DC. (*re.*) stemless. Britain. .... 6. 8. H. ♃. Eng. bot. 1081.
β álba. DC. (*wh.*) *white-flowered.* ———— .... —— H. ♃.
2 pumílio. DC. (*pu.*) dense. Germany. 1823. —— H. ♃. Stur. deut. 1. fas. 22. t. 11.
3 fimbriàta. DC. (*wh.*) fringe-flower'd. Crete. 1803. 5. 8. H. ♃. Bot. mag. 908.
4 lácera. DC. (*wh.*) jagged-flower'd. Caucasus. 1818. —— H. ♃. ———— 2255.
5 stellàta. DC. (*wh.*) four-leaved. N. America. 1696. 6. 8. H. ♃. ———— 1107.
6 inflàta. DC. (*wh.*) bladder. Britain. .... 5. 9. H. ♃. Eng. bot. 164.
Cucùbalus Bèhen. E.B.
7 linearis. (*wh.*) linear-leaved. Naples. 1820. —— H. ♃. Ten. flor. nap. t. 37.
Cucùbalus angustifòlius. T.N.
8 Willdenówii. (*wh.*) Willdenow's. S. Europe. —— —— H. ♃.
Cucùbalus gláucus. W. en. supp.
9 marítima. E.B. (*wh.*) sea. Britain. .... 8. 9. H. ♃. Eng. bot. 957.
10 fabària. H.K. (*wh.*) thick-leaved. Sicily. 1731. 6. 8. F. ♃. Flor. græc. t. 415.
11 Bèhen. DC. (*bh.*) Cretan. Crete 1713. 6. 7. H. ⊙. ———— t. 416.
12 índica. DC. (*pu.*) Indian. Nepaul. 1824. —— F. ♃.
13 viscaginoídes. DC. (*w.*) Viscago-like. Dahuria. 1826. —— H. ♃.
14 procúmbens. DC. (*wh.*) procumbent. Siberia. 1820. —— H. ♃. Murr. c. gœt. p. 83. t. 2.
15 rubélla. DC. (*re.*) small red. Portugal. 1732. 5. 6. H. ⊙. Flor. græc. t. 426.
16 apètala. DC. (*gr.*) petalless. Spain. 1801. 6. 7. H. ⊙.
17 spergulifòlia. DC. (*wh.*) Spurry-like. Armenia. 1824. —— F. ♃. Desf. cor. Tourn. t. 55.
18 Gypsóphila. DC. (*wh.*) Gypsophila-like. .... —— H. ♃.
19 angustifòlia. DC. (*wh.*) narrow-leaved. Caucasus. 1824. —— H. ♃.
20 Otìtes. DC. (*st.*) Spanish. England. .... 7. 8. H. ♃. Eng. bot. 85.
Cucùbalus Otìtes. E.B.
21 wolgénsis. DC. (*wh.*) Wolga. Wolga. 1821. 6. 8. H. ♃.
22 parviflòra. DC. (*st.*) small-flowered. Hungary. 1796. —— H. ♃.
23 effùsa. DC. (*st.*) spreading. Wolga. 1821. —— H. ♃.
24 sibírica. DC. (*wh.*) Siberian. Siberia. 1773. —— H. ♃. Hall. g. 1. p. 150. ic.
25 multiflora. DC. (*wh.*) many-flowered. Hungary. 1794. —— H. ♂. W. et K. hung. 1. t. 56.
26 ruthénica. DC. (*wh.*) Russian. Russia. 1827. —— H. ♃.

27 staticifòlia. DC.	(*wh.*)	Statice-leaved.	Levant.	1828.	6. 8.	H. ♃.	Flor. græc. t. 434.
28 tatarica. DC.	(*wh.*)	Tartarian.	Tartary.	1769.	——	H. ♃.	
29 gigántea. DC.	(*wh.*)	gigantic.	Africa.	1738.	——	G. ♃.	Flor. græc. t. 432.
30 viscòsa. DC.	(*wh.*)	clammy.	Levant.	1739.	7.	H. ♂.	Tourn. it. 2. p. 361. ic.
31 cònica. DC.	(*fl.*)	corn	England.	....	6. 7.	H. ☉.	Eng. bot. 922.
32 conóidea. DC.	(*pu.*)	conoid.	S. Europe.	1683.	——	H. ☉.	Clus. his. 1. p. 288. f. 2.
33 undulàta. DC.	(*re.*)	wave-leaved.	C. B. S.	1775.	8.	G. ♂.	
34 ánglica. DC.	(*wh.*)	English.	England.	....	6. 7.	H. ☉.	Eng. bot. 1178.
35 lusitànica. DC.	(*fl.*)	Portugal.	Portugal.	1732.	——	H. ☉.	Dill. elt. t. 311. f. 401.
36 tridentàta. DC.	(*ro.*)	three-toothed.	Spain.	1824.	——	H. ☉.	
37 gállica. DC.	(*fl.*)	French.	France.	1683.	5. 6.	H. ☉.	Vail. par. t. 16. f. 12.
38 ocymoídes. DC.	(*pu.*)	Basil-like.	....	1819.	6. 7.	H. ☉.	
39 dísticha. DC.	(*ro.*)	two-ranked.	....	1817.	——	H. ☉.	Schrank. pl. rar. t. 39.
40 cerastoídes. DC.	(*re.*)	Cerastium-like.	S. Europe.	1732.	7. 8.	H. ☉.	Flor. græc. t. 412.
41 quinquevúlnera. DC.	(*p.w.*)	variegated.	England.	....	6. 8.	H. ☉.	Eng. bot. 86.
42 noctúrna. DC.	(*w.re.*)	spiked.	S. Europe.	1683.	——	H. ☉.	Flor. græc. t 408.
43 refléxa. H.K.	(*wh.*)	reflexed.	————	1726.	——	H. ☉.	Magn. h. mons. 171. ic.
44 mutábilis. H.K.	(*bh.*)	changeable.	————	1688.	——	H. ☉.	
45 micropétala. DC.	(*re.*)	small-petaled.	Chili.	1822.	6. 7.	H. ☉.	
46 micrántha. DC.	(*re.*)	small-blossomed.	Portugal.	1825.	——	H. ☉.	
47 hirsùta. Lag.	(*re.*)	hairy.	————	1824.	——	H. ☉.	
sabuletòrum. Lk. *hirsutíssimum.* DC.							
48 canéscens. DC.	(*re.*)	canescent.	Naples.	1823.	6. 8.	H. ♃.	Ten. fl. nap. 1. t. 39.
49 decúmbens. DC.	(*wh.*)	decumbent.	S. Europe.	1828.	——	H. ☉.	
50 diffùsa. DC.	(*fl.*)	diffuse.	Italy.	1825.	——	H. ♃.	
51 divaricàta. F.G.	(*wh.*)	divaricate.	Levant.	1827.	——	H. ♂.	Flor. græc. t. 414.
racemòsa. DC.							
52 dichótoma. DC.	(*wh.*)	dichotomous.	Hungary.	1791.	——	H. ☉.	Flor. græc. t. 413.
53 nyctántha. DC.	(*st.*)	various-leaved.	....	1815.	6. 8.	H. ☉.	
54 trinérvia. DC.	(*re.*)	three-nerved.	Rome.	1821.	——	H. ☉.	
55 bellidifòlia. DC.	(*re.*)	Daisy-leaved.	....	1794.	——	H. ☉.	Jacq. vind. 3. t. 81.
56 vespertìna. DC.	(*re.*)	evening.	Barbary.	1796.	7. 8.	H. ☉.	Swt. br. fl. gar. 58.
bipartìta. Desf. atl. 1. t. 100.							
57 obtusifòlia. DC.	(*re.*)	blunt-leaved.	————	1823.	6. 8.	H. ☉.	
58 coloràta. DC.	(*fl.*)	coloured.	————	1804.	——	H. ☉.	
59 crassifolia. DC.	(*br.*)	thick-leaved.	C. B. S.	1774.	7. 8.	G. ♂.	
60 grácilis. DC.	(*wh.*)	slender.	....	1823.	6. 8.	H. ☉.	
61 jeniseénsis. DC.	(*st.*)	two-coloured.	Siberia.	1817.	6. 7.	H. ♃.	
62 ciliata. DC.	(*pu.*)	ciliated.	Crete.	1804.	——	H. ☉.	
63 pendula. DC.	(*re.*)	pendulous.	Sicily.	1731.	5. 7.	H. ☉.	Bot. mag. 114.
64 longicáulis. DC.	(*wh.*)	long-stalked.	Spain.	1818.	6. 7.	H. ☉.	
65 secundiflòra. DC.	(*wh.*)	side-flowered.	....	——	——	H. ☉.	
66 viscosíssima. DC.	(*wh.*)	viscous.	Naples.	1822.	——	H. ☉.	
67 quadridentàta. DC.	(*w.*)	four-toothed.	Italy.	1783.	5. 6.	H. ☉.	Jacq. aust. 2. t. 120.
Ly'chnis quadridentàta. W.							
68 pusílla. W. en.	(*wh.*)	dwarf.	Hungary.	1804.	6. 7.	H. ♃.	W. et K. hung. 3. t. 212.
69 alpéstris. DC.	(*wh.*)	Austrian.	Austria.	1774.	5. 7.	H. ♃.	Jacq. aust. 1. t. 96.
70 rupéstris. DC.	(*re.*)	rock.	Switzerland.	——	——	H. ♃.	Flor. dan. 4.
71 glaucifòlia. DC.	(*bh.*)	glaucous-leav'd.	Spain.	1823.	——	H. ♃.	
72 inapérta. DC.	(*st.*)	unopen-flow'r'd.	S. Europe.	1732.	6. 7.	H. ♃.	Flor. græc. t. 420.
73 clandéstina. DC.	(*w.p.*)	hidden-flower'd.	C. B. S.	1801.	——	H. ☉.	Jacq. col. sup. t. 3. f. 3.
74 porténsis. W.	(*pu.*)	Oporto.	Portugal.	1759.	7. 8.	H. ☉.	
75 antirrhìna. DC.	(*wh.*)	Snap-dragon.	Virginia.	1732.	6. 7.	H. ☉.	Dill. elt. t. 313. f. 403.
76 geminiflòra. DC.	(*pu.*)	twin-flowered.	....	1816.	....	H. ☉.	
77 flavéscens. DC.	(*st.*)	yellowish.	Hungary.	1804.	——	H. ♃.	W. et K. hung. 2. t. 175.
78 linoídes.	(*st.*)	Flax-like.	....	1817.	7. 8.	H. ☉.	
linifòlia. DC. *non* Flor. græc.							
79 linifòlia. F.G.	(*fl.*)	Flax-leaved.	Levant.	1828.	6. 8.	H. ♃.	Flor. græc. t. 433.
80 crética. DC.	(*re.*)	Candian.	Candia.	1732.	5. 8.	H. ☉.	———— t. 422.
81 sedóides. DC.	(*re.*)	Sedum-like.	Crete.	1804.	6. 8.	H. ☉.	———— t. 425.
82 sícula. Cyr.	(*ro.*)	Sicilian.	S. Europe.	1825.	——	H. ☉.	
divaricàta. DC. *non* Flor. græc.							
83 Saxífraga. DC.	(*w.br.*)	Saxifrage.	S. Europe.	1640.	——	H. ♃.	Lodd. bot. cab. 454.
84 petræ̀a. DC.	(*wh.*)	fine-toothed.	Hungary.	1806.	——	H. ♃.	W. et K. hung. 2. t. 164.
85 campánula. DC.	(*wh.*)	bell-flowered.	Piedmont.	1822.	——	H. ♃.	All. auct. p. 28. t. 1. f. 3.
86 longipétala. DC.	(*g.st.*)	long-petaled.	Aleppo.	——	——	H. ☉.	Flor. græc. t. 419.
87 nùtans. E.B.	(*wh.*)	Nottingham.	England.	....	6. 7.	H. ♃.	Eng. bot. 465.
88 infrácta. W.K.	(*wh.*)	smooth.	Hungary.	1816.	——	H. ♃.	W. et K. hung. 3. t. 213.

90 saxátilis. B.M.	(*st.*)	stone.	Siberia.	1800.	6. 7.	H. ♃.	Bot. mag. 689.
91 lívida. W. en.	(*w.br.*)	livid.	Carniola.	1816.	——	H. ♃.	
92 ténuis. DC.	(*st.*)	slender.	Siberia.	——	——	H. ♃.	
93 quadrífida. DC.	(*wh.*)	four-cleft.	Italy.	1820.	——	H. ♃.	
94 viridiflòra. DC.	(*gr.*)	green-flowered.	Portugal.	1739.	——	H. ♂.	Herm. parad. t. 199.
95 chlorántha. DC.	(*st.*)	pale-flowered.	Germany.	1732.	6. 8.	H. ♃.	Dill. elt. t. 316. f. 408.
96 psammìtis. DC.	(*st.*)	sulphur-colour'd.	....	1820.	——	H. ♃.	
97 nicæénsis. DC.	(*w.p.*)	villous-clammy.	S. Europe.	——	——	H. ⊙.	All. ped. t. 44. f. 2.
98 cathólica. DC.	(*wh.*)	panicled.	Italy.	1711.	7. 9.	H. ♃.	Jacq. vind. 1. t. 59.
99 Mussìni. H.	(*wh.*)	Mussin's.	Russia.	1824.	——	H. ♃.	
100 diversifòlia. DC.	(*wh.*)	various-leaved.	....	1820.	——	H. ⊙.	
101 rèpens. DC.	(*wh.*)	creeping.	Siberia.	1806.	7. 8.	H. ♃.	
102 gláuca. DC.	(*pu.*)	glaucous-leav'd.	....	1824.	——	H. ⊙.	
103 tenuifòlia. DC.	(*pu.*)	slender-leaved.	Dahuria.	1820.	——	H. ⊙.	
104 virgínica. DC.	(*re.*)	Virginian.	Virginia.	1783.	5. 8.	H. ♃.	Pluk. alm. t. 203. f. 1.
105 Catesb'æi. DC.	(*cr.*)	Catesby's.	Carolina.	....	——	H. ♃.	Catesb. car. 54. ic.
106 strícta. DC.	(*re.*)	upright.	Spain.	1802.	6. 7.	H. ⊙.	
107 Muscípula. DC.	(*re.*)	Spanish.	———	1596.	7. 8.	H. ⊙.	Clus. hist. 1. p. 289. f. 1.
108 noctiflòra. DC.	(*fl.*)	night-flowering.	England.	....	7.	H. ⊙.	Eng. bot. 291.
109 ornàta. DC.	(*cr.*)	dark-coloured.	C. B. S.	1775.	5. 9.	G. ♂.	Bot. mag. 382.
110 córsica. DC.	(*st.*)	Corsican.	Corsica.	1825.	6. 8.	H. ♃.	Bocc. mus. t. 54.
111 ægyptìaca. DC.	(*fl.*)	Egyptian.	Egypt.	1800.	7. 8.	G. ♂.	
112 sericea. DC.	(*fl.*)	silky.	Corsica.	1801.	6. 8.	H. ⊙.	All. ped. t. 79. f. 3.
113 pícta. DC.	(*w.p.*)	painted.	Mt. Carmel.	1818.	7. 9.	H. ⊙.	Swt. br. fl. gar. 92.
114 bícolor. DC.	(*w.p.*)	two-coloured.	France.	——	6. 8.	H. ⊙.	DC. ic. gall. rar. t. 42.
115 lævigàta. F.G.	(*re.*)	smooth.	Greece.	1828.	7. 9.	H. ⊙.	Flor. græc. t. 418.
116 reticulàta. DC.	(*re.*)	netted-flow'r'd.	Barbary.	1804.	7. 8.	H. ⊙.	Desf. atl. 1. t. 99.
117 pensylvànica. DC.	(*re.*)	Pensylvanian.	N. America.	1806.	6. 9.	H. ♃.	Bot. reg. 247.
incarnàta. Bot. cab. 41.							
118 vallèsia. DC.	(*pu.*)	hard-rooted.	Switzerland.	1765.	6. 8.	H. ♃.	All. ped. t. 23. f. 2.
119 fruticòsa. DC.	(*fl.*)	shrubby.	Sicily.	1629.	6. 7.	F. ♄.	Flor. græc. t. 428.
120 cáspica. DC.	(*st.*)	Caspian.	Caucasus.	1823.	——	H. ♄.	
121 am'œna. DC.	(*wh.*)	Tartarian.	Tartary.	1779.	7.	H. ♃.	
122 supìna. DC.	(*wh.*)	trailing.	Caucasus.	1804.	6. 8.	H. ♃.	Bot. mag. 1997.
123 depréssa. DC.	(*wh.*)	depressed.	Iberia.	1823.	——	H. ♃.	
124 paradoxa. DC.	(*wh.*)	Dover.	England.	....	7.	H. ♃.	Jacq. vind. 3. t. 84.
125 chloræfòlia. DC.	(*wh.*)	Armenian.	Armenia.	1796.	8. 9.	H. ♃.	Bot. mag. 807.
126 itálica. DC.	(*wh.*)	Italian.	Italy.	1759.	5. 6.	H. ♂.	Jacq. obs. 4. t. 79.
Cucùbalus itálicus. L.							
127 polyphy'lla. DC.	(*w.p.*)	many-leaved.	Austria.	....	6. 7.	H. ♃.	Clus. hist. 1. t. 290.
128 nemoràlis. DC.	(*wh.*)	wood.	Hungary.	1823.	——	H. ♃.	W. et K. hung. 3. t. 249.
129 longiflòra. DC.	(*wh.*)	long-flowered.	———	1793.	7. 9.	H. ♃.	——— 1. p. 7. t. 8.
130 bupleuroides. DC.	(*fl.*)	spear-leaved.	Persia.	1801.	6. 7.	H. ♃.	Tourn. it. t. 154.
131 pilòsa. S.S.	(*wh.*)	pilose.	Hungary.	1825.	6. 9.	H. ♃.	W. et K. hung. 3. t. 248.
mollissima. W.K.							
132 mollíssima. S.S.	(*wh.*)	velvet.	Italy.	1739.	7. 9.	H. ♃.	
Cucùbalus mollíssimus. L.							
133 inclùsa. H.	(*wh.*)	included.	....	1828.	6. 9.	H. ⊙.	
134 pínguis. H.	(*wh.*)	fleshy-leaved.	C. Spartel.	1825.	——	H. ⊙.	
135 táurica. F.	(*wh.*)	Taurian.	Tauria.	1828.	——	H. ♃.	
136 dahùrica. F.	(*wh.*)	Dahurian.	Dahuria.	1827.	——	H. ♃.	
137 scàbra. F.	(*wh.*)	rough.	Siberia.	——	——	H. ♃.	
138 marginàta. Sct.	(*wh.*)	margined.	....	1825.	——	H. ♃.	
139 stenopetala. Led.	(*re.*)	narrow-petaled.	Siberia.	1827.	——	H. ⊙.	
140 saponariæfòlia. Sct.	(*re.*)	Soapwort-ld.	....	1822.	——	H. ⊙.	
141 Persoónii. Sct.	(*wh.*)	Persoon's.	Europe.	——	——	H. ♃.	
142 vesicària.	(*pu.*)	bladdered.	Siberia.	1822.	6. 9.	H. ⊙.	
143 règia. DC.	(*cr.*)	splendid.	N. America.	1811.	5. 8.	G. ♂.	Bot. mag. 1724.
144 Atòcion. DC.	(*pu.*)	Orchis-flow'r'd.	Levant.	1781.	5. 7.	H. ⊙.	Jacq. vind. 3. t. 32.
orchídea. H.K. Flor. græc. t. 427.							
145 pseùdo-Atòcion. DC.	(*pu.*)	Atlantic.	Atlantic.	1818.	——	H. ⊙.	
146 Arméria. DC.	(*re.*)	Lobel's.	England.	....	7. 9.	H. ⊙.	Eng. bot. 1398.
α rùbra.	(*re.*)	*red-flowered.*	———	....	——	H. ⊙.	
β álba.	(*wh.*)	*white-flowered.*	———	....	——	H. ⊙.	
147 congésta. DC.	(*re.*)	close-flowered.	Greece.	1816.	——	H. ⊙.	
148 compácta. DC.	(*re.*)	compact.	Russia.	1816.	——	H. ♂.	
VISC'ARIA. Rœhl.		VISC'ARIA.	Decandria Pentagynia. L.				
1 vulgàris. Rœhl.	(*re.*)	common.	Britain.	....	5. 7.	H. ♃.	Eng. bot. 788.
Ly'chnis viscària. L.							
β flòre-plèno.	(*re.*)	*double-red.*	———	....	——	H. ♃.	

2 albiflòra.	(*wh.*)	white-flowered.	....	....	5. 7.	H. ♃.	
LY'CHNIS. DC.		LY'CHNIS.	Decandria Pentagynia. L.				
1 chalcedònica. DC.	(*sc.*)	scarlet.	Japan.	1596.	6. 7.	H. ♃.	Bot. mag. 257.
β albiflòra.	(*wh.*)	*white-flowered.*	——	....	——	H. ♃.	
γ flòre plèno.	(*sc.*)	*double-flowered.*	——	....	——	H. ♃.	
2 Flós-jòvis. DC.	(*re.*)	umbelled.	Piedmont.	1726.	7.	H. ♃.	Bot. mag. 398.
Agrostémma Flós-jòvis. B.M.							
3 coronàta. W.	(*sc.*)	Chinese.	China.	1774.	6. 9.	H. ♃.	Bot. mag. 223.
grandiflòra. Jacq. ic. rar. t. 84.							
4 fúlgens. DC.	(*sc.*)	brilliant.	Siberia.	1819.	7. 9.	H. ♃.	Bot. mag. 2104.
5 Cœ'li-ròsa. DC.	(*fl.*)	smooth-leaved.	Levant.	1713.	7. 8.	H. ⊙.	Bot. mag. 295.
Agrostémma Cœ'li-ròsa. L.							
AGROSTE'MMA.		AGROSTE'MMA.	Decandria Pentagynia. L.				
1 apétala. DC.	(*wh.*)	petalless.	Lapland.	1819.	6. 8.	H. ♃.	Lin. fl. lap. t. 12. f. 1
Ly'chnis apétala. L.							
2 brachypétala. L. en.	(*wh.*)	short-petal'd.	....	——	——	H. ♃.	
3 sylvéstris. DC.	(*re.*)	red-flowered.	Britain.	....	6. 9.	H. ♃.	Eng. bot. 1579.
Ly'chnis diòica rùbra. E. B.							
β flòre-plèno.		*double-flowered.*	....	....	——	H. ♃.	
4 diòica. DC.	(*wh.*)	white-flowered.	Britain.	....	——	H. ♃.	Eng. bot. 1580.
Ly'chnis diòica álba. E.B. *vespertìna.* Sibthorp.							
β flòre-plèno.	(*wh.*)	*double-white.*	....	....	——	H. ♃.	
γ viridiflora.	(*gr.*)	*double-green.*	....	....	——	H. ♃.	
5 diclìnis. DC.	(*ro.*)	long-pedicled.	Spain.	1824.	——	H. ♃.	
6 l'æta. DC.	(*re.*)	small.	Portugal.	1778.	7.	H. ⊙.	
7 córsica. DC.	(*re.*)	Corsican.	Corsica.	1824.	6. 8.	H. ⊙.	
8 pyrenàica. DC.	(*re.*)	Pyrenean.	Pyrenees.	1820.	4. 6.	H. ♃.	
9 suècica. L.C.	(*re.*)	Swedish.	Sweden.	....	6. 9.	H. ♃.	Lodd. bot. cab. 881.
10 alpìna. DC.	(*re.*)	Alpine.	Scotland.	....	4. 5.	H. ♃.	Eng. bot. 2254.
Ly'chnis alpìna. E.B.							
11 Flóscùculi. DC.	(*re.*)	Ragged Robin.	Britain.	....	6. 9.	H. ♃.	Eng. bot. 573.
Ly'chnis Flóscùculi. L.							
β flòre-plèno.	(*re.*)	*double.*	....	....	——	H. ♃.	
12 coronària. DC.	(*ro.*)	Rose Campion.	Italy.	1596.	——	H. ♃.	
α rùbra.	(*ro.*)	*red-flowered.*	——	——	——	H. ♃.	Bot. mag. 24.
β álba.	(*wh.*)	*white.flowered.*	....	——	——	H. ♃.	
γ plèna.	(*ro.*)	*double-flowered.*	....	....	——	H. ♃.	
GITH'AGO. Desf.		CORN-COCKLE.	Decandria Pentagynia. L.				
1 segetum. Desf.	(*re.*)	common.	Britain.	....	6. 8.	H. ⊙.	Eng. bot. 741.
Agrostémma Githàgo. L.							
2 nicæénsis.	(*wh.*)	long-calyxed.	Italy.	1794.	——	H. ⊙.	
VEL'EZIA. DC.		VEL'EZIA.	Pentandria Digynia. L.				
1 rígida. DC.	(*re.*)	rigid.	Spain.	1683.	7.	H. ⊙.	Flor. græc. t. 390.
2 quadridentàta. F.G.	(*r.*)	four-toothed.	Levant.	1826.	——	F. ♃.	—— t. 391.
DR'YPIS. DC.		DR'YPIS.	Pentandria Trigynia. L.				
spinòsa. DC.	(*bh.*)	prickly.	Italy.	1775.	6. 7.	H. ♄.	Bot. mag. 2216.

Tribus II. *ALSINEÆ.* DC.

ORT'EGIA. DC.		ORT'EGIA.	Triandria Monogynia. L.				
1 hispánica. DC.	(*gr.*)	Spanish.	Spain.	1768.	6. 7.	H. ♃.	
2 dichótoma. DC.	(*gr.*)	forked.	Italy.	1781.	8. 9.	H. ♃.	All ped. t. 4. f. 1.
hispánica. Cav. ic. 1. t. 47.							
BUFF'ONIA. DC.		BUFF'ONIA.	Tetrandria Digynia. L.				
1 ánnua. DC.	(*wh.*)	annual.	England.	....	6.	H. ⊙.	Eng. bot. 1313.
tenuifòlia. E.B.							
2 perénnis. DC.	(*wh.*)	perennial.	France.	1820.	6. 7.	H. ♃.	Lam. ill. t. 87. f. 2.
SAG'INA. DC.		PEARL-WORT.	Tetrandria Tetragynia. L.				
1 procúmbens. DC.	(*gr.*)	procumbent.	Britain.	....	5. 9.	H. ⊙.	Eng. bot. 880.
β multipléx.	(*wh.*)	*double-flowered.*	——	....	——	H. ♃.	
2 cerastoídes DC.	(*wh.*)	Cerastium like.	——	....	6. 7.	H. ⊙.	Eng. bot. 166.
Cerástium tetràndrum H. K.							
3 filifórmis. DC.	(*wh.*)	filiform.	Pyrenees.	1824.	——	H. ⊙.	
4 marítima. E.B.	(*wh.*)	sea-side.	Ireland.	....	5. 8.	H. ⊙.	Eng. bot. 2195.
5 strícta. DC.	(*wh.*)	straight.	N. Europe.	1828.	——	H. ⊙.	
6 apétala. DC.	(*wh.*)	small-flowered.	Britain.	....	5. 6.	H. ⊙.	Eng. bot. 881.

MŒ'NCHIA. P.S. MŒ'NCHIA. Tetrandria Tetragynia. L.
gláuca. P.S. (*wh.*) glaucous. Britain. .... 4. 5. H. ⊙. Eng. bot. 609.
quaternélla. L. en. *Sagìna erécta.* E.B.
MŒHRI'NGIA. DC. MŒHRI'NGIA. Octandria Digynia. L.
1 muscòsa. DC. (*wh.*) mossy. S. Europe. 1775. 6. 7. H. ♃. Lam. ill. t. 314.
2 sedifòlia. DC. (*wh.*) Stonecrop-ld. ——— 1823. —— H. ♃. W.ber.m.1818.t.3.f.23.
ELA'TINE. DC. WATERWORT. Octandria Tetragynia. L.
1 Hydropìper. DC. (*wh.*) small. Britain. .... 7. 8. H. ♃. Eng. bot. 955.
2 hexándra. DC. (*wh.*) hexandrous. France. .... —— H. ♃. DC. ic. rar. 1. t. 43. f. 1.
BE'RGIA. DC. BE'RGIA. Decandria Pentagynia. L.
verticillàta. DC. (*wh.*) whorl-leaved. E. Indies. 1826. 7. 10. H. ⊙. Delil. fl. æg. t. 26. f. 1.
MOLL'UGO. W. MOLL'UGO. Triandria Trigynia. L.
1 verticillàta. DC. (*wh.*) whorled. Virginia. 1748. 6. 8. H. ⊙. Ehret pict. t. 6.
2 Schránkii. DC. (*wh.*) Schrank's. Brazil. 1821. 7. 9. H. ⊙. Schrank pl. rar. 64. ic.
dichótoma. Schk.
3 Línkii. DC. (*wh.*) Link's. ——— —— —— H. ⊙.
triphy'lla. L. en. *non* Lour.
PHARN'ACEUM. W PHARN'ACEUM. Pentandria Trigynia. L.
1 Mollùgo. R.S. (*wh.*) LadiesBedstraw. E. Indies. 1752. 7. 8. S. ⊙. Burm. ind. t. 5. f. 4.
Mollùgo Spérgula. DC.
2 cordifòlium. R.S. (*wh.*) heart-leaved. C. B. S. 1820. —— H. ⊙. Jacq. h. schœn. t. 349.
3 dichótomum. R.S. (*wh.*) forked. ——— 1783. —— H. ⊙.
4 Cerviàna. R.S. (*wh.*) umbelled. Russia. 1771. 6. 7. H. ⊙. Lam. ill. t. 214. f. 1.
5 bellidifòlium. R.S. (*w.*) Daisy-leaved. W. Indies. 1826. —— H. ⊙. Plum. amer. t. 21. f. 1.
HOLO'STEUM. DC. HOLO'STEUM. Triandria Trigynia. L.
1 umbellàtum. DC. (*wh.*) umbelled. England. .... 7. 8. H. ⊙. Flor. dan. t. 1204.
β ròseum. (*ro.*) *rose-coloured.* ——— .... —— H. ⊙. Eng. bot. 27.
2 diándrum. DC. (*wh.*) diandrous. Jamaica 1822. 6. 8. S. ⊙. Swartz ic. t. 7.
SPERG'ULA. DC. SPURREY. Decandria Pentagynia. L.
1 arvénsis. DC. (*wh.*) rough-seeded. Britain. .... 7. 8. H. ⊙. Eng. bot. 1535.
2 pentándra. DC. (*wh.*) smooth-seeded. England. .... 6. 7. H. ⊙. ——— 1536.
3 pállida. DC. (*pa.*) pale. C. B. S. .... 7. 9. G. ♃.
4 nodòsa. DC. (*wh.*) knotted. Britain. .... 7. 8. H. ♃. Eng. bot. 694
5 laricìna. DC. (*wh.*) Larch-leaved. Siberia. 1823. 6. 8. H. ♃. Sm. ic. ined. 1. t. 18.
6 saginoídes. DC. (*wh.*) Sagina like Scotland. .... —— H. ♃. Eng. bot. 2105.
7 subulàta. DC. (*wh.*) awl-leaved. Britain. .... 6. 7. H. ♃. ——— 1082.
8 glábra. DC. (*wh.*) smooth. Europe. 1824. —— H. ♃. All. ped. t. 64. f. 1.
LA'RBREA. DC. LA'RBREA. Decandria Trigynia. L.
aquática. DC. (*wh.*) water. Britain. .... 7. 8. H.*w.*♃. Eng. bot. t. 1074.
Stellària aquática. Poll. pal. n. 422. *uliginòsa.* E.B. *alsìne* Hoff. germ. 1. t. 5.
DRYM'ARIA DC. DRYM'ARIA. Pentandria Trigynia. L.
1 cordàta. DC. (*wh.*) heart-leaved. W. Indies. 1814. 7. 9. H. ⊙. Lam. ill. t. 51. f. 2.
Holósteum cordàtum. L.
2 divaricàta. DC. (*wh.*) divaricate. Lima. 1823. —— H. ⊙.
STELL'ARIA. DC. STICHWORT. Decandria Trigynia. L.
1 némorum. DC. (*wh.*) wood. Britain. .... 4. 6. H. ♃. Eng. bot. 92.
2 latifòlia. DC. (*wh.*) broad-leaved. Germany. 1816. 6. 8. H. ⊙.
3 mèdia. DC. (*wh.*) Chickweed. Britain. .... 1. 12. H. ⊙. Eng. bot. 537.
Alsìne mèdia. L.
4 dichótoma. DC. (*wh.*) forked. Siberia. 1774. 6. 8. H. ♃. Smith ic. ined. 1. t. 14.
5 bulbòsa. DC. (*wh.*) bulbiferous. Carinthia. 1820. —— H. ♃. Jacq. ic. rar. 3. t. 468.
6 víscida. DC. (*wh.*) viscous. Tauria. 1804. —— H. ⊙. W. et K. hung. 1. t. 22.
Cerástium anómalum. W.
7 sabulòsa. DC. (*wh.*) gravel. Persia. 1828. —— H. ⊙.
8 Holóstea. DC. (*wh.*) greater. Britain. .... 4. 6. H. ♃. Eng. bot. 511.
9 Laxmánni. DC. (*wh.*) Laxmann's. Siberia. 1827. —— H. ♃.
10 velutìna. DC. (*wh.*) velvetty. Siberia. 1824. 6. 8. H. ♃.
11 gramínea. DC. (*wh.*) grass-leaved. Britain. .... 5. 8. H. ♃. Eng. bot. 803.
12 gláuca. DC. (*wh.*) glaucous marsh. ——— .... —— H. ♃. ——— 825.
13 cerastoídes. DC. (*wh.*) Alpine. Scotland. .... —— H. ♃. Eng. bot. 911.
14 multicáulis, W. (*wh.*) many-stalked. Carinthia. 1815. 6. 8. H. ♃. Jacq. coll. 1. t. 19.
cerastoídes. Jacq. coll.
15 arenària. DC. (*wh.*) sand. Spain. 1799. 6. 8. H. ⊙.
16 scapígera. DC. (*wh.*) naked-stalked. Scotland. .... 6. 7. H. ♃. Eng. bot. 1269.
17 Fischeriàna. DC. (*wh.*) Fischer's. Siberia. 1825. —— H. ♃.
18 Edwárdsii. B. (*wh.*) Edwards's. N. America. 1827. —— H. ♃.
nítida. Hook.
19 humifùsa. DC. (*wh.*) trailing. N. Europe. 1825. —— H. ⊙. Flor. dan. t. 978.
20 l'æta. S.S. (*wh.*) short-petaled. N. America. 1827. —— H. ⊙.

21	grácilis. S.S.	(*wh.*)	slender.	N. America.	1827.	6. 7.	H. ⊙.	
22	muràlis. DC.	(*wh.*)	wall.	Crete.	1821.	——	H. ⊙.	
23	élegans. DC.	(*wh.*)	elegant.	Siberia.	1823.	——	H. ♃.	
24	lóngipes. DC.	(*wh.*)	long pedicled.	N. America.	1821.	——	H. ♃.	
25	mollugínea.	(*wh.*)	Mollugo-like.	New Spain.	1825.	7. 9.	H. ⊙.	
	Alsìne mollugínea. Lag. *Arenària? mollugìnea.* DC.							
	ARENARIA. DC.		SANDWORT.	Decandria Trigynia. L.				
1	segetàlis. DC.	(*wh.*)	hedge.	France.	1805.	6. 10.	H. ⊙.	Vail. par. t. 3. f. 3.
	Alsìne segetàlis. L.							
2	rùbra. DC.	(*re.*)	purple.	Britain.	....	6. 8.	H. ⊙.	Eng. bot. 852.
3	marìna. L.	(*re.*)	sea-side.	———	....	——	H. ♃.	
4	salìna. DC.	(*re.*)	salt marsh.	Bohemia.	1820.	——	H. ⊙.	
5	rubélla. F.L.	(*re.*)	red-flowered.	England.	....	——	H. ⊙.	Flor. lond. t.
6	glutinòsa. DC.	(*pu.*)	glutinous.	Astracan.	1828.	——	H. ⊙.	
7	mèdia. DC.	(*re.*)	marine.	Britain.	....	——	H. ♃.	Eng. bot. 958.
	marìna. E.B. non L.							
8	glandulòsa. Jac.	(*wh.*)	glandular.	Europe.	1820.	6. 7.	H. ⊙.	Jacq. Schœnb. 3. t. 355.
9	canadénsis. DC.	(*re.*)	Canadian.	N. America.	1812.	——	H. ⊙.	
10	graminifòlia. DC.	(*wh.*)	grass-leaved.	Caucasus.	1815.	——	H. ♃.	Schrad. h. gœtt. t. 5.
	β *glabérrima.* DC.	(*w.*)	*smooth.*	———	——	——	H. ♃.	
11	longifòlia. DC.	(*wh.*)	long-leaved.	Volga.	1820.	——	H. ♃.	Gmel. sib. 4. t. 63. f. 2.
12	dahùrica. DC.	(*wh.*)	Dahurian.	Dahuria.	1824.	——	H. ♃.	
13	nardifòlia. F.	(*wh.*)	Nardus-leaved.	Siberia.	1827.	——	H. ♃.	
14	otitoídes. DC.	(*wh.*)	dense-flowered.	———	——	——	H. ♃.	
15	rígida. DC.	(*wh.*)	rigid.	———	——	——	H. ♃.	
16	formòsa. DC.	(*wh.*)	handsome.	Dahuria.	——	——	H. ♃.	
17	Gmelíni. DC.	(*wh.*)	Gmelin's.	———	1827.	——	H. ♃.	Gmel. sib. 4. t. 61. f. 1.
18	cephalòtes. DC.	(*wh.*)	close-headed.	Tauria.	1820.	——	H. ♃.	
19	imbricàta. DC.	(*wh.*)	imbricate.	Caucasus.	1821.	——	H. ♃.	
20	juniperìna. DC.	(*wh.*)	Juniper-leaved.	Armenia.	1800.	——	H. ♃.	Sm. ic. ined. t 35.
21	strícta. DC.	(*wh.*)	upright.	N. America.	1812.	5. 6.	H. ♃.	
22	pròcera. DC.	(*wh.*)	tall.	....	1827.	——	H. ♃.	
23	laricifòlia. L.	(*wh.*)	Larch-leaved.	Switzerland.	....	8.	H. ♃.	Jacq. aust. 3. t. 272.
24	rostràta. W. en.	(*wh.*)	beaked.	Hungary.	1816.	6. 7.	H. ♃.	
25	striàta. W.	(*wh.*)	striated.	Switzerland.	1683.	——	H. ♃.	Vill. dauph. 4. t. 47. f. 6.
26	árctica. DC.	(*wh.*)	small.	Siberia.	1807.	——	H. ♃.	
27	austrìaca. DC.	(*wh.*)	Austrian.	Austria.	1793.	6. 9.	H. ♃.	Jacq. aust. 3. t. 270.
28	Villársii. P.S.	(*wh.*)	Villars's.	Dauphiny.	1826.	——	H. ♃.	Vill. delph. 3. t. 47.
29	grandiflòra. DC.	(*wh.*)	great-flowered.	Switzerland.	1783.	——	H. ♃.	All ped. n. 1711. t. 10. f. 1.
30	triflòra. L.	(*wh.*)	three-flowered.	———	1816.	——	H. ♃.	Cav. ic. 3. t. 249. f. 2.
31	Hélmii. DC.	(*wh.*)	long-peduncled.	Siberia.	1824.	——	H. ♃.	
32	macrocárpa. DC.	(*wh.*)	large-capsuled.	N. America.	1828.	——	H. ♃.	
33	hírta. DC.	(*re.*)	hairy.	Greenland.	1827.	——	H. ♃.	Flor. dan. t. 1646.
34	vérna. DC.	(*wh.*)	vernal.	Britain.	....	5 8.	H. ♃.	Eng. bot. 512.
35	cæspitòsa. DC.	(*wh.*)	tufted.	S. Europe.	1824.	——	H. ♃.	
36	ramosíssima. DC.	(*wh.*)	branching.	Hungary.	1820.	——	H. ♃.	
37	saxátilis. DC.	(*wh.*)	rock.	Germany.	1732.	7 8.	H. ♃.	Barr. ic. t. 580.
38	péndula. DC.	(*wh.*)	pendulous.	Hungary.	1816.	6. 7.	H. ♃.	W. et K. hung. 1. t. 87.
39	tenuifòlia. DC.	(*wh.*)	fine-leaved.	England.	....	——	H. ⊙.	Eng. bot. 219.
40	villòsa. DC.	(*wh.*)	villous.	Siberia.	1828.	——	H. ♃.	
41	calycìna. DC.	(*wh.*)	large-calyxed.	Barbary.	1824.	——	H. ⊙.	
42	triándra. DC.	(*wh.*)	triandrous.	....	——	——	H. ⊙.	Schrank. h. mon. t. 30.
43	mediterrànea. DC.	(*w.*)	Mediterranean.	S. Europe.	1821.	——	H. ⊙.	
44	recúrva. DC.	(*wh.*)	recurved-leav'd.	———	——	——	H. ♃.	Jacq. coll. 2. t. 16. f. 1.
45	hirsùta. DC.	(*wh.*)	hairy.	Tauria.	1824.	——	H. ♃.	
46	setàcea. DC.	(*wh.*)	bristly.	France.	1820.	——	H. ♃.	
47	fasciculàta. DC.	(*wh.*)	dense-flowered.	S. Europe.	1815.	——	H. ⊙.	Flor. græc. t. 442.
48	fastigiàta. E.B.	(*wh.*)	fastigiate.	Scotland.	....	6. 7.	H. ⊙.	Eng. bot. 1744.
49	filifòlia. DC.	(*wh.*)	thread-leaved.	Arabia.	——	——	H. ♃.	Vahl. symb. 1. t. 12.
50	mucronàta. DC.	(*wh.*)	mucronate.	———	1777.	——	H. ⊙.	
51	uliginòsa. DC.	(*wh.*)	marsh.	Switzerland.	1824.	——	H. ♃.	DC. ic. gall. t. 46.
52	polygonóides. DC.	(*wh.*)	procumbent.	———	——	——	H. ♃.	Jacq. coll. 1. t. 15.
53	marginàta. DC.	(*wh.*)	margined.	Siberia.	1824.	——	H. ♃.	
54	canéscens. DC.	(*wh.*)	canescent.	....	1828.	——	H. ♃.	
	macrocárpa. H. non Ph.							
55	tetraquètra. DC.	(*wh.*)	square-stalked.	S. France.	1731.	8.	H. ♃.	All. ped. 2 t. 89. f. 1.
56	Giesèkii. DC.	(*wh.*)	Gieseke's.	Greenland.	1827.	——	H. ♃.	Horn. fl. dan t. 1518.
57	lanceolàta. DC.	(*wh.*)	lanceolate.	Piedmont.	1820.	——	H. ♃.	All. fl. ped. t. 26. f. 5.
58	cherlerióides. P.S.	(*wh.*)	cherleria-leav'd.	S. France.	....	7. 8.	H. ♃.	Vill. delph. 4. t. 47. f. 1.

59 montàna. DC.	(*wh.*)	mountain.	Spain.	1800.	4. 7.	H. ♃.	Bot. mag. 1118.
60 serpyllifòlia. DC.	(*wh.*)	Thyme-leaved.	Britain.	....	6. 7.	H. ⊙.	Eng. bot. 923.
61 ciliàta. DC.	(*wh.*)	fringed.	Ireland.	....	3. 8.	H. ♃.	———— 1745.
62 multicáulis. W.	(*wh.*)	many-stalked.	Europe.	1794.	7. 8.	H. ♃.	Jacq. coll. 1. t. 17. f. 1.
63 trinérvia. DC.	(*wh.*)	three-nerved.	Britain.	....	5. 7.	H. ⊙.	Eng. bot. 1483.
64 norvègica. DC.	(*wh.*)	Norway.	Norway.	1828.	6. 7.	H. ♃.	Flor. dan. t. 1269.
65 baleàrica. DC.	(*wh.*)	Majorca.	Majorca.	1787.	3. 8.	H. ♃.	L. Her. stirp. 1. t. 15.
66 biflòra. DC.	(*wh.*)	two-flowered.	S. Europe.	1823.	——	H. ♃.	All. ped. t. 64. f. 3.
67 procúmbens. DC.	(*re.*)	procumbent.	Egypt.	1801.	6. 8.	F. ♃.	Vahl. symb. 1. t. 33.
ADEN'ARIUM. DC.		SEA CHICKWEED.	Decandria Tri-Pentagynia.				
peploídes. DC.	(*wh.*)	common.	Britain.	....	6. 7.	H. ♃.	Eng. bot. 189.
Arenària peploídes. E.B.							
CERA'STIUM. DC.		MOUSE-EAR CHICKWEED.	Decandria Pentagynia. L.				
1 nemoràle. DC.	(*wh.*)	wood.	Caucasus.	1823.	5 6.	H. ⊙.	
2 perfoliàtum. DC.	(*wh.*)	perfoliate.	Greece.	1725.	6. 7.	H. ⊙.	Dill. elt. t. 217. f. 284.
3 caucásicum. DC.	(*wh.*)	Caucasean.	Caucasus.	1825.	——	H. ⊙.	
4 dahùricum. DC.	(*wh.*)	glaucous.	Dahuria.	1815.	5. 9.	H. ♃.	Schrank. h. mon. t. 75.
amplexicáule. B.M. 1789.							
5 máximum. DC.	(*wh.*)	greatest.	Siberia.	1792.	6. 7.	H. ⊙.	Gmel. sib. 4. t. 62. f. 2.
6 dichótomum. DC.	(*wh.*)	forked.	Spain.	1725.	——	H. ⊙.	
7 ruderàle. DC.	(*wh.*)	pendulous-fruit.	Caucasus.	1824.	——	H. ⊙.	
8 táuricum. DC.	(*wh.*)	Taurian.	Tauria.	——	——	H. ⊙.	
9 vulgàtum. DC.	(*wh.*)	common.	Britain.	....	4. 6.	H. ⊙.	Eng. bot. 789.
10 viscòsum. DC.	(*wh.*)	clammy.	————	....	4. 9.	H. ♃.	———— 790.
11 barbulàtum. L. en.	(*wh.*)	bearded.	S. Europe.	1821.	——	H. ⊙.	
12 holosteoídes. DC.	(*wh.*)	Holosteum like.	Sweden.	——	——	H. ♃.	
13 pentándrum. DC.	(*wh.*)	five-stamened.	Spain.	1820.	——	H. ⊙.	
14 semidecándrum. DC.	(*wh.*)	least.	Britain.	....	3. 4.	H. ⊙.	Eng. bot. 1630.
15 brachypétalum. DC.	(*wh.*)	small-flowered.	France.	1816.	4. 5.	H. ♃.	DC. ic. gall. t. 44.
16 diffùsum. DC.	(*wh.*)	spreading.	....	....	4. 9.	H. ♃.	
17 serpyllifòlium. W. en.	(*wh.*)	Thyme-ld.	S. Europe.	1824.	5. 6.	H. ♃.	
18 mánticum. DC.	(*wh.*)	long peduncl'd.	Hungary.	1801.	6. 7.	H. ⊙.	W. et K. hung. 1. t. 96.
19 campanulàtum. DC.	(*wh.*)	early-flowd.	Italy.	1826.	4. 6.	H. ⊙.	Viv. ann. bot. 1. p. 2. t. 1.
pr'æcox. T.N.							
20 grandiflòrum. DC.	(*wh.*)	large-flowered.	————	1820.	6. 7.	H. ♃.	W. et K. hung. 2. t. 168.
21 argénteum. M.B.	(*wh.*)	silvery-leaved.	Iberia.	1821.	——	H. ♃.	
22 tomentòsum. DC.	(*wh.*)	white-leaved.	S. Europe.	1648.	——	H. ♃.	Col. phyt. ed. 1744. t. 31.
23 Bieberstéinii. DC.	(*wh.*)	Bieberstein's.	Tauria.	1820.	——	H. ♃.	DC. pl. rar. gen. t. 11.
24 lanàtum. DC.	(*wh.*)	woolly.	Pyrenees.	——	——	H. ♃.	
25 alpìnum. DC.	(*wh.*)	Alpine.	Britain.	....	——	H. ♃.	Eng. bot. 472.
26 ovàtum. DC.	(*wh.*)	oval-leaved.	Carinthia.	1816.	——	H. ♃.	
27 latifòlium. DC.	(*wh.*)	broad-leaved.	Britain.	....	——	H. ♃.	Eng. bot. 473.
28 aquáticum. E.B.	(*wh.*)	water.	————	....	——	H. *w.* ♃.	———— 538.
29 sylváticum. DC.	(*wh.*)	wood.	Hungary.	1826.	——	H. ♃.	W. et K. hung. 1. t. 97.
30 arvénse. DC.	(*wh.*)	field.	————	....	5. 8.	H. ♃.	Eng. bot. 93.
31 rèpens. W.	(*wh.*)	creeping.	S. Europe.	1759.	5. 7.	H. ♃.	Lam. ill. t. 392. f. 2.
32 stríctum. W.	(*wh.*)	upright.	Austria.	1793.	——	H. ♃.	Scop. carn. t. 19. f. 1.
33 suffruticòsum. W.	(*wh.*)	suffruticose.	S. Europe.	1796.	——	H. ♃.	
34 pensylvànicum. DC.	(*wh.*)	Pensylvanian.	N. America.	1823.	——	H. ♃.	
35 pubéscens. DC.	(*wh.*)	pubescent.	————	1822.	——	H. ♃.	
36 matrénse. L. en.	(*wh.*)	Matram.	Hungary.	1823.	——	H. ♃.	
37 dióicum. H.K.	(*wh.*)	Spanish.	Spain.	1766.	6. 7.	H. ♃.	
38 pilòsum. L. en.	(*wh.*)	hairy-leaved.	....	1823.	——	H. ♃.	
39 Sprengèlii. DC.	(*wh.*)	Sprengel's.	....	——	——	H. ♂.	
40 Tenoreànum. DC.	(*wh.*)	Tenore's.	Italy.	1825.	——	H. ⊙.	
41 Samniànum. DC.	(*wh.*)	Samnian.	————	——	——	H. ♃.	
42 hirsùtum. DC.	(*wh.*)	hairy.	————	——	——	H. ♃.	
43 glandulòsum. Lk.	(*wh.*)	glandular.	....	1827.	——	H. ♃.	
44 inflàtum. Lk.	(*wh.*)	inflated.	....	——	——	H. ⊙.	
CHERL'ERIA. DC.		CHERL'ERIA.	Decandria Trigynia. L.				
sedoídes. DC.	(*wh.*)	dwarf.	Scotland.	....	7. 8.	H. ♃.	Eng. bot. 1212.

ORDO XXVII.

LINEÆ. *DC. prodr.* 1. *p.* 423.

L'INUM. DC.		FLAX.	Pentandria Pentagynia. L.				
1 gállicum. DC.	(*ye.*)	annual yellow.	France.	1777.	7. 8.	H. ⊙.	Flor. græc. t. 303.
2 áureum. DC.	(*ye.*)	golden.	Hungary.	1823.	——	H. ⊙.	W. et K. hung. 2. t. 177.
3 tènue. DC.	(*ye.*)	slender.	Barbary.	1826.	——	H. ⊙.	
4 lutèolum. DC.	(*ye.*)	yellowish.	Tauria.	——	——	H. ⊙.	Buxb. cent. 5. t. 59.
5 nodiflòrum. DC.	(*ye.*)	knotted.	Italy.	1759.	——	H. ♃.	Flor. græc. t. 307.
6 stríctum. DC.	(*ye.*)	upright.	S. Europe.	——	5. 7.	H. ♂.	—————— t. 304.
7 rígidum. DC.	(*st.*)	stiff-leaved.	Missouri.	1807.	7.	H. ⊙.	
8 virginiànum. DC.	(*ye.*)	Virginian.	N. America.	——	——	H. ♂.	
9 mexicànum. DC.	(*ye.*)	Mexican.	Mexico.	1827.	8. 11.	F. ♃.	
10 dahùricum. DC.	(*ye.*)	Dahurian.	Dahuria.	1822.	6. 8.	H. ♃.	
11 marítimum. DC.	(*ye.*)	sea.	S. Europe.	1596.	7. 8.	H. ♃.	Jacq. vind. 2. t. 154.
12 arbòreum. R.S.	(*ye.*)	shrubby.	Candia.	1788.	5. 8.	G. ♄.	Flor. græc. t. 305.
13 campanulàtum. R.S.	(*y.*)	glaucous-leav'd.	S. Europe.	1795.	——	H. ♄.	Lob. ic. 414.
14 capitàtum. R.S.	(*ye.*)	headed.	Hungary.	1826.	6. 10.	H. ♃.	
15 flàvum. W.	(*ye.*)	yellow.	S. Europe.	1793.	——	H. ♃.	Bot. mag. 312.
16 trig'ynum. DC.	(*ye.*)	three-styled.	E. Indies.	1799.	10. 3.	G. ♄.	—————— 1100.
17 africànum. DC.	(*ye.*)	African.	C. B. S.	1771.	6. 7.	G. ♄.	—————— 403.
18 quadrifòlium. DC.	(*ye.*)	four-leaved.	——————	1787.	5. 6.	G. ♄.	—————— 431.
19 viscòsum. DC.	(*pi.*)	viscous.	S. Europe.	1821.	6. 8.	H. ♃.	Scop. carn. n. 383. t. 11.
20 hypericifòlium. P.L.	(*pi.*)	Mallow-flow'r'd	Caucasus.	1807.	6. 7.	H. ♃.	Bot. mag. 1048.
venústum. A.R. 477.							
21 hirsùtum. W.	(*bl.*)	hairy.	Austria.	1759.	7. 8.	H. ♃.	Jacq. aust. 1. t. 31.
22 ascyrifòlium. B.M.	(*w.b.*)	blue and white.	Portugal.	1800.	——	H. ♃.	Bot. mag. 1087.
23 nervòsum. DC.	(*bl.*)	nervose.	Hungary.	1820.	——	H. ♃.	W. et K. hung. 2. t. 105.
24 narbonénse. DC.	(*bl.*)	Narbonne.	S. France.	1759.	5. 7.	F. ♃.	Barr. ic. t. 1007.
25 usitatíssimum. DC.	(*bl.*)	common.	Britain.	....	6. 7.	H. ⊙.	Eng. bot. 1357.
26 marginàtum. DC.	(*bl.*)	white-margin'd.	....	1823.	——	H. ♃.	
angustifòlium. W. en. *non* E.B.							
27 refléxum. DC.	(*bl.*)	reflexed-leav'd.	S. Europe.	1777.	7.	H. ♃.	
28 squamulòsum. DC.	(*bl.*)	scaly.	Tauria.	1821.	6. 8.	H. ♃.	
29 diffùsum. DC.	(*bl.*)	spreading.	....	1823.	——	H. ♃.	
30 angustifòlium. DC.	(*bl.*)	narrow-leaved.	England.	....	7.	H. ♃.	Eng. bot. 381.
31 sibíricum. DC.	(*bl.*)	Siberian.	Siberia.	1775.	6. 7.	H. ♃.	Mill. ic. t. 166. f. 2.
austrìacum. B.M. 1086.							
32 Lewísii. Ph.	(*bl.*)	Lewis's.	N. America.	1824.	——	H. ♃.	Bot. reg. 1163.
33 ánglicum. DC.	(*bl.*)	English.	England.	....	6. 8.	H. ♃.	Eng. bot. 40.
perénne. E.B.							
β *procúmbens.* DC.	(*bl.*)	*procumbent.*	——————	....	——	H. ♃.	
34 austrìacum. DC.	(*bl.*)	Austrian.	Austria.	1775.	6. 7.	H. ♃.	Jacq. aust. t. 418.
35 montànum. DC.	(*bl.*)	mountain.	Switzerland.	....	7. 8.	H. ♃.	Scop. carn. ed. 2. t. 11.
36 alpìnum. DC.	(*bl.*)	Alpine.	Austria.	1739.	——	H. ♃.	Swt. br. fl. gar. t. 17.
37 tenuifòlium. DC.	(*li.*)	slender-leaved.	Europe.	1759.	6. 7.	H. ♃.	Jacq. aust. 3. t. 215.
38 salsoloídes. DC.	(*fl.*)	Salt-wort-like.	S. France.	1823.	7. 8.	H. ♄.	
39 suffruticòsum. DC.	(*fl.*)	Spanish.	Spain.	1759.	8.	F. ♄.	Cav. ic. 2. t. 108.
40 cathárticum. DC.	(*wh.*)	purging.	Britain.	....	6. 8.	H. ⊙.	Eng. bot. 382.
RAD'IOLA. DC.		RAD'IOLA.	Tetrandria Tetragynia.				
linoídes. DC.	(*wh.*)	All-seed.	Britain.	....	7. 8.	H. ⊙.	Eng. bot. 893.
millegràna. E.B. *Lìnum Radìola.* L.							

ORDO XXVIII.

MALVACEÆ. *DC. prodr.* 1. *p.* 429.

MA'LOPE. DC.		MA'LOPE.	Monadelphia Polyandria. L.				
1 malacoídes. DC.	(*pu.*)	Barbary.	Barbary.	1710.	6. 7.	G. ♂.	Cav. diss. 2. t. 27. f. 1.
2 trífida. DC.	(*pu.*)	trifid.	——————	1808.	7. 8.	H. ⊙.	Swt. br. fl. gar. t. 153.

MA'LVA. DC. MALLOW. Monadelphia Polyandria. L.

1	tricúspidàta. DC.	(ye.)	Jamaica.	W. Indies.	1726.	7. 8.	S. ♂.	Cav. diss. 2. t. 22. f. 2.
	coromandeliàna. W. *americàna*. Cav.							
2	americàna. DC.	(ye.)	American.	——	1756.	6. 7.	H. ⊙.	
3	scàbra. DC.	(ye.)	rough-stemmed.	Peru.	1798.	——	G. ♄.	Cav. diss. 5. t. 138. f. 1.
	scopària. Jacq. ic. 139.							
4	scopària. DC.	(ye.)	Birch-leaved.	——	1782.	8. 9.	G. ♄.	L. her. stirp. t. 27.
5	borbònica. DC.	(ye.)	Bourbon.	Mauritius.	1816.	7. 8.	S. ♄.	
6	polystàchya. DC.	(ye.)	many-spiked.	Peru.	1798.	——	G. ♄.	Cav. dis. 5. t. 138. f. 3.
7	spicàta. DC.	(ye.)	simple-spiked.	Jamaica.	1726.	9. 10.	S. ♄.	—— 2. t. 20. f. 4.
8	ovàta. L. en.	(ye.)	oval-leaved.	Brazil.	....	——	S. ♄.	—— 2. t. 20. f. 2.
9	subhastàta. DC.	(ye.)	subhastate.	S. America.	1821.	7. 10.	S. ♄.	—— 2. t. 21. f. 3.
10	tomentòsa. DC.	(ye.)	tomentose.	India.	1820.	——	S. ♄.	Pluk. alm. t. 356. f. 1.?
11	waltherifòlia. DC.	(ye.)	Waltheria-ld.	Java.	1824.	....	S. ♃.	
12	trachelifòlia. DC.	(ye.)	Trachelium-ld.	....	——	7. 8.	H. ⊙.	
13	gangética. DC.	(ye.)	Indian.	E. Indies.	——	——	H. ⊙.	Pluk. t. 74. f. 6.
14	lepròsa. DC.	(wh.)	leprous.	Cuba.	1815.	5. 7.	S. ♃.	
15	Sherardiàna. DC.	(re.)	Sherard's.	Bithynia.	....	——	S. ♃.	Cav. dis. 2. t. 26. f. 4.
16	crética. DC.	(fl.)	Cretan.	Crete.	1828.	——	H. ⊙.	—— 5. t. 138. f. 2.
17	hispánica. DC.	(ro.)	Spanish.	Spain.	1710.	7.	H. ⊙.	Desf. atl. 2. t. 170.
18	stipulàcea. DC.	(pu.)	long-stipuled.	——	1815.	6. 8.	H. ⊙.	Cav. dis. 2. t. 15. f. 2.
19	ægy'ptia. DC.	(bh.)	Egyptian.	Egypt.	1739.	6. 7.	H. ⊙.	—— 2. t. 17. f. 1.
20	trífida. DC.	(bl.)	large-flowered.	Spain.	1815.	——	H. ⊙.	—— 5. t. 137. f. 2.
21	Tournefortiàna. DC.	(r.)	Tournefort's.	——	1759.	7. 8.	H. ⊙.	—— 2. t. 17. f. 3.
22	Alcèa. DC.	(pi.)	Vervain.	Europe.	1597.	7. 10.	H. ♃.	Bot. mag. 2297.
23	Morènii. S.S.	(pi.)	Moreni's.	Italy.	1824.	——	H. ♃.	—— 2793.
	alceoídes. Ten. fl. nap. t. 64.							
24	moschàta. DC.	(pi.)	musk.	Britain.	....	7. 8.	H. ♃.	Eng. bot. 754.
	β *laciniàta*. DC.	(wh.)	*white-flowered*.	....	....	——	H. ♃.	Bot. mag. 2298.
25	hirsùta. T.	(pi.)	hairy.	Sicily.	1825.	——	H. ⊙.	
26	fastigiàta. DC.	(re.)	fastigiate.	Spain.	1828.	6. 8.	H. ♂.	Cav. dis. 2. t. 23. f. 2.
27	mauritiàna. DC.	(pu.)	Ivy-leaved.	S. Europe.	1768.	6. 7.	H. ⊙.	Swt. br. fl. gar. 81.
28	excélsa. Fr.	(pu.)	tall.	Sicily.	1828.	——	H. ♃.	
29	sylvéstris. DC.	(pu.)	common.	Britain.	....	5. 10.	H. ♃.	Eng. bot. 671.
30	vulgàris. T.	(pu.)	prostrate.	Italy.	1827.	——	H. ♂.	
31	Henníngii. DC.	(pu.)	Henning's.	Moscow.	1823.	6. 8.	H. ♃.	
32	rotundifòlia. DC.	(wh.)	round-leaved.	Britain.	....	6. 9.	H. ⊙.	Eng. bot. 1092.
	β *pusílla*. E.B.	(wh.)	*dwarf*.	——	....	——	H. ⊙.	—— 241
33	brasiliénsis. DC.	....	Brazilian.	Brazil.	1820.	——	S. ♃.	
34	nicæénsis. DC.	(w.p.)	Italian.	Italy.	1823.	——	H. ⊙.	Cav. dis. 2. t. 25. f. 1.
35	microcárpa. DC.	(fl.)	small-fruited.	Egypt.	——	——	H. ♃.	
36	parviflòra. DC.	(ro.)	small-flowered.	Barbary.	1779.	6. 7.	H. ♃.	Cav. dis. 2. t. 26. f. 1.
37	verticillàta. DC.	(w.p.)	whorl-flowered.	China.	1683.	——	H. ⊙.	Jacq. vind. 1. t. 40.
38	mareótica. DC.	(fl.)	headed.	Egypt.	1828.	6. 10.	H. ⊙.	
39	críspa. DC.	(bh.)	curled.	Syria.	1573.	6. 8.	H. ⊙.	Cav. dis. 2. t. 23. f. 1.
40	flexuòsa. DC.	(re.)	flexuose.	....	1823.	——	H. ⊙.	
41	amœ'na. DC.	(pi.)	lobe-leaved.	C. B. S.	1796.	4. 8.	G. ♄.	Bot. mag. 1998.
	calycìna. B.R. 297.							
42	calycìna. DC.	(pi.)	large-calyxed.	——	1812.	——	G. ♄.	Cav. dis. 2. t. 23. f. 4.
43	virgàta. DC.	(fl.)	twiggy.	——	1727.	5. 7.	G. ♄.	—— 2. t. 18. f. 2.
44	capénsis. DC.	(w.r.)	Cape.	——	1713.	1. 12.	G. ♄.	Bot. reg. 295.
45	balsámica. DC.	(re.)	balsamic.	——	1800.	5. 9.	G. ♄.	Jacq. ic. 1. t. 140.
46	oxyacanthoides. DC.	(w.r.)	Hawthorn-ld.	——	1825.	——	G. ♄.	
47	tridactylìtes. DC.	(wh.)	reflex-flowered.	——	1791.	6. 8.	G. ♄.	Cav. dis. 2. t. 21. f. 2.
	refléxa. A.R. t. 135.							
48	divaricàta. DC.	(wh.)	straddling.	——	1792.	6. 9.	G. ♄.	And. bot. rep. 182.
49	retùsa. DC.	(y.bh.)	blunt-leaved.	——	1803.	3. 5.	G. ♄.	Cav. dis. 2. t. 21. f. 1.
50	fràgrans. DC.	(pu.)	fragrant.	——	1759.	5. 7.	G. ♄.	Bot. reg. 296.
51	strícta. DC.	(w.r.)	upright.	——	1805.	5. 8.	G. ♄.	Jacq. shœnb. 3. t. 294.
52	bryonifòlia. DC.	(re.)	Bryony-leaved.	——	1731.	7. 8.	G. ♄.	Wend. h. her. 1. t. 4.
53	grossulariæfòlia. DC.	(r.)	Gooseberry-ld.	——	1732.	5. 9.	G. ♄.	Bot. reg. 561.
54	aspérrima. DC.	(wh.)	roughest.	——	1796.	——	G. ♄.	Jacq. schœnb. 2. t. 139.
55	láctea. DC.	(wh.)	panicled white.	Mexico.	1780.	1. 2.	G. ♄.	Cav. ic. 1. t. 30.
	vitifòlia. Cav.							
56	capitàta. DC.	(pu.)	various-leaved.	Peru.	1798.	11. 12.	G. ♄.	Cav. dis. 5. t. 137. f. 1.
57	miniàta. DC.	(re.)	painted.	S. America.	——	4. 5.	G. ♄.	Cav. ic. 3. t. 278.
58	operculàta. DC.	(pu.)	lid capsuled.	Peru.	1795.	7. 8.	G. ♄.	Cav. dis. 2. t. 35. f. 1.
59	peruviàna. DC.	(pu.)	Peruvian.	——	1759.	6. 8.	H. ⊙.	Jacq. vind. 2. t. 156.
60	liménsis. DC.	(bl.)	blue-flowered.	——	1768.	——	H. ⊙.	—— 2. t. 140.

61 itálica. DC.	(*re.*)	Italian.	Italy.	1826.	6. 8.	H. ♃.	
62 anómala. L. en.	(*re.*)	anomalous.	C. B. S.	——	——	G. ♃.	Pl. sel. hort. ber. t. 22.
63 umbellàta. DC.	(*cr.*)	umbelled.	S. America.	1814.	1. 3.	G. ♄.	Lod. Bot. cab. 222.
64 ròsea. DC.	(*ro.*)	rose-coloured.	Mexico.	1828.	——	F. ♄.	Flor. mex. ic. ined.
65 abutilóides. DC.	(*pu.*)	Bahama.	Bahama Isl.	1725.	6. 9.	G. ♄.	Jacq. schœnb. 3. t. 293.
66 obtusilòba. B.M.	(*pu.*)	blunt-lobed.	Chili.	1824.	7. 8.	F. ♄.	Bot. mag. 2787.
67 élegans. DC.	(*y.re.*)	elegant.	C. B. S.	1791.	5. 8.	G. ♄.	Cav. dis. 2. t. 16. f. 1.
68 angustifòlia. DC.	(*li.*)	narrow-leaved.	Mexico.	1780.	8.	G. ♄.	Bot. mag. 2839.
69 caroliniàna. DC.	(*re.*)	Carolina.	Carolina.	1723.	6. 7.	H. ☉.	Cav. dis. 2. t. 15. f. 1.
70 pròstràta. DC.	(*sc.*)	trailing.	Brazil.	1806.	5. 7.	G. ♃.	Bot. mag. 2515.
71 decúmbens. DC.	(*re.*)	decumbent.	S. America.	1815.	6. 8.	G. ♃.	
KITAIB'ELIA. DC.		KITAIB'ELIA.	Monadelphia Polyandria. L.				
vitifòlia. DC.	(*wh.*)	Vine-leaved.	Hungary.	1801.	7. 9.	H. ♃.	Bot. mag. 821.
ALTH'ÆA. DC.		ALTH'ÆA.	Monadelphia Polyandria. L.				
1 officinàlis. DC.	(*fl.*)	Marsh-Mallow.	Britain.	....	7. 9.	H. ♃.	Eng. bot. 147.
2 taurinénsis. DC.	(*pu.*)	intermediate.	S. Europe.	1823.	——	H. ♃.	
3 narbonénsis. DC.	(*pu.*)	Narbonne.	————	1780.	8. 9.	H. ♃.	Jacq. ic. 1. t. 138.
4 cannabìna. DC.	(*ro.*)	Hemp-leaved.	————	1597.	6. 7.	H. ♃.	Jacq. aust. 2. t. 101.
5 hirsùta. DC.	(*bh.*)	hairy.	England.	1683.	——	H. ☉.	———— 2. t. 170.
6 Ludwígii. DC.	(*wh.*)	Ludwig's.	Sicily.	1791.	——	H. ☉.	Cav. ic. 5. t. 423.
7 leucántha. OT.	(*wh.*)	white-flowered.	Mt. Sinai.	1826.	6. 8.	H. ♂.	
8 nudiflòra. F.	(*ro.*)	naked-flowered.	Altay.	1824.	6. 10.	H. ♂.	
9 Froloviàna. F.	(*pu.*)	Siberian.	Siberia.	1826.	6. 8.	H. ♂.	
10 glabrifòlia. Led.	(*pu.*)	smooth-leaved.	————	1827.	——	H. ♂.	
11 acáulis. DC.	(*p.ye.*)	stemless.	Aleppo.	1680.	——	H. ☉.	Cav. dis. 2. t. 27. f. 3.
12 carib'æa. DC.	(*ro.*)	West Indian.	W. Indies.	1816.	3. 4.	S. ♂.	Bot. mag. 1916.
13 striàta. DC.	(*li.*)	streaked.	....	1820.	6. 8.	H. ♂.	
14 pállida. DC.	(*li.*)	pale-flowered.	Hungary.	1805.	——	H. ♂.	
15 ròsea. DC.	(*va.*)	Hollyhock.	China.	1573.	7. 9.	H. ♂.	Cav. dis. 2. t. 28. f. 1.
β *bilòba.* DC.	(*va.*)	*two-lobed.*	....	....	——	H. ♂.	
16 sinénsis. DC.	(*v.*)	China.	————	....	——	H. ☉.	Cav. dis. 2. t. 29. f. 3.
17 flexuòsa. DC.	(*pu.*)	flexuose.	E. Indies.	1803.	6. 8.	S. ♃.	Bot. mag. 892.
18 ficifòlia. DC.	(*ye.*)	Fig leaved.	Levant.	1597.	6. 9.	H. ♂.	Cav. dis. 2. t. 28. f. 2.
LAVAT'ERA. DC.		LAVAT'ERA.	Monadelphia Polyandria. L.				
1 triméstris. DC.	(*li.w.*)	common annual.	S. Europe.	1633.	7. 9.	H. ☉.	Bot. mag. 109.
2 phœnícea. DC.	(*cr.*)	scarlet.	Madeira.	1823.	6. 9.	G. ♄.	Vent. malm. t. 120.
3 acerifòlia. DC.	(*li.*)	Maple-leaved.	Teneriffe.	——	——	G. ♄.	Lois. h. amat. t. 322.
4 híspida. DC.	(*ro.*)	hispid.	Algiers.	1804.	6. 7.	G. ♄.	Desf. atl. 2. t. 171.
5 O'lbia. DC.	(*pu.*)	downy-leaved.	France.	1570.	6. 10.	G. ♄.	Cav. dis. 2. t. 32. f. 2.
6 unguiculàta. DC.	(*pu.*)	clawed.	Greece.	1807.	7. 9.	G. ♄.	
7 mìcans. DC.	(*pu.*)	glittering.	Spain.	1796.	6. 7.	G. ♄.	Moris. ox. t. 17. f. 9.
8 lusitànica. DC.	(*pu.*)	Portugal.	Portugal.	1731.	8. 9.	G. ♄.	
9 flàva. DC.	(*ye.*)	yellow.	Sicily.	1823.	6. 8.	H. ♃.	Desf. atl. 2. t. 172.
10 plebéia. B.M.	(*li.*)	vulgar.	N. S. W.	1820.	——	G. ♃.	Bot. mag. 2269.
Weinmanniàna. Bes.							
11 thuringìaca. DC.	(*pu.*)	large-flowered.	Germany.	1731.	7. 9.	H. ♃.	Bot. mag. 517.
12 biénnis. DC.	(*pu.*)	biennial.	Caucasus.	1823.	5. 8.	H. ♂.	
13 punctàta. DC.	(*bh.*)	spotted-stalked.	Italy.	1800.	7. 9.	H. ☉.	
14 marítima. DC.	(*wh.*)	sea side.	S. Europe.	1597.	4. 6.	G. ♄.	Cav. dis 2. t. 32. f. 3.
15 trilòba. DC.	(*li.*)	three lobed.	Spain.	1759.	6. 7.	G. ♄.	Bot. mag. 2226.
16 arbòrea. DC.	(*li.*)	Tree Mallow.	Britain.	....	7. 10.	H. ♂.	Eng. bot. 1841.
17 neapolitàna. DC.	(*pu.*)	Neapolitan.	Naples.	1823.	——	H. ♃.	
18 crética. DC.	(*pu.*)	Cretan.	Crete.	1723.	7. 9.	H. ☉.	Cav. dis. 2. t. 32. f. 1.
19 lanceolàta. W.en.	(*pu.*)	spear-leaved.	....	1823.	——	H. ☉.	
20 sylvéstris. DC.	(*pu.*)	wood.	Portugal.	——	——	H. ☉.	
21 ambígua. DC.	(*pu.*)	ambiguous.	Naples.	1824.	——	H. ☉.	
22 móllis. Ehr.	(*pu.*)	soft.	Egypt.	1825.	7. 10.	H. ☉.	
MALA'CHRA. DC.		MALA'CHRA.	Monadelphia Polyandria. L.				
1 rotundifòlia. DC.	(*ro.*)	round-leaved.	Brazil.	1823.	6. 9.	S. ☉.	Schrank. h. mon. t. 56
2 capitàta. DC.	(*ye.*)	headed.	W. Indies.	1759.	8. 9.	S. ☉.	Cav. dis. 2. t. 33. f. 1. 2.
3 fasciàta. DC.	(*ro.*)	fasciate.	Caracas.	1820.	——	S. ☉.	Jacq. ic. 3. t. 548.
4 trilòba. DC.	(*wh.*)	three-lobed.	....	——	——	S. ☉.	
5 radiàta. DC.	(*pu.*)	rayed.	St. Domingo.	1794.	7. 9.	S. ♂.	Cav. dis. 2. t. 33. f. 3.
6 alceæfòlia. DC.	(*ye.*)	Hollyhock-ld.	Caracas.	1805.	8. 9.	S. ☉.	Jacq. ic. 3. t. 549.
7 heptaphy'lla. DC.	(*ro.*)	many-flower'd.	Brazil.	1820.	6. 10.	S. ♃.	
fasciàta. B. reg. 467.		*non* DC.					
8 palmàta. DC.	(*ye.*)	palmate.	....	1825.	——	S. ☉.	
UR'ENA. DC.		UR'ENA.	Monadelphia Polyandria. L.				
1 lobàta. DC.	(*ro.*)	angular-leav'd.	China.	1731.	6. 7.	S. ♄.	Dill. elt. t. 319. f. 412.

2 scabriúscula. DC.	(*ro.*)	roughish.	India.	1820.	6. 7.	S. ♄.	Cav. dis. 6. t. 185. f. 1.
3 repánda. DC.	(*ro.*)	netted-leaved.	———	——	——	S. ♄.	
4 multífida. DC.	(*ye.*)	multifid.	Mauritius.	1817.	10. 1.	S. ♄.	Cav. dis. 6. t. 184. f. 2.
5 tricúspis. DC.	(*ye.*)	three-pointed.	———	1820.	6. 9.	S. ♄.	——— 6. t. 183. f. 1.
6 reticulàta. DC.	(*ye.*)	netted.	St. Domingo.	1816.	——	S. ♄.	——— 6. t. 183. f. 2.
americàna. L. *non* Sm.							
7 subtrilòba. DC.	(*re.*)	various-leaved.	Brazil.	1823.	——	S. ☉.	Schrank. h. mon. t. 79.
8 Swártzii. DC.	(*re.*)	Swartz's.	Surinam.	....	——	S. ♄.	
sinuàta. Sw. *americàna.* Sm.							
9 sinuàta. DC.	(*re.*)	cut-leaved.	E. Indies.	1759.	7. 8.	S. ♄.	Cav. dis. 6. t. 185. f. 2.
10 muricàta. DC.	(*re.*)	muricate-fruit'd	———	1823.	——	S. ♄.	
PAV'ONIA. DC.		PAV'ONIA. Monadelphia Polyandria. L.					
1 spìnifex. DC.	(*ye.*)	prickly-seeded.	W. Indies.	1778.	7. 8.	S. ♄.	Bot. reg. 339.
2 Typhàlea. DC.	(*wh.*)	globular-headed.	———	1823.	——	S. ♄.	Cav. dis. 2. t. 197.
3 ùrens. DC.	(*ro.*)	stinging.	Mauritius.	1801.	——	S. ♄.	Jacq. ic. 3. t. 522.
4 coccìnea. DC.	(*sc.*)	scarlet.	St. Domingo.	1816.	——	S. ♄.	Cav. dis. 3. t. 47. f. 1.
5 Columélla. DC.	(*pu.*)	angular-leaved.	Bourbon.	1807.	——	S. ♄.	——— 3. t. 48. f. 3.
6 parviflòra. DC.	(*st.*)	small-flowered.	....	1820.	——	S. ♄.	
7 racemòsa. DC.	(*ye.*)	naked-racemed.	Jamaica.	——	——	S. ♄.	Cav. dis. 3. t. 46. f. 1.
spicàta. Cav.							
8 præmórsa. DC.	(*ye.*)	bitten-leaved.	C. B. S.	1774.	6. 8.	G. ♄.	Bot. mag. 436.
9 paniculàta. DC.	(*ye.*)	panicled.	Caracas.	1823.	——	S. ♄.	Cav. dis. 3. t. 46. f. 2.
10 odoràta. DC.	(*re.*)	fragrant.	E. Indies.	1807.	——	S. ♄.	
11 sidoídes. DC.	(*re.*)	Sida like.	———	1820.	——	S. ♂.	
12 zeylánica. DC.	(*bh.*)	Ceylon.	———	1790.	7. 9.	S. ☉.	Burm. ind. t. 48. f. 2.
13 cancellàta. DC.	(*re.*)	arrow-leaved.	Surinam.	1823.	——	S. ☉.	Lin. sup. t. 311.
MALVAVI'SCUS. DC.		ACH'ANIA. Monadelphia Polyandria. L.					
1 arbòreus. DC.	(*sc.*)	smooth-leaved.	Jamaica.	1714.	1. 12.	S. ♄.	Cav. dis. 3. t. 48. f. 1.
Achània malvavíscus. W.							
2 móllis. DC.	(*sc.*)	soft-leaved.	Mexico.	1780.	——	S. ♄.	Bot. reg. 11.
Achània móllis. B.R.							
3 pilòsa. DC.	(*sc.*)	hairy.	Jamaica.	1780.	10. 11.	S. ♄.	Lodd. bot. cab. 829.
LEBRET'ONIA. DC.		LEBRET'ONIA. Monadelphia Polyandria. L.					
coccìnea. DC.	(*sc.*)	scarlet.	Brazil.	1823.	6. 8.	S. ♄.	Schrank. h. mon. t. 90.
HIBI'SCUS. DC.		HIBI'SCUS. Monadelphia Polyandria. L.					
1 liliiflòrus. DC.	(*va.*)	Lily-flowered.	Bourbon.	1822.	8. 12.	S. ♄.	Cav. dis. 3. t. 57. f. 1.
2 Boryànus. DC.	(*wh.*)	Bory St. Vincent's.	———	1826.	6. 10.	S. ♄.	
3 pedunculàtus. DC.	(*pi.*)	long-peduncled.	C. B. S.	1812.	5. 12.	S. ♄.	Bot. reg. 231.
4 Lámpas. DC.	(*st.*)	three-pointed.	E. Indies.	1806.	10. 2.	S. ♄.	Cav. dis. 3. t. 56. f. 2.
5 membranàceus. DC.	(*y.*)	leafy-calyxed.	....	1816.	....	S. ♄.	——— 3. t. 57. f. 2.
6 spiràlis. DC.	(*y.fl.*)	spiral.	Mexico.	1826.	6. 9.	G. ♄.	Cavan. ic. 2. t. 162.
7 tubulòsus. DC.	(*y.p.*)	tubular.	E. Indies.	1796.	7. 10.	S. ♂.	Cav. dis. 3. t. 68. f. 2.
8 pentacárpos. DC.	(*bh.*)	angular-fruited.	Venice.	1752.	——	G. ♃.	Jacq. ic. 1. t. 143.
9 virgínicus. DC.	(*ro.*)	Virginian.	Virginia.	1798.	——	H. ♃.	——— 1. t. 142.
10 pentaspérmus. DC.	(*ye.*)	five-seeded.	Jamaica.	1825.	——	S. ♄.	
11 Mánihot. DC.	(*y.p.*)	palmated.	E. Indies.	1712.	7. 9.	S. ♄.	Bot. mag. 1702.
12 palmàtus. C.	(*ye.p.*)	palmate-leav'd.	S. America.	....	——	S. ♄.	Cav. dis. 3. t. 63. f. 1.
13 ficúlneus. DC.	(*wh.*)	Fig-leaved.	Ceylon.	1732.	6. 7.	S. ♄.	——— 3. t. 52. f. 2.
sinuàtus. C.							
14 ficulneoídes. B.R.	(*st.*)	Fig-like.	S. America.	1824.	6. 8.	S. ♄.	Bot. reg. 938.
15 tetraphy'llus. DC.	(*ye.*)	four-leaved.	E. Indies.	1807.	7. 10.	S. ☉.	
16 platanifòlius.	(*ye.*)	Plane-leaved.	W. Indies.	1820.	——	S. ♄.	Pl. sel. h. ber. t. 1.
acerifòlius. DC. *non* P.L. *Pavònia platanifòlia.* L. en.							
17 Ròsa-sinénsis. DC.	(*sc.*)	China Rose.	E. Indies.	1731.	1. 12.	S. ♄.	Bot. mag. 158.
β rùbro-plènus.	(*sc.*)	*double-red.*	———	——	——	S. ♄.	Lod. bot. cab. 995.
γ cárneo-plènus.	(*sa.*)	*double-salmon.*	———	——	——	S. ♄.	
δ variegàtus.	(*sc.w.*)	*double-striped.*	———	——	——	S. ♄.	Lod. bot. cab. 963.
ε flávo-plènus.	(*co.*)	*double-buff.*	———	——	——	S. ♄.	——— 513.
ζ lùteo-plènus.	(*ye.*)	*double-yellow.*	———	1822.	——	S. ♄.	——— 932.
18 acerifòlius. P.L.	(*va.*)	Maple-leaved.	China.	1798.	3. 6.	G. ♄.	Sal. par. lond. 22.
19 syrìacus. DC.	(*va.*)	Althæa frutex.	Syria.	1596.	8. 9.	H. ♄.	Bot. mag. 83.
α purpùreus.	(*pu.*)	*purple-flowered.*	———	——	——	H. ♄.	
β rùber.	(*re.*)	*red-flowered.*	———	——	——	H. ♄.	
γ álbus.	(*wh.*)	*white-flowered.*	———	——	——	H. ♄.	
δ variegàtus.	(*w.p.*)	*striped-flowered.*	———	——	——	H. ♄.	
ε marginàtus.	(*pu.*)	*variegated.*	———	——	——	H. ♄.	
ζ álbo-plènus.	(*wh.*)	*double-white.*	———	——	——	H. ♄.	
η purpùreo-plènus.	(*p.*)	*double-purple.*	———	——	——	H. ♄.	
20 prùriens. DC.	(*y.pu.*)	stinging.	E. Indies.	1804.	7. 9.	S. ♂.	Andr. reposit. 498.

21	suratténsis. DC.	(*y.p.*)	prickly-stalk'd.	E. Indies.	1731.	7. 9.	S. ⊙.	Bot. mag. 1356.
22	radiàtus. DC.	(*st.p.*)	rayed.	———	1790.	6. 8.	S. ⊙.	——— 1911.
23	furcàtus. DC.	(*ye.p.*)	forked.	———	1816.	7. 9.	S. ⊙.	
24	unidéns. B.R.	(*y.p.*)	one-toothed.	Brazil.	1824.	11. 4.	S. ♄.	Bot. reg. 878.
25	scàber. DC.	(*st.*)	scabrous.	Carolina.	1810.	7. 9.	G.*w.*♃.	
26	strigòsus. B.R.	(*ro.*)	strigose.	S. America.	1822.	1. 12.	S. ♄.	Bot. reg. 860.
27	bifùrcatus. DC.	(*pu.*)	two-forked.	Brazil.	1826.	——	S. ♄.	Cav. dis. 3. t. 51. f. 1.
28	trilòbus. DC.	(*co.*)	three-lobed.	W. Indies.	1824.	——	S. ♄.	——— 3. t. 53. f. 2.
29	diversifòlius. DC.	(*y.p.*)	different-leav'd.	E. Indies.	1798.	6. 7.	S. ♄.	Jacq. ic. 3. t. 551.
30	heterophy'llus. DC.	(*w.p.*)	various-leaved.	N. S. W.	1808.	8. 9.	G. ♄.	Bot. reg. 29.
31	cannabìnus. DC.	(*y.p.*)	Hemp-leaved.	E. Indies.	1759.	6. 7.	S. ♂.	Roxb. cor. 2. t. 190.
32	vitifòlius. DC.	(*y.p.*)	Vine-leaved.	———	1690.	7. 10.	S. ♂.	Rheed. mal. 6. t. 46.
33	obtusifòlius. DC.	(*y.p.*)	blunt-leaved.	———	1820.	——	S. ♂.	
34	Bámmia. L.en.	(*y.p.*)	Egyptian.	Egypt.	....	6. 7.	S. ⊙.	Tozz. mus. d. fir. 2. t. 7.
35	esculéntus. L.en.	(*y.p.*)	eatable.	India.	1692.	——	S. ⊙.	
36	longifòlius. DC.	(*ye.*)	long-leaved.	E. Indies.	1820.	7. 9.	S. ♂.	
37	moscheùtos. DC.	(*w.p.*)	swamp.	N. America.	....	8. 10.	H.*w.*♃.	Swt. br. fl. gar. 286.
38	palústris. DC.	(*pu.*)	marsh.	———	1759.	7. 9.	H.*w.*♃.	Cav. dis. 3. t. 65. f. 2.
39	ròseus. DC.	(*ro.*)	rose-colour'd.	S. Europe.	1824.	8. 10.	H.*w.*♃.	Swt. br. fl. gar. 277.
40	aquáticus. DC.	(*wh.*)	water.	Etruria.	1820.	7. 9.	H.*w.*♃.	
41	incànus. DC.	(*st.*)	hoary.	Carolina.	1806.	9.	G.*w.*♃.	Wend. h. her. 4. t. 24.
42	militàris. DC.	(*pu.*)	smooth.	Louisiana.	1804.	8. 9.	G.*w.*♃.	Bot. mag. 2385.
43	speciòsus. DC.	(*cr.*)	superb.	Carolina.	1778.	6. 9.	G.*w.*♃.	——— 360.
44	grandiflòrus. DC.	(*fl.*)	great-flower'd.	Georgia.	1823.	——	G.*w.*♃.	
45	ferrugíneus. DC.	(*re.*)	rusty	Madagascar.	1820.	——	S. ♄.	Cav. dis. 3. t. 60. f. 1.
46	lunarifòlius. DC.	(*st.p.*)	Lunaria-leav'd.	E. Indies.	....	——	S. ♄.	
47	æthiòpicus. DC.	(*ye.*)	dwf. wedge-ld.	C. B. S.	1774.	8.	G. ♃.	Cav. dis. 3. t. 61. f. 1.
48	mutàbilis. DC.	(*va.*)	changeable.	E. Indies.	1690.	10. 12.	S. ♄.	Bot. reg. 589.
	β *flòre-plèno.*	(*va.*)	*double-flowered.*	———	....	——	S. ♄.	
49	clypeàtus. DC.	(*r.y.*)	shield-capsuled.	Jamaica.	1759.	7. 8.	S. ♄.	Cav. dis. 3. t. 58. f. 1.
50	Abelmóschus. DC.	(*ye.*)	Musk Okro.	India.	1640.	7. 9.	S. ♄.	——— 3. t. 62. f. 2.
51	eriocárpus. DC.	(*ye.*)	woolly-capsul'd.	E. Indies.	1820.	——	S. ♄.	
52	phœníceus. B.R.	(*cr.*)	crimson-flowd.	S. America.	1796.	——	S. ♄.	Bot. reg. 230.
53	hírtus. L.	(*sc.*)	hairy.	E. Indies.	....	2. 10.	S. ♄.	Cav. dis. 3. t. 67. f. 3.
54	Ròsa malabárica.	(*sc.*)	Malabar Rose.	———	1818.	8. 10.	S. ♄.	Bot. reg. 337.
55	rhombifòlius. DC.	(*wh.*)	rhomb-leaved.	———	1823.	——	S. ♄.	Cav. dis. 3. t. 69. f. 3.
56	gossypìnus. DC.	(*re.*)	cottony.	C. B. S.	1818.	——	G. ♄.	
57	micránthus. DC.	(*ye.*)	small-flowered.	E. Indies.	1794.	——	S. ♄.	Cav. dis. 3. t. 66. f. 1.
58	Triònum. DC.	(*st.p.*)	BladderKetmia.	Italy.	1596.	6. 9.	H. ⊙.	Bot. mag. 209.
59	hispidus. B.R.	(*y.p.*)	hispid.	C. B. S.	....	——	G. ♄.	Bot. reg. 806.
60	africànus. R.B.	(*st.p.*)	African.	Africa.	1826.	6. 10.	H. ⊙.	
61	Richardsòni.	(*st.*)	Richardson's.	N. S. W.	1822.	——	G. ♄.	Bot. reg. 875.
62	vesicàrius. DC.	(*st.p.*)	African.	Africa.	1713.	7. 8.	H. ⊙.	Cav. dis. 3. t. 62. f. 2.
63	Sabdaríffa. DC.	(*y.ro.*)	Indian.	E. Indies.	1596.	6. 9.	S. ♂.	Bonp. nav. t. 29.
64	digitàtus. DC.	(*y.r.*)	fingered.	Brazil.	1818.	——	S. ⊙.	Cav. dis. 3. t. 70. f. 2.
	β *Kerriànus.* DC.	(*w.r.*)	*Ker's.*	Rio Janeiro.	——	——	S. ⊙.	Bot. reg. 608.
65	tricúspis. DC.	(*ye.*)	sharp-pointed.	Society Isl.	1820.	——	S. ♄.	Cav. dis. 3. t. 55. f. 2.
66	racemòsus. B.R.	(*y.p.*)	racemed.	Nepaul.	1824.	8. 2.	G. ♄.	Bot. reg. 917.
67	tiliàceus. DC.	(*ye.p.*)	Lime-tree-ld.	E. Indies.	1739.	7. 8.	S. ♄.	——— 232.
68	elàtus. DC.	(*y.p.*)	tall.	Jamaica.	1790.	——	S. ♄.	
69	abutiloídes. DC.	(*y.p.*)	smooth-leaved.	S. America.	1820.	....	S. ♄.	Sloan. jam. 1. t. 234. f. 4.
70	flavéscens.	(*ye.*)	flavescent.	Brazil.	1822.	6. 9.	S. ♄.	Cav. dis. 3. t. 70. f. 3.
71	borbònicus. DC.	(*ye.*)	Bourbon.	I. Bourbon.	1826.	——	S. ♄.	
72	Macleànus. BF.	(*ye.*)	Macleay's.	W. Indies.	1827.	8. 9.	S. ♄.	
73	macrophy'llus. DC.	(*ye.*)	large-leaved.	E. Indies.	1820.	....	S. ♄.	
74	collìnus. H.B.	(*ye.*)	hill.	———	1822.	....	S. ♄.	
75	cancellàtus. H.B.	(*ye.*)	netted.	Nepaul.	1823.	10. 2.	S. ♂.	
76	truncàtus. H.B.	(*ye.*)	truncate.	E. Indies.	——	8. 10.	S. ⊙.	
77	scándens. H.B.	(*ye.*)	climbing.	———	1822.	....	S. ♄.	
78	stríctus. H.B.	(*ye.*)	upright.	———	1809.	8. 10.	S. ⊙.	
79	gangéticus. L.en.	(*ye.*)	dwarf.	———	1804.	5. 10.	S. ♄.	
80	setòsus. H.B.	(*ye.*)	bristly.	———	1820.	....	S. ♄.	
81	tortuòsus. H.B.	(*ye.*)	twisted.	———	1808.	6. 9.	S. ♄.	
82	pentaphy'llus. H.B.	(*y.*)	five-leaved.	China.	1825.	7. 9.	S. ♂.	
83	púngens. H.B.	(*ye.*)	pungent.	Nepaul.	——	....	G. ♃.	
84	fràgrans. H.B.	(*ye.*)	fragrant.	E. Indies.	1826.	....	S. ♄.	
85	Patersònii. DC.	(*fl.*)	Norfolk Island.	Norfolk Isl.	1792.	6. 8.	G. ♄.	Andr. rep. 286.

Lagun'æa Patersònia. Bot. mag. 769. *squàmea.* Vent. malm. t. 42.

THESP'ESIA. DC.		THESP'ESIA. Monadelphia Polyandria. L.					
popúlnea. DC.	(*y.p.*)	Poplar-leaved.	E. Indies.	1770.	....	S. ♄.	Cav. dis. 3. t. 56. f. 1.
Hibíscus popúlneus. W.							
GOSSY'PIUM. DC.		COTTON-TREE. Monadelphia Polyandria. L.					
1 herbàceum. DC.	(*y.p.*)	common.	E. Indies.	1594.	7.	S. ♂.	Roxb. corom. t. 269.
2 índicum. DC.	(*y.p.*)	Indian.	———	....	——	S. ♂.	Cav. dis. 6. t. 169.
3 arbòreum. DC.	(*y.p.*)	tree.	———	1694.	7. 8.	S. ♄.	——— 6. t. 195.
4 vitifòlium. DC.	(*y.p.*)	Vine-leaved.	———	1805.	——	S. ♄.	——— 6. t. 166.
5 hirsùtum. DC.	(*ye.*)	hairy.	S. America.	1731.	——	S. ♂.	——— 6. t. 167.
6 eglandulòsum. DC.	(*ye.*)	glandless.	....	....	——	S. ♂.	
7 religiòsum. DC.	(*bh.*)	spotted-bark'd.	E. Indies.	1777.	——	S. ♄.	Cav. dis. 6. t. 164. f. 1.
8 latifòlium. DC.	(*ye.p.*)	broad-leaved.	....	....	——	S. ♄.	Murr. c. gœt. p. 32. t. 1.
9 barbadénse. DC.	(*y.p.*)	Barbadoes.	Barbadoes.	1759.	8. 10.	S. ♂.	Bot. reg. 84.
10 obtusifòlium. H.B.	(*y.p.*)	blunt-leaved.	E. Indies.	1822.	7. 8.	S. ♄.	
REDO'UTEA. DC.		REDO'UTEA. Monadelphia Polyandria. L.					
heterophy'lla. DC.	(*st.p.*)	various-leaved.	St. Thomas.	1824.	7. 9.	S. ⊙.	Vent. cels. t. 11.
LOPI'MIA. DC.		LOPI'MIA. Monadelphia Polyandria. L.					
malacophy'lla. DC.	(*sc.*)	Mallow-leaved.	Brazil.	1824.	....	S. ♄.	Pl. sel. h. ber. t. 30.
Sìda malacophy'lla. L. en.							
PAL'AVIA. DC.		PAL'AVIA. Monadelphia Polyandria. L.					
1 malvæfòlia. DC.	(*pu.*)	Mallow-leaved.	Lima.	1794.	6. 8.	H. ⊙.	Cav. dis. 1. t. 11. f. 4.
2 moschàta. DC.	(*pu.*)	musk-scented.	———	1822.	——	H. ⊙.	——— 1. t. 11. f. 5.
CRIST'ARIA. Ph.		CRIST'ARIA. Monadelphia Polyandria. L.					
? coccínea. Ph.	(*sc.*)	scarlet.	Missouri.	1811.	7. 9.	H. ♃.	Bot. mag. 1673.
Málva coccínea. Nut. *Sìda? coccínea.* DC.							
AN'ODA. DC.		AN'ODA. Monadelphia Polyandria. L.					
1 hastàta. DC.	(*bl.wh.*)	halberd-leav'd.	Mexico.	1799.	6. 7.	S. ⊙.	Bot. rep. 588.
2 trilòba. DC.	(*pu.*)	three-lobed.	———	1720.	7. 9.	G. ⊙.	Cav. dis. 1. t. 10. f. 3.
Sìda cristàta. W. *non* B.M.							
3 Dilleniàna. DC.	(*ro.*)	Dillenius's.	———	1725.	6. 11.	G. ⊙.	Cav. dis. 1. t. 11. f. 1.
Sìda cristàta. B.M. 330. *non* W.							
4 triangulàris. DC.	(*pu.*)	triangular-ld.	———	1824.	7. 9.	G. ⊙.	
5 incarnàta. DC.	(*fl.*)	flesh-color'd.	———	1825.	——	G. ♂.	
6 acerifòlia. DC.	(*bl.*)	Maple-leaved.	———	1809.	1. 12.	S. ♂.	Bot. mag. 1541.
Sìda hastàta. B.M. *non* W. *centrota.* S.S.							
7 parviflòra. DC.	(*st.*)	small-flowered.	New Spain.	1823.	6. 10.	G. ⊙.	Cav. ic. 5. t. 431.
PERI'PTERA. DC.		PERI'PTERA. Monadelphia Polyandria. L.					
punícea. DC.	(*cr.*)	crimson.	New Spain.	1813.	6. 9.	S. ♄.	Bot. mag. 1644.
Sìda períptera. B.M.		*Anòda punícea.* Lag.					
NUTTA'LLIA. B.A.		NUTTA'LLIA. Monadelphia Polyandria.					
1 digitàta. B.A.	(*d.p.*)	digitate-leaved.	N. America.	1824.	7. 10.	H. ♃.	Swt. br. fl. gar. 129.
2 pedàta. B.A.	(*d.pu.*)	pedate-leaved.	———	——	——	H. ♃.	Hook. ex. flor. 172.
S'IDA. Kth.		S'IDA. Monadelphia Polyandria. L.					
1 linifòlia. DC.	(*wh.*)	Flax-leaved.	S. America.	1822.	6. 9.	S. ♄.	Cav. dis. 1. t. 2. f. 1.
2 angustifòlia. DC.	(*ye.*)	narrow-leaved.	Bourbon.	1726.	7. 9.	S. ♄.	——— 1. t. 2. f. 2.
3 lineàris. DC.	(*ye.*)	linear-leaved.	New Spain.	1820.	6. 10.	G. ♄.	Cav. ic. 4. t. 312. f. 1.
4 vimínea. L. en.	(*ye.*)	slender.	Brazil.	1823.	——	S. ♄.	
5 spinòsa. DC.	(*ye.*)	prickly.	E. Indies.	1680.	——	S. ♂.	Cav. dis. 1. t. 1. f. 9.
6 acùta. DC.	(*ye.*)	acute-leaved.	———	1827.	——	S. ♄.	——— 1. t. 2. f. 3.
7 álba. DC.	(*wh.*)	white-flowered.	———	1732.	6. 7.	S. ⊙.	Dill. elt. 2. t. 171. f. 210.
8 stipulàta. DC.	(*ye.*)	large-stipuled.	Mauritius.	1820.	....	S. ♄.	Cav. dis. 1. t. 3. f. 10.
9 jamaicénsis. DC.	(*ye.*)	Jamaica.	Jamaica.	1818.	6. 9.	S. ♄.	——— 1. t. 2. f. 5.
10 bracteolàta. DC.	(*ye.*)	bracteolate.	S. America.	——	——	S. ♄.	
11 carpinifòlia. DC.	(*ye.*)	Hornbeam-ld.	Canaries.	1774.	7. 9.	G. ♄.	Cav. dis. 5. t. 134. f. 1.
12 carpinoídes. DC.	(*ye.*)	Hornbeam-like.	....	....	——	S. ♄.	
13 semicrenàta. L. en.	(*y.*)	semicrenate.	Manila.	1826.	——	S. ♃.	
14 acrántha. L. en.	(*ye.*)	close-flower'd.	Brazil.	——	——	S. ♄.	
15 spiræifòlia. L. en.	(*ye.*)	Spiræa-leaved.	....	....	——	S. ♄.	
16 Schránkii. DC.	(*ye.*)	Schrank's.	Brazil.	1824.	——	S. ♃.	
brasíla. L. en.							
17 betulìna. H.	(*ye.*)	Birch-leaved.	S. America.	1823.	——	S. ♄.	
18 eròsa. DC.	(*ye.*)	jagged.	Brazil.	——	——	S. ♃.	
19 ciliàris. DC.	(*cr.*)	ciliated.	St. Domingo.	1798.	——	S. ♄.	Cav. dis. 1. t. 3. f. 9.
20 alnifòlia. DC.	(*ye.*)	Alder-leaved.	E. Indies.	1732.	——	S. ♂.	Dill. elt. t. 172. f. 211.
21 sessiliflòra. B.M.	(*ye.*)	sessile-flower'd.	S. America.	1826.	11. 2.	S. ♄.	Bot. mag. 2857.
22 frutéscens. DC.	(*ye.*)	frutescent.	Mauritius.	1824.	6. 9.	S. ♄.	Cav. dis. 1. t. 10. f. 1.
23 bícolor. DC.	(*r.y.*)	two-colour'd.	New Spain.	1826.	——	S. ♄.	Cav. ic. 4. t. 311.
24 canariénsis. DC.	(*wh.*)	Canary Tea.	Canaries.	....	——	S. ♄.	Cav. dis. 1. t. 3. f. 8.
álba. Cav. *non* L.							

25 rhombifòlia. DC.	(*ye.*)	rhomboid-ld.	Brazil.	1732.	6. 8.	S. ♄.	Cav. dis. 1. t. 3. f. 12.
26 rhombóidea. DC.	(*ye.*)	rhomboid.	E. Indies.	1818.	——	S. ♄.	
27 retùsa. DC.	(*ye.*)	retuse-leaved.	————	——	——	S. ♄.	Cav. dis. 1. t. 3. f. 4.
28 recìsa. DC.	(*ye.*)	notched.	Brazil.	1823.	——	S. ♃.	
29 calyxhymènia. DC.	(*st.*)	New Holland.	N. Holland.	1820.	——	G. ♄.	
30 globiflòra. B.M.	(*st.*)	globe-flowered.	Mauritius.	1825.	10. 12.	S. ♄.	Bot. mag. 2821.
31 maculàta. DC.	(*ye.r.*)	corky.	St. Domingo.	1798.	....	S. ♄.	Cav. dis. 1. t. 3. f. 7.
suberòsa. L. her. stirp. t. 54.							
32 acuminàta. DC.	(*ye.*)	acuminate.	St. Domingo.	1805.	7. 9.	S. ♄.	
33 pilòsa. DC.	(*ye.*)	pilose.	W. Indies.	1793.	——	S. ♂.	Cav. dis. 1. p. 9. t. 8.
34 hùmilis. DC.	(*ye.*)	dwarf.	E. Indies.	1800.	7. 8.	S. ♂.	Cav. dis. 5. t. 134. f. 2.
uniloculàris. L'Herit. stirp. 1. t. 53. bis.							
35 rotundifòlia. DC.	(*pu.*)	round-leaved.	Bourbon.	1826.	——	S. ⊙.	Cav. dis. 1. t. 3. f. 6.
36 cordifòlia. DC.	(*ye.*)	heart-leaved.	E. Indies.	1732.	6. 9.	S. ⊙.	———— 1. t. 3. f. 2.
37 multiflòra. DC.	(*ye.*)	many-flowered.	Brazil.	1825.	——	S. ♄.	———— 1. t. 3. f. 3.
38 argùta. DC.	(*ye.*)	sharp-leaved.	W. Indies.	1732.	7. 8.	S. ♄.	
39 althæifòlia. DC.	(*co.*)	Althæa-leaved.	————	1818.	——	S. ♄.	Sloan. his. 1. t. 136. f. 2.
40 ùrens. DC.	(*ye.*)	stinging.	————	1781.	——	S. ♄.	Cav. dis. 1. t. 2. f. 7.
41 verticillàta. DC.	(*ye.*)	whorl-flowered.	Brazil.	1823.	——	S. ♄.	———— 1. t. 1. f. 12.
42 paniculàta. DC.	(*d.pu.*)	dark purple.	S. America.	1795.	7. 9.	S. ♄.	———— 1. t. 12. f. 5.
atrosanguínea. Jacq. ic. 1. t. 136.							
43 fruticòsa. DC.	(*wh.*)	shrubby.	Jamaica.	....	6. 10.	S. ♄.	
44 villòsa. DC.	(*ye.*)	villous.	S. America.	....	——	S. ♄.	
45 verruculàta. DC.	(*ye.*)	wart-stem'd.	Brazil.	1827.	5. 9.	S. ♄.	
argùta. F. in L. en. *non* Swartz.							
46 gravèolens. DC.	(*ye.*)	strong-scented.	E. Indies.	1820.	6. 9.	S. ♃.	
47 purpuráscens. DC.	(*pu.*)	purplish.	Brazil.	——	7. 9.	S. ♄.	
48 pàtens. DC.	(*pu.*)	spreading.	Abyssinia.	1806.	——	S. ♂.	Andr. rep. 571.
49 conférta. DC.	(*st.*)	close-flowered.	Brazil.	1826.	——	S. ♄.	
50 hirsùta. DC.	(*wh.*)	hairy.	....	....	——	S. ♄.	
51 lasióstega. DC.	(*ye.*)	woolly.	Brazil.	1823.	——	S. ♄.	
52 pentacárpos. DC.	(*ye.*)	five-capsuled.	E. Indies.	1823.	8. 10.	S. ⊙.	
53 trilòba. DC.	(*wh.*)	three-lobed.	C. B. S.	1794.	——	G. ♂.	Jacq. schœn. 2. t. 142.
54 ricinoídes. DC.	(*bh.*)	Ricinus-like.	Peru.	1826.	7. 10.	H. ⊙.	L'Her. stirp. 1. t. 56.
palmàta. Cav. dis. 1. t. 3. f. 3.							
55 jatrophoídes. DC.	(*vi.*)	Jatropha-like.	Peru.	1787.	8.	S. ♂.	L. Her. stirp. 1. t. 56.
palmàta. Jacq. ic. rar. 3. t. 547.							
56 malvæflòra. B.R.	(*bh.*)	Mallow-flow'd.	Columbia.	1826.	9. 11.	H. ♃.	Bot. reg. 1036.
57 Nap'æa. DC.	(*wh.*)	smooth.	Virginia.	1748.	8. 9.	H. ♃.	Bot. mag. 2193.
Nap'æa læ'vis. L.							
58 dioíca. DC.	(*wh.*)	rough.	Virginia.	1759.	8. 9.	H. ♃.	Cav. dis. 5. t. 132. f. 2.
BASTA'RDIA. Kth.		BASTA'RDIA.	Monadelphia	Polyandria. L.			
1 viscòsa. K.S.	(*ye.*)	viscous.	S. America.	1822.	7. 8.	S. ♄.	Sloan. his. 1. t. 139. f. 4.
Sìda viscòsa. DC.							
2 fœ'tida.	(*ye.*)	stinking.	Peru.	1795.	7. 8.	S. ⊙.	L. Her. stirp. 1. t. 53.
Sìda fœ'tida. DC. *Sìda viscòsa.* L.H.							
GA'YA. Kth.		GA'YA.	Monadelphia Polyandria. L.				
1 occidentàlis. H.S.	(*vi.*)	downy.	America.	1732.	7. 8.	S. ⊙.	Cav. dis. 1. t. 4. f. 3.
Sìda occidentàlis. DC.							
2 hermannioídes. K.S.	(*y.*)	Hermannia-like.	Mexico.	1826.	4. 9.	G. ♄.	H. et B. n. gen. 5. t. 475.
3 dísticha.	(*ye.*)	two-ranked.	New Spain.	1820.	6. 9.	G. ♄.	Cav. ic. 5. t. 432.
Sìda dísticha. Cav.							
4 nùtans.	(*ye.*)	nodding.	Peru.	1825.	——	G. ♄.	L. Her. stirp. 1. t. 57.
Sìda nùtans. DC. *calyptràta.* Cav.							
AB'UTILON. Kth.		AB'UTILON.	Monadelphia Polyandria. L.				
1 periplocifòlium.	(*st.*)	Periploca-ld.	India.	1691.	7. 8.	S ♂.	Cav. dis. 1. t. 5. f. 2.
Sìda periplocifòlia. DC.							
2 ferrugíneum. K.S.	(*ye.*)	ferruginous.	Peru.	1822.	7. 8.	S. ♄.	
3 hernandioídes.	(*ye.*)	Hernandia-ld.	Hispaniola.	1798.	....	S. ♄.	L. Her. stirp. 1. t. 58.
4 luciànum.	(*wh.*)	St. Lucian.	St. Lucia.	1822.	....	S. ♄.	
Sìda luciàna. DC.							
5 Lechenaultiànum.	(*ye.*)	Lechenault's.	E. Indies.	1820.	....	S. ♄.	
Sìda Lechenaultiàna. DC.							
6 nudiflòrum.	(*ye.*)	naked-flower'd.	Peru.	1731.	5. 6.	S. ♄.	L. Her. stirp. 1. t. 59.
7 polyánthon.	(*ye.*)	many-flowered.	....	1822.	——	S. ♄.	
Sìda polyántha. DC.							
8 contráctum.	(*ye.*)	contracted.	Madagascar.	1826.	5. 7.	S. ♄.	
Sìda contrácta. L. en.							
9 áuritum.	(*co.*)	eared-stipuled.	Java.	1820.	12. 3.	S. ♄.	Bot. mag. 2495.
Sìda áu'ita. B.M.							

10 triquètrum.	(*ye.*)	triangular.	W. Indies.	1775.	7. 8.	S. ♄.	Jacq. vind. 2. t. 118.
11 incànum.	(*ye.*)	hoary.	Sandwich Isl.	1821.	——	S. ♄.	
Sìda incàna. L. en.							
12 pulchéllum.	(*wh.*)	neat.	N. Holland.	——	——	S. ♄.	Bonpl. nav. t. 2.
13 umbellàtum.	(*ye.*)	umbelled.	Jamaica.	1788.	7. 9.	S. ♂.	Jacq. vind. t. 56.
14 gigánteum.	(*ye.*)	gigantic.	Caracas.	1823.	....	S. ♄.	Jacq. schœnb. 2. t. 141.
15 refléxum.	(*re.*)	reflex-flowered.	Peru.	1799.	7. 8.	S. ♄.	Cav. dis. 1. t. 7.
16 críspum.	(*wh.*)	curled.	America.	1726.	——	H. ⊙.	——— 1. t. 7. f. 1.
17 virgàtum.	(*ye.*)	twiggy.	Peru.	1820.	——	S. ♄.	Cav. ic. 1. t. 73.
18 arbòreum.	(*pu.*)	great-flowered.	———	1772.	——	S. ♄.	L. her. stirp. 1. t. 63.
19 Mauritiànum.	(*ye.*)	hairy-capsuled.	Mauritius.	1789.	7. 9.	S. ♄.	Jacq. ic. 1. t. 137.
Sìda planiflòra. Cav. dis. 1. t. 7. f. 4.							
20 tiliæfòlium.	(*ye.*)	Lime-tree-ld.	China.	1820.	——	S. ⊙.	Jacq. f. ecl. 1. t. 35.
21 mùticum.		awnless.	Egypt.	1825.	8. 10.	H. ⊙.	
Sìda mùtica. DC.							
22 Avicénnæ. G.	(*ye.*)	broad-leaved.	S. France.	1596.	6. 8.	H. ⊙.	Houtt. syst. 8. t. 61.
Sìda Abùtilon. DC.							
23 americànum.	(*ye.*)	American.	Jamaica.	1730.	6. 8.	S. ⊙.	
Sìda americàna. DC.							
24 asiàticum.	(*ye.*)	small-flowered.	E. Indies.	1768.	7. 8.	S. ⊙.	Cav. dis. 1. t. 7. f. 2.
25 populifòlium.	(*ye.*)	Poplar-leaved.	———	1796.	——	S. ⊙.	——— 1. t. 7. f. 9.
26 hírtum.	(*ye.*)	hairy.	———	1820.	——	S. ⊙.	Rumph. amb. 4. t. 10?
Sìda hírta. DC.							
27 mólle.	(*ye.*)	large-leaved.	Peru.	1817.	10.12.	S. ♄.	Bot. mag. 2759.
Sìda móllis. DC. *grandifòlia.* Bot. reg. 360.							
28 mollíssimum.	(*st.*)	soft-leaved.	Peru.	1789.	6. 7.	S. ♂.	Cav. dis. 2. t. 14. f. 1.
29 permólle.	(*ye.*)	woolly-leaved.	———	1818.	——	S. ♄.	
Sìda permóllis. DC.							
30 cornùtum.	(*ye.*)	horned.	S. America.	1827.	——	S. ♄.	
Sìda cornùta. W. enum.							
31 índicum.	(*ye.*)	Indian.	E. Indies.	1731.	7. 8.	S. ⊙.	Cav. dis. 1. t. 7. f. 10.
32 vesicàrium.	(*st.*)	bladdered.	Mexico.	1796.	——	S. ♄.	——— 2. t. 14. f. 3.
33 gláucum.	(*ye.*)	glaucous.	Senegal.	1822.	——	S. ♄.	Cav. ic. 1. t. 11.
34 álbidum.	(*ye.*)	white-leaved.	Canaries.	——	——	G. ♄.	
Sìda álbida. DC.							
35 mollicòmum.	(*ye.*)	soft.	....	1804.	7. 9.	S. ♄.	
Sìda serícea. Cav? *Sìda mollicòma.* DC.							
36 Sonneratiànum.	(*ye.*)	Sonnerat's.	C. B. S.	1806.	6. 7.	G. ♂.	Cav. dis. 1. t. 6. f. 4.
LAGUN'EA. DC.		LAGUN'EA. Monadelphia Polyandria. L.					
1 lobàta. DC.	(*wh.*)	Maple-leaved.	Bourbon.	1787.	7. 8.	S. ⊙.	Cav. dis. 5. t. 136. f. 1.
2 sinuàta. DC.	(*wh.*)	sinuate.	....	....	——	S. ⊙.	

ORDO XXIX.

BOMBACEÆ. *DC. prodr.* 1. *p.* 475.

HELI'CTERES. DC.		SCREW-TREE. Monadelphia Dodecandria. L.					
1 Isòra. DC.	(*re.*)	great-fruited.	E. Indies.	1733.	6. 7.	S. ♄.	Bot. mag. 2061.
2 baruénsis. DC.	(*wh.*)	small-fruited.	W. Indies.	1739.	9. 10.	S. ♄.	Jacq. amer. t. 149.
3 jamaicénsis. DC.	(*wh.*)	broad-leaved.	———	——	6. 7.	S. ♄.	Jacq. vind. 2. t. 143.
4 guazumæfòlia. DC.	(*pu.*)	Guazuma-ld.	S. America.	1820.	6. 8.	S. ♄.	
5 verbascifòlia. DC.	(*re.*)	Verbascum-ld.	Brazil.	——	——	S. ♄.	Lodd. bot. cab. 504.
6 ferruginàta. DC.	(*ye.*)	yellow-flow'r'd.	———	——	——	S. ♄.	
7 lanceolàta. DC.	(*pu.*)	spear-leaved.	E. Indies.	1823.	....	S. ♄.	
8 spicàta. H.B.	(*pu.*)	spike-flowered.	———	——	....	S. ♄.	
MYR'ODIA. DC.		MYR'ODIA. Monadelphia Dodecandria. L.					
turbinàta. DC.	(*wh.*)	short-flowered.	W. Indies.	1793.	....	S. ♄.	Fl. mex. ic. ined.
PLAGIA'NTHUS. DC.		PLAGIA'NTHUS. Monadelphia Dodecandria.					
divaricàtus. DC.	....	small-leaved.	NewZealand.	1821.	....	G. ♄.	Forst. gen. t. 43.
ADANS'ONIA. DC.		SOUR-GOURD. Monadelphia Polyandria. L.					
digitàta. DC.	(*wh.*)	palmate.	Senegal.	1724.	....	S. 5.	Bot. mag. 2791-2792.

CAROLI'NEA. DC.		CAROLI'NEA.	Monadelphia Polyandria. L.				
1 princéps. DC.	(re.)	digitated.	W Indies.	1787.	....	S. ♄.	Aub. gui. 2. t. 291-2.
Páchira aquática. A.							
2 insígnis. DC.	(sc.)	great-flowered.	———	1796.	....	S. ♄.	Lodd. bot. cab. 1004.
3 robústa. L.C.	(re.)	robust.	S. America.	1825.	....	S. ♄.	
4 fastuòsa. DC.	(cr.)	crimson.	New Spain.	——	....	S. ♄.	Hern. mex. 68. ic.
5 mìnor. DC.	(re.)	lesser.	Guiana.	1798.	7. 8.	S. ♄.	Bot. mag. 1412.
6 álba. B.C.	(wh.)	white-flowered.	Brazil.	....	——	S. ♄.	Lod. bot. cab. 752.
BO'MBAX. DC.		SILK-COTTON-TREE.	Monadelphia Polyandria.				
1 Céiba. DC.	(ro.)	five-leaved.	S. America.	1692.	....	S. ♄.	Jacq. am. t. 176. f. 1.
quinàtum. Jacq.							
2 septenàtum. DC.	....	seven-leaved.	———	1699.	....	S. ♄.	
heptaphy'llum. L. *non* Cav.							
3 malabáricum. DC.	(sc.)	Malabar.	E. Indies.	....	....	S. ♄.	Roxb. cor. 3. t. 247.
heptaphy'llum. Roxb. *non* L.							
4 globòsum. DC.	(pa.)	globe-fruited.	Guiana.	1824.	....	S. ♄.	Aubl. guian. 2. t. 281.
5 buonopozénse. DC.	(cr.)	African.	Africa.	1822.	....	S. ♄.	Beauv. fl. ow. 2. t. 83.
ERIODE'NDRON. DC.		ERIODE'NDRON.	Monadelphia Dodecandria.				
1 leianthèrum. DC.	(wh.)	Brazilian.	Brazil.	1822.	....	S. ♄.	Cav. dis. 5. t. 152. f. 1.
Bómbax eriánthos. Cav.							
2 anfractuòsum. DC.	(st.)	twisted-anther'd	....	1739.	....	S. ♄.	Cav. dis. 5. t. 151.
Bómbax pentándrum. L.							
α *índicum.* DC.	(ye.)	*East Indian.*	E. Indies.	——	....	S. ♄.	Rheed. mal. 3. t. 49. 51.
β *carib'æum.* DC.	(ro.)	*West Indian.*	W. Indies.	——	....	S. ♄.	Jacq. am. t. 176. f. 70.
γ *africànum.* DC.	(cr.)	*African.*	Africa.	1823.	....	S. ♄.	
OCHR'OMA. DC.		OCHR'OMA.	Monadelphia Pentandria. L.				
1 Lagòpus. DC.	(wh.)	downy-leaved.	Jamaica.	1802.	....	S. ♄.	Sw. fl. 2. t. 23.
2 tomentòsa. DC.	(wh.)	woolly-leaved.	S. America.	1816.	....	S. ♄.	
CHEIROST'EMON. DC.		CHEIROST'EMON.	Monadelphia Decandria.				
platanoides. DC.	(wh.)	plane-leaved.	New Spain.	1820.	11. 1.	S. ♄.	H. et B. pl. æq. 1. t. 24.
Chiranthodéndron. Larr. dis. ic.							

ORDO XXX.

BUTTNERIACEÆ. *Kunth Synops.* 3. *p.* 266.

Tribus I. *STERCULIEÆ.* DC. *prodr.* 1. *p.* 481.

STERC'ULIA. DC.		STERC'ULIA.	Monœcia Monadelphia. L.				
1 Balánghas. DC.	(st.)	many-seeded.	E. Indies.	1787.	6. 9.	S. ♄.	Rheed. mal. 1. t. 49.
2 nòbilis. DC.	(st.)	few-seeded.	China.	——	——	S. ♄.	Bot. reg. 185.
Balánghas. B.R. *Southwéllia nòbilis.* Par. Lond. 69.							
3 acuminàta. DC.	(ye.)	taper-pointed.	Africa.	1822.	....	S. ♄.	Beau. fl. ow. 1. t. 24.
4 grandiflòra. DC.	(ye.)	large-flowered.	E. Indies.	1820.	....	S. ♄.	
5 coccínea. DC.	(re.)	scarlet-fruited.	———	——	——	S. ♄.	
6 angustifòlia. DC.	(ye.)	narrow-leaved.	Népaul.	1823.	....	S. ♄.	
7 guttàta. DC.	(fl.)	spotted.	E. Indies.	1819.	....	S. ♄.	Rheed. mal. 4. t. 61.
8 heterophy'lla. DC.	(ye.)	various-leaved.	Africa.	1824.	....	S. ♄.	Beauv. fl. ow. 1. t. 40.
9 Ivìra. DC.	(fl.)	hairy capsuled.	W. Indies.	1723.	....	S. ♄.	Aub. gui. 2. t. 279.
crinìta. Cav. dis. 5. t. 162.							
10 macrophy'lla. DC.	(ye.)	large-leaved.	E. Indies.	1822.	....	S. ♄.	
11 platanifòlia. DC.	(gr.)	Plane-leaved.	China.	1757.	7.	G. ♄.	Cav. dis. 5. t. 145.
12 coloràta. DC.	(re.)	scarlet.	E. Indies.	1818.	....	S. ♄.	Roxb. cor. 1. t. 25.
13 ùrens. DC.	(gr.)	stinging.	———	1793.	6. 9.	S. ♄.	——— 1. t. 24.
14 villòsa. DC.	....	villous-leaved.	———	1818.	....	S. ♄.	
15 fœ'tida. DC.	(fl.)	fetid.	———	1690.	....	S. ♄.	Cav. dis. 5. t. 145.
HERITIE'RA. DC.		LOOKING-GLASS PLANT.	Diœcia Monadelphia.				
1 littoràlis. DC.	(re.)	large-leaved.	E. Indies.	1780.	....	S. ♄.	Rheed. mal. 6. t. 21.
2 mìnor. DC.	....	small-leaved.	———	1816.	....	S. ♄.	Gært. fr. 2. t. 98. f. 2.

Tribus II. *BUTTNERIEÆ.*

THEOBR'OMA. DC.		CHOCOLATE-NUT. Polyadelphia Decandria. L.					
1 Cacào. DC.	(*ye.*)	smooth-leaved.	S. America.	1739.	6. 9.	S. ħ.	Lodd. bot. cab. 545.
2 guianénsis. DC.	(*ye.*)	woolly-leaved.	Guiana.	1803.	....	S. ♄.	Aub. gui. 2. t. 275.
3 carib'æa.	(*ye.*)	Caribean.	W. Indies.	1821.	....	S. ħ.	
4 bícolor. DC.	(*re.*)	two-coloured.	S. America.	1820.	....	S. ♄.	H. et B. pl. eq. 1. t. 30.
ABR'OMA. DC.		ABR'OMA. Polyadelphia Dodecandria. L.					
1 augústa. DC.	(*d.pu.*)	smooth-stalked.	E. Indies.	1770.	6. 9.	S. ♄.	Bot. reg. 518.
2 fastuòsa. DC.	(*d.pu.*)	rough-stalked.	N. Holland.	1800.	——	S. ♄.	Sal. par. lond. 102.
3 móllis. DC.	(*d.pu.*)	soft.	Moluccas.	....	——	S. ♄.	
GUAZ'UMA. DC.		BASTARD CEDAR. Polyadelphia Dodecandria. L.					
1 ulmifòlia. DC.	(*ye.*)	Elm-leaved.	W. Indies.	1739.	8. 9.	S. ħ.	Trew ehret. t. 76.
Bubròma Guazùma. W.							
2 tomentòsa. DC.	(*ye.*)	tomentose.	S. America.	....	——	S. ♄.	
3 polybòtrya. DC.	(*ye.*)	velvetty.	W. Indies.	....	——	S. ♄.	Cav. ic. 3. t. 299.
COMMERS'ONIA. DC.		COMMERS'ONIA. Pentandria Pentagynia. L.					
1 echinàta. DC.	(*wh.*)	oval-leaved.	Moluccas.	1806.	6. 7.	S. ħ.	Rumph. amb. 3. t. 119.
2 platyph'ylla. DC.	(*wh.*)	broad-leaved.	————	——	——	S. ♄.	And. bot. rep. 519.
RULI'NGIA. Br.		RULI'NGIA. Pentandria Pentagynia. L.					
1 dasyphy'lla.	(*wh.*)	hairy-leaved.	N. Holland.	1808.	4. 7.	G. ħ.	Bot. rep. 603.
Commersònia dasyphy'lla. A.R. *Byttnèria dasyphy'lla.* DC.							
2 pannòsa. B.M.	(*wh.*)	cloth-leaved.	N. Holland.	1819.	4. 7.	G. ħ.	Bot. mag. 2191.
3 hermanniæfòlia. DC.	(*w.*)	Hermannia-l^d.	————	——	——	G. ♄.	
BUTTN'ERIA. L.		BUTTN'ERIA. Pentandria Monogynia. L.					
1 herbàcea. DC.	(*wh.*)	herbaceous.	E. Indies.	1820.	4. 7.	S. ♃.	Roxb. cor. 1. t. 29.
2 grandifòlia. DC.	(*wh.*)	large-leaved.	————	——	——	S. ♄.	
3 catalpæfòlia. DC.	(*wh.*)	Catalpa-leav'd.	Caracas.	1823.	....	S. ħ.	Jacq. schœnb. 1. t. 46.
4 cordata. DC.	(*wh.*)	heart-leaved.	Peru.	——	....	S. ♄.	Cav. dis. 5. t. 150.
5 microphylla. DC.	(*va.*)	small-leaved.	S. America.	1816.	....	S. ħ.	———— 5. t. 148. f. 2.
6 scàbra. DC.	(*wh.*)	rough.	————	1793.	7. 8.	S. ♄.	———— 5. t. 148. f. 1.
AY'ENIA. DC.		AY'ENIA. Pentandria Monogynia.					
1 pusílla. DC.	(*pu.*)	small.	Jamaica.	1756.	7. 9.	S. ♂.	Mill. ic. t. 118.
2 lævigàta. DC.	(*cr.*)	smooth.	————	....	——	S. ♄.	
KLEINH'OVIA. DC.		KLEINH'OVIA. Monadelphia Dodecandria.					
Hóspita. DC.	(*pu.*)	purple-flower'd.	Moluccas.	1800.	7. 9.	S. ♄.	Cav. dis. 5. t. 146.

Tribus III. *LASIOPETALEÆ.* DC.

SERI'NGIA. DC.		SERI'NGIA. Pentandria Monogynia.					
platyphy'lla. DC.	(*wh.*)	broad-leaved.	N. Holland.	1802.	4. 7.	G. ħ.	Gay dis. p. 13. t. 1. 2.
Lasiopétalum arboréscens. H.K.							
LASIOPE'TALUM. DC.		LASIOPE'TALUM. Pentandria Monogynia.					
1 ferrugíneum. DC.	(*wh.*)	rusty.	N. Holland.	1791.	4. 7.	G. ħ.	Bot. mag. 1766.
2 parviflòrum. DC.	(*wh.*)	small-flowered.	————	1810.	——	G. ♄.	Linn. trans. 10. t. 19. f. 2.
THOM'ASIA. DC.		THOM'ASIA. Pentandria Monogynia.					
1 purpùrea. DC.	(*pu.*)	purple.	N. Holland.	1803.	4. 7.	G. ♄.	Bot. mag. 1755.
Lasiopétalum purpùreum. H.K. *coccineum.* Donn. cant.							
2 foliòsa. DC.	(*wh.*)	leafy.	N. Holland.	1822.	——	G. ♄.	Gay dis. p. 24. t. 7.
3 solanàcea. DC.	(*wh.*)	Solanum-like.	————	1803.	——	G. ♄.	———— p. 26. t. 6.
Lasiopétalum solanàceum. B.M. 1486.							
4 quercifòlia. DC.	(*pu.*)	Oak-leaved.	————	——	——	G. ħ.	Bot. mag. 1485.

Tribus IV. *HERMANNIEÆ.* DC.

MEL'OCHIA. DC.		MEL'OCHIA. Monadelphia Pentandria. DC.					
1 pyramidàta. DC.	(*vi.*)	pyramidal.	Brazil.	1768.	7. 8.	S. ♄.	Jacq. vind. 1. t. 30.
2 tomentòsa. DC.	(*vi.*)	downy.	W. Indies.	——	5. 6.	S. ♄.	Cav. dis. 6. t. 172. f. 2
3 parvifòlia. DC.	(*wh.*)	small-leaved.	Caracas.	1823.	....	S. ħ.	

RIEDLE'IA. DC.	RIEDLE'IA.	Monadelphia Pentandria. L.				
1 velutìna. DC.	soft-leaved.	E. Indies.	1820.	6. 8.	S. ♄.	
2 depréssa. DC. (st.)	depressed.	Havannah.	——	——	S. ♄.	
3 supìna. DC.	procumbent.	E. Indies.	1823.	——	S. ♂.	Pluk. alm. t. 232. f. 4.
4 corchorifòlia. DC. (bh.)	Corchorus-l^d^.	————	1732.	7. 8.	S. ⊙.	Dill. elt. t. 176. f. 217
Melòchia corchorifòlia. L.						
5 nodiflòra. DC. (wh.)	knot flowered.	S. America.	1823.	——	S. ♄.	Sloan. his. t. 135. f. 2.
Mougeòtia nodiflòra. Kth. *Melochia nodiflora.* Sw.						
6 caracasàna. DC. (wh.)	Caracas.	Caracas.	1823.	——	S ♄.	Jacq. ic. 3. t. 507.
Mougeòtia caracasàna. Kth. *Melòchia caracasàna.* Jacq.						
7 serràta. DC. (pu.)	saw-leaved.	St. Domingo.	1827.	6. 8.	S. ♃.	Vent. choix. t. 37.
WALTH'ERIA. DC.	WALTH'ERIA.	Monadelphia Pentandria. L.				
1 americàna. DC. (ye.)	American.	S. America.	1691.	5. 10.	S. ♄.	Jacq. ic. 1. t. 130.
2 glàbra. DC. (ye.)	smooth.	W. Indies.	1823.	——	S. ♃.	Schrank. h. mon. t. 55.
lævis. L. en.						
3 índica. DC. (ye.)	Indian.	E. Indies.	1759.	6. 8.	S. ♄.	Burm. zeyl. t. 68.
4 ellíptica. DC. (ye.)	woolly-leaved.	————	1812.	——	S. ♄.	Cav. dis. 6. t. 171. f. 2.
5 microphy'lla. DC. (ye.)	small-leaved.	————	1823.	——	S. ♄.	———— 6. t. 170. f. 2.
HERMA'NNIA. DC.	HERMA'NNIA.	Monadelphia Pentandria. L.				
1 multífida. DC. (ye.)	multifid-leaved.	C. B. S.	1825.	5. 8.	G. ♄.	
2 halicácaba. DC. (ye.)	bladder-calyx'd.	————	1820.	——	G. ♄.	
3 comosa. DC. (ye.)	tufted.	————	——	1. 12.	G. ♄.	
4 althæifòlia. DC. (ye.)	Althæa-leaved.	————	1728.	3. 7.	G. ♄.	Bot. mag. 307.
5 althæoídes. L. en. (ye.)	Althæa-like	————	....	——	G. ♄.	
6 plicàta. DC. (ye.)	plaited-leaved.	————	1774.	11. 12.	G. ♄.	Jacq. schœnb. 2. t. 213.
7 cándicans. DC. (ye.)	white-leaved.	————	——	4. 6.	G. ♄.	———— 1. t. 117.
8 móllis. DC. (ye.)	soft-leaved.	————	1818.	——	G. ♄.	
9 decúmbens. DC. (ye.)	decumbent.	————	——	——	G. ♄.	
10 hyssopifòlia. DC. (ye.)	Hyssop-leaved.	————	1725.	——	G. ♄.	Cav. dis. 6. t. 181. f. 3.
11 trifoliàta. DC. (ye.)	three-leaved.	————	1752.	5. 8.	G. ♄.	———— 6. t. 182. f. 2.
12 triphy'lla. DC. (st.)	three-parted.	————	1820.	——	G. ♄.	———— 6. t. 178. f. 3.
13 glandulòsa. DC. (ye.)	glandular.	————	——	——	G. ♄.	
14 frágrans. DC. (ye.)	fragrant.	————	——	——	G. ♄.	
15 disermæfòlia. DC. (ye.)	simple-flow'r'd.	————	1795.	3. 4.	G. ♄.	Jacq. schœnb. 1. t. 121
16 dísticha. DC. (ye.)	round-leaved.	————	1789.	5. 8.	G. ♄.	S. et Wend. sert. h. t. 10.
rotundifòlia. Jacq. schœn. 3. t. 121.						
17 melochioídes. DC. (ye.)	Melochia-like.	————	1818.	——	G. ♄.	
18 bryonifòlia. DC. (ye.)	Bryony-leaved.	————	——	——	G. ♄.	
19 salvifòlia. DC. (ye.)	Sage-leaved.	————	1795.	4. 6.	G. ♄.	Cav. dis. 6. t. 180. f. 2.
20 mìcans. DC. (y. or.)	glittering.	————	1790.	5. 8.	G. ♄.	S. et Wend. s. han. t. 5.
latifòlia. Jacq. schœn. 3. t. 119.						
21 involucràta. DC. (st.)	involucred.	————	1794.	5. 6.	G. ♄.	Cav. dis. 6. t. 177. f. 1.
22 scordifòlia. DC. (or. y.)	Germander-l^d^.	————	——	4. 11.	G. ♄.	Jacq. schœn. 1. t. 120.
23 denudàta. DC. (ye.)	smooth.	————	1774.	5. 7.	G. ♄.	———— 1. t. 122.
24 alnifòlia. DC. (ye.)	Alder-leaved.	————	1728.	2. 5.	G. ♄.	Bot. mag. 299.
25 cuneifòlia. DC. (ye.)	wedge-leaved.	————	1791.	8. 9.	G. ♄.	Jacq. schœn. 1. t. 124.
26 holoserícea. DC. (ye.)	velvet-leaved.	————	1792.	5. 6.	G. ♄.	———— 3. t. 292.
27 hirsùta. DC. (ye.)	hairy-branch'd.	————	1790.	——	G. ♄.	S. et Wend. s. h. 1. t. 4.
28 scàbra. DC. (ye.)	rough-leaved.	————	1789.	3. 4.	G. ♄.	Jacq. schœn. 1. t. 127.
29 multiflòra. DC. (ye.)	many-flowered.	————	1791.	3. 5.	G. ♄.	———— 1. t. 128.
30 flámmea. DC. (re.)	flame-flowered.	————	1794.	1. 12.	G. ♄.	Bot. mag. 1349.
31 angulàris. DC. (or.)	angular-calyx'd.	————	1791.	4. 5.	G. ♄.	Jacq. schœn. 1. t. 126.
32 trifurcàta. DC. (re.)	three-forked.	————	1789.	4. 7.	G. ♄.	———— 1. t. 125.
33 odoràta. DC. (re.)	sweet-scented.	————	1780.	2. 10.	G. ♄.	
34 lavandulæfòlia. DC. (y.)	Lavender-l^d^.	————	1732.	5. 9.	G. ♄.	Bot. mag. 304.
35 velutina. DC. (ye.)	mucronate.	————	1818.	——	G. ♄.	
36 filifòlia. DC. (or.)	thread-leaved.	————	1816.	5. 8.	G. ♄.	Jacq. schœn. 1. t. 123
37 procúmbens. DC. (st.)	procumbent.	————	1792.	5. 6.	G. ♄.	Cav. dis. 6. t. 177. f. 2.
38 tenuifòlia. DC. (ye.)	slender-leaved.	————	....	6. 7.	G. ♄.	Bot. mag. 1348.
39 incísa. DC. (ye.)	cut-leaved.	————	1806.	——	G. ♄.	
40 pulveràta. DC. (or.)	powdered.	————	1800.	5. 8.	G. ♄.	And. bot. rep. 164.
41 argéntea. DC. (or.)	silvery-leaved.	————	1818.	——	G. ♄.	
42 coronopifòlia. DC. (ye.)	Buckshorn-l^d^.	————	——	6. 8.	G. ♄.	
43 hispidula. S. S. (ye.)	bristly.		1823.	3. 8.	G. ♄.	
MAHE'RNIA. DC.	MAHE'RNIA.	Monadelphia Pentandria.				
1 verticillàta. DC. (ye.)	whorl-leaved.	C. B. S.	1820.	6. 8.	G. ♄.	Cav. dis. 6. t. 176. f. 1.
2 resedæfòlia. DC. (re.)	Reseda-leaved.	————	1818.	——	G. ♄.	
3 bipinnàta. DC. (re.)	winged-leaved.	————	1752.	——	G. ♄.	Bot. mag. 277.
pinnàta. B. M.						

4 incìsa. DC.	(*st.*)	cut-leaved.	C. B. S.	1792.	7. 9.	G. ♄.	Bot. mag. 353.
5 diffùsa. DC.	(*ye.*)	diffuse.	———	1774.	6. 8.	G. ♄.	Lodd. bot. cab. 187.
6 heterophy'lla. DC.	(*ye.*)	various-leaved.	———	1731.	4. 5.	G. ♄.	Cav. dis. 6. t. 178. f. 1.
Hermánnia grossulariæfòlia. L.							
7 seselifòlia. DC.	(*ye.*)	Seseli-leaved.	———	1818.	6. 8.	G. ♄.	
8 pulchélla. DC.	(*re.*)	neat.	———	1792.	7. 9.	G. ♄.	Cav. dis. 6. t. 177. f. 3.
9 glabràta. DC.	(*ye.*)	sweet-scented.	———	1789.	4. 5.	G. ♄.	Jacq. schœnb. 1. t. 53.
odoràta. A.R. 85.							
10 vesicària. DC.	(*ye.*)	inflated.	———	1818.	——	G. ♄.	Cav. dis. 6. t. 181. f. 2.
11 grandiflòra.	(*sc.*)	great-flowered.	———	1791.	5. 8.	G. ♄.	
Hermánnia grandiflòra. H.K.							
12 Burchéllii.	(*sc.*)	Burchell's.	———	1818.	——	G. ♄.	Bot. reg. 224.
grandiflòra. B.R. *non* H.K.							
13 biserràta. DC.	(*ye.*)	double-tooth'd.	———	1825.	——	G. ♄.	Cav. dis. 6. t 200. f. 2.

Tribus V. *DOMBEYACEÆ.* DC.

RUI'ZIA. DC.		RUI'ZIA. Monadelphia Polyandria. L.					
1 lobàta. DC.	(*ye.*)	lobed-leaved.	Bourbon.	1820.	....	S. ♄.	Cav. dis. 3. t. 36. f. 1.
2 variábilis. DC.	(*bh.*)	various-leaved.	———	1792.	5.	S. ♄.	Jacq. schœn. 3. t. 295.
PENTAP'ETES. DC.		PENTAP'ETES. Monadelphia Dodecandria. L.					
1 phœnícea. DC.	(*cr.*)	scarlet-flower'd.	India.	1690.	7. 8.	S. ♂.	Bot. reg. 575.
2 ovàta. DC.	(*sn.*)	oval-leaved.	New Spain.	1805.	6. 9.	S. ♂.	Cav. ic. 5. t. 433.
Brotèra ovàta. Cav.							
ASS'ONIA. DC.		ASS'ONIA. Monadelphia Dodecandria.					
1 populnea. DC.	(*wh.*)	Poplar-leaved.	Bourbon.	1822.	....	S. ♄.	Cav. dis. 3. t. 42. f. 3.
2 viburnoídes. DC.	(*wh.*)	Viburnum-l^d^.	———	——	....	S. ♄.	
DOMB'EYA. DC.		DOMB'EYA. Monadelphia Dodecandria.					
1 cordifòlia. DC.	(*wh.*)	heart-leaved.	E. Indies.	1820.	....	S. ♄.	
2 ferrugínea. DC.	(*wh.*)	rusty.	Mauritius.	1815.	....	S. ♄.	Cav. dis. 3. t. 42. f. 2.
3 ovàta. DC.	(*wh.*)	oval-leaved.	Bourbon.	1822.	....	S. ♄.	——— 3. t. 41. f. 2.
MELH'ANIA. DC.		MELH'ANIA. Monadelphia Pentandria.					
1 Erythróxylon. DC.	(*wh.*)	red-wood.	St. Helena.	1722.	5. 8.	S. ♄.	Bot. mag. 1000.
2 Melanóxylon. DC.	(*wh.*)	black-wood.	———	1800.	7. 8	S. ♄.	Pluk. mant. 6. t. 333. f. 5.
3 Burchéllii. DC.	(*wh.*)	Burchell's.	C. B. S.	1818.	——	S. ♄.	
PTEROSPE'RMUM. DC.		PTEROSPE'RMUM. Monadelphia Dodecandria.					
1 acerifòlium. DC.	(*wh.*)	Maple-leaved.	E. Indies.	1790.	7. 9.	S. ♄.	Bot. mag. 620.
2 suberifòlium. DC.	(*wh.*)	various-leaved.	———	1783.	9. 10.	S. ♄.	——— 1526.
3 lanceæfòlium. DC.	(*w.*)	lance-leaved.	———	1818.	——	S. ♄.	
4 canéscens. H.B.	(*wh.*)	canescent.	Ceylon.	——	——	S. ♄.	
5 semisagittàtum. DC.	(*w.*)	fringe-bracted.	E. Indies.	——	——	S. ♄.	
ASTRAP'ÆA. DC.		ASTRAP'ÆA. Monadelphia Dodecandria.					
1 Wallíchii. DC.	(*sc.*)	Wallich's.	Mauritius.	1820.	12. 1.	S. ♄.	Bot. reg. 691.
2 viscòsa.		viscous.	Madagascar.	1823.	....	S. ♄.	
3 tiliæfòlia.	(*wh.*)	Lime-leaved.	Bourbon.	1824.	....	S. ♄.	
KY'DIA. DC.		KY'DIA. Monadelphia Dodecandria.					
1 calycìna. DC.	(*wh.*)	calycine.	E. Indies.	1818.	....	S. ♄.	Roxb. cor. 3. t. 215.
2 fratérna. DC.	(*wh.*)	long-stamen'd.	———	1823.	....	S. ♄.	——— 3. t. 216.

Tribus VI. *WALLICHIEÆ.* DC.

ERIOL'ÆNA. DC.	ERIOL'ÆNA. Monadelphia Polyandria.					
Wallíchii. DC.	Wallich's.	E. Indies.	1823.	....	S. ♄.	DC. m. mus. 10. p. 102. ic.

ORDO XXXI.

TILIACEÆ. *DC. prodr.* 1. *p.* 503.

ENT'ELEA. B.M.		CORK-TREE.	Polyandria Monogynia.				
arboréscens. B.M.	(*wh.*)	New Zealand.	N. Zealand.	1821.	5. 6.	G. ♄.	Bot. mag. 2480.
SPARRMA'NNIA. DC.		SPARRMA'NNIA.	Polyandria Monogynia. L.				
1 africàna. DC.	(*wh.*)	African.	C. B. S.	1790.	3. 7.	G. ħ.	Bot. mag. 516.
2 rugòsa. Bhi.	(*wh.*)	rugged-leaved.	———	1825.	3. 6.	G. ♄.	
HELIOCA'RPUS. DC.		HELIOCA'RPUS.	Dodecandria Monogynia.				
americàna. DC.	(*wh.*)	American.	Vera Cruz.	1733.	....	S. ♄.	Trew ehret. t. 45.
CO'RCHORUS. DC.		CO'RCHORUS.	Polyandria Monogynia. L.				
1 siliquòsus. DC.	(*ye.*)	Germander-ld.	S. America.	1732.	6. 8.	S. ♃.	Jacq. vind. 3. t. 59.
2 foliòsus. S.S.	(*ye.*)	leafy.	Cuba.	1827.	——	S. ♃.	
3 hírtus. DC.	(*ye.*)	hairy.	S. America.	1820.	——	S. ♃.	Plum. ed. B. t. 103. f. 2.
4 pilolòbus. DC.	(*ye.*)	hairy-podded.	———	——	——	S. ♃.	
5 americànus. Sck.	(*ye.*)	American.	S. America.	1825.	——	S. ħ.	
6 triloculàris. DC.	(*ye.*)	three-celled.	Arabia.	1790.	7. 8.	S. ⊙.	Jacq. vind. 2. t. 173.
7 olitòrius. DC.	(*ye.*)	bristly-leaved.	India.	1640.	6. 8.	S. ⊙.	Bot. mag. 2810.
8 æ'stuans. DC.	(*ye.*)	Hornbeam-ld.	S. America.	1731.	6. 7.	S. ⊙.	Jacq. vind. 1. t. 85.
9 acutángulus. DC.	(*ye.*)	acute-angled.	India.	1816.	——	S. ♄.	Pluk. phyt. t. 44. f. 1.
10 capsulàris. DC.	(*ye.*)	heart-leaved.	E. Indies.	1725.	——	S. ⊙.	Jacq. f. eclog. 2. t. 120.
11 hirsùtus. DC.	(*ye.*)	woolly-capsul'd.	S. America.	1752.	——	S. ♄.	Jacq. vind. 3. t. 59.
TRIUMFE'TTA. DC.		TRIUMFE'TTA.	Dodecandria Monogynia. L.				
1 Láppula. DC.	(*ye.*)	prickly-seeded.	W. Indies.	1739.	7. 8.	S. ♄.	Plum. ic. t. 255.
2 heterophy'lla. DC.	(*ye.*)	various-leaved.	———	1823.	——	S. ħ.	Pluk. t. 425. f. 3.
3 rubricáulis. DC.	(*ye.*)	red-stem'd.	Caracas.	1826.	——	S. ħ.	
4 suborbiculàta. DC.	(*ye.*)	subrotund.	E. Indies.	....	——	S. ħ.	
5 glandulòsa. DC.	(*ye.*)	gland-toothed.	Mauritius.	....	——	S. ħ.	Lam. ill. t. 400. f. 1.
6 trichoclàda. DC.	(*ye.*)	oval-leaved.	Nepaul.	1823.	——	S. ♂.	
7 oblongàta. DC.	(*ye.*)	oblong-leaved.	———	——	——	S. ♂.	
8 ánnua. DC.	(*ye.*)	annual.	Java.	1760.	8. 9.	S. ⊙.	Bot. mag. 2296.
9 angulàta. DC.	(*ye.*)	angular-leaved.	E. Indies.	1739.	6. 7.	S. ħ.	Lam. ill. t. 400. f. 2.
10 semitrilòba. DC.	(*ye.*)	Mallow-leaved.	W. Indies.	1773.	——	S. ♄.	Jacq. vind. 3. t. 76.
11 havanénsis. DC.	(*ye.*)	Havannah.	Cuba.	1825.	——	S. ♄.	
12 micropétala. B.R.	(*ye.*)	small-petal'd.	E. Indies.	———	4. 8.	S. ħ.	Bot. reg. 1058.
13 petiolàris. B.R.	(*ye.*)	petiolate.	Peru.	1819.	5. 6.	S. ♄.	Lindl. collect. t. 20.
rhomboídea. L. col. *non* Jacq.							
14 rhomboídea. DC.	(*ye.*)	rhomboid.	S. America.	1819.	5. 6.	S. ♄.	Jacq. amer. t. 90.
15 althæoídes. DC.	(*ye.*)	Althæa-leaved.	Cayenne.	——	——	S. ♄.	
macrophy'lla. Vahl. ecl. 2. p. 34.							
16 grandiflòra. DC.	(*ye.*)	large-flowered.	W. Indies.	1810.	——	S. ♄.	
17 triloculàris. DC.	(*ye.*)	three-celled.	E. Indies.	1817.	10. 4.	S. ♄.	
GRE'WIA. DC.		GRE'WIA.	Polyandria Monogynia. L.				
1 tomentòsa. DC.		tomentose.	Java.	1818.	....	S. ♄.	Juss. an. 2. t. 49. f. 1.
2 hirsùta. DC.		hairy.	———	1816.	7. 9.	S. ħ.	
3 bracteàta. DC.	(*gr.*)	bracted.	E. Indies.	1818.	——	S. ħ.	
4 Mallocócca. DC.		rough-fruited.	Tongatabu.	1792.	8. 9.	S. ♄.	Forst. gen. t. 39.
Mallocócca crenàta. Forst.							
5 oppositifòlia. DC.		sharp-leaved.	Nepaul.	1818.	....	S. ♄.	
6 umbellàta. DC.		umbel-flowered.	E. Indies.	——	....	S. ħ.	
7 bícolor. DC.		two-coloured.	Senegal.	1822.	....	S. ħ.	Juss. ann. t. 50. f. 2.
8 Róthii. DC.		Roth's.	E. Indies.	1819.	....	S. ♄.	
9 salvifòlia. DC.	(*ye.*)	Sage-leaved.	———	——	....	S. ♄.	
10 flàva. DC.	(*ye.*)	yellow.	C. B. S.	1818.	——	G. ħ.	
11 ovalifòlia. DC.		oval-leaved.	E. Indies.	——	....	S. ħ.	
12 Micròcos. DC.	(*st.*)	panicled.	———	1779.	....	S. ♄.	Rheed. mal. 1. t. 56.
Micròcos paniculàta. L.							
13 paniculàta. DC.		villous-leaved.	———	1820.	....	S. ħ.	
14 orientàlis. DC.	(*wh.*)	oriental.	———	1767.	7. 8.	S. ♄.	Rheed mal. 5. t. 46.
15 pilòsa. DC.	(*wh.*)	pilose.	———	1804.	——	S. ħ.	
16 glandulòsa. DC.		glandular.	Mauritius.	1827.	....	S. ħ.	Juss. an. 2. t. 48. f. 1.
17 serrulàta. DC.		serrulate.	E. Indies.	1818.	....	S. ♄.	
18 occidentàlis. DC.	(*pu.*)	Elm-leaved.	C. B. S.	1690.	7. 9.	G. ♄.	Bot. mag. 422.
19 asiàtica.	(*li.*)	Asiatic.	E. Indies.	1792.	7. 8.	S. ♄.	Sonn. voy. 2. t. 138.

20 subin'æqualis. DC. .... oblique-leaved. E. Indies. 1816. .... S. ♄.
21 villífera. .... wool-bearing. ——— 1826. .... S. ♄.
villòsa. Roth. *nec aliorum.*
22 áspera. DC. .... rough-leaved. ——— 1818. .... S. ♄.
23 tiliæfòlia. DC. ...... Lime-leaved. ——— 1812. .... S. ♄.
24 villòsa. DC. .... villous. ——— 1816. .... S. ♄.
25 therebinthinàcea. DC. strong-scented. .... —— .... S. ♄.
26 obtusifòlia. DC. ...... blunt-leaved. C. B. S. 1818. .... G. ♄.
27 ulmifòlia. H.B. ...... Chinese. China. —— .... G. ♄.
28 carpinifòlia. DC. (*pu.*) Hornbeam-ld. Africa. 1822. .... S. ♄. Beauv. fl. ow. 2. t.
29 megalocárpa. B.O.(*re*) black-fruited. ——— —— .... S. ♄. ——— 2. t. 102.
BROWNLO'WIA. R. BROWNLO'WIA. Polyandria Monogynia.
elàta. R. (*ye.*) tall. E. Indies. 1812. .... S. ♄. Roxb. cor. 3. t. 265.
T'ILIA. DC. LIME-TREE. Polyandria Monogynia. L.
1 microphy'lla. DC. (*wh.*) small-leaved. Britain. .... 8. 9. H. 5. Eng. bot. 1705.
parvifòlia. E.B.
2 intermèdia. DC. (*wh.*) intermediate. ——— .... —— H. 5. Hayn. et Sv. bot. t. 40.
europ'æa. Eng. bot. 610.
β *laciniàta.* (*wh.*) jagged-leaved. ——— .... 6. 8. H. 5.
3 corallìna. E.F. (*wh.*) coral-twigg'd. Britain. .... 7. H. 5.
4 rùbra. DC. (*wh.*) red-twigged. Europe. .... 6. 8. H. 5.
5 platyphy'lla. DC. (*wh.*) broad-leav'd. Britain. .... —— H. 5. Vent. dis. p. 6. t. 1. f. 2.
grandifòlia. E.F.
6 glábra. DC. (*wh.*) broad-leaved. N. America. 1752. 6. 7. H. 5. Vent. dis. p. 9. t. 2.
americàna. L.
7 laxiflòra. DC. (*wh.*) loose-flowered. ——— 1820. —— H. 5.
8 pubéscens. DC. (*wh.*) pubescent. ——— 1726. 7. 8. H. 5. Vent. dis. p. 10. t. 3.
β *leptophy'lla.* DC. (*w.*) *narrow-leaved.* ——— —— —— H. 5.
9 heterophy'lla. DC. (*wh.*) white-leaved. ——— .... 6. 8. H. 5. Vent. dis. p. 16. t. 5.
álba. Mich. arb. 3. t. 2.
10 argéntea. DC. (*wh.*) silver-leaved. Hungary. 1767. —— H. 5. Vent. dis. p. 13. t. 4.
álba. W. et K. hung. 1. t. 3. *rotundifòlia.* Vent.
11 petiolàris. DC. (*wh.*) long-petioled. .... 1824. .... H. 5.
MUNTI'NGIA. DC. MUNTI'NGIA. Polyandria Monogynia. L.
Calabùra. DC. (*wh.*) Jamaica. Jamaica. 1690. 6. 7. S. ♄. Jacq. amer. t. 107.
APE'IBA. DC. APE'IBA. Polyandria Monogynia. L.
1 Tibóurbou. DC. (*ye.*) hairy. S. America. 1756. .... S. ♄. Aub. gui. 1. t. 213.
Aublètia Tibóurbou. W.
2 Petóumo. DC. (*ye.*) hoary. ——— 1817. .... S. ♄. Aub. gui. 1. t. 215.
3 áspera. DC. (*ye.*) prickly-capsul'd. Cayenne. 1792. .... S. ♄. ——— 1. t. 216.
Aublètia áspera. W.
4 glábra. DC. (*gr.*) smooth. ——— 1817. .... S. ♄. Aub. gui. 1. t. 214.
Aublètia l'ævis. W.
SL'OANEA. DC. SL'OANEA. Polyandria Monogynia. L.
1 dentàta. DC. (*ye.*) Chestnut-leav'd. S. America. 1752. .... S. ♄. Plum. ic. 244.
2 sinemariénsis. DC. (*ye.*) round-leaved. W. Indies. 1820. .... S. ♄. Aub. gui. 1. t. 212.
GYROST'EMON. DC. GYROST'EMON. Diœcia Polyandria.
ramulòsum. DC. .... branching. N. Holland. 1820. .... G. ♄. Desf. m. mus. 6. t. 6.
BE'RRYA. R.C. BE'RRYA. Polyandria Monogynia.
ammonílla. R.C. (*st.*) Ceylon. Ceylon. 1818. .... S. ♄. Roxb. corom. 3. t. 264.

ORDO XXXII.

ELÆOCARPEÆ. *DC. prodr.* 1. *p.* 519.

ELÆOCA'RPUS. DC. ELÆOCA'RPUS. Dodecandria Monogynia.
1 serràtus. DC. (*pu.*) saw-leaved. E. Indies. 1774. .... S. ♄. Burm. zeyl. t. 40.
2 cyáneus. DC. (*wh.*) blue-fruited. N. Holland. 1803. 6. 8. G. ♄. Bot. mag. 1737.
reticulàtus. B.R. 657.
ACER'ATIUM. DC. ACER'ATIUM. Dodecandria Monogynia.
oppositifòlium. DC. opposite-leaved. Amboyna. 1818. .... S. ♄.

DIC'ERA. DC.		Dic'era. Dodecandria Monogynia.					
dentàta. DC.	(*wh.*)	tooth-leaved.	N. Zealand.	1820.	6. 8.	G. ♄.	Forst. gen. t. 40.
FRIE'SIA. DC.		Frie'sia. Dodecandria Monogynia.					
pedunculàris. DC.	(*wh.*)	peduncled.	Van Diem.	1818.	....	G. ♄.	Lab. n. holl. 2. t. 155.
Elæocárpus pedunculàris. Lab.							

ORDO XXXIII.

CHLENACEÆ. *DC. prodr.* 1. *p.* 521.

† *Genus Chlenaceis affine.*

HUG'ONIA. DC.		Hug'onia. Monadelphia Decandria.					
1 My'stax. DC.	(*ye.*)	entire-leaved.	E. Indies.	1818.	6. 9.	S. ƀ.	Rheed. mal. 2. t. 29.
2 serràta. DC.	(*ye.*)	saw-leaved.	Mauritius.	1820.	——	S. ƀ.	Cav. dis. 3. t. 73. f. 1.

ORDO XXXIV.

TERNSTRŒMIACEÆ. *DC. prodr.* 1. *p.* 523.

Tribus I. *TERNSTRŒMIEÆ.* DC.

TERNSTR'ŒMIA. DC.		Ternstr'œmia. Polyandria Monogynia.					
1 pedunculàris. DC.	(*w.*)	long-peduncled.	W. Indies.	1818.	6. 8.	S. ♄.	
2 punctàta. DC.	(*wh.*)	spotted-leaved.	Guiana.	1824.	....	S. ♄.	Aubl. gui. 1. t. 228.
3 dentàta. DC.	(*ye.*)	toothed.	———	1826.	....	S. ♄.	——— 1. t. 227.
4 venòsa. DC.	(*wh.*)	veined-leaved.	Brazil.	1820.	6. 8.	S. ♄.	
5 serràta. H. B.	(*wh.*)	saw-leaved.	Silhet.	——	....	S. ♄.	

Tribus II. *FREZIEREÆ.* DC.

CL'EYERA. DC.		Cl'eyera. Polyandria Monogynia.					
japónica. DC.	(*wh.*)	Japan.	Japan.	1822.	....	S. ƀ.	Kæmp. am. 5. p. 774. ic.
FREZIE'RA. DC.		Frezie'ra. Polyandria Monogynia.					
thæoídes. DC.	(*wh.*)	Tea-like.	Jamaica.	1818.	....	S. ƀ.	
E'URYA. DC.		E'urya. Polygamia Monœcia.					
1 chinénsis. DC.	(*wh.*)	China.	China.	1818.	1. 3.	G. ƀ.	Lodd. bot. cab. 1213.
2 multiflòra. DC.	(*wh.*)	many-flowered.	Nepaul.	1823.	....	S. ƀ.	
LETTS'OMIA. DC.		Letts'omia. Polyandria Monogynia.					
tomentòsa. DC.	....	tomentose.	Peru.	1823.	....	S. ♄.	Ruiz et P. prod. t. 14.

Tribus III. *SAURAUJEÆ* DC.

SAURA'UJA. DC.		Saura'uja. Polyandria Pentagynia.					
1 excélsa. DC.	(*wh.*)	tall.	Caracas.	1824.	....	S. ♄.	Willd. n. act. ber. 3. t. 4.
2 napalénsis. DC.	(*wh.*)	Nepaul.	Nepaul.	——	....	S. ♄.	

Tribus IV. *LAPLACEÆ.* DC.

WITTELSBA'CHIA. M.N. WITTELSBA'CHIA. Polyandria Monogynia.
1 Gossy'pium. M.N. (*ye.*) large-flowered. E. Indies. 1822. .... S. ♄. Sonn. voy. 2. t. 133.
Cochlospérmum Gossy'pium. DC. *Bombax Gossy'pium.* Cav. dis. 5. t. 157.
2 vitifòlia. M.N. (*ye.*) Vine-leaved S. America. 1820. .... S. ♄. Fl. mex. ic. ined.
Cochlospérmum serratifòlium. DC. *hibiscoídes.* K.S. *Bombax vitifòlium.* W.en.
3 insígnis. M.N. (*ye.*) spot-flowered. Brazil. 1821. .... S. ♄. Mart. nov. gen. 1. t. 55.

Tribus V? *GORDONIEÆ.* DC.

MALACHODE'NDRON. DC. MALACHODE'NDRON. Monadelphia Polyandria. L.
ovàtum. DC. (*wh.*) oval-leaved. N. America. 1785. 8. 9. H. ♄. Bot. reg. 1104.
Stuártia pentagy'nia. Ex. bot. t. 101.
STUA'RTIA. W. STUA'RTIA. Monadelphia Polyandria. L.
virgínica. DC. (*wh.*) Virginian. N. America. 1742. 5. 8. H. ♄. Cav. dis. 5. t. 159. f. 2.
Malachodéndron. L. *marylándica.* A.R. t. 73.
GORD'ONIA. DC. GORD'ONIA. Monadelphia Polyandria. L.
1 Lasiánthus. DC. (*wh.*) smooth. N. America. 1739. 8. 11. F. ♄. Bot. mag. 668.
2 hæmatóxylon. DC. (*w.*) red-wooded. Jamaica. 1820. .... S. ♄.
3 pubéscens. DC. (*wh.*) pubescent. N. America. 1774. 8. 11. F. ♄. Vent. malm. t. 1.
4 Franklìni. L'Her. (*w.*) Franklin's. ——— ——— ——— F. ♄.
POLYSP'ORA. N.L.F. POLYSP'ORA. Monadelphia Polyandria.
axillàris. N.L.F. (*wh.*) axillary-flow'r'd E. Indies. 1816. 11. 3. S. ♄. Bot. reg. 349.
Camèllia axillàris. B.R. *et* B.M. 2047.

ORDO XXXV.

CAMELLIEÆ. *DC. prodr.* 1. *p.* 529.

CAME'LLIA. Ker. CAME'LLIA. Monadelphia Polyandria. L.
1 Thèa. L. en. (*wh.*) Bohea Tea. China. 1768. 11. 2. G. ♄. Lob. bot. cab. 226.
2 víridis. L. en. (*wh.*) green Tea. ——— ——— 10. 12. G. ♄. ——— 227.
Thèa chinénsis. B.M. 998.
3 euryoídes. B.R. (*wh.*) small-flower'd. ——— 1824. 1. 5. G. ♄. Bot. reg. 983.
4 oleifera. B.R. (*wh.*) oleiferous. ——— 1819. 1. 3. G. ♄. ——— 942.
5 Sesánqua. DC. (*wh.*) Lady Banks's. ——— 1811. 11. 1. G. ♄. Thunb. fl. jap. t. 30.
β semidùplex. (*wh.*) *semidouble.* ——— ——— ——— G. ♄. Bot. reg. 12.
γ flòre-plèno. (*wh.*) *double-flowered.* ——— 1823. ——— G. ♄. ——— 1091.
6 maliflòra. B.R. (*ro.*) pink-flowered. ——— 1819. 1. 4. G. ♄. ——— 547.
Sesanqua β rosea. Bot. reg. 547.
7 Kissi. DC. (*wh.*) Nepaul. Nepaul. 1823. .... G. ♄.
8 reticulàta. B.R. (*sc.*) Captain Rawes'. China. 1822. 2. 4. G. ♄. Bot. reg. 1078.
9 japónica. DC. (*va.*) common. ——— 1739. 12. 5. G. ♄. Cav. dis. 6. t. 160.
1 *rùbra.* (*sc.*) *single-red.* ——— ——— ——— G. ♄. Bot. mag. 42.
2 *álba.* (*wh.*) *single-white.* ——— .... ——— G. ♄. Bot. reg. 353.
3 *semidùplex.* (*re.*) *semidouble red.* ——— .... ——— G. ♄. Andr. rep. 559.
4 *rùbro-plèna.* (*re.*) *double red.* ——— .... ——— G. ♄. ——— 199.
5 *cárnea.* (*pi.*) *Middlemist's.* ——— .... ——— G. ♄. Lod. bot. cab. 455.
6 *myrtifòlia.* (*pi.*) *Myrtle-leaved.* ——— .... ——— G. ♄. Bot. mag. 1670.
involuta. *Bot. reg.* 633.
7 *hexangulàris.* (*re.*) *six sided.* ——— .... ——— G. ♄.
8 *atrorùbens.* (*sc.*) *Loddiges' red.* ——— .... ——— G. ♄. Lod. bot. cab. 170.
9 *anemoniflòra.* (*cr.*) *red Waratah.* ——— .... ——— G. ♄. Bot. mag. 1654.
10 *dianthiflòra.* (*v.*) *Carnation War.* Knights. .... ——— G. ♄. ——— 2577.
11 *blánda.* (*bh.*) *blush Waratah.* China. .... ——— G. ♄.
12 *pompònia.* (*bh.*) *Kew-blush.* ——— .... ——— G. ♄. Bot. reg. 22.

13 *pæoniflòra.*	(*pi.*)	*Pæony-flowered.*	China.	....	12. 5.	G. ♄.	Lod. bot. cab. 238.
14 *Welbánkii.*	(*wh.*)	*Welbank's.*	——	....	——	G. ♄.	——— 1198.
15 *lùteo-álba.*	(*st.*)	*pale yellow.*	——	....	——	G. ♄.	Bot. reg. 708.
16 *flavéscens.*	(*st.*)	*buff.*	——	....	——	G. ♄.	——— 112.
17 *álbo-plèna.*	(*wh.*)	*double white.*	——	....	——	G. ♄.	Andr. rep. 25.
18 *fimbriàta.*	(*wh.*)	*fringed white.*	——	....	1. 4.	G. ♄.	
19 *imbricàta.*	(*re.*)	*imbricate-petal'd.*	——	....	——	G. ♄.	
20 *variegàta.*	(*v.*)	*double striped.*	——	....	——	G. ♄.	Andr. rep. 91.
21 *crassinérvis.*	(*re.*)	*thick nerved.*	——	....	——	G. ♄.	
22 *expánsa.*	(*re.*)	*expanded.*	——	....	——	G. ♄.	
23 *conchiflòra.*	(*re.*)	*shell-flowered.*	——	....	——	G. ♄.	
24 *rubricáulis.*	(*sc.*)	*Lady Ad'l. Campbell's*	——	....	——	G. ♄.	
25 *longifòlia.*	(*re.*)	*long-leaved.*	Malcolm's.	....	——	G. ♄.	
26 *Chandlèri.*	(*v.*)	*striped Waratah.*	Chandler's.	1822.	12. 5.	G. ♄.	Chandl. camel. t. 1. 2.
versicolor. *Bot. reg.* 887.							
27 *Aitòni.*	(*sc.*)	*large single-red.*	——	——	——	G. ♄.	Chandl. camel. t. 3.
28 *althæiflora.*	(*re.*)	*Hollyhock-fl^d.*	——	——	——	G. ♄.	——— t. 4.
29 *corallìna.*	(*re.*)	*coral-flowered.*	——	——	——	G. ♄.	——— t. 5.
30 *insígnis.*	(*re.*)	*splendid.*	——	——	——	G. ♄.	——— t. 6.
31 *flòrida.*	(*re.*)	*cluster-flowered.*	——	——	——	G. ♄.	——— t. 7.
32 *anemonæflòra. alba.*	(*w.*)	*white Anemone-fl^d.*	——	——	——	G. ♄.	——— t. 8.
33 *Róssii.*	(*re.*)	*Ross's.*	Ross's.	——	——	G. ♄.	
34 *Wiltòniæ.*	(*va.*)	*Mrs. Wilton's.*	Knight's.	——	——	G. ♄.	
35 *Elphinstoniàna.*	(*re.*)	*Miss Elphinstone's.*	——	——	——	G. ♄.	
36 *rosæflòra.*	(*re.*)	*Rose-flowered.*	——	——	——	G. ♄.	
37 *Cliveàna.*	(*re.*)	*Lady H. Clive's.*	——	——	——	G. ♄.	
38 *heterophy'lla.*	(*re.*)	*various-leaved.*	——	——	——	G. ♄.	
39 *Egertòniæ.*	(*re.*)	*Mrs. Egerton's.*	——	——	——	G. ♄.	
40 *Goussòniæ.*	(*ro.*)	*Mrs. Goussone's.*	——	——	——	G. ♄.	
41 *Barnabìæ.*	(*pi.*)	*Mrs. Barnaby's.*	——	——	——	G. ♄.	
42 *radiàta.*	(*re.*)	*rayed single.*	——	——	——	G. ♄.	
43 *supìna.*	(*re.*)	*supine.*	——	——	——	G. ♄.	
44 *Hibbértii.*	(*re.*)	*Mr. Hibbert's.*	——	——	——	G. ♄.	
45 *Perc'yæ.*	(*re.*)	*D's. of Northumberland's*	——	——	——	G. ♄.	
46 *Woódsii.*	(*ro.*)	*Mr. Woods's.*	Chandler's.	——	——	G. ♄.	
47 *argéntea.*	(*re.w.*)	*silvery-cup'd.*	——	——	——	G. ♄.	
48 *basílica.*	(*re.*)	*princely.*	——	——	——	G. ♄.	
49 *princéps.*	(*sc.*)	*double-carmine.*	——	——	——	G. ♄.	
50 *sanguínea.*	(*cr.*)	*single-crimson.*	——	——	——	G. ♄.	
51 *serícea.*	(*ro.*)	*double-silky.*	——	——	——	G. ♄.	
52 *Byrónii.*	(*ro.*)	*Lord Byron's.*	——	——	——	G. ♄.	
53 *carnéscens.*	(*fl.*)	*single pale-red.*	——	——	——	G. ♄.	
54 *exímia.*	(*re.*)	*fine double-red.*	——	——	——	G. ♄.	
55 *bícolor.*	(*ro.w.*)	*single rosy & white.*	——	——	——	G. ♄.	
56 *purpuráscens.*	(*re.*)	*single dark-red.*	——	——	——	G. ♄.	
57 *gigántea.*	(*ro.*)	*large double.*	——	——	——	G. ♄.	
58 *lùcida.*	(*re.*)	*double bright-red.*	——	——	——	G. ♄.	
59 *ornàta.*	(*pi.*)	*pink Waratah.*	——	——	——	G. ♄.	
60 *papaveràcea.*	(*re.*)	*Poppy-flower'd.*	Loddiges.	——	——	G. ♄.	
61 *rotundifòlia.*	(*re.*)	*round-leaved.*	——	——	——	G. ♄.	
62 *Spofforthiàna.*	(*wh.*)	*red-lined white.*	Spofforth.	——	——	G. ♄.	
63 *Herbérti.*	(*ro.*)	*Mr. Herbert's.*	——	——	——	G. ♄.	
Spofforthiàna. rosea. *Herbert Mss.*			——	——	——	G. ♄.	
64 *Haylóckii.*	(*wh.*)	*Haylock's.*	——	——	——	G. ♄.	
65 *reflexa.*	(*cr.*)	*single reflex-petaled.*	——	——	——	G. ♄.	

ORDO XXXVI.

OLACINEÆ. *DC. prodr.* 1. *p.* 531.

OLA'X. DC. OLA'X. Triandria Monogynia. L.

1 scándens. DC.	(*wh.*)	climbing.	E. Indies.	1820.	1. 12.	S. ♄. ◡.	Roxb. cor. 2. t. 102.
2 imbricàta. DC.	(*wh.*)	imbricate.	——	——	....	S. ♄. ◡.	

SPERMA'XYRUM. DC.		SPERMA'XYRUM. Triandria Monogynia.					
stríctum. DC.	(*wh.*)	upright.	N. S. W.	1820.	....	G. ħ.	
XIM'ENIA. DC.		XIM'ENIA. Octandria Monogynia. L.					
1 americàna. DC.	(*wh.*)	spring.	W. Indies.	1759.	6. 9.	S. ♄.	Jacq. am. pict. t. 107.
2 inérmis. DC.	(*wh.*)	unarmed.	———	1820.	....	S. ħ.	

ORDO XXXVII.

AURANTIACEÆ. *DC. prodr.* 1. *p.* 535.

ATALA'NTIA. DC.		ATALA'NTIA. Monadelphia Octandria.					
monophy'lla. DC.	(*wh.*)	simple-leaved.	E. Indies.	1777.	6. 8.	S. ħ.	Roxb. cor. 2. t. 83.
Limònia monophy'lla. R.							
TRIPH'ASIA. DC.		TRIPH'ASIA. Hexandria Monogynia.					
trifoliàta. DC.	(*wh.*)	three-leaved.	China.	1787.	7. 8.	S. ħ.	Jacq. ic. rar. t. 463.
Limònia trifoliàta. And. rep. 143.							
LIM'ONIA. DC.		LIM'ONIA. Decandria Monogynia. L.					
1 acidíssima. DC.	(*wh.*)	sour-fruited.	E. Indies.	1824.	4. 8.	S. ♄.	Rumph. amb. 2. t. 43.
2 crenulàta. DC.	(*wh.*)	crenulate.	———	1808.	5. 7.	S. ♄.	Roxb. cor. 1. t. 86.
3 ambígua. DC.	(*wh.*)	ambiguous.	....	....	——	S. ♄.	
4 scándens. H.B.	(*wh.*)	climbing.	E. Indies.	1820.	....	S. ♄.	
CO'OKIA. DC.		CO'OKIA. Decandria Monogynia. L.					
púnctata. DC.	(*wh.*)	Wampee tree.	China.	1795.	5. 7.	S. ħ.	Jacq. schœnb. 1. t. 101.
MURRA'YA. DC.		MURRA'YA. Decandria Monogynia. L.					
1 exótica. DC.	(*wh.*)	Ash-leaved.	E. Indies.	1771.	8 9.	S. ♄.	Bot. reg. 434.
2 paniculàta. DC.	(*wh.*)	panicled.	———	1821.	4. 5.	S. ħ.	Hook. ex. flor. 134.
AGLA'IA. DC.		AGLA'IA. Pentandria Monogynia.					
odoràta. DC.	(*st.*)	sweet-scented.	China.	1810.	2. 5.	S. ħ.	Rumph. amb. 5. t. 18.
BE'RGERA. DC.		BE'RGERA. Decandria Monogynia. L.					
1 K'œnigii. DC.	(*gr.*)	Kœnig's.	E. Indies.	1818.	3. 6.	S. ♄.	Lodd. bot. cab. 1019.
2 integérrima. DC.	(*gr.*)	entire-leaved.	———	1820.	....	S. ♄.	
CLAUS'ENIA. DC.		CLAUS'ENIA. Octandria Monogynia.					
pentaphy'lla. DC.	(*wh.*)	five-leaved.	E. Indies.	1823.	....	S. ħ.	
GLYCO'SMIS. DC.		GLYCO'SMIS. Decandria Monogynia.					
1 arbòrea. DC.	(*wh.*)	tree.	E. Indies.	1796.	5. 8.	S. ħ.	Roxb. cor. 1. t. 85.
Limònia arbòrea. R.C.							
2 citrifòlia. H.T.	(*wh.*)	Citron-leaved.	China.	1820.	2. 6.	G. ħ.	Bot. mag. 2416.
Limònia citrifòlia. DC. *parviflora.* B.M.							
3 pentaphy'lla. DC.	(*wh.*)	five-leaved.	E. Indies.	1790.	6. 7.	S. ħ.	Roxb. cor. 1. t. 84.
Limònia pentaphy'lla. R.C.							
FER'ONIA. DC.		ELEPHANT APPLE. Decandria Monogynia.					
elephántum. DC.	(*bh.*)	Indian.	E. Indies.	1804.	....	S. ♄.	Roxb. cor. 2. t. 141.
ÆGLE. DC.		BENGAL QUINCE. Polyandria Monogynia.					
1 Mármelos. DC.	(*wh.*)	thorny.	E. Indies.	1759.	....	S. ħ.	Roxb. cor. 2. t. 143.
CI'TRUS. DC.		ORANGE-TREE. Polyadelphia Polyandria. L.					
1 Médica. DC.	(*wh.*)	Lemon-tree.	Asia.	1648.	5. 7.	G. ħ.	Ris. an. mus. 20. t. 2. f. 2.
2 Limétta. DC.	(*wh.*)	Bergamotte.	———	....	——	G. ♄.	——— 20. t. 2. f. 1.
3 Limònum. DC.	(*wh.*)	Citron.	———	....	——	G. ħ.	Ferr. hesp. t. 247. &c.
4 ácida. H.B.	(*wh.*)	Lime.	———	....	——	G. ħ.	
5 Aurántium. DC.	(*wh.*)	common.	———	1595.	——	G. ♄.	Ris. an. mus. 20. t. 1. f. 1. 2.
6 vulgàris. DC.	(*wh.*)	Myrtle-leaved.	———	....	——	G. ♄.	Bot. reg. t. 346.
7 decumàna. DC.	(*wh.*)	Shaddock.	India.	1724.	——	G. ħ.	Rumph. am. 2. t. 24. f. 2.
8 hy'strix. DC.	(*wh.*)	very spiny.	———	———	——	G. ♄.	——— 2. t. 28?
9 japónica. DC.	(*wh.*)	small-fruited.	Japan.	1815.	——	G. ħ.	Th. ic. jap. t. 15.
10 nòbilis. DC.	(*wh.*)	Mandarin.	China.	1805.	——	G. ♄.	Andr. rep. 608.
β minor.	(*wh.*)	Tangerine.	———	———	——	G. ħ.	Bot. reg. 211.
11 Margarìta. DC.	(*wh.*)	sweet Lemon.	———	1820.	——	G. ħ.	
12 madurénsis. DC.	(*wh.*)	spineless.	China.	1816.	——	G. ħ.	Rumph. amb. 2. t. 31.
13 angulàta. DC.	(*wh.*)	angular-fruited.	E. Indies.	....	——	G. ħ.	——— 2. t. 32.
14 buxifòlia. DC.	(*wh.*)	retuse-leaved.	China.	....	——	G. ħ.	

ORDO XXXVIII.

HYPERICINEÆ. *DC. prodr.* 1. *p.* 541.

Subordo I. HYPERICINEÆ VERÆ. DC.

Tribus I. *VISMIEÆ.* DC.

HARO'NGA. DC.		HARO'NGA. Polyadelphia Dodecandria.					
1 madagascariénsis. DC.	(*y.*)	Madagascar.	Madagascar.	1824.	....	S. ♄.	Lam. ill. t. 645.
Arungàna paniculàta. P.S.							
VI'SMIA. DC.		VI'SMIA. Polyadelphia Polyandria.					
1 sessilifòlia. DC.	(*ye.*)	sessile-leaved.	Guiana.	1826.	....	S. ♄.	Aubl. guian. 2. t. 312. f. 2.
2 guianénsis. DC.	(*ye.*)	Guiana.	———	1828.	....	S. ♄.	——— 2. t. 311.
3 brasiliénsis. DC.	(*ye.*)	Brasilian.	Brazil.	1822.	5. 6.	S. ♄.	Chois. p. hyp. 35. t. 2.
4 guineénsis. DC.	(*ye.*)	African.	Sierra Leon.	1823.	——	S. ♄.	Lin. amœn. 8. t. 8. f. 1.

Tribus II. *HYPERICEÆ.* DC.

ANDROS'ÆMUM. DC.		TUTSAN. Monadelphia Polyandria.					
officinale. DC.	(*ye.*)	common.	Britain.	....	7. 9.	H. ♄.	Eng. bot. 1225.
HYPE'RICUM. DC.		ST. JOHN'S WORT. Polyadelphia Polyandria. L.					
1 elàtum. DC.	(*ye.*)	tall.	N. America.	1762.	7. 8.	H. ♄.	Wats. dend. brit. t. 85.
2 frondòsum. DC.	(*ye.*)	green.	———	1806.	——	H. ♄.	
3 amœ'num. DC.	(*ye.*)	pretty.	Carolina.	1812.	——	H. ♄.	
4 grandifòlium. DC.	(*ye.*)	great-leaved.	Teneriffe.	1818.	——	G. ♄.	Chois. pr. hyp. t. 3.
5 hircìnum. DC.	(*ye.*)	stinking.	S. Europe.	1640.	7. 9.	H. ♄.	Wats. dend. brit. t. 86.
β mìnus. W.D.	(*ye.*)	*smaller.*	———	——	——	H. ♄.	——— t. 87.
6 foliòsum. DC.	(*ye.*)	shining.	Azores.	1778.	8.	G. ♄.	
7 floribúndum. DC.	(*ye.*)	many-flowered.	Madeira.	1779.	——	G. ♄.	Com. hort. 2. t. 68.
8 oly'mpicum. DC.	(*ye.*)	Olympic.	Levant.	1706.	7. 9.	H. ♄.	Exot. bot. 2. t. 96.
9 canariénse. DC.	(*ye.*)	Canary.	Canaries.	1699.	——	G. ♄.	
10 chinénse. DC.	(*ye.*)	Chinese.	China.	1753.	3. 9.	G. ♄.	Bot. mag. 334.
monógynum. B.M.							
11 cordifòlium. DC.	(*ye.*)	heart-leaved.	Nepaul.	1822.	6. 9.	G. ♄.	
12 pyramidàtum. DC.	(*ye*)	pyramidal.	Canada.	1759.	7. 8.	H. ♃.	Vent. malm. t. 118.
13 A'scyron. DC.	(*ye.*)	Siberian.	Siberia.	1774.	6. 9.	H. ♃.	Gmel. sib. 4. t. 69.
14 ascyroídes. DC.	(*ye.*)	large-capsuled.	N. America.	1812.	6. 7.	H. ♃.	
15 pátulum. DC.	(*ye.*)	spreading.	Japan.	1820.	7. 9.	G. ♄.	Thunb. jap. ic. 17.
16 Uràlum. D.P.	(*ye.*)	Nepaul.	Nepaul.	——	——	F. ♄.	Bot. mag. 2375.
17 Kalmiànum. DC.	(*ye.*)	Kalmia-leaved.	Virginia.	1759.	6. 7.	H. ♄.	
18 calycìnum. DC.	(*ye.*)	large-flowered.	Ireland.	....	6. 9.	H. ♄.	Eng. bot. 2017.
19 baleàricum. DC.	(*ye.*)	warted.	Majorca.	1714.	3. 9.	G. ♄.	Bot. mag. 137.
20 virgínicum. DC.	(*re.*)	Virginian.	N. America.	1800.	7. 9.	H. ♃.	Andr. rep. 552.
Elòdea campanulàta. Ph.							
21 angulòsum. DC.	(*ye.*)	tooth'd-flow'r'd.	N. America.	1812.	6. 7.	H. ♃.	
22 punctàtum. DC.	(*ye.*)	spotted.	———	1789.	7. 9.	F. ♃.	
carymbòsum. W. *maculàtum.* Walt.							
23 dolabrifórme. DC.	(*ye.*)	hatchet-like.	———	1818.	——	H. ♃.	
24 densiflòrum. DC.	(*ye.*)	dense-flowered.	Virginia.	1812.	6. 8.	F. ♄.	
25 rosmarinifòlium. DC.	(*y.*)	Rosemary-ld.	Carolina.	——	——	F. ♄.	
26 virgàtum. DC.	(*ye.*)	twiggy.	N. America.	1823.	6. 8.	H. ♃.	
27 prolíficum. DC.	(*ye.*)	prolific.	———	1758.	——	H. ♄.	Wats. dend. brit. t. 88.
Kalmiànum. Maund. bot. gard. 194. t. 49. *nec aliorum. foliosum.* Jacq. schœnb. 3. t. 299. *non* Ait.							
28 gláucum. DC.	(*ye.*)	glaucous.	N. America.	1812.	7. 8.	H. ♃.	
29 lævigàtum. H.K.	(*ye.*)	smooth.	———	1772.	——	H. ♃.	
30 nudiflòrum. DC.	(*ye.*)	naked-panicled.	———	1811.	8. 10.	H. ♄.	
31 sphærocárpon. DC.	(*y.*)	round-capsuled.	———	1827.	6. 8.	H. ♃.	

32 quadrángulum. W.	(*ye.*)	square-stalked.	Britain.	....	7. 8.	H. ♃.	Eng. bot. 370.
33 dùbium. E.B.	(*ye.*)	imperforate.	————	....	——	H. ♃.	———— 296.
34 maculàtum. All.	(*ye.*)	spot-flowered.	Europe.	....	——	H. ♃.	All. ped. t. 83. f. 1.
35 undulàtum. W.en.	(*y.*)	wave-leaved.	Barbary.	1802.	——	F. ♃.	
36 attenuàtum. DC.	(*ye.*)	attenuated.	Dahuria.	1828.	——	H. ♃.	Chois. pr. hyp. p. 47. t. 6.
37 rèpens. DC.	(*ye.*)	creeping.	S. Europe.	1827.	——	H. ♃.	
38 japónicum. DC.	(*ye.*)	Japan.	Japan.	1820.	——	G. ♃.	Thunb. jap. t. 31.
39 críspum. DC.	(*ye.*)	curl-leaved.	Greece.	1688.	——	F. ♃.	Bocc. mus. 2. t. 12.
40 pilòsum. DC.	(*ye.*)	pilose.	N. America.	1818.	——	H. ♄.	Pluk. alm. t. 245. f. 6.
setòsum. H.K.							
41 símplex. DC.	(*ye.*)	simple-stalked.	————	——	——	H. ☉.	Pluk. alm. t. 421. f. 3.
42 heterophy'llum. DC.	(*y.*)	various-leaved.	Persia.	1812.	——	G. ♄.	Vent. cels. t. 68.
43 ægyptìacum. DC.	(*ye.*)	Egyptian.	Egypt.	1787.	6. 7.	G. ♄.	Bot. reg. 196.
44 humifùsum. DC.	(*ye.*)	trailing.	Britain.	....	7. 8.	H. ♃.	Eng. bot. 1226.
45 pusíllum. DC.	(*ye.*)	small.	N. S. W.	1818.	——	G. ♃.	Lab. nov. holl. 2. t. 175.
46 involutum. DC.	(*ye.*)	involute-fl^d.	————	1822.	——	G. ♃.	———— 2. t. 174.
47 perforàtum. DC.	(*ye.*)	perforate.	Britain.	....	——	H. ♃.	Eng. bot. 295.
48 veronénse. Sck.	(*ye.*)	Verona.	Italy.	1820.	——	H. ♃.	
49 quinquenérvium. DC.	(*ye.*)	small-fl^d.	N. America.	1759.	6. 9.	F. ♃.	
mùtilum. W.							
50 canadénse. DC.	(*wh.*)	Canadian.	Canada.	1770.	7. 9.	H. ♃.	
51 perfoliàtum. W.	(*wh.*)	perfoliate.	Italy.	1785.	5. 6.	F. ♃.	
52 axillàre. DC.	(*ye.*)	clustered.	N. America.	1806.	7.	F. ♄.	
fasciculàtum. W.							
53 Elòdes. DC.	(*ye.*)	marsh.	Britain.	....	7. 8.	H. ♃.	Eng. bot. 109.
54 tomentòsum. DC.	(*ye.*)	woolly.	S. Europe.	1648.	7. 9.	F. ♃.	Moris. his. 2. s. 5. t. 6. n. 5.
55 lanuginòsum. DC.	(*ye.*)	downy.	Levant.	1822.	——	F. ♃.	
56 hirsùtum. DC.	(*ye.*)	hairy.	Britain.	....	6. 7.	H. ♃.	Eng. bot. 1156.
57 nummulàrium. DC.	(*ye.*)	Money-wort.	S. Europe.	1818.	——	H. ♃.	Lam. ill. t. 643.
58 élegans. DC.	(*ye.*)	elegant.	Sibéria.	——	7. 9.	H. ♃.	Spr. fl. hal. n. 864. t. 9.
Kohliànum. Spr.							
59 glandulòsum. DC.	(*ye.*)	glandular.	Madeira.	1777.	5. 8.	G. ♄.	
60 refléxum. DC.	(*ye.*)	reflex-leaved.	Teneriffe.	1778.	6. 9.	G. ♄.	
61 púlchrum. DC.	(*ye.*)	small upright.	Britain.	....	7. 8.	H. ♃.	Eng. bot. 1227.
62 elodeoídes. DC.	(*ye.*)	close panicled.	Nepaul.	1823.	....	G. ♃.	
nervòsum. D.P.							
63 barbàtum. DC.	(*ye.*)	bearded.	Scotland.	....	6. 10.	H. ♃.	Eng. bot. 1985.
64 montànum. DC.	(*ye.*)	mountain.	Britain.	....	7. 8.	H. ♃.	———— 371.
65 fimbriàtum. DC.	(*ye.*)	fimbriate.	Pyrenees.	1818.	——	H. ♃.	Vill. delph. 3. t. 44.
Richèri. Vill.							
66 æthiòpicum. DC.	(*ye.*)	Ethiopian.	C. B. S.	1817.	——	G. ♄.	
67 ciliàtum. DC.	(*ye.*)	fringed.	S. Europe.	1739.	7.	F. ♃.	Boc. mus. 2. t. 127.
68 serpyllifòlium. DC.	(*ye.*)	Thyme-leaved.	Levant.	1688.	7. 8.	F. ♄.	Mor. his. 2. s. 5. t. 6. f. 2.
69 triplinérve. DC.	(*ye.*)	triply-nerved.	N. America.	1823.	7. 9.	H. ♃.	Vent. cels. t. 58.
70 empetrifòlium. DC.	(*y.*)	Crowberry-l^d.	S. Europe.	....	6. 8.	H. ♄.	Wats. dend. brit. 141.
71 Còris. DC.	(*ye.*)	Coris-leaved.	Levant.	1640.	5. 9.	G. ♄.	Bot. mag. 178.
72 ericoídes. DC.	(*ye.*)	Heath-leaved.	S. Europe.	1818.	——	G. ♄.	Cav. ic. 2. t. 122.
73 mexicànum. DC.	(*ye.*)	Mexican.	Mexico.	1826.	....	G. ♄.	Linn. amœn. 8. t. 8. f. 2.
74 fasciculàtum. DC.	(*ye.*)	Aspalathus-like.	N. America.	1811.	6. 8.	F. ♄.	
aspalathoídes. W.							
75 verticillàtum. DC.	(*ye.*)	whorl-leaved.	C. B. S.	1784.	——	G. ♃.	
76 cochinchinénse. DC.	(*sc.*)	Cochinchina.	Cochinchina.	1822.	6. 8.	G. ♄.	
A'SCYRUM. DC.		A'SCYRUM. Monadelphia Polyandria.					
1 pùmilum. DC.	(*ye.*)	dwarf.	Georgia.	1806.	6. 8.	G ♃.	
2 Crúx A'ndreæ. DC.	(*y.*)	St. Andrew's-cr.	N. America.	1759.	7.	G. ♄.	Burm. am. 146. t. 152.
3 hypericoídes. DC.	(*ye.*)	Hypericum-like.	————	——	7. 9.	G. ♄.	———— t. 152. f. 1.
4 stáns. DC.	(*ye.*)	large-flowered.	————	1806.	——	G. ♄.	
5 amplexicáule. DC.	(*ye.*)	stem-clasping.	————	——	——	G. ♄.	
SAR'OTHRA. W.		SAR'OTHRA. Pentandria Trigynia.					
gentianoídes. W.	(*ye.*)	Gentian-like.	N. America.	1768.	6. 10.	H. ☉.	Pluk. mant. t. 342. f. 2.
hypericoídes. N. *Hypericum Sarothra.* M.							

Subordo II. *HYPERICINEÆ ANOMALÆ.* DC.

CARPODO'NTOS. DC.		CARPODO'NTOS. Polyandria Pentagynia.					
lùcida. DC.	(*wh.*)	shining.	V. Diem. Is.	1820.	....	G. ♄.	Lab. voy. 2. t. 18.

ORDO XXXIX.

GUTTIFERÆ. *DC. prodr.* 1. *p.* 557.

Tribus I. *CLUSIEÆ.* DC.

MAH'UREA. DC.		MAH'UREA. Polyandria Monogynia.					
palústris. DC.	(*pu.*)	marsh.	Trinidad.	1820.	....	S. ♄.	Aubl. guian. 1. t. 222.
M'ARILA. DC.		M'ARILA. Monadelphia Polyandria.					
racemòsa. DC.	......	clustered.	W. Indies.	1827.	....	S. ♄.	
GODO'YA. DC.		GODO'YA. Polyandria Monogynia.					
gemmiflòra. M.N.	(*ye.*)	gem-flowered.	Brazil.	1824.	....	S. ♄.	Mart. nov. gen. 3. t. 74.
CL'USIA. DC.		BALSAM-TREE. Polygamia Monœcia. L.					
1 ròsea. DC.	(*ro.*)	Rose-coloured.	America.	1692.	7. 8.	S. ♄.	Cates. car. 2. t. 99.
2 álba. DC.	(*wh.*)	white-flowered.	S. America.	1752.	——	S. ♄.	Jacq. amer. t. 166.
3 flàva.	(*ye.*)	yellow-flowered.	Jamaica.	1759.	9.	S. ♄.	Bot. rep. 223.
4 venòsa. DC.	(*wh.*)	veiny-leaved.	S. America.	1733.	....	S. ♄.	Plum. ic. 87. f. 2.
5 nemoròsa. DC.	(*w.pu.*)	wood.	W. Indies.	1826.	....	S. ♄.	
HAV'ETIA. Kth.		HAV'ETIA. Diœcia Tetrandria.					
laurifòlia. Kth.	(*wh.*)	Laurel-leaved.	S. America.	1820.	....	S. ♄.	H. et B. n. g. am. 5. t. 462.
Clùsia tetrándra. W.							

Tribus II. *GARCINIEÆ.* DC.

MARIA'LVA. DC.		MARIA'LVA. Monadelphia Polyandria.					
guianénsis. DC.	(*gr.*)	Guiana.	Guiana.	1827.	....	S. ♄.	Aubl. guian. 2. t. 364.
MICRA'NTHERA. DC.		MICRA'NTHERA. Diœcia Polyandria.					
clusiæfòlia. DC.	....	Clusia-leaved.	Cayenne.	1823.	....	S. ♄.	Chois. m. h. P. 1. t. 11. 12.
GARCI'NIA. DC.		GARCI'NIA. Polyandria Monogynia.					
1 Mangostana. DC.	(*ro.*)	Mangosteen.	Java.	1789.	....	S. ♄.	Lod. bot. cab. 845.
2 cornea. DC.	(*ye.*)	nodding-flow'd.	E. Indies.	1817.	....	S. ♄.	Rumph. amb. 3. t. 30.
3 Cambogia. DC.	(*ye.*)	yellow-flower'd.	———	1822.	....	S. ♄.	Rheed. mal. 1. t. 24.
4 Cowa. DC.	(*ye.*)	polygamous.	———	——	....	S. ♄.	

Tribus III. *CALOPHYLLEÆ.* DC.

MAMM'EA. DC.		MAMMEE-TREE. Polyandria Monogynia. L.					
1 americàna. DC.	(*wh.*)	America.	S. America.	1737.	....	S. ♄.	Jacq. amer. t. 182. f. 32.
2 africàna. G.D.	(*wh.*)	African.	Sierra Leon.	1820.	....	S. ♄.	
XANTHOCH'YMUS. DC.		XANTHOCH'YMUS. Polyadelphia Polyandria.					
1 pictòrius. H.B.	(*wh.*)	painters.	E. Indies.	1796.	....	S. ♄.	Roxb. cor. 2. t. 196.
2 dúlcis. H.B.	(*st.*)	sweet.	———	1820.	....	S. ♄.	——— 3. t. 270.
3 ovalifòlius. H.B.	....	oval-leaved.	———	——	....	S. ♄.	
4 lùteus. L.C.	(*ye.*)	yellow.	E. Indies.	1823.	....	S. ♄.	
5 macrophy'llus. L.C.	..	large leaved.	———	——	....	S. ♄.	
6 purpùreus. L.C.	(*pu.*)	purple.	———	——	....	S. ♄.	
CALOPHY'LLUM. DC.		CALOPHY'LLUM. Polyandria Monogynia. L.					
1 Inophy'llum. DC.	(*wh.*)	sweet-scented.	E. Indies.	1793.	....	S. ♄.	Rheed. mal. 4. t. 38.
2 Tacamaháca. DC.	(*wh.*)	acute-leaved.	Bourbon.	1822.	....	S. ♄.	Pluk. alm. t. 147. f. 3.
3 Cálaba. DC.	(*wh.*)	Calaba-tree.	W. Indies.	1780.	....	S. ♄.	Jacq. amer. t. 165.
4 spùrium. DC.	(*wh.*)	spurious.	E. Indies.	——	....	S. ♄.	Rheed. mal. 4. t. 39.
Cálaba. L.							
PENTADE'SMA. H.T.		BUTTER-TREE. Polyandria Monogynia.					
butyràcea. H.T.	(*ye.*)	African.	Sierra Leon.	1821.	....	S. ♄.	

Tribus IV. *SYMPHONIEÆ.* DC.

CANE'LLA. DC.		CANELLA. Dodecandria Monogynia.					
1 álba. DC.	(*pu.*)	broad-leaved.	W. Indies.	1735.	....	S. ♄.	Linn. trans. 1. t. 8.
2 laurifòlia. L.C.	(*pu.*)	narrow-leaved.	—	....	....	S. ♄.	
MORONOB'EA. DC.		MORONOB'EA. Monadelphia Pentandria.					
coccínea. DC.	(*sc.*)	scarlet.	Guiana.	1825.	....	S. ♄.	Aubl. guian. 2. t. 313.
Symphònia globulífera. L.							
SING'ANA. DC.		SING'ANA. Polyandria Monogynia.					
guianénsis. DC.	(*wh.*)	Guiana.	Guiana.	1827.	....	S. ♄.◡.	Aubl.guian.sup.2.t.230.
Sternbéckia lateriflòra. W.							

ORDO XL.

MARCGRAVIACEÆ. *DC. prodr.* 1. *p.* 565.

Subordo I. *MARCGRAVIEÆ.* DC.

ANTHOL'OMA. DC.		ANTHOL'OMA. Polyandria Monogynia.					
montàna. DC.	(*ye.*)	mountain.	N.Caledonia.	1820.	....	S. ♄.	Labill. voy. t. 41.
MARCGR'AVIA. DC.		MARCGR'AVIA. Polyandria Monogynia. L.					
1 umbellàta. DC.	(*gr.*)	umbelled.	W. Indies.	1792.	6. 8.	S. ♄.	Hook. exot. flor. 160.
2 coriàcea. DC.	(*gr.*)	leathery-leav'd.	Guiana.	1820.	—	S. ♄.	
3 pícta. DC.	(*gr.*)	painted-leav'd.	....	1827.	....	S. ♄.	

Subordo II. *NORANTEÆ.* DC.

NORA'NTEA. DC.		NORA'NTEA. Polyandria Monogynia. L.					
1 guianénsis. DC.	(*pu.*)	Guiana.	Guiana.	1827.	....	S. ♄.	Aubl. guian. 1. t. 220.
2 brasiliénsis. DC.	....	Brasilian.	Brazil.	1820.	....	S. ♄.	
3 índica.	....	Indian.	Mauritius.	1822.	....	S. ♄.	
RUY'SCHIA. DC.		RUY'SCHIA. Pentandria Monogynia.					
1 Souroubèa. DC.	(*st.*)	Guiana.	Guiana.	1826.	....	S. ♄.	Aubl. guian. 1. t. 97.
2 clusiæfòlia. DC.	(*pu.*)	Clusia-leaved.	W. Indies.	1823.	....	S. ♄.	Jacq. amer. t. 51. f. 2.

ORDO XLI.

HIPPOCRATEACEÆ. *DC. prodr.* 1. *p.* 567.

HIPPOCR'ATEA. DC.		HIPPOCR'ATEA. Triandria Monogynia. L.					
1 obcordàta. DC.	(*st.*)	obcordate.	W. Indies.	1819.	....	S. ♄.◡.	Lam. ill. 1. t. 28. f. 1.
scándens. Jacq. amer. t. 9.							
2 ovàta. DC.	(*ye.*)	oval-leaved.	S. America.	1739.	....	S. ♄.◡.	Lam. ill. 1. t. 28. f. 2.
volùbilis. L.							
3 índica. DC.	(*ye.*)	Indian.	E. Indies.	1818.	....	S. ♄.◡.	Roxb. cor. 2. t. 130.
4 obtusifòlia. DC.	(*st.*)	blunt-leaved.	—	1820.	....	S. ♄.◡.	
5 arbòrea. DC.	(*st.*)	tree.	—	1822.	....	S. ♄.◡.	Roxb. cor. 3. t. 205.
A'NTHODON. DC.		A'NTHODON. Triandria Monogynia.					
1 paniculàtum. DC.	....	panicled.	Rio Janeiro.	1818.	....	S. ♄.◡.	
2 ellípticum. DC.	....	elliptic-leaved.	—	—	....	S. ♄.◡.	
TONSE'LLA. DC.		TONSE'LLA. Triandria Monogynia.					
1 pyrifórmis. H.T.	(*gr.*)	pear-fruited.	Sierra Leon.	1822.	....	S. ♄.	
2 scàbra. R.S.	(*gr.*)	rough.	Trinidad.	1824	....	S. ♄.◡.	Aubl. guian. 1. t. 10.
Tontèlea scándens. Aub.							

JO'HNIA. F.I. Jo'hnia. Triandria Monogynia. F. I.
1 salacioídes. F.I. (or.) entire-leaved. E. Indies. 1822. .... S. ♄.◡.
2 coromandeliàna. F.I. saw-leaved. ——— —— .... S. ♄.◡.
TRIG'ONIA. DC. Trig'onia. Monadelphia Pent-Decandria.
1 villòsa. DC. (y.re.) villous. Cayenne. 1820. .... S. ♄. Aub. gui. 1. t. 149.
2 móllis. DC. .... soft-leaved. Brazil. 1823. .... S. ♄.
3 lævis. DC. (wh.) smooth. Guiana. 1826. .... S. ♄. Aub. gui. 1. t. 150.

ORDO XLII.

ERYTHROXYLEÆ. *DC. prodr.* 1. *p.* 573.

ERYTHRO'XYLUM. DC. Erythro'xylum. Decandria Trigynia. L.
1 hypericifòlium. DC. (w.) Hypericum-ld. Mauritius. 1818. .... S. ♄. Cav. dis. 8. t. 230.
2 havanense. DC. (wh.) Havannah. W. Indies. 1822. .... S. ♄. Jacq. amer. t. 87. f. 2.
3 laurifolium. DC. (wh.) Laurel-leaved. Mauritius. —— .... S. ♄. Cav. dis. 8. t. 226.
4 areolàtum. DC. (wh.) Carthagena. Carthagena. 1826. .... S. ♄. Jacq. amer. t. 187. f. 1.
carthagenénse. Jacq.
S'ETHIA. Kth. S'ethia. Decandria Monogynia.
índica. DC. (ye.) Indian. E. Indies. 1823. .... S. ♄. Roxb. cor. 1. t. 88.
Erythróxylon monog'ynum. R.C.

ORDO XLIII.

MALPIGHIACEÆ. *DC. prodr.* 1. *p.* 577.

Tribus I. *MALPIGHIEÆ.* DC.

MALPI'GHIA. DC. Malpi'ghia. Decandria Trigynia. L.
1 fucàta. DC. (li.) painted. W. Indies. 1814. 1. 12. S. ♄. Bot. reg. 189.
macrophy'lla. Colla hort. ripul. t. 11. *non* Juss.
2 ùrens. DC. (li.) stinging. S. America. 1737. 7. 10. S. ♄. Bot. reg. 96.
3 angustifòlia. DC. (li.) narrow-leaved. W. Indies. —— 7. 8. S. ♄. Lod. bot. cab. 321.
4 aquifòlia. DC. (li.) Holly-leaved. S. America. 1759. 8. 9. S. ♄. ——— 1079.
5 coccífera. DC. (li.) kermes-oak-ld. ——— 1733. —— S. ♄. Bot. reg. 568.
6 glábra. DC. (pu.) smooth-leaved. ——— 1757. 3. 7. S. ♄. Bot. mag. 813.
7 biflòra. DC. (ye.) two-flowered. ——— .... —— S. ♄. Cav. dis. 8. t. 234. f. 2.
punicifòlia. Cav.
8 punicifòlia. DC. (pu.) Punica-leaved. ——— 1690. —— S. ♄. Plum. ic. t. 166. f. 2.
9 nitida. DC. (pu.) glossy-leaved. S. America. 1733. 3. 8. S. ♄.
10 incàna. DC. (ro.) hoary. Campeachy. —— —— S. ♄.
11 ? faginea. DC. (ye.) Beech-leaved. S. America. 1822. .... S. ♄.
12 ? dùbia. DC. (ye.) doubtful. W. Indies. 1824. .... S. ♄. Cav. dis. 8. t. 242.
BYRSON'IMA. DC. Byrson'ima. Decandria Trigynia. L.
1 verbascifòlia. DC. (ye.) Mullein-leaved. Guiana. 1818. 3. 8. S. ♄. Aub. gui. 1. t. 184.
2 nervòsa. DC. (ye.) prominent-nd. Brazil. —— —— S. ♄.
3 altíssima. DC. (wh.) tall. Guiana. 1820. .... S. ♄. Aub. gui. 1. t. 181.
4 crassifòlia. DC. (ye.) thick-leaved. ——— 1793. 8. S. ♄. ——— 1. t. 182.
Malpíghia crassifòlia. W.
5 chrysophy'lla. DC. (ye.) golden-leaved. S. America. 1822. .... S. ♄.
6 spicàta. DC. (ye.) spiked. Brazil. 1818. 5. 8. S. ♄. Cav. dis. 8. t. 237.
7 lùcida. DC. (re.) wedge-leaved. W. Indies. 1759. —— S. ♄. Bot. mag. 2462.
Malpíghia lùcida. W.
8 lævigàta. DC. (st.) smooth. Cayenne. 1826. .... S. ♄.
9 élegans. DC. (li.) elegant. Guiana. 1827. .... S. ♄.
10 dénsa. DC. (re.) dense-flower'd. ——— 1825. .... S. ♄.
11 coriàcea. DC. (wh.) leathery-leaved. Jamaica. 1814. —— S. ♄. Sloan. his. 2. t. 163. f. 1.

12 pállida. DC. (ye.) pale-leaved. W. Indies. 1820. 5. 8. S. ♄.
13? volùbilis. DC. (ye.) twining. ——— 1793. 8. 9. S. ♄. Bot. mag. 809.
Malpíghia volùbilis. B.M.
BUNCH'OSIA. DC. BUNCH'OSIA. Decandria Monogynia.
1 glandulòsa. DC. (ye.) 2-glanded. Antilles. 1804. 3. 5. S. ♄. Cav. dis. 8. t. 239. f. 2.
Malpíghia glandulòsa. Cav.
2 glandulífera. DC. (ye.) 4-glanded. Caracas. 1806. —— S. ♄. Jacq. coll. 5. f. 3.
Malpíghia glandulòsa. Jacq. ic. 3. t. 469.
3 cornifòlia. DC. (wh.) Cornus-leaved. S. America. 1824. .... S. ♄.
4 polystàchya. DC. (ye.) many-spiked. W. Indies. 1806. 3. 5. S. ♄. And. bot. rep. 604.
5 mèdia. DC. (ye.) intermediate. ——— 1790. —— S. ♄.
Malpíghia mèdia. H.K.
6 tuberculàta. DC. (ye.) tuberculate. Caracas. 1823. .... S. ♄. Jacq. schœn. 1. t. 104.
7 argéntea. DC. (ye.) silvery. ——— 1820. .... S. ♄. Jacq. frag. t. 83.
8 nítida. DC. (ye.) glossy-leaved. W. Indies. 1733. 3. 8. S. ♄. Cav. dis. 8. t. 239. f. 1.
9 canéscens. DC. ...... downy-leaved. ——— 1742. .... S. ♄.
10? paniculàta. DC. (pu.) panicled. ——— 1733. .... S. ♄.
GALPH'IMIA. DC. GALPH'IMIA. Decandria Trigynia.
1 hirsùta. DC. (ye.) hairy. Mexico. 1824. 3. 6. S. ♄.
2 gláuca. DC. (ye.) glaucous-leav'd. ——— 1826. .... S. ♄. Cav. ic. 5. t. 489.
3 glandulòsa. DC. (ye.) glandular. ——— 1824. 4. 6. S. ♄. ——— 6. t. 563.
THRYA'LLIS. L. THRYA'LLIS. Decandria Trigynia.
? brachystachys. B.R. (y.) short-spiked. Brazil. 1823. 9. 11. S. ♄.⌣. Bot. reg. 1162.

Tribus II. *HYPTAGEÆ.* DC.

HIPT'AGE. DC. HIPT'AGE. Decandria Monogynia.
1 Madablòta. DC. (wh.) clustered. E. Indies. 1796. 3. 4. S. ♄.⌣. Roxb. cor. 1. t. 18.
Gærtnèra racemòsa. A.R. 600.
2 obtusifòlia. DC. (wh.) blunt-leaved. China. 1822. .... S. ♄.⌣.
Gærtnèra obtusifòlia. H.B.
GAUDICHA'UDIA. DC. GAUDICHA'UDIA. Pentandria Monogynia.
cynanchoídes. DC. (ye.) Cynanchum-like. Mexico. 1824. .... S. ♄.⌣. H.etB.n.g.am. 5. t. 445.

Tribus III. *BANISTERIEÆ.* DC.

HIR'ÆA. DC. HIR'ÆA. Decandria Trigynia. L.
1 odoràta. DC. .... sweet-scented. Guinea. 1822. .... S. ♄.⌣.
2 glaucéscens. L.C. .... glaucescent. .... S. ♄.⌣.
3 nùtans. DC. ...... nodding. E. Indies. 1818. 5. 8. S. ♄.⌣.
4 índica. DC. (wh.) Indian. ——— —— —— S. ♄.⌣. Roxb. cor. 2. t. 160.
TRIO'PTERIS. DC. TRIO'PTERIS. Decandria Trigynia. L.
1 serícea. L.C. .... silky. S. America. 1820. .... S. ♄.⌣.
2 jamaicénsis. DC. (ye.) Jamaica. Jamaica. 1818. 5. 9. S. ♄.⌣.
3 lùcida. K.S. (pi.) glossy. Cuba. 1822. .... S. ♄.⌣. H.et.B.n.g.am.5.t.451.
TETRA'PTERIS. DC. TETRA'PTERIS. Decandria Trigynia. L.
1 buxifòlia. DC. (ye.) Box-leaved. St. Domingo. .... 5. 8. S. ♄.⌣. Cav. dis. 9. t. 262. f. 1.
2 acapulcénsis. K.S. (ye.) Acapulco. Mexico. 1824. .... S. ♄.⌣.
3 acutifòlia. DC. (ye.) acute-leaved. Cayenne. 1826. .... S. ♄.⌣. Cav. dis. 9. t. 261.
4 citrifolia. DC. (ye.) Citron-leaved. Jamaica. 1819. —— S. ♄. ——— 9. t. 260.
5 discolor. DC. (ye.) two-coloured. Guiana. 1827. .... S. ♄.⌣.
BANIST'ERIA. DC. BANIST'ERIA. Decandria Trigynia. L.
1 auriculàta. DC. (ye.) eared. Brazil. 1818. .... S. ♄.⌣. Cav. dis. 9. t. 255.
2 ciliàta. DC. (or.) ciliated. ——— 1796. .... S. ♄.⌣. ——— 9. t 254.
3 spléndens. DC. (ye.) shining. S. America. 1812. .... S. ♄.⌣. ——— 9. t. 253.
heterophy'lla. W. *fúlgens.* Cav.
4 dichótoma. DC. (ye.) forked. S. America. .... S. ♄.⌣. Cav. dis. 9. t. 256.
5 emarginàta. DC. (ye.) emarginate. W. Indies. 1826. .... S. ♄.⌣. ——— 9. t. 249.
6 serícea. DC. (ye.) silky. Brazil. 1810. .... S. ♄.⌣. ——— 9. t. 258.
7 tomentòsa. DC. (ye.) woolly-leaved. Antilles. 1818. 5. 8. S. ♄.⌣.
8 tiliæfòlia. DC. (pu.) Lime-leaved. Java. 1820. .... S. ♄.⌣. Vent. choix. t. 50.

9 sinemariénsis. DC.	(*ye.*)	prickly-leaved.	Guiana.	1824.	....	S. ♄.⌣.	Aubl. guian. 1. t. 185.
10 ferrugìnea. DC.	(*bh.*)	ferruginous.	Brazil.	1818.	....	S. ♄.⌣.	Cav. dis. 9. t. 248.
11 periplocæfòlia. DC.	(*ye.*)	Periploca-ld.	Porto Rico.	——	....	S. ♄.⌣.	
12 laurifòlia. DC.	(*ye.*)	Laurel-leaved.	Jamaica.	1733.	7. 8.	S. ♄.⌣.	Bot. reg. 937.
13 fúlgens. DC.	(*ye.*)	fulgent.	W. Indies.	1759.	....	S. ♄.⌣.	
14 ovàta. DC.	(*bh.*)	oval-leaved.	St. Domingo.	1819.	7. 8.	S. ♄.⌣.	Vent. choix. t. 51. A.
HETERO'PTERIS. DC.		HETERO'PTERIS.	Decandria Trigynia. L.				
1 purpùrea. DC.	(*pu.*)	purple.	W. Indies.	1759.	....	S. ♄.⌣.	Cav. dis. 9. t. 246. f. 1.
2 parvifòlia. DC.	(*ye.*)	small-leaved.	St. Thomas.	1822.	....	S. ♄.⌣.	Vent. choix. t. 51.
3 brachiàta. DC.	....	cross-branched.	S. America.	1759.	....	S. ♄.⌣.	
4 chrysophy'lla. DC.	(*or.*)	Star-apple-ld.	Brazil.	1793.	....	S. ♄.⌣.	Jacq. schœn. 1. t. 105.
5 cœrùlea. DC.	(*bl.*)	blue-flowered.	W. Indies.	1822.	....	S. ♄.⌣.	Cav. dis. 9. t. 243.
6 floribúnda. DC.	(*bl.*)	many-flowered.	Mexico.	1826.	....	S. ♄.⌣.	
7 nítida. DC.	....	glossy.	Brazil.	1809.	....	S. ♄.⌣.	Cav. dis. 9. t. 244.
8 appendiculàta. DC.	..	long-pointed.	W. Indies.	1823.	....	S. ♄.	

ORDO XLIV.

ACERINEÆ. *DC. prodr.* 1. *p.* 593.

ACER. DC.		MAPLE.	Polygamia Monœcia. L.				
1 oblóngum.	(*gr.*)	oblong-leaved.	Nepaul.	1820.	....	H. 5.	
2 tatáricum. DC.	(*gr.*)	Tartarian.	Tartary.	1759.	5. 6.	H. 5.	Pall. fl. ros. t. 3.
3 striàtum. DC.	(*st.*)	striped-barked.	N. America.	1755.	——	H. 5.	Mich. f. arb. 2. t. 17.
pensylvánicum. L.							
4 Pseùdo-Plátanus.	(*st.*)	Sycamore.	Britain.	....	4. 5.	H. 5.	Eng. bot. 303.
β variegàtum.	(*st.*)	*variegated.*	———	....	——	H. 5.	Duham. arb. 1. t. 36.
γ subobtùsum. DC.	(*st.*)	*blunt-leaved.*	———	....	——	H. 5.	
5 spicàtum. DC.	(*st.*)	mountain.	N. America.	1750.	——	H. 5.	Schm. arb. 1. t. 11.
montànum. H.K.							
6 hy'bridum. DC.	(*gr.*)	hybrid.	Hybrid.	....	——	H. 5.	Lodd. bot. cab. 1221.
7 macrophy'llum. DC.	(*st.*)	large-leaved.	Columbia.	1827.	....	H. 5.	
8 campéstre. DC.	(*gr.*)	common.	Britain.	....	5. 6.	H. 5.	Eng. bot. 304.
9 austrìacum. L.en.	(*gr.*)	Austrian.	Austria.	....	——	H. 5.	Tratt. arch. 1. n. 6. ic.
10 Opàlus. W.	(*st.*)	Italian.	Italy.	1752.	——	H. 5.	
11 opalifòlium. L. en.	(*st.*)	Opalus-leaved.	S. Europe.	....	——	H. 5.	Tratt. arch. 1. n. 13. ic.
12 obtusàtum. DC.	(*st.*)	Hungarian.	Hungary.	1818.	——	H. 5.	——— 1. n. 14. ic.
13 neapolitànum. T.S.	(*st.*)	Neapolitan.	Naples.	1820.	5. 6.	H. 5.	Ten. at. ac. neap. 1819. ic.
14 créticum. DC.	(*gr.*)	Cretan.	Levant.	1752.	——	H. 5.	Tratt. arch. 1. n. 19. ic.
15 monspessulànum. DC.	(*gr.*)	Montpelier.	France.	1739.	5.	H. 5.	——— 1. n. 20. ic.
16 ibéricum. DC.	(*gr.*)	Iberian.	Iberia.	1826.	....	H. 5.	
17 heterophy'llum. DC.	(*gr.*)	evergreen.	Levant.	1759.	5. 6.	H. 5.	Willd. arb. 10. t. 1. f 1.
sempervìrens. L.							
18 barbàtum. DC.	(*gr.*)	bearded.	N. America.	1812.	4. 5.	H. 5.	
19 Lobèlii. T.S.	(*st.*)	Lobel's.	Italy.	1826.	....	H. 5.	
20 platanoídes. DC.	(*st.*)	Norway.	Europe.	1683.	5. 7.	H. 5.	Schm. arb. 1. t. 3. 4.
β laciniàtum.	(*st.*)	*cut-leaved.*	———	———	——	H. 5.	——— 1. t. 5.
21 saccharìnum. DC.	(*st.*)	Sugar.	N. America.	1735.	4. 5.	H. 5.	Mich. f. arb. 2. t. 15.
22 nìgrum. DC.	(*st.*)	black-sugar.	———	1812.	——	H. 5.	——— 2. t. 16.
23 circinnàtum. DC.	....	round-leaved.	Columbia.	1827.	....	H. 5.	
24 eriocárpum. DC.	(*st.*)	Sir C. Wager's.	N. America.	1725.	——	H. 5.	Mich. f. arb. 2. t. 13.
dasycárpum. W.							
25 rùbrum. DC.	(*re.*)	red-flowered.	———	1656.	——	H. ♄.	Mich. f. arb. 2. t. 14.
26 palmàtum. DC.	(*st.*)	palmate.	China.	1820.	....	H. ♄.	Tratt. arch. 1. n. 17. ic.
NEGU'NDO. DC.		NEGU'NDO.	Diœcia Pentandria.				
fraxinifòlium. DC.	(*st.*)	Ash-leaved.	N. America.	1688.	4.	H. 5.	Mich. f. arb. 2. t. 16.
Acer Negúndo. L.							

ORDO XLV.

HIPPOCASTANEÆ. *DC. prodr.* 1. *p.* 597.

ÆʹSCULUS. DC.		HORSE-CHESTNUT. Heptandria Monogynia. L.					
1 Hippocástanum. DC.	(*wh.*)	common.	Asia.	1629.	4. 5.	H. 5.	Schm. arb. 1. t. 38.
β *variegàta.*	(*wh.*)	*striped-leaved.*	——	....	——	H. 5.	
γ *flòre plèno.*	(*wh.*)	double-flowered.	....	....	——	H. 5.	
2 rubicúnda. DC.	(*re.*)	red-flowered.	....	....	5. 6.	H. 5.	Lod. bot. cab. 1242.
cárnea. W.D. 121 *non* B. reg.							
3 cárnea. B.R.	(*fl.*)	flesh-coloured.	....	....	5. 6.	H. 5.	Bot. reg. 993.
4 glábra. DC.	(*st.*)	smooth-leaved.	N. America.	1812.	——	H. 5.	
5 ohioénsis. DC.	(*wh.*)	Ohio.	——	1818.	——	H. 5.	
6 pállida. DC.	(*st.*)	pale-flowered.	——	1812.	6.	H. 5.	
PʹAVIA.		BUCK'S-EYE-TREE. Heptandria Monogynia.					
1 macrostàchya. DC.	(*w.*)	long-spiked.	N. America.	1786.	7. 8.	H. ♄.	Jacq. ecl. 1. t. 9.
Æʹsculus parviflòra. H.K.							
2 rùbra. DC.	(*re.*)	red-flowered.	N. America.	1711.	5. 6.	H. 5.	Duham. arb. 2. t. 19.
Æʹsculus Pàvia. Lod. bot. cab. 1257.							
β *argùta.* B.R.	(*fl.*)	*sharp-toothed.*	——	....	——	H. 5.	Bot. reg. 993.
γ *sublaciniata.* W.D.	(*re.*)	*deep-toothed.*	——	....	——	H. 5.	Wats. dend. brit. 120.
3 hyʹbrida. DC.	(*fl.*)	variegated-fld.	——	1812.	——	H. 5.	
4 díscolor.	(*fl.*)	dwarf.	——	——	——	H. ♄.	Bot. reg. 310.
Æʹsculus díscolor. Ph. *non hyʹbrida.* DC.							
5 hùmilis. B.R.	(*re.*)	dwarf.	——	....	——	H. ♄.	Bot. reg. 1018.
6 neglécta. B.R.	(*st.*)	neglected.	——	....	——	H. 5.	—— 1009.
7 flàva. DC.	(*st.*)	yellow-flowered.	——	1764.	——	H. 5.	Lod. bot. cab. 1280.

ORDO XLVI.

RHIZOBOLEÆ. *DC. prodr.* 1. *p.* 599.

CARʹYOCAR. DC.		BUTTER-NUT. Polyandria Tetragynia.					
1 nucíferum. DC.	(*re.*)	nut-bearing.	S. America.	1826.	....	S. ♄.	Bot. mag. 2728.
2 glábrum. DC.	......	smooth-leaved.	Guiana.	1820.	....	S. ♄.	Aub. gui. 2. t. 240.
3 tomentòsum. DC.	(*wh.*)	woolly-leaved.	——	——	....	S. ♄.	—— 1. t. 239.

ORDO XLVII.

SAPINDACEÆ. *DC. prodr.* 1. *p.* 601.

Tribus I. *PAULLINIEÆ.* DC.

CARDIOSPEʹRMUM. L.		HEART-SEED. Octandria Trigynia. L.					
1 Halicácabum. DC.	(*w.*)	smooth-leaved.	India.	1594.	6. 8.	S. ☉.⌣.	Bot. mag. 1049.
2 grandiflòrum. DC.	(*wh.*)	large-flowered.	Jamaica.	1823.	....	S. ♄.⌣.	
3 coluteoídes. DC.	(*wh.*)	Colutea-like.	Caracas.	——	....	S. ♃.⌣.	
4 Coríndum. L.	(*wh.*)	Parsley-leaved.	Brazil.	1750.	7. 8.	S. ☉.⌣.	
5 pubéscens. DC.	(*wh.*)	pubescent.	N. Spain.	1820.	——	S. ♄.⌣.	
URVILLʹÆA. Kth.		URVILLʹÆA. Octandria Monogynia.					
ulmàcea. Kth.	(*wh.*)	Elm-leaved.	Caracas.	1823.	....	S. ♄.⌣.	H. et B. n. gen. 5. t. 440.
SERIʹANA. Kth.		SERIʹANA. Octandria Trigynia. L.					
1 sinuàta. DC.	(*wh.*)	sinuate-leaved.	S. America.	....	7. 9.	S. ♄.⌣.	Sch. ac. hafn. 3. t. 12. f. 1.

2 divaricàta. DC. (*wh.*) divaricate. W. Indies. 1823. .... S. ♄.⌣ Sch. ac. hafn.3.t.12.f.2.
3 caracasàna. DC. (*wh.*) tooth-leaved. Caracas. 1816. 7. 9. S. ♄.⌣ Jacq. schœn. 1. t. 99.
4 mexicàna. DC. (*wh.*) Mexican. Mexico. 1825. .... S. ♄.⌣. Sch. ac. hafn.3.t.11.f.3.
5 triternàta. DC. (*wh.*) Supple-Jack. W. Indies. 1739. .... S. ♄.⌣. Plum. ed. Burm. t. 112.
Paullínia polyphy'lla. L.

PAULLI'NIA. DC. PAULLI'NIA. Octandria Trigynia. L.
1 pinnàta. DC. (*wh.*) winged-leaved. S. America. 1752. 7. 9. S. ♄.⌣. Jacq. obs. 3. t. 62. f. 12.
2 Vespertílio. DC. (*wh.*) Bat-winged. W. Indies. 1823. —— S. ♄.⌣. Sch. act. hafn.3.t.11.f.1.
3 meliæfòlia. DC. (*wh.*) Melia-leaved. Brazil. 1818. —— S. ♄.⌣. Hook. ex. flor. 110.
4 curassávica. DC. (*wh.*) shining-leaved. S. America. 1739. 6. 8. S. ♄.⌣. Jacq. obs. 3. t. 61. f. 1.
5 barbadénsis. DC. (*wh.*) Barbadoes. W. Indies. 1786. .... S. ♄.⌣. ———— 3. t. 61. f. 9.
6 polyphy'lla. DC. (*wh.*) many-leaved. S. America. 1739. 6. 8. S. ♄.⌣. Pluk. alm. t. 168. f. 5?
7 carthagenénsis. DC. (*wh.*) biternate. Trinidad. 1821. —— S. ♄.⌣. Jacq. obs. 3. t. 61. f. 6.
8 carib'æa. DC. (*wh.*) Caribean. Caribees. .... S. ♄.⌣. Jacq. obs. 3. t. 62. f. 7.
9 híspida. DC. (*wh.*) hispid. Caracas. 1823. .... S. ♄.⌣. Jacq. schœn. 3. t. 268.
10 cauliflòra. DC. (*wh.*) stem-flowering. ———— 1821. 6. 8. S. ♄.⌣. Jacq. ic. rar. 3. t. 450.

Tribus II. *SAPINDEÆ.* DC.

SAPI'NDUS. DC. SOAP BERRY. Octandria Trigynia. L.
1 Saponària. DC. (*wh.*) common W. Indies. 1697. 7. 9. S. ♄. Comm. hort. 1. t. 94.
2 marginàtus. DC. (*wh.*) margined. Carolina. .... —— G. ♄.
3 rígidus. DC. (*wh.*) Ash-leaved. S. America. 1759. —— S. ♄. Pluk. alm. t. 217. f. 7.
4 senegalénsis. DC. (*wh.*) Senegal. Senegal. 1823. .... S. ♄.
5 laurifòlius. DC. (*wh.*) Laurel-leaved. E. Indies. 1818. .... S. ♄. Rheed mal. 4. t. 19.
6 longifòlius. DC. (*wh.*) long-leaved. ———— —— .... S. ♄.
7 emarginàtus. DC. (*wh.*) emarginate. ———— 1820. .... S. ♄.
8 rubiginòsus. DC. (*wh.*) rusty-leaved. ———— —— .... S. ♄. Roxb. cor. 1. t. 62.
9 ? índicus. DC. (*wh.*) various-leaved. ———— —— .... S. ♄.

BL'IGHIA. DC. AKEE-TREE. Octandria Monogynia.
sápida. DC. (*wh.*) Ash-leaved. Africa. 1723. .... S. ♄. Kon. an. bot. 2. t. 16. 17.

MATA'YBA. DC. MATA'YBA. Octandria Monogynia.
guianénsis. DC. (*wh.*) Ash-leaved. Guiana. 1803. 7. 9. S. ♄. Aub. gui. 1. t. 128.
Ephièlis fraxínea. W.

SCHMI'DELIA. DC. SCHMI'DELIA. Octandria Monogynia. L.
1 serràta. DC. (*wh.*) saw-leaved. E. Indies. 1804. 6. 8. S. ♄. Roxb. cor. 1. t. 61.
Ornítrophe serràta. W. ————
2 racemòsa. DC. (*wh.*) clustered. ———— 1818. ..... S. ♄. Burm. ind. t. 32. f. 1.
3 Comínia. DC. (*wh.*) yellow-berried. Jamaica. 1759. 6. 9. S. ♄. Sloan. h. 2. t. 208 f. 1.
Ornítrophe Comínia. W. *Rhús Comínia.* L.

EUPH'ORIA. DC. EUPH'ORIA. Octandria Monogynia. L.
1 Lítchi. DC. (*wh.*) Lee-Chee. China. 1786. 5. 6. S. ♄. Lam. ill. t. 306.
Dimocárpus Lítchi. W. *Scytàlia chinénsis.* G. fr. t. 42. f. 3.
2 Longàna. DC. (*wh.*) loose-panicled. China. —— —— S. ♄. Buchoz. ic. col. t. 99.
3 verticillàta. B.R. (*bh.*) whorl-flowered. Moluccas. 1822. 2. 3. S. ♄. Bot. reg. 1059.
4 Nephèlium. DC. (*wh.*) Rambootan. E. Indies. 1809. .... S. ♄. Lam. ill. t. 764.
Nephèlium lappàceum. W. *Scytàlia Rambootán.* H.B.

CUP'ANIA. DC. CUP'ANIA. Octandria Monogynia.
1 tomentòsa. DC. (*wh.*) woolly-leaved. W. Indies. 1818. 5. 7. S. ♄.
Trigònis tomentòsa. Jacq. am. 102.
2 glábra. DC. (*wh.*) smooth-leaved. ———— —— —— S. ♄.
3 Saponària. DC. (*wh.*) Soap-wort-ld. ———— 1820. .... S. ♄.

MOLIN'ÆA. J. MOLIN'ÆA. Octandria Monogynia.
canéscens. H.B. (*wh.*) canescent. E. Indies. 1817. 5. 6. S. ♄.

COSSI'GNIA. DC. COSSI'GNIA. Hexandria Monogynia.
borbònica. DC. (*wh.*) golden-leaved. Bourbon. 1811. .... S. ♄. Lam. ill. t. 256.
pinnàta. Lam. *Ruízia aúrea.* Hort.

MELIC'OCCA. DC. HONEY-BERRY. Octandria Monogynia. L.
1 bìjuga. DC. (*wh.*) winged-leaved. Jamaica. 1778. .... S. ♄. Jacq. amer. 103. t. 72.
2 olivæfórmis. DC. (*wh.*) large-leaved. S. America. 1818. .... S. ♄.

Tribus III. *DODONÆACEÆ.* DC.

KOELREUT'ERIA. DC.		KOELREUT'ERIA.	Octandria Monogynia. L.					
paniculàta. DC.	(*ye.*)	panicled.	China.	1763.	7. 8.	H.	♄.	Bot. reg. 320.
DODON'ÆA. DC.		DODON'ÆA.	Octandria Monogynia. L.					
1 viscòsa. DC.	(*gr.*)	clammy.	S. America.	1690.	6. 7.	S.	♄.	Plum. ic. t. 247. f. 2.
2 jamaicénsis. DC.	(*gr.*)	Jamaica.	Jamaica.	——	——	S.	♄.	Cav. ic. t. 327.
viscòsa. Cav. *angustifòlia.* Sw.								
3 Burmanniàna. DC.	(*gr.*)	Burman's.	E. Indies.	1758.	5. 8.	S.	ħ.	Burm. zeyl. t. 23.
angustifòlia. H.B.								
4 microcárpa. DC.	(*gr.*)	small-fruited.	Bourbon.	1818.	——	S.	♄.	Lam. ill. t. 304. f. 2.
5 attenuàta. F.T.	(*gr.*)	attenuated.	N. S. W.	1824.	2. 6.	G.	♄.	Bot. mag. 2860.
6 salicifòlia. DC.	(*gr.*)	Willow-leaved.	C. B. S.	1758.	——	G.	ħ.	
angustifòlia. Lam.								
7 dioíca. DC.	(*gr.*)	diœcious.	E. Indies.	....	——	S.	♄.	
8 trìquetra. DC.	(*gr.*)	three-sided.	N. S. W.	1790.	6. 8.	G.	♄.	Andr. rep. 230.
9 oblongifòlia. L.en.	(*gr.*)	oblong-leaved.	————	1816.	——	G.	♄.	Bot. reg. 1051.
10 cuneàta. DC.	(*gr.*)	wedge-leaved.	————	——	——	G.	♄.	Linn. trans. 11. t. 19.
11 asplenifòlia. DC.	(*gr.*)	Asplenium-ld.	————	——	——	G.	♄.	———— 11. t. 20.
12 filifórmis. DC.	(*gr.*)	slender-leaved.	————	1820.	——	G.	♄.	
13 angustíssima. DC.	(*gr.*)	narrow-leaved.	————	——	——	G.	♄.	
14 elæagnoídes. DC.	(*gr.*)	scaly leaved.	W. Indies.	1819.	....	S.	ħ.	
15 ? pinnàta. DC.	(*gr.*)	winged-leaved.	N. S. W.	1810.	....	G.	♄.	

ORDO XLVIII.

MELIACEÆ. *DC. prodr.* 1. *p.* 619.

Tribus I. *MELIEÆ.* DC.

AIT'ONIA. W.		AIT'ONIA.	Monadelphia Octandria.					
capénsis. W.	(*pi.*)	Cape.	C. B. S.	1774.	5. 9.	G.	♄.	Bot. mag. 173.
TURR'ÆA. DC.		TURR'ÆA.	Decandria Monogynia. L.					
1 vìrens. DC.	(*ye.*)	green.	E. Indies.	1822.	....	S.	♄.	Sm. ic. ined. 1. t. 10.
2 rígida. DC.	(*li.*)	rigid.	Mauritius.	1820.	....	S.	ħ.	Vent. choix. t. 48.
QUIVI'SIA. DC.		QUIVI'SIA.	Octandria Monogynia.					
heterophy'lla. DC.	(*bh.*)	various-leaved.	Mauritius.	1821.	....	S.	ħ.	Cav. dis. 7. t. 213.
SAND'ORICUM. DC.		SAND'ORICUM.	Decandria Monogynia.					
índicum. DC.	(*st.*)	Indian.	E. Indies.	1818.	....	S.	♄.	Roxb. corom. 3. t. 261.
M'ELIA. DC.		BEAD-TREE.	Decandria Monogynia. L.					
1 Azedarách. DC.	(*li.*)	common.	Syria.	1656.	6. 8.	G.	ħ.	Bot. mag. 1066.
2 supérba. H.B.	(*li.*)	superb.	E. Indies.	1821.	1. 2.	S.	♄.	
3 robústa. H.B.	(*li.*)	robust.	————	——	....	S.	♄.	
4 sempervìrens. DC.	(*li.*)	evergreen.	Jamaica.	1656.	6. 9.	S.	♄.	Bot. reg. 643.
5 compòsita. DC.	(*li.*)	compound-ld.	E. Indies.	1818.	1. 3.	S.	♄.	
6 Azadiráchta. DC.	(*li.*)	Ash-leaved.	————	1759.	6. 8.	S.	ħ.	Cav. dis. 7. t. 208.
7 austràlis.	(*li.*)	New Holland.	N. S. W.	1824.	....	G.	♄.	

Tribus II. *TRICHILIEÆ.* DC.

TRICHI'LIA. DC.		TRICHI'LIA.	Decandria Monogynia. L.					
1 hírta. DC.	(*wh.*)	hairy.	Jamaica.	1820.	6. 7.	S.	ħ.	Sloan. h. 2. t. 220. f. 1.
2 spondioídes. DC.	(*st.*)	Spondias-like.	————	——	——	S.	ħ.	Jacq. schœn. 1. t. 102.
3 havanénsis. DC.	(*st.*)	smooth.	W. Indies.	1794.	——	S.	ħ.	Jacq. amer. t. 175. f. 38.
glábra. L.								
4 odoràta. A.R.	(*gr.*)	sweet-scented.	————	1801.	——	S.	♄.	Andr. rep. 637.
5 terminàlis. DC.	(*wh.*)	terminal.	Jamaica.	1825.	....	S.	♄.	
6 glandulòsa. DC.	(*gr.*)	glandular.	N. S. W.	1821.	....	G.	♄.	
7 trifoliàta. DC.	(*wh.*)	three-leaved.	S. America.	1828.	....	S.	♄.	Jacq. amer. 129. t. 82.

EKEBE'RGIA. DC. Ekebe'rgia. Decandria Monogynia.
1 capénsis. DC. (*wh.*) Cape. C. B. S. 1789. 7. 8. G. ♄. Lam. ill. t. 358.
2 índica. H.B. (*wh.*) Indian. E. Indies. 1826. .... S. ♄.
GUA'REA. DC. Gua'rea. Octandria Monogynia. L.
1 trichilioídes. L. (*wh.*) Ash-leaved. S. America. 1752. 5. 6. S. ♄. Jacq. am. t. 176. f. 37.
grandifòlia. DC. *Mèlia Guára.* Jacq.
2 Swártzii. (*wh.*) Swartz's. W. Indies. 1822. .... S. ♄. Sloan. h. 2. t. 170. f. 1.
trichilioídes. Sw.
3 ramiflòra. DC. (*li.*) branch-flow'r'd. Porto Rico. —— .... S. ♄. Vent. choix. t. 41.
HE'YNEA. DC. He'ynea. Decandria Monogynia.
1 tríjuga. DC. (*wh.*) Walnut-like. Nepaul. 1812. 9. 10. S. ♄. Bot. mag. 1738.
2 quínquejuga. H.B. (*w.*) five-paired. Moluccas. 1821. .... S. ♄.
AMOO'RA. R.C. Amoo'ra. Hexandria Monogynia.
cucullàta. R.C. (*ye.*) hooded. E. Indies. 1824. .... S. ♄. Roxb. corom. 3. t. 258.

ORDO XLIX.

CEDRELEÆ. *Brown gen. rem.* 64.

CEDR'ELA. DC. Bastard Cedar. Pentandria Monogynia. L.
1 odoràta. DC. (*st.*) fragrant. W. Indies. 1739. 7. 9. S. ♄. Lam. ill. t. 137.
2 Toóna. DC. (*st.*) Nepaul. Nepaul. 1821. .... S. ♄. Roxb. cor. 3. t. 238.
3 velutìna. DC. ...... velvetty. .... S. ♄.
SWIET'ENIA. DC. Mahogany-tree. Decandria Monogynia. L.
1 Mahógoni. DC. (*wh.*) common. W. Indies. 1734. .... S. ♄. Cav. dis. 7. t. 209.
2 febrifùga. DC. (*wh.*) Febrifuge. E. Indies. 1796. .... S. ♄. Roxb. cor. 1. t. 17.
CHLORO'XYLON. DC. Chloro'xylon. Decandria Monogynia.
Swietènia. DC. (*ye.*) Satin-wood. E. Indies. 1822. .... S. ♄. Roxb. cor. 1. t. 64.
Swietènia chloróxylon. R.C.
FLINDE'RSIA. DC. Flinde'rsia. Decandria Monogynia.
austràlis. DC. (*wh.*) New Holland. N. Holland. 1821. .... G. ♄. Brown gen. rem. t. 1.
CARA'PA. DC. Cara'pa. Decandria Monogynia.
1 pròcera. DC. (*st.*) tall. W. Indies. .... S. ♄.
2 guineénsis. (*st.*) glossy-leaved. Sierra Leon. 1793. .... S. ♄.
Afzèlia spléndens. Hort.
3 guianénsis. DC. (*st.*) Guiana. Guiana. 1820. .... S. ♄. Aubl. guian. 2. s. t. 387.

ORDO L.

AMPELIDEÆ. *DC. prodr.* 1. *p.* 627.

Tribus I. *VINIFERÆ* seu *SARMENTACEÆ.* DC.

CI'SSUS. DC. Ci'ssus. Tetrandria Monogynia. L.
1 vitigínea. DC. (*re.*) vine-leaved. India. 1772. 7. 8. S. ♄.⌣. Pluk. mant. t. 337. f. 4.
2 adnàta. DC. (*gr.*) pointed-leaved. E. Indies. 1818. —— S. ♄.⌣.
3 gláuca. DC. (*gr.*) glaucous. —— —— —— S. ♄.⌣.
4 c'æsia. DC. (*gr.*) grey. Guinea. 1822. 5. 6. S. ♄.⌣.
5 rèpens. DC. (*pu.*) creeping. E. Indies. 1821. —— S. ♄.⌣. Rheed. mal. 7. t. 48.
6 puncticulòsa. DC. (*gr.*) strumous. Cayenne. 1818. .... S. ♄.⌣.
7 uvífera. DC. (*gr.*) berry-bearing. Guinea. 1822. 5. 6. S. ♄.⌣.
8 tiliàcea. DC. (*gr.*) Lime-leaved. Mexico. 1825. .... G. ♄.⌣.
9 quadrangulàris. DC. (*w.*) square-stalked. E. Indies. 1790. .... S. ♄.⌣. Rumph. amb. 5. t. 44. f. 2.
10 sicyoídes. DC. (*gr.*) naked-leaved. Jamaica. 1768. 7. 8. S. ♄.⌣. Jacq. amer. 22. t. 15.
11 smilacìna. DC. (*gr.*) Smilax-like. S. America. 1826. .... S. ♄.⌣.
12 ovàta. DC. (*gr.*) oval-leaved. ———— 1822. .... S. ♄.⌣. Brown. jam. t. 4. f. 1. 2.
13 antárctica. DC. (*gr.*) Kanguru Vine. N. S. W. 1790. 6. 8. G. ♄.⌣. Bot. mag. 2488.

14 glandulòsa. DC.	(*gr.*)	glandular.	Arabia.	1824.	6. 7.	G.	♃.◡.	
15 capénsis. DC.	(*gr.*)	Cape.	C. B. S.	1792.	——	G.	♄.◡.	
16 ácida. DC.	(*gr.*)	acid.	Jamaica.	1692.	——	S.	♄.◡.	Jacq. schœn. 1. t. 33.
17 carnòsa. DC.	(*wh.*)	flat-stemmed.	E. Indies.	1818.	....	S.	♄.◡.	Rheed. mal. 7. t. 9.
18 trifoliàta. DC.	(*re.*)	three-leaved.	Jamaica.	1739.	6. 8.	S.	♄.◡.	Sloan.jam.t.142.f.5.6.?
19 pauciflòra. DC.	(*gr.*)	few-flowered.	C. B. S.	1818.	5. 6.	G.	♄.◡.	
20 alàta. DC.	(*re.*)	winged-stalked.	Jamaica.	——	——	S.	♄.◡.	Sloan. jam. 1. t. 144.
21 angustifòlia. DC.	(*st.*)	narrow-leaved.	E. Indies.	1818.	....	S.	♄.◡.	
22 cirrhòsa. DC.	(*gr.*)	tendrilled.	C. B. S.	1822.	5. 6.	G.	♄.◡.	
23 quinàta. DC.	(*gr.*)	wedge-leaved.	————	1790.	7.	G.	♄.◡.	
24 pentaphy'lla. DC.	(*gr.*)	Japan.	Japan.	——	4. 9.	G.	♄.◡.	
25 quinquefòlia. B.M.	(*gr.*)	five-leaved.	Brazil.	——	8. 9.	S.	♄.◡.	Bot. mag. 2443.
26 diversifòlia. DC.	(*gr.*)	various-leaved.	....	——	....	S.	♄.◡.	
heterophy'lla. L.en.*non* DC.								
27 elongàta. DC.	(*gr.*)	long-branched.	E. Indies.	1818.	....	S.	♄.◡.	
28 auriculàta. DC.	(*gr.*)	ear-stipuled.	————	——	....	S.	♄.◡.	
29 orientàlis. DC.	(*gr.*)	oriental.	Levant.	——	6. 8.	G.	♄.◡.	Lam. ill. t. 84. f. 2.

AMPELO'PSIS. DC. AMPELO'PSIS. Pentandria Monogynia. L.

1 cordàta. DC.	(*gr.*)	heart-leaved.	N. America.	1803.	6. 8.	H.	♄.◡.	
Císsus Ampelópsis. P.S.								
2 hederàcea. DC.	(*gr.*)	Virginian creeper.	————	1629.	——	H.	♄.◡.	Corn. canad. t. 100.
Hèdera quinquefòlia. L. *Císsus hederàcea.* P.S.								
3 hirsùta. DC.	(*gr.*)	hairy.	————	1806.	——	H.	♄.◡.	
4 bipinnàta. DC.	(*gr.*)	Pepper Vine.	————	1700.	7. 9.	H.	♄.◡.	Pluk. mant. t. 412. f.2.
Vìtis arbòrea. W. *Císsus stáns.* P.S.								

V'ITIS. DC. VINE. Pentandria Monogynia. L.

1 vinífera. DC.	(*gr.*)	common Grape.	Asia.	....	6. 7.	H.	♄.◡.	Jacq. ic. 1. t. 50.
2 laciniòsa. DC.	(*gr.*)	Parsley-leaved.	....	1648.	——	H.	♄.◡.	Schm. ic. 34. t. 8.
3 dentàta. DC.	(*gr.*)	toothed.	....	1818.	....	H.	♄.◡.	
4 Wallíchii. DC.	(*gr.*)	Wallich's.	Nepaul.	——	....	H.	♄.◡.	
5 índica. DC.	(*g.pu.*)	Indian.	E. Indies.	1692.	....	S.	♄.◡.	Rheed. mal. 7. t. 6.
6 lanàta. F.I.	(*gr.*)	woolly.	————	1824.	....	S.	♄.◡.	
7 latifòlia. F.I.	(*br.*)	broad-leaved.	————	——	....	S.	♄.◡.	Rheed. mal. 7. t. 7.
8 carib'æa. DC.	(*st.*)	West Indian.	W. Indies.	....	....	S.	♄.◡.	Sloan. his. 2. t. 210. f. 4.
9 Labrúsca. DC.	(*gr.*)	downy-leaved.	N. America.	1656.	5. 6.	H	♄.◡.	Jacq. schœn. t. 426.
10 blánda.	(*gr.*)	white-berried.	————	——	——	H.	♄.◡.	
11 æstívalis. DC.	(*gr.*)	Fox Grape.	————	——	——	H.	♄.◡.	Jacq. schœn. t. 425.
vulpìna. W.								
12 cordifòlia. DC.	(*gr.*)	Winter Grape.	————	1806.	....	H.	♄.◡.	
13 ripària. DC.	(*gr.*)	sweet-scented.	————	——	5. 6.	H.	♄.◡.	Bot. mag. 2429.
14 rotundifòlia. DC.	(*gr.*)	Bullet Grape.	————	——	——	H.	♄.◡.	
15 palmàta. DC.	(*gr.*)	palmate-leaved.	————	1820.	——	H.	♄.◡.	

Tribus II. *LEEACEÆ.* DC.

LE'EA. DC. LE'EA. Pentandria Monogynia. L.

1 sambucina. DC.	(*st.*)	Elder-leaved.	E. Indies.	1790.	....	S.	♄.	Cav. dis. 7. t. 218.
2 críspa. DC.	(*wh.*)	fringe-stalked.	C. B. S.	1676.	8. 10.	S.	♃.	Bot. rep. t. 355.
pinnàta. A.R.								
3 æquàta. DC.	(*gr.*)	round-stalked.	E. Indies.	1777.	....	S.	♄.	
4 hírta. DC.	(*gr.*)	hairy.	————	1816.	....	S.	♄.	
5 robusta. F.I.	(*gr.*)	robust.	————	1820.	....	S.	♄.	
6 macrophy'lla. DC.	(*wh.*)	large-leaved.	————	1806.	....	S.	♃.	

ORDO LI.

GERANIACEÆ. *DC. prodr.* 1. *p.* 637.

Tribus I. *GERANIEÆ.* S.G. 1. p. VII.

SARCOCA'ULON. DC.		SARCOCA'ULON.	Monadelphia Dodecandria.				
1 L'Heritíeri.	(*ye.*)	L. Heritier's.	C. B. S.	1790.	5. 6.	G. ♄.	L. Herit. ger. t. 42.
Monsònia L'Heritíeri. DC. *Monsonia spinosa.* L. Her.							
2 Burmánni. DC.	(*ye.*)	Burman's.	C. B. S.	1790.	7.	G. ♄.	Burm. afr. t. 31.
Gerànium spinòsum. Cav. dis. 4. t. 75. f. 2.							
MONS'ONIA. S.G.		MONS'ONIA.	Monadelphia Dodecandria.				
1 ovàta. DC.	(*st.*)	oval-leaved.	C. B. S.	1774.	8.	G. ♂.	Cav. dis. 4. t. 113. f. 1.
2 lobàta. DC.	(*st. bh.*)	broad-leaved.	——	——	4. 5.	G. ♃.	Sweet ger. 3. t. 273.
3 pilòsa. W. en.	(*bh.*)	hairy.	——	1778.	7. 8.	G. ♃.	—— 2. t. 199.
fìlia. A.R. 276. *nec aliorum.*							
4 speciòsa. L.	(*va.*)	large-flowered.	——	1774.	4. 5.	G. ♃.	
α rùbra.	(*ro.*)	*red-flowered.*	——	——	12. 2.	G. ♃.	Bot. mag. 73.
β pállida.	(*st.*)	*pale-flowered.*	——	——	4. 5.	G. ♃.	Sweet ger. 1. t. 77.
GER'ANIUM. DC.		CRANES-BILL.	Monadelphia Decandria.				
1 sibíricum. DC.	(*wh.*)	Siberian.	Siberia.	1758.	6. 7.	H. ♃.	Jacq. vind. 1. t. 19.
2 sanguíneum. DC.	(*cr.*)	bloody.	Britain.	....	6. 9.	H. ♃.	Eng. bot. 272.
3 prostràtum. L. en	(*std.*)	Lancashire.	——	....	——	H. ♃.	Cav. dis. 4. t. 76. f. 3.
4 multífidum. S.G.	(*li.*)	multifid.	C. B. S.	1817.	——	G. ♃.	Sweet ger. 3. t. 245.
5 íncànum. DC.	(*li.*)	hoary.	——	1701.	5. 7.	G. ♃.	Cav. dis. 4. t. 82. f. 2.
6 canéscens. DC.	(*li.*)	silky-leaved.	——	1787.	——	G. ♃.	L. Herit. ger. t. 38.
7 argénteum. DC.	(*bh.*)	silvery-leaved.	S. Europe.	1699.	6. 7.	H. ♃.	Sweet ger. 1. t. 59.
8 cinéreum. DC.	(*vi.*)	grey.	Pyrenees.	....	6. 8.	H. ♃.	Cav. dis. 4. t. 89. f. 1.
vàrium. L'Her. ger. t. 37.							
9 anemonefòlium. DC.	(*p.*)	Anemone-ld.	Canaries.	1778.	5. 8.	G. ♄.	Sweet ger. 3. t. 244.
10 macrorhìzon. DC.	(*re.*)	long-rooted.	Italy.	1576.	5. 6.	H. ♄.	—— 3. t. 271.
11 tuberòsum. DC.	(*re.*)	tuberous-rooted.	——	1596.	5. 8.	H. ♃.	—— 2. t. 155.
12 ibéricum. DC.	(*bl.*)	Iberian.	Iberia.	1802.	6. 9.	H. ♃	—— 1. t. 84.
13 nodòsum. DC.	(*std.*)	knotted.	England.	....	5. 10.	H. ♃.	Eng. bot. 1091.
14 angulàtum. DC.	(*std.*)	angular-stalk'd.	....	1789.	5. 6.	H. ♃.	Bot. mag. 203.
15 Lambérti. S.G.	(*li.*)	Mr. Lambert's.	Nepaul.	1825.	8. 10.	H. ♃.	Sweet ger. 4. t. 338.
16 Wallichiànum. S.G.	(*p.*)	Wallich's.	——	1820.	6. 10.	H. ♃.	—— 1. t. 90.
17 Vlassoviànum. DC.	(*st.*)	Vlassof's.	Siberia.	——	5. 8.	H. ♃.	—— 3. t. 228.
18 striàtum. DC.	(*std.*)	streaked.	Italy.	1629.	5. 10.	H. ♃.	Bot. mag. 55.
19 refléxum. DC.	(*re*)	reflex-flowered.	——	1758.	5. 6.	H. ♃.	Cav. dis. 4. t. 81. f. 1.
20 ph'æum. DC.	(*br.*)	dusky.	England.	....	4. 6.	H. ♃.	Eng. bot. 322.
21 fúscum. W.	(*da.*)	brown.	S. Europe.	1759.	7.	H. ♃.	
22 lívidum. W.	(*li.*)	wrinkle-leaved.	Switzerland.	1775.	6. 7.	H. ♃.	Sweet ger. 3. t. 268.
23 erióstemon. DC.	(*bl.*)	woolly-stamen'd	Dahuria.	1822.	7. 9.	H. ♃.	—— 2. t. 197.
α cœrùleum. S.G.	(*bl.*)	*bright blue.*	——	——	——	H. ♃.	—— 2. t. 197. f. a.
β pállidum. S.G.	(*pa.*)	*pale blue.*	Nepaul.	——	——	H. ♃.	—— 2. t. 197. f. b.
24 sylváticum. DC.	(*pu.*)	wood.	Britain.	....	5. 6.	H. ♃.	Eng. bot. 121.
25 batrachioídes. L en.	(*b.*)	hairy.	Europe.	....	——	H. ♃.	Cav. dis. 4. t. 85. f. 1.
26 praténse. DC.	(*bl.*)	Crowfoot-leav'd	Britain.	....	5. 7.	H. ♃.	Eng. bot. 404.
α cœrùleum.	(*bl.*)	*blue-flowered.*	——	....	——	H. ♃.	
β variegàtum.	(*v.*)	*variegated.*	——	....	——	H. ♃.	
γ álbum.	(*wh.*)	*white-flowered.*	——	....	——	H. ♃.	
δ flòre plèno.	(*bl.*)	*double-blue.*	——	....	——	H. ♃.	
27 lóngipes. DC.	(*li.*)	long-peduncled.	....	1824.	——	H. ♃.	
Londèsii, L. en.							
28 maculàtum. DC.	(*bh.*)	spotted.	N. America.	1732.	5. 8.	H. ♃.	Sweet ger. 4. t. 332.
29 collìnum. DC.	(*re.*)	hill.	Tauria.	1815.	——	H. ♃.	
30 palústre. DC.	(*re.*)	marsh.	Europe.	1732.	——	H. ♃.	Sweet ger. 1. t. 3.
31 aconitifòlium. DC.	(*wh.*)	Aconite-leaved.	Switzerland.	1775.	5. 6.	H. ♃.	L. Herit. ger. t. 40.
32 pilòsum. DC.	(*li.*)	pointed anth'r'd	N. Zealand.	1821.	5. 8.	F. ♃.	Sweet ger. 2. t. 119.
33 parviflòrum. DC.	(*li.*)	small-flowered.	Van Diem.	1816.	——	G. ♃.	
34 nepalénse. S.G.	(*li.*)	Nepaul.	Nepaul.	1819.	——	H. ♃.	Sweet ger. 1. t. 12.
35 cristàtum. DC.	(*li.*)	crested.	Iberia.	1823.	——	H. ♃.	Stev. m. mosc. 4. t. 5.
albànum. M.B.							

36 pyrenàicum. DC.	(*li.*)	mountain.	Britain.	....	5. 8.	H. ♃.	Eng. bot. 405.
37 umbròsum. P.S.	(*li.*)	naked-stalked.	Hungary.	1804.	6. 9.	H. ♃.	W.etKit.hung.2.t.124.
38 mólle. DC.	(*li.*)	Dove's-foot.	Britain.	....	4. 8.	H. ⊙.	Eng. bot. 778.
39 pusíllum. DC.	(*li.*)	small-flowered.	England.	....	6. 9.	H. ⊙.	——— 385.
40 rotundifòlium. DC.	(*li.*)	round-leaved.	———	....	6. 7.	H. ⊙.	——— 157.
41 columbìnum. DC.	(*li.*)	long-stalked.	Britain.	....	——	H. ⊙.	——— 259.
42 disséctum. DC.	(*li.*)	jagged-leaved.	———	....	5. 7.	H. ⊙.	——— 753.
43 caroliniànum. DC.	(*st.*)	veined petaled.	N. America.	1725.	7. 8.	H. ⊙.	Cav. dis. 4. t. 84. f. 1.
44 lanuginòsum. W.en.	(*p.*)	woolly.	———	....	——	H. ⊙.	Jacq. schœn. 2. t. 140.
45 villòsum. DC.	(*li.*)	villous.	Naples.	1824.	——	H. ⊙.	
46 bohêmicum. DC.	(*bl.*)	Bohemian.	Bohemia.	1683.	6. 8.	H. ⊙.	Sweet ger. ser. 2. t.
47 divaricàtum. DC.	(*li.*)	straddling.	Hungary.	1799.	7. 8.	H. ⊙.	W. etKit.hung.2.t.123.
48 lùcidum. DC.	(*li.*)	shining.	Britain.	....	5. 8.	H. ⊙.	Eng. bot. 75.
49 Briceànum.	(*wh.*)	Miss Brice's.	Bristol.	....	4. 10.	H. ⊙.	
50 Robertiànum. DC.	(*pu.*)	Herb-Robert.	Britain.	....	——	H. ⊙.	Eng. bot. 1486.
51 purpùreum. W.	(*pu.*)	purple.	———	....	——	H. ⊙.	Vill. dauph. 3. t. 40.
ER'ODIUM. DC.		HERON'S-BILL.	Monadelphia Pentandria.				
1 petr'æum. DC.	(*pu.*)	rock.	S. Europe.	1640.	6. 7.	H. ♃.	Gouan. ill. t. 21. f. 1.
2 glandulòsum. DC.	(*pu.*)	glandulous.	Spain.	1798.	——	H. ♃.	Cav. dis. 5. t. 125. f. 2.
3 Stephaniànum. D.C.	(*bl.*)	multifid.	Dahuria.	1820.	——	H. ♃.	
4 alpìnum. DC.	(*bl.*)	Alpine.	Italy.	1814.	5. 8.	H. ♃.	L'Herit. ger. t. 3.
5 crassifòlium. DC.	(*li.*)	thick-leaved.	Cyprus.	1788.	3. 8.	G. ♄.	Sweet ger. 2. t. 111.
6 laciniàtum. DC.	(*li.*)	laciniated.	Crete.	1794.	5. 8.	G. ♃.	Cav. dis. 4. t. 113. f. 3.
7 hírtum. DC.	(*li.*)	hairy.	Egypt.	1823.	——	G. ♃.	Jacq. f. ecl. 1. t. 58.
8 Cicònium. DC.	(*li.*)	long-beaked.	S. Europe.	1711.	6. 7.	H. ⊙.	Jacq. vind. 1. t. 18.
9 cicutàrium. E.B.	(*pi.*)	Hemlock-leav'd	Britain.	....	4. 9.	H. ⊙.	Eng. bot. 1768.
10 pimpinellæfòlium.	(*pi.*)	Pimpinella-ld.	Europe.	1823.	——	H. ⊙.	Cav. dis. 4. t. 95. f. 2.
11 bipinnàtum. W.	(*pi.*)	Numidian.	Numidia.	1803.	5. 6.	H. ♂.	——— 5. t. 126. f. 3.
12 romànum. W.	(*pi.*)	Roman.	Rome.	1724.	——	H. ♃.	Bot. mag. 377.
13 caucalifòlium. S.G.	(*pi.*)	Caucalis-leaved.	S. Europe.	1816.	6. 8.	H. ♃.	Sweet ger. t. 6.
14 moschàtum. DC.	(*li.*)	musky.	England.	....	5. 7.	H. ⊙.	Eng. bot. 902.
15 bòtrys. DC.	(*bl.*)	large-blue.	Italy.	1819.	——	H. ⊙.	Cav. dis. 4. t. 90. f. 2.
16 múrcicum. DC.	(*bl.*)	small-blue.	Murcia.	——	——	H. ⊙.	——— 5. t. 126. f. 1.
17 gruìnum. DC.	(*bl.*)	broad-leaved.	Crete.	1596.	6. 7.	H. ⊙.	——— 4. t. 88. f. 2.
18 serotìnum. DC.	(*bl.*)	late-flowering.	Tyre.	1816.	6. 9.	H. ♃.	Stev.m.ac.P.3.t.15.f.2.
multicáule. S.G. t. 137.							
19 chìum. DC.	(*bl.*)	Levant.	Levant	1724.	6. 7.	H. ⊙.	Cav. dis. 4. t. 92. f. 1.
20 hymenòdes. DC.	(*li.*)	three-leaved.	Barbary.	1789.	1. 12.	H. ♃.	Sweet ger. 1. t. 23.
21 Gussònii. S.G.	(*li.*)	Gussone's.	Naples.	1822.	5. 8.	H. ♃.	——— 2. t. 200.
22 malachoídes. DC.	(*pu.*)	Mallow-leaved.	S. Europe.	1596.	5. 7.	H. ⊙.	Cav. dis. 4. t. 91. f. 1.
23 ribifòlium. W.	(*pu.*)	Currant-leaved.	C. B. S.	....	6. 8.	H. ⊙.	Jacq. ic. rar. 3. t. 509.
24 incarnàtum. DC.	(*fl.*)	flesh-coloured.	———	1787.	5. 7.	G. ♄.	Sweet ger. 1. t. 94.
25 glaucophy'llum. DC.	(*li.*)	glaucous-leav'd.	Egypt.	1732.	7. 8.	H. ♃.	——— 3. t. 283.
26 malopoídes. DC.	(*li.*)	velvetty-leaved.	Sicily.	1818.	5. 8.	H. ♃.	Cav. dis. 4. t. 90. f. 1.
27 littòreum. DC.	(*li.*)	smooth.	Narbonne.	——	——	H. ♃.	
28 marítimum. DC.	(*li.*)	sea-side.	England.	....	5. 9.	H. ♃.	Eng. bot. 646.
29 Reichárdi. DC.	(*wh.*)	dwarf.	Minorca.	1783.	4. 9.	H. ♃.	Bot. mag. 18.
chamædryoídes. W. *Gerànium Reichárdi*. B.M.							

Tribus II. *GRIELEÆ*. S.G. p. VII.

GRI'ELUM. W.		GRI'ELUM.	Decandria Pentagynia.				
1 tenuifòlium. W.	(*ye.*)	large-flowered.	C. B. S.	1790.	4. 6.	G. ♄.	Sweet ger. 2. t. 171.
2 laciniàtum. G.	(*ye.*)	jagged-leaved.	———	1825.	——	G. ♄.	——— 4. t. 306.

Tribus III. *PELARGONIEÆ*. S.G. p. VIII.

PHYMATA'NTHUS. S.G.		WART-FLOWER.	Monadelphia Pentandria.				
1 trícolor. S.G.	(*w.pu.*)	three-coloured.	C. B. S.	1791.	6. 10.	G. ♄.	Sweet ger. 1. t. 43.
Pelargònium trícolor. B.M. 240.							
2 elàtus. S.G.	(*w.pu.*)	tall.	———	1795.	——	G. ♄.	Sweet ger. 1. t. 96.

3 villòsus.	(*w.pu.*)	villous.	Hybrid.	1823.	6. 10.	G. ♄.	
4 grandiflòrus.	(*w.pu.*)	large-flowered.	——	——	——	G. ♄.	
5 latifòlius. C.C.	(*w.pu.*)	broad-leaved.	——	——	——	G. ♄.	
CAMP'YLIA. S.G.		CAMP'YLIA.	Monadelphia Pentandria.				
1 carinàta. S.G.	(*w.pu.*)	keeled-stipuled.	C. B. S.	1810.	4. 9.	G. ♄.	Sweet ger. 1. t. 21.
2 variegàta. S.G.	(*v.*)	variegated-fl'd.	Hybrid.	1823.	——	G. ♄.	—— 3. t. 266.
3 holosericea. S.G.	(*da.*)	silky-leaved.	——	1820.	——	G. ♄.	—— 1. t. 75.
4 blattària. S.G.	(*li.*)	downy-leaved.	C. B. S.	1790.	6. 9.	G. ♄.	—— 1. t. 88.
5 élegans. S.G.	(*bh.*)	elegant.	Hybrid.	1822.	——	G. ♄.	—— 3. t. 222.
6 verbasciflòra. S.G.	(*li.*)	Mullein-fl'd.	C. B. S.	1811.	——	G. ♄.	—— 2. t. 157.
7 càna. S.G.	(*li.*)	hoary-leaved.	——	1794.	7. 12.	G. ♄.	—— 2. t. 114.
8 erióstemon. S.G.	(*wh.*)	velvet-leaved.	——	——	3. 6.	G. ♄.	Jacq. schœn. 2. t. 132.
Pelargònium erióstemon. Jacq.							
9 laciniàta. S.G.	(*d.re.*)	jagged-leaved.	Hybrid.	1827.	4. 10.	G. ♄.	Swt. ger. ser. 2. t. 1.
10 coronopifòlia.	(*li.*)	Buck's-horn-ld.	C. B. S.	1791.	6. 10.	G. ♄.	Jacq. ic. rar. 3. t. 526.
11 ? œnothèræ.	(*li.*)	œnothera-like.	——	1812.	4. 8.	G. ♃.	—— 3. t. 525.
Pelargònium œnothèræ. Jacq.							
OTI'DIA. S.G.		OTI'DIA.	Monadelphia Pentandria.				
1 carnòsa. S.G.	(*wh.*)	fleshy-stalked.	C. B. S.	1724.	6. 8.	G. ♄.	Swt. ger. 1. t. 98.
2 ceratophy'lla. S.G.	(*wh.*)	horn-leaved.	Africa.	1786.	5. 7.	G. ♄.	Bot. mag. 315.
Pelargònium ceratophy'llum. B.M.							
3 dasycaùlon. S.G.	(*wh.*)	thick-stemmed.	Africa.	1795.	7. 12.	G. ♄.	Bot. mag. 2029.
4 láxa. S.G.	(*wh.*)	loose-panicled.	C. B. S.	1821.	6. 9.	G. ♄.	Sweet ger. 2. t. 196.
5 crithmifòlia. S.G.	(*wh.*)	Samphire-ld.	——	1790.	4. 9.	G. ♄.	—— 4. t. 354.
6 altérnans. S.G.	(*wh.*)	Parsley-leaved.	——	1791.	5. 8.	G. ♄.	—— 3. t. 286.
GRENVI'LLEA. S.G.		GRENVI'LLEA.	Monadelphia Tetrandria.				
conspícua. S.G.		conspicuous.	Africa.	1810.	7. 11.	G. ♃.	Sweet ger. 3. t. 262. f. 2.
Gerànium Grenvílliæ. A. G. C. ic.							
H'OAREA. S.G.		H'OAREA.	Monadelphia Pentandria.				
1 violæflòra. S.G.	(*wh.*)	white violet-fld.	C. B. S.	1821.	5. 8.	G. ♃.	Sweet ger 2. t. 123.
2 nívea. S.G.	(*wh.*)	snowy white.	——	——	——	G. ♃.	—— 2. t. 182.
3 bubonifòlia.†	(*wh.*)	Bubon-leaved.	——	1800.	3. 7.	G. ♃.	And. bot. rep. 328.
4 Leeàna. S.G.	(*wh.*)	Mr. Lee's.	——	1823.	6. 8.	G. ♃.	Sweet ger. 3. t. 323.
5 laciniàta.†	(*v.*)	jag-leaved.	——	1800.	5. 6.	G. ♃.	And bot. rep. 131.
6 pennifórmis.†	(*ye.*)	winged.	——	——	——	G. ♃.	—— 269.
7 purpuráscens †	(*pu.*)	purplish.	——	——	——	G. ♃.	—— 204.
8 incrassàta.†	(*std.*)	fleshy-leaved.	——	1801.	6. 7.	G. ♃.	Bot. mag. 761.
9 carinàta. S.G.	(*ye.*)	boat-flowered.	——	1812.	5. 8.	G. ♃.	Sweet ger. 2. t. 135.
10 corydaliflòra. S.G.	(*ye.*)	Fumitory-fld.	——	——	——	G. ♃.	—— 1. t. 18.
11 rapàcea.	(*pi.*)	Turnep-rooted.	——	1788.	4. 6.	G. ♃.	And. rep. 239.
Gerànium Selìnum. A.R.							
12 setòsa. S.G.	(*bh.*)	bristle-pointed.	——	1818.	4. 8.	G. ♃.	Sweet ger. 1. t. 38.
13 pilòsa.	(*pu.*)	hairy.	——	1801.	5. 7.	G. ♃.	And. bot. rep. 259.
Gerànium pilòsum. var. 1. A.G. ic.							
14 bícolor.		two-coloured.	——	——	3. 8.	G. ♃.	Andr. geran. c. ic.
Gerànium pilòsum. var. 2. A.G. ic.							
15 blánda.	(*bh.*)	blush-flowered.	——	——	——	G. ♃.	Andr. geran. c. ic.
Gerànium pilòsum. var. 3. A.G. ic.							
16 congésta. S.G.	(*sp.*)	crowded-fld.	——	1820.	5. 10.	G. ♃.	Sweet ger. 3. t. 202.
17 retùsa. S.G.	(*da.*)	notched-petal'd.	Hybrid.	1824.	——	G. ♃.	—— 4. t. 307.
18 selinifòlia. S.G.	(*d.pu.*)	Milk-parsley-ld.	——	1821.	3. 8.	G. ♃.	—— 2. t. 159.
19 radicàta. S.G.	(*ye.*)	fleshy-fringed-ld.	C. B. S.	1802.	5. 8.	G. ♃.	—— 1. t. 174.
20 ciliàta.	(*wh.*)	ciliated.	——	1795.	4. 6.	G. ♃.	L'Herit. ger. t. 7.
21 ovalifòlia. S.G.	(*wh.*)	oval-leaved.	——	1820.	4. 8.	G. ♃.	Sweet ger. 2. t. 106.
22 reticulàta. S.G.	(*std.*)	netted-petaled.	——	——	——	G. ♃.	—— 1. t. 91.
23 nervifòlia.	(*wh.*)	nerved-leaved.	——	——	5. 8.	G. ♃.	Jacq. ic. rar. 3. t. 517.
24 triphy'lla.	(*fl.cr.*)	three-leaved.	——	——	——	G. ♃.	—— 3. t. 515.
25 lùtea.†	(*ye.*)	small-yellow.	——	1800.	——	G. ♃.	And. bot. rep. t. 123.
26 nummularifòlia.	(*ye.*)	money-wort-ld.	——	1801.	——	G. ♃.	Sal. par. lond. t. 23.
27 integrifòlia. C.C.	(*w.r.*)	entire-leaved.	Hybrid.	1821.	4. 8.	G. ♃.	
28 fuscàta. C.C.	(*br.*)	brown-flowered.	——	——	——	G. ♃.	
29 nígricans. C.C.	(*bl.*)	blackish.	——	1822.	——	G. ♃.	
30 mágica. C.C.	(*bl.*)	dark-flowered.	——	——	——	G. ♃.	
31 àtra. S.G.	(*da.*)	dark brown.	C. B. S.	1793.	——	G. ♃.	Sweet ger. 1. t. 72.
32 dioíca. C.C.	(*bl.*)	diœcious.	——	1795.	6. 8.	G. ♃.	And. bot. rep. 209.
Gerànium melanánthum. A.R.							

Those marked thus (†) are only known to us by Andrews's Figures.

33 undulæflòra. S.G. (*da.*) waved-petaled. C. B. S. 1821. 4. 8. G. ♃. Sweet ger. 3. t. 263.
34 melanántha. S.G. (*bl.*) black-flowered. ——— 1790. —— G. ♃. ——— 1. t. 73.
35 vària. S.G. (*da.*) various-leaved. Hybrid. 1821. —— G. ♃. ——— 2. t. 166.
36 Colvíllii. S.G. (*cr.*) Colvill's. ——— 1823. —— G. ♃. ——— 3. t. 260.
37 Jenkinsòni. C.C. (*sc.*) Jenkinson's. ——— —— —— G. ♃.
38 dilùta (*bh.*) diluted. ——— 1821. —— G. ♃.
39 replicàta. (*re.*) replicate. ——— —— —— G. ♃.
40 rubéscens. (*re.*) reddish. ——— —— —— G. ♃.
41 intermíxta. (*re.*) intermixed. ——— 1822. —— G. ♃.
42 capitàta. (*bh.*) headed. ——— —— —— G. ♃.
43 supérba. (*sc.*) superb. ——— —— —— G. ♃.
44 villòsa. (*sc.*) villous. ——— 1821. —— G. ♃.
45 unguiculàta. (*re.*) clawed. ——— —— 5. 8. G. ♃.
46 tenuifòlia. (*bh.*) slender-leaved. ——— —— —— G. ♃.
47 marginàta. (*bh.*) margined. ——— 1823. —— G. ♃.
48 atrosanguínea. S.G. (*cr.*) dark-crimson. ——— 1822. 4. 8. G. ♃. Sweet ger. 2. t. 151.
49 élegans. S.G. (*cr.*) elegant. ——— —— —— G. ♃. ——— 2. t. 132.
Pelargònium Sweetiànum. DC.
50 coluteæfòlia. S.G. (*cr.*) BladderSenna-ld. ——— 1823. —— G. ♃. Sweet ger. 4. t. 311.
51 orobifòlia. S.G. (*st.pu.*) Orobus-leav'd. ——— —— —— G. ♃. ——— 4. t. 304.
52 coccínea. S.G. (*sc.*) scarlet. ——— —— —— G. ♃. ——— 4. t. 398.
53 recurviflòra. (*sc.*) recurv'd-petal'd. ——— 1821. —— G. ♃.
54 sanguinolénta. (*cr.*) bloody-flower'd. ——— 1823. —— G. ♃.
55 pulchélla. (*re.*) neat. ——— —— —— G. ♃.
56 am'œna. (*re.*) delightful. ——— —— —— G. ♃.
57 pátens. (*re.*) spreading. ——— 1822. —— G. ♃.
58 lilacìna. (*li.*) Lilac-coloured. ——— —— —— G. ♃.
59 venòsa. S.G. (*std.*) veined-petaled. ——— 1823. —— G. ♃. Sweet ger. 3. t. 209.
60 literàta. (*std.*) lettered. ——— —— —— G. ♃.
61 labyrínthica. S.G. (*std.*) labyrinth-fld. ——— —— —— G. ♃. Sweet ger. 3. t. 276.
62 ròsea. S.G. (*ro.*) rose-coloured. C. B. S. 1792. —— G. ♃. ——— 3. t. 262. f. 1.
Geranium ròseum. A.R. 173. *Pelargònium condensàtum.* P.S.
63 hedysarifòlia. S.G. (*cr.*) Hedysarum-ld. Hybrid. 1823. —— G. ♃. Sweet ger. 4. t. 355.
64 galegifòlia. (*cr.*) Galega-leaved. ——— —— —— G. ♃.
65 sisymbriifòlia. S.G. (*cr.*) Water-rocket-ld. ——— —— —— G. ♃. Sweet ger. 3. t. 292.
66 hirsùta. (*bh.*) various-ld. hairy. C. B. S. 1788. 3. 5. G. ♃. Cav. dis. 4. t. 101. f. 2.

DIMA'CRIA. S.G. DIMA'CRIA. Monadelphia Pentandria.

1 pinnàta. S.G. (*bh.*) wing-leaved. C. B. S. 1779. 4. 6. G. ♃. Sweet ger. 1. t. 46.
Pelargònium viciæfòlium. DC.
2 bipartìta. S.G. (*st.*) forked-leaved. Hybrid. 1821. 4. 8. G. ♃. Sweet ger. 2. t. 142.
3 foliolòsa. S.G. (*st.*) leafy. C. B. S. 1800. —— G. ♃. And. bot. rep. 311.
Geranium pinnàtum. A.R.
4 astragalifòlia. S.G. (*st.*) Astragalus-ld. ——— 1788. 6. 8. G. ♃. Sweet ger. 2. t. 103.
5 Smithiàna. S.G. (*re.*) Mr. E. D. Smith's. Hybrid. 1823. 4. 8. G. ♃. ——— 4. t. 358.
6 tenélla.† (*bh.*) tender. ——— 1802. —— G. ♃. Andr. geran. c. ic.
Geranium tenéllum. A.G.
7 imbùta. (*bh.*) stained-petaled. Hybrid. 1822. 4. 8. G. ♃.
8 fúlgens. C.C. (*cr.*) shining-scarlet. ——— 1821. —— G. ♃.
9 Andrèwsii.† (*bh.p.*) Andrews's. C. B. S. 1802. —— G. ♃. Andr. geran. c. ic.
Geranium heterophy'llum. A.G. *non* Jacq.
10 élegans. S.G. (*sc.*) elegant. Hybrid. 1822. —— G. ♃. Sweet ger. 3. t. 202.
11 coronillæfòlia.† (*co.*) Coronilla-leav'd. C. B. S. 1795. 6. 7. G. ♃. And. bot. rep. 305.
12 apiifòlia.† (*bh.*) Parsley-leaved. ——— 1802. 4. 8. G. ♃. Andr. geran. c. ic.
Geranium apiifòlium. A.G. *non.* Jacq.
13 auriculàta. S.G. (*w.pu.*) eared-leaved. C. B. S. .... 5. 8. G. ♃. Sweet ger. 4. t. 395.
Pelargònium ciliàtum. Jacq. ic. 3. t. 519.
14 sulphùrea. S.G. (*st.*) sulphur-color'd. Hybrid. 1821. —— G. ♃. Sweet ger. 2. t. 163.
15 cárnea. (*fl.*) flesh-coloured. C. B. S. 1812. 4. 6. G. ♃. Jacq. ic. rar. 3. t. 512.
16 floribúnda.† (*v.*) many-flowered. ——— 1795. 3. 5. G. ♃. And. bot. rep. 420.
17 fissifòlia.† (*bh.*) cloven-leaved. ——— —— 4. 8. G. ♃. ——— 378.
18 barbàta. S.G. (*fl.*) bearded. ——— 1790. 5. 8. G. ♃. Sweet ger. 4. t. 391.
19 recurvàta. S.G. (*wh.*) recurved. ——— —— —— G. ♃. And. rep. 323.
Geranium barbàtum. v. minor. A.R.
20 aristàta. S.G. (*wh.*) awned. ——— —— —— G. ♃. Andr. rep. 366.
Geranium barbàtum. v. undulàtum. A.R.
21 setígera. S.G. (*wh.*) bristle-bearing. ——— —— —— G. ♃. Andr. bot. rep. 303.
Geranium barbàtum. A.R. *nec aliorum.*
22 revolùta.† (*pu.*) revolute. ——— 1800. 7. 8. G. ♃. Andr. rep. 354.
23 oxalidifòlia.† (*st.*) Oxalis-leaved. ——— 1801. 5. 8. G. ♃. ——— 300.

24	refléxa.†	(*wh.*)	reflex-leaved.	C. B. S.	1800.	6. 7.	G. ♃.	Andr. rep. 224.
25	rumicifòlia. S.G.	(*ye.*)	Dock-leaved.	——	1825.	5. 8.	G. ♃.	Sweet ger. 4. t. 318.
26	depréssa. S.G.	(*st.*)	depressed.	——	1812.	——	G. ♃.	—— 3. t. 290.
27	longiflòra. S G.	(*st.*)	long-flowered.	——	——	——	G. ♃.	Jacq. ic. rar. 3. t 521.
28	longifòlia. S.G.	(*ro.*)	long-leaved.	——	——	——	G ♃.	—— 3. t. 518.
29	lanceolàta. S.G.	(*bh.*)	spear-leaved.	——	1826.	——	G. ♃.	Sweet ger. 4. t. 387.
30	undulàta.†	(*wh.*)	wave-flowered.	——	1795.	5. 7.	G. ♃.	Andr. rep. 292.
31	spathulàta.†	(*wh.*)	spatula-leaved.	——	——	4. 6.	G. ♃.	—— 152.
32	affínis.†	(*wh.*)	fring'd-spat.-l^d.	——	——	——	G. ♃.	—— 282.
33	virgínea.†	(*wh.*)	virgin.	——	——	5. 7.	G. ♃.	—— 317.
34	radiàta.†	(*ye.*)	ray-leaved.	——	1801.	6. 8.	G. ♃.	—— 222.
35	lineàris.†	(*st.*)	linear-petaled.	——	1800.	——	G. ♃.	—— 193.
36	punctàta.†	(*st.*)	dotted-flowered.	——	1794.	4. 6.	G. ♃.	—— 60.
	SEYMO'URIA. S.G.		SEYMO'URIA.	Monadelphia Pentandria.				
1	asarifòlia. S.G.	(*da.*)	Asarum-leaved.	C. B. S.	1821.	11. 2.	G. ♃.	Sweet ger. 3. t. 206.
2	L'Heritiéri. S.G.	(*pu.*)	L'Heritier's.	——	1795.	4. 5.	G. ♃.	L'Herit. ger. t. 43.
	Pelargònium dipétalum. L'Her.							
	PELARG'ONIUM. S.G.		STORK'S-BILL.	Monadelphia Heptandria.				
1	amœ'num. S.G.	(*sa.*)	delightful.	Hybrid.	1821.	4. 8.	G. ♃.	Sweet ger. 2. t. 121
2	dimacriæflòrum. S.G.	(*sc.*)	Dimacria-fl^d.	——	1822.	——	G. ♃.	—— 3. t. 220.
3	múndulum. S.G.	(*sc.*)	spruce.	——	1823.	3. 10.	G. ♃.	—— 3. t. 288.
4	ríngens. S.G.	(*sc.*)	ringent-flow'rd.	——	——	——	G. ♃.	—— 3. t. 256.
5	concàvum. S.G.	(*re.*)	concave-petaled.	——	——	——	G. ♃.	—— 3. t. 237.
6	concláusum. S.G	(*d.re.*)	shut-petaled.	——	——	4. 10.	G. ♄.	—— 4. t. 305.
7	hoareæflòrum S.G.	(*sa.*)	Hoarea-flow'r'd.	——	1821.	——	G. ♃.	—— 2. t. 133.
8	cláusum. C.C.	(*re.*)	closed petaled.	——	1822.	——	G. ♃.	
9	pàtens. S.G.	(*sc.*)	spreading-fl^d.	——	1820.	——	G. ♄.	Sweet ger. 2. t. 125.
10	cruéntum. S.G.	(*cr.*)	blood-red.	——	1822.	——	G. ♃.	—— 2. t. 170.
11	áulicum.	(*cr.*)	courtly.	——	1823.	5. 10.	G. ♃.	
12	Avroniànum. S.G.	(*sc.*)	Avron's.	——	——	——	G. ♃.	Sweet ger. 3. t. 394.
13	ligulàtum. S.G.	(*sc.*)	ligulate-petaled.	——	1823.	4. 9.	G. ♄.	—— 4. t. 301.
14	puníceum. C.C.	(*cr.*)	puniceous.	——	——	——	G. ♃.	
15	pusíllum. C.C.	(*sc.*)	dwarf.	——	——	——	G. ♃.	
16	pygm'æum. C.C.	(*sc.*)	small.	——	——	5. 8.	G. ♃.	
17	venulòsum.	(*sc.*)	veined-petaled.	——	1821.	——	G. ♃.	
18	intertéxtum. S.G.	(*sc.*)	interwoven.	——	1822.	4. 8.	G. ♃.	Sweet ger. 2. t. 185.
19	ácidum. S.G.	(*cr.*)	sour-leaved.	——	1823.	5. 9.	G. ♃.	—— 3. t. 261.
20	pállens. S.G.	(*st.*)	cream coloured.	C. B. S.	1800.	——	G. ♄.	—— 2. t. 148.
21	caryophyllàceum.	(*bh.*)	Clove-scented.	Hybrid.	1822.	——	G. ♄.	—— 4. t. 347.
22	paradòxum.	(*wh.*)	paradoxical.	——	——	——	G. ♄.	
23	multiradiàtum. S.G.	(*da.*)	many-rayed.	C. B. S.	1818.	6. 10.	G. ♄.	Sweet ger. 2. t. 145.
24	rutàceum. S.G.	(*da.*)	Rue-like.	Hybrid.	1823.	5. 10.	G. ♄.	—— 3. t. 279.
25	pedunculatùm. S.G.	(*da.*)	long-peduncled.	——	——	——	G. ♄.	—— 4. t. 346.
26	œnanthifòlium. S.G.	(*d.cr.*)	Dropwort-l^d.	——	——	6. 10.	G. ♄.	—— ser. 2. t. 13.
27	sphondyliifòlium.	(*d.cr.*)	Sphondylium-l^d	Hybrid.	1822.	4. 8.	G. ♃.	—— 3. t. 246.
28	chærophy'llum. S.G.	(*cr.*)	Cow-Parsley-l^d	——	——	——	G. ♄.	—— 3. t. 257.
29	nítidum. S.G.	(*sc.*)	glossy.	——	——	——	G. ♄.	—— 3. t. 298.
30	sanguíneum. S.G.	(*cr.*)	crimson.	C. B. S ?	1819.	——	G. ♄.	—— 1. t. 76.
31	árdens. S.G.	(*sc.*)	glowing.	Hybrid.	1810.	3. 8.	G. ♄.	—— 1. t. 45.
32	confertifòlium.	(*sc.*)	close-leaved.	——	——	——	G. ♄.	—— 3. t. 297.
	Gerànium árdens minor. A.G.							
33	pulchéllum. B.M.	(*v.*)	nonesuch.	C. B. S.	1795.	3. 5.	G. ♄.	Sweet ger. 1. t. 31.
34	jonquillìnum. S.G.	(*pi.*)	Jonquil-scent'd.	Hybrid.	1822.	5. 10.	G. ♄.	—— 3. t. 241.
35	seléctum. S.G.	(*d.pu.*)	select.	——	——	——	G. ♃.	—— 2. t. 190.
36	imbricàtum. S.G.	(*v.*)	imbricate-pet'd.	C. B. S ?	1800.	5. 8.	G. ♄.	—— 1. t. 65.
37	bícolor. S.G.	(*da.*)	two-coloured.	....	1778.	7. 8.	G. ♄.	Bot. mag. 201.
	α átrum. S.G.	(*da.*)	*dark-edged.*	....	——	——	G. ♄.	Sweet ger. 1. t. 97. f. 1.
	β pállidum. S.G.	(*pu.*)	*pale-edged.*	....	——	——	G. ♄.	—— 1. t. 97. f. 2.
38	quinquevúlnerum.	(*da.*)	dark-flower'd.	Hybrid.	1796.	5. 10.	G. ♄.	—— 2. t. 161.
39	Blandfordiànum.	(*bh.*)	hoary-leaved.	——	1812.	1. 12	G. ♄.	—— 2. t. 101.
40	millefoliàtum. S.G.	(*v.*)	Milfoil-leaved.	C. B. S.	1819.	6. 10.	G. ♃.	—— 3. t. 230.
41	flàvum. DC.	(*st.*)	pale yellow.	——	1724.	7. 9.	G. ♃.	—— 3. t. 254.
42	scábridum. C.C.	(*br.*)	rough-edged.	——	1822	——	G. ♃.	
43	tríste. DC.	(*da.*)	night-smelling.	——	1632.	5. 10.	G. ♃.	Delaun. h. am. t. 27.
44	filipendulifòlium.	(*br.*)	Dropwort-l^d.	——	1812.	——	G. ♃.	Sweet ger. 1. t. 85.
45	pentastíctum.	(*bh.*)	five-marked.	Hybrid.	1824.	——	G. ♃.	—— ser. 2. t. 25.
46	heracleifòlium. S.G.	(*br.*)	Heracleum-l^d.	C. B. S.	1820.	——	G. ♃.	—— 3. t. 211.
47	lobàtum. DC.	(*da.*)	Cow-parsnep-l^d.	——	1710.	——	G. ♃.	—— 1. t. 51.
48	salebròsum. S.G.	(*d.pu.*)	rugged-leaved.	Hybrid.	1824.	——	G. ♃.	—— 4. t. 309.

49 Lawranceànum. s.g.	(d.pu.)	Mrs. Kearse's.	——	1824.	5. 10.	G. ♃.	Swt. ger. ser. 2. t. 23.
50 orbiculàtum.	(da.)	orbicular-ld.	Hybrid	1823.	——	G. ♃.	
51 pulveruléntum. s.g.	(sp.)	powdered-ld	C. B. S.	1822.	——	G. ♃.	Sweet ger. 3. t. 218.
52 pedicellàtum. s.g.	(v.)	long-pedicled.	——	——	——	G. ♃.	—— 3. t. 250.
53 lùridum. s.g.	(co.)	lurid.	——	1811.	7. 12.	G. ♃.	—— 3. t. 281.
Geranium lùridum. A.G. ic.							
54 rapæfòlium.	(da.)	Turnep-leaved.	Hybrid.	1821.	5. 10.	G. ♃.	
55 glauciifòlium. s.g.	(da.)	Glaucium-ld.	——	——	——	G. ħ.	Sweet ger. 2. t. 179.
56 gibbòsum. DC.	(st.)	knotted-stalk'd	C. B. S.	1712.	——	G. ♄.	—— 1. t. 61.
57 mutábile. s.g.	(bh.)	changeable-cold	Hybrid.	1822.	7. 10.	G. h.	—— 3. t. 213.
58 vespertìnum. s.g.	(pu.)	evening-scent'd	——	——	8. 12.	G. ♄.	—— 3. t. 239.
59 sinapifòlium.	(pi.)	Mustard-leav'd.	Hybrid.	——	6. 10.	G. h.	
60 brassicàtum.	(st.)	Cabbage-leav'd.	——	——	——	G. ♄.	
61 adscéndens.	(pa.)	ascending.	——	——	——	G. h.	
62 circumfléxum.	(pa.)	bending.	——	——	——	G. ♄.	
63 campyliæflòrum. s.g.	(li.)	Campylia-fld	——	1823.	——	G. ♄.	Sweet ger. 3. t. 251.
64 renifórme. DC.	(pu.)	kidney-leaved.	C. B. S.	1791.	1. 12.	G. ♄.	—— 1. t. 48.
65 flexuòsum. s.g.	(sc.)	bent-stalked.	Hybrid.	1821.	3. 10.	G. h.	—— 2. t. 180.
66 sæpeflòrens. s.g.	(pi.)	frequent-flow'rg	——	1801.	——	G. h.	—— 1. t. 58.
67 cómptum. s.g.	(pi.)	decked.	——	1821.	——	G. h.	—— 3. t. 255.
68 echinàtum. DC.	(wh.)	prickly-stalked.	C. B. S.	1789.	6. 10.	G. h.	—— 1. t. 54.
69 armàtum. s.g.	(d.pu.)	long-spined.	——	——	3. 6.	G. ♄.	—— 3. t. 214.
70 intermèdium. c.c.	(li.)	intermediate.	Hybrid.	1821.	6. 10.	G. h.	
71 particéps. s.g.	(li.)	participant.	——	1806.	5. 10.	G. ♄.	Sweet ger. 1. t. 49.
72 cortusæfòlium. DC.	(wh.)	Cortusa-leavd	Africa.	1786.	6. 10.	G. ħ.	—— 1. t. 14.
73 eréctum. s.g.	(pu.)	upright.	Hybrid.	1822.	——	G. ħ.	—— 2. t. 187.
74 Stapletòni. s.g.	(pu.)	Miss Stapleton's.	——	——	——	G. ♄.	—— 2. t. 212.
75 crassicáule. s.g.	(wh.)	Cowslip-scent'd.	Africa.	1786.	——	G. ♄.	L'Herit. ger. t. 26.
α *álbum.* s.g.	(wh.)	*white-flowered.*	——	——	——	G. ♄.	Sweet ger. 2. t. 192. f. a.
β *maculàtum.* s.g.	(sp.)	*spotted-flowered*	——	——	——	G. ħ.	—— 2. t. 192. f. b.
76 Victoriàna. s.g.	(sc.)	Princess Victoria's.	Hybrid.	1826.	——	G. ħ.	—— ser. 2. t. 7.
77 anómalum. s.g.	(pi.)	anomalous.	——	1822.	4. 8.	G. h.	—— 4. t. 315.
78 tenuifòlium. DC.	(pi.)	slender-leaved.	C. B. S.	1768.	5. 7.	G. ħ.	L'Herit. ger. t. 12.
79 hírtum. DC.	(pi.)	hairy.	——	——	3. 9.	G. ħ.	Sweet ger. 2. t. 113.
80 anthriscifòlium. s.g.	(sc.)	Anthriscus-ld.	Hybrid.	1823.	7. 9.	G. ħ.	—— 3. t. 233.
81 magnistipulàtum.	(sc.)	large-stipuled.	——	1825.	5. 10.	G. ħ.	—— 4. t. 313.
82 fúlgidum. DC.	(sc.)	Celandine-ld.	C. B. S.	1723.	3. 10.	G. h.	—— 1. t. 69.
83 aurantìacum. s.g.	(or.)	orange-colour'd.	Hybrid.	1821.	4. 8.	G. h.	—— 2. t. 198.
α *undulàtum.* s.g.	(or.)	*wave-leaved.*	——	——	——	G. h.	—— 2. t. 198. f. a.
β *planifòlium.* s.g.	(or.)	*flat-leaved.*	——	——	——	G. h.	—— 2. t. 198. f. b.
84 ardéscens. s.g.	(cr.)	burnished.	——	1822.	——	G. ♄	—— 3. t. 231.
85 torrefáctum. s.g.	(da.)	burnt-petaled.	——	——	——	G. h.	—— 3. t. 243.
86 Pottèri. s.g.	(st.)	Potter's.	——	——	6. 9.	G. h.	—— 2. t. 147.
87 luculéntum.	(sc.)	bright-flowered.	——	——	——	G. ♄.	
88 diléctum.	(sc.da.)	chosen.	——	——	4. 9.	G. ħ.	
89 Loudoniànum. s.g.	(d.sc.)	Mr. Loudon's.	——	1827.	5. 9.	G. ħ.	Sweet ger. ser. 2. t. 17.
90 eratìnum. s.g.	(sc.)	lovely.	——	——	——	G. ħ.	—— ser. 2. t. 30.
91 nànum. s.g.	(sc.)	pigmy.	——	1820.	——	G. ħ.	—— 2. t. 102.
92 incúrvum. s.g.	(sc.)	incurved-petal'd	——	1821.	——	G. h.	—— 3. t 249.
93 ignéscens. s.g.	(sc.)	fiery-flowered.	——	1812.	——	G. ♄.	—— 1. t. 2. & 55.
α *màjor.*	(sc.)	*large-flowered.*	——	1818.	——	G. ♄.	—— 1. t. 2.
β *coccíneum.*	(sc.)	*scarlet-flowered.*	——	1822.	——	G. ♄.	
γ *spléndens.*	(sc.)	*splendid.*	——	——	——	G. ♄.	
δ *supérbum.*	(sc.)	*superb.*	——	——	——	G. h.	
ε *stérile.* s.g.	(sc.)	*barren-anthered.*	——	1812.	——	G. ♄.	Sweet ger. 1. t. 55.
94 laxiflòrum. s.g.	(sc.)	spreading-umbd.	——	1822.	——	G. h.	—— 3. t. 216.
95 concrètum. s.g.	(sc.)	compounded.	——	1827.	4. 9.	G. ħ.	—— ser. 2. t. 24.
96 Leghkéckæ. s.g.	(sc.)	Mrs. Legh Keck's	——	1826.	——	G. h.	—— 4. t. 377.
97 Charlwoòdii. s.g.	(d.sc.)	Charlwood's.	——	——	——	G. h.	—— 4. t. 380.
98 incomparabile.	(d.sc.)	incomparable.	——	1822.	——	G. h.	
99 eriocáulon. s.g.	(sc.)	woolly-stalked.	——	1824.	——	G. ♄.	Sweet ger. 4. t. 357.
100 quadriflòrum. s.g.	(or.)	four-flowered.	——	——	——	G. ħ.	—— 4. t. 321.
101 chelidoniifòlium. s.g.	(or.)	Chelidonium-ld	——	1825.	——	G. ♄.	—— 4. t. 341.
102 chenopodiifòlium. s.g.	(sc.)	Goosefoot-ld	——	——	——	G. ♄.	—— 4. t. 328.
103 latidentàtum. s.g.	(sc.)	broad-toothed.	——	1827.	——	G. ♄.	—— ser. 2. t. 28.
104 variifòlium. s.g.	(sc.)	various-leaved.	——	1823.	——	G. ♄.	—— 3. t. 280.
105 diversilòbum. s.g.	(re.)	various-lobed.	——	1825.	——	G. h.	—— 4. t. 361.
106 Southcoteànum. s.g.	(sc.)	Miss Southcote's	——	1826.	——	G. ♄.	—— 4. t. 348.
107 insígnitum. s.g.	(d.sc.)	marked-flower'd.	——	1823.	——	G. ħ.	—— 3. t. 300.

108	plectophy'llum. S.G.	(d.sc.)	plaited-leaved.	Hybrid.	1826.	4. 9.	G.	♄.	Sweet ger. ser. 2. t. 41.
109	floccòsum. S.G.	(cr.)	nappy.	———	1821.	——	G.	♄.	——— 2. t. 129.
110	rubicúndum.	(re.)	rubicund.	———	1822.	——	G.	♄.	
111	hæmánthon.	(cr.)	blood-coloured.	———	1823.	——	G.	♄.	
112	multiflòrum.	(sc.)	many-flowered.	———	——	——	G.	♄.	Sweet ger. 4. t. 396.
113	exìle.	(sc.)	thin-branched.	———	1822.	——	G.	♄.	
114	graciléscens.	(sc.)	slender-stalked.	———	——	——	G.	♄.	
115	spléndidum.	(cr.)	splendid.	———	——	——	G.	♄.	
116	consímile.	(sc.)	likened.	———	1822.	——	G.	♄.	
117	Broughtòniæ. S.G.	(sc.)	Lady Broughton's	———	——	——	G.	♄.	Sweet ger. 2. t. 181.
118	Bakeriànum. S.G.	(sc.)	Mrs. Jenkinson's	———	——	——	G.	♄.	——— 3. t. 240.
119	áltum. S.G.	(re.)	tall upright.	———	1827.	——	G.	♄.	——— ser. 2. t. 26.
120	Debúrghæ. S.G.	(sc)	Mrs. De Burgh's	———	——	——	G.	♄.	——— ser. 2. t. 21.
121	Cummíngiæ. S.G.	(re.)	Ly. Gordon Cumming's	———	——	——	G.	♄.	——— ser. 2. t. 35.
122	rubéscens. S.G.	(re.)	Lady Liverpool's	———	1819.	——	G.	♄.	——— 1. t. 30.
123	Boscawèniæ. C.C.	(re.)	Mrs. Boscawen's.	———	——	——	G.	♄.	
124	pavonìnum. S.G.	(re.)	peacock-spott'd	———	——	——	G.	♄.	Sweet ger. 1. t. 40.
125	carnéscens. S.G.	(sa.)	pale red.	———	1822.	——	G.	♄.	——— 4. t. 388.
126	ramígerum. S.G.	(sc.)	branching-veined	———	1826.	1. 12.	G.	♄.	——— 4. t. 352.
127	fláccidum. S.G.	(d.re.)	flaccid-petal'd.	———	——	4. 9.	G.	♄.	——— 4. t. 337.
128	Smíthii. S.G.	(re.)	Smith's.	———	1819.	——	G.	♄.	——— 2. t. 110.
129	spectábile. S.G.	(sc.)	showy.	———	1821.	——	G.	♄.	——— 2. t. 136.
	α maculàtum. S.G.	(sc.)	*spot-flowered.*	———	——	——	G.	♄.	——— 2. t. 136. f. 1.
	β atroròbens. S.G.	(d.re.)	*dark-red.*	———	——	——	G.	♄.	
	γ striàtum. S.G.	(sc.)	*streak-flowered.*	———	——	——	G.	♄.	Sweet ger. 2. t. 136. f. 2.
	δ recúrvum. S.G.	(sc.)	*recurved-petaled.*	———	——	——	G.	♄.	——— 2. t. 136. f. 3.
	ε purpuráscens.	(pu.)	*purplish.*	———	——	——	G.	♄.	
130	Tibbitsiànum. S.G.	(sc.)	Mr. Tibbits'.	———	——	——	G.	♄.	Sweet ger. 2. t. 158.
131	flàgrans. S.G.	(sc.)	burning.	———	1827.	——	G.	♄.	——— ser. 2. t. 16.
132	malachræfòlium. S.G.	(re.)	Malachra-ld.	———	——	——	G.	♄.	——— ser. 2. t. 2.
133	magnifòlium. S.G.	(re.)	very large-ld.	———	——	——	G.	♄.	——— ser. 2. t. 10.
134	Lacòniæ. S.G.	(d.sc.)	Lady Lacon's.	———	1826.	——	G.	♄.	——— 4. t. 374.
135	Francísii. S.G.	(re.)	Mr. Francis's.	———	1824.	5. 10.	G.	♄.	——— 4. t. 349.
136	cartilagíneum. S.G.	(re.)	horny-toothed.	———	1824.	——	G.	♄.	——— 4. t. 382.
137	dilùtum. S.G.	(re.)	stained-petal'd.	———	1823.	5. 9.	G.	♄.	——— 3. t. 293.
138	basílicum. S.G.	(sc.)	princely.	———	1824.	——	G.	♄.	——— 4. t. 360.
139	Hùmei.	(d.re.)	Sir A. Hume's.	———	——	4. 10.	G.	♄.	——— 4. t. 324.
140	erióphorum.	(sc.)	wool-bearing.	———	1823.	——	G.	♄.	
141	irregulàre.	(sc.)	irregular-ld.	———	——	——	G.	♄.	
142	flámmeum. C.C.	(sc.)	flame-coloured.	———	1822.	——	G.	♄.	
143	affine. S.G.	(sc.)	related.	———	——	——	G.	♄.	Sweet ger. 3. t. 277.
144	Colvíllii. S.G.	(sc.)	Colvill's.	———	1820.	——	G.	♄.	——— 1. t. 86.
145	Vallèti. C.C.	(sc.)	Vallet's.	———	1822.	——	G	♄.	
146	heteromállum. S.G.	(re.)	soft-leaved.	———	1823.	——	G.	♄.	Sweet ger. ser. 2. t. 32.
147	hibiscifòlium.	(re.)	Hibiscus-ld.	———	——	——	G.	♄.	
148	atroròbens. C.C.	(re.)	dark-red.	———	1822.	——	G.	♄.	
149	Allènii. S.G.	(d.re.)	Allen's.	———	1823.	——	G.	♄.	Sweet ger. 3. t. 229.
150	melanchòlicum.	(d.re.)	melancholy.	———	1827.	——	G.	♄.	
151	Daveyànum. S.G.	(d.re.)	Davey's.	———	1819.	——	G.	♄.	Sweet ger. 1. t. 32.
152	phœníceum. S.G.	(d.re.)	reddish purple.	———	1824.	4. 12.	G.	♄.	——— 3. t. 207.
153	heterotrìchum. S.G.	(d.re.)	various-haired.	———	1823.	4. 10	G.	♄.	——— 4. t. 303.
154	Burnettiànum. S.G.	(d.re.)	Miss Burnett's.	———	1826.	——	G.	♄.	——— 4. t. 369.
155	imperiàle. S.G.	(d.re.)	imperial.	———	——	——	G.	♄.	——— 4. t. 365.
156	saturàtum. S.G.	(d.re.)	saturated.	———	1827.	——	G.	♄.	——— ser. 2. t. 37.
157	règium. S.G.	(sc.)	kingly.	———	1826.	——	G.	♄.	——— 4. t. 368.
158	Hesilrígeæ. S.G.	(pu.)	Lady Hesilrige's.	———	1824.	——	G.	♄.	——— 4. t. 319.
159	acetabulòsum. S.G.	(sa.)	saucer-leaved.	———	1827.	——	G.	♄.	——— ser. 2. t. 44.
160	Knìpeæ. S.G.	(li.)	Mrs. Knipe's	———	1826.	——	G.	♄.	——— 4. t. 372.
161	calycìnum. S.G.	(li.)	large-calyxed.	———	1815.	——	G.	♄.	——— 1. t. 81.
	α vèrum. S.G.	(li.)	*true.*	———	——	——	G.	♄.	——— 1. t. 81. f. a.
	β maculàtum. S.G.	(li.)	*large-spotted.*	———	1820.	——	G.	♄.	——— 1. t. 81. f. b.
	γ angulàtum. S.G.	(li.)	*angular-leaved.*	———	——	——	G.	♄.	——— 1. t. 81. f. c.
162	planiflòrum. C.C.	(li.)	flat-flowered.	———	——	——	G.	♄.	
163	Dennisiànum. S.G.	(pu.)	Dennis's.	———	1819.	——	G.	♄.	Sweet ger. 1. t. 20.
164	Barringtònii. DC.	(pu.)	Barrington's	C. B. S.	....	——	G.	♄.	
165	elàtum. C.C.	(pu.)	tall purple.	Hybrid.	1812.	——	G.	♄.	
166	Tormánni. L.en.	(pu.)	Torman's.	———	1821.	——	G.	♄.	
167	Chandlèri. C.C.	(pu.)	Chandler's.	———	1814.	——	G.	♄.	
168	crenulàtum. S.G.	(pu.)	crenulate-leav'd.	———	1821.	——	G.	♄.	Sweet ger. 2. t. 162.

169	megalostíctum. S.G.	(*d.pu.*)	large-marked.	Hybrid.	1824.	4. 10.	G. ♄.	Sweet ger. ser. 2. t. 8.
170	poculifòlium. S.G.	(*pu.*)	cup-leaved.	——	1826.	——	G. ♄.	—— ser. 2. t. 31.
171	Foljàmbeæ. S.G.	(*r.pu.*)	Mrs. Foljambe's	——	1825.	——	G. ♄.	—— 4. t. 312.
172	Robinsòni. S.G.	(*bh.*)	Robinson's.	——	1822.	——	G. ♄.	—— 2. t. 150.
173	Most'ynæ. S.G.	(*pu.*)	Mrs. Mostyn's.	——	1816.	——	G. ♄.	—— 1. t. 10.
174	cucullàtum. DC.	(*pu.*)	hooded-leaved.	C. B. S.	1690.	——	G. ♄.	Cav. dis. 4. t. 106. f. 1.
	β striatiflòrum.	(*pu.*)	*Prince Regent.*	——	1810.	——	G. ♄.	
	γ màjor. C.C.	(*pu.*)	*Royal George.*	——	1812.	——	G. ♄.	
	δ grandiflòrum. C.C.	(*pu.*)	*great-flowered.*	——	1818.	——	G. ♄.	
	ε parvifòlium. C.C.	(*pu.*)	*smaller-leaved.*	——	——	——	G. ♄.	
175	cochleàtum. W. en.	(*wh.*)	concave-l[d].	——	....	——	G. ♄.	Wendl. col. 2. t. 64.
	concavifòlium. Wendl.							
176	rùbens. W. en.	(*re.*)	red-flowered.	——	....	——	G. ♄.	Wendl. col. 2. t. 54.
177	robústum. C.C.	(*pu.*)	robust.	Hybrid.	1821.	——	G. ♄.	
178	incarnàtum. S.G.	(*fl.*)	pale flesh-col[d].	——	——	——	G. ♄.	Sweet ger. 4. t. 308.
179	lasiocáulon. S.G.	(*bh.*)	villous-stalked.	——	1826.	——	G. ♄.	—— 4. t. 364.
180	veníferum. S.G.	(*std.*)	veined-petal'd.	——	1825.	——	G. ♄.	—— 4. t. 322.
181	foliòsum. S.G.	(*r.pu.*)	leafy.	——	1826.	——	G. ♄.	—— 4. t. 340.
182	anisodónton. S.G.	(*pu.*)	unequal-toothed.	——	1825.	——	G. ♄.	—— 4. t. 399.
183	malacophy'llum. S.G.	(*pu.*)	soft-leaved.	——	——	——	G. ♄.	—— 4. t. 397.
184	angulòsum. DC.	(*pu.*)	Marshmallow-l.	C. B. S.	1724.	7. 9.	G. ♄.	Cav. dis. 4. t. 112. f. 2.
185	cardiifòlium. S.G.	(*pu.*)	Cockle-shell-l[d].	——	1816.	4. 9.	G. ♄.	Sweet ger. 1. t. 15.
186	fasciculàtum. C.C.	(*pu.*)	Banbury.	Hybrid.	1803.	5. 10.	G. ♄.	Andr. geran. c. ic.
	Geranium speciòsum var. flore purpureo. A.G.							
187	rugòsum. C.C.	(*fl.*)	wrinkled-leav'd.	C. B. S.	....	——	G. ♄.	Andr. geran. c. ic.
188	speciòsum. DC.	(*wh.*)	specious.	——	1794.	——	G. ♄.	
189	formòsum. S.G.	(*li.*)	variegated-fl[d].	Hybrid.	....	——	G. ♄.	Sweet ger. 2. t. 120.
190	solùbile. S.G.	(*pu.*)	dissolvible-col'[d].	——	1818.	——	G. ♄.	—— 1. t. 24.
191	multinérve. S.G.	(*pu.*)	many-nerved.	——	——	——	G. ♄.	—— 1. t. 17.
192	tyrianthìnum. S.G.	(*pu.*)	royal purple.	——	1820.	——	G. ♄.	—— 2. t. 183.
193	albinotàtum. S.G.	(*pu.*)	white-marked.	——	1824.	——	G. ♄.	—— 4. t. 359.
194	æmulum. S.G.	(*pu.*)	rival.	——	——	——	G. ♄.	—— 2. t. 160.
195	accèdens. C.C.	(*pu.*)	approached.	——	——	——	G. ♄.	
196	villòsum. S.G.	(*pu.*)	villous.	——	——	——	G. ♄.	Sweet ger. 1. t. 100.
197	flóridum. S.G.	(*li.*)	abundant-fl[d].	——	1816.	6. 10.	G. ♄.	—— 1. t. 41.
198	púlchrum. S.G.	(*pu.*)	gay.	——	1820.	5. 10.	G. ♄.	—— 2. t. 107.
199	Watsòni. S.G.	(*pu.*)	Watson's.	——	1812.	7. 11.	G. ♄.	—— 2. t. 130.
200	fusciflòrum. S.G.	(*pu.*)	brown-marked.	——	1822.	4. 10.	G. ♄.	—— 3. t. 210.
	fuscàtum. S.G. *errata.*							
201	coarctàtum. S.G.	(*pu.*)	close-leaved.	——	1814.	6. 10.	G. ♄.	—— 1. t. 70.
202	paucidentàtum. S.G.	(*pu.*)	distant-tooth'd.	——	1821.	5. 10.	G. ♄.	—— 2. t. 186.
203	nervòsum. S.G.	(*pu.*)	prominent-nerv'd.	——	1818.	——	G. ♄.	—— 1. t. 47.
204	exornàtum.	(*pu.*)	adorned.	——	1823.	——	G. ♄.	—— 4. t. 381.
205	Yoúngii. S.G.	(*wh.*)	Young's.	——	1820.	——	G. ♄.	—— 2. t. 131.
206	Jenkinsòni. S.G.	(*d.bh.*)	Mr. Jenkinson's.	——	——	——	G. ♄.	—— 2. t. 154.
207	nùbilum. S.G.	(*d.bh.*)	clouded.	——	1827.	——	G. ♄.	—— ser. 2. t. 20.
208	Brównii. S.G.	(*d.bh.*)	Brown's.	——	——	——	G. ♄.	—— 2. t. 146.
209	diffórme. S.G.	(*bh.*)	various-leaved.	——	——	——	G. ♄.	—— 2. t. 105.
210	Obrieniànum. S.G.	(*bh.*)	Miss O'Brien's.	——	——	——	G. ♄.	—— ser. 2. t. 11.
211	melanostíctum. S.G.	(*li.*)	dark-spotted.	——	——	——	G. ♄.	—— ser. 2. t. 6.
212	vestiflùum. S.G.	(*wh.*)	clothed.	——	——	——	G. ♄.	—— ser. 2. t. 27.
213	pùrum. S.G.	(*wh.*)	pure white.	——	1824.	——	G. ♄.	—— 4. t. 334.
214	scìtulum. S.G.	(*wh.*)	neat white.	——	1826.	——	G. ♄.	—— 4. t. 390.
215	involucràtum. S.G.	(*va.*)	large-bracted.	——	....	——	G. ♄.	
	α máximum. S.G.	(*w.*)	*largest-flowered.*	——	1818.	——	G. ♄.	Sweet ger. 1. t. 33.
	β álbidum. S.G.	(*wh.*)	*white-flowered.*	——	1817.	——	G. ♄.	
	γ incarnàtum. S.G.	(*fl.*)	*commander in chief.*	——	1815.	——	G. ♄.	
	δ intermèdium. S.G.	(*bh.*)	*high-admiral.*	——	1817.	——	G. ♄.	
	ε lilacìnum. S.G.	(*li.*)	*Lilac-coloured.*	——	——	——	G. ♄.	
	ζ ròseum. S.G.	(*ro.*)	*rosy-coloured.*	——	——	——	G. ♄.	
	η coriàceum. C.C.	(*w.*)	*leathery.*	——	1820.	——	G. ♄.	
	ϑ maculàtum. C.C.	(*wh.*)	*coronation.*	——	——	——	G. ♄.	
216	polytríchum. S.G.	(*w.*)	many-haired.	——	1823.	——	G. ♄.	Sweet ger. 3. t. 274.
217	Baileyànum. S.G.	(*w.*)	Bailey's.	——	1819.	——	G. ♄.	—— 1. t. 87.
218	Josephínæ. Lee.	(*wh.*)	Josephine's.	——	1822.	——	G. ♄.	
219	formosíssimum. S.G.	(*w.*)	superb-white.	C. B. S.	1794.	——	G. ♄.	Sweet ger. 3. t. 215.
	α álbum. S.G.	(*wh.*)	*white-flowered.*	——	——	——	G. ♄.	—— 3. t. 215. f. a.
	β lineàtum. S.G.	(*wh.*)	*lined-flowered.*	——	——	——	G. ♄.	—— 3. t. 215. f. b.
220	Bóyleæ. S.G.	(*wh.*)	Countess of Cork's.	Hybrid.	1818.	——	G. ♄.	—— 1. t. 50.

221	pannifòlium. S.G.	(w.)	cloth-leaved.	Hybrid.	1818.	5. 10.	G. ♄.	Sweet ger. 1. t. 9.
222	biflòrum. S.G.	(wh.)	two-flowered.	——	1822.	——	G. ♄.	—— 3. t. 287.
223	cándidum. S.G.	(wh.)	fair-flowered.	——	1818.	4. 9.	G. ♄.	—— 2. t. 128.
224	blándum. S.G.	(bh.)	blush-flowered.	——	1816.	——	G. ♄.	—— 1. t. 4.
225	aceroídes. S.G.	(fl.)	Acer-leaved.	——	1822.	——	G. ♄.	—— 3. t. 242.
226	Gardnèriæ. C.C.	(wh.)	Lady Gardner's.	——	1817.	——	G. ♄.	
227	ramulòsum. S.G.	(d.w.)	small-branch'd.	——	1820.	——	G. ♄.	Sweet ger. 2. t. 177.
228	maculàtum.	(d.bh.)	spot-flowered.	C. B. S.	1805.	——	G. ♄.	Andr. ger. c. ic.
229	Hammerslèiæ. S.G.	(b.)	Mrs. Hammersley's.	Hybrid.	1822.	——	G. ♄.	Sweet ger. 3. t. 225.
230	pulchérrimum. S.G.	(pu.)	beautiful.	——	1819.	5. 10.	G. ♄.	—— 2. t. 134.
231	Rollisòni. S.G.	(d.ro.)	Rollison's.	——	1826.	——	G. ♄.	—— 4. t. 371.
232	eriosépalon. S.G.	(bh.)	woolly-calyxed.	——	——	——	G. ♄.	—— 4. t. 375.
233	veniflòrum. S.G.	(std.)	veined-flowered.	——	1822.	——	G. ♄.	—— 3. t. 258.
234	sidæfòlium. W.en.	(li.)	Sida-leaved.	——	....	4. 8.	G. ♄.	
235	acerifòlium. DC.	(li.)	Maple-leaved.	C. B. S.	1784.	4. 6.	G. ♄.	L'Herit. ger. t. 21.
236	cuneàtum. DC.	(pu.)	wedge-leaved.	Hybrid.	....	5. 8.	G. ♄.	
237	Balbisiànum. DC.	(pu.)	Balbis's.	——	....	——	G. ♄.	
238	gloriòsum. L.en.	(pu.)	renowned.	——	....	——	G. ♄.	
239	Beaufortiànum. S.G.	(li.)	Beaufort's.	C. B. S.?	....	——	G. ♄.	Sweet ger. 2. t. 138.
240	principíssæ. S.G.	(pu.)	P's. Charlotte's.	——	1810.	5. 10.	G. ♄.	—— 2. t. 139.
241	purpùreum.	(pu.)	purple-flowered.	——	1800.	——	G. ♄.	Andr. geran. c. ic.
242	versícolor. S.G.	(pu.)	various-colour'd	Hybrid.	1819.	——	G. ♄.	Sweet ger. 1. t. 78.
243	pectinifòlium. S.G.	(li.)	Scallop-shell-ld.	——	1820.	——	G. ♄.	—— 1. t. 66.
244	Thynnèæ. S.G.	(pu.)	Lady Bath's.	——	1815.	——	G. ♄.	—— 1. t. 74.
245	obtusifòlium. S.G.	(li.)	blunt-leaved.	——	——	——	G. ♄.	—— 1. t. 25.
246	Beadòniæ. S.G.	(pu.)	Mrs. Beadon's.	——	——	——	G. ♄.	—— 2. t. 191.
247	adulterìnum. DC.	(pu.)	hoary-trifid-ld.	——	1785.	4. 6.	G. ♄.	—— 1. t. 22.
248	dilatilòbum. S.G.	(bh.)	dilated-lobed.	——	1826.	——	G. ♄.	—— 4. t. 378.
249	gráphicum. S.G.	(pu.)	written-petal'd.	——	——	——	G. ♄.	—— ser. 2. t. 12.
250	obtusilòbum. S.G.	(pu.)	blunt-lobed.	——	1806.	5. 10.	G. ♄.	—— 1. t. 8.
251	Comptòniæ. S.G.	(li.)	Ly. Northampton's	——	1818.	——	G. ♄.	—— 2. t. 122.
252	afflùens. S.G.	(li.)	numerous-fld.	——	1821.	——	G. ♄.	—— 2. t. 194.
253	Saundérsii. S.G.	(pu.)	Saunders's.	——	——	——	G. ♄.	—— 3. t. 205.
254	atropurpùreum. S.G.	(pu.)	dark purple.	——	1822.	——	G. ♄.	—— 2. t. 152.
255	corúscans. S.G.	(d.re.)	glittering.	——	1821.	——	G. ♄.	—— 2. t. 173.
256	concínnum. S.G.	(re.)	comely.	——	1819.	——	G. ♄.	—— 2. t. 108.
257	Seymóuriæ. S.G.	(pu.)	Mrs. Seymour's.	——	——	——	G. ♄.	—— 1. t. 37.
258	Lousadiànum. S.G.	(bh.)	Miss Lousada's.	——	——	——	G. ♄.	—— 1. t. 44.
259	lépidum. S.G.	(li.)	pretty.	——	1822.	——	G. ♄.	—— 2. t. 156.
260	inscríptum. S.G.	(std.)	scribbled.	——	——	——	G. ♄.	—— 2. t. 193.
261	labyrínthicum.	(std.)	labyrinth-mark'd.	——	——	——	G. ♄.	
262	planifòlium. S.G.	(li.d.)	flat-leaved.	——	1823.	——	G. ♄.	Sweet ger. 3. t. 219.
263	incanéscens. S.G.	(pu.)	whitish-leaved.	——	——	6. 12.	G. ♄.	—— 3. t. 203.
264	notàtum. S.G.	(pu.)	marked-petal'd.	——	1821.	5. 10.	G. ♄.	—— 3. t. 208.
265	undulæflòrum. S.G.	(bh.)	waved-flower'd	——	1824.	——	G. ♄.	—— 4. t. 329.
266	lineàtum. S.G.	(st.d.)	striped-flower'd.	——	1810.	——	G. ♄.	—— 1. t. 16.
267	dumòsum. S.G.	(pu.)	bushy.	——	1816.	——	G. ♄.	—— 1. t. 19.
268	Scarboròviæ. S.G.	(d.li.)	L. Scarborough's	——	1817.	——	G. ♄.	—— 2. t. 117.
269	pustulòsum. S.G.	(bh.)	blistered-leav'd.	——	1816.	5. 10.	G. ♄.	—— 1. t. 11.
270	bellùlum. S.G.	(pu.)	neat.	——	——	——	G. ♄.	—— 1. t. 60.
	moschàtum. A.G. C. ic.							
271	Fairlièæ. S.G.	(ro.)	Mrs. Fairlie's.	——	1821.	——	G. ♄.	Sweet ger. 2. t. 178.
272	scintíllans. S.G.	(sa.)	sparkling.	——	1812.	——	G. ♄.	—— 1. t. 28.
273	platypètalon. S.G.	(w.)	broad-petaled.	——	——	——	G. ♄.	—— 2. t. 116.
	α rígidum. S.G.	(wh.)	*rigid.*	——	——	——	G. ♄.	—— 2. t. 116. f. a.
	β hirsùtum. S.G.	(w.)	*hairy.*	——	——	——	G. ♄.	—— 2. t. 116. f. b.
274	optábile. S.G.	(wh.)	desirable.	——	——	——	G. ♄.	—— 1. t. 62.
275	ornàtum. S.G.	(wh.)	ornate.	——	1817.	——	G. ♄.	—— 1. t. 39.
276	modéstum. S.G.	(bh.)	modest.	——	1822.	——	G. ♄.	—— 3. t. 204.
277	rhodoléntum. S.G.	(bh.)	rose-smelling.	——	1824.	——	G. ♄.	—— 3. t. 291.
278	picturàtum. S.G.	(d.bh.)	pictured.	——	1827.	——	G. ♄.	—— ser. 2. t. 34.
279	papyràceum. S.G.	(w.)	paper-white.	——		——	G. ♄.	—— ser. 2. t. 22.
280	Newshamiànum. S.G.	(bh.)	Miss Newsham's.	——	1821.	——	G. ♄.	—— 2. t. 144.
281	Cosmiànum. S.G.	(w.)	perfumed.	——		——	G. ♄.	—— 2. t. 189.
282	Brightiànum. S.G.	(w.)	Miss Bright's.	——	1823.	——	G. ♄.	—— 3. t. 227.
283	depéndens. S.G.	(wh.)	pendent-petal'd.	——	——	——	G. ♄.	—— 2. t. 195.
284	Mattocksiànum. S.G.	(wh.)	Mrs. Sweet's.	——	——	——	G. ♄.	—— 3. t. 234.
285	venústum. S.G.	(bh.)	comely-flower'd.	——	1822.	——	G. ♄.	—— 2. t. 167.
286	tínctum. S.G.	(w.pu.)	stained	——	1826.	——	G. ♄.	—— ser. 2. t. 29.

287	obscùrum. S.G.	(*wh.d.*)	darkened-pet'd.	——	1818.	——	G.	♄.	Sweet ger. 1. t. 89.
288	atrofúscum. S.G.	(*d.pu.*)	dark-brown.	——	——	——	G.	♄.	——— 1. t. 82.
289	acutilòbum. S.G.	(*wh.*)	acute-lobed.	——	1822.	——	G.	♄.	——— 2. t. 184.
290	compár. C.C.	(*wh.p.*)	partner.	——	1818.	——	G.	♄.	
291	Hoareànum. S.G.	(*bh.p.*)	fair Rosamond.	——	——	——	G.	♄.	Sweet ger. 1. t. 80.
292	Lambérti. S.G.	(*pu.*)	) Lambert's.	——	——	——	G.	♄.	——— 2. t. 104.
293	recurvàtum. S.G.	(*li.*)	recurved-petd.	——	——	——	G.	♄.	——— 3. t. 223.
294	divérgens. S.G.	(*li.*)	spreading-petal'd	——	1820.	——	G.	♄.	——— ser. 2. t. 39.
295	penicillàtum. W.en.	(*wh.*)	pencilled.	——	1794.	——	G.	♄.	Willd. h. ber. 1. t. 37.
296	dentàtum. A.G.	(*re.*)	toothed.	——	....	——	G.	♄.	Andr. geran. c. ic.
297	floréscens. H.G.	(*wh.*)	triangular-ld.	——	1818.	——	G.	♄.	
298	Sabìni. H.S.	(*pu.*)	Sabine's.	——	1821.	——	G.	♄.	
299	betulìnum. DC.	(*pu.*)	Birch-leaved.	C. B. S.	1759.	——	G.	♄.	Bot. mag. 148.
300	alnifòlium. W.en.	(*fl.*)	Alder-leaved.	Hybrid.	....	——	G.	♄.	
301	betulæfòlium. Sch.	(*bh.*)	Birch-like.	——	....	——	G.	♄.	
302	Willdenówii. S.	(*bh.*)	Willdenow's.	——	....	——	G.	♄.	
303	hermannifòlium. DC.	(*bh.*)	Hermannia-ld.	C. B. S.	....	4. 6.	G.	♄.	Jacq. ic. rar. 3. t. 545.
304	rígidum. DC.	(*wh.*)	rigid.	——	1790.	5. 8.	G.	♄.	
305	ribifòlium. DC.	(*wh.*)	Currant-leaved	——	1798.	——	G.	♄.	Jacq. ic. rar. 3. t. 538.
306	odoríferum.	(*wh.*)	odoriferous.	——	....	——	G.	♄.	Andr. geran. c. ic.
	Geranium scábrum. A.G. *nec aliorum.*								
307	jocúndum. H.G.	(*li.*)	pleasant.	Hybrid.	1818.	——	G.	♄.	
308	Broadlèiæ. C.C.	(*pu.*)	Mrs. Broadley's.	——	——	——	G.	♄.	Andr. geran. c. ic.
309	propínquum. C.C.	(*li.*)	neighbouring.	——	1818.	——	G.	♄.	
310	decòrum. S.G.	(*li.*)	neat-flowered.	——	——	——	G.	♄.	Sweet ger. ser. 2. t. 15.
311	delicàtum. H.G.	(*li.*)	delicate.	——	——	——	G.	♄.	——— 3. t. 267.
312	semitrilòbum. DC.	(*fl.*)	three-lobed.	C. B. S.	1800.	4. 7.	G.	♄.	Jacq. schœn. 2. t. 139.
313	Irbyànum. C.C.	(*ro.*)	Irby's.	Hybrid.	....	——	G.	♄.	Andr. geran. c. ic.
314	zingiberìnum. S.G.	(*bh.*)	Ginger-scented.	——	1826.	——	G.	♄.	Sweet ger. ser. 2. t. 42.
315	míxtum. S.G.	(*bh.*)	mixed.	——	1819.	5. 8.	G.	♄.	——— 1. t. 71.
316	my'rtillum. H.S.	(*pu.*)	Myrtle-scented.	——	1820.	——	G.	♄.	
317	rùbidum.	(*pu.*)	red-purple.	C. B. S.	1805.	——	G.	♄.	Andr. geran. c. ic.
	Geranium rùbens. A. G. *non* Wendl.								
318	spùrium. W.en.	(*pu.*)	spurious.	Hybrid.	....	——	G.	♄.	
319	nóthum. W.en.	(*pu.*)	mungrel.	——	....	——	G.	♄.	
320	críspum. DC.	(*li.*)	curl-leaved.	C. B. S.	1774.	6. 11.	G.	♄.	Sweet ger 4. t. 383.
321	gràtum. W.en.	(*bh.*)	grateful-scent'd.	Hybrid.	....	4. 8.	G.	♄.	
322	consanguíneum. W. en.	(*pu.*)	kindred.	——	....	——	G.	♄.	
323	pállidum. W.en.	(*pa.*)	pale-flowered.	——	....	——	G.	♄.	
324	citriodòrum.	(*wh.*)	Citron-scented.	C. B. S.	....	——	G.	♄.	Andr. geran. c. ic.
325	uniflòrum. DC.	(*fl.*)	one-flowered.	——	....	——	G.	♄.	
326	limònium. S.G.	(*pu.*)	Lemon-scented.	Hybrid.	1823.	——	G.	♄.	Sweet ger. 3. t. 278.
327	notábile. H.G.	(*li.*)	Hoare's Harlequin.	——	1822.	——	G.	♄.	
328	scábrum. DC.	(*li.*)	rough wedge-ld.	C. B. S.	1775.	——	G.	♄.	L'Herit. ger. t. 31.
329	tricuspidàtum. DC.	(*wh.*)	three-point'd.	——	1780.	——	G.	♄.	——— t. 30.
330	Perryànum. C.C.	(*wh.*)	Perry's.	Hybrid.	1816.	——	G.	♄.	
331	glaucoídes. W.C.	(*wh.*)	glaucum-like.	——	——	——	G.	♄.	Wendl. coll. ic.
332	gláucum. DC.	(*wh.*)	glaucous-leaved.	C. B. S.	1775.	6. 8.	G.	♄.	Sweet ger. 1. t. 57.
333	lanceolàtum.	(*wh.*)	lance-leaved.	——	——	——	G.	♄.	Andr. geran. c. ic.
334	diversifòlium. W.C.	(*wh.*)	different-ld.	——	1794.	——	G.	♄.	Wendl. coll. ic.
335	cuspidàtum. DC.	(*wh.*)	sharp-pointed.	——	....	——	G.	♄.	
336	soròrium. DC.	(*wh.*)	sister.	Hybrid.	....	——	G.	♄.	
337	Tankervilliæ. H.G.	(*wh.*)	Lady Tankerville's	——	....	——	G.	♄.	
338	acutifòlium.	(*wh.*)	acute-leaved.	——	1821.	——	G.	♄.	
339	trifurcàtum. C.C.	(*wh.*)	three-forked.	——	....	——	G.	♄.	
340	lævigàtum. DC.	(*wh.*)	glaucous trifid-l.	C. B. S.	....	——	G.	♄.	Cav. dis. 4. t. 121. f. 1.
341	trifoliatum. S.G.	(*wh.*)	trifoliate-leaved.	——	1799.	——	G.	♄.	Sweet ger. 3. t. 294.
	oxyphy'llum. DC. *Geranium trifoliàtum.* A.G. ic.								
342	pátulum. DC.	(*ro.*)	spreading.	——	1812.	4. 7.	G.	♄.	Jacq. ic. rar. 3. t. 541.
343	hepaticifòlium. DC.	(*li.*)	Hepatica-ld.	——	1791.	5. 8.	G.	♄.	
344	mitellæfòlium.	(*wh.*)	Mitella-leaved.	——	1791.	5. 10.	G.	♄.	Andr. geran. c. ic.
	Geranium hepaticifòlium. A.G. *nec aliorum.*								
345	fuscàtum. DC.	(*ro.*)	dark-marked.	——	1812.	——	G.	♄.	Jacq. ic. rar. 3. t. 540.
346	saniculæfòlium. DC.	(*pu.*)	Sanicle-leav'd	——	1806.	6. 9.	G.	♄.	——— 3. t. 539.
	cortusæfòlium. Jacq. *non* L. Her.								
347	crenæflòrum C.C.	(*w.*)	notch'd-petaled.	Hybrid.	1817.	——	G.	♄.	
348	sceleràtum. C.C.	(*wh.*)	Celery-leaved.	——	1819.	——	G.	♄.	
349	viridifòlium. S.G.	(*sa.*)	bright-green-ld.	——	1825.	5. 9.	G.	♄.	Sweet ger. 4. t. 331.
350	expánsum.	(*re.*)	expanded.	——	1823.	5. 10.	G.	♄.	

351	multidentàtum. (*re.*) many-toothed.	Hybrid.	1823.	5. 10.	G. ♄.	
352	grandidentàtum. S.G. large-toothed.	———	——	——	G. ♄.	Sweet ger. 3. t. 217.
353	Belladónna. S.G.(*bh.*) painted-lady.	———	——	——	G. ♄.	——— 3. t. 270.
354	triúmphans. C.C. (*bh.*) triumphant.	———	1818.	——	G. ♄.	
355	serratifòlium. S.G.(*bh.*) saw-leaved.	———	1822.	——	G. ♄.	Sweet ger. 3. t. 221.
356	variegàtum. DC.(*wh.*) variegated-fld.	C. B. S.	1812.	——	G. ♄.	Cav. dis. 4. t. 118. f. 3.
357	amplíssimum. W.C.(*wh.*) stately.	Hybrid.	1796.	——	G. ♄.	Wendl. coll. ic.
	Gerànium grandiflòrum. var. A.G. C. ic. *Formanni.* Hort.					
358	grandiflòrum. DC.(*wh.*) great-flower'd.	C. B. S.	1794.	——	G. ♄.	Sweet ger. 1. t. 29.
359	Parmentiéri. C.C.(*wh.*) Parmentier's.	Hybrid.	1822.	——	G. ♄.	
360	delphinifòlium. DC.(*vi.*) Larkspur-ld.	———	....	——	G. ♄.	
361	gossypiifòlium. (*li.*) Cotton-tree-ld.	———	1805.	5. 10.	G. ♄.	Andr. ger. c. ic.
	Gerànium grandiflòrum. var. *carnea.* A.G.					
362	eléctum. S.G. (*wh.*) elect.	Hybrid.	1804.	——	G. ♄.	Sweet ger. 3. t. 238.
363	asperifòlium. S.G.(*re.*) rough-leaved.	———	1807.	——	G. ♄.	——— 2. t. 169.
364	Curtisiànum. DC.(*wh.*) Curtis's.	———	....	——	G. ♄.	
365	nòbile. DC. (*li.*) noble.	———	1822.	——	G. ♄.	
366	opulifòlium. S.G. (*sa.*) Guelder-Rose-ld.	———	1819.	——	G. ♄.	Sweet ger. 1. t. 53.
367	malvàceum. (*bh.*) Mallow-like.	———	....	——	G. ♄.	
368	hæmastíctum. S.G.(*bh.*) blood-spotted.	———	1827.	——	G. ♄.	Sweet ger. ser. 2. t. 18.
369	psilophy'llum. S.G.(*re.*) smooth-leaved.	———	1825.	——	G. ♄.	——— 4. t. 356.
370	lùcidum. S.G. (*bh.*) glossy-leaved.	———	1826.	——	G. ♄.	——— 4. t. 373.
371	Princeànum. S.G.(*wh.*) Prince's.	———	1827.	——	G. ♄.	——— 4. t. 386.
372	clàrum. S.G. (*wh.*) clear white.	———	1826.	——	G. ♄.	——— 4. t. 366.
373	recurvifòlium. S.G.(*wh.*) recurve-leav'd.	———	1825.	——	G. ♄.	——— 4. t. 343.
374	bryoniæfòlium. S.G.(*bh.*) Bryony-leav'd.	———	1824.	——	G. ♄.	——— 4. t. 320.
375	Richiànum. S.G.(*bh.*) Miss Rich's.	———	1826.	——	G. ♄.	——— 4. t. 370.
376	platanifòlium. S.G.(*li.*) Plane-leaved.	———	1824.	——	G. ♄.	——— 4. t. 326.
377	verecúndum. S.G.(*pi.*) blushing.	———	——	——	G. ♄.	——— 4. t. 316.
378	macránthon. S.G.(*wh.*) large-flowered.	———	1821.	——	G. ♄.	Sweet ger. 1. t. 83.
379	Coúttsiæ. S.G. (*sa.*) Mrs. Coutts's.	———	1822.	——	G. ♄.	——— 3. t. 269.
380	rùtilum. C.C. (*sa.*) light red.	———	1821.	——	G. ♄.	
381	schizophy'llum. S.G.(*sa.*) jagged.	———	1822.	——	G. ♄.	Sweet ger. 3. t. 289.
382	Barnardiànum. S.G.(*sa.*) Mr. Barnard's.	———	1820.	——	G. ♄.	——— 2. t. 127.
383	mucronàtum. S.G.(*d.re.*) mucronate.	———	1823.	——	G. ♄.	——— 3. t. 275.
384	Gurneyànum. S.G.(*d.re.*) Mr. Gurney's.	———	1826.	——	G. ♄.	——— 4. t. 393.
385	argùtum. S.G. (*sc.*) sharp-toothed.	———	1824.	——	G. ♄.	——— 4. t. 344.
386	latifòlium. S.G. (*re.*) broad-leaved.	———	——	——	G. ♄.	——— 4. t. 335.
387	megalánthum. S.G.(*re.*) grand-flowered.	———	1826.	——	G. ♄.	——— ser. 2. t. 4.
388	megàleion. S.G. (*sc.*) magnificent.	———	——	——	G. ♄.	——— ser. 2. t. 5.
389	cratægifòlium. S.G.(*re.*) Hawthorn-ld.	———	1824.	——	G. ♄.	——— ser. 2. t. 19.
390	coilophy'llum. (*re.*) hollow-leaved.	———	1827.	——	G. ♄.	——— ser. 2. t. 33.
391	Clintòniæ. S.G. (*re.*) Ds. of Newcastle's.	———	——	——	G. ♄.	——— 4. t. 392.
392	campylosépalon. S.G.(*pu.*) reflex-calyxed.	———	1826.	——	G. ♄.	——— 4. t. 379.
393	Nàirnii. S.G. (*d.re.*) Nairn's.	———	1825.	——	G. ♄.	——— 4. t. 376.
394	rhodánthum. S.G.(*ro.*) rose-colour'd.	———	1824.	——	G. ♄.	——— 3. t. 282.
395	rhodopétalon. S.G.(*ro.*) rosy-petal'd.	———	1826.	——	G. ♄.	——— ser. 2. t. 14.
396	acutidentàtum. S.G.(*sa.*) acute-toothed.	———	1827.	——	G. ♄.	——— ser. 2. t. 40.
397	Mílleri. (*re.*) Mr. Miller's.	———	1824.	——	G. ♄.	
398	conchiifòlium. (*li.*) shell-leaved.	———	1820.	——	G. ♄.	
399	striàtum. S.G. (*bh.*) streak-flowered.	———	1818.	3. 8.	G. ♄.	Sweet ger. 1. t. 1.
400	rigéscens. S.G. (*li.*) stiff-leaved.	———	——	4. 9.	G. ♄.	——— 2. t. 112.
401	eriophy'llum. S.G.(*pu.*) woolly round-ld.	———	——	5. 10.	G. ♄.	——— 2. t. 141.
402	intermíxtum. (*li.*) intermixed.	———	——	4. 8.	G. ♄.	
403	abutiloídes. S.G.(*pu.*) Abutilon-like.	———	1820.	4. 9.	G. ♄.	Sweet ger. ser. 2. t. 3.
404	cordàtum. DC. (*pu.*) heart-leaved.	C. B. S.	1774.	3. 8.	G. ♄.	——— 1. t. 67.
405	conduplicàtum. DC.(*pu.*) curl'd-heart-ld.	Hybrid.	....	——	G. ♄.	
406	rubro-cínctum. DC.(*pu.*) red-edged.	———	....	——	G. ♄.	
407	Góweri. S.G. (*pi.*) Mr. Levison Gower's.	———	1825.	5. 8.	G. ♄.	Sweet ger. 4. t. 333.
408	papilionàceum. DC.(*pu.*) Butterfly.	C. B. S.	1724.	4. 7.	G. ♄.	——— 1. t. 27.
409	tomentòsum. DC.(*wh.*) Pennyroyal.	———	1790.	——	G. ♄.	——— 2. t. 168.
410	vitifòlium. DC. (*li.*) Vine-leaved.	———	1724.	4. 8.	G. ♄.	Cav. dis. 4. t. 111. f. 2.
411	capitàtum. DC. (*li.*) Rose-scented.	———	1690.	——	G. ♄.	——— 4. t. 105. f. 1.
412	cóncolor. S.G. (*re.*) self-coloured.	Hybrid.	1820.	——	G. ♄.	Sweet ger. 2. t. 140.
	β *màjor.* S.G. (*cr.*) *larger.*	———	——	——	G. ♄.	——— 2. t. 140. f. b.
413	congéstum. S.G. (*pu.*) close-headed.	———	1825.	——	G. ♄.	——— 4. t. 325.
414	calamistràtum. S.G.(*pu.*) curled-lobed.	———	1828.	——	G. ♄.	——— ser. 2. t. 36.
415	gravèolens. DC. (*li.*) Odour of Rose.	C. B. S.	1774.	3. 7.	G. ♄.	L'Herit. ger. t. 17.
	β *variegàtum.* (*li.*) *striped-leaved.*		....	——	G. ♄.	Andr. geran. c. ic.
	Geranium capitàtum. var. A.G.					

416	asperum. DC. (*fl.*)	rough multifid.	C. B. S.	1795.	5. 9.	G. ♄.	Roth. abh. p. 51. t. 10.
417	Rádula. DC. (*li.*)	rasp-leaved.	———	1774.	3. 7.	G. ♄.	L'Herit. ger. t. 16.
418	balsàmeum. DC. (*fl.*)	balsamic.	———	1790.	5. 9.	G. ♄.	Jacq. ic. rar. 3. t. 53.
419	Vandèsiæ. S.G. (*bh.*)	Comtesse de Vandes'.	Hybrid.	1818.	——	G. ♄.	Sweet ger. 1. t. 7.
420	híspidum. DC. (*wh.*)	hispid.	C. B. S.	1790.	——	G. ♄.	Cav. dis. 4. t. 110. f. 1.
421	callíston. S.G. (*sa.*)	graceful.	———	1827.	——	G. ♄.	Sweet ger. ser. 2. t. 9.
422	anthemifòlium. (*re.*)	Anthemis-ld.	Hybrid.	1822.	——	G. ♄.	
423	pyrethriifòlium. S.G. (*sc.*)	Feverfew-ld.	———	1821.	——	G. ♄.	Sweet ger. 2. t. 153.
424	tanacetifòlium. S.G. (*re.*)	Tansy-leaved.	———	1824.	——	G. ♄.	——— 4. t. 336.
425	sphærocéphalon. S.G. (*sc.*)	round-headed.	———	——	——	G. ♄.	——— 4. t. 317.
426	Moreànum. S.G. (*sc.*)	More's victory.	———	1823.	——	G. ♄.	——— 3. t. 285.
427	obcordàtum. S.G. (*sc.*)	obcordate-petal'd.	———	——	——	G. ♄.	——— 4. t. 310.
428	cuneiflòrum. S.G. (*sc.*)	wedge-petal'd.	———	1825.	——	G. ♄.	——— 4. t. 330.
429	chrysanthemifòlium. (*sc.*)	Chrysanthm.-ld.	———	1821.	4. 10.	G. ♄.	——— 2. t. 124.
430	Dobreeànum. S.G. (*sc.*)	Mrs. Dobree's.	———	1818.	——	G. ♄.	——— 3. t. 253.
431	Bisshóppæ. S.G. (*sc.*)	Mrs. Bisshopp's.	———	1822.	——	G. ♄.	——— 3. t. 272.
432	signàtum. S.G. (*re.*)	marked-leaved.	———	——	——	G. ♄.	——— 3. t. 265.
433	Spìnii. C.R. (*bh.d.*)	Spin's.	———	....	——	G. ♄.	——— 4. t. 362.
434	quercifòlium. DC. (*bh.*)	Oak-leaved.	C. B. S.	1774.	3. 8.	G. ♄.	L'Herit. ger. t. 14.
	bipinnatífidum. (*bh.*)	*bipinnatifid.*	———	——	——	G. ♄.	——— t. 15.
435	tortuòsum. S.G. (*li.*)	twisted-petal'd.	———	1826.	——	G. ♄.	Sweet ger. ser. 2. t. 43.
436	exímium. S.G. (*pi.*)	select.	Hybrid.	1818.	4. 8.	G. ♄.	——— 1. t. 26.
437	Esséxiæ. C.C. (*pi.*)	Lady Essex's.	———	——	——	G. ♄.	
438	oblàtum. S.G. (*pi.*)	oblate-leaved.	———	——	——	G. ♄.	Sweet ger. 1. t. 35.
439	Murryànum. S.G. (*bh.*)	Ld. Jas. Murray's.	———	——	——	G. ♄.	——— 2. t. 164.
440	calocéphalon. S.G. (*bh.*)	pretty-headed.	———	1820.	——	G. ♄.	——— 3. t. 201.
441	obovàtum. S.G. (*li.*)	obovate-petal'd.	———	1825.	——	G. ♄.	——— 4. t. 367.
442	dilatàtum. S.G. (*li.*)	dilated-leav'd.	———	1824.	——	G. ♄.	——— 4. t. 314.
443	augústum. DC. (*bh.*)	august.	———	1809.	——	G. ♄.	Andr. geran. c. ic.
	β *Brùcii.* C.C. (*bh.*)	*Bruce's.*	———	....	——	G. ♄.	
444	Breesiànum. S.G. (*pi.*)	Breese's.	———	1818.	——	G. ♄.	Sweet ger. 1. t. 64.
445	Husseyànum. S.G. (*pu.*)	Lady Hussey's.	———	——	——	G. ♄.	——— 1. t. 92.
446	Wellsiànum. S.G. (*sc.*)	Mr. Wells's.	———	1822.	3. 8.	G. ♄.	——— 1. t. 175.
447	Kíngii. S.G. (*sc.*)	Mr. King's.	———	——	3. 12.	G. ♄.	——— 3. t. 248.
448	Scóttii. S.G. (*sc.*)	Sir C. Scott's.	———	——	3. 9.	G. ♄.	——— 3. t. 264.
449	Pálkii. S.G. (*cr.*)	Mr. Palk's.	———	——	1. 12.	G. ♄.	——— 3. t. 224.
450	rotundilòbum. S.G. (*sc.*)	round-lobed.	———	1823.	3. 9.	G. ♄.	——— 3. t. 252.
451	latilòbum. S.G. (*sc.*)	broad-lobed.	———	——	——	G. ♄.	——— 3. t. 236.
452	Russelliànum. S.G. (*sc.*)	Russell's.	———	1826.	4. 9.	G. ♄.	——— 4. t. 385.
453	translùcens. S.G. (*sc.*)	transparent.	———	1824.	——	G. ♄.	——— 4. t. 400.
454	Barclayànum. S.G. (*sc.*)	Mr. Barclay's.	———	——	——	G. ♄.	——— 4. t. 384.
455	Harewoòdiæ. S.G. (*d.re.*)	Lady Harewood's.	———	1826.	——	G. ♄.	——— 4. t. 389.
456	Stewártii. S.G. (*da.sc.*)	Mr. Stewart's.	———	1825.	——	G. ♄.	——— 4. t. 353.
457	lasiophy'llum. S.G. (*sa.*)	woolly-divided-ld.	———	——	——	G. ♄.	——— 3. t. 296.
458	Lechiànum. DC. (*bh.*)	acute-notched.	———	1820.	——	G. ♄.	
459	jatrophæfòlium. DC. (*fl.*)	Jatropha-ld.	———	——	——	G. ♄.	
460	denticulàtum. DC. (*li.*)	tooth-leaved.	C. B. S.	1789.	5. 8.	G. ♄.	Sweet ger. 2. t. 109.
461	viscosíssimum. S.G. (*bh.*)	viscous.	———	1820.	3. 8.	G. ♄.	——— 2. t. 118.
	α *álbum.* (*wh.*)	*white-flowered.*	———	——	——	G. ♄.	
	β *cárneum.* (*fl.*)	*flesh-coloured.*	———	——	——	G. ♄.	
462	carbasìnum. S.G. (*sa.*)	linen-flower'd.	Hybrid.	1827.	——	G. ♄.	Sweet ger. ser. 2. t. 38.
463	glutinòsum. DC. (*fl.*)	clammy.	C. B. S.	1777.	5. 6.	G. ♄.	L'Herit. ger. t. 20.
	β *nigréscens.* (*fl.*)	*black-marked.*	———	——	——	G. ♄.	Bot. mag. 143.
464	cynosbatifòlium. DC. (*li.*)	Cynosbati-ld.	Hybrid.	....	4. 8.	G. ♄.	Sweet ger. 3. t. 259.
	Geranium oxoniénse. A.G. ic.						
465	verbenæfòlium. S.G. (*li.*)	Vervain-ld.	Hybrid.	1818.	——	G. ♄.	Sweet ger. 2. t. 149.
	piperàtum. DC?						
466	melíssinum. S.G. (*da.*)	Balm-scented.	Hybrid.	——	——	G. ♄.	Sweet ger. 1. t. 5.
467	volatiflòrum. S.G. (*sc.*)	flying-flower'd.	———	1823.	——	G. ♄.	——— 3. t. 284.
468	brèvipes. (*sc.*)	short-peduncled.	———	1824.	——	G. ♄.	
469	ternàtum. DC. (*li.*)	ternate-leaved.	C. B. S.	1789.	5. 8.	G. ♄.	Sweet ger. 2. t. 165.
470	patentíssimum. W.C. (*li.*)	most-spreading.	———	1822.	——	G. ♄.	Wendl. coll. ic.
471	spinòsum. DC. (*st.*)	thorny.	———	1795.	5. 6.	G. ♄.	Paters. it. t. ad. p. 67.
472	tripartìtum. S.G. (*st.*)	brittle-stalked.	———	1789.	5. 8.	G. ♄.	Sweet ger. 2. t. 115.
	Geranium frágile. A.R. 37.						
473	incìsum. DC. (*st.*)	cut-leaved.	———	1791.	——	G. ♄.	Sweet ger. 1. t. 93.
474	canéscens. S.G. (*bh.*)	canescent.	———	....	——	G. ♄.	
475	abrotanifòlium. DC. (*bh.*)	Southernwood-ld.	———	1791.	——	G. ♄.	Jacq. schœn. 2. t. 136.
476	artemisiæfòlium. DC.	Wormwood-ld.	———	1817.	——	G. ♄.	

477	fràgrans. DC.	(*wh.*)	Nutmeg-scent'd.	Hybrid.	....	5. 10.	G. ħ.	Sweet ger. 2. t. 172.
478	odoratíssimum. DC.	(*wh.*)	sweet-scent'd.	C. B. S.	1724.	——	G. ħ.	—— 3. t. 299.
479	exstipulàtum. DC.	(*bh.*)	soft trifid-l^d^.	——	1779.	——	G. ♄.	L'Herit. ger. t. 35.
	Geranium suavéolens. A.G. ic.							
480	disséctum. S.G.	(*wh.*)	dissected-l^d^.	Hybrid.	1822.	5. 11.	G. ħ.	Sweet ger. 3. t. 247.
481	glomeràtum. DC.	(*w.*)	close-headed.	N. S. W.	1792.	5. 8.	G. ħ.	Jacq. fl. ecl. 1. t. 98.
	australe. S.G. 68. *non* DC.							
482	austràle. DC.	(*wh.*)	New Holland.	——	——	——	G. ♃.	Jacq. fl. ecl. 1. t. 100.
483	inodòrum. DC.	(*w.pu.*)	scentless.	——	1796.	3. 8.	G. ♃.	Sweet ger. 1. t. 56.
484	columbìnum. DC.	(*pu.*)	Doves-foot.	C. B. S.	1795.	6. 10.	G. ♃.	Jacq. schœn. 2. t. 133.
485	procúmbens. DC.	(*bh.*)	procumbent.	——	1801.	4. 5.	G. ♃.	Andr. bot. rep. 234.
486	humifùsum. DC.	(*bh.*)	trailing.	Canaries.	——	4. 10.	H. ⊙.	Sweet ger 1. t. 42.
487	chamædryfòlium. DC.	(*wh.*)	Chamædrys-l^d^.	C. B. S.	1812.	5. 6.	G. ♃.	Jacq. ic. rar. 3. t. 523.
488	althæoídes. DC.	(*wh.*)	Althæa-leaved.	——	1724.	4. 6.	G. ♂.	L'Herit. ger. t. 10.
489	grossularioídes. DC.	(*fl.*)	Gooseberry-l^d^.	——	1731.	4. 8.	G. ♃.	Cav. dis. 4. t. 119. f. 2.
490	ancéps. DC.	(*pu.*)	flat-stalked.	——	1788.	5. 7.	G. ♃.	Jacq. col. 4. t. 22. f. 3.
491	parviflòrum. A.G.	(*pu.*)	small-flowered.	——	1800.	——	G. ♃.	Andr. geran. c. ic.
492	acugnáticum. L.T.	(*re.*)	Tristan d'Acuna.	T. d'Acunna.	1818.	——	G. ♃.	
493	senecioídes. DC.	(*wh.*)	Groundsel-l^d^.	C. B. S.	1775.	6. 7.	G. ⊙.	Sweet ger. 4. t. 327.
494	alchimilloídes. DC.	(*wh.*)	mantle-leaved.	——	1693.	5. 10.	G. ♃.	Cav. dis. 4. t. 98. f. 1.
495	Heritièri. Jacq.	(*wh.*)	L'Heritier's.	——	1820.	——	G. ♃.	
496	tabulàre. DC.	(*wh.*)	rough-stalked.	——	1775.	5. 8.	G. ħ.	L'Herit. ger. t. 9.
497	ovàle. DC.	(*ro.*)	oval-leaved.	——	1774.	——	G. ħ.	Sweet ger. 3. t. 235.
498	élegans. DC.	(*wh. pu.*)	elegant.	——	1795.	——	G. ħ.	And. repos. 98.
	α *màjor.* S.G.	(*wh.*)	*smaller-flowered.*	——	——	——	G. ħ.	Sweet ger. 1. t. 36. f. a.
	β *mìnor.* S.G.	(*v.*)	*larger-flowered.*	——	——	——	G. ħ.	—— 1. t. 36. f. b.
499	schizopétalum. S.G.	(*y.br.*)	divided-petal'd.	——	1821.	6. 10.	G. ħ.	—— 3. t. 232.
500	pinguifòlium. S.G.	(*bh.*)	greasy leaved.	Hybrid?	....	5. 10.	G. ♄.	—— 1. t. 52.
501	latéripes. DC.	(*va.*)	Ivy-leaved.	C. B. S.	1787.	——	G. ♄.⌣.	L'Herit. ger. t. 24.
	α *viridifòlium.*	(*li.*)	*green-leaved.*	——	——	——	G. ♄.⌣.	Andr. geran. c. ic.
	β *zonàtum.*	(*li.*)	*Horseshoe-mark'd.*	——	——	——	G. ♄.⌣.	
	γ *ròseum.*	(*ro.*)	*rose-coloured.*	——	——	——	G. ♄.⌣.	
	δ *albomarginàtum.* DC.		*white-margined.*	——	——	——	G. ♄.⌣.	Andr. geran. c. ic.
502	peltàtum. DC.	(*li.*)	peltate-leaved.	——	1701.	——	G. ♄.⌣.	Bot. mag. 20.
	β *variegàtum.* A.G.	(*li.*)	*variegated-l^d^.*	——	——	——	G. ♄.⌣.	Andr. geran. c. ic.
503	scutàtum. S.G.	(*wh.*)	shield-leaved.	——	1819.	——	G. ♄.⌣.	Sweet ger. 1. t. 95.
	JENKINS'ONIA. S.G.		JENKINS'ONIA.	Monadelphia Heptandria.				
1	tetragòna. S.G.	(*li.*)	square-stalked.	C. B. S.	1774.	5. 8.	G. ♄.⌣	Sweet ger. 1. t. 99.
	β *variegàta.*	(*li.*)	*variegated.*	——	——	——	G. ♄.⌣.	—— 1. t. 99.
2	quinàta. S.G.	(*st.*)	quinate-leaved.	——	1793.	——	G. ♄.	—— 1. t. 79.
	Pelargònium quinàtum. B.M. 547. *Geranium præmórsum.* A.R. 150.							
3	péndula. S.G.	(*pu.*)	pendulous-branch'd.	——	....	5. 8.	G. ♄.⌣.	Sweet ger. 2. t 188.
4	longicáulis. S.G.	(*bh.*)	long-stalked.	——	....	——	G. ♄.⌣.	Jacq. ic. rar. 3. t. 533.
	Pelargònium longicáule. Jacq.							
5	anemonefòlia. S.G.	(*re.*)	Anemone-l^d^.	——	....	——	G. ♄.⌣.	Jacq. ic. rar. 3. t. 535.
6	caucalifòlia. S.G.	(*bh.*)	Caucalus-leav'd.	——	1812.	3. 9.	G. ♃.	—— 3. t. 529.
7	multicáulis. S.G.	(*vi.*)	many-stalked.	——	1802.	6. 8.	G. ♃.	—— 3. t. 534.
8	lácera. S.G.	(*bh.*)	torn-leaved.	——	1731.	——	G. ♃.	—— 3. t. 532.
9	Synnòti. S.G.	(*pu.*)	Synnot's.	——	1825.	——	G. ♃.	Sweet ger. 4. t. 342.
10	coriandrifòlia. S.G.	(*bh.*)	Coriander-l .	——	1724.	3. 9.	G. ♃.	—— 1. t. 34.
	Pelargònium coriandrifòlium. S.G.							
11	myrrhifòlia. S.G.	(*li.*)	Myrrh-leaved.	——	1696.	5. 8.	G. ħ.	Jacq. ic. rar. 3. t. 531.
	Pelargonium betònicum. Jacq.							
12	bullàta. S.G.	(*wh.*)	blistered.	——	....	——	G. ħ.	Jacq. ic. rar. 3. t. 530.
13	canariénsis.	(*wh.*)	Canary.	Canaries.	1802.	5. 9.	G. ♂.	Willd. hort. ber. t. 17.
	Pelargònium canariénse. W.							
	CIC'ONIUM. S.G.		CIC'ONIUM.	Monadelphia Heptandria.				
1	acetòsum. S.G.	(*sa.*)	Sorrel-leaved.	C. B. S.	1710.	5. 9.	G. ♄.	Andr. geran. c. ic.
2	scándens. S.G.	(*ro.*)	climbing.	——	1800.	——	G. ♄.	
3	stenopétalum. S.G.	(*sc.*)	narrow-petal'd.	——	——	——	G. ♄.	
4	leptopétalum.	(*sc.*)	slender-petal'd.	——	——	——	G. ♄.⌣.	Andr. geran. c. ic.
	Geranium stenopétalum. A.G. *nec aliorum.*							
5	pùmilum. S.G.	(*sc.*)	dwarf.	C. B. S.	——	——	G. ħ.	Andr. geran. c. ic.
	Geranium zonàle minima. A.G.							
6	laterítium. S.G.	(*sc.*)	brick-coloured.	——	——	——	G. ♄.	Jacq. ecl. 1. t. 97. f. ult.
7	malvæfòlium.	(*li.*)	Mallow-leaved.	——	1812.	——	G. ħ.	—— 1. t. 97.
	Pelargònium malvæfòlium. Jacq.							
8	heterogàmum. S.G.	(*li.*)	six-stamened.	Hybrid.	1786.	——	G. ♄.	L'Herit. geran. t. 18.
9	álbidum. C.C.	(*wh.*)	white-flowered.	C. B. S.	....	——	G. ♄.	

10 zonàle. S.G.	(*va.*)	horse-shoe.	C. B. S.	1710.	4. 12.	G. ♄.	Cav. dis. 4. t. 98. f. 2.
α lilacìnum.	(*li.*)	*lilac-coloured.*	———	——	——	G. ♄.	
β coccíneum.	(*sc.*)	*scarlet.*	———	——	——	G. ♄.	
γ crystallìnum.	(*sc.*)	*coral-stalked.*	———	——	——	G. ♄.	
δ marginàtum.	(*sc.*)	*white-edged.*	———	——	——	G. ♄.	
11 reticulàtum. S.G.	(*sc.*)	netted-veined.	Hybrid.	1810.	5. 10.	G. ♄.	Sweet ger. 2. t. 143.
12 Fothergíllii. S.G.	(*sc.*)	Fothergill's.	C. B. S.	....	——	G. ♄.	——— 3. t. 226.
α coccíneum.	(*sc.*)	*scarlet.*	———	——	——	G. ♄.	——— 3. t. 226. a.
β purpùreum.	(*pu.*)	*purple.*	———	——	——	G. ♄.	——— 3. t. 226. b.
13 mónstrum. S.G.	(*pu.*)	cluster-leaved.	Hybrid	1784.	7. 8.	G. ♄.	——— 1. t. 13.
14 æqualiflòrum. C.C.	(*sc.*)	victory.	———	1819.	5. 10.	G. ♄.	
15 hy'bridum. S.G.	(*sc.*)	hybrid.	C. B. S.	1732.	——	G. ♄.	Sweet ger. 1. t. 63.
16 oxyphy'llum. C.C.	(*wh.*)	sour-leaved.	Hybrid.	....	——	G. ♄.	Andr. geran. c. ic.
Gerànium miniàtum. v. album. A.G.							
17 hùmile.	(*sc.*)	low-spreading.	C. B. S.	....	——	G. ♄.	
18 crenàtum. S.G.	(*sc.*)	scolloped-ld.	———	....	——	G. ♄.	Sweet ger. 4. t. 345.
19 bracteòsum.	(*sc.*)	large-bracted.	Hybrid.	....	——	G. ♄.	
Pelargònium bracteòsum. DC.							
20 Bentinckiànum.	(*sc.*)	Bentinck's.	C. B. S.	....	——	G. ♄.	Sweet ger. 4. t. 350.
Pelargònium Bentinckiànum. DC. *Gerànium crenatum ; var. mollifoliata.* A.G. c. ic.							
21 fúlgens. S.G.	(*sc.*)	Basilisk.	Hybrid.	1823.	——	G. ♄.	Sweet ger. 4. t. 339.
22 glabrifòlium. S.G.	(*sc.*)	smooth-leaved.	———	1825.	——	G. ♄.	——— 4. t. 363.
23 cerìnum. S.G.	(*pi.*)	wax-flowered.	Hybrid ?	....	——	G. ♄.	——— 2. t. 176.
24 inquínans. S.G.	(*sc.*)	staining-gland'd.	C. B. S.	1714.	5. 9.	G. ♄.	Andr. Geran. c. ic.
25 micránthum. S.G.	(*sc.*)	small-flowered.	———	....	——	G. ♄.	Sweet ger. 3. t. 295.
ISOPE'TALUM. S.G.		ISOPE'TALUM.	Monadelphia Pentandria.				
1 Cotyledònis. S.G.	(*wh.*)	Hollyhock-ld.	St. Helena.	1765.	5. 8.	G. ♄.	Sweet ger. 2. t. 126.
Pelargònium Cotyledònis. W.							
2 ? díscipes. C.C.	......	central-stalked.	———	1808.	....	G. ♄.	

ORDO LII.

TROPÆOLEÆ. *DC. prodr.* 1. *p.* 683.

TROP'ÆOLUM. DC.		INDIAN-CRESS.	Octandria Monogynia. L.				
1 mìnus. DC.	(*ye.*)	small.	Peru.	1596.	6. 10.	H. ⊙.	Bot. mag. 98.
β flòre plèno.	(*ye.*)	*double-flowered.*	———	——	——	G. ♃.	
2 màjus. DC.	(*ye.*)	great.	———	1686.	——	H. ⊙.	Bot. mag. 23.
β flòre plèno.	(*ye.*)	*double-flowered.*	———	——	——	G. ♃.	
3 hy'bridum. DC.	(*ye.*)	hybrid.	Hybrid.	....	——	G. ♃.	Berg. act. holm. 32. t. 1.
4 adúncum. DC.	(*ye.*)	fringe-flower'd.	Peru.	1775.	——	G. ⊙.	Bot. reg. t. 718.
peregrìnum. B.R. *non* L.							
5 pinnatum. DC.	(*ye.*)	pinnate-flower'd.	———	....	——	G. ♃.	Andr. reposit. t. 535.
6 tricolòrum. B.F.G.	(*or.*)	three-coloured.	Chile.	1828.	——	H. ♃.	Sweet br. fl. gar. 3. t. 270.

ORDO LIII.

BALSAMINEÆ. *DC. prodr.* 1. *p.* 685.

BALSAM'INA. DC.		BALSAM.	Pentandria Pentagynia.				
1 hortênsis. DC.	(*v.*)	garden.	E. Indies.	1596.	7. 10.	S. ⊙.	Blackw. t. 583.
Impàtiens Balsamìna. L.							
2 coccìnea. DC.	(*sc.*)	glandular-ld.	———	1808.	6. 9.	S. ⊙.	Bot. mag. 1256.
3 mysoréensis. DC.	(*re.*)	Mysore.	———	1823.	——	S. ⊙.	
4 bífida. DC.	(*re.*)	bifid-spurred.	C. B. S.	1818.	7. 10.	H. ⊙.	
5 capénsis. DC.	(*re.*)	Cape.	———	——	——	H. ⊙.	

6 cochleàta. DC.	(*sc.*)	shell-spurred.	China.	1826.	——	H. ⊙.	
7 chinénsis. DC.	(*pu.*)	Chinese.	China.	1823.	——	H. ⊙.	
IMP'ATIENS. DC.		TOUCH ME NOT.	Pentandria Monogynia. L.				
1 nàtans. DC.	(*ye.*)	floating.	E. Indies.	1823.	6. 10.	S.*w.*⊙.	
2 biflòra. W.	(*or.*)	two-flowered.	N. America.	....	——	H. ⊙.	Sweet br. fl. gar. t. 43.
fúlva. DC.							
3 pállida. DC.	(*ye.*)	pale-flowered.	————	1812.	——	H. ⊙.	
4 Nòli-tángere. DC.	(*ye.*)	common.	England.	....	——	H. ⊙.	Eng. bot. 937.
5 parviflòra. DC.	(*ye.*)	small-flowered.	Russia.	1828.	——	H. ⊙.	
6 tripétala. DC.	(*ye.*)	three-petaled.	E. Indies.	1823.	——	S. ⊙.	

ORDO LIV.

OXALIDEÆ. *DC. prodr.* 1. *p.* 689.

AVERRH'OA. DC.		AVERRH'OA.	Pent-Decandria Pentagynia. L.				
1 Carambòla. DC.	(*st.*)	Carambola-tree.	E. Indies.	1793.	....	S. ♄.	Rumph. amb. 1. t. 35.
2 Bilímbi. DC.	(*st.*)	Bilimbi-tree.	————	1791.	5. 7.	S. ♄.	———— 1. t. 36.
BIO'PHYTUM. DC.		BIO'PHYTUM.	Decandria Pentagynia.				
sensitìvum. DC.	(*ye.*)	sensitive.	E. Indies.	1824.	5. 7.	S. ♃.	Jacq. ox. n. 21. t. 78. f. 4.
O'xalis sensitìva. L.							
O'XALIS. DC.		WOOD-SORREL.	Decandria Pentagynia. L.				
1 Plumiéri. DC.	(*ye.*)	Plumier's.	S. America.	1822.	4. 6.	S. ♄.	Bot. reg. 810.
2 Barreliéri. DC.	(*bh.*)	Barrelier's.	————	——	——	S. ♄.	Jacq. ox. n. 4. t. 3.
3 fruticòsa. DC.	(*ye.*)	frutescent.	Rio Janeiro.	——	——	S. ♄.	
4 perénnans. DC.	(*br.y.*)	perennial.	N. S. W.	....	5. 9.	G. ♃.	
5 Dillènii. DC.	(*ye.*)	Dillenius's.	America.	1798.	——	H. ⊙.	Dill. elth. 2. t. 221.
6 flòrida. S.P.	(*ye.*)	prostrate.	————	....	——	H. ⊙.	
7 strícta. DC.	(*ye.*)	upright.	N. America.	1658.	6. 10.	H. ♃.	Jacq. ox. n. 9. t. 4.
8 corniculàta. DC.	(*ye.*)	procumbent.	Britain.	....	5. 10.	H. ♃.	Eng. bot. 1726.
9 Lyòni. DC.	(*ye.*)	Lyons'.	N. America.	1812.	——	H. ♃.	
10 microphy'lla. DC.	(*ye.*)	small-leaved.	N. Holland.	....	——	H. ⊙.	
rùbens. H.M.							
11 rèpens. DC.	(*ye.*)	creeping-stalk'd.	C. B. S.	1793.	3. 4.	G. ♃.	Jacq. ox. n. 11. t. 78. f. 1.
12 ròsea. DC.	(*ro.*)	rose-coloured.	Chile.	1826.	3. 6.	G. ♃.	Bot. mag. 2830.
floribunda. B.R. 1123.							
13 Símsii.	(*cr.*)	Sims's.	————	1822.	4. 6.	G. ♃.	Bot. mag. 2415.
ròsea. Sims Bot. mag. *nec aliorum.*							
14 lateriflòra. DC.	(*pu.*)	side-flowering.	C. B. S.	1826.	3. 6.	G. ♃.	Jacq. Schœnb. 2. t. 204.
15 macróstylis. DC.	(*pu.*)	long-styled.	————	1793.	10. 11.	G. b.	Jacq. ox. n. 22. t. 9.
16 tubiflòra. DC.	(*vi.*)	tube-flowered.	————	1790.	——	G. b.	———— n. 23. t. 10.
17 canéscens. DC.	(*li.*)	canescent.	————	1822.	——	G. b.	———— n. 24. t. 11.
18 secùnda. DC.	(*li.*)	secund.	————	1790.	——	G. b.	———— n. 68. t. 12.
19 hírta. DC.	(*vi.*)	hairy-stalked.	————	1787.	1. 3.	G. b.	———— n. 26. t. 13
20 hirtélla. DC.	(*li.*)	tooth-stamen'd.	————	1820.	——	G. b.	———— n. 27. t. 14.
21 multiflòra. DC.	(*vi.*)	many-flowered.	————	1789.	2. 3.	G. b.	———— n. 28. t. 15.
22 fúlgida. B.R.	(*cr.*)	crimson.	————	....	——	G. b.	Bot. reg. 1073.
23 rubélla. DC.	(*vi.*)	branching red.	————	1791.	9. 11.	G. b.	Bot. mag. 1031.
24 rosàcea. DC.	(*vi.*)	rose-coloured.	————	1793.	10. 2.	G. b.	———— 1698.
25 virgínea. DC.	(*wh.*)	virgin.	————	1822.	11. 12.	G. b.	Jacq. schœnb. 3. t. 275.
26 reptátrix. DC.	(*li.*)	creeping-rooted.	————	1795.	——	G. ♃.	Jacq. ox. n. 33. t. 20.
27 incarnàta. DC.	(*bh.*)	flesh-coloured.	————	1739.	4. 6.	G. b.	Jacq. vind. t. 71.
28 dísticha. DC.	(*ye.*)	distichous.	————	1822.	——	G. b.	Jacq. ox. n. 31. t. 18.
29 venòsa. DC.	(*vi.*)	veined-flower'd.	————	1821.	——	G. b.	
30 cùprea. B.C.	(*co.*)	copper-colour'd.	————	1822.	——	G. b.	Lodd. bot. cab. 824.
31 tetraphy'lla. DC.	(*pu.*)	four-leaved.	Mexico.	——	——	H. b.	———— t. 790.
32 serícea. DC.	(*ye.*)	silky-leaved.	C. B. S.	1794.	——	G. b.	Jacq. ox. n. 13. t. 77. f. 1.
33 carnòsa. B.R.	(*ye.*)	fleshy.	Chile.	1825.	4. 9.	H. ♃.	Bot. reg. 1063.
34 bipunctàta. Gr.	(*li.*)	two-spotted.	Brazil.	——	4. 6.	G. ♃.	Bot. mag. 2781.
35 violacea. DC.	(*vi.*)	violet-coloured.	N. America.	1772.	5. 6.	H. b.	Jacq. vind. t. 180.
36 Déppii.	(*li.*)	Depp's.	Mexico.	1827.	6. 8.	G. b.	
37 Martiàna. Zuc.	(*li.*)	Martius's.	Brazil.	1828.	5. 9.	S. ♃.	
38 palústris. S.S.	(*li.*)	marsh.	————	——	5. 12.	S. ♃.	
papilionacea. Zuc.							

39 caprìna DC.	(*vi.*)	goats-foot.	C. B. S.	1757.	3. 6.	G. b.	Jacq. ox. t. 76. f. 1.
40 cérnua. DC.	(*ye.*)	drooping-yellow.	———	——	2. 5.	G. b.	——— n. 16. t. 6.
caprìna. B.M. 237. *nec aliorum.*							
β flòre plèno.	(*ye.*)	*double-flowered.*	———	——	——	G. b.	
41 compréssa. DC.	(*ye.*)	compressed.	———	1794.	12. 1.	G. b.	Jacq. ox. n. 19. t. 78. f. 3.
42 dentàta. DC.	(*li.*)	toothed.	———	1793.	11. 12.	G. b.	——— n. 17. t. 7.
43 lívida. DC.	(*fl.*)	livid.	———	——	10. 11.	G. b.	——— n. 18. t. 8.
44 purpuràta. DC.	(*bh.*)	purplish-leav'd.	———	1820.	——	G. b.	Jacq. schœnb. t. 356.
45 macrophy'lla. DC.	(*vi.*)	large-leaved.	———	——	——	G. b.	
46 lobàta. DC.	(*ye.*)	lobed.	Chile.	1821.	4. 6.	H. b.	Bot. mag. 2386.
47 ténera. B.R.	(*li.*)	tender.	Brazil.	1826.	5. 6.	S. ♃.	Bot. reg. 1046.
48 monophylla. DC.	(*bh.*)	simple-leaved.	C. B. S.	1774.	10. 11.	G. b.	Jacq ox. n. 35. t. 79. f. 3.
49 lépida. DC.	(*wh.*)	pretty.	———	1822.	10. 2.	G. b.	——— n. 34. t. 21.
50 rostràta. DC.	(*bh.*)	beaked.	———	1795.	——	G. b.	——— n. 36. t. 22.
51 críspa. DC.	(*wh.*)	curled.	———	1793.	——	G. b.	——— n. 37. t. 23.
52 leporìna. DC.	(*std.*)	Hare's-eared.	———	1795.	10. 11.	G. b.	——— n. 30. t. 25.
53 asinìna. DC.	(*ye.*)	asses-eared.	———	1792.	11. 12.	G. b.	——— n. 38. t. 24.
54 lanceæfòlia. DC.	(*ye.*)	spear-leaved.	———	1795.	10. 11.	G. b.	——— n. 40. t. 26.
55 fabæfòlia. DC.	(*ye.*)	Bean-leaved.	———	1794.	——	G. b.	——— n. 41. t. 27.
56 laburnifòlia. DC.	(*ye.*)	Laburnum.l[d].	———	1793.	9. 10.	G. b.	——— n. 42. t. 28.
57 sanguínea. DC.	(*ye.*)	bloody-leaved.	———	1795.	10. 12.	G. b.	——— n. 43. t. 29.
58 trícolor. DC.	(*std.*)	three-coloured.	———	1794.	——	G. b.	——— n. 63. t. 47.
59 ciliàris. DC.	(*pu.*)	ciliated.	———	1793.	——	G. b.	——— n. 45. t. 30.
60 arcuàta. DC.	(*vi.*)	gland-covered.	———	1795.	——	G. b.	——— n. 46. t. 31.
61 fláccida. DC.	(*re.wh.*)	flaccid.	———	1812.	——	G. b.	——— n. 66. t. 51.
62 ferruginàta. DC.	(*wh.*)	rust-spotted.	———	1822.	——	G. b.	Jacq. schœnb. 3. t. 274.
63 ambígua. DC.	(*v.*)	ambiguous.	———	1790.	9. 12.	G. b.	Jacq. ox. n. 59. t. 43.
64 undulàta. DC.	(*wh.*)	wave-leaved.	———	1795.	10. 11.	G. b.	——— n. 60. t. 44.
65 glandulòsa. DC.	(*wh.*)	glandulous.	———	1822.	——	G. b.	——— n. 62. t. 46.
66 fuscàta. DC.	(*wh.*)	brown-spotted.	———	1795.	5. 6.	G. b.	——— n. 61. t. 45.
67 sulphùrea. DC.	(*st.*)	sulphur-colour'd.	———	——	10. 11.	G. b.	——— n. 77. t. 63.
68 breviscàpa. DC.	(*wh.*)	short-stalked.	———	1823.	9. 2.	G. b.	——— n. 72. t. 58.
69 speciòsa. DC.	(*pu.*)	specious.	———	1690.	9. 11.	G. b.	——— n. 74. t. 60.
70 rigídula. Jac.	(*wh.*)	rigid.	———	1822.	——	G. b.	——— n. 73. t. 59.
71 variábilis. DC.	(*v*)	variable.	———	1795.	10. 12.	G. b.	——— n. 67.
α longiscàpa. DC.	(*wh.*)	*white-flowered.*	———	——	——	G. b.	——— t. 52.
β rùbra. DC.	(*re.*)	*red-flowered.*	———	——	——	G. b.	——— t. 53.
72 grandiflòra. W.	(*wh.*)	great-flowered.	———	1810.	——	G. b.	——— n. 68. t. 54.
β Símsii. DC.	(*wh.*)	*Sims's.*	———	——	——	G. b.	Bot. mag. 1683.
73 láxula. Jac.	(*wh.*)	loose.	———	1820.	——	G. b.	Jacq. ox. n. 71. t. 57.
74 purpùrea. DC.	(*pu.*)	purple.	———	1812.	——	G. b.	——— n. 70. t. 56.
75 convéxula. DC.	(*or.*)	convex-leaved.	———	1789.	11. 1.	G. b.	——— n. 69. t. 55.
76 lævigàta. DC.	(*pu.*)	smooth.	....	1818.	6. 8.	H. ⊙.	
77 punctàta. DC.	(*bh.*)	spot-leaved.	C. B. S.	1823.	10. 2.	G. b.	Jacq. ox. n. 82. t. 66.
78 strumòsa. DC.	(*wh.*)	swollen-styled.	———	1821.	——	G. b.	——— n. 79. t. 64.
79 marginàta. DC.	(*wh.*)	green-margin'd.	———	1812.	9. 12.	G. b.	——— n. 85. t. 68.
80 pulchélla. DC.	(*bh.*)	beautiful.	———	1795.	10. 11.	G. b.	——— n. 86. t. 69.
81 obtùsa. DC.	(*re.*)	blunt-leaved.	———	1812.	——	G. b.	——— n. 83. t. 79. f. 1.
82 lanàta. DC.	(*wh.*)	woolly-leaved.	———	1791.	——	G. b.	——— n. 81. t. 77. f. 2.
83 lutèola. DC.	(*ye.*)	yellowish.	———	1822.	——	G. b.	——— n. 80. t. 65.
84 fállax. DC.	(*ye.*)	related.	———	1825.	——	G. b.	——— n. 84. t. 67.
85 Acetosélla. DC.	(*wh.*)	common.	Britain.	....	4. 6.	H. ♃.	Eng. bot. 762.
86 americàna. DC.	(*wh.*)	American.	N. America.	....	——	H. ♃.	
87 tenélla. DC.	(*vi.*)	slender.	C. B. S.	1793.	——	G. b.	Jacq. ox. n. 32. t. 19.
88 nàtans. DC.	(*wh.*)	floating.	———	1795.	9. 12.	G. *w.* b.	——— n. 78. t. 79. f. 2.
89 cruentàta. DC.	(*pu.*)	bloody.	———	1826.	——	G. b.	Jacq. f. ecl. 1. t. 45.
90 filicáulis. DC.	(*vi.*)	bilobed-leaved.	———	1815.	9. 10.	G. b.	Jacq. schœnb. 2. t. 205.
91 bífida. DC.	(*vi.*)	cloven-leaved.	———	1791.	——	G. b.	Jacq. ox. n. 89. t. 79. f. 4.
92 bifúrca. B.C.	(*re.*)	forked-leaved.	———	1825.	6. 9.	G. b.	Lodd. bot. cab. 1056.
93 cuneàta. DC.	(*wh.*)	wedge-leaved.	———	1818.	——	G. b.	Jacq. ox. n. 55. t. 40.
94 cuneifòlia. DC.	(*wh.*)	wedge-shaped.	———	1793.	4. 5.	G. b.	——— n. 56. t. 41.
95 pusílla. DC.	(*bh.*)	small.	———	1826.	——	G. b.	——— n. 57. t. 42.
96 lineàris. DC.	(*vi.*)	linear-leaved.	———	1795.	9. 11.	G. b.	——— n. 47. t. 32.
97 reclinàta. DC.	(*sc.*)	reclining.	———	——	——	G. b.	——— n. 49. t. 34.
98 grácilis. DC.	(*re.*)	slender-stalked.	———	....	——	G. b.	——— n. 48. t. 33.
99 miniàta. DC.	(*re.*)	miniated.	———	1818.	——	G. b.	——— n. 50. t. 35.
100 glábra. DC.	(*pu.*)	smooth.	———	1795.	5. 6.	G. b.	——— n. 58. t. 76. f. 3.
101 versícolor. DC.	(*std.*)	striped-flower'd.	———	1774.	1. 3.	G. b.	Bot. mag. 155.
102 elongàta. DC.	(*std.*)	elongated.	———	1791.	9. 10.	G. b.	Jacq. ox. n. 52. t. 37.

103 am'œna. L.en.	(*ro.*)	handsome.	C. B. S.	1791.	9. 10.	G. b.	Jacq. schœnb. t. 206.
104 tenuifòlia. DC.	(*std.*)	fine-leaved.	———	1790.	10. 11.	G. b.	Lodd. bot. cab. t. 712.
105 polyphy'lla. DC.	(*ro.*)	many-leaved.	———	1791.	9. 1.	G. b.	Jacq. ox. n. 59. t. 39.
106 filifòlia. DC.	(*ro.*)	slender-leaved.	———	1818.	——	G. b.	Jacq. schœn. 3. t. 273.
107 pentaphy'lla. DC.	(*li.*)	five-leaved.	———	1800.	11. 2.	G. b.	Bot. mag. 1549.
108 lupinifòlia. DC.	(*ye.*)	Lupine-leaved.	———	1791.	10. 11.	G. b.	Jacq. ox. n. 92. t. 72.
109 flàva. DC.	(*ye.*)	narrow-leaved.	———	1775.	3. 4.	G. b.	Bot. reg. 117.
110 pectinàta. DC.	(*ye.*)	pectinated.	———	1790.	9. 11.	G. b.	Jacq. ox. n. 95. t. 75.
111 flabellifolia. DC.	(*ye.*)	fan-leaved.	———	1789.	——	G. b.	——— n. 94. t. 74.
112 tomentòsa. DC.	(*wh.*)	downy-leaved.	———	1791.	4. 5.	G. b.	——— n. 96. t. 81.

ORDO LV.

ZYGOPHYLLEÆ. *DC. prodr.* 1. *p.* 703.

TR'IBULUS. DC.		CALTROPS. Decandria Monogynia. L.					
1 cistoìdes. DC.	(*ye.*)	Cistus-like.	S. America.	1752.	7.	S. ♃.	Bot. reg. 791.
2 terréstris. DC.	(*ye.*)	small.	S. Europe.	1596.	6. 7.	H. ⊙.	Lam. ill. t. 346. f. 1.
3 subinérmis. F.	(*ye.*)	hairy.	Thibet.	1824.	6. 9.	H. ⊙.	
4 maximus. DC.	(*st.*)	great.	Jamaica.	1728.	6. 7.	S. ⊙.	Jacq. ic. rar. t. 462.
5 trijugàtus. DC.	(*ye.*)	Nuttall's.	America.	1823.	——	H. ⊙.	
FAG'ONIA. DC.		FAG'ONIA. Decandria Monogynia. L.					
1 crética. DC.	(*pu.*)	Cretan.	Candia.	1739.	6. 8.	H. ⊙.	Bot. mag. 241.
2 aràbica. DC.	(*pu.*)	Arabian.	Arabia.	1759.	——	G. ♄.	
LA'RREA. DC.		LA'RREA. Decandria Monogynia.					
nítida. DC.	(*ye.*)	glossy-leaved.	BuenosAyres.	1823.	6. 8.	G. ♄.	Cav. ic. 6. t. 559.
ROEP'ERIA. J.R.		ROEP'ERIA. Decandria Monogynia.					
fabagifòlia. J.R.	(*ye.*)	Fabago-leaved.	N. Holland.	1822.	6. 8.	G. ♄.	Juss. Rut. t. 2. f. 3.
ZYGOPHY'LLUM. DC.		BEAN-CAPER. Decandria Monogynia. L.					
1 cordifòlium. DC.	(*ye.*)	heart-leaved.	C. B. S.	1774.	10.	G. ħ.	
2 macrópterum. Led.	(*ye.*)	large-winged.	Siberia.	1827.	7. 9.	H. ♃.	
3 Fabàgo. DC.	(*st.*)	common.	Syria.	1596.	——	H. ♃.	Lam. ill. t. 345. f. 1.
4 fœ'tidum. DC.	(*ye.*)	fœtid.	C. B. S.	1790.	6. 8.	G. ♄.	Schr. s. han. p. 17. t. 9.
β *insuáve.* B.M.	(*ye.*)	*unpleasant.*	———	——	——	G. ♄.	Bot. mag. 372.
5 maculàtum. DC.	(*ye.*)	spotted.	C. B. S.	1782.	10. 11.	G. ♄.	
6 microphy'llum. DC.	(*ye.*)	small-leaved.	———	1820.	——	G. ♄.	
7 Morgsàna. DC.	(*ye.*)	four-leaved.	———	1732.	5. 9.	G. ♄.	Dill. elth. t. 116. f. 141.
8 sessilifòlium. DC.	(*ye.*)	sessile-leaved.	———	1713.	7. 8.	G. ♄.	——— t. 116. f. 142.
9 spinòsum. DC.	(*ye.*)	spiny.	———	1818.	——	G. ♄.	
10 coccíneum. DC.	(*sc.*)	scarlet.	Egypt.	——	——	G. ♄.	Forsk. des. 87. ic. t. 11.
11 álbum. DC.	(*wh.*)	white.	———	1770.	10. 11.	G. ♄.	DC. pl. gras. t. 154.
12 prostràtum. DC.	(*ye.*)	hairy-jointed.	C. B. S.	1818.	——	G. ♃.	
13 tridentàtum. DC.	(*ye.*)	three-toothed.	Mexico.	1827.	9. 11.	G. ♄.	Flor. mex. ic. ined.
GUA'IACUM. DC.		LIGNUM VITÆ. Decandria Monogynia. L.					
1 officinàle. DC.	(*bl.*)	common.	W. Indies.	1694.	7. 9.	S. ħ.	Lam. ill. t. 342.
2 verticàle. DC.	(*bl.*)	vertical-petal'd.	New Spain.	1823.	....	S. ħ.	
3 arbòreum. DC.	(*ye.*)	tree.	S. America.	——	....	S. ħ.	Jacq. amer. 130. t. 83.
Zygophy'llum arbòreum. Jac.							
PORLIE'RA. DC.		PORLIE'RA. Octandria Monogynia.					
hygrométrica. DC.	(*st.*)	pinnate-leaved.	Peru.	1820.	....	S. ħ.	Fl. per. v. 4. ined. t. 343.

§ 2. *ZYGOPHYLLEÆ SPURIÆ ALTERNIFOLIÆ.* DC.

MELIAN'THUS. DC.		HONEY-FLOWER. Didynamia Angiospermia. L.					
1 màjor. DC.	(*br.*)	great.	C. B. S.	1688.	5. 7.	G. ħ.	Bot. reg. 45.
2 mìnor. DC.	(*br.*)	small.	———	1696.	8.	G. ħ.	Bot. mag. 301.
3 comòsus. DC.	(*br.*)	tufted.	———	1818.	——	G. ħ.	Comm. rar. t. 4.
BALAN'ITES. DC.		BALAN'ITES. Decandria Monogynia.					
ægyptíaca. DC.	(*wh.*)	Egyptian.	Africa.	1822.	....	S. ħ.	Del. fl. eg. t. 28. f. 1.
Ximènia ægyptíaca. L.							

ORDO LVI.

RUTACEÆ. *DC. prodr.* 1. *p.* 709.

Tribus I. *DIOSMEÆ.* DC.

R'UTA. DC.		RUE. Decandria Monogynia. L.						
1 pinnàta. DC.	(*ye.*)	winged-leaved.	Canaries.	1780.	3. 8.	G.	♄.	Bot. reg. 307.
2 montàna. DC.	(*ye.*)	mountain.	S. Europe.	1596.	8. 9.	H.	♃.	Jacq. ic. rar. 1. t. 76.
3 gravèolens. DC.	(*ye.*)	common.	———	1562.	6. 9.	H.	♄.	Duham. arb. 2. t. 61.
4 bracteòsa. DC.	(*ye.*)	large-bracted.	Sicily.	1823.	——	H.	♄.	
5 divaricàta. DC.	(*ye.*)	spreading.	Italy.	——	——	H.	♄.	Ten. fl. nap. 1. t. 36.
6 angustifòlia. DC.	(*ye.*)	narrow-leaved.	S. France.	1722.	——	G.	♄.	Bot. mag. 2311.
7 macrophy'lla. DC.	(*ye.*)	long-leaved.	Africa.	——	——	G.	♄.	——— 2018.
8 albiflòra. D.P.	(*wh.*)	white-flowered.	Nepaul.	1820.	9. 10.	G.	♄.	Hook. ex. fl. t. 79.
APLOPHY'LLUM. J.R.		APLOPHY'LLUM. Decandria Monogynia.						
1 patavìna. J.R.	(*ye.*)	Italian.	Italy.	1820.	6. 9.	H.	♃.	Mich. gen. t. 19.
2 pubéscens. J.R.	(*ye.*)	pubescent.	Spain.	1816.	5. 8.	H.	♃.	
3 tuberculàta. J.R.	(*ye.*)	tubercled.	Egypt.	1824.	6. 8.	H.	♃.	
4 villòsa. J.R.	(*ye.*)	villous.	Caucasus.	1828.	....	H.	♃.	Buxb. cent. 2. t. 28. f. 2.
5 linifòlia. J.R.	(*ye.*)	Flax leaved.	Spain.	1752.	6. 9.	H.	♄.	Andr. rep. 565.
6 suavèolens. J.R.	(*ye.*)	Primrose-scent^d.	Tauria.	....	——	H.	♄.	Bot. mag. 2254.
7 thesioídes. F.	(*ye.*)	Thesium-like.	Caspia.	1828.	....	H.	♃.	
8 dahùrica. J.R.	(*st.*)	Milk-wort-l^d.	Dahuria.	1816.	7. 8.	H.	♃.	Gmel. sib. 4. t. 68. f. 2.
Péganum daùricum. β. L. *Rùta dahùrica.* DC.								
PE'GANUM. DC.		SYRIAN-RUE. Dodecandria Monogynia. L.						
Hármala. DC.	(*wh.*)	common.	Levant.	1570.	7. 8.	H.	♃.	Lam. ill. t. 401.
DICTA'MNUS. DC.		FRAXINE'LLA. Decandria Monogynia. L.						
1 Fraxinélla. L.en.	(*re.*)	red-flowered.	S Europe.	1596.	5. 7.	H.	♃.	
2 álbus. L.en.	(*wh.*)	white-flowered.	———	——	——	H.	♃.	Jacq. aust. 5. t. 428.
CALODE'NDRON. DC.		CALODE'NDRON. Pentandria Monogynia.						
capénse. DC.	(*bh.*)	Cape.	C. B. S.	1789.	....	G.	♄.	Lam. j. h. nat. 56. t. 3.
ADENA'NDRA. W.en.		ADENA'NDRA. Pentandria Monogynia. L.						
1 uniflòra. W.en.	(*wh.*)	one-flowered.	C. B. S.	1775.	4. 7.	G.	♄.	Bot. mag. 273.
2 acuminàta.	(*wh.*)	acute-leaved.	———	1812.	——	G.	♄.	Lodd. bot. cab. 493.
Diósma acuminàta. B.C. *non* DC.								
3 am'œna.	(*bh.*)	charming.	———	1798.	——	G.	♄.	Bot. reg. 553.
Diósma am'œna. B.C. 161.								
4 speciòsa.	(*wh.*)	umbel-flower'd.	———	1790.	——	G.	♄.	Bot. mag. 1271.
umbellàta. W.en. *Diosma speciosa.* B.M.								
5 fràgrans. R.S.	(*re.*)	red-flowered.	———	1812.	5. 7.	G.	♄.	Bot. mag. 1519.
6 villòsa. R.S.	(*bh.*)	villous.	———	——	——	G.	♄.	
Diósma villòsa. DC. *non* Hort. *tomentòsa.* Hort.								
7 marginàta. R.S.	(*bh.*)	marginate.	———	1806.	4. 6.	G.	♄.	
8 lineàris. B.D.	(*wh.*)	linear-leaved.	———	1800.	3. 7.	G.	♄.	
BARO'SMA. W. en.		BARO'SMA. Pentandria Monogynia. L.						
1 serratifòlia. W.en.	(*wh.*	saw-leaved.	C. B. S.	1789.	3. 6.	G.	♄.	Bot. mag. 456.
2 odoràta. R.S.	(*wh.*)	odoriferous.	———	1790.	3. 5.	G.	♄.	Wend. coll. 1. t. 15.
Diósma latifòlia. B.C. 290.								
3 latifòlia. R.S.	(*wh.*)	broad-leaved.	———	——	——	G.	♄.	Andr. rep. t. 33?
4 crenàta. B.D.	(*wh.*)	crenated.	———	1774.	1. 5.	G.	♄.	Lodd. bot. cab. t. 404.
5 betulìna. B.D.	(*wh.*)	wedge-leaved.	———	1790.	5. 7.	G.	♄.	
orbiculàris. Hort. *Diósma betulìna.* DC.								
6 ovàta. B.D.	(*wh.*)	oval-leaved.	———	——	2. 9.	G.	♄.	Bot. mag. 1616.
Diósma ovàta. B.M.								
7 pulchélla. B.D.	(*pu.*)	blunt-leaved.	———	1787.	2. 9.	G.	♄.	Bot. mag. 1357.
Diósmapulchélla. B.M. *Búcco pulchélla.* R.S.								
8 oblónga. B.D.	(*wh.*)	oblong-leaved.	———	1821.	4. 8.	G.	♄.	
9 dioíca. B.D.	(*pi.*)	dioecious.	———	1816.	——	G.	♄.	Bot. reg. 502.
Diósma dioíca. B.R. *linifolia.* B.C. 400.								
10 angustifolia. B.D.	(*bh.*)	narrow-leaved.	———	1820.	——	G.	♄.	
11 fœtidíssima. B.D.	(*bh.*)	strong-scented.	———	1818.	——	G.	♄.	

MACRO'STYLIS. B.D. MACRO'STYLIS. Pentandria Monogynia.
1 lanceolàta. B.D. (*li.*) spear-leaved. C. B. S. 1824. 4. 6. G. ♄.
Diósma barbàta. S.S.
2 barbígera. B.D. (*li.*) beard-bearing. ——— 1826. —— G. ♄.
3 obtùsa. B.D. (*li.*) blunt-leaved. ——— 1824. —— G. ♄.
4 squarrosa. B.D. (*li.*) squarrose. ——— 1821. —— G. ♄.
EUCH'ÆTIS. B.D. EUCH'ÆTIS. Pentandria Monogynia.
glomeràta. B.D. (*wh.*) close-flowered. C. B. S. 1821. 4. 6. G. h. B. et Wend. dios. t. A. f. 1.
ACMAD'ENIA. B.D. ACMAD'ENIA. Pentandria Monogynia.
1 juniperìna. B.D. (*wh.*) Juniper-leaved. C. B. S. 1818. 4. 6. G. h.
2 obtusàta. B.D. (*re.*) blunt-leaved. ——— —— —— G. ♄.
Diósma obtusàta. R.S.
3 lævigàta. B.D. (*wh.*) smooth. ——— .... —— G. ♄. Thunb. voy. 4. t. 5.
Diósma tetragona. Th. non L.
4 púngens. B.D. (*wh.*) pungent. ——— 1823. —— G. h.
5 tetragòna. B.D. (*bh.*) four-sided. ——— 1720. 6. 8. G. ♄.
COLEON'EMA. B.D. COLEON'EMA. Pentandria Monogynia.
álba. B.D. (*wh.*) white-flowered. C. B. S. 1800. 3. 6. G. ♄.
Diósma álba. DC. *Adenándra álba.* R.S.
DIO'SMA. W. en. DIO'SMA. Pentandria Monogynia. L.
1 oppositifòlia. DC. (*wh.*) opposite-leav'd. C. B. S. 1752. 3. 7. G. ♄.
2 succulénta. DC. (*wh.*) succulent-ld. ——— .... 4. 6. G. ♄. Wendl. coll. 1. t. 1.
3 cupressìna. DC. (*wh.*) Cypress-like. ——— 1790. 6. 7. G. ♄. Lodd. bot. cab. t. 303.
4 ramosíssima. B.D. (*wh.*) branching. ——— .... 4. 8. G. h.
cupressìna. Lam.
5 Meyeriàna. S.S. (*wh.*) Meyer's. ——— 1826. —— G. ♄.
6 teretifòlia. L. en. (*wh.*) round-leaved. ——— 1818. —— G. ♄.
7 pectinàta. DC. (*wh.*) pectinated. ——— 1812. 4. 6. G. h.
8 subulàta. DC. (*wh.*) awl-leaved. ——— —— 3. 6. G. h. Wendl. coll. 1. t. 8.
9 rùbra. DC. (*bh.*) Heath-leaved. ——— 1752. 2. 5. G. h. Bot. reg. 563.
ericifòlia. Andr. rep. t. 451.
10 hirsùta. DC. (*wh.*) hairy-leaved. ——— 1731. —— G. h. Wendl. coll. 1. t. 27.
11 ambígua. B.D. (*wh.*) ambiguous. ——— 1824. —— G. h.
12 ericoídes. DC. (*wh.*) Heath-like. ——— 1756. 3. 7. G. ♄. Mill. ic. t. 124. f. 2.
13 longifòlia. DC. (*wh.*) long-leaved. ——— .... —— G. ♄. Wendl. coll. 1. t. 19.
AGATHO'SMA. W. en. AGATHO'SMA. Pentandria Monogynia.
1 híspida. S.S. (*wh.*) rough-leaved. C. B. S. 1786. 6. 8. G. ♄.
2 pátula. S.S. (*li.*) spreading. ——— 1821. —— G. ♄.
3 brevifòlia. S.S. (*li.*) short-leaved. ——— 1816. —— G. ♄.
4 tenuíssima. S.S. (*li.*) slender-leaved. ——— .... 2. 5. G. ♄.
5 vìrgata. S.S. (*wh.*) twiggy. ——— 1816. 4. 6. G. ♄.
6 eréctа. B.D. (*wh.*) upright. ——— —— —— G. ♄. Wendl. coll. p. 17. t. 13.
7 bruniàdes. S.S. (*li.*) Brunia-like. ——— —— 6. 8. G. ♄.
8 Thunbergiàna. S.S. (*li.*) Thunberg's. ——— .... 4. 6. G. ♄.
Diósma ciliàta. Th. *non* L.
9 linifòlia. S.S. (*li.*) Flax-leaved. ——— 1821. —— G. ♄.
10 prolífera. S.S. (*wh.*) proliferous. ——— .... 4. 8. G. h. Wendl. coll. 3. t. 77.
11 ciliàta. S.S. (*bh.*) ciliated. ——— 1774. —— G. ♄. Bot. reg. 366.
12 Cerefòlium. S.S. (*wh.*) Chervil-scented. ——— 1790. —— G. ♄. Vent. malm. t. 93.
13 refléxa. L. en. (*li.*) reflexed. ——— 1816. 6. 8. G. ♄.
14 hírta. B.R. (*pu.*) hairy. ——— 1794. 5. 7. G. ♄. Bot. reg. 369.
Diósma Ventenatiàna. S.S. *Búcco Ventenatiàna.* R.S.
15 láxa. S.S. (*li.*) loose-branched. ——— 1816. —— G. ♄.
16 villòsa. W. en. (*li.*) villous. ——— 1786. 6. 8. G. ♄. Wendl. coll. 1. t. 2.
Búcco villòsa. R.S. *Diósma Wendlandiàna.* DC.
17 glábrata. B.D. (*li.*) smooth. ——— 1821. —— G. ♄.
18 hy'brida. S.S. (*li.*) hybrid. ——— —— —— G. ♄.
19 rugòsa. L. en. (*wh.*) rugged-leaved. ——— 1790. 4. 8. G. h.
20 móllis. B.D. (*wh.*) soft. ——— 1824. —— G. ♄.
21 obtùsa. DC. (*li.*) blunt-leaved. ——— 1774. —— G. ♄. Lodd. bot. cab. 210.
Diósma ciliata. B.C. *non* L.
α ovàta. (*li.*) *oval-leaved.* ——— —— —— G. ♄. Wendl. coll. 1. t. 13.
β oblónga. (*li.*) *oblong-leaved.* ——— —— —— G. ♄. ——— 1. t. 14.
22 lanceolàta. (*li.*) lance-leaved. ——— —— —— G. ♄. Bot. reg. 366.
23 orbiculàris. S.S. (*wh.*) round-leaved. ——— 1790. —— G. ♄.
24 ambígua. DC. (*wh.*) doubtful. ——— .... —— G. ♄. Lodd. bot. cab. 461.
25 imbricàta. W. en. (*li.*) imbricated. ——— 1774. —— G. ♄. Wendl. coll. 1. t. 9.
26 acuminàta. W. en. (*pa.*) taper-pointed. ——— 1812. —— G. ♄. ——— 1. t. 28.
Diósma cordàta. Hort.
27 obtusàta. (*li.*) obtuse-leaved. ——— 1816. —— G. h. Wendl. coll. 3. p. 7. t. 76.

EMPLE'URUM. DC.		EMPLE'URUM.	Monœcia Tetrandria.					
serrulàtum. DC.	(st.)	saw-leaved.	C. B. S.	1774.	6. 7.	G.	♄.	Sm. exot. bot. 2. t. 63.
CORR'EA. DC.		CORR'EA.	Octandria Monogynia.					
1 álba. DC.	(wh.)	white.	N. S. W.	1793.	4. 7.	G.	♄.	Andr. bot. rep. 18.
2 rùfa. DC.	(wh.)	rusty-leaved.	———	1819.	——	G.	♄.	Labill. voy. 2. t. 17.
3 pulchélla. S.F.A.	(pi.)	pretty.	N. Holland.	1824.	11. 5.	G.	♄.	Sweet fl. aust. t. 1.
4 speciòsa. DC.	(re.)	red-flowered.	N. S. W.	1806.	——	G.	♄.	Bot. reg. 26.
5 vìrens. DC.	(gr.)	green-flowered.	———	1800.	——	G.	♄.	Sm. exot. bot. 2. t. 72.
PHEB'ALIUM. DC.		PHEB'ALIUM.	Decandria Monogynia.					
1 squammulòsum. DC.	(wh.)	scaly.	N. S. W.	1820.	4. 6.	G.	♄.	Vent. malm. t. 102.
2 elàtum. F.T.	(wh.)	tall.	———	1823.	4. 8.	G.	♄.	
3 áureum. F.T.	(ye.)	golden.	———	——	——	G.	♄.	
4 lachnoídes. F.T.	(wh.)	Lachnæa-like.	———	——	——	G.	♄.	
5 lineàre. ψ.	(wh.)	linear-leaved.	N. Holland.	1825.	....	G.	♄.	
6 salicifòlium. ψ.	(wh.)	Willow-leaved.	———	——	....	G.	♄.	
CR'OWEA. DC.		CR'OWEA.	Decandria Monogynia.					
salígna. DC.	(pi.)	Willow-leaved.	N. S. W.	1790.	7. 12.	G.	♄.	Bot. mag. 989.
ERIO'STEMON. DC.		ERIO'STEMON.	Decandria Monogynia.					
1 buxifòlium. DC.	(li.)	Box-leaved.	N. S. W.	1822.	4. 6.	G.	♄.	
2 salicifòlium. DC.	(re.)	Willow-leaved.	———	——	——	G.	♄.	Bot. mag. 2854.
3 cuspidàtum. F.T.	(wh.)	sharp-pointed.	———	1823.	4. 8.	G.	♄.	Lodd. bot. cab. 1247.
4 obovàle. F.T.	(li.)	oboval-leaved.	———	——	——	G.	♄.	
5 lanceolàtum. DC.	(li.)	spear-leaved.	———	——	——	G.	♄.	
6 linearifòlium. DC.	(wh.)	linear-leaved.	———	——	——	G.	♄.	
7 squámmeum. DC.	(wh.)	scaly.	N. Holland.	1822.	——	G.	♄.	Lab. nov. hol. 1. t. 141.
8 ericæfòlium. M.C.	(li.)	Heath-leaved.	———	1825.	——	G.	♄.	
PHILOTH'ECA. L.T.		PHILOTH'ECA.	Monadelphia Decandria.					
australis. L.T.	(bh.)	Salsola-leaved.	N. S. W.	1822.	——	G.	♄.	Linn. trans. 11. t. 21.
BOR'ONIA. DC.		BOR'ONIA.	Octandria Monogynia.					
1 pinnàta. DC.	(re.)	Hawthorn-scent'd.	N. S. W	1794.	2. 5.	G.	♄.	Bot. mag. 1763.
2 tetrándra. DC.	(re.)	tetrandrous.	N. Holland.	1824.	....	G.	♄.	Lab. nov. hol. 1. t. 125.
3 anemonæfòlia. F.T.	(re.	Anemone-leav'd.	N. S. W.	——	5. 8.	G.	♄.	
4 alàta. DC.	(re	winged-leaved.	N. Holland.	——	4 8.	G.	♄.	Sweet fl. aust. t. 48.
5 serrulàta. DC.	(ro.)	Rose-scented.	N. S. W.	1816.	2. 5.	G.	♄.	Bot. reg. 842.
6 denticulàta. DC.	(re.)	tooth-leaved.	N. Holland.	1823.	4. 8.	G.	♄.	———1000.
7 parviflòra. DC.	(re.)	small-flowered.	N. S. W.	1826.	——	G.	♄.	Smith's tracts. t. 6.
8 polygalæfòlia. DC.	(re.)	Polygala-ld.	———	1824.	....	G.	♄.	——— t. 7.
9 ledifòlia. DC.	(re.)	Ledum-leaved.	———	——	....	G.	♄.	
CYMINO'SMA. DC.		CYMINO'SMA.	Octandria Monogynia.					
1 pedunculàta. DC.	(wh.)	peduncled.	E. Indies.	1800.	....	G.	♄.	Vahl. symb. 3. t. 61.
Jambolífera pedunculàta. W.								
2 odoràta. DC.	(wh.)	sweet-scented.	China.	1818.	....	G.	♄.	
ZIE'RIA. DC.		ZIE'RIA.	Tetrandria Monogynia.					
1 lanceolàta. DC.	(wh.)	spear-leaved.	N. S. W.	1808.	4. 7.	G.	♄.	Bot. mag. 1395.
Smíthii. Bot. rep. 606.								
2 macrophy'lla. DC.	(wh.)	large-leaved.	———	1820.	——	G.	♄.	
3 lævigàta. DC.	(wh.)	smooth-leaved.	———	1822.	——	G.	♄.	
4 octándra.	(gr.)	octandrous.	———	1825.	4. 6.	G.	♄.	
5 obcordàta. F.T.	(wh.)	obcordate.	———	1824.	——	G.	♄.	
6 revolùta. F.T.	(wh.)	revolute.	———	——	3. 5.	G.	♄.	
7 microphy'lla. DC.	(wh.)	small-leaved.	———	——	——	G.	♄.	
8 pilòsa. DC.	(wh.)	hairy.	———	——	——	G.	♄.	Linn. trans. 10. t. 17. f. 2.
9 hirsùta. DC.	(wh.)	hirsute.	———	——	——	G.	♄.	
MELIC'OPE. DC.		MELIC'OPE.	Octandria Monogynia.					
ternàta. DC.	....	ternate-leaved.	N. Zealand.	1822.	....	G.	♄.	Lam. ill. t. 294.
EV'ODIA. DC.		EV'ODIA.	Tetrandria Monogynia.					
triphy'lla. DC.	(wh.)	three-leaved.	Pulo Penang.	1820.	....	S.	♄.	Rumph. amb. 2. t. 62.
Fagàra triphy'lla. F.I.								
ZANTHO'XYLUM. DC.		ZANTHO'XYLUM.	Tetrandria Monogynia et Diœcia Pentandria.					
1 tragòdes. DC.	(wh.)	prickly-leaved.	St. Domingo.	1759.	8. 9.	S.	♄.	Jacq. amer. 21. t. 14.
Fagàra tragòdes. W.								
2 affíne. K.S.	(wh.)	related.	Mexico.	1826.	....	G.	♄.	
3 Pteròta. DC.	(wh.)	Lentiscus-ld.	W. Indies.	1768.	——	S.	♄.	Br. jam. 146. t. 5. f. 1.
Fagàra lentiscifòlia.		W. en.						
4 piperìtum. DC.	(wh.)	Ash-leaved.	Japan.	1773.	9.	G.	♄.	Kæmpf. am. t. 893.
5 Avicénnæ. DC.	(wh.)	short-panicled.	China.	1823.	....	G.	♄.	Lob. ic. 2. t. 133. f. 2.
6 heterophy'llum. DC.	(wh.)	various-leav'd.	Bourbon.	——	....	S.	♄.	
7 emarginàtum. DC.	(wh.)	notch-leaved.	Jamaica.	1739.	....	S.	♄.	Sloan. j. 2. t. 168. f. 4.
8 acuminàtum. DC.	(wh.)	taper-pointed.	———	1818.	....	S.	♄.	

9 spinòsum. DC.	(*wh.*) spiny.	Jamaica.	1824.	....	S. ♄.	
10 tricárpum. DC.	(*gr.*) Ash-leaved.	N. America.	1806.	....	H. ♄.	Cat. car. 1. t. 26.
11 fraxíneum. DC.	(*gr.*) common.	————	1759.	3. 4.	H. ♄.	Duham. arb. 1. t. 97.
12 mìte. DC.	(*gr.*) soft-leaved.	————	1812.	——	H. ♄.	
13 hermaphrodìtum. DC.	(*wh.*) hermaphrodite.	Guiana.	1823.	....	S. ♄.	Aubl. gui. 1. t. 30.
14 juglandifòlium. DC.	(*wh.*) Walnut-leaved.	W. Indies.	1822.	....	S. ♄.	Pluk. t. 239. f. 6.
15 ClàvaHérculis. L.	(*wh.*) clubbed.	————	1739.	4. 5.	S. ♄.	—— t. 239. f. 4.
16 aromáticum. DC.	(*wh.*) aromatic.	St. Domingo.	1823.	....	S. ♄.	Jacq. f. ecl. 1. t. 70.
17 armàtum. DC.	(*wh.*) armed.	E. Indies.	1816.	....	S. ♄.	
18 nítidum. DC.	(*wh.*) shining-leaved.	China.	1822.	2. 3.	G. ♄.	Bot. mag. 2558.
19 sapindoídes. DC.	(*wh.*) Licca-tree.	Jamaica.	....	....	G. ♄.	Br. jam. 207. t. 20. f. 2.
Sapíndus spinòsus. L.						
SPIRANTH'ERA. DC.	SPIRANTH'ERA. Pentandria Monogynia.					
odoratíssima. DC.	(*bh.*) Jasmine-scent'd.	Brazil.	1823.	....	S. ♄.	N.etMart.ac.b.xi.t.31.
Terpnánthus jasminiodòrus. Nees.						

Tribus II. *CUSPARIEÆ.* DC.

MONNIE'RA. DC.	MONNIE'RA. Monadelphia Diandria.					
trifòlia. DC.	(*wh.*) three-leaved.	Guiana.	1792.	7. 8.	S. ⊙.	Aub. gui. 2. t. 293.
TIC'OREA. DC.	TICO REA. Pentandria Monogynia.					
1 fœ'tida. DC.	(*wh.*) strong-scented.	Guiana.	1825.	....	S. ♄.	Aubl. gui. 2. t. 277.
2 jasminiflòra. DC.	(*wh.*) Jasmine-fl'd.	Brazil.	1827.	....	S. ♄.	N.etMa.n.ac.b.11.t.18.
GALIP'EA. DC.	GALIP'EA. Diandria Monogynia.					
trifoliàta. DC.	(*gr.*) three-leaved.	Guiana.	1803.	....	S. ♄.	Aub. gui. 2. t. 2o9.

ORDO LVII.

SIMARUBEÆ. *DC. prodr.* 1. *p.* 733.

QUA'SSIA. DC.	QUA'SSIA. Decandria Monogynia. L.					
amára. DC.	(*re.*) bitter.	Guiana.	1790.	6. 7.	S. ♄.	Bot. mag. 497.
SIMAR'UBA. DC.	SIMAR'UBA. Monœcia Decandria.					
1 officinàlis. DC.	(*wh.*) officinal.	Guiana.	1787.	....	S. ♄.	Aub. gui. 2. t. 331-2.
Quássia Simarùba. L. *Simarùba amára.* Aub.						
2 gláuca. DC.	(*wh.*) glaucous-leav'd.	Cuba.	1824.	....	S. ♄.	
3 excélsa. DC.	(*wh.*) tall.	Jamaica.	1822.	....	S. ♄.	Sw.act.holm.1788.t.8.
Quássia excélsa. Sw.						
SIM'ABA. DC.	SIM'ABA. Decandria Monogynia.					
guianénsis. DC.	(*wh.*) Guiana.	Guiana.	1822.	....	S. ♄.	Aub. gui. 1. t. 153.
Zwíngera amára. W.						

ORDO LVIII.

OCHNACEÆ. *DC. prodr.* 1. *p.* 735.

O'CHNA. DC.	O'CHNA. Polyandria Monogynia. L.					
1 squarròsa. R.C.	(*ye.*) blunt-leaved.	E. Indies.	1790.	7. 8.	S. ♄.	Roxb. cor. 1. t. 89.
2 lùcida. DC.	(*pu.*) acute-leaved.	————	1818.	——	S. ♄.	Lam. ill. t. 472. f. 1.
3 nítida. DC.	(*wh.*) shining.	C. B. S.	1816.	....	S. ♄.	DC.ann.m.17.n.3.t.2.
4 multiflòra. DC.	(*ye.*) many-flowered.	SierraLeon.	1822.	....	S. ♄.	———— n. 4. t. 3.
5 mauritiàna. DC.	(*ye.*) Mauritius.	Mauritius.	——	....	S. ♄.	———— n. 8. t. 5.
6 atropurpùrea. DC.	(*ye.*) purple-calyxed.	C. B. S.	1816.	....	S. ♄.	Pluk. alm. t. 263. f. 1. 2.

GO'MPHIA. DC.		Go'mphia.	Decandria Monogynia. W.				
1 obtusifòlia. DC.	(*ye.*)	blunt-leaved.	Madagascar.	1803.	....	S. ♄.	DC. ann. mus. n. 4. t. 8.
O'chna lævigàta. Vahl.							
2 guianénsis. DC.	(*ye.*)	Guiana.	Guiana.	1825.	....	S. ♄.	DC. ann. mus. n. 7. t. 9.
3 Jabotápita. DC.	(*ye.*)	spear-leaved.	W. Indies.	1818.	....	S. ♄.	Lam. ill. 472. f. 2.
4 nítida. DC.	(*ye.*)	glossy-leaved.	———	1803.	....	S. ♄.	DC. an. mus. n. 13. t. 13.
5 laurifòlia. DC.	(*ye.*)	laurel-leaved.	———	1822.	....	S. ♄.	——— n. 15. t. 15.
6 mexicàna. DC.	(*ye.*)	Mexican.	Mexico.	1825.	....	G. ♄.	H. et B. pl. æq. 2. t. 74.
WALK'ERA. DC.		Walk'era.	Pentandria Monogynia.				
serràta. DC.	(*ye.*)	saw-leaved.	E. Indies.	1824.	....	S. ♄.	Rheed. mal. 5. t. 48.

ORDO LIX.

CORIARIEÆ. *DC. prodr.* 1. *p.* 739.

CORI'ARIA. DC.		Cori'aria.	Mono-Diœcia Decandria.				
1 myrtifòlia. DC.	(*gr.*)	Myrtle-leaved.	S. Europe.	1629.	5. 8.	H. ♄.	Duham. arb. 1. t. 73.
2 sarmentòsa. DC.	(*gr.*)	sarmentose.	N. Zealand.	1820.	7. 9.	G. ♃.	Bot. mag. 2470.

** Subclassis II. *CALYCIFLORÆ.*

ORDO LX.

CELASTRINEÆ. *Brown gen. rem. p.* 22.

Tribus I. *STAPHYLEACEÆ.* DC. *prodr.* 2. *p.* 2.

STAPHYL'EA. DC.		Bladder-nut.	Pentandria Trigynia. L.				
1 trifòlia. DC.	(*wh.*)	three-leaved.	N. America.	1640.	5. 6.	H. ♄.	Schmidt. arb. t. 80.
2 Bumálda. DC.	(*wh.*)	Japan.	Japan.	1804.	6. 9.	G. ♄.	
Bumálda trifòlia. Th.							
3 pinnata. DC.	(*wh.*)	five-leaved.	England.	....	4. 6.	H. ♄.	Eng. bot. 1560.
TURPI'NIA. DC.		Turpi'nia.	Polygamia Diœcia. DC.				
paniculàta. DC.	(*wh.*)	panicled.	W. Indies.	1826.	....	S. ♄.	Vent. choix. t. 31.
Dalry'mplea domingénsis. S.S.							
DALRY'MPLEA. R.C.		Dalry'mplea.	Pentandria Monogynia. F.I.				
pomífera. R.C.	(*wh.*)	Apple-bearing.	E. Indies.	1823.	....	S. ♄.	Roxb. corom. 3. t. 279.
Turpínia pomìfera. DC.							

Tribus II. *EUONYMEÆ.* DC.

EUO'NYMUS. DC.		Spindle-tree.	Pentandria Monogynia. L.				
1 japónicus. DC.	(*wh.*)	Japan	Japan.	1804.	6. 8.	G. ♄.	Kæmpf. ic. t. 8.
2 europ'æus. DC.	(*wh.*)	common.	Britain.	....	5. 7.	H. ♄.	Eng. bot. 362.
3 verrucòsus. DC.	(*pu.*)	warted.	Austria.	1763.	5. 6.	H. ♄.	Schmidt arb. t. 72.
4 latifòlius. DC.	(*gr.*)	broad-leaved.	S. Europe.	1730.	5. 7.	H. ♄.	——— t. 74.
5 atropurpùreus. DC.	(*d.pu.*)	dark-purple.	N. America.	1756.	——	H. ♄.	Jacq. vind. 2. t. 120.
6 americànus. DC.	(*st.*)	evergreen.	———	1683.	6. 7.	H. ♄.	Schmidt arb. t. 75.
7 sarmentòsus. N.	(*st.*)	sarmentose.	———	1823.	——	H. ♄.	
8 obovàtus. N.	(*gr.*)	obovate-leaved.	———	——	——	H. ♄.	

9 angustifòlius. DC.	(*st.*)	narrow-leaved.	Georgia.	1806.	——	H. ħ.	
10 lùcidus. D.P.	(*wh.*)	shining-leaved.	Nepaul.	1820.	6. 7.	H. ħ.	
11 micránthus. D.P.	(*wh.*)	small-flowered.	———	——	....	H. ħ.	
12 chinénsis. DC.	(*wh.*)	Chinese.	China.	1822.	5. 6.	H. ħ.⌣.	
13 lácerus. D.P.	(*wh.*)	torn.	Nepaul.	1820.	6. 7.	H. ħ.	
14 grandiflòrus. F.I.	(*wh.*)	large-flowered.	———	1823.	——	H. ħ.	
15 echinàtus. F.I.	(*st.*)	spinous-fruited.	———	——	4. 5.	H. ħ.⌣.	Bot. mag. 2767.
CELA'STRUS. DC.		STAFF-TREE.	Pentandria Monogynia. L.				
1 oleoídes. DC.	(*wh.*)	Olive-leaved.	C. B. S.	1823.	5. 6.	G. ♄.	
2 pterocárpus. DC.	(*wh.*)	wing-capsuled.	———	1816.	5. 8.	G. ♄.	Burm. afr. t. 97. f. 1.
3 rostràtus. DC.	(*wh.*)	beaked.	———	1821.	——	G. ♄.	
4 tricuspidàtus. DC.	(*wh.*)	three-pointed.	———	1816.	5. 6.	G. ♄.	Buchoz. dec. 6. t. 3.
Cassìne lævigàta. Lam. *'Olea capénsis.* Buchoz.							
5 lùcidus. DC.	(*wh.*)	shining.	C. B. S.	1722.	4. 9.	G. ♄.	L'Her. stirp. 1. t. 25.
6 macrocárpus. DC.	(*wh.*)	large-capsuled.	Peru.	1826.	....	G. ♄.	Flor. per. 3. t.230. f.16.
7 bullàtus. DC.	(*wh.*)	scarlet-fruited.	Virginia.	1759.	7.	H. ♄.⌣.	Pluk. alm. t. 28. f. 5.
8 scándens. DC.	(*wh.*)	climbing.	N. America.	1736.	5. 6.	H. ♄.⌣.	Duham. arb. 1. t. 95.
9 punctàtus. DC.	(*wh.*)	spotted-stalked.	Japan.	1816.	....	G. ♄.⌣.	
10 nùtans. F.I.	(*wh.*)	drooping-fl^d.	E. Indies.	1824.	....	S. ♄.⌣.	
11 cassinoídes. DC.	(*wh.*)	crenated.	Canaries.	1779.	8. 9.	G. ♄.	L'Her. sert. ang. t. 10.
12 lycioídes. DC.	(*wh.*)	Box-thorn-like.	———	1821.	——	G. ♄.	
13 cérnuus. DC.	(*wh.*)	drooping.	C. B. S.	1815.	5. 6.	G. ♄.	
14 myrtifòlius. DC.	(*wh.*)	Myrtle-leaved.	Jamaica.	....	6. 8.	S. ♄.	Sloan. jam. 2. t. 193. f.1.
15 tetragònus. DC.	(*wh.*)	four-sided.	C. B. S.	1816.	——	G. ♄.	
16 ilicifòlius. DC.	(*wh.*)	Holly-leaved.	Brazil.	1824.	....	S. ♄.	
17 ilicìnus. DC.	(*wh.*)	Ilex-like.	C. B. S.	1816.	5. 8.	G. ♄.	
18 retùsus. DC.	(*wh.*)	notched.	Peru.	1823.	——	G. ♄.	Flor. per. 3. t. 229. f. a.
emarginàtus. F.P. *non* W.							
19 lineàris. DC.	(*wh.*)	linear-leaved.	C. B. S.	1823.	5. 8.	G. ♄.	
20 emarginàtus. DC.	(*st.*)	emarginate.	E. Indies.	1820.	....	S. ♄.	
21 buxifòlius. DC.	(*wh.*)	Box-leaved.	C. B. S.	1752.	5. 8.	G. ♄.	Bot. mag. 2114.
22 cymòsus. B.M.	(*wh.*)	cyme-flowered.	———	1815.	6. 8.	G. ♄.	——— t. 2070.
23 multiflòrus. DC.	(*wh.*)	many-flowered.	———	1816.	5. 8.	G. ♄.	
24 pyracánthus. DC.	(*wh.*)	red-fruited.	———	1752.	——	G. ♄.	Bot. mag. 1167.
25 flexuòsus. DC.	(*wh.*)	flexuose.	———	1816.	——	G. ♄.	
26 montànus. DC.	(*wh.*)	mountain.	E. Indies.	1824.	....	S. ♄.	
PLECTR'ONIA. W.		PLECTR'ONIA.	Pentandria Monogynia.				
ventòsa. W.	(*wh.*)	corymbed.	C. B. S.	1816.	....	G. ♄.	Burm. afr. t. 94?
corymbòsa. P.S. *Celástrus? Plectrònia.* DC.							
MA'YTENUS. DC.		MAYTENUS.	Pentandria Monogynia.				
1 octogònus. DC.	(*gr.*)	angular-leaved.	Peru.	1786.	10. 11.	G. ♄.	
Celástrus octogònus. W. *Senàcia octogòna.* Lam.							
2 chilénsis. DC.	(*re.*)	Chile.	Chile.	1824.	....	G. ♄.	Feuill. obs. 3. t. 27.
3 verticillàtus. DC.	(*gr.*)	whorled.	Peru.	1823.	9. 11.	G. ♄.	Flor. per. 3. t. 229. B.
ELÆODE'NDRON. DC.		OLIVE-WOOD.	Pentandria Monogynia.				
1 orientàle. DC.	(*wh.*)	taper-leaved.	Madagascar.	1771.	....	S. ♄.	Jacq. ic. 1. t 48.
índicum. Gært. fruct. 1. t. 57. *Rubéntia olivìna.* J.							
2 gláucum. DC.	(*st.*)	glaucous.	Ceylon.	1823.	....	S. ♄.	Rottb. n. a. hav. 2. t. 4. f. 1.
3 xylocárpum. DC.	(*wh.*)	leathery-leaved.	Antilles.	1816.	7. 8.	S. ♄.	Vent. choix. t. 23.
Cassìne xylocárpa. Vent.							
4 austràle. DC.	(*wh.*)	thick-leaved.	N. S. W.	1796.	6. 8.	G. ♄.	Vent. malm. t. 117.
Portenschlágia austràlis. Tratt. arch. t. 250. *Lamárckia dentàta.* Hort.							
5 integrifolium.	(*wh.*)	entire-leaved.	N. S. W.	——	——	G. ♄.	Tratt. arch. t. 284.
Portenschlágia integrifolia. Tratt.							
6? cròceum. DC.	(*wh.*)	Cape Holly.	C. B. S.	1794.	5. 6.	G. ♄.	
'Ilex cròcea. W.							
PTELI'DIUM. DC.		PTELI'DIUM.	Tetrandria Monogynia.				
ovàtum. DC.	(*gr.*)	oval-leaved.	Madagascar.	1825.	....	S. ♄.	Pet. Th. veg. afr. 1. t. 2.

Tribus III. *AQUIFOLIACEÆ.* DC. *prod.* 2. *p.* 11.

CASSINE. DC.		CASSINE.	Pentandria Trigynia. L.				
1 Maurocènia. DC.	(*wh.*)	Hottentot-cherry.	C. B. S.	1690.	7. 8.	G. ♄.	Dill. elt. t. 121. f. 147.
2 capénsis. DC.	(*wh.*)	Cape Phillyrea.	———	1629.	——	G. ♄.	——— t. 236. f. 305.
3 Colpoón. W.	(*wh.*)	Colpoon-tree.	———	1791.	——	G. ♄.	Burm. afr. t. 86.

4 bárbara. R.S.	(*wh.*)	quadrangular.	C. B. S.	1816.	——	G. ♄.	
5 æthiòpica. R.S.	(*wh.*)	Ethiopian.	———	....	——	G. ♄.	
6 oppositifòlia. DC.	(*wh.*)	opposite-leaved.		....	....	S. ♄.	
7 excélsa. F.I.	(*wh.*)	tall.	Nepaul.	1823.	....	G. ♄.	
HART'OGIA. DC.		HARTOGIA. Tetrandria Monogynia.					
capénsis. DC.	(*wh.*)	Cape.	C. B. S.	1816.	....	G. ♄.	Thunb. n. gen. 5. p. 55. ic.
Schrèbera schinoìdes. Th. prodr. t. 2. *Elæodéndron schinoìdes.* S.S.							
CURTI'SIA. DC.		CURTI'SIA. Tetrandria Monogynia.					
fagínea. DC.	(*wh.*)	Hassagay-tree.	C. B. S.	1775.	....	G. ♄.	Burm. afr. 235. t. 82.
MYGI'NDA. DC.		MYGINDA. Tetrandria Tetragynia.					
1 Uragòga. DC.	(*pu.*)	saw-leaved.	S. America.	1790.	8. 9.	S. ♄.	Jacq. amer. 24. t. 16.
2 myrtifòlia. DC.	(*wh.*)	Myrtle-leaved.	N. America.	1818.	5. 8.	H. ♄.	
Ilex myrsinites. Ph.							
3 Rhácoma. DC.	(*re.*)	blunt-leaved.	Jamaica.	1798.	....	S. ♄.	Jacq. ic. 2. t. 311.
4 latifòlia. DC.	(*wh.*)	broad-leaved.	W. Indies.	1795.	4. 5.	S. ♄.	
ILEX. DC.		HOLLY. Tetrandria Tetragynia. L.					
1 Aquifòlium. DC.	(*wh.*)	common.	Britain.	....	4. 6.	H. ♄.	Eng. bot. 496.
β *heterophy'lla.*	(*wh.*)	*various-leaved.*	———	....	——	H. ♄.	
γ *crassifòlia.*	(*wh.*)	*thick-leaved.*	———	....	——	H. ♄.	
δ *feróx.*	(*wh.*)	*hedgehog.*	———	....	——	H. ♄.	
ε *echinàta.*	(*wh.*)	*striped hedgehog.*	———	....	——	H. ♄.	
ζ *flàva.*	(*wh.*)	*yellow-berried.*	———	....	——	H. ♄.	
η *senéscens.*	(*wh.*)	*spineless.*	———	....	——	H. ♄.	
ϑ *álbo-marginàta.*	(*w.*)	*white-edged.*	———	....	——	H. ♄.	
ι *áureo-marginàta.*	(*w.*)	*gold-edged.*	———	....	——	H. ♄.	
κ *pícta.*	(*wh.*)	*painted.*	———	....	——	H. ♄.	
2 recúrva. L.en.	(*wh.*)	recurved-spined		....	——	H. ♄.	
3 baleàrica. DC.	(*wh.*)	Minorca.	Minorca.	1815.	5. 6.	H. ♄.	
4 Peràdo. DC.	(*bh.*)	thick-leaved.	Madeira.	1760.	4. 5.	F. ♄.	Lodd. bot. cab. 549.
β *obtùsa.* DC.	(*bh.*)	*blunt-leaved.*	———	....	——	F. ♄.	Duham. ed. nov. 1. t. 2.
5 canariénsis. DC.	(*wh.*)	Canary.	Canaries.	1818.	5. 6.	F. ♄.	
6 opàca. DC.	(*wh.*)	Carolina.	Carolina.	1744.	——	F. ♄.	Wats. dend. brit. t. 3.
7 laxiflòra. DC.	(*wh.*)	loose-flowered.	———	1811.	——	F ♄.	
8 chinénsis. DC.	(*wh.*)	Chinese.	China.	1810.	4. 5.	F. ♄.	Bot. mag. 2043.
9 Dahoòn. DC.	(*wh.*)	Dahoon.	Carolina.	1726.	5. 6.	F. ♄.	Willd. hort. ber. 1. t. 31.
β *laurifòlia.* DC.	(*wh.*)	*large-leaved.*	———	....	——	F. ♄.	Wats. dend. brit. t. 114.
10 Cassíne. DC.	(*wh.*)	Cassine-like.	———	1726.	8.	F. ♄.	Catesb. car. 1. t. 31.
11 angustifòlia. DC.	(*wh.*)	narrow-leaved.	N. America.	1806.	5. 6.	F. ♄.	Wats. dend. brit. t. 4.
12 vomitòria. DC.	(*wh.*)	South-sea Tea.	Florida.	1700.	6. 7.	F. ♄.	Jacq. ic. 2. t. 310.
13 latifòlia. DC.	(*wh.*)	broad-leaved.	Japan.	1824.	....	F. ♄.	
14 serràta. DC.	(*wh.*)	saw-leaved.	———	——	....	F. ♄.	
15 myrtifòlia. DC.	(*wh.*)	Myrtle-leaved.	W. Indies.	1806.	7. 8.	S. ♄.	
16 salicifòlia. DC.	(*wh.*)	Willow-leaved.	Mauritius.	1818.	4. 5.	S. ♄.	Jacq. col. 5. t. 2. f. 2.
Burglària lùcida. Wendl.							
17 Macoucóua. DC.	(*wh.*)	acuminate.	Trinidad.	1826.	....	S. ♄.	Aubl. gui. 1. t. 34.
18 paraguariénsis. DC.	(*w.*)	Paraguay Tea.	Brazil.	1823.	....	S. ♄.	Lamb. pin. ic.
PR'INOS. DC.		WINTER BERRY. Hexandria Monogynia. L.					
1 decíduus. DC.	(*wh.*)	deciduous.	N. America.	1760.	6. 7.	H. ♄.	Wats. dend. brit. 115.
Ilex prinoìdes. R.S.							
2 ambíguus. DC.	(*wh.*)	Carolina.	———	1812.	7. 8.	H. ♄.	Wats. dend. t. 29.
3 verticillàtus. DC.	(*wh.*)	whorled.	———	1736.	——	H. ♄.	——— t. 30.
padifòlius. W. en.							
4 lævigàtus. DC.	(*wh.*)	smooth.	———	1812.	——	H. ♄.	Wats. dend. t. 28.
5 lanceolàtus. DC.	(*wh.*)	scarlet berried.	———	1811.	6. 7.	H. ♄.	
6 prunifòlius. DF.	(*wh.*)	Plum-leaved.	———	1820.	——	H. ♄.	
7 gláber. DC.	(*wh.*)	ever-green.	———	1759.	7. 8.	H. ♄.	Wats. dend. brit. t. 27.
8 atomàrius. DC.	(*wh.*)	atomiferous.	Georgia.	1820.	6. 7.	H. ♄.	
9 coriàceus. DC.	(*wh.*)	leathery-leaved.	———	——	——	H. ♄.	
10 lùcidus. DC.	(*wh.*)	shining-leaved.		1778.	——	F. ♄.	
NEMOPA'NTHES. DC.		NEMOPA'NTHES. Polygamia Diœcia.					
canadénsis. DC.	(*wh.*)	Canadian.	N. America.	1812.	6. 7.	H. ♄.	DC. pl. rar. gen. t. 3.
Ilex canadensis. Mich. fl. amer. 2. t. 49.							

ORDO LXI.

RHAMNEÆ. *DC. prodr.* 2. *p.* 19.

ZI'ZYPHUS. R.S.		ZI'ZYPHUS. Pentandria Monogynia. R.S.					
1 Lòtus. R.S.	(*st.*)	Barbary.	Africa.	1731.	....	G. ♄.	Desf. act. ac. 1788. t. 21.
2 mucronàta. W.en.	(*st.*)	mucronate.	C. B. S.	1820.	....	G. ♄.	
bubalìna. R.S.							
3 sinénsis. R.S.	(*wh.*)	Chinese.	China.	1818.	5. 6.	H. ♄.	
4 Oenóplia. R.S.	(*st.*)	oblique-leaved.	Ceylon.	....	....	S. ♄.	Burm. zeyl. 131. t. 61.
5 Jujùba. R.S.	(*st.*)	blunt-leaved.	E. Indies.	1759.	4. 5.	S. ♄.	Rheed. mal. 4. t. 41.
6 soròria. R.S.	(*ye.*)	sister.	———	1823.	....	S. ♄.	
7 álbens. F.I.	(*wh.*)	white-fruited.	China.	1822.	....	G. ♄.	
8 Spìna Chrísti. R.S.	(*st.*)	Christ's-thorn.	Egypt.	....	4. 5.	G. ♄.	Pluk. alm. t. 197. f. 3.
9 Napèca. R.S.	(*wh.*)	spotted-fruited.	Ceylon.	1816.	....	S. ♄.	——— t. 216. f. 6.
10 vulgàris. R.S.	(*ye.*)	common.	S. Europe.	1640.	8. 9.	H. ♄.	Pall. ross. 2. t. 59.
11 nítida. F.I.	(*ye.*)	shining.	China.	1822.	....	G. ♄.	
12 Xylop'yrus. R.S.	(*ye.*)	cork-barked.	E. Indies.	———	....	S. ♄.	
13 Caracútta. F.I.	(*ye.*)	villous-branch'd.	———	1820.	....	S. ♄.	
14 microphy'lla. F.I.	(*ye.*)	small-leaved.	———	1824.	....	S. ♄.	Pluk. alm. t. 197. f. 2.
15 incúrva. F.I.	(*gr.*)	incurved.	Nepaul.	1820.	5. 6.	F. ♄.	
16 flexuòsa. F.I.	(*ye.*)	flexuose.	———	———	....	F. ♄.	
PALI'URUS. DC.		CHRIST'S-THORN. Pentandria Trigynia.					
1 aculeàtus. DC.	(*ye.*)	common.	S. Europe.	1596.	6. 7.	H. ♄.	Bot. mag. 1893.
austràlis. R.S. *vulgàris.* D.P. *Zízyphus Paliùrus.* B.M.							
2 virgàtus. DC.	(*ye.*)	twiggy.	Nepaul.	1819.	8. 9.	H. ♄.	Bot. mag. t. 2535.
COND'ALIA.		COND'ALIA. Pentandria Monogynia.					
microphy'lla. DC.	(*gr.*)	small-leaved.	Chili.	1824.	....	G. ♄.	Cav. ic. 6. t. 525.
BERCH'EMIA. DC.		BERCH'EMIA. Pentandria Monogynia.					
1 volùbilis. DC.	(*gr.*)	climbing.	Carolina.	1714.	6. 7.	H. ♄.⌣.	Jacq. ic. 2. t. 336.
Rhámnus volùbilis. L. *Zízyphus volùbilis.* W. *Oenoplìa volubìlis.* R.S.							
2 lineàta. DC.	(*wh.*)	lined.	China.	1804.	....	G. ♄.	Osb. it. 219. t. 7.
3 floribúnda. Bro.	(*wh.*)	many-flowered.	Nepaul.	1827.	....	G. ♄.⌣.	Brong. m. ram. pl. 2. f. 1.
SAGER'ETIA. Bro.		SAGER'ETIA. Pentandria Monogynia.					
1 theèzans. Bro.	(*gr.*)	Tea-like.	China.	....	9. 12.	G. ♄.	
Rhámnus theèzans. L.							
2 hamòsa. Bro.	(*gr.*)	hook-thorned.	Nepaul.	1826.	....	G. ♄.⌣.	
Zízyphus hamósa. F.I.							
RHA'MNUS. DC.		BUCKTHORN. Pentandria Monogynia. L.					
1 Alatérnus. DC.	(*ye.*)	Alaternus.	S. Europe.	1629.	4. 6.	H. ♄.	Duham. ed. nov. 3. t. 14.
α latifòlius.	(*ye.*)	*broad-leaved.*	———	....	———	H. ♄.	
β baleàricus. DC.	(*ye.*)	*round-leaved.*	———	....	———	H. ♄.	
γ hispánicus. DC.	(*ye.*)	*Spanish.*	———	....	———	H. ♄.	
δ vulgàris. DC.	(*ye.*)	*oval saw-leaved.*	———	....	———	H. ♄.	
ε glàber.	(*ye.*)	*smooth-leaved.*	———	....	———	H. ♄.	
ζ integrifòlius.	(*ye.*)	*entire-leaved.*	———	....	———	H. ♄.	
η maculàtus.	(*ye.*)	*spot-leaved.*	———	....	———	H. ♄.	
ϑ áureus.	(*ye.*)	*gold-striped.*	———	....	———	H. ♄.	
ι argénteus.	(*ye.*)	*silver-striped.*	———	....	———	H. ♄.	
2 Clùsii. W.en.	(*gr.*)	narrow-leaved.	———	1629.	———	H. ♄.	Clus. hist. 1. p. 50.
3 hy'bridus. DC.	(*gr.*)	hybrid.	Hybrid.	....	5. 6.	H. ♄.	L'Her. sert. t. 5.
4 glandulòsa. DC.	(*gr.*)	Madeira.	Canaries.	1785.	6. 7.	G. ♄.	Vent. malm. t. 34.
5 integrifòlius. DC.	(*gr.*)	Teneriffe.	———	1820.	———	G. ♄.	Nees hor. phys. t. 22.
coriàceus. Nees.							
6 longifòlius. DC.	(*gr.*)	long-leaved.	N. America?	1822.	5. 6.	H. ♄.	
7 prinoídes. DC.	(*gr.*)	Prinos-leaved.	C. B. S.	1778.	8. 9.	G. ♄.	L'Her. sert. ang. t. 9.
8 celtifòlius. DC.	(*gr.*)	Celtis-leaved.	———	....	———	G. ♄.	Burm. afr. 242. t. 88.
9 cathárticus. DC.	(*gr.*)	purging.	England.	....	5. 6.	H. ♄.	Eng. bot. 1629.
10 virgàtus. DC.	(*ye.*)	slender.	Nepaul.	1819.	4. 5.	H. ♄.	
11 tinctòrius. DC.	(*st.*)	dying.	Hungary.	1823.	5. 6.	H. ♄.	Pl. rar. hung. 3. t. 255.
12 infectòrius. DC.	(*ye.*)	yellow-berried.	S. Europe.	1683.	6. 7.	H. ♄.	Clus. hist. 1. p. 111. ic.
13 saxátilis. DC.	(*gr.*)	rock.	———	1752.	5. 6.	H. ♄.	Jacq. aust. 1. t. 53.
14 oleoídes. DC.	(*gr.*)	Olive-leaved.	Spain.	———	6. 7.	H. ♄.	
15 buxifòlius. DC.	(*gr.*)	Box-leaved.	S. Europe.	1820.	———	H. ♄.	
16 pubéscens. DC.	(*gr.*)	pubescent.	S. France.	1820.	6. 7.	H. ♄.	
17 crenulàtus. DC.	(*gr.*)	crenulate.	Teneriffe.	1778.	3. 4.	G. ♄.	

18 lycioídes. DC.	(*gr.*)	Box-thorn-like.	Spain.	1752.	9. 12.	H. ♄.	Cavan. ic. 2. t. 182.
β latifòlius.	(*gr.*)	*broader-leaved.*	———	——	——	H. ♄.	
19 erythróxylon. DC.	(*ye.*)	red-wooded.	Siberia.	1822.	4. 5.	H. ♄.	Pall. ross. 2. t. 62.
20 microphy'llus. DC.	(*ye.*)	small-leaved.	Mexico.	——	——	G. ♄.	H.etB.n.gen.a.7.t.616.
21 pùmilus. L.	(*bh.*)	dwarf.	Europe.	1752.	7.	H. ♄.	Scop. carn. ed. 2. t. 5.
β rupéstris. Vill.	(*bh.*)	*small.*	Dauphiny.	——	——	H. ♄.	
22 pusíllus. T.P.	(*gr.*)	little.	Naples.	1825.	——	H. ♄.	
23 valentìnus. W.	(*ye.*)	Spanish.	Spain.	——	6. 7.	H. ♄.	Cavan. ic. 2. t. 181.
pùmilus. C. *non* W.							
24 Wulfènii. S.S.	(*ye.*)	Wulfen's.	Tergestum.	——	——	H. ♄.	Jacq. coll. 2. t. 11.
pùmilus. Wulf. *non* L.							
25 dahùricus. DC.	(*gr.*)	Dahurian.	Dahuria.	1826.	....	H. ♄.	Pall. ross. 2. t. 61.
26 alnifòlius. DC.	(*gr.*)	Alder-leaved.	N. America.	1778.	5. 6.	H. ♄.	
27 franguloídes. R.S.	(*gr.*)	Frangula-like.	———	——	——	H. ♄.	
28 Purshiànus. DC.	(*gr.*)	Pursh's.	———	1826.	——	H. ♄.	
alnifòlius. Ph. *nec aliorum.*							
29 alpìnus. DC.	(*ye.pu.*)	Alpine.	Switzerland.	1752.	——	H. ♄.	Lodd. bot. cab. 1077.
30 caroliniànus. DC.	(*gr.*)	Carolina.	N. America.	1824.	5. 7.	H. ♄.	
31 Frángula. DC.	(*gr.*)	berry-bearing Alder.	Britain.	....	4. 5.	H. ♄.	Eng. bot. 250.
32 latifòlius. DC.	(*ye.*)	broad-leaved.	Azores.	1778.	6. 7.	H. ♄.	Wats. dend. brit. t. 11.
33 lanceolàtus. Ph.	(*gr.*)	spear-leaved.	N. America.	1812.	4. 5.	H. ♄.	
34? mystacìnus. R.S.	(*gr.*)	wiry.	Abyssinia.	1775.	11.	S. ♄.	
35? tetragonus. DC.	(*gr.*)	square-branch'd.	C. B. S.	1816.	6. 8.	G. ♄.	Th. act. gor. 1812. c. ic.
SC'UTIA. Bro.		SC'UTIA. Pentandria Monogynia.					
1 índica. Bro.	(*wh.*)	Indian.	E. Indies.	1824.	....	S. ♄.	Gært. fruct. 2. t. 106.
Rhámnus circumscíssus. L. *Ceanòthus circumscíssus.* DC.							
RETANI'LLA. Bro.		RETANI'LLA. Pentandria Monogynia.					
obcordàta. Bro.	(*ye.*)	obcordate-ld.	Peru.	1822.	5. 7.	G. ♄.	Vent. hort. cels. t. 92.
Collétia obcordàta. V.							
COLLE'TIA. Bro.		COLLE'TIA. Pentandria Monogynia.					
1 spinòsa. DC.	(*ye.*)	spiny.	Peru.	1823.	....	G. ♄.	Lam. ill. 2. t. 129.
2 serratifòlia. DC.	(*ye.*)	saw-leaved.	———	——	5. 8.	G. ♄.	Vent. choix. t. 15.
3 pubéscens. Bro.	(*ye.*)	pubescent.	N. Holland.	1824.	....	G. ♄.	
HOV'ENIA. DC.		HOV'ENIA. Pentandria Monogynia.					
1 dúlcis. DC.	(*gr.*)	smooth-leaved.	China.	1812.	4. 6.	G. ♄.	Lam. ill. t. 131.
acérba. Bot. reg. 501.							
2 inæquàlis. DC.	(*gr.*)	pubescent-ld.	Nepaul.	1823.	....	G. ♄.	
dulcìs. D.P. *et* F.I. *nec aliòrum.*							
COLUBR'INA. Bro.		COLUBR'INA. Pentandria Monogynia.					
1 ferruginòsa. Bro.	(*st.*)	Bahama red-wood.	Bahama.	1726.	6.	S. ♄.	Jacq. vind. 3. t. 50.
Rhámnus colubrìnus. L. *Ceanòthus colubrìnus.* DC.							
2 reclinàta. Bro.	(*wh.*)	oval-leaved.	Jamaica.	1758.	8.	S. ♄.	Brown jam. t. 20. f. 2.
Ceanòthus reclinàtus. DC. *Rhámnus ellípticus.* H.K.							
3 triflòra. Bro.	(*st.*)	three-flowered.	Mexico.	1826.	....	G. ♄.	
4 cubénsis. Bro.	(*st.*)	Cuba.	Cuba.	1824.	....	S. ♄.	Jacq. vind. 3. t. 49.
5 asiática. Bro.	(*st.*)	Asiatic.	Asia.	1691.	7. 8.	S. ♄.	Cavan. ic. 5. t. 440. f. 1.
Ceanòthus asiáticus. DC.							
CEAN'OTHUS. Bro.		CEAN'OTHUS. Pentandria Monogynia. L.					
1 azùreus. DC.	(*bl.*)	blue.	New Spain.	1818.	4. 5.	G. ♄.	Bot. reg. 291.
cærùleus. Lod. bot. cab. 110. *bicolor.* R.S.							
2 tardiflòrus. DC.	(*wh.*)	late-flowering.	N. America.	1812.	9. 10.	H. ♄.	
3 americànus. DC.	(*wh.*)	New Jersey Tea.	———	1713.	7. 8.	H. ♄.	Bot. mag. 1479.
4 ovàtus. DC.	(*wh.*)	smooth-leaved.	———	....	——	H. ♄.	
5 perénnis. DC.	(*wh.*)	perennial.	———	1822.	6. 7.	H. ♃.	
6 intermèdius. DC.	(*wh.*)	intermediate.	———	1812.	——	H. ♄.	
7 collìnus. D.	(*wh.*)	hill.	———	1827.	——	H. ♄.	
8 sanguíneus. DC.	(*wh.*)	bloody twigged.	———	1812.	5. 7.	H. ♄.	
9 microphy'llus. DC.	(*wh.*)	small-leaved.	———	1806.	6. 7.	H. ♄.	
10 buxifòlius. DC.	(*wh.*)	Box-leaved.	Mexico.	1824.	4. 6.	G. ♄.	H.etB.n.gen.a.7.t.615.
11 macrocárpus. DC.	(*st.*)	large-fruited.	New Spain.	1823.	6. 7.	G. ♄.	Cavan. ic. 3. t. 276.
12 nepalénsis. F.I.	(*st.*)	Nepaul.	Nepaul.	——	——	H. ♄.	
WILLEM'ETIA. Bro.		WILLEM'ETIA. Pentandria Monogynia.					
africàna. Bro.	(*st.*)	African.	C. B. S.	1712.	3. 4.	G. ♄.	Pluk. phyt. 126. f. 1.
Ceanòthus africànus. DC.							
POMADE'RRIS. DC.		POMADE'RRIS. Pentandria Monogynia.					
1 globulòsa. Bro.	(*st.*)	round-headed.	N. Holland.	1803.	4. 5.	G. ♄.	Lab. nov. hol. 1. t. 85.
Ceanòthus globulòsus. Lab.							
2 spathulàta. Bro.	(*st.*)	spathulate-leav'd.	———	1824.	——	G. ♄.	Lab. nov. 1. t. 84.
3 díscolor. DC.	(*ye.*)	two-coloured.	———	1803.	——	G. ♄.	Swt. fl. aust. t. 41.
Ceanòthus díscolor. Vent. malm. t. 58.							

4 intermèdia. DC. (*st.*) intermediate. N. S. W. 1818. 4. 6. G. ♄.
5 acuminàta. L.en. (*wh.*) loose-flowered. N. Holland. 1816. 4. 7. G. ♄.
6 ellíptica. DC. (*ye.*) elliptic-leaved. ——— 1805. —— G. ♄. Lab. nov. holl. 1. t. 86.
7 lanígera. (*ye.*) ferruginous. ——— 1806. —— G. ♄. Andr. rep. 569.
Ceanòthus laníger. A.R.
8 Símsii. (*st.*) Sims's. N. S. W. 1821. —— G. ♄. Bot. mag. 1823.
lanígera. Sims bot. mag. *non* Andr.
9 viridirùfa. Sieb. (*st.*) brown green. N. S. W. —— —— G. ♄.
10 ledifòlia. F.T. (*ye.*) Ledum-leaved. ——— 1824. —— G. ♄.
11 andromedæfòlia. F.T. (*ye.*) Andromeda-l^d ——— —— G. ♄.
12 phylliræoídes. DC. (*ye.*) Phillyrea-leaved. ——— 1803. —— G. ♄.
13 apétala. DC. (*br.*) petalless. N. Holland. —— 5. 6. G. ♄. Lab. nov. holl. 1. t. 87.
14 áspera. DC. (*br.*) rough. N. S. W. 1818. 6. 7. G. ♄.
15 ligustrìna. DC. (*wh.*) Privet-like. ——— 1826. —— G. ♄.
16 phylicifòlia. DC. (*wh.*) Phylica-leaved. N. Holland. 1810. 3. 5. G. ♄. Lodd. bot. cab. 120.

CRYPTA'NDRA. DC. CRYPTA'NDRA. Pentandria Monogynia. R.S.
1 ericifòlia. DC. (*pa.*) Heath-leaved. N. S. W. 1823. 5. 7. G. ♄. Linn. trans. 10 t. 18. f. 1.
2 amàra. DC. (*pa.*) bitter. ——— —— 4. 6. G. ♄. ——— 10. t. 18. f. 2.
3 spinéscens. DC. (*pa.*) spiny. ——— 1824. —— G. ♄.

VENTIL'AGO. DC. VENTIL'AGO. Pentandria Monogynia. R.S.
maderaspatàna. DC. (*gr.*) Madras. E. Indies. 1822. .... S. ♄.⌣. Roxb. corom. 1. t. 76.

GOUA'NIA. DC. GOUA'NIA. Polygamia Monœcia. L.
1 integrifòlia. DC. .... entire-leaved. S. America? 1824. .... S. ♄.⌣.
2 domingénsis. DC. (*gr.*) Chaw-stick. W. Indies. 1739. .... S. ♄.⌣. Jacq. amer. t. 179. f. 40.
3 cordifòlia. DC. (*ye.*) heart-leaved. Brazil. 1822. .... S. ♄.⌣.
4 tomentòsa. DC. (*ye.*) woolly-leaved. W. Indies. 1823. .... S. ♄.⌣. Gært. f. carp. 3. t. 183. f. 1
5 mauritiàna. DC. (*br.*) rusty. Mauritius. —— .... S. ♄.⌣.
6 tiliæfòlia. DC. (*gr.*) Lime-leaved. Bourbon. 1810. 4. 6. S. ♄.⌣.
7 leptostàchya. DC. (*gr.*) slender-spiked. E. Indies. —— —— S. ♄.⌣. Roxb. corom. 1. t. 98.
tiliæfòlia. R.C. *non* Lam.
8 Retinària. DC. (*gr.*) oval-leaved. Mauritius. 1824. .... S. ♄.⌣. Gært. fruct. 2. t. 120. f. 4.

TRICHOCE'PHALUS. Bro. TRICHOCE'PHALUS. Pentandria Monogynia.
1 stipulàris. Bro. (*wh.*) stipuled. C. B. S. 1786. 5. 9. G. ♄. Wendl. coll. t. 32.
Phy'lica stipulàris. DC.
2 spicàta. Bro. (*wh.*) spiked. ——— 1774. 11. 12. G. ♄. Lam. ill. t. 127. f. 3.
Phy'lica spicàta. L. *non* B.M.

PHY'LICA. Bro. PHY'LICA. Pentandria Monogynia. L.
1 parviflòra. R.S. (*wh.*) small-flower'd. C. B. S. 1790. 4. 7. G. ♄.
2 ericoídes. R.S. (*wh.*) Heath-leaved. ——— —— 9. 4. G. ♄. Bot. mag. 224.
3 glábrata. R.S. (*wh.*) smooth-leaved. ——— 1820. 3. 5. G. ♄.
4 aceròsa. R.S. (*wh.*) needle-leaved. ——— 1819. 4. 7. G. ♄. Spr. ber. mag. 8. t. 8 f. 2.
5 lanceolàta. R.S. (*wh.*) spear-leaved. ——— 1790. 4. 5. G. ♄.
6 erióphora. R.S. (*wh.*) pale-flowered. ——— 1774. 11. 3. G. ♄.
7 callòsa. R.S. (*wh.*) callous-leaved. ——— —— 3. 4. G. ♄.
8 imbérbis. R.S. (*wh.*) beardless. ——— 1823. 5. 9. G. ♄. Breyn. cent. 18. t. 7.
9 secúnda. R.S. (*wh.*) secund. ——— 1812. —— G. ♄.
10 ? australis. L.en. (*wh.*) southern. ——— —— 3. 5. G. ♄.
11 pùmila. DC. (*wh.*) dwarf. ——— 1820. 4. 6. G. ♄.
12 bícolor. R.S. (*wh.*) two-coloured. ——— 1823. 4. 5. G. ♄.
13 spicàta. B.M. (*wh.*) close-spiked. ——— —— 8. 12. G. ♄. Bot. mag. 2704.
14 hirsùta. R.S. (*wh.*) hairy. ——— 1820. 5. 9. G. ♄.
15 pìnea. R.S. (*wh.*) Pine-like. ——— 1824. —— G. ♄.
16 villòsa. R.S. (*wh.*) villous. ——— 1790. —— G. ♄.
17 papillòsa. R.S. (*wh.*) papillose. ——— 1820. 3. 5. G. ♄.
18 rosmarinifòlia. R.S. (*w.*) Rosemary-l^d. ——— 1815. 5. 8. G. ♄. Lodd. bot. cab 849.
19 cylíndrica. R.S. (*wh.*) cylindrical. ——— .... 4. 8. G. ♄. Wendl. coll. 1. t. 7.
20 excélsa. R.S. (*wh.*) tall. ——— 1820. 3. 5. G. ♄. ——— 3. t. 4.
21 horizontàlis. DC. (*wh.*) spreading. ——— 1800. 6. 8. G. ♄. Spr. ber. mag. 8. f. 7.
plumòsa. Spr. *non* L.
22 squarròsa. DC. (*wh.*) squarrose. ——— 1820. 3. 5. G. ♄. Lodd. bot. cab. 36.
23 Commelìni. R.S. (*wh.*) Commelin's. ——— 1800. 5. 8. G. ♄. Comm. præl. 63. t. 13.
24 plumòsa. R.S. (*wh.*) feathered. ——— 1752. 3. 5. G. ♄. Lam. ill. t. 127. f. 4.
25 capitàta. R.S. (*wh.*) headed. ——— 1774. —— G. ♄. Bot. reg. 711.
26 reclinàta. R.S. (*wh.*) reclined. ——— 1820. —— G. ♄. Wendl. coll. 2. t. 56.
27 pedicellàta. DC. (*wh.*) pedicled. ——— —— —— G. ♄.
28 atràta. R.S. (*wh.*) dark-haired. ——— —— 3. 5. G. ♄.

SOULA'NGIA. Bro. SOULA'NGIA. Pentandria Monogynia.
1 axillàris. Bro. (*wh.*) axillary-flower'd. C. B. S. 1812. 5. 6. G. ♄. Spr. ber. mag. 8. t. 8. f. 4.
2 thymifòlia. Bro. (*wh.*) Thyme-leaved. Antarctic. 1824. —— S. ♄. Vent. malm. t. 57.

3 oleæfòlia. Bro. (*wh.*) Olive-leaved. C. B. S. .... 1. 4. G. ♄. Lodd. bot. cab. t. 323.
Phy'lica spicàta. Lod. bot. cab. *nec aliorum.*
4 paniculàta. Bro. (*wh.*) panicled. —— 1812. 4. 6. G. ♄.
Phy'lica paniculàta. DC. *myrtifòlia.* Poir.
5 buxifòlia. Bro. (*wh.*) Box-leaved. —— 1759. 5. 9. G. ♄. Lodd. bot. cab. 848.
Phy'lica buxifòlia. Wend. coll. 1. t. 26.
6 cordàta. Bro. (*wh.*) heart-leaved. —— 1789. 5. 6. G. ♄. Comm. præl. 62. t. 12.
7 orientàlis. (*wh.*) oriental. —— 1820. —— G. ♄. Herb. amat. t. 283.
Phylica orientàlis. Herb. amat. t. 283.

---

† *Genera ad Rhamneas accedentia.*

SCHÆFF'ERIA. DC. SCHÆFF'ERIA. Diœcia Tetrandria.
frutéscens. DC. (*wh.*) white-flowered. W. Indies. 1793. .... S. ♄. Swartz.fl.ind.1.t.7.f.A.
β *buxifòlia.* DC. (*wh.*) *Box-leaved.* —— .... S. ♄. Sloan.hist.2.t.209.f.1.
OLI'NIA. DC. OLI'NIA. Pentandria Monogynia.
cymòsa. DC. (*wh.*) sweet-scented. C. B. S. 1812. 4. 6. G. ♄.
Sideróxylum cymòsum. L.

ORDO LXII.

BRUNIACEÆ. *DC. prodr.* 2. *p.* 43.

BERZ'ELIA. Bro. BERZ'ELIA. Pentandria Monogynia.
1 abrotanoídes.Bro.(*w.*) Southernwood-like. C. B. S. 1787. 5. 7. G. ♄. Wendl. coll. t. 45.
Brùnia abrotanoídes. Lodd. bot. cab. 355.
2 lanuginòsa. Bro. (*wh.*) woolly. —— 1774. 6. 8. G. ♄. Wendl. coll. 1. t. 11.
Brùnia lanuginòsa. DC.
BR'UNIA. R.S. BR'UNIA. Pentandria Monogynia. L.
1 l'ævis. R.S. (*wh.*) smooth. C. B. S. 1823. 5. 7. G. ♄.
2 nodiflòra. R.S. (*wh.*) imbricated. —— 1786. 7. 8. G. ♄. Breyn. cent. 22. t. 10.
3 deùsta. R.S. (*wh.*) black-tipped. —— 1804. 6. 8. G. ♄.
4 comòsa. R.S. (*wh.*) tufted. —— 1823. —— G. ♄.
5 verticillata. R.S. (*wh.*) whorled. —— 1794. —— G. ♄.
6 virgàta. Bro. (*wh.*) twiggy. —— .... —— G. ♄.
7 squarròsa. R.S. (*wh.*) squarrose. —— 1818. —— G. ♄.
8 alopecuroídes.R.S.(*w.*) Fox-tail. —— 1816. 5. 7. G. ♄.
9 láxa. R.S. (*wh.*) loose-branched. —— —— —— G. ♄.
10 plumòsa. R.S. (*wh.*) feathered. —— 1822. —— G. ♄.
11 supérba. R.S. (*wh.*) superb. —— 1791. —— G. ♄.
12 fragarioídes. R.S.(*wh.*) Strawberry-like. —— 1794. 5. 7. G. ♄.
13 ciliàta. R.S. (*wh.*) fringed. —— 1812. —— G. ♄.
14 ericoídes. R.S. (*wh.*) Heath-leaved. —— —— —— G. ♄. Wendl. coll. II. t. 57.
15 arachnoídea. R.S.(*wh.*) cobweb. —— 1810. —— G. ♄. —— t. 62.
16 formòsa. R.S. (*wh.*) handsome. —— 1812. —— G. ♄.
17 macrocéphala. S.S.(*w.*) large-headed. —— —— —— G. ♄.
18 racemòsa. Bro. (*wh.*) clustered. —— 1790. 5. 9. G. ♄.
Phy'lica racemòsa. L.
19 pinifòlia. Bro. (*wh.*) Pine-leaved. —— 1789. 5. 6. G. ♄. Brong. brun. t. 1. f. 2.
Phy'lica pinifòlia. L.
20 imbricàta. (*wh.*) imbricated. —— 1801. 8. 11. G. ♄.
Phy'lica imbricàta. Th.
RASP'ALIA. Bro. RASP'ALIA. Pentandria Digynia.
microphy'lla. Bro.(*w.*) small-leaved. C. B. S. 1804. 6. 8. G. ♄. Brong. brun. t. 3. f. 1.
Brùnia microphy'lla. Th.
STA'AVIA. DC. STA'AVIA. Pentandria Monogynia.
1 radiàta. DC. (*wh.*) rayed. C. B. S. 1787. 5. 6. G. ♄. Brong. brun. t. 2. f. 2.
2 glutinòsa. DC. (*wh.*) clammy. —— 1793. 4. 5. G. ♄. Lodd. bot. cab. 852.

3 nùda. Bro. (*wh.*) naked. C. B. S. 1821. 4. 5. G. ♄.
4 ciliàta. Bro. (*wh.*) fringed. ——— 1812. 5. 7. G. ♄.
NEB'ELIA. Neck. Neb'elia. Pentandria Digynia.
1 paleàcea. (*wh.*) chaffy. C B. S. 1791. 6. 8. G. ♄. Wendl. coll. t. 21.
Berárdia. Bro. *non* P.S. *Brùnia paleàcea.* R.S.
2 affínis. (*wh.*) related. C. B. S. 1816. —— G. ♄.
Berárdia affínis. Bro.
3 phylicoídes. (*wh.*) Phylica-leaved. ——— 1822. —— G. ♄.
Berárdia phylicoídes. Bro. *Brùnia phylicoides.* Th.
LINC'ONIA. L. Linc'onia. Pentandria Digynia.
1 alopecuroídea. L. (*wh.*) Fox-tail. C. B. S. 1824. 6. 8. G. ♄. Swa. in. ber. m. 1810. t. 4.
2 thymifòlia. DC. (*wh.*) Thyme-leaved. ——— 1825. —— G. ♄. ——— p. 86. t. 4.
3 cuspidàta. DC. (*wh.*) cuspidate. ——— —— —— G. ♄. ——1811. p. 284. t. 7. f 1.
AUDOUI'NIA. Bro. Audoui'nia. Pentandria Monogynia.
capitàta. Bro. (*wh.*) headed. C. B. S. 1790. 4. 6. G. ♄.
Diósma capitàta. DC.
TITTMA'NNIA. Bro. Tittma'nnia. Pentandria Monogynia.
lateriflòra. Bro. (*wh.*) side-flowering. C. B. S 1812. 4. 6. G. ♄. Brong. brun. t. 4. f. 2.
THA'MNEA. Bro. Tha'mnea. Pentandria Monogynia.
uniflòra. Bro. (*wh.*) one-flowered. C. B. S. 1810. 4. 6. G. ♄. Brong. brun. t. 4. f. 3.

ORDO LXIII.

SAMYDEÆ. *DC. prodr.* 2. *p.* 47.

SAM'YDA. DC. Sam'yda. Decandria Monogynia. S.S.
1 glábrata. DC. (*wh.*) smooth. Jamaica. 1818. 5. 8. S. ♄.
2 villòsa. DC. (*wh.*) villous. ——— —— —— S. ♄.
3 decúrrens. S.S. (*wh.*) decurrent. Brazil. 1822. —— S. ♄.
4 pubéscens. S.S. (*wh.*) pubscent. W. Indies. 1793. —— S. ♄.
5 serrulàta. DC. (*wh.*) saw-leaved. ——— 1723. 7. 8. S. ♄. Jacq. col. 2. t. 17. f. 1.
6 spinulòsa. DC. (*wh.*) spiny. St. Thomas. 1826. .... S. ♄. Vent. choix. t. 43.
7 macrocárpa. DC. (*wh.*) large-fruited. Mexico. —— .... S. ♄.
8 ròsea. B M. (*ro.*) rose-coloured. W. Indies. 1793. 6. 7. S. ♄. Bot. mag. 550.
serrulàta. A.R. 202. *nec aliorum.*
9 nítida. DC. (*wh.*) glossy-leaved. Jamaica. —— —— S. ♄. Br. jam. 217. t. 23. f. 3.
10 macrophy'lla. DC. (*gr.*) large-leaved. E. Indies. 1823. .... S. ♄.
CASE'ARIA. DC. Case'aria. Decandria Monogynia. S.S.
1 glomeràta. DC. (*wh.*) cluster-flower'd. E. Indies. 1824. 5. 7. S. ♄.
2 ovàta. DC. (*gr.*) oval-leaved. ——— —— —— S. ♄. Rheed. mal. 4. t. 49.
3 ramiflòra. DC. (*wh.*) branch-flower'd. Guiana. —— —— S. ♄. Aubl. guian. 1. t. 127.
4 nítida. DC. (*wh.*) shining. Carthagena. —— —— S. ♄. Act. helv. 8. p. 58. ic.
5 sylvéstris. DC. (*wh.*) wood. Jamaica. 1823. .... S. ♄.
6 parviflòra. DC. (*wh.*) small-flowered W. Indies. 1818. 5. 7. S. ♄. Sloan. hist. 2. t. 211. f. 2.
7 serrulàta. DC. (*wh.*) serrrulate. ——— —— —— S. ♄.
8 glábra. H.B. (*wh.*) smooth. E. Indies. 1820. —— S. ♄.
9 hirsùta. DC. (*wh.*) hairy. W. Indies. —— .... S. ♄.
10 zizyphoídes. K.S. (*wh.*) Zizyphus-like. Caracas. 1824. .... S. ♄.
11 argùta. K.S. .... sharp-toothed. Mexico. —— .... S. ♄.

ORDO LXIV.

HOMALINEÆ. *DC. prodr.* 2. *p.* 53.

HOM'ALIUM. DC. Hom'alium. Polyadelphia Polyandria.
racemòsum. DC. (*wh.*) racemed. W. Indies. 1818. 5. 7. S. ♄. Bot. reg. 519.
BLACKWE'LLIA. DC. Blackwe'llia. Dodecandria Pentagynia.
1 paniculàta. DC. (*wh.*) panicled. Bourbon. 1818. 5. 7. S. ♄.

2 gláuca. DC. (wh.) glaucous. Mauritius. 1823. .... S. ♄. Vent. choix. t. 55.
3 nepalensis. DC. (wh.) Nepaul. Nepaul. 1824. .... S. ♄.
4 axillàris. DC. (wh.) axillary-flower'd. Madagascar. 1823. .... S. ♄. Lam. ill. t. 412. f. 1.
5 spiràlis. DC. (wh.) spiral. Pegu. 1820. .... S. ♄.
6 fagìfolia. (wh.) Beech-leaved. China. 1824. 5. G. ♄.
ASTRA'NTHUS. DC. STAR-FLOWER. Heptandria Tetragynia.
cochinchinénsis. (wh.) Chinese. China. 1821. 6. 7. S. ♄. Bot. mag. 2659.
ARISTOT'ELIA. DC. ARISTOTE'LIA. Dodecandria Monogynia. L.
Máqui. DC. (wh.) shining-leaved. Chile. 1773. 4. 5. H. ♄. Wats. dend. brit. t. 44.

ORDO LXV.

CHAILLETIACEÆ. *DC. prodr.* 2. *p.* 57.

CHAILL'ETIA. DC. CHAILL'ETIA. Pentandria Di-Trigynia.
1 toxicària. D.D. (wh.) poisonous. Sierra Leone. 1822. .... S. ♄.
2 erécta. D.D. (wh.) upright. ——— ——— .... S. ♄.

ORDO LXVI.

AQUILARINEÆ. *DC. prodr.* 2. *p.* 59.

AQUIL'ARIA. DC. AQUIL'ARIA. Decandria Monogynia.
Agallòcha. H.B. .... Aloe-wood. E. Indies. 1824. .... S. ♄.

ORDO LXVII.

CASSUVIEÆ. *R. Brown cong. p.* 12.

ANACA'RDIUM. DC. CASHEW-NUT. Enneandria Monogynia. L.
1 occidentàle. DC. (y.gr.) common. India. 1599. .... S. ♄. Catesb. car. 3. t. 9.
Cassùvium pomíferum. Lam. ill. 3. t. 322.
2 dùbium. H.B. (gr.) doubtful. E. Indies. 1824. .... S. ♄.
SEMECA'RPUS. DC. SEMECA'RPUS. Pentandria Trigynia. L.
Anacárdium. DC. (y.gr.) common. E. Indies. 1820. .... S. ♄.
α angustifòlium. DC. (y.gr.) *long-leaved.* ——— ——— .... S. ♄. Rumph. amb. 1. t. 70.
β cuneifòlium. DC. (y.gr.) *wedge-leaved.* ——— ——— .... S. ♄.
γ obtusiúsculum. DC. (y.gr.) *broad-leaved.* ——— ——— .... S. ♄. Roxb. corom. 1. t. 12.
MANGI'FERA. DC. MANGO-TREE. Pentandria Monogynia.
1 índica. DC. (ye.) common. E. Indies. 1690. 6. 9. S. ♄. Andr. rep. 425.
2 oppositifòlia. F.I. (ye.) opposite-leaved. ——— 1823. .... S. ♄.
BUCHANA'NIA. R.C. BUCHANA'NIA. Decandria Monogynia.
1 latifòlia. DC. (wh.) broad-leaved. E. Indies. 1820. .... S. ♄.
2 angustifòlia. DC. (wh.) narrow-leaved. ——— ——— .... S. ♄. Roxb. corom. 3. t. 262.
Mangífera axillàris. Lam. *Cambessèdea.* K.S.
RHU'S. R.S. SUMA'CH. Pentandria Trigynia. L.
1 Coriària. R.S. (gr.) Elm-leaved. S. Europe. 1596. 7. H. ♄. Flor. græc. t. 290.
2 typhina. R.S. (pu.) Virginian. N. America. 1629. 7. 8. H. ♄. Wats. dend. t. 17. mas.
α arborescens. R.S. (pu.) *arborescent.* ——— ——— ——— H. ♄. ——— t. 18. fem.
β frutescens. R.S. (pu.) *frutescent.* ——— ——— ——— H. ♄.

3 javánica. R.S.	(*wh.*)	Java.	Java.	1799.	7. 9.	S. ♄.	
4 semialàta. R.S.	(*wh.*)	Service-leaved.	Macao.	1780.	——	G. ♄.	Murr. com. got. VI. t. 3.
5 viridiflòra. R.S.	(*gr.*)	green-flowered.	N. America.	....	7.	H. ♄.	
6 glábra. R.S.	(*gr.*)	smooth.	———	1726.	7. 9.	H. ♄.	Wats. dend. t. 15.
α hermaphrodíta. DC.	(*gr.*)	*hermaphrodite.*	———	——	——	H. ♄.	Dill. elt. t. 243.
β dioíca. DC.	(*gr.*)	*dioecious.*		——	——	H. ♄.	Lam. ill. t. 207. f. 1.
7 élegans. R.S.	(*sc.*)	scarlet-flowered.	———	——	——	H. ♄.	Wats. dend. t. 16.
8 Vérnix. R.S.	(*gr.*)	varnish.	———	1713.	——	H. ♄.	——— t. 19.
9 vernicífera. DC.	(*gr.*)	Chinese varnish.	China.	....	——	G. ♄.	Kæmp. amœn. t. 792.
Rhus Vernix. Th. *non* Ph.							
10 A'mela. D.P.	(*gr.*)	large-leaved.	Nepaul.	1823.	....	G. ♄.	
11 Wallíchii.	(*gr.*)	Wallich's.	———	——	....	G. ♄.	
juglandifòlia. D.P. *non* Kth.							
12 fraxinifòlia. D.P.	(*gr.*)	Ash-leaved.	———	——	——	G. ♄.	
13 acuminàta. DC.	(*gr.*)	pointed-leaved.	———	——	....	G. ♄.	
14 succedànea. R.S.	(*gr.*)	red Lac.	China.	1768.	6.	G. ♄.	Kæmp. amœn. t. 795.
15 juglandifòlia. K.S.	(*gr.*)	Walnut-leaved.	S. America.	1826.	....	G. ♄.	H.etB.n.ge.a.7.t.603-4.
16 copallìna. R.S.	(*gr.*)	Lentiscus-lea'd.	N. America.	1688.	8. 9.	H. ♄.	Jacq. schœn. 3. t. 341.
17 alàta. R.S.	(*gr.*)	winged-leaved.	C. B. S.	1818.	....	G. ♄.	
18 pauciflòra. R.S.	(*gr.*)	few-flowered.	———	——	....	G. ♄.	
19 pùmila. R.S.	(*gr.*)	dwarf poisonous.	N. America.	1806.	7.	H. ♄.	
20 digitàta. R.S.	(*gr.*)	fingered.	C. B. S.	1818.	....	G. ♄.◡.	
21 oblíqua. R.S.	(*gr.*)	oblique-leaved.	———	——	....	G. ♄.	
22 schinoídes. R.S.	(*gr.*)	Schinus-like.	Brazil.	1825.	....	S. ♄.	
23 pentaphy'lla. R.S.	(*ye.*)	various-leaved.	Barbary.	1816.	....	G. ♄.	Desf. atl. 1. t. 77.
24 chinénsis. Mill.	(*gr.*)	Chinese.	China.	1740.	....	G. ♄.	
25 rádicans. DC.	(*wh.*)	rooting poison-oak.	N. America.	1724.	6. 7.	H. ♄.◡.	Bot. mag. 1806.
26 Toxicodéndron. DC.	(*w.*)	poison-oak.	———	1640.	——	H. ♄.◡.	Schkuhr. hand. 1. t. 82.
27 cirrhiflòra. R.S.	(*gr.*)	tendril-flower'd.	C. B. S.	1825.	....	G. ♄.◡.	
28 Thunbergiàna. R.S.	(*gr.*)	Thunberg's.	———	1697.	7. 8.	G. ♄.	
29 gláuca. R.S.	(*gr.*)	glaucous.	———	1818.	——	G. ♄.	
30 tridentàta. R.S.	(*gr.*)	three-toothed.	———	——	....	G. ♄.◡.	
31 crenàta. R.S.	(*gr.*)	crenated.	———	....	6. 7.	G. ♄.	
32 dentàta. R.S.	(*gr.*)	rough-stalked.	———	1798.	——	G. ♄.	
33 sinuàta. R.S.	(*gr.*)	sinuated.	———	1818.	——	G. ♄.	
34 cuneifòlia. R.S.	(*gr.*)	wedge-leaved.	———	1816.	——	G. ♄.	
35 dimidiàta. R.S.	(*gr.*)	variable-leaved.	———	——	——	G. ♄.	
36 dissécta. R.S.	(*gr*)	dissected-leaved.	———	——	——	G. ♄.	
37 incísa. R.S.	(*gr.*)	cut-leaved.	———	1789.	——	G. ♄.	
38 tomentòsa. R.S.	(*gr.*)	woolly-leaved.	———	1691.	——	G. ♄.	Comm. hort. ams. 1. t. 92.
39 álbida. R.S.	(*gr.*)	white-barked.	Barbary.	1824.	....	F. ♄.	
40 villòsa. R.S.	(*gr.*)	villous-leaved.	C. B. S.	1714.	7.	G. ♄.	Pluk. alm. t. 219. f. 8.
41 spicàta. R.S.	(*gr.*)	spiked.	———	1816.	——	G. ♄.	
42 pubéscens. R.S.	(*gr.*)	pubescent.	———	——	——	G. ♄.	
43 atomària. DC.	(*gr.*)	velvetty.	———	1820.	6. 8.	G. ♄.	Jacq. schœnb. 3. t. 343.
44 parviflòra. DC.	(*gr.*)	small-flowered.	Nepaul.	1824.	....	G. ♄.	
45 mucronàta. R.S.	(*gr.*)	mucronate.	C. B. S.	1825.	....	G. ♄.	
46 viminàlis. R.S.	(*gr.*)	Willow-leaved.	———	1774.	6. 7.	G. ♄.	Jacq. schœn. 3. t. 344.
47 eròsa. R.S.	(*gr.*)	erose-leaved.	———	1816.	——	G. ♄.	
48 láncea. R.S.	(*gr.*)	lance-leaved.	———	——	——	G. ♄.	
49 angustifòlia. R.S.	(*wh.*)	narrow-leaved.	———	1714.	——	G. ♄.	Pluk. alm. t. 219. f. 6.
50 rosmarinifòlia. R.S.	(*gr.*)	Rosemary-leaved.	———	1800.	——	G. ♄.	Burm. afr. t. 91. f. 1.
51 undulàta. R.S.	(*gr.*)	wave-leaved.	———	1818.	——	G. ♄.	Jacq. schœn. 3. t. 346.
52 lævigàta. R.S.	(*gr.*)	polished-leaved.	———	1758.	——	G. ♄.	——— 3. t. 345.
elongàta. Jacq.							
53 ellíptica. R.S.	(*gr.*)	elliptic-leaved.	———	1816.	——	G. ♄.	
54 pendulìna. R.S.	(*gr.*)	pendulous.	———	——	——	G. ♄.	
55 lùcida. DC.	(*gr.*)	shining-leaved.	———	1697.	7. 8.	G. ♄.	Comm. hort. 1. t. 93.
56 Cavanillèsii. DC.	(*wh.*)	Cavanilles'.	Mexico.	....	——	G. ♄.	Cavan. ic. 2. t. 132.
lùcidum. C.I. *non* DC.							
57 Burmánni. DC.	(*gr.*)	Burman's.	C. B. S.	1697.	7. 8.	G. ♄.	Burm. afr. t. 91. f. 2.
lùcidum. α. H.K.							
58 concínna. B.T.	(*gr.*)	pretty.	———	1816.	——	G. ♄.	
59 nervòsa. R.S.	(*gr.*)	nerved-leaved.	———	....	——	G. ♄.	
60 pyroídes. B.T.	(*gr.*)	Pyrus-like.	———	1816.	——	G. ♄.	
61 rígida. DC.	(*gr.*)	rigid.	———	——	——	G. ♄.	
62 tridáctylis. B.T.	(*gr.*)	three-fingered.	———	1816.	....	G. ♄.	
63 serræfòlia. B.T.	(*gr.*)	saw-leaved	———	——	——	G. ♄.	
64 dioíca. DC.	(*ye.*)	dioecious.	Mogador	1823.	——	F. ♄.	
oxyacanthoídes. P.S.							

65 oxyacántha. DC.	(*ye.*)	Hawthorn-like.	Magador.	1823.	....	F. ♄.	
66 lobàta. DC.	(*gr.*)	lobed-leaved.	Teneriffe.	1824.	....	F. ♄.	
67 heterophy'lla. DC.	(*gr.*)	various-leaved.	....	1816.	6. 8.	F. ♄.	
68 zizyphina. R.S.	(*gr.*)	Zizyphus-like.	Sicily.	——	....	G. ♄.	
69 Cotinus. R.S.	(*gr.*)	Venice.	S. Europe.	1656.	6. 7.	H. ♄.	Jacq. aust. 3. t. 210.
LOB'ADIUM. Raf.		LOB'ADIUM.	Pentandria Trigynia.				
1 aromáticum.	(*gr.*)	aromatic.	N. America.	1772.	5.	H. ♄.	Turp. ann. mus. 5. t. 30.
Rhús aromática. R.S. *Schmáltzia aromática.* Desv.							
2 suavèolens.	(*gr.*)	sweet.	N. America.	1759.	——	H. ♄.	
Schmáltzia suavèolens. Desv. *Rhús suavèolens.* R.S.							

ORDO LXVIII.

TEREBINTHACEÆ. *Brown.*

PIST'ACIA. DC.		PIST'ACHIA-TREE.	Diœcia Pentandria. L.				
1 vèra. DC.	(*st.*)	true.	Levant.	1570.	4. 5	H. ♄.	Duham. ed. nov. 4. t. 17.
Mas. P. *trifolia*. W.	(*st.*)	*male-flowered.*	————	——	——	H. ♄.	Bocc. mus. t. 93.
Fem. P. *vera*. W.	(*gr.*)	*female.*	————	——	——	H. ♄.	Rauw. it. 72. t. 9.
2 reticulàta. W.	(*st.*)	netted-leaved.	————	1752.	——	H. ♄.	
3 Terebínthus. DC.	(*st.*)	turpentine-tree.	S. Europe.	1656.	6. 7.	H. ♄.	Duham. arb. v. 2. t. 87.
β *sphærocárpa*. DC.	(*st.*)	*round-fruited.*	Levant.	——	——	H. ♄.	J. Bauh. hist. 1. p. 278. ic.
4 atlántica. DC.	(*st.*)	Atlantic.	Barbary.	1790.	....	G. ♄.	
5 mexicàna. DC.	(*st.*)	Mexican.	Mexico.	1825.	....	G. ♄.	H. et B. n. ge. am. 7. t. 608.
6 fagaroídes. DC.	(*st.*)	Fagara-like.	....	——	....	G. ♄.	
7 Lentíscus. DC.	(*st.*)	Mastick-tree.	S. Europe.	1664.	5. 6.	H. ♄.	Bot. mag. 1967.
β *massiliénsis*. Mill.	(*st.*)	*narrow-leaved.*	————	——	——	H. ♄.	
ASTR'ONIUM. DC.		ASTR'ONIUM.	Diœcia Pentandria.				
gravèolens. DC.	(*re.*)	strong-scented.	S. America.	1826.	....	S. ♄.	Jacq. amer. t. 181. f. 96.
COMOCLA'DIA. DC.		MAIDEN-PLUM.	Triandria Monogynia. L.				
1 ilicifòlia. DC.	(*pu.*)	Holly-leaved.	Caribees.	1789.	7. 9.	S. ♄.	Lam. ill. t. 27. f. 2.
2 dentàta. DC.	(*pu.*)	tooth-leaved.	W. Indies.	1790.	——	S. ♄.	Jacq. amer. t. 173. f. 4.
3 propínqua. DC.	(*pu.*)	related.	Cuba.	1826.	——	S. ♄.	
4 integrifòlia. DC.	(*pu.*)	entire-leaved.	Jamaica.	1778.	——	S. ♄.	Sloan. his. 2. t. 222. f. 1.
PICRA'MNIA. DC.		PICRA'MNIA.	Diœcia Pentandria. L.				
1 Antidésma. DC.	(*st.*)	Ash-leaved.	Jamaica.	1793.	6. 8.	S. ♄.	Sloan. his. 2. t. 208. f. 2.
2 pentándra. DC.	(*st.*)	pentandrous.	W. Indies.	1822.	——	S. ♄.	
SCH'INUS. DC.		SCH'INUS.	Diœcia Decandria. L.				
1 Mólle. DC.	(*wh.*)	Mastick-tree.	Peru.	1597.	7. 8.	G. ♄.	Mill. ic. 2. t. 246.
2 terebinthifòlius. DC.	(*w.*)	Terebinthus-ld.	Brazil.	1822.	....	S. ♄.	
DUVA'UA. DC.		DUVA'UA.	Monœcia Octandria.				
1 depéndens. DC.	(*wh.*)	simple-leaved.	Chile.	1790.	6. 7.	H. ♄.	Cavan. ic. 3. t. 239.
Am'yris polygàma. S.S. *Schìnus depéndens.* H.K.							
2? dentàta. DC.	(*wh.*)	tooth-leaved.	Owhyhee.	1795.	5. 7.	G. ♄.	Andr. rep. 620.
Schìnus dentàta. A.R.							
MA'URIA. K.S.		MA'URIA.	Octo-Decandria Monogynia.				
1 simplicifolia. K.S.	(*bh.*)	simple-leaved.	Peru.	1822.	....	G. ♄.	H. et B. n. gen. a. 7. t. 605.
2 heterophy'lla. K.S.	(*bh.*)	various-leaved.	————	——	....	G. ♄.	———— 7. t. 606.

ORDO LXIX.

AMYRIDEÆ. *Kth. gen. tereb. p.* 21.

A'MYRIS. W.		A'MYRIS.	Octandria Monogynia. L.				
1 marítima. DC.	(*wh.*)	sea-side.	Jamaica.	1824.	6. 7.	S. ♄.	
2 floridàna. DC.	(*wh.*)	Florida.	Florida.	1826.	——	G. ♄.	
3 brasiliénsis. S.S.	(*wh.*)	Brazilian.	Brazil.	1822.	——	S. ♄.	

4 sylvática. DC.	(*wh.*)	wood.	Carthagena.	1793.	6. 7.	S. ♄.	Jacq. amer. pict. t. 188.
5 Plumiéri. DC.	(*wh.*)	Plumier's.	Antilles.	....	——	S. ♄.	Plum. ed. Burm. t. 100.
6 toxífera. DC.	(*wh.*)	poisonous.	W. Indies.	1800.	——	S. ♄.	Catesb. car. 1. t. 40.
balsamífera. L.							
7 pinnàta. DC.	(*wh.*)	pinnate.	S. America.	1824.	——	S. ♄.	H. et B. n.gen. 7.t.610.
8 Agallòcha. H B.	(*wh.*)	hill.	E. Indies.	1820.	....	S. ħ.	
9 acuminàta. H.B.	(*wh.*)	acuminate.	———	1822.	....	S. ♄.	
10 heptaphy'lla. H.B.	(*wh.*)	seven-leaved.	———	——	....	S. ♄.	
11 nàna. H.B.	(*wh.*)	dwarf.	———	1824.	....	S. ħ.	
12 punctàta. H.B.	(*wh.*)	dotted.	———	1818.	....	S. ħ.	

ORDO LXX.

SPONDIACEÆ. *Kth. gen. tereb. p.* 30.

SPO'NDIAS. DC.		Hog-Plum.	Decandria Pentagynia. L.				
1 Mómbin. L.	(*re.*)	purple-fruited.	W. Indies.	1817.	6. 8.	S. ♄.	Jacq. amer. t. 88.
purpùrea. DC. *Myrobálanus.* Jacq. *non* L.							
2 Myrobálanus. L.	(*wh.*)	yellow-fruited.	W. Indies.	1739.	——	S. ♄.	Sloan. hist. t. 219. f. 1. 2.
lùtea. DC. *Mómbin.* Jacq. *non* L.							
3 mangífera. DC.	......	Mango-bearing.	E. Indies.	1818.	....	S. ħ.	
Mangífera pinnàta. L.							
4 dúlcis. DC.	......	Otaheite-apple.	Society Isles.	1793.	6. 8.	S. ħ.	Sonn. it. 2. t. 123.
5 axillàris. H.B.	......	axillary.	Nepaul.	1824.	....	S. ħ.	
6 acuminàta. H.B.	....	taper-pointed.	Malabar.	——	....	S. ħ.	
POUPA'RTIA. DC.		Poupa'rtia.	Decandria Pentagynia.				
borbònica. DC.	(*pu.*)	Bourbon.	Bourbon.	1824.	....	S. ħ.	

ORDO LXXI.

BURSERACEÆ. *Kth. gen. tereb. p.* 14.

BOSWE'LLIA. DC.		Boswe'llia.	Decandria Monogynia.				
1 glábra. DC.	(*wh.*)	smooth.	Coromandel.	1824.	....	S. ♄.	Roxb. corom. 3. t. 207.
2 hirsùta. DC.	(*wh.*)	hairy.	Amboyna.	——	....	S. ħ.	Rumph amb. 2. t. 51.
3 serràta. DC.	(*wh.*)	Olibanum-tree.	E. Indies.	1816.	....	S. ħ.	Asiat. res. 9. p 377. ic.
ELA'PHRIUM. K.S.		Ela'phrium.	Octandria Monogynia.				
1 glábrum.	(*st.*)	smooth.	Carthagena.	1818.	....	S. ħ.	Jacq. amer. t. 71. f. 4.
Fagàra Eláphrium. W.							
2 copallíferum. DC.	(*st.*)	pubescent.	Mexico.	1824.	....	S. ħ.	Hern. mex. 45. f. 1.
BALSAMODE'NDRON. DC.		Balsam-tree.	Octandria Monogynia.				
gileadénse. DC.	(*gr.*)	blunt-leaved.	Arabia.	1827.	....	S. ħ.	Vahl. symb. 1. t. 11.
BURS'ERA. DC.		Burs'era.	Polygamia Diœcia. L.				
1 gummífera. DC.	(*wh.*)	Jamaica Birch.	W. Indies.	1690.	....	S. ħ.	Jacq. amer. t. 65.
2 serràta. Rox.	(*wh.*)	saw-leaved.	E. Indies.	1818.	....	S. ♄.	
3 simplicifòlia. DC.	(*wh.*)	simple-leaved.	Jamaica.	1824.	....	S. ♄.	
COLOPH'ONIA. DC.		Coloph'onia.	Hexandria Monogynia.				
mauritiàna. DC	(*pu.*)	panicled.	Mauritius.	1826.	....	S. ♄.	
CAN'ARIUM. DC.		Can'arium.	Diœcia Hexandria.				
1 commùne. DC.	(*wh.*)	common.	Moluccas.	1821.	....	S. ħ.	Rumph. amb. 2. t. 47.
2 microcárpum. DC.	(*w.*)	small-fruited.	———	1826.	....	S. ħ.	——— 2. t. 54.
3 Pimèla. DC	(*wh.*)	Chinese.	China.	1823.	....	S. ♄.	Ann. bot. 1. t. 7. f. 2.
GAR'UGA. DC.		Gar'uga.	Decandria Monogynia.				
1 pinnàta. DC.	(*ye.*)	winged-leaved.	E. Indies.	1808.	....	S. ♄.	Roxb. cor. 3. t. 208.
2 madagascariénsis. DC.	(*ye.*)	Madagascar.	Madagascar.	1824.	....	S. ♄.	

ORDO LXXII.

PTELEACEÆ. *Kth. gen. tereb. p.* 22.

PT'ELEA. DC.		SHRUBBY TREFOIL. Tetrandria Monogynia. L.					
trifoliàta. DC.	(*wh.*)	three-leaved.	N. America.	1704.	6. 7.	H. ♄.	Schmidt arb. 2. t. 76.
TODD'ALIA. DC.		TODD'ALIA. Pentandria Monogynia. L.					
1 aculeàta. DC.	(*wh.*)	prickly.	E. Indies.	1790.	6. 8.	S. ♄.	Rheed. mal. 5. t. 41.
asiática. R.S. *Scopòlia*		*aculeàta.* W.					
2 paniculàta. DC.	(*wh.*)	panicled.	Mauritius.	1824.	....	S. ♄.	Lam. ill. t. 139. f. 2.
inérmis. P.S. *Scopòlia*		*paniculàta.* S.S.					
CNE'ORUM. DC.		WIDOW-WAIL. Triandria Monogynia. L.					
1 tricóccum. DC.	(*ye.*)	smooth.	S. Europe.	1793.	4. 9.	F. ♄.	Lam. ill. t. 27.
2 pulveruléntum. DC.	(*y.*)	powdered.	Teneriffe.	1820.	——	F. ♄.	Vent. cels. t. 77.
SPATH'ELIA. DC.		SPATH'ELIA. Pentandria Ti:gynia. L.					
simpléx. DC.	(*re.*)	Sumach-leaved.	Jamaica.	1778.	4. 6.	S. ♄.	Bot. reg. 670.

ORDO LXXIII.

CONNARACEÆ. *R. Brown cong. p.* 12.

CO'NNARUS. L.		CO'NNARUS. Monadelphia Decandria.					
1 paniculàtus. H.B.	(*wh.*)	panicled.	Chittagong.	1824.	....	S. ♄.	
2 nítidus. H.B.	(*wh.*)	glossy.	Silhet.	——	....	S. ♄.	
ROU'REA. Aub.		ROU'REA. Monadelphia Decandria.					
frutéscens. Aub.	(*wh.*)	pubescent.	Guiana.	1820.	....	S. ♄.	Aub. gui. 1. t. 187.
OMPHAL'OBIUM. DC.		OMPHAL'OBIUM. Monadelphia Decandria.					
1 índicum. DC.	(*wh.*)	Indian.	E. Indies.	....	....	S. ♄.	Gært. fruct. 1. t. 46.
2 africànum. DC.	(*wh.*)	African.	Sierra Leon.	1822.	....	S. ♄.	Cav. dis. 7. t. 221.
EURYC'OMA. DC.		EURYC'OMA. Pentandria Monogynia.					
longifòlia. DC.	(*pu.*)	long-leaved.	E. Indies.	1826.	....	S. ♄.	
CNE'STIS. DC.		CNE'STIS. Decandria Monogynia.					
1 glábra. DC.	(*pu.*)	smooth.	Mauritius.	1824.	....	S. ♄.	Lam. ill. t. 387. f. 1.
2 polyphy'lla. DC.	(*pu.*)	many-leaved.	Madagascar.	——	....	S. ♄.	———— t. 387. f. 2.
3 monadélpha. DC.	....	monadelphous.	E. Indies.	1826.	....	S. ♄.	
BRUNE'LLIA. DC.		BRUNE'LLIA. Dodecandria Monogynia.					
inérmis. DC.	......	various-leaved.	Peru.	1823.	....	S. ♄.	
BR'UCEA. DC.		BR'UCEA. Diœcia Tetrandria.					
1 ferrugínea. W.	(*st.*)	ferruginous.	Abyssinia.	1775.	4. 5.	S. ♄.	Bruce's tr. 5. p. 69. ic.
antidysentérica. DC.							
2 grácilis. DC.	(*st.*)	slender.	E. Indies.	....	....	S. ♄.	
3 sumatràna. DC.	(*pu.*)	Sumatra.	Sumatra.	1822.	4. 6.	S. ♄.	
TETRA'DIUM. DC.		TETRA'DIUM. Tetrandria Monogynia.					
trichótomum. DC.	....	trichotomous.	China.	1822.	....	S. ♄.	
Brùcea trichótoma. S.S.							
AILA'NTHUS. DC.		AILA'NTHUS. Polygamia Monœcia.					
1 glandulòsa. DC.	(*st.*)	Chinese.	China.	1751.	8.	H. 5.	Wats. dend. brit. t. 104.
2 excélsa. DC.	(*st.*)	Indian.	E. Indies.	1800.	....	S. 5.	Roxb. corom. 1. t. 23.

ORDO LXXIV.

LEGUMINOSÆ. *DC. prodr.* 2. *p.* 93.

Subordo I. *PAPILIONACEÆ.* DC. l. c. p. 94.

Tribus I. SOPHOREÆ. DC. l. c. p. 94.

MYROSPE'RMUM. K.S.		BALSAM-SEED. Decandria Monogynia.					
frutéscens. K.S.	(*bh.*)	shrubby.	Caracas.	1824.	....	S. ♄.	H.et B.n.g. 6. t. 570-71.
MYROXYLUM. K.S.		BALSAM-WOOD. Decandria Monogynia.					
1 pubéscens. K.S.	(*wh.*)	pubescent.	Carthagena.	1820.	....	S. ♄.	Lamb. cinch. 92. t. 1.
2 peruíferum. K.S.	(*wh.*)	Balsam of Peru.	Peru.	1824.	....	S. ♄.	Lam. ill. t. 341. f. 1.
3 toluíferum. K.S.	(*wh.*)	Balsam of Tolu.	S. America.	....	....	S. ♄.	Woodv. m. bot. 3. t. 193.
S'OPHORA. DC.		S'OPHORA. Decandria Monogynia. L.					
1 japónica. DC.	(*wh.*)	Japanese.	Japan.	1753.	8. 9.	H. 5.	Andr. reposit. 585.
β pendula. L.C.	(*wh.*)	*pendulous.*	———	....	——	H. 5.	
γ variegàta. L.C.	(*wh.*)	*striped-leaved.*	———	....	——	H. 5.	
2 chinénsis. L.C.	(*wh.*)	Chinese.	China.	....	——	H. 5.	
3 gláuca. DC.	(*pu.*)	glaucous.	E. Indies.	1818.	....	S. ♄.	
4 tomentòsa. DC.	(*ye.*)	downy.	———	1690.	....	S. ♄.	Lam. ill. t. 325. f. 2.
5 occidentàlis. W.	(*ye.*)	West Indian.	W. Indies.	....	....	S. ♄.	Trew. ehr. t. 59.
6 crassifòlia. DC.	(*ye.*)	thick-leaved.	Senegal.	1824.	....	S. ♄.	
7 secundiflòra. DC.	(*bl.*)	side-flowering.	New Spain.	1818.	....	G. ♄.	Cavan. ic. 5. t. 401.
Virgília secundiflòra. C.I. *Broussonètia secundiflòra.* Ort. dec. 5. t. 7.							
8 macrocárpa. DC.	(*ye.*)	large-podded.	Chile.	1822.	4. 5.	F. ♄.	Lodd. bot. cab. 1125.
9 littoràlis. DC.	(*ye.*)	sea-side.	Brazil.	1824.	....	S. ♄.	Pluk. alm. t. 104. f. 3.
10 robústa. H.B.	......	robust.	Silhet.	1820.	....	S. ♄.	
11 velutìna. B.R.		velvetty.	Nepaul.	1824.	6. 7.	G. ♄.	Bot. reg. 1185.
12 flavéscens. DC.	(*ye.*)	pale-yellow.	Siberia.	1785.	5. 7.	H. ♃.	
PSEUDOS'OPHORA.		PSEUDOS'OPHORA. Diadelphia Decandria.					
1 alopecuroídes. DC.	(*wh.*)	Fox-tail.	Levant.	1731.	7. 8.	H. ♃.	Dill. elth. f. 136.
2 serícea. N.	(*wh.*)	silky.	Missouri.	1811.	6. 7.	H. ♃.	
Astrágalus carnòsus. Ph.							
EDWA'RDSIA. DC.		EDWA'RDSIA. Decandria Monogynia.					
1 grandiflòra. DC.	(*ye.*)	large-flowered.	N. Zealand.	1772.	3. 6.	F. ♄.	Bot. mag. 167.
Sòphora tetráptera. B.M.							
2 chrysophy'lla. DC.	(*ye.*)	golden-leaved.	———	———	——	F. ♄.	Bot. reg. 738.
3 microphy'lla. DC.	(*ye.*)	small-leaved.	———	1772.	——	F. ♄.	Bot. mag. 1442.
4 mínima. L.C.	(*ye.*)	smallest.	———	1816.	——	F. ♄.	
ORM'OSIA. DC.		ORM'OSIA. Decandria Monogynia.					
1 dasycárpa. DC.	(*pu.*)	smooth-leaved.	W. Indies.	1793.	6. 7.	S. ♄.	Linn. trans. 10. t. 26.
Sòphora monospérma. Swartz.							
2 coccínea. DC.	(*pu.*)	scarlet-seeded.	Guiana.	1826.	....	S. ♄.	Linn. trans. 10. t. 25.
VIRGI'LIA. DC.		VIRGI'LIA. Decandria Monogynia.					
1 capénsis. DC.	(*bh.*)	Cape.	C. B. S.	1767.	7. 8.	G. ♄.	Bot. mag. 1590.
2 intrùsa. H.K.	(*ye.*)	small-flowered.	———	1790.	5. 8.	G. ♄.	——— t. 2617.
Sòphora sylvática. Burchell.							
3 áurea. DC.	(*ye.*)	golden.	Abyssinia.	1777.	7.	G. ♄.	L'Herit. st. nov. 1. t. 75.
4 lùtea. DC.	(*ye.*)	American.	N. America.	1812.	6. 7.	G. ♄.	Mich. fil. arb. 3. t. 3.
MACROTR'OPIS. DC.		MACROTR'OPIS. Decandria Monogynia.					
1 f'œtida. DC.	(*wh.*)	strong-scented.	China.	1822.	4. 5.	F. ♄.	
2 inodòra. DC.	(*wh.*)	scentless.	———	1824.	——	F. ♄.	
ANAG'YRIS. DC.		BEAN TREFOIL. Decandria Monogynia. L.					
1 f'œtida. DC.	(*ye.*)	strong-scented.	Spain.	1570.	4. 5.	F. ♄.	Lodd. bot. cab. t. 740.
β gláuca. DC.	(*ye.*)	*glaucous-leaved.*	S. Europe.	1824.	——	F. ♄.	
2 latifòlia. DC.	(*ye.*)	broad-leaved.	Teneriffe.	1815.	——	F. ♄.	
PIPTA'NTHUS. B.F.G.		PIPTA'NTHUS. Decandria Monogynia.					
nepalénsis. B.F.G.	(*ye.*)	Laburnum l^d^.	Nepaul.	1819.	5. 6.	H. ♄.	Swt. br. fl. gar. 264.
Thermópsis napaulénsis. DC. *laburnifòlia.* D.P. *Baptísia nepalénsis.* Hook. ex. flor. 131.							

THERMO'PSIS. DC. THERMO'PSIS. Decandria Monogynia.
1 rhombifòlia. DC. (ye.) rhomb-leaved. Louisiana. 1811. 6. 7. H. ♃.
Cy'tisus rhombifòlius. Ph.
2 fabàcea. DC. (ye.) Bean-leaved. Kamtschatka. 1824. —— H. ♃. Pall. astr. t. 90. f. 2.
3 lanceolàta. DC. (ye.) spear-leaved. Siberia. 1776. —— H. ♃. Bot. mag. 1389.
Sòphora lupinoídes. Pall. astr. t. 89. *Podaly'ria lupinoídes.* B.M.
4 corgonénsis. DC. (ye.) alpine. Altay. 1824. —— H. ♃. Pall. astr. t. 90. f. 1.
Sòphora alpìna. Pall. *Podaly'ria alpìna.* W. enum.
BAPTI'SIA. DC. BAPTI'SIA. Decandria Monogynia.
1 perfoliàta. DC. (ye.) perfoliate. Carolina. 1732. 8. F. ♃. Lodd. bot. cab. 1104.
Crotalària perfoliàta. L. *Ráfnia perfoliàta.* W.
2 exaltàta. B.F.G. (bl.) tall upright. N. America. 1812. 6. 7. H. ♃. Swt. br. fl. gar. 97.
3 auriculàta. (bl.) eared. —— —— —— H. ♃.
4 austràlis. DC. (bl.) common. —— 1758. —— H. ♃. Bot. mag. 509.
Sòphora austràlis. B.M. *Podaly'ria austràlis.* Vent. cels. t. 56.
5 confùsa. (bl.) persistent-bract'd. N. America. 1812. —— H. ♃.
6 villòsa. DC. (st.) villous. —— 1811. —— H. ♃.
7 versícolor. B.C. (va.) various-colour'd. —— 1824. 6. 8. H. ♃. Lodd. bot. cab. 1144.
8 álba. DC. (wh.) white-flowered. —— 1724. 6. 7. H. ♃. Bot. mag. 1177.
9 tinctòria. DC. (ye.) dyer's. —— 1759. 7. 8. H. ♃. —— t. 1099.
Sòphora tinctòria. L. *Podaly'ria tinctòria.* B.M.
CYCL'OPIA. DC. CYCL'OPIA. Decandria Monogynia.
1 genistoídes. DC. (ye.) Genista like. C. B. S. 1787. 7. 8. G. ♄. Bot. mag. 1259.
Ibbetsònia genistoídes. B.M. *Gompholòbium maculàtum.* A.R. 427.
2 galioídes. DC. (ye.) Galium-like. C. B. S. 1818. 7. 8. G. ♄. Pluk. alm. t. 413. f. 4.
3 latifòlia. DC. (ye.) broad-leaved. —— —— —— G. ♄.
PODALY'RIA. DC. PODALY'RIA. Decandria Monogynia.
1 Burchéllii. DC. (pu.) Burchell's. C. B. S. 1818. 5. 7. G. ♄.
2 hirsùta. DC. (pu.) hairy. —— 1774. 7. 8. G. ♄.
3 cuneifòlia. DC. (wh.) wedge-leaved. —— 1804. 5. 8. G. ♄. Vent. cels. t. 99.
4 oleæfòlia. P.L. (pu.) Olive-leaved. —— —— —— G. ♄. Sal. par. lond. 114.
5 serícea. DC. (pu) silky. —— 1778. 10. 1. G. ♄. Bot. mag. 1923.
6 myrtillifòlia. DC. (pu.) Myrtle-leaved. —— 1795. 4. 7. G. ♄.
7 buxifòlia. DC. (pu.) Box-leaved. —— 1790. 5. 10. G. ♄. Bot. reg. 869.
8 styracifòlia. DC. (pu.) Storax-leaved. —— 1792. 5. 7. G. ♄. Bot. mag. 1580.
9 calyptràta. W. (pu.) one-flowered. —— —— 4. 5. G. ♄.
10 corúscans. Rch. (pu.) glittering. —— 1822. —— G. ♄.
11 gláuca. DC. (pu.) glaucous-leaved. —— —— —— G. ♄. Lam. ill. t. 327. f. 4.
12 argéntea. DC. (li.) silvery-leaved. —— 1780. 2. 6. G. ♄. Sal. par. lond. t. 7.
biflòra. Bot. mag. t. 753.
13 liparioídes. DC. (li.) Liparia-like. —— 1824. 4. 6. G. ♄.
14 subbiflòra. DC. (pu.) netted-leaved. —— —— —— G. ♄.
15 cordàta. DC. (pu.) heart-leaved. —— 1794. 5. 7. G. ♄.
CHORIZ'EMA. DC. CHORIZ'EMA. Decandria Monogynia.
1 ilicifòlia. DC. (sc.) Holly-leaved. N. Holland. 1803. 3. 10. G. ♄. Labill. voy. 1. t. 21.
2 nàna. DC. (sc.) dwarf prickly. —— —— —— G. ♄. Bot. mag. 1032.
3 rhómbea. DC. (sc.) various-leaved. —— —— 4. 6. G. ♄. Swt. flor. aust. t. 40.
4 Henchmánni. B.R. (sc.) Henchmann's. —— 1825. 5. 9. G. ♄. Bot. reg. 986.
5 platylobioídes. DC. (ye.) Platylobium-like. —— —— —— G. ♄.
PODOL'OBIUM. DC. PODOL'OBIUM. Decandria Monogynia.
1 trilobàtum. DC. (ye.) three-lobed. N. S. W. 1791. 4. 7. G. ♄. Bot. mag. 1477.
Chorozèma trilobàtum. L.T. *Pulten'æa ilicifòlia* A.R. 320.
2 staurophy'llum. DC. (y.) spiny-leaved. N. S. W. 1821. 3. 5. G. ♄. Bot. reg. 959.
3 scándens. DC. (ye.) climbing. —— 1825. 4. 6. G. ♄.◡.
Chorozèma scándens. L.T.
OXYL'OBIUM. DC. OXYL'OBIUM. Decandria Monogynia.
1 obtusifòlium. S.F.A. (sc.) blunt-leaved. N. Holland. 1825. 4. 6. G. ♄. Swt. flor. aust. t. 5.
2 arboréscens. DC. (ye.) arborescent. V. Diem. Isl. 1805. 5. 9. G. ♄. Bot. reg. 392.
3 retùsum. B.R. (or.) retuse-leaved. N. Holland. 1822. 4. 6. G. ♄. —— t. 913.
4 ellípticum. DC. (ye.) elliptic-leaved. V. Diem. Isl. 1805. 5. 9. G. ♄. Lab. nov. hol. 1. t. 135.
5 cordifòlium. DC. (sc.) heart-leaved. N. S. W. 1807. 4. 9. G. ♄. Andr. rep. 492.
6 spinòsum. DC. (ye.) spiny. N. Holland. 1823. 5. 8. G. ♄.
7 Pultene'æ. DC. (ye.) wood. N. S. W. —— 4. 6. G. ♄.
CALLIST'ACHYS. DC. CALLIST'ACHYS. Decandria Monogynia.
1 lanceolàta. DC. (ye.) spear-leaved. N. Holland. 1815. 5. 8. G. ♄. Bot. reg. 216.
2 ovàta. DC. (ye.) oval-leaved. —— —— —— G. ♄. Bot. mag. 1925.
BRACHYS'EMA. DC. BRACHYS'EMA. Decandria Monogynia.
1 latifòlium. DC. (sc.) broad-leaved. N. Holland. 1803. 4. 7. G. ♄.◡. Bot. reg. 118.
2 undulàtum. DC. (ye.) wave-leaved. —— 1820. 2. 7. G. ♄.◡. —— t. 642.

GOMPHOL'OBIUM. DC. GOMPHOL'OBIUM. Decandria Monogynia.
1 barbígerum. DC. (ye.) bearded-flow'd. N. S. W. 1824. 5. 8. G. ♄.
2 grandiflòrum. DC. (ye.) large-flowered. ——— 1803. 3. 9. G. ♄. Bot. reg. 484.
3 virgàtum. DC. (ye.) twiggy. ——— 1824. 5. 8. G. ♄.
4 latifòlium. DC. (ye.) broad-leaved. ——— 1803. 3. 9. G. ♄. Labill. nov. hol. t. 133.
fimbriàtum. Sm. ex. bot. t. 58. *psoraleæfòlium.* Sal. par. lond. t. 6.
5 glaucéscens. F.T. (ye.) glaucescent. N. S. W. 1821. 4. 6. G. ♄.
6 marginàtum. DC. (ye.) small-flower'd. N. Holland. 1803. 3. 9. G. ♄.
7 pedunculàre. DC. (ye.) long-stalk'd. N. S. W. 1825. 4. 7. G. ♄.
8 tetrathecoídes. DC. (ye.) Tetratheca-like. ——— —— —— G. ♄.
9 polymórphum. DC. (sc.) variable. N. Holland. 1803. 3. 8. G. ♄. ◡. Bot. mag. 1533.
10 tomentòsum. DC. (ye.) tomentose. ——— —— 4. 7. G. ♄. Labill. nov. hol. 1. t. 134.
11 glábratum. DC. (ye.) smooth. ——— 1824. —— G. ♄.
12 venústum. DC. (pu.) purple-flowered. ——— 1803. —— G. ♄.
13 pinnàtum. DC. (ye.) wing-leaved. N. S. W. 1822. 5. 6. G. ♂.
BURT'ONIA. DC. BURT'ONIA. Decandria Monogynia.
1 scábra. DC. (pu.) rough-leaved. N. Holland. 1803. 5. 7. G. ♄.
2 mìnor. DC. (ye.) hairy-stalked. N. S. W. 1812. 3. 8. G. ♄.
3 sessilifòlia. DC. (ye.) sessile-leaved. ——— 1825. —— G. ♄.
JACKS'ONIA. DC. JACKS'ONIA. Decandria Monogynia.
1 scopària. DC. (ye.) Broom-like. N. S. W. 1803. 6. 8. G. h. Lodd. bot. cab. 427.
2 spinòsa. DC. (ye.) spiny. N. Holland. —— 4. 9. G. ♄. Labill. nov. hol. 1. t. 136.
3 hórrida. DC. (ye.) horrid. ——— 1825. —— G. h.
4 furcellàta. DC. (ye.) forked. N. S. W. 1822. 4. 6. G. h. Bonpl. nav. t. 11.
5 reticulata. DC. (ye.) netted. ——— —— —— G. h.
Davièsia reticulàta. L.T.
VIMIN'ARIA. DC. VIMIN'ARIA. Decandria Monogynia.
denudàta. DC. (ye.) half-naked. N. S. W. 1789. 6. 9. G. ♄. Bot. mag. 1190.
SPHÆROL'OBIUM. DC. SPHÆROL'OBIUM. Decandria Monogynia.
1 vimíneum. DC. (ye.) yellow-flower'd. N. S. W. 1802. 5. 8. G. ♄. Bot. mag. 969.
2 mèdium. DC. (re.) red-flowered. N. Holland. 1803. 6. 8. G. h.
A'OTUS. DC. A'OTUS. Decandria Monogynia.
1 villòsa. DC. (ye.) villous. V. Diem. Isl. 1790. 4. 6. G. ♄. Bot. mag. 949.
2 virgàta. DC. (ye.) slender. N. Holland. 1820. 4. 8. G. ♄.
DILLW'YNIA. DC. DILLW'YNIA. Decandria Monogynia.
1 floribúnda. DC. (ye.) close-flowered. N. S. W. 1794. 4. 7. G. ♄. Sm. exot. bot. t. 26.
ericifòlia. Bot. mag. 1545. *nec aliorum.*
2 ericifòlia. DC. (ye.) Heath-leaved. ——— —— 3. 7. G. ♄. ——— t. 25.
3 juniperìna. B.C. (ye.) Juniper-leaved. ——— 1818. 6. 7. G. ♄. Lodd. bot. cab. 401.
4 glabérrima. DC. (ye.) smooth. ——— 1800. 3. 7. G. ♄. Bot. mag. 944.
5 parvifòlia. DC. (ye.) small-leaved. ——— —— —— G. ♄. ——— t. 1527.
6 acicularis. DC. (ye.) needle-leaved. ——— 1822. 5. 7. G. ♄.
7 rùdis. DC. (ye.) villous-branched. ——— 1812. —— G. ♄.
8 teretifòlia. Sie. (ye.) round-leaved. ——— 1825. —— G. ♄.
9 hispídula. Sie. (ye.) hispid. ——— —— —— G. ♄.
10 tenuifòlia. DC. (ye.) slender-leaved. ——— 1820. —— G. ♄.
11 phylicoídes. F.T. (ye.) Phylica-leaved. ——— 1824. 5. 8. G. ♄.
12 serícea. F.T. (ye.) silky. ——— —— —— G. ♄.
13 cineráscens. DC. (ye.) ash-coloured. V. Diem. Isl. 1818. 6. 7. G. ♄. Bot. mag. 2247.
EUTA'XIA. DC. EUTA'XIA. Decandria Monogynia.
1 myrtifòlia. DC. (ye.) Myrtle-leaved. N. Holland. 1803. 3. 6. G. h. Bot. mag. 1274.
Dillw'ynia obovàta. B.M.
2 púngens. S.F.A. (ye.) pungent-leaved. ——— 1825. 6. 8. G. h. Swt. flor. aust. t. 28.
SCLEROTHA'MNUS. DC. SCLEROTHA'MNUS. Decandria Monogynia.
mícrophy'llus. DC. (ye.) small-leaved. N. Holland. 1803. 5. 6. G. ♄.
GASTROL'OBIUM. DC. GASTROL'OBIUM. Decandria Monogynia.
bilòbum. DC. (ye.) two-lobed. N. Holland. 1803. 3. 5. G. h. Bot. reg. 411.
EUCH'ILUS. DC. EUCH'ILUS. Decandria Monogynia.
obcordàtus. DC. (ye.) heart-leaved. N. Holland. 1803. 3. 6. G. ♄. Bot. reg. 403.
PULTEN'ÆA. DC. PULTEN'ÆA. Decandria Monogynia.
1 daphnoídes. DC. (ye.) Daphne-leaved. N. S. W. 1792. 6. 7. G. h. Bot. mag. 1394.
2 obcordàta. DC. (ye.) heart-leaved. V. Diem. Isl. 1808. 5. 7. G. h. Andr. rep. t. 574.
3 bilòba. DC. (ye.) two-lobed. N. Holland. 1818. —— G. h. Bot. mag. t. 2091.
4 scábra. H.K. (y.pu.) rough-leaved. N. S. W. 1803. —— G. ♄.
5 incurvàta. F.T. (ye.) incurved-ld. ——— 1825. —— G. ♄.
6 canéscens. F.T. (ye.) canescent ——— 1824. 4. 7. G. ♄.
7 polifòlia. F.T. (ye.) Poley-leaved. ——— —— —— G. ♄.
8 argéntea. F.T. (ye.) silvery. ——— 1826. —— G. ♄.
9 ferrugínea. DC. (ye.) ferruginous. ——— 1818. —— G. ♄. Linn. trans. 11. t. 23.

10 racemulòsa. DC.	(*ye.*)	racemed.	N. S. W.	1821.	4. 7.	G. ♄.	
11 parviflòra. DC.	(*ye.*)	small-flowered.	——	1824.	5. 7.	G. ♄.	
12 procúmbens. F.T.	(*ye.*)	procumbent.	——	——	——	G. ♄.	
13 strícta. DC.	(*ye.*)	upright.	——	1803.	4. 6.	G. ħ.	Bot. mag. 1588.
14 ellíptica. DC.	(*ye.*)	elliptic-leaved.	——	1818.	4. 7.	G. ħ.	Linn. trans. 11. t. 24.
15 plumòsa. DC.	(*ye.*)	feathered.	——	1822.	5. 7.	G. ħ.	
16 thymifòlia. DC.	(*ye.*)	Thyme-leaved.	——	——	——	G. ħ.	
17 hypolámpra. DC.	(*ye.*)	whorl-branched.	——	——	——	G. ħ.	
18 villífera. DC.	(*ye.*)	wool-bearing.	——	1824.	——	G. ħ.	
19 polygalifòlia. DC.	(*ye.*)	Milkwort-ld.	——	1818.	4. 7.	G. ħ.	Swt. flor. aust. t. 37.
20 fléxilis. DC.	(*ye.*)	fragrant.	——	1801.	4. 6.	G. ħ.	—— t. 35.
21 euchìla. DC.	(*ye.*)	wedge-leaved.	——	1824.	5. 7.	G. ħ.	
22 retùsa. DC.	(*ye.*)	blunt.	——	1789.	4. 5.	G. ħ.	Bot. reg. 378.
23 capitellàta. DC.	(*ye.*)	small-headed.	——	1825.	——	G. ħ.	
24 linophy'lla. DC.	(*ye.*)	Flax-leaved.	——	1789.	5. 7.	G. ħ.	Schrad. s. han. 3. t. 18.
25 microphy'lla. DC.	(*ye.*)	small-leaved.	——	1823.	——	G. ħ.	
26 stipulàris. DC.	(*ye.*)	scaly.	——	1792.	4. 6.	G. ♄.	Bot. mag. 475.
27 paleàcea. DC.	(*ye.*)	chaffy.	——	1789.	5. 7.	G. ♄.	Lodd. bot. cab. t. 291.
28 aristàta. DC.	(*ye.*)	awned-calyxed.	——	1823.	——	G. ♄.	
29 echinùla. DC.	(*ye.*)	harsh-leaved.	——	——	——	G. ♄.	
30 vestìta. DC.	(*ye.*)	awned.	N. Holland.	1803.	4. 7.	G. ♄.	
31 cándida. B.C.	(*ye.*)	white-leaved.	——	1824.	——	G. ♄.	Lod. bot. cab. 1236.
32 tenuifòlia. DC.	(*ye.*)	slender-leaved.	——	1818.	——	G. ♄.	Bot. mag. 2086.
33 villòsa. DC.	(*ye.*)	villous.	N. S. W.	1790.	4. 5.	G. ħ.	—— t. 967.
34 pedunculàta. B.M.	(*ye.*)	pedunculated.	——	1824.	5. 6.	G. ħ.	Bot. mag. 2859.
35 erióphora. L.C.	(*ye.*)	wool-bearing.	——	1794.	4. 7.	G. ħ.	
36 incarnàta. M.C.	(*fl.*)	flesh-coloured.	N. Holland.	1824.	——	G. ħ.	
37 rosmarinifòlia. ψ.	(*ye.*)	Rosemary-leav'd.	——	——	——	G. ħ.	
38 áspera. DC.	(*ye.*)	rough.	N. S. W.	——	5. 8.	G. ħ.	
39 comòsa. DC.	(*ye.*)	tufted.	——	——	——	G. ♄.	
40 squarròsa. DC.	(*ye.*)	squarrose.	——	——	——	G. ħ.	
41 phylicoídes. DC.	(*ye.*)	Phylica-like.	——	——	——	G. ħ.	
DAVI'ESIA. DC.		DAVI'ESIA. Decandria Monogynia.					
1 latifòlia. DC.	(*ye.*)	broad-leaved.	N. S. W.	1803.	5. 8.	G. ħ.	Bot. mag. 1757.
2 corymbòsa. DC.	(*ye.*)	corymb-flower'd.	——	1804.	——	G. ħ.	
3 mimosoídes. H.K.	(*ye.*)	green-leaved.	——	1800.	——	G. ħ.	Andr. reposit. 526.
corymbòsa. A.R. *non* Smith.							
4 gláuca. B.C.	(*ye.*)	glaucous-leaved.	——	1812.	——	G. ħ.	Lodd bot. cab. 43.
mimosoídes. Bot. mag. 1957. *non* H.K.							
5 umbellulàta. DC.	(*ye.*)	small-umbelled.	——	1822.	4. 8.	G. ħ.	
6 incrassàta. DC.	(*ye.*)	thick-leaved.	——	1820.	5. 8.	G. ħ.	
7 aciculàris. DC.	(*ye.*)	needle-leaved.	——	1804.	6. 7.	G. ♄.	
8 ulicìna. DC.	(*ye.*)	Furze-like.	N. S. W.	1792.	4. 8.	G. ♄.	Andr. reposit. 304.
9 púngens. M.C.	(*ye.*)	pungent.	N. Holland.	1825.	——	G. ♄.	
10 squarròsa. DC.	(*ye.*)	squarrose.	N. S. W.	1822.	5. 7.	G. ħ.	
11 cordàta. DC.	(*ye.*)	heart-leaved.	N. Holland.	1824.	——	G. ħ.	Bot. reg. 1005.
12 alàta. DC.	(*ye.*)	winged.	——	1820.	4. 6.	G. ♄.	—— t. 728.
MIRB'ELIA. DC.		MIRB'ELIA. Decandria Monogynia.					
1 reticulàta. DC.	(*pu.*)	netted-leaved.	N. S. W.	1792.	5. 8.	G. ♄.	Bot. mag. 1211.
2 púngens. M.C.	(*pu.*)	pungent-leav'd.	N. Holland.	1824.	——	G. ♄.	
3 speciòsa. DC.	(*pu.*)	beautiful.	N. S. W.	——	——	G. ♄.	Swt. flor. aust. t. 34.
4 grandiflòra. B.M.	(*y.re.*)	large-flowered.	——	1823.	——	G. ♄.	Bot. mag. 2771.
5 Baxtèri. M.C.	....	Baxter's.	N. Holland.	1824.	....	G. ħ.	
6 dilatàta. DC.	(*pu.*)	lobed-leaved.	——	1803.	5. 8.	G. ħ.	Bot. reg. 1041.

Tribus II. *LOTEÆ.* DC. prodr. 2. p. 115.

Subtribus I. GENISTEÆ. *DC.* l. c. 115.

H'OVEA. DC.		H'OVEA. Monadelphia Decandria.					
1 rosmarinifòlia. F.T.	(*bl.*)	Rosemary-ld.	N. S. W.	1824.	3. 6.	G. ħ.	
2 longifòlia. DC.	(*bl.*)	long-leaved.	——	1805.	3. 7.	G. ħ.	Bot. reg. 614.
3 lineàris. DC.	(*bl.*)	linear-leaved.	——	1796.	——	G. ♄.	—— t. 463.
4 lanceolàta. DC.	(*bl.*)	spear-leaved.	——	1805.	——	G. ♄.	Bot. mag. 1624.

5 ellíptica. DC. (*bl.*) elliptic-leaved. N. S. W. 1817. 3. 7. G. ♄.
Poirétia ellíptica. L.T.
6 purpùrea. S.F.A. (*pu.*) purple-flowered. ——— 1824. 2. 5. G. ♄. Swt. flor. aust. t. 13.
7 latifòlia. DC. (*bl.*) broad-leaved. ——— 1817. 3. 6. G. ♄. Lodd. bot. cab. 30.
8 Celsi. DC. (*bl.*) Cels's. N. Holland. 1818. 3. 7. G. ♄. Bot. reg. 280.
PLAGIOL'OBIUM. S.F.A. PLAGIOL'OBIUM. Diadelphia Decandria.
1 chorizemæfòlium. S.F.A. (*bl.*) Chorizema-l^d. N. Holland. 1824. 2. 5. G. ♄. Swt. flor. aust. t. 2.
2 ilicifòlium. S.F.A. (*bl.*) Holly-leaved. ——— 1827. —— G. ♄. ——— t. 2. f. 10-11.
PLATYL'OBIUM. DC. FLAT-PEA. Monadelphia Decandria.
1 formòsum. DC. (*ye.*) large-flowered. N. S. W. 1790. 6. 8. G. ♄. Bot. mag. 469.
2 parviflòrum. DC. (*ye.*) smaller-flower'd. ——— 1792. 5. 9. G. ♄. ——— t. 1520.
3 ovàtum. DC. (*ye.*) oval-leaved. ——— 1820. 5. 7. G. ♄.
4 triangulàre. DC. (*ye.*) triangular-l^d. V. Diem. Isl. 1805. 5. 9. G. ♄. Bot. mag. 1508.
BOSSI'ÆA. DC. BOSSI'ÆA. Monadelphia Decandria.
1 scolopéndria. DC. (*ye.*) Plank-plant. N. S. W. 1792. 5. 7. G. ♄. Vent. malm. 55.
Platylòbium Scolopéndrium. Andr. rep. 191.
2 ensàta. DC. (*ye.*) sword-stemmed. N. S. W. 1822. —— G. ♄. Swt. flor. aust. t. 51.
rùfa. Lod. bot. cab. 1119. *non* DC.
3 rùfa. DC. (*re.*) red-flowered. N. Holland. 1803. 6. 9. G. ♄.
4 linophy'lla. DC. (*ye.*) narrow-leaved. ——— —— 7. 9. G. ♄. Bot. mag. 2491.
5 heterophy'lla. DC. (*ye.*) various-leaved. N. S. W. 1792. 4. 12. G. ♄. Vent. cels. t. 7.
lanceolàta. B.M. 1144. *Platylòbium lanceolàtum et ovàtum.* A.R. 205 et 276.
6 rotundifòlia. DC. (*ye.*) round-leaved. N. S. W. 1822. 5. 7. G. ♄.
7 rhombifòlia. DC. (*ye.*) rhomb-leaved. ——— —— 4. 6. G. ♄. Swt. flor. aust. t. 9.
lenticulàris. Lod. bot. cab. 1238. *non* DC.
8 microphy'lla. DC. (*ye.*) small-leaved. ——— 1803. 5. 7. G. ♄. Lodd. bot. cab. t. 656.
9 lenticulàris. DC. (*ye.*) orbicular-leaved. ——— 1822. —— G. ♄.
10 foliòsa. F.T. (*ye.*) leafy. ——— 1823. 4. 7. G. ♄.
11 buxifòlia. F.T. (*ye.*) Box-leaved. ——— —— —— G. ♄.
12 prostràta. DC. (*ye.*) prostrate. ——— 1803. —— G. ♄. Bot. mag. 1493.
13 cordifòlia. S.F.A. (*y.pu.*) heart-leaved. N. Holland. 1824. —— G. ♄. Swt. flor. aust. t. 20.
14 cinèrea. DC. (*y.pu.*) sharp-leaved. V. Diem. Isl. 1803. 5. 7. G. ♄. Bot. reg. 306.
GO'ODIA. DC. GO'ODIA. Monadelphia Decandria.
1 lotifòlia. DC. (*ye.*) Lotus-leaved. V. Diem. Isl. 1793. 4. 7. G. ♄. Sal. par. lond. 41.
2 subpubéscens. (*ye.*) hairy-leaved. ——— 1812. —— G. ♄.
3 pubéscens. DC. (*y.pu.*) downy. ——— 1805. —— G. ♄. Bot. mag. 1310.
4 retùsa. M.C. (*ye.*) retuse-leaved. N. Holland. 1824. —— G. ♄.
TEPHROTHA'MNUS. TEPHROTHA'MNUS. Monadelphia Decandria.
tomentòsus. (*ye.*) woolly. C. B. S. 1798. 7. 8. G. ♄. And. rep. 237.
Cy'tisus tomentòsus. A.R. *Goòdia? polyspérma.* DC.
SCO'TTIA. H.K. SCO'TTIA. Monadelphia Decandria.
1 dentàta. DC. (*sc.*) broad-leaved. N. Holland. 1803. 1. 6. G. ♄.
2 trapezifórmis. M.C. .. trapeziform. ——— 1825. .... G. ♄.
3 angustifòlia. M.C. (*br.*) narrow-leaved. ——— —— 1. 6. G. ♄.
TEMPLET'ONIA. DC. TEMPLET'ONIA. Monadelphia Decandria.
1 retùsa. DC. (*sc.*) green-leaved. N. Holland. 1803. 3. 7. G. ♄. Bot. reg. 383.
2 gláuca. DC. (*sc.*) glaucous-leaved. ——— —— G. ♄. Bot. mag. 2088.
RA'FNIA. DC. RA'FNIA. Monadelphia Decandria.
1 cordàta. DC. (*ye.*) heart-leaved. C. B. S. 1821. 5. 7. G. ♄.
2 ellíptica. DC. (*ye.*) elliptic-leaved. ——— 1816. —— G. ♄.
3 cuneifòlia. DC. (*y.pu.*) wedge-leaved. ——— 1820. —— G. ♂.
4 triflóra. DC. (*ye.*) three-flowered. ——— 1786. —— G. ♂. Bot. mag. 482.
Crotalària triflòra. L. *Borbònia cordàta.* And. rep. t. 31. *nec aliorum.*
5 láncea. DC. (*ye.*) lance-leaved. C. B. S. 1816. 5. 7. G. ♂. Th. ac. holm. 1800. t. 4.
Œdmánnia láncea. Th.
6 opposìta. DC. (*ye.*) opposite-leaved. ——— 1825. —— G. ♄.
7 angulàta. DC. (*ye.*) angular-stalked. ——— 1816. —— G. ♄.
8 filifòlia. DC. (*ye.*) narrow-leaved. ——— —— —— G. ♄.
VASC'OA. DC. VASC'OA. Monadelphia Decandria.
1 amplexicáulis. DC. (*ye.*) stem-clasping. C. B. S. 1812. 6. 7. G. ♄.
Crotalària amplexicáulis. L. *Ráfnia amplexicáulis.* Th.
2 perfoliàta. DC. (*ye.*) perfoliate. C. B. S. 1812. 7. 8. G. ♄. Seba. thes. 1. t. 24. f. 5.
Borbònia perfoliàta. Th. *Crotalária amplexicáulis.* Lam.
BORB'ONIA. DC. BORB'ONIA. Monadelphia Decandria.
1 barbàta. DC. (*ye.*) bearded. C. B. S. 1816. 6. 8. G. ♄. Lam. ill. t. 610. f. 2.
2 trinérvia. DC. (*ye.*) three-nerved. ——— 1759. —— G. ♄. Pluk. alm. t. 297. f. 4.
3 lanceolàta. DC. (*ye.*) many-nerved. ——— 1752. —— G. ♄. Jacq. schœnb. 2. t. 217.
4 cordàta. DC. (*ye.*) heart-leaved. ——— 1759. —— G. ♄. —— ——— 2. t. 218.
5 ruscifòlia. DC. (*ye.*) Ruscus-leaved. ——— 1816. —— G. ♄. Bot. mag. 2128.

6 parviflòra. DC.	(*ye.*)	small-flowered.	C. B. S.	1821.	6. 8.	G. ♄.	
7 crenàta. DC.	(*ye.*)	notch-leaved.	———	1774.	——	G. ♄.	Bot. mag. 274.
8 ciliàta. DC.	(*ye*)	fringed.	———	1816.	——	G. ♄.	Houtt. syst. 8. t. 62 f.2.
ACHYR'ONIA. DC.		ACHYR'ONIA.	Diadelphia Decandria.				
villòsa. DC.	(*ye.*)	villous.	N. Holland.	1810.	5. 8.	G. ♄.	Wendl. hort. her. 1. t. 12.
LIP'ARIA. DC.		LIP'ARIA.	Diadelphia Decandria.				
sphærica. DC.	(*ye.*)	globe-headed.	C. B. S.	1794.	7. 8.	G. ♄.	Bot. mag. 1241.
PRIESTLE'YA. DC.		PRIESTLE'YA	Diadelphia Decandria.				
1 myrtifolia. DC.	(*ye.*)	Myrtle-leaved.	C. B. S.	1825.	4. 7.	G. ♄.	DC. Leg. mem. t. 29.
2 hirsùta. DC.	(*ye.*)	shaggy-stemmed.	———	1792.	12. 4.	G. ♄.	Bot. reg. 8.
Lipària hirsùta. B.R.							
3 lævigàta. DC.	(*ye.*)	polished.	———	1799.	7. 8.	G. ♄.	DC. Leg. mem. t. 30.
Borbònia lævigàta. Lod. bot. cab. 247.							
4 capitàta. DC.	(*ye.*)	headed.	———	1812.	3. 6.	G. ♄.	
5 graminifòlia. DC.	(*ye*)	narrow-leaved.	———	1800.	4. 7.	G. ♄.	
6 téres. DC.	(*ye.*)	smooth-stalked.	———	1816.	——	G. ♄.	
7 ericæfòlia. DC.	(*w.pu.*)	Heath-leaved.	———	1812.	4. 8.	G. ♄.	DC. Leg. mem. t. 31.
8 sericea. DC.	(*ye.*)	silky-leaved.	———	1794.	——	G. ♄.	Pluk. t. 388. f. 3.
9 axillàris. DC.	(*ye.*)	axillary-flower'd.	———	1824.	——	G. ♄.	DC. Leg. mem. t. 32.
10 ellíptica. DC.	(*ye.*)	elliptic-leaved.	———	1816.	——	G. ♄.	——— t. 33.
11 villòsa. DC.	(*ye.*)	villous-leaved.	———	1774.	——	G. ♄.	Seb. thes. 1. t. 24. f. 2.
12 vestíta. DC.	(*ye.*)	concave-leaved.	———	1800.	5. 6.	G. ♄.	Bot. mag. 2223.
Lipària vestíta. B.M. *Lipària villòsa.* A.R. 382. *nec aliorum.*							
13 tomentòsa. DC.	(*ye.*)	downy.	C. B. S.	1812.	5. 8.	G. ♄.	
14 umbellífera. DC.	(*ye.*)	umbel-flowered.	———	1825.	6. 7.	G. ♄.	
HA'LLIA. DC.		HA'LLIA.	Monadelphia Decandria.				
1 alàta. DC.	(*pu.*)	wing-stalked.	C. B. S.	1818.	6. 8.	G. ♄.	
2 fláccida. DC.	(*pu.*)	long-leaved	———	1789.	6. 9.	G. ♂.	
3 angustifòlia. DC.	(*pu.*)	narrow-leaved.	———	1816.	——	G. ♂.	
4 cordàta. DC.	(*pu.*)	heart-leaved.	———	1787.	——	G. ♄.	Jacq. schœnb. 3. t. 296.
Glycìne monophy'lla. Jacq. *Hedy'sarum cordàtum.* Th. n. ac. ups. 6. t. 1.							
5 Asarìna. DC.	(*pu.*)	Asarum-like.	C. B. S.	1816.	6. 9.	G. ♄.	
6 imbricàta. DC.	(*pu.*)	imbricated.	———	1812.	——	G. ♄.	Bot. mag. 1850. et 2596.
HEYLA'NDIA. DC.		HEYLA'NDIA.	Monadelphia Decandria.				
hebecárpa. DC.	(*ye.*)	hairy-podded.	Ceylon.	1820.	6. 8.	S. ♄.	DC. Leg. mem. t. 34.
CROTAL'ARIA. DC.		CROTAL'ARIA.	Monadelphia Decandria.				
1 alàta. DC.	(*st.*)	winged-stalked.	Nepaul.	1818.	4. 8.	S. ♄.	
2 platycárpa. DC.	(*ye.*)	flat-podded.	N. America.	1824.	7. 9.	H. ⊙.	
3 Púrshii. DC.	(*ye.*)	Pursh's.	———	1827.	——	H. ⊙.	Pluk. alm. t. 277. f. 2.
lævigàta. Ph. *non* Lam.							
4 parviflòra. DC.	(*ye.*)	small-flowered.	———	1823.	6. 9.	H. ⊙.	Pluk. alm. t. 169. f. 6 ?
5 sagittàlis. DC.	(*ye.*)	Virginian.	———	1731.	——	H. ⊙.	Herm. lugd. p. 203. ic.
6 ovàlis. DC.	(*ye.*)	oval-leaved.	———	1827.	——	H. ⊙.	
7 rubiginòsa. DC.	(*ye.*)	rusty.	E. Indies.	1807.	——	S. ⊙.	
8 verrucòsa. DC.	(*bl.*)	blue-flowered.	———	1731.	——	S. ⊙.	Andr. reposit. 308.
cærùlea. Jacq. ic. t. 144. *angulòsa.* Cav. ic. 4. t. 321.							
9 semperflòrens. DC.	(*ye.*)	ever-blowing.	E. Indies.	1816.	3. 9.	S. ♄.	Vent. hort. cels. t. 17.
10 retùsa. DC.	(*ye.*)	wedge-leaved.	———	1731.	5. 9.	S. ⊙.	Bot. reg. 253.
11 Lechenáultii. DC.	(*ye.*)	Lechenault's.	———	1826.	——	S. ⊙.	
12 spectábilis. DC.	(*pu.*)	showy.	———	1824.	——	S. ⊙.	
13 pulchérrima. DC.	(*ye.*)	Mysore.	———	1814.	——	S. ♃.	Bot. mag. 2027.
14 bengalénsis. DC.	(*ye.*)	Bengal.	———	1806.	——	S. ⊙.	Pluk. alm. t. 169. f. 5.
15 júncea. DC.	(*ye.*)	rush-stalked.	———	1700.	6. 9.	S. ⊙.	Bot. mag. 490.
16 tenuifòlia. DC.	(*ye.*)	slender-leaved.	———	1818.	——	S. ⊙.	
17 fenestràta. DC.	(*ye.*)	window-calyxed.	———	1816.	——	S. ⊙.	Bot. mag. 1933.
18 sericea. DC.	(*ye.*)	silky.	———	1807.	——	S. ⊙.	
19 técta. DC.	(*ye.*)	striped-flower'd.	———	1824.	6. 10.	S. ⊙.	
20 Burmánni. DC.	(*ye.*)	Burmann's.	———	———	——	S. ⊙.	Burm. ind. t. 48. f. 1.
21 hirsùta. DC.	(*ye.*)	hairy-podded.	———	1823.	——	S. ⊙.	
22 mysorénsis. DC.	(*ye.*)	Mysore.	———	1826.	——	S. ⊙.	
23 montàna. DC.	(*ye.*)	mountain.	———	1816.	——	S. ♃.	
24 álbida. DC.	(*wh.*)	white-flowered.	———	1824.	——	S. ⊙.	
25 paniculàta. DC.	(*ye.*)	panicled.	———	1807.	6. 8.	S. ♄.	
26 púlchra. DC.	(*ye.*)	short-podded.	———	1800.	3. 5.	S. ♄.	Andr. reposit. t. 601.
27 Nòvæ-Hollándiæ. DC.	(*pu.*)	NewHolland.	N. Holland.	1820.	6. 8.	S. ♄.	
28 paulìna. DC.	(*ye.*)	St. Paul.	Brazil.	———	7. 9.	S. ♄.	Schr. pl. rar. h. mon. t. 88.
29 breviflòra. DC.	(*ye.*)	short-flowered.	———	1824.	——	S. ♄.	
30 bifària. DC.	(*ye.*)	two-ranked.	E. Indies.	———	——	S. ♄.	
31 dichótoma. DC.	(*ye.*)	forked.	———	———	——	S. ♄.	

32 diffùsa. DC.	(*ye*)	diffuse.	....	1818.	6. 8.	S. ♃.	
33 biflòra. DC.	(*ye.*)	two-flowered.	E. Indies.	1790.	——	S. ⊙.	Petiv. gaz. t. 30. f. 10.
34 nàna. DC.	(*ye.*)	dwarf.	————	——	——	S. ⊙.	Burm. ind. t. 48. f. 2.
35 bialàta. DC.	(*ye.*)	two-winged.	————	1818.	——	S. ⊙.	
36 tetragòna. DC.	(*ye.*)	square-stalked.	Nepaul.	1806.	8. 9.	S. ♂.	Andr. rep. t. 593.
37 linifòlia. DC.	(*ye.*)	Flax-leaved.	E. Indies.	1823.	6. 7.	S. ⊙.	
38 pellìta. DC.	(*ye.*)	villous-podded.	Jamaica.	1822.	7. 9.	S. ħ.	
39 virgúltalis.	(*ye.*)	twiggy.	C. B. S.	1816.	——	G. ħ.	
40 acuminàta. DC.	(*ye.*)	taper-pointed.	————	——	——	G. ħ.	
41 spartioídes. DC.	(*ye.*)	Broom-like.	————	——	——	G. ħ.	
42 thebaíca. DC.	(*ye.*)	stripe-flowered.	Egypt.	1828.	7. 8.	G. ħ.	Delil. fl. eg. t. 37. f. 1.
43 nummulària. DC.	(*ye.*)	money-wort-l^d.	E. Indies.	——	——	S. ♃.	
44 tuberòsa. DC.	(*pu.*)	tuberous-rooted.	Nepaul.	1816.	6. 7.	S. ♃.	
45 anthylloídes. DC.	(*ye.*)	large-calyxed.	Java.	1789.	8. 9.	S. ♄.	
46 calycìna. DC.	(*st.*)	hispid-calyxed.	E. Indies.	1818.	7. 9.	S. ♃.	Schr. pl. rar. mon. t. 12.
47 nepalénsis. L.en.	(*bl.*)	Nepaul.	Nepaul.	1821.	6. 8.	S. ⊙.	
48 Roxburghiàna. DC.	(*ye.*)	Roxburgh's.	E. Indies.	1823.	7. 9.	S. ♃.	
49 chinénsis. DC.	(*ye.*)	Chinese.	China.	1820.	6. 7.	S. ♄.	
50 hírta. DC.	(*ye.*)	hairy.	E. Indies.	1816.	6. 8.	S. ⊙.	Mart. acad. mun. 6. t. F.
51 prostràta. DC.	(*wh.*)	prostrate.	————	1804.	——	S. ♃.	———— t. E.
52 stipulàcea. H.B.	(*ye.*)	large-stipuled.	————	1823.	——	S. ♃.	
53 arboréscens. DC.	(*ye.*)	spreading.	C. B. S.	1774.	6. 10.	G. ♄.	Jacq. vind. 3. t. 64.
incanéscens. W. *capensis.* Th.							
54 túrgida. DC.	(*ye.*)	turgid-podded.	....	1820.	——	G. ħ.	Herb. amat. t. 238.
55 laburnifòlia. DC.	(*ye.*)	Laburnum-l^d.	E. Indies.	1739.	7. 9.	S. ♂.	Rheed. mal. 9. t. 27.
56 péndula. DC.	(*ye.*)	pendulous-pod'^d.	W. Indies.	1820.	——	S. ħ.	
57 Brównei. DC.	(*ye.*)	Browne's.	Jamaica.	——	——	S. ♄.	
58 bracteàta. DC.	(*ye.*)	large-bracted.	E. Indies.	1818.	5. 9.	S. ♄.	
59 cytisoídes. DC.	(*ye.*)	Cytisus-like.	Nepaul.	1823.	6. 8.	S. ♄.	
60 foliòsa. DC.	(*ye.*)	leafy.	E. Indies.	1826.	——	S. ♄.	
61 orixénsis. DC.	(*ye.*)	minute-flower'd.	————	1816.	7. 8.	S. ⊙.	Mart. acad. mun. 6. t. H.
62 virgàta. DC.	(*ye.*)	slender.	————	1824.	——	S. ⊙.	———— 6. t. G.
63 purpuráscens. DC.	(*pu.*)	purplish.	Mauritius.	1823.	8. 10.	S. ⊙.	
64 incàna. DC.	(*ye.*)	hoary.	W. Indies.	1714.	6. 7.	S. ⊙.	Bot. reg. 377.
65 mìcans. DC.	(*ye.*)	glittering.	E. Indies.	1823.	——	S. ♃.	
66 curtàta. DC.	(*ye.*)	short-keeled.	....	1820.	——	S. ⊙.	
67 vitellìna. DC.	(*ye.*)	yellow-twigged.	Brazil.	1819.	5. 7.	S. ♄.	Bot. reg. 447.
68 medicagínea. DC.	(*ye.*)	Trefoil-leaved.	E. Indies.	1818.	6. 7.	S. ♄.	Rott. n. act. nat. c. 4. t. 5.
69 Grahámi.	(*ye.*)	Graham's.	Mexico.	1824.	9. 12.	S. ħ.	Bot. mag. 2714.
dichótoma B.M. *non* DC							
70 Bojèri.	(*ye.*)	Bojer's.	Madagascar.	1828.	....	S. ħ.	
dichótoma. Bojer. *nec aliorum.*							
71 trichótoma. Boj.	(*ye.*)	three-forked	————	——	....	S ħ.	
72 stríćta. DC.	(*ye.*)	upright.	E. Indies.	1826.	8. 10.	S. ⊙.	
73 purpùrea. DC.	(*pu.*)	purple-flowered.	C. B. S.	1790.	3. 5.	G. ♄.	Bot. reg. 128.
74 microphy'lla. DC.	(*ye.*)	small-leaved.	Arabia.	1823.	8. 10.	H. ⊙.	
75 micrántha. DC.	(*ye.*)	small-blossom'd.	Ceylon.	1820.	6. 7.	S. ⊙.	
76 pállida. DC.	(*st.*)	pale-flowered.	Africa.	1775.	——	S. ⊙.	
77 argéntea. DC.	(*ye.*)	silvery-leaved.	C. B. S.	1823.	——	G. ♄.	Jacq. schœnb. 2. t. 220.
78 pulchélla. DC.	(*ye.*)	large-flowered.	————	1800.	——	G. ♄.	Bot. mag. 1699.
79 Saltiàna. A.R.	(*ye.*)	Salt's.	Abyssinia.	1810.	6. 7.	S. ♄.	Andr. reposit. 648.
80 angustifòlia. DC.	(*st.*)	narrow-leaved.	C. B. S.	1815.	5. 9.	G. ħ.	Jacq. schœnb. 2. t. 219.
81 villòsa. DC.	(*ye.*)	villous.	————	1823.	——	G. ♃.	
82 latifòlia.	(*ye.*)	broad-leaved.	E. Indies.	1825.	——	S. ħ.	
83 ellíptica. H.B.	(*ye.*)	elliptic-leaved.	China.	1823.	7. 9.	G. ħ.	
84 lanceolàta. H.B.	(*ye.*)	spear-leaved.	W. Indies.	1806.	——	S. ħ.	
85 lotifòlia. DC.	(*ye.*)	Lotus-leaved.	————	1732.	6. 7.	S. ♄.	Dill. elth. t. 102. f. 131.
86 axillàris. DC.	(*ye.*)	axillary-fl^d.	Guinea.	1781.	7. 8.	S. ⊙.	
87 quinquefòlia. DC.	(*ye.*)	five-leaved.	E. Indies.	1792.	6. 7.	S. ⊙.	Rheed. mal. 9. t. 28.
HYPOCALY'PTUS. DC.		HYPOCALY'PTUS. Monadelphia Decandria.					
obcordàtus. DC.	(*pu.*)	obcordate-leav'd.	C. B. S.	1790.	4. 6.	G. ħ.	Lodd. bot. cab. 1158.
Crotalària cordifòlia. B.C.							
VIBO'RGIA. DC.		VIBO'RGIA. Monadelphia Decandria.					
1 obcordàta. DC.	(*ye.*)	many-flower'd.	C. B. S.	1812.	7. 9.	G. ♄.	Lodd. bot. cab. 509.
Crotalària floribúnda. B.C. *Anthy'llis cuneàta.* Hort.							
2 serícea. DC.	(*ye.*)	silky.	C. B. S.	1780.	——	G. ♄.	
LODDIG'ESIA. DC.		LODDIG'ESIA. Monadelphia Decandria.					
oxalidifòlia. DC.	(*ro.*)	Oxalis-leaved.	C. B. S.	1802.	5. 9.	G. ħ.	Bot. mag. 965.
DICH'ILUS. DC.		DICH'ILUS. Monadelphia Decandria.					
lebeckoídes. DC.	(*st.*)	Lebeckia-like.	C. B. S.	1816.	6. 8.	G. ħ.	DC. leg. mem. 6. t. 35.

LEBE'CKIA. DC.		LEBE'CKIA.	Monadelphia Decandria.				
1 subnùda. DC.	(ye.)	few-leaved.	C. B. S.	1823.	4. 6.	G. ♄.	
2 contaminàta. DC.	(ye.)	narrow-leaved.	——	1787.	——	G. ♄.	
3 sepiària. DC.	(ye.)	filiform-leaved.	——	1816.	——	G. ♄.	Pluk. alm. t. 424. f. 1.
4 subternàta. DC.	(ye.)	various-leaved.	——	——	——	G. ♄.	
5 serícea. DC.	(ye.)	silky-leaved.	——	1774.	——	G. ♄.	
6 cytisoídes. DC.	(ye.)	Cytisus-like.	——	——	——	G. ♄.	Com. hort. amst. 2. t. 107.
SARCOPHY'LLUM. DC.		SARCOPHY'LLUM.	Monadelphia Decandria.				
carnósum. DC.	(ye.)	fleshy-leaved.	C. B. S.	1812.	5. 8.	G. ♄.	Bot. mag. 2502.
ASPA'LATHUS. DC.		ASPA'LATHUS.	Monadelphia Decandria.				
1 spinòsa. DC.	(ye.)	spiny.	C. B. S.	1823.	5. 8.	G. ♄.	Breyn. cent. t. 26.
2 aculeàta. DC.	(ye.)	yellow-spined.	——	——	——	G. ♄.	
3 acuminàta. DC.	(ye.)	taper-pointed.	——	1816.	——	G. ♄.	Lam. ill. t. 620. f. 4.
4 laricifòlia. DC.	(ye.)	Larch-leaved.	——	1825.	——	G. ♄.	
5 capitàta. DC.	(ye.)	headed.	——	1823.	6. 7.	G. ♄.	Lam. ill. t. 620. f. 2.
6 astroítes. DC.	(ye.)	spiny-bracted.	——	1825.	5. 8.	G. ♄.	Seba. thes. t. 24. f. 6.
7 chenopòda. DC.	(or.)	cluster-leaved.	——	1759.	7. 8.	G. ♄.	Bot. mag. 2225.
8 álbens. DC.	(wh.)	white-leaved.	——	1774.	——	G. ♄.	
9 argyr'æa. DC.	(wh.)	silvery.	——	1818.	——	G. ♄.	
10 hy'strix. DC.	(wh.)	silky-spined.	——	1824.	——	G. ♄.	Lam. ill. t. 620. f. 1.
11 thymifòlia. DC.	(ye.)	Thyme-leaved.	——	1825.	5. 8.	G. ♄.	Pluk. alm. t. 413. f. 1.
12 ericifòlia. DC.	(ye.)	Heath-leaved.	——	1780.	7. 8.	G. ♄.	—— t. 413. f. 6.
13 híspida. DC.	(ye.)	hispid.	——	1823.	——	G. ♄.	
14 asparagoídes. DC.	(ye.)	Asparagus-like.	——	1812.	——	G. ♄.	
15 multiflòra. DC.	(ye.)	many-flowered.	——	1818.	——	G. ♄.	
16 uniflora. DC.	(ye.)	one-flowered.	——	1812.	——	G. ♄.	Pluk. mant. t. 414. f. 7.
17 carnòsa. DC.	(ye.)	fleshy-leaved.	——	1795.	5. 8.	G. ♄.	Bot. mag. 1289.
18 pínguis. DC.	(ye.)	succulent-leaved.	——	1818.	7. 8.	G. ♄.	
19 crassifòlia. DC.	(ye.)	thick-leaved.	——	1800.	——	G. ♄.	Andr. reposit. t. 351.
20 aff ínis. DC.	(ye.)	likened.	——	1816.	——	G. ♄.	
21 genistoídes. DC.	(ye.)	Genista-like.	——	1825.	6. 8.	G. ♄.	
22 squarròsa. DC.	(ye.)	squarrose.	——	——	——	G. ♄.	
23 galioídes. DC.	(ye.)	Galium-like.	——	1816.	7. 8.	G. ♄.	
24 pìnea. DC.	(ye.)	Pine-leaved.	——	——	——	G. ♄.	
25 abietìna. DC.	(ye.)	Fir-leaved.	——	1826.	——	G. ♄.	
26 araneòsa. DC.	(pa.)	cobweb.	——	1795.	6. 7.	G. ♄.	Bot. mag. 829.
27 globòsa. A.R.	(ye.)	globe-headed.	——	1802.	——	G. ♄.	Andr. reposit. 302.
28 ciliaris. DC.	(ye.)	ciliated.	——	1790.	7. 8.	G. ♄.	Bot. mag. 2233.
29 quinquefòlia. DC.	(br.)	five-leaved.	——	1816.	——	G. ♄.	
30 cándicans. DC.	(ye.)	white.	——	1774.	——	G. ♄.	
31 heterophy'lla. DC.	(ye.)	various-leaved.	——	1828.	——	G. ♄.	
32 argéntea. DC.	(ye.)	silvery-leaved.	——	1759.	——	G. ♄.	
33 subulàta. DC.	(ye.)	awl-leaved	——	1789.	——	G. ♄.	
34 serícea. DC.	(ye.)	silky.	——	1816.	——	G. ♄.	
35 callòsa. DC.	(ye.)	oval-spiked.	——	1812.	——	G. ♄.	Bot. mag. 2329.
36 fusca. DC.	(ye.)	brown.	——	1816.	——	G. ♄.	
37 mucronàta. DC.	(ye.)	thorny-branch'd.	——	1796.	6. 7.	G. ♄.	
38 pedunculàta. DC.	(ye.)	small-leaved.	——	1775.	7. 8.	G. ♄.	Bot. mag. 344.
39 bracteata. DC.	(ye.)	bracteate.	——	1816.	——	G. ♄.	
ULEX. DC.		FURZE.	Monadelphia Decandria.				
1 europ'æus. DC.	(ye)	common.	Britain.	....	3. 7.	H. ♄.	Eng. bot 742.
2 nànus. DC.	(ye.)	dwarf.	——	....	1. 12.	H. ♄.	—— t. 743.
3 provinciàlis. DC.	(ye.)	Provence.	S. Europe.	1815.	6. 8.	H. ♄.	Loisel. not. 105. t. 6. f. 2.
STAURACA'NTHUS. DC.		STAURACA'NTHUS.	Monadelphia Decandria.				
aphy'llus. DC.	(ye.)	leafless.	Portugal.	1815.	6. 8.	H. ♄.	
SPA'RTIUM. DC.		SPANISH BROOM.	Monadelphia Decandria.				
júnceum. DC.	(ye.)	common.	S. Europe.	1548.	7. 9.	H. ♄.	Bot mag. 85.
Spartiánthus júnceus. L. en.							
β flòre plèno.	(ye.)	*double-flowered.*	——	....	——	H. ♄.	
GENI'STA. DC.		GENI'STA.	Monadelphia Decandria.				
1 parviflòra. DC.	(ye.)	small-flowered.	Levant.	1739.	5. 7.	H. ♄.	Venten. cels. t. 87.
Spártium parviflòrum. Vent.							
2 clavàta. DC.	(ye.)	clavate.	Barbary.	1812.	5. 8.	F. ♄.	Venten. hort. cels. t. 17.
Spártium seríceum. Vent. *non* Ait.							
3 canariénsis. DC.	(ye.)	Canary.	Canaries.	1656.	5. 9.	F. ♄.	Bot. reg. t. 217.
4 cándicans. DC.	(ye.)	hoary.	Spain.	1735.	4. 7.	H. ♄.	Wats. dend. brit. t. 80.
5 pàtens. DC.	(ye.)	spreading.	——	1825.	5. 7.	H. ♄.	Cavan. ic. 2. t. 176.
6 linifòlia. DC.	(ye.)	Flax-leaved.	——	1739.	1. 6.	F. ♄.	Bot. mag. 442.
Spártium linifòlium. Desf. atl. 2. t. 181.							

7	triquètra. DC.	(*ye.*)	triangular.	Corsica.	1770.	5. 6.	H. ♄.	Wats. dend. brit. t. 79
8	bracteolàta. DC.	(*ye.*)	bracteolate.	....	1826.	——	H. ♄.	
9	microphy'lla. DC.	(*ye.*)	small-leaved.	Canaries.	1824.	——	F. ♄.	
10	umbellàta. DC.	(*ye.*)	umbelled.	Barbary.	1799.	4. 6.	F. ♄.	Desf. atl. 2. t. 180.
	Spártium umbellàtum. Desf.							
11	radiàta. DC.	(*ye.*)	rayed.	Italy.	1758.	5. 7.	H. ♄.	Bot. mag. 2260.
	Spártium radiàtum. B.M. Mill. ic. t. 249. f. 1.							
12	hórrida. DC.	(*ye.*)	large-spined.	Pyrenees.	1826.	5. 7.	H. ♄.	Gilib. b. prat. 2. p. 239. ic.
13	lusitànica. DC.	(*ye.*)	Portugal.	Portugal.	1771.	3. 5.	H. ♄.	Andr. reposit. t. 419.
14	Lobèlii. DC.	(*ye.*)	Lobel's.	S. Europe.	1826.	4. 6.	H. ♄.	Lob. adv. p. 409. icon.
15	fèrox. DC.	(*ye.*)	sharp-spined.	Barbary.	1800.	6. 7.	F. ♄.	Desf. atl. 2. t. 182.
16	Cupàni. DC.	(*ye.*)	Cupani's.	Naples.	1826.	——	H. ♄.	Cup pan. ed. 1. v. 2. t. 233.
17	triacánthos. DC.	(*ye.*)	three-spined.	Portugal.	1821.	5. 6.	H. ♄.	Brot. phyt. 130. t. 54.
18	cuspidòsa. DC.	(*ye.*)	sharp-pointed.	C. B. S.	1816.	——	G. ♄.	
19	tricuspidàta. DC.	(*ye.*)	three-pointed.	Algiers.	1826.	——	F. ♄.	Desf. alt. 2. t. 183.
20	gibraltárica. DC.	(*ye.*)	Gibraltar.	Gibraltar.	——	——	F. ♄.	
21	sylvéstris. DC.	(*ye.*)	wood.	Hungary.	1824.	6. 8.	H. ♄.	Jacq. ic. rar. t. 557.
	hispánica. Jacq. *non* L.							
22	ægyptìaca. DC.	(*ye.*)	Egyptian.	Egypt.	1828.	——	H. ♄.	
23	falcàta. DC.	(*ye.*)	falcate-podded.	Portugal.	1821.	5. 7.	H. ♄.	Brot. phyt. 133. t. 55.
24	córsica. DC.	(*ye.*)	Corsican.	Corsica.	1826.	——	H. ♄.	
25	Scórpius. DC.	(*ye.*)	Scorpion.	S. Europe.	1570.	3. 4.	H. ♄.	Wats. dend. brit. t. 78.
26	hispánica. DC.	(*ye.*)	dwarf prickly.	Spain.	1759.	6. 8.	H. ♄.	Cavan. ic. 3. t. 211.
27	ánglica. DC.	(*ye.*)	petty whin.	Britain.	....	5. 8.	H. ♄.	Eng. bot. 132.
28	germánica. DC.	(*ye.*)	German.	Germany.	1773.	6. 8.	H. ♄.	Fuchs. hist. 220. ic.
29	púrgans. DC.	(*ye.*)	purging.	S. France.	1768.	6. 7.	H. ♄.	Bull. herb. t. 115.
30	cinèrea. DC.	(*ye.*)	ash-coloured.	————	1818.	5. 7.	H. ♄.	Wats. dend. brit. t. 76.
31	virgàta. DC.	(*ye.*)	long-twigged.	Madeira.	1777.	3. 6.	F. ♄.	Bot. mag. 2265.
32	congésta. DC.	(*ye.*)	close-branched.	Teneriffe.	....	4. 7.	F. ♄.	
33	serícea. DC.	(*ye.*)	silky-leaved.	Austria.	1812.	5. 6.	H. ♄.	Jacq. ic. rar. 3. t. 556.
34	aphy'lla. DC.	(*vi.*)	violet-coloured.	Volga.	1818.	6. 7.	H. ♄.	Pall. it. 3. app. t. 5. f. 2.
	Spártium aphy'llum. L. en.							
35	monospérma. DC.	(*wh.*)	single-seeded.	S. Europe.	1690.	——	F. ♄.	Bot. mag. 683.
	Spártium monospérmum. B.M.							
36	sphærocárpa. DC.	(*ye.*)	round-podded.	————	1731.	——	F. ♄.	Renealm. spec. t. 33.
37	æthnénsis. DC.	(*ye.*)	three-seeded.	Mt. Ætna.	1822.	6. 8.	F. ♄.	Bot. mag. 2674.
	Spartium ætnense. B.M.							
38	tetragòna. DC.	(*ye.*)	four-sided.	Poland.	1826.	——	H. ♄.	
39	depréssa. DC.	(*ye.*)	depressed.	Tauria.	1828.	....	H. ♄.	
40	scariòsa. DC.	(*ye.*)	scariose.	Italy.	1820.	6. 8.	H. ♄.	Viv. frag. ital. 1. t. 8.
41	anxántica. DC.	(*ye.*)	smooth.	————	——	——	H. ♄.	
42	tinctòria. DC.	(*ye.*)	Dyer's green-weed.	Britain.	....	——	H. ♄.	Eng. bot. 44.
	β *latifòlia.* DC.	(*ye.*)	*broad-leaved.*	Switzerland.	1824.	——	H. ♄.	
	γ *hirsùta.* DC.	(*ye.*)	*villous.*	————	....	——	H. ♄.	
	δ *praténsis.* DC.	(*ye.*)	*meadow.*	Italy.	....	——	H. ♄.	
43	sibírica. DC.	(*ye.*)	Siberian.	Siberia.	1785.	6. 8.	H. ♄.	Jacq. vind. 2. t. 190.
44	polygalæfòlia. DC.	(*ye.*)	Milkwort-leav'd.	Portugal.	1826.	——	H. ♄.	
45	flòrida. DC.	(*ye.*)	Spanish.	Spain.	1752.	——	H. ♄.	
46	mántica. DC.	(*ye.*)	prostrate.	Italy.	1824.	——	H. ♄.	Poll. fl. veron. 2. t. 4. f. 7.
47	ovàta. DC.	(*ye.*)	oval-leaved.	Hungary.	1816.	——	H. ♄.	Wats. dend. brit. t. 77.
48	pátula. DC.	(*ye.*)	spreading.	Siberia.	1828.	....	H. ♄.	
49	triangulàris. DC.	(*ye.*)	three-sided.	Hungary.	1815.	5. 6.	H. ♄.	W. et K. hung. 2. t. 153.
50	sagittàlis. DC.	(*ye.*)	jointed.	Germany.	1570.	——	H. ♄.	Jacq. aust. 3. t. 209.
51	diffùsa. DC.	(*ye.*)	diffuse.	Italy.	1816.	——	H. ♄.	Jacq. ic. rar. 3. t. 555.
	Spàrtium decúmbens. Jacq. *non* Ait.							
52	prostràta. DC.	(*ye.*)	trailing.	Italy.	1826.	——	H. ♄.	Reyn. mem. 1. p. 211. ic.
53	procúmbens. DC.	(*ye.*)	procumbent.	Hungary.	1816.	6. 8.	H. ♄.	Bot. reg. 1150.
54	pilòsa. DC.	(*ye.*)	hairy green-weed.	England.	....	5. 6.	H. ♄.	Eng. bot. 44.
55	micrántha. DC.	(*ye.*)	small-blossom'd.	Spain.	1824.	——	H. ♄.	Orteg. dec. 6. t. 10. f. 1.
56	pilocárpa. DC.	(*ye.*)	hairy-podded.	....	1820.	5. 7.	H. ♄.	
	CY'TISUS. DC.		CY'TISUS. Monadelphia Decandria.					
1	nubigènus. DC.	(*wh.*)	cluster-flower'd.	Teneriffe.	1779.	5. 8.	H. ♄.	
2	álbus DC.	(*wh.*)	white-flowered.	Portugal.	1752.	6. 7.	H. ♄.	Duham. arb. t. 23.
	Spártium multiflòrum. W.							
	β *incarnàtus* B.C.		*flesh-coloured.*	————	1823.	——	H. ♄.	Lodd. bot. cab. 1052.
3	Labúrnum. DC.	(*ye.*)	Laburnum.	Europe.	1596.	5. 6.	H. ♄.	Bot. mag. 176.
4	alpìnus. DC.	(*ye.*)	Scotch Laburnum.	Hungary.	....	6. 7.	H. ♄.	W. et K. hung. 3. t. 260.
5	nígricans. DC.	(*ye.*)	black-rooted.	Austria.	1730.	——	H. ♄.	Bot. reg. 802.
6	sessilifòlius. DC.	(*ye.*)	common.	Italy.	1629.	5. 6.	H. ♄.	Bot. mag. 255.

7 triflòrus. DC.	(*ye.*)	three-flowered.	Spain.	1640.	6. 7.	H. ♄.	Clus. hist. 1. p. 94. f. 3.
8 móllis. DC.	(*ye.*)	soft-leaved.	....	1820.	——	H. ♄.	
9 pátens. DC.	(*ye.*)	woolly-podded.	Portugal.	1752.	——	H. ♄.	
10 grandiflòrus. DC.	(*ye.*)	great-flowered.	————	1816.	——	H. ♄.	
11 scopàrius. DC.	(*ye.*)	commonBroom.	Britain.	....	4. 6.	H. ♄.	Eng. bot. 1330.
Spártium scopàrium. E.B.							
12 spinòsus. DC.	(*ye.*)	prickly.	S. Europe.	1596.	6. 7.	F. ♄.	Lobel. ic. 2. t. 95.
13 lanìgerus. DC.	(*ye.*)	villous.	————	1822.	——	F. ♄.	
Spártium lanígerum. Df. *Calycótome villòsa.* L.en.							
14 prolíferus. DC.	(*wh.*)	silky.	Canaries.	1779.	4. 5.	G. ♄.	Bot. reg. 121.
15 leucánthus. DC.	(*wh.*)	cream-colour'd.	Hungary.	1806.	6. 7.	H. ♄.	Bot. mag. 1438.
16 álbidus. DC.	(*wh.*)	whitish.	S. Europe.	1824.	——	F. ♄.	
17 purpùreus. DC.	(*pu.*)	purple-flower'd.	Austria.	1792.	5. 8.	H. ♄.	Bot. mag. 1176.
β álbus. DC.	(*wh.*)	*white-flowered.*	————	————	——	H. ♄.	
18 biflòrus. DC.	(*ye.*)	two-flowered.	Hungary.	1760.	5. 7.	H. ♄.	Bot. reg. 308.
19 elongàtus. DC.	(*ye.*)	long-branched.	————	1804.	——	H. ♄.	Wats. dend. brit. 82.
20 multiflòrus. B.R.	(*ye.*)	many-flowered.	————	1824.	——	H. ♄.	Bot. reg. 1191.
21 falcàtus. DC.	(*ye.*)	sickle-shaped.	————	1816.	6. 8.	H. ♄.	W. et K. hung. 3. t. 238.
22 austrìacus. DC.	(*ye.*)	Austrian.	Austria.	1741.	6. 9.	H. ♄.	Jacq. aust. 1. t. 21.
23 supìnus. DC.	(*ye.*)	trailing.	S. Europe.	1755.	5. 8.	H. ♄.	———— 1. t. 20.
24 hirsùtus. DC.	(*ye.*)	hairy.	————	1739.	6. 8.	H. ♄.	Jacq obs. 4. t. 96.
25 serotìnus. DC.	(*ye.*)	late-flowering.	Hungary.	1826.	——	H. ♄.	
26 capitàtus. DC.	(*ye.*)	cluster-flower'd.	Austria.	1774.	6. 7.	H. ♄.	Jacq. aust. t. 33.
27 ciliàtus. DC.	(*ye.*)	fringed.	Mt.Carpathian.	1824.	——	H. ♄.	
28 polytrìchus. DC.	(*ye.*)	many-haired.	Tauria.	1822.	——	H. ♄.	
29 argénteus. DC.	(*ye.*)	silver-leaved.	France.	1739.	7. 9.	H. ♄.	Lobel. ic. 2. p. 41. f. 2.
30 calycìnus. DC.	(*ye.*)	dwarf.	Levant.	1816.	6. 8.	H. ♄.	Lodd. bot. cab. t. 673.
nànus. Wats. dend. brit. t. 81.							
31 orientàlis. DC.	(*ye.*)	oriental.	————	....	5. 7.	G. ♄.	
32 canéscens. DC.	(*ye.*)	hoary.	....	1826.	——	H. ♄.	
33 pròcerus. DC.	(*ye.*)	tall.	Portugal.	1816.	6. 7.	H. ♄.	
ADENOCA'RPUS. DC.		ADENOCA'RPUS. Monadelphia Decandria.					
1 hispánicus. DC.	(*ye.*)	Spanish.	Spain.	1824.	5. 7.	F. ♄.	
2 intermèdius. DC.	(*ye.*)	intermediate.	S. Europe.	....	7. 8.	H. ♄.	Clus. hist. 1. p. 94. f. 1.
3 parvifòlius. DC.	(*ye.*)	small-leaved.	France.	....	4. 7.	H. ♄.	
4 telonénsis. DC.	(*ye.*)	loose-flowered.	S. Europe.	1824.	5. 7.	H. ♄.	Du.arb.ed.n.5.t.47.f.2.
Cy'tisus divaricàtus. β. Bot. mag. 1387 ?							
5 frankenioídes. DC.	(*ye.*)	Frankenia-like.	Teneriffe.	1815.	——	G. ♄.	
Genísta viscòsa. W. *Adenocárpus anag'yrus.* S.S.							
6 foliolòsus. DC.	(*ye.*)	leafy.	Canaries.	1770.	7. 8.	F. ♄.	Bot. mag. 426.
ON'ONIS. DC.		REST-HARROW. Monadelphia Decandria.					
1 críspa. DC.	(*ye.*)	curl-leaved.	Spain.	1739.	6. 8.	F. ♄.	Rœm. arch. 1. p. 3. t. 1.
2 hispánica. DC.	(*ye.*)	Spanish.	————	1799.	5. 9.	F. ♄.	Bot. mag. 2450.
3 vaginàlis. DC.	(*ye.*)	sheathed.	Egypt.	1815.	6. 8.	F. ♄.	Vent. cels. t. 32.
4 Nátrix. DC.	(*ye.*)	yellow-shrubby.	S. Europe.	1683.	5. 9.	F. ♄.	Bot. mag. 329.
5 pícta. DC.	(*std.*)	painted-flower'd.	Barbary.	1825.	6. 8.	H. ♃.	Desf. atl. 2. t. 137.
6 pínguis. W.	(*ye.*)	greasy.	————	1739.	7. 8.	F. ♄.	
7 longifòlia. DC.	(*ye.*)	long-leaved.	Canaries.	1816.	5. 6.	F. ♄.	
8 ramosíssima. DC.	(*ye.*)	branching.	Barbary.	1825.	7. 8.	H. ♄.	Desf. atl. 2. t. 186.
9 arenària. DC.	(*ye.*)	sand.	S. France.	———	——	H. ♄.	
10 biflòra. DC.	(*ye.*)	two-flowered.	Barbary.	1824.	——	H. ⊙.	
11 viscòsa. DC.	(*ye.*)	clammy.	S. France.	1750.	——	H. ⊙.	Barrel. ic. t. 1239.
12 brachycárpa. DC.	(*ye.*)	short-podded.	Spain.	1824.	——	H. ⊙.	
13 brevifòlia. DC.	(*ye.*)	short-leaved.	S. Europe.	....	——	H. ⊙.	
14 sícula. DC.	(*ye.*)	Sicilian.	Sicily.	1822.	6. 8.	H. ⊙.	
15 pubéscens. DC.	(*ye.*)	downy.	S. Europe.	1680.	——	H. ⊙.	Brot. phyt. 140. t. 58.
arthropòdia. Brot.							
16 ornithopodioídes. DC.	(*ye.*)	Bird's-foot.	Sicily.	1713.	——	H. ⊙.	Cavan. ic. 2. t. 192.
17 rotundifòlia. DC.	(*re.*)	round-leaved.	Switzerland.	1570.	5. 7.	H. ♄.	Bot. mag. 335.
18 fruticòsa. DC.	(*re.*)	shrubby.	S. Europe.	1680.	5. 7.	H. ♄.	———— t. 317.
19 tridentàta. DC.	(*re.*)	three-toothed.	Spain.	1752.	6. 8.	F. ♄.	Cavan. ic. 2. t. 152.
20 angustíssima. DC.	(*re.*)	narrow-leaved.	————	1826.	——	F. ♄.	
21 cenísia. DC.	(*re.*)	crested.	Italy.	1759.	——	H. ♃.	Barrel. ic. t. 1104.
22 fœtída. DC.	(*re.*)	strong-scented.	Barbary.	1804.	7. 10.	H. ⊙.	
23 geminiflòra. DC.	(*re.*)	twin-flowered.	Spain.	1825.	8. 10.	H. ⊙.	
24 Schóuwii. DC.	(*re.*)	Schouw's.	Italy.	1828.	7. 10.	H. ⊙.	
25 reclinàta. DC.	(*re.*)	reclined.	S. Europe.	1800.	6. 8.	H. ⊙.	Barrel. ic. t 761.
26 móllis. DC.	(*re.*)	soft.	Spain.	1824.	——	H. ⊙.	
27 Cherlèri. DC.	(*re.*)	dwarf.	S. Europe.	1771.	6. 7.	H. ⊙.	Bauh. hist. 2. p. 394. f. 2

28 péndula. DC. (re.) pendulous. Barbary. 1826. 7. 9. H. ⊙. Desf. atl. 2. t. 191.
29 Siebèri. DC. (re.) Sieber's. Crete. —— —— H. ⊙.
30 arboréscens. DC. (re.) tree. Barbary. 1825. 6. 8. F. ♄. Desf. atl. 2. t. 193.
31 altíssima. DC. (re.) stinking. Italy. 1596. 5. 8. H. ♃. All. ped. t. 41. f. 1.
hircìna. Jacq. vind. t. 93.
32 procúrrens. DC. (re.) rooting-branch'd. Europe. .... 6. 8. H. ħ. Fuchs. hist. 60. ic.
33 répens. W. (re.) creeping. Britain. .... —— H. ♃. Dill. elth. t. 25. f. 28.
34 spinòsa. DC. (re.) Cammock. ———— .... —— H. ħ. Eng. bot. 682.
arvénsis. E.B.
35 antiquorum. W. (re.) tall. S. Europe. 1790. —— H. ♃.
36 mitíssima. DC. (re.) cluster-flower'd. Portugal. 1732. 6. 7. H. ⊙. Dill. elth. t. 24. f. 27.
37 diffùsa. DC. (re.) diffuse. Italy. 1824. 7. 8. H. ⊙.
38 serràta. DC. (pu.wh.) serrate. Egypt. 1828. 6. 10. H. ⊙.
39 hírta. DC. (bl.) hairy. Levant. —— —— H. ♃.
40 alopecuroídes. DC. (re.) Fox-tail. Portugal. 1696. 7. 8. H. ⊙. Schkuhr. handb. t. 194.
41 álba. DC. (wh.) white-flowered. Naples. 1826. —— H. ⊙.
42 oligophy'lla. DC. (re.w.) few-leaved. ———— 1824. —— H. ⊙. Ten. fl. nap. 2. t. 67.
43 capitàta. DC. (ye.) capitate. Spain. 1818. 6. 7. H. ♃. Cavan. ic. 2. t. 159. f. 2.
44 striàta. DC. (ye.) striated. S. France. 1814. —— H. ♃.
45 Colúmnæ. DC. (ye.) small-flowered. Europe. 1732. —— H. ♃. All. p. 1. n. 1166. t. 20. f. 3.
minutíssima. Jacq. aust. t. 240. *parviflora.* P.S.
46 minutíssima. DC. (ye.) pointed-calyx'd. S. Europe. 1818. 6. 8. H. ⊙. Cavan. ic. 2. t. 153.
barbàta. Cav. *saxátilis.* Lam.
47 variegàta. DC. (ye.) variegated. ———— 1784. 7. 8. H. ⊙. Desf. atl. 2. t. 185.
48 pállida. DC. (pa.) pale. .... 1824. —— H. ⊙.
49 emarginàta. Boj. .... emarginate. Mauritius. —— .... S. ♄.
50 villòsa. DC. (ye.) villous. C. B. S. 1825. 7. 9. G. ħ.
51 decúmbens. DC. (ye.) decumbent. ———— 1822. —— G. ♃.
52 geminàta. DC. (ye.) twin-podded. ———— 1787. —— G. ♃.
53 umbellàta DC. (ye.) umbel-flowered. ———— 1818. —— G. ♃.
54 glábra. DC. (ye.) smooth-leaved. ———— 1816. —— G. ħ.
55 capénsis. DC. (pu.) Cape annual. ———— 1823. —— H. ⊙.
56 cérnua. DC. (ye.) hanging-podded. ———— 1774. —— G. ♄.

REQUIE'NIA. DC. REQUIE'NIA. Monadelphia Decandria.
1 obcordàta. DC. (ye.) obcordate. Senegal. 1824. 6. 8. S. ♄. DC. leg. mem. t. 37.
2 sphærospérma. DC. (ye.) round-seeded. C. B. S. 1816. 4. 7. G. ♄. ———— t. 38.

ANTHY'LLIS. DC. KIDNEY VETCH. Monadelphia Decandria.
1 Gerárdi. DC. (ro.) Gerard's. S. Europe. 1806. 6. 8. H. ♃. Ger. gallopr. t. 18.
2 onobrychioídes. DC. (y.) Saintfoin-like. Spain. 1818. —— H. ♃. Cavan. ic. 2. t. 150.
3 cytisoídes. DC. (ye.) downy-leaved. ———— 1731. 4. 6. F. ♄. Barrel. ic. 1182.
4 genístæ. DC. (ye.) Genista-like. ———— 1824. —— F. ♄.
5 Hermánniæ. DC. (ye.) Lavender-ld. Levant. 1739. 4. 7. F. ♄. Bot. mag. 2576.
6 aspálathi. DC. (ye.) spiny. ———— —— —— F. ♄.
7 erinàcea. DC. (bl.) prickly. Spain. 1759. 4. 6. H. ħ. Bot. mag. 676.
8 hy'strix. (wh.) porcupine. S. Europe. .... —— H. ♄.
9 Bárba-Jòvis. (st.) Jupiter's-beard. ———— 1640. 3. 5. G. ♄. Bot. mag. 1927.
10 heterophy'lla. DC. (v.) various-leaved. ———— 1768. 6. 7. G. ♄.
11 serícea. DC. (ye.) silky. Spain. 1786. —— F. ħ.
12 montàna. DC. (ro.) mountain. Europe. 1759. —— H. ♃. Swt. br. fl. gar. 1. t. 79.
13 Vulnerària. DC. (ye.) common. Britain. .... 5. 8. H. ♃. Eng. bot. 104.
14 rústica. Mil. (wh.) white-flower'd. .... —— H. ♃.
15 Dillénii. Sct. (sc.) red-flowered. S. Europe. .... —— H. ♃. Dill. elth. t. 320. f. 413.
16 polyphy'lla. Kit. (ye.) many-leaved. Hungary. 1827. —— H. ♃.
17 tetraphy'lla. DC. (st.) four-leaved. S. Europe. 1640. 7. 8. H. ⊙. Bot. mag. 108.
18 cornícina. DC. (st.) horned. Spain. 1759. —— H. ⊙. Cavan. ic. 1. t. 39. f. 2.
19 hamòsa. DC. (st.) hooked. Barbary. 1820. —— H. ⊙.
20 lotoídes. DC. (st.) Lotus-like. Spain. 1739. 6. 7. H. ⊙. Cavan. ic. 1. t. 40.
21 chilénsis. DC. (st.) Chile. Chile. 1824. —— G. ♃.
Lòtus subpinnàtus. Lag.

Subtribus II. *TRIFOLIEÆ.* DC. prodr. 2. p. 171.

MEDIC'AGO. DC. MEDICK. Diadelphia Decandria.
1 circinnàta DC. (ye) kidney-podded. Italy. 1640. 7. 8. H. ⊙.
Hymenocárpos circinnàtus. L. en.

2	nummulària. DC.	(*ye.*)	money-wort.	S. Europe.	....	——	H. ⊙.	Gært. sem. 2. t.155.f.7.
	circinnàta. G. *Hymenocárpos nummulàrius.* L. en.							
3	radiàta. DC.	(*ye.*)	ray-podded.	Italy.	1629.	6. 7.	H ⊙.	Moris. his. s.2. t.15. f. 3.
4	rupéstris. DC.	(*ye.*)	rock.	Tauria.	1827.	6. 8.	H. ♃.	
5	brachycárpa. DC.	(*ye.*)	short-podded.	Tiflin.	1818.	7. 8.	H. ⊙.	
6	lupulìna. DC.	(*ye.*)	nonesuch.	Britain.	....	5. 8.	H. ♂.	Eng. bot. 971.
7	cretàcea. DC.	(*ye.*)	suffrutescent.	Tauria.	1805.	7. 8.	H. ♄.	
8	falcàta. DC.	(*ye.*)	sickle-podded.	Britain.	....	——	H. ♃.	Eng. bot. 1016.
9	procúmbens. DC.	(*ye.*)	procumbent.	Poland.	1827.	——	H. ♃.	
10	cancellàta. DC.	(*ye.*)	netted-podded.	Caucasus.	1828.	——	H. ♃.	
11	suffruticòsa. DC.	(*vi.ye.*)	frutescent.	Pyrenees.	1818.	6. 8.	H. ♄.	
12	arbòrea. DC.	(*ye.*)	moon trefoil.	Italy.	1596.	5. 11.	H. ♄.	Flor græc. 767.
13	sibírica. DC.	(*ye.*)	Siberian.	Siberia.	1824.	6. 9.	H. ⊙.	
14	mèdia. DC.	(*bl.ye.*)	pale-flowered.	Europe.	....	6. 7.	H. ♃.	
15	satìva. DC.	(*bl.*)	Lucerne.	Britain.	....	——	H. ♃.	Eng. bot. 1749.
	β *versicolor.* DC.	(*bl.y.*)	*various-colour'd.*	———	....	——	H. ♃.	
16	mexicàna.	(*bl.*)	Mexican.	Mexico.	1824.	5. 8.	F. ♄.	
17	prostràta. DC.	(*ye.*)	prostrate.	Hungary.	1793.	7. 8.	H. ♃.	Jacq. vind. 1. t. 89.
18	glomeràta. DC.	(*ye.*)	clustered.	Italy.	1818.	——	H. ♃.	
19	glutinòsa. DC.	(*vi.*)	glutinous.	Caucasus.	——	——	H. ♃.	
20	obscùra. DC.	(*ye.*)	doubtful.	S. France.	1734.	——	H. ⊙.	Retz. obs. fasc. 1. t. 1.
21	l'ævis. DC.	(*ye.*)	smooth.	Tunis.	1816.	6. 8.	H. ⊙.	
	Hèlix. W.							
22	orbiculàris. DC.	(*ye.*)	flat-podded.	S. Europe.	1688.	7. 8.	H. ⊙.	Gært. sem. 2. t. 155.
23	marginàta. DC.	(*ye.*)	margined.	———	1816.	——	H. ⊙.	Moris. his. s. 2. t. 15. f. 2.
24	applanàta. DC.	(*ye.*)	obovate-leaved.	....	1824.	7. 9.	H. ⊙.	
25	scutellàta. DC.	(*ye.*)	snail-podded.	S. Europe.	1562.	6. 8.	H. ⊙.	Gært. sem. 2. t. 155. f. 7.
26	rugòsa. DC.	(*ye.*)	elegant.	———	1680.	7. 8.	H. ⊙.	Moris. his. s. 2. t. 15. f. 4.
	élegans. W.							
27	saxátilis. DC.	(*ye.*)	stone.	Tauria.	1828.	6. 8.	H. ♃.	
28	tornàta. DC.	(*ye.*)	smooth-podded.	S. Europe.	1658.	——	H. ⊙.	
29	turbinàta. DC.	(*ye.*)	Turban.	———	1680.	——	H. ⊙.	Moris. his. s. 2. t. 15. f. 5.
30	tuberculàta. DC.	(*ye.*)	wart-podded.	———	1658	——	H. ⊙.	Bauh. hist. 2. f. 1.
31	striàta. DC.	(*std.*)	stripe-podded.	———	1828.	——	H. ⊙.	
32	apiculàta. DC.	(*ye.*)	tufted.	Italy.	1800.	——	H. ⊙.	
33	coronàta. DC.	(*ye.*)	crowned.	S. Europe.	1660.	——	H. ⊙.	Moris. his. s. 2. t. 15. f. 16.
34	catalònica. DC.	(*ye.*)	Catalonian.	Catalonia.	1816.	——	H. ⊙.	Schrank. h. mon. t. 28.
35	denticulàta. DC.	(*ye.*)	toothed.	S. France.	1800.	——	H. ⊙.	
36	flexuòsa. DC.	(*ye.*)	flexuose.	Naples.	1824.	——	H. ⊙.	
37	spinulòsa. DC.	(*ye.*)	prickly-podded.	S. Europe.	1820.	——	H. ⊙.	
38	pubéscens. DC.	(*ye.*)	pubescent.	Montpelier.	1823.	——	H. ⊙.	
39	terebéllum. DC.	(*ye.*)	short-spined.	S. Europe.	1798.	——	H. ⊙.	
40	marìna. DC.	(*ye.*)	sea-side.	———	1596.	——	H. ⊙.	Cavan. ic. 2. t. 130.
41	tentaculàta. DC.	(*ye.*)	distichous-spin'd.	———	1824.	——	H. ⊙.	Gært. sem. 2. t. 155. f. 7?
42	Hornemanniàna. DC.	(*ye.*)	Hornemann's.	Morocco.	1823.	——	H. ⊙.	
43	littoràlis. DC.	(*ye.*)	shore.	S. Europe.	1827.	——	H. ♃	J. Bauh. his. 2. p. 385. ic.
44	lappàcea. DC.	(*ye.*)	bur-podded.	———	——	——	H. ⊙.	
45	pentac'ycla. DC.	(*ye.*)	five-whorled.	———	....	——	H. ⊙.	
46	hístrix. DC.	(*ye.*)	porcupine.	Naples.	1824.	——	H. ⊙.	
47	disciférmis. DC.	(*ye.*)	hook-spined.	Montpelier.	1826.	——	H. ⊙.	
48	carstiénsis. DC.	(*ye.*)	creeping-rooted.	Carinthia.	1789.	6. 9.	H. ♃.	Bot. mag. 909.
49	nìgra. DC.	(*ye.*)	black.	France.	——	6. 8.	H. ⊙.	Moris. his. s. 2. t. 15. f. 19.
50	tribuloídes. DC.	(*ye.*)	Caltrops-like.	S. Europe.	1739.	——	H. ⊙.	
51	cylindràcea. DC.	(*ye.*)	cylinder-podded	....	1824.	——	H. ⊙.	
52	græca. DC.	(*ye.*)	Grecian.	Greece.	1804.	——	H. ⊙.	
53	mínima. DC.	(*ye.*)	least.	England.	....	——	H. ⊙.	Flor. dan. t. 211.
54	arenària. DC.	(*ye.*)	sand.	Italy.	1824.	——	H. ⊙.	
55	muricoléptis. DC.	(*ye.*)	arch-spined.	Sicily.	——	——	H. ⊙.	
56	uncinàta. DC.	(*ye.*)	hooked.	S. Europe.	1820.	——	H. ⊙.	
57	dístans. DC.	(*ye.*)	distant.	———	——	——	H. ⊙.	Moris. his. s. 2. t. 15. f. 21.
58	aculeàta. DC.	(*ye.*)	prickly.	....	1802.	——	H. ⊙.	
59	maculàta. DC.	(*ye.*)	heart.	England.	....	5. 6.	H. ⊙.	Eng. bot. 1616.
	polymórpha. E.B.							
60	Gerárdi. DC.	(*ye.*)	Gerard's.	Hungary.	1816.	6. 8.	H. ⊙.	Moris. his. s. 2. t. 15. f. 18.
	villòsa. DC. fl. fr. 4. p. 545.							
61	agréstis. DC.	(*ye.*)	field.	Italy.	1824.	——	H. ⊙.	
62	rigídula. DC.	(*ye.*)	thorny-podded.	S. Europe.	1730.	——	H. ⊙.	
63	muricàta. DC.	(*ye.*)	prickly.	England.	....	5. 8.	H. ⊙.	Vaill. par. t. 33. f. 7.
64	sphærocárpos. DC.	(*ye.*)	round-podded.	Italy.	1826.	6. 8.	H. ⊙.	Sebast. pl. rom. p. 15 t. 3.

65 mùrex. DC.	(*ye.*)	Burdock-podded.	......	1802.	6. 8.	H. ⊙.	
66 laciniàta. DC.	(*ye.*)	cut-leaved.	S. Europe.	1683.	——	H. ⊙.	Breyn. cent. 81. t. 34.
67 Tenoreàna. DC.	(*ye.*)	Tenore's.	Naples.	1827.	——	H. ⊙.	
68 granaténsis. DC.	(*ye.*)	Spanish.	Spain.	1816.	——	H. ⊙.	Jacq. coll.sup.t.15. f. 2.
69 intertéxta. DC.	(*ye.*)	hedgehog.	S. Europe.	1629.	——	H. ⊙.	Moris. his. s. 2.t.15. f. 8.
70 ciliàris. DC.	(*ye.*)	fringed.	France.	1686.	7. 8.	H. ⊙.	———— s. 2. t. 15. f. 7.
71 èchinus. DC.	(*ye.*)	Sea-egg.	S. Europe.	1820.	6. 8.	H. ⊙.	———— s. 2. t. 15. f.9.
72 ægagróphila. DC.	(*ye.*)	goats.	....	1824.	7. 9.	H. ⊙.	
73 càspica. L.en.	(*ye.*)	Caspian.	Caucasus.	1820.	6. 8.	H. ⊙.	
74 ovàta. H.E.	(*ye.*)	oval-leaved.	S. Europe.	1818.	——	H. ⊙.	
75 strumària. H.E.	(*ye.*)	strumous.	————	——	——	H. ⊙.	
76 mícrodon. E.	(*ye.*)	small-toothed.	Egypt.	1826.	——	H. ⊙.	
77 pedunculàta. E.	(*ye.*)	peduncled.	————	——	——	H. ⊙.	
TRIGONE'LLA. DC.		**Fenugreek.**	Diadelphia Decandria.				
1 cœrùlea. DC.	(*bl.*)	blue.	Switzerland.	1562.	7. 9.	H. ⊙.	Bot. mag. 2283.
Trifòlium cœrùleum. B.M. *Melilòtus cœrùlea.* Lam.							
2 Besseriàna. DC.	(*bl.*)	Besser's.	Poland.	....	——	H. ⊙.	
3 marítima. DC.	(*ye.*)	sea-side.	Egypt.	1827.	——	H. ⊙.	
4 uncinàta. DC.	(*st.*)	hooked.	Iberia.	1823.	6. 9.	H. ⊙.	
Melilòtus hamòsum. M.B. *uncinàta.* Bess.							
5 littoràlis. DC.	(*ye.*)	shore.	Sicily.	1826.	——	H. ⊙.	
6 callicèras. DC.	(*ye.*)	rostrate.	Iberia.	1821.	——	H. ⊙.	DC. hort. gen. 2. ic.
oxyrhy'nchus. F.							
7 prostràta. DC.	(*ye.*)	prostrate.	S. France.	....	——	H. ⊙.	J.Bauh.his.2.p.365.f.2.
8 gladiàta. DC.	(*wh.*)	sword-podded.	Tauria.	1829.	——	H. ⊙.	
9 Fœnum-græ̀cum. DC.	(*st.*)	common.	S. France.	1597.	——	H. ⊙.	Schk. s. han. 2. t. 211.
10 spinòsa. DC.	(*ye.*)	thorny.	Candia.	1710.	7. 9.	H. ⊙.	Breyn.cent.79.t.33.f.1.
11 striàta. DC.	(*ye.*)	streaked.	Abyssinia?	1824.	6. 9.	H. ⊙.	
12 hamòsa. DC.	(*ye.*)	hook-podded.	Egypt.	1640.	7. 8.	H. ⊙.	Alp. ægypt. t. 124.
13 flexuòsa. DC.	(*st.*)	flexuose.	————	1818.	6. 8.	H. ⊙.	
14 Fischeriàna. DC.	(*ye.*)	Fischer's.	Tiflin.	——	——	H. ⊙.	
flexuòsa. M.B. *non Delil.*							
15 anguìna. DC.	(*ye.*)	snake-podded.	Egypt.	1828.	——	H. ⊙.	Delil. fl.egypt.t.110.f.2.
16 ténuis. DC.	(*ye.*)	slender.	Tiflin.	1818.	——	H. ⊙.	
17 elongàta. W. en.	(*ye.*)	elongated.	....	1823.	——	H. ⊙.	
18 cancellàta. DC.	(*ye.*)	netted-podded.	....	1818.	——	H. ⊙.	
19 monspelìaca. DC.	(*ye.*)	Montpelier.	Montpelier.	1710.	——	H. ⊙.	W. et K. hung. 2. t. 142.
20 parviflòra. DC.	(*ye.*)	small-flowered.	Switzerland.	1824.	——	H. ⊙.	
21 pinnatífida. DC.	(*ye.*)	pinnatifid.	Spain.	1801.	——	H. ⊙.	Cavan. ic. 1. t. 38.
mèdia. Delil.							
22 polycèrata. DC.	(*ye.*)	broad-leaved.	S. Europe.	1640.	7. 9.	H. ⊙.	
23 ægyptìaca. DC.	(*ye.*)	Egyptian.	Egypt.	1823.	6. 8.	H. ⊙.	
24 ornithory'nchos. DC.	(*ye.*)	beak-podded.	Russia.	1825.	6. 9.	H. ⊙.	
25 crassifòlia. DC.	(*ye.*)	thick-leaved.	Egypt.	1828.	6. 8.	H. ⊙.	
26 ruthènica. DC.	(*ye.*)	small.	Siberia.	1741.	5. 7.	H. ♃.	Lodd. bot. cab. 1391.
27 laciniàta. DC.	(*ye.*)	laciniate.	Egypt.	1823.	7. 9.	H. ⊙.	
28 platycárpos. DC.	(*ye.*)	round-leaved.	Siberia.	1741.	6. 9.	H. ♂.	Gmel. sib. 4. t. 9.
29 hy'brida. DC.	(*ye.*)	hybrid.	France.	1806.	——	H. ♃.	DC. ic. rar. 1. t. 29.
30 ornithopodioìdes. DC.	(*re.*)	Bird's-foot.	Britain.	....	——	H. ⊙.	Eng. bot. t. 1047.
Trifòlium ornithopodioìdes. E.B. *Melilòtus ornithopodioìdes.* P.S.							
31 corniculàta. DC.	(*ye.*)	horse-shoe.	S. Europe.	1597.	6. 7.	H. ⊙.	Moris. his. s. 2.t.16.f.11.
32 esculénta. DC.	(*ye.*)	esculent.	E. Indies.	1815.	6. 9.	S. ⊙.	
POCO'CKIA. DC.		**Poco'ckia.**	Diadelphia Decandria.				
crética. DC.	(*ye.*)	Cretan.	Candia.	1713.	6. 9.	H. ⊙.	Moris. his. s. 2. t. 14. f. 3.
Trifòlium créticum. L. *Melilòtus crética.* P.S.							
MELIL'OTUS. DC.		**Melilot.**	Diadelphia Decandria.				
1 Kochiàna. DC.	(*ye.*)	Koch's.	Germany.	1816.	6. 9.	H. ♂.	
2 dentàta. DC.	(*ye.*)	toothed.	Hungary.	1802.	6. 8.	H. ♃.	W. et K. hung. 1. t. 42.
3 lineàris. DC.	(*ye.*)	linear-leaved.	Spain.	1816.	——	H. ⊙.	
4 ruthènica. DC.	(*wh.*)	Russian.	Sarepta.	1828.	——	H. ⊙.	
5 melanospérma. DC.	(*ye.*)	black-seeded.	Tauria.	1825.	——	H. ♂.	
6 officinàlis. DC.	(*ye.*)	common.	Britain.	....	6. 9.	H. ⊙.	Eng. bot. 1340.
7 palústris. DC.	(*ye.*)	marsh.	Hungary.	1824.	——	H. ♂.	W. et K. hung. 3. t. 266.
8 altíssima. DC.	(*wh.*)	tallest.	S. Europe.	1818.	——	H. ⊙.	
9 leucántha. DC.	(*wh.*)	white-flowered.	Europe.	....	——	H. ♂.	Sturm. deuts. 1. f. K.
vulgàris. W. en.							
10 macrorhìza. DC.	(*ye.*)	long-rooted.	Hungary.	1801.	7. 9.	H. ♃.	W. et K. hung. 1. t. 26.
11 parviflòra. DC.	(*ye.*)	small-flowered.	S. Europe.	1810.	6. 8.	H. ⊙.	
12 rugulòsa. W. en.	(*wh.*)	rugged-podded.	E. Indies.	——	——	H. ⊙.	

13 índica. W. en.	(*ye.*)	Indian.	E. Indies.	1680.	6 8.	H. ⊙.	Pluk. alm. t. 45. f. 4.
14 Báumeti. H.	(*ye.*)	one-seeded.	....	1818.	——	H. ⊙.	
15 mínima. DC.	(*wh.*)	small-flowered.	E. Indies.	1810.	8. 10.	H. ⊙.	
16 segetàlis. DC.	(*ye.*)	hedge.	Spain.	1824.	6. 8.	H. ⊙.	
17 polónica. DC.	(*ye.*)	Polish.	Poland.	1778.	——	H. ⊙.	
18 táurica. DC.	(*wh.*)	Taurian.	Tauria.	1823.	6. 9.	H. ⊙.	
19 itálica. DC.	(*ye.*)	Italian.	Italy.	1596.	——	H. ⊙.	Camer. hort. t. 29.
20 rotundifòlia. T.N.	(*ye.*)	round-leaved.	————	1824.	——	H. ⊙.	
21 grácilis. DC.	(*ye.*)	slender.	France.	——	6. 8.	H. ⊙.	
22 neapolitàna. T.P.	(*ye.*)	Neapolitan.	Naples.	1827.	——	H. ⊙.	
23 pállida. DC.	(*ye.*)	pale-podded.	Poland.	1828.	——	H. ♂.	
24 suavèolens. DC.	(*ye.*)	sweet-scented.	Dahuria.	1826.	——	H. ♃.	
25 hirsùta. Bal.	(*ye.*)	hairy.	....	——	——	H. ⊙.	
26 procúmbens. H. Pr.	(*y.*)	procumbent.	....	1829.	——	H. ⊙.	
27 arvénsis. DC.	(*ye.*)	corn-field.	Germany.	1822.	6. 9.	H. ⊙.	
28 petitpierriàna. W.	(*wh.*)	rough-podded.	————	1816.	——	H. ♂.	
29 Besseriàna. DC.	(*ye.*)	Besser's.	Tauria.	1826.	6. 8.	H. ♂.	
imbricàta. Bes.							
30 plicàta. F.	(*ye.*)	plaited.	————	1823.	——	H. ⊙.	
31 messanénsis. DC.	(*ye.*)	Sicilian.	Sicily.	1680.	——	H. ⊙.	
32 sulcàta. DC.	(*ye.*)	furrowed.	Barbary.	1798.	——	H. ⊙.	
mauritánica. W. enum.							

TRIFOLIUM. DC. TREFOIL. Diadelphia Decandria. L.

1 angustifòlium. DC.	(*ro.*)	narrow-leaved.	S. Europe.	1640.	6. 8.	H. ⊙.	Barrel. ic. t. 698.
2 intermèdium. DC.	(*wh.*)	intermediate.	Italy.	1829.	——	H. ⊙.	
3 purpùreum. DC.	(*pu.*)	purple.	Montpelier.	1826.	——	H. ⊙.	Loisel. fl gall. t. 14.
4 rùbens. DC.	(*re.*)	long-spiked.	S. Europe.	1633.	6. 9.	H. ♃.	Jacq. aust. 4. t. 385.
5 cœruléscens. DC.	(*pu.*)	bluish.	Tauria.	1827.	——	H. ⊙.	
6 incarnàtum. DC.	(*sc.*)	scarlet.	Italy.	1596.	7. 9.	H. ⊙.	Bot. mag. 328.
7 Molineri. W.	(*bh.*)	blush-flowered.	————	1823.	——	H. ⊙.	
8 arvénse. DC.	(*wh.*)	Hare's-foot.	Britain.	....	7. 8.	H. ⊙.	Engl. bot. v. 14. t. 944.
9 grácile. P.S.	(*re.*)	slender.	France.	1818.	——	H. ⊙.	Barrel. ic. 901.
10 ligústicum. DC.	(*re.*)	awned.	Portugal.	1820.	6. 8.	H. ⊙.	Balb. act. ac. it. 1. f. 2.
arrectisètum. Brot. phyt. t. 63. f. 1. *aristàtum.* H.							
11 divaricàtum. DC.	(*wh.*)	spreading.	....	1827.	——	H. ⊙.	
12 geméllum. DC.	(*wh.*)	double-headed.	Spain.	1825.	——	H. ⊙.	
13 phleoídes. DC.	(*re.*)	Phleum-spiked.	————	——	——	H. ⊙.	
14 lappàceum. DC.	(*st.*)	burr-spiked.	S. Europe.	1787.	——	H. ⊙.	Barrel. ic. 871.
15 echinàtum. DC.	(*wh.*)	bristly.	Caucasus.	1823.	——	H. ⊙.	
16 erinàceum. DC.	(*wh.*)	villous-leaved.	Iberia.	1827.	——	H. ⊙.	
17 malacánthum. DC.	(*wh.*)	soft-headed.	....	1824.	——	H. ⊙.	
18 sylváticum. DC.	(*pu.*)	wood.	S. France.	1826.	——	H. ♃.	
19 Bocccòni. DC.	(*re.*)	Boccone's.	S. Europe.	1825.	7. 8.	H. ⊙.	Brot. phyt. t. 63. f. 2.
semiglábrum. Brot. *collìnum.* Poir.							
20 striàtum. DC.	(*pu.*)	soft-knotted.	Britain.	....	6. 7.	H. ⊙.	Eng. bot. v. 26. t. 1843.
21 tenuiflòrum. DC.	(*pa.*)	slender-flower'd.	Naples.	1824.	6. 8.	H. ⊙.	
22 scábrum. DC.	(*wh.*)	rough.	Britain.	....	5. 7.	H. ⊙.	Eng. bot. v. 13. t. 903.
23 marítimum. DC.	(*li.*)	sea-shore.	————	....	6. 8.	H. ⊙.	———— v. 4. t. 220.
rígidum. Savi. fl. pis. 2. p. 159. t. 1. f. 1.							
24 supìnum. DC.	(*bh.*)	spreading-branch'd.	S. Europe.	1827.	——	H. ⊙.	Savi. obs. trif. p. 46. f. 2.
25 constantinopolitànum. DC.	(*ye.*)	Turkish.	Constantinople.	——	——	H. ⊙.	
26 cínctum. DC.	(*st.*)	girded.	Montpelier.	1820.	——	H. ⊙.	
27 alexandrìnum. DC.	(*st.*)	Egyptian.	Egypt.	1798.	——	H. ⊙.	
28 ochroleùcum. DC.	(*st.*)	sulphur-color'd.	Britain.	....	——	H. ♃.	Eng. bot. v. 17. t. 1224.
29 tricocéphalum. DC.	(*st.*)	hairy-headed.	Caucasus.	1829.	——	H. ♃.	
30 canéscens. DC.	(*wh.*)	gray.	————	1803.	5. 7.	H. ♃.	Bot. mag. 1168.
31 pannònicum. DC.	(*wh.*)	Hungarian.	Hungary.	1752.	6. 8.	H. ♃.	Jacq. obs. 2. t. 42.
32 barbàtum. DC.	(*ye.*)	bearded.	Montpelier.	1827.	——	H. ♃.	
33 oly'mpicum. B.M.	(*st.*)	long-flowered.	Levant.	1810.	——	H. ♃.	Bot. mag. 2790.
34 armènium. DC.	(*st.*)	Armenian.	Armenia.	1816.	5. 7.	H. ♃.	
35 squarròsum. DC.	(*wh.*)	various-leaved.	Spain.	1640.	7. 8.	H. ⊙.	Moris. his. s. 2. t. 13. f. 1.
β *flavicans.* DC.	(*st.*)	*yellowish.*	————	....	——	H. ⊙.	Savi. obs. tri. p. 65. t. 3.
36 álbidum. DC.	(*st.*)	whitish.	....	1796.	——	H. ⊙.	
37 cònicum. DC.	(*st.*)	conical.	....	1823.	6. 8.	H. ⊙.	
38 Kitaibeliànum. DC.	(*st.*)	Kitaibel's.	Hungary.	1828.	——	H. ⊙.	
39 elongàtum. DC.	(*re.*)	long-flowered.	Galatia.	1810.	——	H. ♃.	
40 alpéstre. DC.	(*pu.*)	oval-spiked.	Europe.	1789.	——	H. ♃.	Bot. mag. 2779.
41 mèdium. DC.	(*pu.*)	Cow-grass.	Britain.	....	7. 8.	H. ♃.	Eng. bot. v. 3. t. 190.
42 bracteàtum. DC.	(*pu.*)	large-bracted.	Morocco.	1804.	6. 8.	H. ♂.	

43 expánsum. DC. (*pu.*) expanded. .... 1804. 6. 7. H. ♃.
44 praténse. DC. (*pu.*) common clover. Britain. .... 5. 9. H. ♃. Eng. bot. v. 25. t. 1770.
α perénne. (*pu.*) *perennial.* ———— .... —— H. ♃. Sincl. gr. wob. 221.
β satívum. (*pu.*) *cultivated.* ———— .... —— H. ♂.
45 vaginàtum. Schl. (*ye.*) sheathed. Switzerland. 1824. —— H. ♃.
praténse β flávicans. Ser. in DC. prodr. 2. p. 195. *ochroleùcum.* γ. DC.
46 microphy'llum. DC. (*p.*) small-leaved. France. 1824. 5. 9. H. ♃.
47 nòricum. DC. (*st.*) shaggy. Carinthia. 1823. 6. 7. H. ♃. Sturm deuts. 1. fas. 16.
48 pensylvànicum. DC. (*p.*) Buffalo-clover. N. America. 1811. 6. 9. H. ♃.
49 pállidum. DC. (*wh.*) pale. Hungary. 1803. 6. 8. H. ☉. W. et K. hung. 1. t. 36.
50 diffùsum. DC. (*pu.*) diffuse. ———— 1801. —— H. ☉. ———— 1. t. 50.
51 hírtum. DC. (*pu.*) hairy. S. Europe. 1806. —— H. ☉. Desf. atl. 2. t. 209. f. 1.
híspidum. Desf.
52 píctum. W. (*pu.*) painted-leaved. ———— —— —— H. ☉.
53 pauciflòrum. DC. (*pu.*) few-flowered. Archipelago. 1828. —— H. ♂.
54 Cherlèri. DC. (*st.*) hispid-headed. S. Europe. 1750. 5. 7. H. ☉. Barrel. ic. 859.
55 globòsum. DC. (*wh.*) globular-headed. Levant. 1713. 6. 8. H. ☉.
56 clypeàtum. DC. (*wh.*) oriental. ———— 1711. —— H. ☉. Alpin. exot. t. 306.
57 stellàtum. DC. (*ro.*) star-headed. England. .... 7. 8. H. ☉. Eng. bot. v. 22. t. 1545.
58 leucánthum. DC. (*wh.*) white-flowered. Tauria. 1823. 6. 8. H. ☉.
59 Waldsteiniànum. (*pu.*) Waldstein's. Hungary. 1826. —— H. ☉. W. et K. hung. 3. t. 269.
refléxum. W. K. *non* L.
60 saxátile. DC. (*wh.*) rock. Switzerland. 1816. 6. 8. H. ♂. All ped. t. 59. f. 3.
61 suffocàtum. DC. (*wh.*) suffocated. England. .... —— H. ☉. Eng. bot. v. 15. t. 1049.
62 glomeràtum. DC. (*ro.*) round-headed. ———— .... 6. 7. H. ☉. ———— v. 15. t. 1063.
63 parviflòrum. DC. (*li.*) small-flowered. Hungary. 1805. 7. 8. H. ☉. Sturm deuts. 1. fasc. 15.
stríctum. Sturm.
64 stríctum. DC. (*bh.*) upright. S. Europe. —— —— H. ☉. W. et K. hung. 1. t. 37.
65 rèpens. DC. (*wh.*) white clover. Britain. .... 5. 9. H. ♃. Eng. bot v. 25. t. 1769.
β pentaphyllum. P. S. (*w.*) *purple 5-leaved.* ———— .... —— H. ♃.
66 anómalum. DC. (*wh.*) anomalous. .... 1827. —— H. ♃. Schrank pl. rar. t. 47.
67 pallèscens. DC. (*st.*) pale. Hungary. 1804. 6. 8. H. ♃. Sturm. deuts. fl. 1. fas. 15.
68 cæspitòsum. DC. (*wh.*) turfy. Switzerland. 1815. —— H. ♃. Vill. delph. 3. t. 41.
69 angulàtum. DC. (*bh.*) angular. Hungary. 1803. —— H. ☉. W. et K. hung 1. t. 27.
70 suavèolens. DC. (*pu.*) sweet-scented. S. Europe. 1818. —— H. ☉. W. hort. berol. t. 108.
71 hy'bridum. DC. (*wh.*) hybrid. ———— 1805. —— H. ☉. Sturm deuts. fl. 1. fas. 15.
72 Micheliànum. DC. (*st.*) Michel's. Italy. 1815. —— H. ☉. Mich. n. gen. t. 25. f. 25.
73 élegans. DC. (*li.*) elegant. S. Europe. 1818. —— H. ♃. Savi. fl. pi. 2. t. 1. f. 2.
formòsum. Savi. Vail. par. t. 22. f. 1.
74 refléxum. DC. (*pu.*) reflexed. Virginia. 1794. —— H. ♃.
75 montànum. DC. (*wh.*) mountain. Europe. 1786. 7. 8. H. ♃. Flor. dan. t. 1172.
76 rupéstre. DC. (*wh.*) flat-headed. Italy. 1824. —— H. ♃.
77 Prutetiànum. DC. (*re.*) soft hairy. ———— 1828. —— H. ♃.
78 subterràneum. DC. (*w.*) subterraneous. England. .... 5. 6. H. ☉. Eng. bot. v. 15. t. 1048.
79 vesiculòsum. DC. (*re.*) recurved. Hungary. 1805. 6. 8. H. ☉. W. et K. hung. 2. t. 165.
recúrvum. W. K.
80 spumòsum. DC. (*pu.*) bladdered. France. 1771. —— H. ☉. Cupan. panp. t. 69. f. 1.
81 resupinàtum. DC. (*pu.*) resupinate. Germany. 1713. —— H. ☉. Sturm deuts. fl. 1. fas. 16.
82 fragíferum. DC. (*li.*) Strawberry. England. .... 7. 8. H. ♃. Eng. bot. v. 15. t. 1050.
83 tomentòsum. DC. (*ye.*) woolly. S. Europe. 1640. 6. 8. H. ☉. Barrel. ic. 864.
84 physòdes. DC. (*li.*) swollen-calyx'd. Iberia. 1829. —— H. ♃.
85 alàtum. DC. (*li.*) netted-calyx'd. Sicily. 1820. 7. 8. H. ☉. Cupan. panp. t. 97. f. 1?
Cupàni. Tineo.
86 uniflòrum. DC. (*re.*) one-flowered. S. Europe. 1822. 6. 8. H. ♃. Buxb. cent. 3. t. 31. f. 1.
β Sternbergiànum. DC. (*wh.*) *Sternberg's.* ———— —— H. ♃. ———— 3. p. 18. t. 31. f. 2.
87 exímium. DC. (*pu*) choice. Dahuria. 1828. —— H. ♃.
β albiflòrum. DC. (*wh.*) *white-flowered.* Altay. —— —— H. ♃.
88 alpìnum. DC. (*re. wh.*) alpine. Italy. 1775. —— H. ♃. Sturm deut. fl. 1. fasc. 15.
89 involucràtum. DC. (*re.*) involucred. Mexico. 1825. —— H. ☉. Kth. pl. legum. t. 53.
90 Gussòni. DC. (*pu.*) Gussone's. Sicily. 1822. —— H. ♃. Bonan. sic. t. 245.
91 fimbriàtum. B. R. (*re.*) fringed. N. America. 1826. 9. 10. H. ♃. Bot. reg. 1070.
92 tridentàtum. B. R. (*re.*) three-toothed. ———— —— 8. 10. H. ♃.
93 cyathíferum. B. R. (*re.*) cup-like. ———— —— —— H. ♃.
94 bádium. DC. (*br.*) villous-stalked. Pyrenees. .... 6. 8. H. ♃. Sturm deuts. fl. 1. fas. 16.
95 agràrium. DC. (*ye.*) golden. Europe. 1815. —— H. ☉. ———— fl. 1. fasc. 16.
96 spadíceum. DC. (*ye.*) bay-coloured. ———— 1778. —— H. ♃. Bot. mag. 557.
97 decípiens. DC. (*ye.*) decipient. .... 1827. 6. 8. H. ☉.
98 speciòsum. DC. (*br.*) large-flowered. Candia. 1752. —— H. ☉. Labill. dec. 5. t. 10.
comòsum. Labil.

99	procúmbens. DC.	(*st.*)	Hop.	Britain.	....	6. 7.	H. ⊙.	Eng. bot. v. 14. t. 945.
100	parisiénse. DC.		Paris.	France.	....	——	H. ⊙.	Sturm deuts. 1. fas. 16.
	pàtens. Sturm.							
101	Sebastiàni. DC.	(*ye.*)	Sebastian's.	Italy.	1823.	——	H. ⊙.	Sebas. pl. rom. fas. 2. t. 4.
102	flavéscens. DC.	(*ye.*)	yellowish.	Sicily.	1828.	6. 8.	H. ⊙.	
103	mìnus. E.B.	(*ye.*)	lesser yellow.	Britain.	....	——	H. ⊙.	Eng. bot. v. 18. t. 1256.
104	filifórme. E.B.	(*ye.*)	slender yellow.	———	....	——	H. ⊙.	——— v. 18. t. 1257.
105	comòsum. W.	(*wh.*)	tufted.	N. America.	1798.	——	H. ♃.	
106	tenuifòlium. DC.	(*wh.*)	slender-leaved.	Naples.	1824.	——	H. ⊙.	
107	capénse. DC.	(*st.*)	Cape.	C. B. S.	1825.	——	G. ♃.	
108	congéstum. DC.	(*wh.*)	close-headed.	Sicily.	1829.	——	H. ⊙.	
109	formòsum. DC.	(*pi.*)	handsome.	Isl. Melo.	1828.	——	H. ♂.	
110	lasiocéphalum. DC.	(*w.*)	woolly-headed.	C. B. S.	1825.	——	H. ⊙.	
111	villòsum. Pr.	(*wh.*)	villous.	Sicily.	1818.	5. 7.	H. ⊙.	
112	plicàtum Pr.	(*br.*)	plaited.	———	1824.	6. 8.	H. ⊙.	
113	Ehrenbérgii. H ber.	(*w.*)	Ehrenberg's.	Egypt.	1826.	——	H. ⊙.	
114	libanáticum. E.	(*wh.*)	Lebanon.	Mt. Lebanon.	1827.	——	H. ⊙.	
115	albicéps. E.	(*wh.*)	white-headed.	Egypt.	1829.	——	H. ⊙.	
	LUPINA'STER. Mh.		LUPINA'STER.	Diadelphia Decandria.				
1	pentaphy'llus. Mh.	(*re.*)	five-leaved.	Siberia.	1741.	7. 8.	H. ♃.	Bot. mag. 879.
	Trifòlium Lupinaster. B.M. *Péntaphy'llon Lupinaster.* P.S.							
2	álbens. F.	(*wh.*)	white-flowered.	Volga.	1827.	——	H. ♃.	
3	purpuráscens. F.	(*pu.*)	purple.	———	1821.	6. 8.	H. ♃.	
4	cœruléscens. F.	(*pu.*)	blue purple.	———	1826.	——	H. ♃.	
	DORY'CNIUM. DC.		DORY'CNIUM.	Diadelphia Decandria.				
1	réctum. DC.	(*ro.*)	upright.	S. Europe.	1640.	6. 8.	H. ♄.	Barrel. ic. t. 544.
	Lòtus réctus. W.							
2	latifòlium. DC.	(*wh.*)	broad-leaved.	Iberia.	1818.	6. 9.	H. ♄.	
3	hirsùtum. DC.	(*ro.*)	hairy.	S. Europe.	1683.	——	H. ♄.	Bot. mag. 336.
	Lòtus hirsùtus. B.M.							
4	seríceum.	(*bh.*)	silky.	———	1820.	——	H. ♄.	
	Lòtus seríceus. DC. cat. h. monsp. p. 122.							
5	argénteum. DC.	(*ye.*)	silvery.	Egypt.	1825.	——	H. ♄.	Delil. fl. æg. 113. t. 40.
6	parviflòrum. DC.	(*ye.*)	small-flowered.	S. Europe.	1824.	7. 8.	H. ⊙.	Desf. atl. 2. t. 211.
7	subbiflòrum. DC.	(*ye.*)	two-flowered.	Spain.	1820.	6. 8.	H. ⊙.	
8	herbàceum. DC.	(*wh.*)	herbaceous.	S. Europe.	1802.	6. 9.	H. ♃.	Vill. dauph. 3. t. 41.
9	suffruticòsum. DC.	(*wh.*)	suffrutescent.	———	1640.	7. 9.	F. ♄.	Lobel. ic. 2. p. 51. f. 1. 2.
10	microphy'llum.	(*ro.*)	small-leaved.	C. B. S.	1826.	——	G. ♄.	Bot. mag. 2808.
	Lòtus microphy'llus. B.M. 2808.							
	L'OTUS. DC.		BIRD'S-FOOT TREFOIL.	Diadelphia Decandria.				
1	èdulis. DC.	(*ye.*)	esculent.	S. Europe.	1710.	7. 8.	H. ⊙.	Cavan. ic. 2. t. 157.
2	ornithopodioídes.	(*ye.*)	claw-podded.	———	1683.	6. 8.	H. ⊙.	——— 2. t. 163.
3	peregrìnus. DC.	(*ye.*)	flat-podded.	———	1713.	7. 8.	H. ⊙.	
4	índicus. DC.	(*ye.*)	Indian.	E. Indies.	....	——	H. ⊙.	Pluk. phyt. t. 200. f. 7.
5	tetraphy'llus. DC.	(*ye.*)	four-leaved.	Archipelago.	1828.	——	H. ⊙.	Viv. fl. lib. t. 17. f. 13.
6	flexuòsus. DC.	(*ye.*)	flexuose.	....	1820.	6. 8.	H. ⊙.	
7	gláucus. DC.	(*ye.*)	glaucous.	Madeira.	1777.	——	G. ♂.	
8	sessilifòlius. DC.	(*ye.*)	sessile leaved.	Canaries.	1816.	6. 9.	G. ♃.	
9	anthylloídes. DC.	(*ye.*)	Anthyllis-like.	C. B. S.	1812.	——	G. ♄.	Venten. malm. t. 92.
10	atropurpùreus. DC.	(*da.*)	dark-purple.	Cap. Verd Isl.	1823.	3. 9.	G. ♄.	
11	jacob'æus. DC.	(*da.*)	dark-flowered.	———	1714.	1. 12.	G. ♄.	Bot. mag. 79.
12	arenàrius. DC.	(*ye.*)	sand.	Portugal.	1824.	6. 8.	H. ⊙.	
13	créticus. DC.	(*ye.*)	silver-leaved.	Levant.	1680.	6. 9.	G. ♄.	Cavan. ic. 2. t. 156.
	β *vàrians.* DC.	(*ye.*)	*variable.*	———	1829.	——	G. ♄.	
14	cytisoídes. DC.	(*ye.*)	downy.	S. Europe.	1752.	7. 8.	H. ♃.	Barrel. ic. t. 1031.
15	Dioscóridis. DC.	(*ye.*)	Dioscorides's.	Crete.	1658.	6. 8.	H. ⊙.	All. ped. 1. t. 59. f. 1.
16	Gebèlia. DC.	(*ro.*)	Aleppo.	Aleppo.	1823.	5. 9.	F. ♃.	Vent. h. cels. t. 57.
17	arábicus. DC.	(*re.*)	red-flowered.	Arabia.	1773.	7. 9.	F. ♃.	Jacq. vind. t. 155.
18	austràlis. DC.	(*re.*)	New Holland.	N. S. W.	1803.	5. 9.	G. ♄.	Bot. mag. 1365.
19	Candolleànus. ed. 1.	(*re.*)	De Candolle's.	———	1812.	6. 9.	G. ♄.	
	austràlis β *angustifoliola.* DC.							
20	álbidus. B.C.	(*wh.*)	white-flowered.	———	1823.	——	G. ♄.	Lodd. bot. cab. 1063
21	obovàtus. ed. 1.	(*ye.*)	glaucous obovate.	———	1803.	4. 9.	G. ♃.	
22	decúmbens. DC.	(*ye.*)	decumbent.	France.	1820.	5. 7.	H. ♃.	
23	Forstèri.	(*ye.*)	Forster's.	Britain.	....	——	H. ♃.	
	decúmbens. E.F. *non* DC.							
24	pilosíssimus. DC.	(*ye.*)	hairy.	S. France.	1826.	6. 8.	H. ⊙.	
25	híspidus. DC.	(*ye.*)	hispid.	S. Europe.	1820.	——	H. ⊙.	Loisel. fl. gal. t. 16.
26	suavèolens. DC.	(*ye.*)	fragrant.	———	1804.	——	H. ♃.	

27	angustíssimus. DC.	(*ye.*)	narrow-podded.	S. Europe.	....	5. 7.	H. ♂.	Bauh. hist. 2. p. 356. f. 2.
28	diffùsus. DC.	(*ye.*)	slender-podded.	England.	....	——	H. ⊙.	Eng. bot. v. 13. t. 925.
29	ciliàtus. DC.	(*ye.*)	ciliated.	Italy.	1820.	6. 8.	H. ⊙.	
30	grácilis. DC.	(*ye.*)	slender.	Hungary.	1818.	5. 7.	H. ⊙.	W. et K. hung. 3. t. 229.
31	lanuginòsus. DC.	(*ro.*)	drooping-fl[d].	Levant.	1818.	7. 8.	G. ♃.	
32	conimbricénsis. DC.	(*bh.*)	Portuguese.	Portugal.	1800.	6. 8.	H. ⊙.	Brot. phyt. t. 53.
33	glaberrimus. DC.	(*bh.*)	smooth.	....	1820.	——	H. ⊙.	
34	odoràtus. DC.	(*ye.*)	sweet-scented.	Barbary.	1804.	——	H. ♃.	Bot. mag. 1233.
35	corniculàtus. E.B.	(*ye.*)	common.	Britain.	....	——	H. ♃.	Eng. bot. v. 30. t. 2090.
36	màjor. E.B.	(*ye.*)	large hairy.	————	....	——	H. ♃.	———— v. 30. t. 2091.
37	tenùis. W.K.	(*ye.*)	slender.	Hungary.	1816.	——	H. ♃.	
	deprèssus. W. enum. *humifùsus.* W. en. *ex.* Link enum.							
38	pedunculàtus. DC.	(*ye.*)	long-peduncled.	Spain.	1814.	——	H. ♃.	Cavan. ic. 2. t. 164.
39	palústris. DC.	(*ye.*)	marsh.	Levant.	1824.	——	H. ♃.	
	TETRAGONO'LOBUS. DC. WINGED PEA. Diadelphia Decandria.							
1	purpùreus. DC.	(*pu.*)	purple-flower'd.	S. Europe.	1596.	6. 8.	H. ⊙.	Bot. mag. 151.
	Lòtus tetragonólobus. B.M.							
2	biflòrus. DC.	(*ye.*)	two-flowered.	Sicily.	1754.	——	H. ⊙.	Desf. atl. t. 210.
3	conjugàtus. DC.	(*pu.*)	twin-podded.	Montpelier.	1759.	——	H. ⊙.	
4	siliquòsus. L. en.	(*ye.*)	square-podded.	S. Europe.	1683.	7. 8.	H. ♃.	Jacq. aust. 4. t. 361.
5	marítimus. L. en.	(*ye.*)	sea-side.	————	——	——	H. ♃.	Flor. dan. 800.

Subtribus III. *CLITORIÆ.* DC. prodr. p. 216.

	PSOR'ALEA. DC. PSOR'ALEA. Diadelphia Decandria. L.							
1	odoratíssima. DC.	(*bl.*)	sweet-scented.	C. B. S.	1795.	5. 10.	G. ♄.	Jacq. schœnb. 2. t. 229.
	pínnata. Andr. rep. 474. *nec aliorum.*							
2	arbòrea. B.M.	(*bl.*)	tree.	————	1814.	——	G. ♄.	Bot. mag. 2090.
3	pinnàta. DC.	(*bl.*)	winged-leaved.	————	1690.	5. 7.	G. ♄.	Herm. lugd. 273. ic.
4	lævigàta. DC.	(*pu.*)	smooth.	————	1824.	——	G. ♄.	
5	verrucòsa. DC.	(*bl.*)	warted.	————	1774.	5. 8.	G. ♄.	Jacq. schœnb. 2. t. 226.
	angustifòlia. Jacq.							
6	filifórmis. DC.	(*bl.*)	slender.	————	1821.	——	G. ♄.	
7	axillàris. DC.	(*bl.*)	axillary-flower'd.	————	1816.	6. 7.	G. ♄.	
8	tenuifòlia. DC.	(*bl.*)	slender-leaved.	————	1793.	——	G. ♄.	Jacq. schœnb. 2. t. 225.
9	fasciculàris. DC.	(*bl.*)	fascicle-flowered.	————	1816.	——	G. ♄.	
10	multicáulis. DC.	(*v.*)	many-stalked.	————	1793.	8. 10.	G. ♄.	Jacq. schœnb. 2. t. 230.
11	aphy'lla. DC.	(*bl.*)	leafless.	————	1790.	6. 8.	G. ♄.	Bot. mag. 1727.
12	lathyrifòlia. DC.	(*bl.*)	Lathyrus-leaved.	————	1823.	——	G. ♄.	Balb. h. taur. p. 25. ic.
13	decúmbens. DC.	(*bl.*)	decumbent.	————	1774.	4. 6.	G. ♄.	Lodd. bot. cab. 3. t. 282.
14	répens. DC.	(*bl.*)	creeping	————	——	7. 8.	G. ♄.	
15	hírta. DC.	(*bl.*)	hairy.	————	1713.	5. 9.	G. ♄.	
16	Jacquiniàna.	(*bl.*)	Jacquin's.	————	——	——	G. ♄.	Jacq. schœnb. 2. t. 228.
	hírta. Jacq. *nec aliorum.*							
17	aculeàta. DC.	(*bl.*)	prickle-leaved.	————	1774.	——	G. ♄.	Andr. reposit. 146.
18	bracteata. DC.	(*bl.*)	oval-spiked.	————	1731.	6. 8.	G. ♄.	Bot. mag. 446.
19	involucràta. DC.	(*bl.*)	large-bracted.	————	1824.	7. 8.	G. ♄.	
20	spicàta. DC.	(*bl.wh.*)	long-spiked.	————	1774.	——	G. ♄.	Andr. reposit. 411.
21	Stáchydis. DC.	(*bl.*)	Stachy's-like.	————	1793.	4. 6.	G. ♄.	
22	striàta. DC.	(*bl.*)	streaked.	————	1816.	6. 8.	G. ♄.	
23	argéntea. DC.	(*bl.*)	silvery.	————	——	——	G. ♄.	
24	tomentòsa. DC.	(*bl.*)	woolly.	————	1825.	——	G. ♄.	
25	capitàta. DC.	(*bl.*)	headed.	————	1793.	——	G. ♄.	
26	corylifòlia. DC.	(*vi.*)	Hazel-leaved.	Arabia.	1739.	6. 7.	S. ♂.	Bot. mag. 665.
27	virgàta. DC.	(*bl.*)	twiggy.	Florida.	1826.	6. 8.	H. ♃.	
28	arenària. DC.	(*bl.*)	sand.	Missouri.	1823.	——	H. ♃.	
	lanceolàta et ellíptica. Ph. *ex* Nutt.							
29	palæstìna. DC.	(*bl.*)	Palestine.	Levant.	1771.	4. 9.	F. ♃.	Jacq. vind. 2. t. 184.
30	bituminòsa. DC.	(*pu.*)	bituminous.	Italy.	1570.	——	F. ♄.	Schkuhr hand. 2. t. 210.
31	serícea. DC.	(*pu.*)	long-peduncled.	C. B. S.	1815.	1. 12.	G. ♄.	Bot. reg. 223.
	pedunculàta. B.R.							
32	Mutísii. DC.	(*pu*)	Mutis's.	Mexico.	1828.	6. 8.	G. ♄.	Kth. pl. legum. t. 54.
33	esculénta. DC.	(*bl.*)	Bread-root.	Missouri.	1811.	——	H. ♃.	Pursh. amer. 2. t. 22.

34	cuspidàta. DC.	(*bl.*)	large-rooted.	Louisiana.	1811.	6. 8.	H. ♃.	
	macrorhìza. Fraser.							
35	incàna. DC.	(*bl.*)	silky.	Missouri.	1813.	——	H. ♃.	
	argophy'lla. Ph.							
36	lupinélla. DC.	(*pu.*)	narrow-leaved.	Carolina.	1812.	——	H. ♃.	
37	tenuiflòra. DC.	(*pu.*)	slender-flower'd.	Missouri.	1826.	——	H. ♃.	
38	longifòlia. DC.	(*wh.*)	long-leaved.	Louisiana.	1811.	——	H. ♃.	
39	melilotoídes. DC.	(*bl.*)	Melilot-like.	Carolina.	1814.	——	F. ♃.	Bot. reg. 454.
40	Onobr'ychis. DC.	(*pu.*)	rough-podded.	Louisiana.	1812.	——	H. ♃.	———— t. 453.
41	glandulòsa. DC.	(*v.*)	glandular.	Peru.	1770.	5. 9.	H. ♄.	Swt. br. fl. gar. 3. t. 296.
42	pubéscens. DC.	(*bl.*)	pubescent.	Lima.	1823.	——	G. ♂.	Bot. reg. 968.
43	canéscens. DC.	(*bl.*)	canescent.	Carolina.	——	6. 8.	F. ♃.	
44	divaricàta. DC.	(*bl.*)	divaricate.	Peru.	1820.	5. 9.	G. ♄.	
45	dentàta. DC.	(*wh.pu.*)	tooth-leaved.	Madeira.	1816.	——	F. ♄.	Lobel. ic. 2. t. 31. f. 1.
	americàna. Jacq. schœnb. t. 227.							
	CARMICH'ÆLIA. B.R.		CARMICH'ÆLIA. Diadelphia Decandria.					
	austràlis. B.R.	(*li.*)	New Zealand.	New Zealand.	1822.	3. 5.	G. ♄.	Bot. reg. 912.
	Lòtus arbòreus. Forster.							
	INDIGO'FERA. DC.		INDIGO. Diadelphia Decandria. L.					
1	linifòlia. DC.	(*wh.*)	Flax-leaved.	E. Indies.	1792.	7. 8.	S. ☉.	Roxb. cor. 2. t. 194.
2	paniculàta. DC.	(*re.*)	panicled.	Sierra Leon.	1824.	——	S. ♄.	
3	filifòlia. DC.	(*re.*)	filiform-leaved.	C. B. S.	1812.	7. 10.	G. ♄.	
4	echinàta. DC.	(*re.*)	prickly-podded.	E. Indies.	1824.	6. 9.	S. ♃.	
5	cordifòlia. DC.	(*pu.*)	heart-leaved.	————	1826.	——	S. ♄.	
6	monophy'lla. DC.	(*pu.*)	simple-leaved.	N. S. W.	——	——	G. ♄.	
7	diphy'lla. DC.	(*ro.*)	two-leaved.	Africa.	1824.	7. 10.	S. ☉.	
8	subulàta. DC.	(*li.*)	awl-stipuled.	W. Indies.	——	——	S. ☉.	
9	Lechenáultii. DC.	(*pu.*)	Lechenault's.	Bengal.	1826.	7. 10.	S. ☉.	
10	trifoliàta. DC.	(*re.*)	trifoliate.	E. Indies.	1818.	7. 8.	S. ♃.	
11	multicáulis. DC.	(*re.*)	many-stalked.	Nepaul.	1824.	7. 10.	S. ☉.	
12	glandulòsa. DC.	(*re.*)	glandular.	E. Indies.	1818.	5. 7.	S. ♃.	
13	biflòra. DC	(*pu.*)	two-flowered.	————	1826.	——	S. ♄.	
14	denudàta. DC.	(*pu.*)	smooth-leaved.	C. B. S.	1790.	——	G. ♄.	Jacq. schœnb. 2. t. 233.
15	am'œna. DC.	(*sc.*)	scarlet-flowered.	————	1774.	3. 5.	G. ♄.	Bot. reg. 300.
16	rígida. DC.	(*re.*)	rigid.	E. Indies.	1820.	5. 8.	S. ♄.	
17	virgàta. DC.	(*re.*)	twiggy.	Nepaul.	1823.	7. 8.	S. ♄.	
18	argéntea. DC.	(*re.*)	silvery.	Egypt.	1776.	——	S. ♄.	L'Herit. strip. t. 79.
19	tinctòria. DC.	(*re.*)	East Indian.	E. Indies.	1731.	——	S. ♄.	Rheed. mal. 1. t. 54.
20	brachycárpa. DC.	(*re.*)	short-podded.	W. Indies.	....	——	S. ♄.	
21	A'nil. DC.	(*re.*)	West Indian.	————	1731.	——	S. ♄.	Lam. ill. t. 626. f. 2.
22	cœrùlea. H.B.	(*pu.*)	blue-flowered.	E. Indies.	1820.	6. 7.	S. ♄.	
23	leptostàchya. DC.	(*pu.*)	slender-spiked.	————	1827.	6. 8.	S. ♄.	
24	atropurpùrea. DC.	(*pu.*)	dark-purple.	Nepaul.	1820.	7. 8.	S. ♄.	
25	cylíndrica. DC.	(*ro.*)	cylindric-podd'd.	C. B. S.	1822.	6. 8.	G. ♄.	
26	júncea. DC.	(*ro.*)	rushy.	————	....	7. 10.	G. ♄.	Herb. amat. t. 227.
	aphylla. L. enum. *Lebéckia contaminàta*. B.R. 104. *nùda*. B.M. 2214.							
27	austràlis. DC.	(*li.*)	round-stemm'd.	N. S. W.	1790.	3. 6.	G. ♄.	Bot. reg. 386.
28	angulàta. B.R.	(*li.*)	angular-stalked.	————	1824.	——	G. ♄.	———— t. 991.
29	macrostàchya. DC.	(*ro.*)	long-spiked.	China.	1822.	4. 6.	G. ♄.	Vent. malm. t. 44.
30	frutéscens. DC.	(*ro.*)	frutescent.	C. B. S.	1823.	7. 9.	G. ♄.	
31	divaricàta. DC.	(*ru.*)	spreading.	....	1824.	——	G. ♄.	Jacq. schœnb. t. 365.
32	mucronàta. DC.	(*re.*)	mucronate.	Jamaica.	——	——	S. ♄.	
33	ellíptica. H.B.	(*re*)	elliptic-leaved.	Bengal.	1820.	7. 8.	S. ♄.	
34	pulchélla. H.B.	(*re.*)	pretty.	E. Indies.	1823.	——	S. ♄.	
35	arbòrea. H.B.	(*re.*)	tree.	————	——	....	S. ♄.	
36	purpuráscens. H.B.	(*p.*)	purplish.	————	1827.	7. 8.	S. ♄.	
37	violácea. H.B.	(*vi.*)	violet-coloured.	————	1825.	——	S. ♄.	
38	uncinàta. H.B.	(*re.*)	hooked.	————	——	——	S. ♄.	
39	circinnàta. Rox.	(*re.*)	rounded.	————	1822.	——	S. ♄.	
40	laterítia. DC.	(*re.*)	brick-coloured.	Guinea.	——	——	S. ☉.	Jacq. ic. t. 359.
	hirsùta. Jacq. *non* Linn.							
41	viscòsa. DC.	(*re.*)	clammy.	E. Indies.	1806.	6. 8.	S. ☉.	Wend. sert. han. 2. t. 12.
42	dendroídes. DC.	(*ro.*)	tree-like.	Guinea.	1822.	——	S. ☉.	Jacq. ic. 3. t. 571.
43	polyphy'lla. DC.	(*re.*)	many-flowered.	Nepaul.	1825.	4. 8.	G. ♃.	
44	heterotrícha. DC.	(*ro.*)	various-haired.	C B. S.	1816.	——	G. ♄.	
45	hirsùta. DC.	(*pu.*)	hairy-leaved.	E. Indies.	1739.	7. 9.	S. ♄.	Burm. zeyl. 37. t. 14.
46	endecaphy'lla. DC.	(*re.*)	eleven-leaved.	Guinea.	1822.	——	S. ☉.	Bot. reg. 789.
47	altérnans. DC.	(*re.*)	alternate-leav'd.	C. B. S.	1816.	——	G. ♄.	
48	fràgrans. DC.	(*re.*)	fragrant.	E. Indies.	1822.	——	S. ♃.	

49 pusílla. DC.	(*re.*)	small.	Madagascar.	1822.	7. 9.	S. ħ.	
50 glábra. DC.	(*re.*)	smooth.	E. Indies.	1824.	——	S. ☉.	Pluk. alm. t. 166. f. 1.
51 enneaphy'lla. DC.	(*pu.*)	trailing.	——	1776.	——	S. ☉.	—— t. 166. f. 2.
52 cytisoídes. DC.	(*ro.*)	angular-stem'd.	C. B. S.	1774.	——	G. ♄.	Bot. mag. 742.
53 lotoídes. DC.	(*ro.*)	Lotus-like.	——	1816.	——	G. ♄.	Comm. hort. 2. t. 84?
54 angustifòlia. DC.	(*pu.*)	narrow-leaved.	——	1774.	6. 10.	G. ♄.	Bot. mag. 465.
55 aspalathoídes. DC.	(*re.*)	Aspalathus-like.	Ceylon.	1823.	4. 6.	S. ♄.	
56 filifórmis. DC.	(*ro.*)	filiform.	C. B. S.	1816.	——	G. ♄.	
57 alopecuroídes. DC.	(*ro.*)	Fox-tail.	——	1825.	——	G. ♄.	
58 coriàcea. DC.	(*ro.*)	leathery-leaved.	——	1774.	7. 8.	G. ħ.	
59 sarmentòsa. DC.	(*ro.*)	dwarf.	——	1786.	6. 8.	G. ħ.	
60 microphy'lla. P.S.	(*ro.*)	small-leaved.	——	1812.	1. 12.	G. ħ.	
61 digitàta. DC.	(*ro.*)	finger-leaved.	——	1820.	4. 8.	G. ♄.	
62 sessilifòlia. DC.	(*ro.*)	sessile-leaved.	——	1816.	——	G. ♄.	
63 spinòsa. DC.	(*re.*)	spiny.	Arabia.	1822.	——	S. ♄.	
64 cándicans. DC.	(*ro.*)	white-leaved.	C. B. S.	1774.	5. 9.	G. ♄.	Bot. mag. 198.
65 psoraloídes. DC.	(*ro.*)	Psoralea-like.	——	1758.	7. 8.	G. ♄.	—— t. 476.
66 stipulàris. DC.	(*ro.*)	stipuled.	——	1816.	6. 7.	G. ħ.	
67 incàna. DC.	(*ro.*)	hoary.	——	1812.	5. 7.	G. ħ.	Bot. reg. 957.
68 tríta. DC.	(*re.*)	oval-leaved.	E. Indies.	1802.	6. 8.	S. ♂.	
69 procúmbens. DC.	(*re.*)	procumbent.	C. B. S.	1816.	5. 7.	G. ħ.	
70 prostràta. DC.	(*re.*)	prostrate.	E. Indies.	1828.	8. 10.	S. ☉.	
CLIT'ORIA. DC.		CLIT'ORIA.	Diadelphia Decandria.	L.			
1 lascívia. Boj.	....	Madagascar.	Madagascar.	1826.	....	S. ♄.◡.	
2 heterophy'lla. DC.	(*bl.*)	various-leaved.	E. Indies.	1812.	7. 8.	S. ☉.◡.	Bot. mag. 2111.
3 Ternàtea. DC.	(*bl.wh.*)	wing-leaved.	——	1739.	5. 9.	S. ♄.◡.	Rheed. mal. 8. t. 38.
α cœrùlea.	(*bl.*)	*blue-flowered.*	——	——	——	S. ♄.◡.	Bot. mag. 1542.
β álba.	(*wh.*)	*white-flowered.*	——	....	——	S. ♄.◡.	
4 mariàna. DC.	(*bl.*)	Maryland.	N. America.	1759.	8. 9.	H. ♃.◡.	
5 mexicàna. DC.	(*pu.*)	Mexican.	Mexico.	1823.	6. 8.	G. ♃.◡.	
6 virginiàna. DC.	(*pu.*)	Virginian.	America.	1732.	7. 8.	G. ♃.◡.	Dill. elt. t. 76.
calcarígera. Sal. par. lond. t. 51.							
7 occidentàlis.	(*pu.*)	West Indian.	Trinidad.	1824.	6. 8.	S. ♄.◡.	Bot. reg. 1047.
virginiàna. B.R. 1047. *non* Linn.							
8 brasiliàna. DC.	(*bl.*)	Brazilian.	Brazil.	1759.	——	S. ♃.◡.	Breyn. cent. t. 32.
9 Plumiéri. DC.	(*pu.*)	Plumier's.	W. Indies.	1815.	8. 11.	S. ♄.◡.	Bot. reg. 268.
10 erécta. H.B.	....	erect.	S. America.	1822.	....	S. ♄.	
11 Berteriàna. DC.	(*ye.*)	Bertero's.	St. Domingo.	1824.	6. 8.	S. ♃.◡.	
12 arboréscens. DC.	....	tree.	Trinidad.	1804.	....	S. ♄.	
NEUROCA'RPUM. DC.		NEUROCA'RPUM.	Diadelphia Decandria.				
1 guianénse. DC.	(*pu.*)	Guiana.	Guiana.	1826.	6. 8.	S. ♃.	Aubl. guian. 2. t. 305.
2 falcàtum. DC.	(*pu.*)	sickle-podded.	St. Domingo.	——	——	S. ♄.◡.	Plum. spec. 8. t. 85.
COLOG'ANIA. DC.		COLOG'ANIA.	Diadelphia Decandria.				
1 intermèdia. DC.	(*fl.*)	intermediate.	Mexico.	1824.	6. 8.	G. ♃.◡.	
2 angustifòlia. DC.	(*vi.*)	narrow-leaved.	——	——	——	G. ♃.◡.	Kth. mim. 209. t. 58.
3 Broussonnétii. DC.	(*vi.*)	Broussonnet's.	....	1820.	——	S. ♃.◡.	
GALA'CTIA. DC.		GALA'CTIA.	Diadelphia Decandria.				
1 péndula. DC.	(*ro.*)	pendulous.	Jamaica.	1794.	7. 8.	S. ♄.◡.	Bot. reg. 269.
2 serícea. DC.	(*pu.*)	silky.	Isl. Bourbon.	1824.	——	S. ♄.◡.	
3 cubénsis. DC.	(*ro.*)	Cuba.	Cuba.	1826.	——	S. ♃.◡.	
4 móllis. DC.	(*pu.*)	soft.	Carolina.	....	——	G. ♃.◡.	Dill. elth. 1. f. 170.
5 glabélla. DC.	(*re.*)	smooth.	——	1826.	6. 8.	G. ♃.◡.	
VILMOR'INIA. DC.		VILMORI'NIA.	Diadelphia Decandria.				
multiflòra. DC.	(*pu.*)	many-flowered.	W. Indies.	1823.	7. 8.	S. ♄.	
Clitòria multiflòra. Swartz.							
COLL'ÆA. DC.		COLL'ÆA.	Monadelphia Pent-Decandria.				
1 speciòsa. DC.	(*pu.*)	pentandrous.	Peru.	1824.	6. 8.	G. ♄.	DC. leg. me. p. 245. t. 40.
2 trinérvia. DC.	(*pu.*)	decandrous.	E. Indies.	1826.	——	S. ♄.	—— p. 247. t. 41.
OTO'PTERA. DC.		OTO'PTERA.	Monadelphia Decandria.				
Burchéllii. DC.	(*pu.*)	Burchell's.	C. B. S.	1816.	5. 7.	G. ♄.◡.	DC. leg. mem. 250. t. 42.
PUER'ARIA. DC.		PUER'ARIA.	Monadelphia Decandria.				
1 tuberòsa. DC.	(*ye.*)	tuberous-root'd.	E. Indies.	1806.	6. 8.	S. ♄.◡.	
Hed'ysarum tuberòsum. W.							
2 Wallíchii. DC.	(*ye.*)	Wallich's.	Nepaul.	1826.	4. 6.	G. ♄.◡.	DC. leg. mem. 254. t. 43.
DUM'ASIA. DC.		DUM'ASIA.	Diadelphia Decandria.				
1 villòsa. DC.	(*ye.*)	villous.	Nepaul.	1824.	8. 12.	G. ♄.◡.	DC. leg. mem. 257. t. 44.
2 pubéscens. DC.	(*ye.*)	pubescent.	——	——	——	G. ♄.◡.	Bot. reg. 962.
GL'YCINE. DC.		GL'YCINE.	Diadelphia Decandria.	L.			
1 clandestìna. DC.	(*bl.*)	hidden-flower'd.	N. S. W.	1810.	4. 8.	G. ♄.◡.	

2 dèbilis. DC. (ye.) hairy. E. Indies. 1778. 6. 7. S. ♃.‿.
3 mínima. DC. (bl.) small. N. S. W. 1812. 4. 10. G. ♃.‿.
4 striàta. DC. (li.) streaked. S. America. 1823. 7. 9. S. ♄.‿. Jacq. vind. 1. t. 76.
5 parviflòra. DC. (re.) small-flowered. E. Indies. 1812. 7. 8. S. ♂.‿.
6 senegaléensis. DC. (ye.) Senegal. Senegal. 1822. —— S. ♃.‿.
7 heterophy'lla. DC. (ye.) various-leaved. C. B. S. 1816. 6. 8. G. ♃.‿.
8 argéntea. DC. (ye.) silvery-leaved. ———— —— —— G. ♃.‿.
9 secúnda. DC. (ye.) side-flowering. ———— 1825. 6. 9. G. ♃.‿.
CHÆTOCALYX. DC. CHÆTOCALYX. Diadelphia Decandria.
vincentìna. DC. (ye.) St. Vincent. W. Indies. 1823. 6. 8. S. ♄.‿. Bot. reg. 799.
Gl'ycine vincentìna. B.R. 799.

Subtribus IV. *GALEGEÆ.* DC. prodr. 2. p. 243.

PETALOST'EMUM. DC. PETALOST'EMUM. Monadelphia Pentandria.
1 cándidum. DC. (wh.) white. N. America. 1811. 7. 8. H. ♃. Mich. amer. 2. t. 37. f. 1.
2 cárneum. DC. (li.) flesh-coloured. Florida. —— —— H. ♃.
3 violàceum. DC. (pu.) purple-flower'd. Missouri. —— 7. 9. H. ♃. Bot. mag. 1707.
4 villòsum. DC. (ro.) villous. ———— 1826. —— H. ♃.
5 corymbòsum. DC. (wh.) corymbose. Carolina. 1806. —— H. ♃.
Dàlea Kuhnistèra. W. *Kuhnistéra caroliné nsis.* Lam. *Cylipògon virgàtum.* Rafin.
D'ALEA. DC. D'ALEA. Monadelphia Decandria.
1 laxiflòra. DC. (wh.) loose-flowered. Louisiana. 1811. 7. 8. H. ♃.
2 áurea. DC. (ye.) golden. ———— —— —— H. ♃.
3 alopecuroídes. DC. (w.) Fox-tail. ———— 1812. —— H. ⊙. Mich. amer. 2. t. 38.
Dàlea Linn'æi. Mich.
4 Cliffortiàna. W. (bl.) Vera Cruz. Vera Cruz. 1737. 7. 8. H. ⊙. Linn. h. cliff. t. 22.
5 Lagòpus. DC. (vi.) downy-spiked. Mexico. 1780. 10. 11. G. ⊙. Cavan. ic. 1. t. 86.
6 serícea. DC. (d.vi.) silky. ———— 1825. 9. 11. G. ♃. Kth. mim. 166. t. 48.
grácilis. Kth.
7 tuberculàta. DC. (pu.) tubercled. ———— —— —— G. ♄.
8 citriodòra. DC. (v.) leafy. New Spain. 1780. 10. 11. G. ⊙. Cavan. ic. 3. t. 271.
Psoràlea citriodòra. Cav. *foliolòsa.* H.K.
9 lùtea. DC. (ye.) yellow. Mexico. 1823. 7. 9. G. ♃. Cavan. ic. 4. t. 325.
10 mutábilis. DC. (v.) changeable. ———— 1821. 9. 10. G. ♄. Bot. mag. 2486.
bícolor. Hook. ex. flor. 43.
11 nùtans. DC. (vi.) nodding. ———— 1824. 7. 9. G. ♃. Cavan. ic. 3. t. 201.
12 reclinàta. DC. (vi.) reclining. ———— —— 5. 8. G. ♃.
13 tomentòsa. DC. (vi.) woolly. ———— 1823. 7. 9. G. ♃. Cavan. ic. 3. t. 240.
14 prostràta. DC. (ye.) prostrate. ———— 1826. —— G. ♃.
15 mucronàta. DC. (wh.) mucronate. ———— —— —— G. ♃.
16 Thouíni. DC. (vi.) Thouin's. ———— 1824. 6. 9. H. ⊙. Schrank hort. mon. t. 9.
17 phymatòdes. DC. (vi.) obovate-leaved. Caracas. —— 7. 9. G. ♄. Jacq. ic. 3. t. 563.
18 enneaphy'lla. DC. (re.) nine-leaved. W. Indies. 1772. —— S. ♄.
GLYCYRRH'IZA. DC. LIQUORICE. Diadelphia Decandria. L.
1 glábra. DC. (pu.) common. S. Europe. 1562. 7. 9. H. ♃. Lam. ill. t. 625. f. 2.
2 glandulìfera. DC. (pu.) glandular. Hungary. 1805. 6. 8. H. ♃. W. et K. hung. 1. t. 21.
3 lepidòta. DC. (wh.) scaly. Missouri. 1811. 7. 8. H. ♃. Bot. mag. 2150.
4 fœ'tida. DC. (ye.) strong-scented. Barbary. 1823. 6. 9. H. ♃. Desf. atl. 2. t. 199.
5 echinàta. DC. (pu.) prickly-headed. Italy. 1596. —— H. ♃. Bot. mag. 2154.
6 aspérrima. DC. (li.) rough. Siberia. 1795. 7. 8. H. ♃. Pall. it. app. t. M. f. 3.
7 uralénsis. DC. (li.) Siberian. ———— 1826. —— H. ♃.
8 hirsùta. DC. (li.) hairy. Levant. 1739. —— H. ♃.
GAL'EGA. DC. GOAT'S-RUE. Monadelphia Decandria.
1 officinàlis. DC. (pu.) officinal. Spain. 1568. 6. 9. H. ♃. Schk. hand. 2. t. 208.
α purpùrea. (pu.) *purple-flowered.* ———— —— —— H. ♃.
β álba. (wh.) *white-flowered.* ———— —— —— H. ♃.
2 pérsica. DC. (wh.) Persian. Persia. 1816. —— H. ♃. Swt. br. fl. gar. 3. t. 244.
3 bilòba. B.F.G. (pu.) two-lobed. .... 1823. —— H. ♃. ———— 2. t. 159.
4 orientàlis. DC. (bl.) oriental. Levant. 1801. —— H. ♃. Bot. reg. 326.
TEPHR'OSIA. DC. TEPHR'OSIA. Mono-Diadelphia Decandria.
1 suberòsa. DC. (ro.) cork-barked. E. Indies. 1818. 5. 8. S. ♄.
Robínia suberòsa. H.B
2 fruticòsa. DC. (ro.) frutescent. ———— 1820. .... S. ♄.‿.

3 sophoroídes. DC.	(wh.)	Sophora-like.	Nepaul.	1826.	....	G. ♄.	
4 brachystàchya. DC.	(ro.)	short-spiked.	——	——	....	G. ♄.	
5 cándida. DC.	(ro.)	white-leaved.	E. Indies.	1810.	5. 8.	S. ♄.	
Robínia cándida. H.B.							
6 toxicària. DC.	(bh.)	Fish poison.	S. America.	1791.	——	S. ♄.	Plum. ic. t. 135.
7 virginiàna. DC.	(re.)	Virginian.	Virginia.	1765.	6. 8.	H. ♃.	Pluk. alm. t. 23. f. 2.
8 hispídula. DC.	(li.)	bristle-podded.	——	1826.	——	H. ♃.	
9 ochroleùca. DC.	(st.)	straw-colour'd.	W. Indies.	1799.	——	S. ♄.	Jacq. ic. t. 150.
10 carib'æa. DC.	(re.wh.)	various-colour'd.	Caribees.	1786.	——	S. ♄.	Jacq. amer. t. 125.
11 granditlòra. DC.	(re.)	large-flowered.	C. B. S.	1774.	6. 9.	I. ♄.	Bot. reg. 769.
Galèga grandiflòra. B.R. *Galèga ròsea.* Lam.							
12 villòsa. DC.	(pu.)	villous.	E. Indies.	1779.	6. 8.	S. ♃.	Burm. zeyl. t. 33.
13 purpùrea. DC.	(pu.)	purple.	——	1768.	7. 8.	S. ♃.	—— t. 32.
Galèga purpùrea. L.							
14 lanceæfòlia. DC.	(st.)	lance-leaved.	——	1824.	——	S. ♄.	
15 argéntea. DC.	(pu.)	silvery.	——	1826.	....	S. ♃.	Pluk. alm. t. 52. f. 1.
16 piscatória. DC.	(pu.)	woolly.	——	1778.	6. 8.	S. ♄.	
17 leucántha. DC.	(wh.)	white-flowered.	Mexico.	1826.	6. 10.	G. ♄.	H. et B. n. gen. a. 6. t. 577.
18 coronillæfòlia. DC.	(st.)	Coronilla-leav'd.	Bourbon.	——	....	S. ♄.	
19 capénsis. DC.	(re.)	Cape.	C. B. S.	1823.	5. 9.	G. ♄.	Jacq. ic. t. 574.
20 cinèrea. DC.	(vi.)	cinereous.	W. Indies.	1824.	6. 9.	S. ⊙.	—— t. 575.
21 hypargy'rea. DC.	(pu.)	various-leaved.	E. Indies.	1825.	——	S. ♄.	
heterophy'lla. F.							
22 stricta. DC.	(pu.)	straight-podded.	C. B. S.	1774.	5. 6.	I. ♄.	Jacq. schœnb. t. 236.
23 lineàris. DC.	(pu.)	linear-leaved.	Guinea.	1822.	....	S. ♄.	
24 pállens. DC.	(li.)	pale-flowered.	C. B. S.	1787.	6. 8.	I. ♄.	
25 apollínea. DC.	(bl.)	Egyptian.	Egypt.	1822.	6. 7.	F. ♃.	Delil. fl. eg. t. 53. f. 5.
26 tricolor.	(bl.)	three-coloured.	N. Holland.	1824.	7. 9.	G. ♃.	Hook. ex. flor. t. 185.
Galèga tricolor. H.E.F.							
27 Barclayàna.	(pu.)	Mr. Barclay's.	Madagascar.	1823.	5. 10.	S. ♄.◡.	Hook. ex. flor. t. 188.
Dalbérgia Barcláyii. H.E.F.							
28 Telfáirii.	(pu.)	Mr. Telfair's.	——	——	——	S. ♄.◡.	
29 Coloníla. L.T.	(pu.)	Colonil.	E. Indies.	1818.	6. 8.	S. ♄.	
30 racemòsa.	(pu.)	racemed.	——	1822.	——	S. ♄.◡.	
Robínia racemòsa. H.B.							
31 pentaphy'lla.	(pu.)	five-leaved.	——	1824.	——	S. ♄.	
Galèga pentaphy'lla. H.B.							
32 biflòra. DC.	(pu.)	two-flowered.	....	1826.	6. 9.	G. ♄.	
33 capitulàta. DC.	(pu.)	headed.	Owyhee.	——	——	S. ♃.	
34 incàna.	(pu.)	hoary.	E. Indies.	1820.	——	S. ♄.	
Galèga incàna. H.B.							
35 Heyniàna.	(bl.)	Heyne's.	——	1807.	7. 9.	S. ♄.	
Galèga Heyniàna. H.B.							
36 serícea. DC.	(pu.)	silky.	C. B. S.	1824.	——	I. ♄.	
37 filifòlia. DC.	(re.)	filiform-leaved.	——	——	——	I. ♄.	
AMO'RPHA. DC.		BASTARD INDIGO. Monadelphia Decandria.					
1 fruticòsa. DC.	(bl.)	shrubby.	Carolina.	1724.	6. 8.	H. ♄.	Bot. reg. 427.
2 croceolanàta. W.D.	(pu.)	yellow-haired.	——	1812.	——	H. ♄.	Wats. dend. brit. t. 139.
3 fràgrans. B.F.G.	(pu.)	fragrant.	——	——	——	H. ♄.	Swt. br. fl. gar. t. 241.
nàna. Bot. mag. t. 2112. *nec aliorum.*							
4 emarginàta. L.C.	(bl.)	emarginate.	——	——	——	H. ♄.	
fruticòsa β emarginàta. Ph.							
5 Lewísii. L.C.	(bl.)	Lewis's.	——	——	——	H. ♄.	
fruticòsa γ angustifòlia. Ph.							
6 glábra. DC.	(pu.)	smooth.	——	——	——	H. ♄.	
7 nàna. DC.	(pu.)	dwarf.	Missouri.	1811.	7. 8.	H. ♄.	
microphy'lla. Ph.							
8 herbàcea. DC.	(pu.)	pubescent.	Carolina.	1803.	6. 8.	H. ♄.	Lodd. bot. cab. 689
pubéscens. Ph.							
9 canéscens. DC.	(bl.)	canescent.	Missouri.	1812.	7. 8.	H. ♄.	
NISS'OLIA. DC.		NISS'OLIA. Mono-Diadelphia Decandria.					
1 fruticòsa. DC.	(st.)	shrubby.	S. America.	1766.	7. 11.	S. ♄.◡.	Jacq. vind. t. 167.
2 racemòsa. DC.	(st.)	racemed.	W. Indies.	....	6. 8.	S. ♄.◡.	
3 glábrata. DC.	(wh.)	smooth-leaved.	S. America.	1818.	7. 11.	S. ♄.	
4 robiniæfòlia. DC.	....	Robinia-leaved.	St. Vincent.	1824.	....	S. ♄.	
5 micróptera. DC.	(ye.)	small-winged.	....		5. 8.	S. ♄.	
6 retùsa. DC.	(ye.)	retuse-leaved.	S. America.	1823.	....	S. ♄.◡.	
MULLERA. DC.		MULLERA. Monadelphia Octo-Decandria.					
moniliformis. DC.	....	bracelet.	Guiana.	1792.	....	S. ♄.	Aubl. guian. 4. t. 356.

LONCHOCA'RPUS. DC.		LONCHOCA'RPUS. Mono-Diadelphia Decandria.					
1 domingénsis. DC.	(re.)	oval-leaved.	St. Domingo.	1759.	....	S. ħ.	
2 violàceus. DC.	(vi.)	violet.	W. Indies.	——	5. 8.	S. ♄.	Jacq. amer. t. 177. f. 49.
Robínia violàcea. Jacq.							
3 formosiànus. DC.	(vi.)	Formosa.	Africa.	1822.	....	S. ħ.	Beauv. fl. owar. 2. t. 76.
4 seríceus. DC.	(pu.)	silky.	W. Indies.	1823.	....	S. ♄.	
5 pyxidàrius. DC.	(pu.)	Cuba.	Cuba.	1822.	....	S. ♄.	
6 latifòlius. DC.	(pu.)	broad-leaved.	S. America.	1814.	....	S. ħ.	
Amerímnum latifòlium. W.							
7 pubéscens. DC.	(pu.)	pubescent.	Caracas.	1820.	....	S. ♄.	
8 ròseus. DC.	(ro.)	rose-coloured.	S. America.	....	....	S. ħ.	
9 sèpium. DC.	(ro.)	hedge.	W. Indies.	1820.	....	S. ♄.	Jacq. amer. t. 129. f. 101.
10 Swartzii. DC.	(ro.)	Swartz's.	————	——	....	S. ♄.	
Robínia sèpium. Sw.							
11 Nicóu. DC.	(pu.)	climbing.	Guiana.	1826.	....	S. ♄.⌣.	Aubl. guian. 2. t. 308.
Robínia Nicóu. Aubl. *scándens.* W.							
ROBI'NIA. DC.		ROBI'NIA. Diadelphia Decandria. L.					
1 Pseudacàcia. DC.	(wh.)	com. or Locust Tree.	N. Amer.	1640.	5. 6.	H. 5.	Schmidt arb. 1. t. 32.
2 inérmis. DC.	(wh.)	spineless.	....	....	——	H. 5.	
spectábilis. L.C.							
3 macrophy'lla. L.C.	(ro.)	long-leaved.	Siberia.	1797.	6. 7.	H. 5.	
4 sophoræfòlia. L.C.	(wh.)	Sophora-leaved.	Hybrid.	1811.	——	H. 5.	
5 amorphæfòlia. L.en.	(w.)	Amorpha-leav'd.	————	1820.	——	H. 5.	
6 umbraculìfera. DC.	(w.)	Parasol.	————	1811.	....	H. 5.	
Robínia inérmis. Dum. Cours. *non* DC.							
7 dùbia. DC.	(bh.)	doubtful.	————	——	....	H. 5.	
8 strícta. L.C.	(wh.)	upright.	————	....	....	H. 5.	
9 tortuòsa. DC.	(wh.)	twisting.	————	....	6. 7.	H. 5.	
10 monstròsa. L.C.	(wh.)	monstrous.	————	....	——	H. 5.	
11 péndula. L.en.	(wh.)	pendulous.	————	....	——	H. 5.	
12 viscòsa. DC.	(bh.)	clammy.	N. America.	1797.	6. 8.	H. 5.	Venten. cels. t. 4.
glutinòsa. Bot. mag. 580.							
13 híspida. DC.	(ro.)	Rose Acacia.	Carolina.	1743.	5. 9.	H. ħ.	Bot. mag. 311.
14 ròsea. Ell.	(ro.)	smooth-branch'd.	————	1812.	——	H. ♄.	
15 ? guineénsis. W.en.	(ro.)	Guinea.	Guinea.	1821.	....	S. ♄.	
Cy'tisus híspidus. W.							
16? purpùrea. L.en.	(pu.)	purple.	....	1816.	4. 6.	S. ♄.	
POIT'ÆA. DC.		POIT'ÆA. Diadelphia Decandria.					
galegoídes. DC.	(ro.)	Galega-like.	St. Domingo.	1826.	....	S. ħ.	Venten. choix. t. 36.
COURS'ETIA. DC.		COURS'ETIA. Diadelphia Decandria.					
1 tomentòsa. DC.	(ye.)	woolly.	Peru.	1824.	6. 8.	G. ħ.	Cavan. ic. t. 84.
Láthyrus fruticòsus. Cav. *O'robus tomentòsus.* Df. *Vícia fruticòsa.* W.							
2 virgàta. DC.	(ye.)	slender.	Trinidad.	1820.	——	S. ♄.	Cavan. ic. t. 293.
Æschynómene virgàta. Cav.							
SESBA'NIA. DC.		SESBA'NIA. Diadelphia Decandria.					
1 ægyptíaca. DC.	(ye.)	Egyptian.	Egypt.	1680.	7. 8.	S. ♂.	Alp. ægypt. 81. t. 82.
2 occidentàlis. DC.	(ye.)	West Indian.	W. Indies.	1823.	6. 8.	S. ♄.	Plum. ed Bur. t. 125. f. 1.
3 aculeàta. DC.	(ye.)	prickly.	E. Indies.	1690.	7. 10.	S. ⊙.	Jacq. ic. t. 564.
4 cannabìna. DC.	(ye.)	Hemp-like.	————	1800.	——	S. ⊙.	
5 affínis. DC.	(ye.)	likened.	————	1822.	5. 8.	S. ♂.	
6 pícta. DC.	(std.)	painted.	W. Indies.	——	4. 9.	S. ħ.	Bot. reg. 873.
7 macrocárpa. DC.	(st.)	long-podded.	Louisiana.	——	7. 10.	H. ⊙.	Bart. fl. amer. 1. t. 28.
8 paludòsa. DC.	(ye.)	marsh.	E. Indies.	1810.	——	S. ⊙.	
9 grácilis. DC.	(ye.)	slender.	....	1824.	——	S. ⊙.	
10 serícea. DC.	(ye.)	silky.	E. Indies.	1818.	——	S. ⊙.	
11 procúmbens. H.B.	(ye.)	procumbent.	————	——	——	S. ⊙.	
12 uliginòsa.	(ye.)	swamp.	————	——	——	S. ⊙.	
Æschynómene uliginòsa. H.B.							
AGA'TI. DC.		AGA'TI. Diadelphia Decandria.					
1 grandiflòra. DC.	(bh.)	great-flowered.	E. Indies.	1768.	7. 8.	S. ħ.	Rheed. mal. 1. t. 51.
Sesbánia grandiflòra. P.S. *Æschynómene grandiflòra.* L. *Coronílla grandiflòra.* W.							
2 coccínea. DC.	(sc.)	red-flowered.	E. Indies.	1824.	7. 8.	S. ♄.	Rumph. amb. 1. t. 77.
GLOTTI'DIUM. DC.		GLOTTI'DIUM. Diadelphia Decandria.					
floridànum. DC.	(ye.)	flat-podded.	Florida.	1822.	7. 8.	S. ⊙.	Jacq. ic. 1. t. 148.
Sesbánia platycárpa. P.S. *Robínia vesicària.* Jacq. *Phàca floridàna.* W.							
PISCI'DIA. DC.		JAMAICA DOGWOOD. Monadelphia Decandria.					
1 Erythrìna. DC.	(v.)	oval-leaved.	W. Indies.	1690.	....	S. ♄.	Sloan. jam. 2. t. 176. f. 4-5.
2 carthagenénsis. DC.	(v.)	obovate-leaved.	————	....	....	S. ♄.	Plum. ed Bur. t. 133. f. 2.

DAUBENT'ONIA. DC.	DAUBENT'ONIA. Diadelphia Decandria.					
1 punícea. DC.	(*pu.*) red-flowered.	Mexico.	1824.	....	S. ♄.	Cavan. ic. 4. t. 316.
2 longifòlia. DC.	(*ye.*) long-leaved.	New Spain.	1828.	....	S. ♄.	——— 4. t. 315.
CORYNE'LLA. DC.	CORYNE'LLA. Diadelphia Decandria.					
polyántha. DC.	(*pu.*) many-flowered.	W. Indies	1824.	....	S. ♄.	
Robínia polyántha. Swartz.						
CARAG'ANA. DC.	CARAG'ANA. Diadelphia Decandria.					
1 Altagàna. DC.	(*ye.*) flat-podded.	Dahuria.	1789.	4. 6.	H. ♄.	L'Herit. stirp. t. 76.
2 microphy'lla. DC.	(*ye.*) small-leaved.	Siberia.	1816.	——	H. ♄.	Pall. ross. t. 42. f. 1-2.
3 arenària. L.C.	(*ye.*) sand.	———	1802.	6. 7.	H. ♄.	Bot. mag. 18.6.
Robínia Caragàna. β. arenària. B.M.						
4 arboréscens. DC.	(*ye.*) common.	———	1752.	4. 6.	H. ♄.	Schmidt arb. 1. t. 33.
Robínia Caragàna. L.						
5 Chamlàgu. DC.	(*ye.*) shining.	China.	1773.	5. 6.	H. ♄.	L'Herit. stirp. t. 77.
6 frutéscens. DC.	(*ye.*) shrubby.	Siberia.	1752.	4. 6.	H. ♄.	Swt. br. fl. gar. 3. t. 227.
7 móllis. M.B.	(*ye.*) soft-leaved.	Tauria.	1822.	——	H. ♄.	
8 grandiflòra. DC.	(*ye.*) large-flowered.	Iberia.	—	——	H. ♄.	
9 pygm'æa. DC.	(*ye.*) pigmy.	Siberia.	1751.	——	H. ♄.	Bot. reg. 1021.
10 mongòlica F.	(*ye.*) Mongolian.	Tartary.	1826.	——	H. ♄.	
11 Redówskii. F.	(*ye.*) Redowski's.	Siberia.	1822.	——	H. ♄.	DC. leg. mem. t. 11. f. 45.
12 spinòsa. DC.	(*ye.*) thorny.	———	1775.	——	H. ♄.	Pall. ross. 1. t. 44.
13 tragacanthoídes. DC.	(*y.*) Goat's-thorn-like.	———	1816.	——	H. ♄.	Pall. astr. t. 86.
14 jubàta. DC.	(*bh.*) bearded.	———	1796.	——	H. ♄.	——— p. 113. t. 85.
HALIMODE'NDRON. DC.	SALT-TREE. Diadelphia Decandria.					
1 argénteum. DC.	(*pu.*) silvery-leaved.	Iberia.	1779.	5. 6.	H. ♄.	Bot. mag. 1016.
2 triflòrum.	(*pu.*) smooth-leaved.	———	....	——	H. ♄.	L'Herit. stirp. n. 162.
Robínia triflòra. L'Herit.						
CALOPHA'CA. DC.	CALOPHA'CA. Diadelphia Decandria.					
wolgàrica. DC.	(*ye.*) wing-leaved.	Siberia.	1786.	5. 6.	H. ♄.	Wats. dend. brit. 83.
Cy'tisus wolgàricus. L. *pinnàtus.* Pall. ross. 1. t. 47.						
COLUT'EA. DC.	BLADDER SENNA. Diadelphia Decandria.					
1 arboréscens. DC.	(*ye.*) common.	S. Europe.	1568.	6. 8.	H. ♄.	Bot. mag. 81.
2 cruènta. DC.	(*re.*) red-flowered.	Levant.	1710.	——	H. ♄.	Schmidt arb. t. 119.
3 Haléppica. DC.	(*ye.*) Pocock's.	———	1752.	5. 10.	H. ♄.	——— t. 120.
Pocóckii. H.K. *Istria.* Mill. ic. t. 100.						
4 mèdia. DC.	(*ye.*) smaller.	....	....	6. 8.	H. ♄.	Wats. dend. brit. t. 140.
5 nepalénsis. DC.	(*ye.*) Nepaul.	Nepaul.	1822.	8. 9.	H. ♄.	Bot. mag. 2622.
SPHÆROPH'YSA. DC.	SPHÆROPH'YSA. Diadelphia Decandria.					
1 Salsùla. DC.	(*re.*) salt-field.	Dahuria.	1829.	....	H. ♄.	Pall. itin. 4. t. 9. f. 1.
2 cáspica. DC.	(*pu.*) Caspian.	Caucasus.	——	....	H. ♄.	
Colutèa cáspica. M.B. *Phàca Sálsula.* Bieb. casp. 210.						
SWAINS'ONA. DC.	SWAINS'ONA. Diadelphia Decandria.					
1 galegifòlia. DC.	(*sc.*) red-flowered.	N. S. W.	1800.	7. 8.	G. ħ.	Bot. mag. 792.
2 coronillæfòlia. DC.	(*pu.*) purple-flowered.	———	1802.	——	G. ħ.	——— t. 1725.
3 lessertiæfòlia. DC.	(*pu.*) small-flowered.	N. Holland.	1822.	——	G. ħ.	
4 astragalifòlia. C.C.	(*wh.*) white-flowered.	N. S. W.	1802.	——	G. ħ.	Bot. reg. 994.
galegifòlia. var. *albiflòra.* B.R.						
LESSE'RTIA. DC.	LESSE'RTIA. Diadelphia Decandria.					
1 ánnua. DC.	(*pu.*) annual.	C. B. S.	1731.	6. 7.	G. ⊙.	Hook. exot. flor. t. 84.
2 diffùsa. DC.	(*pu.*) procumbent.	———	1792.	7. 8.	G. ⊙.	Jacq. ic. 3. t. 576.
Galéga dùbia. Jacq.						
3 fruticòsa. B.R.	(*pu.*) frutescent.	———	1826.	——	G. ⊙.	Bot. reg. 970.
4 perénnans. DC.	(*ro.*) perennial.	———	1753.	——	G. ♃.	Jacq. vind. 3. t. 3.
Colutèa perénnans. Jacq. *fistulòsa.* Hort.						
5 púlchra. DC.	(*ro.*) pretty.	———	1817.	5. 8.	G. ⊙.	Bot. mag. 2064.
6 falcifórmis. DC.	(*pu.*) sickle-podded.	———	1825.	——	G. ♂.	DC. leg. mem. t. 46.
7 annulàris. DC.	(*pu.*) ring-podded.	———	1816.	——	G. ⊙.	
8 brachystàchya. DC.	(*pu.*) short-spiked.	———	——	——	G. ħ.	
9 prostràta. DC.	(*pu.*) prostrate.	———	1820.	——	G. ⊙.	
10 virgàta. L.O.	(*pu.*) slender.	———	1828.	6. 8.	G. ⊙.	
11 vesicària. DC.	(*pu.*) bladder-podded.	———	1825.	——	G. ♂.	
12 tomentòsa. DC.	(*pu.*) woolly.	———	1822.	——	G. ♃.	
13 procúmbens. DC.	(*pu.*) procumbent.	———	....	——	G. ħ.	
SUTHERLA'NDIA. DC.	SUTHERLA'NDIA. Diadelphia Decandria.					
1 frutéscens. DC.	(*sc.*) frutescent.	C. B. S.	1683.	6. 9.	F. ħ.	Bot. mag. 181.
Colutèa frutéscens. B.M.						
β canéscens.	(*sc.*) *hoary-ld. dwarf.*	———	1816.	——	F. ħ.	
2 microphy'lla. DC.	(*sc.*) small-leaved.	———	——	——	F. ħ.	

Subtribus V. *ASTRAGALEÆ.* DC. prodr. 2. p. 273.

PHA'CA. DC.		BASTARD VETCH. Diadelphia Decandria.					
1 b'ætica. DC.	(*wh.*)	hairy.	Spain.	1640.	7. 8.	H. ♃.	Moris. his. s. 2. t. 8. f. 1.
2 frígida. DC.	(*st.*)	small.	Austria.	1795.	——	H. ♃.	Jacq. aust. t. 166.
3 alpìna. DC.	(*st.*)	Alpine.	————	1759.	——	H. ♃.	Jacq. ic. 1. t. 151.
4 membranàcea. F.	(*st.*)	membranous.	Dahuria.	1823.	——	H. ♃.	
5 lùtea. Led.	(*ye.*)	yellow.	Siberia.	1827.	——	H. ♃.	
6 cæspitòsa. DC.	(*st.*)	tufted.	Missouri.	1826.	——	H. ♃.	
Astrágalus triphy'llus. Ph.							
7 densifòlia. DC.	(*re.*)	dense-leaved.	California.	1822.	——	H. ♃.	
8 glábra. DC.	(*v.*)	smooth.	S. Europe.	——	——	H. ♃.	
9 austràlis. DC.	(*pa.*)	trailing.	————	1779.	5. 6.	H. ♃.	Lodd. bot. cab. 490.
10 astragalìna. DC.	(*v.*)	procumbent.	Europe.	1771.	6. 7.	H. ♃.	———— t. 429.
11 lappònica. DC.	(*v.*)	Lapland.	Lapland.	1824.	7. 8.	H. ♃.	Flor. dan. t. 51.
12 oroboídes. DC.	(*bl.*)	Orobus-like.	N. Europe.	1826.	——	H. ♃.	———— 8. t. 1396.
13 triangulàris. F.	(*bl.*)	triangular.	Siberia.	1824.	——	H. ♃.	
14 exaltàta. F.	....	tall.	Altay.	1828.	....	H. ♃.	
15 arenària. DC.	(*pu.*)	sand.	Siberia.	1796.	7. 8.	H. ♃.	Pall. it. 4. t. 19. f. 3-4.
OXY'TROPIS. DC.		OXY'TROPIS. Diadelphia Decandria.					
1 montàna. DC.	(*bl.*)	mountain.	Europe.	1531.	7. 8.	H. ♃.	Bot. mag. 843.
Astrágalus montànus. B.M.							
2 cyánea. DC.	(*bl.*)	bright blue.	Iberia.	1826.	——	H. ♃.	
3 sórdida. DC.	(*pa.*)	sordid.	N. Europe.	....	5. 8.	H. ♃.	Flor. dan. t. 1041.
4 argyr'æa. DC.	(*pu.*)	silvery.	Altay.	1828.	——	H. ♃.	
5 uraléns is. DC.	(*pu.*)	villous.	Siberia.	1800.	6. 8.	H. ♃.	Pall. ast. t. 42. f. A.
6 serícea. DC.	(*pu.*)	hairy mountain.	Scotland.	....	5. 8.	H. ♃.	Eng. bot. t. 466.
7 cœrùlea. DC.	(*bl.*)	blue-flowered.	Siberia.	1827.	6. 8.	H. ♃.	Pall. ast. t. 52.
8 árctica. DC.	(*pu.*)	northern.	Melvill. Isl.	——	——	H. ♃.	
9 argentàta. DC.	(*pa.*)	silver-leaved.	Siberia.	——	——	H. ♃.	Pall. ast. t. 48.
10 ambígua. DC.	(*pu.*)	ambiguous.	————	1824.	5. 8.	H. ♃.	———— t. 43.
11 setòsa. DC.	(*pu.*)	pointed-calyxed.	————	1828.	6. 8.	H. ♃.	———— t. 44.
12 Lambérti. DC.	(*pu.*)	Mr. Lambert's.	Missouri.	1811.	8. 9.	H. ♃.	Bot. mag. 2147.
13 grandiflòra. DC.	(*ro.*)	large-flowered.	Siberia.	1824.	5. 8.	H. ♃.	Pall. astr. t. 46.
14 cándicans. DC.	(*pa.*)	hoary.	————	1827.	6. 8.	H. ♃.	———— t. 49.
15 songàrica. DC.	(*vi.*)	spear-podded.	————	1824.	5. 8.	H. ♃.	———— t. 51.
16 sylvática. DC.	(*pu.*)	wood.	————	1820.	——	H. ♃.	———— t. 78.
17 longiróstra. DC.	(*pu.*)	long-beaked.	————	1824.	6. 8.	H. ♃.	DC. astr. n. 17. t. 5.
Gmelìni. F.							
18 breviróstra. DC.	(*pu.*)	short-beaked.	Altay.	1802.	7. 9.	H. ♃.	DC. astr. n. 19. t. 6.
Astrágalus altáicus. Pall. astr. t. 45.							
19 filifórmis. DC.	(*bl.*)	filiform.	Dahuria.	1824.	7. 8.	H. ♃.	DC. astr. n. 16. t. 4.
20 nigréscens. DC.	(*bl.*)	black-haired.	Siberia.	1828.	——	H. ♃.	Pall. astr. t. 53.
21 campéstris. DC.	(*st.*)	field.	Scotland.	....	——	H. ♃.	Eng. bot. v. 36. t. 2522.
Astrágalus campéstris. E.B.							
22 sulphùrea. F.	(*st.*)	sulphur-color'd.	Siberia.	1827.	7. 8.	H. ♃.	
23 microphy'lla. DC.	(*pu.*)	small-leaved.	————	1823.	——	H. ♃.	Pall. ast. t. 76.
24 pùmila. DC.	(*bl.*)	small.	————	1828.	——	H. ♃.	
25 verticillàris. DC.	(*pu.*)	whorl-leaved.	————	1824.	——	H. ♃.	Amm. ruth. t. 19. f. 1.
26 myriophy'lla. DC.	(*va.*)	many-leaved.	————	——	——	H. ♃.	Pall. astr. t. 71.
27 oxyphy'lla. DC.	(*pu.*)	sharp-leaved.	————	1823.	7. 9.	H. ♃.	———— t. 74.
28 prostràta. DC.	(*pu.*)	prostrate.	————	1820.	6. 9.	H. ♃.	———— t. 72.
Astrágalus dahùricus. Pall. ast. *Phàca prostràta.* L.							
29 pilòsa. DC.	(*st.*)	hairy.	Siberia.	1732.	6. 8.	H. ♃.	Bot. mag. 2483.
30 floribúnda. DC.	(*pu.*)	many-flowered.	————	1827.	5. 8.	H. ♂.	Pall. astr. t. 37.
31 glábra. DC.	(*vi.*)	smooth.	Dahuria.	——	——	H. ♃.	DC. astr. n. 31. t. 8.
32 defléxa. DC.	(*vi.*)	small-flowered.	Siberia.	1800.	6. 8.	H. ♃.	L'Herit. stirp. t. 80.
Astrágalus retrofléxus. Pall. astr. t. 27. *hìans.* Jacq. ic. t. 252. *parviflòrus.* Lam.							
33 Fischèri. DC.	(*bl.*)	Fischer's.	Siberia.	1828.	——	H. ♃.	
34 vaginàta. DC.	(*bl.*)	sheathed.	————	——	——	H. ♃.	
35 tenélla. DC.	(*bl.*)	slender.	————	——	——	H. ♃.	
36 dichóptera. DC.	(*bl.*)	pubescent.	————	1815.	——	H. ♃.	
37 longicúspis. Led.	(*pu.*)	long-pointed.	————	1827.	——	H. ♃.	
ASTRA'GALUS. DC.		MILK-VETCH. Diadelphia Decandria. L.					
1 hypoglóttis. DC.	(*pu.*)	purple mountain.	Britain.	....	6. 7.	H. ♃.	Eng. bot. v. 4. t. 274.
2 dasyglottis. DC.	(*pu.*)	creeping-root'd.	Siberia.	1828.	——	H. ♃.	

3	purpùreus. DC.	(*pu.*)	purple.	S. France.	1824.	6. 7.	H. ♃.	DC. astr. n. 17. t. 12.
4	sínicus. L.	(*ro.*)	Chinese annual.	China.	1763.	7. 9.	H. ⊙.	Bot. mag. 1350.
	lotoídes. DC.							
5	onobrychioídes. DC.	(*pu.*)	Saintfoin-like.	Persia.	1828.	——	H. ♄.	DC. astr. n. 39. t. 16.
	canéscens. DC. ast. *cephalòtes.* Pall. astr. t. 24.							
6	leontìnus. DC.	(*pu.*)	Lion's-tail.	Austria.	1816.	5. 7.	H. ♃.	Lodd. bot. cab. 432.
7	Doniànus. DC.	(*pu.*)	Mr. D. Don's.	Nepaul.	1821.	8. 9.	H. ♃.	
	macrorhìzus. D.P. *non* Cav.							
8	arenàrius. DC.	(*bl.*)	sand.	N. Europe.	1798.	6. 7.	H. ♃.	Retz. obs. 3. t. 3.
9	bayonénsis. DC.	(*pu.*)	woolly-podded.	S. France.	1828.	——	H. ♃.	
10	macrópterus. DC.	(*pu.*)	large-winged.	Siberia.	1827.	——	H. ♃.	
11	taùricus. DC.	(*pu.*)	Taurian.	Tauria.	1824.	7. 8.	H. ♃.	Pall. astr. t. 38.
12	austrìacus. DC.	(*pu.*)	Austrian.	Austria.	1640.	6. 7.	H. ♃.	Jacq. aust. 2. t. 195.
13	olópterus. DC.	(*pa.*)	long-racemed.	Altay.	1827.	——	H. ♃.	
14	melilotoídes. DC.	(*pu.*)	Melilot-like.	Siberia.	1785.	——	H. ♃.	Pall. astr. t. 41.
15	sulcàtus. DC.	(*pu.*)	furrowed.	——	——	7. 8.	H. ♃.	Jacq. vind. 3. t. 40.
16	grácilis. DC.	(*li.*)	slender.	Missouri.	1812.	6. 7.	H. ♃.	
17	subulàtus. DC.	(*pu.*)	awl-podded.	Tauria.	1806.	5. 8.	H. ♃.	Pall. astr. t. 20.
18	ceratoídes. DC.	(*pu.*)	triangular-pod'd.	Altay.	1827.	——	H. ♃.	———— t. 20. A.
19	virgàtus. DC.	(*pu.*)	twiggy.	Siberia.	1806.	——	H. ♄.	———— t. 18.
	vàrius. Gmel. itin. 2. t. 17.							
20	brachylòbus. DC.	(*pu.*)	short-podded.	————	1827.	——	H. ♄.	
21	dahùricus. DC.	(*pu.*)	Dahurian.	Dahuria.	1824.	——	H. ♃.	Pall. it. 3. ap. t. w. f. 1.
22	Steveniànus. DC.	(*pu.*)	Steven's.	Iberia.	1829.	....	H. ♃.	
23	tenuifòlius. W.	(*pu.*)	fine-leaved.	Siberia.	1780.	7. 8.	H. ♃.	Swt. br. fl. gar. t. 73.
	linearifòlius. DC.							
24	Onobr'ychis. DC.	(*pu.*)	purple-spiked.	Europe.	1640.	6. 8.	H. ♃.	Jacq. aust. 1. t. 38.
25	fláccidus. DC.	(*wh.*)	flaccid.	Iberia	1826.	——	H. ♃.	
26	vimíneus. DC.	(*pu.*)	twiggy.	Siberia.	1824.	7. 8.	H. ♄.	Pall. astr. t. 21.
27	fruticòsus. DC.	(*pu.*)	woody	————	1804.	6. 8.	H. ♄.	———— t. 19.
28	adsúrgens. DC.	(*bl.*)	assurgent.	————	1824.	——	H. ♃.	———— t. 31.
29	Laxmánni. DC.	(*pa.*)	Laxmann's.	————	1804.	——	H. ♃.	Jacq. vind. 3. t. 37.
30	missouriénsis. DC.	(*pu.*)	Missouri.	Missouri.	1826.	6. 8.	H. ♃.	
31	caryocárpus. DC.	(*vi.*)	swelled-podded.	Louisiana.	1811.	——	H. ♃.	Bot. reg. 176.
32	pentaglóttis. DC.	(*pu.*)	rough Spanish.	Spain.	1739.	6. 7.	H. ⊙.	Cavan. ic. 2. t. 188.
33	Gláux. DC.	(*pu.*)	small Spanish.	————	1596.	——	H. ⊙.	Clus. hist. 2. p. 242. ic.
34	oxyglóttis. DC.	(*pu.*)	sharp-podded.	Tauria.	1824.	6. 9.	H. ⊙.	
35	psiloglóttis. DC.	(*pu.*)	smooth-podded.	Astracan.	1829.	— –	H. ⊙.	
36	cruciàtus. DC.	(*pu.*)	oblong-leaved.	Iberia.	1823.	——	H. ⊙.	
37	Stélla. DC.	(*pu.*)	star-podded.	S. Europe.	1658.	——	H. ⊙.	Lobel. ic. 2. t. 83.
38	tribuloídes. DC.	(*pu.*)	Caltrops-like.	Egypt.	1828.	——	H. ⊙.	
39	sesàmeus. DC.	(*pu.*)	Sesama.	S. Europe.	1616.	——	H. ⊙.	Garid. prov. t. 12.
40	vesicàrius. DC.	(*pa.*)	white Italian.	Europe.	1737.	7. 8.	H. ♃.	Vill. dauph. t. 42. f. 1.
	álbidus. W. et K. hung. 1. t. 40. *dealbàtus.* Pall. astr. t. 23. f. 1.							
41	albicáulis. DC.	(*wh.*)	white-stalked.	Caucasus.	1803.	7. 8.	H. ♃.	DC. astr. n. 59. t. 21.
	dealbàtus Pall. ast. t. 23. f. 2 et 3. *Oxy'tropis dealbàta.* P.S.							
42	Nuttalliànus. DC.	(*bl.*)	Nuttall's.	N. America.	1826.	6. 8.	H. ♃.	
	micránthus. N. *non* Desv.							
43	cymbæcárpos. DC.	(*pa.*)	boat-podded.	Portugal.	1800.	——	H. ⊙.	Brot. phyt. t. 59.
44	striátellus. DC.	(*bl.*)	striated.	Tauria.	1829.	——	H. ⊙.	Bieb. cent. 1. t. 20.
45	annulàris. DC.	(*ro.*)	ring-podded.	Egypt.	1800.	——	H. ⊙.	DC. astr. app. n. 6. t. 7.
46	hispídulus. DC.	(*pa.*)	rough-podded.	————	1828.	——	H. ⊙.	
47	scorpioídes. DC.	(*bl.*)	blue-flowered.	Spain.	1816.	6. 9.	H. ⊙.	
	canaliculàtus. W. enum. ex Link enum. 2. p. 254.							
48	mareóticus. DC.	(*li.*)	Alexandrian.	Egypt.	1828.	——	H. ⊙.	Delil. fl. eg. t. 39. f. 3.
49	corrugàtus. DC.	(*pu.*)	rugged-podded.	————		——	H. ⊙.	
50	reticulàtus. DC.	(*pa.*)	netted.	Volga.		——	H. ⊙.	
51	contortuplicàtus. DC.	(*st.*)	wave-podd'd.	Siberia.	1764.	7. 8.	H. ⊙.	Pall. astr. t. 79.
52	triméstris. DC.	(*st.*)	Egyptian.	Egypt.	1739.	——	H. ⊙.	Jacq. vind. 2. t. 174.
53	ægicèras. DC.	(*st.*)	Goat's-horn.	....	1816.	6. 8.	H. ⊙.	
54	brachycèras. DC.	(*st.*)	short-podded.	Persia.	1828.	——	H. ⊙.	
55	hamòsus. DC.	(*st.*)	hook-podded.	Spain.	1633.	——	H. ⊙.	Moris. ox. s. 2. t. 9. f. 10.
	bucèras. W. enum. ex Link enum.							
56	epiglóttis. DC.	(*st.*)	heart-podded.	S. Europe.	1737.	——	H. ⊙.	Herm. lugd. t. 77.
57	triangulàris. DC.	(*st.*)	triangular.	....	1828.	——	H. ⊙.	
58	bæ̀ticus. DC.	(*st.*)	slender-podded.	S. Europe.	1759.	——	H. ⊙.	Munt. phyt. t. 110.
59	capitàtus. W.	(*st.*)	headed.	Levant.	——	7. 8.	H. ♃.	
60	microphy'llus. DC.	(*st.*)	small-leaved.	Siberia.	1773.	6. 7.	H. ♃.	
61	semibiloculàris. DC.	(*st.*)	semibilocular.	————	1804.	6. 8.	H. ♃.	DC. astr. n. 64. t. 23.

62 réptans. DC.	(*wh.*)	creeping.	Mexico.	1829.	6. 8.	F. ♃.	Willd. hort. ber. 2. t. 88.
63 odoràtus. DC.	(*st.*)	sweet-scented.	Levant.	1816.	——	H. ♃.	DC. astr. n. 67. t. 24.
64 stipulàtus. DC.	(*st.*)	large-stipuled.	Nepaul.	1821.	8. 9.	H. ♃.	Bot. mag. 2380.
65 glycyphy'llos. DC.	(*st.*)	sweet-leaved.	Britain.	....	6. 7.	H. ♃.	Eng. bot. 3. t 203.
66 depréssus. DC.	(*st.*)	depressed.	Europe.	1772.	5. 6.	H. ♃.	Lodd. bot. cab. 680.
67 leucoph'æus. DC.	(*wh.*)	dwarf-white.	....	1776.	5. 8.	H. ♃.	—— t. 111.
68 Cìcer. DC.	(*st.*)	bladder-podd'd.	Europe.	1570.	6. 7.	H. ♃.	Jacq. aust. t. 251.
69 uliginòsus. DC.	(*st.*)	marsh.	Siberia.	1752.	6. 8.	H. ♃.	Pall. astr. t. 26.
70 micránthus. DC.	(*st.*)	small-blossom'd.	....	1824.	——	H. ♃.	
71 canadénsis. DC.	(*st.*)	woolly.	N. America.	1732.	6. 7.	H. ♃.	Lodd. bot. cab. 372.
72 caroliniànus. DC.	(*st.*)	Carolina.	——	——	——	H. ♃.	Dill. elth. t. 39. f. 45.
73 falcàtus. DC.	(*st.*)	sickle-podded.	Siberia.	1737.	——	H. ♃.	DC. astr. n. 70. t. 26.
virèscens. H.K.							
74 falcifórmis. DC.	(*st.*)	sickle-like.	Barbary.	....	——	H. ♃.	Desf. atl. 2. t. 207.
75 asper. DC.	(*st.*)	rough Astracan.	Siberia.	1796.	7. 8.	H. ♃.	Jacq. ic. 1. t. 33.
chloránthus. Pall. astr. t. 25.							
76 chinénsis. DC.	(*st.*)	upright.	Chinese.	1795.	6. 8.	H. ♃.	Linn. f. decad. 1. t. 3.
77 galegifórmis. DC.	(*st.*)	Goat's-rue-like.	Siberia.	1729.	——	H. ♃.	Pall. astr. t. 29.
78 racemòsus. DC.	(*wh.*)	racemed.	Louisiana.	1827.	6. 8.	H. ♃.	
galegoídes. N.							
79 alopècias. DC.	(*ye.*)	axillary-headed.	Siberia.	1823.	——	H. ♃.	Pall. astr. p. 12. t. 9.
80 alopecuroídes. DC.	(*ye.*)	Foxtail-like.	——	1737.	——	H. ♃.	—— t. 8.
81 narbonénsis. DC.	(*ye.*)	French.	S. Europe.	1780.	——	H. ♃.	—— t. 10.
82 pónticus. DC.	(*ye.*)	villous-calyxed.	Tauria.	1824.	——	H. ♃.	—— t. 11.
83 vulpìnus. DC.	(*ye.*)	Fox-tail.	Siberia.	1815.	——	H. ♃.	—— t. 7.
alopecuroídes. Pall. astr. t. 7.							
84 macrocéphalus. DC.	(*ye.*)	large-headed.	Persia.	1829.	——	H. ♃.	
85 christiànus. DC.	(*st.*)	great-yellow.	Armenia.	1737.	7. 8.	H. ♃.	Tourn. it. 2. t. 254.
86 tomentòsus. DC.	(*st.*)	downy-leaved.	Egypt.	1800.	——	H. ♃.	DC. astr. n. 80. t. 29.
87 Siebèri. DC.	(*st.*)	Sieber's.	——	1828.	——	H. ♃.	
88 Bárba-Jòvis. DC.	(*pu.*)	silvery-leaved.	Armenia.	....	——	H. ♃.	
Tragacántha argéntea. Mill.							
89 caucásicus. DC.	(*wh.*)	Caucasean.	Caucasus.	1822.	——	H. ♄.	Pall. astr. p. 2. t. 2.
90 breviflòrus. DC.	(*pu.*)	short-flowered.	Armenia.	1826.	6. 8.	H. ♄.	Lodd. bot. cab. 1388.
91 arnacántha. DC.	(*li.*)	prickly-leaved.	Tauria.	1640.	——	H. ♄.	Pall. astr. p. 1. t. 1.
Potèrium. Pall. ast. *nec aliorum.*							
92 sículus. DC.	(*bh.*)	Sicilian.	Sicily.	1827.	——	H. ♄.	
93 Potèrium. DC.	(*wh.*)	small goat's-thorn.	S. Europe.	1640.	——	H. ♄.	Morris. ox. s. 2. t. 13. f. 2.
94 massiliénsis. DC.	(*wh.*)	sea-side.	——	1827.	——	H. ♄.	Duhain. arb. 2. t. 100.
Tragacántha. Pall. astr. t. 4. f. 1. 2.							
95 aristàtus. DC.	(*wh.*)	awned.	Pyrenees.	1791.	5. 7.	H. ♄.	Garid. aix. t. 104.
96 lagùrus. DC.	(*ye.*)	hare's-tail'd.	Armenia.	1829.	——	H. ♄.	DC. astr. n. 104. t. 36.
97 armàtus. W.	(*ye.*)	armed.	Barbary.	....	6. 7.	H. ♄.	Desf. atl. 2. t. 194.
Anthy'llis tragacanthoídes. Desf.							
98 halicácabus. DC.	(*ye.*)	bladdered.	Armenia.	1806.	——	H. ♃.	Schreb. dec. p. 5. t. 3.
Phàca vesicària. Schreb.							
99 laguroídes. DC.	(*pu.*)	Hare's-tail-like.	Siberia.	1829.	....	H. ♃.	Pall. astr. t. 16.
100 ammod'ytes. DC.	(*wh.*)	silky-leaved.	——	1823.	6. 7.	H. ♃.	—— p. 7. t. 5.
101 dasyánthus. DC.	(*ye.*)	woolly-headed.	——	1828.	——	H. ♃.	—— p. 79. t. 65.
eriocéphalus. W. et K. hung. 1. t. 46.							
102 caprìnus. DC.	(*ye.*)	scented.	Barbary.	1683.	——	H. ♃.	Moris. ox. s. 2. t. 24. f. 3.
103 longiflòrus. DC.	(*ye.*)	long-flowered.	Tartary.	——	——	H. ♃.	Pall. astr. t. 60.
104 utríger. DC.	(*ye.*)	turgid-podded.	——	1825.	——	H. ♃.	—— t. 61.
105 exscàpus. DC.	(*ye.*)	hairy-podded.	Hungary.	1787.	5. 7.	H. ♃.	Jacq. ic. 3. t. 561.
106 diffùsus. DC.	(*st.*)	diffuse.	Tauria.	1824.	6. 7.	H. ♃.	Pall. astr. t. 68.
107 buchtorménsis. DC.	(*y.*)	pointed-podd'd.	Siberia.	1826.	——	H. ♃.	—— t. 62. f. A.
108 monspessulànus. DC.	(*pu.*)	Montpelier.	S. Europe.	1710.	7. 8.	H. ♃.	Bot. mag. 375.
109 incànus. DC.	(*pu.*)	hoary.	——	1759.	6. 7.	H. ♃.	Magn. monsp. p. 32. ic.
110 macrorhìzus. DC.	(*pu.*)	long-rooted.	Spain.	1821.	6. 8.	H. ♃.	Cavan. ic. 2. t. 133.
111 brachycárpus. DC.	(*pu.*)	short-podded.	Siberia.	1820.	——	H. ♃.	Bot. mag. 2335.
112 physòdes. DC.	(*pu.*)	inflated	——	1759.	——	H. ♃.	DC. astr. n. 127. t. 48.
113 Pallàsii. DC.	(*pu.*)	Pallas's.	Caucasus.	1828.	——	H. ♃.	Pall. astr. t. 59.
114 testiculàtus. DC.	(*bh.*)	woolly-podded.	——	1824.	——	H. ♃.	—— t. 67.
115 rupifràgus. DC.	(*pu.*)	dry rock.	Tauria.	1828.	——	H. ♃.	—— t. 70.
116 redúncus. DC.	(*pu.*)	crook-podded.	Volga.	——	——	H. ♃.	—— p. 109. t. 82.
117 chlorostàchys. H.C.	(*st.*)	straw-colour'd.	N. America.	1826.	9. 10.	H. ♃.	
118 uncàtus. DC.	(*wh.*)	hooked.	Aleppo.	1768.	7. 8.	H. ♃.	
119 ràdicans. DC.	(*pu.*)	rooting-branch'd.	....	1828.	——	H. ♃.	

BISE'RRULA. DC. Hatchet-Vetch. Diadelphia Decandria. L.
Pelecìnus. DC. (*bl.*) common. S. Europe. 1640. 7. 9. H. ⊙. Tourn. inst. 417. t. 234.

Tribus III. HEDYSAREÆ. *DC. prodr.* 2. *p.* 307.

Subtribus I. *CORONILLEÆ.* DC. loc. cit. p. 308.

SCORPI'URUS. DC. Caterpillar. Diadelphia Decandria. L.
1 lævigàta. DC. (*ye.*) smooth. Levant. 1820. 6. 8. H. ⊙. Flor. græc. t. 718.
2 muricàta. DC. (*ye.*) two-flowered. S. Europe. 1640. —— H. ⊙. Moris. ox. s. 2. t. 11. f. 4.
3 sulcàta. DC. (*ye.*) three-flowered. ——— 1596. —— H. ⊙. Desf. atl. 2. f. 4.
4 acutifòlia. DC. (*ye.*) acute-leaved. Corsica. 1828. —— H. ⊙. Viv. fl. lyb. t. 19. f. 4.
5 vermiculàta. DC. (*ye.*) common. S. Europe. 1621. —— H. ⊙. Moris. ox. s. 2. t. 11. f. 3.
CORONI'LLA. DC. Coroni'lla. Diadelphia Decandria. L.
1 E'merus. DC. (*ye.*) Scorpion Senna. France. 1596. 4. 6. H. ♄. Bot. mag. 445.
2 júncea. DC. (*ye.*) rush-like. ——— 1656. 6. 8. F. ♄. Bot. reg. 820.
3 valentína. L. (*ye.*) nine-leaved. S. Europe. 1596. 2. 6. F. ♄. Bot. mag. 185.
stipulàris. Lam.
4 pentaphy'lla. DC. (*ye.*) five-leaved. Algiers. .... 3. 6. F. ♄. Mill. ic. t. 289. f. 2?
5 argéntea. DC. (*ye*) silvery-leaved. Crete. 1664. 5. 6. F. ♄. —— 2. t. 289. f. 3.
6 gláuca. DC. (*ye.*) seven-leaved. France. 1722. 9. 5. F. ♄. Bot. mag. t. 13.
7 viminàlis. DC. (*bh.*) slender. Mogador. 1798. 5. 11. G. ♄. Sal. par. lond. t. 13.
8 mínima. DC. (*ye.*) least. S. Europe. 1658. 7. 8. H. ♃. Bot. mag. 2179.
9 coronàta. DC. (*ye.*) large-headed. ——— 1776. 6. 8. H. ♃. Clus. hist. 1. p. 98. f. 2.
10 montàna. DC. (*ye.*) mountain. Europe. 1816. 6. 9. H. ♃. Scop. carn ed. 2. t. 41.
coronàta. B.M. 907. *non* L.
11 ibérica. DC. (*ye.*) Iberian. Iberia. 1818. 7. 10. H. ♃. Swt. br. fl. gar. t. 25.
12 parviflòra. DC. (*ye.*) small-flowered. Crete. 1828. —— F. ♃.
13 crética. DC. (*v.*) Cretan. ——— 1731. 6. 8 H. ⊙. Jacq. vind. 1. t. 25.
14 vària. DC. (*pu.*) purple. Europe. 1597. 6. 11. H. ♃. Bot. mag. 258.
ASTROL'OBIUM. DC. Astrol'obium. Diadelphia Decandria.
1 ebracteàtum. DC. (*ye.*) round-podded. Portugal. .... 5. 6. H. ⊙. Dalech. his. 1. 487. f. 1.
Ornithòpus lævigàtus. Sm. *nudiflòrus.* Lag. *ebracteàtus.* Brot.
2 dùrum. DC. (*ye.*) perennial. S. Europe. 1805. 6. 7. H. ♃. Cavan. ic. 1. t. 41.
3 repándum. DC. (*ye.*) repand. Barbary. —— —— H. ⊙. Lam. ill. t. 631. f. 4.
4 scorpioídes. DC. (*ye.*) Purslane-leav'd. S. Europe. 1596. —— H. ⊙. Cavan. ic. 1. t. 37.
Ornithòpus scorpioìdes. L. *trifoliàtus.* Lam.
ORNITH'OPUS. DC. Bird's-foot. Diadelphia Decandria. L.
1 compréssus. DC. (*ye.*) hairy. S. Europe. 1730. 6. 7. H. ⊙. Berg. phyt. t. 191.
2 intermèdius. RO. (*ro.*) intermediate. Portugal. 1816. —— H. ⊙.
satìvus. Brot. fl. lus. 2. p. 160.
3 perpusíllus. DC. (*r.*) common. Britain. .... 5. 6. H. ⊙. Eng. bot. v. 6. t. 369.
HIPPOCR'EPIS. DC. Horseshoe-Vetch. Diadelphia Decandria. L.
1 baleàrica. DC. (*ye.*) shrubby. Minorca. 1776. 5. 6. F. ♄. Bot. mag. 427.
2 helvética. (*ye.*) Swiss. Switzerland. 1822. 5. 8. H. ♃.
3 comòsa. DC. (*ye.*) tufted. England. .... 4. 8. H. ♃. Eng. bot. 1. t. 31.
4 gláuca. DC. (*ye.*) glaucous. Naples. 1825. 5. 8. H. ♃. Ten. fl. neap. 2. t. 69.
5 syrìaca. E. (*ye.*) Syrian. Syria. 1826. 6. 8. H. ⊙.
6 scábra. DC. (*ye.*) rough. Spain. —— —— H. ⊙.
7 multisiliquòsa. DC. (*ye.*) many-podded. S. Europe. 1683. —— H. ⊙. Schkuhr hand. 2. t. 206.
8 monocárpa. DC. (*ye.*) one-podded. Caucasus. 1825. —— H. ⊙.
9 unisiliquòsa. DC. (*ye.*) single-podded. Italy. 1570. 6. 7. H. ⊙. Lam. ill. t. 630. f. 3.
10 ciliàta. DC. (*ye.*) ciliated. Spain. 1823. 6. 8. H. ⊙.
11 biflòra. DC. (*ye.*) two-flowered. .... 1824. —— H. ⊙.
SECURI'GERA. DC. Hatchet-Vetch. Diadelphia Decandria.
Coronílla. DC. (*ye.*) common. S. Europe. 1562. 7. 8. . ⊙. Lam. ill. t. 629.

Subtribus II. *EUHEDISAREÆ.* DC. prodr. 2. p. 313.

PICT'ETIA. DC. Pict'etia. Diadelphia Decandria.
aristàta. DC. (*ye.*) awned. St. Domingo. 1820. 4. 6. S. ♄. DC. leg. mem. t. 47. f. 5.
Æschynómene aristàta. Jacq. schœnb. 2. t. 237.

ORMOCA'RPUM. DC. ORMOCA'RPUM. Diadelphia Decandria.
sennoídes. DC. (*ye.*) Senna-like. E. Indies. 1818. 6.7. S. ♄.
Hedy'sarum sennoídes. W.
POIR'ETIA. DC. POIR'ETIA. Monadelphia Decandria.
scándens. DC. (*ye.*) climbing. St. Domingo. 1826. .... S. ♄.⌣. Venten. choix. t. 42.
Turpínia punctàta. P.S. *Gly'cine punctàta.* Lam. ill. t. 609.
TETRAN'EMA. TETRAN'EMA. Monadelphia Decandria.
nùtans. (*li.*) drooping-fl^d^. E. Indies. 1823. 5.8. S. ♄. Bot. mag. 2867.
Hedy'sarum nùtans. Gr. *Desmòdium nùtans.* B.M.
ZO'RNIA. DC. ZO'RNIA. Monadelphia Decandria.
1 angustifòlia. DC. (*ye.*) narrow-leaved. E. Indies. 1733. 7.8. S. ⊙. Rheed. mal. 9. t. 82.
diphy'lla. P.S. *Hedy'sarum diphy'llum.* L.
2 reticulàta. DC. (*ye.*) netted-bracted. W. Indies. .... —— S. ⊙.
3 thymifòlia. DC. (*ye.*) Thyme-leaved. Mexico. 1826. —— G. ♄.
4 zeylonénsis. DC. (*ye.*) Ceylon. Ceylon. 1733. —— S. ⊙. Burm. zeyl. t. 50. f. 1.
conjugàta. Sm. *Hedy'sarum conjugàtum.* W.
5 dyctiocárpa. DC. (*ye.*) netted-podded. N. Holland. 1824. —— H. ⊙.
6 tetraphy'lla. DC. (*ye.*) four-leaved. Carolina. 1823. —— H. ♃. Mich. amer. 2. t. 41.
7 capénsis. DC. (*ye.*) Cape. C. B. S. 1824. 5.6. G. ♃. Thunb. n. ac. ups. 6. t. 3.
STYLOSA'NTHES. DC. STYLOSA'NTHES. Monadelphia Decandria.
1 procúmbens. DC. (*ye.*) procumbent. Jamaica. 1820. 6.8. S. ♄. Swar. ac. h. 1789. t. 11. f. 1
2 viscòsa. DC. (*ye.*) viscous. ———— —— —— S. ♄. ———— t. 9. f. 2.
3 mucronàta. DC. (*ye.*) mucronate. E. Indies. 1818. —— S. ♄. Burm. zeyl. t. 106. f. 2.
Hedy'sarum hamàtum. Burm. *A'rachis fruticòsa.* Retz.
4 elatior. DC. (*ye.*) tall. Carolina. 1823. 7.8. H. ♃. Pluk. alm. t. 447. f. 7.
5 guianénsis. DC. (*ye.*) Guiana. Guiana. 1826. —— S. ⊙. Aubl. guian. 2. t. 309.
ADE'SMIA. DC. ADE'SMIA. Decandria Monogynia.
1 muricàta. DC. (*ye.*) prickly-podded. Patagonia. 1793. 6.7. G. ♃. Jacq. ic. 3. t. 568.
Hedy'sarum muricàtum. Jacq. *Patagònium Hedysaroídes.* Schk.
2 hispídula. DC. (*ye.*) hispid. Peru. 1826. 6.8. G. ♃. DC. leg. mem. t. 48.
3 pappòsa. DC. (*ye.*) feathery-podd'd. Chili. 1823. —— G. ♃.
ÆSCHYNO'MENE. DC. ÆSCHYNO'MENE. Diadelphia Decandria. L.
1 áspera. DC. (*ye.*) rough-stemm'd. E. Indies. 1759. 6.7. S. ⊙. Breyn. cent. 51. t. 52.
2 sensitìva. DC. (*wh.*) sensitive. W. Indies. 1733. 5.8. S. ♄. Plum. ic. t. 139.
3 índica. DC. (*ye.*) Indian. E. Indies. 1799. 6.7. S. ⊙. Rheed. mal. 9. t. 18.
4 americàna. DC. (*ye.*) hairy. Jamaica. 1732. 7.8. S. ⊙. Sloan. jam. 1. t. 118. f. 3.
5 pátula. DC. (*ye.*) spreading. Mauritius. 1826. 6.8. S. ♄.
6 subviscòsa. DC. (*v.*) glandular. E. Indies. 1820. 9.10. S. ⊙.
7 híspida. DC. (*ye.*) hispid. N. America. 1803. 8.9. H. ⊙. Bart. fl. amer. t. 29.
8 micrántha. DC. (*ye.*) small-flowered. Madagascar. 1826. 6.8. S. ♃.
9 viscídula. DC. (*ye.*) viscous. N. America. 1818. 7.8. H. ⊙.
10 falcàta. DC. (*ye.*) sickle-podded. Brazil. 1825. —— S. ⊙.
11 péndula. Boj. (*ye.*) pendulous. Madagascar. —— —— S. ⊙.
SMI'THIA. DC. SMI'THIA. Diadelphia Decandria.
1 sensitìva. DC. (*ye.*) sensitive. E. Indies. 1785. 7.10. S. ⊙. Sal. par. lond. t. 92.
2 conférta. DC. (*ye.*) close-flowered. N. Holland. 1825. —— S. ⊙.
3 áspera. H.B. (*ye.*) rough. E. Indies. 1812. —— S. ⊙.
LO'UREA. DC. LO'UREA. Diadelphia Decandria.
1 Vespertiliònis. DC. (*w.*) bat-winged. Cochinchina. 1780. 7.8. S. ♂. Jacq. ic. 3. t. 566.
Hedy'sarum Vespertiliònis. Jacq. *Chrístia lunàta.* Mœnch.
2 renifórmis. DC. (*pu.*) kidney-leaved. China. 1826. —— S. ♂.
UR'ARIA. DC. UR'ARIA. Diadelphia Decandria.
1 pícta. DC. (*pu.*) painted-leaved. Guinea. 1788. 6.8. S. ♂. Jacq. ic. 3. t. 567.
Hedy'sarum píctum. Jacq.
2 crinìta. DC. (*br.*) crook-podded. E. Indies. 1780. 7.9. S. ♄. Burm. ind. t. 56.
3 lagopoídes. DC. (*wh.*) Harefoot-like. ———— 1790. —— S. ♄. ———— t. 53. f. 2.
Hedy'sarum lagopodioídes. L. *Lespedèza lagopodioídes.* P. S.
4 Lágopus. DC. (*pu.*) Hare's-foot. Nepaul. 1823. 6.8. S. ♄.
5 hamòsa. (*wh.*) hooked. E. Indies. —— 6 7. S. ♃.
Doódia hamòsa et Hedy'sarum hamòsum. H.B.
6 alopecuroídes. (*wh.*) Fox-tail. E. Indies. —— —— S. ♃.
Doódia alopecuroídes et Hedy'sarum alopecuroídes. H.B.
7 lagocéphala. DC. (*ye.*) hairy. Brazil. —— 7.9. S. ♃.
NICOLS'ONIA. DC. NICOLS'ONIA. Diadelphia Decandria.
barbàta. DC. (*pu.*) bearded. W. Indies. 1826. 6.8. S. ♃.
DESM'ODIUM. DC. DESM'ODIUM. Diadelphia Decandria.
1 dùbium. B.R. doubtful. Nepaul. 1823. 7.8. S. ♄. Bot. reg. 967.
2 umbellàtum. DC. (*wh.*) umbel-flower'd. E. Indies. 1801. 6.7. S. ♄. Jacq. schœnb. 3. t. 297.
3 auriculàtum. DC. (*pu.*) eared. Timor. 1825. —— S. ♄.

4	triquètrum. DC.	(*pu.*)	triangular.	E. Indies.	1802.	7. 8.	S. ♂.	Burm. ind. t. 52. f. 2.
5	pseudotriquètrum. DC.	(*pu.*)	three-sided.	Nepaul.	1818.	——	S. ♂.	
6	alàtum. DC.	(*pu.*)	winged-stalked.	E. Indies.	1809.	——	S. ♄.	
7	g'yrans. DC.	(*re.*)	moving plant.	————	1775.	7. 9.	S. ♂.	Jacq. ic. t. 565.
8	gyroídes. DC.	(*li.*)	gyrans-like.	————	1818.	——	S. ♄.	
9	gangéticum. DC.	(*li.*)	oval-leaved.	————	1762.	——	S. ♄.	Burm. zeyl. t. 49.
10	maculàtum. DC.	(*li.*)	spotted.	————	1732.	6. 8.	S. ☉.	Dill. elth. 1. t. 141. f. 168.
	Hedy'sarum maculàtum. L. *Æschynómene maculàta.* Poir.							
11	sagittàtum. DC.	(*re.*)	arrow-leaved.	————	1807.	——	S. ♃.	
12	angustifòlium. DC.	(*pu.*)	narrow-leaved.	Mexico.	1825.	——	S. ♄.	
13	latifòlium. DC.	(*bl.*)	broad-leaved.	E. Indies.	1800.	7. 8.	S. ♄.	Bot. reg. t. 356.
	Hedy'sarum latifòlium. B.R.							
14	velutìnum. DC.	(*vi.*)	velvetty.	Trinidad.	1820.	——	S. ♄.	
15	lasiocárpum. DC.	(*ro.*)	woolly-podded.	Guinea.	1822.	7. 9.	S. ♄.	Beauv. fl. ow. 1. t. 18.
16	canadénse. DC.	(*pu.*)	Canadian.	N. America.	1640.	——	H. ♃.	Corn. canad. t. 45.
17	canéscens. DC.	(*pu.*)	canescent.	————	1733.	——	H. ♃.	
18	marylándicum. DC.	(*p.*)	Maryland.	————	1725.	7. 10.	H. ♃.	Dill. elth. t. 144. f. 171.
19	obtùsum. DC.	(*vi.*)	obtuse.	————	1805.	7. 8.	H. ♃.	
20	ciliàre. DC.	(*vi.*)	fringed.	————	1823.	——	H. ♃.	
21	viridiflòrum. DC.	(*pu.*)	green-flower'd.	————	1787.	7. 9.	H. ♃.	Pluk. alm. t. 308. f. 5.
22	lævigàtum. DC.	(*pu.*)	smooth-stalk'd.	————	1826.	——	H. ♃.	
23	bracteòsum. DC.	(*pu.*)	bracted.	————	1806.	7. 8.	H. ♃.	
24	cuspidàtum. DC.	(*pu.*)	sharp-pointed.	————	——	——	H. ♃.	
25	glábellum. DC.	(*pu.*)	smooth.	————	1824.	7. 9.	H. ♃.	
26	paniculàtum. DC.	(*pu.*)	panicled.	————	1781.	7. 8.	H. ♃.	Pluk. alm. t. 432. f. 6.
27	acuminàtum. DC.	(*pu.*)	glutinous.	————	1805.	——	H. ♃.	
	Hedy'sarum glutinòsum. W. *acuminàtum.* Ph.							
28	nudiflòrum. DC.	(*pu.*)	naked-flowered.	————	1823.	——	H. ♃.	
29	pauciflòrum. DC.	(*wh.*)	small-flowered.	————	1828.	——	H. ♃.	
30	rotundifòlium. DC.	(*pu.*)	round-leaved.	————	1733.	——	H. ♃.	
31	Aparìnes. DC.	(*li.*)	hairy-stalked.	Mexico.	1820.	7. 9.	S. ♄.	
32	cinèreum. DC.	(*pu.*)	white-branched.	————	1826.	——	S. ♄.	H. et B. n. pl. am. 6. t. 599.
33	uncinàtum. DC.	(*vi.*)	uncinate.	Caracas.	1823.	6. 7.	S. ♄.	Jacq. schœnb. 3. t. 208.
34	incànum. DC.	(*pu.*)	hoary.	W. Indies.	1820.	7. 8.	S. ♄.	Plum. ed. Bur. t. 149. f. 1.
35	erythrinæfòlium. DC.	(*wh.*)	Erythrina-ld.	Trinidad.	——	——	S. ♄.	
36	supìnum. DC.	(*pu.*)	trailing.	Jamaica.	——	——	S. ♄.	Sloan. jam. 1. t. 118. f. 2.
37	tortuòsum. DC.	(*pu.*)	twisted-podded.	————	1781.	——	S. ♄.	———— 1. t. 116. f. 2.
38	adscéndens. DC.	(*pu.*)	ascending.	————	1820.	7. 9.	S. ♄.	H. et B. n. pl. am. 6. t. 597.
39	mexicànum.	(*bl.*)	Mexican.	Mexico.	1820.	11. 5.	S. ♄.	Bot. reg. 815.
	Hedy'sarum adscéndens β? cœrùleum. DC.							
40	trigònum. DC.	(*li.*)	trigonal.	Jamaica.	1733.	7. 9.	S. ♄.	
41	spiràle. DC.	(*wh.*)	spiral-podded.	————	....	....	S. ♄. ⌣.	
42	ásperum. DC.	(*pa.*)	harsh.	....	1820.	6. 8.	S. ♄.	
43	cæspitòsum. DC.	(*bh.*)	tufted.	Mauritius.	——	——	S. ♃.	
44	mauritiànum. DC.	(*wh.*)	creeping-rooted.	————	1822.	7. 9.	S. ♃.	
45	Scálpe. DC.	(*pu.*)	repand-leaved.	Madagascar.	1823.	7. 8.	S. ♄.	
	Hedy'sarum repándum. Poir. *non* Vahl.							
46	diversifòlium. DC.	(*li.*)	different-leav'd.	————	1820.	——	S. ♄.	
47	spectábile.	(*pu.*)	showy.	Zanzibar.	1825.	....	S. ♄.	
	Hedy'sarum spectábile. Boj.							
48	triflòrum. DC.	(*pu.*)	three-flowered.	India.	1796.	6. 7.	S. ♃.	Burm. ind. t. 54. f. 2.
49	heterophy'llum. DC.	(*pu.*)	various-leav'd.	Ceylon.	1820.	——	S. ♃.	Burm. zeyl. t. 54. f. 1.
50	parvifòlium. DC.	(*pu.*)	slender.	Nepaul.	1823.	7. 9.	S. ♄.	
	Hedy'sarum tenéllum. D.P. *non* Kth.							
51	Roxbúrghii.	(*pu.*)	Roxburgh's.	E. Indies.	1825.	——	S. ♄.	
	diffùsum. DC. p. 335. *non* p. 336.							
52	floribúndum. D.P.	(*pu.*)	abundant-fld.	Nepaul.	1823.	——	S. ♄.	
53	retùsum. D.P.	(*pu.*)	retuse-leaved.	————	——	——	S. ♄.	
54	confértum. DC.	(*li.*)	close-flowered.	————	1826.	——	S. ♄.	
55	sambuénse. D.P.	(*li.*)	villous-leaved.	————	——	——	S. ♄.	
56	concínnum. DC.	(*li.*)	neat.	————	1823.	——	S. ♄.	
57	trichocaúlum. DC.	(*vi.*)	hairy-stalked.	————	1827.	——	S. ♄.	
58	Leschenaúltii. DC.	(*li.*)	Leschenault's.	E. Indies.	——	——	S. ♄.	
59	capitàtum. DC.	(*pu.*)	headed.	Ceylon.	....	7. 8.	S. ♄.	Burm. ind. t. 54. f. 1.
	Hedy'sarum cònicum. Poir.							
60	láxum. DC.	(*li.*)	loose.	Nepaul.	1828.	——	S. ♄. ⌣.	
61	tomentòsum. DC.	(*wh.*)	woolly.	China.	1782.	6. 7.	G. ♃.	
62	rèpens. DC.	(*pu.*)	creeping.	W. Indies.	1820.	——	S. ♄.	
63	malacophy'llum. DC.	(*li.*)	soft-leaved.	Manila.	1822.	——	S. ♃.	

64 grandifòlium. DC. (*li.*) large-leaved. E. Indies. 1822. 6. 7. S. ♄.
65 serotìnum. DC. (*pu.*) late-flowering. N. America. 1806. 8. 10. H. ♃.
66 cephalòtes. H.B. (*li.*) close-headed. E. Indies. 1825. 6. 8. S. ♄.
67 altíssimum. (*pa.*) tallest. —— —— .... S. ♄.
68 heptaphy'llum. (*li.*) seven-leaved. —— 1820. 6. 8. S. ♄.
69 abyssínicum. DC. (*li.*) Abyssinian. Abyssinia. 1802. 7. S. ♃.
70 bracteàtum. H.B. (*li.*) bracted. E. Indies. 1820. 7. 8. S. ♄.
71 rhomboídeum. (*pa.*) rhomboid. —— —— 7. 9. S. ♄.
Hedy'sarum rhombifòlium. H.B. non Elliot.
72 orbiculàtum. H.B. (*li.*) round-leaved. —— 1818. —— S. ♄. ⌣.
73 recurvàtum. H.B. (*li.*) recurved. —— 1824. —— S. ♄.
74 divérgens. H.B. (*pa.*) divergent. —— —— —— S. ♄.
75 purpùreum. H.B. (*pu.*) purple-flowered. —— 1818. 6. 8. S. ♃.
76 pátens. H.B. (*li.*) spreading —— 1824. —— S. ♃.
77 arbòreum. .... tree. —— 1818. .... S. ♄.
Hedy'sarum arbòreum. H.B.
78 lácteum. Boj. (*st.*) milk-coloured. Zanzibar. 1825. 10. 11. S. ♂.
79 Helsinbérgii. Boj. .... Helsinberg's. Madagascar. —— .... S. ♄.

DICE'RMA. DC. DICE'RMA. Diadelphia Decandria.
1 pulchéllum. DC. (*ye.*) neat. E. Indies. 1799. 7. 8. S. ♄. Burm. zeyl. t. 52.
Hedy'sarum pulchéllum. W. *Zórnia pulchélla.* P.S. *Phyllòdium pulchéllum.* Desv.
2 élegans. DC. (*ye.*) elegant. China. 1826. 7. 8. G. ♄.
3 biarticulàtum. DC. (*ye.*) two-jointed. E. Indies. 1808. —— S. ♄. Burm. zeyl. t. 50. f. 2.
Hedy'sarum biarticulàtum. L.

HEDY'SARUM. DC. HEDY'SARUM. Diadelphia Decandria. L.
1 grandiflòrum. DC. (*st.*) large-flowered. Iberia. 1820. 6. 8. H. ♃. Pall. itin. 2. p. 743. t. Y.
seríceum. M.B. *Astrágalus grandiflòrus.* L.
2 argénteum. DC. (*pu.*) silvery. Siberia. 1827. —— H. ♃. Gmel. sib. 4. t. 13.
3 cándidum. DC. (*pu.*) silver-leaved. —— 1796. 7. 8. H. ♃.
4 spléndens. DC. (*pa.*) glossy-leaved. —— 1828. —— H. ♃.
5 pállidum. DC. (*pa.*) pale-flowered. Barbary. 1811. —— H. ♃.
6 coronàrium. DC. (*sc.*) F. Honeysuckle. Italy. 1596. 6. 8. H. ♂. Gært. fruct. 2. t. 155.
α coccíneum. (*sc.*) *scarlet-flowered.* —— —— —— H. ♂.
β álbum. (*wh.*) *white-flowered.* —— —— —— H. ♂.
7 hùmile. DC. (*pu.*) dwarf. Spain. 1640. 7. 8. H. ♃. J. Bauh. his. 2. p. 336. f. 1.
8 spinosíssimum. DC. (*pu.*) thorny. —— 1731. —— H. ☉. Pluk. alm. t. 50. f. 2.
9 capitàtum. DC. (*ro.*) headed. Barbary. .... —— H. ♃. Bot. mag. 1251.
10 flexuòsum. DC. (*pu.*) wave-podded. Asia. 1680. —— H. ☉. Schk. hand. 2. t. 207.
11 cárneum. DC. (*fl.*) flesh-coloured. Siberia. 1818. —— H. ♃. Lodd. bot. cab. 312.
12 elongàtum. F. (*pu.*) long-stalked. Russia. 1823. 6. 8. H. ♃. —— t. 1401.
13 lasiocárpum. DC. (*pu.*) woolly-podded. Altay. 1824. —— H. ♃.
14 taúricum. DC. (*ro.*) Taurian. Tauria. 1804. 7. 8. H. ♃. Bieb. cent. 2. t. 85.
ròseum. Bot. mag. t. 996. *non* DC.
15 ròseum. DC. (*ro.*) rose-coloured. Siberia. 1803. —— H. ♃.
16 altaícum. DC. (*pu.*) Altay. —— 1828. —— H. ♃.
17 venústum. DC. (*pu.*) handsome. Altay. —— —— H. ♃.
18 fruticòsum. DC. (*li.*) shrubby. Siberia. 1782. 6. 7. H. ♄. Pall. it. 3. . 5. f. 1.
19 ibéricum. DC. (*li.*) Iberian. Iberia. 1824. 6. 8. H. ♃.
20 Razoumowiànum. DC. (*li.*) net-podded. Scythia. 1828. —— H. ♃.
21 cretáceum. DC. (*pu.*) slender-spiked. Volga. 1818. —— H. ♃.
22 consanguíneum. DC. (*pu.*) related. Altay. —— —— H. ♃.
23 obscùrum. DC. (*pu.*) creeping-rooted. Europe. 1640. 7. 8. H. ♃. Bot. mag. 282.
24 brachisèma. DC. (*pu.*) short-lipped. Altay. 1820. —— H. ♃. Gmel. sib. 4. t. 10.
25 sibíricum. DC. (*pu.*) Siberian. Siberia. 1798. 6. 7. H. ♃. Lam. ill. t. 628. f. 3.
26 alpìnum. W. (*pu.*) Alpine. —— —— —— H. ♃. Bot. mag. 2213.
27 caucásicum. DC. (*pu.*) Caucasean. Caucasus. 1824. 6. 8. H. ♃.
28 argyrophy'llum. F. (*pu.*) silvery-leaved. Altay. 1827. —— H. ♃.

ONOBRY'CHIS. DC. SAINTFOIN. Diadelphia Decandria.
1 satìva. DC. (*ro.*) common. Britain. .... 6. 7. H. ♃. Eng. bot. v. 2. t. 96.
Hedy'sarum Onobry'chis. Jacq. aust. t. 352.
2 tanàitica. F. (*ro.*) spreading. Siberia. 1823. —— H. ♃.
3 carpáthica. DC. (*ro.*) Carpathian. Carp. Mount. 1818. —— H. ♃.
4 incurvàta. DC (*ro.*) incurved. .... 1826. —— H. ♃.
5 montàna. DC. (*pu.*) mountain. Pyrenees. 1816. —— H. ♃.
6 conférta. DC. (*pu.*) close-flowered. Caucasus. 1827. 6. 8. H. ♃.
7 procúmbens. DC. (*ro.*) procumbent. Iberia. 1823. —— H. ♃.
8 supìna. DC. (*ro.*) trailing. S. Europe. 1816. —— H. ♃.
9 glábra. DC. (*ro.*) smooth. Tauria. 1829. —— H. ♃.
10 arenària. DC. (*ro.*) sand. Hungary. 1818. —— H. ♃.

11 mareótica. DC.	(*ro.*)	villous-calyxed.	Siberia.	1829.	6. 8.	H. ♃.	
12 álba. DC.	(*wh.*)	white-flowered.	Hungary.	1804.	7. 8.	H. ♃.	W. et K. hung. 2. t. 111.
13 petr'æa. DC.	(*wh.*)	stone.	Caucasus.	1818.	6. 8.	H. ♃.	
14 grácilis. DC.	(*li.*)	slender.	Poland.	1828.	——	H. ♃.	
15 saxátilis. DC.	(*wh.*)	rock.	S. France.	1790.	——	H. ♃.	All. ped. 1191. t. 19. f. 1.
16 Cáput-gálli. DC.	(*vi.*)	Cock's-head.	S. Europe.	1731.	7. 8.	H. ♃.	Lobel. ic. 2. p. 81. f. 1.
17 Crísta-gálli. DC.	(*vi.*)	Cock's-comb.	————	1710.	6. 8.	H. ⊙.	Gært. fruct. 2. t. 148.
18 foveolàta. DC.	(*vi.*)	Sicilian.	Sicily.	1828.	——	H. ⊙.	
19 æquidentàta. DC.	(*re.*)	equal-toothed.	Levant.	1823.	——	H. ⊙.	Desv. j. bot. 1. t. 23.
20 Pallàsii. DC.	(*st.*)	Pallas's.	Tauria.	——	——	H. ♃.	Bieb. pl. ros. 1. t. 35.
Buxbaumiàna. S.S. *Hedy'sarum Pallàsii.* W. Buxb. cent. 2. t. 42.							
21 radiàta. DC.	(*st.*)	rayed.	Caucasus.	1823.	7. 9.	H. ♃.	Desf. an. mus. 12. t. 13.
Hedy'sarum Buxbaúmii. M.B.							
22 ptolemàica. DC.	(*ye.*)	striped-flower'd.	Egypt.	1829.	——	H. ♃.	Delil. fl. eg. t. 39. f. 1.
ELEI'OTIS. DC.		ELEI'OTIS.	Diadelphia Decandria.				
1 monophy'lla. DC.	(*li.*)	simple-leaved.	E. Indies.	1820.	4. 8.	S. ♃.	Burm. ind. t. 50. f. 2.
2 soròria. DC.	(*bh.*)	ternate-leaved.	————	——	——	S. ♃.	Desv. j. bot. 3. t. 6. f. 31.
Hedy'sarum soròrium. L.							
LESPED'EZA. DC.		LESPED'EZA.	Diadelphia Decandria.				
1 júncea. DC.	(*wh.*)	slender-branch'd.	Siberia.	1776.	7. 8.	G. ħ.	Linn. f. dec. 1. t. 4.
Hedy'sarum júnceum. L. *Anthy'llis cuneàta.* Dum. Cours.							
2 reticulàta. DC.	(*pu.*)	netted.	N. America.	1823.	——	H. ♃.	
3 sessiliflòra. DC.	(*pu.*)	sessile-flowered.	————	....		H. ♃.	
4 eriocárpa. DC.	(*vi.*)	woolly-podded.	Nepaul.	1812.	7. 9.	G. ħ.	
5 frutéscens. DC.	(*pu.*)	frutescent.	Carolina.	1739.	——	G. ħ.	Jacq. vind. 3. t. 89.
6 capitáta. DC.	(*li.*)	headed.	N. America.	1789.	6. 8.	H. ♃.	
7 angustifòlia. DC.	(*li.*)	narrow-leaved.	Carolina.	1827.	——	H. ♃.	
8 Stùvei. DC.	(*pu.*)	Stuve's.	New Jersey	1823.	7. 9.	H. ⊙.	
9 polystàchya. DC.	(*wh.*)	hairy.	N. America.	1789.	6. 8.	H. ♃.	Mich. amer. 2. t. 40.
Hedy'sarum hírtum. W.							
10 villòsa. DC.	(*wh.*)	villous.	————	1812.	——	H. ♃.	
11 violàcea. DC.	(*vi.*)	violet-colour'd.	————	1789.	7. 8.	H. ♃.	
12 divérgens. DC.	(*vi.*)	divergent.	————	1827.	——	H. ♃.	
13 procúmbens. DC.	(*pu.*)	procumbent.	————	1812.	——	H. ♃.	Mich. amer. 2. t. 39.
14 prostràta. DC.	(*pu.*)	prostrate.	————	1818.	——	H. ♃.	
E'BENUS. DC.		EBE'NUS.	Monadelphia Decandria.				
1 crética. DC.	(*bh.*)	Cretan.	Crete.	1737.	6. 7.	F. ♄.	Bot. mag. 1092.
Anthy'llis crética. Lam.							
2 pinnàta. DC.	(*pi.*)	wing-leaved.	Barbary.	1786.	7. 8.	G. ♂.	Desf. ac. par. 1. t. 3.
Anthy'llis serícea. W.							
3 Sibthórpii. DC.	(*pi.*)	Sibthorp's.	Greece.	1826.	——	G. ♂.	DC. leg. mem. t. 53.
FLEMI'NGIA. DC.		FLEMI'NGIA.	Diadelphia Decandria.				
1 strícta. DC.	(*re.*)	straight.	E. Indies.	1798.	7. 9.	S. ♄.	Roxb. corom. 3. t. 248.
2 semialàta. DC.	(*ro.*)	winged.	Nepaul.	1805.	7. 8.	S. ♄.	———— 3. t. 249.
3 congésta. DC.	(*re.*)	crowded spiked.	E. Indies.	1802.	7. 9.	S. ♄.	
4 nàna. DC.	(*re.*)	dwarf.	————	1804.	8. 9.	S. ♄.	
5 lineàta. DC.	(*re.*)	branch-spiked.	————	1793.	7. 8.	S. ♄.	Burm. ind. t. 53. f. 1.
6 angustifòlia. DC.	(*re.*)	narrow-leaved.	————	1812.	7. 9.	S. ♄.	
7 procúmbens. DC.	(*re.*)	procumbent.	————	1816.	8.	S. ♄.	
8 prostràta. DC.	(*re.*)	prostrate.	China.	1820.	7. 9.	S. ♄.	
9 strobilífera. DC.	(*wh.*)	Beech-leaved.	E. Indies.	1787.	——	S. ♄.	Bot. reg. 617.
Hedy'sarum strobilíferum. L. *Zórnia strobilífera.* P.S.							

Subtribus III. *ALHAGEÆ.* DC. *prodr.* 2. *p.* 352.

ALH'AGI. DC.		ALH'AGI.	Diadelphia Decandria.				
1 mauròrum. DC.	(*re.*)	Levant.	Levant.	1714.	7. 8.	G. ħ.	Tourn. cor. 54. t. 489.
Hedy'sarum Alhàgi. L. *mannífera.* Dv. *Mánna hebraica.* D.D.							
2 camelòrum. DC.	(*re.*)	Caspian.	Caucasus.	1824.	——	H. ♃.	
ALSICA'RPUS. DC.		ALSICA'RPUS.	Diadelphia Decandria.				
1 bupleurifòlius. DC.	(*li.*)	Bupleurum-l[d].	E. Indies.	1793.	7. 8.	S. ♂.	Bot. mag. 1722.
Hedy'sarum bupleurifòlium. Roxb. corom. 2. t. 194.							

2 vaginàlis. DC.	(*pu.*)	sheathed.	E. Indies.	1790.	7. 8.	S. ⊙.	Burm. zeyl. t. 49. f. 1.
3 nummularifòlius. DC.	(*pu.*)	moneywort.	——	1777.	7. 9.	S. ⊙.	Petiv. gaz. t. 26 f. 1.
4 monilífer. DC.	(*pu.*)	neck-lace.	——	1824.	——	S. ♃.	Burm. ind. t. 52. f. 3.
5 glumàceus. DC.	(*vi.*)	rugged-podded.	Arabia.	——	——	G. ♃.	
6 styracifòlius. DC.	(*pu.*)	Storax-leaved.	E. Indies.	1796.	——	S. ♄.	
Hedy'sarum styracifòlium. L.							

Tribus IV. *VICIEÆ.* DC. prodr. 2. p. 353.

C'ICER. DC.		Chick-Pea. Diadelphia Decandria. L.					
1 arietìnum. DC.		common.	S. Europe.	1548.	7. 8.	H. ⊙.	Bot. mag. 2274.
2 songáricum. DC.		large-flowered.	Persia.	1828.	——	H. ⊙.	DC. leg. mem. t. 54.
F'ABA. DC.		Bean. Diadelphia Decandria.					
vulgàris. DC.	(*wh.bk.*)	common.	Egypt.	....	5. 8.	H. ⊙.	Tourn. inst. t. 222.
α èquina.	(*wh.bk.*)	*Horse.*	——	....	——	H. ⊙.	
β pr'æcox.	(*wh.bk.*)	*Mazagon.*	——	....	——	H. ⊙.	
γ ensifórmis.	(*wh.bk.*)	*long-pod.*	——	....	——	H. ⊙.	
δ albiflòra.	(*wh.*)	*white-blossomed.*	——	....	——	H. ⊙.	
ε purpùrea.	(*pu.*)	*purple-flowered.*	——	....	——	H. ⊙.	
ζ macrospérma.	(*wh.bk.*)	*broad Windsor.*	——	....	——	H. ⊙.	
η chlorospérma.	(*wh.bk.*)	*green Windsor.*	——	....	——	H. ⊙.	
VI'CIA. DC.		Vetch. Diadelphia Decandria.					
1 pisifórmis. DC.	(*st.*)	Pea-shaped.	S. Europe.	1739.	7. 8.	H. ♃.	Jacq. aust. 4. t. 364.
2 am'œna. DC.	(*pu.*)	beautiful.	Siberia.	1829.	——	H. ♃.	Gmel. sib. 4. n. 11. t. 3 ?
3 caroliniàna. DC.	(*wh.*)	Carolina.	Carolina.	——	——	H. ♃.	
4 dumetòrum. DC.	(*vi.*)	great wood.	France.	1752.	5. 7.	H. ♃.	Sturm deuts. 1. fas. 31.
5 dentàta. F.	(*vi.*)	toothed.	Russia.	1818.	6. 7.	H. ♃.	
6 sylvatìca. DC.	(*wh.vi.*)	wood.	Britain.	....	7. 8.	H. ♃.	Eng. bot. 2. t. 79.
7 americàna. DC.	(*vi.*)	American.	N. America.	1818.	——	H. ♃.	
8 variegàta. DC.	(*v.*)	variegated.	Caucasus.	1829.	——	H. ♃.	Desf. cor. p. 86. t. 66.
9 purpùrea. DC.	(*pu.*)	purple.	Tauria.	1827.	——	H. ♃.	
10 alpéstris. DC.	(*pu.*)	alpine.	——	1828.	——	H. ♃.	
11 cassùbica. DC.	(*vi.*)	Cassubian.	Germany.	1711.	6. 7.	H. ♃.	Sturm deuts. 1. fas. 31.
12 abbreviàta. DC.	(*vi.*)	abbreviated.	Caucasus.	1818.	——	H. ♃.	
13 cinèrea. DC.	(*pa.bl.*)	ash-coloured.	Tauria.	1829.	——	H. ♃.	
14 Crácca. DC.	(*vi.*)	tufted.	Britain.	....	7. 9.	H. ♃.	Eng. bot. 17. t. 1168.
15 Bivònea. DC.	(*ro.*)	Bivoni's.	Sicily.	1828.	——	H. ♃.	
16 Gerárdi. DC.	(*vi.*)	Gerard's.	S. Europe.	1711.	6. 7.	H. ♃.	Jacq. aust. t. 229.
17 pseudocrácca. DC.	(*vi.*)	slender-leaved.	Italy.	1824.	7. 9.	H. ⊙.	
tenuifòlia. Tenore. *non* Roth.							
18 microphy'lla. DC.	(*li*)	small-leaved.	Archipelago.	1828.	6. 8.	H. ♃.	
19 polyphy'lla. DC.	(*li.*)	many-leaved.	Algiers.	1823.	——	H. ♃.	
20 tenuifòlia. DC.	(*vi.*)	slender-leaved.	Germany.	1799.	6. 7.	H. ♃.	Sturm deuts. 1. fas. 31.
21 consentìna. DC.	(*st.*)	straw-coloured.	Naples.	1824.	——	H. ♃.	
22 capénsis. DC.	(*pu.*)	Cape.	C. B. S.	1802.	6. 8.	G. ♃.	
23 canéscens. DC.	(*bl.*)	hoary.	Libanus.	1800.	——	H. ⊙.	Labil. dec. syr. 1. t. 7.
24 ochroleùca. DC.	(*st.*)	sulphur-color'd.	Italy.	1824.	——	H. ♃.	
25 punctàta. DC.	(*pu.*)	spot-leaved.	Switzerland.	1826.	——	H. ⊙.	
26 onobrychioídes. DC.	(*pu.*)	Saintfoin-like.	Europe.	1789.	——	H. ⊙.	Sturm deuts. 1. fas. 31.
27 atropurpùrea. DC.	(*d.pu.*)	dark-purple.	Algiers.	1815.	——	H. ⊙.	Bot. reg. t. 871.
28 benghalénsis. DC.	(*sc.*)	Bengal.	E. Indies.	1792.	7. 9.	H. ⊙.	Herm. lugd. t. 625.
29 perénnis. DC.	(*pu.*)	perennial.	S. Europe.	1820.	6. 8.	H. ♃.	
30 longifòlia. DC.	(*st.*)	long-leaved.	Syria.	1824.	——	H. ♃.	
31 argéntea. DC.	(*ro.*)	silvery.	Pyrenees.	1829.	——	H. ♃.	
32 villòsa. DC.	(*pu.*)	villous.	Germany.	1815.	——	H. ⊙.	Sturm deuts. 1. fas. 31.
33 dispérma. DC.	(*li.*)	two-seeded.	S. France.	1826.	——	H. ⊙.	
34 biénnis. DC.	(*vi.*)	biennial.	Siberia.	1753.	7. 9.	H. ♂.	Gmel. sib. 4. t. 2.
35 Nissoliàna. DC.	(*re.*)	red-flowered.	Levant.	1773.	6. 7.	H. ⊙.	
36 pellùcida. DC.	(*pu.*)	pellucid.	C. B. S.	1812.	6. 8.	G. ♃.	Jacq. schœnb. 2. t. 222.
37 biflòra. DC.	(*bl.*)	two-flower'd.	Algiers.	1801.	——	H. ⊙.	Desf. atl. 2. t. 197.
38 calcaràta. DC.	(*pu.*)	spurred.	——	1790.	7. 9.	H. ⊙.	
monántha. W.							
39 syrìaca. DC.	(*vi.*)	Syrian.	Syria.	1822.	——	H. ⊙.	
40 pimpinelloídes. DC.	(*pu.*)	Pimpinella-like.	Rome.	——	6. 9.	H. ⊙.	

41 satìva. DC.	(*pu.*) common.	Britain.	....	5. 7.	H. ⊙.	Eng. bot. 5. t. 334.
α nemoràlis. P.S.	(*pu.*) *wood.*	———	....	——	H. ⊙.	Sturm deuts. 1. fas. 31.
β segetàlis. DC.	(*pu.*) *hedge.*	———	....	——	H. ⊙.	Eng. bot. 5. t. 334.
42 leucospérma. L.en.	(*wh.*) white-seeded.	....	....	——	H. ⊙.	
43 angustifòlia. W.en.	(*pu.*) narrow-leaved.	Europe.	....	——	H. ⊙.	Sturm deuts. 1. fas. 31.
44 megalospérma. W.en.	(*pu.*) large-seeded.	Tauria.	1798.	6. 8.	H. ⊙.	
45 incìsa. DC.	(*pu.*) cut-leaved.	———	1829.	——	H. ⊙.	
46 rèpens. DC.	(*pu.*) creeping.	Archipelago.	1828.	——	H. ♃.	
47 canadénsis. DC.	(*li.*) Canadian.	Canada.	——	——	H. ⊙.	
48 globòsa. DC.	(*bl.*) globular.	....	1804.	——	H. ⊙.	
49 cordàta. DC.	(*pu.*) heart-leaved.	Germany.	1828.	——	H. ⊙.	Sturm deuts. 1. fas. 32.
50 pilòsa. DC.	(*pu.*) pilose.	Tauria.	1820.	——	H. ⊙.	
51 peregrìna. DC.	(*pu.*) broad-podded.	S. Europe.	1779.	7. 8.	H. ⊙.	Sturm deuts. 1. fas. 32.
52 Michaúxii. DC.	(*wh.*) Michaux's.	....	1803.	——	H. ⊙.	
53 amphicárpa. DC.	(*pu.*) subterraneous.	S. France.	1815.	——	H. ⊙.	Ger. m. an. 6. v.3. p.344.
54 lathyroídes. DC.	(*pu.*) spring.	Britain.	....	4. 6.	H. ⊙.	Eng. bot. 1. t. 30.
55 lævigàta. DC.	(*br.*) smooth-podded.	England.	....	7. 9.	H. ♃.	——— 7. t. 483.
56 dùbia. DC.	(*pu.*) doubtful.	....	1827.	——	H. ♃.	
57 lùtea. DC.	(*ye.*) yellow-flowered.	Britain.	....	6. 8.	H. ♃.	Eng. bot 7. t. 481.
58 hírta. P.S.	(*wh.*) hairy.	France.	1820.	——	H. ♃.	
59 hy'brida. DC.	(*st.*) hairy-flower'd.	England.	....	——	H. ♃.	Eng. bot. 7. t. 482.
60 spuria. DC.	(*st.*) spurious.	Sicily.	1826.	——	H. ♃.	
61 grandiflòra. DC.	(*ye.*) large-flowered.	Switzerland.	1823.	——	H. ♂.	Scop. fl. carn. 2. t. 42.
62 sórdida. DC.	(*ye.*) large-yellow.	Hungary.	1802.	——	H. ⊙.	W. et K. hung. 2. t. 134.
63 Biebersteínii. DC.	(*ye.*) Bieberstein's.	Poland.	1823.	——	H. ♃.	
64 trícolor. DC.	(*ye.pu.*) three-coloured.	Naples.	1827.	——	H. ⊙.	Seb. et Maur. pr. rom. t. 4.
65 sèpium. DC.	(*pu.*) bush.	Britain.	....	5. 6.	H. ♃.	Eng. bot. 22. t. 1515.
66 pannònica. DC.	(*wh.*) Hungarian.	Hungary.	1658.	6. 7.	H. ⊙.	Jacq. aust. 1. t. 34.
67 pyrenaìca. DC.	(*pu.*) Pyrenean.	Pyrenees.	1828.	——	H. ⊙.	DC. ic. rar. 1. t. 33.
68 striàta. M.B.	(*std.*) striped.	Tauria.	1824.	——	H. ⊙.	
69 truncàtula. DC.	(*st.*) truncate-leav'd.	Caucasus.	1828.	6. 8.	H. ⊙.	
70 narbonénsis. DC.	(*d.pu.*) broad-leaved.	France.	1596.	——	H. ⊙.	Roth. abhand. 10. t. 2.
71 serratifòlia. W.	(*pu.*) saw-leaved.	Hungary.	1723.	——	H. ⊙.	Sturm fl. germ. 1. fas. 32.
72 platycárpos. DC.	(*pu.*) flat-podded.	Germany.	....	——	H. ⊙.	Roth. abhand. 10. t. 1.
73 bícolor. DC.	(*v.*) two-coloured.	....	1818.	——	H. ♃.	
E'RVUM. DC.	TARE. Diadelphia Decandria. L.					
1 Léns. DC.	(*pa.*) Lentil.	France.	1548.	5. 7.	H. ⊙.	Sturm fl. germ. 1. fas. 32.
Cìcer Léns. W. *Léns esculénta.* Mœnch.						
2 nígricans. DC.	(*pa.*) dark-seeded.	Tauria.	1823.	6. 7.	H. ⊙.	
lentoídes. T.P.						
3 lentícula. DC.	(*bl.*) one-flowered.	Carinthia.	1824.	6. 8.	H. ⊙.	Sturm fl. germ. 1. fas. 32.
soloniénse. Schrank pl. rar. t. 48. *uniflòrum.* Seb. rom. pl. fas. 2. t. 4.						
4 hirsùtum. DC.	(*pu.wh.*) hairy.	Britain.	....	6. 7.	H. ⊙.	Eng. bot. 14. t. 970.
5 dispérmum. S.S.	(*li.*) two-seeded.	E. Indies.	1824.	6. 8.	H. ⊙.	
6 camelòrum. S.S.	(*li.*) broad-podded.	....	——	——	H. ⊙.	
7 péndulum. S.S.	(*li.*) pendulous-podd'd.	E. Indies.	1828.	——	H. ⊙.	
8 Ervília. DC.	(*pu.*) officinal.	S. Europe.	1596.	6. 8.	H. ⊙.	Sturm fl. germ. 1. fas. 32.
Ervília satìva. L. en. *Vícia Ervília.* W.						
9 monánthos. DC.	(*li.*) single-flowered.	———	1790.	7. 8.	H. ⊙.	Sturm fl. ge. 1. fas. 32. ic.
Vícia articulàta. W. *Láthyrus monánthos.* W. en.						
10 agrigentìnum. DC.	(*li.*) Sicilian.	Sicily.	1827.	——	H. ⊙.	
11 tetraspérmum. DC.	(*li.*) four-seeded.	Britain.	....	6. 8.	H. ⊙.	Eng. bot. 17. t. 1223.
12 grácile. DC.	(*li.*) slender.	S. Europe.	1823.	——	H. ⊙.	Lois. fl. gal. t. 12.
tenuíssimum. P.S. *longifòlium* T.N. *Vícia laxiflòra.* Brot. phyt. t. 54.						
13 pubéscens. DC.	(*li.*) pubescent.	S. Europe.	1828.	6. 8.	H. ⊙.	
14 Loiseleùrii. DC.	(*li.*) Loiseleur's.	Tauria.	——	——	H. ⊙.	
15 diphy'llum. DC.	(*li.*) two-leaved.	....	——	——	H. ⊙.	
PI'SUM. DC.	PEA. Diadelphia Decandria. L.					
1 satìvum. DC.	(*wh.*) cultivated.	S. Europe.	....	5. 10.	H. ⊙.	Lam. ill. t. 633.
α saccharàtum. DC.	(*wh.*) *sugar.*	———	....	——	H. ⊙.	Tourn. inst. t. 215.
β macrocárpon. DC.	(*wh.*) *marrowfat.*	———	....	——	H. ⊙.	
γ víride.	(*wh.*) *green marrowfat.*	———	....	——	H. ⊙.	
δ rugòsum.	(*wh.*) *Knight's marrowfat.*	———	....	——	H. ⊙.	
ε chlorospérmum.	(*wh.*) *green rouncival.*	———	....	——	H. ⊙.	
ζ cæruléscens.	(*wh.*) *blue Prussian.*	———	....	——	H. ⊙.	
η prolíficum.	(*wh.*) *prolific.*	———	....	——	H. ⊙.	
θ pr'æcox.	(*wh.*) *early Charlton.*	———	....	——	H. ⊙.	
ι uniflòrum.	(*wh.*) *early frame.*	———	....	——	H. ⊙.	

κ *hùmile.*	(*wh.*)	*dwarf.*	S. Europe.	....	——	H. ⊙.	
λ *umbellàtum.* DC.	(*pi.*)	*rose or crown.*	——	....	——	H. ⊙.	Tabern. icon. t. 495.
μ *quadràtum.* DC.	(*pu.*)	*grey or swine's.*	——	....	——	H. ⊙.	
2 elàtum. DC.	(*pa.pu.*)	upright.	Iberia.	1827.	6. 10.	H. ⊙.	
3 arvénse. DC.	(*pu.*)	field.	Europe.	....	6. 9.	H. ⊙.	Moris. his. 2. s. 2. t. 1. f. 4.
4 marítimum. DC.	(*pu.*)	sea.	England.	....	7. 8.	H. ♃.	Eng. bot. 15. t. 1046.
5 americànum. DC.	(*pu.*)	American.	America.	....	——	H. ⊙.	
6 Jomárdi. DC.	(*wh.*)	angular-stalk'd.	Egypt.	1818.	6. 9.	H. ⊙.	
7 thebàicum. DC.	(*wh.*)	short-peduncled.	....	——	——	H. ⊙.	
LA'THYRUS. DC.		LA'THYRUS.	Diadelphia Decandria. L.				
1 sylvéstris. DC.	(*pu.*)	wood.	Britain.	....	7. 8.	H. ♃.	Eng. bot. v. 12. t. 805.
2 intermèdius. DC.	(*ro.*)	intermediate.	N. Europe.	1818.	7. 8.	H. ♃.	Flor. dan. t. 785.
3 magellànicus. DC.	(*pu.*)	LordAnson'sPea.	Cape Horn.	1744.	——	H. ♃.	
4 latifòlius. DC.	(*ro.*)	everlasting Pea.	England.	....	7. 9.	H. ♃.	Eng. bot. v. 16. t. 1108.
β *albiflòrus.*	(*wh.*)	*white-flowered.*	....	....	——	H. ♃.	
5 rotundifòlius. DC.	(*ro.*)	round-leaved.	Tauria.	1822.	——	H. ♃.	Bieb. cent. 1. t. 22.
6 praténsis. DC.	(*ye.*)	meadow.	Britain.	....	6. 8.	H. ♃.	Eng. bot. v. 10. t. 670.
7 tuberòsus. DC.	(*ro.*)	tuberous-rooted.	Europe.	1596.	7. 8.	H. ♃.	Bot. mag. 111.
8 ròseus. DC.	(*ro.*)	rose-coloured.	Iberia.	1828.	——	H. ♃.	
9 pisifórmis. DC.	(*std.*)	Siberian.	Siberia.	1759.	6. 8.	H. ♃.	Linn. fil. dec. t. 20.
10 mutábilis. B.F.G.	(*std.*)	changeable.	——	1825.	——	H. ♃.	Swt. br. fl. gar. t. 194.
11 venòsus. DC.	(*pu.*)	veined.	N. America.	1823.	7. 8.	H. ♃.	
12 califórnicus. B.R.	(*pu.*)	Californian.	California.	1827.	6. 10.	H. ♃.	Bot. reg. 1144.
13 palústris. DC.	(*pu.*)	marsh.	Britain.	....	7. 8.	H. ♃.	Eng. bot. v. 3. t. 169.
14 myrtifòlius. DC.	(*pu.*)	Myrtle-leaved.	N. America.	1823.	——	H. ♃.	
15 polymórphus. DC.	(*pu.*)	variable.	Missouri.	——	——	H. ♃.	
16 heterophy'llus. DC.	(*li.*)	various-leaved.	Europe.	1731.	7. 9.	H. ♃.	J. Bauh. his. 2. p. 304. f. 1.
17 stipulàceus. DC.	(*pu.*)	large-stipuled.	N. America.	1827.	——	H. ♃.	
18 Aphàca. DC.	(*st.*)	yellow vetchling.	England.	....	6. 7.	H. ⊙.	Eng. bot. v. 17. t. 1167.
19 Nissòlia. DC.	(*cr.*)	crimson.	——	....	5. 6.	H. ⊙.	—— v. 2. t. 112.
20 inconspícuus. DC.	(*pa.*)	small-flowered.	Levant.	1739.	7. 8.	H. ⊙.	
21 sph'æricus. DC.	(*sc.*)	scarlet.	S. Europe.	1800.	6. 7.	H. ⊙.	DC. ic. rar. 1. t. 32.
coccìneus. P.S. *axillàris.* Lam. *inconspícuus.* Jacq. vind. t. 86. *O'robus coccíneus.* Lodd. b. cab. t. 883.							
22 micránthus. DC.	(*pu.*)	small-blossom'd.	S. France.	....	6. 7.	H. ⊙.	
23 angulàtus. DC.	(*pu.*)	angular-seeded.	S. Europe.	1683.	——	H. ⊙.	Buxb. cent. 3. t. 42. f. 2.
24 longepedunculàtus. DC.	(*pu.*)	long-stalked.	....	1824.	6. 9.	H. ⊙.	
25 leptophy'llus. DC.	(*pu.*)	slender-leaved.	Caucasus.	1828.	——	H. ⊙.	
26 setifòlius. DC.	(*ro.*)	bristle-leaved.	S. Europe.	1739.	6. 7.	H. ⊙.	J. Bauh. 2. p. 308. ic.
27 amphicárpos. DC.	(*li.*)	Earth Pea.	Levant.	1680.	6. 9.	H. ⊙.	Swt. br. fl. gar. 3. t. 236.
28 satìvus. DC.	(*bl.wh.*)	chickling vetch.	S. Europe.	1640.	6. 8.	H. ⊙.	Jacq. f. eclog. t. 116.
α *cærùleus.*	(*bl.*)	*blue-flowered.*	——	——	——	H. ⊙.	Bot. mag. 115.
β *álbus.*	(*wh.*)	*white-flowered.*	——	——	——	H. ⊙.	
29 Cícera. DC.	(*pu.*)	flat-podded.	——	1633.	——	H. ⊙.	Jacq. f. eclog. t. 115.
30 ánnuus. DC.	(*ye.*)	yellow annual.	——	1621.	——	H. ⊙.	Buxb. cent. 3. t. 42. f. 1.
31 hirsùtus. DC.	(*pu.*)	rough-podded.	England.	....	7. 8.	H. ⊙.	Eng. bot. v. 18. t. 1255.
32 hírtus. DC.	(*pu.*)	hairy.	....	1824.	7. 9.	H. ⊙.	
33 odoràtus. DC.	(*va.*)	Sweet Pea.	Sicily.	1700.	6. 9.	H. ⊙.	Bot. mag. t. 60.
α *purpùreus.* DC.	(*pu.*)	*dark-purple.*	——	——	——	H. ⊙.	—— t. 60.
β *rùber.*	(*re.*)	*red-flowered.*	....	——	——	H. ⊙.	
γ *ròseus.* DC.	(*ro.*)	*painted lady.*	Ceylon.	——	——	H. ⊙.	
δ *álbus.*	(*wh.*)	*white-flowered.*	——	——	——	H. ⊙.	
34 grandiflòrus. DC.	(*pu.*)	large-flowered.	Italy.	1814.	——	H. ♃.	Bot. mag. t. 1938.
35 tingitànus. DC.	(*pu.*)	Tangier Pea.	Barbary.	1680.	——	H. ⊙.	—— t. 100.
36 bithy'nicus. DC.	(*vi.*)	tumid-podded.	S. Europe.	1816.	6. 8.	H. ⊙.	All. ped. 1. t. 26. f. 2.
tùmidus. W. *túrgidus.* Lam. *Vícia bithy'nica.* L.							
37 incúrvus. DC.	(*pu.*)	curve-podded.	Siberia.	1802.	7. 8.	H. ♃.	Roth abh. 13. t. 4
38 alàtus. DC.	(*pu.*)	wing-stalked.	Italy.	1829.	——	H. ⊙.	
39 Cly'menum. DC.	(*re.bl.*)	various-flower'd.	Levant.	1713.	6. 7.	H. ⊙.	Pluk. alm. t. 114. f. 6
40 articulàtus. DC.	(*ro.wh.*)	joint-podded.	S. Europe.	1640.	7. 9.	H. ⊙.	Bot. mag. 258.
41 auriculàtus. Bert.	(*pu.*)	eared.	Italy.	1820.	6. 8.	H. ⊙.	
42 spùrius. DC.	(*pu.*)	bastard.	....	1815.	——	H. ⊙.	
43 tenuifòlius. DC.	(*bl.*)	slender-leaved.	Algiers.	1824.	6. 7.	H. ⊙.	
44 'Ochrus. DC.	(*ye.*)	pale-yellow.	S. Europe.	1633.	——	H. ⊙.	Schk. hand. 2. t. 280.
'Ochrus pállida. P.S. *Pìsum 'Ochrus.* L.							
45 purpùreus. DC.	(*pu.*)	purple-flower'd.	Levant.	1826.	6. 8.	H. ♃.	Desf. cor. p. 81. t. 61.
46 cornùtus. DC.	(*pu.*)	horned.	....	1824.	——	H. ⊙.	
47 aphy'llus.	(*pu.*)	leafless.	Smyrna.	1828.	7. 8.	H. ⊙.	
PLATYST'YLIS. B.F.G.		PLATYST'YLIS.	Diadelphia Decandria.				
1 cyànea. B.F.G.	(*bl.*)	blue-flowered.	Iberia.	1826.	7. 8.	H. ♃.	Swt. br. fl. gar. 3. t. 239.

2 sessilifòlius. B.F.G.	(pu.)	sessile-leaved.	Levant.	1824.	6. 8.	H. ♃.	Buxb. cent. 2. t. 38.
O'robus sessilifòlius. Bot. Mag. 2796. *digitàtus.* M.B.							
O'ROBUS. DC.		BITTER VETCH. Diadelphia Decandria. L.					
1 hirsùtus. DC.	(li.)	hairy.	Levant.	1818.	5. 6.	H. ♃.	Bot. mag. t. 2345.
2 formòsus. DC.	(pu.)	handsome.	Caucasus.	1824.	——	H. ♃.	
3 lathyroídes. DC.	(bl.)	upright.	Siberia.	1758.	6. 7.	H. ♃.	Bot. mag. t. 2098.
4 vérnus. DC.	(pu.)	spring.	Europe.	1629.	3. 4.	H. ♃.	———— t. 521.
5 venòsus. DC.	(bl.)	veined.	Siberia.	1818.	5. 6.	H. ♃.	Gmel. fl. sib. 4. t. 5.
6 alpéstris. DC.	(pu.)	alpine.	Hungary.	1818.	——	H. ♃.	W. et K. hung. 2. t. 126.
7 multiflòrus. DC.	(li.)	many-flowered.	Italy.	1824.	5. 7.	H. ♃.	
8 variegàtus. DC.	(pu.)	variegated.	Naples.	1818.	5. 6.	H. ♃.	Ten. fl. nap. t. 68.
9 vicioídes. DC.	(pu.)	Vetch-like.	Hungary.	——	6. 8.	H. ♃.	W. et K. hung. 3. t. 242.
Vícia oroboídes. W.K.							
10 lùteus. DC.	(ye.)	yellow.	Siberia.	1759.	6. 7.	H. ♃.	Lodd. bot. cab. 783.
11 lævigàtus. DC.	(ye.)	smooth.	Hungary.	1818.	——	H. ♃.	W. et K. hung. 3. t. 243.
12 aurántius. DC.	(ye.)	orange-color'd.	Iberia.	1824.	6. 8.	H. ♃.	
13 sylváticus. DC.	(std.)	wood.	Britain.	....	6. 7.	H. ♃.	Eng. bot. v. 8. t. 518.
14 ochroleùcus. DC.	(st.)	straw-colour'd.	Hungary.	1816.	——	H. ♃.	W. et K. hung. 2. t. 118.
15 nìger. DC.	(pu.)	black-rooted.	Europe.	1596.	5. 7.	H. ♃.	Bot. mag. t. 2261.
16 eréctus. DC.	(pu.)	upright.	....	1820.	6. 7.	H. ♃.	
17 tuberòsus. DC.	(pu.)	tuberous-rooted.	Britain.	....	5. 6.	H. ♃.	Eng. bot. v. 17. t. 1153.
18 pyrenàicus. L.	(pu.)	Pyrenean.	Pyrenees.	1699.	——	H. ♃.	Pluk. alm. t. 210. f. 2.
19 tenuifòlius. Roth.	(pu.)	slender-leaved.	Germany.	1818.	5. 6.	H. ♃.	
20 vàrius. DC.	(v.)	parti-coloured.	Italy.	1759.	——	H. ♃.	Bot. mag. t. 675.
21 angustifòlius. DC.	(v.)	narrow-leaved.	Siberia.	1766.	——	H. ♃.	Gmel. sib. 4. p. 14. t. 5.
22 canéscens. DC.	(bl.wh.)	canescent.	S. Europe.	1823.	——	H. ♃.	
23 palléscens. M.B.	(wh.)	pale-flowered.	Tauria.	1824.	——	H. ♃.	
24 àlbus. DC.	(wh.)	white-flowered.	Austria.	1794.	——	H. ♃.	Swt. br. fl. gar. 1. t. 22.
pannònicus. Jacq. aust. 1. t. 39.							
25 atropurpùreus. DC.	(d.pu.)	dark-purple.	Algiers.	1826.	6. 7.	H. ♃.	Desf. atl. 2. t. 196.
26 Fischèri. B.F.G.	(d.pu.)	Fischer's.	Siberia?	1827.	——	H. ♃.	Swt. br. fl. gar. 3. t. 289.
27 saxátilis. DC.	(pu.)	rock.	S. France.	1823.	6. 8.	H. ⊙.	Venten. cels. t. 94.
28 longifòlius. N.	(li.)	long-leaved.	Missouri.	1827.	——	H. ♃.	

Tribus V. *PHASEOLEÆ.* DC. prodr. 2. p. 381.

ABRUS. DC.		WILD LIQUORICE. Monadelphia Enneandria.					
precatòrius. DC.	(ro.)	common.	India.	1680.	3. 5.	S. ♄.◡.	Rheed. mal. 3. t. 39.
α erythrospérma.	(ro.)	*red-seeded.*	————	——	——	S. ♄.◡.	Rumph. amb. 5. t. 32.
β melanospérma.	(ro.)	*black-seeded.*	————	——	——	S. ♄.◡.	
SWE'ETIA. DC.		SWE'ETIA. Diadelphia Decandria.					
1 longifòlia. DC.	(pu.)	long-leaved.	S. America.	1824.	6. 7.	S. ħ.◡.	Jacq. ic. 3. t. 572.
Gálega longifòlia. Jacq. *Tephròsia longifòlia.* P.S.							
2 filifórmis. DC.	(pu.)	slender.	————	1826.	——	S. ♄.◡.	Jacq. ic. 3. t. 573.
3 lignòsa. DC.	(pu.)	woody.	Trinidad.	1820.	....	S. ♄.◡.	
Gl'ycine lignòsa. P.S.							
MACRA'NTHUS. DC.		MACRA'NTHUS. Diadelphia Decandria.					
cochinchinénsis. DC.	(wh.)	eatable-podd'd.	Cochinchina.	1826.	6. 9.	H. ⊙.◡.	
WEST'ONIA. S.S.		WEST'ONIA. Monadelphia Decandria.					
trifoliàta. S.S.	(st.re.)	trifoliate.	....	1820.	6. 8.	S. ⊙.	
Róthia trifoliàta. DC. *Gl'ycine humifùsa.* W. en.							
TERA'MNUS. DC.		TERA'MNUS. Monadelphia Pentandria.					
1 volùbilis. DC.	(re.)	twining.	Jamaica.	1822.	7. 8.	S. ♄.◡.	
2 uncinàtus. DC.	(re.)	hooked.	————	——	——	S. ♄.◡.	Plum. ed. Burm. t. 221.
AMPHICARP'ÆA. DC.		AMPHICARP'ÆA. Diadelphia Decandria.					
1 monoíca. DC.	(pa.)	pale-flowered.	N. America.	1781.	8. 10.	H. ⊙.◡.	Rœm. arch. 1. p. 3. t. 2.
2 comòsa.	(bl.)	tufted.	————	1812.	7. 9.	H. ♃.◡.	
Gl'ycine comòsa. Ph.							
3 sarmentòsa. DC.	(pa.)	sarmentose.	————	1805.	——	H. ⊙.◡.	Schk. bot. an. 12. t. 2.
Gl'ycine sarmentòsa. Ph. *monoíca.* Schkuhr.							
KENN'EDYA. DC.		KENN'EDYA. Diadelphia Decandria.					
1 rubicúnda. DC.	(cr.)	dingy-flowered.	N. S. W.	1788.	3. 8.	G. ♄.◡.	Bot. mag. 208.
2 prostràta. DC.	(sc.)	scarlet one-fl^d.	————	1790.	——	G. ħ.◡.	———— t. 270.
Gl'ycine coccínea. B.M.							
3 coccínea. DC.	(sc.)	many-flowered.	N. Holland.	1803.	5. 8.	G. ♄.◡.	Swt. fl. aust. t. 23.
4 Comptoniàna. DC.	(pu.)	Comptonian.	————	——	3. 6.	G. ♄.◡.	Bot. reg. 298.
Gl'ycine Comptoniàna. B.R.							

5 heterophy'lla. ψ. (pu.) various-leaved. H. Holland. 1826. .... G. ♄.⌣.
6 inophy'lla. ψ. (pu.) few-leaved. ——— 1825. .... G. ♄.⌣.
7 dilatàta. ψ. (pu) dilated-leaved. ——— —— .... G. ♄.⌣.
8 monophy'lla. (pu.) simple-leaved. N. S. W. 1790. 3. 8. G. ♄.⌣. Vent. malm. t. 106.
Gl'ycine bimaculàta. Bot. mag. t. 263.
9 ovàta. DC. (pu.) oval-leaved. ——— 1820. —— G. ♄.⌣. Bot. mag. 2169.
cordàta. Bot. reg. 944.

RHYNCH'OSIA. DC. RHYNCH'OSIA. Diadelphia Decandria.
1 renifórmis. DC. (ye.) kidney-leaved. Carolina. 1806. 7. 8. F. ♃.
2 diffórmis. DC. (ye.) downy climbing. ——— 1732. 6. 9. F. ♃.⌣. Dill. elth. t. 26. f. 29.
Gl'ycine tomentòsa β volùbilis. M.
3 erécta. DC. (br.) upright. ——— .... —— F. ♃.
4 carib'æa. DC. (ye.) trailing. W. Indies. 1742. 9. 10. S. ♄.⌣. Bot. reg. t. 275.
Gly'cine carib'æa. B.R.
5 móllis. DC. (ye.) soft-leaved. Guinea. 1823. 6. 7. S. ♄.⌣. Hook. ex. flor. 201.
Gl'ycine móllis. H.E.F.
6 mínima. DC. (ye.) slender-stalked. W. Indies. —— —— S. ♄.⌣. Jacq. obs. 1. t. 22.
7 volùbilis. DC. (ye.) black-podded. China. 1826. —— S. ♄.⌣.
8 phaseoloídes. DC. (ye.) Abrus-seeded. W. Indies. 1820. —— S. ♄.⌣. Bot. mag. t. 2284.
9 precatòria. DC. (ye.) Abrus-like. S. America. —— —— S. ♄.⌣.
10 macrophy'lla. DC. (ye.) large-leaved. Cuba. 1826. —— S. ♄.⌣.
11 reticulàta. DC. (ye.) netted. Jamaica. 1779. 7. 9. S. ♄.⌣.
12 rhombifòlia. DC. (ye.) rhomb-leaved. E. Indies. 1815. 5. 8. S. ♄.⌣.
13 medicagínea. DC. (ye.) Medick-like. Ceylon. 1826. —— S. ♄.⌣. Burm. zeyl. 188. t. 88.
14 suavèolens. DC. (ye.) sweet-scented. E. Indies. 1816. —— S. ♄.
15 víscida. DC. (ye.) viscous. ——— 1823. 6. 7. S. ♄.⌣.
16 viscòsa. DC. (ye.) clammy. ——— 1824. —— S. ♄.⌣.
17 Fridericiàna. DC. (ye.) blunt-leaved. Manilla. 1826. —— S. ♄.⌣.
18 scaraboeoídes. DC. (ye.) rough-stemmed. E. Indies. —— —— S. ⊙.⌣. Pluk. alm. t. 52. f. 3.
19 biflòra. DC. (ye.) two-flowered. E. Indies. 1827. 6. 8. S. ⊙.⌣.
20 angustifòlia. DC. (ye.) narrow-leaved. C. B. S. 1795. —— G. ♄.⌣. Jacq. schœn. 2. t. 231.
21 Tótta. DC. (ye.) smooth-stemm'd. ——— 1819. —— G. ♄.⌣.

FAG'ELIA. DC. FAG'ELIA. Diadelphia Decandria.
bituminòsa. DC. (ye.) clammy. C. B. S. 1774. 4. 9. G. ♄.⌣. Bot. reg. t. 261.
Gl'ycine bituminòsa. B.R.

WIST'ERIA. DC. WIST'ERIA. Diadelphia Decandria.
1 frutéscens. DC. (bl.) handsome. N. America. 1724. 6. 9. H. ♄.⌣. Bot. mag. t. 2103.
speciòsa. N. *Gl'ycine frutéscens.* B.M. *Apios frutéscens.* Ph.
2 chinénsis. DC. (p.bl.) Chinese. China. 1818. 5. 7. H. ♄.⌣. Swt. br. fl. gar. 3. t. 211.
Gl'ycine chinénsis. Bot. mag. 2083. *sinénsis.* Bot. reg. 650.

APIOS. DC. APIOS. Diadelphia Decandria.
tuberòsa. DC. (ro.) tuberous-rooted. N. America. 1640. 8. 9. H. ♃.⌣. Bot. mag. t. 1198.
Gl'ycine 'Apios. B.M.

AMPH'ODUS. B.R. AMPH'ODUS. Diadelphia Decandria.
ovàtus. B.R. (pu.) oval-leaved. Trinidad. 1824. 4. 6. S. ♄.⌣. Bot. reg. t. 1101.

PHAS'EOLUS. DC. KIDNEY BEAN. Diadelphia Decandria. L.
1 Caracálla. DC. (pu.) Snail-flower. E. Indies. 1690. 8. 9. S. ♃.⌣. Bot. reg. t. 341.
2 tuberòsus. DC. (ye.) tuberous-rooted. China. 1825. —— S. ♃.⌣.
3 perénnis. DC. (pu.) perennial. Carolina. 1823. 7. 9. F. ♃.⌣.
4 macrostàchyus. DC. (pu.) long-spiked. N. America. 1824. —— H. ♃.⌣.
5 alàtus. DC. (pu.) winged. Carolina. 1732. 7. 8. H. ⊙. Dill. elth. t. 235. f. 303.
6 multiflòrus. DC. (sc.) scarlet-runner. S. America. 1633. 6. 10 H. ♃. Schk. hand. 2. t. 199. a.
α coccíneus. DC. (sc.) *scarlet-flowered.* ——— —— —— H. ♃.⌣. Moris. his. 52. t. 5. f. 4.
β albiflòrus. DC. (wh.) *white-flowered.* ——— .... —— H. ♃.⌣.
7 vulgàris. DC. (wh.li.) common. E. Indies. 1597. 5. 9. H. ⊙.⌣. Lobel. ic. 2. p. 59. ic.
8 compréssus. DC. (wh.) white-runner. .... 6. 9. H. ⊙.⌣. Savi. mem. 3. t. 10. f. 20.
9 oblóngus. DC. (li.) oblong-seeded. .... —— H. ⊙.⌣. —— 3. p. 17. t. 10. f. 14.
10 saponàceus. DC. (wh.) spotted-white. .... —— H. ⊙.⌣. —— 3. p. 19. t. 10. f. 15.
11 tùmidus. DC. (wh.) thick-seeded. .... —— H. ⊙.⌣. —— 3. p. 19. t. 10. f. 16.
12 hæmatocárpus. DC. (li.) red-podded. .... —— H. ⊙.⌣. —— 3. p. 20. t. 10. f. 17.
13 sph'æricus. DC. (li.) roundish-seeded. .... —— H. ⊙.⌣. —— 3. p. 20. t. 10. f. 18.
14 gonospérmus. DC. (wh.) angular-seeded. .... —— H. ⊙.⌣. —— 3. p. 21. t. 10. f. 19.
15 deràsus. DC. (gr.wh.) narrow-podded. Brazil. 1823. —— H. ⊙.⌣. Schk. h. mon. 1. t. 89.
16 lunàtus. DC. (gr.) cimeter-podded. E. Indies. 1779. —— S. ⊙.⌣. Hout. syst. 8. t. 63. f. 1.
17 inam'œnus. DC. (gr.wh.) various-colour'd. Africa. 1794. 7. 8. H. ⊙.⌣. Jacq. vind. t. 66.
18 Xuarèsii. DC. (gr.wh.) pointed-leaved. .... 1823. 6. 9. H. ⊙.⌣.
19 adenánthus. DC. (wh.bl.) painted-flower'd. Guiana. 1821. —— S. ⊙.⌣.
20 macrocárpus. DC. (wh.) long-podded. .... —— H. ⊙.⌣.
21 calcaràtus. H.B. (li.) spurred. E. Indies. 1820. 7. 8. S. ♂.⌣.

22 dolichoídes. H.B.	(*pu.*)	Dolichos-like.	E. Indies.	1820.	7. 8.	S. ♃.◡.	
23 mínimus. H.B.	(*li.*)	small.	China.	1823.	——	H. ⊙.◡.	
24 angustifòlius. H.B.	(*li.*)	narrow-leaved.	————	1818.	6. 8.	H. ⊙.◡.	
25 microspérmus. DC.	(*pu.*)	small-seeded.	Cuba.	1821.	6. 9.	S. ⊙.	
26 heterophy'llus. DC.	(*re.*)	different-leaved.	Mexico.	1818.	7. 8.	H. ⊙.◡.	
27 aconitifòlius. DC.	(*ye.*)	Aconite-leaved.	E. Indies.	1731.	6. 8.	S. ⊙.◡.	Jacq. obs. 3. t. 52.
28 sublobàtus. H.B.	(*li.*)	lobed-leaved.	————	1820.	——	S. ⊙.◡.	
29 subtrilòbus. L.en.	(*ye.*)	acute-lobed.	Brazil.	1818.	——	S. ⊙.◡.	
30 diversifòlius. DC.	(*pu.*)	various-leaved.	N. America.	1806.	6. 8.	H. ⊙.	
31 trilòbus. DC.	(*pu.*)	three-lobed.	E. Indies.	1777.	7. 8.	S. ⊙.	Burm. ind. t. 50. f. 1.
32 angulòsus. DC.	(*bh.*)	angular-stalked.	S. America.	1818.	7. 8.	S. ⊙.◡.	
33 farinòsus. DC.	(*re.*)	mealy.	E. Indies.	1759.	——	S. ♃.◡.	Niss. act. par. 1730. t. 42.
34 stipulàris. DC.	(*ye.*)	large-stipuled.	Peru.	1805.	——	H. ⊙.	
35 mesoleùcus. L.en.	(*wh.*)	Brazilian.	Brazil.	1818.	——	H. ⊙.◡.	
36 hélvolus. DC.	(*ro.*)	pale red.	Carolina.	1732.	——	H. ⊙.◡.	Dill. elth. t. 233. f. 300.
37 vexillàtus. DC.	(*bh.*)	sweet-scented.	————	——	——	H. ⊙.◡.	Jacq. vind. 2. t. 102.
38 Múngo. DC.	(*ye.*)	small-fruited.	E. Indies.	1790.	6. 7.	H. ⊙.	
39 hírtus. W.	(*ye.*)	hairy.	C. B. S.	1818.	7. 8.	H. ⊙.◡.	
40 Hernandèsii. DC.	(*ye.*)	Hernandez.	Mexico.	1824.	——	H. ⊙.	Hern. mex. 887. ic.
41 chrysánthos. DC.	(*ye.*)	golden-flowered.	....	....	——	H. ⊙.	
42 radiàtus. DC.	(*bh.*)	rayed.	China.	1732.	6. 7.	H. ⊙.	Dill. elth. t. 235. f. 305.
43 scáber. DC.	(*st.*)	rough-podded.	....	....	6. 9.	H. ⊙.	Moris. ox. s. 2. t. 5. f. 8.
44 lathyroídes. DC.	(*re.*)	Lathyrus-like.	Jamaica.	1786.	7. 8.	S. ⊙.	Sloan. hist. 1. t. 116. f. 1.
45 semieréctus. DC.	(*re.*)	dark-red.	S. America.	1732.	——	S. ⊙.◡.	Bot. reg. t. 743.
46 violàceus. DC.	(*vi.wh.*)	violet-flowered.	Africa ?	....	6. 9.	H. ⊙.	
47 Máx. DC.	(*re.*)	hairy-podded.	E. Indies.	1758.	6. 7.	S. ⊙.	Rumph. amb. 5. t. 140.
48 gláber. H.B.	(*li.*)	smooth.	Mauritius.	1823.	7. 8.	S. ⊙.	
49 toròsus. H.B.	(*li.*)	twisted-podded.	Nepaul.	——	——	H. ⊙.	
S'OJA. DC.		Soy. Diadelphia Decandria.					
híspida. DC.	(*vi.*)	hispid.	E. Indies.	1790.	7. 8.	S. ⊙.	Jacq. ic. 1. t. 145.
Dólichos Soja. Jacq. *Sòja japónica.* Savi.							
β *pállida.* DC.	(*ye.*)	*pale-flowered.*	————	....	——	S. ⊙.	
DO'LICHOS. DC.		Do'lichos. Diadelphia Decandria. L.					
1 lignòsus DC.	(*ro.*)	woody.	E. Indies.	1776.	7. 8.	S. ♄.◡.	Smith spicil. t. 21.
β? *falcifórmis.*	(*ro.*)	*falcate-podded.*	C. B. S.	——	——	G. ♄.◡.	Bot. mag. t. 382.
2 Jacquíni. DC.	(*wh.*)	Jacquin's.	W. Indies.	1822.	7. 9.	S. ♄.◡.	
3 spùrius. DC.	(*pu.*)	spurious.	Guiana.	1826.	——	S. ♄.◡.	
4 tuberòsus. DC.	(*li.*)	tuberous-rooted.	Martinica.	——	——	S. ♄.◡.	Plum. ed. Burm. t. 220.
5 articulàtus. DC.	(*pu.*)	jointed.	St. Domingo.	1828.	——	S. ♄.◡.	
6 hirsùtus. DC.	(*pu.*)	hirsute.	China.	1802.	6. 7.	G. ♄.◡.	Kæmpf. ic. t. 41.
7 pubéscens. DC.	(*st.*)	pubescent.	S. America.	1826.	7. 9.	G. ⊙.◡.	
8 pilòsus. DC.	(*li.*)	hairy-podded.	E. Indies.	1790.	7. 8.	S. ⊙.◡.	
9 tetraspérmus. DC.	(*li.*)	four-seeded.	————	1816.	5. 8.	S. ♂.◡.	
10 capénsis. DC.	(*ye.*)	Cape.	C. B. S.	1818.	——	G. ♃.◡.	
11 falcàtus. DC.	(*li.*)	falcate-podded.	E. Indies.	1820.	7. 8.	S. ♃.◡.	
12 heterophy'llus. DC.	(*pu.*)	various-leaved.	Canaries.	1827.	6. 9.	H. ⊙.◡.	
13 lutèus. DC.	(*ye.*)	yellow-flowered.	Jamaica.	1812.	7. 8.	S. ♃.◡.	
14 biflòrus. DC.	(*li.*)	two-flowered.	E. Indies.	1776.	——	S. ⊙.	Pluk. alm. t. 213. f. 4.
15 Catiáng. DC.	(*pu.*)	small-podded.	————	1793.	——	S. ⊙.	
16 monáchalis. DC.	(*pu.*)	hybrid.	Portugal.	1826.	——	H. ⊙.	
17 sinénsis. DC.	(*fl.*)	Chinese.	China.	1776.	——	H. ⊙.◡.	Rumph. amb. 5. t. 134.
18 nilóticus. DC.	(*st.*)	Egyptian.	Egypt.	1828.	——	H. ⊙.◡.	Delil. fl. eg. t. 38. f. 1.
19 Lùbia. DC.	(*bl.wh.*)	cultivated.	————	——	——	H. ⊙.	
20 vexillàtus. DC.	(*st.*)	hispid.	S. America.	....	——	H. ⊙.◡.	Dill. elt. 2. f. 302.
21 lobàtus. DC.	(*li.*)	lobe-leaved.	C. B. S.	1818.	6. 7.	G. ♃.◡.	Hout. syst. 8. t. 64. f. 1.
22 viròsus. H.B.	(*li.*)	virose.	E. Indies.	——	5. 8.	S. ♂.◡.	Rheed. mal. 8. t. 45.
23 gangéticus. H.B.	(*pu.*)	Ganges.	Bengal.	1824.	6. 7.	S. ♃.◡.	
24 glutinòsus. H.B.	(*pu.*)	glutinous.	E. Indies.	1820.	——	S. ♃.◡.	
25 unguiculàtus. DC.	(*li.*)	Bird's-foot.	W. Indies.	1780.	——	S. ⊙.◡.	Jacq. vind. 1. t. 23.
26 tranquebáricus. DC.	(*vi.*)	Tranquebar.	E. Indies.	1801.	6. 7.	S. ⊙.◡.	Jacq. vind. 3. t. 70.
27 melanophthálmus. DC.	(*vi.*)	black-circled.	....	1822.	7. 9.	H. ⊙.◡.	
28 sesquipedàlis. DC.	(*wh.*)	long-podded.	W. Indies.	1781.	8. 9.	S. ⊙.◡.	Jacq. vind. 1. t. 67.
29 Brunélli. DC.	(*wh.*)	branching.	....	1826.	6. 9.	S. ⊙.	
30 sphærospérmos. DC.	(*st.*)	round-seeded.	S. America.	1823.	7. 8.	S. ⊙.	Sloan. jam. 1. t. 117.
31 reticulàtus. DC.	(*li.*)	netted-leaved.	N. S. W.	1781.	6. 7.	G. ♄.◡.	
32 phaseoloídes. H.B.	(*li.*)	KidneyBean-like.	China.	1823.	7. 8.	G. ♃.◡.	
33 frutéscens. D.P.	(*pu.*)	frutescent.	Nepaul.	1824.	——	G. ♄.	
34 incànus. Wal.	(*pu.*)	hoary.	E. Indies.	1825.	——	S. ⊙.	

VI'GNA. DC. Vi'gna. Diadelphia Decandria.
1 glábra. DC. (ye.) smooth. America. 1805. 7. 8. H. ⊙.⌣. Jacq. vind. 1. t. 90.
Dólichos lutèolus. Jacq.
2 villòsa. DC. (ye.) villous. Chile. 1826. —— H. ⊙.⌣.
LA'BLAB. DC. La'blab. Diadelphia Decandria.
1 vulgàris. DC. (pu.) common. Egypt. 1694. 6. 7. S. ⊙. Bot. mag. t. 896.
Dólichos Láblab. B.M.
2 purpùreus. (pu.) purple. E. Indies. 1790. 5. 10. S. ♃.⌣. Exot. bot. 2. t. 74.
Dólichos purpùreus. Bot. reg. t. 830.
3 benghalénsis. (wh.) Bengal. Bengal. 1820. 7. 8. S. ♃.⌣. Jacq. vind. 2. t. 124.
4 nankínicus. DC. (wh.) Chinese. China. 1821. —— H. ⊙.⌣. Savi. dis. 1821. f. 8. d-g.
5 leucocárpus. DC. (wh.) white-podded. E. Indies. —— —— S. ⊙.⌣. —————— f. 9. a-d.
6 cultràtus. DC. (wh.) sharp-edged. Japan. 1818. 6. 7. H. ⊙.⌣. Banks ic. kæmpf. t. 25.
7 microcárpus. DC. (pu.) short-podded. E. Indies. 1824. —— S. ⊙.⌣. Rumph. am. 5. t. 141. f. 1.
8 perénnans. DC. (wh.) perennial. China. 1823. 7. 8. H. ⊙.⌣. —————— 5. t. 137.
Dólichos álbus. W.
PACHYRH'IZUS. DC. Pachyrh'izus. Diadelphia Decandria.
1 angulàtus. DC. (vi.) angular-leaved. E. Indies. 1781. 7. 8. S. ♄.⌣. Rumph. amb. 5. t. 132.
Dólichos bulbòsus. L.
2 trilòbus. DC. (pu.) three-lobed. China. 1825. 6. 8. S. ♄.⌣.
PAROCH'ETUS. D.P. Paroch'etus. Diadelphia Decandria.
1 commùnis. D.P. (li.) common. Nepaul. 1827. 6. 8. G. ♃.
2 màjor. D.P. (li.) larger. —————— —— —— G. ♃.
DIOCL'EA. DC. Diocl'ea. Diadelphia Decandria.
Jacquiniàna. DC. (re.) Jacquin's. Martinica. 1826. 6. 9. S. ♄.⌣. Jacq. amer. t. 123.
Dólichos rùber. Jacq.
PSOPHOCA'RPUS. DC. Psophoca'rpus. Diadelphia Decandria.
tetragonólobus. DC. (bl.) square-podded. Mauritius. 1816. 9. 11. S. ⊙. Rumph. amb. 5. t. 133.
Dólichos tetragonólobus. L.
CANAV'ALIA. DC. Canav'alia. Monadelphia Decandria.
1 obtusifòlia. DC. (pu.) blunt-leaved. E. Indies. 1818. 7. 8. S. ♄.⌣. Rheed. mal. 8. t. 43.
Dólichos rotundifòlius. Vahl.
β emarginàta. DC. (pu.) *emarginate.* —————— .... —— S. ♄.⌣. Jacq. schœn. 2. t. 221.
2 gladiàta. DC. (bh.) sabre-podded. —————— 1801. 6. 7. S. ♄.⌣. Jacq. ic. 3. t. 560.
Dólichos gladiàtus. W. *Malócchia gladiàta.* Savi.
3 ensifórmis. DC. (pu.) cimeter-podded. E. Indies. 1778. 7. 8. S. ⊙. Jacq. ic. 3. t. 559.
4 ròsea. DC. (ro.) rose-colour'd. Jamaica. 1812. —— S. ♄.
5 bonariénsis. B.R. (pu.) Buenos Ayres. Buenos Ayres. 1826. 6. 10. G. ♄.⌣. Bot. reg. t. 1199.
MUC'UNA. DC. Cow-Itch. Diadelphia Decandria.
1 ùrens. DC. (st.) broad-podded. W. Indies. 1691. 6. 7. S. ♄.⌣. Jacq. amer. t. 182. f. 84.
Dólichos ùrens. L. *Stizolòbium ùrens.* P.S.
2 altíssima. DC. (li.) tallest. Martinica. 1779. 6. 8. S. ♄.⌣. Jacq. amer. t. 182. f. 85.
3 prùriens. DC. (vi.) common. India. 1680. —— S. ♄.⌣. Rheed. mal. 8. t. 85.
Dólichos prùriens. L. *Stizolòbium prùriens.* P.S. *Carpopògon prùriens.* H.B.
4 gigántea. DC. (pa.) gigantic. E. Indies. 1815. 7. 9. S. ♄.⌣. Rheed. mal. 8. t. 36.
5 ellíptica. DC. (li.) elliptic-leaved. Peru. 1828. .... G. ♄.⌣.
6 mìtis. DC. (li.) oblique-leaved. —————— 1823. 7. 9. G. ♄.⌣.
Negrètia mìtis. R.P.
7 comòsa. DC. (wh.) tufted. Guiana. 1827. —— S. ♄.⌣.
8 monospérma. DC. (li.) one-seeded. E. Indies. 1818. —— S. ♄.⌣.
Carpopògon monospérmum. H.B.
9 atropurpùrea. DC. (d.pu.) dark-purple. —————— 1820. —— S. ♄.⌣.
10 anguínea. (pu.) snake-like. —————— 1822. .... S. ♄.⌣.
Carpopògon anguíneum. Roxb.
11 nívea. DC. (wh.) snowy-white. —————— 1820. 7. 9. S. ⊙.⌣.
12 imbricàta. DC. (pu.) imbricated. —————— 1815. —— S. ♄.⌣.
13 bracteàta. DC. (pu.) bracteate. —————— 1822. —— S. ♄.⌣.
14 càpitàta. (pu.) headed. —————— —— 8. 11. S. ⊙.⌣.
Carpopògon capitàtum. H.B.
PHYLLOL'OBIUM. DC. Phyllol'obium. Diadelphia Decandria. S.S.
1 chinénse. DC. (wh.) Chinese. China. 1818. 6. 8. G. ♃.
2 zanzibarénse. Boj. .. Zanzibar. Zanzibar. 1825. .... S. ♄.
CAJ'ANUS. DC. Pigeon Pea. Diadelphia Decandria.
1 bícolor. DC. (ye.pu.) two-coloured. E. Indies. 1687. 7. 8. S. ♄. Jacq. vind. 2. t. 119.
Cy'tisus pseùdo-cájan. Jacq.
2 flàvus. DC. (ye.) yellow-flowered. —————— —— —— S. ♄. Jacq. obs. 1. t. 1.
Cy'tisus Cájan. L.
LUP'INUS. DC. Lupine. Monadelphia Decandria.
1 álbus. DC. (wh.) white. Levant. 1596. 7. 8. H. ⊙. Blackw. herb. t 282.

2	Térmis. DC.	(*wh.bl.*)	Egyptian.	Egypt.	1802.	6. 7.	H. ⊙.	
3	vàrius. DC.	(*bl.*)	variable.	S. Europe.	1596.	7. 9.	H. ⊙.	
4	hirsùtus. DC.	(*bl.*)	blue-flowered.	—	1629.	—	H. ⊙.	J. Bauh. his. 2. p. 289. ic.
5	bracteolàris. DC.	(*bl.*)	bracteolate.	Monte Video.	1828.	—	H. ⊙.	
6	pilòsus. DC.	(*ro.*)	rose.	S. Europe.	1710.	—	H. ⊙.	
7	angustifòlius. DC.	(*bl.*)	narrow-leaved.	—	1686.	—	H. ⊙.	Knor. del. 2. t. L. 7.
8	linifòlius. DC.	(*bl.*)	Flax-leaved.	....	1790.	—	H. ⊙.	Roth. abhand. 14. t. 5.
9	lùteus. DC.	(*ye.*)	yellow.	S. Europe.	1596.	—	H. ⊙.	Bot. mag. t. 140.
	β sulphùreus.	(*st.*)	*straw-coloured.*	—	—	—	H. ⊙.	
10	microcárpus. DC.	(*bl.*)	small-podded.	Chile.	1822.	4. 6.	H. ⊙.	Bot. mag. t 2413.
11	micránthus. B.R.	(*pu.*)	small-flowered.	Columbia.	1827.	5. 7.	H. ⊙.	Bot. reg. t. 1251.
12	áridus. B.R.	(*bl.pu.*)	arid.	—	—	6. 9.	H. ♂.	— t. 1242.
13	bícolor. B.R.	(*bl.wh.*)	two-coloured.	—	—	5. 10.	H. ♃.	— t. 1109.
14	lépidus. B.R.	(*bl.li.*)	lively.	—	—	7. 10.	H. ♃.	— t. 1149.
15	laxiflòrus. B.R.	(*bl.pi.*)	lax-flowered.	—	—	8. 10.	H. ♃.	— t. 1140.
16	littoràlis. B.R.	(*pu.bl.*)	shore.	—	—	6. 10.	H. ♃.	— t. 1198.
17	ornàtus. B.R.	(*bl.*)	adorned.	—	—	5. 11.	H. ♃.	— t. 1216.
18	plumòsus. B.R.	(*bl.*)	feathered.	—	—	—	H. ♃.	— t. 1217.
19	Sabìni. D.	(*bl.*)	Sabine's.	—	—	6. 10.	H. ♃.	
20	mexicanus. DC.	(*bl.*)	Mexican.	Mexico.	1819.	7. 9.	F. ♃.	Bot. reg. t. 457.
21	perénnis. DC.	(*bl.*)	smooth perenial.	N. America.	1658.	5. 8.	F. ♃.	Bot. mag. t. 202.
	β albiflòrus.	(*wh.*)	*white-flowered.*	—	1817.	—	F. ♃.	
22	polyphy'llus. B.R.	(*bl.*)	many-leaved.	Columbia.	1827.	—	H. ♃.	Bot. reg. t. 1096.
23	leucophy'llus. B.R.	(*pu.*)	white-leaved.	—	—	6. 8.	H. ♃.	— t. 1124.
24	nootkaténsis. DC.	(*bl.*)	hairy perennial.	Nootka sound.	1794.	—	H. ♃.	— t. 1311.
	β variábilis.	(*va.*)	*variable coloured.*	—	—	—	H. ♃.	— t 2136.
25	variegàtus. F.	(*v.*)	variegated.	Silka.	1828.	—	H. ♃.	
26	argyr'æus. DC.	(*bl.*)	silvery-leaved.	....	—	—	F. ♃.	
27	arbústus. B.R.	(*pu.*)	half-shrubby.	California.	1827.	6. 9.	F. ħ.	Bot. reg. 1230.
28	canaliculàtus. B.F.G.	(*bl.*)	channel-l[d].	BuenosAyres.	1828.	6. 9.	H. ħ.	Swt. br. fl. gar. t. 283.
29	versícolor. B.F.G.	(*v.*)	various-colour'd.	Mexico.	1828.	—	H. ħ.	— ser. 2. t. 12.
30	tomentòsus. DC.	(*va.*)	woolly.	Peru.	1826.	6. 10.	H. ħ.	Swt. br. fl. gar. t. 261.
31	arbòreus. DC.	(*ye.*)	tree.	S. America.	1793.	7. 9.	H. ♄.	Bot. mag. t. 682.
32	mutábilis. B.F.G.	(*wh.y.*)	changeable.	Begota.	1825.	—	F. ♄.	Swt. br. fl. gar. t. 130.
33	villòsus. DC.	(*pu.*)	simple-leaved.	Carolina.	1787.	—	F. ♃.	Pursh. fl. amer. 2. t. 21.

CYLI'STA. DC. CYLI'STA. Diadelphia Decandria.

1	scariòsa. DC.	(*bh.*)	rough.	E. Indies.	1806.	4. 6.	S. ♄. ◡.	Roxb. corom. 1. t. 92.
2	tomentòsa. DC.	(*ye.*)	tomentose.	—	1818.	—	S. ♄. ◡.	— t. 221.
3	albiflòra. DC.	(*wh.*)	white-flowered.	Mauritius.	....	4. 5.	S. ♄. ◡.	Bot. mag. t. 1859.
4	villòsa. DC.	(*wh.*)	hairy.	C. B. S.	1776.	4. 6.	G. ♄. ◡.	Andr. rep. t. 446.
	Dólichos hírtus. A.R.							

ERYTHR'INA. DC. CORAL-TREE. Diadelphia Decandria. L.

1	herbàcea. DC.	(*sc.*)	herbaceous.	Carolina.	1724.	6. 9.	G. ħ.	Bot. mag. t. 877.
2	cárnea. DC.	(*fl.*)	flesh-coloured.	Vera Cruz.	1733.	5. 6.	G. ħ.	Trew. ehr. 2. t. 8.
	β pubéscens.	(*fl.*)	*pubescent-leaved.*	—	—	—	G. ħ.	Bot. reg. t. 389.
3	Corallodéndron. DC.	(*sc.*)	common.	W. Indies.	1690.	—	S. ♄.	Comm. h. ams. 1. t. 108.
4	enneándra. DC.	(*sc.*)	nine-anthered.	....	....	—	S. ♄.	Jacq. schœn. 4. t. 466.
	velutìna. Jacq. *non* W.							
5	macrophy'lla. DC.	(*sc.*)	large-leaved.	....	1816.	—	S. ♄.	
6	mìtis. DC.	(*sc.*)	smooth-stem'd.	Caracas.	1818.	6. 8.	S. ♄.	Jacq. schœn. 2. t. 216.
7	poíanthes. DC.	(*sc.*)	crowded-flower'd.	....	1826.	4. 6.	S. ♄.	Bot. reg. t. 1246.
8	secundiflòra. DC.	(*sc.*)	side-flowering.	Brazil.	1818.	—	S. ♄.	Linn. trans. 14. t. 12.
9	speciòsa. DC.	(*sc.*)	large-flowered.	W. Indies.	1805.	5. 10.	S. ♄.	Andr. rep. t. 443.
10	umbròsa. DC.	(*sc.*)	deltoid-leaved.	Caracas.	1824.	....	S. ♄.	
11	Humeàna. S.S.	(*sc.*)	Sir A. Hume's.	C. B. S.	1818.	5. 6.	S. ħ.	Bot. mag. t. 2431.
	cáffra. Bot. reg. t. 736. *non Thunb.*							
12	cáffra. Th.	(*pu.*)	Cape-tree.	—	1816.	....	S. 5.	
13	fúlgens. DC.	(*sc.*)	fulgent.	E. Indies.	1801.	5. 10.	S. ♄.	
14	arboréscens. DC.	(*cr.*)	arborescent.	Nepaul.	1818.	....	S. ♄.	Roxb. corom. 3. t. 219.
15	índica. DC.	(*sc.*)	Indian.	E. Indies.	1814.	5. 6.	S. ♄.	Rheed. mal. 6. t. 7.
16	pícta. DC.	(*cr.*)	prickly-leaved.	—	1696.	5. 10.	S. ♄.	Rumph. amb. 2. t. 77.
17	aculeatíssima. Df.	(*sc.*)	very prickly.	....	1816.	—	S. ♄.	
18	velutìna. DC.	(*sc.*)	velvetty.	Caracas.	1824.	....	S. ♄.	
19	fúsca. DC.	(*br.re.*)	red-brown.	China.	1800.	....	S. ♄.	Rumph. amb. 2. t. 78.
20	laurifòlia. Jacq.	(*sc.*)	Laurel-leaved.	Brazil.	....	5. 10.	H. ħ.	Swt. br. fl. gar. t. 142.
21	crísta-gálli. DC.	(*sc.*)	Cockscomb.	—	1771.	6. 10.	H. ħ.	— t. 1214.
22	nervòsa. DC.	(*sc.*)	nerved-leaved.	—	1824.	—	S. ♄.	
23	glaùca. DC.	(*sn.*)	glaucous.	Caracas.	—	....	S. ♄.	
24	piscidioídes. DC.	(*sc.*)	Piscidia-leaved.	E. Indies.	1800.	....	S. ♄.	

25 incàna. W. (sc.) hoary. .... 1820. 6. 8. S. ♄.
26 lithospérma. W.H. .... stony-seeded. N. Holland. 1824. .... S. ♄.
27 arbòrea. D.L. .... tree. Trinidad. 1822. .... S. ♄.
28 strícta. H.B. (sc.) upright. E. Indies. 1816. 4. 6. S. ♄.
29 ovalifòlia. H.B. .... oval-leaved. ———— .... S. ♄.
30 suberòsa. H.B. .... cork-barked. ———— 1818. .... S. ♄.
31 resupinàta. H.B. .... resupinate. ———— 1822. .... S. ♄.
32 abyssínica. DC. .... Abyssinian. Africa. 1820. .... S. ♄.
RUDO'LPHIA. DC. RUDO'LPHIA. Diadelphia Decandria.
1 volùbilis. DC. (sc.) twining. Portorico. 1820. .... S. ♄.◡. Vahl ecl. 3. t. 30.
2 portoricénsis. (sc.) Portorico. ———— 1820. .... S. ♄.◡.
Erythrìna portoricénsis. Df.
3 dùbia. DC. (sc.) doubtful. S. America. 1815. .... S. ♄.◡. H. et B. n. gen. 6. t. 591.
B'UTEA. DC. B'UTEA. Diadelphia Decandria.
1 frondòsa. DC. (sc.) downy-branch'd. E. Indies. 1796. .... S. ♄. Roxb. corom. 1. t. 21.
2 supérba. DC. (sc.) smooth-branch'd. ———— 1798. .... S. ♄. ———— v. 1. t. 22.
3 parviflòra. DC. (sc.) small-flowered. Coromandel. 1820. .... S. ♄.

Tribus VI. *DALBERGIEÆ.* DC. prod. 2. p. 415.

DE'RRIS. DC. DE'RRIS. Monadelphia Decandria.
trifoliàta. DC. (wh.) three-leaved. China. 1826. 4. 6. G. ♄.◡.
PONG'AMIA. DC. PONG'AMIA. Monadelphia Decandria.
1 glàbra. DC. (wh.) smooth-leaved. E. Indies. 1699. 6. 8. S. ♄. Vent. malm. t. 28.
2 ellíptica. (wh.) elliptic-leaved. ———— 1820. —— S. ♄.◡.
Galedùpa ellíptica. H.B.
3 chinénsis. DC. (ye.) Chinese. China. 1827. —— S. ♄.
4 uliginòsa. DC. .... marsh. E. Indies. —— .... S. ♄.◡.
5 piscìdia. .... Piscidia-leaved. ———— 1816. .... S. ♄.
Galedùpa piscídia. H.B.
6 marginàta. H.B. .... margined. ———— 1824. .... S. ♄.◡.
DALBE'RGIA. DC. DALBE'RGIA. Mono-Diadelphia Decandria.
1 ougeinénsis. DC. (wh.) Hindoo. E. Indies. 1828. .... S. ♄.
2 latifòlia. DC. (wh.) broad-leaved. ———— 1811. .... S. ♄. Roxb. corom. 2. t. 113.
3 Sissoò. DC. (wh.) large. ———— 1823. .... S. ♄.
4 rubiginòsa. DC. (wh.) rusty. ———— 1811. 6. 8. S. ♄.◡. Roxb. corom. 2. t. 115.
5 robùsta. DC. (wh.) robust. ———— —— .... S. ♄.
6 scándens. DC. (ro.) scandent. ———— —— 5. 7. S. ♄.◡. Roxb. corom. 2. t. 192.
7 volùbilis. DC. (wh.) twining. ———— 1820. —— S. ♄.◡. ————. v. 2. t. 191.
8 frondòsa. DC. (wh.) frondose. ———— 1818. .... S. ♄.
9 paniculàta. DC. (wh.) panicled. ———— 1811. .... S. ♄. Roxb. corom. 2. t. 114.
10 emarginàta. H.B. (wh.) emarginate. ———— 1823. .... S. ♄.
11 zeylànica. H.B. (wh.) Ceylon. Ceylon. 1816. .... S. ♄.
12 marginàta. H.B. (wh.) margined. E. Indies. 1820. 5. 7. S. ♄.◡.
13 tamarindifòlia. H.B. (wh.) Tamarind-leav'd. ———— —— .... S. ♄.◡.
14 stipulàcea. H.B. (wh.) stipuled. ———— —— 6. 8. S. ♄.
15 rimòsa. H.B. (wh.) rimose. ———— 1824. .... S. ♄.
16 alàta. H.B. (wh.) winged. ———— 1823. .... S. ♄.
17 Crówei. DC. (wh.) Silhet. ———— —— .... S. ♄.
PTEROCA'RPUS. DC. PTEROCA'RPUS. Mono-Diadelphia Decandria.
1 Dràco. DC. (ye.) officinal. S. America. 1818. .... S. ♄. Jacq. amer. t. 183. f. 92.
officinàlis. Jacq. *hemíptera.* Gært.
2 suberòsus. DC. (ye.) corky. Guiana. 1826. .... S. ♄. Aubl. guian. 2. t. 299.
3 Marsùpium. DC. (st.) leathery-leaved. E. Indies. 1811. .... S. ♄. Roxb. corom. 2. t. 116.
4 dalbergioídes. DC. (ye.) Andaman red-wood. —— 1820. .... S. ♄.
5 índicus. DC. (ye.) Indian. ———— 1818. .... S. ♄.
6 santalìnus. DC. (ye.) red sandal wood. ———— 1800. .... S. ♄.
7 santalinoídes. DC. (ye.) African. Sierra Leon. 1826. .... S. ♄.
8 flàvus. DC. (ye.) yellow-barked. China. —— .... S. ♄. Rumph. amb. 3. t. 117.
9 scándens. DC. (vi.) climbing. Caracas. 1822. 6. 8. S. ♄.◡.
Amerímnum scándens. W.
DREPANOCA'RPUS. DREPANOCA'RPUS. Monadelphia Decandria.
lunàtus. DC. (wh.) crescent-podded. S. America. 1792. .... S. ♄. Plu. ed. Burm. t. 201. f. 2.
Pterocárpus lunàtus. L. *Pterocárpus ápterus.* Gært. fruct. t. 156.

ECASTAPHY'LLUM. DC.		ECASTAPHY'LLUM. Diadelphia Octo-Decandria.					
Brównei. DC.	(*wh.*)	Browne's.	W. Indies.	1733.	6. 8.	S. ♄.◡.	Brown. jam. t. 32. f. 1.
Pterocárpus Ecastaphy'llum. L.							
AMERI'MNUM. DC.		AMERI'MNUM. Monadelphia Decandria.					
Brównei. DC.	(*wh.*)	Browne's.	W. Indies.	1793.	6. 8.	S. ♄.	Brown.jam.288.t.32.f.3
BR'YA. DC.		BR'YA. Monadelphia Decandria.					
E'benus. DC.	(*ye.*)	Jamaica Ebony.	W. Indies.	1713.	7. 8.	S. ♄.	Brown. jam. t. 31. f. 2.

Subordo II.—*Tribus* VII. SWARTZIEÆ. DC. prodr. 2. p. 422.

SWA'RTZIA. DC.		SWA'RTZIA Monadelphia Dec-Polyandria.					
1 simplicifòlia. DC.	(*ye.*)	simple-leaved.	W. Indies.	1822.	....	S. ♄.	
Rittera símplex. Vahl. *Póssira símplex.* Swartz.							
2 grandiflòra. DC.	(*ye.*)	great-flowered.	Trinidad.	1822.	....	S. ♄.	Vahl. pl. am. dec. 1. t. 9.
3 pinnàta. DC.	(*ye.*)	wing-leaved.	———	1820.	....	S. ♄.	Corr.an.mus. 9. t. 24. f. 2.
4 tomentòsa. DC.	(*ye.*)	woolly.	Cayenne.	1826.	....	S. ♄.	Aubl. guian. 2. t. 307.
BA'PHIA. DC.		BA'PHIA. Decandria Monogynia.					
nítida. DC.	(*wh.*)	glossy.	Sierra Leone.	....	5. 9.	S. ♄.	Lodd. bot. cab. t. 367.

Subordo III.—*Tribus* VIII. MIMOSEÆ. DC. prodr. 2. p. 424.

ENT'ADA. DC.		ENT'ADA. Polygamia Monœcia.					
1 Gigalòbium. DC.	(*wh.*)	West Indian.	W. Indies.	1780.	4. 6.	S. ♄.◡.	
2 Purs'ætha. DC.	(*wh.*)	East Indian.	E. Indies.	——	——	S. ♄.◡.	Rumph. amb. 5. t. 4.
Parràna. S.S. *Acàcia scándens.* W. *Mimòsa scándens.* L.							
3 monostàchya. DC.	(*wh.*)	single-spiked.	———	——	——	S. ♄.◡.	Rheed. mal. 9. t. 77.
Rheèdii. S.S. *Mimòsa Entàda.* W.							
4 polystàchya. DC.	(*wh.*)	many-spiked.	W. Indies.	1816.	4. 9.	S. ♄.◡.	DC. leg. mem. t. 61—62.
MIM'OSA. DC.		SENSITIVE-PLANT. Polygamia Monœcia. L.					
1 sensitìva. DC.	(*ro.*)	true.	Brazil.	1648.	4. 9.	S. ♄.	Botan. regist. t. 25.
2 obtusifòlia. DC.	(*ro.*)	blunt-leaved.	———	1816.	——	S. ♄.	
3 strigòsa. DC.	(*ro.*)	strigose.	S. America.	1822.	——	S. ♄.	
4 floribúnda. DC.	(*ro.*)	many-flowered.	Cumana.	1824.	6. 7.	S. ♄.	
5 vìva. DC.	(*ro.*)	lively.	Jamaica.	1739.	7. 9.	S. ♃.	Sloan. his. 2. t. 182. f. 7.
6 polycárpa. DC.	(*ro.*)	many-podded.	Peru.	1816.	——	S. ♄.	Kth. mim. p. 8. t. 3.
7 pùdica. DC.	(*ro.*)	humble-plant.	Brazil.	1638.	4. 9.	S. ♄.	Bot. reg. v. 11. t. 941.
8 pudibúnda. DC.	(*ro.*)	fringe-stipuled.	———	1822.	——	S. ♄.	
9 polydáctyla. DC.	(*ro.*)	many-fingered.	Demerara.	1824.	——	S. ♄.	Kth. mim. p. 14. t. 5.
10 hamàta. DC.	(*ro.*)	hooked.	E. Indies.	1820.	4. 9.	S. ♄.	
11 intermèdia. DC.	(*ro.*)	intermediate.	Caracas.	1825.	——	S. ♄.	Kth. mim. p. 16. t. 6.
12 víscida. DC.	(*ro.*)	viscid	Brazil.	——	——	S. ♄.	
13 canéscens. DC.	(*wh.*)	canescent.	Guinea.	1826.	——	S. ♄.	
14 híspida. DC.	(*wh.*)	hispid.	Caracas.	1824.	6. 9.	S. ♄.	
15 ciliàta. DC.	(*wh.*)	ciliated.	Brazil.	——	——	S. ♄.	
16 polyacántha. DC	(*wh.*)	many-spined.	Africa.	1828.	——	S. ♄.	Bruce itin. 5. t. 7.
17 asperàta. DC.	(*wh.*)	various spined.	W. Indies.	1822.	6. 7.	S. ♄.	DC. leg. mem. t. 63.
18 pìgra. L.	(*wh.*)	straight-spined.	Vera Cruz.	1733.	——	S. ♄.	Mill. ic. 2. t. 182. f. 3.
19 Sicària. DC.	(*wh.*)	subulate-spined.	Brazil.	1828.	——	S. ♄.	
20 rubicaúlis. DC.	(*st.*)	Bramble-stalk'd.	E. Indies.	1799.	6. 7.	S. ♄.	Roxb. corom. 2. t. 200.
octándra. R.C. *spinosíliqua.* Rottler. *Rottleri.* S.S.							
21 Ceratònia. DC.	(*wh.*)	round-leaved.	St. Domingo.	1800.	....	S. ♄.	Plum. ed. Burm. 17. t. 8.
Acàcia Ceratònia. W.							
22 cásta. DC.	(*wh.*)	chaste.	E. Indies.	1741.	6. 8.	S. ♄.	Comm. hort. 1. t. 28.
23 angulàta. S.S.	(*wh.*)	angular-stalked.	Brazil.	1826.	——	S. ♄.	
24 latispinòsa. S.S.	(*ye.*)	broad-spined.	Madagascar.	1822.	——	S. ♄.	
25 Barclayàna.	....	Mr. Barclay's.	———	——	....	S. ♄.	
26 brasiliènsis. S.S.	(*wh.*)	Brazilian.	Brazil.	1824.	6. 8.	S. ♄.	
27 abstérgens. S.S.	(*ro.*)	spotted-stalked.	E. Indies.	1818.	——	S. ♄.	
28 ferrugìnea. S.S.	(*ro.*)	ferruginous.	———	——	——	S. ♄.	

GAGNEB'INA. DC. GAGNEB'INA. Decandria Monogynia.
1 tamariscìna. DC. (*st.*) Tamarisk-like. Mauritius. 1824. .... S. ♄. DC. leg. mem. t. 64. f. B.
Acàcia tamariscìna. W. *Mimòsa tamariscìna.* Lam. Pluk. t. 329. f. 3.
2 axillàris. DC. (*st.*) axillary-spiked. Mauritius. 1822. .... S. ♄. DC. leg. mem. t. 64. f. A.
ERYTHROPHL'EUM. B. RED-WATER-TREE. Decandria Monogynia.
leonénse. G.D. .... Sierra Leone. Sierra Leone. 1822. .... S. ♄.
PA'RKIA. B. PA'RKIA. Monadelphia Decandria.
africàna. B. (*cr.*) African. Africa. 1822. 6. 8. S. ♄. Beauv. fl. ow. 2. t. 90.
I'nga biglobòsa. DC. *Mimòsa biglobòsa.* Jacq. amer. t. 179. f. 87.
I'NGA. DC. I'NGA. Polygamia Monœcia.
1 quassiæfòlia. DC. (*wh.*) Quassia-leaved. Brazil. 1820. .... S. ♄.
2 pilosiúscula. DC. (*wh.*) glossy-leaved. Cayenne. 1827. .... S. ♄. Vahl. ecl. 3. t. 24.
Mimòsa lùcida. Vahl.
3 setífera. DC. (*wh.*) setiferous. Guiana. 1824. .... S. ♄.
4 álba. DC. (*wh.*) white. Cayenne. 1804. 7. 8. S. ♄.
5 Feuillèi. DC. (*wh.*) Feuillee's. Peru. 1824. —— S. ♄. Feuil. obs. 3. p. 2. t. 19.
6 fastuòsa. DC. (*re.*) villous-flower'd. Caracas. 1820. .... S. ♄. Jacq. fragm. t. 10.
Mimòsa fastuòsa. Jacq.
7 sapindoídes. DC. (*wh.*) Soap-berry-like. ———— 1826. .... S. ♄.
8 spùria. DC. (*wh.*) spurious. Cumana. 1823. 7. 8. S. ♄. Kth. mim. 39. t. 12.
9 vèra. DC. (*wh.*) sweet-fruited. S. America. 1739. —— S. ♄. Sloan. his. t. 183. f. 1.
Mimòsa I'nga. L.
10 rhoifòlia. DC. (*wh.*) Rhus-leaved. Brazil. 1815. —— S. ♄.
11 Bourgòni. DC. (*st.*) Beech-leaved. W. Indies. 1752. —— S. ♄. Aubl. guian. 2. t. 358.
marginàta. W. *Mimòsa fagifòlia.* L. *non* Jacq.
12 marginàta. DC. .... margined. Caracas. —— —— S. ♄.
13 laurìna. DC. (*wh.*) Laurel-leaved. ———— 1768. 3. 5. S. ♄.
14 punctàta. DC. (*wh.*) white-spotted. ———— 1818. —— S. ♄. Jacq. amer. t. 164.
Mimòsa fagifòlia. Jacq. *non* L.
15 spléndens. DC. (*wh.*) shining-leaved. ———— 1825. .... S. ♄.
16 nodòsa. DC. (*wh.*) knobbed. Ceylon. 1690. 4. 6. S. ♄. Pluk. alm. t. 211. f. 5.
17 microphy'lla. DC (*wh.*) small-leaved. Cumana. 1815. .... S. ♄.
18 dúlcis. DC. (*wh.*) Sappan fruit. E. Indies. 1800. 5. 7. S. ♄. Roxb. corom. 1. t. 99.
19 U'nguis-càti. DC. (*wh.*) four-leaved. W. Indies. 1690. 4. 6. S. ♄. Jacq. schœn. 2. t. 34.
20 mellífera. DC. (*wh.*) honey-bearing. Arabia. 1826. .... S. ♄.
21 ligustrìna. DC. (*wh.*) Privet-leaved. Caracas. —— .... S. ♄. Jacq. frag. t. 32. f. 5.
22 fœtida. DC. (*wh.*) strong-smelling. W. Indies. 1818. .... S. ♄. Jacq. schœn. 3. t. 390.
23 tergémina. DC. (*pu.*) trifoliolate. Martinica. —— 4. 6. S. ♄. Jacq. amer. t. 177. f. 81.
24 emarginàta. DC. (*ro.*) emarginate. Mexico. 1825. .... S. ♄. Kth. mim. 54. t. 17.
25 latifòlia. DC. (*ro.*) broad-leaved. ———— 1768. 3. 5. S. ♄. Plum. ed. Burm. t. 9.
26 circinàlis. DC. (*sc.*) spiral-podded. W. Indies. 1726. 4. 6. S. ♄. Catesb. car. 2. t. 97.
27 purpùrea. DC. (*pu.*) Soldier-wood. ———— 1733. 3. 5. S. ♄. Bot. reg. t. 129.
Mimòsa purpùrea. Andr. reposit. t. 372.
28 xylocárpa. DC. (*wh.*) woody-podded. E. Indies. 1822. 4. 7. S. ♄. Roxb. corom. 1. t. 100.
29 bigémina. DC. (*wh.*) two-paired. ———— 1823. —— S. ♄. Rheed mal. 6. t. 12.
30 Jirínga. DC. (*wh.*) twisted-podded. ———— 1828. .... S. ♄.
31 comòsa. DC. (*wh.*) tufted. Jamaica. 1820. 4. 6. S. ♄.
32 cyclocárpa. DC. (*wh.*) shell-podded. Caracas. 1826. .... S. ♄. Jacq. fragm. t. 34. f. 1.
33 Samán. DC. .... linear-podded. Jamaica. —— .... S. ♄. ———— t. 9.
34 corcondiàna. DC. (*wh.*) Coromandel. E. Indies. 1818. .... S. ♄.
35 pulchérrima. Cerv. (*pu.*) handsome. Mexico. 1822. .... S. ♄.
ANNESLE'IA. P.L. ANNESLE'IA. Monadelphia Polyandria.
1 grandiflòra. (*cr.*) large-flowered. Mexico. 1769. 6. 9. S. ♄. Andr. reposit. t. 592.
Mimòsa grandiflòra. A.R. *Acàcia grandiflòra.* W. *I'nga anómala.* Kth. mim. t. 22.
2 Houstòni. (*cr.*) Houstoun's. Vera Cruz. 1729. 9. 11. S. ♄. Bot. reg. t. 98.
falcifòlia. Sal. par. lond. 64. *Acàcia Houstòni.* W. *Inga? Houstòni.* DC.
SCHRA'NKIA. DC. SCHRA'NKIA. Polygamia Monœcia.
1 aculeàta. DC. (*ro.*) prickly. Vera Cruz. 1733. 7. 8. S. ♃. Mill. ic. 1. t. 182. f. 1.
2 leptocárpa. DC. (*ro.*) slender-podd'd. St. Domingo. 1827. —— S. ♃.
3 uncinàta. DC. (*ro.*) hooked. N. America. 1789. —— H. ♃. Vent. choix. t. 28.
DARLINGT'ONIA. DC. DARLINGT'ONIA. Pentandria Monogynia.
1 brachylòba. DC. (*wh.*) short-podded. N. America. 1803. 9. 10. H. ♃. DC. leg. mem. t. 66.
2 glandulòsa. DC. (*wh.*) glandular. ———— 1806. —— F. ♄. Vent. choix. t. 27.
Mimòsa glandulòsa. M.
DESMA'NTHUS. DC. DESMA'NTHUS. Polygamia Monœcia.
1 lacústris. DC. (*st.*) marsh. S. America. 1823. 6. 8. S. *w.* ⊙. H. et B. pl. æq. 1. t. 16.
2 nàtans. DC. (*ye.*) floating. China. 1800. 7. 9. S. *w.* ⊙. Andr. reposit. t. 629.
3 polyphy'llus. DC. (*wh.*) many-leaved. .... 1826. —— S. *w.* ⊙.
4 triquètrus. DC. (*wh.*) triquetrous. E. Indies. 1823. —— S. *w.* ⊙.

5 plènus. DC. (*ye.*) double-yellow. Vera Cruz. 1733. 7. 9. S.*w.*⊙. Mill. ic. t. 182. f. 2.
6 depréssus. DC. (*wh.*) depressed. W. Indies. 1828. .... S. ♄. Kth. mim. p. 115. t. 35.
7 diffùsus. DC. (*wh.*) prostrate. ———— 1731. 7. 9. S. ♄. Pluk. alm. t. 307. f. 3.
8 virgàtus. DC. (*wh.*) twiggy. E. Indies. 1774. —— S. ♄. Jacq. vind. 1. t. 80.
9 stríctus. DC. (*wh.*) upright. Caribees. 1827. —— S. ♄.
10 punctàtus. DC. (*wh.*) spotted-stalked. Jamaica. 1636. —— S.♂. Comm. hort. 1. t. 31.
11 divérgens. DC. (*ye.*) prickly. Abyssinia. 1816. 6. 7. S. ♄. Bruce itin. 5. t. 6.
12 callistáchys. DC. (*ro.ye.*) pretty-spiked. .... 1820. —— S. ♄.
13 cinèreus. DC. (*ro.ye.*) ash-coloured. E. Indies. 1739. —— S. ♄. Roxb. corom. 2. t. 174.
Mimòsa cinèrea. R.C.
14 nùtans. DC. (*ro.ye.*) nodding. Senegal. 1826. —— S. ♄.

ADENANTH'ERA. DC. BASTARD FLOWER-FENCE. Decandria Monogynia.
1 pavonìna. DC. (*st.*) pale-pellow. E. Indies. 1759. 5. 8. S. ♄. Rumph. amb. 3. t. 109.
2 falcàta. DC. (*ye.*) woolly-leaved. ———— 1812. —— S. ♄. ———— 3. t. 111.

PR'OSOPIS. DC. PR'OSOPIS. Polygamia Monœcia.
1 spicígera. DC. (*ye.*) eatable-podded. E. Indies. 1812. 6. 7. S. ♄. Roxb. corom. 1. t. 63.
2 Adenanthèra. DC. (*ye.*) prickly. Coromandel. 1821. .... S. ♄.
Adenanthèra aculeàta. H.B.
3 hórrida. DC. (*wh.*) long-spined. S. America. 1826. .... S. ♄. Kth. mim. p. 106. t. 33.
4 dúlcis. DC. (*wh.*) sweet-podded. New Spain. 1824. .... S. ♄. ———— p. 110. t. 34.
5 siliquástrum. DC. (*wh.*) falcate-podded. Chile. 1827. .... S. ♄.
6 cumanénsis. DC. (*wh.*) Cumana. Cumana. 1824. .... S. ♄.
7 domingénsis. DC. (*wh.*) St. Domingo. St. Domingo. 1826. .... S. ♄.
8 juliflòra. DC. (*ye.*) long-spiked. Jamaica. 1793. 5. 6. S. ♄.
Acàcia juliflòra. W. *Mimòsa juliflòra.* Swartz.

LAGONY'CHIUM. DC. LAGONY'CHIUM. Decandria Monogynia.
Stephaniànum. DC. Stephan's. Levant. 1821. .... H. ♄. Breyn. cent. 1. t. 56. f. 4.

AC'ACIA. DC. AC'ACIA. Polygamia Monœcia.
1 alàta. DC. (*ye.*) wing-stalked. N. Holland. 1803. 4. 7. G. ♄. Bot. reg. 396.
2 dolabrifórmis. DC. (*ye.*) hatchet-leaved. ———— 1818. 4. 6. G. ♄.
3 decípiens. DC. (*ye.*) paradoxical. ———— 1803. 3. 6. G. ♄. Bot. mag. t. 1745.
4 biflòra. DC. (*ye.*) two-flowered. ———— —— —— G. ♄. Wendl. dis. n. 3. t. 2.
5 hastulàta. DC. (*ye.*) hastulate. ———— 1824. 4. 6. G. ♄.
6 nervòsa. DC. (*ye.*) nervose. ———— —— —— G. ♄.
7 paradóxa. DC. (*ye.*) wave-leaved. ———— 1818. 3. 6. G. ♄. Wendl. dis. n. 4. t. 3.
undulàta. Bot. reg. t. 843.
8 ornithóphora. S.F.A. (*ye.*) Bird-leaved. ———— 1825. 4. 6. G. ♄. Swt. fl. aust. t. 24.
9 armàta. DC. (*ye.*) armed. ———— 1803. —— G. ♄. Bot. mag. t. 1653.
10 hy'brida. (*ye.*) Whitley's mule. Hybrid. 1822. 3. 6. G. ♄. Lodd. bot. cab. 1342.
11 genistifòlia. DC. (*ye.*) Genista-leaved. N. Holland. 1820. —— G. ♄.
12 juniperìna. DC. (*st.*) Juniper-leaved. N. S. W. 1790. —— G. ♄. Vent. malm. t. 64.
ulicifòlia. Wendl. coll. 2. t. 6. *Mimòsa ulicìna.* Sal. prodr.
13 Bròwnii. DC. (*st.*) Brown's. N. S. W. 1796. —— G. ♄.
acicularis. H.K. *non* W.
14 echínula. DC. (*st.*) pungent. ———— 1824. —— G. ♄.
15 pugionifórmis. DC. (*st.*) needle-leaved. ———— 1803. —— G. ♄. Wendl. dis. 26. t. 9.
16 asparagoídes. F.T. (*st.*) Asparagus-like. ———— 1824. —— G. ♄.
17 Cunninghámii. (*ye.*) Cunningham's. ———— 1825. —— G. ♄.
taxifòlia. F.T. *non* W.
18 diffùsa. DC. (*ye.*) spreading. ———— 1822. 3. 6. G. ♄. Bot. reg. t. 634.
prostràta. Lodd. bot. cab. 631.
19 sulcàta. DC. (*ye.*) furrowed-leaved. N. Holland. 1803. 4. 8. G. ♄. Bot. reg. v. 11. t. 928.
20 canaliculàta. (*ye.*) glaucous-furrow'd. ———— 1823. 5. 8. I. ♄.
21 salígna. DC. (*ye.*) Willow-leaved. ———— 1820. 3. 6. G. ♄. Labil. n. hol. 2. t. 235.
22 emarginàta. DC. (*st.*) emarginate. ———— 1816. —— G. ♄.
23 strícta. DC. (*ye.*) upright. N. S. W. 1790. 2. 5. G. ♄. Bot. mag. t. 1121.
24 lepròsa. DC. (*ye.*) leprous. ———— 1824. 3. 6. G. ♄.
25 dodoneæfòlia. DC. (*ye.*) clammy. ———— 1818. 4. 6. G. ♄. Wendl. dis. n. 19. t. 7.
viscòsa. Wendl.
26 verniciflùa. F.T. (*ye.*) varnished. ———— 1819. 3. 6. G. ♄.
27 lanígera. F.T. (*ye.*) woolly-stem'd. ———— 1824. —— G. ♄. Bot. mag. t. 2922.
virgàta. Lodd. bot. cab. 1246.
28 multinérvia. DC. (*ye.*) many-nerved. N. Holland. —— —— G. ♄.
29 eglandulòsa. DC. (*ye.*) glandless. ———— —— —— G. ♄.
30 hispídula. DC. (*ye.*) little harsh. N. S. W. 1794. 4. 5. G. ♄. Sm. nov. holl. t. 16.
31 cochleàris. DC. (*ye.*) spoon-leaved. N. Holland. 1820. 3. 6. G. ♄. Labill. n. holl. 2. t. 234.
32 coriàcea. DC. (*ye.*) leathery. ———— 1824. —— G. ♄.
33 ancéps. DC. (*ye.*) flat-edged. ———— —— —— G. ♄.
34 laurifòlia. DC. (*ye.*) Laurel-leaved. S. Sea Isl. 1775. —— S. ♄.

35	elongàta. DC.	(*ye.*)	elongated.	N. S. W.	1824.	3. 6.	G. ♄.	
36	trinervàta. DC.	(*ye.*)	three-nerved.	———	1823.	4. 6.	G. ♄.	
37	calamifòlia. DC.	(*ye.*)	quill-leaved.	———	1822.	1. 12.	G. ħ.	Bot. reg. t. 839.
38	quadrilateràlis. DC.	(*ye.*)	tetragonal-leav'd	———	——	4. 6.	G. ħ.	
39	falcàta. DC.	(*st.*)	sickle-leaved.	———	1790.	5. 6.	G. ħ.	Lodd. bot. cab. 1115.
40	falcifórmis. DC.	(*st.*)	sickle-shaped.	———	1824.	4. 6.	G. ♄.	
41	pennirérvis. DC.	(*wh.*)	feather-nerved.	———	1822.	——	G. ħ.	Bot. mag. t. 2754.
	impréssa. Bot. reg. v. 13. t. 1115! Lodd. bot. cab. 1319.							
42	platyphy'lla.	(*st.*)	broad-leaved.	N. Holland.	——	——	I. ♄.	
43	melanòxylon. DC.	(*wh.*)	black-wooded.	V. Diem. Isl.	1808.	——	G. ♄.	Bot. mag. t. 1659.
44	heterophy'lla. DC.	(*ye.*)	various-leaved.	Bourbon.	1820.	....	S. ħ.	
45	am'œna. DC.	(*st.*)	handsome.	N. S. W.	1824.	4. 6.	G. ♄.	Wendl. dis. 16. t. 4.
46	myrtifòlia. DC.	(*st.*)	Myrtle-leaved.	———	1789.	2. 5.	G. ♄.	Swt. flor. aust. t. 49.
47	vestìta. DC.	(*ye.*)	clothed.	N. Holland.	1820.	3. 6.	G. ♄.	Bot. reg. t. 698.
48	marginàta. DC.	(*ye.*)	thick margined.	N. S. W.	1803.	4. 6.	G. ♄.	Wendl. dis. 19. t. 5. A.
49	pyrifòlia. DC.	(*ye.*)	Pear-leaved.	———	1824.	——	G. ♄.	
50	binervàta. DC.	(*ye.*)	two-nerved.	———	——	——	G. ħ.	
51	bivenòsa. DC.	(*ye.*)	two-veined.	———	——	——	G. ħ.	
52	buxifòlia. F.T.	(*ye.*)	Box-leaved.	———	——	3. 5.	G. ħ.	
53	lunàta. DC.	(*ye.*)	lunate-leaved.	V. Diem. Isl.	1816.	3. 6.	G. ♄.	Swt. flor. aust. t. 42.
	β *brevifòlia*. B.C.	(*ye.*)	*short-leaved*.	———	——	——	G. ħ.	Lodd. bot. cab. t. 1235.
54	obtusàta. DC.	(*ye.*)	obtuse.	N. S. W.	1824.	4. 6.	G. ħ.	
55	crassiúscula. DC.	(*st.*)	thick-leaved.	———	1803.	2. 6.	G. ♄.	Wendl. dis. 20. t. 8.
56	suavèolens. DC.	(*st.*)	sweet-scented.	———	1790.	——	G. ħ.	Lodd. bot. cab. t. 730.
	β *platycárpa*. DC.	(*st.*)	*flat-podded*.	———	——	——	G. ♄.	
57	angustifòlia. DC.	(*st.*)	narrow-leaved.	———	——	——	G. ħ.	Lodd. bot. cab. t. 768.
	Mimòsa angustifòlia. Jacq. schœnb. 3. t. 391.							
58	linifòlia. DC.	(*wh.*)	Flax-leaved	———	——	——	G. ♄.	Bot. mag. t. 2168.
59	abietìna. DC.	(*wh.*)	Fir-leaved.	———	——	3. 6.	G. ♄.	
60	subulàta. DC.	(*ye.*)	awl-leaved.	N. Holland.	1824.	4. 6.	G. ħ.	Bonpl. nav. t. 45.
61	spinulosa.	(*ye.*)	small-spined.	———	1822.	3. 6.	I. ħ.	
62	cordifòlia. ψ.	(*ye.*)	heart-leaved.	———	1824.	4. 6.	I. ♄.	
63	dillwyniæfòlia. ψ.	(*st.*)	Dillwynia-leaved.	———	——	——	G. ħ.	
64	thegonocárpa. ψ.	(*ye.*)	angular-podded.	———	——	——	G. ♄.	
65	rìgens. ψ.	(*ye.*)	stiff.	———	——	——	G. ♄.	
66	Cyclòpis. ψ.	(*ye.*)	Cyclopia-like.	———	——	——	G. ħ.	
67	muconulàta. ψ.	(*ye.*)	sharp-pointed.	———	——	——	G. ħ.	
68	oleifòlia. ψ.	(*ye.*)	Olive-leaved.	———	——	——	G. ħ.	
69	péndula. ψ.	(*wh.*)	pendulous.	———	——	——	G. ħ.	
70	leucophy'lla. C.C.	(*ye.*)	white-leaved.	———	1820.	3. 6.	I. ħ.	
71	taxifòlia. DC.	(*ye.*)	Yew-leaved.	China.	——	——	G. ħ.	
72	Oxycèdrus. DC.	(*ye.*)	sharp-leaved.	N. S. W.	1824.	4. 6.	G. ħ.	Swt. flor. aust. t. 6.
	taxifòlia. Lodd. bot. cab. t. 1225. *non* W.							
73	verticillàta. DC.	(*ye.*)	whorl-leaved.	V. Diem. Isl.	1780.	3. 5.	G. ħ.	Bot. mag. t. 110.
	α *glàbra*. DC.	(*ye.*)	*smooth-leaved*.	———	——	——	G. ħ.	
	β *angústa*. DC.	(*ye.*)	*narrow-leaved*.	———	——	——	G. ħ.	Vent. malm. t. 63.
	γ *latifòlia*. DC.	(*ye.*)	*broad-leaved*.	———	——	——	G. ħ.	Wendl. coll. 1. t. 30.
74	lineàris. B.M.	(*wh.*)	linear-leaved.	N. S. W.	1812.	8. 12.	G. ħ.	Bot. mag. t. 2156.
75	longíssima. W.A.	(*ye.*)	long-linear-ld.	———	——	——	G. ħ.	Wendl. dis. 31. t. 11.
76	uncinàta. C.C.	(*ye.*)	hooked-leaved.	N. Holland.	1823.	9. 1.	G. ħ.	
77	mucronàta. DC.	(*wh.*)	mucronate.	N. S. W.	1812.	——	G. ħ.	Bot. mag. t. 2747.
78	floribúnda. DC.	(*wh.*)	many-flowered.	———	1796.	5. 6.	G. ♄.	Vent. choix. t. 13.
79	doratòxylon. F.T.	(*ye.*)	spear-wood.	———	1824.	3. 6.	G. ♄.	
80	obtusifòlia. F.T.	(*ye.*)	blunt-leaved.	———	——	4. 6.	G. ♄.	
81	homomálla. DC.	(*ye.*)	downy-leaved.	———	——	——	G. ♄.	Wendl. dis. 34. t. 13.
82	longifòlia. DC.	(*ye.*)	long-leaved.	———	1792.	3. 6.	G. ♄.	Bot. mag. t. 1827.
	β *latifòlia*.	(*ye.*)	*broader-leaved*.	———	——	——	G. ♄.	——— t. 2166.
83	intertéxta. DC.	(*ye.*)	netted-veined.	———	1824.	4. 6.	G. ♄.	
84	glaucéscens. DC.	(*ye.*)	glaucescent.	———	1790.	2. 6.	G. ħ.	Willd. hort. ber. t. 101.
85	leucolòbia.	(*ye.*)	white-podded.	———	1824.	3. 6.	G. ♄.	
	dealbàta. F.T. *non* Link.							
86	cineráscens. DC.	(*ye.*)	greyish.	———	——	——	G. ħ.	
87	subcœrùlea. B.R.	(*ye.*)	bluish-leaved.	N. Holland.	——	4. 6.	I. ♄.	Bot. reg. t. 1075.
88	rùbida. F.T.	(*st.*)	red-margined.	N. S. W.	1824.	——	G. ħ.	
89	Sophòræ. DC.	(*ye.*)	Sophora-podd'd.	V. Diem. Isl.	1805.	——	G. ♄.	Labil. n. holl. 2. t. 237.
90	heteromálla. C.C.	(*ye.*)	broad woolly-ld.	N. Holland.	1822.	——	I. ♄.	
91	coronillæfòlia. DC.	(*ye.*)	Coronilla-leaved.	Mogador.	1826.	4. 5.	G. ħ.	
92	strumbulífera. DC.	(*ye.*)	twisted-podded.	Peru.	1823.	4. 7.	G. ħ.	
93	reticulàta. DC.	(*ye.*)	netted.	C. B. S.	1816.	4. 6.	G. ħ.	Pluk. alm. t. 123. f. 2.

94	pulchélla. DC.	(*ye.*)	zigzag spiny.	N. Holland.	1803.	4. 7.	G. ♄.	Lodd. bot. cab. t. 212.
95	hispidíssima. DC.	(*ye.*)	hispid.	———	1824.	4. 6.	G. ♄.	
96	detìnens. DC.	(*ye.*)	detaining.	C. B. S.	1816.	6. 10.	I. ƀ.	
97	viridiràmis. DC.	(*ye.*)	green-branched.	———	——	4. 6.	I. ♄.	
98	cassioídes. DC.	(*ye.*)	Cassia-like.	S. America.	1822.	——	S. ƀ.	
99	díptera. DC.	(*ye.*)	two-winged.	———	——	....	S. ♄.	
100	lophanthoídes. DC.	(*st.*)	crested.	Jamaica.	1826.	5. 6.	S. ♄.	
101	lophántha. DC.	(*st.*)	two-spiked.	N. Holland.	1803.	11. 4.	G. ƀ.	Bot. reg. t. 361.
	Mimòsa élegans. Andr. reposit. 563.							
102	guianénsis. DC.	(*wh.*)	Guiana.	Guiana.	1803.	....	S. ƀ.	Aubl. guian. 2. t. 357.
103	elephantìna. B.T.	(*ye.*)	Elephant.	C. B. S.	1816.	....	I. ƀ.	
104	pulchérrima. DC.	(*re.*)	beautiful.	Brazil.	1826.	....	S. ♄.	
105	aspidioídes. DC.	(*wh.*)	Aspidium-leav'd.	Guiana.	1827.	....	S. ♄.	
106	dealbàta. DC.	(*ye.*)	white-leaved.	V. Diem. Isl.	1818.	3. 6.	G. ƀ.	Bot. reg. t. 371.
	decúrrens β móllis. B.R.							
107	Wallichiàna. DC.	(*ye.*)	Wallich's.	E. Indies.	1827.	....	S. ƀ.	
108	adenanthèra. H.B.	...	perennial.	———	1825.	....	S. ♃.	
109	ferrugínea. DC.	(*wh.*)	ferruginous.	———	——	....	S. ♄.	
110	Súndra. DC.	(*ye.*)	hook-spined.	———	1789.	5. 6.	S. ♄.	Roxb. corom. 3. t. 225.
111	Sùma. H.B.	(*ye.*)	Shumee-tree.	———	1822.	——	S. ƀ.	
112	Catèchu. DC.	(*ye.*)	medicinal.	———	1790.	——	S. ƀ.	Roxb. corom. 2. t. 175.
113	fasciculàta. DC.	(*wh.*)	fascicled.	Mexico.	1828.	....	S. ƀ.	Kth. mim. 75. t. 23.
114	cáffra. DC.	(*ye.*)	Hottentot.	C. B. S.	1816.	5. 6.	I. ♄.	
115	Senegal. DC.	(*wh.*)	Senegal.	Africa.	1823.	....	S. ♄.	
116	cornígera. DC.	(*ye.*)	Cuckold-tree.	S. America.	1692.	5. 6.	S. ♄.	
	α americàna. DC.	(*ye.*)	*brown-spined.*	———	——	——	S. ♄.	Comm. hort. 1. t. 107.
	β índica. DC.	(*ye.*)	*yellow-spined.*	E. Indies.	——	——	S. ƀ.	Seba thes. 1. t. 70. f. 13.
117	latrònum. DC.	(*wh.*)	white-spined.	———	1824.	——	S. ƀ.	
118	Séyal. DC.	(*ye.*)	Egyptian.	Egypt.	1828.	....	G. ♄.	Delil. fl. eg. t. 52. f. 2.
119	hórrida. DC.	(*ye.*)	common-cape.	C. B. S.	1823.	3. 5.	I. ♄.	Pluk. t. 121. f. 4.
	Mimòsa leucacántha. Jacq. *capénsis.* B.T.							
120	ebúrnea. DC.	(*ye.*)	ivory-thorned.	E. Indies.	1792.	4. 6.	S. ♄.	Roxb. corom. 2. t. 199.
121	tortuòsa. DC.	(*st.*)	twisted.	Jamaica.	1822.	....	S. ƀ.	
122	Burmanniàna. DC.	(*ye.*)	Burmann's.	E. Indies.	1824.	....	S. ♄.	
123	vèra. DC.	(*ye.*)	Egyptian-thorn.	Egypt.	1596.	7. 8.	I. ♄.	Vest egypt. t. 8.
	Mimòsa nilótica. L.							
124	arábica. DC.	(*ye.*)	Gum Arabic.	Arabia.	....	6. 8.	S. ƀ.	Roxb. corom. 2. t. 149.
125	hebeclàda. DC.	(*ye.*)	stoloniferous.	C. B. S.	1816.	4. 6.	I. ♄.	
	stolonífera. B.T.							
126	mauroceàna. DC.	(*wh.*)	Marocca.	Marocca.	....	——	G. ƀ.	
127	farnesiàna. DC.	(*ye.*)	Sponge-tree.	St. Domingo.	1656.	6. 8.	I. ƀ.	Duham. ar. ed. n. 2. t. 28.
128	álba. DC.	(*wh.*)	white.	E. Indies.	1828.	....	S. ♄.	
129	álbicans. DC.	(*wh.*)	whitish.	Mexico.	1827.	....	S. ♄.	Kth. mim. 87. t. 27.
130	leucophl'æa. DC.	(*st.*)	panicled.	E. Indies.	1812.	4. 6.	S. ♄.	Roxb. corom. 2. t. 150.
131	hæmatóxylon. DC.	(*ye.*)	hoary.	C. B. S.	1816.	....	S. ƀ.	
	atomiphy'lla. B.T.							
132	tomentòsa. DC.	(*st.*)	tomentose.	E. Indies.	——	5. 7.	S. ♄.	
133	índica. DC.	(*ye.*)	Indian.	———	1824.	....	S. ♄.	
134	flexuòsa. DC.	(*ye.*)	flexuose.	Cumana.	1822.	....	S. ♄.	
135	peruviàna. DC.	(*ye.*)	Peruvian.	Peru.	——	....	S. ♄.	
136	brachyacántha. DC.	...	short-spined.	S. America.	1816.	....	S. ƀ.	
137	ciliàta. DC.	....	fringed.	———	1823.	....	S. ƀ.	
138	aciculàris. DC.	....	needle-spined.	———	——	....	S. ♄.	
139	Giráffæ. DC.	(*ye.*)	Camelopard's.	C. B. S.	1816.	....	I. ♄.	
140	rhodacántha. DC.	(*ye.*)	red-spined.	....	——	....	S. ♄.	
141	heteracántha. DC.	(*ye.*)	various-spined.	C. B. S.	——	....	I. ♄.	
142	Bancroftiàna. DC.	(*st.*)	Bancroft's.	Jamaica.	1826.	....	S. ƀ.	
143	polycéphala. DC.	(*wh.*)	many-headed.	Mauritius.	——	....	S. ♄.	
144	subinérmis. DC.	(*ye.*)	scarcely spiny.	Jamaica.	——	....	S. ƀ.	
145	èdulis. W. en.	....	eatable.	S. America.	1822.	....	S. ƀ.	
146	acanthocárpa. DC.	(*bh.*)	prickly-podd'd.	New Spain.	1818.	4. 7.	F. ƀ.	
	Mimòsa aculeaticárpa. P.S.							
147	acantholòba. DC.	(*st.*)	prickly-fruited.	S. America.	1822.	....	S. ♄.	
148	concínna. DC.	(*bh.*)	neat.	E. Indies.	1794.	6. 8.	S. ƀ.	
149	tamarindifòlia. DC.	(*wh.*)	Tamarind-l[d].	W. Indies.	1774.	....	S. ƀ.	Jacq. schœnb. 3. t. 396.
150	I'ntsia. DC.	(*st.*)	angular-stalked.	E. Indies.	1778.	6. 8.	S. ♄.	Rheed. mal. 6. t. 4.
151	pennàta. DC.	(*ye.*)	fine-leaved.	———	1773.	——	S. ♄.	Burm. zeyl. 2. t. 1.
152	c'æsia. DC.	(*wh.*)	gray.	———	——	——	S. ♄.	Pluk. mant. 1. t. 330. f. 3.
153	centrophy'lla. DC.	(*wh.*)	prickly-leaved.	Jamaica.	1826.	——	S. ƀ.	

154	Courantiàna. DC.	(*wh.*)	Courant's.	....	1824.	6. 8.	S.	♄.	
155	arróphula. DC.	(*wh.*)	Nepaul.	Nepaul.	1823.	....	G.	♄.	
156	scándens. DC.	....	climbing.	Brazil.	1826.	....	S.	♄.	
157	sarmentòsa. DC.	....	trailing.	....	——	....	S.	♄.	
158	trichòdes. DC.	(*wh.*)	warted-stalked.	Caracas.	1824.	6. 8.	S.	♄.	Jacq. schœnb. 3. t. 394.
159	formòsa. DC.	(*wh.*)	handsome.	Mexico.	1826.	....	S.	♄.	Kth. mim. t. 32.
160	nìgricans. DC.	(*st.*)	unequal-winged.	N. Holland.	1803.	5. 7.	G.	♄.	Bot. mag. t. 2188.
161	strigòsa. DC.	(*ye.*)	ciliate-winged.	————	——	3. 6.	G.	♄.	
	ciliàta. H.K. *nec aliorum.*								
162	vàga. DC.	(*st.*)	Brazilian.	Brazil.	1823.	....	S.	♄.	Marcg. bras. p. 111. f. 1.
163	Lébbeck. DC.	(*pa.*)	Egyptian.	Egypt.	....	4. 6.	G.	♄.	Pluk. mant. 2. t. 331. f. 1.
164	pròcera. DC.	(*pa.*)	lofty.	Coromandel.	1824.	....	S.	♄.	Roxb. corom. 2. t. 121.
165	odoratíssima. DC.	(*pa.*)	fragrant.	————	1790.	4. 6.	S.	♄.	———— 2. t. 120.
166	speciòsa. DC.	(*wh.*)	splendid.	E. Indies.	1742.	8. 9.	S.	♄.	Jacq. ic. 1. t. 198.
167	glaùca. DC.	(*wh.*)	glaucous.	S. America.	1690.	6. 9.	S.	♄.	Trew ehret. t. 46.
168	leucocéphala. DC.	(*wh.*)	white-headed.	————	1818.	4. 6.	S.	♄.	
169	latisíliqua. DC.	(*wh.*)	broad-podded.	W. Indies.	1777.	3. 6.	S.	♄.	Plum. ed. Burm. t. 6.
170	portoricénsis. DC.	(*wh.*)	Porto Rico.	Porto Rico.	1820.	....	S.	♄.	Jacq. ic. 3. t. 633.
171	caracasàna. DC.	(*pu.*)	Caracas.	Caracas.	1823.	....	S.	♄.	——— 3. t. 632.
172	Lambertiàna. DC.	(*pu.*)	Mr. Lambert's.	Mexico.	1820.	7. 9.	G.	♄.	Bot. reg. t. 721.
173	tetragòna. DC.	(*wh.*)	square-stalked.	Caracas.	1822.	....	S.	♄.	
174	quadrangulàris. DC.	(*wh.*)	quadrangular	....	1816.	5. 7.	S.	♄.	Bot. mag. t. 2651.
175	díscolor. DC.	(*ye.*)	two-coloured.	N. S. W.	1788.	3. 6.	G.	♄.	———— t. 1750.
	Mimòsa discolor. Andr. reposit. t. 235. *Mimòsa botrycéphala.* Vent. cels. t. 1.								
176	angulàta. DC.	(*ye.*)	angular.	N. Holland.	1824.	4. 6.	G.	♄.	
177	pubéscens. DC.	(*ye.*)	hairy-stemmed.	N. S. W.	1790.	3. 6.	G.	♄.	Bot. mag. t. 1263.
178	filicìna. DC.	(*wh.*)	fern-leaved.	Mexico.	1825.	4. 6.	G.	♄.	Cavan. ic. 1. t. 78.
179	villòsa. DC.	(*wh.*)	villous.	Jamaica.	1826.	——	S.	♄.	
180	frondòsa. DC.	(*wh.*)	frondose.	E. Indies.	1816.	....	S.	♄.	
181	venústa. W. en.	(*ye.*)	charming.	S. America.	——	....	S.	♄.	
182	stipulàta. DC.	(*wh.*)	stipuled.	Bengal.	1820.	5. 6.	S.	♄.	
183	divaricàta. DC.	(*wh.*)	straddling.	E. Indies.	——	——	S.	♄.	Jacq. schœnb. t. 395.
184	umbellulífera. DC.	(*wh.*)	small-umbell'd.	Mexico.	1825.	——	G.	♄.	Kth. mim. t. 31.
185	arbòrea. DC.	(*bh.*)	tree.	Jamaica.	1768.	——	S.	♄.	Sloan. his. 2. t. 182. f. 1. 2.
186	amára. DC.	(*wh.*)	bitter.	Coromandel.	1823.	4. 6.	S.	♄.	Roxb. corom. 2. t. 122.
187	Nemù. DC.	(*bh.*)	Japan tree.	Japan.	——	....	G.	♄.	Banks ic. Kæmpf. t. 19.
188	Julibríssin. DC.	(*bh.*)	smooth tree.	Levant.	1745.	8. 9.	H.	♄.	Scop. insub. 1. t. 8.
189	polyphy'lla. DC.	(*wh.*)	many-leaved.	St. Martha.	1826.	6. 9.	S.	♄.	
190	decúrrens. DC.	(*ye.*)	decurrent.	N. S. W.	1790.	5. 7.	G.	♄.	Vent. malm. t. 61.
191	mollíssima. DC.	(*ye.*)	soft-leaved.	————	1818.	4. 6.	G.	♄.	Swt. flor. aust. t. 12.
192	esculénta. DC.	(*wh.*)	eatable.	New Spain.	1829.	....	S.	♄.	Flor. mexic. ic. ined.
193	peregrìna. DC.	(*wh.*)	white-flowered.	S. America.	1780.	7. 8.	S.	♄.	Kth. mim t. 30.
194	gràta. DC.	(*wh.*)	grateful.	Brazil.	1828.	....	S.	♄.	
195	microphy'lla. DC.	(*wh.*)	small-leaved.	Caracas.	1826.	....	S.	♄.	
196	chrysostàchys.	(*ye.*)	golden-spiked.	Mauritius.	1823.	....	S.	♄.	
197	lùcida. H.B.	....	glossy.	E. Indies.	1816.	....	S.	♄.	
198	Siríssa. H.B.	(*wh.*)	Shireesh.	————	——	....	S.	♄.	
199	elàta. H.B.	(*wh.*)	tall.	————	1820.	....	S.	♄.	
200	Kalkòra. H.B.	(*wh.*)	large.	————	——	....	S.	♄.	
201	Smithiàna. H.B.	....	Smith's.	Silhet.	1823.	....	S.	♄.	
202	semicordàta. H.B.	(*wh.*)	semicordate.	E. Indies.	——	....	S.	♄.	
203	fruticòsa. H.B.	(*wh.*)	frutescent.	China.	1820.	5. 6.	G.	♄.	
204	pedunculàta. H.B.	(*wh.*)	peduncled.	E. Indies.	1822.	....	S.	♄.	
205	triquètra. H.B.	(*wh.*)	triquetrous.	Coromandel.	1824.	....	S.	♄.	
206	robústa. B.T.	(*ye.*)	robust.	C. B. S.	1816.	....	I.	♄.	
207	litakunénsis. B.T.	(*ye.*)	Litakun.	————	——	....	I.	♄.	
208	guilandinæfòlia.	(*gr.*)	Guilandina-ld.	N. Holland.	1818.	7. 9.	I.	♄.	

Subordo IV. CÆSALPINEÆ.

Tribus V. *GEOFFREÆ.* DC. prodr. 2. p. 473.

	ARACHIS. DC.		EARTH-NUT.	Diadelphia Decandria. L					
1	hypog'æa. DC.	(*ye.*)	American.	S. America.	1712.	5. 6.	H.	⊙.	Trew ehret. t. 3. f. 3.
2	africàna. W.	(*ye.*)	African.	Africa.	1824.	——	H.	⊙.	

VOANDZE'IA. DC.		VOANDZE'IA. Diadelphia Decandria.					
subterrànea. DC.	(ye.)	subterraneous.	Madagascar.	1820.	6. 8.	S. ☉.	DC. leg. me. t. 20. f. 106.
Gly'cine subterrànea. Linn. dec. 37. t. 17.							
PERA'LTEA. DC.		PERA'LTEA. Diadelphia Decandria.					
lupinoídes. DC.	(pu.)	Lupine-like.	New Spain.	1827.	9. 11.	F. ♄.	H. et B. n. gen. 6. t. 589.
BRONGNIA'RTIA. DC.		BRONGNIA'RTIA. Diadelphia Decandria.					
podalyrioídes. DC.	(fl.)	Podalyria-like.	New Spain.	1827.	9. 11.	F. ♄.	H. et B. n. gen. 6. t. 588.
AND'IRA. DC.		AND'IRA. Diadelphia Decandria.					
1 racemòsa. DC.	(pu.)	racemed.	Brazil.	1824.	....	S. ♄.	Aubl. gui. sup. t. 373.
2 inérmis. DC.	(pu.)	bastard cabbage-tree.	Jamaica.	1778.	....	S. ♄.	Phil. trans. 1777. t. 10.
GEOFFRO'YA. DC.		GEOFFRO'YA. Diadelphia Decandria.					
1 spinòsa. DC.	(st.)	spiny.	S. America.	1823.	....	S. ♄.	Jacq. amer. t. 180. f. 62.
2 Bredemeyèri. DC.	(ye.)	striated.	Caracas.	1824.	....	S. ♄.	
Robínia striàta. W.							
3 violàcea. DC.	(vi.)	violet.	Guiana.	1827.	....	S. ♄.	Aubl. guian. t. 301.
BRO'WNEA. DC.		BRO'WNEA. Monadelphia Dec-Dodecandria.					
1 Ròsa. DC.	(ro.)	mountain Rose.	Trinidad.	1816.	....	S. ♄.	Ber. act. ang. 1771. t. 8-9.
speciòsa. Sieber.							
2 coccínea. DC.	(sc.)	scarlet-flower'd.	W. Indies.	1793.	7. 9.	S. ♄.	Jacq. amer. t. 121.
3 latifòlia. DC.	(ro.)	broad-leaved.	Caracas.	1826.	....	S. ♄.	Jacq. fragm. t. 17.
4 racemòsa. DC.	(ro.)	racemed.	———		....	S. ♄.	——— t. 18-19.
5 grandicéps. DC.	(ro.)	close-headed.	———	1828.	....	S. ♄.	——— t. 22-23.
DI'PTERIX. DC.		TONQUIN BEAN. Monadelphia Octo-Decandria.					
odoràta. DC.	(pu.)	sweet-scented.	Guiana.	1793.	....	S. ♄.	Aubl. guian. 3. t. 296.
Coumaroùna odoràta. Aubl. *Baryósma Tónga.* Gært. fruct. 2. t. 93.							

Tribus X. *CASSIEÆ.* DC. prodr. 2. p. 478.

MORI'NGA. DC.		HORSE-RADISH-TREE. Decandria Monogynia.					
1 pterigospérma. DC.	(st.)	wing-seeded.	E. Indies.	1759.	....	S. ♄.	Jacq. ic. 3. t. 461.
oleífera. Lam. *Zeylánica.* P.S. *Hyperanthèra Morínga.* Vahl. *Guilandìna Morínga.* L.							
2 polygòna. DC.	(st.)	many-sided.		1822.	....	S. ♄.	
GLEDI'TSCHIA. DC.		GLEDI'TSCHIA. Polygamia Diœcia. L.					
1 triacánthos. DC.	(gr.)	three-thorned.	N. America.	1700.	6. 8.	H. 5.	Wats. dend. brit. 2. t. 138.
β inérmis. DC.	(gr.)	*spineless.*	———	———	———	H. 5.	DC. leg. mem. t. 22. f. 109
2 brachycárpa. DC.	(gr.)	curved-spined.	———	....	———	H. 5.	
3 monospérma. DC.	(gr.)	single-seeded.	———	1723.	———	H. 5.	Mich. f. arb. 3. t. 10.
4 sinénsis. DC.	(gr.)	Chinese.	China.	1774.	———	H. 5.	DC. leg. mem. 1. t. 1.
hórrida. W.							
5 macracántha. DC.	(gr.)	large-spined.	———	....	———	H. 5.	
6 fèrox. DC.	(gr.)	flat-spined.	———	....	———	H. 5.	
orientàlis. Bosc.							
7 cáspica. DC.	(gr.)	Caspian.	Caucasus.	1822.	———	H. ♄.	
8 micracántha. L.C.	(gr.)	small-spined.	....	1824.	———	H. ♄.	
9 índica. DC.	(gr.)	Indian.	E. Indies.	1812.	....	S. ♄.	
10 l'ævis. L.C.	(gr.)	smooth.	....	1824.	....	H. ♄.	
11 latisíliqua. L.C.	(gr.)	broad-podded.	....	———	....	H. 5.	
GYMNOCLA'DUS. DC.		KENTUCKY COFFEE-TREE. Diœcia Decandria.					
canadénsis. DC.	(wh.)	Canadian.	N. America.	1748.	5. 6.	H. 5.	Mich. amer. 2. t. 51.
Guilandìna dioica. L.							
GUILAND'INA. DC.		NICKER-TREE. Decandria Monogynia. L.					
1 Bónduc. L.T.	(ye.)	Sumatra.	Sumatra.	1640.	....	S. ♄.	Rumph. amb. 5. t. 48.
2 Bonducélla. L.T.	(ye.)	common.	E. Indies.	———	....	S. ♄.	Schrank. h. mon. t. 68.
3 microphy'lla. DC.	....	small-leaved.	———	———	....	S. ♄.	Rumph. amb. 5. t. 49. f. 2?
COULT'ERIA. DC.		COULT'ERIA. Decandria Monogynia.					
1 tinctòria. DC.	(ye.)	dyer's.	Peru.	1818.	....	S. ♄.	H. et B. n. ge. am. 6. t. 569.
Poinciàna Tàra. R.P. *Cæsalpínia pectinàta.* Cavan.							
2 mexicàna. DC.	(ye.)	Mexican.	New Spain.	1826.	....	S. ♄.	Flor. mexic. ic. ined.
CÆSALPI'NIA. DC.		BRASILE'TTO. Decandria Monogynia. L.					
1 Nùga. DC.	(ye.)	acute-leaved.	Amboyna.	1801.	4. 6.	S. ♄.	Rumph. amb. 5. t. 50.
2 paniculàta. DC.	(ye.)	panicled.	E. Indies.	1812.	....	S. ♄. ◡.	Rheed. mal. 6. t. 19.
3 brasiliénsis. DC.	(ye.)	Brazilian.	Brazil.	1818.	4. 6.	S. ♄.	
4 Sappán. DC.	(ye.)	narrow-leaved.	E. Indies.	1773.	———	S. ♄.	Roxb. corom. 1. t. 16.

5 dig'yna. DC. (*ye.*) digynous. E. Indies. 1828. .... S. ♄. Will.n.ac.1803.v.4.t.3.
6 mimosoídes. DC. (*ye.*) Mimosa-leaved. ——— 1806. 5. 7. S. ♄. Rheed. mal. 6. t. 8.
7 bijùga. DC. (*ye.*) broad-leaved. Jamaica. 1770. 6. 9. S. ♄. Sloan. j. 2. t. 181. f. 2. 3.
8 bahaménsis. DC. (*wh.*) emarginate. Bahama. 1823. .... S. ♄. Catesb. car. 2. t. 51.
9 Crísta. DC. (*ye.*) oval-leaved. Jamaica. .... 4. 6. S. ♄. Plum. gen. 28. t. 9.
10 glandulòsa. DC. (*ye.*) glandular. St. Domingo. 1828. .... S. ♄.
11 Cacaláco. DC. (*ye.*) round-leaved. Mexico. 1826. .... S. ♄. H. et B. pl. æq. 2. t. 137.
12 coriària. DC. (*ye.*) leathery. W. Indies. —— .... S. ♄. Kth. mim. t. 45.
13 lebbekkoídes. DC. (*ye.*) Lebbek-podd'd. China. 1812. .... S. ♄.
14 punctàta. DC. (*ye.*) dotted-leaved. Brazil. 1818. .... S. ♄.
15 pròcera. S.S. (*ye.*) tall. Cuba. 1822. .... S. ♄.
16 cassioídes. DC. (*ye.*) Cassia-like. S. America. 1823. .... S. ♄.
17 mucronàta. DC. (*ye.*) mucronate. Brazil. —— .... S. ♄.
18 olæospérma. H.B. (*ye.*) oil-seeded. E. Indies. 1800. 5. 6. S. ♄ ◡.
19 cucullàta. H.B. (*ye.*) hooded. ——— 1822. .... S. ♄ ◡.
20 sepiària. H.B. (*ye.*) Mysore-thorn. ——— 1806. 6. 9. S. ♄ ◡.
21 resupinàta. H.B. (*ye.*) resupinate. ——— 1828. .... S. ♄ ◡.
22 enneaphy'lla. H.B. (*ye.*) nine-leaved. ——— 1812. .... S. ♄ ◡.
23 chinénsis. H.B. (*ye.*) Chinese. China. —— .... S. ♄ ◡.
24 tortuòsa. H.B. (*ye.*) twisting. E. Indies. 1822. .... S. ♄ ◡.
25 sumatràna. H.B. (*ye.*) Sumatra. ——— —— .... S. ♄ ◡.

POINCI'ANA. DC. FLOWER FENCE. Decandria Monogynia. L.
1 règia. B.M. (*sc.*) superb. Madagascar. 1827. .... S. ♄. Bot. mag. t. 2884.
2 pulchérrima. DC. (*or.*) Barbadoes. E. Indies. 1691. 6. 9. S. ♄. ——— t. 995.
3 elàta. DC. (*ye.*) fringe-petal'd. ——— 1778. —— S. ♄.
Cæsalpínia elàta. S.S.

HOFFMANSE'GGIA. DC. HOFFMANSE'GGIA. Decandria Monogynia.
1 falcària. DC. (*ye.*) sickle-podded. Chile. 1806. 7. 8. F. ♃. Cavan. ic. t. 392.
2 prostràta. DC. (*ye.*) prostrate. Lima. 1824. —— F. ♃.

MELANOSTI'CTA. DC. MELANOSTI'CTA. Decandria Monogynia.
Burchéllii. DC. (*ye.*) Burchell's. C. B. S. 1816. .... G. ♄. DC. leg. mem. t. 69.

POM'ARIA. DC. POM'ARIA. Decandria Monogynia.
glandulòsa. DC. (*ye.*) glandular. New Spain. 1826. 5. 6. G. ♄. Cavan. ic. 5. t. 402.

HÆMATO'XYLON. D. LOGWOOD. Decandria Monogynia. L.
campechiànum. DC. (*ye.*) common. S. America. 1724. .... S. ♄. Catesb. car. 2. t. 66.

PARKINS'ONIA. DC. PARKINS'ONIA. Decandria Monogynia. L.
aculeàta. DC. (*ye.*) prickly. W. Indies. 1739. 6. 9. S. ♄. Jacq. amer. t. 80.

CA'DIA. DC. CA'DIA. Decandria Monogynia.
vària. DC. (*bh.*) various-colour'd. Arabia. 1755. 7. 10. S. ♄. Picc. hort. pan. 9. ic.
purpùrea. W. *Spændóncea tamarindifòlia.* Desf.

CERAT'ONIA. DC. CAROB-TREE. Polygamia Diœcia.
Síliqua. DC. (*gr.*) St. John's bread. Levant. 1570. 9. 10. G. ♄. Andr. reposit. 567.

HARDWI'CKIA. DC. HARDWI'CKIA. Oct-Decandria Monogynia.
1 binàta. DC. (*ye.*) binate-leaved. E. Indies. 1824. .... S. ♄. Roxb. corom. t. 209.
2 pinnàta. DC. (*ye.*) pinnate-leaved. ——— 1826. .... S. ♄.

JON'ESIA. DC. JON'ESIA. Octandria Monogynia.
Asòca. DC. (*or.*) winged-leaved. E. Indies. 1796. 4. 6. S. ♄. Rheed. mal. 5. t. 59.
pinnàta. W. *Sarâca índica.* L.

TACHIG'ALIA. DC. TACHIG'ALIA. Decandria Monogynia.
paniculàta. DC. (*ye.*) panicled. Guiana. 1827. .... S. ♄. Aubl. guian. 1. t. 143. f 1.

MOLDENHAW'ERA. DC. MOLDENHAW'ERA. Decandria Monogynia.
floribúnda. DC. (*ye.*) many-flowered. Brazil. 1828. .... S. ♄. Neuw. reis. v. 2. ic.

TAMARI'NDUS. DC. TAMARIND-TREE. Monadelphia Triandria.
1 índica. DC. (*ye.*) East Indian. E. Indies. 1633. 6. 7. S. ♄. Rheed. mal. 1. t. 23.
2 occidentàlis. DC. (*ye.*) West Indian. W. Indies. —— —— S. ♄. Jacq. amer. t. 10.

CA'SSIA. DC. CA'SSIA. Decandria Monogynia. L.
1 brasiliàna. DC. (*ye.*) large. Brazil. 1822. .... S. ♄. Breyn. cent. t. 21.
grândis. L. *móllis.* Vahl. *Cathartocárpus grándis.* P.S.
2 ferrugínea. DC. (*ye.*) ferruginous. Brazil. 1828. .... S. ♄.
3 Roxbúrghii. DC. (*ye.*) Roxburgh's. Coromandel. 1823. 6. 7. S. ♄.
marginàtus. H.B. *non* W.
4 trinitàtis. DC. (*ye.*) Trinidad. Trinidad. 1824. .... S. ♄.
5 javánica. DC. (*fl.*) Java. Java. 1779. 6. 7. S. ♄. Rumph. amb. 2. t. 22.
6 Fístula. DC. (*ye.*) purging. E. Indies. 1731. —— S. ♄. Woodv. med. bot. t. 163.
7 rhombifòlia. H.B. (*ye.*) rhomb-leaved. ——— 1820. —— S. ♄.
8 nodòsa. H.B. (*ye.*) knotted. ——— 1824. —— S. ♄.
9 conspícua. G.D. (*ye.*) drumstick-tree. Sierra Leon. 1822. .... S. ♄.
10 bacillàris. DC. (*ye.*) smooth. Surinam. 1818. —— S. ♄. Houst. rel. t. 17.
Cathartocárpus Bacillus. Bot. Reg. t. 881.

11 speciòsa. DC.	(*ye.*) handsome.	Brazil.	1828.	6. 8.	S ♄.	
12 melanocárpa. DC.	(*ye.*) black-podded.	Jamaica.	1825.	——	S. ♄.	
13 corymbòsa. DC.	(*ye.*) corymbose.	Buenos Ayres.	1796.	7. 8.	S. ♄.	Bot. mag. t. 633.
14 floribúnda. DC.	(*ye.*) many-flowered.	New Spain.	1822.	5. 8.	S. ♄.	
15 lævigàta. DC.	(*ye.*) smooth.	——	1811.	——	S. ♄.	
16 péndula. DC.	(*ye.*) pendulous.	S. America.	1820.	....	S. ♄.	
17 toròsa. DC.	(*ye.*) twisted.	China.	1807.	5. 7.	G. ♄.	Jacq. ic. 1. t. 73.
chinénsis. Jacq. *non* Lam.						
18 chinénsis. DC.	(*ye.*) Chinese.	——	——	——	G. ♄.	
19 gigántea. DC.	(*ye.*) gigantic.	Jamaica.	1825.	....	S. ♄.	
20 coromandeliàna. DC.	(*or.*) Coromandel.	Coromandel.	1822.	5. 7.	S. ♄.	Jacq. fragm. t. 100.
21 Sòphera. DC.	(*ye.*) round-podded.	E. Indies.	1653.	7. 9.	S. ♄.	Burm. zeyl. t. 98.
22 alàta. DC.	(*ye.*) winged-podded.	S. America.	1731.	....	S. ♄.	Jacq. obs. 2. t. 45. f. 2.
23 Rumphiàna. DC.	(*ye.*) Rumphius's.	E. Indies.	1822.	....	S. ♄.	Rumph. amb. 7. t. 18.
24 bracteàta. DC.	(*ye.*) bracteate.	Surinam.	——	....	S. ♄.	
25 obovàta. DC.	(*ye.*) obovate-leaved.	Egypt.	1640.	7. 9.	G. ⊙.	Jacq. f. ecl. 1. t. 87.
Sénna. Lam. ill. t. 332. f. 2.						
26 lanceolàta. DC.	(*ye.*) spear-leaved.	——	1759.	6. 8.	G. ⊙.	Tabern. ic. 507.
orientàlis. P.S. *acutifòlia.* Delil. ill. eg. 75. t. 27. f. 1.						
27 ligustrinoídes. DC.	(*ye.*) Privet-like.	Arabia.	1829.	——	G. ♄.	
28 Browniàna. DC.	(*ye.*) Brown's.	Mexico.	1828.	....	S. ♄.	Kth. mim. t. 41.
29 cuspidàta. W. en.	(*ye.*) sharp-pointed.	S. America.	1820.	5. 7.	S. ♄.	
30 serícea. DC.	(*ye.*) silky-leaved.	——	1731.	5. 8.	S. ⊙.	Jacq. ic. 3. t. 459.
sensitiva. Jacq.						
31 obtusifòlia. DC.	(*ye.*) blunt-leaved.	Jamaica.	1732.	7. 8.	S. ⊙.	Dill. elth. t. 62. f. 72.
32 hùmilis. DC.	(*ye.*) dwarf.	S. America.	....	——	S. ⊙.	Plum. ed. Burm. t. 76. f. 2
33 Tòra. DC.	(*st.*) oval-leaved.	E. Indies.	1693.	——	S. ⊙.	Dill. elth. t. 63. f. 73.
34 toroídes. H.B.	(*st.*) Mysore.	——	1812.	——	S. ⊙.	
35 ciliàris. DC.	(*ye.*) fringe-stipuled.	——	——	——	S. ♂.	
36 Tágera. DC.	(*ye.*) long-podded.	——	1803.	——	S. ♂.	Rheed. mal. 2. t. 53.
37 venústula. DC.	(*ye.*) pretty.	Cumana.	1825.	——	S. ♄.	
38 aversiflòra. B.M.	(*ye.*) contrary-flowr'd.	Brazil.	1823.	4. 12.	S. ♄.	Bot. mag. t. 2638.
39 vimínea. DC.	(*ye.*) twiggy.	W. Indies.	1786.	6. 7.	S. ♄.	Sloan. jam. 2. t. 180. f. 67.
40 chrysotrícha. DC.	(*ye.*) golden-podded.	Guiana.	1828.	——	S. ♄.	Coll. mon. p. 99. t. 13.
41 Apoucouìta. DC.	(*ye.*) acuminate.	——	1823.	——	S. ♄.	Aubl. gui. 1. t. 146.
acuminàta. W. *non* Mœnch.						
42 sennoídes. DC.	(*ye.*) Senna-leaved.	E. Indies.	1808.	7. 8.	S. ♄.	Jacq. ic. 1. t. 70.
43 bicapsulàris. DC.	(*ye.*) six-leaved.	W. Indies.	1730.	5. 6.	S. ♄.	Bot. mag. t. 633.
44 mollíssima. DC.	(*ye.*) soft-leaved.	S. America.	1816.	7. 8.	S. ♄.	
45 artemisioídes. DC.	(*ye.*) Wormwood-like.	N. S. W.	1823.	4. 7.	G. ♄.	
46 glutinòsa. DC.	(*ye.*) glutinous.	N. Holland.	——	——	G. ♄.	
47 oxyadèna. DC.	(*ye.*) acute-glanded.	Jamaica.	1820.	——	S. ♄.	
48 díspar. DC.	(*ye.*) various-leaved.	S. America.	1822.	5. 8.	S. ♄.	
49 acapulcénsis. DC.	(*ye.*) Acapulco.	Mexico.	1824.	——	S. ♄.	
50 sulfùrea. DC.	(*st.*) straw-coloured.	E. Indies.	1800.	——	S. ♄.	Rheed. mal. 6. t. 9-10.
arboréscens. Vahl. *non* Miller.						
51 coluteoídes. DC.	(*ye.*) Colutea-like.	....	....	——	S. ♄.	Coll. mon. 102. t. 12.
52 Barclayàna. S.F.A.	(*ye.*) Mr. Barclay's.	N. Holland.	1824.	6. 10.	G. ♄.	Swt. flor. aust. t. 32.
53 biflòra. DC.	(*ye.*) two-flowered.	W. Indies.	1766.	12. 4.	S. ♄.	Bot. mag. t. 810.
54 tomentòsa. DC.	(*st.*) tomentose.	S. America.	1822.	6. 8.	S. ♄.	Jacq. ic. 1. t. 72.
multiglandulòsa. Jacq.						
55 mexicàna. DC.	(*ye.*) Mexican.	Mexico.	1824.	——	S. ♄.	Jacq. schœnb. t. 203.
56 stipulàcea. DC.	(*ye.*) large-stipuled.	Chile.	1786.	——	S. ♄.	Feuill. per. 3. t. 42.
57 auriculàta. DC.	(*ye.*) eared.	E. Indies.	1777.	7. 9.	S. ♄.	Pluk. alm. t. 314. f. 4.
58 frondòsa. DC.	(*ye.*) smooth shrubby.	W. Indies.	1769.	3. 4.	S. ♄.	Jacq. ic. 1. t. 74.
59 austràlis. B.M.	(*ye.*) New Holland.	N. Holland.	1820.	5. 12.	I. ♄.	Bot. mag. t. 2676.
60 polyphy'lla. DC.	(*ye.*) many-leaved.	Portorico.	1828.	6. 9.	S. ♄.	Jacq. ic. t. 460.
61 marginàta. DC.	(*ye.*) white-edged.	Surinam.	1822.	——	S. ♄.	
62 ægyptìaca. DC.	(*ye.*) Egyptian.	Egypt.	1823.	7. 9.	G. ⊙.	
63 esculénta. H.B.	(*ye.*) esculent.	E. Indies.	——	——	S. ⊙.	
64 occidentàlis. DC.	(*ye.*) occidental.	W. Indies.	1759.	3. 8.	S. ♄.	Bot. reg. t. 83.
65 purpùrea. DC.	(*ye.*) purple-stalked.	E. Indies.	1823.	6. 11.	S. ♄.	—— t. 856.
66 hirsùta. DC.	(*ye.*) hairy.	S. America.	1778.	7. 9.	S. ♄.	Jacq. schœnb. t. 270.
caracasàna. Jacq.						
67 lineàris. DC.	(*ye.*) linear-podded.	Carolina.	1828.	5. 7.	F. ♃.	
68 pátula. DC.	(*ye.*) shining.	W. Indies.	1778.	8. 9.	S. ♄.	
69 pubéscens. DC.	(*ye.*) pubescent.	....	1818.	5. 8.	S. ♄.	Jacq. fragm. t. 57.
70 sulcàta. DC.	(*ye.*) furrowed.	S. America.	1826.	6. 8.	S. ♄.	Coll. mon. p. 110. t. 6.
71 ruscifòlia. DC.	(*ye.*) Ruscus-leaved.	Caracas.	1816.	5. 7.	G. ♄.	Jacq. ic. 1. t. 71.

72 ligustrìna. DC.	(*ye.*)	Privet-leaved.	America.	1726.	6. 8.	S. ♄.	Bot. reg. t. 109.
73 robinioídes. DC.	(*ye.*)	Robinia-like.	S. America.	1820.	——	S. ♄.	
74 marylándica. DC.	(*ye.*)	Maryland.	N. America.	1723.	8. 10.	H. ♃.	Schkuhr hand. 1. t. 113.
75 latifòlia. DC.	(*ye.*)	broad-leaved.	Trinidad.	1826.	6. 8.	S. ♄.	
76 arboréscens. DC.	(*ye.*)	arborescent.	New Spain.	....	——	S. ♄.	
77 opàca. Gr.	(*or.*)	dull-leaved.	S. America.	1825.	8. 12.	S. ♄.	
78 pulchélla. Boj.	(*ye.*)	pretty.	Zanzibar.	1825.	7. 9.	S. ♄.	
79 Millèri. DC.	(*ye.*)	Miller's.	Jamaica.	.....	6. 8.	S. ♄.	
80 Houstoniàna. DC.	(*ye.*)	Houstoun's.	———	....	7. 10.	S. ⊙.	
81 Plumiéri. DC.	(*ye.*)	Plumier's.	Guadaloupe.	1822.	——	S. ♄.	Plum. ed. Burm. t. 77.
planisíliqua. Lam. *non* Linn.							
82 hírta. DC.	(*ye.*)	hairy-branched.	S. America.	1822.	6. 9.	S. ♄.	
83 sumatràna. DC.	(*ye.*)	Sumatra.	Sumatra.	1823.	——	S. ♄.	
84 formòsa.	(*ye.*)	beautiful.	E. Indies.	——	——	S. ♄.	
speciòsa. H.B. *non* Schrader.							
85 emarginàta. DC.	(*ye.*)	notch-leaved.	Jamaica.	1759.	5. 6.	S. ♄.	Sloan. his. 2. t. 180. f. 1-4.
86 atomària. DC.	(*ye.*)	woolly-leaved.	S. America.	1822.	3. 8.	S. ♄.	
87 reticulàta. DC.	(*ye.*)	netted-leaved.	Brazil.	——	4. 6.	S. ♄.	
88 spectábilis. DC.	(*ye.*)	showy.	Caracas.	1823.	5. 7.	S. ♄.	Coll. mon. p. 115. t. 7.
89 setígera. DC.	(*ye.*)	seta-bearing.	E. Indies.	1826.	——	S. ♄.	
90 montàna. DC.	(*ye.*)	mountain.	———	1822.	——	S. ♄.	
91 A'bsus. DC.	(*sn.*)	four-leaved.	Ceylon.	1777.	6. 7.	S. ⊙.	Burm. zeyl. t. 97.
92 híspida. DC.	(*ye.*)	hispid.	Cayenne.	1826.	——	S. ⊙.	
93 diphy'lla. DC.	(*ye.*)	two-leaved.	W. Indies.	1781.	5. 7.	S. ♂.	Cavan. ic. 5. t. 600. f. 1.
94 bifoliolàta. DC.	(*ye.*)	two-leafleted.	Brazil.	1825.	——	S. ♄.	Coll. mon. 120. t. 9. f. B.
95 exígua. H.B.	(*ye.*)	dwarf.	Bengal.	1820.	8. 10.	S. ⊙.	
96 grácilis. DC.	(*ye.*)	slender.	S. America.	1816.	6. 8.	S. ♄.	Kth. mim. t. 36.
97 uniflòra. DC.	(*re.*)	one-flowered.	Brazil.	1824.	——	S. ♄.	
98 lineàta. DC.	(*ye.*)	lined.	Jamaica.	1820.	5. 7.	S. ♄.	
99 prostràta. DC.	(*ye.*)	prostrate.	S. America.	1823.	7. 9.	S. ♄.	
100 triflòra. DC.	(*ye.*)	three-flowered.	....	1825.	——	S. ⊙.	Jacq. schœnb. t. 480.
101 Burmánni. DC.	(*ye.*)	Burmann's.	C. B. S.	1816.	——	G. ♄.	
102 virgàta. DC.	(*ye.*)	slender-twigg'd.	Jamaica.	1820.	5. 8.	S. ♄.	
103 níctitans. DC.	(*ye.*)	Virginian.	N. America.	1800.	7.	G. ♂.	Linn. hort. cliff. t. 36.
104 chamæcrísta. DC.	(*ye.*)	dwarf.	America.	1699.	6. 9.	G. ♂.	Bot. mag. t. 157.
105 chamæcristoídes. DC.	(*ye.*)	large-flow[d].	Vera Cruz.	....	——	S. ♄.	
106 stricta. Sck.	(*ye.*)	straight.	Jamaica.	....	——	S. ♂.	Schk. hort. mon. t. 34.
107 glandulòsa. DC.	(*ye.*)	glandular.	W. Indies.	1822.	——	S. ♂.	
108 mimosoídes. DC.	(*ye.*)	Mimosa-leaved.	E. Indies.	1806.	——	S. ⊙.	
109 capénsis. DC.	(*ye.*)	Cape.	C. B. S.	1816.	4. 6.	G. ♄.	Coll. mon. 124. t. 19.
110 pùmila. DC.	(*ye.*)	small.	E. Indies.	1820.	7. 10.	S. ⊙.	Pluk. alm. t. 120. f. 1.
111 procúmbens. DC.	(*ye.*)	procumbent.	N. America.	1806.	6. 9.	H. ⊙.	Comm. petrop. t. 11.
112 Wallichiàna. DC.	(*ye.*)	fine-leaved.	Nepaul.	1812.	——	S. ⊙.	
dimidiàta. D.P.							
113 ténera.	(*ye.*)	small-branched.	E. Indies.	1818.	——	S. ♂.	
tenélla. H.B. *non* Kth.							
114 microphy'lla. DC.	(*ye.*)	small-leaved.	W. Indies.	1822.	——	S. ⊙.	
115 Hornemánni. DC.	(*ye.*)	Hornemann's.	....	1827.	——	S. ⊙.	
venòsa. Horn. *non* Zucc.							
AFZ'ELIA. DC.		Afz'elia.	Octandria Monogynia.				
africàna. DC.	(*cr.*)	African.	Sierra Leone.	1821.	....	S. ♄.	
SCHO'TIA. DC.		Scho'tia.	Decandria Monogynia.				
1 speciòsa. DC.	(*cr.*)	superb.	C. B. S.	1759.	7. 12.	I. ♄.	Andr. reposit. t. 348.
2 tamarindifòlia. DC.	(*cr.*)	Tamarind-leav'd.	———	1795.	5. 9.	I. ♄.	Bot. mag. t. 1153.
3 stipulàta. DC.	(*cr.*)	falcate-stipuled.	———	1794.	——	I. ♄.	
4 alàta. DC.	(*re.*)	winged.	———	1816.	6. 7.	I. ♄.	
5 simplicifòlia. DC.	(*re.*)	simple-leaved.	———	——	——	I. ♄.	
6 latifòlia. DC.	(*re.*)	broad-leaved.	———	——	5. 8.	I. ♄.	Hook. ex. flor. t. 159.
Omphalòbium Schòtia. S.S. Jacq. ecl. ic.							
COPAI'FERA. DC.		Balsam of Capevi.	Decandria Monogynia. L.				
1 officinàlis. DC.	(*st.*)	officinal.	S. America.	1774.	....	S. ♄.	H. et B. n. gen. 7. t. 659.
2 guianénsis. DC.	(*st.*)	Guiana.	Guiana.	1826.	....	S. ♄.	
3 coriàcea. DC.	(*st.*)	leathery-leaved.	Brazil.	1824.	....	S. ♄.	
CYNOM'ETRA. DC.		Cynom'etra.	Decandria Monogynia. L.				
1 cauliflòra. DC.	(*re.*)	stem-flowered.	E. Indies.	1804.	....	S. ♄.	Rumph. amb. 1. t. 62.
2 ramiflòra. DC.	(*re.*)	branch-flowered.	———	1826.	....	S. ♄.	——— 1. t. 63.
3 polyándra. DC.	(*re.*)	many stamen'd.	———	1822.	....	S. ♄.	Roxb. corom. t. 286.
PARIV'OA. DC.		Pariv'oa.	Diadelphia Decandria.				
grandiflòra. DC.	(*pu.*)	large-flowered.	Cayenne.	1816.	....	S. ♄.	Aubl. guian. 2. t. 303
Dimórpha grandiflòra. Linn. trans. 9. t. 20.							

O'UTEA. DC. O'UTEA. Tri-Tetrandria Monogynia.
1 guianénsis. DC. (*wh.*) Guiana. Guiana. 1827. .... S. ♄. Aubl. guian. 1. t. 9.
2 bijùga. DC. (*wh.*) two-paired. E. Indies. 1823. .... S. ♄. Linn. trans. 12. t. 17.
Macrolòbium bijùgum. Linn. trans.

VOU'APA. DC. VOU'APA. Triandria Monogynia.
bifòlia. DC. .... two-leaved. Guiana. 1823. .... S. ♄. Aubl. guian. 1. t. 7.

HYMEN'ÆA. DC. LOCUST-TREE. Decandria Monogynia.
1 Coùrbaril. DC. (*wh.*) leathery-leaved. S. America. 1688. 6. 8. S. ♄. Lam. ill. t. 330. f. 1.
2 Candolleàna. DC. (*wh.*) Decandolle's. Mexico. 1825. .... S. ♄. H. et B. n. gen. 6. t. 566.
3 verrucòsa. DC. (*wh.*) wart-podded. Madagascar. 1808. 6. 7. S. ♄. Lam. ill. t. 330. f. 2.

BAUHI'NIA. DC. BAUHI'NIA. Monadelphia v. Decandria Monogynia.
1 divaricàta. DC. (*wh.*) dwarf. W. Indies. 1742. 6. 9. S. ♄. Linn. hort. cliff. t. 15.
2 Lamarckiàna. DC. (*wh.*) Lamarck's. S. America. .... —— S. ♄.
3 spathàcea. DC. (*wh.*) spathaceous. New Spain. 1825. —— S. ♄.
4 subrotundifòlia. DC. (*w.*) round-leaved. S. America. 1816. 4. 6. S. ♄. Cavan. ic. 5. t. 406.
5 aurìta. DC. (*wh.*) long-eared. Jamaica. 1756. 9. 10. S. ♄. Miller ic. t. 61.
6 porrécta. DC. (*wh.*) smooth-leaved. W. Indies. 1737. 7. 8. S. ♄. Bot. mag. t. 1708.
7 latifòlia. DC. (*wh.*) broad-leaved. New Spain. 1825. .... S. ♄. Cavan. ic. t. 405.
8 cándida. DC. (*wh.*) white-flowered. E. Indies. 1777. 5. 6. S. ♄.
9 acuminàta. DC. (*wh.*) acute-leaved. ———— 1808. 7. 8. S. ♄. Rheed. mal. 1. t. 34.
10 grandiflòra. DC. (*wh.*) large-flowered. Peru. 1824. .... S. ♄.
11 aculeàta. DC. (*wh.*) prickly-stalked. W. Indies. 1737. 6. 8. S. ♄. Plum. ed. Bur. t. 44. f. 1.
12 forficàta. DC. (*wh.*) forked. Brazil. 1824. —— S. ♄.
13 emarginàta. DC. (*wh.*) emarginate. Carthagena. .... —— S. ♄.
14 rufèscens. DC. .... rufous. Africa. 1816. .... S. ♄.
15 parviflòra. DC. (*wh.*) small-flowered. E. Indies. 1808. .... S. ♄.
16 U'ngula. DC. (*wh.*) villous-leaved. Caracas. 1827. .... S. ♄. Jacq. fragm. t. 15. f. 1.
aculeàta. Jacq. amer. t. 177. f. 2.
17 variegàta. DC. (*v.*) variegated-fl[d]. E. Indies. 1690. 6. 7. S. ♄. Rheed. mal. 1. t. 32.
18 tomentòsa. DC. (*st.*) tomentose. ———— 1808. .... S. ♄. Burm. zeyl. t. 18.
19 pubéscens. DC. (*wh.*) pubescent. W. Indies. 1824. .... S. ♄.
20 multinérvia. DC. (*wh.*) many-nerved. Caracas. 1825. .... S. ♄.
21 racemòsa. DC. .... large-leaved. E. Indies. 1790. .... S. ♄.◡. Vahl. symb. 3. t. 62.
22 purpùrea. DC. (*pu.*) purple. ———— 1778. .... S. ♄. Rheed. mal. 1. t. 33.
23 retùsa. DC. (*wh.*) retuse-leaved. ———— 1822. .... S. ♄.
24 corymbòsa. DC. (*wh.*) corymbose. China. 1816. .... S. ♄.◡. DC. leg. mem. t. 70.
25 anguìna. DC. (*wh.*) snake. E. Indies. 1818. .... S. ♄.◡. Roxb. corom. 3. t. 285.
scándens. L. *excl. synon. Rumph.* Rheed. mal. 8. t. 30-31.
26 Língua. DC. (*st.*) tongue-leaved. E. Indies. 1799. .... S. ♄.◡. Rumph. amb. 5. t. 1.
scándens. L. *excl. syn. Rheed.*
27 coccínea. DC. (*sc.*) scarlet-flowered. China. 1826. .... S. ♄.◡.
28 guianénsis. DC. (*wh.*) Guiana. Guiana. 1824. .... S. ♄.◡. Aubl. guian. 1. t. 145.
29 heterophy'lla. DC. (*wh.*) various-leaved. Caracas. 1824. .... S. ♄. Kth. mim. t. 46.
30 cumanénsis. DC. (*wh.*) Cumana. Cumana. —— 7. 8. S. ♄.◡. Bot. reg. t. 1133.
31 anatómica. L. en. .... jagged-leaved. S. America. 1816. .... S. ♄.
32 triándra. H.B. (*wh.*) triandrous. E. Indies. 1823. .... S. ♄.
33 semibífida. H.B. (*wh.*) half-cleft. ———— 1818. .... S. ♄.◡.
34 malabárica. H.B. .... Malabar. Malabar. 1823. .... S. ♄.
35 ferrugínea. H.B. .... ferruginous. E. Indies. 1822. .... S. ♄.◡.
36 cordifòlia. H.B. .... heart-leaved. ———— 1828. .... S. ♄.◡.
37 spicàta. .... spiked. ———— 1820. .... S. ♄.
38 speciòsa. L.C. .... handsome. .... S. ♄.
39 microphy'lla. L.C. .... small-leaved. .... S. ♄.
40 esculénta. B.T. (*wh.*) eatable-rooted. C. B. S. 1816. —— I. ♃.

CE'RCIS. DC. JUDAS-TREE. Decandria Monogynia. L.
1 Siliquástrum. DC. (*ro.*) European. S. Europe. 1596. 5. 6. H. ♄. Bot. mag. t. 1138.
2 canadénsis. DC. (*ro.*) American. N. America. 1730. —— H. ♄. Miller ic. t. 2.

CR'UDYA. DC. CR'UDYA. Octo-Decandria Monogynia.
1 spicàta. DC. (*wh.*) spiked. Guiana. 1825. .... S. ♄. Aubl. guian. 1. t. 147.
Apalatòa spicàta. Aubl.
2 aromática. DC. (*gr.*) aromatic. ———— 1828. .... S. ♄. Aubl. guian. 1. t. 148.
Touchiròa aromática. Aubl.

COD'ARIUM. DC. VELVET TAMARIND. Diandria Monogynia.
1 acutifòlium. DC. (*re.*) acute-leaved. Sierra Leone. 1800. .... S. ♄. Rœm. arch. 1. t. 6.
nítidum. Vahl. *Diàlium guineénse.* W.
2 obtusifòlium. DC. (*re.*) blunt-leaved. ———— —— .... S. ♄.

Tribus X. *DETARIEÆ.* DC. prodr. 2. p. 521.

DET'ARIUM. DC.		DET'ARIUM.	Decandria Monogynia.				
senegalénsis. DC.	(*gr.*)	Senegal.	Senegal.	1822.	....	S. ♄.	

ORDO LXXV.

CHRYSOBALANEÆ. *Brown cong. p.* 14.

CHRYSOBA'LANUS. DC.		COCOA PLUM.	Icosandria Monogynia. L.				
1 Icáco. DC.	(*wh.*)	West Indian.	W. Indies.	1752.	4. 6.	S. ♄.	Jacq. amer. t. 94.
2 ellípticus. DC.	(*wh.*)	Pigeon plum.	Sierra Leone.	1822.	....	S. ♄.	
3 oblongifòlius. DC.	(*wh.*)	American.	Georgia.	1812.	4. 5.	G. ♄.	
PARIN'ARIUM. DC.		PARIN'ARIUM.	Dodecandria Monogynia.				
1 excélsum. G.D.	(*wh.*)	rough plum.	Sierra Leone.	1822.	....	S. ♄.	Loud. encycl. pl. f. 5072.
2 macrophy'llum. G.D.	(*w.*)	gingerbread plum.	———	——	....	S. ♄.	
GRANG'ERIA. DC.		GRANG'ERIA.	Dodecandria Monogynia.				
borbònica. DC.	(*wh.*)	Bourbon.	Bourbon.	1822.	....	S. ♄.	Lam. ill. t. 427.
HIRTE'LLA. DC.		HIRTE'LLA.	Tri-Dodecandria Monogynia.				
1 triándra. DC.	(*wh.*)	triandrous.	W. Indies.	1782.	....	S. ♄.	Jacq. amer. t. 8.
americàna. Jacq. *non* Aubl. *paniculàta.* R.S.							
2 glandulòsa. DC.	(*wh.*)	glandular.	Brazil.	1827.	....	S. ♄.	Spreng. anl. t. 7. f. 1-4.
3 racemòsa. DC.	(*bl.*)	racemed.	Guiana.	1782.	....	S. ♄.	Aubl. guian. 1. t. 98.
americàna. Aubl. *non* Jacq.							
4 oblongifòlia. DC.	(*ro.*)	oblong-leaved.	New Spain.	1826.	....	S. ♄.	
5 nemoròsa. DC.	(*ye.*)	wood.	Brazil.	1825.	....	S. ♄.	

ORDO LXXVI.

AMYGDALINÆ. *D. Don prodr. flor. nepal. p.* 239.

AMY'GDALUS. K.S.		ALMOND.	Icosandria Monogynia. L.				
1 nàna. W.	(*pi.*)	common dwarf.	Russia.	1683.	3. 4.	H. ♄.	Bot. mag. t. 161.
2 geórgica. Df.	(*pi.*)	Georgian.	Georgia.	1820.	——	H. ♄.	
3 campéstris. Bes.	(*pi.*)	Besser's.	Poland.	1829.	——	H. ♄.	
Besseriàna. Schott.							
4 orientàlis. W.	(*ro.*)	silvery-leaved.	Levant.	1756.	——	F. ♄.	
argéntea. Lam.							
5 incàna. W.	(*ro.*)	woolly.	Caucasus.	....	——	H. ♄.	Pall. ross. 1. t. 7.
6 prostràta. K.S.	(*pi.*)	prostrate.	Crete.	1802.	5.	H. ♄.	Bot. reg. t. 136.
Prùnus prostràta. B.R.							
7 pùmila. W.	(*pi.*)	double dwarf.	Africa.	1683.	5. 6.	H. ♄.	Bot. mag. t. 2176.
Prùnus sinénsis. P.S. Prùnus japónica, B.R. is a very different plant, and a true Cérasus.							
8 sibírica. L.C.	(*pi.*)	Siberian.	Siberia.	1820.	3. 4.	H. ♄.	
9 commùnis. DC.	(*pi.*)	common.	Barbary.	1548.	——	H. 5.	
α amara. DC.	(*pi.*)	*bitter.*	———	——	——	H. 5.	
β dúlcis. DC.	(*bh.*)	*sweet.*	———	——	——	H. 5.	Lam. ill. t. 430. f. 2.
γ frágilis. DC.	(*bh.*)	*tender.*	———	....	——	H. 5.	Nois. jard. fr. t. 3. f. 2.
δ macrocarpa. DC.	(*pi.*)	*large-fruited.*	....	....	——	H. 5.	Bot. reg. t. 1160.
ε persicoídes. DC.	(*pi.*)	*Almond Peach.*	Hybrid.	....	——	H. 5.	Nois. jard. fr. t. 3. f. 1.
10 cochinchinénsis. DC.	(*pi.*)	Chinese.	Cochinchina.	1825.	——	H. 5.	
11 Pérsica. W.	(*pi.*)	common Peach.	Persia.	1562.	4. 5.	H. 5.	Nois. j. fr. 1-16. 22-35. ic.
β flòre plèno.	(*pi.*)	*double-blossomed.*	....	....	——	H. 5.	
δ nàna.	(*pi.*)	*dwarf.*	China.	1826.	——	H. ♄.	
ε nectarìna.	(*pi.*)	*Nectarine.*	Persia.	1562.	——	H. 5.	Nois. jard. fr. t. 20-21.

ARMEN'IACA. DC.		APRICOT.	Icosandria Monogynia.				
1 vulgàris. DC.	(*wh.*)	common.	Levant.	1548.	2. 4.	H. 5.	Lam. ill. t. 431.
α *ovalifòlia.* DC.	(*wh.*)	*oval-leaved.*	——	——	——	H. 5.	Nois. jar. fr. t. 1-2. f. 1-2.
β *cordifòlia.* DC.	(*wh.*)	*heart-leaved.*	——	——	——	H. 5.	—— t. 1. f. 3. t. 2. f. 3.
2 dasycárpa. DC.	(*wh.*)	black-fruited.	....	....	3. 4.	H. 5.	Bot. reg. t. 1243.
Prùnus dasycárpa. B.R.							
β *persicifòlia.* DC.	(*wh.*)	*Peach-leaved.*	....	....	——	H. ♄.	Du. ar. ed. no. 5. t. 52. f. 1.
3 sibírica. DC.	(*pi.*)	Siberian.	Siberia.	1788.	4.	H. ♄.	Pall. ross. 1. t. 8.
4 brigantìaca. DC.	(*wh.*)	spiny.	S. Europe.	....	——	H. ♄.	Duh. arb. ed. n. 5. t. 59.
PR'UNUS. DC.		PLUM-TREE.	Icosandria Monogynia. L.				
1 spinòsa. DC.	(*wh.*)	Sloe-tree.	Britain.	....	3. 4.	H. ♄.	Eng. bot. v. 12. t. 842.
α *vulgàris.* DC.	(*wh.*)	*common.*	——	....	——	H. ♄.	Duham. ed. n. 5. t. 54. f. 1.
β *microcárpa.* DC.	(*wh.*)	*small-fruited.*	——	....	——	H. ♄.	
γ *macrocárpa.* DC.	(*wh.*)	*large-fruited.*	——	....	——	H. ♄.	
δ *ovàta.* DC.	(*wh.*)	*oval-leaved.*	——	....	——	H. ♄.	Blackw. herb. t. 494.
2 insitítia. DC.	(*wh.*)	Bullace-tree.	Britain.	....	4.	H. ♄.	Eng. bot. v. 12. t. 841.
3 cándicans. DC.	(*wh.*)	snowy.	....	1818.	4. 5.	H. ♄.	Bot. reg. t. 1135.
4 Cocomília. DC.	(*wh.*)	twin-flowered.	Calabria.	1826.	——	H. ♄.	
5 marítima. DC.	(*wh.*)	sea-side.	N. America.	1800.	——	H. ♄.	
6 pubéscens. DC.	(*wh.*)	pubescent.	....	1819.	——	H. ♄.	
7 diváricata. DC.	(*wh.*)	spreading.	Caucasus.	1828.	....	H. ♄.	
8 cerasífera. S.S.	(*wh.*)	Myrobalan.	N. America.	1629.	4. 5.	H. 5.	Duham. arb. 2. t. 2. f. 15.
9 pyramidàlis. DC.	(*wh.*)	pyramidal.	Europe.	....	4.	H. 5.	—— 2. t. 15—20.
10 doméstica. DC.	(*wh.*)	common.	England.	....	——	H. 5.	Eng. bot. v. 25. t. 1783.
α *armenioídes.* DC.	(*wh.*)	*Apricot-like.*	....	....	——	H. 5.	Duham. arb. 2. t. 13—14.
β *Claudiàna.* DC.	(*wh.*)	*Reine Claude.*	....	....	——	H. 5.	—— 2. t. 11—12.
γ *damascèna.* DC.	(*wh.*)	*Damson.*	....	....	——	H. 5.	—— 2. n. 13. t. 5.
δ *turonénsis.* DC.	(*wh.*)	*Monsieur hatif, &c.*	....	....	——	H. 5.	—— 2. n. 16. t. 20. f. 1.
ε *Juliàna.* DC.	(*wh.*)	*Saint Julien.*	....	....	——	H. 5.	Duh. ed. nov. 5. t. 54. f. 2.
ζ *Catharìnea.* DC.	(*wh.*)	*Saint Catherine.*	....	....	——	H. 5.	Duham. arb. 2. f. 1.
η *Aubertiàna.* DC.	(*wh.*)	*Imperatrice.*	....	....	——	H. 5.	Nois. jar. fr. t. 57-58. f. 4.
C'ERASUS. DC.		CHERRY TREE.	Icosandria Monogynia.				
1 ávium. DC.	(*wh.*)	small-fruited.	Britain.	....	4. 5.	H. 5.	Blackw. t. 425.
α *sylvéstris.*	(*wh.*)	*wood.*	——	....	——	H. 5.	
β *macrocárpa.* DC.	(*wh.*)	*large-fruited.*	....	....	——	H. 5.	Du. ed. no. 1. n. 2. t. 4. f. D.
γ *pállida.* DC.	(*wh.*)	*pale-fruited.*	....	....	——	H. 5.	—— 5. n. 3. t. 4. f. B. C.
δ *multipléx.* DC.	(*wh.*)	*double-blossomed.*	....	....	——	H. 5.	
2 duracìna. DC.	(*wh.*)	hard-fruited.	Europe.	....	——	H. 5.	
α *cordígera.* DC.	(*wh.*)	*Bigarreaux.*	....	....	——	H. 5.	Nois. jard. fr. t. 6. f. 1.
β *obtusàta.* DC.	(*wh.*)	*blunt-fruited.*	....	....	——	H. 5.	Duh. ed. no. 5. t. 18. f. B.
γ *mamillàris.* DC.	(*wh.*)	*large-leaved.*	....	....	——	H. 5.	—— 5. n. 12. t. 16. f. A.
3 juliàna. DC.	(*wh.*)	early.	S. Europe.	....	——	H. 5.	—— 5. n. 5. t. 15. f. A.
β *Heaumiàna.* DC.	(*wh.*)	*long-leaved.*	——	....	——	H. 5.	—— n. 29. t. 19.
γ *péndula.* DC.	(*wh.*)	*pendulous-branch'd.*	——	....	——	H. 5.	
4 Caproniàna. DC.	(*wh.*)	common.	——	....	——	H. 5.	
α *Montmorencyàna.* DC.		*May Duke, &c.*	——	....	——	H. 5.	Duh. ed. no. 5. n. 14. t. 7.
β *palléscens.* DC.	(*wh.*)	*pale-fruited.*	——	....	——	H. 5.	—— n. 18. t. 11.
γ *gobétta.* DC.	(*wh.*)	*Kentish.*	——	....	——	H. 5.	—— t. 12. f. A.
δ *polyg'yna.* DC.	(*wh.*)	*polygynous.*	——	....	——	H. 5.	Duham. arb. 4. t. 3.
ε *multipléx.* DC.	(*wh.*)	*semi-double.*	....	....	——	H. 5.	—— p. 173. t. 5.
ϑ *persiciflòra.* DC.	(*wh.*)	*double-flowered.*	....	....	——	H. 5.	
κ *variegàta.* DC.	(*wh.*)	*striped-leaved.*	....	....	——	H. 5.	
θ *griotta.* DC.	(*wh.*)	*globular-fruited.*	....	....	——	H. 5.	Duh. ed. nov. n. 21. t. 14.
ι *cordígera.* DC.	(*wh.*)	*red-heart.*	....	....	——	H. 5.	Duham. arb. 4. t. 16. f. 1.
5 semperflòrens. DC.	(*wh.*)	late-fruited.	N. America.	....	5. 6.	H. ♄.	Wats. dend. brit. 131.
6 chamæcérasus. DC.	(*wh.*)	bastard.	Germany.	1597.	5.	H. ♄.	Jacq. ic. 1. t. 90.
7 persicifòlia. DC.	(*wh.*)	Peach-leaved.	N. America.	1818.	4. 5.	H. ♄.	
8 púmila. DC.	(*wh.*)	dwarf.	——	1756.	5.	H. ♄.	Mill. ic. t. 89. f. 2.
9 pygm'æa. DC.	(*wh.*)	small.	——	1818.	——	H. ♄.	
10 nígra. DC.	(*wh.*)	black-fruited.	Canada.	1773.	5. 6.	H. ♄.	Bot. mag. t. 1117.
11 boreàlis. DC.	(*wh.*)	northern.	N. America.	1818.	——	H. ♄.	
12 depréssa. DC.	(*wh.*)	depressed.	——	1805.	5.	H. ♄.	
13 Susquehánnæ. W. en.	(*wh.*)	glaucous-ld.	N. America.	1800.	——	H. ♄.	
14 hyemàlis. DC.	(*wh.*)	mountain.	——	1805.	——	H. ♄.	
15 Chícasa. DC.	(*wh.*)	Chicasaw.	——	1806.	4. 5.	H. ♄.	
16 pubéscens. DC.	(*wh.*)	pubescent.	——	1822.	5.	H. ♄.	
17 japónica. DC.	(*pi.*)	Japan.	Japan.	1810.	3. 5.	H. ♄.	B[illegible]g. t. 27.
18 serrulàta. H.C.	(*bh.*)	saw-leaved.	China.	1822.	4. 5.	H. ♄.	
19 salicìna. H.C.	(*bh.*)	Willow-like.	——	——	——	H. ♄.	

20	pseùdo-cérasus. HC.	(*wh.*) Chinese.	China.	1820.	4. 5.	F. ♄.	Bot. reg. t. 800.
	paniculàta. B.R. *non* Thunb.						
21	pensylvánica. DC.	(*wh.*) Pensylvanian.	N. America.	1773.	5.	H. ♄.	Willd. arb. t. 3. f. 3.
22	Maháleb. DC.	(*wh.*) perfumed.	S. Europe.	1714.	4. 5.	H. ♄.	Jacq. aust. 3. t. 227.
23	Pádus. DC.	(*wh.*) Bird.	Britain.	....	——	H. ♄.	Eng. bot. v. 20. t. 1383.
24	rùbra. W.	(*wh.*) red-fruited.	England.	....	——	H. ♄.	Willd. arb. t. 4. f. 2.
25	virginiàna. DC.	(*wh.*) Virginian.	N. America.	1724.	5. 6.	H. ♄.	——— t. 5. f. 1.
26	serotìna. DC.	(*wh.*) American Bird.	———	1629.	——	H. ♄.	Wats. dend. brit. 48.
27	occidentàlis. DC.	(*wh.*) West Indian.	Jamaica.	1784.	12. 1.	S. ♄.	Catesb. car. 2. t. 94?
28	lusitánica. DC.	(*wh.*) Portugal Laurel.	Portugal.	1648.	6.	H. ħ.	Mill. ic. t. 196. f. 1.
29	Híxa. W. en.	(*wh.*) Teneriffe.	Teneriffe.	1824.	——	F. ħ.	
30	Laurocérasus. DC.	(*wh.*) Laurel.	Levant.	1629.	4. 5.	H. ħ.	Blackw. herb. t. 512.
31	caroliniàna. DC.	(*wh.*) evergreen.	Carolina.	1759.	5.	H. ħ.	

ORDO LXXVII.

POMACEÆ. *D. Don prodr. flor. nepal. p.* 236.

	MESPILUS. K.S.	MESPILUS. Icosandria Di-Pentagynia.					
1	germánica. DC.	(*wh.*) Medlar.	England.	....	5. 7.	H. 5.	Eng. bot. v. 22. t. 1523.
	α sylvéstris. DC.	(*wh.*) *wild.*	———	....	——	H. 5.	
	β stricta. DC.	(*wh.*) *upright.*	....	....	——	H. 5.	
	γ diffùsa. DC.	(*wh.*) *Dutch.*	....	....	——	H. 5.	Duham. arb. fr. 1. t. 3.
2	grandiflòra. S.S.	(*wh.*) large-flow red.	S. Europe.	1800.	5. 6.	H. 5.	Exot. bot. 1. t. 18.
	Smíthii. DC.						
3	Pyracántha. S.S.	(*wh.*) evergreen thorn.	———	1629.	5.	H. ħ.	Schmidt arb. t. 90.
4	crenulàta. D.P.	(*wh.*) crenulate.	Nepaul.	1818.	6. 7.	H. ħ.	
5	mexicàna. DC.	(*wh.*) Mexican.	Mexico.	1824.	5. 6.	F. ħ.	
6	Crús-gálli. L.	(*wh.*) Cockspur-thorn.	N. America.	1691.	5. 6.	H. 5.	Wats. dend. brit. t. 56.
7	pyracanthifòlia.	(*wh.*) Pyracantha-leaved.	———	——	——	H. 5.	
	lùcida. Dum-Cours. bot. cult. ed. 2. v. 5. p. 448. *non* Miller.						
8	salicifolia. L.C.	(*wh.*) Willow-leaved.	———	——	——	H. 5.	
9	lineàris. DF.	(*wh.*) linear-leaved.	———	——	——	H. 5.	
10	nàna. Dum-C.	(*wh.*) dwarf.	———	1826.	——	H. ♄.	
11	spinosíssima. L.C.	(*wh.*) long-spined.	S. Europe.	....	——	H. ♄.	Lodd. bot. cab. 1100.
12	punctàta. W. en.	(*wh.*) spotted-fruited.	N. America.	1746.	——	H. 5.	Wats. dend. brit. t. 57.
	α rùbra. Ph.	(*wh.*) red-fruited.	———	——	——	H. ♄.	
	β aùrea. Ph.	(*wh.*) *yellow-fruited.*	———	——	——	H. ♄.	
13	ellíptica. W. en.	(*wh.*) oval-leaved.	———	1765.	5.	H. 5.	
14	prunifòlia. S.S.	(*wh.*) Plum-leaved.	———	1812.	5. 6.	H. 5.	
15	prunellifòlia. DC.	(*wh.*) Sloe-leaved.	———	1824.	——	H. ♄.	
16	latifòlia. Poir.	(*wh.*) broad-leaved.	N. America.	....	——	H. 5.	
17	parvifòlia. W. en.	(*wh.*) Gooseberry-leav'd.	———	1704.	——	H. ♄.	Wats. dend. brit. 65.
	xanthocárpos. S.S. *Crat'ægus parvifòlia.* DC. *tomentòsa.* L. Trew. ehr. t. 17.						
18	ovalifòlia. DC.	(*wh.*) oval-leaved.	———	....	——	H. ♄.	
19	víridis. DC.	(*wh.*) smooth green.	———	....	——	H. ♄.	
20	pauciflòra. Poir.	(*wh.*) few-flowered.	Switzerland.	1818.	——	H. ♄.	
21	lùcida. DC.	(*wh.*) glossy-leaved.	———	....	——	H. ♄.	
22	apiifòlia. S.S.	(*wh.*) Parsley-leaved.	———	1812.	——	H. ♄.	
23	pyrifòlia. S.S.	(*wh.*) Pear-leaved.	———	1765.	6.	H. 5.	Wats. dend. brit. t. 61.
24	glandulòsa. S.S.	(*wh.*) hollow-leaved.	N. America.	1750.	5. 6.	H. 5.	——— brit. t. 58.
	Crat'ægus sanguínea. Pall. ross. 1. t. 11.						
25	spathulàta. S.S.	(*wh.*) spatula-leaved.	———	1806.	——	H. 5.	Lodd. bot. cab. t. 1261.
26	coccínea. S.S.	(*wh.*) scarlet-fruited.	———	1683.	4. 5.	H. 5.	Wats. dend. brit. t. 62.
27	purpùrea. W.D.	(*wh.*) purple-branch'd.	———	....	——	H. 5.	——— t. 60.
28	cordàta. S.S.	(*wh.*) Maple-leaved.	———	1738.	5. 6.	H. 5.	——— t. 63.
	Crat'ægus cordàta. Bot. reg. t. 1151.						
29	lobàta. Poir.	(*wh.*) lobed-leaved.	———	1818.	——	H. 5.	
30	florentìna. Be.	(*wh.*) Florentine.	Florence.	1824.	——	H. 5.	
31	nìgra. S.S.	(*wh.*) black-fruited.	Hungary.	1818.	——	H. 5.	Lodd. bot. cab. t. 1021.

32 pentágyna. S.S.	(*wh.*)	five-styled.	Hungary.	1818.	5. 6.	H. 5.	
33 flàva. W. en.	(*wh.*)	yellow-berried.	N. America.	1724.	——	H. 5.	Wats. dend. brit. t. 59.
34 físsa. Poir.	(*wh.*)	cloven-leaved.	———	1818.	——	H. 5.	
35 Oliveriàna. DC.	(*wh.*)	Oliver's.	Levant.	——	——	H. 5.	
36 Poiretiàna. DC.	(*wh.*)	Poiret's.	N. America.	....	——	H. 5.	
lineàris. Poir. *non* P.S.							
37 dissécta. Dum.	(*wh.*)	cut-leaved.	Persia.	1828.	....	H. 5.	
38 Oxyacántha. S.S.	(*wh.*)	Hawthorn.	Britain.	....	5. 6.	H. 5.	Eng. bot. v. 35. t. 2504.
Crat'ægus Oxyacántha. DC.							
β major.	(*wh.*)	*great-fruited.*	———	....	——	H. 5.	
γ aùrea.	(*wh.*)	*yellow-berried.*	———	....	——	H. 5.	
39 pr'æcox.	(*wh.*)	Glastonbury.	....	....	3. 5.	H. 5.	
40 oxyacanthoídes. B.R.		Hawthorn-like.	S. Europe.	....	5. 6.	H. 5.	Bot. reg. t. 1128.
β multiplèx.	(*wh.*)	*double-flowered.*	———	....	——	H. 5.	
41 élegans. Poir.	(*ro.*)	*rose-coloured.*	———	....	——	H. 5.	Blackw. herb. t. 149.
β rubérrima.	(*re.*)	*dark-red.*	———	1827.	——	H. 5.	
42 eriocárpa. Lind.	(*wh.*)	woolly-fruited.	Britain.	....	——	H. 5.	
43 monógyna. Pal.	(*wh.*)	one-styled.	Siberia.	....	——	H. 5.	Pall. ross. 1. t. 12.
44 heterophy'lla. DC.	(*wh.*)	various-leaved.	Levant?	1818.	——	H. 5.	Bot. reg. t. 1161.
45 Azaròlus. S.S.	(*wh.*)	Azarole.	S. Europe.	1640.	——	H. 5.	Andr. reposit. t. 579.
46 Arònia. W. en.	(*wh.*)	Aronian.	Levant.	....	——	H. 5.	Pocock tr. t. 85.
47 orientàlis. Poir.	(*wh.*)	oriental.	———	1824.	——	H. 5.	
48 tanacetifòlia. A.R.	(*wh.*)	Tansy-leaved.	———	1789.	——	H. ♄.	Andr. reposit. t. 591.
49 odoratíssima. A.R.	(*wh.*)	sweet-scented.	Crimea.	....	——	H. ♄.	——— t. 590.
50 Celsiàna. W. en.	(*wh.*)	Cels's.	Levant.	1822.	——	H. ♄.	
51 laciniàta. DC.	(*wh.*)	jagged.	Sicily.	——	——	H. ♄.	
52 melanocárpa. S.S.	(*wh.*)	black-berried.	Tauria.	1818.	——	H. ♄.	
53 pectinàta. L. en.	(*wh.*)	pectinated.	Persia.	1824.	——	H. ♄.	
54 flabellàta. L.C.	(*wh.*)	fan-leaved.	....	1823.	——	H. ♄.	
55 pruinòsa. Lk.	(*wh.*)	frosted	....	1826.	——	H. ♄.	
56 succulénta. Scr.	(*wh.*)	succulent.	....	——	——	H. ♄.	
57 subvillòsa. Scr.	(*wh.*)	villous.	....	1827.	——	H. ♄.	
58 punícea. F.	....	crimson.	Dahuria.	——	....	H. ♄.	
RAPHIOL'EPIS. DC.		RAPHIOL'EPIS.	Icosandria Digynia.				
1 índica. DC.	(*wh.*)	Indian.	China.	1806.	5. 8.	G. ♄.	Bot. mag. t. 1726.
Crat'ægus índica. B.M.							
2 phæostèmon. DC.	(*wh.*)	red-stamen'd.	———	1818.	——	G. ♄.	Bot. reg. t. 468.
índica. Bot. reg. *non* Bot. mag.							
3 rùbra. DC.	(*wh.*)	red-stemmed.	———	1806.	——	G. ♄.	Lindl. coll. t. 3.
4 salicifòlia. DC.	(*wh.*)	Willow-leaved.	———	1820.	——	G. ♄.	Bot. reg. t. 652.
PHOTI'NIA. DC.		PHOTI'NIA.	Icosandria Digynia.				
1 serrulàta. DC.	(*wh.*)	smooth-leaved.	China.	1804.	4. 7.	H. ♄.	Bot. mag. t. 2105.
Crat'ægus glábra. B.M.							
2 arbutifòlia. DC.	(*wh.*)	Arbutus-leaved.	California.	1796.	6. 8.	H. ♄.	Bot. reg. t. 491.
3 integrifòlia. DC.	(*wh.*)	entire-leaved.	Nepaul.	1820.	5. 6.	H. ♄.	
P'yrus integérrima. D.P.							
4 dùbia. DC.	(*wh.*)	doubtful.	———	1822.	——	H. ♄.	Linn. trans. 13. t. 10.
Méspilus tinctòria. D.P. *Méspilus bengalénsis.* Roxb.							
ERIOB'OTRYA. DC.		LOQUAT.	Icosandria Pentagynia.				
1 japónica. DC.	(*wh.*)	Japan.	Japan.	1787.	9. 10.	H. ♄.	Bot. reg. t. 365.
Méspilus japónica. Vent. malm. t. 19.							
2 ellíptica. DC.	(*wh.*)	elliptic-leaved.	Nepaul.	1822.	....	H. ♄.	
Méspilus Cuìla. D.P.							
COTONEA'STER. DC.		COTONEA'STER.	Icosandria Di-Pentagynia.				
1 vulgàris. DC.	(*wh.*)	common.	Europe.	1656.	4. 5.	H. ♄.	Flor. dan. t. 112.
Méspilus Cotoneáster. L.							
2 affínis. DC.	(*bh.*)	likened.	Nepaul.	1820.	6. 7.	H. ♄.	
3 tomentòsa. DC.	(*wh.*)	Quince-leaved.	Europe.	1759.	4. 5.	H. ♄.	Wats. dend. brit. t. 55.
Méspilus eriocárpa. W.D.							
4 denticulàta. K.S.	(*wh.*)	tooth-leaved.	Mexico.	1826.	....	F. ♄.	H.etB.n.ge.am.6.t.556.
5 laxiflòra. J.F.	(*wh.*)	loose-flowered.	Nepaul.	1820.	5. 7.	H. ♄.	
6 frígida. B.R.	(*wh.*)	mountain.	———	1823.	4. 5.	H. ♄.	Bot. reg. t. 1229.
7 acuminàta. DC.	(*bh.*)	pointed-leaved.	———	1820.	4. 6.	H. ♄.	Linn. trans. v. 13. t. 9.
Méspilus acuminàta. Lodd. bot. cab. t. 919.							
8 rotundifòlia. B.R.	(*wh.*)	round-leaved.	———	——	——	H. ♄.	Bot. reg. t. 1187.
microphy'lla β 'Uva úrsi. B.R.							
9 microphy'lla. B.R.	(*wh.*)	small-leaved.	———	——	——	H. ♄.	Bot. reg. t. 1114.
10 buxifòlia. B.R.	(*wh.*)	Box-leaved.	———	1824.	——	H. ♄.	

AMELA'NCHIER. DC. AMELA'NCHIER. Icosandria Pentagynia.
1 vulgàris. DC. (*wh.*) common. S. Europe. 1596. 4. 5. H. ♄. Schmidt. arb. t. 85.
Méspilus Amelánchier. Jacq. aust. t. 300. *P'yrus Amelánchier.* W.
2 Botryàpium. DC. (*wh.*) snowy. N. America. 1746. —— H. ♄. Schmidt. arb. t. 84.
Méspilus canadénsis. L. *P'yrus Botryàpium.* W. *Arònia Botryàpium.* P.S.
3 ovàlis. DC. (*wh.*) oval-leaved. ———— 1800. —— H. ♄.
4 sanguínea. DC. (*wh.*) red-branched. ———— 1812. —— H. ♄. Bot. reg. t. 1171.
AR'ONIA. P.S. AR'ONIA. Icosandria. Di-Pentagynia.
1 arbutifòlia. N. (*wh.*) red-berried. N. America. 1700. 5. 6. H. ♄. Mill. ic. t. 109.
pyrifòlia. P.S. *P'yrus arbutifòlia.* Ph.
2 grandifòlia. (*wh.*) large-leaved. ———— .... —— H. ♄. Bot. reg. t. 1154.
P'yrus grandifòlia. B.R.
3 floribúnda. (*wh.*) many-flowered. ———— .... —— H. ♄. Bot. reg. t. 1006.
4 xanthocárpa. L.C. (*wh.*) yellow-fruited. ———— 1820. —— H. ♄.
5 melanocárpa. N. (*wh.*) black-fruited. ———— 1700. —— H. ♄. Schmidt. arb. t. 86.
6 chamæméspilus. P.S. (*wh.*) Bastard Quince. Pyrenees. 1683. —— H. ♄. ———— t. 87.
Méspilus chamæméspilus. L. *Crat'ægus chamæméspilus.* Jacq. aust. t. 231.
P'YRUS. DC. PEAR & APPLE. Icosandria Di-Pentagynia.
1 commùnis. DC. (*wh.*) common Pear. England. .... 4. H. 5. Eng. bot. v. 25. t. 1784.
α *'Achras.* DC. (*wh.*) *pointed-leaved.* ———— .... —— H. 5.
β *pyráster.* DC. (*wh.*) *round-leaved.* ———— .... —— H. 5. Gært. fr. t. 87. f. 2.
γ *satìva.* DC. (*wh.*) *cultivated.* ———— .... —— H. 5.
2 Bollwylleriàna. DC. (*wh.*) woolly-leaved. Germany. 1786. 4. 5. H. 5. Lodd. bot. cab. t. 1009.
3 salvifòlia. DC. (*wh.*) Sage-leaved. France. 1824. —— H. 5.
4 amygdalifórmis. DC. (*wh.*) Almond-like. ———— 1820. —— H 5.
5 elæagnifòlia. DC. (*wh.*) Oleaster-leaved. Tauria. 1818. —— H. 5.
6 salicifòlia. DC. (*wh.*) Willow-leaved. Siberia. 1780. 5. 6. H. 5. Bot. reg. t. 514.
7 crenàta. D.P. (*wh.*) crenated. Nepaul. 1820. —— H. 5.
8 sinàica. DC. (*wh.*) Mount Sinai. Levant. —— 5. H. 5. Wats. dend. brit. t. 49.
pérsica. P.S.
9 nivàlis. DC. (*wh.*) white-leaved. Austria. .... —— H. 5. Jacq. aust. 2. t. 107.
10 Páshia. D.P. (*wh.*) rough-fruited. Nepaul. 1825. 5. 6. H. 5.
11 Nússia. D.P. (*wh.*) small-fruited. ———— 1823. —— H. 5.
12 sinénsis. B.R. (*wh.*) Chinese. China. —— —— H. 5. Bot. reg. t. 1248.
13 acérba. DC. (*bh.*) sour. Europe. .... 4. 5. H. 5. Flor. dan. t. 1101.
14 Màlus. DC. (*bh.*) common Apple. Britain. .... —— H. 5. Eng. bot. v. 3. t. 179.
15 pr'æcox. (*bh.*) Paradise. Russia. 1784. 4. H. 5.
16 dioíca. DC. (*st.*) linear petal'd. .... 4. 5. H. 5.
17 astracànica. DC. (*bh.*) Astracan. Astracan. 1824. —— H. 5.
18 spectábilis. DC. (*pi.*) Chinese Apple. China. 1780. 5. H. 5. Bot. mag. t. 267.
19 prunifòlia. DC. (*bh.*) Siberian Crab. Siberia. 1758. 4. 5. H. 5. Mill. ic. 2. t. 269.
20 tomentòsa. DC. (*bh.*) woolly. ———— 1824. —— H. 5.
21 rubicúnda. DC. (*bh.*) red-fruited. .... 1820. —— H. 5.
22 baccàta. DC. (*bh.*) small Crab. Siberia. 1784. —— H. 5. Pall. fl. ross. t. 10.
23 coronària. DC. (*pi.*) sweet-scented. Virginia. 1724. 5. H. 5. Bot. reg. t. 651.
24 angustifòlia. DC. (*bh.*) narrow-leaved. N. America. 1750. —— H. ♄. Wats. dend. brit. 132.
25 pùmila. Mil. (*bh.*) dwarf evergreen......... .... 4. 5. H. ♄.
26 'Aria. DC. (*wh.*) white-beam. Britain. .... 5. 6. H. 5. Eng. bot. v. 26. t. 1858.
27 alpìna. W. en. (*wh.*) Alpine. Europe. 1820. —— H. 5. Duham. ed. nov. 4. t. 34.
longifòlia. Duham.
28 intermèdia. DC. (*wh.*) Swedish. Sweden. 1789. 4. 5. H. 5. Flor. dan. 301.
29 latifòlia. P.S. (*wh.*) broad-leaved. France. .... 5. 6. H. 5. Duham. arb. 1. t. 80.
30 èdulis. W. en. (*wh.*) eatable-fruited. Europe. 1816. —— H. 5. Wats. dend. brit. t. 52.
31 lanàta. D.P. (*wh.*) woolly. Nepaul. 1820. —— H. 5.
32 torminàlis. DC. (*wh.*) Wild Service. England. .... 4. 5. H. 5. Eng. bot. v. 5. t. 298.
33 auriculàta. DC. (*wh.*) eared-leaved. Egypt. 1828. .... H. 5.
34 pinnatífida. DC. (*wh.*) pinnatifid. England. .... 5. 6. H. 5. Eng bot. v. 33. t. 23-31.
35 arbúscula. Poir. (*wh.*) little tree. Germany. .... —— H. 5.
36 microcárpa. DC. (*wh.*) small-berried. N. America. .... —— H. 5.
37 aucupària. DC. (*wh.*) Mountain Ash. Britain. .... —— H. 5. Eng. bot. v. 5. t. 337.
38 americàna. DC. (*wh.*) purple-berried. Canada. 1782. —— H. 5. Wats. dend. brit. t. 54.
39 sorbifòlia. W.D. (*wh.*) Service-leaved. .... 5. H. 5. ———— t. 53
40 Sórbus. DC. (*wh.*) Service-tree. England. .... 5. 6. H. 5. Eng. bot. v. 5. t. 350.
doméstica. E.B. *Sórbus doméstica.* L.
41 lanuginòsa. DC. (*wh.*) woolly-leaved. Hungary. 1820. —— H. 5.
42 spùria. DC. (*bh.*) spurious. Hybrid. .... —— H. 5. Bot. reg. t. 1196.
hy'brida. Mœnch. Weis. t. 6. *Sórbus spuria.* P.S.

CYD'ONIA. DC.		QUINCE. Icosandria Pentagynia.					
1 vulgàris. DC.	(wh.)	common.	S. Europe.	1573.	5. 6.	H. 5.	Jacq. aust. 4. t. 342.
α malifórmis. DC.	(wh.)	*Apple-fruited.*	——	——	——	H. 5.	
β lusitànica. DC.	(wh.)	*large-fruited.*	——	——	——	H. 5.	Duha. arb. fr. 1. p. 195. ic.
γ oblónga. DC.	(wh.)	*oblong-fruited.*	——	——	——	H. 5.	Blackw. herb. t. 137.
2 sinénsis. DC.	(wh.)	Chinese.	China.	1815.	4. 5.	H. 5.	Bot. reg. t. 905.
3 japónica. DC.	(sc.)	Japan.	Japan.	1796.	1. 12.	H. ♄.	Bot. mag. t. 692.
α rùbra.	(sc.)	*red-flowered.*	——	——	——	H. ♄.	
P'yrus japónica. B.M. *Màlus japónica.* A.R. 462.							
β álba.	(wh.)	*white-flowered.*	——	1812.	——	H. ♄.	Lodd. bot. cab. t. 541.

ORDO LXXVIII.

ROSACEÆ. *D. Don prodr. fl. nepal. p.* 229.

Tribus I. *ROSEÆ.* DC. prodr. 2. p. 596.

R'OSA. DC.		ROSE. Icosandria Polygynia. L.					
1 berberifòlia. W.	(ye.)	Berberry-ld.	Persia.	1790.	6. 8.	H. ♄.	Redout. ros. 1. 27. t. 2.
simplicifòlia. P.L. 101. *Lo'wea berberifòlia.* Bot. reg. 1261.							
2 fèrox. L.R.	(pu.)	hedgehog.	Caucasus.	1796.	6. 8.	H. ♄.	Bot. reg. t. 420.
3 kamchática. L.R.	(pu.)	Kamtschatka.	Kamtschatka.	1802.	7. 8.	H. ♄.	—— t. 419.
β nitens. B.R.	(pu.)	*glossy.*	——	1820.	——	H. ♄.	—— t. 824.
4 involucràta. L.R.	(wh.)	floral-leaved.	China.	1814.	5. 10.	H. ♄⌣.	—— t. 739.
5 clinophy'lla. R.R.	(wh.)	naked-flowered.	——	1820.	——	H. ♄⌣.	Redout. ros. 1. t. 10.
6 bracteàta. L.R.	(wh.)	Macartney.	——	1795.	6. 10.	H. ♄⌣.	—— 1. t. 6.
β scabriúscula. L.R.	(w.)	*rough Macartney.*	——	——	——	H. ♄⌣.	Bot. mag. 1377.
7 nítida. W. en.	(ro.)	dwarf Labrador.	N. America.	1773.	7. 8.	H. ♄⌣.	Lindl. ros. t. 2.
8 ràpa. R.R.	(ro.)	double Burnet.	——	1726.	——	H. ♄.	Redout. ros. 1. 7. t. 2.
9 lùcida. W.	(re.)	single Burnet.	——	1773.	——	H. ♄.	Miss Lawr. ros. t. 24.
10 Lindlèyi. S.S.	(ro.)	spreading.	——	1726.	——	H. ♄.	Lindl. ros. t. 3.
láxa. L.R. *non* Retz.							
11 parviflòra. W.	(bh.)	small-flowered.	——	1724.	——	H. ♄.	
α Single Pensylvanian	(bh.)	....		——	——	H. ♄.	Miss Lawr. ros. t. 68.
β Double Pensylvanian	(bh.)	....		——	——	H. ♄.	—— t. 66.
12 Woòdsii. L.R.	(pi.)	Woods's.	——	1815.	6. 7.	H. ♄.	
13 gemélla. S.S.	(re.)	twin-flowering.	——	1800.	7. 8.	H. ♄.	
14 Lyònii. S.S.	(ro)	Lyon's.	Tennassee.	1812.	——	H. ♄.	
15 flórida. L.C.	(re.)	flowering.	N. America.	——	6. 7.	H. ♄.	
16 carolìna. L.R.	(re.)	Carolina.	——	1726.	——	H. ♄.	Lindl. ros. t. 4.
17 pensylvànica. L. en.	(re.)	Pensylvanian.	——	——	——	H. ♄.	
18 blánda. L.R.	(re.)	Hudson's Bay.	——	1773.	5. 8.	H. ♄.	Miss Lawr. ros. t. 27.
19 fraxinifòlia. L.R.	(re.)	Ash-leaved.	Newfoundland.	—	4. 6.	H. ♄.	Bot. reg. 458.
β Red Alpine Rose.	(re.)		....		——	H. ♄.	Miss Lawr. ros. t. 75.
20 cinnamòmea. E.B.	(ro.)	Single Cinnamon.	England.	....	5.	H. ♄.	Eng. bot. v. 34. t. 2388.
β flòre semiplèno.	(ro.)	*semidouble.*	——	....	——	H. ♄.	
γ fœcundissima.	(ro.)	*double Cinnamon.*	Europe.	....	4. 5.	H. ♄.	Lindl. ros. t. 5.
21 majális. L.R.	(ro.)	dwarf Cinnamon.	S. Europe.	....	——	H. ♄.	Flor. dan. t. 688.
22 frutetòrum. S.S.	(bh.)	short peduncl'd.	Volhynia.	1820.	5. 6.	H. ♄.	
23 ambígua. S.S.	(bh.)	ambiguous.	....		——	H. ♄.	
24 macrophy'lla. L.R.	(ro.)	long-leaved.	Nepaul.	1822.	....	H. ♄.	Lindl. ros. t. 6.
25 americàna. L. en.	(ro.)	American.	America.	——	....	H. ♄.	
26 alpìna. L.R.	(re.)	Alpine.	Europe.	1683.	6. 7.	H. ♄.	Bot. reg. 424.
α Alpine Rose.	(re.)	*common.*	——	——	——	H. ♄.	Miss Lawr. ros. t. 30.
β pyrenàica. L.R.	(re.)	*Pyrenean.*	——	——	——	H. ♄.	Jacq. schœnb. 4. t. 416.
γ pendulìna. L.R.	(re.)	*pendulous.*	——	1726.	——	H. ♄.	Miss Lawr. ros. t. 9.
27 pygm'æa. S.S.	(ro.)	pigmy.	Tauria.	....	——	H. ♄.	
28 rubélla. E.B.	(re.)	long-fruited.	England.	....	7.	H. ♄.	Eng. bot. v. 36. t. 2521.
β melanocárpa. L.R.	(re.)	*black-fruited.*	——	....	——	H. ♄.	

29	muricàta. L.en.	(*ro.*)	muricate.	....	1823.	6. 7.	H. ♄.	
30	balsámica. S.S.	(*ro.*)	balsamic.	....	——	——	H. ♄.	
31	stricta. L.R.	(*ro.*)	upright.	N. America.	1726.	——	H. ♄.	Lindl. ros. t. 7.
	Upright Carolina.	(*ro.*)		———	——	——	H. ♄.	Miss Lawr. ros. t. 36.
32	suávis. S.S.	(*ro.*)	sweet.	....	1817.	——	H. ♄.	
33	aciculàris. L.R.	(*ro.*)	grey Siberian.	Siberia.	....	3. 4.	H. ♄.	Lindl. ros. t. 8.
34	pauciflòra. Ly.	(*ro.*)	few-flowered.	———	....	——	H. ♄.	
35	sulphùrea. W.	(*st.*)	double yellow.	Levant.	1629.	7.	H. ♄.	Bot. reg. 46.
36	lutéscens. Ph.	(*st.*)	yellowish.	Siberia.	1780.	5. 6.	H. ♄.	Lindl. ros. t. 9.
	híspida. Bot. mag. t. 1570.							
37	vimínea. L.R.	(*ro.*)	twiggy.	Russia.	1826.	5. 6.	H. ♄.	
38	spinosíssima. E.B.	(*wh.*)	Scotch.	Britain.	....	6. 7.	H. ♄.	Eng. bot. v. 3. t. 187.
	β revérsa. L.R.	(*wh.*)	*reversed prickle.*	———	....	——	H. ♄.	Bot. reg. 431.
	γ striped-flowered Scotch.	(*std.*)		———	....	——	H. ♄.	Miss Lawr. ros. t. 15.
	δ Red Scotch.	(*re.*)		———	....	——	H. ♄.	——— t. 62.
	ε Double Scotch.	(*wh.*)		———	....	——	H. ♄.	——— t. 63.
	ζ Tall Scotch.	(*wh.*)		———	....	——	H. ♄.	——— t. 19.
	η Marbled Scotch.	(*pu.*) (*wh.*)		———	....	——	H. ♄.	——— t. 78.
	ϑ sanguisorbifòlia. L.R.	(*wh.*)	*Sanguisorba-ld.*	....	....	——	H. ♄.	
39	altàica. W.en.	(*wh.*)	Pallas's.	Siberia.	....	——	H. ♄.	Pall. ros. 2. t. 75.
40	ochroleùca. L.en.	(*st.*)	sulphur-color'd.	———	1814.	——	H. ♄.	
41	melanocárpa. S.S.	(*wh.*)	small black fruited.	S. Europe.	1823.	——	H. ♄.	
42	grandiflòra. L.R.	(*wh.*)	large flowered.	Siberia.	....	5. 6.	H. ♄.	Bot. reg. 888.
43	myriacántha. DC.	(*wh.*)	numerous-spin'd.	S. France.	1820.	——	H. ♄.	Lindl. ros. t. 10.
44	involùta. E.B.	(*bh.*)	Dr. Walker's.	Hebrides.	....	6. 7.	H. ♄.	Eng. bot. v. 29. t. 2068.
45	revérsa. L.R.	(*bh.*)	reversed.	Hungary.	1817.	——	H. ♄.	W. et K. hung. 3. t. 264.
46	Biebersteínii. Ly.	(*ro.*)	Bieberstein's.	Caucasus.	1821.	5.	H. ♄.	
47	Doniàna. L.T.	(*wh.*)	Mr. Don's.	Britain.	....	6. 7.	H. ♄.	
48	grácilis. L.T.	(*ro.*)	slender.	———	....	7. 8.	H. ♄.	Eng. bot. v. 9. t. 583.
	villòsa. E.B. *non* L.							
49	Sabini. L.T.	(*ro.*)	Mr. Sabine's.	———	....	——	H. ♄.	
50	marginàta. L.R.	(*ro.*)	margined.	Germany.	1823.	——	H. ♄.	
51	corúscans. L.en.	(*ro.*)	glittering.	....	——	——	H. ♄.	
52	damascèna. L.R.	(*re.*)	red damask.	Syria?	1575.	6. 7.	H. ♄.	Miss Lawr. ros. t. 38.
53	centifòlia. L.R.	(*ro.*)	Provins.	Caucasus.	1596.	——	H. ♄.	——— t. 8.
	β muscòsa. L.R.	(*ro.*)	*single moss.*	....	1724.	——	H. ♄.	Bot. reg. 53.
	γ multipléx.	(*ro.*)	*double moss.*	....	——	——	H. ♄.	Bot. mag. t. 69.
	δ albiflòra.	(*wh.*)	*white moss.*	....	——	——	H. ♄.	Bot. reg. t. 102.
	ε pomponia. L.R.	(*pi.*)	*Pompone.*	....	....	——	H. ♄.	Redout. ros. 1. 65. t. 21.
	ζ bipinnata. L.R.	(*pi.*)	*bipinnate.*	Caucasus.	....	——	H. ♄.	——— 2. 11. t. 4.
54	gállica. L.R.	(*re.*)	officinal.	Europe.	1596.	——	H. ♄.	Bot. reg. 448.
	β pùmila. L.R.	(*re.*)	*dwarf.*	Austria.	1773.	——	H. ♄.	Jacq. aust. 2. t. 198.
55	pulchélla. S.S.	(*re.*)	neat.	France.	....	——	H. ♄.	
56	parvifòlia. L.R.	(*pu.*)	Burgundy.	Europe.	....	——	H. ♄.	Bot. reg. t. 452.
57	turbinàta. L.R.	(*ro.*)	Frankfort.	———	1629.	——	H. ♄.	Miss Lawr. ros. t. 69.
58	villòsa. L.T.	(*bh.*)	villous.	Britain.	....	6.	H. ♄.	Eng. bot. v. 35. t. 2459.
	móllis. E.B.							
	α single apple bearing.	(*bh.*)	———	———	....	——	H. ♄.	Miss Lawr. ros. t. 33.
	β cærùlea. L.T.	(*bh.*)	*blue-stalked.*	———	....	——	H. ♄.	
	γ concavifòlia. L.T.	(*bh.*)	*concave-leaved.*	Scotland.	....	——	H. ♄.	
	δ suberécta. L.T.	(*bh.*)	*suberect.*	Yorkshire.	....	——	H. ♄.	
	ε double apple bearing.	(*pi.*)		———	....	——	H. ♄.	
59	scabriúscula. L.T.	(*bh.*)	roughish.	Britain.	....	——	H. ♄.	Miss Lawr. ros. t. 29.
60	heterophy'lla. L.T.	(*bh.*)	various-leaved.	Scotland.	....	——	H. ♄.	Eng. bot. t. 1896.
61	tomentòsa. L.T.	(*bh.*)	downy-ld. dog.	Britain.	....	——	H. ♄.	Eng. bot. v. 14. t. 990.
	β sylvéstris. L.T.	(*bh.*)	*wood.*	———	....	——	H. ♄.	
	γ canéscens. L.T.	(*bh.*)	*canescent.*	———	....	——	H. ♄.	
	δ incàna. L.T.	(*bh.*)	*hoary.*	———	....	——	H. ♄.	
	ε resinòsa. L.R.	(*re.*)	*resinous.*	Ireland.	....	——	H. ♄.	
62	nùda. L.T.	(*bh.*)	naked.	Westmoreland.	..	——	H. ♄.	
63	álba. L.R.	(*wh.*)	single white.	Europe.	1597.	6. 7.	H. ♄.	Miss Lawr. ros. t. 37.
64	montàna. S.S.	(*bh.*)	mountain.	Tauria.	1820.	——	H. ♄.	
65	hibérnica. L.T.	(*pi.*)	Irish.	Ireland.	....	6. 11.	H. ♄.	Eng. bot. v. 31. t. 2196.
66	lùtea. L.R.	(*ye.*)	yellow.	Germany.	1596.	6.	H. ♄.	Bot. mag. 363.
67	punícea. Mil.	(*re. ye.*)	red and yellow.	Persia.	——	——	H. ♄.	——— t. 1077.
68	Sherárdi. Sm.	(*bh.*)	Sherard's.	Britain.	....	6. 7.	H. ♄.	
69	Dicksoniàna. L.B.	(*bh.*)	Dickson's.	Ireland.	....	——	H. ♄.	
70	rubiginòsa. E.B.	(*ro.*)	sweet briar.	Britain.	....	——	H. ♄.	Eng. bot. v. 14. t. 991.
71	micrántha. L.T.	(*ro.*)	small.	England.	....	——	H. ♄.	——— v. 35. t. 2490.

72	sèpium. DC.	(*bh.*) hedge.	France.	....	——	H. ♄.	
73	Borrèri. L.T.	(*bh.*) Borrer's.	Britain.	....	——	H. ♄.	Eng. bot. v. 36. t. 2579.
	dumetòrum. E.B. *nec aliorum.*						
74	umbellàta. DC.	(*ro.*) umbel-flowered.	———	....	——	H. ♄.	
75	adenophy'lla. S.S.	(*bh.*) glandulous.	....	1816.	——	H. ♄.	
76	pruinòsa. L.C.	(*ro.*) frosted.	....	1817.	5. 6.	H. ♄.	
	pulverulénta. L.R. *non* M.B.						
77	glutinòsa. L.R.	(*bh.*) glutinous.	Levant.	1821.	6. 7.	H. ♄.	Cupan. panp. ed. 1. t. 61.
	rubiginòsa crética. Red. ros. 125. t. 47.						
78	Montezùmæ. K.S.	(*ro.*) Mexican.	Mexico.	1823.	....	H. ♄.	Redout. ros. 1. t. 16.
79	rèpens. Scop.	(*bh.*) creeping.	....	1825.	6. 7.	H. ♄.	
80	uncinàta. Bes.	(*bh.*) hooked.	Poland.	1826.	——	H. ♄.	
81	Boreykiàna. Bes.	(*bh.*) Boreyki's.	....	1827.	——	H. ♄.	
82	Fischeriàna. Bes.	(*bh.*) Fischer's.	Russia.	1825.	——	H. ♄.	
83	gorenkénsis. Bes.	(*bh.*) Gorenki.	———	——	——	H. ♄.	
84	guttensteiniénsis. Bes.	(*bh.*) Guttenstein.	Poland.	1822.	——	H. ♄.	
85	Kosinskiàna. Bes.	(*bh.*) Koscinszki.	Europe.	1825.	——	H. ♄.	
86	polyphy'lla. Bes.	(*bh.*) many-leaved.	....	1827.	——	H. ♄.	
87	terebinthinàcea. Bes.	(*bh.*) Turpentine.	Germany.	1825.	——	H. ♄.	
88	Redowskiàna. Bes.	(*bh.*) Redowski's.	....	1828.	——	H. ♄.	
89	uncinélla. Bes.	(*bh.*) small-hooked.	Europe.	1825.	——	H. ♄.	
90	dimórpha. Bes.	(*bh.*) variable.	Volhynia.	1821.	——	H. ♄.	
91	caryophyllàcea. Bes.	(*bh.*) Clove-scent'd.	Poland.	1825.	——	H. ♄.	
92	glaùca. Bes.	(*bh.*) glaucous.	———	——	——	H. ♄.	
93	nitídula. Bes.	(*bh.*) glossy.	Europe.	——	——	H. ♄.	
94	floribúnda. Bes.	(*bh.*) abundant-fl'd.	———	1828.	——	H. ♄.	
95	solstitiàlis. Bes.	(*bh.*) summer.	Poland.	1821.	——	H. ♄.	
96	acùta. F.	(*bh.*) acute-leaved.	Caucasus.	——	——	H. ♄.	
97	purpuráscens. F.	(*pu.*) purplish.	Tanaim.	——	——	H. ♄.	
98	rupíncola. F.	(*bh.*) rocky.	Altay.	——	——	H. ♄.	
99	saxátilis. M.B.	(*bh.*) rock.	Tauria.	1825.	——	H. ♄.	
100	caucásica. M.B.	(*bh.*) Caucasean.	Caucasus.	1793.	5. 6.	H. ♄.	Lindl. ros. t. 11.
101	c'æsia. L.T.	(*bh.*) grey-leaved.	Scotland.	....	7.	H. ♄.	Eng. bot. v. 33. t. 2367.
102	sarmentàcea. L.T.	(*ro.*) sarmentose.	Britain.	....	6. 7.	H. ♄.	
	β *nitens.* L.T.	(*ro.*) *glossy-leaved.*	———	....	——	H. ♄.	
103	bractéscens. L.T.	(*bh.*) long-bracted.	Lancashire.	....	——	H. ♄.	
104	dumetòrum. L.T.	(*bh.*) bushy.	England.	....	6.	H. ♄.	
105	Forstèri. Sm.	(*bh.*) Forster's.	———	....	——	H. ♄.	
106	collìna. L.T.	(*bh.*) hill.	———	....	——	H. ♄.	Jacq. aust. 2. t. 197.
107	canìna. L.T.	(*bh.*) dog or hep.	Britain.	....	6. 7.	H. ♄.	Eng. bot. v. 14. t. 992.
	β *cera.* L.T.	(*bh.*) *waxy.*	———	....	——	H. ♄.	
	γ *glandulifera.* L.T.	(*bh.*) *glandular.*	———	....	——	H. ♄.	
	δ *simpliciuscula.* L.T.	(*bh.*) *simple-calyxed.*	———	....	——	H. ♄.	
108	surculòsa. L.T.	(*bh.*) surculose.	———	....	——	H. ♄.	
109	leucántha. L.en.	(*wh.*) white-flowered.	Siberia.	1824.	....	H. ♄.	
110	rubrifòlia. L.R.	(*re.*) reddish-leaved.	S. Europe.	....	5. 6.	H. ♄.	Bot. reg. t. 430.
111	serícea. L.R.	(*bh.*) silky-leaved.	Nepaul.	1823.	....	H. ♄.	Lindl. ros. t. 12.
112	microphy'lla. L.R.	(*re.*) small-leaved.	China.	1822.	6. 8.	H. ♄.	Bot. reg. t. 919.
113	longifòlia. W.	(*bh.*) Willow-leaved.	———	....	2. 10.	F. ♄.	Redout. ros. 2. 27. t. 12.
114	índica. W.	(*ro.*) common China.	———	1789.	1. 12.	H. ♄.	Miss Lawr. ros. t. 26.
	β *centifòlia.*	(*ro.*) *large-double.*	———	....	——	H. ♄.	
	γ *purpùrea.*	(*pu.*) *purple China.*	———	....	——	H. ♄.	
	δ *subálba.*	(*bh.*) *whitish China.*	———	....	——	H. ♄.	
	ε *mèdia.*	(*ro.*) *middle China.*	———	....	——	H. ♄.	
	ζ *mìnor.*	(*ro.*) *small China.*	———	....	——	H. ♄.	Andr. ros. ic.
	η *fràgrans.*	(*pu.*) *animating.*	———	....	——	H. ♄.	
115	odoràta.	(*pa.*) sweet-scented.	———	1810.	2. 12.	H. ♄.	Bot. reg. 804.
	índica odoràta. A.R. *índica fràgrans.* R.R. 1. 115. t. 42.						
	β *flavéscens.*	(*st.*) *yellow China.*	China.	1823.	——	H. ♄.	
116	semperflòrens. L.R.	(*cr.*) red China.	———	1789.	1. 12.	F. ♄.	Exot. bot. 2. t. 91.
	α *diversifòlia.* V.	(*re.*) *single.*	———	....	——	F. ♄.	Vent. cels. t. 35.
	β *atrorùbens.*	(*cr.*) *double red.*	———	1789.	....	F. ♄.	Bot. mag. 284.
	γ *respléndens.*	(*cr.*) *resplendent.*	———	....	——	F. ♄.	
	δ *fràgrans.*	(*re.*) *fragrant.*	———	1820.	——	F. ♄.	
	ε *Terneáuxi.*	(*re.*) *Terneaux's.*	———	——	——	F. ♄.	
	ζ *Bichónii.*	(*re.*) *Bichon's.*	———	——	——	F. ♄.	
117	Lawranceàna. L.R.	(*bh.*) Miss Lawrance's.	———	1810.	——	F. ♄.	Bot. reg. t. 538.
	β *flore plèno.*	(*bh.*) *double-flowered.*	———	1824.	——	F. ♄.	

118	syst'yla. L.T.	(*bh.*)	clustered dog.	Britain.	....	6. 7.	H.	♄.	Eng. bot. v. 27. t. 1895.
	β *lanceolata.* L.R.	(*bh.*)	spear-leaved.	Ireland.	....	——	H.	♄.	
	γ *Monsoniæ.* L.R.	(*bh.*)	Lady Monson's.	England.	....	——	H.	♄.	
119	arvénsis. L.T.	(*wh.*)	field.	Britain.	....	——	H.	♄.	Eng. bot. v. 3. t. 188.
	β *scándens.*	(*wh.*)	*Ayrshire.*	———	....	——	H.	♄.◡.	Bot. mag. 2054.
120	geminàta. Rau.	(*bh.*)	Swiss.	Switzerland.	....	——	H.	♄.	
121	sempervìrens. W.	(*wh.*)	evergreen.	S. Europe.	1629.	6. 8.	H.	♄.◡.	Bot. reg. t. 465.
122	Boursoùlti.	(*ro.*)	Boursoult's.	Hybrid.	1821.	4. 7.	H.	♄.◡.	
123	hyacinthìna.	(*ro.*)	Hyacinth-scented.	———	——	——	H.	♄.◡.	
124	Grevíllii.	(*ro.bh.*)	Greville's.	China.	....	——	H.	♄.◡.	
125	Roxbúrghii.	(*ro.*)	Roxburgh's.	———	....	——	H.	♄.◡.	
126	multiflòra. W.	(*va.*)	many-flowered.	———	1804.	6. 8.	H.	♄.◡.	Bot. reg. 425.
	α *ròsea.*	(*ro.*)	*common.*	———	——	——	H.	♄.◡.	Bot. mag. t. 1059.
	β *albiflòra.*	(*wh.*)	white-flowered.	———	1825.	——	H.	♄.◡.	
127	Brunònii. L.R.	(*wh.*)	Mr. Brown's.	Nepaul.	1820.	——	H.	♄.	Lindl. ros. t. 14.
128	moschàta. L.R.	(*wh.*)	musk-scented.	Africa.	1596.	7. 9.	H.	♄.	Miss Lawr. ros. t. 64.
	β *flòre plèno.*	(*wh.*)	*double musk.*	———	——	——	H.	♄.	——— t. 53.
129	nívea. Dup.	(*wh.*)	snow white.	Hybrid.	1823.	——	H.	♄.	Bot. reg. 861.
130	Noisettiàna. R.R.	(*bh.*)	Noisette's.	———	1816.	5. 11.	H.	♄.	
	β *grandiflòra.*	(*bh.*)	*large-flowered.*	———	1827.	——	H.	♄.	
	γ *Fraserì.*	(*bh.*)	*Fraser's.*	———	1815.	——	H.	♄.	
	δ *purpùrea.*	(*pu.*)	*purple.*	———	1823.	——	H.	♄.	
131	rubifòlia. L.R.	(*re.*)	bramble-leav'd.	N. America.	1800.	7. 9.	H.	♄.	
	β *fenestràlis.* L.R.	(*re.*)	*thin-leaved.*	———		——	H.	♄.	Lindl. ros. t. 15.
132	lævigàta. L.R.	(*wh.*)	smooth.	Georgia.	1826.	——	H.	♄.◡.	Bot. mag. t. 2847.
133	sínica. L.R.	(*wh.*)	shining-leaved.	China.	1759.	6. 8.	F.	♄.	Lindl. ros. t. 16.
134	Bánksiæ. L.R.	(*wh.*)	Lady Banks's.	———	1807.	——	H.	♄.◡.	Bot. reg. t. 397.
	α *albiflòra.*	(*wh.*)	white-flowered.	———	——	——	H.	♄.◡.	Bot. mag. t. 1954.
	β *lùtea.*	(*ye.*)	*yellow-flowered.*	———	1823.	——	H.	♄.◡.	——— v. 13. t. 1105.

GARDEN VARIETIES.

I. *spinosissima.*

1 Aberdeen.
2 Aberfoil.
3 alba.
4 Alloa.
5 Ancram.
6 Anderson's double lady's [blush.
7 Arbroath.
8 Argyll.
9 Arrochar.
10 Aurora borealis.
11 Balloch.
12 Banff.
13 Bannochburn.
14 Borisdale.
15 Bass.
16 Bengloe.
17 Ben Lomond.
18 Ben More.
19 Berwick.
20 bicolor.
21 Biggar.
22 Birnam.
23 Blair Athol.
24 Boharm.
25 Borthwick.
26 Buchan.
27 Bute.
28 Caithness.
29 Calder.
30 Callender.
31 Campsie.
32 carnescens.
33 Carron.
34 Cheviot.
35 ciphieri.

GARDEN VARIETIES.

36 Clydesdale.
37 Cromarty.
38 Dalkeith.
39 Dalrymple.
40 Dornock.
41 double blush Princess.
42 double blush Provins.
43 double crimson.
44 double crimson mottled.
45 double dark marbled.
46 double dark red.
47 double lady's blush.
48 double light marbled.
49 double light purple.
50 double pink.
51 double purple.
52 double rose blush.
53 Douglas.
54 Dumbarton.
55 Dumblane.
56 Dumfernline.
57 Dumfries.
58 Dunbar.
59 Duncrieff.
60 Dundee.
61 Dunglass.
62 Dunkeld.
63 Dunlop.
64 Dunmore.
65 Dutch double blush.
66 dwarf bicolor.
67 Dysart.
68 Eden.
69 Elgin.
70 Etterick.
71 Fairy.
72 Falkirk.

GARDEN VARIETIES.

73 Falkland.
74 Falla.
75 Fife.
76 flavescens.
77 Forfar.
78 Forth.
79 Fort William.
80 fulgens.
81 Furness.
82 Galloway.
83 Glasgow.
84 Glenco.
85 Glenfallach.
86 Glengarry.
87 Gourock.
88 Grahamston.
89 Grampian.
90 Greenock.
91 Haddington.
92 Halkirk.
93 Hamilton.
94 Hawick.
95 Hawthorndean.
96 Hector.
97 humilis.
98 Huntly.
99 incarnata.
100 Inverary.
101 Inverness.
102 Invermay.
103 Janus.
104 Jedburgh.
105 Jura.
106 Keith.
107 Kelso.
108 Kilmarnock.
109 Kincardine.

GARDEN VARIETIES.

110 Kinnaird.
111 Kinross.
112 Kircaldy.
113 Kirkwall.
114 Lanark.
115 Laxford.
116 Leith.
117 Leslie.
118 Lismore.
119 Lochaber.
120 Lochaird.
121 Loch fine.
122 Lochleven.
123 Lochlomond.
124 Lothian.
125 Maidenkirk.
126 Melrose.
127 Moncrieff.
128 Monteith.
129 Montgomery.
130 Montrose.
131 Mount Stuart.
132 Mull.
133 Moray.
134 mutabilis.
135 Nevis.
136 Northumberland.
137 Paisley.
138 penicillata.
139 Pentland.
140 perpetual.
141 picta. L.B.C. 687.
142 Proteus.
143 red, double light.
144 Roberton.
145 Rosslyn.
146 Rothsay.
147 rubicunda.
148 Selkirk.
149 Shetland.
150 Sky.
151 small double light purple.
152 small double white.
153 small double yellow.
154 Staffa.
155 Stirling.
156 Strathmore.
157 striata.
158 Stronsa.
159 Sutherland.
160 Tiviotdale.
161 Tarbet.
162 Thornhill.
163 Thurso.
164 Tranent.
165 true double red.
166 two-coloured large double.
167 two-coloured small double.
168 variegata.
169 venulosa glabra.
170 venulosa hispida.
171 white, large double.
172 white, large semidouble.
173 white, small double.
174 Whitley's double white.
175 yellow, globe double.
176 yellow, large double.
177 yellow, pale double.

GARDEN VARIETIES.

II. *gallica.*

178 admirable.
179 aigle noir.
180 aimable beaute.
181 albanian.
182 Amaranth.
183 Antwerp.
184 atlas.
185 beaute rouge.
186 beaute supreme.
187 belle aurore.
188 belle cramoisie.
189 belle herminie.
190 belle pourpre.
191 belle violette.
192 Bijou.
193 Bishop. M.L.R. 20.
194 black frizzled.
195 blue.
196 blue and purple.
197 blush hundred-leaved.
198 bouquet rouge royale.
199 bright purple.
200 brunette.
201 Brussels.
202 Buonaparte.
203 burning coal.
204 cardinal M.L.R. 59.
205 carmine.
206 carmine, brillante.
207 carmine, proliferous.
208 carnation.
209 Catalonian.
210 champion.
211 chancellor.
212 changeable.
213 cherry.
214 Clementine.
215 coquette.
216 couleur de feu.
217 cramoisie, grand.
218 crown.
219 Cupid.
220 damask black.
221 dark marbled.
222 dark violet.
223 delicious.
224 dingy.
225 double velvet.
226 double marbled. M.L.R. 57.
227 Duc de Guiche.
228 Duchess d'Orleans.
229 dutch crimson.
230 dutch tree.
231 dutch 100 leaved.
232 early Ranunculus.
233 enchanter.
234 enfant de France.
235 Eucharis.
236 Fanny Bias.
237 favourite purple.
238 fiery.
239 Flanders.
240 Flemish.
241 formidable.
242 fringed. M.L.R. 82.
243 garnet.
244 gay.

GARDEN VARIETIES.

245 giant. M.L.R. 49.
246 gloria mundi.
247 gloria mundi flore pallido.
248 gloria mundi pourpre.
249 granaat appel.
250 grand monarque.
251 grand purple.
252 grand Sultan.
253 Henry IV.
254 Hervy.
255 imperatrice.
256 incomparable.
257 infernal.
258 invincible.
259 Italian.
260 Josephine.
261 Junon.
262 king.
263 La Dauphine.
264 la grand belle pourpre.
265 Leyden.
266 light purple.
267 Lisbon. M.L.R. 67.
268 lively.
269 L'ombre agreable.
270 L'ombre superbe.
271 lurid.
272 maiden.
273 Majorca.
274 Malabar.
275 Malta.
276 manteau royal.
277 marbled.
278 marbled, grand.
279 Margaret.
280 matchless.
281 Mauve.
282 mignonne.
283 mignonne, blush.
284 mignonne, dark.
285 mignonne, favorite.
286 mignonne, incised.
287 mignonne, red.
288 mignonne, semidouble.
289 mignonne, striped.
290 mirabelle.
291 Mogul.
292 Montaubon.
293 Morocco.
294 mottled, black.
295 mundi. M.L.R. 13.
296 natalie.
297 negrette.
298 negro.
299 ninon de l'Enclos.
300 nonpareil.
301 nonsuch.
302 noir de Hollande.
303 Normandy.
304 officinal.
305 officinal blush.
306 officinal carmine.
307 Orleans.
308 ornement de parade.
309 Pæstana.
310 paradise.
311 paragon.
312 pavot.

GARDEN VARIETIES.

313 perruque.
314 petite 100 leav'd. M.L.R. 55.
315 petite panachèe.
316 phœnix.
317 plicate. M.L.R. t. 87.
318 Pluto. M.L.R. 32.
319 Pomona.
320 pompadour.
321 pompone bizard.
322 Poniatowsky.
323 poppy.
324 porcelaine.
325 Portland.
326 pourpre bouquet.
327 pourpre charmante.
328 pourpre de Tyr.
329 pourprée, grand belle.
330 pourprée, point.
331 pourpre velours.
332 prince.
333 princess.
334 prince William V.
335 prolific.
336 Pronville.
337 Proserpine.
338 Provins pulmonaire.
339 purple blue.
340 purple crimson.
341 pyramidal.
342 queen.
343 Ranunculus.
344 red and violet.
345 Roi de France.
346 Roi de pourpres.
347 royal crimson.
348 royal red.
349 royal purple.
350 royal virgin. M.L.R. 7.
351 sable.
352 Saint John's.
353 Sanspareil.
354 sceptre.
355 semidouble velvet.
356 shell. M.L.R. 89.
357 single velvet.—51.
358 Singleton's 100 leaved.—85.
359 Spanish.
360 stadtholder.
361 Stepney. M.L.R. 46.
362 striped nosegay.
363 striped velvet.
364 superb red.
365 Sultan. M.L.R. 35.
366 Trafalgar.
367 triumphant.
368 Tuscany. B.R. v. 6. t. 448.
369 two-colored.
370 Venetian.
371 victory.
372 violette and rouge.

III. *centifolia.*

373 aurora.
374 Belle d'Aunay.
375 Blandford, M.L.R. 21.
376 black mottled.
377 blush royal. M.L.R. 47.

GARDEN VARIETIES.

378 blush cabbage.
379 Bourbon.
380 bright crumpled.
381 carmine.
382 Centfeuilles Anemone.
383 Centfeuilles de Bruxelles.
384 Centfeuilles de Hesse.
385 Centfeuilles gaufrèe.
386 chamois.
387 cluster.
388 Constance.
389 Cumberland.
390 dragon.
391 Duchesse d'Angoulème.
392 Duchesse de Berri.
393 early 100 leaved.
394 Elysian.
395 emperor.
396 grand chamois.
397 grand marbled.
398 grand Provins.
399 Louis XVIII.
400 Malta.
401 mère gryone.
402 mignonne, scarlet.
403 mignonne, crimson.
404 mignonne, purple.
405 mottled purple.
406 Neapolitan.
407 one-sided.
408 Œillet.
409 pencilled.
410 petite Hollande.
411 Persian.
412 pompone, gros.
413 pourprée aimable.
414 pourprée favorite.
415 pourprée violette.
416 Provins, single. R.R. 1. t. 1.
417 Provins, common. M.L.R. 8.
418 Provins, cabbage.
419 Provins, blush cabbage.
420 Provins, scarlet, M.L.R. 22.
421 Provins, Childings.—43.
422 Provins, blush.—1.
423 Provins, white.—4.
424 Provins, Shailer's.—76.
425 Provins, Colvill's.
426 Provins, damask.
427 Provins, dwarf.
428 Provins, invincible.
429 Provins, Dutch.
430 Provins, imperial.
431 Provins, royal.
432 Provins, early.
433 Provins, semidouble.
434 red Belgic.
435 rouge superbe.
436 sans petales.
437 souchet.
438 Spong's.
439 superb carmine.
440 surpassante.
441 Syren.
442 Trianon, Belle de
443 Versailles.
444 Vilmorin.

GARDEN VARIETIES.

* *centifolia* muscosa.
445 single moss.
446 common moss.
447 blush moss.
448 dark moss.
449 mottled moss.
450 prolific moss.
451 royal moss.
452 scarlet moss.
453 semidouble moss.
454 striped moss.
455 Shailer's white moss.
456 Bristol white moss.
† *centifolia* pomponia.
457 de Meax. M.L.R. 31.
458 de Rheims.—71.
459 dwarf Bagshot.
460 mignonne charmante.
461 mossy de Meaux.
462 mossy Pompone.
463 Pompone. M.L.R. 50.
464 proliferous Pompone.
465 St. Francis. M.L.R. 88.
466 small provins.

IV. *damascena.*

467 argentea.
468 Aurora.
469 bifera carnea.
470 bifera de Naples.
471 bifera grandiflora.
472 Belgique carnèe.
473 Belgique violette.
474 belle couronnee.
475 blush Belgic. M.L.R. 70.
476 blush monthly.—18.
477 blush damask.—52.
478 Brunswick.
479 couronnèe petite.
480 damas argente.
481 damas pourpre.
482 early blush.
483 Egyptian.
484 Emperor.
485 favorite mignonne.
486 felicitè.
487 Goliath. M.L.R. t. 80.
488 gracieuse.
489 grande monarque.
490 great royal. M.L.R. 58.
491 imperial blush. M.L.R. 90.
492 incomparable.
493 lesser Belgic.
494 pale cluster.
495 paragon.
496 Parnassus.
497 Pæstana.
498 petite 100-leaved.
499 perpetual.
500 prolific.
501 quatre saisons.
502 quatre saisons blanche.
503 quatre saisons, flesh-colored.
504 quatre saisons Francois.
505 quatre saisons panaché.
506 quatre saisons pompone.
507 quatre saisons sans épines.
508 quatre saisons, semidouble.

GARDEN VARIETIES.

509 red damask. M.L.R. 38.
510 red monthly. M.L.R. 5.
511 red Belgic.—83.
512 rouge agathe.
513 Swiss.
514 valiant.
515 Watson's blush.
516 white damask.
517 white monthly. M.L.R. 17.
518 York and Lancaster.—10.
519 Zealand.

V. *alba.*

520 agate.
521 agate magnifique.
522 agate Marie Louise.
523 agate nouvelle.
524 agate prolifere.
525 agate royale.
526 agate superb.
527 agate thalie.
528 belle aurore.
529 belle Henriette.
530 blanche à cœur vert.
531 blanche de Belgique.
532 bouquet blanc.
533 cœlestial.
534 cluster maiden's blush. [M.L.R. 23.
535 double white blush.
536 Duc d'York.
537 Eliza.
538 feuille fermèe.
539 full double white. M.L.R. 25.
540 grand cuisse de nymphe.
541 great maiden's blush. [M.L.R. 230.
542 Joanne d'Arc.
543 Moraga la favourite.
544 muscat rouge.
545 nova cœlestis.
546 nova plena.
547 petite cuisse de nymphe.
548 rosea.
549 semidouble white.
550 Simonville.
551 small maiden's blush.
552 spineless virgin.
553 thornless virgin.
554 triangularis.

VI. *rubiginosa.*

555 Clementine sweet briar.
556 cluster blush sweet briar.
557 double mossy sweet briar. [M.L.R. 72.
558 double red sweet briar.—61.
559 double rosy sweet briar.
560 double tree sweet briar.
561 eglantine sweet briar.
562 eglanteria alba.
563 evergreen sweet briar.
564 large blush sweet briar.
565 maiden sweet briar.
566 Manning's sweet briar. [M.L.R. 41.
567 monstrous sweet briar.

GARDEN VARIETIES.

568 petite hessoise.
569 royal sweet briar. [M.L.R. 74.
570 scarlet sweet briar.
571 semidouble dwarf.
572 semidouble white sweet briar. [M.L.R. 65.
573 Zabeth.

VII. *indica* et *semperflorens.*

574 atronigra.
575 Bengale a bouquet.
576 Bengale à flore panachè.
577 Bengale blanche.
578 carnescens.
579 Chiffonnèe.
580 Cérise éclatante.
581 cucullàta.
582 elegant.
583 flòrida.
584 gigántea.
585 lie de vin.
586 lùcida.
587 màjor.
588 monstròsa.
589 moonshine.
590 nigra.
591 sanguínea.
592 sans épines.
593 Thisbe.
594 Veloutèe.

VIII. *various garden Roses.*

595 abundant.
596 abyla.
597 Achaia.
598 Achilles.
599 Adelina.
600 admirable.
601 Adonis.
602 African black.
603 agate japanese.
604 aglace noir.
605 aimable violette.
606 Ajax.
607 aland.
608 alba feuille fermee.
609 alba bifera.
610 Albanian.
611 alba nova pleno.
612 alba nova cœlestis.
613 albo-rosea.
614 alba triangularis.
615 aline.
616 Amazon.
617 amour constant.
618 Andalusian.
619 anemoneflora.
620 Arcadian.
621 arcaro.
622 archbishop.
623 ardoise.
624 Armenian.
625 armida.
626 Assyrian.
627 Athenian.
628 augusta.

GARDEN VARIETIES.

629 aurelia.
630 Baden.
631 Balbec.
632 beaute frappante.
633 beaute insurmontable.
634 beaute panache.
635 beaute renommee.
636 beaute sanspareille.
637 beaute tendre.
638 belladonna.
639 belle actrise.
640 belle aimable.
641 belle aurore descemet.
642 belle brune.
643 belle cataline.
644 belle cerisie.
645 belle d'aunay.
646 belle du jour.
647 belle flamande.
648 belle forme.
649 belle galathee.
650 belle hative.
651 belle incarnate.
652 belle inconnue.
653 belle matilde.
654 belle mahæca.
655 belle parade.
656 belle Tigridia.
657 belle zulme.
658 belle rouge.
659 belle sans flatterie.
660 belle sanspareille.
661 belle therese.
662 belle vue.
663 bellona.
664 Berkshire.
665 bizard royal.
666 black purple.
667 blanche superbe.
668 blondine.
669 blush Italian.
670 blush velvet.
671 bold.
672 Bonifacius.
673 bouquet blanc.
674 bouquet panache.
675 bouquet parfait.
676 bouquet superb.
677 Bourbon.
678 Brabant.
679 bracelet d'amour.
680 brigitte.
681 brilliant.
682 brune brillante.
683 brunette aimable.
684 brunette superbe.
685 bucephalus.
686 bullata.
687 cabinet.
688 Calypso.
689 camellote.
690 cannabina.
691 canopus.
692 Carthaginian.
693 Castile.
694 Cecropia.
695 Ceres.

GARDEN VARIETIES.

696 cervalis.
697 chance.
698 charmante pourpre.
699 chausse.
700 chiron.
701 cicris pale rose.
702 Cleopatra.
703 comtesse de Genlis.
704 cornelian.
705 cossack.
706 couleur a la mode.
707 couleur de cendre.
708 couleur de cerise.
709 couleur excellent.
710 coupee.
711 couronne.
712 couronne des roses.
713 couronne imperiale.
714 cramoisie eblouissant.
715 cramoisie elegant.
716 cramoisie fonce.
717 cramoisie imperial.
718 cramoisie incomparable.
719 cramoisie minor.
720 cramoisie nouveau.
721 cramoisie panache.
722 cramoisie sanspareil.
723 cramoisie triomphant.
724 crimson.
725 crimson perpetual.
726 crown imperial.
727 crown purple.
728 cybele.
729 danæ.
730 Danish.
731 dark mignonne.
732 dark mottled.
733 dark shell.
734 darling.
735 de bengal gros violet
736 Dejanira.
737 delicate.
738 Derby.
739 descemets.
740 diadem.
741 dianthiflora.
742 Dickson's duchess.
743 Dido.
744 don des dames.
745 Dione.
746 discolor.
747 Dorothee.
748 double apple bearing.
[M.L.R. 29.
749 double dog.—60.
750 double mimms.
751 double Pæony.
752 double purple.
753 double red thornless.
754 double white thornless.
755 duc de Brabant.
756 Duchess.
757 duchess d'Oldenburgh.
758 duke of Clarence.
759 Durham.
760 Dutch Cinnamon.
761 Dutch velvet.

GARDEN VARIETIES.

762 dwarf burnet-leaved.
763 dwarf Chester.
764 dwarf prolific.
765 early red.
766 erato.
767 Etna.
768 euphrosyne.
769 evratina.
770 ex alba rosea.
771 fair maid.
772 favonius.
773 favorite.
774 favorite des dames.
775 fausse unique.
776 felicie.
777 felicite.
778 Felix.
779 feu amoureux.
780 feuille chiffonee.
781 feuille crenulee.
782 feuille de chene.
783 feuille de frene.
784 few-spined.
785 fine panachee.
786 fine crimson.
787 fine purple.
688 flavia pourpre.
789 flesh-coloured.
790 flesh-coloured four seasons.
791 fleur de parade.
792 fleur de pomme.
793 fleur et feuille marbree.
794 fleur panachee.
795 flora.
796 flore nigricante.
797 florentine.
798 flore rubra.
799 floribunda.
800 florio.
801 flos puniceus.
802 folie de corse.
803 four seasons semidouble.
804 franconian.
805 Frankfort.
806 Frankfort agate.
807 freckled.
808 frilled.
809 frizzled.
810 gallica foliacea.
811 garnett.
812 Gascony.
813 Genoese.
814 Genoese velvet.
815 gienvieve.
816 globe.
817 gloomy.
818 gloria rubrorum.
819 gloriosa.
820 glowing.
821 grand Alexandre.
822 grande Angloise.
823 grand Cæsar.
824 grand centifolia.
825 grande chausse.
826 grand cramoisi.
827 grande foncee.
828 grand Mogul.

GARDEN VARIETIES.

829 grand pivoine.
830 grande pompadour.
831 grand pyramidal.
832 grand rouge.
833 grand triumphant.
834 grandesse royale.
835 grandeur royal.
836 great Mogul.
837 grey.
838 grisdeline.
839 grisdeline sans epines.
840 Hainault.
841 Haarlem.
842 harpagus.
843 Helena.
844 Hebe.
845 hero.
846 Hertford.
847 Hildesheim.
848 Hollandia magniflora.
849 hortulanus.
850 hundred leaved mignonne.
851 hyacinth.
852 hybrida perfecta.
853 hybride nouvelle.
854 iconium.
855 idumea.
856 illustrious.
857 imperatrice.
858 imperatrice de France.
859 Indian queen.
860 inermis.
861 iris noir.
862 Isabelle.
863 Isis.
864 jaqueline.
865 jeanne d'arc.
866 Jersey.
867 jessica.
868 jolie.
869 Julian.
870 king Agrippa.
871 king of Mexico.
872 la belle elise.
873 la belle rosine.
874 l'abondance.
875 la constance.
876 la coquette.
877 la Diane.
878 l'admirable.
879 la fidele.
880 la gracieuse.
881 la grandesse.
882 l'aimable de Stors.
883 lake.
884 la maculee.
885 la magnifique.
886 la majesteuse.
887 la mignarde.
888 l'amitie.
889 la moderne.
890 Lancaster.
891 lanckman.
892 la negresse.
893 la parfaite.
894 la Parisienne.
895 la petite lisette.

GARDEN VARIETIES.

896 la plus belle.
897 la plus rouge.
898 la pomme grenade.
899 la precieuse pourpre.
900 la reine.
901 la reine des roses.
902 large platter.
903 large perfect.
904 la rose de Medicis.
905 la rose d'Ispahan.
906 la royale.
907 la superbe.
908 la superbe en brun.
909 la tendresse.
910 la victorieuse.
911 la virginité.
912 leafy.
913 Leander.
914 le dauphin.
915 Lee's perpetual.
916 le grandeur.
917 le grand lowendael.
918 le grand suwarrow.
919 le grand visir.
920 Leipsick.
921 les alliées.
922 le violet triomphant.
923 light 100 leaved.
924 l'impayable.
925 l'importante.
926 l'obscurité.
927 Lodona.
928 Lodoiska.
929 l'ombre panaché.
930 l'ombre sanspareille.
931 London pride.
932 Louis 18th.
933 Lubeck.
934 Ludolph.
935 Ludovicus.
936 Luxemburg.
937 Lyonnoise.
938 lystra.
939 madder.
940 magdalen.
941 maiden blush.
942 mantle.
943 marble apple bearing.
944 Margaretta.
945 Marie anne.
946 marquis de la Romana.
947 Marseilles.
948 maxima.
949 mere gigon.
950 merveilleuse.
951 Mexican.
952 Milanese.
953 Minerva.
954 miroir des dames.
955 mon bijou.
956 monstrosa.
957 monstrous cluster.
958 monstrous 100 ld. M.L.R. 73.
959 Montpelier.
960 morin.
961 morning star.
962 mottled.

GARDEN VARIETIES.

963 mourning.
964 multiplex.
965 Napolean.
966 Napoli di romana.
967 Narbonne.
968 Nassau.
969 negre panaché.
970 Neptune.
971 new mottled.
972 new Provins.
973 nigritienne.
974 Niobe.
975 noblesse.
976 noir foncé.
977 non plus ultra.
978 nouvelle favorite.
979 nouvelle de Provins.
980 nouvelle pivoine.
981 Oldenburg.
982 Oliver.
983 Olympic.
984 omphale.
985 orient.
986 orion.
987 ornement de park.
988 ornement de violet.
989 oublie des François.
990 painted.
991 painted Belgic.
992 Palestine.
993 panachée admirable.
994 panachée nouvelle.
995 panachée sanspareille.
996 panachée superbe.
997 Pannonian.
998 paradise.
999 Paris 100-leaved.
1000 Parnassus.
1001 parson.
1002 party coloured.
1003 Patmos.
1004 Palmyra.
1005 Paulina.
1006 Pearson's gigantic.
1007 Pegasus.
1008 pencilled mignonne.
1009 pencilled Provins.
1010 Penelope.
1011 perfecta.
1012 perle de l'orient.
1013 perpetuelle rouge.
1014 Peruvian.
1015 petit Cæsar.
1016 petit cramoisi.
1017 petit favorite.
1018 petit mignon.
1019 pierian.
1020 pilgrim.
1021 pink velvet.
1022 plicata purpurea.
1023 Pomeranian.
1024 pomegranate.
1025 pomona.
1026 pompone.
2027 pompone blanc.
1028 pompone bifera.
1029 pompone quatre saisons.

GARDEN VARIETIES.

1030 pompone varin.
1031 Pope's cluster.
1032 portlandica grandiflora.
1033 portlandica perpetua.
1034 Portobella.
1035 Portugal.
1036 pourpre agreable.
1037 pourpre de Paris.
1038 pourpre sans epines.
1039 pourpre de Vienne.
1040 pourpre imperial.
1041 pourpre incomparable.
1042 pourpre obscur.
1043 pourpre sans defaut.
1044 pourpre sanspareil.
1045 pourpre superb.
1046 predominant.
1047 Presburg.
1048 prince.
1049 prince d'Aremberg.
1050 princely.
1051 prince regent.
1052 princess Charlotte.
1053 princesse noble.
1054 professor.
1055 purple crown.
1056 purple imperial.
1057 purple velvet.
1058 pyramide pourpre.
1059 quatre saisons blanche.
1060 quatre saisons François.
1061 quatre saisons panaché.
1062 quatre saisons sans epines.
1063 queen's Provins.
1064 ragged robin.
1065 Raphael.
1066 Ratisbon.
1067 ravenna.
1068 red cluster.
1069 red crown.
1070 red Provins.
1071 red velvet.
1072 refulgent.
1073 regina florum.
1074 reine Caroline.
1075 reine d'Hongrie.
1076 renoncule nouvelle.
1077 rex rubrorum.
1078 riche en fleurs.
1079 roi des negres.
1080 roi de Rome.
1081 roi de Maroc.
1082 roi des pays bas.
1083 Roman.
1084 Rosabel.
1085 Rosanna.
1086 rose agreeable.
1087 rose bouquet.
1088 rose de Ceres.
1089 rose des dames.
1090 rose du roi.
1091 rose d'Orleans.
1092 rose du prince.
1093 rose la mode.
1094 rose lee.
1095 rose pivoine.
1096 rose prolifere.

Garden Varieties.

1097 rouge agreable.
1098 rouge bien vif.
1099 rouge eclatant.
1100 rouge formidable.
1101 rouge frappant.
1102 rouge luisant.
1103 rouge panaché.
1104 rouge sans epines.
1105 rouge semidouble.
1106 rouge vegetable.
1107 royal bouquet.
1108 royal carmine.
1109 royal gabina.
1110 royal mantle.
1111 royal viscous.
1112 rubrispina.
1113 rubiginosa umbellata.
1114 ruby.
1115 sable.
1116 St. Antoine.
1117 St. Catharine.
1118 St. George.
1119 St. Germain.
1120 St. Gothard.
1121 St. Jago.
1122 St. Mark.
1123 St. Patrick.
1124 sans defaut.
1125 sans petales.
1126 Sarmatian.
1127 scarlet Brabant.
1128 sceptre.
1129 scone purple.
1130 semidouble hep.
1131 semidouble mignonne.
1132 semidouble monthly.
1133 semidouble new.
1134 seigneur d'Ærtzelaar.
1135 septum cerise.
1136 Seville.
1137 Sheffield.
1138 Shylock.
1139 Silenus.
1140 Silesian.
1141 shining semidouble.
1142 simplicité.
1143 singuliere agate.

Garden Varieties.

1144 Sirius.
1145 soliditas.
1146 Southampton.
1147 Spartan.
1148 spineless red.
1149 spiral.
1150 stæbon.
1151 standard.
1152 state.
1153 Stephanus.
1154 striped monthly.
1155 striped Provins.
1156 sugar loaf.
1157 sulphurea.
1158 sulphurea minor.
1159 superb.
1160 superb amaranth.
1161 superbe brune.
1162 superbe pyramide.
1163 surpasse Singleton.
1164 surpasse tout.
1165 Swiss.
1166 Syrian.
1167 Tangiers.
1168 ten-leaved.
1169 temple d'Apollon.
1170 tête de mort.
1171 Theseus.
1172 Tigris.
1173 tricolor.
1174 tomentosa alba.
1175 toujours.
1176 tous les mois gris.
1177 transparent.
1178 transparent nouvelle.
1179 Trebonius.
1180 tree burnet-leav'd.
1181 tree pæony.
1182 tresorier.
1183 triomphe.
1184 triomphe des dames.
1185 triomphe royale.
1186 Triton.
1187 turban.
1188 two-coloured 100-leaved.
1189 two-coloured mignonne.
1190 vacuna.

Garden Varieties.

1191 vagrant.
1192 velours cramoisie.
1193 veloute.
1194 venerable.
1195 venetatus.
1196 ventome noir.
1197 Venus.
1198 venusta.
1199 veritas.
1200 vermillion.
1201 vesta.
1202 Vesuvius.
1203 Victoria.
1204 Vidua.
1205 villosa nouvelle.
1206 Vilmorin.
1207 violette.
1208 violette agreable.
1209 violette aimable.
1210 violette brillante.
1211 violette curieuse.
1212 violette foncée.
1213 violette maculée.
1214 violette nouvelle.
1215 violette sanspareille.
1216 violette superbe.
1217 violette superieure.
1218 virgin's.
1219 virginale.
1220 virbilia.
1221 virgo cramoisie.
1222 ulterio.
1223 ultramarine.
1224 umbrella.
1225 unique rouge.
1226 Waterloo.
1227 Watson's blush.
1228 Watson's white.
1229 Wellington.
1230 woolly-leaved.
1231 white damask.
1232 white Pompone.
1233 York.
1234 Yorkshire blush.
1235 Yorkshire Provins.
1236 zatre.
1237 Zenobia.

Subordo II. *POTENTILLÆ*. Juss. gen. 337.

R'UBUS. L.		Bramble. Icosandria Polygynia. L.					
1 rosæfòlius. DC.	(*wh.*)	Rose-leaved.	Mauritius.	1811.	10. 4.	G. ♄.	Smith ic. 3. t. 60.
2 eglantèria. DC.	(*wh.*)	Eglantine.	N. Holland.	1824.	5. 6.	G. ♄.	
3 coronàrius.	(*wh.*)	double-flowered.	China.	1811.	10. 4.	H. ♄.	Bot. mag. t. 1783.
rosæfòlius β coronarius. B.M.							
4 pinnàtus. DC.	(*wh.*)	pinnate.	St. Helena.	1789.	6. 7.	G. ♄.	
5 grácilis. DC.	(*wh.*)	slender.	E. Indies.	1826.	——	S. ♄.	
6 micránthus. D.P.	(*re.*)	small-flowered.	Nepaul.	1818.	6. 8.	H. ♄.	Bot. reg. t. 854.
pauciflòrus. B.R.							
7 dístans. D.P.	(*re.*)	distant-leaved.	————	———	——	H. ♄.	
8 apétalus. DC.	(*gr.*)	petalless.	Mauritius.	1823.	——	S. ♄.	
9 macropòdus. DC.	(*wh.*)	large-footed.	N. Holland.	1822.	——	G. ♄.	
10 strigòsus. DC.	(*wh.*)	strigose.	N. America	....	6. 7.	H. ♄.	

11	pedunculòsus. D.P.	(*wh.*)	long-peduncled.	Nepaul.	1820.	7. 8.	H. ♄.	
12	occidentàlis. DC.	(*wh.*)	American raspberry.	N.Amer.	1696.	5. 6.	H. ♄.	Dill. elth. t. 247. f. 319.
13	ásper. DC.	(*wh.*)	rough.	Nepaul.	1822.	6. 8.	H. ♄.	
14	Id'æus. DC.	(*wh.*)	Raspberry.	Britain.	....	5. 6.	H. ♄.	Eng. bot. v. 34. t. 2442.
15	laciniàtus. DC.	(*bh.*)	jagged-leaved.	....	....	6. 9.	H. ♄.	Willd. hort. ber. t. 82.
16	c'æsius. DC.	(*wh.*)	Dewberry.	Britain.	....	6. 8.	H. ♄.	Eng. bot. v. 12. t. 826.
17	foliolòsus. D.P.	(*li.*)	leafy.	Nepaul.	1823.	7. 8.	H. ♄.	
18	flagellàris. DC.	(*wh.*)	shining-leaved.	N. America.	1789.	6. 7.	H. ♄.	
19	inérmis. W. en.	(*wh.*)	smooth.	———	1805.	——	H. ♄.	
20	corylifòlius. E.B.	(*wh.*)	Hazel-leaved.	Britain.	....	7.	H. ♄.	Eng. bot. v. 12. t. 827.
21	agréstis. DC.	(*wh.*)	field.	Hungary.	1818.	7. 8.	H. ♄.	W. et K. hung. 3. t. 268.
22	spectábilis. DC.	(*pu.*)	showy.	Columbia.	1827.	6. 8.	H. ♄.	Ph. fl. amer. 1. t. 16.
23	plicàtus. W.N.	(*wh.*)	plaited-leaved.	Britain.	....	7. 8.	H. ♄.	Wei. et N. rub. germ. t. 1.
24	affínis. W.N.	(*wh.*)	related.	———	....	6. 9.	H. ♄.	——— p. 22. t. 3 et 36.
25	ulmifòlius. DC.	(*ro.*)	Elm-leaved.	Gibraltar.	1824.	7. 8.	H. ♄.	
26	Linkiànus. DC.	(*wh.*)	double-flowered.	....	....	6. 9.	H. ♄.	
27	álbicans. DC.	(*wh.*)	white-leaved.	Hungary.	1818.	——	H. ♄.	
28	fruticòsus. DC.	(*fl.*)	Blackberry.	Britain.	....	——	H. ♄.	Eng. bot. v. 10. t. 715.
	β *pompònius*. DC.	(*bh.*)	*semidouble.*	....	....	——	H. ♄.	
	γ *leucocárpus*. DC.	(*bh.*)	*white-berried.*	....	....	——	H. ♄.	
	δ *inermis*. DC.	(*bh.*)	*smooth-stemm'd.*	....	....	——	H. ♄.	
29	tomentòsus. W.N.	(*wh.*)	woolly-leaved.	Germany.	1816.	6. 8.	H. ♄.	Wei. et N. rub. germ. t. 8.
30	sánctus. DC.	(*pk.*)	holy.	Levant.	1823.	——	H. ♄.	Schreb. dec. t. 8.
31	canéscens. DC.	(*bh.*)	canescent.	Piedmont.	1826.	——	H. ♄.	
32	collìnus. DC.	(*bh.*)	hill.	Montpelier.	1820.	7. 8.	H. ♄.	Noc. et Balb. fl. tic. t. 9.
33	fastigiàtus. W.N.	(*wh.*)	close-flowered.	Britain.	....	——	H. ♄.	Wei. et N. rub. germ. t. 2.
34	suberéctus. E.B.	(*wh.*)	upright.	———	....	6. 9.	H. ♄.	Eng. bot. v. 36. t. 2572.
35	cordifòlius. W.N.	(*wh.*)	heart-leaved.	———	....	7. 8.	H. ♄.	Wei. et N. rub. germ. t. 5.
36	Ménkii. S.S.	(*wh.*)	velvetty.	Germany.	1816.	——	H. ♄.	
37	Schlechtendàlii. S.S.	(*w.*)	Schlechtendale's.	———	——	——	H. ♄.	
38	tiliæfòlius. S.S.	(*wh.*)	Lime-tree-ld.	———	——	——	H. ♄.	
39	Schleichèri. S.S.	(*wh.*)	Schleicher's.	———	1824.	——	H. ♄.	
40	hórridus. S.S.	(*wh.*)	very prickly.	———	——	——	H. ♄.	
41	leucostàchys. S.S.	(*wh.*)	white-spiked.	Britain.	....	——	H. ♄.	
42	rhamnifòlius. W.N.	(*wh.*)	Rhamnus-leaved.	———	....	——	H. ♄.	Wei. et N. rub. germ. t. 6.
43	abruptus. L.B.	(*wh.*)	abrupt.	———	....	——	H. ♄.	
44	nítidus. W.N.	(*bh.*)	glossy-leaved.	Britain.	....	7. 8.	H. ♄.	Wei. et N. rub. germ. t. 4.
45	argùtus. DC.	(*wh.*)	sharp-toothed.	N. America.	1824.	6. 7.	H. ♄.	
46	cuneifòlius. DC.	(*wh.*)	wedge-leaved.	———	1811.	6. 8.	H. ♄.	
47	sanguinoléntus. DC.	(*w.*)	red-wooded.	Mauritius.	1823.	——	S. ♄.	
48	villòsus. DC.	(*wh.*)	shaggy.	N. America.	1777.	7. 8.	H. ♄.	
49	hírtus. W.K.	(*wh.*)	hairy.	Britain.	....	——	H. ♄.	W. et K. hung. 2. t. 141.
50	vulpìnus. Df. *Sprengèlii*. W.N.	(*re.*)	fox-tail.	Germany.	1816.	——	H. ♄.	Flor. dan. t. 1165.
51	jamaicénsis. DC.	(*wh.*)	Jamaica.	Jamaica.	1820.	——	S. ♄.	Sloan. jam. 2. t. 213. f. 1.
52	híspidus. DC.	(*wh.*)	bristly.	Canada.	1768.	——	H. ♄.	
53	triviàlis. Ph.	(*bh.*)	American dewberry.	N.Amer.	1789.	6. 7.	H. ♄.	
54	macrophy'llus. W.N.	(*w.*)	large-leaved.	Wales.	....	7. 8.	H. ♄.	
55	vulgàris. W.N.	(*bh.*)	common.	Britain.	....	——	H. ♄.	
56	diversifòlius. L.B.	(*bh.*)	different-leaved.	———	....	——	H. ♄.	
57	díscolor. W.N.	(*bh.*)	two-coloured.	———	....	——	H. ♄.	
58	fusco-ater. W.N.	(*bh.*)	dark-brown.	———	....	——	H. ♄.	
59	pállidus. W.N.	(*bh.*)	pale-leaved.	———	....	——	H. ♄.	
60	Kœhlèri. W.N.	(*bh.*)	Kœhler's.	———	....	——	H. ♄.	
61	rùdis. W.N.	(*bh.*)	way-side.	———	....	——	H. ♄.	
62	rubricaùlis. W.N.	(*bh.*)	red-stalked.	Germany.	1816.	——	H. ♄.	
63	echinàtus. L.B.	(*bh.*)	hedgehog.	Britain.	....	——	H. ♄.	
64	dumetòrum. W.N.	(*wh.*)	bushy.	———	....	——	H. ♄.	
65	tetraphy'llus. W. en.	(*w.*)	four-leaved.	N. America.	1812.	6. 7.	H. ♄.	
66	parvifòlius. DC.	(*pu.*)	small-leaved.	China.	1818.	6. 8.	H. ♄.	Bot. reg. t. 496.
67	canádénsis. DC.	(*wh.*)	purple-stalked.	N. America.	1811.	6. 7.	H. ♄.	
68	saxátilis. DC.	(*wh.*)	stone.	Britain.	....	——	H. ♃.	Eng. bot. v. 32. t. 2233.
69	triflòrus. Rich.	(*wh.*)	three-flowered.	N. America.	....	——	H. ♃.	
70	árcticus. DC.	(*cr.*)	dwarf crimson.	Scotland.	....	5. 8.	H. ♃.	Eng. bot. v. 23. t. 1585.
71	pistillàtus. Ph. *ucaùlis*. DC.	(*ro.*)	close-styled.	Labrador.	1802.	6. 7.	H. ♃.	Sm. exot. bot. 2. t. 86.
72	obovàlis. DC.	(*wh.*)	oboval-leaved.	N. America.	1823.	5. 7.	H. ♃.	
73	parviflòrus. DC.	(*wh.*)	Vine-leaved.	———	1824.	——	H. ♃.	
74	stellàtus. DC.	(*ro.*)	star-flowered.	———	1823.	6. 7.	H. ♃.	Sm. ic. fasc. 3. t. 64.

75 Chamæmòrus. DC.	(*wh.*)	Cloud-berry.	Britain.	....	6. 7.	H. ♃.		Eng. bot. v. 10. t. 716.
76 flòridus. DC.	(*wh.*)	large-flowered.	N. America.	1827.	6. 8.	H. ♄.		
77 odoràtus. DC.	(*re.*)	flowering raspberry.	——	1700.	6. 9.	H. ♄.		Bot. mag. t. 323.
78 refléxus. DC.	(*wh.*)	reflexed.	China.	1818.	6. 7.	G. ♄ ⌣.		Bot. reg. t. 461.
79 Hamiltoniànus. DC.	(*wh.*)	Hamilton's.	Nepaul.	——	——	H. ♄.		
rugòsus. D.P. *non* Smith.								
80 moluccànus. DC.	(*wh.*)	Moluccas.	Amboyna.	1810.	——	S. ♄.		Rumph. amb. 5. t. 47. f. 2.
81 rugòsus. DC.	(*wh.*)	rugose.	S. America.	1824.	——	S. ♄.		
82 tiliàceus. DC.	(*wh.*)	Lime-tree-ld.	Nepaul.	——	——	H. ♄.		
83 paniculàtus. DC.	(*wh.*)	panicled.	————	1826.	7. 9.	H. ♄.		
84 betulìnus. DC.	(*wh.*)	Birch-leaved.	————	1825.	——	H. ♄.		
DALIBA'RDA. DC.		DALIBA'RDA.	Icosandria Pent-Decagynia.					
1 rèpens. DC.	(*wh.*)	Violet-leaved.	N. America.	1768.	5. 6.	H. ♃.		Mich. amer. 1. t. 27.
violæoídes. M.								
FRAG'ARIA. DC.		STRAWBERRY.	Icosandria Polygynia.					
1 monophy'lla. S.S.		simple-leaved.	....	1773.	5. 6.	H. ♃.		Bot. mag. t. 63.
2 vésca. DC.	(*wh.*)	wood.	Britain.	....	4. 6.	H. ♃.		Eng. bot. v. 22. t. 1524.
α sylvéstris. DC.	(*wh.*)	*wild.*	————	....	——	H. ♃.		
β semperflòrens. DC.	(*wh.*)	*ever-bearing.*	....	....	4. 12.	H. ♃.		Nois. jard. fr. t. 11. f. 2.
γ álba.	(*wh.*)	*white-fruited.*	....	....	——	H. ♃.		
δ mìnor. DC.	(*wh.*)	*small English.*	Britain.	....	——	H. ♃.		
ε horténsis. DC.	(*wh.*)	*garden.*	....	....	——	H. ♃.		
ζ efflágellis. DC.	(*wh.*)	*short-shooted.*	....	....	——	H. ♃.		
η multipléx. DC.	(*wh.*)	*double-blossomed.*	....	....	4. 6.	H. ♃.		Blackw. herb. t. 77. f. 3.
θ botryfórmis. DC.	(*wh.*)	*crown-flowered.*	....	....	——	H. ♃.		
ι muricàta. DC.	(*gr.*)	*monstrous.*	....	....	——	H. ♃.		
3 calycìna. DC.	(*wh.*)	large-calyxed.	Britain.	....	——	H. ♃.		
4 collìna. DC.	(*wh.*)	Alpine.	Germany.	1768.	4. 11.	H. ♃.		
α rùbra.	(*wh.*)	*red Alpine.*	————	——	——	H. ♃.		
β álba.	(*wh.*)	*white Alpine.*	————	——	——	H. ♃.		
5 platanoídes. DC.	(*bh.*)	Plane-leaved.	N. America.	....	4. 6.	H. ♃.		
6 Majaùfea. DC.	(*wh.*)	plaited-leaved.	France.	....	——	H. ♃.		
α bífera. DC.	(*wh.*)	*double-bearing.*	————	....	——	H. ♃.		
β dùbia. DC.	(*wh.*)	*flat-fruited.*	————	....	——	H. ♃.		
7 Breslíngea. DC.	(*wh.*)	leathery-leaved.	————	....	——	H. ♃.		
α abórtiva. DC.	(*wh.*)	*sterile.*	————	....	——	H. ♃.		
β nìgra. DC.	(*wh.*)	*black-fruited.*	....	....	——	H. ♃.		
γ péndula. DC.	(*wh.*)	*pear-fruited.*	....	....	——	H. ♃.		
δ híspida. DC.	(*wh.*)	*Longchamp.*	France.	....	——	H. ♃.		
ε víridis. DC.	(*wh.*)	*green-fruited.*	....	....	——	H. ♃.		Nois jard. fr. t. 13. f. 2.
ζ praténsis. DC.	(*wh.*)	*field.*	France.	....	——	H. ♃.		
8 elàtior. DC.	(*wh.*)	tall.	Europe.	....	——	H. ♃.		
9 moschàta. Ehr.	(*wh.*)	Hautboy.	England.	....	5. 6.	H. ♃		Eng. bot. v. 31. t. 2197.
elàtior. E.B.								
β dioíca. DC.	(*wh.*)	*diœcious.*	————	....	——	H. ♃.		
10 virginiàna. DC.	(*wh.*)	scarlet.	N. America.	1629.	4. 6.	H. ♃.		Nois. jard. fr. t. 12. f. 2.
β hy'brida. DC.	(*wh.*)	*Keen's seedling.*	Hybrid.	....	——	H. ♃.		
11 chilénsis. DC.	(*wh.*)	Chile.	S. America.	1727.	5. 6.	H. ♃.		Dill. elth. t. 20. f. 140.
12 grandiflòra. DC.	(*wh.*)	Pine.	Surinam.	1759.	---	H. ♃.		Mill. ic. t. 288.
β ananássa. DC.	(*wh.*)	*Pine apple.*	....	....	——	H. ♃.		Nois. jard. fr. t. 14. f. 2.
γ calyculàta. DC.	(*wh.*)	*Bath scarlet*	....	....	——	H. ♃.		——— t. 11. f. 1. t. 14. f. 1.
δ tíncta. DC.	(*wh.*)	*red-fleshed.*	....	....	——	H. ♃.		
13 canadénsis. DC.	(*wh.*)	Canadian.	Canada.	....	——	H. ♃.		
14 índica. DC.	(*ye.*)	yellow-flower'd.	Nepaul.	1805.	5. 10.	H. ♃.		Bot. reg. t. 61.
Duchésnea fragarioídes. Smith Linn. trans. 10. p. 373.								
COMARO'PSIS. DC.		COMARO'PSIS.	Icosandria Pent-Octagynia.					
1 fragarioídes. DC.	(*ye.*)	strawberry-like.	N. America.	1803.	5. 6.	H. ♃.		Mich. amer. 1. t. 28.
Dalibárda fragarioídes. M.								
2 Doniàna. DC.	(*ye.*)	Don's.	————	——	——	H. ♃.		Bot. mag. t. 1567.
Dalibárda fragarioídes. B.M. *non* Mich. *Waldsteínia Doniàna.* Tratt.								
3 pedàta. DC.	(*wh.*)	pedate.	————	1827.	——	H. ♃.		Smith. ic. ined. t. 63.
Rùbus pedàtus. Sm. *Dalibárda pedàta.* S.S.								
HORK'ELIA. B.M.		HORK'ELIA.	Decandria Polygynia.					
congésta. B.M.	(*wh.*)	tufted-flowered.	N. America.	1827.	7. 9.	H. ♃.		Bot. mag. t. 2880.
WALDST'EINIA. DC.		WALDST'EINIA.	Icosandria Digynia.					
geoídes. DC.	(*ye.*)	Avens-leaved.	Hungary.	1804.	6. 7.	H. ♃.		Lodd. bot. cab. 402.
COL'URIA. B.		COL'URIA.	Icosandria Polygynia.					
potentilloídes. B.	(*or.*)	Siberian.	Siberia.	1780.	6. 7.	H. ♃.		Jacq. vind. 3. t. 68.
Gèum potentilloídes. W. *Dr'yas geoídes.* Pall. *Sievérsia geoídes.* S.S.								

SIEVE'RSIA. B.		SIEVE'RSIA.	Icosandria Polygynia.				
1 Péckii. B.M.	(*ye.*)	Peck's.	N. America.	1827.	6. 7.	H. ♃.	Bot. mag. t. 2863.
2 radiàta.	(*ye.*)	radiate.	———	1822.	— —	H. ♃.	
Gèum radiàtum. Ph.							
3 triflòra. B.M	(*ye.bh.*)	three-flowered.	———	1827.	——	H. ♃.	Bot. mag. t. 2858.
4 montàna. S.S.	(*ye.*)	mountain.	Austria.	1597.	5. 9.	H. ♃.	Jacq. aust. 4. t. 373.
β *minor.*	(*ye.*)	*smaller mountain.*	Alps.	——	——	H. ♃.	Barrel. rar. t. 399.
5 glaciàlis. S.S.	(*ye.*)	arctic.	Siberia.	1827.	— —	H. ♃.	
6 Róssii. B.	(*ye.*)	Ross's.	Melville Isl.	——	——	H. ♃.	Brown ch.mel.p.18.t.C.
7 anemonoídes. S.S.	(*wh.*)	Anemone-like.	Kamtschatka.	1824.	——	H. ♃.	
8 réptans. S.S.	(*ye.*)	creeping.	Switzerland.	1775.	6. 8.	H. ♃.	Jacq. aust. 5. ap. t. 22.
G'EUM. B.		AVENS.	Icosandria Polygynia. L.				
1 agrimonoídes. Ph.	(*wh.*)	Agrimony-leaved.	N. America.	1811.	6. 7.	H. ♃.	Murr. com. gœt. 5. t. 2?
2 híspidum. S.S.	(*wh.*)	hispid.	Sweden.	1828.	——	H. ♃.	Reich. ic. cent. t. 3.
3 stríctum. W.	(*ye.*)	upright.	N. America.	1778.	5. 7.	H. ♃.	Jacq. ic. 1. t. 93.
aléppicum. Jacq.							
4 intermèdium. W.	(*ye.*)	intermediate.	Europe.	1794.	5. 8.	H. ♃.	W. hort. ber. 1. t. 69.
5 hy'bridum. Jacq.	(*ye.*)	hybrid.	———	....	——	H. ♃.	Jacq. ic. 1. t. 94.
6 rivàle. W.	(*ye.bh.*)	water.	Britain.	....	5. 9.	H. ♃.	Eng. bot. 2. t. 106.
7 pyrenàicum. DC.	(*ye.*)	Pyrenean.	Pyrenees.	1804.	6. 7.	H. ♃.	Lam. ill. t. 44.
8 brachypétalum. DC.	(*ye.bh.*)	short-petal'd.	......	1826.	5. 8.	H. ♃.	
9 sylváticum. DC.	(*ye.*)	wood.	S. Europe.	1820.	6. 7.	H. ♃.	
altánticum. Desf. atl. 1. p. 401. *biflòrum.* Brot.							
10 Thomasiànum. DC.	(*ye.*)	Thomas's.	Pyrenees.	1828.	——	H. ♃.	
11 Quéllyon. B.F.G.	(*sc.*)	scarlet Chile.	Chile.	1826.	5. 10.	H. ♃.	Swt. br. fl. gar. t. 292.
coccíneum. Bot. reg. t. 1088. *non* Flor. græc.							
β *grandiflòrum.*	(*sc.*)	*large-flowered.*	———	1829.	——	H. ♃.	
12 macrophy'llum. DC.	(*ye.*)	large-leaved.	Kamtschatka.	1804.	6. 7.	H. ♃.	
13 canadénse. DC.	(*ye.*)	Canadian.	Canada.	1778.	5. 8.	H. ♃.	Murr. com gœt. 5. t. 4.
14 heterophy'llum. DC.	(*wh.*)	various-leaved.	....	....	——	H. ♃.	
15 Besseriànum.	(*ye.*)	Besser's.	Volhynia.	1828.	——	H. ♃.	
intermèdium. Besser *non* Willd.							
16 hirsùtum. L.en.	(*st.*)	hairy.	N. America.	1818.	5. 7.	H. ♃.	
17 virginiànum. DC.	(*wh.*)	Virginian.	———	....	7. 8.	H. ♃.	Herm. parad. t. 111.
18 álbum. DC.	(*wh.*)	white-flowered.	———	1730.	——	H. ♃.	Jacq. vind. 2. t. 175.
canadénse. Jacq. *non* Murr.							
19 urbànum. DC.	(*ye.*)	common.	Britain.	....	5. 8.	H. ♃.	Eng. bot. v. 20. t. 1400.
20 ranunculoídes. DC.	(*ye.*)	Ranunculus-like	....	1827.	——	H. ♃.	
DR'YAS. DC.		DR'YAS.	Icosandria Polygynia. L.				
1 octopétala. DC.	(*wh.*)	mountain.	Britain.	....	7. 8.	H. ♃.	Eng. bot. v. 7. t. 451.
2 integrifòlia. DC.	(*wh.*)	entire-leaved.	Greenland.	1824.	4. 8.	H. ♃.	Hook. ex. flor. t. 220.
POTENTI'LLA. DC.		CINQUEFOIL.	Icosandria Polygynia. L.				
1 nívea. DC.	(*ye.*)	snowy-leaved.	N. Europe.	1816.	6. 8.	H. ♃.	Lodd. bot. cab. 460.
2 angustifòlia. DC.	(*ye.*)	narrow-leaved.	Siberia.	1826.	——	H. ♃.	Lehm. pot. t. 19.
3 macrántha. DC.	(*ye.*)	great-flowered.	———	1829.	5. 8.	H. ♃.	
4 uniflòra. DC.	(*ye.*)	single-flowered.	Dahuria.	1827.	——	H. ♃.	Lehm. pot. t. 18.
5 mínima. DC.	(*ye.*)	small.	Switzerland.	1816.	4. 6.	H. ♃.	Lodd. bot. cab. 480.
6 Brauneàna. N.P.	(*ye.*)	Braune's.	———	——	——	H. ♃.	Nestl. pot. t. 10. f. 4.
7 glaciàlis. DC.	(*ye.*)	very hairy.	———	——	5. 7.	H. ♃.	Ser. mus. helv. 1. t. 7.
8 frígida. DC.	(*ye.*)	frigid.	Dauphiny.	——	——	H. ♃.	Nestl. pot. t. 10. f. 3.
9 grandiflòra. DC.	(*ye.*)	large-flowered.	Siberia.	1640.	6. 7.	H. ♃.	Bot. mag. t. 75.
10 villòsa. DC.	(*ye.*)	villous.	N. America.	1827.	——	H. ♃.	Lehm. pot. t. 16.
11 norvègica. DC.	(*ye.*)	Norwegian.	N. Europe.	1764.	6. 8.	H. ♂.	Flor. dan. t. 171.
12 Morisòni. DC.	(*wh.*)	Morison's.	N. America.	1680.	7. 8.	H. ⊙.	Nestl. pot. t. 9. f. 1.
monspeliénsis. L. Moris. his. s. 2. t. 20. f. 2. *hirsuta.* M.							
13 hirsùta. DC.	(*ye.*)	hairy-leaved.	Greenland.	1820.	6. 8.	H. ♃.	Flor. dan. t. 1390.
14 subacaùlis. DC.	(*ye.*)	short-stalked.	S. France.	——	5. 7.	H. ♃.	Scop. carn. 1. t. 22.
velutìna. L.P.							
15 Tormentílla DC.	(*ye.*)	Septfoil.	Britain.	....	5. 10.	H. ♃.	Eng. bot. v. 12. t. 863.
Tormentílla officinàlis. E.B. *erécta.* W.							
16 nemoràlis. N.P.	(*ye.*)	creeping-wood.	———	....	6. 8.	H. ♃.	Eng. bot. v. 12. t. 864.
Tormentílla réptans. E.B.							
17 umbròsa. DC.	(*go.*)	shaded.	Tauria.	1821.	——	H. ♃.	
18 réptans. DC.	(*ye.*)	common.	Britain.	....	——	H. ♃.	Eng. bot. v. 12. t. 862.
19 bithy'nica. DC.	(*ye.*)	Bithynian.	Bithynia.	1822.	——	H. ♃.	
20 símplex. DC.	(*ye.*)	sarmentose.	N. America.	1804.	5. 8.	H. ♃.	
sarmentòsa. W.en. *caroliniàna.* Poiret.							
21 canadénsis. DC.	(*ye.*)	Canadian.	Canada.	1818.	——	H. ♃.	Nestl. pot. t. 10. f. 1.
22 vérna. E.B.	(*ye.*)	spring.	Britain.	....	3. 5.	H. ♃.	Eng. bot. v. 1. t. 37.

23 cinèrea. DC.	(*ye.*)	ash-coloured.	S. Europe.	1820.	4. 6.	H. ♃.	Jacq. ic. 3. t. 491.
subacaùlis. Jacq. *non.* L.							
24 pátula. DC.	(*ye.*)	spreading.	Hungary.	1806.	6. 8.	H. ♃.	W. et K. hung. 2. t. 199.
25 opàca. DC.	(*ye.*)	opaque-leaved.	Scotland.	....	——	H. ♃.	Ser. mus. helv. 1. t 4. f. 1.
26 alpéstris. E.F.	(*ye.*)	alpine.	————	....	——	H. ♃.	———— v. 8. t. 561.
aùrea. E.B. *nec aliorum.*							
27 aùrea. DC.	(*go.*)	golden.	Alps Europe.	....	——	H. ♃.	Ser. mus. helv. t. 8. f. a. b.
28 salisburgénsis. Jac.	(*ye.*)	broad-lobed.	Switzerland.	1825.	——	H. ♃.	Jacq. ic. 3. t. 490.
29 Hallèri. DC.	(*ye.*)	Haller's.	————	....	——	H. ♃.	Hall. hist. t. 21. f. 4.
30 Thomàsii. DC.	(*ye.*)	silky-haired.	Italy.	1825.	——	H. ♃.	Ten. fl. nap. 1. t. 44.
31 calábra. T.N.	(*ye.*)	Calabrian.	————	——	——	H. ♃.	———— 1. p. 287. t. 45.
32 impólita. Wah.	(*ye.*)	unpolished.	Carpathian.	1818.	——	H. ♃.	
33 Gunthèri. L.P.	(*ye.*)	Gunther's.	Bohemia.	1823.	——	H. ♃.	Lehm. pot. 97. t. 10.
34 argéntea. E.B.	(*ye.*)	silvery-leaved.	Britain.	....	——	H. ♃.	Eng. bot. v. 2. t. 89.
35 collìna. DC.	(*ye.*)	hill.	Germany.	1823.	——	H. ♃.	
36 intermèdia. DC.	(*ye.*)	intermediate.	Scotland.	....	——	H. ♃.	Eng. bot. v. 35. t. 2449.
opàca. E.B. *non* L.							
37 elongàta. Bes.	(*ye.*)	long-stalked.	Russia.	1827.	——	H. ♃.	
38 incìsa. DC.	(*ye.*)	jagged-leaved.	....	1818.	5. 8.	H. ♃.	
39 canéscens. DC.	(*ye.*)	canescent.	S. Europe.	1822.	6. 8.	H. ♃.	Tausch. h. can. 1. t. 10.
40 thuringìaca. DC.	(*ye*)	decumbent.	Germany.	——	——	H. ♃.	
41 stipulàris. DC.	(*ye.*)	large-stipuled.	Siberia.	1807.	——	H. ♃.	Gmel. fl. bad. 3. n. 38. f. 2.
42 hírta. DC.	(*ye.*)	hairy.	S. Europe.	1725.	5. 9.	H. ♃.	Lehm. pot. t. 8.
43 pedàta. N.P.	(*ye.*)	pedate-leaved.	Italy.	1818.	——	H. ♃.	Nestl. pot. 44. t. 7.
44 astracànica. S.S.	(*ye.*)	Astracan.	Levant.	1787.	6. 8.	H. ♃.	Jacq. ic. 1. t. 92.
45 taùrica. S.S.	(*ye.*)	Taurian.	Tauria.	1824.	——	H. ♃.	Lehm. pot. 90. t. 9.
46 récta. S.S.	(*ye.*)	upright.	S. Europe.	1648.	——	H. ♃.	Nestl. pot. 42. t. 6.
47 obscùra. S.S.	(*ye.*)	obscure.	Siberia.	1816.	6. 7.	H. ♃.	Noc. et Balb. fl. tic. t. 10.
48 laciniòsa. S.S.	(*ye.*)	jagged.	Hungary.	1818.	——	H. ♃.	Lehm. pot. 86. t. 9.
49 atrosanguínea. DC.	(*cr.*)	dark crimson.	Nepaul.	1820.	6. 9.	H. ♃.	Swt. br. fl. gar. t. 124.
50 Russelliàna. B.F.G.	(*sc.*)	Russell's.	Hybrid.	1827.	——	H. ♃.	———— v. 3. t. 279.
51 formòsa. D.P.	(*ro.*)	handsome.	Nepaul.	1820.	5. 9.	H. ♃.	———— v. 2. t. 136.
nepalénsis. Hook. ex. flor. t. 88.							
52 ruthènica. DC.	(*ye.*)	Russian.	Siberia.	1799.	5. 7.	H. ♃.	Moris. his. s. 2. t. 20. f. 2.
53 diffùsa. DC.	(*ye.*)	diffuse.	————	1825.	——	H. ♃.	
54 chrysántha. L.en.	(*go.*)	golden-flowered.	....	1821.	——	H. ♃.	
55 declinàta. H.H.	(*ye.*)	declining.	....	1827.	6. 8.	H. ♃.	
56 pùmila. S.S.	(*ye.*)	dwarf.	N. America.	1818.	5. 6.	H. ♃.	
57 Weinmánnii. L.C.	(*ye.*)	Weinmann's.	Thuringia.	1822.	6. 8.	H. ♃.	Lodd. bot. cab. t. 706.
58 ornithopodioídes.	(*ye.*)	Bird's-foot.	....	1827.	——	H. ♃.	
59 lanuginòsa. F.	(*ye.*)	woolly.	Caucasus.	1821.	——	H. ♃.	
60 Sieversiàna. Ot.	(*ye.*)	Sievers's.	....	1829.	——	H. ♃.	
61 normàlis. F.	(*ye.*)	normal.	Siberia.	1827.	——	H. ♃.	
62 divaricàta. S.S.	(*ye.*)	divaricate.	Corsica.	1824.	——	H. ♃.	
63 fruticòsa. DC.	(*ye.*)	shrubby.	England.	....	6. 8.	H. ♄.	Eng. bot. 2. t. 88.
64 dahùrica. N.P.	(*ye.*)	Dahurian.	Dahuria.	1824.	——	H. ♄.	Nestl. pot. 31. t. 1.
65 floribúnda. Ph.	(*ye.*)	cluster-flower'd.	N. America.	1811.	5. 9.	H. ♄.	Wats. dend. brit. t. 70.
66 bifúrca. DC.	(*ye.*)	bifid-leaved.	Siberia.	1773.	6. 7.	H. ♃.	Gmel. it. 1. t. 37. f. 1.
67 supìna. DC.	(*ye.*)	trailing.	Europe.	1696.	7. 8.	H. ♃.	Jacq. aust. 5. t. 406.
68 pimpinelloídes. DC.	(*ye.*)	Burnet-leaved.	Levant.	1758.	6. 8.	H. ♃.	Buxb. cent. 1. t. 48.
69 geoídes. DC.	(*ye.*)	Avens-like.	Tauria.	1820.	——	H. ♃.	
70 Sprengeliàna. DC.	(*ye.*)	Sprengel's.	Siberia.	1824.	——	H. ♃.	Lehm. pot. 49. t. 3.
71 viscòsa. DC.	(*ye.*)	clammy.	————	1797.	7. 8.	H. ♃.	Nestl. pot. t. 36.
híspida. N.P. *non* W.							
72 pensylvànica. DC.	(*ye.*)	Pensylvanian.	N. America.	1725.	6. 8.	H. ♃.	Jacq. vind. 2. t. 189.
73 pectinàta. F.	(*ye.*)	pectinated.	Siberia.	1824.	——	H. ♃.	
74 argùta. Ph.	(*ye.*)	sharp-toothed.	N. America.	1818.	5. 7.	H. ♃.	
75 multífida. DC.	(*ye.*)	cut-leaved.	Siberia.	1759.	——	H. ♃.	Ser. mus. helv. 1. t. 8.
76 agrimonioídes. S.S.	(*ye.*)	Agrimony-l[d].	Caucasus.	1823.	6. 8.	H. ♃.	
77 verticillàris. DC.	(*ye.*)	whorled.	Dahuria.	——	——	H. ♃.	
78 serícea. DC.	(*ye.*)	silky.	Siberia.	1780.	5. 6.	H. ♃.	Lehm. pot. 66. t. 6.
79 pulchélla. DC.	(*ye.*)	neat.	Melville Isl.	1827.	5. 7.	H. ♃.	
80 Egèdii. DC.	(*ye.*)	smooth-leaved.	Greenland.	——	——	H. ♃.	Flor. dan. t. 1578.
81 anserìna. DC.	(*ye.*)	wild Tansey.	Britain.	....	5. 9.	H. ♃.	Eng. bot. 12. t. 861.
β geminiflòra. DC.	(*ye.*)	*twin-flowered.*	————	....	——	H. ♃.	Moris. his. s. 2. t. 20. f. 4.
γ cóncolor. DC.	(*ye.*)	*silky-leaved.*	————	....	——	H. ♃.	
82 cándicans. DC.	(*ye.*)	hoary-leaved.	America.	1826.	——	H. ♃.	Nestl. pot. 34. t. 3. f. 2.
83 spléndens. DC.	(*ye.*)	glossy-leaved.	Nepaul.	1818.	7. 9.	H. ♃.	Swt. br. fl. gar. 2. t. 150.
striàta. S.S.							

84 Còmarum. DC.	(*pu.*)	marsh.	Britain.	....	6. 8.	H. ♃.	Eng. bot. v. 3. t. 172.
Còmarum palústre. E.B.							
85 Salesòvii. DC.	(*ye.*)	white shrubby.	Siberia.	1823.	——	H. ♄.	Lehm. pot. 35. t. 1.
86 glábra. DC.	(*wh.*)	smooth white.	———	1818.	6. 10.	H. ♄.	Lodd. bot. cab. t. 914.
87 rupéstris. DC.	(*wh.*)	rock.	England.	....	6. 8.	H. ♃.	Eng. bot. v. 29. t. 2058.
88 fragarioídes. DC.	(*wh.*)	Strawberry-like.	Siberia.	1773.	5. 7.	H. ♃.	Lehm. pot. 50. t. 4.
89 álba. DC.	(*wh.*)	white-flowered.	Wales.	....	6. 9.	H. ♃.	Eng. bot. v. 20. t. 1384.
90 cauléscens. DC.	(*wh.*)	caulescent.	Austria.	1759.	5. 7.	H. ♃.	Jacq. aust. t. 220.
91 Clusiàna. S.S.	(*wh.*)	Clusius's.	S. Europe.	1806.	7. 8.	H. ♃.	Bot. mag. t. 1327.
92 lupinoídes. S.S.	(*wh.*)	close-flowered.	Pyrenees.	1739.	6. 7.	H. ♃.	Lodd. bot. cab. t. 654.
93 valdèria. DC.	(*st.*)	crowded-flower'd.	S. Europe.	1818.	6. 8.	H. ♃.	All. fl. ped. t. 24. f. 1.
94 nítida. DC.	(*bh.*)	glossy.	———	1798.	——	H. ♃.	Jacq. aust. ap. t. 25.
95 apennìna. DC.	(*wh.*)	Apennine.	Italy.	1825.	——	H. ♃.	Ten. fl. nap. 1. t. 46.
96 tridentàta. DC.	(*wh.*)	trifid-leaved.	Scotland.	....	5. 7.	H. ♃.	Eng. bot. v. 34. t. 2389.
97 micrántha. DC.	(*wh.*)	small-blossom'd.	S. Europe.	1816.	6. 8.	H. ♃.	Ser. mus. helv. 1. t. 5.
98 Fragària. DC.	(*wh.*)	sterile strawberry.	Britain.	....	3. 6.	H. ♃.	Eng. bot. v. 25. t. 1785.
Fragària stérilis. E.B.		*Còmarum fragarioídes.* W. enum.					
99 emarginàta. Ph.	(*wh.*)	emarginate.	Labrador.	1828.	——	H. ♃.	
100 Jamesoniàna. DC.	(*ye.*)	Jameson's.	Greenland.	1822.	6. 7.	H. ♃.	
SIBBA'LDIA. DC.		SIBBA'LDIA.	Pentandria Pentagynia.				
1 procúmbens. DC.	(*ye.*)	procumbent.	Britain.	....	6. 8.	H. ♃.	Eng. bot. v. 13. t. 897.
2 parviflòra. DC.	(*ye.*)	small-flowered.	Levant.	....	——	H. ♃.	
3 erécta. DC.	(*ye.*)	upright.	Siberia.	1806.	——	H. ♃.	Amm. ruth. 112. t. 15.
4 grandiflòra. DC.	(*ye.*)	large-flowered.	Dahuria.	1828.	——	H. ♃.	
5 polygyna. DC.	(*ye.*)	many-styled.	Siberia.	1824.	——	H. ♃.	
AGRIM'ONIA. DC.		AGRIMONY.	Dodecandria Digynia. L.				
1 Eupatòria. DC.	(*ye.*)	common.	Britain.	....	6. 8.	H. ♃.	Eng. bot. v. 19. t. 1335.
2 nepalénsis. DC.	(*ye.*)	Nepaul.	Nepaul.	1820.	——	H. ♃.	
3 odoràta. DC.	(*ye.*)	sweet-scented.	Italy.	1640.	7. 8.	H. ♃.	
4 dahùrica. DC.	(*ye.*)	Dahurian.	Dahuria.	1824.	6. 8.	H. ♃.	
5 pilòsa. DC.	(*ye.*)	hairy.	Siberia.	1820.	——	H. ♃.	
6 parviflòra. DC.	(*ye.*)	small-flowered.	N. America.	1766.	6. 7.	H. ♃.	
7 suavèolens. DC.	(*st.*)	fragrant.	———	1820.	——	H. ♃.	
8 striàta. DC.	(*wh.*)	striped-flowered.	———	1812.	6. 8.	H. ♃.	
9 rèpens. DC.	(*ye.*)	creeping-rooted.	Levant.	1737.	7. 9.	H. ♃.	
AREM'ONIA. DC.		AREM'ONIA.	Pent-Decandria Digynia.				
agrimonoídes. DC.	(*ye.*)	Agrimony-like.	Italy.	1739.	6. 8.	H. ♃.	Col. ecphr. 1. t. 144.
Spallanzània agrimonoídes. S.S.		*Agrimònia agrimonoídes.* L.					

Subordo III. *SANGUISORBEÆ.* DC. prodr. v. 2. p. 588.

ALCHIMILLA. DC.		LADIES MANTLE.	Di-Tetrandria Monogynia. L.				
1 capénsis. DC.	(*gr.*)	Cape.	C. B. S.	1818.	4. 6.	G. ♃.	Lam. ill. t. 86. f. 2.
2 vulgàris. DC.	(*gr.*)	common.	Britain.	....	6. 8.	H. ♃.	Eng. bot. v. 9. t. 597.
β *variegàta.*	(*gr.*)	*variegated-leav'd.*	———	....	——	H. ♃.	
γ *glábra.* DC.	(*gr.*)	*smooth.*	Pyrenees.	....	——	H. ♃.	
3 hy'brida. DC.	(*gr.*)	mountain.	Saxony.	....	——	H. ♃.	Pluk. t. 240. f. 2.
montàna. W. en.							
4 pubéscens. DC.	(*gr.*)	pubescent.	Caucasus.	1813.	——	H. ♃.	Willd. hort. ber. 2. t. 79.
5 alpìna. DC.	(*gr.*)	silvery.	Britain.	....	7.	H. ♃.	Eng. bot. v. 4. t. 244.
6 serícea. DC.	(*gr.*)	silky.	Caucasus.	1813.	6. 8.	H. ♃.	
7 pentaphy'lla. DC.	(*gr.*)	five-leaved.	Switzerland.	1784.	7.	H. ♃.	Bocc. mus. p. 18. t. 1.
8 sibbaldiæfòlia. DC.	(*gr.*)	Sibbaldia-l[d].	Mexico.	1823.	——	F. ♃.	H. et B. n. g. am. 6. t. 561.
9 arvénsis. DC.	(*gr.*)	Parsley piert.	Britain.	....	4. 6.	H. ⊙.	Eng. bot. v. 15. t. 1011.
Aphànes. W. *Aphànes arvénsis.* L.							
10 cornucopioídes. DC.	(*gr.*)	Spanish.	Spain.	1825.	4. 7.	H. ⊙.	
CEPHAL'OTUS. DC.		CEPHAL'OTUS.	Dodecandria Hexagynia.				
folliculàris. DC.	(*wh.*)	pitcher-plant.	N. Holland.	1822.	4. 6.	G. *w.* ♃.	Lab. nov. holl. 2. t. 145.
MARGYRICA'RPUS. DC.		MARGYRICA'RPUS.	Diandria Monogynia.				
setòsus. DC.	(*gr.*)	bristle-pointed.	S. America.	1829.	....	F. ♄.	Flor. peruv. 1. t. 8. f. d.
AC'ÆNA. Kth.		AC'ÆNA.	Tri-Pentandria Monogynia.				
1 elongàta. DC.	(*gr.*)	long-spiked.	Mexico.	1827.	4. 8.	F. ♄.	
2 latebròsa. DC.	(*gr.*)	hairy.	C. B. S.	1774.	4. 6.	G. ♃.	Gært. fruct. 1. t. 32.

3 pinnatífida. DC.	(*gr.*)	pinnatifid.	Chili.	1823.	4. 6.	G. ♃.	Fl. per. 1. t. 104. f. 1. b.
4 serícea. DC.	(*gr.*)	silky.	Port Desire.	....	——	G. ♃.	Jacq. f. ecl. 1. t. 55.
ANCI'STRUM. Kth.		ANCI'STRUM.	Di-Pentandria Monogynia.				
1 sanguisórbæ. L.	(*gr.*)	Burnet-leaved.	N. Zealand.	1796.	6.	F. ♃.	Gært. fruct. t. 32.
2 ovalifòlium. DC.	(*gr.*)	oval-leaved.	Peru.	1802.	5. 6.	F. ♃.	Vent. h. cels. t. 5.
rèpens. Vent.							
3 argénteum. K.S.	(*gr.*)	silvery-leaved.	S. America.	1823.	——	F. ♄.	Fl. per. 1. t. 103. f. b.
4 adscéndens. DC.	(*gr.*)	ascending.	Magellan.	——	——	H. ♃.	
5 lævigàtum. DC.	(*gr.*)	smooth	———	1790.	6. 8.	H. ♃.	
6 magellánicum. DC.	(*gr.*)	Magellan.	———	....	——	H. ♃.	Lam. ill. t. 22. f. 2.
7 lùcidum. Lam.	(*gr.*)	shining.	Falkland Isl.	1777.	5. 6.	H. ♃.	——— t. 22. f. 3.
SANGUISO'RBA. DC.		GREAT-BURNET.	Tetrandria Monogynia. L.				
1 officinàlis. DC.	(*pu.*)	officinal.	Britain.	....	6. 8.	H. ♃.	Eng. bot. v. 19. t. 1312.
β *auriculàta.* DC.	(*pu.*)	*eared.*	Italy.	....	——	H. ♃.	Bocc. mus. 19. t. 9.
2 cárnea. DC.	(*pi.*)	flesh-coloured.	....	1818.	——	H. ♃.	
3 rùbra. Sck.	(*re.*)	red-spiked.	Dahuria.	1824.	——	H. ♃.	Schrank. h. mon. t. 69.
4 præcox. Bes.	(*pi.*)	early-flowered.	Siberia.	1827.	5. 7.	H. ♃.	
5 angustifòlia. F.	(*pi.*)	narrow-leaved.	———	1820.	6. 8.	H. ♃.	
6 tenuifòlia. DC.	(*re.*)	slender-leaved.	Dahuria.	——	——	H. ♃.	
7 íngrica.	(*wh.*)	glossy-leaved.	....	1829.	——	H. ♃.	
8 mauritànica. DC.	(*gr.*)	Barbary.	Algiers.	....	——	H. ♃.	Moris. ox. s. 8. t. 18. f. 4.
9 mèdia. DC.	(*wh.*)	short-spiked.	Canada.	1785.	——	H. ♃.	——— s. 8. t. 18. f. 8.
10 canadénsis. DC.	(*wh.*)	Canadian.	———	1633.	——	H. ♃.	——— s. 8. t. 18. f. 12.
POTE'RIUM. DC.		BURNET.	Monœcia Polyandria. L.				
1 spinòsum. DC.	(*wh.*)	prickly shrubby.	Levant.	1595.	4. 8.	G. ♄.	Moris. ox. s. 8. t. 18 f. 5.
β *críspum.* DC.	(*wh.*)	*curled-leaved.*	———	....	——	G. ♄.	
γ *inérme.* DC.	(*wh.*)	*spineless.*	———	....	——	G. ♄.	
2 caudàtum. DC.	(*wh.*)	shrubby.	Canaries.	1779.	1. 4.	G. ħ.	Bot. mag. 2341.
3 Sanguisórba. DC.	(*pu.*)	common.	England.	....	7.	H. ♃.	Eng. bot. v. 12. t. 860.
4 agrimonifòlium. DC.	(*pu.*)	Agrimony-ld.	Spain.	1818.	6. 7.	H. ♃.	
5 hy'bridum. W.	(*pu.*)	sweet.	France.	1683.	——	H. ♃.	Barrel. ic. t. 632.
6 polygàmum. DC.	(*pu.*)	Hungarian.	Hungary.	1803.	7. 8.	H. ♃.	W. et K. hung. 2. t. 198.
7 villòsum. DC.	(*pu.*)	villous.	Levant.	1829.	——	H. ♃.	Flor. græc. t. 942.
8 verrucòsum. E.	(*pu.*)	warted.	Egypt.	1827.	——	H. ♃.	
CERCOCA'RPUS. DC.		CERCOCA'RPUS.	Icosandria Monogynia.				
fothergilloídes. DC.	(*pu.*)	Fothergilla-like.	Mexico.	1828.	....	G. ♄.	H. et B. n. gen. 6. t. 559.
CLIFFO'RTIA. DC.		CLIFFO'RTIA.	Diœcia Polyandria. L.				
1 ilicifòlia. DC.	(*st.*)	Holly-leaved.	C. B. S.	1714.	5. 9.	G. ħ.	Linn. hort. cliff. t. 30.
2 cordifòlia. DC.	(*st.*)	heart-leaved.	———	1816.	——	G. ♄.	
3 tridentàta. DC.	(*st.*)	three-toothed.	———	1752.	6. 7.	G. ħ.	
4 ruscifòlia. DC.	(*st.*)	Ruscus-leaved.	———	——	——	G. ♄.	Linn. hort. cliff. t. 31.
arachnoídea. B.C. 260.							
5 gramínea. DC.	(*st.*)	grass-leaved.	———	1825.	——	G. ♄.	
6 ferrugínea. DC.	(*st.*)	ferruginous.	———	1823.	——	G. ħ.	
7 cuneàta. DC.	(*st.*)	wedge-leaved.	———	1787.	4.	G. ħ.	
8 strobilífera. DC.	(*st.*)	cone-bearing.	———	1816.	5. 7.	G. ħ.	Pluk. alm. t. 275. f. 2.
9 ericæfòlia. DC.	(*st.*)	heath-leaved.	———	1799.	7. 9.	G. ♄.	
10 sarmentòsa. DC.	(*st.*)	twiggy.	———	1795.	6. 8.	G. ♄.	
11 falcàta. DC.	(*st.*)	falcate-leaved.	———	1816.	——	G. ♄.	
12 ternàta. DC.	(*st.*)	ternate-leaved.	———	——	——	G. ♄.	Linn. hort. cliff. t. 32.
13 trifoliàta. DC.	(*st.*)	three-leaved.	———	1752.	4. 7.	G. ♄.	Pluk. alm. t. 319. f. 4.
14 obcordàta. DC.	(*st.*)	obcordate.	———	1790.	6. 8.	G. ħ.	
15 oblíqua. DC.	(*st.*)	oblique-leaved.	———	1816.	——	G. ♄.	
16 dentàta. DC.	(*st.*)	toothed.	———	——	——	G. ♄.	
17 crenàta. DC.	(*st.*)	notched-leaved.	———	1791.	7. 8.	G. ♄.	
18 pulchella. DC.	(*st.*)	beautiful.	———	1795.	4. 5.	G. ħ.	
19 cinerea. DC.	(*st.*)	cinereous.	———	1800.	6. 7.	G. ♄.	

ORDO LXXIX.

SPIRÆACEÆ. *D. Don prodr. flor. nepal. p.* 227.

PU'RSHIA. DC.		PU'RSHIA.	Icosandria Monogynia.				
tridentàta. DC.	(*ye.*)	three-toothed.	Columbia.	1827.	....	H. ♄.	Pursh. fl. amer. 1. t. 15.
Tigàrea tridentàta. Ph.							

KE'RRIA. DC.		KE'RRIA. Icosandria Pent-Octagynia.					
japónica. DC.	(*ye.*)	Japanese.	Japan.	1804.	4. 9.	H. ♄.	
β flòre plèno.	(*ye.*)	*double-flowered.*	———	——	——	H. ♄.	Andr. reposit. t. 587.
Córchorus japónicus. Bot. mag. t. 1296.							
SPIR'ÆA. DC.		SPIR'ÆA. Icosandria Pentagynia. L.					
1 opulifòlia. DC.	(*wh.*)	Guelder-rose-ld.	N. America.	1690.	6. 7.	H. ♄.	Schmidt arb. 1. t. 52.
2 capitàta. DC.	(*wh.*)	headed.	Columbia.	1827.	——	H. ♄.	
3 ulmifòlia. DC.	(*wh.*)	Elm-leaved.	Siberia.	1790.	——	H. ♄.	Jacq. vind. 2. t. 140.
4 flexuòsa. DC.	(*wh.*)	flexuose.	———	1820.	——	H. ♄.	Camb. mon. spir. t. 26.
5 bélla. DC.	(*ro.*)	red-flowered.	Nepaul.	1818.	6. 9.	H. ♄.	Bot. mag. t. 2426.
6 chamædrifòlia. DC.	(*wh.*)	Germander-ld.	Siberia.	1780.	6. 7.	H. ♄.	Bot. reg. t. 1222.
7 mèdia. Ph.	(*wh.*)	middle.	Canada.	1827.	——	H. ♄.	
8 oblongifòlia. W. K.	(*wh.*)	oblong-leaved.	Hungary.	1815.	5. 7.	H. ♄.	W. et K. hung. 3. t. 235.
9 càna. DC.	(*wh.*)	hoary-leaved.	———	1824.	6. 7.	H. ♄.	——— 3. t. 227.
10 trilobàta. DC.	(*wh.*)	three-lobed.	Siberia.	1801.	5. 6.	H. ♄.	Pall. ross. 1. t. 17.
11 alpìna. DC.	(*wh.*)	Alpine.	———	1806.	6. 7.	H. ♄.	——— 1. t. 20.
12 thalictroídes. DC.	(*wh.*)	meadow-rue-ld.	———	1790.	5. 6.	H. ♄.	——— 1. t. 18.
13 pikowiénsis. DC.	(*wh.*)	Pikow.	Poland.	1824.	——	H. ♄.	
14 hypericifòlia. W.	(*wh.*)	Hypericum-ld.	N. America.	1640.	4. 5.	H. ♄.	Schmidt arb. 1. t. 26.
15 uralénsis.	(*wh.*)	Fischer's.	Siberia.	1822.	——	H. ♄.	
crenàta. F. *non* W. *hypericifòlia β uralénsis.* DC.							
16 acutifòlia. W. en.	(*wh.*)	acute-leaved.	Siberia.	——	——	H. ♄.	
17 crenàta. L.	(*wh.*)	crenated.	———	1739.	——	H. ♄.	
18 cratægifòlia. L. en.	(*wh.*)	Hawthorn-leav'd.	......	1824.	5. 6.	H. ♄.	
19 obovàta. W. K.	(*wh.*)	obovate leaved.	Hungary.	1816.	——	H. ♄.	
20 savránica. Bes.	(*wh.*)	small-flowered.	Poland.	1826.	——	H. ♄.	
21 Besseriàna.	(*wh.*)	Besser's.	———	——	——	H. ♄.	
crenàta. Bes. *non* L.							
22 canéscens. DC.	(*wh.*)	canescent.	Nepaul.	1825.	6. 8.	H. ♄.	
23 ceanothifòlia. DC.	(*wh.*)	Ceanothus-ld.	....	1820.	5. 6.	H. ♄.	
24 corymbòsa. DC.	(*wh.*)	corymbose.	Virginia.	1812.	——	H. ♄.	Lodd. bot. cab. t. 671.
25 vacciniifòlia. D. P.	(*wh.*)	Vaccinium-ld.	Nepaul.	1824.	6. 8.	H. ♄.	——— t. 1403.
26 betulæfòlia. DC.	(*wh.*)	Birch-leaved.	Siberia.	1812.	6. 7.	H. ♄.	Pall. ross. 1. t. 16.
27 lævigàta. DC.	(*wh.*)	smooth-leaved.	———	1774.	4. 6.	H. ♄.	Schmidt arb. 1. t. 49.
28 salicifòlia. DC.	(*ro.*)	Willow-leaved.	Britain.	....	6. 8.	H. ♄.	Eng. bot. v. 21. t. 1468.
29 álba. W. D.	(*wh.*)	white-flowered.	....	....	——	H. ♄.	Wats. dend. brit. t. 133.
30 carpinifòlia. W. en.	(*bh.*)	broad-leaved.	N. America.	....	7. 8.	H. ♄.	——— t. 66.
31 tomentòsa. DC.	(*ro.*)	tomentose.	———	1736.	7. 9.	H. ♄.	Schmidt arb. 1. t. 51.
32 ariæfòlia. DC.	(*ro.*)	Aria-leaved	———	1827.	6. 8.	H. ♄.	
33 discolor. DC.	(*wh.*)	two-coloured.	———	——	——	H. ♄.	
34 sorbifòlia. DC.	(*wh.*)	Service-leaved.	Siberia.	1759.	7. 9.	H. ♄.	Schmidt arb. 1. t. 58.
35 grandiflòra.	(*wh.*)	large-flowered.	Kamtschatka.	1817.	7. 8.	H. ♄.	Pall. ros. 1. t. 25.
sorbifòlia β alpìna. DC.							
36 Arúncus. DC.	(*wh.*)	Goat's-beard.	Siberia.	1633.	6. 7.	H. ♃.	Pall. ros. 1. t. 26.
37 americàna.	(*wh.*)	glossy-leaved.	N. America.	1825.	——	H. ♃.	
38 Ulmària. DC.	(*wh.*)	Meadow-sweet.	Britain.	....	6. 10.	H. ♃.	Eng. bot. v. 14. t. 960.
α tomentòsa. DC.	(*wh.*)	*tomentose.*	———	....	——	H. ♃.	
β múltiplex. DC.	(*wh.*)	*double-flowered.*	———	....	——	H. ♃.	
γ variegàta. DC.	(*wh.*)	*variegated-leav'd.*	———	....	——	H. ♃.	
39 lobàta. DC.	(*ro.*)	lobed-leaved.	N. America.	1765.	7. 8.	H. ♃.	Jacq. vind. 1. t. 88.
40 palmàta. DC.	(*wh.*)	palmated.	China.	1823.	——	H. ♃.	
41 digitàta. DC.	(*ro.*)	digitate.	Siberia.	——	——	H. ♃.	Pall. ros. 1. t. 27.
42 Filipéndula. DC.	(*wh.*)	Dropwort.	Britain.	....	6. 10.	H. ♃.	Eng. bot. v. 4. t. 284.
β múltiplex. DC.	(*wh.*)	*double-flower'd.*	———	....	——	H. ♃.	
GILL'ENIA. DC.		GILL'ENIA. Dodecandria Pentagynia.					
1 trifoliàta. DC.	(*bh.*)	three-leaved.	N. America.	1713.	6. 7.	H. ♃.	Bot. mag. t. 489.
Spir'æa trifoliàta. B. M.							
2 stipulàcea. DC.	(*bh.*)	large-stipuled.	———	1805.	——	H. ♃.	Bart. med. bot. t. 6.

ORDO LXXX.

CALYCANTHEÆ. *DC. prodr.* 3. *p.* 1.

CALYCA'NTHUS. DC.		ALLSPICE-TREE. Icosandria Polygynia. L.					
1 flòridus. DC.	(*d. pu.*)	Carolina.	Carolina.	1726.	5. 8.	H. ♄.	Bot. mag. t. 503.

2 prunifòlius. (*d.pu.*) Plum-leaved. Carolina. 1822. 5. 8. H. ♄.
3 glaùcus. DC. (*d.pu.*) glaucous-leaved. ——— 1806. —— H. ♄. Guimp. ab. holz. t. 5.
fértilis. Bot. reg. t. 404.
4 oblongifòlius. N.(*d.pu.*) oblong-leaved. ——— 1812. —— H. ♄.
5 lævigàtus. DC. (*d.pu.*) smooth-leaved. ——— 1806. —— H. ♄. Bot. reg. t. 481.
CHIMONA'NTHUS. DC. JAPAN ALLSPICE. Icosandria Polygynia.
fràgrans. DC. (*ye.*) fragrant. Japan. 1766. 12. 3. H. ♄. Bot. mag. t. 466.
Calycánthus pr'æcox. B.M.
α verus. (*ye.pu.*) *common.* ——— —— —— H. ♄. Hort. Kew. v. 2. t. 10.
β lùteus. (*ye.*) *yellow-flowered.* ——— 1814. —— H. ♄.
γ grandiflòrus. B.R.(*ye.*) *large-flowered.* ——— 1812. —— H. ♄. Bot. reg. t. 451.

ORDO LXXXI.

GRANATEÆ. *D.Don in Jameson Edinb. phil. journ. jul.* 1826. *p.* 134.

P'UNICA. DC. POMEGRANATE. Icosandria Monogynia. L.
1 Granàtum. DC. (*or.*) common. S. Europe. 1548. 6. 9. H. ♄. Bot. mag. t. 1832.
α rùbrum. DC. (*or.*) *red-flowered.* ——— —— —— H. ♄. Schkuhr handb. t. 131.
β flòre plèno. DC. (*or.*) *double red.* ——— .... 6. 8. H. ♄. Trew ehret. t. 71. f. 2.
γ flavéscens. (*ye.*) *yellow-flowered.* China. 1810. —— F. ♄.
δ albéscens. (*wh.*) *white-flowered.* ——— —— —— F. ♄. Andr. reposit. t. 96.
ε múltiplex. (*wh.*) *double white.* ——— .... —— F. ♄.
2 nàna. DC. (*or.*) dwarf. S. America. 1723. 7. 9. F. ♄. Bot. mag. t. 634.

ORDO LXXXII.

MEMECYLEÆ. *DC. prodr.* 3. *p.* 5.

MEM'ECYLON. DC. MEM'ECYLON. Octandria Monogynia. L.
1 angulàtum. DC. (*pu.*) square-stalked. Mauritius. 1826. 5. 7. S. ♄.
2 capitellàtum. DC. (*pu.*) headed. Ceylon. 1796. 7. 8. S. ♄. Burm. zeyl. t. 30.
3 èdule. DC. (*pu.*) eatable-fruited. E. Indies. 1823. —— S. ♄. Roxb. corom. 1. t. 82.
4 gránde. DC. (*bl.*) tree. ——— 1824. .... S. ♄. Rheed. mal. 2. t. 14.
MOURI'RIA. DC. MOURI'RIA. Decandria Monogynia.
1 myrtilloídes. DC. (*wh.*) sessile-leaved. Jamaica. 1818. 6. 8. S. ♄. Sloan. his. 2. t. 87. f. 3.
Petalòma myrtilloídes. Swartz fl. ind. occ. 2. p. 833. t. 14.
2 guianénsis. DC. (*ye.*) Guiana. Guiana. 1825. —— S. ♄. Aubl. guian. 1. t. 180.
3 grandiflòra. DC. (*wh.*) great flowered. Brazil. 1822. .... S. ♄.
4 alternifòlia. H.B. .... alternate-leav'd. E. Indies. 1823. .... S. ♄.

ORDO LXXXIII.

COMBRETACEÆ. *DC. prodr.* 3. *p.* 9.

Tribus I. *TERMINALIEÆ.* DC. prodr. 3. p. 9.

BUC'IDA. DC. OLIVE-BARK-TREE. Decandria Monogynia. L.
1 Bùceras. DC. (*wh.*) Jamaica. Jamaica. 1793. 7. 10. S. ♄. Bot. reg. t. 907.
2 capitàta. DC. (*wh.*) headed. W. Indies. 1826. .... S. ♄. Vahl. ecl. 1. t. 8.
TERMIN'ALIA. DC. TERMIN'ALIA. Polygamia Monœcia. L.
1 Taniboùca. DC. (*wh.*) Guiana. Guiana. 1826. .... S. ♄. Aubl. guian. 1. t. 178.
2 angustifòlia. DC. (*wh.*) narrow-leaved. E. Indies. 1692. 5. 7. S. ♄. Jacq. vind. 3. t. 100.
Catáppa Benzoin. Gært. fruct. 2. p. 206. t. 127.

3 mauritiàna. DC. (*wh.*) Mauritius. Mauritius. 1822. .... S. ♄. Lam. ill. t. 848. f. 2.
4 Catáppa. DC. (*wh.*) large-leaved. E. Indies. 1778. 5. 7. S. ♄. Jacq. ic. 1. t. 197.
5 subcordàta. K.S. (*wh.*) heart-leaved. W. Indies. 1816. .... S. ♄.
6 moluccàna. DC. (*wh.*) smooth-leaved. Moluccas. ——— .... S. ♄. Rumph. amb. 1. t. 68.
7 glábrata. DC. (*wh.*) smooth-leaved. Society Isles. 1822. .... S. ♄. Spreng. antiq. bot. t. 2.
8 latìfolia. DC. (*wh.*) broad-leaved. W. Indies. 1778. 6. 8. S. ♄.
9 bengalénsis. DC. (*wh.*) Bengal. E. Indies. 1826. .... S. ♄.
10 Bellerìca. DC. (*wh.*) fetid-flowered. ——— 1818. .... S. ♄. Roxb. corom. 2. t. 198.
11 Chébula. DC. (*wh.*) oval-leaved. ——— 1796. 6. 8. S. ♄. ——— 2. t. 197.
12 citrìna. DC. (*st.*) Citron-leaved. ——— 1823. .... S. ♄. Gært. fruct. 2. t. 97.
13 pròcera. DC. (*wh.*) lofty. ——— 1816. .... S. ♄. Roxb. corom. 3. t. 224.
14 gangética. H.B. (*wh.*) Ganges. ——— 1823. .... S. ♄.
15 biticària. H.B. (*wh.*) Silhet. ——— 1824. .... S. ♄.
16 arbúscula. DC. (*wh.*) small tree. Jamaica. 1822. .... S. ♄.
17 serícea. DC. (*wh.*) silky-leaved. C. B. S. 1816. .... I. ♄.
18 Fatr'æa. DC. (*wh.*) Madagascar. Madagascar. 1826. .... S. ♄.

PENTA'PTERA. DC. PENTA'PTERA. Polygamia Monœcia.
1 tomentòsa. DC. (*wh.*) woolly. E. Indies. 1824. .... S. ♄. DC. diss. t. 1.
2 paniculàta. DC. (*wh.*) panicled. ——— 1826. .... S. ♄.
3 Arjùna. DC. (*wh.*) glaucous-leaved. ——— 1825. .... S. ♄. DC. diss. t. 2.
4 bialàta. H B. (*wh.*) two-winged. ——— ——— .... S. ♄.

GET'ONIA. DC. GET'ONIA. Decandria Monogynia.
1 floribúnda. DC. (*wh.*) many-flowered. E. Indies. 1815. 5. 7. S. ♄⌣. Roxb. corom. 1. t. 87.
2 nùtans. DC. (*wh.*) nodding. ——— 1823. ——— S. ♄⌣.

CONOCA'RPUS. DC. BUTTON-TREE. Pent-Decandria Monogynia.
1 erécta. W. (*wh.*) upright. Jamaica. 1752. 6. 8. S. ♄. Jacq. amer. t. 52. f. 1.
2 procúmbens. W. (*wh.*) procumbent. Cuba. 1730. ——— S. ♄. ——— t. 52. f. 2.
3 acuminàta. DC. (*wh.*) pointed-leaved. E. Indies. 1822. .... S. ♄.
4 latifòlia. DC. (*wh.*) broad-leaved. ——— 1816. .... S. ♄.

LAGUNCUL'ARIA. DC. LAGUNCUL'ARIA. Decandria Monogynia.
racemòsa. DC. (*gr.*) racemed. W. Indies. 1825. .... S. ♄. Jacq. amer. t. 53.

POI'VREA. DC. POI'VREA. Decandria Monogynia.
1 comòsa. (*cr.*) tufted. Sierra Leone. 1821. 5. 8. S. ♄⌣. Bot. reg. t. 1165.
Combrètum comòsum. Bot. reg.
2 intermèdia. G.D. (*cr.*) intermediate. ——— ——— .... S. ♄⌣.
3 alternifòlia. DC. (*wh.*) alternate-leav'd. S. America. 1826. .... S. ♄⌣. Jacq. am. pic. t. 260. f. 27.
Combrètum decándrum. Jacq.
4 coccínea. DC. (*sc.*) scarlet-flower'd. Madagascar. 1818. 3. 10. S. ♄⌣. Lam. ill. t. 282. f. 2.
Combrètum purpùreum. Bot. reg. 429. *Cristària coccínea.* Sonn. itin. 2. t. 140. *non* Pursh.
5 Roxbúrghii. DC. (*wh.*) Roxburgh's. E. Indies. 1822. .... S. ♄⌣. Roxb. corom. 1. t. 59.
Combrètum decándrum. Roxb. *non* Jacq.

Tribus II. *COMBRETEÆ.* DC. prodr. 3. p. 18.

COMBR'ETUM. DC. COMBR'ETUM. Octandria Monogynia. L.
1 secúndum. G.D. (*ye.sc.*) side-flowering. W. Indies. 1822. .... S. ♄⌣. Jacq. am. pic. t. 260. f. 26.
2 élegans. DC. (*re.*) elegant. S. America. 1823. .... S. ♄⌣. Aubl. gui. 1. t. 137.
3 formòsum. G.D. (*ye.re.*) beautiful. Brazil. 1824. .... S. ♄⌣.
4 grandiflòrum. G.D. (*sc.*) large-flowered. Sierra Leone. 1821. .... S. ♄⌣.
5 farinòsum. G.D. (*or.*) mealy-leaved. Mexico. 1823. .... G. ♄⌣.
6 láxum. G.D. (*wh.*) loose-racemed. S. America. 1822. .... S. ♄⌣. Lam. ill. t. 282. f. 1.
7 pulchéllum. G.D. (*sc.*) pretty. Brazil. 1823. .... S. ♄⌣.
8 nànum. G.D. (*wh.*) dwarf. Nepaul. 1826. .... G. ♄.
9 micránthum. G.D. (*sc.*) small-flowered. Sierra Leone. 1828. .... S. ♄.
parviflòrum. Reichenb. hort. bot. t. 62.

CACO'UCIA. DC. CACO'UCIA. Dodecandria Monogynia.
coccínea. DC. (*sc.*) scarlet. Guiana. 1815. .... S. ♄⌣. Aubl. guian. 1. t. 179.
Schousb'æa coccínea. W.

QUISQU'ALIS. DC. QUISQU'ALIS. Decandria Monogynia. L.
1 índica. DC. (*cr.*) Indian. China. 1815. 5. 8. S. ♄⌣. Bot. reg. t. 492.
2 glábra. DC. (*cr.*) smooth. E. Indies. 1825. ——— S. ♄⌣. Burm. ind. t. 28. f. 2.
3 pubéscens. S.S. .... pubescent. China. 1815. .... S. ♄⌣.
4 villòsa. S.S. .... villous. Pegu. 1818. .... S. ♄⌣.

ORDO LXXXIV.

VOCHYSIEÆ. *DC. prodr.* 3. *p.* 25.

VOCHY'SIA. DC.		VOCHY'SIA. Monandria Monogynia.					
1 guianénsis. DC.	(*ye.*)	Guiana.	Guiana.	1822.	....	S. ♄.	Aubl. guian. 1. t. 6.
Cuculària excélsa. W.							
2 tomentòsa. DC.	(*ye.*)	woolly.	———	1825.	....	S. ♄.	

ORDO LXXXV.

RHIZOPHOREÆ. *DC. prodr.* 3. *p.* 31.

OLISBEA. DC.		OLISBEA. Decandria Monogynia.					
rhizophoræfòlia. DC...		Rhizophora-ld.	W. Indies.	1825.	....	S. ♄.	
RHIZO'PHORA. DC.		RHIZO'PHORA. Octandria Monogynia.					
Mángle. DC.	(*ye.*)	common.	India.	1818.	....	S. ♄.	Jacq. amer. t. 89.
CARA'LLIA. DC.		CARA'LLIA. Dodecandria Monogynia.					
lùcida. DC.	(*ye.*)	glossy.	E. Indies.	1826.	....	S. ♄.	Roxb. corom. 3. t. 211.

ORDO LXXXVI.

ONAGRARIEÆ. *DC. prodr. v.* 3. *p.* 35.

Tribus I. *MONTINIEÆ.* DC. prodr. loc. cit.

MONTI'NIA. DC.		MONTI'NIA. Diœcia Tetrandria. L.					
ácris. DC.	(*wh.*)	Sea purslane-ld.	C. B. S.	1774.	7. 8.	G. ♄.	Smith spicil. t. 15.
caryophyllàcea. Thunb.							

Tribus II. *FUCHSIEÆ.* DC. prod. v. 3. p. 36.

FUCHSIA. DC.		FUCHSIA. Octandria Monogynia. L.					
1 microphy'lla. DC.	(*re.*)	small-leaved.	Mexico.	1827.	6. 9.	F. ♄.	Swt. br. fl. gar. s. 2. t. 16.
2 lycioídes. DC.	(*re.*)	Box-thorn-ld.	Chile.	1796.	4. 10.	G. ♄.	Bot. mag. t. 1024.
3 parviflòra. DC.	(*re.*)	small-flowered.	Mexico.	1824.	6. 9.	F. ♄.	Bot. reg. t. 1048.
4 arboréscens. DC.	(*ro.*)	arborescent.	———	1823.	6. 10.	G. ♄.	——— t. 943.
5 grácilis. DC.	(*sc.*)	slender.	———	1822.	5. 10.	F. ♄.	——— t. 847.
β multiflòra. DC.	(*sc.*)	*many-flowered.*	———	1824.	——	F. ♄.	——— t. 1052.
6 macrostémma. DC.	(*sc.*)	large-crowned.	Chile.	1823.	——	G. ♄.	Flor. per. 3. t. 324. f. b.
tenélla. DC.	(*sc.*)	*weak.*	———	1822.	——	G. ♄.	
7 cònica. B.R.	(*sc.*)	conical-tubed.	———	1824.	——	F. ♄.	Bot. reg. t. 1062.
8 virgàta.	(*sc.*)	Colvill's.	Mexico.	1825.	——	F. ♄.	
9 coccínea. DC.	(*sc.*)	scarlet.	Chile.	1788.	——	F. ♄.	Bot. mag. t. 97.
10 excorticàta. DC.	(*gr, re.*)	changeable.	N. Zealand.	1821.	——	G. ♄.	Bot. reg. t. 857.

Tribus III. *ONAGREÆ.* DC. prodr. v. 3. p. 40.

CHAMÆN'ERION. Tourn.		CHAMÆN'ERION. Octandria Monogynia.					
1 spicàtum. G.B.	(*re.*)	spiked.	England.	....	7. 8.	H. ♃.	Eng. bot. v. 28. t. 1947.
Epilòbium angustifòlium. E.B.							
β albiflòrum.	(*wh.*)	*white-flowered.*	———	....	——	H. ♃.	

2 speciòsum. L.C.	(*re.*)	handsome.	Europe.	....	7. 8.	H. ♃.	
Epilòbium angustifòlium β latum. DC. *latifòlium.* Schmidt.							
3 americànum.	(*re.*)	American.	N. America.	1825.	——	H. ♃.	
4 rosmarinifòlium. DC.	(*re.*)	Rosemary-l^{d}.	Europe.	1775.	6. 8.	H. ♃.	Reich. ic. 4. t. 341.
Epilòbium angustíssimum. Bot. mag. t. 76.							
5 Hallèri. Retz.	(*re.*)	Haller's.	Switzerland.	1798.	——	H. ♃.	———— 4. t. 342.
Epilòbium Dodonæ̀i. Allioni. *rosmarinifòlium β alpinum.* DC.							
6 latifòlium.	(*li.*)	broad-leaved.	Europe.	1779.	7. 8.	H. ♃.	Sal. par. lond. t. 58.
halimifòlium. P.L.							
EPIL'OBIUM. DC.		WILLOW-HERB. Octandria Monogynia. L.					
1 alpìnum. DC.	(*ro.*)	alpine.	Britain.	....	6. 7.	H. ♃.	Eng. bot. v. 28. t. 2001.
2 origanifòlium. DC.	(*ro.*)	Marjoram-leav'd.	————	....	7. 8.	H. ♃.	———— v. 28. t. 2000.
alsinifòlium. E.B.							
3 ròseum. E.B.	(*ro.*)	rose-coloured.	————	....	——	H. ♃.	———— v. 10. t. 693.
4 alpéstre. S.S.	(*ro.*)	simple-stalked.	Switzerland.	....	——	H. ♃.	
5 montànum. DC.	(*li.*)	mountain.	Britain.	....	——	H. ♃.	Eng. bot. v. 17. t. 1177.
6 nùtans. DC.	(*bh.*)	nodding.	Bohemia.	1827.	——	H. ♃.	Reich. ic. 2. t. 197.
7 Hornemánni. DC.	(*li.*)	Hornemann's.	Finland.	——	——	H. ♃.	———— p. 73. t. 180.
nùtans. Horn. *nec aliorum.*							
8 coloràtum. DC.	(*ro.*)	pink-flowered.	N. America.	1805.	——	H. ♃.	
9 dahùricum. DC.	(*wh.*)	Dahurian.	Dahuria.	1816.	——	H. ⊙.	
10 hirsùtum. DC.	(*pu.*)	Codlins & cream.	Britain.	....	——	H. ♃.	Eng. bot. v. 12. t. 838.
11 tomentòsum. DC.	(*pu.*)	woolly.	Levant.	1821.	——	H. ♃.	Vent. hort. cels. t. 90.
12 villòsum. DC.	(*pu.*)	villous.	C. B. S.	1799.	——	G. ♃.	
13 cylíndricum. DC.	(*ro.*)	round-stemmed.	Nepaul.	1827.	——	H. ♃.	
14 palústre. DC.	(*ro.*)	round-stalked.	Britain.	....	——	H. ♃.	Eng. bot. v. 5. t. 346.
15 rivulàre. DC.	(*ro.*)	rivulet.	N. Europe.	1828.	——	H. ♃.	Reich. ic. 2. t. 170.
16 parviflòrum. DC.	(*pu.*)	small-flowered.	Britain.	....	——	H. ♃.	Eng. bot. v. 12. t. 795.
17 tetragònum. DC.	(*pu.*)	square-stemmed.	————	....	——	H. ♃.	———— v. 28. t. 1948.
18 obscùrum. P.S.	(*pu.*)	obscure.	Europe.	1820.	——	H. ♃.	Reich. ic. 2. t. 199.
19 canéscens. H.H.	(*ro.*)	canescent.	....	1826.	——	H. ♃.	
GA'URA. DC.		GA'URA. Octandria Monogynia. L.					
1 biénnis. DC.	(*bh.*)	biennial.	N. America.	1762.	8. 10.	H. ♂.	Bot. mag. t. 389.
2 angustifòlia. DC.	(*fl.*)	narrow-leaved.	Carolina.	1816.	7. 9.	F. ♂.	Jacq. ic. 3. t. 457.
fruticòsa. Jacq. *ex* DC. prodr.							
3 sinuàta. DC.	(*fl.*)	sinuate-leaved.	Arkanza.	1826.	——	H. ♂.	
4 tripétala. DC.	(*bh.*)	three-petaled.	Mexico.	1804.	8. 10.	H. ⊙.	Cavan. ic. 4. t. 396. f. 1.
5 linifòlia. DC.	(*wh.*)	Flax-leaved.	Arkanza.	1826.	7. 9.	H. ♂.	
6 coccínea. DC.	(*ro.*)	pale-scarlet.	Louisiana.	1811.	8. 10.	H. ♃.	
7 mutábilis. DC.	(*ye.*)	changeable-col'd.	N. Spain.	1795.	7. 9.	F. ♂.	Bot. mag. t. 388.
8 œnotheriflòra. DC.	(*ye.*)	canescent.	....	1816.	——	F. ♂.	
ŒNOTH'ERA. DC.		EVENING PRIMROSE. Octandria Monogynia. L.					
1 dentàta. DC.	(*ye.*)	toothed.	Chile.	1818.	6. 9.	H. ⊙.	Lindl. coll. t. 10.
2 micrántha. S.S.	(*ye.*)	small-blossom'd.	California.	1820.	——	H. ⊙.	
hírta. L. enum.							
3 cheiranthifòlia. DC.	(*ye.*)	Wallflower-l^{d}.	Chile.	——	——	F. ♄.	Bot. reg. t. 1040.
4 biénnis. DC.	(*ye.*)	common.	N. America.	1629.	——	H. ♂.	Eng. bot. v. 22. t. 1534.
5 elàta. DC.	(*ye.*)	tall upright.	Mexico.	1827.	7. 10.	F. ♂.	
6 grandiflòra. DC.	(*ye.*)	large-flowered.	N. America.	1778.	6. 8.	H. ⊙.	Bot. mag. t. 2068.
7 suavéolens. DC.	(*ye.*)	fragrant.	————	....	——	H. ♂.	
8 Simsiàna. DC.	(*ye.*)	Sims's.	Mexico.	1816.	7. 10.	F. ♂.	Bot. mag. t. 1974.
corymbòsa. B.M. *non* Lamarck.							
9 macrocárpa. DC.	(*ye.*)	broad-leaved.	N. America.	1811.	6. 8.	H. ♃.	Swt. br. fl. gar. t. 5.
10 missouriénsis. B.M.	(*ye.*)	narrow-leaved.	————	——	——	H. ♃.	Bot. mag. t. 1592.
alàta. N. *non* Rafin.							
11 salicifòlia. DC.	(*ye.*)	Willow-leaved.	....	1818.	6. 9.	H. ⊙.	
12 parviflòra. DC.	(*ye.*)	small-flowered.	N. America.	1757.	6. 8.	H. ♂.	Meerb. ic. 1. t. 34.
13 cruciàta. N.	(*st.*)	cross-petaled.	————	1821.	6. 10.	H. ♂.	
14 muricàta. DC.	(*ye.*)	prickly-stalked.	N. America.	1789.	7. 9.	H. ♂.	Murr. com. goet. 6. t. 1.
15 noctúrna. DC.	(*ye.*)	night-smelling.	C. B. S.	1790.	4. 8.	G. ♂.	Jacq. ic. 3. t. 455.
16 longiflòra. DC.	(*ye.*)	long-flowered.	BuenosAyres.	1776.	7. 9.	H. ♂.	Bot. mag. t. 365.
17 villòsa. DC.	(*ye.*)	villous.	C. B. S.	1791.	——	G. ♂.	
18 mollíssima. DC.	(*ye.*)	soft wave-l^{d}.	BuenosAyres.	1732.	6. 10.	H. ♂.	Schkuhr hand. 1. t. 105.
19 odoràta. DC.	(*ye.*)	sweet-scented.	Patagonia.	1790.	4. 8.	H. ♃.	Bot. mag. t. 2403.
α glaucéscens. DC.	(*ye.*)	*glaucescent.*	————	————	——	H. ♃.	Jacq. ic. 3. t. 456.
β viréscens. DC.	(*ye.*)	*purple-stalked.*	————	————	——	H. ♃.	Bot. reg. t. 147.
20 striàta. DC.	(*ye.*)	striped-seeded.	....	1816.	6. 9.	H. ⊙.	
21 sinuàta. DC.	(*ye.*)	scollop-leaved.	N. America.	1770.	7. 8.	H. ⊙.	Murr. com. gœt. 5. t. 9.
22 tenélla. DC.	(*pu.*)	slender.	Chile.	1822.	6. 9.	H. ⊙.	Swt. br. fl. gar. t. 167.

23 tenuifòlia. DC.	(*pu.*)	slender-leaved.	Chile.	1829.	6. 9.	H. ☉.	Cavan. ic. 4. t. 397.
24 vimínea. B.M.	(*pu.*)	twiggy.	N. America.	1827.	——	H. ☉.	Bot. reg. t. 1220.
25 decúmbens. D.	(*pu.*)	decumbent.	——	——	——	H. ☉.	—— t. 1221.
26 quadrivúlnera. D.	(*pu.*)	four-spotted.	——	——	——	H. ☉.	—— t. 1119.
27 Lindleyàna. D.	(*pu.*)	Lindley's.	——	——	——	H. ☉.	Swt. br. fl. gar. s. 2. t. 19.
28 ròseo-álba. DC.	(*wh.ro.*)	rosy and white.	America.	1826.	——	H. ☉.	—— t. 268.
29 purpùrea. DC.	(*pu.*)	purple-flower'd.	N. America.	1794.	5. 9.	H. ☉.	Bot. mag. t. 352.
30 Romanzòvii. DC.	(*pu.*)	Romanzoff's.	——	1819.	——	H. ☉.	Bot. reg. t. 662.
31 corymbòsa. DC.	(*ye.*)	corymbose.	....	1826.	6. 9.	H. ♂.	
32 spectábilis. H.H.	(*ye.*)	showy.	Mexico.	1822.	——	F. ♂.	
33 gauroídes. DC.	(*ye.*)	Gaura-like.	Baltimore.	1816.	7. 9.	H. ♂.	
34 mèdia. DC.	(*ye.*)	intermediate.	N. America.	1821.	6. 9.	H. ♃.	
35 serrulàta. DC.	(*ye.*)	saw-leaved.	——	——	6. 10.	H. ♃.	Swt. br. fl. gar. t. 133.
36 strícta. Led.	(*ye.*)	straight.	....	1822.	——	H. ☉.	
37 am'œna. Leh.	(*ye.*)	delicate.	....	1826.	——	H. ☉.	
38 nervòsa. H.H.	(*ye.*)	nerved-leaved.	....	——	——	H. ♂.	
39 cæspitòsa. DC.	(*wh.*)	tufted.	N. America.	1811.	6. 9.	H. ♃.	Bot. mag. t. 1593.
40 acaùlis. DC.	(*wh.*)	short-stalked.	Chile.	1822.	——	H. ♃.	Bot. reg. t. 763.
41 taraxacifòlia. B.F.G.	(*w.*)	Dandelion-l[d].	——	1823.	——	H. ♃.	Swt. br. fl. gar. t. 294.
grandiflòra. Flor. peruv. 3. t. 318. f. 6. *nec aliorum.*							
42 trilòba. DC.	(*st.*)	root-flowering.	N. America.	1821.	6. 10.	H. ☉.	Bot. mag. t. 2566.
rhizocárpa. S.S.							
43 virgàta. DC.	(*pu.*)	slender-twigg'd.	Peru.	1823.	6. 9.	H. ☉.	Flor. per. 29. t. 315.
44 pinnatífida. DC.	(*ye.*)	pinnatifid.	Mexico.	1825.	——	F. ♃.	
45 tetraptera. DC.	(*wh.*)	four-winged.	——	1796.	——	H. ☉.	Bot. mag. t. 468.
46 latiflòra. DC.	(*wh.*)	broad-flowered.	——	1826.	——	H. ♃.	Flor. mex. ined. t. 376.
47 glaùca. DC.	(*ye.*)	glaucous-leav'd.	N. America.	1812.	5. 10.	H. ♃.	Bot. mag. t. 1606.
48 fruticòsa. DC.	(*ye.*)	frutescent.	——	1737.	6. 8.	H. ♃.	—— t. 332.
49 ambígua. DC.	(*ye.*)	ambiguous.	——	....	....	H. ♃.	
50 serotìna. B.F.G.	(*ye.*)	late-flowered.	——	1820.	7. 10.	H. ♄.	Swt. br. fl. gar. t. 184.
51 tetragòna. DC.	(*ye.*)	blunt-leaved.	——	....	6. 8.	H. ♃.	
52 Fraséri. DC.	(*ye.*)	Fraser's.	——	1811.	5. 10.	H. ♃.	Bot. mag. t. 1674.
53 pùmila. DC.	(*ye.*)	dwarf.	——	1757.	5. 9.	H. ♃.	—— t. 355.
54 ròsea. DC.	(*ro.*)	rose-coloured.	Peru.	1783.	——	F. ♃.	—— t. 347.
55 lineàris. DC.	(*ye.*)	linear-leaved.	N. America.	....	6. 8.	H. ♃.	
56 speciòsa. DC.	(*wh.*)	handsome.	Louisiana.	1821.	——	H. ♃.	Swt. br. fl. gar. t. 253.
57 pállida. B.R.	(*bh.*)	pale.	N. America.	1827.	——	H. ♃.	Bot. reg. t. 1142.
58 albicaùlis. DC.	(*wh.*)	white-stalked.	——	1811.	5. 9.	H. ☉.	
pinnatífida. N. *non* Kth.							
59 Nuttállii.	(*wh.*)	Nuttall's.	N. America.	——	5. 8.	H. ♃.	
albicaùlis. N. *non* Pursh.							
60 canadénsis. DC.	(*ye.*)	Canadian.	Canada.	1825.	6. 9.	H. ☉.	
CLA'RKIA. Ph.		CLA'RKIA. Octandria Monogynia.					
pulchélla. Ph.	(*ro.*)	beautiful.	N. America.	1827.	5. 10.	H. ☉.	Bot. reg. t. 1100.
β albiflòra.	(*wh.*)	*white-flowered.*	——	——	——	H. ☉.	
γ integrilòba.	(*ro.*)	*entire-lobed.*	——	——	——	H. ☉.	

Tribus IV. *JUSSIEÆ.* DC. prodr. v. 3. p. 52.

JUSSI'ÆA. DC.		JUSSI'ÆA. Octo-Dec-Dodecandria Monogynia.					
1 variábilis. DC.	(*ye.*)	variable.	Guiana.	1820.	7. 9.	S.*w.* ♄.	
2 leptocárpa. DC.	(*ye.*)	narrow-podded.	Missisippi.	1824.	——	H. ☉.	
3 pilòsa. DC.	(*ye.*)	hairy.	Caracas.	1829.	——	S.*w.* ♃.	H. et B. n. gen. 6. t. 532.
4 grandiflòra. DC.	(*ye.*)	great-flowered.	Carolina.	1812.	7. 10.	G.*w.* ♃.	Bot. mag. t. 2122.
5 rèpens. DC.	(*ye.*)	creeping.	E. Indies.	1816.	——	S.*w.* ♃.	Rheed. mal. 2. t. 51.
6 Swartziàna. DC.	(*ye.*)	Swartz's.	W. Indies.	1826.	——	S.*w.* ♃.	
7 inclinàta. DC.	(*ye.*)	inclined.	Surinam.	1820.	——	S.*w.* ☉.	
8 acuminàta. DC.	(*ye.*)	acuminate.	Jamaica.	——	——	S. ☉.	
9 linifòlia. DC.	(*ye.*)	Flax-leaved.	W. Indies.	——	——	S. ☉.	
10 erécta. DC.	(*ye.*)	upright.	S. America.	1739.	——	S.*w.* ♂.	Gært. fruct. t. 31.
11 frutéscens. DC.	(*ye.*)	frutescent.	....	1829.	4. 6.	S.*w.* ♃.	
12 longifòlia. DC.	(*ye.*)	long-leaved.	Brazil.	1828.	6. 9.	S.*w.* ♃.	DC. pl. rar. gen. 1. t. 4.
13 decúrrens. DC.	(*ye.*)	decurrent.	N. America.	——	——	H.*w.* ♃.	Abb. ins. t. 40.
14 ovalifòlia. DC.	(*ye.*)	oval-leaved.	Madagascar.	1823.	7. 9.	S.*w.* ♃.	Bot. mag. t. 2530.
15 Burmànni. DC.	(*ye.*)	Burmann's.	E. Indies.	1808.	8. 9.	S.*w.* ♄.	Andr. reposit. 621.
suffruticòsa. A.R.							

16 villòsa. DC.	(*ye.*)	villous.	E. Indies.	1826.	8. 9.	S.*w.*♄.	Rheed. mal. 2. t. 50.
17 octonérvia. DC.	(*ye.*)	eight-nerved.	S. America.	....	7. 9.	S.*w.*♃.	Lam. ill. t. 280. f. 1.
Œnothèra octoválvis. Jacq. amer. t. 70.							
18 octofíla. DC.	(*ye.*)	eight-veined.	Caribees.	....	——	S.*w.*♃.	Plum.edBur.t.175.f.1.
19 scábra. DC.	(*ye.*)	rough.	Brazil.	1816.	——	S.*w.*♃.	
20 hírta. DC.	(*ye.*)	hairy.	S. America.	1823.	——	S.*w.*♄.	Plum.edBur.t.174.f.2.
LUDWI'GIA. DC.		Ludwi'gia. Tetrandria Monogynia. L.					
1 jussiæoídes. DC.	(*ye.*)	Jussiæa-like.	Mauritius.	1826.	7. 9.	S.*w.*♃.	
2 parviflòra. DC.	(*ye.*)	small-flowered.	E. Indies.	1812.	——	S. ⊙.	
ISNA'RDIA. DC.		Isna'rdia. Tetrandria Monogynia. L.					
1 alternifòlia. DC.	(*ye.*)	alternate-leav'd.	N. America.	1752.	6. 7.	H.*w.*♃.	Lam. ill. t. 77.
Ludwígia alternifòlia. L. *macrocárpa.* M.							
2 hirsùta. DC.	(*ye.*)	hairy.	————	1812.	6. 8.	H.*w.*♃.	
3 palústris. DC.	(*gr.*)	marsh.	Europe.	1776.	7. 8.	H.*w.*♃.	Schk. handb. 1. t. 25.
4 nítida.	(*gr.*)	American.	America.	....	——	H.*w.*♃.	
Ludwígia nítida. M. *apétala.* Walter. *palústris.* Torrey.							

Tribus V. *CIRCÆEÆ.* DC. prodr. v. 3. p. 61.

LOP'EZIA. DC.		Lop'ezia. Monandria Monogynia.					
1 racemòsa. DC.	(*pu.*)	smooth.	Mexico.	1792.	8. 10.	G. ♂.	Bot. mag. t. 254.
2 cordàta. DC.	(*pu.*)	heart-leaved.	————	1826.	7. 10.	H. ⊙.	
3 coronàta. DC.	(*pu.*)	coronet-flower'd.	————	1805.	——	H. ⊙.	Swt. br. fl. gar. t. 108.
4 oppositifòlia. DC.	(*pu.*)	opposite-leaved.	————	1825.	——	H. ⊙.	
5 integrifòlia. DC.	(*pu.*)	entire-leaved.	————	——	——	H. ⊙.	
6 miniàta. DC.	(*pu.*)	frutescent.	————	1824.	9. 11.	G. ♄.	Jacq. f. ecl. t. 109.
7 hirsùta. DC.	(*pu.*)	hairy.	————	1796.	——	G. ♂.	Jacq. coll. 5. t. 15. f. 4.
8 pùmila. DC.	(*pu.*)	dwarf.	————	1825.	7. 9.	H. ⊙.	
CIRC'ÆA. DC.		Enchanter's Nightshade. Diandria Monogynia.					
1 lutetiàna. DC.	(*wh.*)	common.	Britain.	....	6. 8.	H. ♃.	Eng. bot. v. 15. t. 1056.
2 canadénsis. L.	(*wh.*)	smooth-stalked.	N. America.	....	——	H. ♃.	
3 alpìna. DC.	(*wh.*)	mountain.	Britain.	....	6. 9.	H. ♃.	Eng. bot. v. 15. t. 1057.

Tribus VI. *HYDROCARYEÆ.* DC. prodr. v. 3. p. 63.

TR'APA. DC.		Water-caltrops. Tetrandria Monogynia. L.					
1 nàtans. DC.	(*wh.*)	floating.	S. Europe.	1781.	6. 8.	H.*w.*⊙.	Bot. reg. t. 259.
2 quadrispinòsa. DC.	(*wh.*)	four-spined.	E. Indies.	1823.	——	S.*w.*⊙.	
3 bispinòsa. DC.	(*wh.*)	two-spined.	————	1822.	6. 9.	S.*w.*⊙.	Roxb. corom. t. 234.
4 bicórnis. DC.	(*wh.*)	two-horned.	China.	1790.	——	G.*w.*♃.	Braam ic. chin. t. 22.

ORDO LXXXVII.

HALORAGEÆ. *DC. prodr. v. 3. p. 65.*

Tribus I. *CERCODIANÆ.* DC. prodr. v. 3. p. 65.

SERPI'CULA. DC.		Serpi'cula. Monœcia Tetrandria. L.					
rèpens. DC.	(*gr.*)	creeping.	C. B. S.	1789.	7. 8.	G. ♃.	Berg. cap. t. 5. f. 10.
GONIOCA'RPUS. DC.		Gonioca'rpus. Octandria Tetragynia.					
1 micránthus. DC.	(*wh.*)	small-flowered.	Japan.	1805.	6. 9.	H. ⊙.	Thunb. fl. jap. t. 15.
2 scáber. DC.	(*wh.*)	rough-leaved.	China.	1820.	——	H. ⊙.	Ann. bot. 1. t. 12. f. 6.
3 teucrioídes. DC.	(*wh.*)	Teucrium-like.	N. S. W.	1826.	——	H. ⊙.	
HALOR'AGIS. DC.		Halor'agis. Octandria Di-Tetragynia.					
1 prostràta. DC.	(*gr.*)	prostrate.	N. S. W.	1818.	4. 9.	G. ♄.	
2 dig'yna. DC.	(*gr.*)	two-styled.	N. Holland.	1826.	——	G. ♄.	Labil. n. holl. 1. t. 129.

CERC'ODIA. DC.		Cerc'odia. Octandria Tetragynia.					
1 erécta. dc.	(*gr.*)	upright.	N. Zealand.	1772.	4. 9.	G. ♄.	Jacq. ic. 1. t. 69.
Haloràgis alàta. Jacq. *Cercòdia.* h.k. *Tetragònia ivæfòlia.* l.							
PROSERPIN'ACA. DC.		Proserpin'aca. Triandria Trigynia.					
1 palústris. dc.	(*wh.*)	marsh.	N. America.	1822.	7. 8.	H.*w.*⊙.	Jus. an. mus. 3. t. 30. f. 1.
2 pectinàta. dc.	(*wh.*)	pectinated.	———	——	——	H.*w.*⊙.	Lam. ill. 1. t. 50. f. 1.
MYRIOPHY'LLUM.		Water Milfoil. Monœcia Tetr-Octandria.					
1 spicàtum. dc.	(*ro.*)	spiked.	Britain.	....	7. 8.	H.*w.*♃.	Eng. bot. v. 2. t. 83.
2 pectinàtum. dc.	(*ro.*)	pectinated.	Europe.	1828.	——	H.*w.*♃.	Flor. dan. t. 1046.
verticillàtum. f.d.							
3 verticillàtum. dc.	(*ro.*)	verticillate.	England.	....	——	H.*w.*♃.	Eng. bot. v. 4. t. 218.

Tribus II. *CALLITRICHINEÆ.* DC. prodr. v. 3. p. 70.

CALLI'TRICHE. DC.		Water Starwort. Monandria Digynia. L.					
1 vérna. dc.	(*wh.*)	vernal.	Britain.	....	4. 10.	H.*w.*⊙.	Eng. bot. v. 11. t. 722.
2 autumnàlis. dc.	(*wh.*)	autumnal.	———	....	6. 10.	H.*w.*⊙.	Gmel. sib. 3. t. 1.

Tribus III. *HIPPURIDEÆ.* DC. prodr. v. 3. p. 71.

HIPP'URIS. DC.		Mare's-tail. Monandria Monogynia. L.					
vulgàris. dc.	(*gr.*)	common.	Britain.	....	5.	H.*w.*♃.	Eng. bot. v. 11. t. 763.

ORDO LXXXVIII.

CERATOPHYLLEÆ. *Gray brit. pl. arr. 2. p. 554.*

CERATOPHY'LLUM. DC.		Hornwort. Monœcia Polyandria. L.					
1 demérsum. dc.	(*gr.*)	spiny-fruited.	Britain.	....	7. 9.	H.*w.*♃.	Eng. bot. v. 14. t. 947.
2 submérsum. dc.	(*gr.*)	naked-fruited.	———	....	9.	H.*w.*♃.	——— v. 10. t. 679.

ORDO LXXXIX.

LYTHRARIEÆ. *DC. prodr. v. 3. p. 75.*

Tribus I. *SALICARIEÆ.* DC. prodr. v. 3. p. 75.

PE'PLIS. DC.		Water Purslane. Hexandria Monogynia. L.					
Pórtula. dc.	(*re.*)	common.	Britain.	....	7. 9.	H. ⊙.	Eng. bot. v. 17. t. 1211.
AMMA'NNIA. DC.		Amma'nnia. Tetr-Heptandria Monogynia.					
1 ægyptìaca. dc.	(*gr.*)	Egyptian.	Egypt.	1818.	7. 8.	H.*w.*⊙.	Willd. hort. ber. t. 6.
2 vesicatòria. dc.	(*ye.*)	bladdered.	E. Indies.	1825.	——	S.*w.*⊙.	Pluk. alm. t. 136. f. 22.
3 occidentàlis. dc.	(*gr.*)	occidental.	Portorico.	1829.	——	S.*w.*⊙.	
4 cáspica. dc.	(*gr.*)	Caspian.	Astracan.	1818.	——	H.*w.*⊙.	
5 ramòsior. dc.	(*pu.*)	branching.	Virginia.	1759.	——	H.*w.*⊙.	Bocc. mus. t. 104.
6 latifòlia. dc.	(*wh.*)	broad-leaved.	W. Indies.	1733.	——	S.*w.*⊙.	Sloan. his. 1. t. 7. f. 4.
7 hùmilis. dc.	(*ro.*)	small.	N. America.	1823.	——	H.*w.*⊙.	
8 baccífera. r.s.	(*ye.*)	berry-bearing.	China.	1820.	——	H.*w.*⊙.	Burm. ind. t. 15. f. 3.
9 verticillàta. dc.	(*pu.*)	whorled.	———	———	——	H.*w.*⊙.	Lam. ill. t. 77. f. 3.
10 multiflòra. dc.	(*re.*)	many-flowered.	E. Indies.	1818.	——	S.*w.*⊙.	
11 dèbilis. dc.	(*li.*)	cluster-flower'd.	———	1778.	——	S.*w.*⊙.	
12 diffùsa. dc.	(*wh.*)	spreading.	———	1818.	——	S.*w.*⊙.	
13 rotundifòlia. dc.	(*fl.*)	round-leaved.	Nepaul.	——	——	H.*w.*⊙.	
14 pentándra. dc.	(*wh.*)	pentandrous.	E. Indies.	1823.	——	S.*w.*⊙.	

15 sanguinolénta. DC. (cr.) bloody. Jamaica. 1803. 7. 8. S.w.⊙.
16 octándra. DC. (cr.) octandrous. E. Indies. 1818. —— S.w.⊙. Roxb. corom. 2. t. 133.
17 auriculàta. DC. (pu.) eared. Egypt. 1820. —— H.w.⊙. Willd. h. ber. 1. t. 7.
18 rùbra. DC. (re.) red-flowered. Nepaul. 1822. —— H.w.⊙.
19 pállida. DC. (pa.) pale-flowered. .... 1828. —— H.w.⊙.
20 víridis. DC. (gr.) green. .... 1818. —— H.w.⊙.
21 Wormskiòldii. F. .... Wormskiold's. Siberia. 1828. .... H.w.⊙.

L'YTHRUM. DC. L'YTHRUM. Di-Dodecandria Monogynia.
1 thymifòlium. DC. (li.) Thyme-leaved. S. Europe. 1820. 7. 9. H. ⊙. Barrel. ic. 773. f. 2.
2 thesioídes DC. (li.) Thesium-like. Caucasus. 1828. —— H. ⊙.
3 lineàre. DC. (wh.) linear-leaved. N. America. 1812. 7. 8. H. ♃.
4 alàtum. DC. (pu.) winged-stalked. ———— 1812. 5. 11. F. ♃. Bot. mag. t. 1812.
5 hyssopifòlium. DC. (li.) Hyssop-leaved. England. .... 8. 9. H. ⊙. Eng. bot. v. 5. t. 292.
6 lanceolàtum. DC. (pu.) spear-leaved. Carolina. 1822. 7. 9. H. ♃.
7 álbum. DC. (wh.) white-flowered. Mexico. 1827. —— H. ♄.
8 Græfferi. DC. (pu.) Græffer's. S. Europe. 1825. —— H. ⊙.
9 Salicària. DC. (pu.) common. Britain. .... 7. 9. H.w. ♃. Eng. bot. v. 15. t. 1061.
10 pubéscens. (pu.) pubescent. N. America. .... 7. 8. H. ♃.
Salicària β pubéscens. Ph.
11 diffùsum. B.F.G. (pu.) spreading. ———— 1822. 7. 9. H. ♃. Swt. br. fl. gar. t. 149.
12 virgàtum. DC. (ro.) slender-branch'd. Europe. 1776. 6. 9. H. ♃. Bot. mag. t. 1003.

C'UPHEA. DC. C'UPHEA. Dodecandria Monogynia.
1 circæoídes. DC. (vi.) Circæa-like. S. America. 1819. 8. 10. H. ⊙. Bot. mag. t. 2201.
2 spicàta. DC. (ro.) spiked. Peru. —— —— H. ⊙. Cavan. icon. 4. t. 381.
3 Melvílla. DC. (sc.gr.) scarlet & green. Guiana. 1822. 6. 10. S. ♄. Bot. reg. t. 852.
4 viscosíssima. DC. (li.) clammy. America. 1776. 7. 10. H. ⊙. Swt. br. fl. gar. t. 60.
5 procúmbens. DC. (li.) procumbent. Mexico. 1816. 6. 9. H. ⊙. Bot. reg. t. 182.
6 lanceolàta. DC. (li.) spear-leaved. ———— 1796. 5. 8. G. ♂.
7 æquipètala. DC. (vi.) equal-flowered. ———— 1828. 6. 9. H. ⊙. Cavan. ic. 4. t. 382. f. 2.
8 virgàta. DC. (vi.) twiggy. ———— —— —— H. ⊙. ———— t. 382. f. 1.
9 decándra. DC. (pu.) ten-stamened. Jamaica. 1789. 6. 10. S. ♄.
10 parviflòra. DC. (ro.) small-flowered. Demerara. 1824. 10. 12. S. ♄. Hook. ex. flor. t. 161.
11 Símsii. ed. 1. (li.) Sims's. Trinidad. 1823. 8. 10. S. ♄. Bot. mag. t. 2580.
serpyllifòlia. B.M. *non* Kth. *Trinitatis.* DC.
12 multiflòra. DC. (vi.) many-flowered. ———— 1822. —— S. ♄. Lodd. bot. cab. t. 808.

ACISANTH'ERA. DC. ACISANTH'ERA. Decandria Monogynia.
quadràta. DC. (pi.) four-sided. Jamaica. 1804. .... S. ♃. Brown. jam. t. 22. f. 1.

HE'IMIA. DC. HE'IMIA. Dodecandria Monogynia.
1 salicifòlia. DC. (ye.) Willow-leaved. Mexico. 1821. 6. 9. F. ♄. Swt. br. fl. gar. t. 281.
2 myrtifòlia. Ot. (ye.) Myrtle-leaved. ———— 1827. —— F. ♄.

DEC'ODON. DC. DEC'ODON. Decandria Monogynia.
verticillàtum. DC. (pu.) whorl-flowered. N. America. 1759. 7. 9. H.w. ♃.
L'ythrum verticillàtum. L. *Nes'æa verticillàta.* Kth.

NES'ÆA. DC. NES'ÆA. Oct-Dodecandria Monogynia.
triflòra. DC. (bl.) three-flowered. Mauritius. 1802. 8. 9. S.w. ♃.
L'ythrum triflòrum. L.

LAWS'ONIA. DC. LAWS'ONIA. Octandria Monogynia. L.
1 spinòsa. W. (wh.) prickly. E. Indies. 1759. 6. 9. S. ♄. Rheed. mal. 1. t. 40.
2 inérmis. W. (wh.) smooth. Egypt. —— —— G. ♄. Rauw. ic. 60. t. 7.

GIN'ORIA. DC. GIN'ORIA. Dodecandria Monogynia.
americàna. DC. (bl.) American. Cuba. 1827. .... S. ♄. Jacq. amer. t 91.

GRI'SLEA. DC. GRI'SLEA. Oct-Dodecandria Monogynia.
1 secúnda. DC. (re.) secund. Caracas. 1824. 6. 8. S. ♄.
2 tomentòsa. DC. (sc.) downy. E. Indies. 1804. 5. 12. S. ♄. Bot. reg. t. 30.
Woodfòrdia floribùnda. Sal. par. lond. t. 42. *L'ythrum fruticòsum.* L.

Tribus II. *LAGERSTRŒMIEÆ.* DC. prodr. v. 3. p. 92.

LAGERSTR'ŒMIA. DC. LAGERSTR'ŒMIA. Icosandria Monogynia.
1 índica. DC. (re.) Indian. China. 1759. 8. 10. F. ♄. Bot. mag. t. 405.
β carnea. (fl.) *flesh-coloured.* ———— 1816. —— F. ♄.
γ álba. (wh.) *white-flowered.* ———— —— —— F. ♄.
2 parviflòra. DC. (wh.) small-flowered. E. Indies. —— —— S. ♄. Roxb. corom. 1. t. 66.
3 speciòsa. DC. (ro.) beautiful. China. 1826. .... S. ♄. Munch. haus. 5. t. 356.
4 grandiflòra. DC. (ro.) large-flowered. E. Indies. 1809. .... S. ♄.
5 regìnæ. DC. (ro.) oblong-leaved. ———— 1792. .... S. ♄. Roxb. corom. 1. t. 651.

ORDO XC.

TAMARISCINEÆ. *DC. prodr. v.* 3. *p.* 95.

TA'MARIX. DC.		TAMARISK. Tetr-Pentandria Trigynia.					
1 gállica. DC.	(*fl.*)	red-wooded.	England.	....	7. 8.	H. ♄.	Eng. bot. v. 19. t. 1318.
2 Pallàsii. DC.	(*fl.*)	Pallas's.	Caucasus.	1827.	——	H. ♄.	Pall. fl. ross. t. 77.
pentándra. Pall. *gállica.* M.B. *nec aliorum.*							
3 índica. DC.	(*bh.*)	Indian.	E. Indies.	....	6. 8.	S. ħ.	
4 dioíca. DC.	(*bh.*)	diœcious.	————	1812.	——	S. ♄.	
5 articulàta. DC.	(*pi.*)	jointed.	————	1816.	....	S. ♄.	Vahl. symb. 2. t. 32.
6 tetrándra. DC.	(*wh.*)	tetrandrous.	Tauria.	1821.	7. 8.	H. ħ.	
MYRIC'ARIA. DC.		MYRIC'ARIA. Monadelphia Decandria.					
1 germánica. DC.	(*pi.*)	green-wooded.	Germany.	1582.	6. 9.	H. ħ.	Flor. dan. t. 234.
Támarix germánica. L.							
2 dahùrica. DC.	(*pi.*)	Dahurian.	Dahuria.	1827.	....	H. ♄.	

ORDO XCI.

MELASTOMACEÆ. *DC. prodr. v.* 3. *p.* 99.

Subordo I. *MELASTOMEÆ.—Tribus* I. LAVOISIEREÆ. *DC. loc. cit.* 100.

MERI'ANIA. DC.		MERI'ANIA. Decandria Monogynia.					
1 leucántha. DC.	(*wh.*)	white-flowered.	Jamaica.	1825.	6. 8.	S. ♄.	
2 ròsea. DC.	(*ro.*)	mountain rose.	————	——	——	S. ♄.	Tussac fl. ant. t. 6.
3 purpùrea. DC.	(*pu.*)	purple.	————	——	——	S. ħ.	Swart. fl. ind. oc. 2. t. 15.

Tribus II. RHEXIEÆ. *DC. prodr.* 3. *p.* 114.

SPENN'ERA. DC.		SPENN'ERA. Octo-Decandria Monogynia.					
1 paludòsa. DC.	(*ro.*)	marsh.	Brazil.	1825.	6. 8.	S. ♃.	
2 frágilis. DC.	(*wh.*)	brittle.	Trinidad.	1822.	——	S. ♄.	
3 aquática. DC.	(*ro.*)	marsh.	S. America.	1793.	3. 9.	S. ♄.	Aubl. guian. 1. t. 169.
Rhéxia aquática. Bonpl. rhex. t. 40. *Melástoma aquática.* Aubl.							
4 pendulifòlia. DC.	(*ro.*)	hanging-leaved.	Guiana.	1826.	——	S. ♄.	Bonpl. nav. t. 26.
5 glandulòsa. DC.	(*wh.*)	glandulous.	————	1824.	——	S. ♄.	———— 70. t. 27.
MICROLI'CIA. DC.		MICROLI'CIA. Decandria Monogynia.					
1 brevifòlia. DC.	(*pu.*)	short-leaved.	Guiana.	1823.	7. 10.	S. ⊙.	Aubl. guian. 1. t. 155. f. b.
Melástoma triválvis. Aubl. *Rhéxia triválvis.* Vahl.							
2 biválvis. DC.	(*pu.*)	two-valved.	Trinidad.	1822.	6. 8.	S. ⊙.	Aubl. guian. 1. t. 155. f. a.
3 recúrva. DC.	(*pu.*)	recurved-leav'd.	————	1820.	——	S. ♃.	
RHE'XIA. DC.		RHE'XIA. Octandria Monogynia. L.					
1 marianà. DC.	(*li.*)	Maryland.	N. America.	1759.	7. 9.	H. ♃.	Pluk. mant. t. 428. f. 1.
α purpùrea. DC.	(*li.*)	*purplish.*	————	——	——	H. ♃.	
β rubélla. DC.	(*re.*)	*reddish.*	————	——	——	H. ♃.	Swt. br. fl. gar. t. 41.
2 virgínica. DC.	(*pu.*)	Virginian.	————	——	6. 8.	H. ♃.	Bot. mag. t. 968.
3 ciliòsa. DC.	(*pu.*)	ciliated.	————	1812.	8. 10.	H. ♃.	Swt. br. fl. gar. t. 298.
4 angustifòlia. DC.	(*wh.*)	narrow-leaved.	————	——	7. 9.	H. ♃.	
OXYSP'ORA. DC.		OXYSP'ORA. Octandria Monogynia.					
paniculàta. DC.	(*wh.*)	panicled.	Nepaul.	1826.	6. 8.	S. ♄.	
Arthrostémma paniculàtum. D.P.							

Tribus III. OSBECKIÆ. *DC. prodr.* 3. *p.* 127.

LASIA'NDRA. DC. LASIA'NDRA. Decandria Monogynia.
1 Fontanesiàna. DC. (*pu.*) rough-leaved. Brazil. 1818. 4. 8. S. ♄. Bonpl. rhex. t. 36.
Melástoma granulòsa. B. reg. 671. *Rhéxia Fontanèsii.* Bonpl.
2 Langsdorfiàna. DC. (*pu.*) Langsdorf's. Brazil. 1821. —— S. ♄. Bonpl. rhex. t. 51.
3 argéntea. DC. (*bl.*) silvery-leaved. ———— 1816. —— S. ♄. ———— t. 12.
Rhéxia holosericea. Bot. reg. t. 323. *Pleròma holoseríceum.* Don.
CHÆTOGA'STRA. DC. CHÆTOGA'STRA. Decandria Monogynia.
1 lanceolàta. DC. (*wh.*) spear-leaved. Trinidad. 1827. 1. 5. S. ♄. Bot. mag. t. 2836.
2 hypericoídes. DC. (*wh.*) Hypericum-like. ———— 1820. 8. 12. S. ⊙. Aubl. guian. 1. t. 168.
Rhéxia villosíssima. Bonpl. rhex. t. 31. *Melástoma villòsa.* Aubl.
3 ciliàris. DC. (*pu.*) fringed. S. America. 1822. 4. 7. S. ♃. Vent. choix. t. 34.
Meriània ciliàris. Vent.
4 grácilis. DC. (*ro.*) slender. Brazil. 1824. —— S. ♃. Bonpl. rhex. t. 52.
ARTHROSTE'MMA. DC. ARTHROSTE'MMA. Octandria Monogynia.
1 ladanoídes. DC. (*li.*) bristly-stalked. Trinidad. 1820. 6. 9. S. ⊙. Bonpl. rhex. t. 27.
Rhéxia ladanoídes. Bonpl. *Melástoma trichótoma.* W.
2 villòsum. DC. (*vi.*) villous. Trinidad. —— —— S. ♄. Aubl. guian. 1. t. 129. f. 1.
3 versícolor. DC. (*bh.*) changeable-col'd. Brazil. 1825. 5. 9. S. ♄. Bot. reg. t. 1066.
Rhéxia versícolor. B.R.
OSBE'CKIA. DC. OSBE'CKIA. Octo-Decandria Monogynia.
1 urceolàris. DC. (*pu.*) urceolate. Brazil. 1825. 4. 9. S. ♄.
2 Símsii. DC. (*pu.*) Sims's. Mauritius. 1820. —— S. ♄. Bot. mag. t. 2235.
Melástoma osbeckioídes. B.M.
3 glomeràta. DC. (*ro.*) cluster-flower'd. S. America. 1821. 12. 4. S. ♄. Bot. mag. t. 2838.
β albiflòra DC. (*wh.*) *white-flowered.* Brazil. —— —— S. ♄. Lodd. bot. cab. t. 334.
Rhéxia glomeràta. B.C.
4 chinénsis. DC. (*ro.*) Chinese. China. 1816. 3. 10. S. ♄. Bot. reg. t. 542.
5 zeylánica. DC. (*wh.*) Ceylon. Ceylon. 1749. 5. 9. S. ♄. ———— t. 565.
6 angustifòlia. DC. (*ro.*) narrow-leaved. Nepaul. 1826. —— S. ♄.
7 nepalénsis. H.E.F. (*ro.*) handsome. ———— 1820. —— S. ♄. Hook. ex. flor. t. 31.
speciòsa. D.P.
8 stellàta. DC. (*li.*) starry-haired. ———— 1818. —— S. ♄. Bot. reg. t. 674.
9 ternifòlia. DC. (*li.*) whorl-leaved. ———— 1825. —— S. ♄.
10 octándra. DC. (*ro.*) octandrous. Ceylon. 1815. 4. 8. S. ♄.
Melástoma octándra. L.
11 decúmbens. DC. (*ro.*) decumbent. Africa. 1821. —— S. ♃. Beauv. fl. ow. t. 41.
Melástoma decúmbens. Beauv.
TIBOUCH'INA. DC. TIBOUCH'INA. Decandria Monogynia.
áspera. DC. (*pu.*) rough. Guiana. 1820. 4. 6. S. ♄. Aubl. guian. 1. t. 177.
MELA'STOMA. DC. MELA'STOMA. Decandria Monogynia. L.
1 affìne. DC. (*ro.*) scaly-stalked. E. Indies. 1816. 1. 8. S. ♄.
2 malabáthricum. DC. (*ro.*) Malabar. ———— 1793. —— S. ♄. Bot. reg. t. 672.
3 ásperum. DC. (*ro.*) rough. ———— 1815. 8. 11. S. ♄. Rheed. mal. 4. t. 43.
4 macrocárpum. DC. (*ro.*) large-fruited. ———— 1793. 1. 8. S. ♄. Bot. mag. t. 529.
malabáthrica. B.M. *nec aliorum.*
5 sanguíneum. DC. (*li.*) red-veined. Sunda. St. 1818. 4. 12. S. ♄. Bot. mag. t. 2241.
6 rubrolimbàtum. DC. (*wh.*) red-limbed. E. Indies. 1823. —— S. ♄.
7 cándidum. DC. (*li.*) white-leaved. China. 1820. 4. 8. S. ♄.
8 normàle. DC. (*li.*) Nepaul. Nepaul. 1819. 2. 6. S. ♄. Lodd. bot. cab. t. 707.
nepalénsis. B.C.
9 Wallíchii. DC. (*li.*) Wallich's. ———— 1824. —— S. ♄.
10 decémfidum. DC. (*pu.*) ten-cleft. E. Indies. 1826. 1. 8. S. ♄.
11 cymòsum. DC. (*li.*) cymose. Sierra Leone. 1792. 4. 8. S. ♄. Vent. h. malm. t. 14.
corymbòsum. Bot. mag. t. 984.
12 elongàtum. DC. (*li.*) tuberous-rooted. ———— 1822. .... S. ♃.
13 pulverulèntum. DC. (*re.*) powdered. Sumatra. —— .... S. ♄.
PLER'OMA. DC. PLER'OMA. Decandria Monogynia.
1 vimíneum. DC. (*vi.*) slender. Brazil. 1820. 4. 8. S. ♄. Bot. reg. t. 664.
Rhéxia vimínea. B.R.
2 heteromállum. DC. (*bl.*) woolly-leaved. ———— —— —— S. ♄. Bot. reg t. 644.
Melástoma heteromálla.
DIPLOST'EGIUM. DC. DIPLOST'EGIUM. Decandria Monogynia.
canéscens. DC. (*ro.*) hoary-stemm'd. Brazil. 1820. 4. 8. S. ♄.
ACI'OTIS. DC. ACI'OTIS. Octandria Monogynia.
díscolor. DC. (*bh.*) two-coloured. W. Indies. 1793. 3. 9. S. ♃.

Tribus IV. MICONIEÆ. *DC. prodr. v.* 3. *p.* 152.

CLID'EMIA. DC.		CLID'EMIA. Decandria Monogynia.					
1 aggregàta. DC.	(*wh.*)	clustered.	Peru.	1820.	7. 10.	S. ♄.	
2 hírta. DC.	(*wh.*)	hairy.	Jamaica.	1740.	3. 6.	S. ♄.	Bot. mag. t. 1971.
Melástoma hírta. B.M.							
3 crenàta. DC.	(*wh.*)	crenated.	W. Indies.	1820.	——	S. ♄.	Vahl. ic. pl.am. 2. t. 22.
4 élegans. DC.	(*wh.*)	elegant.	Trinidad.	1822.	——	S. ♄.	Aubl. guian. 1. t. 167.
5 biserràta. DC.	(*pu.*)	doubly serrate.	Brazil.	1825.	——	S. ♄.	
6 spicàta. DC.	(*wh.*)	spiked.	Trinidad.	1820.	——	S. ♄.	Aubl. guian. 1. t. 165.
7 strigillòsa. DC.	(*wh.*)	strigose.	Jamaica.	1825.	——	S. ♄.	
8 agréstis. DC.	(*wh.*)	field.	Cayenne.	1822.	4. 8.	S. ♄.	Aubl. guian. 1. t. 166.
9 lanàta. DC.	(*wh.*)	sweet-scented.	Trinidad.	——	——	S. ♄.	
TOC'OCA. DC.		TOC'OCA. Decandria Monogynia.					
guianénsis. DC.	(*ro.*)	Guiana.	Guiana.	1822.	4. 7.	S. ♄.	Aubl. guian. 1. t. 174.
Melástoma physíphora. Vahl.							
MAI'ETA. DC.		MAI'ETA. Decandria Monogynia.					
guianénsis. DC.	(*ro.*)	Guiana.	Guiana.	1824.	....	S. ♄.	Aubl. guian. 1. t. 176.
OSS'ÆA. DC.		OSS'ÆA. Octandria Monogynia.					
purpuráscens. DC.	(*pu.*)	purplish.	Jamaica.	1822.	3. 6.	S. ♄.	
Melástoma purpuráscens. Swartz.							
SAGR'ÆA. DC.		SAGR'ÆA. Octandria Monogynia.					
1 sessiliflòra. DC.	(*re.*)	sessile-flowered.	Guiana.	1793.	4. 8.	S. ♄.	Vahl. ic. am. t. 19.
Melástoma rùbra. Bonpl. mel. t. 39.							
2 hirtélla. DC.	(*wh.*)	hispid-fruited.	Jamaica.	1824.	——	S. ♄.	
3 umbròsa. DC.	(*wh*)	shaded.	W. Indies.	——	3. 7.	S. ♄.	
4 pilòsa. DC.	(*wh.*)	pilose.	Jamaica.	1828.	——	S. ♄.	
5 microphy'lla. DC.	(*wh.*)	small-leaved.	————	1820.	5. 6.	S. ♄.	
6 hirsùta. DC.	(*wh.*)	purple-haired.	————	1823.	——	S. ♄.	
TETRAZY'GIA. DC.		TETRAZY'GIA. Tetr-Octandria Monogynia.					
1 tetrándra. DC.	(*wh.*)	tetrandrous.	Jamaica.	1815.	3. 6.	S. ♄.	
Melástoma tetrándra. Swartz.							
2 díscolor. DC.	(*wh.*)	two-coloured.	W. Indies.	1793.	5. 7.	S. ♄.	Bonpl. mel. t. 34.
Melástoma díscolor. Jacq. amer. t. 84.							
3 angustifòlia. DC.	(*wh.*)	narrow-leaved.	————	1823.	——	S. ♄.	Bonpl. mel. t. 10.
HETEROTRI'CHUM. DC.		HETEROTRI'CHUM. Decandria Monogynia.					
1 níveum. DC.	(*bh.*)	snowy-white.	St. Domingo.	1820.	5. 10.	S. ♄.	
Melástoma nívea. Lam.							
2 pátens. DC.	(*bh.*)	spreading.	————	1825.	4. 8.	S. ♄.	
CONOST'EGIA. DC.		CONOST'EGIA. Dec-Dodecandria Monogynia.					
1 pròcera. DC.	(*wh.*)	tall.	Jamaica.	1822.	....	S. ♄.	Bonpl. mel. t. 51.
2 Balbisiàna. DC.	(*wh.*)	Balbis's.	————	1825.	....	S. ♄.	
3 semicrenàta. DC.	(*wh.*)	semicrenate.	W. Indies.	1823.	4. 6.	S. ♄.	Bonpl. mel. t. 41.
DIPLOCH'ITA. DC.		DIPLOCH'ITA. Octo-Decandria Monogynia.					
1 Fothergílla. DC.	(*wh.*)	Fothergill's.	Trinidad.	1815.	....	S. ♄.	Bonpl. mel. t. 32.
Melástoma Fothergílla. s.s. *Fothergílla mirábilis.* Aubl. gui. 1. t. 175.							
2 Swartziàna. DC.	(*ro.*)	Swartz's.	Jamaica.	1815.	4. 6.	S. ♄.	Bonpl. mel. t. 33.
Melástoma Tamònea. Swartz.							
3 macrophy'lla. DC.	.....	large-leaved.	Mexico.	1820.	....	S. ♄.	
L'OREYA. DC.		L'OREYA. Decandria Monogynia.					
arboréscens. DC.	(*wh.*)	arborescent.	Guiana.	1822.	....	S. ♄.	Aubl. guian. 1. t. 163.
Melástoma arboréscens. Aubl.							
MIC'ONIA. DC.		MIC'ONIA. Decandria Monogynia.					
1 racemòsa. DC.	(*ro.*)	racemed.	Trinidad.	1822.	4. 8.	S. ♄.	Bonpl. mel. t. 27.
Melástoma racemòsa. Aubl. guian. t. 156.							
2 purpuráscens. DC.	(*wh.*)	purple-berried.	Guiana.	1804.	3. 6.	S. ♄.	Aubl. guian. t. 154.
Melástoma purpùrea. W.							
3 ciliàta. DC.	(*wh.*)	fringed.	————	1824.	4. 7.	S. ♄.	Bonpl. mel. t. 26.
4 holoserícea. DC.	(*wh.*)	velvetty.	W. Indies.	1816.	——	S. ♄.	———— t. 23-24.
5 argéntea. DC.	(*wh.*)	silvery-leaved.	————	1827.	——	S. ♄.	Sloan. jam. t. 196. f. 1.
6 impetiolàris. DC.	(*wh.*)	great-leaved.	————	1822.	——	S. ♄.	Bonpl. mel. t. 29.
Melástoma macrophy'lla. Lam.							
7 Lambertiàna. DC.	(*wh.*)	Mr. Lambert's.	————	1824.	——	S. ♄.	
8 caudàta. DC.	(*ro.*)	tailed-leaved.	S. America.	1822.	5. 8.	S. ♄.	Bonpl. mel. t. 7.
9 pyramidàlis. DC.	(*wh.*)	pyramidal.	Trinidad.	——	——	S. ♄.	Pluk. alm. t. 159. f. 1.
Melástoma pyramidàlis. Lam.							

10 Acinodéndron.	(*wh.*)	oval-leaved.	Jamaica.	1804.	5. 8.	S. ♄.	Plum. ic. 142. f. 2.
Melástoma Acinodéndron. L.							
11 lævigàta. DC.	(*wh.*)	smooth.	Caribees.	1815.	4. 8.	S. ♄.	Bot. reg. t. 363.
12 prasìna. DC.	(*wh.*)	green.	Jamaica.	1824.	——	S. ♄.	
13 ? ecostàta.	(*pu.*)	ribless.	———	1793.	5. 6.	S. ♄.	
Melástoma ecostata. H.K.							
BL'AKEA. DC.		BL'AKEA.	Dodecandria Monogynia. L.				
1 trinérvia. DC.	(*ro.*)	three-ribbed.	Jamaica.	1789.	6. 7.	S. ♄.	Bot. mag. t. 451.
2 quinquenérvis. DC.	(*fl.*)	five-nerved.	Guiana.	1820.	——	S. ♄.	Aubl. guian. 1. t. 210.
triplinérvia. L.							

ORDO XCII.

ALANGIEÆ. *DC. prodr.* 3. *p.* 203.

ALA'NGIUM. DC.		ALA'NGIUM.	Icosandria Monogynia.				
1 decapètalum. DC.	(*wh.*)	Sage-leaved.	E. Indies.	1779.	5. 8.	S. ♄.	Rheed. mal. 4. t. 17.
Gre'wia salvifòlia. L.							
2 hexapétalum. DC.	(*wh.*)	six-petaled.	———	1822.	——	S. ♄.	Rheed. mal. 4. t. 26.

ORDO XCIII.

PHILADELPHEÆ. *DC. prodr. v.* 3. *p.* 205.

PHILADE'LPHUS. DC.		SYRINGA.	Icosandria Monogynia. L.				
1 coronàrius. DC.	(*wh.*)	common.	S. Europe.	1596.	5. 6.	H. ♄.	Bot. mag. t. 391.
β *variegàtus.*	(*wh.*)	*striped-leaved.*	———	....	——	H. ♄.	
γ *multipléx.*	(*wh.*)	*double-flowered.*	———	....	——	H. ♄.	
2 nànus. Mil.	(*wh.*)	dwarf.	....	....	6. 7.	H. ♄.	
3 Zeyhèri. DC.	(*wh.*)	Zeyher's.	N. America.	1812.	——	H. ♄.	Schrad. diss. c. ic.
4 floribúndus. DC.	(*wh.*)	many-flowered.	———	1816.	——	H. ♄.	——— c. ic.
5 verrucòsus. DC.	(*wh.*)	warted.	———	1812.	——	H. ♄.	——— c. ic.
grandiflòrus. Bot. reg. t. 570. *non* W.							
6 latifòlius. DC.	(*wh.*)	broad-leaved.	———	1820.	——	H. ♄.	Lois. herb. amat. t. 208.
7 grandiflòrus. DC.	(*wh.*)	large-flowered.	———	1816.	——	H. ♄.	Swt. br. fl. gar. s. 2. t. 8.
8 speciòsus. DC.	(*wh.*)	beautiful.	———	1820.	——	H. ♄.	Schrad. diss. c. ic.
9 láxus. DC.	(*wh.*)	loose-branched.	———	1816.	——	H. ♄.	——— c. ic.
10 hirsùtus. DC.	(*wh.*)	slender hairy.	———	——	——	H. ♄.	Wats. dend. brit. t. 47.
11 inodòrus. DC.	(*wh.*)	entire-leaved.	Carolina.	1738.	——	H. ♄.	Bot. mag. t. 1478.
DECUM'ARIA. DC.		DECUM'ARIA.	Dodecandria Monogynia. L.				
1 bárbara. Ph.	(*wh.*)	smaller.	Carolina.	1785.	7. 8.	H. ♄. ⌣.	
2 sarmentòsa. Ph.	(*wh.*)	larger.	———	1758.	——	H. ♄. ⌣.	Bosc. act. par. 1. t. 13.

ORDO XCIV.

MYRTACEÆ. *DC. prodr. v.* 3. *p.* 207.

Tribus I. *CHAMÆLAUCIEÆ.* DC. prodr. 3. p. 208.

CA'LYTHRIX. DC.		CA'LYTHRIX.	Icosandria Monogynia.				
1 glábra. DC.	(*wh.*)	smooth.	N. S. W.	1818.	4. 6.	G. ♄.	Bot. reg. t. 409.
2 ericoídes. F.T.	(*wh.*)	Heath-like.	———	1823.	——	G. ♄.	
3 scábra. DC.	(*wh.*)	rough-leaved.	———	1824.	——	G. ♄.	

DARWI'NIA. L.T. DARWI'NIA. Decandria Monogynia.
1 fascicularis. L.T. (wh.) fascicled. N. S. W. 1826. 4. 6. G. ♄. Linn. trans. v. 11. t. 22.
2 taxifòlia. F.T. (wh.) Yew-leaved. ——— 1827. —— G. ♄.
VERTICO'RDIA. DC. VERTICO'RDIA. Decandria Monogynia.
1 Fontanèsii. DC. (wh.) Desfontaines'. N. Holland. 1826. 4. 6. G. ♄. Desf. mem. mus. 5. t. 4.
Chamelaùcium plumòsum. Df.
2 Bro'wnii. DC. (wh.) Brown's. ——— —— —— G. ♄. Desf. mem. mus. 5. t. 19.
CHAMÆLA'UCIUM. DC. CHAMÆLA'UCIUM. Decandria Monogynia.
ciliàtum. DC. (wh.) fringed. N. Holland. 1825. 4. 6. G. ♄. Df. mem. mus. 5. t. 3. f. B.
GENETY'LLIS. DC. GENETY'LLIS. Icosandria Monogynia.
diosmoídes. DC. (wh.) Diosma-like. N. Holland. 1827. 4. 6. G. ♄. DC. dict. cl. v. 11. c. ic.
PILE'ANTHUS. DC. PILE'ANTHUS. Icosandria Monogynia.
Limàcis. DC. (wh.) clubbed-leaved. N. Holland. 1824. 4. 7. G. ♄. Lab. n. holl. 2. t. 149.

Tribus II. *LEPTOSPERMEÆ.* DC. prodr. 3. p. 209.

Subtribus I. MELALEUCEÆ. Stamina polyadelpha. *DC. loc. cit. p.* 210.

ASTA'RTEA. DC. ASTA'RTEA. Polyadelphia Icosandria.
fasciculàris. DC. (wh.) fascicled. V. Diem. Isl. 1824. 4. 6. G. ♄. Lab. n. holl. 2. t. 170.
Melaleùca fasciculàris. Labill.
TRIST'ANIA. DC. TRIST'ANIA. Polyadelphia Icosandria.
1 neriifòlia. DC. (ye.) Oleander-leav'd. N. S. W. 1804. 7. 9. G. ♄. Bot. mag. t. 1058.
Melaleùca neriifòlia. B.M. *salicifòlia.* Andr. reposit. t. 485.
2 persicæfòlia. F.T. (ye.) Peach-leaved. N. S. W. 1824. —— G. ♄.
3 laurìna. DC. (ye.) Laurel-leaved. ——— 1798. —— G. ♄.
4 depréssa. DC. (ye.) depressed. ——— 1820. —— G. ♄.
5 álbens. DC. (ye.) white-leaved. ——— 1822. —— G. ♄.
6 conférta. DC. (ye.) Pittosporum-l[d]. ——— 1805. —— G. ♄.
7 suavèolens. DC. (ye.) sweet-scented. N. Holland. 1826. .... G. ♄. Gært. fruct. 1. t. 35.
BEAUFO'RTIA. DC. BEAUFO'RTIA. Polyadelphia Icosandria.
1 decussàta. DC. (sc.) splendid. N. Holland. 1803. 3. 8. G. ♄. Bot. reg. t. 18.
2 spársa. DC. (sc.) alternate-leav'd. ——— —— —— G. ♄.
CALOTHA'MNUS. DC. CALOTHA'MNUS. Polyadelphia Icosandria.
1 quadrífida. DC. (sc.) four-cleft. N. Holland. 1803. 1. 12. G. ♄. Bot. mag. t. 1506.
2 clavata. M.C. (sc) club-leaved. ——— 1826. —— G. ♄. Lodd. bot. cab. t. 1447.
3 villòsa. DC. (sc.) hairy 5-cleft. ——— 1803. —— G. ♄. Bot. reg. t. 1099.
4 grácilis. DC. (sc.) long-leaved. ——— —— —— G. ♄.
MELALE'UCA. DC. MELALE'UCA. Polyadelphia Icosandria.
1 Leucadéndron. DC. (wh.) aromatic. E. Indies. 1796. .... S. ♄. Rumph. amb. 2. t. 16.
2 minòr. DC. (wh.) Cajuputi-tree. ——— 1816. .... S. ♄. Rumph. amb. 2. t. 17. f. 1.
3 viridiflòra. DC. (gr.) green-flowered. N. S. W. 1798. 4. 6. G. ♄. Cavan. ic. t. 333.
Metrosidèros quinquenérvia. Cavan.
4 paludòsa. DC. (re.) long-leaved red. N. Holland. 1803. 7. 9. G. ♄.
5 globífera. DC. (wh.) globe-fruited. ——— —— —— G. ♄.
6 diosmifòlia. DC. (gr.) Diosma-leaved. ——— 1794. 6. 7. G. ♄. Andr. reposit. t. 476.
7 styphelioídes. DC. (wh.) Styphelia-leav'd. N. S. W. 1793. 4. 6. G. ♄.
8 genistifòlia. DC. (wh.) Genista-leaved. ——— —— —— G. ♄. Sm. exot. bot. 1. t. 55.
9 lanceolàta. DC. (st.) spear-leaved. N. Holland. 1812. 5. 6. G. ♄. Otto hort. ber. 36.
10 striàta. DC. (wh.) rigid-leaved. ——— 1803. —— G. ♄. Lab. n. holl. 2. t. 165.
11 thymoídes. DC. (ye.) yellow-spear leav'd. ——— —— —— G. ♄. ——— 2. t. 167.
12 squámea. DC. (pu.) scaly-branched. V. Diem. Isl. 1805. 6. 8. G. ♄. Bot. reg. t. 477.
13 nodòsa. DC. (wh.) Whin-leaved. N. S. W. 1790. —— G. ♄. Sm. exot. bot. 1. t. 35.
14 tenuifòlia. DC. (wh.) slender-leaved. ——— 1824. —— G. ♄.
15 ericifòlia. DC. (wh.) Heath-leaved. ——— 1788. 7. 9. G. ♄. Sm. exot. bot. 1. t. 34.
16 erubéscens. DC. (li.) Lilac-coloured. ——— 1820. —— G. ♄. Otto hort. ber. 37.
17 armillàris. DC. (wh.) pale-flowered. ——— 1788. 6. 8. G. ♄. Vent. malm. t. 76.
ericifòlia. Andr. rep. t. 175. *non* Sm. *Metrosidèros armillàris.* G. Cav. ic. t. 335.
18 uncinàta. DC. (re.) hook-leaved. N. Holland. 1803. —— G. ♄.
19 scábra. DC. (pu.) rough-leaved. ——— —— —— G. ♄. Swt. flor. aust. t. 10.
20 juniperoídes. DC. (st.) Juniper-like. N. S. W. 1826. —— G. ♄.
21 pulchélla. DC. (li.) pendulous. N. Holland. 1803. 6. 9. G. ♄. Lodd. bot. cab. t. 200.
22 thymifòlia. DC. (pu.) Thyme-leaved. N. S. W. 1792. —— G. ♄. Sm. exot. bot. 1. t. 36.
coronàta. Andr. rep. t. 278. *gnidiæfòlia.* Vent. malm. t. 7.
23 decussàta. DC. (li.) cross-leaved. N. Holland. 1803. 6. 9. G. ♄. Bot. mag. t. 2268.
24 tetragòna. DC. (li.) four-sided. ——— 1824. —— G. ♄.

25 cuticulàris. DC.	(*pu.*)	solitary-flower'd.	N. Holland.	1824.	6. 9.	G. ♄.	Lab. n. holl. 2. t. 171.
26 fúlgens. DC.	(*sc.*)	splendid.	———	1803.	6. 8.	G. ♄.	Bot. reg. t. 103.
27 linariifòlia. DC.	(*wh.*)	Toad-flax-leav'd.	N. S. W.	1793.	5. 7.	G. ♄.	Sm. exot. bot. 1. t. 56.
28 abietìna. DC.	(*wh.*)	Fir-leaved.	N. Holland.	1824.	——	G. ħ.	
29 hypericifòlia. DC.	(*sc.*)	Hypericum-ld.	N. S. W.	1792.	7. 9.	G. ♄.	Andr. reposit. t. 200.
30 ellíptica. DC.	(*cr.*)	elliptic-leaved.	N. Holland.	1824.	——	G. ħ.	Lab. n. holl. 2. t 173.
31 squarròsa. DC.	(*wh.*)	Myrtle-leaved.	N. S. W.	1792.	——	G. ♄.	Bot. mag. t. 1935.
myrtifòlia. Vent. malm. t. 47.							
32 gibbòsa. DC.	(*li.*)	gibbous.	V. Diem. Isl.	1822.	6. 8.	G. ♄.	Lab. n. holl. 2. t. 172.
33 imbricàta. L.en.	(*wh.*)	imbricated.	N. Holland.	1803.	5. 6.	G. ♄.	
34 sprengelioídes. DC.	(*wh.*)	Sprengelia-like.	N. S. W.	1824.	——	G. ♄.	
35 calycìna. DC.	(*wh.*)	permanent-cup'd.	N. Holland.	1803.	6. 8.	G. ♄.	
36 dénsa. DC.	(*pu.*)	dense-leaved.	———	———	——	G. ħ.	
37 incàna. DC.	(*st.*)	hoary.	N. S. W.	1812.	——	G. ♄.	Bot. reg. t. 410.
canéscens. Otto hort. ber. p. 87. Abbild. t. 81.							
EUDE'SMIA. DC.		EUDE'SMIA. Polyadelphia Icosandria.					
tetragòna. DC.	(*wh.*)	square-stemm'd.	N. Holland.	1824.	6. 8.	G. ♄.	Swt. flor. aust. t. 21.

Subtribus II. EULEPTOSPERMEÆ. Stamina libera. *DC. prodr.* 3. *p.* 216.

EUCALY'PTUS. DC.		EUCALY'PTUS. Icosandria Monogynia.					
1 cornùta. DC.	(*wh.*)	horned.	N. Holland.	1803.	3. 6.	G. ħ.	Lab. voy. 1. t. 20.
2 tereticórnis. DC.	(*wh.*)	long-horned.	N. S. W.	1804.	4. 7.	G. ♄.	
3 resinífera. DC.	(*wh.*)	red gum-tree.	———	1788.	——	G. ♄.	Andr. reposit. t. 400.
4 longifòlia. DC.	(*wh.*)	long-leaved.	———	1818.	——	G. ħ.	Link et Ot. abb. t. 45.
5 robústa. DC.	(*wh.*)	brown gum-tree.	———	1794.	8. 9.	G. ħ.	Smith nov. holl. t. 13.
6 rostràta. Cav.	(*wh.*)	beaked.	———	1804.	——	G. ♄.	Cavan. ic. 4. t. 342.
7 marginàta. DC.	(*wh.*)	thick-edged.	———	1794.	4. 7.	G. ħ.	
8 incrassàta. DC.	(*wh.*)	thick-leaved.	N. Holland.	1820.	——	G. ♄.	Lab. nov. holl. 2. t 150.
9 persicifòlia. DC.	(*wh.*)	Peach-leaved.	N. S. W.	1812.	——	G. ħ.	Lodd. bot. cab. t. 501.
10 punctàta. DC.	(*wh.*)	black-dotted.	———	1810.	——	G. ħ.	
11 acérvula. DC.	(*wh.*)	flat-peduncled.	———	1818.	——	G. ♄.	
12 virgàta. DC.	(*wh.*)	slender-twigged.	———	1820.	5. 8.	G. ♄.	
13 micrántha. DC.	(*wh.*)	small-flowered.	———	1824.	——	G. ♄.	
14 stellulàta. DC.	(*wh.*)	starry.	———	1816.	——	G. ħ.	
15 oblónga. DC.	(*wh.*)	oblong-leaved.	———	1819.	6. 8.	G. ħ.	
16 viminàlis. DC.	(*wh.*)	slender.	N. Holland.	1816.	——	G. ♄.	Lab. nov. holl. 2. t. 151.
17 capitellàta. DC.	(*wh.*)	small-headed.	N. S. W.	1804.	4. 7.	G. ♄.	White's voy. 226. ic.
18 triántha. L.en.	(*wh.*)	three-flowered.	———	1820.	4. 6.	G. ♄.	
19 salígna. DC.	(*wh.*)	Willow-like.	———	1804.	4. 7.	G. ħ.	
20 angustifòlia. L.en.	(*wh.*)	narrow-leaved.	———	1816.	——	G. ħ.	
21 microphy'lla. DC.	(*wh.*)	small-leaved.	———	——	——	G. ♄.	
22 stenophy'lla. DC.	(*wh.*)	slender-leaved.	———	——	——	G. ♄.	
23 myrtifòlia. DC.	(*wh.*)	Myrtle-leaved.	———	1812.	4. 6.	G. ħ.	
24 elongàta. DC.	(*wh.*)	taper-pointed.	———	1820.	——	G. ♄.	
25 mèdia. DC.	(*wh.*)	acuminate.	———	——	——	G. ħ.	
26 reticulàta. DC.	(*wh.*)	netted-leaved.	———	1816.	....	G. ♄.	
27 umbellàta. DC.	(*wh.*)	umbel-flowered.	———	1820.	5. 8.	G. ♄.	
28 ovàta. DC.	(*wh.*)	oval-leaved.	N. Holland.	——	——	G. ħ.	Lab. nov. holl. 2. t. 153.
mucronàta. L. enum.							
29 scábra. DC.	(*wh.*)	rough-leaved.	N. S. W.	1803.	——	G. ♄.	
30 pilulàris. DC.	(*wh.*)	small-fruited.	———	1804.	2. 5.	G. ħ.	
31 radiàta. DC.	(*wh.*)	rayed.	———	1824.	3. 6.	G. ħ.	
32 strícta. DC.	(*wh.*)	straight.	———	——	....	G. ħ.	
33 hæmastómma. DC.	(*wh.*)	red-mouthed.	———	1803.	4. 7.	G. ħ.	White's voy. 226. c. ic.
racemòsa. Cavan. ic. 4. n. 377.							
34 ligustrìna. DC.	(*wh.*)	Privet-like.	N. S. W.	1824.	....	G. ♄.	
35 amygdalìna. DC.	(*wh.*)	Almond-leaved.	N. Holland.	1822.	....	G. ħ.	Lab. nov. holl. 2. t. 154.
36 ambígua. DC.	(*wh.*)	ambiguous.	———	1824.	....	G. ♄.	
37 Lindleyàna. DC.	(*wh.*)	Lindley's.	N. S. W.	1816.	4. 6.	G. ♄.	Bot. reg. t. 947.
longifòlia. B.R. *non* Link.							
38 botryoídes. DC.	(*wh.*)	clustered.	———	1803.	4. 7.	G. ħ.	Cavan. ic. 4. t. 341.
platypòdos. Cavan.							
39 piperìta. DC.	(*wh.*)	Peppermint-tree.	———	1788.	——	G. ħ.	Reich. gart. mag. t. 42.
40 pállens. DC.	(*wh.*)	white-leaved.	———	1822.	——	G. ħ.	
41 oblíqua. DC.	(*wh.*)	oblique-leaved.	V. Diem. Isl.	1774.	4. 8.	G. ħ.	Sal. par. lond. t. 15.

42 corymbòsa. DC. (*wh.*) corymbose-fl^d. N. S. W. 1788. 4. 8. G. ♄. Cavan. ic. 4. t. 340.
43 paniculàta. DC. (*wh.*) panicled. ——— 1801. 6. 8. G. ♄.
44 cneorifòlia. DC. (*wh.*) Willow-wail-l^d. ——— 1824. —— G. ♄.
45 obtusifòlia. DC. (*wh.*) blunt-leaved. ——— —— .... G. ♄.
46 Cunninghámi. (*wh.*) Cunningham's. ——— 1825. .... G. ♄.
microphy'lla. F.T. *non* Link.
47 gomphocéphala. DC. (*w.*) persistent-headed. ——— 1824. .... G. ♄.
48 Glòbulus. DC. (*wh.*) large tree. V. Diem. Isl. 1804. 5. 6. G. ♄. Lab. voy. 1. t. 13.
pulverulénta. L. enum. *non* B.M. *glaùca.* DC.
49 diversifòlia. DC. (*wh.*) various-leaved. N. S. W. 1810. 5. 8. G. ♄. Bonpl. nav. 1. t. 13.
50 perfoliàta. C.C. (*wh.*) perfoliate glaucous. N. Holland. 1803. 4. 6. G. ♄.
51 cordàta. DC. (*wh.*) heart-leaved. V. Diem. Isl. 1816. 11. 3. G. ♄. Lab. nov. holl. 2. t. 152.
52 pulverulénta. DC. (*wh.*) powdered. N. S. W. —— —— G. ♄. Bot. mag. t. 2087.
53 purpuráscens. DC. (*wh.*) purple-branch'd. ——— 1804. .... G. ♄.
54 tuberculàta. DC. (*wh.*) warted-branch'd. ——— 1816. .... G. ♄.
55 rígida. DC. (*wh.*) rigid-leaved. ——— —— .... G. ♄.
56 hypericifòlia. (*wh.*) Hypericum-leaved. ——— 1804. .... G. ♄.

ANGO'PHORA. DC. ANGO'PHORA. Icosandria Monogynia.
1 cordifòlia. DC. (*wh.*) hispid. N. S. W. 1789. 5. 8. G. ♄. Bot. mag. t. 1960.
Metrosidèros híspida. Sm. exot. bot. 1. t. 42. *hirsùta.* Andr. rep. t. 281. *anómala.* Vent. malm. t. 2.
2 floribúnda. (*wh.*) many-flowered. N. S. W. 1788. —— G. ♄.
Metrosidèros floribúnda. L.T. *non* Vent.
3 intermèdia. DC. (*wh.*) intermediate. ——— 1820. —— G. ♄.
4 lanceolàta. DC. (*wh.*) ribbed. ——— 1816. —— G. ♄. Cavan. ic. 4. t. 337.
Metrosidèros costàta. Gært. fruct. 1. t. 34. f. 2.

CALLIST'EMON. DC. CALLIST'EMON. Icosandria Monogynia.
1 pinifòlium. DC. (*gr.*) Pine-tree-leav'd. N.S.W. 1806. 6. 8. G. ♄. Wendl. coll. 1. t. 16.
2 viridiflòrum. DC. (*gr.*) green-flowered. ——— 1820. —— G. ♄. Bot. mag. t. 2602.
ruscifòlium. C.C. *Metrosidèros viridiflòra.* B.M.
3 lophánthum. S.F.A. (*st.*) crested. ——— 1816. —— G. ♄. Swt. flor. aust. t. 29.
4 leptostàchyum. S.F.A. (*st.*) slender-spik'd. ——— —— —— G. ♄.
5 salígnum. DC. (*st.*) Willow-leaved. ——— 1788. 5 7. G. ♄. Bonpl. nav. t. 4.
Metrosidèros salígna. Vent. cels. t 70. *non* Bot. mag.
6 pállidum. DC. (*st.*) pale-leaved. N. S. W. 1816. —— G. ♄. Bonpl. nav. p. 101. t. 41.
7 Siebèri. DC. (*st.*) Sieber's. ——— 1820. —— G. ♄.
8 rígidum. DC. (*sc.*) rigid-leaved. ——— 1800. 3. 8. G. ♄. Bot. reg. t. 393.
9 lineàre. DC. (*sc.*) linear. ——— 1788. 6. 7. G. ♄. Sert. hann. t. 11.
10 rugulòsum. DC. (*sc.*) tubercle-leaved. ——— 1820. —— G. ♄.
11 scábrum. B.C. (*sc.*) rough-leaved. N. Holland. 1824. —— G. ♄. Lodd. bot. cab. t. 1288.
12 linearifòlium. DC. (*sc.*) linear-leaved. ——— 1820. —— G. ♄.
13 microphy'llum. M.C. (*sc.*) small-leaved. ——— 1825. —— G. ♄.
14 formòsum. M.C. (*sc.*) handsome. ——— —— —— G. ♄.
15 lanceolàtum. DC. (*sc.*) spear-leaved. N. S. W. 1788. 11. 6. G. ♄. Bot. mag. t. 260.
Metrosidèros citrìna. B.M. *lanceolàta.* L.T.
β *semperflòrens.* B.C. (*sc.*) *ever-blowing.* ——— —— —— G. ♄. Lodd. bot. cab. t. 523.
16 marginàtum. DC. (*sc.*) thick-margined. ——— 1816. 6. 8. G. ♄. Cavan. ic. 4. t. 332.
17 glaùcum. (*sc.*) splendid. N. Holland. 1803. 3. 6. G. ♄. Bonpl. nav. t. 34.
Metrosidèros glaùca. Bonp. *speciòsa.* Bot. mag. t. 1761.
18 hy'bridum. DC. (*sc.*) hybrid. N. S. W. 1824. 4. 6. G. ♄.

METROSID'EROS. DC. METROSID'EROS. Icosandria Monogynia.
1 vèra. DC. (*gr.*) Ironwood. E. Indies. 1815. 5. 8. S. ♄. Lindl. coll. t. 18.
2 flòrida. DC. (*st.*) many-flowered. N. Zealand. 1820. —— G. ♄.
3 glomulífera. DC. (*st.*) cluster-flower'd. N. S. W. 1805. —— G. ♄.
4 umbellàta. DC. (*re.*) umbel-flowered. ——— 1822. —— G. ♄. Cavan. ic. 4. t. 337.
5 angustifòlia. DC. (*wh.*) narrow-leaved. C. B. S. 1787. —— G. ♄. Hout. pfl. sys. 3. t. 25. f. 2.

STENOSPE'RMUM. STENOSPE'RMUM. Icosandria Monogynia.
1 capitàtum. (*li.*) headed. N. S. W. 1824. 4. 6. G. ♄. Reich. hort. bot. 1. t. 84.
Metrosidèros capitàta. DC.
2 corifòlium. (*wh.*) Coris-leaved. ——— 1791. 6. 7. G. ♄. Vent. malm. t. 46.
Leptospérmum ambíguum. Sm. exot. bot. t. 59. *Metrosidèros corifòlia.* Vent.

AG'ONIS. DC. AG'ONIS. Icosandria Monogynia.
1 marginàta. DC. (*wh.*) white margined. N. Holland. 1816. 6. 7. G. ♄. Lab. nov. holl. 2. t. 148.
Leptospérmum marginàtum. Labill.
2 flexuòsa. DC. (*wh.*) flexuose. ——— 1812. —— G. ♄. Colla h. rip. ap. 1. t. 2.
Leptospérmum flexuòsum. L. enum. *Metrosidèros flexuòsa.* W. enum.
3 salígna. (*wh.*) Willow-like. N. S. W. 1812. 6. 7. G. ♄.
4 ovalifòlia. (*wh.*) oval-leaved. N. Holland. —— —— G. ♄.
Fabrícia myrtifòlia. Hort. *nec aliorum.*
5 linearifòlia. DC. linear-leaved. ——— 1824. .... G. ♄.

6 ellíptica.	(*wh.*)	elliptic-leaved.	N. Holland.	1824.	....	G. ♄.	
Billóttia ellíptica. M.C.							

LEPTOSPE'RMUM. DC. SOUTH SEA MYRTLE. Icosandria Monogynia.

1 emarginàtum. DC.		emarginate.	N. S. W.	1816.	6. 7.	G. ♄.	
2 seríceum. DC.	(*wh.*)	silky-leaved.	V. Diem. Isl.	1820.	——	G. ♄.	Lab. nov. holl. 2. t. 147.
3 tuberculàtum. DC.	(*wh.*)	tuberculate.	N. S. W.	1816.	——	G. ♄.	Cavan. ic. 4. t. 330.
stellàtum. Cavan.							
4 grandifòlium. DC.	(*wh.*)	large-leaved.	——	1803.	——	G. ♄.	Bot. mag. t. 1810.
5 lanígerum. DC.	(*wh.*)	woolly-leaved.	——	1774.	——	G. ♄.	Lodd. bot. cab. t. 1192.
6 pubéscens. W.	(*wh.*)	hoary-leaved.	——	——	——	G. ♄.	
7 trinérve. DC.	(*wh.*)	three-nerved.	——	1812.	——	G. ♄.	White itin. 229. c. ic.
8 grandiflòrum. B.C.	(*wh.*)	large-flowered.	——	1816.	——	G. ♄.	Lodd. bot. cab. t. 514.
9 scopàrium. DC.	(*wh.*)	N. Zealand Tea.	N. Zealand.	1772.	——	G. ♄.	Andr. rep. t. 622.
10 squarròsum. G.	(*wh.*)	squarrose.	N. S. W.	——	——	G. ♄.	Forst. itin. 1. t. 22.
11 floribúndum.	(*wh.*)	many-flowered.	——	——	——	G. ♄.	Rœm. et Ust. mag. 7. t. 2.
12 rubricaùle. L.en.	(*wh.*)	red-stemmed.	——	1812.	——	G. ♄.	
13 flavéscens. DC.	(*wh.*)	yellowish.	——	1787.	5. 7.	G. ♄.	Bot. mag. t. 2695.
Melaleùca Thèa. Wendl. sert. han. 1. t. 14.							
14 gnidiæfòlium. DC.	(*wh.*)	Gnidia-leaved.	——	1824.	——	G. ♄.	
15 thymifòlium. F.T.	(*wh.*)	Thyme-leaved.	——	1823.	——	G. ♄.	
16 multicaùle. F.T.	(*wh.*)	many-branched.	——	——	——	G. ♄.	
17 porophy'llum. DC.	(*wh.*)	dotted-leaved.	——	——	——	G. ♄.	Cav. ic. 4. t. 330. f. 2.
18 parvifòlium. DC.	(*wh.*)	small-leaved.	——	1795.	——	G. ♄.	
19 myrtifòlium. DC.	(*wh.*)	Myrtle-leaved.	——	1824.	6. 8.	G. ♄.	
20 obovàtum. S.F.A.	(*wh.*)	obovate-leaved.	——	1823.	——	G. ♄.	Swt. flor. aust. t. 36.
21 attenuàtum. DC.	(*wh.*)	fine-branched.	——	1795.	——	G. ♄.	
22 multiflorum. DC.	(*wh.*)	abundant-flower'd.	——	1820.	——	G. ♄.	Cav. ic. 4. t. 331. f. 1
23 juniperìnum. DC.	(*wh.*)	Juniper-leaved.	——	1790.	——	G. ♄.	Vent. malm. 89.
24 baccàtum. DC.	(*wh.*)	berry-fruited.	——	1790.	6. 7.	G. ♄.	Cavan. ic. 4. t. 331. f. 2.
25 arachnoídeum. DC.	(*w.*)	cobweb.	——	1795.	5. 7.	G. ♄.	Gært. fruct. 1. t. 35. f. 3.
26 triloculàre. DC.	(*wh.*)	trilocular.	——	1816.	——	G. ♄.	Vent. malm. t. 84.
27 dùbium. DC.	(*wh.*)	doubtful.	——	1824.	6. 7.	G. ♄.	
28 péndulum. DC.	(*wh.*)	pendulous.	——	——	——	G. ♄.	

FABRI'CIA. DC. FABRI'CIA. Icosandria Monogynia.

1 myrtifòlia. DC.	(*wh.*)	Myrtle-leaved.	N. S. W.	1803.	5. 6.	G. ♄.	Gært. fruct. 1. t. 35. f. 4.
2 lævigàta. DC.	(*wh.*)	smooth-leaved.	——	1788.	——	G. ♄.	Bot. mag. t. 1304.
3 serícea.	(*wh.*)	silky-leaved.	N. Holland.	1803.	....	G. ♄.	
4 stricta. B.C.	(*wh.*)	upright.	N. S. W.	1822.	5. 7.	G. ♄.	Lodd. bot. cab. t. 1219.

BÆ'CKEA. DC. BÆ'CKEA. Pent-Dodecandria Monogynia.

1 frutéscens. DC.	(*wh.*)	Chinese.	China.	1806.	9. 1.	G. ♄.	Bot. mag. t. 2802.
2 grácilis. F.T.	(*wh.*)	slender-branch'd.	N.S.W.	1824.	6. 9.	G. ♄.	
3 virgàta. DC.	(*wh.*)	slender-twiggy.	——	1806.	8. 10.	G. ♄.	Bot. mag. t. 2127.
4 linifòlia. DC.	(*wh.*)	Flax-leaved.	——	1820.	6. 8.	G. ♄.	Linn. trans. 8. t. 12.
5 diffùsa. DC.	(*wh.*)	spreading.	——	1824.	——	G. ♄.	
6 densifòlia. DC.	(*wh.*)	close-leaved.	——	1812.	——	G. ♄.	
7 pulchélla. DC.	(*wh.*)	neat.	——	1825.	——	G. ♄.	
8 diosmifòlia. DC.	(*wh.*)	Diosma-leaved.	——	1823.	——	G. ♄.	
9 camphoràta. DC.	(*wh.*)	Camphor-scented.	——	1803.	8. 10.	G. ♄.	Bot. mag. t. 2694.
10 microphy'lla. DC.	(*wh.*)	small-leaved.	——	1812.	——	G. ♄.	
11 brevifòlia. DC.	(*wh.*)	short-leaved.	——	1824.	——	G. ♄.	
12 ramosíssima. DC.	(*wh.*)	branching.	——	1825.	7. 9.	G. ♄.	
13 crenulàta. DC.	(*wh.*)	notch leaved.	——	1822.	——	G. ♄.	
14 diosmoídes. DC.	(*wh.*)	Diosma-like.	——	1825.	——	G. ♄.	

Tribus III. *MYRTEÆ.* DC. prodr. v. 3. p. 230.

SONNER'ATIA. DC. SONNER'ATIA. Icosandria Monogynia. L.

1 ácida. DC.	(*re.*)	acid-fruited.	E. Indies.	1822.	....	S. ♄.	Rheed. mal. 3. t. 40.
2 álba. DC.	(*wh.*)	white-flowered.	——	1824.	....	S. ♄.	Rumph. amb. 3. t. 73.
3 apètala. DC.	(*gr.*)	petalless.	——	1826.	....	S. ♄.	Sym. emb. to Ava. 3. t. 25.

NEL'ITRIS. DC. NEL'ITRIS. Icosandria Monogynia.

paniculàta. DC.	(*wh.*)	panicled.	Moluccas.	1820.	....	S. ♄.	

CAMPOMAN'ESIA. DC. CAMPOMAN'ESIA. Icosandria Monogynia.

lineatifòlia. DC.	(*wh.*)	yellow-fruited.	Peru.	1824.	4. 6.	G. ♄.	Flor. per. prod. t. 13.

PSI'DIUM. DC. GUAVA. Icosandria Monogynia. L.

1 pùmilum. DC.	(*wh.*)	dwarf.	E. Indies.	1824.	——	S. ♄.	Rumph. amb. 1. t. 49.
2 aromáticum. DC.	(*wh.*)	aromatic.	W. Indies.	1779.	6. 8.	S. ♄.	Aubl. guian. 1. t. 191.

3 montànum. DC.	(*wh.*)	mountain.	W. Indies.	1779.	5. 7.	S. ♄.	
4 pyríferum. DC.	(*wh.*)	Pear-fruited.	——	1656.	6. 8.	S. ♄.	Trew. ehr. t. 43.
5 pomíferum. DC.	(*wh.*)	Apple-fruited.	——	1692.	——	S. ♄.	Rumph. amb. 1. t. 48.
6 sapidíssimum. S.S.	(*wh.*)	pleasant.	....	1823.	——	S. ♄.	Jacq. schœnb. 3. t. 366.
7 guineénse. DC.	(*wh.*)	Guinea.	Guinea.	——	——	S. ♄.	
8 polycárpon. DC.	(*wh.*)	many-fruited.	Trinidad.	1809.	4. 8.	S. ♄.	Linn. trans. 11. t. 17.
9 Aráca. DC.	(*wh.*)	egg-fruited.	Brazil.	1820.	5. 7.	S. ♄.	Raddi mem. 1821. t. 1.
10 fluviátile. DC.	(*wh.*)	Guiana.	Guiana.	1823.	——	S. ♄.	
guianénse. P.S.							
11 oligospérmum. DC.	(*wh.*)	few-seeded.	Brazil.	——	——	S. ♄.	
12 Cattleiànum. DC.	(*wh.*)	Mr. Cattley's.	——	1816.	1. 12.	S. ♄.	Lindl. coll. t. 16.
13 cordàtum. DC.	(*wh.*)	spice.	W. Indies.	1811.	5. 7.	S. ♄.	Bot. mag. t. 1779.
14 nìgrum. DC.	(*wh.*)	black-fruited.	China.	....	——	S. ♄.	
15 rùbrum. DC.	(*wh.*)	red-fruited.	——	1820.	——	S. ♄.	
16 chinénse. L.C.	(*wh.*)	yellow Chinese.	——	1818.	——	S. ♄.	
17 fràgrans. L.C.	(*wh.*)	fragrant.	....	....	4. 8.	S. ♄.	
18 myrtifòlia. L.C.	(*wh.*)	Myrtle-leaved.	....	....	——	S. ♄.	
19 latifòlium. DC.	(*wh.*)	broad-leaved.	....	1820.	6. 8.	S. ♄.	
20 índicum. DC.	(*wh.*)	Indian.	E. Indies.	1824.	——	S. ♄.	
JOSSI'NIA. DC.		JOSSI'NIA. Icosandria Monogynia.					
1 mespiloídes. DC.	(*wh.*)	Mespilus-like.	Isl. Bourbon.	1826.	....	S. ♄.	
2 ellíptica. DC.	(*wh.*)	elliptic-leaved.	Mauritius.	1824.	....	S. ♄.	
Eugènia ellíptica. Lam. *non* Smith.							
3 buxifòlia. DC.	(*wh.*)	Box-leaved.	Isl. Bourbon.	1822.	....	S. ♄.	
MY'RTUS. DC.		MYRTLE. Icosandria Monogynia. L.					
1 commùnis. DC.	(*wh.*)	common.	S. Europe.	1597.	7. 8.	F. ♄.	Lam. ill. t. 410.
A. *melanocárpa.* DC.	(*w.*)	*black-berried.*	——	——	——	F. ♄.	
a romana. DC.	(*wh.*)	*broad-leaved.*	——	——	——	F. ♄.	Mill. ic. t. 184. f. 1.
β multipléx.	(*wh.*)	*double-blossomed.*	——	....	——	F. ♄.	
B. *leucocárpa.* DC.	(*wh.*)	*white-fruited.*	Levant.	....	——	F. ♄.	
2 tarentìna.	(*wh.*)	Box-leaved.	S. Europe.	1597.	——	F. ♄.	
3 itálica.	(*wh.*)	upright Italian.	——	——	——	F. ♄.	
β variegàta.	(*wh.*)	*striped-leaved.*	——	....	——	F. ♄.	
γ maculàta.	(*wh.*)	*blotch-leaved.*	——	....	——	F. ♄.	
4 b'œtica.	(*wh.*)	Orange-leaved.	——	1597.	——	F. ♄.	Blackw. t. 114.
5 lusitánica.	(*wh.*)	Portugal.	Portugal.	——	——	F. ♄.	Clus. hist. 1. p. 66. f. 1.
6 bélgica.	(*wh.*)	sharp-leaved.	S. Europe.	——	——	F. ♄.	
7 mucronàta.	(*wh.*)	small-leaved.	——	——	——	F. ♄.	
8 tomentòsa. DC.	(*ro.*)	woolly-leaved.	China.	1776.	5. 9.	S. ♄.	Bot. mag. t. 250.
9 tenuifòlia. DC.	(*wh.*)	slender-leaved.	N. S. W.	1824.	——	G. ♄.	
10 lùcida. DC.	(*wh.*)	shining.	Surinam.	1793.	4. 8.	S. ♄.	
MY'RCIA. DC.		MY'RCIA. Icosandria Monogynia.					
1 punctàta. DC.	(*wh.*)	punctated.	Santa Cruz.	1825.	....	S. ♄.	
2 coriàcea. DC.	(*wh.*)	Sumach-leaved.	Hispaniola.	1759.	5. 7.	S. ♄.	Plum. ic. t. 208. f. 2.
3 àcris. DC.	(*wh.*)	broad-ld. Allspice.	W. Indies.	——	——	S. ♄.	Pluk. alm. t. 155. f. 3.
4 pimentoídes. DC.	(*wh.*)	Allspice-like.	——	——	——	S. ♄.	Lodd. bot. cab. t. 178.
My'rtus Piménta. β latifòlia. Lodd. bot. cab. t. 178.							
5 divaricàta. DC.	(*wh.*)	spreading-flow'd.	W. Indies.	1820.	——	S. ♄.	Plum. ic. t. 208. f. 1.
6 soròria. DC.	(*wh.*)	sister.	Trinidad.	1822.	——	S. ♄.	
7 spléndens. DC.	(*wh.*)	glossy-leaved.	Hispaniola.	——	——	S. ♄.	Jacq. coll. 2. t. 4.
Eugènia periplocæfòlia. Jacq.							
8 Mìni.	(*wh.*)	small-fruited.	Guiana.	1803.	——	S. ♄.	Aubl. guian. 1. t. 197.
Eugènia Mìni. Aubl.							
9 crassinérvia. DC.	(*wh.*)	thick-nerved.	——	1780.	——	S. ♄.	
10 bracteàta. DC.	(*wh.*)	leafy-bracted.	Brazil.	1824.	——	S. ♄.	
11 pseùdo-Mìni. DC.	(*wh.*)	Brazilian.	——	1822.	——	S. ♄.	
CALYPTRA'NTHES. DC.		CALYPTRA'NTHES. Icosandria Monogynia.					
1 rígida. DC.	(*wh.*)	rigid-leaved.	Jamaica.	1822.	....	S. ♄.	
2 Chytracùlia. DC.	(*wh.*)	forked-panicled.	——	1778.	3. 5.	S. ♄.	Brown. jam. t. 37. f. 2.
3 Zuzy'gium. DC.	(*wh.*)	oval-leaved.	——	——	5. 7.	S. ♄.	—— 240. t. 7. f. 2.
SYZY'GIUM. DC.		SYZY'GIUM. Icosandria Monogynia.					
1 obovàtum. DC.	(*wh.*)	obovate-leaved.	Mauritius.	1822.	....	S. ♄.	
2 paniculàtum. DC.	(*wh.*)	panicled.	——	1821.	4. 6.	S. ♄.	
Eugènia paniculàta. Lam.							
3 glomeràtum. DC.	(*wh.*)	cluster-flowered.	——	1824.	....	S. ♄.	
4 Jambolàna. DC.	(*wh.*)	Jambolana-tree.	E. Indies.	1796.	5. 7.	S. ♄.	Rumph. amb. 1. t. 42.
Calyptránthes Jambolàna. W. *Eugènia Jambolàna.* Lam.							
5 inophy'llum. DC.	(*wh.*)	equal-leaved.	E. Indies.	1826.	....	S. ♄.	

P 2

6 caryophyllifòlium. DC. (*wh.*) Clove-leav'd. E. Indies. 1822. 5. 7. S. ♄. Rumph. amb. 1. t. 41.
Eugènia caryophyllifòlia. Lam. *Calyptránthes caryophyllifòlia.* W.
7 venòsum. DC. (*wh.*) veined-leaved. Nepaul. 1824. .... S. ♄.
8 fruticósum. DC. (*wh.*) shrubby. E. Indies. —— .... S. ♄.
9 zeylánicum. DC. (*wh.*) Ceylon. ——— 1798. 6. 7. S. ♄. Herm. zeyl. 435. ic.

CARYOPHY'LLUS. DC. CLOVE-TREE. Icosandria Monogynia. L.
aromáticus. DC. (*wh.*) aromatic. Moluccas. 1797. .... S. ♄. Bot. mag. t. 2749-2750.

ACM'ENA. DC. ACM'EMA. Icosandria Monogynia.
ellíptica. DC. (*wh.*) elliptic-leaved. N. S. W. 1790. 5. 9. G. ♄. Bot. mag. t. 1872.
Eugènia ellíptica. B.M. *My'rtus Smíthii.* S.S. *Metrosidèros floribúnda.* Vent. malm. t. 75.

EUG'ENIA. DC. EUG'ENIA. Icosandria Monogynia.
1 Michèlii. DC. (*wh.*) one-flowered. Brazil. 1759. 1. 3. S. ♄. Bot. mag. t. 473.
uniflòra. W. *Plìnia pedunculàta.* B.M.
2 ligustrìna. DC. (*wh.*) Privet-leaved. Caribees. 1798. 8. 9. S. ♄.
3 balsámica. DC. (*wh.*) balsamic. W. Indies. 1816. 6. 7. S. ♄. Jacq. frag. t. 45. f. 2.
4 angustifòlia. DC. (*wh.*) narrow-leaved. St. Domingo. 1825. .... S. ♄. Plum. ed Bur. t. 207. f. 2.
5 obscùra. DC. (*wh.*) obscure. Brazil. 1824. 10. 11. S. ♄. Bot. reg. t. 1044.
My'rtus obscùra. B.R.
6 alpìna. DC. (*wh.*) alpine. Jamaica. 1827. 6. 8. S. ♄.
7 latifòlia. DC. (*wh.*) broad-leaved. Guiana. 1793. —— S. ♄. Aubl. guian. 1. t. 199.
8 Lambertiàna. DC. (*wh.*) Mr. Lambert's. W. Indies. 1794. .... S. ♄.
9 Parkeriàna. DC (*wh.*) Parker's. Trinidad. —— .... S. ♄.
10 Roxbúrghii. DC. (*wh.*) Roxburgh's. E. Indies. 1798. 6. 7. S. ♄. Andr. reposit. t. 619?
zeylànica. Roxb. *non* W.
11 dísticha. DC. (*wh.*) globe-berried. Jamaica. 1793. 4. 7. S. ♄. Lindl. coll. t. 19.
My'rtus dísticha. L. coll.
12 glabràta. DC. (*wh.*) smooth. W. Indies. 1825. —— S. ♄.
13 axillàris. DC. (*wh.*) axillary. Jamaica. 1793. 9. 10. S. ♄.
14 floribúnda. DC. (*wh.*) many-flowered. Santa Cruz. 1822. 4. 6. S. ♄.
15 buxifòlia. DC. (*wh.*) Box-leaved. W. Indies. —— —— S. ♄.
16 baruénsis. DC. (*wh.*) many-flowered. ——— .... —— S. ♄. Jacq. ic. 3. t. 486.
17 dumòsa. DC. (*wh.*) bushy. ——— 1793. 6. 7. S. ♄.
18 biflòra. DC. (*wh.*) two-flowered. Jamaica. 1759. 4. 5. S. ♄. Brown jam. t. 25. f. 3.
My'rtus biflòra. W.
19 Sieberiàna. DC. (*wh.*) Sieber's. Trinidad. 1824. —— S. ♄.
20 trinérvia. DC. (*wh.*) three-nerved. N. S. W. 1823. .... G. ♄.
My'rtus trinérvia. L.T.
21 trinitàtis. DC. (*wh.*) Trinidad. Trinidad. —— .... S. ♄.
22 Grégii. DC. (*wh.*) Greg's. Dominica. 1776. 6. 7. S. ♄. Gært. fruct. 1. t. 33.
Gréggia aromática. G. *My'rtus Grégii.* W.
23 frágrans. DC. (*wh.*) sweet-scented. Jamaica. 1790. 4. 5. S. ♄. Bot. mag. t. 1242.
24 virgultòsa. DC. (*wh.*) twiggy. W. Indies. 1787. 7. 8. S. ♄.
My'rtus virgultòsa. W.
25 corymbòsa. DC. (*wh.*) corymbose. E. Indies. 1824. .... S. ♄. Rheed. mal. 5. t. 27.
26 pulchélla. H.B. (*wh.*) pretty. ——— 1822. .... S. ♄.
27 glandulífera. H.B. (*wh.*) glandular. ——— 1823. .... S. ♄.
28 obtusifòlia. H.B. (*wh.*) blunt-leaved. ——— .... S. ♄.
29 cerasoídes. H.B. (*wh.*) Cherry-fruited. ——— .... S. ♄.
30 Piménta. DC. (*wh.*) Allspice. W. Indies. 1723. 5. 7. S. ♄. Bot. mag. t. 1236.
My'rtus Piménta. W.

JAMB'OSA. DC. JAMB'OSA. Icosandria Monogynia.
1 vulgàris. DC. (*st.*) narrow-leaved. E. Indies. 1768. 2. 7. S. ♄. Bot. mag. t. 1696.
Eugenìa Jámbos. B.M.
2 venòsa. DC. (*bh.*) veined-leaved. Madagascar. 1824. .... S. ♄.
3 macrophy'lla. DC. (*wh.*) large-leaved. E. Indies. 1822. .... S. ♄.
4 macrocárpa. (*wh.*) large-fruited. ——— —— .... S. ♄.
Eugènia macrocárpa. H.B.
5 purpuráscens. DC. (*pu.*) purple-flower'd. W. Indies. 1768. 5. 8. S. ♄. Andr. reposit. t. 458.
Eugènia malaccénsis. Smith exot. bot. t. 61.
6 malaccénsis. DC. (*wh.*) Malay Apple. E. Indies. —— —— S. ♄. Rheed. mal. 1. t. 18.
7 amplexicaùlis. DC. (*gr.*) stem-clasping. Sumatra. 1820. 3. 6. S. ♄. Bot. reg. t. 1033.
8 áquea. DC. (*ro.*) watery. Moluccas. 1819. —— S. ♄. Rumph. amb. 1. t. 38. f. 2.
Eugènia áquea. H.B.
9 acuminàta. (*gr.*) taper-pointed. Amboyna. 1816. 5. 6. S. ♄.
Eugènia acuminàta. H.B.
10 laurifòlia. DC. (*wh.*) Laurel-leaved. E. Indies. 1824. .... S. ♄.
11 ternifòlia. (*wh.*) three-leaved. ——— 1822. —— S. ♄.
Eugènia ternifòlia. H.B.

12 austràlis. DC. (*wh.*) Myrtle-leaved. N. Holland. 1812. 5. 9. G. ♄. Bot. mag. t. 2230.
Eugènia austràlis. Colla hort. rip. app. t. 8. *Eugènia myrtifòlia.* Bot. reg. t. 627.

Tribus IV. *BARRINGTONIEÆ.* DC. prodr. v. 3. p. 288.

BARRINGT'ONIA. DC.	BARRINGT'ONIA. Monadelphia Polyandria. L.						
1 speciòsa. DC. (*pu.wh.*)	beautiful.	E. Indies.	1786.	....	S. ♄.	Rumph. amb. 3. t. 114.
2 racemòsa. DC. (*wh.*)	racemed.	Malabar.	1822.	....	S. ♄.	Rheed. mal. 4. t. 6.
Eugènia racemòsa. L.						
STRAV'ADIUM. DC.	STRAV'ADIUM. Monadelphia Polyandria.					
1 álbum. DC. (*wh.*)	white-flowered.	Moluccas.	1822.	....	S. ♄.	Rumph. amb. 3. t. 116.
2 rùbrum. DC. (*re.*)	red-flowered.	———	——	....	S. ♄.	——— 3. t. 115.
Barringtònia acutángula. G. *Eugènia acutángula.* L.						
GUST'AVIA. DC.	GUST'AVIA. Monadelphia Polyandria. L.					
1 augústa. DC. (*wh.*)	large-flowered.	Guiana.	1794.	....	S. ♄.	Aubl. guian. 1. t. 192.
2 fastuòsa. DC. (*wh.*)	six-petaled.	———	1824.	....	S. ♄.	——— 1. t. 193.

Tribus V. *LECYTHIDEÆ.* DC. prodr. v. 3. p. 290.

LEC'YTHIS. DC.	LEC'YTHIS. Monadelphia Polyandria.					
1 Ollària. DC. (*wh.*)	sessile-leaved.	Brazil.	1820.	....	S. ♄.	
2 mìnor. DC. (*wh.*)	smaller.	Carthagena.	1823.	....	S. ♄.	Jacq. amer. t. 109.
3 grandiflòra. DC. (*ro.*)	large-flowered.	Cayenne.	1822.	....	S. ♄.	Au.gu.2.t.283.284.285.
4 Idatìmon. DC. (*fl.*)	flesh-coloured.	Guiana.	——	....	S. ♄.	——— 2. t. 289.
5 amára. DC. (*ye.*)	bitter-seeded.	———	——	....	S. ♄.	———2.t.286et.t.285.f.1.
6 parviflòra. DC. (*st.*)	small-flowered.	———	——	....	S. ♄.	——— 2. t. 287.
BERTHOLL'ETIA. DC.	BERTHOLL'ETIA. Monadelphia Polyandria.					
excélsa. DC. (*st.*)	lofty.	Brazil.	1824.	....	S. ♄.	H. et B. pl. equ. 1. t. 36.
COUROUP'ITA. DC.	COUROUP'ITA. Monadelphia Polyandria.					
guianénsis. DC. (*bh.*)	Guiana.	Guiana.	1820.	....	S. ♄.	Aubl. guian. 2. t. 282.
Lec'ythis bracteàta. W. *Pèkea Couroupìta.* Juss.						

MYRTACEÆ DUBIÆ. DC. prodr. v. 3. p. 394.

C'AREYA. R.C.	C'AREYA. Monadelphia Polyandria.					
1 sph'ærica. H.B. (*wh.*)	sphærical.	E. Indies.	1823.	....	S. ♄.	
2 arbòrea. R.C. (*gr.bh.*)	tree.	———	——	....	S. ♄.	Roxb. corom. 3. t. 218.
3 herbàcea. R.C. (*re.*)	herbaceous.	———	1808.	7. 8.	S. ♃.	——— 3. t. 217.
GR'IAS. DC.	ANCHOVY PEAR. Polyandria Monogynia.					
cauliflòra. DC. (*wh.*)	stem-flowering.	Jamaica.	....	....	S. ♄.	Sloan.his.2.t.217.f.1.2.
MA'RLEA. R.C.	MA'RLEA. Icosandria Monogynia.					
begonifòlia. R.C. (*wh.*)	Begonia-leaved.	E. Indies.	1824.	....	S. ♄.	Roxb. corom. 3. t. 283.

ORDO XCV.

CUCURBITACEÆ. *DC. prodr. v.* 3. *p* 297.

Tribus II. *CUCURBITEÆ.* DC. prod. 3. p. 299.

LAGEN'ARIA. DC.	GOURD. Monœcia Monadelphia.					
1 vulgàris. DC. (*wh.*)	bottle.	India.	1597.	7. 9.	H. ⊙.⌣.	
α Gòurda. DC. (*wh.*)	*common.*	———	——	——	H. ⊙.⌣.	Moris. his. s. 1. t. 5. f. 1.
β Cougoùrda. DC. (*wh.*)	*Cow-gourd.*	China.	——	——	H. ⊙.⌣.	Rumph. amb. 5. t. 144.
γ depréssa. DC. (*wh.*)	*flat-fruited.*	India.	——	——	H. ⊙.⌣.	
δ turbinàta. DC. (*wh.*)	*top-fruited.*	———	——	——	H. ⊙.⌣.	Moris. his. s. 1. t. 5. f. 2.
ε clavàta. DC. (*wh.*)	*club-fruited.*	———	——	——	H. ⊙.⌣.	——— sect. 1. t. 5. f. 3.

2 vittàta. DC.	(*wh.*)	ribbon.	E. Indies.	....	7. 9.	H. ⊙.◡.	
3 idolátrica. DC.	(*wh.*)	Pear-fruited.	———	....	——	H. ⊙.◡.	
CUCUMIS. DC.		CUCUMBER.	Monœcia Monadelphia. L.				
1 Mèlo. DC.	(*st.*)	Melon.	Tartary.	1570.	5. 9.	F. ⊙.	Blackw. herb. t. 329.
2 deliciòsus. DC.	(*st.*)	delicious.	E. Indies.	1818.	——	F. ⊙.	
3 satìvus. DC.	(*ye.*)	common.	———	1573.	7. 9.	H. ⊙.	Blackw. herb. t. 4.
α víridis. DC.	(*ye.*)	*green-fruited.*	———	——	——	H. ⊙.	
β flavus. DC.	(*ye*)	*pale-fruited.*	———	——	——	H. ⊙.	
γ álbus. DC.	(*ye.*)	*white Turkey.*	———	——	——	H. ⊙.	
δ variegàtus. DC.	(*ye.*)	*striped-fruited.*	———	——	——	H. ⊙.	
ε fastigiàtus. DC.	(*ye.*)	*clustered.*	———	——	——	H. ⊙.	
4 flexuòsus. DC.	(*ye.*)	flexuous snake.	———	1597.	5. 9.	F. ⊙.	Lob. stirp. p. 363. f. 2.
β refléxus. DC.	(*ye.*)	*angular-leaved.*	———	....	——	F. ⊙.	
5 anguìnus. W.	(*ye.*)	snake-fruited.	———	1820.	7. 9.	F. ⊙.	
6 jamaicénsis. DC.	(*ye.*)	round-fruited.	Jamaica.	1818.	——	F. ⊙.	
7 Châte. DC.	(*ye.*)	hairy.	Levant.	1759.	6. 7.	H. ⊙.	Alp. ægypt. t. 117.
8 Dudàim. DC.	(*ye.*)	Apple-fruited.	———	1705.	7. 8.	H. ⊙.	Andr. reposit. t. 548.
9 prophetàrum. DC.	(*ye.*)	globe-fruited.	———	1777.	6. 9.	F. ⊙.	Jacq. vind. 1. t. 9.
10 africànus. DC.	(*ye.*)	African.	C. B. S.	1816.	——	G. ♃.◡.	Herm. parad. t. 134.
11 Angùria. DC.	(*ye.*)	round prickly.	Jamaica.	1692.	7. 9.	H. ⊙.	Mill. ic. 1. t. 33.
12 Citrúllus. DC.	(*ye.*)	Water Melon.	Africa.	1597.	5. 9.	F. ⊙.	Rumph. amb. 5. t. 146.
α Pásteca. DC.	(*ye.*)	*firm-fruited.*	———	——	——	F. ⊙.	
β Jàce. DC.	(*ye.*)	*juicy-fruited.*	———	——	——	F. ⊙.	
13 pubéscens. DC.	(*ye.*)	pubescent.	....	1815.	6. 9.	F. ⊙.	
14 maculàtus. DC.	(*ye.*)	spotted-fruited.	....	....	——	F. ⊙.	Ser. dis. t. 3.
15 Colocy'nthis. DC.	(*ye.*)	bitter.	Japan.	1551.	5. 8.	F. ⊙.	Blackw. herb. t. 441.
16 perénnis. DC.	(*ye.*)	perennial.	N. America.	1824.	6. 8.	F. ⊙.	
17 coronàrius.	(*ye.*)	crowned.	....	1827.	——	F. ⊙.	
18 Momórdica. H.B.	(*ye.*)	Momordica-like.	E. Indies.	1812.	7. 9.	F. ⊙.	
19 utilitíssimus. H.B.	(*ye.*)	useful.	———	1820.	——	F. ⊙.	
20 turbinàtus. H.B.	(*ye.*)	top-shaped.	———	——	——	F. ⊙.	
21 maderaspatànus. W.	(*ye.*)	Madras.	———	1805.	7. 8.	F. ⊙.	Pluk. alm. t. 170. f. 2.
Bryònia maderaspatàna. DC.							
22 integrifòlius. H.B.	(*ye.*)	entire-leaved.	———	1825.	——	F. ⊙.◡.	
LU'FFA. DC.		LU'FFA.	Pentandria Monogynia.				
1 fœtida. DC.	(*ye*)	strong-scented.	E. Indies.	1812.	6. 10.	F. ⊙.◡.	Bot. mag. t. 1638.
2 acutángula. DC.	(*ye.*)	acute-angled.	———	1692.	6. 9.	F. ⊙.◡.	Jacq. vind. 3. t. 73-74.
Cùcumis acutángulus. Jacq.							
3 Plukenetiàna. DC.	(*ye.*)	Plukenet's.	———	....	——	F. ⊙.◡.	Pluk. alm. t. 172. f. 1.
4 clavàta. H.B.	(*ye.*)	erect.	———	1820.	6. 10.	F. ⊙.	
5 amára. H.B.	(*ye.*)	bitter.	———	——	——	F. ⊙.◡.	
6 gravèolens. H.B.	(*ye.*)	strong-smelling.	———	1823.	——	F. ⊙.◡.	
7 racemòsa. H.B.	(*ye.*)	racemed.	———	——	——	F. ⊙.◡.	
8 ægyptìaca. DC.	(*ye.*)	Egyptian.	Egypt.	1739.	7. 8.	F. ⊙.	Alp. ægyp. t. 58.
Momórdica Lúffa. L.							
BENINC'ASA. DC.		BENINC'ASA.	Polygamia Monœcia.				
cerífera. DC.	(*ye.*)	musky.	E. Indies.	....	7. 9.	F. ⊙.	Savi. mem. c. 1818. p. 6.
BRY'ONIA. DC.		BRYONY.	Monœcia Monadelphia. W.				
1 rostràta. DC.	(*st.*)	rostrate.	E. Indies.	1825.	7. 9.	S. ⊙.◡.	
2 scábra. DC.	(*st.*)	globe-fruited.	C. B. S.	1774.	9. 10.	G. ♃.◡.	
3 verrucòsa. DC.	(*st.*)	warted.	Canaries.	1779.	6. 10.	G. ♃.◡.	
4 scabràta. DC.	(*st.*)	roughish.	E. Indies.	1826.	——	S. ♃.◡.	
5 punctàta. DC.	(*st.*)	dotted.	C. B. S.	1816.	——	G. ♃.◡.	
6 cordàta. DC.	(*st.*)	cordate.	———	——	——	G. ♃.◡.	
7 grándis. DC.	(*re.*)	great-flowered.	E. Indies.	1783.	5. 8.	S. ♃.◡.	Rump. amb. 5. t. 166. f. 1.
8 umbellàta. DC.	(*st.*)	umbel-flowered.	———	1825.	——	S. ♃.◡.	Rheed. mal. 8. t. 26?
9 hederæfòlia. DC.	(*st.*)	Ivy-leaved.	Teneriffe.	——	——	G. ♃.◡.	Jacq. fragm. 73. t. 113.
10 epig'æa. DC.	(*st.*)	leathery-leaved.	E. Indies.	1815.	——	G. ♃.◡.	
11 scabrélla. DC.	(*st.*)	rough-seeded.	———	1781.	6. 8.	S. ⊙.◡.	
12 latebròsa. DC.	(*st.*)	hairy.	Canaries.	1779.	——	G. ♃.◡.	
13 trilobàta. DC.	(*st.*)	three-lobed.	C. B. S.	1816.	——	G. ⊙.◡.	
14 álba. DC.	(*wh.*)	black-fruited.	Europe.	1807.	——	H. ♃.◡.	Lam. ill. t. 796.
15 dioíca. DC.	(*st.*)	red-fruited.	Britain.	....	5. 9.	H. ♃.◡.	Eng. bot. v. 7. t. 439.
16 nítida. DC.	(*st.*)	glossy.	....	1824.	——	S. ♃.◡.	
17 crética. DC.	(*st.*)	Cretan.	Candia.	1759.	7. 9.	G. ♃.◡.	Desf. an. mus. 12. t. 17.
18 quinquelòba. DC.	(*st.*)	five-lobed.	C. B. S.	....	6. 10.	G. ♃.◡.	Bot. reg. t. 82.
19 ficifòlia. DC.	(*st.*)	Fig-leaved.	Buenos Ayres.	1726.	——	G. ♃.◡.	Dill. elth. t. 50. f. 58.
20 variegàta. DC.	(*st.*)	variegated.	America.	....	——	S. ♃.◡.	
21 racemòsa. DC.	(*st.*)	racemed.	Trinidad.	1820.	——	S. ♃.◡.	Plum. amer. 83. t. 97.

22 pinnatífida. DC.	(*st.*)	pinnatifid.	C. B. S.	1816.	7. 8.	G. ♃.◡.	
23 divìsa. W. en.	(*st.*)	divided-leaved.	....	1818.	6. 10.	G. ♃.◡.	
24 palmàta. DC.	(*st*)	palmated.	Ceylon.	1778.	7. 8.	S. ♃.◡.	
25 Garcìni. DC.	(*st.*)	round-lobed.	———	1818.	——	S. ♃.◡.	Burm. ind. t. 57. f. 3.
26 laciniòsa. DC.	(*st.*)	laciniated.	E. Indies.	1710.	——	S. ♃.◡.	Herm. lugd. 95. t. 97.
27 africàna. DC.	(*st.*)	African.	C. B. S.	1759.	7. 8.	G. ♃.◡.	Herm. parad. t. 708.
28 dissécta. DC.	(*st.*)	smooth-leaved.	———	1710.	——	G. ♃.◡.	
29 tenélla. H.B.	(*st.*)	slender.	E. Indies.	1820.	——	S. ♃.◡.	
30 Wallichiàna. DC.	(*st.*)	filiform.	Nepaul.	——	——	S. ♃.◡.	
filifórmis. H.B.							
SI'CYOS. DC.		SINGLE-SEEDED CUCUMBER.	Monœcia Monadelphia. W.				
1 angulàtus. DC.	(*ye.*)	angular-leaved.	N. America.	1710.	7. 9.	H. ⊙.◡.	Dill. elth. t. 51. f. 59.
2 parviflòrus. DC.	(*ye.*)	small-flowered.	Quito.	1823.	——	H. ⊙.◡.	
3 microphy'llus. DC.	(*ye.*)	small-leaved.	Mexico.	——	——	H. ⊙.◡.	
4 vitifòlius. DC.	(*ye.*)	Vine-leaved.	....	——	——	H. ⊙.◡.	
5 laciniàtus. DC.	(*ye.*)	jagged-leaved.	S. America.	1827.	——	F. ⊙.◡.	Plum. ed Burm. t. 243.
ELAT'ERIUM. DC.		ELAT'ERIUM.	Monœcia Monadelphia.				
1 geméllum. DC.	(*wh.*)	twin-tendrilled.	Mexico.	1826.	6. 9.	H. ⊙.◡.	
2 carthaginénse. DC.	(*wh.*)	Carthagenian.	Caracas.	1824.	——	F. ⊙.◡.	Jacq. am. pict. t. 232.
3 tamnoídes. DC.	(*st.*)	Tamus-like.	Mexico.	1818.	——	H. ⊙.◡.	
MOMO'RDICA. DC.		MOMO'RDICA.	Monœcia Monadelphia. W.				
1 balsamìna. DC.	(*ye.*)	Balsam Apple.	E. Indies.	1568.	6. 7.	F. ⊙.◡.	Blackw. herb. 6. t. 539.
2 muricàta. DC.	(*ye.*)	tubercled.	———	....	——	F. ⊙.◡.	Rheed. mal. 8. t. 10.
3 Charántia. DC.	(*ye.*)	hairy.	———	1710.	——	F. ⊙.◡.	Bot. mag. t. 2455.
β *abbreviàta.* DC.	(*ye.*)	*short-fruited.*	———	——	——	F. ⊙.◡.	
4 cylíndrica. DC.	(*ye.*)	cylindrical.	China.	1816.	——	F. ⊙.◡.	
5 operculàta. DC.	(*ye.*)	rough-fruited.	W. Indies.	1731.	6. 9.	F. ⊙.◡.	Comm. rar. t. 22.
6 Elatèrium. DC.	(*st.*)	squirting cucumber.	S. Europe.	1548.	——	H. ♃.	Bot. mag. t. 1914.
7 dioíca. DC.	(*ye.*)	diœcious.	E. Indies.	1820.	7. 8.	S. ♃.◡.	Rheed. mal. 8. t. 18.
8 monadélpha. H.B.	(*ye.*)	monadelphous.	———	1812.	——	S. ♃.◡.	
9 míxta. H.B.	(*ye.*)	mixed.	———	1820.	——	S. ♃.◡.	
10 umbellàta. H.B.	(*ye.*)	umbel-flowered.	———	1818.	——	S. ♃.◡.	
11 tubiflòra. H.B.	(*ye.*)	tube-flowered.	———	1824.	7. 10.	F. ⊙.◡.	
S'ECHIUM. DC.		CHOCHO.	Monœcia Monadelphia.				
1 èdule. DC.	(*ye.*)	eatable.	W. Indies.	1823.	7. 9.	F. ⊙.◡.	Jacq. amer. t. 163.
2 americànum. DC.	(*ye.*)	American.	Jamaica.	1825.	——	F. ⊙.◡.	
3 palmàtum. DC.	(*st.*)	palmated.	Mexico.	1827.	9. 11.	F. ⊙.◡.	Moc. et Ses. fl. mex. ic.
MELO'THRIA. DC.		MELO'THRIA.	Monœcia Triandria.				
1 péndula. DC.	(*ye.*)	pendulous.	America.	1752.	6. 7.	H. ⊙.◡.	Plum. ic. t. 66. f. 2.
2 fœ'tida. DC.	(*ye.*)	strong-scented.	Guinea.	1825.	——	S. ♃.◡.	Jacq. ic. 3. t. 624.
Trichosánthes fœtidíssima. Jacq.							
TRICHOSA'NTHES. DC.		SNAKE-GOURD.	Monœcia Monadelphia. W.				
1 anguìna. DC.	(*wh.*)	common.	China.	1755.	5. 8.	F. ⊙.◡.	Bot. mag. t. 722.
2 colubrìna. DC.	(*wh.*)	panicled.	....	1820.	——	F. ⊙.◡.	Jacq. f. eclog. t. 128.
3 dioíca. H.B.	(*wh.*)	diœcious.	E. Indies.	——	5. 6.	S. ♃.◡.	
4 cucumerìna. DC.	(*wh.*)	Cucumber-like.	———	1804.	6. 8.	F. ⊙.◡.	Rheed. mal. 8. t. 15.
5 lobàta. H.B.	(*wh.*)	lobed-leaved.	———	1812.	——	F. ⊙.◡.	
6 cordàta. H.B.	(*wh.*)	heart-leaved.	———	1823.	——	S. ♃.◡.	
7 palmàta. H.B.	(*wh.*)	palmated.	———	1825.	——	S. ♄.◡.	
8 corniculàta. DC.	(*wh.*)	horn-gourd.	W. Indies.	1810.	6. 7.	S. ♃.◡.	Bot. mag. t. 2703.
tuberòsa. B.M. *Ceratosánthes tuberòsa.* S.S.							
JOLI'FFIA. DC.		JOLI'FFIA.	Diœcia Monadelphia.				
africàna. DC.	(*pu.*)	African.	Zanzibar.	1825.	6. 8.	S. ♄.◡.	Bot. mag. t. 2681.
♀ *fœ'mina.*	(*pu.*)	*female.*	———	——	——	S. ♄.◡.	——— t. 2681.
Feuill'æa pedàta. B.M.							
♂ *máscula.*	(*pu.*)	*male.*	———	1827.	——	S. ♄.◡.	Bot. mag. t. 2751 et 2752.
Telfaíria pedàta. Hook. bot. mag. t. 2751.							
CUCU'RBITA. DC.		PUMPKIN.	Monœcia Monadelphia. W.				
1 máxima. DC.	(*ye.*)	Elephant.	....	....	7. 10.	H. ⊙.◡.	Tourn. inst. n. 2. t. 34.
α *Potíro.* DC.	(*ye.*)	*yellow-fruited.*	....	....	——	H. ⊙.◡.	
β *víridis.* DC.	(*ye.*)	*green-fruited.*	....	....	——	H. ⊙.◡.	
γ *Coùrgero.* DC.	(*ye.*)	*small-fruited.*	....	....	——	H. ⊙.◡.	
2 Melopèpo. DC.	(*ye.*)	Squash.	....	1597.	5. 9.	H. ⊙.◡.	Moris. hist. 1. s. 1. t. 8.
3 moschàta. DC.	(*ye.*)	musk Melon.	W. Indies.	——	6. 9.	F. ⊙.◡.	Dalech. hist. 616. f. 3.
4 villòsa. DC.	(*ye.*)	woolly-fruited.	E. Indies.	1818.	——	F. ⊙.◡.	
5 Pèpo. DC.	(*ye.*)	common.	Levant.	1570.	——	H. ⊙.◡.	
α *subrotúnda.* DC.	(*ye.*)	*round-fruited.*	———	——	——	H. ⊙.◡.	
β *oblonga.* DC.	(*ye.*)	*oblong-fruited.*	———	——	——	H. ⊙.◡.	
6 farinòsa. DC.	(*ye.*)	mealy-fruited.	E. Indies.	1822.	——	F. ⊙.◡.	

7 verrucòsa. DC.	(*ye.*)	warted.	....	1658.	6. 9.	H. ⊙.‿.	Bauh. hist. 2. p. 222. ic.
8 subverrucòsa. DC.	(*ye.*)	oblong-fruited.	....	....	——	H. ⊙.‿.	
9 aurántia. DC.	(*ye.*)	orange-gourd.	....	....	——	H. ⊙.‿.	
α orangìna. DC.	(*ye.*)	*true orange.*	....	....	——	H. ⊙.‿.	
β colocynthoìdes. DC	(*ye.*)	*variegated-fruit'd.*	....	....	——	H. ⊙.‿.	
10 ovífera. DC.	(*ye.*)	egg-shaped.	Astracan.	....	——	H. ⊙.‿.	Dod. pempt. 670. f. 1.
α pyrifórmis. DC.	(*ye.*)	*pear-fruited.*	————	....	——	H. ⊙.‿.	Ser. diss. t. 1.
β subglobòsa. DC.	(*ye.*)	*white-striped.*	————	....	——	H. ⊙.‿.	
γ grìsea. DC.	(*ye.*)	*white-spotted.*	————	....	——	H. ⊙.‿.	
11 fœtidíssima. DC.	(*ye.*)	strong-scented.	Mexico.	1825.	——	H. ⊙.‿.	
12 marsupiàlis. Jacq.	(*ye.*)	bag-bearing.	....	1821.	——	H. ⊙.‿.	
13 dipsàcea. E.	(*ye.*)	Teasel-like.	Egypt.	1827.	——	H. ⊙.‿.	
ANG'URIA. DC.		ANG'URIA.	Monœcia Diandria. W.				
1 trilobàta. DC.	(*re.*)	three-lobed.	Martinico.	1793.	6. 7.	S. ♃.‿.	Jacq. am. pict. t. 234.
2 pedatisécta. DC.	(*re.*)	cut-leaved.	Peru.	1827.	7. 10.	H. ⊙.‿.	Feuil per. 1. t. 41.
3 pedàta. DC.	(*re.*)	pedate-leaved.	St. Domingo.	——	6. 8.	S. ♃.‿.	Jacq. am. pict. t. 155.

GENERA CUCURBITACEIS AFFINIA.

GRON'OVIA. DC.		GRON'OVIA.	Pentandria Monogynia. L.				
scándens. DC.	(*st.*)	climbing.	Jamaica.	1731.	6. 7.	S. ♃.‿.	Jacq. ic. 2. t. 338.
CA'RICA. S.S.		PAPAW-TREE.	Diœcia Decandria. L.				
1 Papáya. S.S.	(*st.*)	common.	India.	1690.	7. 8.	S. ♄.	Bot. reg. t. 459.
2 pyrifórmis. S.S.	(*ro.*)	Pear-fruited.	Peru.	1825.	....	S. ♄.	Feuil. journ. 3. t. 39.
3 citrifórmis. S.S.	(*st.*)	Citron-fruited.	Guinea.	——	....	S. ♄.	
4 cauliflòra. S.S.	(*st.*)	stem-flowering.	Caracas.	1806.	6. 8.	S. ♄.	Jacq. schœnb. 3. t. 311.
5 microcárpa. S.S.	(*st.*)	small-fruited.	————	——	——	S. ♄.	———— 3. t. 309. 310.
6 monoíca. Df.	(*st.*)	monœcious.	————	1816.	——	S. ♄.	
7 digitàta. S.S.	(*st.*)	spiny.	Guiana.	1822.	....	S. ♄.	Aubl. guian. 2. t. 246.
spinòsa. W.							

ORDO XCVI.

PASSIFLOREÆ. *DC. prodr. v. 3. p.* 321.

Tribus I. *PAROPSIEÆ.* DC. prodr. v. 3. p. 322.

SMEATHMA'NNIA. L.T.		SMEATHMA'NNIA.	Polyandria Pentagynia.				
lævigàta. L.T.	(*wh.*)	smooth.	Sierra Leone.	1822.	....	S. ♄.	

Tribus II. *PASSIFLOREÆ VERÆ.* DC. prodr. v. 3. p. 322.

PASSIFL'ORA. DC.		PASSION-FLOWER.	Monadelphia Pentandria.				
1 holoserícea. DC.	(*st.*)	silky-leaved.	Vera Cruz.	1733.	5. 8.	S. ♄.‿.	Bot. reg. t. 59.
2 pállida. DC.	(*gr.*)	pale-flowered.	W. Indies.	1820.	5. 10.	S. ♄.‿.	———— t. 660.
3 cùprea. DC.	(*co.*)	copper-colour'd.	Bahama Isl.	1724.	7. 8.	S. ♄.‿.	Jacq. ic. 3. t. 606.
4 multiflòra. DC.	(*wh.*)	many-flowered.	Vera Cruz.	1731.	6. 9.	S. ♄.‿.	Cavan. diss. 10. t. 272.
5 pubéscens. DC.	(*wh.*)	pubescent.	Caracas.	1825.	——	S. ♄.‿.	
6 mexicàna. DC.	(*wh.*)	Mexican.	Mexico.	1812.	——	S. ♄.‿.	Jus. an. mus. 6. t. 38. f. 2.
7 normàlis. DC.	(*cr.*)	linear-lobed.	Jamaica.	1771.	5. 7.	S. ♄.‿.	
8 angustifòlia. DC.	(*gr.*)	narrow-leaved.	W. Indies.	1773.	6. 9.	S. ♄.‿.	Bot. reg. t. 188.
heterophy'lla. Jacq. schœn. 2. t. 181.							
9 maculàta. DC.	(*wh.*)	painted-leaved.	N. America.	1812.	——	G. ♄.‿.	Pluk. t. 210. f. 3.
10 grácilis. DC.	(*wh.*)	slender.	————	1823.	6. 10.	S. ⊙.‿.	Bot. reg. t. 870.
11 lutèa. DC.	(*st.*)	pale yellow.	————	1714.	5. 8.	F. ♃.‿.	———— t. 79.
12 mínima. DC.	(*gr.*)	small.	S. America.	1690.	7. 9.	S. ♄.‿.	———— t. 144.
13 hirsùta. DC.	(*gr.*)	hairy-stalked.	W. Indies.	1778.	8. 10.	S. ♄.‿.	Plum. amer. t. 88.
14 suberòsa. DC.	(*wh.*)	cork-barked.	————	1759.	6. 9.	S. ♄.‿.	Sm. exot. bot. 1. t. 28.

15 peltàta. DC. (*st.*) peltate. W. Indies. 1778. 8. 10. S. ♄.‿. Bot. reg. t. 507.
16 hederàcea. DC. (*wh.*) Ivy-leaved. Antilles. .... 6. 7. S. ♄.‿. Plum. amer. t. 84.
17 pannòsa. DC. (*fl.*) cloth-leaved. W. Indies. 1824. 6. 9. S. ♄.‿.
18 cuneàta. W.en. (*wh.*) wedge-leaved. S. America. 1820. 6. 10. S. ♄.‿.
19 multifórmis. DC. (*wh.*) multiform. Caracas. 1825. —— S. ♄.‿. Jacq. frag. t. 67. f. 1.
caracasàna. W. en.
20 heterophy'lla. DC. (*wh.*) various-leaved. St. Domingo. 1817. —— S. ♄.‿. Plum.ed.Bur.t.139.f.1.
21 rubricaùlis. Jac. (*re.*) red-stalked. S. America. 1821. 6. 8. S. ♄.‿.
22 perfoliàta. DC. (*pu.*) perfoliate. W. Indies. 1806. 7. 9. S. ♄.‿. Bot. reg. t. 78.
23 rùbra. DC. (*fl.*) red-fruited. ———— 1731. 4. 9. S. ♄.‿. ———— t. 95.
24 capsulàris. DC. (*wh.*) angular-fruited. ———— 1826. —— S. ♄.‿. Bot. mag. t. 2868?
25 biflòra. DC. (*wh.*) crescent-leaved. ———— 1733. —— S. ♄.‿. Sm. ic. pict. 1. t. 1.
lunàta. Sm. *non* Juss. *Vespertìlio.* Miss Lawr. pass. t. 8. *nec aliorum.*
β? acutilòba. (*wh.*) *acute-lobed.* W. Indies. —— —— S. ♄.‿. Bot. reg. t. 577.
26 Vespertìlio. DC. (*wh.*) bat-winged. ———— 1732. 5. 6. S. ♄.‿. Dill. elth. t. 137. f. 164.
27 Maximiliàna. DC. (*wh.*) two-coloured. Brazil. 1820. 6. 10. S. ♄.‿. Bot. reg. t. 597.
Vespertìlio. B.R. *nec aliorum. discolor.* Lodd. bot. cab. t. 565.
28 oblongàta. DC. (*wh.*) oblong-leaved. Jamaica. 1822. —— S. ♄.‿.
29 tuberòsa. DC. (*pa.*) tuberous-root'd. S. America. 1784. —— S. ♄.‿. Bot. reg. t. 432.
punctàta. Lodd. bot. cab. t. 101.
30 rotundifòlia. DC. (*wh.*) round-leaved. W. Indies. 1779. 5. 8. S. ♄.‿. Cavan. dis. t. 290.
31 punctàta. DC. (*st.*) dotted-leaved. Peru. 1784. 5. 6. S. ♄.‿. ———— 6. t. 269.
32 serratifòlia. DC. (*pu.*) notch-leaved. W. Indies. 1731. 5. 10. S. ♄.‿. Bot mag. t. 651.
33 coccìnea. DC. (*sc.*) scarlet-flower'd. Guiana. 1823. .... S. ♄.‿. Aubl. guian. 3. t. 324.
34 mucronàta. DC. (*v.*) mucronate. Brazil. 1822. .... S. ♄.‿. Cavan. diss. t. 282.
35 malifórmis. DC. (*v.*) Apple-fruited. W. Indies. 1731. 7. 11. S. ♄.‿. Bot. reg. t. 94.
36 tiliæfòlia. DC. (*wh.*) Lime-tree-l'd. Peru. 1822. .... S. ♄.‿. Feuill. obs. 2. t. 12.
37 ligularis. DC. (*v.*) taper-pointed. ———— —— 4. 8. S. ♄.‿. Juss. an. mus. 6. t. 40.
38 quadrangulàris. DC. (*v.*) square-stalked. Jamaica. 1768. 8. 10. S. ♄.‿. Bot. reg. t. 14.
39 alàta. DC. (*v.*) wing-stalked. W. Indies. 1772. 4. 8. S. ♄.‿. Bot. mag. t. 66.
40 álbida. DC. (*wh.*) white-flowered. Brazil. 1816. 7. 11. S. ♄.‿. Bot. reg. t. 677.
41 angulàta. Cerv. (*wh.*) angular-stalked. Mexico. 1823. —— G. ♄.‿.
42 laurifòlia. DC. (*v.*) Laurel-leaved. W. Indies. 1690. 6. 8. S. ♄.‿. Bot. reg. t. 13.
43 tinifòlia. DC. (*v.*) Tinus-leaved. ———— 1822. —— S. ♄.‿. Juss. an. mus. 6. t. 41. f. 2.
44 racemòsa. DC. (*sc.*) racemed. Brazil. 1816. 1. 12. S. ♄.‿. Bot. reg. t. 285.
45 sanguìnea. DC. (*cr.*) crimson. Hybrid. 1820. —— G. ♄.‿.
46 cœrùleo-racemòsa. (*pu.*) Whitley's mule. ———— —— —— G. ♄.‿. Lodd. bot. cab. t. 573.
α quinquelobàta. H.T. (*pu.*) *five-lobed.* ———— —— —— G. ♄.‿. Hort. trans. 4. t. 9.
β trilobàta. H.T. (*pu.*) *three-lobed.* ———— —— —— G. ♄.‿.
γ racemífera. (*pu.*) *raceme-bearing.* ———— —— —— G. ♄.‿.
47 alàto-cœrùlea. DC. (*ro.*) Masters's mule. Hybrid. 1823. 4. 8. G. ♄.‿. Bot. reg. t. 848.
48 stipulàta. DC. (*wh.*) glaucous-leaved. Cayenne. 1779. 8. 9. S. ♄.‿. Aubl. guian. 2. t. 325.
glaùca. Bot. reg. t. 88. *non* Humb.
49 incarnàta. DC. (*bh.*) flesh-coloured. N. America. 1629. 7. 9. H. ♄.‿. Miss Lawr. pass. ic.
β? integrilòba. DC. (*bh.*) *entire-lobed.* ———— 1812. —— H. ♄.‿. Bot. reg. t. 332.
50 èdulis. DC. (*bh.*) eatable-fruited. Brazil. 1816. 6. 10. G. ♄.‿. Bot. mag. t. 1989.
incarnàta. β. Bot. reg. t. 152.
51 vitifòlia. DC. (*wh.*) Vine-leaved. S. America. 1824. .... S. ♄.‿.
52 filamentòsa. DC. (*v.*) filamentose. ———— 1819. 7. 10. G. ♄.‿. Cavan. diss. 10. t 294.
53 palmàta. DC. (*v.*) palmate-leaved. Brazil. —— —— F. ♄.‿. Lodd. bot. cab. t. 97.
54 Colvíllii. DC. (*v.*) Colvill's mule. Hybrid. 1823. —— H. ♄.‿. Swt. br. fl. gar. t. 126.
55 cœrùlea. DC. (*wh.bl.*) common. Brazil. 1699. —— H. ♄.‿. Bot. reg. t. 488.
β glaucophy'lla. (*wh.bl.*) glaucous-leaved. ———— .... —— H. ♄.‿.
γ angustifòlia. (*wh.bl.*) narrow long-l'd. ———— .... —— H. ♄.‿.
56 chinénsis. (*wh.bl.*) Chinese. China. 1819. 6. 10. G. ♄.‿.
cœrùlea Lour. *nec aliorum.*
57 serrata. DC. (*pa.*) saw-leaved. Martinica. 1800. .... S. ♄.‿. Cavan. diss. 10. t. 296.
58 digitàta. R.P. (*pa.*) digitate. Peru. 1824. .... S. ♄.‿.
59 pedàta. DC. (*pa.*) curl-flowered. W. Indies. 1781. .... S. ♄.‿. Plum. am. t. 81.
60 picturàta. DC. (*v.*) pictured. Brazil. 1823. 6. 9. S. ♄.‿. Bot. reg. t. 673.
61 hibiscifòlia. (*wh.bl.*) Hibiscus-leav'd. Caribees. 1825. —— S. ♄.‿. Plum. am. t. 86.
62 fœtida. (*wh.bl.*) stinking. W. Indies. 1731. 7. 9. S. ♂.‿. Bot. mag. t. 2619.
63 hircìna. ed. 1. (*wh.bl.*) strong-scented. Brazil. .... 5. 10. S. ♄.‿. Bot. reg. t. 321.
fœtida. Bot. reg. *nec aliorum. hirsùta.* Bot. cab. 138. *nec aliorum.*
64 ciliàta. DC. (*wh.ro.*) ciliated. Jamaica. 1783. 7. 9. S. ♄.‿. Bot. mag. t. 288.

DISE'MMA. DC. DISE'MMA. Monadelphia Pentandria.
1 Herbertiàna. DC. (*st.*) L'd.Caernarvon's. N. Holland. 1822. 6. 10. G. ♄.‿. Bot. reg. t. 737.
Passiflòra Herbertiàna. B.R.
2 adianthifòlia. DC. (*or.*) Adiantum-l'd. Norfolk Isl. 1792. —— G. ♄.‿. Bot. reg. t. 233.
Passiflòra aurántia. Andr. reposit. t. 295. *Passiflòra glábra.* Wendl. coll. 1. t. 17.

MURUC'UIA. DC. Muruc'uia. Monadelphia Pentandria.
ocellàta. DC. (*cr.*) ocellated. W. Indies. 1739. 7. 8. S. ♄.⌣. Bot. reg. t. 574.
Passiflòra Murucùja. B.R. Lin. am. 1. t. 10. f. 10.
TACS'ONIA. DC. Tacs'onia. Monadelphia Pentandria.
1 pinnatistìpula. DC. (*ro.*) feather-stipuled. Chile. 1826. .... S. ♄.⌣. Cavan. ic. 5. t. 428.
2 pedunculàris. DC. (*ro.*) long-peduncled. Peru. 1815. .... S. ♄.⌣. ——— 5. t. 426.
Passiflòra pedunculàris. Cav.
MODE'CCA. DC. Mode'cca. Diœcia Monadelphia.
1 lobàta. DC. (*gr.*) lobe-leaved. Sierra Leone. 1816. 6. 9. S. ♄.⌣. Bot. reg. t. 433.
2 trilobàta. H.B. (*gr.*) three-lobed. E. Indies. 1823. —— S. ♄.⌣. Roxb. corom. 3. ic.
3 tuberòsa. H.B. (*gr.*) tuberous-rooted. ——— 1822. —— S. ♄.⌣. Rumph. amb. 8. t. 20.

ORDO XCVII.

LOASEÆ. *DC. prodr. v.* 3. *p.* 339.

BART'ONIA. DC. Bart'onia. Polyandria Monogynia.
1 ornàta. DC. (*wh.*) naked-seeded. Missouri. 1811. 7. 9. F. ♂. Bot. mag. t. 1487.
decapétala. B.M.
2 nuda. DC. (*wh.*) winged-seeded. ——— —— —— F. ♃.
BLUMENBA'CHIA. DC. Blumenba'chia. Polyandria Monogynia.
insígnis. DC. (*wh.*) elegant. Chile. 1826. 7. 10. H. ⊙. Swt. br. fl. gar. t. 171.
Loàsa palmàta. Treviranus. *pátula.* Gr.
SCYPHA'NTHUS. B.F.G. Cup-Flower. Polyandria Monogynia.
élegans. B.F.G. (*ye.*) elegant. Chile. 1824. 5. 9. F. ♄.⌣. Swt. br. fl. gar. t. 238.
LO'ASA. DC. Lo'asa. Polyandria Monogynia.
1 bryoniæfòlia. DC. (*ye.*) Bryony-leaved. Chile. 1828. 6. 9. H. ⊙. Schrad. pl. rar. gœt. ic.
2 trícolor. DC. (*ye.*) three-coloured. ——— 1822. 8. 10. H. ⊙. Bot. reg. t. 667.
3 nítida. DC. (*ye.*) glossy-leaved. Peru. —— 7. 10. H. ⊙. Hook. exot. bot. t. 83.
4 Plàcei. H.T. (*ye.*) Place's. Chile. —— —— H. ⊙. Bot. reg. t. 785.
acanthifòlia. B.R. *non* Lamarck.
5 grandiflòra. DC. (*ye.*) large-flowered. Peru. 1825. —— H. ⊙. Juss. an. mus. 5. t. 4. f. 2.
MENTZ'ELIA. DC. Mentz'elia. Polyandria Monogynia.
1 áspera. DC. (*or.*) rough. W. Indies. 1773. 7. 9. S. ⊙. Plum. ed. Bur. t. 174. f. 1.
2 oligospérma. DC. (*or.*) few-seeded. Louisiana. 1812. 5. 7. F. ♃. Bot. mag. t. 1760.
3 híspida. DC. (*or.*) hispid. Mexico. 1824. 6. 9. F. ♃. Cavan. ic. 1. t. 70.

ORDO XCVIII.

TURNERACEÆ. *DC. prodr. v.* 3. *p.* 345.

TURN'ERA. DC. Turn'era. Pentandria Trigynia. L.
1 ulmifòlia. DC. (*ye.*) Elm-leaved. Jamaica. 1733. 6. 9. S. ♄. Linn. h. cliff. t. 10.
2 angustifòlia. B.M. (*ye.*) narrow-leaved. ——— —— 4. 9. S. ♄. Bot. mag. t. 281.
3 cuneifórmis. DC. (*ye.*) wedge-leaved. Brazil. 1821. —— S. ♄.
4 trioniflòra. DC. (*ye.*) Ketmia-flow'd. Trinidad. 1812. —— S. ♄. Bot. mag. t. 2106.
élegans. R.S.
5 acùta. DC. (*ye.*) acute-leaved. Jamaica. 1826. —— S. ♄.
6 Pumílea. DC. (*ye.*) nettle-leaved. ——— 1796. 8. 9. S. ⊙. Sloan. hist. t. 127. f. 6.
7 rupéstris. DC. (*ye.*) rock. Guiana. 1826. 6. 9. S. ♄. Aubl. guian. 1. t. 113. f. 1.
8 guianénsis. DC. (*ye.*) Guiana. ——— —— —— S. ⊙. ——— 1. t. 114.
9 cistoídes. DC. (*ye.*) Cistus-like. America. 1774. 6. 10. H. ⊙. Plum. ic. t. 150. f. 1.
10 racemòsa. DC. (*ye.*) raceme-flower'd. St. Domingo. 1789. 7. 10. S. ⊙. Jacq. vind. 3. t. 94.
PIRIQU'ETA. DC. Piriqu'eta. Pentandria Trigynia.
villòsa. DC. (*ye.*) villous. Guiana. 1826. 7. 10. S. ⊙. Aubl. guian. 1. t. 117.

ORDO XCIX.

PORTULACEÆ. *DC. prodr. v.* 3. *p.* 351.

TRIA'NTHEMA. DC.	TRIA'NTHEMA. Pent-Decandria Mono-Trigynia.					
1 decándra. DC.	(*gr.*) trailing.	E. Indies.	1762.	7. 8.	S. ⊙.	Burm. ind. t. 31. f. 3.
2 triquètra. DC.	(*gr.*) triquetrous.	———	1826.	——	S. ⊙.	
3 obcordàta. H.B.	(*gr.*) obcordate.	———	1820.	——	S. ⊙.	
4 monógyna. DC.	(*gr.*) Purslane-leav'd.	Jamaica.	1716.	——	S. ⊙.	DC. plant. gras. t. 109.
PORTUL'ACA. DC.	PURSLANE. Oct-Dodecandria Monogynia.					
1 oleràcea. H.S.	(*ye.*) small.	Europe.	1562.	6. 8.	H. ⊙.	DC. plant. gras. t. 123.
2 satìva. H.S.	(*ye.*) garden.	America.	1652.	6. 9.	H. ⊙.	
β aùrea. H.S.	(*ye.*) *golden.*	———	——	——	H. ⊙.	
3 parvifòlia. H.S.	(*ye.*) small-leaved.	Jamaica.	1799.	8. 10.	H. ⊙.	
4 foliòsa. DC.	(*ye.*) leafy.	Guinea.	1822.	6. 8.	S. ♃.	Bot. reg. t. 793.
5 involucràta. H.R.	(*ye.*) leafy-flowered.	....	1819.	8. 9.	S. ♄.	
6 marginàta. DC.	(*ye.*) red-margined.	Caracas.	1822.	6. 10.	H. ⊙.	
7 halimoìdes. DC.	(*ye.*) Halimus-leaved.	Jamaica.	1823.	8. 9.	S. ⊙.	Sloan. hist. 1. t. 129. f. 3.
8 mucronàta. DC.	(*ye.*) mucronate.	....	1827.	6. 9.	S. ⊙.	
9 quadrífida. DC.	(*ye.*) quadrifid.	Asia.	1773.	8. 9.	S. ⊙.	Jacq. coll. 2. t. 17. f. 4.
10 meridiàna. H.S.	(*ye.*) noon-day.	E. Indies.	1791.	5. 6.	S. ⊙.	Rheed. mal. 10. t. 31.
11 grandiflòra. B.M.	(*va.*) large-flowered.	Chile.	1828.	6. 8.	F. ♃.	Bot. mag. t. 2885.
a purpùrea.	(*pu.*) *purple-flowered.*	———	——	——	F. ♃.	——— t. 2885. f. 2.
β aurantìaca.	(*or.*) *orange-coloured.*	———	——	——	F. ♃.	——— t. 2885. f. 1.
12 pilòsa. DC.	(*pu.*) hairy.	W. Indies.	1690.	——	S. ♄.	Bot. reg. t. 792.
13 setàcea. H.S.	(*pu.*) bristly.	———	——	——	S. ⊙.	Pluk. phyt. t. 246. f 6.
ANACA'MPSEROS. DC.	ANACA'MPSEROS. Dodecandria Monogynia.					
1 telephiástrum. DC.	(*re.*) round-leaved.	C. B. S.	1732.	7. 9.	D.I. ♄.	DC. plant. gras. t. 3.
Portulàca Anacámpseros. L. *Talìnum Anacámpseros.* W. *Rulíngia Anacámpseros.* H.S.						
2 arachnoìdes. DC.	(*bh.*) cobweb.	C. B. S.	1790.	7. 9.	D.I. ħ.	Bot. mag. t. 1368.
3 vàrians.	(*re.*) various-leaved.	———	1812.	——	D.I. ♄.	
Rulíngia vàrians. H.S.S.						
4 rùbens. DC.	(*re.*) red-leaved.	———	1796.	——	D.I. ♄.	
5 filamentòsa. DC.	(*ro.*) thready.	———	1795.	——	D.I. ħ.	Bot. mag. t. 1367.
6 lanceolàta. DC.	(*ro.*) spear-leaved.	———	1796.	——	D.I. ♃.	
7 angustifòlia. DC.	(*ro.*) narrow-leaved.	———	1819.	——	D.I. ♄.	
8 rufèscens. DC.	(*bh.*) rufescent.	———	1818.	——	D.I. ♄.	
9 trigòna. DC.	(*bh.*) trigonal-leaved.	———	——	——	D.I. ♄.	Burm. afr. t. 30. f. 2.
10 lanígera. DC.	(*ro.*) woolly.	———	——	——	D.I. ♄.	
11 intermèdia. H.P.	(*ro.*) intermediate.	———	1823.	——	D.I. ♄.	
12 polyphy'lla. DC.	(*ro.*) many-leaved.	———	——	——	D.I. ♄.	
Rulíngia polyphy'lla. H.S.S.						
TAL'INUM. DC.	TAL'INUM. Dodecandria Monogynia.					
1 teretifòlium. DC.	(*ro.*) round-leaved.	N. America.	1820.	7. 8.	H. ♃.	Lodd. bot. cab. t. 819.
2 crassifòlium. DC.	(*ro.*) thick-leaved.	Caribees.	1800.	8. 9.	S. ♄.	Jacq. vind. 3. t. 52.
3 fruticòsum. W.	(*wh.*) white-flowered.	S. America.	....	——	S. ♄.	Comm. hort. 1. t. 4.
4 triangulàre. DC.	(*ye.*) triangular.	W. Indies.	1739.	——	S. ♄.	Jacq. obs. 1. t. 23.
5 pátens. DC.	(*re.*) spreading-fl'd.	S. America.	1776.	8. 10.	S. ♂.	Jacq. vind. 2. t. 151.
paniculàtum. Gært. fruct. 2. t. 128. *Portulàca paniculàta.* Jacq. amer. *non* R.P.						
6 refléxum. DC.	(*ye.*) reflex-flowered.	S. America.	1800.	8. 10.	S. ♂.	Bot. mag. t. 1543.
7 cuneifòlium. DC.	(*re.*) wedge-leaved.	Arabia.	1823.	——	S. ♄.	
8 lineàre. K.S.	(*ye.*) linear-leaved.	Mexico.	1827.	——	H. ⊙.	
9 purpùreum. F.	(*pu.*) purple.	....	——	——	H. ⊙.	
CALANDRI'NIA. DC.	CALANDRI'NIA. Dodecandria Monogynia.					
1 paniculàta. DC.	(*ro.*) panicled.	S. America.	1812.	8. 10.	H. ⊙.	
2 pilosiúscula. DC.	(*ro.*) ciliated.	Chile.	1823.	——	H. ⊙.	Hook. ex. flor. 1. t. 82.
Talìnum ciliàtum. Hooker *non* R.P. ex DC.						
3 Phacospérma. DC.	(*ro.*) Peruvian.	Peru.	1827.	——	G. ♂.	
4 caulèscens. DC.	(*ro.*) caulescent.	Mexico.	——	——	H. ⊙.	H. et B. n. ge. am. 6. t. 526.
5 compréssa. DC.	(*ro.*) compressed.	Chile.	1826.	——	H. ⊙.	
β adscéndens. Otto.	(*ro.*) *ascending*	———	——	——	H. ⊙.	
6 glaùca. DC.	(*ro.*) glaucous-leaved.	———	1827.	——	D.G. ♄.	
7 grandiflòra. B.R.	(*ro.*) large-flowered.	———	1826.	6. 10.	D.G. ♄.	Bot. reg. t. 1194.
8 Andre'wsii.	(*ro.*) Andrews's.	W. Indies.	1812.	8. 10.	D.G. ♄.	Andr. rep. t. 253.
Talìnum pátens. A.R. *nec aliorum.*						
9 Lockhárti.	(*ro.*) Lockhart's.	Trinidad.	1825.	6. 9.	D.G. ♄.	
Talìnum grandiflòrum. Lockhart.						

PORTULAC'ARIA. DC.		PURSLANE-TREE. Pentandria Trigynia.					
A'fra. DC.	(*ro.*)	African.	Africa.	1732.	6. 9.	D.G. ♄.	DC. plant. gras. t. 132.
CLAYT'ONIA. DC.		CLAYT'ONIA. Pentandria Monogynia. L.					
1 perfoliàta. DC.	(*wh.*)	small-flowered.	N. America.	1794.	5. 8.	H. ☉.	Bot. mag. t. 1336.
cubénsis of Bonp.an.mus.7.t.6. is a very different species,according to G.Don,who saw it growing wild.							
2 alsinoídes. DC.	(*wh.*)	chickweed-like.	Nootka-sound.	1794.	3. 6.	H. ☉.	Bot. mag. t. 1309.
3 lanceolàta. DC.	(*wh.*)	spear-leaved.	N. America.	1812.	3. 5.	H. ♃.	Pursh. fl. amer. 1. t. 3.
4 sibírica. DC.	(*ro.*)	Siberian.	Siberia.	1768.	5. 8.	H. ☉.	Swt. br. fl. gar. t. 16.
5 acutifòlia. DC.	(*wh.*)	acute-leaved.	———	1827.	....	H. ♃.	
6 Vestiàna. F.	(*ro.*)	Vest's.	Altay.	——	....	H. ♃.	
7 unalaschkénsis. R.S.	(*wh.*)	Unalaschka.	Unalaschka.	1824.	3. 6.	H. ☉.	
8 virginiàna. L.	(*wh.*)	notch-petal'd.	N. America.	1748.	3. 8.	H. ♃.	Pluk. alm. t. 102. f. 3.
9 acutiflòra.	(*wh.*)	acute-petal'd.	———	....	——	H. ♃.	Bot. mag. t. 941.
virgínica. B.M. *non* L. *virginiàna a. acutiflòra.* DC.							
10 grandiflòra. B.F.G.	(*pi.*)	large-flowered.	N. America.	....	——	H. ♃.	Swt. br. fl. gar. t. 216.
11 caroliniàna. M.	(*pi.*)	spatula-leaved.	———	1789.	——	H. ♃.	——— v. 3. t. 208.
MO'NTIA. DC.		WATER CHICKWEED. Triandria Trigynia. L.					
fontàna. DC.	(*wh.*)	common.	Britain.	....	4. 6.	H. ☉.	Eng. bot. v. 17. t. 1206.
GINGI'NSIA. DC.		GINGI'NSIA. Pentandria Trigynia.					
1 brevicaùlis. DC.	(*pi.*)	short-stalked.	C. B. S.	1816.	5. 6.	G. ♄.	DC. m.s.his.n.pa.v.4.ic.
2 elongàta. DC.	(*pi.*)	long-peduncled.	———	1795.	——	G. ♄.	Andr. rep. t. 329?
Pharnàceum incànum. L. *lineàre.* Andr. *non alior.*							
3 aurántia. DC.	(*or.*)	orange-flowered.	C. B. S.	——	——	G. ♄.	Andr. rep. t. 326.
4 álbens. DC.	(*wh.*)	white-flowered.	———	——	——	G. ♄.	——— t. 329?
5 conférta. DC.	(*wh.*)	close-leaved.	———	1782.	5. 10.	G. ♄.	Bot. mag. t. 1883.
Pharnàceum incànum. B.M. *non* L.							
L'IMEUM. W.		L'IMEUM. Heptandria Digynia. L.					
africànum. W.	(*wh.*)	African.	C. B. S.	1774.	6. 7.	G. ♃.	

ORDO C.

PARONYCHIEÆ. *DC. prodr. v.* 3. *p.* 365.

Tribus I. *TELEPHIEÆ.* DC. prodr. v. 3. p. 366.

TEL'EPHIUM. DC.		ORPINE. Pentandria Trigynia. L.					
Imperàti. DC.	(*wh.*)	true.	S. Europe.	1658.	6. 8.	H. ♃.	Schkuhr. handb.1.t.85.
CORRIGI'OLA. DC.		STRAPWORT. Pentandria Trigynia. L.					
1 telephiifòlia. DC.	(*wh.*)	Orpine-leaved.	Spain.	1822.	6. 8.	H. ♃.	
2 littoràlis. DC.	(*wh.*)	sand.	England.	....	7. 8.	H. ♃.	Eng. bot. v. 10. t. 668.

Tribus II. *ILLECEBREÆ.* DC. prodr. v. 3. p. 367.

HERNI'ARIA. DC.		RUPTURE-WORT. Pentandria Digynia. L.					
1 cinèrea. DC.	(*gr.*)	cinereous.	Montpelier.	1816.	7. 8.	H. ☉.	
2 glábra. DC.	(*gr.*)	smooth.	England.	....	——	H. ♃.	Eng. bot. v. 3. t. 206.
3 hirsùta. DC.	(*gr.*)	hairy.	———	....	——	H. ♃.	——— v. 20. t. 1379.
4 incàna. DC.	(*gr.wh.*)	hoary.	S. Europe.	1816.	5. 8.	H. ♃.	Pluk. alm. t. 53. f. 3.
5 Bessèri. DC.	(*gr.*)	Besser's.	Tauria.	1824.	——	H. ♃.	
6 alpìna. DC.	(*gr.*)	alpine.	S. Europe.	1816.	7. 8.	H. ♃.	
7 fruticòsa. DC.	(*gr.*)	frutescent.	Spain.	1814.	5. 8.	H. ♄.	
8 polygonoídes. DC.	(*gr.*)	Knot-grass-like.	S. Europe.	1752.	——	F. ♄.	Cavan. ic. 2. t. 137.
ANY'CHIA. DC.		ANY'CHIA. Pentandria Monogynia.					
1 dichótoma. DC.	(*gr.*)	forked.	N. America.	1806.	5. 8.	H. ♂.	Ort. dec. t. 15. f. 2.
Quèria canadénsis. L.							
2 capillàcea. DC.		smooth-stalked.	———	1824.	——	H. ♂.	
ILLE'CEBRUM. DC.		KNOT-GRASS. Pentandria Monogynia. L.					
verticillàtum. DC.	(*wh.*)	whorled.	England.	....	7. 8.	H. ♃.	Eng. bot. v. 13. t. 895.
CHÆTONY'CHIA. DC.		CHÆTONY'CHIA. Pentandria Monogynia.					
cymòsa. DC.	(*wh.*)	cymose.	S. Europe.	1817.	6. 7.	H. ♃.	Flor. Græc. t. 245.
Illecèbrum cymòsum. L.							
PARONY'CHIA. DC.		PARONY'CHIA. Pentandria Monogynia.					
1 echinàta. DC.	(*wh.*)	bristly.	S. Europe.	1817.	6. 7.	H. ♃.	Schrad. journ. 1801. t. 4.

2 bonariénsis. DC.	(*wh.*)	Buenos Ayres.	BuenosAyres.	1826.	8. 10.	H. ⊙.	
brasiliàna. Lam. dict. 5. p. 23.							
3 arábica. DC.	(*wh.*)	Arabian.	Arabia.	1828.	7. 10.	H. ⊙.	Descr. de l'Egyp. t. 18.
4 polygonifòlia. DC.	(*wh.*)	Polygonum-l[d].	Dauphiny.	1816.	5. 8.	H. ♃.	Vill. delph. 2. t. 16.
5 argéntea. DC.	(*wh.*)	silvery.	S. Europe.	1640.	——	H. ♃.	Flor. græc. t. 246.
hispánica. S.S. *Illécebrum Parony'chia.* L.							
6 itálica. R.S.	(*wh.*)	Italian.	Italy.	1816.	——	H. ♃.	Vill. delph. 2. p. 410. ic.
7 capitàta. DC.	(*wh.*)	capitate.	Spain.	1683.	6. 8.	H. ♃.	Flor. græc. t. 247.
8 serpyllifòlia. DC.	(*wh.*)	Thyme-leaved.	S. Europe.	1820.	——	H. ♃.	Schrad. journ. 1801. t. 4.
9 nívea. DC.	(*wh.*)	villous-leaved.	————	1812.	——	H. ♃.	Barrel. ic. t. 687.
10 tenuifòlia. L.en.	(*wh.*)	slender-leaved.	....	1822.	——	H. ⊙.	
11 alsinefòlia. R.S.	(*wh.*)	Chickweed-leav'd.	Spain.	....	——	H. ⊙.	Scop. del. ins. t. 13.
14 pubéscens. R.S.	(*wh.*)	pubescent.	S. France.	1818.	——	H. ♃.	Schra. jour. 1801. p. 412.

Tribus III. *POLYCARPÆÆ.* DC. prodr. v. 3. p. 373.

POLYCARP'ÆA. DC. POLYCARP'ÆA. Pentandria Monogynia.

1 gnaphalòdes. DC.	(*wh.*)	woolly-leaved.	Canaries.	1818.	6. 8.	F. ♄.	
Hàgea gnaphalòdes. P.S. *Móllia gnaphalòdes.* S.S. *Laháya gnaphalòdes.* R.S.							
2 latifòlia. DC.	(*wh.*)	broad-leaved.	Teneriffe.	1816.	7. 8.	F. ♄.	Schrank pl. rar. t. 29.
3 Teneríffæ. DC.	(*wh.*)	forked.	————	1779.	——	H. ⊙.	W. hort. ber. 1. t. 11.
Móllia diffùsa. W. *Laháya diffùsa.* Schult. syst.							
4 aristàta. DC.	(*wh.*)	bearded.	Canaries.	1780.	6. 7.	F. ♄.	
5 memphítica. DC.	(*wh.*)	Egyptian.	Egypt.	1828.	7. 9.	H. ⊙.	Delil. fl. æg. t. 24. f. 2.
6 carnòsa. DC.	(*wh.*)	fleshy-leaved.	Canaries.	1816.	6. 8.	F. ♄.	
7 Smíthii. DC.	(*wh.*)	Smith's.	————	——	——	F. ♄.	

ORT'EGIA. DC. ORT'EGIA. Triandria Monogynia. L.

1 hispánica. DC.	(*gr.*)	Spanish.	Spain.	1768.	6. 7.	H. ♃.	Cavan. ic. 1. t. 47.
2 dichótoma. DC.	(*gr.*)	forked.	Italy.	1781.	8. 9.	H. ♃.	All. ac. taur. 3. t. 4. f. 1.

POLYCA'RPON. DC. ALL-SEED. Tri-Pentandria Trigynia.

1 tetraphy'llum. DC.	(*wh.*)	four-leaved.	England	....	5. 8.	H. ⊙.	Eng. bot. v. 15. t. 1031.
2 alsinefòlium. DC.	(*wh.*)	Chickweed-l[d].	S. Europe.	....	——	H. ⊙.	Bocc. sic. p. 71. t. 38.
Laháya alsinefòlia. S.							
3 peploídes. DC.	(*wh.*)	Peplis-like.	Sicily.	1818.	6. 8.	H. ♃.	
Laháya polycarpoídes. S. *Móllia polycárpon.* S.S.							

Tribus IV. *POLLICHIEÆ.* DC. prodr. v. 3. p. 377.

POLLI'CHIA. DC. POLLI'CHIA. Monandria Monogynia.

campéstris. DC.	(*wh.*)	whorl-leaved.	C. B. S.	1780.	8. 10.	G. ♂.	Smith spicil. 1. t. 1.

Tribus V. *SCLERANTHEÆ.* DC. prodr. v. 3. p. 377.

MNI'ARUM. DC. MNI'ARUM. Monandria Digynia.

1 biflòrum. DC.	(*gr.*)	two-flowered.	N. Holland.	1818.	5. 7.	G. ♃.	Forst. gen. 1. t. 1.

SCLER'ANTHUS. DC. KNAWEL. Decandria Digynia. L.

1 ánnuus. DC.	(*gr.*)	annual.	Britain.	....	7. 8.	H. ⊙.	Eng. bot. v. 5. t. 351.
2 perénnis. DC.	(*gr.*)	perennial.	————	....	8. 9.	H. ♃.	———— v. 5. t. 352.

Tribus VI. *QUERIACEÆ.* DC. prodr. v. 3. p. 379.

QU'ERIA. DC. QU'ERIA. Triandria Trigynia. L.

hispánica. DC.	(*gr.*)	Spanish.	Spain.	1800.	5. 9.	H. ⊙.	Quer. fl. esp. 6. t. 15. f. 2.

Tribus VII. *MINUARTIEÆ.* DC. prodr. v. 3. p. 379.

MINUA'RTÍA. DC. MINUA'RTIA. Tri-Decandria Trigynia.

1 dichótoma. DC.	(*wh.*)	forked.	Spain.	1771.	6. 7.	H. ⊙.	Lœfl. it. p. 121. t. 1. f. 3.

2 montàna. DC.	(*wh.*)	mountain.	Spain.	1806.	6. 7.	H. ☉.	Lœfl. it. p. 122. t. 1. f. 4.
3 campéstris. DC.	(*wh.*)	field.	———	——	——	H. ☉.	Act. stock. 1758. t. 1. f. 3.
LŒFLI'NGIA. DC.		LŒFLI'NGIA.	Triandria Monogynia. L.				
hispánica. DC.	(*wh.*)	Spanish.	Spain.	1770.	6. 7.	H. ☉.	Cavan. ic. 1. t. 94.

ORDO CI.

CRASSULACEÆ. *DC. prodr. v.* 3. *p.* 381.

Tribus I. *CRASSULEÆ.* DC. prodr. v. 3. p. 381.

TILL'ÆA. DC.		TILL'ÆA.	Tetrandria Tetragynia. L.				
1 muscòsa. DC.	(*wh.*)	mossy.	England.	....	5. 7.	H. ☉.	Eng. bot. v. 2. t. 116.
2 moschàta. DC.	(*wh.*)	musk-scented.	Magellan.	1794.	5. 11.	H. ☉.	
Crássula moschàta. Forster.							
3 verticillàris. DC.	(*wh.*)	whorled.	N. S. W.	1823.	6. 9.	H. ☉.	
BULLIA'RDA. DC.		BULLIA'RDA.	Tetrandria Tetragynia.				
1 Vaillántii. DC.	(*wh.*)	Vaillant's.	France.	1825.	6. 8.	H. *w.* ☉.	DC. plant. gras. t. 74.
2 aquática. DC.	(*wh.*)	water.	N. Europe.	1816.	——	H. *w.* ☉.	
DASYST'EMON. DC.		DASYST'EMON.	Pentandria Pentagynia.				
calycìnum. DC.	(*wh.*)	leafy-calyxed.	N. Holland.	1823.	6. 9.	H. ☉.	
SE'PTAS. DC.		SE'PTAS.	Heptandria Heptagynia. L.				
1 capénsis. DC.	(*wh.*)	Cape.	C. B. S.	1774.	8. 9.	D.G. ♃.	Andr. rep. t. 98.
2 globífera. B.M.	(*wh.*)	globe-flowering.	———	1809.	3. 5.	D.G. ♃.	Bot. mag. t. 1472.
3 Umbélla. DC.	(*wh.*)	skreen.	———	1800.	7. 8.	D.G. ♃.	Jacq. ic. t. 352.
Crássulu Umbélla. Jacq.							
CRA'SSULA. H.R.		CRA'SSULA.	Pentandria Pentagynia. L.				
1 arboréscens. DC.	(*pi.*)	arborescent.	C. B. S.	1730.	5. 6.	D.G. ♄.	Bot. mag. t. 384.
Cotylèdon. B.M.							
2 portulàcea. DC.	(*ro.*)	oblique-leaved.	———	1759.	4. 5.	D.G. ♄.	DC. plant. gras. t. 79.
oblíqua. H.S.							
3 láctea. H.S.	(*wh.*)	milk white.	———	1774.	9. 2.	D.G. ♄.	Bot. mag. t. 1771.
4 telephioídes. H.R.	(*ro.*)	Orpine-like.	———	1820.	7. 8.	D.G. ♄.	
5 ramòsa. H.S.	(*ro.*)	branching.	———	1774.	——	D.G. ♄.	
6 revólvens. H.P.	(*wh.*)	recurved-leaved.	———	1820.	8. 10.	D.G. ♄.	
7 tetragòna. H.S.	(*wh.*)	square-leaved.	———	1711.	——	D.G. ♄.	DC. plant. gras. t. 19.
8 biplanàta. H.P.	(*wh.*)	flat-leaved.	———	1820.	——	D.G. ♄.	
9 acutifòlia. H.S.	(*wh.*)	acute-leaved.	———	1795.	9. 11.	D.G. ♄.	DC. plant. gras. t. 2.
10 bibracteàta. H.P.	(*wh.*)	double-bracted.	———	1820.	8. 10.	D.G. ♄.	
α minor. H.P.	(*wh.*)	*smaller.*	———	——	——	D.G. ♄.	
β major. H.P.	(*wh.*)	*larger.*	———	——	——	D.G. ♄.	
11 filicaùlis. H.P.	(*wh.*)	thready-stem'd.	———	——	——	D.G. ♄.	
12 scábra. H.R.	(*wh.*)	rough-leaved.	———	1730.	6. 7.	D.G. ♄.	Dill. elth. t. 99. f. 117.
13 scabrélla. H.R.	(*wh.*)	patent-leaved.	———	1820.	6. 9.	D.G. ♄.	
14 squamulòsa. H.R.	(*wh.*)	small-scaly.	———	1817.	——	D.G. ♄.	
15 bullulàta. H.S.	(*ye.*)	rough yellow.	———	1800.	6. 7.	D.G. ♄.	Mart. cent. t. 24.
16 pruinòsa. DC.	(*wh.*)	frosted.	———	1821.	——	D.G. ♄.	
17 columnàris. H.S.	(*wh.*)	columnar.	———	1789.	7. 10.	D.G. ♄.	Burm. afr. t. 9. f. 2.
18 lycopodioídes. DC.	(*wh. pu.*)	whip-cord.	———	1760.	6. 7.	D.G. ♄.	
imbricàta. H.K.							
19 ericoídes. H.P.	(*wh.*)	large whip-cord.	———	1820.	9. 10.	D.G. ♄.	
20 vestìta. DC.	(*ye.*)	powdered-leav'd.	———	....	6. 10.	D.G. ♄.	
21 argéntea. H.S.	(*wh.*)	silvery-leaved.	———	1820.	7. 9.	D.G. ♄.	
22 perfóssa. DC.	(*wh.*)	threaded.	———	1785.	8. 9.	D.G. ♄.	DC. plant. gras. t. 25.
perfilàta. H.S. Scop. del. ins. 3. t. 6. c.							
23 perforàta. DC.	(*wh.*)	perfoliate.	———	——	——	D.G. ♄.	
24 marginàlis. DC.	(*wh.*)	red-margined.	———	1774.	7. 8.	D.G. ♄.	Jacq. schœnb. 4. t. 471.
25 pellùcida. H.S.	(*bh.*)	pellucid.	———	1732.	6. 9.	D.G. ♄.	Dill. elth. t. 100. f. 119.
26 centauroídes. DC.	(*ro.*)	Centaury-flow'd.	———	1774.	5. 7.	D.G. ♄.	Bot. mag. t. 1765.
27 cordàta. DC.	(*ro.*)	heart-leaved.	———	——	5. 8.	D.G. ♄.	DC. plant. gras. 2. t. 121.
28 spathulàta. DC.	(*ro.*)	notch-leaved.	———	——	7. 9.	D.G. ♄.	——— 1. t. 49.
29 orbiculàris. H.S.	(*bh.*)	starry.	———	1731.	——	D.G. ♃.	——— 1. t. 43.
30 rosulàris. H.R.	(*wh.*)	great starry.	———	1819.	——	D.G. ♃.	
31 glomeràta. H.S.	(*wh.*)	rough clustered.	———	1774.	8. 10.	H. ☉.	DC. plant. gras. 1. t. 57.

32 glábra. H.R. (*wh.*) smooth-cluster'd. C. B. S. 1774. 6. 10. H. ⊙.
33 expánsa. H.S. (*wh.*) awl-leaved. ——— —— 6. 8. H. ⊙.
34 diffùsa. H.S. (*wh.*) diffuse. ——— —— —— H. ⊙.
35 subulàta. H.S. (*wh.*) awl-leaved. ——— 1820. —— H. ⊙. Petiv. gaz. t. 89. f. 8.
36 lineolàta. H.S. (*wh.*) channelled. ——— 1774. —— D.G. ♂.
37 biconvéxa. H.S. (*wh.*) doubly convex. ——— 1800. 8. 9. D.G. ♄.
38 pulchélla. H.R. (*ye.*) pretty. ——— 1778. 7. 8. H. ⊙.
39 spársa. H.S. (*st.*) alternate-leaved. ——— 1774. 6. 9. H. ⊙.

PURG'OSEA. H.P. PURG'OSEA. Pentandria Pentagynia.

1 linguæfòlia. H.R. (*wh.*) tongue-leaved. C. B. S. 1803. 8. 9. D.G. ♃.
2 ciliàta. DC. (*st.*) ciliated. ——— 1732. 7. 8. D.G. ♃. DC. plant. gras. t. 7.
3 concínna. (*wh.*) small neat. ——— 1820. —— D.G. ♄.
Crássula concínna. H.R.
4 ligulifòlia. (*wh.*) latchet-leaved. ——— —— —— D.G. ♄.
Crássula ligulifòlia. H.R.
5 concinnélla. DC. elegant. ——— 1822. —— D.G. ♄.
6 tomentòsa. H.R. (*wh.*) tomentose. ——— 1790. 4. 5. D.G. ♄.
7 cotyledònis. DC. (*wh.*) Cotyledon-like. ——— 1800. 7. 9. D.G. ♃.
8 conspícua. (*wh.*) conspicuous. ——— 1820. —— D.G. ♃.
Crássula conspícua. H.R.
9 corymbulòsa. DC. (*wh.*) impress-dotted. ——— 1818. 9. 11. D.G. ♃. Ot. et Lk. abb. t. 16.
pertùsa. H.R. *Crássula corymbulòsa.* L.en.
10 pertúsula. H.P. (*wh.*) small impress-dot'd. ——— 1822. —— D.G. ♃.
11 capitellàta. H.R. (*wh.*) square-spiked. ——— 1774. 7. 8. D.G. ♂.
12 thyrsiflòra. H.R. (*wh.*) thyrse-flowered. ——— 1824. —— D.G. ♂.
13 obovàta. H.R. (*wh.*) obovate-leaved. ——— 1818. —— D.G. ♂.
14 aloídes. H.R. (*bh.*) Aloe-like. ——— 1774. 6. 8. D.G. ♂.
15 punctàta. DC. (*wh.*) dotted. ——— 1759. 4. 8. D.G. ♄.
16 ramuliflòra. DC. (*wh.*) branch-flowered. ——— 1821. 6. 7. D.G. ♄. Ot. et Lk. abb. t. 17.
17 turrìta. H.R. (*bh.*) tower-like. ——— 1818. 9. 11. D.G. ♂. Jacq. schœnb. 1. t. 52.
α álba. (*wh.*) *white-flowered.* ——— —— —— D.G. ♂.
β ròsea. H.S.S. *rose-coloured.* ——— —— —— D.G. ♂.

GLOB'ULEA. DC. GLOB'ULEA. Pentandria Pentagynia.

1 cultràta. DC. (*wh.*) sharp-leaved. C. B. S. 1732. 7. 8. D.G. ♄. Bot. mag. t. 1940.
2 rádicans. DC. (*wh.*) rooting-branch'd. ——— 1823. 12. 3. D.G. ♄.
3 atropurpùrea. DC. (*wh.*) purple-leaved. ——— 1823. 7. 8. D.G. ♄.
4 Língua. DC. (*wh.*) tongue-leaved. ——— —— 5. 6. D.G. ♃.
5 língula. DC. (*wh.*) lesser-tongue. ——— —— —— D.G. ♃.
6 capitàta. DC. (*wh.*) large grey-leav'd. ——— 1819. 7. 8. D.G. ♃.
7 obvallàta. DC. (*wh.*) Houseleek-leav'd. ——— 1795. —— D.G. ♄. DC. plant. gras. t. 61.
Crássula obvallàta. L. Trew. pl. rar. 1. t. 11.
8 canéscens. DC. (*wh.*) canescent. ——— 1800. —— D.G. ♃.
9 nudicaùlis. DC. (*wh.*) naked-stalked. ——— 1732. 5. 9. D.G. ♃. DC. plant. gras. t. 133.
10 sulcàta. DC. (*wh.*) sulcate. ——— 1820. —— D.G. ♃.
11 impréssa. DC. (*wh.*) impressed. ——— 1822. 7. 8. D.G. ♃. Comm. præl. t. 29.
β mìnor. H.P. (*wh.*) *lesser.* ——— —— —— D G. ♃.
12 paniculàta. DC. (*wh.*) panicled. ——— 1823. —— D.G. ♃.
13 híspida. DC. (*wh.*) hispid. ——— —— 10. 11. D.G. ♄.
14 mesembryanthoídes. DC. (*wh.*) hedgehog. ——— 1822. 7. 9. D.G. ♄.
β áltior. (*wh.*) *taller.* ——— —— —— D.G. ♄.
15 móllis. DC. (*wh.*) soft. ——— 1774. —— D.G. ♄.
16 subincàna. DC. (*wh.*) hoary frutescent. ——— 1822. —— D.G. ♄.
α decumbens. H.P. (*wh.*) *decumbent.* ——— —— —— D.G. ♄.
β erécta. H.P. (*wh.*) *erect.* ——— —— —— D.G. ♄.

CURTO'GYNE. H.R. CURTO'GYNE. Pentandria Pentagynia.

1 dejécta. DC. (*wh.*) twisted-branch'd. C B. S. 1821. 6. 7. D.G. ♄. Jacq. schœnb. t. 423.
Crássula dejécta. Jacq.
2 undulàta. H.R. (*wh.*) lesser-waved. ——— 1797. 8. 11. D.G. ♄. Lodd. bot. cab. t. 584.
3 undàta. H.R. (*wh.*) greater-waved. ——— 1818. —— D.G. ♄.

GRAMMA'NTHES. DC. GRAMMA'NTHES. Pentandria Pentagynia.

1 chloræflòra. DC. (*ye.*) Chlora-flower'd. C. B. S. 1774. 6. 8. H. ⊙. Herm. lugd. t. 553.
Crássula dichòtoma. L. *Vauánthes chloræflòra.* H.R.
2 retrofléxa. (*or.*) orange-flower'd. ——— 1788. —— H. ⊙.
Crássula retrofléxa. H.S.
3 gentianoídes. DC. (*ye.*) Gentian-flower'd. ——— 1816. —— H. ⊙. Pluk. mant. t. 415. f. 6.

R'OCHEA. DC. R'OCHEA. Pentandria Pentagynia.

1 falcàta. DC. (*or.*) falcated. C. B. S. 1785. 6. 9. D.G. ♄. DC. plant. gras. t. 103.
Crássula falcàta. Bot. mag. t. 2035. *oblíqua.* Andr. reposit. t. 414.
β mìnor. H.P. (*sc.*) *smaller.* C. B. S. 1821. 6. 9. D.G. ♄.

2 perfoliàta. DC. (va.) great perfoliate. C. B. S. 1725. 7. 8. D.G. ♄. Mill. ic. t. 108.
α coccínea. DC. (sc.) *scarlet.* ——— —— —— D.G. ♄.
β álba. DC. (wh.) *white-flowered.* ——— —— —— D.G. ♄. DC. plant. gras. t. 13.
3 albiflòra. DC. (wh.) white-flowered. ——— 1820. 7. 9. D.G. ♄. Bot. mag. t. 2391.
Crássula albiflòra. B.M.

KALOSA'NTHES. H.R. KALOSA'NTHES. Pentandria Pentagynia.
1 coccínea. H.R. (sc.) scarlet. C. B. S. 1710. 6. 8. D.G. ♄. Bot. mag. t. 495.
Crássula coccínea. B.M. *Ròchea coccínea.* DC. plant. gras. t. 1.
β álba. H.R. (wh.) *white-flowered.* C. B. S. 1710. 6. 8. D.G. ♄. Bradl. succ. t. 50.
2 mèdia. H.R. (std.) great changeable. ——— 1820. —— D.G. ♄.
3 versícolor. H.R. (std.) various-coloured. ——— 1816. —— D.G. ♄. Bot. reg. t. 320.
Crássula versícolor. B.R.
4 bícolor. H.R. (ye.sc.) two-coloured. ——— 1820. —— D.G. ♄.
5 capitàta. (wh.) headed. ——— 1822. —— D.G. ♄. Lodd. bot. cab. t. 1029.
Crássula capitàta. L.C. *nec aliorum.*
6 odoratíssima. H.R. (st.) sweet-scented. ——— 1793. 6. 7. D.G. ♄. Andr. rep. t. 26.
Crássula odoratíssima. Jacq. schœnb. t. 434. *Dietríchia odoratíssima.* Tratt.
7 flàva. H.S. (ye.) yellow-flower'd. C. B. S. 1802. 6. 9. D.G. ♄. Burm. afr. t. 23. f. 3.
8 cymòsa. H.R. (wh.) cymose. ——— 1800. —— D.G. ♄.
9 biconvexa. H.P. (wh.) convex-leaved. ——— 1822. —— D.G. ♄.
10 jasmínea. H.R. (wh.) Jasmine-flower'd. ——— 1812. 6. 8. D.G. ♄. Bot. mag. t. 2178.
Crássula jasmínea. B.M. *Ròchea jasmínea.* DC.

KALANCH'OE. DC. KALANCH'OE. Octandria Tetragynia.
1 spathulàta. DC. (ye.) spathulate. China. 1820. 7. 9. D.G. ♄. DC. plant. gras. t. 65.
2 ægyptìaca. DC. (or.) Egyptian. Egypt. 1820. 7. 9. D.G. ♄. DC. plant. gras. t. 64.
3 rotundifòlia. H.P. (ye.) round-leaved. C. B. S. —— —— D.G. ♄.
4 acutiflòra. H.P. (wh.) white-flowered. E. Indies. 1806. —— D.S. ♄. Andr. reposit. t. 560.
Vèria acutiflòra. A.R.
5 crenàta. DC. (ye.) great notched. Sierra Leone. 1793. —— D.S. ♄. Andr. reposit. t. 21.
Cotylèdon crenàtum. Bot. mag. t. 1436. *Vèria crenàta.* A.R.
6 laciniàta. DC. (ye.) cut-leaved. E. Indies. 1731. —— D.S. ♄. DC. plant. gras. t. 100.
7 ceratophy'lla. DC. (st.) buck's-horn-l'd. China. 1820. 9. 1. D.G. ♄.

BRYOPHY'LLUM. DC. BRYOPHY'LLUM. Octandria Tetragynia.
calycìnum. DC. (br.) large-cupped. Mauritius. 1800. 4. 7. D.G. ♄. Bot. mag. t. 1409.

COTYL'EDON. DC. NAVELWORT. Decandria Pentagynia. L.
1 undulàta. H.S.S. (ye.) wave-leaved. C. B. S. 1818. 7. 8. D.G. ♄.
2 orbiculàta. H.S. (ye.bh.) round-leaved. ——— 1789. —— D.G. ♄.
3 ovàta. H.S. (ye.bh.) ovate-leaved. ——— —— —— D.G. ♄. Bot. mag. t. 321.
orbiculàta. B.M. *orbiculàta β obovàta.* DC. pl. gras. t. 76.
4 oblónga. H.S. (ye.bh.) oblong-leaved. C. B. S. 1690. 7. 10. D.G. ♄.
5 elàta. H.SS. (ye.bl.) tall powdered. ——— 1816. 6. 8. D.G. ♄.
6 crassifòlia. H.P. .... thick-leaved. ——— 1823. .... D.G. ♄.
7 víridis. H.P. .... green-leaved. ——— —— .... D.G. ♄.
8 ramòsa. H.S.S. (ye.bh.) branching. ——— 1768. 7. 10. D.G. ♄. Moris. his. s. 12. t. 7. f. 39.
9 ramosíssima. H.S.S. (ye.) cluster-branch'd. ——— 1816. —— D.G. ♄.
10 corúscans. H.S.S. (or.) sparkling. ——— 1818. 3. 5. D.G. ♄. Bot. mag. t. 2601.
canalifòlia. H.P.
11 ungulàta. DC. (pu.) ungulate. ——— 1822. 5. 8. D.G. ♄. Bur. afr. dec. 3. t. 22. f. 1.
12 papillàris. DC. (or.) cross-leaved. ——— 1813. —— D.G. ♄. Bot. reg. t. 915.
decussàta. Bot. mag. t. 2518.
13 tricuspidàta H.P. (or.) trident-leaved. ——— 1820. —— D.G. ♄.
14 purpùrea. DC. (pu.) purple. ——— 1813. 7. 8. D.G. ♄. Burm. afr. t. 20. f. 1.
15 rotundifòlia. H.P. (ye.bh.) round-leav'd. ——— 1826. —— D.G. ♄.
16 cristàta. H.P. (pu.) Coxcomb-leaved. ——— 1825. 8. 10. D.G. ♃.
17 clavifòlia. H.P. (pu.) club-leaved. ——— —— —— D.G. ♄.
18 curviflòra. DC. (ye.bh.) curve-flower'd. ——— —— —— D.G. ♄. Bot. mag. t. 2044.
19 tuberculòsa. DC. (or.) tubercled. ——— 1822. —— D.G. ♄. DC. plant. gras. 1. t. 86.
20 cacalioides. DC. (ye.) Cacalia-like. ——— 1820. 5. 6. D.G. ♄.
21 ventricòsa. DC. (gr.) bellied-tubed. ——— 1825. 6. 8. D.G. ♄. Burm. afr. d. 3. t. 21. f. 1.
22 spùria. DC. (ye.) narrow-leaved. ——— 1731. 7. 8. D.G. ♄. ——— t. 19. f. 1.
23 fasciculàris. DC. (gr.re.) clustered. ——— 1759. 7. 9. D.G. ♄. ——— t. 18.
tardiflòrum. Bonpl. nav. t. 37.
24 triflòra. DC. (wh.pu.) three-flowered. ——— 1821. 7. 19. D.G. ♄.
25 maculàta. DC. (v.) spotted-leaved. ——— 1818. 5. 7. D.G. ♄.
26 rhombifòlia. H.P. (ye.) rhomb-leaved. ——— 1823. —— D.G. ♄.
27 jasminiflòra. DC. (r.) Jasmine-flower'd. ——— 1818. 6. 8. D.G. ♄.
28 hemisph'ærica. DC. (r.) thick-leaved. ——— 1731. —— D.G. ♄. DC. plant. gras. 1. t. 87.
29 mamillàris. DC. (r.) mamillary. ——— 1818. 5. 7. D.G. ♄.
30 cuneàta. DC. (ye.) large pubescent. ——— 1813. 6. 8. D.G. ♄.

31 caryophyllàcea. DC. (*re.*) Clove-flowered. C. B. S. 1826. 6. 8. D.G. ♄. Burm. dec. 2. t. 17.
32 mucronàta. DC. (*ye.*) mucronate. ——— 1818. —— D.G. ♄. Burm. afr. t. 19. f. 2.
33 canaliculàta. H.S.S. (*v.*) channel-leav'd. ——— 1817. —— D.G. ♄.
34 grácilis. H.S.S. (*ye.*) slender. ——— 1800. 5. 8. D.G. ♄.
35 dichótoma. H.R. (*gr.*) fork-spined. ——— 1818. —— D.G. ♄.
36 cuneifórmis. H.P. .. wedge-leaved. C. B. S. 1823. .... D.G. ♄.
37 interjécta. H.P. .... fleshy-stemmed. ——— —— .... D.G. ♄.

PISTORI'NIA. DC. Pistori'nia. Decandria Pentagynia.
hispánica. DC. (*re.*) Spanish. Spain. 1796. 6. 7. H. ♂. DC. plant. grass. t. 122.
Cotylèdon hispanica. Lœfl. it. p. 77. t. 1.

UMBI'LICUS. DC. Pennywort. Decandria Pentagynia.
1 pendulìnus. DC. (*gr.*) drooping-flower'd. Britain. .... 6. 9. H. ♃. Eng. bot. v. 5. t. 325.
Cotylèdon Umbílicus. E.B.
2 eréctus. DC. (*ye.*) upright-flower'd. England. .... —— H. ♃. Eng. bot. v. 22. t. 1522.
Cotylèdon lùteum. E.B.
3 horizontàlis. DC. (*st.*) horizontal. Sicily. 1828. —— H. ♃.

OROST'ACHYS. F. Orost'achys. Decandria Pentagynia.
1 serràta. (*pu.*) notch-leaved. Crete. 1732. 6. 7. F. ♃. Dill. elth. t. 95. f. 112.
Cotylèdon serràta. H.S.
2 spinòsa. (*ye.*) spiny. Siberia. .... —— F. ♃. Murr. com. gœt. 7. t. 5.
chlorántha. F. *Cotylèdon spinòsum.* L. Gmel. sib. 4. t. 67. f. 2.
3 malachophy'lla. F. (*wh.*) soft-leaved. Dahuria. 1815. —— H. ☉. Pall. it. 3. ap. t. G. f. 1.
4 thyrsiflòra. F. (*bh.*) thyrse-flowered. ——— 1820. —— H. ☉.

ECHEV'ERIA. DC. Echev'eria. Decandria Pentagynia.
1 coccínea. DC. (*sc.*) scarlet. Mexico. 1816. 8. 10. D.G. ♄. Bot. mag. t. 2572.
Cotylèdon coccínea. B.M.
2 grandifòlia. H.P. (*or.*) large-leaved. ——— 1826. —— D.G. ♄. Swt. br. fl. gar. t. 275.
3 gibbiflòra. DC. (*or.*) gibbous-flower'd. ——— —— —— D.G. ♄. Bot. reg. t. 1247.
4 cæspitòsa. DC. (*ye.*) tongue-leaved. California. 1796. 6. 8. D.F. ♃. Jacq. f. ecl. 1. t. 17.
Cotylèdon linguifórmis. H.K. *Sèdum Cotylèdon.* Jacq.

RHOD'IOLA. L. Rose-root. Diœcia Octandria. L.
1 ròsea. L. (*ye.*) common. Britain. .... 5. 7. H. ♃. Eng. bot. v. 8. t. 508.
Sèdum Rhodìola. DC. plant. gras. t. 143.
2 sibírica. F. (*ye.*) Siberian. Siberia. 1826. —— H. ♃.

S'EDUM. DC. Stonecrop. Decandria Pentagynia.
1 Aizòon. DC. (*ye.*) yellow Orpine. Siberia. 1757. 7. 9. H. ♃. DC. plant. gras. t. 101.
2 hy'bridum. DC. (*st.*) Germander-leaved. ——— 1766. 5. 7. H. ♃. Murr. com. gœt. 6. t. 5.
3 involucràtum. DC. (*wh.*) involucred. Caucasus. 1822. —— H. ♃.
4 verticillàtum. DC. (*pu.*) whorl-leaved. Kamtschatka. .... 7. 9. H. ♃. Linn. amœn. 2. t. 4. f. 14.
5 latifòlium. DC. (*st.*) broad-leaved. Europe. 1794. —— H. ♃. Clus. hist. 2. p. 66. f. 1.
6 máximum. H.S. (*pu.*) great purple. Spain. —— —— H. ♃.
7 álbicans. H.S. (*wh.*) great white. Europe. —— —— H. ♃.
8 argùtum. H.R. (*pu.*) lance-toothed. M.Carpathian. 1820. —— H. ♃.
9 paucidéns. H.R. (*wh.*) few-toothed. Europe. —— —— H. ♃.
10 Telèphium. E.B. (*pu.*) common Orpine. Britain. .... —— H. ♃. Eng. bot. v. 19. t. 1319.
11 triphy'llum. (*pu.*) three-leaved. Portugal. 1794. —— H. ♃.
Anacámpseros triphy'lla. H.S.
12 purpùreum. H.S. (*pu.*) purple-stalked. ——— —— —— H. ♃. DC. plant. gras. t. 92.
13 telephioídes. DC. (*pu.*) Rhodiola-like. N. America. 1810. 7. 8. H. ♃.
14 spùrium. DC. (*pu.*) greater-fringed. Caucasus. —— 7. 9. H. ♃. Bot. mag. t. 2370.
15 ròseum. DC. (*ro.*) rose-coloured. ——— 1827. —— H. ♃.
16 dentàtum. DC. (*li.*) large-toothed. .... 1810. —— H. ♃.
17 pectinàtum. DC. (*wh.*) pectinated. .... 1819. —— H. ♃.
18 lívidum. DC. (*wh.*) livid-leaved. .... 1820. —— H. ♃.
19 cruciàtum. DC. (*wh.*) cross leaved. S. Europe. 1816. 6. 7. H. ♃. Lodd. bot. cab. t. 464.
monregalénse. Balb. misc. p. 23. t. 6.
20 ciliàre. H.S. (*wh.*) fringed. Caucasus. 1810. 7. 9. H. ♃. Bot. mag. t. 1807.
oppositifòlium. B.M. *Anacámpseros ciliàris.* H.S.
21 ternàtum. DC. (*bh.*) dwarf 3-leaved. N. America. 1789. 7. 8. H. ♃. Bot. reg. t. 142.
22 púlchrum. DC. (*pu.*) pretty. ——— 1824. —— H. ♃.
23 populifòlium. DC. (*wh.*) Poplar-leaved. Siberia. 1780. —— H. ♄. Bot. mag. t. 211.
24 Notarjánni. DC. (*wh.*) few-flowered. Naples. 1824. —— H. ♄. Ten. flor. nap. 1. t. 40.
25 Anacámpseros. DC. (*pu.*) evergreen. France. 1596. 7. 8. H. ♃. Bot. mag. t. 118.
Anacámpseros sempervìrens. H.S.
26 divaricàtum. W. (*li.*) spreading. Madeira. 1777. 6. 7. G. ♄.
27 Sempervìvum. DC. (*re.*) houseleek-leav'd. Iberia. 1823. 6. 8. H. ♃. Bot. mag. t. 2474.
28 deltoídeum. DC. (*pu.*) deltoid-leaved. Naples. 1826. —— H. ☉.
29 stellàtum. DC. (*bh.*) starry. S. Europe. 1640. 6. 7. H. ☉. Camer. hort. 7. t. 2.
30 Cep'æa. DC. (*wh.*) panicled. France. —— 7. 8. H. ♂. Clus. hist. 2. p. 68. ic.
31 spathulàtum. W.K. (*wh.*) spathulate. Hungary. 1820. —— H. ♂. W. et K. hung. 2. t. 104.

32	tetraphy'llum. F.G.	(*wh.*) four-leaved.	Levant.	....	7. 8.	H. ♂.	
33	cœrùleum. DC.	(*pu.*) blue pubescent.	Tunis.	1821.	7. 9.	H. ♃.	Bot. reg. t. 520.
34	ánglicum. DC.	(*wh.re.*) English.	England.	....	7. 8.	H. ♃.	Eng. bot. v. 3. t. 171.
	β microphy'llum. H.R.	(*wh.*) *small-leaved.*	———	....	——	H. ♃.	
35	oblóngum. H.R.	(*wh.*) oblong-leaved.	———	....	——	H. ♃.	
36	atrátum. DC.	(*wh.*) dark annual.	Italy.	1795.	7. 9.	H. ⊙.	DC. plant. gras. t. 120.
37	glaùcum. DC.	(*wh.*) glaucous.	Hungary.	1816.	——	H. ♂.	W. et K. hung. 2. t. 181.
38	sexfìdum. DC.	(*wh.*) six-cleft.	Caucasus.	1818.	——	H. ⊙.	
39	pállens. H.R.	(*wh.*) annual glaucous.	....	....	7. 9.	H. ⊙.	
40	villòsum. DC.	(*ro.*) hairy.	Britain.	....	6. 7.	H. ♃.	Eng. bot. v. 6. t. 394.
41	pállidum. M.B.	(*ro.*) pale red.	Tauria.	1820.	7. 9.	H. ⊙.	
42	rùbens. DC.	(*re.*) annual red.	S. Europe.	1759.	5. 8.	H. ⊙.	DC. plant. gras. t. 55.
	Crássula rùbens. DC.						
43	cæspitòsum. DC.	(*re.*) tufted.	———	1788.	7. 8.	H. ⊙.	Cavan. ic. t. 69. f. 2.
	Crássula cæspitòsa. Cav. *Crássula Magnòlii.* DC. *verticillàris.* L?						
44	dasyphy'llum. DC.	(*wh.*) thick-leaved.	England.	....	6. 7.	H. ♃.	Eng. bot. v. 10. t. 656.
45	hispánicum. DC.	(*wh.*) Spanish.	Spain.	1732.	——	H. ♃.	Jacq. aust. 5. ap. t. 47.
46	álbum. DC.	(*wh.*) white.	England.	....	6. 8.	H. ♃.	Eng. bot. v. 22. t. 1578.
47	altàicum. F.	.... Altay.	Altay.	1827.	....	H. ♃.	
48	quinquéfidum. H.R.	(*gr.wh.*) five-cleft.	....	....	6. 8.	H. ♃.	
49	hexapétalum. H.R.	(*ye.*) six-petal'd.	....	....	——	H. ♃.	
50	quadrífidum. DC.	(*ye.*) four-cleft.	Dahuria.	1800.	7. 8.	H. ♃.	Pall. it. 3. a. t. P. f. 1.
51	nùdum. DC.	(*ye.*) naked-branch'd.	Madeira.	1777.	——	G. ♄.	
52	ochroleùcum. H.P.	(*st.*) tall sharp-cup'd.	Greece.	....	——	F. ♃.	
53	altíssimum. DC.	(*st.*) tall blunt-cup'd.	Italy.	1769.	——	H. ♃.	DC. plant. gras. t. 116.
	Jacquini. H.P. *Sempervìvum sedifórme.* Jacq. vind. 1. t. 81.						
54	cæruléscens. H.P.	(*st.*) great-straw-color'd.	....	....	——	H. ♃.	
55	refléxum. H.P.	(*ye.*) reflex-leaved.	Britain.	....	6. 7.	H. ♃.	Eng. bot. v. 10. t. 695.
56	septangulàre. H.P.	(*ye.*) seven-angled.	Europe.	....	7. 8.	H. ♃.	
57	albéscens. H.R.	(*ye.*) white-leaved.	England.	....	6. 7.	H. ♃.	Eng. bot. v. 35. t. 2477.
	glaùcum. E.B. *non* W.K.						
58	mìnus. H.P.	(*ye.*) lesser short-leaved.	......	....	——	H. ♃.	DC. plant. gras. t. 116.
	refléxum. DC.						
59	spiràle. H.P.	(*ye.*) spiral-leaved.	Europe.	....	——	H. ♃.	
60	sexangulàre. DC.	(*ye.*) insipid.	England.	....	——	H. ♃.	Eng. bot. v. 28. t. 1946.
61	àcre. DC.	(*ye.*) biting.	Britain.	....	——	H. ♃.	——— v. 12. t. 839.
62	recurvàtum. H.P.	(*ye.*) slender recurv'd.	....	1820.	——	H. ♃.	
63	anopétalum. DC.	(*st.*) broader-leaved.	S. France.	1815.	——	H. ♃.	Bauh. icon. 3. 428.
	β aurantiacum. H.P.	(*or.*) *orange-colour'd.*	———	——	——	H. ♃.	
64	vìrens. H.P.	(*ye.*) deep green.	Portugal.	1774.	——	H. ♃.	
65	virídulum. H.P.	(*ye.*) small-green.	....	1826.	6. 8.	H. ♃.	
66	rupéstre. H.P.	(*ye.*) expanding glaucous.	England.	....	——	H. ♃.	Eng. bot. v. 3. t. 170.
67	Forsteriànum. H.P.	(*ye.*) Forster's.	Wales.	....	7. 8.	H. ♃.	——— v. 26. t. 1802.
68	saxátile. DC.	(*ye.*) rock.	S. Europe.	1739.	7. 9.	H. ⊙.	DC. plant. gras. t. 119.
69	rèpens. DC.	(*ye.*) creeping.	Pyrenees.	1829.	——	H. ⊙.	Vill. dauph. 3. t. 45.
	Guettàrdi. Villars.						
70	magellénse. DC.	(*ye.*) Neapolitan.	Naples.	1826.	——	H. ⊙.	
	MONA'NTHES. H.R.	MONA'NTHES.	Dodecandria Heptagynia.				
	polyphy'lla. H.R.	(*pu.*) many-leaved.	Canaries.	1777.	7. 9.	D.G. ♃.	Bot. mag. t. 93.
	Sempervìvum monánthes. B.M.						
	SEMPERVI'VUM. DC.	HOUSELEEK.	Dodecandria Dodecagynia. L.				
1	tortuòsum. DC.	(*ye.*) twisted.	Canaries.	1779.	7. 8.	D.G. ♄.	Bot. mag. t. 296.
2	villòsum. DC.	(*ye.*) small villous.	Madeira.	1777.	6. 7.	D.G. ♄.	
3	ciliàtum. DC.	(*wh.*) ciliated.	Teneriffe.	1815.	——	D.G. ♄.	
4	glandulòsum. DC.	(*ye.*) gland-edged.	Madeira.	1777.	3. 5.	D.G. ♄.	
5	glutinòsum. DC.	(*ye.*) glutinous.	———	——	7. 8.	D.G. ♄.	Bot. reg. t. 278.
6	úrbicum. DC.	(*ye.*) wedge-leaved.	Canaries.	1824.	——	D.G. ħ.	
7	retùsum. H.P.	(*ye.*) retuse-leaved.	Teneriffe.	——	——	D.G. ħ.	
8	arbòreum. DC.	(*ye.*) tree.	Levant.	1640.	12. 3.	D.G. ħ.	Bot. reg. t. 99.
	β variegàtum. H.R.	(*ye*) *variegated.*	———	....	——	D.G. ħ.	
	γ atropurpùreum. H.R.	(*ye.*) *dark purple-leav'd.*	———	....	——	D.G. ħ.	
9	frutéscens. H.P.	(*ye.*) frutescent.	Teneriffe.	1824.	3. 5.	D.G. ħ.	
10	tabulæfórme. DC.	(*ye.*) table-shaped.	Madeira.	1815.	7. 9.	D.G. ♄.	Lodd. bot. cab. t. 1328.
11	canariénse. DC.	(*wh.*) Canary Island.	Canaries.	1699.	6. 7.	D.G. ♄.	DC. plant. gras. t. 141.
12	Smíthii. DC.	(*ye.*) Smidt's.	———	1815.	8. 10.	D.G. ♄.	Bot. mag. t. 1980.
13	barbàtum. DC.	(*ye.*) bearded.	Madeira.	——	6. 7.	D.G. ♄.	
	lineolàre. H.S.S. *spathulàtum.* Horn.						
14	hy'bridum. H.R.	(*ye.*) hybrid.	Hybrid.	....	——	D.G. ♄.	
15	cæspitòsum. DC.	(*ye.*) latchet-leaved.	Canaries.	1815.	9. 4.	D.G. ♄.	Bot. mag. t. 1978.
	ciliàtum. B.M. *non* W. *barbàtum.* H.R. non Smith. *ciliàre.* H.S.S.						

16 láxum. H.R.	(*ye.*)	loose-flowered.	Canaries.	1820.	7. 10.	D.G. ♂.	
17 dodrantàle. DC.	(*st.*)	smooth-leaved.	———	1815.	7. 8.	D.G. ♂.	
18 aùreum. DC.	(*ye.*)	chalice-form.	Teneriffe.	——	6. 7.	D.G. ♃.	Bot. reg. t. 892.
calyciförme. B.R. Lodd. bot. cab. 1368.							
19 punctàtum. DC.	(*ye.*)	dotted-leaved.	Canaries.	1824.	7. 9.	D.G. ♂.	
20 stellàtum. DC.	(*ye.*)	annual villous.	Madeira.	1790.	7. 10.	H. ⊙.	Bot. mag. t. 1809.
villòsum. B.M. *non* H.S.							
21 dichótomum. DC.	(*ye.*)	forked.	Canaries.	1824.	6. 8.	D.G. ♂.	DC. jard. gen. t. 21.
22 africànum. DC.	(*ye.*)	African.	C. B. S.	1768.	....	D.G. ♃.	
23 hírtum. DC.	(*st.*)	hairy.	Italy.	....	6. 7.	H. ♃.	DC. plant. gras. t. 107.
sobolíferum. B.M. t. 1457 ex DC. *globíferum.* H.R. *non* DC.							
24 globíferum. DC.	(*ye.*)	great-flowered.	Germany.	1731.	——	H. ♃.	Bot. mag. t. 507 et 2115.
25 tectòrum. DC.	(*pu.*)	common.	Britain.	....	6. 9.	H. ♃.	Eng. bot. v. 19. t. 1320.
26 flagellifórme. DC.	(*re.*)	whip-lash.	Siberia.	1818.	6. 8.	H. ♃.	
27 montànum. DC.	(*re.*)	mountain.	Switzerland.	1752.	——	H. ♃.	DC. plant. gras. t. 105.
28 arachnoídeum. DC.	(*pu.*)	Cobweb.	Italy.	1699.	——	F. ♃.	Bot. mag. t. 68.

Tribus II. *PENTHOREÆ.*

PENTH'ORUM. DC. PENTH'ORUM. Decandria Pentagynia. L.

sedoídes. DC.	(*st.*)	American.	Virginia.	1768.	7. 8.	H. ♃.	Act. ups. 1744. t. 2.

LEWI'SIA. Ph. LEWI'SIA. Dodecandria Trigynia.

redivìva. Ph.	(*wh.*)	white-flowered.	N. America.	1827.	7. 8.	H. ♃.

ORDO CII.

FICOIDEÆ. *DC. prodr. v.* 3. *p.* 415.

Tribus I. *MESEMBRYANTHEMEÆ.*

MESEMBRYA'NTHEMUM. FIG-MARYGOLD. Icosandria Pentagynia. L.

1 minùtum. H.R.	(*re.*)	small-white.	C. B. S.	1795.	9. 11.	D.I. ♃.	Bot. mag. t. 1376.
2 mínimum. H.R.	(*st.*)	small chequer'd.	———	1766.	9. 12.	D.I. ♃.	Pet. gaz. t. 39. f. 3.
3 perpusíllum. H.R.	(*st.*)	small exserted.	———	1819.	——	D.I. ♃.	
4 obcordéllum. H.R.	(*wh.*)	glaucous chequer'd.	———	1776.	10. 2.	D.I. ♃.	Bot. mag. t. 1647.
5 obconéllum. H.R.	(*wh.*)	green chequer'd.	———	1794.	——	D.I. ♃.	
6 ficifórme. H.R.	....	fig-shaped.	———	1819.	....	D.I. ♃.	
7 truncatéllum. H.R.	(*st.*)	large dotted.	———	1795.	10. 2.	D.I. ♃.	
8 fibulifórme. H.R.	....	cloth button.	———	——	11.	D.I. ♃.	
9 uvæfórme. H.R.	....	Grape-like.	———	1820.	....	D.I. ♃.	Burm. dec. t. 10. f. 2?
10 nucifórme. H.R.	....	nut-shaped.	———	1790.	....	D.I. ♃.	
11 testiculàre. H.R.	(*wh.*)	broad short white.	———	1774.	11.	D.I. ♃.	
12 octophy'llum. H.R.	(*ye.*)	narrow short white.	———	——	——	D.I. ♃.	Bot. mag. t. 1573.
testiculàre. B.M.							
β *longiúsculum.*	(*ye.*)	*longer-leaved.*	———	——	——	D.I. ♃.	
13 testiculàtum. DC.	(*ro.*)	rosy.	———	——	——	D.I. ♃.	Jacq. frag. 73. t. 12. f. 2.
14 obtùsum. H.R.	(*re.*)	obtuse cloven.	———	1792.	3. 4.	D.I. ♃.	
15 físsum. H.R.	....	cloven white.	———	1776.	....	D.I. ♃.	
16 digitifórme. H.R.	(*wh.*)	finger-formed.	———	1775.	4. 6.	D.I. ♃.	
digitàtum. H.K?							
17 magnipúnctum. H.R.	(*ye.*)	great dotted club.	———	1816.	9. 11.	D.I. ♃.	
α *sesquiunciàle.* H.R.	(*ye.*)	*thickest-leaved.*	———	——	——	D.I. ♃.	
β *unciàle.* H.R.	(*ye.*)	*thinner-leaved.*	———	——	——	D.I. ♃.	
18 nòbile. H.P.	(*ye.*)	noble.	———	——	11.	D.I. ♃.	
19 cànum. H.R.	(*ye.*)	hoary club-leav'd.	———	1795.	....	D.I. ♃.	
20 aloìdes. H.R.	(*ye.*)	Aloe-like.	———	1816.	9. 11.	D.I. ♃.	
21 canìnum. H.R.	(*ye.*)	dog-chop.	———	1717.	8. 10.	D.G. ♄.	DC. plant. gras. t. 95.
22 vulpìnum. H.R.	(*ye.*)	fox-chop.	———	1795.	——	D.G. ♄.	
23 hy'bridum. H.R.	(*ye.*)	hybrid.	Hybrid.	....	——	D.G. ♃.	
24 álbidum. H.R.	(*ye.*)	white.	C. B. S.	1714.	7. 8.	D.G. ♃.	Bot. mag. t. 1824.

25	albinòtum. H.P.	(*ye.*)	keeled white dott'd.	C.B.S.	1824.	7. 9.	D.G. ♃.	
26	albipúnctum. H.P.	(*ye.*)	white-spotted.	———	——	8. 9.	D.G. ♃.	
27	tigrìnum. H.P.	(*ye.*)	tiger-chop.	———	1790.	9. 11.	D.G. ♃.	Bot. reg. t. 260.
28	felìnum. H.P.	(*ye.*)	cat-chop.	———	1730.	8. 11.	D.G. ♃.	DC. plant. gras. t. 152.
29	lupìnum. H.P.	(*ye.*)	wolf's chop.	———	1822.	6. 8.	D.G. ♃.	
30	agnìnum. H.P.	(*ye.*)	lamb's chop.	———	——	5. 6.	D.G. ♃.	
31	mustelìnum. H.P.	(*ye.*)	weasel-chop.	———	1818.	8. 11.	D.G. ♃.	
32	musculìnum. H.P.	(*ye.*)	small trailing.	———	——	6. 8.	D.G. ♃.	
33	erminìnum. H.P.	(*ye.*)	Ermine-chop.	———	1823.	5. 6.	D.G. ♃.	
	β *màjus.* H.P.	(*ye.*)	*red-tipped.*	———	——	——	D.G. ♃.	
34	murìnum. H.P.	(*ye.*)	mouse-chop.	———	1790.	9.	D.G. ♃.	
35	dolabrifórme. H.R.	(*ye.*)	hatchet-leaved.	———	1705.	5. 11.	D.G. ♄.	Bot. mag. t. 32.
36	cárinans. H.R.	(*ye.*)	scymetar-leav'd.	———	1820.	....	D.G. ♃.	
37	scapíger. H.P.	(*ye.*)	great-scaped.	———	1822.	8. 9.	D.G. ♃.	
38	denticulàtum. H.R.	(*st.*)	toothed-beak'd.	———	1793.	4.	D.G. ♃.	
	α *cànum.* H.R.	(*st.*)	*white-leaved.*	———	——	——	D.G. ♃.	
	β *glaùcum.* H.R.	(*st.*)	*glaucous-leaved.*	———	——	——	D.G. ♃.	
	γ *candidíssimum.* H.R.	(*st.*)	*whitest.*	———	——	——	D.G. ♃.	
39	robústum. H.R.	(*ye.*)	robust-beaked.	———	1795.	....	D.G. ♃.	
40	compáctum. H.R.	(*ye.*)	compact.	———	1780.	11.	D.G. ♄.	
41	quadrífidum. H.R.	(*ye.*)	quadrifid.	———	1795.	——	D.G. ♄.	
42	bífidum. H.R.	(*ye.*)	bifid beaked.	———	——	——	D.G. ♃.	
43	bibracteàtum. H.R.	(*ye.*)	double-bracted.	———	1803.	4. 11.	D.G. ♄.	
44	purpuráscens. DC.	(*ye.*)	purple-bracted.	———	1826.	——	D.G. ♃.	
45	rostràtum. H.R.	(*ye.*)	heron-beaked.	———	1732.	4. 5.	D.G. ♃.	Dill. elth. f. 229.
	tuberculàtum. H.R.	(*ye.*)	*tuberculate.*	———	....	——	D.G. ♃.	
46	ramulòsum. H.R.	(*ye.*)	small-beaked.	———	1803.	3. 11.	D.G. ♄.	
47	pisifòrme. H.R.	....	pea-shaped.	———	1796.	....	D.I. ♃.	
48	monilifòrme. H.R.	(*wh.*)	necklace.	———	1791.	3. 4.	D.I. ♃.	
49	scalpràtum. H.R.	(*ye.*)	great tongue.	———	1714.	8. 10.	D.G. ♃.	Dill. elth. t. 183. f. 224.
50	fràgrans. H.R.	(*ye.*)	sweet-scented.	———	1820.	——	D.G. ♃.	Ott. et Lk. abb. t. 43.
51	grandiflórum. H.P.	(*ye.*)	great-blistered.	———	1824.	——	D.G. ♃.	
52	præpíngue. H.R.	(*ye.*)	soft-tongue.	———	1792.	——	D.G. ♃.	
53	mèdium. H.R.	(*ye.*)	large-tongue.	———	1816.	——	D.G. ♃.	
54	cultràtum. H.R.	(*ye.*)	long cultrated.	———	1820.	——	D.G. ♃.	
55	lùcidum. H.R.	(*ye.*)	high-polished.	———	1768.	——	D.G. ♃.	Dill. elth. f. 227.
56	adscéndens. H.R.	(*ye.*)	long-peduncled.	———	1805.	8. 11.	D.G. ♃.	
57	pustulàtum. H.R.	(*ye.*)	blistered tongue.	———	1816.	——	D.G. ♃.	
58	lóngum. H.R.	(*ye.*)	long-thin-leaved.	———	1725.	8. 10.	D.G. ♃.	
	α *fláccidum.* H.R.	(*ye.*)	*long-flaccid.*	———	——	——	D.G. ♃.	Bot. mag. t. 1866.
	depréssum. B.M. *nec aliorum.*							
	β *declìve.* H.R.	(*ye.*)	*lesser erect-edged.*	———	——	——	D.G. ♃.	
	γ *angústius.* H.R.	(*ye.*)	*narrow-leaved.*	———	——	——	D.G. ♃.	
	δ *purpuráscens.* H.R.	(*ye.*)	*purple green.*	———	——	——	D.G. ♃.	
	ε *uncàtum.* H.R.	(*ye.*)	*leaden green.*	———	1819.	——	D.G. ♃.	
	ζ *attóllens.* H.R.	(*ye.*)	*narrow dark.*	———	——	——	D.G. ♃.	
59	linguæfórme. H.R.	(*ye.*)	common tongue.	———	1732.	11. 3.	D.G. ♃.	Dill. elth. f. 226.
	β *rufèscens.* H.R.	(*ye.*)	*reddish green.*	———	1804.	....	D.G. ♃.	
	γ *subcruciàtum.* H.R.	(*ye.*)	*subcruciate.*	———	1820.	8. 11.	D.G. ♃.	
	δ *prostràtum.* H.R	(*ye.*)	*prostrate.*	———	——	——	D.G. ♃.	
	ε *assúrgens.* H.R.	(*ye.*)	*assurgent.*	———	1819.	——	D.G. ♃.	
60	làtum. H.R.	(*ye.*)	broad tongue.	———	1732.	——	D.G. ♃.	Dill. elth. f. 225.
	β *brève.* H.R.	(*ye.*)	*short flat.*	———	1802.	——	D.G. ♃.	
61	depréssum. H.R.	(*ye.*)	small recurving-ld.	———	1795.	9. 11.	D.G. ♃.	Dill. elth. f. 226?
	α *pállens.* H.R.	(*ye.*)	*light green.*	———	——	——	D.G. ♃.	
	β *lívidum.* H.R.	(*ye.*)	*rufous green.*	———	1819.	——	D.G. ♃.	
62	cruciàtum. H.R.	(*ye.*)	soft cruciate.	———	1792.	5. 11.	D.G. ♄.	
63	taurìnum. H.R.	(*ye.*)	Bull's-horn.	———	1795.	9. 11.	D.G. ♄.	
64	Sálmii. H.R.	(*ye.*)	Prince de Salm's.	———	1818.	——	D.G. ♃.	
	α *decussàtum.* H.R.	(*ye.*)	*decussate.*	———	——	——	D.G. ♃.	
	β *semicruciàtum.* H.R.	(*ye.*)	*semicruciate.*	———	1820.	——	D.G. ♃.	
	γ *angustifòlium.* H.R.	(*ye.*)	*narrow-leaved.*	———	——	——	D.G. ♃.	
65	surréctum. H.R.	(*ye.*)	thick acute tongue.	———	1819.	——	D.G. ♃.	
66	heterophy'llum. H.R.	(*ye.*)	various-leaved.	———	1795.	——	D.G. ♃.	
67	angùstum. H.R.	(*ye.*)	narrow tongue.	———	1790.	3. 10.	D.G. ♃.	
	β *pállidum.* H.R.	(*ye.*)	*paler-leaved.*	———	1819.	....	D.G. ♃.	
	γ *heterophy'llum.* H.K.	(*ye.*)	*semcylindric-ld.*	———	....	8. 10.	D.G. ♃.	Andr. reposit. 540.
68	diffórme. H.R.	(*ye.*)	difformed.	———	1732.	——	D.G. ♃.	Dill. elth. f. 242.
	β *brevicaùle.* H.R.	(*ye.*)	*short-stalked.*	———	——	——	D.G. ♄.	

69	bidentàtum.	H.R.	(*ye.*)	double-toothed.	C. B. S.	1816.	8. 10.	D.G.♃.	
	α mìnus.	H.R.	(*ye.*)	*smaller.*	——	——	——	D.G.♃.	
	β màjus.	H.R.	(*ye.*)	*larger.*	——	——	——	D.G.♃.	
70	bigibberàtum.	H.P.	(*ye.*)	lesser difformed.	——	1824.	——	D.G.♄.	
71	semicylíndricum.	H.R.	(*ye.*)	small difformed.	——	1732.	3. 11.	D.G.♄.	Dill. elth. t. 194. f. 241.
72	gibbòsum.	H.R.	(*re.*)	great pale gibbous.	——	1780.	1. 4.	D.G.♄.	
73	lùteo-víride.	H.R.	(*re.*)	long pale gibbous.	——	1795.	——	D.G.♄.	
74	pervíride.	H.R.	(*re.*)	dark green gibbous.	——	1792.	1. 5.	D.G.♄.	
75	pubéscens.	H.R.	(*re.*)	hoary gibbous.	——	——	——	D.G.♄.	
76	calamifórme	H.R.	(*wh.*)	white fl^d^. quill.	——	1717.	7. 9.	D.G.♄.	DC. plant. gras. t. 5.
77	obsubulàtum.	H.R.	....	reversed quill.	——	1796.	....	D.G.♄.	
78	cylíndricum.	H.R.	(*re.*)	cylindric quill.	——	1792.	9. 2.	D.G.♄.	
79	teretifòlium.	H.R.	(*re.*)	round quill.	——	1794.	——	D.G.♄.	
80	teretiúsculum.	H.R.	(*re.*)	roundish quill.	——	——	——	D.G.♄.	
81	bellidiflòrum.	H.R.	(*re.*)	daisy-flowered.	——	1717.	6. 8.	D.G.♄.	Dill. elth. t.224. f. 233.
	α glaucéscens.	H.R.	(*re.*)	*small glaucous.*	——	——	——	D.G.♄.	
	β víride.	H.R.	(*ro.*)	*pea green.*	——	1822.	——	D.G.♄.	
82	subulàtum.	DC.	(*std.*)	great green.	——	1768.	6. 8.	D.G.♄.	DC. plant. gras. t. 41.
83	acùtum.	H.R.	(*pu.*)	great awl-leaved.	——	1793.	11. 4.	D.G.♃.	
84	punctàtum.	H.R.	(*pu.*)	middle-spotted.	——	....	....	D.G.♃.	
85	diminùtum.	H.R.	(*pu.*)	small awl-leaved.	——	1789.	4.	D.G.♃.	
	β cauliculàtum.	H.R.	(*pu.*)	*stalked.*	——	1817.	——	D.G.♃.	
86	lòreum.	H.R.	(*st.*)	leathery-stalk'd.	——	1732.	9.	D.G.♄.	Dill. elth. t. 200. f. 255.
	β congéstum.	H.R.	(*st.*)	*crowded-leaved.*	——	1805.	....	D.G.♄.	
87	diversifòlium.	H.R.	(*st.*)	short horn-leav'd.	——	1726.	3. 10.	D.G.♄.	Dill. elth. t. 198. f. 252.
	α màjus.	H.R.	(*std.*)	*larger.*	——	1819.	——	D.G.♄.	
	β glaùcius.	H.R.	(*st.*)	*glaucescent.*	——	1726.	——	D.G.♄.	
	γ brevifòlium.	H.R.	(*st.*)	*short-leaved.*	——	1819.	8.	D.G.♄.	
	δ l'æte-vìrens.	H.R.	(*st.*)	*pale green.*	——	——	——	D.G.♄.	
	ε átro-vìrens.	H.R.	(*st.*)	*dark green.*	——	——	8. 9.	D.G.♄.	
88	decípiens.	H.R.	(*st.*)	middle horn-l^d^.	——	1820.	....	D.G.♄.	
89	dùbium.	H.R.	(*st.or.*)	round stalked.	——	1800.	11. 5.	D.G.♄.	Bradl. succ. d. 4. t. 40.
90	corniculatum.	H.R.	(*std.*)	long-horn-leav'd.	——	1732.	3. 5.	D.G.♄.	DC. plant. gras. t. 108.
91	purpùreo-álbum.	H.P.	(*std.*)	purple & white.	——	1824.	7. 10.	D.G.♄.	
92	procúmbens.	H.R.	(*v.*)	purple-jointed.	——	1820.	....	D.G.♄.	
93	tricolòrum.	H.R.	(*st.re.*)	three-coloured.	——	1794.	10.	D.G.♄.	
94	pugionifórme.	H.R.	(*ye.*)	great dagger-l^d^.	——	1714.	7. 9.	D.G.♄.	DC. plant. gras. t. 82.
	β cárneum.	H.R.	(*ye.*)	*flesh-coloured.*	——	....	——	D.G.♄.	
	γ purpùreum	H.R.	(*pu.*)	*purple.*	——	....	——	D.G.♄.	
	δ biénne.	H.R.	(*ye.*)	*biennial.*	——	....	——	D.G.♂.	
95	capitàtum.	H.R.	(*ye.*)	lesser dagger-l^d^.	——	1717.	7. 9.	D.G.♄.	Bot. reg. t. 494.
	β ramígerum.	H.R.	(*ye.*)	*branching.*	——	1816.	——	D.G.♄.	
96	brevicàule.	H.R.	(*ye.*)	dwarf dagger-l^d^.	——	——	——	D.G.♄.	
97	corúscans.	H.R.	(*ye.*)	glittering dagger.	——	1813.	——	D.G.♄.	
98	elongàtum.	H.R.	(*ye.*)	tuberous long-l^d^.	——	1793.	5.	D.G.♄.	DC. plant. gras. t. 72.
	α grandiflòrum.		(*ye.*)	*large-flowered.*	——	——	——	D.G.♄.	
	β mìnus.	H.R.	(*ye.*)	*lesser.*	——	——	——	D.G.♄.	Bot. reg. t. 493.
	γ fusifórme.	H.R.	(*ye.*)	*fusiform.*	——	——	——	D.G.♄.	Breyn. cent. t. 80.
99	válidum.	H.P.	(*ro.*)	robust light green.	——	1823.	——	D.G.♄.	
100	geminiflòrum.	H.R.	(*ro.*)	small pale.	——	1819.	——	D.G.♄.	
101	símile.	H.R.	(*ro.*)	short-jointed.	——	——	5. 7.	D.G.♄.	
102	láxum.	H.R.	(*ro.*)	long-jointed.	——	——	——	D.G.♄.	
103	sarmentòsum.	H.R.	(*ro.*)	sarmentose.	——	1805.	4. 5.	D.G.♄.	
104	rigidicàule.	H.R.	(*ro.*)	stiff-stemmed.	——	1819.	——	D.G.♄.	
105	Schóllii.	H.R.	(*ro.*)	Scholl's.	——	1818.	——	D.G.♄.	Jacq. frag. t. 51. f. 2.
	adúncum. Jacq. *non* H. *recúrvum.* H.S.S.								
106	filamentòsum.	H.R.	(*ro.*)	thready.	——	1732.	3. 4.	D.G.♄.	Dill. elth. t. 212. f. 273.
107	serrulàtum.	H.R.	(*ro.*)	rosy-stalked.	——	1795.	11. 12.	D.G.♄.	
	α glaùcum.	H.R.	(*ro.*)	*glaucous.*	——	——	——	D.G.♄.	
	β viridius.	H.R.	(*ro.*)	*greenish.*	——	——	——	D.G.♄.	
108	rubricaùle.	H.R.	(*ro.*)	deep red-stalk'd.	——	1802.	12. 2.	D.G.♄.	
	α gracílius.	H.R.	(*ro.*)	*slenderest.*	——	——	——	D.G.♄.	
	β dénsius.	H.R.	(*ro.*)	*crowded.*	——	....	——	D.G.♄.	
	γ subvìrens.	H.R.	(*ro.*)	*tall-green.*	——	....	——	D.G.♄.	
109	acinacifórme.	H.R.	(*ro.*)	great scymetar-l^d^.	——	1714.	8. 9.	D.G.♄.	Dill. elth. f. 270. 271.
	α brève.	H.R.	(*ro.*)	*short-branched.*	——	——	——	D.G.♄.	Andr. reposit. 508.
	β lóngum.	H.R.	(*ro.*)	*longer-branched.*	——	——	——	D.G.♄.	
110	lævigàtum.	H.R.	(*ro.*)	smooth scymetar.	——	1802.	——	D.G.♄.	

111	rùbro-cínctum. H.R.	(ro.) red-edged.	C. B. S.	1811.	8. 9.	D.G. ♄.	
	α crássum. H.R.	(ro.) *thick-leaved.*	——	——	....	D.G. ♄.	
	β compréssum. H.R.	(ro.) *compressed.*	——	....	....	D.G. ♄.	
	γ ténerum. H.R.	(ro.) *tender.*	——	....	....	D.I. ♄.	
112	subalàtum. H.R.	(ro.) wing-stalked.	——	1802.	....	D.G. ♄.	
113	èdule. H.R.	(ye.) Hottentot's-fig.	——	1690.	7. 8.	D.G. ♄.	Dill. elth. t. 212. f. 272.
114	dimidiàtum. H.R.	(ro.) lesser eatable.	——	1811.	——	D.G. ♄.	DC. plant. gras. t. 89.
115	glaucéscens. H.R.	(ro.) glaucescent.	N. Holland.	1804.	7.	D.G. ♄.	
116	abbreviàtum. H.P.	.... short-jointed.	N. S. W.	1825.	....	D.G. ♄.	
117	Róssi. H.R.	(ro.) Ross's.	V. Diem. Isl.	1821.	....	D.G. ♄.	
118	viréscens. H.R.	(ro.) virescent.	N. Holland.	1804.	6. 8.	D.G. ♄.	
119	æquilateràle. H.R.	(ro.) equal angled.	——	1791.	——	D.G. ♄.	
	β decag'ynum. DC.	(ro.) *ten-styled.*	——	——	——	D.G. ♄.	
120	vìrens. H.R.	(ro.) smooth-edged green.	C. B. S.	1821.	7. 8.	D.G. ♄.	
121	réptans. H.R.	(ro.) creeping.	——	1774.	6. 8.	D.G. ♄.	
122	débile. H.P.	(ro.) smooth weak.	——	1821.	——	D.G. ♄.	
123	austràle. H.R.	(bh.) New Zealand.	N. Zealand.	1773.	——	D.G. ♄.	
124	crassifòlium. H.R.	(ro.) thick-leaved.	C. B. S.	1727.	5. 8.	D.G. ♄.	Dill. elth. t. 208. f. 257.
125	clavellàtum. H.R.	(re.) club-l^d. creeping.	N. Holland.	1803.	6. 7.	D.G. ♄.	
	β aggregàtum. H.R.	(re.) *cluster branched.*	——	——	——	D.G. ♄.	
	γ mìnus. H.R.	(re.) *shorter-branched.*	——	——	——	D.G. ♄.	
126	forficàtum. H.R.	(re.) scissors-leaved.	C. B. S.	1758.	9. 10.	D.G. ♄.	Jacq. vind. 1. t. 26.
127	geminàtum. H.R.	(re.) white twin-shoot'd.	——	1792.	....	D.G. ♄.	
128	marginàtum. H.R.	(re.) white-edged.	——	1793.	....	D.G. ♄.	
129	rostéllum. H.R.	(wh. re.) little prostrate beak.	——	1820.	8. 10.	D.G. ♄.	
130	perfoliàtum. H.R.	(ro.) great-hooked.	——	1714.	6. 8.	D.G. ♄.	Dill. elth. t. 192. f. 240.
	α tricánthum. H.R.	(ro.) *three-thorned.*	——	——	——	D.G. ♄.	Bradl. suc. 5. t. 46.
	β monacánthum. H.R.	(ro.) *one-thorned.*	——	——	——	D.G. ♄.	Bradl. suc. dec. 3. f. 26.
131	uncinéllum. H.R.	(li.) small-hooked.	——	1819.	——	D.G. ♄.	Dill. elth. f. 239.
132	uncinàtum. H.R.	(ro.) greenish-hooked.	——	1725.	8.	D.G. ♄.	DC. plant. gras. t. 54.
133	semidentátum. H.R.	(ro.) slender-hooked.	——	1818.	6. 8.	D.G. ♄.	
134	ùnidens. H.P.	(li.) dwarf uncinate.	——	1824.	——	D.G. ♄.	
135	víride. H.R.	(li.) deep green.	——	1792.	7.	D.G. ♄.	
136	acutángulum. H.P.	(wh.) sharp-angled.	——	1820.	6. 8.	D.G. ♄.	
137	cúrtum. H.R.	(wh.) short-sheathed.	——	1811.	——	D.G. ♄.	
	β màjus. H.R.	(wh.) *larger.*	——	——	——	D.G. ♄.	
138	hamàtum. W.	(wh.) hook'd sheathing.	——	——	——	D.G. ♄.	
139	vaginàtum. H.R.	(wh.) rough sheathing.	——	1802.	7. 8.	D.G. ♄.	
	β parviflòrum. H.R.	(wh.) *smaller-flowered.*	——	——	——	D.G. ♄.	
140	parviflòrum. H.R.	(wh.) small-flowered.	——	1800.	8.	D.G. ♄.	
141	rígidum. H.R.	(wh.) rigid.	——	1793.	——	D.G. ♄.	
142	tenéllum. H.R.	(wh.) least perfoliate.	——	1792.	——	D.G. ♄.	
143	imbricàtum. H.R.	(wh.) imbricated.	——	——	7.	D.G. ♄.	
	α mìnus. H.R.	(wh.) *smallest.*	——	——	——	D.G. ♄.	
	β màjus. H.R.	(wh.) *blunt crowded.*	——	——	——	D.G. ♄.	
	γ víride. H.R.	(wh.) *dull green.*	——	——	——	D.G. ♄.	
144	multiflòrum. H.R.	(wh.) many-flowered.	——	——	7. 9.	D.G. ♄.	Pluk. phyt. t. 117. f. 1.
	α màjus. H.R.	(wh.) *larger white.*	——	——	——	D.G. ♄.	
	β mìnus. H.R.	(wh.) *lesser white.*	——	——	——	D.G. ♄.	
	γ rùbrum. H.R.	(re.) *red-flowered.*	——	1816.	——	D.G. ♄.	
	δ pàtens. H.R.	(wh.) *single flowered.*	——	——	——	D.G. ♄.	
	ε nìtens. H.R.	(wh.) *bright-green.*	——	....	——	D.G. ♄.	
145	umbellàtum. H.R.	(wh.) umbel-flowered.	——	1727.	6. 9.	D.G. ♄.	Dill. elth. t. 208. f. 266.
	β apétalum. H.R.	(wh.) *petalless.*	——	——	——	D.G. ♄.	
146	tumídulum. H.R.	(re.) tumid.	——	1802.	3.	D.G. ♄.	
	β mìnus. H.R.	(re.) *lesser.*	——	——	——	D.G. ♄.	
147	foliòsum. H.R.	(re.) many-leaved.	——	——	9.	D.G. ♄.	
148	lineolàtum. H.R.	(re.) tumid sheathed.	——	1819.	6. 8.	D.G. ♄.	
	β mìnus. H.R.	(re.) *lesser.*	——	——	——	D.G. ♄.	
	γ nìtens. H.R.	(re.) *glittering.*	——	——	——	D.G. ♄.	
149	serràtum. H.R.	(ye.) saw-keeled.	——	1707.	6. 7.	D.G. ♄.	Dill. elth. t. 192. f. 238.
150	lácerum. H.R.	(re.) ragged-keeled.	——	1792.	6.	D.G. ♄.	
151	heteropétalum. H.R.	(ye.) various-petal'd.	——	1794.	5. 8.	D.G. ♄.	
	α albiflòrum.	(wh.) *white-flowered.*	——	——	——	D.G. ♄.	
	β ròseum.	(ro.) *rose-coloured.*	——	——	——	D.G. ♄.	
152	glaucìnum. H.R.	(li.) glaucine.	——	1818.	——	D.G. ♄.	
	α angústum. H.R.	(li.) *slenderest-leaved.*	——	——	——	D.G. ♄.	
	β crássum. H.R.	(li.) *thicker-leaved.*	——	——	——	D.G. ♄.	

153 mutábile. H.R. (*wh.ro.*) changeable. C. B. S. 1792. 7. 9. D.G. ♄. DC. plant. gras. t. 60.
filamentòsum. DC.
154 inclaùdens. H.R. (*pu.*) open-flowered. —— 1805. 6. 9. D.G. ♄. Andr. reposit. 388.
155 cauléscens. H.R. (*li.*) smooth delta-l^d. —— 1731. 5. 7. D.G. ♄. Dill. elth. f. 243-4.
156 deltoídes. H.R. (*li.*) great delta-leaved. —— —— 5. D.G. ♄. DC. plant. gras. 53.
157 muricàtum. H.R. (*li.*) small delta-leaved. —— —— —— D.G. ♄. Dill. elth. t. 195. f. 27.
β mìnus. H.R. (*li.*) *smaller-leaved.* —— .... —— D.G. ♄.
158 microphy'llum. H.R. (*ro.*) small-green-leav'd. —— 1795. —— D.G. ♄.
159 mucronàtum. H.R. (*ro.*) bristle-pointed. —— 1794. 6. 7. D.G. ♄.
160 pygm'æum. H.R. (*ro.*) pigmy connate. —— 1820. —— D.G. ♄.
161 pulchéllum. DC. (*ro.*) small grey-leaved. —— 1793. 4. D.G. ♄.
canéscens. H.R.
β revolùtum. H.R. (*ro.*) *revolving-branched.* —— 1819. —— D.G. ♄.
162 máximum. H.R. (*ro.*) great crescent-l^d. —— 1787. 3. 8. D.G. ♄.
163 lunàtum. H.R. (*ro.*) small crescent-l^d. —— .... 4. 6. D.G. ♄.
164 falcàtum. H.R. (*ro.*) small sickle-leav'd. —— 1727. 6. 8. D.G. ♄. Dill. elth. f. 275-6.
165 decúmbens. H.R. (*li.*) dwarf decumbent. —— 1759. 5. 10. D.G. ♄.
166 incúrvum. H.R. (*li.*) incurved. —— 1802. 6. D.G. ♄.
β dilàtans. H.R. (*li.*) *dilated.* —— 1805. 9. 10. D.G. ♄.
γ pallìdius. H.R. (*li.*) *pale flowering.* —— —— —— D.G. ♄.
δ densifòlium. H.R. (*li.*) *dense lesser-leav'd.* —— 1819. —— D.G. ♄.
ε ròseum. H.R. (*ro.*) *rosy.* —— —— —— D.G. ♄.
167 confértum. H.R. (*li.*) crowded-leaved. —— 1805. —— D.G. ♄.
168 falcifórme. H.R. (*li.*) sickle-shaped. —— —— 7. 8. D.G. ♄.
169 glomeràtum. H.R. (*re.*) clustered. —— 1732. 6. 8. D.G. ♄. Dill. elth. t. 213. f. 274.
170 infléxum. H.R. (*re.*) inflex-leaved. —— 1818. —— D.G. ♄.
171 scábrum. H.R. (*re.*) rough. —— 1731. 7. D.G. ♄. Dill. elth. t. 197. f. 251.
α pállidum. H.R. (*li.*) *pale-bloomed.* —— —— —— D.G. ♄.
β purpùreum. H.R. (*pu.*) *purple.* —— —— —— D.G. ♄.
172 versícolor. H.R. (*wh.re.*) changeable rough. —— 1795. 5. 8. D.G. ♄.
173 retrofléxum. H.R. (*re.*) reflex-flowered. —— 1794. 5. 10. D.G. ♄.
174 imbricàns. H.R. (*re.*) imbricate-leaved. —— 1818. —— D.G. ♄.
175 defléxum. H.R. (*re.*) smooth deflexed. —— 1774. 7. 10. D.G. ♄.
176 polyánthon. H.R. (*pu.*) copious-flowered. —— 1803. 8. D.G. ♄.
177 polyphy'llum. H.R. (*re.*) crowded-leaved. —— 1819. 6. 10. D.G. ♄.
178 violàceum. H.R. (*vi.*) deep violaceous. —— 1820. 10. D.G. ♄. DC. plant. gras. t. 84.
179 emarginàtum. H.R. (*vi.*) notch-petaled. —— 1732. 6. 8. D.G. ♄. Dill. elth. t. 197. f. 250.
180 dilatàtum. H.R. (*wh.*) gibbous-keeled. —— 1795. 7. D.G. ♄.
181 virgàtum. H.R. (*li.*) slender-twigged. —— 1793. 2. 4. D.G. ♄.
182 bracteàtum. H.R. (*re.*) double-bracted. —— 1774. 7. 10. D.G. ♄.
183 ancéps. H.R. (*li.*) broad-bracted. —— 1811. 9. 10. D.G. ♄.
β pállidum. H.R. (*li.*) *paler-flowered.* —— —— —— D.G. ♄.
184 grácile. H.R. (*li.*) slender bracteate. —— 1794. 8. 11. D.G. ♄.
stellàtum. H.S. *non* DC.
β gracílius. H.R. (*li.*) *pale flat-bracted.* —— .... —— D.G. ♄.
185 radiàtum. H.R. (*re.*) radiate. —— 1732. —— D.G. ♄. Dill. elth. t. 197. f. 249.
186 compréssum. H.R. (*re.*) rough compressed. —— 1792. 7. 9. D.G. ♄.
187 pátulum. H.R. (*li.*) paler-flowered. —— 1811. 10. 11. D.G. ♄.
188 ásperum. H.R. (*ro.*) large rough-leav'd. —— 1818. 7. 9. D.G. ♄.
α màjus. H.R. (*ro.*) *broader-leaved.* —— —— —— D.G. ♄.
β cæruléscens. H.R. (*ro.*) *bluish-leaved.* —— 1820. —— D.G. ♄.
189 formòsum. H.R. (*re.*) white-eyed red. —— —— 8. 9. D.G. ♄.
190 am'œna. DC. (*re.*) beautiful. —— 1825. —— D.G. ♄.
191 spectábile. H.R. (*pu.*) glaucous showy. —— 1787. 5. 8. D.G. ♄. Bot. mag. t. 396.
192 conspícuum. H.R. (*re.*) dark-red showy. —— 1806. 9. 10. D.G. ♄.
193 turbinàtum. DC. (*re.*) top-shaped. —— 1826. —— D.G. ♄. Jacq. vind. t. 476.
194 blándum. H.R. (*bh.*) fair-flowered. —— 1816. 6. 12. D.G. ♄. Bot. reg. t. 582.
195 curviflòrum. H.R. (*bh.*) incurving white. —— 1818. —— D.G. ♄.
196 àureum. H.R. (*or.*) great orange. —— 1750. 3. 10. D.G. ♄. Bot. mag. t. 262.
197 aurántium. H.R. (*or.*) glaucous orange. —— 1793. 6. 8. D.G. ♄.
198 glaùcum. H.R. (*ye.*) great gold-flower'd. —— 1696. 6. 7. D.G. ♄. DC. plant. gras. t. 146.
199 cymbifòlium. H.P. (*ye.*) greater boat-l^d. —— 1823. .... D.G. ♄.
200 stríctum. H.R. (*ye.*) upright boat-l^d. —— 1795. .... D.G. ♄.
201 mucroníferum. H.P. (*ye.*) mucronate boat-l^d. —— 1823. .... D.G. ♄.
202 cymbifórme. H.R. (*re.*) small boat-leav'd. —— 1793. .... D.G. ♄.
203 granifórme. H.R. (*ye.*) grain-leaved. —— 1727. 9. 10. D.G. ♄. Bradl. succ. 2. t. 20.
204 mólle. H.R. (*re.*) soft. —— 1774. 10. D.G. ♄.

205	coccíneum. H.R.	(sc.) scarlet-flower'd.	C. B. S.	1696.	5. 9.	D.G.♄.	Bot. mag. t. 59.
	α obtùsum. H.R.	(sc.) *blunt-calyxed.*	——	——	——	D.G.♄.	DC. plant. grass. t. 83.
	β acùtius. H.R.	(sc.pu.) *acute-calyxed.*	——	....	——	D.G.♄.	
	γ mìnus. H.R.	(sc.) *lesser.*	——	....	——	D.G.♄.	
206	bicolòrum. H.R.	(sc.ye.) two-coloured.	——	1732.	——	D.G.♄.	
	α erectum. H.R.	(sc.ye.) *tall upright.*	——	——	——	D.G.♄.	Dill. elth. t. 202. f. 258.
	β patulum. H.R.	(sc.ye.) *spreading.*	——	....	——	D.G.♄.	
	γ mìnus. H.R.	(sc.ye.) *dwarf spreading.*	——	....	——	D.G.♄.	
207	inæquàle. H.R.	(ye.) unequal-cupped.	——	1716.	——	D.G.♄.	Bradl. succ. 1. t. 7.
208	tenuifòlium. H.R.	(sc.) slender-leaved.	——	1700.	6. 9.	D.G.♄.	Dill. elth. t. 201. f. 256.
	β eréctum. H.R.	(sc.) *upright slender-ld.*	——	....	——	D.G.♄.	DC. plant. gras. t. 82.
209	variábile. H.R.	(ye.) changeable-fld.	——	1796.	6. 8.	D.G.♄.	
	α dénsius. H.R.	(ye.) *rough close-leaved.*	——	——	——	D.G.♄.	
	β lǽvius. H.R.	(ye.) *smoother distant-ld.*	——	——	——	D.G.♄.	
210	spinifórme. H.R.	(li.) thorn-shaped.	——	1793.	9. 10.	D.G.♄.	
	β subadúncum. H.R.	(li.) *greater hook-ld.*	——	1810.	10. 11.	D.G.♄.	
211	curvifòlium. H.R.	(li.) great curve-ld.	——	1799.	10.	D.G.♄.	
212	flexifòlium. H.R.	(li.) lesser curve-ld.	——	1818.	——	D.G.♄.	
	α pervíride. H.R.	(li.) *dark-green.*	——	——	——	D.G.♄.	
	β lǽte vìrens. H.R.	(li.) *apple-green.*	——	...	——	D.G.♄.	
213	inconspícuum. H.P.	(li.) inconspicuous.	——	1824.	6. 9.	D.G.♄.	
214	adúncum. H.R.	(li.) small curve-leav'd.	——	1795.	2. 3.	D.G.♄.	
215	filicaùle. H.R.	(li.) thready-stemmed.	——	1800.	9.	D.G.♄.	
216	spinòsum. H.R.	(pu.) thorny.	——	1714.	6. 9.	D.G.♄.	Dill. elth. t. 208. f. 265.
217	prodúctum. H.P.	(ro.) great-cupped.	——	1820.	6. 8.	D.G.♄.	
218	lépidum. H.P.	(wh.) pretty.	——	1824.	7. 9.	D.G.♄.	
219	stipulàceum. H.R.	(re.) tall, stipuled.	——	1723.	5. 8.	D.G.♄.	Dill. elth. f. 267 & 268.
220	corallìnum. H.R.	(ro.) coral-stemmed.	——	1819.	——	D.G.♄.	
221	Hawórthii. H.R.	(ro.) Haworth's.	——	1793.	6. 1.	D.G.♄.	
222	lǽve. H.R.	(pu.) blunt white-leav'd.	——	1774.	7. 9.	D.G.♄.	
223	veruculàtum. H.R.	(ye.) long spit-leaved.	——	1731.	5. 6.	D.G.♄.	
	α Dillènii. H.R.	(ye.) *Dillenius's.*	——	——	——	D.G.♄.	Dill. elth. f. 259.
	β Candóllii. H.R.	(ye.) *cluster spit-leaved.*	——	——	——	D.G.♄.	DC. plant. gras. t. 36.
224	purpùro-cròceum. H.R.	purple saffron-fld.	——	1780.	8. 10.	D.G.♄.	
225	flàvo-cròceum. H.P.	yellow and saffron.	——	1816.	5. 9.	D.G.♄.	
226	lùteum. H.P.	(ye.) upright yellow.	——	1822.	6. 8.	D.G.♄.	
227	lutèolum. H.P.	(ye.) small yellow.	——	——	——	D.G.♄.	
228	cròceum. H.R.	(sn.) saffron-flowered.	——	1816.	10. 11.	D.G.♄.	Jacq.frag.p.17.t.11.f.2.
229	crystallìnum. H.R.	(wh.) diamond.	——	1727.	5. 8.	D.G.♂.	DC. plant. gras. 128.
230	glaciàle. H.R.	(wh.) Ice-plant.	Greece.	1787.	——	G.⊙.	
231	pinnatífidum. H.R.	(ye.) jagged-leaved.	C. B. S.	1774.	5. 10.	G.⊙.	Bot. mag. t. 67.
232	sessiliflòrum. H.R.	(ye.) small ice-plant.	——	——	7. 8.	G.⊙.	
	α lùteum. H.R.	(ye.) *yellow-flowered.*	——	——	——	G.⊙.	
	β álbum. H.R.	(wh.) *white-flowered.*	——	1816.	——	G.⊙.	
233	humifùsum. H.R.	(wh.) narrow-ld. ice-plant.	——	1774.	——	D.G.♄.	
234	clandestìnum. H.P.	(wh.) minute-flower'd.	——	1823.	5. 8.	D.G.♄.	
235	Aitòni. H.R.	(ro.) Aiton's.	——	——	6. 10.	D.G.♂.	Jacq. vind. 3. t. 7.
236	lanceolàtum. H.R.	(wh.) spear-leaved.	——	1795.	5. 8.	G.⊙.	
	α álbum. H.R.	(wh.) *white-flowered.*	——	——	——	G.⊙.	
	β ròseum. H.R.	(ro.) *rosy.*	——	1812.	——	G.⊙.	
237	cordifòlium. H.R.	(re.) heart-leaved.	——	1774.	5. 9.	D.G.♃.	DC. plant. gras. t. 102.
238	pomeridiànum. H.R.	(ye.) great yellow.	——	——	7. 9.	G.⊙.	Bot. mag. t. 540.
	β Andre'wsii. H.R.	(ye.) *Andrews's.*	——	....	——	G.⊙.	Andr. reposit. t. 57.
239	Candóllii. H.R.	(ye.) Decandolle's.	——	....	——	G.⊙.	DC. plant. gras. t. 135.
	helianthoìdes. DC. *nec aliorum.*						
240	pilòsum. H.R.	(ye.) hairy-yellow.	——	1800.	6. 8.	G.⊙.	Breyn. cent. t. 79.
241	calendulàceum. H.R.	(y.) hispid narrow-cup'd.	——	1819.	8.	G.⊙.	
242	fláccidum. DC.	(ye.) flaccid.	——	1826.	——	D.G.♂.	
243	helianthoídes. H.R.	(ye.) spatula-leaved.	——	1774.	8. 10.	G.⊙.	
244	cuneifòlium. DC.	(pu.) transparent.	——	——	7.	G.⊙.	Jacq. ic. 3. t. 288.
	límpidum. H.R.						
245	pyrop'æum. H.S.S.	(v.) annual 3-colour'd.	——	1820.	4. 7.	G.⊙.	
	α rùbidum.	(pu.) *the ruby.*	——	——	——	G.⊙.	Bot. mag. t. 2144.
	trícolor. W. hort. ber. t. 22.						
	β ròseum. H.R.	(ro.) *the rosy.*	——	——	——	G.⊙.	
	γ álbum. H.R.	(bh.) *blush white.*	——	1819.	——	G.⊙.	
246	villòsum. H.R.	(bh.) villous.	——	1759.	7.	D.G.♄.	

247	cadùcum. H.R. (*wh.*) deciduous.	C. B. S.	1774.	7. 8.	G.⊙.	
248	apétalum. H.R. (*wh.*) dwarf spreading.	——	——	——	G.⊙.	Jacq. vind. 3. t. 6.
249	nodiflòrum. H.R. (*wh.*) knot-flowered.	Egypt.	1739.	8. 10.	G.⊙.	DC. plant. gras. t. 88.
250	ciliàtum. H.R. (*wh.*) ciliated.	C. B. S.	1774.	....	D.G.♄.	
251	geniculiflòrum. H.R. (*ye.*) joint-flower'd.	Egypt.	1727.	7. 9.	D.G.♄.	DC. plant. gras. t. 17.
252	Tripòlium. H.R. (*wh.*) Aster-leaved.	C. B. S.	1700.	6. 10.	D.G.♂.	Dill. elth. t. 179. f. 220.
253	expánsum. H.R. (*st.*) houseleek-leav'd.	——	1705.	7. 8.	D.G.♄.	—— f. 223.
	tortuòsum. DC. plant. gras. t. 94. *non* L.					
254	vàrians. H.R. (*wh.st.*) varying.	——	1706.	7. 10.	D.G.♄.	Pet. gaz. t. 78. f. 10.
255	tortuòsum. H.R. (*st.*) twisting-leaved.	——	1705.	6. 10.	D.G.♄.	Dill. elth. t. 181. f. 222.
256	pállens. H.R. (*wh.*) pale-flowered.	——	1774.	7. 8.	D.G.♄.	DC. plant. gras. t. 47.
	expánsum. DC. *non* L.					
257	crassicaùle. H.P. (*ye.*) thick-stalked.	——	1822.	5.	D.G.♄.	
258	loràtum. H.R. (*wh.*) glaucous strap-ld.	——	1819.	——	D.G.♄.	
259	relaxàtum. H.R. (*re.*) livid strap-leav'd.	——	....	——	D.G.♄.	
260	anatómicum. H.R. (*wh.*) skeleton-leav'd.	——	1803.	7. 8.	D.G.♄.	
	β *frágile.* H.R. (*wh.*) *brittle.*	——	——	——	D.G.♄.	
261	réctum. H.R. (*wh.*) ugright egg-leav'd.	——	1819.	——	D.G.♄.	
262	crassuloídes. H.R. (*bh.*) Crassula-like.	——	——	7. 8.	D.G.♄.	
263	incómptum. H.R. (*wh.*) persistent cup'd.	——	1816.	——	D.G.♄.	
264	spléndens. H.R. (*wh.*) shining white fld.	——	1716.	6. 8.	D.G.♄.	DC. plant. gras t. 35.
265	flexuòsum. H.R. (*bh.*) flexuose-leaved.	——	1795.	7. 8.	D.G.♄.	
266	acuminàtum. H.P. (*wh.*) long-cupped.	——	1820.	8. 9.	D.G.♄.	
267	albicaùle. H.P. (*wh.*) white-twigged.	——	1823.	8. 10.	D.G.♄.	
268	sulcàtum. H.R. (*wh.*) acute-cupped.	——	1819.	——	D.G.♄.	
269	fastigiàtum. H.R. (*bi.*) level-topped.	——	1794.	7. 9.	D.G.♄.	
	β *refléxum.* H.R. (*br.*) *reflex-leaved.*	——	1792.	8. 9.	D.G.♄.	
270	umbelliflòrum. H.R... umbel-flowered.	——	1825.	——	D.G.♄.	
271	longistỳlum. DC. (*bh.*) long-styled.	——	1822.	——	D.G.♄.	
	pállens. Jacq. schœnb. 3. t. 279. *non* Ait. *palléscens.* H.R.					
	β *purpuráscens.* DC. (*pu.*) *purplish.*	C. B. S.	——	——	D.G.♄.	
272	micránthon. H.R. (*wh.*) small-blossom'd.	——	1804.	——	D.G.♄.	Jacq. schœnb. 3. t. 278.
	parviflòrum. Jacq. *non* H.R.					
273	júnceum. H.R. (*ro.*) rushy.	——	1800.	8. 10.	D.G.♄.	
274	granulicaùle. H.P. (*ro.*) shagreen-stem'd.	——	1821.	6. 8.	D.G.♄.	
275	ténue. H.R. (*wh.*) slender rushy.	——	1819.	——	D.G.♄.	
276	spinulíferum. H.R. (*st.*) spinulescent.	——	1794.	6. 10.	D.G.♄.	
277	gróssum. H.R. (*bh.*) gouty.	——	1774.	8. 10.	D.G.♄.	
278	salmòneum. H.R. (*ro.*) salmon-colour'd.	——	1819.	——	D.G.♄.	
279	canaliculàtum. H.R. (*ye.*) channel-leav'd.	——	1794.	7. 10.	D.G.♄.	
280	viridiflòrum. H.R. (*gr.*) green-flowered.	——	1774.	7. 11.	D.G.♄.	Bot. mag. t. 326.
281	tenuiflòrum. H.R. (*re.*) slender-flower'd.	——	1824.	....	D.G.♄.	Jacq. frag. 52. f. 3.
282	longispìnulum. H.P. (*st.*) rosy-cupp'd yellow.	——	1821.	8. 11.	D.G.♄.	
283	nítidum. H.R. (*ye.*) shrubby crystalline.	——	1790.	7. 10.	D.G.♄.	DC. plant. gras. t. 129.
	brachiàtum. DC. *nec aliorum.*					
284	subincànum. H.P. (*wh.*) soft-leav'd white.	——	1821.	8. 9.	D.G.♄.	
285	brachiàtum. H.R. (*ye.*) three-forked.	——	1774.	6. 8.	D.G.♄.	
286	testàceum. H.R. (*sn.*) tile-coloured.	——	1816.	——	D.G.♄.	
287	macrorhìzum. H.P. (*wh.*) large-rooted.	——	1823.	6. 9.	D.G.♄.	
288	tuberòsum. H.R. (*co.*) tuberous shrubby.	——	1714.	6. 10.	D.G.♄.	DC. plant. gras. t. 78.
	β *mìnus.* H.R. (*bl.*) *lesser-leaved.*	——	——	——	D.G.♄.	
289	noctiflòrum. H.R. (*ro.wh.*) night-flowering.	——	——	6. 8.	D.G.♄.	DC. plant. gras. t. 10.
	α *phœníceum.* H.R. (*ro.*) *deep rosy.*	——	——	——	D.G.♄.	Dill. elth. f. 262.
	β *stramíneum.* H.R. (*st.*) *straw-coloured.*	——	——	——	D.G.♄.	—— f. 263.
	γ *elàtum.* H.R. (*wh.ro.*) *tall-rosy.*	——	1819.	——	D.G.♄.	
290	fúlvum. H.R. (*br.wh.*) grey-barked.	——	1732.	——	D.G.♄.	
291	clavàtum. DC. (*wh.*) clubbed.	——	....	——	D.G.♄.	Jacq. schœnb. t. 108.
	defoliàtum. H.R.					
292	horizontàle. H.R. (*st.*) spreading-leav'd.	——	1795.	——	D.G.♄.	
293	speciòsum. H.R. (*cr.*) glittering orange.	——	1793.	5. 10.	D.G.♄.	
294	mìcans. H.R. (*sc.*) dark orange.	——	1704.	5. 8.	D.G.♄.	Bot. mag. t. 448.
295	maculàtum. H.R. (*sc.*) spotted-stalked.	——	1792.	——	D.G.♄.	
296	flàvum. H.R. (*go.*) small-glittering.	——	1820.	8.	D.G.♄.	
297	oblíquum. H.R. (*re.*) bright afternoon flg.	——	1819.	——	D.G.♄.	
298	parvifòlium. H.R. (*pu.*) small triquetrous.	——	——	6. 10.	D.G.♄.	
299	brevifòlium. H.R. (*re.*) short-leaved.	——	1777.	——	D.G.♄.	Jacq. vind. t. 477.
	erigerifòlium. Jacq.					
300	subglobòsum. H.R. (*re.*) globular.	——	1795.	——	D.G.♄.	
301	pulverulèntum. H.R. (*re.*) powdered.	——	1792.	5.	D.G ♄.	

302 híspidum. H.R. (*pu.*) hispid-stalked. C. B. S. 1704. 5. 10. D.G.♄. DC. plant. gras. t. 66.
α Dillènii. H.R. (*pu.*) *the bushy.* ——— ——— ——— D.G.♄. Dill. elth. f. 278.
β platypétalum. H.R. *tall.* ——— 1820. ——— D.G.♄.
303 hirtéllum. H.R. (*re.*) dwarf bristly. ——— 1792. 5. 11. D.G.♄.
304 cándens. H.R. (*wh.*) trailing hispid. ——— 1820. ——— D.G.♄.
305 subcompréssum. H.P. (*vi.*) upright twiggy. ——— 1824. 6. 8. D.G.♄.
β mìnus. H.P. (*vi.*) *lesser.* ——— ——— ——— D.G.♄.
306 floribúndum. H.R. (*fl.*) crowded-flower'd. ——— 1704. 5. 10. D.G.♄. Dill. elth. t. 214. f. 280.
307 calycìnum. H.R. (*wh.*) hispid long-cup'd. ——— 1819. ——— D.G.♄.
308 striàtum. H.R. (*std.*) striped bristly. ——— 1727. ——— D.G.♄. Dill. elth. t. 215. f. 281.
α ròseum. H.R. (*std.*) *rosy.* ——— ——— ——— D.G.♄. ——— t. 215. f. 281.
β pállens. H.R. (*std.*) *pale-flowered.* ——— ——— ——— D.G.♄. DC. plant. gras. t. 132.
309 attenuàtum. H.R. (*v.*) slender hispid. ——— 1820. ——— G. ♄.
310 tuberculàtum. DC. (*v.*) hispid-leaved. ——— 1814. ——— G. ♄.
hispifòlium. H.R.
α fastigiàtum. H.R. (*v.*) *level-topped.* ——— ——— ——— G. ♄.
β ròseum. H.R. (*ro.*) *rosy-flowered.* ——— ——— ——— G. ♄.
311 echinàtum. H.R. (*ye.*) hedge-hog. ——— 1774. 7. 10. G. ♄. DC. plant. gras. t. 24.
α lùteum. (*ye.*) *yellow.* ——— ——— ——— G. ♄.
β álbum. (*wh.*) *white-flowered.* ——— ——— ——— G. ♄.
312 strumòsum. H.R. (*st.*) tuberous hedge-hog. ——— 1820. ——— G. ♄.
313 barbàtum. H.P. (*pu.*) bearded bushy. ——— 1793. 5. 10. G. ♄. Bot. mag. t. 70.
314 stellígerum. H.P. (*pu.*) procumbent bearded. ——— 1705. 6. 8. G. ♄.
315 bulbòsum. H.P. (*re.*) bearded tuberous. ——— 1820. ——— G. ♄.
316 intónsum. H.P. (*re.*) black bearded. ——— ——— 7. G. ♄.
α rubicúndum. H.P. (*re.*) *red-flowered.* ——— ——— ——— G. ♄.
β álbum. H.P. (*bh.*) *white-flowered.* ——— ——— ——— G. ♄.
317 stellàtum. H.P. (*pu.*) small-bearded. ——— 1716. 9. 10. G. ♄. DC. plant. gras. t. 29.
318 dénsum. H.P. (*pu.*) dwarf-bearded. ——— 1732. 5. 8. G. ♄. Bot. mag. t. 1220.

HYMENO'GYNE. H.R. HYMENO'GYNE. Icosandria Dodecagynia.
glábra. H.R. (*st.*) smooth. C. B. S. 1787. 7. 8. G. ⊙.
Mesembryánthemum glábrum. H.S. *non* A.R.

TETRAG'ONIA. W. TETRAG'ONIA. Icosandria Pentagynia. L.
1 expánsa. H.S. (*gr.*) horned. N. Zealand. 1772. 8. 9. H. ⊙. DC. plant. gras. t. 114.
2 crystallìna. H.S. (*gr.*) diamond. Peru. 1788. 6. 12. H. ⊙. ——— t. 34.
3 echinàta. H.S. (*gr.*) hedgehog. C. B. S. 1774. 5. 8. G. ♂. ——— t. 113.
4 hirsùta. DC. (*gr.*) hairy. ——— 1823. ——— G. ♂.
5 fruticòsa. H.S. (*gr.*) shrubby. ——— 1712. 7. 9. G. ♄. Mill. ic. 2. t. 263. f. 2.
6 lineàris. H.R. (*gr.*) linear-leaved. ——— 1819. 9. G. ♄.
7 obovàta. H.R. (*gr.*) obovate-leaved. ——— ——— .... G. ♄.
8 decúmbens. H.S. (*gr.*) trailing. ——— 1758. 7. 9. G. ♄. DC. plant. gras. t. 23.
9 Tetrápteris. H.S. (*gr.*) wing-seeded. ——— 1795. ——— G. ♄.
10 spicàta. H.S. (*gr.*) spiked. ——— ——— ——— G. ♃.
11 herbàcea. H.S. (*gr.*) tuberous-rooted. ——— 1752. 6. 7. G. ♃. Comm. hort. 2. t. 102.
12 purpùrea. Ot. (*pu.*) purple. .... 1825. ——— H. ⊙.

SES'UVIUM. DC. SES'UVIUM. Icosandria Tri-Pentagynia.
1 pedunculàtum. P.S. (*pu.*) peduncled. W. Indies. 1692. 6. 8. D.S. ♃. Pluk. phyt. t. 216. f. 1.
2 sessìle. P.S. (*pu.*) sessile-flowered. ——— ——— ——— D.S. ♃. DC. plant. gras. t. 9.
Aizoòn canariénse. Andr. rep. t. 201. *nec aliorum.*
3 longifòlium. DC. (*pu.*) long-leaved. S. America. 1816. 7. 8. D.S. ♃.
4 revolutifòlium. DC. (*pu.*) revolute-leaved. Cuba. .... ——— D.S. ♃. Bot. mag. t. 1701.
5 rèpens. DC. (*pu.*) creeping. E. Indies. 1816. ——— S. ⊙. Rumph. amb. 5. t. 72. f. 1.
6 microphy'llum. DC. (*pu.*) small-leaved. Havannah. 1823. ——— D.S. ♃.

AIZO'ON. DC. AIZO'ON. Icosandria Pentagynia. L.
1 canariénse. DC. (*ye.*) Purslane-leav'd. Canaries. 1731. 7. 8. D.G. ♃. DC. plant. gras. t. 136.
2 glinoídes. DC. (*ro.*) hairy. C. B. S. 1774. 6. 8. D.G. ♃.
3 hispánicum. DC. (*wh.*) Spanish. Spain. 1728. 7. 8. H. ⊙. DC. plant. gras. t. 30.
4 tomentòsum. DC. (*ye.*) woolly. C. B. S. 1824. ——— D.G. ♄.
5 stellàtum. DC. (*ye.*) starry. ——— ——— ——— D.G. ♄.
6 perfoliàtum. DC. (*ye.*) perfoliate. ——— 1816. 6. 8. D.G. ♄.
7 sarmentòsum. DC. (*ye.*) trailing. ——— ——— ——— D.G. ♄. Burm. afr. t. 26. f. 2.
8 lanceolàtum. W. (*ye.*) spear-leaved. ——— 1752. 8. D.G. ♂.

GL'INUS. DC. GL'INUS. Dodecandria Pentagynia. L.
lotoídes. DC. (*ye.*) hairy. S. Europe. 1788. 7. 8. H. ⊙. Bocc. sic. 21. t. 11.

Tribus II. *REAUMURIEÆ.*

REAUM'URIA. DC. REAUM'URIA. Icosandria Pentagynia.
1 vermiculàta. DC. (*wh.*) white-flowered. Persia. 1829. .... G. ♄. Lob. ic. t. 380.

2 hypericoídes. DC.	(pu.)	purple-flowered.	Syria.	1800.	7. 10.	G. ♄.	Bot. mag. t. 2057.
linifòlia. Sal. par. lond. t. 18.							
NITR'ARIA. DC.		NITR'ARIA. Dodecandria Trigynia.					
1 Schobèri. L.	(wh.)	thick-leaved.	Siberia.	1778.	5. 8.	H. ħ.	Andr. rep. t. 529.
2 sibírica. F.	(wh.)	Siberian.	———	1826.	....	H. ħ.	Gmel. sib. 2. t. 98.
3 cáspica. W.en.	(wh.)	Caspian.	Caspia.	1822.	5. 8.	H. ħ.	Pall. ross. t. 50. f. B.
4 tridentàta. DC.	(wh.)	tridentate.	Africa.	——	——	H. ħ.	

ORDO CIII.

CACTEÆ. *DC. prodr. v.* 3. *p.* 457.

Tribus I. *OPUNTIACEÆ.* DC. prodr. v. 3. p. 458.

MAMMILL'ARIA. H.S.		MAMMILL'ARIA. Icosandria Monogynia.					
1 magnimámma. DC.	(re.)	large-beaded.	Mexico.	1823.	6. 8.	D.S. ♄.	
2 coronària. DC.	(sc.)	the great.	———	1820.	——	D.S. ♄.	Orteg. dec. t. 16.
3 díscolor. DC.	(wh.re.)	two-coloured.	S. America.	1800.	——	D.S. ♄.	DC. diss. t. 2. f. 2.
Cáctus Spinii. Colla. *depréssus*. DC. cat. h. monsp. p. 84. *pseudo-mammillàris*. Salm Dyck.							
4 lanífera. DC.	(re.)	wool-bearing.	Mexico.	1823.	6. 8.	D.S. ♄.	DC. diss. t. 4.
5 flavéscens. DC.	(re.)	yellow-spined.	S. America.	1811.	——	D.S. ♄.	
6 stramínea. H.S.S.	(re.)	straw-coloured.	———	——	——	D.S. ♄.	
7 prolífera. DC.	(wh.)	white-spined.	———	1688.	7. 8.	D.S. ♄.	
8 símplex. DC.	(wh.)	small red-spin'd.	W. Indies.	——	——	D.S. ♄.	DC. plant. gras. t. 3.
Cáctus mammillàris. L.							
9 missouriénsis. L.C.	(wh.)	Missouri.	Missouri.	1813.	6. 8.	D.G. ♄.	
Cáctus mammillàris. Nuttall. *nec aliorum*.							
10 vivípara. DC.	(re.)	viviparous.	———	——	——	D.G. ♄.	
11 parvimámma. DC.	....	small-beaded.	S. America.	1817.	....	D.S. ♄.	
Cáctus microthèle. S.S.							
12 geminispìna. DC.	(re.)	twin-spined.	Mexico.	1823.	6. 8.	D.S. ♄.	DC. diss. t. 3.
13 glomeràta. DC.	(re.)	tufted.	St. Domingo.	1825.	....	D.S. ♄.	Plum. ed Bur. t. 201. f. 1.
14 pusílla. DC.	(bh.)	hoary starry.	S. America.	1816.	6. 9.	D.S. ♄.	DC. diss. t. 2. f. 1.
Cáctus stellàtus. Lodd. bot. cab. t. 79.							
15 Helícteres. DC.	(ro.)	simple obovate.	Mexico.	1827.	——	D.S. ♄.	DC. diss. t. 5.
16 cònica. H.S.S.	....	cone-beaded.	....	1808.	....	D.S. ♄.	
17 coccínea. Gil.	(sc.)	scarlet-flower'd.	Chile.	1827.	....	D.S. ♄.	
18 speciòsa. Gil.	(re.)	handsome.	———	——	....	D.S. ♄.	
19 fuscàta. Ot.	....	brown.	S. America.	——	....	D.S. ♄.	
20 canéscens. Ot.	....	canescent.	———	——	....	D.S. ♄.	
21 cæspitòsa. Ot.	....	tufted.	———	——	....	D.S. ♄.	
22 chrysántha. Ot.	(ye.)	yellow-flowered.	———	——	....	D.S. ♄.	
MELOCA'CTUS. DC.		MELON THISTLE. Icosandria Monogynia.					
1 commùnis. DC.	(re.)	Turk's-cap.	W. Indies.	1688.	7. 8.	D.S. ♄.	DC. plant. gras. t. 112.
β *oblóngus*. DC.	(re.)	*oblong*.	———	——	——	D.S. ♄.	
γ *macrocéphalus*.		*large-beaded*.	———	——	——	D.S. ♄.	
2 macracánthus. DC.	(re.)	large-spined.	S. America.	1820.	——	D.S. ♄.	Lk. et Ot. dis. t. 12.
3 pyramidàlis. DC.	....	pyramidal.	Curassao.	1824.	....	D.S. ♄.	——— p. 10. t. 25.
4 placentifórmis. DC.	(re.)	black-spined.	Brazil.	——	7. 9.	D.S. ♄.	Besl. h. eyst. 4. ord. f. 1.
Cáctus Melocáctus. Besl. *non* L. *Melocáctus Beslèri*. Lk. et Ot. dis. p. 11. t. 21.							
5 Langsdórfii. DC.	(ye.)	Langsdorf's.	Brazil.	1826.	7. 9.	D.S. ♄.	
6 Sellòwii. DC.	....	Sellow's.	Monte Video.	——	....	D.S. ♄.	Lk. et Ot. dis. t. 22.
7 polyacánthus. DC.	....	many-spined.	Brazil.	——	....	D.S. ♄.	——— p. 13. t. 16. f. 1.
ECHINOCA'CTUS. DC.		ECHINOCA'CTUS. Icosandria Monogynia.					
1 gibbòsus. DC.	(wh.)	gibbous.	Jamaica.	1808.	6. 8.	D.S. ♄.	Bot. reg. t. 137.
Cáctus gibbòsus. B.R.							
2 latispìnus. H.P.	(pu.)	flat-spined.	Mexico.	1823.	....	D.S. ♄.	Lk. et Ot. dis. t. 14.
cornígerus. DC. dis. t. 7 ?							
3 crispàtus. DC.	(pu.)	curl-ribbed.	———	1826.	....	D.S. ♄.	DC. diss. t. 8.
4 obvallàtus. DC.	(pu.)	guarded.	———	——	....	D.S. ♄.	——— t. 9.
5 tenuispìnus. DC.	(ye.)	slender-spined.	Brazil.	——	6. 8.	D.S. ♄.	Lk. et Ot. dis. t. 19. f. 1.
6 melocactifórmis. DC.	(bh.)	Melocactus-like.	Mexico.	——	....	D.S. ♄.	DC. diss. t. 10.
7 tephracánthus. DC.	....	ashy-spined.	Brazil.	1825.	....	D.S. ♄.	Lk. et Ot. dis. t. 14. f. 2.
8 recúrvus. DC.	....	recurve-spined.	Mexico.	1768.	....	D.S. ♄.	
Cáctus nóbilis. L.							

9	intórtus. DC.	....	twisted.	Antigua.	1768.	....	D.S.♄.	
10	Salmiànus. DC.	....	Salm-Dyck's.	Curassao.	1824.	....	D.S.♄.	
11	tuberculàtus. DC.	....	tubercled.	Mexico.	1826.	....	D.S.♄.	Lk. et Ot. dis. t. 26.
12	gladiàtus. DC.	....	sword-spined.	———	———	....	D.S.♄.	——— p. 17. t. 17.
13	subulíferus. DC.	....	awl-spined.	———	1824.	....	D.S.♄.	——— p. 16. t. 27.
14	depréssus. DC.	....	depressed.	S. America.	1780.	....	D.S.♄.	
15	orthacánthus. DC.	....	straight-spined.	Monte Video.	1828.	....	D.G.♄.	Lk. et Ot. dis. t. 18.
16	acuàtus. DC.	....	sharp-ribbed.	———	———	....	D.G.♄.	——— p. 15. t. 23.
17	parvispìnus. DC.	....	small-spined.	S. America.	1815.	....	D.S.♄.	
18	intricàtus. DC.	....	intricate.	Monte Video.	1828.	....	D.S.♄.	Lk. et Ot. dis. t. 24.
19	meonacánthus. DC.	....	oblong.	Jamaica.	....	....	D.S.♄.	——— p. 19. t. 15.
20	inflàtus. Gil.	....	inflated.	Chile.	1828.	....	D.S.♄.	
21	erinàceus.	....	Hedgehog.	....	1818.	....	D.S.♄.	
	Cáctus erinàceus. H.S.S.							
	C'EREUS. DC.		C'EREUS.	Icosandria Monogynia.				
1	multangulàris. DC.	....	many angled.	S. America.	1815.	....	D.S.♄.	
2	redúctus. DC.	....	removed.	Mexico.	1796.	....	D.S.♄.	
	Cáctus nóbilis. H.S.							
3	Scòpa. DC.	....	setaceous.	Brazil.	1826.	....	D.S.♄.	
4	sénilis. DC.	....	grey-headed.	Mexico.	1823.	....	D.S.♄.	
	Cáctus sénilis. H.P.							
5	peruviànus. DC.	(*re.wh.*)	great Peruvian.	Peru.	1728.	8. 10.	D.S.♄.	DC. plant. gras. t. 58.
6	monstròsus. DC.	(*re.wh.*)	irregular.	S. America.	1816.	——	D.S.♄.	DC. diss. t. 11.
	Cáctus abnórmis. W. enum. sup. 31.							
7	monoclònos. DC.	(*ro.wh.*)	obcordate-petal'd.	Caribees.	....	6. 8.	D.S.♄.	Plum. ed. Burm. t. 191.
8	fimbriàtus. DC.	(*ro.*)	fringe-petal'd.	St. Domingo.	1826.	——	D.S.♄.	——— t. 195. f. 1.
9	hexagònus. H.S.	(*ro.wh.*)	great 6-angled.	Surinam.	1690.	7. 8.	D.S.♄.	Bradl. succ. t. 1.
10	heptàgonus. H.S.	(*wh.*)	great 7-angled.	W. Indies.	1728.	——	D.S.♄.	
11	Hístrix. DC.	....	Porcupine.	....	1818.	....	D.S.♄.	
12	ebúrneus. DC.	....	ivory-thorned.	Curassao.	——	....	D.S.♄.	Bradl. succ. t. 10.
	β *monstròsus.* DC.	....	*monstrous.*	Hybrid.	....	....	D.S.♄.	
13	grìseus. H.S.	(*wh.*)	grey-spined.	S. America.	1809.	9. 10.	D.S.♄.	
14	crenulàtus. DC.	....	crenulate.	———	1822.	....	D.S.♄.	
15	albispìnus. DC.	....	white-spined.	———	——	....	D.S.♄.	
16	regàlis. DC.	....	the royal.	———	1816.	....	D.S.♄.	
17	stríctus. DC.	....	straight.	———	1822.	....	D.S.♄.	
18	Hawórthii. DC.	....	Haworth's.	Caribees.	1811.	....	D.S.♄.	
	Cáctus Hawórthii. S.S. *Cèreus nóbilis.* H.S.							
19	aùreus. DC.	....	gold-spined.	S. America.	1825.	....	D.S.♄.	
20	nìger. DC.	(*wh.*)	black.	———	1820.	....	D.S.♄.	
21	fulvispinòsus. DC.	....	tawny-spined.	———	1795.	....	D.S.♄.	
22	flavispìnus. DC.	....	yellow-spined.	———	1822.	....	D.S.♄.	
23	chiloénsis. DC.	....	ten-angled.	Chile.	1825.	....	D.S.♄.	
24	euphorbioídes. DC.	....	Euphorbia-like.	S. America.	1815.	....	D.S.♄.	
25	Royèni. DC.	(*wh.*)	Royen's 9-angled.	W. Indies.	1728.	6. 9.	D.S.♄.	DC. plant. gras. t. 143.
26	lanuginòsus. DC.	(*wh.*)	woolly.	———	1690.	7. 8.	D.S.♄.	Herm. parad. t. 115.
27	repándus. DC.	(*wh.*)	waved short-spin'd.	———	1728.	8. 10.	D.S.♄.	Bot. reg. t. 336.
28	subrepándus. DC.	(*wh.*)	waved long-spin'd.	———	1817.	——	D.S.♄.	
29	polygònus. DC.	(*wh.*)	many-sided.	St. Domingo.	1826.	....	D.S.♄.	Plum. ed. Burm. t. 196.
30	rosàceus. Ot.	(*ro.*)	rosy.	....	1826.	....	D.S.♄.	
31	Déppii. Ot.	....	Deppi's.	....	——	....	D.S.♄.	
32	incrustàtus. Ot.	....	incrusted.	....	——	....	D.S.♄.	
33	grándis. DC.	....	straddling-spin'd.	......	1816.	....	D.S.♄.	
34	tetragònus. DC.	(*wh.*)	great 4-angled.	S. America.	1710.	7. 8.	D.S.♄.	
35	paniculàtus. DC.	(*wh.re.*)	panicled.	St. Domingo.	1827.	....	D.S.♄.	Plum. ed. Burm. t. 192.
36	obtùsus. DC.	....	blunt 3-angled.	Brazil.	1819.	....	D.S.♄.	
37	serpentìnus. DC.	(*pu.wh.*	white-spined.	S. America.	1817.	7. 8.	D.S.♄.	Lk. et Ott. abb. t. 91.
38	ambíguus. DC.	(*pu.wh.*)	ambiguous.	....	1827.	——	D.S.♄.	
39	flagellifórmis. DC.	(*ro.*)	creeping.	Peru.	1690.	3. 6.	D.G.♄.	Bot. mag. t. 17.
40	caudàtus. Gil.	....	tailed.	Chile.	1828.	....	D.G.♄.	
41	horizontàlis. Gil.	....	horizontal.	———	——	....	D.G.♄.	
42	grandiflòrus. DC.	(*ye.wh.*)	night-flowered.	Jamaica.	1700.	6. 8.	D.S.♄.	Andr. reposit. t. 508.
43	Colvíllii.	....	Colvill's hybrid.	Hybrid.	1825.	....	D.S.♄.	
44	pentagònus. DC.	(*wh.*)	five-angled.	S. America.	1769.	7. 8.	D.S.♄.	
45	réptans. H.S.S.	....	creeping.	———	1813.	....	D.S.♄.	
46	albisetòsus. H.S.S.	....	white-spin'd trailg.	St. Domingo	1816.	....	D.S.♄.	
47	rádicans. DC.	....	rooting-branch'd.	S. America.	1827.	....	D.S.♄.	
48	hùmilis. DC.	....	slender dwarf.	....	——	....	D.S.♄.	
49	quadrangulàris. DC.	(*wh.*)	quadrangular.	S. America.	1809.	7. 9.	D.S.♄.	Plum. ed. Bur. t. 199. f. 1.

50 trípteris. DC. .... broad-lobed. .... 1827. .... D.S.♄.
51 bifróns. H.S.S .... rooting 4-angular. St. Domingo. 1818. .... D.S.♄.
52 speciosíssimus. DC. (sc.pu.) beautiful. Mexico. 1816. 6. 9. D.G.♄. Bot. reg. t. 486.
Cáctus speciosíssimus. B.R. *speciòsus.* Colla hort. rip. t. 10.
53 triangulàris. DC. (wh.) great triangular. W. Indies. 1690. 7. 8. D.S.♄. Bot. mag. t. 1884.
54 trigònus. DC. (wh.) small triangular. ——— 1809. —— D.S.♄. Plum. ed. Bur. t. 200. f. 2.
55 triquèter. DC. .... least triangular. S. America. 1794. .... D.S.♄.
56 coccíneus. DC. (sc.) scarlet-flowered. Brazil. 1828. .... D.S.♄.
57 Smithiànus. (sc.) Smith's hybrid. Hybrid. 1824. 6. 9. D.G.♄.
58 exténsus. DC. .... extended. .... 1826. .... D.S.♄.
59 squamulòsus. DC. (ye.) scaly. Brazil. —— .... D.S.♄.
60 prismáticus. DC. .... prismatic. .... —— .... D.S.♄.
61 setáceus. DC. .... bristly. Brazil. 1828. .... D.S.♄.
62 myosurus. DC. .... mouse-tail. ——— —— .... D.S.♄.
63 tènuis. DC. (ro.) slender-branched. ——— 1826. .... D.S.♄.
64 tenuispìnus. H.P. .... slender-spined. Mexico. —— .... D.S.♄.
65 phyllanthoídes. DC. (pi.) rosy-flowered. ——— 1816. 5. 8. D.G.♄. Bot. mag. t. 2092.
Cáctus speciósus. Bot. reg. t. 304. *Epiphy'llum phyllanthoídes.* H.R.
66 Jenkinsòni. (sc.) Mr. Jenkinson's. Hybrid. 1824. 6. 9. D.G.♄.
67 Ackermánni. H.R. (sc.) Ackermann's. Mexico. 1824. 5. 7. D.G.♄.
68 truncàtus. DC. (cr.) truncated. Brazil. 1818. 7. 11. D.S.♄. Bot. reg. t. 696.
Cáctus truncàtus. Bot. mag. t. 2092. *Epiphy'llum truncàtum.* H.P.
69 alàtus. DC. (gr.wh.) wing-stemmed. Jamaica. 1818. 8. 11. D.S.♄. Bot. mag. t. 2820.
70 Hookèri. H.P. (wh.) Hooker's. S. America. .... 7. 8. D.S.♄. ——— t. 2692.
Cáctus Phyllánthus. B.M. *nec aliorum.* *Epiphy'llum Hookèri.* H.P.
71 Phyllánthus. DC. (wh.) long-tubed. S. America. 1710. 6. 7. D.S.♄. DC. plant. gras. t. 145.
72 grácilis. DC. .... slender. ——— 1826. .... D.S.♄.
73 imbricàtus. H.R. .... imbricated. ——— 1820. .... D.S.♄.
74 cylíndricus. H.S. .... cylindrical. Peru. 1799. .... D.G.♄.

OPU'NTIA. DC. INDIAN-FIG. Icosandria Monogynia.
1 foliòsa. DC. (ye.) greater Pin-pillow. S. America. 1805. 6. 7. D.S.♄.
pusílla. H.S. *non* DC.
2 curassávica. DC. (ye.) Pin-pillow. Curassao. 1690. —— D.S.♄. Bradl. succ. t. 4.
α màjor. H.S. (ye.) *larger.* ——— —— —— D.S.♄.
β mèdia. H.S. (ye.) *middle.* ——— —— —— D.S.♄.
γ mìnor. H.S. (ye.) *smaller.* ——— —— —— D.S.♄.
δ lónga. H.R. (ye.) *longer.* Brazil. 1820. —— D.S.♄.
3 frágilis. DC. (ye.) brittle Pin-pillow. Missouri. 1814. .... D.F.♄.
4 pusílla. DC. (ye.) small Pin-pillow. S. America. 1826. .... D.S.♄.
5 polyacántha. H.S.S. (st.) many-spined. Missouri. 1814. 7. 9. D.F.♄.
missouriénsis. DC. *Cáctus feróx.* N.
6 mèdia. H.S.S. (ye.) lesser many-spined. ——— —— —— D.F.♄.
7 fèrox. DC. (ye.) ferocious. S. America. 1817. .... D.S.♄.
8 spinosíssima. DC. .... cluster-spined. Jamaica. 1732. 7. D.S.♄.
9 hùmilis. H.S. (ye.) humble. W. Indies. 1795. 7. 8. D.S.♄.
hórrida. DC.
10 Dillènii. DC. (ye.) Dillenius's. S. America. 1731. 9. 10. D.S.♄. Bot. reg. t. 255.
Cáctus Dillènii. B.R. Dill. elth. 2. t. 296. f. 382.
11 Tùna. DC. (ye.) yellow-spined. ——— —— —— D.S.♄. Dill. elth. t. 295. f. 380.
12 polyántha. DC. (ye.) many-flowered. ——— 1811. 7. 9. D.S.♄. Bot. mag. t. 2691.
13 monacántha. DC. (ye.) one-spined. ——— 1816. —— D.S.♄. DC. pla. gras. n. 137. t. 2.
14 elàtior. DC. (ye.) great black-spin'd. ——— 1731. 7. 8. D.S.♄. Dill. elth. t. 204. f. 379.
15 nígricans. DC. (ye.) lesser black-spined. ——— 1795. —— D.S.♄. Bot. mag. t. 1557.
Cáctus Tùna γ nigricàns. B.M.
16 máxima. DC. (ye.) great oblong. S. America. 1768. 7. 9. D.S.♄.
17 triacántha. DC. (ye.) three-spined. ——— 1796. —— D.S.♄.
18 decumàna. H.R. .... great. ——— 1820. .... D.S.♄.
19 elongàta. H.S.S. .... various-spined. ——— 1816. .... D.S.♄.
20 cochinillífera. DC. (re.) Cochineal-fig. ——— 1688. 7. 9. D.S.♄. Andr. rep. t. 523.
Cáctus cochinillífer. Bot. mag. t. 2741-2742.
21 Fìcus-I'ndica. DC. (ye.) Indian Fig. ——— 1731. 7. 8. D.S.♄.
22 pseùdo-Tùna. DC. (ye.) bastard. ——— .... —— D.S.♄.
23 tomentòsa. DC. .... woolly-branch'd. ——— 1824. .... D.S.♄.
24 serícea. Gil. .... silky. Chile. 1827. .... D.S.♄.
25 lanceolàta. DC. (ye.) spear-shaped. ——— 1796. 7. 9. D.S.♄.
26 tuberculàta. DC. (ye.) tubercled. ——— 1818. —— D.S.♄.
27 inérmis. DC. (ye.) upright spineless. ——— 1796. —— D.S.♄. DC. plant. gras. n. 138. ic.
stricta. H.S.
28 Amycl'æa. DC. (ye.) Neapolitan. Naples. 1825. —— D.F.♄.

29	subinérmis. L.en.	(ye.)	nearly smooth	S. America.	1824.	7. 9.	D.S. ♄.	
30	rubéscens. DC.	....	red-stemmed.	Brazil.	1828.	....	D.S. ♄.	
31	crássa. DC.	(ye.)	thick-lobed.	Mexico.	1817.	6. 9.	D.G. ♄.	
32	longispìna. Ot.	(ye.)	long-spined.	S. America.	1824.	....	D.S. ♄.	
33	leucántha. Ot.	(wh.)	white-flowered.	———	——	....	D.S. ♄.	
34	pulvinàta. DC.	....	pulvinate.	Mexico.	1827.	....	D.S. ♄.	
35	Hernandèsii. DC.	(re.)	Hernandez's.	———	——	....	D.G. ♄.	Hern. mex. 78. ic.
36	vulgàris. DC.	(ye.)	common.	America.	1596.	7. 9.	D.F. ♄.	Mill. dict. t. 191.
	Cáctus Opúntia. Bot. mag. t. 2393.							
	β mìnor.	(ye.)	*lesser.*	N. America.	1814.	——	D.F. ♄.	
	Cáctus Opúntia. N.							
37	brasiliénsis. DC.	(gr.)	round-stemm'd.	Brazil.	1817.	....	D.S. ♄.	Pis. bras. p. 100. f. 2.
	PERE'SKIA. DC.		PERE'SKIA. Icosandria Monogynia.					
1	aculeàta. DC.	(wh.)	Barbadoes Goosebʏ.	W. Indies.	1696.	10. 11.	S. ♄.	Dill. elth. t. 227. f. 294.
	Cáctus Peréskia. L.							
2	longispìna. H.S.S.	(wh.)	long-spined.	S. America.	1808.	....	S. ♄.	
3	portulacæfòlia. DC.	(pu.)	Purslane-leav'd.	W. Indies.	1820.	....	S. ♄.	Plum. ed Bur. t. 197. f. 1.
4	grandifòlia. DC.	....	great-leaved.	Brazil.	1818.	....	S. ♄.	

Tribus II. *RHIPSALIDEÆ.* DC. prodr. v. 3. p. 475.

	RHIPS'ALIS. DC.		RHIPS'ALIS. Dodecandria Monogynia.					
1	Cass'ytha. DC.	(wh.)	pendulous.	W. Indies.	1758.	7. 10.	S. ♄.	Gært. sem. 1. t. 28. f. 1.
2	Hookeriàna. DC.	(wh.)	four-petal'd.	———	....	——	S. ♄.	Hook. ex. flor. t. 2.
3	fasciculàta. DC.	(wh.)	cluster-branch'd.	S. America.	1800.	——	S. ♄.	
4	parasìtica. DC.	(wh.)	parasitical.	W. Indies.	——	——	S. ♄.	Plum. ed Bur. t. 197. f. 2.
5	funàlis. DC.	(wh.)	great-flowered.	S. America.	1816.	7. 8.	S. ♄.	Bot. mag. t. 2740.
	grandiflòrus. B.M. *Cáctus funàlis.* S.S.							
6	salicornioídes. DC.	(ye.)	Glass-wort-like.	W. Indies.	1818.	5. 8.	D.S. ♄.	Bot. mag. t. 2461.
7	mesembryanthoídes. DC.		Fig Marygold-like.	———	——	——	D.S. ♄.	

ORDO CIV.

GROSSULARIEÆ. *DC. prodr. v. 3. p. 477.*

	R'IBES. DC.		CURRANT. Pentandria Monogynia. L.					
1	oxyacanthoídes. DC.	(st.)	Hawthorn-leav'd.	N. America.	1705.	4. 5.	H. ♄.	Dill. elth. t. 139. f. 166.
2	fèrox. DC.	(st.)	ferocious.	California.	1827.	——	H. ♄.	
3	lacústre. DC.	(br.)	swamp.	N. America.	1812.	——	H. ♄.	Lodd. bot. cab. t. 884.
4	Menzièsii. DC.	(cr.)	Menzies'.	———	1827.	——	H. ♄.	
5	speciòsum. DC.	(pu.)	beautiful.	———	——	——	H. ♄.	
6	ùva-críspa. R.S.	(st.)	smooth Gooseberry.	England.	....	——	H. ♄.	Eng. bot. v. 29. t. 2057.
7	reclinàtum. R.S.	(st.)	procumbent.	Germany.	1683.	——	H. ♄.	
8	Grossulària. R.S.	(st.)	rough Gooseberry.	England.	....	——	H. ♄.	Eng. bot. v. 12. t. 1292.
9	caucásicum. DC.	(st.)	Caucasian.	Caucasus.	1820.	——	H. ♄.	
10	setòsum. B.R.	(st.)	bristly.	Missouri.	1827.	5.	H. ♄.	Bot. reg. t. 1237.
11	hirtéllum. DC.	(st.)	hairy.	N. America.	1812.	4. 5.	H. ♄.	
12	grácile. DC.	(st.)	slender.	———	——	——	H. ♄.	
13	Cynósbati. DC.	(gr.)	prickly-fruited.	Canada.	1759.	4.	H. ♄.	Schmidt arb. t. 98.
14	triflòrum. DC.	(br.)	three-flowered.	N. America.	1812.	4. 5.	H. ♄.	Lodd. bot. cab. t. 1094.
15	Dikúscha. F.	(st.)	Fischer's.	Siberia.	1821.	——	H. ♄.	
16	Stakòri. F.	(st.)	Siberian.	———	——	——	H. ♄.	
17	orientàle. DC.	(st.)	oriental.	Syria.	1820.	——	H. ♄.	
18	saxátile. DC.	(br.)	rock.	Siberia.	1828.	——	H. ♄.	
19	diacántha. DC.	(st.)	two-spined.	———	1781.	5. 6.	H. ♄.	Pall. ross. 2. t. 66.
20	alpìnum. DC.	(st.)	Alpine.	Britain.	....	4. 5.	H. ♄.	Eng. bot. v. 10. t. 704.
	α stérile. DC.	(st.)	*barren.*	———	....	——	H. ♄.	
	β baccíferum. DC.	(st.)	*berry-bearing.*	———	....	——	H. ♄.	
21	resinòsum. DC.	(st.)	clammy.	N. America.	1800.	——	H. ♄.	Bot. mag. t. 1583.
22	procúmbens. DC.	(br.)	trailing.	Dahuria.	1804.	5. 6.	H. ♄.	Pall. ross. 2. t. 65.
23	multiflòrum. DC.	(st.)	many-flowered.	Hungary.	1818.	5.	H. ♄	Bot. mag. t. 2368.
24	spicàtum. DC.	(br.)	acid.	England.	....	——	H. ♄.	Eng. bot. v. 18. t. 1290.

25 rùbrum. DC.	(*st.*)	common red.	Britain.	....	4. 5.	H. ♄.	Eng. bot. v. 18. t. 1289.
α sylvéstre. DC.	(*st.*)	*wild.*	———	....	——	H. ♄.	
β horténse. DC.	(*st.*)	*cultivated.*	....	....	——	H. ♄.	
γ cárneum. DC.	(*st.*)	*pale-fruited.*	....	....	——	H. ♄.	
δ variegàtum. DC.	(*st.*)	*stripe-fruited.*	....	....	——	H. ♄.	
ε álbum. DC.	(*st.*)	*white-fruited.*	....	....	——	H. ♄.	Berl. mem. t. 2. f. 15.
26 petr'æum. DC.	(*re.*)	rock.	England.	....	5. 6.	H. ♄.	Eng. bot. v. 10. t. 705.
27 cèreum. B.R.	(*wh.*)	waxy.	Columbia.	1827.	4. 5.	H. ♄.	Bot. reg. t. 1263.
28 atropurpùreum. Led.	...	dark-purple.	Siberia.	——	....	H. ♄.	
29 carpáthicum. R.S.	(*st.*)	Carpathian.	Carpathian.M.	1816.	4. 5.	H. ♄.	
30 tríste. DC.	(*re.ye.*)	dark-coloured.	Siberia.	1818.	——	H. ♄.	
31 nìgrum. DC.	(*wh.*)	black-fruited.	Britain.	....	——	H. ♄.	Eng. bot. v. 18. t. 1291.
32 trífidum. DC.	(*pu.*)	trifid-calyxed.	N. America.	1828.	——	H. ♄.	
33 glandulòsum. DC.	(*st.*)	glandular.	Chile.	1820.	5. 6.	H. ♄.	Flor. per. t. 232. f. b.
34 prostràtum. DC.	(*pu.*)	prostrate.	N. America.	1777.	4. 5.	H. ♄.	L'Herit. stirp. 1. t. 2.
glandulòsum. H.K. *non* R.P.							
35 Biebersteínii. DC.	(*st.*)	Bieberstein's.	Caucasus.	1820.	——	H. ♄.	
caucásicum. M.B. *nec aliorum.*							
36 rìgens. DC.	(*st.*)	rigid.	N. America.	1812.	——	H. ♄.	
37 albinérvium. DC.	(*st.*)	white-nerved.	Canada.	1828.	——	H. ♄.	
38 laxiflòrum. DC.	(*st.*)	loose-flowered.	N. America.	1812.	——	H. ♄.	
39 viscosíssimum. DC.	(*ye.*)	viscous.	———	1827.	——	H. ♄.	
40 sanguíneum. DC.	(*cr.*)	crimson.	———	1820.	5. 6.	H. ♄.	
41 flòridum. DC.	(*st.*)	Pensylvanian.	———	1729.	4. 5.	H. ♄.	Dill. elth. t. 244. f. 315.
42 aùreum. DC.	(*ye.*)	golden.	Missouri.	1812.	4. 6.	H. ♄.	Bot. reg. t. 125.
43 longiflòrum. L.C.	(*ye.*)	long-flowered.	N. America.	1818.	——	H. ♄.	
44 tenuiflòrum. B.R.	(*ye.*)	slender-flower'd.	———	1824.	——	H. ♄.	Bot. reg. t. 1274.
α melanocárpa.	(*ye.*)	*black-fruited.*	———	——	——	H. ♄.	
β chrysocárpa.	(*ye.*)	*yellow-fruited.*	———	——	——	H. ♄.	

ORDO CV.

ESCALLONEÆ. *Kth. synops. v.* 3. *p.* 364.

ESCALL'ONIA. L.		ESCALL'ONIA.	Pentandria Monogynia. L.				
1 rùbra. R.S.	(*re.*)	red-flowered.	Chile.	1826.	8. 10.	G. ♄.	Bot. mag. t. 2890.
Stereóxylon rùbrum. Flor. per. v. 3. t. 236. f. b.							
2 glandulòsa. B.C.	(*wh.*)	glandular.	Chile.	——	——	G. ♄.	Lodd. bot. cab. t. 1291.
3 floribúnda. K.S.	(*wh.*)	many-flowered.	New Granada.	1827.	....	G. ♄.	
4 caracasàna. K.S.	(*wh.*)	Caracas.	Caracas.	——	....	S. ♄.	
ANO'PTERUS. B.P.		ANO'PTERUS.	Hexandria Monogynia.				
glandulòsa. B.P.	(*wh.*)	glandular.	V. Diem. Isl.	1823.	....	G. ♄.	Lab. nov. holl. 1. t. 112.

ORDO CVI.

SAXIFRAGEÆ. *Juss. gen.* 308.

SAXI'FRAGA. L.		SAXIFRAGE.	Decandria Digynia. L.				
1 crassifòlia. L.T.	(*ro.*)	thick-leaved.	Siberia.	1765.	3. 5.	H. ♃.	Bot. mag. t. 126.
Bergènia crassifòlia. Mœn. *Megàsea crassifòlia.* H.R.							
2 cordifòlia. L.T.	(*ro.*)	heart-leaved.	———	1779.	——	H. ♃.	
3 mèdia.	(*ro.*)	intermediate.	———	1819.	——	H. ♃.	
Megàsea mèdia. H.R.							
4 ligulàta. L.T.	(*wh.*)	fringed-leaved.	Nepaul.	1818.	——	H. ♃.	Swt. br. fl. gar. 59.
Megàsea ciliàta. H.R.							
5 Gèum. L.T.	(*sp.*)	kidney-leaved.	Ireland.	....	6. 7.	H. ♃.	Eng. bot. v. 22. t. 1561.
6 hirsùta. L.T.	(*sp.*)	hirsute.	———	....	5. 6.	H. ♃.	——— v. 33. t. 2322.
7 polìta. L.en.	(*sp.*)	polished.	———	....	——	H. ♃.	
Robertsònia polìta. H.S.							

8 dentàta. L.en. (*sp.*) toothed. Ireland. .... 5. 6. H. ♃.
Robertsònia dentàta. H.S.
9 crenàta. (*sp.*) crenated. .... 1790. —— H. ♃.
Robertsònia crenàta. H.S.
10 sphæroídea. (*sp.*) sphæroid. Pyrenees? 1800. —— H. ♃.
Robertsònia sphæroídea. H.S.
11 punctàta. L. (*sp.*) dotted-flower'd. Siberia. 1699. —— H. ♃.
12 serràta. L.en. (*sp.*) sawed-leaved. Ireland. .... —— H. ♃.
Robertsònia serràta. H.S.
13 umbròsa. L.T. (*sp.*) London pride. Britain. .... 4. 6. H. ♃. Eng. bot. v. 10. t. 633.
14 cuneifòlia. L.T. (*sp.*) wedge-leaved. Switzerland. 1768. 5. 6. H. ♃. W. et K. hung. 1. t. 44.
15 hy'brida. L.en. (*sp.*) hybrid. .... 1820. —— H. ♃. Sternb. sax. 17. t. 8. f. 3.
16 stellàris. L.T. (*wh.*) starry. Britain. .... 6. 7. H. ♃. Eng. bot. v. 3. t. 167.
δ *Schleichèri.* L.T. (*wh.*) *Schleicher's.* Switzerland. 1819. —— H. ♃.
Spatulària stellàris. γ *depilàta.* H.R.
17 leucanthemifòlia. (*wh.*) Stock-leav'd. N. America. 1812. —— H. ♃. Lap. pyr. sax. t. 25.
18 sarmentòsa. L.T. (*wh.*) sarmentose. China. 1771. —— F. ♃. Bot. mag. t. 92.
Díptera sarmentòsa. Bork. *Ligulària sarmentòsa.* H.R.
19 cuscutæfòrmis. B.C. (*wh.*) dodder-like. China. 1815. 2. 7. F. ♃. Lodd. bot. cab. t. 186.
Ligulària mìnor. H.R.
20 eròsa. L.T. (*wh.*) jagged-leaved. N. America. 1812. 5. 6. H. ♃.
Auláxis nùda. H.R.
β *hirsùta.* L.T. (*wh.*) *hairy.* ——— —— —— H. ♃.
Auláxis micranthifòlia. H.R.
21 rotundifòlia. L.T. (*wh.*) round-leaved. Austria. 1596. —— H. ♃. Bot. mag. t. 424.
Miscopétalum rotundifòlium. H.R.
β *repánda.* L.T. (*wh.*) *repand.* ——— .... —— H. ♃. Sternb. saxif. t. 5.
22 granulàta. L.T. (*wh.*) grain-rooted. Britain. .... 5. H. ♃. Eng. bot. v. 7. t. 500.
β *multipléx.* (*wh.*) *double-flowered.* ——— .... —— H. ♃.
23 bulbífera. L.T. (*wh.*) bulbiferous. S. Europe. .... 6. 7. H. ♃. Sternb. sax. t. 12. f. 1.
24 cérnua. L.T. (*wh.*) drooping. Scotland. .... 7. H. ♃. Eng. bot. v. 10. t. 664.
Lobària cérnua. H.R.
25 sibírica. L.T. (*wh.*) Siberian. Siberia. 1802. 5. 6. H. ♃. Sternb. sax. p. 23. t. 25.
26 rivulàris. L.T. (*wh.*) Alpine brook. Scotland. .... 6. 7. H. ⊙. Eng. bot. v. 32. t. 2275.
Lobària rivulàris. H.R.
27 hederàcea. L.T. (*wh.*) Ivy-leaved. Levant. 1752. 7. H. ⊙. Flor. græc. t. 379.
28 Hírculus. L.T. (*ye.*) yellow marsh. England. .... 8. H. ♃. Eng. bot. v. 15. t. 1009.
Hírculus ranunculoídes. H.R.
29 flagellàris. L.T. (*ye.*) flagelliferous. Baffin's Bay. 1820. —— H. ♃. Sternb. sax. p. 25. t. 6.
30 aizoídes. L.T. (*ye.*) smaller mountain. Britain. .... 7. 8. H. ♃. ——— t. 8. f. 1.
β *autumnàlis.* (*ye.*) *larger mountain.* ——— .... —— H. ♃. Eng. bot. v. 1. t. 39.
Leptàsea aizoídes. H.R.
31 bronchiàlis. L.T. (*st.*) spiny-fringed, Siberia. 1819. —— H. ♃. Gmel. sib. 4. t. 65. f. 2.
32 tenélla. L.T. (*wh.*) slender. Carinthia. 1824. —— H. ♃. Jacq. coll. 3. t. 17.
33 juniperìna. L.T. (*ye.*) Juniper-like. Caucasus. 1826. —— H. ♃. Sternb. sax. p. 31. t. 10.
34 áspera. L.T. (*st.*) rough. Switzerland. 1752. 8. H. ♃. Jacq. aust. 5. t. 31.
Ciliària áspera. H.R.
35 bryoìdes. L.T. (*st.*) thread-moss like. ——— —— 6. 7. H. ♃. Jacq. misc. 2. t. 5. f. 1.
36 hieracifòlia. L.T. (*br.*) Hawkweed-ld. Hungary. 1789. 5. 6. H. ♃. W. et K. hung. 1. t. 18.
37 pensylvánica. L.T. (*gr.*) Pensylvanian. N. America. 1732. —— H. ♃. Dill. elt. t. 253. f. 328.
β *glábra.* (*gr.*) *smoother-leaved.* ——— —— —— H. ♃.
38 semipubéscens. L.T. (*st.*) semipubescent. ——— .... —— H. ♃.
Micránthes semipubéscens. H.S.
39 virginiénsis. L.T. (*wh.*) Virginian. ——— 1790. 5. 7. H. ♃. Bot. mag. t. 1664.
Dermàsea pilòsa. H.R.
40 congésta. H.S.L. (*wh.*) close-headed. ——— 1812. —— H. ♃.
nivàlis. Ph. *nec aliorum.*
41 nivàlis. L.T. (*wh.*) clustered Alpine. Britain. .... 6. 7. H. ♃. Eng. bot. v. 7. t. 440.
Dermasea nivàlis. H.R.
42 melaleùca. S.S. (*wh.*) Fischer's. Siberia. 1826. —— H. ♃.
43 longiscàpa. L.T. (*wh.*) long-scaped. ——— 1824. 5. 7. H. ♃.
44 davùrica. L.T. (*wh.*) Davurian. Davuria. —— —— H. ♃.
45 Cotyledon. L.T. (*wh.*) pyramidal. Alps Europe. 1596. 5. 7. H. ♃. Flor. dan. t. 241.
Chondròsea pyramidális. H.R.
46 intermèdia. (*wh.*) related. .... 1808. 7. H. ♃
Chondròsea intermèdia. H.R.
47 lingulàta. L.T. (*sp.*) tongue-leaved. Pyrenees. .... 5. 7. H. ♃.
longifòlia. S.S.
β *mèdia.* L.T. (*sp.*) *middle.* ——— .... —— H. ♃. Sternb. sax. p. 1. t. 1. a.

48	crustàta. L.en.	(*sp.*)	crusted.	Pyrenees.	....	5. 7.	H. ♃.	Sternb. sax. t. 1. b.
49	Aizoòn. L.T.	(*sp.*)	large margined.	——	1731.	——	H. ♃.	Jacq. aust. t. 438.
50	récta. P.S.	(*sp.*)	straight-leaved.	Alps Europe.	....	——	H. ♃.	Lap. pyr. sax. t. 15.
	intácta. L.T. W. hort. ber. 2. t. 75. *Chondròsea Aizoòn.* H.R.							
	β *mìnor.* H.R.	(*sp.*)	*lesser.*	Alps Europe.	....	——	H. ♃.	
	γ *parviflòs.* H.R.	(*sp.*)	*smaller-flowered.*	——	....	——	H. ♃.	
51	mutàta. L.T.	(*ye.*)	saffron-colour'd.	Switzerland.	1779.	6. 7.	H. ♃.	Bot. mag. t. 351.
52	aretioídes. L.T.	(*ye.*)	Aretia-like.	——	1825.	——	H. ♃.	Lap. pyr. sax. t. 13.
53	Burseriàna. L.T.	(*st.*)	Burser's.	Italy.	——	5. 6.	H. ♃.	Jacq. misc. 1. t. 17. f. 3.
54	retùsa. L.T.	(*ro.*)	purple-flower'd.	Pyrenees.	....	3. 4.	H. ♃.	Lap. pyr. sax. t. 18.
55	oppositifòlia. L.T.	(*re.*)	opposite-leaved.	Britain.	....	——	H. ♃.	Eng. bot. v. 1. t. 9.
	Antiphy'lla cærùlea. H.R.							
56	biflòra. L.T.	(*re.*)	two-fl^d purple.	Switzerland.	1825.	——	H. ♃.	Lap. pyr. sax. t. 17.
57	cæ'sia. L.T.	(*st.*)	grey.	——	1752.	5. 6.	H. ♃.	Jacq. aust. 4. t. 374.
58	diapensioídes. L.T.	(*wh.*)	Diapensia-like.	——	1828.	——	H. ♃.	Flor. ped. app. p. 21. t. 3.
59	compácta. S.S.	(*wh.*)	compact.	——	——	——	H. ♃.	
60	androsàcea. L.T.	(*wh.*)	Androsace-leav'd.	——	1792.	——	H. ♃.	Jacq. aust. 4. t. 389.
61	sedioídes. L.T.	(*st.*)	Stonecrop-like.	Europe.	1825.	——	H. ♃.	Jacq. misc. 2. t. 21. f. 22.
62	ténera. L.T.	(*st.*)	flat-leaved.	Switzerland.	——	5. 6.	H. ♃.	Sternb. sax. t. 9. f. 4.
	planifòlia. S.S.							
63	geranioídes. L.S.	(*wh.*)	cranesbill-leav'd.	Pyrenees.	1770.	4. 5.	H. ♃.	Lap. pyr. sax. t. 43.
64	irrígua. L.T.	(*wh.*)	crowfoot-leav'd.	Siberia.	1820.	——	H. ♃.	M.B. cent. pl. rar. t. 73.
65	pedatífida. L.T.	(*wh.*)	pedatifid.	Scotland.	....	5. 6.	H. ♃.	Eng. bot. v. 32. t. 2278.
	β *mìnor.* H.R.	(*wh.*)	*smaller villous.*	——	....	——	H. ♃.	
	γ *ladanífera.* L.T.	(*wh.*)	*ladaniferous.*	Pyrenees.	....	——	H. ♃	Lap. sax. pyr. t. 42.
66	quinquéfida. H.R.	(*wh.*)	quinquefid.	Scotland.	....	——	H. ♃.	
67	ceratophy'lla. L.T.	(*wh.*)	shining-calyx'd.	Spain.	1804.	——	H. ♃.	Bot. mag. t. 1651.
68	ajugifòlia. L.T.	(*wh.*)	Bugle-leaved.	Pyrenees.	1770.	6. 7.	H. ♃.	Lap. pyr. sax. t. 31.
69	affínis. L.T.	(*wh.*)	affined.	Ireland.	....	——	H. ♃.	
70	pentadáctyla. L.T.	(*wh.*)	five-fingered.	Pyrenees.	1815.	5. 6.	H. ♃.	Lap. pyr. sax. t. 40.
71	latífida. L.T.	(*wh.*)	broad-cleft.	Spain.	1826.	——	H. ♃.	
72	decípiens. L.T.	(*wh.*)	palmate.	Wales.	....	——	H. ♃.	Sternb. sax. p. 55. t. 23.
	palmàta. Eng. bot. v. 7. t. 455. *cæspitòsa.* Flor. dan. t. 71.							
73	hírta. L.T.	(*wh.*)	hairy.	Ireland.	....	——	H. ♃.	Eng. bot. v. 32. t. 2291.
74	platypétala. L.T.	(*wh.*)	broad-petaled.	Scotland.	....	6.	H. ♃.	—— v. 32. t. 2276.
75	Schraderi. S.S.	(*wh.*)	Schrader's.	Siberia.	1826.	——	H. ♃.	
76	incurvifòlia. L.T.	(*wh.*)	incurved-leav'd.	Ireland.	....	——	H. ♃.	
77	denudàta. L.T.	(*wh.*)	half-naked.	Scotland.	....	——	H. ♃.	
78	Sternbérgii. L.T.	(*wh.*)	Sternberg's.	Germany.		5. 6.	H. ♃.	Sternb. sax. t. 24.
79	pulchélla. L.T.	(*st.*)	pretty.	——	1818.	——	H. ♃.	
80	tridentàta. L.T.	(*wh.*)	three-toothed.	....	....	——	H. ♃.	
81	cæspitòsa. L.T.	(*st.*)	tufted.	Wales.	....	——	H. ♃.	Eng. bot. v. 12. t. 794.
82	exaràta. L.T.	(*wh.*)	furrowed.	Pyrenees.	1818.	——	H. ♃.	Vill. delph. 4. t. 45.
	nervòsa. Lap. pyr. sax. t. 39.							
83	pedemontàna. L.T.	(*wh.*)	Piedmont.	Piedmont.	1824.	——	H. ♃.	All. ped. t. 21. f. 6.
	cymòsa. Pl. rar. hung. t. 88. *heterophy'lla.* Sternb. sax. t. 20. f. 1-2.							
84	moschàta. L.T.	(*ye.*)	musky.	Piedmont.	....	5. 6.	H. ♃.	Jacq. misc. 2. t. 21. f. 2.
85	muscoídes. L.T.	(*st.*)	Moss-like.	England.	....	——	H. ♃.	Lap. pyr. sax. t. 35.
	Muscària muscoídes. H.R.							
86	pygmæ'a. L.T.	(*st.*)	pigmy.	Scotland.	....	——	H. ♃.	Eng. bot. v. 33. t. 2214.
	moschàta. E.B. *non* Jacq. *Muscària cæspitòsa.* H.R.							
87	tridacty'lites. E.B.	(*bh.*)	Rue-leaved.	Britain.	....	3. 5.	H. ⊙.	Eng. bot. v. 7. t. 501.
	Tridacty'lites ánnua. H.R.							
88	controvérsa. S.S.	(*wh.*)	confused.	N. Europe.	....	——	H. ⊙.	Sternb. sax. t. 16.
89	petræ'a. L.T.	(*wh.*)	rock.	Carinthia.	1752.	4. 5.	H. ⊙.	Jacq. ic. 1. t. 81.
90	adscéndens. L.T.	(*wh.*)	ascending.	Pyrenees.	——	5. 6.	H. ♃.	Sternb. sax. t. 19. f 1.2.
	aquática. Lap. pyr. sax. t. 28.							
91	hypnoídes. L.T.	(*wh.*)	mossy.	Britain.	....	4. 6.	H. ♃.	Eng. bot. v. 7. t. 454.
	β *viscòsa.* L.T.	(*wh.*)	*viscous.*	——	....	——	H. ♃.	
	γ *angustifòlia.* L.T.	(*wh.*)	*narrow-leaved.*	——	....	——	H. ♃.	
	δ *muscòsa.* L.T.	(*wh.*)	*tufted.*	——	....	——	H. ♃.	
	ε *pulchélla.* L.T.	(*wh.*)	*larger.*	——	....	——	H. ♃.	
	ζ *spatulàta.* H.R.	(*wh.*)	*spatulate.*	——	....	——	H. ♃.	
	η *latífida.* H.R.	(*wh.*)	*broad-cleft.*	——	....	——	H. ♃.	
	θ *rhodopétala.* H.R.	(*bh.*)	*rosy-petuled.*	——	....	——	H. ♃.	
	ι *recúrvula.* H.R.	(*wh.*)	*recurving-cup'd.*	——	....	——	H. ♃.	
	κ *aristàta.* H.R.	(*wh.*)	*awned.*	——	....	——	H. ♃.	
	λ *septífida.* H.R.	(*wh.*)	*seven-cleft.*	——	....	——	H. ♃.	
92	condensàta. L.T.	(*wh.*)	close.	Scotland.	....	——	H. ♃.	Gmel. fl. bad. t. 3.

93 elongélla. L.T.	(*wh.*)	long peduncled.	Scotland.	....	5. 6.	H. ♃.	**Eng. bot. v. 32. t. 2277.**
94 leptophy'lla. L.T.	(*wh.*)	narrow-leaved.	Wales.	....	——	H. ♃.	
β *angustifida.* L.T.	(*wh.*)	*narrow-lobed.*	————	....	——	H. ♃.	
95 lætevìrens. L.T.	(*wh.*)	light-green.	Scotland.	....	——	H. ♃.	
96 lanceolàta. H.R.	(*wh.*)	spear-leaved.	Europe.	1808.	——	H. ♃.	
obtùsa. H.R.	(*wh.*)	*blunt-leaved.*	————	1820.	——	H. ♃.	
97 altífida. H.R.	(*wh.*)	deep-cleft.	————	....	——	H. ♃.	
98 viscòsa. H.R.	(*wh.*)	viscous.	————	....	——	H. ♃.	
99 quinquedèns. H.R.	(*wh.*)	five-toothed.	————	....	——	H. ♃.	
100 Hawórthii. H.S.L.	(*wh.*)	Haworth's.	————	....	——	H. ♃.	
hírta. H.R. *non* E.B.							
101 æstivàlis. F.	(*wh.*)	summer.	Altay.	1821.	5. 7.	H. ♃.	
HEUCH'ERA. L.		ALLUM-ROOT.	Pentandria Digynia. L.				
1 americàna. L.	(*gr.*)	viscid.	N. America.	1656.	5. 7.	H. ♃.	**Schkuhr hand. t. 58.**
víscida. Ph.							
2 pubéscens. Ph.	(*bh.*)	pubescent.	————	1812.	——	H. ♃.	
3 híspida. Ph.	(*gr.*)	hispid.	————	1820.	——	H. ♃.	
4 villòsa. M.	(*gr.*)	villous.	————	1812.	——	H. ♃.	
5 cauléscens. Ph.	(*wh.*)	caulescent.	————	....	——	H. ♃.	
6 glábra. R.S.	(*gr.*)	smooth.	Altay.	1827.	——	H. ♃.	
TIARE'LLA. L.		TIARE'LLA.	Decandria Digynia. L.				
1 cordifòlia. Ph.	(*wh.*)	heart-leaved.	N. America.	1731.	4. 5.	H. ♃.	**Bot. mag. t. 1589.**
2 Menzièsii. Ph.	(*wh.*)	leafy-stemmed.	————	1812.	——	H. ♃.	
3 trifoliàta. Ph.	(*wh.*)	three-leaved.	N. America.	1826.	——	H. ♃.	
4 polyphy'lla. D.P.	(*wh.*)	many-leaved.	Nepaul.	1818.	——	H. ♃.	
ASTI'LBE. D.P.		ASTI'LBE.	Oct-Decandria Digynia.				
1 decándra. D.P.	(*st.*)	decandrous.	Carolina.	1812.	5. 6.	F. ♃.	**Vent. malm. t. 54.**
Tiarélla biternàta. V.							
2 rivulàris. D.P.	(*st.*)	rivulet.	Nepaul.	1825.	——	H. ♃.	
TELL'IMA. B.		TELL'IMA.	Decandria Digynia.				
grandiflòra. B.R.	(*st. re.*)	large-flowered.	N. America.	1826.	4. 5.	H. ♃.	**Bot. reg. t. 1178.**
Mitélla grandiflòra. Ph.							
MITE'LLA. L.		MITE'LLA.	Decandria Digynia. L.				
1 diphy'lla. W.	(*wh.*)	two-leaved.	N. America.	1731.	4. 5.	H. ♃.	**Bot. reg. t. 166**
2 cordifòlia. W.	(*wh.*)	heart-leaved.	————	1812.	——	H. ♃.	**Lam. ill. t. 373. f. 3.**
3 renifórmis. Ph.	(*wh.*)	kidney-leaved.	————	1758.	6. 8.	H. ♃.	**———— t. 373. f. 2.**
nùda. W.							
4 prostràta. M.	(*wh.*)	prostrate.	————	1812.	5. 6.	H. ♃.	
5 pentándra. B.M.	(*wh.*)	pentandrous.	————	1827.	6. 7.	H. ♃.	**Bot. mag. t. 2933.**
CHRYSOSPL'ENIUM. L.		GOLDEN SAXIFRAGE.	Oct-Decandria Digynia.				
1 alternifòlium. W.	(*ye.*)	alternate-leav'd.	Britain.	....	4. 5.	H. ♃.	**Eng. bot. v. 1. t. 54.**
2 oppositifòlium. W.	(*ye.*)	opposite-leaved.	————	....	——	H. ♃.	**———— v. 7. t. 490.**
3 nepalénse. D.P.	(*ye.*)	Nepaul.	Nepaul.	1824.	——	H. ♃.	
ADO'XA. L.		MOSCHATEL.	Oct-Decandria Tetragynia.				
Moschatéllina. E.B.	(*st.*)	tuberous.	Britain.	....	3. 5	H. ♃.	**Eng. bot. v. 7. t. 453.**
HYDRA'NGEA. L.		HYDRA'NGEA.	Decandria Digynia. L.				
1 arboréscens. W.	(*wh.*)	shrubby.	Virginia.	1736.	7. 8.	H. ♄.	**Bot. mag. t. 437.**
vulgàris. Ph.							
2 cordàta. Ph.	(*wh.*)	heart-leaved.	Carolina.	1806.	——	H. ♄.	**Wats. dend. brit. t. 42.**
3 radiàta. W.	(*wh.*)	white-leaved.	————	1786.	——	H. ♄.	**Lam. ill. t. 370. f. 2.**
nívea. Wats. dend. brit. t. 43.							
4 heteromálla. D.P.	(*wh.*)	woolly-leaved.	Nepaul.	1823.	....	H. ♄.	
5 quercifòlia. W.	(*wh.*)	Oak-leaved.	Florida.	1803.	6. 9.	H. ♄.	**Bot. mag. t. 975.**
6 horténsis. W.	(*pi.*)	changeable.	China.	1788.	4. 9.	H. ♄.	**———— t. 438.**
β *cærùlea.*	(*bl.*)	*blue-flowered.*	————	——	——	H. ♄.	

ORDO CVII.

CUNONIACEÆ. *Brown.*

WEINMA'NNIA. L.		WEINMA'NNIA.	Octandria Digynia. L.				
1 glábra. L.	(*wh.*)	smooth.	W. Indies.	1818.	5. 7.	S. ♄.	**Lam. ill. t. 313. f. 1.**
2 hírta. W.	(*wh.*)	hairy.	————	——	——	S. ♄.	
3 ellíptica. K.S.	(*wh.*)	elliptic-leaved.	Peru.	1823.	....	G. ♄.	
4 ovàta. K.S.	(*wh.*)	oval-leaved.	————	——	....	G. ♄.	**Cav. ic. 6. t. 566.**

CUN'ONIA. L. Cun'onia. Decandria Digynia. L.
capénsis. w. (*wh.*) Cape. C. B. S. 1816. 5. 7. G. ♄. Bot. reg. t. 828.
CALLIC'OMA. A.R. Callic'oma. Dodecandria Digynia.
serratifòlia. A.R. (*st.*) saw-leaved. N. S. W. 1793. 5. 8. G. ♄. Andr. reposit. t. 566.
CERATOP'ETALUM. Sm. Ceratop'etalum. Decandria Monogynia.
gummíferum. Sm.(*ye.*) gum-bearing. N. Holland. 1820. .... G. ♄. Sm. n. holl. 1. t. 3.
BAU'ERA. Sal. Bau'era. Polyandria Digynia. H.K.
1 rubiæfòlia. Sal. (*re.*) Madder-leaved. N. S. W. 1793. 7. 12. G. ♄. Sal. ann. bot. 1. t. 10.
rubioídes. A.R. 198. B.M. 715.
2 hùmilis. L.en. (*re.*) dwarf. ——— 1804. —— G. ♄. Lodd. bot. cab. t. 1197.

ORDO CVIII.

UMBELLIFERÆ. *Juss. gen.* 218.

Subordo I. *ORTHOSPERMEÆ.* DC.

Tribus I. Hydrocotylineæ.—Subtribus I. *Hydrocotyleæ.* DC.

HYDROCO'TYLE. S.S. Marsh-pennywort. Pentandria Digynia. L.
1 vulgàris. s.s. (*bh.*) common. Britain. .... 5. 6. H.*w.*♃. Eng. bot. v. 11. t. 751.
2 umbellàta. s.s. (*wh.*) umbelled. N. America. 1795. 7. 8. H.*w.*♃. Spreng. umb. 1. t. 1.
3 bonariénsis. s.s. (*ye.*) Buenos Ayres. Buenos Ayres. 1828. —— F.*w.*♃. Cavan. icon. 5. t. 488.
4 americàna. s.s. (*gr.*) American. N. America. 1790. 5. 8. H.*w.*♃. Spreng.um.p.3.t.2.f.3.
5 villòsa. s.s. (*gr.*) villous. C. B. S. 1795. 8. G.*w.*♃.
6 nepalénsis. H.E.F. (*gr.*) Nepaul. Nepaul. 1818. 2. 6. G.*w.*♃. Hook. ex. flor. t. 30.
híspida. D.P.
7 nitídula. s.s. (*gr.*) glossy. Java. 1821. —— S.*w.*♃. Hook. ex. flor. t. 29.
8 canariénsis. (*gr.*) Canary. Canaries. 1828. 4. 6. G.*w.*♃.
9 prolífera. Ot. (*gr.*) proliferous. Brazil. 1824. —— S.*w.*♃.
10 repánda. s.s. (*wh.*) Pilewort-leav'd. N. America. 1806. 6. 7. G.*w.*♃. Spreng. umb. t. 2. f. 4.
11 asiática. s.s. (*wh.*) thick-leaved. Asia. 1690. 7. 8. G.*w.*♃. Rheed. mal. 10. t. 46.
12 sibthorpioídes. s.s. (*wh.*) Sibthorpia-like. Mauritius. 1818. —— S.*w.*♃.
ERIG'ENIA. N. Erig'enia. Pentandria Digynia.
bulbòsa. N. (*wh.*) bulbous. N. America. 1825. 3. 5. H. ♃.
DIDI'SCUS. DC. Didi'scus. Pentandria Digynia.
1 incísus. DC. (*wh.*) cut-leaved. N. S. W. 1820. 6. 8. H. ⊙. Linn. trans. 10. t. 21. f. 2.
Trachymène incísa. L.T.
2 azùreus. DC. (*bl.*) blue-flowered. ——— 1827. 7. 9. H. ⊙. Bot. mag. t. 2875.
cærùleus. B.M. *Trachymène cærùlea.* Bot. reg. t. 1225.
TRACHYM'ENE. DC. Trachym'ene. Pentandria Digynia.
1 compréssa. s.s. (*st.*) flat-stalked. N. Holland. 1824. 5. 6. G. ♃. Lab. n. holl. 1. t. 101.
2 lineàris. s.s. (*ye.*) linear-leaved. N. S. W. —— —— G. ♄. Cavan. icon. 5. t. 485.
Azorélla linearifòlia. Cav.
3 ovàta. s.s. (*st.*) oval-leaved. ——— 1827. —— G. ♄. Lab. n. holl. 1. t. 100.
Azorélla ovàta. Labill.
ASTROTRI'CHA. DC. Astrotri'cha. Pentandria Digynia.
floccòsa. DC. (*wh.*) woolly. N. S. W. 1825. 5. 7. G. ♄.
XANTH'OSIA. DC. Xanth'osia. Pentandria Digynia.
1 pilòsa. L.T. (*wh.*) hairy. N. S. W. 1826. —— G. ♃. Linn. trans. 10. t. 22. f. 1.
BOWL'ESIA. S.S. Bowl'esia. Pentandria Digynia.
ténera. s.s. (*wh.*) slender. Monte Video. 1827. 6. 8. H. ⊙.

Subtribus II. *Mulineæ.* DC.

SPANA'NTHE. S.S. Spananthe. Pentandria Digynia. S.S.
paniculàta. s.s. (*wh.*) panicled. Caracas. 1795. 7. 8. S. ♂. Jacq. ic. 2. t. 350.
Hydrocótyle Spanánthe. W.
POZ'OA. Poz'oa. Pentandria Digynia.
coriàcea. s.s. (*wh.*) leathery-leaved. Peru. 1827. 7. 9. F. ♃.

Tribus II. Saniculeæ. DC.

ACTIN'OTUS. Lab. Actin'otus. Pentandria Digynia.
Heliánthi. Lab. (*wh.*) larger. N. S. W. 1821. 6. 12. G. ♄. Bot. reg. t. 654.
Eriocàlia màjor. Ex. bot. t. 80.

SANI'CULA. S.S.		SANICLE. Pentandria Digynia. L.					
1 europ'æa. s.s.	(*wh.*)	wood.	Britain.	....	6. 7.	H. ♃.	Eng. bot. v. 2. t. 98.
2 canadénsis. s.s.	(*wh.*)	Canadian.	Canada.	1800.	——	H. ♃.	
3 marilándica. s.s.	(*wh.*)	Maryland.	N. America.	1765.	——	H. ♃.	Jacq. ic. 2. t. 348.
HACQU'ETIA. DC.		HACQU'ETIA. Pentandria Digynia.					
Epipáctis. DC.	(*ye.*)	dwarf.	S. Europe.	1818.	5. 7.	H. ♃.	Jacq. aust. 5. t. 11.
Astrántia Epipáctis. W. *Dóndia Epipáctis.* S.S.							
ASTRA'NTIA. S.S.		ASTRA'NTIA. Pentandria Digynia. L.					
1 màjor. W.	(*bh.*)	great black.	Alps Europe.	1596.	5. 9.	H. ♃.	Exot. bot. 2. t. 76.
2 carniòlica. W.	(*bh.*)	Carniolian.	Carniola.	1812.	5. 6.	H. ♃.	Jacq. aust. 5. app. t. 10.
3 caucàsica. R.S.	(*bh.*)	Caucasean.	Caucasus.	1824.	——	H. ♃.	
intermèdia. M.B.							
4 máxima. Pall.	(*bh.*)	Hellebore-leav'd.	———	1804.	6. 7.	H. ♃.	Bot. mag. t. 1553.
helleborifòlia. P.L. 60. *heterophy'lla.* M.B.							
5 mìnor. W.	(*wh.*)	small.	Switzerland.	1686.	5 6.	H. ♃.	Sm. exot. bot. 2. t. 77.
ERY'NGIUM. S.S.		ERY'NGO. Pentandria Digynia. L.					
1 aquáticum. s.s.	(*wh.*)	marsh.	N. America.	1699.	7. 9.	H. ♃.	Bot. reg. t. 372.
2 longifòlium. s.s.	(*wh.*)	long-leaved.	Mexico.	1812.	——	F. ♃.	Cav. ic. 6. t. 555.
3 gramíneum. s.s.	(*bl.*)	grass-leaved.	New Spain.	1825.	7. 10.	F. ♃.	Laroch. eryng. t. 26.
4 bromelifòlium. s.s.	(*st.*)	Bromelia-leav'd.	———	1822.	——	F. ♃.	——— t. 28.
5 monocéphalum. s.s.	(*pu.*)	one-headed.	Mexico.	1824.	——	F. ♃.	Cavan. icon. 6. t. 553.
6 ebracteàtum. s.s.	(*wh.*)	bractless.	Monte Video.	1810.	7. 9.	F. ♃.	Laroch. eryng. t. 32.
7 ternàtum. s.s.	(*pu.*)	ternate-leaved.	Crete.	....	——	F. ♃.	Moris. his. s. 7. t. 36. f. 24.
8 plànum. s.s.	(*bl.*)	flat-leaved.	Europe.	1596.	——	H. ♃.	Jacq. aust. 4. t. 391.
9 tricuspidàtum. s.s.	(*pu.*)	trifid.	Spain.	1699.	9.	H. ♂.	Laroch. eryng. t. 9.
10 asperifòlium. s.s.	(*wh.*)	rough-leaved.	....	1824.	6. 8.	H. ♃.	——— t. 11.
11 dichótomum. s.s.	(*bl.*)	forked.	S. Europe.	——	7. 9.	H. ♃.	Desf. atl. 1. t. 55.
12 cœrùleum. M.B.	(*bl.*)	blue.	Caucasus.	1816.	——	H. ♃.	
13 gigánteum. s.s.	(*bl.*)	gigantic.	Armenia.	1823.	——	H. ♃.	
14 alpìnum. s.s.	(*bl.*)	Alpine.	Switzerland.	1597.	——	H. ♃.	Bot. mag. t. 922.
15 spìna-álba. DC.	(*wh.*)	white-spined.	S. Europe.	1816.	——	H. ♃.	Vill. delph. t. 17.
16 virgàtum. s.s.	(*bl.*)	oval-leaved.	N. America.	1810.	6. 7.	H. ♃.	Laroch. eryng. t. 20.
17 grácile. s.s.	(*ye.*)	slender.	New Spain.	1818.	7. 9.	F. ♃.	——— t. 24.
18 Baldwìni. s.s.	(*bl.*)	small-headed.	Carolina.	1824.	——	H. ♃.	
19 aquifòlium. s.s.	(*bl.*)	holly-leaved.	Spain.	1827.	——	H. ♃.	Laroch. eryng. t. 10.
20 corniculàtum. s.s.	(*bl.*)	horned.	Portugal.	1803.	6. 8.	H. ♃.	Bot. mag. t. 1427.
21 tènue. s.s.	(*bl.*)	slender annual.	———	....	——	H. ⊙.	Moris. sect. 7. t. 35. f. 15.
22 pusíllum. s.s.	(*bl.*)	dwarf.	Spain.	1640.	——	H. ♃.	Laroch. eryng. t. 16.
23 virginiànum. s.s.	(*pa.*)	Virginian.	Virginia.	....	7. 9.	H. ♃.	——— t. 19.
24 serràtum. s.s.	(*wh.*)	serrated.	New Spain.	1823.	8. 10.	F. ♃.	Cav. ic. 6. t. 55.
25 Carlìnæ. s.s.	(*bl.*)	Carlina-like.	———	1827.	——	F. ♃.	Laroch. eryng. t. 23.
26 f'œtidum. s.s.	(*wh.*)	stinking.	W. Indies.	1714.	——	S. ♂.	Sloan. jam. t. 156. f. 3 4.
27 marítimum. s.s.	(*bl.*)	sea Holly.	Britain.	....	7. 10.	H. ♃.	Eng. bot. v. 10. t. 718.
28 trìquètrum. s.s.	(*bl.*)	triquetrous.	Sicily.	1827.	——	H. ♃.	Desf. atl. 1. t. 54.
29 campéstre. s.s.	(*wh.*)	field.	Britain.	....	7. 8.	H. ♃.	Eng. bot. v. 1. t. 57.
30 Bourgàti. s.s.	(*bl.*)	cut-leaved.	S. France.	1731.	6. 8.	H. ♃.	Gouan. ill. 7. t. 3.
31 amethy'stinum. s.s.	(*bl.*)	amethystine.	Styria.	1648.	7. 8.	H. ♃.	W. et K. hung. 3. t. 215.
32 dilatàtum. s.s.	(*bl.*)	dilated.	Portugal.	....	——	H. ♃.	Laroch. eryng. t. 4.
33 crinìtum. s.s.	(*bl.*)	fringe-leaved.	Spain.	1826.	——	H. ♃.	
34 comòsum. s.s.	(*pa.*)	tufted.	New Spain.	1818.	——	F. ♃.	Laroch. eryng. t. 7.
35 azùreum. R.S.	(*bl.*)	azure blue.	....	——	——	H. ♃.	

Tribus III. AMMINEÆ. DC.

R'UMIA. H.U.		R'UMIA. Pentandria Digynia.					
1 taùrica. H.U.	(*wh.*)	Taurian.	Tauria.	1820.	7. 8.	H. ♃.	Hoff. umb. p. 173. f. 17.
Cáchrys taùrica. S.S.							
2 sibírica. F.	(*wh.*)	Siberian.	Siberia.	1826.	——	H. ♃.	
3 microcárpa. H.U.	(*wh.*)	small-fruited.	Caucasus.	1827.	——	H. ♃.	Moris. his. s. 9. t. 1. f. 3.
4 athamanthoídes. H.U.	(*wh.*)	Athamanta-like.	Siberia.	——	——	H. ♃.	
5 seseloídes. H.U.	(*wh.*)	Seseli-like.	Tauria.	——	——	H. ♃.	Hoff. umb. ed. 2. t. tit. f. 4.
CIC'UTA. S.S.		COWBANE. Pentandria Digynia. L.					
1 viròsa. s.s.	(*wh.*)	long-leaved.	Britain.	....	7.	H. ♃.	Eng. bot. v. 7. t. 479.
2 davùrica. s.s.	(*wh.*)	Daurian.	Siberia.	1824.	——	H. ♃.	Gmel. sib. 1. t. 47.
3 maculàta. s.s.	(*wh.*)	spotted.	N. America.	1759.	7. 8.	H. ♃.	Pluk. alm. t. 76. f. 1.
4 bulbífera. s.s.	(*wh.*)	bulb-bearing.	———	1823.	——	H. ♃.	
5 sinénsis. R.S.	(*wh.*)	Chinese.	China.	1826.	——	H. ♃.	

Z'IZIA. DC. Z'IZIA. Pentandria Digynia.
aùrea. DC. (*ye.*) golden. N. America. 1699. 5. 6. H. ♃.
Smy'rnium aùreum. W. *Tháspium aùreum.* N. *Sìson aùreus.* S.S.
APIUM. DC. CELERY. Pentandria Digynia. L.
1 gravéolens. S.S. (*wh.*) common. Britain. .... 6. 8. H. ♂. Eng. bot. v. 17. t. 1210.
α fistulòsum. (*wh.*) *hollow-stalked.* ——— .... —— H. ♂.
β sólidum. (*wh.*) *solid-stalked.* ——— .... —— H. ♂.
γ rùbrum. (*wh.*) *red-stalked.* ——— .... —— H. ♂.
δ napàceum. Mill. (*wh.*) *Turnep-rooted.* .... —— H. ♂.
PETROSEL'INUM. DC. PARSLEY. Pentandria Digynia. L.
1 satìvum. K.U. (*gr.*) common. Sardinia. 1548. 6. 7. H. ♂. Moris. sect. 9. t. 8. f. 2.
β críspum. (*gr.*) *curled-leaved.* ——— .... —— H. ♂. ——— 9. t. 8. f. 3.
2 romànum. Zuc. (*gr.*) large. Greece. .... —— H. ♂.
3 fractophy'llum. Lag. (*st.*) broken-leaved. .... 1825. —— H. ♂.
4 prostràtum. S.S. (*gr.*) prostrate. V. Diem. Isl. 1818. —— G. ♂. Vent. malm. t. 81.
5 peregrìnum. K.U. (*wh.*) Lovage-like. Portugal. 1633. 6. 7. H. ♂. Jacq. vind. 3. t. 18.
Ligústicum peregrìnum. Jacq. *Sìson peregrìnum.* R.S.
6 ségetum. K.U. (*bh.*) corn. England. .... 8. H. ⊙. Eng. bot. v. 4. t. 228.
Sìson ségetum. E.B.
TR'INIA. H.U. TR'INIA. Pentandria Digynia. L. en.
1 Hoffmánni. M.B. (*wh.*) Hoffman's. England. .... 5. 6. H. ♃. Eng. bot. v. 17. t. 1209.
Pimpinélla dioica. E.B.
2 Henníngii. M.B. (*wh.*) Henning's. Hungary. 1803. 6. 8. H. ♃.
Pimpinélla glaùca. W.
3 Kitaibèlii. M.B. (*wh.*) Kitaibel's. ——— 1827. —— H. ♃. W. et K. hung. 1. t. 72.
4 ramosíssima. F. (*wh.*) branching. Siberia. —— —— H. ♃.
HELOSC'IADIUM. K.U. HELOSC'IADIUM. Pentandria Digynia.
1 lateriflòrum. K.U. (*wh.*) narrow-leaved. S. Europe. 1824. 7. 8. H. ⊙. Jacq. vind. t. 200.
Sìson Ammi. Jacq. *Pimpinélla lateriflòra.* L. enum.
2 inundàtum. K.U. (*wh.*) inundated. Britain. .... 5. 6. H. *w.* ⊙. Eng. bot. v. 4. t. 227.
Sìson inundàtum. E.B. *Siùm inundàtum.* E.F. *Mèum inundàtum.* S.S.
3 nodiflòrum. K.U. (*wh.*) procumbent. Britain. .... 7. 8. H. *w.* ♃. Eng. bot. v. 9. t. 639.
Siùm nodiflòrum. E.B.
4 rèpens. K.U. (*wh.*) creeping. ——— .... 6. H. *w.* ♃. Eng. bot. v. 20. t. 1431.
5 bulbòsum. K.U. (*wh.*) bulb-bearing. Switzerland. 1826. 6. 8. H. *w.* ♃. Thor. jo. bot. 1. t. 7. f. 1-2.
DISCOPLE'URA. DC. DISCOPLE'URA. Pentandria Digynia.
Nuttállii. DC. (*wh.*) slender-leaved. N. America. 1812. 7. 8. H. ⊙.
A'mmi capillàceum. M. *Sìson capillàceus.* S.S.
LEPTOCA'ULIS. DC. LEPTOCA'ULIS. Pentandria Digynia.
divaricàtus. DC. (*ro. wh.*) spreading. N. America. 1812. 6. 8. H. ⊙.
Æthùsa divaricàta. N. *A'mmi divaricàtum.* P.S.
PTYCH'OTIS. K.U. PTYCH'OTIS. Pentandria Digynia.
1 heterophy'lla. K.U. (*wh.*) Coriander-leav'd. Pyrenees. 1778. 7. 8. H. ♂. Jacq. vind. 2. t. 198.
Mèum heterophy'llum. S.S. *Æthùsa Bùnius.* W.
2 ammoídes. K.U. (*wh.*) Milfoil-leaved. S. Europe. 1759. 6. 7. H. ⊙. Jacq. vind. 1. t. 52.
Séseli ammoídes. S.S.
3 verticillàta. DC. (*wh.*) whorled. ——— 1825. —— H. ⊙.
4 cóptica. DC. (*wh.*) Egyptian. Egypt. 1773. —— H. ⊙. Jacq. vind. 2. t. 196.
Trachyspérmum cópticum. L. enum. *A'mmi cópticum.* L. *Bùnium cópticum.* R.S.
FALC'ARIA. DC. FALC'ARIA. Pentandria Digynia.
agréstis. (*wh.*) decurrent. Europe. 1726. 7. 8. H. ♃. Jacq. aust. 3. t. 257.
Siùm Falcària. R.S. *Drepanophy'llum agréste.* K.U.
S'ISON. S.S. S'ISON. Pentandria Pentagynia. L.
1 Amòmum. E.B. (*wh.*) hedge. Britain. .... 7. 8. H. ⊙. Eng. bot. v. 14. t. 954.
SCHU'LTZIA. S.S. SCHU'LTZIA. Pentandria Digynia.
crinìta. S.S. (*wh.*) slender-leaved. Altay. 1828. 6. 7. H. ♂. Pall. act. pet. 1779. 2. t. 7.
A'MMI. S.S. BISHOP'S WEED. Pentandria Digynia. L.
1 màjus. S.S. (*wh.*) great. S. Europe. 1551. 6. 7. H. ⊙. Flor. græc. t. 273.
2 Bœbèri. L. en. (*wh.*) Bœber's. Crete. 1818. 7. 8. H. ♃.
cicutæfòlium. R.S.
3 Visnàga. S.S. (*wh.*) Toothpick. S. Europe. 1596. 6. 8. H. ⊙. Gært. fr. 1. t. 107.
Visnàga daucoídes. G.
4 glaucifòlium. S.S. (*wh.*) glaucous-leav'd. France. 1816. —— H. ♂. Lapeyr. pyr. t. 75.
ÆGOP'ODIUM. L. GOUT-WEED. Pentandria Digynia. L.
Podagrària. E.B. (*wh.*) common. Britain. .... 5. 7. H. ♃. Eng. bot. v. 14. t. 940.
Sìson Podagrària. S.S.
C'ARUM. K.U. CARAWAY. Pentandria Digynia. L.
1 Cárvi. S.S. (*wh.*) common. Britain. .... 6. H. ♂. Eng. bot. v. 21. t. 1503.
Fœnículum Cárvi. L. en.

2 Bulbocástanum.K.U.(*wh.*)Earth-nut. Europe. 1797. 6. 7. H. ♃. Flor. dan. t. 220.
Bùnium Bulbocástanum. W. *My'rrhis Bùnium.* R.S.
3 flexuòsum. (*wh.*) common. Britain. .... 5. 6. H. ♃. Eng. bot. v. 14. t. 988.
Bùnium flexuòsum. E.B.
4 verticillàtum.K.U.(*wh.*)whorl-leaved. ———— .... 7. 8. H. ♃. Eng. bot. v. 6. t. 395.
Siùm verticillàtum. E. B.

B'UNIUM. DC. B'UNIUM. Pentandria Digynia. L.
1 peucedanoídes.M.B.(*st.*) Sulphur-wort. Caucasus. 1823. 7. 8. H. ♃.
2 lùteum. M.B. (*ye.*) yellow. Volga. 1827. —— H. ♃.
3 viréscens. DC. (*gr.*) greenish-flower'd.Europe. 1826. —— H. ♃.
Siùm viréscens. R.S.
4 acaùle. M.B. (*ye.*) stalkless. Caucasus. 1821. 6. 8. H. ♂.
A'mmi acaùle. S.S.

CRYPTOT'ÆNIA. DC. CRYPTOT'ÆNIA. Pentandria Digynia.
1 canadénse. DC. (*wh.*) Canadian. N. America. 1699. 7. 8. H. ♃. Moris. his. s. 9. t. 11.
Sìson canadénse. W. *My'rrhis canadénsis.* S.S.
2 Thomàsii. DC. (*wh.*) Thomas's. Italy. 1827. —— H. ♃.

PIMPINE'LLA. K.U. BURNET SAXIFRAGE. Pentandria Digynia. L.
1 mágna. S.S. (*wh.*) great. England. .... 6. 8. H. ♃. Eng. bot. v. 6. t. 408.
2 mèdia. P.S. (*wh.*) middle. Germany. .... —— H. ♃. Hoff. deut. fl. t. 397.
orientàlis. Jacq. aust. t. 397.
3 dissécta. W. (*wh.*) cut-leaved. France. .... —— H. ♃. Retz. obs. 3. t. 2.
4 rotundifòlia.M.B.(*wh.*) round-leaved. Caucasus. 1816. —— H. ♂. Spr.Anl.ed.2.t.4.f.1.4.
Sìson rotundifòlius. S.S.
5 Saxífraga. W. (*wh.*) common. Britain. .... —— H. ♃. Eng. bot. v. 6. t. 407.
6 Tràgium. DC. (*wh.*) Columna's. S. Europe. 1824. 6. 8. H. ♃. Column. phyt. t. 17.
Tràgium Colúmnæ. S.S.
7 peregrìna. W. (*wh.*) nodding. Italy. 1640. —— H. ♃. Jacq. vind. 2. t. 131.
8 aromática. M.B. (*wh.*) aromatic. Caucasus. 1824. —— H. ♂.
9 bubonoídes. P.S.(*wh.*) Bubon-like. Portugal. 1814. 6. 8. H. ♃.
Tràgium Brotèri. R.S.
10 villòsa. Scho. (*wh.*) villous. Barbary. 1824. —— H. ♃.
Tràgium Schousb'œi. R.S.
11 taùrica. DC. (*wh.*) Taurian. Tauria. 1824. 6. 8. H. ♃.
Ledebùria taùrica. L.en. *Tràgium taùricum.* S.S.
12 nìgra. W.en. (*wh.*) black-rooted. Germany. 1683. —— H. ♃. Moris. sect. 9. t. 5.
13 lateriflòra. L.en.(*wh.*) side-flowering. Hungary. 1816. —— H. ♃.
14 dichótoma. W. (*wh.*) dichotomous. Spain. 1798. —— H. ♃.
15 A'nisum. L. (*wh.*) Anise. Egypt. 1551. 6. 8. H. ⊙. Moris. his. s. 9. t. 9.
Síson A'nisum. R.S. *A'nisum vulgàre.* Gaert.

S'IUM. K.U. WATER-PARSNEP. Pentandria Digynia. L.
1 latifòlium. S.S. (*wh.*) broad-leaved. Britain. .... 7. 8. H.*w.*♃. Eng. bot. v. 3. t. 204.
2 lancifòlium. S.S. (*wh.*) spear-leaved. Siberia. 1823. —— H.*w.*♃. Gmel. sib. 1. t. 47.
3 angustifòlium.S.S.(*wh.*) narrow-leaved. Britain. .... —— H.*w.*♃. Eng. bot. v. 2. t. 139.
4 Sísarum. S.S. (*wh.*) Skirret. China. 1548. —— H. ♃. Schkuhr. hand. 1. t. 69.

BUPLE'URUM. S.S. HARE'S-EAR. Pentandria Digynia. L.
1 coriàceum. S.S. (*st.*) thick-leaved. Gibraltar. 1784. 7. 8. G. ♄. L'Herit. st. nov. t. 67.
2 fruticòsum. S.S. (*st.*) shrubby. S. Europe. 1596. —— H. ♄. Flor. græc. t. 263.
3 fruticéscens. S.S. (*st.*) frutescent. Spain. 1752. 8. 9. H. ♄. Cav. ic. 2. t. 106.
4 plantagíneum. S.S.(*st.*) plantain-leaved. Barbary. 1809. —— G. ♄. Desf. atl. 1. t. 57.
β *canéscens.* P.S. (*st.*) *hoary.* ———— —— —— G. ♄.
5 spinòsum. S.S. (*st.*) thorny. Spain. 1752. —— F. ♄. Gouan. ill. 8. t. 2. f. 3.
6 diffórme. S.S. (*st.*) various-leaved. C. B. S. .... —— G. ♄. Burm. afr. t. 71. f. 1.
7 rotundifòlium.S.S.(*ye.*) round-leaved. England. .... 6. 7. H. ⊙. Eng. bot. v. 2. t. 99.
8 protráctum. S.S. (*st.*) protracted. Portugal. 1824. —— H. ⊙.
9 heterophy'llum.S.S.(*st.*)various-leaved. Egypt. —— —— H. ⊙.
10 aùreum. S.S. (*ye.*) golden. Siberia. —— —— H. ♃.
11 longifòlium. S.S. (*st.*) long-leaved. Switzerland. 1713. 5. 7. H. ♃. Moris. sect. 9. t. 11.
12 pyrenàicum. S.S. (*st.*) Pyrenean. Pyrenees. 1814. —— H. ♃. Gouan. ill. t. 4.
13 angulòsum. W. (*st.*) angular-leaved. Switzerland. 1759. —— H. ♃. Moris. sect. 9. t. 12.
14 ranunculoídes. W.(*st.*) Crowfoot-like. Pyrenees. 1790. 7. 8. H. ♃. Park. theatr. f. 7.
15 caricifòlium. W. (*st.*) Carex-leaved. S. France. 1824. —— H. ♃.
16 Burserìanum.W.en. (*st.*) Burser's. ———— 1820. —— H. ♃.
17 petr'æum. S.S. (*st.*) alpine. S. Europe. 1823. —— H. ♃. Rich. de Bellev. t. 206.
18 sibiricum. M.B. (*st.*) Siberian. Siberia. 1826. —— H. ♃.
19 graminifòlium.S.S.(*st.*) Grass-leaved. Switzerland. 1768. 5. 7. H. ♃. Jacq. ic. 1. t. 56.
20 stellàtum. S.S. (*st.*) starry. ———— 1775. —— H. ♃. Hall. helv. t. 18.
21 falcàtum. S.S. (*st.*) twisted-stalk'd. Germany. 1739. 5. 9. H. ♃. Jacq. aust. 2. t. 158.

22 scorzoneræfòlium. W.en.(*st.*) Scorzonera-l[d]. Germany. 1818. 6. 8. H. ♃.
23 rígidum. R.S. (*st.*) rigid. S. France. 1739. 5. 9. H. ♃. Moris. sect. 9. t. 12.
24 nùdum. S.S. (*st.*) naked-stalked. C. B. S. 1778. 10. G. ♃.
25 gibraltáricum. Lag. (*st.*) Gibraltar. Gibraltar. 1825. —— G. ♃.
26 glaùcum. DC. (*ye.*) glaucous. S. France. 1828. —— H. ⊙.
27 Odontìtes. S.S. (*bh.*) narrow-leaved. England. .... —— H. ⊙. Eng. bot. v. 35. t. 2468.
Odontìtes lutéola. R.S.
28 júnceum. S.S. (*ye.*) linear-leaved. S. Europe. 1722. 7. 8. H. ⊙. Moris. sect. 9. t. 12.
29 Gerárdi. R.S. (*ye.*) Gerard's. ———— 1804. —— H. ⊙. Jacq. aust. t. 256.
30 baldénse. R.S. (*ye.*) channel-leav'd. M. Baldo. 1818. —— H. ⊙. W. et K. hung. 3. t. 257.
exaltàtum. P.S.
31 lancifòlium. H.H. (*ye.*) lance-leaved. .... 1824. 6. 9. H. ⊙.
32 semicompòsitum. S.S. (*ye.*) dwarf. Spain. 1778. —— H. ⊙. Flor. græc. t. 261.
Odontìtes semicompòsita. R.S.
33 grácile. M.B. (*ye.*) slender-twigged. Tauria. 1827 —— H. ⊙.
34 tenuíssimum. S.S. (*ye.*) slender. England. .... 7. 8. H. ⊙. Eng. bot. v. 7. t. 478.
35 trífidum. S.S. (*ye.*) trifid-umbel'd. Italy. 1824. 5. 9. H. ♃.

HETEROMO'RPHA. DC. HETEROMO'RPHA. Pentandria Digynia.
arboréscens. DC. (*st.*) arborescent. C. B. S. 1816. 7. 8. G. ♄.
Bupleùrum arboréscens. S.S. *Tenòria arboréscens.* R.S.

Tribus IV. SESELINEÆ. DC.

LICHTENSTEI'NIA. DC. LICHTENSTEI'NIA. Pentandria Digynia.
inèbrians. DC. (*wh.*) various-leaved. C. B. S. 1816. 8. 9. G. ♃.
Œnánthe inèbrians. S.S.

ŒNA'NTHE. S.S. WATER-DROPWORT. Pentandria Digynia. L.
1 fistulòsa. S.S. (*bh.*) common. Britain. .... 6. 8. H. ♃. Eng. bot. v. 6. t. 363.
2 tenuifòlia. S.S. (*wh.*) slender-leaved. C. B. S. 1816. 6. 8. G. ♃.
3 peucedanifòlia. S.S. (*wh.*) round-rooted. S. Europe. .... —— H.*w.* ♃. Pollich. palat. t. 3.
4 silaifòlia. S.S. (*wh.*) oblong-rooted. England. .... —— H.*w.* ♃. Eng. bot. v. 5. t. 348.
peucedanifòlia. E.B. *non.* S.S.
5 gymnorhìza. S.S. (*wh.*) naked rooted. N. Europe. 1824. —— H.*w.* ♃.
6 pimpinelloídes. S.S. (*bh.*) Burnet Saxifrage. England. .... —— H.*w.* ♃. Eng. bot. v. 5. t. 347.
7 prolífera. S.S. (*wh.*) proliferous. S. Europe. 1739. —— H.*w.* ♃. Jacq. vind. 3. t. 62.
8 globulòsa. S.S. (*wh.*) globe-headed. Portugal. 1710. —— H. ♂. Gouan. ill. 18. t. 9.
9 crocàta. S.S. (*wh.*) poisonous. Britain. .... —— H.*w.* ♃. Eng. bot. v. 33. t. 2313.
10 Phellándrium. S.S. (*wh.*) water Hemlock. ———— .... 6. 7. H.*w.* ♂. ———— v. 10. t. 684.
Phellándrium aquáticum. E.B.
11 apiifòlia. S.S. (*wh.*) Parsley-leaved. Portugal. 1806. 6. 8. H. ♃.
12 austràlis. R.S. (*wh.*) southern. S. Europe. 1824. —— H. ♃.
13 involucràta. Cav. (*wh.*) involucred. Spain. 1820. —— H. ♃.
14 am'œna. Leh. (*wh.*) handsome. .... 1827. —— H ♃.

ANESORH'IZA. DC. ANESORH'IZA. Pentandria Digynia.
interrúpta. (*wh.*) interrupted. C. B. S. 1816. 6. 8. G. ♃.
Œnánthe interrúpta. S.S.

SCLEROSC'IADIUM. DC. SCLEROSC'IADIUM. Pentandria Digynia.
rìgens. DC. (*wh.*) rigid-stalked. C. B. S. 1787. 6. 8. G. ♄.
Cònium rìgens. W. *Bùnium rìgens.* S.S.

ÆTH'USA. R.S. FOOLS-PARSLEY. Pentandria Digynia. L.
1 Cynápium. W. (*wh.*) common. Britain. .... 7. 9. H. ⊙. Eng. bot. v. 17. t. 1192.
2 cynapioídes. M.B. (*wh.*) Caucasean. Caucasus. 1820. —— H. ♃. Hoff. umb. ed. 2. t. tit. f. 9.
3 elàta. L.en. (*wh.*) tall. Siberia. ———— —— H. ⊙.
4 fátua. R.S. (*wh.*) fine-leaved. .... 1781. —— H. ♃.

FŒNI'CULUM. G. FENNEL. Pentandria Digynia.
1 vulgàre. G. (*ye.*) common. Britain. .... 7. 8. H. ♃. Eng. bot. v. 17. t. 1208.
Anèthum Fœnículum. E.B. *Mèum Fœnículum.* S.S.
2 dúlce. DC. (*ye.*) sweet. S. Europe. 1824. —— H. ⊙. DC. hort. m. ined. t. 96.
3 piperìtum. (*ye.*) peppery. Italy. ———— —— H. ♂.
Mèum piperìtum. R.S. *Anèthum piperìtum.* Bert.
4 lùteum. F. (*ye.*) yellow. Persia. 1829. —— H. ♂.

KUNDMA'NNIA. DC. KUNDMA'NNIA. Pentandria Digynia.
pastinacæfòlia. (*ye.*) Parsnep-leaved. Sicily. 1686. 7. 8. H. ♃. Jacq. vind. 2. t. 133.
Brignòlia pastinacæfòlia. K.U. *Siùm sìculum.* R.S.

DEVE'RRA. DC. DEVE'RRA. Pentandria Digynia.
1 tortùosa. DC. (*wh.*) twisting. Levant. 1828. .... G. ♄. Desf. atl. 1. t. 73.
Bùbon tortuòsum. Df.

2 aphy'lla. DC.	(*wh.*)	leafless.	C. B. S.	1816.	....	G. ♄.	
Bùbon aphy'llus. Schlecht.							
SE'SELI. S.S.		MEADOW SAXIFRAGE. Pentandria Digynia. S.S.					
1 elàtum. S.S.	(*wh.*)	tall.	Austria.	1710.	7. 8.	H. ♃.	Gouan. ill. 16. t. 8.
2 vàrium. S.S.	(*wh.*)	variable.	....	1818.	——	H. ♃.	
3 glaùcum. S.S.	(*pu.wh.*)	glaucous.	France.	1759.	——	H. ♃.	Jacq. aust. 2. t. 144.
4 taùricum. S.S.	(*wh.*)	Taurian.	Tauria.	1818.	6. 8.	H. ♃.	
5 montànum. S.S.	(*wh.*)	mountain.	Italy.	1658.	6. 7.	H. ♃.	Jacq. vind. 2. t. 129.
6 ánnuum. S.S.	(*bh.*)	annual.	Europe.	1816.	7. 8.	H. ♂.	Jacq. aust. t. 55.
Cùrum símplex. W.							
7 campéstre. S.S.	(*wh.*)	field.	Poland.	1827.	——	H. ♃.	
8 pimpinelloídes. S.S.	(*wh.*)	nodding-flower'd.	S. Europe.	1796.	——	H. ♃.	
9 Hippomárathrum. S.S.	(*bh.*)	various-leav'd.	Austria.	1656.	——	H. ♃.	Jacq. aust. 2. t. 143.
10 tortuòsum. S.S.	(*wh.*)	crooked.	S. Europe.	1597.	10.	H. ♃.	Moris. sect. 9. t. 6.
11 leucospérmum. S.S.	(*wh.*)	white-seeded.	Hungary.	1805.	7.	H. ♃.	W. et K. hung. t. 89.
12 prolíferum. S.S.	(*wh.*)	prolific.	....	1823.	5. 8.	H. ♃.	
13 chærophylloídes. S.S.	(*wh.*)	Chervil-like.	C. B. S.	1816.	——	G. ♃.	
14 Boccòni. S.S.	(*wh.*)	Boccone's.	Sicily.	1826.	——	H. ♃.	Bocc. sic. s. 27. 28.
15 divaricàtum. S.S.	(*ye.*)	shining-leaved.	N. America.	1812.	6. 7.	H. ♃.	Bot. mag. t. 1742.
16 striàtum. S.S.	(*wh.*)	striped-stalked.	C. B. S.	1816.	7. 8.	G. ♃.	
17 grácile. S.S.	(*ye.*)	slender.	Hungary.	1805.	6. 7.	H. ♃.	W. et K. hung. 2. t. 117.
18 peucedanifòlium. K.U.	(*wh.*)	smooth-stalk'd.	Hungary.	1826.	7. 9.	H. ♃.	———— t. 146.
19 petr'æum. K.U.	(*wh.*)	glaucous.	Tauria.	1824.	7. 8.	H. ♃.	
Bùbon glaùcus. S.S.							
20 cuneifòlium. M.B.	(*wh.*)	wedge-leaved.	Sicily.	1823.	7. 9.	H. ♃.	Moris. his. s. 9. t. 7.
21 gummíferum. K.U.	(*bh.*)	gum-bearing.	Crimea.	1710.	——	H. ♂.	Sm. exot. bot. t. 20.
22 dichótomum. M.B.	(*wh.*)	forked.	Tauria.	1816.	——	H. ♂.	Bot. mag. t. 2073.
23 eriocéphalum.	(*wh.*)	woolly-headed.	Siberia.	1827.	——	H. ♂.	
24 córsicum. L.en.	(*wh.*)	dwarf annual.	Corsica.	1825.	——	H. ⊙.	
25 Pallàsii. Bess.	(*wh.*)	Pallas's.	Siberia.	1823.	——	H. ♃.	
26 Lobeliànum. Bess.	(*wh.*)	Lobel's.	....	——	——	H. ♃.	
LIBAN'OTIS. DC.		LIBAN'OTIS. Pentandria Digynia.					
1 buchtorménsis. DC.	(*wh.*)	Siberian.	Siberia.	1822.	7. 8.	H. ♃.	
Bùbon buchtorménsis. R.S. *Athamánta rígida.* Horn.							
2 vulgàris. DC.	(*wh.*)	mountain.	England.	....	——	H. ♃.	Eng. bot. v. 2. t. 138.
Athamántu Libanòtis. E.B.							
CENOL'OPHIUM. DC.		CENOL'OPHIUM. Pentandria Digynia.					
Fischèri. DC.	(*wh.*)	Fischer's.	Siberia.	1820.	6. 8.	H. ♃.	
Cnídium Fischèri. S.S. *Angélica Fischèri.* R.S. *Ligústicum Fischèri.* L. enum.							
CNI'DIUM. S.S.		CNI'DIUM. Pentandria Digynia. S.S.					
1 Monniéri. S.S.	(*wh.*)	annual.	S. Europe.	1771.	7. 8.	H. ⊙.	Jacq. vind. 1. t. 62.
Selìnum Monniéri. Jacq.							
2 apioídes. H.U.	(*wh.*)	Parsley-leaved.	Europe.	1791.	6. 7.	H. ♃.	Jacq. aust. 5. ap. t. 44.
Ligústicum apioìdes. R.S. *silaifòlium.* Jacq.							
3 venòsum. DC.	(*wh.*)	veined.	Europe.	1818.	7. 8.	H. ♃.	
Séseli dùbium. R.S.							
THA'SPIUM. N.		THA'SPIUM. Pentandria Digynia.					
1 actæifòlium. N.	(*ye.*)	Actæa-leaved.	N. America.	1824.	7. 8.	H. ♃.	
Ligústicum actæifòlium. M.							
2 barbinòde. N.	(*ye.*)	bearded-jointed.	———	1759.	6. 7.	H. ♃.	
Thápsia trifòliata. R.S.							
TROCHISCA'NTHES. K.U.		TROCHISCA'NTHES. Pentandria Digynia.					
nodiflòrus. K.U.	(*bh.*)	knot-flowered.	Apennines.	1818.	6. 7.	H. ♃.	All. ped. t. 72.
Ligústicum nodiflòrum. S.S. *verticillàtum.* W. et K. hung. 2. t. 171.							
ATHAMA'NTA. S.S.		SPIGNELL. Pentandria Digynia. L.					
1 macedònica. S.S.	(*wh.*)	Macedonian.	Greece.	1596.	6. 8.	G. ♂.	Lobel. hist. 406. ic. 708.
Bùbon macedònicum. L.							
2 condensàta. W.	(*wh.*)	close-headed.	Siberia.	1773.	7. 9.	H. ♃.	Gmel. sib. 1. t. 44.
3 sibírica. W.	(*st.*)	Siberian.	———	1771.	7. 8.	H. ♃.	Lam. ill. t. 194. f. 1.
4 sícula. S.S.	(*wh.*)	Flixweed-leav'd.	Sicily.	1686.	6. 7.	H. ♃.	Ten. fl. nap. 1. t. 25.
5 Matthìoli. S.S.	(*wh.*)	fine-leaved.	Carniola.	1802.	——	H. ♃.	Jacq. ic. rar. 1. t. 57.
6 stricta. Led.	(*wh.*)	upright.	Siberia.	1827.	——	H. ♃.	
7 creténsis. S.S.	(*wh.*)	candy-carrot.	Austria.	1596.	——	H. ♃.	Jacq. aust. 1. t. 62.
8 incàna. S.S.	(*wh.*)	hoary.	Siberia.	1802.	7. 8.	H. ♃.	
9 ánnua. W.	(*wh.*)	annual.	Candia.	1731.	6. 7.	H. ⊙.	
LIGU'STICUM. S.S.		LIGU'STICUM. Pentandria Digynia. L.					
1 scóticum. S.S.	(*bh.*)	Scotch.	Britain.	....	——	H. ♃.	Eng. bot. v. 17. t. 1207.
2 obtusifòlium. S.S.	(*wh.*)	blunt-leaved.	Tangiers.	1824.	7. 8.	H. ♃.	
3 athamantoídes. S.S.	(*st.*)	Carrot-leaved.	Pyrenees.	1734.	——	H. ♃.	Scop. carn. t. 10.
A'mmi daucifòlium. W. *Athamánta pyrenàica.* Jacq. vind. 2. t. 197.							

4 alàtum. s.s. (*ro.wh.*) winged-stalked. Caucasus. 1823. 7. 8. H. ♃.
5 símplex. s.s. (*wh.pu.*) shining. Switzerland. 1775. 7. H. ♂. Jacq. misc. 2. t. 2.
Laserpítium símplex. Jacq. *lùcidum.* H.K.
6 ferulàceum. s.s. (*wh.*) Fennel-leaved. Levant. 1752. 6. H. ♃. Allion. ped. t. 60.
Laserpítium daùricum. Jacq. vind. 3. t. 38.
7 cándicans. s.s. (*wh.*) pale. .... 1780. 7. 8. H. ♃.
8 pyrenàicum. K.U. (*wh.*) Pyrenean. Pyrenees. 1739. 6. 7. H. ♃. Gouan. ill. t. 7. f. 2.
Cnídium pyren'æum. s.s. *Séseli aristàtum.* H.K.
9 Seguiérii. K.U. (*wh.*) Fennel-leaved. Italy. 1774. 7. 8. H. ♃. Jacq. vind. 1. t. 61.
Selìnum Seguiérii. Jacq. *Imperatòria Seguiérii.* s.s.
10 Fontanèsii. s.s. (*wh.*) fine-leaved. Algiers. .... 6. 7. H. ♃. Desf. atl. 1. t. 71.
Laserpítium peucedanoídes. Desf.
11 cuneifòlium. Led. (*wh.*) wedge-leaved. Siberia. 1827. —— H. ♃.

SIL'AUS. K.U. SIL'AUS. Pentandria Digynia.
1 praténsis. K.U. (*st.*) meadow. England. .... 6. 8. H. ♃. Eng. bot. v. 30. t. 2142.
Peucedànum Silàus. E.B. *Cnídium Silàus.* s.s.
2 Matthìoli. K.U. (*ye.*) late-flowering. Italy. 1824. 7. 8. H. ♃.
Peucedànum Matthìoli. s.s.
3 alpéstris. K.U. (*st.*) Alpine. France. 1739. 6. 7. H. ♃.

WALLRO'THIA. DC. WALLRO'THIA. Pentandria Digynia.
1 alpìna. DC. (*wh.*) alpine. Hungary. 1827. 6. 8. H. ♃. W. et K. hung. 2. t. 182.
tuberòsa. R.S. *Bùnium alpìnum.* W.K.
2 tenuifòlia. DC. (*wh.*) slender-leaved. Pyrenees. 1826. —— H. ♃. DC. mem. umb. t. 1. f. A.

M'EUM. S.S. BAWD-MONEY. Pentandria Digynia. S.S.
1 athamánticum. s.s. (*wh.*) common. Britain. .... 4. 6. H. ♃. Eng. bot. v. 32. t. 2249.
2 Mutellìna. s.s. (*re.*) Alpine. Austria. 1774. 7. 8. H. ♃. Jacq. aust. t. 56.
Phellándrium Mutellìna. L.
3 sibirícum. s.s. (*ye.*) Siberian. Siberia. 1820. —— H. ♃.

CONIOSEL'INUM. H.U. CONIOSEL'INUM. Pentandria Digynia.
1 íngricum. F. (*wh.*) Ingrian. Russia. 1827. 7. 8. H. ♃.
2 tatáricum. F. (*wh.*) Tartarian. Tartary. 1825. —— H. ♃.
Angélica Fischèri. R.S.

CRI'THMUM. L.en. SAMPHIRE. Pentandria Digynia. L.
1 Libanòtis. L.en. (*ye.*) smooth-seeded. Sicily. 1570. 7. 8. H. ♃. Schk. handb. 1. t. 65.
Cáchrys Libanòtis. s.s.
2 marítimum. W. (*st.*) sea-side. Britain. .... 7. 9. H. ♃. Eng. bot. v. 12. t. 819.

Tribus V. ANGELICEÆ. DC.

LEVI'STICUM. K.U. LOVAGE. Pentandria Digynia.
officinàle. K.U. (*ye.*) common. Italy. 1596. 6. 7. H. ♃. Schk. handb. t. 68.
Ligústicum Levísticum. s.s.

SEL'INUM. L. MILK-PARSLEY. Pentandria Digynia. L.
1 Carvifòlium. W. (*bh.*) Caraway-leav'd. Austria. 1774. 7. 8. H. ♃. Jacq. aust. 1. t. 16.

OSTE'RICUM. H.U. OSTE'RICUM. Pentandria Digynia.
praténse. H.U. (*wh.*) field. Galicia. 1824. 7. 8. H. ♃.
Angélica praténsis. M.B.

ANGE'LICA. K.U. ANGE'LICA. Pentandria Digynia. L.
1 sylvéstris. s.s. (*wh.*) wild. Britain. .... 6. 8. H. ♃. Eng. bot. v. 16. t. 1128.
2 Razoùlii. s.s. (*bh.*) decurrent-l[d]. Pyrenees. 1816. —— H. ♃. Gouan. ill. 13. t. 6.
3 montàna. K.U. (*wh.*) mountain. Switzerland. —— —— H. ♃.
4 flavéscens. Bess. (*st.*) yellowish. Pyrenees. —— —— H. ♃.
5 pyren'æa. s.s. (*wh.*) Pyrenean. ——— 1818. —— H. ♃. Gouan. ill. p. 11. t. 5.
6 lùcida. s.s. (*br.*) shining. Canada. 1640. —— H. ♂. Jacq. vind. 3. t. 24.
7 chærophy'llea. s.s. (*wh.*) Chervil-leaved. Livonia. 1827. —— H. ♃.
8 Wormskióldii. F. (*wh.*) Wormskiold's. Kamtschatka. —— —— H. ♃.
9 lancifòlia. s.s. (*wh.*) lance-leaved. .... 1818. —— H. ♃.
Selìnum angelicástrum. L. enum.

ARCHANGE'LICA. H.U. ARCHANGE'LICA. Pentandria Digynia.
1 officinàlis. H.U. (*wh.*) garden. England. .... 6. 8. H. ♂. Eng. bot. v. 36. t. 2561.
Angélica Archangélica. E.B.
2 atropurpùrea. H.U. (*pu.*) dark purple. Canada. 1759. 7. 8. H. ♃. Cornut. canad. t. 199.

Tribus VI. PEUCEDANEÆ. DC.

OP'OPANAX. K.U. OP'OPANAX. Pentandria Digynia.
Chirònium. K.U. rough. S. Europe. 1640. 6. 7. H. ♃. W. et K. hung. 3. t. 211.
Pastinàca Opòpanax. W. *Ferùla Opòpanax.* s.s.

	FER'ULA. S.S.		GIANT-FENNEL. Pentandria Digynia. L.					
1	lævigàta. s.s.	(ye.)	smooth.	C. B. S.	1774.	12. 3.	G. ♄.	
	Bùbon lævigàtum. W.							
2	nudicaùlis. s.s.	(ye.)	naked-stalked.	Sicily.	1825.	——	H. ♃.	
3	A'sa f'œtida. s.s.	(st.)	Asa fœtida.	Persia.	1818.	7. 8.	H. ♃.	
4	commùnis. s.s.	(ye.)	common.	S. Europe.	1597.	6. 7.	H. ♃.	Moris. s. 9. t. 15 f. 3.
5	glaùca. s.s.	(ye.)	glaucous.	Italy.	1596.	——	H. ♃.	
6	sylvática. s.s.	(ye.)	wood.	Poland.	1827.	——	H. ♃.	
7	campéstris. s.s.	(ye.)	field.	Tauria.	——	——	H. ♃.	
8	nodiflòra. s.s.	(ye.)	knot-flowered.	S. Europe.	1596.	——	H. ♃.	Desf. atl. 1. t. 67.
	sulcàta. Desf.							
9	Ferulàgo. s.s.	(ye.)	Fennel-leaved.	————	——	——	H. ♃.	Jacq. aust. 5. ap. t. 5.
	nodiflòra. Jacq. *non* s.s.							
10	thyrsiflòra. s.s.	(st.)	thyrse-flower'd.	Crete.	1823.	——	F. ♃.	Flor. græc. t. 280.
11	tingitàna. s.s.	(ye.)	Tangier.	Barbary.	1680.	——	H. ♂.	Herm. parad. t. 165.
12	stricta. s.s.	(ye.)	upright.	C. B. S.	1816.	——	G. ♂.	
13	nùda. s.s.	(ye.)	naked.	Siberia.	1827.	——	H. ♃.	
14	cáspica. s.s.	(ye.)	Caspian.	Caspia.	1824.	——	H. ♃.	Spreng. umb. t. 7. f. 14.
15	orientàlis. s.s.	(ye.)	eastern.	Levant.	1759.	7. 8.	H. ♃.	Tourn. it. 3. t. 239.
16	pérsica. s.s.	(ye.)	stinking.	Persia.	1782.	——	H. ♃.	Andr. rep. t. 558.
17	tatárica. F.	(ye.)	Tartarian.	Tartary.	1828.	——	H. ♃.	
18	capillàris. s.s.	(ye.)	capillary.	Portugal.	——	——	H. ♃.	
19	desertòrum. Led.	(ye.)	desert.	Siberia.	1827.	——	H. ♃.	
20	songàrica. R.S.	(ye.)	Songarian.	————	——	——	H. ♃.	
	ERIOSYNA'PHE. DC.		ERIOSYNA'PHE. Pentandria Digynia.					
	longifòlia. DC.	(ye.)	long-leaved.	Wolga.	1823.	7. 8.	H. ♃.	DC.mem.umb.pl.1.f.E.
	Ferùla longifòlia. F.							
	PÆLI'MBIA. DC.		PÆLI'MBIA. Pentandria Digynia.					
	Chabr'æi. DC.	(bh.)	fine-leaved.	Europe.	1791.	7. 8.	H. ♃.	Jacq. aust. 1. t. 72.
	Selìnum Chabr'æi. Jacq. *Oreoselìnum Chabr'æi.* M.B. *Imperatòria Chabr'æi.* R.S.							
	PEUCED'ANUM. S.S.		SULPHURWORT. Pentandria Digynia. L.					
1	officinàle. s.s.	(ye.)	officinal.	England.	....	5. 7.	H. ♃.	Eng. bot. v. 25. t. 1767.
2	paniculàtum. R.S.	(ye.)	panicled.	Corsica.	1827.	——	H. ♃.	
3	alsáticum. W.	(wh.)	small-headed.	Austria.	1774.	6. 8.	H. ♃.	Jacq. aust. 1. t. 70.
4	gállicum. s.s.	(wh.)	French.	France.	1816.	——	H. ♃.	Gmel. sib. 1. t. 41.
5	itálicum. P.S.	(ye.)	yellow-flower'd.	Italy.	....	5. 7.	H. ♃.	
6	ruthénicum. R.S.	(ye.)	Russian.	Russia.	1824.	6. 8.	H. ♃.	
7	tàuricum. R.S.	(ye.)	Taurian.	Tauria.	1828.	——	H. ♃.	
8	longifòlium. R.S.	(ye.)	long-leaved.	Hungary.	1826.	——	H. ♃.	W. et K. hung. t. 251.
9	arenàrium. s.s.	(ye.)	sand.	————	1816.	6. 7.	H. ♃.	———— 1. t. 20.
10	Morisòni. R.S.	(ye.)	Morison's.	————	——	——	H. ♃.	Moris. his. 3. s. 9. t. 15. f. 1.
11	nodòsum. s.s.	(wh.)	knotted-stalked.	Crete.	....	——	H. ♃.	Moris. his. s. 9. t. 17.
12	aùreum. s.s.	(ye.)	golden.	Canaries.	1779.	6.	G. ♂.	Bot. reg. t. 559.
13	tenuifòlium. s.s.	(ye.)	slender-leaved.	C. B. S.	1816.	6. 8.	G. ♂.	
14	iseténse. s.s.	(st.)	straw-coloured.	Siberia.	1828.	——	H. ♂.	Gmel. sib. 1. t. 42-43.
15	sibíricum. W.	(ye.)	Siberian.	————	1804.	6. 7.	H. ♃.	W. et K. hung. 1. t. 20.
16	palústre. K.U.	(wh.)	marsh.	Britain.	....	7. 8.	H. ♃.	Eng. bot. v. 4. t. 229.
	Selìnum palústre. E.B. *Thysselìnum palústre.* H.U.							
17	baicalénse. K.U.	(wh.)	Baical.	Siberia.	1824.	——	H. ♃.	
18	Cervària. K.U.	(wh.pu.)	large-leaved.	Europe.	1597.	——	H. ♃.	Jacq. aust. 1. t. 69.
	Athamánta Cervària. Jacq. *Ligústicum Cervària.* s.s.							
19	Oreoselìnum. K.U.	(re.wh.)	bending.	Europe.	1780.	7. 9.	H. ♃.	Jacq. aust. 1. t. 68.
	Athamánta Oreoselìnum. Jacq. *Oreoselìnum legítimum.* M.B.							
20	montànum. K.U.	(wh.)	mountain.	Switzerland.	1816.	7. 8.	H. ♃.	
21	involucràtum. K.U.	(wh.)	involucred.	S. France.	1824.	——	H. ♃.	
	Selìnum peucedanoìdes. R.S. *Oreoselìnum peucedanoìdes.* H.U.							
22	austrìacum. K.U.	(wh.)	Austrian.	Austria.	1804.	——	H. ♃.	Jacq. aust. 1. t. 71.
	Selìnum austrìacum. Jacq. *Oreoselìnum austrìacum.* H.U.							
23	rablénse. K.U.	(wh.)	linear-leaved.	N. Europe.	1816.	7. 8.	H. ♃.	Flor. dan. t. 1330.
24	latifòlium.	(wh.)	broad-leaved.	Caucasus.	——	——	H. ♃.	
	Oreoselìnum latifòlium. M.B.							
25	polymórphum. Lag.	(wh.)	variable.	Russia.	1824.	——	H. ♃.	
26	Bellárdi. Lag.	(wh.)	Bellardi's.	N. Europe.	——	——	H. ♃.	
27	élegans.	(wh.)	elegant.	————	——	——	H. ♃.	
	Oreoselìnum élegans. H.U.							
28	podòlicum. H.U.	(wh.)	Polish.	Poland.	1819.	——	H. ♃.	
29	nígricans. H.U.	(wh.)	black-rooted.	Switzerland.	1823.	——	H. ♃.	
30	caucásicum.	(st.)	Caucasean.	Caucasus.	1824.	——	H. ♃.	
31	Ostrùthium. K.U.	(wh.)	Masterwort.	Scotland.	....	5. 7.	H. ♃.	Eng. bot. v. 20. t. 1380.
	Imperatòria Ostrùthium. Eng. bot.							

32 imperatorioídes. K.U. (*wh.*) narrow-leav'd. Alps, Europe. 1818. —— H. ♃. Moris. his. s. 9. t 4.
Imperatòria angustifòlia. S.S. *Selìnum imperatorioídes.* L.en.

IMPERAT'ORIA. DC. Masterwort. Pentandria Digynia. L.
verticillàris. S.S. (*wh.*) whorl-flowered. Italy. 1683. 7. H. ♃. Jacq. vind. 2. t. 130.
Angélica verticillàris. W.

CALLIS'ACE. F. Callis'ace. Pentandria Digynia.
davùrica. F. (*wh.*) Daurian. Dauria. 1820. 7. 8. H. ♃.
Thysselìnum davùricum. S.S.

B'UBON. K.U. B'ubon. Pentandria Digynia.
Galbànum. K.U. (*ye.*) Lovage-leaved. C. B. S. 1696. 6. 8. G. ♄. Jacq. vind. 3. t. 36.
Selìnum Galbànum. R.S.

AN'ETHUM. L.en. Dill. Pentandria Digynia. L.
1 gravèolens. W. (*ye.*) common. Spain. 1570. 6. 7. H. ♂. Hoff. umb. gen. t. 1. f. 13.
Pastinàca Anèthum. S.S.
2 ségetum. L.en. (*ye.*) Portugal. Portugal. 1796. —— H. ☉. Jacq. vind. t. 132.
3 Sówa. H.B. (*st.*) Indian. E. Indies. 1810. —— H. ☉.

CAPNOPHY'LLUM. S.S. Capnophy'llum. Pentandria Digynia. S.S.
africànum. S.S. (*wh.*) Rue-leaved. C. B. S. 1759. 6. 9. H. ☉. Jacq. vind. t. 194.
Cònium africànum. W. *Rùmia capénsis.* L.en.

ARCHEM'ORA. DC. Archem'ora. Pentandria Digynia.
rígida. DC. (*wh.*) rigid. Virginia. 1774. 7. 8. H. ♃. Moris. dis. s. 9. t. 7. f. 1.
Siùm rígidius. W. *Œnánthe rígida.* N. *Pastinàca rígida.* S.S.

PASTIN'ACA. S.S. Parsnep. Pentandria Digynia. L.
1 lùcida. S.S. (*ye.*) shining-leaved. S. Europe. 1771. 6. 7. H. ♂. Jacq. vind. 2. t. 199.
2 satìva. W.en. (*ye.*) cultivated. .... 7. H. ♂. Flor. dan. t. 1206.
3 opàca. W.en. (*ye.*) wild. Britain. .... —— H. ♂. Eng. bot. v. 8. t. 556.
satìva β arvénsis. P.S.
4 taraxacifòlia. F. (*ye.*) Dandelion-ld. Russia. 1822. —— H. ♃.
5 gravèolens. S.S. (*ye.*) strong-scented. Tauria. 1819. 7. 8. H. ♃. Hoff. umb. t. 1. B. f. 6. c.
Malabaíla gravèolens. H.U.
6 panacifòlia. F. (*ye.*) Panax-leaved. Tauria. 1826. 7. 8. H. ♃.
7 pimpinellifòlia. S.S. (*ye.*) BurnetSaxifrage. Caucasus. 1824. —— H. ♂. —— p. 209. t. 1. B. f. 6. a. b.
8 umbròsa. F. (*ye.*) shady. ———— 1821. —— H. ♃.
9 dissécta. S.S. (*ye.*) cut-leaved. Levant. 1816. 6. 7. H. ♂. Venten. cels. t. 78.
10 nudicaùlis. S.S. (*ye.*) naked-stalked. N. America. 1816. —— H. ♃.
Smy'rnium nudicaùle. Ph.
11 divaricàta. Desf. (*ye.*) spreading. .... 1824. —— H. ♃.

ASTYD'AMIA. DC. Astyd'amia. Pentandria Digynia.
canariénsis. DC. (*ye.*) cut-leaved. Canaries. 1780. 7. G. ♂. DC. dis. umb. t. 1. f. D.
Bupleùrum canariénse. S.S. *Críthmum latifòlium.* L. *Tenòria canariénsis.* R.S.

HERA'CLEUM. S.S. Cow-parsnep. Pentandria Digynia. L.
1 Sphondy'lium. W. (*wh.*) common. Britain. .... 5. 6. H. ♃. Eng. bot. v. 14. t. 939.
2 flavéscens. W. (*wh.*) yellowish. Austria. 1789. 6. 7. H. ♃. Jacq. aust. 2. t. 173.
3 élegans. W. (*wh.*) rough-leaved. ———— 1800. 5. 6. H. ♃. ———— 2. t. 175.
4 angustifòlium. W. (*wh.*) narrow-leaved. ———— .... —— H. ♃. ———— 2. t. 174.
5 longifòlium. Jacq. (*wh.*) long-leaved. ———— .... —— H. ♃. ———— 2. t. 374.
6 sibíricum. S.S. (*st.*) Siberian. Siberia. 1768. —— H. ♃. Gmel. sib. 1. t. 50.
7 Panáces. S.S. (*wh.*) Fig-leaved. ———— 1596. 7. 8. H. ♃. Lobel. ic. 701.
8 pyrenàicum. W. (*wh.*) Pyrenean. Pyrenees. 1804. —— H. ♃.
9 amplifòlium. Lap. (*wh.*) large-leaved. ———— —— —— H. ♃. Lap. fl. pyr. t. 79.
10 gummíferum. W. (*wh.*) gum-bearing. ———— 1816. —— H. ♃. Wild. hort. ber. t. 53. 54.
11 speciòsum. H.U. (*wh.*) handsome. ———— —— —— H. ♃.
12 ásperum. M.B. (*wh.*) rough. Caucasus. —— —— H. ♃.
13 cordàtum. Pr. (*wh.*) heart-leaved. Sicily. 1825. —— H. ♃.
14 villòsum. S.S. (*wh.*) villous. Siberia. 1816. —— H. ♃.
15 grandiflòrum. Stev. (*w.*) large-flowered. Caucasus. 1818. —— H. ♃.
16 decípiens. H.U. (*wh.*) deceiving. ———— —— —— H. ♃.
17 marginàtum. F. (*wh.*) margined. Siberia. 1816. —— H. ♃.
18 pubéscens. M.B. (*wh.*) pubescent. Caucasus. 1824. —— H. ♃.
19 caucásicum. R.S. (*wh.*) Caucasean. ———— —— —— H. ♃.
20 alpìnum. S.S. (*wh.*) Alpine. Switzerland. 1739. 6. 7. H. ♃. Barrel. ic. 55.
21 austrìacum. S.S. (*wh.*) Austrian. Austria. 1816. —— H. ♃. Jacq. aust. t. 61.
Tord'ylium siifòlium. W.
22 ligusticifòlium. S.S. (*wh.*) Lovage-leaved. Tauria. 1824. —— H. ♂.
23 subvillòsum. F. (*wh.*) subvillous. Siberia. 1820. —— H. ♃.
24 gigánteum. F. (*wh.*) gigantic. ———— —— —— H. ♃.
25 básvicum. F. (*wh.*) stately. ———— —— —— H. ♃.
26 laciniàtum. F. (*wh.*) jagged. ———— 1824. —— H. ♃.
27 taùricum. F. (*wh.*) Taurian. Tauria. 1823. —— H. ♃.

28 cáspicum. F. (*wh.*) Caspian. ——— Caucasus. 1828. 6. 7. H. ♃.
29 lanàtum. M. (*wh.*) woolly. N. America. ——— ——— H. ♃.
30 pùmilum. DC. (*wh.*) dwarf. Dauphiny. 1800. 5. 7. H. ♃. Vill. delph. 2. t. 14.
Peucedànum Vocontiòrum. S.S.
ZOSI'MIA. H.U. ZOSI'MIA. Pentandria Digynia.
orientàlis. H.U. (*ye.*) Wormwood-l^d. Persia. 1816. 7. 8. H. ⊙. Vent. choix. t. 22.
Herácleum absinthifòlium. S.S. *Tordy'lium absinthifòlium.* P.S.

Tribus VII. TORDYLINEÆ. DC.

HASSELQUI'STIA. S.S. HASSELQUI'STIA. Pentandria Digynia. L.
1 ægyptìaca. S.S. (*wh.*) Egyptian. Egypt. 1768. 7. H. ⊙. Jacq. vind. 1. t. 37.
2 cordàta. S.S. (*wh.*) heart-leaved. .... 1787. ——— H. ⊙. ——— 2. t. 193.
TORDY'LIUM. K.U. HARTWORT. Pentandria Digynia. L.
1 syrìacum. S.S. (*wh.*) Syrian. Syria. 1597. 7. H. ⊙. Jacq. vind. 1. t. 54.
2 cordàtum. E. (*wh.*) heart-leaved. ——— 1826. ——— H. ⊙.
3 máximum. S.S. (*wh.*) great. England. .... 6. 7. H. ⊙. Eng. bot. v. 17. t. 1173.
4 lusitánicum. W.en. (*wh.*) Portugal. Portugal. 1820. ——— H. ⊙.
CONDYLOCA'RPUS. K.U. CONDYLOCA'RPUS. Pentandria Monogynia.
1 officinàlis. K.U. (*bh.*) officinal. England. .... 7. H. ⊙. Eng. bot. v. 34. t. 2440.
Tordy'lium officinàle. S.S.
2 ápulus. K.U. (*wh.*) small. Italy. 1739. ——— H. ⊙. Jacq. vind. 1. t. 53.

Tribus VIII. SILERINEÆ. DC.

KRUB'ERA. H.U. KRUB'ERA. Pentandria Digynia.
leptophy'lla. K.U. slender-leaved. Barbary. 1800. 6. 7. H. ⊙. Desf. atl. 1. t. 66.
Cònium dichótomum. Df. *Cáchrys dichótoma.* R.S. *Tordy'lium peregrìnum.* L.
AGASY'LLIS. S.U. AGASY'LLIS. Pentandria Digynia.
1 caucásica. S.U. (*wh.*) Caucasean. Caucasus. 1823. 5. 7. H. ♃.
Silér caucásicum. S.S. *Cáchrys latifòlia.* M.B.
2 sálsa. S.U. (*st.*) fine-leaved. Siberia. 1804. 7. 8. H. ♃. Pall. act. pet. 1779. t. 8.
Silér sálsum. R.S. *Sìson sálsum.* L.
SILE'R. K.U. SILE'R. Pentandria Digynia.
trilòbum. K.U. (*wh.*) Columbine-leav'd. Austria. 1796. 5. 7. H. ♃. Jacq. aust. 2. t. 147.
aquilegifòlium. R.S. *Laserpìtium aquilegifòlium.* W.

Tribus IX. CUMINEÆ. DC.

CU'MINUM. S.S. CUMIN. Pentandria Digynia. L.
Cy'minum. S.S. (*wh.*) officinal. Egypt. 1594. 6. 7. H. ⊙. Cav. ic. 4. t. 360.

Tribus X. THAPSIEÆ. DC.

THA'PSIA. S.S. DEADLY CARROT. Pentandria Digynia. L.
1 villòsa. S.S. (*ye.*) villous. S. Europe. 1710. 6. 7. H. ♃. Moris. s. 9. t. 18. f. 3.
2 fœtida. S.S. (*ye.*) stinking. Spain. 1596. 7. 8. H. ♃. ——— t. 18. f. 7.
3 gummífera. S.S. (*wh.*) gum-bearing. S. Europe. 1824. ——— H. ♃. Desf. atl. 1. t. 72.
Laserpìtium gummíferum. Desf.
4 Asclèpium. S.S. (*ye.*) oriental. Levant. .... ——— H. ♃. Flor. græc. t. 286.
5 gargánica. S.S. (*ye.*) Garganian. Barbary. 1683. ——— H. ♃. Gouan. ill. 18. t. 10.
6 altíssima. R.S. (*ye.*) tall. .... ——— H. ♃.
CYMO'PTERUS. DC. CYMO'PTERUS. Pentandria Digynia.
glomeràta. (*wh.*) clustered. N. America. 1824. 5. 6. H. ♃.
Selìnum acaùle. Ph.
LASERPI'TIUM. DC. LASERWORT. Pentandria Digynia. L.
1 latifòlium. S.S. (*wh.*) broad-leaved. Europe. 1640. 6. 7. H. ♃. Jacq. aust t. 146.
2 Libanòtis. S.S. (*pu.*) pubescent. S. Europe. 1824. ——— H. ♃. Lob. ic. 705.
3 marginàtum. S.S. (*st.*) purple-margin'd. Hungary. 1818. ——— H. ♃. W. et K. hung. 2. t 192.
4 Silér. S.S. (*wh.*) mountain. ——— 1640. ——— H. ♃. ——— 2. t. 145.

5 alpìnum. K.U.	(*wh.*)	alpine.	S. Europe.	1640.	5. 7.	H. ♃.	W. et K. hung. 3. t. 253.
trilòbum. S.S.							
6 angustifòlium. W.	(*wh.*)	narrow-leaved.	S. Europe.	1738.	6. 7.	H. ♃.	Moris. s. 9. t. 19. f. 9.
7 peucedanoídes. W.	(*wh.*)	fine-leaved.	Italy.	....	——	H. ♃.	Jacq. ic. 2. t. 350.
8 resinòsum. S.S.	(*ye.*)	resinous.	Sicily.	1825.	——	H. ♃.	
9 hirsùtum. S.S.	(*wh.*)	hairy.	Switzerland.	1759.	——	H. ♃.	Hall. stirp. t. 19.
10 Archangélica. S.S.	(*wh.*)	spotted-stalked.	Hungary.	1818.	——	H. ♃.	Jacq. ic. 1. t. 58.
11 pilòsum. S.S.	(*st.*)	sulphur-color'd.	Caucasus.	1759.	——	H. ♃.	
12 pruthènicum. S.S.	(*wh.*)	Prussian.	Germany.	——	7. 8.	H. ♃.	Jacq. aust. 2. t. 153.
13 híspidum. M.B.	(*bh.*)	hispid.	Caucasus.	1818.	——	H. ♃.	
14 Athamántæ. S.S.	(*wh.*)	branching.	Siberia.	——	——	H. ♂.	
15 gállicum. S.S.	(*wh.*)	French.	S. Europe.	1683.	6. 7.	H. ♃.	Pluk. phyt. t. 198. f. 6.
16 ferulàceum. W.	(*wh.*)	Fennel-leaved.	Levant.	1752.	——	H. ♃.	Tourn. it. 2. t. 121.
17 praténse. Led.	(*wh.*)	field.	Siberia.	1827.	——	H. ♃.	
18 laxiflòrum. F.	(*wh.*)	loose-flowered.	————	1828.	——	H. ♃.	
LOPHOSC'IADIUM. DC.		LOPHOSC'IADIUM. Pentandria Digynia.					
millefòlium. DC.	(*ye.*)	milfoil-leaved.	Levant.	1823.	7. 8.	H. ♃.	DC. mem. umb. t. 2. f. P.
Ferùla meoídes. S.S.							
MELANOSEL'INUM. H.U.		BLACK-PARSLEY. Pentandria Digynia.					
decípiens. H.U.	(*wh.*)	shrubby.	Madeira.	1785.	6. 7.	G. ♄.	Wend. s. han. 3. t. 13.

Tribus XI. DAUCINEÆ. DC.

ART'EDIA. S.S.		ART'EDIA. Pentandria Digynia. L.					
squamàta. S.S.	(*wh.*)	Fennel-leaved.	Levant.	1740.	7.	H. ⊙.	Lam. ill. t. 193.
ORLA'YA. H.U.		ORLA'YA. Pentandria Digynia.					
1 grandiflòra. H.U.	(*wh.*)	great-flowered.	S. Europe.	1648.	7. 8.	H. ⊙.	Jacq. aust. 1. t. 54.
Caùcalis grandiflòra. S.S.							
2 marítima. K.U.	(*bh.*)	sea-side.	————	1825.	——	H. ⊙.	Cavan. icon. 2. t. 101.
Caùcalis marítimus. S.S.							
3 platycárpos. K.U.	(*wh.*)	broad-seeded.	————	1800.	——	H. ⊙.	Jacq. vind. 3. t. 10.
DA'UCUS. S.S.		CARROT. Pentandria Digynia. L.					
1 Caròta. W.	(*wh.*)	wild.	Britain.	....	6. 7.	H. ♂.	Eng. bot. v. 17. t. 1174.
β *lùtea.*	(*wh.*)	*yellow-rooted, garden.*	————	....	——	H. ♂.	
γ *aurántia.*	(*wh.*)	*long-orange.*	————	....	——	H. ♂.	
δ *pr'æcox.*	(*wh.*)	*early horn.*	————	....	——	H. ♂.	
2 mauritànicus. W.	(*wh.*)	fine-leaved.	Spain.	1768.	——	H. ♂.	All. ped. 2. t. 61. f. 1.
3 polygàmus. L. en.	(*wh.*)	polygamous.	————	....	——	H. ♂.	Jacq. vind. 3. t. 78.
4 pusíllus. R.S.	(*wh.*)	small.	N. America.	1812.	——	H. ♂.	
5 marítimus. E.B.	(*ro.*)	sea-side.	England.	....	——	H. ♂.	Eng. bot. v. 36. t. 2560.
6 gummífer. S.S.	(*ro.*)	gum-bearing.	Italy.	1722.	——	H. ♂.	Moris. sect. 9. t. 13.
7 prolífer. S.S.	(*wh.*)	proliferous.	Sicily.	1828.	——	H. ♂.	
8 híspidus. S.S.	(*bh.*)	hispid.	Barbary.	1804.	——	H. ♂.	Desf. atl. t. 63.
9 sículus. S.S.	(*bh.*)	Sicilian.	————	——	——	H. ♂.	
10 Gingídium. S.S.	(*ro.*)	shining-leaved.	France.	1722.	——	H. ♂.	Moris. s. 9. t. 13. f. 4.
lùcidus. W.							
11 muricàtus. S.S.	(*bh.*)	prickly-seeded	Barbary.	1683.	——	H. ⊙.	Moris. s. 9. t. 14. f. 4.
12 littoràlis. S.S.	(*wh.*)	sea-shore.	Greece.	1823.	——	H. ⊙.	Flor. græc. t. 271.
13 montevidénsis. Ot.	(*wh.*)	Monte Video.	Monte Video.	1826.	——	H. ⊙.	
14 parviflórus. S.S.	(*wh.*)	small-flowered.	Barbary.	1823.	——	H. ♂.	Desf. atl. 1. t. 68.
15 crinìtus. K.U.	(*wh.*)	whorl-leaved.	————	1804.	6. 7.	H. ♃.	———— 1. t. 62.
Tórilis crinìta. S.S.							
16 pulchérrimus. DC.	(*wh.*)	beautiful.	Caucasus.	1816.	——	H. ♂.	
Caùcalis pulchérrima. M.B.							
17 orientàlis. DC.	(*wh.*)	oriental.	————	1699.	——	H. ♂.	Buxb. cent. 3. t. 28.
Caùcalis orientàlis. R.S.							
18 Marschalliànus. DC.	(*wh.*)	Marschall's.	Odessa.	1824.	——	H. ♂.	
Caùcalis littoràlis. M.B. *Cachrys litoràlis.* S.S.							

Subordo II. *CAMPYLOSPERMEÆ.* DC.

Tribus XII. CAUCALINEÆ. DC.

CA'UCALIS. S.S.		BUR-PARSLEY. Pentandria Digynia. L.					
1 daucoídes. S.S.	(*ro.*)	small.	England.	....	7. 8.	H. ⊙.	Eng. bot. v. 3. t. 197.

2 mauritànica. S.S.	(*wh.*)	Barbary.	Barbary.	1823.	7. 8.	H. ⊙.	
3 leptophy'lla. S.S.	(*bh.*)	fine-leaved.	Europe.	1739.	——	H. ⊙.	Jacq. vind. t. 195.
4 pùmila. W.	(*bh.*)	dwarf.	————	1640.	——	H. ⊙.	Ger. gallop. n. 4. t. 10.
5 híspida. S.S.	(*st.*)	hispid.	S. Europe.	1825.	——	H. ⊙.	
6 hùmilis. Lag.	(*wh.*)	low.	————	——	——	H. ⊙.	
TURG'ENIA. DC.		TURG'ENIA.	Pentandria Digynia.				
latifòlia. DC.	(*bh.*)	broad-leaved.	England.	....	7. 8.	H. ⊙.	Eng. bot. v. 3. t. 198.
Caùcalis latifòlia. E.B.							
TOR'ILIS. S.S.		TOR'ILIS.	Pentandria Digynia. S.S.				
1 Anthríscus. S.S.	(*bh.*)	upright.	Britain.	....	7. 8.	H. ⊙.	Eng. bot. v. 14. t. 987.
2 inféſta. E.F.	(*wh.*)	spreading.	————	....	——	H. ⊙.	———— v. 19. t. 1314.
helvética. S.S. *arvénsis.* L.en. *Caùcalis inféſta.* E.B.							
3 neglécta. R.S.	(*wh.*)	confused.	Austria.	....	——	H. ⊙.	Jacq. aust. t. 29.
4 ucrànica. S.S.	(*wh.*)	Ukraine.	Ukraine.	1828.	——	H. ⊙.	
5 nodòsa. S.S.	(*wh.*)	knotted.	Britain.	....	5. 7.	H. ⊙.	Eng. bot. v. 3. t. 199.
6 trichospérma. S.S.	(*fl.*)	hairy-seeded.	Egypt.	1818.	——	H. ⊙.	

Tribus XIII. SCANDICINEÆ. DC.

SCA'NDIX. DC.		CICELY.	Pentandria Digynia. L.				
1 Pécten. R.S.	(*wh.*)	Venus's-comb.	Britain.	....	6. 7.	H. ⊙.	Eng. bot. v. 20. t. 1397.
2 austràlis. R.S.	(*wh.*)	southern.	S. Europe.	1713.	5. 7.	H. ⊙.	Flor. græc. t. 285.
W'ylia austràlis. H.U.							
3 falcàta. R.S.	(*wh.*)	radiated.	Tauria.	——	——	H. ⊙.	Hoff. umb. t. 2. f. 2.
W'ylia ràdians. H.U.							
4 grandiflòra. R.S.	(*wh.*)	large-flowered.	————	1827.	6. 7.	H. ⊙.	Hoff.umb.p.15.t.2.f.3.
5 brachycárpa. S.S.	(*wh.*)	short-seeded.	Sicily.	1829.	——	H. ⊙.	
6 ibèrica. R.S.	(*wh.*)	Iberian.	Iberia.	1828.	——	H. ⊙.	Hoff. umb. t. 2. f. 4.
7 pinnatífida. R.S.	(*wh.*)	cut-leaved.	Persia.	1805.	5. 7.	H. ⊙.	Venten. cels. t. 14.
8 parviflòra. R.S.	(*wh.*)	small-flowered.	Tauria.	1827.	——	H. ⊙.	
ANTHRI'SCUS. S.S.		ROUGH CHERVIL.	Pentandria Digynia. S.S.				
1 vulgàris. S.S.	(*wh.*)	common.	Britain.	....	5. 6.	H. ⊙.	Eng. bot. v. 12. t. 818.
Scándix anthríscus. E.B.							
2 trichospérma. R.S.	(*wh.*)	hairy-seeded.	Persia.	1827.	——	H. ⊙.	
3 nemoròsa. S.S.	(*wh.*)	wood.	Caucasus.	1823.	7. 8.	H. ♃.	Hoff.umb.2.t.1.B.f.19.
4 fumarioídes. S.S.	(*wh.*)	Fumitory-like.	Hungary.	1818.	——	H. ♂.	W. et K. hung. 3. t. 224.
5 bulbòsa. DC.	(*wh.*)	bulbous-rooted.	Europe.	1726.	6. 7.	H. ♂.	Jacq. aust. 1. t. 63.
My'rrhis bulbòsa. S.S. *Chærophy'llum bulbòsum.* Jacq.							
6 sylvéstris. K.U.	(*wh.*)	smooth.	Britain.	....	4. 5.	H. ♃.	Eng. bot. v. 11. t. 752.
Chærophy'llum sylvéstre. E.B.							
7 Cerefòlium. K.U.	(*wh.*)	true Chervil.	————	....	6. 7.	H. ⊙.	Eng. bot. v. 18. t. 1268.
Scándix cerefòlium. E.B. *Chærophy'llum satìvum.* S.S.							
8 capénsis. S.S.	(*wh.*)	Cape.	C. B. S.	1816.	——	G. ♃.	
CHÆROPHY'LLUM. S.S.		CHERVIL.	Pentandria Digynia. L.				
1 aromáticum. K.U.	(*wh.*)	aromatic.	Scotland.	....	6. 8.	H. ♃.	Jacq. aust. 2. t. 150.
My'rrhis aromática. S.S.							
2 angelicæfòlium. M.B.	(*wh.*)	Angelica-leav'd.	Iberia.	1828.	6. 8.	H. ♃.	
3 hirsùtum. W.	(*wh.*)	hairy-leaved.	Switzerland.	1759.	6. 7.	H. ♃.	Jacq. aust. 2. t. 148.
4 cicutàrium. L.en.	(*bh.*)	Cicuta-like.	S. Europe.	1816.	——	H. ♃.	
5 aùreum. L.	(*st.*)	golden.	Scotland.	....	6.	H. ♃.	Eng. bot. v. 30. t. 2103.
6 temuléntum. E.B.	(*wh.*)	rough.	Britain.	....	7. 8.	H. ♂.	———— v. 22. t. 1521.
7 monogònum. L.en.	(*wh.*)	one-angled.	Hungary.	1820.	4. 6.	H. ♃.	
8 angulàtum. Kit.	(*wh.*)	angular.	————	——	——	H. ♃.	
9 hùmile. M.B.	(*wh.*)	dwarf.	....	1825.	——	H. ♃.	
Anthríscus hùmilis. Besser. *My'rrhis hùmilis.* R.S.							
10 nítidum. Wahl.	(*wh.*)	glossy-seeded.	Carpathian-m.	1820.	——	H. ♃.	
11 Biebersteínii. Lag.	(*wh.*)	Bieberstein's.	Caucasus.	1824.	5. 6.	H. ♃.	
12 nodòsum. K.U.	(*wh.*)	knotted.	Sicily.	1656.	5. 6.	H. ♃.	Jacq. vind. 3. t. 25.
Scándix nodòsa. Jacq. *Anthríscus nodòsa.* S.S.							
13 procúmbens. Ph.	(*wh.*)	procumbent.	N. America.	1699.	7. 8.	H. ⊙.	Moris.his.s.9.t.11.f.ult.
14 ròseum. M.B.	(*ro.*)	rose-coloured.	Iberia.	1824.	6. 7.	H. ♂.	Hoff. umb. t. 1. B. f. 33.
15 coloràtum. W.	(*st.*)	color'd-bract'd.	Illyria.	1806.	7. 8.	H. ♂.	
16 tenuifòlium. Poir.	(*wh.*)	slender-leaved.	Morocco.	——	——	F. ♂.	Jacq. vind. t. 51.

SPHALLEROCA'RPUS. DC. SPHALLEROCA'RPUS. Pentandria Digynia.
grácilis. DC. (*wh.*) slender. Siberia. 1827. 6. 8. H. ♂. DC. dis. umb. t. 2. f. N.
My'rrhis grácilis. S.S. *Chærophy'llum Cy'minum.* F.
MOLOPOSPE'RMUM. K.U. MOLOPOSPE'RMUM. Pentandria Digynia.
peloponnesìacum. K.U.(*pa.*) Hemlock-l[d]. Switzerland. 1596. 5. 7. H. ♃. Jacq. aust. app. t. 13.
Ligústicum peloponesìacum. S.S.
MY'RRHIS. S.S. MYRRH. Pentandria Digynia. S.S.
odoràta. S.S. (*wh.*) sweet-scented. Britain. .... 5. 6. H. ♃. Eng. bot. v. 10. t. 697.
OSMORH'IZA. DC. OSMORH'IZA. Pentandria Digynia.
Claytòni. DC. (*wh.*) sweet-rooted. N. America. 1806. 7. 8. H. ♃. Spreng. umb. t. 3. f. 6.
My'rrhis Claytòni. S.S. *Uraspérmum hirsutum.* Bigelow. *Claytòni.* N.

Tribus XIV. SMYRNEÆ. DC.

LAG'ŒCIA. L. WILD CUMIN. Pentandria Monogynia. L.
cuminoídes. S.S.(*gr.ye.*) common. Levant. 1640. 6. 7. H. ☉. Flor. græc. t. 243.
OLIV'ERIA. V. OLIV'ERIA. Pentandria Digynia. S.S.
decúmbens. V. (*bh.*) Thyme-scented. Bagdad. 1816. 5. 7. H. ☉. Venten. cels. t. 21.
ECHINO'PHORA. S.S. SEA PARSNEP. Pentandria Digynia. L.
1 spinòsa. S.S. (*bh.*) prickly. England. .... 7. H. ♃. Eng. bot. v. 34. t. 2413.
2 tenuifòlia. S.S. (*bh.*) fine-leaved. Apulia. 1731. 7. 8. H. ♃. Flor. græc. t. 266.
3 trichophy'lla. S.S. (*bh.*) slender-leaved. Armenia. 1829. —— H. ♃.
ARCT'OPUS. S.S. ARCT'OPUS. Polygamia Diœcia. L.
echinàtus. S.S. (*wh.*) rough. C. B. S. 1774. 6. 9. G. ♃. Bot. reg. t. 705.
CA'CHRYS. L.en. CA'CHRYS. Pentandria Digynia. L.
1 odontálgica. S.S. (*ye.*) Pallas's. Siberia. 1824. 7. 8. H. ♃. Pall. reis. ap. 3. t. 9.
2 Morisòni. S.S. (*ye.*) Morison's. S. Europe. 1710. —— H. ♃. Moris. umb. t. 3. f. 1.
PERL'EBIA. DC. PERLE'BIA. Pentandria Digynia.
triquètra. DC. (*wh.*) winged. Levant. 1816. 6. 7. H. ♃. Vent. cels. t. 79.
Laserpítium triquètrum. V. *Cáchrys triquètra.* S.S.
HIPPOMAR'ATHRUM. HIPPOMAR'ATHRUM. Pentandria Digynia.
sículum. L.en. (*ye.*) hairy-seeded. Sicily. 1640. 7. 8. H. ♃. Moris. umb. t. 3. f. 2.
Cáchrys sícula. S.S.
MAGYD'ARIS. DC. MAGYD'ARIS. Pentandria Digynia.
panacìna. DC. (*wh.*) Parsnep-leav'd. Sicily. 1752. 7. 8. H. ♃. Bocc. sic. 1. t. 1.
Cáchrys panacifòlia. W. *Athamánta panacifòlia.* R.S.
HE'RMAS. S.S. HE'RMAS. Polygamia Monœcia. W.
1 gigántea. S.S. (*pu.*) gigantic. C. B. S. 1794. 6. 7. G. ♃. Thunb. act. pet. 14. t. 11.
2 villòsa. S.S. (*bh.*) villous. ———— 1795. —— G. ♃. Burm. afr. t. 71. f. 2
depauperàta. L.
3 ciliàta. S.S. (*wh.*) ciliated. ———— 1816. —— G. ♃. Burm. t. 72. f. 1.
4 capitàta. S.S. (*wh.*) headed. ———— —— 6. 8. G. ♃. Thun. n. act. pet. 14. t. 12.
5 quinquedentàta. S.S.(*wh.*) toothed. ———— 1821. —— G. ♃. ———— 14. p. 533. t. 12.
C'ONIUM. H.U. HEMLOCK. Pentandria Digynia. L.
1 maculàtum. S.S. (*wh.*) common. Britain. .... 6. 7. H. ♂. Eng. bot. v. 17. t. 1191.
2 croáticum. W.K. (*wh.*) Hungarian. Hungary. .... —— H. ♂.
ARRAC'ACIA. DC. ARRAC'ACIA. Pentandria Digynia.
esculénta. DC. (*wh.*) esculent. S. America. 1822. 4. 5. F. ♃. Hook. ex. flor. t. 152.
Cònium Arracácha. H.E.F.
PLEUROSPE'RMUM. S.S. PLEUROSPE'RMUM. Pentandria Digynia. S.S.
austrìacum. S.S. (*wh.*) Austrian. Austria. 1596. 6. 8. H. ♃. Jacq. aust. 2. t. 151.
Ligústicum austrìacum. W.
SMY'RNIUM. S.S. ALEXANDERS. Pentandria Digynia. L.
1 Olusàtrum. S.S. (*st.*) common. Britain. .... 5. 6. H. ♂. Eng. bot. v. 4. t. 230.
2 perfoliàtum. W. (*st.*) perfoliate. Italy. 1596. 5. H. ♃. Lobel. ic. 709.
Dodon'æi. S.S.
3 Dioscóridis. S.S. (*ye.*) Dioscorides. S. Europe. .... —— H. ♂. Flor. græc. t 289.
4 apiifòlium. S.S. (*ye.*) Smallage-leav'd. Candia. 1731. 5. 7. H. ♃.
5 nudicáule. S.S. (*ye.*) naked-stalked. Caucasus. 1820. —— H. ♃. Spreng. umb. t. 4. f. 7
6 cicutàrium. S.S. (*ye.*) Cicuta-like. ———— 1829. —— H. ♃.
7 cordàtum. S.S. (*ye.*) heart-leaved. N. America. 1759. 6. 7. H. ♄.
8 integérrimum. Ph.(*ye.*) entire-leaved. N. America. 1759. —— H. ♃.
PHYSOSPERMUM. S.S. PHYSOSPERMUM. Pentandria Digynia. S.S.
commutàtum. S.S.(*wh.*) Cornish. England. .... 7. 8. H. ♃. Eng. bot. v. 10. t. 683.
Ligústicum cornubiénse. E.B.

Subordo III. *CÆLOSPERMEÆ.* DC.

Tribus XV. CORIANDREÆ. DC.

BIF'ORIS. S.S.		BIF'ORIS. Pentandria Digynia. S.S.					
1 flosculòsa. s.s.	(*wh.*)	flosculose.	S. Europe.	1640.	6. 7.	H. ☉.	Moris. sect. 9. t. 11.
Còrion testiculàtum. L.en. *Coriándrum testiculàtum.* L.							
2 ràdians. s.s.	(*wh.*)	rayed.	Tauria.	....	——	H. ☉.	Hoff.umb.ed.2.t.tit.f.2.
ATR'EMA. DC.		ATR'EMA. Pentandria Digynia.					
americàna. DC.	(*wh.*)	American.	N. America.	1827.	6. 7.	H. ☉.	DC. mem. umb. t. 18.
Coriándrum americànum. N.							
CORIA'NDRUM. S.S.		CORIANDER. Pentandria Digynia. L.					
sativum. s.s.	(*bh.*)	common.	England.	....	6.	H. ☉.	Eng. bot. v. 1. t. 67.

ORDO CIX.

ARALIACEÆ. *Juss. gen. p.* 217.

P'ANAX. L.		P'ANAX. Polygamia Diœcia. L.					
1 quinquefòlium. W.	(*gr.*)	Ginseng.	N. America.	1740.	6.	H. ♃.	Bot. mag. t. 1333.
2 trifòlium. W.	(*wh.*)	lesser.	———	1759.	5. 6.	H. ♃.	——— t. 1334.
pusílla. a. B.M.							
3 fruticòsum. W.	(*gr.*)	shrubby.	Ternate.	1800.	8. 9.	S. ♄.	Andr. rep. t. 595.
4 digitàtum. H.B.	(*gr.*)	fingered.	E. Indies.	1818.	....	S. ♄.	
5 arboreùm. W.	(*gr.*)	tree.	N. Zealand.	1822.	....	G. ♄.	
6 chrysophyllum. R.S.	(*gr.*)	golden-leaved.	Guiana.	——	....	S. ♄.	Aub. gui. 2. t. 360.
7 conchifòlium. H.B.	(*gr.*)	shell-leaved.	———	1820.	....	S. ♄.	Rumph. amb. 4. t. 31.
8 attenuàtum. W.	(*gr.*)	attenuated.	W. Indies.	1825.	....	S. ♄.	
9 aculeatum. W.	(*gr.*)	prickly.	China.	1773.	11.	S. ♄.	Jacq. ic. 3. t. 634.
CUSS'ONIA. W.		CUSS'ONIA. Pentandria Digynia. L.					
1 thyrsiflòra. W.	(*gr.*)	thyrse-flower'd.	C. B. S.	1795.	8. 10.	G. ♄.	Th. nov. act. ups.3.t.12.
2 tríptera. C.R.	(*gr.*)	tripterous.	———	1816.	....	G. ♄.	Colla hort. rip. t. 26.
3 spicàta. W.	(*gr.*)	spike-flowered.	———	1789.	8. 10.	G. ♄.	Th. nov. act. ups.3.t.13.
SCIODAPHY'LLUM. S.S.		SCIODAPHY'LLUM. Pentandria Pentagynia. S.S.					
1 pentándrum. s.s.	(*bh.*)	pentandrous.	Peru.	1828.	....	G. ♄.	Flor. peruv. t. 311.
2 acuminàtum. s.s.	(*ye.*)	acuminate-petal'd.	———	1818.	....	S. ♄.⌣.	——— t. 310.
Actinophy'llum acuminàtum. R.P.							
3 cònicum. s.s.	(*bh.*)	conical-flower'd.	———	——	....	S. ♄.⌣.	Flor. peruv. p. 75. t. 309.
4 pedicellàtum. s.s.	(*bh.*)	pedicelled.	———	——	....	S. ♄.⌣.	——— t. 308.
Actinophy'llum pedicellàtum. R.P.							
5 angulàtum. s.s.	(*ye.*)	tree.	———	1824.	....	G. ♄.	
6 Brównii. s.s.	(*gr.*)	upright.	W. Indies.	1793.	6. 8.	S. ♄.	Brown jam. t. 19. f. 1. 2.
Aràlia Sciodaphy'llum. W.							
AR'ALIA. D.P.		AR'ALIA. Pentandria Pentagynia. L.					
1 pentaphy'lla. s.s.	(*gr.*)	five-ld prickly.	China.	1822.	....	G. ♄.	
2 spinòsa. s.s.	(*wh.*)	Angelica-tree.	Virginia.	1688.	9.	H. ♄.	Comm. hort. 1. t. 47.
3 racemòsa. s.s.	(*gr.*)	berry-bearing.	N. America.	1658.	6. 9.	H. ♃.	Schkuhr. hand. 1. t. 86.
4 nudicaùlis. s.s.	(*wh.*)	naked-stalked.	———	1731.	6. 7.	H. ♃.	Pluk. alm. t. 238. f. 5.
5 híspida. s.s.	(*wh.*)	hispid.	———	1799.	——	H. ♄.	Bot. mag. t. 1085.
6 pubéscens. DC.	(*wh.*)	pubescent.	New Spain.	1819.	....	G. ♄.	DC. mons. ined. t. 45.
GAST'ONIA. J.		GAST'ONIA. Dodecandria Monogynia.					
palmàta. H.B.	(*wh.*)	palmate-leaved.	E. Indies.	1818.	3. 4.	S. ♄.	Bot. reg. t. 894.
H'EDERA. D.P.		IVY. Pentandria Mono-Pentagynia.					
1 umbraculífera.	(*wh.*)	large.	E. Indies.	1818.	3. 4.	S. ♄.	
Aràlia umbraculífera. H.B.							
2 digitàta. H.B.	(*wh.*)	fingered.	———	——	——	S. ♄.	

3 aculeàta. D.P.	(*wh.*)	prickly Nepaul.	Nepaul.	1816.	....	F. ♄.	
4 frà̀grans. D.P.	(*wh.*)	fragrant.	——	——	....	F. ♄.	
5 cochleàta.	(*wh.*)	shell-leaved.	E. Indies.	1820.	....	S. ♄.	Rumph. amb. 4. t. 31.
Aràlia cochleàta. R.S.							
6 arbòrea. R.S.	(*wh.*)	tree.	Jamaica.	——	....	S. ♄.	Jacq. schœnb. t. 51.
7 capitàta. R.S.	(*gr.*)	cluster-flower'd.	W. Indies.	1777.	8.	S. ♄.	Smith ic. pict. 4.
8 canariénsis. R.S.	(*gr.*)	Irish.	Canaries.	1807.	9. 10.	H. ♄.◡.	Willd. ber. m. 2. t. 5. f. 1.
9 Hèlix. R.S.	(*gr.*)	common.	Britain.	....	——	H. ♄.◡.	Eng. bot. v. 18. t. 1267.
β *poética.* R.S.	(*gr.*)	*Poet's.*	——	....	——	H. ♄.◡.	
γ *álbo-variegàta.*	(*gr.*)	*white-variegated.*	——	....	——	H. ♄.◡.	
δ *flàvo-variegàta.*	(*gr.*)	*yellow-variegated.*	——	....	——	H. ♄.◡.	

ORDO CX.

CAPRIFOLIACEÆ. *Juss. gen. p.* 210.

CO'RNUS. L.		DOGWOOD. Tetrandria Monogynia. L.					
1 alternifòlia. R.S.	(*wh.*)	alternate-leav'd.	N. America.	1760.	9.	H. ♄.	Schm. arb. 2. t. 70.
2 paniculàta. R.S.	(*wh.*)	panicled.	——	1758.	6. 7.	H. ♄.	—— 2. t. 68.
3 stricta. R.S.	(*wh.*)	upright.	——	——	——	H. ♄.	—— 2. t. 67.
β *variegàta.*	(*wh.*)	*variegated-leav'd.*	——	——	——	H. ♄.	
4 oblónga. F.I.	(*bh.*)	oblong-leaved.	Nepaul.	1818.	7. 8.	H. ♄.	
5 macrophy'lla. F.I.	(*wh.*)	large-leaved.	——	1827.	——	H. ♄.	
6 circinàta. R.S.	(*wh.*)	Pensylvanian.	N. America.	1784.	——	H. ♄.	Schm. arb. 2. t. 69.
7 álba. R.S.	(*wh.*)	white berried.	Siberia.	1741.	6. 9.	H. ♄.	—— 2. t. 65.
8 sibírica. L.C.	(*wh.*)	Siberian.	——	....	——	H. ♄.	
9 serícea. R.S.	(*wh.*)	blue-berried.	N. America.	1683.	8.	H. ♄.	Schm. arb. 2. t. 64.
10 sanguínea. R.S.	(*wh.*)	common.	Britain.	....	6. 7.	H. ♄.	Eng. bot. v. 4. t. 249.
β *variegàta.*	(*wh.*)	*variegated.*	——	....	——	H. ♄.	
11 máscula. R.S.	(*ye.*)	CornelianCherry.	Europe.	1596.	2. 4.	H. ♄.	Flor. græc. t. 151.
β *variegàta.*	(*ye.*)	*variegated.*	——	——	——	H. ♄.	
12 flòrida. R.S.	(*ye.*)	great-flowered.	N. America.	1731.	4. 5.	H. ♄.	Bot. mag. t. 526.
13 capitàta. F.I.	(*ye.*)	headed.	Nepaul.	1820.	——	H. ♄.	
14 canadénsis. R.S.	(*wh.*)	Canadian.	Canada.	1774.	6. 8.	H. ♃.	Bot. mag. t. 880.
15 suècica. R.S.	(*wh.*)	dwarf.	Britain.	....	4.	H. ♃.	Eng. bot. v. 5. t. 310.
SAMB'UCUS. L.		ELDER. Pentandria Trigynia. L.					
1 Ebùlus. R.S.	(*fl.*)	herbaceous.	Britain.	....	6. 7.	H. ♃.	Eng. bot. v. 7. t. 475.
2 hùmilis. Mill.	(*fl.*)	dwarf.	....	....	——	H. ♃.	
3 chinénsis. H.T.	(*fl.*)	Chinese.	China.	1825.	——	H. ♃.	
4 nìgra. R.S.	(*wh.*)	common.	Britain.	....	5. 7.	H. ♄.	Eng. bot. v. 7. t. 476.
β *variegàta.*	(*wh.*)	*variegated.*	——	....	——	H. ♄.	
γ *aùrea.*	(*wh.*)	*golden-leaved.*	——	....	——	H. ♄.	
δ *pulverulénta.*	(*wh.*)	*powdered-leaved.*	——	....	——	H. ♄.	
ε *monstròsa.*	(*wh.*)	*deformed.*	——	....	——	H. ♄.	
ζ *viréscens.*	(*wh.*)	*green-fruited.*	——	....	——	H. ♄.	
η *álbida.*	(*wh.*)	*white-fruited.*	——	....	——	H. ♄.	
5 laciniàta. L.en.	(*wh.*)	Parsley-leaved.	....	....	——	H. ♄.	Schm. arb. t. 144.
6 hy'brida. Jac.	(*wh.*)	hybrid.	....	1826.	——	H. ♄.	
7 rotundifòlia. L.C.	(*wh.*)	round-leaved.	Wiltshire.	....	——	H. ♄.	
8 canadénsis. R.S.	(*wh.*)	Canadian.	N. America.	1761.	6. 8.	H. ♄.	Schm. arb. t. 142.
9 racemòsa. R.S.	(*st.*)	red-berried.	S. Europe.	1596.	5. 6.	H. ♄.	Jacq. ic. 1. t. 59.
β *purpùrea.*	(*br.*)	*purple-flowered.*	——	——	——	H. ♄.	
10 pùbens. R.S.	(*st.*)	pubescent.	N. America.	1812.	——	H. ♄.	
VIBU'RNUM. L.		VIBU'RNUM. Pentandria Trigynia. L.					
1 mólle. R.S.	(*wh.*)	soft.	N. America.	1812.	6. 7.	H. ♄.	
2 acerifòlium. R.S.	(*wh.*)	Maple-leaved.	——	1736.	——	H. ♄.	Wats. dend. brit. t. 118.
3 èdule. R.S.	(*wh.*)	eatable-fruited.	——	1812.	7.	H. ♄.	
4 Oxycóccos. R.S.	(*wh.*)	Cranberry-like.	——	....	——	H. ♄.	
5 O'pulus. R.S.	(*wh.*)	Guelder Rose.	Britain.	....	5. 6.	H. ♄.	Eng. bot. v. 5. t. 332.
β *ròseum.* R.S.	(*wh.*)	*Snowball tree.*	....	....	——	H. ♄.	Schm. arb. t. 174.
γ *variegàtum.*	(*wh.*)	*variegated-leav'd.*	....	....	——	H. ♄.	
6 davùricum. R.S.	(*wh.*)	Siberian.	Dahuria.	1785.	6. 7.	H. ♄.	Pall. ross. 1. t. 38.
7 lantanoídes. R.S.	(*wh.*)	Lantana-like.	N. America.	....	5. 6.	H. ♄.	
8 Lantàna. R.S.	(*wh.*)	wayfaring-tree.	Britain.	....	——	H. ♄.	Eng. bot. v. 5. t. 331.
9 nervòsum. D.P.	(*wh.*)	nerved-leaved.	Nepaul.	1826.	——	H. ♄.	

10 Lentàgo. R.S. (*wh.*) tree. N. America. 1761. 7. H. ♄. Wats. dend. brit. t. 21.
11 squamàtum. R.S. (*wh.*) scaly. ——— 1812. 6. 7. H. ♄. ——— t. 24.
12 pyrifòlium. R.S. (*wh.*) pear-leaved. ——— .... 5. 6. H. ♄. ——— t. 22.
13 prunifòlium. R.S. (*wh.*) plum-leaved. ——— 1731. —— H. ♄. ——— t. 23.
14 pubéscens. R.S. (*wh.*) pubescent. ——— 1736. 6. 7. H. ♄.
15 dentàtum. R.S. (*wh.*) tooth-leaved. ——— —— —— H. ♄. Wats. dend. brit. t. 25.
16 lævigàtum. R.S. (*wh.*) Cassioberry bush. ——— 1724. 7. 8. H. ♄. Mill. ic. 1. t. 83. f. 1.
17 nítidum. R.S. (*wh.*) shining-leaved. ——— 1758. 5. 6. H. ♄.
18 cassinoídes. R.S. (*wh.*) thick-leaved. ——— 1761. 6. 7. H. ♄.
19 obovàtum. R.S. (*wh.*) obovate-leaved. ——— 1812. 5. 6. H. ♄.
20 odoratíssimum. B.R. (*w.*) sweet-scented. China. 1810. —— G. ♄. Bot. reg. t. 456.
chinénse. Zeyh.
21 nùdum. R.S. (*wh.*) oval-leaved. N. America. 1752. —— H. ♄. Wats. dend. brit. t. 20.
22 cylíndricum. D.P. (*wh.*) tube-flowered. Nepaul. 1826. 6. 8. H. ♄.
23 villòsum. R.S. (*wh.*) villous. Jamaica. 1820. 4. 6. S. ♄.
24 rugòsum. R.S. (*wh.*) rugged-leaved. Canaries. 1796. 4. 5. H. ♄. Bot. reg. t. 376.
rígidum. Vent. malm. t. 98.
25 stríctum. L.en. (*wh.*) upright Laurestine. S. Europe. 1596. 3. 5. H. ♄.
β *virgàtum.* (*wh.*) *twiggy.* ——— —— —— H. ♄.
26 Tìnus. R.S. (*wh.*) common Laurestine. —— —— 12. 4. H. ♄. Bot. mag. t. 38.
27 lùcidum. R.S. (*bh.*) shining Laurestine. ——— —— 2. 5. H. ♄. Clus. hist. 1. 49. ic.

SYMPH'ORIA. S.S. St. Peter's-wort. Pentandria Monogynia.
1 glomeràta. S.S. (*bh.*) common. N. America. 1730. 8. 9. H. ♄. Schmidt arb. t. 115.
2 racemòsa. S.S. (*bh.*) Snowberry. ——— 1812. 7. 9. H. ♄. Wats. dend. brit. t. 7.

DIERVI'LLA. J. Diervi'lla. Pentandria Monogynia.
hùmilis. P.S. (*ye.*) yellow-flower'd. N. America. 1739. 6. 7. H. ♄. Bot. mag. t. 1796.
lùtea Ph. *Tournefórtii.* R.S. *canadénsis.* S.S. *Lonícera Diervílla.* L.

LONI'CERA. R.S. Fly-Honeysuckle. Pentandria Monogynia. L.
1 alpigèna. S.S. (*re.*) red-berried. Switzerland. 1596. 4. 5. H. ♄. Schmidt arb. t. 112.
2 microphy'lla. S.S. (*re.*) small-leaved. Siberia. 1818. 5. 6. H. ♄.
3 sibírica. R.S. (*re.*) Siberian. ——— .... —— H. ♄.
4 orientàlis. R.S. (*wh.*) oriental. Levant. 1820. —— H. ♄.
caucásica. Pall.
5 puníceа. B.M. (*cr.*) crimson-flower'd........ —— 4. 5. H. ♄. Bot. mag. t. 2469.
6 cœrùlea. R.S. (*st.*) blue-berried. Switzerland. 1629. 3. 4. H. ♄. ——— t. 1965.
β *altaíca.* F. (*st.*) *Altay.* *Altay.* 1827. —— H. ♄.
7 canadénsis. R.S. (*st.*) Canadian. Canada. 1812. 4. 5. H. ♄.
8 involucràta. S.S. (*ye.*) involucred. N. America. 1824. 5. 6. H. ♄. Bot. reg. t. 1179.
9 ibérica. R.S. (*br.*) Iberian. Iberia. 1818. 5. 6. H. ♄.
10 villòsa. R.S. (*st.*) villous. Canada. —— 4. 5. H. ♄.
Xylósteum villòsum. M.
11 Xylósteum. R.S. (*wh.*) fly. England .... 6. 7. H. ♄. Eng. bot. v. 13. t. 916.
12 canéscens. R.S. (*wh.*) canescent. Barbary. 1820. —— H. ♄. Desf. atl. 1. t. 52.
biflòra. Desf.
13 ciliàta. R.S. (*bh.*) ciliated. N. America. 1824. 5. 6. H. ♄.
β *álba.* R.S. (*wh.*) *white berried.* ——— —— —— H. ♄.
14 tatárica. R.S. (*wh.*) Tartarian. Tartary. 1752. 4. 5. H. ♄. Pall. ros. 1. t. 36.
α *álba.* (*wh.*) *white-flowered.* ——— —— —— H. ♄.
β *rùbra.* (*pi.*) *red-flowered.* ——— —— —— H. ♄. Bot. reg. t. 31.
15 nìgra. R.S. (*pu.*) black-berried. Switzerland. 1597. 3. 4. H. ♄. Schmidt arb. t. 110.
16 flexuòsa. R.S. (*wh.*) flexuose. China. 1806. 7. 9. H. ♄.
17 quadrifòlia. R.S. (*wh.*) four-leaved. Siberia. 1824. —— H. ♄.
18 pyrenàica. R.S. (*wh.*) Pyrenean. Pyrenees. 1739. 5. H. ♄. Magn. hort. t. 209.
19 híspida. R.S. (*wh.*) hispid-branch'd. Dahuria. 1821. —— H. ♄.
20 ligustrìna. D.P. (*st.*) Privet-leaved. Nepaul. 1825. 4. 5. H. ♄.

NINTO'OA. Ninto'oa. Pentandria Monogynia.
1 longiflòra. (*st.*) long-flowered. China. 1816. 5. 10. F. ♄.⌣. Bot. reg. t. 1233.
Caprifòlium longiflòrum. B.R. *japónicum.* D.P.
2 confùsa. (*st.*) pale-flowered. China. 1806. 7. 10. F. ♄.⌣. Andr. reposit. t. 583.
Nin-too. Kæmpf. amœn. 5. p. 785. *Lonícera japónica.* Bot. reg. t. 70. *non Thunb.*
3 japónica. Th. (*re.*) red-flowered. China. 1814. 4. 11. H. ♄.⌣. Wats. dend. brit. t. 117.
Lonícera chinénsis. W.D. *Lonícera flexuòsa.* Bot. reg. v. 9. t. 712. *nec aliorum.*

CAPRIF'OLIUM. R.S. Honeysuckle. Pentandria Monogynia.
1 itálicum. R.S. (*bh.*) early. England. .... 4. 6. H. ♄.⌣. Eng. bot. v. 12. t. 799.
álbum. (*wh.*) *white-flowered.* ——— .... —— H. ♄.⌣. Jacq. aust. t. 357.
Lonícera Caprifòlium. E.B.
β *rùbrum.* (*re.*) *early red.* Italy. .... —— H. ♄.⌣. Schmidt arb. t. 106.
2 etrúscum. R.S. (*wh.pu.*) Italian. ——— 1824. —— H. ♄. San. Viag. al. m. t. 1.
3 baleàricum. R.S. (*st.*) red-barked. S. Europe. —— —— H. ♄.⌣.

4 dioícum. R.S. (*re.*) small-flowered. N. America. 1766. 6. 7. H. ♄.◡. Bot. reg. t. 138.
parviflòrum. Ph? *bracteòsum.* M? *Lonícera mèdia.* Murr. com. got. 1776. t. 3.
5 gràtum. R.S. (*re.*) evergreen. N. America. 1730. 6. 8. H. ♄.◡. Hort. angl. n. 10. t. 8.
6 flàvum. B.M. (*ye.*) golden-flowered. ——— 1810. 5. 7. H. ♄.◡. Bot. mag. t. 1318.
Frasèri. Ph. *Lonícera flava.* B.M.
7 Douglásii. H.T. (*ye.*) Douglas's. Canada. 1824. 5. 7. H. ♄.◡.
8 pubéscens. H.E.F.(*ye.*) pubescent. N. America. 1820. —— H. ♄.◡. Hook. ex. flor. t. 27.
-*Lonícera Góldii.* S.S.
9 sempervìrens. R.S.(*sc.*) Trumpet. ——— 1656. 5. 8. H. ♄. Bot. mag. t. 781.
10 oblongifòlium. (*sc.*) oblong-leaved. Carolina. .... —— F. ♄. ——— t. 1753.
Lonícera sempervìrens. β minor. B.M.
11 ciliòsum. R.S. (*ye.*) fringed. N. America. 1827. —— H. ♄.◡.
12 impléxum. R.S. (*pa.*) glaucous-leaved. Minorca. 1772. 6. 9. H. ♄. Bot. mag. t. 640.
13 Pericly'menum. R.S.(*st.*) Woodbine. Britain. .... 5. 7. H. ♄.◡. Eng. bot. v. 12. t. 800.
α vulgàris. H.K. (*st.*) *common.* ——— .... —— H. ♄.◡. Schmidt arb. t. 107.
β serotìnum. H.K.(*re.*) *late red.* ——— .... 6. 8. H. ♄.◡. ——— t. 108.
γ bélgicum. H.K. (*bh.*) *dutch.* ——— .... 5. 7. H. ♄. Hort. angl. n. 5. t. 6.
δ quercifòlium. H.K.(*st.*) *Oak-leaved.* Britain. .... —— H. ♄.◡.
ε variegàtum. H.K.(*st.*) *variegated-leav'd.* .... —— H. ♄.◡.
TRIO'STEUM. R.S. FEVERWORT. Pentandria Monogynia. L.
1 perfoliàtum. R.S. (*pu.*) perfoliate. N. America. 1730. 6. 7. H. ♃.
2 angustifòlium. R.S.(*ye.*) narrow-leaved. ——— 1699. —— H. ♃. Pluk. alm. t. 104. f. 2.
LINN'ÆA. W. LINN'ÆA. Didynamia Angiospermia. L.
boreàlis. W. (*bh.*) two-flowered. Britain. .... 5. 8. H. ♃. Eng. bot. v. 7. t. 433.

ORDO CXI.

LORANTHEÆ. *Juss. ann. mus.* 12. *p.* 292.

LORA'NTHUS. L. LORA'NTHUS. Hexandria Monogynia.
europ'æus. S.S. (*st.*) European. Europe. 1824. .... H. ♄. Jacq. aust. t. 30.
VI'SCUM. L. MISSELTOE. Diœcia Tetrandria. L.
álbum. E.B. (*gr.*) common. England. .... 5. H. ♄. Eng. bot. v. 21. t. 1470.

ORDO CXII.

CHLORANTHEÆ. *R. Brown.*

CHLORA'NTHUS. R.S. CHLORA'NTHUS. Tetrandria Monogynia. R.S.
1 inconspícuus. R.S. (*st.*) trailing. China. 1781. 5. 9. G. ♄. L'Her. sert. angl. t. 2.
2 eréctus. H.S.L. (*wh.*) upright. ——— 1809. 6. 8. G. ♄.
elàtior. L.enum.
3 odoríferus. (*ye.*) sweet-scented. ——— 1820. —— G. ♄.
Creòdus odorífer. Loureiro.
4 monostàchys. S.S. (*st.*) single-spiked. ——— 1819. 2. 6. G. ♃. Bot. mag. t. 2190.
5 monánder. B. (*ye.*) single-stamen'd. ——— 1817. 1. 12. G. ♃.

ORDO CXIII.

LOPHIREÆ. *G. Don.*

LO'PHIRA. S.S. SCRUBBY OAK. Icosandria Monogynia. S.S.
alàta. S.S. (*gr.*) winged. SierraLeone. 1822. .... S. ♄.

ORDO CXIV.

RUBIACEÆ. *Juss. gen.* 196.

Sect. I. GUETTARDEÆ. *Kunth synops.* 3. *p.* 67.

GUETTA'RDA. S.S.		GUETTA'RDA.	Monœcia Tetr-Pent-Hexandria.				
1 argéntea. S.S.	(*wh.*)	silvery-leaved.	Jamaica.	1826.	....	S. ♄.	Brown. jam. t. 20. f. 1.
2 hirsùta. S.S.	(*wh.*)	hairy.	Peru.	1822.	....	S. ♄.	Flor. per. 2. t. 145. f.a.
Laugèria hirsùta. R.P.							
3 tomentòsa. S.S.	(*wh.*)	tomentose.	Jamaica.	——	....	S. ♄.	
4 lùcida. S.S.	(*wh.*)	glossy.	————	1818.	....	S. ♄.	Vahl. symb. 3. t. 57.
Laugèria lùcida. Sw.							
5 odoràta. R.S.	(*br.*)	sweet-scented.	Havannah.	1827.	....	S. ♄.	Jacq.am.pic.t.259.f.16.
6 scábra. S.S.	(*wh.*)	rough.	W. Indies.	——	....	S. ♄.	Vent. choix. 1. t. 1.
Matthìola scábra. L.							
7 rugòsa. S.S.	(*wh.*)	rough-leaved.	————	1793.	....	S. ♄.	
8 ellíptica. R.S.	....	elliptic-leaved.	Jamaica.	1825.	....	S. ♄.	
9 speciòsa. S.S.	(*wh.*)	great-flowered.	E. Indies.	1771.	....	S. ♄.	Rheed. mal. 4. t. 47.48.
NONAT'ELIA. R.S.		NONAT'ELIA.	Pentandria Monogynia.				
1 racemòsa. R.S.	(*wh.*)	racemed.	Trinidad.	1825.	....	S. ♄.	Aubl. guian. 1. t. 72.
2 officinàlis. R.S.	(*wh.*)	officinal.	W. Indies.	1824.	....	S. ♄.	———— 1. t. 70. f. 1.
Psychòtria involucràta. W.							
VANGU'ERIA. J.		VANGU'ERIA.	Pentandria Monogynia. S.S.				
1 èdulis. S.S.	(*gr.*)	eatable.	E. Indies.	1809.	....	S. ♄.	Lam. ill. t. 159.
2 spinòsa. F.I.	(*gr.*)	spiny.	————	1818.	....	S. ♄.	
ERI'THALIS. S.S.		ERI'THALIS.	Pentandria Monogynia. L.				
fruticòsa. S.S.	(*wh.*)	shrubby.	Jamaica.	1793.	7. 8.	S. ♄.	Jacq. am. p. t. 260. f. 20.
odorífera. R.S.							

Sect. II. HAMELIACEÆ. *Kth. synops.* 3. *p.* 63.

HAM'ELIA. S.S.		HAM'ELIA.	Pentandria Monogynia. L.				
1 axillàris. S.S.	(*ye.*)	axillary-flower'd.	Jamaica.	1820.	7. 8.	S. ♄.	
2 ventricòsa. S.S.	(*ye.*)	large-flowered.	W. Indies.	1778.	9. 11.	S. ♄.	Bot. reg. 1195.
grandiflòra. par. lond. t. 55.							
3 chrysántha. S.S.	(*ye.*)	yellow-flower'd.	————	1820.	——	S. ♄.	Jacq. ic. 2. t. 335.
4 pátens. S.S.	(*sc.*)	spreading.	————	1752.	7. 8.	S. ♄.	Exot. bot. 1. t. 24.
5 sphærocárpa. P.S.	(*or.*)	round-fruited.	Peru.	1811.	——	S. ♄.	Flor. per. t. 221. f. b.
ISE'RTIA. S.S.		ISE'RTIA.	Hexandria Monogynia. S.S.				
coccínea. S.S.	(*sc.*)	scarlet-flower'd.	S. America.	1820.	....	S. ♄.	Aubl. guian. 1. t. 123.
SABI'CEA. R.S		SABI'CEA.	Pentandria Monogynia.				
1 áspera. R.S.	(*wh.*)	rough.	Guiana.	1824.	....	S. ♄.⌣.	Aubl. guian. 1. t. 76.
2 hírta. R.S.	(*wh.*)	hairy.	Jamaica.	1825.	....	S. ♄.⌣.	

Sect. III. GARDENIEÆ. *Sal. par. lond.* 46.

MUSS'ÆNDA. R.S.		MUSS'ÆNDA.	Pentandria Monogynia. L.				
1 corymbòsa. F.I.	(*or.*)	corymbose.	E. Indies.	1827.	5. 8.	S. ♄.	
2 frondòsa. R.S.	(*ye.*)	frondose.	————	1815.	5. 10.	S. ♄.	Bot. reg. t. 517.
3 glábra. R.S.	(*ye.*)	smooth.	————	1820.	——	S. ♄.	
4 macrophy'lla. F.I.	(*or.*)	large-leaved.	Nepaul.	1827.	....	S. ♄.	
5 pubéscens. R.S.	(*ye.*)	pubescent.	China.	1805.	5. 9.	G. ♄.	Bot. mag. t. 2099.
6 arcuàta. R.S.	(*ye.*)	arched.	Isle France.	1822.	....	S. ♄.	
LUC'ULIA. B.F.G.		LUC'ULIA.	Pentandria Monogynia.				
gratíssima. B.F.G.	(*ro.*)	sweet-scented.	Nepaul.	1818.	10. 2.	G. ♄.	Swt br. fl. gar. t. 145.
Cinchòna gratíssima. F.I. *Muss'ænda Lucùlia.* D.P.							
RA'NDIA. R.S.		RA'NDIA.	Pentandria Monogynia. R.S.				
1 aculeàta. R.S.	(*wh.*)	prickly.	W. Indies.	1733.	5. 6.	S. ♄.	Bot. mag. t. 1841.
Gardènia Rándia. W.							

2 obovàta. R.S.	(wh.)	obovate-leaved.	Peru.	1818.	5. 6.	S. ♄.	Flor. per. 2. t. 220. f. b.
3 rotundifòlia. R.S.	(wh.)	round-leaved.	———	——	——	S. ♄.	
4 parvifòlia. R.S.	(wh.)	small-leaved.	Jamaica.	1825.	5. 7.	S. ♄.	Sloan. hist. 2. t. 207. f. 1.
5 macrántha. R.S.	(wh.)	large-flowered.	Sierra Leon.	1796.	8. 9.	S. ♄.	Sal. par. lond. t. 93.
longiflòra. P.L. *non* R.S.							
6 racemòsa. H.B.	(wh.)	racemed.	E. Indies.	1818.	....	S. ♄.	
7 longiflòra. R.S.	(wh.)	long-flowered.	———	——	....	S. ♄.	Lam. ill. t. 156. f. 3.
Gardènia multiflòra. W.							
8 strícta. F.I.	(wh.)	upright.	———	1824.	5. 7.	S. ♄.	
Macrocnèmum stríctum. R.S.							
9 hórrida. R.S.	(wh.)	large-spined.	China.	1825.	——	S. ♄.	
10 sinénsis. R.S.	(wh.)	Chinese.	———	——	——	S. ♄.	
POSOQU'ERIA. F.I.		POSOQU'ERIA.	Pentandria Monogynia.				
1 rígida. F.I.	(wh.)	rigid.	Nepaul.	1825.	....	S. ♄.	
2 floribúnda. F.I.	(wh.)	many-flowered.	E. Indies.	——	....	S. ♄.	
3 fasciculàta. F.I.	(wh.)	fascicle-flower'd.	———	1824.	5. 6.	S. ♄.	
4 fràgrans. F.I.	(wh.)	fragrant.	———	1822.	4. 6.	S. ♄.	
5 longispìna. F.I.	(wh.)	long-spined.	———	1812.	——	S. ♄.	
Gardènia longispìna. S.S.							
6 nùtans. F.I.	(wh.)	drooping.	———	1818.	7. 9.	S. ♄.	
7 dumetòrum. F.I.	(wh.)	bushy.	———	1779.	——	S. ♄.	Roxb. corom. 2. t. 136.
Gardènia dumetòrum. W. *Cánthium coronàtum.* Lam.							
8 uliginòsa. F.I.	(wh.)	marsh.	E. Indies.	1802.	——	S. ♄.	Roxb. corom. 2. t. 135.
9 spinòsa.	(wh.)	spiny.	China.	1804.	——	S. ♄.	Thunb. gar. n. 7. t. 2. f. 4.
Gardènia spinòsa. Th. *Cánthium chinénse.* P.S.							
10 longiflòra. R.S.	(wh.)	long-flowered.	Guiana.	1820.	....	S. ♄.	Aubl. guian. 1. t. 51.
GARD'ENIA. R.S.		GARD'ENIA.	Pentandria Monogynia. L.				
1 am'œna. S.S.	(re.wh.)	crimson-tipped.	China.	1810.	6. 8.	S. ♄.	Bot. mag. t. 1904.
2 armàta. S.S.	(wh.)	armed.	W. Indies.	1813.	....	S. ♄.	Jacq. amer. pict. t. 71.
Muss'ænda spinòsa. Jacq.							
3 tetracántha. S.S.	(wh.)	four-thorned.	New Spain.	1818.	....	S. ♄.	
4 propínqua. B.R.	(wh.)	strong-spined.	....	1816.	5. 8.	S. ♄.	Bot. reg. t. 975.
5 montàna. F.I.	(wh.)	mountain.	E. Indies.	1818.	6. 8.	S. ♄.	
6 campanulàta. F.I.	(st.)	bell-flowered.	———	1812.	——	S. ♄.	
7 micrántha. R.S.	(wh.)	small-flowered.	China.	1806.	——	S. ♄.	Thunb. gar. n. 8. t. 1. f. 2.
8 flòrida. S.S.	(wh.)	Cape Jasmine.	———	1754.	7. 10.	S. ♄.	Bot. reg. t. 449.
β *flòre-plèno.*	(wh.)	*double-flowered.*	———	——	——	S. ♄.	Ehret. pict. t. 15.
9 angustifòlia. B.C.	(wh.)	narrow-leaved.	———	1820.	——	S. ♄.	Lodd. bot. cab. 512.
10 rádicans. S.S.	(wh.)	rooting.	———	1804.	3. 6.	G. ♄.	Bot. reg. t. 73.
11 grandiflòra. R.S.	(wh.)	large-flowered.	———	1816.	——	S. ♄.	Bot. mag. t. 2627.
flòrida var. *ovalifòlia.* Bot. mag. t. 2627.							
12 costàta. F.I.	(wh.)	ribbed.	E. Indies.	1825.	....	S. ♄.	
13 latifòlia. S.S.	(wh.)	broad-leaved.	———	1787.	....	S. ♄.	Roxb. cor. 2. t. 134.
14 lùcida. F.I.	(wh.)	glossy-leaved.	———	1820.	....	S. ♄.	
15 pubéscens. R.S.	(wh.)	pubescent.	———	1825.	5. 7.	S. ♄.	
16 Thunbérgia. S.S.	(wh.)	starry.	C. B. S.	1773.	1. 3.	S. ♄.	Bot. mag. t. 1004.
17 Rothmánnia. S.S.	(sp.)	spotted-flower'd.	———	1774.	7.	G. ♄.	——— t. 690.
OXYA'NTHUS. DC.		OXYA'NTHUS.	Pentandria Monogynia.				
speciòsus. DC.	(wh.)	tube-flowered.	Sierra Leon.	1789.	7. 12.	S. ♄.	Bot. mag. t. 1992.
Gardènia tubiflòra. A.R. 183.							
GEN'IPA. R.S.		GENIP-TREE.	Pentandria Monogynia.				
1 oblongifòlia. R.S.	(st.)	oblong-leaved.	Peru.	1823.	4. 6.	S. ♄.	Flor. per. 2. t. 220. f. a.
2 americàna. R.S.	(wh.)	American.	S. America.	1779.	5. 7.	S. ♄.	Plum. ic. t. 136.
3 Meriànæ. R.S.	(wh.)	hairy.	Cayenne.	1800.	——	S. ♄.	Merian. sur. t. 43.
Duroìa eriópila. L.							
4 èdulis. R.S.	(wh.)	eatable.	———	1825.	——	S. ♄.	
5 esculénta. R.S.	(wh.)	Cherry-fruited.	China.	——	——	S. ♄.	
CA'NTHIUM. F.I.		CA'NTHIUM.	Pentandria Monogynia.				
1 angustifòlium. F.I.	(st.)	narrow-leaved.	E. Indies.	1826.	5. 7.	S. ♄.	
2 parvifòlium. F.I.	(wh.)	small-leaved.	———	1824.	——	S. ♄.	
3 parviflòrum. F.I.	(ye.)	small-flowered.	———	——	——	S. ♄.	Roxb. corom. 1. t. 51.
Wébera tetrándra. W.							
4 dídymum. F.I.	(wh.)	cyme-flowered.	———	1811.	——	S. ♄.	Gært. carp. 3. t. 196.
Wébera cymòsa. W.							
AZ'IMA. Lam.		AZ'IMA.	Tetrandria Monogynia.				
tetracántha. P.S.	(wh.)	four-spined.	E. Indies.	1758.	7.	S. ♄.	Seb. thes. 1. t. 13. f. 1.
Monètia barlerioídes. L'Herit. stirp. nov. 1. t. 1.							
WE'BERA. S.S.		WE'BERA.	Pentandria Monogynia.				
1 corymbòsa. S.S.	(wh.)	corymbose.	E. Indies.	1759.	4. 6.	S. ♄.	Bot. reg. t. 119.

2 macrophy'lla. F.I.	(wh.)	large-leaved.	E. Indies.	1824.	5. 7.	S. ♄.	
3 odoràta. F.I.	(wh.)	sweet-scented.	———	1826.	——	S. ♄.	
BURCHE'LLIA. B.R.		BURCHE'LLIA. Pentandria Monogynia.					
1 capénsis. B.R.	(sc.)	broad-leaved.	C. B. S.	1818.	3. 6.	G. ♄.	Bot. reg. t. 466.
2 parviflòra. B.R.	(or.)	smaller-flower'd.	———	——	——	G. ♄.	——— t. 891.
bubalìna. B.M. 2339.							
CATESB'ÆA. S.S.		LILY-THORN. Tetrandria Monogynia. L.					
1 spinòsa. S.S.	(wh.)	spiny.	Providence I.	1726.	5. 9.	S. ♄.	Bot. mag. t. 131.
2 latifòlia. B.R.	(wh.)	broad-leaved.	W. Indies.	1818.	9. 11.	S. ♄.	Bot. reg. t. 858.
3 parviflòra. S.S.	(wh.)	small-flowered.	———	1810.	7. 10.	S. ♄.	Sloan. h. 2. t. 207. f. 1.
FERN'ELIA. J.		FERN'ELIA. Tetrandria Monogynia.					
1 obovàta. R.S.	(wh.)	obovate-leaved.	Isl. France.	1816.	....	S. ♄.	Lam. ill. t. 67. f. 1.
2 biflòra. R.S.	(wh.)	two-flowered.	Bourbon.	——	....	S. ♄.	
Coccocy'pselum buxifòlium. S.S. *biflòrum.* W							
3 uniflòra. R.S.	(wh.)	one-flowered.	Mauritius.	——	....	S. ♄.	
buxifòlia. β. Lamarck.							
COCCOCY'PSELUM. S.S.		COCCOCY'PSELUM. Tetrandria Monogynia. S.S.					
rèpens. S.S.	(li.)	creeping.	W. Indies.	1793.	5.	S. ♃.	Brown. jam. t. 6. f. 1.

Sect. IV. CINCHONEÆ. *Kth. synops.* 3. *p.* 47.

PORTLA'NDIA. S.S.		PORTLA'NDIA. Pentandria Monogynia. L.					
1 grandiflòra. R.S.	(wh.)	great-flowered.	Jamaica.	1775.	6. 8.	S. ♄.	Sm. ic. pict. 1. t. 6.
2 coccínea. R.S.	(sc.)	scarlet-flowered.	———	1812.	....	S. ♄.	
COUT'AREA. S.S.		COUT'AREA. Hexandria Monogynia. S.S.					
speciòsa. S.S.	(sc.)	Laurel-leaved.	Guiana.	1803.	....	S. ♄.	Aubl. guian. 1. t. 122.
Portlándia hexándra. Jacq. am. pict. t. 65.							
EXOSTE'MMA. R.S.		EXOSTE'MMA. Pentandria Monogynia. R.S.					
1 carib'æum. R.S.	(wh.)	Caribean.	W. Indies.	1780.	6. 9.	S. ♄.	Andr. reposit. t. 481.
Cinchòna carib'æa. W.							
2 floribúndum. R.S.	(bh.)	many-flowered.	———	1794.	....	S. ♄.	Lamb. cinch. 27. t. 7.
PINCKN'EYA. M.		PINCKN'EYA. Pentandria Monogynia.					
pùbens. M.	(bh.)	pubescent.	Georgia.	1786.	6. 7.	F. ♄.	Mich. amer. t. 13.
HYMENODI'CTYON. F.I.		HYMENODI'CTYON. Pentandria Monogynia.					
1 thyrsiflòrum. Wall.	(st.)	thyrse-flower'd.	E. Indies.	1816.	....	S. ♄.	
2 excélsum. Wall.	(wh.)	tall.	———	——	....	S. ♄.	Roxb. cor. 2. t. 106.
Cinchòna excélsa. R.S.							
HI'LLIA. S.S.		HI'LLIA. Tetr-Hexandria Monogynia. S.S.					
1 longiflòra. S.S.	(wh.)	long-flowered.	W. Indies.	1789.	2. 3.	S. ♄.	Bot. mag. t. 721.
2 tetrándra. S.S.	(wh.)	mountain.	Jamaica.	1793.	6. 7.	S. ♄.	
MACROCN'EMUM. L.		MACROCN'EMUM. Pentandria Monogynia. L.					
1 jamaicénse. R.S.	(wh.)	Jamaica.	Jamaica.	1806.	....	S. ♄.	Swartz obs. t. 3. f. 1.
2 speciòsum. S.S.	(li.)	handsome.	Caracas.	1828.	....	S. ♄.	Jacq. schœnb. 1. t. 43.
Muss'ænda speciòsa. R.S.							
3 coccíneum. S.S.	(sc.)	scarlet.	Trinidad.	1825.	....	S. ♄.	
Muss'ænda coccínea. R.S.							
RONDEL'ETIA. R.S.		RONDEL'ETIA. Pentandria Monogynia. L.					
1 americàna. R.S.	(wh.)	American.	W. Indies.	1752.	8.	S. ♄.	Plum. ic. t. 242. f. 1.
2 lævigàta. R.S.	(wh.)	smooth-leaved.	———	1790.	7. 8.	S. ♄.	
3 trifoliàta. R.S.	(re.)	three-leaved.	Jamaica.	....	——	S. ♄.	Jacq. amer. pict. t. 62.
4 paniculàta. F.I.	(wh.)	panicled.	E. Indies.	1820.	——	S. ♄.	
5 tinctòria. F.I.	(wh.)	dyer's.	———	1825.	....	S. ♄.	
6 racemòsa. R.S.	(wh.)	racemed.	W. Indies.	1818.	——	S. ♄.	Brown. jam. 143. t. 2. f. 3.
7 tomentòsa. R.S.	(wh.)	downy-leaved.	———	——	——	S. ♄.	
8 hírta. R.S.	(br.ye.)	hairy.	———	1776.	6. 8.	S. ♄.	Lodd. bot. cab. t. 350.

Sect. V. HEDYOTIDEÆ. *Kth. synops.* 3. *p.* 40.

HEDY'OTIS. R.S.		HEDY'OTIS. Tetrandria Monogynia. L.					
1 diffùsa. R.S.	(wh.)	spreading.	E. Indies.	1818.	7. 9.	S. ☉.	
2 herbàcea. R.S.	(wh.)	herbaceous.	———	1816.	6. 9.	S. ☉.	Rheed. mal. 10. t. 35.
3 capénsis. R.S.	(wh.)	Cape.	C. B. S.	1824.	7. 10.	H. ☉.	
4 umbellàta. R.S.	(wh.)	Indian Madder.	E. Indies.	1792.	7. 8.	S. ♃.	Roxb. cor. 1. t. 3.
Oldenlándia corymbòsa L.							

5 cymòsa. R.S. (*wh.*) Hyssop-leaved. Jamaica. 1739. 6. 10. S. ⊙. Ehret pic. t. 2. f. 1. et t. 4.
Oldenlándia corymbòsa. W.
6 ramosíssima. R.S. (*wh.*) branching. .... 1816. —— S. ⊙.
7 biflòra. R.S. (*pu.*) two-flowered. E. Indies. —— —— S. ⊙.
8 láctea. R.S. (*st.*) milk-coloured. ———— —— —— S. ⊙.
9 campanuliflòra. B.M. (*pu.*) bell-flow'r'd. Brazil. 1825. 2. 10. S. ♃.

MANE'TTIA. R.S. MANE'TTIA. Tetrandria Monogynia. L.
1 Lygístum. R.S. (*re.*) veined-leaved. W. Indies. 1822. 2. 4. S. ♄.◡. Brown. jam. t. 3. f. 2.
2 coccínea. R.S. (*sc.*) scarlet. Trinidad. 1806. 12. 4. S. ♄.◡. Bot. reg. t. 693.

BOUVA'RDIA. R.S. BOUVA'RDIA. Tetrandria Monogynia.
1 triphy'lla. R.S. (*sc.*) blunt-leaved. Mexico. 1794. 4. 11. G. ♄. Sal. par. lond. t. 88.
2 Jacquínii. K.S. sharp-leaved. ———— —— —— G. ♄. Jacq. schœnb. 1. t. 257.
Ixòra americàna. Jacq. *Bouvárdia triphy'lla.* Bot. reg. t. 107.
3 versícolor. S.S. (*re.ye.*) various-color'd. S. America. 1814. 7. 1. S. ♄. Bot. reg. t. 245.

DENTE'LLA. W. DENTE'LLA. Pentandria Monogynia. W.
rèpens. W. (*wh.*) creeping. N. Holland. 1802. 7. H. ⊙. Rumph. am. 6. t. 170. f. 4.

SPERMADI'CTYON. B.R. SPERMADI'CTYON. Pentandria Monogynia.
1 suavèolens. B.R. (*wh.*) sweet-scented. E. Indies. 1816. 8. 12. S. ♄. Bot. reg. t. 348.
Hamiltònia suavèolens. F.I.
2 azùreum. B.R. (*bl.*) blue-flowered. Nepaul. 1827. 1. 3. S. ♄. Bot. reg. t. 1235.

Sect. VI. CEPHALANTHEÆ. *Kth. synops.* 3. *p.* 37.

NA'UCLEA. R.S. NA'UCLEA. Pentandria Monogynia. L.
1 Adìna. Sm. (*wh.*) Myrtle-leaved. China. 1804. 7. 8. G. ♄. Bot. reg. t. 895.
Adìna globiflòra. Sal. par. lond. t. 115.
2 adinoídes. S.S. (*wh.*) Adina-like. ———— 1825. —— G. ♄.
3 orientàlis. R.S. (*ye.*) oriental. E. Indies. 1800. .... S. ♄.
4 Cadámba. F.I. (*or.*) broad-leaved. ———— 1823. .... S. ♄. Rheed. mal. 3. t. 33.
5 undulàta. F.I. (*ye.*) wave-leaved. ———— —— .... S. ♄.
6 parvifòlia. F.I. (*ye.*) small-leaved. ———— 1825. —— S. ♄. Roxb. corom. 1. t. 52.
7 cordifòlia. R.S. (*st.*) heart-leaved. ———— —— .... S. ♄. ———— 1. t. 53.
8 purpùrea. F.I. (*pu.*) purple-flowered. ———— 1827. .... S. ♄. ———— 1. t. 54.

UNC'ARIA. F.I. UNC'ARIA. Pentandria Monogynia.
Gámbier. F.I. (*pi.*) Gambier-tree. E. Indies. 1826. .... S. ♄.◡. Linn. trans. v. 9. t. 22.

SARCOCE'PHALUS. Afz. SARCOCE'PHALUS. Pentandria Monogynia.
esculéntus. Afz. (*wh.*) eatable-fruited. Sierra Leone. 1822. 6. 8. S. ♄. Hort. trans. 5. t. 18.

CEPHALA'NTHUS. L. BUTTON-WOOD. Tetrandria Monogynia. L.
occidentàlis. L. (*wh.*) American. N. America. 1735. 8. S. ♄. Schmidt arb. 1. t. 45.

MORI'NDA. L. MORI'NDA. Pentandria Monogynia. L.
1 umbellàta. R.S. (*wh.*) umbelled. E. Indies. 1809. 6. 8. S. ♄.
2 citrifòlia. R.S. (*wh.*) broad-leaved. Otaheite. 1793. —— S. ♄. Rheed. mal. 1. t. 52.
3 tinctòria. F.I. (*wh.*) dyer's. ———— 1826. .... S. ♄.
4 bracteàta. F.I. (*wh.*) bracted. E. Indies. 1816. 5. 7. S. ♄.
5 angustifòlia. F.I. (*wh.*) narrow-leaved. ———— —— —— S. ♄. Roxb. corom. 3. t. 237.
6 Róyoc. R.S. (*wh.*) Laurel-leaved. W. Indies. 1793. 7. 10. S. ♄. Jacq. vind. 1. t. 16.

Sect. VII. COFFEACEÆ. *Kth. synops.* 3. *p.* 20.

MITCHE'LLA. L. MITCHE'LLA. Tetrandria Monogynia. L.
rèpens. D. (*wh.*) creeping. N. America. 1761. 6. H. ♄. Cat. car. 1. t. 20.

PAVE'TTA. R.S. PAVE'TTA. Tetrandria Monogynia. L.
1 índica. R.S. (*wh.*) Indian. E. Indies. 1791. 8. 10. S. ♄. Burm. ind t. 13. f. 3.
2 tomentòsa. R.S. (*wh.*) woolly. ———— 1824. —— S. ♄.
3 cáffra. R.S. (*wh.*) Cape. C. B. S. 1816. —— G. ♄.
4 pentándra. R.S. (*wh.*) pentandrous. W. Indies. 1823. —— S. ♄. Plum. ic. t. 156. f. 1.
5 arenòsa. R.S. (*wh.*) tubercled. China. .... 3. 10. S. ♄. Bot. reg. t. 198.
índica. B.R. *nec aliorum.*

IX'ORA. L. IX'ORA. Tetrandria Monogynia. L.
1 grandiflòra. B.R. (*sc.*) great-flowered. E. Indies. 1814. 7. 9. S. ♄. Bot. reg. t. 154.
coccínea. L. *non* B.M.

2 Bandhùca. F.I. (*sc.*) stem-clasping. E. Indies. 1814. 1. 4. S. ♄. Bot. reg. t. 513.
3 fúlgens. F.I. (*sc.*) glossy. ——— 1820. .... S. ♄. Rumph. amb. 4. t. 46.
longifòlia. Sm. *lanceolàta.* Lam.
4 flámmea. S.P. (*or.*) flame-coloured. China. 1690. 4. 8. S. ♄. Bot. mag. t. 169.
coccínea. B.M. *strícta.* F.I. *speciòsa.* W. enum.
5 crocàta. B.R. (*sn.*) saffron-colour'd. ——— 1820. —— S. ♄. Bot. reg. t. 782.
6 incarnàta. B.C. (*fl.*) flesh-coloured. ——— —— —— S. ♄. Lodd. bot. cab. 1048.
7 blánda. B.R. (*bh.*) charming. ——— 1812. —— S. ♄. Bot. reg. t. 100.
8 álba. W. (*wh.*) white. E. Indies. 1768. 6. S. ♄. Rheed. mal. 2. t. 14.
9 cuneifòlia. F.I. (*wh.*) wedge-leaved. ——— 1820. —— S. ♄. Bot. reg. t. 648.
10 brachiàta. F.I. (*wh.*) brachiate-panicled. ——— —— —— S. ♄.
11 Pavétta. A.R. (*wh.*) fragrant. ——— .... —— S. ♄. Andrews's reposit. t. 78.
12 acuminàta. F.I. (*wh.*) tapering-leaved. ——— 1824. 3. 5. S. ♄.
13 parviflòra. F.I. (*wh.*) small-flowered. ——— 1800. 8. 10 S. ♄. Vahl symb. 3. t. 52.
14 barbàta. F.I. (*wh.*) bearded. ——— 1822. 6. 8. S. ♄. Bot. mag. t. 2505.
15 undulàta. F.I. (*wh.*) wave-leaved. ——— 1818. —— S. ♄.
16 ròsea. F.I. (*ro.*) rose-coloured. ——— 1820. 4. 8. S. ♄. Bot. reg. t. 540.
17 obovàta. R.S. (*cr.*) obovate-leaved. ——— 1810. —— S. ♄.
purpùrea. F. ?

CHOM'ELIA. R.S. CHOM'ELIA. Tetrandria Monogynia. W.
1 spinòsa. R.S. (*wh.*) spiny. W. Indies. 1793. .... S. ♄. Jacq. amer. 18. t. 13.
2 fasciculàta. R.S. (*wh.*) fascicled. Jamaica. 1824. 4. 6. S. ♄.

SIDERODE'NDRON. S.S. IRON-TREE. Tetrandria Monogynia. W.
triflòrum. S.S. (*ro.wh.*) three-flowered. W. Indies. 1793. .... S. ♄. Jacq. amer. 19. t. 175. f. 9.

TETRAM'ERIUM. G. TETRAM'ERIUM. Tetrandria Monogynia. S.S.
1 odoratíssimum. S.S. (*wh.*) sweet-scented. W. Indies. 1793. 4. 6. S. ♄. Jacq. amer. t. 47.
Coffèa occidentàlis. W. *Ixòra americàna.* Sw.
2 paniculàtum. S.S. (*wh.*) panicled. Guiana. 1822. .... S. ♄. Aub. guian. t. 58.

COFF'EA. L. COFFEE-TREE. Pentandria Monogynia. L.
arábica. W. (*wh.*) Arabian. Yemen. 1696. 8. 11. S. ♄. Bot. mag. t. 1303.

CEPHA'ELIS. S.S. CEPHA'ELIS. Pentandria Monogynia. L.
1 elàta. S.S. (*pu.*) tall. Jamaica. 1793. .... S. ♄.
2 tomentòsa. R.S. (*pu.*) tomentose. Trinidad. 1820. 4. 6. S. ♄.
3 punícea. R.S. (*pu.*) puniceous. Jamaica. 1825. —— S. ♄.
4 violàcea. R.S. (*vi.*) violet-coloured. W. Indies. 1824. —— S. ♄. Aubl. guian. 1. t. 60.
Tapogòmea violàcea. Aubl.
5 axillàris. S.S. (*pu.*) axillary-flow'r'd. Brazil. 1816. 4. 5. S. ♄. Lind. coll. t. 21.
calycìna. L. col.
6 glábra. R.S. (*bl.*) smooth. Trinidad. 1820. 5. 8. S. ♄. Aubl. guian. 1. t. 63.
7 álba. R.S. (*bh.*) white. Guiana. 1824. —— S. ♄. ——— 1. t. 62. f. 1.
8 involucràta. R.S. (*wh.*) involucred. ——— —— .... S. ♄. ——— 1. t. 64.
Carapíchea guianénsis. Aubl.
9 pedunculàta. P.L. (*wh.*) long-peduncled. Sierra Leon. .... 2. 4. S. ♄. Sal. par. lond. t. 103.

PSYCH'OTRIA. R.S. PSYCH'OTRIA. Pentandria Monogynia. L.
1 asiática. R.S. (*wh.*) Indian. E. Indies. 1806. 4. 5. S. ♄. Jacq. amer. p. t. 260. f. 19.
2 glabràta. R.S. (*wh.*) smooth. Jamaica. 1822. —— S. ♄.
3 laurifòlia. R.S. (*wh.*) Laurel-leaved. W. Indies. 1825. 5. 8. S. ♄.
4 f'œtens. R.S. (*wh.*) strong-scented. Jamaica. —— —— S. ♄.
5 citrifòlia. R.S. (*wh.*) Citron-leaved. W. Indies. 1793. 1. 12. S. ♄.
6 ellíptica. B.R. (*wh.*) elliptic-leaved. Brazil. 1820. 2. 6. S. ♄. Bot. reg. t. 607.
7 myrtiphy'llum. R.S. (*wh.*) Myrtle-leaved. Jamaica. 1823. 2. 5. S. ♄. Sloan. his. 2. t. 209. f. 2.
8 parasìtica. R.S. (*wh.*) parasitic. W. Indies. 1802. 5. 8. S. ♄. Jacq. amer. t. 51. f. 1.
9 brachiàta. R.S. (*wh.*) cross-branched. ——— 1793. .... S. ♄.
10 pubéscens. R.S. (*wh.*) pubescent. Jamaica. 1812. .... S. ♄.
11 corymbòsa. R.S. (*pu.*) corymbose. ——— —— 3. 4. S. ♄.
12 pedunculàta. R.S. (*st.*) pedunculate. ——— 1824. 5. 6. S. ♄.
13 undàta. R.S. (*wh.*) wave-leaved. Bahama. 1818. 4. 6. S. ♄. Jacq. schœnb. 3. t. 260.
14 Reèvesii. F.I. (*gr.*) Reeves's. China. 1823. —— G. ♄.

GEO'PHILA. D.P. GEO'PHILA. Pentandria Monogynia.
renifórmis. D.P. (*wh.*) kidney-leaved. W. Indies. 1793. 4. 6. S. ♃. Jacq. amer. t. 46
Psychòtria herbàcea. W.

CHIOCO'CCA. W. SNOWBERRY. Pentandria Monogynia. L.
racemòsa. W. (*wh.*) cluster-flower'd. Jamaica. 1729. 2. S. ♄. Andr. reposit. t. 284.

ERN'ODEA. S.S. ERN'ODEA. Tetrandria Monogynia. W.
montàna. S.S. (*pu.*) mountain. Sicily. 1820. 6. 7. H. ♄. Flor. græc. t. 143.
Aspérula calábrica. L.

SERI'SSA. S.S. SERI'SSA. Pentandria Monogynia. W.
f'œtida. S.S. (*wh.*) Japanese. Japan. 1787. 5. 9. G. ♄. Bot. mag. t. 361.
β *múltiplex.* (*wh.*) *double-flowered.* ——— —— —— G. ♄.

PLOC'AMA. W.		PLOC'AMA.	Pentandria Monogynia. W.				
péndula. w.	(*wh.*)	pendulous.	Canaries.	1779.	....	G. ♄.	
PÆD'ERIA. W.		PÆD'ERIA.	Pentandria Monogynia. L.				
1 fœtida. w.	(*pu.*)	stinking.	China.	1806.	....	G. ♄.‿.	Icon. Kæmpf. 9.
2 Língun. Boj.	....	tongue.	Madagascar.	1823.	....	S. ♄.‿.	

Sect. VIII. SPERMACOCEÆ. *Kth. synops.* 3. *p.* 14.

RICHARDS'ONIA. Kth.		RICHARDS'ONIA.	Hexandria Monogynia. L.				
1 scábra. B.F.G.	(*wh.*)	rough.	S. America.	1812.	9. 10.	H. ⊙.	Swt. br. fl. gar. t. 91.
2 pilòsa. K.S.	(*wh.*)	hairy.	Lima.	1824.	7. 9.	H. ⊙.	Flor. peruv. 3. t. 279. f. b.
PHY'LLIS. W.		BASTARD HARE'S-EAR.	Pentandria Digynia. L.				
Nòbla. w.	(*br.*)	Canary.	Canaries.	1699.	6. 7.	G. ♄.	Dill. elth. t. 299. f. 386.
DI'ODIA. S.S.		DI'ODIA.	Tetrandria Monogynia. W.				
1 virgínica. S.S.	(*wh.*)	Virginian.	Virginia.	1820.	6. 8.	H. ♃.	Jacq. ic. 1. t. 29.
2 sarmentòsa. S.S.	(*wh.*)	sarmentose.	W. Indies.	1821.	——	S. ħ.	
3 verticillàta. S.S.	(*wh.*)	whorl-leaved.	Santa Cruz.	——	....	S. ♃.	
BIGEL'OVIA. S.S.		BIGEL'OVIA.	Tetrandria Monogynia. S.S.				
1 verticillàta. S.S.	(*wh.*)	whorl-flowered.	Africa.	1732.	6. 8.	S. ħ.	Dill. elth. t. 277. f. 358.
Spermacòce verticillàta. L.							
2 commutàta. S.S.	(*wh.*)	starry-headed.	W. Indies.	1818.	——	S. ♃.	
3 stricta. S.S.	(*wh.*)	upright.	E. Indies.	1821.	——	S. ⊙.	
Spermacòce strícta. L.							
SPERMAC'OCE. S.S.		BUTTON-WEED.	Tetrandria Monogynia. L.				
1 glábra. S.S.	(*wh.*)	smooth.	N. America.	1823.	7. 8.	H. ⊙.	
2 diodìna. S.S.	(*wh.*)	Diodia-like.	————	1820.	——	H. ⊙.	
3 híspida. S.S.	(*vi.*)	bristly.	E. Indies.	1781.	8. 9.	S. ⊙.	Burm. zeyl. t. 20. f. 3.
4 spinòsa. S.S.	(*wh.*)	spiny.	W. Indies.	1825.	——	S. ⊙.	Brown. jam. p. 141. 1.
5 latifòlia. S.S.	(*wh.*)	broad-leaved.	Guiana.	1803.	7.	S. ♂.	Aub. gui. t. 19. f. 1.
6 cornifòlia. S.S.	(*wh.*)	Dogwood-leav'd.	Brazil.	1820.	8. 9.	S. ⊙.	
7 stylòsa. S.S.	(*wh.*)	long-styled.	Manila.	——	——	S. ⊙.	
8 suffruticòsa. S.S.	(*bh.*)	suffruticose.	....	1823.	7. 8.	S. ħ.	Jacq. schœn. t. 322.
9 rádicans. S.S.	(*wh.*)	rooting.	Guiana.	1803.	——	S. ♃.	Aub. gui. 1. t. 20. f. 4.
10 áspera. S.S.	(*wh.*)	rough.	W. Indies.	1822.	——	S. ⊙.	———— t. 22. f. 6.
11 ténuior. S.S.	(*bh.*)	slender.	————	1732.	6. 8.	H. ⊙.	Sch. hand. 1. t. 22.
12 lanceolàta. L.en.	(*wh.*)	spear-leaved.	————	1818.	——	H. ⊙.	
13 hírta. S.S.	(*wh.*)	hairy.	Jamaica.	1822.	7. 9.	H. ⊙.	Jacq. ic. t. 308.
14 villòsa. S.S.	(*wh.*)	villous.	————	1823.	——	S. ⊙.	
15 scábra. S.S.	(*wh.*)	rough.	E. Indies.	1816.	——	S. ⊙.	Rheed. mal. 9. t. 76?
16 capitàta. S.S.	(*wh.*)	headed.	Peru.	1826.	6. 8.	G. ♃.	Flor. peruv. 1. t. 91. f. B.
17 rùbra. S.S.	(*pi.*)	Cross-wort.	S. America.	1760.	7. 8.	S. ♂.	Jacq. schœn. 3. t. 256.
strigòsa. B.M. 1558.							
18 Fischèri. S.S.	(*wh.*)	Fischer's.	Jamaica.	1820.	——	S. ⊙.	
19 sexangulàris. S.S.	(*bl.*)	six-angled.	W. Indies.	1822.	——	S. ⊙.	Aubl. guian. 1. t. 20. f. 3.
20 involucràta. R.S.	(*wh.*)	involucred.	Carolina.	1812.	——	H. ⊙.	
21 Roxbúrghii. S.S.	(*wh.*)	Roxburgh's.	E. Indies.	1818.	——	S. ♂.	
lævis. F.I. *non* Lam.							
KNO'XIA. S.S.		KNO'XIA.	Tetrandria Monogynia. W.				
1 corymbòsa. S.S.	(*wh.*)	Sumatra.	E. Indies.	1818.	4. 6.	S. ħ.	
Spermacòce sumatrénsis. F.I.							
2 umbellàta. S.S.	(*wh.*)	umbel-flower'd.	Nepaul.	1820.	——	S. ħ.	
Spermacòce tères. F.I.							

Sect. IX. GALEÆ. *Kth. synops.* 3. *p.* 10.

ANTHOSPE'RMUM. W.		AMBER-TREE.	Diœcia Tetrandria. L.				
æthiòpicum. S.S.	(*gr.*)	Ethiopian.	C. B. S.	1692.	6. 7.	G. ♄.	Pluk. alm. t. 183. f. 1.
CRUCIANE'LLA. R.S.		CROSS-WORT.	Tetrandria Monogynia. L.				
1 angustifòlia. R.S.	(*st.*)	narrow-leaved.	S. Europe.	1658.	6. 7.	H. ⊙.	Sm. exot. bot. 2. t. 109.
2 ciliàta. R.S.	(*st.*)	ciliated.	Levant.	1805.	7. 8.	H. ⊙.	
3 pátula. R.S.	(*ye.*)	spreading.	Spain.	1798.	6. 7.	H. ⊙.	
4 latifòlia. R.S.	(*gr.*)	broad-leaved.	S. Europe.	1633.	——	H. ⊙.	Flor. græc. t. 139.
5 ægyptìaca. R.S.	(*st.*)	Egyptian.	Egypt.	1800.	——	H. ⊙.	
6 monspelìaca. R.S.	(*st.*)	Montpelier.	S. France.	1791.	7. 8.	H. ⊙.	Flor. græc. t. 140.

7	marítima. R.S.	(*ye.*)	sea-side.	S. France.	1640.	7. 8.	G. ♄.	Sabb. hort. 2. t. 13.
8	americàna. R.S.	(*bl.*)	American.	Vera Cruz.	....	——	S. ♄.	
9	pubéscens. R.S.	(*pu.*)	pubescent.	Candia.	1799.	7. 8.	G. ♃.	
10	áspera. M.B.	(*st.*)	rough.	Iberia.	1827.	6. 8.	H. ♃.	
11	glomeràta. M.B.	(*st.*)	clustered.	————	1828.	——	H. ♃.	
12	molluginoídes. R.S.	(*wh.*)	Mollugo-like.	Caucasus.	1800.	——	H. ♃.	Buxb. cent. 2. t. 30. f. 1.
13	híspida. R.S.	(*pu.*)	hispid.	Vera Cruz.	....	——	S. ♃.	
	SHERA'RDIA. R.S.		SHERA'RDIA.	Tetrandria Monogynia. L.				
	arvénsis. R.S.	(*li.*)	little.	Britain.	....	4. 9.	H. ⊙.	Eng. bot. v. 13. t. 891.
	R'UBIA. R.S.		MADDER.	Tetrandria Monogynia. L.				
1	tinctòrum. R.S.	(*gr.*)	dyer's.	S. Europe.	1596.	6.	H. ♃.	Flor. græc. t. 141.
2	spléndens. R.S.	(*ye.*)	glossy.	Portugal.	1812.	6. 8.	H. ♄.	
3	galioídes. R.S.	(*gr.*)	Galium-like.	Canaries.	1816.	——	F. ♄.	
4	peregrìna. R.S.	(*ye.*)	wild.	England.	....	7.	H. ♃.	Eng. bot. v. 12. t. 851.
5	lùcida. R.S.	(*ye.*)	shining.	Majorca.	1762.	——	F. ♄.	Flor. græc. t. 142.
6	fruticòsa. R.S.	(*ye.*)	prickly-leaved.	Canaries.	1779.	9.	F. ♄.	Jacq. ic. 1. t. 25.
7	Boccòni. R.S.	(*bh.*)	Boccone's.	Italy.	1823.	——	H. ♃.	Bocc. mus. t. 75. f. 2-3.
8	angustifòlia. R.S.	(*st.*)	narrow-leaved.	S. Europe.	1772.	7. 8.	F. ♄.	Lam. ill. t. 60. f. 2.
9	cordifòlia. R.S.	(*wh.*)	heart-leaved.	Siberia.	1783.	7.	H. ♃.	Pall. it. 3. ap. t. L. f. 1.
	ASPE'RULA. R.S.		WOODRUFF.	Tetrandria Monogynia. L.				
1	lævigàta. R.S.	(*wh.*)	shining.	S. Europe.	1775.	6.	H. ♃.	Moris. his. 3. t. 21. f. 4.
2	taurìna. R.S.	(*li.*)	broad-leaved.	Italy.	1739.	4. 6.	H. ♃.	Lob. ic. 800. f. 1.
3	longiflòra. R.S.	(*li.*)	long-flowered.	Hungary.	1818.	——	H. ♃.	W. et K. hung. 2 t. 150.
4	aristàta. R.S.	(*st.*)	awned.	S. Europe.	1823.	——	H. ♃.	
5	supìna. R.S.	(*bh.*)	trailing.	Caucasus.	——	6. 7.	H. ♃.	
6	brevifòlia. R.S.	(*li.*)	short-leaved.	Levant.	1818.	——	H. ♄.	
7	cynánchica. R.S.	(*bh.*)	small.	England.	....	7.	H. ♃.	Eng. bot. v. 1. t. 33.
8	alpìna. S.S.	(*bh.*)	Alpine.	Caucasus.	1825.	——	H. ♃.	
9	crassifòlia. R.S.	(*wh.*)	thick-leaved.	Levant.	1775.	6.	H. ♃.	
10	scábra. S.S.	(*wh.*)	rough.	Italy.	1824.	——	H. ♃.	
11	tinctòria. R.S.	(*bh.*)	narrow-leaved.	Europe.	1764.	6. 7.	H. ♃.	Tabern. ic. 151.
12	montàna. R.S.	(*bh.*)	mountain.	Hungary.	1801.	——	H. ♃.	
13	arvénsis. R.S.	(*bh.*)	field.	Europe.	1596.	7.	H. ⊙.	Lobel. ic. t. 801. f. 2.
14	arcadiénsis. S.S.	(*ro.*)	Arcadian.	Arcadia.	1819.	5.	H. ♃.	Bot. mag. t. 2146.
15	tomentòsa. T.N.	(*pu.*)	woolly.	Italy.	1826.	——	H. ♃.	Ten. fl. nap. 1. t. 9.
16	hírta. R.S.	(*pu.*)	thick-rooted.	Pyrenees.	——	5. 6.	H. ♃.	Ram. phil. t. 9. f. 1. 2.
17	hirsùta. R.S.	(*ro.*)	hairy creeping.	S. Europe.	1820.	——	H. ♃.	
	rèpens. Brot.							
18	incàna. R.S.	(*pu.*)	hoary.	Crete.	1823.	6. 7.	H. ♃.	Flor. græc. t. 119.
19	galioídes. S.S.	(*wh.*)	glaucous.	S. Europe.	1710.	6. 9.	H. ♃.	Bocc. mus. 2. t. 116.
	Gàlium glaùcum. L.							
20	humifùsa. S.S.	(*wh.*)	spreading.	Caucasus.	1825.	——	H. ♃.	
21	longifòlia. R.S.	(*wh.*)	long-leaved.	Greece.	1823.	——	H. ♃.	Flor. græc. t. 118.
22	Aparìne. R.S.	(*wh.*)	marsh.	Crete.	——	——	H. ♃.	———— t. 117.
	rivàlis. Sm. *Gálium uliginòsum.* Pall.							
23	odoràta. R.S.	(*wh.*)	sweet-scented.	Britain.	....	5. 6.	H. ♃.	Eng. bot. 755.
24	tyràica. Bes.	(*wh.*)	Tyraican.	Levant.	1826.	——	H. ♃.	
	VALA'NTIA. S.S.		VALA'NTIA.	Tetrandria Monogynia. S.S.				
1	muràlis. S.S.	(*st.*)	wall.	S. Europe.	1739.	5. 7.	H. ⊙.	Colum. Ecphr. t. 297.
2	híspida. S.S.	(*wh.*)	bristly.	————	1768.	——	H. ⊙.	Gært. fr. C. 2. t. 24. f. 1.
3	filifórmis. S.S.	(*wh.*)	least.	Canaries.	1780.	7. 9.	H. ⊙.	
	GA'LIUM. S.S.		BEDSTRAW.	Tetrandria Monogynia. L.				
1	cruciàtum. S.S.	(*ye.*)	Crosswort.	Britain.	....	5. 6.	H. ♃.	Eng. bot. v. 2. t. 143.
	Valántia cruciàta. W.							
2	chersonénse. R.S.	(*ye.*)	fringed.	Tauria.	1829.	5. 7.	H. ♃.	
3	taùricum. R.S.	(*ye.*)	fragrant.	————	1820.	——	H. ♃.	Bot. mag. t. 2670.
	Valántia taùrica. B.M.							
4	pedemontànum. S.S.	(*ye.*)	Piedmont.	Hungary.	1799.	7.	H. ⊙.	W. et K. hung. 1. t. 33.
5	Bauhìni. S.S.	(*ye.*)	Bauhin's.	Europe.	1731.	7. 8.	H. ♃.	———— 1. t. 32.
	Valántia glábra. W.							
6	Scopoliànum. R.S.	(*ye.*)	Scopoli's.	————	——	——	H. ♃.	Scop. c. e. 2. 1. n. 144. t. 2.
	vérnum. Scop.							
7	Hallèri. R.S.	(*st.*)	Haller's.	Switzerland.	1823.	5. 6.	H. ♃.	
8	articulàtum. S.S.	(*ye.*)	jointed.	Egypt.	1752.	7. 8.	H. ⊙.	Lam. ill. t. 843. f. 3.
9	rubioídes. S.S.	(*st.*)	Madder-leaved.	S. Europe.	1775.	7.	H. ♃.	Buxb. cent. 2. t. 29.
10	diffùsum. S.S.	(*st.*)	spreading.	————	1820.	——	H. ♃.	
11	tatáricum. S.S.	(*wh.*)	Tartarian.	Tartary.	——	6. 8.	H. ♃.	
12	valantioídes. S.S.	(*wh.*)	Valantia-like.	Caucasus.	1823.	——	H. ♃.	
13	palústre. S.S.	(*wh.*)	water.	Britain.	....	——	H. *w.* ♃.	Eng. bot. v. 26. t. 1857.

14	satureiæfòlium. S.S.	(*wh.*)	Savory-leaved.	Caucasus.	1818.	7.	H.*w.*♃.	
15	trífidum. S.S.	(*wh.*)	3-fid-flowered.	N. Europe.	———	———	H.*w.*♃.	Flor. dan. t. 48.
16	fruticòsum. S.S.	(*wh.*)	frutescent.	Crete.	1823.	7. 8.	F. ♄.	
17	helòdes. R.S.	(*wh.*)	weak-branched.	Portugal.	1827.	———	H.*w.*♃.	
18	Claytòni. R.S.	(*wh.*)	Clayton's.	N. America.	———	———	H.*w.*♃.	
19	Witheríngii. S.S.	(*wh.*)	rough.	England.	....	6. 7.	H. ♃.	Eng. bot. v. 31. t. 2206.
20	glábrum. S.S.	(*wh.*)	smooth.	C. B. S.	1816.	———	G. ♃.	
21	uliginòsum. S.S.	(*wh.*)	marsh.	Britain.	....	7. 8.	H.*w.*♃.	Eng. bot. v. 28. t. 1972.
22	divaricàtum. S.S.	(*wh.*)	divaricate.	S. France.	1820.	———	H. ⊙.	DC. ic. gal. rar. t. 24.
23	campéstre. S.S.	(*st.*)	field.	Barbary.	1818.	8. 9.	H. ⊙.	
24	austrìacum. R.S.	(*wh.*)	Austrian.	Europe.	1804.	6. 7.	H. ♃.	Jacq. aust. t. 80.
25	Boccòni. R.S.	(*bh.*)	Boccone's.	———	1801.	5. 6.	H. ♃.	Bocc. mus. t. 101.
26	oblíquum. R.S.	(*wh.*)	oblique-leaved.	S. France.	....	———	H. ♃.	Vil. delph. 2. n. 9. t. 8.
27	pusíllum. E.B.	(*wh.*)	least.	England.	....	7. 8.	H. ♃.	Eng. bot. v. 2 t. 74.
28	alpéstre. R.S.	(*wh.*)	smooth-leaved.	S. Europe.	1824.	———	H. ♃.	Vill. delph. 2. t. 7. bis.
29	tenuifòlium. R.S.	(*wh.*)	slender-leaved.	———	1820.	6. 7.	H. ♃.	
30	rùbrum. S.S.	(*re.*)	red.	Italy.	1597.	———	H. ♃.	Ger. herb. 967. f. 3.
31	tenuíssimum. S.S.	(*br.*)	slenderest.	Caucasus.	1818.	7. 9.	H. ⊙.	
32	vèrum. S.S.	(*ye.*)	cheese-rennet.	Britain.	....	———	H. ♃.	Eng. bot. v. 10. t. 660.
33	mutábile. R.S.	(*st.*)	changeable-color'd.	Poland.	1828.	7. 10.	H. ♃.	
34	vèro-símile. R.S.	(*ye.*)	large yellow.	....	1825.	7. 9.	H. ♃.	
35	caucásicum. R.S.	(*ye.*)	Caucasean.	Caucasus.	1820.	———	H. ♃.	
36	tyrolénse. R.S.	(*wh.*)	Tyrolese.	Tyrol.	1801.	7.	H. ♃.	
37	Mollùgo. R.S.	(*wh.*)	great hedge.	Britain.	....	7. 8.	H. ♃.	Eng. bot. v. 24. t. 1673.
38	elàtum. P.S.	(*wh.*)	tall.	France.	1827.	———	H. ♃.	
39	campanulàtum. S.S.	(*wh.*)	bell-flowered.	Europe.	1710.	6. 9.	H. ♃.	Jacq. aust. 1. t. 81.
	glaùcum. Jacq. *non* L.							
40	eréctum. S.S.	(*wh.*)	upright.	Britain.	....	6. 7.	H. ♃.	Eng. bot. v. 29. t. 2067.
41	lùcidum. R.S.	(*wh.*)	glossy.	S. Europe.	1820.	———	H. ♃.	All. ped. 1. t. 77. f. 2.
42	linifòlium. S.S.	(*wh.*)	Flax-leaved.	———	1759.	———	H. ♃.	Barr. ic. t. 356 et 583.
43	rígidum. S.S.	(*wh.*)	rigid.	Azores.	1778.	———	H. ♃.	
44	sylváticum. S.S.	(*wh.*)	wood.	S. Europe.	1658.	7. 8.	H. ♃.	
45	cinèreum. S.S.	(*wh.*)	cinereous.	Scotland.	....	8.	H. ♃.	All. ped. t. 77. f. 4.
46	aristàtum. S.S.	(*wh.*)	awned.	———	....	7. 8.	H. ♃.	Barr. ic. t. 356.
47	purpùreum. S.S.	(*pu.*)	purple.	Switzerland.	1731.	6. 7.	H. ♃.	Noc. et B. fl. tic. 1. t. 3.
48	saxátile. S.S.	(*bh.*)	smooth heath.	Britain.	....	7. 8.	H. ♃.	Eng. bot. v. 12. t. 815.
49	helvéticum. S.S.	(*wh.*)	Swiss.	Switzerland.	1823.	———	H. ♃.	J. m. ac. p. 1714. t. 15. f. 1.
50	baldénse. S.S.	(*wh.*)	tufted.	Mount Baldo.	1825.	———	H. ♃.	
51	tricórne. E.B.	(*st.*)	three-flowered.	Britain.	....	7.	H. ⊙.	Eng. bot. v. 23. t. 1641.
52	spùrium. E.B.	(*gr.*)	spurious.	———	....	6. 7	H. ♃.	——— v. 26. t. 1871.
53	verrucòsum. E.B.	(*st.*)	warted.	———	....	6. 8.	H. ⊙.	——— v. 31. t. 2173.
	saccharàtum. R.S. *Valántia Aparìne.* W.							
54	Cucullària. S.S.	(*ye.*)	hooded.	Levant.	1780.	5. 6.	H. ⊙.	Buxb. cent. 1. t. 19. f. 2.
55	boreàle. E.B.	(*wh.*)	cross-leaved.	Britain.	....	7.	H. ♃.	Eng. bot. v. 2. t. 105.
56	septentrionàle. R.S.	(*wh.*)	northern.	N. America.	1827.	———	H. ♃.	
57	brachyphy'llum. R.S.	(*wh.*)	short-leaved.	Caucasus.	———	———	H. ♃.	
58	rotundifòlium. S.S.	(*wh.*)	round-leaved.	Europe.	....	6. 7.	H. ♃.	Jacq. aust. t. 94.
59	ellípticum. S.S.	(*wh.*)	elliptic-leaved.	Sicily.	1820.	———	H. ♃.	Barr. ic. t. 21. f. 4.
60	pilòsum. R.S.	(*wh.*)	hairy.	N. America.	1778.	———	H. ♃.	
61	suavèolens. R.S.	(*wh.*)	sweet-scented.	N. Europe.	1827.	———	H. ♃.	
62	refléxum. R.S.	(*wh.*)	reflexed.	Tauria.	1818.	———	H. ♃.	
63	humifùsum. R.S.	(*wh.*)	trailing.	———	1820.	———	H. ♃.	
64	marítimum. S.S.	(*pu.*)	sea-side.	S. Europe.	———	7. 8.	H. ♃.	
65	triflòrum. R.S.	(*wh.*)	three-flowered.	N. America.	1827.	———	H. ♃.	Willd. hort. ber. t. 66.
66	ánglicum. E.B.	(*st.*)	English.	England.	....	6. 7.	H. ⊙.	Eng. bot. v. 6. t. 384.
67	parvifòlium. R.S.	(*wh.*)	small-leaved.	Switzerland.	1825.	———	H. ⊙.	
68	parisiénse. R.S.	(*st.*)	Paris.	France.	1818.	———	H. ⊙.	
69	muràle. S.S.	(*st.*)	wall.	Italy.	1805.	6. 8.	H. ⊙.	All. ped. 34. t. 77. f. 1.
	Sherárdia muràlis. W.							
70	verticillàtum. R.S.	(*wh.*)	whorled.	France.	1829.	———	H. ⊙.	Flor. græc. t. 116.
71	gr'æcum. S.S.	(*pu.*)	Candian.	Candia.	1798.	6. 7.	H. ♃.	——— t. 136.
72	capillàre. R.S.	(*wh.*)	woolly-fruited.	Spain.	1828.	———	H. ⊙.	Cavan. icon. 2. t. 191. f. 1.
73	Aparìne. E.B.	(*wh.*)	Cleavers.	Britain.	....	5. 9.	H. ⊙.	Eng. bot. v. 12. t. 316.
74	inféstum. R.S.	(*st.*)	infesting.	Europe.	1818.	———	H. ⊙.	W. et K. hung. 3. t. 202.
75	scabérrimum. R.S.	(*st.*)	roughest.	Egypt.	———	6. 9.	H. ⊙.	

ORDO CXVI.

OPERCULARIEÆ. *Juss. ann. mus.* 4. *p.* 418.

OPERCUL'ARIA. G.		OPERCUL'ARIA.	Tetrandria Monogynia. W.				
1 umbellàta. G.	(*wh.*)	umbel-flower'd.	N. Holland.	1818.	6. 7.	G. ♃.	Gært. fr. 1. t. 24. f. 4.
2 Lamarckiàna. R.S.	(*wh.*)	Lamarck's.	N. S. W	1825.	7. 9.	H. ⊙.	Lam. ill. t. 58. f. 1.
3 áspera. S.S.	(*wh.*)	rough.	N. Zealand.	1790.	6. 7.	G. ♃.	Gært. fr. 1. t. 24. f. 4.
4 híspida. S.S.	(*wh.*)	hispid.	N. S. W.	——	——	G. ♃.	Juss. an. mus. 4. t. 70. f. 1.
áspera. J.							
5 sessiliflòra. R.S.	(*wh.*)	sessile-flow'r'd.	——	1816.	——	G. ♃.	Juss. an. mus 4. t. 70. f. 2.
6 ocymifòlia. R.S.	(*wh.*)	Basil-leaved.	——	1820.	——	G. ♃.	
CRYPTOSPE'RMUM. P.S.		CRYPTOSPE'RMUM.	Tetrandria Monogynia. P.S.				
Yoúngii. P.S.	(*gr.re.*)	Young's.	N. S. W.	1793.	7. 8.	G. ♃	Linn. trans. 3. t. 5.
Operculària paleàta. R.S.							

ORDO CXVII.

VALERIANEÆ. *DC. fl. fr. ed.* 3. *v.* 4. *p.* 418.

CENTRA'NTHUS. DC.		CENTRA'NTHUS.	Monandria Monogynia. R.S.				
1 rùber. DC.	(*re.*)	common.	England.	....	6. 10.	H. ♃.	Eng. bot. v. 22. t. 1531.
Valeriàna rùbra. E.B.							
α coccíneus.	(*sc.*)	*scarlet.*	——	....	——	H. ♃.	
β cárneus.	(*fl.*)	*flesh-coloured.*	——	....	——	H. ♃.	
γ álbus.	(*wh.*)	*white-flowered.*	——	....	——	H. ♃.	
2 intermèdius.	(*re.*)	spear-leaved.	Europe.	....	——	H. ♃.	
3 angustifòlius. DC.	(*re.*)	narrow-leaved.	——	1731.	5. 7.	H. ♃.	Flor. græc. 1. t. 29.
4 Calcítrapa. R.S.	(*pi.*)	cut-leaved.	Portugal.	1570.	——	H. ⊙.	—— 1. t. 30.
VALERI'ANA. DC.		VALERIAN.	Triandria Monogynia. L.				
1 dioíca. E.B.	(*bh.*)	diœcious.	Britain.	....	5. 7.	H. ♃.	Eng. bot. v. 9. t. 628.
2 Phù. R.S.	(*wh.*)	garden.	Germany.	1597.	——	H. ♃.	Plenck. t. 28.
3 officinàlis. E.B.	(*bh.*)	great wild.	Britain.	....	6. 7.	H. ♃.	Eng. bot. v. 10. t. 698.
4 tenuifòlia. Vahl.	(*bh.*)	linear-leaved.	Germany.	....	——	H. ♃.	Sturm. fl. germ. fas. 9.
5 sambucifòlia. R.S.	(*bh.*)	Elder-leaved.	Sweden.	1818.	——	H. ♃.	
6 cardámines. M.B.	(*bh.*)	Ladiessmock-l^d.	Caucasus.	1823.	6. 7.	H. ♃.	
7 sisymbriifòlia. R.S.	(*pi.*)	water-rocket-l^d.	Levant.	1820.	——	H. ♃.	
8 montàna. R.S.	(*bh.*)	mountain.	Switzerland.	1748.	——	H. ♃.	Lodd. bot. cab. 317.
9 trípteris. R.S.	(*bh.*)	three-leaved.	——	1752.	3. 5.	H. ♃.	Jacq. aust. 3. t. 268.
10 pyrenàica. E.B.	(*ro.*)	heart-leaved.	Scotland.	....	5. 6.	H. ♃.	Eng. bot. v. 23. t. 1591.
11 alliariæfòlia. R.S.	(*wh.*)	Alliaria-leaved.	Levant.	1825.	5. 8.	H. ♃.	Buxb. cent. 2. t. 11.
12 tuberòsa. R.S.	(*ro.*)	tuberous-root'd.	S. Europe.	1629.	——	H. ♃.	Lapeyr. pyr. t. 53.
13 elongàta. R.S.	(*st.*)	elongated.	Austria.	1812.	6. 7.	H. ♃.	Jacq. ic. 3. t. 219.
14 supìna. R.S.	(*ro.*)	creeping-root'd.	Germany.	1818.	——	H. ♃.	Jacq. misc. 2. t. 17. f. 2.
15 saxátilis. R.S.	(*wh.*)	rock.	——	1748.	7.	H. ♃.	Jacq. aust. 3. t. 267.
16 saliúnca. R.S.	(*bh.*)	headed-flower'd.	Italy.	1824.	6. 7.	H. ♃.	All. ped. 1. t. 70. f. 1.
17 céltica. R.S.	(*st.*)	raceme-flower'd.	Switzerland.	1748.	——	H ♃.	Jacq. coll. 1. t. 1.
18 capénsis. R.S.	(*bh.*)	Cape.	C. B. S.	1816.	——	F. ♃.	
PATRI'NIA. J.		PATRI'NIA.	Tetrandria Monogynia. R.S.				
1 sibírica. J.	(*ye.*)	Siberian.	Siberia.	1759.	5. 6.	H. ♂.	Bot. mag. t. 2325.
Valeriàna sibírica. L. *ruthènica.* B.M.							
2 coronàta. S.S.	(*ye.*)	crowned.	Altay.	1826.	5. 7.	H. ♂.	
3 rupéstris. J.	(*re.*)	rock.	Siberia.	1801.	——	H. ♂.	Bot. mag. t. 714.
Valeriàna sibírica. B.M.							
4 intermèdia. R.S.	(*ye.*)	intermediate.	——	1823.	——	H. ♂.	
5 nudiúscula. S.S.	(*ye.*)	naked-stemm'd.	Altay.	1826.	6. 8.	H. ♂.	
6 scabiosæfòlia. S.S.	(*ye.*)	Scabious-leav'd.	Dahuria.	——	——	H. ♃.	Swt. br. fl. gar. t. 154.
7 serratulæfòlia. F.	(*ye.*)	Serratula-leav'd.	——	1826.	——	H. ♃.	
F'EDIA. DC.		F'EDIA.	Diandria Monogynia.				
Cornucòpiæ. R.S.	(*pu.*)	purple.	S. Europe.	1596.	5. 8.	H. ⊙.	Bot. reg. t. 155.
Valeriàna cornucòpiæ. B.R.							

VALERIANE'LLA. DC. Valeriane'lla. Triandria Monogynia.

1	echinàta. DC.	(*li.*)	prickly-capsul'd.	S. Europe.	1807.	7. 8.	H. ⊙.	Reich. ic. C. 1. t. 68.
2	olitòria. DC.	(*bh.*)	Lamb's Lettuce.	Britain.	....	4. 8.	H. ⊙.	Eng. bot. v. 12. t. 811.
	Valeriàna locústa. E.B. *Fèdia olitòria.* R.S. *Fèdia locústa.* Reich. ic. t. 60.							
3	carinàta. DC.	(*bh.*)	keeled.	France.	....	——	H. ⊙.	Reich. ic. C. 1. t. 61.
4	dentàta. DC.	(*bh.*)	oval-fruited.	Britain.	....	——	H. ⊙.	Eng. bot. v. 20. t. 1370.
	Valeriàna dentàta. E.B. *Fèdia dentàta.* Reich. ic. 1. t. 62.							
5	discoídea. DF.	(*bh.*)	discoid.	Italy.	1731.	4. 7.	H. ⊙.	Reich. ic. C. 2. t. 116.
6	rotàta. R.S.	(*bh.*)	rotate.	————	1828.	——	H. ⊙.	———— 1. t. 67.
7	vesicària. DF.	(*bh.*)	bladdered.	Candia.	1739.	——	H. ⊙.	Flor. græc. 1. t. 34.
8	hamàta. DC.	(*bh.*)	hooked.	France.	1825.	——	H. ⊙.	
9	coronàta. DC.	(*bh.*)	prickly seeded.	Portugal.	1731.	——	H. ⊙.	Dufres. val. t. 3. f. 2.
10	radiàta. DF.	(*bh.*)	rayed.	N. America.	1818.	——	H. ⊙.	———— t. 3. n. 7.
11	aurícula. DC.	(*bh.*)	eared.	Montpelier.	——	——	H. ⊙.	Reich. ic. C. 1. t. 63.
12	pùmila. DC.	(*re.*)	small.	S. Europe.	1823.	——	H. ⊙.	———— 2. t. 113.
13	míxta. DF.	(*bh.*)	mixed.	————	1828.	——	H. ⊙.	Dufr. val. t. 3. n. 6.
14	campanulàta. Biv.	(*bh.*)	bell-flowered.	Italy.	1829.	——	H. ⊙.	
15	eriocárpa. DC.	(*bh.*)	woolly-capsul'd.	France.	1823.	——	H. ⊙.	Reich. ic. C. 1. t. 65.
16	truncàta. S.S.	(*bh.*)	truncate.	Crete.	1829.	——	H. ⊙.	———— 2. t. 115.
17	microcárpa. S.S.	(*bh.*)	small-capsuled.	S. Europe.	1828.	——	H. ⊙.	———— 2. t. 114.
18	chenopodiifòlia. R.S.	(*bh.*)	Goosefoot-l[d].	N. America.	....	——	H. ⊙.	
19	uncinàta. DF.	(*bh.*)	uncinate.	Tauria.	1818.	——	H. ⊙.	Reich. ic. C. 1. t. 69.
20	dasycárpa. L.en.	(*bh.*)	thick-capsuled.	————	——	——	H. ⊙.	
21	muricàta. M.B.	(*bh.*)	muricate.	Caucasus.	1828.	——	H. ⊙.	
22	tridentàta. L.en.	(*bh.*)	three-toothed.	Tauria.	1823.	——	H. ⊙.	Reich. ic. C. 1. t. 64.
23	angulòsa. DC.	(*bh.*)	angular.	....	1826.	——	H. ⊙.	
24	congésta. B.R.	(*ro.*)	close-headed.	N. America.	——	5. 6.	H. ⊙.	Bot. reg. t. 1094.

ORDO CXVIII.

DIPSACEÆ. *Juss. gen.* 194.

MOR'INA. R.S. Mor'ina. Diandria Monogynia. L.

	pérsica. R.S.	(*re.*)	Persian.	Persia.	1740.	7. 8.	G. ♃.	Flor. græc. 1. t. 28.

DI'PSACUS. S.S. Teasel. Tetrandria Monogynia. L.

1	fullònum. S.S.	(*li.*)	manured.	Britain.	....	7.	H. ♂.	Eng. bot. v. 29. t. 2080.
2	sylvéstris. S.S.	(*li.*)	wild.	————	....	——	H. ♂.	———— v. 15. t. 1032.
3	laciniàtus. S.S.	(*li.*)	cut-leaved.	Germany.	1683.	7. 8.	H. ♂.	Jacq. aust. 5. t. 403.
4	divaricàtus. S.	(*li.*)	spreading-leav'd.	Sicily.	1827.	——	H. ♂.	
5	Gmelìni. S.S.	(*bl.*)	Gmelin's.	Caucasus.	1820.	——	H. ♂.	Jacq. f. ecl. t. 106.
6	fèrox. S.S.	(*bh.*)	fierce.	Corsica.	——	——	H. ♂.	———— t. 107.
7	pilòsus. S.S.	(*wh.*)	small.	Britain.	....	8.	H. ♂.	Eng. bot. v. 13. t. 877.
8	inérmis. F.I.	(*st.*)	spineless.	Nepaul.	1820.	8. 10.	H. ♂.	

SCABI'OSA. S.S. Scabious. Tetrandria Monogynia. L.

1	arvénsis. S.S.	(*li.*)	field.	Britain.	....	7. 10.	H. ♃.	Eng. bot. v. 10. t. 659.
	Trichèra arvénsis. R.S.							
2	campéstris. Bes.	(*li.*)	pasture.	Poland.	1824.	——	H. ♃.	
3	sylvática. S.S.	(*re.*)	broad-leaved.	Austria.	1633.	7.	H. ♃.	Jacq. aust. 4. t. 362.
4	montàna. S.S.	(*bl.*)	mountain.	Caucasus.	1823.	——	H. ♃.	Jacq. f. eclog. t. 60.
5	longifòlia. S.S.	(*li.*)	long-leaved.	Hungary.	1802.	7. 8.	H. ♃.	W. et K. hung. 1. t. 5.
6	pubéscens. S.S.	(*st.*)	pubescent.	————	1820.	——	H. ♃.	
7	revérsa. S.S.	(*li.*)	reversed.	Siberia.	1829.	——	H. ♃.	
8	ciliàta. S.S.	(*st.*)	ciliated.	Hungary.	1820.	——	H. ♃.	
9	hy'brida. S.S.	(*li.*)	hybrid.	S. Europe.	——	6. 9.	H. ⊙.	
10	integrifòlia. S.S.	(*pu.*)	entire-leaved.	————	1748.	6. 8.	H. ⊙.	
11	Hornemánni. S.S.	(*li.*)	Horneman's.	....	1825.	——	H. ♃.	
12	austràlis. S.S.	(*ye.*)	creeping-root'd.	Italy.	1823.	——	H. ♃.	

SUCC'ISA. S.S. Succ'isa. Tetrandria Monogynia. S.S.

1	praténsis. S.S.	(*bl.*)	Devil's-bit.	Britain.	....	8. 10.	H. ♃.	Eng. bot. v. 13. t. 878.
	Scabiòsa Succìsa. L.							
2	glabràta.	(*bl.*)	smooth-leaved.	Austria.	1823.	7. 9.	H. ♃.	
	Scabiòsa glabràta. R.S.							
3	syrìaca. S.S.	(*wh.*)	Syrian.	Syria.	1633.	7.	H. ♃.	Flor. græc. t. 105.
4	Vaillánti. R.S.	(*bl.*)	Vaillant's.	Levant.	1826.	——	H. ⊙.	
5	transylvànica. S.S.	(*li.*)	Transylvanian.	Hungary.	1699.	——	H. ⊙.	Jacq. vind. 2. t. 111.

6 jóppica. s.s. (*st.*) Joppa. Joppa. 1828. 7. H. ⊙. Reich. H. bot. t. 15.
7 uralénsis. s.s. (*st.*) Uralian. Siberia. 1789. 7. 8. H. ⊙. Murr. c. got. 1782. t. 4.
8 tatárica. s.s. (*ye.*) Tartarian. Russia. 1759. 6. 8. H. ♃. Lin. act. ups. 1744. t. 1.
9 alpìna. s.s. (*st.*) Alpine. Switzerland. 1570. 6. 7. H. ♃. Moris. s. 6. t. 13. f. 10.
Cephalària alpìna. R.S. *Scabiòsa alpìna.* W.
10 leucántha. s.s. (*wh.*) white flowered. S. Europe. 1633. 9. 10. H. ♃. Ger. emac. f. 8.
11 centauroídes. s.s. (*st.*) Centaury-like. ——— 1816. 7. 8. H. ♃.
12 lævigàta. s.s. (*st.*) smooth. Hungary. 1823. —— H. ♃. W. et K. hung. 3. t. 230.
13 corniculàta. (*st.*) horned. ——— 1801. —— H. ♃. ——— 1. t. 13.
Cephalària corniculàta. R.S. *Scabiòsa corniculàta.* W.K.
14 cretàcea. M.B. (*st.*) leathery-leaved. Tauria. 1828. —— H. ♃.
15 rígida. s.s. (*wh.*) rough-leaved. C. B. S. 1731. 7. G. ♄.
16 attenuàta. (*wh.*) narrów-leaved. ——— 1774. 7. 9. G. ♄.
trífida. s.s. *Scabiòsa attenuàta.* L.

ASTEROCE'PHALUS. S.S. ASTEROCE'PHALUS. Tetrandria Monogynia. S.S.

1 prolífer. s.s. (*st.*) prolific. Egypt. 1683. 7. 9. H. ⊙. Flor. græc. t. 107.
Scabiòsa prolífera. L.
2 graminifòlius. s.s. (*bl.*) grass-leaved. Switzerland. —— 6. H. ♃. Bot. reg. t. 835.
Scabiòsa graminifòlia. B.R.
3 créticus. s.s. (*li.*) Cretan. Crete. 1596. 6. 10. G. ♄. Moris. s. 6. t. 15. f. 31.
4 limoniifòlius. s.s. (*li.*) Limonium-leav'd. Sicily. .... 6. 8. H. ♄.
5 Massòni. (*li.*) Masson's. Azores. 1779. —— H. ♃.
lùcidus. s.s. *Scabiòsa lùcida.* H.K. *Massòni.* H.S.L. *nítens.* R.S.
6 dichótomus. s.s. (*re.*) forked. Sicily. 1804. 6. 8. H. ⊙. Bocc. mus. t. 120.
7 africànus. s.s. (*wh.*) African. Africa. 1690. 7. 10. G. ♄. Herm. parad. t. 219.
8 palæstìnus. s.s. (*st.*) Palestine. Palestine. 1771. 7. 8. H. ⊙. Jacq. vind. 1. t. 96.
9 lùcidus. (*li.*) shining. France. 1800. 6. 9. H. ♃.
Scabiòsa lùcida. R.S.
10 nòricus. R.S. (*li.*) simple-stalked. Caucasus. 1822. —— H. ♃.
11 stríctus. s.s. (*li.*) upright. Hungary. 1827. —— H. ♃. W. et K. hung. 2. t. 138.
12 silenifòlius. s.s. (*li.*) Silene-leaved. ——— 1823. 7. 8. H. ♃. ——— 2. t. 157.
13 canéscens. s.s. (*li.*) canescent. ——— 1802. —— H. ♃. ——— 1. t. 53.
14 suavèolens. s. (*li.*) sweet-scented. France. 1827. —— H. ♃.
15 setífer. s.s. (*wh.*) setiferous. ——— 1820. —— H. ♃.
16 lyràtus. s.s. (*li.*) lyrate-leaved. Turkey. 1799. —— G. ♂.
17 micránthus. s.s. (*ro.*) small-flowered. Armenia. 1823. —— H. ♃. Desf. an. mus. 11. t. 25.
18 Biebersteínii. (*li.*) Bieberstein's. Caucasus. —— —— H. ⊙.
Scabiòsa Biebersteínii. R.S. *sícula.* M.B.
19 sículus. s.s. (*li.*) Sicilian. Sicily. 1783. —— H. ⊙. Jacq. vind. 1. t. 15.
20 am'œnus. s.s. (*ro.*) glossy. .... 1823. —— H. ♃.
21 stellàtus. s.s. (*li.*) starry. Spain. 1596. —— H. ⊙. Moris. s. 6. t. 15. f. 39.
22 ceratophy'llus. R.S. (*li.*) stag's-horn-leav'd. Naples. 1825. —— H. ♂. Ten. fl. nap. 1. t. 8.
23 rotàtus. s.s. (*li.*) rotate. Iberia. 1827. —— H. ⊙.
24 diffùsus. s.s. (*li.*) diffuse. Teneriffe. 1825. 6. 8. F. ♃. Reich. hort. t. 16.
25 atropurpùreus s.s. (*pu.*) sweet. E. Indies. 1629. 7. 9. H. ♂. Bot. mag. t. 247.
Scabiòsa atropurpùrea. B.M.
β cárneus. (*fl.*) *flesh-coloured.* ——— —— —— H. ♂.
γ ròseus. (*ro.*) *rose-coloured.* ——— —— —— H. ♂.
δ variegàtus. (*v.*) *variegated.* ——— —— —— H. ♂.
ε álbus. (*wh.*) *white-flowered.* ——— —— —— H. ♂.
ζ prolífer. (*d.pu.*) *large-flowered.* ——— 1814. —— H. ♂. Moris. s. 6. t. 14. f. 2.
26 caucásicus. s.s. (*wh.*) Caucasean. Caucasus. 1803. 7. 8. H. ♃. Jacq. fragm. t. 125. f. 1.
27 connàtus. H.H. (*bh.*) connate-leaved. ——— 1823. —— H. ♃.
28 élegans. R.S. (*vi.*) elegant. ——— 1803. 7. 8. H. ♃. Bot. mag. t. 886.
Scabiòsa caucásica. B.M. *élegans.* R.S.
29 holoseríceus. s.s. (*bl.*) silky. Pyrenees. —— —— H. ♃.
30 pyrenàicus. R.S. (*li.*) Pyrenean. ——— 1823. —— H. ♃.
31 tomentòsus. s.s. (*vi.*) woolly. Spain. 1826. —— H. ♃. Cavan. icon. 2. t. 183.
32 Webbiànus. s.s. (*wh.*) Webb's. Levant. 1820. —— H. ♃. Bot. reg. t. 717.
Scabiòsa Webbiàna. B.R.
33 crenàtus. s.s. (*fl.*) crenated. Sicily. 1824. 7. 8. H. ♃. Cyril. pl. rar. 1. t. 3.
34 rutæfòlius. R.S. (*fl.*) Rue-leaved. ——— 1804. 6. 8. H. ♃. Bocc. sic. t. 40. f. 3.
35 urceolàtus. s.s. (*st.*) jagged. Barbary. —— 7. 8. H. ♃. ——— t. 52.
36 argénteus. s.s. (*wh.li.*) silvery. Levant. 1713. 6. 10. H. ♃. Flor. græc. t. 108.
37 marítimus. s.s. (*li.*) sea-side. S. Enrope. 1818. 7. 9. H. ⊙. Bauh. hist. 3. p. 7. f. 2.
38 ucrànicus. s.s. (*wh.*) Ukraine. Ukraine. 1795. 9. H. ♃. Gmel. sib. 2. t. 87.
39 Columbària. s.s. (*li.*) fine-leaved. Britain. .... 7. 8 H. ♃. Eng. bot. v. 19. t. 1311.
40 grandiflòrus. R.S. (*st.*) great-flowered. Barbary. 1804. 6. 9. H. ♃. Scop. del. ins 3. t. 14.
41 móllis. L.en. (*li.*) soft. .... 1827. —— H. ♃.
Scabiòsa capillàta. R.S.

42	altíssimus. s.s.	(*li.*)	tall.	C. B. S.	1816.	6. 9.	G. ♄.	Jacq. vind. 2. t. 185.
43	mollíssimus. s.s.	(*bl.*)	soft.	S. Europe.	——	7. 8.	H. ♃.	
44	pilòsus. s.s.	(*bh.*)	hairy.	Spain.	1826.	——	H. ♃.	
45	ochroleùcus. s.s.	(*st.*)	pale-flowered.	Germany.	1597.	——	H. ♃.	Jacq. aust. 5. t. 439.
46	gramúntius.s.s.	(*wh.li.*)	cut-leaved.	S. Europe.	——	——	H. ♃.	Gerard herb. 582. f. 2.
47	iseténsis. s.s.	(*wh.li.*)	Siberian.	Siberia.	1801.	——	H. ♃.	Gmel. sib. 2. t. 88. f. 1.
48	bannáticus. s.s.	(*fl.*)	Hungarian.	Hungary.	1800.	——	H. ♃.	W. et K. hung. 1. t. 12.
49	Colúmnæ. Ten.	(*li.*)	Columna's.	Italy.	1823.	——	H. ♃.	Ten. fl. nap. 1. t. 7.
50	Scopòlii. R.S.	(*st.*)	Scopoli's.	S. Europe.	1822.	——	H. ♃.	
51	comòsus. R.S.	(*vi.*)	tufted.	Dahuria.	1828.	——	H. ♃.	
52	agréstis. S.S.	(*li.*)	field.	Hungary.	——	——	H. ♃.	W. et K. hung. 3. t. 204.
53	símplex. s.s.	(*wh.*)	simple-stalked.	Barbary.	....	——	H. ♃.	Desf. atl. 1. t. 39. f. 1.
54	incànus. R.S.	(*li.*)	hoary.	....	——	——	H. ♃.	
55	dahùricus. F.	(*li.*)	Dahurian.	Dahuria.	1824.	——	H. ♃.	

PTEROCE'PHALUS. S.S. PTEROCE'PHALUS. Tetrandria Monogynia. S.S.

1	plumòsus. s.s.	(*li.*)	feathered.	Greece.	1824.	7. 8.	H. ⊙.	Flor. græc. 1. t. 3.
	Scabiòsa plumòsa. Sm. *Knaùtia plumòsa.* L.							
2	Vailllántii. s.s.	(*wh.*)	Vaillant's.	S. Europe.	1739.	——	H. ⊙.	
	Scabiòsa pappòsa. L.							

KNA'UTIA. R.S. KNA'UTIA. Tetrandria Monogynia. L.

1	orientàlis. R.S.	(*re.*)	red-flowered.	Levant.	1713.	6. 9.	H. ⊙.	Schk. handb. 1. t. 22.
2	propóntica. R.S.	(*pu.*)	purple-flowered.	————	1768.	——	H. ♂.	Lam. ill. t. 58.

ORDO CXIX.

CALYCEREÆ. *Brown in Linn. trans. v.* 12. *p.* 132.

ACICA'RPHA. L.T. ACICA'RPHA. Syngenesia Necessaria. P.S.

spathulàta. L.T.	(*ye.*)	spathulate-leav'd.	Brazil.	1816.	6. 9.	G. ♃.	

ORDO CXX.

COMPOSITÆ. *Adanson fam.* 2. 103.

Subordo I. *CICHORACEÆ.* Kth. synops. 2. p. 352.

Tribus I. HIERACEÆ. *D. Don in Edinb. philos. journ.*

HIER'ACIUM. W. HAWKWEED. Syngenesia Polygamia Æqualis.

1	alpéstre. W.	(*ye.*)	smooth-leaved.	Austria.	1823.	5. 7.	H. ♃.	Jacq. aust. 2. t. 191.
2	rupéstre. W.	(*ye.*)	rock.	Switzerland.	——	——	H. ♃.	All. auct. t. 1. f. 2.
3	alpìnum. W.	(*ye.*)	Alpine.	Britain.	....	7. 8.	H. ♃.	Eng. bot. v. 16. t. 1110.
4	pùmilum. W.	(*ye.*)	small.	Germany.	1804.	——	H. ♃.	
5	glabràtum. W.	(*ye.*)	smooth.	Switzerland.	1823.	6. 8.	H. ♃.	
6	bulbòsum. W.	(*ye.*)	bulbous-rooted.	S. Europe.	——	——	H. ♃.	
7	Pilosélla. W.	(*st.*)	Mouse-ear.	Britain.	....	5. 7.	H. ♃.	Eng. bot. v. 16. t. 1093.
8	dùbium. W.	(*st.*)	branching.	————	....	7. 8.	H. ♃.	———— v. 33. t. 2332.
9	Auricula. W.	(*go.*)	umbell'd mouse-ear.	England.	....	——	H. ♃.	———— v. 33. t. 2368.
10	fállax. W.en.	(*go.*)	hairy spear-leav'd.	......	1816.	——	H. ♃.	
11	angustifòlium. W.	(*ye.*)	narrow-leaved.	Mt. Cenis.	1822.	——	H. ♃.	
12	florentìnum. W.	(*ye.*)	Florentine.	Germany.	1796.	——	H. ♃.	Bauhin pin. t. 67.
13	præáltum. s.s.	(*ye.*)	tall.	Europe.	1824.	——	H. ♃.	
14	cymòsum. W.	(*ye.*)	small-flowered.	————	1739.	5. 6.	H. ♃.	Moris. s. 7. t. 8. f. 10.
15	piloselloídes. P.S.	(*ye.*)	Mouse-ear like.	France.	1823.	5. 7.	H. ♃.	Vill. delph. 3. p. 100. t. 27.
16	staticifòlium. W.	(*ye.*)	thrift-leaved.	Europe.	1804.	6. 7.	H. ♃.	———— 3. p. 116. t. 27.
17	præmórsum. W.	(*ye.*)	præmorse.	Siberia.	1820.	——	H. ♃.	Gmel. sib. 2. t. 13.
18	succisæfòlium. P.S.	(*ye.*)	Succisa-leaved.	Piedmont.	1822.	——	H. ♃.	
19	flagellàre. W.en.	(*ye.*)	creeping.	....	1816.	5. 7.	H. ♃.	
20	bifúrcum. M.B.	(*ye.*)	two-forked.	S. Europe.	1822.	——	H. ♃.	
21	bífidum. W.en.	(*ye.*)	bifid.	Hungary.	——	6. 7.	H. ♃.	

22	incarnàtum. W.	(*fl.*)	flesh-coloured.	Carniola.	1815.	6. 7.	H. ♃.	Jacq. ic. t. 578.
23	aurantìacum. W.	(*or.*)	orange-colour'd.	Scotland.	....	——	H. ♃.	Eng. bot. v. 21. t. 1469.
24	venòsum. W.	(*ye.*)	red-veined.	N. America.	1790.	——	H. ♃.	
25	Gronòvii. W.	(*ye.*)	Gronovius's.	————	1798.	——	H. ♃.	
26	Gmelìni. W.	(*ye.*)	Gmelin's.	Siberia.	——	——	H. ♃.	Gmel. sib. 2. t. 8. f. 2.
27	Bauhìni. S.S.	(*ye.*)	Bauhin's.	Switzerland.	1820.	——	H. ♃.	
28	Lawsòni. W.	(*st.*)	Lawson's.	Britain.	....	——	H. ♃.	Eng. bot. v. 29. t. 2083.
29	collìnum. W.	(*ye.*)	hill.	Europe.	1824.	6. 8.	H. ♃.	
30	Gochnàti. S.S.	(*ye.*)	Gochnat's.	————	——	——	H. ♃.	
31	paniculàtum. W.	(*ye.*)	panicled.	Canada.	1800.	——	H. ♃.	
32	chondrilloídes. W.	(*ye.*)	Gum-succory.	Austria.	1640.	——	H. ♃.	Jacq. aust. 5. t. 429.
33	hùmile. W.	(*ye.*)	low.	S. Europe.	1816.	——	H. ♃.	———— t. 189.
	pùmilum. Jacq. *Jacquìni.* Vill. delph. 3. t. 28.							
34	calcàreum. H.H.	(*ye.*)	spurred.	S. Europe.	1823.	7. 8.	H. ♃.	
35	sonchifòlium. M.B.	(*ye.*)	Sonchus-leaved.	Caucasus.	1828.	——	H. ♃.	
36	píctum. P.S.	(*ye.*)	painted-leaved.	Switzerland.	1823.	——	H. ♃.	
37	nigréscens. W.	(*ye.*)	dark-coloured.	....	1801.	7. 8.	H. ♃.	Willd. hort. ber. t. 10.
38	prunellæfòlium. W.	(*ye.*)	Prunella-leav'd.	Switzerland.	1823.	6. 7.	H. ♃.	All. ped. t. 15. f. 2.
39	montànum. W.	(*ye.*)	mountain.	S. Europe.	1775.	——	H. ♃.	Jacq. aust. 2. t. 190.
40	hy'bridum. S.S.	(*ye.*)	hybrid.	Switzerland.	1825.	——	H. ♃.	
41	porrifòlium. W.	(*ye.*)	Leek-leaved.	Austria.	1640.	7. 8.	H. ♃.	Jacq. aust. 3. t. 286.
42	saxátile. W.	(*ye.*)	rock.	————	1801.	——	H. ♃.	Jacq. ic. 1. t. 163.
43	glaùcum. W.	(*ye.*)	glaucous.	S. Europe.	1807.	6. 7.	H. ♃.	All. ped. t. 81. f. 1.
44	mólle. W.	(*go.*)	soft-leaved.	Scotland.	....	7. 8.	H. ♃.	Eng. bot. v. 31. t. 2210.
45	croáticum. W.K.	(*ye.*)	Hungarian.	Hungary.	1820.	——	H. ♃.	
46	Sternbérgii. L.en.	(*ye.*)	Sternberg's.	————	——	——	H. ♃.	
47	palléscens. S.S.	(*st.*)	pale.	————	1815.	6. 7.	H. ♃.	W. et K. hung. 3. t. 217.
48	muròrum. E.B.	(*go.*)	wall.	Britain.	....	7.	H. ♃.	Eng. bot. v. 29. t. 2082.
49	maculàtum. E.B.	(*ye.*)	stained-leaved.	England.	....	7. 8.	H. ♃.	———— v. 30. t. 2121.
50	ramòsum. W.	(*ye.*)	branching.	Hungary.	1805.	——	H. ♃.	
51	incísum. S.S.	(*ye.*)	cut-leaved.	————	1823.	——	H. ♃.	
52	parvifòlium. W.	(*ye.*)	small-leaved.	Austria.	——	——	H. ♃.	
53	cydoniæfòlium. S.S.	(*ye.*)	Quince-leaved.	France.	1816.	——	H. ♃.	
54	denudàtum. S.S.	(*ye.*)	half-naked.	Pyrenees.	1828.	——	H. ♃.	
55	sylváticum. E.B.	(*ye.*)	wood.	Britain.	....	8.	H. ♃.	Eng. bot. v. 29. t. 2031.
56	pulmonàrium. E.B.	(*go*)	lungwort.	Scotland.	....	7. 8.	H. ♃.	———— v. 33. t. 2307.
57	lapsanoídes. W.	(*ye.*)	Lapsana-like.	Pyrenees.	1812.	——	H. ♃.	Gouan. ill. t. 21. f. 3.
58	lyràtum. W.	(*ye.*)	lyre-leaved.	Siberia.	1777.	——	H. ♃.	Gmel. sib. 2. t. 9.
59	cerinthoídes. E.B.	(*ye.*)	honeywort.	Scotland.	....	8.	H. ♃.	Eng. bot. v. 34. t. 2378.
60	flexuòsum. W.	(*ye.*)	bending-stalk'd.	Hungary.	1804.	7. 8.	H. ♃.	
61	amplexicaùle. W.	(*ye.*)	heart-leaved.	Pyrenees.	1739.	——	H. ♃.	All. ped. 1. t. 15. f. 1.
62	grandiflòrum. W.	(*ye.*)	great-flowered.	Europe.	1791.	——	H. ♃.	W. et K. hung. 1. t. 99.
63	intybàceum. W.	(*ye.*)	endive-leaved.	————	1794.	——	H. ♃.	Jacq. aust. ap. t. 43.
64	Hallèri. W.	(*ye.*)	Haller's.	————	1802.	——	H. ♃.	Vill. dauph. 3. t. 26.
65	eriophy'llum. L.en.	(*ye.*)	villous.	————	1816.	——	H. ♃.	
66	villòsum. E.B.	(*st.*)	shaggy alpine.	Scotland.	....	——	H. ♃.	Eng. bot. v. 34. t. 2379.
67	Schradèri. DC.	(*ye.*)	Schrader's.	Switzerland.	1823.	——	H. ♃.	
68	elongàtum. Lap.	(*ye.*)	elongated.	————	1824.	——	H. ♃.	
69	pilocéphalum. L.en.	(*ye.*)	hairy-headed.	Europe.	1816.	——	H. ♃.	
70	trichocéphalum. W.	(*ye.*)	bristly-headed.	————	1823.	——	H. ♃.	
71	prostràtum. S.S.	(*ye.*)	prostrate.	S. Europe.	——	——	H. ♃.	
72	calyculàtum. H.H.	(*ye.*)	long-scaled.	————	1826.	——	H. ♃.	
73	longifòlium. H.H.	(*ye.*)	long-leaved.	....	1823.	——	H. ♃.	
74	speciosíssimum. S.S.	(*ye.*)	handsome.	S. Europe.	——	——	H. ♃.	
	speciòsum. H.H.							
75	ciliàtum. W.	(*ye.*)	fringed.	Crete.	1823.	——	H. ♃.	
76	Millèri. L.en.	(*ye.*)	Miller's.	....	....	——	H. ♃.	
77	lanàtum. W.	(*ye.*)	woolly.	Hungary.	——	——	H. ♃.	W. et K. hung. 2. t. 127.
78	andryaloídes. P.S.	(*ye.*)	Andryala-like.	France.	1818.	——	H. ♃.	Vill. delph. 3. t. 29.
79	echioídes. W.	(*ye.*)	viper's-bugloss.	Hungary.	1802.	——	H. ♃.	W. et K. hung. 1. t. 85.
80	verruculàtum. L.en.	(*ye.*)	warted.	....	1823.	——	H. ♃.	
81	undulàtum. W.	(*ye.*)	wave-leaved.	Spain.	1778.	——	H. ♃.	
82	glutinòsum. W.	(*ye.*)	clammy.	S. Europe.	1796.	——	H. ⊙.	
83	Kalmii. W.	(*ye.*)	Kalm's.	Pensylvania.	1794.	8.	H. ♃.	
84	fasciculàtum. Ph.	(*ye.*)	bundled.	Canada.	....	7. 8.	H. ♃.	
85	racemòsum. W.	(*ye.*)	racemed.	Hungary.	1818.	——	H. ♃.	W. et K. hung. 2. t. 193.
86	foliòsum. W.	(*ye.*)	leafy.	————	1805.	——	H. ♃.	———— 2. t. 145.
87	corymbòsum. P.S.	(*ye.*)	corymbose.	....	1820.	——	H. ♃	
88	sabaùdum. W.	(*go.*)	frutescent.	Britain.	....	——	H. ♃.	Eng. bot. v. 5. t. 349.

89 umbellàtum. w. (*ye.*) umbel-flowered. Britain. .... 8. 9. H. ♃. Eng. bot. v. 25. t. 1771.
90 prenanthoides. E.B. (*ye.*) rough-bordered. Scotland. .... 6. 9. H. ♃. ——— v. 32. t. 2235.
91 latifòlium. s.s. (*ye.*) broad-leaved. Pyrenees. 1823. —— H. ♃.
92 denticulàtum. E.B. (*ye.*) small-toothed. Scotland. .... 7. 8. H. ♃. Eng. bot. v. 30. t. 2122.
93 lævigàtum. w. (*ye.*) smooth. .... 1804. 8. 9. H. ♃. Willd. hort. ber. t. 16.
94 canéscens. L.en. (*ye.*) canescent. .... 1820. 7. 8. H. ♃.
95 auriculàtum. s.s. (*ye.*) ear-leaved. .... 1826. —— H. ♃.
96 boreàle. s.s. (*ye.*) northern. N. Europe. 1828. —— H. ♃.
97 bracteolàtum. L.en. (*ye.*) bracted. S. Europe. 1823. —— H. ♃.
98 coronopifòlium. L.en. (*ye.*) Buckshorn-l^d^. .... —— —— H. ♃.
99 parviflòrum. Schl. (*ye.*) small-flowered. Switzerland. —— —— H. ♃.
100 ochroleùcum. Schl. (*st.*) sulphur-color'd. ——— 1821. —— H. ♃.
101 ovàtum. Schl. (*ye.*) oval-leaved. ——— 1820. —— H. ♃.
102 compòsitum. s.s. (*ye.*) compound. Pyrenees. 1823. —— H. ♃.
103 crassifòlium. s. (*ye.*) thick-leaved. .... —— —— H. ♃.
104 rotundàtum. s.s. (*ye.*) round-leaved. Hungary. —— —— H. ♃.
105 sudéticum. St. (*ye.*) Swedish. Sweden. —— —— H. ♃.
106 virgàtum. Ph. (*ye.*) twiggy. N. America. 1816. —— H. ♃.
107 viréscens. Schl. (*gr.*) greenish. Switzerland. 1820. —— H. ♃.
108 stoloníferum. Viv. (*ye.*) stoloniferous. Levant. 1823. —— H. ♃.
109 verbascifòlium. P.S. (*ye.*) Mullein-leav'd. S. Europe. 1732. 5. 6. H. ♃. Mill. ic. 1. t. 146. f. 1.
Andr'yala lanàta. s.s.

HAPALOST'EPHIUM. D.P. HAPALOST'EPHIUM. Syngenesia Polygamia Æqualis.
1 paludòsum. D.P. (*ye.*) Succory-leaved. Britain. .... 7. 8. H. ♃. Eng. bot. t. 1094.
Hieràcium paludòsum. E.B.
2 pyrenàicum. D.P. (*ye.*) Pyrenean. Pyrenees. 1723. —— H. ♃.
Hieràcium blattarioìdes. P.S.
3 pilòsum. D.P. (*ye.*) hairy. ——— —— —— H. ♃. Herm. parad. t. 184.
4 austriàcum. (*ye.*) Austrian. Austria. —— —— H. ♃. Jacq. aust. t. 441.
5 sibíricum. D.P. (*ye.*) Siberian. Siberia. 1755. —— H. ♃. Gmel. sib. 2. t. 10.
Crèpis sibírica. L. *Hieràcium sibíricum.* W.
6 grandiflòrum. (*ye.*) large-flowered. Idria. 1824. 6. 8. H. ♃.
Crèpis grandiflòra. L.

LAGO'SERIS. L.en. LAGO'SERIS. Syngenesia Polygamia Æqualis.
1 bursifòlia. L.en. (*ye.*) pinnatifid. Sicily. 1823. 7. 8. H. ♃. Bocc. m. 2. t. 106 et 112.
Crèpis bursifòlia. L.
2 versícolor. L.en. (*ye.*) various-color'd. Dauria. —— —— H. ♃.
3 cérnua. L.en. (*ye.*) nodding. Calabria. —— 6. 8. H. ⊙.
4 nemausénsis. (*ye.*) Palestine. S. Europe. 1794. 6. 7. H. ⊙. Vill. delph. 3. t. 26.
Crepis nemausénsis. W.
5 leontodontoìdes. L.en. (*ye.*) Dandelion-like. Italy. 1804. 7. 8. H. ♂.
6 raphanifòlia. L.en. (*ye.*) Radish-leaved. .... 1816. 6. 7. H. ♃.
7 taurinénsis. L.en (*ye.*) bladdered. S. Europe. 1823. 5. 7. H. ⊙.
8 intybàcea. L.en. (*ye.*) Succory-leaved. Portugal. 1816. 6. 7. H. ♂.

CR'EPIS. S.S. CR'EPIS. Syngenesia Polygamia Æqualis. L.
1 nudicaùlis. s.s. (*ye.*) naked-stalked. Crete. 1828. 6. 8. H. ♃.
2 rìgens. s.s. (*ye.*) bristle-leaved. Azores. 1778. 7. 8. F. ♃.
3 ramosíssima. s.s. (*ye.*) branched. Archipelago. 1828. —— F. ♃.
4 vìrens. w. (*ye.*) green. Switzerland. 1796. 6. 7. H. ⊙. Lobel. ic. 229. f. 2.
5 bannática. s.s. (*ye.*) Hungarian. Hungary. 1823. 7. 8. H. ⊙.
6 parviflòra. s.s. (*ye.*) small-flowered. Levant. 1829. —— H. ⊙.
7 rhagadioloìdes. s.s. (*ye.*) Spanish. Spain. 1818. 6. 7. H. ⊙. Jacq. schœnb. 2. t. 144.
Pìcris Rhagadìolus. P.S.
8 áspera. s.s. (*ye.*) rough. Sicily. 1797. 7. 8. H. ⊙. Flor. græc. t. 804.
9 Sprengeriàna. s.s. (*ye.*) Portugal. Portugal. 1783. 6. 7. H. ⊙. Moris. his. 7. t. 5. f. 17.
Pìcris Sprengeriàna. P.S. *Hieràcium Sprengeriànum.* L.
10 hieracioìdes. s.s. (*ye.*) hawkweed-like. Hungary. 1816. 7. 8. H. ♃. W. et K. hung. 1. t. 17.
11 rígida. s.s. (*ye.*) rigid. ——— 1805. 5. 7. H. ♃. ——— 1. t. 19.
12 púlchra. s.s. (*ye.*) hawkweed-leav'd. Scotland. .... 6. 9. H. ⊙. Eng. bot. v. 33. t. 2325.
13 nàna. s.s. (*ye.*) dwarf. N. America. 1827. —— H. ♃.
14 heterospérma. s.s. (*ye.*) various-seeded. Germany. 1824. —— H. ⊙.
15 Dioscóridis. s.s. (*ye.*) Dioscorides'. France. 1772. 6. 7. H. ⊙. Flor. græc. t. 808.
16 lodomiriénsis. s.s. (*ye.*) rough-headed. Germany. 1827. 6. 9. H. ⊙.
Gmelìni. s.
17 neglécta. s.s. (*ye.*) neglected. ——— .... —— H. ⊙.
18 scábra. s.s. (*ye.*) rough. S. Europe. 1829. —— H. ♂.
19 agréstis. s.s. (*ye.*) field. Hungary. 1801. 7. 8. H. ⊙.
20 corymbòsa. s.s. (*ye.*) corymbose. Italy. 1828. —— H. ♂.
21 cérnua. s.s. (*ye.*) nodding-flower'd. Naples. 1824. —— H. ♂.

22 tectòrum. E.B.	(*ye.*)	smooth.	Britain.	....	6. 9.	H. ⊙.	Eng. bot. v. 16. t. 1111.
23 Lachenàlii. DC.	(*ye.*)	entire-leaved.	Europe.	....	6. 9.	H. ⊙.	
24 cinèrea. P.S.	(*ye.*)	red-stalked.	Britain.	....	——	H. ⊙.	
25 taurinénsis. W.	(*ye.*)	early-flowered.	Europe.	....	5. 6.	H. ⊙.	Balb. misc. p. 37. t. 9.
26 biénnis. S.S.	(*ye.*)	biennial.	Britain.	....	6. 7.	H. ♂.	Eng. bot. v. 3. t. 149.
27 fœtida. E.B.	(*st.*)	strong-scented.	————	....	——	H. ♂.	———— v. 6. t. 406.
28 gravèolens.	(*ye.*)	strong-smelling.	....	1824.	7. 9.	H. ⊙.	
29 pulcherrima. F.	(*ye.*)	elegant.	Siberia.	1823.	6. 8.	H. ⊙.	
30 latifòlia. P.S.	(*ye.*)	broad-leaved.	Italy.	1816.	——	H. ⊙.	
31 pinnatífida. S.S.	(*ye.*)	pinnatifid.	S. Europe.	1820.	4. 6.	H. ⊙.	
32 lácera. S.S.	(*ye.*)	torn-leaved.	Naples.	1826.	6. 7.	H. ⊙.	
BEHRI'NIA. Sieb.		BEHRI'NIA. Syngenesia Polygamia Æqualis.					
chondrilloídes. Sie.	(*ye.*)	Chondrilla-like.	Carniola.	1829.	6. 7.	H. ⊙.	
Crèpis chondrilloídes. S.S. *Wibèlia chondrilloídes.* Hoppe.							
PRENA'NTHES. D.D.		PRENA'NTHES. Syngenesia Polygamia Æqualis. L.					
1 purpùrea. W.	(*pu.*)	purple-flower'd.	Germany.	1658.	7. 9.	H. ♃.	Jacq. aust. 4. t. 317.
2 híspida. S.S.	(*ye.*)	hispid.	Siberia.	1823.	6. 8.	H. ♃.	
3 diversifòlia. S.S.	(*ye.*)	various-leaved.	————	1827.	——	H. ♂.	
HARPAL'YCE. D.D.		HARPAL'YCE. Syngenesia Polygamia Æqualis.					
1 altíssima. D.D.	(*ye.*)	tallest.	N. America.	1696.	7. 8.	H. ♃.	Pluk. alm. t. 317. f. 2.
Prenánthes altíssima. W.							
2 cordàta. D.D.	(*ye.*)	heart-leaved.	————	1816.	——	H. ♃.	
3 virgàta. D.D.	(*li.*)	tall slender.	————	1823.	8. 9.	H. ♃.	
4 álba. D.D.	(*wh.*)	white-flowered.	N. America.	1762.	7. 8.	H. ♃.	Bot. mag. t. 1079.
Prenánthes álba. B.M. *suávis.* Sal. par. lond. t. 85.							
5 rubicúnda. D.D.	(*fl.*)	rubicund.	————	1823.	8. 10.	H. ♃.	
6 serpentària. D.D.	(*pu.*)	Lion's-foot.	————	——	——	H. ♃.	Pursh am. sept. 2. t. 24.
7 racemòsa. D.D.	(*pu.*)	racemed.	Canada.	1822.	——	H. ♃.	
CHORI'SMA. D.D.		CHORI'SMA. Syngenesia Polygamia Æqualis.					
rèpens. D.D.	(*ye.*)	creeping.	Siberia.	1828.	7. 9.	H. ♃.	Linn. am. ac. 2. t. 4. f. 23.
Prenánthes rèpens. L.							
ANDR'YALA. W.		ANDR'YALA. Syngenesia Polygamia Æqualis.					
1 cheiranthifòlia. W.	(*ye.*)	various-leaved.	Madeira.	1777.	5. 10.	F. ♃.	L'Herit. stirp. 35. t. 18.
2 nígricans. W.	(*ye.*)	dark-headed.	Barbary.	1804.	6. 8.	H. ⊙.	
3 pinnatífida. W.	(*ye.*)	wing-leaved.	Madeira.	1778.	7. 8.	F. ♂.	
4 crithmifòlia. W.	(*ye.*)	Samphire-leav'd.	————	——	6. 8.	F. ♂.	
5 ragusìna. W.	(*ye.*)	downy.	Archipelago.	1753.	——	F. ♃.	Mill. ic. 1. t. 146. f. 1.
6 incàna. DC.	(*ye.*)	hoary.	Pyrenees.	1826.	——	H. ♃.	
RO'THIA. W.		RO'THIA. Syngenesia Polygamia Æqualis.					
1 runcinàta. W.	(*ye.*)	hoary.	S. Europe.	1711.	7. 8.	H. ♂.	
2 cheiranthifòlia. W.	(*ye.*)	Stock-leaved.	Spain.	1768.	——	H. ⊙.	
3 andryaloídes. W.	(*ye.*)	Spanish.	————	1810.	8.	H. ⊙.	Gært. sem. 2. t. 174.
Andr'yala Róthia. P.S.							
LAPS'ANA. D.D.		NIPPLE-WORT. Syngenesia Polygamia Æqualis.					
1 commùnis. W.	(*ye.*)	common.	Britain.	....	6. 9.	H. ⊙.	Eng. bot. v. 12. t. 844.
2 lyràta. W.en.	(*ye.*)	lyrate.	Caucasus.	1816.	7. 8.	H. ♃.	
grandiflòra. M.B.							
3 intermèdia. M.B.	(*ye.*)	intermediate.	————	1823.	——	H. ⊙.	
4 pubéscens. Zey.	(*ye.*)	pubescent.	Europe.	1820.	——	H. ⊙.	
5 críspa. W.en.	(*ye.*)	curled.	....	1799.	——	H. ⊙.	

Tribus II. TARAXACEÆ. *D. Don.*

LEO'NTODON. W.		DANDELION. Syngenesia Polygamia Æqualis. L.					
1 Taráxacum. W.	(*ye.*)	common.	Britain.	....	4. 7.	H. ♃.	Eng. bot. v. 8. t. 510.
2 corniculàtus. H.H.	(*ye.*)	horned.	Europe.	1827.	——	H. ♃.	
3 bessarábicus. F.	(*ye.*)	small-flowered.	S. Europe.	1823.	6. 8.	H. ♃.	
4 alpìnus. S.S.	(*ye.*)	Alpine.	————	1827.	——	H. ♃.	
5 caucásicus. M.B.	(*ye.*)	Caucasean.	Caucasus.	——	——	H. ♃.	
6 serotìnus. W.	(*ye.*)	late-flowered.	Hungary	1816.	7. 9.	H. ♃.	W. et K. hung. 2. t. 114.
7 glaucéscens. S.S.	(*ye.*)	glaucescent.	Caucasus.	1820.	4. 7.	H. ♃.	
8 obovàtus. S.S.	(*ye.*)	obovate-leaved.	Spain.	1805.	7.	H. ♃.	
9 palùstris. E.B.	(*ye.*)	marsh.	Britain.	....	6. 7.	H. ♃.	Eng. bot. v. 8. t. 553.
10 nígricans. W.K.	(*ye.*)	dark-leaved.	Hungary.	1823.	7.	H. ♃.	
11 lævigàtus. S.S.	(*ye.*)	smooth.	Spain.	——	——	H. ♃.	
APA'RGIA. W.		APA'RGIA. Syngenesia Polygamia Æqualis.					
1 áspera. S.S.	(*ye.*)	hairy.	Hungary.	1805.	6. 7.	H ♃.	W. et K. hung. 2. t. 110.
2 saxátilis. S.S.	(*ye.*)	rock.	Naples.	1827.	——	H. ♃.	

Name	Colour	English name	Native of	Year	Months	Habit	Figure
3 híspida. s.s.	(*ye.*)	rough.	Britain.	....	7.	H. ♃.	Eng. bot. v. 8. t. 554.
Hedy'pnois híspida. E. B. *Leóntodon híspidum.* L.							
4 dùbia. w.	(*ye.*)	tooth-leaved.	Germany.	——	——	H. ♃.	
5 críspa. s.s.	(*ye.*)	curled.	France.	1803.	7. 8.	H. ♃.	Vill. dauph. 3. t. 25.
6 fasciculàta. s.s.	(*ye.*)	bunch-rooted.	Naples.	1828.	——	H. ♃.	
7 hispánica. s.s.	(*ye.*)	Spanish.	Spain.	1825.	——	H. ♃.	Cavan. icon. t. 149.
8 strigòsa. M.B.	(*ye.*)	strigose.	Caucasus.	1828.	——	H. ♃.	
9 caucásica. M.B.	(*ye.*)	Caucasean.	———	1823.	——	H. ♃.	
10 coronopifòlia. s.s.	(*ye.*)	buck's-horn-l^d.	Barbary.	1820.	——	H. ♃.	Desf. atl. 2. t. 214.
11 incàna. s.s.	(*ye.*)	hoary.	Europe.	1784.	5. 6.	H. ♃.	Jacq. aust. 3. t. 287.
12 alpìna. s.s.	(*ye.*)	alpine.	Austria.	1816.	6. 8.	H. ♃.	——— 1. t. 93.
13 Villársii. s.s.	(*ye.*)	Villars's.	France.	1818.	——	H. ♃.	Vill.dauph.3.p.82.t.25.
14 Taráxaci. s.s.	(*ye.*)	Dandelion-like.	Britain.	....	7. 9.	H. ♃.	Eng. bot. v. 16. t. 1109.
Hedy'pnois Taráxaci. E.B. *Hieràcium Taráxaci.* L. *Pìcris Taráxaci.* All.							
15 aurantìaca. w.	(*or.*)	orange-colored.	Hungary.	1816.	5. 6.	H. ♃.	
16 cròcea. w.	(*sn.*)	saffron-colored.	———	1820.	——	H. ♃.	
17 hástilis. s.s.	(*ye.*)	shining-leaved.	S. Europe.	1796.	7. 8.	H. ♃.	Jacq. aust. 2. t. 164.
OPORI'NIA. D.D.		OPORI'NIA. Syngenesia Polygamia Æqualis.					
autumnàlis. D.D.	(*ye.*)	autumnal.	Britain.	....	8. 9.	H. ♃.	Eng. bot. v. 12. t. 830.
Hedy'pnois autumnàlis. E.B. *Apárgia autumnàlis.* W.							
THRI'NCIA. W.		THRI'NCIA. Syngenesia Polygamia Æqualis.					
1 tuberòsa. DC.	(*ye.*)	tuberous-rooted.	France.	1683.	5. 7.	H. ♃.	Flor. græc. t. 797.
Apárgia tuberòsa. W. *Leóntodon tuberòsum.* L.							
2 glábra. Schl.	(*ye.*)	smooth.	Switzerland.	1823.	7. 8.	H. ♃.	
3 hírta. w.	(*ye.*)	simple-haired.	Britain.	....	——	H. ♃.	Eng. bot. v. 8. t. 555.
pilocályx. Lag ? *Hedy'pnois hírta.* E.B.							
4 nudicályx. Lag.	(*ye.*)	naked-calyxed.	Spain.	1823.	——	H. ♃.	
5 híspida. s.s.	(*ye.*)	hispid.	S. Europe.	1815.	6. 8.	H. ⊙.	Vill. dauph. 3. t. 25.
6 maroccàna. P.S.	(*ye.*)	Morocco.	Morocco.	1799.	7. 8.	H. ⊙.	
7 pygm'æa. P.S.	(*ye.*)	small.	Madeira.	....	——	H. ⊙.	
Hyóseris pygm'æa. W.							
CALLI'OPEA. D.D.		CALLI'OPEA. Syngenesia Polygamia Æqualis.					
aùrea. D.D.	(*go.*)	golden-flower'd.	Italy.	1769.	5. 7.	H. ♃.	Jacq. aust. 3. t. 297.
Leóntodon aùreum. Jacq. *Hieràcium aùreum.* W.							
ÆTH'ONIA. D.D.		ÆTH'ONIA. Syngenesia Polygamia Æqualis.					
1 fruticòsa. D.D.	(*ye.*)	shrubby.	Madeira.	1785.	6. 7.	G. ♄.	
Hieràcium fruticosum. H.K.							
2 filifórmis. D.D.	(*ye.*)	fine-leaved.	———	1777.	——	G. ♂.	
Crèpis filifórmis. W.							
3 lagopòda. D.D.		hare's-foot.	Canaries.	1817.	——	G. ♂.	
TO'LPIS. S.S.		TO'LPIS. Syngenesia Polygamia Æqualis.					
1 barbàta. s.s.	(*ye.*)	purple-eyed.	France.	1620.	6. 9.	H. ⊙.	Bot. mag. t. 35.
Crèpis barbàta. B.M. *Drepània barbàta.* DC.							
2 umbellàta. s.s.	(*ye.*)	umbelled.	Italy.	1818.	6. 7.	H. ⊙.	
3 quadriaristàta. Biv.	(*ye.*)	four-awned.	———	1824.	——	H. ⊙.	
4 virgàta. s.s.	(*ye.*)	twiggy.	S. Europe.	1826.	——	H. ⊙.	
5 altíssima. s.s.	(*ye.*)	tall.	———	1818.	——	H. ⊙.	
Drepània ambígua. DC. *Crèpis ambígua.* Balbis.							
6 coronopifòlia. s.s.	(*ye.*)	fleshy-leaved.	Madeira.	1777.	8. 9.	H. ⊙.	Desf. act. par. 38. t. 9.
Crèpis coronopifòlia. W. *succulénta.* H.K.							
HEDY'PNOIS. W.		HEDY'PNOIS. Syngenesia Polygamia Æqualis.					
1 monspeliénsis. w.	(*ye.*)	branching.	S. Europe.	1683.	6. 7.	H. ⊙.	
2 crètica. w.	(*ye.*)	Cretan.	Crete.	1731.	——	H. ⊙.	Flor. græc. t. 813.
3 coronopifòlia. T.C.	(*ye.*)	buck's-horn-l^d.	Italy.	1823.	——	H. ⊙.	
4 tubæfórmis. T.C.	(*ye.*)	thick-tubed.	———	——	——	H. ⊙.	
5 mauritánica. w.	(*ye.*)	Barbary.	Barbary.	——	——	H. ⊙.	
6 rhagadioloídes. w.	(*ye.*)	Nipple-wort.	S. Europe.	1773.	7. 8.	H. ⊙.	
7 pérsica. M.B.	(*ye.*)	Persian.	....	1822.	6. 9.	H. ⊙.	
8 péndula. w.	(*ye.*)	pendulous.	Barbary.	1818.	——	H. ⊙.	
9 aculeàta. Jac.	(*ye.*)	prickly.	....	1829.	——	H. ⊙.	
RHAGAD'IOLUS. W.		RHAGAD'IOLUS. Syngenesia Polygamia Æqualis.					
1 stellàtus. w.	(*ye.*)	starry.	S. Europe.	1633.	6. 7.	H. ⊙.	Flor. græc. t. 817.
2 èdulis. w.	(*ye.*)	heart-leaved.	Levant.	——	——	H. ⊙.	——— t. 818.
Lapsàna Rhagadìolus. L.							
KŒLPI'NIA. Pall.		KŒLPI'NIA. Syngenesia Polygamia Æqualis.					
lineàris. Pall.	(*ye.*)	linear-leaved.	Levant.	1788.	7. 8.	H. ⊙	Flor. græc. t. 819.
Lapsàna Kœlpínia. L. *Rhagadìolus Kœlpínia.* W.							
ARNO'SERIS. G.		ARNO'SERIS. Syngenesia Polygamia Æqualis.					
1 fœ'tida. L.en.	(*ye.*)	stinking.	Italy.	1722.	7. 8.	H. ♃.	W. et K. hung. 1. t. 49.
Lapsàna fœ'tida. W. *Hyóseris fœ'tida.* P.S.							

2 mínima. L.en.	(st.) least.	Britain.	....	5. 7. H. ⊙.		Eng. bot. v. 2. t. 95.
Hyóseris mínima. E.B. *Lapsàna pusílla.* W.						
HYO'SERIS. W.	Hyo'seris. Syngenesia Polygamia Æqualis.					
1 arenària. W.	(ye.) sand.	Morocco.	1800.	7. 8. H. ⊙.		
2 híspida. W.	(ye.) hispid.	————	1779.	—— H. ⊙.		
3 scábra. W.	(ye.) rugged.	Sicily.	1789.	—— H. ⊙.		Lam. ill. t. 654. f. 1.
4 lùcida. W.	(ye.) shining.	Levant.	1770.	6. 8. H. ♃.		Schm. ic. t. 39. 41.
5 radiàta. W.	(ye.) starry.	S. Europe.	1640.	—— H. ♃.		Pluk. alm. t. 37. f. 2.
ZACI'NTHA. W.	Zaci'ntha. Syngenesia Polygamia Æqualis.					
verrucòsa. W.	(ye.) warted.	S. Europe.	1633.	6. 7. H. ⊙.		Gært. sem. 2. t. 157. f. 7.
Lapsàna Zacy'ntha. L.						
TRO'XIMON. D.D.	Tro'ximon. Syngenesia Polygamia Æqualis.					
1 glaùcum. Ph.	(ye.) glaucous-leaved.	N. America.	1811.	7. 8. H. ♃.		Bot. mag. t. 1667.
2 cuspidàtum. Ph.	(ye.) cuspidate.	————	——	4. 6. H. ♃.		
marginàtum. N.						
CY'NTHIA. D.D.	Cy'nthia. Syngenesia Polygamia Æqualis.					
amplexicaùlis. D.D.	(go.) stem-clasping.	N. America.	1799.	7. 8. H. ♃.		
Hyóseris amplexicaùlis. M. *Krígia amplexicaùlis.* N. *Tróximon virgínicum.* Ph.						
KRI'GIA. W.	Kri'gia. Syngenesia Polygamia Æqualis.					
virgínica. W.	(ye.) Virginian.	N. America.	1811.	5. 7. H. ⊙.		
P'ICRIS. W.	P'icris. Syngenesia Polygamia Æqualis.					
1 hieracioídes. W.	(ye.) hawkweed-like.	England.	....	7. 8. H. ⊙.		Eng. bot. v. 3. t. 196.
2 strigòsa. M.B.	(ye.) strigose.	Caucasus.	1823.	—— H. ♃.		
3 rígida. S.S.	(ye.) rigid.	Tauria.	1829.	—— H. ♃.		
4 híspida. H.K.	(ye.) hispid.	Levant.	1789.	—— H. ♃.		
5 altíssima. S.S.	(ye.) tall.	Egypt.	1825.	—— H. ⊙.		
6 pauciflòra. W.	(ye.) few-flowered	S. France.	1818.	6. 8. H. ⊙.		Gært. fr. 2. t. 158. f. 2.
7 pilòsa. S.S.	(ye.) hairy.	Egypt.	1829.	—— H. ⊙.		
8 asplenioídes. W.	(ye.) Spleenwort-ld.	Barbary.	1803.	7. 8. H. ⊙.		L'Herit. stirp. t. 82.
9 dahùrica. F.	(ye.) Dahurian.	Dahuria.	1820.	—— H. ⊙.		
10 kamtschática. Led.	(ye.) Kamtschatka.	Kamtschatka.	1823.	—— H. ⊙.		
HELMI'NTHIA. J.	Ox-tongue. Syngenesia Polygamia Æqualis.					
echioídes. W.	(ye.) bristly.	Britain.	....	6. 7. H. ⊙.		Eng. bot. v. 14. t. 972.
Pìcris echioídes. E.B.						

Tribus III. Hypochœrideæ. *D. Don in Edinb. phil. journ.*

HYPOCH'ŒRIS. S.S.	Cat's-ear. Syngenesia Polygamia Æqualis. L.					
1 helvética. S.S.	(ye.) one-flowered.	Switzerland.	1799.	6. 7. H. ♃.		Jacq. ic. 1. t. 165.
2 híspida. S.S.	(ye.) bristly.	S. Europe.	1804.	7. 8. H. ♃.		Willd. hort. ber. 1. t. 46.
3 arachnìtes. Biv.	(ye.) Cobweb.	Italy.	——	—— H. ⊙.		
4 mínima. S.S.	(ye.) least.	Barbary.	1797.	—— H. ⊙.		Flor. græc. t. 816.
5 glábra. S.S.	(st.) smooth.	Britain.	....	—— H. ⊙.		Eng. bot. v. 8. t. 575.
ACHYRO'PHORUS. G.	Achyro'phorus. Syngenesia Polygamia Æqualis.					
1 maculàtus. D.D.	(ye.) spotted.	England.	....	6. 7. H. ♃.		Eng. bot. v. 4. t. 225.
2 radicàtus. D.D.	(ye.) long-rooted.	Britain.	....	6. 9. H. ♃.		———— v. 12. t. 831.
Hypoch'æris radicàta. E.B.						
3 Balbísii. W.en.	(ye.) Balbis's.	S. Europe.	1822.	—— H. ♃.		
ROBE'RTIA. DC.	Robe'rtia. Syngenesia Polygamia Æqualis.					
1 taraxacoídes. DC.	(ye.) Dandelion-like.	Corsica.	1824.	7. 8. H. ♃.		Lois. gal. 530. t. 28.
2 pinnatífida. S.S.	(ye.) pinnatifid.	Naples.	1827.	—— H. ♃.		
Hypoch'æris pinnatífida. Tenore.						
SER'IOLA. S.S.	Ser'iola. Syngenesia Polygamia Æqualis. L.					
1 ùrens. S.S.	(ye.) stinging.	S. Europe.	1773.	7. 8. H. ⊙.		Schmid. ic. t. 32.
2 æthnénsis. W.	(ye.) rough.	Italy.	1763.	—— H. ⊙.		Jacq. obs. 4. t. 79.
3 glaùca. S.S.	(ye.) glaucous.	Sicily.	1829.	—— H. ⊙.		
4 rubéscens. S.S.	(li.) reddish.	————	——	—— H. ⊙.		
5 álbicans. S.S.	(ye.) whitish.	————	——	—— H. ⊙.		
6 creténsis. S.S.	(ye.) Cretan.	Crete.	1820.	—— H. ⊙.		
7 Alliàtæ. L.en.	(ye.) wood.	Sicily.	1823.	—— H. ⊙.		Biv. cent. pl. 2. t. 7.
8 apargioídes. Trev.	(ye.) Apargia-like.	S. Europe.	1829.	—— H. ⊙.		
AGE'NORA. D.D.	Age'nora. Syngenesia Polygamia Æqualis.					
lævigàta. D.D.	(ye.) smooth.	Candia.	1772.	7. 8. H. ⊙.		Desf. atl. 2. t. 216.
Serìola lævigàta. W. *Rodígia lævigàta.* S.S.						
RODI'GIA. S.S.	Rodi'gia. Syngenesia Polygamia Æqualis.					
commutàta. S.S.	(ye.) Poppy-leaved.	Caucasus.	1823.	6. 7. H. ⊙.		
Crèpis rhæadifòlia. M.B.						
SOLDEVI'LLA. S.S.	Soldevi'lla. Syngenesia Polygamia Æqualis.					
setòsa. S.S.	(ye.) bristly.	Spain.	1826.	7. 8. H. ⊙.		

Tribus IV. Lactuceæ. *D. Don in Edinb. phil. journ.*

LACT'UCA. S.S. Lettuce. Syngenesia Polygamia Æqualis. L.
1 satìva. w. (*ye.*) garden. .... 1562. 6. 9. H. ⊙. Rob. ic. pl. c. ic.
2 críspa. w. (*ye.*) curled. .... 1570. 6. 8. H. ⊙.
3 palmàta. w. (*ye.*) jagged-leaved. .... 1683. —— H. ⊙.
lacìniàta. Roth.
4 intybàcea. w. (*ye.*) Endive-leaved. S. America. 1781. —— H. ⊙. Jacq. ic. 1. t. 162.
5 quercìna. w. (*ye.*) Oak-leaved. Europe. 1686. 5. 7. H. ⊙.
6 stricta. w. (*ye.*) upright. Hungary. 1805. 6. 7. H. ♂. W. et K. hung. 1. t. 48.
7 elongàta. w. (*st.*) elongated. Pensylvania. —— —— H. ♂.
8 cichoriifòlia. DC. (*ye.*) Succory-leaved. France. 1818. —— H. ♃.
9 Scarìola. w. (*st.*) prickly. Britain. .... 7. 8. H. ♂. Eng. bot. v. 4. t. 268.
10 maculàta. L.en. (*st.*) spot-leaved. Europe. .... —— H. ♂.
11 viròsa. E.B. (*st.*) strong-scented. Britain. .... 7. 9. H. ♂. Eng. bot. v. 28. t. 1957.
12 augustàna. w. (*st.*) entire-leaved. Italy. 1791. 7. 8. H. ⊙. All. ped. 1. t. 52. f. 1.
13 altíssima. M.B. (*ye.*) tall. Caucasus. 1818. —— H. ♃.
14 sagittàta. W.K. (*ye.*) arrow-leaved. Hungary. 1805. —— H. ♂. W. et K. hung. 1. t. 1.
15 salígna. w. (*ye.*) least. England. .... 7. 8. H. ♂. Eng. bot. v. 10. t. 707.
16 muràlis. D.D. (*st.*) wall. Britain. .... —— H. ♃. ———— v. 7. t. 457.
Prenánthes muràlis. E.B.
17 vimínea. L.en. (*st.*) rushy-twigged. Austria. 1789. —— H. ♂. Jacq. aust. 1. t. 9.
Prenánthes vimínea. Jacq.
18 villòsa. S.S. (*bl.*) villous. .... 1825. —— H. ♂. Jacq. schœnb. 3. t. 367.
19 tuberòsa. S.S. (*ye.*) tuberous. .... 1820. —— H. ♃. Jacq. vind. 1. t. 57.
20 índica. S.S. (*ye.*) Indian. E. Indies. 1784. 8. 10. H. ⊙.
21 ? spectábilis. F. .... showy. Persia. 1829. .... H. ♂?
22 pállida. D.P. (*st.*) pale-flowered. Canada. 1704. 7. 9. H. ♃. Rob. ic. 148 et 151.
Sónchus pállidus. w.
23 segusiàna. S.S. (*pu.*) various-leaved. Italy. 1823. 7. 8. H. ⊙.
24 tenérrima. S.S. (*pu.*) purple-flower'd. S. Europe. 1815. —— H. ♃.
25 perénnis. S.S. (*bl.*) perennial. Germany. 1596. 6. 8. H. ♃. Mill. ic. 2. t. 157.

CHONDRI'LLA. W. Gum Succory. Syngenesia Polygamia Æqualis. L.
1 júncea. w. (*ye.*) common. France. 1633. 9. 10. H. ♃. Jacq aust. 5. t. 427.
2 gramínea. M.B. (*ye.*) grass-leaved. Caucasus. 1824. 7. 9. H. ♃.
3 latifòlia. M.B. (*ye.*) broad-leaved. ——— —— —— H. ♂.

AGATHY'RSUS. D.D. Agathy'rsus. Syngenesia Polygamia Æqualis.
1 alpìnus. D.D. (*bl.*) Lapland. Lapland. 1804. 7. 8. H. ♃. Smith ic. ined. t. 21.
Sónchus lappónicus. w.
2 cœrùleus. D.D. (*bl.*) blue-flowered. Scotland. .... —— H. ♃. Eng. bot. v. 34. t. 2425.
Sónchus alpìnus. W. *non* L.
3 Plumíeri. D.D. (*bl.*) Plumier's. Pyrenees. 1794. —— H. ♃.
4 floridànus. D.D. (*bl.*) small-flowered. N. America. 1713. —— H. ♂.
5 acuminàtus. (*bl.*) taper-pointed. ——— 1812. 8. 9. H. ♂.
Sónchus acuminàtus. Ph.
6 macrophy'llus. (*bl.*) large-leaved. N. America. 1820. 7. 8. H. ♃.
Sónchus macrophy'llus. Ph.
7 spicàtus. D.D. (*pa.*) whitish-flower'd. ——— —— —— H. ♂.
Sónchus leucoph'æus. W. *Prenánthes autumnàlis.* Walter.
8 prenanthoídes. D.D. (*bl.*) Prenanthes-like. Caucasus. 1827. —— H. ♃.
9 cacaliæfòlius. D.D. (*bl.*) Cacalia-leaved. Iberia. 1826. —— H. ♃.
10 sibíricus. D.D. (*bl.*) Siberian. Siberia. 1759. 7. 9. H. ♃. Gmel. sib. 2. f. 3.
Sónchus sibíricus. W. *Lactùca sibírica.* D.P.
11 tatáricus. D.D. (*bl.*) Tartarian. Tartary. 1784. 6. 7. H. ♃.
12 pulchéllus. D.D. (*bl.*) entire-leaved. Missouri. 1812. —— H. ♃.
Sónchus pulchéllus. Ph. *Lactùca integrifòlia.* N.
13 sonchifòlia. (*bl.*) Sonchus-leaved. Levant. 1824. —— H. ♂.
Lactùca sonchifòlia. w.

LYGODE'SMIA. D.D. Lygode'smia. Syngenesia Polygamia Æqualis.
júncea. D.D. (*ro.*) rush-like. Missouri. 1827. 5. 6. H. ♃.
Prenánthes júncea. Ph.

ATALA'NTHUS. D.D. Atala'nthus. Syngenesia Polygamia Æqualis.
1 pinnàtus. D.D. (*ye.*) pinnated. Canaries. 1823. 6. 7. F. ♄.
Prenánthes pinnàta. w.
2 arbòreus. (*ye.*) arborescent. ——— —— 6. 8. F. ♄.
Prenánthes arbòrea. W. enum.
3 spinòsus. D.D. (*ye.*) prickly. Barbary. 1640. 3. 5. F. ♄. Park. theatr. 804. f. 7.

SO'NCHUS. W.		Sow Thistle.	Syngenesia Polygamia Æqualis. L.				
1 squarròsus. DC.	(ye.)	shrubby.	Madeira.	1777.	4. 7.	G. ♄.	
fruticòsus. L. *non* Jacq.							
2 Jacquíni. DC.	(ye.)	Jacquin's.	——	——	——	G. ♄.	Jacq. ic. 1. t. 161.
fruticòsus. Jacq. *non* L.							
3 lævigàtus. L.en.	(ye.)	smooth.	——	1816.	——	G. ♄.	
4 lyràtus. L.en.	(ye.)	lyre-leaved.	——	——	——	G. ♄.	
5 hyoserifòlius. H.H.	(ye.)	Hyoseris-leaved.	——	1818.	——	G. ♄.	
6 radicàtus. S.S.	(ye.)	long-rooted.	Canaries.	1780.	7. 8.	G. ♄.	
7 gummífer. S.S.	(ye.)	gummy.	——	1816.	——	G. ♄.	
8 divaricàtus. L.en.	(ye.)	spiny-leaved.	....	1828.	4. 6.	G. ♄.	
9 pinnàtus. S.S.	(ye.)	wing-leaved.	Madeira.	1777.	4. 7.	G. ♄.	
10 marítimus. S.S.	(ye.)	sea-side.	Canaries.	1748.	7. 9.	H. ♃.	All. ped. 1. t. 16. f. 2.
11 oleràceus. S.S.	(ye.)	common-smooth.	Britain.	....	6. 12.	H. ⊙.	Eng. bot. v. 12. t. 843.
l'ævis. L.enum. *oleràceus α. lævis.* W.enum.							
12 lácerus. W.en.	(ye.)	torn-leaved.	Britain.	....	——	H. ⊙.	
13 ásper. S.S.	(ye.)	rough-leaved.	——	....	——	H. ⊙.	Blackw. ic. t. 30.
oleràceus γ. asper. L.							
14 uliginòsus. S.S.	(ye.)	marsh.	Caucasus.	1823.	7. 9.	H. ♃.	
15 palústris. S.S.	(ye.)	tall marsh.	Britain.	....	——	H. ♃.	Eng. bot. v. 13. t. 935.
16 arvénsis. S.S.	(ye.)	corn-field.	——	....	——	H. ♃.	—— v. 10. t. 674.
17 hantoniénsis.	(ye.)	smooth-headed.	Hampshire.	....	——	H. ♃.	
18 tenérrimus. S.S.	(ye.)	clammy.	S. Europe.	1691.	7. 8.	H. ⊙.	Flor. græc. t. 790.
19 chondrilloídes. S.S.	(ye.)	Chondrilla-like.	Sicily.	1826.	——	H. ♃.	
20 caucásicus. F.	(ye.)	Caucasean.	Caucasus.	1820.	——	H. ♃.	
21 flexuòsus. Led.	(ye.)	flexuose.	——	1822.	——	H. ♃.	
22 chinénsis. F.	(ye.)	Chinese.	China.	1820.	——	H. ♃.	
23 tènax. F.	(ye.)	tough.	Siberia.	1824.	——	H. ♃.	
24 longifòlius. Trev.	(ye.)	long-leaved.	....	——	——	H. ♃.	
BARKHA'USIA. DC.		Barkha'usia.	Syngenesia Polygamia Æqualis.				
1 muricàta. S.S.	(ye.)	bristly-headed.	S. Europe.	1825.	6. 8.	H. ♂.	
2 bellidifòlia. DC.	(ye.)	Daisy leaved.	Corsica.	1826.	——	H. ⊙.	
3 Suffreniàna. DC.	(ye.)	mealy-headed.	France.	1825.	——	H. ⊙.	
4 spathulàta. S.S.	(ye.)	spathulate.	Sicily.	1829.	——	H. ⊙.	
5 alpìna. S.S.	(ye.)	alpine.	Italy.	1739.	7. 8.	H. ⊙.	Gmel. sib. 2. t. 5.
Crèpis alpìna. W.							
6 apargioídes. S.S.	(ye.)	Apargia-like.	Germany.	1826.	——	H. ♃.	Jacq. aust. 3. t. 293.
7 vesicària. S.S.	(ye.)	bladdered.	Crete.	1828.	——	H. ♃.	
8 hyemàlis. S.S.	(ye.)	winter.	Sicily.	1829.	8. 11.	H. ♃.	
9 purpùrea. S.S.	(pu.)	purple.	——	1827.	7. 9.	H. ♃.	
10 rùbra. S.S.	(li.)	lilac.	Italy.	1632.	6. 8.	H. ⊙.	Flor. græc. t. 801.
11 nicæénsis. S.S.	(ye.)	gland-headed.	S. Europe.	1824.	——	H. ⊙.	
12 híspida. S.S.	(ye.)	hispid.	——	1798.	——	H. ⊙.	W. et K. hung. 1. t. 43.
13 setòsa. DC.	(ye.)	setose.	Switzerland.	1824.	——	H. ⊙.	
14 púngens. P.S.	(ye.)	pungent.	....	1826.	——	H. ⊙.	
MYO'SERIS. L.en.		Myo'seris.	Syngenesia Polygamia Æqualis.				
purpùrea. L.en.	(pu.)	purple.	Tauria.	1823.	6. 7.	H. ⊙.	
Lagóseris taùrica. M.B. *Crèpis purpùrea.* M.B.							

Tribus V. Scorzonereæ. *D. Don in Edinb. phil. journ.*

SCORZON'ERA. W.		Viper's-grass.	Syngenesia Polygamia Æqualis.				
1 tomentòsa. W.	(ye.)	white.	Armenia.	1789.	6. 7.	H. ♃.	
2 villòsa. S.S.	(ye.)	villous.	Levant.	1821.	6. 8.	H. ♃.	Scop. carn. t. 46.
3 hirsùta. S.S.	(ye.)	hairy.	S. Europe.	1823.	——	H. ♃.	All. ped. t. 31. f. 33.
4 hùmilis. W.	(ye.)	dwarf.	Europe.	1597.	8.	H. ♃.	
5 austrìaca. W.en.	(ye.)	Austrian.	Austria.	....	——	H. ♃.	Jacq. aust. 1. t. 36.
hùmilis. Jacq. *non* W.							
6 plantaginifòlia. Sch.	(ye.)	Plantain-leav'd.	Switzerland.	1816.	——	H. ♃.	
7 hispánica. W.	(ye.)	garden.	Spain.	1576.	6. 9.	H. ♃.	Lam. ill. t. 647. f. 5.
8 glastifòlia. W.	(ye.)	Woad-leaved.	Germany.	1816.	——	H. ♃.	
9 taùrica. M.B.	(ye.)	Taurian.	Tauria.	1820.	——	H. ♃.	
10 rumicifòlia. L.en.	(ye.)	Dock-leaved.	....	1822.	——	H. ♃.	
11 caricifòlia. W.	(ye.)	Carex-leaved.	Siberia.	1805.	——	H. ♃.	Pall. it. 3. ap. t. J. 1. f. 1.
12 parviflòra. W.	(ye.)	small-flowered.	Austria.	1816.	——	H. ♃.	Jacq. aust. 4. t. 305.
13 móllis. M.B.	(ye.)	soft.	Volga.	1829.	——	H. ♃.	
14 strícta. S.S.	(ye.)	upright.	——	1821.	——	H. ♃.	
villòsa. M.B. *nec* Scop.							

15 lanàta. M.B.	(*ye.*)	woolly.	Iberia.	1823.	6. 7	H. ♃.	
16 purpùrea. W.	(*pu.*)	purple-flower'd.	Austria.	1759.	5. 6.	H. ♃.	Jacq. aust. 1. t. 35.
17 angustifòlia. W.	(*ye.*)	narrow-leaved.	S. Europe.	——	6. 8.	H. ♃.	W. et K. hung. 2. t. 122.
18 ensifòlia. M.B.	(*ye.*)	sword-leaved.	Caucasus.	1823.	——	H. ♃.	
19 pusílla. S.S.	(*st.*)	small.	Volga.	1827.	——	H. ♃.	Pall. it. 2. ap. t. L.
20 tuberòsa. S.S.	(*ye.*)	tuberous-rooted.	———	1825.	——	H. ♃.	Pall. it. 3. ap. t. Y. f. 3.
21 críspa. S.S.	(*ye.*)	curled.	Tauria.	1821.	——	H. ♃.	
PODOSPE'RMUM. DC.		PODOSPE'RMUM. Syngenesia Polygamia Æqualis.					
1 calcitrapifòlium. DC.	(*y.*)	Calcitrapa-leav'd.	Levant.	1823.	6. 8.	H. ♃.	Barr. ic. 800.
2 resedifòlium. DC.	(*ye.*)	spreading.	Spain.	1729.	6. 7.	H. ♂.	Bocc. sic. 13. t. 7.
3 laciniàtum. DC.	(*ye.*)	cut-leaved.	S. Europe.	1640.	——	H. ♂.	Jacq. aust. 4. t. 356.
4 octangulàre. DC.	(*ye.*)	8-angled.	———	1820.	——	H. ♂.	
5 pùmilum.	(*ye.*)	small.	Spain.	1822.	7. 9.	H. ⊙.	Cav. ic. 2. t. 121. f. 2.
Scorzonèra pùmila. W.							
6 aspérrimum.	(*ye.*)	roughest.	Galatia.	——	8. 9.	H. ♃.	
Scorzònera aspérrima. W.							
7 ásperum.	(*ye.*)	rough.	Levant.	——	7. 8.	H. ♃.	Desf. an. mus. 1. t. 9.
Scorzonèra áspera. Desf. *stylòsa.* P.S.							
8 subulàtum. DC.	(*ye.*)	grass-leaved.	Portugal.	1759.	——	H. ♃.	Jacq. obs. 4. t. 100.
Scorzonèra graminifòlia. F.G. 784.							
9 muricàtum. DC.	(*ye.*)	bristly.	Piedmont.	1818.	——	H. ♃.	
10 taraxacifòlium.	(*ye.*)	Dandelion-leav'd.	Bohemia.	1801.	6. 8.	H. ♃.	Jacq. ic. 1. t. 106.
Scorzonèra taraxacifòlia. W.							
LASIOSPE'RMUM. F.		LASIOSPE'RMUM. Syngenesia Polygamia Æqualis.					
angustifòlium. F.	(*ye.*)	narrow-leaved.	Caucasus.	1825.	6. 8.	H. ♃.	
Scorzonèra eriospérma. M.B.							
PICRI'DIUM. P.S.		PICRI'DIUM. Syngenesia Polygamia Æqualis.					
1 vulgàre. P.S.	(*ye.*)	various-leaved.	France.	1773.	6. 8.	H. ⊙.	All. ped. 1. t. 16. f. 1.
Scorzonèra picroídes. L. *Sònchus picroídes.* W.							
2 tingitànum. P.S.	(*ye.*)	Tangier.	Barbary.	1713.	6. 9.	H. ⊙.	Bot. mag. t. 142.
Scorzonèra tingitàna. B.M. *Sónchus tingitànus.* W.							
3 hispánicum. V.	(*ye.*)	Spanish.	Spain.	1816.	——	H. ⊙.	Jacq. schœnb. 2. t. 143.
4 álbidum. P.S.	(*ye.*)	pale-flowered.	France.	1781.	7. 10.	H. ♃.	Jacq. ic. 1. t. 164.
Crèpis álbida. W.							
ARNOP'OGON. W.		SHEEP'S-BEARD. Syngenesia Polygamia Æqualis.					
1 Dalechámpii. W.	(*ye.*)	great-flowered.	S. Europe.	1739.	6. 10.	H. ♃.	Bot. mag. t. 1623.
2 picroídes. W.	(*ye.*)	prickly-cupp'd.	———	1683.	7. 8.	H. ⊙.	Flor. græc. t. 781.
3 ásper. W.	(*ye.*)	rough.	Montpelier.	1774.	——	H. ⊙.	——— t. 782.
4 capénse. W.	(*ye.*)	Cape.	C. B. S.	1824.	6. 8.	G. ♂.	Jacq. ic. 3. t. 577.
TRAGOP'OGON. W.		GOAT'S-BEARD. Syngenesia Polygamia Æqualis.					
1 praténsis. W.	(*ye.*)	yellow.	Britain.	....	5. 6.	H. ♂.	Eng. bot. v. 7. t. 434.
2 mutábilis. W.	(*va.*)	changeable.	Siberia.	1816.	——	H. ♂.	Jacq. ic. 1. t. 157.
3 undulàtus. W.	(*st.*)	wave-leaved.	Crimea.	1790.	——	H. ♂.	——— 1. t. 158.
4 parviflòrus. L.en.	(*ye.*)	small-flowered.	....	1822.	——	H. ♂.	
5 orientàlis. W.	(*ye.*)	oriental.	Levant.	1787.	6. 7.	H. ♂.	
6 cànus. W.K.	(*ye.*)	hoary	Hungary.	1822.	——	H. ♂.	
7 màjor. W.	(*ye.*)	great.	Austria.	1788.	5. 6.	H. ♂.	Jacq. aust. 1. t. 29.
8 campéstris. S.S.	(*ye.*)	field.	Poland.	1828.	——	H. ♂.	
9 ròseus. S.S.	(*ro.*)	red-flowered.	Volga.	1827.	——	H. ♂.	
10 porrifòlius. W.	(*pu.*)	purple.	England.	....	——	H. ♂.	Eng. bot. v. 9. p. 638.
11 floccòsus. W.K.	(*ye.*)	woolly.	Hungary.	1816.	——	H. ♂.	W. et K. hung. 2. t. 112.
12 dùbius. L.en.	(*ye.*)	doubtful.	....	1822.	——	H. ♂.	
13 crocifòlius. W.	(*vi.*)	Crocus-leaved.	Italy.	1739.	6. 7.	H. ♂.	Flor. græc. t. 779.
14 angustifòlius. W.	(*vi.*)	narrow-leaved.	———	1816.	——	H. ♂.	
15 villòsus. W.	(*ye.*)	hairy.	Spain.	1794.	5. 6.	H. ♂.	
16 rhodánthus.	(*ro.*)	rose-coloured.	Hungary.	1807.	7.	H ♃.	W. et K. hung. 2. t. 121.
Scorzonèra ròsea. W.K.							
17 hùmilis. F.	....	dwarf.	Caucasus.	1829.	....	H. ♂.	
18 phæopáppus. F.	....	glossy-chaff'd.	Persia.	——	....	H. ♂.	
GEROP'OGON. W.		OLD MAN'S-BEARD. Syngenesia Polygamia Æqualis.					
1 gláber. W.	(*pi.*)	smooth.	Italy.	1704.	7. 8.	H. ⊙.	Bot. mag. t. 479.
2 hirsùtus. W.	(*pi.*)	hirsute.	———	1759.	——	H. ⊙.	Col. ecphr. 1. t. 231.
3 calyculàtus. W.	(*pi.*)	perennial.	———	1774.	6. 8.	H. ♃.	Jacq. vind. 2. t. 106.

Tribus VI. CICHOREÆ. *D. Don in Edinb. phil. journ.*

CICH'ORIUM. S.S.		SUCCORY. Syngenesia Polygamia Æqualis. L.					
1 I'ntybus. S.S.	(*bl.*)	common.	Britain.	....	6. 10.	H. ♃.	Eng. bot. v. 8. t. 539.

2 divaricàtum. s.s.	(*bl.*)	branching.	Barbary.	1798.	7. 9.	H. ⊙.	
3 spinòsum. s.s.	(*bl.*)	prickly.	Candia.	1633.	7. 8.	F. ♄.	Flor. græc. t. 823.
4 pùmilum. s.s.	(*bl.*)	dwarf.	Levant.	1790.	——	H. ⊙.	———— t. 821.
5 Endívia. s.s.	(*bl.*)	Endive.	E. Indies.	1548.	——	H. ⊙.	
α latifòlia.	(*bl.*)	*broad-leaved.*	————	———	——	H. ⊙.	
β críspa.	(*bl.*)	*curled-leaved.*	————	———	——	H. ⊙.	
γ leucophy'lla.	(*bl.*)	*white-leaved.*	————	———	——	H. ⊙.	
SCO'LYMUS. S.S.		GOLDEN THISTLE. Syngenesia Polygamia Æqualis. L.					
1 maculàtus. s.s.	(*ye.*)	annual.	S. Europe.	1633.	7. 8.	H. ⊙.	Flor. græc. t. 824.
2 hispánicus. s.s.	(*ye.*)	Spanish.	————	1658.	7. 9.	H. ♃.	———— t. 825.
3 grandiflòrus. s.s.	(*ye.*)	great-flowered.	Barbary.	1820.	——	H. ♃.	Desf. atl. 2. t. 218.
CATANA'NCHE. S.S.		CATANA'NCHE. Syngenesia Polygamia Æqualis. L.					
1 lùtea. s.s.	(*ye.*)	yellow.	Candia.	1640.	6. 7.	H. ⊙.	Flor. græc. t. 821.
2 cœrùlea. s.s.	(*bl.*)	blue-flowered.	S. Europe.	1596.	7. 10.	H. ♃.	Bot. mag. t. 293.

SUBORDO II. *LABIATIFLORÆ.* *Lagasca et DC. in ann. mus.*

L'ERIA. DC.		L'ERIA. Syngenesia Polygamia Superflua.					
1 nùtans. DC.	(*wh.*)	drooping-flower'd.	W. Indies.	1793.	6. 7.	S. ♂.	Plum. ic. t. 41. f. 1.
Tussilàgo nùtans. W.							
2 álbicans. DC.	(*wh.*)	woolly-leaved.	Jamaica.	1821.	——	S. ⊙.	
CHAPT'ALIA. V.		CHAPT'ALIA. Syngenesia Polygamia Necessaria.					
tomentòsa. V.	(*bh.*)	woolly.	N. America.	1806.	5. 6.	H. ♃.	Bot. mag. t. 2257.
Tussilàgo integrifòlia. M.							
ONO'SERIS. DC.		ONO'SERIS. Syngenesia Polygamia Æqualis.					
1 purpùrea. K.S.	(*pu.*)	purple.	S. America.	1827.	7. 9.	F. ♃.	Smith. ic. ined. t. 65.
Atráctylis purpuràta. Smith.							
2 mexicàna. K.S.	(*wh.*)	Mexican.	Mexico.	1828.	....	F. ♄.	Smith. ic. ined. 3. t. 66.
CHÆTANTH'ERA. R.P.		CHÆTANTH'ERA. Syngenesia Polygamia Superflua.					
ciliàta. R.P.	(*ye.*)	ciliated.	Chile.	1822.	7. 8.	F. ♃.	
MUTI'SIA. L.		MUTI'SIA. Syngenesia Polygamia Æqualis. L.					
speciòsa. B.M.	(*pu.*)	handsome.	Brazil.	1825.	9. 11.	G. ♄. ◡.	Bot. mag. t. 2705.
arachnoìdea. Martius.							
CHUQUIR'AGA. J.		CHUQUIR'AGA. Syngenesia Polygamia Æqualis.					
insígnis. K.S.	(*ye.*)	large-flowered.	S. America.	1824.	....	G. ♄.	Lam. ill. t. 691.
Johánnia insígnis. W.							
DUMERI'LIA. DC.		DUMERI'LIA. Syngenesia Polygamia Æqualis.					
paniculàta. DC.	(*pu.*)	panicled.	Chile.	1825.	7. 9.	H. ♃.	DC. ann. mus. 19. t. 7.
BAC'AZIA. R.P.		BAC'AZIA. Syngenesia Polygamia Æqualis.					
spinòsa. R.P.	(*pi.*)	spiny.	Peru.	1825.	4. 5.	F. ♄.	
TRIPTI'LION. R.P.		TRIPTI'LION. Syngenesia Polygamia Æqualis.					
cordifòlium. B.R.	(*wh.*)	heart-leaved.	Peru.	1823.	6. 8.	H. ⊙.	Bot. reg. t. 853.
LEUC'ERIA. Lag.		LEUC'ERIA. Syngenesia Polygamia Superflua.					
senecioídes. D.D.	(*wh.*)	Ragwort-like.	Chile.	1822.	8. 10.	F. ♃.	Hook. ex. flor. t. 101.
Tríxis senecioìdes. Hooker.							
GASTROCA'RPHA. D.D.		GASTROCA'RPHA. Syngenesia Polygamia Æqualis.					
runcinàta. D.D.	(*wh.*)	Turnep-leaved.	Chile.	1826.	8. 10.	F. ♂.	Swt. br. fl. gar. t. 229.
PERDICIUM. DC.		PERDICIUM. Syngenesia Polygamia Superflua.					
1 Anándria. DC.	(*wh.*)	Siberian.	Siberia.	1759.	8.	H. ♃.	Gmel. sib. 2. t. 68. f. 1.
2 lyràtum.	(*wh.*)	lyrate.	————	1825.	6. 8.	H. ♃.	———— t. 67. f. 2.
Tussilàgo lyràta. W. *Chaptàlia lyràta.* S.S.							
TRI'XIS. DC.		TRI'XIS. Syngenesia Polygamia Superflua.					
1 frutéscens. S.S.	(*wh.*)	frutescent.	Jamaica.	1826.	2. 4.	S. ♄.	Lam. ill. t. 677. f. 2.
Perdícium radiàle. L.							
2 divaricàta. D.D.	(*wh.*)	divaricate.	Brazil.	1824.	4. 6.	S. ♄.	Bot. mag. t. 2765.
auriculàta. B.M. *Perdícium brasiliénse.* Gr.							

SUBORDO III. *CARDUACEÆ.* *Don prodr. p.* 166.

DIV. I. CARDUACEÆ VERÆ. *Kth. synops.* 2. *p.* 363.

A'RCTIUM. L.		BURDOCK. Syngenesia Polygamia Æqualis.					
1 Láppa. W.	(*pu.*)	smooth-headed.	Britain.	....	7. 8.	H. ♂.	Eng. bot. v. 18. t. 1228.

2 minus. M.B.	(*pu.*)	smaller.	Europe.	....	7. 8.	H. ♂.	
Láppa minor. DC.							
3 Bardana. W.	(*pu.*)	woolly-headed.	Britain.	....	——	H. ♂.	Eng. bot. v. 35. t. 2478.
tomentòsum. P.S.							
SERRA'TULA. D.P.		SAWWORT. Syngenesia Polygamia Æqualis.					
1 pulchélla. B.M.	(*pu.*)	purple-scaled.	Siberia.	1823.	6. 7.	H. ♃.	Bot. mag. t. 2589.
2 depréssa. L.T.	(*pu.wh.*)	depressed.	Caucasus.	1820.	——	H. ♃.	Linn. trans. v. 11. t. 38.
3 élegans. L.T.	(*ro.*)	elegant.	———	——	——	H. ♃.	——— v. 11. t. 37.
4 alpìna. E.B.	(*pu.*)	Alpine.	Britain.	....	7. 8.	H. ♃.	Eng. bot. v. 9. t. 599.
Saussùrea alpìna. DC.							
5 angustifòlia. W.	(*pu.*)	narrow-leaved.	Siberia.	1816.	——	H. ♃.	Gmel. sib. 2. t. 33.
6 díscolor. W.	(*pu.wh.*)	two-coloured.	Europe.	——	——	H. ♃.	
Saussùrea díscolor. DC.							
β *lapathifòlia.* DC.		*Dock-leaved.*	———	——	——	H. ♃.	Moris. s. 7. t. 22. f. 1.
7 serràta.	(*pu.*)	saw-leaved.	———	——	——	H. ♃.	
Saussùrea serràta. DC.							
8 salicifòlia. W.	(*re.*)	Willow-leaved.	Siberia.	1796.	——	H. ♃.	Gmel. sib. 2. t. 27.
9 alàta.	(*pu.*)	winged.	———	1823.	——	H. ♃.	
Saussùrea alàta. DC. *non Serrátula alàta.* W.en.							
10 sálsa. M.B.	(*pu.*)	salt.	Caucasus.	1824.	——	H. ♃.	
Heterotrìchum sálsum. M.B. *Saussùrea sálsa.* S.S.							
11 lyràta. F.	(*pu.*)	lyrate-leaved.	Dahuria.	1827.	——	H. ♃.	
12 Gmelìni. Led.	(*pu.*)	Gmelin's.	Siberia.	——	——	H. ♃.	
13 elongàta. DC.	(*pu.*)	elongated.	———	1820.	——	H. ♃.	
14 nitídula. Led.	(*pu.*)	glossy.	———	1827.	——	H. ♃.	
15 pycnocéphala. Led.	(*pu.*)	large-headed.	———	——	——	H. ♃.	
16 rígida. Led.	(*pu.*)	rigid.	———	——	——	H. ♃.	
17 pygmæ̀a. Jac.	(*pu.*)	pigmy.	Austria.	1816.	——	H. ♃.	Jacq. aust. t. 440.
SYNCA'RPHA. DC.		SYNCA'RPHA. Syngenesia Polygamia Æqualis.					
gnaphalòdes. DC.	(*or.*)	squarrose.	C. B. S.	1815.	7. 9.	G. ♄.	Pluk. alm. t. 302. f. 3.
Stæhelìna gnaphaloídes. L. *Leysèra squarròsa.* Th.							
CA'RDUUS. D.P.		THISTLE. Syngenesia Polygamia Æqualis.					
1 alàtus. B.F.G.	(*pu.*)	winged-stalked.	....	1812.	6. 8.	H. ♂.	Swt. br. fl. gar. t. 103.
2 tinctòrius. D.P.	(*pu.*)	common.	Britain.	....	7. 10.	H. ♃.	Eng. bot. v. 1. t. 38.
Serrátula tinctòria. E.B.							
3 quinquefòlius. D.P.	(*pu.*)	five-leaved.	Persia.	1804.	7. 8.	H. ♃.	Bot. mag. t. 1871.
Serrátula quinquefòlia. B.M.							
4 coronàtus. D.P.	(*pu.*)	lyre-leaved.	Italy.	1739.	——	H. ♃.	Gmel. sib. 2. t. 20.
Serrátula coronàta. W.							
5 nítidus.	(*pu.*)	glossy.	Siberia.	1821.	——	H. ♃.	
Serrátula nítida. F.							
6 multiflòrus. L.	(*pu.*)	Volga.	———	——	——	H. ♃	Gmel. sib. 2. t. 28.
7 linearifòlius. DC.	(*pu.*)	linear-leaved.	———	1827.	——	H. ♃.	
8 Pìcris.	(*pu.*)	scariose.	Caucasus.	1822.	——	H. ♃.	
Centaùrea Pìcris. W. *Serrátula Pìcris.* S.S.							
9 centauroídes. D.P.	(*pu.*)	Centaury-like.	Siberia.	1804.	——	H. ♃.	Gmel. sib. 2. t. 40.
10 radiàtus. W.K.	(*pu.*)	rayed.	Hungary.	1800.	——	H. ♂.	W. et K. hung. 1. t. 11.
11 polyclònos. W.	(*pu.*)	Siberian.	Siberia.	1823.	——	H. ♃.	
12 argùtus.	(*pu.*)	sharply-tooth'd.	———	——	——	H. ♃.	
Serrátula argùta. L.en.							
13 cyanoídes. W.	(*pu.*)	blue-bottle-leav'd.	———	1778.	——	H. ♃.	Gmel. sib. 2. t. 15.
14 símplex.	(*pu.*)	one-flowered.	Caucasus.	1817.	——	H. ♃.	Bot. mag. t. 2482.
móllis. W. *Serrátula símplex.* DC.							
15 xeranthemoídes.	(*pu.*)	everlasting-fl[d].	———	1804.	——	H. ♃.	Gmel. sib. 2. t. 47. f. 1.
Serrátula xeranthemoídes. M.B. *Centaùrea radiàta.* W.							
16 lycopiifòlius. W.	(*pu.*)	horehound-leaved.	France.	1823.	——	H. ♃.	Vill. delph. 3. t. 19.
17 heterophy'llus. S.S.	(*pu.*)	various-leaved.	France.	1824.	——	H. ♃.	
18 stæchadifòlius.	(*pu.*)	woolly-headed.	Tauria.	1820.	——	H. ♃.	
Serrátula stæchadifòlia. S.S.							
19 leucógraphus. W.	(*pu.*)	white-spotted.	Italy.	1752.	6. 7.	H. ⊙.	Jacq. vind. 3. t. 23.
20 arábicus. W.	(*pu.*)	Arabian.	Arabia.	1789.	7. 8.	H. ⊙.	Jacq. ic. 1. t. 166.
21 nùtans. W.	(*pu.*)	musk.	Britain.	....	——	H. ⊙.	Eng. bot. v. 16. t. 1112.
22 carlinoídes. W.en.	(*pu.*)	Pyrenean.	Pyrenees.	1784.	——	H. ♂.	Gouan. ill. t. 23.
Carlìna pyrenàica. L.							
23 carlinæfòlius. W.	(*pu.*)	Carline-leaved.	———	1804.	——	H. ♂.	
24 acanthoídes. W.	(*li.*)	welted.	Britain.	....	6. 7.	H. ⊙.	Eng. bot. v. 14. t. 973.
25 hamulòsus. W.	(*pu.*)	hooked.	Hungary.	1802.	——	H ⊙.	
26 tenuiflòrus. E.B.	(*li.*)	slender-flower'd.	Britain.	....	——	H. ⊙.	Eng. bot. v. 6. t. 412.
27 críspus. W.	(*li.*)	curled.	Europe.	1804.	7. 8.	H. ♂.	Flor. dan. t. 621.

28	cándicans. W.	(*pu.*)	hoary.	Hungary.	1805.	7. 8.	H. ♂.	W. et K. hung. 1. t. 83.
29	personàtus. W.	(*pu.*)	cut-leaved.	Austria.	1776.	——	H. ♂.	Jacq. aust. 4. t. 348.
30	polyánthemus. W.	(*pu.*)	many-flowered.	Rome.	1739.	6. 7.	H. ♂.	Triumf. obs. t. 103.
31	paniculàtus. W.	(*pu.*)	panicled.	S. Europe.	1781.	——	H. ♃.	
32	pycnocéphalus. W.	(*pu.*)	Italian.	——	1739.	7. 9.	H. ♃.	Jacq. vind. 1. t. 34.
33	onopordioídes. M.B.	(*pu.*)	Onopordum-like.	Iberia.	1824.	7. 8.	H. ♃.	
34	argentàtus. W.	(*pu.*)	white Egyptian.	Egypt.	1789.	6. 8.	H. ⊙.	Jacq. vind. 2. t. 192.
35	arctioídes. W.	(*pu.*)	pinnated.	Hungary.	1804.	7. 8.	H. ♃.	
36	dùbius. S.S.	(*pu.*)	doubtful.	...?....	1827.	——	H. ♃.	
37	seminùdus. S.S.	(*pu.*)	half-naked.	Caucasus.	——	——	H. ♃.	
38	uncinàtus. S.S.	(*pu.*)	crook-pointed.	Volga.	——	——	H. ♃.	
39	alpéstris. W.	(*pu.*)	alpine.	Hungary.	1805.	7. 8.	H. ♃.	
40	defloràtus. W.	(*pu.*)	different leav'd.	Austria.	1570.	7. 9.	H. ♃.	Jacq. aust. 1. t. 89.
41	corymbòsus. S.S.	(*pu.*)	corymb-flower'd.	Naples.	1827.	——	H. ♃.	
42	cinèreus. S.S.	(*pu.*)	gray.	Caucasus.	——	——	H. ♃.	
43	collìnus. S.S.	(*pu.*)	hill.	Hungary.	1828.	——	H. ♃.	
44	Argemòne. DC.	(*pu.*)	Argemone-leav'd.	Pyrenees.	1816.	——	H. ♃.	
45	parviflòrus. W.	(*pu.*)	small-flowered.	S. Europe.	1781.	6. 7.	H. ♃.	
46	crassifòlius. W.en.	(*pu.*)	thick-leaved.	....	1805.	7.	H. ♃.	
47	sumànus. L.en.	(*pu.*)	Pollini's.	Italy.	1823.	7. 8.	H. ♃.	
48	nítidus. W.	(*pu.*)	glossy.	Hungary.	1806.	——	H. ♃.	W. et K. hung. 1. t. 52.
49	nudicaùlis. P.S.	(*pu.*)	naked-stalked.	S. Europe.	1739.	——	H. ♃.	Cav. ic. 3. t. 226.
	cerinthoídes. W. *glaùcus.* Cav. *Centaùrea nudicaùlis.* L. *Serrátula nudicaùlis.* DC.							
50	clavulàtus. S.S.	(*pu.*)	clubbed.	Canaries.	1827.	——	F. ♂.	
51	álbidus. S.S.	(*pu.*)	white-leaved.	Caucasus.	1829.	——	H. ♃.	
52	argyròa. S.S.	(*pu.*)	silvery.	Sicily.	1827.	——	H. ♃.	
53	orientàlis. M.B.	(*pu.*)	oriental.	Iberia.	1804.	7.	H. ♃.	
54	salìnus. M.B.	(*pu.*)	narrow-leaved.	Siberia.	——	——	H. ♃.	
55	volgénsis. S.S.	(*pu.*)	Volga.	Volga.	——	——	H. ♃.	
56	leucánthus. S.S.	(*wh.*)	white-flowered.	Spain.	1825.	——	H. ⊙.	Cavan. icon. 2. t. 165.
57	atriplicifòlius. S.S.	(*pu.*)	Orach-leaved.	Russia.	1823.	——	H. ♃.	
	SI'LYBUM. G.		SI'LYBUM. Syngenesia Polygamia Æqualis.					
1	marianum. G.	(*pu.*)	milk-thistle.	Britain.	....	7.	H. ♂.	Eng. bot. v. 14. t. 976.
	Cárduus mariànus. E.B.							
2	cérnuum. G.	(*wh.*)	nodding.	Siberia.	1755.	6. 7.	H. ♃.	Mill. ic. 2. t. 248.
	Cnìcus cérnuus. W.							
	HALOCHA'RIS. F.		HALOCHA'RIS. Syngenesia Polygamia Æqualis.					
1	altáica. F.	(*pu.*)	Altay.	Altay.	1827.	7. 8.	H. ♃.	
2	carthamoídes. F.	(*pu.*)	Carthamus-like.	Siberia.	1823.	——	H. ♃.	
	Cnìcus carthamoídes. W. *Leùzea? carthamoídes.* DC. *Cnìcus salìnus.* M.B.							
	CI'RSIUM. G.		HORSE-THISTLE. Syngenesia Polygamia Æqualis.					
1	palústre. DC.	(*li.wh.*)	marsh.	Britain.	....	7. 8.	H. ♂.	Eng. bot. v. 14. t. 974.
	Cárduus palústris. E.B. *Cnìcus palústris.* W.							
2	elàtum. L.en.	(*pu.*)	tall.	....	1823.	——	H. ♃.	
3	uliginòsum. M.B.	(*pu.*)	swamp.	Caucasus.	1820.	——	H. ♃.	
4	cànum. M.B.	(*pu.*)	hoary.	Austria.	1633.	——	H. ♃.	Jacq. aust. 1. t. 42-43.
5	Acárna. DC.	(*pu.*)	yellow.	Spain.	1683.	7. 9.	H. ⊙.	Cavan. ic. 1. t. 53.
6	pyrenàicum. DC.	(*pu.*)	Pyrenean.	Pyrenees.	1818.	7. 8.	H. ♃.	Jacq. obs. 4. t. 95.
7	monspessulànum. DC.	(*pu.*)	Montpelier.	Montpelier.	1596.	6. 7.	H. ♃.	
8	pannònicum. L.en.	(*pu.*)	Hungarian.	Hungary.	1816.	7. 9.	H. ♃.	Jacq. aust. 3. t. 5.
9	stríctum. S.S.	(*pu.*)	straight.	Naples.	1820.	——	H. ♃.	
10	lanceolàtum. DC.	(*pu.*)	common.	Britain.	....	6. 9.	H. ♃.	Eng. bot. v. 2. t. 107.
	Cárduus lanceolàtus. E.B. *Cnìcus lanceolàtus.* W.							
11	desertòrum. L.en.	(*pu.*)	desert.	Siberia.	1823.	7. 8.	H. ♃.	
12	cichoràceum. S.S.	(*pu.*)	Succory-like.	Naples.	1824.	——	H. ♃.	
13	itálicum. S.S.	(*pu.*)	Italian.	Italy.	1826.	——	H. ♃.	
14	sículum. S.S.	(*pu.*)	Sicily.	Sicily.	1828.	——	H. ♃.	
15	negléctum. S.S.	(*pu.*)	neglected.	Siberia.	1829.	——	H. ♃.	
16	púngens. S.S.	(*pu.*)	pungent.	Levant.	——	——	H. ♃.	
17	paniculàtum. S.S.	(*pu.*)	panicled.	Pyrenees.	1827.	——	H. ♃.	
18	créticum. S.S.	(*wh.*)	Cretan.	Crete.	——	——	H. ♃.	
19	ciliàtum. L.en.	(*pu.*)	fringed.	Siberia.	1787.	8.	H. ♃.	Murr. com. got. 6. t. 5.
20	serrulàtum. M.B.	(*pu.*)	serrulate.	Tauria.	1820.	7. 8.	H. ♃.	
21	laniflòrum. M.B.	(*pu.*)	woolly-flowered.	——	1823.	——	H. ♃.	
22	arachnoídeum. M.B.	(*pu.*)	cobweb.	——	——	——	H. ♃.	
23	strigòsum. M.B.	(*pu.*)	strigose.	Caucasus.	1818.	——	H. ♂.	
24	hórridum. M.B.	(*pu.*)	very spiny.	——	1820.	——	H. ♃.	
25	mùnitum. M.B.	(*pu.*)	armed.	——	1816.	——	H. ♃.	
26	obvallàtum. M.B.	(*pu.*)	crowded-flower'd.	——	——	——	H. ♃.	

27 scleránthon. M.B. (*wh.*) hard-flowered. Caucasus. 1824. 7. 8. H. ♂.
carlinoídes. F. *Carlìna Echinus.* M.B.
28 echinàtum. S.S. (*pu.*) hedgehog. Barbary. —— —— H. ♃.
29 erióphorum. DC. (*pu.*) woolly-headed. Britain. .... —— H. ♂. Eng. bot. v. 6. t. 386.
Cárduus erióphorus. E.B. *Cnìcus erióphorus.* W.
30 fèrox. DC. (*wh.*) fierce. S. Europe. 1683. —— H. ♂. All. ped. 1. t. 50.
31 díscolor. S.S. (*pu.*) two-coloured. N. America. 1803. —— H. ♂.
32 inérme. L.en. (*pu.*) unarmed. Pyrenees. 1818. —— H. ♃.
33 cynaroídes. S.S. (*pu.*) Artichoke-like. Crete. 1828. —— H. ♃.
34 altíssimum. S.S. (*pu.*) giant. N. America. 1726. 8. 9. H. ♃. Dill. elth. t. 69. f. 80
35 praténse. (*pu.*) meadow. Britain. .... 6. 7. H. ♃. Eng. bot. v. 3. t. 177.
ánglicum. DC. *Cárduus praténsis.* E.B. *Cnìcus praténsis.* W.
36 heterophy'llum. DC. (*pu.*) melancholy. ———— .... 7. 8. H. ♃. ———— v. 10. t. 675.
37 helenioídes. L.en. (*pu.*) Elecampane-ld. Siberia. 1804. —— H. ♃.
38 serratuloídes. S.S. (*pu.*) sawwort-like. ———— 1752. 6. 10. H. ♃. Jacq. aust. 2. t. 127.
39 heteromállum. (*pu.*) Nepaul. Nepaul. 1820. —— H. ♂.
Cnìcus heteromállus. D.P. *Cárduus cándicans.* Wal.
40 arvénse. DC. (*li.*) corn or way. Britain. .... 7. H. ♃. Eng. bot. v. 14. t. 975.
Serrátula arvénsis. W. *Cárduus arvénsis.* E.B. *Cnìcus arvénsis.* Ph.
41 rivulàre. S.S. (*pu.*) river. Hungary. 1804. 7. 8. H. ♃. Jacq. aust. 1. t. 91.
tricephalòdes. DC. *Cnìcus rivulàris.* W.
β *salisburgénse.* *smaller.* ———— —— —— H. ♃.
42 ambíguum. DC. (*pu.*) ambiguous. S. Europe. 1823. —— H. ♃.
43 orgyàle. S.S. (*pu.*) tall. .... —— —— H. ♃.
44 carniòlicum. S.S. (*wh.*) Carniola. Carniola. 1792. —— H. ♃. Scop. carn. 1005. t. 54.
45 pauciflòrum. S.S. (*pu.*) few-flowered. Hungary. 1816. —— H. ♂. W. et K. hung. 2. t. 161.
46 tatáricum. S.S. (*wh.*) Tartarian. Siberia. 1775. 7. 8. H. ♃. Jacq. aust. 1. t. 90.
47 rìgens. S.S. (*wh.*) upright alpine. Switzerland. —— —— H. ♃. La Chenal. act. h. 4. t. 16.
48 oleràceum. DC. (*wh.*) pale-flowered. Europe. 1570. —— H. ♃. Flor. dan. 860.
49 Erisíthales. L.en. (*st.*) clammy. France. 1752. 6. 8. H. ♃. Jacq. aust. 4. t. 310.
glutinòsum. DC. *Cnìcus Erisíthales.* W.
50 ochroleùcum. DC. (*st.*) pale yellow. Switzerland. 1801. 7. H. ♃.
51 mìte. S.S. (*pu.*) soft. Siberia. 1818. 7. 8. H. ☉.
52 fimbriàtum. S.S. (*pu.*) fringed. Caucasus. 1829. —— H. ♃.
53 mèdium. S.S. (*pu.*) middle. Italy. —— —— H. ♃.
54 níveum. S.S. (*pu.*) snowy-leaved. Sicily. 1828. —— H. ♃.
55 dealbàtum. S.S. (*pu.*) mealy-leaved. Caucasus. —— —— H. ♃.
56 echinocéphalum. S.S. (*pu.*) spiny-headed. ———— 1823. —— H. ♃.
57 lappàceum. M.B. (*pu.*) Burdock-like. ———— 1826. —— H. ♃.
58 pinnatífidum. S.S. (*pu.*) pinnatifid. Spain. —— —— H. ♃.
59 bulbòsum. DC. (*pu.*) tuberous. England. .... 8. 10. H. ♃. Eng. bot. v. 36. t. 2562.
Cnìcus tuberòsus. E.B. *Cárduus tuberòsus.* L.
60 setòsum. M.B. (*pu.*) setose. Tauria. 1823. 7. 8. H. ♃.
61 acaùle. DC. (*pu.*) dwarf. Britain. .... —— H. ♃. Eng. bot. v. 3. t. 161.
Cárduus acaùlis. E.B. *Cnìcus acaùlis.* W.
β *cauléscens.* P.S. (*pu.*) *stalked.* ———— .... —— H. ♃.
62 Casabònæ. DC. (*pu.*) fish-bone. S. Europe. 1714. 6. 8. F. ♂. Schmidt ic. t. 51-52.
63 stellàtum. DC. (*pu.*) starry. Italy. 1665. 6. 7. H. ☉. Flor. græc. t. 830.
64 Diacánthon. S.S. (*pu.*) two-spined. Barbary. 1800. —— F. ♂. Lab. ic. pl. syr. 1. t. 3.
65 áfrum. S.S. (*pu.*) African. ———— 1802. —— F. ♂. Jacq. schœnb. 2. t. 145.
66 squarròsum. S.S. (*pu.*) squarrose. Siberia. 1829. —— H. ♃.
67 incànum. S.S. (*pu.*) hoary. Caucasus. —— —— H. ♃.
68 syrìacum. L.en. (*wh.*) Syrian. Levant. 1771. 7. 8. H. ☉. Flor. græc. t. 831.
69 spinosíssimum. S.S. (*wh.*) feathery-headed. Switzerland. 1759. 6. 8. H. ♃. Bot. mag. t. 1366.
70 tatáricum. S.S. (*wh.*) Tartarian. Tartary. 1829. —— H. ♃.
71 Bertolònii. S.S. (*wh.*) Bertoloni's. Italy. 1828. —— H. ♃.
72 b'æticum. S.S. (*wh.*) Baetian. Spain. 1825. —— H. ♃.
73 semipectinàtum. S.S. (*w.*) semipectinate. Siberia. 1828. —— H. ♃.
74 Gmelìni. F. (*pu.*) Gmelin's. Baical. 1829. —— H. ♃.
75 arábicum. F. (*pu.*) Arabian. Arabia. —— —— H. ☉.
76 pygm'æum. DC. (*pu.*) pigmy. Austria. 1816. —— H. ♃. Jacq. aust. t. 440.
Serrátula pygm'æa. W.

ERYTHROL'ÆNA. ERYTHROL'ÆNA. Syngenesia Polygamia Æqualis.
conspícua. B.F.G. (*sc.*) conspicuous. Mexico. 1824. 9. 10. F. ♂. Swt. br. fl. gar. t. 134.

ONOPO'RDUM. DC. COTTON-THISTLE. Syngenesia Polygamia Æqualis.
1 Acánthium. E.B. (*pu.*) woolly. Britain. .... 7. 8. H. ♂. Eng. bot. v. 14. t. 977.
2 taùricum. W. (*pu.*) Taurian. Tauria. 1800. —— H. ♂.
3 vìrens. DC. (*pu.*) green. S. Europe. 1822. —— H. ♂.
4 macrocánthum. W. (*pu.*) long-spined. Barbary. 1798. —— H. ♂. Schousb. maroc. t. 5.

5 illy'ricum. w.	(*pu.*)	Illyrian.	S. Europe.	1648.	7. 8.	H. ♂.	Jacq. vind. 2. t. 148.
6 arábicum. w.	(*pu.*)	Arabian.	———	1686.	7.	H. ♂.	——— 2. t. 149.
7 elàtum. Sm.	(*pu.*)	tall.	Greece.	1818.	7. 8.	H. ♂.	
8 deltoídes. w.	(*pu.*)	Siberian.	Siberia.	1784.	8.	H. ♃.	
9 gr'æcum. w.	(*pu.*)	Grecian.	Greece.	1799.	6. 7.	H. ♃.	Gouan. ill. t. 25.
10 cynaroídes. L.en.	(*pu.*)	Cynara-like.	Caucasus.	1823.	7. 8.	H. ♃.	
Cárthamus cynaroìdes. M.B.							
11 viscòsum. Schr.	(*pu.*)	viscous.	Europe.	1821.	——	H. ♂.	
12 pyrenàicum. DC.	(*pu.*)	Pyrenean.	Pyrenees.	1818.	——	H. ♃.	
13 acaùlon. w.	(*pu.*)	dwarf.	....	1739.	——	H. ♂.	Jacq. ic. 1. t. 167.
BERA'RDIA. P.S.		BERA'RDIA.	Syngenesia Polygamia Æqualis.				
subacaùlis. P.S.	(*pu.*)	round-leaved.	Italy.	1791.	7. 8.	H. ♃.	Vill. dauph. 3. t. 22.
A'rctium lanuginòsum. DC. *Onopórdum rotundifòlium.* W.							
ACIL'EPIS. D.P.		ACIL'EPIS.	Syngenesia Polygamia Æqualis.				
squarròsa. D.P.	(*pu.*)	squarrose.	Nepaul.	1821.	7. 8.	H. ♃.	
CYN'ARA. W.		ARTICHOKE.	Syngenesia Polygamia Æqualis.				
1 Cardúnculus. w.	(*bl.*)	Cardoon.	Candia.	1658.	8. 9.	H. ♃.	Tabern. ic. 696.
2 Scólymus. w.	(*bl.*)	common.	S. Europe.	1548.	——	H. ♃.	Blackw. t. 458.
3 spinosíssima. S.S.	(*bl.*)	very spiny.	Sicily.	1824.	——	H. ♃.	
4 hórrida. w.	(*bl.*)	very prickly.	Madeira.	1778.	——	G. ♃.	Flor. græc. t. 834.
5 hùmilis. w.	(*bl.*)	dwarf.	Spain.	1613.	7. 8.	H. ♃.	Pluk. alm. t. 81. f. 2.
6 córsica. Viv.	(*bl.*)	Corsican.	Corsica.	1825.	——	H. ♃.	
7 py'gmæa. w.	(*bl.*)	small.	Spain.	1824.	——	H. ♃.	
CESTR'INUS. Cass.		CESTR'INUS.	Syngenesia Polygamia Æqualis.				
carthamoídes. Cass.	(*bl.*)	Carthamus-like.	Barbary.	1799.	7.	H. ♃.	Desf. atl. 2. t. 223.
Cynàra acaùlis. W. *Serrátula acaùlis.* DC.							
CARL'INA. W.		CARLINE-THISTLE.	Syngenesia Polygamia Æqualis.				
1 acaùlis. w.	(*wh.*)	dwarf.	Italy.	1640.	6.	H. ♃.	Knorr. thes. 2. t. c. 1.
β caulèscens. P.S.	(*wh.*)	*stalked.*	———	——	——	H. ♃.	
2 acanthifòlia. w.	(*wh.*)	Acanthus-leav'd.	Pyrenees.	1818.	6. 7.	H. ♃.	Allion. ped. t. 51.
3 símplex. P.S.	(*wh.*)	single-flowered.	Hungary.	1816.	——	H. ♃.	W. et K. hung. 2. t. 152.
4 aggregàta. w.	(*wh.*)	clustered.	———	1804.	6. 9.	H. ♃.	
5 lyràta. w.	(*wh.*)	lyre-leaved.	C. B. S.	1816.	——	G. ♂.	
6 lanàta. w.	(*ro.*)	woolly.	S. Europe.	1683.	6. 7.	H. ⊙.	Garid. aix. t. 21.
7 corymbòsa. w.	(*ye.*)	corymbed.	———	1640.	7. 8.	H. ♃.	Flor. græc. t. 837.
8 vulgàris. w.	(*bh.*)	common.	Britain.	....	5. 9.	H. ♂.	Eng. bot. v. 16. t. 1144.
9 racemòsa. w.	(*ye.*)	racemed.	Spain.	1658.	6. 8.	H. ⊙.	
10 sícula. S.S.	(*wh.*)	Sicilian.	Sicily.	1828.	——	H. ♃.	
11 Biebersteínii. Ot.	(*wh.*)	Bieberstein's.	Caucasus.	1821.	——	H. ♃.	
STOB'ÆA. W.		STOB'ÆA.	Syngenesia Polygamia Æqualis.				
1 pinnàta. w.	(*ye.*)	Carthamus-like.	C. B. S.	1812.	1. 12.	G. ♄.	Bot. mag. t. 1788.
2 glomeràta. S.S.	(*ye.*)	clustered.	———	1816.	7. 8.	G. ♃.	
Cynàra glomeràta. Thunberg.							
3 atractyloídes. S.S.	(*ye.*)	Atractylis-like.	———	1823.	——	G. ♃.	
CA'RTHAMUS. S.S.		BASTARD SAFFRON.	Syngenesia Polygamia Æqualis.				
1 tinctòrius. w.	(*or.*)	common.	Egypt.	1551.	6. 7.	H. ⊙.	Bot. reg. t. 170.
2 oxyacántha. S.S.	(*ye.*)	sharp-spined.	Caucasus.	1829.	——	H. ⊙.	
HERACA'NTHA. L.en.		HERACA'NTHA.	Syngenesia Polygamia Æqualis.				
1 lanàta. L.en.	(*ye.*)	woolly.	S. Europe.	1596.	7. 8.	H. ⊙.	Bot. mag. t. 2143.
Cárthamus lanàtus. B.M. *Centaùrea lanàta.* DC.							
2 crética. L.en.	(*wh.*)	Cretan.	Candia.	1781.	6. 7.	H. ⊙.	
Cárthamus créticus. W.							
3 taùrica. L.en.	(*ye.*)	Taurian.	Tauria.	1818.	——	H. ⊙.	
4 glaùca.	(*pu.*)	glaucous.	Iberia.	——	——	H. ⊙.	
Cárthamus glaùcus. M.B.							
ONOBR'OMA. G.		ONOBR'OMA.	Syngenesia Polygamia Æqualis.				
1 tingitàna.	(*bl.*)	Tangier.	Barbary.	1759.	6. 7.	H. ♃.	Cavan. ic. 2. t. 128.
Cárthamus tingitànus. W.							
2 cœrùlea. L.en.	(*bl.*)	blue-flowered.	Spain.	1640.	——	H. ♃.	Bot. mag. t. 2293.
Cárthamus cœrùleus. W.							
3 arboréscens. S.S.	(*ye.*)	tree.	Spain.	1731.	7. 8.	F. ♄.	
Cárthamus arboréscens. W.							
CARDUNCE'LLUS. DC.		CARDUNCE'LLUS.	Syngenesia Polygamia Æqualis.				
1 vulgàris. DC.	(*bl.*)	common.	France.	1776.	5. 6.	H. ♃.	Lob. ic. 2. p. 20.
Cárthamus Carduncéllus. W.							
2 mitíssimus. DC.	(*bl.*)	small.	———	1734.	6. 7.	H. ♃.	
CARLOW'IZIA. DC.		CARLOW'IZIA.	Syngenesia Polygamia Æqualis.				
salicifòlia. DC.	(*st.*)	Willow-leaved.	Madeira.	1784.	8.	G. ♄.	
Cárthamus salicifòlius. W. *Onobròma salicifòlium.* L.en.							

ATRA'CTYLIS. W.		Atra'ctylis. Syngenesia Polygamia Æqualis.					
hùmilis. w.	(*wh.*)	dwarf.	Spain.	1759.	5. 7.	H. ♃.	Cav. ic. 1. t. 54.
ACA'RNA. W.		Aca'rna. Syngenesia Polygamia Æqualis.					
1 gummífera. w.	(*pu.*)	gummy-rooted.	S. Europe.	1640.	6. 8.	H. ♃.	Cavan. ic. 3. t. 228.
2 serratuloídes. s.s.	(*pu.*)	Sawwort-like.	Palestine.	1827.	——	H. ♃.	
3 comòsa. s.s.	(*pu.*)	tufted.	————	——	——	H. ♃.	
4 cancellàta. w.	(*bl.*)	netted.	S. Europe.	——	6. 7.	H. ☉.	Lam. ill. t. 662. f. 1.
Atráctylis cancellàta. P.S.							
STOK'ESIA. W.		Stok'esia. Syngenesia Polygamia Æqualis.					
cyánea. w.	(*bl.*)	blue-flowered.	Carolina.	1766.	8.	G. ♃.	L'Herit. sert. 27. t. 38.
STÆHEL'INA. W.		Stæhel'ina. Syngenesia Polygamia Æqualis.					
1 dùbia. w.	(*pu.*)	Rosemary-leav'd.	S. Europe.	1640.	6. 7.	H. ♄.	Lam. ill. t. 666. f. 4.
2 arboréscens. w.	(*re.*)	Storax-leaved.	Candia.	1739.	7. 9.	F. ♄.	Flor. græc. t. 345.
3 chamæpeùce. w.	(*re.*)	Pine-leaved.	————	1640.	7. 11.	F. ♄.	———— t. 347.
PTER'ONIA. W.		Pter'onia. Syngenesia Polygamia Æqualis.					
1 camphoràta. w.	(*ye.*)	aromatic.	C. B. S.	1774.	6. 7.	G. ♄.	Pluk. mant. t. 345. f. 2.
2 strícta. w.	(*ye.*)	cluster-flowered.	————	——	4. 6.	G. ♄.	
3 echinàta. w.	(*ye.*)	prickly-cupped.	————	1820.	——	G. ♄.	
4 flexicaùlis. w.	(*ye.*)	bending-stalk'd.	————	1812.	6. 8.	G. ♄.	
5 fasciculàta. w.	(*ye.*)	cluster-flowered.	————	1816.	——	G. ♄.	
6 pállens. w.	(*pa.*)	pale.	————	——	——	G. ♄.	
7 oppositifòlia. w.	(*ye.*)	opposite-leaved.	————	1774.	7.	G. ♄.	Lam. ill. t. 667. f. 2.
8 glomeràta. w.	(*ye.*)	crowded.	————	1823.	6. 8.	G. ♄.	
9 scariòsa. w.	(*ye.*)	window-cupped.	————	1815.	——	G. ♄.	
Z'ŒGEA. W.		Z'œgea. Syngenesia Polygamia Frustranea.					
Leptaùrea. w.	(*ye.*)	yellow-flower'd.	Levant.	1779.	7. 8.	H. ☉.	Jacq. ic. 1. t. 177.
LE'UZEA. DC.		Le'uzea. Syngenesia Polygamia Frustranea.					
1 conífera. DC.	(*pu.*)	cone.	S. Europe.	1683.	6. 9.	H. ♃.	DC. ann. mus. 16. t. 14.
Centaùrea conífera. w.							
2 altàica. L.en.	(*pu.*)	large-flowered.	Altay.	1818.	——	H. ♃.	
3 dahùrica. F.	(*pu.*)	Dahurian.	Dahuria.	1821.	——	H. ♃.	
4 austràlis. D.D.	(*pu.*)	New Holland.	N. Holland.	1821.	8. 9.	H. ♃.	
GALA'CTITES. DC.		Gala'ctites. Syngenesia Polygamia Frustranea.					
tomentòsa. DC.		woolly.	S. Europe.	1738.	7.	H. ☉.	DC. ann. mus. 16. t. 16.
Centaùrea galáctites. w.							
CRUP'INA. Cas.		Starry Scabious. Syngenesia Polygamia Frustranea.					
1 vulgàris. Cas.	(*re.*)	black-seeded.	Italy.	1596.	6. 7.	H. ☉.	Colum. ecph. 1. t. 34.
Centaùrea Crupìna. w.							
2 crupinoídes. P.S.	(*sn.*)	saffron-colour'd.	Barbary.	1826.	——	H. ☉.	
3 Líppii. P.S.		Lippi's.	Egypt.	1739.	——	H. ☉.	Isn. act. par. 1719. t. 10.
Centaùrea Líppii. w.							
4 arenària. P.S.	(*pu.*)	sand.	Volga.	1821.	——	H. ☉.	
CENTA'UREA. W.		Centaury. Syngenesia Polygamia Frustranea.					
1 moschàta. w.	(*li.*)	sweet sultan.	Persia.	1629.	7. 10.	H. ☉.	Knor. thes. 2. t. C. 4.
2 glaùca. w.	(*pu.*)	glaucous.	Caucasus.	1805.	6. 7.	H. ☉.	
3 suavèolens. w.	(*ye.*)	yellow sultan.	Levant.	1683.	7. 10.	H. ☉.	Swt. brit. fl. gar. 1. t. 51.
4 alpìna. w.	(*ye.*)	Alpine.	Italy.	1640.	7. 8.	H. ♃.	Cornut. can. 69. t. 70.
5 Centaùrium. w.	(*bl.*)	great.	————	1596.	——	H. ♃.	
6 ruthènica. w.	(*st.*)	Russian.	Russia.	1806.	——	H. ♃.	Gmel. sib. 2. t. 41.
7 phry'gia. w.	(*pu.*)	feathery-cupp'd.	Switzerland.	1633.	6. 10.	H. ♃.	Flor. dan. t. 520.
8 rivulàris. P.S.	(*br.*)	river-side.	Portugal.	1812.	7. 9.	H. ♃.	
9 salicifòlia. w.	(*pu.*)	Willow-leaved.	Caucasus.	1816.	——	H. ♃.	
10 austrìaca. w.	(*pu.*)	Austrian.	Austria.	1815.	6. 10.	H. ♃.	
11 pectinàta. w.	(*pu.*)	pectinated.	France.	1727.	7. 10.	H. ♃.	
12 uniflòra. w.	(*pu.*)	one-flowered.	S. Europe.	1824.	6. 8.	H. ♃.	Bocc. mus. 2. t. 2.
13 flosculòsa. w.	(*pu.*)	flosculose.	Italy.	1817.	——	H. ♃.	
14 nervòsa. w.	(*pu.*)	nerved.	S. Europe.	1815.	6. 9.	H. ♃.	
15 linifòlia. P.S.	(*pu.*)	Flax-leaved.	————	1826.	——	H. ♃.	
16 capillàta. P.S.	(*pu.*)	hairy-headed.	Volga.	1827.	——	H. ♃.	
17 tricocéphala. w.	(*pu.*)	downy cupped.	Siberia.	1805.	7. 8.	H. ♃.	Gmel. sib. 2. t. 45. f. 1. 2.
18 hyssopifòlia. w.	(*pu.*)	Hyssop-leaved.	Spain.	1812.	——	F. ♄.	Barrel. ic. 306.
19 coronopifòlia. w.	(*ye.*)	Buck's-horn-ld.	Levant.	1739.	6. 7.	H. ☉.	
20 procúmbens. P.S.	(*pu.*)	procumbent.	Italy.	1821.	——	H. ♃.	
21 nìgra. w.	(*pu.*)	black knapweed.	Britain.	....	5. 8.	H. ♃.	Eng. bot. v. 4. t. 278.
22 praténsis. P.S.	(*pu.*)	field.	France.	1824.	——	H. ♃.	
23 nigréscens. w.	(*pu.*)	dark.	Hungary.	1805.	6. 8.	H. ♃.	
24 decípiens. P.S.	(*pu.*)	deceived.	France.	1826.	——	H. ♃.	
25 Triumfétti. w.	(*re.*)	Cenisian.	Mt. Cenis.	——	——	H. ♃.	
26 montàna. w.	(*bl.*)	mountain.	Austria.	1596.	——	H. ♃.	Bot. mag. t. 77.

27	seusàna. P.S.		silvery-fringed.	S. Europe.	1816.	6. 8.	H. ♃.	Barrel. ic. 389.
	axillàris. W. *variegàta.* Lam.							
28	strícta. W.K.	(*bl.*)	upright.	Hungary.	1826.	——	H. ♃.	
29	depréssa. M.B.	(*bl.*)	depressed.	Caucasus.	——	——	H. ♃.	
30	lingulàta. S.S.	(*bl.*)	tongue-leaved.	Spain.	1825.	——	H. ♃.	
31	spathulàta. S.S.	(*bl.*)	spathulate-leav'd.	Naples.	1827.	——	H. ♃.	
32	Cyánus. W.	(*bl.va.*)	blue-bottle.	Britain.	....	——	H. ⊙.	Eng. bot. v. 4. t. 277.
	Cyánus ségetum. Mœnch.							
33	americàna. N.	(*li.*)	American.	N. America.	1823.	7. 10.	H. ⊙.	Swt. br. fl. gar. ic.
34	ochroleùca. W.	(*st.*)	Caucasean.	Caucasus.	1801.	7. 8.	H. ⊙.	Bot. mag. t. 1175.
35	africàna. W.	(*ye.*)	African.	Africa.	1825.	——	F. ♃.	
36	cheiranthifòlia. W.	(*ye.*)	Stock-leaved.	Caucasus.	1829.	——	H. ♃.	
37	ovìna. W.	(*ye.*)	small-flowered.	————	1802.	7. 11.	H. ♃.	
38	paniculàta. W.	(*pu.*)	panicled.	Europe.	1640.	7. 8.	H. ♂.	Jacq. aust. 4. t. 320.
39	intybàcea. P.S.	(*re.*)	Succory-leaved.	S. Europe.	1778.	7. 9.	H. ♃.	
40	leucántha. Lam.	(*wh.*)	white-blossom'd.	————	1826.	——	H. ♃.	Barrel. ic. 359.
41	spinòsa. W.	(*wh.*)	prickly-branch'd.	Candia.	1640.	——	F. ♄.	Bot. mag. t. 2493.
42	ragusìna. W.	(*ye.*)	white-leaved.	————	1710.	6. 7.	G. ♄.	———— t. 494.
43	argéntea. W.	(*ye.*)	silver-leaved.	————	1739.	7. 8.	F. ♄.	Barrel. ic. t. 218.
44	Cinerària. W.	(*pu.*)	hoary-leaved.	Italy.	1710.	——	H. ♃.	Moris. s. 7. t. 26. f. 20.
45	cinèrea. W.	(*pu.*)	gray.	————	——	6. 7.	H. ♃.	Jacq. vind. 1. t. 92.
46	dealbàta. M.B.	(*pu.*)	mealy.	Caucasus.	1804.	7. 8.	H. ♃.	
47	cicutæfòlia. H.H.	(*pu.*)	Cicuta-leaved.	Europe.	1824.	——	H. ♃.	
48	maculòsa. P.S.	(*pu.*)	spotted-cupped.	Siberia.	1816.	——	H. ♃.	Gmel. sib. 2. t. 44. f. 1. 2.
49	Marschalliàna. S.S.	(*pu.*)	Marschall's.	Caucasus.	1829.	——	H. ♃.	
50	declinàta. M.B.	(*pu.*)	declining.	————	1827.	——	H. ♃.	
51	leucophy'lla. M.B.	(*pu.*)	white-leaved.	————	1824.	——	H. ♃.	
52	trinérvia. W.	(*pu.*)	three-nerved.	————	1828.	——	H. ♃.	
53	dissécta. T.N.	(*pu.*)	cut-leaved.	Italy.	1824.	——	H. ♃.	
54	sempervìrens. W.	(*ye.*)	evergreen.	Spain.	1683.	——	H. ♃.	Bocc. sic. t. 39. f. 3.
55	coriàcea. W.	(*re.*)	leathery-leav'd.	Hungary.	1804.	6. 7.	H. ♃.	W. et K. hung. 2. t. 195.
56	Scabiòsa. W.	(*pu.*)	greater-knapweed.	Britain.	....	6. 8.	H. ♃.	Eng. bot. v. 1. t. 56.
57	tatárica. W.	(*ye.*)	Tartarian.	Tartary.	1801.	7. 8.	H. ♃.	
58	St'œbe. W.	(*pu.*)	wing-leaved.	Austria?	1759.	6. 7.	H. ♃.	
59	Fischèri. W.en.	(*pu.*)	Fischer's.	Siberia.	1816.	——	H. ♃.	
60	macrocéphala. W.	(*st.*)	large-headed.	Caucasus.	1805.	6. 8.	H. ♃.	Bot. mag. t. 1248.
61	atropurpùrea. W.	(*pu.*)	dark purple.	Hungary.	1802.	——	H. ♃.	W. et K. hung. 2. t. 116.
62	calocéphala. W.en.	(*ye.*)	smooth-stalked.	Levant.	1816.	——	H. ♃.	
63	canariénsis. W.en.	(*wh.*)	Canary.	Teneriffe.	1815.	——	G. ♄.	
64	orientàlis. W.	(*ye.*)	oriental.	Levant.	1759.	7. 8.	H. ♃.	
65	pulchérrima. W.	(*pu.*)	beautiful.	Caucasus.	1823.	——	H. ♃.	
66	arachnoídea. P.S.	(*ye.*)	cobweb.	Italy.	——	——	H. ♃.	Viv. fl. it. fr. t. 8.
67	sibírica. W.	(*pu.*)	Siberian.	Siberia.	1782.	——	H. ♃.	Gmel. sib. 2. t. 42. f. 2.
68	dilùta. W.	(*li.*)	pale-flowered.	S. Europe.	1781.	——	H. ♃.	
69	alàta. W.	(*ye.*)	winged-stalked.	Tartary.	——	8. 9.	H. ♃.	Venten. cels. t. 80.
70	elongàta. S.S.	(*li.*)	elongated.	Barbary.	1822.	——	H. ♃.	
71	Bèhen. W.	(*ye.*)	saw-leaved.	Levant.	1797.	7. 8.	H. ⊙.	
72	répens. W.	(*ye.*)	creeping.	————	1739.	6. 8.	H. ♃.	
73	Jacea. W.	(*pu.*)	brown knapweed.	England.	....	7. 9.	H. ♃.	Eng. bot. v. 24. t. 1678.
74	decúmbens. P.S.	(*pu.*)	decumbent.	France.	1815.	——	H. ♃.	
75	amára. W.	(*pu.*)	bitter.	Italy.	....	7. 8.	H. ♃.	Bocc. mus. 31. t. 17.
76	álba. W.	(*pu.*)	white.	Spain.	1597.	6. 9.	H. ♃.	
77	spléndens. W.	(*pu.*)	shining.	————	——	7. 8.	H. ♂.	
78	nítens. W.	(*pu.*)	glittering.	Caucasus.	1823.	——	H. ⊙.	Buxb. cent. 2. t. 15. f. 1.
79	tagàna. W.	(*pu.*)	Portugal.	Portugal.	1640.	——	H. ♃.	Brot. phyt. lus. t. 3.
80	babylònica. W.	(*pu.*)	Babylonian.	Levant.	1710.	6. 9.	H. ♃.	Alp. exot. t. 282.
81	glastifòlia. W.	(*ye.*)	woad-leaved.	Siberia.	1731.	——	H. ♃.	Bot. mag. t. 62.
82	parviflòra. S.S.	(*pu.*)	small-flowered.	S. Europe.	1823.	——	H. ♃.	
83	multífida. P.S.	(*st.*)	multifid.	————	1824.	——	H. ♃.	
84	sórdida. S.S.	(*pu.ye.*)	sordid.	....	1817.	——	H. ♃.	
85	rígida. S.S.	(*pu.*)	rigid.	....	1824.	——	H. ♃.	
86	Barrelíeri. S.S.	(*pu.*)	Barrelier's.	Hungary.	1826.	——	H. ♃.	Barrel. ic. 310.
87	stereophy'lla. S.S.	(*pu.*)	variable-leaved.	Poland.	1828.	——	H. ♃.	
88	spinulòsa. S.S.	(*pu.*)	small-spined.	Hungary.	1825.	——	H. ♃.	
89	tenuifòlia. Leh.	(*pu.*)	slender-leaved.	Siberia.	1829.	——	H. ♃.	
90	subspinòsa. Leh.	(*pu.*)	somewhat spiny.	————	——	——	H. ♃.	
91	jacobeæfòlia. S.S.	(*ye.*)	Jacobea-leaved.	....	——	——	H. ♃.	
92	refléxa. S.S.	(*ye.*)	crook-spined.	Iberia.	1801.	7. 8.	H. ♃.	
93	centauroídes. S.S.	(*ye.*)	lyre-leaved.	S. Europe.	1739.	5. 7.	H. ♃.	Column. ecphr. 1. t. 35.

94 collìna. F.S.	(ye.)	hill.	S. Europe.	1596.	6. 7.	H. ♃.	
95 pubéscens. S.S.	(ye.)	downy.	....	1804.	7. 8.	H. ♃.	
96 nicæénsis. s.s.	(ye.)	Italian.	Italy.	1827.	——	H. ♃.	
97 rupéstris. s.s.	(ye.)	rock.	————	1804.	——	H. ♃.	Column.ecph.1.t.et.f.2.
98 hy'brida. s.s.	(pu.ye.)	hybrid.	————	1821.	——	H. ♂.	
99 sabulòsa. s.s.	(wh.)	gravel.	Siberia.	1828.	——	H. ♃.	
100 Crocody'lium.s.s.	(bh.)	blush-flowered.	Levant.	1777.	——	H. ⊙.	Barrel. rar. t. 503.
101 aùrea. s.s.	(go.)	great golden.	S. Europe.	1758.	7. 9.	H. ♃.	Bot. mag. t. 421.
102 scopària. s.s.	(pu.)	broom.	Levant.	1829.	——	H. ♃.	
103 Verùtum. s.s.	(ye.)	dwarf.	————	1780.	8. 9.	H. ⊙.	Jacq. ic. t. 178.
104 muricàta. s.s.	(pu.)	muricated.	Spain.	1621.	7. 8.	H. ⊙.	
105 pubígera. P.S.	(pu.)	pubescent.	S. Europe.	1828.	——	H. ⊙.	
106 peregrìna. s.s.	(pu.)	soft-leaved.	————	1749.	——	H. ⊙.	
107 cichoràcea. s.s.	(pu.)	Succory-like.	Italy.	....	7. 9.	H. ♂.	Till. pis. p. 84. t. 27.
108 salmántica.s.s.	(wh.pu.)	ragwort-leaved.	S. Europe.	1596.	7. 8.	H. ♂.	Jacq. vind. 1. t. 64.
109 sonchifòlia. s.s.	(pu.)	Sow-thistle-ld.	————	1780.	8. 10.	H. ⊙.	Pluk. phyt. 39. f. 1.
110 Séridis. s.s.	(pu.)	purple-flower'd.	Spain.	1686.	6. 8.	H. ♃.	Pluk. alm. t. 38. f. 1.
111 napifòlia. s.s.	(pu)	Turnep-leaved.	Candia.	1691.	7. 9.	H. ⊙.	Herm. parad. t. 189.
112 Zanònii. s.s.	(pu.)	Zanoni's.	————	1822.	——	H. ♂.	
113 Isnárdi. s.s.	(li.)	Jersey.	Jersey.	....	7. 8.	H. ♃.	Eng. bot. v. 32. t. 2256.
114 áspera. s.s.	(pu.)	rough.	S. Europe.	1772.	6. 10.	H. ⊙.	Bocc. mus. 35. t. 26.
115 sphærocéphala.s.s.	(pu.)	globe-headed.	————	1683.	6. 8.	H. ⊙.	Bot. mag. t. 2551.
116 auriculàta. P.S.	(pu.)	eared.	————	1816.	7. 8.	H. ⊙.	
117 polyacántha. s.s.	(pu.)	many-spined.	Portugal.	1804.	——	H. ⊙.	
118 prolífera. P.S.	(st.)	proliferous.	Egypt.	1816.	——	H. ⊙.	Vent. hort. cels. t. 16.
straminea. W. hort. ber. t. 26.							
119 ibérica. s.s.	(pu.)	Iberian.	Iberia.	1821.	——	H. ⊙.	
120 fèrox. s.s.	(ye.)	hedgehog.	Barbary.	1790.	7. 9.	F. ♃.	Desf. atl. 2. t. 242.
121 solstítialis. s.s.	(ye.)	St. Barnaby's.	England.	....	7. 8.	H. ⊙.	Eng. bot. v. 4. t. 243.
122 meliténsis. s.s.	(ye.)	cluster-headed.	Malta.	1710.	——	H. ⊙.	Bocc. sic. t. 35.
123 A'dami. s.s.	(ye.)	Adam's.	Siberia.	1804.	——	H. ♃.	
124 sulphùrea. W.en.	(st.)	straw-coloured.	....	1815.	——	H. ⊙.	
125 sícula. s.s.	(st.)	Sicilian.	Sicily	1710.	——	H. ♃.	Bocc. sic. t. 8. f. 1.
126 Stevènii. s.s.	(ye.)	Steven's.	Caucasus.	1826.	——	H. ⊙.	
127 cruénta. W.en.	(pu.)	obovate-leaved.	S. Europe.	1816.	6. 8.	H. ♃.	
128 romàna. L.	(pu.)	Roman.	Italy.	1739.	7. 9.	H. ♂.	Barrel. rar. t. 504.
129 benedícta. s.s.	(st.)	blessed thistle.	Spain.	1548.	6. 10.	H. ⊙.	Zorn. ic. t. 122.
130 pullàta. s.s.	(br.)	sad-coloured.	S. Europe.	1714.	6. 7.	H. ♂.	Mill. ic. 2. t. 152. f. 2.
131 eriópbora. s.s.	(ye.)	woolly-headed.	Portugal.	1714.	6. 10.	H. ⊙.	
132 ápula. s.s.	(ye.)	Apulian.	Sicily.	1829.	6. 8.	H. ⊙.	Column. ecphr. 1. t. 31.
133 Calcítrapa. s.s.	(li.)	Star-thistle.	Britain.	....	——	H. ⊙.	Eng. bot. v. 2. t. 125.
134 calcitrapoídes.s.s.	(li.)	Phœnician.	Levant.	1683.	6. 7.	H. ⊙.	
135 ægyptiaca. s.s.	(wh.)	Egyptian.	Egypt.	1790.	6. 9.	H. ⊙.	
136 balsamìta. W.	(ye.)	Balsam-scented.	Armenia.	1824.	7. 8.	H. ♃.	
137 lanàta. DC.	(ye.)	hoary.	Switzerland.	1825.	——	H. ⊙.	
138 crética. s.s.	(wh.)	Cretan.	Candia.	1827.	——	H. ⊙.	
139 bullàta. Leh.	....	blistered.	....	1829.	....	H. ♃?	
140 ucrànica. Ot.	....	Ukraine.	Ukraine.	——	....	H. ♃?	
RHAPO'NTICA. DC.		RHAPO'NTICA.	Syngenesia Polygamia Frustranea.				
1 scariòsa. DC.	(pu.)	common.	Switzerland.	1640.	7. 8.	H. ♃.	Bot. mag. t. 1752.
Centaùrea Rhapóntica. B.M.							
β *lyràta.* DC.	(pu.)	*lyrate.*	————	....	——	H. ♃.	
2 uniflòra. DC.		one-flowered.	Siberia.	1796.	——	H. ♃.	Gmel. sib. 2. t. 38.
Cnìcus uniflòrus. W.							
3 doronicæfòlia. F.	(pu.)	Doronicum-leav'd.	————	1825.	6. 8.	H. ♃.	
4 dahùrica. F.	(pu.)	Dahurian.	Dahuria.	1829.	——	H. ♃.	
CARDOP'ATUM. J.		CARDOP'ATUM.	Syngenesia Polygamia Segregata.				
corymbòsum. P.S.	(bl.)	umbelled.	S. Europe.	1640.	6. 7.	H. ♃.	Moris. s. 7. t. 33. f. 17.
Bròtera corymbòsa. W.							

Div. II. ECHINOPSIDEÆ. *Kth. synops.* 2. *p.* 364.

ROLA'NDRA. W.		ROLA'NDRA.	Syngenesia Polygamia Segregata. L.				
argéntea. W.	(wh.)	silver-leaved.	W. Indies.	1714.	7.	S. ħ.	Sloan. jam. 1. t. 7. f. 3.
ECHINO'PS. W.		GLOBE-THISTLE.	Syngenesia Polygamia Segregata. L.				
1 sphærocéphalus.W.	(wh.)	great.	Austria.	1596.	7. 8.	H. ♃.	Flor. græc. t. 923.
β *exaltàtus.*Schr.	(wh.)	*tall.*	————	1820.	....	H. ♃.	

2 spinòsus. W.	(wh.)	thorny-headed.	Egypt.	1597.	7. 8.	H. ♃.	Flor. græc. t. 924.
3 paniculàtus. B.R.	(bl.)	panicled.	Spain.	1815.	——	H. ♂.	Bot. reg. t. 356.
4 hórridus. P.S.	(wh.)	horrid.	Persia.	1817.	——	H. ♃.	
5 viscòsus. M.B.	(wh.)	viscous.	Caucasus.	1823.	——	H. ♂.	
6 taùricus. W.en.	(bl.)	Taurian.	Tauria.	1816.	——	H. ♃.	
7 dahùricus. F.	(bl.)	Dahurian.	Dahuria.	1827.	——	H. ♃.	
8 ruthénicus. M.B.	(bl.)	Russian.	Russia.	1819.	——	H. ♃.	
9 Rìtro. W.	(bl.)	small.	Europe.	1570.	7. 9.	H. ♃.	Bot. mag. t. 932.
10 tenuifòlius. F.	(bl.)	slender-leaved.	Russia.	1823.	——	H. ♃.	
11 stríctus. B.M.	(bl.)	upright.	————	1822.	8. 10.	H. ♂?	Bot. mag. t. 2457.
12 pérsicus. F.	(wh.)	Persian.	Persia.	1821.	——	H. ♃.	
13 bannáticus. W.K.	(wh.)	Hungarian.	Hungary.	1828.	——	H. ♃.	
14 lanuginòsus. W.	(bl.)	woolly.	Levant.	1736.	6. 7.	H. ♃.	Flor. græc. t. 926.
15 hùmilis. M.B.	(bl.)	low.	Siberia.	1826.	6. 8.	H. ♃.	
16 strigòsus. W.	(wh.)	annual.	Spain.	1729.	7. 9.	H. ⊙.	Bot. mag. t. 2109.
GUND'ELIA. W.		GUND'ELIA. Syngenesia Polygamia Segregata. L.					
Tournefórtii. W.	(pu.)	Tournefort's.	Levant.	1739.	6. 8.	H. ♃.	Mill. ic. t. 287.
SPHÆRA'NTHUS. W.		SPHÆRA'NTHUS. Syngenesia Polygamia Segregata. L.					
1 móllis. H.B.	(bl.)	soft.	E. Indies.	1818.	7. 9.	S. ⊙.	
2 índicus. W.	(pu.)	Indian.	————	1699.	8. 12.	S. ♃.	Burm. zeyl. t. 94. f. 3.
3 cochinchinénsis. W.	(bl.)	Cochinchina.	Cochinchina.	1826.	7. 8.	H. ⊙.	
4 hírtus. W.	(bl.)	hairy.	Manila.	1823.	——	G. ⊙.	Lam. ill. t. 718. f. 1.
5 africànus. W.	(bl.)	African.	C. B. S.	1759.	7. 8.	G. ⊙.	Pluk. mant. t. 108. f. 7.
ANGIA'NTHUS. W.C.		ANGIA'NTHUS. Syngenesia Polygamia Segregata.					
tomentòsus. W.C.	(ye.)	woolly.	N. Holland.	1803.	7. 8.	G. ♃.	Wendl. coll. 2. t. 48.
Cassínia aùrea. H.K. *non* L.T.							
BROT'ERA. P.S.		BROT'ERA. Syngenesia Polygamia Segregata. P.S.					
trinervàta. P.S.	(ye.)	three-nerved.	S. America.	1799.	7. 8.	H. ⊙.	Schr. b. jour. 180. 2. t. 5.
Contrayérva. Spr. *Nauenbúrgia trinervàta.* W.							
ŒD'ERA. W.		ŒD'ERA. Syngenesia Polygamia Segregata. L.					
1 prolífera. W.	(ye.)	proliferous.	C. B. S.	1789.	5. 6.	G. ħ.	Bot. mag. t. 1637.
2 alìena. W.	(ye.)	linear-leaved.	————	1821.	5. 7.	G. ħ.	Jacq. schœnb. 2. t. 154.
ELEPHA'NTOPUS. W.		ELEPHANT'S-FOOT. Syngenesia Polygamia Segregata. L.					
1 scàber. W.	(re.)	rough-leaved.	E. Indies.	1695.	6. 9.	S. ♃.	Rheed. mal. 10. t. 7.
2 caroliniànus. W.	(re.)	Carolina.	America.	1732.	7. 9.	G. ♃.	Dill. elth. t. 106. f. 126.
3 tomentòsus. W.	(re.)	woolly.	W. Indies.	1733.	7. 8.	S. ♃.	
4 móllis. K.S.	(wh.)	soft-leaved.	Caracas.	1825.	——	S. ♃.	
5 nudiflòrus. W.	(re.)	naked-flower'd.	W. Indies.	1822.	——	S. ħ.	
6 spicàtus. W.	(re.)	spiked.	————	1821.	——	S. ħ.	Sloan. hist. 1. t. 150. f. 3. 4.
7 angustifòlius. W.	(re.)	narrow-leaved.	————	1823.	——	S. ħ.	———— t. 148. f. 4.
ST'ŒBE. W.		ST'ŒBE. Syngenesia Polygamia Segregata. L.					
1 æthiòpica. W.	(wh.)	Juniper-leaved.	C. B. S.	1759.	8.	G. ♄.	
Seríphium æthiòpicum. P.S.							
2 ericoídes. W.	(wh.)	Heath-like.	————	1816.	7. 9.	G. ♄.	Moris. s. 7. t. 18. f. 12.
3 gomphrenoídes. W.	(w.)	Gomphrena-like.	————	————	——	G. ħ.	Houtt. syst. 4. t. 34. f. 1.
4 gnaphaloídes. W.	(wh.)	Gnaphalium-like.	————	1824.	——	G. ♄.	Burm. afr. t. 77. f. 1.
5 cinèrea. W.	(wh.)	Heath-leaved.	————	1774.	——	G. ♄.	Pluk. mant. t. 297. f. 1.
Seríphium cinèreum. P.S.							
6 alopecuroídes. W.	(wh.)	Fox-tail-like.	————	1824.	——	G. ♄.	Pluk. alm. t. 445. f. 2.
7 reflèxa. W.	(wh.)	reflex-leaved.	————	1816.	——	G. ♄.	
CÆS'ULIA. Roxb.		CÆS'ULIA. Syngenesia Polygamia Segregata. L.T.					
1 axillàris. W.	(wh.)	axillary-flower'd.	E. Indies.	1804.	7. 9.	S. ♃.	Andr. reposit. t. 431.
2 lancifòlia.	(wh.)	spear-leaved.	————	1820.	——	S. ♃.	
LAGA'SCA. Cav.		LAGA'SCA. Syngenesia Polygamia Segregata. L.T.					
móllis. W.	(wh.)	soft.	S. America.	1815.	6. 9.	G. ⊙.	Bot. mag. t. 1804.
NO'CCA. W.		NO'CCA. Syngenesia Polygamia Segregata.					
1 rígida. C.I.	(re.)	red-flowered.	Mexico.	1823.	8. 10.	F. ♄.	Swt. br. fl. gar. s. 2. t. 26.
Lagáscea rùbra. H. et B. pl. amer. 4. t. 311.							
2 suavèolens.	(wh.)	sweet-scented.	Mexico.	——	....	F. ħ.	
Lagáscea suavèolens. K.S.							
3 latifòlia. B.F.G.	(wh.)	broad-leaved.	————	——	....	F. ♄.	Swt. br. fl. gar. t. 215.

Div. III. VERNONIACEÆ. *Kth. synops.* 2. *p.* 367. *et* ASTEREÆ. *l. c. p.* 399.

AMPHE'REPHIS. K.S.		AMPHE'REPHIS. Syngenesia Polygamia Æqualis.					
intermèdia. L.en.	(bl.)	intermediate.	Brazil.	1822.	7. 9.	H. ⊙.	Swt. br. fl. gar. t. 225.

ASCARI'CIDA. Cass.		ASCARI'CIDA. Syngenesia Polygamia Æqualis.					
anthelmíntica.	(*li.*)	Indian.	E. Indies.	1770.	8. 9.	S. ♂.	Rheed. mal. 2. t. 24.
índica. Cass. *Vernònia anthelmíntica.* W. *Con'yza anthelmíntica.* L.							
VERN'ONIA. W.		VERN'ONIA. Syngenesia Polygamia Æqualis. W.					
1 noveboracénsis. W.	(*pu.*)	long-leaved.	N. America.	1710.	9. 11.	H. ♃.	Dill. elt. t. 263. f. 342.
2 præálta. W.	(*pu.*)	tall.	———	1732.	——	H. ♃.	——— t. 264. f. 343.
3 angustifòlia. Ph.	(*pu.*)	narrow-leaved.	———	1817.	——	H. ♃.	
4 glaùca. W.	(*pu.*)	glaucous-leaved.	———	1710.	——	H. ♃.	Dill. elt. t. 262. f. 341.
oligophy'lla. M.							
5 altíssima. N.	(*pu.*)	tallest.	———	1823.	——	H. ♃.	
6 panduràta. L.en.	(*pu.*)	fiddle form.	———	——	——	H. ♃.	
7 serratuloídes. K.S.	(*pu.*)	Sawwort-like.	Mexico.	1824.	——	H. ♃.	
8 axilliflòra. Ot.	(*pi.*)	axillary-flower'd.	———	1829.	9. 10.	F. ♃.	
9 centriflòra. Lk.	(*pu.*)	centre-flower'd.	Brazil.	1826.	——	S. ♄.	
10 serícea. B.R.	(*li.*)	silky.	———	1818.	9. 1.	S. ♄.	Bot. reg. t. 522.
11 arboréscens. P.S.	(*pu.*)	tree.	Jamaica.	1733.	11. 12.	S. ♄.	Plum. sp. 10. t. 130. f. 2.
Con'yza arboréscens. W.							
12 fruticòsa. P.S.	(*pu.*)	shrubby.	S. America.	1818.	——	S. ♄.	Plum. ic. 95. f. 1.
Con'yza fruticòsa. L.							
13 flexuòsa. B.M.	(*pu.*)	flexuose-branch'd.	Brazil.	1823.	9. 10.	S. ♄.	Bot. mag. t. 2477.
LI'ATRIS. W.		LI'ATRIS. Syngenesia Polygamia Æqualis. W.					
1 spicàta. W.	(*pu.*)	long-spiked.	N. America.	1732.	8. 10.	H. ♃.	Swt. br. fl. gar. t. 49.
β *pùmila.* B.C.	(*pu.*)	*dwarf.*	———	....	——	H. ♃.	Lodd. bot. cab. t. 147.
2 pycnostàchya. M.	(*pu.*)	pubescent.	———	1732.	——	H. ♃.	Dill. elt. t. 72. f. 83.
3 heterophy'lla. Ph.	(*pu.*)	various-leaved.	———	1790.	7. 8.	H. ♃.	
4 intermèdia. B.R.	(*pu.*)	intermediate.	———	1823.	8. 9.	H. ♃.	Bot. reg. t. 948.
5 cylindràcea. M.	(*pu.*)	cylindrical-cup'd.	———	1811.	10.	H. ♃.	
6 pilòsa. Ph.	(*pu.*)	hairy-leaved.	———	1783.	——	H. ♃.	Bot. reg. t. 595.
7 grácilis. Ph.	(*pu.*)	smooth-leaved.	———	1812.	——	H. ♃.	
8 turbinàta. B.F.G.	(*pu.*)	turbinate-cup'd.	———	1823.	——	H. ♃.	Swt. br. fl. gar. ic.
9 élegans. W.	(*pu.*)	hairy-cup'd.	———	1787.	9. 10.	H. ♃.	Bot. reg. t. 267.
10 sphæroídea. M.	(*pu.*)	globular-cup'd.	———	1817.	8. 10.	H. ♃.	Swt. br. fl. gar. t. 87.
11 scariòsa. W.	(*pu.*)	scariose-cup'd.	———	1739.	9. 10.	H. ♃.	Bot. reg. t. 590.
12 squarròsa. W.	(*pu.*)	rough-cup'd.	———	1732.	——	H. ♃.	Swt. br. fl. gar. t. 44.
13 paniculàta. W.	(*pu.*)	panicled.	———	1826.	8. 10.	H. ♃.	
14 odoratíssima. W.	(*pu.*)	sweet-scented.	Carolina.	1786.	——	F. ♃.	Andr. reposit. t. 633.
15 corymbòsa. W.	(*pu.*)	corymbose.	N. America.	1825.	——	F. ♃.	
BRACHYL'ÆNA. L.T.		BRACHYL'ÆNA. Syngenesia Polygamia Superflua.					
neriifòlia. L.T.	(*st.*)	Oleander-leav'd.	C. B. S.	1752.	8. 11.	G. ♄.	
Báccharis neriifòlia. W.							
BA'CCHARIS. W.		PLOWMAN'S-SPIKENARD. Syngenesia Polygamia Superflua.					
1 ivæfòlia. W.	(*wh.*)	Peruvian.	America.	1696.	7. 8.	G. ♄.	Schk. handb. 3. t. 244.
2 angustifòlia. M.	(*wh.*)	narrow-leaved.	N. America.	1812.	7. 9.	G. ♄.	
3 glomeruliflòra. M.	(*wh.*)	close-flowered.	———	1817.	8. 10.	H. ♄.	
4 halimifòlia. W.	(*wh.*)	Groundsel-tree.	———	1683.	10. 11.	H. ♄.	Schmidt arb. t. 82.
5 Dioscóridis. W.	(*wh.*)	Dioscorides'.	Levant.	1822.	7. 9.	G. ♄.	Rauw. it. 4. t. 54.
6 scopària. P.S.	(*wh.*)	broom.	Jamaica.	——	——	S. ♄.	Brown. jam. t. 34. f. 4.
Calea scopària. L.							
7 móllis. K.S.	(*wh.*)	soft.	Quito.	1824.	5. 6.	F. ♄.	
8 genistelloídes. K.S.	(*w.*)	wing-stemmed.	Brazil.	1829.	7. 10.	G. ♄.	
alàta. Hort. *Pterocladis genistelloídes.* Lamb. herb.							
9 parviflòra. P.S.	(*wh.*)	small-flowered.	Peru.	1822.	....	S. ♄.	
10 lycopodioídes. P.S.	(*wh.*)	Club-moss-like.	Mauritius.	1828.	....	S. ♄.	
Con'yza lycopodioídes. Lam.							
CON'YZA. W.		FLEA-BANE. Syngenesia Polygamia Superflua. L.					
1 squarròsa. W.	(*ye.*)	great.	Britain.	....	7. 8.	H. ♂.	Eng. bot. v. 17. t. 1195.
2 marylándica. Ph.	(*pu.*)	Maryland.	N. America.	....	8. 10.	H. ☉.	Dill. elt. t. 88. f. 104.
3 camphoràta. Ph.	(*pu.*)	Camphor-scent'd.	———	1704.	——	H. ♃.	——— t. 89. f. 105.
Báccharis fœtida. W.							
4 pátula. W.	(*pu. ye.*)	spreading.	China.	1758.	7. 9.	H. ☉.	Mill. ic. 2. t. 247.
5 axillàris. W.	(*re.*)	axillary.	Mauritius.	1824.	——	H. ☉.	
6 bifróns. W.	(*st.*)	oval-leaved.	N. America.	1739.	8. 9.	H. ♃.	Pluk. alm. t. 87. f. 4.
7 purpuráscens. W.	(*li.*)	purplish.	Jamaica.	1822.	——	S. ☉.	Sloan. hist. 1. t. 159. f. 1.
8 cándida. L.	(*ye.*)	woolly.	Candia.	1714.	6. 7.	F. ♄.	Flor. græc. t. 865.
verbascifòlia. W.							
9 limonifòlia. F.G.	(*ye.*)	white-leaved.	———	....	——	F. ♄.	Barr. ic. t. 217.
cándida. W. *non* L.							
10 balsamífera. W.	(*re.*)	balsamiferous.	China.	1818.	——	S. ♃.	Rumph. amb. 6. t. 24. f. 1.
11 chinénsis. W.	(*bl.*)	Chinese.	———	1796.	7. 8.	S. ♂.	——— 6. t. 14. f. 2.

12 cinèrea. w.	(*pu.*)	grey.	E. Indies.	....	7. 8.	S. ⊙.	Rumph.amb.6.t.14.f.1.
13 fastigiàta. W.en.	(*pu.*)	clustered.	Senegal.	1820.	——	H. ⊙.	
14 prolífera. w.	(*wh.*)	proliferous.	Java.	1819.	7. 9.	S. ♄.	Rumph.am.5.t.104.f.1.
15 hirsùta. w.	(*ye.pu.*)	shaggy.	China.	1767.	8. 9.	G. ♂.	
16 ægyptìaca. w.	(*ye.*)	Egyptian.	Egypt.	1778.	7. 9.	H. ⊙.	Jacq. vind. 3. t. 19.
17 chilénsis. s.s.	(*ye.*)	Chile.	Chile.	1816.	8. 10.	H. ⊙.	
18 mauritiàna. Boj.	(*pu.*)	Mauritius.	Mauritius.	1821.	——	S. ⊙.	
19 Hubértii. Boj.	(*pu.*)	Hubert's.	Madagascar.	1824.	7. 10.	S. ⊙.	
20 viscòsa. Boj.	(*pu.*)	viscous.	Mauritius.	——	8. 10.	S. ⊙.	
21 pùmila. s.s.	(*ye.*)	dwarf.	Crete.	1826.	7. 9.	H. ⊙.	
22 pinnatífida. R.	(*pu.*)	pinnatifid.	E. Indies.	1816.	——	S. ♄.	
23 Gouàni. w.	(*ye.*)	Gouan's.	Canaries.	1772.	7. 8.	G. ♂.	Jacq. vind. 3. t. 79.
24 sícula. w.	(*ye.*)	red-stalked.	Sicily.	1779.	8. 9.	H. ⊙.	Bocc. sic. t. 31. f. 4.
25 aùrita. w.	(*wh.*)	eared.	E. Indies.	1818.	——	S. ⊙.	
26 fœtida. w.	(*wh.re.*)	stinking.	Africa.	1724.	——	G. ♃.	Mill. ic. 2. t. 233.
27 amœna. L.en.	(*pu.*)	pretty.	——	1816.	——	S. ♃.	
28 glomeràta. L.en.	(*pu.*)	crowded.	....	1824.	——	S. ♃.	
29 spatulàta. L.en.	(*bl.*)	spatulate.	....	——	7. 9.	S. ♃.	
30 adnàta. K.S.	(*pu.*)	decurrent.	Mexico.	——	——	S. ♄.	
Báccharis adnàta. W.en.							
31 thapsoídes. M.B.	(*pu.*)	Thapsus-leav'd.	Caspian Sea.	1806.	——	H. ♃.	
32 sórdida. w.	(*br.*)	small-flowered.	S. Europe.	1570.	——	F. ♄.	Barr. ic. t. 368.
33 saxátilis. w.	(*br.*)	stone.	——	1640.	7. 8.	F. ♄.	Schk. handb. 3. t. 241.
34 rupéstris. w.	(*ye.*)	rock.	Arabia.	1790.	——	G. ♄.	Schmidel ic. t. 36.
35 Tenòrii. L.en.	(*br.*)	Tenore's.	S. Europe.	1824.	....	F. ♄.	
36 serícea. w.	(*ye.*)	snowy.	Canaries.	1779.	....	G. ♄.	
37 geminiflòra. L.en.	(*br.*)	twin-flowered.	S. Europe.	1824.	....	F. ♄.	
38 inuloídes. w.	(*pu.*)	cluster-flower'd.	Teneriffe.	1780.	7. 8.	G. ♄.	Jacq. ic. 1. t. 171.
39 odoràta. w.	(*pu.*)	sweet-scented.	India.	1759.	6. 8.	S. ♄.	Plum. ic. t. 97.
40 incìsa. w.	(*pu.*)	cut-leaved.	C. B. S.	1774.	——	G. ♄.	
41 carolinénsis. w.	(*pu.*)	Carolina.	Carolina.	1823.	——	G. ♄.	Jacq. ic. 3. t. 185.
42 virgàta. w.	(*pu.*)	wing-stalked.	America.	1783.	8. 9.	G. ♃.	Sloan.hist.1.t.152.f.5.
43 rugòsa. H.K.	(*br.*)	St. Helena.	St. Helena.	1772.	11.	G. ♄.	
CARP'ESIUM. W.		CARP'ESIUM. Syngenesia Polygamia Superflua. L.					
1 cérnuum. w.	(*ye.*)	drooping.	Austria.	1739.	7. 8.	H. ♃.	Jacq. aust. 3. t. 204.
2 abrotanoídes. w.	(*ye.*)	Southernwood like.	China.	1768.	——	H. ♃.	Osbeck it. t. 10.
3 torulòsum. F.	(*ye.*)	twisted.	Siberia.	1823.	6. 8.	H. ♃.	
4 Wulfeniànum. Schreb.	(*ye.*)	Wulfen's.	....	1829.	——	H. ♂.	
I'NULA. W.		I'NULA. Syngenesia Polygamia Superflua. L.					
1 Helènium. w.	(*ye.*)	Elecampane.	Britain.	....	7. 8.	H. ♃.	Eng. bot. v. 22. t. 1546.
2 O'culus-Chrísti. w.	(*ye.*)	hoary.	Austria.	1759.	7. 9.	H. ♃.	Jacq. aust. 3. t. 223.
3 británnica. w.	(*ye.*)	creeping-rooted.	Germany.	——	——	H. ♃.	Flor. dan. t. 413.
4 undulàta. w.	(*ye.*)	wave-leaved.	Egypt.	1739.	7. 10.	H. ⊙.	
5 índica. w.	(*ye.*)	Indian.	E. Indies.	——	——	S. ⊙.	Burm. zeyl. t. 55. f. 2.
6 paludòsa. s.s.	(*ye.*)	marsh.	Portugal.	1828.	——	H. ♃.	
7 áspera. s.s.	(*ye.*)	rough.	....	——	——	H. ♃.	
8 campéstris. s.s.	(*ye.*)	field.	Poland.	1827.	——	H. ♃.	
9 squarròsa. w.	(*ye.*)	net-leaved.	S. Europe.	1768.	7. 9.	H. ♃.	Flor. græc. t. 875.
10 viscòsa. w.	(*ye.*)	clammy.	Barbary.	1633.	7. 8.	G. ♃.	Jacq. vind. 2. t. 165.
Erígeron viscòsum. Jacq.							
11 tuberòsa. P.S.	(*ye.*)	tuberous-root'd.	S. Europe.	1640.	——	H. ♃.	Moris. s. 7. t. 19. f. 20.
Erígeron tuberòsum. W.							
12 salicìna. w.	(*ye.*)	willow-leaved.	N. Europe.	1648.	8. 9.	H. ♃.	Flor. dan. t. 786.
13 Bubònium. w.	(*ye.*)	Austrian.	Austria.	1801.	7. 9.	H. ♃.	Jacq. aust. 5. ap. t. 19.
14 hírta. w.	(*ye.*)	hairy.	——	1759.	6. 9.	H. ♃.	—— 4. t. 358.
15 glandulòsa. w.	(*ye.*)	glandular.	Georgia.	1804.	7. 8.	H. ♃.	Bot. mag. t. 1907.
16 grandiflòra. w.	(*ye.*)	large-flowered.	Caucasus.	1825.	——	H. ♃.	
17 móllis. L.en.	(*ye.*)	soft.	....	1823.	——	H. ♃.	
18 suavèolens. w.	(*ye.*)	woolly-leaved.	S. Europe.	1758.	6. 8.	H. ♃.	Jacq. vind. 3. t. 51.
19 Vaillántii. w.	(*ye.*)	Vaillant's.	France.	1739.	——	H. ♃.	Hall. helv. n. 73. t. 2.
20 germánica. w.	(*ye.*)	German.	Germany.	1759.	6. 7.	H. ♃.	Jacq. aust. 2. t. 134.
21 ensifòlia. w.	(*ye.*)	sword-leaved.	Austria.	1793.	7. 9.	H. ♃.	—— 2. t. 162.
22 saxátilis. s.s.	(*ye.*)	rock.	S. Europe.	1825.	——	H. ♃.	
Erígeron glutinòsum. L.							
23 crithmoídes. L.	(*ye.*)	Samphire-leav'd.	England.	....	8. 9.	H. ♃.	Eng. bot. v. 1. t. 68.
24 provinciàlis. w.	(*ye.*)	oval-leaved.	France.	1778.	7. 8.	H. ♃.	
25 montàna. w.	(*ye.*)	mountain.	S. Europe.	1759.	——	H. ♃.	Garid. aix. t. 10.
26 bifróns. w.	(*ye.*)	Italian.	——	1713.	6. 8.	H. ♃.	Herm. parad. t. 127.
27 thapsoídes. L.en.	(*ye.*)	Mullein-like.	....	1823.	——	H. ♃.	

28 verbascifòlia. s.s.	(ye.)	Mullein-leaved.	Caucasus.	1828.	6. 8.	H. ♃.	
29 calycìna. s.s.	(ye.)	reflex-scaled.	Italy.	1829.	——	H. ♃.	
30 saturejoìdes. w.	(ye.)	Savory-leaved.	Vera Cruz.	1733.	....	S. ♃.	Reliq. houst. 8. t. 19.
31 f'œtida. w.	(ye.)	stinking.	Malta.	1688.	6. 8.	H. ⊙.	Bocc. sic. 26. t. 13.
32 odòra. w.	(ye.)	sweet-scented.	S. Europe.	1824.	——	H. ♃.	Moris. s. 7. t. 21. f. 6.
33 hy'brida. s.s.	(ye.)	hybrid.	Poland.	1828.	——	H. ♃.	
34 glábra. M.B.	(ye.)	smooth.	Caucasus.	1826.	——	H. ♃.	
35 cáspica. F.	(ye.)	Caspian.	Volga.	——	——	H. ♃.	
PULIC'ARIA. G.		FLEAWORT. Syngenesia Polygamia Superflua.					
1 vulgàris. G.	(ye.)	small.	England.	....	8. 9.	H. ⊙.	Gært. sem. 2. t. 173. f. 7.
I'nula Pulicària. Eng. bot. v. 17. t. 1196.							
2 villòsa. L.en.	(ye.)	villous.	....	1821.	——	H. ⊙.	
3 arábica. L.en.	(ye.)	Arabian.	Arabia.	1823.	——	F. ♃.	Pluk. alm. t. 149. f. 4.
4 dysentérica. L.en.	(ye.)	meadow.	England.	....	——	H. ♃.	Eng. bot. v. 16. t. 1115.
I'nula dysentérica. E.B.							
CHRYSO'PSIS. N.		CHRYSO'PSIS. Syngenesia Polygamia Superflua.					
1 gossypìna. N.	(ye.)	cottony.	N. America.	1823.	6. 8.	H. ♃.	
I'nula gossypìna. M.							
2 trichophy'lla. N.	(ye.)	hairy-leaved.	———	1827.	——	H. ♃.	
3 marianà. N.	(ye.)	American.	———	1742.	7. 8.	H. ♃.	Mill. ic. 1. t. 57. f. 1.
IX'ODIA. H.K.		IX'ODIA. Syngenesia Polygamia Æqualis.					
achillæoìdes. H.K.	(wh.)	Milfoil-like.	N. Holland.	1803.	3. 9.	G. ♄.	Bot. mag. t. 1534.
H'UMEA. Sm.		H'UMEA. Syngenesia Polygamia Æqualis.					
élegans. H.K.	(re.)	elegant.	N. S. W.	1800.	6. 10.	G. ♂.	Exot. bot. 1. t. 1.
CASSI'NIA. L.T.		CASSI'NIA. Syngenesia Polygamia Æqualis.					
1 spectábilis. L.T.	(st.)	Humea-like.	N. Holland.	1821.	5. 10.	G. ♂.	Bot. reg. t. 678.
2 affínis. L.T.	(st.)	related.	N. S. W.	1823.	4. 6.	G. ♄.	
3 aùrea. L.T.	(ye.)	golden.	———	1821.	——	G. ♄.	Bot. reg. t. 764.
4 longifòlia. L.T.	(st.)	long-leaved.	———	——	——	G. ♄.	
5 denticulàta. L.T.	(st.)	tooth-leaved.	———	1826.	6. 8.	G. ♄.	
6 leptophy'lla. L.T.	(wh.)	small-leaved.	N. Zealand.	1821.	4. 6.	G. ♄.	
OZOTHA'MNUS. L.T.		OZOTHA'MNUS. Syngenesia Polygamia Æqualis.					
1 cinéreus. L.T.	(ye.)	rusty.	V. Diem. Isl.	1820.	4. 8.	G. ♄.	Lab. nov. hol. v. 2. t. 182.
Chrysócoma cinèrea. Lab.							
2 rosmarinifòlius. L.T.	(ye.)	Rosemary-leav'd.	———	1822.	——	G. ♄.	Lab. nov. hol. v. 2. t. 181.
Eupatòrium rosmarinifòlium. Lab.							
3 ferrugìneus. L.T.	(ye.)	ferruginous.	———	——	——	G. ♄.	Lab. nov. hol. v. 2. t. 180.
Eupatòrium ferrugìneum. Lab.							
AMM'OBIUM. Br.		AMM'OBIUM. Syngenesia Polygamia Æqualis.					
alàtum. B.M.	(wh.)	winged-stalk'd.	N. S. W.	1823.	8. 9.	F. ♃.	Swt. br. fl. gar. t. 48.
ANTENN'ARIA. L.T.		ANTENN'ARIA. Syngenesia Polygamia Superflua. L.					
1 dioíca. L.T.	(ro.)	red-flowered.	Britain.	....	5. 7.	H. ♃.	Eng. bot. v. 4. t. 267.
Gnaphàlium dioícum. E.B.							
2 alpìna. L.T.	(wh.)	Alpine.	Alps Europe.	1775.	6. 7.	H. ♃.	
3 carpática. L.T.	(wh.)	Carpathian.	Carp. mount.	——	——	H. ♃.	Flor. dan. t. 332.
Gnaphàlium alpìnum. W. *non* L.							
4 plantaginifòlia. L.T.	(wh.)	Plantain-leav'd.	Virginia.	1759.	6. 7.	H. ♃.	Pluk. alm. t. 348. f. 9.
Gnaphàlium plantaginifòlium. W.							
5 margaritàcea. L.T.	(wh.)	pearly.	England.	....	7. 9.	H. ♃.	Eng. bot. v. 29. t. 2018.
6 triplinérvis. D.P.	(wh.)	triple nerved.	Nepaul.	1823.	6. 9.	F. ♃.	Bot. mag. t. 2468.
7 contórta. D.P.	(wh.)	twisted-leaved.	———	1818.	——	F. ♃.	Bot. reg. t. 605.
LEONTOP'ODIUM. L.T.		LION'S-FOOT. Syngenesia Polygamia Superflua. L.					
1 alpìnum. D.D.	(wh.)	Swiss.	Switzerland.	1776.	6. 7.	H. ♃.	Bot. mag. t. 1958.
Gnaphàlium Leontopòdium. B.M.							
2 sibíricum. D.D.	(wh.)	Siberian.	Siberia.	1829.	——	H. ♃.	
GNAPH'ALIUM. L.T.		EVERLASTING. Syngenesia Polygamia Superflua. L.					
1 lùteo-álbum. W.	(st.)	Jersey.	Jersey.	....	7. 8.	H. ⊙.	Eng. bot. v. 14. t. 1002.
2 albéscens. W.	(wh.)	white Jamaica.	Jamaica.	1793.	....	S. ♄.	
3 sanguíneum. W.	(re.)	bloody.	Egypt.	1768.	5. 7.	H. ♃.	Rauw. it. 285. t. 37.
4 undulàtum. W.	(wh.)	wave-leaved.	Africa.	1732.	6. 9.	H. ⊙.	Dill. elth. t. 108. f. 130.
5 obtusifòlium. W.	(wh.)	blunt-leaved.	N. America.	1699.	7. 9.	H. ⊙.	——— t. 108. f. 131.
6 purpùreum. W.	(pu.)	purple-flower'd.	S. America.	1732.	6. 9.	G. ♂.	——— t. 109. f. 132.
7 involucràtum. W.	(gr.)	involucred.	N. Zealand.	1823.	7. 8.	G. ♃.	Bot. mag. t. 2582.
8 americànum. W.	(br.)	Jamaica.	Jamaica.	1815.	....	S. ♃.	
9 pensylvánicum. W.	(st.)	American.	N. America.	——	7. 9.	H. ⊙.	
10 sylváticum. W.	(st.)	highland.	Britain.	....	8. 9.	H. ♃.	Eng. bot v. 13. t. 913.
11 réctum. W.	(br.)	upright wood.	———	....	——	H. ♃.	——— v. 2. t. 124.
12 coarctàtum. W.	(gr.)	crowded.	Chili.	1820.	——	H. ♃.	
13 polycéphalum. M.	(st.)	many-headed.	N. America.	1826.	——	H. ⊙.	

14 stachydifòlium. S.S.	(*br.*) Stachy's-leaved.	Monte Video.	1826.	8. 9.	H. ⊙.	
15 supìnum. W.	(*br.*) dwarf.	Scotland.	....	6. 7.	H. ♃.	Eng. bot. v. 17. t. 1193
16 pusíllum. W.	(*br.*) small.	Sweden.	1818.	——	H. ♃.	Krock. siles. t. 41.
17 sph'æricum. W. en.	(*br.*) sphærical.	S. Europe.	——	——	H. ♃.	
18 cephaloídeum. W.	(*br.*) headed.	————	——	——	H. ⊙.	
19 uliginòsum. W.	(*st.*) marsh.	Britain.	....	8. 9.	H. ⊙.	Eng. bot. v. 17. t. 1194
20 germánicum. W.	(*st.*) common.	————	....	6. 8.	H. ⊙.	———— v. 14. t. 948.
21 gállicum. W.	(*st.*) narrow-leaved.	England.	....	——	H. ⊙.	———— v. 33. t. 2369
22 montànum. W.	(*st.*) mountain.	Europe.	1820.	——	H. ⊙.	
23 pyramidàtum. W.	(*st.*) pyramidal.	S. Europe.	1779.	——	H. ⊙.	
24 mínimum. W.	(*st.*) least.	Britain.	....	7. 8.	H. ⊙.	Eng. bot. v. 17. t. 1157.
25 arvénse. W.	(*st.*) corn.	Europe.	1804.	——	H. ⊙.	
26 austràle. F.	(*st.*) southern.	S. Europe.	1823.	——	H. ⊙.	
27 Sellówii. Lk.	(*st.*) Sellow's.	Buenos Ayres.	1828.	——	H. ⊙.	
28 alàtum. K.S.	(*st.*) wing-stalked.	Quito.	1829.	——	G. ♃.	
29 pompeyànum. E.	(*st.*) Egyptian.	Egypt.	——	——	H. ⊙.	
30 Lágopus. S.S.	(*wh.*) Hare's-foot.	Siberia.	1820.	——	H. ⊙.	
METAL'ASIA. L.T.	METAL'ASIA.	Syngenesia Polygamia Superflua.				
1 umbellàta. D.D.	(*ro.*) umbelled.	C. B. S.	1816.	5. 7.	G. ♄.	
2 muricàta. D.D.	(*ro.*) muricate.	————	——	6. 9.	G. ♄.	
3 divérgens. D.D.	(*wh.*) spreading.	————	——	——	G. ♄.	Burm. afr. t. 72. f. 2.
Gnaphàlium divérgens. W.						
4 fastigiàta.	(*wh.*) close-flowered.	————	1812.	5. 8.	G. ♄.	Petiv. gaz. 12. t. 7. f. 3.
Gnaphàlium fastigiàtum. W.						
5 aùrea. D.D.	(*ye.*) golden.	————	1816.	——	G. ♄.	
6 seriphioídes. D.D.	(*wh.*) Seriphium-like.	————	1825.	——	G. ♄.	
7 mucronàta. L.T.	(*ye.*) mucronate.	————	1824.	——	G. ♄.	Burm. afr. t. 66. f. 3.
8 púngens. D.D.	(*wh.*) pungent.	————	1815.	——	G. ♄.	
9 uniflòra. D.D.	(*wh.*) single-flowered.	————	1816.	——	G. ♄.	
10 phylicoídes. D.D.	(*wh.*) Phylica-leaved.	————	——	——	G. ♄.	
SPIRAL'EPIS. D.D.	SPIRAL'EPIS.	Syngenesia Polygamia Superflua.				
1 squarròsa. D.D.	(*pu.wh.*) squarrose.	C. B. S.	1816.	7. 10.	G. ♃.	Jacq. frag. t. 3. f. 4.
Gnaphàlium squarròsum. Jacq.						
2 glomeràta. D.D.	(*wh.*) cluster-flowered.	————	1774.	3. 9.	G. ♃.	
3 tíncta. D.D.	(*bh.*) stained.	————	1816.	——	G. ♃.	
4? declinàta.	(*wh.*) creeping.	————	1787.	7. 9.	G. ♃.	
Gnaphàlium declinàtum. W.						
PETALA'CTE. D.D.	PETALA'CTE.	Syngenesia Polygamia Superflua.				
1 coronàta. D.D.	(*wh.*) crowned.	C. B. S.	1816.	5. 7.	G. ♄.	Burm. afr. t. 69. f. 3.
Gnaphàlium coronàtum. W.						
2 bícolor. D.D.	(*pu.wh.*) two-coloured.	————	——	——	G. ♄.	
PENTATA'XIS. D.D.	PENTATA'XIS.	Syngenesia Polygamia Superflua.				
micrántha. D.D.	(*wh.*) small-flowered.	C. B. S.	1821.	5. 7.	G. ♄.	
ASTE'LMA. Br.	ASTE'LMA.	Syngenesia Polygamia Superflua.				
1 exímium. B.R.	(*re.*) giant.	C. B. S.	1793.	6. 9.	G. ♄.	Bot. reg. t. 532.
Gnaphàlium exímium. A.R. 654.						
2 milleflòrum. D.D.	(*wh.*) many-flowered.	————	1802.	6. 9.	G. ♄.	
Gnaphàlium milleflòrum. W.						
3 Stæhelìna. D.D.	(*ye.*) keel-leaved.	————	1801.	12. 3.	G. ♄.	Andr. reposit. t. 428.
4 crassifòlium. D.D.	(*ye.*) thick-leaved.	————	1816.	3. 5.	G. ♄.	
5 proteoídes. D.D.	(*wh.*) Protea-like.	————	1821.	——	G. ♄.	
6 speciosíssimum. D.D.	(*wh.*) showy.	————	1691.	7. 9.	G. ♄.	Andr. reposit. t. 51.
Xeránthemum speciosíssimum. A.R. *Elichr'ysum speciosíssimum.* W.						
7 lineàre. D.D.	(*wh.*) linear-leaved.	C. B. S.	1816.	7. 9.	G. ♄.	
8 spiràlis. D.D.	(*re.wh.*) spiral-leaved.	————	1801.	7. 10.	G. ♄.	Andr. reposit. t. 262.
Elichr'ysum spiràle. W.						
9 variegàtum. D.D.	(*v.*) large globular-fld.	————	——	5. 6.	G. ♄.	
10 imbricàtum. D.D.	(*ro.*) imbricated.	————	1816.	——	G. ♄.	
Elichr'ysum imbricàtum. W. *Xeránthemum imbricàtum.* L.						
11 canéscens. D.D.	(*ro.*) elegant.	C. B. S.	1794.	8. 4.	G. ♄.	Bot. mag. t. 420.
12 modéstum. Sie.	(*fl.*) modest.	————	1824.	6. 7.	G. ♄.	Bot. mag. t. 2710.
Gnaphàlium modéstum. B.M.						
HELICHR'YSUM. P.S.	HELICHR'YSUM.	Syngenesia Polygamia Superflua.				
1 grandiflòrum. D.D.	(*w.*) great-flowered.	C. B. S.	1731.	6. 8.	G. ♄.	Burm. afr. t. 76. f. 1.
Gnaphàlium grandiflòrum. W.						
2 frùticans. D.D.	(*wh.*) frutescent.	————	1779.	——	G. ♄.	Bot. mag. t. 1802.
Astélma frùticans. B.R. 726. *Gnaphàlium frùticans.* B.M. *grandiflòrum.* A.R. 489.						
3 críspum. D.D.	(*re.*) curled.	C. B. S.	1809.	——	G. ♄.	
4 arbòreum. D.D.	(*re.*) shrubby.	————	1770.	2. 8.	G. ♄.	

5 congéstum. D.D. (*fl.*) close-headed. C. B. S. 1791. 5. 6. G. ♄. Bot. reg. t. 243.
Gnaphàlium congéstum. B.R.
6 cárneum. P.S. (*fl.*) flesh-coloured. ——— 1816. 4. 6. G. ♄.
7 pátulum. D.D. (*wh.*) spreading. ——— 1771. 8. 1. G. ♄.
8 discolòrum. D.D. (*wh.fl.*) two-coloured. ——— 1815. 5. 8. G. ♄. Burm. afr. t. 97. f. 4.
9 cephalòtes. D.D. (*wh.re.*) large-headed. ——— 1789. 11. 1. G. ♄. Pluk. phyt. t. 410. f. 2.
10? diosmæfòlium. (*wh.*) Diosma-leaved. ——— 1812. 3. 8. G. ♄. Vent. malm. t 74.
Gnaphàlium diosmæfòlium. V.
11 ericoídes. D.D. (*pa.*) Heath-leaved. ——— 1774. —— G. ♄. Bot. mag. t. 435.
Gnaphàlium ericoídes. B.M. *non Helichr'ysum ericoides.* P.S.
12 teretifòlium. D.D. (*wh.*) round-leaved. C. B. S. 1812. —— G. ♄. Burm. afr. t. 77. f. 2.
13? tephròdes. (*wh.*) slender-leaved. ——— 1816. —— G. ♄.
Gnaphàlium tephròdes. L.en.
14 acuminàtum. (*wh.*) taper-pointed. N. Holland. —— —— G. ♄.
Gnaphàlium acuminàtum. L.en.
15? lasiocaùlon. (*wh.*) woolly-stalked. C. B. S. 1823. 6. 9. G. ♄.
Gnaphàlium lasiocaùlon. L.en.
16 St'œchas. D.D. (*ye.*) common shrubby. S. Europe. 1629. 6. 10. H. ♄. Barrel. ic. 410.
17 angustifòlium. (*ye.*) narrow-leaved. ——— 1818. —— H. ♄.
Gnaphàlium angustifòlium. P.S.
18 cònicum. (*ye.*) conical. ——— 1824. —— H. ♄.
Gnaphàlium cònicum. L.en.
19 ignéscens. D.D. (*re.*) red-flowered. C. B. S. 1731. 6. 10. G. ♄.
20 microphy'llum. D.D. (*ye.*) small-leaved. Crete. 1823. —— G. ♄.
21? dasyánthum. (*ye.*) hairy-flowered. C. B. S. 1812. —— G. ♄.
Gnaphàlium dasyánthum. W.en.
22 crassifòlium. D.D. (*ye.*) thick-leaved. S. Europe. 1774. 7. 9 G. ♄.
23 marítimum. D.D. (*ye.*) sea-side. C. B. S. 1772. 6. 8. G. ♄. Burm. afr. t. 77. f. 2.
24 apiculàtum. D.D. (*ye.*) New Holland. V. Diem. Isl. 1804. 1. 12. G. ♄. Bot. reg. t. 240.
Gnaphàlium apiculàtum. B.R.
25 f'œtidum. D.D. (*ye.*) strong-scented. C. B. S. 1692. 6. 9. G. ♂. Bot. mag. t. 1987.
Gnaphàlium f'œtidum. B.M.
26 cylíndricum. D.D. (*ye.*) cylindrical. ——— 1780. 6. 8. G. ♃. Pluk. phyt. 298. f. 4.
27 orientàle. D.D. (*ye.*) eastern. Africa. 1629. 4. 8. G. ♄. Comm. hort. 2. t. 55.
28 arenàrium. D.D. (*ye.*) sand. Europe. 1739. 7. 9. H. ♃. Bot. mag. t. 2159.
29 gravèolens. (*ye.*) strong-smelling. Tauria. 1819. —— H. ♃.
Gnaphàlium gravèolens. M.B.
30 odoratíssimum. (*ye.*) sweet-scented. C. B. S. 1691. 4. 8. G. ♄. Mill. ic. 1. t. 131. f. 2.
31 candidíssimum. D.D. (*ye.*) white-leaved. Caucasus. 1819. 7. 9. H. ♃.
32 rùtilans. D.D. (*re.*) shining-flower'd. C. B. S. 1731. 6. 7. G. ♃. Dill. elt. t. 107. f. 127.
33 cymòsum. D.D. (*ye.wh.*) branching. Africa. —— 4 8. G. ♄. ——— t. 107. f. 128.
34 helianthemifòlium. D. (*wh.*) Sun-rose-leav'd. C. B. S. 1774. 7. 10. G ♄. Volck. norib. t. 194.
35 bracteàtum. D.D. (*ye.*) wave-leaved. N. Holland. 1799. —— H. ⊙. Andr. reposit. 375
chrysánthum. P.S. *Xeránthemum bracteàtum.* V. M. t. 2. *lùcidum.* Maunde.
36 fúlgidum. P.S. (*ye.*) great yellow. C. B. S. 1774. 2. 10. G. ♃. Bot. mag. t. 414.
37 dealbàtum. P.S. (*bh.*) herbaceous. V. Diem. Isl. 1812. 1. 12. G. ♃. Lab. nov. hol. 2. t. 190.
38 paniculàtum. P.S. (*wh.*) corymb-flower'd. C. B. S. 1800. 6. 9. G. ♄. Burm. afr. t. 67. f. 1.
argénteum. A.R. 552. *nec aliorum.*
39 argénteum. P.S. (*wh.*) silvery. ——— —— —— G. ♄.
40 fràgrans. A.R. (*ro.*) fragrant. ——— 1803. 7. 9. G. ♄. Andr. reposit. t. 313.
Elichr'ysum fràgrans. A.R.
41 retórtum. W. (*wh.*) trailing. ——— 1732. 7. 8. G. ♄. Dill. elth. t. 322. f. 415.
42? incànum. B.M. (*ro.wh.*) hoary-leaved. V. Diem. Isl. 1828. 5. 6. G. ♃. Bot. mag. t. 2881.
43 rígidum. A.R. (*wh.*) rigid-leaved. C. B. S. 1801. —— G. ♄. Andr. reposit. t. 387.
44 herbàceum. A.R. (*ye.*) shining-flowered. ——— 1802. 7. 9. G. ♃. ——— t. 487.
spléndens. B.M. 1773.

LEUCOSTE'MMA. D.D. LEUCOSTE'MMA. Syngenesia Polygamia Superflua.
1 vestìtum. D.D. (*wh.*) upright. C. B. S. 1774. 7. 9. G. ♄. Burm. afr. t. 66. f. 1.
Elichr'ysum vestìtum. W.
2 lingulàtum. D.D. (*wh.*) tongue-leaved. ——— 1812. —— G. ♄.

PHŒNO'COMA. D.D. PHŒNO'COMA. Syngenesia Polygamia Superflua.
prolífera. D.D. (*pu.*) proliferous. C. B. S. 1789. 5. 11. G. ♄. Bot. reg. t. 21.
Elichr'ysum prolíferum. B.R. *Xeránthemum prolíferum.* L.

APHELE'XIS. D.D. APHELE'XIS. Syngenesia Polygamia Superflua.
1 hùmilis. D.D. (*pu.*) dwarf. C. B. S. 1812. 6. 9. G. ♄. Andr. reposit. t. 664.
Elichr'ysum spectábile. B.C. 59. *pinifòlium.* P.S.
2 sesamoídes. D.D. (*pu.*) superb. C. B. S. 1739. 4. 6. G. ♄. Bot. mag. t. 425.
Elichr'ysum sesamoídes. W. *Xeránthemum sesamoídes.* B.M.

3 fasciculàta. D.D. (vu.) bundle-leaved. C. B. S. 1799. 3. 9. G. ♄. Andr. reposit. t. 242.
α lutéscens. (st.) *yellowish.* ——— .... —— G. ♄.
β versicolor. (pu.wh.) *variegated-flower'd.*——— .... —— G. ♄. Breyn. ic. t. 16. f. 2.
γ alba. (wh.) *white-flowered.* ——— .... —— G. ♄. Andr. reposit. t. 279.
δ rubra. (sc.) *red-flowered.* ——— .... —— G. ♄. ——— ——— t. 650.
4 ericoídes. (bh.) filiform. ——— 1796. —— G. ♄. Lam. ill. t. 693. f. 2.
filifórmis. D.D. *Helichr'ysum ericoides.* P.S.

XERA'NTHEMUM. D.D. XERA'NTHEMUM. Syngenesia Polygamia Superflua. L.
1 ánnuum. D.D. (pu.wh.) common annual. S. Europe. 1570. 7. 8. H. ⊙. Jacq. aust. 4. t. 388.
α ròseum. (ro.) *rose-coloured.* ——— —— —— H. ⊙.
β album. (wh.) *white-flowered.* ——— —— —— H. ⊙.
2 inapértum. D.D. (pu.) small-flowered. ——— 1620. —— H. ⊙. Moris. his. s. 6. t. 12. f. 1.

CHARDI'NIA. D.D. CHARDI'NIA. Syngenesia Polygamia Superflua.
orientàlis. D.D. (wh.) oriental. Levant. 1713. 7. 8. H. ⊙.
xeranthemoídes. Desf. *Xeránthemum orientàle.* W.

FIL'AGO. W. COTTON-ROSE. Syngenesia Polygamia Necessaria. W.
1 pygm'æa. W. (br.) pigmy. S. Europe. 1629. 7. 8. H. ⊙. Cavan. ic. 1. t. 36.
'Evax pygm'æa. G.
2 asterisciflòra. (br.) simple-stalked. Spain. 1824. —— H. ⊙. Lam. ill. t. 694. f. 2.
'Evax asterisciflòra. P.S.

MICR'OPUS. W. MICR'OPUS. Syngenesia Polygamia Necessaria. W.
1 supìnus. W. (pa.) trailing. S. Europe. 1710. 6. 9. H. ⊙. Schk. hand. 3. t. 267.
2 eréctus. W. (pa.) upright. ——— 1683. —— H. ⊙. Lœfl. hisp. t 1. f. 3.

ERI'GERON. W. ERI'GERON. Syngenesia Polygamia Superflua. L.
1 gravèolens. W. (ye.) strong-smelling. S. Europe. 1633. 7. 8. H. ⊙. Ger. emac. 481. f. 2.
2 compòsitum. Ph. (li.) daisy-flowered. N. America. 1811. —— H. ♃.
3 caroliniànum. W. (ye.) Hyssop-leaved. ——— 1727. —— H. ♃. Dill. elt. t. 306. f. 394.
4 canadénse. W. (wh.) Canada. England. .... 8. 9. H. ⊙. Eng. bot. v. 29. t. 2019.
5 bonariénse. W. (wh.) buck's-horn. S. America. 1732. 7. 8. H. ⊙. Dill. elt. t. 257. f. 334.
6 linifòlium. W. (wh.) flax-leaved. ——— .... —— H. ⊙.
7 lævigàtum. L.en. (wh.) smooth. Cayenne. 1823. —— S. ♂.
8 delphinifòlium. W. (wh.) Larkspur-leav'd. S. America. 1816. —— S. ♂.
9 strigòsum. W. (wh.) strigose. N. America. 1818. —— H. ♂.
10 heterophy'llum. W. (wh.) various-leaved. ——— 1640. 7. 9. H. ♂. Flor. græc. t. 486.
A'ster ánnuus. L. *Diplopáppus dùbius.* Cas.
11 chinénse. W. (wh.) Chinese. China. 1818. —— H. ⊙. Jacq. schœnb. 3. t. 303.
12 jamaicénse. W. (li.) Jamaica. Jamaica. 1820. —— S. ⊙. Swartz obs. t. 8. f. 2.
13 pubéscens. K.S. (wh.) pubescent. Mexico. 1827. 7. 11. H. ♃.
14 involucràtum. Lk. .... involucred. Brazil. 1828. .... S. ♃.
15 armerifòlium. F. (pu.) Thrift-leaved. Baical. 1829. 6. 8. H. ♃.
16 nudicaùle. M. (wh.) naked-stalked. N. America. 1812. 7. H. ♃.
17 philadélphicum. W. (pu.) spreading. ——— 1778. 7. 8. H. ♃.
18 purpùreum. W. (pu.) purple. Hudson'sBay. 1776. —— H. ♃.
19 bellidifòlium. W. (li.) Plantain-leav'd. N. America. 1790. —— H. ♃.
serpentàrium. Hort.
20 glabéllum. N. (pu.) smoothish. ——— 1827. 8. 12. H. ♃. Bot. mag. t. 2923.
21 podòlicum. S.S. (pu.) Polish. Poland. —— 7. 9. H. ♃.
22 Lehmánni. S.S. (li.) Lehman's. .... 1828. —— H. ♃.
23 glaùcum. B.R. (li.) glaucous-leav'd. S. America. 1812. 1. 12. G. ♄. Bot. reg. t. 10.
24 chilénse. D.P. (wh.) Chili. Chili. 1816. 7. 9. H. ⊙.
Con'yza chilénsis. L.en.
25 rivulàre. S.S. (bh.) rivulet. Trinidad. 1821. —— S. ⊙.
26 Villársii. W. (pu.) Villars's. Piedmont. 1804. 7. 8. H. ♃. Bot. reg. t. 583.
átticum. Vil.
27 rupéstre. Schl. (li.) rock. Switzerland. 1818. —— H. ♃.
28 àcre. W. (li.) common. Britain. .... —— H. ♂. Eng. bot. v. 17. t. 1158.
29 alpìnum. W. (li.) Alpine. Scotland. .... 7. H. ♃. ——— v. 7. t. 464.
30 uniflòrum. W. (li.) dwarf. ——— .... 8. 9. H. ♃. ——— v. 34. t. 2416.
31 pùmilum. N. (wh.) dwarf. N. America. 1827. —— H. ♃.
32 humile. Gr. (wh.) low. ——— —— 9. 10. H. ♄.
33 ásperum. N. (wh.) rough. ——— 1828. 8. 9. H. ♃.
34 gramíneum. W. (pu.) grass-leaved. Siberia. 1824. —— H. ♃. Gmel. sib. 2. t. 76. f. 2.
35 caucásicum. M.B. (pu.) Caucasean. Caucasus. —— 7. 8. H. ♃.
36 asteroídes. L.en. (wh.) Starwort-like. .... 1812. —— H. ♃.

JAS'ONIA. Cas. JAS'ONIA. Syngenesia Polygamia Superflua.
longifòlia. Cas. (ye.) long-leaved. N. America. 1818. 7. 8. H. ♃.
Erígeron longifòlium. P.S.

CAL'OTIS. B.R. CAL'OTIS. Syngenesia Polygamia Superflua.
cuneifòlia. B.R. (bl.) wedge-leaved. N. Holland. 1819. 7. 8. G. ♄. Bot. reg. t. 504.

DIPLOSTE'PHIUM. K.S. DIPLOSTE'PHIUM. Syngenesia Polygamia Superflua.

1	linariifòlium.	(*pu.*)	toad-flax-leav'd.	N. America.	1699.	9. 10.	H. ♄.	Lodd. bot. cab. t. 6.
	A'ster pulchérrimus. L.B.C. *linàriifòlius.* W. *Chrysópsis linariifòlia.* N.							
2	rígidum.	(*vi.*)	stiff-leaved.	———	1759.	8. 10.	H. ♃.	Pluk. alm. t. 14. f. 7.
	A'ster rígidus. Ph. *nemoràlis.* W. *non* H.K.							
3	linifòlium.	(*wh.*)	Flax-leaved.	N. America.	1739.	7. 8.	H. ♄.	
	Chrysópsis linifòlia. N. *As'ter linifòlius.* W.							
4	álbum. N.	(*wh.*)	white-flowered.	Missouri.	1824.	8. 9.	H. ♃.	
5	hùmile.	(*wh.*)	low.	N. America.	1699.	8. 10.	H. ♃.	Willd. hort. ber. t. 67.
	A'ster hùmilis. W. *Chrysópsis hùmilis.* N.							
6	cornifòlium. W.	(*wh.*)	Dogwood-leaved.	———	1811.	——	H. ♃.	
7	amygdalìnum.	(*wh.*)	Almond-leaved.	———	1759.	7. 9.	H. ♃.	Hoff. phyt. blat. t. 8. f. 2.
	A'ster umbellàtus. W. *Chrysópsis umbellàta.* N.							

A'STER. W. STARWORT. Syngenesia Polygamia Superflua. L.

1	reflexus. W.	(*re.wh.*)	reflexed-leav'd.	C. B. S.	1759.	2. 9.	G. ♄.	Bot. mag. t. 884.
2	tomentòsus. W.	(*wh.*)	tooth-leaved.	N. S. W.	1793.	5. 7.	G. ♄.	Wend. s. han. p. 8. t. 24.
	dentàtus. Andr. rep. t. 61.							
3	stellulàtus. P.S.		starry-haired.	V. Diem. Isl.	1823.	——	G. ♄.	Lab. nov. hol. 2. t. 196.
4	phlogopáppus. P.S.	(*vi.*)	glossy-haired.	———	1827.	——	G. ♄.	——— 2. t. 195.
5	myrsinoídes. P.S.	(*wh.*)	Myrsine-leaved.	———	1824.	——	G. ♄.	——— p. 53. t. 202.
6	erubéscens. S.S.	(*bh.*)	blush-flowered.	N. S. W.	1826.	——	G. ♄.	
7	argophy'llus. P.S.	(*wh.*)	musk-scented.	V. Diem. Isl.	1804.	——	G. ♄.	Bot. mag. t 1563.
8	liràtus. B.M.	(*wh.*)	fluted-stemm'd.	N. S. W.	1812.	——	G. ♄.	——— t. 1509.
9	seríceus. W.	(*bl.*)	silky-leaved.	Missouri.	1802.	5. 11.	G. ♄.	Venten. cels. t. 33.
10	Cymbalàriæ. W.	(*wh.*)	Ivy-leaved.	C. B. S.	1786.	——	G. ♄.	Venten. malm. t. 95.
11	pappòsus. S.S.	(*vi.*)	cluster-leaved.	———	1812.	——	G. ♄.	
12	taxifòlius. S.S.	(*li.*)	Yew-leaved.	———	1824.	——	G. ♄.	
13	angustifòlius. W.	(*bl.*)	narrow-leaved.	———	1804.	5. 7.	G. ♄.	Jacq. schœnb. 3. t. 370.
14	villòsus. W.	(*li.*)	villous.	———	1812.	——	G. ♄.	
15	obtusàtus. W.	(*li.*)	obtuse-leaved.	———	1793.	——	G. ♄.	
16	fruticulòsus. W.	(*li.*)	shrubby.	———	1759.	3. 7.	G. ♄.	Jacq. frag. 9. t. 5. f. 4.
17	filifòlius. V.	(*wh.*)	thread-leaved.	———	1812.	——	G. ♄.	Venten. malm. t. 82.
18	aculeàtus. P.S.	(*wh.*)	prickly-leaved.	V. Diem. Isl.	1818.	——	G. ♄.	Lab. nov. hol. 2. t. 200.
	exasperàtus. L. enum.							
19	tenéllus. W.	(*vi.*)	slender.	C. B. S.	1769.	4. 10.	H. ⊙.	Bot. mag. t. 33.
20	alpìnus. W.	(*pu.*)	Alpine.	Alps Europe.	1658.	5. 8.	H. ♃.	——— t 199.
21	lusitànus. S.S.	(*pu.*)	Portugal.	Portugal.	1820.	8. 10.	H. ♃.	
22	salsuginòsus. S.S.	(*vi.*)	salt plain.	N. America.	1827.	5. 6.	H. ♃.	Bot. mag. t. 2942.
23	pulchéllus. W.	(*re.*)	narrow-rayed.	Armenia.	1818.	——	H. ♃.	
24	caucásicus. W.	(*pu.*)	Caucasean.	Caucasus.	1804.	7. 8.	H. ♃.	
25	alwarténsis. B.M.	(*pu.*)	large-flowered.	———	....	——	H. ♃.	Bot. mag. t. 2321.
26	diversifòlius. M.	(*li.*)	various-leaved.	N. America.	1811.	7. 9.	H. ♃.	
27	macrophy'llus. W.	(*bl.*)	large-leaved.	———	1739.	——	H. ♃.	
28	corymbòsus. W.	(*wh.*)	corymbed.	———	1765.	9.	H. ♃.	
29	cordifòlius. W.	(*bh.*)	heart-leaved.	———	1759.	7. 8.	H. ♃.	
30	paniculàtus. W.	(*li.*)	panicled.	———	1640.	9. 10.	H. ♃.	Corn. canad. t. 65.
31	sagittifòlius. W.	(*bl.*)	arrow-leaved.	———	1818.	7. 9.	H. ♃.	
32	undulàtus. W.	(*pu.*)	wave-leaved.	———	1699.	8. 10.	H. ♃.	Herm. parad. t. 96.
33	præcox. W.en.	(*li.*)	early-flowering.	———	1800.	7. 8.	H. ♃.	
34	pállens. W.en.	(*li.*)	pale-flowered.	———	....	9. 10.	H. ♃.	
35	conyzoídes. W.	(*wh.*)	Conyza-like.	———	1778.	8. 9.	H. ♃.	
36	altàicus. W.en.	(*pu.*)	dwarf.	Siberia.	1804.	5. 8.	H. ♃.	
	pùmilus. F.							
37	nemoràlis. H.K.	(*li.*)	wood.	N. America.	1778.	8. 9.	H. ♃.	
	ledifòlius. Ph.							
38	hyssopifòlius. W.	(*bh.*)	Hyssop-leaved.	———	1683.	9. 10.	H. ♃.	Dodart. mem. t. 60.
39	punctàtus. W.	(*vi.*)	dotted.	Hungary.	1815.	8. 9.	H. ♃.	W. et K. hung. 2. t. 109.
40	ácris. W.	(*bl.*)	acrid.	S. Europe.	1731.	——	H. ♃.	Pluk. alm. t. 271. f. 3.
41	cànus. W.	(*bl.*)	hoary-leaved.	Hungary.	1816.	——	H. ♃.	W. et K. hung. 1. t. 30.
42	solidaginoídes. W.	(*wh.*)	Solidago-like.	N. America.	1699.	——	H. ♃.	Pluk. alm. t. 79. f. 2.
43	subulàtus. N.	(*li.*)	awl-leaved.	———	1823.	8. 10.	H. ⊙.	
44	pilòsus. W.	(*wh.*)	hairy.	———	1812.	——	H. ♃.	
45	foliolòsus. W.	(*wh.*)	leafy.	———	1732.	10.	H. ♃.	Dill. elth. t. 35. f. 39.
46	tenuifòlius. W.	(*wh.*)	slender-leaved.	———	1725.	9. 10.	H. ♃.	Pluk. alm. t. 78. f. 5.
47	dumòsus. W.	(*wh.*)	bushy.	———	1734.	——	H. ♃.	Herm. parad. 95.
	α violàceus. H.K.	(*vi.*)	*violet-flowered.*	———	——	——	H. ♃.	
	β álbus. H.K.	(*wh.*)	*white-flowered.*	———	——	——	H. ♃.	
48	ericoídes. W.	(*wh.*)	heath-leaved.	———	1758.	9.	H. ♃.	
49	multiflòrus. W.	(*wh.*)	many-flowered.	———	1732.	9. 10.	H. ♃.	Dill. elth. t. 36. f. 40.
50	ciliàtus. W.	(*wh.*)	ciliated.	———	....	——	H. ♃.	

51	canéscens. Ph.	(*li.*)	canescent.	N. America.	1812.	——	H. ♂.	
	biénnis. N.							
52	pauciflòrus. N.	(*wh.*)	few-flowered.	N. America.	1825.	8. 9.	H. ♃.	
53	sparsiflòrus. W.	(*wh.*)	scattered-flower'd.	——	1798.	9. 11.	H. ♃.	
54	coridifòlius. W.	(*bl.*)	Coris leaved.	——	....	8. 11.	H. ♃.	
55	squarròsus. W.	(*bl.*)	ragged.	——	1801.	——	H. ♃.	
56	argénteus. H.K.	(*bl.*)	silver-leaved.	——	——	7. 9.	H. ♃.	
57	cóncolor. W.	(*pu.*)	self-coloured.	——	1759.	8. 11.	H. ♃.	
58	reticulàtus. Ph.	(*wh.*)	netted-leaved.	——	1812.	——	H. ♃.	
59	salicifòlius. W.	(*pu.*)	Willow-leaved.	——	1760.	9. 10.	H. ♃.	Rob. ic. 307.
60	æstìvus. W.	(*bl.*)	summer.	——	1776.	7. 8.	H. ♃.	
61	pannònicus. W.	(*vi.*)	Hungarian.	Hungary.	1815.	——	H. ♃.	Jacq. vind. 1. t. 8.
62	Amèllus. W.	(*bl.*)	Italian.	Italy.	1596.	7. 9.	H. ♃.	Bot. reg. t. 340.
	α latifòlius.	(*bl.*)	*broad-leaved.*	——		——	H. ♃.	
	β angustifòlius.	(*bl.*)	*narrow-leaved.*	——		——	H. ♃.	
63	ibèricus. M.B.	(*bl.*)	Iberian.	Iberia.	1827.	7. 9.	H. ♃.	
64	nòvæ ángliæ. W.	(*bl.*)	New England.	N. America.	1710.	9. 10.	H. ♃.	Herm. parad. 98.
	β rùbra.	(*re.*)	*red-flowered.*	——	1812.	——	H. ♃.	Bot. reg. 183. f. *inf.*
65	cyáneus. N.	(*bl.*)	bright-blue.	——	1789.	——	H. ♃.	Hoff. phyt. 1. t. B. f. 1.
66	glaùcus. N.A.	(*bl.*)	glaucous-leaved.	....	1823.	8. 10.	H. ♃.	
67	paludòsus. W.	(*bl.*)	marsh.	N. America.	1784.	——	H. ♃.	
68	grandiflòrus. W.	(*bl.*)	great-flowered.	——	1720.	10. 11.	H. ♃.	Bot. reg. t. 273.
69	caroliniànus. W.	(*pu.*)	tall.	Carolina.	....	8. 9.	H. ♄.	
70	oblongifòlius. N.	(*li.*)	oblong-leaved.	Missouri.	1823.	——	H. ♃.	
71	phlogifòlius. W.	(*vi.*)	Phlox-leaved.	N. America.	1797.	7. 10.	H. ♃.	
72	pátens. W.	(*bl.*)	spreading.	——	1773.	9. 11.	H. ♃.	Swt. brit. fl. gar. t. 234.
73	Tripòlium. W.	(*bl.*)	sea-side.	Britain.	....	8. 9.	H. ♃.	Eng. bot. v. 2. t. 87.
74	Schrebèri. N.A.	(*wh.*)	Schreber's.	....	1827.	8. 10.	H. ♃.	
75	salígnus. W.	(*wh.*)	Sallow-leaved.	Germany.	1815.	——	H. ♃.	
76	gravèolens. N.	(*bl.*)	strong-scented.	Arkansa.	1825.	10. 11.	H. ♃.	
77	puníceus. W.	(*re.*)	red-stalked.	N. America.	1710.	7. 10.	H. ♃.	Herm. lugd. t. 651.
78	híspidus. W.	(*wh.*)	rough-stalked.	China.	1804.	9. 10.	H. ♃.	Kæmpf. ic. t. 29.
79	dentàtus. Th.	(*bl.*)	tooth-leaved.	C. B. S.	1825.	7. 9.	G. ♃.	
80	serràtus. Th.	(*bl.*)	saw-leaved.	——	——	——	G. ♃.	
81	tatáricus. W.	(*bl.*)	Tartarian.	Siberia.	1818.	——	H. ♃.	
82	pyren'æus. S.S.	(*vi.*)	Pyrenean.	Pyrenees.	1768.	6. 9.	H. ♃.	
83	sibíricus. W.	(*pu.*)	Siberian.	Siberia.	1768.	——	H. ♃.	Gmel. sib. 2. t. 80. f. 1.
84	incìsus. F.	(*li.*)	cut-leaved.	——	1816.	6. 9.	H. ♃.	
85	amelloídes. Bes.	(*vi.*)	Amellus-like.	Poland.	1827.	7. 9.	H. ♃.	
86	élegans. W.	(*bl.*)	elegant.	....	1790.	8. 10.	H. ♃.	
87	amplexicaùlis. W.	(*bl.*)	stem-clasping.	N. America.	....	9. 11.	H. ♃.	
88	adulterìnus. W.	(*re.*)	bastard.	——	....	8. 10.	H. ♃.	
89	lævigàtus. W.	(*li.*)	smooth-stem'd.	——	1794.	9. 11.	H. ♃.	
90	longifòlius. P.S.	(*bl.*)	long-leaved.	——	1798.	10.	H. ♃.	Moris. s. 7. t. 22. f. 26.
91	versícolor. W.	(*wh.re.*)	various-color'd.	——	1790.	8. 9.	H. ♃.	
92	mutábilis. W.	(*pu.*)	changeable.	——	1710.	9. 10.	H. ♃.	Herm. lugd. t. 67.
93	l'ævis. W.	(*li.*)	smooth.	——	1753.	——	H. ♃.	
94	concínnus. W.en.	(*li.*)	neat.	——	1800.	——	H. ♃.	
95	floribúndus. W.	(*li.*)	abundant-flower'd.	——	....	——	H. ♃.	
96	nòvi bélgii. W.	(*li.*)	New-York.	——	1710.	——	H. ♃.	Herm. lugd. t. 69.
97	fi'rmus. L.en.	(*li.*)	firm.	——	1818.	——	H. ♃.	
98	bellidiflòrus. W.en.	(*li.*)	daisy-flowered.	——	....	——	H. ♃.	
99	spectábilis. W.	(*bl.*)	showy.	——	1777.	8. 9.	H. ♃.	
100	serotìnus. W.	(*bl.*)	late-flowering.	——	....	9. 11.	H. ♃.	
101	tardiflòrus. W.	(*li.*)	spear-leaved.	——	1775.	7. 9.	H. ♃.	
102	blándus. Ph.	(*bh.*)	charming.	——	1800.	10. 11.	H. ♃.	
103	acuminatus. Ph.	(*wh.*)	acuminate.	——	1806.	8. 10.	H. ♃.	Bot. mag. t. 2707.
104	éminens. W.en.	(*re.*)	eminent.	——	....	9. 11.	H. ♃.	
105	pubéscens. N.A.	(*re.*)	pubescent.	——	1827.	——	H. ♃.	
106	Rádula. W.	(*wh.*)	rasp-leaved.	——	1785.	——	H. ♃.	
107	prenanthoídes. W.	(*bl.*)	hairy-branch'd.	——	1817.	——	H. ♃.	
108	stríctus. Ph.	(*vi.*)	upright dwarf.	——	1806.	——	H. ♃.	
109	Tradescánti. W.	(*wh.*)	Michaelmas daisy.	——	1633.	7. 9.	H. ♃.	Moris. s. 7. t. 21. f. 42.
110	vimíneus. W.	(*bl.*)	slender.	——	1818.	——	H. ♃.	
111	recurvàtus. W.	(*li.*)	recurved.	——	1800.	8. 9.	H. ♃.	
112	expánsus. N.A.	(*bl.*)	expanded.	——	1827.	——	H. ♃.	
113	fastigiàtus. Led.	(*bl.*)	clustered.	Siberia.	——	——	H. ♃.	
114	láxus. W.en.	(*wh.*)	loose-stalked.	N. America.	....	9. 11.	H. ♃.	
115	símplex. W.en.	(*wh.*)	simple-stalked.	——	....	8. 10.	H. ♃.	

116 virgíneus. N.A.	(*wh.*)	virgin.	N. America.	1826.	8. 10.	H. ♃.	
117 polyphy'llus. W.en.	(*w.*)	many-leaved.	———	....	——	H. ♃.	
118 júnceus. w.	(*fl.*)	slender-stalked.	———	1758.	9. 10.	H. ♃.	
119 lanceolàtus. w.	(*wh.*)	lanceolate.	———	1811.	8. 11.	H. ♃.	
120 dracunculoídes. w.	(*w.*)	Tarragon-like.	———	——	9. 11.	H. ♃.	
121 frágilis. w.	(*wh.*)	brittle.	———	1800.	9.	H. ♃.	
122 cárneus. N.A.	(*fl.*)	flesh-coloured.	———	1827.	——	H. ♃.	
123 pinifòlius. L.en.	(*wh.*)	Pine-leaved.	———	1823.	——	H. ♃.	
124 laxiflòrus. L.en.	(*li.*)	loose-flowered.	———	....	8. 10.	H. ♃.	
125 mìser. w.	(*wh.*)	meagre-flower'd.	———	1759.	9. 10.	H. ♃.	
126 divérgens. w.	(*wh.*)	spreading downy.	———	1758.	——	H. ♃.	
127 diffùsus. w.	(*wh.*)	diffuse.	———	1777.	——	H. ♃.	
128 scopàrius. N.A.	(*wh.*)	broom-like.	———	1827.	——	H. ♃.	
129 péndulus. w.	(*wh.*)	pendulous.	———	1758.	——	H. ♃.	
130 parviflòrus. L.en.	(*w.*)	small-flowered.	———	1822.	——	H. ♃.	
131 rigídulus. L.en.	(*re.*)	stiffish.	———	——	——	H. ♃.	
132 myrtifòlius. L.en.	(*w.*)	Myrtle-leaved.	———	1819.	8. 9.	H. ♃.	
133 abbreviàtus. N.A.	(*bl.*)	short-rayed.	———	1827.	——	H. ♃.	
134 robústus. N.A.	(*bl.*)	robust.	....	——	——	H. ♃.	
CALLIST'EMA. Cas.		CHINA ASTER.	Syngenesia Polygamia Superflua.				
hortênsis. Cas.	(*va.*)	garden.	China.	1731.	7. 9.	H. ⊙.	Dill. elth. t. 34. f. 38.
α cœrùlea.	(*bl.*)	*blue-flowered.*	———	——	——	H. ⊙.	
β rùbra.	(*re.*)	*red-flowered.*	———	——	——	H. ⊙.	
γ álba.	(*wh.*)	*white-flowered.*	———	——	——	H. ⊙.	
δ variegàta.	(*bl.wh.*)	*variegated.*	———	——	——	H. ⊙.	
ε versícolor.	(*re.wh.*)	*red and white.*	———	——	——	H. ⊙.	
ζ multipléx.	(*va.*)	*double-flowered.*	———	——	——	H. ⊙.	
η brachyántha.	(*va.*)	*bonnet.*	———	——	——	H. ⊙.	
SOLID'AGO. W.		GOLDEN-ROD.	Syngenesia Polygamia Superflua. L.				
1 canadénsis. w.	(*ye.*)	Canadian.	N. America.	1648.	7. 9.	H. ♃.	Schk. hand. 3. f. 246.
2 pròcera. w.	(*ye.*)	great.	———	1758.	9. 10.	H. ♃.	
3 serotìna. w.	(*ye.*)	upright smooth.	———	——	7. 8.	H. ♃.	
4 gigántea. w.	(*ye.*)	gigantic.	———	——	8. 9.	H. ♃.	
5 ciliàris. w.	(*ye.*)	ciliated.	———	1811.	——	H. ♃.	
6 refléxa. w.	(*ye.*)	hanging-leaved.	———	1758.	8 9.	H. ♃.	
7 fràgrans. W.en.	(*ye.*)	fragrant.	———	....	7. 9.	H. ♃	
8 lateriflòra. w.	(*ye.*)	lateral-flowered.	———	1758.	8. 9.	H. ♃.	
9 glábra. s.s.	(*ye.*)	smooth.	———	....	——	H. ♃.	
10 áspera. w.	(*ye.*)	rough-leaved.	———	1732.	9.	H. ♃.	Dill. elth. t. 305. f. 392.
11 altíssima. w.	(*ye.*)	tall.	———	1686.	8. 9.	H. ♃.	Mart. cent. 14.
12 recurvàta. W.en.	(*ye.*)	recurved.	———	1814.	——	H. ♃.	
13 rugòsa. Ph.	(*ye.*)	wrinkle-leaved.	———	1732.	——	H. ♃.	Dill. elth. t. 308. f. 396.
14 villòsa. Ph.	(*ye.*)	villous.	———	——	——	H. ♃.	
15 scábra. w.	(*ye.*)	rough.	———	1811.	——	H. ♃.	
16 pyramidàta. Ph.	(*ye.*)	pyramidal.	———	1812.	8. 9.	H. ♃.	
17 nemoràlis. w.	(*ye.*)	woolly-stalked.	———	1769.	9.	H. ♃.	
18 pátula. w.	(*ye.*)	spreading.	———	1805.	9. 10.	H. ♃.	
19 ulmifòlia. w.	(*ye.*)	Elm-leaved.	———	——	8. 10.	H. ♃.	
20 argùta. w.	(*ye.*)	sharp-notched.	———	1758.	7. 8.	H. ♃.	
21 júncea. w.	(*ye.*)	rush-stalked.	———	1769.	8. 9.	H. ♃.	
22 ellíptica. w.	(*ye.*)	oval-leaved.	———	1759.	——	H. ♃.	
23 asperàta. Ph.	(*ye.*)	roughish-leaved.	Canada.	1825.	——	H. ♃.	
24 odòra. w.	(*ye.*)	sweet-smelling.	N. America.	1699.	8. 10.	H. ♃.	Pluk. alm. t. 116. f. 6.
25 retrórsa. M.	(*ye.*)	reflex-leaved.	———	1824.	——	H. ♃.	
26 lævigàta. N.	(*ye.*)	fleshy-leaved.	———	——	10. 11.	H. ♃.	
27 mexicàna. w.	(*ye.*)	Mexican.	America.	1683.	7. 10.	H. ♃.	Dodart. act 4. t. 219.
28 sempervìrens. N.	(*ye.*)	evergreen.	N. America.	1699.	9. 10.	H. ♃.	Corn. canad. t. 169.
29 speciòsa. N.	(*ye.*)	handsome.	———	1812.	——	H. ♃.	
30 multiflòra. P.S.	(*ye.*)	many-flowered.	———	....	——	H. ♃.	
31 pauciflosculòsa. M.	(*ye.*)	slender-flower'd.	———	1811.	8. 10.	H. ♃.	
32 bícolor. w.	(*wh.ye.*)	two-coloured.	———	1759.	8. 9.	H. ♃.	Pluk. alm. t. 114. f. 8.
A'ster bícolor. Nees.							
33 petiolàris. w.	(*ye.*)	late-flowered.	———	1758.	10. 12.	H. ♃.	
34 strícta. w.	(*ye.*)	Willow-leaved.	———	——	9.	H. ♃.	
35 hùmilis. Ph.	(*ye.*)	dwarf.	———	1811.	7. 8.	H. ♃.	
36 virgàta. M.	(*ye.*)	slender.	———	1827.	——	H. ♃.	
37 lívida. W.en.	(*ye.*)	purple-stemmed.	———	1812.	——	H. ♃.	
38 hírta. W.en.	(*ye.*)	hairy.	———	....	9. 10.	H. ♃.	
39 lithospermifòlia. Ph.	(*ye.*)	Gromwell-leav'd.	———	1811.	8. 10.	H. ♃.	

40 híspida. w.	(*ye.*)	hispid.	N. America.	1820.	8. 10.	H. ♃.	
41 c'æsia. w.	(*ye.*)	Maryland.	———	1732.	9. 10.	H. ♃.	
42 flexicaùlis. w.	(*ye.*)	crook-stalked.	———	1725.	9.	H. ♃.	Pluk. alm. t. 235. f. 3.
43 latifòlia. L.	(*ye.*)	broad-leaved.	———	——	——	H. ♃.	——— t. 235. f. 4.
44 ambígua. w.	(*ye.*)	angular-stalked.	....	1759.	7. 8.	H. ♃.	
45 macrophy'lla. Ph.	(*ye.*)	large-leaved.	N. America.	....	8. 10.	H. ♃.	
46 glomeràta. M.	(*ye.*)	close-flowered.	———	1820.	——	H. ♃.	
47 squarròsa. N.	(*ye.*)	squarrose-cup'd.	———	1827.	——	H. ♃.	
48 alpéstris. W.K.	(*ye.*)	Alpine.	Europe.	——	7. 9.	H. ♃.	
49 Virgaùrea. w.	(*ye.*)	common.	Britain.	....	——	H. ♃.	Eng. bot. v. 5. t. 301.
50 cámbrica. w.	(*ye.*)	Welsh.	Wales.	....	7. 8.	H. ♃.	Dill. elth. t. 306. f. 393.
51 asiática. F.	(*ye.*)	Asiatic.	N. Asia.	1824.	——	H. ♃.	
52 axillàris. Ph.	(*ye.*)	axillary.	N. America.	1811.	8. 10.	H. ♃.	
53 eréctа. Ph.	(*ye.*)	upright.	———	....	——	H. ♃.	
54 vimínea. w.	(*ye.*)	twiggy.	———	1759.	9.	H. ♃.	
55 angustifòlia. N.	(*ye.*)	narrow-leaved.	———	1824.	——	H. ♃.	
56 littoràlis. S.S.	(*ye.*)	sea-side.	Italy.	1827.	7. 9.	H. ♃.	
57 decúrrens. S.S.	(*ye.*)	decurrent.	China.	1823.	8. 10.	H. ♃.	
58 pubérula. N.	(*ye.*)	pubescent.	N. America.	——	——	H. ♃.	
59 minùta. w.	(*ye.*)	least.	Pyrenees.	1772.	7. 8.	H. ♃.	Lodd. bot. cab. t. 189.
60 multiradiàta. w.	(*ye.*)	Labrador.	Labrador.	1776.	——	H. ♃.	
61 elàta. Ph.	(*ye.*)	tall hairy.	N. America.	1811.	8. 10.	H. ♃.	
62 rígida. w.	(*ye.*)	hard-leaved.	———	1710.	9.	H. ♃.	Herm. parad. 243.
EUTH'AMIA. N.		EUTH'AMIA. Syngenesia Polygamia Superflua.					
1 graminifòlia. N.	(*ye.*)	grass-leaved.	N. America.	1758.	8. 10.	H. ♃.	Bot. mag. t. 2546.
Chrysócoma graminifòlia. L. *Solidàgo lanceolàta.* w.							
2 tenuifòlia. N.	(*ye.*)	slender-leaved.	N. America.	——	9. 10.	H. ♃.	
BRACH'YRIS. N.		BRACH'YRIS. Syngenesia Polygamia Superflua.					
Euthàmiæ. N.	(*ye.*)	strong-scented.	Missouri.	1827.	8. 10.	H. ♃.	
Solidàgo Saróthræ. Ph. *Brachyach'yris Euthàmiæ.* S.S.							
CHRYSO'COMA. W.		GOLDY-LOCKS. Syngenesia Polygamia Æqualis. L.					
1 comaùrea. w.	(*ye.*)	great shrubby.	C. B. S.	1731.	6. 8.	G. ♄.	Bot. mag. t. 1972.
2 pátula. w.	(*ye.*)	spreading.	———	1816.	——	G. ħ.	
3 cérnua. w.	(*ye.*)	small shrubby.	———	1712.	5. 9.	G. ♄.	Comm. hort. 2. t. 45.
4 ciliàris. w.	(*ye.*)	Heath-leaved.	———	1759.	7. 10.	G. ♄.	——— 2. t. 48.
5 nívea. w.	(*ye.*)	woolly-leaved.	———	1820.	5. 9.	G. ♄.	Jacq. schœn. 2. t. 147.
tomentòsa. Jacq. *non* L.							
6 scábra. w.	(*ye.*)	rugged.	———	1732.	8. 9.	G. ♄.	Dill. elth. t. 88. f. 103.
7 denticulàta. w.	(*ye.*)	toothed.	....	....	——	G. ħ.	Jacq. schœn. 3. t. 368.
8 Linos'yris. w.	(*ye.*)	flax-leaved.	Europe.	....	9. 10.	H. ♃.	Eng. bot. v. 35. t. 2505.
9 dracunculoídes. w.	(*ye.*)	Siberian.	Siberia.	....	——	H. ♃.	
10 biflòra. w.	(*ye.*)	two-flowered.	———	1741.	8. 9.	H. ♃.	Gmel. sib. 2. t. 82. f. 1.
11 gravèolens. N.	(*ye.*)	Rue-scented.	Missouri.	1822.	——	H. ♃.	
12 nudàta. M.	(*ye.*)	naked.	N. America.	1819.	——	H. ♃.	
13 virgàta. N.	(*ye.*)	slender.	———	——	——	H. ♃.	
14 villòsa. w.	(*ye.*)	villous.	Hungary.	1799.	——	H. ♃.	W. et K. hung. 1. t. 58.
BOLT'ONIA. W.		BOLT'ONIA. Syngenesia Polygamia Superflua. L.					
1 asteroídes. w.	(*wh.*)	Starwort-flower'd.	N. America.	1758.	8. 10.	H. ♃.	Bot. mag. t. 2554.
2 glastifòlia. w.	(*pu.*)	glaucous-leaved.	———	——	9.	H. ♃.	——— t. 2381.
AME'LLUS. W.		AME'LLUS. Syngenesia Polygamia Superflua.					
Lychnìtis. w.	(*bl.*)	Cape.	C. B. S.	1768.	6. 7.	G. ♄.	Bot. reg. t. 586.
SIDERA'NTHUS. Fras.		SIDERA'NTHUS. Syngenesia Polygamia Superflua.					
1 villòsus.	(*ye.*)	villous.	Missouri.	1811.	8. 9.	H. ♃.	
Siderànthus integrifòlius. Fras. *Améllus villòsus.* Ph.							
2 spinulòsus.	(*ye.*)	spiny.	———	——	——	H. ♃.	
Améllus spinulòsus. Ph. *Siderànthus pinnatífidus.* Fras.							
STA'RKEA. W.		STA'RKEA. Syngenesia Polygamia Superflua.					
1 umbellàta. w.	(*ye.*)	umbel-flower'd.	Jamaica.	1768.	6. 7.	S. ♃.	Lam. ill. t. 682. f. 2.
2 igniària. D.D.	(*ye.*)	close-flowered.	S. America.	1823.	6. 8.	F. ♃.	H. et B. pl. æq. 2. t. 112.
Hierba de Santa Maria. incolarum. *Andromáchia igniària.* K.S.							
GRIND'ELIA. W.en.		GRIND'ELIA. Syngenesia Polygamia Superflua. W.					
1 glutinòsa. Dun.	(*ye.*)	glutinous.	Mexico.	1803.	1. 12.	F. ♄.	Bot. reg. t. 187.
Dónia glutinòsa. B.R. *Dorónicum glutinòsum.* w.							
2 inuloídes. W.en.	(*ye.*)	Inula-like.	Mexico.	1815.	6. 9.	F. ♄.	Bot. reg. t. 248.
3 pulchélla. S.S.	(*ye.*)	pretty.	Chile.	1827.	——	F. ♄.	
4 nítida. Bern.	(*ye.*)	glossy.	Mexico.	1825.	——	F. ♄.	
5 coronopifòlia. Leh.	(*ye.*)	Buckshorn-leav'd.	———	1826.	——	F. ♄.	
6 angustifòlia. B.R.	(*ye.*)	narrow-leaved.	———	1822.	——	F. ♄.	Bot. reg. t. 781.
7 spatulàta. L.en.	(*ye.*)	spatulate-leav'd.	———	——	——	F. ♄.	

8 squarròsa. Dun.	(*ye.*)	Snake-headed.	Missouri.	1811.	7. 9.	F. ♃.	Bot. mag. t. 1706.
9 ciliàta. s.s.	(*ye.*)	ciliated.	Arkansa.	1821.	9. 10.	H. ♃.	Hook. ex. flor. t. 45.
Dónia ciliàta. H.E.F.							
10 villòsa. D.	(*ye.*)	villous.	Columbia.	1827.	8. 10.	H. ♃.	
PODOL'EPIS. Lab.		PODOL'EPIS. Syngenesia Polygamia Superflua.					
1 rugàta. H.K.	(*ye.*)	wrinkle-scaled.	N. Holland.	1803.	7. 8.	G. ♃.	Lab. nov. holl. 2. t. 208.
2 acuminàta. H.K.	(*ye.*)	sharp-scaled.	N. S. W.	——	5. 8.	G. ♃.	Bot. mag. t. 956.
Scàlia jaceoídes. B.M.							
3 grácilis. B.F.G.	(*li.*)	slender.	N. Holland.	1827.	8. 10.	H. ♃.	Swt. br. fl. gar. t. 285.
GERB'ERIA. Cass.		GERB'ERIA. Syngenesia Polygamia Superflua.					
1 crenàtá. B.R.	(*li.*)	crenated.	C. B. S.	1820.	4. 8.	G. ♃.	Bot. reg. t. 855.
2 asplenifòlia. s.s.	(*ye.pu.*)	Spleenwort-leaved.	——	1822.	4. 6.	G. ♃.	Burm. afr. 155. t. 26. f. 1.
A'rnica Gerbèra. L. *Dorónicum asplenifòlium.* Lam.							
A'RNICA. W.		A'RNICA. Syngenesia Polygamia Superflua.					
1 montàna. W.	(*ye.*)	mountain.	Europe.	1731.	7. 8.	H. ♃.	Bot. mag. t. 1749.
2 angustifòlia. s.s.	(*ye.*)	narrow-leaved.	Greenland.	1827.	6. 8.	H. ♃.	
3 fúlgens. Ph.	(*ye.*)	glossy-flowered.	Missouri.	1826.	——	H. ♃.	
4 scorpioídes. W.	(*ye.*)	alternate-leaved.	Austria.	1710.	——	H. ♃.	Jacq. aust. 4. t. 349.
5 glaciàlis. W.	(*ye.*)	alpine.	Germany.	1816.	7. 8.	H. ♃.	Jacq. ic. 3. t. 586.
6 Dorónicum. W.	(*ye.*)	Clusius's.	Austria.	——	——	H. ♃.	Jacq. aust. 1. t. 92.
Clùsii. All. t. 17. f. 1. 2.							
7 lanígera. T.N.	(*ye.*)	woolly.	Italy.	1827.	——	H. ♃.	
8 marítima. L.	(*ye.*)	sea-side.	Kamtschatka.	1821.	——	H. ♃.	
BELLIDIA'STRUM. Cas.		BELLIDIA'STRUM. Syngenesia Polygamia Superflua.					
Michèlii. Cas.	(*wh.ro.*)	Micheli's.	Austria.	1570.	6. 8.	H. ♃.	Bot. mag. t. 1196.
A'rnica Bellidiástrum. B.M. *Dorónicum Bellidiástrum.* L.							
DIPLO'COMA. D.D.		DIPLO'COMA. Syngenesia Polygamia Superflua.					
villòsa. D.D.	(*ye.*)	villous.	Mexico.	1827.	8. 10.	F. ♃.	Swt. br. fl. gar. t. 246.
N'EJA. D.D.		N'EJA. Syngenesia Polygamia Superflua.					
grácilis. D.D.	(*ye.*)	slender.	Mexico.	1828.	9. 10.	F. ♄.	Swt. br. fl. gar. ser. 2.
DORO'NICUM. W.		LEOPARD'S-BANE. Syngenesia Polygamia Superflua.					W.
1 Pardaliánches. W.	(*ye.*)	great.	Britain.	....	5.	H. ♃.	Eng. bot. v. 9. t. 630.
2 macrophy'llum. s.s.	(*ye.*)	large-leaved.	Europe.	1821.	——	H. ♃.	
3 scorpioídes. W.	(*ye.*)	mountain.	Germany.	——	4. 6.	H. ♃.	
4 Colúmnæ. T.S.	(*ye.*)	Columna's.	Naples.	1826.	——	H. ♃.	Column. ecphr. 2. t. 46.
5 caucásicum. M.B.	(*ye.*)	Caucasean.	Caucasus.	1815.	——	H. ♃.	
orientàle. W.en.							
6 altàicum. W.	(*ye.*)	Siberian.	Siberia.	1783.	6. 8.	H. ♃.	Pall. a. pet. 1779. t. 16.
7 dentàtum. L.en.	(*ye.*)	toothed.	....	1825.	5. 6.	H. ♃.	
8 plantagíneum. W.	(*ye*)	Plantain-leav'd.	S. Europe.	1570.	——	H. ♃.	
9 austrìacum. W.	(*ye.*)	Austrian.	Austria.	1816.	——	H. ♃.	Jacq. aust. t. 130.
LEPTOSTE'LMA. D.D.		LEPTOSTE'LMA. Syngenesia Polygamia Superflua.					
máxima. D.D.	(*wh.*)	gigantic.	Mexico.	1828.	8. 9.	H. ♃	Swt. br. fl. gar. s. 2. t. 38.
BE'LLIS. W.		DAISY. Syngenesia Polygamia Superflua. L.					
1 sylvéstris. W.	(*wh.*)	large Portugal.	Portugal.	1797.	5. 7.	H. ♃.	Bot. mag. t. 2511.
2 perénnis. W.	(*wh.*)	common.	Britain.	....	3. 8.	H. ♃.	Eng. bot. v. 6. t. 424.
β horténsis.	(*wh.*)	*large double.*	....	....	——	H. ♃.	Bot. mag. t. 228.
γ fistulòsa.	(*re.*)	*red quilled.*	....	....	——	H. ♃.	
δ tubulòsa.	(*wh.*)	*white quilled.*	....	....	——	H. ♃.	
ε prolífera.	(*wh.re.*)	*Hen & Chicken.*	....	....	——	H. ♃.	
3 hy'brida. Ten.	(*wh.*)	Italian.	Italy.	1824.	4. 8.	H. ♃.	
4 ánnua. W.	(*wh.*)	annual.	S. Europe.	1759.	3. 7.	H. ☉.	Bot. mag. t. 2174
BE'LLIUM. W.		BE'LLIUM. Syngenesia Polygamia Superflua. L.					
1 bellidioídes. W.	(*wh.*)	small.	Italy.	1796.	6. 9.	H. ☉.	Lam. ill. t. 684.
2 minùtum. W.	(*wh.*)	dwarf.	Levant.	1772.	6. 10.	H. ♃.	Schreb. a. ups. 1. t. 5. f. 2.
PSI'ADIA. W.		PSI'ADIA. Syngenesia Polygamia Necessaria. W.					
glutinòsa. W.	(*ye.*)	glutinous.	Mauritius.	1796.	6. 8.	S. ♄.	Jacq. schœn. 2. t. 152.

Subordo IV. *EUPATOREÆ.* *Kth. synops.* 2. *p.* 408.

K'UHNIA. W.		K'UHNIA. Syngenesia Polygamia Æqualis.					
1 eupatorioídes. W.	(*wh.*)	Eupatorium-like.	N. America.	1812.	6. 8.	H. ♃.	Linn. fil. dec. t. 11.
Critònia eupatorioídes. G.							
2 Critònia. W.	(*st.*)	glandular-leaved.	——	1816.	——	H. ♃.	Gært. carp. 2. t. 174. f. 7.
Critònia Kùhnia. G.							
3 rosmarinifòlia. s.s.	....	Rosemary-leaved.	Cuba.	1827.	....	S. ♃.	

EUPAT'ORIUM. W. EUPAT'ORIUM. Syngenesia Polygamia Æqualis. L.

1	Dàlea. w.	(*wh.*)	shrubby.	Jamaica.	1773.	8.	S. ♄.	Jacq. schœn. 2. t. 146.
2	parviflòrum. w.	(*wh.*)	small-flowered.	———	1823.	——	S. ♄.	
3	triflòrum. w.	(*wh.*)	three-flowered.	Trinidad.	1821.	7. 8.	S. ♄.⌣.	Aub. gui. 2. t. 314.
4	macrophy'llum.w.	(*li.*)	large-leaved.	S. America.	1824.	....	S. ♃.	Plum. spec. 10. t. 129.
5	populifòlium.K.S.	(*wh.*)	Poplar-leaved.	Mexico.	——	....	S. ♄.	
6	canéscens. w.	(*wh.*)	canescent.	S. America.	1822.	6. 8.	S. ♄.	Vahl symb. 3. t. 73.
7	veronicæfòlium.K.S.	(*w.*)	Speedwell-leav'd.	Mexico.	1825.	....	S. ♄.	H. et B. v. 4. t. 341.
8	fœniculàceum. w.	(*st.*)	Fennel-leaved.	N. America.	1807.	6. 9.	H ♃.	
9	coronopifòlium.w.	(*wh.*)	buckshorn-leaved.	———	1819.	8. 9.	H. ♃.	
10	hyssopifòlium.w.	(*wh.*)	Hyssop-leaved.	———	1699.	——	H. ♃.	Dill. elth. t. 115. f. 140.
11	sessilifòlium. w.	(*wh.*)	sessile-leaved.	———	1777.	9. 10.	H. ♃.	
12	truncàtum. w.	(*wh.*)	truncate-leaved.	———	1822.	8. 10.	H. ♃.	
13	álbum. w.	(*wh.*)	white-flowered.	———	——	——	H. ♃.	
14	lanceolàtum. w.	(*wh.*)	spear-leaved.	———	1819.	8. 11.	H. ♃.	
15	teucrifòlium. w.	(*wh.*)	Teucrium-leav'd.	———	1816.	——	H. ♃.	Willd. hort. ber. t. 32.
16	rotundifòlium.w.	(*wh.*)	round-leaved.	———	1699.	7. 8.	H. ♃.	Pluk. alm. t. 88. f. 4.
17	pubéscens. w.	(*wh.*)	pubescent.	———	1816.	8. 10.	H. ♃.	
18	ceanothifòlium.w.	(*w.*)	Ceanothus-like.	———	1812.	8. 11.	H. ♃.	
19	altíssimum. w.	(*wh.*)	tall.	———	1699.	9. 10.	H. ♃.	Jacq. vind. 2. t. 164.
20	trifoliàtum. w.	(*wh.*)	three-leaved.	———	1768.	8. 10.	H. ♃.	
21	cannabìnum. w.	(*pu.*)	Hemp agrimony.	Britain.	....	7. 10.	H. ♃.	Eng. bot. v. 6. t. 428.
22	syrìacum. w.	(*ro.*)	Syrian.	Syria.	1807.	7. 9.	H. ♃.	Jacq. ic. 1. t. 170.
23	diffùsum. S.S.	(*wh.*)	spreading.	S. America.	1821.	6. 8.	S. ♃.	
24	purpùreum. w.	(*pu.*)	purple-stalked.	N. America.	1640.	9. 10.	H. ♃.	Corn. canad. t. 72.
25	maculàtum. w.	(*pu.*)	spotted-stalked.	———	1656.	8. 9.	H. ♃.	Herm. parad. t. 158.
26	punctàtum. Ph.	(*pu.*)	punctated.	———	1815.	——	H. ♃.	
27	verticillàtum. w.	(*pu.*)	whorl-leaved.	———	1811.	——	H. ♃.	
28	perfoliàtum. w.	(*wh.*)	Feverwort.	———	1699.	8. 10.	H. ♃.	Pluk. alm. t. 87. f. 6.
29	salviæfòlium.B.M.	(*wh.*)	Sage-leaved.	———	1812.	——	H. ♃.	Bot. mag. t. 2010.
30	lamiifòlium.L.en.	(*wh.*)	Lamium-leaved.	———	1818.	——	H. ♃.	
31	paniculàtum.Mil.	(*wh.*)	panicled.	———	....	——	H. ♃.	
32	melissoídes. w.	(*wh.*)	Balm-leaved.	———	1821.	8. 10.	H. ♃.	
33	aromáticum. w.	(*wh.*)	aromatic.	———	1739.	7. 8.	H. ♃.	Pluk. alm. t. 88. f. 3.
34	urticæfòlium. w.	(*wh.*)	Nettle-leaved.	S. America.	1803.	——	G. ♃.	Smith ined. t. 68.
35	ageratoídes. w.	(*wh.*)	Ageratum-like.	N. America.	1640.	8. 10.	H. ♃.	Corn. canad. t. 21.
36	deltoídeum. w.	(*pi.*)	deltoid-leaved.	....	1818.	——	S. ♃.	Jacq. schœn. 3. t. 369.
37	scándens. L.en.	(*ye.*)	climbing.	S. America.	1822.	6. 8.	S. ♄.⌣.	
38	squarròsum. w.	(*st.*)	squarrose.	Mexico.	1824.	——	S. ♃.	Cavan. ic. 1. t. 98.
39	atriplicifòlium. w.	(*li.*)	Orache-leaved.	W. Indies.	1820.	....	S. ♄.	
40	sinuàtum. P.S.	(*wh.*)	sinuate.	———	1824.	6. 8.	S. ♄.	Plum. sp. 10. ic. 128. f. 1.
41	odoràtum. w.	(*wh.*)	sweet-scented.	Jamaica.	1752.	8. 10.	S. ♄.	Pluk. alm. t. 177. f. 3.
42	Ayapàna. v.	(*pu.*)	medicinal.	Brazil.	1821.	——	S. ♄.	Vent. malm. p. 3. t. 3.
43	triplinérve. P.S.	....	triply-nerved.	Santa Cruz.	——	——	S. ♄.	
44	ivæfòlium. w.	(*wh.*)	Iva-leaved.	Jamaica.	1794.	6. 7.	S. ♃.	
45	stœchadifòlium.L.	(*vi.*)	Stœchas-leav'd.	NewGranada.	1826.	....	S. ♄.	Sm. ic. ined. t. 69.
46	myosotifòlium. w.	(*bl.*)	Myosotis-leav'd.	S. America.	1824.	6. 8.	S. ♂.	Jacq. ic. 3. t. 282.
47	iresinoídes. K.S.	(*wh.*)	Iresine-like.	———	——	——	S. ♃.	H. et B. v. 4. t. 340.
48	pulchéllum. K.S.	(*vi.*)	pretty.	Mexico.	——	——	S. ♄.	——— v. 4. t. 345.
49	xalapénse. K.S.	(*wh.*)	six-angled.	———	——	——	S. ♄.	
50	repándum. w.	(*wh.*)	repand-leaved.	Antilles.	1822.	——	S. ♄.	Plum. ic. 130. f. 1.
51	ásperum. H.B.	(*wh.*)	rough.	E. Indies.	1820.	6. 8.	S. ♄.	

MIK'ANIA. W. MIK'ANIA. Syngenesia Polygamia Æqualis. W.

1	Houstòni. w.	(*wh.*)	Houstoun's.	Jamaica.	1733.	7. 8.	S. ♄.	
2	hastàta. w.	(*wh.*)	halbert-leaved.	———	....	——	S. ♄.	Brown jam. t. 34. f. 3.
3	scándens. w.	(*wh.*)	climbing.	N. America.	1714.	8. 9.	H. ♃.⌣.	Jacq. ic. 1. t. 169.
4	Guáco. K.S.	(*wh.*)	medicinal.	S. America.	1823.	....	S. ♃.⌣.	H. et B. pl. 2. t. 105.
5	amára. w.	(*wh.*)	bitter.	Trinidad.	1824.	....	S. ♄.⌣.	Aubl. gui. 2. t. 315.
6	opífera. Mart.	(*wh.*)	snakewort.	Brazil.	1822.	....	S. ♄.⌣.	
7	nummulària.D.D.	(*wh.*)	moneywort.	Mexico.	1826.	6. 8.	S. ♃.⌣.	

ST'EVIA. W. ST'EVIA. Syngenesia Polygamia Æqualis. W.

1	connàta. L.G.	(*wh.*)	connate-leaved.	New Spain.	1825.	8. 10.	F. ♄.	
2	subpubéscens.L.G.	(*w.*)	downy-branch'd.	———	——	——	F. ♄.	
3	coriàcea. L.G.	(*wh.*)	leather-leaved.	Peru.	1826.	——	F. ♄.	
4	lùcida. L.G.	(*wh.*)	glossy-leaved.	New Spain.	1822.	——	F. ♄.	
5	salicifòlia. w.	(*wh.*)	Willow-leaved.	Mexico.	1803.	——	F. ♄.	Cavan. ic. 4. t. 354.
6	angustifòlia. K.S.	(*wh.*)	narrow-leaved.	———	——	——	F. ♄.⌣.	
7	trífida. L.G.	(*wh.*)	trifid-leaved.	New Spain.	1828.	——	F. ♃.	
8	micrántha. L.G.	(*wh.*)	small-flowered.	———	1825.	8. 10.	F. ♃.	

9 incanéscens. L.G.	(*wh.*)	villous-stalked.	New Spain.	1825.	8. 10.	F. ♃.	
10 pubéscens. L.G.	(*pu.*)	pubescent.	Mexico.	1823.	——	F. ♃.	
11 tomentosa. K.S.	(*ro.*)	tomentose.	————	1825.	——	F. ♃.	
12 monardæfòlia. K.S.	(*ro.*)	Monarda-leav'd.	————	——	8. 9.	F. ♃.	
13 fastigiàta. K.S.	(*wh.*)	crowded-flower'd.	New Spain.	1828.	——	F. ♃.	
14 adenóphora. L.G.	(*wh.*)	glandular.	Chile.	1822.	——	F. ♃.	
15 enarthrotrìcha. L.G.	(*w.*)	jointed haired.	New Spain.	1828.	——	F. ♃.	
16 ovata. L.G.	(*wh.*)	oval-leaved.	Mexico.	1816.	8. 9.	F. ♃.	
17 suboctoaristàta. L.G.	(*w.*)	8-awned.	Peru.	1824.	——	F. ♃.	
18 paniculàta. L.G.	(*wh.*)	panicled.	New Spain.	——	——	F. ♃.	
19 suavèolens. L.G.	(*wh.*)	sweet-scented.	————	1827.	——	F. ♃.	
20 rhombifòlia. K.S.	(*wh.*)	rhomb-leaved.	Mexico.	1826.	8. 10.	F. ♃.	
21 nepetæfòlia. K.S.	(*wh.*)	catmint-leaved.	————	——	——	F. ♃.	
22 Eupatòria. W.	(*ro.wh.*)	entire-leaved.	————	1798.	7. 9.	F. ♃.	Bot. mag. t. 1849.
23 lanceolàta. L.G.	(*bh.*)	spear-leaved.	————	1824.	——	F. ♃.	
24 hyssopifòlia. B.M.	(*wh.*)	Hyssop-leaved.	————	....	——	F. ♃.	Bot. mag. t. 1861.
25 ivæfolia. W.en.	(*wh.*)	Iva-leaved.	————	1816.	7. 9.	F. ♃.	
26 serràta. W.	(*wh.*)	saw-leaved.	————	1799.	——	F. ♃.	Jacq. schœn. 3. t. 300.
27 virgàta. K.S.	(*wh.*)	slender.	————	1824.	——	F. ♃.	
28 purpùrea. W.en.	(*pu.*)	purple-flowered.	————	1812.	8. 9.	H. ♃.	Bot. reg. t. 93.
Eupatòria. B.R. *nec aliorum.*							
29 víscida. K.S.	(*ro.*)	rose-coloured.	————	1824.	9. 10.	F. ♃.	
30 pilòsa. L.G.	(*ro.*)	hairy.	New Spain.	1825.	8. 10.	F. ♃.	
31 microphy'lla. K.S.	(*bh.*)	small-leaved.	Mexico.	1828.	——	F. ♃.	
PALAFO'XIA. Lag.		PALAFO'XIA.	Syngenesia Polygamia Æqualis.				
1 lineàris. L.G.	(*wh.*)	linear-leaved.	Mexico.	1816.	8. 10.	F. ♃.	Bot. mag. t. 2132.
Stèvia lineàris. W. *Ageràtum lineàre.* Cav. ic. 3. t. 205.							
2 callòsa. D.D.	(*wh.*)	callous.	Arkansa.	1824.	6. 8.	H. ⊙.	
Stèvia callòsa. N.							
AGER'ATUM. W.		AGER'ATUM.	Syngenesia Polygamia Æqualis. L.				
1 stríctum. B.M.	(*wh.*)	upright.	Nepaul.	1820.	9. 11.	H. ⊙.	Bot. mag. t. 2410.
2 conyzoídes. W.	(*bl.*)	hairy.	America.	1714.	7. 8.	H. ⊙.	Schk. hand. 3. t. 238.
3 cœrùleum. P.S.	(*bl.*)	blue-flowered.	S. America.	1820.	——	H. ⊙.	
4 mexicànum. B.M.	(*bl.*)	Mexican.	Mexico.	1823.	8. 10.	H ⊙.	Swt. br. fl. gar. t. 89.
5 latifòlium. W.	(*wh.*)	broad-leaved.	Peru.	1800.	7. 8.	H. ⊙.	Cavan. ic. 4. t. 357.
6 Houstoniànum. Mil.	(*bl.*)	Houstoun's.	S. America.	....	6. 8.	S. ⊙.	
7 paniculàtum. Scr.	(*bl.*)	panicled.	————	1827.	9. 10.	H. ⊙.	
8 álbum. W.	(*wh.*)	white-flowered.	————	1822.	6. 9.	H. ⊙.	
hírtum. Lam.							
9 marítimum. K.S.	(*bl.*)	sea-side.	Cuba.	1826.	——	S. ♃.	
10 cordifòlium. H.B.	(*wh.*)	heart-leaved.	Bengal.	1820.	8. 10.	S. ⊙.	
11 aquáticum. H.B.	(*wh.*)	water.	E. Indies.	——	——	S. ⊙.	
CŒLEST'INA. Cas.		CŒLEST'INA.	Syngenesia Polygamia Æqualis.				
1 suffruticòsa.	(*bl.*)	suffruticose.	S. America.	....	7. 10.	G. ♄.	Bot. mag. t. 1730.
Ageràtum cœlestìnum. B.M.							
2 ageratoídes. K.S.	(*bl.*)	Ageratum-like.	New Spain.	1824.	——	G. ♃.	
3 cœrùlea. Cas.	(*bl.*)	blue-flowered.	N. America.	1732.	7. 11.	H. ♃.	Dill. elth. t. 114. f. 139.
Eupatòrium cœlestìnum. W.							
AL'OMIA. K.S.		AL'OMIA.	Syngenesia Polygamia Æqualis.				
ageratoídes. K.S.	(*wh.*)	white-flowered.	New Spain.	1823.	7. 9.	G. ♃.	H. et B. 4. t. 354.
PIQU'ERIA. W.		PIQU'ERIA.	Syngenesia Polygamia Æqualis. W.				
1 trinérvia. W.	(*wh.*)	three-nerved.	Mexico.	1798.	7. 8.	G. ♃.	Cavan. ic. 3. t. 235.
2 pilòsa. K.S.	(*wh.*)	hairy.	————	1824.	——	G. ♃.	
3 artemisioídes. K.S.	(*wh.*)	Wormwood-like.	Quito.	——	8. 12.	G. ♃.	
SELL'OA. Spr.		SELL'OA.	Syngenesia Polygamia Æqualis.				
glutinòsa. B.R.	(*ye.*)	clammy.	Brazil.	1815.	1. 5.	G. ♄.	Bot. reg. t. 462.
LEYS'ERA. W.		LEYS'ERA.	Syngenesia Polygamia Superflua. W.				
1 ciliàta. W.	(*ye.*)	ciliated.	C. B. S.	1816.	7. 8.	G. ♃.	
2 gnaphalòdes. W.	(*or.*)	woolly.	————	1774.	7. 9.	G. ♄.	Jacq. ic. 3. t. 588.

SUBORDO V. *JACOBEÆ.* *Kth. synops.* 2. *p.* 440.

KLEI'NIA. W.		KLEI'NIA.	Syngenesia Polygamia Æqualis. W.				
1 ruderàlis. W.	(*gr.ye.*)	glaucous-leav'd.	S. America.	1824.	7. 10.	S. ⊙.	Jacq. amer. t. 127.
2 Porophy'llum. W.	(*gr.ye.*)	perforated.	————	1699.	6. 10.	S. ⊙.	Cavan. ic. 3. t. 222.
3 viridiflòra. K.S.	(*gr.*)	green-flowered.	Mexico.	1823.	——	S. ♄.	

4 tagetoídes. K.S. (*gr.*) Tagetes-like. Mexico. 1823. 6. 10. H. ⊙.
5 suffruticòsa. W. (*pu.*) suffruticose. S. America. 1825. —— G. ♄. Cavan. ic. 3. t. 257.
6 coloràta. K.S. (*vi.*) violet-colour'd. Mexico. —— —— H. ⊙.

CAC'ALIA. W. CAC'ALIA. Syngenesia Polygamia Æqualis. L.
1 papillàris. W. (*wh.*) rough-stalked. C. B. S. 1727. .... G. ħ. Dill. elth. t. 55. f. 63.
2 Anteuphórbium. W. (*st.*) oval-leaved. ———— 1596. 2. 3. G. ħ. ———— t. 55. f. 2. 3.
3 Kleínia. W. (*wh.*) Oleander-leaved. Canaries. 1732. 9. 10. G. ħ. DC. plant. gras. t. 12.
4 Ficoídes. W. (*st.*) flat-leaved. C. B. S. 1710. 6. 11. G. ħ. ———— t. 90.
5 rèpens. W. (*st.*) glaucous-leaved. ———— 1759. 6. 10. G. ♄. ———— t. 42.
6 cylíndrica. DC. (*ye.*) cylindrical. Africa. 1818. —— G. ħ. ———— t. ic.
7 carnòsa. W. (*st.*) narrow-leaved. C. B. S. 1757. 6. G. ħ.
8 canéscens. W. (*st.*) woolly-leaved. ———— 1795. .... G. ♄.
Hawórthii. H.S.L. *Kleínia tomentòsa.* H.S. *non* C. *tomentòsa.* L.
9 longifòlia. H.R. ...... long-leaved. .... 1820. .... S. ħ.
pugionifórmis. L.en.
10 articulàta. W. (*wh.*) jointed. C. B. S. 1775. 9. 11. G. ħ. DC. plant. gras. t. 18.
laciniàta. Jacq. ic. 1. t. 168.
11 tomentòsa. W. (*ye.*) tomentose. ———— 1795. —— G. ħ.
12 reticulàta. W. (*ye.*) netted. Bourbon. 1823. .... G. ħ.
13 appendiculàta. W. (*ye.*) appendaged. Teneriffe. 1815. 6. 9. G. ħ.
14 salicìna. Lab. (*ye.*) willow-leaved. V. Diem. Isl. 1820. 4. 6. G. ħ. Bot. reg. t. 923.
15 bícolor. W. (*or.*) two-coloured. E. Indies. 1804. 9. 5. S. ♄. ———— t. 110.
16 ovàlis. B.R. (*ye.*) oval-leaved. ———— .... S. ♄. ———— t. 102.
17 sonchifòlia. W. (*re.*) Sonchus-leaved. ———— 1768. 7. S. ⊙. Rheed. mal. 10. t. 68.
18 coccínea. B.M. (*sc.*) scarlet-flower'd. S. America. 1799. 6. 7. H. ⊙. Bot. mag. t. 564.
19 sarracènica. W. (*st.*) creeping-rooted. France. 1772. 8. 10. H. ♃.
20 runcinàta. K.S. (*ro.*) runcinate. Mexico. 1825. .... H. ♃.
21 hastàta. W. (*wh.*) spear-leaved. Siberia. 1780. 8. 10. H. ♃. Gmel. sib. 2. t. 66.
22 rhombifòlia. W.en. (*st.*) rhomb-leaved. ———— 1816. —— H. ♃.
23 macrophy'lla. M.B. (*st.*) large-leaved. Caucasus. 1828. —— H. ♃.
24 suavèolens. W. (*wh.*) sweet-scented. N. America. 1752. —— H. ♃.
25 atriplicifòlia. W. (*wh.*) Orache-leaved. ———— 1699. 8. H. ♃. Pluk. alm. t. 101. f. 2.
26 cordifòlia. K.S. (*ye.*) heart-leaved. Mexico. 1825. 7. 9. F. ♃.
27 renifórmis. W. (*wh.*) kidney-leaved. N. America. 1801. 7. 8. H ♃.
28 tuberòsa. N. (*wh.*) tuberous-rooted. ———— 1812. —— H. ♃.
29 pinnàta. W.en. (*ye.*) pinnate-leaved. Iberia. 1816. —— H. ♃.
30 peltàta. K.S. (*gr.*) peltated. Mexico. 1825. —— G. ♃. H. et Bonpl. 4. t. 361.
31 scándens. W. (*pi.*) climbing. C. B. S. 1774. 4. G. ♃.

EMI'LEA. Cas. EMI'LEA. Syngenesia Polygamia Æqualis.
flámmea. Cas. (*or.pu.*) flame-flowered. Java. 1823. 6. 9. S. ⊙.
Cacàlia sagittàta. W.

ADENO'STYLES. Cas. ADENO'STYLES. Syngenesia Polygamia Æqualis.
1 candidíssima. Cas. (*pu.*) hoary. S. France. 1828. 7. 9. H. ♃.
Cacàlia tomentòsa. Vill. *leucophy'lla.* W.
2 álbida. Cas. (*pi.*) white-leaved. Austria. 1739. 7. 8. H. ♃. Jacq. aust. 3. t. 235.
Cacàlia álbifrons. Jacq.
3 víridis. Cas. (*li.*) alpine. ———— —— —— H. ♃. Jacq. aust. 3. t. 234.
Cacàlia alpìna. Jacq.

TUSSIL'AGO. W. COLTSFOOT. Syngenesia Polygamia Superflua. L.
1 alpìna. W. (*pu.*) Alpine. Austria. 1710. 3. 5. H. ♃. Bot. mag. t. 84.
2 díscolor. W. (*pu.*) two-coloured. ———— 1633. 4. 5. H. ♃. Jacq. aust. 3. t. 247.
3 sylvéstris. W. (*pu.*) wood. ———— 1816. —— H. ♃. ———— 5. app. t. 12.
4 Fárfara. W. (*ye.*) common. Britain. .... 3. 4. H. ♃. Eng. bot. v. 6. t. 429.
β variegàta. (*ye.*) *variegated-leav'd.* ———— .... —— H. ♃.
5 frígida. W. (*pa.*) Lapland. Lapland. 1710. 5. H. ♃. Flor. dan. t. 61.
6 fràgrans. W. (*bh.*) fragrant. Italy. 1806. 1. 3. H. ♃. Bot. mag. t. 1388.
7 álba. W. (*wh.*) white butter-bur. Europe. 1683. 1. 4. H. ♃. Flor. dan. t. 524.
8 corymbòsa. S.S. (*wh.*) corymbose. Arctic regions. 1827. —— H. ♃.
9 nívea. W. (*wh.*) downy-leaved. Switzerland. 1713. 4. H. ♃. Retz. obs. 2. t. 3.
paradóxa. Retz.
10 Petasìtes. E.B. (*bh.*) common butter bur. Britain. .... 3. 4. H. ♃. Eng. bot. v. 6. t. 431.
11 hy'brida. E.B. (*bh.*) large-leaved. ———— .... —— H. ♃. ———— v. 6. t. 430.
12 spùria. W. (*wh.*) lobe-leaved. Germany. 1790. —— H. ♃. Retz. obs. 1. t. 2.
paradóxa. Roth.
13 palmàta. W. (*wh.*) cut-leaved. Labrador. 1778. 4. H. ♃. Hort. Kew. v. 3. t. 11.

SEN'ECIO. W. GROUNDSEL. Syngenesia Polygamia Superflua. W.
1 reclinàtus. W. (*ye.*) grass-leaved. C. B. S. 1774. 6. 8. G. ♂. Jacq. ic. 1. t. 174.
2 hieracifòlius. W. (*ye.*) hawkweed-leav'd. N. America. 1699. 8. H. ⊙. Herm. parad. t. 226.
3 purpùreus. W. (*pu.*) purple. C. B. S. 1774. 7. 9. G. ♃. Jacq. ic. 3. t. 580.

4	cérnuus. W.	(*pu.*)	drooping.	E. Indies.	1780.	7. 8.	H. ⊙.	Jacq. vind. 3. t. 98.
5	erubéscens. W.	(*bh.*)	blush-coloured.	C. B. S.	1774.	6. 10.	G. ♂.	
6	hæmatophy'llus. W.en.	(*ye.*)	purple leaved.	———	1789.	——	G. ♄.	
7	divaricàtus. W.	(*ye.*)	straddling.	China.	1801.	7.	G. ♂.	
8	croáticus. W.	(*ye.*)	Croatian.	Hungary.	1805.	7. 8.	H. ♃.	W. et K. hung. 2. t. 143.
9	Pseùdo-Chìna. W.	(*or.*)	Chinese.	China.	1732.	6. 8.	G. ♃.	Dill. elth. t. 258. f. 335.
10	japónicus. W.	(*ye.*)	jagged-leaved.	Japan.	1774.	8.	H. ♃.	
11	vulgàris. W.	(*ye.*)	common.	Britain.	....	1. 12.	H. ♃.	Eng. bot. v. 11. t. 747.
12	lanuginòsus. S.S.	(*ye.*)	woolly.	....	1827.	4. 8.	H. ⊙.	
13	valerianæfòlius. S.S.	(*ye.*)	Valerian-leav'd.	Brazil.	1826.	6. 10.	F. ⊙.	
14	peucedanifòlius. W.	(*w.*)	Peucedanum-leav'd.	C. B. S.	1816.	6. 8.	G. h.	Jacq. ic. 3. t. 581.
	Cacàlia peucedanifòlia. Jacq.							
15	arábicus. W.	(*ye.*)	Arabian.	Egypt.	1804.	7. 8.	H. ♂.	
16	verbenæfòlius. W.	(*ye.*)	Vervain-leav'd.	———	1803.	6. 7.	H. ⊙.	Jacq. vind. 1. t. 3.
17	glomeràtus. L.en.	(*ye.*)	close-flowered.	N. Holland.	1816.	8. 10.	H. ⊙.	
18	cacalioìdes. L.en.	(*ye.*)	Cacalia-like.	Brazil.	1820.	——	H. ⊙.	
19	triflòrus. W.	(*ye.*)	three-flowered.	Egypt.	1776.	7. 9.	H. ⊙.	
20	ægy'ptius. W.	(*ye.*)	Egyptian.	———	1771.	7. 8.	H. ⊙.	
21	crassifòlius. W.	(*ye.*)	thick-leaved.	S. Europe.	1815.	——	H. ⊙.	
22	lívidus. W.	(*ye.*)	livid.	Britain.	....	——	H. ⊙.	Eng. bot. v. 35. t. 2515.
23	telephifòlius. W.	(*ye.*)	Telephium-leav'd.	C. B. S.	1823.	——	H. ⊙.	Jacq. frag. 1. t. 1. f. 3.
24	trilòbus. W.	(*ye.*)	three-lobed.	Spain.	1728.	6. 8.	H. ⊙.	
25	cineráscens. W.	(*ye.*)	gray.	C. B. S.	1774.	5. 7.	G. h.	Jacq. schœn. 2. t. 150.
26	lilacìnus. S.S.	(*li.*)	lilac-coloured.	———	1826.	6. 12.	G. ♄.	Link et Ott. abb. ic.
27	viscòsus. W.	(*ye.*)	stinking.	Britain.	....	6. 10.	H. ⊙.	Eng. bot. v. 1. t. 32.
28	sylváticus. W.	(*st.*)	mountain.	———	....	7. 8.	H. ⊙.	——— v. 11. t. 748.
29	coronopifòlius. W.	(*ye.*)	Buck'shorn-leav'd.	Spain.	1821.	——	H. ⊙.	
30	nebrodénsis. W.	(*ye.*)	Nebrodes.	S. Europe.	1704.	6. 8.	H. ⊙.	Barrel. rar. 401.
31	glaùcus. W.	(*ye.*)	sea-green.	Egypt.	1739.	——	H. ⊙.	
32	Marmòræ. S.S.	(*ye.*)	Sardinian.	Sardinia.	1829.	4. 5.	H. ♂.	
33	hastàtus. W.	(*ye.*)	halbert-leaved.	C. B. S.	1722.	5. 8.	G. ♃.	Dill. elth. t. 152. f. 184.
34	pubígerus. W.	(*ye.*)	runcinate-leav'd.	———	1816.	——	G. ♄.	Moris. s. 7. t. 18. f. 32.
35	vernàlis. W.	(*ye.*)	spring.	Hungary.	1803.	4. 6.	H. ⊙.	W. et K. hung. 1. t. 24.
36	montànus. W.	(*ye.*)	smooth mountain.	Germany.	1819.	——	H. ⊙.	
37	rupéstris. W.	(*ye.*)	rock.	Hungary.	1805.	6. 7.	H. ♃.	W. et K. hung. 2. t. 128.
38	laciniàta. Bert.	(*ye.*)	jagged.	Italy.	1822.	5. 8.	H. ♂.	
39	dentàtus. W.	(*ye.*)	toothed-leaved.	C. B. S.	1823.	——	G. ♃.	
40	squamòsus. W.	(*ye.*)	scaly.	———	1816.	——	G. ♃.	
41	venústus. W.	(*re.*)	wing-leaved.	———	1774.	5. 11.	G. ♄.	Bot. reg. t. 901.
42	élegans. W.	(*re.wh.*)	elegant.	———	1700.	6. 8.	H. ⊙.	
	α rùber.	(*re.*)	*red-flowered.*	———	——	——	H. ⊙.	
	β álbus.	(*wh.*)	*white-flowered.*	———	——	——	H. ⊙.	
	γ plèno-álbus.	(*wh.*)	*double-white.*	———	——	4. 11.	G. ♄.	
	δ plèno-rùber.	(*re.*)	*double red.*	———	——	——	G. ♄.	Bot. mag. t. 238.
43	speciòsus. W.	(*re.*)	red-flowered.	China.	1789.	7. 8.	G. ♃.	Bot. reg. t. 41.
44	squálidus. W.	(*ye.*)	inelegant.	England.	....	6. 10.	H. ⊙.	Eng. bot. v. 9. t. 600.
45	erucæfòlius. W.	(*ye.*)	Eruca-leaved.	Europe.	1816.	7. 8.	H. ♃.	
46	chrysanthemifòlius.	(*ye.*)	Chrysanthemum-l[d].	Sicily.	1823.	7. 9.	H. ♃.	Bocc. sic. t. 67.
47	uniflòrus. W.	(*ye.*)	one-flowered.	S. Europe.	1799.	7. 8.	H. ♃.	All. ped. t. 17. f. 8.
48	incànus. W.	(*ye.*)	downy.	———	1759.	——	H. ♃.	Pluk. alm. t. 39. f. 6.
49	abrotanifòlius. W.	(*ye.*)	Southernwood-leav'd.	———	1640.	7. 10.	H. ♃.	Jacq. aust. 1. t. 79.
50	artemisiæfòlius. P.S.	(*ye.*)	Wormwood-leav'd.	France.	1820.	——	H. ♃.	
51	præáltus. S.S.	(*ye.*)	lofty.	Italy.	1827.	7. 8.	H. ♃.	
52	Othónnæ. M.B.	(*ye.*)	Othonna-like.	Caucasus.	1822.	——	H. ♃.	
53	Marschalliànus. S.S.	(*ye.*)	Bieberstein's.	Ukraine.	1826.	——	H. ♃.	
	arenàrius. M.B. *non* Thunb.							
54	grandiflòrus. S.S.	(*ye.*)	large-flowered.	C. B. S.	1824.	6. 8.	G. ♄.	Pluk. mant. t. 422. f. 5.
55	glaucéscens. S.S.	(*ye.*)	glaucescent.	....	1826.	6. 10.	H. ⊙.	
56	carniòlicus. S.S.	(*ye.*)	Carniolian.	Carniola.	1827.	7. 8.	H. ♃.	
57	adonidifòlius. Lois.	(*ye.*)	Adonis-leaved.	....	1826.	7. 10.	H. ♃.	
58	dahùricus. F.	(*ye.*)	Dahurian.	Dahuria.	1821.	——	H. ♃.	
59	umbellàtus. S.S.	(*ye.pu.*)	umbel-flowered.	C. B. S.	1822.	6. 8.	G. ♄.	
60	canadénsis. W.	(*ye.*)	Canadian.	N. America.	1818.	——	H. ♃.	
61	delphinifòlius. W.	(*ye.*)	Larkspur-leav'd.	Sicily.	1823.	——	H. ♃.	Vahl symb. 2. t. 43.
62	tenuifòlius. W.	(*ye.*)	slender-leaved.	Britain.	....	7. 8.	H. ♃.	Eng. bot. v. 8. t. 574.
63	Jacob'æa. W.	(*ye.*)	common-ragwort.	———	....	——	H. ♃.	——— v. 16. t. 1130.
64	erráticus. S.S.	(*ye.*)	water.	S. Europe.	1826.	5. 7.	H. ♃.	
65	aquáticus. W.	(*ye.*)	marsh.	———	....	——	H. ♃.	Eng. bot. v. 16. t. 1131.
66	aùreus. W.	(*ye.*)	golden.	N. America.	1758.	7. 8.	H. ♃.	

67 lyratifòlius. s.s.	(*ye.*)	lyrate-leaved.	S. Europe.	1826.	5. 6.	H. ♃.	
68 fràgrans. s.s.	(*ye.*)	fragrant.	Siberia.	——	6. 8.	H. ♃.	
69 Balsamìtæ. w.	(*ye.*)	various-leaved.	N. America.	1819.	——	H. ♃.	
70 obovàtus. w.	(*ye.*)	obovate-leaved.	————	——	——	H. ♃.	
71 leucophy'llus. DC.	(*ye.*)	white-leaved.	S. Europe.	1821.	——	H. ♃.	
72 rosmarinifòlius. w.	(*ye.*)	Rosemary-leav'd.	C. B. S.	....	7. 8.	G. ħ.	Jacq. ic. 3. t. 587.
73 ásper. w.	(*ye.*)	rough.	————	1774.	——	G. ħ.	
74 linifòlius. w.	(*ye.*)	Flax-leaved.	Spain.	1817.	——	H. ♃.	Barrel. rar. t. 802.
75 rigéscens. w.	(*ye.*)	stiff-leaved.	C. B. S.	1815.	——	G. ħ.	Jacq. coll. 5. t. 6. f. 1.
76 nervàtus. s.s.	(*pu.*)	nerved-leaved.	————	1824.	7. 8.	G. ♄.	
77 paludòsus. w.	(*ye.*)	Bird's-tongue.	England.	....	6. 8.	H. ♃.	Eng. bot. v. 10. t. 650.
78 persicæfòlius. DC.	(*ye.*)	peach-leaved.	Pyrenees.	1817.	——	H. ♃.	Ram. bul. phil. t. 11. f. 3.
79 nemorénsis. w.	(*ye.*)	branching.	Austria.	1785.	7. 8.	H. ♃.	Jacq. aust. 2. t. 184.
80 sarracènicus. w.	(*ye.*)	creeping-root'd.	Britain.	....	7. 10.	H. ♃.	Eng. bot. v. 31. t. 2211.
81 ovàtus. w.	(*ye.*)	oval-leaved.	Germany.	1817.	——	H. ♃.	
82 coriàceus. w.	(*ye.*)	leathery-leaved.	Levant.	1728.	7. 8.	H. ♃.	Dill. elth. t. 105. f. 125.
83 Dòria. w.	(*ye.*)	broad-leaved.	Austria.	1570.	7. 9.	H. ♃.	Jacq. aust. 2. t. 185.
84 macrophy'llus. M.B.	(*ye.*)	large-leaved.	Caucasus.	1817.	——	H. ♃.	
85 Dorónicum. w.	(*ye.*)	Leopard's-bane.	S. Europe.	1705.	——	H. ♃.	Jacq. aust. 5. t. ap. 45.
86 umbròsus. W.K.	(*ye.*)	ear-leaved.	Hungary.	1815.	——	H. ♃.	
87 Barreliéri. w.	(*ye.*)	Barrelier's.	Pyrenees.	1824.	——	H. ♃.	Barrel. ic. 146.
88 lánceus. w.	(*ye.*)	spear-leaved.	C. B. S.	1774.	7. 10.	G. ħ.	
89 oporìnus. w.	(*ye.*)	Jacquin's.	————	....	——	G. ♃.	Jacq. schœn. 3. t. 304.
90 longifòlius. w.	(*ye.*)	long-leaved.	————	1775.	8. 11.	G. ħ.	Comm. hort. 2. t. 71.
91 halimifòlius. w.	(*ye.*)	succulent-leaved.	————	1723.	7.	G. ħ.	Dill. elth. t. 104. f. 124.
92 ilicifòlius. w.	(*ye.*)	Ilex-leaved.	————	1731.	6. 7.	G. ħ.	Comm. rar. t. 42.
93 rígidus. w.	(*ye.*)	hard-leaved.	————	1704.	6. 9.	G. ħ.	Comm. hort. 2. t. 75.
94 solidaginoídes. w.	(*ye.*)	Golden-rod-like.	————	1819.	——	G. ħ.	Hout. syst. 4. t. 32. f. 1.
95 vérnus. Biv.	(*ye.*)	vernal.	Italy.	——	——	H. ♃.	
96 cordifòlius. s.s.	(*ye.*)	heart-leaved.	C. B. S.	1827.	7. 9.	G. ♃.	
97 arenàrius. s.s.	(*ye.*)	sand.	————	1826.	——	G. ♃.	
98 spicàtus.	(*ye.*)	spike-flowered.	....	1828.	——	H. ♃.	
ADENOTRI'CHA. B.R.		ADENOTRI'CHA.	Syngenesia Polygamia Superflua.				
amplexicaùlis. B.R.	(*ye.*)	stem-clasping.	Chile.	1826.	5. 7.	F. ♃.	Bot reg. t. 1190.
SENECI'LLIS. G.		SENECI'LLIS.	Syngenesia Polygamia Superflua.				
glaùca. G.	(*ye.*)	glaucous.	Siberia.	1790.	6. 7.	H. ♂.	Gmel. sib. 2. t. 74.
Cinerària glaùca. w.							
CINER'ARIA. W.		CINER'ARIA.	Syngenesia Polygamia Superflua. L.				
1 americàna. w.	(*ye.*)	American.	S. America.	1823.	....	S. ħ.	
2 incàna. P.S.	(*ye.*)	hoary.	Jamaica.	——	6. 8.	S. ħ.	
3 díscolor. P.S.	(*ye.*)	two-coloured.	————	1822.	——	S. ħ.	
4 lùcida. P.S.	(*ye.*)	glossy.	W. Indies.	——	——	S. ħ.	Plum. 2. ic. 154.
5 glabràta. s.s.	(*ye.*)	smooth.	————	1827.	4. 7.	S. ħ.	
6 geifòlia. w.	(*ye.*)	kidney-leaved.	C. B. S.	1710.	4. 8.	G. ♄.	Comm. hort. 2. t. 73.
7 canéscens. L.en.	(*ye.*)	small-flowered.	————	1790.	——	G. ♄.	Bot. mag. t. 1990.
8 aùrita. w.	(*pu.*)	purple-flower'd.	Madeira.	1777.	6. 7.	G. ♄.	———— t. 1786.
9 láctea. W.en.	(*wh.*)	milk-coloured.	....	1816.	——	G. ♄.	
10 lanàta. w.	(*pu.*)	woolly.	Canaries.	1780.	5. 9.	G. ♄.	Bot. mag. t. 53.
11 hy'brida. W.en.	(*ye.*)	hybrid.	Hybrid.	....	2. 5.	G. ♄.	
12 elàtior. L.en.	(*wh.*)	tall.	....	....	——	G. ♄.	
13 cruénta. w.	(*pu.*)	purple-leaved.	Canaries.	1777.	——	G. ♃.	Bot. mag. t. 406.
aùrita. A.R. 24. *nec aliorum.*							
14 pulchélla.	(*re.*)	bright red-flower'd.	————	1818.	——	G. ♃.	
15 populifòlia. H.K.	(*re.*)	Poplar-leaved.	————	1780.	6. 9.	G. ♄.	Venten. malm. t. 100.
16 hypoleùca. s.s.	(*pu.*)	snowy-leaved.	C. B. S.	1828.	4. 7.	G. ♄.	
17 venústa. s.s.	(*re.*)	charming.	————	——	——	G. ♃.	
18 lobàta. w.	(*ye.*)	lobed.	————	1774.	6. 8.	G. ♄.	
19 malvæfòlia. w.	(*li.*)	Mallow-leaved.	Azores.	1777.	8.	G. ♃.	
20 petasìtes. B.M.	(*ye.*)	butter-bur-leav'd.	Mexico.	1812.	12. 2.	G. ħ.	Bot. mag. t. 1536.
21 pr'æcox. w.	(*ye.*)	early-flowering.	————	1824.	2. 4.	G. ħ.	DC. pl. rar. gen. t. 7.
22 sibírica. w.	(*ye.*)	Siberian.	Siberia.	1784.	6. 8.	H. ♃.	Bot. mag. t. 1869.
23 speciòsa. L.en.	(*ye.*)	large-flowered.	————	1818.	——	H. ♃.	Bot. reg. t. 812.
24 caucásica. M.B.	(*ye.*)	Caucasean.	Caucasus.	1820.	——	H. ♃.	
25 gigántea. H.K.	(*bh.*)	gigantic.	Cape Horn.	1801.	7. 8.	H. ♃.	Sm. exot. bot. 2. t. 65.
26 l'ævis. s.s.	(*ye.*)	smooth.	C. B. S.	1828.	4. 7.	G. ♃.	
27 crenàta. s.s.	(*st.*)	crenate.	————	——	——	G. ♃.	
28 palústris. w.	(*ye.*)	marsh.	England.	....	——	H. ♃.	Eng. bot. v. 3. t. 151.
29 aurantìaca. w.	(*or.*)	orange-colour'd.	Switzerland.	1818.	5. 7.	H. ♃.	Swt. br. fl. gar. t. 256.
30 campéstris. w.	(*ye.*)	mountain.	N. Europe.	....	——	H. ♂.	Jacq. aust. t. 180.

31	integrifòlia. E.B.	(ye.)	entire-leaved.	England.	....	5. 7.	H. ♃.	Eng. bot. v. 3. t. 152.
32	longifòlia. W.	(ye.)	long-leaved.	Austria.	1792.	6. 8.	H. ♃.	Jacq. aust. 2. t. 181.
33	críspa. W.	(ye.)	curled.	———	1818.	——	H. ♃.	——— 2. t. 178.
34	cordifòlia. W.	(ye.)	heart-leaved.	———	1739.	7. 8.	H. ♃.	——— 2. t. 176.
35	mèdia. W.K.	(ye.)	intermediate.	Hungary.	1818.	——	H. ♃.	
36	alpìna. W.	(ye.)	Alpine.	Austria.	1683.	6. 8.	H. ♃.	Jacq. aust. 2. t. 177.
37	aùrea. W.	(ye.)	golden.	Siberia.	1825.	——	H. ♃.	
38	rivulàris. S.S.	(ye.)	rivulet.	Hungary.	1822.	——	H. ♃.	
39	crassifòlia. S.S.	(ye.)	thick-leaved.	Carinthia.	1825.	——	H. ♃.	
40	alpéstris. S.S.	(ye.)	rough-leaved.	N. Europe.	1826.	——	H. ♃.	
41	spathulæfòlia. S.S.	(ye.)	spathulate-leav'd.	Germany.	1820.	——	H. ♃.	
42	pappòsa. S.S.	(ye.)	long-chaffed.	Gallicia.	1828.	——	H. ♃.	
43	parviflòra. S.S.	(ye.)	small-flowered.	Caucasus.	1827.	——	H. ♃.	
44	congésta. S.S.	(ye.)	close-headed.	Melville Isl.	——	5. 6.	H. ♃.	
45	taraxacifòlia. S.S.	(ye.)	Dandelion-leav'd.	Caucasus.	1824.	6. 8.	H. ♃.	
46	exsquàmata.	(ye.)	scaleless.	Portugal.	1827.	——	H. ⊙.	
	Senècio exsquàmeus. Brot.							
47	frígida. S.S.	(ye.)	frigid.	N. America.	1827.	5. 6.	H. ♃.	
48	scapiflòra. S.S.	(ye.)	long-peduncled.	C. B. S.	1828.	——	G. ♃.	
49	gibbòsa. S.S.	(ye.)	gibbous-headed.	Sicily.	——	6. 8.	H. ♄.	
50	Aitoniàna. S.S.	(ye.)	Aiton's.	....	....	——	G. ♄.	
51	marítima. W.	(ye.)	sea ragwort.	S. Europe.	1633.	7. 9.	F. ħ.	Flor. græc. t. 871.
52	bícolor. W.	(ye.)	shining-leaved.	....	1820.	——	F. ħ.	
53	acanthifòlia. S.S.	(ye.)	Acanthus-leav'd.	Levant.	1826.	——	G. ♄.	
54	thyrsoídea. F.	(ye.)	thyrse-flower'd.	Armenia.	1827.	——	H. ♃.	
55	angulàta.	(ye.)	angular-leaved.	Mexico.	——	12. 1.	G. ♃.	
56	canadénsis.	(ye.)	Canadian.	Canada.	1739.	6. 8.	H. ♃.	
57	linifòlia. W.	(ye.)	flax-leaved.	C. B. S.	....	——	G. ħ.	Jacq. schœn. 3. t. 308.
58	angustifòlia. K.S.	(ye.)	narrow-leaved.	Mexico.	1825.	....	G. ħ.	
59	humifùsa. W.	(ye.)	trailing.	C. B. S.	1704.	7. 8.	G. ♃.	
60	viscòsa. W.	(wh.)	clammy.	———	1774.	6. 8.	G. ♂.	Jacq. frag. 12. t. 7. f. 2
61	vestìta. H.R.	(ye.)	clothed.	———	1824.	5. 6.	G. ♃.	
	AGATH'ÆA. Cas.		AGATH'ÆA.	Syngenesia Polygamia Superflua.				
	cœléstis. Cas.	(bl.)	Cape-Aster.	C. B. S.	1753.	2. 9.	G. ♄.	Bot. mag. t. 249.
	Cinerària amelloídes. W.							
	KAULFU'SSIA. Nees.		KAULFU'SSIA.	Syngenesia Polygamia Superflua.				
1	amelloídes. Nees.	(bl.)	blue-flowered.	C. B. S.	1819.	6. 7.	H. ⊙.	Bot. reg. t. 490.
2	ciliàta. S.S.	(st.)	fringed.	———	1822.	6. 9.	H. ⊙.	
	TAG'ETES. W.		TAG'ETES.	Syngenesia Polygamia Superflua. L.				
1	lùcida. W.	(ye.)	sweet-scented.	Mexico.	1798.	7. 11.	F. ♃.	Bot. mag. t. 740.
2	flòrida. B.F.G.	(ye.)	gay-flowering.	———	1827.	8. 11.	F. ♃.	Swt. br. fl. gar. s. 2. t. 35.
3	pátula. W.	(pu. ye.)	French Marygold.	———	1573.	7. 10.	H. ⊙.	Bot. mag. t. 150.
4	corymbòsa. B.F.G.	(va.)	corymbus-flower'd.	———	1825.	9. 11.	H. ⊙.	Swt. br. fl. gar. t. 151.
	α purpùrea.	(pu.)	*purple-flowered.*	———	——	——	H. ⊙.	
	β lùtea.	(ye.)	*yellow-flowered.*	———	——	——	H. ⊙.	
5	erécta. W.	(ye.)	African Marygold.	———	1596.	6. 9.	H. ⊙.	Lam. ill. t. 684.
6	tenuifòlia. W.	(ye.)	fine-leaved.	———	1797.	7. 10.	H. ⊙.	Swt. br. fl. gar. t. 141.
7	subvillòsa. LG.	(ye.)	villous-leaved.	———	1825.	8. 11.	H. ⊙.	
8	minùta. W.	(ye.)	small-flowered.	Chili.	1728.	8. 10.	H. ⊙.	Dill. elth. t. 280. f. 362.
9	clandestìna. LG.	(ye.)	small-rayed.	Mexico.	1822.	——	H. ⊙.	
	coronopifòlia. W. en.							
10	micrántha. W.	(ye.)	solitary-flower'd.	———	——	——	H. ⊙.	Cavan. ic. 4. t. 352.
11	glandulòsa. L. en.	(ye.)	glandular.	———	1825.	——	H. ⊙.	
12	glandulífera. Sck.	(ye.)	gland-bearing.	———	1826.	9. 11.	H. ⊙.	
	BŒB'ERA. W.		BŒB'ERA.	Syngenesia Polygamia Superflua.				
1	glandulòsa. N.	(ye.)	glandular.	N. America.	1823.	8. 9.	H. ⊙.	Venten. cels. t. 36.
	chrysanthemoídes. W. en. *Tagètes pappòsa.* V. *Dyssòdia glandulòsa.* Cav.							
2	Porophy'llum. K.S.	(ye.)	Mexican.	Mexico.	——	——	H. ⊙.	
	Dyssòdia Porophy'llum. W. en.							
3	fastigiàta. K.S.	(ye.)	clustered.	———	1827.	9. 11.	H. ⊙.	
4	divaricàta.	(ye.)	spreading.	———	1828.	——	H. ⊙.	
	Dyssòdia divaricàta. Rich.							
5	pubéscens. S.S.	(ye.)	pubescent.	New Spain.	1826.	8. 10.	H. ⊙.	Cavan. icon. 3. t. 212.
	Dyssòdia pubéscens. LG. *A'ster pinnàtus.* Cavan.							
	OTHO'NNA. W.		RAGWORT.	Syngenesia Polygamia Necessaria. L.				
1	flabellifòlia. B.C.	(ye.)	fan-leaved.	C. B. S.	1822.	7. 9.	G. ♄.	Lodd. bot. cab. t. 726.
2	pinnatífida. P.S.	(ye.)	pinnatifid.	———	1823.	——	G. ♃.	
3	pinnàta. W.	(ye.)	wing-leaved.	———	1759.	4. 6.	G. ♃.	Bot. mag. t. 768.
4	ciliàta. W.	(ye.)	fringed-leaved.	———	1824.	5. 6.	G. ♃.	
5	Tagètes. W.	(ye.)	slender-stalked.	———	1823.	7. 9.	H. ⊙.	

X

6 pectinàta. w. (*ye.*) Wormwood-leav'd. C. B. S. 1731. 4. 6 G. ♄. Bot. mag. t. 306.
7 Athanàsiæ. w. (*ye.*) Athanasia-like. —— 1795. 11. 12. G. ♄. Jacq. schœn. 2. t. 242.
8 abrotanifòlia. w. (*ye.*) Southernwood-leav'd. —— 1692. 1. 5. G. ♄. Bot. reg. t. 108.
9 digitàta. w. (*ye.*) finger-leaved. —— 1822. 6. 9. G. ♃.
10 trifurcàta. w. (*ye.*) three-forked. —— —— —— G. ♄.
11 trífida. w. (*ye.*) trifid-leaved. —— 1823. —— G. ♄.
12 retrofrácta. w. (*ye.*) axillary-flower'd. —— 1812. —— G. ♄. Jacq. schœn. 3. t. 376.
13 coronopifòlia. w. (*ye.*) Buckshorn-leav'd. —— 1731. 7. 9. G. ♄. Comm. hort. 2. t. 70.
14 cheirifòlia. w. (*ye.*) Stock-leaved. Barbary. 1752. 4. 6. H. ♄. Bot. reg. t. 266.
15 crassifòlia. w. (*ye.*) thick-leaved. C. B. S. 1710. 9. 10. G. ♄. Mill. ic. 2. t. 245. f. 2.
16 denticulàta. w. (*ye.*) tooth-leaved. —— 1774. 4. 7. G. ♄. Bot. mag. t. 1979.
17 heterophy'lla. w. (*ye.*) various-leaved. —— 1812. —— G. ♃.
18 Língua. w. (*ye.*) tongue-leaved. —— 1787. 5. 9. G. ♃. Jacq. schœn. 2. t. 238.
19 filicaùlis. w. (*ye.*) Yam-rooted. —— 1791. 4. 5. G. ♃. —— 2. t. 241.
20 bulbòsa. w. (*ye*) bulbous. —— 1774. 5. 6. G. ♃. Breyn. cent. t. 66.
21 capillàris. w. (*ye.*) capillary-stalked. —— 1822. 6. 9. H. ⊙.
22 parviflòra. w. (*ye.*) small-flowered. —— 1704. 7. 8. G. ♄. Volk. norib. t. 226.
23 perfoliàta. H.K. (*ye.*) perfoliate. —— 1789. 5. 7. G. ♃. Jacq. schœn. t. 240.
amplexicaùlis. B.M. 1312. *non* Thunb.
24 ericoídes. w. (*ye.*) Heath-leaved. —— 1815. 7. 8. G. ♄.
25 linifòlia. w. (*ye.*) Flax-leaved. —— 1825. —— G. ♄.
26 tenuíssima. w. (*ye.*) fine-leaved. —— 1759. 4. 7. G. ♄. Jacq. schœn. 2. t. 239.
27 frutéscens. w. (*ye.*) frutescent. —— 1816. 7. 8. G. ♄. Comm. hort. 2. t. 70.
28 arboréscens. w. (*ye.*) tree. —— 1723. —— G. ♄. Dill. elth. t. 103. f. 123.
29 cacalioídes. w. (*ye.*) tuberous. —— 1774. 5. 9. G. ♄.
30 amplexicaùlis. w. (*ye.*) stem-clasping. —— 1826. —— G. ♄.

Subordo VI. *HELIANTHEÆ.* *Kth. synops.* 2. *p.* 463.

HYMENOPA'PPUS. J. HYMENOPA'PPUS. Syngenesia Polygamia Æqualis.
1 pedàtus. LG. (*wh.*) pedate-leaved. Mexico. 1803. 7. 9. G. ⊙. Cavan. ic. 4. t. 356.
Stèvia pedàta. B.M. 2040. *Florestìna pedàta.* Cas.
2 scabiòsæus. Ph. (*wh.*) Scabious-like. Missouri. 1812. 6. 8. H. ⊙. Jour. de hist. nat. n. 1. ic.
3 tenuifòlius. Ph. (*wh.*) slender-leaved. —— 1811. —— F. ♂.
SCHK'UHRIA. W. SCHK'UHRIA. Syngenesia Polygamia Superflua. L.
abrotanoídes. w. (*wh.*) Wormwood-leav'd. Mexico. 1798. 7. 9. H. ⊙. Schk. hand. 3. t. 250. b.
PE'CTIS. W. PE'CTIS. Syngenesia Polygamia Superflua. L.
1 humifùsa. w. (*ye.*) trailing. W. Indies. 1822. 7. 9. S. ⊙. Plum. ic. t. 95. f. 2.
2 prostrata. w. (*ye.*) prostrate. New Spain. 1824. —— H. ⊙. Cavan. ic. 4. t. 324.
3 ciliàris. w. (*ye.*) ciliated. Hispaniola. 1793. 7. S. ⊙. Plum. ic. 151. f. 2.
4 linifòlia. w. (*ye.*) Flax-leaved. Jamaica. 1732. 7. 8. S. ⊙. Sloan. hist. 1. t. 149. f. 3.
5 canéscens. K.S. (*ye.*) canescent. Mexico. 1827. 8. 9. H. ⊙. H. et B. n. gen. 4. t. 393.
6 punctàta. w. (*ye.*) dotted. Hispaniola. 1824. 6. 8. S. ⊙. Jacq. amer. t. 126.
HETEROSPE'RMUM. HETEROSPE'RMUM. Syngenesia Polygamia Superflua. W.
pinnàtum. w. (*ye.*) wing-leaved. New Spain. 1799. 8. 9. H. ⊙. Cavan. ic. 3. t. 267.
MELANANTH'ERA. M. MELANANTH'ERA. Syngenesia Polygamia Æqualis. P.S.
1 Linn'æi. K.S. (*li.*) Linnæus's. S. America. 1799. 6. 8. S. ♂. Jacq. ic. 3. t. 583.
Bìdens nívea. L. *Càlea áspera.* Jacq.
2 hastàta. M. (*wh.*) hastate-leaved. N. America. 1732. —— G. ♃. Dill. elth. t. 47. f. 55.
β panduràta. Ph. (*wh.*) *fiddle-leaved.* —— —— —— G. ♃. —— t. 46. f. 54.
MARSHA'LLIA. W. MARSHA'LLIA. Syngenesia Polygamia Æqualis. W.
1 lanceolàta. Ph. (*li.*) spear-leaved. Carolina. 1812. 6. 7. F. ♃.
2 latifòlia. Ph. (*li.*) broad-leaved. —— 1806. —— F. ♃. Mich. amer. 2. t. 43.
Persoònia latifòlia. M. *Trattenickia latifòlia.* P.S.
PLATY'PTERIS. K.S. PLATY'PTERIS. Syngenesia Polygamia Æqualis.
crocàta. K.S. (*sn.*) saffron-colour'd. Mexico. 1812. 1. 3. S. ♃. Bot. mag. t. 1627.
Spilánthes crocàta. B.M.
SPILA'NTHES. W. SPILA'NTHES. Syngenesia Polygamia Æqualis. L.
1 tenélla. K.S. (*st.*) slender. Caracas. 1823. 8. 10. G. ⊙.
2 Pseùdo-Acmélla. w. (*ye.*) spear-leaved. Ceylon. 1768. 7. S. ⊙. Pluk. alm. t. 159. f. 4.
3 uliginòsa. w. (*ye.*) marsh. Jamaica. 1820. 7. 9. S. ⊙.
4 exasperàta. w. (*wh.*) rough-stalked. Caracas. 1825. —— S. ⊙. Jacq. ic. 3. t. 584.
5 álba. w. (*wh.*) white-flowered. Peru. 1783. 6. 7. H. ⊙. L'Herit. stirp. 7. t. 4.
6 oleràcea. w. (*ye.*) esculent. E. Indies. 1770. 7. 9. S. ♂. Jacq. vind. 2. t. 135.
7 pállida. F. (*st.*) pale-flowered. Jamaica. 1821. —— S. ⊙.
8 multiflòra. Ot. (*ye.*) many-flowered. .... 1829. —— S. ⊙.

ACME'LLA. P.S. ACME'LLA. Syngenesia Polygamia Superflua. P.S.
1 mauritiàna. P.S. (*ye.*) Balm-leaved. Mauritius. 1768. 7. 8. S. ☉. Rumph. amb. 6. t. 65.
*Spilánthes Acméll*a. W.
2 rèpens. P.S. (*ye.*) creeping. N. America. 1819. —— H. ☉.
3 occidentàlis. P.S. (*ye.*) oval-leaved. S. America. 1825. —— H. ☉.
4 buphthalmoídes. P.S. (*ye.*) Ox-eye-leaved. ——— 1798. 7. 9. H. ☉. Jacq. schœnb. 2. t. 151.
A'nthemis buphthalmoídes. Jacq.
LAV'ENIA. W. LAV'ENIA. Syngenesia Polygamia Æqualis.
erécta. W. (*ye.*) upright. E. Indies. 1739. 7. 9. S. ☉. Burm. zeyl. t. 42.
SA'LMEA. DC. SA'LMEA. Syngenesia Polygamia Æqualis.
1 scándens. DC. (*wh.*) climbing. S. America. 1815. 4. 6. S. ♄.◡. Bot. mag. t. 2062.
Càlea Améllus. L.
2 hirsùta. DC. (*wh.*) hairy. ——— —— —— S. ♄.◡.
PETR'OBIUM. L.T. PETR'OBIUM. Syngenesia Polygamia Æqualis.
arbòreum. (*st.*) tree. St. Helena. 1816. .... S. ♄.
Spilánthus arbòreus. Forst.
C'ALEA. L.T. C'ALEA. Syngenesia Polygamia Æqualis. L.
1 jamaicénsis. L.L. (*li.*) purple-flower'd. W. Indies. 1739. 6. 7. S. ♃. Sloan. jam. 1. t. 151. f. 3.
2 martinicénsis. (*li.*) Martinico. Martinico. 1821. .... S. ♃.
3 cordifòlia. L.T. (*ye.li.*) heart-leaved. Jamaica. 1822. 6. 7. S. ♃.
CALEA'CTE. L.T. CALEA'CTE. Syngenesia Polygamia Æqualis.
1 urticifòlia. L.T. (*ye.*) Nettle-leaved. Vera Cruz. .... S. ♄.
Solidàgo urticæfòlia. Mill.
2 pinnatífida. L.T. (*ye.*) pinnatifid. Brazil. 1816. 6. 8. S. ♄.
ISOCA'RPHA. L.T. ISOCA'RPHA. Syngenesia Polygamia Æqualis.
oppositifòlia. (*wh.*) opposite-leaved. W. Indies. 1739. 7. 8. S. ♂.
Càlea oppositifòlia. W.
NEUROL'ÆNA. L.T. NEUROL'ÆNA. Syngenesia Polygamia Æqualis.
lobàta. (*ye.*) halberd-weed. W. Indies. 1733. 6. 7. S. ♄. Bot. mag. t. 1734.
Càlea lobàta. B.M.
M'ADIA. W. M'ADIA. Syngenesia Polygamia Superflua. W.
1 viscòsa. W. (*ye.*) viscous. Chili. 1794. 7. 8. H. ☉. Bot. mag. t. 2574.
2 mellòsa. W. (*ye.*) stem-clasping. ——— 1824. —— H. ☉.
TETRAGONOTH'ECA. TETRAGONOTH'ECA. Syngenesia Polygamia Superflua. W.
helianthoídes. W. (*ye.*) Sunflower-like. Virginia. 1726. 8. 10. H. ♃. Schk. hand. 3. t. 263.
XIMEN'ESIA. W. XIMEN'ESIA. Syngenesia Polygamia Superflua. W.
1 heterophy'lla. K.S. (*ye.*) various-leaved. Mexico. 1827. 8. 10. F. ♃.
2 encelioídes. W. (*ye.*) oval-leaved. ——— 1795. 6. 11. G. ♂. Cavan. ic. 2. t. 178.
3 f'œtida. K.S. (*ye.*) fœtid. ——— 1799. 7. 8. H. ☉. ——— 1. t. 77.
Coreópsis f'œtida. Cav. *Símsia ficifòlia*. P.S.
4 cordàta. K.S. (*ye.*) heart-leaved. Mexico. —— —— F. ♃.
HEL'ENIUM. W. HEL'ENIUM. Syngenesia Polygamia Superflua. W.
1 autumnàle. W. (*ye.*) smooth. N. America. 1729. 8. 10. H. ♃. Schk. hand. 3. t. 250.
2 pubéscens. W. (*ye.*) downy. ——— 1776. 8. 9. H. ♃.
3 quadripartìtum. L.en. (*ye.*) four-cleft. .... 1825. —— G. ♂.
4 quadridentàtum. W. (*ye.*) four-toothed. Louisiana. 1790. 5. 10. H. ♃. Bot. reg. t. 598.
Rudbéckia alàta. Jacq. ic. t. 193.
5 pùmilum. W.en. (*ye.*) dwarf. N. America. 1819. —— H. ♃.
6 mexicànum. K.S. (*ye.*) Mexican. Mexico. 1823. 8. 10. H. ♃.
ZI'NNIA. W. ZI'NNIA. Syngenesia Polygamia Superflua. W.
1 pauciflòra. W. (*ye.*) yellow-flower'd. Peru. 1753. 7. 8. H. ☉. Mill. ic. 1. t. 64.
2 multiflòra. W. (*re.*) red-flowered. N. America. 1770. 6. 10. H. ☉. Bot. mag. t. 149.
3 hy'brida. B.M. (*sc.*) hybrid. Mexico. 1818. 6. 9. H. ☉. ——— t. 2123.
4 verticillàta. W. (*re.*) whorl-leaved. ——— 1789. 7. 8. H. ☉. Andr. repos. t. 189.
5 violàcea. C.I. (*pu.*) purple-flowered. ——— 1796. 7. 10. H. ☉. ——— t. 55.
élegans. Bot. mag. t. 527.
β *coccínea*. B.R. (*sc.*) *scarlet-flowered*. ——— 1829. —— H. ☉. Bot. reg. t. 1294.
6 tenuiflòra. W. (*re.*) slender-flower'd. ——— 1799. 7. 8. H. ☉. Bot. mag. t. 555.
7 angustifòlia. K.S. (*re.*) narrow-leaved. ——— 1825. —— H. ☉.
RELH'ANIA. W. RELH'ANIA. Syngenesia Polygamia Superflua. W.
1 squarròsa. W. (*ye.*) cross-leaved. C. B. S. 1774. 5. 6. G. ♄.
2 genistæfolia. W. (*ye.*) Genista-leaved. ——— 1823. 5. 7. G. ♄.
3 lateriflòra. W. (*ye.*) side-flowering. ——— —— —— G. ♄.
4 púngens. W. (*ye.*) pungent-leaved. ——— 1820. 6. 8. G. ♄. Bot. reg. t. 587.
5 pedunculàta. W. (*ye.*) dwarf. ——— 1824. 5. 8. G. ♄.
6 paleàcea. W. (*ye.*) chaffy. ——— 1825. —— G. ♄. Gært. fruct. 2. t. 173. f. 9.
Leysèra paleàcea. Gært.
ATHRI'XIA. B.R. ATHRI'XIA. Syngenesia Polygamia Superflua.
capénsis. B.R. (*li.*) Cape. C. B. S. 1823. 6. 8. G. ♃. Bot. reg. t. 681.

MATA'XA. S.S. Mata'xa. Syngenesia Polygamia Superflua.
capénsis. s.s. (*ye.*) fleshy-leaved. C. B. S. 1824. 6. 8. G. ♄.
Lasiospérmum radiàtum. Treviranus.
LONGCHA'MPIA. W. Longcha'mpia. Syngenesia Polygamia Superflua. W.
capillifòlia. w. (*st.*) Leysera-like. Africa. 1824. 6. 8. H. ⊙.
Gnaphàlium leyseroìdes. Df. *Péctis discoídea.* H.H. *Leysèra capillifòlia.* Spr.
SANVIT'ALIA. W. Sanvit'alia. Syngenesia Polygamia Superflua. W.
procúmbens. w. (*ye.*) trailing. Mexico. 1798. 7. 8. H. ⊙. Bot. reg. t. 707.
TR'IDAX. L. Tr'idax. Syngenesia Polygamia Superflua. L.
procúmbens. L. (*st.*) procumbent. Mexico. 1804. 7. 8. H. ⊙.
Balbísia elongàta. w.
COLUME'LLIA. W. Colume'llia. Syngenesia Polygamia Superflua. W.
biénnis. w. (*ye.*) biennial. C. B. S. 1820. 7. 9. G. ♂. Jacq. schœn. 3. t. 301.
Nestlèra biénnis. s.s.
MEY'ERA. L.T. Mey'era. Syngenesia Polygamia Superflua.
séssilis. s.s. (*ye.*) sessile-flowered. W. Indies. .... 7. 8. S. ♃.
ECLI'PTA. W. Ecli'pta. Syngenesia Polygamia Superflua. L.
1 erécta. w. (*wh.*) upright. S. America. 1690. 7. 9. H. ⊙. Dill. elth. t. 113. f. 137.
2 punctàta. w. (*wh.*) dotted. W. Indies. 1822. —— S. ⊙. Jacq. amer. pict. t. 197.
3 prostràta. w. (*wh.*). trailing. E. Indies. 1732. —— S. ⊙. Dill. elth. t. 113. f. 138.
4 procúmbens. M. (*wh.*) procumbent. Carolina. 1819. —— H. ⊙.
5 lineàris. Ot. (*wh.*) linear-leaved. .... 1825. —— S. ⊙.
CHRYSANTHE'LLUM. Chrysanthe'llum. Syngenesia Polygamia Superflua. P.S.
procúmbens. P.S. (*ye.*) procumbent. W. Indies. 1768. 6. 7. S. ⊙. Swartz obs. 314. t. 8. f. 1.
Verbesìna mùtica. w.
SIEGESBE'CKIA. W. Siegesbe'ckia. Syngenesia Polygamia Superflua. L.
1 orientàlis. w. (*ye.*) oriental. E. Indies. 1730. 8. 10. H. ⊙. Schk. hand. 3. t. 256.
2 triangulàris. C.I. (*ye.*) triangular-leav'd. Mexico. .... —— H. ⊙. Cav. ic. 3. t. 253.
3 ibérica. w. (*ye.*) Iberian. Iberia. 1823. —— H. ⊙. Buxb. cent. 3. t. 52.
4 droseroídes. B.F.G. (*ye.*) Sundew-like. Mexico. 1825. —— F. ♃. Swt. br. fl. gar. t. 203.
5 flosculòsa. w. (*ye.*) small-flowered. Peru. 1784. 6. 7. H. ⊙. L'Herit. stirp. t. 19.
VERBES'INA. W. Verbes'ina. Syngenesia Polygamia Superflua. L.
1 alàta. w. (*or.*) wing-stalked. S. America. 1699. 5. 10. G. ♃. Bot. mag. t. 1716.
2 chinénsis. w. (*or.*) Chinese. China. 1814. —— G. ♃.
3 virgínica. w. (*wh.*) white-flowered. N. America. 1812. 7. 9. H. ♃.
4 virgàta. w. (*ye.*) slender. Mexico. 1825. —— H. ♃. Cav. ic. 3. t. 275.
5 gigántea. w. (*wh.*) tree. W. Indies. 1758. —— S. ♄. Jacq. ic. 1. t. 175.
6 Boswéllia. L. (*ye.*) eatable. E. Indies. 1818. —— S. ⊙.
7 salicifòlia. K.S. (*ye.*) Willow-leaved. Mexico. 1824. 10. 11. G. ♃.
8 tridentàta. s.s. (*ye.*) three-toothed. W. Indies. 1759. 6. 8. S. ♄.
9 atriplicifòlia. s.s. (*st.*) Atriplex-leav'd. Trinidad. 1822 9. 11. S. ♄. Colla hort. rip. t. 31.
10 buphthalmoídes. s.s. (*ye.*) procumbent. Monte Video. 1827. 6. 8. F. ♄. Link et Ott. abb. t. 49.
11 serràta. w. (*ye.*) saw-leaved. Mexico. 1803. 7. 10. G. ♃. Cav. ic. 3. t. 214.
12 Siegesbéckia. w. (*ye.*) American. Virginia. 1731. 10. 11. H. ♃.
13 dichótoma. w. (*ye.*) forked. E. Indies. 1789. 6. 7. S ⊙. Murr. c. got. 1779. t. 4.
14 satìva. B.M. (*ye.*) oil-seed. ———— 1806. 8. 9. S. ⊙. Bot. mag. t. 1017.
15 biflòra. w. (*ye.*) two-flowered. ———— 1818. 7. 9. S. ⊙. Rheed. mal. 19. t. 40.
16 calendulàcea. w. (*ye.*) Marygold-flower'd. Ceylon. 1739. —— S. ⊙. Burm. zeyl. t. 22. f. 1.
17 fruticòsa. w. (*ye.*) shrubby. W. Indies. 1759. 6. 8. S. ♄. Plum. ic. t. 52.
18 pinnatífida. w. (*st.*) pinnatifid. Mexico. 1824. 8. 10. F. ♃. Jacq. schœn. 3. t. 305.
SYNEDRE'LLA. P.S. Synedre'lla. Syngenesia Polygamia Superflua. P.S.
nodiflòra. P.S. (*ye.*) sessile-flowered. W. Indies. 1726. 6. 7. S. ⊙. Dill. elth. t. 45. f. 53.
ENC'ELIA. J. Enc'elia. Syngenesia Polygamia Superflua. P.S.
1 canéscens. P.S. (*ye.*) canescent. Mexico. 1786. 7. 9. G. ♄. Bot. reg. t. 909.
Pallàsia halimifòlia. w. *Coreópsis liménsis.* Jacq. ic. 3. t. 594.
2 halimifòlia. P.S. (*ye.*) great-flowered. Mexico. 1825. —— G. ♄. Cavan. ic. 3. t. 216.
Pallàsia grandiflòra. w.
GALINS'OGEA. K.S. Galins'ogea. Syngenesia Polygamia Superflua. W.
1 trilobàta. w. (*ye.*) three-lobed. Peru. 1797. 8. 11. H. ⊙. Swt. br. fl. gar. t. 56.
2 balbisioídes. K.S. (*ye.*) Mexican. Mexico. 1825. —— H. ⊙. H. et B. 4. t. 386.
PTILOSTE'PHIUM. K.S. Ptiloste'phium. Syngenesia Polygamia Superflua.
1 coronopifòlium. K.S. (*ye.*) Buckshorn-leav'd. Mexico. 1823. 8. 10. H. ⊙. H. et B. 4. t. 387.
2 trifidum. K.S. (*ye.*) trifid-leaved. ———— —— —— H. ⊙. ———— 4. t. 388.
WIBO'RGIA. K.S. Wibo'rgia. Syngenesia Polygamia Superflua.
parviflòra. K.S. (*wh.*) small-flowered. S. America. 1796. 6. 8. H. ⊙. Cavan. icon. 3. t. 281.
Galinsògea parviflòra. w. *Wibórgia Acméllа.* Roth.
PASC'ALIA. W. Pasc'alia. Syngenesia Polygamia Superflua. W.
glaùca. w. (*ye.*) glaucous-leav'd. Chili. 1799. 6. 8. H. ♃. Andr. reposit. t. 549.

HELIO'PSIS. P.S. Helio'psis. Syngenesia Polygamia Superflua. P.S.
1 canéscens. K.S. (*ye.*) hoary-leaved. S. America. 1820. 8. 10. H. ♃. Bot. reg. t. 592.
2 l'ævis. P.S. (*ye.*) smooth. Mexico. 1714. 7. 10. H. ♃.
3 helianthoídes. (*ye.*) roughish. N. America. —— —— H. ♃. L'Herit. stirp. t. 45.
Buphthálmum helianthoídes. L.
4 scábra. S.S. (*ye.*) rough-leaved. ——— .... —— H. ♃.
DIOM'EDEA. Cas. Diom'edea. Syngenesia Polygamia Superflua.
1 glabràta. K.S. (*ye.*) smooth. S. America. 1699. 5. 9. G. ♄. Dill. elth. t. 38. f. 43.
Buphthálmum arborésens. W.
2 bidentàta. S.S. (*ye.*) two-toothed. Carolina. 1696. 6. 8. G. ♄. Catesb. carol. 1. t. 93.
Buphthálmum frutéscens. Ph.—Dill. elth. t. 28. f. 4.
3 argéntea. K.S. (*ye.*) silvery-leaved. Cuba. 1823. —— G. ♄.
BUPHTHA'LMUM. Ox-eye. Syngenesia Polygamia Superflua. L.
1 stenophy'llum. S.S. (*ye.*) linear-leaved. Canaries. 1816. 5. 7. G. ♄.
2 élegans. S.S. (*ye.*) elegant. ——— —— —— G. ♄.
3 lævigàtum. S.S. (*ye.*) smooth-leaved. Teneriffe. 1824. —— G. ♄.
4 flosculòsum. S.S. (*ye.*) rayless. Mesopotamia. 1827. —— G. ♄. Vent. h. cels. t. 25.
5 seríceum. W. (*ye.*) silky. Canaries. 1779. 5. 7. G. ♄. Bot. mag. t. 1836.
6 spinòsum. W. (*ye.*) prickly. Spain. 1570. 6. 9. H. ⊙. Barrel. icon. 551.
7 aquáticum. W. (*ye.*) sweet-scented. S. Europe. 1731. 7. 8. H. ⊙. Breyn. cent. t. 77.
8 marítimum. W. (*ye.*) sea. Sicily. 1640. 7. 9. H. ♃. Bocc. mus. t. 129.
9 salicifòlium. W. (*ye.*) Willow-leaved. Austria. 1759. 6. 10. H. ♃. Jacq. aust. 4. t. 370.
10 grandiflòrum. W. (*ye.*) great-flowered. ——— 1722. —— H. ♃. Moris. s. 6. t. 7. f. 52.
11 speciosíssimum. S.S. (*ye.*) handsome. Italy. 1828. —— H. ♃. Arduin. spec. 1. t. 12.
12 cordifòlium. W. (*ye.*) heart-leaved. Hungary. 1739. 6. 8. H. ♃. W. et K. hung. 2. t. 113.
13 procúmbens. Df. (*ye.*) procumbent. .... 1820. —— H. ♃.
ZEXM'ENIA. D.D. Zexm'enia. Syngenesia Polygamia Necessaria.
tagetiflòra. D.D. (*ye.*) Tagetes-flowered. Mexico. 1828. 8. 10. F. ♄.
WED'ELIA. K.S. Wed'elia. Syngenesia Polygamia Necessaria. W.
1 acapulcénsis. K.S. (*ye.*) Acapulco. Mexico. 1823. 6. 8. S. ♄.
2 híspida. K.S. (*ye.*) hispid. New Spain. 1819. —— S. ♄. Bot. reg. t. 543.
3 aùrea. D.D. (*ye.*) golden-flower'd. Mexico. 1828. 8. 10. F. ♄.
4 radiòsa. B.R. (*ye.*) rayed. S. America. 1822. —— S. ♄. Bot. reg. t. 610.
GYMNOL'OMA. K.S. Gymnol'oma. Syngenesia Polygamia Necessaria.
1 triplinérve. K.S. (*ye.*) triple-nerved. S. America. 1825. 6. 9. S. ♄.
2 maculàtum. B.R. (*ye.*) spotted-stalked. ——— 1822. —— S. ♄. Bot. reg. t. 662.
ACTINOM'ERIS. N. Actinom'eris. Syngenesia Polygamia Frustranea.
1 squarròsa. N. (*ye.*) squarrose. N. America. 1640. 9. 11. H. ♃. Jacq. vind. 2. t. 110.
Coreópsis alternifòlia. Jacq. *Verbesìna Coreópsis.* Ph.
2 pròcera. N. (*ye.*) tall. N. America. 1765. 9. 10. H. ♃.
Coreópsis pròcera. H.K.
β *álba.* N. (*wh.*) *white-flowered.* ——— —— —— H. ♃.
3 helianthoídes. N. (*ye.*) Sunflower-like. Louisiana. 1825. —— H. ♃.
4 alàta. N. (*ye.*) wing-stalked. Mexico. 1803. 9. 11. H. ♃. Cavan. ic. 3. t. 260.
Coreópsis alàta. W.
5 ovàta. N. (*ye.*) oval-leaved. ——— 1829. —— H. ♃. Cavan. ic. 3. t. 280.
HELIA'NTHUS. W. Sunflower. Syngenesia Polygamia Frustranea. L.
1 ánnuus. W. (*ye.*) annual. S. America. 1596. 6. 10. H. ⊙.
2 índicus. W. (*ye.*) dwarf annual. Egypt. 1785. —— H. ⊙. Tabern. ic. 764.
3 lenticulàris. B.R. (*ye.*) freckled. N. America. 1827. 8. 9. H. ⊙. Bot. reg. t. 1265.
4 petiolàris. N. (*ye.*) long-petioled. Arkansa. 1826. 7. 11. H. ⊙. Swt. br. fl. gar. ser. 2.
ásper. Roth.
5 dentàtus. W. (*ye.*) tooth-leaved. Mexico. 1798. 9. 11. F. ♃. Cavan. ic. 3. t. 220.
6 multiflòrus. W. (*ye.*) many-flowered. N. America. 1597. 8. 10. H. ♃. Bot. mag. t. 227.
β *plènus.* (*ye.*) *double-flowered.* ——— —— —— H. ♃.
7 tuberòsus. W. (*ye.*) Jerusalem Artichoke. Brazil. 1617. 9. 10. H. ♃. Jacq. vind. 2. t. 161.
8 pubéscens. B.R. (*ye.*) pubescent. N. America. 1795. 7. 10. H. ♃. Bot. reg. t. 524.
9 atrorùbens. W. (*ye.*) dark eyed. ——— 1732. —— H. ♃. Dill. elth. t. 91. f. 110.
10 diffùsus. B.M. (*ye.*) spreading. ——— 1812. —— H. ♃. Bot. mag. t. 2020.
11 lætiflòrus. P.S. (*ye.*) pale-flowered. ——— .... —— H. ♃.
12 divaricàtus. W. (*ye.*) divaricate. ——— 1759. 8. 10. H. ♃. Moris. s. 6. t. 7. f. 66.
13 trachelifòlius. W. (*ye.*) Trachelium-leav'd. ——— 1818. —— H. ♃.
14 longifòlius. Ph. (*ye.*) long-leaved. Georgia. 1812. —— H. ♃.
15 pauciflòrus. N. (*ye.*) few-flowered. Louisiana. 1823. —— H. ♃.
16 gigánteus. W. (*ye.*) gigantic. N. America. 1714. 9. 10. H. ♃.
17 altíssimus. W. (*ye.*) tallest. ——— 1731. 7. 9. H. ♃. Jacq. vind. 2. t. 160.
18 strumòsus. W. (*ye.*) Carrot-rooted. ——— 1710. —— H. ♃. Bocc. sic. t. 27. f. 4.
19 prostràtus. W. (*ye.*) rough. ——— 1800. —— H. ♃.
20 decapétalus. W. (*ye.*) ten-rayed. ——— 1759. 8. 11. H. ♃. Rob. ic. 235.

21 móllis. w. (ye.) soft. N. America. 1805. 7. 10. H. ♃.
22 macrophy'llus. w. (ye.) large-leaved. ——— 1800. 8. 10. H. ♃. Willd. hort. ber. t. 70.
23 angustifòlius. w. (ye.) narrow-leaved. ——— 1789. 9. 10. H. ♃. Bot. mag. t. 2051.
24 excélsus. w. (ye.) lofty. Mexico. 1824. —— H. ♃. Cav. ic. 3. t. 219.
gigánteus. Cav. *non* w.
25 lineàris. w. (ye.) linear-leaved ——— 1818. 8. 10. H. ♃. Bot. reg. t. 523.
26 parviflòrus. K.S. (ye.) small-flowered. ——— 1824. —— H. ♃. H. et B. 4. t. 378.
27 cornifòlius. K.S. (ye.) Dogwood-leaved. ——— —— —— H. ♃.
28 trilobàtus. L.en. (ye.) three-lobed. ——— 1823. —— H. ♃.

VIGUI'ERA. K.S. VIGUI'ERA. Syngenesia Polygamia Frustranea.
helianthoídes. K.S. (ye.) Sunflower-like. Cuba. 1825. 6. 9. S. ♃. H. et B. 4. t. 379.

GAILLA'RDIA. W. GAILLA'RDIA. Syngenesia Polygamia Frustranea. W.
1 bícolor. w. (ye. re.) two-coloured. Carolina. 1787. 7. 10. H. ♃. Bot. mag. t. 1602.
2 aristàta. Ph. (ye.) long-awned. N. America. 1812. —— H. ♃. Bot. reg. t. 1186.

RUDBE'CKIA. D.D. RUDBE'CKIA. Syngenesia Polygamia Frustranea. L.
1 pinnàta. Ph. (ye.) fragrant. N. America. 1803. 8. 9. H. ♃. Swt. br. fl. gar. t. 146.
2 digitàta. Ph. (ye.) narrow-jagged-leav'd. ——— 1759. —— H. ♃. Moris. s. 6. t. 6. f. 54.
3 laciniàta. Ph. (ye.) broad jagged-leav'd. ——— 1640. 7. 9. H. ♃. ——— s. 6. t. 6. f. 53.
4 columnàris. Ph. (ye.) high-crowned. ——— 1811. 8. 9. H. ♃. Bot. mag. t. 1601.
5 lævigàta. Ph. (ye.) smooth. Georgia. 1812. 7. 8. H. ♃.
6 amplexifòlia. Ph. (ye.) stem-clasping. Louisiana. 1793. 7. 9. H. ⊙. Jacq. icon. 3. t. 592.

CENTROCA'RPHA. D.D. CENTROCA'RPHA. Syngenesia Polygamia Frustranea.
1 subtomentòsa. Ph. (ye.) downy-lobed. N. America. 1802. 8. 9. H. ♃.
2 trilòba. D.D. (ye.) three-lobed. ——— 1699. 7. 10. H. ♃. Bot. reg. t. 525.
Rudbéckia trilòba. B.R.
3 hírta. D.D. (ye.) great hairy. ——— 1714. 6. 11. H. ♃. Swt. br. fl. gar. t. 82.
Rudbéckia hírta. B.F.G.
4 fúlgida. D.D. (ye.) small hairy. ——— 1760. 7. 8. H. ♃. Bot. mag. t. 1996.
5 acutifòlia. B.F.G. (ye.) pointed-leaved. ——— 1822. 7. 11. H. ♃. Swt. br. fl. gar. s. 2.
6 spathulàta. w. (ye.) spathulate-leav'd. ——— 1825. 8. 11. H. ♃.
7 grácilis. N. (ye.) slender hairy. ——— —— —— H. ♃.

ECHIN'ACEA. D.D. ECHIN'ACEA. Syngenesia Polygamia Frustranea.
1 purpùrea. D.D. (pu.) reflexed-rayed. N. America. 1699. 7. 10. H. ♃. Bot. mag. t. 2.
Rudbéckia purpùrea. B.M.
2 serotìna. D.D. (li.) hispid-stalked. ——— 1816. 8. 10. H. ♃. Swt. br. fl. gar. t. 4.
Rudbéckia serotìna. B.F.G.
3 napifòlia. (ro.) Rape-leaved. New Spain. 1824. 7. 10. F. ♃.
Rudbéckia napifòlia. K.S.
4 heterophy'lla. D.D. (li.) various-leaved. Mexico. 1828. 9. 11. H. ♃. Swt. br. fl. gar. s. 2. t. 32.

CO'SMOS. C.I. CO'SMOS. Syngenesia Polygamia Frustranea. W.
1 sulphùreus. (st.) Southernwood-leav'd. Mexico. 1799. 7. 8. H. ⊙. Jacq. ic. 3. t. 595.
2 tenéllus. K.S. (ro.) slender. Mexico. 1825. 10. 11. G. ⊙.
3 bipinnàtus. C.I. (pu.) purple-flower'd. ——— 1799. —— G. ⊙. Bot. mag. t. 1535.
4 crithmifòlius. K.S. (ro.) Samphire-leav'd. ——— 1825. —— G. ⊙.
5 parviflòrus. w. (wh.) white-flowered. ——— 1800. —— H. ⊙. Jacq. schœn. 3. t. 374.

GEORG'INA. W. GEORG'INA. Syngenesia Polygamia Superfluo—Frustranea.
1 Cervantèsii. LG. (sc.) Cervantes'. Mexico. 1820. 8. 10. H. ♃. Swt. br. fl. gar. s. 2. t. 22.
β *Auròra.* S.F.G. (sc.) *double-flowered.* ——— 1828. —— H. ♃. Swt. flor. guid. t. 117.
2 crocàta. LG. (sc.) saffron-coloured. ——— 1818. —— H. ♃. Swt. br. fl. gar. t. 282.
3 coccínea. w. (sc.) scarlet-flowered. ——— 1802. —— H. ♃. Cavan. icon. t. 266.
4 ròsea. C.I. (ro.) bipinnate-leaved. ——— —— 7. 10. H. ♃. ——— t. 265.
5 astrantiæflòra. (pu.) Astrantia-flower'd. ——— 1812. —— H. ♃.
6 variábilis. w. (va.) variable. ——— 1789. 6. 11. H. ♃. Cavan. icon. 1. t. 80.
Dáhlia pinnàta. C.I. *supérflua.* H.K.
α *purpùrea.* (pu.) *purple-flowered.* ——— —— —— H. ♃. Ann. mus. 3. t. 37. f. 1.
β *lilacìna.* (li.) *lilac-coloured.* ——— —— —— H. ♃. Sal. par. lond. t. 16.
γ *pállida.* (bh.) *blush-flowered.* ——— —— —— H. ♃.
δ *álba.* (wh.) *white-flowered.* ——— —— —— H. ♃.
ε *flàva.* (ye.) *yellow-flowered.* ——— —— —— H. ♃.
ζ *cùprea.* (co.) *copper-coloured.* ——— —— —— H. ♃.
η *rùbra.* (re.) *red-flowered.* ——— —— —— H. ♃. Bot. mag. t. 1885. A.
ϑ *coccínea.* (sc.) *scarlet-flowered.* ——— —— —— H. ♃.
ι *punícea.* (cr.) *crimson-flowered.* ——— —— —— H. ♃. Bot. reg. t. 55.
κ *atrosanguínea.* (d.cr.) *dark crimson.* ——— —— —— H. ♃.
λ *atropurpùrea.* (d.pu.) *dark purple.* ——— —— —— H. ♃.
μ *nígra.* (da.) *dark-coloured.* ——— —— —— H. ♃.
ν *variegàta.* (v.) *variegated-flower'd.* ——— —— —— H. ♃.
ξ *nàna.* (li.) *double dwarf lilac.* ——— —— —— H. ♃. Andr. reposit. t. 483.
ο *supérba.* (pu.) *double purple.* ——— .... —— H. ♃. Bot. mag. t. 1885. B.

π Junoniàna. (d.pu.) Juno. Mexico. .... 6. 11. H. ♃. Swt. flor. guid. t. 102.
ρ imperiòsa. (d.pu.) Dennis's imperial. ——— 1827. —— H. ♃. ———— v. 1. t. 71.
σ stellàris. (sc.) morning star. ——— 1826. —— H. ♃. ———— v. 1. t. 65.
τ sphærocéphala.(d.cr.)globe-flowered. ——— 1827. —— H. ♃. ———— v. 2. t. 115.
υ Belladónna. (pi.) painted Lady. ——— —— —— H. ♃. ———— v. 2. t. 110.
The varieties of the present species are almost endless.

CALLIO'PSIS. S.S. CALLIO'PSIS. Syngenesia Polygamia Frustranea.
1 tinctòria. N. (ye.pu.) two-coloured. Arkansa. 1823. 6. 10. H. ⊙. Swt. br. fl. gar. t. 72.
bícolor. S.S. *Coreópsis tinctòria.* N.
2 palmàta. S.S. (ye.) palmate-leaved. Louisiana. —— 8. 10. H. ⊙.
3 ròsea. S.S. (ro.) rose-coloured. N. America. —— —— H. ♃.
Coreópsis ròsea. N.

COREO'PSIS. W. COREO'PSIS. Syngenesia Polygamia Frustranea. L.
1 grandiflòra. N. (ye.) large-flowered. N. America. 1826. 8. 9. H. ♃. Swt. br. fl. gar. t. 175.
2 verticillàta. W. (ye.) whorl-leaved. ——— 1759. 7. 10. H. ♃. Bot. mag. t. 156.
3 tenuifòlia. W. (ye.) slender-leaved. ——— 1780. 7. 8. H. ♃. Pluk. man. t. 341. f. 4.
4 trichospérma. M. (ye.) pinnate leaved. Carolina. 1822. 8. 10. G. ♂.
5 chrysántha. W. (ye.) Angelica-leav'd. W. Indies. 1752. 7. 9. S. ♂. Plum. ic. 53. f. 1.
6 aùrea. W. (ye.) Hemp-leaved. N. America. 1785. 8. 9. H. ♃. Bot. reg. t. 1228.
7 trípteris. Ph. (ye.) three-leaved. ——— 1737. 8. 10. H. ♃. Moris. s. 6. t. 3. f. 41.
8 díscolor. L.en. (ye.) two-coloured. ——— 1818. —— H. ♃.
9 senifòlia. Ph. (ye.) sessile-leaved. ——— 1812. —— H. ♃.
10 amplexicaùlis. K.S. (ye.) stem-clasping. .... 1806. 7. 8. H. ♃.
Símsia amplexicaùlis. P.S.
11 álba. W. (wh.) climbing. Jamaica. 1699. 6. 7. S. ♃.◡. Herm. parad. 124.
12 incìsa. B.R. (ye.) jagged-leaved. W. Indies. .... 9. 12. S. ♃.◡. Bot. reg. t. 7.
13 réptans. W. (ye.) trailing. ——— 1792. 7. 9. S. ⊙. Smith spic. t. 22.
14 auriculàta. W. (ye.) ear-leaved. N. America. 1699. 8. 10. H. ♃. Schk. hand. 3. t. 260.
15 latifòlia. W. (ye.) broad-leaved. ——— 1786. 8. 9. H. ♃.
16 argùta. Ph. (ye.) sharp-notched. Carolina. .... —— H. ♃.
17 lanceolàta. W. (ye.) spear-leaved. ——— 1724. 7. 9. H. ♃. Swt. br. flr. gar. t. 10.
18 crassifòlia. W. (ye.) thick-leaved. ——— 1786. 8. 10. H. ♃.
19 angustifòlia. W. (ye.) narrow-leaved. N. America. 1778. 6. 8. H. ♃.

B'IDENS. K.S. B'IDENS. Syngenesia Polygamia Frustranea.
1 nodiflòra. W. (ye.) sessile-flowered. E. Indies. 1732. 7. 8. S. ⊙. Dill. elth. t. 44. f. 52.
2 tripartìta. W. (ye.) trifid. Britain. .... 7. 9. H. ⊙. Eng. bot. t. 1113.
3 cérnua. W (ye.) nodding. ——— .... —— H. ⊙. ——— t. 1114.
4 chrysanthemoídes. W. (ye.) large yellow. N. America. 1827. 8. 10. H. ⊙.
5 argùta. K.S. (ye.) sharp-toothed. Mexico. 1825. —— H. ♃.
6 triplinérvia. K.S. (ye.) triply-nerved. ——— —— —— H. ♃.
7 heterophy'lla. W. (ye.) various-leaved. ——— 1803. 8. 9. F. ♃. Orteg. dec. 8. t. 12.
8 luxùrians. W.en. (ye.) luxuriant. ——— 1823. —— F. ♃.
9 frondòsa. W. (ye.) smooth-stalked. N. America. 1710. 7. 8. H. ⊙. Moris. his. s. 6. t. 5. f. 21.
10 foliòsa. W.en. (ye.) leafy. .... 1823. —— H. ⊙.
11 connàta. W. (ye.) connate. N. America. —— —— H. ⊙.
12 leucántha. W. (wh.) white-flowered. S. America. .... —— H. ⊙.
13 striàta. B.F.G. (std.) stripe-flowered. Mexico. 1827. 8. 11. H. ⊙. Swt. br. fl. gar. t. 237.
14 chinénsis. W. (wh.) Chinese. China. 1801. 6. 7. H. ⊙. Rumph. amb. 6. t. 15. f. 2.
15 diversifòlia. W. (ye.) different-leav'd. S. America. 1811. 9. 10. H. ⊙. Bot. mag. t. 1689.
Cósmea lùtea. B.M. *Coreópsis diversifòlia.* Jacq.
16 pilòsa. W. (wh.) hairy. N. America. 1732. 7. H. ⊙. Dill. elth. t. 43. f. 51.
17 sambucifòlia. W. (sc.) Elder-leaved. S. America. 1801. 7. 8. H. ♃. Cavan. ic. 3. t. 229.
18 odoràta. W. (wh.) sweet-scented. Mexico. 1823. —— H. ⊙. ——— 1. t. 13.
19 grandiflòra. S.S. (wh.) large-flowered. S. America. 1825. 8. 11. H. ⊙.
20 bipinnàta. W. (ye.) Hemlock-leav'd. N. America. 1687. —— H. ⊙. Herm. parad. t. 123.
21 parviflòra. S.S. (ye.) small-flowered. Siberia. 1823. —— H. ⊙.
22 bullàta. W. (ye.) rough-leaved. Italy. 1759. —— H. ⊙. Arduin. spec. 2. t. 18.
23 refléxa. L.en. (ye.) reflexed. Mexico. 1824. —— G. ♃.
24 crithmifòlia. K.S. (ye.) Samphire-leav'd. Quito. —— 9. 10. F. ♃.
25 angustifòlia. K.S. (or.) narrow-leaved. New Spain. —— —— H. ♃.
26 pròcera. B.R. (ye.) tall. S. America. 1818. —— H. ♃. Bot. reg. t. 684.
27 ferulæfòlia. (ye.) Fennel-leaved. Mexico. 1799. 10. 11. F. ♃. Jacq. schœnb. 3. t. 373.
Coreópsis ferulæfòlia. Jacq.
28 ciliàta. (ye.) fringed. Mexico. 1827. 8. 10. H. ⊙.
29 chrysántha. Ort. (ye.) golden-flowered. ——— —— —— H. ⊙.
30 coronàta. F. (ye.) crowned. .... 1829. —— H. ⊙.
31 serrulàta. Df. (ye.) serrulate. .... —— —— H. ⊙.

OSM'ITES. W. OSM'ITES. Syngenesia Polygamia Frustranea. W.
1 bellidiástrum. W. (wh.) Daisy-like. C. B. S. 1816. 4. 8. G. ♄.
2 asteriscoídes. W. (wh.) starry. ——— 1823. 4. 7. G. ♄. Burm. afr. t. 58. f. 1.

3 camphorìna. w.	(*wh.*)	Camphor-scent'd.	C. B. S.	1794.	4. 7.	G. ♄.	Seb. mus. 1. t. 90. f. 2.
4 dentàta. w.	(*wh.*)	tooth-leaved.	——	1820.	——	G. ♄.	
SCLEROCA'RPUS. W.		Scleroca'rpus.	Syngenesia Polygamia Frustranea. W.				
africànus. w.	(*ye.*)	African.	Guinea.	1812.	7. 8.	G. ⊙.	Jacq. ic. 1. t. 176.
CULL'UMIA. H.K.		Cull'umia.	Syngenesia Polygamia Frustranea. H.K.				
1 ciliàris. h.k.	(*ye.*)	ciliated.	C. B. S.	1774.	5. 6.	G. ♄.	Burm. afr. t. 54. f. 1.
2 setòsa. h.k.	(*ye.*)	recurved smooth-leav'd.	——	1780.	6. 8.	G. ♄.	Comm. hort. 2. t. 28.
3 squarròsa. h.k.	(*ye.*)	recurved-awl-leaved.	——	1786.	——	G. ♄.	Th. act. hafn. 3. t. 5.
BERCKH'EYA. H.K.		Berckh'eya.	Syngenesia Polygamia Frustranea. H.K.				
1 cynaroìdes. w.	(*ye.*)	Artichoke-cup'd.	C. B. S.	1789.	6.	G. ♃.	
2 obovàta. w.	(*ye.*)	smooth shrubby.	——	1794.	6. 8.	G. ♄.	Hout. n. h. 6. t. 34. f. 2.
3 incàna. w.	(*ye.*)	hoary.	——	1739.	7. 8.	G. ♄.	Jacq. ic. 3. t. 591.
4 cuneàta. w.	(*ye.*)	wedge-leaved.	——	1812.	6. 8.	G. ♄.	Th. act. hafn. 3. t. 10.
5 palmàta. w.	(*ye.*)	palmated.	——	1800.	——	G. ♄.	—— 3. t. 13.
6 grandiflòra. w.	(*ye.*)	large-flowered.	——	1812.	——	G. ♄.	Bot. mag. t. 1844.
7 uniflòra. w.	(*ye.*)	single-flowered.	——	1815.	——	G. ♄.	Th. act. hafn. 3. t. 7.
8 cérnua. h.k.	(*ye.*)	drooping-flower'd.	——	1774.	5. 7.	G. ♂.	Meerb. ic. 1. t. 40.
DIDE'LTA. H.K.		Dide'lta.	Syngenesia Polygamia Frustranea. H.K.				
1 carnòsum. h.k.	(*ye.*)	alternate-leav'd.	C. B. S.	1774.	6. 7.	G. ♄.	L'Her. stirp. t. 28.
2 spinòsum. h.k.	(*ye.*)	opposite-leaved.	——	——	——	G. ♄.	Wendl. obs. t. 4. f. 32.
GORT'ERIA. H.K.		Gort'eria.	Syngenesia Polygamia Frustranea. H.K.				
personàta. h.k.	(*ye.*)	procumbent.	C. B. S.	1774.	7. 8.	G. ⊙.	Jacq. coll. 4. t. 21. f. 1.
GAZ'ANIA. H.K.		Gaz'ania.	Syngenesia Polygamia Frustranea. H.K.				
1 uniflòra. b.m.	(*ye.*)	golden.	C. B. S.	1816.	7. 8.	G. ♄.	Bot. mag. t. 2270.
2 rigens. h.k.	(*ye.ve.*)	great-flowered.	——	1755.	5. 9.	G. ♄.	—— t. 90.
3 heterophy'lla. w.	(*ye.ve.*)	various-leaved.	——	1812.	——	G. ♃.	Willd. hort. ber. t. 97.
4 Pavònia. h.k.	(*ye.ve.*)	Peacock.	——	1804.	6. 7.	G. ♃.	Bot. reg. t. 35.
5 subulàta. h.k.	(*ye.*)	awl-leaved.	——	1792.	7. 8.	G. ♃.	
CRYPTOSTE'MMA. H.K.		Cryptoste'mma.	Syngenesia Polygamia Frustranea.				
1 calendulàceum. h.k.	(*ye.*)	Marygold-flow'r'd.	C. B. S.	1752.	6. 8.	H. ⊙.	Bot. mag. t. 2252.
2 hypochondrìacum. h.k.	(*ye.*)	divided-rayed.	——	1731.	7. 8.	H. ⊙.	
3 runcinàtum. h.k.	(*ye.*)	Dandelion-leaved.	——	1794.	——	H. ⊙.	
ARCTOTH'ECA. W.		Arctoth'eca.	Syngenesia Polygamia Frustranea. H.K.				
1 rèpens. w.	(*ye.*)	creeping.	C. B. S.	1793.	7. 8.	G. ♃.	Jacq. schœn. 3. t. 306.
2 hírta. L.en.	(*ye.*)	hairy.	——	1818.	7. 9.	G. ♃.	
SPHENO'GYNE. H.K.		Spheno'gyne.	Syngenesia Polygamia Frustranea. H.K.				
1 anthemoìdes. h.k.	(*ye.*)	white-crowned.	C. B. S.	1774.	7. 9.	H. ⊙.	Bot. mag. t. 544.
2 crithmifòlia. h.k.	(*ye.*)	Samphire-leav'd.	——	1768.	4. 8.	G. ♄.	Burm. afr. t. 65. f. 1.
3 scariòsa. h.k.	(*ye.*)	scaly-cupped.	——	1774.	——	G. ♄.	
4 abrotanifòlia. h.k.	(*ye.*)	Southernwood-leav'd.	——	1789.	5. 8.	G. ♄.	
5 dentàta. h.k.	(*ye.*)	small-leaved.	——	1787.	6. 7.	G. ♄.	Burm. afr. t. 64.
6 pilífera. b.r.	(*ye.*)	hair toothed.	——	1820.	——	G. ♄.	Bot. reg. t. 604.
7 serràta. s.s.	(*ye.*)	saw-leaved.	——	1826.	——	G. ♄.	
8 fœniculàcea.	(*ye.*)	Fennel-leaved.	——	1825.	5. 9.	H. ⊙.	Jacq. schœn. 2. t. 156.
Arctòtis fœniculàcea. Jacq.							
9 leucanthemoìdes.	(*ye.*)	Ox-eye Daisy-leav'd.	——	——	——	H. ⊙.	—— 2. t. 164.
Arctòtis leucanthemoìdes. Jacq.							
10 odoràta. h.k.	(*ye.*)	smooth-seeded.	——	1774.	4. 6.	G. ♄.	
TITH'ONIA. W.		Tith'onia.	Syngenesia Polygamia Frustranea. W.				
1 tagetiflòra. w.	(*or.*)	Tagetes-flower'd.	W. Indies.	1821.	8. 9.	S. ♃.	Bot. reg. t. 591.
2 tubæfórmis. d.d.	(*ye.*)	tube-flowered.	Mexico.	1799.	9. 11.	H. ⊙.	Jacq. schœn. 3. t. 375.
Heliánthus tubæfórmis. Jacq.							
ARCT'OTIS. H.K.		Arct'otis.	Syngenesia Polygamia Necessaria. L.				
1 acaùlis. w.	(*ye.*)	dwarf.	C. B. S.	1759.	4. 7.	G. ♃.	Bot. reg. t. 122.
2 trícolor. w.	(*pu.wh.*)	three-coloured.	——	1794.	5. 7.	G. ♃.	—— t. 131.
3 undulàta. w.	(*or.*)	wave-leaved.	——	1795.	4. 6.	G. ♃.	Jacq. schœn. 2. t. 160.
4 grandiflòra. w.	(*or.cr.*)	great-flowered.	——	1774.	3. 5.	G. ♄.	—— 3. t. 378.
5 Massoniàna. s.s.	(*st.re.*)	Masson's.	——	——	——	G. ♂.	
grandiflòra. h.k. *non* Jacq.							
6 speciòsa. w.	(*pu.ye.*)	showy.	——	1812.	6. 8.	G. ♃.	Bot. mag. t. 2182.
7 glaucophy'lla. w.	(*or.pi.*)	Sea-green-leav'd.	——	1794.	5. 8.	G. ♃.	Jacq. schœn. 2. t. 170.
8 plantagìnea. w.	(*ye.pu.*)	Plantain-leaved.	——	1768.	6. 8.	G. ♃.	
9 argéntea. w.	(*v.*)	silver-leaved.	——	1774.	8.	G. ♂.	
10 ròsea. w.	(*ro.*)	Rose.	——	1793.	7. 9.	G. ♃.	Jacq. schœn. 2. t. 162.
11 decúmbens. w.	(*wh.pi.*)	decumbent.	——	1790.	——	G. ♃.	—— 3. t. 381.
12 angustifòlia. w.	(*or.*)	narrow-leaved.	——	1739.	——	G. ♃.	
13 fláccida. w.	(*wh.re.*)	bending-stalked.	——	1794.	5. 7.	G. ⊙.	Jacq. schœn. 2. t. 163.
14 virgàta. w.	(*ye.pu.*)	slender-twigged.	——	1816.	7. 9.	H. ⊙.	—— 3. t. 307.
15 flámmea. j.f.	(*or.*)	flame-coloured.	——	1822.	——	G. ♃.	Jacq. frag. t. 47. f. 1.

16 am'œna. J.F. (wh.pu.) handsome. C. B. S. 1822. 7. 9. G. ♃. Jacq. frag. t. 118.
17 paniculàta.w.(wh.cr.) panicled. —— —— —— G. ♄. Jacq. schœn. 3. t. 380.
18 decúrrens.w. (wh.pi.) decurrent. —— —— 6. 7. G. ♄. —— 2. t. 165.
19 melanocìcla.w.(wh.re.)various-colour'd. —— 1812. —— G. ♄.
20 réptans. w. (st.br.) creeping. —— 1795. 7. 9. G. ♃. Jacq. schœn. 3. t. 382.
21 auriculàta.w. (ye.pu.) ear-leaved. —— —— 6. 8. G. ♄. —— 2. t. 169.
22 fastuòsa. w. (or.cr.) Orange-flower'd. —— —— 5. 7. G. ♂. —— 2. t. 166.
23 spinulòsa. w. (or.bk.) thorny-leaved. —— —— 5. 8. G. ☉. —— 2. t. 167.
24 maculàta. w. (st.pu.) spotted. —— 1812. —— G. ♄. Bot. reg. t. 130.
25 áspera. B.R. (wh.ro.) broad rough-leav'd. —— 1710. 7. 9. G. ♄. —— t. 34.
26 elàtior. w. (ye.pu.) tall. —— 1816. —— G. ♄. Jacq. schœn. 2. t. 172.
27 arboréscens.w.(wh.pi.)arborescent. —— —— —— G. ♄. —— 2. t. 171.
28 aurèola. B.R. (or.) narrow rough-leav'd. —— 1710. —— G. ♄. Bot. reg. t. 32.
29 revolùta. w. (ye.bk.) revolute-leaved. —— 1824. 8. 10. H. ☉. Jacq. schœn. 2. t. 173.
30 cùprea. w. (co.pu.) copper-coloured. —— —— 7. 8. G. ♄. —— 2. t. 176.
31 squarròsa. w. (or pu.) squarrose. —— 1825. —— G. ♄. —— 2. t. 177.
32 Cinerària. w. (ye.or.) black-eyed. —— —— —— G. ♃. —— 2. t. 174.
33 bícolor. W.en. (wh.re.)two-coloured. —— 1812. —— G. ♃.
34 glabràta. w. (ye.pu.) smooth-leaved. —— 1816. 7. 9. G. ♄. Jacq. schœn. 2. t. 175.
35 purpùrea. W.en.(pu.) purple. —— 1827. —— G. ♄.

CALE'NDULA. W. MARYGOLD. Syngenesia Polygamia Necessaria. L.
1 arvénsis. w. (ye.) field. Europe. 1597. 6. 9. H. ☉. Ger. herb. 603. f. 10.
2 sícula. W.en. (ye.) Sicilian. Sicily. 1816. —— H. ☉.
3 stellàta. w. (ye.) starry. Barbary. 1795. —— H. ☉. Schk. hand 3. t. 265.
4 officinàlis. w. (ye.) common. S. Europe. 1573. —— H. ☉.
β plèna. (ye.) *double-flowered.* —— —— —— H. ☉.
5 ægyptìaca. P.S. (ye.) Egyptian. Egypt. 1820. —— H. ☉.
6 suffruticòsa. w. (ye.) suffrutescent. S. Europe. 1823. —— G. ♄.
7 denticulàta. S.S. (ye.) toothed. Barbary. 1821. 6. 8. F. ♄.
8 sáncta. w. (st.) pale-flowered. Levant. 1731. 5. 9. H. ☉.
9 incàna. w. (ye.) hoary. Barbary. 1796. 6. 8. H. ☉. Desf. atl. 2. t. 245.
tomentòsa. Df. *non* L.
10 pluviàlis. w. (wh.pu.) small Cape. C. B. S. 1699. —— H. ☉. Mill. ic. t. 75. f. 2.
11 hy'brida. w. (wh.pu.) great Cape. —— 1752. 6. 7. H. ☉. Swt. br. fl. gar. t. 39.
12 nudicaù'is. w. (wh.) naked-stalked. —— 1731. 6. 8. H. ☉. Comm. hort. 2. t. 33.
13 graminifòlia.w.(wh.pu.)Grass-leaved. —— —— 5. 9. G. ♃. Bot. reg. t. 289.
14 Trâgus. w. (pu.wh.) bending-stalked. —— 1774. 5. 6. G. ♄. Bot. mag. t. 408.
β flàccida. V. (or.) *flaccid.* —— —— —— G. ♄. Bot. reg. t. 28.
15 viscòsa. H.K. (or.) viscous. —— 1790. 6. 9. G. ♄. Andr. reposit. t. 412.
Arctòtis glutinòsa. B.M. 1343.
16 dentàta. A.R. (wh.) tooth-leaved. —— 1790. 5. 6. G. ♄. Andr. reposit. c. ic.
17 oppositifòlia. w. (ye.) glaucous-leaved. —— 1774. 8. G. ♄.
18 fruticòsa. w. (wh.) shrubby. —— 1752. 6. 7. G. ♄. Mill. ic. 2. t. 283.
19 chrysanthemifòlia. v. (ye.)large-flower'd. —— 1790. 3. 8. G. ♄. Bot. reg. t. 40.
20 arboréscens. w. (ye.) rough-leaved. —— 1774. 12. G. ♄. Jacq. ic. 3. t. 596.
21 muricàta. w. (ye.) muricate. —— 1816. —— G. ♄.

POLY'MNIA. W. POLY'MNIA. Syngenesia Polygamia Necessaria. L.
1 canadénsis. w. (ye.) Canadian. N. America. 1768. 7. 8. H. ♃. Lin. am. ac. 3. t. 1. f. 5.
2 Uvedàlia. w. (ye.) broad-leaved. —— 1699. 8. 10. H. ♃. Moris. s. 6. t. 7. f. 55.
3 maculàta. C.I. (ye.) spotted-stalked. Mexico. 1823. 9. 11. H. ♃. Cav. ic. 3. t. 227.
4 abyssínica. w. (ye.) upright. Africa. 1775. 4. 5. S. ♂.

MELAMP'ODIUM. W. MELAMP'ODIUM. Syngenesia Polygamia Necessaria. W.
1 americànum. w.(ye.) American. Vera Cruz. 1733. 8. 10. S. ☉. Reliq. Houst. 9. t. 21.
2 hùmile. w. (ye.) dwarf. Jamaica. 1782. 6. 10. S. ☉.
3 híspidum. K.S. (ye.) hispid. New Spain. 1825. 9. 11. H. ☉. H. et B. 4. t. 399.
4 perfoliàtum. K.S.(ye.) perfoliate. —— 1796. 8. 10. H. ☉. Cavan. ic. 1. t. 15.
Alcìna perfoliàta. Cav. *Wedèlia perfoliàta.* w.
5 longifòlium.W.en.(ye.)long-leaved. —— 1820. —— H. ☉.
6 ovatifòlium. R.I. (ye.) oval-leaved. S. America. 1828. —— H. ☉.
Wedèlia ovatifòlium. w. *Alcìna ovalifòlia.* LG.
7 seríceum. S.S. (ye.) silky-leaved. New Spain. 1825. —— H. ☉.

MILL'ERIA. L. MILL'ERIA. Syngenesia Polygamia Necessaria. L.
1 quinqueflòra. w.(ye.) five flowered. Vera Cruz. 1731. 7. 10. H. ☉. Cavan. ic. 1. t. 82.
2 biflòra. w. (ye.) two-flowered. S. America. 1730. —— H. ☉. Mart. dec. 47. f. 1.

FLAV'ERIA. J. FLAV'ERIA. Syngenesia Polygamia Necessaria. P.S.
1 repánda. LG. (ye.) repand. S. America. 1799. 7. 9. H. ☉.
2 Contrayérba.K.S.(ye.)Peruvian. Peru. 1794. 8. 10. S. ♂. Cavan. ic. 1. t. 4.
3 angustifòlia. K.S.(ye.) narrow-leaved. Mexico. 1825. —— H. ♃. —— 3. t. 223.

4 lineàris. LG.	(*ye.*)	linear-leaved.	S. America.	1825.	8. 10.	S. ⊙.	
5 marítima. K.S.	(*ye.*)	sea-side.	Cuba.	1822.	——	S. ⊙.	
BALTIM'ORA. W.		BALTIM'ORA.	Syngenesia Polygamia Necessaria. L.				
1 recta. W.	(*ye.*)	upright.	Vera Cruz.	1699.	6. 7.	H. ⊙.	Schk. ha. 3. t. 261. C.
2 álba. P.S.	(*wh.*)	white.	....	1824.	——	H. ⊙.	
ERIO'COMA. K.S.		ERIO'COMA.	Syngenesia Polygamia Frustranea.				
1 floribúnda. K.S.	(*wh.*)	many-flowered.	Mexico.	1828.	9. 11.	F. ♄.	H. et B. 4. t. 396.
2 frágrans. D.D.	(*wh.*)	sweet-scented.	——	——	——	F. ♄.	Swt. br. fl. gar. s. 2.
SI'LPHIUM. W.		SI'LPHIUM.	Syngenesia Polygamia Necessaria. L.				
1 laciniàtum. W.	(*ye.*)	jagged-leaved.	N. America.	1781.	7. 9.	H. ⊙.	Lin. fil. fas. 1. t. 3.
2 compósitum. W.	(*ye.*)	scollop-leaved.	——	1789.	——	H. ♃.	
3 therebinthinàceum.	(*ye.*)	broad-leaved.	——	1765.	8. 9.	H. ♃.	Jacq. vind. 1. t. 43.
4 perfoliàtum. W.	(*ye.*)	perfoliate.	——	1766.	7. 10.	H. ♃.	
5 conjúnctum. W.en.	(*ye.*)	conjoined.	——	....	——	H. ♃.	
6 connàtum. W.	(*ye.*)	round-stalked.	——	1765.	——	H. ♃.	
7 Asteríscus. W.	(*ye.*)	hairy-stalked.	——	1732.	7. 9.	H. ♃.	Dill. elth. t. 37. f. 42.
8 scábrum. N.	(*ye.*)	rough-leaved.	——	1823.	——	H. ♃.	
9 pùmilum. W.	(*ye.*)	dwarf woolly.	——	——	——	H. ♃.	
10 trifoliàtum. W.	(*ye.*)	three-leaved.	——	1755.	7. 10.	H. ♃.	Moris. s. 6. t. 3. f. 68.
11 ternàtum. W.	(*ye.*)	various-leaved.	——	1806.	——	H. ♃.	
12 atropurpùreum. W.	(*ye.*)	purple-stalked.	——	1812.	——	H. ♃.	
OSTEOSPE'RMUM.		OSTEOSPE'RMUM.	Syngenesia Polygamia Necessaria. L.				
1 spinòsum. H.K.	(*ye.*)	rough-leaved.	C. B. S.	1700.	2. 10.	G. ♄.	Com. hort. 2. t. 43.
2 spinéscens. H.K.	(*ye.*)	smooth-leaved.	——	1793.	3. 6.	G. ♄.	Jacq. schœn. 3. t. 377.
3 pisíferum. W.	(*ye.*)	smooth.	——	1757.	3. 5.	G. ♄.	Mill. ic. 2. t. 194. f. 1.
4 monilíferum. W.	(*ye.*)	Poplar-leaved.	——	1714.	7. 8.	G. ♄.	Dill. elth. t. 68. f. 79.
5 hirsùtum. W.	(*ye.*)	hairy.	——	1822.	——	G. ♄.	
6 ciliàtum. W.	(*ye.*)	fringed.	——	1818.	——	G. ♄.	
7 imbricàtum. W.	(*ye.*)	imbricated.	——	1824.	——	G. ♄.	
8 calendulàceum. W.	(*ye.*)	Marygold-like.	——	1823.	——	G. ♄.	
9 ilicifòlium. W.	(*ye.*)	Holly-leaved.	——	1816.	——	G. ♄.	Burm. afr. 172. t. 62.
10 rígidum. W.	(*ye.*)	rigid.	——	1774.	4. 7.	G. ♄.	
11 cœrùleum. W.	(*bl.*)	blue-flowered.	——	——	6. 9.	G. ♄.	Jacq. ic. 1. t. 179.
12 perfoliàtum. W.	(*ye.*)	perfoliate.	——	1816.	5. 8.	G. ♄.	
13 níveum. W.	(*ye.*)	white-leaved.	——	1820.	——	G. ♄.	
14 corymbòsum. W.	(*ye.*)	corymbed.	——	——	——	G. ♄.	
15 incànum. W.	(*ye.*)	hoary-leaved.	——	1816.	7. 8.	G. ♄.	
16 polygaloídes. W.	(*ye.*)	Milkwort-leaved.	——	1759.	——	G. ♄.	Pluk. mant. t. 382.

Subordo VII. *AMBROSIACEÆ. Link enum.* 2. *p.* 366.

PARTH'ENIUM. W.		PARTH'ENIUM.	Syngenesia Polygamia Necessaria. L.				
1 integrifòlium. W.	(*wh.*)	entire-leaved.	Virginia.	1661.	6. 10.	H. ♃.	Willd. hort. ber. t. 4
2 incànum. K.S.	(*wh.*)	hoary.	Mexico.	1824.	——	H. ⊙.	H. et B. 4. t. 391.
3 Hysteróphorus. W.	(*wh.*)	cut-leaved.	W. Indies.	1728.	7. 10.	H. ⊙.	Bot. mag. t. 2275.
Argyroch'æta bipinnatífida. Cavan. icon. 4. t. 378.							
'IVA. W.		'IVA.	Syngenesia Polygamia Necessaria. L.				
1 ciliàta. W.	(*ye.*)	ciliated.	N. America.	1820.	7. 8.	H. ⊙.	
2 ánnua. W.	(*ye.*)	annual.	S. America.	1768.	——	H. ⊙.	Schmidel ic. t. 16.
3 xanthiifòlia. N.	(*ye.*)	Xanthium-leaved.	Missouri.	1827.	——	H. ⊙.	
4 axillàris. Ph.	(*ye.*)	axillary-flowered.	——	——	5. 6.	H. ♃.	
5 frutéscens. W.	(*ye.*)	shrubby.	N. America.	1711.	8.	H. ♄.	Pluk. alm. t. 27. f. 1.
AMBR'OSIA. W.		AMBR'OSIA.	Monœcia Pentandria. L.				
1 integrifòlia. W.	(*gr.*)	entire-leaved.	N. America.	1816.	7. 9.	H. ⊙.	
2 trífida. W.	(*gr.*)	trifid-leaved.	——	1699.	——	H. ⊙.	Moris. s. 6. t. 1. f. 4.
3 elàtior. W.	(*gr.*)	tall.	——	1696.	7. 8.	H. ⊙.	Herm. lugdb. t. 35.
4 artemisiæfòlia. W.	(*gr.*)	Mugwort-leaved.	——	1759.	——	H. ⊙.	
5 paniculàta. W.	(*gr.*)	panicled.	——	1811.	7. 9.	H. ⊙.	Pluk. alm. t. 10. f. 5.
6 marítima. W.	(*gr.*)	sea.	Italy.	1570.	7. 8.	H. ⊙.	Schk. hand. 3. t. 292.
7 cumanénsis. K.S.	(*gr.*)	Cumana.	S. America.	1823.	9. 11.	G. ♃.	
8 peruviàna. W.	(*gr.*)	Peruvian.	Peru.	1821.	——	G. ♃.	
XA'NTHIUM. W.		XA'NTHIUM.	Monœcia Pentandria. L.				
1 Strumàrium. W.	(*gr.*)	small burdock.	England.	....	7. 9.	H. ⊙.	Eng. bot. v. 36. t. 2544.
2 orientàle. W.	(*gr.*)	oriental.	China.	1685.	——	H. ⊙.	Linn. dec. 33. t. 17.
3 canadénse. S.S.	(*gr.*)	Canadian.	Canada.	——	——	H. ⊙.	
4 echinàtum. W.	(*gr.*)	hedgehog.	....	1818.	——	H. ⊙.	

5 cathárticum. K.S. (*gr.*) divided-leaved. S. America. 1825. 9. 10. H. ⊙.
6 spinòsum. W. (*gr.*) spiny. S. Europe. 1713. 7. 9. H. ⊙. Herm. parad. 246.
FRANS'ERIA. W. Frans'eria. Monœcia Pentandria. W.
1 artemisioídes. W. (*gr.*) Mugwort-leaved. Peru. 1759. 7. 9. G. ♄. Willd. hort. ber. t. 2.
2 ambrosioídes. W. (*gr.*) Ambrosia-leaved. Mexico. 1796. —— G. ♄. Cavan. ic. 2. t. 200.

Subordo VIII. *ANTHEMIDEÆ.* *Kth. synops.* 2. *p.* 513.

CEPHALO'PHORA. W. Cephalo'phora. Syngenesia Polygamia Æqualis. W.
glaùca. W. (*wh.*) glaucous. Chili. 1798. 7. 8. H. ⊙. Cavan. ic. 6. t. 599.
Gr'æmia aromática. Hook. ex. flor. 189.
ETH'ULIA. W. Eth'ulia. Syngenesia Polygamia Æqualis. W.
1 conyzoídes. W. (*li.*) panicled. India. 1776. 7. 8. S. ⊙. Linn. fil. dec. t. 1.
2 divaricàta. W. (*pu.*) spreading. ——— 1815. —— S. ⊙. Lam. ill. t. 699.
Epáltes divaricàta. Cas.
3 ramòsa. H.B. (*pu.*) branching. ——— 1818. —— S. ⊙.
4 brasiliénsis. L.en. (*pu.*) Brazilian. Brazil. 1822. —— S. ♃.
SPARGANO'PHORUS. G. Spargano'phorus. Syngenesia Polygamia Æqualis.
1 Vaillántii. P.S. (*pu.*) Vaillant's. E. Indies. 1824. 6. 8. S. ♃. Gært. fr. 2. t. 165. f. 4.
Ethùlia Sparganóphorus. L.
2 Strùchium. P.S. (*pu.*) axillary-flower'd. W. Indies. 1826. —— S. ♃. Brown. jam. t. 34. f. 2.
3 verticillàtus. M. (*pu.*) whorl-leaved. N. America. 1827. 8. 10. H.*w.* ♃.
TARCHONA'NTHUS. W. African Fleabane. Syngenesia Polygamia Æqualis. L.
1 camphoràtus. W. (*ye.*) strong-scented. C. B. S. 1690. 6. 10. G. ♄. Lam. ill. t. 671.
2 dentàtus. P.S. (*ye.*) scentless. ——— 1816. —— G. ♄.
3 ellípticus. P.S. (*ye.*) elliptic-leaved. ——— 1823. —— G. ♄.
PODA'NTHUS. LG. Poda'nthus. Syngenesia Polygamia Æqualis.
1 míliqui. H.C. (*ye.*) Chilian. Chile. 1826. 7. 9. F. ♄.
2 gràtus. D.D. (*ye.*) grateful. ——— 1825. —— F. ♄. Hor. phys. ber. 75. t. 16.
Euxènia gràta. S.S.
OTA'NTHUS. L.en. Ota'nthus. Syngenesia Polygamia Æqualis.
marítimus. L.en. (*ye.*) sea-side. England. .... 7. 9. H. ♄. Eng. bot. v. 2. t. 141.
Santolìna marítima. E.B.
SANTOL'INA. W. Lavender-Cotton. Syngenesia Polygamia Æqualis. L.
1 Chámæ-Cyparíssus. W. (*ye.*) common. S. Europe. 1573. 7. H. ♄. Lam. ill. t. 671. f. 3.
2 ericoídes. P.S. (*ye.*) Heath-like. ——— 1827. 7. 9. H. ♄.
3 squarròsa. W. (*ye.*) hoary. ——— 1570. 7. 8. H. ♄. Moris. s. 6. t. 3. f. 17.
4 víridis. W. (*ye.*) dark-green. ——— 1727. 7. H. ♄.
5 rosmarinifòlia. W. (*ye.*) Rosemary-leaved. ——— 1683. 7. 9. H. ♄. Sm. exot. bot. 2. t. 62.
6 pectinàta. S.S. (*ye.*) pectinate. Spain. 1825. —— H. ♄.
7 viscòsa. S.S. (*ye.*) viscous. ——— —— —— H. ♄.
8 crithmifòlia. W.en. (*ye.*) Samphire-leaved. S. Europe. 1823. —— H. ♃.
9 pinnàta. Viv. (*wh.*) wing-leaved. ——— 1791. —— F. ♄.
10 anthemoídes. W. (*ye.*) Chamomile-leaved. ——— 1727. 7. 8. H. ♃. Flor. græc. t. 854.
11 alpìna. L. (*ye.*) alpine. ——— 1798. 7. 9. H. ♃. Barr. rar. t. 522.
ATHAN'ASIA. W. Athan'asia. Syngenesia Polygamia Æqualis. L.
1 crenàta. W. (*ye.*) crenated. C. B. S. 1816. 6. 8. G. ♄.
2 punctàta. W. (*ye.*) punctated. ——— 1822. 5. 7. G. ♄. Petiv. gaz. t. 81. f. 6.
3 capitàta. W. (*ye.*) hairy. ——— 1774. 1. 3. G. ♄. Moris. s. 6. t. 3. f. 48.
4 pubéscens. W. (*ye.*) villous-leaved. ——— 1768. 6. 8. G. ♄. Comm. hort. 2. t. 47.
5 canéscens. W. (*ye.*) canescent. ——— 1822. —— G. ♄. Cavan. ic. 1. t. 3.
lanuginòsa. Cav.
6 cuneifòlia. P.S. (*ye.*) wedge-leaved. ——— 1816. 7. 9. G. ♄. Lam. ill. t. 670. f. 3.
7 ánnua. W. (*ye.*) annual. Barbary. 1686. 7. 8. H. ⊙. Bot. mag. t. 2276.
8 dentàta. W. (*ye.*) tooth-leaved. C. B. S. 1759. —— G. ♄. Comm. rar. t. 41.
9 trifurcàta. W. (*ye.*) trifid-leaved. ——— 1710. —— G. ♄. Comm. hort. 2. t. 49.
10 virgàta. W. (*ye.*) twiggy. ——— 1815. —— G. ♄. Jacq. schœn. 2. t. 148.
11 tomentòsa. W. (*ye.*) Lavender-leaved. ——— 1774. 5. 6. G. ♄. Lam. ill. t. 670. f. 1.
cinèrea. L.
12 filifórmis. W. (*ye.*) fine-leaved. ——— 1787. 8. G. ♄.
13 crithmifòlia. W. (*ye.*) Samphire-leav'd. ——— 1723. 7. 8. G. ♄. Comm. hort. 2. t. 50.
14 parviflòra. W. (*ye.*) small-flowered. ——— 1731. 4. G. ♄. Jacq. schœn. 2. t. 149.
15 tricúspis. P.S. (*ye.*) three-pointed. ——— 1816. —— G. ♄.
16 pectinàta. W. (*ye.*) pectinated. ——— 1774. 5. 6. G. ♄.
17 pinnàta. W. (*ye.*) pinnated. ——— 1824. 6. 8. G. ♄. Lam. ill. t. 670. f. 4.
BALSAM'ITA. W. Costmary. Syngenesia Polygamia Æqualis. W.
1 virgàta. W. (*ye.*) twiggy. Italy. 1791. 6. 7. H. ♃. Jacq. obs. 4. t. 81.

2 ageratifòlia. w. (ye.) Ageratum-leav'd. Candia. 1605. 6. 10. G. ♄. Alp. exot. t. 326.
3 vulgàris. w. (ye.) common. Italy. 1568. 8. 9. H. ♃. Schk. hand. 3. t. 240.
4 ánnua. DC. (ye.) annual. S. Europe. 1629. 7. 8. H. ⊙. Mill. ic. 2. t. 227. f. 1.
Tanacètum ánnuum. w.

PE'NTZIA. Th. PE'NTZIA. Syngenesia Polygamia Æqualis. W.
flabellifórmis. w. (ye.) fan-leaved. C. B. S. 1774. 5. 8. G. ♄. Bot. mag. t. 212.
Tanacètum flabellifórme. B.M.

TANAC'ETUM. W. TANSY. Syngenesia Polygamia Superflua. L.
1 vestìtum. w. (ye.) imbricated. C. B. S. 1816. 8. 9. G. ♄.
2 linifòlium. w. (ye.) Flax-leaved. ——— 1774. 8. G. ♄.
3 suffruticòsum. w. (ye.) shrubby. ——— 1751. 5. 9. G. ♄. Comm. hort. 2. t. 100.
4 canariénse. DC. (ye.) Canary. Canaries. 1816. —— G. ♄.
5 incànum. s.s. (ye.) hoary. Levant. 1827. —— H. ♃.
6 sibíricum. w. (ye.) Siberian. Siberia. 1823. —— H. ♃. Gmel. sib. 2. t. 65. f. 2.
7 argénteum. w. (ye.) silvery. Levant. 1812. —— H. ♃.
8 grandiflòrum. w. (ye.) great-flowered. C. B. S. 1825. —— G. ♃.
9 myriophy'llum. w. (ye.) many-leaved. Levant. 1818. —— H. ♃.
Achillèa bipinnàta. L.
10 pauciflòrum. s.s. (ye.) few-flowered. N. America. 1827. —— H. ♃.
11 vulgàre. w. (ye.) common. Britain. .... —— H. ♃. Eng. bot. v. 18. t. 1229.
β críspum. (ye.) *curled.* .... —— H. ♃.

ARTEMI'SIA. W. WORMWOOD. Syngenesia Polygamia Superflua. L.
1 judàica. w. (ye.) Judean. Levant. 1688. 8. G. ♄. Pluk. alm. t. 73. f. 2.
2 valentìna. w. (ye.gr.) Spanish. Spain. 1739. 7. 8. F. ♄. Barr. ic. t. 485.
3 arragonénsis. w. (ye.gr.) snowy-white. ——— 1816. —— F. ♄. Asso arrag. t. 8. f. 1.
4 salsoloídes. s.s. (ye.gr.) saltwort-like. Siberia. 1827. —— F. ♄.
5 subcanéscens. w. (ye.gr.) hoary-leav'd. S. Europe. .... —— F. ♄.
6 Abrotànum. w. (ye.gr.) Southernwood. ——— 1548. 8. 10. H. ♄. Blackw. t. 555.
7 viridifòlia. s.s. (ye.gr.) green-leaved. Siberia. 1828. —— H. ♄.
8 pròcera. w. (ye.gr.) tall. S. Europe. 1816. —— F. ♄. Lobel. ic. 768.
paniculàta. Lam.
9 tenuifòlia. w. (ye.gr.) slender-leaved. .... 1732. 9. 12. G. ♄. Dill. elth. t. 33. f. 37.
10 áfra. w. (wh.gr.) Cape. C. B. S. 1816. —— G. ♄.
11 arboréscens. w. (ye.gr.) tree. Levant. 1640. 6. 8. H. ♄. Flor. græc. t. 856.
12 argéntea. w. (ye.) silvery. Madeira. 1777. 6. 7. G. ♄.
13 lednicénsis. s.s. (ye gr.) linear-lobed. Hungary. 1829. —— H. ♄.
14 pauciflòra. s.s. (ye.) few-flowered. Siberia. 1824. —— H. ♄.
15 glaciàlis. w. (ye.gr.) silky. Switzerland. 1739. 7. 8. H. ♃. Jacq. aust. 5. t. ap. 35.
16 rupéstris. w. (br.) stone. N. Europe. .... —— H. ♄. All. ped. t. 9. f. 1.
17 mutellìna. w. (ye.gr.) Alpine. Alps of Europ. 1815. —— H. ♃. Vill. daup. 3. t. 35.
18 spléndens. s.s. (ye.) glossy. Caucasus. 1827. —— H. ♃.
19 furcàta. s.s. (br.) forked. Siberia. 1829. —— H. ♃.
20 lanàta. s.s. (wh.ye.) woolly. S. Europe. 1827. —— H. ♃.
21 alpìna. M.B. (ye.gr.) alpine. Caucasus. ——— —— H. ♃.
22 jeniseénsis. s.s. (ye.gr.) leafy-panicled. Siberia. 1824. —— H. ♃.
23 pedemontàna. L.en. (br.) Swiss. Switzerland. 1823. —— H. ♃.
24 caucásica. w. (ye.gr.) Caucasian. Caucasus. 1804. 6. 7. H. ♃.
25 spicàta. w. (br.) spiked. Switzerland. 1790. —— H. ♃. Jacq. aust. 5. t. ap. 34.
26 norvègica. s.s. (br.) Norwegian. Norway. 1823. —— H. ♃. Flor. dan. t. 801.
rupéstris. Flor. dan. *nec aliorum.*
27 grœnlándica. s.s. (br.) Greenland. Greenland. 1827. —— H. ♃.
28 ludoviciàna. N. (br.) St. Louis. Missouri. ——— —— H. ♃.
29 desertòrum. s.s. (br.) desert. Tartary. 1817. 7. 9. H. ♃.
leucanthemifòlia. Hort.
30 pectinàta. w. (br.) comb-leaved. Dauria. 1806. —— H. ⊙. Pall. it. 3. t. Hh. f. 2.
31 tanacetifòlia. w. (br.) Tansy-leaved. Siberia. 1768. 7. 8. H. ♂. All. ped. 1. t. 10. f. 3.
32 laciniàta. w. (gr.) jagged. Dauria. 1823. —— H. ♃.
33 insípida. w. (br.) insipid. S. Europe. ——— —— H. ♃. Vill. delph. 3. t. 35.
34 mollíssima. D.P. (ye.) soft-leaved. Nepaul. ——— 8. 11. H. ♃.
35 potentillæfòlia. s.s. (gr.st.) Cinquefoil-leav'd. Siberia. 1827. 7. 9. H. ♃.
36 armenìaca. s.s. (gr.ye.) Armenian. Armenia. 1828. —— H. ♃.
37 canéscens. s.s. (gr.ye.) canescent. ——— ——— —— H. ♃.
38 Santònica. w. (wh.gr.) Tartarian. Siberia. 1596. 9. 11. H. ♄. Gmel. sib. 2. t. 51.
39 scopària. w. (wh.gr.) besom. Hungary. 1796. 7. 9. H. ⊙. W. et K. hung. 1. t. 65.
40 campéstris. w. (br.) field. England. .... 8. H. ♃. Eng. bot. v. 5. t. 338.
41 herbàcea. w. (ye.gr.) herbaceous. Siberia. 1817. —— H. ♃.
42 Marschalliàna. s.s. (gr.ye.) M. Bieberstein's. ——— 1800. 8. 9. H. ♃.
inodòra. M.B. *pauciflòra.* w.
43 virgàta. s.s. (br.) twiggy. N. America. 1827. 7. 9. H. ♃.

44 palústris. w. (*ye.*) marsh. Siberia. 1801. 7. 8. H. ⊙. Gmel. sib. 2. t. 55.
45 neglécta. W.en. (*gr.ye.*) neglected. ——— 1815. —— H. ♃.
46 camphoràta. w. (*ye.gr.*) Camphor-scent'd. S. Europe. 1820. —— H. ♄. Lob. ic. t. 769. f. 1.
47 álbida. W.en. (*st.*) white-leaved. Siberia. —— —— H. ♃.
48 hùmilis. W.en. (*ye.gr.*) dwarfish. Carniola. .... 8. 10. H. ♄.
49 crithmifòlia. w. (*br.*) Samphire-leav'd. Portugal. 1739. —— H. ♄.
50 saxátilis. w. (*wh.*) rock. Hungary. 1816. 6. 8. H. ♃.
51 glaùca. w. (*gr.*) glaucous. Siberia. 1806. —— H. ♃.
52 nùtans. w. (*br.gr.*) nodding. Tartary. 1819. 7. 9. H. ♃.
53 monog'yna. w. (*ye.*) one-styled. Hungary. 1816. —— H. ♃. W. et K. hung. 1. t. 75.
54 palmàta. w. (*gr.ye.*) palmated. S. Europe. 1739. 6. 7. G. ♄.
55 nívea. W.en. (*gr.ye.*) snowy. Siberia. 1815. —— H. ♃.
56 marítima. w. (*br.*) drooping-flower'd. Britain. .... 8. 9. H. ♃. Eng. bot. v. 24. t. 1706.
57 gállica. w. (*br.*) upright-flower'd. ——— .... —— H. ♃. ——— t. 1001.
58 salìna. w. (*wh.gr.*) sea-side. Germany. 1817. —— H. ♃. Spreng. halens. t. 12.
59 fràgrans. w. (*st.*) Lavender-leav'd. Armenia. 1739. 6. 7. H. ♃.
60 austrìaca. w. (*br.*) Austrian. Austria. 1597. 8. 10. H. ♃. Jacq. aust. 1. t. 100.
61 frígida. s.s. (*gr.ye.*) frigid. Siberia. —— —— H. ♃.
62 vallesìaca. w. (*st.*) downy. Italy. 1739. 7. 8. H. ♃.
63 taùrica. w. (*wh.gr.*) Taurian. Tauria. 1824. —— H. ♃.
64 orientàlis. w. (*br.*) oriental. Armenia. .... 8. 10. H. ♃.
65 serícea. w. (*wh.*) silky-leaved. ——— 1796. 6. 7. H. ♃. Gmel. sib. t. 64. f. 1.
66 rèpens. w. (*br.*) creeping. ——— 1805. —— H. ♃.
67 ramòsa. s.s. (*st.*) branching. Canaries. 1816. 6. 8. G. ♄.
68 bal-amìta. W.en. (*ye.*) Balsam-scented. .... —— —— H. ♃.
69 póntica. w. (*ye.*) Roman. Austria. 1570. 9. H. ♃. Jacq. aust. 1. t. 99.
70 chamæmelifòlia. w. (*br.*) Chamomile-leav'd. S. Europe. 1739. 7. 8. F. ♃. Vill. dauph. 3. t. 35.
71 ánnua. w. (*wh.gr.*) annual. Siberia. 1741. —— H. ⊙. Am. ruth. t. 196. f. 23.
72 biénnis. Ph. (*ye.gr.*) biennial. Missouri. 1804. —— H. ♂. Bot. mag. t. 2472.
73 Absínthium. w. (*ye.*) common. Britain. .... 7. 10. H. ♃. Eng. bot. v. 18. t. 1230.
74 Sieversiàna. w. (*br.gr.*) Sievers's. Siberia. 1800. 7. 8. H. ♂.
75 vulgàris. w. (*pu.*) Mugwort. Britain. .... 8. 9. H. ♃. Eng. bot. v. 14. t. 978.
76 procúmbens. DC. (*ye gr.*) procumbent. Siberia. 1823. —— H. ♃.
77 incìsa. W.en. (*ye pu.*) cut-leaved. .... —— —— H. ♃.
78 fasciculàta. M.B. (*ye.pu.*) fascicled. Iberia. 1820. —— H. ♃.
79 élegans. F. (*ye.*) elegant. China. —— 7. 9. H. ♃.
80 camtschática. F. (*gr.st.*) Kamtschatka. Kamtschatka. 1827. 8. 10. H. ♃.
81 sacròrum. Led. (*gr.st.*) sacred. Siberia. 1828. 7. 9. H. ♃.
82 Mertensiàna. Wb. (*gr.st.*) Mertens's. Hungary. 1826. —— H. ♃.
83 índica. w. (*ye.gr.*) Indian. E. Indies. 1796. 9. 10. F. ♃. Rheed. mal. 10. t. 45.
84 integrifòlia. w. (*ye.gr.*) entire-leaved. Siberia. 1759. 7. 8. H. ♃. Gm. sib. 2. t. 48. f. 1. 2.
85 japónica. w. (*wh.*) Japanese. Japan. 1804. 10. 11. H. ♃.
86 dracunculoídes. Ph. (*gr.*) Tarragon-like. N. America. 1811. 8. 10. H. ♃.
cérnua. N. *nùtans.* Fraser.
87 serràta. N. (*gr.*) saw-leaved. Missouri. 1827. —— H. ♃.
88 gnaphalioídes. N. (*br.*) Cudweeed-like. ——— —— —— H. ♃.
89 cæruléscens. w. (*ye.*) blueish. England. .... 8. 10. H. ♄. Eng. bot. v. 34. t. 2426.
90 inodòra. w. (*ye.*) inodorous. Siberia. 1548. 7. 8. H. ♃. Gm. sib. 2. t. 59&60. f. 1.
91 Dracúnculus. w. (*wh.gr.*) Tarragon. S. Europe. —— —— H. ♃.
92 chinénsis. w. (*ye.gr.*) Moxa. China. 1816. —— H. ♄. Gmel. sib. 2. t. 61. f. 1. 2.

ERIOCE'PHALUS. W. WOOLHEAD. Syngenesia Polygamia Necessaria. L.
1 africànus. w. (*ye.*) cluster-leaved. C. B. S. 1732. 1. 4. G. ♄. Bot. mag. t. 833.
frutéscens. H.K.
2 decussàtus. B.T. (*ye.*) cross-leaved. ——— 1816. 3. 6. G. ♄.
3 purpùreus. B.T. (*pu.*) purple. ——— —— —— G. ♄.
4 racemòsus. w. (*ye.*) silver-leaved. ——— 1739. 3. 4. G. ♄.
5 spinéscens. B.T. (*ye.*) spiny. ——— 1816. —— G. ♄.

HI'PPIA. W. HI'PPIA. Syngenesia Polygamia Necessaria. L.
1 frutéscens. w. (*ye.*) shrubby. C. B. S. 1710. 2. 8. G. ♄. Bot. mag. t. 1855.
2 integrifòlia. w. (*ye.*) annual. E. Indies. 1777. 7. 8. S. ⊙. Hout. n. hist. t. 67. f. 2.

SOL'IVA. L.T. SOL'IVA. Syngenesia Polygamia Necessaria.
1 anthemifòlia. (*gr.*) Chamomile-leav'd. S. America. 1812. 4. 12. H. ⊙. J. ann. mus. t. 61. f. 1.
Gymnóstyles anthemifòlia. P.S.
2 pterospérma. J. (*gr.*) wing-seeded. Brazil. 1827. 5. 9. H. ⊙.
3 stolonífera. L.T. (*gr.*) stoloniferous. Portugal? 1818. 4. 12. H. ⊙.
Híppia stolonífera. P.S. *Gymnóstyles lusitánica.* S.S.
4 minùta. (*ye.*) fringe-leaved. S. America. 1825. —— H. ⊙.
Híppia minùta. L. *Gymnóstyles minùta.* S.S.
5 pedicellàta. P.S. (*gr.*) pedicelled. S. America. —— —— H. ⊙.

LIDBE'CKIA. W.		LIDBE'CKIA.	Syngenesia Polygamia Superflua. W.				
1 lobàta. W.	(ye.)	lobe-leaved.	C. B. S.	1816.	5. 6.	G. ♄.	Lam. ill. t. 701. f. 3.
Lancísia lobàta. P.S. *Còtula quinquelòba.* L.							
2 pectinàta. W.	(ye.)	silver-leaved.	C. B. S.	1774.	5. 6.	G. ♄.	Berg. cap. 306. t. 5. f. 9.
C'ENIA. J.		C'ENIA.	Syngenesia Polygamia Superflua.				
turbinàta. P.S.	(wh.re.)	turbinated.	C. B. S.	1713.	7. 8.	H. ⊙.	Lam. ill. t. 701. f. 1.
Lidbéckia turbinàta. W. *Còtula turbinàta.* L.							
C'OTULA. W.		C'OTULA.	Syngenesia Polygamia Superflua. W.				
1 anthemoídes. W.	(ye.)	Anthemis-like.	St. Helena.	1696.	7. 8.	S. ⊙.	Dill. elth. t. 23. f. 25.
2 aùrea. W.	(ye.)	golden.	S. Europe.	1823.	7. 9.	H. ⊙.	
3 nudicaùlis. W.	(ye.)	naked-stalked.	C. B. S.	1816.	——	G. ♃.	Houtt. syst. t. 69. f. 4.
4 coronopifòlia. W.	(ye.)	Buck's Horn.	————	1683.	——	G. ⊙.	Lam. ill. t. 700. f. 1.
5 umbellàta. W.	(ye.)	umbel-flowered.	————	1822.	——	H. ⊙.	
6 viscòsa. W.	(wh.)	clammy.	Vera Cruz.	1739.	8.	S. ♃.	
7 tanacetifòlia. W.	(ye.)	Tansy-leaved.	C. B. S.	1780.	6. 8.	H. ⊙.	Pluk. man. t. 430. f. 7.
8 sericea. W.	(ye.)	silky.	————	1816.	——	G. ♃.	
9 sphæránthus. L. en.	(ye.)	round-headed.	Congo.	1820.	——	S. ♃.	
GRA'NGEA. J.		GRA'NGEA.	Syngenesia Polygamia Superflua.				
1 cuneifòlia. Lk.	(br.ye.)	wedge-leaved.	China.	1816.	7. 9.	H. ⊙.	Lam. ill. t. 699. f. 2.
2 mínima. W.	(ye.)	least.	————	1768.	——	H. ⊙.	Burm. ind. t. 58. f. 3.
3 decúmbens. Df.	(ye.)	decumbent.	N. Holland.	1815.	——	H. ⊙.	
4 maderaspatàna. J.	(ye.)	Madras.	E. Indies.	1780.	7. 8.	S. ⊙.	Lam. ill. t. 699. f. 3.
Còtula maderaspatàna. W.							
5 bícolor. Df.	(ye.)	two-coloured.	————	1804.	——	S. ⊙.	Lam. ill. t. 699. f. 1.
Còtula bícolor. W. *latifòlia.* P.S.							
6 cinèrea. L. en.	(ye.)	grey.	Egypt.	1825.	——	H. ⊙.	
ANAC'YCLUS. W.		ANAC'YCLUS.	Syngenesia Polygamia Superflua. W.				
1 créticus. W.	(wh.)	trailing.	Candia.	1759.	6. 8.	H. ⊙.	Ann. mus. 11. t. 22.
2 orientàlis. W.	(wh.)	oriental.	Levant.	1731.	——	H. ⊙.	Boer. lugd. 1. t. 110.
3 aùreus. W.	(ye.)	golden-flower'd.	————	1570.	——	H. ⊙.	Lam. ill. t. 700. f. 2.
4 radiàtus. L. en.	(ye.)	rayed.	————	1823.	——	H. ⊙.	
5 alexandrìnus. W.	(wh.)	Egyptian.	Egypt.	1828.	——	H. ⊙.	
6 clavàtus. P.S.	(wh.)	clavate.	Barbary.	1823.	——	H. ⊙.	
7 Pyrèthrum. L. en.	(w.)	Pellitory of Spain.	S. Europe.	1570.	6. 7.	H. ♃.	Bot. mag. t. 462.
A'nthemis Pyrèthrum. W.							
8 valentìnus. W.	(ye.)	fine-leaved.	Spain.	1656.	——	H. ⊙.	Schk. han. 3. t. 254. b.
9 purpuráscens. P.S.	(ye.cr.)	two-coloured.	S. Europe.	1829.	——	H. ⊙.	
10 officinàrum. Hayn.	(wh.)	officinal.	....	1828.	——	H. ⊙.	
11 inflàtus. Leh.	(wh.)	inflated.	Siberia.	1829.	——	H. ⊙.	
12 tomentòsus. Leh.	(wh.)	woolly.	————	——	——	H. ⊙.	
13 divaricàtus. Cav.	(wh.)	spreading.	————	1824.	——	H. ⊙.	
ZALUZ'ANIA. P.S.		ZALUZ'ANIA.	Syngenesia Polygamia Superflua.				
trilòba. P.S.	(ye.)	three-lobed.	Mexico.	1798.	7. 9.	F. ♄.	
A'nthemis trilòba. W.							
FERDINA'NDIA. LG.		FERDINA'NDIA.	Syngenesia Polygamia Superflua.				
1 augústa. LG.	(ye.)	august.	Mexico.	1816.	8. 10.	F. ♄.	
2 integrifòlia. D.D.	(ye.)	entire-leaved.	————	1828.	——	F. ♄.	
A'NTHEMIS. W.		CHAMOMILE.	Syngenesia Polygamia Superflua. L.				
1 rigéscens. W. en.	(wh.)	rigid.	Caucasus.	1805.	7. 9.	H. ♃.	W. hort. ber. 1. t. 62.
nigricans. Hort.							
2 Còta. W.	(wh.)	Venetian.	Italy.	1714.	7. 8.	H. ⊙.	Pluk. alm. t. 17. f. 5.
3 altíssima. W.	(wh.)	tall.	S. Europe.	1731.	8.	H. ⊙.	Flor. græc. t. 881.
4 punctàta. W.	(wh.)	dotted.	Levant.	1825.	7. 9.	G. ♄.	Desf. atl. 2. t. 139.
5 marítima. W.	(wh.)	sea.	England.	....	7. 8.	H. ⊙.	Eng. bot. v. 33. t. 2370.
6 tomentòsa. W.	(wh.)	downy.	Levant.	1795.	7. 10.	H. ♃.	
7 ruthénica. M.B.	(wh.)	Russian.	Tauria.	1823.	——	H. ⊙.	
8 fruticulòsa. M.B.	(wh.)	frutescent.	————	——	——	H. ♄.	Barrel. ic. t. 451.
9 pubéscens. W.	(wh.)	pubescent.	S. Europe.	1803.	7. 8.	H. ♃.	
10 biaristàta. S.S.	(wh.)	two-awned.	S. France.	1822.	——	H. ♃.	
11 revolùta. S.S.	(wh.)	revolute.	Canaries.	1816.	8. 10.	G. ♄.	
12 secundiràmea. S.S.	(w.)	secund-branch'd.	Sicily.	1824.	7. 9.	H. ⊙.	
13 míxta. W.	(wh.)	simple-leaved.	France.	1731.	7. 8.	H. ⊙.	Mich. gen. t. 30. f. 1.
14 Kitaibèlii.	(wh.)	Kitaibel's.	Hungary.	1807.	——	H. ♃.	
15 coronopifòlia. W.	(wh.)	Buckshorn-leav'd.	Spain.	——	——	H. ♃.	
16 austràlis. W.	(wh.)	procumbent.	S. France.	1818.	——	H. ♃.	
17 ibèrica. M.B.	(wh.)	Iberian.	Iberia.	1829.	——	H. ♃.	
18 Barrelièri. S.S.	(wh.)	Barrelier's	Naples.	——	8. 10.	H. ⊙.	
19 saxátilis. DC.	(wh.)	rock.	S. France.	1807.	7. 8.	H. ♃.	
20 alpìna. W.	(wh.)	Alpine.	Europe.	——	——	H. ♃.	Jacq. aust. 6. ap. t. 30.

21	carpática. w.	(wh.)	Carpathian.	Carpathian-m.	1818.	7. 8.	H. ♃.	
22	corymbòsa. w.	(wh.)	corymbose.	Austria.	1821.	——	H. ♃.	
23	retùsa. s.s.	(wh.)	retuse.	....	1824.	——	H. ♃.	
24	ramòsa. s.s.	(wh.)	branching.	....	——	9. 11.	H. ♃.	
25	mucronulàta. s.s.	(wh.)	mucronulate.	Italy.	1829.	8. 10.	H. ⊙.	
26	Chamomílla. W.	(wh.)	various-leaved.	S. Europe.	——	——	H. ♃.	
27	chìa. w.	(wh.)	cut-leaved.	Chio.	1731.	6. 10.	H. ⊙.	
28	nòbilis. w.	(wh.)	common.	Britain.	....	7. 9.	H. ♃.	Eng. bot. v. 14. t. 908.
29	arvénsis. w.	(wh.)	corn.	————	....	6. 8.	H. ♂.	———— v. 9. t. 602.
30	austrìaca. w.	(wh.)	Austrian.	Austria.	1759.	5. 8.	H. ⊙.	Jacq. aust. 5. t. 444.
31	Còtula. w.	(wh.)	stinking.	Britain.	....	6. 9.	H. ⊙.	Eng. bot. v. 25. t. 1772.
32	fállax. W.en.	(wh.)	false Chamomile.	————	——	——	H. ⊙.	
33	fuscàta. w.	(wh.)	brown-scaled.	Portugal.	1805.	7. 8.	H. ⊙.	
34	peregrina. w.	(wh.)	close-branched.	Piedmont.	1817.	——	H. ⊙.	
35	candidíssima. s.s.	(w.)	silvery-leaved.	Caucasus.	1812.	——	H. ♃.	
36	montàna. w.	(wh.)	mountain.	Italy.	1759.	7. 10.	H. ♃.	Ger. prov. t. 8.
37	incrassàta. s.s.	(wh.)	thick-stalked.	S. France.	1823.	——	H. ⊙.	
38	pedunculàta. s.s.	(wh.)	large-stalked.	Portugal.	——	8. 10.	H. ⊙.	
39	petr'æa. L.en.	(wh.)	long-stalked.	Naples.	——	——	H. ⊙.	
40	apiifòlia. B.R.	(wh.)	Parsley-leaved.	....	1764.	8. 9.	H. ♃.	Bot. reg. t. 527.
41	minuta. Leh.	(wh.)	minute.	....	1829.	——	H. ⊙.	
42	suffruticòsa. Leh.	(wh.)	suffrutescent.	....	——	6. 9.	H. ♄.	
43	globòsa. w.	(ye.)	globe.	Naples.	——	8. 9.	G. ♃.	Jacq. schœn. 3. t. 371.
44	valentìna. w.	(ye.)	purple-stalked.	S. Europe.	1596.	7. 8.	H. ⊙.	Breyn. cent. t. 75.
45	tinctòria. w.	(ye.)	Ox Eye.	Britain.	....	6. 11.	H. ♃.	Eng. bot. v. 21. t. 1472.
46	discoídea. w	(ye.)	saw-leaved.	Italy.	1800.	7. 8.	H. ⊙.	
47	Marschalliàna. w.	(ye.)	M. Bieberstein's.	Caucasus.	1828.	——	H. ♃.	
48	Rudolphiàna. M.B.	(ye.)	Rudolph's.	————	——	——	H. ♃.	
CLADA'NTHUS. Cas.			CLADA'NTHUS.	Syngenesia Polygamia Superflua.				
	arábicus. Cas.	(ye.)	Arabian.	Barbary.	1759.	7. 8.	H. ⊙.	Smith spic. 9. t. 10.
	A'nthemis arábica. w.							
MATRIC'ARIA. W.			MATRIC'ARIA.	Syngenesia Polygamia Superflua. L.				
1	suavèolens. w.	(wh.)	sweet.	Europe.	1781.	6. 8.	H. ⊙.	
2	Chamomílla. w.	(wh.)	Wild Chamomile.	Britain.	....	5. 7.	H. ⊙.	Eng. bot. v. 18. t. 1232.
3	capénsis. w.	(wh.)	Cape.	C. B. S.	1699.	7. 9.	G. ♂.	Seb. thes. 1. t. 16. f. 2.
4	pusílla. W.en.	(wh.)	small.	....	....	——	H. ⊙.	
ACHILL'EA. W.			MILFOIL.	Syngenesia Polygamia Superflua. L.				
1	lingulàta. w.	(wh.)	tongue-leaved.	Hungary.	1815.	7. 8.	H. ♃.	W. et K. hung. 1. t. 2.
2	Hérba-ròta. w.	(wh.)	Herbarota.	France.	1640.	6. 7.	H. ♃.	All. ped. 1. t. 9. f. 3.
3	grandiflòra. M.B.	(wh.)	great-flowered.	Caucasus.	1815.	7. 8.	H. ♃.	
4	Ptármica. w.	(wh.)	Sneezewort.	Britain.	....	7. 11.	H. ♃.	Eng. bot. v. 11. t. 757.
	β flòre plèno.	(wh.)	*double-flowered.*	————	....	——	H. ♃.	
5	cristàta. w.	(wh.)	slender-branch'd.	Italy.	1784.	7. 8.	H. ♃.	
6	Ageràtum. w.	(ye.)	Sweet Maudlin.	S. Europe.	1570.	8. 10.	H. ♃.	
7	decolòrans. W.en.	(st.)	pale yellow.	England.	1798.	6. 8.	H. ♃.	Eng. bot. v. 36. t. 2531.
	serràta. E.B. *nec aliorum.*							
8	speciòsa. W.en.	(wh.)	spear-leaved.	....	1804.	7. 9.	H. ♃.	
9	biserràta. M.B.	(wh.)	two-toothed.	Iberia.	1825.	——	H. ♃.	
10	salicifòlia. F.	(wh.)	Willow-leaved.	Siberia.	1827.	——	H. ♃.	
11	dentífera. Jac.	(wh.)	tooth-leaved.	....	1824.	——	H. ♃.	
12	alpìna. w.	(wh.)	Alpine.	Siberia.	1731.	7. 11.	H. ♃.	Bocc. mus. 144. t. 101.
13	serràta. w.	(wh.)	saw-leaved.	Switzerland.	1686.	8. 9.	H. ♃.	
14	coronopifòlia. w.	(st.)	buckshorn-leav'd.	Levant.	1816.	——	H. ♃.	
15	Clavénnæ. w.	(wh.)	silver-leaved.	Austria.	1656.	6. 7.	H. ♃.	Bot. mag. t. 1287.
16	Thomasiàna. DC.	(wh.)	Thomas's.	S. France.	1828.	——	H. ♃.	
17	impátiens. w.	(wh.)	impatient.	Siberia.	1759.	6. 9.	H. ♃.	Gmel. sib. 2. t. 83. f. 1.
18	pectinàta. w.	(st.)	comb-leaved.	Hungary.	1801.	8. 9.	H. ♃.	W. et K. hung. 1. t. 34.
	ochroleùca. W.K. *non* w.							
19	Gérberi. w.	(ye.)	various-leaved.	Siberia.	1816.	——	H. ♃.	Gmel. sib. 2. t. 83. f. 2.
20	pilòsa. w.	(ye.)	hairy.	....	1826.	——	H. ♃.	
21	squarròsa. w.	(wh.)	rough-headed.	....	1775.	7. 8.	H. ♃.	
22	falcàta. w.	(wh.)	sickle-leaved.	Levant.	1739.	6. 9.	H. ♃.	Lam. ill. t. 683. f. 3.
23	santolinoídes. LG.	(w.)	Santolina-like.	Spain.	1824.	——	H. ♃.	
24	tenuifòlia. w.	(ye.)	slender-leaved.	Levant.	1733.	6. 8.	H. ♄.	
25	Santolìna. w.	(ye.)	Lavender-cotton-leav'd.	—	1759.	——	H. ♃.	Flor. græc. t. 891.
26	álbida. s.s.	(st.)	white woolly.	....	1819.	6. 10.	H. ♃.	
27	anthemoídes. w.	(ye.)	Chamomile-like.	....	....	——	H. ♃.	
28	decúmbens. s.s.	(ye.)	decumbent.	Kamtschatka.	1800.	6. 8.	H. ♃.	
29	atràta. w.	(wh.)	black-cupped.	Austria.	1596.	7. 9.	H. ♃.	Jacq. aust. 1. t. 77.

30	moschàta. W.	(*wh.*)	musk.	Italy.	1775.	6. 7.	H. ♃.	Jacq. aust. 5. t. ap. 33
31	nàna. W.	(*wh.*)	dwarf.	———	1759.	6. 8.	H. ♃.	All. ped. 1. t. 9. f. 3.
32	crética. W.	(*wh.*)	Cretan.	Candia.	1739.	7. 8.	H. ♃.	Bocc. mus. t. 34.
33	ægyptìaca. W.	(*ye.*)	Egyptian.	Levant.	1640.	7. 9.	G. ♃.	Flor. græc. t. 892.
34	macrophy'lla. W.	(*wh.*)	large-leaved.	Italy.	1710.	7. 8.	H. ♃.	Triumf. obs. t. 23.
35	holoserícea. F.G.	(*wh.*)	silky-leaved.	Greece.	1823.	7. 9.	H. ♃.	
36	punctàta. T.N.	(*st.*)	dotted.	Naples.	1828.	——	H. ♃.	
37	leptophy'lla. M.B.	(*ye.*)	narrow-leaved.	Tauria.	1823.	——	H. ♃.	
38	taùrica. M.B.	(*st.*)	Taurian.	———	1829.	——	H. ♃.	
39	aùrea. W.	(*go.*)	golden-flower'd.	Levant.	1739.	6. 9.	H. ♃.	
40	Eupatòrium. W.	(*go.*)	Caspian.	Caspian Shores.	1803.	7. 8.	H. ♃.	Buxb. cent. 2. t. 19.
41	filipendulìna. S.S.	(*go.*)	short-rayed.	———	1812.	——	H. ♃.	
42	compácta. W.	(*ye.*)	compact.	....	——	——	H. ♃.	
43	pubéscens. W.	(*st.*)	downy.	Levant.	1739.	6. 9.	H. ♃.	
44	crithmifòlia. W.	(*wh.*)	Samphire-leav'd.	Hungary.	1804.	7. 8.	H. ♃.	W. et K. hung. 1. t. 66.
45	tanacetifòlia. W.	(*ro.*)	Tansy-leaved.	Switzerland.	1658.	——	H. ♃.	Moris. s. 6. t. 11. f. 14.
46	dístans. W.	(*wh.*)	branching.	Italy.	1804.	——	H. ♃.	All. ped. t. 53. f. 1.
47	lanàta. W. en.	(*wh.*)	woolly.	....	— —	——	H. ♃.	
48	mágna. W.	(*wh.*)	great.	S. Europe.	1683.	6. 11.	H. ♃.	Flor. græc. t. 896.
49	Millefòlium. W.	(*wh.*)	Yarrow.	Britain.	....	6. 10.	H. ♃.	Eng. bot. v. 758.
	β rùbra.	(*re.*)	*red-flowered.*	———	....	——	H. ♃.	
50	asplenifòlia. P.S.	(*ro.*)	Rose-coloured.	N. America.	1803.	6. 8.	H. ♃.	Vent. cels. t. 93.
51	micrántha W.	(*st.*)	small-flowered.	Levant.	1805.	6. 10.	H. ♃.	
52	tomentòsa. W.	(*ye.*)	tomentose.	S. Europe.	1648.	5. 10.	H. ♃.	Bot. mag. t. 498.
53	ochroleùca. W.	(*st.*)	cream-coloured.	....	1804.	7. 9.	H. ♃.	
54	montàna. L. en.	(*wh.*)	mountain.	Switzerland.	1798.	——	H. ♃.	
55	bannática. W.K.	(*wh.*)	Hungarian.	Hungary.	1823.	——	H. ♃.	
56	microphy'lla. W.	(*wh.*)	small-leaved.	Spain.	1800.	——	H. ♃.	Barr. ic. 1114.
57	ligústica. W.	(*wh.*)	Ligurian.	Italy.	1791.	6. 8.	H. ♃.	Flor. græc. t. 897.
58	nòbilis. W.	(*wh.*)	showy.	Germany.	1640.	——	H. ♃.	Schk. hand. 3. t. 255.
59	myriophy'lla. S.S.	(*wh.*)	many-leaved.	....	1798.	7. 9.	H. ♃.	
60	odoràta. W.	(*wh.*)	sweet-scented.	Spain.	1729.	6. 8.	H. ♃.	Jacq. col. 1. t. 21.
61	setàcea. W.	(*wh.*)	bristly.	Hungary.	1805.	5. 6.	H. ♃.	W. et K. hung. 7. t. 80.
62	abrotanifòlia. W.	(*ye.*)	Southernwood-leav'd.	Levant.	1739.	6. 8.	H. ♃.	
63	chamæmelifòlia. DC.	(*wh.*)	Chamomile-ld.	Pyrenees.	1820.	——	H. ♃.	
64	mongòlica. F.	(*wh.*)	Mongolic.	Tartary.	1823.	——	H. ♃.	
65	sibírica. Df.	(*wh.*)	Siberian.	Siberia.	1800.	——	H. ♃.	
66	ròsea. W.K.	(*ro.*)	rosy.	Hungary.	1803.	——	H. ♃.	
67	dùbia. F.	(*wh.*)	doubtful.	Siberia.	1800.	——	H. ♃.	
68	heterophy'lla. S.	(*wh.*)	different-leaved.	....	1804.	7. 9.	H. ♃.	
69	intermèdia. S.	(*wh.*)	intermediate.	....	1824.	——	H. ♃.	
70	paradóxa. L.C.	(*wh.*)	uncertain.	....	1822.	——	H. ♃.	
	CHRYSA'NTHEMUM. W.		CHRYSA'NTHEMUM.	Syngenesia Polygamia Superflua. L.				
1	atrátum. W.	(*wh.*)	fleshy-leaved.	Austria.	1731.	7. 8.	H. ♃.	
2	Leucánthemum. W.		Ox-eye Daisy.	Britain.	....	6. 7.	H. ♃.	Eng. bot. v. 9. t. 601.
3	montànum. W.	(*wh.*)	mountain.	France.	1759.	——	H. ♃.	Jacq. obs. 4. t. 91.
4	heterophy'llum. W.	(*wh.*)	various-leaved.	Switzerland.	1806.	7. 8.	H. ♃.	
5	sylvéstre. W. en.	(*wh.*)	wood.	Portugal.	1804.	6. 7.	H. ♃.	
6	hy'bridum. Gus.	(*st.*)	hybrid.	Sicily.	1827.	——	H. ♃.	
7	paludòsum. W.	(*wh.*)	marsh.	Barbary.	1817.	8. 9.	H. ♃.	Desf. atl. 2. t. 238.
8	rotundifòlium. W.	(*wh.*)	round-leaved.	Hungary.	1819.	7. 8.	H. ♃.	
9	mexicànum. K.S.	(*wh.*)	Mexican.	Mexico.	1827.	8. 11.	H. ♃.	
10	graminifòlium. W.	(*wh.*)	Grass-leaved.	Montpelier.	1739.	5. 7.	H. ♃.	Jacq. obs. 4. t. 92.
11	integrifòlium. S.S.	(*wh.*)	entire-leaved.	N. America.	1827.	6. 8.	H. ♃.	
12	perpusíllum. S.S.	(*wh.*)	small.	Corsica.	——	——	H. ♃.	
13	anómalum. P.S.	(*wh.*)	anomalous.	Spain.	1824.	7. 9.	H. ♄.	
14	daucifòlium. P.S.	(*wh.*)	carrot-leaved.	....	——	——	H. ♃.	
15	monspeliénse. W.	(*wh.pu.*)	Montpelier.	———	1739.	6. 9.	H. ♃.	Jacq. obs. 4. t. 93.
16	Achille'æ. W.	(*wh.*)	Milfoil-leaved.	Italy.	1775.	6. 8.	H. ♃.	Mich. gen. 34. t. 29.
17	argénteum. W.	(*wh.*)	silver-leaved.	Levant.	1731.	7. 8.	H. ♃.	
18	árcticum. W.	(*bh.*)	northern.	Kamtschatka.	1801.	6. 8.	H. ♃.	Willd. hort. ber. t. 33.
19	carinàtum. W.	(*wh.pu.*)	three-coloured.	Barbary.	1796.	7. 10.	H. ☉.	Bot. mag. t. 508.
20	pùmilum. W. en.	(*wh.*)	small.	C. B. S.	1806.	——	H. ☉.	
21	coronàrium. W.	(*wh.ye.*)	garden.	Sicily.	1629.	——	H. ☉.	Lam. ill. t. 678. f. 6.
	α álbum.	(*wh.*)	*white-flowered.*	———	——	——	H. ☉.	
	β aùreum.	(*ye.*)	*golden.*	———	——	——	H. ☉.	
22	ségetum. W.	(*ye.*)	corn.	Britain.	....	6. 8.	H. ☉.	Eng. bot. v. 8. t. 540.
23	Mycònis. W.	(*ye*)	tongue-leaved.	Italy.	1775.	7. 8.	H. ☉.	Jacq. obs. 4. t. 94.
24	itálicum. W.	(*ye.*)	Italian.	———	1796.	6. 7.	H. ♃.	

25 tripartìtum.B.F.G.(*ye.*)three-parted. China. 1820. 10. 12. F.♃. Swt. br. fl. gar. t. 193.
26 índicum. P.S. (*ye.*) Indian. ——— 1819. —— G.♄. Hort. trans. v. 4. pl. 12.
27 sinénse. L.T. (*va.*) variable chinese. ——— 1790. —— H.♄.
A'nthemis artemisiæfòlia. W. *Chrysánthemum índicum.* B.M. *non* L.
β purpùreum. (*pu.*) *old purple.* ——— —— —— H.♄. Bot. mag. t. 327.
γ variábile. (*wh.*) *changeable white.* ——— 1802? —— H.♄. ——— t. 2042.
δ tubulòsum álbum. (*wh.*) *quilled white.* ——— 1808. —— H.♄. Bot. reg. t. 4.
ε supérbum. (*wh.*) *superb white.* ——— 1817. —— H.♄. ——— t. 455.
ζ tessellàtum. (*wh.*) *tasselled white.* ——— 1816. —— H.♄.
η tubulòsum lùteum. (*ye.*) *quilled yellow.* ——— 1802. —— H.♄.
ϑ sulphùreum. (*st.*) *straw-coloured.* ——— —— —— H.♄.
ι aùreum. (*go.*) *golden yellow.* ——— —— —— H.♄. Bot. reg. t. 4. f. sup.
κ díscolor. (*li.*) *large lilac.* ——— 1808. —— H.♄.
λ lilacìnum. (*pi.*) *pink or lilac.* ——— 1798. —— H.♄.
μ cùpreum. (*co.*) *buff or copper-colour'd.* ——— —— —— H.♄.
ν fúlvum. (*br.*) *Spanish brown.* ——— 1806. —— H.♄.
ξ flámmeum. (*ye.*) *quilled flame yellow.* ——— 1819. —— H.♄. Hort. trans. v. 4. pl. 14.
ο tubulòsum ròseum. (*li.*) *quilled pink.* ——— —— —— H.♄. Bot. reg. t. 616.
π atropurpùreum. (*cr.*) *early crimson.* ——— 1820. —— H.♄. Hort. trans. v. 5. pl. 3. inf.
ρ aurantìacum. (*or.*) *large quilled orange.* ——— —— —— H.♄. ——— v. 5. pl. 3. f. sup.
σ expánsum. (*li.*) *expanded light purple.* ——— —— —— H.♄. Bot. mag. t. 2556.
τ purpuráscens. (*li.*) *quilled light purple.* ——— —— —— H.♄.
υ involùtum. (*li.*) *curled lilac.* ——— —— —— H.♄. Swt. br. fl. gar. t. 7.
φ fasciculàtum. (*ye.*) *superb clustered yellow.* —— —— —— H.♄. ——— t. 14.
χ tubulòsum cárneum. *semidouble quilled pink.* —— —— —— H.♄. Hort. tr. v. 5. pl. 17*. inf.
ψ álbum semidùplex. *semidouble quilled white.* —— —— H.♄.
ω tubulòsum aurántium. *semidouble quilled orange.* —— —— H.♄. Hort. tr. v. 5. pl. 17**. inf.
αα serotìnum. (*p. pu.*) *late pale purple.* ——— —— —— H.♄.
ββ salmòneum. (*sa.*) *quilled salmon-colour'd.* —— —— —— H.♄. Hort. t. v. 5. pl. 17**. sup.
γγ párvulum. (*ye.*) *small yellow.* ——— —— —— H.♄. Hort. t. 5. pl. 17**. f. sup.
δδ papyràceum. (*wh.*) *paper white.* ——— 1821. —— H.♄.
εε pállidum. (*li.*) *late pale pink.* ——— 1822. —— H.♄.
ζζ chrysocòmum. (*ye.*) *tasselled yellow.* ——— 1824. —— H.♃.
ηη Waratáh. (*ye.*) *yellow Waratah.* ——— —— —— H.♃.
ϑϑ Sabìni. (*ye.*) *golden Lotus.* ——— —— —— H.♃.
ιι chryseídes. (*ye.*) *double Indian yellow.* ——— —— —— H.♃.
κκ Párkii. (*ye.*) *Park's small yellow.* ——— —— —— H.♃.
λλ pállens. (*p. or.*) *semidouble quilled pale orange.* —— —— H.♃.
μμ stramíneum. (*st.*) *pale buff.* ——— —— —— H.♃.
νν mutábile. (*st.*) *changeable pale buff.* ——— —— —— H.♃.
ξξ bícolor. (*v.*) *two-coloured incurved.* ——— —— —— H.♃.
οο versícolor. (*re.*) *two-coloured red.* ——— —— —— H.♃.
ππ stellàtum. (*pu.*) *starry purple.* ——— —— —— H.♃.
ρρ ornàtum. (*li.*) *tasselled lilac.* ——— —— —— H.♃.
σσ fulvéscens. (*pu.*) *brown purple.* ——— —— —— H.♃.
ττ verecúndum. (*bh.*) *early blush.* ——— —— —— H.♃.
υυ blándum. (*bh.*) *blush.* ——— —— —— H.♃.
φφ leucánthum. (*wh.*) *double Indian white.* ——— —— —— H.♃.

CENTROSPE'RMUM. S.S. CENTROSPE'RMUM. Syngenesia Polygamia Superflua.
chrysánthemum.s.s.(*ye.*)yellow-flower'd.Spain. 1821. 8. 10. H.⊙.
Heteránthemis hírta. Schot.

PYR'ETHRUM. W. FEVERFEW. Syngenesia Polygamia Superflua. W.
1 fœniculàceum.W.en.(*wh.*)Fennel-leav'd.Teneriffe. 1815. 1. 12. G.♄. Bot. reg. t. 272.
2 crithmifòlium.W.en.(*wh.*)Samphire-leav'd. ——— —— —— G.♄.
3 anethifòlium.W.en. (*wh.*)Dill-leaved. ——— —— —— G.♄.
4 frutéscens.W.en.(*wh.*) shrubby. Canaries. 1699. —— G.♄.
5 coronopifòlium.W.en.(*wh.*)Buckshorn-leav'd. —— .... —— G.♄.
6 grandiflòrum.W.en.(*wh.*)great-flowered. ——— 1815. —— G.♄.
7 speciòsum.W.en.(*wh.*) large-flowered. ——— —— —— G.♄.
8 Broussonètii.s.s.(*wh.*) Broussonet's. ——— 1829. 7. 10. G.♄.
9 pinnatífidum.L.en.(*wh.*)cut-leaved. Madeira. 1778. 5. 8. G.♄.
Chrysánthemum pinnatífidum. S.S.
10 ptarmicæfòlium.w.(*wh.*)Sneezewort-leav'd.Caucasus. 1803. 7. 8. H.♃.
11 serotìnum. w. (*wh.*) creeping-rooted.N. America. 1731. 9. 10. H.♃. Jacq. obs. 4. t. 90.
12 uliginòsum. w. (*wh.*) marsh. Hungary. 1816 7. 9. H.♃. Bot. mag. t. 2706.
13 alpìnum. w. (*wh.*) Alpine. Switzerland. 1759. 7. 8. H.♃.
14 saxátile. B.C. (*wh.*) rock. ——— 1816. —— H.♃. Lodd. bot. cab. t. 126.
15 Barrelièri. Lk. (*wh.*) Barrelier's. Alps Europe. .... —— H.♃.
16 Hallèri. W.en. (*wh.*) Haller's. Switzerland. 1820. —— H.♃. Barrel. ic. 458. f. 2.

17	ceratophylloídes. W.	(*wh.*) Stags-horn-ld.	Piedmont.	1808.	6. 7.	H. ♃.	All. ped. t. 37. f. 1.
	Chrysánthemum ceratophylloídes. All.						
18	simplicifòlium. W.	(*wh.*) simple-leaved.	S. America.	1825.	8. 10.	H. ⊙.	
19	máximum. DC.	(*wh.*) broad-leaved.	Pyrenees.	1816.	5. 8.	H. ♃.	
	latifòlium. W.						
20	Balsamìta. W.	(*wh.*) various-leaved.	Levant.	1779.	——	H. ♃.	Jacq. obs. 4. t. 89.
21	oppositifòlium. S.S.	(*wh.*) opposite-leaved.	Portugal.	1828.	7. 9.	H. ♃.	
22	tomentòsum. S.S.	(*wh.*) woolly.	Corsica.	1829.	——	H. ♃.	
23	palústre. W.	(*wh.*) marshy.	Armenia.	1824.	7. 10.	H. ⊙.	
24	macrophy'llum. W.	(*wh.*) large-leaved.	Hungary.	1803.	——	H. ♃.	W. et K. hung. 1. t. 94.
25	corymbòsum. W.	(*wh.*) mountain.	Germany.	1596.	6. 8.	H. ♃.	Jacq. aust. 4. t. 379.
26	pulveruléntum. W. en.	(*wh.*) powdered.	Iberia.	1806.	——	H. ♃.	
27	cinerarifòlium. S.S.	(*wh.*) Ragwort-leav'd.	Dalmatia.	1824.	——	H. ♃.	
28	Parthènium. W.	(*wh.*) common.	Britain.	....	6. 9.	H. ♃.	Eng. bot. v. 18. t. 1231.
	β *flòre plèno.*	(*wh.*) *double-flowered.*	————	....	——	H. ♃.	
29	parthenifòlium. W.	(*wh.*) Parthenium-leav'd.	Caucasus.	1804.	6. 7.	H. ♃.	Vent. cels. t. 43.
30	caucásicum. W.	(*wh.*) Caucasian.	————	——	7. 8.	H. ♃.	
31	tenuifòlium. T.N.	(*wh.*) slender-leaved.	Naples.	1806.	——	H. ♃.	
32	seríceum. M.B.	(*wh.*) silky.	Iberia.	1824.	——	H. ♃.	
33	inodòrum. W.	(*wh.*) scentless.	Britain.	....	8. 9.	H. ⊙.	Eng. bot. v. 10. t. 676.
34	marítimum. W.	(*wh.*) sea.	————	....	6. 10.	H. ♃.	———— v. 14. t. 979.
35	parviflòrum. W.	(*wh.*) small-flowered.	Caucasus.	——	6. 7.	H. ⊙.	
36	præcox. M.B.	(*wh.*) early-flowered.	————	1827.	5. 7.	H. ♃.	
37	incànum. Led.	(*wh.*) hoary.	Siberia.	——	7. 8.	H. ♃.	
38	glàbrum. LG.	(*wh.*) smooth.	Spain.	1825.	——	H. ♃.	
39	leptophy'llum. S.S.	(*wh.*) narrow-leaved.	Caucasus.	1829.	——	H. ♃.	
40	diversifòlium. H.E.F.	(*wh.*) different-leaved.	N. Holland.	1824.	5. 7.	G. ♄.	Hook. ex. flor. t. 215.
41	cárneum. M.B.	(*pi.*) flesh-coloured.	Caucasus.	1804.	8. 9.	H. ♃.	Bot. mag. t. 1080.
	Chrysánthemum coccíneum. B.M.						
42	ròseum. M.B.	(*ro.*) rose-coloured.	————	1818.	——	H. ♃.	
	Chrysánthemum coccíneum. W.						
43	achillæfòlium. M.B.	(*ye.*) Milfoil-like.	Cumana.	1824.	6. 7.	H. ⊙.	
44	millefoliàtum. W.	(*ye.*) Milfoil-leaved.	Siberia.	1731.	5. 9.	H. ♃.	Mill. ic. 1. t. 9.
45	bipinnàtum. W.	(*ye.*) wing-leaved.	————	1796.	6. 7.	H. ♃.	Gmel. sib. 2. t. 85. f. 1.
46	índicum. H.K.	(*ye.*) Indian.	E. Indies.	1810.	6. 9.	H. ⊙.	Bot. mag. t. 1521.

ERIOPHY'LLUM. LG. ERIOPHY'LLUM. Syngenesia Polygamia Superflua.

1	lanàtum.	(*ye.*) woolly.	N. America.	1827.	6. 8.	H. ♃.	Bot. reg. t. 1167.
	cæspitòsum. B.R. *Actinélla lanàta.* Ph. *Trichophy'llum lanàtum.* N. *Helènium lanàtum.* S.S.						
2	oppositifolium. N.	(*ye.*) opposite-leaved.	N. America.	1826.	——	H. ♃.	
3	trolliifòlium. LG.	(*ye.*) Trollius-leaved.	New Spain.	1828.	8. 10.	F. ♃.	

ORDO CXXI.

LOBELIACEÆ. *Juss. ann. mus.* 18. *p.* 1.

LOB'ELIA. R.S. LOB'ELIA. Pentandria Monogynia. R.S.

1	lineàris. R.S.	(*bl.*) linear-leaved.	C. B. S.	1791.	5. 8.	G. ♄.	
2	setàcea. R.S.	(*bl.*) bristle-leaved.	————	1816.	——	G. ♃.	Act. Gorenk. 1811. f. 11.
3	commutàta. R.S.	(*bl.*) notch-leaved.	————	——	——	G. ♄.	
4	símplex. R.S.	(*bl.*) simple-stalked.	————	1794.	——	G. ♂.	
5	unidentàta. R.S.	(*d.pu.*) single-toothed.	————	——	——	G. ♃.	Bot. mag. t. 1484.
6	pinifòlia. R.S.	(*bl.*) Pine-leaved.	————	1752.	——	G. ♃.	Andr. reposit. t. 240.
7	Dortmánna. R.S.	(*bl.*) water.	Britain.	....	7. 8.	H. *w.* ♃.	Eng. bot. v. 2. t. 140.
8	paludòsa. R.S.	(*bl.*) marsh.	N. America.	1823.	——	H. *w.* ♃.	
9	Tùpa. R.S.	(*sc.*) Mullein-leav'd.	Peru.	1824.	8. 10.	F. ♃.	Swt. br. fl. gar. t. 284.
10	salicifòlia. H.S.L.	(*or.*) Willow-leaved.	Chili.	1794.	6. 8.	G. ♄.	Bot. mag. t. 1325.
	Tùpa. H.K. *non* L. *gigántea.* B.M. *non.* Cav. *argùta.* B.R. t. 973.						
11	Kálmii. R.S.	(*bl.*) Kalm's.	N. America.	1820.	7. 9.	H. ♂.	Bot. mag. t. 2238.
12	Nuttállii. R.S.	(*bl.*) Nuttall's.	————	——	——	H. ♃.	
	grácilis. N. *nec aliorum.*						
13	pauciflòra. K.S.	(*bl.*) few-flowered.	Mexico.	1824.	——	F. ♃.	
14	chinénsis. R.S.	(*p.bl.*) Chinese.	China.	1822.	——	H. ⊙.	
15	bellidifòlia. R.S.	(*bl.*) Daisy-leaved.	C. B. S.	1790.	5. 8.	G. ♃.	
16	rhizoph'yta. R.S.	(*wh.*) rooting branched.	————	1814.	——	G. ♃.	Bot. mag. t. 2519.

17 decúmbens. B.M.	(bl.)	decumbent.	C. B. S.	1820.	5. 8.	G. ♃.	Bot. mag t. 2277.
18 triquètra. R.S.	(bl.)	triangular.	——	1774.	——	G. ♃.	
19 cinèrea. R.S.	(wh.)	gray.	——	1824.	6. 10.	H. ⊙.	
20 fenestràlis. R.S.	(bl.)	windowed.	Mexico.	1826.	9. 11.	H. ⊙.	Cavan. icon. 6. t. 512. f. 1.
21 gigántea. R.S.	(or.)	gigantic.	S. America.	1828.	8. 11.	S. ℏ.	—— 6. t. 513.
22 longiflòra. R.S.	(wh.)	long-flowered.	Jamaica.	1752.	5. 8.	S. ♃.	Bot. mag. t. 2563.
23 tomentòsa. R.S.	(bl.)	tomentose.	C. B. S.	1816.	——	G. ♃.	
24 secúnda. R.S.	(wh.)	side-flowering.	——	1790.	——	G. ♃.	
25 trialàta. D.P.	(bl.)	three-winged.	Nepaul.	1820.	——	G. ♃.	Hook. ex. flor. t. 44.
micrántha. H.E.F. *non* K.S.							
26 alàta. B.P.	(bl.)	winged-stalked.	N. S. W.	1804.	——	G. ♃.	Lab. nov. hol. 1. t. 72.
27 acuminàta. R.S.	(st.)	taper-pointed.	Jamaica.	1822.	6. 9.	S. ♄.	Sloan. his. 1. t. 95. f. 2.
28 nicotianæfòlia. R.S.	(st.)	Tobacco-leaved.	E. Indies.	——	——	S. ♃.	
29 circiifòlia. R.S.	(st.)	Cirsium-leaved.	W. Indies.	1825.	8. 10.	S. ♃.	Plum. sp. v. 4. t. 116.
30 persicifòlia. R.S.	(pu.)	Peach-leaved.	——	1818.	6. 9.	S. ♃.	
31 Cavanillesiàna. R.S.	(re.)	Cavanilles'.	New Spain.	1825.	——	S. ♃.	Cavan. ic. 6. t. 518.
persicifòlia. Cav. *non* Lam.							
32 Claytoniàna. R.S.	(p.bl.)	Clayton's.	N. America.	1799.	6. 8.	H. ♃.	Willd. hort. ber. t. 30.
goodenioìdes. W. *pállida.* Roth.							
33 assúrgens. R.S.	(pu.)	purple.	W. Indies.	1787.	6. 10.	S. ♄.	Andr. reposit. t. 553.
34 pyramidàlis. D.P.	(pu.)	pyramidal.	Nepaul.	1819.	7. 10.	G. ♃.	Bot. mag. t. 2387.
35 fúlgens. R.S.	(cr.)	fulgent.	Mexico.	1809.	5. 9.	F. ♃.	Bot. reg. t. 165.
36 spléndens. R.S.	(sc.)	splendid.	——	1814.	——	F. ♃.	—— t. 60.
37 cardinàlis. R.S.	(sc.)	Cardinal flower.	Virginia.	1629.	——	F. ♃.	Bot. mag. t. 320.
38 syphilítica. R.S.	(bl.)	blue cardinal.	——	1665.	8. 10.	H. ♃.	Bot. reg. t. 537.
39 débilis. R.S.	(bl.)	feeble.	C. B. S.	1774.	4. 8.	H. ⊙.	
40 grácilis. B.P.	(bl.)	slender.	N. S. W.	1801.	7. 10.	G. ⊙.	Bot. mag. t. 741.
41 purpuráscens. B.P.	(pu.)	purplish.	——	1809.	6. 8.	G. ♃.	
42 surinaménsis. R.S.	(bh.)	Surinam.	W. Indies.	1786.	1. 7.	S. ♄.	Bot. mag. t. 225.
β *rùbra.*	(re.)	*red-flowered.*	Brazil.	1820.	12. 4.	S. ♄.	Lodd. bot. cab. t. 749.
43 Cliffortiàna. S.S.	(pu.)	purple-flowered.	S. America.	1733.	7. 8.	H. ⊙.	Lin. h. clif. 426. t. 26.
44 Michaùxii. N.	(pu.)	Michaux's.	N. America.	....	——	H. ⊙.	
Cliffortiàna. M. *non* L.							
45 inflàta. R.S.	(bl.)	bladder-podded.	——	1759.	——	H. ⊙.	Swt. br. fl. gar. t. 99.
46 ùrens. R.S.	(bl.)	acrid.	England.	....	6. 7.	H. ♃.	Eng. bot. v. 14. t. 953.
47 pubérula. R.S.	(bl.)	downy.	N. America.	1819.	——	H. ♃.	
48 am'œna. R.S.	(bl.)	bright blue.	——	1812.	6. 8.	H. ♃.	Ann. mus. 18. t. 1. n. 1.
49 minùta. R.S.	(bh.)	small.	C. B. S.	1772.	6. 9.	G. ♃.	Bot. mag. t. 2590.
50 mínima. B.M.	(wh.)	least.	——	....	6. 10.	G. ♃.	—— t. 2077.
51 Lauréntia. R.S.	(bl.)	Italian.	Italy.	1778.	7. 8.	H. ⊙.	Mich. gen. 18. t. 14.
52 tenélla. R.S.	(vi.)	slender-leaved.	Sicily.	1820.	6. 8.	H. ⊙.	
53 campanuloídes. R.S.	(bl.)	campanulate.	Japan.	1819.	7. 9.	G. ♃.	Bot. reg. t. 733.
54 Erìnus. R.S.	(bl.)	ascending.	C. B. S.	1752.	6. 9.	H. ♃.	Bot. mag. t. 901.
55 Bre'ynii. R.S.	(bl.)	Breynius's.	——	1823.	——	G. ♃.	Breyn. cent. t 89.
56 erinoídes. R.S.	(bl.)	trailing.	——	1759.	6. 8.	G. ♂.	Herm. lugd. t. 109.
57 bícolor. R.S.	(bl.wh.)	spotted.	——	1795.	——	H. ⊙.	Bot. mag. t. 514.
58 Schránkii.	(bl.)	Schrank's.	——	1824.	——	G. ♃.	Schrank. h. mon. t. 31.
secúnda. Sck. *non* Thunb. *decúmbens.* L. enum. *non* B.M.							
59 corymbòsa. B.M.	(wh.pu.)	corymbose.	C. B. S.	1825.	——	G. ♃.	Bot. mag. t. 2693.
60 ilicifòlia. B.M.	(wh.pu.)	Holly-leaved.	——	1815.	5. 9.	G. ♃.	—— t. 1896.
61 ánceps. R.S.	(bl.)	flat-edged.	——	1824.	7. 10.	H. ⊙.	
62 pubéscens. R.S.	(wh.)	downy-leaved.	——	1780.	6. 8.	G. ♃.	Jacq. schœn. 2. t. 178.
63 hirsùta. R.S.	(bl.)	blue-flower'd hairy.	——	1759.	5. 9.	G. ♃.	Burm. afr. t. 40. f. 2.
64 zeylánica. R.S.	(pu.)	Ceylon.	E. Indies.	1820.	——	S. ♃.	
65 lùtea. R.S.	(ye.)	yellow.	C. B. S.	1774.	6. 7.	G. ♃.	Bot. mag. t. 1319.
66 variifòlia. B.M.	(ye.)	various-leaved.	——	1812.	——	G. ♃.	—— t. 1692.
67 coronopifòlia. R.S.	(bl.)	Buckshorn-leaved.	——	1752.	7. 8.	G. ♃.	—— t. 644.
68 cœrùlea. B.M.	(bl.)	blue-flowered.	——	1823.	5. 8.	G. ♄.	—— t. 2701.
69 Thunbérgii.	(bl.)	upright-raceme-flower'd.		1812.	7. 10.	G. ♃.	
coronopifòlia. Th. *non* B.M.							
70 crenàta. R.S.	(bl.)	notched-leaved.	——	1794.	4. 5.	G. ♃.	
71 Símsii.	(bl.)	long-peduncled.	——	1819.	8. 11.	G. ♃.	Bot. mag. t. 2251.
pedunculàta. B.M. *non* B.P.							
72 verbascifòlia. Sm.	(pu.)	Mullein-leaved.	Nepaul.	1822.	5. 7.	G. ♃.	
73 racemòsa. B.M.	(gr.)	racemed.	W. Indies.	1818.	7. 9.	S. ♄.	Bot. mag. t. 2137.
74 críspa. Gr.	(bl.)	curled-leaved.	Mexico.	1825.	5. 6.	F. ♃.	
75 heteromàlla. Jac.	(bl.)	woolly-leaved.	....	1829.	....	G. ♃.	
76 móllis. Gr.	(pu.)	soft.	Dominica.	1828.	5. 8.	S. ♃.	
77 rugulòsa. Gr.	(wh.)	rugged.	New Zealand.	——	——	G. ♃.	

78 cóncolor. B.P.	(*bl.*)	self-coloured.	N. S. W.	1828.	5. 6.	G. ♃.	
79 inundàta. B.P.	(*bl.*)	inundated.	———	1822.	5. 8.	G.*w.* ♃.	
80 pedunculàta. B.P.	(*bl.*)	long-stalked.	———	——	——	G. ♃.	
81 dentàta. B.P.	(*bl.*)	toothed.	———	1819.	——	G. ♃.	Cavan. ic. 6. t. 522.
82 gibbòsa. B.P.	(*bl.*)	gibbous.	———	1824.	——	G. ♃.	Lab. nov. hol. 1. t. 71.
83 senecioídes. B.M.	(*bl.*)	Groundsel-like.	———	——	7. 10.	G. ♄.	Bot. mag. t. 2702.
Isotòma axillàris. Bot. reg. t. 964.							
MONO'PSIS. Sal.		Mono'psis.	Pentandria Monogynia.				
1 conspícua. H.T.	(*bl.*)	conspicuous.	C. B. S.	1812.	7. 8.	H. ⊙.	Bot. mag. t. 1499.
Lobèlia spéculum. B.M.							
2 inconspícua. H.T.	(*bl.*)	inconspicuous.	———	....	——	H. ⊙.	
CLINT'ONIA. B.R.		Clint'onia.	Pentandria Monogynia.				
élegans. B.R.	(*bl.*)	elegant.	Columbia.	1827.	6. 9.	H. ⊙.	Bot. reg. t. 1241.
C'YPHIA. R.S.		C'yphia.	Pentandria Monogynia. W.				
1 volùbilis. R.S.	(*pi.*)	climbing.	C. B. S.	1795.	6. 8.	S. ♃. ◡.	
2 digitàta. R.S.	(*li.*)	finger-leaved.	———	1820.	7. 10.	G. ♃.	
3 incìsa. R.S.	(*bh.*)	cut-leaved.	———	——	——	G. ♃.	
4 bulbòsa. R.S.	(*pi.*)	bulb-bearing.	———	1791.	8. 9.	S. ♃.	Burm. afr. t. 38. f. 2.
5 Phyteùma. R.S.	(*pi.*)	Phyteuma-leav'd.	———	1820.	2. 5.	S. ♃.	Bot. reg. t. 625.
6 Cardámines. R.S.	(*bh.*)	Cardamine-leaved.	———	——	——	S. ♃.	

ORDO CXXII.

STYLIDEÆ. *Brown prodr.* 1. *p.* 565.

STYLI'DIUM. B.P.		Styli'dium.	Gynandria Diandria.				
1 graminifòlium. B.P.	(*ro.*)	grass-leaved.	N. S. W.	1803.	4. 8.	G. ♃.	Bot. reg. t. 90.
2 lineàre. B.P.	(*li.*)	linear-leaved.	———	1820.	——	G. ♃.	Sm. exot. bot. 2. t. 67.
Ventenàtia mìnor. Ex. b.							
3 scándens. B.P.	(*wh.*)	climbing.	N. Holland.	1803.	7. 8.	G. ♃. ◡.	
4 fruticòsum. B.P.	(*wh.*)	frutescent.	———	——	5. 10.	G. ♄.	Sal. par. lond. t. 77.
glandulòsum. P.L.							
5 laricifòlium. B.R.	(*ro.*)	Larch-leaved.	N. S. W.	1818.	——	G. ♄.	Bot. reg. t. 550.
tenuifòlium. B.M. 2249.							
6 adnàtum. B.P.	(*bh.*)	adnate.	N. Holland.	1824.	——	G. ♄.	Bot. reg. t. 914.

ORDO CXXIII.

GOODENOVIÆ. *Brown prodr.* 1. *p.* 573.

GOOD'ENIA. B.P.		Good'enia.	Pentandria Monogynia. W.				
1 bellidifòlia. B.P.	(*ye.*)	Daisy-leaved.	N. S. W.	1823.	6. 9.	G. ♃.	
2 stellígera. B.P.	(*ye.*)	starry-haired.	———	1825.	——	G. ♃.	
3 paniculàta. B.P.	(*ye.*)	panicled.	———	——	——	G. ♃.	Cavan. ic. 6. t. 507.
4 grácilis. B.P.	(*ye.*)	slender.	N. Holland.	1824.	5. 8.	G. ♃.	Lodd. bot. cab. t. 1032.
5 decúrrens. B.P.	(*ye.*)	decurrent.	N. S. W.	1825.	——	G. ♃.	
6 ovàta. B.P.	(*ye.*)	oval-leaved.	———	1793.	6. 10.	G. ♄.	Andr. reposit. t. 68.
7 grandiflòra. B.P.	(*ye.*)	large-flowered.	———	1802.	6. 8.	G. ♄.	Bot. mag. t. 890.
8 hederàcea. B.P.	(*ye.*)	Ivy-leaved.	———	1823.	——	G. ♃.	
E'UTHALES. B.P.		E'uthales.	Pentandria Monogynia. R.S.				
trinérvis. B.P.	(*ye.pu.*)	three-nerved.	N. Holland.	1803.	5. 9.	G. ♃.	Bot. mag. t. 1137.
Goodènia tenélla. A.R. 466. *Velleia trinérvis.* Lab. n. hol. t. 77.							
VELLE'IA. B.P.		Velle'ia.	Pentandria Monogynia. R.S.				
1 paradóxa. B.P.	(*ye.*)	paradoxical.	N. S. W.	1823.	6. 8.	G. ♃.	Bot. reg. t. 971.
2 lyràta. B.P.	(*ye.*)	lyrate-leaved.	———	1819.	4. 6.	G. ♃.	——— t. 551.
3 spathulàta. B.P.	(*ye.*)	spathulate.	———	1823.	——	G. ♃.	
LECHENA'ULTIA. B.P.		Lechena'ultia.	Pentandria Monogynia. R.S.				
1 formòsa. B.P.	(*sc.*)	handsome.	N. Holland.	1823.	6. 9.	G. ♄.	Swt. flor. aust. t. 26.
2 oblàta. S.F.A.	(*ye.*)	oblate-petaled.	———	——	——	G. ♄.	——— t. 46.
SC'ÆVOLA. B.P.		Sc'ævola.	Pentandria Monogynia. R.S.				
1 Kœnígii. R.S.	(*wh.*)	Konig's.	E. Indies.	1818.	8. 10.	S. ♄.	Bot. mag. t. 2732.

2 Plumièri. R S.	(*wh.*)	Plumier's.	W. Indies.	1724.	8. 10.	S. ♄.	Jacq. amer. t. 179. f. 88.
Lobèlia Plumièri. Jacq.							
3 ivæfòlia. Lht.	(*wh.*)	Iva-leaved.	Trinidad.	1820.	——	S. ♄.	
4 Taccàda. F.I.	(*wh.*)	Indian.	E. Indies.	1818.	——	S. ♄.	Gært. fruct. 1. t. 25.
5 crassifòlia. B.P.	(*bl.*)	thick-leaved.	N. Holland.	1805.	8. 10.	G. ♄.	Lab. nov. hol. 1. t. 79.
6 cuneifórmis. B.P.	(*bl.*)	wedge-leaved.	V. Diem. Isl.	1823.	——	G. ♃.	—————— 1. t. 80.
7 microcárpa. B.P.	(*bl.*)	small-fruited.	N. S. W.	1790.	5. 9.	G. ♃.	Cavan. ic. 6. t. 509.
Goodènia lævigàta. B.M. 287.							
8 suavèolens. B.P.	(*bl.*)	sweet-scented.	———	1798.	——	G. ♃.	Andr. reposit. t. 22.
Goodènia calendulàcea. A.R.							
9 híspida. B.P.	(*li.*)	branching.	N. S. W.	1820.	5. 8.	G. ♃.	Cavan. ic. 6. t. 510.
Goodènia ramosíssima. Sm. new holl. 15. t. 5.							
DAMPI'ERA. B.P.		DAMPI'ERA.	Pentandria Monogynia. R.S.				
1 undulàta. B.P.	(*bl.*)	wave-leaved.	N. S. W.	1824.	5. 8.	G. ♄.	
2 ovalifòlia. B.P.	(*pu.*)	oval leaved.	———	1823.	5. 9.	G. ♄.	
3 purpùrea. B.P.	(*pu.*)	purple-flowered.	———	1825.	——	G. ♄.	
4 strícta. B.P.	(*bl.*)	upright.	N. S. W.	1814.	6. 8.	G. ♃.	

ORDO CXXIV.

CAMPANULACEÆ. *Juss. gen.* 163.

CANAR'INA. W.		CANAR'INA.	Hexandria Monogynia. W.				
Campánula. W.	(*or.*)	Canary.	Canaries.	1696.	1. 3.	G. ♃.	Bot. mag. t. 444.
MICHA'UXIA. W.		MICHA'UXIA.	Octandria Monogynia. W.				
1 campanuloídes. W.	(*li.*)	rough-leaved.	Levant.	1787.	6. 8.	G. ♂.	Bot. mag. t. 219.
2 decándra. F.		decandrous.	Persia.	1829.	....	H. ♂?	
LIGHTFO'OTIA. R.S.		LIGHTFO'OTIA.	Pentandria Monogynia. R.S.				
1 oxycoccoídes. R.S.	(*pa.*)	lance-leaved.	C. B. S.	1787.	7.	G. ♄.	Sm. exot. bot. 2. t. 69.
2 tenélla. B.C.	(*pa.*)	slender.	———	1824.	6. 8.	H. ♄.	Lodd bot. cab. t. 1038.
3 lanceolàta. L.en.	(*w.bl.*)	spear-leaved.	———	1826.	——	G. ♃.	
4 muscòsa. L.en.	(*wh.bl.*)	moss-like.	———	——	——	G. ♃.	
5 subulàta. R.S.	(*bl.*)	awl-leaved.	———	——	——	G. ♃.	L'Herit. s. angl. 4. t. 5.
ADENO'PHORA. F.		ADENO'PHORA.	Pentandria Monogynia.				
1 verticillàta. F.	(*bl.*)	whorled.	Dahuria.	1783.	6. 8.	H. ♃.	Pall. it. 3. t. G. f. 1.
2 marsupiiflòra. F.	(*bl.*)	bellied.	———	1815.	7. 8.	H. ♃.	Bot. reg. t. 149.
Campánula coronàta. B.R.							
3 Gmelìni. F.	(*bl.*)	Gmelin's.	———	1823.	5. 8.	H. ♃.	Gmel. sib. 3. t. 33.
4 coronopifòlia. F.	(*bl.*)	buckshorn-leav'd.	———	1822.	——	H. ♃.	Swt. br. fl. gar. t. 104.
5 denticulàta. F.	(*bl.*)	toothed.	———	——	5. 7.	H. ♃.	———— t. 116.
Campánula tricuspidàta. R.S.							
6 latifòlia. F.	(*bl.*)	broad-leaved.	Siberia.	1827.	——	H. ♃.	
Campánula pereskiæfòlia. R.S.							
7 Lamárckii. F.	(*bl.*)	Lamarck's.	———	1826.	——	H. ♃.	Gmel. sib. 3. n. 18. t. 26.
8 stylòsa. F.	(*bl.*)	long-styled.	———	1820.	——	H. ♃.	———— 3. n. 20. t. 27.
9 reticulàta. F.	(*bl.*)	netted.	———	1829.	——	H. ♃.	
10 commùnis. F.	(*p.bl.*)	common.	———	1784.	5. 9.	H. ♃.	Jacq. schœnb. 3. t. 335.
Campánula liliifòlia. B.R.							
β *suavèolens.* R.S.	(*p.bl.*)	*sweet-scented.*	Hungary.	1818.	——	H. ♃.	Bot. reg. t. 236.
γ *hy'brida.* R.S.	(*p.bl.*)	*upright-flower'd.*		——	——	H. ♃.	
11 intermèdia.	(*p.bl.*)	intermediate.	Siberia.	1820.	——	H. ♃.	
Campánula intermèdia. R.S.							
WAHLENBE'RGIA. Sc.		WAHLENBE'RGIA.	Pentandria Monogynia.				
1 grandiflòra. Sc.	(*bl.*)	great-flowered.	Siberia.	1782.	6. 8.	H. ♃.	Bot. mag. t. 252.
Campánula grandiflòra. B.M.							
2 péndula. Sc.	(*bh.*)	pendulous.	Canaries.	1777.	7. 8.	H. ⊙.	
Campánula lobelioídes. R.S.							
3 elongàta. Sc.	(*p.bl.*)	elongated.	....	1816.	——	H. ⊙.	
CAMPA'NULA. R.S.		BELL-FLOWER.	Pentandria Monogynia. L.				
1 cenísia. R.S.	(*bl.*)	ciliated.	Switzerland.	1775.	6. 7.	H. ♃.	All. ped. 1. t. 6. f. 2.
2 uniflòra. R.S.	(*bl.*)	single-flowered.	Lapland.	1827.	— —	H. ♃.	Linn. fl. lap. t. 9. f. 5. 6.
3 Kitaibeliàna. R.S.	(*bl.*)	Kitaibel's.	Hungary.	1828.	——	H. ♃.	
4 Bellárdi. R.S.	(*bl.*)	Bellardi's.	Italy.	1813.	——	H. ♃.	All. ped. t. 85. f. 5.
5 Stevènii. M.B.	(*bl.*)	Steven's.	Siberia.	1828.	——	H. ♃.	
6 púlla. R.S.	(*d.pu.*)	dark-flowered.	Austria.	1779.	——	H. ♃.	Bot. mag. t. 2492.

7 Zóysii. R.S.	(*bl.*)	blunt-leaved.	Carniola.	1813.	6. 8.	H. ♃.	Jacq. ic. 2. t. 334.
8 diffùsa. R.S.	(*bl.*)	diffuse.	Italy.	1818.	——	H. ♃.	
β frágilis. L.en.	(*bl.*)	*brittle.*	——	——	——	H. ♃.	Cyr. pl. rar. n. t. 11. f. 2.
9 carpáthica. R.S.	(*bl.*)	Carpathian.	Carp. Alps.	1774.	——	H. ♃.	Bot. mag. t. 117.
10 dasycárpa. R.S.	(*bl.*)	fleshy-podded.	Hungary.	1827.	——	H. ♃.	
11 rotundifòlia. R.S.	(*bl.*)	round-leaved.	Britain.	....	——	H. ♃.	Eng. bot. v. 13. t. 866.
α cærùlea.	(*bl.*)	*blue-flowered.*	——	....	——	H. ♃.	
β albiflòra.	(*wh.*)	*white-flowered.*	——	....	——	H. ♃.	
12 excìsa. R.S.	(*bl.*)	cut-flowered.	Switzerland.	1816.	——	H. ♃.	Lodd. bot. cab. t. 561.
13 pusílla. R.S.	(*bl.wh.*)	small.	Europe Alps.	....	——	H. ♃.	Scheuch. it. 4. f. 4.
14 pùmila. R.S.	(*bl.wh.*)	dwarf.	Switzerland.	....	——	H. ♃.	Bot. mag. t. 512.
α cærùlea.	(*bl.*)	*blue-flowered.*	——	....	——	H. ♃.	
β álba.	(*wh.*)	*white-flowered.*	——	....	——	H. ♃.	
15 cæspitòsa. R.S.	(*bl.*)	tufted.	Carniola.	1818.	——	H. ♃.	Scop. carn. ed. 2. t. 4.
16 pubéscens. R.S.	(*bl.*)	pubescent.	Bohemia.	1813.	——	H. ♃.	
17 grácilis. B.M.	(*bl.*)	slender.	N. S. W.	1794.	4. 8.	G. ♂.	Bot. mag. t. 691.
18 erécta.	(*bl.*)	upright.	——	——	——	G. ♂.	Sm. exot. bot. t. 45.
stricta. Sm. *non* L.							
19 littoràlis. Lab.	(*bl.*)	sea-side.	V. Diem. Isl.	1820.	——	G. ♂.	Lab. nov. hol. 1. t. 70.
20 capillàris. B.C.	(*bl.*)	capillary.	N. Holland.	1824.	——	G. ♂.	Lodd. bot. cab. t. 1406.
21 quadrífida. B.P.	(*bl.*)	quadrifid.	N. S. W.	1822.	——	G. ♂.	
22 saxícola. B.P.	(*bl.*)	small rock.	V. Diem. Isl.	1825.	5. 10.	G. ♂.	
23 Waldsteiniàna. R.S.	(*vi.*)	Waldstein's.	Hungary.	——	6. 8.	H. ♃.	W. et K. hung. t. 136.
24 lanceolàta. R.S.	(*bl.*)	spear-leaved.	Pyrenees.	1827.	——	H. ♃.	All. ped. t. 47. f. 2.
25 divaricàta. R.S.	(*bl.*)	divaricate.	N. America.	——	——	H. ♃.	
26 virgàta. R.S.	(*bl.*)	twiggy.	——	1823.	——	H. ♃.	
27 linifòlia. R.S.	(*bl.*)	Flax-leaved.	Sweden.	1816.	——	H. ♃.	Vill. delph. 1. t. 10.
28 Scheuchzèri. S.S.	(*bl.*)	Scheuchzer's.	Alps Europe.	1813.	——	H. ♃.	Lodd. bot cab. t. 485.
29 pátula. R.S.	(*bl.*)	spreading.	Britain.	....	7. 8.	H. ♂.	Eng. bot. v. 1. t. 42.
30 neglécta. R.S.	(*bl.*)	neglected.	Europe.	1818.	——	H. ♃.	
crenàta. L.en. *ucrànica*. Hort.							
31 Rapúnculus. R.S.	(*bl.*)	Rampion.	Britain.	....	——	H. ♃.	Eng. bot. v. 4. t. 283.
32 persicifòlia. R.S.	(*va.*)	Peach-leaved.	Europe.	1596.	7. 9.	H. ♃.	Flor. dan. t. 1087.
α cærùlea.	(*bl.*)	*blue-flowered.*	Scotland.	....	——	H. ♃.	Flor. græc. t. 205.
β álba.	(*wh.*)	*white-flowered.*	Europe.	....	——	H. ♃.	
γ máxima. B.M.	(*bl.*)	*large-flowered.*	——	....	——	H. ♃.	Bot. mag. t. 397.
δ álbo-plèna.	(*wh.*)	*double white.*	——	....	——	H. ♃.	
ε cærùleo-plèna.	(*bl.*)	*double blue.*	——	....	——	H. ♃.	
33 pyramidàlis. R.S.	(*bl.*)	pyramidal.	Carniola.	1596.	——	H. ♃.	
α cærùlea.	(*bl.*)	*common blue.*	——	——	——	H. ♃.	
β albiflòra.	(*wh.*)	*white flowered.*	——	1816.	——	H. ♃.	
34 versícolor. R.S.	(*bl.*)	various-color'd.	Greece.	1788.	——	H. ♃.	Andr. reposit. t. 396.
planiflòra. W.en. *non* Lam. *Willdenowiàna*. R.S.							
35 oblíqua. R.S.	(*bl.*)	oblique-flower'd.	N. Europe.	1813.	6. 7.	H. ♂.	Jacq. schœn. 3. t. 336.
36 americàna. R.S.	(*bl.*)	American.	N. America.	1763.	7.	H. ♂.	
37 nítida. H.K.	(*bl.wh.*)	glossy.	——	1731.	7. 8.	H. ♃.	Dodart. mem. 4. t. 113.
α cærùlea.	(*bl.*)	*blue-flowered.*	——	——	——	H. ♃.	
β álba.	(*wh.*)	*white-flowered.*	——	——	——	H. ♃.	
38 aùrea. R.S.	(*ye.*)	golden-flower'd.	Madeira.	1777.	7. 9.	G. ♄.	Bot. reg. t. 57.
α latifòlia. R.S.	(*ye.*)	*broad-leaved.*	——	——	——	G. ♄.	Duham. ed. nov. t. 41.
β angustifòlia. R.S.	(*ye.*)	*narrow-leaved.*	——	——	——	G. ♄.	Jacq. schœn. 4. t. 472.
39 rhomboídea. R.S.	(*bl.*)	Germander-ld.	Switzerland.	1775.	7.	H. ♃.	Barrel. ic. 567.
40 asteroídes. R.S.	(*bl.*)	star-flowered.	....	1824.	——	H. ♃.	
41 Alpìni. R.S.	(*bl.*)	Alpinus'.	Italy.	1823.	7. 9.	H. ♃.	
42 Ottoniàna. R.S.	(*bl.*)	Otto's.	C. B. S.	1822.	5. 7.	G. ♄.	
43 undulàta. R.S.	(*bl.*)	waved.	——	——	——	G. ♃.	
44 latifòlia. R.S.	(*bl.*)	giant.	Britain.	....	7.	H. ♃.	Eng. bot. v. 5. t. 302.
α cærùlea.	(*bl.*)	*blue-flowered.*	——	....	——	H. ♃.	
β álba.	(*wh.*)	*white-flowered.*	——	....	——	H. ♃.	Flor. dan. t. 85.
45 macránthа. L.en.	(*bl.*)	great-flowered.	Russia.	1822.	6. 7.	H. ♃.	Bot. mag. t. 2553.
46 eriocárpa. R.S.	(*bl.*)	woolly-capsul'd.	Caucasus.	1824.	——	H. ♃.	
47 urticifòlia. R.S.	(*bl.*)	Nettle-leaved.	Germany.	1800.	8.	H. ♃.	
β flòre plèno.	(*wh.*)	*double white.*	——	....	——	H. ♃.	
48 Trachèlium. R.S.	(*bl.*)	throatwort.	Britain.	....	6. 8.	H. ♃.	Eng. bot. v. 1. t. 12.
α cærùlea.	(*bl.*)	*blue-flowered.*	——	....	——	H. ♃.	
β álba.	(*wh.*)	*white-flowered.*	——	....	——	H. ♃.	
γ múltiplex.	(*bl.wh.*)	*double-flowered.*	——	....	——	H. ♃.	
49 rapunculoídes. R.S.	(*bl.*)	creeping.	England.	....	6. 7.	H. ♃.	Eng. bot. v. 20. t. 1369.
infundibulifórmis. Bot. mag. t. 2632.							

50	ucrànica. R.S.	(*bl.*)	Ukraine.	Ukraine.	1819.	6. 7.	H. ♃.	
51	trachelioídes. R.S.	(*bl.*)	throatwort-like.	Caucasus.	1818.	——	H. ♃.	
52	bononiénsis. R.S.	(*bl.*)	panicled.	Italy.	1773.	8. 9.	H. ♃.	Moris. s. 5. t. 4. f. 38.
53	símplex. DC.	(*bl.*)	simple-stalked.	————	1823.	7. 8.	H. ♃.	
54	lychnìtis. H.H.	(*bl.*)	Lychnis-like.	....	——	——	H. ♃.	
55	obliquifòlia. R.S.	(*bl.*)	oblique-leaved.	Italy.	1820.	——	H. ♃.	Ten. fl. neap. t. 17.
56	trichocalycìna. R.S.	(*bl.*)	long-pointed-calyx'd.	——	1824.	——	H. ♃.	———— t. 16.
57	ruthènica. R.S.	(*bl.*)	Russian.	Russia.	1815.	6. 8.	H. ♃.	Bot. mag. t. 2653.
58	graminifòlia. R.S.	(*vi.*)	grass-leaved.	Hungary.	1823.	——	H. ♃.	W. et K. hung. 2. t. 154.
59	tenuifòlia. R.S.	(*vi.*)	slender-leaved.	————	1820.	5. 9.	H. ♃.	———— t. 155.
60	glomeràta. R.S.	(*bl.*)	clustered.	Britain.	....	——	H. ♃.	Eng. bot. v. 2. t. 90.
	β *albiflòra.*	(*wh.*)	*white-flowered.*	————	....	——	H. ♃.	
61	ellíptica. R.S.	(*bl.*)	elliptic-leaved.	Hungary.	1819.	——	H. ♃.	Bocc. mus. 70. t. 58.
62	cephalántha. F.	(*p.bl.*)	pale-flowered.	Siberia.	1824.	——	H. ♃.	
63	farinòsa. Bes.	(*bl.*)	mealy.	Poland.	——	——	H. ♃.	
64	aggregàta. R.S.	(*bl.*)	crowded-flower'd.	Bavaria.	——	——	H. ♃.	
65	speciòsa. R.S.	(*bl.*)	specious.	Siberia.	——	——	H. ♃.	Bot. mag. t. 2649.
	glomeràta β dahùrica. Bot. reg. 620.							
66	nicæénsis. R.S.	(*bl.*)	Nice.	Italy.	1820.	——	H. ♃.	All. ped. t. 39. f. 1.
67	foliòsa. R.S.	(*bl.*)	leafy.	————	——	6. 9.	H. ♃.	Ten. fl. neap. t. 18.
68	petr'æa. R.S.	(*bl.*)	woolly-leaved.	Mt. Baldo.	1818.	——	H. ♃.	
69	Rainèri. R.S.	(*bl.*)	Rainer's.	Italy.	1825.	——	H. ♃.	
70	Cervicària. R.S.	(*bl.*)	wave-leaved.	Germany.	1768.	7.	H. ♂.	Flor. dan. t. 787.
71	cervicarioídes. R.S.	(*bl.*)	intermediate.	Italy.	1819.	——	H. ♃.	
72	multiflòra. R.S.	(*vi.*)	many-flowered.	Hungary.	1814.	6. 7.	H. ♂.	W. et K. hung. t. 263.
	macrostàchya. W.en.							
73	collìna. R.S.	(*bl.*)	Sage-leaved.	Caucasus.	1803.	7. 8.	H. ♃.	Bot. mag. t. 927.
74	azùrea. R.S.	(*bl.*)	azure.	Switzerland.	1778.	6. 7.	H. ♃.	———— t. 551.
75	Lòrei. R.S.	(*bl.*)	Pollini's.	Mt. Baldo.	1819.	6. 9.	H. ⊙.	———— t. 2581.
76	lactiflòra. R.S.	(*st.*)	milk-coloured.	Caucasus.	1814.	7. 9.	H. ♃.	Bot. reg. t. 241.
77	thyrsoídea. R.S.	(*st.*)	long-spiked.	Switzerland.	1785.	6. 8.	H. ♂.	Bot. mag. t. 1290.
78	peregrìna. R.S.	(*bl.*)	rough-leaved.	C. B. S.	1794.	——	G. ♂.	———— t. 1257.
79	tomentòsa. R.S.	(*wh.*)	tomentose.	Levant.	1820.	6. 9.	H. ♃.	
80	aparinoídes. R.S.	(*p.bl.*)	Aparine-like.	N. America.	1825.	7. 9.	H. ⊙.	
81	cérnua. R.S.	(*wh.*)	nodding-flower'd.	C. B. S.	1804.	——	G. ♂.	
82	capénsis. R.S.	(*vi.*)	Cape.	————	1803.	——	G. ⊙.	Bot. mag. t. 782.
	Roélla decúrrens. And. rep. t. 238.							
83	paniculàta. R.S.	(*wh.*)	panicled.	C. B. S.	1822.	6. 8.	G. ♄.	
84	spathulàta. F.G.	(*bl.*)	spathulate.	Greece.	1824.	——	G. ♃.	Flor. græc. t. 203.
85	velutìna. R.S.	(*p.bl.*)	velvetty.	Gibraltar.	——	——	F. ♃.	Desf. atl. 1. t. 51.
86	microphy'lla. R.S.	(*p.bl.*)	small-leaved.	Spain.	1820.	——	H. ♃.	
87	Alliòni. R.S.	(*p.bl.*)	Allioni's.	S. Europe.	1818.	——	H. ♃.	All. ped. n. 418. t. 6. f. 3.
88	liguláris. R.S.	(*bl.*)	strap-leaved.	Alps.	1825.	——	H. ♃.	
89	barbàta. R.S.	(*p.bl.*)	bearded.	Italy.	1752.	——	H. ♃.	Bot. mag. t. 1258.
90	punctàta. R.S.	(*wh.*)	dotted-flower'd.	Siberia.	1813.	5. 6.	H. ♃.	———— t. 1723.
91	Mèdium. R.S.	(*bl.*)	Canterbury-bell.	Germany.	1597.	6. 9.	H. ♂.	Knor. delic. 1. t. 9.
	β *álbida.*	(*wh.*)	*white-flowered.*	————	....	——	H. ♂.	Garid. aix. p. 76. t. 18.
92	longifòlia. R.S.	(*bl.*)	long-leaved.	Pyrenees.	1824.	——	H. ♂.	Lapeyr. pyren. t. 6.
93	betonicifòlia. R.S.	(*bl.*)	Betony-leaved.	Olympus.	1823.	——	H. ♂.	Flor. græc. t. 210.
94	dichótoma. R.S.	(*bl.*)	forked.	Levant.	1827.	7. 10.	H. ⊙.	Swt. br. fl. gar. t. 280.
95	spicàta. R.S.	(*bl.*)	spiked.	Switzerland.	1786.	7.	H. ♂.	All. ped. 1. t. 46. f. 2.
96	alpìna. R.S.	(*pa.bl.*)	Alpine.	————	1779.	——	H. ♃.	Bot. mag. t. 957.
97	móllis. R.S.	(*bl.*)	soft.	Syria.	1788.	5. 8.	G. ♃.	———— t. 404.
98	saxátilis. R.S.	(*bl.*)	rock.	Candia.	1768.	——	G. ♃.	Barrel. ic. t. 813.
99	alliariæfòlia. R.S.	(*wh.*)	Alliaria-leaved.	Caucasus.	1803.	8. 9.	H. ♃.	Sal. par. lond. t. 26.
	macrophy'lla. B.M. t. 912.							
100	sarmática. B.R.	(*pa.bl.*)	gum-bearing.	Siberia.	——	6. 8.	H. ♃.	Bot. reg. t. 237.
	gummífera. R.S. *betonicæfòlia.* M.B.							
101	lamiifòlia. R.S.	(*st.*)	Lamium-leaved.	Caucasus.	1818.	——	H. ♃.	Buxb. cent. 5. t. 18.
102	péndula. R.S.	(*st.*)	pendulous.	————	1824.	——	H. ♃.	
103	sibírica. R.S.	(*bl.*)	Siberian.	Siberia.	1783.	7. 9.	H. ♃.	Bot. mag. t. 659.
104	divérgens. R.S.	(*bl.*)	divergent.	Hungary.	1814.	6. 7.	H. ♂.	W. et K. hung. t. 258.
	spathulàta. W.K.							
105	violæfòlia. R.S.	(*wh.*)	Violet-leaved.	Siberia.	1827.	6. 8.	H. ♃.	
106	A'dami. R.S.	(*bl.*)	Adam's.	Caucasus.	1829.	——	H. ♃.	
107	Pallasiàna. R.S.	(*bl.*)	Pallas's.	Siberia.	1827.	——	H. ♃.	
108	lingulàta. R.S.	(*vi.*)	tongue-leaved.	Hungary.	1804.	7. 8.	H. ♃.	W. et K. hung. 1. t. 64.
109	caucásica. R.S.	(*bl.*)	Caucasian.	Caucasus.	——	——	H. ♃.	
110	laciniàta. R.S.	(*bl.*)	jagged-leaved.	Greece.	1788.	5. 8.	G. ♃.	Andr. reposit. t. 385.

111 lyràta. R.S. (*bl.*) lyre-leaved. Levant. .... 6. 8. H. ♃. Moris. 2. v. 5. t. 3. f. 51.
112 rupéstris. R.S. (*bl.*) various-leaved. ——— 1827. —— H. ♂. Flor. græc. t. 213.
113 strícta. R.S. (*bl.*) straight. Egypt. —— —— H. ♃. Tour.an.mu.p.137.t.13.
114 cichoràcea. R.S. (*bl.*) Succory-leaved. Greece. 1824. —— F. ♂. Flor. græc. t. 209.
115 capitàta. R.S. (*bl.*) headed. ——— 1768. 6. 7. H. ♃. Bot. mag. t. 811.
116 lanuginòsa. R.S. (*bl.*) woolly. .... 1824. —— H. ♂.
117 Erìnus. R.S. (*pa.bl.*) forked. S. Europe. 1768. 7. 8. H. ⊙. Moris. s. 5. t. 3. f. 25.
118 drabifòlia. R.S. (*bl.*) Draba-leaved. Athens. 1823. —— H. ⊙. Flor. græc. t. 215.
119 erinoídes. R.S. (*bl.*) Erinus-like. Africa. —— —— H. ⊙. Herm. lugd. t. 111?
120 Elátines. R.S. (*pa.pu.*) trailing-stalked. S. Europe. —— 7. 9. H. ♃. All. ped. n.422. t. 7. f. 1.
121 hederàcea. R.S. (*pa.bl.*) Ivy-leaved. England. .... 5. 6. H. ♃. Eng. bot. v. 2. t. 73.
122 arvática. R.S. (*bl.*) Spanish. Spain. 1825. —— H. ♃. Pluk. phyt. t. 23. f. 1.
123 hispídula. R.S. (*bl.wh.*) hispid-calyxed. C. B. S. 1816. 7. 9. H. ⊙. Comm. hort. 2. t. 37.

PRISMATOCA'RPUS. L'H. VENUS-LOOKING-GLASS. Pentandria Monogynia.

1 fruticòsus. L'H. (*bl.*) shrubby. C. B. S. 1787. 8. G. ♄.
2 Hookèri. (*wh.*) Hooker's. ——— 1823. —— G. ♄. Bot mag. t. 2733.
Campánula Prismatocárpus. B.M. *nec* L'Her.
3 nítidus. L'H. (*bl.*) shining. ——— —— 6. 8. H. ⊙. L'Her. sert. ang. 2. t. 3.
Campánula Prismatocárpus. W.
4 hispídula. R.S. (*bl.wh.*) bristly. ——— 1822. —— G. ⊙. Comm. hort. 2. t. 37.
5 interrúptus. L'H. (*bl.*) interrupted. ——— 1820. —— G. ⊙.
6 Spéculum. L'H. (*bl.*) common. S. Europe. 1596. 5. 8. H. ⊙. Bot. mag. t. 102.
β *álbus.* (*wh.*) *white-flowered.* ——— —— —— H. ⊙.
7 hirsùtus. T.N. (*bl.*) hairy. Italy. 1822. —— H. ⊙. Ten. fl. neap. t. 19.
8 falcàtus. T.N. (*bl.*) falcate. ——— —— —— H. ⊙. ——— t. 20.
9 hy'bridus. L'H. (*bl.*) corn. England. .... —— H. ⊙. Eng. bot. v. 6. t. 375.
10 pentagònius. L'H. (*bl.*) five-angled. Turkey. 1686. —— H. ⊙. Bot. reg. t. 56.
11 perfoliàtus. (*bl.*) perfoliate. N. America. 1680. —— H. ⊙. Moris. s. 5. t. 2. f. 23.
Campánula perfoliàta. R.S. *amplexicaùlis.* M.

ROE'LLA. R.S. ROE'LLA. Pentandria Monogynia. L.

1 ciliàta. R.S. (*bl.wh.*) ciliate. C. B. S. 1774. 6. 9. G. ♄. Bot. mag. t. 378.
2 pedunculàta. R.S. (*bl.*) peduncled. ——— 1827. —— G. ♄.
3 filifórmis. R.S. (*bl.*) filiform. ——— 1816. —— G. ♄. Lam. ill. t. 123. f. 2.
4 decúrrens. R.S. (*bl.*) decurrent. ——— 1787. 7. 9. H. ⊙. L'Her. sert. ang. 4. t. 6.
5 squarròsa. R.S. (*bl.*) trailing. ——— —— 7. 8. G. ♃.
6 muscòsa. R.S. (*bl.*) moss-like. ——— 1802. 8. 10. H. ⊙.
7 spicàta. R.S. (*wh.*) spiked. ——— 1824. —— G. ♄.

PHYTE'UMA. R.S. PHYTE'UMA. Pentandria Monogynia. L.

1 pauciflòrum. R.S. (*bl.*) few-flowered. Germany. 1823. 6. 8. H. ♃.
2 globulariæfòlium. R.S. (*bl.*) Globularia-leav'd. S. Europe. —— —— H. ♃.
3 Scheuchzèri. R.S. (*bl.*) Scheuchzer's. Switzerland. 1813. 5. 6. H. ♃. Bot. mag. t. 1797.
4 scorzonerifòlium. R.S. (*bl.*) Scorzonera-leav'd. Europe. 1817. —— H. ♃. ——— t. 2271.
5 Michèlii. R.S. (*bl.*) Micheli's. Pyrenees. 1822. 6. 7. H. ♃. All. ped. n. 427. t. 7. f. 3.
6 hemisph'æricum. R.S. (*bl.*) linear-leaved. Switzerland. 1752. 7. H. ♃. Jacq. ic. 2. t. 333.
7 comòsum. R.S. (*da.bl.*) tufted. Austria. —— 6. 7. H. ♃. Jacq. aust. ap. t. 50.
8 orbiculàre. R.S. (*bl.*) round-headed. England. .... 6. 8. H. ♃. Eng. bot. v. 2. t. 142.
9 Charmèlii. R.S. (*bl.*) long-bracted. Pyrenees. 1819. —— H. ♃. Vill. delph. 2. t. 11. f. 3. A.
10 Siebèrii. S.S. (*bl.*) Sieber's. Levant. 1827. 7. 8. H. ♃.
11 inæquàtum. R.S. (*bl.*) unequal-leaved. Hungary. 1826. 6. 7. H. ♃.
12 cordàtum. B.M. (*bl.*) heart-leaved. S. Europe. 1804. 7. 8. H. ♃. Bot. mag. t. 1466.
13 ellípticum. R.S. (*bl.*) elliptic-leaved. Switzerland. 1820. —— H. ♃. Vill. delph. 2. t. 11. f. 2.
14 nìgrum. R.S. (*da.pu.*) black-flowered. Bohemia. —— —— H. ♃.
15 betonicifòlium. R.S. (*pa.*) Betony-leaved. Switzerland. 1816. 6. 8. H. ♃. Bot. mag. t. 2066.
16 spicàtum. R.S. (*st.*) spiked. Europe. 1597. —— H. ♃. Flor. dan. t. 362.
17 Hallèri. R.S. (*pa.vi.*) Haller's. Switzerland. 1819. —— H. ♃.
β *ovàtum.* (*pa.*) *oval-spiked.* ——— 1814. —— H. ♃.
18 stylòsum. Bes. (*bl.*) long-styled. Poland. 1827. —— H. ♃.
19 stríctum. B.M. (*bl.*) upright. S. Europe. 1819. —— H. ♃. Bot. mag. t. 2145.
20 virgàtum. R.S. (*bl.*) slender-branched. Levant. 1823. 7. 9. H. ♃. Lab. syr. 2. p. 11. t. 6.
21 campanuloídes. R.S. (*bl.*) Campanula-like. Caucasus. 1804. 6. 8. H. ♃. Bot. mag. t. 1015.
22 canéscens. R.S. (*li.*) canescent. Hungary. —— —— H. ♃. W. et K. hung. 1. t. 14.
23 pinnàtum. R.S. (*bl.*) winged-leaved. Candia. 1640. —— G. ♃. Venten. cels. t. 52.

TRACH'ELIUM. R.S. THROATWORT. Pentandria Monogynia. L.

1 cœrùleum. R.S. (*bl.*) blue. Italy. 1640. 7. 9. H. ♂. Bot. reg. t. 72.
2 diffùsum. R.S. (*bl.*) shrubby. C. B.S. 1787. —— G. ♄.

JAS'IONE. R.S. SHEEP'S SCABIOUS. Pentandria Monogynia. W.

1 montàna. R.S. (*bl.*) mountain. Britain. .... 6. 7. H. ♃. Eng. bot. v. 13. t. 882.
2 perénnis. R.S. (*bl.*) perennial. France. 1787. —— H. ♃. Bot. reg. t. 505.
3 foliòsa. S.S. (*bl.*) leafy. S. Spain. 1826. —— H. ♃.

ORDO CXXV.

GESNEREÆ. *Rich. et Juss. ann. mus.* 5. *p.* 428.

GLOXI'NIA. S.S.		GLOXI'NIA.	Didynamia Angiospermia. L.				
1 hirsùta. B.R.	(*bl.*)	hairy.	Brazil.	1825.	6. 8.	S. ♃.	Bot. reg. t. 1004.
2 speciòsa. B.R.	(*bl.*)	many-flowered.	———	1815.	6. 11.	S. ♃.	——— t. 213.
β álba.	(*wh.*)	*white-flowered.*		1822.	——	S. ♃.	
3 cauléscens. B.R.	(*vi.*)	caulescent.	Pernambuca.	1820.	8. 10.	S. ♃.	Bot. reg. t. 1127.
4 maculàta. S.S.	(*pu.*)	spotted-stalked.	S. America.	1739.	7. 10.	S. ♃.	Bot. mag t. 1191.
SINNI'NGIA. Nees.		SINNI'NGIA.	Didynamia Angiospermia.				
1 guttàta. B.R.	(*gr.ye.*)	spotted-throated.	Brazil.	1825.	6. 8.	S. ♄.	Bot. reg. t. 1134.
2 Hellèri. B.R.	(*gr.ye.*)	Heller's.	———	1824.	——	S. ♄.	——— t. 997.
3 velutìna. B.R.	(*gr.ye.*)	velvetty.	———	——	——	S. ♄.	Lodd. bot. cab. t. 1398.
4 villòsa. B.R.	(*gr.ye.*)	shaggy.	———	1826.	——	S. ♄.	Bot. reg. t. 1134.
CODONO'PHORA. B.R.		CODONO'PHORA.	Didynamia Angiospermia.				
1 grandiflòra. B.R.	(*gr.*)	large-flowered.	Brazil.	1816.	5. 6.	S. ♄.	Bot. reg. t. 428.
Gesnèria prasinàta. Bot. reg.							
2 lanceolàta. B.R.	(*gr.*)	spear-leaved.	———	1826.	——	S. ♄.	
GESN'ERIA. L.		GESN'ERIA.	Didynamia Angiospermia. L.				
1 tomentòsa. S.S.	(*gr.pu.*)	woolly.	S. America.	1752.	11. 6.	S. ♄.	Bot. mag. t. 1023.
2 scábra. S.S.	(*pu.*)	rough.	Jamaica.	1822.	6. 8.	S. ♄.	
3 corymbòsa. S.S.	(*ye.*)	corymbose.	———	——	——	S. ♄.	
4 Douglásii. B.R.	(*sp.*)	Douglas's.	Brazil.	1825.	5. 8.	S. ♄.	Bot. reg. t. 1110.
verticillàta. Bot. mag. t. 2776. *non* Cavan.							
5 aggregàta. S.S.	(*sc.*)	clustered.	———	1816.	6. 10.	S. ♃.	Bot. reg. t. 329.
6 pendulìna. B.R.	(*sc.*)	drooping-flowered.	———	1825.	6. 9.	S. ♃.	——— t. 1032.
7 rùtila. B.R.	(*sc.*)	brilliant.	———	1826.	6. 10.	S. ♃.	——— t. 1158.
β atrosanguínea. B.R.	(*cr.*)	*dark-crimson.*	———	——	——	S. ♃.	——— t. 1279.
8 latifòlia. M.N.	(*sc.*)	long-spiked.	———	1825.	7. 10.	S. ♃.	——— t. 1202.
macrostàchya. B.R.							
9 tubiflòra. C.I.	(*sc.*)	tube-flowered.	S. America.	1815.	2. 6.	S. ♃.	Cavan. ic. 6. t. 584.
10 bulbòsa. S.S.	(*sc.*)	bulbous-rooted.	Brazil.	1816.	5. 8.	S. ♃.	Bot. reg. t. 343.
11 acaùlis. S.S.	(*sc.*)	stemless.	Jamaica.	1793.	——	S. ♃.	Sloan. jam. 1. t. 102. f. 1.
BESL'ERIA. L.		BESL'ERIA.	Didynamia Angiospermia. L.				
1 cristàta. S.S.	(*ye.*)	crested.	W. Indies.	1739.	6. 8.	S. ♄.	Jacq. amer. t. 119.
2 coccìnea. S.S.	(*ye.*)	scarlet-calyxed.	Guiana.	1824.	5. 8.	S. ♄ ◡.	Aub. gui. 2. t. 255.
3 dichrus. S.S.	(*ye.re.*)	two-coloured.	Brazil.	1823.	——	S. ♄.	
4 pulchélla. S.S.	(*or.*)	large-flowered.	W. Indies.	1806.	——	S. ♄.	Bot. mag. t. 1146.
5 melittifòlia. S.S.	(*pu.*)	Balm-leaved.	Guiana.	1739.	6. 7.	S. ♄.	Sm. exot. bot. 1. t. 54.
6 serrulàta. S.S.	(*st.*)	saw-leaved.	W. Indies.	1806.	——	S. ♄.	Jacq. schœn. 3. t. 290.
7 lùtea. S.S.	(*ye.*)	yellow-flowered.	Guiana.	1739.	7. 8.	S. ♄.	Plum. ic. 49. f. 1.
8 grandifòlia. S.S.	(*ye.*)	large-leaved.	Brazil.	1823.	——	S. ♄.	
9 hirtélla. S.S.	(*ye.*)	hairy-leaved.	———	——	——	S. ♄.	
10 incarnàta. S.S.	(*fl.*)	flesh-coloured.	Guiana.	——	——	S. ♄.	Aub. gui. 2. t. 256.

ORDO CXXVI.

VACCINIEÆ. *DC. theor. ed.* 1. *p.* 216.

VACCI'NIUM. W.		WHORTLE-BERRY.	Oct-Decandria Monogynia.				
1 Myrtíllus. W.	(*bh.*)	Blea-berry.	Britain.	....	4. 6.	H. ♄.	Eng. bot. 7. t. 456.
2 stamíneum. Ph.	(*wh.*)	long-stamened.	N. America.	1772.	5. 6.	H. ♄.	Pluk. alm. t. 339. f. 3.
3 álbum. Ph.	(*wh.*)	white-flowered.	———	——	——	H. ♄.	Andr. reposit. t. 263.
stamíneum. A.R. *non* Ph.							
4 pállidum. W.	(*wh.*)	pale.	———	——	——	H. ♄.	
5 uliginòsum. W.	(*ro.*)	great Bilberry.	Britain.	....	4. 5.	H. ♄.	Eng. bot. v. 9. t. 581.
6 diffùsum. W.	(*bh.*)	tree.	N. America.	1765.	5. 11.	H. ♄.	Bot. mag. t. 1607.
arbòreum. M.							
7 dumòsum. B.M.	(*bh.*)	bushy.	———	1774.	5. 6.	H. ♄.	Bot. mag. t. 1106.
hirtéllum. H.K.							
β hùmile. W.D.	(*wh.*)	*dwarf.*	———	——	——	H. ♄.	Wats. dend. brit. t. 32.

8 frondòsum. W.	(*wh.*)	blunt-leaved.	N. America.	1761.	5. 6.	H. ♄.	Andr. reposit. t. 140.
9 venústum. H.K.	(*wh.*)	red-twigged.	——	1770.	——	H. ♄.	
10 resinòsum. W.	(*st.*)	clammy.	——	1772.	——	H. ♄.	Wang. am. t. 30. f. 69.
β rubéscens.	(*re.*)	*red-flowered.*	——	——	——	H. ♄.	Bot. mag. t. 1288.
11 parviflòrum. A.R.	(*re.*)	small-flowered.	——	1804.	——	H. ♄.	Andr. reposit. t. 125.
12 corymbòsum. W.	(*wh.*)	corymbose.	——	1806.	6. 7.	H. ♄.	Wats. dend. brit. t. 123.
13 am`œnum. W.	(*wh.*)	broad-leaved.	——	1765.	5. 6.	H. ♄.	Bot. reg. t. 400.
14 canadénse. S.S.	(*wh.*)	Canadian.	——	1827.	5. 7.	H. ♄.	
15 virgàtum. W.	(*pa.re.*)	twiggy.	——	1767.	4. 5.	H. ♄.	Andr. reposit. t. 181.
16 Watsòni.	(*bh.*)	Watson's.	——	1812.	——	H. ♄.	Wats. dend. brit. t. 34.
virgàtum β angustifòlium. W.D.							
17 mariànum. W.D.	(*wh.*)	Maryland.	——	——	5. 6.	H. ♄.	Wats. dend. brit. t. 124.
18 grandiflòrum. W.D.	(*wh.*)	great-flowered.	——	——	7. 8.	H. ♄.	—— t. 125. A.
19 elongàtum. W.D.	(*bh.*)	elongated.	——	——	——	H. ♄.	—— t. 125. B.
20 minutiflòrum. W.D.	(*bh.*)	minute-flowered.	——	——	——	H. ♄.	—— t. 125. C.
21 glábrum. W.D.	(*re.*)	smooth.	——	——	——	H. ♄.	—— t. 125. D.
22 mucronàtum. L.	(*wh.*)	mucronate.	——	....	5. 6.	H. ♄.	
23 fuscàtum. W.	(*re.*)	cluster-flowered.	——	1770.	——	H. ♄.	Bot. reg. t. 302.
formòsum. A.R. 97.							
β angustifòlium. Ph.	(*re.*)	*narrow-leaved.*	——	——	——	H. ♄.	
24 galèzans. M.	(*st.*)	Gale-leaved.	——	1806.	——	H. ♄.	
25 ligustrìnum. W.	(*pu.re.*)	Privet-leaved.	——	....	——	H. ♄.	
26 tenéllum. W.	(*ro.*)	Pensylvanian.	——	1772.	——	H. ♄.	Wats. dend. brit. t. 35.
pensylvánicum. M.							
27 angustifòlium. W.	(*wh.*)	narrow-leaved.	——	1776.	4. 5.	H. ♄.	
myrtilloídes. M.							
28 Arctostáphylos. W.	(*re.*)	oriental.	Levant.	1777.	6. 8.	H. ħ.	Bot. mag. t. 974.
29 maderénse. S.S.	(*re.*)	Madeira.	Madeira.	1810.	7. 2.	H. ħ.	Andr. reposit. t. 30.
Arctostáphylos. A.R.							
30 meridionàle. W.	(*re.*)	Jamaica.	Jamaica.	1778.	3. 6.	G. ħ.	
31 salicìnum. S.S.	(*bh.*)	Willow-like.	N. America.	1827.	6. 8.	H. ħ.	
32 crassifòlium. S.S.	(*pi.*)	thick-leaved.	Carolina.	1787.	6. 7.	F. ħ.	Bot. mag. t. 1152.
33 myrtifòlium. M.	(*pi.*)	Myrtle-leaved.	——	1812.	5. 7.	F. ħ.	
34 nítidum. S.S.	(*pi.*)	glossy-leaved.	——	1794.	5. 6.	F. ħ.	Andr. rep. t. 480.
35 decúmbens. B.M.	(*pi.*)	decumbent.	——	——	——	F. ħ.	Bot. mag. t. 1550.
36 myrsinìtes. M.	(*pu.*)	Myrsine-leaved.	——	....	——	F. ħ.	
37 buxifòlium. S.S.	(*bh.*)	Box-leaved.	N. America.	1794.	——	H. ħ.	Bot. mag. t. 928.
38 Vìtis Id`æa. W.	(*bh.*)	Cow-berry.	Britain.	....	4. 6.	H. ħ.	Eng. bot. v. 9. t. 598.
α mìnor.	(*bh.*)	*smaller.*	——	....	——	H. ħ.	Lodd. bot. cab. t. 1023.
β màjor.	(*ro.*)	*larger.*	N. America.	....	——	H. ħ.	—— t. 616.
39 ovàtum. Ph.	(*bh.*)	oval-leaved.	——	1827.	5. 7.	H. ħ.	
40 obtùsum. Ph.	(*bh.*)	blunt-creeping.	——	——	——	H. ħ.	
41 halleriæfòlium. L.C.	(*wh.*)	Halleria-leaved.	——	....	5. 6.	H. ♄.	

OXYCO'CCUS. P.S. CRANBERRY. Octandria Monogynia.

1 palústris. P.S.	(*pi.*)	common.	Britain.	....	5. 6.	H. ħ.	Eng. bot. v. 5. t. 319.
Vaccínium Oxycóccos. E.B. *Schöllera Oxycóccos.* Roth.							
2 macrocárpus. Ph.	(*pi.*)	large-fruited.	N. America.	1760.	5. 8.	H. ħ.	Wats. dend. brit. t. 122.
3 eréctus. Ph.	(*ro.*)	upright.	——	1806.	6. 8.	H. ♄.	—— t. 31.
Vaccínium erythrocárpum. M.							
4 hispídulus. P.S.	(*pi.*)	Thyme-leaved.	——	1776.	5. 7.	H. ħ.	Mich. amer. 1. t. 23.
Vaccínium hispídulum. W. *Arbùtus thymifòlia.* Ait. *Gaulthèria serpyllifòlia.* Ph.							

ORDO CXXVII.

ERICEÆ. *Don Prodr. p.* 148.

Sect. I. *ERICEÆ VERÆ.*

ARCTOSTA'PHYLOS. K.S. ARCTOSTA'PHYLOS. Decandria Monogynia.

1 `Uva-úrsi. S.S.	(*bh.*)	Bear-berry.	Britain.	....	4. 6.	H. ħ.	Eng. bot. v. 10. t. 714.
A'rbutus `Uva-úrsi.* E.B. *Mairània `Uva-úrsi. Desv.							
2 alpìna. S.S.	(*pi.*)	Alpine.	Scotland.	....	4. 5.	H. ħ.	Eng. bot. v. 29. t. 2030.

A'RBUTUS. K.S.		STRAWBERRY-TREE.	Decandria Monogynia. L.				
1 Unèdo. s.s.	(*wh.re.*)	common.	Ireland.	....	9. 12.	H. ♄.	Eng. bot. v. 34. t. 2377.
β *rùbra.*	(*re.*)	*red-flowered.*	———	....	——	H. ♄.	Lodd. bot. cab. t. 123.
γ *flòre plèno.*	(*wh.*)	*double-flowered.*	———	....	——	H. ♄.	
δ *schizopétala.*	(*wh.*)	*jagged-petaled.*	———	....	——	H. ♄.	
ε *integrifòlia.*	(*wh.*)	*entire-leaved.*	———	....	——	H. ♄.	
ζ *críspa.*	(*wh.*)	*curled-leaved.*	———	....	——	H. ♄.	
η *salicifòlia.*	(*wh.*)	*Willow-leaved.*	———	....	——	H. ♄.	
2 canariénsis. s.s.	(*ro.*)	long-leaved.	Canaries.	1796.	5. 6.	G. ♄.	Bot. mag. t. 1577.
3 hy'brida. s.s.	(*wh.*)	hybrid.	Hybrid.	....	——	H. ♄.	Bot. reg. t. 619.
serratifòlia. B.C. 580. *andrachnoídes.* L.en.							
4 Millèri.	....	Miller's.	———	1825.	....	H. ♄.	
5 Andráchne. s.s.	(*wh.*)	oriental.	Levant.	1724.	3. 4.	F. ♄.	Bot. reg. t. 113.
6 pròcera. D.	(*wh.*)	tall.	N. America.	1827.	....	H. ♄.	
7 Menzièsii. Ph.	(*wh.*)	Mr. Menzies's.	———	——	....	H. ♄.	
8 tomentòsa. Ph.	(*wh.*)	woolly.	———	——	....	H. ♄.	
9 laurifòlia. s.s.	(*wh.*)	Laurel-leaved.	Mexico.	1825.	....	F. ♄.	
10 phillyreæfòlia. s.s.	(*wh.*)	Phillyrea-leaved.	Peru.	1812.	....	G. ♄.	
11 ? pùmila. s.s.	(*wh.*)	dwarf.	Magellan.	1824.	....	G. ♄.	
GAULTH'ERIA. S.S.		GAULTH'ERIA.	Decandria Monogynia. L.				
1 procúmbens. s.s.	(*bh.*)	trailing.	N. America.	1762.	7. 9.	H. ♄.	Andr. reposit. t. 116.
2 fràgrantíssima. s.s.	(*wh.*)	fragrant.	Nepaul.	1824.	....	H. ♄.	Wal. as. res. 13. p. 297. ic.
3 Shállon. Ph.	(*bh.*)	Shallon.	N. America.	1827.	5. 6.	H. ♄.	Lodd. bot. cab. t. 1732.
ENKIA'NTHUS. B.M.		ENKIA'NTHUS.	Decandria Monogynia.				
1 quinqueflòrus. B.R.	(*pi.*)	Canton.	China.	1812.	10. 2.	G. ♄.	Bot. reg. t. 884.
2 reticulàtus. B.R.	(*bh.*)	netted-leaved.	———	1820.	——	G. ♄.	——— t. 885.
ANDRO'MEDA. N.		ANDRO'MEDA.	Decandria Monogynia. L.				
1 hypnoídes. W.	(*wh.*)	Moss-like.	Lapland.	1798.	6. 7.	H. ♄.	Bot. mag. t. 2936.
2 polifòlia. W.	(*ro.*)	wild rosemary.	Britain.	....	——	H. ♄.	Eng. bot. v. 10. t. 713.
3 glaucophy'lla. L.en.	(*wh.*)	glaucous-leaved.	N. America.	....	5. 9.	H. ♄.	
4 rosmarinifòlia. Ph.	(*bh.*)	Rosemary-leaved.	———	....	——	H. ♄.	Pall. ross. 2. t. 70. f. B.
5 subulàta. P.S.	(*ro.*)	awl-leaved.	Europe.	....	——	H. ♄.	
6 angustifòlia. Ph.	(*wh.*)	narrow-leaved.	N. America.	....	2. 4.	H. ♄.	
críspa. Link enum.							
7 calyculàta. W.	(*wh.*)	various-leaved.	Russia.	1748.	2. 4.	H. ♄.	Pall. ross. 2. t. 72. f. 1.
α *ventricòsa.* H.K.	(*wh.*)	*globe-flowered.*	———	——	——	H. ♄.	Bot. mag. t. 1286.
β *nàna.* B.C.	(*wh.*)	*dwarf.*	....	....	——	H. ♄.	Lodd. bot cab. t. 862.
γ *latifòlia.* H.K.	(*wh.*)	*broad-leaved.*	Newfoundland.	....	——	H. ♄.	——— ——— t. 530.
8 japónica. W.	(*wh.*)	Japan.	Japan.	1806.	5. 6.	H. ♄.	Thunb. jap. 181. t. 22.
9 ovalifòlia. D.P.	(*bh.*)	oval-leaved.	Nepaul.	1823.	——	G. ♄.	Wall. as. res. 13. p. 39. ic.
10 buxifòlia. P.S.	(*cr.*)	Box-leaved.	Mauritius.	1822.	4. 6.	G. ♄.	Bot. mag. t. 2660.
11 fasciculàta. W.	(*wh.*)	bundle-flower'd.	Jamaica.	——	——	G. ♄.	
12 jamaicénsis. W.	(*wh.*)	Jamaica.	———	1793.	——	G. ♄.	
13 marìana. W.	(*bh.*)	Maryland.	N. America.	1736.	6. 7.	H. ♄.	Pluk. mant. t. 448. f. 6.
α *ovàlis.*	(*bh.*)	*oval-leaved.*	———	——	——	H. ♄.	Bot. mag. t. 1579.
β *oblónga.*	(*bh.*)	*oblong-leaved.*	———	——	——	H. ♄.	
14 speciòsa. Ph.	(*wh.*)	large-flowered.	Carolina.	1800.	6. 9.	H. ♄.	
α *nítida.* Ph.	(*wh.*)	*smooth-leaved.*	———	——	——	H. ♄.	Bot. mag. t. 970.
cassinæfòlia. B.M.							
β *glaùca.* W.D.	(*wh.*)	*glaucous-leaved.*	———	——	——	H. ♄.	Wats. dend. brit. t. 26.
Andrómeda dealbàta. Bot. reg. t. 1010.							
γ *pulverulénta.* Ph.	(*wh.*)	*mealy-leaved.*	———	——	——	H. ♄.	Bot. mag. t. 667.
15 coriàcea. W.	(*pi.*)	thick-leaved.	N. America.	1765.	6. 8.	H. ♄.	——— t. 1095.
nítida. M. *lùcida.* Lam. *mariána.* Jacq. ic. 3. t. 79. *nec aliorum.*							
16 acuminàta. W.	(*wh.*)	acute-leaved.	N. America.	——	8.	H. ♄.	Sm. exot. bot. 2. t. 89.
17 Catesb'æi. W.	(*wh.*)	spiny-leaved.	———	1793.	6. 7.	H. ♄.	Bot. mag. t. 1955.
spinulòsa. Ph.							
18 axillàris. W.	(*wh.*)	axil-flowering.	———	1765.	5. 8.	H. ♄.	Duham. ar. ed. nov. t. 39.
β *longifòlia.*	(*wh.*)	*long-leaved.*	———	——	——	H. ♄.	Bot. mag. t. 2357.
19 floribúnda. Ph.	(*wh.*)	many-flowered.	Georgia.	1812.	5. 6.	H. ♄.	——— t. 1566.
20 racemòsa. W.	(*wh.*)	branching.	N. America.	1736.	7.	H. ♄.	
21 spicàta. W.D.	(*wh.*)	spiked.	———	——	6. 7.	H. ♄.	Wats. dend. brit. t. 36.
22 arbòrea. W.	(*wh.*)	Sorrel-tree.	———	1752.	7. 9.	H. ♄.	Bot. mag. t. 905.
LY'ONIA. N.		LY'ONIA.	Decandria Monogynia. N.				
1 ferrugínea. N.	(*wh.*)	rusty-leaved.	Georgia.	1784.	6. 7.	G. ♄.	Venten. malm. t. 80.
Andrómeda ferrugínea. W.							
2 rígida. N.	(*wh.*)	rigid.	———	1774.	4. 5.	G. ♄.	Lodd. bot. cab. t. 430.
3 salicifòlia. W.D.	(*wh.*)	Willow-leaved.	N. America.	....	6. 7.	H. ♄.	Wats. dend. brit. t. 38.
4 paniculàta. N.	(*wh.*)	panicled.	———	1748.	5. 6.	H. ♄.	——— ——— t. 37.

5 capreæfòlia. W.D.	(*wh.*)	tendril-leaved.	N. America.	1812.	7.	H. ♄.	Wats. dend. brit. t. 127
6 multiflòra. W.D.	(*wh.*)	many-flowered.	———	———	———	H. ♄.	——— t. 128.
7 frondòsa. N.	(*wh.*)	bristly-flowered.	———	1806.	5. 6.	H. ♄.	
MYLOC'ARYUM. W.en.		BUCKWHEAT-TREE.	Decandria Monogynia. W.				
ligustrìnum. Ph.	(*wh.*)	Privet-leaved.	Georgia.	1786.	5. 6.	G. ♄.	Bot. mag. t. 1625.
Cliftònia ligustrìna. S.S.							
CLE'THRA. W.		CLE'THRA.	Decandria Monogynia. W.				
1 alnifòlia. S.S.	(*wh.*)	Alder-leaved.	N. America.	1731.	8. 10.	H. ♄.	Lam. ill. t. 369.
2 tomentòsa. S.S.	(*wh.*)	woolly-leaved.	———	———	———	H. ♄.	Wats. dend. brit. t. 39.
3 scábra. S.S.	(*wh.*)	rough-leaved.	Georgia.	1806.	———	H. ♄.	
4 acuminàta. S.S.	(*wh.*)	acute-leaved.	Carolina.	———	———	H. ♄.	
5 paniculàta. S.S.	(*wh.*)	panicled.	N. America.	1770.	———	H. ♄.	
6 arbòrea. S.S.	(*wh.*)	tree.	Madeira.	1784.	———	G. ♄.	Bot. mag. t. 1057.
β *variegàta.*	(*wh.*)	*variegated.*	———	———	———	G. ♄.	
γ *mìnor.*	(*wh.*)	*dwarf.*	———	....	———	G. ♄.	
7 ferrugìnea. L.en.	(*wh.*)	ferruginous.	S. America.	1818.	———	G. ♄.	
CYRI'LLA. L.		CYRI'LLA.	Pentandria Monogynia. L.				
racemiflòra. L.	(*wh.*)	Carolina.	Carolina.	1765.	6. 8.	F. ♄.	Bot. mag. t. 2456.
caroliniàna. P.S. *Itea Cyrílla.* W.							
BROSS'ÆA. L.		BROSS'ÆA.	Pentandria Monogynia. L.				
coccínea. L.	(*sc.*)	scarlet.	S. America.	....	....	S. ♄.	Lam. ill. t. 111. f. 3.
BL'ÆRIA. W.		BL'ÆRIA.	Tetrandria Monogynia. W.				
1 ericoídes. S.S.	(*pu.*)	heath-leaved.	C. B. S.	1774.	8. 10.	G. ♄.	Wendl. coll. 1. t. 25.
2 muscòsa. S.S.	(*wh.*)	Moss-leaved.	———	———	6. 8.	G. ♄.	
3 fasciculàta. S.S.	(*wh.*)	close-headed.	———	1812.	———	G. ♄.	
4 paucifòlia. S.S.	(*fl.*)	few-leaved.	———	———	———	G. ♄.	Wendl. coll. 2. t. 43.
5 articulàta. S.S.	(*bh.*)	jointed.	———	1795.	5. 6.	G. ♄.	——— 2. t. 44.
6 scábra. S.S.	(*wh.*)	rough.	———	1806.	———	G. ♄.	——— 1. t. 31.
7 depréssa. R.S.	(*pu.*)	depressed.	———	1816.	———	G. ♄.	
8 purpùrea. S.S.	(*pu.*)	purple-flowered.	———	1791.	———	G. ♄.	
9 dumòsa. S.S.	(*ro.*)	bushy.	———	1806.	6. 8.	G. ♄.	Wendl. coll. 2. t. 38.
10 glabélla. S.S.	(*wh.*)	smooth.	———	1816.	———	G. ♄.	Seb. thes. 1. t. 20. f. 2.
11 ciliàris. S.S.	(*wh.*)	fringed.	———	1795.	———	G. ♄.	Wendl. coll. 2. t. 49.
SYMPI'EZA. R.S.		SYMPI'EZA.	Tetrandria Monogynia. R.S.				
capitellàta. R.S.	(*pu.*)	headed.	C. B. S.	1812.	6. 8.	G. ♄.	Wendl. coll. 2. t. 37.
Bl'æria bracteàta. Wend.							
CALL'UNA. Sal.		CALL'UNA.	Octandria Monogynia.				
vulgàris. W.en.	(*pi.*)	common.	Britain.	....	6. 8.	H. ♄.	Eng. bot. v. 15. t. 1013.
Erìca vulgàris. E.B.							
α *cárnea.*	(*pi.*)	*flesh-coloured.*	———	....	———	H. ♄.	
β *álba.*	(*wh.*)	*white-flowered.*	———	....	———	H. ♄.	
γ *múltiplex.*	(*pi.*)	*double-flowered.*	———	....	———	H. ♄.	
ER'ICA. W.		HEATH.	Octandria Monogynia. L.				
1 absinthoídes. H.K.	(*pu.*)	Wormwood-like.	C. B. S.	1792.	3. 6.	G. ♄.	
2 act'æa. L.en.	(*wh.*)	cluster-leaved.	———	1810.	———	G. ♄.	
3 acuminàta. A.H.	(*pi.*)	pointed-leaved.	———	1798.	5. 6.	G. ♄.	Andr. heath. v. 3.
β *pállida.*	(*pa.*)	*pale-flowered.*	———	———	———	G. ♄.	
4 acùta. A.H.	(*re.*)	pointed-cupped.	———	1799.	5. 8.	G. ♄.	Andr. heath. v. 2.
5 acutangulàris. L.C.	(*re.*)	acute-angled.	———	1823.	———	G. ♄.	
6 adenóphora. S.S.	(*wh.*)	gland-haired.	———	1810.	———	G. ♄.	
7 aggregàta. S.S.	(*li.*)	clustered.	———	———	———	G. ♄.	Wendl. er. f. 13. n. 5.
8 Aitoniàna. H.K.	(*wh.re.*)	Aiton's.	———	1790.	6. 9.	G. ♄.	Bot. mag. t. 429.
9 álbens. H.K.	(*wh.*)	pallid.	———	1789.	3. 8.	G. ♄.	——— t. 440.
10 álbida. L.C.	(*wh.*)	white-flowered.	———	1823.	5. 8.	G. ♄.	
11 alopecuroídes. S.S.	(*li.*)	fox-tail.	———	1812.	———	G. ♄.	Lodd. bot. cab. t. 874.
12 ambígua. S.S.	(*pi.*)	ambiguous.	———	1798.	———	G. ♄.	Andr. heath. v. 1.
cylíndrica. A.H. *non* Thunb. *hy'brida.* Hort.							
13 am'œna. H.K.	(*pu.*)	feathery.	C. B. S.	1795.	3. 7.	G. ♄.	Wend. eric. 17. p. 73. ic.
plumòsa. A.H.							
14 ampullàcea. H.K.	(*bh.*)	flask.	———	1790.	6. 8.	G. ♄.	Bot. mag. t. 303.
15 andromedæflòra. H.K.	(*pi.*)	Andromeda-fld.	———	1803.	3. 6.	G. ♄.	——— t. 1250.
16 anthìna. S.S.	(*pu.*)	full-flowering.	———	1811.	———	G. ♄.	
flòrida. Hort. *non* W.							
17 apérta. S.S.	(*re.*)	open-flowered.	———	———	———	G. ♄.	
18 Aphànes. S.S.	(*wh.*)	Parsley-piert.	———	1820.	———	G. ♄.	
19 arbòrea. S.S.	(*wh.*)	tree.	S. Europe.	1658.	2. 6.	G. ♄.	
β *stylòsa.* P.S.	(*wh.*)	*long-styled.*	———	———	———	G. ♄.	
20 arbúscula. B.C.	(*re.*)	little-tree.	C. B. S.	1818.	4. 8.	G. ♄.	Lodd. bot. cab. t. 843.
21 arctàta. L.C.	(*wh.*)	arched.	———	———	———	G. ♄.	

22	Archeriàna. H.K.	(*re.*)	Lady Archer's.	C. B. S.	1796.	8. 11.	G. ♄.	Andr. heath. v. 2.
23	árdens. H.K.	(*sc*)	ardent.	———	1800.	4. 6.	G. ♄.	Bot. reg. t. 115.
24	argentiflòra. S.S.	(*wh.*)	silvery-flowered.	———	1816.	——	G. ♄.	Andr. heath. v. 4.
25	aristàta. H.K.	(*cr.*)	awned.	———	1801.	3. 8.	G. ♄.	Bot. mag. t. 1249.
26	assúrgens. L.en.	(*wh.*)	assurgent.	———	1810.	5. 8.	G. ♄.	
27	aùrea. H.K.	(*ye.*)	golden.	———	1799.	7. 9.	G. ♄.	Andr. heath. v. 2.
28	auriculàris. L.T.	(*re.*)	eared.	———	1810.	3. 7.	G. ♄.	
29	austràlis. H.K.	(*re.*)	Spanish.	Spain.	1769.	4. 7.	H. ♄.	Andr. heath. v. 3.
30	axillàris. W.	(*pu.*)	axillary.	C. B. S.	1798.	5. 7.	G. ♄.	
31	azaleæfòlia. L.T.	(*li.*)	Azalea-leaved.	———	——	5. 8.	G. ♄.	
32	báccans. H.K.	(*li.*)	Arbutus-flower'd.	———	1774.	4. 7.	G. ♄.	Bot. mag. t. 358.
33	Bandoniàna. L.C.	(*bh.*)	Lord Bandon's.	———	1816.	5. 8.	G. ♄.	Andr. heath. v. 4.
34	Bánksii. H.K.	(*st.*)	Banks's.	———	1787.	2. 7.	G. ♄.	——— v. 1.
35	barbàta. H.K.	(*wh.*)	bearded.	———	1799.	5. 8.	G. ♄.	——— v. 2.
36	Beaumóntiæ. A.H.	(*wh.*)	Mrs. Beaumont's.	———	1820.	6. 8.	G. ♄.	——— v. 4.
37	bélla. S.S.	(*pu.*)	pretty.	———			G. ♄.	
38	Bergiàna. H.K.	(*re.*)	Bergius's.	———	1787.	4. 8.	G. ♄.	Lodd. bot. cab. t. 939.
	quadriflòra. A.H. v. 2.							
39	bícolor. W.	(*re.gr.*)	two-coloured.	———	1790.	8. 3.	G. ♄.	
	calathiflòra. L.T.							
40	biflòra. B.C.	(*wh.*)	two-flowered.	———	1820.	5. 8.	G. ♄.	Lodd. bot. cab. t. 683.
41	blánda. H.K.	(*pi.*)	charming.	———	1800.	4. 9.	G. ♄.	Andr. heath. v. 3.
42	Blandfordiàna. H.K.	(*ye.*)	Blandford's.	———	1803.	3. 6.	G. ♄.	Bot. mag. t. 1793.
43	Bonplandiàna. B.M.	(*wh.*)	Bonpland's.	———	1816.	——	G. ♄.	——— t. 2126.
44	borboniæfòlia. L.T.	(*wh.*)	Borbonia-leaved.	———			G. ♄.	
45	Bowieàna. B.C.	(*wh.*)	Bowie's.	———	——	3. 9.	G. ♄.	Lodd. bot. cab. t. 842.
	Bauèria. Andr. heath. v. 4.							
46	brachiàlis. L.T.	(*re.*)	armed.	———	1792.	——	G. ♄.	
47	bracteàta. W.	(*re.*)	red-bracted.	———	1800.	5. 7.	G. ♄.	
48	bracteolàris. P.S.	(*wh.*)	bracteate.	———	——	3. 7.	G. ♄.	
49	brevifòlia. L.T.	(*wh.*)	short-leaved.	———	....	8. 1.	G. ♄.	
50	Broadleyàna. A.H.	(*pi.*)	Broadley's.	———	1810.	5. 7.	G. ♄.	Andr. heath. v. 3.
51	bruniàdes. W.	(*wh.*)	Brunia-like.	———	1790.	4. 7.	G. ♄.	
52	bucciniflòra. B.M.	(*li.wh.*)	trumpet-flowered.	———	1816.	5. 9.	G. ♄.	Bot. mag. t. 2465.
53	c'æsia. L.T.	(*wh.*)	grey.	———	1795.	2. 5.	G. ♄.	
54	cáffra. A.H.	(*wh.*)	sweet-scented.	———	1774.	8. 10.	G. ♄.	Lodd. bot. cab. t. 196.
55	caledònica. S.S.	(*ro.*)	Caledonian.	———	1816.	5. 9.	G. ♄.	
56	callòsa. S.S.	(*li.*)	callous-leaved.	———	1799.	6. 9.	G. ♄.	Andr. heath. v. 3.
	canaliculàta β mìnor. A.H.							
57	calycìna. W.	(*pi.*)	calycine.	———	——	——	G. ♄.	
58	calyculàta. P.S.	(*br.*)	rusty-flowered.	———	1800.	5. 7.	G. ♄.	
59	campanulàta. H.K.	(*ye.*)	bell-flowered.	———	1791.	4. 8.	G. ♄.	Andr. heath. v. 1.
60	campylophy'lla. S.S.	(*li.*)	hook-leaved.	———	1802.	3. 5.	G. ♄.	
	incúrva. A.H.							
61	canaliculàta. H.K.	(*li.*)	channelled.	———	1799.	8. 2.	G. ♄.	Andr. heath. v. 3.
62	canéscens. H.K.	(*li.*)	canescent.	———	1790.	5. 8.	G. ♄.	——— v. 2.
	eriocéphala. A.H. *non* Lam.							
63	càpax. L.T.	(*re.wh.*)	capacious.	———	1806.	——	G. ♄.	
64	capitàta. H.K.	(*st.*)	downy-headed.	———	1774.	3. 7.	G. ♄.	Andr. heath. v. 1.
65	carduifòlia. L.T.	(*li.*)	Carduus-leaved.	———	1806.	5. 7.	G. ♄.	
66	carinàta. B.C.	(*pi.*)	keeled.	———	——	6. 9.	G. ♄.	Lodd. bot. cab. t. 1071.
67	cárnea. L.	(*pi.*)	early dwarf.	Germany.	1763.	1. 8.	G. ♄.	Jacq. aust. t. 32.
68	carnìula. B.C.	(*fl.*)	flesh-coloured.	C. B. S.	1816.	5. 8.	G. ♄.	Lodd. bot. cab. t 926.
69	catervæfòlia. L.T.	(*li.*)	huddled-leaved.	———	1790.	4. 7.	G. ♄.	
70	Celsiàna. L.C.	(*re.*)	Cels's.	———	....	——	G. ♄.	
71	cephalòtes. W.	(*pu.*)	purple-headed.	———	1812.	3. 7.	G. ♄.	
72	cerinthoídes. H.K.	(*sc.*)	honeywort-flower'd.	———	1774.	5. 9.	G. ♄.	Bot. mag. t. 220.
	α glabriúscula.	(*sc.*)	*smoothish.*	———	——	——	G. ♄.	
	β híspida.	(*sc.*)	*bristly.*	———	——	——	G. ♄.	
73	cérnua. H.K.	(*bh.*)	drooping-flowered.	———	1791.	8. 12.	G. ♄.	Mont. ac. ups. 2. t. 9. f. 3.
74	cerviciflòra. L.T.	(*ye.*)	Stag's-horn-flower'd.	———	....	3. 7.	G. ♄.	
	inapérta. Hort.							
75	chlamydiflòra. L.T.	(*re.*)	cloaked-flower'd.	———	1801.	5. 9.	G. ♄.	
76	ciliàris. W.	(*pi.*)	ciliated.	Portugal.	1759.	7. 9.	H. ♄.	Bot. mag. t. 484.
77	ciliciiflòra. L.T.	(*wh.*)	fringe-flower'd.	C. B. S.	....	5. 8.	G. ♄.	
78	cineráscens. W.	(*pu.*)	greyish.	———	....	——	G. ♄.	
79	cinèrea. W.	(*pu.*)	fine-leaved.	Britain.	....	6. 9.	H. ♄.	Eng. bot. v. 15. t. 1015.
	α atropurpùrea.	(*d.pu.*)	*dark-purple.*	———	....	——	H. ♄.	Lodd. bot. cab. t. 1409.
	β rùbra.	(*re.*)	*red-flowered.*	———	....	——	H. ♄.	

	γ cárnea.	(*fl.*)	*flesh-coloured.*	Britain.	....	6. 9.	H. ♄.	
	δ álba.	(*wh.*)	*white-flowered.*	——	....	——	H. ♄.	
80	cistifòlia. L.en.	(*wh.*)	Cistus-leaved.	C. B. S.	1799.	5. 8.	G. ♄.	Andr. heath. v. 2.
	barbàta β minor. A.H.							
81	clavæflòra. H.K.	(*gr.*)	club-flowered.	——	1779.	8. 10.	G. ♄.	Andr. heath. v. 2.
	sessiliflòra. A.H. *non* L.							
82	clavàta. S.S.	(*gr.*)	clubbed.	——	1812.	——	G. ♄.	Andr. heath. ic.
83	Clintòniæ. Lee.	(*ro.*)	Lady Clinton's.	——	1816.	5. 8.	G. ♄.	
84	coarctàta. S.S.	(*li.*)	close.	——	....	——	G. ♄.	Wendl. eric. f. 19. n. 1.
85	coccìnea. H.K.	(*sc.*)	scarlet-flowered.	——	1783.	1. 12.	G. ♄.	Andr. heath. v. 1.
86	cólorans. B.R.	(*wh.*)	changing-coloured.	——	1812.	1. 6.	G. ♄.	Bot. reg. t. 601.
87	comòsa. H.K.	(*re.wh.*)	tufted.	——	1787.	4. 8.	G. ♄.	Ic. hort. kew. t. 18.
	α rùbra.	(*re.*)	*red-flowered.*	——	——	——	G. ♄.	Wendl. eric. 12. p. 7. ic.
	β álba.	(*wh.*)	*white-flowered.*	——	——	——	G. ♄.	Andr. heath. v. 2.
88	complanàta. S.S.	(*re.*)	flattened.	——	....	——	G. ♄.	
89	Comptoniàna. S.S.	(*re.*)	Lady Northampton's.	——	1820.	——	G. ♄.	Andr. heath. v. 4.
90	concàva. B.C.	(*pu.*)	concave-flowered.	——	1808.	2. 5.	G. ♄.	Lodd. bot. cab. t. 134.
91	concínna. H.K.	(*bh.*)	blush.	——	1773.	7. 10.	G. ♄.	Andr. heath. v. 2.
92	cóncolor. L.C.	(*wh.*)	self-coloured.	——	....	3. 6.	G. ♄.	
93	conférta. H.K.	(*wh.*)	crowded-flowered.	——	1800.	10. 2.	G. ♄.	Andr. heath. v. 2.
94	confertifòlia. S.S.	(*re.*)	crowded-leaved.	——	....	5. 8.	G. ♄.	
95	congésta. S.S.	(*wh.*)	close-headed.	——	1812.	——	G. ♄.	Wendl. eric. f. 17. n. 5.
96	cònica. L.C.	(*re.*)	conical.	——	1820.	——	G. ♄.	Lodd. bot. cab. t. 1179.
97	conspícua. H.K.	(*ye.*)	conspicuous.	——	1774.	——	G. ♄.	Andr. heath. v. 2.
98	constántia. Lee.	(*re.*)	constant-flowered.	——	1810.	——	G. ♄.	
99	cordàta. H.K.	(*wh.*)	heart-leaved.	——	1799.	4. 7.	G. ♄.	Andr. heath. v. 3.
100	corifòlia. H.K.	(*bh.*)	Coris-leaved.	——	1774.	8. 12.	G. ♄.	Bot. mag. t. 422.
	articulàris. B.M. *calycìna.* A.H.							
101	cory'dalis. L.T.	(*re.*)	winged anthered.	——	1812.	2. 5.	G. ♄.	
102	costàta. H.K.	(*re.*)	ribbed-flowered.	——	1795.	2. 6.	G. ♄.	Andr. heath. v. 1.
	β supérba. A.H.	(*bh.*)	*superb.*	——	1820.	4. 5.	G. ♄.	—— v. 4.
103	Coventryàna. B.C.	(*ro.*)	Lord Coventry's.	——	1808.	5. 6.	G. ♄.	Lodd. bot. cab. t. 423.
104	crassifòlia. A.H.	(*li.*)	thick-leaved.	——	1826.	——	G. ♄.	Andr. heath. v. 4.
105	crinìta. B.C.	(*wh.*)	long-haired.	——	1825.	——	G. ♄.	Bot. cab. t. 1432.
106	cristæflòra. L.T.	(*li.*)	crested.	——	1803.	——	G. ♄.	
107	crossòta. S.S.	(*wh.*)	fringed-calyxed.	——	1820.	——	G. ♄.	
108	crucifórmis. S.S.	(*ye.*)	cross-like.	——	1818.	——	G. ♄.	Andr. heath. v. 4.
109	cruénta. H.K.	(*cr.*)	bloody-flowered.	——	1774.	9. 5.	G. ♄.	—— v. 1.
110	cùbica. H.K.	(*pu.*)	cube-flowered.	——	1790.	4. 7.	G. ♄.	—— v. 1.
	β minor. B.C.	(*pu.*)	*lesser cube-flowered.*	——	——	——	G. ♄.	Lodd. bot. cab. t. 972.
111	cumuliflòra. L.T.	(*wh.*)	heap-flowered.	——	1801.	5. 9.	G. ♄.	
112	curviflòra. H.K.	(*ye.*)	curve-flowered.	——	1774.	7. 10.	G. ♄.	Andr. heath. v. 1.
113	curviróstris. L.T.	(*li.*)	curve-styled.	——	1790.	——	G. ♄.	
	declinàta. Hort.							
114	Cushiniàna. Lee.	(*re.*)	Cushin's.	——	1816.	——	G. ♄.	
115	cuspidígera. L.T.	(*li.*)	cuspidate.	——	——	2. 5.	G. ♄.	
116	cylíndrica. Th.	(*wh.*)	white cylindrical.	——	1810.	3. 5.	G. ♄.	
117	cyrillæflòra. L.T.	(*li.*)	Cyrilla-flowered.	——	1800.	2. 8.	G. ♄.	
118	daphnæflòra. B.C.	(*bh.*)	Daphne-flowered.	——	1791.	4. 5.	G. ♄.	Lodd. bot. cab. t. 543.
119	daphnoídes. B.C.	(*ro.*)	Daphne-like.	——	——	——	G. ♄.	—— t. 154.
120	decípiens. S.S.	(*fl.*)	slender woolly.	——	1822.	——	G. ♄.	
121	decólorans. W.	(*ro.wh.*)	discoloured.	——	1812.	——	G. ♄.	
122	decòra. H.K.	(*li.*)	graceful.	——	1790.	11. 1.	G. ♄.	Andr. heath. v. 3.
123	defléxa. L.C.	(*wh.*)	deflexed.	——	1812.	5. 8.	G. ♄.	
124	demíssa. L.C.	(*re.*)	low.	——	1810.	——	G. ♄.	
125	dénsa. S.S.	(*bh.*)	dense.	——	——	——	G. ♄.	Andr. heath. ic.
126	densifòlia. W.	(*re.*)	dense-leaved.	——	1811.	3. 8.	G. ♄.	
127	denticulàta. W.	(*wh.*)	tooth-calyxed.	——	——	——	G. ♄.	Lodd. bot. cab. t. 1090.
	fastigiàta. A.H. *nec aliorum.*							
128	depréssa. H.K.	(*wh.*)	depressed.	——	1789.	6. 8.	G. ♄.	Andr. heath. v. 2.
	rupéstris. A.H.							
129	dianthifòlia. L.T.	(*re.*)	pink-leaved.	——	1796.	3. 8.	G. ♄.	
130	diaphàna. S.S.	(*fl.*)	transparent.	——	1800.	5. 8.	G. ♄.	Andr. heath. v. 4.
	transpàrens. B.C. 177. *non* Th.							
131	dìchrus. S.S.	(*re.gr.*)	Andrew's 2-color'd.	——	1790.	8. 5.	G. ♄.	Andr. heath. v. 4.
	bícolor. A.H. *non* W.							
132	Dickensoniàna. L.C.	(*va.*)	Mr. Dickenson's.	——	1809.	——	G. ♄.	
	α álba.	(*wh.*)	*white-flowered.*	——	——	——	G. ♄.	
	β rùbra.	(*re.*)	*red-flowered.*	——	——	——	G. ♄.	

133	diosmæfòlia. L.T.	(*li.*)	Diosma-leaved.	C. B. S.	1792.	3. 7.	G. ♄.	
134	diotæflòra. L.T.	(*wh.*)	Diotis-flowered.	———	1795.	3. 8.	G. ♄.	
135	díscolor. H.K.	(*bh.*)	different-coloured.	———	1788.	10. 3.	G. ♄.	Andr. heath. v. 1.
136	dístans. S.S.	(*vi.*)	distant.	———	1816.	——	G. ♄.	
137	Donniàna. Lee.	(*re.*)	Donn's.	———	1812.	5. 9.	G. ♄.	
138	droseroídes. A.H.	(*pu.*)	Sundew-like.	———	——	7. 10.	G. ♄.	Andr. heath. v. 4.
	droseroides β mìnor. A.H.							
139	dumòsa. A.H.	(*re.*)	bushy.	———	——	——	G. ♄.	Andr. heath. v. 4.
140	echiiflòra. H.K.	(*pi.*)	Echium-flowered.	———	1798.	2. 7.	G. ♄.	——— v. 3.
	β purpùrea. A.H.	(*pu.*)	*purple-flowered.*	———	1812.	——	G. ♄.	——— v. 4.
141	elàta. H.K.	(*ye.*)	tall.	———	1790.	7. 9.	G. ♄.	——— v. 2.
142	élegans. H.K.	(*pi.*)	elegant.	———	1799.	11. 5.	G. ♄.	Bot. mag. t. 960.
143	elongàta. B.C.	(*wh.*)	elongated.	———	1820.	3. 6.	G. ♄.	Lodd. bot. cab. t. 738.
144	emarginàta. A.H.	(*wh.*)	hairy cupped.	———	1800.	6. 10.	G. ♄.	Andr. heath. ic.
145	embothriifòlia. L.T.	(*re.*)	Embothrium-leav'd.	———	——	2. 5.	G. ♄.	
146	empetrifòlia. H.K.	(*re.*)	crowberry-leaved.	———	1774.	4. 7.	G. ♄.	Bot. mag. t. 447.
147	empetroídes. H.K.	(*pu.*)	close-flowered.	———	1788.	5. 8.	G. ♄.	Andr. heath. v. 2.
148	epistòmia. L.C.	(*wh.*)	spiggot.	———	1800.	——	G. ♄.	Lodd. bot. cab. t. 1186.
149	equìsetifòlia. L.T.	(*re.*)	Equisetum-leaved.	———	——	——	G. ♄.	
	articulàris. Hort.							
150	eriocéphala. B.C.	(*wh.*)	woolly-headed.	———	1816.	6. 8.	G. ♄.	Lodd. bot. cab. t. 1270.
151	eròsa. B.C.	(*bh.*)	gnawed-petal'd.	———	——	——	G. ♄.	——— t. 133.
152	erubéscens. H.K.	(*bh.*)	reddish-flowered.	———	1800.	3. 8.	G. ♄.	Andr. heath. v. 3.
153	Eweràna. H.K.	(*cr.*)	Ewer's.	———	1793.	7. 11.	G. ♄.	——— v. 2.
	α glábra.	(*cr.*)	*smooth-leaved.*	———	——	——	G. ♄.	
	β longiflòra.	(*cr.*)	*long-flowered.*	———	——	——	G. ♄.	
	γ pilòsa.	(*cr.*)	*hairy-leaved.*	———	——	——	G. ♄.	
154	exígua. L.T.	(*pu.*)	small downy.	———	1790.	3. 9.	G. ♄.	
155	exímia. L.C.	(*ro.*)	choice.	———	1811.	——	G. ♄.	Lodd. bot. cab. t. 1105.
156	exsérta. L.C.	(*wh.*)	exserted.	———	1820.	——	G. ♄.	
157	expánsa. L.C.	(*re.*)	expanded.	———	1811.	——	G. ♄.	
158	exprómta. S.S.	(*re.*)	woolly-branched.	———	——	——	G. ♄.	
159	exsùdans. B.C.	(*pi.*)	sweating.	———	1810.	8. 10.	G. ♄.	Lodd. bot. cab. t. 287.
160	exsúrgens. H.K.	(*or.re.*)	quiver-formed.	———	1792.	1. 12.	G. ♄.	Andr. heath. v. 1.
161	fábrilis. L.T.	(*re.*)	tiled-leaved.	———	1791.	4. 8.	G. ♄.	
162	fállax. L.T.	(*wh.*)	viscous-flower'd.	———	——	11. 3.	G. ♄.	
163	fasciculàris. H.K.	(*ro.gr.*)	cluster-flower'd.	———	1787.	2. 7.	G. ♄.	Andr. heath. v. 1.
	coronàta. A.H. *radiiflòra.* L.T.							
164	fastigiàta. W.	(*bh.*)	crowded.	———	1792.	2. 7.	G. ♄.	Andr. heath. v. 1.
	Walkèri. A.H. v. 1. *nec* v. 4.							
165	faústa. L.T.	(*re.*)	bristly-leaved.	———	1795.	4. 6.	G. ♄.	
166	ferrugìnea. H.K.	(*cr.bh.*)	rusty.	———	1798.	5. 7.	G. ♄.	Andr. heath. v. 3.
167	fésta. L.T.	(*bh.*)	angular-stalked.	———	....	3. 5.	G. ♄.	
168	fibula. L.en.	(*li.*)	button.	———	1812.	5. 6.	G. ♄.	
169	filamentòsa. H.K.	(*re.*)	long-peduncled.	———	1800.	1. 12.	G. ♄.	Bot. reg. t. 6.
170	filifórmis. L.T.	(*re.*)	filiform.	———	....	2. 5.	G. ♄.	
171	fimbriàta. A.H.	(*re.*)	fringed.	———	1800.	3. 7.	G. ♄.	Andr. heath. ic.
172	finitìma. L.C.	(*pu.*)	related.	———	....	——	G. ♄.	
173	fistulæflòra. L.T.	(*wh.*)	white slender-flower'd.	—	1800.	7. 11.	G. ♄.	Andr. heath. v. 3.
	tenuiflòra β álba. A.H. *Cliffordiàna.* B.C. 34.							
174	fláccida. L.en.	(*wh.*)	flaccid.	———	1810.	——	G. ♄.	
175	flagellàris. L.en.	(*st.*)	whipcord.	———	——	——	G. ♄.	
176	flagellifórmis. A.H.	(*re.*)	whipcord-like.	———	1812.	——	G. ♄.	Andr. heath. v. 4.
177	flámmea. H.K.	(*ye.*)	flame-flowered.	———	1798.	10. 5.	G. ♄.	——— v. 2.
	bìbax. L.T.							
178	flàva. H.K.	(*ye.*)	3-leaved yellow.	———	1795.	9. 4.	G. ♄.	Andr. heath. v. 2.
	β imbricàta. B.M.	(*ye.*)	*imbricated.*	———	——	——	G. ♄.	Bot. mag. t. 1815.
179	flexicaùlis. H.K.	(*re.*)	crook-stalked.	———	1800.	5. 1.	G. ♄.	Andr. heath. v. 2.
	glandulòsa. A.H. *nec aliorum.*							
180	flexuòsa. H.K.	(*wh.*)	zigzag.	———	1792.	4. 7.	G. ♄.	Andr. heath. v. 1.
181	floribúnda. B.C.	(*li.*)	many-flowered.	———	1800.	11. 3.	G. ♄.	Lodd. bot. cab. t. 176.
182	flòrida. H.K.	(*pu.*)	florid.	———	1803.	5. 8.	G. ♄.	Thunb. eric. n. 64. t. 6.
183	foliàcea. A.H.	(*ye.*)	foliaceous.	———	1822.	5. 7.	G. ♄.	Andr. heath. v. 4.
184	folliculàris. H.K.	(*ye.*)	yellow pencilled.	———	1794.	2. 7.	G. ♄.	——— v. 1.
	Petiveriàna. A.H.							
185	formòsa. W.	(*va.*)	handsome.	———	1795.	3. 8.	G. ♄.	Thunb. eric. n. 82. t. 3.
	α álba.	(*wh.*)	*white-flowered. reflexa alba.* Hort.	—	——	——	G. ♄.	Andr. heath. v. 4.
	β rùbra.	(*re.*)	*red-flowered. reflexa rubra.* Hort.	—	——	——	G. ♄.	——— v. 4.
186	fràgrans. H.K.	(*pu.*)	fragrant.	C. B. S.	1803.	3. 7.	G. ♄.	Bot. mag. t. 2181.

187	fùgax. L.T.	(*pu.*)	fugacious.	C. B. S.	1800.	3. 5.	G. ♄.	
188	furfuròsa. H.K.	(*re.*)	columnar-threaded.	——	1789.	8. 12.	G. ♄.	Andr. heath. v. 1.
	monadélpha. A.H. *non* B.M.							
189	gélida. H.K.	(*gr.*)	green verticillate.	——	1790.	4. 7.	G. ♄.	Andr. heath. v. 2.
	gílva. W. *alveiflòra.* L.T.							
190	gemmífera. B.M.	(*or.*)	jewel.	——	1802.	7. 10.	G. ♄.	Bot. mag. t 2266.
191	genistæfòlia. L.T.	(*re.*)	Genista-leaved.	——	....	10. 5.	G. ♄.	
192	glábra. L.en.	(*wh.*)	glabrous.	——	1802.	5. 8.	G. ♄.	
193	glandulòsa. Th.	(*cr.*)	glandular-haired.	——	1801.	——	G. ♄.	
194	glaùca. H.K.	(*pu.*)	glaucous-leaved.	——	1792.	——	G. ♄.	Bot. mag. t. 580.
195	globòsa. W.	(*bh.*)	globular-flowered.	——	1789.	7. 9.	G. ♄.	Andr. heath. v. 4.
196	glomeràta. S.S.	(*li.*)	glomerate.	——	1812.	——	G. ♄.	—— v. 4.
197	glomiflòra. L.T.	(*wh.*)	round-flowered.	——	....	4. 9.	G. ♄.	
198	glutinòsa. H.K.	(*pu.*)	glutinous.	——	1787.	7. 10.	G. ♄.	Icon. hort. kew. t. 17.
	droseroídes. A.H.							
199	gnaphalódes. W.	(*wh.*)	cudweed-like.	——	1812.	2. 4.	G. ♄.	Pluk. m.68. t. 346. f.11
200	grácilis. H.K.	(*re.*)	gracile.	——	1794.	2. 6.	G. ♄.	Wend. eric. 8. p. 9. ic.
201	grandiflòra. H.K.	(*ye.*)	great-flowered.	——	1775.	5. 9.	G. ♄.	Bot. mag. t. 189.
202	grandinòsa. B.C.	(*wh.*)	hail-flowered.	——	1820.	——	G. ♄.	Lodd. bot. cab. t. 627.
203	halicácaba. H.K.	(*wh.*)	bladder-flowered.	——	1780.	5. 8.	G. ♄.	Andr. heath. v. 2.
204	helianthemifòlia. L.T.		plain-leaved.	——	1796.	6. 9.	G. ♄.	
205	herbàcea. L.	(*fl.*)	trailing.	S. Europe.	1763.	1. 8.	H. ♄.	Bot. mag. t. 471.
206	Hibbertiàna. H.K.	(*cr.gr.*)	Hibbert's.	C. B. S.	1800.	6. 9.	G. ♄.	Andr. heath. v. 3.
207	hirsùta. B.C.	(*wh.pu.*)	hairy.	——	1812.	5. 8.	G. ♄.	Lodd. bot. cab. t. 754.
208	hírta. W.	(*re.*)	hairy-leaved.	——	1795.	4. 7.	G. ♄.	Thunb. eric. n. 56. t. 2.
209	hirtiflòra. B.M.	(*li.*)	hairy-flowered.	——	1790.	——	G. ♄.	Bot. mag. t. 481.
210	híspida. S.S.	(*re.*)	hispid.	——	1791.	7. 9.	G. ♄.	Andr. heath. ic.
211	hispídula. W.	(*pu.*)	bristly-stemmed.	——	——	6. 8.	G. ♄.	
212	holosericea. L.T.	(*wh.*)	velvetty.	——	....	2. 7.	G. ♄.	
213	horizontàlis. H.K.	(*wh.*)	horizontal-leaved.	——	1800.	7. 9.	G. ♄.	Andr. heath. v. 2.
214	Humeàna. B.C.	(*pi.*)	Sir A. Hume's.	——	1808.	5. 7.	G. ♄.	Lodd. bot. cab. t. 389.
215	humifùsa. L.T.	(*pi.*)	low.	——	....	6. 7.	G. ♄.	
216	hyacinthoídes. H.K.	(*pi.*)	Hyacinth-flower'd.	——	1798.	6. 8.	G. ♄.	Andr. heath. v. 3.
217	hyssopifòlia. L.T.	(*re.*)	Hyssop-leaved.	——	1800.	5. 9.	G. ♄.	
218	ignéscens. H.K.	(*or.sc.*)	fiery.	——	1792.	3. 7.	G. ♄.	Andr. heath. v. 2.
219	imbecílla.	(*re.*)	feeble.	——	1793.	6. 7.	G. ♄.	
	grácilis. L.T. *nec aliorum.*							
220	imbricàta. W.	(*pu.*)	imbricated.	——	1786.	5. 8.	G. ♄.	Lodd. bot. cab. t. 1243.
221	imperiàlis. S.S.	(*ro.*)	imperial.	——	1816.	——	G. ♄.	Andr. heath. v. 4.
222	incàna. S.S.	(*wh.re.*)	hoary.	——	——	——	G. ♄.	Wend. er. f. 18. n. 4. ic.
223	incarnàta. W.	(*fl.*)	flesh-coloured.	——	1791.	——	G. ♄.	Andr. heath. v. 1.
224	incúrva. W.	(*wh.*)	incurved.	——	——	——	G. ♄.	
225	inflàta. W.	(*ro.*)	inflated.	——	1800.	5. 9.	G. ♄.	Thunb. dis. n. 67. t. 2. f. 2.
226	infundibulifórmis.	(*ro.wh.*)	funnel-flower'd.	——	1812.	——	G. ♄.	Lodd. bot. cab. t. 589.
227	intertéxta. B.C.	(*wh.*)	interwoven.	——	1810.	——	G. ♄.	—— t. 1034.
228	intervallàris. L.T.	(*bh.*)	distant-leaved.	——	——	5. 11.	G. ♄.	
229	Irbyàna. A.H.	(*bh.*)	Irby's.	——	1800.	6. 10.	G. ♄.	Andr. heath. v. 3.
230	jasminiflòra. H.K.	(*wh.*)	Jasmine-flower'd.	——	1794.	——	G. ♄.	—— v. 1.
231	jubàta. S.S.	(*re.*)	crest-flowered.	——	1816.	——	G. ♄.	
232	Juliàna. S.S.	(*re.*)	Julian's.	——	1812.	5. 10.	G. ♄.	Lodd. bot. cab. t. 799.
233	labiàlis. L.T.	(*pi.*)	lipped.	——	——	——	G. ♄.	
234	lachneæfòlia. H.K.	(*pu.*)	Lachnea-leaved.	——	1793.	5. 7.	G. ♄.	Andr. heath. v. 3.
235	lactiflòra. B.C.	(*st.*)	milk-coloured.	——	1816.	——	G. ♄.	Lodd. bot. cab. t. 901.
	trícolor. S.S.							
236	l'ævis. A.H.	(*wh.*)	smooth.	——	1790.	3. 6.	G. ♄.	Andr. heath. v. 3.
237	Lambertiàna. H.K.	(*wh.*)	Lambert's.	——	1800.	5. 8.	G. ♄.	—— v. 2.
238	lanàta. W.	(*sn.*)	woolly.	——	1775.	2. 5.	G. ♄.	Wend. eric. f. 5. n. 2.
239	lanceolàta. P.S.	(*wh.*)	spear-leaved.	——	1791.	6. 12.	G. ♄.	—— fasc. 8. p. 13. ic.
240	lanuginòsa. H.K.	(*br.*)	large brown-flower'd.	——	1803.	9. 1.	G. ♄.	Andr. heath. v. 3.
241	laricìna. S.S.	(*pi.*)	Larch-leaved.	——	1824.	6. 8.	G. ♄.	
242	lasiophy'lla. S.S.	(*li.*)	downy-leaved.	——	1816.	5. 8.	G. ♄.	
243	lateràlis. H.K.	(*fl.*)	side-flowered.	——	1791.	3. 7.	G. ♄.	Andr. heath. v. 1.
244	latifòlia. H.K.	(*re.*)	broad-leaved.	——	1800.	5. 8.	G. ♄.	—— v. 2.
245	lavandulæfòlia. L.T.	(*re.*)	Lavender-leaved.	——	1795.	——	G. ♄.	
246	Lawsòni. B.M.	(*re.*)	red slender-flower'd.	——	1802.	4. 7.	G. ♄.	Bot. mag. t. 1720.
247	láxa. H.K.	(*li.*)	loose-flowered.	——	1800.	9. 2.	G. ♄.	Andr. heath. v. 3.
248	Leeàna. H.K.	(*ye.*)	Lee's.	——	1788.	8. 1.	G. ♄.	—— v. 1.
249	leptocárpha. S.S.	(*re.*)	slender chaffed.	——	1824.	4. 7.	G. ♄.	
250	leucántha. L.en.	(*wh.*)	white-blossomed.	——	1803.	1. 5.	G. ♄.	Andr. heath. v. 2.
	leucanthèra. A.H. *non* L.							

251	leucanthèra. W.	(*wh.*)	white-tipped.	C. B. S.	1803.	2. 6.	G. ♄.	
252	Linnæàna. H.K.	(*pu.wh.*)	Linnæus's.	———	1790.	1. 5.	G. ♄.	Andr. heath. v. 2.
	Linnæoìdes. A.H.							
	β *supérba.* A.H.	(*pu.wh.*)	*superb.*	———	1822.	——	G. ♄.	Andr. heath. v. 4.
253	longiflora. B.C.	(*ye.*)	long-flowered.	———	....	4. 8.	G. ♄.	Lodd. bot. cab. t. 983.
254	longifòlia. W.	(*re.*)	long-leaved.	———	1787.	2. 7.	G. ♄.	Wend. eric. t. 3.
	pìnea. Wend.							
255	longipedunculàta.	(*li.*)	long-stalked.	———	1818.	3. 8.	G. ♄.	Lodd. bot. cab. t. 103.
256	lùcida. H.K.	(*pi.*)	lucid.	———	1800.	4. 1.	G. ♄.	Andr. heath. v. 2.
257	lùtea. H.K.	(*ye.*)	yellow.	———	1774.	2. 5.	G. ♄.	——— v. 1.
258	lyrígera. L.T.	(*re.*)	lyre-bearing.	———	1790.	——	G. ♄.	
259	magnífica. S.S.	(*ro.*)	magnificent.	———	1816.	4. 8.	G. ♄.	Andr. heath. v. 4.
260	mammòsa. W.	(*pu.*)	nipple.	———	1762.	7. 10.	G. ♄.	——— v. 1.
	α *purpùrea.*	(*pu.*)	*purple-flowered.*	———	——	——	G. ♄.	Lodd. bot. cab. t. 125.
	β *pállida.*	(*li.*)	*pale-flowered.*	———	——	——	G. ♄.	——— t. 951.
261	margaritàcea. H.K.	(*wh.*)	pearl-flowered.	———	1775.	5. 9.	G. ♄.	Andr. heath. v. 1.
262	marifòlia. H.K.	(*wh.*)	Marum-leaved.	———	1773.	5. 7.	G. ♄.	——— v. 1.
263	Massòni. H.K.	(*or.gr.*)	Masson's.	———	1787.	7. 10.	G. ♄.	Bot. mag. t. 336.
264	mediterrànea. W.	(*re.*)	Mediterranean.	Portugal.	1648.	3. 5.	H. ♄.	——— t. 471.
265	melanthèra. W.	(*li.*)	dark-anthered.	C. B. S.	1803.	6. 8.	G. ♄.	Lodd. bot. cab. t. 867.
266	melástoma. A.H.	(*ye.*)	black-anthered.	———	1795.	4. 7.	G. ♄.	——— t. 333.
267	mellífera. S.S.	(*pu.*)	honey-bearing.	———	1816.	——	G. ♄.	
268	metulæflòra. H.K.	(*re.*)	nine-pin.	———	1798.	6. 8.	G. ♄.	Bot. mag. t. 612.
269	microphy'lla. L.C.	(*re.*)	small-leaved.	———	....	——	G. ♄.	
270	micróstoma. S.S.	(*re.*)	aggregate-flower'd.	———	1816.	——	G. ♄.	
271	minutæflòra. A.R.	(*pu.*)	minute-flowered.	———	1822.	5. 7.	G. ♄.	Andr. heath. v. 4.
272	mirábilis. A.H.	(*pu.*)	wonderful.	———	——	3. 8.	G. ♄.	——— v. 4.
273	modésta. L.T.	(*bh.*)	modest.	———	....	——	G. ♄.	
274	molleàris. P.S.	(*pu.*)	soft-leaved.	———	1803.	4. 10.	G. ♄.	
275	móllis. S.S.	(*pu.*)	soft.	———	1816.	——	G. ♄.	Andr. heath. v. 4.
276	mollíssima. L.C.	(*wh.*)	softest.	———	....	——	G. ♄.	
277	monadélpha. B.M.	(*wh.*)	monadelphous.	———	1789.	5. 7.	G. ♄.	Bot. mag. t. 1370.
	Bánksia β *purpùrea.* A.H.							
278	Monsòniæ. H.K.	(*wh.*)	Lady Monson's.	———	1787.	4. 9.	G. ♄.	Bot. mag. t. 1915.
279	montàna. L.C.	(*li.*)	mountain.	———	1816.	——	G. ♄.	
280	moschàta. A.H.	(*li.*)	musky.	———	1805.	5. 7.	G. ♄.	Andr. heath. v. 4.
281	mucòsa. H.K.	(*pu.*)	mucous.	———	1787.	2. 4.	G. ♄.	——— v. 1.
282	mùcosoídes. L.C.	(*li.*)	mucous-like.	———	1800.	3. 4.	G. ♄.	Lodd. bot. cab. t. 1212.
283	mucronàta. S.S.	(*re.*)	mucronate.	———	1812.	4. 8.	G. ♄.	Andr. heath. v. 4.
284	multiflòra. H.K.	(*bh.*)	many-flowered.	France.	1731.	6. 11.	G. ♄.	——— v. 2.
285	múnda. L.T.	(*bh.*)	neat-flowered.	C. B. S.	....	5. 7.	G. ♄.	
286	múndula. B.C.	(*pi.*)	neat.	———	1816.	3. 8.	G. ♄.	Lodd. bot. cab. t. 114.
287	Muscàri. H.K.	(*st.*)	musk.	———	1790.	——	G. ♄.	Andr. heath. v. 1.
288	mutábilis. H.K.	(*pu.*)	mutable.	———	1798.	2. 10.	G. ♄.	——— v. 3.
289	nàna. L.T.	(*ye.*)	dwarf.	———	1792.	5. 8.	G. ♄.	——— v. 2.
	depréssa. A.H.							
290	nidiflòra. L.T.	(*bh.*)	nest-flowered.	———	....	10. 5.	G. ♄.	
291	nidulària. B.C.	(*pi.*)	nest-like.	———	1816.	5. 8.	G. ♄.	Lodd. bot. cab. t. 764.
292	nìgricans. S.S.	(*wh.*)	black-stamened.	———	——	——	G. ♄.	
293	nigrìta. H.K.	(*wh.*)	black-tipped.	———	1790.	3. 7.	G. ♄.	Andr. heath. v. 1.
294	nítida. H.K.	(*wh.*)	shining.	———	1800.	7. 10.	G. ♄.	——— v. 3.
295	nivàlis. A.H.	(*wh.*)	snowy-flowered.	———	1820.	5. 6.	G. ♄.	——— v. 4.
296	nívea. S.S.	(*wh.*)	snowy.	———	1816.	5. 9.	G. ♄.	
297	Nivèni. H.K.	(*re.*)	Niven's.	———	1799.	2. 7.	G. ♄.	Andr. heath. v. 2.
298	nolæflòra. L.T.	(*re.*)	bell-shaped.	———	....	2. 5.	G. ♄.	
299	notábilis. S.S.	(*re.*)	notable.	———	1816.	3. 8.	G. ♄.	
300	nudiflòra. W.	(*ye.*)	small-bracted.	———	1783.	7. 8.	G. ♄.	
301	obbàta. H.K.	(*re.wh.*)	bottle.	———	1796.	4. 7.	G. ♄.	Andr. heath. v. 2.
302	obcordàta. S.S.	(*wh.*)	obcordate.	———	1812.	——	G. ♄.	
303	oblíqua. H.K.	(*pu.*)	oblique-leaved.	———	1789.	8. 10.	G. ♄.	Andr. heath. v. 1.
304	obtùsa. B.C.	(*re.*)	obtuse.	———	1816.	5. 8.	G. ♄.	Lodd. bot. cab. t. 1027.
305	octogòna. L.C.	(*re.*)	eight-angled.	———	——	——	G. ♄.	
306	odoràta. H.K.	(*wh.*)	perfumed.	———	1804.	4. 7.	G. ♄.	Andr. heath. v. 3.
307	ollùla. S.S.	(*ro.*)	close smooth-leav'd.	———	1812.	——	G. ♄.	——— v. 4.
308	onosmæflòra. L.T.	(*ye.*)	Onosma-flowered.	———	1789.	9. 5.	G. ♄.	——— v. 1.
	glutinòsa. A.H. *viscòsa.* Wend.							
309	oppositifòlia. A.H.	(*wh.*)	opposite-leaved.	———	1804.	3. 5.	G. ♄.	Andr. heath. c. ic.
	α *álba.*	(*wh.*)	*white-flowered.*	———	——	——	G. ♄.	Lodd. bot. cab. t. 1343.
	β *rùbra.* B.C.	(*re.*)	*red-flowered.*	———	——	——	G. ♄.	——— t. 1060.

Z

310	ostrìna. B.C.	(sc.)	scarlet-flowered.	C. B. S.	1824.	3. 6.	G. ♄.	Lodd. bot. cab. t. 1218.
311	ovàta. B.C.	(re.)	ovate.	——	1791.	2. 5.	G. ♄.	—— t. 417.
312	oxycoccifòlia. L.T.	(bh.)	Cranberry-leaved.	——	——	——	G. ♄.	
313	pállens. S.S.	(st.)	pale.	——	1812.	3. 8.	G. ♄.	Andr. heath. v. 4.
314	pállida. B.C.	(pu.)	pallid.	——	——	——	G. ♄.	Lodd. bot. cab. t. 1355.
315	palliiflòra. L.T.	(pa.)	pale-flowered.	——	....	2. 8.	G. ♄.	
316	palústris. H.K.	(bh.)	marsh.	——	1799.	5. 10.	G. ♄.	Andr. heath. v. 2.
317	paniculàta. W.	(pu.)	panicled.	——	1774.	2. 8.	G. ♄.	Lodd. bot. cab. t. 1194.
318	pannòsa. L.T.	(wh.)	cloth-flowered.	——	....	2. 6.	G. ♄.	
319	párilis. L.T.	(re.)	matched.	——	1789.	5. 8.	G. ♄.	
320	Parmentièri. B.C.	(wh.)	Parmentier's.	——	1816.	——	G. ♄.	Lodd. bot. cab. t. 197.
	β ròsea. L.C.	(ro.)	*rose-coloured.*	——	——	——	G. ♄.	
321	parviflòra. L.T.	(pu.)	small-fld. downy.	——	1790.	3. 9.	G. ♄.	
322	Passerìna. W.	(wh.)	sparrow-wort.	——	1800.	11. 5.	G. ♄.	Petiv. gaz. t. 3. f. 7.
323	pátens. A.H.	(pu.)	spreading.	——	——	3. 6.	G. ♄.	Andr. heath. v. 3.
324	Patersòni. H.K.	(ye.)	Paterson's.	——	1791.	3. 8.	G. ♄.	—— v. 1.
325	pavettæflòra. L.T.	(wh.)	Pavetta-flowered.	——	....	5. 8.	G. ♄.	
326	pectinifòlia. L.T.	(bh.)	pectinated.	——	1800.	11. 6.	G. ♄.	
327	pedunculàta. S.S.	(re.)	peduncled.	——	1810.	5. 9.	G. ♄.	
328	pellùcida. A.H.	(wh.)	pellucid.	——	1800.	10. 6.	G. ♄.	Andr. heath. v. 3.
	β rùbra. A.H.	(re.)	*red-flowered.*	——	1806.	——	G. ♄.	—— v. 4.
329	peltàta. A.H.	(gr.pu.)	peltate-stigma'd.	——	1804.	4. 8.	G. ♄.	—— v. 4.
330	péndula. W.	(pu.)	pendulous.	——	1791.	7. 8.	G. ♄.	Lodd. bot. cab. t. 902.
331	penicillàta. A.H.	(re.)	pencilled.	——	1774.	4. 7.	G. ♄.	Andr. heath. v. 2.
332	penicilliflòra. L.T.	(wh.)	white-pencilled.	——	1792.	5. 8.	G. ♄.	Wend. eric. 4. p. 5. ic.
333	periplocæfòlia. L.T.	(li.)	Periploca-leaved.	——	——	8. 12.	G. ♄.	
334	persolùta. H.K.	(re.)	garland.	——	1774.	2. 5.	G. ♄.	Bot. mag. t. 342.
335	perspícua. H.K.	(wh.)	clear-flowered.	——	1790.	3. 6.	G. ♄.	Wend. eric. 1. p. 7. ic.
	β Linnæ'a. A.H. v. 2.							
336	petiolàta. H.K.	(wh.)	rosemary-leaved.	——	1774.	3. 7.	G. ♄.	Andr. heath. v. 3.
337	Petiveriàna. H.K.	(ye.)	Petiver's.	——	——	——	G. ♄.	Lam. ill. t. 288. f. 3.
338	Pezìza. B.C.	(wh.)	button-flowered.	——	....	5. 8.	G. ♄.	Lodd. bot. cab. t. 265.
339	phylicoídes. W.	(pu.)	Phylica-like.	——	1800.	4. 7.	G. ♄.	
340	physòdes. H.K.	(wh.)	sticky.	——	1788.	3. 7.	G. ♄.	Bot. mag. t. 443.
341	pícta. L.C.	(wh.ye.)	painted.	——	1800.	——	G. ♄.	
342	pilòsa. S.S.	(gr.)	pilose.	——	——	——	G. ♄.	Lodd. bot. cab. t. 606.
343	pilulífera. W.	(pu.)	ball-bearing.	——	1789.	4. 5.	G. ♄.	
344	pìnea. S.S.	(wh.)	Pine-like.	——	1790.	8. 12.	G. ♄.	
	β purpùrea. L.C.	(pu.)	*purple.*	——	——	——	G. ♄.	Lodd. bot. cab. t. 1259.
345	pinifòlia. A.H.	(re.)	Pine-leaved.	——	1790.	6. 12.	G. ♄.	Andr. heath. v. 1.
	pityophy'lla. S.S.							
	a coccínea. A.H.	(sc.)	*scarlet.*	——	——	——	G. ♄.	Andr. heath. v. 1.
	β díscolor. A.H.	(wh.pu.)	*two-coloured.*	——	1820.	——	G. ♄.	—— v. 4.
346	planifòlia. A.H.	(re.)	flat-leaved.	——	——	——	G. ♄.	—— v. 3.
347	Plukenetiàna. H.K.	(sc.)	Plukenet's.	——	1774.	4. 7.	G. ♄.	—— v. 1.
	β pállida. L.C.	(pa.)	*pale-flowered.*	——	——	——	G. ♄.	
	γ álbens. A.H.	(wh.)	*white-flowered.*	——	——	——	G. ♄.	Andr. heath. v. 4.
348	poly'tricha.	(sc.gr.)	many-haired.	——	1804.	7. 10.	G. ♄.	Lodd. bot. cab. t. 1116.
	hírta. Lod. bot. cab. *nec aliorum pilòsa.* Hort.							
349	pr'æcox. L.C.	(li.)	early-dwarf.	——	1805.	2. 5.	G. ♄.	Lodd. bot. cab. t. 1413.
350	præ'gnans. A.H.	(bh.)	swelled.	——	1796.	5. 7.	G. ♄.	Andr. heath. v. 3.
	β coccìnea.	(sc.)	*scarlet-flowered.*	——	——	——	G. ♄.	—— v. 4.
351	præ'stans. S.S.	(bh.)	excelling.	——	——	——	G. ♄.	—— v. 4.
352	primuloídes. A.H.	(re.wh.)	Cowslip-flower'd.	——	1802.	4. 7.	G. ♄.	Bot. mag. t. 1548.
353	prínceps. H.K.	(pi.)	fine red.	——	1800.	——	G. ♄.	Andr. heath. v. 2.
354	pròcera. S.S.	(ye.)	lofty.	——	1791.	——	G. ♄.	
355	procúmbens. S.S.	(re.)	procumbent.	——	1816.	——	G. ♄.	
356	propéndens. H.K.	(pu.)	pendent.	——	1800.	7. 8.	G. ♄.	Bot. mag. t. 2140.
357	protrùdens. L.en.	(wh.)	protruding.	——	1816.	5. 8.	G. ♄.	
358	pubéscens. H.K.	(li.)	pale downy.	——	1790.	2. 12.	G. ♄.	
359	pubígera. L.T.	(li.)	pubescent-flowered.	——	1792.	1. 5.	G. ♄.	
360	pudibúnda. L.T.	(bh.)	nodding.	——	1812.	3. 8.	G. ♄.	Wend. eric. f. 3. p. 5. ic.
	nùtans. Wendland.							
361	pulchélla. W.	(re.)	neat.	——	——	——	G. ♄.	Thunb. dis. n. 24. t. 4. f. 2.
362	pulchérrima.	(ro.wh.)	beautiful.	——	1820.	7. 10.	G. ♄.	Andr. heath. v. 4.
	jasminiflòra β mìnor. A.H. v. 4. *Irbyàna.* Hort. *non* Andr.							
363	pulverulénta. L.C.	(wh.)	powdered.	——	——	3. 8.	G. ♄.	
364	pulviniformis. L.T.	(bh.)	cushion.	——	....	2. 5.	G. ♄.	
365	pùmila. S.S.	(li.)	short.	——	1812.	3. 8.	G. ♄.	Andr. heath. v. 4.

366	pùra. B.C.	(*wh.*) clear.	C. B. S.	1812.	5. 8.	G. ♄.	Lodd. bot. cab. t. 72.
367	purpùrea. H.K.	(*pu.*) purple-flowered.	——	1789.	1. 12.	G. ♄.	Andr. heath. v. 2.
368	pusílla. L.T.	(*bh.*) low.	——	....	6. 7.	G. ♄.	
369	pyramidàlis. H.K.	(*bh.*) pyramidal.	——	1787.	2. 5.	G. ♄.	Bot. mag. t. 366.
370	pyramidifórmis. S.S.	(*bh.*) pyramid-formed.	——	1818.	——	G. ♄.	
371	pyrolæflòra. L.T.	(*wh.*) Pyrola-flowered.	——	1790.	5. 7.	G. ♄.	Andr. heath. v. 4.
	andromedæflòra β álba. A.H. v. 4.						
372	quadræflòra. L.T.	(*re.*) four-flowered.	——	——	3. 8.	G. ♄.	
373	quadrangulàris. A.H.	(*fl.wh.*) square-tubed.	——	1812.	5. 8.	G. ♄.	Andr. heath. v. 4.
374	radiàta. H.K.	(*re.*) rayed.	——	1798.	8. 11.	G. ♄.	—— v. 1.
	β díscolor. A.H.	(*re.wh.*) *two-coloured.*	——	1820.	——	G. ♄.	—— v. 4.
375	racemòsa. W.	(*fl.*) racemed.	——	1795.	4. 5.	G. ♄.	
376	racemífera. H.K.	(*re.*) compact-flowered.	——	1803.	4. 6.	G. ♄.	Andr. heath. v. 2.
377	ramentàcea. H.K.	(*pu.*) slender-branched.	——	1786.	7. 12.	G. ♄.	—— v. 1.
378	ramulòsa. Viv.	(*li.*) Italian.	Italy.	1826.	5. 8.	H. ♄.	
379	recurvàta. B.C.	(*wh.*) recurved.	C. B. S.	1812.	——	G. ♄.	Lodd. bot. cab. t. 1093.
380	refléxa. S.S.	(*pu.wh.*) reflexed.	——	....		G. ♄.	
381	refúlgens. A.H.	(*ro.gr.*) refulgent.	——	1816.	5. 9.	G. ♄.	Andr. heath. v. 4.
382	regérminans. W.	(*li.*) clustered-flower'd.	——	1791.	5. 8.	G. ♄.	
383	resinòsa. H.K.	(*or.*) varnished.	——	1803.	——	G. ♄.	Bot. mag. t. 1139.
	vèrnix. A.H. v. 3.						
384	retórta. H.K.	(*ro.*) recurved-leaved.	——	1787.	——	G. ♄.	Bot. mag. t. 362.
385	retrofléxa. H.K.	(*re.*) jointed.	——	——	7. 9.	G. ♄.	Andr. heath. v. 1.
	pulchélla. A.H. *non* Th. *articulàris.* Th. *non* L.						
386	rígida. B.C.	(*re.*) rigid.	——	1820.	5. 8.	G. ♄.	Lodd. bot. cab. t. 1286.
387	rigidifòlia. S.S.	(*re.*) rigid-leaved.	——	1816.	6. 8.	G. ♄.	
388	Rollisòni.	(*bh.*) Rollison's.	——	1823.	5. 8.	G. ♄.	Andr. heath. v. 4.
	exsúrgens β hy'brida. A.H. v. 4.						
389	ròsea. H.K.	(*ro.*) rose-coloured.	——	1798.	6. 10.	G. ♄.	Andr. heath. v. 2.
390	rubélla. B.M.	(*re.*) reddish.	——	1812.	5. 8.	G. ♄.	Bot. mag. t. 2165.
391	rùbens. H.K.	(*re.*) red-flowered.	——	1798.	6. 9.	G. ♄.	
392	rùbida. B.C.	(*re.*) ruby.	——	1820.	——	G. ♄.	Lodd. bot. cab. t. 1166.
393	rubrosépala.	(*re.wh.*) red-calyxed.	——	1825.	——	G. ♄.	Andr. heath. v. 4.
	rùber-càlyx. A.H. v. 4.						
394	rugòsa. S.S.	(*ye.pi.*) wrinkled.	——	1812.	——	G. ♄.	Andr. heath. v. 4.
395	Russelliàna. B.C.	(*ro.*) Duke of Bedford's.	——	1824.	——	G. ♄.	Lodd. bot. cab. t. 1013.
396	Sainsburyàna. S.S.	(*ro.*) Sainsbury's.	——	1804.	7. 9.	G. ♄.	Andr. heath. v. 4.
397	sálax. L.T.	(*re.*) broad-stigma'd.	——	1796.	5. 9.	G. ♄.	
398	Salisburyàna. S.S.	(*ro.*) Salisbury's.	——	1815.	——	G. ♄.	Andr. heath. v. 4.
399	sanguínea. B.C.	(*cr.*) bloody.	——	....	3. 8.	G. ♄.	Lodd. bot. cab. t. 36.
400	sanguinolénta. B.M.	(*pu.*) dark-flowered.	——	1810.	——	G. ♄.	Bot. mag. t. 2263.
	pygm'æa. Andr. heath. v. 4.						
401	Savileàna.	(*re.*) Savile's.	——	1800.	6. 7.	G. ♄.	Lodd. bot. cab. t. 96.
402	scabriúscula. B.C.	(*wh.*) roughish.	——	——	——	G. ♄.	—— t. 517.
403	scariòsa. B.C.	(*li.*) scariose.	——	——	——	G. ♄.	—— t. 477.
404	Schóllii. B.C.	(*pu.*) Scholl's.	——	1790.	5. 9.	G. ♄.	—— t. 538.
405	scopària. W.	(*gr.*) small green-fld.	S. Europe.	1770.	4. 5.	F. ♄.	Linn. eric. n. 14. f. fl.
406	Sebàna. H.K.	(*re.*) Seba's.	C. B. S.	1774.	3. 6.	G. ♄.	Andr. heath. v. 1.
	α fúsca. Lod.	(*br.*) *brown-flowered.*	——	——	——	G. ♄.	
	β lùtea. Lod.	(*ye.*) *yellow-flowered.*	——	——	——	G. ♄.	Lodd. bot. cab. t. 266.
	γ mìnor. Lod.	(*or.*) *smaller.*	——	——	——	G. ♄.	
407	selaginifòlia. L.T.	(*re.*) Selago-leaved.	——	1801.	2. 6.	G. ♄.	
408	serpyllifòlia. B.C.	(*li.*) Serpyllum-leaved.	——	1812.	5. 8.	G. ♄.	Lodd. bot. cab. t. 744.
409	serratifòlia. H.K.	(*ye.*) saw-leaved.	——	1790.	8. 12.	G. ♄.	Andr. heath. v. 1.
410	serrulàta. L.C.	(*ye.*) sawed.	——	1814.	6. 8.	G. ♄.	
411	setàcea. A.H.	(*bh.*) bristly-leaved.	——	1796.	2. 8.	G. ♄.	Andr. heath. v. 1.
412	sexfària. H.K.	(*wh.*) six-angled.	——	1774.	5. 8.	G. ♄.	—— v. 2.
413	Shannoniàna. A.H.	(*bh.*) Earl Shannon's.	——	1800.	6. 7.	G. ♄.	Lodd. bot. cab. t. 168.
414	sicæfòlia. L.T.	(*re.*) dagger-leaved.	——	1791.	2. 7.	G. ♄.	
415	simpliciflòra. H.K.	(*y.re.*) single-flowered.	——	1774.	3. 7.	G. ♄.	Wend. eric. 17. p. 69. ic.
416	Smithiàna. L.C.	(*re.*) Smith's.	——	1810.	——	G. ♄.	
417	socciflòra. L.T.	(*gr.*) green-pencilled.	——	1799.	4. 5.	G. ♄.	Andr. heath. v. 1.
	Sebàna víridis. A.H.						
418	Solándri. H.K.	(*re.*) Solander's.	——	1800.	3. 9.	G. ♄.	Andr. heath. v. 2.
419	sórdida. H.K.	(*br.*) sordid.	——	1790.	8. 5.	G. ♄.	—— v. 1.
	laniflòra. Wend.						
420	Sparrmánni. H.K.	(*ye.*) Sparrman's.	——	1794.	9. 3.	G. ♄.	Andr. heath. v. 3.
	áspera. A.H. *hystriciflòra.* L.T.						
421	spársa. L.C.	(*li.*) scattered.	——	1800.	3. 8.	G. ♄.	Lodd. bot. cab. t. 1467.

422	speciòsa.	H.K.	(*or.re.*)	specious.	C. B. S.	1800.	6. 9.	G. ♄.	Andr. heath. v. 2.
423	spicàta.	H.K.	(*gr.*)	spiked.	———	1789.	1. 12.	G. ♄.	——— v. 1.
424	spiràlis.	L.C.	(*wh.*)	spiral.	———	1816.	5. 8.	G. ♄.	
425	spléndens.	W.	(*sc.*)	splendid.	———	1792.	4. 9.	G. ♄.	Wend. eric. 8. p. 5. ic.
426	spumòsa.	H.K.	(*st.*)	spumous.	———	1786.	5. 8.	G. ♄.	Lodd. bot. cab. t. 566.
427	spùria.	H.K.	(*pi.*)	spurious.	———	1796.	4. 8.	G. ♄.	Andr. heath. v. 1.
428	squamæflòra.	L.T.	(*pi.*)	scaly-flowered.	———	....	5. 11.	G. ♄.	
429	squamòsa.	H.K.	(*li.*)	scaly-cupped.	———	1794.	4. 6.	G. ♄.	Andr. heath. v. 3.
430	squarròsa.	L.T.	(*pu.*)	squarrose.	———	....	3. 8.	G. ♄.	
431	stagnàlis.	L.T.	(*bh.*)	stagnant.	———	1790.	3. 6.	G. ♄.	
432	stamínea.	A.H.	(*st.*)	reflex-stamened.	———	1799.	6. 9.	G. ♄.	Andr. heath. v. 3.
433	stellàta.	B.C.	(*wh.*)	starry.	———	1806.	——	G. ♄.	Lodd. bot. cab. t. 893.
	solandroìdes. Andr. heath. v. 4.								
434	stellífera.	S.S.	(*bh.*)	star-bearing.	———	——	4. 8.	G. ♄.	Andr. heath. v. 4.
435	stenántha.		(*bh.*)	slender-blush-flower'd.	—	1820.	5. 8.	G. ♄.	——— v. 4.
	tenuiflòra γ cárnea. A.H. v. 4.								
436	strícta.	H.K.	(*li.*)	straight-branch'd.	S. Europe.	1765.	8. 11.	G. ♄.	Andr. heath. v. 2.
437	strigòsa.	H.K.	(*li.*)	dwarf downy.	C. B. S.	1775.	3. 8.	G. ♄.	
438	struthiolæflòra.	L.C.	(*wh.*)	Struthiola-flower'd.	—	1812.	5. 8.	G. ♄.	
439	stylàris.	S.S.	(*re.*)	prominent-styled.	———	——	——	G. ♄.	
440	stylòsa.	L.T.	(*li.*)	long-styled.	———	1789.	1. 8.	G. ♄.	
441	suavèolens.	L.C.	(*li.*)	sweet.	———	1800.	5. 11.	G. ♄.	Lodd. bot. cab. t. 24.
442	subulàta.	S.S.	(*re.*)	awl-leaved.	———	1818.	6. 8.	G. ♄.	
443	sulphùrea.	B.M.	(*st.*)	sulphur-coloured.	———	1812.	3. 5.	G. ♄.	Bot. mag. t. 1984.
444	supérba.		(*re.*)	superb.	———	1795.	6. 9.	G. ♄.	Andr. heath. c. ic.
	formòsa. A.H. *nec aliorum.*								
	α coccínea.		(*sc.*)	*scarlet.*	———	——	——	G. ♄.	Andr. heath. v. 1.
	β ròsea.		(*ro.*)	*rose-coloured.*	———	——	——	G. ♄.	——— v. 2.
	γ bícolor.		(*ye.gr.*)	*two-coloured.*	———	——	——	G. ♄.	——— v. 4.
445	Swainsoniàna.	S.S.	(*pi.*)	Swainson's.	———	1810.	6. 7.	G. ♄.	——— v. 4.
446	tardiflòra.	L.T.	(*li.*)	pubescent.	———	1790.	3. 8.	G. ♄.	Bot. mag. t. 480.
	pubéscens. B.M. *non* L.								
447	taxifòlia.	H.K.	(*bh.*)	yew-leaved.	———	1788.	7. 11.	G. ♄.	Andr. heath. v. 1.
448	tegulæfòlia.	L.T.	(*wh.*)	tiled.	———	....	1. 5.	G. ♄.	
449	Templèæ.	A.H.	(*ro.*)	Lady Temple's.	———	1820.	5. 8.	G. ♄.	Andr. heath. v. 4.
450	tenélla.	H.K.	(*re.*)	delicate.	———	1791.	8. 5.	G. ♄.	——— v. 2.
451	tenuiflòra.	A.H.	(*st.*)	slender yellow-flower'd.	—	1800.	4. 6.	G. ♄.	——— v. 3.
452	tenuifòlia.	H.K.	(*re.*)	slender-leaved.	———	1794.	4. 8.	G. ♄.	Seb. mus. 1. t. 73. f. 6.
453	ténuis.	L.T.	(*wh.*)	slender.	———	1790.	8. 2.	G. ♄.	
454	tenuíssima.	S.S.	(*re.*)	very-slender.	———	1803.	2. 8.	G. ♄.	Wend. eric. 6. p. 9. ic.
455	tetragòna.	H.K.	(*ye.*)	square-flowered.	———	1789.	7. 9.	G. ♄.	Andr. heath. v. 3.
456	Tétralix.	H.K.	(*li.*)	cross-leaved.	Britain.	....	6. 8.	H. ♄.	Eng. bot. v. 15. t. 1014.
	α rùbra.		(*re.*)	*red-flowered.*	———	....	——	H. ♄.	
	β cárnea.		(*li.*)	*flesh-coloured.*	———	....	——	H. ♄.	
	γ álba.		(*wh.*)	*white-flowered.*	———	....	——	H. ♄.	
457	teucriifòlia.	S.S.	(*pu.*)	Teucrium-leaved.	C. B. S.	1812.	5. 9.	G. ♄.	
458	thalictriflòra.	S.S.	(*wh.*)	meadow-rue-flower'd.	—	1810.	——	G. ♄.	Lodd. bot. cab. t. 1294.
459	Thunbérgii.	H.K.	(*or.*)	Thunberg's.	———	1794.	5. 8.	G. ♄.	Bot. mag. t. 1214.
460	thymifòlia.	H.K.	(*re.*)	Thyme-leaved.	———	1789.	——	G. ♄.	Andr. heath. v. 2.
461	tiaræflòra.	H.K.	(*re.*)	turban-flower'd.	———	1800.	——	G. ♄.	——— v. 3.
462	togàta.	B.M.	(*re.*)	large-cupped.	———	1812.	6. 8.	G. ♄.	Bot. mag. t. 1626.
463	tomentòsa.	L.T.	(*wh.*)	woolly-flower'd.	———	....	2. 8.	G. ♄.	
464	tortuòsa.	L.C.	(*wh.*)	twisted.	———	1816.	3. 6.	G. ♄.	
465	tótta.	S.S.	(*wh.*)	bristly-branched.	———	1822.	——	G. ♄.	
466	tragulífera.	L.T.	(*li.*)	broad-anthered.	———	....	5. 8.	G. ♄.	
467	translùcens.	Lee.	(*bh.*)	translucent.	———	1797.	——	G. ♄.	
468	transpàrens.	Th.	(*bh.*)	transparent.	———	1800.	——	G. ♄.	
469	trìceps.	B.C.	(*wh.*)	three-headed.	———	1809.	——	G. ♄.	Lodd. bot. cab. t. 962.
470	trícolor.	H.S.L.	(*or.r.gr.*)	three-coloured.	———	1803.	6. 9.	G. ♄.	Andr. heath. v. 3.
	aristàta minor. A.H. *non trícolor.* S.S.								
471	triflòra.	H.K.	(*wh.*)	three-flowered.	———	1774.	3. 6.	G. ♄.	Wend. eric. 12. p. 13.
472	triphy'lla.	L.en.	(*re.ye.*)	three-leaved.	———	1812.	——	G. ♄.	
473	triúmphans.	B.C.	(*wh.*)	triumphant.	———	——	5. 10.	G. ♄.	Lodd. bot. cab. t. 257.
474	tróssula.	B.C.	(*wh.pi.*)	dapper.	———	——	——	G. ♄.	——— t. 668.
475	tuberculàris.	L.T.	(*pu.*)	tubercled.	———	1790.	2. 8.	G. ♄.	
476	tubiflòra.	H.K.	(*pi.*)	tube-flowered.	———	1775.	4. 7.	G. ♄.	Andr. heath. v. 1.
477	tubiúscula.	B.C.	(*re.*)	small-tubed.	———	1822.	——	G. ♄.	Lodd. bot. cab. t. 1157.
478	tùmida.	B.R.	(*sc.*)	tumid.	———	1812.	5. 9.	G. ♄.	Bot. reg. t. 65.
479	turbiniflòra.	L.T.	(*wh.*)	top-flowered.	———	1793.	10. 2.	G. ♄.	

480	túrgida. L.en.	(*pi.*)	turgid.	C. B. S.	1800.	3. 6.	G. ♄.	Andr. heath. v. 4.
	taxifòlia β màjor. A.H. v. 4.							
481	turrígera. L.T.	(*li.*)	Cypress.	———	1796.	6. 9.	G. ♄.	
	cupressìna. Hort.							
482	umbellàta. H.K.	(*re.*)	umbelled.	Portugal.	1782.	5. 7.	G. ♄.	Andr. heath. v. 2.
483	undulàta. A.H.	(*cr.*)	crumple-flower'd.	C. B. S.	1824.	5. 8.	G. ♄.	——— v. 4.
484	ùnica. S.S.	(*bh.*)	one-flowered.	———	1812.	5. 7.	G. ♄.	——— v. 4.
	pedunculàta. A.H. *nec aliorum.*							
485	urceolàris. H.K.	(*wh.*)	pitcher-flowered.	———	1778.	——	G. ♄.	Icon. hort. kew. t. 16.
486	ursìna. Lee.	(*wh.*)	shaggy.	———	1812.	4. 8.	G. ♄.	
487	vàgans. H.K.	(*bh.*)	Cornish.	Cornwall.	....	7. 8.	H. ♄.	Eng. bot. v. 1. t. 3.
	β álba.	(*wh.*)	*white-flowering.*	———	....	——	H. ♄.	
488	vària. B.C.	(*va.*)	variable.	C. B. S.	1820.	5. 8.	G. ♄.	Lodd. bot. cab. t. 1325.
489	velitàris. L.T.	(*bh.*)	javelin-flowered.	———	1790.	1. 6.	G. ♄.	
490	velleriflòra. H.K.	(*bh.*)	woolly-flowered.	———	1774.	2. 6.	G. ♄.	Andr. heath. v. 1.
	bruniàdes. A.H.							
491	ventricòsa. H.K.	(*bh.*)	porcelain.	———	1787.	4. 9.	G. ♄.	Bot. mag. t. 350.
492	ventròsa.	(*wh.*)	tun-bellied.	———	1822.	5. 8.	G. ♄.	Andr. heath. v. 4.
	obbàta β umbellàta. A.H. v. 4.							
493	verecúnda. L.T.	(*bh.*)	blushing.	———	1791.	4. 9.	G. ♄.	Andr. heath. v. 1.
	cérnua. A.H. *non* Thunb.							
494	verniciflua. L.T.	(*wh.*)	viscous-leaved.	———	1804.	3. 9.	G. ♄.	
495	versícolor. H.K.	(*sc.or.*)	various-colored.	———	1790.	11. 5.	G. ♄.	Andr. heath. v. 1.
496	verticillàta. A.H.	(*sc.*)	verticillate.	———	1774.	7. 10.	G. ♄.	——— v. 1.
497	vesiculàris. L.T.	(*bh.*)	bladdered.	———	1796.	2. 8.	G. ♄.	
498	vestíflua. L.T.	(*re.*)	clothed.	———	——	3. 6.	G. ♄.	
499	vestìta. H.K.	(*va.*)	tremulous.	———	1789.	1. 12.	G. ♄.	
	α álba. H.K.	(*wh.*)	*white.*	———	——	——	G. ♄.	Andr. heath. v. 1.
	β incarnàta. H.K.	(*fl.*)	*flesh-coloured.*	———	1789.	1. 12.	G. ♄.	——— v. 2.
	γ purpùrea. H.K.	(*pu.*)	*purple-flowered.*	———	——	——	G. ♄.	——— v. 1.
	δ ròsea. H.K.	(*ro.*)	*rose-coloured.*	———	——	——	G. ♄.	——— v. 2.
	ε fúlgida. H.K.	(*sc.*)	*bright red.*	———	——	——	G. ♄.	——— v. 2.
	ζ coccínea. H.K.	(*sc.*)	*scarlet.*	———	——	——	G. ♄.	——— v. 1.
	η lùtea. H.K.	(*ye.*)	*yellow.*	———	——	——	G. ♄.	——— v. 3.
500	villòsa. H.K.	(*wh.*)	villous.	———	1800.	2. 6.	G. ♄.	——— v. 3.
501	virgàta. S.S.	(*wh.*)	twiggy.	———	1824.	——	G. ♄.	
502	viridéscens. B.C.	(*gr.*)	greenish.	———	....	1. 6.	G. ♄.	Lodd. bot. cab. t. 233.
503	viridiflòra. A.H.	(*gr.*)	green-flowered.	———	1820.	5. 8.	G. ♄.	Andr. heath. v. 4.
504	viridipurpùrea. W.	(*gr.*)	green & purple.	Portugal.	....	5. 8.	F. ♄.	Linn. eric. n. 9. f. fl.
505	víridis. H.K.	(*gr.*)	green-flowered.	C. B. S.	1800.	——	G. ♄.	Andr. heath. v. 2.
506	víscaria. H.K.	(*li.*)	clammy-flower'd.	———	1774.	3. 7.	G. ♄.	Icon. hort. kew. t. 1.
507	Walkeriàna.	(*re.*)	Walker's.	———	1806.	6. 8.	G. ♄.	Andr. heath. v. 4.
	Walkèria β rùbra. A.H. v. 4.							
508	xeranthemifòlia. L.T.		Xeranthemum-ld.	———	1812.	5. 8.	G. ♄.	
509	Zeyhèri. S.S.	(*li.*)	Zeyher's.	———	1824.	——	G. ♄.	
	MENZI'ESIA Sm.		MENZI'ESIA.	Octandria Monogynia.				
1	ferrugìnea. S.S.	(*co.*)	ferruginous.	N. America.	1811.	5. 6.	H. ♄.	Sm. ic. ined. 1. t. 56.
2	globulàris. S.S.	(*co.*)	globular-flower'd.	———	1806.	——	H. ♄.	Sal. par. lond. t. 44.
	ferrugínea β. B.M. 1571.							
3	polifòlia. S.S.	(*pu.*)	Irish.	Ireland.	....	6. 9.	H. ♄.	Eng. bot. v. 1. t. 35.
	Erìca Daboècia. E.B.							
	β nàna.	(*pu.*)	*dwarf.*	....	....	——	H. ♄.	
4	cœrùlea. S.S.	(*pu.*)	yew-leaved.	Scotland.	....	6. 7.	H. ♄.	Eng. bot. v. 35. t. 2169.
	P'YROLA. S.S.		WINTER-GREEN.	Decandria Monogynia. L.				
1	uniflòra. S.S.	(*wh.*)	single-flower'd.	Britain.	....	6. 7.	H. ♃.	Eng. bot. t. 146.
2	secúnda. S.S.	(*wh.*)	serrated.	———	....	——	H. ♃.	——— t. 517.
3	mìnor. S.S.	(*bh.*)	lesser.	———	....	——	H. ♃.	——— t. 158.
	β ròsea. E.B.	(*ro.*)	*rose-coloured.*	England.	....	7. 8.	H. ♃.	——— t. 2543.
4	mèdia. S.S.	(*bh.*)	intermediate.	———	....	6. 7.	H. ♃.	——— t. 945.
5	dentàta. D.D.	(*st.*)	toothed.	N. America.	1827.	——	H. ♃.	
6	ellíptica. D.D.	(*wh.*)	elliptic-leaved.	———	1826.	——	H. ♃.	
7	occidentàlis. D.D.	(*st.*)	western.	———	1827.	——	H. ♃.	
8	rotundifòlia. S.S.	(*st.*)	round-leaved.	Britain.	....	——	H. ♃.	Eng. bot. t. 213.
9	americàna.	(*wh.*)	American.	N. America.	....	——	H. ♃.	
	rotundifòlia. Ph. *non* Eng. bot.							
10	asarifòlia. S.S.	(*st.*)	Asarum-leaved.	———	1822.	——	H. ♃.	
11	chlorántha. S.S.	(*st.*)	cream-colour'd.	———	——	——	H. ♃.	
	CHIMA'PHILA. Ph.		CHIMA'PHILA.	Decandria Monogynia.				
1	umbellàta. S.S.	(*wh.re.*)	umbelled.	N. America.	1752.	6. 8.	H. ♃.	Bot. mag. t. 778.
	corymbòsa. Ph. *P'yrola umbellàta.* L.							

2 maculàta. S.S.	(*wh.*)	spotted-leaved.	N. America.	1752.	6. 8.	H. ♃.	Bot. mag. t. 897.
DIAPE'NSIA. W.		DIAPE'NSIA.	Pentandria Monogynia. L.				
lappónica. W.	(*wh.*)	obtuse-leaved.	Lapland.	1801.	2. 3.	H. ♃.	Bot. mag. t. 1108.
PYXIDANTH'ERA. M.		PYXIDANTH'ERA.	Pentandria Monogynia.				
barbulàta. M.	(*wh.*)	barbed.	Carolina.	1806.	7.	G. ♃.	Mich. amer. t. 17.
Diapénsia cuneifòlia. Sal. par. lond. t. 104.							

Sect. II. *MONOTROPEÆ.* *Don prodr. p.* 151.

MONO'TROPA. N.		YELLOW BIRD'S-NEST.	Decandria Monogynia. L.				
1 uniflòra. N.	(*st.*)	one-flowered.	N. America.	1822.	6. 8.	H. ♃.	Hook. ex. flor. t. 85.
2 Hypopítys. E.B.	(*st.*)	common.	Britain.	....	6. 7.	H. ♃.	Eng. bot. v. 1. t. 69.
Hypopíthys europ'æa. N.							

Sect. III. *RHODORACEÆ.* *Don prodr. p.* 152.

KA'LMIA. W.		KA'LMIA.	Decandria Monogynia. L.				
1 latifòlia. W.	(*pi.wh.*)	broad-leaved.	N. America.	1734.	5. 7.	H. ♄.	Bot. mag. t. 175.
2 angustifòlia. W.	(*re.*)	narrow-leaved.	——	1736.	——	H. ♄.	—— t. 331.
β variegàta.	(*li.*)	*variegated.*	——	——	——	H. ♄.	
γ mínima.	(*re.*)	*least.*	——	....	——	H. ♄.	
δ nàna.	(*re.*)	*dwarf.*	——	....	——	H. ♄.	
ε pùmila.	(*re.*)	*small.*	——	....	——	H. ♄.	
ζ ròsea.	(*ro.*)	*rose-coloured.*	——	....	——	H. ♄.	
η rùbra. B.C.	(*re.*)	*red-flowered.*	——	....	——	H. ♄.	Lodd. bot. cab. t. 502.
ϑ ovàta. Ph.	(*re.*)	*oval-leaved.*	New Jersey.	....	——	H. ♄.	
3 glàuca. W.	(*re.*)	glaucous.	N. America.	1767.	4. 5.	H. ♄.	Bot. mag. t. 177.
4 rosmarinifòlia. Ph.	(*re.*)	Rosemary-leaved.	——	1812.	——	H. ♄.	
5 hirsùta. W.	(*ro.*)	hairy.	——	1786.	8. 9.	H. ♄.	Bot. mag. t. 138.
EPIG'ÆA. W.		EPIG'ÆA.	Decandria Monogynia. L.				
rèpens. W.	(*wh.*)	creeping.	N. America.	1736.	7. 8.	H. ♄.	Bot. reg. t. 201.
RHODODE'NDRON. W.		RHODODE'NDRON.	Pent-Decandria Monogynia. L.				
1 máximum. W.	(*bh.*)	large.	N. America.	1736.	6. 8.	H. ♄.	Bot. mag. t. 951.
2 álbum. Ph.	(*wh.*)	white-flowered.	——	1811.	——	H. ♄.	
3 purpùreum. Ph.	(*pu.*)	purple-flowered.	——	....	——	H. ♄.	
4 pónticum. W.	(*va.*)	common.	Gibraltar.	1763.	5. 6.	H. ♄.	Bot. mag. t. 650.
α purpùreum.	(*pu.*)	*purple.*	——	——	——	H. ♄.	
β ròseum.	(*ro.*)	*rose-coloured.*	——	....	——	H. ♄.	
γ álbum.	(*wh.*)	*white.*	——	....	——	H. ♄.	
δ variegàtum.	(*li.*)	*variegated.*	——	....	——	H. ♄.	
ε marginàtum.	(*li.*)	*silver-edged.*	——	....	——	H. ♄.	
ζ múltiplex.	(*pu.*)	*double-blossom'd.*	——	....	——	H. ♄.	
5 punctàtum. W.	(*li.*)	dotted-leaved.	N. America.	1786.	6. 8.	H. ♄.	Andr. reposit. t. 36.
α mìnus. W.D.	(*li.*)	*smallest.*	——	——	——	H. ♄.	Wats. dend. brit. t. 162.
β màjor. B.R.	(*li.*)	*larger.*	——	1812.	——	H. ♄.	Bot. reg. t. 37.
6 obtùsum. W.D.	(*li.*)	blunt-leaved.	Hybrid.	....	——	H. ♄.	Wats. dend. brit. t. 162.
7 azaleoídes. S.S.	(*li.*)	Azalea-like.	——	....	——	H. ħ.	Andr. reposit. t. 379.
pónticum β subdecíduum. A.R.							
8 hy'bridum. B.R.	(*li.*)	Herbert's hybrid.	——	....	——	H. ♄.	Bot. reg. t. 195.
9 Goweniànum. B.F.G.	(*pu.*)	Mr. Gowen's.	——	1825.	——	H. ♄.	Swt. br. fl. gar. t. 263.
10 Smíthii.		Smith's.	Arboreo-pont.	1826.	....	H. ♄.	
11 Carnarvòni		Lord Carnarvon's.	Arb-catawb.	——	....	H. ♄.	
12 catawbiénse. M.	(*ro.*)	Catawba.	N. America.	1809.	6. 8.	H. ♄.	Bot. mag. t. 1671.
13 arbòreum. Sm.	(*sc.*)	tree.	India.	1817.	4. 6.	F. ♄.	Swt. br. fl. gar. t. 250.
β ròseum.	(*ro.*)	*rose-coloured.*	——	——	——	F. ♄.	Bot. reg. t. 1240.
γ níveum.	(*wh.*)	*snowy white.*	——	——	——	F. ♄.	
14 campanulàtum. D.P.	(*ro.*)	bell-flowered.	Nepaul.	1824.	....	F. ♄.	
15 anthopògon. D.P.	(*ro.*)	bearded-flower'd.	——	——	....	G. ♄.	
16 cinnamòmeum. Wal.	(*ro.*)	cinnamon-color'd.	——	1826.	....	G. ♄.	
17 setòsum. D.P.	(*pu.*)	bristly.	——	1820.	....	G. ♄.	
18 daùricum. W.	(*pu.*)	Daurian.	Siberia.	1780.	2. 4.	H. ħ.	Bot. mag. t. 636.
β altaícum. F.	(*li.*)	*Altay.*	——	1824.	——	H. ħ.	
γ atrovìrens. B.R.	(*d.pu.*)	*dark green.*	——	....	——	H. ♄.	Bot. reg. t. 194.

19	myrtifòlium. B.C.	(*pu.*)	Myrtle-leaved.	Hybrid.	....	6. 8.	H. ♄.	Lodd. bot. cab. t. 908.
20	ferrugíneum. W.	(*ro.*)	rusty-leaved.	Switzerland.	1752.	5. 7.	H. ♄.	Jacq. aust. 3. t. 255.
21	hirsùtum. W.	(*ro.*)	hairy-leaved.	——	1656.	——	H. ♄.	Bot. mag. t. 1853.
22	chrysánthum. W.	(*ye.*)	yellow.	Siberia.	1796.	6. 7.	H. ♄.	Sal. par. lond. t. 80.
	officinàle. P.L.							
23	caucásicum. W.	(*ro.wh.*)	Caucasean.	Caucasus.	1803.	5. 7.	H. ♄.	Bot. mag. t. 1145.
24	Catesb'æi. L.C.	(*ro.*)	Catesby's.	N. America.	....	——	H. ♄.	
25	camtscháticum. W.	(*pi.*)	Kamtschatka.	Kamtschatka.	1802.	....	H. ♄.	Pall. ros. 1. t. 33.
26	chamæcístus. W.	(*pi.*)	Thyme-leaved.	Siberia.	1786.	5. 6.	H. ♄.	Bot. mag. t. 488.
27	lappónicum. S.S.	(*ro.*)	Lapland.	Lapland.	1827.	....	H. ♄.	Flor. dan. t. 966.
	Azàlea lappónica. L.							
28	Rhodòra.	(*pi.*)	Canadian.	Canada.	1767.	4. 5.	H. ♄.	Bot. mag. t. 474.
	Rhodòra canadénsis. W.							
29	índicum.	(*va.*)	Indian.	China.	1808.	3. 5.	G. ♄.	
	Azàlea índica. B.M.							
	α puníceum.	(*pi.*)	*red-flowered.*	——	——	——	G. ♄.	Bot. mag. t. 1480.
	β phœníceum.	(*pu.*)	*purple-flowered.*	——	1824.	——	G. ♄.	—— t. 2667.
	γ Smíthii.	(*li.*)	*Smith's.*	Hybrid.	1826.	4. 7.	G. ♄.	
	δ purpùreum.	(*pu.*)	*double-purple.*	China.	1819.	3. 5.	G. ♄.	Bot. mag. t. 2509.
	ε lùteum.	(*ye.*)	*double yellow.*	——	1826.	——	G. ♄.	
	ζ álbum.	(*wh.*)	*white-flowered.*	——	1819.	3. 6.	G. ♄.	Bot. reg. t. 811.
	Azàlea ledifòlia. Bot. mag. t. 2901.							
30	sinénse. B.F.G.	(*ye.*)	Chinese.	——	1824.	——	H. ♄.	
	α flámmeum. B.F.G.	(*or.*)	*flame-coloured.*	——	——	——	H. ♄.	Lodd. bot. cab. t. 885.
	β flavéscens. B.F.G.	(*ye.*)	*pale yellow.*	——	1826.	——	H. ♄.	Swt. br. fl. gar. t. 290.
31	lùteum.	(*ye.*)	yellow.	Turkey.	1793.	5. 6.	H. ♄.	Bot. mag. t. 433.
	Azàleu póntica. B.M.							
	β cùpreum.	(*co.*)	*copper-coloured.*	——	....	——	H. ♄.	
	γ glaùcum.	(*pa.*)	*glaucous.*	——	....	——	H. ♄.	
	δ pállidum.	(*pa.*)	*pale.*	——	....	——	H. ♄.	
	ε albiflòrum.	(*wh.*)	*white-flowered.*	——	....	——	H. ♄.	Bot. mag. t. 2383.
32	trícolor.	(*ye.ro.wh.*)	three-coloured.	Hybrid.	1816.	——	H. ♄.	
33	Mortèrii. B.F.G.	(*fl.ye.*)	Morter's.	——	1828.	——	H. ♄.	Swt. br. fl. gar. s. 2. t. 10.
	α cárneum.	(*fl.*)	*flesh-coloured.*	——	——	——	H. ♄.	
	β præ'stans.	(*co.bh.*)	*handsome.*	——	——	——	H. ♄.	
34	calendulàceum.	(*ye.*)	yellow American.	N. America.	1806.	——	H. ♄.	
	Azàlea calendulàcea. M.							
	α cròceum.	(*sn.*)	*saffron-coloured.*	——	——	——	H. ♄.	Bot. mag. t. 1721.
	β chrysoléctum.	(*go.*)	*golden.*	——	....	——	H. ♄.	
	γ grandiflòrum.	(*or.*)	*large-flowered.*	——	....	——	H. ♄.	
	δ cùpreum.	(*co.*)	*copper-coloured.*	——	....	——	H. ♄.	
	ε spléndens.	(*co.ye.*)	*shining.*	——	....	——	H. ♄.	
	ζ flámmeum.	(*co.ye.*)	*flame-coloured.*	——	1812.	——	H. ♄.	Bot. reg. t. 145.
	η triúmphans.	(*co.ye.*)	*triumphant.*	——	——	——	H. ♄.	
	ϑ ignéscens.	(*co.ye.*)	*fiery-flowered.*	——	——	——	H. ♄.	
35	canéscens. M.	(*ro.*)	canescent.	——	1812.	——	H. ♄.	
36	speciòsum.	(*va.*)	large-calyxed.	——	....	——	H. ♄.	Wats. dend. brit. t. 116.
	α màjor.	(*sc.*)	*large scarlet.*	——	....	——	H. ♄.	Lodd. bot. cab. 624.
	β críspum.	(*sc.*)	*curled.*	——	....	——	H. ♄.	
	γ aurántium.	(*or.*)	*orange.*	——	....	——	H. ♄.	
	δ oblíquum.	(*re.*)	*oblique-leaved.*	——	....	——	H. ♄.	
	ε prunifòlium.	(*re.*)	*Plum-leaved.*	——	....	——	H. ♄.	
	ζ cucullàtum.	(*re.*)	*hollow-leaved.*	——	....	——	H. ♄.	
	η revolùtum.	(*re.*)	*revolute.*	——	....	——	H. ♄.	
	ϑ ciliàtum.	(*re.*)	*fringed.*	——	....	——	H. ♄.	
	ι acutifòlium.	(*re.*)	*acute-leaved.*	——	....	——	H. ♄.	
	κ undulàtum.	(*re.*)	*wave-leaved.*	——	....	——	H. ♄.	
	λ tortulifòlium.	(*re.*)	*twisted-leaved.*	——	....	——	H. ♄.	
37	nudiflòrum.	(*va.*)	small calyxed.	——	1734.	——	H. ♄.	
	Azàlea nudiflòra. W.							
	α álbum.	(*wh.*)	*early white.*	——	——	——	H. ♄.	
	β álbo-plènum.	(*wh.*)	*double white.*	——	....	——	H. ♄.	
	γ blándum.	(*bh.*)	*blush-flowered.*	——	....	——	H. ♄.	
	δ cárneum.	(*fl.*)	*pale red.*	——	1734.	——	H. ♄.	Bot. reg. t. 120.
	ε caroliniànum.	(*fl.*)	*Carolina.*	——	....	——	H. ♄.	
	ζ coccíneum.	(*sc.*)	*scarlet.*	——	....	——	H. ♄.	Bot. mag. t. 180.
	η corymbòsum.	(*wh.*)	*corymb-flowered.*	——	....	——	H. ♄.	
	ϑ críspum.	(*bh.*)	*curled.*	——	....	——	H. ♄.	

	ι cumulàtum.	(*wh.*)	*bundled.*	N.America.	....	5. 6.	H. ♄.	
	κ díscolor.	(*re.wh.*)	*two-coloured.*	———	....	——	H. ♄.	
	λ fastigiàtum.	(*bh.*)	*crowded.*	———	....	——	H. ♄.	
	μ flòridum.	(*bh.*)	*abundant-flower'd.*	———	....	——	H. ♄.	
	ν globòsum.	(*bh.*)	*globose.*	———	....	——	H. ♄.	
	ξ glomeràtum.	(*bh.*)	*clustered.*	———	....	——	H. ♄.	
	ο incànum.	(*wh.*)	*hoary-leaved.*	———	....	——	H. ♄.	
	π incarnàtum.	(*fl.*)	*flesh-coloured.*	———	....	——	H. ♄.	
	ρ mirábile.	(*re.*)	*wonderful.*	———	....	——	H. ♄.	
	σ montànum.	(*fl.*)	*mountain.*	———	....	——	H. ♄.	
	τ pállidum.	(*li.wh.*)	*pale-flowered.*	———	....	——	H. ♄.	
	υ paludòsum.	(*li.wh.*)	*marsh.*	———	....	——	H. ♄.	
	φ papilionàceum.	(*fl.*)	*Butterfly.*	———	....	——	H. ♄.	
	χ partìtum.	(*fl.wh.*)	*five-parted.*	———	....	——	H. ♄.	
	ψ parviflòrum.	(*bh.*)	*small-flowered.*	———	....	——	H. ♄.	
	ω prolíferum.	(*bh.*)	*proliferous.*	———	....	——	H. ♄.	
	αα pùmilum.	(*bh.*)	*dwarf.*	———	....	——	H. ♄.	
	ββ purpuráscens.	(*fl.*)	*purplish.*	———	....	——	H. ♄.	
	γγ purpùreum.	(*pu.*)	*purple.*	———	....	——	H. ♄.	
	δδ purpùreo-plènum.	(*pu.*)	*double purple.*	———	....	——	H. ♄.	
	εε ròseum.	(*ro.*)	*rosy.*	———	....	——	H. ♄.	
	ζζ rubérrimum.	(*re.*)	*dark-red.*	———	....	——	H. ♄.	
	ηη rubéscens.	(*pi.*)	*reddish.*	———	....	——	H. ♄.	
	θθ rubicúndum.	(*re.*)	*rubicund.*	———	....	——	H. ♄.	
	ιι rùbrum. B.C.	(*re.*)	*red-flowered.*	———	....	——	H. ♄.	Lodd. bot. cab. t. 51.
	κκ rùbro-plènum.	(*re.*)	*double-red.*	———	....	——	H. ♄.	
	λλ rùfum.	(*re.*)	*rufous.*	———	....	——	H. ♄.	
	μμ rùtilans.	(*re.*)	*deep red.*	———	....	——	H. ♄.	
	νν semidùplex.	(*wh.*)	*semidouble.*	———	....	——	H. ♄.	
	ξξ stamíneum.	(*wh.*)	*long-stamened.*	———	....	——	H. ♄.	
	οο stellàtum.	(*wh.*)	*starry.*	———	....	——	H. ♄.	
	ππ trícolor.	(*wh.re.*)	*three-coloured.*	———	....	——	H. ♄.	
	ρρ variábile.	(*re.wh.*)	*variable.*	———	....	——	H. ♄.	
	σσ variegàtum.	(*std.*)	*variegated.*	———	....	——	H. ♄.	
	ττ versícolor.	(*re.wh.*)	*various-coloured.*	———	....	——	H. ♄.	
	υυ violàceum.	(*vi.*)	*violet.*	———	....	——	H. ♄.	
38	bícolor.	(*wh.re.*)	two-coloured.	———	1734.	——	H. ♄.	Trew. ehret. t. 48.
	Azàlea bícolor. Ph.							
39	viscòsum.	(*wh.*)	viscid.	—— ——	——	7. 8.	H. ♄.	Meerb. ic. 2. t. 9.
	Azàlea viscòsa. Ph.							
	α odoràtum.	(*wh.*)	*common white.*	———	——	——	H. ♄.	
	β críspum.	(*wh.*)	*curled.*	———	——	——	H. ♄.	
	γ dealbàtum.	(*wh.*)	*powdered.*	———	——	——	H. ♄.	
	δ vittàtum.	(*re.wh.*)	*striped-flowered.*	———	——	——	H. ♄.	
	ε penicillàtum.	(*re.wh.*)	*pencilled.*	———	——	——	H. ♄.	
	ζ variegàtum.	(*re.wh.*)	*variegated.*	———	——	——	H. ♄.	
	η rubéscens.	(*re.*)	*reddish-flowered.*	———	——	——	H. ♄.	
	θ pubéscens.	(*re.wh.*)	*pubescent.*	———	——	——	H. ♄.	Lodd. bot. cab. t. 441.
	ι físsum.	(*re.wh.*)	*narrow-petaled.*	———	——	——	H. ♄.	
40	arboréscens.	(*re.*)	arborescent.	———	1818.	5. 7.	H. ♄.	
	Azàlea arboréscens. Ph.							
41	nítidum.	(*re.wh.*)	glossy-leaved.	———	1812.	6. 8.	H. ♄.	Bot. reg. t. 414.
	Azàlea nítida. B.R.							
42	glaùcum.	(*wh.*)	glaucous dwarf.	———	1784.	6.	H. ♄.	Wats. dend. brit. t. 5.
	Azàlea glaùca. Ph.							
43	híspidum.	(*wh.*)	tall glaucous.	———	——	6. 8.	H. ♄.	Wats. dend. brit. t. 6.
	Azàlea híspida. Ph.							
AZ'ALEA. D.P.			AZ'ALEA. Pentandria Monogynia. R.S.					
	procúmbens. E.B.	(*ro.*)	trailing.	Britain.	....	4. 5.	H. ♄.	Eng. bot. v. 13. t. 865.
	Loiseleùria procúmbens. R.S. *Chamælèdon procúmbens.* L.en.							
AMMYRS'INE. Ph.			AMMYRS'INE. Decandria Monogynia. Ph.					
1	buxifòlia. Ph.	(*wh.*)	upright.	N. America.	1736.	4. 5.	H. ♄.	Bot. reg. t. 531.
	Lèdum buxifòlium. W. *Leiophy'llum thymifòlium.* S.S.							
2	Lyòni.	(*wh.*)	spreading.	Carolina.	1812.	——	H. ♄.	
L'EDUM. Ph.			L'EDUM. Decandria Monogynia. L.					
1	palústre. W.	(*wh.*)	marsh.	Europe.	1762.	4. 5.	H. ♄.	Flor. dan. t. 1031.
2	decúmbens. L.C.	(*wh.*)	decumbent.	Hudson's Bay.	——	——	H. ♄.	
3	canadénse. L.C.	(*wh.*)	dwarf.	Canada.	1824.	——	H. ♄.	Lodd. bot. cab. t. 1049.
4	latifòlium. W.	(*wh.*)	broad-leaved.	N. America.	1763.	——	H. ♄.	Jacq. ic. 3. t. 464.

BEF'ARIA. L. Bef'aria. Dodecandria Monogynia. L.
racemòsa. s.s. (*bh.*) sweet-scented. Florida. 1810. 6. 7. G. ♄. Vent. cels. t. 51.
paniculàta. M. *Bejària racemòsa.* Ph.
'ITEA. L. 'Itea. Pentandria Monogynia. L.
virgínica. L. (*wh.*) Virginian. N. America. 1744. 6. 8. H. ♄. Wats. dend. brit. t. 12.

ORDO CXXVIII.

GALACINEÆ. *D. Don in Edinb. phil. journ. Dec.* 1828. *p.* 53.

G'ALAX. W. G'alax. Pentandria Monogynia. W.
aphy'lla. W. (*bh.*) heart-leaved. N. America. 1786. 6. 7. H. ♃. Bot. mag. t. 754.
cordifòlia. H.K. *Blandfórdia cordàta.* Andr. rep. t. 343.

ORDO CXXIX.

EPACRIDEÆ. *Brown prodr.* 1. *p.* 535.

Sect I. *EPACRIDEÆ VERÆ.*

DRACOPHY'LLUM. B.P. Dracophy'llum. Pentandria Monogynia.
1 secúndum. B.P. (*wh.*) side-flowering. N. S. W. 1824. 4. 6. G. ♄.
2 longifòlium. R.S. (*wh.*) long-leaved. N. Zealand. —— —— G. ♄. Fors.ch.gen.t.10.f.i-m.
SPHENO'TOMA. S.F.A. Spheno'toma. Pentandria Monogynia.
grácilis. S.F.A. (*wh.*) slender. N. Holland. 1825. 5. 10. G. ♄. Swt. flor. aust. t. 44.
SPRENG'ELIA. Sm. Spreng'elia. Pentandria Monogynia.
incarnàta. B.P. (*pi.*) flesh-coloured. N. S. W. 1793. 4. 6. G. ♄. Bot. mag. t. 1719.
PONCEL'ETIA. B.P. Poncel'etia. Pentandria Monogynia.
sprengelioídes. B.P. (*w.*) marsh. N. S. W. 1826. 4. 6. G. ♄.
ANDERS'ONIA. B.P. Anders'onia. Pentandria Monogynia.
sprengelioídes. B.P. (*w.*) Sprengelia-like. N. Holland. 1803. 3. 7. G. ♄. Bot. mag. t. 1645.
COSM'ELIA. B.P. Cosm'elia. Pentandria Monogynia.
rùbra. B.P. (*re.*) red-flowered. N. Holland. 1826. 4. 6. G. ♄.
LYSIN'EMA. B.P. Lysin'ema. Pentandria Monogynia.
1 pentapétalum. B.P. (*w.*) five-petaled. N. Holland. 1826. 4. 6. G. ♄.
2 conspícuum. B.P. (*wh.*) conspicuous. —— 1824. .... G. ♄.
3 púngens. B.P. (*wh.*) pungent. N. S. W. 1804. 3. 6. G. ♄. Cav. ic. 4. t. 346.
Epácris attenuàta. Lodd. bot. cab. t. 38.
4 ròseum. B.C. (*ro.*) rose-coloured. —— 1804. —— G. ♄. Lodd. bot. cab. t. 863.
Epácris ròsea. B.C. *púngens.* Bot. mag. t. 1199.
EPA'CRIS. B.P. Epa'cris. Pentandria Monogynia.
1 purpuráscens. B.P. (*li.*) rigid-leaved. N. S. W. 1803. 2. 6. G. ♄. Bot. mag. t. 844.
púngens. B.M.
α lilacìna. (*li.*) *pale-flowered.* —— —— —— G. ♄. Lodd. bot. cab. t. 237.
β rùbra. (*re.*) *red-flowered.* —— —— —— G. ♄. —— t. 876.
2 pulchélla. B.P. (*wh.*) sweet-scented. —— 1804. 3. 6. G. ♄. —— t. 170.
3 microphy'lla. B.P. (*wh.*) small-leaved. —— 1817. —— G. ♄.
4 grandiflòra. B.P. (*cr.*) crimson. —— 1803. 1. 6. G. ♄. Bot. mag. t. 982.
5 ruscifòlia. B.P. (*wh.*) Ruscus-leaved. V. Diem. Isl. 1824. 3. 5. G. ♄.
6 impréssa. B.P. (*ro.*) impressed. —— —— 2. 5. G. ♄. Swt. flor. aust. t. 4.
7 spársa. B.P. (*wh.*) scattered-flower'd. N. S. W. 1825. 3. 5. G. ♄.
8 obtusifòlia. B.P. (*wh.*) blunt-leaved. —— 1804. 4. 6. G. ♄. Sm. exot. bot. 1. t. 40.
9 heteronèma. B.P. (*wh.*) hairy-branched. V. Diem. Isl. 1824. —— G. ♄. Labill. n. hol. 1. t. 56.
10 paludòsa. B.P. (*wh.*) marsh. N. S. W. —— —— G. ♄. Lodd. bot. cab. t. 1226.
11 apiculàta. F.T. (*wh.*) pointed. —— 1823. 5. 8. G. ♄.
12 onosmæflòra. F.T. (*re.*) Onosma-flowered. —— —— —— G. ♄.
13 exsérta. B.P. (*wh.*) exserted. V. Diem. Isl. 1812. 4. 6. G. ♄.
14 mucronulàta. B.P. (*wh.*) short-pointed. —— 1824. —— G. ♄.

Sect. II. *STYPHELIÆ.*

TROCHOCA'RPA. B.P.		TROCHOCA'RPA.	Pentandria Monogynia.				
laurìna. B.P.	(*wh.*)	Laurel-leaved.	N. S. W.	1820.	....	G. ♄.	Linn. trans. 8. t. 9.
Styphèlia cornifòlia. L.T.							
ACRO'TRICHE. B.P.		ACRO'TRICHE.	Pentandria Monogynia.				
1 cordàta. B.P.	(*wh.*)	heart-leaved.	N. Holland.	1825.	4. 6.	G. ♄.	Labill. n. hol. 1. t. 63.
2 ovalifòlia. B.P.	(*wh.*)	oval-leaved.	———	1824.	——	G. ♄.	
3 serrulàta. B.P.	(*wh.*)	saw-leaved.	V. Diem. Isl.	1816.	——	G. ♄.	Labill. n. hol. 1. t. 62.
Styphèlia serrulàta. Labil.							
4 divaricàta. B.P.	(*wh.*)	divaricate.	N. S. W.	1822.	——	G. ♄.	
MONO'TOCA. B.P.		MONO'TOCA.	Pentandria Monogynia.				
1 scopària. B.P.	(*wh.*)	broom.	N. S. W.	1823.	4. 7.	G. ♄.	Smith new holl. 43.
Styphèlia scopària. Smith.							
2 lineàta. B.P.	(*wh.*)	lineated.	V. Diem. Isl.	1804.	5. 8.	G. ♄.	Lab. nov. hol. 1. t. 61.
Styphèlia glaùca. Labill.							
3 álbens. B.P.	(*wh.*)	white-leaved.	N. S. W.	1823.	——	G. ♄.	
4 ellíptica. B.P.	(*wh.*)	elliptic-leaved.	———	1802.	——	G. ♄.	Smith new holl. 49.
LEUCOP'OGON. B.P.		LEUCOP'OGON.	Pentandria Monogynia.				
1 lanceolàtus. B.P.	(*wh.*)	small-flowered.	N. S. W.	1790.	5. 8.	G. ♄.	Swt. flor. aust. t. 47.
Styphèlia parviflòra. Andr. reposit. t. 287. *gnídium.* Vent. malm. t. 23.							
2 Richèi. B.P.	(*wh.*)	oblong-leaved.	N. S. W.	1824.	——	G. ♄.	Lab. nov. hol. 1. t. 60.
3 mùticus. B.P.	(*wh.*)	awnless.	———	1826.	——	G. ♄.	
4 ericoídes. B.P.	(*wh.*)	heath-leaved.	———	1815.	4. 7.	G. ♄.	Smith new holl. 48.
Epácris spùria. Cavan. ic. 4. t. 347. f. 1.							
5 virgàtus. B.P.	(*wh.*)	twiggy.	———	———	——	G. ♄.	Lab. nov. hol. 1. t. 64.
6 collìnus. B.P.	(*wh.*)	hill.	V. Diem. Isl.	1822.	5. 6.	G. ♄.	——— 1. t. 65.
7 amplexicaùlis. B.P.	(*w.*)	stem-clasping.	N. S. W.	1815.	4. 8.	G. ♄.	Linn. trans. 8. t. 8.
8 microphy'llus. B.P.	(*w.*)	small-leaved.	———	1818.	——	G. ♄.	Cavan. ic. 4. t. 349. f. 2.
Perojòa microphy'lla. Cav.							
9 striàtus. B.P.	(*wh.*)	striated.	N. Holland.	1824.	——	G. ♄.	
10 attenuàtus. F.T.	(*wh.*)	attenuated.	N. S. W.	———	——	G. ♄.	
11 biflòrus. B.P.	(*wh.*)	two-flowered.	———	1826.	——	G. ♄.	
12 setíger. B.P.	(*wh.*)	bristle-pointed.	———	1824.	——	G. ♄.	
13 appréssus. B.P.	(*wh.*)	pressed-leaved.	———	———	4. 6.	G. ♄.	
14 juniperìnus. B.P.	(*wh.*)	Juniper-leaved.	———	1804.	——	G. ♄.	Lodd. bot. cab. t. 447.
15 interrúptus. B.P.	(*wh.*)	interrupted.	N. Holland.	1825.	3. 6.	G. ♄.	——— t. 1451.
16 polystàchyus. B.P.	(*wh.*)	many-spiked.	———	———	6. 7.	G. ♄.	——— t. 1436.
LISSA'NTHE. B.P.		LISSA'NTHE.	Pentandria Monogynia.				
1 sápida. B.P.	(*wh.*)	recurved-flowered.	N. S. W.	1823.	11. 1.	G. ♄.	Bot. reg. t. 1275.
2 subulàta. B.P.	(*wh.*)	awl-leaved.	———	——	4. 7.	G. ♄.	
3 strigòsa. B.P.	(*wh.*)	strigose.	———	——	——	G. ♄.	Sm. new holl. 48.
4 dàphnoídes. B.P.	(*wh.*)	Daphne-leaved.	V. Diem. Isl.	1818.	6. 7.	G. ♄.	Lodd. bot. cab. t. 466.
5 ciliàta. B.P.	(*wh.*)	fringed.	———	1823.	——	G. ♄.	
CYATH'ODES. B.P.		CYATH'ODES.	Pentandria Monogynia.				
1 glaùca. B.P.	(*wh.*)	glaucous.	V. Diem. Isl.	1818.	4. 6.	G. ♄.	Lab. nov. holl. 1. t. 81.
2 oxycèdrus. B.P.	(*wh.*)	linear-leaved.	———	1822.	5. 8.	G. ♄.	——— 1. t. 69.
3 abietìna. B.P.	(*wh.*)	Fir-like.	———	1824.	8. 12.	G. ♄.	——— 1. t. 68.
MELI'CHRUS. B.P.		MELI'CHRUS.	Pentandria Monogynia.				
1 rotàtus. B.P.	(*re.*)	rotate-flowered.	N. S. W.	1822.	5. 8.	G. ♄.	Cavan. ic. 4. t. 349. f. 1.
Ventenàtia procúmbens. Cav.							
2 mèdius. F.T.	(*re.*)	intermediate.	———	1823.	——	G. ♄.	
STENANTH'ERA. B.P.		STENANTH'ERA.	Pentandria Monogynia.				
pinifòlia. B.P.	(*sc.*)	Pine-leaved.	N. S. W.	1811.	5. 7.	G. ♄.	Bot. reg. t. 218.
ASTROL'OMA. B.P.		ASTROL'OMA.	Pentandria Monogynia.				
1 humifùsum. B.P.	(*sc.*)	Juniper-leaved.	N. S. W.	1807.	5. 10.	G. ♄.	Bot. mag. t. 1439.
2 denticulàtum. B.P.	(*re.*)	toothed.	———	1825.	——	G. ♄.	
STYPH'ELIA. B.P.		STYPH'ELIA.	Pentandria Monogynia.				
1 tubiflòra. B.P.	(*cr.*)	crimson.	N. S. W.	1802.	5. 8.	G. ♄.	Sm. new holl. 45. t. 14.
2 triflòra. B.P.	(*cr. gr.*)	three-flowered.	———	1796.	——	G. ♄.	Bot. mag. t. 1297.
3 viridiflòra. B.P.	(*gr.*)	green-flowered.	———	1791.	4. 6.	G. ♄.	Swt. flor. aust. t. 50.
4 latifòlia. B.P.	(*gr.*)	broad-leaved.	———	1823.	——	G. ♄.	
5 adscéndens. B.P.	(*gr.*)	ascending.	V. Diem. Isl.	1822.	——	G. ♄.	
6 l'æta. B.P.	(*gr.*)	fruitful.	N. S. W.	——	——	G. ♄.	
7 longifòlia. B.P.	(*gr.*)	long-leaved.	———	1807.	——	G. ♄.	Bot. reg. t. 24.

ORDO CXXX.

PENÆACEÆ.

PEN'ÆA. R.S.		PEN'ÆA. Tetrandria Monogynia. R.S.					
1 myrtoídes. W.	(*ye.*)	Myrtle-like.	C. B. S.	1816.	5. 7.	G. ♄.	
2 fruticulòsa. R.S.	(*re.*)	blunt-leaved.	———	1822.	——	G. ♄.	
3 mucronàta. R.S.	(*re.*)	sharp-pointed.	———	1787.	——	G. ♄.	Vent. malm. t. 87.
4 squamòsa. R.S.	(*re.*)	scaly.	———	——	——	G. ♄.	Bot. reg. t. 106.
5 imbricàta. B.M.	(*re.*)	imbricate.	———	1823.	4. 6.	G. ♄.	Bot. mag. t. 2809.
6 acùta. R.S.	(*re.*)	acute-leaved.	———	1824.	——	G. ♄.	
7 Sarcocólla. R.S.	(*ye.*)	rhomb-leaved.	———	1816.	5. 7.	G. ♄.	Th.m.n.c.ber.1.t.3.f.1.
8 marginàta. R.S.	(*wh.*)	thick-edged.	———	——	——	G. ♄.	

SUBCLASSIS III. *COROLLIFLORÆ.* DC.

ORDO CXXXI.

SYMPLOCINEÆ. *Don Prodr. p.* 144.

SY'MPLOCOS. L'H.		SY'MPLOCOS. Polyadelphia Polyandria. L.					
1 coccínea. K.S.	(*sc.*)	scarlet.	Mexico.	1825.	....	S. ♄.	H. et B. pl. æq. 1. t. 52.
2 tinctòria. W.	(*ye.*)	Laurel-leaved.	Carolina.	1780.	4. 6.	F. ♄.	Catesb. car. 1. t. 54.
Hòpea tinctòria. Ph.							
3 sínica. B.R.	(*wh.*)	Chinese.	China.	1822.	4. 6.	F. ♄.	Bot. reg. t. 710.
4 cratægoídes. D.P.	(*wh.*)	Hawthorn-like.	Nepaul.	1824.	....	G. ♄.	
SCHO'EPFIA. D.P.		SCHO'EPFIA. Pentandria Monogynia.					
fràgrans. D.P.	(*ye.*)	fragrant.	Nepaul.	1827.	....	G. ♄.	

ORDO CXXXII.

STYRACINEÆ. *Kth. synops.* 2. *p.* 315.

ST'YRAX. W.		STORAX. Decandria Monogynia. L.					
1 officinàle. W.	(*wh.*)	officinal.	Italy.	1597.	7.	H. ♄.	Andr. reposit. t. 631.
2 grandifòlium. W.	(*wh.*)	great-leaved.	N. America.	1765.	——	H. ♄.	Wats. dend. brit. t. 129.
3 pulveruléntum. M.	(*wh.*)	powdered.	———	1794.	6. 7.	H. ♄.	——— t. 41.
4 glábrum. N.	(*wh.*)	smooth.	———	1765.	7. 8.	H. ♄.	——— t. 40.
lævigàtum. Bot. mag. t. 921.							
HAL'ESIA. W.		SNOWDROP-TREE. Dodecandria Monogynia. L.					
1 tetráptera. W.	(*wh.*)	four-winged.	Carolina.	1756.	4. 5.	H. ♄.	Bot. mag. t. 910.
2 parviflòra. B.R.	(*wh.*)	small-flowered.	———	1812.	5.	H. ♄.	Bot. reg. t. 952.
3 díptera. W.	(*wh.*)	two-winged.	N. America.	1758.	4. 5.	H. ♄.	Lodd. bot. cab. t. 1172.

ORDO CXXXIII.

MYRSINEÆ. *Brown prodr.* 532.

M'ÆSA. J.		M'ÆSA. Pentandria Monogynia.					
1 índica. D.P.	(*wh.*)	Indian.	E. Indies.	1812.	2. 6.	S. ♄.	Bot. mag. t. 2052.
Bœobòtrys índica. F.I.							
2 tomentòsa. D.P.	(*wh.*)	tomentose.	Nepaul.	1818.	——	S. ♄.	
3 argéntea. F.I.	(*wh.*)	silvery-leaved.	E. Indies.	——	——	S. ♄.	
4 macrophy'lla. F.I.	(*wh.*)	large-leaved.	———	1823.	——	S. ♄.	

JACQUI'NIA. W. JACQUI'NIA. Pentandria Monogynia. W.
1 arbòrea. W. (*wh.*) blunt-leaved. W. Indies. 1820. .... S. ħ.
2 armillàris. W. (*wh.*) armed-leaved. ——— 1768. 4. 6. S. ħ. Jacq. amer. pict. t. 56.
3 aurantìaca. H.K. (*or.*) orange-color'd. Sandwich Isl. 1796. 4. 9. S. ♄. Bot. mag. t. 1639.
4 ruscifòlia. W. (*or.*) Ruscus-leaved. S. America. 1729. —— S. ♄. Jacq. amer. pict. t. 57.
5 macrocárpa. C.I. (*or.*) large capsuled. ——— 1820. —— S. ♄. Cavan. ic. 5. t. 483.
ARDI'SIA. W. ARDI'SIA. Pentandria Monogynia. W.
1 acuminàta. W. (*pi.*) acuminate. Guiana. 1803. 7. 8. S. ♄. Bot. mag. t. 1678.
Myrsìne Icacòrea. R.S.
2 tinifòlia. R.S. (*pi.*) Tinus-leaved. Jamaica. 1824. —— S. ♄. Sloan. hist. 2. t. 105.
3 solanàcea. W. (*ro.*) Nightshade-like. E. Indies. 1798. 6. 9. S. ♄. Bot. mag. t. 1677.
4 paniculàta. F.I. (*ro.*) panicled. ——— 1818. 3. 8. S. ♄. Bot. reg. t. 638.
5 coloràta. F.I. (*ro.*) red-flowered. ——— 1816. —— S. ♄. Lodd. bot. cab. t. 465.
6 umbellàta. F.I. (*pi.*) umbel-flowered. ——— 1809. 6. 9. S. ♄. ——— t. 531.
littoràlis. Andr. repos. t. 630.
7 lanceolàta. F.I. (*pi.*) leathery-leaved. ——— 1818. —— S. ♄.
8 élegans. A.R. (*pi.*) elegant. ——— 1809. 7. 9. S. ħ. Andr. reposit. t. 623.
crenàta. F.I.
9 punctàta. B.R. (*pi.*) spotted-flowered. China. 1823. 6. 9. S. ♄. Bot. reg. t. 827.
10 lentiginòsa. B.R. (*wh.*) freckled. ——— 1809. 3. 12. G. ħ. ——— t. 533.
crenulàta. B.C. 2. *non* Vent. *crenàta.* Bot. mag. t. 1950.
11 glandulòsa. F.I. (*wh.*) glandular. E. Indies. 1827. —— S. ♄.
12 macrocárpa. F.I. (*pi.*) large-fruited. Nepaul. 1824. 5. 9. S. ♄.
13 complanàta. F.I. (*pi.*) pubescent. Pinang. —— —— S. ♄.
14 thyrsiflòra. D.P. (*pi.*) thyrse-flower'd. Nepaul. —— 4. 8. S. ♄.
15 pyramidàlis. Cav. (*pi.*) pyramidal. Santa Cruz. 1818. 4. 9. S. ħ. Lodd. bot. cab. t. 448.
16 excélsa. W. (*gr.*) Laurel-leaved. Madeira. 1784. 7. 8. G. ♄. Gært. sem. 1. t. 77. f. 1.
My'rsine Heberdènia. R.S.
17 coriàcea. R.S. (*ro.*) leathery-leaved. W. Indies. 1822. 4. 8. S. ♄.
18 serrulàta. W. (*pi.*) saw-leaved. ——— 1820. 7. 8. S. ♄. Plum. ic. 80.
19 crenulàta. W. (*re.*) crenulated. ——— 1809. 6. 9. S. ♄. Venten. choix. t. 5.
20 lateriflòra. W. (*pi.*) side-flowering. ——— 1793. —— S. ♄.
21 canaliculàta. W. (*pi.*) channelled. NewGranada. 1824. 4. 8. S. ♄. Lodd. bot. cab. t. 1083.
EMB'ELIA. L. EMB'ELIA. Pentandria Monogynia. L.
1 robústa. F.I. (*gr.wh.*) robust. E. Indies. 1823. .... S. ħ.
2 floribúnda. F.I. (*ye.gr.*) sweet-scented. Nepaul. 1826. .... S. ♄.⌣.
MYRS'INE. B.P. MYRS'INE. Polygamia Diœcia. L.
1 africàna. S.S. (*gr.pu.*) African. C. B. S. 1691. 3. 5. G. ♄. Comm. hort. 1. t. 64.
2 retùsa. S.S. (*gr.pu.*) retuse-leaved. Azores. 1778. 6. G. ħ. Venten. cels. t. 86.
3 rotundifòlia. R.S. (*w.pu.*) round-leaved. C. B. S. 1816. —— G. ħ. Pluk. alm. t. 18. f. 5.
4 semiserràta. F.I. (*bh.*) semiserrate. Nepaul. 1822. .... G. ħ.
5 capitellàta. F.I. (*gr.*) axillary headed. ——— —— .... G. ♄.
6 bifària. F.I. (*bh.*) bifarious. ——— —— .... G. ♄.
7 subspinòsa. D.P. (*wh.*) spiny-leaved. ——— 1823. .... G. ♄.
8 coriàcea. B.P. (*wh.*) leathery-leaved. Jamaica. 1822. .... S. ♄.
Samára coriàcea. Sw.
9 Samára. S.S. (*wh.*) oval-leaved. C. B. S. 1770. 11. 2. G. ♄.
Samára pentándra. W.
10 melanóphleos. S.S. (*wh.*) Laurel-leaved. ——— 1783. 3. 6. G. ħ. Jacq. vind. 1. t. 71.
Sideróxylon melanóphleum. W.
11 mìtis. S.S. (*wh.*) pointed-leaved. ——— .... 6. 9. G. ♄. Bot. mag. t. 1858.
Sideróxylon mìte. B.M. *Scleróxylum mìte.* W. en. *Manglílla Milleriàna.* P.S.
12 venulòsa. S.S. (*wh.*) veined-leaved. ——— 1816. —— G. ♄.
13 canariénse. S.S. (*wh.*) Canary. Canaries. —— —— G. ♄.
14 variábilis. B.P. (*wh.*) variable. N. S. W. 1820. —— G. ħ.
ÆGIC'ERAS. G. ÆGIC'ERAS. Pentandria Monogynia.
fràgrans. B.P. (*wh.*) fragrant. N. Holland. 1824. .... G. ħ. Kon. an. bot. 1. p. 129. ic.

ORDO CXXXIV.

SAPOTEÆ. *Brown prodr.* 1. *p.* 528.

INOCA'RPUS. W. OTAHEITE-CHESNUT. Decandria Monogynia. W.
èdulis. W. (*wh.*) eatable. South-sea Isl. 1793. .... S. ♄. Roxb. corom. 3. t. 263.
BUM'ELIA. S.S. BUM'ELIA. Pentandria Monogynia. W.
1 lycioídes. S.S. (*wh.*) Boxthorn-leav'd. N. America. 1758. 8. H. ♄. Duham. arb. 2. t. 68.

2 oblongifòlia. N.	(*wh.*)	oblong-leaved.	Missisippi.	1818.	7. 8.	H. ♄.	
3 tènax. R.S.	(*wh.*)	silvery-leaved.	Carolina.	1765.	——	H. ♄.	Wats. dend. brit. t. 10.
chrysophylloídes. W.D. *Sideróxylon tènax.* L.							
4 lanuginòsa. R.S.	(*wh.*)	woolly-leaved.	————	——	——	H. ♄.	
5 reclinàta. S.S.	(*wh.*)	reclined.	————	1806.	——	H. ♄.	Venten. choix. t. 22.
6 strigòsa. S.S.	(*wh.*)	strigose.	....	....	——	G. ♄.	
Sideróxylon strigòsum. W.en.							
7 nìgra. S.S.	(*wh.*)	black.	W. Indies.	1806.	6. 8.	S. ♄.	
8 salicifòlia. S.S.	(*wh.*)	Willow-leaved.	S. America.	1758.	——	S. ♄.	Sloan. hist. 2. t. 206. f. 2.
9 fœtidíssima. S.S.	(*wh.*)	fetid.	W. Indies.	....	5. 7.	S. ♄.	
10 rotundifòlia. S.S.	(*wh.*)	round-leaved.	————	1816.	——	S. ♄.	
SERSALI'SIA. B.P.		SERSALI'SIA.	Pentandria Monogynia.				
serícea. B.P.	(*wh.*)	silky-leaved.	N. Holland.	1772.	5. 6.	G. ♄.	
Sideróxylon seríceum. W.							
SIDERO'XYLON. W.		IRON-WOOD.	Pentandria Monogynia. L.				
1 inérme. W.	(*wh.*)	smooth.	C. B. S.	1692.	7. 8.	G. ♄.	Lam. ill. 2. t. 120. f. 1.
2 tomentòsum. W.	(*wh.*)	tomentose.	E. Indies.	1818.	....	S. ♄.	Roxb. corom. 1. t. 28.
ARG'ANIA. R.S.		ARG'ANIA.	Pentandria Monogynia.				
Sideróxylum. R.S.	(*ye.*)	spiny.	Morocco.	1711.	7. 8.	H ♄.	Comm. hort. 1. t. 83.
Sideróxylum spinòsum. L. *Elæodéndron A'rgan.* W.							
CHRYSOPHY'LLUM. W.		STAR-APPLE.	Pentandria Monogynia. L.				
1 Cainìto. W.	(*wh.*)	broad-leaved.	W. Indies.	1737.	5. 6.	S. ♄.	Jacq. amer. t. 37. f. 1.
β *jamaicénse.*	(*wh.*)	*purple-fruited.*	————	——	——	S. ♄.	Jacq. am. pict. t. 52.
γ *cœrùleum.*	(*wh.*)	*blue-fruited.*	————	——	——	S. ♄.	Jacq. amer. t. 37.
δ *microphy'llum.*	(*wh.*)	*small-leaved.*	————	——	——	S. ♄.	Jacq. amer. pict. t. 53.
2 monopyrènum. W.	(*wh.*)	acute-leaved.	————	1816.	——	S. ♄.	Burm. amer. t. 69.
3 argénteum. W.	(*wh.*)	silvery.	Martinico.	1758.	——	S. ♄.	Jacq. amer. pict. t. 51.
4 macrophy'llum. G.D.	(*wh.*)	large-leaved.	Sierra Leone.	1822.	....	S. ♄.	
5 obovàtum. G.D.	(*wh.*)	obovate.	————	——	....	S. ♄.	
6 glábrum. W.	(*wh.*)	smooth-leaved.	Martinico.	1818.	4. 6.	S. ♄.	Jacq. amer. pict. t. 55.
NYCTERISI'TION. R.P.		NYCTERISI'TION.	Pentandria Monogynia.				
ferrugíneum. R.P.	(*wh.*)	ferruginous.	S. America.	1823.	....	S. ♄.	Flor. peruv. 2. t. 187.
A'CHRAS. J.		SAPOTA.	Hexandria Monogynia. S.S.				
1 Sapòta. W.	(*wh.*)	common.	S. America.	1731.	4. 6.	S. ♄.	Jacq. amer. 57. t. 41.
β *Zapotílla.* Jacq.	(*wh.*)	*Naseberry-tree.*	————	——	——	S. ♄.	———— 57. t. 41. b.
2 austràlis. B.P.	(*wh.*)	New Holland.	N. S. W.	1823.	....	G. ♄.	
LUC'UMA. J.		LUC'UMA.	Pentandria Monogynia.				
1 mammòsum. K.S.	(*wh.*)	Mammee.	S. America.	1739.	5. 8.	S. ♄.	Jacq. amer. t. 182. f. 19.
2 Bonplándii. K.S.	(*wh.*)	Bonpland's.	Havannah.	1822.	....	S. ♄.	
3 obovàtum. K.S.	(*wh.*)	obovate-leaved.	Peru.	——	....	S. ♄.	Flor. peruv. 3. t. 239.
4 salicifòlium. K.S.	(*wh.*)	Willow-leaved.	Mexico.	1823.	....	S. ♄.	
MI'MUSOPS. W.		MI'MUSOPS.	Octandria Monogynia. L.				
1 Eléngi. S.S.	(*wh.*)	pointed-leaved.	E. Indies.	1796.	5. 7.	S. ♄.	Roxb. cor. 1. t. 14.
2 dissécta. S.S.	(*wh.*)	oval-leaved.	South-sea Isl.	1804.	——	S. ♄.	
3 hexándra. S.S.	(*wh.*)	hexandrous.	E. Indies.	1816.	——	S. ♄.	Roxb. cor. 1. t. 15.
4 Kaùki. S.S.	(*wh.*)	obtuse-leaved.	————	1796.	5. 8.	S. ♄.	Rumph. amb. 3. t. 8.
IMBRIC'ARIA. J.		IMBRIC'ARIA.	Octandria Monogynia.				
borbònica. G.	(*wh.*)	Bourbon.	Bourbon.	1820.	....	S. ♄.	
BA'SSIA. L.		BA'SSIA.	Dodecandria Monogynia. L.				
1 longifòlia. W.	(*wh.*)	long-leaved.	E. Indies.	1811.	....	S. ♄.	Lam. ill. t. 398.
2 latifòlia. W.	(*wh.*)	broad-leaved.	————	1799.	....	S. ♄.	Roxb. cor. 1. t. 19.
3 butyràcea. D.P.	(*wh.*)	butter-tree.	Nepaul.	1823.	....	S. ♄.	

ORDO CXXXV.

EBENACEÆ. *Brown prodr.* 1. *p.* 524.

M'ABA. J.		M'ABA.	Diœcia Hexandria. W.				
buxifòlia. P.S.	(*wh.*)	Box-leaved.	E. Indies.	1810.	....	S. ♄.	Roxb. cor. 1. t. 45.
Ferreòla buxifòlia. Roxb.							
CARGI'LLIA. B.P.		CARGI'LLIA.	Polygamia Diœcia.				
austràlis. B.P.	(*gr.*)	southern.	N. S. W.	1816.	....	G. ♄.	
DIO'SPYROS. S.S.		DATE PLUM.	Polygamia Diœcia. W.				
1 virginiàna. S.S.	(*p.ye.*)	Persimon.	N. America.	1629.	6. 7.	H. ♄.	Wats. dend. brit. t. 146.
2 pubéscens. S.S.	(*pa.ye.*)	pubescent.	————	1812.	4. 5.	H. ♄.	

3 reticulàta. s.s.	(*wh.*)	netted.	Mauritius.	1823.	4. 5.	S. ♄.	
4 Ebenáster. s.s.	(*wh.*)	Ebenus-like.	E. Indies.	1792.	6. 8.	S. ♄.	Rumph. amb. 1. t. 6.
5 melanóxylon. s.s.	(*wh.*)	black-wooded.	———	1817.	——	S. ♄.	Roxb. cor. 1. t. 46.
6 E'benum. s.s.	(*wh.*)	smooth.	———	1792.	——	S. ♄.	Rotb. act. hafn. 2. t. 5.
7 montàna. s.s.	(*wh.*)	mountain.	———	1819.	——	S. ♄.	Roxb. cor. 1. t. 48.
8 obovàta. s.s.	(*wh.*)	four-seeded.	W. Indies.	1796.	5. 7.	S. ♄.	Jacq. schœn. 3. t. 312.
9 sylvática. s.s.	(*wh.*)	wood.	E. Indies.	1812.	——	S. ♄.	Roxb. cor. 1. t. 47.
10 Embryópteris. s.s.	(*gr.*)	glutiniferous.	———	1796.	5. 8.	S. ♄.	Bot. reg. t. 499.
Embryópteris glutinífera. B.C. 1. t. 70.							
11 lobàta. s.s.	(*wh.*)	lobed-fruited.	China.	1822.	——	G. ♄.	
12 vaccinioídes. L.C.	(*wh.*)	Vaccinium-like.	———	——	——	G. ♄.	Hook. ex. flor. t. 139.
13 èdulis. L.C.	(*wh.*)	eatable.	E. Indies.	——	——	S. ♄.	
14 Lòtus. s.s.	(*pa.ye.*)	European.	Italy.	1596.	6. 7.	H. ♄.	Mill. ic. t. 116.
15 Kàki. s.s.	(*wh.*)	Japan.	Japan.	1789.	——	G. ♄.	Kæmp. amœn. t. 806.
16 hirsùta. s.s.	(*wh.*)	hairy.	Ceylon.	1816.	——	S. ♄.	
17 chloróxylon. s.s.	(*wh.*)	cluster-flowered.	E. Indies.	1822.	....	S. ♄.	Roxb. cor. 1. t. 49.
18 cordifòlia. s.s.	(*wh.*)	heart-leaved.	———	1794.	5. 7.	S. ♄.	——— 1. t. 50.
ROY'ENA. W.		Roy'ena. Decandria Digynia. L.					
1 lùcida. W.	(*wh.*)	shining-leaved.	C. B. S.	1690.	5. 6.	G. ♄.	Lam. ill. t. 370. f. 1.
2 villòsa. W.	(*wh.*)	heart-leaved.	———	1774.	6. 7.	G. ♄.	
3 pállens. W.	(*wh.*)	pale.	———	1789.	——	G. ♄.	
4 myrtifòlia. W.en.	(*wh.*)	Myrtle-leaved.	———	1812.	——	G. ♄.	
5 glábra. W.	(*wh.*)	smooth.	———	1731.	7. 9.	G. ♄.	Comm. hort. 1. t. 65.
6 pubéscens. W.en.	(*wh.*)	pubescent.	———	1752.	——	G. ♄.	Bot. reg. t. 500.
7 hirsùta. W.en.	(*wh.*)	hairy-leaved.	———	——	——	G. ♄.	Lam. ill. t. 370. f. 2.
8 angustifòlia. W.	(*wh.*)	willow-leaved.	———	1789.	6. 7.	G. ♄.	
9 latifòlia. W.en.	(*wh.*)	broad-leaved.	———	1816.	——	G. ♄.	
10 ambígua. V.	(*wh.*)	obovate-leaved.	———	1815.	6. 7.	G. ♄.	Vent. malm. t. 17.
11 polyándra. W.	(*wh.*)	oval-leaved.	———	1774.	....	G. ♄.	
VI'SNEA. W.		Vi'snea. Dodecandria Trigynia. W.					
Mocanèra. W.	(*wh.*)	Canary.	Canaries.	1815.	8. 12.	G. ♄.	

ORDO CXXXVI.

OLEINÆ. *Brown prodr.* 1. *p.* 522.

FRA'XINUS. W.		Ash-tree. Polygamia Diœcia. L.					
1 excélsior. W.	(*gr.*)	common.	Britain.	....	4. 5.	H. 5.	Eng. bot. v. 24. t. 1692.
β argéntea.	(*gr.*)	*silver-striped.*	———	....	——	H. 5.	
γ péndula.	(*gr.*)	*weeping.*	———	....	——	H. 5.	
δ horizontàlis.	(*gr.*)	*horizontal.*	———	....	——	H. 5.	
ε jáspidea.	(*gr.*)	*coloured branch'd.*	———	....	——	H. 5.	
ζ eròsa.	(*gr.*)	*erose-leaved.*	———	....	——	H. 5.	
2 aùrea. W.en.	(*gr.*)	yellow barked.	....	....	....	H. 5.	
3 atrovìrens. P.S.	(*gr.*)	green curled-leav'd.	....	....	....	H. 5.	
4 nàna. W.en.	(*gr.*)	dwarf.	....	....	....	H. ♄.	
5 verrucòsa. L.en.	(*gr.*)	warted.	....	....	4. 5.	H. 5.	
6 heterophy'lla. s.s.	(*gr.*)	various-leaved.	England.	....	——	H. 5.	Eng. bot. v. 35. t. 2476.
simplicifòlia. E.B.							
7 polemonifòlia. R.S.	(*gr.*)	Polemonium-ld.	N. America.	....	——	H. 5.	
8 parvifòlia. s.s.	(*gr.*)	small-leaved.	Levant.	1816.	——	H. 5.	
9 lentiscifòlia. s.s.	(*gr.*)	Lentiscus-leaved.	Aleppo.	1710.	5. 6.	H. 5.	Pluk. alm. t. 182. f. 4.
10 argéntea. s.s.	(*gr.*)	silver-leaved.	Corsica.	1825.	....	H. 5.	
11 sambucifòlia. s.s.	(*gr.*)	Elder-leaved.	N. America.	1800.	....	H. 5.	
12 oxycárpa. s.s.	(*gr.*)	sharp-leaved.	Tauria.	1815.	5. 6.	H. 5.	
oxyphy'lla. M.B.							
13 pállida. s.s.	(*gr.*)	pale.	N. America.	1818.	....	H. 5.	
14 rùfa. s.s.	(*gr.*)	rufous.	———	1822.	....	H. 5.	
15 fúsca. s.s.	(*gr.*)	brown branched.	———	——	....	H. 5.	
16 nìgra. s.s.	(*gr.*)	black-twigged.	———	1818.	4. 6.	H. 5.	
17 ellíptica. s.s.	(*gr.*)	elliptic-leaved.	———	1824.	....	H. 5.	
18 ovàta. s.s.	(*gr.*)	oval-leaved.	———	——	....	H. 5.	
19 Richárdi. s.s.	(*gr.*)	Richard's.	———	——	....	H. 5.	
20 álba. s.s.	(*gr.*)	grey branched.	———	1825.	....	H. 5.	
21 cinèrea. s.s.	(*gr.*)	cinereous.	———	1818.	....	H. 5.	

22 víridis. s.s.	(*gr.*)	green-branched.	N. America.	1816.	....	H.5.	
23 longifòlia. s.s.	(*gr.*)	long-leaved.	———	1825.	....	H.5.	
24 tamariscifòlia. R.S.	(*gr.*)	Tamarisk-leaved.	Levant.	....	——	H.5.	
25 mexicàna.	(*gr.*)	Mexican.	Mexico.	1825.	....	H.5.	
26 rubicúnda. s.s.	(*gr.*)	red-veined.	N. America.	——	....	H.5.	
27 pulverulénta. s.s.	(*gr.*)	powdered.	———	——	....	H.5.	
28 míxta. s.s.	(*gr.*)	mixed.	———	——	....	H.5.	
29 expánsa. s.s.	(*gr.*)	expanded	———	——	....	H.5.	
30 platycárpa. s.s.	(*gr.*)	flat-seeded.	———	1816.	4. 6.	H.5.	Catesb. car. t. 80.
31 quadrangulàta. s.s.	(*gr.*)	four-sided.	———	1811.	——	H.5.	Mich. f. arb. ic.
32 epíptera. s.s.	(*gr.*)	dotted-stalked.	———	——	——	H.5.	——— t. 33.
díscolor. M.							
33 láncea. Bosc.	(*gr.*)	lance-leaved.	———	1823.	——	H.5.	
34 pannòsa. s.s.	(*gr.*)	cloth-leaved.	———	1818.	——	H.5.	
35 pubéscens. s.s.	(*gr.*)	pubescent.	———	1783.	5.	H.5.	Du Roi. ed. 2. t. 1.
36 cúrvidens. R.S.	(*gr.*)	curved-toothed.	Carolina.	1811.	4. 5.	H.5.	
37 caroliniàna. s.s.	(*gr.*)	Carolina.	N. America.	——	4. 6.	H.5.	
38 juglandifòlia. s.s.	(*gr.*)	Walnut-leaved.	———	1783.	——	H.5.	
39 acuminàta. s.s.	(*gr.*)	taper-pointed.	———	1723.	——	H.5.	
40 americàna. s.s.	(*gr.*)	white.	———	——	——	H.5.	
O'RNUS. P.S.		FLOWERING ASH.	Diandria Monogynia. P.S.				
1 americàna. Ph.	(*wh.*)	American.	N. America.	1812.	5. 6.	H.5.	
2 europ'æa. P.S.	(*wh.*)	European.	Italy.	1710.	——	H.5.	Flor. græc. 1. t. 4.
Fráxinus O'rnus. W.							
3 rotundifòlia. P.S.	(*wh.*)	manna.	———	1697.	4. 5.	H.5.	
4 striàta.	(*wh.*)	streak-barked.	....	1818.	——	H.5.	
Fràxinus striàta. s.s.							
5 floribúnda.	(*wh.*)	many-flowered.	Nepaul.	——	....	G.5.	
Fráxinus floribúnda. F.I.							
CHIONA'NTHUS. S.S.		FRINGE-TREE.	Diandria Monogynia. L.				
1 virgínica. W.	(*wh.*)	smooth-leaved.	N. America.	1736.	5. 7.	H.5.	Catesb. car. 1. t. 68.
2 marítima.	(*wh.*)	pubescent-leaved.	———	——	——	H.5.	Wats. dend. brit. t. 1.
virgínica β marítima. Ph. *virgínica β angustifòlia.* W.D.							
3 axillàris. B.P.	(*wh.*)	axil-flowering.	N. Holland.	1810.	....	G. ♄.	
LINOCI'ERA. S.S.		LINOCI'ERA.	Diandria Monogynia. S.S.				
1 ligustrìna. s.s.	(*wh.*)	Privet-like.	W. Indies.	1820.	....	S. ♄.	
2 cotinifòlia. s.s.	(*wh.*)	Cotinus-leaved.	Ceylon.	1818.	....	S. ♄.	
3 compácta. s.s.	(*wh.*)	Caribean.	W. Indies.	1793.	....	S. ♄.	Jacq. coll. 2. t. 6. f. 1.
FONTAN'ESIA. S.S.		FONTAN'ESIA.	Diandria Monogynia. W.				
phillyræoídes. s.s.	(*wh.*)	Phillyrea-leaved.	Syria.	1787.	6. 8.	H. ♄.	Wats. dend. brit. t. 2.
NOTEL'ÆA. B.P.		NOTEL'ÆA.	Diandria Monogynia. S.S.				
1 longifòlia. B.P.	(*wh.*)	long-leaved.	N. S. W.	1790.	3. 6.	G. ♄.	Venten. choix. t. 25.
'Olea apetàla. A.R. 316.							
2 ovàta. B.P.	(*wh.*)	oval-leaved.	———	1818.	——	G. ♄.	
3 ligustrìna. B.P.	(*wh.*)	Privet-leaved.	V. Diem. Isl.	1807.	7. 8.	G. ♄.	
4 rígida. R.S.	(*wh.*)	rigid-leaved.	———	1820.	——	G. ♄.	
'OLEA. B.P.		OLIVE.	Diandria Monogynia. L.				
1 europ'æa. s.s.	(*wh.*)	European.	S. Europe.	1570.	6. 8.	G. ♄.	
α Oleáster. R.S.	(*wh.*)	*spiny-branched.*	Portugal.	....	——	G. ♄.	Flor. græc. 1. t. 3.
β longifòlia. R.S.	(*wh.*)	*long-leaved.*	S. Europe.	....	——	G. ♄.	Lodd. bot. cab. t. 456.
γ ferrugínea. R.S.	(*wh.*)	*ferruginous.*	———	....	——	G. ♄.	
δ latifòlia. R.S.	(*wh.*)	*broad-leaved.*	———	....	——	G. ♄.	
ε oblíqua. R.S.	(*wh.*)	*oblique-leaved.*	———	....	——	G. ♄.	
ζ buxifòlia. R.S.	(*wh.*)	*Box-leaved.*	———	....	——	G. ♄.	
2 verrucòsa. s.s.	(*wh.*)	warted.	C. B. S.	1814.	——	G. ♄.	
3 americàna. s.s.	(*wh.*)	American.	N. America.	1758.	6.	G. ♄.	Catesb. car. 1. t. 61.
4 fràgrans. s.s.	(*wh.*)	fragrant.	China.	1771.	6. 8.	G. ♄.	Bot. mag. t. 1552.
5 láncea. s.s.	(*wh.*)	lance-leaved.	Mauritius.	1818.	——	S. ♄.	
6 dioíca. F.I.	(*wh.*)	diœcious.	E. Indies.	——	....	S. ♄.	
7 capénsis. s.s.	(*wh.*)	Cape.	C. B. S.	1730.	6. 9.	G. ♄.	Bot. reg. t. 613.
8 undulàta. s.s.	(*wh.*)	wave-leaved.	———	——	——	G. ♄.	Lodd. bot. cab. t. 379.
9 emarginàta. R.S.	(*wh.*)	emarginate.	Mauritius.	1816.	....	S. ♄.	
10 excélsa. s.s.	(*wh.*)	Laurel-leaved.	Madeira.	1784.	5. 6.	G. ♄.	Jacq. schœn. 3. t. 251.
11 cérnua. s.s.	(*wh.*)	drooping-flower'd.	Mauritius.	1816.	....	S. ♄.	
12 paniculàta. B.P.	(*wh.*)	panicled.	N. S. W.	1823.	4. 6.	G. ♄.	
13 Roxbúrghii.	(*wh.*)	Roxburgh's.	E. Indies.	1828.	....	S. ♄.	
paniculàta. F.I. *non* B.P.							
14 robústa.	(*wh.*)	large Phillyrea.	———	1818.	....	S. ♄.	
Phillyrèa robústa. F.I.							

15 chinénsis.	(*wh.*)	Chinese.	China.	1820.	4. 6.	H. ħ.	
Phillyrèa paniculàta. F.I.							
16 angustifòlia.	(*wh.*)	narrow-ld. Phillyrea.	S. Europe.	1597.	5. 6.	H. ħ.	Lam. ill. t. 8. f. 3.
Phillyrèa angustifòlia. S.S.							
α *lanceolàta.*	(*wh.*)	*spear-leaved.*	———	——	——	H. ħ.	
β *rosmarinifòlia.*	(*wh.*)	*Rosemary-leaved.*	———	——	——	H. ħ.	
γ *brachiàta.*	(*wh.*)	*cross-branched.*	———	——	——	H. ħ.	
17 mèdia. R.S.	(*wh.*)	twiggy Phillyrea.	———	——	——	H. ħ.	
β *buxifòlia.* R.S.	(*wh.*)	*Box-leaved.*	———	——	——	H. ħ.	
18 ligustrifòlia. R.S.	(*wh.*)	Privet-leaved.	———	——	——	H. ħ.	
19 péndula. R.S.	(*wh.*)	pendulous.	———	——	——	H. ħ.	
20 oleæfòlia. R.S.	(*wh.*)	Olive-leaved Phillyrea.	———	——	——	H. ħ.	Pluk. alm. t. 310. f. 5.
21 l'ævis. R.S.	(*wh.*)	smooth-leaved.	———	——	——	H. ħ.	Duham. arb. t. 125.
22 latifòlia. R.S.	(*wh.*)	broad-leaved	———	——	——	H. ħ.	Flor. græc. t. 2.
23 oblíqua. R.S.	(*wh.*)	oblique-leaved.	———	——	——	H. ħ.	
24 spinòsa. R.S.	(*wh.*)	Holly-leaved.	———	——	——	H. ħ.	
Phillyrèa ilicifòlia. W. en.							
LIGU'STRUM. W.		PRIVET. Diandria Monogynia. L.					
1 vulgàre. R.S.	(*wh.*)	common.	Britain.	....	6. 7.	H. ♄.	Eng. bot. v. 11. t. 764.
β *variegàtum.*	(*wh.*)	*striped-leaved.*	———	....	——	H. ♄.	
γ *angustifòlium.*	(*wh.*)	*narrow-leaved.*	———	....	——	H. ♄.	
δ *leucocárpum.*	(*wh.*)	*white-berried.*	———	....	——	H. ♄.	
ε *itálicum.* Mill.	(*wh.*)	*evergreen.*	Italy.	....	——	H. ħ.	
2 lùcidum. R.S.	(*wh.*)	Wax-tree.	China.	1794.	6. 9.	F. ħ.	Wats. dend. brit. t. 137.
SYRI'NGA. R.S.		LILAC. Diandria Monogynia. L.					
1 vulgàris. W.	(*va.*)	common.	Persia.	1597.	5.	H. ♄.	Schk. hand. 1. t. 2.
α *lilacìna.*	(*li.*)	*pale-flowered.*	———	——	——	H. ♄.	Bot. mag. t. 183.
β *purpùrea.*	(*pu.*)	*purple-flowered.*	———	——	——	H. ♄.	
γ *violàcea.*	(*vi.*)	*violet.*	———	——	——	H. ♄.	
δ *álba.*	(*wh.*)	*white-flowered.*	———	——	——	H. ♄.	
ε *variegàta.*	(*li.*)	*variegated-leaved.*	———	——	——	H. ♄.	
2 chinénsis. W.	(*pu.*)	Siberian.	China.	1795.	——	H. ♄.	
3 pérsica. W.	(*li.*)	Persian.	Persia.	1640.	——	H. ♄.	Bot. mag. t. 486.
α *lilacìna.*	(*li.*)	*pale-flowered.*	———	——	——	H. ♄.	
β *álba.*	(*wh.*)	*white-flowered.*	———	——	——	H. ♄.	
γ *laciniàta.*	(*wh.*)	*cut-leaved.*	———	——	——	H. ♄.	Schmidt arb. 2. t. 79.

ORDO CXXXVII.

JASMINEÆ. *Brown prodr.* 1. *p.* 520.

NYCTA'NTHES. W.		NYCTA'NTHES. Diandria Monogynia. L.					
árbor trístis. W.	(*wh. ye.*)	square-stalked.	E. Indies.	1781.	6. 9	S. ♄.	Bot. reg. t. 399.
JASM'INUM. W.		JASMINE. Diandria Monogynia. L.					
1 Sámbac. S.S.	(*wh.*)	single Arabian.	E. Indies.	1665.	1. 12.	S. ♄.⌣.	Bot. reg. t. 1.
β *múltiplex.*	(*wh.*)	*semidouble.*	———	1700.	——	S. ♄.⌣.	Andr. reposit. t. 497.
γ *trifoliàtum.*	(*wh.*)	*Tuscan.*	———	1730.	——	S. ♄.	Bot. mag. t. 1785.
2 undulàtum. S.S.	(*wh.*)	wave-leaved.	———	1812.	2. 6.	S. ♄.⌣.	Bot. reg. t. 436.
3 hirsùtum. S.S.	(*wh.*)	hairy Indian.	———	1759.	5. 12.	S. ♄.	Sm. ex. bot. 2. t. 118.
multiflòrum. And. rep. t. 436.							
4 angustifòlium. F.I.	(*wh.*)	narrow-leaved.	———	1816.	1. 12.	S. ♄.⌣.	Rheed. mal. 6. t. 53.
5 laurifòlium. F.I.	(*wh.*)	Laurel-leaved.	———	——	——	S. ♄.⌣.	Bot. reg. t. 521.
6 trinérve. S.S.	(*wh.*)	three-nerved.	Java.	1804.	4. 6.	S. ♄.⌣.	——— t. 918.
7 glaùcum. S.S.	(*wh.*)	glaucous.	C. B. S.	1774.	6. 8.	S. ♄.⌣.	Sal. stirp. rar. t. 8.
8 bracteàtum. S.S.	(*wh.*)	bracteate.	Sumatra.	1820.	——	S. ♄.⌣.	
9 latifòlium. S.S.	(*wh.*)	broad-leaved.	E. Indies.	1817.	——	S. ♄.⌣.	
10 arboréscens. S.S.	(*wh.*)	arborescent.	———	1824.	....	S. ♄.	
11 scándens. S.S.	(*wh.*)	scandent.	———	——	....	S. ♄.⌣.	
12 simplicifòlium. S.S.	(*wh.*)	simple-leaved.	Society Isl.	....	3. 6.	S. ♄.⌣.	Bot. mag. t. 980.
13 elongàtum. S.S.	(*wh.*)	long-branched.	E. Indies.	1823.	....	S. ♄.⌣.	Berg. act. an. 1772. t. 11.
14 grácile. S.S.	(*wh.*)	slender.	Norfolk Isl.	1791.	1. 12.	G. ♄.⌣.	Andr. reposit. t. 127.
15 acuminàtum. S.S.	(*wh.*)	acuminate.	N. Holland.	1820.	6. 9.	G. ♄.⌣.	Bot. reg. t. 1296?
16 volùbile. R.S.	(*wh.*)	twisting.	C. B. S.	1816.	4. 7.	G. ♄.⌣.	Jacq. schœn. 3. t. 321.
17 auriculàtum. S.S.	(*wh.*)	auriculate.	E. Indies.	1790.	5. 9.	S. ♄.⌣.	Bot. reg. t. 264.
18 paniculàtum. S.S.	(*wh.*)	variable-leaved.	China.	1812.	2. 12.	S. ♄.⌣.	——— t. 690.

19 fléxile. s.s.	(*wh.*)	acute 3-leaved.	E. Indies.	1820.	4. 8.	S. ♄.◡.	
20 tortuòsum. s.s.	(*wh.*)	twisting.	C. B. S.	——	——	S. ♄.◡.	Jacq. schœn. 1. t. 490.
fléxile. Jacq. *non* W.							
21 campanulàtum.s.s.	(*wh.*)	bell-shaped calyx'd.	....	——	——	S. ħ.◡.	
22 capénse. s.s.	(*wh.*)	Cape.	C. B. S.	1816.	——	G. ħ.	
23 azòricum. s.s.	(*wh.*)	white Azorian.	Madeira.	1724.	4. 11.	G. ♄.	Bot. reg. t. 89.
24 dispérmum. s.s.	(*wh.*)	two-seeded.	Nepaul.	1825.	....	G. ħ.◡.	
25 frùticans. s.s.	(*ye.*)	common yellow.	S. Europe.	1570.	4. 10.	H. ħ.	Bot. mag. t. 461.
26 hùmile. s.s.	(*ye.*)	Italian.	———	1656.	6. 9.	H. ♄.	Bot. reg. t. 350.
27 odoratíssimum.s.s.	(*ye.*)	yellow Azorian.	Madeira.	——	5. 11.	G. ♄.	Bot. mag. t. 285.
28 revolùtum. s.s.	(*ye.*)	yellow Nepaul.	Nepaul.	1812.	3. 10.	H. ħ.	Bot. reg. t. 178.
chrysánthemum. F.I.							
29 pubígerum. D.P.	(*ye.*)	pubescent.	———	1828.	——	H. ♄.◡.	
30 officinàle. s.s.	(*wh.*)	common white.	E. Indies.	1548.	6. 10.	H. ♄.◡.	Bot. mag. t. 31.
β *leucophy'llum.*	(*wh.*)	*white-edged.*	———	....	——	H. ħ.◡.	
γ *chrysophy'llum.*	(*wh.*	*yellow-edged.*	———	....	——	H. ħ.◡.	
δ *múltiplex.*	(*wh.*)	*double-blossomed.*	———	1823.	——	H. ♄.◡.	
31 grandiflòrum. s.s.	(*wh.*)	Catalonian.	———	1629.	——	G. ħ.◡.	Bot. reg. t. 91.

ORDO CXXXVIII.

APOCINEÆ. *Brown prodr.* 1. *p.* 465.

ALY'XIA. B.P.		ALY'XIA.	Pentandria Monogynia.				
Richardsòni.	(*wh.*)	Richardson's.	N. Holland.	1823.	4. 6.	S. ♄.	
RAUWO'LFIA. S.S.		RAUWO'LFIA.	Pentandria Monogynia. W.				
1 nítida. s.s.	(*wh.*)	shining.	S. America.	1752.	6. 9.	S. ħ.	Lodd. bot. cab. t. 339.
2 ternifòlia. s.s.	(*wh.*)	three-leaved.	———	1822.	——	S. ħ.	Bot. mag. t. 2440.
3 canéscens. s.s.	(*fl.*)	hoary.	Jamaica.	1739.	——	S. ♄.	Jacq. amer. pict. f. 17.
4 tomentòsa. s.s.	(*wh.*)	woolly.	S. America.	1823.	——	S. ħ.	——— t. 46.
5 spinòsa. R.S.	(*st.*)	spiny.	Peru.	1827.	——	S. ♄.	Cavan. icon. 6. t. 526.
VALL'ESIA. R.P.		VALL'ESIA.	Pentandria Monogynia.				
cymbæfòlia. R.S.	(*wh.*)	boat-leaved.	S. America.	1816.	6. 8.	S. ♄.	Cavan. ic. 3. t. 297.
Rauwólfia glábra. W.							
OPHIO'XYLON. W.		OPHIO'XYLON.	Polygamia Monœcia. W.				
serpentìnum. W.	(*wh.*)	serpent-wood.	E. Indies.	1690.	5. 6.	S. ♄.	Bot. mag. t. 784.
STRY'CHNOS. S.S.		STRY'CHNOS.	Pentandria Monogynia. W.				
1 Núx vómica. s.s.	(*wh.*)	Poison-nut.	E. Indies.	1778.	....	S. ♄.	Roxb. cor. 1. t. 4.
2 colubrìna. s.s.	(*gr.ye.*)	muricate-fruited.	———	1818.	5. 8.	S. ♄.	Rheed. mal. 8. t. 24.
3 potatòrum. s.s.	(*wh.*)	clearing-nut.	———	1794.	4. 6.	S. ♄.	Roxb. cor. 1. t. 5.
4 madagascariénsis.s.s.	(*wh.*)	Madagascar.	Madagascar.	1818.	....	S. ħ.	
5 spinòsa. s.s.	(*gr.ye.*)	spiny.	———	1824.	....	S. ħ.	
6 axillàris. s.s.	(*wh.*)	axillary-flowered.	E. Indies.	1818.	5. 7.	S. ♄.	Linn. trans. v. 12. t. 15.
THEOPHRA'STA. S.S.		THEOPHRA'STA.	Pentandria Monogynia.				
1 longifòlia. R.S.	(*or.*)	long-leaved.	Caracas.	1827.	....	S. ♄.	Jacq. schœn. 1. t. 116.
2 macrophy'lla. R.S.	....	large-leaved.	Brazil.	——	....	S. ♄.	
3 americàna. R.S.	....	American.	S. America.	——	....	S. ♄.	Plum. icon. 126.
4 Jussi'æi. s.s.	(*wh.*)	Jussieu's.	———	1818.	1. 4.	S. ♄.	Lindl. coll. t. 26.
CLAVI'JA. R.P.		CLAVI'JA.	Pentandria Monogynia.				
macrophy'lla. R.P.	(*wh.*)	long-leaved.	S. America.	1816.	....	S. ♄.	Flor. peruv. gen. t. 30.
FAGR'ÆA. W.		FAGR'ÆA.	Pentandria Monogynia.				
1 zeylánica. W.	(*wh.*)	Ceylon.	Ceylon.	1824.	4. 7.	S. ħ.	Thunb.ac.hol.1782.t.4.
2 fràgrans. F.I.	(*wh.*)	fragrant.	China.	1826.	5. 7.	S. ♄.	
CARI'SSA. L.		CARI'SSA.	Pentandria Monogynia. L.				
1 Carándas. s.s.	(*wh.*)	Jasmine-flower'd.	E. Indies.	1790.	7. 8.	S. ♄.	Roxb. cor. 1. t. 77.
2 spinàrum. s.s.	(*wh.*)	Box-leaved.	———	1809.	8. 12.	S. ħ.	Lodd. bot. cab. t. 162.
3 ovàta. B.P.	(*wh.*)	oval-leaved.	N. Holland.	1817.	6. 9.	G. ħ.	
4 Xylopìcron. s.s.	(*wh.*)	nerved-leaved.	Mauritius.	1820.	——	S. ♄.	Aub. Thou. obs. p. 24. ic.
ARDU'INA. L.		ARDU'INA.	Pentandria Monogynia. L.				
bispinòsa. s.s.	(*wh.*)	two-spined.	C. B. S.	1760.	3. 8.	G. ħ.	Mill. ic. 2. t. 300.
Caríssa Arduìna. P.S.							
GELS'EMIUM. J.		GELS'EMIUM.	Pentandria Monogynia.				
sempervìrens. P.S.	(*ye.*)	evergreen.	N. America.	1640.	6. 7.	G. ħ.◡.	Catesb. car. 1. t. 53.
nítidum. R.S. *Bignònia sempervìrens.* W.							

CE'RBERA. W.		Ce'rbera.	Pentandria Monogynia. L.				
1 Ahoùai. s.s.	(*ye.*)	oval-leaved.	Brazil.	1739.	6. 7.	S. ♄.	Bot. mag. t. 737.
2 ovàta. s.s.	(*ye.*)	ovate-leaved.	N. Spain.	....	——	S. ♄.	Cavan. ic. 3. t. 270.
3 fruticòsa. s.s.	(*re.*)	purple-flowered.	Pegu.	1817.	3. 9.	S. ♄.	Bot. reg. t. 391.
4 Mánghas. l.t.	(*wh.*)	blunt-leaved.	E. Indies.	1759.	4. 8.	S. ♄.	Burm. zeyl. t. 70. f. 1.
5 Odállam. l.t.	(*wh.re.*)	spear-leaved.	———	——	6. 9.	S. ♄.	Bot. mag. t. 1845.
Mánghas. b.m. *non* l.t.							
6 lactària. l.t.	(*wh.*)	Milk-tree.	———	....	——	S. ♄.	Rumph. amb. 2. t. 81.
Mánghas. G. 2. t. 123 et 124. f. 1. *non* l.t.							
7 Tánghin. b.m.	(*bh.re.*)	poison-nut.	Madagascar.	1826.	....	S. ♄.	Bot. mag. t. 2968.
8 laurifòlia. b.c.	(*wh.*)	Laurel-leaved.	E. Indies.	1819.	6. 9.	S. ♄.	Lodd. bot. cab. t. 989.
9 maculàta. s.s.	(*wh.re.*)	wave-leaved.	Bourbon.	1782.	6. 7.	S. ♄.	Andr. reposit. t. 130.
10 dichótoma. b.c.	(*wh.*)	forked.	E. Indies.	1827.	——	S. ♄.	Lodd. bot. cab. t. 1516.
11 thevetioídes. s.s.	(*ye.*)	revolute-margin'd.	N. Spain.	1820.	....	S. ♄.	
12 Thevètia. s.s.	(*ye.*)	linear-leaved.	S. America.	1735.	....	S. ♄.	Jacq. amer. 48. t. 34.
13 cuneifòlia. k.s.	(*ye.*)	wedge-leaved.	Mexico.	1826.	4. 6.	S. ♄.	
OCHR'OSIA. J.		Ochr'osia.	Pentandria Monogynia. R.S.				
borbònica. r.s.	(*st.*)	Bourbon.	Bourbon.	1823.	....	S. ♄.	
Cérbera borbònica. s.s. *Ophióxylon Ochròsia.* p.s.							
DISSOL'ENA. R.S.		Dissol'ena.	Pentandria Monogynia. R.S.				
verticillàta. r.s.	(*wh.*)	whorl-leaved.	China.	1812.	....	S. ♄.	
ALLAMA'NDA. S.S.		Allama'nda.	Pentandria Monogynia. L.				
1 cathártica. s.s.	(*ye.*)	Willow-leaved.	Guiana.	1785.	6. 7.	S. ♄.	Bot. mag. t. 338.
2? verticillàta. s.s.	....	whorl-leaved.	E. Indies.	1812.	....	S. ♄.	
Nèrium tinctòrium. Hort.							
VI'NCA. R.S.		Periwinkle.	Pentandria Monogynia. L.				
1 herbàcea. r.s.	(*bl.*)	herbaceous.	Hungary.	1816.	6. 7.	H. ♃.	Bot. reg. t. 301.
2 mìnor. r.s.	(*bl.*)	lesser.	Britain.	....	3. 9.	H. ♄.	Eng. bot. v. 13. t. 917.
α cærùlea.	(*bl.*)	*blue-flowered.*	———	....	——	H. ♄.	
β álba.	(*wh.*)	*white-flowered.*	———	....	——	H. ♄.	
γ atropurpùrea.	(*pu.*)	*dark-purple.*	———	....	——	H. ♄.	
δ múltiplex.	(*pu.*)	*double-flowered.*	———	....	——	H. ♄.	
ε variegàta.	(*wh.*)	*variegated.*	———	....	——	H. ♄.	
3 màjor. r.s.	(*bl.*)	greater.	England.	....	——	H. ♄.	Eng. bot. v. 8. t. 514.
β múltiplex.	(*bl.*)	*double-blue.*	———	....	——	H. ♄.	
4 pusílla. r.s.	(*st.*)	small-flowered.	E. Indies.	1778.	8.	S. ⊙.	Rheed. mal. 9. t. 33.
parviflòra. w.							
5 ròsea. r.s.	(*ro.*)	Madagascar.	———	1756.	3. 10.	S. ♄.	Bot. mag. t. 248.
β álba.	(*wh.*)	*white-flowered.*	———	——	——	S. ♄.	
γ ocellàta.	(*wh.re.*)	*red-eyed.*	———	——	——	S. ♄.	
WRI'GHTIA. S.S.		Wri'ghtia.	Pentandria Monogynia. S.S.				
1 antidysentérica. s.s.	(*wh.*)	oval-leaved.	E. Indies.	1778.	....	S. ♄.	Burm. zeyl. t. 77.
2 zeylánica. s.s.	(*wh.*)	spear-leaved.	———	....	....	S. ♄.	——— t. 12. f. 2.
3 dùbia. s.s.	(*sc.*)	doubtful.	———	....	5. 8.	S. ♄.	Bot. mag. t. 1646.
Camerària dùbia. b.m.							
4 coccínea. b.m.	(*sc.*)	scarlet.	———	1820.	6. 8.	S. ♄.	Bot. mag. t. 2696.
Nèrium coccíneum. Bot. cab. t. 894.							
5 tinctòria. s.s.	(*wh.*)	dyer's.	———	1812.	——	S. ♄.	Bot. reg. t. 933.
STROPHA'NTHUS. DC.		Stropha'nthus.	Pentandria Monogynia. P.S.				
1 dichótomus. dc.	(*ro.*)	forked.	E. Indies.	1816.	2. 10.	S. ♄.	Burm. ind. 68. t. 26.
2 divérgens. Gr.	(*ye.*)	spreading.	China.	——	——	S. ♄.	Bot. reg. t. 469.
dichótomus. b.r. *non* dc.							
3 sarmentòsus. dc.	(*re.*)	sarmentose.	Sierra Leon.	1822.	....	S. ♄.	Desf. an. mus. 1. t. 27.
N'ERIUM. R.S.		Oleander.	Pentandria Monogynia. L.				
1 Oleánder. r.s.	(*ro.*)	common.	S. Europe.	1596.	6. 10.	G. ♄.	Flor. græc. t. 248.
β álba.	(*wh.*)	*white-flowered.*	———	——	——	G. ♄.	Lodd. bot. cab. t. 700.
γ spléndens.	(*ro.*)	*double-hybrid.*	Hybrid.	1815.	——	G. ♄.	Bot. reg. t. 74.
2 odòrum. r.s.	(*ro.*)	sweet-scented.	E. Indies.	1683.	6. 8.	S. ♄.	Rheed. mal. 9. t. 2.
β cárneum.	(*fl.*)	*flesh-coloured.*	———	1816.	——	S. ♄.	Bot. mag. t. 2032.
γ flòre plèno.	(*ro.*)	*double-flowered.*	———	....	——	S. ♄.	Rheed. mal. 9. t. 1.
3 flavéscens. r.s.	(*st.*)	yellowish.	....	1816.	——	S. ♄.	
TABERNÆMONT'ANA.		Tabernæmont'ana.	Pentandria Monogynia. L.				
1 citrifòlia. r.s.	(*st.*)	Citron-leaved.	Jamaica.	1734.	5. 8.	S. ♄.	Plum. ic. t. 248. f. 2.
2 álba. r.s.	(*wh.*)	white-flowered.	W. Indies.	1780.	——	S. ♄.	Jacq. amer. t. 175. f. 13.
citrifòlia. Jacq. *non* l.							
3 laurifòlia. r.s.	(*st.*)	Laurel-leaved.	———	1768.	——	S. ♄.	Bot. reg. t. 716.
4 cymòsa. r.s.	(*wh.*)	cyme-flowered.	S. America.	1820.	——	S. ♄.	Jacq. amer. t. 181. f. 14.
5 amygdalifòlia. r.s.	(*wh.*)	Peach-leaved.	———	1780.	4. 8.	S. ♄.	Bot. reg. t. 338.
6 díscolor. r.s.	(*st.*)	two-coloured.	Jamaica.	1822.	——	S. ♄.	

7 persicariæfòlia. R.S.	(st.)	Persicaria-leav'd.	Mauritius.	1826.	4. 8.	S. ♄.	Jacq. icon. rar. t. 320.
8 arcuàta. R.S.	(st.)	arch'd follicled.	Peru.	1820.	——	S. ♄.	Flor. peruv. 2. t. 143.
9 críspa. L.T.	(wh.)	wave-leaved.	E. Indies.	1818.	5. 10.	S. ♄.	Rheed. mal. 1. t. 46.
10 coronària. L.T.	(wh.)	broad-leaved.	———	1710.	6. 10.	S. ♄.	Lodd. bot. cab. t. 406.
β *flòre plèno.*	(wh.)	*double-flowered.*	———	——	——	S. ♄.	Bot. mag. t. 1865.
Nèrium coronàrium. B.M.							
11 densiflòra. B.R.	(wh.)	close-flowered.	Ceylon.	1824.	6. 7.	S. ♄.	Bot. reg. t. 1273.
12 undulàta. R.S.	(st.)	waved.	Trinidad.	——	——	S. ♄.	
13 odoràta. S.S.	(ye.)	yellow-flowered.	Cayenne.	1793.	10. 11.	S. ♄.	Aub. guian. 1. t. 102.
Camerària lùtea. W. *Camerària Tamaquarìna.* Aublet.							
CAMER'ARIA. S.S.		BASTARD MANGENEEL.	Pentandria Monogynia. L.				
1 latifòlia. S.S.	(wh.)	broad-leaved.	Havannah.	1733.	8.	S. ♄.	Andr. reposit. t. 261.
2 angustifòlia. S.S.	(wh.)	narrow-leaved.	S. America.	1752.	9.	S. ♄.	Plum. ic. t. 72. f. 2.
AMS'ONIA. R.S.		AMSONIA.	Pentandria Monogynia. P.S.				
1 latifòlia. R.S.	(bl.)	broad-leaved.	N. America.	1759.	5. 6.	H. ♃.	Bot. reg. t. 151.
2 salicifòlia. R.S.	(pa.bl.)	Willow-leaved.	———	1812.	——	H. ♃.	Bot. mag. t. 1873.
3 angustifòlia. R.S.	(bl.)	hairy-stalked.	———	1774.	——	H. ♃.	Venten. choix. t. 29.
PLUM'ERIA. S.S.		PLUM'ERIA.	Pentandria Monogynia. L.				
1 rùbra. R.S.	(re.)	red-flowered.	Jamaica.	1690.	7. 8.	S. ♄.	Bot. reg. t. 780.
2 purpùrea. R.S.	(pu.)	purple.	Peru.	1820.	——	S. ♄.	Flor. peruv. 2. t. 137.
3 acuminàta. H.K.	(wh.)	acute-leaved.	E. Indies.	1790.	6. 9.	S. ♄.	Bot. reg. t. 114.
acutifòlia. R.S.							
4 obtùsa. R.S.	(wh.)	blunt-leaved.	W. Indies.	1733.	7. 8.	S. ♄.	Catesb. car. 1. t. 93.
5 álba. R.S.	(wh.)	white.	Jamaica.	——	——	S. ♄.	Jacq. amer. t. 174. f. 2.
6 incarnàta. R.P.	(fl.)	flesh-coloured.	Peru.	....	——	S. ♄.	Flor. peruv. t. 138.
7 carinàta. R.P.	(wh.re.)	keel-leaved.	———	1824.	——	S. ♄.	
8 bícolor. R.P.	(ye.wh.)	two-coloured.	S. America.	1815.	6. 9.	S. ♄.	Bot. reg. t. 480.
9 trícolor. R.P.	(wh.re.)	three-coloured.	———	——	——	S. ♄.	——— t. 510.
10 lùtea. R.P.	(ye.)	yellow-flowered.	———	——	——	S. ♄.	Flor. peruv. 2. t. 142.
11 aurántia. H.S.S.	(or.)	orange-flowered.	———	1816.	——	S. ♄.	
12 pùdica. R.S.	(st.)	wax-flowered.	———	——	7. 8.	S. ♄.	
13 longifòlia. R.S.	....	long-leaved.	Madagascar.	1816.	....	S. ♄.	
14 tuberculàta. B.C.	(wh.)	tubercled.	S. America.	1820.	6. 8.	S. ♄.	Lodd. bot. cab. t. 681.
15 tenuifòlia. L.C.	....	slender-leaved.	———	....	....	S. ♄.	
16 Northiàna. L.C.	....	North's.	———	....	....	S. ♄.	
17 mexicàna. B.C.	(wh.)	Mexican.	Mexico.	....	6. 9.	S. ♄.	Lodd. bot. cab. t. 1024.
18 Cowàni. D.D.	(wh.)	Cowan's.	———	1824.	——	S. ♄.	
19 Blandfordiàna. L.C.	....	Blandford's.	S. America.	....	....	S. ♄.	
20 nívea. L.C.	(wh.)	snowy white.	———	....	6. 8.	S. ♄.	
21 conspícua. L.C.	....	conspicuous.	———	....	....	S. ♄.	
22 macrophy'lla. L.C.	....	large-leaved.	———	....	....	S. ♄.	
23 leucántha. L.C.	(wh.)	white-blossomed.	———	....	....	S. ♄.	
24 parvifòlia. H.S.S.	(wh.)	small-leaved.	W. Indies.	1813.	7. 8.	S. ♄.	
PREST'ONIA. S.S.		PREST'ONIA.	Pentandria Monogynia. S.S.				
1 tomentòsa. S.S.	(ye.)	tomentose.	Brazil.	1820.	5. 8.	S. ♄. ◡.	
2 glabràta. S.S.	(ye.)	smooth.	S. America.	1823.	——	S. ♄ ◡.	
CRYPTOL'EPIS. B.		CRYPTOL'EPIS.	Pentandria Monogynia.				
élegans. L.C.	....	elegant.	E. Indies.	1823.	....	S. ♄. ◡.	
ALST'ONIA. S.S.		ALST'ONIA.	Pentandria Monogynia. S.S.				
1 scholàris. S.S.	(wh.)	whorl-leaved.	E. Indies.	1816.	....	S. ♄.	Rheed. mal. 1. t. 45.
2 venenàta. S.S.	(wh.)	poisonous.	———	1823.	....	S. ♄.	
THENA'RDIA. K.S.		THENA'RDIA.	Pentandria Monogynia.				
floribúnda. K.S.	(gr.)	abundant-flowered.	Mexico.	1823.	....	S. ♄. ◡	H. et B. 3. t. 240.
ECH'ITES. R.S.		ECH'ITES.	Pentandria Monogynia. L.				
1 longiflòra. S.S.	(wh.)	long-flowered.	Brazil.	1816.	5. 8.	S. ♄. ◡.	
2 bispinòsa. S.S.	(fl.)	two-spined.	C. B. S.	1795.	7. 11.	S. ♄.	
3 succulénta. S.S.	(re.wh.)	succulent.	———	1820.	4. 6.	S. ♄.	Jacq. fragm. t. 117.
tuberòsa. H.S.S.							
4 rèpens. R.S.	(re.)	creeping.	St. Domingo.	1824.	6. 8.	S. ♄.	Jacq. am. pict. t. 35.
5 biflòra. S.S.	(wh.)	twin-flowered.	W. Indies.	1793.	7. 8.	S. ♄. ◡.	Jacq. amer. 30. t. 21.
6 umbellàta. S.S.	(wh.)	umbelled.	Jamaica.	1733.	7.	S. ♄. ◡.	——— 30. t. 22.
7 toròsa. S.S.	(ye.)	climbing.	———	1778.	6. 8.	S. ♄. ◡.	——— 33. t. 27.
8 antidysentérica. S.S.	(pi.)	medicinal.	E. Indies.	1821.	....	S. ♄.	
9 pubéscens. L.T.	(wh.)	pubescent.	———	1822.	....	S. ♄.	Rheed. mal. 1. t. 49.?
10 suberécta. S.S.	(ye.)	Savanna-flower.	W. Indies.	1759.	6. 8.	S. ♄. ◡.	Bot. mag. t. 1064.
11 caryophyllàta. S.S.	(st.)	Clove-scented.	E. Indies.	1812.	9. 11.	S. ♄. ◡.	——— t. 1919.
12 diffórmis. S.S.	(ye.)	deformed.	Carolina.	1806.	6. 7.	S. ♄. ◡.	
13 grandiflòra. R.S.	(pi.)	large-flowered.	E. Indies.	1823.	....	S. ♄.	
macrántha. S.S.							

14 reticulàta. s.s. (*wh.*) netted-leaved. E. Indies. 1818. 6. 7. G. ♄.◡.
15 Héynii. s.s. (*wh.*) Heyne's. ——— ——— ——— S. ♄.◡.
16 frutéscens. H.B. (*wh.*) frutescent. ——— 1816. ——— S. ♄.◡.
17 paniculàta. H.B. (*wh.*) panicled. ——— ——— ——— S. ♄.◡.
BEAUMO'NTIA. B.R. Beaumo'ntia. Pentandria Monogynia.
1 grandiflòra. B.R. (*wh.*) oval-leaved. E. Indies. 1812. 5. 7. S. ♄.◡. Bot. reg. t. 911.
2 longifòlia. L.C. (*wh.*) long-leaved. ——— 1818. .... S. ♄.◡.
HÆMADI'CTYON. H.T. Hæmadi'ctyon. Pentandria Monogynia.
venòsum. H.T. (*ye.*) red-veined. W. Indies. 1820. 4. 8. S. ♄.◡. Bot. mag. t. 2473.
Echìtes nùtans. B.M.
VALL'ARIS. S.S. Vall'aris. Pentandria Monogynia. S.S.
Pergulànus. s.s. (*wh.*) smooth-leaved. E. Indies. 1818. .... S. ♄.◡. Rumph. am. 5. t. 29.
Pergulària glábra. W. *Emerícia Pergulària.* R.S.
ICHNOCA'RPUS. S.S. Ichnoca'rpus. Pentandria Monogynia. S.S.
frutéscens. s.s. (*pu.*) shrubby. E. Indies. 1759. 7. 8. S. ♄. Burm. zeyl. t. 12. f. 1.
PARSO'NSIA. R.S. Parso'nsia. Pentandria Monogynia. R.S.
1 corymbòsa. R.S. (*re.*) corymb-flowered. W. Indies. 1820. 6. 8. S. ♄.◡. Jacq. amer. pict. t. 37.
Echìtes corymbòsa. Jacq.
2 floribúnda. R.S. (*wh.*) many-flowered. Jamaica. ——— ——— S. ♄.
Echìtes floribúnda. W.
WILLUGHBE'IA. S.S. Willughbe'ia. PentandriaMonogynia. S.S.
èdulis. s.s. (*bh.*) eatable. E. Indies. 1818. 6. 8. S. ♄.◡. Roxb. corom. 3. t. 280.
LYO'NSIA. B.P. Lyo'nsia. Pentandria Monogynia. S.S.
stramínea. B.P. (*st.*) climbing. N. S. W. 1820. 6. 8. S. ♄.◡.
MELOD'INUS. R.S. Melod'inus. Pentandria Mono-Digynia.
1 scándens. R.S. (*wh.*) climbing. N. Caledonia. 1775. .... S. ♄.◡. Lam. ill. t. 179.
2 parvifòlius. L.C. (*wh.*) small-leaved. E. Indies. .... S. ♄.
3 undulàtus. L.C. (*wh.*) wave-leaved. ——— .... S. ♄.
4 monog'ynus. B.R. (*wh.*) one-styled. ——— 1816. 4. 8. S. ♄.◡. Bot. reg. t. 834.
CARPOD'INUS. D.D. Pishamin. Pentandria Digynia.
1 dúlcis. D.D. (*wh.*) sweet. Sierra Leone. 1823. .... S. ♄.◡.
2 ácidus. D.D. (*wh.*) sour. ——— — — .... S. ♄.◡.
APO'CYNUM. S.S. Dog's-bane. Pentandria Monogynia. R.S.
1 androsæmifòlium. L.C. (*bh.*) Tutsan-leaved. N. America. 1688. 7. 9. H. ♃. Bot. mag. t. 280.
2 cannabìnum. s.s. (*gr.*) Hemp-like. ——— 1699. ——— H. ♃. Moris. s. 15. t. 3. f. 14.
3 hypericifòlium. s.s. (*wh.*) Hypericum-leav'd. ——— 1758. 6. 7. H. ♃. Jacq. vind. 3. t. 66.
4 venètum. s.s. (*re.*) Venetian. Adriatic Isl. 1690. ——— H. ♃. Lob. ic. 372. f. 1. 2.
β *álbum.* (*wh.*) *white-flowered.* ——— .... ——— H. ♃.

ORDO CXXXIX.

ASCLEPIADEÆ. *Brown prodr.* 1. *p.* 458.

PERIPL'OCA. R.S. Peripl'oca. Pentandria Digynia. L.
1 gr'æca. R.S. (*pu.*) Virginian-silk. Syria. 1597. 7. 8. H. ♄.◡. Bot. reg. t. 803.
2 mauritiàna. R.S. (*pu.*) long-leaved. Mauritius. 1823. .... S. ♄.◡. Rheed. mal. 9. t. 11.
3 lævigàta. R.S. (*ye.*) smooth. Canaries. 1779. 6. 8. G. ♄. Cavan. ic. 3. t. 217.
punicæfòlia. Cav. icon.
HEMIDE'SMUS. R.S. Hemide'smus. Pentandria Digynia. R.S.
índicus. R.S. (*gr.*) Indian. Ceylon. 1796. .... S. ♄. Burm. zeyl. t. 83. f. 1.
CRYPTOST'EGIA. B.R. Cryptost'egia. Pentandria Digynia. S.S.
grandiflòra. B.R. (*ro.*) large-flowered. E. Indies. 1812. 6. 8. S. ♄.◡. Bot. reg. t. 435.
SECAM'ONE. R.S. Secam'one. Pentandria Digynia. R.S.
1 emética. R.S. (*wh.*) narrow-leaved. India. 1816. .... S. ♄. Willd. phyt. 1. t. 5. f. 2.
2 ægyptìaca. H.K. (*wh.*) Egyptian. Egypt. 1752. 7. G. ♄.◡. Alp. ægyp. t. 134.
Alpìni. R.S. *Periplòca Secamòne.* W.
DUVA'LIA. H.S. Duva'lia. Pentandria Digynia. H.S.
1 reclinàta. H.S. (*pu.ye.*) reclined. C. B. S. 1795. 8. 10. D.S. ♄. Bot. mag. t. 1397.
2 élegans. H.S. (*d.pu.*) elegant. ——— ——— ——— D.S. ♄. ——— t. 1184.
Stapèlia élegans. B.M.
3 hirtèlla. s.s. (*d.pu.*) hairy. ——— ——— ——— D.S. ♄. Jacq. stap. c. ic.
Stapèlia hirtèlla. Jacq.
4 cæspitòsa. H.S. (*d.pu.*) tufted. ——— 1790. 8. 11. D.S. ♄. Mass. stap. 20. t. 29.

5 radiàta. H.S. (*d.pu.*) radiated.	C. B. S.	1795.	9. 11.	D.S.♄.	Bot. mag. t. 619.
6 replicàta. S.S. (*br.pu.*) replicate.	———	1806.	8. 10.	D.S.♄.	Jacq. stap. c. ic.
Stapèlia replicàta. Jacq.					
7 mastòdes. S.S. (*d.pu.*) teat.	———	1821.	——	D.S.♄.	Jacq. stap. c. ic.
8 Jacquiniàna. R.S.(*d.pu.*)Jacquin's.	———	1802.	7. 9.	D.S.♄.	——— t. 37.
Stapèlia radiàta. Jacq. *non* B.M.					
9 tuberculàta. H.S.S.(*d.p.*)tubercled.	———	1774.	7. 11.	D.S.♄.	
10 lævigàta. H.S.S. (*d.p.*) smooth.	———	1808.	——	D.S.♄.	
11 glomeràta. H.S.S. (*d.p.*) clustered.	———	——	——	D.S.♄.	
12 compácta. H.S.S. (*d.p.*) compact.	———	1800.	8. 11.	D.S.♄.	

PECTIN'ARIA. H.S.S. PECTIN'ARIA. Pentandria Digynia. H.S.

1 mammillàris. (*br.re.*) prickly.	C. B. S.	1774.	6. 8.	D.S.♄.	Burm. afr. p. 27. t. 11.
Stapèlia mammillàris. W.					
2 articulàta. H.S.S. (*d.r.*) jointed.	———	——	8. 9.	D.S.♄.	Mass. stap. 20. t. 30.
Stapèlia articulàta. Mas.					

O'RBEA. H.S. O'RBEA. Pentandria Digynia. H.S.

1 maculòsa. H.S. (*y.r.*) spotted.	C. B. S.	1804.	6. 9.	D.S.♄.	Bot. mag. t. 1833.
2 míxta. H.S. (*vi.ye.*) mixed.	———	1795.	——	D.S.♄.	Mass. stap. 23. t. 38.
Stapèlia míxta. Mas.					
3 quinquenérvia. H.S.(*s.p.*)five-nerved.	———	1810.	10. 12.	D.S.♄.	
4 trisúlca. J.S. (*ye.pu.*) trisulcate.	———	1806.	——	D.S.♄.	Jacq. stap. c. ic.
5 bisúlca. H.S. (*s.p.*) bisulcate.	———	——	——	D.S.♄.	
6 variegàta. H.S. (*s.p.*) variegated.	———	1727.	6. 10.	D.S.♄.	Moris. s. 15. t. 3. f. 4.
Stapèlia variegàta. L. *non* B.M.					
7 Curtísii. H.S. (*s.p.*) Curtis's.	———	1787.	——	D.S.♄.	Bot. mag. t. 26.
Stapèlia variegàta. B.M. *non* L.					
8 lépida. H.S.S. (*st.cr.*) pretty.	———	1805.	7. 8.	D.S.♄.	Jacq. stap. c. ic.
9 orbiculàris. H.S.S. (*br.*) large orbed.	———	1799.	7. 11.	D.S.♄.	Andr. reposit. t. 448.
10 bufònia. H.S. (*y.bk.*) toad-like.	———	1806.	8. 10.	D.S.♄.	Bot. mag. t. 1676.
Stapèlia Bufònis. B.M.					
11 clypeàta. H.S.S. (*y.p.*) shield-flowered.	———	1799.	6. 11.	D.S.♄.	Jacq. stap. c. ic.
12 planiflòra. H.S.S.(*ye.bk.*)flat-flowered.	———	1805.	7. 11.	D.S.♄.	Lodd. bot. cab. t. 191.
13 inodòra. H.S.S. (*s.p.*) scentless.	———	1788.	6. 8.	D.S.♄.	
14 rugòsa. S.S. (*gr.br.*) rugged.	———	1805.	5. 8.	D.S.♄.	Jacq. stap. c. ic.
15 Wendlandiàna. R.S.(*s.br.*)Wendland's.	———	——	6. 9.	D.S.♄.	Wendl. coll. 2. t. 52.
Stapèlia rugòsa. Wend. *non* Jacq.					
16 marginàta. R.S. (*s.vi.*) margined.	———	1805.	——	D.S.♄.	
17 conspurcàta. R.S.(*s.vi.*) dirtied.	———	1795.	6. 10.	D.S.♄.	Jacq. stap. c. ic.
18 normàlis. R.S. (*s.p.*) sulphur-coloured.	———	1821.	——	D.S.♄.	Bot. reg. t. 755.
19 marmoràta. R.S. (*s.da.*) marbled.	———	——	——	D.S.♄.	
20 retùsa. H.S. (*s.da.*) retuse.	———	1808.	9. 10.	D.S.♄.	
21 anguínea. H.S. (*y.da.*) snake.	———	1806.	8. 9.	D.S.♄.	Lodd. bot. cab. t. 828.
22 pícta. H.S. (*st.d.p.*) painted.	———	1799.	6. 9.	D.S.♄.	Bot. mag. t. 1169.
23 Woodfordiàna. H.S.(*s.p.*)Woodford's.	———	1805.	——	D.S.♄.	

STAP'ELIA. H.S. CARRION-FLOWER. Pentandria Digynia. L.

1 grandiflòra. R.S. (*d.p.*) great-flowered.	C. B. S.	1795.	9. 12.	D.S.♄.	Mass. stap. 13. t. 11.
2 spectábilis. H.S. (*d.p.*) showy.	———	1802.	——	D.S.♄.	Bot. mag. t. 585.
grandiflòra. B.M. *non* Mas.					
3 ambígua. R.S. (*re.pu.*) ambiguous.	———	1795.	6. 11.	D.S.♄.	Mass. stap. 13. t. 12.
β *fúlva.* (*br.*) *brown-flowered.*	———	——	——	D.S.♄.	Jacq. stap. c. ic.
ambígua varietas. Jacq.					
4 soròria. R.S. (*re.pu.*) sister.	———	1797.	6. 8.	D.S.♄.	Mass. stap. 23. t. 39.
5 elongàta. (*br.*) long-branched.	———	——	——	D.S.♄.	Jacq. stap. c. ic.
soròria varietas. Jacq.					
6 lunàta. (*pu.*) lunate-spotted.	———	——	——	D.S.♄.	Jacq. stap. c. ic.
soròria varietas alia. Jacq.					
7 pátula. R.S. (*d.p.*) spreading.	———	....	——	D.S.♄.	Jacq. stap. c. ic.
soròria. Jacq. *non* Mas.					
8 defléxa. B.M. (*ye.re.*) reflexed.	———	1806.	——	D.S.♄.	Bot. mag. t. 1890.
refléxa. H.S. *Gonostèmon defléxum.* S.S.					
9 lùcida. R.S. (*pu.*) glossy.	———	1821.	——	D.S.♄.	
10 breviróstris. W.en.(*br.*)short-beaked.	———	....	——	D.S.♄.	
11 Massòni. H.S. (*d.p.*) Masson's.	———	1808.	——	D.S.♄.	
12 Astèrias. R.S. (*d.p.ye.*) Starfish.	———	1795.	5. 11.	D.S.♄.	Bot. mag. t. 536.
13 stellàris. H.S.S. (*br.p.*) starry.	———	1804.	8. 10.	D.S.♄.	Jacq. stap. c. ic.
14 hirsùta. R.S (*s.d.vi.*) hairy.	———	1710.	6. 8.	D.S.♄.	Jacq. misc. 1. t. 3.
β *atra.* B.R. (*da.pu.*) *dark-flowered.*	———	——	——	D.S.♄.	Bot. reg. t. 756.
15 lanífera. H.S.S. (*d.br.*) woolly-flowered.	———	1814.	6. 9.	D.S.♄.	

16 flavicomàta. H.S.S. (*br.*) yellow-haired. C. B. S. 1814. 6. 9. D.S.♄.
17 hamàta. R.S. (*pu.re.*) hooked. ——— 1805. —— D.S.♄. Lodd. bot. cab. t. 242.
18 comàta. R.S. (*d.p.ye.*) very-hairy. ——— 1821. —— D.S.♄.
19 multiflòra. DC. (*pu.re.*) many-flowered. ——— —— —— D.S.♄.
20 rùfa. R.S. (*bl.re.*) rusty brown. ——— 1795. 6. 11. D.S.♄. Lodd. bot. cab. t. 239.
21 pulvinàta. R.S. (*r.p.*) cushioned. ——— —— —— D.S.♄. Bot. mag. t. 1240.
22 fissiróstris. R.S. (*gr.pu.*) cloven-beaked. ——— 1821. —— D.S.♄. Jacq. stap. c. ic.
23 concínna. R.S. (*br.*) spruce. ——— 1795. 6. 8. D.S.♄. Mass. stap. 15. t. 18.
24 glandulífera. R.S. (*y.br.*) hairy-glanded. ——— —— —— D.S.♄.
25 hispídula. R.S. (*br.*) hispid. ——— 1821. —— D.S.♄.
26 glanduliflòra. R.S. (*st.*) gland-flowered. ——— 1795. 8. 11. D.S.♄. Mass. stap. 16. t. 19.
27 acuminàta. R.S. (*d.p.*) acuminated. ——— —— 7. 9. D.S.♄. ——— 15. t. 17.
28? cordàta. H.S. .... heart-shaped. ——— 1805. 6. 8. D.S.♄.
29? canéscens. H.S. .... canescent. ——— 1795. —— D.S.♄.
30? ophiúncula. H.S. .... serpent-like. ——— 1805. —— D.S.♄.

TRIDE'NTEA. H.S. TRIDE'NTEA. Pentandria Digynia. H.S.

1 gemmiflòra. H.S. (*bk.*) gem-flowered. C. B. S. 1795. 10. 11. D.S.♄. Mass. stap. 14. t. 15.
2 st'ygia. H.S. (*bk.*) Stygian. ——— 1800. 8. 10. D.S.♄. Bot. mag. t. 1839.
Stapèlia gemmiflòra. β. B.M.
3 moschàta. H.S.S. (*d.p.*) musk-scented. ——— 1799. 9. 11. D.S.♄. Lodd. bot. cab. t. 1051.
Stapèlia hircòsa. Jacq. stap. c. ic.
4 vétula. H.S. (*d.p.*) purple smooth. ——— 1793. 5. 11. D.S.♄. Mass. stap. 15. t. 16.
5 Símsii. H.S. (*d.p.*) Sims's. ——— 1800. —— D.S.♄. Bot. mag. t. 1234.
Stapèlia vétula. B.M. *non* Mass.
6 paniculàta. R.S. (*re.*) panicled. ——— 1805. 6. 9. D.S.♄.
7 juvéncula. S.S. (*d.br.*) short-flowered. ——— .... —— D.S.♄. Jacq. stap. c. ic.
8 depréssa. R.S. (*re.p.*) depressed. ——— .... —— D.S.♄. ——— c. ic.

TROMO'TRICHE. H.S. TROMO'TRICHE. Pentandria Digynia. H.S.

1 fuscàta. H.S.S. (*br.*) brown-flowered. C. B. S. 1814. 6. 10. D.S.♄. Jacq. stap. c. ic.
Stapèlia fuscàta. Jacq.
2 glaùca. H.S.S. (*cr.*) large-glaucous. ——— 1799. 6. 11. D.S.♄. Jacq. stap. c. ic.
3 revolùta. H.S. (*cr.*) revolute-flowered. ——— 1790. 6. 9. D.S.♄. Bot. mag. t. 724.
Stapèlia revolùta. B.M.
4 pruinòsa. H.S. (*bh.*) frosted. ——— 1795. 6. 8. D.S.♄. Mass. stap. 24. t. 41.
5 oblíqua. R.S. (*ye.vi.*) oblique-flowered. ——— 1805. 6. 9. D.S.♄.
6 mutábilis. (*gr.br.*) changeable. ——— 1823. —— D.S.♄. Jacq. stap. c. ic.
β *variábilis.* (*st.re.*) *variable.* ——— —— —— D.S.♄. ——— c. ic.
Stapèlia mutábilis varietas. Jacq.
7 ciliàta. (*ye.*) fringed. ——— 1795. 10. 12. D.S.♄. Mass. stap. 9. t. 1.
Stapèlia ciliàta. Mass.

PODA'NTHES. H.S. PODA'NTHES. Pentandria Digynia. H.S.

1 púlchra. H.S. (*st.*) beautiful sulphur. C. B. S. 1800. 8. 10. D.S.♄.
β *màjor.* (*st.d.p.*) *larger.* ——— —— —— D.S.♄. Bot. mag. t. 786.
Stapèlia verrucòsa. B.M. *nec aliorum.*
2 irroràta. H.S. (*st.cr.*) dewy. ——— 1795. 7. 9. D.S.♄. Mass. stap. 12. t. 9.
3 verrucòsa. H.S. (*st.br.*) warted. ——— —— 8. 10. D.S.♄. ——— 11. t. 8.
4 roriflùa. R.S. (*st.cr.*) dew-bearing. ——— 1802. —— D.S.♄. Jacq. stap. c. ic.
Stapèlia roriflùa. Jacq.
5 pulchélla. H.S. (*st.cr.*) beautiful. ——— 1795. 5. 11. D.S.♄. Mass. stap. 22. t. 36.
6 incarnàta. S.S. (*fl.*) flesh-coloured. ——— 1793. 4. 8. D.S.♄. ——— 22. t. 34.

OB'ESIA. H.S. OB'ESIA. Pentandria Digynia. H.S.

1 serrulàta. S.S. (*st.br.*) sawed. C. B. S. 1805. 6. 8. D.S.♄. Jacq. stap. c. ic.
Stapèlia serrulàta. Jacq.
2 geminàta. H.S. (*ye.br.*) twin-flowered. ——— 1795. 5. 11. D.S.♄. Bot. mag. t. 1326.
Stapèlia geminàta. B.M.
3 decòra. H.S. (*ye.pu.*) neat. ——— —— —— D.S.♄. Mass. stap. 19. t. 26.
4? árida. (*ye.br.*) dry. ——— 1795. 8. D.S.♄. ——— 21. t. 33.
Stapèlia árida. Mass.

GONOST'EMON. H.S. GONOST'EMON. Pentandria Digynia. H.S.

1 divaricàtum. H.S. (*fl.*) divaricated. C. B. S. 1793. 6. 11. D.S.♄. Bot. mag. t. 1007.
Stapèlia divaricàta. B.M.
2 stríctum. H.S.S. (*fl.*) upright. ——— 1819. 10. 12. D.S.♄. Bot. mag. t. 2037.
Stapèlia strícta. B.M.
3 pállidum. R.S. (*pa.*) pale-flowered. ——— —— —— D.S.♄. Wendl. coll. 2. t. 51.
Stapèlia pállida. Wendl.

CARUNCUL'ARIA. H.S. CARUNCUL'ARIA. Pentandria Digynia. H.S.

1 Símsii. (*re.bk.*) Sims's. C. B. S. 1790. 6. 11. D.S.♄. Sims bot. mag. t. 793.
Stapèlia pedunculàta. B.M. *pedunculàta Símsii.* Jacq. stap. c. ic.

2 Massòni. (*br.*) Masson's. C. B. S. 1790. 6. 11. D.S.♄. Mass. stap. p. 17. t. 21
Stapèlia pedunculàta. Mass. *pedunculàta Massòni.* Jacq. stap. c. ic.
3 Jacquíni. (*co.*) Jacquin's. ——— —— —— D.S.♄. Jacq. stap. c. ic.
Stapèlia pedunculàta. Jacq.
4 penduliflòra. (*br.*) pendulous-flower'd. —— —— —— D.S.♄. Jacq. stap. c. ic.
Stapèlia pedunculàta alia. Jacq.
5 apérta. (*p.pu.*) open-flowered. ——— 1795. 7. 8. D.S.♄. Mass. stap. 23. t. 37.
Stapèlia apérta. Mass.

PIARA'NTHUS. B. PIARA'NTHUS. Pentandria Digynia. H.S.
1 punctàtus. R.S. (*fl.*) purple dotted. C. B. S. 1795. 7. 11. D.S.♄. Mass. stap. 18. t. 24.
2 púllus. R.S. (*d.p.*) many-flowered. ——— 1774. 8. 9. D.S.♄. Bot. mag. t. 1648.
3 ramòsus. (*bk.*) branched. ——— 1795. 6. 8. D.S.♄. Mass. stap. 21. t. 32.
Stapèlia ramòsa. Mass.
4 parviflòrus. S.S. (*pu.*) small-flowered. ——— —— —— D.S.♄. Mass. stap. 22. t. 35.
5 pilíferus. (*bk.*) hairy-tubercled. ——— 1799. —— D.S.♄. ——— 17. t. 23.

HO'ODIA. HO'ODIA. Pentandria Digynia.
Gordòni. (*br.*) Gordon's. C. B. S. 1796. .... D.S.♄. Mass. stap. 24. t. 40.
Stapèlia Gordòni. Mass.

HUE'RNIA. B. HUE'RNIA. Pentandria Digynia. R.S.
1 reticulàta. H.S. (*d.pu.*) netted. C. B. S. 1793. 7. 8. D.S.♄. Bot. mag. t. 1662.
Stapèlia reticulàta. B.M.
β *defórmis.* J.S.(*wh.pu.*)*deformed.* ——— —— —— D.S.♄. Jacq. stap. c. ic.
2 campanulàta. H.S.(*st.pu.*)bell-shaped. ——— 1795. 7. 10. D.S.♄. Bot. mag. t. 1227.
3 clavígera. H.S.S.(*st.cr.*) club-haired. ——— —— —— D.S.♄. ——— t. 1661.
Stapèlia campanulàta. B.M. *nec aliorum.*
4 venústa. H.S. (*st.re.*) showy. ——— —— 6. 7. D.S.♄. Mass. stap. 10. t. 3.
5 lentiginòsa. H.S.(*st.pu.*) freckled. ——— —— 7. 11. D.S.♄. Bot. mag. t. 506.
6 guttàta. H.S. (*st.cr.*) red-spotted. ——— —— 8. 11. D.S.♄. Mass. stap. 10. t. 4.
7 hùmilis. H.S. (*st.pu.*) humble. ——— —— —— D.S.♄. ——— 10. t. 5.
8 tubàta. H.S. (*st.re.*) tube-flowered. ——— 1805. —— D.S.♄. Jacq. stap. c. ic.
9 duodecímfida. J.S.(*st.re.*)eleven-cleft. ——— —— —— D.S.♄. ——— c. ic.
10 críspa. H.S. (*st.pu.*) curled. ——— —— —— D.S.♄.
11 barbàta. H.S. (*st.pu.*) bearded. ——— 1795. —— D.S.♄. Mass. stap. 11. t. 7.
12 ocellàta. R.S. (*st.pu.*) eyed. ——— 1816. —— D.S.♄.

BRACHYSTE'LMA. B.R. BRACHYSTE'LMA. Pentandria Digynia. S.S.
tuberòsum. B.R. (*pu.*) tuberous-rooted. C. B. S. 1820. 4. 10. S.♃.◡. Bot. reg. t. 722.

CARALL'UMA. H.S. CARALL'UMA. Pentandria Digynia. S.S.
1 adscéndens. H.S.S.(*pu.*) ascending. E. Indies. 1804. 7. 11. D.S.♄. Roxb. cor. 1. t. 30.
2 umbellàta. H.S. (*fl.*) umbelled. ——— —— —— D.S.♄. ——— 3. t. 241.
3 crenulàta. Wa.(*ye.pu.*) crenulate. ——— 1824. 6. 9. D.S.♄. Wall pl. as. rar. t. 7.
4 fimbriàta. Wa. (*pu.*) fringed. ——— —— —— D.S.♄. ——— t. 8.

HO'YA. B.P. HO'YA. Pentandria Digynia. R.S.
1 crassifòlia. H.S. (*wh.*) large-leaved. China. 1817. 10. 1. D.S.♄.◡.
2 carnòsa. H.K. (*bh.*) fleshy-leaved. ——— 1802. 7. 10. D.G.♄.◡. Bot. mag. t. 788.
3 pállida. B.R. (*bh.*) pale. ——— 1820. —— D.S.♄.◡. Bot. reg. t. 951.
acùta. H.R.
4 Póttsii. H.T. (*bh.*) Potts's. ——— 1822. —— D.S.♄.◡.
5 trinérvis. H.T. (*bh.*) three-nerved. ——— 1824. —— D.S.♄.◡.
6 lanceolàta. D.P. (*bh.*) spear-leaved. Nepaul. 1825. —— D.G.♄.◡.
7 viridiflòra. B.P. (*gr.*) green-flowered. E. Indies. —— .... S.♄.◡. Rheed. mal. 9. t. 15.

FISCH'ERIA. DC. FISCH'ERIA. Pentandria Digynia.
scándens. DC. (*gr.ye.*) climbing. S. America. 1826. .... S.♄.◡. DC. hort. mon. ined. t. 67

TYLO'PHORA. B.P. TYLO'PHORA. Pentandria Digynia. R.S.
1 grandiflòra. B.P.(*br.gr.*)great-flowered. N. S. W. 1822. 6. 8. G.♄.◡.
2 barbàta. B.P. (*gr.br.*) bearded. ——— —— —— G.♄.◡.
3 exìlis. L.T. (*pu.*) slender. E. Indies. 1823. —— S.♄.◡. Linn. trans. 12. t. 16.

CEROP'EGIA. R.S. CEROP'EGIA. Pentandria Digynia. R.S.
1 aphy'lla. H.S. (*st.*) leafless. Canaries. 1808. .... S.♄.
2 stapeliæfórmis. H.P.(*br.*)Stapelia-like. C. B. S. 1826. 7. 8. D.S.♄.
3 torulòsa. H.R. .... straddling. ——— 1821. .... D.S.♄.
4 dichótoma. H.R. (*st.*) forked. E. Indies. 1806. 8. D.S.♄.
5 júncea. R.S. (*st.pu.*) rushy. ——— 1820. 4. 6. D.S.♃.◡. Roxb. cor. 1. t. 10.
6 africàna. B.R. (*br.*) African. C. B. S. 1821. 4. 11. D.S.♃.◡. Bot. reg. t. 626.
7 bulbòsa. R.S. (*pu.*) bulbous-rooted. E. Indies. —— —— D.S.♃.◡. Roxb. cor. 1. t. 7.
8 tuberòsa. R.S. (*pu.*) tuberous-rooted. ——— —— —— D.S.♃.◡. ——— 1. t. 9.

PERGUL'ARIA. R.S. PERGUL'ARIA. Pentandria Digynia. R.S.
1 odoratíssima. R.S. (*st.*) large. E. Indies. 1784. 6. 7. S.♄.◡. Bot. reg. t. 412.
2 mìnor. R.S. (*st.*) smaller. ——— 1790. 5. 8. S.♄.◡. Bot. mag. t. 755.
3 sanguinolénta. B.M.(*st.*)bloody-juiced. Sierra Leor 1822. 8. 9. S.♄.◡. ——— t. 2532.

MARSD'ENIA. B.P.	MARSD'ENIA.	Pentandria Digynia. R.S.					
1 suavèolens. B.P.	(*wh.*)	sweet-scented.	N. S. W.	1816.	3. 8.	G. ♄.⌣.	Bot. reg. t. 489.
2 erécta. R.S.	(*wh.*)	upright.	Syria.	1597.	7. 8.	G. ♄.	Jacq. vind. 1. t. 38.
DISCHI'DIA. B.P.	DISCHI'DIA.	Pentandria Digynia. S.S.					
bengalénsis. L.T.	(*st.*)	Bengal.	E. Indies.	1818.	6. 10.	S. ♄.⌣.	Bot. mag. t. 2916.
GYMN'EMA. B.P.	GYMN'EMA.	Pentandria Digynia. R.S.					
1 sylvéstre. R.S.	(*wh.*)	netted-leaved.	Ceylon.	1816.	6. 8.	S. ♄.⌣.	Willd. phyt. 1. t. 5. f. 3.
2 tíngens. S.S.	(*ye.*)	pointed-leaved.	E. Indies.	1823.	——	S. ♄.⌣.	Roxb. corom. 3. t. 239.
Asclèpias tíngens. R.C.							
3 tenacíssimum. S.S.	(*ye.*)	tough.	————	1806.	....	S. ♄.⌣.	Roxb. corom. 3. t. 240.
SARCO'LOBUS. R.S.	SARCO'LOBUS.	Pentandria Digynia. R.S.					
1 globòsus. R.S.	(*wh.*)	globular-fruit'd.	E. Indies.	1823.	....	S. ♄.⌣.	Wall. as. res. 12. t. 4.
2 carinàtus. R.S.	(*gr.ye.*)	keeled-fruited.	————	——	....	S. ♄.⌣.	———— 12. t. 5.
GONO'LOBUS. R.S.	GONO'LOBUS.	Pentandria Digynia. R.S.					
1 marítimus. R.S.	(*gr.p.*)	sea-side.	Caracas.	1823.	6. 10.	S. ♄.⌣.	Bot. reg. t. 931.
2 suberòsus. R.S.	(*pu.*)	cork-barked.	America.	1732.	7. 9.	S. ♄.⌣.	Dill. elt. t. 229. f. 296.
3 crispiflòrus. R.S.	(*wh.gr.*)	curled-flower'd.	S. America.	1741.	7. 8.	S. ♄.⌣.	Plum. ic. t. 216. f. 1.
4 guianénsis. S.S.	(*gr.*)	green-flowered.	Guiana.	1826.	7. 9.	S. ♄.⌣.	
viridiflòrus. R.S. *non* Nutt. *Cynánchum viridiflòrum.* Meyr.							
5 grandiflòrus. R.S.	(*ye.*)	large-flowered.	S. America.	1825.	8. 10.	S. ♄.⌣.	Bot. reg. t. 1053.
6 rostràtus. R.S.	(*gr.*)	beaked-petal'd.	Trinidad.	1823.	——	S. ♄.⌣.	
7 nìger. R.S.	(*bk.*)	black-flowered.	Mexico.	1825.	9. 11.	G. ♃.⌣.	Bot. mag. t. 2799.
Cynánchum nigrum. Cav. ic. 2. t. 159.							
8 uniflòrus. K.S.	(*pu.*)	solitary-flowered.	————	——	——	S. ♄.⌣.	H. et B. 3. t. 238.
9 carolinénsis. R.S.	(*d.p.*)	blunt-flowered.	N. America.	1820.	7. 9.	H. ♃.⌣.	Jacq. ic. 2. t. 342.
10 hirsùtus. R.S.	(*d.pu.*)	hairy.	————	1806.	6. 9.	H. ♃.⌣.	Swt. br. fl. gar. t. 1.
11 macrophy'llus. N.	(*p.pu.*)	large-leaved.	————	1822.	——	H. ♃.⌣.	
12 díscolor. R.S.	(*d.pu.*)	Virginian.	————	1809.	7. 9.	H. ♃.⌣.	Bot. mag. t. 1273.
Cynánchum díscolor. B.M.							
13 oblíquus. R.S.	(*pu.*)	oblique-flower'd.	Carolina.	1818.	——	H. ♃.⌣.	Jacq. ic. 2. t. 341.
14 Baldwyniànus.	(*st.*)	Baldwyn's.	Savannah.	1822.	7. 9.	H. ♃.⌣.	Lodd. bot. cab. t. 365?
hirsùtus. L. bot. cab.? *nec aliorum. caroliniànus.* N. *non* Jacq.							
15 hírtus. R.S.	(*pu.*)	hairy-leaved.	America.	....	——	S. ♄.⌣.	Moris. his. 3. s. 15. t. 3. f. 61
16 undulàtus. R.S.	(*br.pu.*)	wave-leaved.	S. America.	1814.	——	S. ♄.⌣.	Jacq. amer. 85. t. 58.
17 viridiflòrus. N.	(*gr.*)	Nuttall's.	Missisippi.	1822.	——	H. ♃.⌣.	
Nuttalliànus. S.S.							
18 l'ævis. R.S.	(*d.pu.*)	smooth.	————	1806.	6. 7.	H. ♃.⌣.	
19 diademàtus. B.R.	(*st.*)	red-crowned.	Mexico.	1812.	9. 10.	S. ♄.⌣.	Bot. reg. t. 252.
20 barbàtus. R.S.	(*ye.*)	bearded.	————	1826.	——	G. ♄.⌣.	H. et B. n. gen. 3. t. 239.
21 mucronàtus.	(*br.re.*)	sharp-pointed.	Trinidad.	1804.	7. 8.	S. ♄.⌣.	Andr. reposit. t. 515.
Cynánchum mucronàtum. A.R. *non* K.S.							
ENSL'ENIA. N.	ENSL'ENIA.	Pentandria Digynia.					
álbida. N.	(*st.*)	straw-coloured.	Virginia.	1828.	7. 9.	H. ♃.⌣.	
ASCL'EPIAS. R.S.	SWALLOW-WORT.	Pentandria Digynia. R.S.					
1 syrìaca. Lam.	(*li.*)	Syrian.	Astracan.	1629.	7. 8.	H. ♃.	Blackw. t. 521.
Beid el Ossar. Veslingius. Linn. trans. v. 14. p. 239.							
2 obovàta. E.C.	(*pu.*)	obovate-leaved.	Carolina.	1826.	——	H. ♃.	
3 tomentòsa. E.C.	(*pu.*)	woolly.	————	——	——	H. ♃.	
4 exaltàta. L.en.	(*pu.*)	tall.	N. America.	....	——	H. ♃.	
5 phytolaccoídes. R.S.	(*pu.*)	Phytolacca-like.	————	1812.	——	H. ♃.	
6 parviflòra. R.S.	(*wh.*)	small flowered.	————	1774.	7. 10.	H. ♃.	Jacq. f. ecl. t. 28.
7 salicifòlia. B.C.	(*wh.*)	Willow-leaved.	S. America.	1816.	6. 9.	G. ♃.	Lodd. bot. cab. t. 272.
8 angustifòlia. R.S.	(*fl.*)	narrow-leaved.	Mexico.	1824.	——	F. ♃.	
9 curassávica. R.S.	(*or.*)	Curassavian.	S. America.	1692.	6. 10.	S. ♃.	Bot. reg. t. 81.
10 nívea. R.S.	(*wh.*)	Almond-leaved.	N. America.	1730.	7. 9.	H. ♃.	Bot. mag. t. 1181.
11 virgàta. R.S.	(*li.*)	slender.	Mexico.	1825.	——	F. ♃.	
12 incarnàta. R.S.	(*pu.*)	flesh-coloured.	N. America.	1710.	7. 8.	H. ♃.	Bot. reg. t. 250.
13 púlchra. R.S.	(*pu.*)	hairy.	————	——	——	H. ♃.	Swt. br. fl. gar. s. 2. t. 18.
14 am'œna. R.S.	(*pu.*)	oval-leaved.	————	1732.	——	H. ♃.	Dill. elt. t. 27. f. 30.
15 purpuráscens. R.S.	(*pu.*)	purple.	————	——	——	H. ♃.	———— t. 28. f. 31.
16 citrifòlia. R.S.	(*wh.*)	Citron-leaved.	Florida.	1812.	——	F. ♃.	Jacq. icon. t. 345.
17 variegàta. R.S.	(*wh.re.*)	variegated.	————	1597.	——	H. ♃.	Bot. mag. t. 1182.
18 obtusifòlia. R.S.	(*pu.*)	blunt-leaved.	————	1823.	——	H. ♃.	
19 amplexicaùlis. R.S.	(*re.*)	stem-clasping.	————	1812.	——	H. ♃.	
20 laurifòlia. M.	(*pu.*)	Laurel-leaved.	————	1823.	——	H. ♃.	Pluk. amal. t. 358. f. 2.
periplocæfòlia. N.							
21 acuminàta. R.S.	(*wh.pu.*)	taper-pointed.	————	1812.	——	H. ♃.	
22 paupércula. R.S.	(*re.*)	remote-leaved.	————	——	6. 7.	H. ♃.	
23 pedicellàta. R.S.	(*gr.ye.*)	dwarf.	————	——	——	H. ♃.	

24 quadrifòlia. R.S.	(*bh.*) four-leaved.	N. America.	1823.	8. 9.	H. ♃.	Jacq. obs. 2. t. 33.
25 verticillàta. R.S.	(*bh.*) whorl-leaved.	———	1759.	8. 10.	H. ♃.	Swt. br. fl. gar. t. 144.
26 mexicàna. R.S.	(*wh.*) Mexican.	Mexico.	1823.	7. 9.	G. ♄.	Cavan. ic. 1. t. 58.
27 ròsea. K.S.	(*ro.*) rose-coloured.	———	1824.	——	G. ♃.	
28 linifòlia. R.S.	(*wh.*) Flax-leaved.	———	——	——	G. ♄.	
29 Linària. R.S.	(*wh.*) toadflax-leaved.	———	1802.	——	G. ♃.	Cavan. ic. 1. t. 57.
30 rùbra. R.S.	(*re.*) red-flowered.	N. America.	——	——	H. ♃.	
31 tuberòsa. R.S.	(*or.*) tuberous-rooted.	———	1690.	7. 9.	H. ♃.	Bot. reg. t. 76.
32 decúmbens. W.	(*or.*) decumbent.	———	1731.	——	H. ♃.	Swt. br. fl. gar. s. 2. t. 24.
ACER'ATES. E.C.	ACER'ATES. Pentandria Digynia.					
longifòlia. E.C.	(*p.pu.*) long-leaved.	N. America.	1816.	7. 8.	H. ♃.	
Asclèpias longifòlia. R.S. *Gomphocárpus longifòlius.* S.S.						
ANANTH'ERIX. N.	ANANTH'ERIX. Pentandria Digynia. N.					
víridis. N.	(*gr.*) green-flowered.	N. America.	1812.	8. 10.	H. ♃.	
STYLA'NDRA. N.	STYLA'NDRA. Pentandria Digynia. N.					
pùmila. N.	(*st.*) dwarf.	Carolina.	1812.	7. 8.	G. ♃.	
Podostígma pubéscens. S.S. *Asclèpias pedicellàta.* Ph.						
GOMPHOCA'RPUS. R.S.	GOMPHOCA'RPUS. Pentandria Digynia. R.S.					
1 arboréscens. R.S.	(*wh.*) broad-leaved.	C. B. S.	1714.	12.	G. ♄.	Jacq. schœnb. 1. t. 50.
2 críspus. R.S.	(*pu.*) curled-leaved.	———	1774.	7.	G. ♄.	Comm. rar. t. 17.
3 fruticòsus. R.S.	(*wh.*) Willow-leaved.	———	1714.	6. 9.	G. ♄.	Bot. mag. t. 1628.
Asclèpias fruticòsa. W.						
OXYSTE'LMA. R.S.	OXYSTE'LMA. Pentandria Digynia. R.S.					
esculéntum. R.S.	(*wh.ro.*) esculent.	E. Indies.	1816.	....	S. ♃. ⌣.	Roxb. corom. 1. t. 11.
XYSMAL'OBIUM. R.S.	XYSMAL'OBIUM. Pentandria Digynia. R.S.					
1 undulàtum. R.S.	(*gr.*) waved-leaved.	C. B. S.	1783.	7. 8.	G. ♄.	Comm. rar. t. 16.
2 grandiflòrum. R.S.	(*pu.*) large-flowered.	———	1823.	——	G. ♄.	
CALO'TROPIS. R.S.	CALO'TROPIS. Pentandria Digynia. R.S.					
1 pròcera. R.S.	(*wh.*) bell-flowered.	Persia.	1714.	7. 9.	S. ♄.	Andr. reposit. t. 271.
Asclèpias gigántea. A.R.						
2 gigántea. R.S.	(*p.pu.*) curled-flower'd.	E. Indies.	1690.	——	S. ♄.	Bot. reg. t. 58.
OXYPE'TALUM. R.S.	OXYPE'TALUM. Pentandria Digynia.					
1 appendiculàtum. M.N.	appendaged.	Brazil.	1824.	6. 8.	S. ♄. ⌣.	Mart. n. pl. bras. t. 30.
2 Bánksii. R.S.	Banks's.	———	1826.	——	S. ♄. ⌣.	——— t. 29.
CYNA'NCHUM. R.S.	CYNA'NCHUM. Pentandria Digynia. L.					
1 acùtum. R.S.	(*wh.*) acute-leaved.	Europe.	1596.	7.	H. ♃. ⌣.	Flor. græc. t. 250.
2 monspelìacum. R.S.	(*bh.*) Montpelier.	S. Europe.	——	8. 9.	H. ♃. ⌣.	——— t. 251.
3 pauciflòrum. R.S.	(*gr.w.*) few-flowered.	E. Indies.	1822.	5. 8.	S. ♄. ⌣.	Will. phy. 1. n. 23. t. 5. f. 3.
Periplòca tunicàta. W.						
4 pilòsum. R.S.	(*wh.*) hairy.	C. B. S.	1726.	6. 9.	G. ♄. ⌣.	Bot. reg. t. 111.
Periplòca africàna. A.R. 557.						
5 crassifòlium. R.S.	(*gr.w.*) thick-leaved.	———	1816.	....	G. ♄. ⌣.	
6 capénse. R.S.	(*wh.*) Cape.	———	1820.	....	G. ♄. ⌣.	
7 ròseum. R.S.	(*ro.*) rosy.	Dahuria.	1816.	7. 8.	H. ♃.	Gmel. sib. 4. t. 42.
Asclèpias davùrica. W.						
8 nìgrum. R.S.	(*bk.*) black.	S. Europe.	1596.	5. 8.	H. ♃. ⌣.	Bot. mag. t. 2390.
9 mèdium. R.S.	(*br.*) middle.	....	....	——	H. ♃. ⌣.	
10 Vincetóxicum. R.S.	(*w.*) officinal.	Europe.	1596.	——	H. ♃.	Flor. dan. t. 849.
Asclèpias Vincetóxicum. W.						
11 fuscàtum. S.S.	(*st.*) yellow-flowered.	———	——	——	H. ♃.	
12 sibíricum. R.S.	(*st.*) Siberian.	Siberia.	1775.	7. 8.	H. ♃.	Murr. gott. 2. t. 7.
13 fœtidum. R.S.	(*st.*) fetid.	Mexico.	1823.	——	G. ♄. ⌣.	Cavan. ic. 2. t. 158.
14 viridiflòrum. B.M.	(*gr.re.*) green-flowered.	E. Indies.	1814.	10. 12.	S. ♄. ⌣.	Bot. mag. t. 1929.
15 undàtum. A.R.	(*gr.*) wave-leaved.	———	1803.	7. 8.	S. ♄. ⌣.	Andr. reposit. t. 410.
HARRIS'ONIA. B.M.	HARRIS'ONIA. Pentandria Digynia.					
loniceroídes. B.M.	(*cr.*) Honeysuckle.	Brazil.	1824.	9. 10.	S. ♄.	Bot. mag. t. 2699.
D'ÆMIA. R.S.	D'ÆMIA. Pentandria Digynia. R.S.					
1 exténsa. R.S.	(*st.*) smooth-leaved.	E. Indies.	1778.	7. 8.	S. ♄. ⌣.	Jacq. ic. 1. t. 54.
2 bícolor.	(*gr.pu.*) two-coloured.	———	1806.	——	S. ♄. ⌣.	Andr. reposit. t. 562.
Cynánchum bícolor. A.R.						
DIPLOL'EPIS. R.S.	DIPLOL'EPIS. Pentandria Digynia.					
1 vomitòria. H.T.	(*st.*) dull-leaved.	China.	1820.	6. 9.	S. ♄. ⌣.	
2 apiculàta. H.T.	(*st.*) glossy-leaved.	———	——	——	S. ♄. ⌣.	
SARCOSTE'MMA. R.S.	SARCOSTE'MMA. Pentandria Digynia. R.S.					
1 viminàle. R.S.	(*wh.*) twisting.	E. Indies.	1731.	7.	S. ♄. ⌣.	
2 Swartziànum. R.S.	(*wh.*) Swartz's.	W. Indies.	....	....	S. ♄. ⌣.	Sloan. hist. 1. t. 131. f. 1.
EUST'EGIA. R.S.	EUST'EGIA. Pentandria Digynia. R.S.					
hastàta. R.S.	(*gr.wh.*) halberd-leaved.	C. B. S.	1816.	6. 8.	G. ♄.	Spr. n. ent. t. 1. f. 5-10.
Apócynum hastàtum. W.						

METASTE'LMA. R.S.		METASTE'LMA.	Pentandria Digynia.				
parviflòrum. R.S.	(*wh.*)	small-flowered.	W. Indies.	1826.	5. 6.	S. ♄.⌣.	Plum. ic. 215. f. 1.
MICROL'OMA. R.S.		MICROL'OMA.	Pentandria Digynia. R.S.				
1 sagittàtum. R.S.	(*sc.*)	arrow-leaved.	C. B. S.	1775.	7. 8.	G. ♄.⌣.	Jacq. schœn. 1. t. 38.
2 lineàre. R.S.	(*cr.*)	linear-leaved.	———	1816.	——	G. ♄.⌣.	Pluk. man. 17. t. 335. f. 5.
ASTEPHA'NUS. R.S.		ASTEPHA'NUS.	Pentandria Digynia. R.S.				
1 triflòrus. R.S.	(*wh.*)	three-flowered.	C. B. S.	1816.	7. 9.	G. ♃.⌣.	
2 lineàris. R.S.	(*wh.*)	panicled.	———	——	——	G. ♃.⌣.	
3 lanceolàtus. R.S.	(*wh.*)	spear-leaved.	———	1822.	——	G. ♃.⌣.	

ORDO CXL.

GENTIANEÆ. *Brown prodr.* 1. *p.* 449.

GENTIA'NA. R.S.		GENTIAN.	Tetr-Pentandria Digynia.				
1 lùtea. R.S.	(*ye.*)	yellow.	Alps Europe.	1596.	6. 7.	H. ♃.	Mill. ic. t. 139.
2 hy'brida. R.S.	(*ye.re.*)	hybrid.	Switzerland.	1825.	——	H. ♃.	
3 purpùrea. R.S.	(*pu.*)	purple.	———	1768.	——	H. ♃.	Andr. reposit. t. 117.
4 pannònica. R.S.	(*pa.pu.*)	round-petaled.	———	....	——	H. ♃.	Jacq. aust. 2. t. 136.
β albiflòra.	(*wh.*)	*white-flowered.*	———	1826.	——	H. ♃.	
5 punctàta. R.S.	(*st.pu.*)	dotted-flowered.	———	1775.	——	H. ♃.	Jacq. aust. 5. t. ap. 28.
6 campanulàta. R.S.	(*ye.*)	bell-flowered.	———	1824.	——	H. ♃.	——— ap. t. 29.
7 septémfida. R.S.	(*bl.*)	crested.	Levant.	1804.	——	H. ♃.	Bot. mag. t. 1229.
β guttàta.	(*bl.wh.*)	*spotted.*	———	——	——	H. ♃.	——— t. 1410.
8 asclepiadèa. R.S.	(*bl.*)	swallowwort-ld.	Austria.	1629.	7. 8.	H. ♃.	——— t. 1078.
9 cruciàta. R.S.	(*bl.*)	Crosswort.	———	1596.	6. 7.	H. ♃.	Jacq. aust. 4. t. 372.
10 macrophy'lla. R.S.	(*bl.*)	long-leaved.	Siberia.	1796.	7. 8.	H. ♃.	Bot. mag. t. 1414.
11 adscéndens. S.S.	(*bl.*)	ascending.	———	1799.	6. 7.	H. ♃.	——— t. 705.
β decúmbens. R.S.	(*bl.*)	*decumbent.*	———	——	——	H. ♃.	——— t. 723.
12 frígida. R.S.	(*wh.*)	frigid.	Carpathian m.	1816.	——	H. ♃.	Frœl. gent. n. 13. ic.
13 álgida. R.S.	(*st.bl.*)	narrow-leaved.	Siberia.	1808.	——	H. ♃.	Pall. ros. 2. t. 95.
14 gèlida. R.S.	(*st.*)	cream-coloured.	Caucasus.	1816.	——	H. ♃.	
15 Pneumonánthe. R.S.	(*bl.*)	Calathian violet.	England.	....	8. 9.	H. ♃.	Eng. bot. v. 1. t. 20.
β guttàta. B.M.	(*bh.wh.*)	*spotted-flowered.*	....	....	——	H. ♃.	Bot. mag. t. 1101.
16 pseudo-pneumonanthe. R.S.	(*bl.*)	Canadian.	N. America.	1828.	——	H. ♃.	
Pneumonánthe. Ph.							
17 lineàris. R.S.	(*bl.*)	linear-leaved.	———	1816.	——	H. ♃.	
18 triflòra. R.S.	(*p.bl.*)	three-flowered.	Siberia.	1807.	6. 7.	H. ♃.	Pall. ross. 1. t. 93. f. 1.
19 Saponària. B.M.	(*bl.*)	barrel-flower'd.	N. America.	1776.	8. 9.	H. ♃.	Bot. mag. t. 1039.
20 Catesb'æi. N.	(*bl.*)	Catesby's.	———	——	6. 8.	H. ♃.	Andr. reposit. t. 418.
21 intermèdia. B.M.	(*pu.*)	intermediate.	———	1820.	8. 9.	H. ♃.	Bot. mag. t. 2303.
22 incarnàta. B.M.	(*fl.*)	flesh-coloured.	———	1812.	9. 11.	H. ♃.	——— t. 1856.
23 ochroleùca. R.S.	(*st.*)	pale-flowered.	———	1803.	8. 9.	H. ♃.	——— t. 1551.
24 quinqueflòra. R.S.	(*pu.*)	five-flowered.	———	1823.	7. 9.	H. ♂.	
amarelloìdes. M.							
25 aùrea. R.S.	(*wh.ye.*)	large involucred.	Norway.	——	——	H. ♂.	Flor. dan. t. 344.
quinquefòlia. F.D. *involucràta.* Rottb. ac. havn. 10. p. 434. t. 1. f. 2. A.B.							
26 umbellàta. M.B.	(*pa.bl.*)	umbel-flowered.	Caucasus.	1823.	8. 10.	H. ☉.	
27 acaùlis. R.S.	(*bl.*)	dwarf.	Wales.	....	3. 6.	H. ♃.	Eng. bot. v. 23. t. 1594.
28 alpìna. R.S.	(*bl.*)	Alpine.	Switzerland.	1818.	——	H. ♃.	Lodd. bot. cab. t. 476.
29 altàica. R.S.	(*bl.*)	Altay.	Siberia.	1826.	——	H. ♃.	Pall. ross. 2. t. 97. f. 1.
30 pyrenàica. R.S.	(*bl.*)	Pyrenean.	Pyrenees.	1825.	——	H. ♃.	W. et K. hung. t. 207.
31 pùmila. R.S.	(*bl.*)	small.	Switzerland.	——	——	H. ♃.	Jacq. obs. 2. t. 49.
32 vérna. R.S.	(*bl.*)	spring.	England.	....	——	H. ♃.	Eng. bot. v. 7. t. 493.
33 æstìva. R.S.	(*bl.*)	summer.	Germany.	1826.	——	H. ♃.	Rœm. arch. 1. t. 3. f. 8.
34 bavàrica. R.S.	(*bl.*)	Bavarian.	———	1775.	7.	H. ♃.	Vill. delph. 2. t. 10.
35 imbricàta. R.S.	(*bl.*)	imbricated	Switzerland.	1825.	——	H. ♃.	Barrel. ic. 101. f. 2.
36 angulòsa. R.S.	(*bl.*)	angular-calyx'd.	Caucasus.	1826.	4. 6.	H. ♃.	Marsch. cent. rar. 1. t. 47.
37 utriculòsa. R.S.	(*bl.*)	bladdered.	———	——	7. 9.	H. ♂.	W. et K. hung. t. 206.
38 nivàlis. R.S.	(*bl.*)	small alpine.	Scotland.	....	8. 9.	H. ☉.	Eng. bot. v. 13. t. 896.
39 aquática. R.S.	(*bl.*)	water-side.	Dahuria.	1827.	7. 9.	H. ♂.	

40	hùmilis. R.S.	(pa.bl.)	dwarf.	Caucasus.	1827.	7. 9.	H. ♂.	Pall. ross. 2. t. 97. f. 2.
41	uniflòra. R.S.	(vi.)	one-flowered.	M.Carpathian.	1828.	——	H. ♂.	Rœm.ar.1.p.23.t.2.f.4.
42	germánica. R.S.	(vi.)	German.	Germany.	1825.	8. 9.	H. ♂.	Mayrh. fl. mon. t. 278.
43	Amarélla. R.S.	(vi.)	autumnal.	Britain.	....	8. 10.	H. ♂.	Eng. bot. v. 4. t. 236.
44	uliginòsa. R.S.	(bl.)	marsh.	Germany.	1827.	——	H. ♂.	
45	obtusifòlia. R.S.	(bl.st.)	blunt-leaved.	Switzerland.	1828.	——	H. ♂.	Rœm.ar.1.p.22.t.2.f.3.
46	praténsis. R.S.	(vi.)	field.	Siberia.	1824.	——	H. ♂.	
47	caucásica. R.S.	(bl.)	Caucasean.	Caucasus.	1804.	7. 8.	H. ♂.	Bot. mag. t. 1038.
48	campéstris. R.S.	(vi.)	field.	Britain.	....	8. 10.	H. ♂.	Eng. bot. v. 4. t. 237.
49	angustifòlia. R.S.	(bl.)	slender-leaved.	N. America.	1820.	8. 9.	H. ♃.	
50	glaciàlis. R.S.	(bl.)	northern.	N. Europe.	1826.	....	H. ♂.	Sch.rom.arc.1.t.2.f.5.
51	conférta. F.	(bl.)	crowded.	Altay.	1827.	——	H. ♂.	
52	Geblèri. F.	(bl.)	Gebler's.	——	——	——	H. ♂.	Reich. ic. t. 271.
53	rotàta. R.S.	(bl.)	rotate.	Siberia.	——	——	H. ♂.	Pall. ross. 2. t. 89. f. 3.
54	ciliàta. R.S.	(bl.)	fringe-flowered.	S. Europe.	1759.	——	H. ♃.	Jacq. aust. t. 113.
55	fimbriàta. R.S.	(bl.)	fimbriate.	Caucasus.	1824.	7. 9.	H. ♃.	
56	crinìta. R.S.	(bl.)	jagged-flower'd.	N. America.	1804.	7. 10.	H. ♂.	Swt. br. fl. gar. t. 139.
	fimbriàta. Andr. rep. t. 509. *non* W.							
57	barbàta. R.S.	(bl.)	bearded.	Siberia.	....	8. 9.	H. ♂.	Bot. mag. t. 639.
	ciliàta. B.M. *nec aliorum.*							
	SWE'RTIA. R.S.		FELWORT.	Tetr-Pentandria Digynia.				
1	perénnis. R.S.	(br.pu.)	marsh.	England.	....	7. 8.	H. ♃.	Eng. bot. v. 21. t. 1441.
2	Michauxiàna. R.S.	(st.)	American.	N. America.	1824.	——	H. ♂.	
3	dichótoma. R.S.	(gr.re.)	forked.	Siberia.	1827.	7. 9.	H. ♂.	Pall. ross. 2. t. 91.
	CHL'ORA. S.S.		YELLOW-WORT.	Octandria Monogynia. L.				
1	imperfoliàta. S.S.	(ye.)	sessile-leaved.	Italy.	1826.	6. 7.	H. ⊙.	
2	perfoliàta. S.S.	(ye.)	perfoliate.	Britain.	....	——	H. ⊙.	Eng. bot. v. 1. t. 60.
	FRAS'ERA. M.		FRAS'ERA.	Tetrandria Monogynia. R.S.				
	Walteri. M.	(ye.)	Walter's.	N. America.	1795.	7. 8.	H. ♂.	
	carolinénsis. Walt.							
	ERYTHR'ÆA. R.S.		ERYTHR'ÆA.	Pentandria Monogynia. P.S.				
1	Centaùrium. R.S.	(ro.)	common.	Britain.	....	7. 8.	H. ⊙.	Eng. bot. v. 6. t. 417.
2	latifòlia. S.S.	(ro.)	broad-leaved.	Lancashire.	....	——	H. ⊙.	
3	pulchélla. R.S.	(ro.)	dwarf branching.	England.	....	8. 9.	H. ⊙.	Eng. bot. v. 7. t. 458.
4	littoràlis. R.S.	(ro.)	simple dwarf.	Britain.	....	6. 8.	H. ⊙.	——— v. 33. t. 2305.
5	conférta. R.S.	(ro.)	close-flowered.	Spain.	1818.	5. 11.	H. ♃.	
6	spicàta. R.S.	(ro.)	spiked.	S. Europe.	....	——	H. ♂.	Flor. græc. t. 238.
	β *albiflòra.*	(wh.)	*white-flowered.*	——	....	——	H. ♂.	Matth. comm. p. 488. f. 2.
7	austràlis. B.P.	(ro.)	New Holland.	New Holland.	1823.	5. 8.	G. ♂.	
8	uliginòsa. R.S.	(ro.)	marsh.	Hungary.	1825.	6. 8.	H. ♂.	W. et K. pl. hung. t. 238.
9	aggregàta. B.F.G.	(ro.)	cluster-branch'd.	Nepaul.	1824.	3. 11.	H. ♃.	Swt. br. fl. gar. t. 137.
10	marítima. F.G.	(ye.)	sea-side.	S. Europe.	1777.	7. 8.	H. ⊙.	Flor. græc. t. 237.
11	Massòni.	(ye.)	Masson's.	Azores.	——	——	F. ♃.	
	marítima. H.K. *nec aliorum.* vid. obs. in Flor. Græc.							
12	lùtea. R.S.	(ye.)	yellow-flowered.	S. Europe.	....	——	H. ⊙.	Barrel. ic. 468.
	SABB'ATIA. R.S.		SABB'ATIA.	Pentandria Monogynia. R.S.				
1	grácilis. R.S.	(ro.)	slender.	N. America.	....	7. 9.	H. ♂.	Sal. par. lond. t. 32.
2	stellàris. R.S.	(ro.ye.)	starry.	——	1827.	6. 8.	H. ♂.	Bartr. ic. ined. t. 13.
3	angulàris. R.S.	(pu.)	angular-stalked.	——	1823.	7. 8.	H. ♂.	
4	calycòsa. R.S.	(ro.)	dichotomous.	——	1812.	6. 8.	H. ♂.	Bot. mag. t. 1600.
5	chloroídes. R.S.	(ro.)	Chlora-like.	——	1817.	8. 9.	H. ♂.	
6	paniculàta. R.S.	(wh.)	panicled.	——	——	——	H. ♃.	
	VOHI'RIA. R.S.		VOHI'RIA.	Pentandria Monogynia.				
1	ròsea. R.S.	(ro.)	rose-coloured.	Trinidad.	1822.	6. 8.	S. ♃.	Aubl. guian. 1. t. 83. f. 1.
	Lìta ròsea. W.							
2	cœrùlea. R.S.	(bl.)	blue-flowered.	——	1824.	——	S. ♃.	Aubl. guian. 1. t. 83. f. 2.
3	uniflòra. R.S.	(ye.)	one-flowered.	W. Indies.	——	——	S. ♃.	Jacq. amer. 87. t. 60. f. 3.
	CHIR'ONIA. R.S.		CHIR'ONIA.	Pentandria Monogynia. L.				
1	jasminoídes. R.S.	(ro.)	Jasmine-like.	C. B. S.	1812.	4. 7.	G. ♄.	Bot. reg. t. 197.
2	lychnoídes. R.S.	(pu.)	Lychnis-flowered.	——	1816.	——	G. ♄.	
3	nudicaùlis. R.S.	(pu.)	naked-stalked.	——	——	——	G. ♄.	Linn. trans. 7. t. 12. f. 3.
4	linoídes. R.S.	(pi.)	flax-leaved.	——	1787.	7. 9.	G. ♄.	Bot. mag. t. 511.
5	tetragòna. R.S.	(pu.)	four-sided.	——	1822.	——	G. ♄.	Linn. trans. 7. t. 12. f. 2.
6	baccífera. R.S.	(pi.)	berry-bearing.	——	1759.	6. 7.	G. ♄.	Bot. mag. t. 233.
7	angustifòlia. B.M.	(ro.)	narrow-leaved.	——	1800.	6. 8.	G. ♄.	——— t. 818.
8	frutéscens. R.S.	(ro.)	frutescent.	——	1756.	6. 9.	G. ♄.	——— t. 37.
	β *albiflòra.*	(wh.)	*white-flowered.*	——	....	——	G. ♄.	
9	decussàta. R.S.	(ro.)	cross-leaved.	——	1789.	——	G. ♄.	Bot. mag. t. 707.

HI'PPION. S.S.		Hi'ppion.	Pentandria Monogynia. S.S.				
viscòsum. s.s.	(ye.)	clammy.	Canaries.	1781.	6. 7.	G. ♂.	Sm. ic. fas. 3. t. 18.
Exàcum viscòsum. Sm. *Gentiàna viscòsa.* Bot. mag. t. 2135.							
SEB'ÆA. B.P.		Seb'æa.	Tetr-Pentandria Monogynia. S.S.				
1 ovàta. B.P.	(st.)	oval-leaved.	N. S. W.	1820.	8. 10.	G. ☉.	Lab. nov. hol. 1. t. 52.
2 álbens. B.P.	(wh.)	white-flowered.	C. B. S.	——	——	G. ☉.	Burm. afr. t. 74. f. 4.
3 cordàta. B.P.	(ye.)	heart-leaved.	————	1815.	7. 8.	H. ☉.	———— t. 74. f. 5.
CICE'NDIA. Adans.		Cice'ndia.	Tetrandria Monogynia.				
filifórme.	(ye.)	slender-stalk'd.	Britain.	....	——	H. ☉.	Eng. bot. v. 4. t. 235.
Exàcum filifórme. E.B. *Microcàle filifórme.* Lk.							
EX'ACUM. B.P.		Ex'acum.	Tetrandria Monogynia. L.				
tetragònum. D.P.	(ye.)	square-stalked.	Nepaul.	1820.	8. 10.	H. ☉.	
COUTOUB'EA. P.S.		Coutoub'ea.	Tetrandria Monogynia.				
1 ternifòlia. P.S.	(wh.)	three-leaved.	Trinidad.	1820.	6. 10.	S. ☉.	Cavan. icon. 4. t. 328.
2 ramòsa. Aub.	(pu.)	purple-flowered.	————	1823.	——	S. ☉.	Aubl. guian. 1. t. 28.
purpùrea. Lam. ill. t. 79. *Exàcum ramòsum.* R.S.							
MITRASA'CME. B.P.		Mitrasa'cme.	Tetrandria Monogynia. R.S.				
1 polymórpha. B.P.	(wh.)	variable.	N. S. W.	1822.	6. 8.	G. ♃.	
2 canéscens. B.P.	(wh.)	canescent.	————	1820.	——	G. ♃.	
3 serpyllifòlia. B.P.	(wh.)	creeping.	————	1826.	——	G. ♃.	
HOUST'ONIA. R.S.		Houst'onia.	Tetrandria Monogynia. L.				
1 cœrùlea. R.S.	(pa.bl.)	blue-flowered.	N. America.	1785.	5. 8.	H. ♃.	Bot. mag. t. 370.
2 serpyllifòlia. R.S.	(wh.)	Thyme-leaved.	————	1827.	——	H. ♃.	———— t. 2822.
3 tenélla. R.S.	(pu.)	slender.	Carolina.	1812.	——	H. ♃.	
4 albiflòra. N.	(wh.)	white-flowered.	N. America.	1828.	——	H. ♃.	
5 purpùrea. R.S.	(li.)	purple-flowered.	————	1800.	——	H. ♃.	
SPIG'ELIA. R.S.		Worm-grass.	Pentandria Monogynia. L.				
1 Anthélmia. R.S.	(bl.)	annual.	W. Indies.	1759.	7.	S. ☉.	Bot. mag. t. 2359.
2 marylándica. R.S.	(sc.)	perennial.	N. America.	1694.	7. 8.	H. ♃.	———— t. 80.
EUST'OMA. P.L.		Eust'oma.	Pentandria Monogynia.				
silenifòlium. P.L.	(pu.)	Silene-leaved.	Providence Isl.	1804.	7.	S. ♃.	Sal. par. lond. t. 241.
Lisiánthus glaucifòlius. Jacq. ic. 1. t. 33.							
LISIA'NTHUS. R.S.		Lisia'nthus.	Pentandria Monogynia. L.				
1 longifòlius. R.S.	(ye.)	long-leaved.	Jamaica.	1793.	7. 9.	S. ♄.	Bot. reg. t. 880.
2 grandiflòrus. R.S.	(st.)	large-flowered.	Trinidad.	1820.	——	S. ☉.	Aubl. guian. 1. t. 79.
3 exsértus. R.S.	(st.)	oval-leaved.	W. Indies.	1793.	——	S. ♄.	
4 latifòlius. R.S.	(ye.)	broad-leaved.	————	1822.	....	S. ♄.	
5 umbellàtus. R.S.	(ye.)	umbel-flowered.	————	——	....	S. ♄.	
6 cordifòlius. R.S.	(ye.)	heart-leaved.	————	1816.	6. 8.	S. ♄.	Brown. jam. 2. t. 9. f. 2.
7 acutángulus. R.S.	(ye.)	acute-angled.	Peru.	1820.	——	S. ♃.	Flor. per. 2. t. 122. f. a.
8 angustifòlius. K.S.	(gr.)	narrow-leaved.	S. America.	1824.	7. 10.	S. ☉.	
LOG'ANIA. B.P.		Log'ania.	Pentandria Monogynia. R.S.				
1 latifòlia. B.P.	(wh.)	broad-leaved.	N. Holland.	1816.	4. 6.	G. ♄.	Lab. nov. hol. 1. t. 61.
Exàcum vaginàle. Lab.							
2 floribúnda. B.P.	(wh.)	many-flowered.	N. S. W.	1797.	4. 5.	G. ♄.	Andr. reposit. t. 520.
Euósma albiflòra. A.R.							
3 revolùta. B.P.	(wh.)	revolute-leaved.	————	1822.	4. 6.	G. ♄.	
VILLA'RSIA. R.S.		Villa'rsia.	Pentandria Monogynia. R.S.				
1 nymphoídes. R.S.	(ye.)	common.	England.	....	6. 7.	H. w. ♃.	Eng. bot. v. 4. t. 217.
Menyánthes nymphoìdes. E.B.							
2 índica. R.S.	(ye.)	Indian.	E. Indies.	1790.	6. 8.	S. w. ♃.	Bot. mag. t. 658.
3 sarmentòsa. R.S.	(ye.)	sarmentose.	N. S. W.	1806.	——	G. w. ♃.	———— t. 1323.
Menyánthes sarmentòsa. B.M.							
4 lacunòsa. V.	(wh.)	smooth-flowered.	N. America.	1812.	6. 7.	G. w. ♃.	Venten. choix. 9.
5 renifórmis. B.P.	(ye.)	kidney-leaved.	N. S. W.	1820.	——	G. w. ♃.	
6 parnassiifòlia. B.P.	(ye.)	tall.	————	1805.	6. 10.	G. w. ♃.	Bot. mag. t. 1029.
Menyánthes exaltàta. B.M. *Swértia parnassifòlia.* Lab. n. hol. t. 97.							
7 ovàta. R.S.	(ye.)	oval-leaved.	C. B. S.	1786.	5. 7.	G. w. ♃.	Bot. mag. t. 1909.
MENYA'NTHES. R.S.		Buck-bean.	Pentandria Monogynia. L.				
1 trifoliàta. E.B.	(li.)	common.	Britain.	....	6. 7.	H. w. ♃.	Eng. bot. v. 7. t. 495.
2 americàna.	(li.)	American.	N. America.	....	4. 6.	H. w. ♃.	
trifoliàta β minor. P.S.							

ORDO CXLI.

BIGNONIACEÆ. *Brown prodr.* 1. *p.* 470.

CHILO'PSIS. D.D.		CHILO'PSIS.	Didynamia Angiospermia.				
lineàris.	(*ro.*)	linear-leaved.	Mexico.	1825.	....	S. ♄.	Cavan. ic. 3. t. 269.
Bignònia? lineàris. Cav.							
CATA'LPA. J.		CATA'LPA.	Diandria Monogynia. S.S.				
1 syringifòlia. s.s.	(*wh.*)	common.	N. America.	1726.	6. 8.	H. ♄.	Bot. mag. t. 1094.
Bignònia Catálpa. L.							
2 longíssima. s.s.	(*wh.pu.*)	wave-leaved.	W. Indies.	1777.	....	S. ♄.	Plum. ic. t. 57.
3 microphy'lla. s.s.	(*bh.*)	small-leaved.	Hispaniola.	1820.	....	S. ♄.	——— t. 55.
BIGN'ONIA. S.S.		TRUMPET-FLOWER.	Didynamia Angiospermia. L.				
1 U'nguis. s.s.	(*ye.*)	Barbadoes.	W. Indies.	1759.	....	S. ♄.⌣.	Plum. amer. t. 94.
2 stamínea. s.s.	(*ye.pu.*)	long-stamened.	Hispaniola.	1820.	6. 9.	S. ♄.⌣.	Plum. ic. 56. f. 2.
3 capreolàta. s.s.	(*ye pu.*)	four-leaved.	N. America.	1710.	6. 8.	H. ♄.⌣.	Bot. mag. t. 864.
4 pubéscens. s.s.	(*ye.*)	downy.	New Spain.	1759.	6. 7.	S. ♄.⌣.	
5 æquinoctiàlis. s.s.	(*ye.*)	equinoctial.	W. Indies.	1768.	6. 8.	S. ♄.⌣.	Plum. ic. t. 55. f. 1.
6 Chamberla'ynii. B.M.	(*ye.*)	Chamberlayne's.	S. America.	1818.	——	S. ♄.⌣.	Bot. mag. t. 2148.
7 chrysoleùca. s.s.	(*ye.wh.*)	yellow & white.	———	1824.	——	S. ♄.⌣.	
8 alliàcea. s.s.	....	garlick-scented.	W. Indies.	1790.	....	S. ♄.⌣.	
9 spectábilis. s.s.	(*pu.*)	showy.	S. America.	1820.	....	S. ♄.⌣.	
10 picta. K.S.	(*wh.*)	flat-stemmed.	———	——	....	S. ♄.⌣.	
11 oblíqua. K.S.	(*re.*)	oblique-leaved.	Caracas.	1828.	——	S. ♄.⌣.	
12 laurifòlia. s.s.	(*li.*)	Laurel-leaved.	Guiana.	1804.	....	S. ♄.⌣.	
13 lactiflòra. s.s.	(*wh.*)	milk-coloured.	Trinidad.	1822.	4. 6.	S. ♄.⌣.	Vahl. symb. 3. t. 66.
14 villòsa. s.s.	(*pu.*)	villous.	S. America.	1820.	....	S. ♄.⌣.	
15 ophthálmica. A.	(*ye.*)	Anderson's.	St. Vincent.	1812.	....	S. ♄.⌣.	
16 rigéscens. s.s.	(*fl.ye.*)	rigid-leaved.	S. America.	1816.	6. 8.	S. ♄.⌣.	Jacq. schœn. 2. t. 210
17 grandifòlia. B.R.	(*ye.*)	large-leaved.	———	——	4. 7.	S. ♄.⌣.	Bot. reg. t. 418.
18 latifòlia. s.s.	(*ye.*)	broad-leaved.	Cayenne.	1824.	....	S. ♄.⌣.	
19 elongàta. s.s.	(*ye.*)	long-racemed.	S. America.	1820.	....	S. ♄.⌣.	Vahl. ecl. 2. t. 16.
20 mollíssima. K.S.	(*pu.*)	soft-leaved.	Caracas.	1812.	....	S. ♄.⌣.	
21 diversifòlia. K.S.	......	different-leav'd.	Mexico.	1824.	....	S. ♄.⌣.	
22 floribúnda. K.S.	......	abundant-flower'd.	———	——	....	S. ♄.⌣.	
23 cándicans. s.s.	(*pu.*)	white-leaved.	Trinidad.	1822.	9. 12.	S. ♄.⌣.	
24 quadrangulàris. D.L.	...	square-stalked.	———	——	....	S. ♄.⌣.	
25 coloràta. s.s.	(*ye.*)	coloured calyxed.	———	1824.	....	S. ♄.⌣.	
26 crucígera. s.s.	(*or.*)	cross-bearing.	S. America.	1759.	6. 8.	S. ♄.⌣.	Plum. ic. t. 58.
27 Cherère. B.R.	(*or.*)	various-leaved.	Guiana.	1820.	6. 10.	S. ♄.⌣.	Bot. reg. t. 1301.
heterophy'lla. W.							
28 venústa. s.s.	(*or.*)	comely.	S. America.	1816.	9. 12.	S. ♄.⌣.	Bot. reg. t. 249.
29 incarnàta. s.s.	(*ro.*)	flesh-coloured.	Guiana.	1818.	6. 9.	S. ♄.⌣.	Aub. gui. 3. t. 261.
30 móllis. s.s.	(*ye.*)	soft.	———	——	....	S. ♄.⌣.	Vahl. ecl. 2. t. 10.
tomentòsa. Rich. *non* Th.							
31 littoràlis. K.S.	(*re.*)	sea-side.	Mexico.	1824.	....	S. ♄.⌣.	
32 viridiflòra. B.C.	(*gr.*)	green-flowered.	S. America.	1820.	6. 9.	S. ♄.	Lodd. bot. cab. t. 1026.
33 triphy'lla. s.s.	(*wh.*)	three-leaved.	Mexico.	1733.	——	S. ♄.	
34 pállida. B.R.	(*li.*)	pale.	S. America.	1823.	7. 8.	S. ♄.	Bot. reg. t. 965.
35 fluviátilis. s.s.	(*wh.*)	marsh.	———	1818.	——	S. ♄.	Aub. gui. 2. t. 267.
36 Leucóxylon. s.s.	(*pi.*)	white-wooded.	W. Indies.	1759.	6. 7.	S. ♄.	Andr. reposit. t. 43.
37 chrysántha. s.s.	(*ye.*)	golden-flower'd.	Caracas.	1822.	....	S. ♄.	Jacq. schœnb. 2. t. 211.
38 lepidòta. K.S.	(*ro.*)	silver-spotted.	Cuba.	1828.	....	S. ♄.	
39 æsculifòlia. K.S.	(*or.*)	Æsculus-leaved.	Mexico.	1824.	....	S. ♄.	
40 serratifòlia. s.s.	(*ye.*)	saw-leaved.	Trinidad.	1822.	....	S. ♄.	
41 bijùga. s.s.	....	Madagascar.	Madagascar.	——	....	S. ♄.	
42 Clématis. K.S.	(*wh.ye.*)	Clematis-leaved.	Caracas.	——	....	S. ♄.⌣.	
43 chelonoídes. s.s.	(*ye.*)	tree.	E. Indies.	1808.	....	S. ♄.	Rheed. mal. 6. t. 26.
44 variábilis. s.s.	(*gr.ye.*)	variable-leaved.	Caracas.	1820.	6. 8.	S. ♄.⌣.	Jacq. schœnb. 2. t. 212.
45 índica. W.	(*wh.std.*)	Indian.	E. Indies.	1775.	....	S. ♄.	Rheed. mal. 1. t. 43.
46 jasminifòlia. K.S.	....	Jasmine-leaved.	S. America.	1824.	....	S. ♄.⌣.	
47 crenàta. L.C.	....	crenated.	....	1825.	....	S. ♄.⌣.	
48 decípiens. L.C.	......	doubtful.	....	——	....	S. ♄.	
49 lùcida. L.C.	....	glossy.	S. America.	1824.	....	S. ♄.⌣.	
50 álba. W.	(*wh.*)	white-flowered.	Guiana.	1823.	....	S. ♄.⌣.	Aub. gui. 2. t. 266.

51	echinàta. W.	(*bh.*)	bristly-fruited.	Guiana.	1804.	....	S. ♄ ◡.	Aub. gui. 2. t. 264.
52	purpùrea. L.C.	(*pu.*)	purple.	S. America.	1822.	....	S. ♄.	
53	tomentòsa. Th.	....	tomentose.	Japan.	1823.	....	S. ♄.	Kæmpf. amœn. t. 860.
54	stipulàta. H.B.	......	stipuled.	E. Indies.	1825.	....	S. ♄.	
55	undulàta. H.B.	......	wave-leaved.	———	1827.	....	S. ♄.	
56	suavèolens. H.B.	....	sweet-scented.	———	1825.	....	S. ♄.	
57	críspa. H.B.	....	curled.	———	1822.	....	S. ♄.	
58	xylocárpa. H.B.	....	woody-podded.	———	1825.	....	S. ♄.	
59	am'œna	....	handsome.	———	1823.	....	S. ♄.	
60	spicàta. D.L.	....	spiked-flowered.	Trinidad.	1822.	....	S. ♄ ◡.	
61	multijùga. R.	....	many-leaved.	E. Indies.	1826.	....	S. ♄.	
	MILLINGT'ONIA. L.		MILLINGT'ONIA.	Didynamia Angiospermia. L.				
	horténsis. L.	(*wh.*)	cork-barked.	E. Indies.	1818.	....	S. ♄.	Roxb. corom. 3. t. 214.
	Bignònia suberòsa. Roxb.							
	TEC'OMA. J.		TEC'OMA.	Didynamia Angiospermia. S.S.				
1	spléndida. S.S.	(*or.*)	splendid.	Brazil.	1826.	....	S. ♄.	
2	pentaphy'lla. S.S.	(*wh.*)	five-leaved.	Jamaica.	1733.	....	S. ♄.	Marc. bras. t. 118.
3	digitàta. K.S.	....	digitate-leaved.	S. America.	1818.	....	S. ♄.	
4	austràlis. B.P.	(*bh.*)	broad-leaved.	N. S. W.	1793.	4. 7.	G. ♄ ◡.	Bot. mag. t. 865.
	Bignònia Pandòræ. B.M.							
5	meonántha.	(*bh.*)	narrow-leaved.	———	1815.	——	G. ♄ ◡.	
	Bignònia meonántha. L.en.							
6	móllis. K.S.	(*ye.*)	soft-leaved.	Mexico.	1824.	....	S. ♄.	
7	rádicans. S.S.	(*or.*)	Ash-leaved.	N. America.	1640.	7. 8.	H. ♄ ◡.	
	α màjor. H.K.	(*or.*)	*great.*	———	——	——	H. ♄ ◡.	Bot. mag. t. 485.
	β mìnor. H.K.	(*or.*)	*smaller.*	———	——	——	H. ♄ ◡.	Catesb. car. 1. t. 65.
8	grandiflòra.	(*or.*)	large-flowered.	China.	1800.	6. 9.	G. ♄ ◡.	Bot. mag. t. 1398.
	Bignònia grandiflòra. B.M. *Incarvíllea grandiflòra.* S.S.							
9	stáns. S.S.	(*ye.*)	upright-ash-ld.	S. America.	1730.	7. 9.	S. ♄.	Plum. ic. t. 54.
10	incìsa. D.L.	(*ye.*)	cut-leaved.	Trinidad.	1824.	——	S. ♄.	
11	capénsis. B.R.	(*or.*)	Cape.	C. B. S.	——	4. 8.	G. ♄ ◡.	Bot. reg. t. 1117.
	JACARA'NDA. S.S.		JACARA'NDA.	Didynamia Angiospermia. S.S.				
1	tomentòsa. D.D.	(*pu.*)	tomentose.	Brazil.	1824.	....	S. ♄.	
2	mimosifòlia. B.R.	(*bl.*)	Mimosa-leaved.	———	1818.	4. 5.	S. ♄.	Bot. reg. t. 631.
	ovalifòlia. Bot. mag. t. 2327.							
3	bahaménsis. B.M.	(*bl.*)	Bahama.	Bahama Isl.	1724.	7. 8.	S. ♄.	Catesb. car. 1. t. 42.
	Bignònia cœrùlea. W.							
4	Copaìa. D.D.	(*bl.*)	Box-leaved.	Guiana.	1793.	....	S. ♄.	Aub. gui. 2. t. 265.
	Bignònia pròcera. W.							
5	rhombifòlia. S.S.	(*pu.*)	rhomb-leaved.	Surinam.	1823.	....	S. ♄.	
	filicifòlia. D.D.							
	SPATH'ODEA. S.S.		SPATH'ODEA.	Didynamia Angiospermia. S.S.				
1	uncàta. S.S.	(*ye.*)	hooked.	Guiana.	1804.	6. 9.	S. ♄ ◡.	Bot. mag. t. 1511.
	Bignònia uncàta. B.M.							
2	corymbòsa. S.S.	(*ye.*)	corymb-flower'd.	Trinidad.	1823.	....	S. ♄.	
3	Rheèdii. S.S.	(*wh.*)	Rheede's.	E. Indies.	1794.	....	S. ♄.	Roxb. cor. 2. t. 144.
	Bignònia spathàcea. R.C.							
4	Roxbúrghii. S.S.	(*ro.*)	Roxburgh's.	———	1820.	....	S. ♄.	Roxb. cor. v. 2. t. 145.
	Bignònia quadriloculàris. Roxb.							
5	fraxinifòlia. K.S.	(*ye.*)	Ash-leaved.	Caracas.	1821.	....	S. ♄ ◡.	
	Bignònia fraxinifòlia. S.S.							
	AMPHIL'OPHIUM. K.S.		AMPHIL'OPHIUM.	Didynamia Angiospermia. S.S.				
	paniculàtum. K.S.	(*pu.*)	panicled.	S. America.	1788.	....	S. ♄ ◡.	Jacq. amer. t. 116.
	Bignònia paniculàta. Jacq.							
	CALA'MPELIS. D.D.		CALA'MPELIS.	Didynamia Angiospermia.				
	scábra. B.F.G.	(*or.*)	rough-podded.	Chile.	1823.	7. 11.	H. ♄ ◡.	Swt. br. fl. gar. s. 2. t. 30.
	Eccremocárpus scáber. Bot. reg. t. 939.							
	ECCREMOCA'RPUS.		ECCREMOCA'RPUS.	Didynamia Angiospermia. S.S.				
	longiflòrus. K.S.	(*ye.*)	long-flowered.	Peru.	1824.	....	G. ♄ ◡.	H. et B. pl. æq. 1. t. 65.
	FI'ELDIA. Cun.		FI'ELDIA.	Didynamia Angiospermia.				
	austràlis. Cun.	(*gr.st.*)	New Holland.	N. Holland.	1824.	....	G. ♄.	Hook. ex. flor. t. 232.

ORDO CXLII.

COBŒACEÆ. *D. Don in Edinb. phil. journ.*

COB'ŒA. Cav. COB'ŒA. Pentandria Monogynia.
scándens. Cav. (*bl.*) climbing. Mexico. 1792. 5. 10. G. ♄. ◡. Bot. mag. t. 851.

ORDO CXLIII.

PEDALINÆ. *Brown prodr.* 519.

PED'ALIUM. S.S. PED'ALIUM. Didynamia Angiospermia. S.S.
Mùrex. s.s. (*ye.*) prickly-fruited. E. Indies. 1778. 8. 9. S. ⊙. Lam. ill. t. 538.

ORDO CXLIV.

DIDYMOCARPEÆ. *D. Don.*

STREPTOCA'RPUS. B.R. STREPTOCA'RPUS. Diandria Monogynia.
Réxii. B.R. (*bl.*) Cape. C. B. S. 1824. 2. 10. S. ♃. Bot. reg. t. 1173.
Didymocárpus Réxii. Hook. ex. flor. t. 227.

ORDO CXLV.

SESAMEÆ. *Brown.*

TOURR'ETIA. S.S. TOURR'ETIA. Didynamia Angiospermia. S.S.
lappàcea. s.s. (*sc.*) scarlet-flowered. Peru. 1788. 6. 8. S. ⊙. ◡. Sal. stirp. rar. 5. t. 3.
Dombèya lappàcea. Sal.
MARTY'NIA. J. MARTY'NIA. Didynamia Angiospermia. L.
1 diándra. s.s. (*re.*) two-stamened. New Spain. 1731. 7. 8. S. ⊙. Andr. reposit. t. 575.
2 proboscídea. s.s. (*li.*) horn-capsuled. America. 1738. 6. 8. G. ⊙. Bot. mag. t. 1056.
ánnua. L.
3 lùtea. B.R. (*ye.*) yellow. S. America. 1824. 7. 9. H. ⊙. Bot. reg. t. 934.
4 longiflòra. s.s. (*pu.*) long-flowered. C. B. S. 1781. 7. 8. G. ⊙. Meerb. ic. 1. t. 7.
CRANIOL'ARIA. J. CRANIOL'ARIA. Didynamia Angiospermia. L.
ánnua. K.S. (*wh.sp.*) annual. S. America. 1733. 7. 8. S. ⊙. Jacq. amer. t. 110.
Marty'nia Craniolària. W.
SES'AMUM. S.S. OILY-GRAIN. Didynamia Angiospermia. L.
1 orientàle. s.s. (*bh.*) oriental. E. Indies. 1731. 7. 8. S. ⊙. Rheed. mal. 9. t. 54.
2 índicum. s.s. (*bh.*) Indian. ——— ——— ——— S. ⊙. Bot. mag. t. 1788.
3 lùteum. s.s. (*ye.*) yellow. ——— 1825. ——— S. ⊙.

ORDO CXLVI.

POLEMONIACEÆ. *D. Don in Edinb. phil. journ.*

POLEM'ONIUM. S.S. GREEK VALERIAN. Pentandria Monogynia. L.
1 cœrùleum. s.s. (*bl.*) common. Britain. .... 6. H. ♃. Eng. bot. t. 14.
β *álbum.* (*wh.*) *white-flowered.* ——— .... ——— H. ♃.
γ *maculàtum.* s.s. (*sp.*) *spot-flowered.* ——— .... ——— H. ♃.

2 grácile. W.en.	(*wh.*)	slender.	Dahuria.	1818.	6. 7.	H. ♃.	
3 mexicànum. s.s.	(*bl.*)	Mexican.	Mexico.	1817.	4. 6.	H. ♃.	Bot. reg. t. 460.
4 réptans. s.s.	(*bl.*)	creeping.	N. America.	1758.	4. 5.	H. ♃.	Bot. mag. t. 1887.
cærùleum β pilíferum. Bot. reg. t. 1303.							
5 villòsum. G.R.	(*bl.*)	dwarf hairy.	Siberia.	1827.	6. 9.	H. ♃.	Swt. br. fl. gar. t. 266.
6 Richardsòni. B.M.	(*bl.*)	Richardson's.	N. America.	——	——	H. ♃.	Bot. mag. t. 2800.
7 hùmile. B.M.	(*bl.*)	dwarf.	———	——	——	H. ♃.	Bot. reg. t. 1304.
8 sibíricum. D.D.	(*wh.*)	bipinnate.	Siberia.	1800.	6.	H. ♃.	Swt. br. fl. gar. t. 182.
PHLO'X. R.S.		LYCHNIDEA.	Pentandria Monogynia. L.				
1 paniculàta. R.S.	(*pu.*)	panicled.	N. America.	1732.	8. 9.	H. ♃.	Dill. elth. t. 166. f. 203.
β álba.	(*wh.*)	*white-flowered.*	———	1813.	——	H. ♃.	
2 undulàta. R.S.	(*pu.*)	wave-leaved.	———	1759.	7. 8.	H. ♃.	
3 acuminàta. R.S.	(*pu.*)	cross-leaved.	———	1812.	6. 8.	H. ♃.	Bot. mag. t. 1880.
decussàta. Lyon. *latifòlia.* Hort. *non* M.							
4 virgínica. L.C.	(*pu.*)	Virginian.	———	——	——	H. ♃.	
5 intermèdia. L.C.	(*pu.*)	intermediate.	———	....	7. 9.	H. ♃.	
6 penduliflòra. B.F.G.	(*pu.*)	drooping-flower'd.	———	1824.	——	H. ♃.	Swt. br. fl. gar. s. 2.
7 corymbòsa. B.F.G.	(*li.*)	corymb-flower'd.	———	——	——	H. ♃.	
8 acutifòlia. B.F.G.	(*pu.*)	sharp-leaved.	———	1825.	——	H. ♃.	
9 odoràta. B.F.G.	(*re.*)	sweet-scented.	———	1824.	——	H. ♃.	Swt. br. fl. gar. t. 224.
10 scábra. B.F.G.	(*li.*)	rough-leaved.	———	1812.	——	H. ♃.	———— t. 248.
Sickmánni. Lehman.							
11 maculàta. R.S.	(*pu.*)	spotted-stalked.	———	1740.	7. 8.	H. ♃.	Jacq. vind. 2. t. 147.
12 pyramidàlis. R.S.	(*li.*)	pyramidal.	———	1800.	6. 8.	H. ♃.	Swt. br. fl. gar. t. 233.
13 latifòlia. R.S.	(*pu.*)	broad-leaved.	———	1812.	——	H. ♃.	
14 carolìna. R.S.	(*re.pu.*)	rough-stemmed.	———	1728.	7. 9.	H. ♃.	Bot. mag. t. 1344.
15 cordàta. E.C.	(*li.*)	heart-leaved.	Carolina.	1826.	6. 9.	H. ♃.	Swt. br. fl. gar. s. 2. t. 13.
16 excélsa. H.E.	(*pu.*)	tall upright.	N. America.	1824.	——	H. ♃.	
17 Wheeleriàna. B.F.G.	(*pi.*)	Wheeler's.	Hybrid.	——	——	H. ♃.	
18 Shephérdii. H.E.	(*pu.*)	Shepherd's.	———	——	——	H. ♃.	
19 glabérrima. R.S.	(*pu.*)	smooth-stalked.	N. America.	1725.	6. 8.	H. ♃.	Swt. br. fl. gar. s. 2 t. 36.
20 triflòra. R.S.	(*bh.*)	flesh-coloured.	———	1816.	7. 9.	H. ♃.	Swt. brit. fl. gar. t. 29.
cárnea. Bot. mag. t. 2155.							
21 suavèolens. R.S.	(*wh.*)	white-flowered.	———	1766.	7. 8.	H. ♃.	
22 longiflòra. B.F.G.	(*wh.*)	long-flowered.	———	1827.	7. 10.	H. ♃.	Swt. br. fl. gar. s. 2. t. 31.
23 refléxa. B.F.G.	(*d.pu.*)	reflexed-leaved.	———	1824.	6. 9.	H. ♃.	Swt. br. fl. gar. t. 232.
24 suffruticòsa. B.R.	(*d.pu.*)	glossy-leaved.	———	1790.	7. 9.	H. ♃.	Bot. reg. t. 68.
25 Listoniàna. H.E.	(*pu.*)	Liston's.	———	1816.	5. 7.	H. ♃.	
26 ovàta. W.	(*pu.*)	ovate-leaved.	———	1759.	——	H. ♃.	Bot. mag. t. 528.
27 stolonífera. B.M.	(*bl.*)	creeping.	———	1800.	6. 9.	H. ♃.	——— t. 563.
prostràta. H.K. *réptans.* M. *stolonífera* has the right of priority.							
28 pilòsa. B.M.	(*p.bl.*)	hairy-leaved.	N. America.	1759.	5. 6.	H. ♃.	Bot. mag. t. 1307.
aristàta. M.							
29 am'œna. B.M.	(*ro.*)	Fraser's hairy.	———	1809.	6. 7.	H. ♃.	Bot. mag. t. 1308.
pilòsa. β. R.S.							
30 canadénsis. B.F.G.	(*pa.bl.*)	Canadian.	Canada.	1826.	5. 7.	H. ♃.	Swt. br. fl. gar. t. 221.
31 divaricàta. R.S.	(*pa.bl.*)	early-flowering.	N. America.	1746.	4. 6.	H. ♃.	Bot. mag. t. 163.
32 procúmbens. B.F.G.	(*bl.*)	procumbent.	———	1829.	5. 6.	H. ♃.	Swt. br. fl. gar. s. 2. t. 7.
33 subulàta. R.S.	(*li.pu.*)	awl-leaved.	———	1786.	4. 6.	H. ♄.	Bot. mag. t. 411.
34 nivàlis. B.C.	(*wh.*)	snowy white.	———	1820.	——	H. ♄.	Swt. br. fl. gar. t. 185.
35 setàcea. R.S.	(*pi.*)	bristle-leaved.	———	1786.	4. 5.	H. ♄.	Bot. mag. t. 415.
COLL'OMIA. N.		COLL'OMIA.	Pentandria Monogynia.				
1 grandiflòra. D.	(*bf.*)	large-flowered.	N. America.	1827.	6. 7.	H. ⊙.	Bot. reg. t. 1174.
2 lineàris. D.	(*fl.*)	linear-leaved.	———	——	——	H. ⊙.	——— t. 1156.
3 grácilis. D.	(*ro.*)	slender.	———	——	5. 7.	H. ⊙.	Bot. mag. t. 2924.
Gília gracilis. B.M.							
4 heterophy'lla. B.M.	(*ro.*	various-leaved.	———	——	5. 8.	H. ⊙.	Bot. mag. t. 2895.
GI'LIA. R.P.		GI'LIA.	Pentandria Monogynia. S.S.				
1 inconspícua. B.M.	(*bl.*)	small blue.	America.	1793.	9. 11.	H. ⊙.	Bot. mag. t. 2883.
Ipomópsis inconspícua. Sm. exot. bot. 1. t. 14.							
2 capitàta. D.	(*bl.*)	cluster-flowered.	———	1826.	5. 9.	H. ⊙.	Swt. br. fl. gar. t. 287.
IPOMO'PSIS. M.		IPOMO'PSIS.	Pentandria Monogynia.				
1 élegans. Ex.B.	(*sc.*)	elegant.	Carolina.	1726.	8. 9.	G. ♂.	Sm. exot. bot. 1. t. 13.
Cántua coronopifòlia. Andr. rep. t.							
2 pulchélla. D.	(*sc.*)	pretty.	N. America.	1827.	——	G. ♂.	Bot. reg. t. 1281.
élegans. Bot. reg. *non* M.							
HOI'TZIA. J.		HOI'TZIA.	Pentandria Monogynia. S.S.				
1 coccínea. K.S.	(*sc.*)	scarlet.	Mexico.	1826.	5. 7.	G. ♄.	Cavan. ic. 4. t. 365.
2 cœrùlea. K.S.	(*bl.*)	blue-flowered.	———	1827.	——	G. ♄.	——— 4. t. 366.

CALD'ASIA S.S.			CALD'ASIA.	Pentandria Monogynia.				
heterophy'lla. s.s.	(*bl.*)	various-leaved.		New Spain.	1813.	5. 12.	S.⊙.	Bot. reg. t. 92.
Bonplándia geminiflòra. Cav. ic. 6. t. 532.								

ORDO CXLVII.

HYDROLEACEÆ. *Kth. synops.* 2. *p.* 234.

HYDR'OLEA. S.S.		HYDR'OLEA.	Pentandria Digynia. L.				
1 spinòsa. s.s.	(*bl.*)	thorny.	S. America.	1791.	7. 8.	S. ♄.	Bot. reg. t. 566.
2 quadriválvis. s.s.	(*bl.*)	four-valved.	Carolina.	1824.	——	G. ♃.	
WIGA'NDIA. K.S.		WIGA'NDIA.	Pentandria Digynia.				
1 ùrens. K.S.	(*vi.*)	stinging.	Mexico.	1827.	4. 6.	G. ♃.	Flor. peruv. 3. t. 243.
2 caracasàna. K.S.	(*vi.*)	Caracas.	Caracas.	1825.	——	S. ħ.	
N'AMA. S.S.		N'AMA.	Pentandria Digynia. L.				
1 jamaicénsis. s.s.	(*wh.bl.*)	Jamaica.	Jamaica.	1812.	6. 9.	S.⊙.	Lam. ill. t. 184.
2 undulàta. K.S.	(*vi.*)	wave-leaved.	Mexico.	1826.	——	G. ♃.	

ORDO CXLVIII.

CONVOLVULACEÆ. *Brown prodr.* 1. *p.* 481.

? RE'TZIA. R.S.		RE'TZIA.	Pentandria Monogynia. R.S.				
spicàta. R.S.	(*br.*)	spiked.	C. B. S.	....	5. 6.	G. ♄.	Lam. ill. t. 103.
CONVO'LVULUS. B.P.		BINDWEED.	Pentandria Monogynia. L.				
1 arvénsis. R.S.	(*pi.wh.*)	small.	Britain.	....	6. 9.	H. ♃.◡.	Eng. bot. v. 5. t. 312.
2 chinénsis. B.R.	(*pi.*)	Chinese.	China.	1816.	——	H. ♃.◡.	Bot. reg. t. 322.
3 Malcólmi. F.I.	(*wh.*)	Malcolm's.	Persia.	1824.	7. 10.	H. ♃.◡.	
4 emarginàtus. s.s.	(*re.*)	emarginate.	....	——	6. 9.	H. ♃.◡.	
5 bicuspidàtus. s.s.	(*bh.*)	two-pointed.	Davuria.	——	——	H. ♃.◡.	
6 Scammònia. R.S.	(*st.bh.*)	Scammony.	Levant.	1596.	7. 8.	H. ♃.◡.	Flor. græc. t. 192.
7 erubéscens. R.S.	(*bh.*)	maiden blush.	N. S. W.	1803.	7. 9.	G. ♂.◡.	Bot. mag. t. 1067.
8 quinqueflòrus. R.S.	(*bh.*)	five-flowered.	Mauritius.	1826.	——	S. ♄.◡.	
9 pannifòlius. R.S.	(*bl.*)	cloth-leaved.	Canaries.	1805.	6. 9.	G. ♄.◡.	Bot. reg. t. 222.
10 canariénsis. R.S.	(*pi.*)	Canary.	————	1690.	——	G. ♄.◡.	Bot. mag. t. 1228.
11 Massòni. R.S.	(*wh.*)	Masson's.	————	1788.	——	G. ħ.◡.	Bot. reg. t. 133.
Dryándri. S.S. *suffruticòsus.* B.R.							
12 cándicans. R.S.	....	hoary.	E. Indies.	1818.	....	S. ♄.◡.	
13 bonariénsis. R.S.	(*std.*)	stripe-flowered.	BuenosAyres.	1827.	6. 8.	G. ♃.◡.	Cavan. icon. 5. t. 48. f. 2.
14 tiliàceus. R.S.	(*wh.*)	Lime-tree-leaved.	Brazil.	1820.	——	S. ♃.◡.	
15 guianénsis. R.S.	(*bh.*)	Guiana.	Guiana.	——	——	S. ♃.◡.	Aub. gui. 1. t. 52.
16 ciliàtus. R.S.	(*bh.*)	fringed-leaved.	Cayenne.	——	——	S. ♄.◡.	
17 pentánthus. R.S.	(*bl.*)	five-flowered.	W. Indies.	1808.	4. 10.	S. ♄.◡.	Bot. reg. t. 439.
18 farinòsus. R.S.	(*bh.*)	mealy-stalked.	Madeira.	1777.	5. 6.	G. ♄.◡.	Sal. par. lond. t. 45.
19 Hermánniæ. s.s.	(*wh.*)	Peruvian.	Peru.	1799.	8. 9.	G. ♃.◡.	L'Herit. stirp. t. 33.
20 tenuíssimus. R.S.	(*li.*)	slender-leaved.	Levant.	....	6. 10.	G. ♃.◡.	Flor. græc. t. 195.
21 althæoídes. R.S.	(*li.*)	Althæa-leaved.	————	1597.	6. 9.	F. ♃.◡.	———— t. 194.
22 bryoniæfòlius. R.S.	(*pu.*)	Bryony-leaved.	China.	1802.	7. 8.	G. ♃.◡.	Bot. mag. t. 943.
23 itálicus. R.S.	(*pi.*)	Italian.	Italy.	....	——	F. ♃.◡.	Ten. fl. neap. 1. t. 15.
hirsùtus. T.N. *non* M.B.							
24 hirsùtus. M.B.	(*pu.ye.*)	hairy.	Tauria.	1824.	——	H. ♃.◡.	
25 alceifòlius. R.S.	(*st.pu.*)	Hollyhock-leav'd.	C. B. S.	1823.	——	G. ♃.◡.	
26 macrocárpus. R.S.	(*pu.*)	long-fruited.	S. America.	1752.	——	S.⊙.◡.	Burm. amer. t. 91. f. 1.
27 quinquefòlius. R.S.	(*w.pu.*)	five-leaved.	Vera Cruz.	....	——	S.⊙.◡.	Pluk. alm. t. 167. f. 6.
28 gláber. R.S.	(*wh.*)	smooth.	Cayenne.	1806.	....	S. ħ.◡.	Aub. gui. 1. t. 53.
29 tenéllus. R.S.	(*wh.*)	slender.	N. America.	1812.	6. 8.	H.⊙.◡.	Pluk. alm. t. 166. f. 4.
30 aquáticus. N.	(*wh.*)	water-side.	————	——	——	H.⊙.	
31 sículus. R.S.	(*bl.*)	small-flowered.	S. Europe.	1640.	——	H.⊙.	Bot. reg. t. 445.
32 pseudosículus. R.S.	(*wh.*)	long-peduncl'd.	Canaries.	1815.	7. 8.	H.⊙.	———— t. 498.
elongàtus. W.enum.							
33 Impèrati. R.S.	(*st.*)	trailing.	Naples.	1824.	6. 9.	H. ♃.	Cyr. pl. rar. fasc. 1. t. 5.

34 sagittifòlius. F.G.	(bh.)	Sibthorp's.	Samos.	1823.	7. 8.	H.♃.	Flor. græc. t. 193.
35 hírtus. R.S.	(bl.)	hairy-stalked.	E. Indies.	1806.	6. 8.	S.☉.	
36 platycárpos. R.S.	(li.)	flat-podded.	Mexico.	1827.	8. 10.	F.♃.	Cavan. ic. 5. t. 482.
37 multífidus. R.S.	(bh.)	multifid.	C. B. S.	1822.	6. 8.	G.♃.	
38 pérsicus. M.B.	(wh.)	Persian.	Persia.	1829.	——	H.♃.	Gmel. iter. 3. t. 7.
39 trícolor. R.S.	(bl.ye.)	three-coloured.	S. Europe.	1629.	7. 9.	H.☉.	Bot. mag. t. 27.
β albiflòrus.	(wh.)	*white-flowered.*	————	——	——	H.☉.	
40 meonanthus. R.S.	(bl.ye.)	pale blue.	Portugal.	1827.	——	H.☉.	
41 pentapetaloídes. R.S.	(bl.)	Majorca.	Majorca.	1789.	6. 7.	H.☉.	Flor. græc. t. 197.
42 lineàtus. R.S.	(bh.)	dwarf.	S. Europe.	1714.	——	H.♃.	———— t. 199.
43 Gerárdi. R.S.	(bh.)	Gerard's.	————	....	——	H.♃.	Barrel. ic. t. 311.
44 undulàtus. R.S.	(wh.re.)	waved.	————	1816.	6. 10.	H.☉.	Cavan. ic. 3. t. 277. f. 1.
hùmilis. Jacq. col. 4. t. 22. f. 3. *evolvuloídes.* F.G. t. 198. *ciliàtus.* P.S.							
45 lanàtus. R.S.	(wh.)	woolly-leaved.	Levant.	1829.	5. 6.	G.♄.	Flor. græc. t. 202.
46 intermèdius. R.S.	(bh.)	intermediate.	S. Europe.	1825.	6. 7.	H.♃.	
47 Cneòrum. R.S.	(bh.)	silvery-leaved.	Levant.	1640.	5. 9.	G.♄.	Bot. mag. t. 459.
48 saxátilis. R.S.	(bh.)	rock.	S. Europe.	1796.	6. 8.	F.♃.	Bocc. mus. 138. t. 96.
49 flòridus. R.S.	(bh.)	many-flowered.	Teneriffe.	1779.	8. 9.	G.ħ.	Jacq. ic. 1. t. 34.
50 Cantábrica. R.S.	(fl.)	flax-leaved.	S. Europe.	1680.	5. 9.	H.♃.	Jacq. aust. 3. t. 296.
51 salvifòlius. L.en.	(bh.)	Sage-leaved.	Palestine.	1825.	——	H.♃.	
52 lineàris. R.S.	(wh.re.)	narrow-leaved.	S. Europe.	1770.	——	G.ħ.	Bot. mag. t. 289.
53 suffruticòsus. R.S.	(bh.)	suffrutescent.	Levant.	1827.	6. 8.	F.♄.	Desf. atl. 1. t. 48.
54 Dory'cnium. R.S.	(ro.)	silky-leaved.	————	1806.	6. 7.	H.ħ.	Flor. græc. t. 201.
55 holoseríceus. R.S.	(st.)	silky.	Tauria.	1824.	——	H.♃.	
56 scopàrius. R.S.	(wh.)	broom.	Canaries.	1733.	8. 9.	G.ħ.	Venten. choix. t. 24.
57 terréstris. F.	(wh.)	small.	Altay.	1828.	7. 9.	H.♃.	
CALYST'EGIA. B.P.		BEARBIND.	Pentandria Monogynia. R.S.				
1 sèpium. R.S.	(wh.)	great hedge.	Britain.	....	6. 9.	H.♃.◡.	Eng. bot. v. 5. t. 313.
2 sylvéstris. R.S.	(wh.)	wood.	Hungary.	1815.	——	H.♃.◡.	W. et K. hung. t. 261.
Convólvulus sylvéstris. W.en.							
3 inflàta. Df.	(re.)	inflated.	N. America.	....	——	H.♃.◡.	Bot. mag. t. 732.
sèpium. Ph. *Convólvulus sèpium β incarnàta.* B.M.							
4 dahùrica. B.M.	(pi.)	Dahurian.	Dahuria.	1820.	7. 8.	H.♃.◡.	Bot. mag. t. 2609.
5 Catesbiàna. R.S.	(ro.)	Catesby's.	Carolina.	1816.	7. 9.	H.♃.◡.	
6 marginàta. B.P.	(wh.)	margined.	N. S. W.	1824.	7. 8.	F.♃.◡.	
7 stáns. M.	(wh.)	woolly.	N. America.	1818.	6. 7.	H.♃.	
tomentòsa. Ph. *Convólvulus stáns.* M.							
8 spitham'æa. R.S.	(wh.)	small upright.	————	1696.	7. 8.	H.♃.	Hook. ex. flor. t. 97.
9 Soldanélla. R.S.	(pi.)	sea-side.	Britain.	....	6. 7.	H.♃.	Eng. bot. v. 5. t. 314.
Convólvulus Soldanélla. E.B.							
10 renifórmis. B.P.	(pi.)	kidney-leaved.	N. S. W.	1822.	6. 8.	F.♃.	Swt. br. fl. gar. v. 2. t. 181.
11 hederàcea. F.I.	(ro.)	Ivy-leaved.	Nepaul.	1826.	——	F.♃.◡.	
12 Keriàna.	(bl.)	Mr. Ker's.	Mauritius.	1818.	——	S.☉.	Bot. reg. t. 318.
Convòlvulus involucràtus. B.R. *non* Beau. *Convólvulus bícolor.* B.M. 2205. *nec aliorum.*							
IPOM'ŒA. B.P.		IPOM'ŒA.	Pentandria Monogynia. R.S.				
1 Quamóclit. R.S.	(sc.)	wing-leaved.	E. Indies.	1629.	7. 9.	S.☉.◡.	Bot. mag. t. 244.
β albiflòra.	(wh.)	*white-flowered.*	————	....	——	S.☉.◡.	
2 dissécta. R.S.	(wh.)	cut-leaved.	Tropic's.	1813.	6. 9.	S.♃.◡.	Willd. phyt. 1. t. 2. f. 3.
3 tuberòsa. R.S.	(st.pu.)	tuberous-rooted.	W. Indies.	1731.	8. 10.	S.♃.◡.	Bot. reg. t. 768.
4 digitàta. R.S.	(sc.)	fingered.	S. America.	....	——	S.♃.◡.	Plum. ic. 92. f. 1.
5 sinuàta. R.S.	(wh.)	sinuated.	N. America.	1813.	6. 9.	G.♃.◡.	Jacq. vind. t. 159.
dissécta. Ph. *non.* W. *Convólvulus disséctus.* M.							
6 tuberculàta. R.S.	(li.)	tubercled.	Bourbon.	1818.	——	S.ħ.◡.	Jacq. schœn. 2. t. 199.
stipulàcea. Jacq.							
7 caírica.	(pu.)	jagged-leaved.	Egypt.	1680.	——	S.ħ.◡.	Bot. mag. t. 699.
Convólvulus caíricus. B.M. *Ipom'œa palmàta.* R.S.							
8 péndula. B.P.	(pu.)	pendulous-flower'd.	N. S. W.	1808.	5. 10.	G.ħ.◡.	Bot. reg. t. 632.
9 dasyspérma. J.E.	(ye.pu.)	warted.	E. Indies.	1815.	8. 9.	G.☉.◡.	———— t. 86.
tuberculàta. B.R. *non* R.S. *Convólvulus pedàtus.* F.I.							
10 cóptica. R.S.	(wh.)	palmate-leaved.	E. Indies.	1824.	8. 10.	S.☉.◡.	
11 pés-tìgridis. R.S.	(bh.)	palmated.	————	1732.	——	S.☉.◡.	Dill. elth. t. 318. f. 411.
12 platénsis. B.R.	(pu.)	Plata.	S. America.	1817.	6. 9.	S.♃.◡.	Bot. reg. t. 333.
13 setòsa. B.R.	(pu.)	bristly-stalked.	Brazil.	——	8. 10.	S.ħ.◡.	———— t. 335.
14 paniculàta. R.S.	(pu.)	panicled.	E. Indies.	1799.	6. 9.	S.♄.◡.	———— t. 62.
15 insígnis. A.R.	(pu.)	large purple.	————	1814.	——	S.♄.◡.	———— t. 75.
gossypifòlia. R.S.							
16 senegalénsis. R.S.	(pu.)	Senegal.	Senegal.	1822.	6. 10.	S.♄.◡.	
17 spléndens. G.D.	(pu.)	splendid.	Sierra Leon.	——	6. 9.	S.♄.◡.	
18 parviflòra. R.S.	(pu.)	small-flowered.	Santa Cruz.	——	7. 10.	S.☉.◡.	Sloan. hist. 1. t. 97. f. 1.

19 heterophy'lla. R.S. (*bl.*) various-leaved. S. America. 1817. 7. 10. S. ♃. Jacq. frag. t. 42. f. 4.
20 macrorhìzos. R.S. (*sc.*) large-rooted. W. Indies. —— —— S. ♃. Plum. cat.mss.v.2.t.58.
21 umbellàta. R.S. (*sc.*) umbelled. S. America. 1739. 6. 7. S. ♃. Burm. amer. t. 92. f. 2.
22 carolìna. R.S. (*pu.*) Carolina. Carolina. 1732. 7. 8. G. ♃. ◡. Dill. elth. t. 84. f. 98.
23 pulchélla. R.S. (*vi.*) neat. E. Indies. 1824. 5. 8. S. ♃. ◡.
24 venòsa. R.S. (*pu.*) veined-leaved. Mauritius. 1820. 7. 8. S. ♄. ◡.
25 pentaphy'lla. R.S. (*wh.*) five-leaved. W. Indies. 1739. 8. 9. S. ⊙. ◡. Jacq. ic. 2. t. 319.
26 Cavanillèsii. R.S. (*bh.*) Cavanille's. S. America. 1815. —— S. ♄. ◡. Cavan. ic. 3. t. 256.
pentaphy'lla. Cav. *non* Jacq.
27 armàta. R.S. (*re.vi.*) armed-calyxed. Mexico. 1824. 7. 10. S. ⊙. ◡. Cavan. ic. 5. t. 478. f. 2.
muricàta. Cav. *non* Jacq.
28 ternàta. R.S. (*wh.*) ternate-leaved. S. America. —— 6. 8. S. ♄. ◡. Jacq. schœnb. 1. t. 37.
29 coccínea. R.S. (*sc.*) bright-scarlet. W. Indies. 1713. 6. 9. S. ⊙. ◡. Andr. reposit. t. 99.
30 lutèola. R.S. (*or.*) common scarlet. Guatimala. 1759. —— H. ⊙. ◡. Bot. mag. t. 221.
coccínea. B.M. *nec aliorum.*
31 angulàris. R.S. (*cr.*) angular-leaved. E. Indies. 1806. —— S. ⊙. ◡.
phœnícea. F.I.
32 hastígera. K.S. (*fl.*) halbert-bearing. Mexico. 1824. 6. 8. S. ♃. ◡.
33 longiflòra. K.S. (*wh.*) long-flowered. Cuba. 1803. —— S. ♃. ◡.
34 sagittàta. R.S. (*ro.*) sagittate. S. Europe. 1826. 7. 9. H. ♃. ◡. Cavan. icon. 2. t. 107.
35 bòna nóx. R.S. (*wh.*) large white. America. 1773. 6. 9. G. ⊙. ◡. Bot. mag. t. 752.
36 lacunòsa. R.S. (*wh.pu.*) starry. N. America. 1640. 7. 8. H. ⊙. ◡. Dill. elth. t. 87. f. 102.
37 Batàtas. R.S. (*wh.pu.*) sweet potatoe. India. 1597. .... S. ♃. Rheed. mal. 7. t. 50.
Convólvulus Batàtas. W.
38 Turpèthum. R.S. (*wh.*) square-stalked. Ceylon. 1752. 3. 6. S. ♃. ◡. Bot. reg. t. 279.
39 chryseídes. B.R. (*ye.*) golden. China. 1817. 6. 10. S. ♃. ◡. ——— t. 270.
40 ochràcea. B.R. (*ye.*) pale yellow. Benin. 1826. 8. 10. S. ⊙. ◡. ——— t. 1060.
41 Jálapa. (*ro.*) true Jalap. S. America. 1733. 8. 11. S. ♄. ◡. Bot. mag. t. 1572.
Jálapa. β ròsea. B.R. 621. *Convólvulus Jálapa.* B.M.
42 Micháuxii. (*wh.bh.*) Michaux's. Georgia. 1815. —— G. ♃. ◡. Bot. reg. t. 342.
Jálapa. α. B.R. *macrorhìza.* M. *non* R.S.
43 sanguínea. R.S. (*sc.*) blood-flowered. W. Indies. 1812. 2. 11. S. ♄. ◡. Bot. reg. t. 9.
44 repánda. R.S. (*sc.*) scolloped. ——— 1793. 5. 10. S. ♄. ◡. Sal. par. lond. t. 81.
45 vària. R.S. (*bl.*) variable-leaved. S. America. 1816. —— S. ♃. ◡.
Convólvulus pubéscens. S.S.
46 scábra. R.S. (*wh.*) rough. Egypt. 1804. —— H. ⊙. ◡.
47 scábrida. R.S. (*wh.*) roughish. S. America. 1824. —— H. ⊙. ◡.
48 hederàcea. B.R. (*bl.*) deep-lobed. N. America. 1729. 8. 10. H. ⊙. ◡. Bot. reg. t. 85.
barbàta. Roth.
49 cuspidàta. R.S. (*pu.*) sharp-pointed. Peru. 1732. 6. 7. H. ⊙. ◡. Flor. per. 2. t. 119. f. a.
50 Dillènii. R.S. (*bl.*) Dillenius's. Guinea. .... —— S. ⊙. ◡. Dill. elth. t. 81. f. 93.
51 Níl. P.S. (*bl.*) blue. S. America. 1597. 7. 9. S. ⊙. ◡. Bot. mag. t. 188.
52 cœrùlea. B.R. (*bl.*) East Indian. E. Indies. 1815. —— S. ⊙. ◡. Bot. reg. t. 276.
53 cœruléscens. F.I. (*bl.*) light blue. ——— 1820. 6. 8. S. ♃. ◡.
54 pudibúnda. B.R. (*ro.*) modest. W. Indies. 1825. 7. 10. S. ⊙. ◡. Bot. reg. t. 999.
55 trilòba. R.S. (*vi.*) three-lobed. ——— 1752. 6. 7. S. ⊙. ◡.
56 hederifòlia. R.S. (*vi.*) Ivy-leaved. S. America. 1773. 7. H. ⊙. ◡. Plum. ic. t. 93. f. 2.
57 hepaticifòlia. R.S. (*bl.*) Hepatica-leav'd. E. Indies. 1759. 8. 9. S. ⊙. ◡. Burm. ind. t. 20. f. 2.
58 punctàta. R.S. (*vi.*) spotted-calyx'd. S. America. .... 7. 11. S. ⊙. ◡. Dill. elth. t. 83. f. 96.
59 mutábilis. B.R. (*bl.*) changeable. ——— 1812. 5. 9. S. ♄. ◡. Bot. reg. t. 39.
60 acuminàta. R.S. (*pu.*) taper-pointed. ——— .... —— S. ♃. ◡.
61 commutàta. R.S. (*li.*) hairy-capsuled. N. America. .... —— H. ⊙. ◡. Dill. elth. t. 84. f. 98.
carolìnus. Ph. *Convólvulus carolìnus.* M. *non* L.
62 gemélla. R.S. (*wh.*) twin-flowered. E. Indies. 1827. 7. 9. S. ♃. ◡. Burm. ind. t. 21. f. 1.
63 gangética. (*pi.*) Ganges. ——— 1812. 5. 9. S. ♄. ◡.
Convólvulus gangéticus. F.I.
64 semidig'yna. F.I. (*wh.*) cleft-styled. ——— 1826. .... S. ♄. ◡.
65 muricàta. R.S. (*pu.*) rough. ——— 1777. 7. 8. S. ⊙. ◡. Jacq. schœn. 3. t. 323.
bòna nóx β purpuráscens. B.R. 290. *Convólvulus muricàtus.* W.
66 pseudomuricàta. L.en. (*pu.*) rough-stem'd. ——— 1827. 7. 9. S. ♄. ◡.
67 cárnea. R.S. (*fl.*) flesh-coloured. S. America. 1799. 8. 9. S. ⊙. ◡. Jacq. amer. 26. t. 18.
68 involucràta. R.S. (*pu.*) involucred. Sierra Leon. 1822. 8. 10. S. ⊙. ◡. Beauv. fl. d'ow. 2. t. 89.
69 barbígera. B.F.G. (*bl.*) bearded-calyx'd. America. 1824. 7. 10. H. ⊙. ◡. Swt. br. fl. gar. t. 86.
70 purpùrea. R.S. (*pu.*) purple. ——— 1629. 6. 9. H. ⊙. ◡. Bot. mag. t. 113.
β incarnàta. (*fl.*) *flesh-coloured.* ——— —— —— H. ⊙. ◡.
γ leucántha. (*wh.*) *white-flowered.* ——— —— —— H. ⊙. ◡.
δ vària. B.M. (*std.*) *striped-flowered.* ——— —— —— H. ⊙. ◡. Bot. mag. t. 1682.
ε elàtior. B.M. (*wh.bl.*) *spotted-flowered.* ——— —— —— H. ⊙. ◡. ——— t. 1005.
discolor. Jacq.

71	violàcea. R.S.	(*vi.*) purple-flower'd.	S. America.	1732.	8. 9.	S.⊙.⌣.	Plum. ic. t. 93. f. 1.
72	corymbòsa. R.S.	(*wh.*) corymbose.	India.	1823.	——	S.♄.	———— t. 89. f. 2.
73	campanulàta. R.S.	(*pu.w.*) bell-flowered.	E. Indies.	1800.	——	S.⊙.	Rheed. mal. 2. t. 56.
74	polyánthes. R.S.	(*ye.gr.*) umbel-flowered.	W. Indies.	1739.	6. 7.	S.♃.⌣.	Plum. amer. 88. t. 102.
	Convólvulus umbellàtus. W.						
75	multiflòra. R.S.	(*pu.*) many-flowered.	Jamaica.	....	——	S.♃.⌣.	Pluk. phyt. t. 167. f. 1.
76	malabárica. R.S.	(*pu.st.*) Malabar.	E. Indies.	1823.	....	S.♄.⌣.	Rheed. mal. 2. t. 51.
77	solanifòlia. R.S.	(*ro.*) Nightshade-ld.	America.	1759.	7. 8.	S.♂.⌣.	Plum. ic. 94. f. 1.
78	leucántha. R.S.	(*wh.*) white-flowered.	S. America.	1823.	8. 10.	H.⊙.⌣.	Jacq. ic. 2. t. 318.
79	obscùra. R.S.	(*wh.pu.*) hairy.	E. Indies.	1732.	6. 8.	S.⊙.⌣.	Bot. reg. t. 239.
80	fastigiàta.	(*pu.*) crowded.	————	1816.	——	S.♃.⌣.	
	Convólvulus fastigiàtus. F.I.						
81	máxima.	(*pu.re.*) great Ceylon.	Ceylon.	1799.	7. 8.	S.♄.⌣.	Rheed. mal. 2. t. 53.
82	hirsùtula. J.E.	(*wh.*) hairy-stemmed.	S. America.	1828.	——	S.⊙.	Jacq. f. eclog. t. 44.
83	blánda.	(*wh.*) fair-flowered.	————	1820.	....	S.♄.⌣.	
	Convólvulus blándus. F.I.						
84	laurifòlia.	(*ro.*) Laurel-leaved.	————	1822.	5. 8.	S.♄.⌣.	
	Convólvulus laurifòlius. F.I.						
85	sphærocéphala.	(*ro.*) round-headed.	————	1816.	8. 10.	S.⊙.⌣.	
	Convólvulus sphærocéphalus. F.I.						
86	sibírica. R.S.	(*wh.*) Siberian.	Siberia.	1779.	7. 8.	H.⊙.⌣.	Pall. it. 3. p. 723. t. K.
87	Hardwíckii. S.S.	(*wh.*) fringe-calyxed.	E. Indies.	1816.	5. 8.	S.♃.⌣.	
	Convólvulus calycìnus. F.I. *non* Kth.						
88	tamnifòlia. R.S.	(*bl.*) Tamus-leaved.	Carolina.	1732.	7.	H.⊙.⌣.	Dill. elth. t. 318. f. 410.
89	capitàta. R.S.	(*bl.*) headed.	E. Indies.	1823.	7. 10.	S.♃.⌣.	
90	triquètra. R.S.	(*pu.*) triquetrous.	Santa Cruz.	....	——	S.♄.	
91	pilòsa.	(*pi.*) various-ld. hairy.	E. Indies.	1815.	——	S.⊙.⌣.	
	Convólvulus pilòsus. F.I.						
92	denticulàta. B.P.	(*ye.*) denticulate.	————	1778.	7. 8.	S.⊙.⌣.	Bot. reg. t. 317.
	Convólvulus Mèdium. W. *non* L.						
93	bícolor.	(*ye.pu.*) two-coloured.	————	1815.	——	S.♃.⌣.	
	Convólvulus bícolor. F.I.						
94	atropurpùrea. F.I.	(*d.pu.*) dark purple.	Nepaul.	1824.	7. 9.	H.⊙.⌣.	
95	dentàta. F.I.	(*ye.*) toothed-lobed.	E. Indies.	1820.	——	S.♃.⌣.	
96	digitifòlia.	(*pu.*) finger-leaved.	————	1815.	6. 8.	S.♄.⌣.	
	Convólvulus digitàtus. F.I.						
97	polytricha.	(*wh.*) many-haired.	————	1826.	6. 10.	S.⊙.⌣.	
	Convólvulus hirsùtus. F.I.						
98	heptaphylla. F.I.	(*pi.*) seven-leaved.	————	1827.	....	S.♄.⌣.	
99	bogoténsis. K.S.	(*pu.*) Bogota.	S. America.	1820.	5. 8.	S.♄.⌣.	
100	sidæfòlia. K.S.	(*wh.*) Sida-leaved.	————	1812.	——	S.♄.⌣.	
101	albivènia.	(*wh.*) white-veined.	Algoa Bay.	1825.	9. 10.	S.♄.⌣.	Bot. reg. t. 1116.
	Convólvulus albivènius. B.R.						
102	scrobiculàta. B.R.	(*wh.pu.*) furrowed.	Trinidad.	1822.	3. 6.	S.♄.⌣.	Bot. reg. t. 1076.
103	grandiflòra. R.S.	(*wh.*) great-flowered.	E. Indies.	1802.	9.	S.♄.⌣.	Andr. reposit. t. 188.
104	noctilùca. B.R.	(*wh.*) night-flowered.	————	1811.	9. 10.	S.♄.⌣.	Bot. reg. t. 889.
105	latiflòra. R.S.	(*wh.*) broad-flowered.	W. Indies.	....	——	S.♄.⌣.	Jacq. vind. 3. t. 69.
106	filifórmis. R.S.	(*pu.*) filiform.	————	1823.	——	S.♄.⌣.	Jacq. amer. pict. t. 26.
107	salicifòlia. F.I.	(*wh.*) Willow-leaved.	E. Indies.	1816.	5. 7.	S.♂.⌣.	
108	Roxbúrghii.	(*ro.*) Roxburgh's.	————	1817.	——	S.♄.⌣.	
	multiflòra. F.I. *nec aliorum.*						
109	sepiària. F.I.	(*ro.*) hedge.	————	——	——	S.♄.⌣.	Rheed. mal. 11. t. 53.
110	panduràta. B.R.	(*w.re.*) Virginian.	N. America.	1732.	6. 9.	G.♃.⌣.	Bot. reg. t. 588.
	Convólvulus panduràtus. B.M. 1939.						
111	cándicans.	(*wh.*) hoary.	————	....	6. 8.	G.♃.⌣.	Bot. mag. t. 1603.
	Convólvulus cándicans. B.M.						
112	glaucifòlia. R.S.	(*fl.*) glaucous-leaved.	Mexico.	1732.	5. 7.	G.♃.⌣.	Dill. elth. t. 87. f. 101.
113	réptans. R.S.	(*li.*) creeping.	E. Indies.	1806.	——	S.♃.	Rumph. am. 5. t. 155. f. 1.
114	angustifòlia. R.S.	(*ye.*) narrow-leaved.	————	1800.	7. 8.	S.⊙.⌣.	Jacq. ic. rar. t. 317.
115	tridentàta. R.S.	(*pu.st.*) trifid.	————	1778.	——	S.⊙.⌣.	Rheed. mal. 2. t. 65.
116	stáns. R.S.	(*vi.*) Mexican.	Mexico.	1824.	——	S.♃.⌣.	Cavan. icon. 3. t. 250.
117	bignonioídes. B.M.	(*d.pu.*) trumpet-flower'd.	Cayenne.	1823.	——	S.♃.⌣.	Bot. mag. t. 2645.
118	sagittifòlia. B.R.	(*ro.*) arrow-leaved.	Carolina.	1819.	6. 9.	G.♃.⌣.	Bot. reg. t. 437.
	Convólvulus sagittifòlius. R.S.						
119	vitifòlia.	(*ye.*) Vine-leaved.	E. Indies.	1820.	——	S.♄.⌣.	Burm. ind. t. 18. f. 1.
	Convólvulus vitifòlius. R.S.						
120	stipulàcea.	(*wh.*) stipuled.	————	1805.	8. 10.	S.♃.⌣.	
	Convólvulus stipulàceus. F.I.						
121	pentagòna. F.I.	(*wh.*) five-angled.	————	1824.	5. 8.	S.♃.	
122	cæspitòsa. F.I.	(*st.*) tufted.	————	1827.	——	S.♂.	

123 Beladámboe. R.S. (*wh.*) whipcord. E. Indies. 1809. 7. 9. S. ♃. Rheed. mal. 11. t. 58.
Convólvulus flagellifórmis. F.I.
124 renifórmis. F.I. (*ye.*) kidney-leaved. ——— 1825. —— S. ♃.
125 bilobàta. F.I. (*pu.*) two-lobed. ——— 1809. 6. 7. S. ♃. Rumph. amb. 5. t. 159. f. 1
126 brasiliénsis. (*pu.*) three-flowered. Brazil. 1726. —— S. ♃. Plum. amer. 89. t. 104.
Convólvulus brasiliénsis. W.
127 pés-càpræ. (*pu.*) single-flowered. E. Indies. 1770. —— S. ♃. Bot. reg. t. 319.
marítima. B.R. *Convólvulus pés-càpræ.* F.I.
128 arboréscens. (*wh.pu.*) tree. Mexico. 1818. .... S. ♄.

ARGYR'EIA. S.S. ARGYR'EIA. Pentandria Monogynia. S.S.
1 spléndens. (*ro.*) glossy-leaved. E. Indies. 1814. 11. 1. S. ♄. ◡. Bot. mag. t. 2628.
Ipom'œa spléndens. B.M. *Lettsòmia spléndens.* F.I.
2 speciòsa. (*pu.*) broad-leaved. E. Indies. 1778. 7. 9. S. ♄. ◡ Bot. mag. t. 2446.
Ipom'œa speciòsa. B.M. *Convólvulus speciòsus.* W. *Lettsòmia nervòsa.* F.I.
3 argéntea. (*pi.*) silvery-leaved. E. Indies. 1820. .... S. ♄. ◡.
Lettsòmia argéntea. F.I.
4 setòsa. F.I. (*pi.*) bristly-stemmed. ——— ——— .... S. ♄. ◡.
5 cuneàta. B.R. (*d.pu.*) wedge-leaved. ——— 1817. 6. 9. S. ♄. ◡. Bot. reg. t. 661.
Lettsòmia cuneàta. F.I. *Ipom'œa atrosanguínea.* B.M. 2170.
6 cymòsa. (*li.*) cyme-flowered. E. Indies. 1823. .... S. ♄. ◡.
Lettsòmia cymòsa. F.I.
7 pomàcea. F.I. (*pi.*) Apple-fruited. ——— 1828. .... S. ♄. ◡.
8 bòna-nóx. F.I. (*wh.*) night-flowering. ——— 1799. 6. 9. S. ♄. ◡.
9 uniflòra. F.I. (*wh.*) kidney-leaved. ——— 1817. .... S. ♄. ◡.
10 ornàta. F.I. (*wh.*) large white-flower'd. ——— 1824. .... S. ♄.
11 virgàta. .... slender-twigged. ——— 1825. .... S. ♄. ◡.

DIN'ETUS. B.F.G. DIN'ETUS. Pentandria Monogynia.
1 racemòsus. B.F.G. (*wh.*) cluster-flower'd. Nepaul. 1823. 7. 11. H. ⊙. ◡. Swt. br. fl. gar. t. 127.
Poràna racemòsa. F.I.
2 paniculàtus. (*wh.*) panicled. E. Indies. ——— .... S. ♄. ◡. Roxb. corom. 3. t. 235.
Poràna paniculàta. F.I.

POR'ANA. W. POR'ANA. Pentandria Digynia.
volùbilis. W. (*wh.*) climbing. E. Indies. 1823. .... S. ♄. ◡. Burm. ind. t. 21. f. 1.

EVO'LVULUS. S.S. EVO'LVULUS. Pentandria Digynia. R.S.
1 nummulàrius. R.S. (*bl.*) round-leaved. Jamaica. 1816. 10. S. ⊙. Jacq. amer. t. 260. f. 23.
2 gangéticus. R.S. (*bl.*) Ganges. E. Indies. 1823. —— S. ⊙.
3 emarginàtus. R.S. (*bl.*) kidney-leaved. ——— 1816. —— S. ⊙. Burm. ind. t. 30. f. 1.
4 alsinoídes. R.S. (*bl.*) chickweed-leaved. ——— 1733. 6. 7. S. ♃. Rheed. mal. 11. t. 64.
5 hirsùtus. R.S. (*pa.bl.*) hairy. Cumana. 1825. 6. 8. S. ♃. Lam. ill. t. 216. f. 2.
6 lanceolàtus. Poi. (*pa.bl.*) spear-leaved. S. America. 1818. —— S. ♃.
7 linifòlius. R.S. (*bl.*) flax-leaved. Jamaica. 1782. 8. 9. S. ⊙. Brown. jam. t. 10. f. 2.
8 seríceus. R.S. (*wh.*) silky. ——— 1827. 8. 10. S. ⊙.
9 incànus. R.S. (*bl.*) hoary. Peru. 1824. —— S. ♃.
10 Nuttalliànus. R.S. (*ye.*) Nuttall's. Missouri. 1821. 5. 6. H. ♃.
argénteus. N.
11 latifòlius. B.R. (*wh.*) broad-leaved. Brazil. 1819. 6. 7. S. ♃. Bot. reg. t. 401.

CRE'SSA. R.S. CRE'SSA. Pentandria Digynia. L.
crética. R.S. (*wh.*) Cretan. Levant. 1822. 7. 9. H. ⊙. Flor. græc. t. 256.

DICHO'NDRA. R.S. DICHO'NDRA. Pentandria Digynia. L.
1 répens. R.S. (*wh.*) creeping. N. S. W. 1803. 6. 8. G. ♃. Sm. ic. ined. 1. t. 8.
2 rotundifòlia. L.en. (*wh.*) round-leaved. Persia. 1816. —— H. ♃.
3 caroliniàna. R.S. (*gr.*) Carolina. Carolina. 1812. —— G. ♃.
4 serícea. S.S. (*wh.*) silky. W. Indies. 1793. —— S. ♃. Swartz. ic. t. 10.
5 argéntea. R.S. (*wh.*) silvery. New Spain. 1818. —— S. ♃. Willd. hort. ber. t. 81.

FA'LKIA. R.S. FA'LKIA. Pentandria Digynia. L.
répens. R.S. (*wh.*) creeping. C. B. S. 1774. 5. 8. G. ♄. Andr. reposit. t. 257.

CUSC'UTA. R.S. DODDER. Pentandria Digynia. R.S.
1 europ'æa. R.S. (*wh.*) great. England. .... 8. 9. H. ⊙. ◡. Eng. bot. v. 6. t. 378.
2 Epíthymum. R.S. (*wh.*) lesser. Britain. .... 7. 10. H. ♃. ◡. ——— v. 1. t. 55.
3 epilìnum. S.S. (*wh.*) Flax. Germany. 1821. —— H. ⊙. ◡.
4 monog'yna. R.S. (*wh.*) one-styled. Levant. ——— —— H. ⊙. ◡. Flor. græc. t. 257.
5 lupulifórmis. R.S. (*wh.*) rough dotted. Silesia. 1824. —— H. ⊙. ◡. Krock. siles. t. 36.
6 macrocárpa. Led. (*wh.*) large-podded. Siberia. 1827. —— H. ⊙. ◡.
7 chinénsis. R.S. (*wh.*) Chinese. China. 1803. —— G. ⊙. ◡.
8 austràlis. B.P. (*wh.*) New Holland. N. Holland. 1818. —— G. ♃. ◡.
9 chilénsis. B.R. (*wh.*) Chili. Chili. 1821. 1. 12. G. ♃. ◡. Bot. reg. t. 603.
10 odoràta. R.S. (*wh.*) sweet-scented. Lima. 1820. —— H. ♃. ◡.
11 verrucòsa. B.F.G. (*wh.*) warted-calyxed. Nepaul. 1822. 9. 11. H. ♃. ◡. Swt. br. fl. gar. t. 6.
12 Hookèri. (*wh.*) Hooker's. E. Indies. 1823. —— H. ⊙. Hook. ex. flor. t. 150.
refléxa, var. *verrucòsa,* H.E.F. *excl. synon.*

ORDO CXLIX.

BORAGINEÆ. *Brown prodr.* 1. *p.* 492.

TIARI'DIUM. S.S.		TIARI'DIUM.	Pentandria Monogynia. S.S.				
1 índicum. S.S.	(*li.*)	Indian.	W. Indies.	1713.	6. 8.	S.⊙.	Bot. mag. t. 1837.
Heliotròpium índicum. B.M.							
2 velutìnum. R.S.	(*wh.*)	villous-stalked.	E. Indies.	....	——	S.⊙.	Rheed. mal. 10. t. 48.
HELIOTR'OPIUM. S.S.		TURNSOLE.	Pentandria Monogynia. L.				
1 peruviànum. S.S.	(*wh.*)	Peruvian.	Peru.	1757.	5. 9.	G.♄.	Bot. mag. t. 141.
β *hy'bridum.*	(*wh.*)	*More's hybrid.*	Hybrid.	1815.	——	G.♄.	
2 corymbòsum. S.S.	(*bl.*)	large-flowered.	———	1808.	——	G.♄.	Bot. mag. t. 1609.
grandiflòrum. L.en.							
3 maroccànum. S.S.	(*wh.*)	Morocco.	Morocco.	1821.	——	G.♄.	
4 parviflòrum. S.S.	(*wh.*)	small-flowered.	India.	1732.	7. 9.	S.♂.	Dill. elth. t. 146. f. 175.
5 europ'æum. S.S.	(*wh.*)	European.	S. Europe.	1562.	6. 10.	H.⊙.	Jacq. aust. 3. t. 207.
6 commutàtum. R.S.	(*wh.*)	smaller.	———	....	——	H.⊙.	
7 oblongifòlium. L.en.	(*w.*)	oblong-leaved.	———	1827.	——	H.⊙.	
8 suavèolens. S.S.	(*wh.*)	sweet-scented.	Caucasus.	1823.	——	H.⊙.	
9 coromandeliànum. W.	(*w.*)	Coromandel.	E. Indies.	1812.	7. 10.	S.♂.	
10 curassávicum. S.S.	(*wh.*)	glaucous.	S. America.	1731.	6. 7.	S.♂.	Bot. mag. t. 2669.
11 chenopodioídes. W.en.	(*w.*)	smooth-leaved.	———	1820.	——	S.♂.	
12 anisophy'llum. B.G.	(*w.*)	unequal-leaved.	Africa.	1822.	——	S.♂.	Beau. fl. d. ow. 2. t. 94.
13 zeylánicum. R.S.	(*wh.*)	Ceylon.	Ceylon.	1815.	——	S.♂.	Burm. ind. t. 16. f. 2.
14 linifòlium. R.S.	(*wh.*)	flax-leaved.	C. B. S.	——	——	G.♄.	
Myosòtis fruticòsa. W.							
15 demíssum. R.S.	(*wh.*)	dwarf.	S. America.	1752.	5. 6.	S.♄.	Plum. ic. 227. f. 2.
Tournefórtia hùmilis. W.							
16 supìnum. S.S.	(*wh.*)	trailing.	S. Europe.	1640.	6. 7.	H.⊙.	Flor. græc. t. 157.
LITHOSPE'RMUM. S.S.		GROMWELL.	Pentandria Monogynia. L.				
1 fruticòsum. S.S.	(*re.*)	shrubby.	S. Europe.	1683.	5. 6.	H.♄.	Flor. græc. t. 161.
2 prostràtum. S.S.	(*pu.*)	prostrate.	———	1825.	——	H.♄.	Lois. fl. gall. add. 1. t. 4.
3 seríceum. S.S.	(*st.*)	silky.	N. America.	1824.	——	H.♃.	
4 divaricàtum. S.S.	(*st.*)	spreading.	Africa.	1829.	7. 9.	H.♃.	
5 hispídulum. S.S.	(*bl.*)	bristly.	Greece.	1828.	——	F.♄.	
6 arvénse. S.S.	(*st.*)	corn.	Britain.	....	5. 6.	H.⊙.	Eng. bot. v. 2. t. 123.
7 officinàle. S.S.	(*ye.*)	officinal.	———	....	5. 8.	H.♃.	——— v. 2. t. 134.
8 latifòlium. S.S.	(*st.*)	broad-leaved.	America.	1816.	——	H.♃.	
9 dístichum. S.S.	(*st.*)	two-spiked.	Mexico.	1806.	——	G.♃.	Jacq. fragm. t. 48. f. 3.
10 orientàle. S.S.	(*ye.*)	yellow-flower'd.	Levant.	1713.	——	H.♃.	Bot. mag. t. 515.
11 lineàtum. S.S.	(*st.*)	lined.	....	1816.	——	H.♃.	
12 ápulum. S.S.	(*ye.*)	small.	S. Europe.	1768.	6. 7.	H.⊙.	Flor. græc. t. 158.
13 crassifòlium. S.S.	(*st.*)	thick-leaved.	Levant.	1825.	5. 8.	H.♃.	
14 decúmbens. S.S.	(*st.*)	decumbent.	Persia.	——	7. 9.	H.⊙.	Vent. cels. t. 37.
15 villòsum. S.S.	(*pu.*)	villous.	S. Europe.	1820.	6. 7.	H.♃.	
16 tinctòrium. R.S.	(*st.*)	dyer's.	———	1596.	6. 10.	H.♃.	
17 purpùreo-cœrùleum.	(*bl.*)	creeping.	England.	....	5. 6.	H.♃.	Eng. bot. v. 2. t. 117.
18 graminifòlium. S.S.	(*bl.*)	grass-leaved.	S. Europe.	1826.	——	H.♄.	Viv. frag. ital. t. 5.
19 tenuiflòrum. S.S.	(*wh.*)	slender-flower'd.	Egypt.	1796.	——	H.⊙.	Jacq. ic. 2. t. 313.
20 dispérmum. S.S.	(*li.*)	two-seeded.	Spain.	1799.	6. 7.	H.⊙.	Linn. dec. 1. t. 7.
21 davùricum. S.S.	(*bl.*)	Daurian.	Davuria.	1812.	5. 6.	H.♃.	Swt. br. fl. gar. t. 121.
Pulmonària davùrica. B.M. 1743.							
22 denticulàtum. S.S.	(*pu.*)	toothed.	N. America.	1801.	——	H.♃.	
23 sibíricum. S.S.	(*pu.*)	Siberian.	Siberia.	——	6. 7.	H.♃.	Gmel. sib. 4. n. 15. t. 39.
24 marítimum. S.S.	(*bl.*)	sea-side.	Britain.	....	——	H.♃.	Eng. bot. v. 6. t. 368.
Pulmonària marítima. E.B.							
25 parviflòrum.	(*bl.*)	small-flowered.	N. America.	1824.	——	H.♃.	
26 púlchrum. S.S.	(*bl.*)	Virginian.	———	1699.	3. 5.	H.♃.	Bot. mag. t. 160.
Pulmonària virgínica. B.M.							
27 paniculàtum. S.S.	(*pu.*)	panicled.	Hudson's Bay.	1778.	5. 6.	H.♃.	Bot. reg. t. 146.
Pulmonària paniculàta. B.R.							
28 lanceolàtum.	(*bl.*)	margined.	Louisiana.	1813.	——	H.♃.	
Pulmonària lanceolàta. Ph. *marginàta.* N.							
29 simplicíssimum. S.S.	(*bl.*)	simple-stalked.	Siberia.	1829.	——	H.♃.	
BA'TSCHIA. M.		BA'TSCHIA.	Pentandria Monogynia.				
1 Gmelìni. M.	(*ye.*)	Carolina.	Carolina.	1812.	5. 7.	H.♃.	
caroliniàna. R.S. *Lithospérmum hírtum.* S.S.							

2 canéscens. M.	(ye.)	canescent.	N. America.	....	5. 7.	H. ♃.	Mich. amer. 1. t. 14.
3 longiflòra. N.	(ye.)	long-flowered.	Missouri.	1812.	——	H. ♃.	
4 decúmbens. N.	(ye.)	decumbent.	N. America.	1826.	——	H. ♃.	
5 conspícua. B.	(ye.)	conspicuous.	————	1827.	——	H. ♃.	
PULMON'ARIA. S.S.		LUNGWORT.	Pentandria Monogynia. L.				
1 azùrea. S.S.	(bl.)	azure.	Hungary.	1823.	5.	H. ♃.	Moris.his.s.11.t.29.f.5.
2 angustifòlia. S.S.	(pu.)	narrow-leaved.	Britain.	....	4. 5.	H. ♃.	Eng. bot. v. 23. t. 1628.
3 tuberòsa. S.S.	(pu.)	tuberous-rooted.	Hungary.	1824.	5.	H. ♃.	
4 móllis. S.S.	(pu.)	soft-leaved.	————	1816.	——	H. ♃.	Swt. br. fl. gar. ic.
5 officinàlis. R.S.	(pu.)	common.	England.	....	——	H. ♃.	Eng. bot. v. 2. t. 118.
6 saccharàta. L.en.	(wh.)	white-flowered.	Germany.	1816.	——	H. ♃.	Moris. s. 11. t. 29.
7 grandiflòra. R.S.	(pu.)	large-flowered.	....	....	——	H. ♃.	DC. ic. ined. t. 64.
8 oblongàta. Schr.	(pu.)	oblong-leaved.	Germany.	1820.	——	H. ♃.	
ONOSM'ODIUM. M.		ONOSM'ODIUM.	Pentandria Monogynia. R.S.				
1 híspidum. R.S.	(st.)	hispid.	N. America.	1759.	6.	H. ♃.	Moris. s.11. t. 28. f. 3.
2 mólle. R.S.	(wh.)	soft.	————	1812.	6. 8.	H. ♃.	Mich. amer. t. 15.
ONO'SMA. R.S.		ONO'SMA.	Pentandria Monogynia. L.				
1 echioídes. S.S.	(ye.)	hairy.	S. Europe.	1683.	3. 6.	H. ♃.	Flor. græc. t. 172.
2 gigánteum. S.S.	(ye.)	gigantic.	Tauria.	1818.	——	H. ♃.	
3 montànum. R.S.	(ye.)	mountain.	Levant.	1827.	——	H. ♃.	Colum.ecphr p.183.f.2.
4 arenàrium. R.S.	(ye.)	sand.	Hungary.	1804.	3. 7.	H. ♃.	W.et K. hung. 3. t. 279.
5 stellulàtum. W.K.	(ye.)	starry-haired.	————	1816.	——	H. ♃.	———— 2. t. 173.
6 taùricum. B.M.	(go.)	golden-flower'd.	Tauria.	1801.	4. 6.	F. ♃.	Bot. mag. t. 889.
7 tinctòrium. R.S.	(st.)	dyer's.	Volga.	1828.	——	H. ♃.	
8 rupéstre. R.S.	(ye.)	rock.	Caucasus.	1823.	——	H. ♃.	
9 eréctum. R.S.	(ye.)	upright.	Crete.	1828.	——	F. ♃.	Flor. græc. t. 173.
10 orientàle. S.S.	(st.)	oriental.	Levant.	1752.	5. 6.	G. ♃.	
11 divaricàtum. S.S.	(st.)	spreading.	Caucasus.	1818.	4. 6.	H. ♃.	Pall. it. 2. app. t. 50.
12 simplicíssimum. R.S.	(st.)	linear-leaved.	Siberia.	1768.	——	H. ♃.	Bot. mag. t. 2248.
13 trinérvium. S.S.	(st.)	three-nerved.	Mexico.	1825.	——	G. ♃.	
14 seríceum. R.S.	(ye.)	silky-leaved.	Levant.	1752.	6. 7.	F. ♃.	
15 synanthèrum. F.	(ye.)	connected.	Altay.	1827.	——	H. ♃.	
MO'LTKIA. R.S.		MO'LTKIA.	Pentandria Monogynia.				
cœrùlea. R.S.	(bl.)	blue-flowered.	Persia.	1829.	4. 6.	H. ♃.	
SY'MPHYTUM. R.S.		COMFREY.	Pentandria Monogynia. L.				
1 officinàle. R.S.	(wh.)	common.	Britain.	....	5. 7.	H. ♃.	Eng. bot. v. 12. t. 817.
β purpuráscens.	(pu.)	*purplish.*	————	....	——	H. ♃.	
2 bohèmicum. P.S.	(re.)	red-flowered.	Bohemia.	....	——	H. ♃.	
3 caucásicum. R.S.	(bl.)	Caucasean.	Caucasus.	1816.	——	H. ♃.	
4 tuberòsum. R.S.	(wh.)	tuberous-root'd.	Scotland.	....	5. 10.	H. ♃.	Eng. bot. v. 23. t. 1502.
5 bulbòsum. S.S.	....	bulbous-rooted.	Levant.	1829.	5. 7.	H. ♃.	Reich. icon. t. 220.
6 orientàle. R.S.	(wh.)	eastern.	Turkey.	1752.	——	H. ♃.	Bot. mag. t. 1912.
7 bullàtum. S.S.	(wh.)	blistered.	Tauria.	1806.	——	H. ♃.	———— t. 1787.
taùricum. B.M.							
8 cordàtum. W.K.	(st.)	heart-leaved.	Hungary.	1813.	——	H. ♃.	W. et K. hung. 6. t. 7.
9 aspérrimum. R.S.	(bl.)	roughest.	Caucasus.	1799.	5. 9.	H. ♃.	Bot. mag. t. 929.
10 echinàtum. R.S.	(pu.)	bristly.	....	1824.	——	H. ♃.	
11 peregrìnum. S.S.	(wh.)	oblique-leaved.	Poland.	1816.	5. 7.	H. ♃.	
CERI'NTHE. R.S.		HONEYWORT.	Pentandria Monogynia. L.				
1 màjor. R.S.	(ye.pu.)	great.	S. Europe.	1596.	7. 8.	H. ⊙.	Bot. mag. t. 333.
2 áspera. R.S.	(pu.ye.)	rough.	————	1633.	——	H. ⊙.	Flor. græc. t. 170.
3 retórta. F.G.	(pu.ye.)	bent topped.	Levant.	1828.	——	H. ⊙.	———— t. 171.
4 mìnor. R.S.	(ye.)	small.	Austria.	1570.	6. 10.	H. ⊙.	Jacq. aust. 2. t. 124.
5 alpìna. R.S.	(pu.ye.)	alpine.	M.Carpathian.	1827.	——	H. ⊙.	
6 maculàta. R.S.	(ye.pu.)	spotted.	Tauria.	1804.	——	H. ♂.	
ECHIUM. R.S.		VIPER'S-BUGLOSS.	Pentandria Monogynia. L.				
1 fruticòsum. R.S.	(wh.bh.)	frutescent.	C. B. S.	1759.	5. 6.	G. ♄.	Bot. reg. t. 86.
2 híspidum. R.S.	(wh.)	hispid-leaved.	————	1816.	——	G. ♄.	
3 verrucòsum. R.S.	(wh.)	warted.	————	1820.	5. 7.	G. ♄.	
4 lævigàtum. R.S.	(wh.)	smooth-stalked.	————	1774.	6. 7.	G. ♄.	
5 glaucophy'llum. R.S.	(bl.)	glaucous-leaved.	————	1792.	6. 8.	G. ♄.	Jacq. ic. t. 312.
6 glábrum. R.S.	(wh.)	smooth.	————	1791.	5. 6.	G. ♄.	
7 paniculàtum. R.S.	(wh.)	panicled.	————	1815.	——	G. ♄.	
8 grandiflòrum. R.S.	(ro.)	great-flowered.	————	1787.	6. 7.	G. ♄.	Bot. reg. t. 124.
formòsum. P.S.							
9 ferocíssimum. A.R.	(bl.)	prickly-stalked.	————	1794.	——	G. ♄.	Andr. reposit. t. 39.
10 petr'æum. S.S.	(bl.)	rock.	Dalmatia.	1828.	——	H. ♂.	
11 calycìnum. R.S.	(bl.ye.)	large-calyxed.	S. Europe.	1829.	6. 8.	H. ⊙.	Viv. frag. ital. 1. t. 4.

12	spicàtum. R.S.	(*wh.*)	spiked dwarf.	C. B. S.	1799.	3. 5.	G. ♄.	
13	caudàtum. R.S.	(*re.*)	tailed.	————	1815.	6. 9.	G. ♄.	
14	capitàtum. R.S.	(*re.*)	headed.	————	1822.	——	G. ♄.	
15	sphærocephalon. S.S.	(*w.*)	round-headed.	————	1824.	——	G. ♄.	
16	argénteum. R.S.	(*bl.*)	silvery.	————	1789.	6. 7.	G. ♄.	Andr. reposit. t. 154.
17	lasiophy'llum. L.en.	(*w.*)	woolly-leaved.	....	1819.	——	G. ♄.	
18	Swártzii. R.S.	(*bl.*)	Swartz's.	C. B. S.	1816.	——	G. ♄.	
19	incànum. R.S.	(*bl.*)	hoary.	————	——	4. 6.	G. ♄.	
20	strigòsum. R.S.	(*vi.*)	strigose.	————	1819.	6. 7.	G. ♄.	
21	gigánteum. R.S.	(*wh.*)	gigantic.	Canaries.	1779.	7. 11.	G. ♄.	Vent. malm. t. 71.
22	aculeàtum. R.S.	(*st.*)	prickly.	————	1815.	5. 8.	G. ♄.	
23	mólle. R.S.	(*wh.*)	soft-leaved.	————	——	——	G. ♄.	
24	nervòsum. H.K.	(*bl.*)	sinewy-leaved.	Madeira.	1777.	6. 8.	G. ♄.	
25	bìfrons. DC.	(*bh.*)	two-fronted.	Canaries.	1815.	——	G. ♄.	
26	ambíguum. DC.	(*bh.*)	ambiguous.	————	——	5. 7.	G. ♄.	DC. ic. mons. ined. t. 20.
27	viréscens. DC.	(*pa.bl.*)	green-leaved.	————	——	——	G. ♄.	
28	cándicans. R.S.	(*bl.*)	hoary tree.	Madeira.	1777.	5. 6.	G. ♄.	Bot. reg. t. 44.
29	cynoglossoídes. R.S.	(*bl.*)	hound's-tongue-ld.	Canaries.	1816.	——	G. ♄.	
30	densiflòrum. DC.	(*li.*)	dense-flowered.	————	——	——	G. ♄.	DC. ic. m. ined. t. 22.
31	fastuòsum. H.K.	(*bl.*)	fastuous.	————	1779.	4. 8.	G. ♄.	
32	stríctum. R.S.	(*bl.*)	upright.	————	——	5. 12.	G. ♂.	Jacq. schœn. 1. t. 35.
33	lineàtum. R.S.	(*bh.*)	lined.	————	1815.	5. 6.	G. ♄.	Jacq. f. eclog. t. 42.
34	macrophy'llum. R.S.	(*bl.*)	large-leaved.	....	1812.	——	G. ♄.	
35	foliòsum. R.S.	(*wh.*)	leafy.	....	——	——	G. ♄.	
36	símplex. DC.	(*wh.*)	simple-stalked.	Canaries.	——	5. 12.	G. ♂.	DC. ic. m. ined. t. 21.
37	brachyánthum. R.S.	(*w.*)	short-flowered.	Madeira.	1820.	5. 7.	G. ♄.	
38	scábrum. R.S.	(*bl.*)	rough.	C. B. S.	——	——	G. ♄.	
39	plantagíneum. R.S.	(*pu.*)	Plantain-leav'd.	Italy.	1776.	7. 10.	H. ⊙.	Flor. græc. t. 179.
40	lusitánicum. R.S.	(*bl.*)	Portugal.	S. Europe.	1731.	7. 8.	H. ♃.	
41	plantaginoídes. R.S.	(*bl.*)	Plantain-like.	————	1825.	——	H. ⊙.	Jacq. vind. 1. t. 45.
42	prostràtum. R.S.	(*re.*)	prostrate.	Egypt.	1824.	——	H. ♃.	
43	macránthum. R.S.	(*vi.*)	large-flowered.	Barbary.	——	7. 10.	H. ⊙.	Desf. atl. 1. t. 46.
	grandiflòrum. Desf. *non* Andr.							
44	austràle. R.S.	(*re.pu.*)	oval-leaved.	S. Europe.	1824.	——	H. ⊙.	Swt. br. fl. gar. t. 101.
45	Símsii.	(*re.bl.*)	Sims's.	————.	1816.	——	H. ⊙.	Bot. mag. t. 1934.
	crèticum. B.M. *non* Flor. græc.							
46	marítimum. R.S.	(*bl.*)	sea-side.	Italy.	1815.	——	H. ⊙.	Bocc. mus. 2. t. 78.
47	itálicum. R.S.	(*wh.*)	white.	Jersey.	....	7. 8.	H. ♂.	Eng. bot. v. 29. t. 2081.
48	pyrenàicum. R.S.	(*bl.*)	Pyrenean.	Pyrenees.	1815.	——	H. ♂.	
49	aspérrimum. M.B.	(*bl.*)	roughest.	Caucasus.	1826.	——	H. ♂.	
50	thyrsoídeum. R.S.	(*bl.*)	bunch-flowered.	....	1823.	——	H. ♂.	
51	rùbrum. R.S.	(*re.*)	red-flowered.	Hungary.	1791.	——	H. ♃.	Bot. mag. t. 1822.
52	vulgàre. R.S.	(*bl.*)	common.	Britain.	....	——	H. ♂.	Eng. bot. v. 3. t. 181.
	β álbum.	(*wh.*)	*white-flowered.*	————	....	——	H. ♂.	
53	tuberculàtum. R.S.	(*bl.*)	tubercled.	S. Europe.	1826.	——	H. ♂.	
54	violàceum. R.S.	(*vi.bl.*)	violet-flowered.	Austria.	1658.	——	H. ⊙.	
55	parviflòrum. H.K.	(*bl.*)	small-flowered.	Barbary.	1798.	——	H. ⊙.	
	micránthum. R.S.							
56	ténue. S.S.	(*bl.*)	slender.	————	1824.	——	H. ⊙.	
57	Sibthórpii. R.S.	(*ro.*)	Sibthorp's.	Naples.	——	——	H. ♂.	Flor. græc. t. 181.
	híspidum. F.G. nec aliorum. *élegans.* Lehm.							
58	diffùsum. R.S.	(*re.*)	spreading.	Crete.	1827.	7. 9.	H. ⊙.	Flor. græc. t. 182.
59	créticum. R.S.	(*pu.*)	Cretan.	Levant.	1683.	——	H. ⊙.	———— t. 183.
60	orientàle. R.S.	(*pu.*)	oriental.	————	1780.	7. 8.	H. ⊙.	Tourn. it. 2. f. 107.
61	dahùricum. F.	(*bl.*)	Dahurian.	Dahuria.	1827.	——	H. ♂.	
62	angustifòlium. S.S.	(*bl.*)	narrow-leaved.	Spain.	1826.	——	H. ⊙.	
63	Lagascànum. R.S.	(*li.*)	Lagasca's.	————	——	——	H. ♃.	
64	salmánticum. R.S.	(*pu.*)	long-stamened.	————	1824.	——	H. ⊙.	
	N'ONEA. DC.		N'ONEA. Pentandria Monogynia. R.S.					
1	violàcea. DC.	(*vi.*)	violet-coloured.	S. Europe.	1686.	5. 7.	H. ⊙.	Moris. s. 11. t. 26. f. 11.
	Lycópsis vesicària. W.							
2	nígricans. DC.	(*d.pu.*)	black-flowered.	————	....	——	H. ⊙.	Zanon. hist. 56. t. 38.
3	púlla. DC.	(*d.pu.*)	dark-flowered.	Germany.	1648.	6. 7.	H. ♃.	Jacq. aust. 2. t. 188.
4	ciliàta. R.S.	(*ye.*)	fringed.	Levant.	1804.	——	H. ⊙.	
5	ròsea. L.en.	(*ro.*)	rose-coloured.	Caucasus.	1824.	——	H. ⊙.	
	Anchùsa ròsea. M.B. *Lycópsis ròsea.* S.S.							
6	setòsa. R.S.	(*ye.*)	bristly.	Iberia.	1826.	——	H. ⊙.	
7	versícolor.	(*bl.wh.*)	various-colored.	————	1820.	——	H. ⊙.	
	Anchùsa versícolor. R.S.							

8 lùtea. DC.	(*ye.*)	pale yellow.	S. Europe.	1805.	6. 7.	H.☉.	Nocc. hort. tran. t. 3.
Anchùsa lùtea. R.S. *Lycópsis lùtea.* S.S.							
9 pícta.	(*bl.std.*)	painted.	Tauria.	1818.	——	H.☉.	
Anchùsa pícta. M.B. *Lycópsis pícta.* S.S.							
LYCO'PSIS. L.		WILD BUGLOSS. Pentandria Monogynia. L.					
arvénsis. L.	(*bl.*)	common.	Britain.	....	5. 8.	H.☉.	Eng. bot. v. 14. t. 938.
ASPERUGO. L.		GERMAN MADWORT. Pentandria Monogynia. L.					
procumbens. L.	(*bl.*)	procumbent.	Britain.	.....	4. 5.	H.☉.	Eng. bot. v. 10. t. 661.
ANCH'USA. R.S.		BUGLOSS. Pentandria Monogynia. L.					
1 Barrelíeri. R.S.	(*bl.*)	Barrelier's.	Italy.	1815.	6. 7.	H.♃.	Barrel. ic. 333.
2 paniculàta. R.S.	(*bl.*)	panicled.	Levant.	1777.	6. 10.	H.♃.	Flor. græc. t. 163.
3 itálica. B.R.	(*bl.*)	Italian.	S. Europe.	1597.	——	H.♃.	Bot. reg. t. 483.
4 Millèri. R.S.	(*ro.*)	Miller's.	....	.....	——	H.☉.	
5 myosotidiflòra. S.S.	(*bl.*)	Myosotis-flower'd.	Caucasus.	1828.	——	H.♃.	
6 stylòsa. R.S.	(*bl.*)	long-styled.	Tauria.	1824.	——	H.☉.	MB. cent. pl. rar. t. 23.
7 pròcera. S.S.	(*bl.*)	tall.	Poland.	1822.	——	H.♃.	
8 lycopsoídes. L.en.	(*vi.*)	Lycopsis-like.	....	——	——	H.♃.	
9 variegàta. S.S.	(*bl.cr.*)	variegated.	Levant.	1683.	6. 7.	H.☉.	Flor. græc. t. 178.
Lycópsis variegàta. F.G.							
10 ovàta. S.S.	(*bl.*)	oval-leaved.	——	1796.	——	H.☉.	Buxb. cent. 5. t. 30.
Lycópsis orientàlis. R.S.							
11 tinctòria. R.S.	(*bl.*)	dyer's.	——	1596.	6. 10.	H.♃.	Flor. græc. t. 166.
12 parviflòra. R.S.	(*bl.*)	small-flowered.	Levant.	1827.	——	H.☉.	
13 aggregàta. S.S.	(*bl.*)	clustered.	——	——	——	H.☉.	Flor. græc. t. 107.
parviflòra. Flor. græc. *non* W.							
14 híspida. R.S.	(*vi.*)	hispid.	Egypt.	1823.	——	H.♃.	
15 tenélla. R.S.	(*wh.*)	slender.	China.	1816.	——	H.☉.	Jacq. f. ecl. t. 29.
zeylánica. Jacq.							
16 cæspitòsa. F.G.	(*bl.*)	tufted.	Levant.	1828.	——	H.♃.	Flor. græc. t. 169.
17 officinàlis. R.S.	(*vi.*)	common.	Britain.	....	——	H.♃.	Eng. bot. v. 10. t. 662.
18 incarnàta. Schr.	(*fl.*)	flesh-coloured.	Europe.	....	——	H.♃.	
19 undulàta. R.S.	(*pu.*)	wave-leaved.	S. Europe.	1739.	6. 8.	H.♃.	Bot. mag. t. 2119.
20 ochroleùca. R.S.	(*st.*)	pale-flowered.	Caucasus.	1810.	7. 8.	H.♃.	——— t. 1608.
21 angustifòlia. R.S.	(*bl.*)	narrow-leaved.	Europe.	1640.	5. 6.	H.♃.	Flor. græc. t. 164.
22 capénsis. R.S.	(*vi.*)	Cape.	C. B. S.	1800.	7. 8.	G.♂.	Andr. reposit. t. 336.
23 hy'brida. R.S.	(*pu.*)	villous.	Naples.	1818.	7. 10.	H.☉.	Ten. fl. neap. t. 11.
24 Gmelìni. S.S.	(*bl.*)	Gmelin's.	Siberia.	1829.	——	H.♃.	
25 alpéstris. S.S.	(*bl.*)	alpine.	Caucasus.	——	——	H.♃.	
26 Agárdhii. R.S.	(*pu.*)	tubercle-leav'd.	Siberia.	1823.	7. 9.	H.♃.	
27 verrucòsa. R.S.	(*wh.*)	warted.	Egypt.	——	7. 10.	H.☉.	Jacq. vind. 3. t. 21.
Asperùgo ægyptìaca. Jacq.							
28 rupéstris. B.P.	(*bl.*)	rock.	Siberia.	1802.	6. 8.	H.♃.	Pall. it. 3. ap. t. E. f. 3.
Myosòtis rupéstris. W.							
MYOS'OTIS. R.S.		SCORPION-GRASS. Pentandria Monogynia. L.					
1 palústris. E.B.	(*bl.*)	marsh.	Britain.	....	6. 9.	H.*w.*♃.	Eng. bot. v. 28. t. 1973.
scorpioídes. W.							
2 cæspitòsa. S.S.	(*bl.*)	tufted.	——	....	5. 7.	H.*w.*♃.	
3 intermèdia. S.S.	(*bl.*)	intermediate.	——	....	4. 6.	H.♂.	Fl. dan. 583. f. maj.
4 sylvática. R.S.	(*bl.ye.*)	wood.	——	....	6. 8.	H.♃.	Raii syn. t. 9. f. 2.
5 commutàta. R.S.	(*bl.*)	changed.	Europe.	....	——	H.♂.	
6 alpéstris. R.S.	(*bl.*)	rock.	Scotland.	....	——	H.♃.	Eng. bot. v. 36. t. 2559.
rupícola. E.B.							
7 lithospermifòlia. R.S.	(*bl.*)	Gromwell-leav'd.	Caucasus.	1823.	——	H.♃.	
8 suavèolens. R.S.	(*bl.*)	sweet-scented.	Europe.	——	——	H.♃.	
9 serícea. S.S.	(*bl.*)	silky-leaved.	Siberia.	1802.	——	H.♃.	
10 villòsa. S.S.	(*bl.*)	villous.	——	1828.	——	H.♃.	
11 sparsiflòra. R.S.	(*wh.*)	loose-flowered.	Europe.	1818.	——	H.☉.	
12 strícta. R.S.	(*bl.ye.*)	upright.	——	——	——	H.☉.	
13 arvénsis. E.B.	(*bl.*)	corn.	Britain.	....	6. 9.	H.☉.	Eng. bot. v. 36. t. 2558.
14 versícolor. E.B.	(*bl.ye.*)	various-colour'd.	——	....	4. 6.	H.☉.	——— t. 480. f. 1.
15 nàna. R.S.	(*bl.*)	dwarf.	——	....	6. 8.	H.♃.	Hacq. pl. alp. car. t. 2. f. 6.
16 pedunculàris. S.S.	(*bl.*)	peduncled.	Astracan.	1822.	——	H.☉.	
17 clavàta. S.S.	(*bl.*)	clavate.	Siberia.	1829.	——	H.☉.	
18 austràlis. B.P.	(*bl.*)	southern.	N. S. W.	1824.	——	G.☉.	
19 ungulàta. F.	(*bl.*)	clawed.	Siberia.	1822.	——	H.☉.	
ECHINOSPE'RMUM. S.S.		ECHINOSPE'RMUM. Pentandria Monogynia. S.S.					
1 zeylanicum. S.S.	(*bl.*)	Ceylon.	Ceylon.	1826.	6. 9.	S.☉.	
2 virginiànum. S.S.	(*pa.*)	Virginian.	Virginia.	1699.	6. 7.	H.♂.	Moris. s. 11. t. 30. f. 9.
Myosòtis virginiàna. W. *Rochèlia virginiàna.* R.S.							

3 defléxum. s.s. (*bl.*) deflexed. Hungary. 1823. 6. 7. H. ♂. Wahl. a. holm.1810. t.4.
4 Láppula. s.s. (*bl.*) common. Europe. 1656. 4. 8. H.⊙. Flor. dan. t. 692.
5 squarròsum. L.en.(*bl.*) squarrose. Siberia. 1802. —— H.⊙.
pátulum. s.s.
6 Redówskii. L.A. (*bl.*) Redowski's. Russia. 1821. 6. 7. H. ♂.
7 barbàtum. s.s. (*bl.*) bearded. Tauria. 1819. —— H.⊙.
8 marginàtum. s.s. (*bl.*) margined. Astracan. —— —— H.⊙.
9 heteracánthum. Led.(*bl.*)various-spined.Siberia. 1826. —— H.⊙.
10 grandiflòrum.Led.(*bl.*) large-flowered. —— 1827. —— H.⊙.

OMPHAL'ODES. L.A. OMPHAL'ODES. Pentandria Monogynia. L.en.
1 nítida. L.A. (*bl.*) shining. Portugal. 1812. 4. 6. H. ♃. Bot. mag. t. 2529.
Cynoglóssum nítidum. B.M. *Picòtia nítida.* R.S.
2 vérna. L.en. (*bl.*) Comfrey-leav'd. S. Europe. 1633. 3. 5. H. ♃. Bot. mag. t. 7.
Cynoglóssum Omphalòdes. B.M. *Picòtia vérna.* R.S.
3 brassicæfòlia. L. (*wh.*) stem clasping. Spain. 1824. 6. 8. H.⊙. Lehm. m. ber. 8. 2. t. 6.
amplexicàulis. Lehm. *Picòtia brassicæfòlia.* R.S.
4 littoràlis. L.A. (*wh.*) sea-side. France. 1827. —— H.⊙.
5 linifòlia. L.A. (*wh.*) Venus Navelwort. Portugal. 1648. —— H.⊙. Barrel. ic. 1234.
6 lithospermifòlia. L.A.(*bl.*)loose-racemed. Levant. 1828. 4. 6. H. ♃.
7 scorpioídes. L.A. (*bl.*) variable-leaved. Bohemia. 1823. 6. 8. H.⊙. Lehm. m. ber. 8. 2. t. 7.
8 sempervìrens. D.P.(*bl.*) evergreen. Britain. .... 5. 6. H. ♃. Eng. bot. v. 1. t. 45.
Anchùsa sempervìrens. E.B.

RIND'ERA. R.S. RIND'ERA. Pentandria Monogynia. R.S.
1 lævigàta. R.S. (*wh.*) smooth. Siberia. 1824. 6. 7. H. ♃. Pall. ross. 2. t. 88.
tetráspis. Pall. *Cynoglóssum Rindèra.* Pall. *Cynoglóssum lævigàtum.* W.
2 magellénsis. R.S. (*vi.*) woolly. Italy. 1823. 5. 7. H. ♃.
Cynoglóssum magellénse. Tenore.
3 Colúmnæ. R.S. (*pu.*) Columna's. —— 1824. —— H. ♂. Column. ecphr. 1. t. 178.
4 cristàta. R.S. (*pu.*) crested. Armenia. 1829. —— H. ♂.

MA'TTIA. R.S. MA'TTIA. Pentandria Monogynia. R.S.
1 umbellàta. R.S. (*re.ye.*) umbel-flowered. Hungary. 1818. 6. 7. H. ♃. W. et K. hung. 2. t.148.
2 lanàta. R.S. (*ro.*) woolly. Levant. 1800. —— G. ♃. Tourn. an. mus. 10. t.37.
Cynoglóssum lanàtum. W.

CYNOGLO'SSUM. R.S. HOUND'S-TONGUE. Pentandria Monogynia. L.
1 officinàle. R.S. (*ro.*) common. Britain. .... 6. 7. H. ♂. Eng. bot. v. 13. t. 921.
2 Hæ'nkii. R.S. (*pu.*) rough-leaved. Hungary. 1815. —— H. ♂.
3 sylváticum. R.S. (*pu.*) green-leaved. Britain. .... —— H. ♂. Eng. bot. v. 23. t. 1642.
4 Dioscòridis. R.S. (*bl.*) Dioscoride's. Dauphiny. 1818. —— H. ♂.
5 bícolor. R.S. (*wh.re.*) two-coloured. Germany. —— —— H. ♂.
6 píctum. R.S. (*re.bl.*) Madeira. Madeira. 1658. 8. G. ♂. Clus. hist. 2. p. 162. f. 2.
7 lanceolàtum. L.A. (*wh.bl.*)spear-leav'd. Africa. 1806. 7. 8. H.⊙.
8 virgínicum. S.S. (*bl.wh.*) stem-clasping. N. America. 1812. 6. 7. H. ♃.
amplexicàule. M.
9 clandestìnum. R.S.(*vi.*) hidden-flower'd. Barbary. 1823. —— G. ♂. Desf. atl. 1. t. 42.
10 canéscens. R.S. (*bl.*) canescent. E. Indies. 1815. 7. 10. G.⊙. Jacq. schœnb. 4. t. 489.
racemòsum. F.I.
11 apennìnum. R.S. (*vi.*) Apennine. Italy. 1731. 4. 7. H. ♂. Sabb. hort. rom. 2. t. 36.
12 tomentòsum. L.A. (*vi.*) woolly-leaved. —— 1823. 5. 8. H. ♃.
13 magellénse. T.S. (*ro.*) woolly. —— —— 5. 7. H. ♃.
14 holoseríceum. S.S. (*vi.*) silky. Caucasus. 1827. 5. 9. H. ♂.
15 diffùsum. L.A. (*wh.*) spreading. E. Indies. 1820. 7. 10. H.⊙.
16 austràle. B.P. (*vi.*) New Holland. N. S. W. —— 5. 7. G. ♃.
17 cheirifòlium. S.S. (*pu.*) silvery-leaved. Levant. 1596. 6. 7. H. ♂.
Anchùsa lanàta. R.S.

TRICHODE'SMA. B.P. TRICHODE'SMA. Pentandria Monogynia. R.S.
1 índicum. R.S. (*bl.*) Indian. E. Indies. 1759. 6. 10. H.⊙. Plnk. alm. 30. t. 76. f.3.
Boràgo índica. L.
2 africànum. R.S. (*bl.*) African. C. B. S. —— 7. 8. H.⊙. Isn. act. par. 1718. t. 11.
3 zeylánicum. R.S. (*bl.*) Ceylon. E. Indies. 1799. —— S.⊙. Jacq. ic. 2. t. 314.

BOR'AGO. R.S. BORAGE. Pentandria Monogynia. L.
1 officinàlis. R.S. (*bl.*) common. Britain. .... 6. 9. H.⊙. Eng. bot. v. 1. t. 36.
β *albiflòra.* (*wh.*) *white-flowered.* —— .... —— H.⊙.
2 orientàlis. R.S. (*bl.*) oriental. Turkey. 1752. 3. 5. H. ♃. Bot. reg. t. 288.
3 crética. R.S. (*bh.*) Cretan. Greece. 1823. 5. 6. H. ♃. Flor. græc. t. 176.
4 laxiflòra. R.S. (*bl.*) bell-flowered. Corsica. 1813. 5. 8. H. ♂. Bot. mag. t. 1798.
5 crassifòlia. R.S. (*fl.*) thick-leaved. Persia. 1824. —— H. ♃. Venten. cels. t. 100.

TOURNEFO'RTIA. R.S. TOURNEFO'RTIA. Pentandria Monogynia. L.
1 cymòsa. R.S. (*fl.*) broad-leaved. Jamaica. 1777. 7. S. ♄. Jacq. ic. 1. t. 31.
2 fœtidíssima. R.S. (*fl.*) Tobacco-leav'd. —— 1739. 9. S. ♄. Plum. ic. 226. t. 230.

3 bícolor. R.S. (*wh.gr.*) two-coloured. Jamaica. 1812. .... S. ♄.
4 suffruticòsa. R.S. (*wh.*) hoary-leaved. ——— 1759. 6. 9. S. ♄. Sloan. hist. 2. t. 162. f. 4.
5 mutábilis. R.S. (*wh.*) changeable. Java. 1820. —— S. ♄. Venten. choix. t. 3.
6 maculàta. R.S. (*st.*) spotted-fruited. Carthagena. 1828. —— S. ♄.
7 gnaphalòdes. R.S. (*wh.*) Gnaphalium-like. W. Indies. 1820. —— S. ♄. Jacq. amer. t. 173. f. 11.
Heliotròpium gnaphalòdes. W.
8 velutìna. K.S. (*wh.*) velvetty. Mexico. 1826. —— G. ♄. H. et B. am. 3. t. 201.
9 umbellàta. K.S. (*wh.*) umbel-flower'd. ——— ——— —— G. ♄. ——— 3. t. 202.
10 fruticòsa. R.S. (*wh.*) shrubby. Canaries. 1779. 6. 10. G. ♄. Bot. reg. t. 464.
Messerschmídia fruticòsa. W.
11 angustifòlia. R.S. (*pa.bl.*) narrow-leav'd. ——— .... —— G. ♄.
12 argéntea. R.S. (*wh.*) silvery-leaved. E. Indies. 1822. —— S. ♄. Rumph. amb. 4. t. 55.
13 Argùzia. R.S. (*wh.*) herbaceous. Siberia. 1780. —— H. ♃. Gmel. it. 2. t. 27.

MESSERSCHMI'DIA. R.S. MESSERSCHMI'DIA. Pentandria Monogynia. R.S.
1 caracasàna. K.S. (*wh.*) Caracas. Caracas. 1828. 5. 9. S. ♄.
2 hirsutíssima. R.S. (*wh.*) very hairy. S. America. 1818. 5. 8. S. ♄. Sloan. hist. 2. t. 212. f. 1.
Tournefórtia hirsutíssima. W.
3 volùbilis. R.S. (*gr.*) climbing. Jamaica. 1752. 7. 8. S. ♄.⌣. Sloan. hist. 1. t. 143. f. 2.
Tournefórtia volùbilis. W.
4 scándens. R.S. (*gr.*) scandent. S. America. 1816. —— S. ♄.⌣. Flor. peruv. 2. t. 148.
5 laurifòlia. R.S. (*st.*) Laurel-leaved. W. Indies. 1823. .... S. ♄.⌣. Venten. choix. t. 2.

BEURR'ERIA. G. BEURR'ERIA. Pentandria Monogynia. R.S.
1 succulénta. S.S. (*wh.*) oval-leaved. W. Indies. 1758. 5. 7. S. ♄. Jacq. obs. 2. t. 26.
2 exsúcca. S.S. (*wh.*) dry-fruited. ——— 1804. —— S. ♄. Jacq. amer. t. 173. f. 17.

EHR'ETIA. S.S. EHR'ETIA. Pentandria Monogynia. L.
1 serràta. F.I. (*wh.*) saw-leaved. Nepaul. 1812. 6. 9. G. ♄.
2 tinifòlia. S.S. (*wh.*) Tinus-leaved. Jamaica. 1734. 6. 7. S. ♄. Trew Ehret. t. 24.
3 acuminàta. B.P. (*wh.*) acuminate. N. S. W. 1820. —— G. ♄.
4 láxa. S.S. (*wh.gr.*) loose-flowered. Mauritius. 1821. —— S. ♄. Jacq. schœn. 1. t. 41.
5 internòdis. S.S. (*wh.*) axillary-flower'd. ——— ——— —— S. ♄. L'Herit. stirp. 1. t. 24.
6 l'ævis. S.S. (*wh.*) smooth. E. Indies. 1816. .... S. ♄. Roxb. corom. t. 56.
7 buxifòlia. S.S. (*wh.*) Box-leaved. ——— ——— .... S. ♄. ——— 1. t. 57.
8 microphy'lla. S.S. (*wh.*) small-leaved. Nepaul. 1818. 6. 8. S. ♄. Pluk. phyt. t. 31. f. 1.
9 áspera. S.S. (*wh.*) rough-leaved. E. Indies. 1795. —— S. ♄. Roxb. corom. 1. t. 55.
10 divaricàta. DC. (*wh.*) divaricate. Havannah. 1820. —— S. ♄.

CO'RDIA. B.P. CO'RDIA. Pentandria Monogynia. L.
1 latifòlia. F.I. (*wh.*) broad-leaved. E. Indies. 1824. 6. 8. S. ♄. Rumph. amb. 2. t. 75.
2 My'xa. R.S. (*wh.*) smooth-leaved. ——— 1640. .... S. ♄. Rheed. mal. 4. t. 37.
3 monoíca. R.S. (*wh.*) Birch-leaved. ——— 1799. 3. 4. S. ♄. Roxb. corom. 1. t. 58.
4 Collocócca. R.S. (*gr.*) long-leaved. Jamaica. 1759. —— S. ♄. Sloan. hist. 2. t. 203. f. 2.
5 micrántha. R.S. (*gr.*) small-flowered. ——— 1822. .... S. ♄.
6 serràta. R.S. (*wh.*) saw-leaved. E. Indies. 1824. .... S. ♄.
7 tetraphy'lla. R.S. (*wh.*) four-leaved. Trinidad. 1823. 4. 7. S. ♄. Aubl. guian. 1. t. 88.
8 geraschanthoídes. K.S. (*w.*) taper-leaved. Cuba. 1824. —— S. ♄.
9 Geraschánthus. R.S. (*w.*) Spanish Elm. W. Indies. 1789. 5. S. ♄. Jacq. amer. t. 175. f. 16.
10 grándis. F.I. (*wh.*) great. E. Indies. 1818. .... S. ♄.
11 nodòsa. R.S. (*wh.*) hairy. Guiana. 1803. 6. 7. S. ♄. Aub. gui. 1. t. 86.
hirsùta. W. *Collocóccus.* Aub. *non* L.
12 flavéscens. R.S. (*wh.*) yellow-fruited. ——— 1818. —— S. ♄. Aub. gui. 1. t. 89.
13 spinéscens. R.S. (*wh.*) spiny. E. Indies. ——— —— S. ♄.
14 macrophy'lla. R.S. (*wh.*) broad-leaved. W. Indies. 1752. —— S. ♄. Sloan. hist. 2. t. 221. f. 1.
15 campanulàta. F.I. (*ye.*) bell-flowered. E. Indies. 1824. .... S. ♄. Rumph. amb. 2. t. 75.
16 Sebestèna. R.S. (*or.*) rough-leaved. W. Indies. 1728. 7. 8. S. ♄. Bot. mag. t. 794.
17 l'ævis. R.S. (*bh.*) smooth-leaved. Caracas. 1816. 4. 8. S. ♄. Jacq. schœn. 1. t. 40.
18 reticulàta. R.S. (*wh.*) netted-leaved. E. Indies. 1820. —— S. ♄.
angustifòlia. F.I.
19 polygàma. F.I. (*wh.*) polygamous. ——— 1812. 6. 8. S. ♄.
20 dentàta. R.S. (*wh.*) toothed. Caracas. 1823. .... S. ♄.
21 ellíptica. R.S. (*wh.*) elliptic-leaved. W. Indies. 1804. 6. 8. S. ♄.
22 dichótoma. R.S. (*wh.*) forked. N. Holland. 1824. .... S. ♄.
23 scabérrima. K.S. (*wh.*) rough. Peru. ——— .... S. ♄.
24 angustifòlia. R.S. (*wh.*) narrow-leaved. Santa Cruz. 1808. 6. 8. S. ♄.
25 curassávica. R.S. (*wh.*) long-spiked. S. America. 1759. —— S. ♄. Centur. amer. t. 56.
Varrònia curassávica. W.
26 martinicénsis. R.S. (*wh.*) Martinico. Martinico. 1795. 8. 9. S. ♄. Jacq. amer. 41. t. 32.
27 bullàta. R.S. (*wh.*) blister-leaved. W. Indies. 1823. .... S. ♄. Jacq. amer. pict. t. 43.
28 monospérma. R.S. (*wh.*) single-seeded. Caracas. 1820. 5. 7. S. ♄. Jacq. schœn. 1. t. 39.
29 lineàta. R.S. (*wh.*) round-spiked. W. Indies. 1793. 6. 9. S. ♄. Brown. jam. t. 13. f. 2.
Varrònia lineàta. W.

30 globòsa. R.S. (*wh.*) globe-headed. Cuba. 1820. 7. 8. S. ♄. Sloan. his. 2. t. 194. f. 2.
31 mirabiloídes. R.S. (*wh.*) jointed. Hispaniola. 1798. 9. S. ♄. Jacq. amer. 41. t. 33.
32 álba. R.S. (*wh.*) white-flowered. Carthagena. 1818. .... S. ♄. Comm. hort. 1. t. 80.
33 bifurcàta. R.S. (*wh.*) forked. Peru. —— .... S. ♄. Flor. peruv. 2. t. 146. f. a.
Varrònia dichótoma. R.P. *parviflòra.* P.S.
PATAG'ONULA. R.S. PATAG'ONULA. Pentandria Monogynia. R.S.
americàna. R.S. (*wh.*) spear-leaved. S. America. 1732. 6. 8. G. ♄. Dill. elth. t. 226. f. 293.
Córdia Patagònula. W.

ORDO CL.

HYDROPHYLLEÆ. *Brown.*

HYDROPHY'LLUM. L. WATER-LEAF. Pentandria Monogynia. L.
1 virgínicum. L. (*wh.bl.*) Virginian. N. America. 1739. 5. 6. H. ♃. Bot. reg. t. 331.
2 canadénse. L. (*wh.*) Canadian. Canada. 1759. —— H. ♃. —— t. 242.
PHAC'ELIA. B.P. PHAC'ELIA. Pentandria Monogynia. R.S.
1 bipinnatífida. M. (*bl.*) bipinnatifid. N. America. 1827. 5. 6. H. ♃. Mich. amer. 1. t. 16.
2 circinnàta. J.E. (*wh.*) woolly. Magellan. 1816. 6. 10. H. ♃. Jacq. f. ecl. t. 91.
Hydrophy'llum magellánicum. Lam. *Aldèa circinnàta.* W. en.
3 Aldèa. B.P. (*wh.*) various-leaved. Peru. 1820. —— G. ♃. Flor. peruv. 2. t. 14.
Aldèa pinnàta. R.P.
NEMO'PHILA. N. NEMO'PHILA. Pentandria Monogynia. S.S.
1 phacelioídes. N. (*bl.*) Phacelia-like. N. America. 1822. 6. 9. H. ☉. Swt. br. fl. gar. t. 32.
2 paniculàta. S.S. (*pa.bl.*) panicled. —— 1812. 5. 6. H. ♃.
Hydrophy'llum appendiculàtum. M.
ELL'ISIA. R.S. ELL'ISIA. Pentandria Monogynia. W.
1 Nyctelèa. R.S. (*wh.*) cut-leaved. Virginia. 1755. 4. 8. H. ☉. Trew pl. select. t. 99.
2 ambígua. N. (*wh.*) ambiguous. N. America. 1825. 4. 6. H. ☉.
E'UTOCA. B. E'UTOCA. Pentandria Monogynia.
1 multiflòra. B.R. (*pu.*) many-flowered. N. America. 1826. 5. 6. H. ☉. Bot. reg. t. 1180.
2 Franklínii. S.S. (*pu.*) Franklin's. Missisippi. 1827. —— H. ☉.

ORDO CLI.

SOLANEÆ. *Brown prodr.* 1. *p.* 443.

Sect. I. *PERICARPIUM CAPSULARE.*

ANTHOCE'RCIS. B.P. ANTHOCE'RCIS. Didynamia Angiospermia. S.S.
1 littòrea. B.P. (*ye.*) yellow. N. Holland. 1803. 5. 8. G. ♄. Swt. flor. aust. t. 17.
2 álbicans. F.T. (*wh.*) white-flowered. —— 1823. 4. 6. G. ♄. —— t. 16.
3 viscòsa. B.P. (*wh.*) viscous. —— —— —— G. ♄. Bot. mag. t. 2961.
4 ilicifòlia. B.M. (*wh.*) Holly-leaved. —— 1829. .... G. ♄.
CE'LSIA. S.S. CE'LSIA. Didynamia Angiospermia.
1 orientàlis. S.S. (*ye.*) oriental. Levant. 1713. 7. 8. H. ☉. Lam. ill. t. 532.
2 Arctùrus. W. (*ye.*) scollop-leaved. Candia. 1780. 7. 9. F. ♂. Bot. mag. t. 1962.
3 coromandelìna. W. (*ye.*) Coromandel. E. Indies. 1783. 7. 8. S. ☉.
4 crética. S.S. (*ye.pu.*) great-flowered. Crete. 1752. 7. 9. F. ♂. Bot. mag. t. 964.
5 viscòsa. S.S. (*ye.*) clammy. .... 1816. —— F. ☉.
6 betonicæfòlia. S.S. (*ye.*) Betony-leaved. Barbary. 1824. —— F. ♂.
7 sublanàta. B.R. (*ye.*) woolly. .... 1818. —— H. ♃. Bot. reg. t. 438.
8 lanceolàta. S.S. (*ye.*) spear-leaved. Levant. 1816. —— F. ♂. Venten. cels. t. 27.
VERBA'SCUM. L. MULLEIN. Pentandria Monogynia. L.
1 Thápsus. R.S. (*st.*) Shepherd's club. Britain. .... 7. 8. H. ♂. Eng. bot. v. 8. t. 549.
2 elongàtum. W. en. (*wh.*) long-spiked. Europe. 1813. —— H. ♂. Tabern. Kraeut. 956. ic.
3 thapsoídes. R.S. (*ye.*) branching. Portugal. .... —— H. ♂. Schrad. verb. t. 5. f. 2.
4 thapsifórme. R.S. (*ye.*) simple-stalked. Europe. .... —— H. ♂.
5 índicum. Wal. (*ye.*) Indian. Nepaul. 1827. —— F. ♂.
6 crassifòlium. R.S. (*ye.*) thick-leaved. S. Europe. 1816. —— H. ♂. H. et Lk. fl. port. 1. t. 26.
7 cuspidàtum. R.S. (*ye.*) sharp-pointed. Austria. —— —— H. ♂. Schrad. verb. t. 1. f. 1.

8	gossypìnum. M.B.	(*ye.*)	cottony.	Iberia.	1824.	6. 8.	H. ♂.	
9	níveum. R.S.	(*ye.*)	snowy.	Naples.	——	——	H. ♂.	Ten. fl. neap. t. 22.
10	candidíssimum. DC.	(*ye.*)	very white-l'd.	S. Europe.	1825.	——	H. ♂.	
11	densiflòrum. R.S.	(*ye.*)	dense-flowered.	Italy.	——	——	H. ♂.	
12	macránthum. R.S.	(*ye.*)	large-flowered.	Portugal.	1823.	——	H. ♂.	H. et Lk. fl. port. t. 27.
13	austràle. R.S.	(*ye.*)	southern.	S. Europe.	1816.	——	H. ♂.	Schrad. verb. t. 2.
14	phlomoídes. R.S.	(*ye.*)	woolly.	Turkey.	1739.	——	H. ♂.	Flor. græc. t. 224.
15	rugulòsum. W.	(*ye.*)	rugged-leaved.	S. Europe.	1820.	——	H. ♂.	
16	conocárpon. S.S.	(*ye.*)	conical-capsuled.	Sardinia.	1829.	——	H. ♂.	
17	condensàtum. R.S.	(*ye.*)	close-flowered.	Austria.	1818.	——	H. ♂.	
18	nemoròsum. R.S.	(*ye.*)	wood.	————	——	——	H. ♂.	Schrad. verb. t. 3.
19	montànum. R.S.	(*st.*)	mountain.	Switzerland.	1816.	——	H. ♂.	———— t. 1. f. 2.
20	collìnum. R.S.	(*ye.*)	hill.	Germany.	1820.	——	H. ♂.	
21	versiflòrum. R.S.	(*br.re.*)	scatter'd-flower'd.	————	1815.	——	H. ♃.	Schrad. verb. t. 5. f. 1.
22	ramígerum. R.S.	(*ye.*)	small-branched.	————	1824.	——	H. ♂.	
23	Bastárdi. R.S.	(*ye.*)	Bastard's.	S. France.	——	——	H. ♂.	Schrad. verb. t. 4.
24	auriculàtum. F.G.	(*ye.*)	eared-leaved.	Levant.	1826.	——	E. ♂.	
25	sinuàtum. R.S.	(*ye.*)	scollop-leaved.	S. Europe.	1570.	7. 8.	H. ♂.	Flor. græc. t. 225.
26	undulàtum. M.B.	(*ye.*)	waved-leaved.	Caucasus.	1816.	——	H. ♂.	———— t. 227.
27	plicàtum. F.G.	(*ye.*)	plaited-leaved.	Greece.	——	——	F. ♃.	
28	bipinnatífidum. B.M.	(*ye.*)	bipinnatifid.	Tauria.	1813.	——	H. ♂.	Flor. græc. t. 226.
29	pinnatífidum. R.S.	(*ye.*)	pinnatifid.	Archipelago.	1788.	5. 6.	G. ♃.	Bot. mag. t. 1777.
30	Chaíxi. R.S.	(*ye.*)	French.	France.	1816.	7. 8.	H. ♂.	Flor. græc. t. 228.
31	Boerhaàvii. R.S.	(*ye.*)	annual.	S. Europe.	1731.	——	H. ☉.	Vill. delph. 2. t. 13.
32	compáctum. R.S.	(*ye.*)	compact-flower'd.	Tauria.	1824.	——	H. ♂.	Mill. ic. 2. t. 273.
33	gnaphalòdes. M.B.	(*ye.*)	Gnaphalium-like.	————	——	——	H. ♂.	
34	ovalifòlium. B.M.	(*ye.*)	oval-leaved.	Caucasus.	1804.	7. 9.	H. ♃.	Bot. mag. t. 1037.
35	formòsum. B.R.	(*ye.pu.*)	beautiful.	Russia.	1814.	7. 8.	H. ♂.	Bot. reg. t. 558.
36	Lychnìtis. R.S.	(*wh.*)	white.	Britain.	....	6. 8.	H. ♂.	Eng. bot. v. 1. t. 58.
37	austrìacum. R.S.	(*st.*)	Austrian.	Austria.	1819.	——	H. ♂.	
38	pulverulèntum. R.S.	(*ye.*)	powdered.	England.	....	——	H. ♂.	Eng. bot. v. 7. t. 487.
39	majàle. DC.	(*ye.*)	purple-stalked.	Montpelier.	1822.	5. 7.	H. ♂.	
40	Stevènii. F.	(*ye.*)	Steven's.	Siberia.	1820.	6. 8.	H. ♂.	
41	fasciculàtum. E.	(*ye.*)	clustered.	Mount Sinai.	1826.	5. 8.	F. ♃.	
42	æthiòpicum. E.	(*ye.*)	Ethiopian.	————	1825.	——	F. ♃.	
43	pyramidàtum. R.S.	(*ye.*)	pyramidal.	Caucasus.	1804.	7. 8.	H. ♃.	Swt. br. fl. gar. t. 31.
44	ovàtum. R.S.	(*ye.*)	oval-leaved.	Spain.	——	——	H. ♂.	Schrad. h. gott. t. 15.
45	rotundifòlium. R.S.	(*ye.*)	round-leaved.	Naples.	1826.	——	H. ♂.	Ten. fl. neap. t. 23.
46	floccòsum. R.S.	(*ye.*)	wool-bearing.	Hungary.	1805.	6. 7.	H. ♂.	W. et K. hung. 1. t. 79.
47	leptostàchyum. R.S.	(*ye.*)	slender-spiked.	S. France.	1824.	7. 9.	H. ♂.	
48	angustifòlium. R.S.	(*ye.*)	narrow-leaved.	Naples.	1822.	——	H. ♂.	
49	longifòlium. R.S.	(*ye.*)	long-leaved.	————	——	——	H. ♂.	Ten. fl. neap. t. 21.
50	nìgrum. R.S.	(*ye.*)	black-rooted.	England.	....	6. 8.	H. ♃.	Eng. bot. v. 1. t. 59.
51	orientàle. R.S.	(*ye.*)	oriental.	Caucasus.	1820.	——	H. ♃.	
52	Alopecùrus. R.S.	(*ye.*)	fox-tail.	France.	1818.	——	H. ♃.	
53	virgàtum. R.S.	(*ye.*)	slender.	Britain.	....	8.	H. ♂.	Eng. bot. v. 8. t. 550.
54	speciòsum. R.S.	(*ye.*)	splendid.	Austria.	1820.	6. 8.	H. ♃.	Schrad. h. gott. t. 16.
55	grandiflòrum. R.S.	(*ye.*)	great-flowered.	....	1816.	——	H. ♃.	———— t. 15.
56	monspessulànum. S.S.	(*ye.*)	Montpelier.	S. France.	1824.	——	H. ♂.	Schrad. ver. p. 2. t. 2. f. 2.
57	lanàtum. S.S.	(*ye.*)	woolly-leaved.	S. Europe.	1828.	——	H. ♂.	———— pars. 2. t. 2. f. 1.
58	hyoserifòlium. S.S.	(*ye.*)	Hyoseris-leav'd.	Levant.	1829.	——	F. ♃.	———— pars. 2. t. 3. f. 1.
59	chrysèrium. S.S.	(*ye.*)	golden.	Palestine.	1827.	——	F. ♃.	
60	Schottiànum. S.S.	(*ye.*)	Schott's.	Austria.	1828.	——	H. ♂.	Schrad. ver. p. 2. t. 3. f. 2.
61	Blattària. R.S.	(*ye.*)	yellow moth.	Britain.	....	7. 9.	H. ♂.	Eng. bot. v. 6. t. 393.
62	glábrum. Mil.	(*wh.*)	white moth.	....	....	——	H. ♂.	
63	blattarioídes. R.S.	(*ye.*)	moth-like.	S. Europe.	1816.	——	H. ☉.	
64	bannáticum. S.S.	(*ye.*)	Hungarian.	Hungary.	1820.	——	H. ♂.	
65	repándum. R.S.	(*ye.*)	waved.	....	1813.	7. 8.	H. ♂.	
66	hæmorrhoidàle. R.S.	(*ye.*)	Madeira.	Madeira.	1777.	6. 8.	G. ♂.	
67	cùpreum. R.S.	(*co.*)	copper-colour'd.	Hybrid.	1798.	5. 9.	H. ♃.	Bot. mag. t. 1226.
68	phœníceum. R.S.	(*d.pu.*)	purple-flower'd.	S. Europe.	1596.	5. 8.	H. ♃.	———— t. 885.
69	puníceum. R.S.	(*pi.*)	pink-flower'd.	————	1815.	——	H. ♃.	Schrad. h. gott. t. 14.
70	rubiginòsum. R.S.	(*re.br.*)	brown-flower'd.	Hungary.	——	——	H. ♂.	W. et K. hung. 2. t. 197.
71	ferrugíneum. R.S.	(*br.*)	rusty.	S. Europe.	1683.	——	H. ♃.	Trew Ehret. t. 16. f. 1.
72	tríste. R.S.	(*da.br.*)	dark-flowered.	Levant.	1788.	——	H. ♃.	Andr. reposit. t. 162.
	ferrugíneum. A.R. *nec aliorum.*							
73	spectábile. M.B.	(*ye.*)	showy.	Tauria.	1822.	6. 8.	H. ♂.	
74	betonicæfòlium. R.S.	(*ye.*)	Betony-leav'd.	Levant.	1828.	——	F. ♂.	Desf. ann. mus. 11. t. 4.
75	urticæfòlium. R.S.	(*ye.re.*)	spotted-flower'd.	......	1827.	——	H. ♂.	

76 lyràtum. R.S. (*ye.*) lyrate-leaved. Spain. 1816. 6. 8. H. ♂.
77 Osbéckii. R.S. (*gr.ye.*) Osbeck's. ——— 1752. 7. 8. H. ♂. Buxb. cent. 5. t. 32.
78 ceratophy'llum. S.S. (*ye.*) buckshorn-leav'd. Levant. 1829. —— H. ♂. Schrad. verb. p. 2. t. 1. f. 2.
79 spinòsum. R.S. (*ye.*) spiny. Candia. 1824. 5. 8. G. ♄. Flor. græc. t. 229.

RAMO'NDA. P.S. RAMO'NDA. Pentandria Monogynia. P.S.
pyrenàica. P.S. (*pu.*) Borage-leaved. Pyrenees. 1731. 5. H. ♃. Bot. mag. t. 236.
Verbáscum Mycòni. B.M.

HYOSC'YAMUS. R.S. HENBANE. Pentandria Monogynia. L.
1 nìger. R.S. (*ye.bk.*) common. Britain. .... 6. 7. H. ♂. Eng. bot. v. 9. t. 591.
2 agréstis. R.S. (*gr.pu.*) field. Hungary. 1816. 6. 9. H. ⊙. Swt. br. fl. gar. t. 27.
3 pállidus. R.S. (*st.*) pale-flowered. ——— 1815. —— H. ⊙.
4 reticulàtus. R.S. (*re.*) Egyptian. Egypt. 1640. 7. 9. H. ⊙. Camer. hort. 77. t. 22.
5 álbus. R.S. (*st.pu.*) white. Greece. 1570. 7. 8. H. ⊙. Flor. græc. t. 230.
6 Clùsii. F.G. (*wh.*) Clusius's. S. Europe. —— —— H. ⊙. Clus. hist. 2. p. 84. f. 1.
álbus β mìnor. R.S.
7 aùreus. R.S. (*ye.pu.*) golden. Levant. 1640. 3. 10. G. ♄. Bot. mag. t. 87.
8 Seneciònis. R.S. (*ye.pu.*) yellow-flowered. Egypt. 1812. —— G. ♃.
9 canariénsis. B.R. (*ye.pu.*) various-leaved. Canaries. 1816. 1. 12. G. ♂. Bot. reg. t. 180.
10 mùticus. R.S. (*wh.pu.*) three-coloured. Egypt. 1818. 5. 8. G. ♂.
11 Datòra. F.Æ. (*ye.pu.*) Egyptian. ——— 1829. —— F. ♃.
12 pusíllus. R.S. (*ye.bk.*) dwarf. Persia. 1691. 7. H. ⊙. Pluk. alm. t. 37. f. 5.
13 auriculàtus. T.N. (*ye.pu.*) eared. Naples. 1828. —— H. ⊙.
14 micránthus. Led. (*st.pu.*) small-flowered. Siberia. 1829. —— H. ♂.
15 màjor. Mil. (*st.pu.*) large. Levant. .... —— F. ♂.
16 mìnor. Mil. (*ye.*) lesser. ——— —— F. ♂.
17 physaloídes. R.S. (*d.pu.*) purple-flowered. Siberia. 1777. 3. 4. H. ♃. Swt. br. fl. gar. t. 13.
18 orientàlis. R.S. (*li.*) oriental. Iberia. 1815. —— H. ♃. ——— t. 12.

SCOP'OLIA. Jacq. SCOP'OLIA. Pentandria Monogynia. R.S.
carniòlica. Jacq. Nightshade-leav'd. Carniola. 1780. 4. 5. H. ♃. Jacq. obs. 1. t. 20.
atropoídes. R.S. *Hyosc'yamus Scopòlia.* Bot. mag. t. 1126.

NICOTI'ANA. R.S. TOBACCO. Pentandria Monogynia. L.
1 chinénsis. R.S. (*ro.*) Chinese. China. 1815. 6. 8. G. ♄.
2 fruticòsa. R.S. (*ro.*) frutescent. C. B. S. 1699. 7. 9. G. ♄. Mill. icon. t. 185. f. 1.
3 álba. Mil. (*wh.*) white-flowered. Tobago. —— 8. 10. H. ⊙.
4 latíssima. DC. (*ro.*) broad-leaved. America. .... 7. 9. H. ⊙.
macrophy'lla. R.S.
5 Tabácum. R.S. (*ro.*) Virginian. ——— 1570. —— H. ⊙. Blackw. t. 146.
α attenuàta. (*re.*) *slender.* ——— .... —— H. ⊙.
β palléscens. (*bh.*) *pale-flowered.* ——— .... —— H. ⊙.
γ àlipes. (*ro.*) *winged.* ——— .... —— H. ⊙.
δ serotìna. (*ro.*) *late-flowered.* ——— .... —— H. ⊙.
ε grácilipes. (*ro.*) *slender-stalked.* ——— .... —— H. ⊙.
ζ Vérdan. (*ro.*) *Verdan.* ——— .... —— H. ⊙.
η língua. (*ro.*) *tongue-leaved.* ——— .... —— H. ⊙.
6 nepalénsis. Ot. (*ro.*) Nepaul. Nepaul. 1829. —— H. ⊙.
7 brasiliénsis. Ot. (*ro.*) Brazilian. Brazil. 1825. —— H. ⊙.
8 decúrrens. Ag. (*ro.*) decurrent. S. America. 1820. —— H. ⊙.
9 petiolàta. Ag. (*ro.*) petiolate. ——— —— —— H. ⊙.
10 sanguínea. Ot. (*sc.*) scarlet-flowered. Brazil. 1829. —— H. ⊙.
11 angustifòlia. R.S. (*gr.*) narrow-leaved. Chili. 1822. —— H. ⊙. Flor. peruv. 2. t. 130. f. a.
12 viscòsa. R.S. (*gr.*) viscous. BuenosAyres. 1824. —— H. ⊙.
13 pusílla. R.S. (*gr.*) Primrose-leaved. Vera Cruz. 1733. 8. S. ♂. Mill. ic. t. 185. f. 2.
14 glutinòsa. R.S. (*or.*) clammy. Peru. 1759. 7. 9. H. ⊙. Swt. br. fl. gar. t. 107.
15 rústica. R.S. (*gr.*) common. America. 1570. —— H. ⊙. Blackw. t. 437.
α asiática. R.S. (*gr.*) *Asiatic.* .... —— —— H. ⊙.
β brasília. R.S. (*gr.*) *Brazil.* Brazil. —— —— H. ⊙.
γ pùmila. R.S. (*gr.*) *dwarf.* S. America. —— —— H. ⊙.
16 hùmilis. L.en. (*gr.*) lesser. America. .... —— H. ⊙.
17 paniculàta. R.S. (*gr.*) panicled. Peru. 1752. —— H. ⊙. Flor. peruv. 2. t. 129. f. b.
18 cerinthoídes. R.S. (*gr.*) honeywort. America. 1816. —— H. ⊙. Lehm. nicot. t. 2.
19 Langsdórffii. S.S. (*gr.ye.*) Langsdorff's. S. America. 1824. —— H. ⊙. Bot. mag. t. 2555.
20 glaùca. B.M. (*st.*) glaucous. BuenosAyres. 1827. —— H. ⊙. ——— t. 2837.
21 repánda. R.S. (*bh.*) stem-clasping. Havannah. 1822. 8. 10. H. ⊙. ——— t. 2484.
lyràta. K.S.
22 plumbaginifòlia. R.S. curled-leaved. America. 1816. 5. 10. G. ♃. Vivian. elench. t. 5.
críspa. Jacq. fragm. t. 84.
23 pulmonarioídes. K.S. (*wh.*) Pulmonaria-like. S. America. 1828. —— H. ⊙.
24 bonariénsis. R.S. (*wh.*) Buenos Ayres. BuenosAyres. 1821. —— H. ⊙. Lehm. nicot. n. 8. t. 1.
25 suavèolens. R.S. (*wh.*) sweet-scented. N. S. W. 1800. 5. 9. G. ♃. Bot. mag. t. 673.
undulàta. Vent. malm. t. 10. *non* R.P.

26 vincæflòra. s.s.	(*wh.*)	Vinca-flowered.	S. America.	1825.	8. 10.	H.⊙.	
27 dilatàta. L.en.	(*wh.*)	dilated-leaved.	———	——	——	H.⊙.	
28 noctiflòra. B.M.	(*wh.*)	night-flowering.	Andes.	1826.	7. 10.	F.♃.	Swt. br. fl. gar. t. 262.
29 acuminàta. B.M.	(*wh.*)	taper-pointed.	BuenosAyres.	——	——	F.♃.	Bot. mag. t. 2919.
30 nàna. B.R.	(*wh.*)	dwarf.	S. America.	1822.	8. 10.	H.⊙.	Bot. reg. t. 833.
31 quadriválvis. R.S.	(*wh.*)	four-valved.	N. America.	1811.	7. 8.	H.⊙.	Bot. mag. t. 1778.
32 multiválvis. B.R.	(*wh.*)	many-valved.	———	1826.	9. 10.	H.⊙.	Bot. reg. t. 1057.
PET'UNIA. J.		PET'UNIA.	Pentandria Monogynia. R.S.				
nyctaginiflòra. J.	(*wh.*)	large-flowered.	S. America.	1824.	4. 10.	G.♃.	Swt. br. fl. gar. t. 119.
SALPIGLO'SSIS. R.P.		SALPIGLO'SSIS.	Didynamia Angiospermia.				
1 stramínea. H.E.F.	(*st.*)	straw-coloured.	Chile.	1826.	6. 9.	F.♃.	Swt. br. fl. gar. t. 231.
2 pícta. B.F.G.	(*wh.bl.*)	painted-flowered.	———	——	——	F.♃.	——— v. 3. t. 258.
3 atropurpùrea. B.M.	(*d.pu.*)	dark purple.	———	——	——	F.♃.	——— v. 3. t. 271.
4 sinuàta. R.P.	(*cr.*)	crimson.	———	1829.	——	F.♃.	Flor.per.v.5.t.515.ined.
DAT'URA. R.S.		THORN-APPLE.	Pentandria Monogynia. L.				
1 fèrox. R.S.	(*wh.*)	Chinese.	China.	1731.	7. 10.	H.⊙.	Zanon. hist. t. 162.
2 Stramònium. R.S.	(*wh.*)	common.	England.	....	——	H.⊙.	Eng. bot. v. 18. t. 1288.
β canéscens. F.I.	(*st.*)	*canescent.*	Nepaul.	1824.	——	H.⊙.	
3 Tátula. R.S.	(*bl.*)	light blue.	S. America.	1629.	——	H.⊙.	Swt. br. fl. gar. t. 83.
4 muricàta. S.S.	(*wh.*)	muricated.	....	1819.	——	H.⊙.	
5 hy'brida. T.P.	(*wh.*)	hybrid.	Hybrid.	1826.	6. 9.	F.♃.	
6 Mètel. R.S.	(*wh.*)	downy.	Asia.	1596.	6. 10.	H.⊙.	Bot. mag. t. 1440.
7 fruticòsa. L.en.	(*wh.*)	frutescent.	S. America.	1825.	——	S.ħ.	
8 quercifòlia. K.S.	(*li.*)	Oak-leaved.	Mexico.	1824.	——	H.⊙.	
9 fastuòsa. R.S.	(*pu.wh.*)	purple.	Egypt.	1629.	——	H.⊙.	Rumph. am.5.t.243.f.2.
β flòre plèno.	(*pu.wh.*)	*double-flowered.*	———	——	——	G.⊙.	
10 dùbia. R.S.	(*wh.*)	doubtful.	....	——	——	H.⊙.	
11 l'ævis. R.S.	(*wh.*)	smooth-fruited.	Africa.	1780.	——	H.⊙.	Jacq. vind. 3. t. 82.
12 ceratocaùla. R.S.	(*wh.pu.*)	horn-stalked.	Cuba.	1805.	8. 10.	H.⊙.	Jacq. schœn. 3. t. 339.
13 guayaquilénsis. K.S.	(*wh.*)	Guayaquil.	S. America.	1824.	——	H.⊙.	
BRUGMA'NSIA. P.S.		BRUGMA'NSIA.	Pentandria Monogynia. R.S.				
1 cándida. P.S.	(*wh.*)	downy-stalked.	Peru.	1818.	8. 9.	S.♄.	Flor. peruv. 2. t. 128.
Datùra arbòrea. L.							
2 suavèolens.	(*wh.*)	smooth-stalked.	S. America.	1733.	——	S.♄.	
Datùra arbòrea. Hort. *suavèolens.* W.enum.							
VE'STIA. W.		VE'STIA.	Pentandria Monogynia. W.				
lycioídes. W.en.	(*gr.ye.*)	Boxthorn-like.	Chili.	1815.	1. 12.	G.♄.	Bot. reg. t. 299.
? C'ODON. W.		C'ODON.	Decandria Monogynia. L.				
Royèni. W.	(*re.wh.*)	prickly.	C. B. S.	1801.	9. 10.	G.♂.	And. repos. t. 325.

Sect. II. *PERICARPIUM BACCATUM.*

ANIS'ODUS. S.S.		ANIS'ODUS.	Pentandria Monogynia. S.S.				
lùridus. S.S.	(*br.*)	lurid.	Nepaul.	1823.	8. 10.	H.♃.	Swt. br. fl. gar. t. 125.
Whitlèya stramonifòlia B.F.G.							
NECTO'UXIA. K.S.		NECTO'UXIA.	Pentandria Monogynia.				
formòsa. K.S.	(*ye.*)	pretty.	Mexico.	1826.	7. 10.	H.⊙.	H.etB. n.g.am.3.t.193.
A'TROPA. R.S.		A'TROPA.	Pentandria Monogynia. L.				
Belladónna. R.S.	(*br.pu.*)	deadly nightshade.	Britain.	....	6. 7.	H.♃.	Eng. bot. v. 9. t. 592.
SA'RACHA. R.S.		SA'RACHA.	Pentandria Monogynia.				
1 procúmbens. R.P.	(*st.*)	procumbent.	Peru.	1822.	7. 9.	H.♃.	Cavan. ic. 1. t. 72.
A'tropa procúmbens. Cav. *Bellínia procúmbens.* R.S.							
2 umbellàta. DC.	(*st.*)	umbel-flowered.	Peru.	——	——	H.♃.	Swt. br. fl. gar. t. 85.
MANDRAG'ORA. J.		MANDRAKE.	Pentandria Monogynia. R.S.				
1 pr'æcox. B.F.G.	(*st.bh.*)	blister-leaved.	Switzerland.	1818.	3. 4.	H.♃.	Swt. br. fl. gar. t. 198.
2 vernàlis. S.S.	(*bh.*)	large-leaved.	S. Europe.	1562.	5. 6.	H.♃.	
3 autumnàlis. S.S.	(*li.*)	autumn-flowered.	Levant.	——	8. 9.	H.♃.	Flor. græc. t. 232.
NOL'ANA. R.S.		NOL'ANA.	Pentandria Monogynia. L.				
1 prostràta. R.S.	(*bl.*)	trailing.	Peru.	1761.	7. 9.	H.⊙.	Bot. mag. t. 731.
2 tenélla. H.C.	(*bl.*)	slender.	Chile.	1824.	——	F.ħ.	
3 paradóxa. B.R.	(*bl.*)	violet-coloured.	Chili.	1822.	——	H.⊙.	Bot. reg. t. 865.
NICA'NDRA. R.S.		NICA'NDRA.	Pentandria Monogynia. R.S.				
1 physaloídes. R.S.	(*bl.*)	American.	America.	1759.	7. 9.	H.⊙.	Bot. mag. t. 2458.
2 índica. R.S.	(*bl.*)	Indian.	Nepaul.	1823.	——	H.⊙.	Rumph.amb.6.t.26.f.1.
PHY'SALIS. R.S.		WINTER CHERRY.	Pentandria Monogynia. L.				
arboréscens. W.	(*ye.*)	arborescent.	Campeachy.	1733.	6. 8.	S.♄.	Mill. ic. 2. t. 206. f. 2.

2 frutéscens. DC.	(*ye.*) frutescent.	Spain.	1737.	1. 3.	G. ♄.	Cavan. ic. 2. t. 102.
A'tropa frutéscens. R.S.						
3 aristàta. W.	(*ye.*) bearded.	Canaries.	1779.	5. 8.	G. ♄.	
A'tropa aristàta. R.S.						
4 somnífera. R.S.	(*st.*) clustered.	Levant.	1596.	7. 8.	G. ♄.	Flor. græc. t. 233.
β *aristàta.*	(*st.*) *awned.*	Mexico.	1826.	——	G. ♄.	
5 flexuòsa. R.S.	(*gr.wh.*) flexuose.	E. Indies.	1759.	——	G. ♄.	Jacq. f. ecl. t. 23.
6 curassávica. R.S.	(*st.*) Curassavian.	S. America.	1699.	6. 9.	S. ♃.	Pluk. alm. t. 111. f. 5.
7 tomentòsa. R.S.	(*ye.*) woolly.	C. B. S.	1816.	——	G. ♄.	
8 viscòsa. R.S.	(*st.*) clammy.	America.	1732.	7.	H. ♃.	Bot. mag. t. 2625.
9 lanceolàta. R.S.	(*ye.*) spear-leaved.	N. America.	1780.	——	H. ♃.	
10 pensylvánica. R.S.	(*ye.*) Pensylvanian.	———	1726.	7. 9.	H. ♃.	
11 Jacquìni. L.en.	(*ye.*) Jacquin's.	———	....	——	H. ♃.	Jacq. vind. 2. t. 136.
viscòsa. Jacq. *non* L.						
12 Alkekéngi. R.S.	(*wh.*) common.	S. Europe.	1548.	——	H. ♃.	Flor. græc. t. 234.
13 tuberòsa. R.S.	(*ye.pu.*) tuberous-rooted.	....	1816.	....	S. ♃.	
14 chenopodifòlia. R.S.	(*ye.*) goose-foot-leav'd.	S. America.	1798.	7. 8.	S. ♄.	
15 peruviána. R.S.	(*ye.pu.*) Peruvian.	Peru.	1773.	——	S. ♃.	
16 èdulis. B.M.	(*ye.pu.*) eatable.	S. America.	——	——	S. ♃.	Bot. mag. t. 1068.
17 pubéscens. R.S.	(*ye.pu.*) pubescent.	America.	1640.	——	H. ⊙.	Feuil. per. 3. t. 1.
18 micrántha. L.en.	(*ye.*) small-flowered.	———	1822.	——	H. ⊙.	
19 nodòsa. R.S.	(*ye.*) knotted.	———	....	——	H. ⊙.	
20 barbadénsis. R.S.	(*st.*) Barbadoes.	W. Indies.	1798.	——	H. ⊙.	Jacq. ic. 1. t. 39.
21 dùbia. L.en.	(*ye.*) dubious.	Brazil.	1819.	——	H. ⊙.	
22 fœtidíssima. R.S.	(*ye.br.*) stinking.	New Spain.	——	——	H. ⊙.	
23 fœtens. R.S.	(*ye.br.*) strong-scented.	Peru.	1827.	——	H. ♃.	
24 angulàta. R.S.	(*wh.*) angular-branch'd.	India.	1732.	6. 9.	H. ⊙.	Dill. elth. t. 12. f. 12.
25 Rothiàna. R.S.	(*wh.*) Roth's.	E. Indies.	1827.	——	S. ♃.	
26 philadélphica. R.S.	(*ye.*) Philadelphian.	N. America.	1824.	——	H. ⊙.	
27 atriplicifòlia. R.S.	(*ye.*) Atriplex-leaved.	....	1798.	7. 8.	H. ⊙.	Jacq. frag. t. 85. f. a.
chenopodifòlia. W. *non* Lamarck.						
28 mínima. R.S.	(*ye.*) least.	E. Indies.	1759.	——	H. ⊙.	Rheed. mal. 10. t. 70.
29 pruinòsa. R.S.	(*ye.*) hairy annual.	America.	1726.	——	H. ⊙.	Dill. elth. 10. t. 9. f. 9.
30 Lagáscii. R.S.	(*ye.*) Lagasca's.	New Spain.	1824.	——	H. ⊙.	
31 pátula. R.S.	(*wh.*) spreading.	Vera Cruz.	1731.	7. 9.	S. ⊙.	
32 villòsa. R.S.	(*st.*) villous.	———	——	——	S. ⊙.	
33 cordàta. R.S.	(*ye.gr.*) heart-leaved.	———	——	——	S. ⊙.	
34 máxima. R.S.	(*st.*) large annual.	———	——	——	S. ⊙.	
35 abyssínica. Ot.	(*ye.*) Abyssinian.	Abyssinia.	1828.	——	S. ⊙.	
36 prostràta. R.S.	(*bl.*) trailing.	Peru.	1782.	——	H. ⊙.	Andr. reposit. t. 75.
CA'PSICUM. R.S.	CA'PSICUM.	Pentandria Monogynia. L.				
1 ánnuum. R.S.	(*wh.*) annual.	India.	1548.	6. 7.	G. ⊙.	Rheed. mal. 2. t. 35.
2 lóngum. R.S.	(*wh.*) long-fruited.	....	——	——	G. ⊙.	Dodon. pemph. 716. f. 3.
3 cordifórme. R.S.	(*wh.*) heart-shaped.	India.	1731.	——	G. ⊙.	
4 tetragònum. R.S.	(*wh.*) four-sided.	———	——	——	S. ⊙.	
5 angulòsum. R.S.	(*wh.*) angular-fruited.	....	——	——	S. ⊙.	
6 sphæricum. R.S.	(*wh.*) round-fruited.	....	1807.	4. 7.	S. ⊙.	
7 ovàtum. R.S.	(*wh.*) oval-fruited.	....	——	——	S. ♄.	
8 péndulum. R.S.	(*wh.*) pendulous.	....	1804.	——	S. ♄.	
9 lùteum. R.S.	(*wh.*) yellow-fruited.	E. Indies.	1820.	——	S. ♄.	
10 gróssum. R.S.	(*wh.*) various-fruited.	India.	1759.	5. 8.	S. ♄.	Besl. eyst. 2. t. 2. f. 1.
α *globòsum.* R.S.	(*wh.*) *globular-fruited.*	———	——	——	S. ♄.	
β *lùteum.* R.S.	(*wh.*) *yellow oval-fruited.*	———	——	——	S. ♄.	
γ *bífidum.*	(*wh.*) *cleft-fruited.*	———	....	——	S. ♄.	
11 conoídes. R.S.	(*st.*) conical-fruited.	India.	1728.	——	S. ♄.	
12 pyramidàle. R.S.	(*wh.*) pyramidal.	Egypt.	1732.	——	S. ♄.	
13 cerasifórme. R.S.	(*st.*) Cherry-fruited.	W. Indies.	1759.	——	S. ♄.	
14 Millèri. R.S.	(*st.*) Miller's.	New Spain.	——	——	S. ♄.	
cerasifórme. Mill. *non* W.						
15 purpùreum. F.I.	(*pu.*) purple.	E. Indies.	1824.	——	S. ♄.	
16 frutéscens. R.S.	(*wh.*) shrubby.	India.	1656.	6. 9.	S. ♄.	Rheed. mal. 2. t. 56.
β *torulòsum.*	(*wh.*) *twisted.*	———	....	——	S. ♄.	
17 bícolor. R.S.	(*pu.*) purple-flower'd.	W. Indies.	1804.	——	S. ♄.	Bot. mag. t. 1835.
nìgrum. W.en. *violàceum.* DC.						
18 cæruléscens. R.S.	(*pu.*) bluish.	....	1827.	——	S. ♄.	
19 baccàtum. R.S.	(*wh.*) Bird pepper.	W. Indies.	1731.	——	S. ♄.	Sloan. j. 1. t. 146. f. 2.
20 mínimum. F.I.	(*wh.*) small-berried.	E. Indies.	1728.	——	S. ♄.	
21 sinénse. R.S.	(*wh.*) oval Chinese.	China.	1807.	——	S. ♄.	Jacq. vind. 3. t. 67.
22 havanénse. K.S.	(*wh.*) Havannah.	Havannah.	1826.	——	S. ♄.	

23 microcárpum. R.S. (*wh.*) small-fruited. S. America. 1824. 6. 9. S. ♄.
24 micránthum. R.S. (*wh.*) small-flowered. Brazil. —— —— S. ♄.

LYCOPE'RSICUM. R.S. LOVE-APPLE. Pentandria Monogynia. R.S.

1 pimpinellifòlium. R.S. (*ye.*) Pimpinella-ld. S. America. .... 6. 8. S. ♃.
2 regulàre. R.S. (*ye.*) regular-leaved. ———— .... 6. 9. G. ⊙.
3 peruviànum. R.S. (*ye.*) Peruvian. Peru. .... 6. 8. G. ♃. Bot. mag. t. 2814.
Solànum peruviànum. Jacq. icon. t. 327.
4 dentàtum. R.S. (*ye.*) tooth-leaved. S. America. .... —— H. ⊙. Dunal solan. 2. ined. t. 82.
5 Humbóldtii. R.S. (*ye.*) Humboldt's. ———— 1815. —— H. ⊙. Willd. hort. ber. 1. t. 27.
6 pyrifórme. R.S. (*ye.*) Pear-fruited. ———— .... —— H. ⊙. Dunal solan. t. 26.
7 cerasifórme. R.S. (*ye.*) Cherry-fruited. Peru. 1800. 7. 9. H. ⊙. Jacq. vind. 1. t. 11.
Solànum pseudolycopérsicum. Jacq.
α rùber. (*ye.*) *red-fruited.* ———— —— —— H. ⊙.
β lùteum. (*ye.*) *yellow-fruited.* ———— —— —— H. ⊙.
8 esculéntum. R.S. (*ye.*) Tomato. S. America. 1596. —— H. ⊙. Dunal solan. t. 3. f. 3.
α erythrocárpum. (*ye.*) *red-fruited.* ———— —— —— H. ⊙.
β chrysocárpum. (*ye.*) *yellow-fruited.* ———— —— —— H. ⊙.
γ leucocárpum. (*ye.*) *white-fruited.* ———— —— —— H. ⊙.
9 procúmbens. R.S. (*ye.*) procumbent. ———— .... —— H. ⊙.

SOLA'NUM. R.S. NIGHTSHADE. Pentandria Monogynia. L.

1 tuberòsum. R.S. (*bl.wh.*) Potatoe. Peru. 1597. 6. 8. H. ♃. Plenck. off. t. 121.
2 appendiculàtum. R.S. (*wh.*) appendaged. Mexico. 1823. —— S. ♄. ◡. Dunal solan. ed. 2. t. 84.
3 Seaforthiànum. R.S. (*pi.*) Seaforth's. Barbadoes. 1804. 7. 9. S. ♄. ◡. Andr. reposit. t. 504.
4 betàceum. R.S. (*bh.*) Beet-leaved. New Spain. 1803. 6. 7. S. ♄. ———— t. 511.
5 muricàtum. R.S. (*wh.vi.*) warted. Peru. 1785. 7. 8. S. ♄. Feuil. peruv. 2. t. 26.
6 diversifòlium. R.S. (*wh.*) different-leaved. Caracas. 1824. —— S. ♄. Dunal solan. ed. 2. t. 88.
7 Bulbocástanum. R.S. (*wh.*) tuberous-rooted. Mexico. 1826. 6. 9. H. ♃. ———— ed. 2. t. 3.
8 oligánthum. R.S. (*wh.*) climbing. Orinoco. 1824. 7. 8. S. ♄. ◡. ———— ed. 2. t. 90.
9 reclinàtum. R.S. (*bl.*) reclining. Peru. .... —— S. ♄. Lam. ill. t. 145. f. 4.
10 laciniàtum. B.P. (*bl.*) cut-leaved. N. Holland. 1772. —— G. ♄. Bot. mag. t. 349.
α fruticòsum. (*bl.*) *frutescent.* N. S. W. —— —— G. ♄.
β herbàceum. (*bl.*) *herbaceous.* V. Diem. Isl. —— —— H. ⊙.
11 quercifòlium. R.S. (*vi.*) Oak-leaved. Peru. 1787. 6. 7. H. ♃. Feuil. per. 2. t. 15 ?
12 rádicans. R.S. (*li.*) rooting. ———— 1771. 7. 8. G. ♃. Linn. dec. 1. t. 10.
13 corymbòsum. R.S. (*vi.*) corymb-flowered. ———— 1786. —— S. ♃. Jacq. ic. 1. t. 40.
14 Dulcamàra. R.S. (*bl.*) bitter-sweet. Britain. .... 6. 8. H. ♄. ◡. Eng bot. v. 8. t. 565.
β álbum. (*wh.*) *white-flowered.* ———— .... —— H. ♄. ◡.
γ variegàtum. (*pa.*) *variegated.* ———— .... —— H. ♄. ◡.
15 littoràle. L.en. (*bl.*) sea-side. Europe. .... —— H. ♄.
16 Tegòre. R.S. (*gr.*) hairy-stemmed. Guiana. 1820. —— S. ♄. Aubl. guian. 1. t. 84.
17 quiténse. S.S. (*wh.pu.*) Lima. Lima. 1825. —— G. ♄. Bot. mag. t. 2739.
18 macrocárpon. S.S. (*bl.*) large-fruited. Peru. 1759. 5. 9. S. ♄. Mill. ic. 2. t. 294.
19 calycìnum. R.S. (*bl.*) large-calyxed. Mexico. 1825. —— S. ♄. Dunal sol. ed. 2. t. 29.
20 racemiflòrum. R.S. (*bl.*) racemed. S. America. 1818. 7. 9. H. ⊙. Jacq. schœn. 3. t. 333.
scábrum. Jacq. *non* Vahl.
21 æthiòpicum. R.S. (*wh.*) Ethiopian. Ethiopia. 1597. —— H. ⊙. Jacq. vind. 1. t. 12.
22 Zuccagniànum. R.S. (*wh.*) Zuccagni's. .... 1818. —— H. ⊙. Dunal solan. t. 11.
23 pseùdo-cápsicum. (*wh.*) Winter-cherry. Madeira. 1596. 6. 9. G. ♄. Sabb. h. rom. t. 59.
24 nodiflòrum. R.S. (*wh.*) knot-flowered. Guiana. 1815. —— G. ⊙. Jacq. ic. 2. t. 326.
25 Dillènii. R.S. (*wh.*) Dillenius's. Hungary. .... —— H. ⊙. Dill. elth. t. 275. f. 355.
26 guineénse. R.S. (*gr.wh.*) large-berried. Guinea. .... —— H. ⊙. ———— t. 274. f. 354.
27 fistulòsum. R.S. (*wh.*) fistulous-stalked. Mauritius. 1826. —— H. ⊙.
28 nìgrum. R.S. (*wh.*) black-berried. Britain. .... —— H. ⊙. Eng. bot. v. 8. t. 566.
29 oleràceum. R.S. (*wh.*) cultivated. W. Indies. .... —— S. ⊙. Piso lib. 4. c. 50. f. 3.
30 pterocaùle. R.S. (*wh.vi.*) wing-stalked. America. .... —— H. ⊙. Dill. elth. t. 275. f. 256.
31 judàicum. R.S. (*wh.*) rough-stalked. Europe. .... —— H. ⊙.
32 suffruticòsum. R.S. (*wh.*) suffrutescent. Barbary. 1804. 5. 9. G. ♄.
33 quadrangulàre. R.S. (*wh.*) quadrangular. C. B. S. 1812. —— H. ⊙.
34 triangulàre. R.S. (*wh.*) three-sided. E. Indies. .... 6. 9. S. ⊙.
35 rùbrum. R.S. (*wh.*) red-berried. S. America. .... —— H. ⊙.
36 incértum. R.S. (*wh.*) uncertain. E. Indies. 1810. —— H. ⊙.
37 miniàtum. R.S. (*wh.*) pubescent-stalk'd. Europe. .... —— H. ⊙.
38 hùmile. R.S. (*wh.*) small. S. Europe. .... —— H. ⊙.
39 flàvum. R.S. (*st.*) yellow-berried. Hungary. 1816. —— H. ⊙.
ochroleùcum. Dun.
40 villòsum. R.S. (*wh.*) villous annual. S. Europe. .... —— H. ⊙. Dill. elth. t. 274. f. 353.
41 hirsùtum. R.S. (*wh.*) hairy annual. Egypt. .... —— H. ⊙.
42 decemdentàtum. F.I. (*wh.*) ten-toothed. China. 1826. —— H. ⊙.
43 Rúmphii. R.S. (*wh.*) Rumphius's. E. Indies. .... —— H. ⊙. Rumph. amb. 6. t. 26. f. 1.

44	Bessèri. R.S.	(wh.)	Besser's.	America.	1824.	6. 9.	H.⊙.	
45	cestrifòlium. S.S.	(wh.)	Cestrum-leaved.	....	1826.	4. 7.	S. ♄.	
46	longiflòrum. R.S.	(bl.)	long-flowered.	Guiana.	1823.	6. 8.	S. ♄.	Dunal solan. t. 9.
	longifòlium. Dun.							
47	triquètrum. R.S.	(wh.)	triquetrous.	New Spain.	1821.	——	S. ♄.	Cavan. ic. 3. t. 259.
48	bombénse. R.S.	(wh.)	Bomba.	S. America.	1818.	——	S. ♄.	Jacq. f. eclog. t. 24.
49	pubígerum. R.S.	(wh.)	downy.	Mexico.	1823.	——	S. ♄.	Dunal sol. t. 6.
	Cervantèsii. Lag.							
50	pyrifòlium. R.S.	(wh.)	Pear-leaved.	W. Indies.	1824.	6. 8.	S. ♄.	Dun. sol. ed. 2. t. 34.
51	macranthèrum. R.S.	(vi.)	large-anthered.	Mexico.	1826.	——	G. ♄.⌣.	H. et B. gen. am. 3. t. 195.
52	laurifòlium. R.S.	(wh.)	Laurel-leaved.	S. America.	1820.	——	S. ♄.	Dunal sol. t. 8.
53	spirále. F.I.	(wh.)	spiral-racemed.	E. Indies.	——	4. 6.	S. ♄.	
54	Hookeriànum. S.S.	(bl.)	Hooker's.	Mexico.	1825.	6. 9.	G. ♄.	Bot. mag. t. 2708.
	coriàceum. Bot. mag. *nec aliorum.*							
55	dealbàtum. Lind.	(wh.)	white-leaved.	Chile.	1826.	10. 11.	G. ♄.	Bot. mag. t. 2697.
	saponàceum. B.M. *non Dunal.*							
56	salicifòlium. D.L.	(wh.)	Willow-leaved.	Trinidad.	1822.	6. 9.	S. ♄.	
57	verbascifòlium. R.S.	(w.)	Mullein-leaved.	W. Indies.	1749.	6. 8.	S. ♄.	Jacq. vind. 1. t. 13.
58	auriculàtum. R.S.	(vi.)	ear-leaved.	Madagascar.	1773.	....	S. ♄.	Scop. del. 3. t. 8.
59	pubéscens. R.S.	(pu.)	pubescent.	E. Indies.	1816.	6. 8.	S. ♄.	Willd. phyt. 1. t. 3.
60	argénteum. R.S.	(wh.)	silvery.	Brazil.	——	——	S. ♄.	Dunal sol. ed. 2. t. 39.
61	nùdum. K.S.	(wh.)	naked.	Mexico.	1824.	3. 6.	S. ♄.	
62	diphy'llum. R.S.	(wh.)	two-leaved.	W. Indies.	1699.	6. 7.	S. ♄.	Jacq. ic. 2. t. 322.
63	arbòreum. R.S.	(wh.)	tree.	Cumana.	1819.	——	S. ♄.	Dunal sol. ed. 2. t. 198.
64	havanénse. R.S.	(bl.)	Havannah.	W. Indies.	1793.	7. 8.	S. ♄.	Jacq. amer. pict. t. 48.
65	aggregàtum. R.S.	(li.)	crowded.	Guinea.	——	——	S. ♄.	Jacq. ic. 2. t. 323.
66	Plukenètii. R.S.	(wh.)	Plukenet's.	S. America.	....	6. 8.	S. ♄.	Pluk. phyt. t. 227. f. 2.
67	eriocályx. R.S.	(wh.)	woolly-calyxed.	Carthagena.	1731.	5. 8.	S. ♄.	
68	umbellàtum. R.S.	(wh.)	umbelled.	S. America.	——	——	S. ♄.	
69	americànum. R.S.	(pu.)	American.	Virginia.	1727.	7. 9.	H.⊙.	
70	axilliflòrum. R.S.	(bl.)	axil-flowered.	Vera Cruz.	1728.	6. 9.	S. ♄.⌣.	
	scándens. Miller.							
71	memphíticum. R.S.	(wh.)	tooth-stalked.	....	1823.	——	S. ♃.	
72	pérsicum. R.S.	(bl.)	Persian.	Persia.	1829.	6. 9.	H. ♄.⌣.	
73	capsicástrum. Ot.	(wh.)	Capsicum-like.	....	1827.	——	S. ♄.	
74	lycioídes. R.S.	(li.)	spiny.	Peru.	1791.	5. 6.	S. ♄.	Jacq. ic. 1. t. 46.
75	monánthum. R.S.	(fl.)	one-flowered.	New Spain.	1816.	——	S. ♃.	
76	fùgax. R.S.	(wh.ye.)	fugacious.	Caracas.	——	——	S. ♄.	Jacq. ic. 2. t. 324.
77	stellàtum. R.S.	(bl.)	starry.	S. America.	——	——	S. ♄.	———— 2. t. 325.
78	virgàtum. R.S.	(vi.)	twiggy.	W. Indies.	1819.	——	S. ♄.	Dunal sol. ed. 2. t. 40.
79	léntum. R.S.	(vi.)	climbing woolly.	New Spain.	1823.	6. 8.	S. ♄.⌣.	Cavan. ic. 4. t. 308.
80	scándens. R.S.	(pu.)	climbing.	W. Indies.	1822.	....	S. ♄.⌣.	
81	atrosanguíneum. Sck.	(c.)	dark crimson.	....	1827.	....	S. ♄.	
82	elæagnifòlium. R.S.	(bl.)	Elæagnus-leaved.	Chile.	1823.	6. 8.	S. ♄.	Cavan. ic. 3. t. 243.
83	obtusifòlium. K.S.	(bl.)	blunt-leaved.	Mexico.	1826.	——	G. ♄.	Dunal sol. ed. 2. t. 119.
84	stellígerum. R.S.	(li.)	starry-haired.	N. Holland.	1820.	——	S. ♄.	Sm. exot. bot. 2. t. 88.
85	racemòsum. R.S.	(wh.)	wave-leaved.	W. Indies.	1781.	7. 8.	S. ♄.	Jacq. amer. pict. t. 50.
86	ígneum. R.S.	(wh.)	red-spined.	S. America.	1714.	3. 11.	S. ♄.	Jacq. vind. 1. t. 14.
87	subarmàtum. R.S.	(wh.)	partly armed.	————	1823.	——	S. ♄.	
88	bahaménse. R.S.	(wh.vi.)	Bahama.	Bahama.	1732.	6. 7.	S. ♄.	Dill. elth. t. 271. f. 350.
89	hírtum. R.S.	(wh.)	hairy-calyxed.	Trinidad.	1823.	——	S. ♄.	Vahl ic. rar. d. 3. t. 21.
90	tomentòsum. R.S.	(bl.)	tomentose.	S. America.	1662.	——	S. ♄.	Bocc. sic. 8. t. 5.
91	hy'bridum. R.S.	(pa.bl.)	hybrid.	Guinea.	1815.	——	S. ♄.	Jacq. vind. 2. t. 113.
92	coccíneum. R.S.	(bh.)	scarlet-fruited.	....	1820.	——	S. ♄.	Jacq. ic. 1. t. 43.
93	jamaicénse. R.S.	(li.)	Jamaica.	Jamaica.	——	——	S. ♄.	
94	cuneifòlium. R.S.	(wh.)	wedge-leaved.	W. Indies.	....	——	S. ♄.	Dunal sol. t. 22.
95	heterophy'llum. R.S.	(bh.)	variable-leaved.	Guiana.	1823.	7. 9.	S. ♄.	———— ed. 2. t. 62.
96	volùbile. R.S.	(bl.)	twining.	Hispaniola.	——	——	S. ♄.⌣.	Plum. Mss. t. 4. f 32.
97	lanceæfòlium. R.S.	(wh.)	lance-leaved.	S. America.	....	——	S. ♄.⌣.	Jacq. ic. 2. t. 329.
98	coriàceum. R.S.	(bl.)	leathery-leav'd.	Trinidad.	1824.	6. 9.	S. ♄.⌣.	Dun. sol. p. 197. t. 14.
99	bonariénse. R.S.	(wh.)	Buenos Ayres.	BuenosAyres.	1727.	——	H. ♄.	Dill. elth t. 272. f. 351.
100	fastigiàtum. R.S.	(pa.bl.)	fastigiate.	S. America.	1820.	——	S. ♄.	Jacq. f. eclog. t. 6.
101	oporìnum. R.S.	(bl.)	large-leaved.	————	————	——	S. ♄.	Dunal sol. t. 16.
	macrophy'llum. Dun.							
102	subinérme. R.S.	(bl.)	spear-leaved.	W. Indies.	1752.	7. 8.	S. ♄.	Jacq. amer. t. 40. f. 3.
103	lanceolàtum. R.S.	(bl.)	lanceolate.	New Spain.	....	6. 7.	S. ♄.	Cavan. ic. 3. t. 245.
104	astroítes. Jacq.	(bl.)	Aster-like.	S. America.	1823.	——	S. ♄.	Jacq. f. eclog. t. 65.
105	glutinòsum. R.S.	(bl.)	glutinous.	....	1818.	6. 8.	S. ♄.	Dunal sol. ed. 2. t. 54.
106	Bro'wnii. R.S.	(vi.)	Brown's.	N. S. W.	1820.	——	S. ♄.	
	violàceum. B.P. *non* Jacq.							

107 gigánteum. R.S. (*wh.vi.*) tall. C. B. S. 1792. 6. 8. G. ♄. Bot. mag. t. 1921.
108 tórvum. R.S. (*wh.*) curved-prickled. W. Indies. 1816. —— S. ♄.
109 ferrugíneum. R.S. (*wh.*) ferruginous. .... —— —— S. ♄. Jacq. schœn. 3. t. 334.
110 elàtum. L.en. (*bl.*) tall. .... 1824. 6. 9. S. ♄.
111 ochroneùrum. L.en. (*bl.*) rusty-branched. Brazil. —— 6. 8. S. ♄.
112 mólle. K.S. (*wh.*) soft-leaved. Caracas. 1826. —— G. ♄.
113 Lichtensteínii. R.S. .. Lichtenstein's. C. B. S. 1824. .... G. ♄.
114 esculéntum. R.S. (*li.*) eatable. Tropics. 1597. 6. 9. G. ⊙. Dunal sol. t. 3. f. E.
115 ovígerum. R.S. (*pa.bl.*) Egg-plant. Arabia. —— —— G. ⊙. Blackw. t. 549.
α violaceum. (*pa.bl.*) *violet-fruited.* ——— .... —— G. ⊙.
β álbum. (*pa.bl.*) *white-fruited.* ——— .... —— G. ⊙.
γ luteùm. (*pa.bl.*) *yellow-fruited.* ——— .... —— G. ⊙.
δ rúber. (*pa.bl.*) *red-fruited.* ——— .... —— G. ⊙.
116 lóngum. F.I. (*vi.*) long-fruited. E. Indies. 1816. —— S. ♄.
117 insànum. W. (*pu.*) insane. ——— 1815. 8. 10. S. ♄. Pluk. alm. t. 226. f. 3.
118 melanóxylon. S.S. (*vi.*) black-wooded. .... 1825. 7. 10. S. ⊙.
119 undàtum. R.S. (*pa.bl.*) waved. E. Indies. 1818. 6. 9. S. ♄. Rheed. mal. 2. t. 37.
120 fuscàtum. R.S. (*vi.*) brown-branched. America. .... 7. 8. G. ⊙. Jacq. ic. 1. t. 42.
121 sodòmeum. R.S. (*pu.*) black-spined. Africa. 1688. 6. 7. G. ♄. Flor. græc. t. 235.
122 incànum. R.S. (*re.*) hoary. Mauritius. 1823. —— S. ♄. Scop. del. 1. t. 1.
123 índicum. R.S. (*bl.*) Indian. India. 1732. 7. S. ♄. Dill. elth. t. 270. f. 349.
124 coágulans. R.S. (*bl.*) scollop-leaved. Arabia Felix. 1802. —— S. ♄. Jacq. schœn. 4. t. 469.
125 sánctum. R.S. (*bl.pu.*) oblique-leaved. Egypt. —— —— S. ♄.
126 marginàtum. R.S. (*li.*) white margined. Africa. 1775. 6. 9. G. ♄. Bot. mag. t. 1928.
127 congénse. L.en. (*bl.*) Congo. ——— 1815. —— G. ♄.
128 campechiénse. R.S. (*li.*) purple-spined. America. 1732. 7. G. ♄. Dill. elth. t. 268. f. 347.
129 platanifòlium. B.M. (*vi.*) Plane-leaved. S. America. 1822. 7. 9. S. ♄. Bot. mag. t. 2618.
130 myriacánthum. R.S. (*w.*) many-spined. ——— —— —— S. ♄.
131 aculeatíssimum. R.S. (*w.*) very prickly. ——— 1800. —— S. ♄. Jacq. ic. 1. t. 41.
132 spinosíssimum. D.L. (*w.*) very spiny. Trinidad. 1820. —— S. ♄.
133 lívidum. S.S. (*wh.*) livid. Madagascar. 1821. —— S. ♄.
134 lasiocárpum. R.S. (*wh.*) four-celled. E. Indies. 1824. 6. 8. S. ♂. Rheel. mal. 2. t. 35.
hirsùtum. F.I. non Dunal. *quadriloculàre.* S.S.
135 mammòsum. R.S. (*pa.bl.*) nipple. W. Indies. 1699. 7. 8. H. ⊙. Merian. surin. t. 27.
136 ciliàtum. R.S. (*pa.bl.*) ciliated. .... 1816. —— H. ⊙. Dunal sol. t. 18.
137 stramonifòlium. R.S. (*bl.*) broad-leaved. E. Indies. 1778. 6. 9. S. ♄. Jacq. ic. 1. t. 44.
138 flavéscens. R.S. (*wh.*) yellowish. Trinidad. 1822. —— S. ♄.
139 fèrox. R.S. (*li.pu.*) Malabar. Malabar. 1795. 8. 9. S. ♃. Dunal sol. ed. 2. t. 68.
140 crassipétalum. F.I. (*li.*) thick-petaled. Nepaul. 1824. 4. 7. G. ♄.
141 lysimachioídes. F.I. (*w.*) Loosestrife-like. E. Indies. 1823. 6. 9. S. ♃.
142 campanulàtum. B.P. (*pu.*) bell-flowered. N. S. W. 1818. 8. 9. G. ⊙.
143 armàtum. B.P. (*pu.*) armed. ——— 1819. —— S. ♃. Dunal sol. ed. 2. t. 69.
144 pungètium. B.P. (*vi.*) pungent. ——— 1823. —— G. ⊙. ——— ed. 2. t. 70.
145 cinèreum. B.P. (*wh.*) grey. ——— —— —— G. ♃.
146 Millèri. R.S. (*wh.*) Miller's. C. B. S. 1762. 7. 8. G. ♄. Jacq. ic. 2. t. 330.
147 trilobàtum. R.S. (*wh.*) three-lobed. W. Indies. 1759. 8. S. ♄. Pluk. alm. t. 316. f. 5.
148 acetosæfòlium. R.S. (*w.*) Sorrel-leaved. E. Indies. —— —— S. ♄. Burm. ind. t. 22. f. 2.
149 rigéscens. R.S. (*vi.*) rigid. C. B. S. 1815. 6. 8. G. ♄. Jacq. schœn. 1. t. 42.
150 carolinénse. R.S. (*w.bl.*) Carolina. Carolina. 1732. 7. 9. G. ♃. Jacq. ic. 2. t. 331.
151 violàceum. R.S. (*bl.*) violet. E. Indies. 1823. 6. 9. S. ♄. Jacq. fragm. t. 133. f. 1.
152 pyracánthum. B.M. (*pu.*) orange-thorn'd. Madagascar. 1789. 8. 9. S. ♄. Sm. exot. bot. 2. t. 54.
β inérmis. (*pu.*) *smooth-calyxed.* ——— —— —— S. ♄. Bot. mag. t. 2547.
153 virginiànum. R.S. (*vi.*) Virginian. Virginia. 1662. 5. 8. H. ⊙. Dill. elth. t. 267. f. 346.
154 Jacquìni. R.S. (*bl.pu.*) Jacquin's. E. Indies. 1804. 9. 11. S. ⊙. Jacq. ic. 2. t. 332.
155 xanthocárpum. R.S. (*bl.*) yellow-fruited. Ethiopia. 1828. 6. 10. H. ⊙. Wend. s. han. 1. t. 2.
156 Balbísii. R.S. (*wh.*) Balbis's. S. America. 1816. 4. 9. S. ♄. Bot. mag. t. 2568.
decúrrens. B.R. 140. *brancæfòlium.* Jacq. f. ecl. t. 7. *viscòsum.* DC.
157 téctum. R.S. (*ye.*) villous-calyxed. Mexico. 1826. 7. 10. G. ♄. Cavan. icon. 4. t. 309.
158 capénse. R.S. (*wh.*) Cape prickly. C. B. S. 1816. 6. 10. G. ♄.
159 Houstòni. R.S. (*wh.*) Houstoun's. Vera Cruz. 1731. —— S. ♄.
160 Angùrium. R.S. (*ye.*) procumbent. ——— —— —— S. ♄.

NYCT'ERIUM. V. NYCT'ERIUM. Pentandria Monogynia.

1 cordifòlium. V. (*vi.*) Canary. Canaries. 1779. 3. 4. G. ♄. Venten. malm. t. 85.
Solànum vespertílio. R.S.
2 Amazònium. B.M. (*bl.*) Lambert's. Brazil. 1814. 5. 9. S. ♄. Bot. mag. t. 1801.
Solànum Amazònium. Bot. reg. t. 71.
3 cornùtum. L.en. (*ye.*) horned. S. America. 1825. —— S. ♄. Juss. ann. mus. 3. t. 9.
4 lobàtum. (*ye.*) Missouri. Missouri. 1813. 7. 9. H. ⊙. Pursh fl. amer. 1. t. 7.
Solànum heterándrum. Ph. *Androçèra lobàta.* N.

5 rostràtum. L.en. (*ye.*) beaked. Mexico. 1823. 7. 9. H.⊙. Dunal sol. t. 24.
6 Fontanesiànum. (*ye.*) Desfontaine's. Brazil. 1813. —— H.⊙. Bot. reg. t. 177.
7 heterodóxum. L.en. (*bl.*) very prickly. Mexico. 1824. —— H.⊙. Dunal sol. t. 25.

WITHERI'NGIA. R.S. WITHERI'NGIA. Tetr-Pentandria Monogynia.
1 solanàcea. R.S. (*st.*) yellow-flowered. S. America. 1742. 5. 9. S.♃. L'Herit. sert. angl. t. 1.
2 stramonifòlia. K.S. (*st.*) Stramonium-ld. Mexico. 1823. —— S.♄.
3 crassifòlia. R.S. (*li.*) thick-leaved. C. B. S. 1729. —— G.♄. Dill. elth. t. 273. f. 352.
4 montàna. R.S. (*wh.bl.*) mountain. Peru. 1822. —— G.♃. Bot. mag. t. 2768.
5 phyllántha. R.S. (*bl.*) annual. ———— —— 7. 9. H.⊙. Cavan. ic. 4 t. 359. f. 1.
6 pinnatífida. R.S. (*li.*) pinnatifid. ———— —— 6. 8. G.♃. Flor. peruv. 2. t. 170. f. b.

CE'STRUM. R.S. CE'STRUM. Pentandria Monogynia. L.
1 laurifòlium. R.S. (*ye.*) Laurel-leaved. W. Indies. 1691. 5. 8. S.♄. Smith spicil. 2. t. 2.
2 citrifòlium. R.S. (*st.*) Citron-leaved. .... 1816. —— S.♄.
3 macrophy'llum. R.S. (*st.*) large-leaved. W. Indies. 1812. —— S.♄. Venten. choix. t. 18.
4 venenàtum. R.S. (*st.*) poisonous. C. B. S. 1787. 2. 4. G.♄.
5 noctúrnum. R.S. (*gr.ye.*) night-smelling. E. Indies. 1732. 11. S.♄. Dill. elth. t. 153. f. 185.
6 hírtum. R.S. (*gr.wh.*) hairy. Jamaica. 1822. 5. 8. S.♄.
7 Párqui. R.S. (*st.*) Willow-leaved. S. America. 1787. 6. 7. G.♄. Bot. mag. t. 1770.
8 thyrsoídeum. K.S. (*st.*) bunch-flowered. Mexico. 1826. 5. 7. S.♄.
9 suberòsum. R.S. (*gr.*) cork-barked. .... 1815. 6. 7. S.♄. Jacq. schœn. 4. t. 452.
10 auriculàtum. W. (*wh.*) eared-leaved. Peru. 1774. —— S.♄. L'Herit. s. nov. 1. t. 35.
11 exstipulàtum. R.S. (*gr.*) exstipulate. S. America. 1824. —— S.♄.
12 tinctòrium. R.S. (*wh.*) dying. Caracas. 1823. 5. 6. S.♄. Jacq. schœn. 3. t. 332.
13 undulàtum. R.S. (*st.*) waved-leaved. Peru. —— —— S.♄. Flor. peruv. 2. t. 155.
14 multiflòrum. R.S. (*gr.*) many-flowered. S. America. 1825. —— S.♄.
15 paniculàtum. K.S. (*gr.ye.*) panicled. Caracas. 1819. .... S.♄.
16 latifòlium. R.S. (*st.*) broad-leaved. Trinidad. 1822. 5. 6. S.♄.
17 alaternoídes. R.S. (*gr.ye.*) Alaternus-like. S. America. 1820. 4. 6. S.♄. Bot. mag. t. 2929.
18 hirsùtum. R.S. (*st.*) hairy-stalked. .... 1822. 5. 6. S.♄. Jacq. schœn. 3. t. 324.
19 cauliflòrum. R.S. (*wh.*) stem-flowering. Martinica. 1818. —— S.♄. ———— 3. t. 325.
20 salicifòlium. R.S. (*st.*) Sallow-leaved. Caracas. .... 3. 6. S.♄. ———— 3. t. 326.
21 pendulìnum. R.S. (*wh.*) pendulous. ———— 1820. —— S.♄. ———— 3. t. 327.
22 vespertìnum. R.S. (*wh.*) cluster-flowered. W. Indies. 1759. 5. 7. S.♄. ———— 3. t. 328.
23 pállidum. R.S. (*gr.ye.*) pale-flowered. Jamaica. 1732. —— S.♄. Sloan. hist. 2. t. 204. f. 2.
24 fœtidíssimum. R.S. (*ye.*) stinking. ———— .... 5. 8. S.♄. Jacq. schœn. 3. t. 329.
25 bracteàtum. B.M. (*st.*) large-bracted. S. America. 1828. 12. 2. S.♄. Bot. mag. t. 2974.
26 acuminàtum. (*ye.gr.*) tapering-leaved. Mexico. 1824. 7. 10. F.♄. Swt. br. fl. gar. ser. 2.
27 fastigiàtum. R.S. (*wh.*) Honeysuckle. W. Indies. .... 10. 12. S.♄. Bot. mag. t. 1729.
28 odontospérmum. R.S. (*w.*) smooth. S. America. 1793. 7. 9. S.♄. Jacq schœn. 3. t. 331.
29 diúrnum. R.S. (*wh.*) day smelling. W. Indies. 1732. 10. 11. S.♄. Dill. elth. t. 154. f. 186.
30 ròseum. R.S. (*ro.*) rose-coloured. Mexico. 1829. .... G.♄.
31 tomentòsum. R.S. (*st.*) downy. S. America. 1790. 6. 7. S.♄.
32 angustifòlium. B.C. (*st.*) narrow-leaved. ———— 1820. —— S.♄. Lodd. bot. cab. t. 618.

LY'CIUM. R.S. BOX-THORN. Pentandria Monogynia. L.
1 áfrum. R.S. (*vi.*) African. C. B. S. 1712. 6. 7. H.♄. Bot. reg. t. 354.
2 ténue. R.S. (*pu.*) slender. ———— —— —— H.♄.
3 rígidum. R.S. (*vi.*) rigid. ———— 1795. —— H.♄. Trew Ehret. t. 24. f. 1.
carnòsum. P.S. *itálicum* Miller.
4 microphy'llum. R.S. (*vi.*) small-leaved. ———— —— —— H.♄.
5 ruthènicum. R.S. (*ro.*) Russian. Siberia. 1804. —— H.♄. Pall. ros. 1. t. 49.
6 tetrándrum. R.S. (*pu.*) tetrandrous. C. B. S. .... —— H.♄.
7 barbàrum. R.S. (*re.*) Willow-leaved. Barbary. 1696. 5. 8. H.♄.⌣. Wats. dend. brit. t. 9.
8 Sháwii. R.S. (*wh.*) white-flowered. ———— .... —— H.♄. Shaw afric. f. 349.
9 Trewiànum. R.S. (*pu.*) Trew's. China. .... —— H.♄. Trew Ehret. t. 68.
chinénse. P.S. *non* Mill.
10 chinénse. R.S. (*re.*) Chinese. ———— .... —— H.♄. Wats. dend. brit. t. 8.
ovàtum. P.S. *barbàrum* β *chinénse.* W.
11 lanceolàtum. P.S. (*vi.*) spear-leaved. S. Europe. .... —— H.♄. Duham. arb. nov. t. 32.
12 turbinàtum. P.S. (*re.*) top-shaped. China. 1709. —— H.♄. ———— t. 31.
13 cinéreum. R.S. (*pu.*) gray. C. B. S. 1815. 6. 8. H.♄.
14 europ'æum. R.S. (*li.*) European. S. Europe. 1730. 5. 8. H.♄. Flor. græc. t. 236.
15 hórridum. R.S. (*wh.*) succulent-leav'd. C. B. S. 1791. 7. 8. H.♄.
16 boerhaaviæfòlium. (*bh.*) glaucous-leaved. Peru. 1780. 4. 5. G.♄. L'Herit. stirp. t. 83.
17 arboréscens. S.S. (*wh.*) arborescent. ———— 1819. 6. 7. G.♄. Flor. peruv. 2. t. 182. f. a.
A'tropa arboréscens. R.S.
18 caroliniànum. R.S. (*pu.*) Carolina. Carolina. 1806. 7. 9. H.♄.
19 capsulàre. R.S. (*pu.*) capsular. Mexico. 1730. 6. 8. G.♄.

SOLA'NDRA. R.S. SOLA'NDRA. Pentandria Monogynia. R.S.
1 grandiflòra. R.S. (*li.*) large-flowered. Jamaica. 1781. 3. 8. S.♄. Jacq. schœn. 1. t. 45.

2 nítida. R.S. (*st.*) glossy-leaved. Jamaica. 1781. 3. 8. S. ♄. Bot. mag. t. 1874.
grandiflòra. B.M. *non* Jacq.
3 viridiflòra. S.S. (*gr.*) green-flowered. Brazil. 1816. 5. 6. S. ♄. Bot. mag. t. 1948.
4 oppositifolia. ...... opposite-leaved. W. Indies. 1824. .... S. ♄.
BRUNFE'LSIA. W. BRUNFE'LSIA. Didynamia Angiospermia. L.
1 undulàta. W. (*wh.st.*) large-flowered. Jamaica. 1780. 6. 7. S. ♄. Bot. reg. t. 228.
2 americàna. W. (*wh.st.*) American. W. Indies. 1735. —— S. ♄. Bot. mag. t. 393.
α latifòlia. (*wh.st.*) *broad-leaved.* ——— —— —— S. ♄.
β angustifòlia. (*wh.st.*) *narrow-leaved.* ——— —— —— S. ♄.
3 violàcea. B.C. (*wh.st.*) purple-stalked. ——— 1820. —— S. ♄. Lod. bot. cab. t. 792.
4 uniflòra. D.D. (*bl.wh.*) single-flowered. Brazil. 1827. 1. 12. S. ♄. Bot. mag. t. 2829.
Franciscea Hopeàna. B.M.
CRESCE'NTIA. W. CALABASH-TREE. Didynamia Angiospermia. L.
1 Cujète. W. (*li.ye.*) oval-fruited. Jamaica. 1690. .... S. ♄. Jacq. amer. t. 111.
2 acuminàta. K.S. (*li.*) acuminate. Cuba. 1822. .... S. ♄.
3 cucurbitìna. W. (*li.ye.*) round-fruited. W. Indies. 1733. .... S. ♄. Plum. ic. t. 109.
4 alàta. K.S. (*li.*) winged-leaved. Mexico. 1825. .... S. ♄.
TRIPINN'ARIA. S.S. TRIPINN'ARIA. Didynamia Angiospermia.
africàna. S.S. (*re.*) African. Mauritius. 1823. .... S. ♄. Jacq. coll. 3. t. 18.
Crescéntia pinnàta. Jacq. *Tan'æcium pinnàtum.* W.

ORDO CLII.

SCROPHULARINÆ. *Brown prodr.* 1. *p.* 433.

Sect. I. *STAMINA* (4) *ANTHERIFERA.*

BU'DDLEA. S.S. BU'DDLEA. Tetrandria Monogynia. L.
1 madagascariénsis. (*or.*) Madagascar. Madagascar. 1824. 9. 10. S. ♄. Bot. mag. t. 2824.
2 salígna. S.S. (*wh.*) Willow-like. C. B. S. 1816. 8. 9. G. ♄. Jacq. schœn. 1. t. 29.
3 paniculàta. S.S. (*wh.*) panicled. Nepaul. 1823. —— G. ♄.
4 americàna. R.S. (*ye.*) American. W. Indies. 1824. 7. 10. S. ♄. Sloan. hist. 2. t. 173. f. 1.
5 occidentàlis. R.S. (*wh.*) occidental. Peru. 1730. .... G. ♄. Flor. peruv. 1. t. 83.
6 salvifòlia. S.S. (*wh.cr.*) Sage-leaved. C. B. S. 1760. 8. 9. G. ♄. Jacq. schœn. 1. t. 28.
7 Neèmda. S.S. (*wh.*) Nepaul. Nepaul. 1822. —— G. ♄.
8 brasiliénsis. S.S. (*or.*) Brazilian. Brazil. —— 9. 11. S. ♄. Bot. mag. t. 2713.
9 heterophy'lla. B.R. (*ye.*) different-leaved. S. America. 1826. 1. 5. S. ♄. Bot reg. t. 1259.
10 globòsa. S.S. (*ye.*) round-headed. Chile. 1774. 5. 7. H. ♄. Bot. mag. t. 174.
11 connàta. S.S. (*or.*) connate-leaved. Peru. 1826. 5. 6. G. ♄. ——— t. 2853.
12 salicifòlia. S.S. (*ye.*) Willow-leaved. S. America. 1823. .... S. ♄.
13 diversifòlia. R.S. (*wh.*) various-leaved. Java. —— .... S. ♄.
SCOP'ARIA. L. SCOP'ARIA. Tetrandria Monogynia. L.
1 dúlcis. L. (*wh.*) sweet. S. America. 1730. 6. 9. S. ♄. Herm. parad. t. 241.
2 flàva. Ot. (*ye.*) yellow. ——— 1829. —— S. ♄.
CAPR'ARIA. S.S. GOAT-WEED. Didynamia Angiospermia. L.
1 undulàta. S.S. (*ro.*) wave-leaved. C. B. S. 1774. 3. 7. G. ♄. Bot. mag. t. 1556.
2 cuneàta. S.S. (*ro.*) wedge-leaved. S. America. 1759. 6. 8. S. ♄.
3 lanceolàta. S.S. (*ye.*) spear-leaved. C. B. S. 1774. 5. 7. G. ♄. Pl. sel. h. ber. t. 4.
4 biflòra. S.S. (*wh.*) two-flowered. S. America. 1752. 7. 8. S. ♄. Jacq. amer. t. 115.
5 hùmilis. W. (*ro.*) dwarf. E. Indies. 1781. —— S. ⊙.
TE'EDIA. S.S. TE'EDIA. Didynamia Angiospermia. S.S.
1 lùcida. S.S. (*pu.*) shining. C. B. S. 1774. 4. 7. G. ♄. Bot. reg. t. 210.
Caprària lùcida. W. *Borkhausènia lùcida.* Roth.
2 pubéscens. S.S. (*pu.*) pubescent. C. B. S. 1816. 5. 10. G. ♄. Bot. reg. t. 214.
HALL'ERIA. S.S. HALL'ERIA. Didynamia Angiospermia. L.
1 lùcida. S.S. (*sc.*) shining-leaved. C. B. S. 1752. 6. 8. G. ♄. Bot. mag. t. 1744.
2 ellíptica. S.S. (*sc.*) elliptic-leaved. ——— 1816. —— G. ♄. Burm. afr. t. 84. f. 1.
STEM'ODIA. S.S. STEM'ODIA. Didynamia Angiospermia. W.
1 suffruticòsa. S.S. (*bl.*) suffrutescent. S. America. 1823. 5. 8. S. ♄.
2 viscòsa. S.S. (*vi.*) viscous. E. Indies. 1818. 7. 10. S. ⊙. Roxb. Corom. t. 163.
3 verticillàris. S.S. (*pu.*) whorled. Brazil. 1822. —— H. ⊙.
4 parviflòra. H.K. (*wh.*) small-flowered. S. America. 1759. 7. 8. S. ♃.
5 marítima. S.S. (*bl.*) sea-side. Cuba. 1822. —— S. ⊙. Jacq. amer. t. 174. f. 66.

RUSS'ELIA. S.S. Russ'elia. Didynamia Angiospermia. W.
1 floribúnda. K.S. (*sc.*) abundant-flowering. Mexico. 1824. 6. 8. S.♃.
2 multiflòra. S.S. (*sc.*) many-flowered. ——— 1812. —— S.♃. Bot. mag. t. 1528.
TREVIR'ANIA. S.S. Trevir'ania. Didynamia Angiospermia. S.S.
coccínea. S.S. (*sc.*) scarlet. Jamaica. 1778. 8. 10. S.♃. Bot. mag. t. 374.
Cyrílla pulchélla. B.M. *Achimènes coccínea.* P.S.
COLU'MNEA. S.S. Colu'mnea. Didynamia Angiospermia. L.
1 scándens. S.S. (*sc.*) climbing. W. Indies. 1759. 8. 9. S.♄. Bot. reg. t. 805.
2 hirsùta. S.S. (*li.*) hairy. ——— 1780. 8. 11. S.♄. Brown jam. t. 30. f. 3.
3 trifoliàta. L.en. (*bl.*) three-leaved. S. America. 1825. —— S.♄.
4 rùtilans. S.S. (*ye.re.*) red-leaved. Jamaica. 1823. —— S.♄.
5 híspida. S.S. (*sc.*) bristly. ——— 1824. —— S.♄.
MAURA'NDIA. S.S. Maura'ndia. Didynamia Angiospermia. W.
1 semperflòrens. S.S. (*ro.*) red-flowered. Mexico. 1796. 1. 12. G.♄.‿. Bot. mag. t. 460.
2 antirrhiniflòra. S.S. (*bl.*) blue-flowered. ——— 1814. —— G.♄.‿. ——— t. 1643.
3 Barclayàna. B.R. (*bl.*) Mr. Barclay's. ——— 1826. 4. 11. G.♄.‿. Bot. reg. t. 1108.
LOPHOSPE'RMUM. D.D. Lophospe'rmum. Didynamia Angiospermia.
scándens. L.T. (*pu.*) climbing. Mexico. 1830. 6. 8. G.♄.‿.
DODA'RTIA. S.S. Doda'rtia. Didynamia Angiospermia. L.
orientàlis. S.S. (*d.pu.*) oriental. Levant. 1752. 7. 8. H.♃. Swt. br. fl. gar. t. 147.
CYMB'ARIA. S.S. Cymb'aria. Didynamia Angiospermia. L.
davùrica. S.S. (*ye.*) Daurian. Davuria. 1796. 6. 7. H.♃. Amm. ruth. t. 1. f. 2.
COLLI'NSIA. N. Colli'nsia. Didynamia Angiospermia.
1 vérna. N. (*bl.pu.*) large-flowered. N. America. 1827. 4. 11. H.⊙. Swt. br. fl. gar. t. 220.
grandiflòra. Bot. reg. t. 1107.
2 parviflòra. D. (*bl.*) small-flowered. ——— —— 6. 9. H.⊙. Bot. reg. t. 1082.
NEM'ESIA. S.S. Nem'esia. Didynamia Angiospermia. S.S.
1 chamædryfòlia. S.S. (*w.bl.*) Chamædry's-ld. C. B. S. 1787. 4. 9. G.♃.
Antirrhìnum macrocárpum. W.
2 bicórne. S.S. (*wh.bl.*) horned. ——— 1774. 7. 8. G.⊙. Burm. afr. t. 75. f. 3.
3 fœ'tens. S.S. (*wh.bl.*) fœtid. ——— 1798. 4. 9. G.♃. Venten. malm. t. 41.
4 lineàris. S.S. (*sn.ro.*) linear-leaved. ——— 1822. —— G.⊙.
ANARRH'INUM. S.S. Anarrh'inum. Didynamia Angiospermia. S.S.
1 pubéscens. E. (*bl.*) pubescent. Mount Sinai. 1826. 5. 8. H.♃.
2 bellidifòlium. S.S. (*bl.*) Daisy-leaved. France. 1629. 6. 8. H.♂. Bot. mag. t. 2056.
ANTIRRH'INUM. S.S. Snap-dragon. Didynamia Angiospermia. L.
1 màjus. S.S. (*va.*) great. England. .... 6. 8. H.♃. Eng. bot. v. 2. t. 129.
α albiflòrum. (*wh.*) *white-flowered.* ——— .... —— H.♃.
β ròseum. (*ro.*) *rose-coloured.* ——— .... —— H.♃.
γ coccíneum. (*sc.*) *scarlet.* ——— .... —— H.♃.
δ bícolor. (*re.wh.*) *two-coloured.* ——— .... —— H.♃.
ε múltiplex. (*sc.*) *double-flowered.* .... —— H.♃.
2 meonánthum. L.en. (*re.*) smaller-flower'd. S. Europe. 1820. —— H.♃. Lk. fl. port. 1. t. 56.
3 angustifòlium. Urv. (*re.*) narrow-leaved. Italy. 1818. —— H.♃.
4 sículum. S.S. (*st.*) Sicilian. Sicily. 1804. 7. 8. H.♃.
5 Oróntium. S.S. (*bh.*) Calf's-snout. Britain. .... 7. 9. H.⊙. Eng. bot. v. 17. t. 1155.
6 calycìnum. L.en. (*wh.*) long-calyxed. S. Europe. .... —— H.⊙.
7 Asarìna. S.S. (*wh.*) heart-leaved. Italy. 1699. 7. F.♃. Bot. mag. t. 902.
8 mólle. S.S. (*wh.*) soft-leaved. Spain. 1752. 7. 10. F.♃.
LIN'ARIA. S.S. Toad-flax. Didynamia Angiospermia. S.S.
1 fruticòsa. S.S. (*ye.*) shrubby. Barbary. 1822. 5. 6. G.♄. Desf. atl. 2. t. 135.
2 frutéscens. S.S. (*ye.*) frutescent. C. B. S. 1816. 5. 7. G.♄.
Antirrhìnum frutéscens. Th.
3 dalmática. S.S. (*ye.*) Dalmatian. Levant. 1731. 6. 7. G.♄. Buxb. cent. 1. t. 24.
4 scopària. S.S. (*ye.*) broom-like. Teneriffe. 1815. —— G.♄.
5 Cymbalària. S.S. (*bh.*) Ivy-leaved. England. .... 5. 11. H.♃. Eng. bot. v. 7. t. 502.
6 pállida. T.S. (*li.*) pale-flowered. Italy. 1824. —— H.♃.
pubéscens. T.P. *non* P.S.
7 pilòsa. S.S. (*li.*) hairy-leaved. ——— 1800. 6. 9. H.♃. Jacq. obs. 2. t. 48.
8 hepaticæfòlia. S.S. (*pu.*) Hepatica-leav'd. Corsica. 1828. —— F.♃.
9 æquitrilòba. S.S. (*d.pu.*) equal 3-lobed. Sardinia. —— —— F.♃. Bot. mag. t. 2941.
10 Elátine. S.S. (*bl.*) sharp-pointed. England. .... 7. 11. H.⊙. Eng. bot. v. 10. t. 692.
11 cirrhòsa. S.S. (*bl.wh.*) cirrhose. Egypt. 1777. 7. 8. H.⊙. Jacq. vind. 1. t. 82.
12 spùria. S.S. (*ye.vi.*) round-leaved. England. .... 7. 9. H.⊙. Eng. bot. v. 10. t. 691.
Antirrhìnum spùrium. E.B.
13 lanígera. S.S. (*st.*) woolly. S. Europe. 1820. —— H.⊙.
14 dealbàta. S.S. (*ye.*) villous-leaved. Portugal. —— 7. 9. H.⊙.
15 latifòlia. S.S. (*ye.*) broad-leaved. Barbary. 1800. 6. 7. H.⊙. Desf. atl. 2. t. 134.
16 triphy'lla. S.S. (*bl.ye.*) three-leaved. Sicily. 1596. 6. 9. H.⊙. Bot. mag. t. 324.

17	refléxa. s.s.	(*wh.*)	reflex-flowered.	S. Europe.	1824.	6. 9.	H.⊙.	Flor. græc. t. 593.
18	origanifòlia. s.s.	(*bl.*)	Marjoram-leav'd.	——	1785.	——	F.♃.	Barrel. ic. 598.
19	cretàcea. s.s.	(*bl.ye.*)	crowded-flower'd.	Siberia.	1826.	——	H.⊙.	
20	virgàta. s.s.	(*bl.*)	slender-stalked.	Barbary.	1824.	——	H.⊙.	Desf. atl. 2. t. 135.
21	thymifòlia. s.s.	(*ye.*)	Thyme-leaved.	S. Europe.	——	——	H.⊙.	
22	villòsa. s.s.	(*wh.*)	villous.	Spain.	1786.	7. 8.	F.♃.	Barrel. ic. 597.
23	rubrifòlia. s.s.	(*bl.*)	red-leaved.	S. France.	1824.	——	F.♃.	
24	saxátilis. P.S.	(*ye.*)	rock.	S. Europe.	1827.	6. 8.	H.♃.	
25	pubéscens. P.S.	(*ye.*)	pubescent.	——	1824.	6. 9.	H.⊙.	
26	hírta. s.s.	(*ye.*)	shaggy-leaved.	Spain.	1759.	——	H.⊙.	Jacq. ic. 1. t. 117.
27	triornithóphora.	(*pu.*)	three bird's.	Portugal.	1710.	——	F.♃.	Bot. mag. t. 523.
28	parviflòra. s.s.	(*ye.*)	small flowered.	Sardinia.	1829.	——	H.⊙.	Desf. atl. 2. t. 137.
29	albifróns. F.G.	(*wh.*)	white fronted.	Greece.	1828.	——	H.⊙.	Flor. græc. t. 588.
30	bipartìta. s.s.	(*bl.*)	cloven flowered.	Barbary.	1801.	——	H.⊙.	Swt. br. fl. gar. t. 30.
	speciòsa. Jacq. fil. eclog. t. 95.							
31	mìnor. s.s.	(*pu.ye.*)	little erect.	England.	....	6. 11.	H.⊙.	Eng. bot. v. 28. t. 2014.
	Antirrhìnum minus. E.B.							
32	littoràlis. s.s.	(*bl.*)	shore.	Austria.	1824.	——	H.⊙.	
33	chalepénsis. s.s.	(*wh.*)	white-flowered.	Levant.	1680.	6. 7.	H.⊙.	Flor. græc. t. 592.
34	trístis. s.s.	(*bk.br.*)	melancholy.	Spain.	1727.	7. 8.	G.♃.	Bot. mag. t. 74.
35	versícolor. s.s.	(*ye.bl.*)	various-colour'd.	France.	1777.	7. 9.	H.⊙.	Jacq. ic. 1. t. 116.
36	linarioídes. s.s.	(*ye.*)	toadflax-like.	S. Europe.	1816.	5. 8.	H.♃.	
37	micrántha. s.s.	(*ye.*)	small-flowered.	Spain.	1825.	5. 7.	H.⊙.	Flor. græc. t. 587.
38	viscòsa. s.s.	(*br.ye.*)	clammy.	——	1786.	7.	H.⊙.	Bot. mag. t. 368.
39	purpùrea. s.s.	(*pu.*)	purple.	S. Europe.	1648.	7. 9.	H.♃.	Flor. græc. t. 589.
40	arenària. s.s.	(*ye.*)	sand.	Hungary.	1825.	——	H.⊙.	
41	alpìna. s.s.	(*bl.*)	Alpine.	Austria.	1570.	7. 11.	H.♃.	Bot. mag. t. 205.
42	lusitánica. s.s.	(*ye.*)	Portugal.	S. Europe.	1820.	6. 7.	H.♃.	
43	pyrenàica. s.s.	(*ye.*)	Pyrenean.	Pyrenees.	1824.	6. 8.	H.⊙.	
44	Hælàva. F.E.	(*ye.*)	hairy calyxed.	Egypt.	1803.	7. 9.	H.⊙.	
45	pelisseriàna. s.s.	(*vi.*)	violet-coloured.	S. Europe.	1640.	6. 9.	H.⊙.	Flor. græc. t. 591.
46	élegans. P.S.	(*bl.*)	elegant.	Spain.	1812.	——	H.⊙.	
47	répens. H.K.	(*std.*)	creeping rooted.	England.	....	7. 10.	H.♃.	Eng. bot. v. 18. t. 1253.
48	amethy'stina. s.s.	(*bl.*)	bright blue.	S. Europe.	1816.	——	H.⊙.	Cavan. ic. 1. t. 33. f. 1.
	Antirrhìnum bipunctàtum. Cav. *nec aliorum.*							
49	supìna. s.s.	(*ye.*)	trailing.	Spain.	1728.	——	H.⊙.	Flor. græc. t. 595.
50	bipunctàta. s.s.	(*ye.*)	two-spotted.	——	1749.	6. 8.	H.⊙.	
51	spártea. s.s.	(*ye.*)	branching.	——	1772.	6. 10.	H.⊙.	Bot. mag. t. 200.
52	símplex. s.s.	(*bl.*)	upright.	S. Europe.	1816.	7. 8.	H.⊙.	Jacq. ic. 3. t. 499.
53	arvénsis. s.s.	(*ye.*)	corn.	——	....	——	H.⊙.	Flor. græc. t. 590.
54	diffùsa. s.s.	(*ye.bl.*)	spreading.	Portugal.	1824.	——	H.⊙.	
55	multicaùlis. s.s.	(*wh.bl.*)	many-stalked.	Levant.	1728.	5. 7.	H.⊙.	Bocc. sic. t. 19. f. 1.
56	macroùra. L.en.	(*ye.*)	large-spurred.	Caucasus.	1823.	6. 9.	H.♃.	
57	incarnàta. s.s.	(*li.*)	flesh-coloured.	S. Europe.	1826.	——	H.⊙.	
58	reticulàta. s.s.	(*ye.pu.*)	netted-flower'd.	Algiers.	1788.	5. 7.	F.♃.	Smith ic. pict. t. 2.
59	glaùca. s.s.	(*ye.*)	glaucous-leaved.	S. Europe.	1800.	6. 8.	H.⊙.	Buxb. cent. 4. t. 37.
60	ægyptìaca. s.s.	(*ye.*)	Egyptian.	Egypt.	1771.	7.	H.⊙.	
61	silenifòlia. Lk.	(*ye.*)	Silene-leaved.	Portugal.	1820.	7. 9.	H.⊙.	
62	genistifòlia. s.s.	(*ye.*)	Broom-leaved.	Austria.	1704.	7. 8.	H.♃.	Flor. græc. t. 596.
63	linifòlia. s.s.	(*ye.*)	flax-leaved.	Italy.	1824.	——	H.♃.	
64	vulgàris. s.s.	(*ye.*)	common.	Britain.	....	6. 9.	H.♃.	Eng. bot. v. 10. t. 658.
	Antirrhìnum Linària. E.B.							
	β Pelòria.	(*ye.*)	*regular-flowered.*	——	....	——	H.♃.	Eng. bot. v. 4. t. 260.
65	Lœsèlii. s.s.	(*ye.bl.*)	Loesel's.	Russia.	1824.	——	H.⊙.	
66	odòra. M.B.	(*ye.bl.*)	sweet-scented.	Caucasus.	1827.	——	H.⊙.	
67	strícta. s.s.	(*ye.*)	upright.	Levant.	1829.	——	H.⊙.	Flor. græc. t. 594.
68	júncea. s.s.	(*ye.*)	rush-stalked.	Spain.	1780.	7. 8.	H.⊙.	
69	canadénsis. s.s.	(*vi.*)	Canadian.	N. America.	1812.	6. 8.	H.⊙.	Venten. cels. t. 49.
70	acutifòlia. F.	(*ye.bl.*)	sharp-leaved.	Levant.	1826.	——	H.⊙.	
71	fasciculàta. Ot.	(*ye.*)	clustered.	....	1829.	——	H.⊙.	
	SCROPHUL'ARIA. L.		FIGWORT.	Didynamia Angiospermia. L.				
1	marilándica. s.s.	(*g.br.*)	Maryland.	N. America.	1759.	5. 7.	H.♃.	
2	peregrìna. s.s.	(*pu.*)	Nettle-leaved.	Italy.	1640.	6. 8.	H.♂.	Flor. græc. t. 597.
3	nodòsa. s.s.	(*d.pu.*)	knotty-rooted.	Britain.	....	5. 7.	H.♃.	Eng. bot. v. 22. t. 1544.
4	aquàtica. s.s.	(*d.pu.*)	water.	——	....	——	H.♃.	—— v. 12. t. 854.
5	glabràta. s.s.	(*pu.ye.*)	spear-leaved.	Canaries.	1779.	4. 5.	G.♂.	Jacq. schœn. 2. t. 209.
6	biserràta. s.s.	(*pu.ye.*)	double-serrate.	....	1816.	5. 7.	H.♃.	
7	argùta. s.s.	(*re.*)	slender upright.	Canaries.	1778.	——	H.⊙.	
8	vernàlis. s.s.	(*ye.*)	yellow.	Britain.	....	3. 5.	H.♂.	Eng. bot. v. 8. t. 567.

9	Smíthii. L.en.	(*br.ye.*)	Smith's.	Canaries.	1815.	4. 8.	G. ♃.	
10	glandulòsa. s.s.	(*br.ye.*)	glandular.	Hungary.	1806.	6. 9.	H. ♃.	W. et K. hung. t. 214.
11	rugòsa. w.k.	(*br ye.*)	rugged.	———	1816.	——	H. ♃.	
12	Scopòlii. p.s.	(*br.ye.*)	Scopoli's.	Germany.	——	——	H. ♃.	Scop. carn. 2. t. 32.
13	Scorodònia. s.s.	(*pu.*)	balm-leaved.	Britain.	....	7. 8.	H. ♃.	Eng. bot. v. 31. t. 2209.
14	grandidentàta. T.s.	(*pu.*)	great-toothed.	Italy.	1829.	6. 8.	H. ♃.	
15	altàica. s.s.	(*st.*)	white-flowered.	Siberia.	1786.	5. 6.	H. ♃.	Mur. com. got. 4. t. 2.
16	betonicæfòlia. s.s.	(*pu.*)	Betony-leaved.	Spain.	1752.	6. 8.	H. ♃.	Barrel. ic. 274.
17	hirsùta. L.en.	(*pu.ye.*)	hairy.	S. Europe.	1822.	——	H. ♃.	
18	appendiculàta. s.s.	(*pu.*)	heart-leaved.	Morocco.	1805.	7.	H. ♃.	
19	Balbísii. s.s.	(*pu.ye.*)	Balbis's.	Italy.	1823.	6. 8.	H. ♃.	
20	auriculàta. s.s.	(*br.ye.*)	ear-leaved.	Spain.	1772.	7. 8.	H. ♃.	Lobel. ic. 533.
21	micrántha. s.s.	(*pu.*)	small-flowered.	Levant.	1828.	——	H. ♃.	
22	heterophy'lla. s.s.	(*pu.ye.*)	various-leaved.	———	1827.	——	G. ♄.	
23	trifoliàta. s.s.	(*ye.br.*)	three-leaved.	Africa.	1731.	5. 9.	H. ♂.	Jacq. schœn. 3. t. 286.
24	laciniàta. w.k.	(*br.ye.*)	jag-leaved.	Hungary.	1806.	7.	H. ♃.	W. et. K. hung. 2. t. 170.
25	verbenæfòlia. Poir.	(*br.ye.*)	Vervain-leav'd.	....	1816.	7. 8.	H. ♃.	
26	c'æsia. s.s.	(*ye.br.*)	grey.	Levant.	1828.	——	F. ♂.	
27	adscéndens. W.en.	(*br.ye*)	ascending.	....	1816.	——	H. ♃.	
28	variegàta. s.s.	(*pu.ye.*)	variegated.	Caucasus.	——	——	H. ♃.	
29	ebulifòlia. s.s.	(*ye.pu.*)	Elder-leaved.	———	1824.	——	H. ♃.	
30	grandiflòra. DC.	(*br.ye.*)	large-flowered.	....	1816.	——	H. ♃.	
31	frutéscens. s.s.	(*br.ye.*)	shrubby.	Portugal.	1768.	6. 8.	F. ♄.	Herm. lugd. t. 547.
32	rupéstris. s.s.	(*br.ye.*)	rock.	Tauria.	1826.	6. 9.	H. ♃.	
33	orientàlis. s.s.	(*br.ye.*)	oriental.	Levant.	1710.	7. 8.	H. ♃.	Schk. hand. 2. t. 173.
34	melissæfòlia. s.s.	(*br.ye.*)	balm-like.	———	1823.	——	H. ♃.	
35	lanceolàta. s.s.	(*st.*)	spear-leaved.	Pensylvania.	1826.	——	H. ♃.	
36	cretàcea. s.s.	(*ye.*)	narrow-leaved.	Siberia.	1828.	——	H. ♃.	
37	lyràta. s.s.	(*br.pu.*)	lyrate-leaved.	Portugal.	1816.	——	H. ♃.	W. hort. ber. 1. t. 55.
38	mellífera. s.s.	(*br.pu.*)	honey-bearing.	Barbary.	1786.	——	H. ♃.	Desf. atl. 2. t. 143.
39	sambucifòlia. s.s.	(*br.g.*)	Elder-like.	Spain.	1640.	7. 9.	H. ♃.	Mill. ic. 2. t. 231.
40	híspida. s.s.	(*br.ye.*)	hispid.	Levant.	1820.	——	H. ♃.	
41	tanacetifòlia. s.s.	(*br.ye.*)	Tansy-leaved.	Tauria.	1804.	——	H. ♃.	W. hort. ber. 1. t. 56.
42	canìna. s.s.	(*or.re.*)	wing-leaved.	S. Europe.	1683.	6. 8.	H. ♂.	Flor. græc. t. 598.
43	chrysanthemifòlia.	(*br.pu.*)	Chrysanthemum-ld.	Tauria.	1816.	——	H. ♃.	W. hort. ber. 1. t. 59.
44	lùcida. s.s.	(*br.ye.*)	shining-leaved.	Levant.	1596.	——	H. ♃.	——— 1. t. 57.
45	filicifòlia. s.s.	(*d.pu.*)	Fern-leaved.	Crete.	....	——	H. ♃.	Flor. græc. t. 600.
46	multífida. s.s.	(*br.ye.*)	multifid-leaved.	....	1816.	——	H. ♃.	W. hort. ber. 1. t. 58.
47	glaùca. F.G.	(*pu.*)	glaucous.	Greece.	1827.	——	H. ♃.	Flor. græc. t. 599.
48	pubéscens. F.	(*pu.*)	pubescent.	....	——	——	H. ♃.	
49	pállida. F.	(*pa.*)	pale-flowered.	Iberia.	——	——	H. ♃.	
50	obtusifòlia. F.	(*pu.*)	blunt-leaved.	Hungary.	1821.	——	H. ♃.	

DIGITALIS. S.S. FOXGLOVE. Didynamia Angiospermia. L.

1	Scéptrum. s.s.	(*or.*)	Madeira shrubby.	Madeira.	1777.	7. 8.	G. ♄.	Sm. exot. bot. 2. t. 73.
2	canariénsis. s.s.	(*ye.*)	Canary.	Canaries.	1698.	6. 7.	G. ♄.	Bot. reg. t. 48.
3	obscùra. s.s.	(*or.*)	Willow-leaved.	Spain.	1778.	7. 8.	H. ♃.	Lind. digit. t. 26.
4	lùtea. s.s.	(*st.*)	small yellow.	France.	1629.	——	H. ♃.	——— t. 23.
5	hy'brida. Sv.	(*pi.wh.*)	hybrid.	Hybrid.	....	——	H. ♃.	Salv. nou. bull. 3. 337. t. 6.
	lùtea γ Lind. digit. t. 25.							
6	fucàta. P.S.	(*pu.*)	painted.	———	....	——	H. ♃.	
	lùtea δ. Lindl. digit. t. 24.							
7	mèdia. s.s.	(*ye.*)	middle yellow.	Germany.	1816.	——	H. ♃.	
8	micrántha. s.s.	(*st.*)	small-blossom'd.	Switzerland.	1824.	——	H. ♃.	Elm. mon. 46. t. 2.
9	tubiflòra. L.D.	(*st.*)	tube-flowered.	S. Europe.	1816.	——	H. ♃.	Lind. digit. t. 22.
10	lutéscens. L.D.	(*st.*)	pale yellow.	———	——	——	H. ♃.	——— t. 21.
11	purpuráscens. s.s.	(*pi.*)	blush-flowered.	Italy.	1776.	——	H. ♃.	——— t. 20.
	erubéscens. H.K.							
12	rígida. L.D.	(*pu.ye.*)	rigid.	S. Europe.	1824.	——	H. ♃.	Lind. digit. t. 19.
13	viridiflòra. L.D.	(*gr.*)	green-flowered.	Levant.	1827.	——	H. ♃.	——— t. 18.
14	parviflòra. s.s.	(*pu.*)	small-flowered.	....	1798.	6. 8.	H. ♃.	——— t. 17.
15	laciniàta. B.R.	(*ye.*)	cut-leaved.	Malaga.	1826.	——	H. ♃.	Bot. reg. t. 1201.
16	orientàlis. s.s.	(*li.wh.*)	oriental.	Levant.	1818.	7. 8.	H. ♃.	——— t. 554.
17	lanàta. s.s.	(*wh.br.*)	woolly.	Hungary.	1789.	6. 7.	H. ♃.	Swt. br. fl. gar. t. 291.
18	leucoph'æa. s.s.	(*wh.br.*)	broad-lipped.	Greece.	1788.	6. 10.	H. ♃.	Lind. digit. t. 14.
19	aùrea. s.s.	(*or.*)	golden.	———	1815.	7. 8.	H. ♃.	——— t. 13.
20	ferrugínea. s.s.	(*br.*)	iron-coloured.	Italy.	1597.	——	H. ♃.	——— t. 12.
21	sibírica. L.D.	(*ye.re.*)	Siberian.	Siberia.	1820.	——	H. ♃.	——— t. 11.
22	lævigàta. s.s.	(*ye.*)	shining-leaved.	Hungary.	1810.	——	H. ♃.	——— t. 10.
23	fúlva. s.s.	(*so.*)	tawny.	S. Europe.	1820.	6. 8.	H. ♃.	——— t. 9.

24 ambígua. w.	(*ye.*)	greater yellow.	Switzerland.	1596.	7. 8.	H. ♃.	Bot. reg. t. 64.
grandiflòra. s.s.							
25 ochroleùca. s.s.	(*st.*)	sulphur-color'd.	Germany.	1816.	———	H. ♃.	Lind. digit. t. 8.
26 fuscéscens. s.s.	(*br.*)	netted-flower'd.	Hungary.	1824.	———	H. ♃.	
27 mìnor. s.s.	(*pu.*)	dwarf.	Spain.	1789.	6. 7.	H. ♃.	Lind. digit. t. 5. 6.
28 eriostàchya. Bes.	(*pu.*)	woolly spiked.	....	1826.	———	H. ♃.	
29 tomentòsa. s.s.	(*pu.*)	tomentose.	Portugal.	1820.	6. 8.	H. ♃.	Bot. mag. t. 2194.
30 Thápsi. s.s.	(*pu.*)	Mullein.	Spain.	1752.	5. 8.	H. ♃.	Sm. exot. bot. 1. t. 43.
31 purpùrea. s.s.	(*pu.*)	common.	Britain.	....	6. 9.	H. ♂.	Eng. bot. v. 19. t. 1297.
β *álba.*	(*wh.*)	*white-flowered.*	———	....	———	H. ♂.	
CHEL'ONE. S.S.		Chel'one.	Didynamia Angiospermia. L.				
1 glábra. w.	(*wh.*)	white-flowered.	N. America.	1730.	8. 10.	H. ♃.	Trew pl. select. t. 88.
2 oblíqua. w.	(*pu.*)	red-flowered.	———	1752.	———	H. ♃.	Bot. reg. t. 175.
3 Lyòni. s.s.	(*re.*)	Lyon's.	———	1812.	7. 9.	H. ♃.	Swt. br. fl. gar. t. 293.
màjor. Bot. mag. t. 1864.							
4 nemoròsa. D.	(*pu.*)	wood.	N. America.	1827.	7. 9.	H. ♃.	Bot. reg. t. 1211.
5 Scoulèri. D.	(*bl.*)	Scouler's.	Columbia.	———	5. 7.	H. ♃.	——— t. 1277.
6 erianthèra. N.	(*bl.*)	blue-flowered.	Louisiana.	1811.	7. 9.	F. ♃.	Bot. mag. t. 1672.
Pentstèmon glábra. B.M.							
7 barbàta. s.s.	(*sc.*)	scarlet.	Mexico.	1794.	6. 9.	H. ♃.	Bot. reg. t. 116.
8 ròsea. B.F.G.	(*ro.*)	rose-coloured.	———	1825.	———	H. ♄.	Swt. br. fl. gar. t. 230.
Pentstèmon angustifòlium. B.R. t. 1122. *non* Pursh.							
9 campanulàta. s.s.	(*ro.*)	bell-flowered.	Mexico.	1794.	3. 10.	F. ♄.	Bot. mag. t. 1878.
Pentstèmon campanulàta. Jacq. schœn. 3. t. 362.							
10 atropurpùrea. B.F.G.	(*d.pu.*)	dark purple.	Mexico.	1824.	———	H. ♄.	Swt. br. fl. gar. t. 235.
11 pulchélla. B.R.	(*pu.*)	pretty.	———	1825.	5. 10.	H. ♄.	Bot. reg. t. 1138.
12 lævigàta. s.s.	(*pi.*)	smooth.	N. America.	1776.	7. 9.	H. ♃.	Bot. mag. t. 1425.
13 hirsùta. s.s.	(*bl.pi.*)	narrow-ld hairy.	———	1758.	———	H. ♃.	Moris. s. 11. t. 21. f. 3.
14 pubéscens.	(*bl.pu.*)	broad-ld hairy.	———	———	———	H. ♃.	Bot. mag. t. 1424.
Pentstèmon pubéscens. B.M.							
15 grácilis. N.	(*li.*)	slender.	———	1825.	———	H. ♃.	Bot. mag. t. 2945.
16 glaùca. D.	(*li.*)	glaucous.	———	1827.	7. 9.	H. ♃.	Bot. reg. t. 1286.
17 digitàlis. B.F.G.	(*wh.*)	foxglove-flower'd.	Arkansa.	1824.	6. 8.	H. ♃.	Swt. br. fl. gar. t. 120.
Pentstèmon digitàlis. Bot. mag. t. 2587.							
18 cristàta. s.s.	(*vi.*)	crested.	Louisiana.	1811.	7. 9.	F. ♃.	
19 cœrùlea. s.s.	(*bl.*)	slender blue.	———	———	———	F. ♃.	
Pentstèmon angustifòlium. Ph. *non* K.S.							
20 grandiflòra. s.s.	(*pi.wh.*)	great-flowered.	———	———	———	F. ♃.	
21 álbida. s.s.	(*wh.*)	dwarf.	———	———	———	F. ♃.	
22 diffùsa. D.	(*pu.*)	spreading.	Columbia.	1827.	6. 10.	H. ♃.	Bot. reg. t. 1132.
23 ovàta. D.	(*bl.pu.*)	oval-leaved.	N. America.	———	6. 8.	H. ♃.	Bot. mag. t. 2903.
24 venústa. D.	(*pu.*)	handsome.	———	———	7. 9.	H. ♃.	Bot. reg. t. 1309.
25 deústa. D.	(*pu.*)	blasted.	———	———	———	H. ♃.	
26 decussàta. D.	(*pu.*)	cross-leaved.	———	———	———	H. ♃.	
27 glandulòsa. D.	(*pu.*)	glandular.	———	———	———	H. ♃.	Bot. reg. t. 1262.
28 acuminàta. D.	(*pu.*)	pointed-leaved.	———	———	6. 8.	H. ♃.	——— t. 1285.
29 Richardsòni. D.	(*pu.*)	Richardson's.	———	———	7. 10.	H. ♃.	——— t. 1121.
30 attenuàta. D.	(*st.*)	taper-pointed.	———	———	7. 9.	H. ♃.	——— t. 1295.
31 conférta. D.	(*st.*)	crowded-flower'd.	Columbia.	———	———	H. ♃.	——— t. 1260.
32 pruinòsa. D.	(*bl.*)	blue-leaved.	N. America.	———	6. 8.	H. ♃.	——— t. 1280.
33 pròcera. D.	(*vi.*)	tall.	———	———	———	H. ♃.	Bot. mag. t. 2954.
34 speciòsa. D.	(*bl.*)	beautiful.	———	———	6. 9.	H. ♃.	Bot. reg. t. 1270.
35 triphy'lla. D.	(*li.*)	three-leaved.	———	———	7. 9.	H. ♃.	——— t. 1245.
36 serrulàta. D.	(*pu.*)	saw-leaved.	———	———	———	H. ♃.	
37 suffruticòsa. D.	(*pu.*)	suffrutescent.	———	———	———	H. ♄.	
HERPE'STIS. G.		Herpe'stis.	Didynamia Angiospermia. S.S.				
1 strícta. s.s.	(*bl.*)	upright.	Brazil.	1823.	6. 8.	S. ♃.	
2 Monniéria. K.S.	(*wh.*)	thyme-leaved.	S. America.	1772.	7. 9.	S. ♃.	Ehret. pict. t. 14. f. 2.
3 portulacàcea. B.M.	(*w.*)	Purslane-leaved.	———	1823.	———	S. ♃.	Bot. mag. t. 2557.
4 cuneifòlia. Ph.	(*bl.*)	wedge-leaved.	N. America.	1812.	8. 9.	H. ♃.	
5 amplexicaùlis. s.s.	(*bl.*)	stem-clasping.	Carolina.	1822.	6. 8.	H. ♃.	
6 micrántha. Ph.	(*wh.*)	small-flowered.	N. America.	1827.	7. 8.	H. ⊙.	
7 semiserràta. Ot.	(*wh.*)	semiserrate.	Mexico.	1824.	———	G. ♃.	
8 chrysántha. Ot.	(*ye.*)	yellow-flowered.	———	1829.	———	G. ♃.	
M'AZUS. B.P.		M'azus.	Didynamia Angiospermia. S.S.				
rugòsus. B.P.	(*wh.*)	China.	China.	1780.	5. 10.	H. ⊙.	Swt. br. fl. gar. t. 36.
Lindérnia japónica. Th. *Hornemánnia bicolor.* W.en.							
TOR'ENIA. B.P.		Tor'enia.	Didynamia Angiospermia. S.S.				
1 ovàta.	(*wh.*)	oval-leaved.	Manila.	1823.	5. 10.	H. ⊙.	Pl. sel. hort. ber. t. 9.
Hornemánnia ovàta. L.en. *Tittmánnia ovàta.* s.s.							

2 viscòsa.	(*wh.*)	viscous.	E. Indies.	1823.	5. 10.	S.⊙.	
Hornemánnia viscòsa. W.en. *Tittmánnia viscòsa.* S.S.							
LINDE'RNIA. B.P.		LINDE'RNIA.	Didynamia Angiospermia. L.				
Pyxidària. W.	(*pu.*)	European.	S. Europe.	1789.	6. 8.	H.⊙.	Lam. ill. t. 522.
MI'MULUS. L.		MONKEY-FLOWER.	Didynamia Angiospermia. L.				
1 glutinòsus. S.S.	(*ye.*)	orange-colored.	California.	1794.	1. 12.	G. ♄.	Bot. mag. t. 354.
2 ríngens. S.S.	(*li.*)	gaping.	N. America.	1759.	7. 8.	H. ♃.	——— t. 283.
3 alàtus. S.S.	(*li.*)	oval-leaved.	———	1783.	——	H. ♃.	
4 guttàtus. DC.	(*ye.*)	spotted-flowered.	America.	1812.	6. 9.	F. ♃.	Bot. mag. t. 1501.
lùteus. B.M. *non* L.							
5 lùteus. B.R.	(*ye.*)	yellow-flower'd.	Chile.	1825.	——	H. ♃.	Feuill. peruv. 2. t. 34.
β *rivulàris.* B.R.	(*ye.pu.*)	*dark-spotted.*	Columbia.	——	——	H. ♃.	Bot. reg. t. 1030.
6 floribúndus. B.R.	(*ye.*)	many-flowered.	———	1826.	8. 10.	H.⊙.	——— t. 1125.
7 moschàtus. B.R.	(*ye.*)	musk-scented.	———	——	7. 10.	H. ♃.	——— t. 1118.
8 parviflòrus. B.R.	(*st.*)	small-flowered.	Mexico.	1824.	6. 9.	H.⊙.	——— t. 874.
BROWA'LLIA. L.		BROWA'LLIA.	Didynamia Angiospermia. L.				
1 elàta. W.	(*bl.wh.*)	upright.	Peru.	1768.	6. 9.	G.⊙.	Bot. mag. t. 34.
2 demíssa. W.	(*bl.wh.*)	spreading.	S. America.	1735.	——	G.⊙.	——— t. 1136.
ANGEL'ONIA. K.S.		ANGEL'ONIA.	Didynamia Angiospermia. L.				
salicariæfòlia. K.S.	(*bl.*)	Salicaria-leav'd.	S. America.	1818.	7. 10.	S. ♄.	Bot. reg. t. 415.
ALONS'OA. R.P.		ALONS'OA.	Didynamia Angiospermia. H.K.				
1 incisifòlia. R.P.	(*sc.*)	Nettle-leaved.	Chile.	1795.	5. 10.	G. ♄.	Bot. mag. t. 417.
Célsia urticæfòlia. B.M. *Hemímeris urticæfòlia.* W.							
2 acutifòlia. R.P.	(*sc.*)	acute-leaved.	Peru.	1790.	——	G. ♄.	
3 intermèdia. B.C.	(*or.*)	intermediate.	———	....	——	G. ♄.	Lodd. bot. cab. t. 1456.
4 lineàris. H.K.	(*sc.*)	linear-leaved.	———	1790.	——	G. ♄.	Bot. mag. t. 210.
5 caulialàta. R.P.	(*sc.*)	wing-stemmed.	Chile.	1823.	5. 10.	G. ♄.	
DIA'SCIA. L.en.		DIA'SCIA.	Didynamia Angiospermia. S.S.				
Bergiàna. L.en.	(*re.*)	Bergius's.	C. B. S.	1816.	7. 10.	H.⊙.	Pluk. phyt. 320. f. 5.
Hemímeris diffùsa. W.							
VANDE'LLIA. S.S.		VANDE'LLIA.	Didynamia Angiospermia. L.				
diffùsa. S.S.	(*wh.*)	spreading.	S. America.	1824.	7. 9.	S.⊙.	Marc. bras. 1. c. 15. f. 1.
LIMOSE'LLA. L.		MUDWORT.	Didynamia Angiospermia. L.				
aquática. L.	(*fl.*)	water.	Britain.	....	7. 9.	H.⊙.	Eng. bot. v. 5. t. 357.
DISA'NDRA. L.		DISA'NDRA.	Heptandria Monogynia. L.				
prostràta. L.	(*ye.*)	trailing.	Madeira.	1771.	5. 8.	G. ♃.	Bot. mag. t. 218.
SIBTHO'RPIA. L.		SIBTHO'RPIA.	Didynamia Angiospermia. L.				
europ'æa. L.	(*wh.*)	Cornish money-wort.	England.	....	7. 8.	H. ♃.	Eng. bot. v. 10. t. 619.
ER'INUS. S.S.		ER'INUS.	Didynamia Angiospermia. L.				
1 Lychnidèa. S.S.	(*wh.pu.*)	Phlox-flower'd.	C. B. S.	1821.	3. 10.	G. ♄.	Bot. reg. t. 748.
2 frágrans. H.K.	(*wh.pu.*)	fragrant.	———	1776.	5. 6.	G. ♄.	Burm. afr. t. 49. f. 4.
3 alpìnus. P.S.	(*pu.*)	smooth-leaved.	Pyrenees.	1739.	3. 4.	H. ♃.	Bot. mag. t. 310.
4 hispánicus. P.S.	(*pu.*)	hairy-leaved.	Spain.	——	——	H. ♃.	
MAN'ULEA. S.S.		MAN'ULEA.	Didynamia Angiospermia. L.				
1 viscòsa. S.S.	(*re.*)	clammy.	C. B. S.	1774.	6. 11.	G. ♄.	Bot. mag. t. 217.
Buchnèra viscòsa. B.M.							
2 tomentòsa. S.S.	(*or.*)	tomentose.	———	1774.	5. 12.	G. ♄.	Bot. mag. t. 322.
3 angustifòlia. L.en.	(*or.*)	narrow-leaved.	———	1822.	——	G. ♄.	Pl. sel. hort. ber. t. 20.
4 rhynchántha. S.S.	(*or.*)	wedge-leaved.	———	1816.	6. 9.	G. ♄.	
5 oppositifòlia. V.	(*wh.*)	opposite-leaved.	———	....	5. 12.	G. ♄.	Vent. malm. t. 15.
6 villòsa. P.S.	(*wh.*)	villous.	———	1783.	6. 7.	G.⊙.	Burm. afr. t. 50. f. 2.
Buchnèra capénsis. W.							
7 cœrùlea. S.S.	(*bl.*)	blue-flowered.	———	1822.	——	G. ♄.	
8 rùbra. S.S.	(*re.*)	red.	———	1790.	4. 9.	G. ♄.	
9 Cheiránthus. S.S.	(*ye.*)	Wallflower-like.	———	1795.	6. 8.	G. ♂.	Comm. hort. 2. t. 42.
10 argéntea. S.S.	(*wh.*)	silvery.	———	1801.	7. 11.	G.⊙.	
11 pedunculàta. P.S.	(*wh.*)	solitary-flower'd.	———	1790.	6. 11.	G. ♄.	Andr. reposit. t. 84.
12 f'œtida. S.S.	(*wh.*)	stinking.	———	1794.	6. 9.	G.⊙.	——— t. 80.
Buchnèra f'œtida. A.R.							
13 violàcea. S.S.	(*vi.*)	violet-flowered.	———	1816.	——	G. ♄.	
14 cordàta. S.S.	(*wh.*)	trailing heart-ld.	———	——	——	G. ♃.	
BUCHN'ERA. S.S.		BUCHN'ERA.	Didynamia Angiospermia. L.				
americàna. S.S.	(*bl.*)	American.	N. America.	1733.	6. 8.	H. ♃.	
EUPHR'ASIA. W.		EYE-BRIGHT.	Didynamia Angiospermia. L.				
1 linifòlia. S.S.	(*li.*)	flax-leaved.	S. France.	1823.	7. 9.	H.⊙.	
2 lùtea. S.S.	(*ye.*)	yellow.	S. Europe.	1816.	——	H.⊙.	Jacq. aust. t. 398.
3 alpìna. S.S.	(*pu.*)	Alpine.	———	1823.	——	H.⊙.	
4 latifòlia. S.S.	(*re.*)	broad-leaved.	———	1826.	——	H.⊙.	Magn. monsp. t. 94.
5 mínima. P.S.	(*wh.pu.*)	least.	Switzerland.	1828.	——	H.⊙.	

6 tricuspidàta. s.s. (*w.pu.*) three-pointed. Italy. 1828. 6. 9. H.⊙. Pluk. alm. t. 177. f. 1.
7 officinàlis. s.s. (*wh.pu.*) common. Britain. .... 7. 9. H.⊙. Eng. bot. v. 20. t. 1416.

BA'RTSIA. S.S. BA'RTSIA. Didynamia Angiospermia. L.

1 viscòsa. s.s. (*ye.*) yellow viscid. Britain. .... 7. 8. H.⊙. Eng. bot. v. 15. t. 1045.
2 alpìna. s.s. (*vi.*) Alpine. ——— .... —— H.⊙. ——— v. 6. t. 361.
3 Odontìtes. s.s. (*re.*) red-flowered. ——— .... 7. 9. H.⊙. ——— v. 20. t. 1415.
4 Trixàgo. s.s. (*ye.*) simple-stalked. S. Europe. 1820. —— H.⊙. Flor. græc. t. 585.
Trixàgo rhinanthìna. L.enum.
5 latifòlia. F.G. (*re.wh.*) broad-leaved. Greece. 1827. 6. 9. H.⊙. Flor. græc. v. 5. t. 586.

CASTILL'EJA. S.S. CASTILL'EJA. Didynamia Angiosperma. S.S.

1 integrifòlia. s.s. (*wh.*) entire-leaved. S. America. 1825. .... G.♄. Sm. ic. ined. t. 39.
2 toluccénsis. s.s. (*wh.*) procumbent. Mexico. —— .... G.♄.
3 septentrionàlis. B.R. (*st.*) northern. N. America. 1824. 7. 8. H.⊙. Bot. reg. t. 925.
4 pállida. s.s. (*st.*) pale-flowered. Siberia. 1782. 6. 9. H.♃. Gmel. sib. 3. t. 42.
Bártsia pállida. W.
5 coccínea. s.s. (*or.*) vermilion. N. America. 1787. 7. 9. H.⊙. Bot. reg. t. 1136.
Bártsia coccínea. W. *Euchròma coccínea.* N.
6 sessiliflòra. Ph. (*bh.*) sessile-flowered. Louisiana. 1811. 7. 8. H.♃.
grandiflòra s.s. *Euchròma grandiflòrum.* N.

RHINA'NTHUS. W. YELLOW RATTLE. Didynamia Angiospermia. L.

1 Crísta-gálli. W. (*ye.*) Cock's-comb. Britain. .... 6. 8. H.⊙. Eng. bot. v. 10. t. 657.
2 màjor. E.F. (*ye.*) larger. England. .... —— H.⊙.
3 Alecterolòphus. L.en. villous. Germany. 1816. —— H.⊙. Bull. herb. t. 125.
villòsus. P.S.

MELAMP'YRUM. S.S. COW-WHEAT. Didynamia Angiospermia. L.

1 cristàtum. E.B. (*ye.pu.*) crested. England. .... 7. 8. H.⊙. Eng. bot. v. 1. t. 41.
2 arvénse. E.B. (*ye.pu.*) corn-field. ——— .... 6. 7. H.⊙. ——— v. 1. t. 53.
3 nemoròsum. s.s. (*ye.*) coloured-bracted. Europe. 1823. —— H.⊙. Flor. dan. t. 305.
4 praténse. E.B. (*ye.*) common. Britain. .... 7. 8. H.⊙. Eng. bot. v. 2. t. 113.
5 sylváticum. E.B. (*co.*) wood. ——— .... —— H.⊙. ——— v. 12. t. 804.

PEDICUL'ARIS. S.S. LOUSEWORT. Didynamia Angiospermia. L.

1 incarnàta. s.s. (*fl.*) flesh-coloured. Austria. 1796. 6. 7. H.♃. Jacq. aust. 2. t. 140.
2 rostràta. s.s. (*pu.*) beaked. S. Europe. 1823. —— H.♃. ——— t. 205.
3 tuberòsa. s.s. (*ye.*) tuberous. Switzerland. 1799. 7. 8. H.♃. Hall. helv. n. 323. t. 10.
4 ascéndens. Schl. (*ye.*) ascending. ——— 1825. —— H.♃.
5 gyrofléxa. s.s. (*pu.*) bending. ——— —— —— H.♃. Vill. delph. 2. t. 9.
6 atrorùbens. DC. (*d.re.*) dark-red. ——— —— —— H.♃.
7 aspleniifòlia. P.S. (*pu.*) Fern-leaved. ——— 1828. —— H.♃. Willd. bot. zeit. c. ic.
8 uncinàta. s.s. (*fl.ye.*) hooked-flowered. Siberia. 1815. —— H.♃. Gmel. sib. 3. t. 45.
9 compácta. s.s. (*ye.*) close-headed. ——— —— —— H.♃.
10 resupinàta. s.s. (*pu.*) resupinate. ——— 1816. 5. 7. H.♃. Gmel. sib. 3. t. 44.
11 palústris. E.B. (*pu.*) marsh. Britain. .... 6. 7. H.♃. Eng. bot. v. 6. t. 399.
12 euphrasioídes. s.s. (*ye.re.*) eyebright-leaved. Siberia. 1816. – — H.♃. Gmel. sib. 3. t. 43.
13 striàta. s.s. (*ye.cr.*) stripe-flowered. Dahuria. 1826. —— H.♃. Pall. it. 3. a. n. 98. t. R. f. 2.
14 canadénsis. s.s. (*ye.*) Canadian. N. America. 1800. 7. 8. H.♃. Swt. br. fl. gar. t. 37.
15 sylvática. E.B. (*ro.*) common. Britain. .... 5. 7. H.♃. Eng. bot. v. 6. t. 400.
16 myriophy'lla. s.s. (*ye.*) Milfoil-leaved. Dahuria. 1816. —— H.♃. Pall. it. 3. ap. t. 8. f. 1.
17 rùbens. s.s. (*pu.*) red-flowered. ——— 1827. —— H.⊙.
18 sudética. s.s. (*pu.*) Swedish. N. Europe. —— —— H.♃.
19 flámmea. s.s. (*ye.sc.*) upright. ——— 1775. 7. H.♃. Hall. helv. t. 8. f. 3.
20 Oèderi. s.s. (*ye.*) Oeder's. ——— 1827. —— H.♃.
21 versícolor. s.s. (*ye.br.*) various-coloured. Switzerland. 1825. 7. 8. H.♃.
22 achillæfòlia. s.s. (*ye.*) Milfoil-leaved. Siberia. 1827. —— H.♃.
23 comòsa. s.s. (*ye.*) spiked. Italy. 1775. —— H.♃. All. ped. 1. t. 4. f. 1.
24 pállida. s.s. (*st.*) pale yellow. N. America. 1826. —— H.♃.
25 spicàta. s.s. (*pu.*) spike-flowered. Dahuria. 1827. —— H.♃. Pall. it. 3. a. n. 100. t. 8. f. 2.
26 Scéptrum. s.s. (*ye.cr.*) sceptred. Sweden. 1793. —— H.♃. Flor. dan. t. 26.
27 flàva. s.s. (*ye.*) yellow. Siberia. 1828. —— H.♃. Pall. it. 3. a. n. 97. t. R. f. 1.
28 recutìta. s.s. (*pu.*) jagged-leaved. Austria. 1787. —— H.♃. Jacq. aust. 3. t. 258.
29 foliòsa. s.s. (*ye.*) leafy. ——— 1786. —— H.♃. ——— 2. t. 139.
30 verticillàta. s.s. (*pu.*) whorled. ——— 1790. 5. 6. H.♃. ——— 3. t. 206.
31 proboscídea. s.s. .... proboscis-flower'd. Siberia. 1827. .... H.♃.
32 élegans. Led. (*pu.*) elegant. ——— —— 6. 8. H.♃.
33 speciòsa. Led. (*pu.*) beautiful. ——— —— —— H.♃.
34 elàta. s.s. (*pu.*) lofty. ——— 1826. —— H.⊙.
35 exaltàta. Led. (*pu.*) tall. ——— 1827. —— H.♃.
36 ròsea. s.s. (*ro.*) rose-coloured. S. Europe. 1825. 7. 8. H.♃. Jacq. ic. 1. t. 115.

GERA'RDIA. S.S. GERA'RDIA. Didynamia Angiospermia.

1 delphinifòlia. s.s. (*ro.*) Larkspur-leav'd. E. Indies. 1800. 6. 7. S.♃. Roxb. corom. 1. t. 90.

2 purpùrea.s.s.	(*pu.*)	purple.	N. America.	1772.	7. 8.	H. ♂.	Bot. mag. t. 2048.
3 tenuifòlia. s.s.	(*pu.*)	slender-leaved.	——	1812.	——	H. ♂.	Bart. fl. amer. 3. t. 82.
4 marítima. s.s.	(*li.*)	sea-side.	——	1823.	——	H. ♂.	
5 flàva. s.s.	(*ye.*)	yellow.	——	1796.	——	H. ♃.	Pluk. amal. t. 389. f. 3.
6 quercifòlia. s.s.	(*ye.*)	oak-leaved.	——	1812.	——	H. ♃.	Pursh amer. 1. t. 19.
7 Pediculària. s.s.	(*ye.*)	variable.	——	1826.	6. 8.	H. ♂.	Lam. ill. t. 529. f. 2.
SEYM'ERIA. S.S.		SEYM'ERIA.	Didynamia Angiospermia. S.S.				
1 tenuifòlia. s.s.	(*ye.*)	slender-leaved.	Carolina.	....	7. 9.	H.☉.	
Afzèlia cassioìdes. Gmel. *Gerárdia Afzèlia.* M.							
2 pectinàta. s.s.	(*ye.*)	pectinated.	Carolina.	....	——	H.☉.	

Sect. II. *STAMINA* (2) *ANTHERIFERA.*

CALCEOL'ARIA. R.S.		SLIPPERWORT.	Diandria Monogynia. L.				
1 Fothergíllii. R.S.	(*pu.*)	Fothergill's.	Falkland. Isl.	1777.	5. 8.	G. ♃.	Bot. mag. t. 348.
2 plantagínea. R.S.	(*ye.*)	Plantain-leaved.	Chile.	1826.	6. 9.	F. ♃.	—— t. 2805.
3 corymbòsa. R.S.	(*ye.*)	corymb-flower'd.	——	1823.	4. 8.	G. ♃.	Bot. reg. t. 723.
4 arachnoídea. B.M.	(*pu.*)	cobweb.	——	1827.	6. 10.	F. ♃.	Bot. mag. t. 2874.
5 purpùrea. B.M.	(*pu.*)	purple-flowered.	——	1826.	——	F. ♃.	—— t. 2775.
6 petioalàris. R.S.	(*ye.*)	much branched.	——	1827.	6. 9.	G. ♃.	Bot. reg. t. 1214.
floribúnda. B.R. *connàta.* Bot. mag. t. 2876.							
7 ascéndens. B.R.	(*ye.*)	ascending.	——	1826.	——	G. ♄.	Bot. reg. t. 1215.
8 polifòlia. B.M.	(*ye.*)	white-leaved.	——	——	——	G. ♃.	Bot. mag. t. 2897.
9 rugòsa. R.P.	(*ye.*)	rugged-leaved.	——	1823.	6. 8.	G. ♄.	Hook. ex. flor. t. 99.
crenàta. Bot. reg. t. 790.							
10 integrifòlia. R.P.	(*ye.*)	entire-leaved.	——	——	——	G. ♄.	Bot. reg. t. 744.
rugòsa. Bot. mag. t. 2523.							
11 angustifòlia.	(*ye.*)	narrow-leaved.	——	1826.	——	G. ♄.	Bot. reg. t. 1083.
12 thyrsiflòra. B.M.	(*ye.*)	tufted-flowered.	——	1827.	——	G. ♄.	Bot. mag. t. 2915.
13 pinnàta. R.S.	(*st.*)	wing-leaved.	Peru.	1773.	7. 9.	H.☉.	—— t. 41.
14 scabiosæfòlia. R.S.	(*st.*)	Scabious-leaved.	——	1823.	——	G. ♃.	—— t. 2405.
SCHIZA'NTHUS. R.P.		SCHIZA'NTHUS.	Diandria Monogynia.				
1 pinnàtus. H.E.F.	(*li.pu.*)	winged-leaved.	Chile.	1823.	7. 10.	H.☉.	Swt. br. fl. gar. t. 63.
2 pórrigens. H.E.F.	(*wh.pu.*)	spreading.	——	——	——	H.☉.	—— t. 76.
SCHWE'NCKIA. R.S.		SCHWE'NCKIA.	Diandria Monogynia. W.				
1 americàna. R.S.	(*vi.*)	American.	Cayenne.	1781.	8. 9.	S. ♂.	Schw. h. m. haag. t. 1.
2 browallioídes. K.S.	(*gr.pu.*)	Browallia-like.	Caracas.	1824.	——	S. ♃.	H.etB. n. g. am. 2. t. 181.
3 Hilariàna. DC.	(*gr.pu.*)	St. Hilaire's.	Brazil.	1826.	——	S.☉.	DC. pl. rar. genev. t. 10.
BONNA'YA. S.S.		BONNA'YA.	Diandria Monogynia. S.S.				
1 brachiàta. s.s.	(*wh.*)	brachiate.	E. Indies.	1825.	6. 9.	H.☉.	Pl. sel. h. ber. t. 11.
2 brachycárpa. Ot.	(*vi.*)	short-capsuled.	——	1829.	——	S.☉.	
3 veronicæfòlia. s.s.	(*vi.*)	Speedwell-leaved.	——	1798.	——	S. ♂.	Roxb. corom. 2. t. 154.
Gratiòla veronicæfòlia. W.							
4 réptans. s.s.	(*li.*)	creeping.	——	1820.	——	S.☉.	
GRATI'OLA. B.P.		HEDGE-HYSSOP.	Diandria Monogynia. L.				
1 latifòlia. B.P.	(*wh.*)	broad-leaved.	N. S. W.	1822.	6. 8.	G. ♃.	
2 officinàlis. R.S.	(*wh.*)	officinal.	Europe.	1568.	5. 8.	H. ♃.	Flor. dan. t. 363.
3 virgínica. R.S.	(*st.*)	Virginian.	Virginia.	1759.	8.	H. ♃.	
4 aùrea. Ph.	(*ye.*)	golden pert.	N. America.	1826.	7. 9.	H. ♃.	Lodd. bot. cab. t. 1399.
5 quadridentàta. Ph.	(*st.*)	four-toothed.	——	——	6. 8.	H. ♃.	Lam. ill. 1. t. 16. f. 2.
6 acuminàta. Ph.	(*st.*)	sharp-pointed.	——	1828.	7. 8.	H. ♃.	
7 pilòsa. Ph.	(*wh.*)	hairy.	——	1827.	——	H. ♃.	
WULF'ENIA. R.S.		WULFE'NIA.	Diandria Monogynia. R.S.				
carinthìaca. R.S.	(*bl.*)	blue-flowered.	Carinthia.	1817.	6. 9.	H. ♃.	Swt. br. fl. gar. t. 66.
PÆDER'OTA. R.S.		PÆDER'OTA.	Diandria Monogynia. L.				
1 Agèria. R.S.	(*st.*)	sulphur-coloured.	Italy.	1823.	6. 7.	H. ♃.	Jacq. vind. 2. t. 121.
2 Bonaròta. R.S.	(*vi.*)	violet-coloured.	Germany.	1826.	——	H. ♃.	Jacq. aust. 5. t. 39.
LEPTA'NDRA. N.		LEPTA'NDRA.	Diandria Monogynia. N.				
1 virgínica.	(*wh.*)	Virginian.	Virginia.	1714.	7. 9.	H. ♃.	Pluk. alm. t. 70. f. 2.
Verónica virgínica. W.							
2 incarnàta. G.D.	(*li.*)	lilac-flowered.	——	——	——	H. ♃.	
3 sibírica. N.	(*wh.*)	Siberian.	Siberia.	1779.	7. 8.	H. ♃.	Amm. ruth. 20. t. 4.
VERO'NICA. R.S.		SPEEDWELL.	Diandria Monogynia. L.				
1 foliòsa. R.S.	(*bl.*)	leafy.	Hungary.	1805.	7. 9.	H. ♃.	W. et K. hung. 2. t. 102.
2 crenulàta. R.S.	(*bl.*)	notch-flowered.	S. Europe.	1804.	——	H. ♃.	Hoff. phyt. p. 95. t. E. f. 3.
3 marítima. R.S.	(*bl.*)	sea-side.	——	1570.	——	H. ♃.	Schrad. ver. t. 1. f. 1.
4 Hóstii. R.S.	(*bl.*)	Host's.	Italy.	....	——	H. ♃.	Tabern. n. kreut. 2. f. 4.

5 spùria. R.S.	(*bl.*) spurious.	Siberia.	1731.	7. 9.	H. ♃.	Gmel. it. 1. t. 39.
6 paniculàta. R.S.	(*bl.*) panicled.	Tartary.	1797.	6. 7.	H. ♃.	
7 Stephaniàna. R.S.	(*bl.*) Stephan's.	Persia.	1821.	——	H. ♃.	
8 complicàta. R.S.	(*bl.*) bundled.	S. Europe.	1812.	9. 10.	H. ♃.	Hoff. phyt. p. 98. t. E. f. 4.
9 azùrea. R.S.	(*bl.*) azure.	....	1818.	6. 9.	H. ♃.	
10 glábra. R.S.	(*bl.*) smooth.	S. Europe.	1804.	7. 9.	H. ♃.	Schrad. ver. p. 25. t. 1. f. 4.
β albiflòra.	(*wh.*) *white-flowered.*	————	——	——	H. ♃.	
11 brachyphy'lla. R.S.	(*bl.*) short-leaved.	....	1818.	——	H. ♃.	
brevifòlia. L.en.						
12 elàtior. R.S.	(*bl.*) tall.	S. Europe.	1808.	——	H. ♃.	
13 exaltàta.	(*bl.*) lofty.	Siberia.	1816.	——	H. ♃.	
14 falcàta. R.S.	(*bl.*) falcate-leaved.	....	1820.	6. 7.	H. ♃.	
15 acùta. R.S.	(*bl.*) acute-leaved.	....	——	7. 9.	H. ♃.	
16 argùta. R.S.	(*bl.*) sharp-notched.	S. Europe.	1812.	——	H. ♃.	Schrad. ver. n. 7. t. 2. f. 2.
17 nìtens. R.S.	(*bl.*) glossy-leaved.	....	1816.	——	H. ♃.	
18 longibracteàta. R.S.	(*bl.*) long-bracted.	....	——	——	H. ♃.	
β latifòlia.	(*bl.*) *broad-leaved.*	....	——	——	H. ♃.	
19 mèdia. R.S.	(*bl.*) long-spiked.	Germany.	1804.	——	H. ♃.	Schrad. ver. n. 8. t. 1. f. 2.
20 persicifòlia. R.S.	(*bl.*) Peach-leaved.	....	1819.	——	H. ♃.	
21 austràlis. R.S.	(*bl.*) pubescent.	S. Europe.	1812.	——	H. ♃.	Schrad. ver. n. 9. t. 2. f. 3.
22 longifòlia. R.S.	(*bl.*) long-leaved.	————	1731.	——	H. ♃.	———— p. 26. t. 2. f. 1.
23 gróssa. R.S.	(*bl.*) stout.	S. Russia.	1827.	6. 9.	H. ♃.	
24 ambígua. R.S.	(*bl.*) ambiguous.	England.	....	——	H. ♃.	
25 neglécta. R.S.	(*bl.*) canescent.	Siberia.	1797.	7. 9.	H. ♃.	Swt. br. fl. gar. t. 55.
26 angustifòlia. S.S.	(*bl.*) narrow-leaved.	————	1823.	——	H. ♃.	
27 incàna. R.S.	(*bl.*) hoary.	Russia.	1759.	——	H. ♃.	Hoff. c. gott. 15. t. 6.
28 rìgens. R.S.	(*bl.*) rigid.	....	1824.	——	H. ♃.	
29 élegans. R.S.	(*ro.*) elegant.	Italy.	....	——	H. ♃.	
30 spicàta. R.S.	(*bl.*) spiked.	England.	....	——	H. ♃.	Eng. bot. v. 1. t. 2.
31 Clùsii. R.S.	(*bl.*) Clusius's.	N. Europe.	....	——	H. ♃.	Flor. dan. t. 52.
32 menthæfòlia. R.S.	(*bl.*) Mint-leaved.	Austria.	1818.	——	H. ♃.	
33 Barrelíeri. R.S.	(*bl.*) Barrelier's.	S. Europe.	....	——	H. ♃.	Barrel. ic. t. 682.
34 orchidèa. R.S.	(*bl.*) Orchis-like.	Europe.	....	——	H. ♃.	
35 crassifòlia. R.S.	(*vi.*) thick-leaved.	Hungary.	1818.	——	H. ♃.	
36 hy'brida. R.S.	(*bl.*) Welsh.	England.	....	——	H. ♃.	Eng. bot. v. 10. t. 673.
37 confùsa. R.S.	(*bl.*) confused.	....	1819.	6. 7.	H. ♃.	
38 longiflòra. R.S.	(*li.*) long-flowered.	....	1823.	7. 9.	H. ♃.	
39 ruthénica. R.S.	(*bl.*) Russian.	Russia.	——	——	H. ♃.	
40 grándis. R.S.	(*wh.*) large white.	Siberia.	1826.	——	H. ♃.	
41 Pònæ. R.S.	(*bl.*) dwarf.	Pyrenees.	1823.	——	H. ♃.	Gouan. ill. t. 1. f. 1.
42 polystàchya. R.S.	(*bl.*) many-spiked.	....	1819.	——	H. ♃.	
43 præálta. L.C.	(*bl.*) tall upright	....	1817.	6. 8.	H. ♃.	
44 incarnàta. G.D.	(*pi.*) flesh-coloured.	....	....	——	H. ♃.	
45 cárnea. Ræu.	(*pi.*) pale red.	....	....	——	H. ♃.	
46 corymbòsa. G.D.	(*bl.*) corymbose.	....	1817.	6. 7.	H. ♃.	
47 melanchólica. Df.	(*bl.*) melancholy.	....	1820.	——	H. ♃.	
48 leucántha. F.	(*wh.*) white-flowered.	Siberia.	1824.	——	H. ♃.	
49 ascéndens. R.S.	(*bl.*) ascending.	....	1828.	——	H. ♃.	
50 villòsa. R.S.	(*bl.*) villous.	S. Europe.	1800.	——	H. ♃.	Schrad. ver. p. 13. t. 1. f. 3.
51 pinnàta. R.S.	(*bl.*) wing-leaved.	Siberia.	1776.	6. 8.	H. ♃.	Hoff. c. gott. 15. t. 10.
52 incísa. R.S.	(*bl.*) cut-leaved.	————	1779.	——	H. ♃.	
53 laciniàta. R.S.	(*bl.*) jagged-leaved.	————	1780.	——	H. ♃.	Jung. ic. rar. C. 1. f. 2.
54 gentianoídes. F.G.	(*vi.*) Gentian-like.	Levant.	1748.	5. 6.	H. ♃.	Flor. græc. 1. t. 5.
55 gentianifòlia.	(*pa.bl.*) Gentian-leaved.	————	——	——	H. ♃.	Bot. mag. t. 1002.
Verónica gentianoídes. B.M. *non* Flor. græc.						
56 pállida. R.S.	(*pa.*) pale.	————	1816.	——	H. ♃.	
57 bellidioídes. R.S.	(*bl.*) Daisy-leaved.	Switzerland.	1775.	6. 7.	H. ♃.	Hall. hist. t. 15. f. 1.
58 fruticulòsa. R.S.	(*pu.*) flesh-coloured.	Scotland.	....	6. 8.	H. ♄.	Eng. bot. v. 15. t. 1028.
59 saxátilis. R.S.	(*bl.*) blue rock.	————	....	6. 7.	H. ♃.	———— v. 15. t. 1027.
60 crética. R.S.	(*bl.*) Cretan.	Crete.	1819.	5. 6.	H. ♃.	
61 Baumgartènii. R.S.	(*pa.bl.*) Baumgarten's.	Transylvania.	1826.	——	H. ♃.	
62 nummulària. R.S.	(*bl.*) moneywort.	Pyrenees.	1820.	6. 8.	H. ♃.	Gouan. ill. 1. t. 1. f. 2.
63 alpìna. R.S.	(*bl.*) Alpine.	Scotland.	....	5. 6.	H. ♃.	Eng. bot. v. 7. t. 484.
α acutifòlia.	(*bl.*) *acute entire-leav'd.*	Europe.	....	——	H. ♃.	
β pùmila. All.	(*bl.*) *saw-leaved.*	————	....	——	H. ♃.	Allion. ped. t. 22. f. 5.
γ obtusifòlia.	(*bl.*) *blunt entire-leav'd.*	Scotland.	....	——	H. ♃.	Eng. bot. v. 7. t. 484.
δ integrifòlia. W.	(*bl.*) *entire-leaved.*	Europe.	....	——	H. ♃.	Krocker siles. t. 3.
ε heterophy'lla.	(*bl.*) *various-leaved.*	————	....	——	H. ♃.	
ζ rotundifòlia.	(*bl.*) *round-leaved.*	————	....	——	H. ♃.	Braun. salisb. n. 15. t. 2.

64	Wormskiòldii. R.S.	(*bl.*)	Wormskiold's.	Greenland.	1819.	5. 6.	H. ♂.	Bot. mag. t. 2975.
65	grandifòlia. R.S.	(*bl.*)	large-leaved.	Italy.	1828.	——	H. ♃.	
66	depauperàta. R.S.	(*bl.*)	few-flowered.	Hungary.	1819.	——	H. ♃.	W.et K. hung. 3. t. 245
67	serpyllifòlia. R.S.	(*bl.fl.*)	smooth.	Britain.	....	4. 7.	H. ♃.	Eng. bot. v. 15. t. 1075
	α ovàta.	(*bl.fl.*)	*common.*	————	....	——	H. ♃.	
	β tenélla. Schm.	(*bl.*)	*roundish-leaved.*	Europe.	....	——	H. ♃.	
	γ neglécta. Schm.	(*pa.*)	*neglected.*	————	....	——	H. ♃.	
	δ quaternàta.	(*pa.li.*)	*four-leaved.*	————	....	——	H. ♃.	
	ε nummulariæfòlia.	(*pa.*)	*moneywort-leaved.*	————	....	——	H. ♃.	
	ζ humifùsa. L.T.	(*pa.*)	*trailing.*	————	....	——	H. ♃.	
68	hirsùta. R.S.	(*pa.*)	bristly.	Scotland.	....	——	H. ♃.	
	setígera. S.S.							
69	tenélla. R.S.	(*pu.*)	slender.	Pyrenees.	1820.	5. 6.	H. ♃.	Allion. ped. t. 22. f. 1.
70	microphy'lla. R.S.	(*pu.*)	small-leaved.	Hungary.	1824.	——	H. ♃.	
71	parviflòra. R.S.	(*bl.*)	small-flowered.	N. Zealand.	1822.	6. 8.	H. ♃.	
72	undulàta. R.S.	(*wh.*)	wave-leaved.	E. Indies.	1827.	——	G. ♃.	
73	decussàta. R.S.	(*wh.*)	cross-leaved.	Falkland Isl.	1776.	——	F. ♄.	Bot. mag. t. 242.
74	aphy'lla. R.S.	(*bl.*)	naked-stalked.	Italy.	1775.	5.	H. ♃.	Seguier. ver. t. 3. f. 2.
75	Beccabúnga. R.S.	(*bl.*)	brook-lime.	Britain.	....	5. 6.	H.*w.* ♃.	Eng. bot. v. 10. t. 655.
76	Anagállis. R.S.	(*pa.bl.*)	long-leaved water.	————	....	6. 8.	H.*w.* ♃.	———— v. 11. t. 781.
77	scutellàta. R.S.	(*li.*)	marsh.	————	....	——	H.*w.* ♃.	———— v. 11. t. 782.
78	Parmulària. R.S.	(*li.*)	tooth-leav'd marsh.	Austria.	1816.	——	H.*w.* ♃.	Trattin. fl. aust. b. 20.
79	caroliniàna. R.S.	(*bl.*)	Carolina.	Carolina.	1821.	6. 7.	H.*w.* ♃.	
80	thymifòlia. F.G.	(*bl.*)	thyme-leaved.	Crete.	1827.	5. 7.	F. ♃.	Flor. græc. v. 1. t. 6.
81	pectinàta. R.S.	(*bl.*)	pectinated.	Levant.	1820.	6. 8.	H. ♃.	Buxb. C. 1. t. 39. f. 1.
82	orientàlis. R.S.	(*bl.*)	various-leaved.	————	1748.	——	H. ♃.	Sal. stirp. rar. 7. t. 4.
83	trichocárpa. R.S.	(*bl.*)	hairy-capsuled.	————	1820.	——	H. ♃.	
	pilocárpa. L.en.							
84	Jacquìni. R.S.	(*bl.*)	Jacquin's.	Austria.	1748.	——	H. ♃.	Jacq. aust. 4. t. 329.
	austrìaca. Jacq. *non* L.							
85	austrìaca. R.S.	(*bl.*)	Austrian.	————	———	——	H. ♃.	Moris. h. 2. s. 3. t. 23. f. 12.
	polymórpha. W.en.							
86	tenuifòlia. R.S.	(*bl.*)	slender-leaved.	Iberia.	1821.	6. 7.	H. ♃.	
87	pinnatífida. W.	(*bl.*)	pinnatifid.	....	1817.	5. 6.	H. ♃.	
88	multífida. R.S.	(*bl.*)	multifid-leaved.	Siberia.	1748.	6. 8.	H. ♃.	Bot. mag. t. 1679.
89	caucásica. R.S.	(*wh.bl.*)	Caucasean.	Caucasus.	1815.	——	H. ♃.	
90	taùrica. W.	(*bl.*)	Taurian.	Tauria.	———	——	H. ♃.	Lodd. bot. cab. t. 911.
91	Alliònii. R.S.	(*bl.*)	shining-leaved.	S. Europe.	1748.	——	H. ♃.	Allion. ped. 1. t. 46. f. 3.
92	officinàlis. R.S.	(*bl.*)	officinal.	Britain.	....	4. 7.	H. ♃.	Eng. bot. v. 11. t. 765.
	β rígida. Gr.	(*bl.*)	*rigid-leaved.*	Scotland.	....	——	H. ♃.	
	γ albiflòra.	(*wh.*)	*white-flowered.*	Britain.	....	——	H. ♃.	
93	Mulleriàna. R.S.	(*bl.*)	Muller's.	Styria.	1825.	5. 7.	H. ♃.	
94	Tournefórtii. R.S.	(*bl.*)	Tournefort's.	S. France.	1821.	5. 6.	H. ♃.	
95	Chaìxi. R.S.	(*bl.*)	Chaix's.	S. Europe.	1825.	5. 7.	H. ♃.	
96	prostràta. R.S.	(*bl.*)	trailing.	Germany.	1774.	5. 6.	H. ♃.	Bauh. hist. 3. p. 287. ic.
97	pilòsa. R.S.	(*bl.*)	hairy.	Bohemia.	1819.	——	H. ♃.	
98	acutiflòra. R.S.	(*li.*)	acute-flowered.	France.	1821.	——	H. ♃.	
99	plicàta. R.S.	(*bl.*)	plaited-leaved.	Bohemia.	1823.	——	H. ♃.	Pohl tent. fl. b. p. 15. f. 1.
100	micrántha. R.S.	(*wh.pu.*)	soft-leaved.	Portugal.	———	——	H. ♃.	Hoff. f. port. t. 57.
	móllis. Zea.							
101	latifòlia. R.S.	(*bl.*)	broad-leaved.	Austria.	1748.	6. 7.	H. ♃.	Swt. br. fl. gar. t. 23.
102	Teùcrium. R.S.	(*bl.*)	saw-leaved.	Europe.	1596.	6. 8.	H. ♃.	Lodd. bot. cab. t. 425.
103	crinìta. R.S.	(*bl.*)	very hairy.	Hungary.	1826.	——	H. ♃.	
104	lutetiàna. R.S.	(*pa.bl.*)	Portugal.	Portugal.	———	——	H. ♃.	
105	dentàta. R.S.	(*bl.*)	tooth-leaved.	S. Europe.	1816.	——	H. ♃.	
106	Schmídtii. R.S.	(*bl.*)	Schmidt's.	Bohemia.	1797.	6. 7.	H. ♃.	
	paniculàta. W. *non* Pall.							
107	pedunculàris. R.S.	(*bl.*)	long-peduncled.	Caucasus.	1825.	——	H. ♃.	
108	petr'æa. R.S.	(*pa.bl.*)	Caucasean rock.	————	1821.	——	H. ♃.	
109	umbròsa. M.B.	(*bl.*)	shady wood.	Tauria.	1827.	——	H. ♃.	M. B. cent. ros. 1. t. 7.
110	Cham'ædrys. R.S.	(*bl.*)	Germander.	Britain.	....	5. 8.	H. ♃.	Eng. bot. v. 9. t. 623.
111	stolonífera. R.S.	(*bl.*)	creeping.	N. America.	1827.	——	H. ♃.	
112	divaricàta. R.S.	(*bl.*)	divaricate.	Sweden.	1828.	——	H. ♃.	
113	Vestiàna. R.S.	(*bl.*)	Vest's.	....	1820.	——	H. ♃.	
114	máxima. R.S.	(*pa.bl.*)	large.	Iberia.	1824.	——	H. ♃.	Buxb. cent. 1. t. 34.
115	urticæfòlia. R.S.	(*pa.bl.*)	Nettle-leaved.	Austria.	1776.	6. 7.	H. ♃.	Jacq. aust. 1. t. 59.
116	montàna. R.S.	(*pa.*)	mountain.	Britain.	....	7. 8.	H. ♃.	Eng. bot. v. 11. t. 766.
117	plebèia. B.P.	(*bl.*)	plebeian.	N. S. W.	1822.	6. 8.	G. ♃.	
	argùta. R.S.							

118 Bro'wnii. B.P. (bl.) sharp-toothed. N. S. W. 1822. 6. 8. G.♃.
119 grácilis. B.P. (bl.) slender. ——— ——— ——— G.♃.
120 perfoliàta. B.P. (bl.) perfoliate. ——— 1815. 7. 8. G.♃. Bot. mag. t. 1936.
121 labiàta. B.P. (pa.bl.) labiated. N. Holland. 1802. 4. 7. F.♃. ——— t. 1660.
Derwéntia. Andr. reposit. t. 531.
122 mollíssima. R.S. (bl.) softest. .... 1819. 6. 8. H.♃.
123 renifórmis.Ph.(pa.bl.)kidney-leaved. Missouri. 1827. 5. 8. H.♃.
124 Waldsteiniàna.R.S.(bl.)Waldstein's. .... 1819. 6. 8. H.♃.
125 vérna. R.S. (pa.bl.) vernal. Britain. .... 4. 5. H.⊙. Eng. bot. v. 1. t. 25.
126 digitàta. R.S. (bl.) fingered. Spain. 1805. 6. 7. H.⊙. Reich. ic. t. 36.
127 triphy'llos. R.S. (bl.) variable-leaved. Britain. .... 4. 5. H.⊙. Eng. bot. v. 1. t. 26.
128 hederifòlia.R.S.(pa.bl.)Ivy-leaved. ——— .... 3. 10. H.⊙. ——— v. 11. t. 784.
129 Cymbalària. R.S.(wh.)fleshy-leaved. Levant. 1823. 4. 8. H.⊙. Flor. græc. t. 9.
130 peregrìna. R.S. (pa.) knotgrass-leaved.N. Europe. 1680. 5. 6. H.⊙. Flor. dan. t. 407.
131 depréssa. R.S. (bl.) depressed. Hungary. 1826. 4. 8. H.⊙.
132 glaùca. F.G. (bl.) glaucous. Greece. ——— ——— H.⊙. Flor. græc. v. 1. t. 7.
133 am'œna. R.S. (bl.) fine blue. Iberia. 1823. 5. 6. H.⊙. MB. cent. ros. t. 1. f. 18.
134 bilòba. R.S. (wh.) two-lobed. Levant. 1819. 4. 6. H.⊙. Buxb. cent. 1. t. 36.
135 pr'æcox. R.S. (bl.) early-flowered. Europe. ——— 3. 5. H.⊙. Allion. ped. t. 1. f. 1.
136 acinifòlia. R.S.(pa.bl.) Basil-leaved. S. Europe. 1788. 4. 5. H.⊙. Poit.etTurp. par. 1. t. 23.
137 arvénsis. R.S. (bl.) wall. Britain. .... 4. 7. H.⊙. Eng. bot. v. 11. t. 734.
138 agréstis. R.I. (va.) field. ——— .... 1. 12. H.⊙. Reich. ic. C. 3. t. 277.
pulchélla. DC. *non* Bernhard. *versícolor.* Fries.
139 opàca. R.I. (pa.bl.) dull-leaved. Britain. .... ——— H.⊙. Reic.ic.C.3.p.67.t.278.
140 polìta. R.I. (bl.) polished. ——— .... ——— H.⊙. ——— C.3.p.45.t.246.
141 dídyma. T.N. (bl.) toothed-calyxed.Naples. 1827. ——— H.⊙.
142 pulchélla. R.S.(wh.ro.)neat. .... 1823. 3. 6. H.⊙.
143 Buxbaùmii. S.S. (bl.) Buxbaum's. S. Europe. 1820. ——— H.⊙. Ten. fl. nap. t. 1.
144 filifórmis. S.S. (bl.) long-stalked. Levant. 1780. 5. H.⊙. Buxb. cent. 1. t. 40. f. 1.
145 pérsica. S.S. (bl.) Persian. Persia. 1816. 1. 12. H.⊙.
DIPLOPHY'LLUM. S.S. DIPLOPHY'LLUM. Diandria Monogynia. S.S.
veronicæfórme.S.S.(bl.)Speedwell-like. Caucasus. 1813. 4. 6. H.⊙. Linn. trans. v. 11. t. 31.
Verónica crísta gálli. L.T.
HEMI'MERIS. L. HEMI'MERIS. Diandria Monogynia.
montàna. L. (re.) mountain. C. B. S. 1816. 6. 8. G.♃. Lam. ill. t. 532.

ORDO CLIII.

LABIATÆ. *Brown prodr.* 1. *p.* 499.

Tribus I. MENTHOIDEÆ. *Bentham in Bot. reg. fol.* 1282.

LYC'OPUS. R.S. WATER-HOREHOUND. Diandria Monogynia. L.
1 austràlis. B.P. (wh.) New Holland. N. S. W. 1823. 5. 7. G.♃.
2 europ'æus. R.S. (wh.) Europæan. Britain. .... 7. 8. H.♃. Eng. bot. v. 16. t. 1105.
3 intermèdius. (wh.) intermediate. Europe. 1816. ——— H.♃.
4 exaltàtus. R.S. (wh.) tall. Italy. 1739. ——— H.♃. Flor. græc. 1. t. 12.
5 virgínicus. R.S. (wh.) Virginian. N. America. 1760. 8. 9. H.♃.
6 sinuàtus. E.C. (wh.) sinuate-leaved. ——— 1812. ——— H.♃.
ISA'NTHUS. M. ISA'NTHUS. Didynamia Gymnospermia. S.S.
cœrùleus. M. (bl.) blue. N. America. 1732. 6. 8. H.♂. Dill. elt. t. 285. f. 369.
Trichostèma brachiàta. L.
AUDIBE'RTIA. B.R. AUDIBE'RTIA. Didynamia Gymnospermia.
pusílla. B.R. (pu.) small. Corsica. 1829. 5. 8. H.♃.
Th'ymus parviflòrus. Requien.
ME'NTHA. L. MINT. Didynamia Gymnospermia. L.
1 víridis. W. (pu.) spear. Britain. .... 8. H.♃. Eng. bot. v. 34. t. 2424.
2 lævigàta. W.en. (pu.) smooth-leaved. Germany. .... 6. 8. H.♃.
3 balsàmea. W.en. (pu.) Balsam-scent'd. Italy. 1804. 7. 8. H.♃.
4 crispàta. W.en. (pu.) crumpled. Europe. 1807. ——— H.♃.
5 piperìta. E.B. (pu.) pepper. England. .... 8. 9. H.♃. Eng. bot. v. 10. t. 687.
6 nilìaca. S.S. (pu.) Egyptian. Egypt. 1796. 7. 8. H.♃. Jacq. vind. 3. t. 87.
7 tènuis. M. (wh.) slender. N. America. 1826. ——— H.♃.

8 sylvéstris. E.B.	(*pu.*)	wild.	Britain.	....	7. 8.	H.♃.	Eng. bot. v. 10. t. 686.
9 nemoròsa. W.en.	(*pu.*)	wood.	———	....	——	H.♃.	Flor. dan. t. 484.
10 hírta. W.en.	(*pu.*)	hairy.	Europe.	....	——	H.♃.	
11 pubéscens. W.en.	(*pu.*)	pubescent.	———	....	——	H.♃.	
12 Auriculària. W.	(*li.*)	Indian.	E. Indies.	1796.	——	S.♃.	Rumph. amb. 6. t. 16.
13 críspa. W.	(*pu.*)	curled.	Siberia.	1640.	——	H.♃.	
14 pyramidàlis. T.N.	(*pu.*)	pyramidal.	Naples.	1824.	——	H.♃.	
15 undulàta. W.en.	(*pu.*)	wave-leaved.	Europe.	1816.	——	H.♃.	
16 incàna. W.en.	(*pu.*)	hoary.	———	1790.	——	H.♃.	
17 lavandulàcea. W.en.	(*pu.*)	Lavender-leaved.	Spain.	1825.	——	H.♃.	
18 capénsis. W.	(*pu.*)	Cape.	C. B. S.	1816.	——	G.♃.	
19 rotundifòlia. E.B.	(*pu.*)	round-leaved.	Britain.	....	8. 9.	H.♃.	Eng. bot. v. 7. t. 446.
β variegàta.	(*pu.*)	*variegated.*	———	....	——	H.♃.	
20 macrostàchya. T.N.	(*pu.*)	long-spiked.	Naples.	1820.	——	H.♃.	
21 gratíssima. W.	(*pu.*)	oblong-leaved.	Britain.	....	7. 8.	H.♃.	
22 citràta. S.S.	(*pu.*)	Bergamot.	England.	....	——	H.♃.	Eng. bot. v. 15. t. 1025.
odoràta. E.B.							
23 hirsùta. E.B.	(*pu.*)	hairy water.	Britain.	....	7. 9.	H.*w.*♃.	Eng. bot. v. 7. t. 447.
24 acutifòlia. E.B.	(*pu.*)	acute-leaved.	England.	....	9.	H.*w.*♃.	——— v. 34. t. 2415.
25 satìva. E.B.	(*re.*)	tall red.	———	....	8. 9.	H.*w.*♃.	——— v. 7. t. 448.
26 rùbra. E.B.	(*re.*)	common red.	Britain.	....	9.	H.*w.*♃.	——— v. 20. t. 1413.
27 géntilis. E.B.	(*re.*)	bushy red.	———	....	7. 8.	H.*w.*♃.	——— v. 30. t. 2118.
28 grácilis. H.K.	(*re.*)	narrow-leaved.	———	....	——	H.*w.*♃.	——— v. 7. t. 449.
29 palústris. S.M.	(*pu.*)	marsh.	———	....	9.	H.*w.*♃.	
30 villòsa. S.M.	(*pu.*)	horse.	———	....	7. 8.	H.*w.*♃.	
31 praténsis. S.M.	(*pu.*)	meadow.	———	....	——	H.*w.*♃.	Sole's Mints. t. 17.
32 rivàlis. S.M.	(*pu.*)	river.	———	....	——	H.*w.*♃.	
33 paludòsa. S.M.	(*pu.*)	brook.	———	....	9.	H.*w.*♃.	
34 arvénsis. E.B.	(*li.*)	corn.	———	....	7. 9.	H.♃.	Eng. bot. v. 30. t. 2119.
35 pr'æcox. S.M.	(*li.*)	early-flowering.	———	....	6.	H.♃.	Sole's Mints. ic.
36 agréstis. E.B.	(*li.*)	field.	———	....	7. 8.	H.♃.	Eng. bot. v. 30. t. 2120.
37 austrìaca. S.S.	(*pu.*)	Austrian.	Austria.	1818.	——	H.♃.	
38 dentàta. W.	(*pu.*)	toothed.	Germany.	1816.	——	H.♃.	
39 canadénsis. W.	(*pu.*)	Canadian.	N. America.	1801.	——	H.♃.	
40 boreàlis. M.	(*li.*)	northern.	———	1824.	——	H.♃.	
41 élegans. S.S.	(*pu.*)	elegant.	S. Europe.	1828.	——	H.♃.	
42 badénsis. S.S.	(*pu.*)	Baden.	Italy.	——	——	H.♃.	
43 rugòsa. Roth.	(*pu.*)	rugged-leaved.	Germany.	1824.	——	H.♃.	
44 glabràta. S.S.	(*pu.*)	smooth.	Egypt.	1802.	——	H.♃.	
45 Pulègium. S.S.	(*li.*)	Pennyroyal.	Britain.	....	8. 9.	H.♃.	Eng. bot. v. 15. t. 1026.
46 cervìna. S.S.	(*wh.*)	Hyssop-leaved.	S. France.	1648.	6. 8.	H.♃.	Moris. s. 3. t. 7. f. 7.
COLEBRO'OKIA. S.S.		COLEBRO'OKIA.	Didynamia Gymnospermia.				
1 oppositifòlia. D.P.	(*wh.*)	opposite-leav'd.	Nepaul.	1816.	6. 9.	S.♄.	Sm. exot. bot. t. 115.
2 ternifòlia. S.S.	(*wh.*)	three-leaved.	E. Indies.	——	——	S.♄.	
PERI'LLA. W.		PERI'LLA.	Didynamia Gymnospermia. L.				
1 ocymoídes. W.	(*wh.*)	Basil-leaved.	India.	1770.	7. 8.	H.☉.	Arduin. spec. 1. t. 13.
2 fruticòsa. D.P.	(*wh.*)	frutescent.	Nepaul.	1818.		S.♄.	
ELSHO'LTZIA. S.S.		ELSHO'LTZIA.	Didynamia Gymnospermia. W.				
1 cristàta. S.S.	(*li.*)	crested.	Siberia.	1789.	5. 7.	H.☉.	Bot. mag. t. 2560.
2 ocymoídes. S.S.	(*li.*)	Basil-like.	E. Indies.	1823.	——	S.♃.	
APHANOCH'ILUS. B.R.		APHANOCH'ILUS.	Didynamia Gymnospermia.				
1 blándus.	(*wh.*)	agreeable.	Nepaul.	1820.	10. 12.	G.♃.	DC. pl. rar. gen. t. 8.
Méntha blánda. DC.							
2 incìsus. B.R.	(*wh.*)	cut-leaved.	———	1818.	6. 10.	G.♃.	
Méntha blánda. H.C. *non* DC.							
DYSOPHY'LLA. B.R.		DYSOPHY'LLA.	Didynamia Gymnospermia.				
1 quadrifòlia. B.R.	(*li.*)	four-leaved.	Nepaul.	1823.	6. 8.	G.♃.	
2 stellàta. B.R.	(*li.*)	starry-leaved.	———	1816.	——	G.♃.	
3 verticillàta. B.R.	(*li.*)	whorl-leaved.	———	1828.	——	G.♃.	
4 pùmila. B.R.	(*li.*)	dwarf.	———	——	——	G.♃.	Bot. mag. t. 2907.
Méntha verticillàta. B.M. *pùmila.* Graham.							
POGOST'EMON. S.S.		POGOST'EMON.	Didynamia Gymnospermia. S.S.				
plectranthoídes. S.S.	(*w.*)	Plectanthrus-like.	......	1818.	6. 8.	S.♄.	Desf. m. mus. 2. t. 6.

Tribus II. Satureineæ. *Benth. in Bot. reg.*

BYSTROP'OGON. S.S.		Bystrop'ogon.	Didynamia Gymnospermia. S.S.					
1 origanifòlius. s.s.	(*li.*)	Marjoram-leav'd.	Teneriffe.	1815.	7. 8.	G. ♄.		
2 plumòsus. s.s.	(*li.*)	woolly-flower'd.	Canaries.	1779.	6. 7.	G. ♄.		
3 canariénsis. s.s.	(*li.*)	Canary.	———	1714.	6. 8.	G. ♄.	Comm. hort. 2. t. 65.	
4 punctàtus. s.s.	(*li.*)	cluster-flower'd.	Madeira.	1775.	7. 9.	G. ♄.		
PYCNA'NTHEMUM. M.		Pycna'nthemum.	Didynamia Angiospermia. S.S.					
1 linifòlium. s.s.	(*wh.*)	Flax-leaved.	N. America.	1739.	7. 8.	H. ♃.	Herm. parad. t. 218.	
2 lanceolàtum. s.s.	(*wh.*)	spear-leaved.	———	1812.	——	H. ♃.		
3 aristàtum. s.s.	(*wh.*)	awned.	———	1752.	8.	H. ♃.	Mich. amer. 2. t. 33.	
4 verticillàtum. s.s.	(*wh.*)	whorl-leaved.	———	1816.	7. 8.	H. ♃.		
5 Monardélla. s.s.	(*li.*)	Monarda-like.	———	——	——	H. ♃.		
6 incànum. s.s.	(*bh.*)	hoary.	———	1732.	7. 10.	H. ♃.	Dill. elt. t. 74. f. 85.	
SATUR'EJA. S.S.		Savory.	Didynamia Gymnospermia. L.					
1 approximàta. s.s.	(*li.*)	bundle-leaved.	Sicily.	1825.	5. 7.	F. ♄.		
2 spinòsa. F.G.	(*wh.*)	spiny.	Crete.	1827.	——	G. ħ.	Flor. græc. t. 545.	
3 vimínea. s.s.	(*li.*)	Pennyroyal tree.	Jamaica.	1783.	——	S. ♄.		
4 montàna. s.s.	(*li.*)	winter.	S. Europe.	1562.	6. 7.	H. ħ.	Flor. græc. t. 543.	
5 rupéstris. s.s.	(*wh.*)	rock.	Carniola.	1798.	——	H. ♃.	Jacq. ic. 3. t. 494.	
6 obovàta. s.s.	(*li.*)	obovate-leaved.	Spain.	1824.	——	H. ħ.		
7 tenuifòlia. s.s.	(*li.*)	slender-leaved.	Italy.	1820.	——	H. ħ.		
8 Juliàna. s.s.	(*li.*)	linear-leaved.	———	1596.	5. 9.	F. ħ.	Flor. græc. t. 540.	
9 gr'æca. s.s.	(*li.*)	Grecian.	Greece.	1759.	6. 7.	F. ♃.	——— t. 542.	
10 hirsùta. s.s.	(*li.*)	hairy.	Sicily.	1824.	——	F. ♃.		
11 Teneríffæ. s.s.	(*pu.*)	Teneriffe.	Teneriffe.	1815.	——	F. ♃.		
12 horténsis. s.s.	(*wh.*)	summer.	Italy.	1652.	6. 8.	H. ☉.	Lam. ill. t. 504. f. 1.	
13 capitàta. w.	(*li.*)	ciliated.	Levant.	1596.	6. 10.	H. ħ.	Flor. græc. t. 544.	
14 nervòsa. s.s.	(*li.*)	nerved.	Sicily.	1824.	——	H. ♄.		
15 Thy'mbra. s.s.	(*li.*)	whorl-flowered.	Candia.	1640.	5. 7.	G. ♄.	Flor. græc. t. 541.	
16 congésta. s.s.	(*li.*)	close-flowered.	....	1825.	——	G. ♄.		
TH'YMUS. S.S.		Thyme.	Didynamia Gymnospermia. L.					
1 villòsus. s.s.	(*li.*)	hairy.	Portugal.	1759.	6. 7.	F. ħ.	Flor. græc. t. 578.	
2 cephalòtus. s.s.	(*li.*)	great headed.	———	——	7. 8.	H. ħ.	Hof. et Lk. fl. lus. t. 13.	
3 capitàtus. L.en.	(*li.*)	headed.	———	1823.	——	H. ħ.		
4 odoratíssimus. s.s.	(*li.*)	very sweet.	Tauria.	——	——	H. ħ.		
5 acicularis. s.s.	(*li.*)	needle-shaped.	Hungary.	1806.	6. 8.	H. ħ.	W. et K. hung. 2. t. 147.	
6 hirsùtus. s.s.	(*li.*)	hairy.	———	1816.	——	H. ħ.		
7 Z'ygis. s.s.	(*wh.*)	Spanish.	Spain.	1771.	8.	H. ħ.	Flor. græc. t. 574.	
8 Marschalliànus. s.s.	(*li.*)	Marschall's.	Russia.	1815.	6. 8.	H. ħ.		
9 glabréscens. L.en.	(*li.*)	smooth.	S. Europe.	1823.	——	H. ♃.	Hof. et Lk. fl. lus. t. 15.	
10 angustifòlius. s.s.	(*li.*)	narrow-leaved.	Europe.	....	——	H. ħ.		
11 Serpy'llum. s.s.	(*li.*)	wild or mother of.	Britain.	....	——	H. ħ.	Eng. bot. v. 22. t. 1514.	
12 exserens. s.s.	(*li.*)	exserted.	———	....	——	H. ħ.		
13 citriodòrus. L.en.	(*li.*)	Lemon.	Europe.	....	——	H. ħ.		
14 collìnus. s.s.	(*li.*)	hill.	Tauria.	1823.	——	H. ħ.		
15 montanus. W.K.	(*li.*)	mountain.	Hungary.	1800.	6. 7.	H. ħ.	W. et K. hung. 1. t. 71.	
16 nummulàrius. M.B.	(*li.*)	moneywort.	Caucasus.	1818.	——	H. ħ.	Bot. mag. t. 2666.	
17 pannònicus. W.en.	(*li.*)	Hungarian.	Hungary.	——	——	H. ħ.		
18 lanuginòsus. s.s.	(*li.*)	woolly.	Europe.	....	6. 8.	H. ħ.		
19 tomentòsus. s.s.	(*li.*)	tomentose.	Spain.	1816.	——	H. ħ.		
20 álbicans. s.s.	(*li.*)	white-leaved.	Portugal.	——	——	H. ħ.		
21 adscéndens. L.en.	(*li.*)	ascending.	S. Europe.	....	——	H. ħ.		
22 lùcidus. s.s.	(*li.*)	shining-leaved.	....	1816.	——	H. ħ.		
23 vulgàris. s.s.	(*li.*)	common.	S. Europe.	1548.	5. 8.	H. ħ.		
β latifòlius.	(*li.*)	*broad-leaved.*	———	——	——	H. ħ.		
24 Mastichìna. s.s.	(*li.*)	Mastick.	Spain.	1596.	7. 9.	F. ħ.	Blackw. t. 134.	
25 elongàtus. L.en.	(*li.*)	elongated.	....	1816.	——	F. ħ.		
26 gravèolens. F.G.	(*li.*)	strong-scented.	Levant.	1823.	——	F. ħ.	Flor. græc. t. 576.	
27 Tragoríganum. s.s.	(*li.*)	goat's.	Candia.	1640.	5. 6.	F. ħ.	Alp. exot. t. 78.	
28 filifórmis. s.s.	(*li.*)	filiform.	Minorca.	1770.	6. 7.	G. ħ.		
29 fruticulòsus. s.s.	(*li.*)	frutescent.	Sicily.	1823.	——	H. ħ.		
30 Teneríffæ. s.s.	(*li.*)	Teneriffe.	Teneriffe.	1828.	——	F. ħ.		
31 incànus. s.s.	(*wh.*)	hoary.	Levant.	——	——	F. ♃.	Flor. græc. t. 577.	
32 croáticus. P.S.	(*li.*)	oval-leaved.	Hungary.	1802.	7. 8.	H. ħ.	W. et K. hung. t. 156.	
Piperélla. W.K. *non* L.								
33 ericæfòlius. W.en.	(*li.*)	Heath-leaved.	Spain.	1806.	6. 8.	H. ħ.		

ORI'GANUM. S.S.		MARJORAM.	Didynamia Gymnospermia. L.				
1 vulgàre. s.s.	(*ro.*)	common.	Britain.	....	6. 10.	H. ♃.	Eng. bot. v. 16. t. 1143
2 hùmile. s.s.	(*wh.*)	dwarf.	....	1827.	——	H. ♃.	
3 vìrens. s.s.	(*ro.*)	green.	Portugal.	1816.	——	H. ♃.	Hoff. et Lk. fl. p. 1. t. 9
4 heracleóticum. s.s.	(*li.*)	winter, sweet.	S. Europe.	1640.	——	H. ♃.	Lobel. ic. 492.
5 créticum. s.s.	(*li.*)	Cretan.	———	1596.	——	H. ♃.	Schk. hand. 2. t. 164.
6 megastàchyum. L.en.	(*li.*)	large-spiked.	———	1824.	——	H. ♃.	
7 macrostàchyon. s.s.	(*li.*)	long-spiked.	Portugal.	1820.	——	H. ♃.	
8 hírtum. L.en.	(*wh.*)	hairy.	S. Europe.	——	——	H. ♃.	
9 oblongàtum. L.en.	(*wh.*)	oblong-spiked.	———	1823.	——	H. ♃.	
10 smyrn'æum. L.	(*wh.*)	Smyrna.	Smyrna.	1722.	6. 7.	G. ♄.	Flor. græc. t. 571.
11 Onìtes. s.s.	(*wh.*)	pot.	Sicily.	1759.	7. 11.	H. ♃.	——— t. 572.
12 Marù. W.	(*li.*)	Lavender-scented.	Crete.	1822.	——	F. ♃.	Bot. mag. t. 2605.
13 Majoranoídes. W.	(*wh.*)	shrubby sweet.	....	....	——	F. ♄.	
14 Majoràna. s.s.	(*wh.*)	knotted.	Portugal.	1573.	6. 7.	F. ♂.	Moris. s. 11. t. 3. f. 1.
15 índicum. s.s.	(*wh.*)	Indian knotted.	E. Indies.	1820.	——	S. ♂.	
16 salvifòlium. Roth.	(*li.*)	Sage-leaved.	....	....	——	F. ♄.	
17 pallídum. P.S.	(*pa.*)	pale.	Levant.	1821.	——	G. ♄.	
18 stoloníferum. F.	(*li.*)	creeping.	....	1828.	——	H. ♃.	
19 Tournefórtii. s.s.	(*ro.*)	Tournefort's.	Amorgos.	1788.	8. 9.	F. ♄.	Flor. græc. t. 569.
20 sipy'leum. s.s.	(*ro.*)	Mount Sipylus.	Levant.	1699.	6. 9.	F. ♄.	—— —— t. 570.
21 Dictámnus. s.s.	(*li.*)	Dittany of Crete.	Candia.	1551.	——	F. ♄.	Bot. mag. t. 298.
22 ægyptìacum. s.s.	(*li.*)	Egyptian.	Egypt.	1731.	6. 8.	F. ♄.	Alp. ægypt. t. 95.
LOPHA'NTHUS. B.R.		LOPHA'NTHUS.	Didynamia Gymnospermia.				
1 chinénsis. B.R.	(*ye.*)	Mint-leaved.	Siberia.	1752.	8. 9.	H. ♃.	Jacq. vind. 2. t. 182.
Hyssòpus Lophánthus. s.s.							
2 nepetoídes. B.R.	(*ye.*)	square-stalked.	N. America.	1692.	8. 10.	H. ♃.	Jacq. vind. 1. t. 68.
3 scrophularifòlius. B.R.	(*re.*)	Figwort-leaved.	———	1800.	7. 8.	H. ♃.	Herm. parad. t. 106.
4 anisàtus. B.R.	(*bl.*)	Anise-like.	———	1820.	7. 9.	H. ♃.	Bot. reg. t. 1282.
5 urticifòlius. B.R.	(*pu.*)	Nettle-leaved.	———	1827.	——	H. ♃.	
6 multífidus. B.R.	(*wh.*)	multifid.	Siberia.	1797.	7. 8.	H. ♃.	Gmel. sib. 3. t. 55.
Népeta multífida. W.							
HYSS'OPUS. B.R.		HYSSOP.	Didynamia Gymnospermia.				
1 officinàlis. s.s.	(*bl.*)	common.	S. Europe.	1548.	6. 9.	H. ♄.	Jacq. aust. 3. t. 254.
2 orientàlis. W.	(*bl.*)	narrow-leaved.	Caucasus.	1816.	——	H. ♄.	Bot. mag. t. 2299.
angustifòlius. M.B.							
3 díscolor. F.	(*bl.*)	two-coloured.	Siberia.	1818.	——	H. ♄.	
4 septémfidus. E.	....	seven-cleft.	Egypt.	1827.	....	H. ♄.	
5 septemcrenàtus. E	....	seven-notched.	———	1829.	....	H. ♄.	
WESTRI'NGIA. S.S.		WESTRI'NGIA.	Didynamia Gymnospermia. S.S.				
1 rosmarinifórmis. B.P.	(*w.*)	Rosemary like.	N. S. W.	1791.	5. 8.	G. h.	Andr. reposit. t. 214.
2 Dampièri. B.P.	(*wh.*)	Dampier's.	N. Holland.	1803.	5. 7.	G. ♄.	
3 angustifòlia. B.P.	(*wh.*)	narrow-leaved.	V. Diem. Isl.	1823.	——	G. ♄.	
4 longifòlia. B.P.	(*wh.*)	long-leaved.	N. S. W.	1826.	——	G. h.	
Tribus III. AJUGOIDEÆ. *Benth. in Bot. reg.*							
LEUCOSCE'PTRUM. S.S.		LEUCOSCE'PTRUM.	Didynamia Gymnospermia.				
cànum. s.s.	(*wh.*)	white-flowered.	Nepaul.	1825.	7. 9.	F. ♃.	
TE'UCRIUM. S.S.		GERMANDER.	Didynamia Gymnospermia. L.				
1 frùticans. W.	(*bl.*)	narrow-ld. shrubby.	Spain.	1640.	6. 9.	H. ♄.	Flor. græc. t. 527.
2 latifòlium. L.	(*li.*)	broad-ld. shrubby.	———	——	——	H. ♄.	Bot. mag. t. 245.
3 heterophy'llum. s.s.	(*li.*)	various-leaved.	Madeira.	1759.	6. 7.	G. h.	
4 Laxmánni. s.s.	(*ye.pu.*)	Laxmann's.	Siberia.	1800.	6. 8.	H. ♃.	
5 alpéstre. F.G.	(*wh.*)	alpine.	Levant.	1821.	6. 9.	G. ♄.	Flor. græc. t. 538.
6 brevifòlium. s.s.	(*li.*)	short-leaved.	Crete.	1824.	——	H. h.	——— t. 528.
7 asiáticum. s.s.	(*bh.*)	Asiatic.	....	1777.	6. 10.	G. ♄.	Jacq. vind. 3. t. 41.
8 cubénse. s.s.	(*li.*)	Cuba.	Cuba.	1733.	5.	S. ♂.	Jacq. obs. 2. t. 30.
9 Nissoliànum. s.s.	(*pu.*)	Nissole's.	Spain.	1752.	6. 7.	F. ♃.	Moris. s. t. 22. f. 19.
10 lævigàtum. s.s.	(*li.*)	smooth.	BuenosAyres.	1829.	——	G. ♃.	
11 campanulàtum. s.s.	(*bh.*)	bell-flowered.	Levant.	1728.	7. 8.	H. ♃.	
12 règium. s.s.	(*pu.*)	royal.	Spain.	1699.	5. 10.	F. h.	Pluk. alm. t. 65. f. 1.
13 multiflòrum. s.s.	(*pu.*)	many-flowered.	———	1731.	7. 9.	H. ♃.	Bocc. mus. t. 117.
14 lùcidum. s.s.	(*re.*)	shining.	S. Europe.	1730.	6. 9.	H. ♃.	Flor. græc. t. 532.
15 Cham'ædrys. s.s.	(*pu.*)	wall.	England.	....	5. 8.	H. ♃.	Eng. bot. v. 10. t. 680.
16 Scórdium. s.s.	(*pu.*)	water.	———	....	7. 8.	H. *w.* ♃.	——— v. 12. t. 828.

17 scordioídes. s.s.	(*pu.*)	likened.	S. Europe.	1816.	7. 8.	H.♃.	
18 bracteàtum. s.s.	(*pu.*)	bracted.	Barbary.	1828.	——	G.♃.	Desf. atl. 2. t. 120.
19 spinòsum. s.s.	(*wh.*)	thorny.	Spain.	1640.	5. 6.	H.ħ.	Flor. græc. t. 539.
20 Bòtrys. s.s.	(*pu.*)	cut-leaved.	S. Europe.	1633.	7. 9.	H.☉.	Ger. emac. 525. f. 2.
21 trífidum. s.s.	(*re.*)	trifid-leaved.	C. B. S.	1791.	6. 8.	G.♄.	
22 angustíssimum. s.s.	(*re.*)	narrow-leaved.	Spain.	1818.	——	H.♄.	Barrel. rar. t. 1080.
23 pùmilum. s.s.	(*re.*)	small.	———	1816.	7. 8.	F.ħ.	——— t. 1092.
24 Libanìtis. s.s.	(*ye.bh.*)	Spanish.	Spain.	1824.	6. 8.	H.♃.	Cavan. ic. 2. t. 118.
25 thymifòlium. s.s.	(*pu.*)	Thyme-leaved.	———	———	6. 10.	F.ħ.	
26 supìnum. s.s.	(*wh.*)	procumbent.	Austria.	1752.	——	H.ħ.	Jacq. aust. 5. t. 417.
27 montànum. s.s.	(*st.*)	dwarf mountain.	S. Europe.	1710.	7. 10.	F.ħ.	Flor. græc. t. 534.
28 Pòlium. s.s.	(*wh.*)	Poley.	———	1562.	7. 9.	F.ħ.	——— t. 535.
29 cuneifòlium. F.G.	(*wh.*)	wedge-leaved.	Levant.	1828.	6. 8.	F.♄.	——— t. 537.
30 Pseudohyssòpus. s.s.	(*w.*)	Hyssop-leaved.	Italy.	1804.	6. 7.	H.♄.	Col. ecphr. 1. t. 67.
31 capitàtum. s.s.	(*ro.*)	round-headed.	Spain.	1731.	7. 8.	F.♄.	Flor. græc. t. 536.
32 lusitánicum. s.s.	(*wh.*)	Portugal.	Portugal.	1816.	——	F.♄.	
33 pycnophy'llum. s.s.	(*wh.*)	clustered.	Spain.	———	——	H.♃.	Barrel. rar. t. 1096.
34 valentìnum. s.s.	(*wh.*)	Valencia.	———	1826.	——	F.ħ.	——— t. 1048.
35 gnaphalòdes. s.s.	(*re.*)	woolly-calyx'd.	———	———	7. 9.	F.♄.	——— t. 1083.
36 aùreum. s.s.	(*ye.*)	golden Poley.	S. Europe.	1731.	6. 7.	F.♄.	Cavan. ic. 2. t. 117.
37 flavéscens. P.S.	(*ye.*)	yellow Poley.	———	....	7. 9.	F.♄.	Barrel. rar. t. 1073.
38 hyrcànicum. s.s.	(*pu.*)	Betony-leaved.	Persia.	1763.	8. 10.	H.♃.	Bot. mag. t. 2013.
39 inflàtum. s.s.	(*pu.*)	thick-spiked.	Jamaica.	1778.	——	S.♃.	
40 orchioídes. B.R.	(*bh.*)	Orchis-flowered.	Chile.	1826.	——	F.ħ.	Bot. reg. t. 1255.
heterophy'lla. Cavan. icon. 6. t. 577. *non* L'Herit.							
41 virgínicum. s.s.	(*pu.*)	Virginian.	N. America.	1768.	5. 6.	H.♃.	Schkuhr. handb. 160.
42 canadénse. s.s.	(*pu.*)	Canadian.	———	———	8. 9.	H.♃.	
43 Arduìni. s.s.	(*wh.*)	Arduini's.	Levant.	1827.	——	H.♃.	Flor. græc. t. 531.
44 abutiloídes. s.s.	(*ye.*)	Mulberry-leaved.	Madeira.	1777.	4. 5.	G.♄.	Jacq. schœn. 3. t. 358.
45 Scorodònia. s.s.	(*st.*)	wood sage.	Britain.	....	7.	H.♃.	Eng. bot. v. 22. t. 1543.
46 betónicum. s.s.	(*pu.*)	hoary.	Madeira.	1775.	5. 8.	G.♄.	Bot. mag. t. 1114.
47 flàvum. s.s.	(*st.*)	yellow-flowered.	S. Europe.	1640.	7. 9.	H.ħ.	Flor. græc. t. 533.
48 massiliénse. s.s.	(*li.*)	sweet scented.	France.	1731.	6. 7.	H.♄.	Jacq. vind. 1. t. 94.
49 resupinàtum. s.s.	(*st.*)	resupinate.	Barbary.	1801.	7. 8.	H.☉.	Desf. atl. 2. t. 117.
50 Màrum. s.s.	(*pu.*)	Cat-thyme.	Spain.	1640.	7. 9.	F.ħ.	Park. theat. 17. f. 2.
51 subspinòsum. s.s.	(*pu.*)	Minorca.	Minorca.	1816.	——	G.♄.	
52 quadràtulum. s.s.	(*ro.*)	many-branched.	Levant.	1828.	——	F.ħ.	Flor. græc. t. 530.
53 créticum. s.s.	(*ro.*)	Cretan.	Crete.	1825.	——	G.♄.	——— t. 529.
54 pyrenàicum. s.s.	(*pu.*)	Pyrenean.	Pyrenees.	1731.	6. 8.	H.♃.	
55 orientàle. s.s.	(*bl.*)	great-flowered.	Levant.	1752.	7. 8.	H.♃.	Bot. mag. t. 1279.
AMETHY'STEA. R.S.		AMETHY'STEA.	Diandria Monogynia. L.				
cærùlea. R.S.	(*bl.*)	blue-flowered.	Siberia.	1759.	6. 7.	H.☉.	Bot. mag. t. 2448.
TRICHOST'EMA. S.S.		TRICHOST'EMA.	Didynamia Gymnospermia. L.				
1 dichótoma. W.	(*bl.*)	Marjoram-leav'd.	N. America.	1759.	6. 7.	H.☉.	Lam. ill. t. 515.
2 lineàris. s.s.	(*bl.*)	linear-leaved.	———	———	——	H.☉.	
AJ'UGA. S.S.		BUGLE.	Didynamia Gymnospermia. L.				
1 'Iva. s.s.	(*ye.*)	musky.	S. Europe.	1759.	7. 8.	H.☉.	Flor. græc. t. 525.
2 réptans. s.s.	(*bl.*)	common.	Britain.	....	5. 6.	H.♃.	Eng. bot. v. 7. t. 489.
α cærùlea.	(*bl.*)	*blue-flowered.*	———	....	——	H.♃.	
β rùbra.	(*pu.*)	*red-flowered.*	———	....	——	H.♃.	
γ álba.	(*wh.*)	*white-flowered.*	———	....	——	H.♃.	
3 álpina. s.s.	(*bl.*)	Alpine.	England.	....	5. 7.	H.♃.	Eng. bot. v. 7. t. 477.
4 genevénsis. s.s.	(*bl.*)	Geneva.	Switzerland.	1656.	5. 6.	H.♃.	Bull. herb. t. 361.
5 pyramidàlis. s.s.	(*li.*)	pyramidal.	Britain.	....	——	H.♃.	Eng. bot. v. 18. t. 1270.
6 oblongàta. s.s.	(*bl.*)	oblong-leaved.	Caucasus.	1829.	——	H.♃.	
7 orientàlis. s.s.	(*bl.*)	oriental.	Levant.	1732.	——	H.♃.	Swt. br. fl. gar. ser. 2.
8 Chamæpitys. s.s.	(*ye.*)	ground pine.	England.	....	4. 7.	H.☉.	Eng. bot. v. 2. t. 77.
9 Chìa. F.G.	(*ye.*)	Grecian.	Levant.	1828.	——	H.☉.	Flor. græc. t. 524.
10 salicifòlia. F.G.	(*ye.*)	Willow-leaved.	———	———	——	H.☉.	——— t. 526.
ANISOM'ELES. B.P.		ANISOM'ELES.	Didynamia Gymnospermia. S.S.				
1 ovàta. s.s.	(*vi.wh.*)	broad-leaved.	E. Indies.	1783.	7. 8.	S.☉.	Burm. zeyl. t. 71. f. 1.
2 malabárica. s.s.	(*vi.*)	Malabar.	———	1817.	——	S.♄.	Bot. mag. t. 2071.

Tribus IV. MONARDEÆ. *Benth. in Bot. reg.*

MONA'RDA. R.S.		MONA'RDA.	Diandria Monogynia. L.				
1 fistulòsa. R.S.	(*pu.*)	fistulous.	N. America.	1656.	6. 8.	H.♃.	Mill. ic. t. 122. f. 2.
2 mèdia. R.S.	(*da.pu.*)	purple-bracted.	———	....	6. 9.	H.♃.	Swt. br. fl. gar. t. 98.

3 móllis. R.S.	(*pa.pu.*) soft.	N. America.	....	6. 7.	H.♃.	
4 oblongàta. R.S.	(*pa.pu.*) long-leaved.	———	1761.	7. 9.	H.♃.	
5 altíssima. R.S.	(*li.*) tallest.	———	1818.	——	H.♃.	
6 affìnis. L.en.	(*li.*) affined.	———	....	——	H.♃.	
7 menthæfòlia. B.M.	(*pu.*) Mint-leaved.	———	1827.	——	H.♃.	Bot. mag. t. 2958.
8 rugòsa. R.S.	(*wh.*) white-flowered.	———	1761.	——	H.♃.	
9 purpùrea. R.S.	(*pu.*) purple-flowered.	———	1789.	6. 8.	H.♃.	Bot. mag. t. 145.
fistulòsa. var. *crimson.* B.M.						
10 Kalmiàna. R.S.	(*sc.*) pubescent-flower'd.	———	1813.	——	H.♃.	Pursh. amer. 1. t. 1.
11 dídyma. R.S.	(*sc.*) Oswego Tea.	———	1752.	——	H.♃.	Bot. mag. t. 546.
12 Russelliàna. N.	(*wh.re.*) Russell's.	———	1823.	——	H.♃.	Swt. br. fl. gar. t. 166.
13 clinopòdia. Ph.	(*st.*) wild basil-leav'd.	———	1771.	7.	H.♃.	
14 grácilis. R.S.	(*st.*) slender.	———	1824.	7. 8.	H.♃.	
15 punctàta. R.S.	(*ye.re.*) spotted.	———	1714.	6. 10.	F.♃.	Bot. reg. t. 87.
16 citriodòra. R.S.	(*li.*) Citron-scented.	New Spain.	1823.	——	F.♃.	
BLEPHI'LIA. B.R.	BLEPHI'LIA.	Diandria Monogynia.				
1 ciliàta. B.R.	(*bl.*) blue-flowered.	N. America.	1798.	7. 8.	H.♃.	Pluk. alm. t. 164. f. 3.
Monárda ciliàta. R.S.						
2 hirsùta. B.R.	(*li.*) hairy.	———	——	7. 9.	H.♃.	
ZIZI'PHORA. S.S.	ZIZI'PHORA.	Diandria Monogynia. L.				
1 clinopodioídes. R.S.	(*li.*) headed.	Siberia.	1799.	7. 8.	H.☉.	Rud. mem. p. 1810. t. 11.
Cùnila capitàta. L.						
2 acinoídes. W.en.	(*ro.*) Thyme-leaved.	———	1786.	——	H.☉.	
3 mèdia. L.en.	(*li.*) middle.	....	1816.	——	H.ħ.	
4 capitàta. R.S.	(*ro.*) oval-leaved.	Syria.	1752.	——	H.☉.	Flor. græc. 1. t. 13.
5 hispánica. R.S.	(*li.*) Spanish.	Spain.	1759.	6.	H.☉.	Lam. ill. t. 18. f. 1.
6 tenùior. R.S.	(*li.*) spear-leaved.	Levant.	1752.	6. 7.	H.☉.	——— t. 18. f. 2.
7 taùrica. R.S.	(*li.*) Taurian.	Tauria.	1816.	7. 9.	H.☉.	M.B. cent. pl. rar. 2. t. 82.
8 serpyllàcea. R.S.	(*li.*) sweet-scented.	Caucasus.	1803.	7. 8.	H.ħ.	Bot. mag. t. 906?
9 dasyántha. R.S.	(*li.*) hairy-flowered.	Siberia.	——	6. 8.	H.ħ.	——— t. 1093.
Pouschkìni. B.M.						
C'UNILA. N.	C'UNILA.	Diandria Monogynia. L.				
mariàna. N.	(*pu.*) Mint-leaved.	N. America.	1759.	7. 9.	H.♃.	Swt. br. fl. gar. t. 243.
HEDE'OMA. N.	HEDE'OMA.	Diandria Monogynia. P.S.				
1 pulegioídes. N.	(*wh.vi.*) Pennyroyal-ld.	N. America.	1777.	6. 8.	H.☉.	
2 thymoídes. P.S.	(*bh.*) Thyme-leaved.	France.	1699.	——	H.☉.	Moris. s. 11. t. 19. f. 6.
ROSMAR'INUS. R.S.	ROSEMARY.	Diandria Monogynia. L.				
1 officinàlis. R.S.	(*pa.bl.*) common.	S. Europe.	1548.	1. 4.	H.ħ.	Flor. græc. 1. t. 14.
α *angustifòlius*	(*pa.bl.*) *narrow-leaved.*	———	——	——	H.ħ.	
β *latifòlius.*	(*pa.bl.*) *broad-leaved.*	———	——	——	H.ħ.	
γ *argénteus.*	(*pa.bl.*) *silver-striped.*	———	——	——	H.ħ.	
δ *aùreus.*	(*pa.bl.*) *gold-striped.*	———	——	——	H.ħ.	
2 chiléusis. R.S.	(*wh.*) Chili.	Chili.	1795.	7.	G.ħ.	
SYNA'NDRA. N.	SYNA'NDRA.	Didynamia Gymnospermia.				
grandiflòra. N.	(*st.*) great-flowered.	N. America.	1827.	6. 7.	H.♃.	

Tribus V. NEPETEÆ. *Benth. in Bot. reg.*

LEON'OTIS. S.S.	LION'S-EAR.	Didynamia Gymnospermia. S.S.				
1 Leonùrus. S.S.	(*or.*) narrow-leaved.	C. B. S.	1712.	10. 12.	G.ħ.	Bot. mag. t. 478.
Phlòmis Leonùrus. W.						
2 ovàta. S.S.	(*or.*) dwarf shrubby.	———	1713.	6. 7.	G.ħ.	Mill. ic. 2. t. 162. f. 1.
3 intermèdia. S.S.	(*or.*) intermediate.	S. Africa.	1823.	——	G.ħ.	Bot. reg. t. 850.
4 nepetifòlia. S.S.	(*or.*) Catmint-leaved.	E. Indies.	1778.	9. 10.	S.☉.	——— t. 281.
LE'UCAS. B.P.	LE'UCAS.	Didynamia Gymnospermia. S.S.				
1 chinénsis. S.S.	(*wh.*) Chinese.	China.	1822.	6. 8.	S.ħ.	
2 zeylánica. S.S.	(*wh.*) Ceylon.	E. Indies.	1777.	6. 10.	S.☉.	Jacq. ic. 1. t. 111.
3 linifòlia. S.S.	(*wh.*) Flax-leaved.	———	1816.	——	S.☉.	
4 Pluknètii. S.S.	(*wh.*) Pluknet's.	———	1824.	——	S.☉.	
5 cephalòtes. S.S.	(*wh.pu.*) large-headed.	———	1811.	——	S.☉.	
6 nùtans. S.S.	(*wh.*) nodding.	———	1826.	——	S.☉.	
7 hírta. S.S.	(*wh.*) hairy.	———	1824.	——	S.☉.	
8 áspera. S.S.	(*wh.*) rough.	Persia.	1823.	——	S.☉.	
9 índica. S.S.	(*bh.*) Indian	E. Indies.	1789.	7. 8.	S.☉.	
10 urticæfòlia. S.S.	(*wh.*) Nettle-leaved.	———	1810.	——	S.☉.	

11	martinicénsis.	(*wh.*)	West-Indian.	W. Indies.	1781.	7. 8.	S.⊙.	Jacq. ic. 1. t. 110.
	Phlòmis martinicénsis. W.							
12	biflòra. S.S.	(*wh.*)	two-flowered.	E. Indies.	1820.	——	S.⊙.	Burm. zeyl. t. 63. f. 1.
	PHL'OMIS. S.S.		JERUSALEM-SAGE.	Didynamia Gymnospermia. L.				
1	fruticòsa. S.S.	(*ye.*)	shrubby.	Spain.	1596.	6. 7.	H. ♄.	Flor. græc. t. 563.
2	ferrugínea. S.S.	(*ye.*)	ferruginous.	Naples.	....	——	H. ♄.	
3	lanàta. S.S.	(*ye.*)	small shrubby.	S. Europe.	1596.	——	H. ♄.	
4	crética. S.S.	(*ye.*)	Cretan.	Crete.	1823.	——	G. h.	
5	scariòsa. S.S.	(*ye.*)	scariose-bracted.	Italy.	——	——	F. h.	
6	vìrens. S.S.	(*ye.*)	green shrubby.	Levant.	....	——	F. ♄.	
7	microphy'lla. Sie.	(*ye.*)	small-leaved.	————	1827.	——	F. ħ.	
8	purpùrea. S.S.	(*pu.*)	purple.	S. Europe.	1661.	6. 8.	F. ♄.	Smith spicil. 6. t. 3.
9	itálica. S.S.	(*pu.*)	Italian.	Italy.	——	——	F. ♄.	
10	Nissòlii. S.S.	(*ye.*)	Nissole's.	Levant.	1757.	6. 7.	F. ♃.	Mill. ic. 2. t. 204.
11	armeniaca. S.S.	(*ye.*)	Armenian.	Armenia.	1829.	——	F. ħ.	
12	Lychnìtes. S.S.	(*ye.*)	lamp-wick.	S. Europe.	1658.	6. 8.	G. ♄.	Bot. mag. t. 999.
13	floccòsa. B.R.	(*ye.*)	woolly.	Egypt.	1828.	8. 11.	F. ħ.	Bot. reg. t. 1300.
14	salviæfòlia. P.S.	(*ye.re.*)	Sage-leaved.	....	1822.	6. 9.	F. ħ.	Jacq. schœnb. 3. t. 359.
15	agrària. Led.	(*pu.*)	field.	Siberia.	1827.	——	H. ♃.	
16	Sàmia. S.S.	(*pu.*)	Samian.	N. Africa.	1714.	6. 7.	H. ♃.	Flor. græc. t. 564.
17	Russelliàna. Lag.	(*ye.*)	Russell's.	Levant.	1821.	——	H. ♃.	Russ. alep. 2. t. 16.
	lunarifòlia β Russelliàna. B.M. 2542.							
18	Hérba-vénti. S.S.	(*pu.*)	rough-leaved.	S. Europe.	1596.	7. 9.	H. ♃.	
19	púngens. S.S.	(*pu.*)	pungent.	Levant.	1818.	6. 8.	H. ♃.	Swt. br. fl. gar. t. 33.
20	crinìta. S.S.	(*or.*)	white woolly.	Spain.	1829.	——	H. ♃.	Cavan. icon. 3. t. 247.
21	alpìna. S.S.	(*pu.*)	Alpine.	Siberia.	1802.	6. 9.	H. ♃.	Pall. act. pet. 2. t. 13.
22	tuberòsa. S.S.	(*pu.*)	tuberous-root'd.	————	1759.	6. 10.	H. ♃.	Bot. mag. t. 1555.
23	laciniàta. S.S.	(*ye.*)	jagged-leaved.	Levant.	1731.	7.	H. ♃.	Swt. br. fl. gar. t. 24.
	BALL'OTA. S.S.		STINKING HOREHOUND.	Didynamia Gymnospermia. L.				
1	vulgàris. S.S.	(*pu.*)	common.	Europe.	....	7. 9.	H. ♃.	
2	f'œtida. L.en.	(*pu.*)	strong-scented.	Britain.	....	——	H. ♃.	Eng. bot. v. 1. t. 46.
	α nìgra.	(*pu.*)	*purple-flowered.*	————	....	——	H. ♃.	
	β álba.	(*wh.*)	*white-flowered.*	————	....	——	H. ♃.	
	BERING'ERIA. B.R.		BERING'ERIA.	Didynamia Gymnospermia.				
1	cinèrea. B.R.	(*li.*)	cinereous.	S. Europe.	1824.	6. 9.	H. ♃.	
	Marrùbium cinèreum. S.S.							
2	acetabulòsum. B.R.	(*li.*)	saucer-leaved.	Candia.	1676.	6. 8.	F. ♃.	Barrel. icon. 129.
3	pseùdo-dictámnus. B.R.	(*li.*)	shrubby white.	———	1596.	7. 8.	F. ħ.	Flor. græc. t. 562.
	Marrùbium pseùdo-dictámnus. F.G.							
4	africànum. B.R.	(*li.*)	African.	C. B. S.	1710.	7. 9.	G. ♃.	Comm. hort. 2. t. 90.
5	críspum. B.R.	(*li.*)	curled-leaved.	S. Europe.	1714.	7. 8.	H. ♃.	Herm. parad. t. 200.
6	hispánicum. B.R.	(*li.*)	Spanish.	Spain.	——	——	H. ♃.	———— t. 201.
7	hirsùtum. B.R.	(*li.*)	hairy.	S. Europe.	....	6. 7.	H. ♃.	
8	orientàle. B.R.	(*li.*)	oriental.	Levant.	1828.	——	H. ♃.	
	MOLUCCE'LLA. S.S.		MOLUCCA-BALM.	Didynamia Gymnospermia. L.				
1	tuberòsa. S.S.	(*wh.*)	tuberous-root'd.	Tartary.	1796.	7.	H. ♃.	Pall. it. 3. app. t. T.
2	l'ævis. S.S.	(*wh.re.*)	smooth.	Syria.	1570.	7. 8.	H.⊙.	Bot. mag. t. 1852.
	CHASM'ONIA. B.R.		CHASM'ONIA.	Didynamia Gymnospermia.				
	incìsa. B.R.	(*wh.re.*)	spiny.	Levant.	1596.	7. 8.	H.⊙.	Bot. reg. t. 1244.
	Moluccélla spinòsa. Flor. græc. t. 567.							
	LEON'URUS. S.S.		MOTHERWORT.	Didynamia Gymnospermia. L.				
1	Cardìaca. S.S.	(*li.wh.*)	common.	Britain.	....	7. 8.	H. ♃.	Eng. bot. v. 4. t. 286.
2	villòsus. S.S.	(*li.wh.*)	villous.	Tauria.	1824.	——	H. ♃.	
3	críspus. S.S.	(*li.*)	curl-leaved.	Siberia.	1658.	——	H. ♃.	Murr. c. gott. 8. t. 4.
4	supìnus. S.S.	(*wh.*)	procumbent.	————	1816.	6. 8.	H. ♃.	
5	glaucéscens. Led.	(*li.*)	glaucous-leaved.	————	1827.	7. 9.	H. ♃.	
6	tatáricus. S.S.	(*li.*)	Tartarian.	Tartary.	1756.	8. 10.	H. ♂.	Mill. ic. 1. t. 80.
7	sibíricus. S.S.	(*ro.*)	Siberian.	Siberia.	1759.	6. 8.	H. ♂.	Swt. br. fl. gar. t. 204.
8	condensàtus. S.S.	(*li.*)	condensed.	————	1824.	——	H. ♃.	
9	heterophy'llus. B.F.G.	(*ro.*)	various-leav'd.	S. America.	——	——	H.⊙.	Swt. br. fl. gar. t. 197.
10	multífidus. Desf.	(*li.*)	multifid.	....	——	——	H.⊙.	
11	negléctus. Sch.	(*li.*)	neglected.	....	——	——	H. ♃.	
12	altàicus. S.S.	(*li.*)	Altaic.	Altay.	1826.	——	H. ♃.	
13	lanàtus. S.S.	(*st.*)	woolly.	Siberia.	1752.	——	H. ♃.	Gmel. sib. 3. t. 54.
	Ballòta lanàta. W.							
	GALEO'BDOLON. S.S.		DEAD-NETTLE.	Didynamia Angiospermia. S.S.				
	lùteum. E.B.	(*ye.*)	yellow.	Britain.	....	5. 6.	H. ♃.	Eng. bot. v. 11. t. 787.
	GALEO'PSIS. S.S.		HEMP-NETTLE.	Didynamia Gymnospermia. L.				
1	Ládanum. E.B.	(*ro.*)	red.	Britain.	....	7. 9.	H.⊙.	Eng. bot. v. 13. t. 884.

2 angustifòlia. P.S.	(*ro.*)	narrow-leaved.	Europe.	1820.	7. 9.	H.⊙.	
3 canéscens. S.S.	(*li.*)	canescent.	Germany.	1827.	——	H.⊙.	
4 pubéscens. S.S.	(*wh.re.*)	pubescent.	————	1820.	——	H.⊙.	
5 ochroleùca. S.S.	(*ye.*)	downy.	Britain.	....	7. 8.	H.⊙.	Eng. bot. v. 33. t. 2353.
villòsa. E.B.							
6 Tetràhit. E.B.	(*wh.pu.*)	common.	————	....	——	H.⊙.	Eng. bot. v. 3. t. 207.
7 versícolor. E.B.	(*ye.vi.*)	various-colour'd.	————	....	——	H.⊙.	———— v. 10. t. 667.
cannabìna. W.							
L'AMIUM. S.S.		ARCHANGEL.	Didynamia Gymnospermia. L.				
1 longiflòrum. T.N.	(*pu.*)	long-flowered.	S. Europe.	....	4. 9.	H.♃.	Eng. bot. v. 36. t. 2550.
maculàtum. Eng. bot. *non* Flor. Græc.							
2 rugòsum. S.S.	(*pu.*)	rough.	Italy.	1766.	7. 8.	H.♃.	Flor. græc. t. 555.
3 maculàtum. E.B.	(*pu.*)	spotted.	Britain.	....	4. 5.	H.♃.	———— t. 556.
4 álbum. E.B.	(*wh.*)	white.	————	....	4. 9.	H.♃.	Eng. bot. v. 11. t. 768.
5 striàtum. F.G.	(*wh.std.*)	striped-flower'd.	Levant.	1829.	——	H.♃.	Flor. græc. t. 557.
6 bífidum. S.S.	(*wh.*)	bifid-helmetted.	Naples.	1825.	——	H.♃.	Cyr. rar. fasc. 1. t. 7.
7 moschàtum. S.S.	(*wh.*)	musk-scented.	Levant.	1739.	7. 9.	H.♃.	
8 mólle. S.S.	(*wh.*)	Pellitory-leaved.	....	1683.	4. 5.	H.♃.	
9 flexuòsum. T.N.	(*wh.*)	flexuose.	Naples.	1824.	4. 8.	H.♃.	Ten. flor. nap. ic.
10 gargánicum. S.S.	(*pu.*)	woolly.	Italy.	1729.	7. 8.	H.♃.	Sm. exot. bot. 1. t. 48.
11 lævigàtum. S.S.	(*pu.*)	smooth.	————	1711.	3. 10.	H.♃.	Pluk. alm. t. 198. f. 1.
12 purpùreum. S.S.	(*pu.*)	purple.	Britain.	....	2. 8.	H.⊙.	Eng. bot. v. 11. t. 769.
β *álbidum.*	(*wh.*)	*white-flowered.*	————	....	——	H.⊙.	
13 incìsum. S.S.	(*li.*)	cut-leaved.	England.	....	5. 7.	H.⊙.	Eng. bot. v. 27. t. 1933.
14 amplexicàule. S.S.	(*pu.*)	Henbit.	Britain.	....	——	H.⊙.	———— v. 11. t. 770.
15 multífidum. S.S.	(*pu.*)	multifid-leaved.	Levant.	1752.	4. 5.	H.⊙.	Comm. rar. t. 26.
ORV'ALA. DC.		ORV'ALA.	Didynamia Gymnospermia.				
1 lamioìdes. DC.	(*pu.bh.*)	balm-leaved.	Italy.	1596.	4. 7.	H.♃.	Bot. mag. t. 172.
Làmium Orvàla. B.M.							
2 gargànica. DC.	(*pu.*)	Garganian.	————	————	——	H.♃.	
PHYSOST'EGIA. B.R.		PHYSOST'EGIA.	Didynamia Angiospermia.				
1 virginiàna. B.R.	(*cr.*)	Virginian.	N. America.	1683.	7. 9.	H.♃.	Bot. mag. t. 467.
2 speciòsa. B.F.G.	(*ro.*)	beautiful.	————	1820.	——	H.♃.	Swt. br. fl. gar. t. 93.
Dracocéphalum speciòsum. B.F.G.							
3 denticulàta. B.R.	(*ro.*)	toothed.	————	1787.	8. 9.	H.♃.	Bot. mag. t. 214.
Dracocéphalum denticulàtum. B.M.							
4 variegàta. B.R.	(*re.wh.*)	variegated.	————	1812.	——	H.♃.	Vent. cels. t. 44.
SPHA'CELE. B.R.		SPHA'CELE.	Didynamia Gymnospermia.				
Lindlèi. B.R.	(*li.*)	Lindley's.	Valparaiso.	1825.	7. 9.	G.♄.	Bot. reg. t. 1226.
Stáchys Sálviæ. B.R.							
BETO'NICA. S.S.		BETONY.	Didynamia Gymnospermia. L.				
1 officinàlis. S.S.	(*pu.*)	wood.	Britain.	....	7. 8.	H.♃.	Eng. bot. v. 16. t. 1142.
2 strícta. S.S.	(*pu.*)	Danish.	Denmark.	1592.	6. 7.	H.♃.	
3 incàna. S.S.	(*pi.*)	hoary.	Italy.	1759.	——	H.♃.	Bot. mag. t. 2125.
4 orientàlis. S.S.	(*pu.*)	oriental.	Levant.	1737.	——	H.♃.	Lam. ill. t. 507. f. 2.
5 alopecùros. S.S.	(*st.*)	Foxtail.	S. Europe.	1759.	7.	H.♃.	Jacq. aust. 1. t. 78.
6 hirsùta. S.S.	(*pu.*)	hairy.	Italy.	1710.	6. 7.	H.♃.	Mur. com. got. 2. t. 3.
7 grandiflòra. S.S.	(*pu.*)	great-flowered.	Siberia.	1800.	——	H.♃.	Bot. mag. t. 700.
ST'ACHYS. S.S.		HEDGE-NETTLE.	Didynamia Gymnospermia. L.				
1 spinòsa. S.S.	(*wh.*)	thorny.	Candia.	1640.	7.	F.♄.	Flor. græc. t. 559.
2 glutinòsa. S.S.	(*wh.*)	clammy.	————	1729.	6. 7.	H.♄.	Moris. s. 11. t. 4. f. 17.
3 fruticulòsa. S.S.	(*pu.*)	frutescent.	Caucasus.	1825.	——	H.♄.	
4 æthiòpica. S.S.	(*li.*)	Ethiopian.	C. B. S.	1770.	4. 7.	G.♃.	Jacq. obs. 4. t. 77.
5 córsica. S.S.	(*li.*)	Corsican.	Corsica.	1825.	——	H.♃.	
6 angustifòlia. S.S.	(*li.*)	narrow-leaved.	Tauria.	1823.	6. 7.	H.♃.	Swt. br. fl. gar. 2. t. 180.
7 rugòsa. S.S.	(*st.*)	rough.	C. B. S.	1774.	7. 8.	G.♄.	Jacq. ic. 3. t. 493.
8 betonicæfòlia. S.S.	(*ye.*)	Betony-leaved.	Crete.	1812.	6. 7.	H.♃.	
9 stenophy'lla. S.S.	(*ye.*)	linear-leaved.	S. Spain.	1819.	——	F.♄.	
10 orientàlis. S.S.	(*st.*)	oriental.	Levant.	1768.	——	H.♃.	Flor. græc. t. 560.
11 marítima. S.S.	(*ye.*)	sea-side.	S. Europe.	1714.	7.	H.♃.	Jacq. vind. 1. t. 70.
12 ánnua. S.S.	(*wh.pu.*)	annual.	————	1713.	6. 8.	H.⊙.	Jacq. aust. 4. t. 360.
13 hírta. S.S.	(*ye.pu.*)	procumbent.	Spain.	1725.	——	H.♃.	All. ped. 1. t. 2. f. 3.
14 arvénsis. S.S.	(*pu.*)	corn.	Britain.	....	7. 8.	H.⊙.	Eng. bot. v. 17. t. 1154.
15 arenària. S.S.	(*pu.*)	sand.	Barbary.	1824.	6. 8.	H.♃.	Desf. atl. 2. t. 126.
16 diffùsa. Schweig.	(*re.*)	diffuse.	S. Europe.	1804.	——	H.♃.	Bot. mag. t. 1915.
arenària. B.M. *non* Desf. *rùbra.* Donn.							
17 ibérica. B.M.	(*pu.*)	Iberian.	Iberia.	1823.	——	H.♃.	
18 scordifòlia. W.en.	(*pu.*)	wedge-leaved.	....	1816.	7. 8.	H.♃.	
19 áspera. M.	(*pu.*)	harsh-leaved.	N. America.	——	——	H.♃.	

20 híspida. s.s.	(*pu.*)	hispid.	N. America.	1827.	7. 8.	H. ♃.	
21 itálica. Mil.	(*pu.*)	Italian.	Italy.	1747.	——	H. ♂.	
22 Balbísii. L.en.	(*ye.*)	Balbis's.	———	1823.	——	H. ♃.	
23 nepetæfòlia. s.s.	(*pu.*)	Catmint-leaved.	Levant.	1805.	6. 8.	H. ♃.	
24 arábica. s.s.	(*re.*)	Arabian.	Arabia.	1820.	——	H. ⊙.	
25 ambígua. s.s.	(*pu.*)	ambiguous.	Britain.	....	6. 7.	H. ♃.	Eng. bot. v. 30. t. 2089.
26 sylvática. s.s.	(*pu.*)	common.	———	....	7. 8.	H. ♃.	——— v. 6. t. 416.
27 circinàta. s.s.	(*pu.*)	blunt-leaved.	Barbary.	1777.	5. 7.	H. ♃.	L'Herit. s. nov. t. 26.
canariénsis. Jacq. ic. 1. t. 108.							
28 mollíssima. W.en.	(*li.*)	soft-leaved.	Corfu.	1806.	7. 8.	H. ♃.	W. hort. ber. t. 60.
29 pubéscens. T.S.	(*st.*)	pubescent.	Italy.	1826.	——	H. ♃.	
30 decúmbens. P.S.	(*st.*)	decumbent.	S. Europe.	1816.	5. 7.	H. ♃.	
31 coccínea. s.s.	(*sc.*)	scarlet.	S. America.	1798.	6. 8.	F. ♃.	Bot. mag. t. 666.
32 récta. s.s.	(*ye.*)	upright.	S. Europe.	1683.	——	H. ♃.	Jacq. aust. 4. t. 359.
33 heráclea. W.	(*pu.*)	Phlomis-leaved.	Italy.	1816.	6. 7.	H. ♃.	All. ped. t. 84. f. 1.
phlomoídes. W.en.							
34 oblíqua. P.S.	(*st.*)	oblique-leaved.	Hungary.	——	——	H. ♃.	W. et K. hung. t. 133.
35 palústris. s.s.	(*pu.*)	Clown's all-heal.	Britain.	....	8.	H. ♃.	Eng. bot. v. 24. t. 1675.
36 germánica. s.s.	(*pu.*)	downy.	England.	....	7. 8.	H. ♃.	——— v. 12. t. 829.
37 intermèdia. W.	(*pu.*)	oblong-leaved.	Carolina.	1762.	6. 7.	H. ♃.	
38 sibírica. L.en.	(*pu.*)	Siberian.	Siberia.	1822.	6. 8.	H. ♃.	Swt. br. fl. gar. t. 100.
39 polystàchya. s.s.	(*pu.*)	many-spiked.	Naples.	1824.	——	H. ♃.	
40 salviæfòlia. s.s.	(*pu.*)	Sage-leaved.	———	——	——	H. ♃.	
41 crética. s.s.	(*pu.*)	Cretan.	Candia.	1640.	——	F. ♃.	Flor. græc. t. 558.
42 dasyántha. Raf.	(*pu.*)	woolly-flower'd.	Sicily.	1829.	——	H. ♃.	
43 exaltàta. Ot.	(*pu.*)	tall.	....	——	——	H. ♃.	
44 alpìna. s.s.	(*pu.*)	Alpine.	Germany.	1597.	——	H. ♃.	Lap. pyr. 1. t. 8.
45 lanàta. s.s.	(*pu.*)	woolly.	Siberia.	1782.	6. 9.	H. ♃.	Jacq. ic. 1. t. 107.
46 latifòlia. s.s.	(*pu.*)	broad-leaved.	....	1775.	6. 7.	H. ♄.	
47 biénnis. Roth.	(*pu.*)	biennial.	Germany.	1801.	6. 8.	H. ♂.	
48 lusitánica.	(*pu.*)	Portugal.	Portugal.	1800.	——	H. ♃.	
Eriostèmon lusitánicum. Lk.							
ZIET'ENIA. P.S.		ZIET'ENIA. Didynamia Gymnospermia. P.S.					
orientàlis. P.S.	(*li.*)	Lavender-leaved.	Levant.	1816.	5. 7.	H. ♄.	
lavandulifòlia. L.en. *Stàchys lavandulifòlia.* W.							
CHAIT'URUS. B.R.		CHAIT'URUS. Didynamia Gymnospermia.					
marrubiástrum. Ehr.	(*li.*)	small-flowered.	Austria.	1710.	6. 8.	H. ⊙.	Jacq. aust. 5. t. 405.
Leonùrus marrubiastrum. W.							
CRANIO'TOME. B.R.		CRANIO'TOME. Didynamia Gymnospermia.					
versícolor. B.R.	(*re.wh.*)	various-coloured.	Nepaul.	1823.	7. 8.	G. ♄.	
Ajùga furcàta. L.enum. *Anisomèles nepalénsis.* s.s.							
NE'PETA. L.		CATMINT. Didynamia Gymnospermia. L.					
1 serpyllifòlia. s.s.	(*li.*)	Thyme-leaved.	Tauria.	1826.	6. 8.	H. ♃.	
2 parviflòra. s.s.	(*li.*)	small-flowered.	———	1828.	——	H. ♃.	
3 sibírica. s.s.	(*bl.*)	Siberian.	Siberia.	1804.	7.	H. ♃.	
4 grandiflòra. s.s.	(*bl.*)	large-flowered.	Caucasus.	1806.	7. 8.	H. ♃.	
5 melissæfòlia. s.s.	(*bl.*)	Balm-leaved.	Candia.	1752.	——	H. ♃.	
6 coloràta. W.en.	(*vi.*)	coloured.	Caucasus.	1806.	——	H. ♃.	
7 longiflòra. H.K.	(*bl.*)	long-flowered.	Persia.	1802.	6. 8.	H. ♃.	Venten. cels. t. 66.
8 Mussìni. H.K.	(*bl.*)	scollop-leav'd.	Siberia.	1804.	5. 8.	H. ♃.	Bot. mag. t. 923.
longiflòra. B.M. *non* V.							
9 incàna. s.s.	(*bl.*)	hoary.	Levant.	1723.	8.	H. ♃.	
10 gravèolens. s.s.	(*bl.*)	strong-smelling.	S. Europe.	1804.	7. 8.	H. ♃.	All. ped. t. 2. f. 1.
11 amethy'stina. Poir.	(*bl.*)	amethystine.	———	1820.	——	H. ♃.	
12 Nepetélla. s.s.	(*li.*)	small.	———	1758.	——	H. ♃.	
13 teucriifòlia. s.s.	(*bl.*)	Teucrium-leav'd.	Armenia.	1816.	——	H. ♃.	
14 lamiifòlia. s.s.	(*bl.*)	Lamium-leaved.	———	1806.	——	H. ♃.	
15 citriodòra. Lag.	(*wh.*)	Citron-scented.	Spain.	1826.	——	H. ♃.	
16 lophántha. F.	(*bl.*)	crested.	Altay.	1821.	——	H. ♃.	
17 angustifòlia. s.s.	(*wh.*)	narrow-leaved.	Spain.	1798.	6. 7.	H. ♃.	
18 Catària. s.s.	(*wh.*)	common.	Britain.	....	7. 9.	H. ♃.	Eng. bot. v. 2. t. 137.
19 ucrànica. s.s.	(*vi.*)	Ukranian.	Ukraine.	1789.	7. 8.	H. ♃.	
20 nùda. L.	(*li.*)	naked.	S. Europe.	1710.	——	H. ♃.	Flor. græc. t. 547.
21 pannònica. L.	(*li.*)	close-flowered.	Hungary.	1683.	8. 10.	H. ♃.	Jacq. aust. 2. t. 129.
22 violàcea. s.s.	(*vi.*)	violet-coloured.	Spain.	1723.	7. 9.	H. ♃.	Bocc. mus. t. 36.
23 cœrùlea. s.s.	(*bl.*)	blue-flowered.	....	1777.	5. 6.	H. ♃.	
24 latifòlia. s.s.	(*pu.*)	broad-leaved.	Pyrenees.	1818.	7. 8.	H. ♃.	
25 macrùra. s.s.	(*li.*)	long-tailed.	Siberia.	——	——	H. ♃.	
26 críspa. s.s.	(*bl.*)	curl-leaved.	Levant.	1800.	——	H. ♃.	

27 diffùsa. s.s. (*bl.*) spreading. Siberia. 1820. 7. 8. H. ♃.
28 imbricàta. s.s. (*pu.*) imbricated. Spain. 1825. —— H. ♃.
29 teucrioídes. s.s. (*br.*) Teucrium-like. Levant. 1819. —— H. ♃.
30 itálica. w. (*wh.sp.*) Italian. Italy. 1640. 6. 8. H. ♃. Flor. græc. t. 548.
31 marrubioídes. W.en. (*pu.*) horehound-leav'd. —— .... —— H. ♃.
32 multibracteàta. s.s. (*vi.*) many-bracted. Barbary. 1823. —— H. ♃. Desf. atl. 2. t. 123.
33 lanàta. s.s. (*vi.*) woolly. S. Europe. 1774. 5. 6. H. ♃. Jacq. obs. 3. t. 75.
34 tuberòsa. s.s. (*vi.*) tuberous-root'd. Spain. 1683. 6. 8. H. ♃. Barrel. ic. t. 602.
35 reticulàta. s.s. (*li.*) netted. Morocco. 1801. 7. 8. H. ♃. Desf. atl. 2. t. 124.
36 Scordòtis. s.s. (*wh.*) sessile-spiked. Crete. 1823. —— G. ♄. Alp. exot. t. 283.
37 suavèolens. Rœm. (*wh.*) sweet-scented. .... 1804. —— H. ♃.
38 botryoìdes. w. (*wh.*) annual. —— 1779. 6. 7. H. ☉. Cavan. ic. 1. t. 49.
bipinnàta. Cav.

GLE'CHOMA. L. GROUND-IVY. Didynamia Gymnospermia. L.
1 hederàcea. E.B. (*bl.*) common. Britain. .... 3. 5. H. ♃. Eng. bot. v. 12. t. 853.
2 hirsùta. P.S. (*pu.*) hairy. Hungary. .... —— H. ♃. W. et K. hung. t. 119.

HEMIG'ENIA. B.P. HEMIG'ENIA. Didynamia Gymnospermia.
purpùrea. B.P. (*pu.*) purple. N. S. W. 1824. 4. 8. G. ♄.

SIDE'RITIS. S.S. IRONWORT. Didynamia Gymnospermia. L.
1 canariénsis. s.s. (*ye.*) Canary. Canaries. 1697. 5. 8. G. ♄. Jacq. vind. 3. t. 30.
2 cándicans. s.s. (*ye.*) Mullein-leaved. Madeira. 1714. 4. 7. G. ♄. Comm. hort. 2. t. 99.
3 crética. s.s. (*wh.*) Cretan. Crete. 1823. —— G. ♄.
4 montàna. s.s. (*ye.*) mountain. Austria. 1752. 7. 8. H. ☉. Flor. græc. t. 551.
5 élegans. s.s. (*wh.bl.*) dark-flowered. .... 1787. 7. H. ☉. Murr. com. got. 1. t. 4.
6 romàna. s.s. (*wh.*) Roman. Italy. 1759. 6. 8. H. ☉. Flor. græc. t. 552.
7 syrìaca. s.s. (*ye.*) Syrian. Levant. 1597. 6. 9. G. ♃. —— t. 550.
8 taùrica. W.en. (*ye.*) Taurian. Tauria. 1818. —— H. ♄.
9 brùtia. T.S. (*ye.*) woolly. Naples. 1827. —— F. ♄.
10 perfoliàta. s.s. (*ye.*) perfoliate. Levant. 1731. 8. 11. H. ♃.
11 pullùlans. s.s. (*wh.ye.*) brownish. —— 1828. —— F. ♃. Vent. cels. 1. t. 98.
12 leucántha. s.s. (*wh.*) white-flowered. Spain. —— —— F. ♄. Cavan. ic. 2. t. 304.
13 incàna. s.s. (*ye.*) Lavender-leav'd. —— 1752. 7. 8. H. ♄. —— 2. t. 186.
14 angustifòlia. s.s. (*ye.*) narrow-leaved. —— 1825. —— H. ♄.
15 spinòsa. s.s. (*ye.*) spiny. —— .... 6. 9. H. ♃. Cavan. ic. 3. t. 209.
16 ilicifòlia. s.s. (*ye.*) Holly-leaved. Levant. .... —— H. ♃.
17 serràta. s.s. (*ye.*) saw-leaved. Spain. 1818. —— H. ♃.
18 hyssopifòlia. s.s. (*ye.*) Hyssop-leaved. Pyrenees. 1597. 6. 11. H. ♃. Schk. hand. 2. t. 158.
19 f'œtida. L.en. (*ye.*) strong-scented. .... —— H. ♃.
20 pyrenàica. s.s. (*ye.*) Pyrenean. Pyrenees. 1816. —— H. ♃.
21 chamædryfòlia. s.s. (*ye.*) Germander-ld. Spain. —— —— H. ♃. Cavan. ic. 4. t. 301.
22 scordioídes. s.s. (*ye.*) scollop-leaved. France. 1597. 8. 11. H. ♃. Barrel. ic. t. 343.
23 hirsùta. s.s. (*ye.*) hairy. S. Europe. 1731. 6. 7. H. ♃. Cavan. ic. 4. t. 302.
24 crispàta. s.s. (*ye.*) curled-leaved. Gibraltar. 1816. 6. 9. F. ♃.

MARR'UBIUM. S.S. HOREHOUND. Didynamia Gymnospermia. L.
1 Aly'ssum. s.s. (*pu.*) plaited-leaved. Spain. 1597. 7. 8. H. ♃. Ger. herb. 379. f. 1.
2 astracánicum. s.s. (*bl.*) Astracan. Levant. 1816. —— H. ♃. Jacq. ic. 1. t. 109.
3 peregrìnum. w. (*wh.*) Sicilian. Sicily. 1640. 7. 9. H. ♃. Jacq. aust. 2. t. 160.
4 candidíssimum. s.s. (*wh.*) woolly white. Levant. 1732. —— H. ♃. Dill. elt. 274. f. 214.
5 créticum. w. (*wh.*) Cretan. —— 1596. —— H. ♃.
6 aff'ine. L.en. (*pu.*) related. Siberia. 1816. —— H. ♃.
7 supìnum. s.s. (*wh.*) procumbent. S. Europe. 1714. 8. 10. H. ♃. Bocc. mus. 2. t. 96.
8 catariæfòlium. s.s. (*wh.*) Catmint-leav'd. Levant. 1823. 7. 9. H. ♃.
9 velutìnum. F.G. (*ye.*) velvetty. —— 1827. —— F. ♃. Flor. græc. t. 561.
10 vulgàre. s.s. (*wh.*) common. Britain. .... 6. 9. H. ♃. Eng. bot. t. 6. f. 410.

LAVA'NDULA. S.S. LAVENDER. Didynamia Gymnospermia. L.
1 Spìca. s.s. (*bl.*) common. S. Europe. 1568. 7. 9. H. ♄. Schk. hand. 2. t. 157.
β álba. (*wh.*) *white-flowered.* —— —— —— H. ♄.
2 latifòlia. s.s. (*bl.*) broad-leaved. —— —— —— H. ♄.
3 St'œchas. s.s. (*bl.*) French. S. Europe. 1562. 5. 7. G. ♄. Flor. græc. t. 549.
4 víridis. s.s. (*bl.*) Madeira. Madeira. 1777. —— G. ♄. Hoff. et Lk. lus. 1. t. 4.
5 dentàta. s.s. (*bl.*) tooth-leaved. Spain. 1597. 6. 9. G. ♄. Bot. mag. t. 401.
6 pinnàta. s.s. (*bl.*) pinnated. Madeira. 1777. 4. 8. G. ♄. —— t. 400.
7 heterophy'lla. s.s. (*bl.*) various-leaved. .... 1816. —— G. ♄.
8 formòsa. L.en. (*bl.*) handsome. Canaries. —— —— G. ♄.
9 multífida. s.s. (*bl.*) cut-leaved. —— 1597. 7. 9. G. ♂. Lobel. ic. 432.
10 abrotanoídes. s.s. (*bl.*) Southernwood-ld. —— 1699. 6. 9. G. ♄. Comm. rar. t. 27.

DRACOC'EPHALUM. DRAGON'S-HEAD. Didynamia Gymnospermia. L.
1 Ruyschiàna. s.s. (*bl.*) Hyssop-leaved. N. Europe. 1699. 6. 9. H. ♃. Flor. dan. t. 121.
2 austrìacum. s.s. (*bl.*) Austrian. Austria. 1597. 6. 7. H. ♃. Jacq. ic. 1. t. 112.

3 origanoídes. s.s.	(*bl.*)	Marjoram-like.	Siberia.	1827.	6. 7.	H.♃.	
4 botryoídes. s.s.	(*bl.*)	woolly-leaved.	Caucasus.	1823.	——	H.♃.	
5 palmàtum. s.s.	(*bl.*)	palmated.	Siberia.	1815.	6. 8.	H.♃.	
6 canariénse. s.s.	(*li.*)	Balm of Gilead.	Canaries.	1697.	7. 9.	G.♄.	Comm. hort. 2. t. 41.
7 pinnàtum. s.s.	(*bl.*)	wing-leaved.	Siberia.	1827.	——	H.♄.	
8 peregrìnum. s.s.	(*bl.*)	prickly-leaved.	————	1759.	7. 8.	H.♃.	Bot. mag. t. 1084.
9 argunénse. s.s.	(*bl.*)	Fischer's.	————	1822.	7. 9.	H.♃.	Swt. br. fl. gar. t. 47.
10 integrifòlium. Led.	(*bl.*)	entire-leaved.	————	1827.	——	H.♃.	
11 grandiflòrum. s.s.	(*bl.*)	great-flowered.	————	1759.	——	H.☉.	
12 altaiénse. s.s.	(*bl.*)	Betony-leaved.	Georgia.	1787.	——	H.♃.	Swt. br. fl. gar. s. 2.
13 Moldávica. s.s.	(*bl.*)	Moldavian.	Moldavia.	1596.	7. 8.	H.☉.	Lam. ill. t. 513. f. 1.
14 canéscens. s.s.	(*bl.*)	hoary-leaved.	Levant.	1711.	——	H.☉.	Swt. br. fl. gar. t. 38.
15 ibéricum. s.s.	(*bl.*)	Iberian.	Iberia.	1827.	——	H.☉.	
16 peltàtum. s.s.	(*bl.*)	Willow-leaved.	Levant.	1711.	——	H.☉.	Lam. ill. t. 513. f. 2.
17 thymiflòrum. s.s.	(*bl.*)	small-flowered.	Siberia.	1752.	6. 9.	H.☉.	Gmel. sib. 3. t. 50.
18 nùtans. s.s.	(*bl.*)	nodding.	————	1731.	7. 8.	H.♃.	Bot. reg. t. 841.
19 chamædryoídes. s.s.	(*bl.*)	Chamædrys-like........		1828.	——	H.♃.	
20 sibíricum. s.s.	(*bl.*)	Siberian.	Siberia.	1760.	6. 8.	H.♃.	Bot. mag. t. 2185.
HORM'INUM. P.S.		HORM'INUM.	Didynamia Gymnospermia. P.S.				
pyrenàicum. s.s.	(*bl.*)	Pyrenean.	Pyrenees.	1820.	6. 7.	H.♃.	Swt. br. fl. gar. t. 252.
MELI'SSA. S.S.		BALM.	Didynamia Gymnospermia. L.				
1 officinàlis. W.	(*bh.*)	common.	S. Europe.	1573.	6. 10.	H.♃.	
2 hirsùta. H.H.	(*bh.*)	hairy.	————	....	——	H.♃.	
3 cordifòlia. P.S.	(*bh.*)	heart-leaved.	————	....	——	H.♃.	
4 altíssima. s.s.	(*bh.*)	tall.	————	1823.	——	H.♃.	Flor. græc. t. 579.
LEPECHI'NIA. S.S.		LEPECHI'NIA.	Didynamia Gymnospermia. W.				
spicàta. s.s.	(*st.*)	spiked.	Mexico.	1800.	7. 8.	H.♃.	Bot. reg. t. 1292.
THY'MBRA. S.S.		THY'MBRA.	Didynamia Gymnospermia. L.				
1 spicàta. L.	(*re.*)	spiked-flower'd.	Levant.	1699.	6. 7.	G.♄.	Flor. græc. t. 546.
2 verticillàta. L.	(*re.*)	whorl-flowered.	Spain.	1702.	——	G.♄.	
A'CYNOS. P.S.		A'CYNOS.	Didynamia Gymnospermia. P.S.				
1 vulgàris. P.S.	(*vi.*)	Basil-leaved.	Britain.	....	6. 8.	H.☉.	Eng. bot. v. 6. t. 411.
Th'ymus A'cinos. E.B.							
2 gravèolens. L.en.	(*vi.*)	strong-scented.	Tauria.	1823.	——	H.♄.	
3 villòsus. P.S.	(*vi.*)	villous.	Germany.	1817.	——	H.☉.	
4 exíguus. F.G.	(*pu.*)	small.	Levant.	1828.	——	H.☉.	Flor. græc. t. 575.
5 alpìnus. P.S.	(*vi.*)	Alpine.	Austria.	1731.	6. 9.	H.♂.	Jacq. aust. 1. t. 97.
6 purpuráscens. P.S.	(*pu.*)	purple.	Spain.	1820.	——	H.♂.	
7 patavìnus. P.S.	(*pu.*)	Marjoram-leav'd.	S. Europe.	1776.	6. 8.	H.♂.	Jacq. obs. 4. t. 87.
8 rotundifòlius. P.S.	(*pu.*)	round-leaved.	Spain.	1820.	——	H.♄.	
GARD'OQUIA. S.S.		GARD'OQUIA.	Didynamia Gymnospermia.				
1 díscolor. K.S.	(*pu.*)	two-coloured.	Caracas.	1827.	4. 7.	S.♄.	
2 origanoídes. s.s.	(*pu.*)	Marjorum-like.	Trinidad.	1824.	——	S.♄.	
CALAMI'NTHA. Ph.		CALAMINT.	Didynamia Gymnospermia.				
1 marifòlia. P.S.	(*wh.sp.*)	Marum-leaved.	Spain.	1788.	6. 7.	H.♃.	Cavan. ic. 6. t. 576.
Népeta marifòlia. Cav. *Th'ymus marifòlius.* W.en.							
2 crética. P.S.	(*pu.*)	Cretan.	S. Europe.	1596.	——	F.♄.	Barrel. ic. 1166.
3 fruticòsa. P.S.	(*pu.*)	shrubby.	Spain.	1752.	7. 9.	F.♄.	
4 álba. L.en.	(*wh.*)	white.	Hungary.	1818.	——	H.♃.	
5 Népeta. P.S.	(*li.*)	lesser.	England.	....	7. 10.	H.♃.	Eng. bot. v. 20. t. 1414.
Melissa Népeta. L. *Th'ymus Népeta.* E.B.							
6 vulgàris.	(*li.*)	common.	England.	....	7. 8.	H.♃.	Eng. bot. v. 24. t. 1676.
Melissa Calamíntha. L. *Th'ymus Calamíntha.* E.B.							
7 grandiflòra. P.S.	(*ro.*)	great-flowered.	Italy.	1596.	6. 9.	H.♃.	Bot. mag. t. 208.
Melissa grandiflòra. B.M.							
8 caroliniàna. N.	(*li.*)	Carolina.	Carolina.	1804.	6. 7.	F.♄.	Bot. mag. t. 997.
Th'ymus grandiflòrus. B.M.							
CLINOP'ODIUM. S.S.		WILD-BASIL.	Didynamia Gymnospermia. L.				
1 vulgàre. s.s.	(*ro.*)	common.	Britain.	....	6. 8.	H.♃.	Eng. bot. v. 20. t. 1401.
2 atropurpùreum. H.H.	(*pu.*)	dark purple.	S. Europe.	1820.	——	H.♃.	
3 ægyptìacum. s.s.	(*pu.*)	Egyptian.	Egypt.	1759.	——	H.♃.	
MELI'TTIS. S.S.		BASTARD-BALM.	Didynamia Gymnospermia. L.				
1 Melissophy'llum. s.s.	(*pu.*)	common.	England.	....	6. 8.	H.♃.	Eng. bot. v. 9. t. 577.
β *alpìna.*	(*pu.wh.*)	*Alpine.*	Switzerland.	....	——	H.♃.	
2 grandiflòra. s.s.	(*wh.vi.*)	great-flowered.	England.	....	5.	H.♃.	Eng. bot. v. 9. t. 636.
MACBR'IDEA. E.C.		MACBR'IDEA.	Didynamia Gymnospermia.				
púlchra. N.	(*re.std.*)	pretty.	Carolina.	1804.	7. 9.	G.♄.	
Thy'mbra caroliniàna. Walter.							

PRUNE'LLA. S.S.		SELF-HEAL.	Didynamia Gymnospermia. L.				
1 vulgàris. s.s.	(*bl.*)	common.	Britain.	....	7. 8.	H.♃.	Eng. bot. v. 14. t. 961.
β *álba.*	(*wh.*)	*white-flowered.*	———	....	——	H.♃.	
2 austràlis.	(*bl.*)	New Holland.	N. S. W.	1826.	——	F.♃.	
vulgàris. B.P. *nec aliorum.*							
3 pensylvánica. s.s.	(*bl.*)	Pensylvanian.	N. America.	1801.	7. 9.	H.⊙.	W. hort. ber. 1. t. 9.
4 grandiflòra. s.s.	(*bl.*)	great-flowered.	Austria.	1596.	——	H.♃.	Bot. mag. t. 2014.
5 intermèdia. s.s.	(*bl.*)	intermediate.	S. Europe.	1790.	——	H.♃.	——— t. 337.
grandiflòra. B.M.							
6 incìsa. L.en.	(*bl.*)	cut-leaved.	———	....	——	H.♃.	
7 laciniàta. L.	(*ye.*)	yellow-flower'd.	Austria.	1713.	——	H.⊙.	Lam. ill. t. 516. f. 2.
8 hastàta. s.s.	(*bl.*)	halberd-leaved.	Portugal.	....	——	H.♃.	
9 longifòlia. P.S.	(*bl.*)	long-leaved.	France.	....	——	H.♃.	
10 hyssopifòlia. s.s.	(*pu.*)	hyssop-leaved.	S. Europe.	1731.	——	H.♃.	Moris. s. 11. t. 5. f. 7.
CLE'ONIA. S.S.		CLE'ONIA.	Didynamia Gymnospermia. L.				
lusitánica. s.s.	(*pa.bl.*)	sweet-scented.	Portugal.	1710.	6. 7.	H.⊙.	Mill. ic. 1. t. 7C
SCUTELL'ARIA. S.S.		SKULLCAP.	Didynamia Gymnospermia. L.				
1 galericulàta. s.s.	(*bl.*)	common.	Britain.	....	6. 9.	H.♃.	Eng. bot. v. 8. t. 523.
2 hastifòlia. s.s.	(*pu.*)	hastate-leaved.	Germany.	1798.	6. 7.	H.♃.	
3 mìnor. s.s.	(*ro.*)	lesser.	Britain.	....	7. 8.	H.♃.	Eng. bot. v. 8. t. 524.
4 párvula. s.s.	(*bl.*)	slender.	N. America.	1823.	——	H.♃.	Hook. ex. flor. t. 106.
5 havanénsis. s.s.	(*wh.bl.*)	Havannah.	Havannah.	1793.	5. 6.	S.♃.	Jacq. obs. 2. t. 29.
6 hùmilis. B.P.	(*pu.*)	dwarf.	N. S. W.	1822.	6. 8.	G.♃.	
7 grandiflòra. s.s.	(*re.ye.*)	large-flowered.	Siberia.	1804.	7. 8.	H.♃.	Bot. mag. t. 635.
8 lateriflòra. s.s.	(*pu.*)	Virginian.	N. America.	1752.	6. 9.	H.♃.	
9 nervòsa. s.s.	(*bl.*)	nerved-leaved.	———	1827.	——	H.♃.	
10 orientàlis. s.s.	(*ye.*)	yellow-flower'd.	Levant.	1729.	7. 9.	H.♃.	Flor. græc. t. 580.
11 altàica. F.	(*bl.wh.*)	Altay.	Siberia.	1816.	——	H.♃.	Swt. br. fl. gar. t. 45.
12 víscida. s.s.	(*bl.wh.*)	viscid.	Iberia.	1820.	——	H.♃.	
13 variegàta. s.s.	(*st.bl.*)	variegated.	———	——	——	H.♃.	
14 alpìna. s.s.	(*bl.*)	Alpine.	Hungary.	1752.	6. 10.	H.♃.	Swt. br. fl. gar. t. 90.
15 peregrìna. s.s.	(*d.pu.*)	Florentine.	Italy.	1783.	——	H.♃.	Flor. græc. t. 582.
16 altíssima. s.s.	(*d.pu.*)	tall.	Levant.	1731.	7. 8.	H.♃.	
17 pállida. s.s.	(*pa.*)	pale-flowered.	Tauria.	1824.	——	H.♃.	
18 Colúmnæ. s.s.	(*d.pu.*)	heart-leaved.	Italy.	1806.	6. 8.	H.♃.	Swt. br. fl. gar. t. 52.
19 rubicúnda. s.s.	(*re.*)	rubicund.	....	1815.	——	H.♃.	
20 nigréscens. s.s.	(*d.pu.*)	dark.	Crete.	1826.	——	F.♃.	
21 álbida. s.s.	(*wh.*)	hairy white.	Levant.	1771.	6. 7.	H.♃.	Flor. græc. t. 581.
22 lupulìna. s.s.	(*st.*)	Tartarian.	Tartary.	1739.	6. 9.	H.♃.	Schmidel ic. t. 73.
23 integrifòlia. s.s.	(*bl.*)	entire-leaved.	N. America.	1731.	——	H.♃.	Pluk. alm. t. 441. f. 6.
24 hyssopifòlia. W.	(*bl.*)	Hyssop-leaved.	———	——	——	H.♃.	
25 purpuráscens. s.s.	(*pu.*)	purplish.	W. Indies.	1823.	——	S.♃.	
26 caroliniàna. s.s.	(*pu.*)	Carolina.	Carolina.	1811.	——	H.♃.	Lam. ill. t. 515. f. 3.
27 macrántha. F.	(*pu.*)	large-flowered.	Dahuria.	1827.	——	H.♃.	
28 pilòsa. s.s.	(*d.pu.*)	pilose.	N. America.	——	——	H.♃.	
29 serrata. s.s.	(*d.pu.*)	saw-leaved.	———	1800.	——	H.♃.	Andr. reposit. t. 494.
30 scordifòlia. F.	(*pu.wh.*)	Scordium-leaved.	Siberia.	1823.	——	H.♃.	
31 vérna. Bess.	(*wh.pu.*)	vernal.	———	——	5. 6.	H.♃.	
SA'LVIA. R.S.		SAGE.	Diandria Monogynia. L.				
1 pomifera. R.S.	(*li.*)	Apple-bearing.	Candia.	1699.	7. 8.	H.♃.	Flor. græc. t. 15.
2 calycìna. R.S.	(*li.*)	broad-calyxed.	Greece.	1823.	——	G.♄.	——— t. 16.
3 canariénsis. R.S.	(*li.*)	Canary.	Canaries.	1697.	6. 9.	G.h.	Trew. pl. rar. 2. t. 19.
4 aùrea. R.S.	(*br.*)	golden.	C. B. S.	1731.	4. 11.	G.h.	Bot. mag. t. 182.
5 dentàta. R.S.	(*bl.*)	tooth-leaved.	———	1774.	12. 1.	G.h.	
6 Régla. R.S.	(*sc.*)	Regla.	Mexico.	1828.	9. 11.	G.♄.	Cavan. icon. 5. t. 455.
7 spléndens. s.s.	(*sc.*)	splendid.	Brazil.	1823.	8. 3.	S.h.	Bot. reg. t. 687.
8 bracteàta. s.s.	(*li.*)	long-bracted.	Syria.	1788.	6. 8.	G.h.	Bot. mag. t. 2320.
9 pilántha. L.en.	(*bl.*)	hairy-flowered.	....	1823.	6. 9.	G.♄.	
10 incarnàta. R.S.	(*fl.*)	flesh-coloured.	Levant.	1729.	——	G.♃.	Pluk. phyt. t. 194. f. 6.
11 interrúpta. R.S.	(*wh.bl.*)	interrupted.	Barbary.	1825.	5. 6.	H.♄.	Swt. br. fl. gar. t. 169.
12 pinnàta. R.S.	(*pu.*)	winged-leaved.	Levant.	1731.	7.	H.♂.	Boer. lugd. 1. t. 167.
13 ríngens. F.G.	(*li.*)	ringent.	———	1827.	——	F.♃.	Flor. græc. t. 18.
14 rosæfòlia. R.S.	(*pu.*)	Rose-leaved.	———	——	——	G.♄.	Smith ic. ined. 1. t. 5.
15 scabiosæfòlia. R.S.	(*wh.li.*)	Scabious-leav'd.	Peru.	1826.	——	G.♃.	Jacq. f. ecl. t. 3.
16 Habliziàna. R.S.	(*w.re.*)	Taurian.	Tauria.	1795.	8.	H.♄.	Bot. mag. t. 1429.
17 vulnerariæfòlia. R.S.	(*w.bl.*)	Armenian.	Armenia.	1829.	6. 9.	F.♄.	
18 lanceolàta. R.S.	(*bl.*)	lance-leaved.	S. America.	1818.	7. 9.	H.⊙.	Jacq. f. ecl. t. 13.
19 hirsùta. R.S.	(*bl.*)	hirsute.	Mexico.	1801.	5. 6.	H.⊙.	Jacq. schœn. 3. t. 252.
20 leucántha. R.S.	(*wh.*)	white-flowered.	———	1825.	7. 10.	G.♃.	Cavan. ic. 1. t. 24.

21 angustifòlia. R.S.	(*bl.*)	narrow-leaved.	Mexico.	1806.	6. 7.	G.♃.	Cavan. ic. 4. t. 317.
réptans. Jacq. schœn. t. 319.							
22 trichostemmoídes. R.S.	(*bl.*)	Missouri.	Missouri.	1827.	6. 9.	H.⊙.	
23 azùrea. R.S.	(*bl.*)	azure-flowered.	Carolina.	1806.	8. 9.	F.♃.	Bot. mag. t. 1728.
24 bengalénsis. F.I.	(*wh.*)	Bengal.	E. Indies.	1825.	6. 9.	S.♄.	
25 occidentàlis. R.S.	(*wh.*)	West Indian.	W. Indies.	1820.	——	S.♃.	
26 pseudococcínea.	(*sc.*)	pale scarlet.	S. America.	1797.	6. 8.	S.♄.	Jacq. ic. 2. t. 209.
27 amœ̀na. B.M.	(*bl.*)	fine blue.	W. Indies.	1793.	9. 12.	S.♄.	Bot. mag. t. 1294.
Boosiàna. Jacq. f. ecl. 1. t. 47. *amœ̀na.* B.R. 446.							
28 ròsea. R.S.	(*ro.*)	rose-coloured.	E. Indies.	1816.	6. 9.	S.♄.	
29 tubífera. R.S.	(*sc.*)	long-tubed.	Mexico.	1824.	——	S.♃.	Cavan. ic. 1. t. 25.
30 papilionàcea. R.S.	(*bl.*)	butterfly-like.	New Spain.	1825.	7. 10.	S.♄.	——— 4. t. 319.
31 mexicàna. R.S.	(*bl.*)	Mexican.	Mexico.	1724.	5. 7.	G.♄.	——— 1. t. 26.
32 chamædryoídes. R.S.	(*re.*)	Germander-like.	———	1795.	6. 9.	G.♄.	Bot. mag. t. 808.
33 cæ̀sia. R.S.	(*pa.bl.*)	grey.	S. America.	1813.	——	G.♄.	
34 involucràta. R.S.	(*pu.*)	large-bracted.	Mexico.	1825.	8. 11.	G.♄.	Bot. reg. t. 1205.
35 purpùrea. R.S.	(*pu.*)	purple.	———	———	5. 7.	S.♃.	Cavan. ic. 2. t. 166.
36 hispánica. R.S.	(*bl.*)	Spanish.	Spain.	1739.	6. 8.	H.⊙.	Bot. reg. t. 359.
37 præ̀cox. S.S.	(*pu.*)	early-flowered.	N. Africa.	1826.	3. 6.	F.♃.	
38 scordioídes. R.S.	(*bl.*)	glandular.	....	1824.	5. 8.	S.♃.	
39 papillòsa. R.S.	(*wh.*)	papillose.	....	1827.	6. 8.	G.♄.	
40 serotìna. R.S.	(*bl.*)	late-flowering.	Chio.	1803.	8. 10.	G.♄.	Jacq. ic. 1. t. 3.
41 domínica. R.S.	(*wh.*)	Dominican.	W. Indies.	1759.	7.	S.♃.	Swartz.obs.p.18.t.1.f.1.
42 tenélla. R.S.	(*bl.*)	slender.	Jamaica.	1821.	6. 9.	S.⊙.	Swartz. icon. t. 2.
43 tiliæfòlia. R.S.	(*bl.*)	Lime-leaved.	S. America.	1793.	6. 8.	S.♃.	Jacq. schœn. 3. t. 254.
44 polystàchya. R.S.	(*bl.*)	many-spiked.	New Spain.	1822.	9. 12.	G.♄.	Cavan. ic. 1. t. 27.
45 lamiifòlia. R.S.	(*bl.*)	Lamium-leaved.	S.America.	1826.	——	G.♄.	Jacq. schœn. 3. t. 318.
46 micrántha. R.S.	(*wh.*)	small-flowered.	———	1818.	6. 8.	S.♃.	
47 leonuroídes. R.S.	(*sc.*)	shining-leaved.	Peru.	1783.	4. 10.	G.♄.	Lam. ill. t. 20. f. 3.
formòsa. B.M. 376.							
48 fúlgens. R.S.	(*sc.*)	glossy.	Mexico.	1827.	1. 12.	G.♄.	Swt. br. fl. gar. s. 2.
49 coccínea. R.S.	(*sc.*)	scarlet-flowered.	S. America.	1774.	4. 10.	G.♄.	Mirb.an.m.15.t.15.f.8.
50 pulchélla. R.S.	(*sc.*)	pretty.	———	....	8. 12.	G.♄.	Coll. hort. rip. t. 16.
51 amethy̆stina. R.S.	(*bl.*)	amethystine.	———	1824.	6. 9.	S.♄.	Smith ic. ined. 2. t. 27.
52 amaríssima. R.S.	(*bl.*)	bitter.	Mexico.	1803.	7. 8.	G.♃.	Bot. reg. t. 347.
53 glutinòsa. R.S.	(*st.*)	glutinous.	Europe.	1596.	6. 9.	H.♃.	Sabb. hort. rom. t. 3.
54 nubícola. B.F.G.	(*st.*)	Nepaul.	Nepaul.	1824.	8. 10.	H.♃.	Swt. br. fl. gar. t. 140.
55 ægyptìaca. R.S.	(*wh.*)	Egyptian.	Egypt.	1770.	6. 7.	H.⊙.	Jacq. vind. 2. t. 108.
56 crética. R.S.	(*vi.*)	Cretan.	Crete.	1760.	6. 8.	F.♄.	Rivin. monop. t. 128.
57 paniculàta. R.S.	(*pu.*)	panicled.	C. B. S.	1758.	——	G.♄.	Mill. ic. t. 225. f. 1.
58 africàna. R.S.	(*bl.*)	African.	———	1731.	4. 6.	G.♄.	Comm. hort. 2. t. 91.
59 coloràta. R.S.	(*bl.cr.*)	coloured-calyxed.	———	1758.	7. 8.	G.♄.	Mill. ic. t. 225. f. 2.
60 officinàlis. R.S.	(*bl.*)	garden.	S. Europe.	1597.	6. 7.	H.♄.	Lam. ill. t. 20. f. 1.
61 chromática. Hg.	(*bl.*)	variegated.	———	——	——	H.♄.	
62 tenùior. Df.	(*bl.*)	narrow-leaved.	Spain.	——	——	H.♄.	
63 hispanòrum. Lg.	(*bl.*)	Spanish.	———	——	——	H.♄.	
64 lavandulæfòlia. R.S.	(*bl.*)	Lavender-leaved.	———	——	——	H.♄.	
65 rosmarinifòlia. Lg.	(*bl.*)	Rosemary-leaved.	———	——	——	H.♄.	
66 Spielmánni. R.S.	(*bl.*)	Spielman's.	Caucasus.	1813.	——	H.♃.	Jacq. f. ecl. 2. t. 15.
67 phlomoídes. R.S.	(*pu.*)	Mullein-leaved.	Spain.	1805.	5. 6.	H.♂.	Rœm. script. his. t. 1. f. 1.
68 urticifòlia. R.S.	(*bl.*)	Nettle-leaved.	N. America.	1799.	6. 7.	H.♃.	Moris. his. 3. t. 13. f. 31.
69 élegans. R.S.	(*cr.*)	elegant.	Mexico.	1820.	7. 10.	G.♄.	H. et B. pl. am. 2. t. 144.
incarnàta. K.S. *non* W.							
70 lævigàta. K.S.	(*pu.*)	smooth.	———	1826.	——	G.♄.	H.etB.pl.am. 2. t. 147.
71 bullàta. W.en.	(*re.*)	blistered.	Spain.	1804.	7. 8.	H.♃.	Jacq. f. ecl. t. 38.
lusitánica. J.E. *elongàta.* Spr.							
72 rugòsa. R.S.	(*wh.*)	wrinkle-leaved.	C. B. S.	1775.	——	H.♄.	
73 verticillàta. R.S.	(*li.*)	whorled.	Germany.	1658.	6. 11.	H.♃.	Mirb.an.m.15.t.15.f.10.
β *parviflòra.*	(*li.*)	*small-flowered.*	———	——	——	H.♃.	
74 índica. R.S.	(*bl.wh.*)	Indian.	India.	1731.	5. 7.	H.♃.	Bot. mag. t. 395.
75 Tenòrii. R.S.	(*bl.*)	Tenore's.	Italy.	1820.	5. 8.	H.♃.	Swt. br. fl. gar. t. 26.
76 candidíssima. R.S.	(*p.bl.*)	hoary-leaved.	Levant.	1804.	6. 7.	G.♃.	Flor. græc. t. 26.
crassifòlia. F.G. *non* Cav.							
77 verbascifòlia. R.S.	(*wh.ye.*)	Verbascum-ld.	Iberia.	1825.	——	H.♃.	
78 odoràta. R.S.	(*wh.*)	sweet-scented.	Bagdad.	1804.	7.	F.♄.	Jacq. f. ecl. t. 16.
79 compréssa. R.S.	(*wh.*)	compressed.	Levant.	1816.	7. 8.	H.♃.	Venten. cels. t. 59.
80 móllis. R.S.	(*re.*)	soft.	Siberia.	1596.	6. 10.	H.♃.	Jacq. f. ecl. t. 37.
81 grandiflòra. R.S.	(*bl.*)	great-flowered.	S. Europe.	1616.	6. 9.	H.♄.	——— t. 36.
82 crassifòlia. R.S.	(*bl.*)	thick-leaved.	Barbary.	1804.	——	F.♄.	

83 praténsis. R.S.	(*bl.*)	meadow.	England.	....	5. 11.	H. ♃.	Eng. bot. v. 3. t. 153.
84 variegàta. R.S.	(*bl.wh.*)	variegated.	Hungary.	1814.	6. 8.	H. ♃.	
85 hæmatòdes. R.S.	(*bl.*)	bloody-veined.	Italy.	1699.	7. 8.	H. ♃.	Barrel. ic. t. 185.
86 nepetifòlia. R.S.	(*bl.*)	Catmint-leaved.	....	1823.	——	H. ⊙.	
87 aspidophy'lla. R.S.	(*bl.*)	slender-leaved.	Peru.	1824.	7. 10.	H. ⊙.	
88 lineatifòlia. Lg.	(*bl.*)	line-leaved.	Mexico.	1823.	6. 9.	G. ♄.	
89 melissodòra. Lg.	(*bl.*)	balm-scented.	New Spain.	1826.	——	G. ♄.	
90 agglùtinans. Lg.	(*sc.*)	clammy-stalked.	——	1827.	——	G. ♄.	
91 viscòsa. R.S.	(*wh.cr.*)	clammy.	Italy.	1773.	5. 6.	H. ♃.	Jacq. ic. 1. t. 5.
92 disérmas. R.S.	(*wh.*)	long-spiked.	Syria.	——	7.	H. ♃.	Barrel. ic. 187.
93 nùtans. R.S.	(*bl.*)	nodding.	Russia.	1780.	6. 8.	H. ♃.	Bot. mag. t. 2436.
94 betonicæfòlia. R.S.	(*bl.*)	Betony-leaved.	——	1804.	6. 7.	H. ♃.	
95 péndula. R.S.	(*bl.*)	drooping.	——	——	6. 8.	H. ♃.	
β *grandiflòra.*	(*d.bl.*)	*large-flowered.*	——	——	——	H. ♃.	
96 amplexicaùlis. R.S.	(*li.*)	stem-clasping.	Levant.	1813.	7. 9.	H. ♃.	
97 parviflòra. R.S.	(*st.*)	small-flowered.	—— ——	1816.	——	F. ♃.	Lab. syr. dec. 4. t. 7.
98 austrìaca. R.S.	(*st.*)	Austrian.	Austria.	1776.	6. 7.	H. ♃.	Jacq. aust. 2. t. 112.
99 syrìaca. R.S.	(*wh.*)	Syrian.	Levant.	1759.	7.	G. ♄.	Arduin. spec. 1. t. 1.
100 nùbia. R.S.	(*bl.*)	Nubian.	Africa.	1784.	6. 7.	G. ♃.	Murr. c. gott. 1778. t. 3.
101 virgàta. R.S.	(*wh.ro.bl.*)	long-branched.	Armenia.	1758.	7. 11.	H. ♃.	Jacq. schœn. 1. t. 37.
102 campéstris. R.S.	(*bl.*)	field.	Tauria.	1813.	6. 7.	H. ♃.	
103 sylvéstris. R.S.	(*bl.*)	wood.	Germany.	1759.	6. 10.	H. ♃.	Jacq. aust. 3. t. 212.
104 nemoròsa. R.S.	(*va.*)	spear-leaved.	——	1728.	——	H. ♃.	
105 pátula. R.S.	(*bh.*)	spreading.	Portugal.	1805.	5. 7.	H. ♃.	
106 tingitàna. R.S.	(*st.*)	Tangier.	Barbary.	1796.	7.	G. ♄.	Rivin. monop. 1. t. 52.
107 Sclàrea. R.S.	(*bh.*)	common Clary.	Italy.	1562.	7. 9.	H. ♂.	Flor. græc. 1. t. 25.
108 spinòsa. R.S.	(*wh.*)	thorny-calyxed.	Egypt.	1789.	6.	G. ♂.	Jacq. ic. 1. t. 7.
109 Æthiòpis. R.S.	(*wh.*)	woolly.	Austria.	1570.	5. 6.	H. ♂.	Jacq. aust. 3. t. 211.
110 argéntea. R.S.	(*wh.*)	silvery.	Crete.	1759.	5. 7.	H. ♂.	Flor. græc. 1. t. 27.
111 applanàta. R.S.	(*wh.*)	short-spined.	....	1818.	6. 8.	H. ♂.	
112 Hormìnum. R.S.	(*pu.*)	annual Clary.	S. Europe.	1596.	——	H. ⊙.	Flor. græc. 1. t. 20.
α *violàcea.*	(*bl.*)	*purple-topped.*	——	——	——	H. ⊙.	
β *rùbra.*	(*re.*)	*red-topped.*	——	——	——	H. ⊙.	
113 víridis. R.S.	(*li.*)	green topped.	Italy.	1759.	——	H. ⊙.	Flor. græc 1. t. 19.
114 pyramidàlis. R.S.	(*li.*)	pyramidal.	——	1825.	——	H. ♃.	
115 rùbra. S.S.	(*re.*)	red-flowered.	Egypt.	1827.	6. 8.	H. ♃.	
116 Simsiàna. B.R.	(*bl.wh.*)	Sims's.	Russia.	1820.	6. 7.	H. ♃.	Bot. reg. t. 1003.
117 verbenàca. R.S.	(*bl.*)	wild clary.	Britain.	....	6. 10.	H. ♃.	Eng. bot. v. 3. t. 154.
118 oblongàta. J.E.	(*bl.*)	oblong-leaved.	Tauria.	1820.	——	H. ♃.	Jacq. f. ecl. 1. t. 14.
119 Sibthórpii. F.G.	(*li.*)	Sibthorp's.	Levant.	1829.	6. 9.	F. ♃.	Flor. græc. t. 22.
120 trilòba. R.S.	(*vi.*)	three-lobed.	S. Europe.	1596.	6. 7.	H. ♄.	—— 1. t. 17.
121 Clùsii. R.S.	(*wh.*)	Clusius's.	Levant.	1827.	6. 9.	F. ♄.	Jacq. schœn. 2. t. 195.
122 gigántea. R.S.	(*wh.*)	giant.	....	1816.	6. 9.	H. ♂.	
123 scorodoniæfòlia. R.S.		villous.	....	1824.	——	F. ♄.	
124 lyràta. R.S.	(*pu.*)	lyre-leaved.	N. America.	1728.	——	H. ♃.	Moris. s. 11. t. 13. f. 27.
125 abyssínica. R.S.	(*pu.*)	Abyssinian.	Abyssinia.	1775.	6. 7.	G. ♃.	Jacq. ic. 1. t. 6.
126 nilótica. R.S.	(*bl.*)	Nilotic.	Egypt.	1780.	6. 8.	H. ⊙.	Jacq. vind. 3. t. 92.
127 Forskòhlii. R.S.	(*bl.*)	Forskohl's.	Levant.	1800.	——	H. ♃.	Bot. mag. t. 988.
128 napifòlia. R.S.	(*li.*)	Rape-leaved.	Italy.	1776.	6. 7.	H. ♃.	Jacq. vind. 2. t. 262.
129 aurìta. R.S.	(*li.*)	ear-leaved.	C. B. S.	1795.	5. 6.	G. ♄.	
130 bícolor. R.S.	(*bl.ye.*)	two-coloured.	Barbary.	1793.	6. 7.	H. ♃.	Bot. mag. t. 1774.
131 Barrelièri. R.S.	(*bl.*)	Barrelier's.	S. Europe.	1821.	5. 8.	H. ♃.	Barrel. ic. 186.
132 laciniàta. R.S.	(*bl.wh.*)	jagged-leaved.	....	1818.	——	H. ♃.	
133 runcinàta. R.S.	(*bl.*)	runcinated.	C. B. S.	1795.	——	G. ♄.	Jacq. schœn. 1. t. 8.
134 scábra. R.S.	(*pu.*)	rough-leaved.	——	1774.	4. 9.	G. ♄.	
135 eròsa. R.S.	(*st.*)	erose-leaved.	....	1823.	6. 9.	H. ⊙.	
136 polymórpha. R.S.	(*bl.*)	variable.	Portugal.	1821.	——	H. ♃.	Barrel. ic. 220.
137 multífida. F.G.	(*pa.bl.*)	multifid.	S. Europe.	1823.	5. 6.	H. ♃.	Flor. græc. 1. t. 23.
138 clandestìna. R.S.	(*li.*)	cut-leaved.	Italy.	1739.	4. 7.	H. ♂.	—— 1. t. 24.
139 ceratophy'lla. R.S.	(*li.*)	stag's-horn-leav'd.	Persia.	1699.	7. 8.	H. ♂.	Pluk. alm. t. 194. f. 5.
140 ceratophylloídes.	(*li.*)	branching.	Egypt.	1771.	6. 8.	H. ♂.	Arduin. spec. 2. t. 2.
141 prismática. R.S.	(*bl.*)	slender-spiked.	Mexico.	1822.	——	H. ⊙.	
COLLINS'ONIA. R.S.		COLLINS'ONIA.	Diandria Monogynia. L.				
1 canadénsis. R.S.	(*ye.*)	Nettle-leaved.	N. America.	1735.	8. 10.	H. ♃.	Linn. hort. clif. t. 5.
2 scabriúscula. R.S.	(*re.ye.*)	rough-stalked.	Florida.	1776.	7. 9.	H. ♃.	
3 ovàlis. R.S.	(*ye.*)	oval-leaved.	Carolina.	1812.	8.	H. ♃.	
4 tuberòsa. M.	(*ye.*)	tuberous.	——	1806.	——	H. ♃.	
serotìna. R.S.							
5 anisàta. R.S.	(*ye.*)	Anise-scented.	——	——	10.	H. ♃.	Bot. mag. t. 1213.

Tribus VI. Prasieæ. *Benth. in Bot. reg.*

PR'ASIUM. S.S.		Pr'asium.	Didynamia Gymnospermia. L.				
1 màjus. s.s.	(*wh.*)	great.	S. Europe.	1699.	6. 8.	G.♄.	Flor. græc. t. 584.
2 mìnus. s.s.	(*wh.*)	smaller.	———	1752.	——	F.♄.	
PHR'YMA. S.S.		Phr'yma.	Didynamia Gymnospermia. L.				
leptostàchya. s.s.	(*wh.*)	slender-spiked.	N. America.	1802.	8. 9.	H.♃.	Lam. ill. t. 516.

Tribus VII. Ocymoideæ. *Benth. in Bot. reg.*

MOSCHO'SMA. B.R.		Moscho'sma.	Didynamia Gymnospermia.				
1 ocymoìdes. B.R.	(*wh.*)	Basil-like.	....	1823.	6. 8.	S.⊙.	
Lumnítzera ocymoìdes. s.s. *O'cymum polyclàdum.* Link.							
2 polystàchya. B.R.	(*wh.*)	many-spiked.	E. Indies.	1783.	7. 8.	S.⊙.	Murr. c. gott. 3. t. 3.
O'CYMUM. W.		Basil.	Didynamia Gymnospermia. L.				
1 thyrsiflòrum. s.s.	(*wh.*)	thyrse-flowered.	E. Indies.	1806.	7. 8.	S.♂.	Jacq. vind. 3. t. 72.
2 gratíssimum. s.s.	(*wh.*)	shrubby.	———	1752.	——	S.♄.	Jacq. ic. 3. t. 495.
3 Basílicum. w.	(*wh.*)	common sweet.	India.	1548.	——	H.⊙.	Blackw. t. 104.
4 pilòsum. W.en.	(*wh.*)	hairy.	....	1816.	7. 9.	H.⊙.	
5 micránthum. s.s.	(*wh.*)	small-flowered.	....	——	——	S.⊙.	
6 víride. s.s.	(*wh.*)	green.	E. Indies.	——	——	S.♄.	
7 suàve. s.s.	(*wh.*)	sweet-scented.	———	——	——	S.♄.	
8 cànum. B.M.	(*wh.*)	hoary.	China.	1822.	——	H.⊙.	Bot. mag. t. 2452.
9 sánctum. s.s.	(*wh.*)	purple-stalked.	E. Indies.	1758.	——	S.⊙.	Rheed. mal. 10. t. 92.
10 febrifùgum. s.s.	(*wh.*)	Feverwort.	Sierra Leon.	1821.	——	S.⊙.	Bot. reg. t. 753.
11 álbum. s.s.	(*wh.*)	large white-fld.	E. Indies.	1816.	——	S.⊙.	
12 americànum. s.s.	(*wh.*)	American.	S. America.	1789.	7. 8.	S.⊙.	Jacq. vind. 3. t. 86.
13 caryophyllàtum. H.B.	(*w.*)	Clove-scented.	E. Indies.	1817.	——	S.♃.	
14 tenuiflòrum. w.	(*li.*)	slender-spiked.	———	1768.	——	S.♂.	Rumph. amb.5.t.92.f.2.
15 menthoìdes. s.s.	(*wh.*)	Mint-leaved.	———	1783.	——	S.⊙.	
16 inodòrum. Lam.	(*wh.*)	scentless.	———	1823.	——	S.⊙.	
17 cristàtum. H.B.	(*wh.*)	crested.	———	1812.	——	S.⊙.	
18 grandiflòrum. s.s.	(*wh.*)	great-flowered.	Abyssinia.	1802.	9. 10.	G.♄.	L'Herit. s. nov. t. 43.
19 mínimum. s.s.	(*wh.*)	bush.	Ceylon.	1573.	——	H.⊙.	Schk. handb. 2. t. 166.
20 carnòsum. s.s.	(*wh.*)	thick-leaved.	Monte Video.	1828.	6. 8.	H.⊙.	
ORTHOS'IPHON. B.R.		Orthos'iphon.	Didynamia Gymnospermia.				
1 adscéndens. B.R.	(*wh.*)	ascending.	E. Indies.	1828.	6. 8.	S.♃.	
O'cymum adscéndens. w.							
2 ásperum. B.R.	(*wh.*)	rough.	———	1827.	——	S.♃.	
O'cymum ásperum. Roth. *Plectránthus ásper.* s.s.							
C'OLEUS. B.R.		C'oleus.	Didynamia Gymnospermia.				
1 Forskóhlæi. B.R.	(*vi.wh.*)	Forskohl's.	Abyssinia.	1806.	10. 11.	S.♄.	Bot. mag. t. 2036.
Plectránthus Forskohlæi. B.M. *barbàtus.* Andr. reposit. t. 594.							
2 comòsus. B.R.	(*bl.*)	tufted.	Nepaul.	1820.	6. 9.	G.♄.	Bot. mag. t. 2318.
3 scutellarioídes. B.R.	(*bl.w.*)	Scullcap-like.	E. Indies.	1764.	7. 8.	S.⊙.	——— t. 1446.
4 amboínicus. Lour.	(*bl.*)	Amboyna.	———	1826.	——	S.♄.	Rumph. amb. 8. t. 72.
PLECTRA'NTHUS. S.S.		Plectra'nthus.	Didynamia Gymnospermia. S.S.				
1 fruticòsus. s.s.	(*bl.*)	shrubby.	C. B. S.	1774.	6. 9.	G.♄.	L'Herit. stirp. 85. t. 41.
2 galeàtus. s.s.	(*bl.*)	helmetted.	Nepaul.	1820.	7. 8.	S.⊙.	
3 graveòlens. B.P.	(*bl.*)	strong-scented.	N. S. W.	——	——	G.♄.	
4 austràlis. B.P.	(*bl.*)	New Holland.	———	1823.	——	G.⊙.	
5 incànus. s.s.	(*bl.*)	hoary.	Nepaul.	——	——	G.♃.	
6 monachòrum. s.s.	(*bh.*)	monk's basil.	E. Indies.	1796.	——	S.⊙.	
O'cymum monachòrum. w.							
7 móllis. s.s.	(*li.*)	heart-leaved.	———	1781.	9. 10.	S.⊙.	
8 strobilíferus. H.B.	(*bl.*)	thick-spiked.	———	1804.	6. 9.	S.⊙.	Rheed. mal. 10. t. 90.
9 secúndus. H.B.	(*bl.*)	side-flowering.	———	1816.	——	S.⊙.	
10 punctàtus. s.s.	(*wh.bl.*)	dotted.	Africa.	1775.	1. 5.	G.♂.	L'Herit. stirp. 1. t. 41.
11 rotundifòlius. s.s.	(*bl.*)	round-leaved.	Mauritius.	1823.	6. 8.	S.⊙.	
12 parviflòrus. s.s.	(*bl.*)	small-flowered.	S. America.	1805.	6. 9.	G.♄.	W. hort. ber. 1. t. 65.
13 cordifòlius. D.P.	(*bl.*)	heart-leaved.	Nepaul.	1823.	——	G.♄.	
14 rubicúndus. D.P.	(*pu.*)	rubicund.	———	1826.	——	G.♄.	
15 virgàtus. D.P.	(*bl.*)	slender.	———	——	——	G.♄.	
16 coloràtus. D.P.	(*bl.*)	coloured bracted.	———	1823.	——	G.♄.	

17 Coétsa. D.P.	(*bl.*)	canescent.	Nepaul.	1823.	6. 9.	G. ♃.	
18 ternifòlius. D.P.	(*bl.*)	three-leaved.	———	1825.	——	G. ♄.	
19 ternàtus. B.M.	(*bl.*)	Omime-root.	Madagascar.	1820.	11.	S. ♃.	Bot. mag. t. 2460.
GENIO'SPORUM. B.R.	GENIO'SPORUM.	Didynamia Gymnospermia.					
prostràtum. B.R.	(*pa.bl.*)	prostrate.	E. Indies.	1812.	6. 9.	S. ⊙.	
O'cymum prostràtum. W.							
ACROCE'PHALUS. B.R.	ACROCE'PHALUS.	Didynamia Gymnospermia.					
1 capitàtus. B.R.	(*wh.*)	headed.	China.	1806.	7. 9.	H. ⊙.	
Lumnítzera capitàta. S.S. *O'cymum capitellàtum.* W.							
2 villòsus. B.R.	(*wh.*)	villous.	E. Indies.	1817.	7. 8.	S. ♃.	
ANISOCH'ILUS. B.R.	ANISOCH'ILUS.	Didynamia Gymnospermia.					
carnòsus. B.R.	(*li.*)	thick-leaved.	E. Indies.	1788.	6. 7.	S. ♂.	Linn. amœn. ac. 10. t. 3.
Lavándula carnòsa. W. *Plectránthus dùbius.* S.S.							
PYCNOST'ACHYS.	PYCNOST'ACHYS.	Didynamia Gymnospermia.					
cœrùlea. H.E.F.	(*bl.*)	blue-flowered.	Madagascar.	1825.	8. 10.	F. ⊙.	Hook. ex. flor. t. 202.
ÆOLLA'NTHUS. M.P.	ÆOLLA'NTHUS.	Didynamia Gymnospermia.					
suavèolens. M.P.R.	(*wh.*)	sweet-scented.	Brazil.	1829.	7. 10.	S. ⊙.	
HY'PTIS. S.S.	HY'PTIS.	Didynamia Gymnospermia. S.S.					
1 ebracteàta. S.S.	(*bl.*)	small-headed.	W. Indies.	1778.	10. 1.	S. ⊙.	Poit. an. mus. 7. t. 29. f. 2.
Ballòta suavèolens. Jacq. schœnb. 3. t. 42.							
2 radiàta. S.S.	(*wh.*)	Carolina.	Carolina.	1690.	6. 7.	G. ♃.	Poit. an. mus. 7. t. 27. f. 2.
3 capitàta. S.S.	(*wh.*)	Jamaica.	W. Indies.	1714.	——	S. ♂.	Jacq. ic. 1. t. 114.
4 polyánthes. S.S.	(*wh.*)	many-flowered.	S. America.	1819.	——	S. ⊙.	
5 brèvipes. S.S.	(*li.*)	short-peduncled.	W. Indies.	1822.	——	S. ⊙.	
6 recurvàta. S.S.	(*wh.*)	recurved.	Cayenne.	——	——	S. ♃.	Poit. an. mus. t. 4. f. 3.
7 álbida. S.S.	(*bl.*)	white-leaved.	Mexico.	1825.	——	S. ♄.	
8 pérsica. P.S.	(*st.*)	Persian.	Persia.	1800.	7.	G. ♄.	Linn. trans. 6. t. 12.
Brotèra pérsica. L.T.							
9 spicàta. S.S.	(*bl.*)	spiked.	S. America.	1819.	7. 9.	G. ⊙.	Poit. an. mus. t. 28. f. 2.
10 verticillàta. S.S.	(*wh.*)	whorled.	———	1824.	——	S. ⊙.	Jacq. icon. t. 113.
11 Plumièri. K.S.	(*bl.*)	Plumier's.	———	1819.	——	S. ⊙.	Plum. ic. t. 163. f. 1.
12 stachyòdes. S.S.	(*wh.*)	whorl-spiked.	Brazil.	1820.	——	S. ⊙.	
13 pectinàta. S.S.	(*ye.*)	pectinated.	W. Indies.	1776.	12. 1.	S. ♃.	Poit. an. mus. 7. t. 30.
PROSTANTH'ERA. B.P.	PROSTANTHE'RA.	Didynamia Gymnospermia, S.S.					
1 lasiánthos. B.P.	(*wh.*)	villous-flowered.	N. S. W.	1808.	6. 7.	G. ♄.	Bot. reg. t. 143.
2 cœrùlea. B.P.	(*bl.*)	blue-flowered.	———	1824.	4. 7.	G. ♄.	
3 prunelloídes. B.P.	(*pu.*)	Prunella-like.	———	1826.	——	G. ♄.	
4 incísa. B.P.	(*wh.*)	cut-leaved.	———	1825.	——	G. ♄.	
5 lineàris. B.P.	(*wh.*)	linear-leaved.	———	1823.	——	G. ♄.	
6 denticulàta. B.P.	(*wh.*)	tooth-leaved.	———	——	——	G. ♄.	
7 marifòlia. B.P.	(*wh.*)	Marum-leaved.	———	1827.	4. 6.	G. ♄.	
8 saxícola. B.P.	(*wh.*)	rock.	———	1823.	——	G. ♄.	
9 violàcea. B.P.	(*bl.*)	violet-coloured.	———	——	——	G. ♄.	Bot. reg. t. 1072.

ORDO CLIV.

SELAGINEÆ. *Choisy mem. fam. Selag.* 1823. *p.* 19.

POLYC'ENIA. C.S.	POLYC'ENIA.	Didynamia Gymnospermia.					
hebenstretioídes. C.S.	(*w.*)	spike-flowered.	C. B. S.	1816.	8. 10.	G. ⊙.	Chois. sel. t. 2. f. 1.
HEBENSTRE'ITIA. C.S.	HEBENSTRE'ITIA.	Didynamia Angiospermia.					
1 dentàta. W.	(*wh.*)	dwarf tooth-ld.	C. B. S.	1739.	5. 9.	G. ⊙.	Bot. mag. t. 483.
2 integrifòlia. W.	(*ye.*)	golden.	———	1792.	4. 6.	G. ♄.	Andr. reposit. t. 252.
aùrea. Andr.							
3 fruticòsa. B.M.	(*wh.*)	frutescent.	———	1816.	5. 11.	G. ♂.	Bot. mag. t. 1970.
4 albiflòra. L.en.	(*wh.*)	white-flowered.	———	1815.	——	G. ♄.	
5 chamædryfòlia. L.en.		Chamædrys-leav'd.	———	1816.	——	G. ♄.	
6 tenuifòlia. H.H.	(*wh.*)	slender-leaved.	———	1826.	——	G. ♂.	
7 scábra. W.	(*wh.*)	rough-leaved.	———	——	5. 9.	G. ♄.	
8 cordàta. W.	(*wh.re.*)	heart-leaved.	———	1774.	7. 8.	G. ♄.	
DISCHI'SMA. C.S.	DISCHI'SMA.	Didynamia Angiospermia.					
1 ciliàtum. C.S.	(*wh.*)	ciliated.	C. B. S.	1815.	5. 7.	G. ♂.	
2 erinoídes.	(*wh.*)	Erinus-leaved.	———	1816.	5. 11.	G. ♂.	
Hebenstreìtia erinoídes. W.							

3 híspidum. (*wh.*) long-spiked. C. B. S. 1816. 5. 11. G.⊙. Burm. afr.109. t.41. f.1.
Hebenstreìtia híspida. Lam.
4 spicàtum. C.S. (*wh.*) spiked. ——— 1815. 5. 7. G.⊙.
5 capitàtum. C.S. (*ye.*) headed. ——— 1822. —— G.⊙.

AGATHE'LPIS. C.S. AGATHE'LPIS. Diandria Monogynia.
1 angustifòlia. C.S. (*wh.*) narrow-leaved. C. B. S. 1823. 5. 8. G.♄. Burm. afr. t. 47. f. 3.
Eránthemum angustifòlium. L. *Selàgo angustifòlia.* W.
2 parvifòlium. C.S. (*wh.*) small-leaved. C. B. S. 1816. —— G.♄. Lam. ill. t. 17. f. 2.

MI'CRODON. C.S. MI'CRODON. Didynamia Gymnospermia.
1 lùcidum. C.S. (*wh.*) glossy. C. B. S. 1812. 6. 7. G.♄. Venten. malm. t. 26.
2 ovàtum. C.S. (*wh.*) oval-headed. ——— 1774. —— G.♄. Bot. mag. t. 186.
Selàgo ovàta. B.M.

SEL'AGO. C.S. SEL'AGO. Didynamia Gymnospermia. L.
1 micrántha. C.S. (*ye.*) small-flowered. C. B. S. 1824. 5. 7. G.♄.
2 divaricàta. C.S. (*wh.*) spreading. ——— 1816. —— G.♄.
3 fruticòsa. C.S. (*wh.*) frutescent. ——— 1822. 6. 8. G.♄.
4 minutíssima. C.S. (*ye.*) small. ——— 1816. —— G.♄. Chois. sel. t. 3.
5 corymbòsa. C.S. (*wh.*) fine-leaved. ——— 1699. 7. 9. G.♄. Comm. hort. 2. t. 40.
6 cinèrea. C.S. (*vi.*) cinereous. ——— 1816. —— G.♄.
7 híspida. C.S. (*ye.*) hispid. ——— 1823. 6. 9. G.♄.
8 adpréssa. C.S. (*wh.*) close-flowered. ——— 1816. —— G.♄. Chois. sel. t. 4.
9 canéscens. S.S. (*wh.*) canescent. ——— 1812. 7. 11. G.♄.
10 spìnea. L.en. (*wh.*) reflex-leaved. ——— 1816. 7. 8. G.♄.
11 diffùsa. S.S. (*wh.*) diffuse. ——— 1807. —— G.♄.
12 spicàta. L.en. (*wh.*) spiked. ——— 1816. —— G.♄.
13 teretifòlia. L.en. (*wh.*) cylinder-leaved. ——— 1819. —— G.♄.
14 rapunculoìdes. S.S. (*vi.*) Rampion-like. ——— 1790. 7. 11. G.♄. Burm. afr. t. 42. f. 1.
15 fulvomaculàta. L.en. (*vi.*) brown-spotted. ——— 1816. —— G.♄.
16 spùria. S.S. (*vi.*) linear-leaved. ——— 1779. 7. 10. G.♄. Burm. afr. t. 42. f. 3.
17 ramulòsa. L.en. (*wh.*) branching. ——— 1815. 6. 9. G.♄.
18 fasciculàta. S.S. (*bl.*) cluster-flowered. ——— 1774. —— G.♄. Bot. reg. t. 184.
19 heterophy'lla. S.S. (*wh.*) various-leaved. ——— 1829. —— G.♄.
20 polystàchya. S.S. (*wh.*) many-spiked. ——— 1823. 6. 8. G.♄.
21 bracteàta. S.S. (*wh.*) imbricate-bracted. ——— 1816. 6. 9. G.♄.
22 polygaloídes. S.S. (*vi.*) milkwort-like. ——— 1807. 7. 11. G.♄.
23 rotundifòlia. S.S. (*wh.*) round-leaved. ——— 1816. 7. 9. G.♄.
24 pterophy'lla. Ot. .... winged-leaved. ——— 1828. .... G.♄.
25 ciliàta. C.S. (*wh.*) fringed. ——— 1824. —— G.♄. Chois. sel. t. 5.

ORDO CLV.

VERBENACEÆ. *Brown prodr.* 1. *p.* 510.

CLERODE'NDRUM. B.P. CLERODE'NDRUM. Didynamia Angiospermia. L.
1 pubéscens. B.R. (*wh.*) downy-leaved. W. Indies. 1824. 7. 9. S.♄. Bot. reg. t. 1035.
2 ligustrìnum. S.S. (*wh.*) Privet-leaved. Mauritius. 1789. 8. 11. S.♄. Jacq. col. sup. t. 5. f. 1.
3 inérme. S.S. (*wh.*) smooth. E. Indies. 1692. —— S.♄. ——— t. 4. f. 1.
4 angustifòlium. S.S. (*wh.*) narrow-leaved. Jamaica. 1823. —— S.♄. Lam. ill. t. 544. f. 2.
5 coromandelìnum. S.S. (*w.*) Coromandel. E. Indies. 1824. —— S.♄.
6 emirnénse. B.M. (*wh.*) small-flowered. Madagascar. 1823. 6. 12. S.♄.◡. Bot. mag. t. 2925.
floribúndum. Bot. reg. t. 1035. *non* B.P.
7 volùbile. S.S. (*wh.*) climbing. Guinea. 1822. 8. 11. S.♄.◡.
8 buxifòlium. S.S. (*wh.*) Box-leaved. .... 1818. 7. 8. S.♄.
Volkamèria buxifòlia. W.en.
9 heterophy'llum. S.S. (*w.*) various-leaved. Mauritius. 1805. 8. 9. S.♄. Andr. reposit. t. 554.
Volkamèria angustifòlia. A.R.
10 macrophy'llum. B.M. (*st.bl.*) large-leaved. ——— 1822. —— S.♄. Bot. mag. t. 2536.
11 serràtum. S.S. (*wh.*) saw-leaved. E. Indies. —— —— S.♄. Rheed. mal. 4. t. 29.
12 lívidum. B.R. (*wh.*) livid-bracted. China. 1824. 9. 11. S.♄. Bot. reg. t. 945.
13 cérnuum. Wal. (*wh.*) drooping. E. Indies. 1822. .... S.♄.
13 Siphonánthus. S.S. (*wh.*) whorled-leaved. ——— 1796. .... S.♄. Burm. ind. t. 43. f. 1.
Siphonánthus índica et *Ovièda mìtis.* L.
15 verticillàtus. D.P. (*wh.*) four-leaved. Nepaul. 1818. .... S.♄.
16 hastàtum. B.R. (*wh.*) halbert-leaved. E. Indies. 1824. 7. 11. S.♄. Bot. reg. t. 1307.

17	odoràtum. D.P.	(*re.*)	odoriferous.	Nepaul.	1820.	7. 8.	S. ♄.	
18	fœtidum. D.P.	(*wh.*)	strong-scented.	———	——	——	S. ♄.	
19	cordàtum. D.P.	(*wh.*)	heart-leaved.	———	1826.	——	S. ♄.	
20	nùtans. D.P.	(*wh.*)	nodding.	———	——	....	S. ♄.	
21	scándens. S.S.	(*wh.*)	scandent.	Guinea.	1822.	7. 9.	S. ♄. ◡.	
22	trichótomum. S.S.	(*wh.*)	three-forked.	Japan.	1800.	——	G. ♄.	Banks ic. Kæmp. t. 22.
23	fortunàtum. S.S.	(*wh.*)	spear-leaved.	E. Indies.	1784.	7. 8.	S. ♄.	Osb. it. t. 11.
24	attenuàtum. B.P.	(*wh.*)	attenuated.	N. S. W.	1824.	5. 8.	G. ♄.	
25	squamàtum. S.S.	(*sc.*)	scarlet.	China.	1790.	6. 9.	S. ♄.	Jacq. ic. 3. t. 500.
26	tomentòsum. S.S.	(*wh.*)	downy.	N. S. W.	1794.	3. 5.	G. ♄.	Bot. mag. t. 1518.
27	infortunàtum. S.S.	(*wh.*)	long-flowered.	E. Indies.	....	....	S. ♄.	
28	phlomoídes. S.S.	(*wh.*)	Phlomis like.	———	1820.	5. 7.	S. ♄.	Burm. ind. t. 45. f. 1.
29	fràgrans. S.S.	(*wh.*)	fragrant.	China.	1790.	12. 8.	S. ♄.	Venten. malm. t. 70.
	β *múltiplex.*	(*wh.*)	*double-flowered.*	———	——	——	S. ♄.	Bot. mag. t. 1834.
30	viscòsum. S.S.	(*wh.*)	clammy.	E. Indies.	1796.	5. 8.	S. ♄.	——— t. 1805.
31	dentàtum. H.B.	(*wh.*)	tooth-leaved.	———	1826.	——	S. ♄.	
32	neriifòlium. H.B.	(*wh.*)	Oleander-leaved.	———	1824.	——	S. ♄.	
33	paniculàtum. S.S.	(*sc.*)	panicled.	Java.	1809.	7. 10.	S. ♄.	Bot. reg. t. 406.
34	urticifòlium. H.B.	(*wh.*)	Nettle-leaved.	E. Indies.	1822.	——	S. ♄.	
35	violàceum. L.en.	(*vi.*)	violet-coloured.	....	——	....	S. ♄.	
36	japónicum.	(*wh.*)	Japan.	Japan.	1823.	....	S. ♄.	
	Volkamèria japónica. Th.							
37	costàtum. B.P.	(*wh.*)	ribbed-leaved.	N. S. W.	1823.	5. 8.	G. ♄.	
	VOLKAM'ERIA. S.S.		VOLKAM'ERIA.	Didynamia Angiospermia. L.				
	aculeàta. S.S.	(*wh.*)	prickly.	W. Indies.	1739.	8. 10.	S. ♄.	Brown jam. t. 20. f. 2.
	ÆGI'PHILA. R.S.		ÆGI'PHILA.	Tetrandria Monogynia. L.				
1	martinicénsis. R.S.	(*st.*)	Martinico.	W. Indies.	1780.	11. 1.	S. ♄.	Lodd. bot. cab. t. 132.
2	arboréscens. R.S.	(*st.*)	arborescent.	Trinidad.	1820.	....	S. ♄.	Aub. gui. 1. t. 24.
3	l'ævis. R.S.	(*ye.*)	smooth.	Guiana.	1824.	6. 7.	S. ♄.	——— 1. t. 25.
4	elàta. R.S.	(*ye.*)	tall.	W. Indies.	1823.	7. 11.	S. ♄.	Bot. reg. t. 946.
5	f'œtida. R.S.	(*ye.*)	fetid.	———	1800.	6. 7.	S. ♄.	
6	trífida. R.S.	(*wh.*)	trifid-flowered.	Jamaica.	1826.	——	S. ♄.	
7	diffúsa. R.S.	(*st.*)	diffuse.	W. Indies.	1804.	7. 8.	S. ♄.	Andr. reposit. t. 578. f. 1.
8	obovàta. R.S.	(*st.*)	oval-leaved.	———	——	——	S. ♄.	——— t. 578. f. 2.
	CALLICA'RPA. R.S.		CALLICA'RPA.	Tetrandria Monogynia. L.				
1	americàna. R.S.	(*re.*)	American.	Carolina.	1724.	6. 7.	G. ♄.	Catesb. car. 2. t. 47.
2	càna. R.S.	(*li.*)	hoary.	E. Indies.	1799.	——	S. ♄.	
3	lanàta. R.S.	(*pu.*)	woolly.	———	1788.	——	S. ♄.	Gært. fruct. 1. t. 94.
4	incàna. F.I.	(*ro.*)	white-leaved.	———	1816.	6. 10.	S. ♄.	
5	macrophy'lla. R.S.	(*ro.*)	large-leaved.	———	1808.	——	S. ♄.	
6	lanceolària. F.I.	(*pu.*)	lance-leaved.	———	1816.	——	S. ♄.	
7	ferrugìnea. R.S.	(*li.*)	rusty.	Jamaica.	1794.	6. 7.	S. ♄.	
8	reticulàta. R.S.	(*wh.*)	netted-leaved.	———	1822.	——	S. ♄.	
9	longifòlia. R.S.	(*wh.*)	long-leaved.	China.	——	——	S. ♄.	Hook. ex. flor. t. 133.
10	rubélla. B.R.	(*re.*)	red-flowered.	———	——	——	S. ♄.	Bot. reg. t. 883.
11	purpùrea. F.I.	(*pu.*)	purple-flowered.	———	——	7. 9.	S. ♄.	
12	arbòrea. F.I.	(*p.pu.*)	tree.	Nepaul.	——	....	S. ♄.	
	V'ITEX. S.S.		CHASTE-TREE.	Didynamia Angiospermia. L.				
1	ovàta. S.S.	(*bl.*)	oval-leaved.	China.	1796.	7. 8.	S. ♄.	
2	triflòra. S.S.	(*vi.*)	three-flowered.	Cayenne.	1823.	——	S. ♄.	
3	latifòlia. S.S.	(*bl.*)	broad-leaved.	E. Indies.	1820.	——	S. ♄.	Rheed. mal. 5. t. 2.
4	altíssima. S.S.	(*vi.*)	tall.	Ceylon.	1802.	....	S. ♄.	
5	alàta. S.S.	(*vi.*)	winged.	E. Indies.	1818.	6. 8.	S. ♄.	
6	gigántea. K.S.	(*vi.*)	giant.	Guayaquil.	1826.	....	S. ♄.	
7	bícolor. S.S.	(*bl.wh.*)	two-coloured.	E. Indies.	1810.	6. 8.	S. ♄.	
8	trifòlia. S.S.	(*vi.*)	three-leaved.	———	1759.	——	S. ♄.	Bot. mag. t. 2187.
9	Negúndo. S.S.	(*bl.*)	Ash-leaved.	———	....	——	S. ♄.	Rheed. mal. 2. t. 11.
10	umbròsa. S.S.	(*vi.*)	bushy.	W. Indies.	1824.	——	S. ♄.	
11	A'gnus-cástus. W.	(*wh.*)	common.	Sicily.	1570.	9.	H. ♄.	
	β *latifòlia.*	(*wh.*)	*broad-leaved.*	———	——	——	H. ♄.	
12	incìsa. S.S.	(*bl.*)	cut-leaved.	China.	1758.	7. 9.	G. ♄.	Bot. mag. t. 364.
	Negúndo. B.M.							
13	arbòrea. H.B.	(*bl.*)	tree.	E. Indies.	1820.	....	S. ♄.	
14	bignonioídes. K.S.	(*bl.*)	trumpet-flowered.	Caracas.	1826.	....	S. ♄.	
15	salígna. H.B.	(*bl.*)	Willow-leaved.	E. Indies.	1823.	....	S. ♄.	
16	Doniàna.	(*vi.*)	Don's.	Sierra Leon.	1824.	....	S. ♄.	
	umbròsa. H.T. *non* Sw.							
17	heterophy'lla. H.B.	(*bl.*)	various-leaved.	E. Indies.	1820.	....	S. ♄.	
18	capitàta. P.S.	(*bl.*)	headed-flowered.	Trinidad.	1822.	6. 8.	S. ♄.	Vahl. ecl. 2. t. 18.

WALLRO'THIA. S.S. WALLRO'THIA. Didynamia Angiospermia. S.S.
Leucóxylon. s.s. (*pu.*) white-wooded. Ceylon. 1793. .... S. ♄.
Vitex Leucóxylon. L.
CHLOA'NTHES. B.P. CHLOA'NTHES. Didynamia Angiospermia. S.S.
1 st'œchadis. B.P. (*st.*) white-leaved. N. S. W. 1822. 6. 8. G. ♄.
2 glandulòsa. B.P. (*st.*) glandular-leaved. ——— 1824. —— G. ♄.
PRE'MNA. S.S. PRE'MNA. Didynamia Angiospermia. S.S.
1 integrifòlia. s.s. (*wh.*) entire-leaved. E. Indies. 1822. .... S. ♄. Rumph. amb. 3. t. 133.
2 tomentòsa. s.s. (*wh.*) woolly. ——— 1825. .... S. ♄.
3 reticulàta. s.s. (*wh.*) netted-leaved. Jamaica. 1826. .... S. ♄.
4 spinòsa. H.B. (*wh.*) spiny. ——— 1822. .... S. ♄.
5 esculénta. H.B. (*wh.*) esculent. ——— 1824. .... S. ♄.
6 latifòlia. H.B. (*wh.*) broad-leaved. E. Indies. 1827. .... S. ♄.
7 serratifòlia. s.s. (*wh.*) saw-leaved. ——— ——— .... S. ♄.
HOLMSKI'OLDIA. HOLMSKI'OLDIA. Didynamia Angiospermia. S.S.
1 sanguínea. s.s. (*sc.*) scarlet. E. Indies. 1796. 10. 4. S. ♄. Bot. reg. t. 692.
Hastíngia coccínea. Sm. ex. bot. t. 80.
2 scándens. (*pi.*) climbing. ——— 1824. —— S. ♄.◡.
Hastíngia scándens. H.B.
PETR'ÆA. S.S. PETR'ÆA. Didynamia Angiospermia. L.
1 volùbilis. s.s. (*pu.*) climbing. Vera Cruz. 1733. 7. 8. S. ♄.◡. Bot. mag. t. 628.
2 rugòsa. K.S. (*pu.*) rough-leaved. Caracas. 1824. .... S. ♄.
3 eréctа. L.C. (*pu.*) upright. S. America. —— .... S. ♄.
HO'STA. Jacq. HO'STA. Diandria Monogynia.
1 cœrùlea. Jacq. (*bl.*) blue-flowered. S. America. 1733. 6. 9. S. ♄. Bot. reg. t. 1204.
Cornùtia punctàta. B.M. 2611.
2 latifòlia. K.S. (*vi.*) broad-leaved. Mexico. 1824. —— S. ♄.
CORN'UTIA. P.S. CORN'UTIA. Diandria Monogynia. S.S.
pyramidàta. P.S. (*bl.*) pyramidal. W. Indies. 1733. .... S. ♄. Lam. ill. t. 541.
GMEL'INA. S.S. GMEL'INA. Didynamia Angiospermia. L.
1 arbòrea. s.s. (*br.ye.*) tree. E. Indies. 1812. .... S. ♄. Roxb. corom. 3. t. 246.
2 villòsa. s.s. (*ye.*) villous. ——— 1818. .... S. ♄.
3 asiática. s.s. (*ye.*) oval-leaved. ——— 1792. .... S. ♄. Lam. ill. t. 542.
4 parviflòra. P.S. (*ye.*) small-flowered. ——— 1817. .... S. ♄. Roxb. corom. 1. t. 32.
parvifòlia. s.s.
5 sinuàta. L.en. (*ye.*) sinuated. ——— 1824. .... S. ♄.
CITHARE'XYLUM. W. FIDDLE-WOOD. Didynamia Angiospermia. L.
1 caudàtum. s.s. (*wh.*) upright. Jamaica. 1763. .... S. ♄. Jacq. ic. 3. t. 501.
eréctum. Jacq.
2 villòsum. s.s. (*wh.*) hairy-leaved. W. Indies. 1784. .... S. ♄. Jacq. ic. 1. t. 118.
3 subserràtum. s.s. (*wh.*) subserrate. Hispaniola. 1819. .... S. ♄.
4 quadrangulàre. s.s. (*w.*) square-stalked. Jamaica. 1759. 6. 8. S. ♄. Jacq. vind. 1. t. 22.
5 cinéreum. s.s. (*wh.*) ash-coloured. W. Indies. 1739. —— S. ♄. Jacq. amer. t. 118.
6 mólle. s.s. (*wh.*) soft-leaved. .... 1822. —— S. ♄.
7 pentándrum. s.s. (*pu.*) pentandrous. Portorico. 1815. —— S. ♄. Vent. cels. t. 47.
DURA'NTA. S.S. DURA'NTA. Didynamia Angiospermia. L.
1 Plumiéri. B.R. (*bl.*) smooth. S. America. 1733. 6. 10. S. ♄. Bot. reg. t. 244.
2 Ellísia. B.M. (*bl.*) prickly. W. Indies. 1739. 8. S. ♄. Bot. mag. t. 1759.
3 inérmis. Mill. (*bl.*) unarmed. S. America. .... —— S. ♄.
4 xalapénsis. K.S. (*bl.*) Mexican. Mexico. 1822. .... S. ♄.
5 Mutísii. K.S. (*bl.*) Mutis's. S. America. —— 6. 8. S. ♄.
AMAS'ONIA. S.S. AMAS'ONIA. Didynamia Angiospermia. S.S.
punícea. s.s. (*ye.*) puniceous. Trinidad. 1823. .... S. ♄.
LANT'ANA. S.S. LANT'ANA. Didynamia Angiospermia. L.
1 aculeàta. s.s. (*or.*) changeable-color'd. W. Indies. 1692. 4. 11. S. ♄. Bot. mag. t. 96.
2 purpùrea. s.s. (*pu.*) purple-flowered. S. America. 1822. —— S. ♄.
3 scábrida. s.s. (*or.*) rough. W. Indies. 1774. 7. 10. S. ♄.
4 místa. W. (*st.li.*) Nettle-leaved. ——— 1732. 8. 10. S. ♄. Lodd. bot. cab. t. 68.
5 melissæfolia. s.s. (*ye.*) Balm-leaved. ——— —— 7. 9. S. ♄. Dill. elt. t. 57. f. 66.
6 nívea. s.s. (*wh.*) snowy white. E. Indies. 1810. —— S. ♄. Bot. mag. t. 1946.
7 strícta. s.s. (*ro.*) narrow-leaved. Jamaica. 1733. —— S. ♄. Sloan. jam. 2. t. 195. f. 4.
8 Rádula. W. (*pu.*) Rasp-leaved. W. Indies. 1803. —— S. ♄.
9 salviæfòlia. s.s. (*re.*) Sage-leaved. C. B. S. 1823. —— G. ♄. Jacq. schœnb. 3. t. 285.
10 récta. s.s. (*bh.*) upright. Jamaica. 1758. 6. 8. S. ♄. ——— 3. t. 360.
11 involucràta. s.s. (*li.*) round-leaved. W. Indies. 1690. 5. 7. S. ♄. Pluk. alm. t. 114. f. 5.
12 odoràta. s.s. (*wh.*) sweet-scented. ——— 1758. 5. 11. S. ♄. Plum. ic. 71. f. 2.
13 Cámara. s.s. (*ye.*) various-coloured. ——— 1691. 4. 9. S. ♄. Dill. elt. t. 56. f. 65.
14 trifòlia. s.s. (*ro.*) three-leaved. ——— 1733. 6. 9. S. ♄. Bot. mag. t. 1449.
15 ánnua. s.s. (*ro.*) annual. S. America. —— 7. 8. S. ⊙. ——— t. 1022.

16 fucata. s.s. (ro.) painted. Brazil. 1822. 7. 8. S.♄. Bot. reg. t. 798.
17 lavandulàcea.s.s.(wh.) Lavender-like. .... —— —— S.♄.
18 cròcea. L.en. (or.) saffron-coloured. Jamaica. —— —— S.♄. Jacq. schœn. t. 473.
19 violàcea. Desf. (vi.) violet-coloured. S. America. 1824. .... S.♄.
20 brasiliénsis. L.en.(wh.) Brazilian. Brazil. —— 7. 8. S.♄.
21 álba. L.en. (wh.) white-flowered. S. America. .... —— S.♄.
22 hírta. Gr. (wh.) hairy white. ———— 1825. 8. 11. S.♄.
23 hórrida. K.S. (ye.) horrid. Mexico. —— 7. 8. S.♄.
24 Sellòi. L.O. (ro.) Sello's. Monte Video. 1829. 12. 4. S.♄. Bot. mag. t. 2981.

SPIELMA'NNIA. S.S. Spielma'nnia. Didynamia Angiospermia. S.S.
africàna. s.s. (wh.) African. C. B. S. 1710. 2. 11. G.♄. Bot. mag. t. 1899.

TECT'ONA. W. Teak-wood. Pentandria Monogynia. L.
grándis. w. (wh.) great. E. Indies. 1777. .... S.♄. Roxb. corom. 1. t. 6.

STRE'PTIUM. S.S. Stre'ptium. Didynamia Angiospermia. S.S.
ásperum. s.s. (wh.) rough. E. Indies. 1799. 7. 8. S.♃. Roxb. corom. 2. t. 146.
Tórtula áspera w. *Prìva leptostàchya.* P.S.

PR'IVA. S.S. Pr'iva. Didynamia Angiospermia. S.S.
1 lappulàcea. P.S. (pu.) bristly-fruited. S. America. 1822. 6. 8. S.♃. Jacq. obs. 1. t. 24.
echinàta. K.S.
2 mexicàna. P.S. (li.) Mexican. Mexico. 1726. 8. 9. S.♃. Dill. elth. t. 302. f. 389.
híspida. K.S. *Verbèna mexicàna.* L.

TAM'ONEA. S.S. Tam'onea. Diandria Monogynia.
1 verbenàcea. s.s. (bl.) thorny-fruited. W. Indies. 1733. 8. S.⊙. Banks. vel. Hous. t. 2.
curassávica. P.S. *Ghìnia spinòsa.* W.
2 spicàta. K.S. (bl.) spiked. Trinidad. 1824. 8. 10. S.⊙. Aubl. guian. t. 268.

STACHYTA'RPHETA. Bastard Vervain. Diandria Monogynia. S.S.
1 angustifòlia. R.S. (vi.) narrow-leaved. S. America. .... 6. 9. S.⊙. Jacq. obs. 4. t. 86.
2 elàtior. R.S. (vi.) tall. Brazil. 1821. 8. 11. S.⊙.
3 índica. R.S. (vi.) Indian. Ceylon. 1732. 8. 9. S.⊙.
4 urticifòlia. B.M. (bl.) Nettle-leaved. S. America. .... 5. 9. S.♂. Bot. mag. t. 1848.
5 jamaicénsis. R.S. (bl.) Jamaica. W. Indies. 1714. 6. 9. S.♂. ———— t. 1860.
6 hírta. K.S. (vi.) hairy-leaved. New Granada. 1821. 5. 10. S.♄.
7 cayennénsis. s.s. (bl.) Cayenne. Cayenne. 1819. —— S.♄.
8 umbròsa. K.S. (bl.) shade. Cumana. 1829. —— S.♄.
9 prismática. R.S. (bl.) Germander-leav'd. W. Indies. 1699. 5. 6. S.♂. Jacq. ic. 2. t. 208.
10 orùbica. R.S. (vi.) Orubian. S. America. —— 6. 8. G.♂. Ehr. pict. t. 5. f. 1.
11 mutábilis. R.S. (ro.) changeable-flower'd. —— 1801. 3. 9. S.♄. Bot. mag. t. 976.
Verbèna mutábilis. w. *Cymbùrus mutábilis.* Par. lond. t. 49.
12 crassifòlia. R.S. (vi.) thick-leaved. Brazil. 1826. 6. 9. S.♄.
13 Zuccágni. R.S. (ro.vi.) Zuccagni's. .... 1824. —— S.♄.
14 hirsutíssima. L.en.(bl.) hairy. Brazil. 1821. —— S.♃.

ZAP'ANIA. J. Zap'ania. Didynamia Angiospermia. P.S.
1 stœchadifòlia. P.S.(wh.) oval-spiked. W. Indies. 1732. 8. 9. S.♃. Brown jam. t. 3. f. 1.
2 nodiflòra. B.P. (pi.) knot-flowered. America. 1664. 7. 8. G.♃. Flor. græc. t. 553.
Verbèna nodiflòra. L. *Líppia nodiflòra.* s.s.

ALO'YSIA. P.S. Alo'ysia. Didynamia Angiospermia. P.S.
citrodòra. P.S. (li.) Lemon-scented. Chile. 1784. 5. 9. G.♄. Bot. mag. t. 367.
Verbèna triphy'lla. B.M. *Líppia citrodòra.* K.S.

VERB'ENA. S.S. Vervain. Didynamia Angiospermia. S.S.
1 angustifòlia. s.s. (bl.) narrow-leaved. N. America. 1802. 6. 8. H.♃.
rugòsa. w.
2 paniculàta. s.s. (pu.) panicled. ———— 1800. 7. 8. H.♃. Bot. reg. t. 1102.
3 bonariénsis. s.s. (bl.) cluster-flower'd. BuenosAyres. 1732. 7. 10. H.♂. Dill. elt. t. 300. f. 387.
4 caroliniàna. s.s. (pi.) Carolina. N. America. —— 6. 9. H.♃. ———— t. 301. f. 388.
5 urticæfòlia. s.s. (wh.) Nettle-leaved. ———— 1683. 7. 9. H.♃. Rob. ic. 26.
6 strícta. s.s. (bl.) upright. ———— 1802. 7. 8. H.♃. Bot. mag. t. 1976.
7 lasiostàchys. s.s. (bl.) woolly-spiked. California. 1823. —— H.♃.
8 bracteòsa. s.s. (bl.) long-bracted. N. America. 1812. 7. H.♃. Bot. mag. t. 2910.
9 Aublètia. s.s. (ro.) rose. ———— 1774. 6. 8. F.♂. ———— t. 308.
10 Lambérti. B.M. (ro.) Lambert's. Peru. 1816. —— F.♃. ———— t. 2200.
11 chamædryfòlia. P.S.(sc.) scarlet-flower'd. BuenosAyres. 1827. 4. 10. F.♃. Swt. br. fl. gar. ser. 2. t. 9.
Melíndres. Bot. reg. t. 1184. *Erìnus peruviànus.* L.
12 barbàta. Gr. (li.) bearded. Mexico. —— 9. 12. G.♄.
13 hastàta. s.s. (pu.) halberd-leaved. Canada. 1710. —— H.♃. Herm. parad. t. 242.
14 scábra. s.s. (vi.) rough. Mexico. 1822. —— H.♃.
15 alàta. B.F.G. (pu.) winged-stalked. MonteVideo. 1827. 7. 10. F.♃. Swt. br. fl. gar. ser. 2. t. 41.
16 officinàlis. E.B. (li.) common. Britain. .... 6. 9. H.♃. Eng. bot. v. 11. t. 769.
17 soròria. D.P. (li.) sister. Nepaul. 1823. 8. 10. H.♃. Swt. br. fl. gar. t. 202.
18 spùria. s.s. (li.) jagged-leaved. N. America. 1731. 7. 8. H.♂.

19 erinoídes. s.s.	(*ro.*)	Erinus-like.	Peru.	1820.	7. 8.	H.⊙.	Flor. peruv. 1. t.33.f.c.
multífida. R.P.							
20 pulchélla. B.F.G.	(*li.*)	pretty.	BuenosAyres.	1827.	4. 10.	F.♃.	Swt.br.fl.gar.s.2.t.295.
21 canéscens. s.s.	(*vi.*)	canescent.	Mexico.	1820.	7. 8.	H.♃.	H. et B.2. t. 136.
22 supína. s.s.	(*li.*)	trailing.	Spain.	1640.	6. 7.	H.⊙.	Flor. græc. t. 554.
β *hirsùta.* E.	(*li.*)	*hairy.*	Egypt.	1829.	——	H.⊙.	
23 prostràta. H.K.		prostrate.	N. America.	1794.	——	H.♃.	
LI'PPIA. J.		LI'PPIA.	Didynamia Angiospermia.	S.S.			
purpùrea. s.s.	(*pu.*)	purple.	Mexico.	1823.	6. 8.	G.♄.	Jacq. f. eclog. t. 85.
XENOP'OMA. S.S.		XENOP'OMA.	Didynamia Angiospermia.				
obovàtum. s.s.	(*wh.*)	obovate-leaved.	China.	1816.	5. 8.	G.♄.	Coll. hort. rip. t. 25.

ORDO CLVI.

MYOPORINÆ. *Brown prodr. p.* 514.

MYOP'ORUM. B.P.		MYOP'ORUM.	Didynamia Angiospermia.	S.S.			
1 ellípticum. B.P.	(*wh.*)	smooth-leaved.	N. S. W.	1789.	1. 3.	G.♄.	Andr. reposit. t. 283.
Pogònia glábra. A.R.							
2 acuminàtum. B.P.	(*wh.*)	acuminate.	———	1812.	....	G.♄.	
3 montànum. B.P.	(*wh.*)	mountain.	———	1822.	2. 6.	G.♄.	
4 parvifòlium. B.P.	(*wh.*)	small-leaved.	N. Holland.	1803.	3. 9.	G.♄.	Bot. mag. t. 1691.
5 dèbile. B.P.	(*bl.*)	procumbent.	N. S. W.	1793.	5. 8.	G.♄.	——— t. 1830.
6 diffùsum. B.P.	(*bl.*)	diffuse.	N. Holland.	....	8. 2.	G.♄.	
7 tuberculàtum. B.P.	(*wh.*)	tubercled.	———	1803.	5. 8.	G.♄.	
8 viscòsum. B.P.	(*wh.*)	viscid.	———	——	——	G.♄.	
9 crassifòlium. s.s.	(*wh.*)	thick-leaved.	N. Zealand.	1822.	——	G.♄.	
10 oppositifòlium. s.s.	(*wh.*)	opposite-leaved.	N. Holland.	1803.	1. 12.	G.♄.	
STENOCH'ILUS. B.P.		STENOCH'ILUS.	Didynamia Angiospermia.	S.S.			
1 gláber. B.P.	(*sc.*)	smooth-leaved.	N. Holland.	1803.	1. 12.	G.♄.	Bot. mag. t. 1942.
2 maculàtus. B.R.	(*cr.sp.*)	spotted-flowered.	———	1820.	4. 8.	G.♄.	Bot. reg. t. 647.
3 viscòsus. B.M.	(*ye.*)	clammy.	———	1825.	——	G.♄.	Bot. mag. t. 2930.
BO'NTIA. S.S.		BO'NTIA.	Didynamia Angiospermia.	L.			
daphnoídes. s.s.	(*st.*)	Barbadoes.	W. Indies.	1690.	6.	S.♄.	Dill. elth. t. 49. f. 57.
AVICE'NNIA. B.P.		AVICE'NNIA.	Didynamia Angiospermia.	L.			
tomentòsa. B.P.	(*wh.*)	downy-leaved.	India.	1793.	....	S.♄.	Palisot. d'Owar. t. 47.

ORDO CLVII.

ACANTHACEÆ. *Brown prodr.* 1. *p.* 472.

ACA'NTHUS. S.S.		BEARS-BREECH.	Didynamia Angiospermia.	L.			
1 móllis. W.	(*wh.*)	soft-leaved.	Italy.	1548.	7. 9.	H.♃.	Lam. ill. t. 550.
2 nìger. L.en.	(*wh.*)	shining-leaved.	Portugal.	1759.	——	H.♃.	
3 spinòsus. B.M.	(*wh.*)	prickly-leaved.	S. Europe.	1629.	——	H.♃.	Bot. mag. t. 1808.
4 spinosíssimus. P.S.	(*wh.*)	white-spined.	———	——	——	H.♃.	
5 carduifòlius. s.s.	(*wh.*)	Cape.	C. B. S.	1816.	——	G.♃.	
6 ilicifòlius. s.s.	(*wh.*)	Holly-leaved.	E. Indies.	1759.	....	S.♄.	Rheed. mal. 2. t. 48.
BLEPH'ARIS. J.		BLEPH'ARIS.	Didynamia Angiospermia.	P.S.			
1 capénsis. s.s.	(*wh.*)	simple-spined.	C. B. S.	1816.	5. 9.	G.♄.	
2 furcàta. s.s.	(*wh.*)	fork-spined.	———	——	6. 8.	G.♄.	
3 procúmbens. s.s.	(*wh.*)	procumbent.	———	1825.	——	G.♄.	
THUNBE'RGIA. S.S.		THUNBE'RGIA.	Didynamia Angiospermia.	S.S.			
1 grandiflòra. s.s.	(*bl.*)	large-flowered.	E. Indies.	1822.	3. 8.	S.♄⌣.	Bot. reg. t. 495.
2 frágrans. s.s.	(*wh.*)	white Indian.	———	1796.	5. 9.	S.♄⌣.	Bot. mag. t. 1881.
3 cordàta. C.R.	(*wh.*)	heart-leaved.	Brazil.	1823.	3. 8.	S.♄⌣.	Coll. h. rip. t. 21.
4 capénsis. s.s.	(*wh.*)	Cape.	C. B. S.	1816.	——	G.♄⌣.	
5 alàta. B.M.	(*or.*)	winged.	Zanzebar.	1825.	1. 12.	S.♄⌣.	Bot. mag. t. 2591.
6 angulàta. H.E.F.	(*pu.*)	angular-stalked.	Mauritius.	1824.	5. 8.	S.♄⌣.	Hook. ex. flor. t. 166.
7 coccínea. D.P.	(*sc.*)	scarlet-flowered.	Nepaul.	——	6. 2.	S.♄⌣.	——— t. 195.

BARL'ERIA. S.S.		BARL'ERIA.	Didynamia Angiospermia. L.				
1 buxifòlia. S.S.	(*bl.*)	Box-leaved.	E. Indies.	1763.	6. 7.	S. ♄.	Rheed. mal. 2. t. 47.
2 Prionìtis. S.S.	(*ye.*)	thorny.	———	1759.	7. 8.	S. ♄.	——— 9. t. 41.
3 spicàta. H.B.	(*pu.*)	spiked-flowered.	———	1822.	4. 8.	S. ♄.	
4 longifòlia. S.S.	(*vi.*)	long-leaved.	———	1781.	7. 9.	S. ♂.	Pluk alm. t. 133. f. 4.
5 cristàta. S.S.	(*bl.*)	crested.	———	1796.	4. 9.	S. ♄.	Bot. mag. t. 1615.
6 álba. B.C.	(*wh.*)	white-flowered.	———	1818.	——	S. ♄.	Lodd. bot. cab. t. 360.
7 ciliàta. H.B.	(*pu.*)	fringed.	———	1822.	4. 10.	S. ♄.	
8 tomentòsa. S.S.	(*bl.*)	woolly.	———	1825.	——	S. ♄.	
9 flàva. S.S.	(*ye.*)	yellow-flowered.	Arabia.	1816.	——	S. ♄.	Jacq. f. ecl. t. 46.
mítis. B.R. 191. *Justícia flàva*. W. *Eránthemum flàvum*. W.en.							
10 longiflòra. S.S.	(*vi.*)	long-flowered.	E. Indies.	1816.	6. 9.	S. ♄.	Vahl symb. 1. t. 16.
11 purpùrea. B.C.	(*pu.*)	purple.	———	1817.	——	S. ♄.	Lodd. bot. cab. t. 344.
12 dichótoma. H.B.	(*pu.*)	forked.	———	1822.	——	S. ♄.	
13 cœrùlea. H.B.	(*bl.*)	blue-flowered.	———	———	——	S. ♄.	
HYGRO'PHILA. B.P.		HYGRO'PHILA.	Didynamia Angiospermia. S.S.				
ríngens. S.S.	(*li.*)	gaping-flowered.	E. Indies.	1807.	7. 8.	S. ♄.	
GEISSOM'ERIA. B.R.		GEISSOM'ERIA.	Didynamia Angiospermia.				
1 longiflòra. B.R.	(*sc.*)	long-flowered.	Brazil.	1825.	8. 10.	S. ♄.	Bot. reg. t. 1045.
2 fúlgida.	(*sc.*)	bright-flowered.	W. Indies.	1804.	7. 8.	S. ♄.	Andr. reposit. t. 527.
Ruéllia fúlgida. A.R.							
RUE'LLIA. S.S.		RUE'LLIA.	Didynamia Angiospermia. L.				
1 ocymoídes. S.S.	(*bl.*)	Basil-like.	Mexico.	1815.	7. 8.	S. ♄.	Cavan. ic. 5. t. 456.
2 ciliàta. S.S.	(*li.*)	fringed.	E. Indies.	1806.	——	S. ♄.	
3 austràlis. S.S.	(*re.*)	New Holland.	N. S. W.	1820.	6. 9.	G. ♄.	Cavan. ic. 6. t. 586. f. 1.
4 fœtida. S.S.	(*vi.*)	strong-scented.	Mexico.	1824.	——	G. ♄.	
5 biflòra. S.S.	(*bl.*)	two-flowered.	Carolina.	1765.	7.	G. ♃.	
6 violàcea. S.S.	(*bl.*)	violet-flowered.	Guiana.	1822.	6. 9.	S. ♃.	Aub. gui. 2. t. 271.
7 salicifòlia. S.S.	(*vi.*)	Willow-leaved.	E. Indies.	———	——	S. ♄.	
8 ovàta. S.S.	(*bl.*)	oval-leaved.	Mexico.	1800.	7. 8.	S. ♃.	Cavan. ic. 3. t. 254.
9 pátula. S.S.	(*vi.*)	spreading.	E. Indies.	1774.	——	S. ♄.	Jacq. ic. 1. t. 119.
10 Sabiniàna. B.R.	(*vi.*)	Sabine's.	———	1827.	4. 8.	S. ♄.	Bot. reg. t. 1238.
11 brasìla. Mart.	(*vi.*)	Brazil.	Brazil.	1828.	8. 9.	S. ♄.	
12 anisophy'lla. H.E.F.	(*bl.*)	unequal-leaved.	E. Indies.	1823.	9. 4.	S. ♄.	Hook. ex. flor. 191.
persicifòlia. Bot. reg. t. 955							
13 formòsa. S.S.	(*sc.*)	splendid.	Brazil.	1808.	6. 9.	S. ♄.	Bot. mag. t. 1400.
14 macrophy'lla. S.S.	(*re.*)	large-leaved.	S. America.	1820.	——	S. ♄.	Vahl symb. 2. t. 39.
15 tuberòsa. S.S.	(*vi.*)	tuberous-root'd.	Jamaica.	1752.	7. 8.	S. ♃.	Sloan. jam. 1. t. 95. f. 1.
16 strèpens. S.S.	(*pa.bl.*)	whorl-flowered.	N. America.	1726.	——	G. ♃.	Schk. hand. 2. t. 177.
17 láctea. S.S.	(*li.*)	white.	Mexico.	1796.	6. 8.	G. ♃.	Cavan. ic. 3. t. 255.
18 clandestìna. S.S.	(*bl.*)	three-flowered.	Barbadoes.	1728.	7. 8.	S. ♃.	Dill. elth. t. 248. f. 320.
19 paniculàta. S.S.	(*li.*)	panicled.	W. Indies.	1768.	8. 10.	S. ♃.	Bot. reg. t. 585.
20 obovàta. H.B.	(*pu.*)	obovate-leaved.	E. Indies.	1820.	7. 11.	S. ⊙.	
21 undulàta. S.S.	(*vi.*)	wave-leaved.	———	1817.	——	S. ♃.	
22 hírta. S.S.	(*vi.*)	hairy.	———	———	——	S. ♃.	Vahl symb. 3. t. 67.
23 pubéscens. P.S.	(*bl.*)	pubescent.	C. B. S.	1816.	5. 7.	G. ♄.	
24 tetragòna. S.S.	(*bl.*)	four-sided.	Brazil.	1818.	6. 9.	S. ♃.	
25 longifòlia. S.S.	(*pu.*)	long-leaved.	E. Indies.	1821.	——	S. ♄.	
26 cérnua. H.B.	(*vi.*)	cernuous.	———	———	——	S. ♄.	
27 semitetrándra. Jac.		semitetrandrous.	....	1829.	....	S. ♄.	
BL'ECHUM. B.P.		BL'ECHUM.	Didynamia Angiospermia. S.S.				
1 Brównei. S.S.	(*li.*)	dense-spiked.	W. Indies.	1780.	6.	S. ⊙.	Sloan. jam. 1. t. 109. f. 1.
2 laxiflòrum. S.S.	(*wh.*)	loose-flowered.	———	1822.	6. 8.	S. ♃.	
3 angustifòlium. S.S.	(*bl.*)	narrow-leaved.	———	1826.	7. 10.	S. ♃.	
APHELA'NDRA. B.P.		APHELA'NDRA.	Didynamia Angiospermia. S.S.				
cristàta. S.S.	(*sc.*)	dense-spiked.	W. Indies.	1733.	6. 9.	S. ♄.	Bot. mag. t. 1578.
Ruéllia cristàta. A.R. 505. *Justícia pulchérrima*. L.							
CROSSA'NDRA. P.S.		CROSSA'NDRA.	Didynamia Angiospermia.				
undulæfòlia. P.L.	(*sc.*)	wave-leaved.	E. Indies.	1800.	6. 1.	S. ♄.	Bot. reg. t. 69.
Ruéllia infundibulifórmis. A.R. 542. *Harráchia speciòsa*. Jacq. ecl. t. 21.							
PHAYLO'PSIS. W.		PHAYLO'PSIS.	Didynamia Angiospermia. W.				
1 longifòlia. B.M.	(*wh.*)	long-leaved.	Sierra Leon.	1822.	3. 5.	S. ♄.	Bot. mag. t. 2433.
2 imbricàta.	(*wh.*)	imbricated.	Mauritius.	———	——	S. ♄.	
Ætheilèma imbricàta. S.S.							
3 parviflòra. W.	(*wh.*)	small-flowered.	E. Indies.	1825.	4. 8.	S. ♃.	
LEPIDAG'ATHIS. W.		LEPIDAG'ATHIS.	Didynamia Angiospermia. W.				
cristàta. W.	(*pu.*)	crested.	E. Indies.	1817.	5. 8.	S. ♃.	
ELYTR'ARIA. R.S.		ELYTR'ARIA.	Diandria Monogynia. S.S.				
1 crenàta. R.S.	(*wh.*)	notched-leav'd.	E. Indies.	1819.	11. 2.	S. ♃.	Roxb. corom. 2. t. 127.
Justícia acaùlis. F.I.							

2 lyràta. R.S.	(*wh.*)	lyrate-leaved.	E. Indies.	1825.	11. 2.	S. ♃.	
3 virgàta. R.S.	(*wh.*)	twiggy.	Carolina.	1813.	7. 8.	H. ♃.	Mich. amer. 1. t. 1.
4 ramòsa. R.S.	(*bl.*)	branching.	Mexico.	1825.	——	G. ♃.	
5 caulèscens. R.S.	(*wh.*)	caulescent.	Manilla.	1828.	——	G. ♃.	
JUSTI'CIA. R.S.		JUSTI'CIA.	Diandria Monogynia. L.				
1 bicalyculàta. R.S.	(*li.*)	Malabar.	E. Indies.	1785.	8. 10.	S. ⊙.	Cavan. ic. 1. t. 71.
2 híspida. R.S.	(*wh.*)	hispid.	Sierra Leon.	1824.	4. 8.	S. ħ.	
3 Ecbòlium. R.S.	(*pa.bl.*)	long-spiked.	E. Indies.	1759.	3. 8.	S. ħ.	Bot. mag. t. 1847.
4 ventricòsa. B.M.	(*sp.*)	hop-flowered.	China.	1825.	4. 6.	S. ♄.	——— t. 2766.
5 calycotrìcha. S.S.	(*ye.*)	yellow-flower'd.	Brazil.	——	1. 3.	S. ħ.	——— t. 2816.
flavícoma. Bot. reg. t. 1027.							
6 thyrsiflòra. F.I.	(*or.*)	thyrse-flower'd.	E. Indies.	1812.	5. 9.	S. ħ.	
7 vitellìna. F.I.	(*pi.ye.*)	reflex-leaved.	———	1818.	——	S. ħ.	
8 álba. F.I.	(*wh.*)	white-flowered.	———	1816.	4. 8.	S. ♄.	
9 coccínea. R.S.	(*sc.*)	scarlet.	S. America.	1770.	12. 4.	S. ♄.	Bot. mag. t. 432.
10 polyspérma. F.I.	(*bl.*)	many-seeded.	E. Indies.	1821.	8. 10.	S. ⊙.	
11 echioídes. R.S.	(*wh.pu.*)	Echium-like.	———	1812.	6. 9.	S. ⊙.	Rheed. mal. 9. t. 46.
12 quadrífida. R.S.	(*sc.*)	twiggy.	Mexico.	1795.	3. 9.	S. ħ.	Sal. par. lond. t. 50.
virgulàris. P.L.							
13 pùmila. R.S.	(*sc.*)	dwarf.	S. America.	1820.	——	S. ħ.	
14 nìgricans. R.S.	(*wh.re.*)	black-striped.	China.	1819.	6. 9.	G. ♄.	
15 nítida. R.S.	(*wh.sp.*)	glossy.	W. Indies.	1790.	3. 9.	S. ħ.	Andr. reposit. t. 570.
16 bracteolàta. R.S.	(*pu.*)	bracteolate.	Caracas.	1824.	6. 8.	S. ♄.	Jacq. ic. t. 205.
17 pícta. R.S.	(*cr.*)	painted.	E. Indies.	1780.	7. 8.	S. ħ.	Rheed. mal. 6. t. 60.
α álba.	(*cr.*)	*white-marked.*	———	——	——	S. ħ.	Bot. reg. t 1227.
β lùrido-sanguínea.	(*pu*)	*purple-leaved.*	———	1815.	——	S. ħ.	Bot. mag. t. 1870.
18 paniculàta. R.S.	(*ro.*)	panicled.	———	1811.	——	S. ⊙.	Jacq. f. eclog. t. 34.
19 salicìna. R.S.	(*re.*)	Willow-like.	Peru.	1819.	——	S. ♄.	
20 secúnda. R.S.	(*pu.*)	side-flowering.	W. Indies.	1793.	5. 9.	S. ħ.	Bot. mag. t. 2060.
21 geniculàta. B.M.	(*sc.*)	jointed-stalked.	———	1822.	7. 10.	S. ħ.	——— t. 2487.
22 nodòsa. B.M.	(*cr.pu.*)	swoln-jointed.	Brazil.	1826.	8. 10.	S. ħ.	——— t. 2914.
23 oblongàta. L.O.	(*pi.*)	oblong-leaved.	Java.	——	9. 10.	S. ħ.	
24 ciliàris. R.S.	(*wh.*)	ciliated.	Cumana.	1780.	7. 8.	S. ⊙.	Jacq. vind. 2. t. 104.
25 lùcida. R.S.	(*sc.*)	shining-leaved.	W. Indies.	1795.	——	S. ħ.	Bot. mag. t. 1014.
26 salviæflòra. R.S.	(*sc.*)	Sage-flowered.	Mexico.	1823.	——	S. ♄.	
27 Gandarússa. R.S.	(*vi.*)	Willow-leaved.	E. Indies.	1800.	6. 7.	S. ħ.	Bot. reg. t. 635.
28 carthaginénsis. R.S.	(*sc.*)	Carthagenian.	Carthagena.	1792.	——	S. ħ.	——— t. 797.
29 pedunculòsa. R.S.	(*pu.*)	long-peduncled.	N. America.	1759.	7. 8.	H. ♃.	Bot. mag. t. 2367.
americàna. W.							
30 quadrangulàris. B.M.	(*pu.*)	square-stalked.	Mauritius.	——	1. 3.	S. ħ.	Bot. mag. t. 2845.
31 procúmbens. R.S.	(*ro.*)	procumbent.	E. Indies.	1798.	7. 8.	S. ♃.	Pluk. alm. t. 56. f. 3.
32 quinquangulàris. F.I.	(*ro.*)	five-angled.	———	1820.	——	S. ♃.	
33 comàta. R.S.	(*li.*)	Balsam-herb.	Jamaica.	1795.	——	S. ♃.	Sloan. jam. 1. t. 103. f. 2.
34 eustachiàna. R.S.	(*pu.*)	Eustachian.	St. Eustachia.	1799.	8. 9.	S. ħ.	Jacq. amer. pict. t. 5.
35 Keriàna.	(*sc.or.*)	Ker's.	China.	1815.	6. 9.	S. ħ.	Bot. reg. t. 309.
eustachiàna. B.R. Bot. mag. t. 2076. *non* Jacq.							
36 nasùta. R.S.	(*wh.*)	white-flowering.	E. Indies.	1790.	2. 10.	S. ħ.	Bot. mag. t. 325.
37 lanceolària. F.I.	(*ro.*)	lance-leaved.	———	1818.	8. 12.	S. ♄.	
38 speciòsa. F.I.	(*pu.*)	beautiful.	———	1823.	1. 12.	S. ♃.	Bot. mag. t. 2722.
39 Roxburghiàna. R.S.	(*pi.*)	Roxburgh's.	———	1815.	6. 9.	S. ♄.	Rumph. amb. 6. t. 22. f. 1.
tinctòria. F.I. *non* Lour. *báphica.* S.S.							
40 pectoràlis. R.S.	(*pu.*)	Garden Balsam.	W. Indies.	1787.	5. 6.	S. ♃.	Bot. reg. t. 796.
41 periplocæfòlia. R.S.	(*pu.*)	Periploca-leav'd.	S. America.	1799.	6. 8.	S. ♄.	Jacq. coll. s. t. 7. f. 2.
42 furcàta. R.S.	(*vi.*)	forked.	Peru.	1795.	4. 8.	S. ♄.	Bot. mag. t. 430.
peruviàna. B.M. Cavan. ic. 1. t. 28.							
43 lithospermifòlia.	(*vi.*)	Gromwell-leav'd.	———	1796.	——	S. ♄.	Jacq. schœn. 1. t. 4.
44 caracasàna. R.S.	(*vi.*)	Caracas.	Caracas.	1818.	——	S. ♄.	Jacq. ic. 2. t. 206.
45 leucántha. R.S.	(*wh.*)	dirty white.	Brazil.	1824.	——	S. ♄.	
46 Adhátoda. R.S.	(*wh.pu.*)	Malabar-nut.	Ceylon.	1699.	5. 7.	G. ♄.	Bot. mag. t. 861.
47 decussàta. F.I.	(*li.*)	cross-leaved.	E. Indies.	1824.	2. 4.	S. ♄.	
48 Betónica. R.S.	(*wh.*)	Betony-leaved.	———	1739.	——	S. ♄.	Rheed. mal. 2. t. 21.
49 ramosíssima. F.I.	(*wh.*)	branching.	———	1812.	8. 12.	S. ħ.	
50 plumbaginifòlia. R.S.	(*re.*)	Plumbago-leav'd.	......	1819.	6. 9.	S. ♄.	Jacq. f. ecl. t. 12.
51 formòsa. R.S.	(*pu.*)	handsome.	....	——	——	S. ♄.	
52 maculàta. B.C.	(*pu.*)	spotted-leaved.	W. Indies.	1823.	——	S. ħ.	Lodd. bot. cab. t. 626.
53 divaricàta. R.S.	(*wh.*)	divaricate.	C. B. S.	1822.	3. 6.	G. ♄.	
54 pátula. R.S.	(*wh.*)	spreading-branch'd.	——	1824.	——	G. ħ.	
55 hyssopifòlia. R.S.	(*st.*)	Hyssop-leaved.	Canaries.	1690.	3. 8.	G. ħ.	Mill. ic. 9. t. 13.
56 orchioídes. R.S.	(*bh.*)	Broom-leaved.	C. B. S.	1774.	8	G. ♄.	Vent. malm. t. 51.

57 tranquebariénsis. R.S.	(bh.)	Tranquebar.	E. Indies.	1818.	4. 8.	S. ♃.	
58 glábra. F.I.	(ro.ye.)	smooth.	———	1824.	6. 9.	S. ♃.	
59 vestìta. R.S.	(vi.)	clothed.	———	1827.	——	S. ♃.	
DICLI'PTERA. J.		Dicli'ptera.	Diandria Monogynia. R.S.				
1 spinòsa. B.C.	(ye.)	spiny.	Madagascar.	1827.	4. 6.	S. ♄.	Lodd. bot. cab. t. 1244.
2 chinénsis. R.S.	(ro.)	Chinese.	China.	1816.	8. 12.	S. ♄.	Burm. ind. t. 4. 1. 1. ?
3 biválvis. R.S.	(ro.)	two-valved.	E. Indies.	1820.	....	S. ♄.	
4 multiflòra. R.S.	(vi.)	many-flowered.	Quito.	1821.	6. 9.	S. ♄.	Flor. peruv. 1. t. 14. f. b.
5 retùsa. R.S.	(pu.)	retuse.	Santa Cruz.	1816.	7. 9.	S. ♃.	Lodd. bot. cab. t. 724.
6 congésta. K.S.	(pu.)	crowded-flowered.	Mexico.	1825.	——	S. ♃.	
7 resupinàta. R.S.	(vi.)	resupinate.	N. Spain.	1805.	5. 8.	S. ♃.	Cavan. ic. 3. t. 203.
Justícia sexangulàris. Cav. *non* L.							
8 pectinàta. R.S.	(bl.)	pectinated.	E. Indies.	1801.	7. 9.	S. ♃.	Roxb. corom. 2. t. 153.
Justícia parviflòra. W.							
9 rèpens. R.S.	(li.)	creeping.	———	1822.	——	S. ♃.	Roxb. corom. 2. t. 152.
10 brasiliénsis. R.S.	(pu.)	Brazilian.	Brazil.	1824.	——	S. ♃.	
11 scorpioídes. R.S.	(pu.)	Scorpion-grass-ld.	Vera Cruz.	....	——	S. ♄.	Reliqu. Houst. t. 1.
12 sexangulàris. R.S.	(re.)	chickweed-ld.	S. America.	1733.	7.	S. ☉.	Pluk. alm. t. 279. f. 6.
13 assúrgens. R.S.	(sc.)	assurgent.	W. Indies.	1827.	——	S. ♃.	
NELS'ONIA. B.P.		Nels'onia.	Diandria Monogynia.				
lamiifòlia. F.I.	(pu.)	Lamium-leaved.	E. Indies.	1824.	6. 8.	S. ♃.	
HYPOE'STES. R.S.		Hypoe'stes.	Diandria Monogynia. R.S.				
1 purpùrea. R.S.	(pu.)	purple-flower'd.	China.	1822.	6. 8.	S. ♃.	
Justícia purpùrea. L.							
2 verticillàris. R.S.	(bh.)	whorl-flowered.	C. B. S.	1823.	4. 6.	G. ♃.	
3 sérpens. R.S.	(li.)	creeping-stalked.	Mauritius.	1823.	6. 8.	S. ♃.	
4 involucràta. R.S.	(wh.)	involucred.	E. Indies.	1811.	7. 8.	S. ♄.	
ERA'NTHEMUM. R.S.		Era'nthemum.	Diandria Monogynia. R.S.				
1 variábile. B.P.	(pu.)	variable.	N. S. W.	1820.	6. 9.	G. ♄.	
2 montánum. F.I.	(bl.)	mountain.	E. Indies.	1825.	7. 2.	S. ♄.	
3 pulchéllum. F.I.	(bl.)	nervose.	———	1796.	1. 10.	S. ♄.	Andr. reposit. t. 88.
Justícia nervòsa. B.M. 1358.							
4 racemòsum. F.I.	(bh.)	racemed.	———	1826.	7. 2.	S. ♄.	
5 barlerioídes. F.I.	(pa.bl.)	Barleria-like.	———	1824.	——	S. ♄.	
6 stríctum. F.I.	(bl.)	upright.	———	1822.	——	S. ♄.	Bot. reg. t. 867.
7 crenulàtum. B.R.	(li.)	crenulate.	———	1824.	10. 2.	S. ♄.	——— t. 879.
8 bícolor. S.S.	(wh.pu.)	two-coloured.	Philippines.	1802.	5. 8.	S. ♄.	Bot. mag. t. 1423.
Justícia bícolor. B.M.							
9 spinòsum. R.S.	(wh.)	thorny.	W. Indies.	1733.	7. 8.	S. ♄.	Jacq. amer. 2. t. 2. f. 1.
10 acanthóphorum. R.S.	(li.)	spiny.	China.	1822.	——	S. ♄.	
11 ambíguum. R.S.	....	bunch-leaved.	....	1823.	....	S. ♄.	
parviflòrum. L.enum. *nec aliorum.*							

ORDO CLVIII.

OROBANCHEÆ. *Ventenat.*

LATHR'ÆA. S.S.		Toothwort.	Didynamia Angiospermia. L.				
squamária. S.S.	(pu.)	scaly.	Britain.	....	4.	H. ♃.	Eng. bot. v. 1. t. 50.
OROBA'NCHE. S.S.		Broom-rape.	Didynamia Angiospermia. L.				
1 màjor. E.B.	(br.pu.)	greater.	Britain.	....	6. 7.	H. ♃.	Eng. bot. v. 6. t. 421.
2 elàtior. E.B.	(br.)	taller.	———	....	7. 8.	H. ♃.	——— v. 8. t. 568.
3 mìnor. E.B.	(pa.pu.)	smaller.	———	....	——	H. ♃.	——— v. 6. t. 422.
4 rùbra. E.B.	(pu.re.)	red.	Ireland.	....	8.	H. ♃.	——— v. 25. t. 1786.
5 cœrùlea. E.B.	(bl.)	blue.	Britain.	....	7.	H. ♃.	——— v. 6. t. 423.
6 ramòsa. E.B.	(pu.bl.)	branching.	———	....	8. 9.	H. ♃.	——— v. 3. t. 184.

ORDO CLIX.

LENTIBULARIÆ. *Brown prodr.* 1. *p.* 429.

PINGU'ICULA. L.		Butterwort.	Diandria Monogynia. L.				
1 vulgàris. S.S.	(vi.)	common.	Britain.	....	5.	H. w. ♃.	Eng. bot. v. 1. t. 70.
2 alpìna. S.S.	(wh.st.)	Alpine.	Europe.	1794.	4.	H. w. ♃.	Flor. dan. t. 453.

3 grandiflòra. S.S.	(*bl.*)	great-flowered.	Ireland.	....	5. 6.	H.*w*.♃.	Eng. bot. v. 31. t. 2184.
4 lusitánica. S.S.	(*fl.*)	pale.	Britain.	....	6. 7.	H.*w*.♃.	——— v. 3. t. 145.
5 lùtea. B.R.	(*ye.*)	yellow.	Carolina.	1816.	——	H.*w*.♃.	Bot. reg. t. 126.
6 edéntula. H.E.F.	(*ye.*)	toothless.	N. America.	1821.	4. 5.	H.*w*.♃.	Hook. ex. flor. t. 16.
UTRICUL'ARIA. L.		HOODED MILFOIL.	Diandria Monogynia. L.				
1 vulgàris. E.B.	(*ye.*)	common.	Britain.	....	7.	H.♃.	Eng. bot. v. 4. t. 253.
2 intermèdia. E.B.	(*ye.*)	intermediate.	Ireland.	....	——	H.♃.	——— v. 35. t. 2489.
3 minor. E.B.	(*st.*)	lesser.	Britain.	....	——	H.♃.	——— v. 4. t. 254.

ORDO CLX.

PRIMULACEÆ. *Brown prodr. p.* 427.

CY'CLAMEN. L.		CY'CLAMEN.	Pentandria Monogynia. L.				
1 còum. R.S.	(*re.*)	green, round-ld.	S. Europe.	1596.	1. 4.	H.♃.	Bot. mag. t. 4.
2 vérnum. B.F.G.	(*re.*)	spring.	———	1814.	2. 5.	H.♃.	Swt. br. fl. gar. t. 9.
3 europ'æum. R.S.	(*li.*)	round-leaved.	Europe.	1596.	7. 8.	H.♃.	——— t. 176.
4 pérsicum. R.S.	(*wh.*)	Persian.	Cyprus.	1731.	2. 4.	G.♃.	Bot. mag. t. 44.
α inodòrum.	(*wh.pu.*)	*scentless.*	———	——	——	G.♃.	
β odoràtum.	(*wh.pu.*)	*sweet-scented.*	———	——	——	G.♃.	
γ albiflòrum.	(*wh.*)	*pure white.*	———	——	——	G.♃.	
δ laciniàtum. B.R.	(*wh.pu.*)	*jagged.*	———	——	——	G.♃.	Bot. reg. t. 1095.
5 hederæfòlium. E.B.	(*li.*)	Ivy-leaved.	Britain.	....	6. 8.	H.♃.	Eng. bot. v. 8. t. 548.
α purpuráscens.	(*li.*)	*purplish-flower'd.*	———	....	——	H.♃.	
β álbidum.	(*wh.*)	*white-flowered.*	———	....	——	H.♃.	
6 neapolitànum. T.N.	(*re.*)	Neapolitan.	Naples.	1826.	——	H.♃.	Clus. hist. 1. p. 265. ic.
7 repándum. F.G.	(*re.*)	angular-leaved.	Greece.	....	3. 4.	F.♃.	Swt. br. fl. gar. t. 117.
8 latifòlium. F.G.	(*li.*)	broad-leaved.	———	1823.	4. 5.	F.♃.	Flor. græc. t. 185.
DODEC'ATHEON. L.		AMERICAN COWSLIP.	Pentandria Monogynia. L.				
Mèadia. R.S.	(*li.*)	Mead's.	Virginia.	1744.	4. 6.	H.♃.	Bot. mag. t. 12.
β albiflòra.	(*wh.*)	*white-flowered.*	———	1820.	——	H.♃.	
SOLDANE'LLA. R.S.		SOLDANE'LLA.	Pentandria Monogynia. L.				
1 montàna. R.S.	(*bl.*)	mountain.	Bohemia.	1816.	4.	H.♃.	Swt. br. fl. gar. t. 11.
2 alpìna. R.S.	(*bl.*)	Alpine.	Switzerland.	1656.	——	H.♃.	Bot. mag. t. 49.
3 púsilla. B.F.G.	(*vi.*)	lesser.	S. Europe.	1824.	——	H.♃.	Swt. br. fl. gar. s.2. t.48.
4 mínima. R.S.	(*bl.*)	least.	Carp. mount.	1820.	——	H.♃.	——— s.2. t.53.
CORT'USA. L.		BEARS-EAR SANICLE.	Pentandria Monogynia. L.				
Mathìoli. R.S.	(*li.*)	short calyxed.	Austria.	1596.	4. 6.	H.♃.	Bot. mag. t. 987.
PRI'MULA. L.		PRIMROSE.	Pentandria Monogynia. L.				
1 prænìtens. B.R.	(*pi.*)	glossy.	China.	1820.	1. 12.	F.♃.	Bot. reg. t. 539.
sinénsis. Lind. col. *non* Lour.							
α ròsea.	(*ro.*)	*rose-coloured.*	———	——	——	F.♃.	Bot. mag. t. 2564.
β albiflòra. B.F.G.	(*wh.*)	*white-flowered.*	———	——	——	F.♃.	Swt. br. fl. gar. t. 196.
γ dentiflòra.	(*ro.*)	*jagged-flowered.*	———	——	——	F.♃.	Lindl. coll. bot. t. 7.
2 cortusoídes. R.S.	(*re.*)	Cortusa-leaved.	Siberia.	1794.	5. 7.	H.♃.	Bot. mag. t. 399.
3 dentiflòra. A.R.	(*re.*)	tooth-flowered.	———	....	——	H.♃.	Andr. reposit. t. 451.
4 suavèolens. R.S.	(*ye.*)	sweet-scented.	Naples.	1826.	——	H.♃.	Ten. fl. neap. 1. t. 13.
Colúmnæ. T.N.							
5 inflàta. R.S.	(*ye.*)	bellied-calyxed.	Hungary.	1824.	——	H.♃.	Swt. br. fl. gar. s. 2.
6 brevist'yla. R.S.	(*st.*)	short-styled.	France.	....	——	H.♃.	
β versícolor.	(*va.*)	*various-coloured.*	———	....	——	H.♃.	
7 vèris. R.S.	(*ye.*)	Cowslip.	Britain.	....	5. 6.	H.♃.	Eng. bot. v. 1. t. 5.
β rùbra.	(*re.*)	*red cowslip.*	———	....	——	H.♃.	
8 acaùlis. R.S.	(*st.*)	common.	Britain.	....	3. 5.	H.♃.	Eng. bot. v. 1. t. 4.
vulgàris. E.B.							
β álbo-plèna.	(*wh.*)	*double white.*	....	....	——	H.♃.	
γ sulphùreo-plèna.	(*st.*)	*double-sulphur.*	....	....	——	H.♃.	
δ cùpreo-plèna.	(*co.*)	*double-copper.*	....	....	——	H.♃.	
ε salmòneo-plèna.	(*pa.*)	*double-salmon.*	....	....	——	H.♃.	
ζ lilacìno-plèna.	(*li.*)	*double-lilac.*	....	....	——	H.♃.	Bot. mag. t. 229.
η rùbro-plèna.	(*re.*)	*double Scotch.*	....	....	——	H.♃.	
ϑ purpùreo-plèna.	(*d.pu.*)	*double dark purple.*	....	....	——	H.♃.	

9	elàtior. R.S.	(*st.*)	Oxlip.	Britain.	....	3. 5.	H. ♃.	Eng. bot. v. 8. t. 518.
	β polyántha.	(*va.*)	*Polyanthus.*	....	....	——	H. ♃.	
	γ Fletchèri.	(*ve.ye.*)	*Fletcher's defiance.*	....	....	——	H. ♃.	Swt. flor. guid. t. 49.
	δ Burnardiàna.	(*ve.ye.*)	*Burnard's Formosa*	.....	....	——	H. ♃.	—————— t. 97.
	ε Shadiàna.	(*pu.ye.*)	*Shad's telegraph.*	....	....	——	H. ♃.	—————— t. 125.
	ζ calycántha.	(*va.*)	*coloured-calyxed.*	....	....	——	H. ♃.	
10	Perreiniàna. R.S.	(*pu.*)	Perrein's.	Spain.	....	——	H. ♃.	Flugg. an. mus. 12. t. 37.
	Fluggeàna. Lehm. mon. prim. p. 36. t. 2.							
11	Pallàsii. R.S.	(*st.*)	Pallas's.	Altay.	1820.	——	H. ♃.	Bot. reg. t. 896.
12	Aurícula. R.S.	(*va.*)	Auricula.	Switzerland.	1596.	4. 5.	H. ♃.	Jacq. aust. 5. t. 415.
	β lùtea.	(*ye.*)	*yellow-flowered.*	————	———	——	H. ♃.	
	γ calycántha.	(*ye.*)	*yellow-calyxed.*	————	....	——	H. ♃.	
	δ integérrima.	(*va.*)	*entire-leaved.*	————	....	——	H. ♃.	Trattin. tabul. t. 431.
	ε horténsis.	(*va.*)	*garden.*	....	....	——	H. ♃.	—————— t. 432.
	ζ venústa.	(*pu.*)	*Howe's Venus.*	....	....	——	H. ♃.	Swt. flor. guid. t. 84.
	η Redmàni.	(*bl.*)	*Redman's metropolitan.*	..	....	——	H. ♃.	—————— t. 69.
	θ melanoc'ycla.	(*bk.wh.*)	*Wild's black and clear*	....	....	——	H. ♃.	—————— t. 63.
	ι albifòlia.	(*pu.st.*)	*Taylor's glory.*	....	....	——	H. ♃.	—————— t. 3.
	κ Hedgeàna.	(*pu.st.*)	*Hedge's Britannia.*	....	....	——	H. ♃.	—————— t. 22.
	λ dilécta.	(*pu.wh.*)	*Wood's delight.*	....	....	——	H. ♃.	—————— t. 131.
	μ Warrisiàna.	(*ve.gr.*)	*Warris's union.*	....	....	——	H. ♃.	—————— t. 134.
	ν Moreàna.	(*pu.gr.*)	*More's Navarino.*	....	....	——	H. ♃.	—————— t. 58.
	ο Quiragiàna.	(*pu gr.*)	*Burnard's Gen. Quiraga.*		....	——	H. ♃.	—————— t. 129.
	π Hilliàna.	(*pu gr.*)	*Page's Lord Hill.*	......	....	——	H. ♃.	—————— t. 118.
	ρ violàcea.	(*vi.gr.*)	*Moore's Violet.*	....	....	——	H. ♃.	—————— t. 105.
	σ Oldenbúrghii.	(*ve.gr.*)	*Page's Dutchess of Oldenburgh*		....	——	H. ♃.	—————— t. 2.
	τ Alexándri.	(*ve.gr.*)	*Stretch's Alexander.*	....	....	——	H. ♃.	—————— t. 10.
	υ undulàta.	(*ve.gr.*)	*Page's champion.*	......	....	——	H. ♃.	—————— t. 15.
	φ Grìmesii.	(*ve.gr.*)	*Grimes's privateer.*	......	....	——	H. ♃.	—————— t. 29.
	χ Lawrieàna.	(*ve.gr.*)	*Lawrie's Hertfordshire hero.*		....	——	H. ♃.	—————— t. 94.
	ψ indepéndens.	(*bk.gr.*)	*Booth's freedom.*	....	....	——	H. ♃.	—————— t. 91.
	ω obscùra.	(*pu.gr.*)	*Cockup's Eclipse.*	....	....	——	H. ♃.	—————— t. 45.
	αα Pollitiàna.	(*ve.gr.*)	*Pollit's highland boy.*	....	....	——	H. ♃.	—————— t. 52.
	ββ Smíthii.	(*ve.gr.*)	*Smith's Waterloo.*	......	....	——	H. ♃.	—————— t. 76.
	γγ gloriòsa.	(*ve.gr.*)	*Lawrie's glory of Cheshunt.*		....	——	H. ♃.	—————— t. 79.
13	Palinùri. R.S.	(*ye.*)	unequal-bracted.	Naples.	1816.	——	H. ♄.	Swt. br. fl. gar. t. 8.
14	Balbísii. R.S.	(*ye.*)	Balbis's.	Italy.	1826.	——	H. ♃.	Moret. fl. vicent. p. 7. ic.
15	marginàta. B.M.	(*ro.*)	silver-edged.	Switzerland.	1777.	3. 4.	H. ♃.	Bot. mag. t. 191.
	crenàta. R.S.							
16	longiflòra. R.S.	(*li.*)	long-flowered.	————	1818.	4. 6.	H. ♃.	Jacq. aust. 5. t. 46.
17	longifòlia. B.M.	(*li.*)	long-leaved.	Levant.	1790.	4. 5.	H. ♃.	Bot. mag. t. 392.
	auriculàta. R.S.							
18	farinòsa. R.S.	(*re.*)	Bird's-eye.	Britain.	....	4. 7.	H. ♃.	Eng. bot. v. 1. t. 6.
19	scótica. S.S.	(*pu.*)	Scotch bird's-eye.	Scotland.	....	7.	H. ♃.	Hook. fl. lond. t. 133.
20	strícta. R.S.	(*li.*)	Horneman's.	N. Europe.	1822.	6. 7.	H. ♃.	Flor. dan. t. 1385.
	Hornemanniàna. S.S.							
21	davùrica. R.S.	(*ro.*)	Siberian bird's-eye.	Davuria.	1806.	5. 6.	H. ♃.	Bot. mag. t. 1219.
	intermèdia. B.M.							
22	pusílla. S.S.	(*li.*)	American bird's-eye.	N. America.	1819.	——	H. ♃.	Hook. ex. flor. t. 68.
23	altàica. R.S.	(*li.*)	pale-flowered.	Altay.	1826.	3. 5.	H. ♃.	Lehm. prim. t. 5.
24	finmárchica. R.S.	(*vi.*)	Norwegian.	Norway.	1798.	——	H. ♃.	Flor. dan. t. 188.
25	nivàlis. R.S.	(*pu.*)	snowy mountain.	Davuria.	1790.	4. 5.	H. ♃.	Pall. it. 3. app. t. 9. f. 2.
26	viscòsa. R.S.	(*pu.*)	clammy.	Piedmont.	1792.	4.	H. ♃.	Trattin. tabul. t. 434.
27	glutinòsa. R.S.	(*vi.*)	glutinous.	S. Europe.	1820.	3. 6.	H. ♃.	Jacq. aust. t. 26.
28	carniòlica. R.S.	(*vi.*)	elliptic-leaved.	Carniola.	1826.	——	H. ♃.	———— 5. t. 4.
29	glaucéscens. S.S.	(*pu.*)	glaucescent.	Switzerland.	——	——	H. ♃.	Swt. br. fl. gar. t. 254.
30	integrifòlia. R.S.	(*pu.*)	entire-leaved.	Pyrenees.	——	6. 7.	H. ♃.	Bot. mag. t. 942.
31	Símsii.	(*st.*)	Sims's.	Switzerland.	....	4. 6.	H. ♃.	———— t. 1161.
	villòsa β flòre álbo. B.M.							
32	nívea. F.	(*wh.*)	snowy white.	Siberia.	....	——	H. ♃.	
	nivàlis. Hort. *nec aliorum.*							
33	helvética. B.C.	(*pu.*)	Swiss.	Switzerland.	1790.	——	H. ♃.	Lodd. bot. cab. t. 348.
34	pubéscens. R.S.	(*pu.*)	pubescent.	Pyrenees.	1768.	——	H. ♃.	Trattin. tabul. t. 427.
	villòsa. B.M. 14. *non* Jacq.							
35	villòsa. R.S.	(*pu.*)	villous.	S. Europe.	——	——	H. ♃.	Jacq. aust. 5. t. 27.
36	ciliàta. R.S.	(*pu.*)	ciliated.	Switzerland.	....	——	H. ♃.	Swt. br. fl. gar. s. 2.
	decòra. Bot. mag. t. 1922.							
37	mínima. R.S.	(*pu.*)	smallest.	————	1819.	4. 5.	H. ♃.	Bot. reg. t. 581.

38 involucràta. L.O.	(*ye.*)	involucred.	Egypt.	1826.	3. 4.	F. ♃.	Bot. mag. t. 2842.
verticillàta. B.M. *non* Vahl. *nec* Forsk.							
ANDROS'ACE. L.		ANDROS'ACE.	Pentandria Monogynia. L.				
1 máxima. R.S.	(*wh.*)	oval-leaved.	Austria.	1597.	3. 6.	H. ⊙.	Jacq. aust. 4. t. 331.
2 elongàta. R.S.	(*wh.*)	cluster-flowered.	———	1776.	4. 5.	H. ⊙.	——— 4. t. 330.
3 nàna. R.S.	(*wh.*)	dwarf.	S. Europe.	1825.	——	H. ⊙.	
4 septentrionàlis. R.S.	(*wh.*)	tooth-leaved.	Russia.	1755.	——	H. ⊙.	Bot. mag. t. 2021.
5 alismoídes. H.H.	(*wh.*)	Alisma-like.	Siberia.	1827.	——	H. ⊙.	
6 coronopifòlia. A.R.	(*wh.*)	Buckshorn-ld.	———	1806.	6. 9.	H. ♂.	Andr. reposit. t. 647.
lactiflòra. R.S.							
7 filifórmis. R.S.	(*wh.*)	filiform.	———	1823.	——	H. ⊙.	Gmel. sib. 4. t. 44. f. 4.
8 villòsa. R.S.	(*bh.*)	villous.	Pyrenees.	1790.	6. 7.	H. ♃.	Bot. mag. t. 743.
9 chamæjásme. R.S.	(*bh.*)	grass-leaved.	Austria.	1768.	6. 8.	H. ♃.	Lodd. bot. cab. t. 232.
10 brevifòlia. R.S.	(*wh.*)	short-leaved.	Siberia.	1827.	——	H. ♂.	Vill. delph. 2. t. 15.
11 macrocárpa. Led.	(*wh.*)	large-capsuled.	———	——	——	H. ⊙.	
12 láctea. R.S.	(*wh.*)	white-flowered.	———	1752.	——	H. ♃.	Bot. mag. t. 868.
13 cárnea. R.S.	(*ro.*)	awl-leaved.	Switzerland.	1768.	7. 8.	H. ♃.	Lodd. bot. cab. t. 40.
14 obtusifòlia. R.S.	(*bh.*)	blunt-leaved.	———	1820.	——	H. ♃.	All. ped. 1. t. 46. f. 1.
15 albàna. R.S.	(*bh.*)	capitate.	Caucasus.	1827.	——	H. ♃.	Linn. trans. 11. t. 33.
16 acaùlis. Ot.	(*wh.*)	stalkless.	Siberia.	1825.	——	H. ♂.	
AR'ETIA. W.		AR'ETIA.	Pentandria Monogynia. L.				
1 helvética. W.	(*st.*)	imbricated.	Switzerland.	1775.	5. 6.	H. ♃.	Schk. hand. 1. t. 32.
2 argéntea.	(*st.*)	silvery.	———	1826.	——	H. ♃.	Lap. pyr. ic. ined. t. 67.
Androsàce argéntea. R.S.							
3 alpìna. W.	(*pi.*)	linear-leaved.	———	1775.	——	H. ♃.	Jacq. aust. 5. app. t. 18.
4 pubéscens. L.C.	(*pi.*)	pubescent.	———	1820.	——	H. ♃.	Lodd. bot. cab. t. 1273.
5 Vitaliàna. W.	(*ye.*)	grass-leaved.	Pyrenees.	1787.	——	H. ♃.	——— t. 166.
Prímula sedifòlia. P.L. 107. *Androsàce Vitaliàna.* R.S.							
TRIENT'ALIS. W.		WINTER-GREEN.	Heptandria Monogynia. L.				
1 europ'æa. W.	(*wh.*)	oval-leaved.	Britain.	....	5. 6.	H. ♃.	Eng. bot. v. 1. t. 15.
2 americàna. Ph.	(*wh.*)	spear-leaved.	N. America.	1816.	7. 8.	H. ♃.	Bart. flor. amer. 2. t. 48.
C'ORIS. L.		C'ORIS.	Pentandria Monogynia. L.				
monspeliénsis. L.	(*vi.*)	Montpelier.	S. Europe.	1640.	6. 7.	G. ♂.	Bot. reg. t. 536.
HOTT'ONIA. L.		WATER-VIOLET.	Pentandria Monogynia. L.				
palústris. E.B.	(*pi.*)	marsh.	England.	....	7. 8.	H. *w.* ♃.	Eng. bot. v. 6. t. 364.
LYSIMA'CHIA. L.		LOOSE-STRIFE.	Pentandria Monogynia. L.				
1 vulgàris. R.S.	(*ye.*)	common.	Britain.	....	7. 9.	H. ♃.	Eng. bot. v. 11. t. 761.
2 Ephémerum. R.S.	(*wh.*)	Willow-leaved.	Spain.	1730.	——	H. ♃.	Bot. mag. t. 2346.
3 atropurpùrea. R.S.	(*d.pu.*)	dark purple.	Levant.	....	——	H. ⊙.	Flor. græc. t. 187.
4 dùbia. R.S.	(*li.*)	pale purple.	———	1759.	——	H. ♂.	——— t. 188.
5 strícta. R.S.	(*ye.*)	upright.	N. America.	1781.	7. 8.	H. ♃.	Bot. mag. t. 104.
bulbífera. B.M. *racemòsa.* M.							
6 capitàta. R.S.	(*ye.*)	headed.	———	1813.	5. 7.	H. ♃.	
7 thyrsiflòra. R.S.	(*ye.*)	tufted.	England.	....	——	H. ♃.	Eng. bot. v. 3. t. 176.
8 quadrifòlia. R.S.	(*ye.*)	four-leaved.	N. America.	1798.	7. 8.	H. ♃.	Lam. ill. 1. t. 101. f. 2.
hirsùta. M.							
9 longifòlia. R.S.	(*ye.*)	long-leaved.	———	——	——	H. ♃.	Bot. mag. t. 660.
quadriflòra. B.M.							
10 lanceolàta. R.S.	(*ye.*)	spear-leaved.	———	1826.	——	H. ♃.	
11 verticillàta. R.S.	(*ye.*)	whorl-leaved.	Hungary.	1818.	6. 8.	H. ♃.	Swt. br. fl. gar. t. 21.
12 ciliàta. R.S.	(*ye.*)	ciliated.	N. America.	1732.	——	H. ♃.	Wachend. hort. t. 12.
13 punctàta. R.S.	(*ye.*)	punctated.	S. Europe.	1658.	7. 8.	H. ♃.	Jacq. aust. 4. t. 366.
14 angustifòlia. R.S.	(*ye.*)	narrow-leaved.	N. America.	1803.	7. 9.	H. ♃.	
15 hy'brida. R.S.	(*ye.*)	hybrid.	———	1806.	——	H. ♃.	
16 nemòrum. R.S.	(*ye.*)	wood.	Britain.	....	5. 7.	H. ♃.	Eng. bot. v. 8. t. 527.
17 anagalloídes. F.G.	(*ye.*)	Pimpernell-fl'd.	Crete.	1789.	——	F. ♃.	Flor. græc. t. 198.
18 maculàta. B.P.	(*ye.*)	spotted.	N. S. W.	1822.	——	G. ♃.	
19 Nummulària. R.S.	(*ye.*)	Moneywort.	Britain.	....	6. 7.	H. ♃.	Eng. bot. v. 8. t. 528.
LUBI'NIA. S.S.		LUBI'NIA.	Pentandria Monogynia. S.S.				
atropurpùrea. S.S.	(*pu.*)	dark-purple.	C. B. S.	1823.	7. 9.	H. ♃.	Swt. br. fl. gar. s. 2. t. 34.
ASTEROL'INON. Lk.		ASTEROL'INON.	Pentandria Monogynia. L.en.				
stellàtum. L.en.	(*bh.*)	starry.	Italy.	1658.	6.	H. ⊙.	Flor. græc. t. 189.
Lysimáchia Linum-stellàtum. F.G.							
ANAGA'LLIS. L.		PIMPERNELL.	Pentandria Monogynia. L.				
1 arvénsis. E.B.	(*sc.*)	red-flowered.	Britain.	....	6. 9.	H. ⊙.	Eng. bot. v. 8. t. 529.
phœnícea. R.S.							
2 cárnea. Schl.	(*fl.*)	flesh-coloured.	Switzerland.	——	——	H. ⊙.	
3 índica. B.F.G.	(*bl.*)	Indian.	E. Indies.	1824.	——	H. ⊙.	Swt. br. fl. gar. t. 132.
4 cœrùlea. R.S.	(*bl.*)	blue-flowered.	Britain.	....	——	H. ⊙.	Eng. bot. v. 26. t. 1823.

5 collina. R.S.	(sc.) large-flowered.	Morocco.	1803.	5. 7.	G. ♂.	Bot. mag. t. 831.
fruticòsa. B.M. *grandiflòra.* A.R. 367.						
6 latifòlia. R.S.	(bl.) broad-leaved.	Spain.	1759.	——	G. ♂.	Bot. mag. t. 2389.
7 Monélli. R.S.	(bl.) blue Italian.	Italy.	1648.	5. 9.	G. ♃.	———— t. 319.
8 linifòlia. R.S.	(bl.) Flax-leaved.	Portugal.	1796.	6. 7.	G. ♃.	Moris. s. 5. t. 26. f. 3.
9 tenélla. R.S.	(ro.) bog.	Britain.	....	7. 8.	H. ♃.	Eng. bot. v. 8. t. 530.
CENTU'NCULUS. L.	Bastard Pimpernell.	Tetrandria Monogynia. L.				
mínimus. E.B.	(fl.) least.	Britain.	....	6. 7.	H. ☉.	Eng. bot. v. 8. t. 531.
? SAM'OLUS. B.P.	Brook-weed.	Pentandria Monogynia. L.				
1 Valerándi. E.B.	(wh.) common.	Britain.	....	6. 8.	H. ♃.	Eng. bot. v. 10. t. 703.
2 littoràlis. B.P.	(wh.) sea-side.	N. S. W.	1806.	7. 9.	G. ♃.	Lab. n. holl. 1. t. 54.
3 campanuloídes. S.S.	(w.) Campanula like.	C. B. S.	1816.	——	G. ♃.	
Campánula poròsa. W.						

ORDO CLXI.

GLOBULARIÆ. *Lam. et DC. fl. fr. ed. 3. v. 3. p. 427.*

GLOBUL'ARIA. L.	Globul'aria.	Tetrandria Monogynia. L.				
1 longifòlia. W.	(bl.) Willow-leaved.	Madeira.	1775.	7. 8.	G. ♄.	Bot. reg. t. 685.
salicìna. R.S.						
2 A'lypum. R.S.	(pa.bl.) three-toothed.	S. Europe.	1640.	8. 9.	G. ♄.	Garid. aix. t. 42.
β *integrifòlia.*	(pa.bl.) *entire-leaved.*	———	....	——	G. ♄.	Rœm. et Ust. b. m. 7. t. 5.
3 linifòlia. R.S.	(bl.) Flax-leaved.	Spain.	1818.	5. 6.	H. ♃.	
4 vulgàris. R.S.	(bl.) common.	Europe.	1640.	——	H. ♃.	Swt. br. fl. gar. t. 20.
5 spinòsa. R.S.	(bl.) prickly-leaved.	Spain.	——	——	F. ♃.	
6 cordifòlia. R.S.	(bl.) wedge-leaved.	Germany.	1633.	6. 7.	H. ♃.	Swt. br. fl. gar. t. 34.
7 nàna. R.S.	(li.) dwarf.	S. France.	1818.	——	H. ♃.	DC. ic. pl. gal. rar. 1 t. 3.
8 nudicaùlis. R.S.	(bl.) naked-stalked.	Germany.	1629.	——	H. ♃.	Jacq. aust. 3. t. 230.
9 bellidifòlia. R.S.	(bl.) Daisy-leaved.	Italy.	1824.	——	H. ♃.	
10 incanéscens. R.S.	(pu.) hoary.	———	1828.	——	H. ♃.	Viv. frag. it. 1. t. 3.

**** Subclassis IV. *MONOCHLAMYDEÆ.*

ORDO CLXII.

PLUMBAGINEÆ. *Brown prodr. 1. p. 425.*

STA'TICE. T.	Thrift.	Pentandria Pentagynia. L.				
1 vulgàris.	(pa. li.) common.	Europe.	....	6. 8.	H. ♃.	Schk. handb. t. 87.
Armèria vulgàris. R.S.						
2 arenària. P.S.	(li.) sand.	France.	....	5. 8.	H. ♃.	
3 alpìna.	(vi.) flattish-stalked.	Carinthia.	....	——	H. ♃.	
4 littoràlis.	(vi.) ciliate-leaved.	Portugal.	....	——	H. ♃.	
Armèria littoràlis. R.S.						
5 marítima. Mill.	(ro.) sea-side.	Britain.	....	5. 7.	H. ♃.	Eng. bot. v. 4. t. 226.
Armèria. E.B. *Armèria marítima.* R.S.						
6 hùmilis. P.S.	(li.) dwarf.	Portugal.	....	——	H. ♃.	
7 juniperifòlia. W.	(li.) Juniper-leaved.	Spain.	1818.	——	H. ♃.	Quer. hisp. 6. t. 15. f. 1.
8 pinifòlia. Brot.	(li.) Pine-leaved.	Portugal.	....	——	H. ♃.	
9 púngens. Brot.	(li.) pungent-leav'd.	———	....	——	H. ♃.	
10 alliàcea. W.	(wh.) Garlic-leaved.	Spain.	1798.	5. 6.	H. ♃.	Flor. græc. t. 29.
11 scorzonerifòlia.	(li.) spear-leaved.	S. Europe.	1816.	——	H. ♃.	
Armèria scorzonerifòlia. S.S.						

12	lusitánica. P.S.	(sc.)	large-leaved.	S. Europe.	1814.	5. 6.	F. ♃.	
13	latifòlia.	(li.)	broad-leaved.	Portugal.	1740.	5. 7.	H. ♃.	Jacq. vind. 1. t. 42.
	cephalòtes. H.K. *Armèria latifòlia.* R.S.							
14	plantagínea.	(li.)	Plantain-leav'd.	S. Europe.	1816.	5. 6.	H. ♃.	
15	denticulàta. Brot.	(fl.)	tooth-leaved.	Italy.	——	——	H. ♃.	
16	dianthoídes. B.F.G.	(li.)	Pink-like.	S. Europe.	1824.	5. 7.	H. ♃.	Swt. br. fl. gar. ser. 2.
	Armèria dianthoìdes. S.S.							
17	hírta.	(li.)	hairy.	Gibraltar.	1818.	——	H. ♃.	
	Armèria hírta. R.S.							
18	fasciculàta. P.S.	(li.)	bundle-leaved.	Portugal.	....	4. 8.	H. ♄.	Venten. cels. t. 38.
	TAXA'NTHEMA. B.P.		SEA LAVENDER.	Pentandria Pentagynia.				
1	graminifòlia. R.S.	(sc.)	Grass-leaved.	Siberia.	1780.	6. 7.	H. ♃.	
2	Limònium. R.S.	(bl.)	common.	England.	....	5. 8.	H. ♃.	Eng. bot. v. 2. t. 102.
	Státice Limònium. E.B.							
3	elàta. S.S.	(bl.)	tall.	Siberia.	1824.	——	H. ♃.	
4	Gmelini. R.S.	(bl.)	Gmelin's.	——	1796.	6. 8.	H. ♃.	Gmel. sib. 2. t. 90.
5	caroliniàna. R.S.	(bl.)	Carolina.	N. America.	1824.	——	H. ♃.	
6	scopària. R.S.	(bl.)	Broom.	Siberia.	1796.	——	H. ♃.	
7	latifòlia. R.S.	(bl.)	broad-leaved.	——	1791.	5. 7.	H. ♃.	Hoff. c. gott. 1796. t. 1.
8	purpuràta. R.S.	(pu.)	purple.	C. B. S.	1800.	6. 7.	G. ♃.	
9	bellidifòlia. R.S.	(bl.)	Daisy-leaved.	Greece.	1823.	——	G. ♃.	Flor. græc. t. 295.
10	globulariæfòlia. R.S.	(li.)	Globularia-leav'd.	S. Europe.	——	——	F. ♃.	—— t. 296.
11	oleæfòlia. R.S.	(vi.)	Olive-leaved.	Italy.	1688.	5. 8.	H. ♃.	Lobel. ic. p. 295. f. 2.
12	Willdenowiàna. R.S.	(vi.)	Willdenow's.	S. France.	1804.	6. 8.	H. ♃.	Willd. h. ber. t. 63.
	Státice spathulàta. W. *non* Desf.							
13	auriculæfòlia. R.S.	(bl.)	Auricula-leav'd.	Barbary.	1781.	7. 8.	F. ♃.	
14	incàna. B.F.G.	(re.)	hoary.	Levant.	1823.	——	H. ♃.	Swt. br. fl. gar. t. 272.
	α sanguínea.	(cr.)	crimson-flowered.	——	——	——	H. ♃.	—— t. 272. f. 1.
	β ròsea.	(ro.)	rose-coloured.	——	——	——	H. ♃.	—— t. 272. f. 2.
	γ álba.	(wh.)	white-flowered.	——	——	——	H. ♃.	
15	spathulàta. R.S.	(bl.)	spatula-leaved.	Barbary.	1804.	6. 8.	F. ♃.	Bot. mag. t. 1617.
16	cordàta. R.S.	(bl.)	retuse-leaved.	S. Europe.	1752.	5. 7.	F. ♃.	Barrel. ic. 805.
17	emarginàta. R.S.	(bl.)	emarginate.	Gibraltar.	....	——	F. ♃.	
18	cáspia R.S.	(pa. bl.)	Caspian.	Siberia.	....	7. 8.	H. ♃.	Gmel. sib. 2. t. 89. f. 2.
19	scábra. R.S.	(bl.)	rough-branch'd.	C. B. S.	1788.	5. 7.	G. ♃.	
20	tetragòna. R.S.	(bl.)	square-stalked.	——	1816.	——	G. ♃.	
21	reticulàta. R.S.	(bl.)	matted.	England.	....	7. 8.	H. ♃.	Eng. bot. v. 5. t. 328.
	Státice reticulàta. E.B.							
22	dichótoma. R.S.	(bl.)	forked.	Spain.	1818.	——	H. ♃.	Cavan. ic. 1. t. 50.
23	virgàta. R.S.	(bl.)	twiggy.	——	....	——	H. ♃.	
24	vimínea. R.S.	(bl.)	slender.	....	1823.	——	H. ♃.	
25	echioídes. F.G.	(li.)	rough-leaved.	S. Europe.	1752.	——	F. ☉.	Flor. græc. t. 299.
26	ròrida. F.G.	(pa. bl.)	frosted.	Levant.	——	——	F. ♃.	—— t. 298.
	echioìdes. Prodr. fl. græc. *non* L.							
27	palmàris. F.G.	(ro.)	rosy-coloured.	——	1826.	——	F. ♃.	Flor. græc. t. 297.
28	speciòsa. B.F.G.	(re.)	Plantain-leaved.	Russia.	1776.	——	H. ♃.	Swt. br. fl. gar. t. 105.
	Státice speciòsa. Bot. mag. 656.							
29	conspícua. R.S.	(re.)	conspicuous.	——	1804.	——	H. ♃.	Bot. mag. t. 1629.
	Státice conspícua. B.M.							
30	tatárica. B.F.G.	(re.)	Tartarian.	——	1731.	6. 7.	H. ♃.	Swt. br. fl. gar. t. 37.
31	aceròsa. M.B.	(re.)	needle-leaved.	M. Ararat.	1829.	——	F. ♃.	Buxb. cent. 2. p. 18. t. 10.
32	flexuòsa. R.S.	(ro.)	zigzag.	Siberia.	1791.	7. 8.	H. ♃.	Gmel. sib. 2. t. 89. f. 1 ?
33	minùta. R.S.	(li.)	small.	S. Europe.	1658.	6. 7.	H. ♃.	Bocc. sic. t. 13. f. 3.
34	pectinàta. R.S.	(bl.)	triangular-stalk'd.	Canaries.	1780.	9. 10.	F. ♃.	
35	cinèrea. R.S.	(wh.)	ash-coloured.	C. B. S.	1822.	——	G. ♃.	
36	macrophy'lla. S.S.	(bl. w.)	large-leaved.	Canaries.	1816.	7. 9.	F. ♃.	
37	suffruticòsa. R.S.	(li.)	narrow-leaved.	Siberia.	1779.	5. 9.	H. ♄.	Gmel. sib. 2. t. 88. f. 2. 3.
38	monopètala. R.S.	(pu.)	Sicilian shrubby.	Sicily.	1731.	7. 8.	F. ♄.	Bocc. sic. t. 16. 17.
39	cylindrifòlia. R.S.	(vi.)	round-leaved.	N. Africa.	1738.	——	G. ♄.	
40	ferulàcea. R.S.	(ye.)	Fennel-like.	Siberia.	1796.	5.	H. ♄.	Moris. s. 15. t. 1. f. 23.
41	sinuàta. R.S.	(bl. ye.)	scollop-leaved.	Levant.	1629.	5. 9.	F. ♃.	Bot. mag. t. 71.
42	Thouìni. R.S.	(bl. ye.)	Thouin's.	Africa.	1738.	6. 10.	F. ♃.	Mart. cent. t. 84.
43	alàta. R.S.	(ye.)	winged.	....	1806.	——	F. ♃.	
44	spicàta. R.S.	(wh.)	spiked.	Caucasus.	1828.	8. 10.	H. ☉.	Gmel. sib. 2. t. 91. f. 2.
45	ægyptìaca. R.S.	(wh.)	Egyptian.	Egypt.	1816.	6. 10.	F. ♃.	Bot. mag. t. 2363.
46	mucronàta. R.S.	(li.)	curled.	Barbary.	1784.	——	F. ♃.	L'Herit. stirp. t. 13.
	PLUMB'AGO. L.		LEADWORT.	Pentandria Monogynia. L.				
1	europ'æa. R.S.	(vi.)	European.	S. Europe.	1596.	9. 10.	H. ♃.	Bot. mag. t. 2139.
2	lapathifòlia. R.S.	(wh.)	Dock-leaved.	Iberia.	1824.	——	H. ♃.	

3	capénsis. R.S.	(bl.)	blue-flowered.	C. B. S.	1818.	9. 3.	S. ♄.	Bot. reg. t. 417.
4	trístis. H.K.	(br.)	dark-flowered.	———	1792.	5. 6.	G. ♄.	
5	zeylánica. R.S.	(wh.)	Ceylon.	E. Indies.	1731.	4. 9.	S. ♄.	Rheed. mal. 10. t. 8.
6	occidentàlis.	(wh.)	West Indian.	W. Indies.	1817.	——	S. ♄.	
	zeylánica β. R.S. *scándens.* Hort.							
7	mexicàna. K.S.	(vi.)	Mexican.	Mexico.	1829.	4. 8.	S. ♃.	
8	ròsea. R.S.	(ro.)	rose-coloured.	E. Indies.	1777.	3. 7.	S. ♄.	Bot. mag. t. 230.
9	scándens. R.S.	(wh.)	climbing.	W. Indies.	1699.	7. 8.	S. ♄. ⌣.	Jacq. amer. pict. t. 13.
10	rhomboídea. B.M.	(bl.)	rhomboid-leaved	....	1828.	8. 10.	S. ⊙.	Bot. mag. t. 2917.
11	micrántha. F.	(pu.)	small-flowered.	Persia.	1829.	6. 8.	H. ⊙.	

ORDO CLXIII.

PLANTAGINEÆ. *Juss. gen.* 89.

GLA'UX. L.			BLACK SALTWORT.	Pentandria Monogynia. L.				
	marítima. R.S.	(li.)	sea-side.	Britain.	....	5. 6.	H. ♃.	Eng. bot. v. 1. t. 13.
PLANTAGO. L.			PLANTAIN.	Tetrandria Monogynia. L.				
1	màjor. R.S.	(gr.)	greater.	Britain.	....	5. 7.	H. ♃.	Eng. bot. v. 22. t. 1558.
	β ròsea.	(gr.)	*rose-bracted.*	———	....	——	H. ♃.	
2	asiática. R.S.	(li.)	broad-leaved.	Siberia.	1787.	7.	H. ♃.	Gmel. sib. 4. t. 37. ?
3	Cornùti. R.S.	(wh.)	Cornuti's.	S. Europe.	1801.	7. 8.	H. ♃.	Corn. canad. p. 163. ic.
	Gouàni. S.S.							
4	Brùtia. R.S.	(wh.)	Calabrian.	Calabria.	1824.	——	H. ♃.	
5	gentianoídes. S.S.	(wh.)	Gentian-like.	Greece.	1826.	5. 8.	H. ♃.	
6	exaltàta. R.S.	(wh.)	tallest.	....	1824.	7. 8.	H. ♃.	
7	cordàta. R.S.	(wh.)	heart-leaved.	N. America.	1818.	——	H. ♃.	Jacq. f. ecl. t. 72.
8	máxima. R.S.	(li.)	hollow-leaved.	Siberia.	1763.	——	H. ♃.	Jacq. ic. 1. t. 26.
9	crássa. S.S.	(wh.)	succulent.	S. Europe.	1793.	6. 7.	H. ♃.	Jacq. col. sup. 34. t. 16.
	críspa. Jacq.							
10	mèdia. R.S.	(li.)	hoary.	Britain.	....	5. 7.	H. ♃.	Eng. bot. v. 22. t. 1559.
11	camtschática. S.S.	(wh.)	Kamtschatkan.	Kamtschatka.	1819	——	H. ♃.	
12	lanceolàta. R.S.	(br.)	Rib-grass.	Britain.	....	——	H. ♃.	Eng. bot. v. 8. t. 507.
13	sibírica. R.S.	(wh.)	Siberian.	Siberia.	1823.	——	H. ♃.	
14	erióphora. R.S.	(wh.)	woolly-leaved.	Portugal.	1825.	——	H. ♃.	
15	montàna. R.S.	(wh.)	mountain.	S. Europe.	1823.	——	H. ♃.	
16	capénsis. R.S.	(wh.)	Cape.	C. B. S.	1788.	5. 8.	G. ♃.	
17	altíssima. R.S.	(wh.)	tall.	Italy.	1774.	6. 7.	H. ♃.	Jacq. obs. 4. t. 83.
18	Schóttii. R.S.	(br.)	Schott's.	Dalmatia.	1829.	——	H. ♃.	
19	Lagòpus. R.S.	(br.)	round-headed.	Spain.	1683.	——	H. ♃.	Flor. græc. t. 144.
20	eriostàchya. R.S.	(wh.)	woolly-spiked.	Naples.	1825.	7. 8.	H. ⊙.	
21	sphærocéphala. R.S.	(w.)	round-headed.	Austria.	1774.	6. 7.	H. ♃.	Jacq. vind. 2. t. 125.
	alpìna. Jacq. *non* L.							
22	lusitánica. R.S.	(st.)	Portugal.	Portugal.	1781.	7. 8.	H. ♃.	Barrel. ic. t. 745.
23	pseùdo-lusitánica. R.S.		five-nerved.	———	1819.	——	H. ♃.	
24	virgínica. R.S.	(gr.)	Virginian.	N. America.	1688.	6. 9.	H. ⊙.	Moris. s. 8. t. 15. f. 8.
25	depréssa. R.S.	(br.)	depressed.	———	1818.	——	H. ♃.	
26	sparsiflòra. R.S.	(bh.)	scattered-flowered.	———	——	——	H. ⊙.	
27	saxátilis. R.S.	(wh.)	rock.	Iberia.	1820.	6. 8.	H. ♃.	
28	hungàrica. R.S.	(br.)	Hungarian.	Hungary.	1824.	——	H. ♃.	W. et K. hung. 3. t. 203.
29	serícea. R.S.	(wh.)	silky.	———	1820.	——	H. ♃.	——— t. 151.
	argéntea. DC.							
30	álbicans. R.S.	(br.)	woolly.	S. Europe.	1776.	6. 9.	H. ♃.	Flor. græc. t. 145.
31	mexicàna. S.S.	(wh.)	Mexican.	Mexico.	1820.	6. 7.	H. ♃.	
32	patagònica. R.S.	(wh.)	Patagonian.	Patagonia.	1793.	6. 9.	H. ⊙.	Jacq. ic. 2. t. 306.
33	victoriàlis. R.S.	(wh.)	Victoria.	S. Europe.	1824.	——	H. ♃.	
34	holoserícea. R.S.	(wh.)	hairy-scaped.	Switzerland.	——	——	H. ♃.	
35	microcéphala. R.S.	(wh.)	small-headed.	....	....	——	H. ⊙.	
36	Bellárdi. R.S.	(br.)	Bellardi's.	S. Europe.	1797.	——	H. ⊙.	Flor. græc. t. 146.
37	crética. R.S.	(st.)	Cretan.	Candia.	1711.	6. 7.	H. ⊙.	——— t. 147.
38	gramínea. R.S.	(wh.)	grass-leaved.	France.	1804.	——	H. ♃.	Dod. pempt. 108. ic.
39	marítima. R.S.	(gr.)	sea-side.	Britain.	....	7.	H. ♃.	Eng. bot. v. 3. t. 175.
40	sálsa. R.S.	(wh.)	saltwort.	Tauria.	1819.	——	H. ♃.	
41	crassifòlia. R.S.	(br.)	thick-leaved.	Egypt.	1824.	——	H. ♃.	
42	hirsùta. R.S.	(wh.)	hairy.	C. B. S.	1801.	6. 7.	G. ♂.	Jacq. schœn. 3. t. 258.

43 tùmida. L.en.	(*wh.*)	tumid-capsuled.	Chile.	1819.	6. 7.	H.⊙.	
44 alpìna. R.S.	(*li.*)	Alpine.	S. Europe.	1774.	——	H.♃.	
45 tenuiflòra. R.S.	(*br.*)	slender-flowered.	Hungary.	1819.	——	H.⊙.	W. et K. hung. 1. t. 39.
46 bidentàta. R.S.	(*wh.*)	white margined.	Switzerland.	1824.	——	H.♃.	
47 serpentìna. R.S.	(*wh.*)	serpentine.	S. France.	1826.	——	H.♃.	
48 recurvàta. R.S.	(*wh.*)	recurved-leaved.	S. Europe.	1799.	——	H.♂.	Murr. c. gott. 1780. t. 6.
49 Wulfèni. R.S.	(*wh.*)	Wulfen's.	Carinthia.	1802.	——	H.♃.	
50 subulàta. R.S.	(*wh.*)	awl-leaved.	S. Europe.	1596.	7.	H.♃.	Lobel. ic. 429.
51 Holósteum. R.S.	(*wh.*)	Scopoli's.	Carniola.	1825.	7. 8.	H.♃.	Jacq. coll. 1. t. 10.
52 elongàta. R.S.	(*wh.*)	long-spiked.	Louisiana.	1821.	——	H.♃.	
53 humifùsa. Ber.	(*wh.*)	trailing.	....	1825.	——	H.♃.	
54 canéscens. Schr.	(*wh.*)	canescent.	....	1826.	7. 9.	H.⊙.	
55 micrántha. Led.	(*gr.*)	small-flowered.	Siberia.	1829.	6. 8.	H.♃.	
56 Corónopus. R.S.	(*gr.*)	star of the earth.	Britain.	....	4. 9.	H.⊙.	Eng. bot. v. 13. t. 892.
57 Jacquìni. R.S. *Cornùti.* Jacq.	(*wh.*)	Jacquin's.	S. Europe.	1801.	7. 8.	H.♃.	Jacq. ic. 1. t. 27.
58 Serrària. R.S.	(*wh.*)	saw-leaved.	Barbary.	1640.	6. 7.	H.♃.	Moris. s. 8. t. 16. f. 19.
59 macrorhìza. R.S.	(*br.*)	large-rooted.	Morocco.	1798.	7. 8.	H.♃.	Bocc. sic. t. 15. f. 2.
60 Lœflíngii. R.S.	(*br.*)	narrow-leaved.	Spain.	....	——	H.⊙.	Jacq. vind. 2. t. 126.
61 amplexicàulis. R.S.	(*br.*)	stem-clasping.	———	1797.	6. 7.	H.⊙.	Cavan. ic. 2. t. 125.
62 villòsa. R.S.	(*gr.*)	villous.	Germany.	1804.	——	H.⊙.	
63 vaginàta. R.S.	(*li.*)	sheathed.	Canaries.	1816.	——	G.♄.	Venten. cels. t. 29.
64 Cynops. R.S.	(*gr.*)	shrubby.	S. Europe.	1596.	5. 8.	H.♄.	Jacq. fragm. t. 182.
65 áfra. R.S.	(*wh.*)	Barbary.	Barbary.	1640.	6.	F.♄.	Moris. s. 8. t. 17. f. 4.
66 divaricàta. R.S.	(*wh.*)	spreading.	....	1824.	——	H.♃.	
67 squarròsa. R.S.	(*wh.*)	leafy-spiked.	Egypt.	1787.	8. 9.	H.⊙.	Jacq. ic. 1. t. 28.
68 Psy'llium. R.S.	(*bh.*)	Fleawort.	S. Europe.	1562.	7. 8.	H.⊙.	Flor. græc. t. 149.
69 arenària. R.S.	(*wh.*)	sand.	Hungary.	1804.	5. 8.	H.⊙.	W. et K. hung. 1. t. 51.
70 stricta. R.S.	(*bh.*)	upright.	Morocco.	——	7. 8.	H.⊙.	Moris. s. 3. 8. t. 17. f. 2.
71 sícula. R.S.	(*wh.*)	Sicilian.	Sicily.	1820.	7. 10.	H.⊙.	
72 parviflòra. R.S.	(*br.*)	small-flowered.	Barbary.	1826.	——	H.⊙.	
73 Ispághul. R.S.	(*wh.*)	East Indian.	E. Indies.	1819.	7. 8.	S.⊙.	
74 índica. R.S.	(*wh.*)	Indian.	India.	1780.	——	H.⊙.	
75 pùmila. R.S.	(*wh.*)	dwarf.	S. Europe.	1790.	——	H.⊙.	Murr. c. gott. 1778. t. 5.
76 nítida. R.S.	(*bl.*)	glossy.	....	1822.	7. 10.	H.⊙.	
77 brasiliénsis. B.M.	(*wh.*)	Brazilian.	Brazil.	——	——	S.♃.	Bot. mag. t. 2616.
LITTORE'LLA. L.		SHORE-WEED.	Monœcia Tetrandria. L.				
lacústris. E.B.	(*gr.*)	Plantain-leaved.	Britain.	....	6. 9.	H.♃.	Eng. bot. v. 7. t. 468.

ORDO CLXIV.

NYCTAGINEÆ. *Juss. gen.* 90.

OXYBA'PHUS. R.S.		UMBRELLA-WORT.	Triandria Monogynia R.S.				
1 viscòsus. R.S.	(*pu.*)	viscid.	Peru.	1793.	5. 9.	H.♃.	Bot. mag. t. 434.
2 glabrifòlius. R.S.	(*pu.*)	smooth-leaved.	New Spain.	1811.	7. 8.	H.♃.	Cavan. ic. 4. t. 379.
3 ovàtus. R.S.	(*re.*)	oval-leaved.	Peru.	1820.	——	H.♃.	Flor. peruv 1. t. 75. f. b.
4 Cervantèsii. Lag.	(*pu.*)	Cervantes'.	Mexico.	1822.	7. 10.	H.♃.	Swt. br. fl. gar. t. 84.
5 expánsus. R.S.	(*pu.*)	expanded.	Lima.	——	——	H.♃.	Flor. peruv. 1. t. 75. f. a.
6 aggregàtus. R.S.	(*re.*)	aggregate.	New Spain.	1811.	——	H.♃.	Cavan. ic. 5. t. 437.
7 nyctagínea. N. *Allìonia nyctagínea.* M. *Calymènia nyctagínea.* N.	(*re.*)	heart-leaved.	Missouri.	1823.	7. 9.	H.♃.	
8 álbida. N.	(*li.*)	white-leaved.	Carolina.	1824.	——	H.⊙.	
9 pilòsa. N. *Calymènia pilòsa.* N. *Alliònia ovàta.* Ph.	(*re.*)	pilose.	Missouri.	1812.	——	H.⊙.	
10 hirsùta. N.	(*re.*)	hairy.	Louisiana.	——	——	H.♃.	
11 angustifòlia. *Alliònia lineàris.* Ph. *Calymènia angustifòlia.* N.	(*li.*)	narrow-leaved.	———	——	——	H.♃.	
12 decúmbens. N.	(*li.*)	decumbent.	Missouri.	1818.	——	H.♃.	
MIRA'BILIS. R.S.		MARVEL OF PERU.	Pentandria Monogynia. L.				
1 dichótoma. R.S.	(*re.wh.*)	forked.	Mexico.	1640.	7. 8.	G.♃.	Plenck ic. off. t. 139.
α rùbra.	(*re.*)	*red-flowered.*	———	——	——	F.♃.	
β álba.	(*wh.*)	*white-flowered.*	———	——	——	F.♃.	

2 Jalápa. R.S.	(*va.*)	common.	India.	1596.	6. 9.	H. ♃.	Bot. mag. t. 371.
a álba.	(*wh.*)	*white-flowered.*	———	——	——	H. ♃.	
β flàva.	(*ye.*)	*yellow-flowered.*	———	——	——	H. ♃.	
γ rùbra.	(*re.*)	*red-flowered.*	———	——	——	H. ♃.	
δ rùbro-flàva.	(*re.ye.*)	*red and yellow.*	———	——	——	H. ♃.	
ε rùbro-álba.	(*re.wh.*)	*red and white.*	———	——	——	H. ♃.	
3 hy'brida. R.S.	(*re.*)	hybrid.	Hybrid.	1818.	6. 9.	H. ♃.	
4 longiflòra. R.S.	(*wh.*)	long-flowered.	Mexico.	1759.	——	H. ♃.	Sm. exot. bot. 1. t. 23.
5 suavèolens. K.S.	(*wh.*)	sweet-scented.	———	1824.	——	H. ♃.	
ABR'ONIA. R.S.		ABR'ONIA.	Pentandria Monogynia. R.S.				
1 umbellàta. R.S.	(*bh.*)	flesh-coloured.	California.	1823.	6. 8.	H. ♃.	Hook. ex. flor. t. 194.
2 mellífera. B.M.	(*wh.*)	honey-scented.	———	1827.	5. 9.	H. ♃.	Bot. mag. t. 2879.
ALLI'ONIA. K.S.		ALLI'ONIA.	Tetrandria Monogynia. L.				
1 violàcea. K.S.	(*vi.*)	violet-coloured.	Cumana.	1820.	7. 9.	H. ☉.	
2 incarnàta. K.S.	(*bh.*)	flesh-coloured.	———	——	——	H. ☉.	L'Herit. stirp. t. 31.
BOERHA'AVIA. L.		HOGWEED.	Mono-Di-Tetrandria Monogynia.				
1 erécta. R.S.	(*wh.*)	upright.	W. Indies.	1733.	7. 9.	S. ♃.	Jacq. vind. 1. t. 5. 6.
2 procúmbens. F.I.	(*w.re.*)	procumbent.	E. Indies.	....	——	S. ♃.	Rheed. mal. 7. t. 56.
3 diffùsa. R.S.	(*sc.*)	spreading.	W. Indies.	1690.	8. 9.	S. ♃.	
4 adscéndens. R.S.	(*re.*)	ascending.	Guinea.	1824.	6. 9.	S. ♃.	
5 hirsùta. R.S.	(*sc.*)	scarlet-trailing.	Jamaica.	1733.	5. 8.	S. ♄.	Jacq. vind. 1. t. 7.
6 plumbagínea. R.S.	(*ro.*)	Plumbago-leaved.	Spain.	1822.	——	S. ♃.	Cavan. ic. 2. t. 112.
7 díscolor. K.S.	(*bh.*)	two-coloured.	Guayaquil.	1825.	——	S. ♃.	
8 viscòsa. K.S.	(*sc.*)	viscous.	———	——	——	S. ♃.	
9 scándens. R.S.	(*gr.*)	climbing.	Jamaica.	1691.	4. 9.	S. ♄.	Jacq. vind. 1. t. 4.
10 excélsa. R.S.	(*pu.*)	lofty.	S. America.	1820.	——	S. ♄.	
PIS'ONIA. L.		PIS'ONIA.	Heptandria Monogynia. L.				
1 grándis. B.P.	(*wh.*)	superb.	N. Holland.	1805.	....	G. ♄.	
2 hirtélla. K.S.	(*wh.*)	hairy.	Mexico.	1825.	....	G. ♄.	
3 fràgrans. S.S.	(*gr.*)	fragrant.	....	——	....	S. ♄.	
4 macrophy'lla. L.en.	(*w.*)	long-leaved.	....	1820.	....	S. ♄.	
5 obovàta. L.en.	(*gr.*)	obovate-leaved.	S. America.	——	....	S. ♄.	
6 nígricans. S.S.	(*wh.*)	black.	W. Indies.	1806.	5. 7.	S. ♄.	
7 obtusàta. S.S.	(*gr.*)	blunt-leaved.	———	1822.	....	S. ♄.	
8 mexicàna. L.en.	(*ro.*)	Mexican.	Mexico.	1825.	6. 8.	H. ♄.	
Boerhaàvia arboréscens. P.S.							
9 nítida. W.en.	(*li.*)	glossy.	Madagascar.	1824.	....	S. ♄.	
10 aculeàta. S.S.	(*ye.*)	prickly.	Jamaica.	1739.	3. 4.	S. ♄.	Lam. ill. t. 861.
BOLD'OA. S.S.		BOLD'OA.	Triandria Monogynia. S.S.				
lanceolàta. S.S.	(*re.*)	lance-leaved.	Mexico.	1824.	5. 7.	G. ♄.	H. et B. pl. æq. 1. t. 44.
Salpiánthus arenàrius. K.S.							

ORDO CLXV.

AMARANTHACEÆ. *Brown prodr. p.* 413.

AMARA'NTHUS. L.		AMARANTH.	Monœcia Tri-Pentandria. L.				
1 tenuifòlius. W.	(*gr.*)	fine-leaved.	E. Indies.	1801.	7. 9.	H. ☉.	
2 angustifòlius. W.	(*gr.*)	narrow-leaved.	Levant.	1723.	——	H. ☉.	
3 pállidus. S.S.	(*pa.*)	pale.	Tauria.	1823.	——	H. ☉.	
4 persicarioídes. S.S.	(*re.*)	Persicaria-like.	Nepaul.	1818.	9. 11.	H. ☉.	
5 Acroglóchin. S.S.	(*gr.*)	spiny-bracted.	———	1820.	——	H. ☉.	
6 álbus. W.	(*wh.*)	white.	N. America.	1778.	7. 9.	H. ☉.	W. amar. 9. t. 1. f. 2.
7 græcìzans. W.	(*gr.*)	Pellitory-leaved.	———	1759.	——	H. ☉.	——— 8. t. 4. f. 7.
8 poly'gamus. W.	(*gr.*)	hermaphrodite.	E. Indies.	1780.	——	H. ☉.	Rumph. amb. 5. t. 82.
9 Blitum. E.B.	(*gr.*)	wild.	England.	....	6. 8.	H. ☉.	Eng. bot. v. 31. t. 2212.
10 víridis. W.	(*gr.*)	green.	Brazil.	1768.	8. 9.	H. ☉.	W. amar. 18. t. 8. f. 16.
11 campéstris. W.	(*gr.*)	field.	E. Indies.	1818.	——	H. ☉.	
12 polygonoídes. W.	(*gr.*)	spotted-leaved.	Jamaica.	1778.	7. 8.	H. ☉.	W. amar. 11. t. 6. f. 12.
13 melanchólicus. W.	(*pu.*)	melancholy.	E. Indies.	1731.	6. 9.	H. ☉.	——— 15. t. 9. f. 18.
14 trícolor. W.	(*st.*)	three-coloured.	———	1548.	——	G. ☉.	Knor. thes. 2. t. A. 3. 5.
15 bícolor. W.	(*st.*)	two-coloured.	———	1802.	7. 9.	H. ☉.	
16 lineàtus. B.P.	(*gr.*)	line-leaved.	N. Holland.	1828.	——	H. ☉.	
17 gangéticus. W.	(*gr.*)	oval-spiked.	E. Indies.	1778.	——	H. ☉.	W. amar. 16. t. 6. f. 11.
18 mangostànus. W.	(*gr.*)	rhomb-leaved.	———	1801.	——	H. ☉.	——— 13. t. 12.

19	polystàchyus. w.	(gr.)	many-spiked.	E. Indies.	1816.	7. 9.	H.⊙.	
20	trístis. w.	(pu.)	round-headed.	China.	1759.	6. 8.	H.⊙.	W. amar. 21. t. 5. f. 10.
21	inam'œnus. w.	(gr.)	inelegant.	———	1816.	——	H.⊙.	
22	incómptus. W.en.	(gr.)	slovenly.	———	——	——	H.⊙.	
23	lívidus. w.	(br.)	livid.	N. America.	1759.	7. 9.	H.⊙.	W. amar. 20. t. 1. f. 1.
24	oleràceus. w.	(li.)	eatable.	E. Indies.	1764.	7. 8.	H.⊙.	——— 17. t. 5. f. 9.
25	lanceæfòlius. H.B.	(gr.)	lance-leaved.	———	1816.	8. 10.	H.⊙.	
26	atropurpùreus. H.B.	(pu.)	dark-purple.	———	1821.	——	H.⊙.	
27	grácilis. s.s.	(gr.)	oval-leaved.	Guinea.	1806.	7. 8.	H.⊙.	Jacq. ic. 2. t. 344.
	Chenopòdium caudàtum. Jacq.							
28	prostràtus. w.	(gr.)	trailing.	France.	1739.	7. 9.	H.⊙.	
29	scándens. w.	(gr.)	climbing.	S. America.	1796.	7. 8.	S.⊙.	
30	defléxus. w.	(gr.)	bending.	....	1805.	——	H.⊙.	W. amar. 10. t. 10. f. 20.
31	fasciàtus. H.B.	(gr.)	fasciate.	E. Indies.	1820.	8. 10.	H.⊙.	
32	bullàtus. s.s.	(gr.)	blistered.	....	——	7. 9.	H.⊙.	
33	Arardhànus.	(gr.)	Indian.	E. Indies.	1825.	——	G.⊙.	
34	Berchthóldi. H.	(pi.gr.)	Berchthold's.	S. Europe.	1826.	——	H.⊙.	
35	bahiénsis. Lk.	(gr.)	Bahia.	Bahia.	1823.	——	H.⊙.	
36	purpuráscens. Ot.	(pu.)	purplish.	....	1828.	——	H.⊙.	
37	littoràlis. Ot.	(gr.)	sea-side.	....	1823.	——	H.⊙.	
38	rígidus. s.	(gr.)	rigid.	....	——	——	H.⊙.	
39	tortuòsus. H.H.	(gr.)	twisted.	....	——	——	H.⊙.	
40	laxifòlius. Ot.	(gr.)	loose-leaved.	....	——	——	H.⊙.	
41	glomeràtus. Ot.	(gr.)	glomerate.	....	1829.	——	H.⊙.	
42	spicàtus. P.S.	(gr.)	spiked.	France.	....	——	H.⊙.	
43	pùmilus. s.s.	(gr.)	dwarf.	N. America.	——	6. 9.	H.⊙.	
44	curvifòlius. s.s.	(gr.)	curved-leaved.	Caucasus.	1827.	——	H.⊙.	
45	hy'bridus. w.	(re.)	clustered.	———	1656.	——	H.⊙.	W. amar. 26. t. 9. f. 17.
46	hécticus. w.	(pi.)	oval-spiked.	....	1796.	8. 9.	H.⊙.	——— 25. t. 7. f. 13.
47	l'ætus. w.	(re.)	blunt-leaved.	Italy.	1799.	7. 9.	H.⊙.	——— 28. t. 8. f. 15.
48	stríctus. w.	(gr.)	upright.	———	1793.	——	H.⊙.	——— 27. t. 3. f. 5.
49	paniculàtus. w.	(re.)	panicled.	N. America.	1798.	——	H.⊙.	——— 32. t. 2. f. 4.
50	celosioídes. K.S.	(wh.)	Spinach-like.	S. America.	1824.	9. 11.	H.⊙.	
51	caracasànus. K.S.	(wh.)	Caracas.	Caracas.	1821.	——	H.⊙.	
52	retrofléxus. w.	(gr.)	hairy.	N. America.	1759.	7. 9.	H.⊙.	W. amar. 33. t. 11. f. 21.
53	flàvus. w.	(st.)	pale yellow.	India.	——	——	H.⊙.	——— 35. t. 3. f. 6.
54	chlorostàchys. w.	(ye.)	nodding.	———	1796.	——	H.⊙.	——— 34. t. 10. f. 19.
55	parisiénsis. s.s.	(gr.)	Paris.	France.	1819.	——	H.⊙.	
56	hypochondrìacus. w.	(re.)	Prince's feather.	Virginia.	1684.	——	H.⊙.	
57	sanguíneus. w.	(re.)	spreading.	Bahama Isl.	1775.	——	H.⊙.	W. amar. 31. t. 2. f. 3.
58	cruéntus. w.	(cr.)	various-leaved.	China.	1728.	6. 8.	H.⊙.	
59	speciòsus. D.P.	(sc.)	showy.	Nepaul.	1819.	9. 11.	H.⊙.	Bot. mag. t. 2227.
60	caudàtus. w.	(re.)	Love lies bleeding.	E. Indies.	1596.	8. 9.	H.⊙.	
	β *máximus.*	(re.)	*tree.*	———	——	——	H.⊙.	
61	spinòsus. w.	(gr.)	prickly.	W. Indies.	1683.	7. 9.	H.⊙.	W. amar. 38. t. 4. f. 8.
62	frumentàceus. H.B.	(gr.)	cultivated.	E. Indies.	1819.	——	H.⊙.	
	LECANOCA'RPUS. N.E.		LECANOCA'RPUS.	Monœcia Pentandria.				
	nepalénsis. N.E.	(gr.)	Nepaul.	Nepaul.	1816.	8. 10.	H.⊙.	Nees ab Ess. ic. sel. t. 2.
	Amaránthus caulifldrus. Lk.							
	CEL'OSIA. R.S.		COCKS-COMB.	Pentandria Monogynia. L.				
1	argéntea. R.S.	(bh.)	silvery-spiked.	China.	1714.	6. 9.	S.⊙.	Mart. cent. t. 7.
	β *lineàris.*	(bh.)	*narrow-leaved.*	E. Indies.	——	——	S.⊙.	Rheed. mal. 10. t. 39.
2	pyramidàlis. R.S.	(wh.)	pyramidal.	———	——	——	S.⊙.	Burm. ind. 65. t. 25. f. 1.
3	margaritàcea. R.S.	(wh.re.)	pearl-flowered.	India.	....	——	S.⊙.	Rheed. mal. 10. t. 38.
4	cristàta. R.S.	(re.)	common.	Asia.	1570.	——	S.⊙.	Rumph. amb. 5. t. 84.
	α *elàta.*	(re.)	*tall red.*	———	——	——	S.⊙.	Knor. del. 1. t. H. 5. 6.
	β *compácta.*	(re.)	*dwarf red.*	———	——	——	S.⊙.	
	γ *flavéscens.*	(st.)	*yellow.*	———	——	——	S.⊙.	
5	comòsa. R.S.	(re.)	tufted.	E. Indies.	1802.	——	S.⊙.	
6	coccínea. R.S.	(sc.)	scarlet.	China.	1597.	——	S.⊙.	
7	castrénsis. R.S.	(re.)	branched.	E. Indies.	1739.	——	S.⊙.	Barrel. rar. t. 1195.
8	cérnua. A.R.	(sc.)	drooping.	———	1809.	7. 8.	S.⊙.	Andr. reposit. t. 635.
9	Monsòniæ. R.S.	(bh.)	downy.	———	1778.	7. 9.	S.⊙.	Pluk. alm. 11. t. 334. f. 4.
	Achyránthes Monsòniæ. s.s. *Illécebrum Monsòniæ.* L.							
10	nodiflòra. R.S.	(wh.)	knotted.	E. Indies.	1780.	——	S.⊙.	Jacq. vind. 1. t. 98.
	LESTIBUD'ESIA. R.S.		LESTIBUD'ESIA.	Pentandria Monogynia. R.S.				
1	virgàta. R.S.	(gr.)	wave-leaved.	W. Indies.	1815.	8. 10.	S.♄.	Jacq. ic. 2. t. 339.
	Celòsia virgàta. Jacq.							

2 paniculàta. R.S.	(*st.*)	panicled.	Jamaica.	1733.	6. 9.	S. ♂.	Sloan. hist. 1. t. 91. f. 2.
3 trig'yna. R.S.	(*wh.*)	oval-leaved.	Senegal.	1777.	8. 10.	S. ♂.	Jacq. vind. 3. t. 15.
DEERI'NGIA. R.S.		DEERI'NGIA.	Pentandria Trigynia.				
1 índica. S.S.	(*wh.*)	Berry-bearing.	E. Indies.	1804.	——	S. ♂.	
2 celosioídes. B.P.	(*wh.*)	Cockscomb-like.	N. Holland.	1822.	——	S. ♄.	
CHAMISS'OA. K.S.		CHAMISS'OA.	Pentandria Monogynia. R.S.				
altíssima. K.S.	(*st.*)	tall.	S. America.	1819.	5. 6.	S. ♄.⌣.	H. et B. n. gen. 2. t. 125.
CLADOST'ACHYS. D.P.		CLADOST'ACHYS.	Pentandria Monogynia.				
1 frutéscens. D.P.	(*gr.*)	frutescent.	E. Indies.	1777.	8. 11.	S. ħ.	Rumph. amb. 5. t. 83.
Achyránthes muricàta. L. *Chamissòa muricàta.* S.S.							
2 alternifòlia.	(*pu.*)	alternate-leav'd.	E. Indies.	1789.	7. 8.	S. ⊙.	Pluk. alm. t. 260. f. 1.
Achyránthes alternifòlia. W. *Desmoch'æta alternifòlia.* DC.							
ACHYRA'NTHES. R.S.		ACHYRA'NTHES.	Pentandria Monogynia. L.				
1 argéntea. R.S.	(*ro.*)	upright.	Sicily.	1713.	5. 10.	H. ħ.	Flor. græc. t. 244.
2 áspera. R.S.	(*gr.*)	rough.	India.	1751.	——	S. ħ.	Mill. ic. t. 11. f. 2.
3 obtusifòlia. Lam.	(*gr.*)	blunt-leaved.	————	....	——	S. ħ.	
4 críspa. R.S.	(*gr.*)	curl-leaved.	....	1816.	——	S. ħ.	
5 virgàta. R.S.	(*gr.*)	slender.	Portorico.	1823.	——	S. ħ.	
6 fruticòsa. R.S.	(*wh.*)	shrubby.	E. Indies.	1816.	——	S. ♄.	
7 nívea. R.S.	(*wh.*)	white.	Canaries.	1780.	5. 7.	G. ♄.	
8 pubéscens. R.S.	(*wh.*)	pubescent.	....	1824.	——	S. ħ.	
9 brachiàta. R.S.	(*wh.*)	armed.	E. Indies.	1829.	8. 10.	S. ⊙.	Pluk. alm. 2. t. 334. f. 5.
10 globulifera. F.	(*wh.*)	globe-headed.	....	1828.	——	S. ⊙.	
DESMOCH'ÆTA. R.S.		DESMOCH'ÆTA.	Pentandria Monogynia. R.S.				
1 atropurpùrea. DC.	(*pu.*)	dark-purple.	E. Indies.	1759.	8. 10.	S. ħ.	Rheed. mal. 10. t. 59.
Achyránthes lappàcea. L. *Pupàlia lappàcea.* J.							
2 flavéscens. DC.	(*st.*)	yellowish.	E. Indies.	....	——	S. ♃.	DC. h. mons. ined. t. 79.
3 pátula. R.S.	(*wh.*)	spreading.	————	1824.	——	S. ħ.	
4 prostràta. DC.	(*pu.*)	prostrate.	————	1793.	7. 8.	S. ♃.	Rumph. amb. 6. t. 11.
5 sanguinolénta. Lk.	(*gr.*)	bloody-leaved.	————	1824.	——	S. ħ.	———— t. 27. f. 2.
ALTERNANTH'ERA.		ALTERNANTH'ERA.	Pentandria Monogynia. R.S.				
1 séssilis. R.S.	(*wh.*)	sessile-flowered.	E. Indies.	1778.	7. 10.	S. ♂.	Rheed. mal. 10. t. 11.
2 ficoídes. R.S.	(*wh.*)	creeping-stalk'd.	S. America.	1819.	——	S. ♃.	Jacq. amer. pict. t. 90.
Gomphrèna ficoídea. Jacq.							
3 spinòsa. R.S.	(*st.*)	spiny.	....	——	——	S. ♃.	
axillàris. DC. *Achyránthes spinòsa.* Horn.							
4 denticulàta. B.P.	(*wh.*)	tooth-leaved.	N. S. W.	1822.	——	G. ♃.	
5 tenélla. DC.	(*wh.*)	slender.	E. Indies.	1828.	7. 9.	H. ⊙.	
6 Achyrántha. R.S.	(*br.*)	creeping.	BuenosAyres.	1732.	6. 8.	S. ♃.	Mart. pl. bras. 2. t. 152.
BUCHO'LZIA. M.P.		BUCHO'LZIA.	Pentandria Monogynia.				
1 polygonoídes. M.P.	(*w.*)	Persicaria-leaved.	S. America.	1731.	——	G. ♃.	Herm. parad. t. 17.
Illécebrum polygonoídes. W. *Alternanthèra polygonoídes.* R.S.							
2 frutéscens. M.P.	(*wh.*)	frutescent.	Peru.	1822.	——	S. ħ.	L'Herit. stirp. 1. t. 37.
ÆR'UA. J.		ÆR'UA.	Pentandria Monogynia. R.S.				
1 lanàta. R.S.	(*wh.*)	woolly.	E. Indies.	1691.	4. 8.	S. ♂.	Mill. ic. 1. t. 11. f. 1.
Illécebrum lanàtum. L.							
2 javánica. R.S.	(*wh.*)	spear-leaved.	————	1786.	——	S. ♄.	Burm. ind. t. 65. f. 2.
PHILO'XERUS. B.P.		PHILO'XERUS.	Pentandria Monogynia. R.S.				
1 vermiculàris. R.S.	(*wh.*)	creeping.	S. America.	....	——	S. ♃.	Herm. parad. t. 15.
2 aggregàtus. R.S.	(*wh.*)	clustered.	Cumana.	1819.	——	S. ♃.	
MOGIPH'ANES. M.P.		MOGIPH'ANES.	Pentandria Monogynia.				
1 brasiliénsis. M.P.	(*wh.*)	upright.	Brazil.	1790.	7. 10.	S. ħ.	Mart. pl. bras. 2. t. 133.
Gomphrèna brasiliénsis. Jacq. ic. 2. t. 346. *Philóxerus brasiliénsis.* B.P.							
2 stramínea. M.P.	(*st.*)	straw-coloured.	Brazil.	1790.	7. 10.	S. ħ.	Mart. pl. bras. 2. t. 135.
GOMPHR'ENA. R.S.		GLOBE AMARANTH.	Pentandria Monogynia. L.				
1 globòsa. R.S.	(*re.*)	annual.	India.	1714.	5. 10.	S. ⊙.	Bot. mag. t. 2815.
β álba.	(*wh.*)	*white-flowered.*	————	——	——	S. ⊙.	
2 decúmbens. R.S.	(*pu.wh.*)	decumbent.	Mexico.	....	——	S. ⊙.	Jacq. schœn. t. 482.
3 perénnis. R.S.	(*st.*)	perennial.	S. America.	1732.	7. 10.	S. ♃.	Bot. mag. t. 2614.
4 arboréscens. R.S.	(*st.*)	tree.	————	1802.	——	S. ♄.	
5 villòsa. S.S.	(*st.*)	villous.	Monte Video.	1826.	6. 8.	S. ħ.	
6 nígricans. M.P.	(*bk.*)	dark-flowered.	Brazil.	1827.	——	S. ħ.	Mart. pl. bras. 2. t. 113.
7 callòsa. S.	....	callous.	S. America.	1829.	....	S. ħ.	
8 cárnea. Jac.	(*fl.*)	flesh-coloured.	————	——	....	S. ħ.	
OPLOTH'ECA. N.		OPLOTH'ECA.	Pentandria Monogynia.				
1 láctea. M.P.	(*wh.*)	milk-coloured.	S. America.	1818.	7. 10.	S. ħ.	DC. h. mons. ined. t. 93.
2 interrúpta. N.	(*wh.*)	trailing.	W. Indies.	1733.	——	S. ♂.	Jacq. icon. 1. t. 51.
Gomphrèna interrúpta. R.S.							
3 floridàna. N.	(*wh.*)	Florida.	Florida.	1823.	8. 10.	F. ♃.	Bot. mag. t. 2603.

BRAND'ESIA. M.P. BRAND'ESIA. Pentandria Monogynia.
1 pubérula. M.P. (*wh.*) pubescent. Brazil. 1825. 6. 9. S. ♄.
Achyránthes capituliflòra. Colla hort. rip. t. 18.
2 villòsa. M.P. (*wh.*) villous. Brazil. 1826. —— S. ⊙. Mart. bras. 2. t. 128.
3 pórrigens. M.P. (*re.*) crimson-flowered. Peru. 1802. 4. 8. G. ♄. Bot. mag. t. 830.
Achyránthes pórrigens. B.M.
PFA'FFIA. M.P. PFA'FFIA. Pentandria Monogynia.
gnaphaloídes. M.P. (*w.*) Cudweed-like. Brazil. 1822. 6. 8. S. ♄.
Celòsia gnaphaloídes. L.
IRES'INE. S.S. IRES'INE. Diœcia Pentandria. L.
1 celosioídes. S.S. (*wh.*) Florida. America. 1733. 7. 8. F. ♃. Lam. ill. t. 313.
2 diffùsa. S.S. (*wh.*) spreading. S. America. 1812. 7. 9. S. ♃.
3 elongàta. S.S. (*wh.*) long-stalked. ——— 1822. —— S. ♃.
R'OSEA. M.P. R'OSEA. Polygamia Monœcia.
elàtior. M.P. (*st.*) tall. S. America. 1820. 8. 10. S. ⊙. Sloan. hist. 1. t. 90.
Iresìne elàtior. W.

ORDO CLXVI.

PHYTOLACEÆ. *Brown.*

PHYTOLA'CCA. W. PHYTOLA'CCA. Decandria Decagynia. L.
1 octándra. W. (*wh.*) white-flowered. Mexico. 1732. 7. 11. G. ♃. Dill. elt. t. 239. f. 308.
2 abyssínica. W. (*wh.*) African. Africa. 1775. 5. 6. S. ♄. Hoff. c. gœt. t. 2.
3 heptándra. S.S. (*wh.*) 7-anthered. .... 1819. 7. 10. G. ♃.
4 bogoténsis. K.S. (*wh.*) variable anther'd. S. America. 1823. —— G. ♃.
5 decándra. W. (*bh.*) Virginian poke. Virginia. 1615. 8. 9. H. ♃. Bot. mag. t. 931.
6 dodecándra. W. (*bh.*) recurved-leaved......... .... 5. 6. S. ♃.
7 mexicàna. (*wh.*) Mexican. Mexico. 1824. 7. 10. G. ♃. Bot. mag. t. 2633.
icosándra. B.M. *non* W.
8 icosándra. W. (*wh.*) red-stemmed. E. Indies. 1758. 7. 11. S. ♃. Mill. ic. t. 207.
9 dioíca. W. (*wh.*) tree. S. America. 1768. .... S. ♄. L'Herit. st. nov. t. 70.
RIV'INA. R.S. RIV'INA. Tetrandria Monogynia. L.
1 hùmilis. R.S. (*wh.*) downy. W. Indies. 1699. 1. 10. S. ♄. Bot. mag. t. 1781.
2 l'ævis. R.S. (*bh.*) smooth. ——— 1733. 2. 9. S. ♄. Lam. ill. t. 81. f. 2.
3 latifòlia. R.S. (*pu.*) broad-leaved. Madagascar. 1826. 7. 9. S. ⊙.
4 brasiliénsis. R.S. (*gr.*) wave-leaved. Brazil. 1790. 6. 7. S. ♄.
5 purpuráscens. R.S. (*pu.*) purple. W. Indies. 1815. 5. 8. S. ♄. Schrad. gen. nov. t. 5.
6 lanceolàta. L.en. (*wh.*) spear-leaved. Brazil. —— —— S. ♄.
7 octándra. R.S. (*wh.*) climbing. W. Indies. 1752. 5. 6. S. ♄. ◡. Plum. ic. 241.
MICR'OTEA. K.S. MICR'OTEA. Pentandria Digynia. R.S.
débilis. K.S. (*wh.*) slender. S. America. 1824. 7. 9. S. ⊙.
ANCISTROCA'RPUS. K.S. ANCISTROCA'RPUS. Octandria Tetragynia.
maypurénsis. K.S. (*wh.*) branching. S. America. 1822. 5. 6. H. ⊙. H. et B. pl. am. 2. t. 122.
PETIV'ERIA. W. PETIV'ERIA. Heptandria Monogynia. L.
1 alliàcea. W. (*wh.*) Garlic-scented. Jamaica. 1759. 6. 7. S. ♄. Lodd. bot. cab. t. 148.
2 octándra. W. (*wh.*) dwarf. W. Indies. 1737. —— S. ♄. Plum. ic. 213. t. 219.
GIS'EKIA. W. GIS'EKIA. Pentandria Pentagynia. L.
pharnacioídes. W. (*wh.*) trailing. E. Indies. 1783. 6. S. ⊙. Roxb. corom. 2. t. 183.

ORDO CLXVII.

CHENOPODEÆ. *Brown prodr. p.* 405.

BASE'LLA. R.S. MALABAR-NIGHTSHADE. Pentandria Trigynia. L.
1 rùbra. R.S. (*re.*) red-flowered. E. Indies. 1731. 7. 9. S. ♂. ◡. Rheed. mal. 7. t. 24.
2 álba. R.S. (*wh.*) white-flowered. ——— 1688. 7. 11. S. ♂. ◡. Pluk. alm. t. 63. f. 1.
3 lùcida. R.S. (*wh.*) shining. ——— 1802. —— S. ⊙. ◡.
4 cordifòlia. R.S. (*re.*) heart-leaved. ——— —— —— S. ⊙. ◡.
5 tuberòsa. R.S. (*st.*) tuberous-rooted. S. America. 1824. —— S. ♃. ◡.

6 marginàta. R.S. (*wh.*) red-margined. S. America. 1824. 7. 11. S.♃.◡.
7 japónica. R.S. (*bh.*) dwarf. China. 1814. —— G.☉. Burm. ind. t. 39. f. 4.
HABLI'TZIA. M.B. HABLI'TZIA. Pentandria Monogynia.
tamnoídes. M.B. (*wh.*) Tamus-like. Caucasus. 1828. 7. 10. H.♃.◡.
ANRED'ERA. J. ANRED'ERA. Pentandria Trigynia.
spicàta. K.S. (*bh.*) spiked. Cuba. 1749. 7. 9. S.♃.◡. Sloan. hist. 1. t.90. f. 1.
Basélla vesicària. Lam. ill. t. 215. f. 1. *Poly'gonum scándens.* W.
ANABA'SIS. L. ANABA'SIS. Pentandria Digynia. L.
1 aphy'lla. R.S. (*gr.*) leafless. Levant. 1823. 5. 7. F.ħ. Buxb. cent. 1. t. 18.
2 flòrida. M.B. (*wh.*) flowering. Persia. 1829. 7. 11. H.☉. M.B.cent.pl.ros.1.t.17.
3 cretàcea. R.S. (*gr.*) long-rooted. Siberia. 1828. 6. 8. H.♃. Pall.it.1.ap.n.109.t.1.N
4 spinosíssima. R.S.(*gr.*) spiny-branched. Iberia. 1829. —— H.♄. Delil.eg.n.305.t.21.f.2.
5 tamariscifòlia. R.S.(*gr.*) Tamarisk-leaved. Spain. 1752. 6. 7. F.ħ. Cavan. ic. 3. t. 283.
6 foliòsa. S.S. (*gr.*) leafy. Volga. 1828. 7. 10. H.☉. Pall. ill. 2. t. 23.
Salsòla clavifòlia. M.B. *baccífera.* Pall.
SALS'OLA. R.S. SALTWORT. Pentandria Digynia. L.
1 Kàli. R.S. (*fl.*) prickly. Britain. .... 7. 8. H.☉. Eng. bot. v. 9. t. 634.
2 Tràgus. R.S. (*ro.*) smooth. S. Europe. 1816. —— H.☉. Pall. ill. 2. t. 29. f. 2.
3 tamariscìna. R.S. (*br.*) Tamarisk-like. Tauria. 1828. 7. 9. H.☉.
4 crássa. R.S. (*re.wh.*) thick-leaved. Caucasus. 1823. 8. 10. H.☉. Buxb. cent.1. t.14. f. 2.
5 rosàcea. R.S. (*pu.*) rose-coloured. Asia. 1759. 7. 8. H.☉. Schk. hand. 1. t. 57.
6 oppositifòlia. R.S. (*ro.*) opposite-leaved. Sicily. 1829. .... H.ħ.
7 brachiàta. R.S. (*pu.*) armed. Tauria. 1828. 7. 10. H.☉. Pall. ill. 2. t. 22.
8 glaùca. R.S. (*ro.*) glaucous. Armenia. 1829. .... H.ħ. ——— 2. t. 19.
9 vermiculàta. R.S. (*gr.*) suffruticose. Siberia. 1759. 7. 8. H.♄. Gmel. sib. 3. t. 18. f. 2.
10 microphy'lla. R.S. (*re.*) small-leaved. Spain. —— —— H.☉. Cavan. ic. 3. t. 287.
11 spíssa. R.S. (*ro.*) dwarf. Caucasus. 1829. 7. 10. H.☉. Pall. ill. 2. t. 15.
12 verrucòsa. R.S. (*gr.*) warted. ——— 1822. 7. 8. H.♄. ——— 2. t. 16.
13 ericoídes. R.S. (*br.*) Heath-like. ——— 1829. .... H.ħ. ——— 2. t. 14.
14 laniflòra. R.S. (*re.*) woolly-flowered. Siberia. 1797. 6. 8. H.☉. ——— 2. t. 21.
15 satìva. R.S. (*pi.*) cultivated. Spain. 1783. 7. 8. H.♃. Cavan. ic. 3. t. 291.
16 obtusifòlia. Led. (*br.*) blunt-leaved. Siberia. 1827. .... H.☉.
17 marginàta. F. (*gr.*) thick margined. Armenia. 1829. .... H.♄ ?
18 Sòda. R.S. (*wh.*) long fleshy-leav'd. S. Europe. 1683. 7. 8. H.☉. Jacq. vind. 1. t. 68.
K'OCHIA. R.S. K'OCHIA. Pentandria Digynia. R.S.
1 arenària. R.S. (*wh.*) sand. Hungary. 1819. 7. 9. H.☉. W. et K. hung. 1. t. 78.
2 dasyántha. R.S. (*wh.*) woolly-flowered. Caucasus. 1829. —— H.☉. Gmel. jun. it. 1. t. 137.
3 prostràta. R.S. (*gr.*) trailing. S. Europe. 1780. 6. 8. F.♄. Jacq. aust. 3. t. 294.
Salsòla prostràta. Jacq.
4 trig'yna. L.en. (*gr.*) slender-leaved. Spain. 1804. 7. 8. H.☉. Cavan. ic. 3. t. 289.
Salsòla altíssima. Cav. *Chenopòdium trig'ynum.* R.S.
5 hyssopifòlia. R.S.(*gr.re.*) Hyssop-leaved. Siberia. 1801. 6. 8. H.☉. Pall. it. 1. ap. t. H. f. 1.
6 sedoídes. R.S. (*gr.*) Stonecrop-like. Tauria. 1823. —— H.♃. —— 3. t. M. f. 8.
7 erióphora. R.S. (*gr.*) woolly. Spain. 1816. —— H.☉. Trattin. arch. t. 124.
8 scopària. R.S. (*gr.*) Summer Cypress. Greece. 1629. 6. 9. H.♃. Buxb. cent. 1. t. 16.
Chenopòdium Scopària. W.
CHEN'OLEA. L. CHEN'OLEA. Pentandria Monogynia. L.
diffùsa. W. (*wh.*) silky. C. B. S. 1758. 8. 9. G.♄.
Salsòla serícea. H.K. *Kòchia serícea.* R.S. *Chenopòdium seríceum.* S.S.
ENCHYL'ÆNA. B.P. ENCHYL'ÆNA. Pentandria Di-Trigynia.
tomentòsa. B.P. (*wh.*) woolly. N. Holland. 1824. 6. 8. G.♄.
RHAG'ODIA. B.P. RHAG'ODIA. Polygamia Monœcia.
1 Billardíeri. B.P. (*gr.*) narrow-leaved. N. S. W. 1822. 6. 7. G.♄. Lab. nov. hol. 1. t. 96.
2 hastàta. B.P. (*gr.*) halberd-leaved. ——— 1803. —— G.♄. Loud. enc. pl. f. 14270.
3 nùtans. B.P. (*gr.*) nodding-branch'd. V. Diem. Isl. 1824. 7. 9. G.♃.
ACROGLO'CHIN. Schr. ACROGLO'CHIN. Pentandria Digynia.
chenopodioídes. F. (*gr.*) Goose-foot-like. Siberia. 1823. 5. 9. H.☉.
CHENOP'ODIUM. R.S. GOOSE-FOOT. Pentandria Digynia. L.
1 Bònus-Henrìcus. (*gr.*) English Mercury. Britain. .... 5. 8. H.♃. Eng. bot. v. 15. t. 1033.
2 úrbicum. E.B. (*gr.*) upright. ——— .... 8. H.☉. ——— v. 10. t. 717.
3 Atríplicis. R.S. (*cr.*) purple. China. 1780. 8. 9. H.☉. Jacq. vind. 3. t. 80.
4 punctulàtum. R.S. (*re.*) white-spotted. .... 1820. —— H.☉. Scop. del. ins. 1. t. 11.
5 gigánteum. D.P. (*gr.*) gigantic. Nepaul. —— 9. 11. H.☉.
6 rùbrum. R.S. (*fl.*) red. Britain. .... 8. 9. H.☉. Eng. bot. v. 24. t. 1721.
7 chrysomelanospérmum. R.S. black-seeded. Galicia. 1825. —— H.☉.
8 carthagenénse. R.S. (*gr.re.*) Carthage. Carthage. 1824. 6. 8. H.☉.
9 rhombifòlium. R.S. (*gr.*) angular-leaved. N. America. 1807. 7. 9. H.☉.
10 guineénse. W. (*gr.*) Guinea. Guinea. 1790. 8. 9. H.☉. Jacq. ic. 2. t. 345.
11 muràle. E.B. (*gr.*) nettle-leaved. Britain. .... —— H.☉. Eng. bot. v. 24. t. 1722.

12 Quinòa. R.S.	(*gr.*)	green Quinoa.	Chile.	1822.	7. 9.	H.⊙.	Feuil. peruv. t. 10.
β *rùbrum.*	(*re.*)	*red Quinoa.*	———	——	——	H.⊙.	Feuil.per.ed.germ.t.10.
13 serotìnum. R.S.	(*gr.*)	late-flowering.	S. Europe.	....	8. 9.	H.⊙.	
14 ficifòlium. E.B.	(*gr.*)	Fig-leaved.	England.	....	——	H.⊙.	Eng. bot. v. 24. t. 1724.
15 álbum. E.B.	(*gr.*)	white.	Britain.	....	6. 9.	H.⊙.	——— v. 24. t. 1723.
β *subrotúndum.*	(*gr.*)	*round-leaved.*	———	....	——	H.⊙.	Petiv. h. brit. t. 8. f. 4.
γ *víride.* W.	(*gr.*)	*green.*	———	....	——	H.⊙.	
δ *integrifòlium.*	(*gr.*)	*entire-leaved.*	———	....	——	H.⊙.	
ε *crassifòlium.*	(*gr.*)	*thick-leaved.*	———	....	——	H.⊙.	
16 opulifòlium. R.S.	(*gr.*)	Guelder-rose-ld.	S. Europe.	....	——	H.⊙.	Bast. j. bot. 1814. t. 3.
17 hy'bridum. E.B.	(*gr.*)	Maple-leaved.	Britain.	....	8. 9.	H.⊙.	Eng. bot. v. 27. t. 1919.
18 blitoídes. R.S.	(*gr.*)	Blitum-like.	France.	1827.	——	H.⊙.	
19 Bòtrys. R.S.	(*gr.*)	cut-leaved.	S. Europe.	1548.	6. 9.	H.⊙.	Flor. græc. t. 253.
20 f'œtidum. W.en.	(*gr.*)	strong-scented.	....	1823.	6. 8.	H.⊙.	
21 multífidum. R.S.	(*gr.*)	multifid.	BuenosAyres.	1732.	6. 10.	G.♃.	Dill. elt. t. 66. f. 77.
22 botryoídes. E.B.	(*re.*)	many-clustered.	Britain.	....	6. 8.	H.⊙.	Eng. bot. v. 32. t. 2247.
23 ambrosioídes. R.S.	(*gr.*)	Mexican.	Mexico.	1640.	6. 10.	H.⊙.	Moris. s. 5. t. 35. f. 8.
24 suffruticòsum. R.S.	(*gr.*)	suffrutescent.	———	1819.	——	G.♄.	
25 anthelmínticum. W.	(*gr.*)	American.	America.	1732.	7. 8.	G.♄.	Dill. elt. t. 66. f. 76.
26 gravèolens. R.S.	(*gr.*)	strong-scented.	Mexico.	1819.	8. 11.	H.⊙.	
27 incìsum. R.S.	(*gr.*)	jagged.	....	1826.	——	H.⊙.	
28 glaùcum. R.S.	(*gr.*)	Oak-leaved.	England.	....	7. 8.	H.⊙.	Eng. bot. v. 21. t. 1454.
29 crassifòlium. R.S.	(*cr.*)	fleshy-leaved.	....	1820.	——	H.⊙.	
30 humifùsum. R.S.	(*pu.*)	trailing.	....	1824.	——	H.⊙.	
31 sèpium. R.S.	(*re.*)	hedge.	Bohemia.	1822.	——	H.⊙.	
32 radiàtum. R.S.	(*gr.*)	rayed.	N. America.	1803.	6. 8.	H.⊙.	W. hort. ber. 1. t. 28.
Kòchia dentàta. W. *Salsòla platyphy'lla.* M.							
33 ólidum. E.B.	(*gr.*)	stinking.	Britain.	....	7. 8.	H.⊙.	Eng. bot. v. 15. t. 1034.
f'œtidum. R.S. *Vulvària.* W.							
34 polyspérmum. R.S.	(*gr.*)	Allseed.	———	....	——	H.⊙.	Eng. bot. v. 21. t. 1480.
35 acutifòlium. E.B.	(*gr.*)	acute-leaved.	———	....	——	H.⊙.	——— v. 21. t. 1481.
36 marginàtum. R.S.	(*gr.*)	bony-edged.	....	1824.	——	H.⊙.	
37 caudàtum. R.S.	(*gr.*)	long-tailed.	Guinea.	1806.	——	G.⊙.	Jacq. ic. 2. t. 341.
38 lateràle. R.S.	(*gr.*)	oblong-leaved.	....	1781.	8. 9.	S.⊙.	
39 stríctum. R.S.	(*gr.*)	upright.	E. Indies.	——	9. 11.	H.⊙.	
40 aristàtum. R.S.	(*gr.*)	bearded.	Siberia.	1771.	6. 9.	H.⊙.	Gmel. sib. 3. t. 15. f. 1.
41 lanceolàtum. R.S.	(*gr.*)	spear-leaved.	Pensylvania.	1809.	7.	H.⊙.	
42 fruticòsum. R.S.	(*gr.*)	shrubby.	England.	....	8. 9.	H.♄.	Eng. bot. v. 9. t. 635.
Salsòla fruticòsa. E.B.							
43 parvifòlium. R.S.	(*gr.*)	small-leaved.	Caucasus.	1829.	....	H.♄.	Pall. ill. 3. t. 44.
44 altíssimum. R.S.	(*wh.*)	Grass-leaved.	Russia.	1775.	7. 8.	H.⊙.	Schrad. d. hal. t. 1. f. 3
Salsòla altíssima. W. *Suaèda altíssima.* Pall. ill. t. 42.							
45 trig'ynum. R.S.	(*gr.*)	trigynous.	Spain.	——	——	H.⊙.	Cavan. icon. 3. t. 289.
46 hirsùtum. R.S.	(*gr.*)	hairy.	Denmark.	1791.	——	H.⊙.	Flor. dan. t. 187.
Salsòla hirsùta. F.D.							
47 sálsum. R.S.	(*gr.*)	Saltwort.	Astracan.	1782.	8. 9.	H.⊙.	Jacq. vind. 3. t. 83.
48 spicàtum. R.S.	(*gr.*)	spiked.	Spain.	1816.	8. 10.	H.⊙.	Cavan. ic. 3. t. 290.
49 setígerum. R.S.	(*gr.*)	setiferous.	———	——	——	H.⊙.	DC. h. mons. ined. t. 87.
50 marítimum. R.S.	(*gr.*)	sea blite.	Britain.	....	8.	H.⊙.	Eng. bot. v. 9. t. 633.
A'TRIPLEX. R.S.		ORACHE.	Polygamia Monœcia. L.				
1 Hálimus. R.S.	(*wh.*)	tall shrubby.	Spain.	1640.	7. 8.	H.♄.	Park. theat. 724. f. 2.
2 portulacoídes. R.S.	(*gr.*)	dwarf shrubby.	Britain.	....	——	H.♄.	Eng. bot. v. 4. t. 261.
3 verrucíferum. R.S.	(*gr.*)	warted-fruited.	Caucasus.	1827.	——	H.♄.	
4 álbicans. R.S.	(*bh.*)	white.	C. B. S.	1774.	6. 7.	G.♄.	
5 glaùcum. W.	(*gr.*)	glaucous.	S. Europe.	1732.	7. 8.	F.♄.	Dill. elt. t. 40. f. 46.
6 linifòlium. R.S.	(*gr.*)	Flax-leaved.	S. America.	1819.	——	G.♄.	
7 polygàmum. R.S.	(*gr.*)	polygamous.	New Spain.	1826.	——	F.♄.	
8 coriàceum. R.S.	(*gr.*)	leathery.	Egypt.	1828.	6. 9.	F.♄.	
9 ròseum. R.S.	(*gr.*)	rose-like.	S. Europe.	1739.	6. 7.	H.⊙.	Schk. hand. 3. t. 350.
10 foliòsum. R.S.	(*gr.*)	leafy.	Portugal.	1824.	——	H.⊙.	
11 sibíricum. R.S.	(*wh.*)	Siberian.	Siberia.	1783.	7. 8.	H.⊙.	Schk. hand. 3. t. 350.
12 tatáricum. R.S.	(*gr.*)	Tartarian.	Tartary.	1778.	——	H.⊙.	—— 3. p. 539. t. 349.
13 bengalénse. R.S.	(*gr.*)	Bengal.	E. Indies.	....	——	H.⊙.	
14 horténse. R.S.	(*gr.*)	garden.	Tartary.	1548.	——	H.⊙.	Blackw. t. 99 et 552.
β *intermèdium.*	(*re.*)	*intermediate.*	———	——	——	H.⊙.	
γ *rùbrum.*	(*re.*)	*red garden.*	———	——	——	H.⊙.	
15 nìtens. R.S.	(*gr.*)	glossy.	Hungary.	1746.	——	H.⊙.	Schkuhr hand. t. 348.
16 acuminàtum. R.S.	(*gr.*)	acuminate.	———	1820.	——	H.⊙.	W. et K. hung. t. 103.
17 vénetum. R.S.	(*gr.*)	Venetian.	Venice.	——	——	H.⊙.	

18 incànum. R.S.	(*gr.*)	hoary-leaved.	Spain.	1818.	7. 8.	H.⊙.	
19 álbum. R.S.	(*gr.*)	white-leaved.	Italy.	1826.	——	H.⊙.	
20 incìsum. M.B.	(*gr.*)	cut-leaved.	Tauria.	1829.	——	H.⊙.	
21 laciniàtum. R.S.	(*gr.*)	frosted sea.	Britain.	....	——	H.⊙.	Eng. bot. v. 3. t. 165.
22 Besseriànum. R.S.	(*gr.*)	Besser's.	Hungary.	1825.	——	H.⊙.	
23 hastàtum. R.S.	(*gr.*)	halberd-leaved.	Europe.	....	——	H.⊙.	Flor. dan. t. 1286.
24 prostràtum. DC.	(*gr.*)	prostrate.	————	....	——	H.⊙.	
25 obtùsum. R.S.	(*gr.*)	blunt-leaved.	Siberia.	1825.	——	H.⊙.	
26 microspérmum. R.S.	(*w.*)	small-seeded.	Hungary.	1800.	——	H.⊙.	W. et K. hung. t. 250.
27 oblongifòlium. R.S.	(*gr.*)	oblong-leaved.	————	——	——	H.⊙.	———— t. 221.
28 pátulum. R.S.	(*gr.*)	spreading.	Britain.	....	6. 9.	H.⊙.	Eng. bot. v. 13. t. 936.
29 angustifòlium. R.S.	(*gr.*)	narrow-leaved.	————	....	6. 8.	H.⊙.	———— v. 25. t. 1774.
30 eréctum. R.S.	(*gr.*)	upright.	England.	....	8.	H.⊙.	———— v. 31. 2223.
31 littoràle. E.B.	(*gr.*)	grass-leaved.	Britain.	....	8. 9.	H.⊙.	———— v. 10. t. 708.
32 pedunculàtum. R.S.	(*gr.*)	pedunculated.	England.	....	7. 9.	H.⊙.	———— v. 4. t. 232.
33 oppositifòlium. R.S.	(*gr.*)	opposite-leaved.	France.	1827.	——	H.⊙.	
34 verticillàtum. R.S.	(*gr.*)	whorl-leaved.	Spain.	1822.	——	H.⊙.	Cavan. H. R. M. t. 21.
35 sulcàtum. R.S.	(*gr.*)	sulcate.	....	——	——	H.⊙.	
B'ETA. R.S.		BEET. Pentandria Digynia. L.					
1 vulgàris. R.S.	(*gr.*)	common.	S. Europe.	1548.	——	H.♂.	Schk. hand. 1. t. 56.
α víridis.	(*gr.*)	*green.*	————	——	——	H.♂.	Plenck off. t. 169.
β rùbra.	(*gr.*)	*red.*	————	——	——	H.♂.	
γ lùtea.	(*gr.*)	*yellow.*	————	——	——	H.♂.	Kern. abbild. t. 235.
δ macrorhìza.	(*gr.*)	*mangold-wortzel.*	————	....	——	H.♂.	
2 Cìcla. R.S.	(*gr.*)	white.	Portugal.	1570.	——	H.♂.	Plenck off. t. 170.
3 críspa. R.S.	(*gr.*)	curled-leaved.	....	....	——	H.♂.	
4 trig'yna. R.S.	(*wh.*)	Hungarian.	Hungary.	1796.	7. 8.	H.♃.	W. et K. hung. 1. t. 35.
5 marítima. E.B.	(*gr.*)	sea-side.	Britain.	....	6. 8.	H.♃.	Eng. bot. v. 4. t. 285.
6 macrorhìza. R.S.	(*gr.*)	large-rooted.	Caucasus.	1827.	——	H.♃.	
7 foliòsa. E.	(*gr.*)	leafy.	Egypt.	1826.	7. 9.	H.⊙.	
8 pátula. R.S.	(*gr.*)	spreading.	Madeira.	1778.	8. 9.	G.♂.	
SPIN'ACIA. W.		SPINAGE. Diœcia Pentandria. L.					
oleràcea. W.	(*gr.*)	common.	....	1568.	3. 10.	H.⊙.	Schk. hand. 3. t. 324.
α spinòsa. W.	(*gr.*)	*prickly.*	....	——	——	H.⊙.	
β inérmis.	(*gr.*)	*round.*	....	——	——	H.⊙.	
ACN'IDA. W.		VIRGINIAN HEMP. Diœcia Pentandria. L.					
1 cannabìna. W.	(*gr.*)	common.	N. America.	1640.	6. 7.	H.⊙.	
2 ruscocárpa. W.	(*gr.*)	rugged-capsuled.	Virginia.	1824.	——	H.⊙.	Mich. fl. amer. 2. t. 50.
B'OSEA. R.S.		GOLDEN-ROD. Pentandria Digynia. L.					
yervamòra. R.S.	(*br.*)	tree.	Canaries.	1728.	6. 9.	G.♄.	Jacq. f. eclog. t. 25.
CORISPE'RMUM. L.		TICK-SEED. Monandria Digynia. L.					
1 hyssopifòlium. R.S.	(*wh.*)	Hyssop-leaved.	Europe.	1739.	7.	H.⊙.	Flor. græc. 1. t. 1.
2 nítidum. R.S.	(*wh.*)	glossy.	Hungary.	1827.	8. 9.	H.⊙.	
3 Redo'wskii. R.S.	(*wh.*)	Redowski's.	Caucasus.	1828.	——	H.⊙.	
4 spicàtum. R.S.	(*wh.*)	spiked.	....	——	——	H.⊙.	
5 canéscens. R.S.	(*wh.*)	canescent.	Hungary.	1822.	7. 9.	H.⊙.	
6 intermèdium. R.S.	(*wh.*)	intermediate.	Prussia.	——	——	H.⊙.	
7 ténue. L.en.	(*wh.*)	slender.	....	1823.	——	H.⊙.	
8 squarròsum. R.S.	(*wh.*)	rough-spiked.	Russia.	1759.	8. 9.	H.⊙.	Pall. ros. 2. t. 99.
9 Marschállii. R.S.	(*wh.*)	Marschall's.	Caucasus.	1829.	——	H.⊙.	
10 latifòlium. Lk.	(*wh.*)	broad-leaved.	————	1826.	——	H.⊙.	
11 sabulòsum. Led.	(*wh.*)	gravel.	Siberia.	1827.	——	H.⊙.	
BL'ITUM. R.S.		STRAWBERRY-BLITE. Monandria Digynia. L.					
1 capitàtum. R.S.	(*re.*)	berry-headed.	Austria.	1633.	5. 8.	H.⊙.	Poif. et T. fl. p. 1. t. 2.
2 virgàtum. R.S.	(*re.*)	slender.	S. Europe.	1680.	5. 9.	H.⊙.	Bot. mag. t. 276.
3 petiolàre. L.en.	(*re.*)	petiolate.	....	1818.	——	H.⊙.	
4 chenopodioídes. R.S.	(*gr.*)	Goosefoot-like.	....	1826.	——	H.⊙.	
5 marítimum. N.	(*gr.*)	sea-side.	N. America.	1823.	——	H.⊙.	
CERATOCA'RPUS. L.		CERATOCA'RPUS. Monœcia Monandria. L.					
arenàrius. L.	(*gr.*)	sand.	Tartary.	1757.	6. 7.	H.⊙.	Buxb. in. act. pet. 1. t. 9.
SALICO'RNIA. L.		GLASSWORT. Monandria Monogynia. L.					
1 herbàcea. S.S.	(*gr.*)	marsh.	Britain.	....	8. 9.	H.⊙.	Eng. bot. v. 6. t. 415.
2 procúmbens. E.B.	(*gr.*)	procumbent.	England.	....	——	H.⊙.	———— v. 35. t. 2475.
3 rádicans. E.B.	(*gr.*)	rooting.	Britain.	....	9.	H.♃.	———— v. 24. t. 1691.
4 pygm'æa. R.S.	(*gr.*)	small.	Caucasus.	1827.	——	H.⊙.	
5 perénnans. R.S.	(*gr.*)	perennial.	Siberia.	1823.	8. 9.	H.♃.	Pall. it. 1. ap. T. D. f. 1.
6 fruticòsa. E.B.	(*gr.*)	shrubby.	Britain.	....	——	H.♄.	Eng. bot. v. 35. t. 2467.
7 arábica. R.S.	(*gr.*)	Arabian.	Arabia.	1758.	——	G.♄.	Moris. s. 2. t. 33. f. 7.
8 foliòsa. R.S.	(*gr.*)	leafy.	Siberia.	1827.	——	H.♄.	

HALOCN'EMUM. M.B.		HALOCN'EMUM.	Monandria Digynia.				
caspicum. R.S.	(*gr.*)	Caspian.	Caucasus.	1827.	7. 8.	H. ♄.	Pall. it. 1. ap. t. A. f. 2.
POLYCN'EMUM. R.S.		POLYCN'EMUM.	Triandria Monogynia. L.				
1 arvénse. R.S.	(*wh.*)	trailing.	S. Europe.	1640.	7.	H.⊙.	Jacq. aust. 4. t. 365.
2 recúrvum. R.S.	(*wh.*)	recurved-leaved.	Switzerland.	1824.	7. 8.	H.⊙.	
3 sálsum. R.S.	(*gr.*)	saltwort.	Tartary.	1828.	——	H.⊙.	Pall. ill. t. 50.
4 sibíricum. R.S.	(*wh.*)	Siberia.	Siberia.	1829.	——	H.⊙.	——— t. 51.
5 oppositifòlium. R.S.	(*wh.*)	opposite-leaved.	Tartary.	1820.	——	H.⊙.	Pall. it. ap. T. E. f. 2.
6 crassifòlium. R.S.	(*wh.*)	thick-leaved.	Astracan.	1828.	——	H.⊙.	Pall. ill. t. 55.
7 sclerospérmum. R.S.	(*w.*)	hard-seeded.	———	1821.	——	H.⊙.	——— t. 56.
8 pubéscens. Led.	(*wh.*)	pubescent.	Siberia.	1827.	——	H.⊙.	
CAMPHORO'SMA. L.		CAMPHORO'SMA.	Tetrandria Monogynia. L.				
1 monspelìacum. W.	(*wh.*)	hairy.	S. Europe.	1640.	8. 9.	G. ħ.	Schk. hand. 1. t. 26.
perénne. Pall. ill. pl. imp. cogn. t. 57.							
2 ovàtum. R.S.	(*wh.*)	Hungarian.	Hungary.	1823.	——	H.⊙.	W. et K. hung. 1. t. 63.
GAL'ENIA. W.		GAL'ENIA.	Octandria Digynia. L.				
africàna. W.	(*pa.*)	African.	C. B. S.	1752.	6. 8.	G. ♄.	Lam. ill. t. 314.
AX'YRIS. W.		AX'YRIS.	Monœcia Triandria. L.				
1 amaranthoídes. W.	(*gr.*)	simple-spiked.	Siberia.	1758.	6. 7.	H.⊙.	Gmel. sib. t. 2. f. 2.
2 hy'brida. W.	(*gr.*)	bastard.	———	1780.	6. 8.	H.⊙.	——— t. 4. f. 1.
3 prostràta. W.	(*gr.*)	trailing.	———	1798.	7. 8.	H.⊙.	——— t. 4. f. 2.
DI'OTIS. W.		DI'OTIS.	Monœcia Tetrandria. L.				
ceratoídes. W.	(*wh.*)	shrubby.	Siberia.	1780.	8.	H. ♄.	Jacq. ic. 1. t. 189.
Ax'yris ceratoídes. Jacq. *Ceratospérmum pappòsum.* P.S.							

ORDO CLXVIII.

BEGONIACEÆ. *Ventenat. et Don prodr. p.* 223.

BEG'ONIA. L.		BEG'ONIA.	Monœcia Polyandria. L.				
1 nítida. S.S.	(*bh.*)	shining-leaved.	Jamaica.	1777.	5. 12.	S. ♄.	Sal. par. lond. t. 72.
2 dichótoma. S.S.	(*wh.*)	forked.	Caracas.	1800.	7. 8.	S. ♄.	Jacq. ic. 3. t. 619.
3 renifórmis. S.S.	(*wh.*)	kidney-leaved.	Brazil.	1823.	——	S. ♄.	Linn. trans. 1. t. 14. f. 1. 2.
4 spathulàta. S.S.	(*wh.*)	spatula-leaved.	W. Indies.	1813.	8. 10.	S. ħ.	Lodd. bot. cab. t. 17.
5 odoràta. S.S.	(*wh.*)	sweet-scented.	S. America.	1823.	6. 8.	S. ħ.	
6 sanguínea. S.S.	(*pi.*)	bloody.	Brazil.	1828.	6. 10.	S. ♃.	
7 Sellòii. L.O.	(*wh.*)	Sello's.	———	——	——	S. ħ.	
8 undulàta. B.M.	(*wh.*)	wave-leaved.	———	1825.	8. 12.	S. ♄.	Bot. mag. t. 2723.
9 semperflòrens. B.C.	(*w.*)	ever-blowing.	Mexico.	1829.	1. 12.	S. ♄.	Lodd. bot. cab. t. 1439.
10 Hookèri.	(*pi.*)	Hooker's.	....	1827.	——	S. ♄.	Bot. mag. t. 2920.
semperflòrens. B.M. *non.* B.C.							
11 dísticha. S.S.	(*wh.*)	distichous.	S. America.	1820.	6. 8.	S. ħ.	
12 macrophy'lla. S.S.	(*bh.*)	large-leaved.	Jamaica.	1793.	5. 9.	S. ♄.	
13 hirtélla. S.S.	(*wh.*)	fringed.	W. Indies.	1818.	——	S. ħ.	
14 Evansiàna. A.R.	(*pi.*)	Evans's.	China.	1804.	——	S. ♃.	Bot. mag. t. 1473.
díscolor. H.K. *published* 1813. *Evansiàna.* A.R. *published* 1811. B.M. 1812.							
15 pícta. Sm.	(*pi.*)	painted.	Nepaul.	1823.	5. 9.	S. ♃.	Sm. exot. bot. t. 101.
16 palmàta. D.P.	(*pi.*)	palmated.	———	1825.	——	S. ♃.	
17 vitifòlia. L.O.	......	Vine-leaved.	Brazil.	1829.	....	S. ħ.	
18 platanifòlia. L.O.	....	Plane-leaved.	———	——	....	S. ħ.	
19 diversifòlia. B.M.	(*pi.*)	various-leaved.	Mexico.	1827.	9. 12.	S. ħ. ◡.	Bot. mag. t. 2966.
20 villòsa. B.R.	(*wh.*)	villous.	Brazil.	——	8. 10.	S. ħ.	Bot. reg. t. 1252.
21 pauciflòra. B.R.	(*bh.*)	few-flowered.	W. Indies.	1816.	5. 9.	S. ħ.	——— t. 471.
22 pátula. H.S.S.	(*bh.*)	spreading.	Brazil.	1811.	——	S. ħ.	
23 bulbífera. L.O.	(*pi.*)	bulb-bearing.	———	1829.	9. 11.	S. ♃.	
24 monóptera. L.O.	(*pi.*)	one-winged.	———	——	——	S. ♃.	
25 tuberòsa. S.S.	(*pi.*)	tuberous.	Amboyna.	1810.	7. 9.	S. ♃.	Rumph. am. 5. t. 169. f. 2.
26 acutifòlia. S.S.	(*wh.*)	acute-leaved.	W. Indies.	1822.	——	S. ♄.	Sloan. jam. t. 127. f. 1. 2.
27 acuminàta. S.S.	(*wh.*)	pointed-leaved.	Jamaica.	1790.	5. 12.	S. ħ.	Bot. reg. t. 364.
28 lùcida. H.R.	(*wh.*)	polished.	Trinidad.	1820.	——	S.⊙.	
29 argyrostígma. B.R.	(*pi.*)	silver-spotted.	Brazil.	1819.	7. 9.	S. ♄.	Bot. reg. t. 666.
maculàta. S.S.							
30 dipétala. B.M.	(*pi.*)	two-petaled.	E. Indies.	1826.	4. 8.	S. ♄.	Bot. mag. t. 2849.
31 papillòsa. B.M.	(*pi.*)	papillose.	S. America.	1823.	——	S. ♄.	——— t. 2846.

32 insígnis. B.M.	(*pi.*) handsome.	Brazil.	1828.	1. 12.	S. ♄.	Bot. mag. t. 2900.
33 incarnàta. L.O.	(*pi.*) flesh-coloured.	———	1829.	——	S. ♄.	
34 Martiàna. L.O.	(*pi.*) Martius's.	———	——	——	S. ♄.	
35 suavèolens. B.C.	(*wh.*) fragrant.	W. Indies.	1816.	5. 9.	S. ♄.	Lodd. bot. cab. t. 69.
36 hùmilis. H.E.F.	(*wh.*) dwarf.	———	1788.	10.	S. ♂.	Hook. ex. flor. t. 17.
37 hirsùta. S.S.	(*wh.*) shaggy-leaved.	———	1789.	5. 6.	S. ♂.	Aub. gui. 2. t. 348.
38 ulmifòlia. S.S.	(*bh.*) Elm-leaved.	S. America.	1820.	6. 1.	S. ♄.	Hook. ex. flor. t. 57.
39 scándens. S.S.	(*gr.*) climbing.	Jamaica.	1824.	7. 9.	S. ♄. ◡.	Aub. gui. 2. t. 349.

ORDO CLXIX.

POLYGONEÆ. *Brown prodr. p.* 419.

COCCOL'OBA. L.	SEASIDE-GRAPE.	Octandria Trigynia. L.				
1 uvífera. S.S.	(*wh.*) round-leaved.	W. Indies.	1690.	....	S. ♄.	Jacq. amer. t. 73.
2 pubéscens. S.S.	(*wh.*) downy.	———	——	....	S. ♄.	Pluk. phyt. 222. f. 8.
3 excoriàta. S.S.	(*wh.*) oval-leaved.	———	1733.	....	S. ♄.	Plum. ic. t. 146. f. 1.
4 nívea. S.S.	(*st.*) snowy.	Jamaica.	1826.	....	S. ♄.	
5 punctàta. S.S.	(*wh.*) spear-leaved.	W. Indies.	1733.	....	S. ♄.	Jacq. amer. 114. t. 77.
6 laurifòlia. W.en.	(*wh.*) Laurel-leaved.	S. America.	1816.	8. 9.	S. ♄.	
7 diversifòlia. S.S.	(*wh.*) different-leaved.	Hispaniola.	——	——	S. ♄.	Hook. ex. flor. t. 102.
8 obtusifòlia. S.S.	(*wh.*) blunt-leaved.	S. America.	1823.	....	S. ♄.	Jacq. amer. t. 74.
9 microstàchya. S.S.	(*wh.*) small-spiked.	W. Indies.	——	....	S. ♄.	
10 barbadénsis. S.S.	(*wh.*) Barbadoes.	Barbadoes.	1790.	....	S. ♄.	Jacq. obs. 1. t. 8.
11 latifòlia. S.S.	(*wh.*) broad-leaved.	S. America.	1812.	....	S. ♄.	Lam. ill. t. 316. f. 4.
12 acuminàta. K.S.	(*wh.*) sharp-pointed.	———	1824.	....	S. ♄.	
13 longifòlia. L.en.	(*wh.*) long-leaved.	———	1818.	....	S. ♄.	
14 tenuifòlia. S.S.	(*wh.*) small-leaved.	Jamaica.	1824.	....	S. ♄.	Brown. jam. 210. f. 3.
BRUNNI'CHIA. S.S.	BRUNNI'CHIA.	Decandria Trigynia. W.				
cirrhòsa. S.S.	(*gr.*) Carolina.	Carolina.	1787.	....	G. ♄. ◡.	Gært. s. 1. t. 48. f. 2.
POLY'GONUM. L.	PERSICARIA.	Tetr-Octandria Di-Trigynia.				
1 Bistorta. W.	(*pi.*) Snake-weed.	Britain.	....	5. 9.	H. ♃.	Eng. bot. v. 8. t. 509.
2 bistortoídes. Ph.	(*bh.*) Snakeweed-like.	N. America.	1828.	6. 7.	H. ♃.	
3 vivìparum. E.B.	(*gr.*) alpine bistort.	Britain.	....	5. 9.	H. ♃.	Eng. bot. v. 10. t. 669.
4 petiolàtum. D.P.	(*pi.*) petiolate.	Nepaul.	1823.	——	H. ♃.	
5 macrophy'llum. D.P.	(*pu.*) large-leaved.	———	1822.	——	H. ♃.	
6 affi'ne. D.P.	(*ro.*) related.	———	1823.	——	H. ♃.	
7 mìte. P.S.	(*li.*) tasteless.	N. America.	1818.	6. 9.	H. ☉.	
8 hydropiperoídes. Ph.	(*w.*) acrid.	———	1824.	——	H. ☉.	
9 tenuiflòrum. S.S.	(*li.*) slender-flowered.	Sicily.	1826.	——	H. ♃.	
10 scándens. W.	(*pi.*) climbing.	N. America.	1749.	7. 9.	H. ♃. ◡.	Pluk. alm. t. 177. f. 7.
11 Fagop'yrum. W.	(*pi.*) Buck-wheat.	England.	....	7. 8.	H. ☉.	Eng. bot. v. 15. t. 1044.
12 emarginàtum. W.	(*pi.*) Indian buck-wheat.	Nepaul.	1796.	——	H. ☉.	Bot. reg. t. 1065.
13 tatáricum. W.	(*gr.*) Tartarian.	Siberia.	1759.	——	H. ☉.	Gmel. sib. 3. t. 13. f. 1.
14 Convólvulus. W.	(*bh.*) common climbing.	Britain.	....	5. 9.	H. ☉. ◡.	Eng. bot. v. 14. t. 941.
15 dumetòrum. W.	(*li.*) bush.	S. Europe.	1803.	——	H. ☉.	Flor. dan. t. 759.
16 adpréssum. B.P.	(*pi.*) polygamous.	N. S. W.	1822.	——	G. ♄. ◡.	Lab. n. hol. 1. t. 127.
17 sagittàtum. W.	(*wh.*) arrow-leaved.	N. America.	1759.	7. 8.	H. ☉.	Bart. fl. amer. t. 101.
18 chinénse. W.	(*pi.*) Chinese.	China.	1795.	——	H. ☉.	Burm. ind. t. 30. f. 3.
19 pátens. D.P.	(*pi.*) spreading.	Nepaul.	1823.	——	H. ♃.	
20 élegans. S.S.	(*wh.*) elegant.	Naples.	1824.	——	H. ♄.	
21 alpìnum. W.	(*wh.*) Alpine.	Italy.	1816.	——	H. ♃.	Bocc. mus. 2. t. 27.
22 undulàtum. W.	(*wh.*) wave-leaved.	Siberia.	1789.	6. 7.	H. ♃.	Gmel. sib. 3. t. 10.
23 divaricàtum. W.	(*wh.*) divaricated.	———	1759.	7. 8.	H. ♃.	——— 3. t. 11. f. 1.
24 salígnum. W.	(*wh.*) Willow-like.	———	1816.	5. 8.	H. ♃.	
25 acídulum. W.	(*wh.*) narrow-leaved.	———	———	6. 7.	H. ♃.	
26 cymòsum. S.S.	(*wh.*) cyme-flowered.	Naples.	1827.	7. 9.	H. ♃.	
27 aviculàre. E.B.	(*ro.*) Knot-grass.	Britain.	....	4. 10.	H. ☉.	Eng. bot. v. 18. t. 1252.
28 Dryándri. S.S.	(*bh.*) Dryander's.	E. Indies.	1796.	4. 8.	S. ♃.	
élegans. H.K. *nec aliorum.*						
29 arenàrium. W.K.	(*pu.*) sand.	Hungary.	1807.	5. 8.	H. ☉.	W. et K. hung. t. 67.
30 ramòsum. L.en.	(*wh.*) branching.	Siberia.	1822.	6. 8.	H. ♃.	
31 littoràle. L.en.	(*wh.*) sea-side.	S. Europe.	1816.	5. 8.	H. ♄.	
32 Bellárdi. P.S.	(*wh.*) Bellardi's.	———	1822.	——	H. ♃.	All. ped. t. 90. f. 2.
33 pátulum. M.B.	(*wh.*) spreading-branch'd.	Tauria.	——	——	H. ♃.	

34	marítimum. w.	(wh.)	sea-shore.	S. Europe.	1816.	6. 8.	H. ♄.	Flor. græc. t. 363.
35	acetòsum. M.B.	(li.)	Sorrel-leaved.	Astracan.	1823.	——	H. ♃.	
36	eréctum. w.	(wh.)	upright.	N. America.	1792.	7. 8.	H. ♃.	
37	linifòlium. B.A.	(bh.)	Flax-leaved.	————	1824.	6. 8.	H. ⊙.	Bart. fl. amer. 3. t. 95. f. 2.
38	herniarioídes. s.s.	(wh.)	Egyptian.	Egypt.	1822.	——	F. ♄.	
39	pensylvánicum. w.	(ro.)	Pensylvanian.	N. America.	1819.	7. 8.	H. ⊙.	
40	orientàle. w.	(re.)	common.	E. Indies.	1707.	7. 10.	H. ⊙.	Bot. mag. t. 213.
	β álbum.	(wh.)	*white-flowered.*	————	....	——	H. ⊙.	
41	barbàtum. s.s.	(re.)	bearded.	China.	1822.	6. 9.	F. ♃.	Rheed. mal. 12. t. 77.
42	tinctòrium. w.	(re.)	dyer's.	————	1776.	7. 8.	G. ♂.	
43	decípiens. B.P.	(ro.)	decipient.	N. S. W.	1822.	——	G. ♃.	
44	plebèium. B.P.	(bh.)	linear-leaved.	————	——	——	G. ♃.	
45	Persicària.	(li.)	spotted.	Britain.	....	——	H. ⊙.	Eng. bot. v. 11. t. 756.
46	lapathifòlium. s.s.	(gr.)	Dock-leaved.	————	....	7. 10.	H. ⊙.	———— v. 20. t. 1382.
47	incànum. W. en.	(li.)	hoary.	Germany.	1804.	——	H. ⊙.	Schrank. bav. 1. t. 3.
48	persicarioídes. K.S.	(w.)	Mexican.	Mexico.	1825.	——	H. ♃.	
49	Hydropìper. E.B.	(gr.)	Water pepper.	Britain.	....	7. 9.	H. ⊙.	Eng. bot. v. 14. t. 989.
50	glábrum. w.	(gr.)	smooth.	E. Indies.	1822.	——	S. ♃.	
51	mìnus. E.B.	(li.)	slender.	England.	....	——	H. ⊙.	Eng. bot. v. 15. t. 1043.
52	rêpens.	(wh.)	rooting-branched.	Fulham.	....	8. 10.	H. ⊙.	
53	tortuòsum. D.P.	(wh.)	twisting.	Nepaul.	1824.	7. 10.	F. ♃.	
54	equisetifórme. s.s.	(wh.)	Equisetum-like.	Britain.	....	——	H. ⊙.	Flor. græc. t. 364.
55	oxyspérmum. Led.	(w.)	sharp-seeded.	Siberia.	1827.	——	H. ♃.	
56	arifòlium. w.	(bh.)	Arum-leaved.	N. America.	1816.	5. 10.	H. ⊙.	Bart. fl. amer. t. 100.
57	ocreàtum. w.	(bh.)	spear-leaved.	Siberia.	1780.	6. 9.	H. ♃.	Gmel. sib. 3. t. 8.
58	virginiànum. w.	(pa.)	Virginian.	N. America.	1640.	8. 9.	H. ♃.	Park. theat. 857. f. 6.
59	amphíbium. w.	(ro.)	amphibious.	Britain.	....	6. 8.	H. w. ♃.	Eng. bot. v. 7. t. 436.
60	grácile. B.P.	(pi.)	slender-stalked.	N. S. W.	1822.	——	G. ♃.	
61	acetosæfòlium. P.S.	(re.)	Brazilian.	Brazil.	——	——	S. ♄. ◡.	Venten. cels. t. 88.
	TRAGOP'YRUM. M.B.		**TRAGOP'YRUM.**	Octandria Trigynia.				
1	lanceolàtum. M.B.	(pi.)	spear-leaved.	Siberia.	1770.	7. 8.	H. ♄.	Bot. reg. t. 255.
	Poly'gonum frutéscens. B.R.							
2	buxifòlium. M.B.	(pi.)	Box-leaved.	Caucasus.	1800.	——	H. ♄.	Bot. mag. t. 1065.
	Poly'gonum críspulum. B.M.							
	ATRAPHA'XIS. L.		**ATRAPHA'XIS.**	Hexandria Digynia. L.				
1	spinòsa. s.s.	(gr.)	prickly.	Levant.	1732.	8.	G. ♃.	L'Herit. stirp. 1. t. 14.
2	undulàta. s.s.	(gr.)	waved-leaved.	C. B. S.	——	6. 7.	G. ♄.	Dill. elt. t. 32. f. 36.
	OXY'RIA. Sm.		**MOUNTAIN SORREL.**	Hexandria Digynia. Sm.				
	renifórmis. Sm.	(gr.)	kidney-leaved.	Britain.	....	6. 7.	H. ♃.	Eng. bot. v. 13. t. 910.
	Rùmex dig'ynus. E.B.							
	R'UMEX. L.		**DOCK.**	Hexandria Trigynia. L.				
1	Hydrolapáthum. s.s.	(gr.)	great water.	Britain.	....	7. 8.	H. w. ♃.	Eng. bot. v. 30. t. 2104.
	aquáticus. E.B.							
2	Británnica. s.s.	(gr.)	Virginian.	N. America.	....	6. 7.	H. ♃.	
3	máximus. s.s.	(gr.)	great.	Europe.	....	——	H. ♃.	
4	acùtus. E.B.	(gr.)	sharp.	Britain.	....	——	H. ♃.	Eng. bot. v. 11. t. 724.
5	Nemolapáthum. w.	(gr.)	flat-valved.	Europe.	1816.	——	H. ♃.	
6	nemoròsus. W. en.	(gr.)	wood.	————	————	——	H. ♃.	
7	sanguíneus. E.B.	(gr.)	bloody-veined.	England.	....	6. 7.	H. ♃.	Eng. bot. v. 22. t. 1533.
8	condylòdes. M.B.	(gr.)	spreading.	Caucasus.	1823.	——	H. ♃.	
9	críspus. E.B.	(gr.)	curled.	Britain.	....	——	H. ♃.	Eng. bot. v. 28. t. 1998.
10	Patiéntia. w.	(gr.)	Patience.	Italy.	1573.	——	H. ♃.	
11	confértus. W. en.	(gr.)	close-leaved.	————	1816.	——	H. ♃.	
12	brasiliénsis. L. en.	(gr.)	Brazilian.	Brazil.	1823.	——	G. ♃.	
13	verticillàtus. s.s.	(gr.)	whorled.	N. America.	1818.	——	H. ♃.	
14	spathulàtus. P.S.	(gr.)	spatulate-leav'd.	C. B. S.	1824.	8. 2.	G. ♃.	
15	purpùreus. P.S.	(gr.)	purple-veined.	Europe.	....	6. 7.	H. ♃.	
16	lineàris. s.s.	(gr.)	linear-leaved.	C. B. S.	1824.	8. 12.	G. ♃.	
17	salicifòlius. s.s.	(gr.)	Willow-leaved.	California.	1821.	7. 10.	F. ♃.	
18	lævigàtus. s.s.	(gr.)	smooth.	....	1828.	——	H. ♃.	
19	sylvéstris. s.s.	(gr.)	wood.	Germany.	1826.	6. 8.	H. ♃.	
20	obtusifòlius. E.B.	(gr.)	broad-leaved.	Britain.	....	——	H. ♃.	Eng. bot. v. 28. t. 1999
21	nepalénsis. s.s.	(gr.)	Nepalese.	Nepaul.	1823.	——	H. ♃.	
22	palústris. E.B.	(ye.)	yellow marsh.	England.	....	7. 8.	H. ♃.	Eng. bot. v. 27. t. 1932
23	marítimus. E.B.	(ye.)	sea-side.	Britain.	....	——	H. ♃.	———— v. 11. t. 725.
24	divaricàtus. w.	(gr.)	divaricate.	Italy.	1793.	——	H. ♃.	Till. pis. t. 37. f. 2.
25	stríctus. L. en.	(gr.)	upright.	....	1824.	——	H. ♃.	
26	persicarioídes. w.	(gr.)	Persicaria-like.	N. America.	1773.	6. 7.	H. ♃.	
27	dentàtus. w.	(gr.)	dentated.	Egypt.	1782.	7. 8.	H. ⊙.	Dill. elt. t. 158. f. 191.

28 ægyptìacus. w.	(*gr.*)	Egyptian.	Egypt.	1734.	6. 7.	H.⊙.	Till. pis. t. 37. *f*. 1.
29 ucrànicus. s.s.	(*gr.*)	Ukraine.	Ukraine.	1820.	——	H.⊙.	
30 crispàtulus. M.	(*gr.*)	curled-leaved.	N. America.	——	——	H.⊙.	
31 cristàtus. DC.	(*gr.*)	crested.	....	——	——	H.♃.	
32 púlcher. E.B.	(*gr.*)	fiddle.	Britain.	....	6. 8.	H.♃.	Eng. bot. v. 22. t. 1576.
33 aquáticus. s.s.	(*gr.*)	water.	Europe.	1818.	——	H.*w.*♃.	
34 scutàtus. w.	(*gr.*)	French sorrel.	France.	1596.	6. 7.	H.♃.	Jacq. ic. 1. t. 67.
35 hastifòlius. M.B.	(*gr.*)	hastate-leaved.	Caucasus.	1816.	——	H.♃.	
36 vesicàrius. w.	(*gr.*)	bladder.	Africa.	1656.	7. 8.	H.⊙.	Moris. s. 5. t. 28. f. 7.
37 tingitànus. w.	(*gr.*)	Tangier.	Barbary.	1680.	6. 8.	H.♃.	Zanon. hist. 9. t. 6.
38 longifòlius. K.S.	(*gr.*)	long-leaved.	S. America.	1827.	7. 10.	F.♃.	
39 sarcorhìzus. L.en.	(*gr.*)	fleshy-rooted.	C. B. S.	1824.	8. 12.	G.♃.	
40 Burchéllii. s.s.	(*gr.*)	Burchell's.	————	1816.	8. 10.	G.ħ.	
41 Lunària. w.	(*gr.*)	tree.	Canaries.	1698.	6. 7.	G.♄.	Pluk. alm. t. 252. f. 3.
42 orientàlis. Bern.	(*gr.*)	oriental.	Levant.	1821.	7. 9.	H.♃.	
43 conspérsus. Ot.	(*gr.*)	scattered.	....	1824.	——	H.♃.	
44 doméesticus. s.s.	(*gr.*)	various-leaved.	Sweden.	1823.	——	H.♃.	
45 ròseus. w.	(*gr.*)	rose.	Egypt.	1737.	7. 8.	H.⊙.	
46 lácerus. s.s.	(*gr.*)	torn-leaved.	————	1825.	——	H.⊙.	
47 bucephalóphorus.	(*gr.*)	Basil-leaved.	Italy.	1683.	6.	H.⊙.	Cavan. ic. 1. t. 41. f. 1.
48 fimbriàtus. s.s.	(*gr.*)	fringe-valved.	C. B. S.	1816.	7. 9.	G.♃.	
49 Bròwnii. s.s.	(*gr.*)	Brown's fringed.	N. S. W.	1824.	——	G.♃.	
50 Acetòsa. E.B.	(*gr.*)	common Sorrel.	Britain.	....	6. 8.	H.♃.	Eng. bot. v. 2. t. 127.
51 Acetosélla. E.B.	(*st.*)	sheep's sorrel.	————	....	5. 7.	H.♃	———— v. 24. t. 1674.
52 abyssínicus. Jacq.	(*gr.*)	Abyssinian.	Abyssinia.	1823.	——	G.ħ	Jacq. vind. 3. t. 98.
53 tuberòsus. w.	(*gr.*)	tuberous-rooted.	S. Europe.	1752.	6. 8.	H.♃.	
54 arifòlius. s.s.	(*gr.*)	Arum-leaved.	Africa.	1775.	12. 4.	G.ħ.	
55 triangulàris. DC.	(*gr.*)	triangular-leaved.	S. Europe.	1824.	——	F.♃.	
56 amplexicaùlis. s.s.	(*gr.*)	stem-clasping.	Pyrenees.	1816.	6. 8.	H.♃.	
57 luxùrians. s.s.	(*gr.*)	luxuriant.	S. Europe.	....	——	H.♃.	
58 alpìnus. w.	(*gr.*)	Alpine.	France.	1597.	6. 7.	H.♃.	Zorn. ic. 261.
59 frutéscens. s.s.	(*gr.*)	frutescent.	Tristan d'Ac.	1816.	6. 9.	G.ħ.	
60 gigánteus. H.K.	(*gr.*)	tall.	Sandwich Isl.	1796.	6. 8.	G.♃.	
EMEX. S.S.		'EMEX.	Hexandria Trigynia. S.S.				
spinòsus. s.s.	(*gr.*)	spiny-seeded.	Candia.	1656.	6. 7.	H.⊙.	
Rùmex spinòsus. w.							
PODO'PTERUS. K.S.		PODO'PTERUS.	Hexandria Trigynia. S.S.				
mexicànus. K.S.	(*wh.*)	Mexican.	Mexico.	1825.	....	G.♄.	H. et B. pl. am. 2. t. 107.
TRIPL'ARIS. K.S.		TRIPL'ARIS.	Diœcia Enneandria. L.				
americàna. K.S.	(*ro.pu.*)	American.	S. America.	1823.	5. 8.	S.♄.	Jacq. amer. 13. t. 173. f. 5.
RH'EUM. L.		RHUBARB.	Enneandria Trigynia. L.				
1 Rhapónticum. w.	(*wh.*)	common.	Asia.	1573.	5. 6.	H.♃.	Sabb. hort. 1. t. 34.
2 undulàtum. w.	(*wh.*)	wave-leaved.	China.	1734.	——	H.♃.	L. amœn. acad. 3. t. 4.
3 críspum. And.	(*wh.*)	curl-leaved.	....	1820.	——	H.♃.	
4 compáctum. w.	(*wh.*)	thick-leaved.	Tartary.	1758.	——	H.♃.	Mill. ic. 2. t. 218.
5 tatáricum. w.	(*wh.*)	Tartarian.	————	1793.	——	H.♃.	
6 hy'bridum. w.	(*wh.*)	bastard.	Asia.	1778.	——	H.♃.	Murr. c. gott. t. 1.
7 austrìacum. And.	(*wh.*)	Austrian.	Austria.	1819.	——	H.♃.	
8 sibíricum. And.	(*wh.*)	Siberian.	Siberia.	——	——	H.♃.	
9 fenestràtum. F.	(*wh.*)	windowed.	....	1760.	——	H.♃.	
10 rotundifòlium. And.	(*w.*)	round-leaved.	....	....	——	H.♃.	
11 nùtans. F.	(*wh.*)	nodding.	Asia.	1824.	——	H.♃.	
12 Rìbes. w.	(*wh.*)	warted-leaved.	Levant.	1724.	——	H.♃.	Ann. mus. 1. t. 49.
13 palmàtum. w.	(*wh.*)	palmated.	China.	1763.	4. 5.	H.♃.	Linn. fasc. 7. t. 4.
14 australe. D.P.	(*d.pu.*)	Nepaul.	Nepaul.	1826.	——	H.♃.	Swt. br. fl. gar. t. 269.
ERIO'GONUM. M.		ERIO'GONUM.	Enneandria Trigynia. N.				
1 tomentòsum. M.	(*wh.*)	woolly.	Carolina.	1811.	5. 6.	H.♃.	Mich. amer. v. 1. t. 24.
2 flàvum N.	(*ye.*)	ilky.	Missouri.	——	7.	H.♃.	
serícеu.m. Ph.							
3 pauciflòrum. Ph.	(*wh.*)	few-flowered.	Louisiana.	1812.	——	H.♃.	
4 longifòlium. N.	(*wh.*)	long-leaved.	Arkansa.	1824.	——	H.♃.	
CALLI'GONUM. W.		CALLI'GONUM.	Dodecandria Tetragynia. L.				
1 Pallàsia. w.	(*wh.gr.*)	Caspian.	Caspian shores.	1780.	8.	H.ħ.	Pall. ros. 2. t. 77. 78.
2 Pandèri. w.	(*gr.wh.*)	Fischer's.	————	1824.	7. 9.	H.ħ.	
3 comòsum. P.S.	(*gr.wh.*)	tufted.	Persia.	1829.	....	H.ħ.	
KŒNI'GIA. W.		KŒNI'GIA.	Triandria Trigynia. L.				
islándica. w.	(*gr.*)	Iceland.	Iceland.	1773.	4.	H.⊙.	Lam. ill. gen. t. 51.

ORDO CLXX.

LAURINÆ. *Brown prodr. p.* 401.

LA'URUS. W.		LA'URUS. Enneandria Monogynia. L.					
1 nòbilis. W.	(*gr.ye.*)	Sweet Bay.	S. Europe.	1561.	4. 5.	H.ħ.	Flor. græc. t. 365.
β *undulàta.*	(*gr.ye.*)	*wave-leaved.*	———	——	——	H.ħ.	
γ *salicifòlia.*	(*gr.ye.*)	*Willow-leaved.*	———	——	——	H.ħ.	
δ *variegàta.*	(*gr.ye.*)	*variegated.*	———	——	——	H.ħ.	
2 coriàcea. S.S.	(*gr.ye.*)	leathery-leaved.	Jamaica.	1822.	....	S.♄.	
3 salicifòlia. S.S.	(*gr.*)	Willow-leaved.	W. Indies.	——	....	S.♄.	
4 Catesb'æi. P.S.	(*wh.*)	Catesby's.	Florida.	1811.	4. 6.	F.♄.	Catesb. car. 1. t. 28.
5 carolinénsis. M.	(*gr.ye.*)	Carolina.	N. America.	1806.	——	F.♄.	——— 1. t. 63.
α *glábra.*	(*gr.ye.*)	*smooth-leaved.*	———	——	——	F.♄.	
β *pubéscens.*	(*gr.ye.*)	*downy.*	———	——	——	F.♄.	
γ *obtùsa.*	(*gr.ye.*)	*blunt-leaved.*	———	——	——	F.♄.	
6 Borbònia. W.	(*gr.ye.*)	Bourbon.	S. America.	1739.	——	F.♄.	Plum. gen. 4.
7 thyrsiflòra. S.S.	(*gr.ye.*)	bunch-flowered.	Madagascar.	1824.	....	S.♄.	
8 índica. W.	(*gr.ye.*)	royal bay.	Madeira.	1665.	10. 3.	G.♄.	Pluk. alm. t. 304. f. 1.
9 f'œtens. W.	(*gr.ye.*)	Madeira or Til.	———	1760.	——	G.♄.	
10 canariénsis. W.en.	(*st.*)	Canary.	Canaries.	1815.	——	G.♄.	
11 aggregàta. B.M.	(*wh.*)	clustered.	China.	1812.	——	S.♄.	Bot. mag. t. 2497.
12 Chloróxylon. W.	(*gr.ye.*)	Cogwood-tree.	Jamaica.	1778.	....	S.♄.	Brown. jam. t. 7. f. 1.
13 péndula. W.	(*gr.ye.*)	pendulous.	———	1822.	....	S.♄.	
14 glaùca. W.	(*gr.ye.*)	glaucous.	Japan.	1806.	....	G.♄.	
15 æstivàlis. W.	(*gr.ye.*)	veined-leaved.	N. America.	1765.	4. 5.	G.♄.	
16 Benzóin. W.	(*gr.ye.*)	Benjamin-tree.	———	1683.	——	H.♄.	Comm. hort. 1. t. 97.
17 Dióspyrus. P.S.	(*gr.ye.*)	twiggy.	———	1810.	——	H.♄.	Bot. mag. t. 1470.
18 geniculàta. M.	(*gr.ye.*)	flexuose.	———	1759.	——	H.♄.	——— t. 1471.
19 Sássafras. W.	(*gr.ye.*)	Sassafras-tree.	———	1633.	5. 6.	H.♄.	Catesb. car. 1. t. 55.
20 álbida. N.	(*gr.ye.*)	white Sassafras.		....	——	H.♄.	
PE'RSEA. G.		ALLIGATOR-PEAR. Enneandria Monogynia. S.S.					
gratíssima. G.	(*st.*)	common.	W. Indies.	1739.	4. 6.	S.♄.	Bot. reg. t. 1258.
Laùrus Pérsea. W.							
CINNAM'OMUM. B.P.		CINNAMON. Enneandria Monogynia.					
1 Cássia. S.S.	(*gr.*)	pubescent.	E. Indies.	1768.	5. 9.	S.♄.	Andr. reposit. t. 596.
Laùrus Cinnamòmum. A.R. *Pérsea Cássia.* S.S.							
2 vèrum.	(*gr.*)	true.	Ceylon.	1763.	6. 9.	S.♄.	Bot. mag. t. 1636.
Laùrus Cássia. B.M. *Pérsea Cinnamòmum.* S.S.							
3 dúlcis. S.S.	(*gr.wh.*)	sweet.	E. Indies.	1824.	....	S.♄.	
4 nítidum. H.E.F.	(*gr.*)	glossy-leaved.	Sumatra.	——	6. 9.	S.♄.	Hook. ex. flor. t. 176.
5 Bejolghòta.	(*gr.*)	square-branch'd.	E. Indies.	1805.	....	S.♄.	
Laùrus Bejolghòta. L.T. *Laùrus malabáthrica.* H.B?							
6 Culitlàban.	(*gr.*)	three-flowered.	Moluccas.	....	....	S.♄.	Rumph. amb. 2. t. 14.
Laùrus Culitlàban. H.B.							
7 grácile. Hort.	(*gr.*)	slender.	E. Indies.	1805.	....	S.♄.	
8 Camphòra.	(*wh.*)	Camphire-tree.	Japan.	1727.	3. 6.	G.♄.	Bot. mag. t. 2658.
Laùrus Camphòra. B.M. *Pérsea Camfòra.* S.S.							
OC'OTEA. K.S.		OC'OTEA. Dodecandria Monogynia.					
psychotrioídes. K.S.	(*gr.*)	spear-leaved.	Mexico.	1828.	....	G.♄.	
CRYPTOC'ARYA. B.P.		CRYPTOC'ARYA. Dodecandria Monogynia.					
1 glaucéscens. B.P.	(*gr.*)	glaucous-leaved.	N. S. W.	1820.	....	G.♄.	
2 obovàta. B.P.	(*gr.*)	obovate-leaved.	———	1823.	....	G.♄.	
TETRANTH'ERA. B.P.		TETRANTH'ERA. Diœcia Enneandria.					
1 laurifòlia. B.R.	(*gr.wh.*)	Laurel-leaved.	Japan.	1823.	1. 5.	G.♄.	Bot. reg. t. 893.
2 apetàla. B.P.	(*gr.wh.*)	petalless.	E. Indies.	——	——	G.♄.	Roxb. corom. 2. t. 147.
3 dealbàta. B.P.	(*gr.*)	white-leaved.	N. S. W.	1824.	....	G.♄.	
4 involucràta. Rox.	(*gr.*)	involucred.	E. Indies.	1820.	1. 5.	G.♄.	
CASS'YTHA. L.		CASS'YTHA. Dodecandria Monogynia.					
1 filifórmis. L.	(*wh.*)	slender.	C. B. S.	....	5. 8.	G.♃.⌣.	Rumph. amb. 5. t. 184. f. 4
2 glabélla. B.P.	(*wh.*)	smooth.	N. S. W.	1822.	——	G.♃.⌣.	
3 paniculàta. B.P.	(*wh.*)	panicled.	———	1824.	——	G.♃.⌣.	

LAURINIS AFFINE.

GYROCA'RPUS. B.P.		GYROCA'RPUS.	Polygamia Monœcia.				
1 americànus. W.	(*ye.*)	American.	W. Indies.	1816.	....	S. ♄.	Jacq. amer. t. 178. f. 80.
2 asiáticus. W.	(*ye.*)	Asiatic.	E. Indies.	1812.	....	S. ♄.	Roxb. corom. 1. t. 1.
Jacquìni. R.C.							
HERNA'NDIA. L.		JACK IN A BOX.	Monœcia Triandria. L.				
1 sonòra. W.	(*wh.*)	peltate-leaved.	W. Indies.	1693.	....	S. ♄.	Rumph. amb. 2. t. 85.
2 guianénsis. W.	(*wh.*)	Guiana.	Guiana.	1823.	....	S. ♄.	Aub. gui. 2. t. 309.
3 ovígera. W.	(*wh.*)	oval-leaved.	E. Indies.	1820.	....	S. ♄.	Rumph. amb. 3. t. 123.

ORDO CLXXI.

MYRISTICÆÆ. *Brown prodr. p.* 399.

MYRI'STICA. W.		NUTMEG.	Diœcia Monadelphia. L.				
1 moschàta. W.	(*ye.*)	true.	E. Indies.	1795.	....	S. ♄.	Hook. ex. flor. t. 155-156
officinàlis. H.E.F. *aromática.* Lam.							
2 sebífera. S.S.	(*ye.*)	woolly-fruited.	S. America.	1829.	....	S. ♄.	
3 fátua. S.S.	(*ye.*)	tasteless.	Surinam.	1812.	....	S. ♄.	Pluk. alm. t. 250. f. 6.

ORDO CLXXII.

PROTEACEÆ. *Brown Linn. trans.* 10. *p.* 46.

A'ULAX. P.L.		A'ULAX.	Diœcia Tetrandria. H.K.				
1 pinifòlia. L.T.	(*ye.*)	Pine-leaved.	C. B. S.	1780.	7. 9.	G. ♄.	Andr. reposit. t. 76.
Pròtea pinifòlia. A.R.							
2 umbellàta. L.T.	(*ye.*)	umbelled.	———	1774.	6. 8.	G. ♄.	Andr. reposit. t. 248.
LEUCADE'NDRON. L.T.		LEUCADE'NDRON.	Diœcia Tetrandria. H.K.				
1 argénteum. L.T.	(*st.*)	Silver-tree.	C. B. S.	1693.	6. 8.	G. ♄.	Bot. reg. t. 979.
Pròtea argéntea. W.							
2 plumòsum. L.T.	(*ye.*)	feathered.	———	1774.	——	G. ♄.	Thunb. diss. 40. t. 4.
Pròtea parviflòra. Th.							
3 retùsum. L.T.	(*ye.*)	retuse-leaved.	———	1825.	——	G. ♄.	
4 spathulàtum. L.T.	(*ye.*)	spathulate.	———	1812.	6. 8.	G. ♄.	
5 angustàtum. L.T.	(*ye.*)	narrow-leaved.	———	1824.	——	G. ♄.	
6 imbricàtum. L.T.	(*ye.*)	imbricated.	———	1790.	——	G. ♄.	
7 buxifòlium. L.T.	(*ye.*)	Box-leaved.	———	1812.	——	G. ♄.	Wend. h. herr. t. 14?
8 Levisànus. L.T.	(*ye.*)	short-leaved.	———	1774.	4. 6.	G. ♄.	——— t. 1.
9 linifòlium. L.T.	(*ye.*)	Flax-leaved.	———	1809.	——	G. ♄.	Jacq. schœn. 1. t. 26.
10 stellàtum.	(*ye.*)	starry.	———	——	5. 6.	G. ♄.	Bot. mag. t. 881.
Pròtea stellàris. B.M. *squarròsa.* K.P.							
11 fusciflòrum. L.T.	(*br.*)	brown-flowered.	———	1812.	——	G. ♄.	Jacq. schœnb. 1. t. 27.
12 tórtum. L.T.	(*ye.*)	twisted-leaved.	———	1790.	3. 5.	G. ♄.	Bot. reg. t. 826.
13 cinèreum. L.T.	(*ye.*)	gray-leaved.	———	1774.	7. 8.	G. ♄.	
14 corymbòsum. L.T.	(*ye.*)	corymbed.	———	1790.	4. 7.	G. ♄.	Bot. reg. t. 402.
15 decòrum. L.T.	(*ye.*)	decorous.	———		5. 8.	G. ♄.	
16 squarròsum. L.T.	(*ye.*)	squarrose.	———	1812.	——	G. ♄.	Boerh. lug. 2. p. 197. c. ic.
Pròtea strobilìna. L.							
17 cóncolor. L.T.	(*ye.*)	one-coloured.	———	1774.	3. 6.	G. ♄.	Andr. reposit. t. 307.
Pròtea globòsa. B.M. 878. *strobilìna.* Schrad. s. han. 1. t. 1.							
18 grandiflòrum. L.T.	(*ye.*)	great-flowered.	C. B. S.	1789.	4. 6.	G. ♄.	Sal. par. lond. t. 105.
19 ovàle. L.T.	(*ye.*)	oval-leaved.	———	1818.	——	G. ♄.	
20 venòsum. L.T.	(*ye.*)	veined-leaved.	———	1815.	——	G. ♄.	
21 decúrrens. L.T.	(*ye.*)	decurrent.	———	1812.	6. 8.	G. ♄.	
22 glábrum. L.T.	(*ye.*)	smooth.	———	——	3. 6.	G. ♄.	

23 stríctum. L.T.	(*ye.*)	upright.	C. B. S.	1795.	4. 6.	G. ♄.	Sal. par. lond. t. 75.
Euryspérmum salicifòlium. P.L. *Pròtea conífera.* Andr. rep. t. 541.							
24 virgàtum. L.T.	(*ye.*)	slender.	C. B. S.	....	——	G. ♄.	
25 adscéndens. L.T.	(*ye.*)	ascending.	————	1774.	6. 8.	G. ♄.	Pluk. mant. t. 229. f. 6.
26 concínnum. L.T.	(*ye.*)	neat.	————	1800.	——	G. ♄.	
27 salígnum. L.T.	(*ye.*)	Willow-leaved.	————	1774.	4. 6.	G. ♄.	Boer. lugd. 2. p. 204. ic.
28 uliginòsum. L.T.	(*ye.*)	swamp.	————	1795.	——	G. ♄.	Breyn. cent. 21. t. 9.
29 flòridum. L.T.	(*ye.*)	florid.	————	——	——	G. ♄.	Pluk. phyt. t. 229. f. 4.
Pròtea salígna mas. *et* fem. Andr. rep. t. 572 ?							
30 platyspérmum. L.T.	(*ye.*)	flat-seeded.	————	1818.	——	G. ♄.	
31 comòsum. L.T.	(*ye.*)	tufted.	————	1812.	5. 8.	G. ♄.	
32 'æmulum. L.T.	(*ye.*)	related.	————	1789.	6. 9.	G. ♄.	
Pròtea incúrva. Andr. rep. t. 429.							
33 abietìnum. L.T.	(*ye.*)	Pine-leaved.	————	——	7. 9.	G. ♄.	Andr. reposit. t. 461.
Pròtea teretifòlia. Andr. rep. t. 461.							
34 scábrum. L.T.	(*ye.*)	rough.	————	1812.	4. 6.	G. ♄.	
35 seríceum. L.T.	(*ye.*)	silky-leaved.	————	1818.	——	G. ♄.	
36 Globulària. L.T.	(*ye.*)	globe-headed.	————	1810.	——	G. ♄.	Lam. ill. 1. t. 53.
37 pubéscens. L.T.	(*ye.*)	pubescent.	————	1822.	——	G. ♄.	
38 ericifòlium. L.T.	(*ye.*)	Heath-leaved.	————	——	——	G. ♄.	
39 infléxum. L.en.	(*ye.*)	inflexed.	————	1812.	——	G. ♄.	
40 caudàtum. L.en.	(*ye.*)	tailed.	————	——	——	G. ♄.	
41 marginàtum. L.en.	(*ye.*)	villous-margined.	————	1818.	——	G. ♄.	
42 crassifòlium. L.T.	(*ye.*)	thick-leaved.	————	1816.	——	G. ♄.	
43 cartilagíneum. L.T.	(*ye.*)	bony-margined.	————	——	——	G. ♄.	
PETRO'PHILA. L.T.		PETRO'PHILA.	Tetrandria Monogynia. R.S.				
1 teretifòlia. L.T.	(*wh.*)	round-leaved.	N. Holland.	1824.	....	G. ♄.	
2 filifòlia. L.T.	(*wh.*)	thread-leaved.	————	——	....	G. ♄.	
3 aciculàris. L.T.	(*wh.*)	needle-leaved.	————	——	....	G. ♄.	
4 pulchélla. L.T.	(*wh.*)	Fennel-leaved.	N. S. W.	1790.	7. 8.	G. ♄.	Bot. mag. t. 796.
5 pedunculàta. L.T.	(*st.*)	peduncled.	————	1824.	4. 8.	G. ♄.	
6 diversifòlia. L.T.	(*wh.*)	various-leaved.	N. Holland.	1803.	4. 6.	G. ♄.	
7 trífida. L.T.	(*wh.*)	trifid-leaved.	————	1824.	....	G. ♄.	
ISOP'OGON. L.T.		ISOP'OGON.	Tetrandria Monogynia. R.S.				
1 teretifòlius. L.T.	(*wh.*)	round-leaved.	N. Holland	1824.	4. 8.	G. ♄.	
2 anethiifòlius. L.T.	(*wh.*)	Dill-leaved.	N. S. W.	1796.	3. 6.	G. ♄.	Cavan. ic. 6. t. 549.
3 formòsus. L.T.	(*li.*)	handsome.	N. Holland.	1805.	——	G. ♄.	Bot. reg. t. 1288.
4 anemonifòlius. L.T.	(*st.*)	Anemone-leav'd.	N. S. W.	1791.	7. 9.	G. ♄.	Bot. mag. t. 697.
5 divaricàtus. A.R.	(*ye.*)	spreading.	————	——	——	G. ♄.	Andr. rep. t. 465.
6 trilòbus. L.T.	(*wh.*)	three-lobed.	N. Holland.	1803.	5. 6.	G. ♄.	
7 longifòlius. L.T.	(*ye.*)	long-leaved.	————	1820.	——	G. ♄.	Bot. reg. t. 900.
8 propínquus. M.C.	(*ye.*)	related.	————	1824.	....	G. ♄.	
9 attenuàtus. L.T.	(*wh.*)	attenuated.	————	——	5. 6.	G. ♄.	
10 polycéphalus. L.T.	(*wh.*)	many-headed.	————	——	....	G. ♄.	
11 axillàris. L.T.	......	axillary.	————	——	....	G. ♄.	
PR'OTEA. L.T.		PR'OTEA.	Tetrandria Monogynia. L.				
1 cynaroídes. L.T.	(*re.*)	Artichoke-fld.	C. B. S.	1774.	11. 3.	G. ♄.	Bot. mag. t. 770.
2 latifòlia. L.T.	(*sc.*)	ray-flowered.	————	1806.	7. 9.	G. ♄.	———— t. 1717.
α coccínea.	(*sc.*)	*scarlet-flower'd.*	————	——	——	G. ♄.	Andr. reposit. t. 646.
β viridiflòra.	(*gr.*)	*green-flowered.*	————	——	——	G. ♄.	———— t. 646.
3 compácta. L.T.	(*pu.*)	compact.	————	1810.	——	G. ♄.	
4 longiflòra. L.T.	(*st.*)	long-flowered.	————	1795.	1. 4.	G. ♄.	Sm. exot. bot. 2. t. 81.
ochroleùca. Sm. *lacticolor.* Par. lond. t. 27.							
5 coccínea. L.T.	(*sc.*)	scarlet-flowered.	————	1810.	5. 6.	G. ♄.	
6 speciòsa. L.T.	(*wh.br.*)	splendid.	————	1786.	3. 6.	G. ♄.	Knight's prot. c. ic.
7 obtùsa.	(*pa. pi.*)	obtuse.	————	——	——	G. ♄.	Andr. reposit. t. 110.
speciòsa. B.M. 1183. *Erodéndrum obtùsum.* K.P.							
8 macrophy'lla. L.T.	(*wh.*)	large-leaved.	————	1810.	——	G. ♄.	
9 formòsa. L.T.	(*sc.*)	handsome.	————	1789.	5. 6.	G. ♄.	Bot. mag. t. 1713.
coronàta. A.R. 469. *Erodéndrum formòsum.* Par. lond. t. 76.							
10 melaleùca. L.T.	(*bk.*)	black-fringed.	————	1786.	3. 7.	G. ♄.	Andr. reposit. t. 103.
11 Lepidocárpon. L.T.	(*bk.*)	crested.	————	1806.	——	G. ♄.	Wein. phyt. 4. t. 895.
12 neriifòlia. L.T.	(*pu.*)	Oleander-leaved.	————	——	2. 4.	G. ♄.	Bot. reg. t. 208.
13 pulchélla. L.T.	(*pu.*)	wave-leaved.	————	1795.	3. 8.	G. ♄.	———— t. 20.
α ciliàta.	(*pu.bk.*)	*fringe-leaved.*	————	——	——	G. ♄.	Andr. reposit. t. 270.
β glábra.	(*pu.bk.*)	*smooth-leaved.*	————	——	——	G. ♄.	———— t. 277.
γ speciòsa.	(*pu.bk.*)	*specious.*	————	——	——	G. ♄.	———— t. 442.
14 pátens. L.T.	(*st.bk.*)	spreading.	————	1789.	3. 6.	G. ♄.	———— t. 543.
15 magnífica. A.R.	(*pu.bk.*)	magnificent.	————	——	——	G. ♄.	———— t. 438.

16	incómpta. L.T.	(*wh.*)	villous-branched.	C. B. S.	1807.	5. 7.	G. ♄.	
17	longifòlia. L.T.	(*st.bk.*)	long-leaved.	———	1798.	3. 8.	G. ♄.	Bot. reg. t. 47.
18	umbonàlis.	(*st.pu.*)	embossed.	———	——	——	G. ♄.	Andr. reposit. t. 144.
	longifòlia. var. *cono turbinato.* A.R. *Erodéndrum umbondle.* K.P.							
19	ligulæfòlia.	(*st.pu.*)	strap-leaved.	———	1798.	3. 8.	G. ♄.	Andr. reposit. t. 133.
	longifòlia ferruginòso-purpùrea. A.R. *Erodéndrum ligulæfòlium.* K.P.							
20	mellífera. L.T.	(*ro.wh.*)	honey-bearing.	———	1774.	12. 5.	G. ♄.	Bot. mag. t. 346.
	β *álba.*	(*wh.*)	*white-flowered.*	———	——	——	G. ♄.	Andr. reposit. ic.
21	grandiflòra. L.T.	(*ro.w.*)	great-flowered.	———	1787.	5. 6.	G. ♄.	Bot. mag. t. 2447.
	β *marginàta.*	(*ro.wh.*)	*red margined.*	———	——	——	G. ♄.	Bot. reg. t. 569.
22	Scólymus. L.T.	(*ro.wh.*)	small-flowered.	———	1780.	——	G. ♄.	Bot. mag. t. 698.
23	mucronifòlia. L.T.	(*wh.ro.*)	dagger-leaved.	———	1803.	7. 12.	G. ♄.	Andr. rep. t. 500.
24	nàna. L.T.	(*cr.*)	dwarf.	———	1787.	4. 7.	G. ♄.	Sm. exot. bot. 1. t. 44.
	rosàcea. Sm. *acuifòlia.* Par. lond. t. 2.							
25	péndula. L.T.	(*wh.gr.*)	pendulous.	———	1806.	3. 6.	G. ♄.	
26	tènax. L.T.	(*st.*)	tough.	———	1801.	2. 5.	G. ♄.	Sal. par. lond. t. 70.
27	canaliculàta. L.T.	(*pi.*)	channel-leaved.	———	1800.	12. 2.	G. ♄.	Andr. reposit. t. 437.
28	acuminàta. B.M.	(*cr.*)	sharp-pointed.	———	1809.	3. 6.	G. ♄.	Bot. mag. t. 1694.
29	acaùlis. L.T.	(*st.br.*)	short-stalked.	———	1802.	5. 9.	G. ♄.	Wein. phyt. 4. t. 898.
30	glaucophy'lla. P.L.	(*st.br.*)	glaucous-leaved.	———	——	7. 9.	G. ♄.	Sal. par. lond. t. 11.
31	elongàta. L.T.	(*st.br.*)	elongated.	———	1810.	——	G. ♄.	
32	angustàta. L.T.	(*st.*)	narrow-leaved.	———	1809.	——	G. ♄.	
33	revolùta. L.T.	(*wh.*)	revolute-leaved.	———	1818.	——	G. ♄.	
34	tenuifòlia. L.T.	(*wh.br.*)	slender-leaved.	———	1816.	——	G. ♄.	
35	l'ævis. L.T.	(*ro.gr.*)	smooth-leaved.	———	1806.	5. 7.	G. ♄.	Bot. mag. t. 2439.
	longifòlia. Par. lond. t. 37.							
36	scábra. L.T.	(*wh.br.*)	rough-leaved.	———	1809.	——	G. ♄.	
37	répens. L.T.	(*re.wh.*)	creeping.	———	1800.	6. 8.	G. ♄.	Wein. phyt. 4. t. 897.a.
38	lòrea. L.T.	(*wh.*)	thong-leaved.	———	1812.	——	G. ♄.	
39	turbiniflòra. L.T.	(*ro.*)	turfy.	———	1803.	4. 5.	G. ♄.	Sal. par. lond. t. 103.
40	Scolopéndrium. L.T.	(*wh.br.*)	Hart's-tongue.	———	1802.	——	G. ♄.	
41	cordàta. L.T.	(*cr.*)	heart-leaved.	———	1790.	3. 5.	G. ♄.	Andr. reposit. t. 289.
42	amplexicaùlis. L.T.	(*re.*)	stem-clasping.	———	1802.	1. 3.	G. ♄.	Sal. par. lond. t. 67.
43	hùmilis. L.T.	(*pu.*)	low-flowering.	———	——	6. 8.	G. ♄.	Andr. reposit. t. 532.
44	aceròsa. L.T.	(*cr.*)	Pine-leaved.	———	1803.	3. 5.	G. ♄.	Bot. reg. t. 351.
	LEUCOSPE'RMUM. L.T. LEUCOSPE'RMUM. Tetrandria Monogynia. R.S.							
1	lineàre. L.T.	(*ye.*)	linear-leaved.	C. B. S.	1774.	8. 9.	G. ♄.	Thunb. dis. n. 35. t. 4.
2	attenuàtum. L.T.	(*ye.*)	attenuated.	———	1810.	——	G. ♄.	
3	Tóttum. L.T.	(*ye.*)	smooth-bracted.	———	1774.	6. 8.	G. ♄.	
4	mèdium. L.T.	(*ye.*)	oval-leaved.	———	1794.	——	G. ♄.	Andr. reposit. t. 17 ?
	Pròtea formòsa. A.R. ?							
5	ellípticum. L.T.	(*ye.*)	elliptic.	———	1803.	5. 8.	G. ♄.	
6	Conocárpum. L.T.	(*go.*)	many-toothed.	———	1774.	——	G. ♄.	Boer. lugd. 2. p. 196. ic.
7	grandiflòrum. L.T.	(*go.*)	great-flowered.	———	1800.	5. 7.	G. ♄.	Sal. par. lond. t. 116.
	Leucadéndron grandiflòrum. P.L.							
8	pùberum. L.T.	(*ye.*)	downy-leaved.	———	1774.	5. 8.	G. ♄.	
9	buxifòlium. L.T.	(*ye.*)	Box-leaved.	———	1822.	——	G. ♄.	
10	pátulum. L.T.	(*ye.*)	spreading.	———	1816.	——	G. ♄.	
11	spathulàtum. L.T.	(*ye.*)	spathulate.	———	1810.	——	G. ♄.	
12	tomentòsum. K.P.	(*ye.*)	cottony.	———	1789.	8. 9.	G. ♄.	
13	párile. K.P.	(*ye.*)	matched.	———	1790.	——	G. ♄.	
14	cándicans.	(*or.*)	Rose-scented.	———	——	——	G. ♄.	Andr. reposit. t. 294.
	Pròtea cándicans. A.R.							
15	Hypophy'llum. L.T.	(*ye.*)	trifid-leaved.	———	1787.	——	G. ♄.	Pluk. mant. t. 440. f. 3.
16	crinìtum. L.T.	(*ye.li.*)	villous-bracted.	———	1824.	....	G. ♄.	
17	oleæfòlium. L.T.	(*ye.*)	Olive-leaved.	———	1822.	4. 6.	G. ♄.	
18	diffùsum. L.T.	(*ye.*)	diffuse.	———	1823.	——	G. ♄.	
	MIM'ETES. L.T. MIM'ETES. Tetrandria Monogynia. R.S.							
1	hírta. L.T.	(*ye.cr.*)	hairy.	C. B. S.	1774.	6. 8.	G. ♄.	Wein. phyt. 4. t. 899. f. a.
2	capitulàta. L.T.	(*re.*)	small-headed.	———	1322.	——	G. ♄.	
3	pauciflòra. L.T.	(*re.*)	few-flowered.	———	1816.	——	G. ♄.	
4	palústris. K.P.	(*re.*)	marsh.	———	1802.	——	G. ♄.	Boer. lugd. 2. p. 194. ic.
5	cucullàta. L.T.	(*re.*)	three-toothed.	———	1789.	....	G. ♄.	Pluk. alm. t. 304. f. 6.
	Pròtea cucullàta. W.							
6	Hartògii. L.T.	(*re.*)	woolly-leaved.	———	1810.	....	G. ♄.	Wein. phyt. 4. t. 906. a.
7	divaricàta. L.T.	(*wh.*)	divaricate.	———	1795.	6. 9.	G. ♄.	
8	vacciniifòlia.	(*re.wh.*)	Vaccinium-leaved.	———	1800.	——	G. ♄.	
	Diastélla vacciniifòlia. K.P.							
9	purpùrea. L.T.	(*pu.*)	Heath-leaved.	———	1789.	11. 12.	G. ♄.	

SERR'URIA. L.T.			SERR'URIA. Tetrandria Monogynia. R.S.					
1	acrocárpa. L.T.	(*li.*)	sharp-fruited.	C. B. S.	1822.	4. 6.	G. ♄.	
2	elevàta. L.T.	(*li.*)	sandy-hill.	———	1821.	——	G. ♄.	
3	Aitòni. L.T.	(*li.*)	Aiton's.	———	1823.	——	G. ♄.	
4	diffùsa. L.T.	(*li.*)	diffuse.	———	1810.	6. 8.	G. ♄.	
5	abrotanifòlia. K.P.	(*li.*)	Southernwood-leaved.	—	1803.	——	G. ♄.	Andr. reposit. t. 522.
	Pròtea abrotanifòlia hírta. A.R.							
6	millefòlia. K.P.	(*li.*)	thousand-leaved.	———	——	——	G. ♄.	Andr. reposit. t. 337.
7	artemisiæfòlia. K.P.	(*li.*)	Wormwood-leaved.	——	1789.	——	G. ♄.	——— t. 264.
8	pinnàta. L.T.	(*li.*)	slender.	———	1803.	——	G. ♄.	——— t. 512.
9	arenària. L.T.	(*li.*)	sand.	———	——	——	G. ♄.	
10	cyanoídes. L.T.	(*li.*)	trifid-leaved.	———	——	——	G. ♄.	Pluk. mant. t. 345. f. 6.
11	scariòsa. L.T.	(*li.*)	scaly-bracted.	———	1816.	5. 7.	G. ♄.	
12	pedunculàta. L.T.	(*li.*)	woolly-headed.	———	1789.	5. 8.	G. ♄.	Andr. reposit. t. 264.
	Pròtea glomeràta. A.R.							
13	scopària. L.T.	(*li.*)	Broom.	———	1809.	——	G. ♄.	
14	Nivèni. L.T.	(*li.*)	Niven's.	———	1800.	——	G. ♄.	Andr. reposit. t. 349.
	Pròtea decúmbens. A.R. *non* Th.							
15	villòsa. L.T.	(*li.*)	villous.	———	1810.	4. 7.	G. ♄.	
16	fœniculàcea L.T.	(*li.*)	Fennel-leaved.	———	1816.	——	G. ♄.	
17	ciliàta. L.T.	(*li.*)	ciliated.	———	1803.	——	G. ♄.	
18	congésta. L.T.	(*li.*)	crowded.	———	1816.	——	G. ♄.	
19	squarròsa. L.T.	(*li.*)	squarrose.	———	1810.	——	G. ♄.	
20	phylicoídes. L.T.	(*li.*)	Phylica-flowered.	———	1788.	——	G. ♄.	Andr. reposit. t. 507. f. 4.
	Pròtea abrotanifòlia. A.R.							
21	'æmula. L.T.	(*li.*)	grey-branched.	———	1803.	——	G. ♄.	
22	párilis. K.P.	(*li.*)	matched.	———	——	——	G. ♄.	Andr. reposit. t. 507.
23	odoràta.	(*li.*)	sweet-scented.	———	——	——	G. ♄.	——— t. 545.
	Pròtea abrotanifòlia odoràta. A.R.							
24	emarginàta.	(*li.*)	emarginate.	———	1800.	——	G. ♄.	Andr. reposit. t. 536.
	arenària. K.P. *non* Br. *abrotanifòlia mìnor.* A.R.							
25	decúmbens. L.T.	(*li.*)	decumbent.	———	1816.	——	G. ♄.	Thunb. diss. I. t. 1.
26	flagellàris. L.T.	(*li.*)	procumbent.	———	——	——	G. ♄.	
27	glomeràta. L.T.	(*li.*)	many-headed.	———	1789.	——	G. ♄.	Burm. afr. t. 99. f. 2.
28	decípiens. L.T.	(*li.*)	decipient.	———	1806.	——	G. ♄.	
29	Roxbúrghii. L.T.	(*li.*)	Roxburgh's.	———	——	——	G. ♄.	
30	Burmánni. L.T.	(*li.*)	Burmann's.	———	1786.	——	G. ♄.	Burm. afr. t. 99. f. 1.
	Pròtea Serrària. L.							
31	triternàta. L.T.	(*wh.*)	silvery-flowered.	———	1802.	——	G. ♄.	Andr. reposit. t. 447.
	Pròtea argentiflòra. A.R.							
32	elongàta. L.T.	(*li.*)	long-peduncled.	———	1800.	——	G. ♄.	
33	crithmifòlia. L.T.	(*li.*)	Samphire-leaved.	———	1816.	——	G. ♄.	
NIV'ENIA. L.T.			NIV'ENIA. Tetrandria Monogynia. R.S.					
1	Scéptrum. L.T.	(*pu.*)	sceptre-like.	C. B. S.	1790.	5. 6.	G. ♄.	Sparm. a. st. 1777. t. 1.
2	spathulàta. L.T.	(*pu.*)	maiden hair-leaved.	——	——	7. 8.	G. ♄.	Thunb. dis. n. 58. t. 5.
3	parvifòlia. L.T.	(*pu.*)	small-leaved.	———	1810.	——	G. ♄.	Thunb. t. 5. quoad fig.
4	spicàta. L.T.	(*pu.*)	spiked.	———	1786.	6. 8.	G. ♄.	
5	crithmifòlia. L.T.	(*pu.*)	Samphire-leaved.	———	1797.	——	G. ♄.	Andr. reposit. t. 243.
	Pròtea Lagòpus. A.R. *non* Thunb.							
6	mèdia. L.T.	(*bh.*)	middle.	———	1803.	——	G. ♄.	Andr. reposit. t. 234?
	Pròtea spicàta. A.R.?							
7	Lagòpus. L.T.	(*pu.*)	Hare's-foot.	———	1810.	——	G. ♄.	
SOROCE'PHALUS. L.T.			SOROCE'PHALUS. Tetrandria Monogynia. R.S.					
1	imbérbis. L.T.	(*pu.*)	smooth.	C. B. S.	1806.	6. 8.	G. ♄.	
2	spatalloídes. L.T.	(*pu.*)	club-bearing.	———	1803.	——	G. ♄.	
3	tenuifòlia. L.T.	(*pu.*)	slender-leaved.	———	1802.	——	G. ♄.	
4	lanàtus. L.T.	(*pu.*)	woolly.	———	1790.	6. 9.	G. ♄.	Thunb. dis. n. 30. t. 3.
5	imbricàtus. L.T.	(*ro.*)	imbricated.	———	1794.	4. 7.	G. ♄.	Andr. reposit. t. 517.
6	diversifòlius. L.T.	(*pu.*)	various-leaved.	———	1807.	——	G. ♄.	
SPATA'LLA. L.T.			SPATA'LLA. Tetrandria Monogynia. R.S.					
1	pedunculàta. L.T.	(*pu.*)	long-peduncled.	C. B. S.	1822.	4. 8.	G. ♄.	
2	nívea. L.T.	(*li.*)	snowy-bearded.	———	1806.	4. 8.	G. ♄.	
3	ramulòsa. L.T.	(*pu.*)	branching.	———	1812.	——	G. ♄.	
4	bracteàta. L.T.	(*pu.*)	bracteate.	———	——	5. 9.	G. ♄.	
5	prolífera. L.T.	(*pu.*)	proliferous.	———	1800.	6. 8.	G. ♄.	Thunb. dis. n. 27. t. 4.
6	pyramidàlis. L.T.	(*pu.*)	pyramidal.	———	1821.	——	G. ♄.	
7	incúrva. L.T.	(*pu.*)	incurved-leaved.	———	1789.	5. 6.	G. ♄.	Thunb. dis. n. 22. t. 3.
8	caudàta. L.T.	(*pu.*)	tailed.	———	1812.	——	G. ♄.	——— sec. ic. t. 2.
9	Thunbérgii. L.T.	(*pu.*)	Thunberg's.	———	1806.	——	G. ♄.	

ADENA'NTHOS. L.T.		ADENA'NTHOS. Tetrandria Monogynia.					
1 obovàta. L.T.	(li.)	obovate-leaved.	N. Holland.	1824.	....	G. ♄.	Labil. n. hol. 1. t. 37.
2 cuneàta. L.T.	(li.)	wedge-leaved.	———	——	....	G. ♄.	———— v. 1. t. 36.
3 serícea. L.T.	(li.)	silky-leaved.	———	——	....	G. ♄.	———— v. 1. t. 38.
CONOSPE'RMUM. L.T.		CONOSPE'RMUM. Triandria Monogynia.					
1 ellípticum. L.T.	(wh.)	elliptic-leaved.	N. S. W.	1822.	5. 7.	G. ♄.	
2 taxifòlium. L.T.	(st.)	Yew-leaved.	———	1823.	4. 6.	G. ♄.	Bot. mag. t. 2724.
3 ericifòlium. L.T.	(bh.)	Heath-leaved.	———	1820.	——	G. ♄.	——— t. 2850.
4 longifòlium. L.T.	(wh.)	long-leaved.	———	1824.	....	G. ♄.	Sm. exot. bot. 2. t. 82.
5 acinacifòlium. Gr.	(wh.)	scymeter-leaved.	———	1824.	4. 6.	G. ♄.	
6 tenuifòlium. L.T.	(wh.)	slender-leaved.	———	——	——	G. ♄.	
AGAST'ACHYS. L.T.		AGAST'ACHYS. Tetrandria Monogynia. R.S.					
odoràta. L.T.	(ye.)	sweet-scented.	V. Diem. Isl.	1824.	4. 7.	G. ♄.	
PERSO'ONIA. L.T.		PERSO'ONIA. Tetrandria Monogynia. R.S.					
1 teretifòlia. L.T.	(ye.)	round-leaved.	N. Holland.	1824.	5. 6.	G. ♄	
2 pinifòlia. L.T.	(ye.)	Pine-leaved.	N. S. W.	1818.	5. 7.	G. ♄.	Linn. trans. 10. t. 16. f. 1.
3 juniperìna. L.T.	(ye.)	Juniper-leaved.	V. Diem. Isl.	1823.	——	G. ♄.	Lab. n. holl. 1. t. 45.
4 hirsùta. L.T.	(ye.)	hairy.	N. S. W.	1800.	——	G. ♄.	Lodd. bot. cab. t. 327.
5 móllis. L.T.	(ye.)	soft.	———	1818.	——	G. ♄.	
6 lineàris. L.T.	(ye.)	linear-leaved.	———	1794.	7. 8.	G. ♄.	Bot. mag. t. 760.
7 pállida. Gr.	(st.)	pale.	N. S. W.	1824.	9. 10.	G. ♄.	
8 lùcida. L.T.	(ye.)	shining.	———	——	6. 8.	G. ♄.	
9 flexifòlia. L.T.	(ye.)	bending-leaved.	———	1822.	——	G. ♄.	Lodd. bot. cab. 922.
10 chamæpítys. F.T.	(ye.)	ground-pine.	N. S. W.	1824.	——	G. ♄.	
11 scábra. L.T.	(st.)	rough.	N. Holland.	——	——	G. ♄.	
12 nùtans. L.T.	(ye.)	nodding.	N. S. W.	1821.	——	G. ♄.	
13 pruinòsa. M.C.	(ye.)	frosted.	N. Holland.	1824.	——	G. ♄.	
14 lanceolàta. L.T.	(st.)	spear-leaved.	———	1791.	6. 7.	G. ♄.	Andr. reposit. t. 74.
β *latifòlia.* B.P.	(st.)	*broad-leaved.*	———	——	——	G. ♄.	———— t. 280 ?
15 salicìna. L.T.	(st.)	Willow-leaved.	———	1795.	6. 7.	G. ♄.	Cavan. ic. 4. t. 389 ?
16 ferrugínea. L.T.	(ye.)	ferruginous.	———	1818.	——	G. ♄.	Sm. exot. bot. 2. t. 83.
BRAB'EJUM. L.T.		AFRICAN ALMOND. Polygamia Monœcia. L.					
stellatifòlium. L.T.		common.	C. B. S.	1731.	3. 4.	G. ♄.	Hou. n. h. par. 2. t. 6. f. 37.
GREVI'LLEA. L.T.		GREVI'LLEA. Tetrandria Monogynia. R.S.					
1 punícea. L.T.	(sc.)	scarlet-flowered.	N. S. W.	1822.	4. 9.	G. ♄.	Bot. reg. t. 1319.
2 dùbia. L.T.	(ro.)	doubtful.	———	——	——	G. ♄.	
3 serícea. L.T.	(ro.)	silky.	———	1790.	——	G. ♄.	Bot. mag. t. 862.
4 lineàris. L.T.	(li.)	linear-leaved.	———	——	——	G. ♄.	Andr. reposit. t. 272.
α *incarnàta.* B.M.	(pi.)	*flesh-coloured.*	———	——	——	G. ♄.	Bot. mag. t. 2661.
β *álba.* B.C.	(bh.)	*white-flowered.*	———	——	——	G. ♄.	Lodd. bot. cab. t. 858.
5 stricta. L.T.	(li.)	upright.	———	1822.	——	G. ♄.	
6 ripària. L.T.	(li.)	river-side.	———	1791.	4. 5.	G. ♄.	
7 rosmarinifòlia. F.T.	(li.)	Rosemary-leaved.	———	1824.	——	G. ♄.	Swt. flor. aust. t. 30.
8 parviflòra. L.T.	(st.)	small-flowered.	———	1824.	——	G. ♄.	
9 juniperìna. L.T.	(gr. st.)	Juniper-like.	———	1821.	——	G. ♄.	Lodd. bot. cab. t. 1003.
10 concínna. L.T.	(st. ro.)	pretty.	N. Holland.	1824.	3. 8.	G. ♄.	Swt. flor. aust. t. 7.
11 arenària. L.T.	(li.)	sand.	N. S. W.	1803.	4. 9.	G. ♄.	
12 montàna. L.T.	(gr.)	mountain.	———	——	——	G. ♄.	
13 acuminàta. L.T.	(gr.)	acute-leaved.	———	1805.	——	G. ♄.	Swt. flor. aust. t. 55.
14 cinérea. L.T.	(gr.)	grey.	———	1821.	——	G. ♄.	Lodd. bot. cab. t. 857.
15 mucronulàta. L.T.	(gr.)	mucronulate.	———	1809.	——	G. ♄.	Swt. flor. aust. t. 38.
16 Bauèri. L.T.	(bh.)	Bauer's.	———	1823.	4. 8.	G. ♄.	Hook. ex. flor. t. 216.
pubèscens. Hooker, et Lodd. bot. cab. t. 1229.							
17 sulphùrea. F.T.	(st.)	sulphur-coloured.	———	1823.	——	G. ♄.	
18 sphacelàta. L.T.	(br.)	sphacelate.	———	1825.	——	G. ♄.	
19 phylicoídes. L.T.	(br.)	Phylica-like.	———	——	——	G. ♄.	
20 collìna. K.P.	(br.)	hill.	———	1812.	——	G. ♄.	
21 buxifòlia. L.T.	(pu.)	Box-leaved.	———	1790.	2. 9.	G. ♄.	Bot. reg. t. 443.
22 podalyriæfòlia. C.C.	(pu.)	Podalyria-leaved.	N. Holland.	1821.	5. 8.	I. ♄.	
23 podocarpifòlia.	......	Podocarpus-leav'd.	———	1823.	....	I. ♄.	
24 berberifòlia. C.C.	....	Berberry-leaved.	———	——	——	I. ♄.	
25 acanthifòlia. F.T.	(gr.)	Acanthus-leaved.	———	——	——	I. ♄.	Bot. mag. t. 2807.
26 Gaudichàudii. B.	(ye.)	Gaudichaud's.	———	1823.	....	I. ♄.	
27 trifurcàta. C.C.	......	three-forked.	———	——	....	I. ♄.	
28 aspleniifòlia. L.T.	(st.)	Spleenwort-leaved.	N. S. W.	1806.	....	I. ♄.	
29 Flindérsii. M.C.	....	Flinders's.	N. Holland.	1823.	....	I. ♄.	
H'AKEA. L.T.		H'AKEA. Tetrandria Monogynia. R.S.					
1 pugionifórmis. L.T.	(wh.)	dagger-fruited.	N. S. W.	1796.	5. 6.	G. ♄.	Lodd. bot. cab. t. 353.
2 párilis. K.P.	(wh.)	matched.	V. Diem. Isl.	——	——	G. ♄.	

3 epiglóttis. L.T.	(*wh.*)	curved-fruited.	V. Diem. Isl.	1822.	5. 6.	G. ♄.	Lab. nov. hol. 1. t. 40.
4 nodòsa. L.T.	(*wh.*)	knot-podded.	N. Holland.	1824.	——	G. ħ.	
5 fléxilis. L.T.	(*wh.*)	bending.	———	——	——	G. ħ.	
6 oblíqua. L.T.	(*wh.*)	oblique-flowered.	N. Holland.	1803.	6. 9.	G. ♄.	
7 subulàta. ψ.	(*wh.*)	awl-leaved.	———	1824.	——	G. ħ.	
8 petrophiloídes. ψ.	(*wh.*)	Petrophila-like.	———	1825.	——	G. ħ.	
9 lissospérma. L.T.	(*wh.*)	smooth-seeded.	V. Diem. Isl.	1824.	——	G. ħ.	
10 gibbòsa. L.T.	(*st.*)	gibbous-fruited.	N. S. W.	1790.	5. 6.	G. ħ.	Cavan. ic. 6. t. 534.
11 aciculàris. L.T.	(*wh.*)	needle-leaved.	———	——	——	G. ħ.	Venten. malm. t. 111.
Cónchium aciculàre. V.							
12 suavèolens. L.T.	(*wh.*)	sweet-scented.	N. Holland.	1803.	11. 1.	G. ♄.	
13 Lambérti. N.L.F.	(*wh.*)	Lambert's.	———	1823.	——	G. ħ.	
14 microcárpa. L.T.	(*wh.*)	small-fruited.	V. Diem. Isl.	1818.	4. 8.	G. ħ.	Bot. reg. t. 475.
15 trifurcàta. L.T.	(*wh.*)	trifurcate.	N. Holland.	1824.	——	G. ♄.	
16 echinàta. ψ.	(*wh.*)	bristly-podded.	———	1825.	——	G. ♄.	
17 angustifòlia. ψ.	(*wh.*)	narrow-leaved.	———	1824.	——	G. ♄.	
18 lineàris. L.T.	(*wh.*)	linear-leaved.	———	——	——	G. ♄.	Swt. flor. aust. t. 43.
19 flòrida. L.T.	(*wh.*)	many-flowered.	———	1803.	5. 6.	G. ħ.	Bot. mag. t. 2579.
20 ilicifòlia. L.T.	(*wh.*)	Holly-leaved.	———	——	7. 9.	G. ♄.	
21 nítida. L.T.	(*wh.*)	glossy.	———	——	6. 8.	G. ħ.	Bot. mag. t. 2246.
22 amplexicàulis. L.T.	(*wh.*)	smooth-stalked.	———	——	....	G. ħ.	
23 prostràta. L.T.	(*wh.*)	pubescent.	———	——	....	G. ♄.	
24 ceratophy'lla. L.T.	(*wh.*)	horn-leaved.	———	——	....	G. ♄.	
25 acanthophy'lla. L. en.	(*w.*)	Acanthus-leaved.	———	——	....	G. ♄.	
26 undulàta. L.T.	(*wh.*)	wave-leaved.	———	——	6. 8.	G. ♄.	
27 repánda. ψ.	(*wh.*)	repand.	———	1824.	——	G. ħ.	
28 propínqua. F.T.	(*wh.*)	related.	———	1823.	——	G. ♄.	
29 oleifòlia. L.T.	(*wh.*)	Olive-leaved.	———	1794.	6. 7.	G. ♄.	
30 salígna. L.T.	(*wh.*)	Willow-leaved.	———	1791.	3. 7.	G. ♄.	Swt. flor. aust. t. 27.
31 marginàta. L.T.	(*wh.*)	marginate.	———	1824.	4. 7.	G. ♄.	
32 ruscifòlia. L.T.	(*st.*)	Ruscus-leaved.	———	——	——	G. ħ.	Lab. nov. holl. 1. t. 39.
33 cinèrea. L.T.	(*wh.*)	hoary-leaved.	———	1803.	6. 7.	G. ħ.	
34 canéscens. L. en.	(*wh.*)	canescent.	N. S. W.	1800.	6. 8.	G. ♄.	
35 dactyloídes. L.T.	(*wh.*)	nerved-leaved.	———	1790.	——	G. ħ.	Cavan. ic. 6. t. 535.
36 ellíptica. L.T.	(*wh.*)	oval-leaved.	N. Holland.	1794.	——	G. ♄.	
37 ferrugínea. S.F.A.	(*wh.*)	rusty-stemmed.	———	1824.	4. 8.	G. ħ.	Swt. flor. aust. t. 45.
38 cucullàta. ψ.	(*wh.*)	hollow-leaved.	———	——	——	G. ♄.	
39 clavàta. L.T.	(*wh.*)	clavate.	———	1824.	——	G. ♄.	Lab. nov. holl. 1. t. 41.
EMBO'THRIUM.		EMBO'THRIUM.	Tetrandria Monogynia.				
strobilìnum. Lab.	(*st.*)	cone-flowered.	N. Holland.	1824.	3. 6.	G. ♄.	Lab. nov. holl. 2. t. 265.
LAMBE'RTIA. L.T.		LAMBE'RTIA.	Tetrandria Monogynia. R.S.				
1 uniflòra. L.T.	(*re.*)	single-flowered.	N. Holland.	1824.	....	G. ħ.	Bauer ic. ined.
2 inérmis. L.T.	(*or.*)	unarmed.	———	——	....	G. ♄.	——— ic. ined.
3 formòsa. L.T.	(*ro.*)	handsome.	N. S. W.	1788.	6. 8.	G. ħ.	Bot. reg. t. 528.
4 echinàta. L.T.	(*ro.*)	lobe-leaved.	N. Holland.	1824.	....	G. ħ.	Bauer ic. ined.
XYLOM'ELUM. L.T.		WOODEN-PEAR-TREE.	Tetrandria Monogynia. R.S.				
pyrifórme. L.T.	(*wh.*)	prickly-leaved.	N. S. W.	1789.	6. 8.	G. ♄.	Cavan. ic. 6. t. 536.
RHOP'ALA. L.T.		RHOP'ALA.	Tetrandria Monogynia. R.S.				
1 montàna. L.T.	(*st.*)	mountain.	Guiana.	1823.	4. 6.	S. ♄.	Aub. gui. 1. t. 32.
2 nítida. L.T.	(*st.*)	glossy.	———	1826.	....	S. ħ.	Rudge pl. gui. 1. t. 39.
3 dentàta. L.T.	(*st.*)	tooth-leaved.	———	1802.	5. 8.	S. ħ.	
4 sessilifòlia. L.T.	(*st.*)	sessile-leaved.	———	1803.	——	S. ♄.	Rudg. gui. 1. t. 31.
TELOP'EA. L.T.		WARATAH.	Tetrandria Monogynia. R.S.				
speciosíssima. L.T.	(*cr.*)	splendid.	N. S. W.	1789.	5. 7.	G. ħ.	Bot. mag. t. 1128.
Embóthrium speciosíssimum. B.M.							
LOM'ATIA. L.T.		LOM'ATIA.	Tetrandria Monogynia. R.S.				
1 silaifòlia. L.T.	(*wh.*)	cut-leaved.	N. S. W.	1792.	6. 8.	G. ħ.	Bot. mag. t. 1272.
2 tinctòria. L.T.	(*st.*)	various-leaved.	V. Diem. Isl.	1822.	——	G. ħ.	Lab. n. holl. 1. t. 42-43.
3 polymórpha. L.T.	(*st.*)	variable.	———	1823.	——	G. ♄.	
4 ilicifòlia. L.T.	(*st.*)	Holly-leaved.	N. Holland.	1824.	——	G. ħ.	
5 longifòlia. L.T.	(*wh.*)	long-leaved.	N. S. W.	1816.	5. 8.	G. ħ.	Bot. reg. t. 442.
6 dentàta. L.T.	(*st.*)	tooth-leaved.	Chile.	1824.	——	G. ħ.	Flor. peruv. 1. t. 94. a.
STENOCA'RPUS. L.T.		STENOCA'RPUS.	Tetrandria Monogynia. R.S.				
salígnus. L.T.	(*wh.*)	sweet-scented.	N. S. W.	1815.	6. 9.	G. ħ.	Bot. reg. t. 441.
BOTRYC'ERAS. W.		BOTRYC'ERAS.	Tetrandria Monogynia. R.S.				
laurìnum. W.	(*gr.*)	Laurel-like.	N. Holland.	1820.	6. 9.	G. ♄.	W.m.b.1811. t.10. f.10.
BA'NKSIA. L.T.		BA'NKSIA.	Tetrandria Monogynia. R.S.				
1 pulchélla. L.T.	(*st.*)	small-flowered.	N. Holland.	1803.	2. 8.	G. ħ.	
2 sphærocárpa. L.T.	(*st.*)	round-fruited.	———	1803.	....	G. ♄.	

3 nùtans. L.T.	(*st.*) nodding-flowered.	N. Holland.	1803.	4. 6.	G. ♄.	
4 ericifòlia. L.T.	(*st.*) Heath-leaved.	N. S. W.	1788.	1. 12.	G. ♄.	Bot. mag. t. 738.
5 spinulòsa. L.T.	(*st.*) spiny-leaved.	———	——	12. 5.	G. ♄.	Andr. reposit. t. 457.
6 collìna. L.T.	(*st.*) hill.	———	1800.	....	G. ♄.	
7 occidentàlis. L.T.	(*st.*) west-coast.	N. Holland.	1803.	4. 8.	G. ♄.	
8 littoràlis. L.T.	(*ye.*) sea-side.	———	——	4. 7.	G. ♄.	
9 marginàta. L.T.	(*gr.*) various-leaved.	N. S. W.	1804.	5. 8.	G. ♄.	Bot. mag. t. 1947.
10 austràlis. L.T.	(*gr.*) southern.	V. Diem. Isl.	1816.	——	G. ♄.	Bot. reg. t. 787.
marcéscens. Hooker bot. mag. 2803. *nec aliorum.*						
11 insulàris. L.T.	(*gr.*) Island.	———	1823.	....	G. ♄.	
12 Cunninghámii. S.S.	(*st.*) Cunningham's.	N. Holland.	1822.	——	G. ♄.	
13 integrifòlia. L.T.	(*st.*) entire-leaved.	N. S. W.	1788.	5. 8.	G. ♄.	Bot. mag. t. 2770.
14 compár. L.T.	(*st.*) related.	N. Holland.	1823.	....	G. ♄.	
15 verticillàta. L.T.	(*st.*) whorl-leaved.	———	1794.	7. 11.	G. ♄.	Hook. ex. flor. t. 96.
16 coccínea. L.T.	(*sc.*) scarlet-flowered.	———	1803.	....	G. ♄.	Bauer. n. holl. ic.
17 paludòsa. L.T.	(*st.*) marsh.	N. S. W.	1805.	1. 4.	G. ♄.	Bot. reg. t. 697.
18 intermèdia. ψ.	(*st.*) intermediate.	N. Holland.	1824.	....	G. ♄.	
19 oblongifòlia. L.T.	(*st.*) oblong-leaved.	N. S. W.	1788.	5. 8.	G. ♄.	Lodd. bot. cab. t. 241.
20 latifòlia. L.T.	(*st.*) broad-leaved.	———	1802.	——	G. ♄.	Cavan. ic. 6. t. 543.
21 marcéscens. L.T.	(*st.*) short-leaved.	N. Holland.	1794.	1. 12.	G. ♄.	Swt. flor. aust. t. 14.
præmórsa. Andr. rep. t. 258. *aspleniifòlia.* K.P.						
22 attenuàta. L.T.	(*st.*) attenuated, smooth-fld.	——	1794.	1. 10.	G. ♄.	
23 serràta. L.T.	(*st.*) saw-leaved.	N. S. W.	1788.	7. 9.	G. ♄.	Andr. reposit. t. 82.
undulàta. Bot. reg. t. 1316.						
24 æmula. L.T.	(*st.*) deeply-sawed.	———	——	1. 6.	G. ♄.	Bot. reg. t. 688.
25 dentàta. L.T.	(*st.*) tooth-leaved.	N. Holland.	1822.	....	S. ♄.	
26 quercifòlia. L.T.	(*st.*) Oak-leaved.	———	1805.	....	G. ♄.	
27 dryandroídes. S.F.A.	(*br.*) Dryandra-like.	———	1824.	3. 6.	G. ♄.	Swt. flor. aust. t. 56.
28 speciòsa. L.T.	(*wh.*) long-leaved.	———	1805.	5. 8.	G. ♄.	
29 grándis. L.T.	(*st.*) great-flowered.	———	1794.	....	G. ♄.	
30 prostràta. ψ.	(*st.*) prostrate.	———	1824.	....	G. ♄.	
31 rèpens. L.T.	(*st.*) creeping.	———	1803.	....	G. ♄.	Lab. voy. 1. t. 23.
32 ilicifòlia. L.T.	(*st.*) Holly-leaved.	———	1824.	....	G. ♄.	
DRYA'NDRA. L.T.	DRYA'NDRA.	Tetrandria Monogynia. R.S.				
1 floribúnda. L.T.	(*ye.*) many-flowered.	N. Holland.	1803.	1. 12.	G. ♄.	Bot. mag. t. 1581.
2 cuneàta. L.T.	(*st.*) wedge-leaved.	———	——	11. 2.	G. ♄.	
α brevifòlia.	(*st.*) *short-leaved.*	———	——	——	G. ♄.	
β longifòlia.	(*st.*) *long-leaved.*	———	——	——	G. ♄.	
3 armàta. L.T.	(*ye.*) acute-leaved.	———	——	1. 12.	G. ♄.	
4 falcàta. L.T.	(*ye.*) falcate-leaved.	———	1824.	....	G. ♄.	
5 formòsa. L.T.	(*or.*) Apricot-scented.	———	1803.	1. 12.	G. ♄.	Swt. flor. aust. t. 53.
6 mucronulàta. L.T.	(*ye.*) pointed-leaved.	———	1824.	....	G. ♄.	
7 plumòsa. L.T.	(*ye.*) feathered.	———	1803.	1. 8.	G. ♄.	
8 obtùsa. L.T.	(*ye.*) blunt-leaved.	———	——	....	G. ♄.	
9 nívea. L.T.	(*ye.*) white-leaved.	———	——	7. 9.	G. ♄.	Lab. voy. 1. t. 24.
10 longifòlia. L.T.	(*ye.*) long-leaved.	———	1805.	1. 12.	G. ♄.	Swt. flor. aust. t. 3.
11 Baxtèri. ψ.	(*ye.*) Baxter's.	———	1824.	....	G. ♄.	
12 tenuifòlia. L.T.	(*ye.*) slender-leaved.	———	1803.	3. 5.	G. ♄.	
13 nervòsa. S.F.A.	(*ye.*) nerved-leaved.	———	1824.	3. 12.	G. ♄.	Swt. flor. aust. t. 22.
14 pteridifòlia. L.T.	(*ye.*) Pteris-leaved.	———	——	——	G. ♄.	

ORDO CLXXIII.

THYMELÆÆ. *Brown prodr. p.* 358.

DI'RCA. W.	LEATHER-WOOD.	Octandria Monogynia. L.				
palústris. W.	(*ye.*) marsh.	Virginia.	1750.	3. 4.	H. ♄.	Bot. reg. t. 292.
LAGE'TTA. J.	LACE-BARK.	Octandria Monogynia. S.S.				
linteària. J.	(*wh.*) Jamaica.	Jamaica.	1793.	12. 1.	S. ♄.	Lam. ill. t. 289.
DA'PHNE. W.	DA'PHNE.	Octandria Monogynia. L.				
1 Mezèreum. W.	(*re.wh.*) Mezereon.	England.	....	2. 4.	H. ♄.	Eng. bot. v. 20. t. 1381.
α álbum.	(*wh.*) *white-flowered.*	———	....	——	H. ♄.	
β rùbrum.	(*re.*) *red-flowered.*	———	....	——	H. ♄.	
γ autumnàle.	(*re.*) *autumnal-flowered.*	———	....	8. 12.	H. ♄.	

2 Laurèola. W.	(*gr.*)	spurge-laurel.	Britain.	....	1. 3.	H. ♄.	Eng. bot. v. 2. t. 119.
3 póntica. W.	(*gr.ye.*)	Pontic.	Pontus.	1759.	4. 5.	H. ♄.	Bot. mag. t. 1282.
4 tinifòlia. W.	(*st.*)	Bonace-bark.	Jamaica.	1733.	5. 7.	S. ♄.	
5 Gnídium. W.	(*wh.*)	Flax-leaved.	Spain.	1597.	6. 8.	H. ♄.	Lodd. bot. cab. t. 150.
6 odòra. W.	(*bh.*)	sweet-scented.	China.	1771.	12. 3.	G. ♄.	Bot. mag. t. 1587.
7 hy'brida. B.F.G.	(*ro.*)	Dauphin.	Hybrid.	1826.	1. 6.	F. ♄.	Swt. br. fl. gar. t. 200.
8 alpìna. W.	(*wh.*)	Alpine.	Italy.	1759.	5. 7.	H. ♄.	Lodd. bot. cab. t. 66.
9 altàica. W.	(*wh.*)	Altaic.	Siberia.	1796.	4. 5.	H. ♄.	Bot. mag. t. 1875.
10 Cneòrum. W.	(*ro.*)	trailing.	Austria.	1752.	4. 9.	H. ♄.	——— t. 313.
11 striàta. S.S.	(*pi.*)	streaked.	Hungary.	....	——	H. ♄.	
12 collìna. S.S.	(*re.*)	hairy.	Italy.	1752.	1. 6.	H. ♄.	Bot. mag. t. 428.
13 serícea. S.S.	(*li.*)	silky-leaved.	Levant.	1818.	——	H. ♄.	
14 oleoídes. S.S.	(*li.*)	Olive-leaved.	Crete.	1815.	1. 12.	H. ♄.	Bot. mag. t. 1917.
15 napolitàna. B.C.	(*re.*)	Neapolitan.	Italy.	1823.	1. 6.	H. ♄.	Lodd. bot. cab. t. 719.
collìna β neapolitàna. B.R. 822.							
16 Thymel'æa. W.	(*st.*)	smooth-leaved.	Spain.	1815.	2. 4.	H. ♄.	Pluk. alm. t. 229. f. 2.
17 Tárton-ràira. W.	(*st.*)	silvery-leaved.	France.	1640.	5. 7.	H. ♄.	——— t. 318. f. 6.
GNI'DIA. L.		GNI'DIA. Octandria Monogynia. L.					
1 pinifòlia. W.	(*wh.*)	Pine-leaved.	C. B. S.	1768.	5. 6.	G. ♄.	Bot. reg. t. 19.
2 radiàta. B.C.	(*wh.*)	radiate.	———	1818.	——	G. ♄.	Lodd. bot. cab. t. 29.
3 imbérbis. H.K.	(*st.*)	smooth-scaled.	———	1792.	8. 4.	G. ♄.	Bot. mag. t. 1463.
4 símplex. H.K.	(*ye.*)	Flax-leaved.	———	1786.	5. 6.	G. ♄.	——— t. 812.
5 juniperifòlia. W.	(*wh.*)	Juniper-leaved.	———	1822.	——	G. ♄.	
6 capitàta. W.	(*wh.*)	purple-twigged.	———	1788.	6. 7.	G. ♄.	
7 imbricàta. W.	(*st.*)	imbricated.	———	1822.	4. 8.	G. ♄.	Bot. reg. t. 757.
denudàta. B.R.							
8 oppositifòlia. H.K.	(*ye.*)	opposite-leaved.	———	1783.	5. 7.	G. ♄.	Bot. reg. t. 2.
9 tomentòsa. S.S.	(*st.*)	tomentose.	———	1822.	——	G. ♄.	
10 serícea. W.	(*ye.*)	silky.	———	1786.	——	G. ♄.	Andr. reposit. t. 225.
oppositifòlia. A.R. *nec aliorum.*							
LACHN'ÆA. W.		LACHN'ÆA. Octandria Monogynia. L.					
1 buxifòlia. Lam.	(*wh.*)	green box-leaved.	C. B. S.	1800.	5. 7.	G. ♄.	Bot. mag. t. 1657.
2 glaùca. H.K.	(*wh.*)	glaucous.	———	——	——	G. ♄.	——— t. 1658.
3 purpùrea. W.	(*pu.*)	purple-flowered.	———	——	——	G. ♄.	——— t. 1594.
4 eriocéphala. W.	(*wh.*)	woolly-headed.	———	1793.	6. 7.	G. ♄.	——— t. 1295.
5 conglomeràta. W.	(*wh.*)	clustered.	———	1773.	——	G. ♄.	
PASSER'INA. W.		SPARROW-WORT. Octandria Monogynia. L.					
1 uniflòra. W.	(*wh.*)	one-flowered.	C. B. S.	1759.	4. 7.	G. ♄.	Wend. obs. t. 2. f. 18.
2 láxa. W.	(*wh.*)	loose.	———	1804.	6. 7.	G. ♄.	Lodd. bot. cab. t. 755.
3 tenuiflòra. W.	(*st.*)	slender-flowered.	———	1812.	——	G. ♄.	
4 filifórmis. W.	(*st.*)	filiform.	———	1752.	6. 8.	G. ♄.	Wend. obs. t. 2. f. 15.
5 capitàta. W.	(*wh.*)	headed.	———	1789.	6. 10.	G. ♄.	——— t. 2. f. 17.
6 hirsùta. W.	(*st.*)	hairy.	S. Europe.	1759.	——	G. ♄.	Bot. mag. t. 1949.
7 striàta. Lam.	(*wh.*)	striate.	C. B. S.	1822.	——	G. ♄.	
8 ericoídes. W.	(*st.*)	Heath-like.	———	——	——	G. ♄.	
9 ciliàta. W.	(*wh.*)	fringed.	———	1819.	——	G. ♄.	Seb. mus. 2. t. 12. f. 9.
10 grandiflòra. W.	(*wh.*)	great-flowered.	———	1789.	5. 6.	G. ♄.	Bot. mag. t. 292.
11 spicàta. W.	(*wh.*)	spiked.	———	1787.	5. 7.	G. ♄.	Lodd. bot. cab. t. 311.
12 anthylloídes. W.	(*st.*)	Anthyllus-like.	———	1816.	——	G. ♄.	
STELL'ERA. W.		STELL'ERA. Octandria Monogynia. L.					
1 Passerìna. W.	(*ye.*)	Flax-leaved.	S. Europe.	1759.	7. 8.	H. ⊙.	Jacq. ic. 1. t. 68.
Passerìna ánnua. S.S.							
2 altàica. P.S.	(*wh.*)	Altay.	Altay.	1824.	——	H. ⊙.	
3 dichótoma. F.	(*wh.*)	forked.	Dahuria.	——	——	H. ⊙.	
D'AIS. L.		D'AIS. Decandria Monogynia. L.					
cotinifòlia. W.	(*pi.*)	Cotinus-leaved.	C. B. S.	1776.	6. 7.	G. ♄.	Bot. mag. t. 147.
STRUTH'IOLA. L.		STRUTH'IOLA. Tetrandria Monogynia. L.					
1 virgàta. H.K.	(*re.*)	twiggy.	C. B. S.	1779.	4. 8.	G. ♄.	Andr. reposit. t. 139.
2 ciliàta. A.R.	(*st.*)	ciliated.	———	——	——	G. ♄.	——— t. 149.
3 lateriflòra. R.S.	(*wh.*)	side-flowering.	———	1816.	——	G. ♄.	
4 pubéscens. R.S.	(*re.*)	downy.	———	1790.	——	G. ♄.	Bot. mag. t. 1212.
5 longiflòra. R.S.	(*wh.*)	long-flowered.	———	1824.	——	G. ♄.	Lam. ill. t. 78.
6 glábra. R.S.	(*wh.*)	upright smooth.	———	1798.	4. 9.	G. ♄.	Wendl. obs. t. 2. f. 10.
erécta. W. *stricta.* Donn.							
7 juniperìna R.S.	(*wh.*)	drooping.	———	1758.	——	G. ♄.	Bot. mag. t. 222.
erécta. B.M. *nec aliorum.*							
8 angustifòlia. R.S.	(*st.*)	narrow-leaved.	———	1816.	——	G. ♄.	
9 incàna. B.C.	(*ye.*)	hoary.	———	1802.	7. 8.	G. ♄.	Lodd. bot. cab. t. 11.
10 imbricàta. R.S.	(*ye.*)	tiled-leaved.	———	1794.	4. 8.	G. ♄.	Andr. reposit. t. 113.

2 G

11 ovàta. R.S.	(*wh.*) oval-leaved.	C. B. S.	1792.	2. 6.	G. ♄.	Andr. reposit. t. 119.
12 tomentòsa. R.S.	(*wh.*) downy-leaved.	————	1799.	8. 9.	G. ♄.	———— t. 334.
PIMEL'EA. B.P.	PIMEL'EA.	Diandria Monogynia. R.S.				
1 linifòlia. B.P.	(*wh.*) flax-leaved.	N. S. W.	1793.	2. 8.	G. ♄.	Bot. mag. t. 891.
2 paludòsa. B.P.	(*wh.*) marsh.	————	1826.	——	G. ♄.	
3 glaùca. B.P.	(*wh.*) glaucous-leaved.	————	1822.	——	G. ♄.	Linn.trans.10. t.13. f.2.
4 ligustrìna. B.P.	(*wh.*) Privet-like.	————	1823.	——	G. ♄.	Lab. nov. holl. 1. t. 3.
5 decussàta. B.P. *ferrugìnea*. Lab.	(*re.*) cross-leaved.	————	——	4. 8.	G. ♄.	Swt. flor. aust. t. 8.
6 ròsea. B.P.	(*ro.*) rose-coloured.	N. Holland.	1800.	3. 9.	G. ♄.	Bot. mag. t. 1458.
7 linoídes. F.T.	(*wh.*) flax-like.	N. S. W.	1824.	——	G. ♄.	
8 filamentòsa. L.T.	(*wh.*) filamentose.	N. Holland.	1826.	——	G. ♄.	Linn. trans.10. t.14. f.1.
9 pauciflòra. B.P.	(*ye.*) few-flowered.	V. Diem. Isl.	1812.	——	G. ♄.	Lodd. bot. cab. t. 179.
10 humilis. B.P.	(*wh.*) dwarf.	————	1825.	6. 9.	G. ♄.	Bot. reg. t. 1268.
11 clavàta. B.P.	(*ye.*) clubbed.	N. Holland.	1826.	3. 6.	G. ♄.	
12 incàna. B.P.	(*st.*) hoary.	V. Diem. Isl.	——	4. 8.	G. ♄.	
13 drupàcea. B.P.	(*wh.*) drupe-bearing.	————	1820.	——	G. ♄.	Swt. flor. aust. t. 52.
14 spicàta. B.P.	(*wh.*) spiked.	N. S. W.	1824.	——	G. ♄.	Linn. trans.10. t.14. f.2.
? TR'OPHIS. W.	RAMOON-TREE.	Diœcia Tetrandria. W.				
1 americàna. W.	(*gr.*) American.	W. Indies.	1789.	4. 5.	S. ♄.	Brown. jam. t. 37. f. 1.
2 áspera. W.	(*gr.*) rough-leaved.	E. Indies.	1802.	——	S. ♄.	
3 spinòsa. S.S.	(*gr.*) spiny.	————	1826.	....	S. ♄.	

ORDO CLXXIV.

OSYRIDEÆ. *Brown.*

OS'YRIS. W.	POET'S CASSIA.	Diœcia Triandria. L.				
álba. W.	(*wh.*) white.	S. Europe.	1739.	....	G. ♄.	Lam. ill. t. 802.
EXOCA'RPOS. B.P.	EXOCA'RPOS.	Pentandria Monogynia.				
1 cupressifórmis. B.P.	(*wh.*) Cypress-like.	N. Holland.	1820.	....	G. ♄.	Labill. voy. 1. t. 14.
2 humifùsa. B.P.	(*gr.*) low-procumbent.	V. Diem. Isl.	1822.	....	G. ♄.	
3 stricta. B.P.	(*gr.*) upright.	N. S. W.	——	....	G. ♄.	

ORDO CLXXV.

SANTALACEÆ. *Brown prodr. p.* 350.

SA'NTALUM. B.P.	SANDAL-WOOD.	Tetrandria Monogynia. W.				
1 álbum. F.I.	(*ye.pu.*) white.	E. Indies.	1804.	....	S. ♄.	Rumph. amb. 2. t. 11.
2 myrtifòlium. F.I.	(*li.*) Myrtle-leaved.	————	1819.	....	G. ♄.	Roxb. corom. 1. t. 2.
3 obtusifòlium. B.P.	blunt-leaved.	N. S. W.	1823.	....	G. ♄.	
FUS'ANUS. B.P.	FUS'ANUS.	Tetrandria Monogynia.				
compréssus. H.K.	(*gr.*) flat-stalked.	C. B. S.	1776.	....	G. ♄.	Berg. cap. 38. t. 1. f. 1.
LEPTOM'ERIA. B.P.	LEPTOM'ERIA.	Pentandria Monogynia. R.S.				
1 ácida. B.P.	(*wh.*) acid.	N. S. W.	1823.	....	G. ♄.	
2 Billardiéri. B.P. *Thèsium drupàceum*. Lab.	(*wh.*) Labillardiere's.	————	——	....	G. ♄.	Lab. nov. hol. 1. t. 93.
TH'ESIUM. R.S.	THE'SIUM.	Pentandria Monogynia. L.				
1 linophy'llum. R.S.	(*gr.*) common.	England.	....	6. 7.	H. ♃.	Eng. bot. v. 4. t. 247.
2 intermèdium. R.S.	(*wh.*) intermediate.	Germany.	1818.	——	H. ♃.	Schkuhr hand. t. 51.
3 montànum. R.S.	(*wh.*) mountain.	Europe.	....	——	H. ♃.	Hayn. sc. j. t. 119.
4 alpìnum. R.S.	(*wh.*) Alpine.	————	1814.	——	H. ♃.	Jacq. aust. 5. t. 416.
5 ramòsum. R.S.	(*wh.*) branching.	————	1824.	——	H. ♃.	Hayn. sc. j. p. 30. t. 7.
6 ebracteàtum. R.S.	(*gr.*) blunt-leaved.	Germany.	——	——	H. ♃.	—— l. c. t. 7.
7 ? amplexicàule. R.S.	(*gr.*) stem-clasping.	C. B. S.	1787.	——	G. ♄.	
COMA'NDRA. N.	COMA'NDRA.	Pentandria Monogynia. N.				
umbellàta. N. *Thèsium umbellàtum*. W.	(*wh.*) umbelled.	N. America.	1782.	6.	H. ♃.	Pluk. mant. t. 342. f. 1.

HAMILTO'NIA. W.		OIL-NUT.	Polygamia Diœcia. W.				
oleífera. W.	(*gr.ye.*)	Olive-bearing.	N. America.	1800.	....	G. ♄.	Pursh. amer. 1. t. 13.
Pyrulària pùbera. M.							
NY'SSA. R.S.		TUPELO.	Polygamia Diœcia. L.				
1 villòsa. R.S.	(*gr.*)	hairy-leaved.	N. America.	1806.	5.	H. ♄.	Mich. arb. 2. t. 21.
2 aquática. R.S.	(*gr.*)	water.	———	1739.	4. 5.	H. ♄.	——— 2. t. 22.
biflòra. W.							
3 cándicans. R.S.	(*gr.*)	Ogechee-lime.	———	1812.	....	H. ♄.	Mich. arb. 2. t. 20.
4 capitàta. H.K.	(*gr.*)	headed.	Carolina.	1806.	....	H. ♄.	
5 tomentòsa. R.S.	(*gr.*)	tomentose.	———	1812.	....	H. ♄.	Mich. arb. 2. t. 19.
6 denticulàta. R.S.	(*gr.*)	toothed.	N. America.	1795.	5. 6.	H. ♄.	Catesb. car. 1. t. 60.

ORDO CLXXVI.

ELÆAGNEÆ. *Don prodr. p.* 67.

HIPPO'PHAE. L.		SEA-BUCKTHORN.	Diœcia Tetrandria. L.				
1 rhamnoídes. W.	(*st.*)	common.	England.	....	4. 5.	H. ♄.	Eng. bot. t. 425.
2 salicifòlia. D.P.	(*st.*)	Willow-leaved.	Nepaul.	1819.	....	H. ♄.	
SHEPHE'RDIA. N.		SHEPHE'RDIA.	Diœcia Octandria. N.				
1 argéntea. N.	(*st.*)	silvery.	Missouri.	1818.	....	H. ♄.	
Hippóphae argéntea. Ph.							
2 canadénsis. N.	(*st.*)	Canadian.	———	1759.	4. 5.	H. ♄.	
Hippóphae canadénsis. W.							
ELÆA'GNUS. L.		OLEASTER.	Tetrandria Monogynia. L.				
1 angustifòlia. R.S.	(*wh.*)	narrow-leaved.	S. Europe.	1633.	7.	H. ♄.	Bot. reg. t. 1156.
2 songáricus. F.	(*st.*)	Fischer's.	Siberia.	1821.	——	H. ♄.	
3 orientàlis. R.S.	(*wh.*)	oriental.	Levant.	1748.	7. 8.	G. ♄.	Pall. ros. 1. t. 5.
4 spinòsa. R.S.	(*wh.*)	spiny.	Egypt.	1826.	——	F. ♄.	
5 argéntea. Ph.	(*wh.*)	silvery.	N. America.	1813.	——	H. ♄.	Wats. dend. brit. t. 16.
6 arbòrea. F.I.	(*wh.*)	tree.	Nepaul.	1819.	....	H. ♄.	
7 conférta. F.I.	(*st.*)	close-flowered.	———	——	....	H. ♄.⌣.	
8 latifòlia. R.S.	(*wh.*)	broad-leaved.	E. Indies.	1712.	7. 8.	S. ♄.	Burm. zeyl. t. 39. f. 2.
9 triflòra. F.I.	(*wh.*)	three-flowered.	Nepaul.	1825.	....	H. ♄.	——— t. 39. f. 1.
10 umbellàta. D.P.	(*wh.*)	umbelled.	———	1829.	....	H. ♄.	Thunb. jap. t. 14.

ORDO CLXXVII.

ASARINÆ. *Kunth synops.* 1. *p.* 442.

A'SARUM. L.		ASARABACCA.	Dodecandria Monogynia. L.				
1 europ'æum. S.S.	(*pu.*)	common.	England.	....	5.	H. ♃.	Eng. bot. v. 16. t. 1083.
2 canadénse. S.S.	(*pu.*)	Canadian.	Canada.	1713.	4. 7.	H. ♃.	Swt. br. fl. gar. t. 95.
3 virgínicum. W.	(*pu.*)	Virginian.	Virginia.	1759.	4. 5.	H. ♃.	——— t. 18.
4 arifòlium. S.S.	(*pu.*)	Arum-leaved.	Carolina.	1818.	5. 6.	H. ♃.	Hook. ex flor. t. 40.
ARISTOL'OCHIA. W.		BIRTHWORT.	Gynandria Hexandria. L.				
1 bilobàta. W.	(*pu.*)	two-lobed.	W. Indies.	1822.	6. 8.	S. ♄.⌣.	Plum. amer. t. 106.
2 trilobàta. W.	(*pu.*)	three-lobed.	S. America.	1775.	6. 7.	S. ♄.⌣.	
3 ríngens. W.	(*br.*)	gaping.	———	1820.	——	S. ♄.⌣.	Vahl. symb. 2. t. 47.
4 labiòsa. B.R.	(*br.*)	lipped.	———	——	5. 9.	S. ♄.⌣.	Bot. reg. t. 689.
5 pentándra. W.	(*pu.*)	pentandrous.	Cuba.	1828.	....	S. ♄.⌣.	Jacq. ic. pict. t. 224.
6 panduriförmis. W.	(*pu.*)	fiddle-leaved.	Caracas.	1823.	....	S. ♄.⌣.	
7 máxima. W.	(*pu.*)	great.	New Spain.	1759.	7.	S. ♄.⌣.	Jacq. amer. t. 146.
8 Sìpho. W.	(*br.*)	broad-leaved.	N. America.	1763.	6. 7.	H. ♄.⌣.	Bot. mag. t. 534.
9 tomentòsa. B.M.	(*ye.*)	downy.	———	1799.	——	H. ♄.⌣.	——— t. 1369.
10 odoratíssima. W.	(*pu.*)	sweet-scented.	Jamaica.	1737.	7.	S. ♄.⌣.	Sloan. jam. 1. t. 104. f. 1.
11 barbàta. W.	(*pu.*)	bearded.	Caracas.	1796.	6. 8.	S. ♄.⌣.	Jacq. ic. 3. t. 608.
12 bilabiàta. W.	(*pu.*)	two-lipped.	Cuba.	1829.	....	S. ♄.⌣.	
13 índica. W.	(*pu.*)	Indian.	E. Indies.	1780.	6. 7.	S. ♄.⌣.	Rheed. mal. 8. t. 25.
14 acuminàta. W.	(*pu.*)	taper-pointed.	Mauritius.	1820.	....	S. ♄.⌣.	

15	fœtida. K.S.	(*pu.*)	fetid.	Mexico.	1822.	6. 8.	S. ♄.⌣.	H. et B. 2. t. 114.
16	hastàta. K.S.	(*pu.*)	halberd-leaved.	Havannah.	——	....	S. ♄.⌣.	———— 2. t. 116.
17	bœtica. W.	(*pu.*)	Spanish.	Spain.	1596.	5. 6.	H. ♄.⌣.	Moris. s. 12. t. 17. f. 6.
18	glaùca. W.	(*ye.*)	glaucous.	Barbary.	1785.	6. 8.	G. ♄.⌣.	Bot. mag. t. 1115.
19	sempervìrens. W.	(*pu.*)	evergreen.	Candia.	1727.	5. 6.	G. ♄.⌣.	———— t. 1116.
20	lónga. W.	(*pu.*)	long-rooted.	S. Europe.	1548.	6. 10.	H. ♃.	Mill. ic. t. 51. f. 2.
21	microphy'lla. W. en.	(*pu.*)	small-leaved.	....	1826.	——	G. ♃.	
22	Serpentària. W.	(*pu.*)	Snake-root.	N. America.	1632.	6. 7.	H. ♃.	Jacq. schœn. 3. t. 385.
23	sagittàta. Muhl.	(*pu.*)	arrow-leaved.	————	1819.	——	H. ♃.	
24	bracteàta. W.	(*pu.*)	bracteate.	E. Indies.	1793.	7.	S. ♃.	
25	Pistolòchia. W.	(*pu.*)	small.	S. Europe.	1597.	6. 7.	H. ♃.	
26	rotúnda. W.	(*pu.*)	round-rooted.	————	1596.	3. 10.	G. ♃.	
27	pállida. W.	(*pu.*)	pale-flowered.	Italy.	1640.	5. 8.	H. ♃.	Moris. s. 12. t. 18. f. 2.
28	hírta. W.	(*pu.*)	hairy.	Chio.	1759.	5. 6.	G. ♃.	Tourn. it. 1. t. 147.
29	Clematítis. W.	(*ye.*)	common.	England.	....	5. 8.	H. ♃.	Eng. bot. v. 6. t. 398.
30	arboréscens. W.	(*pu. ye.*)	tree.	America.	1737.	6. 7.	G. ♄.	

ORDO CLXXVIII.

CYTINEÆ. *Brogniart.*

NEPE'NTHES. L. PITCHER-PLANT. Diœcia Monadelphia. L.

1	distillatòria. W.	(*gr.*)	cylindrical.	Ceylon.	1789.	4. 5.	S. *w.* ♃.	Burm. zeyl. 42. t. 17.
2	Phyllámphora. W.	(*gr.*)	ventricose.	China.	1822.	6. 8.	S. *w.* ♃.	Bot. mag. t. 2629.

ORDO CLXXIX.

STACKHOUSEÆ. *Brown.*

STACKHO'USIA. Lab. STACKHO'USIA. Pentandria Trigynia.

1	linariifòlia. F.T.	(*st.*)	Linaria-leaved.	N. S. W.	1823.	3. 6.	G. ♄.
2	spathulàta. S.S.	(*wh.*)	spathulate-leaved.	————	1825.	4. 6.	G. ♄.

ORDO CLXXX.

EUPHORBIACEÆ. *Juss. gen.* 384.

MERCURIA'LIS. W. MERCURY. Diœcia Enneandria. L.

1	perénnis. W.	(*gr.*)	perennial.	Britain.	....	4. 5.	H. ♃.	Eng. bot. v. 26. t. 1872.
2	ambígua. W.	(*gr.*)	doubtful.	Spain.	1806.	7. 8.	H. ⊙.	Linn. f. dec. 1. t. 8.
3	ánnua. W.	(*gr.*)	annual.	Britain.	....	7. 9.	H. ⊙.	Eng. bot. v. 8. t. 559.
4	ellíptica. W.	(*gr.*)	oval-leaved.	Portugal.	1802.	5. 7.	G. ♄.	Venten. cels. t. 12.
5	tomentòsa. W.	(*gr.*)	woolly.	Spain.	1640.	7. 9.	H. ♄.	

PEDILA'NTHUS. Neck. SLIPPER-SPURGE. Dodecandria Trigynia. L. en.

1	tithymaloídes. B.R.	(*sc.*)	Myrtle-leaved.	S. America.	1699.	6. 7.	D.S. ♄.	Bot. reg. t. 837.
	myrtifòlius. L. en. *Crepidària myrtifòlia.* H.S.S.							
2	carinàtus.	(*sc.*)	keel-leaved.	W. Indies.	1809.	——	D.S. ♄.	Bot. mag. t. 2514.
	Euphórbia carinàta. B.M. *Crepidària carinàta.* H.S.S.							
3	subcarinàtus. H.R.	(*sc.*)	small-keeled.	W. Indies.	1819.	——	D.S. ♄.	
4	canaliculàtus.	(*sc.*)	channelled-leaved.	————	——	——	D.S. ♄.	Lodd. bot. cab. t. 727.
	Euphórbia canaliculàta. B.C.							
5	padifòlius. H.S.	(*sc.*)	Padus-leaved.	————	1699.	——	D.S. ♄.	Dill. elt. t. 288. f. 372.
6	cordellàtus.	(*sc.*)	small-heart-leaved.	————	1795.	——	D.S. ♄.	
	Crepidària cordellàta. H.S.							

EUPHOR'BIA. W. SPURGE. Dodecandria Trigynia. L.

No.	Name	Colour	English name	Native country	Introduced	Flowering	Habit	Figure
1	uncinàta. DC.	(*gr.ye.*)	twin-spined.	C. B. S.	1795.	6. 8.	D.G. ħ.	DC. plant. gras. t. 151.
	Scolopéndria. H.S. *procúmbens.* Meerb. ic. rar. t. 55. *ex* H.P.							
2	squarròsa. H.P.	(*gr.ye.*)	squarrose.	C. B. S.	1825.	——	D.G. ħ.	
3	trigòna. H.S.	(*gr.ye.*)	upright 3-sided.	E. Indies.	1768.	7. 8.	D.S. ♄.	
4	antiquòrum. H.S.	(*gr.ye.*)	spreading 3-sided.	——	1688.	——	D.S. ♄.	Comm. hort. 1. t. 12.
5	láctea. H.S.	(*gr ye.*)	marbled.	——	1804.	——	D.S. ♄.	
6	grandidéns. H.P.	(*gr.ye.*)	large-toothed.	C. B. S.	1820.	——	D.G. ♄.	
7	canariénsis. W.	(*br.re.*)	Canary.	Canaries.	1697.	3. 4.	D.G. ♄.	DC. plant. gras. t. 140.
8	cærulèscens. H.P.	(*st.*)	blue-stemmed.	C. B. S.	1825.	——	D.G. ♄.	
9	tetragòna. H.P.	(*gr.ye.*)	four-sided.	——	——	——	D.G. ♄.	
10	heptagòna. W.	(*br.re.*)	seven-angled.	——	1731.	7. 11.	D.G. ♄.	Bradl. suc. 2. t. 13.
	α nìgra. H.S.	(*br.re.*)	*black-spined.*	——	——	——	D.G. ♄.	
	β rùbra. H.S.	(*br.re.*)	*red-spined.*	——	——	——	D.G. ♄.	
11	pentagòna. H.P.	(*gr.ye.*)	five-sided.	——	1825.	——	D.G. ♄.	
12	enneagòna. H.S.	(*gr.ye.*)	nine-angled.	——	1790.	7. 10.	D.G. ♄.	
13	mammillàris. W.	(*gr.ye.*)	warty-angled.	——	1759.	7. 8.	D.G. ♄.	Comm. præl. t. 9.
14	cereifórmis. W.	(*gr.ye.*)	naked.	——	1731.	6. 7.	D.G. ♄.	Burm. afr. t. 9. f. 3.
15	officinàrum. W.	(*ye.*)	officinal.	Africa.	1597.	——	D.G. ♄.	DC. plant. gras. t. 77.
16	odontophy'lla. W.en.	(*gr.*)	white-spined.	C. B. S.	1824.	——	D.G. ♄.	
17	stellæspìna. H.P.	(*gr.ye.*)	starry-spined.	——	——	——	D.G. ♄.	
18	polygòna. H.S.	(*gr.ye.*)	many-angled.	——	1790.	7. 9.	D.G. ♄.	Lodd. bot. cab. t. 1334.
19	nereifòlia. W.	(*re.*)	Oleander-leaved.	India.	1690.	6. 7.	D.S. ♄.	DC. plant. gras. t. 46.
20	Hy'strix. W.	(*st.*)	Porcupine.	C. B. S.	1695.	——	D.G. ♄.	Jacq. schœn. 2. t. 207.
	loricàta. P.S. *Treísia Hy'strix.* H.S.							
21	vàrians. H.S.	(*gr.ye.*)	variable-stemm'd.	E. Indies.	1800.	6. 9.	D.S. ♄.	
22	grandifòlia. H.S.	(*gr.ye.*)	great-leaved.	Sierra Leon.	1798.	——	D.S. ♄.	
23	eròsa. W.en.	(*gr.ye.*)	erose.	....	1818.	....	D.S. ♄.	
24	tribuloídes. S.S.	(*wh.*)	Caltrops-like.	Canaries.	1816.	6. 9.	D.G. ♄.	
25	Míllii. Desm.	(*sc.*)	splendid.	Madagascar.	1828.	——	D.S. ♄.	Bot. mag. t. 2902.
	spléndens. B.M. ann. 1829. *Míllii.* Desm. ann. 1826.							
26	cucumerìna. W.	(*gr.*)	Cucumber-like.	C. B. S.	....	5. 9.	D.G. ♄.	Vail. it. t. 5.
27	melofórmis. W.	(*ye.*)	Melon-like.	——	1774.	——	D.G. ♄.	Andr. reposit. t. 617.
28	clàva. W.	(*gr.*)	club.	——	——	3. 12.	D.G. ♄.	Jacq. ic. 1. t. 85.
	Treísia tuberculàta. H.S.S.							
29	Hawórthii.	(*gr.*)	Haworth's.	——	——	——	D.G. ♄.	
	Treísia clàva. H.S. *excl. syn.* W.							
30	bupleurifòlia. W.	(*gr.*)	cone-shaped.	——	1791.	6. 7.	D.G. ♄.	Jacq. schœn. 1. t. 106.
31	Cáput medùsæ. W.	(*gr.*)	great Medusa's head.	Africa.	1731.	8.	D.G. ħ.	Lodd. bot. cab. t. 1315.
	Medùsea màjor. H.S.							
	β hùmilis.	(*gr.*)	*dwarfer.*	——	1768.	——	D.G. ħ.	Burm. afr. t. 8.
32	tessellàta. H.S.	(*gr.*)	chequer'd-M'-head.	....	1788.	——	D.G. ħ.	
33	frúctus-pìni. H.S.	(*gr.*)	small M'-head.	C. B. S.	1731.	——	D.G. ħ.	DC. plant. gras. t. 150.
	β geminàta.	(*gr.*)	*twin-branched.*	——	——	——	D.G. ħ.	Burm. afr. 18. t. 9. f. 1.
34	procúmbens. H.S.	(*gr.*)	least M'-head.	——	1768.	——	D.G. ħ.	—— t. 10. f. 1.
35	pátula. Mill.	(*br.gr.*)	spreading.	——	——	——	D.G. ħ.	
	Dactylánthes pátula. H.S.							
36	anacántha. W.	(*br.gr.*)	scaly.	——	1727.	9. 10.	D.G. ħ.	Bot. mag. t. 2520.
37	globòsa. B.M.	(*br.gr.*)	roundish-jointed.	——	1822.	6. 11.	D.G. ħ.	—— t. 2624.
	Dactylánthes globòsa. H.P.							
38	tuberculàta. W.	(*wh.*)	tuberculate.	——	1805.	6. 8.	D.G. ħ.	Jacq. schœn. 2. t. 208.
39	hamàta. H.S.	(*br.gr.*)	hooked.	——	1795.	——	D.G. ħ.	Burm. afr. t. 6. f. 3.
40	Commelìni. DC.	(*br.gr.*)	Commelin's.	Africa.	....	——	D.S. ♄.	
41	Orníthòpus. W.en.	(*w.*)	Bird's-foot.	——	1816.	——	G. ħ.	Jacq. fragm. t. 120.
42	clandestìna. S.S.	(*gr.*)	clandestine.	——	1824.	——	D.G. ♄.	
43	lophogòna. W.	(*gr.*)	curl-winged.	Madagascar.	——	——	D.S. ♄.	DC. plant. gras. ic.
44	péndula. L.en.	(*gr.*)	pendulous.	....	1808.	——	D.S. ♄.	
	Tithymàlus péndulus. H.							
45	Tirucálli. W.	(*wh.*)	Indian tree.	India.	1690.	....	D.S. ♄.	Rheed. mal. 2. t. 34.
46	mauritánica. W.	(*ye.*)	Barbary.	Africa.	1732.	6. 8.	D.G. ♄.	Dill. elt. t. 289. f. 373.
47	Lamárckii. H.S.L.	(*gr.*)	Lamarck's.	——	1808.	——	D.G. ♄.	
	mauritánica. Lam. *nec aliorum. virgàta.* P.S. *non* W.K.							
48	aphy'lla. W.en.	(*gr.*)	leafless.	Teneriffe.	1815.	——	D.G. ♄.	
49	atropurpùrea. W.en.	(*pu.*)	dark purple.	——	——	——	D.G. ♄.	
50	piscatòria. W.	(*gr.ye.*)	smooth spear-ld.	Canaries.	1777.	——	D.G. ♄.	
51	balsamífera. W.	(*gr.ye.*)	balsam.	——	1779.	——	D.G. ♄.	
52	laurifòlia. P.S.	(*gr.ye.*)	Laurel-leaved.	Peru.	1820.	——	D.G. ♄.	
53	bracteàta. S.S.	(*gr.*)	bracteated.	....	1809.	——	D.S. ♄.	Jacq. schœnb. t. 276.
54	dendroídes. W.	(*ye.*)	tree-like.	Italy.	1768.	——	H. ♄.	Barrel. ic. 910.

55 l'æta. W.	(*st.*)	Mezereon-leaved.	S. Europe.	1758.	6. 8.	D.G. ♄.	Jacq. ic. 1. t. 87.
56 punícea. W.	(*cr.*)	scarlet-flowered.	Jamaica.	1778.	1. 9.	D.S. ♄.	Bot. reg. t. 190.
57 heterophy'lla. W.	(*sc.*)	various-leaved.	W. Indies.	1690.	4. 9.	D.S. ♄.	Plum. ic. t. 251. f. 3.
58 cyathóphora. W.	(*ye.pi.*)	coloured.	S. America.	1806.	7. 8.	D.S. ♄.	Jacq. ic. 3. t. 480.
59 repánda.	(*gr.ye.*)	waved.	E. Indies.	1808.	8.	D.S. ♄.	
Tithymàlus repándus. H.S.							
60 biglandulòsa. H.S.	(*gr.*)	twin-glanded.	Isl. Bourbon.	——	9.	D.S. ♄.	
61 geniculàta. P.S.	(*st.*)	jointed.	Cuba.	1823.	7. 10.	S. ⊙.	Jacq. schœn. 3. t. 277.
prunifòlia. H.K. *Tithymàlus prunifòlius.* H.S.							
62 nudiflòra. S.S.	(*st.*)	naked-flowered.	Jamaica.	1800.	8.	D.S. ♄.	Jacq. ic. 3. t. 470.
63 verticillàta. S.S.	(*gr.*)	whorled.	W. Indies.	1826.	——	D.S. ♄.	
64 cotinifòlia. W.	(*wh.*)	Cotinus-leaved.	S. America.	1690.	7. 8.	D.S. ♄.	Hook. ex. flor. t. 59.
65 petiolàris. H.K.	(*ye.*)	long-stalked.	W. Indies.	1800.	5. 6.	D.S. ♄.	Bot. mag. t. 883.
66 mellífera. W.	(*pu.gr.*)	honey-bearing.	Madeira.	1784.	4. 5.	G. ♄.	—— t. 1305.
67 glabràta. S.S.	(*gr.ye.*)	smooth-leaved.	W. Indies.	1824.	——	D.S. ♄.	
68 toxicària. L.en.	(*gr.*)	poisonous.	....	1825.	——	D.S. ♄.	
69 pícta. W.	(*gr.*)	painted.	S. America.	1789.	5. 7.	S. ♃.	Jacq. ic. 3. t. 477.
70 silenifòlia.	(*gr.ye.*)	Silene-leaved.	C. B. S.	1821.	——	G. ♃.	
Tithymàlus silenifòlius. H.R.							
71 críspa. H.R.	(*gr.ye.*)	curl-leaved.	——	——	——	G. ♃.	
72 tuberòsa. W.	(*gr.ye.*)	tuberous.	Ethiopia.	1800.	10. 12.	G. ♄.	Burm. afr. 9. t. 4.
73 affìnis. DC.	(*gr.ye.*)	affined.	....	1823.	6. 8.	G. ♄.	
74 linarifòlia. W.	(*ye.*)	toad-flax-leaved.	....	1794.	——	S. ♄.	Jacq. ic. 1. t. 86.
75 linifòlia. W.	(*ye.*)	Flax-leaved.	W. Indies.	——	——	S. ♄.	
76 ocymoídea. W.	(*ye.*)	Basil-leaved.	S. America.	1733.	——	S. ⊙.	
77 dentàta. M.	(*ye.*)	dentated.	N. America.	1806.	6. 7.	H. ⊙.	
78 hypericifòlia. W.	(*wh.*)	Hypericum-leaved.	America.	1727.	6. 9.	H. ⊙.	Hook. ex. flor. t. 36.
79 Humbóldtii. W.en.	(*gr.*)	Humboldt's.	S. America.	1809.	7. 10.	S. ♂.	
80 prostràta. W.	(*pu.*)	trailing-red.	W. Indies.	1758.	——	S. ⊙.	
81 ròsea. W.	(*ro.*)	rosy.	E. Indies.	1808.	8.	S. ⊙.	
82 maculàta. W.	(*ye.*)	spotted.	S. America.	1660.	7.	H. ⊙.	Jacq. vind. 2. t. 186.
83 humifusa. W.en.	(*ye.*)	trailing.	....	1824.	6. 8.	H. ⊙.	
84 pilulífera. W.	(*gr.*)	globular.	E. Indies.	1800.	——	S. ♃.	Jacq. ic. 3. t. 478.
85 hyssopifòlia. W.	(*gr.*)	Hyssop-leaved.	W. Indies.	1787.	8. 9.	S. ⊙.	
86 scordifòlia. S.S.	(*wh.*)	Scordium-leaved.	S. America.	1826.	6. 8.	S. ♄.	Jacq. icon. t. 476.
87 tannénsis. S.S.	(*gr.*)	Tanna.	N. Hebrides.	——	——	G. ♄.	
88 thymifòlia. W.	(*wh.*)	Thyme-leaved.	India.	1699.	7. 8.	S. ⊙.	Pluk. alm. t. 113. f. 2.
89 parviflòra. W.	(*wh.*)	small-flowered.	E. Indies.	1818.	——	S. ⊙.	Burm. zeyl. t. 105. f. 2.
90 variegàta. B.M.	(*wh.*)	variegated.	Louisiana.	1811.	6. 8.	H. ⊙.	Bot. mag. t. 1747.
91 canéscens. W.	(*wh.re.*)	canescent.	Spain.	1818.	7. 8.	H. ⊙.	Cavan. ic. 1. t. 63.
92 Chamæs'yce. W.	(*wh.*)	scollop-leaved.	S. Europe.	1752.	——	H. ⊙.	Moris. s. 10. t. 2. f. 19.
93 Péplis. W.	(*ye.*)	purple.	Devonshire.	....	——	H. ⊙.	Eng. bot. v. 28. t. 2002.
94 polygonifòlia. W.	(*st.*)	knotgrass-leaved.	N. America.	1704.	6. 7.	H. ⊙.	Jacq. col. sup. t. 13. f. 3.
95 Ipecacuánhæ. W.	(*gr.*)	Ipecacuanha.	—— ——	1812.	——	H. ♃.	Bot. mag. t. 1494.
96 Péplus. W.	(*gr.*)	petty.	Britain.	....	7. 8.	H. ⊙.	Eng. bot. v. 14. t. 959.
97 peploídes. DC.	(*gr.*)	least.	S. Europe.	1800.	7. 9.	H. ⊙.	
'Esula mínima. H.S.							
98 falcàta. W.	(*ye.*)	sickle-leaved.	——	1699.	6. 8.	H. ⊙.	Jacq. aust. 2. t. 121.
99 acuminàta. M.B.	(*gr.*)	acuminate.	'Iberia.	1823.	——	H. ⊙.	
100 exígua. W.	(*gr.*)	dwarf.	Britain.	....	7.	H. ⊙.	Eng. bot. v. 19. t. 1336.
101 diffùsa. W.	(*gr.ye.*)	diffuse.	Austria.	1798.	7. 8.	H. ⊙.	Jacq. ic. 1. t. 88.
102 undulàta. M.B.	(*wh.*)	undulate.	Caucasus.	1820.	——	H. ♃.	
103 micrántha. W.	(*gr.*)	small-flowered.	Persia.	1803.	7. 9.	H. ⊙.	
104 Lagascàna. S.S.	(*gr.*)	Lagasca's.	Spain.	1827.	7. 10.	H. ⊙.	
105 Láthyris. W.	(*gr.*)	Caper.	England.	....	5. 10.	H. ♂.	Eng. bot. v. 32. t. 2255.
106 terracìna. W.	(*ye.*)	retuse-leaved.	S. Europe.	1818.	7. 10.	H. ⊙.	
107 valentìna. P.S.	(*ye.*)	Spanish.	Spain.	1804.	7. 8.	H. ⊙.	
108 'Apios. W.	(*ye.*)	pear-rooted.	Candia.	1596.	6. 7.	H. ♃.	
109 genistoídes. W.	(*ye.*)	Genista-like.	C. B. S.	1808.	7. 8.	G. ♄.	
110 condensàta. S.S.	(*ye.*)	tufted.	Iberia.	1828.	6. 8.	H. ♃.	
111 spinòsa. W.	(*ye.*)	prickly.	Levant.	1710.	6. 7.	F. ♄.	Wats. dend. brit. t. 45.
112 fruticòsa. L.en.	(*ye.*)	shrubby.	Sicily.	1825.	——	F. ♄.	Biv. pl. sic. cent. 1. 35.
113 epithymoídes. W.	(*ye.*)	broad-leaved.	Austria.	1805.	5. 6.	H. ♃.	Andr. reposit. t. 616.
114 villòsa. W.K.	(*sn.*)	villous.	Hungary.	1818.	6. 9.	H. ♃.	W. et K. hung. t. 93.
115 tenuifòlia. S.S.	(*ye.*)	slender-leaved.	S. Europe.	1827.	——	H. ⊙.	
116 echinocárpa. S.S.	(*ye.*)	prickly-podded.	Crete.	1828.	——	H. ♂.	
117 condylocárpa. M.B.	(*ye.*)	warted-fruited.	Caucasus.	1824.	——	H. ♃.	
118 obtusàta. Ph.	(*ye.*)	blunt-bracted.	N. America.	1826.	——	H. ♃.	
119 merculiàna. Ph.	(*ye.*)	Mercury-like.	——	1824.	——	H. ♃.	

120	marginàta. Ph.	(*wh.*)	white-margined.	N. America.	1816.	6. 9.	H.⊙.	
121	dúlcis. w.	(*pu.*)	sweet.	S. Europe.	1759.	5. 6.	H.♃.	Jacq. aust. 3. t. 213.
122	ambígua. w.k.	(*ye.*)	ambiguous.	Hungary.	1824.	——	H.♃.	W. et K. hung. t. 135.
123	carniòlica. w.	(*ye.*)	Carniolian.	Carniola.	1795.	8.	H.♃.	Scop. carn. t. 21.
124	angulàta. w.	(*ye.*)	angular-stalked.	Austria.	1819.	7. 8.	H.♃.	Jacq. ic. t. 481.
125	nummulariæfòlia.	(*ye.*)	moneywort-leaved.	......	1800.	7.	G.♄.	
126	erythrìna. L.en.	(*gr.*)	purple-margined.	C. B. S.	1816.	6. 8.	G.♃.	
127	neapolitàna. T.N.	(*wh.*)	Neapolitan.	Naples.	1825.	7. 8.	H.♃.	Ten. fl. neap. t. 42.
128	Pithyùsa. w.	(*gr.*)	Juniper-leaved.	S. Europe.	1741.	6. 7.	F.♄.	Bocc. sic. t. 5.
129	portlándica. w.	(*ye.*)	Portland.	Britain.	....	5. 9.	H.♃.	Eng. bot. v. 7. t. 441.
130	saxátilis. w.	(*ye.*)	rock.	Austria.	1816.	6. 8.	H.♃.	Jacq. aust. t. 345.
131	glareòsa. M.B.	(*ye.*)	stony.	Caucasus.	1819.	——	H.♃.	
132	Paràlias. w.	(*ye.*)	sea.	England.	....	7. 9.	H.♃.	Eng. bot. v. 3. t. 195.
	β suffruticòsa.	(*ye.*)	*shrubby sea.*	....	....	——	F.♄.	
133	júncea. w.	(*ye.*)	rushy.	Madeira.	1779.	7. 8.	G.♃.	Jacq. schœn. 1. t. 107.
134	congésta. W.en.	(*ye.*)	fiddle-leaved.	Spain.	1817.	——	H.♃.	
135	aléppica. w.	(*ye.*)	Aleppo.	Crete.	1739.	——	F.♃.	Alp. exot. t. 64.
136	taurinénsis. W.en.	(*ye.*)	Italian.	Italy.	1818.	——	H.♃.	All. ped. 2. t. 83. f. 2.
137	segetàlis. w.	(*ye.*)	corn.	S. Europe.	1699.	——	H.⊙.	Jacq. aust. 5. t. 450.
138	grácilis. s.s.	(*ye.*)	slender.	Ukraine.	1826.	——	H.♃.	
139	cæspitòsa. T.N.	(*ye.*)	tufted.	Naples.	——	——	H.♃.	
140	pìnea. s.s.	(*ye.*)	Pine-leaved.	————	——	——	H.♃.	
	linifòlia. T.N. *non* w.							
141	biumbellàta. DC.	(*ye.*)	double-umbel'd.	Barbary.	1780.	8.	H.⊙.	Poir. it. ed. germ. t. 1.
142	angustifòlia.	(*ye.*)	narrow-leaved.	....	——	7. 8.	H.♃.	
	'Esula angustifòlia. H.S.							
143	multicorymbòsa.	(*ye.*)	many-flowered.	....	1805.	——	H.♃.	
144	provinciàlis. w.	(*ye.*)	linear-leaved.	S. Europe.	1800.	8. 11.	H.⊙.	
145	juncoídes. H.S.	(*ye.*)	Rush-like.	————	——	7. 8.	H.⊙.	
146	helioscòpia. w.	(*ye.*)	Wartwort.	Britain.	....	——	H.⊙.	Eng. bot. v. 13. t. 883.
147	stellulàta. Sn.	(*ye.*)	starry.	Europe.	1823.	——	H.⊙.	
148	bialàta. L.en.	(*ye.*)	two-winged.	....	1819.	——	H.⊙.	
149	serràta. w.	(*ye.*)	narrow notch-ld.	S. Europe.	1710.	7.	G.♃.	Jacq. ic. 3. t. 483.
150	uralénsis. F.	(*ye.*)	Fischer's.	Tauria.	1824.	——	H.♃.	
151	crética. w.	(*ye.*)	Cretan hoary.	Levant.	1768.	——	H.♃.	
152	rìgens.	(*ye.*)	stiff.	....	1795.	7. 8.	H.♄.	
	Galarh'æus rígidus. H.S.							
153	verrucòsa. w.	(*ye.*)	warted.	France.	1800.	6.	H.♃.	Moris. s. 10. t. 3. f. 3.
154	corollàta. w.	(*wh.*)	great-flowered.	N. America.	1803.	7. 9.	H.♃.	Lodd. bot. cab. t. 390.
155	spathulæfòlia. H.S.	(*gr.*)	spatula-leaved.	....	1800.	8.	G.♄.	
156	corallioídes. w.	(*ye.*)	Coral-stalked.	S. Europe.	1752.	6. 9.	H.♃.	
157	móllis. Gm.	(*ye.*)	soft.	————	1823.	——	H.♃.	
158	androsæmifòlia. H.S.	(*ye.*)	Tutsan-leav'd.	Hungary.	1804.	7. 9.	H.♃.	
159	pilòsa. w.	(*ye.*)	hairy.	Siberia.	1758.	5. 8.	H.♃.	Gmel. sib. 2. t. 93.
160	orientàlis. w.	(*st.*)	oriental.	Levant.	1739.	6. 7.	H.♃.	
161	fragífera. L.en.	(*ye.*)	Strawberry-like.	Italy.	1825.	——	H.♃.	
162	virgàta. W.K.	(*ye.*)	twiggy.	Hungary.	1807.	7.	H.♃.	W. et K. hung. t. 162.
163	platyphy'lla. H.S.	(*ye.*)	annual warted.	England.	....	7. 8.	H.⊙.	Jacq. aust. t. 376.
164	stricta. E.B.	(*ye.*)	upright warted.	————	....	——	H.⊙.	Eng. bot. v. 5. t. 333.
165	literàta. w.	(*st.*)	blotch-leaved.	....	1790.	8.	H.⊙.	Jacq. ic. 3. t. 482.
166	crispàta. L.en.	(*ye.*)	crisp-leaved.	....	1819.	7. 9.	H.♃.	
167	pubéscens. s.s.	(*ye.*)	pubescent.	Barbary.	1827.	——	H.♃.	Jacq. f. ecl. t. 66.
168	lutéscens. Led.	(*ye.*)	yellowish.	Siberia.	——	——	H.♃.	
169	Ehrenbérgii.	(*ye.*)	Ehrenberg's.	Egypt.	1826.	6. 9.	H.⊙.	
	heterophy'lla. Ehr. *non* w.							
170	trigonocárpa. F.	(*ye.*)	three-sided podded.	Russia.	1821.	——	H.⊙.	
171	retùsa. C.I.	(*gr.*)	notch-leaved.	S. Europe.	——	——	H.⊙.	Cavan. ic. t. 34. f. 3.
172	'Esula. w.	(*gr.*)	leafy-branched.	Britain.	....	5. 7.	H.♃.	Eng. bot. v. 20. t. 1399.
173	Gerardiàna. w.	(*ye.*)	Gerard's.	Germany.	1801.	7.	H.♃.	Jacq. aust. t. 436.
174	Cyparíssias. w.	(*ye.*)	Cypress.	England.	....	5. 9.	H.♃.	Eng. bot. v. 12. t. 840.
175	nicæénsis. w.	(*ye.*)	sharp-leaved.	Spain.	1809.	——	F.♃.	Jacq. ic. 3. t. 485.
176	myrsinìtes. w.	(*ye.*)	glaucous.	S. Europe.	1570.	4. 6.	H.♃.	
177	rígida. M.B.	(*ye.*)	rigid.	Tauria.	1819.	6. 8.	H.♃.	Bot. reg. t. 274.
178	imbricàta. w.	(*ye.*)	imbricated.	Portugal.	1804.	8. 9.	F.♄.	
179	glaucéscens. W.en.	(*ye.*)	glaucescent.	....	1819.	——	H.♃.	
180	agrària. M.B.	(*ye.*)	field.	Tauria.	——	6. 8.	H.♂.	
181	palústris. w.	(*ye.*)	marsh.	Sweden.	1570.	5. 8.	H.♃.	Flor. dan. t. 866.
182	pállida. w.	(*ye.*)	pale.	Hungary.	1818.	6. 8.	H.♃.	
183	lùcida. W.K.	(*ye.*)	glossy.	————	1820.	——	H.♃.	W. et K. hung. t. 55.

184 pròcera. M.B.	(*ye.*)	tall.	Terek.	1824.	6. 8.	H. ♃.	
185 ceratocárpa. T.N.	(*ye.*)	horn-capsuled.	Naples.	——	——	H. ♃.	Ten. fl. neap. t. 43.
186 emarginàta. W.	(*ye.*)	freckled.	Italy.	1758.	7. 8.	H. ♃.	
187 hibérna. E.B.	(*ye.*)	Irish.	Ireland.	....	5. 6.	H. ♃.	Eng. bot. v. 19. t. 1337.
188 salicifòlia. W.	(*ye.*)	Willow-leaved.	Hungary.	1804.	——	H. ♃.	Lodd. bot. cab. t. 973.
189 sylvática. W.	(*pi.*)	wood.	S. Europe.	1768.	7. 9.	F. ♄.	Jacq. aust. t. 375.
190 amygdaloídes. W.	(*st.*)	Almond-leaved.	England.	....	3. 6.	H. ♃.	Eng. bot. v. 4. t. 256.
191 Characias. W.	(*ye.pu.*)	upright red.	———	....	——	H. ♄.	——— v. 7. t. 442.
192 véneta. L.en.	(*ye.*)	venomous.	Europe.	1824.	——	F. ♄.	
193 Bonplándii.	(*wh.*)	Bonpland's.	N. Spain.	1823.	4. 6.	S. ♃.	
marginàta. K.S. *non* Nutt.							
CI'CCA. L.		CI'CCA.	Monœcia Tetrandria. L.				
dísticha. W.	(*gr.*)	long-leaved.	E. Indies.	1796.	4. 6.	S. ♄.	Jacq. schœn. 2. t. 194.
BU'XUS. W.		BOX-TREE.	Monœcia Tetrandria. L.				
1 chinénsis. L.en.	(*ye.*)	China.	China.	1812.	4. 8.	H. ♄.	
2 baleàrica. W.	(*ye.*)	Minorca.	Minorca.	1780.	7.	H. ♄.	
3 sempervìrens. W.	(*ye.*)	common.	England.	....	4.	H. ♄.	Eng. bot. v. 19. t. 1341.
α arboréscens. Mill.	(*ye.*)	*tree.*	———	....	——	H. ♄.	
† *argéntea.*	(*ye.*)	*silvery-leaved.*	....	....	——	H. ♄.	
‡ *aùrea.*	(*ye.*)	*golden-leaved.*	....	....	——	H. ♄.	
‖ *marginàta.*	(*ye.*)	*margined.*	....	....	——	H. ♄.	
β angustifòlia. Mill.	(*ye.*)	*narrow-leaved.*	....	....	——	H. ♄.	
† *variegàta.*	(*ye.*)	*variegated.*	....	....	——	H. ♄.	
γ myrtifòlia. Mill.	(*ye.*)	*myrtle-leaved.*	....	....	——	H. ♄.	
δ suffruticòsa.	(*ye.*)	*dwarf.*	....	....	——	H. ♄.	
SARCOCO'CCA. B.R.		SARCOCO'CCA.	Monœcia Tetrandria.				
coriàcea.	(*wh.*)	leathery.	Nepaul.	1820.	4. 6.	G. ♄.	Bot. reg. t. 358.
prunifórmis. B.R.		*Pachysándra coriacea.* Hook. ex. flor. t. 148.					
PACHYSA'NDRA. M.		PACHYSA'NDRA.	Monœcia Tetrandria. W.				
1 procúmbens. M.	(*bh.*)	procumbent.	N. America.	1800.	3. 4.	H. ♃.	Bot. reg. t. 33.
ACID'OTON. W.		ACID'OTON.	Monœcia Polyandria.				
ùrens. W.	(*gr.*)	stinging.	Jamaica.	1793.	4. 6.	S. ♄.	Sloan. jam. 1. t. 83. f. 1.
PLUKEN'ETIA. W.		PLUKEN'ETIA.	Monœcia Monadelphia. L.				
1 volùbilis. W.	(*gr.*)	twining.	W. Indies.	1739.	7. 8.	S. ♄. ◡.	Plum. ic. t. 226.
2 corniculàta. S.S.	(*gr.*)	horned.	E. Indies.	1827.	——	S. ♄. ◡.	Rumph. amb. 1. t. 79. f. 2.
DALECHA'MPIA. W.		DALECHA'MPIA.	Monœcia Monadelphia.				
1 scándens. W.	(*gr.*)	climbing.	W. Indies.	1739.	6. 7.	S. ♄. ◡.	Jacq. amer. 252. t. 160.
2 brasiliénsis. W.	(*gr.*)	Brazilian.	Brazil.	1820.	——	S. ♄. ◡.	
3 fimbriàta. K.S.	(*gr.*)	fringed.	Mexico.	1829.	——	S. ♄. ◡.	
ACA'LYPHA. W.		ACA'LYPHA.	Monœcia Monadelphia. W.				
1 diversifòlia. W.	(*gr.*)	various-leaved.	Caracas.	1824.	6. 8.	S. ♄.	
2 hernandifòlia. S.S.	(*gr.*)	Hennandia-leaved.	Jamaica.	1825.	....	S. ♄.	
3 virgínica. W.	(*gr.*)	Virginian.	N. America.	1759.	7. 8.	H. ⊙.	Schk. hand. 3. t. 311.
4 caroliniàna. W.	(*gr.*)	Carolina.	———	1811.	——	H. ⊙.	Lam. ill. t. 780. f. 2.
5 ciliàta. W.	(*gr.*)	ciliated.	E. Indies.	1799.	——	S. ⊙.	Vahl symb. 1. t. 20.
6 réptans. W.	(*gr.*)	creeping.	Jamaica.	1822.	——	S. ♄.	Sloan. hist. 1. t. 82. f. 3.
7 índica. W.	(*gr.*)	Indian.	E. Indies.	1759.	7. 9.	S. ⊙.	Rheed. mal. 10. t. 81.
8 híspida. W.	(*gr.*)	hispid.	———	1820.	——	S. ⊙.	
9 cuspidàta. W.	(*gr.*)	cuspidate.	Caracas.	——	....	S. ♄.	Jacq. schœn. 2. t. 243.
10 rùbra. W.en.	(*re.*)	red.	S. America.	1823.	6. 9.	S. ⊙.	
11 brachystàchya. W.en.	(*gr.*)	short-spiked.	China.	1816.	7. 8.	H. ⊙.	
12 pauciflòra. W.en.	(*gr.*)	few-flowered.	———	——	——	H. ⊙.	
13 alopecuroídea. W.	(*gr.*)	Fox-tail.	Venezuela.	1804.	7. 9.	H. ⊙.	Jacq. ic. 3. t. 620.
14 anemioídes. K.S.	(*gr.*)	Anemia-like.	Mexico.	1829.	....	S. ♄.	
15 purpuráscens. K.S.	(*pu.*)	purple-bracted.	———	1827.	....	S. ♄.	
16 ellíptica. S.S.	(*gr.*)	elliptic-leaved.	Jamaica.	1825.	....	S. ♄.	
17 lævigàta. W.	(*gr.*)	smooth.	———	1819.	——	S. ♄.	
18 virgàta. W.	(*gr.*)	twiggy.	———	1822.	——	S. ♄.	Brown. jam. t. 36. f. 2.
19 macrostàchya. W.	(*gr.*)	long-spiked.	Caracas.	1824.	——	S. ♄.	Jacq. schœn. 2. t. 245.
20 polystàchya. W.	(*gr.*)	many-spiked.	....	——	——	S. ♄.	——— 2. t. 246.
21 scabròsa. W.	(*gr.*)	rough.	Jamaica.	1822.	——	S. ♄.	
22 integrifòlia. S.S.	(*gr.*)	entire-leaved.	Madagascar.	1823.	....	S. ♄.	
OMALA'NTHUS. J.		OMALA'NTHUS.	Monœcia Monadelphia.				
populifòlius. B.M.	(*wh.*)	Poplar-leaved.	N. Holland.	1824.	6. 8.	G. ♄.	Bot. mag. t. 2780.
CAPER'ONIA. Hilar.		CAPER'ONIA.	Monœcia Monadelphia.				
1 castaneæfòlia. S.S.	(*gr.*)	Chestnut-leaved.	S. America.	1828.	7. 8.	S. ⊙.	
2 palústris. S.S.	(*gr.*)	marsh.	Vera Cruz.	1731.	——	S. ⊙.	Mart. dec. 4. t. 88.
Cròton palústris. W.							

CR'OTON. W.		Cr'oton.	Monœcia Monadelphia. L.				
1 variegàtus. w.	(gr.)	variegated.	E. Indies.	1804.	6. 9.	S. ♄.	Rheed mal. 6. t. 61.
α latifòlius. H.B.	(gr.)	*broad-leaved.*	———	——	——	S. ħ.	
β críspus. H.B.	(gr.)	*curl-leaved.*	———	....	——	S. ħ.	Rumph.amb.4.t.26.f.2.
γ mèdius. H.B.	(gr.)	*narrow-leaved.*	———	1820.	——	S. ħ.	——— 4. t. 25.
2 píctus. B.C.	(gr.)	painted-leaved.	———	——	——	S. ♄.	Lodd. bot. cab. t. 870.
3 lineàris. S.S.	(gr.)	Willow-leaved.	W. Indies.	1733.	7.	S. ♄.	Jacq. amer. t. 162. f. 4.
4 Cascarílla. S.S.	(gr.)	Cascarilla.	S. America.	1778.	7. 8.	S. ♄.	Catesb. car. 2. t. 46.
5 marítimus. S.S.	(gr.wh.)	sea-side.	Carolina.	1786.	——	G. ♄.	
6 grácilis. K.S.	(gr.)	slender.	Mexico.	1824.	7. 8.	S. ♄.	
7 glabéllus. S.S.	(gr.wh.)	Laurel-leaved.	Jamaica.	1778.	5. 8.	S. ♄.	Sloan. jam.2. t.174. f.2.
8 hírtus. S.S.	(gr.wh.)	hairy.	Trinidad.	1821.	——	S. ⊙.	L'Herit. stirp. 17. t. 9.
9 argénteus. S.S.	(gr.)	silvery-leaved.	S. America.	1733.	7. 8.	S. ⊙.	
10 Tíglium. W.	(wh.)	purging.	E. Indies.	1796.	8. 9.	S. ♄.	Rheed. mal. 2. t. 33.
11 glandulòsus. S.S.	(gr.)	glandular.	W. Indies.	1818.	——	S. ⊙.	Jacq. ic. 1. t. 194.
12 sublùteus. S.S.	(gr.ye.)	yellowish.	Trinidad.	1822.	——	S. ♄.	Aubl. guian. 2. t. 339.
13 populifòlius. S.S.	(gr.wh.)	Poplar-leaved.	Jamaica.	1824.	——	S. ♄.	
14 tiliæfòlius. P.S.	(gr.wh.)	Lime-tree-leaved.	St. Vincent.	1825.	——	S. ♄.	Plum. mss. 4. t. 123.
15 Elutèria. S.S.	(gr.wh.)	sea-side Balsam.	Jamaica.	1748.	——	S. ♄.	Sloan. hist. 2. t.174. f. 2.
16 lùcidus. S.S.	(gr.wh.)	shining.	———	1822.	7. 9.	S. ħ.	
17 mìcans. S.S.	(gr.ye.)	glittering.	———	1815.	——	S. ♄.	Pluk. alm. t. 220. f. 5.
18 púngens. S.S.	(gr.ye.)	pungent.	Caracas.	1791.	——	S. ♄.	Jacq. ic. 3. t. 622.
19 penicillàtus. S.S.	(gr.)	pencilled.	Cuba.	1799.	7. 8.	S. ħ.	Venten. choix. t. 12.
20 flàvens. S.S.	(ye.gr.)	pale yellow.	W. Indies.	1822.	——	S. ħ.	
21 macrophy'llus. S.S.	(wh.)	large-leaved.	Jamaica.	——	——	S. ♄.	
22 aromáticus. S.S.	(wh.)	aromatic.	Ceylon.	1793.	——	S. ♄.	Rumph. amb. 3. t. 126.
23 hùmilis. S.S.	(wh.gr.)	humble.	Jamaica.	1799.	——	S. ħ.	
24 farinòsus. S.S.	(wh.)	mealy.	Madagascar.	1826.	....	S. ħ.	
25 Astroítes. W.	(gr.wh.)	woolly-leaved.	W. Indies.	1782.	7. 8.	S. ♄.	
26 lobàtus. S.S.	(wh.)	various-leaved.	Vera Cruz.	1730.	——	S. ⊙.	Mart. dec. 5. t. 46.
27 gossypifòlius. S.S.	(wh.)	Cotton-leaved.	Trinidad.	1823.	——	S. ħ.	Vahl. symb. t. 49.
28 tomentòsus. L.en.	....	tomentose.	....	——	....	S. ♄.	
CROZO'PHORA. J.		Crozo'phora.	Monœcia Pentandria.				
1 plicàta. J.	(wh.gr.)	plaited.	E. Indies.	1824.	6. 8.	G. ⊙.	Burm. ind. t. 62. f. 1.
2 tinctòria. J.	(wh.gr.)	officinal.	S. Europe.	1570.	7. 8.	H. ⊙.	Niss. act. par. 1712. t. 17.
Cròton tinctòrium. W.							
3 verbascifòlia. J.	(wh.gr.)	Mullein-leaved.	Greece.	1826.	——	H. ⊙.	
ROTTL'ERA. W.		Rottl'era.	Diœcia Icosandria. W.				
1 brasiliénsis. S.S.	(gr.)	Brazilian.	Brazil.	1820.	....	S. ħ.	
2 tinctòria. S.S.	(gr.)	dyer's.	E. Indies.	1810.	....	S. ♄.	Roxb. corom. 2. t. 168.
GEL'ONIUM. W.		Gel'onium.	Diœcia Icosandria. W.				
1 bifàrium. W.	(gr.)	oval-leaved.	E. Indies.	1793.	6. 8.	S. ♄.	
2 lanceolàtum. H.B.	(gr.)	spear-leaved.	———	1825.	——	S. ħ.	
TRE'WIA. W.		Tre'wia.	Diœcia Polyandria. W.				
nudiflòra. W.	(gr.)	naked-flowered.	E. Indies.	1796.	....	S. ♄.	Rheed. mal. 1. t. 42.
HYÆNA'NCHE. H.K.		Hyæna-poison.	Diœcia Dodecandria.				
globòsa. H.K.	(gr.wh.)	Cape.	C. B. S.	1783.	4. 9.	G. ħ.	Lamb. cinch. 52. t. 10.
E'UCLEA. W.		E'uclea.	Diœcia Dodecandria. W.				
1 racemòsa. W.	(wh.)	round-leaved.	C. B. S.	1722.	11. 12.	G. ħ.	Jacq. fragm. 3. t. 1. f. 5.
2 undulàta. W.	(wh.)	wave-leaved.	———	1794.	6. 10.	G. ħ.	
JA'TROPHA. W.		Physic-nut.	Monœcia Monadelphia. L.				
1 gossypifòlia. W.	(sc.)	Cotton-leaved.	W. Indies.	1690.	5. 8.	S. ħ.	Bot. reg. t. 746.
2 panduræfòlia. W.	(sc.)	fiddle-leaved.	Cuba.	1800.	——	S. ♄.	Bot. mag. t. 604.
3 integérrima. W.	(sc.)	spicy.	———	1809.	——	S. ♄.	——— t. 1464.
4 Cúrcas. W.	(gr.)	angular-leaved.	S. America.	1731.	....	S. ♄.	Jacq. vind. 3. t. 63.
5 multífida. W.	(sc.)	multifid.	———	1696.	6. 8.	S. ħ.	Sal. par. lond. t. 91.
6 coccínea. L.en.	(sc.)	scarlet.	....	1822.	——	S. ♄.	
CNIDO'SCOLUS. P.B.		Cnido'scolus.	Monœcia Monadelphia.				
1 ùrens. P.B.	(wh.)	stinging.	Brazil.	1690.	5. 7.	S. ♄.	Jacq. vind. 1. t. 21.
Játropha ùrens. Jacq.							
2 stimulòsus. Ph.	(wh.)	stimulous.	N. America.	1812.	——	F. ♃.	Pluk. alm. t. 220. f. 3.
Játropha stimulòsa. Ph.							
3 napæifòlia. W.	(wh.)	Napæa-leaved.	Antilles.	1823.	7. 8.	S. ♄.	
4 frágrans. K.S.	(wh.)	fragrant.	Cuba.	1822.	6. 8.	S. ħ.	
5 herbàcea. W.	(wh.)	annual.	Vera Cruz.	1759.	7. 8.	S. ⊙.	Reliq. houst. 6. t. 15.
6 Marcgràvii. P.B.	(wh.)	Marcgrave's.	Brazil.	1823.	——	S. ħ.	Pohl. flor. bras. 1. t. 50.
7 vitifòlius. P.B.	(wh.)	Vine-leaved.	———	——	——	S. ħ.	——— 1. t. 52.
MA'NIHOT. P.B.		Ma'nihot.	Monœcia Decandria.				
1 fœtida. P.B.	(bl.br.)	strong-scented.	Mexico.	1824.	7. 8.	S. ħ.	

2 Jánipha. P.B. (*br.gr.*) Lœfling's. S. America. 1824. 7. 8. S. ♄. Jacq. amer. t. 162. f. 1.
Játropha carthaginénsis. Jacq. *Jánipha.* W. *Jánipha Lœflíngii.* K.S.
3 grácilis. P.B. (*br.gr.*) slender. Brazil. 1822. 7. 8. S. ♄. Pohl. pl. bras. 1. t. 16.
4 anómala. P.B. (*br.*) anomalous. ——— ——— ——— S. ♄. ——— 1. t. 21.
5 pruinòsa. P.B. (*bl.br.*) frosted. ——— 1824. ——— S. ♄. ——— 1. t. 22.
6 digitifórmis. P.B. (*bl.br.*) finger-formed. ——— 1826. ——— S. ♄. ——— 1. t. 27.
7 digitàta. ...... finger-leaved. N. S. W. 1810. .... G. ♄.
Cròton digitàtum. F.
8 diversifòlia. .... different-leaved. N. Holland. 1822. .... G. ♄.
9 tenuifòlia. P.B. (*bl.br.*) slender-leaved. Brazil. ——— 6. 8. S. ♄. Pohl. pl. bras. 1. t. 29.
10 æsculifòlia. K.S. (*bl.br.*) Æsculus-leaved. S. America. 1825. 5. 8. S. ♄. H. et B. 2. t. 109.
11 cannabìna. (*bl.br.*) Cassava. ——— 1739. 7. 8. S. ♄. Sloan. jam. 1. t. 85.
Játropha Mánihot. W. *Jánipha Mánihot.* K.S.
12 caricæfòlia. P.B. (*bl.br.*) Papaw-leaved. Brazil. 1822. ——— S. ♄. Pohl. pl. bras. 1. t. 32.
13 Dalechampiæfòrmis. P.B. (*br.*) Dalechampia-like. — 1818. ——— S. ♄. ——— 1. t. 36.
14 sinuàta. P.B. (*br.*) sinuate-leaved. ——— 1824. ——— S. ♄. ——— 1. t. 41.

ALEUR'ITES. W. ALEUR'ITES. Monœcia Monadelphia. L.
1 trilòba. W. (*wh.*) three-lobed. Society Isles. 1793. .... S. ♄.
2 Ambìnux. P.S. (*wh.*) Molucca. Ceylon. 1803. .... S. ♄.
Cròton moluccànum. W.

RI'CINUS. W. PALMA-CHRISTI. Monœcia Monadelphia. L.
1 commùnis. W. (*re.st.*) Castor-oil plant. E. Indies. 1548. 7. 8. H. ⊙. Lam. ill. t. 792.
2 Kráppa. L.O. (*wh.fl.*) Krappa. .... 1827. 8. 11. H. ⊙.
3 leucocárpus. L.O. (*wh.bh.*) white-capsuled. .... ——— ——— F. ⊙.
4 macrophy'llus. L.O. (*w.r.*) large-leaved. .... ——— ——— F. ⊙.
5 paniculàtus. L.O. (*re.wh.*) panicled. .... 1824. 8. 10. F. ⊙.
6 rùtilans. L.O. (*re.wh.*) red-stalked. .... 1827. ——— F. ⊙.
7 undulàtus. L.O. (*re.wh.*) wave-leaved. .... ——— ——— F. ⊙.
8 víridis. W. (*bh.wh.*) green. E. Indies. 1802. 8. H. ⊙. W. hort. ber. 1. t. 49.
9 africànus. W. (*re.wh.*) African. Africa. ..... 7. 8. G. ♄.
10 lívidus. W. (*re.wh.*) livid-leaved. C. B. S. 1795. ——— G. ♄. Jacq. ic. 1. t. 196.
11 inérmis. W. (*re.st.*) smooth-capsuled. India. 1758. ——— S. ♂. ——— 1. t. 195.
12 armàtus. A.R. (*re.wh.*) rough-capsuled. Malta. 1807. ——— S. ♂. Andr. reposit. t. 430.

MA'PPA. J. MA'PPA. Monœcia Monadelphia.
1 moluccàna. S.S. (*wh.*) Molucca. Moluccas. 1828. .... S. ♄.
2 Tanària. S.S. (*wh.*) scollop-leaved. E. Indies. 1810. 7. 9. S. ♄. Rumph. amb. 3. t. 121.
Rícinus Tanàrius. W.

OMPH'ALEA. W. OMPH'ALEA. Monœcia Monadelphia. L.
triándra. W. (*wh.*) long-leaved. Jamaica. 1763. 6. 7. S. ♄. Lodd. bot. cab. t. 519.

HIPPO'MANE. W. MANCHINEEL-TREE. Monœcia Monadelphia. L.
1 Mancinélla. W. (*gr.*) common. W. Indies. 1690. .... S. ♄. Catesb. car. 2. t. 95.
2 spinòsa. S.S. (*gr.*) spiny. ——— 1820. .... S. ♄. Plum. ic. 171. f. 1.
Sàpium ilicifòlium. W.

S'APIUM. W. S'APIUM. Monœcia Monadelphia. L.
1 aucupàrium. S.S. single-spiked. W. Indies. 1692. .... S. ♄. Jacq. amer. t. 158.
2 Hippómane. S.S. (*gr.*) two-glanded. ——— 1824. .... S. ♄.
3 índicum. W. (*gr.*) Indian. E. Indies. 1817. .... S. ♄.

SIPH'ONIA. W. ELASTIC-GUM-TREE. Monœcia Monadelphia. W.
Cahùchu. W. (*gr.*) common. S. America. 1823. .... S. ♄. Lam. ill. t. 790.
elástica. P.S.

AGYN'EJA. W. AGYN'EJA. Monœcia Monadelphia. W.
1 impùbes. W. (*gr.*) smooth. China. 1820. 7. 9. H. ⊙.
2 pùbera. W. (*gr.*) woolly. ——— 1823. .... S. ♄.

STILLI'NGIA. W. STILLI'NGIA. Monœcia Monadelphia. W.
1 sylvática. W. (*ye.*) wood. Carolina. 1787. 7. 8. G. ♃.
2 ligustrìna. W. (*st.*) Privet-leaved. N. America. 1812. ——— G. ♄.
3 sebífera. W. (*st.*) Tallow-tree. China. 1703. 9. S. ♄. Pluk. amal. t. 390. f. 2.

H'URA. W. SANDBOX-TREE. Monœcia Monadelphia. W.
1 strèpens. W.en. (*gr.*) unequal-toothed. S. America. .... 7. 9. S. ♄.
2 crépitans. W.en. (*gr.*) equal-toothed. ——— 1733. ——— S. ♄. Lam. ill. t. 793.

BRADL'EJA. W. BRADL'EJA. Monœcia Monadelphia. W.
1 sínica. W. (*wh.*) Chinese. China. 1816. 7. 9. S. ♄. Gært. sem. 2. t. 109. f. 1.
2 nítida. H.B. (*wh.*) glossy. E. Indies. 1820. ——— S. ♄.
3 multiloculàris. S.S. (*wh.*) multilocular. ——— ——— ——— S. ♄.

PHYLLA'NTHUS. W. PHYLLA'NTHUS. Monœcia Monadelphia. W.
1 obovàtus. W. (*wh.*) annual. N. America. 1803. 7. 8. H. ⊙.
2 maderaspaténsis. W. (*gr.*) Madras. E. Indies. 1783. ——— S. ♄.
3 grácilis. H.B. (*gr.*) slender. ——— 1820. ——— S. ♃.
4 longifòlius. W. (*gr.*) long-leaved. Isl. Bourbon. 1822. ——— S. ♄. Lam. ill. t. 156. f. 1.

5 lanceolàtus w.	(*gr.*)	spear-leaved.	Isl. Bourbon.	1822.	6. 8.	S. ♄.	
6 nùtans. w.	(*bh.st.*)	nodding.	Jamaica.	——	——	S. ♄.	Jacq. schœn. 2. t. 193.
7 grandifòlius. w.	(*gr.*)	great-leaved.	Portorico.	1771.	——	S. ♄.	
8 viròsus. w.	(*gr.*)	venomous.	E. Indies.	1802.	——	S. ♄.	
9 turbinàtus. B.M.	(*gr.*)	shining-leaved.	China.	....	7. 9.	S. ♄.	Bot. mag. t. 1862.
10 Conámi. w.	(*gr.*)	Brazilian.	Brazil.	1791.	——	S. ♄.	Aub. gui. 2. t. 354.
11 obcordàtus. H.B.	(*gr.*)	obcordate.	E. Indies.	1818.	——	S. ♃.	
12 retùsus. H.B.	(*gr.*)	retuse-leaved.	————	1822.	——	S. ♃.	
13 reticulàtus. B.C.	(*bh.*)	netted.	————	1818.	8. 9.	S. ♄.	Lodd. bot. cab. t. 577.
14 lùcens. w.	(*gr.*)	thick-veined.	China.	1822.	7. 9.	S. ♄.	
15 cantoniénsis. L.en.	(*gr.*)	Canton.	————	1819.	7. 8.	H. ⊙.	
16 cuneàtus. W.en.	(*gr.*)	wedge-leaved.	————	——	——	H. ⊙.	
17 Nirùri. w.	(*gr.*)	Indian.	E. Indies.	1692.	6. 9.	S. ⊙.	Rheed. mal. 10. t. 15.
18 Urinària. w.	(*gr.*)	stinging.	————	1819.	——	S. ⊙.	Rumph. amb. 6. t. 17. f. 2.
19 microphy'llus. K.S.	(*gr.*)	small-leaved.	S. America.	1824.	——	S. ♄.	
20 polyphy'llus. w.	(*gr.*)	many-leaved.	E. Indies.	1805.	7. 9.	S. ♄.	
21 mimosoídes. w.	(*gr.*)	Mimosa-like.	W. Indies.	1820.	——	S. ♄.	Lodd. bot. cab. t. 721.
22 rhamnoídes. w.	(*gr.*)	Rhamnus-like.	E. Indies.	——	——	S. ♄.	
23 stríctus. H.B.	(*gr.*)	upright.	————	1823.	——	S. ♄.	
24 fraxinifòlius. B.C.	(*gr.*)	Ash-leaved.	....	1818.	——	S. ♄.	Lodd. bot. cab. t. 731.
25 juglandifòlius. W.en.	(*gr.*)	Walnut-leaved.	New Spain.	——	6. 9.	S. ♄.	
26 scándens. Roxb.	(*gr.*)	climbing.	E. Indies.	1803.	——	S. ♄.	
E'MBLICA. L.T.		E'MBLICA.	Monœcia Syngenesia. L.T.				
1 officinàlis. L.T.	(*st.*)	officinal.	E. Indies.	1768.	6. 9.	S. ♄.	Lodd. bot. cab. t. 548.
2 racemòsa. S.S.	(*st.*)	racemed.	————	1793.	7. 8.	S. ♄.	
Phyllánthus racemòsus. w.							
KIRGAN'ELIA. J.		KIRGAN'ELIA.	Monœcia Monadelphia.				
élegans. J.	(*gr.*)	elegant.	Mauritius.	1820.	6. 9.	S. ♄.	
Phyllánthus Kirganèlia. w.							
XYLOPHY'LLA. W.		XYLOPHY'LLA.	Monœcia Monadelphia. W.				
1 longifòlia. w.	(*re.*)	long-leaved.	E. Indies.	1816.	1. 12.	S. ♄.	Rumph. amb. 7. t. 12.
Phyllánthus cerámica. P.S.							
2 latifòlia. w.	(*wh.*)	Seaside Laurel.	S. America.	1783.	8. 10.	S. ♄.	Bot. mag. t. 1021.
3 speciòsa.	(*wh.*)	handsome.	Jamaica.	1818.	——	S. ♄.	Jacq. ic. 3. t. 616.
arbúscula. w. *Phyllánthus speciòsus.* Jacq.							
4 angustifòlia. w.	(*wh.*)	narrow-leaved.	————	1789.	7. 8.	S. ♄.	
5 lineàris. P.S.	(*re.*)	linear-leaved.	————	1822.	——	S. ♄.	
6 elongàta. B.C.	(*re.*)	elongated.	————	——	——	S. ♄.	Lodd. bot. cab. t. 1091.
7 montàna. w.	(*or.*)	mountain.	————	1820.	11. 4.	S. ♄.	Bot. mag. t. 2652.
8 falcàta. w.	(*re.*)	falcate-leaved.	Bahama Isl.	1699.	7. 9.	S. ♄.	Bot. reg. t. 373.
9 ramiflòra. w.	(*wh.*)	oval-leaved.	Siberia.	1785.	7. 8.	H. ♄.	Pall. it. 3. t. E. f. 2.
10 obovàta. W.en.	(*wh.*)	obovate-leaved.	————	....	——	S. ♄.	
ANDRA'CHNE. W.		BASTARD ORPINE.	Monœcia Gynandria. L.				
telephioídes. w.	(*wh.*)	annual.	Italy.	1732.	7. 8.	H. ⊙.	Lam. ill. t. 797.
BRID'ELIA. W.		BRID'ELIA.	Polygamia Monœcia. W.				
1 scándens. w.	(*wh.*)	climbing.	E. Indies.	1804.	7. 9.	S. ♄. ⌣.	Roxb. corom. 2. t. 173.
2 montàna. S.S.	(*st.*)	mountain.	————	1825.	——	S. ♄.	———— v. 2. t. 171.
3 fruticòsa. P.S.	(*st.*)	frutescent.	————	1827.	——	S. ♄.	———— v. 2. t. 172.
CLU'YTIA. W.		CLU'YTIA.	Diœcia Gynandria. L.				
1 collìna. w.	(*st.*)	hill.	E. Indies.	1807.	....	S. ♄.	Roxb. corom. 2. t. 169.
2 pátula. w.	(*st.*)	spreading.	————	1812.	....	S. ♄.	———— 2. t. 170.
3 pulchélla. w.	(*st.*)	broad-leaved.	C. B. S.	1739.	1. 6.	G. ♄.	Bot. mag. t. 1945.
4 heterophy'lla. w.	(*st.*)	various-leaved.	————	1816.	2. 4.	G. ♄.	
5 tomentòsa. w.	(*st.*)	tomentose.	————	1812.	4. 6.	G. ♄.	
6 pubéscens. w.	(*st.*)	pubescent.	————	1817.	——	G. ♄.	
7 polifòlia. w.	(*wh.*)	Poley-leaved.	————	1790.	——	G. ♄.	Jacq. schœn. 2. t. 50.
8 ericoídes. w.	(*wh.*)	Heath-leaved.	————	——	——	G. ♄.	Bot. reg. t. 779.
9 daphnoídes. w.	(*wh.*)	Daphne-like.	————	1731.	5. 6.	G. ♄.	W. hort. ber. t. 52.
10 polygonoídes.	(*wh.*)	Polygonum-like.	————	1790.	12. 3.	G. ♄.	———— t. 51.
11 alaternoídes. w.	(*wh.*)	narrow-leaved.	————	1692.	——	G. ♄.	Bot. mag. t. 1321.
EXCŒC'ARIA. W.		EXCŒC'ARIA.	Diœcia Monadelphia. L.				
1 lùcida. S.S.	(*wh.*)	shining-leaved.	Jamaica.	1824.	4. 6.	S. ♄.	Sloan. hist. 2. t. 158. f. 2.
2 glandulòsa. S.S.	(*wh.*)	glandular.	————	——	——	S. ♄.	Rumph. amb. 2. t. 79. 80.
3 Agallòcha. S.S.	(*gr.ye.*)	Indian.	E. Indies.	1823.	....	S. ♄.	
4 serràta. H.K.	(*wh.*)	saw-leaved.	Chile.	1796.	11. 2.	G. ♄.	
AD'ELIA. W.		AD'ELIA.	Diœcia Monadelphia. L.				
1 Bernárdia. w.	(*gr.wh.*)	villous-leaved.	Jamaica.	1768.	7. 8.	S. ♄.	
2 Ricinélla. w.	(*gr.wh.*)	smooth-leaved.	————	——	6. 8.	S. ♄.	
3 Acidòton. w.	(*gr.wh.*)	Box-leaved.	————	——	6. 7.	S. ♄.	

LOURE'IRA. W.		LOURE'IRA. Diœcia Monadelphia. W.					
1 cuneifòlia. w.	(*bh.*)	wedge-leaved.	Mexico.	1824.	6. 8.	G. ♄.	Cavan. ic. 5. t. 429.
Mozínna spathùlata. Ort. dec. 8. t. 13.							
2 glandulòsa. w.	(*bh.*)	glandulous.	———	1799.	——	G. ♄.	Cavan. ic. 5. t. 430.
B'ORYA. W.		B'ORYA. Diœcia Diandria. W.					
1 porulòsa. w.	(*gr.wh.*)	Florida.	Florida.	1806.	7. 8.	H. ♄.	
2 ligustrìna. w.	(*gr.wh.*)	Privet-leaved.	N. America.	1812.	——	H. ♄.	
3 prinoídes. W.en.	(*gr.wh.*)	Prinos-like.	———	1816.	——	H. ♄.	
4 nítida. W.en.	(*gr.wh.*)	glossy-leaved.	———	——	——	H. ♄.	
5 retùsa. W.en.	(*gr.wh.*)	retuse-leaved.	———	1812.	——	H. ♄.	
6 acuminàta. w.	(*gr.wh.*)	pointed.	———	——	——	H. ♄.	Mich. amer. 2. t. 48.
Adèlia acuminàta. M.							
SECURI'NEGA. J.		SECURI'NEGA. Monœcia Pentandria.					
nítida. P.S.	(*wh.*)	OtaheiteMyrtle.	Otaheite.	1793.	3. 8.	S. ♄.	Lindl. collect. t. 9.
TR'AGIA. W.		TR'AGIA. Monœcia Triandria. W.					
1 volùbilis. w.	(*gr.*)	twining.	W. Indies.	1739.	6. 7.	S. ♄.	Trew. pl. rar. 2. t. 15.
2 involucràta. w.	(*gr.*)	involucred.	E. Indies.	1759.	——	S. ⊙.	Jacq. ic. 1. t. 190.
3 mercuriàlis. w.	(*gr.*)	Mercury-like.	———	1818.	——	S. ⊙.	Rheed. mal. 10. t. 82.
4 urticæfòlia. S.S.	(*gr.*)	nettle-leaved.	N.America.	1812.	——	H. ⊙.	
5 hirsùta. D.L.	(*gr.*)	hairy.	Trinidad.	1820.	——	S. ⊙.	
6 ùrens. w.	(*gr.*)	stinging.	Virginia.	1699.	8.	H. ⊙.	Pluk. alm. t. 107. f. 5.
7 cannabìna. w.	(*gr.*)	Hemp-leaved.	———	1699.	6. 7.	S. ♃.	Burm. ind. t. 63. f. 1.
MICROST'ACHYS. J.		MICROST'ACHYS. Monœcia Triandria.					
1 corniculàta.	(*gr.*)	two-horned.	Trinidad.	1820.	6. 7.	S. ⊙.	
Tràgia corniculàta. Vahl. *Cnemidostàchys Váhlii.* S.S.							
2 Chamælèa. J.	(*gr.*)	lance-leaved.	E. Indies.	1793.	——	S. ⊙.	Rheed. mal. 2. t. 34.

ORDO CLXXXI.

EMPETREÆ. *D. Don on the affinities of the Empetreæ. p.* 3.

EMP'ETRUM. D.D.		EMP'ETRUM. Diœcia Triandria. L.					
nìgrum. w.	(*gr.*)	black-berried.	Britain.	....	4. 5.	H. ♄.	Eng. bot. v. 8. t. 526.
COR'EMA. D.D.		COR'EMA. Diœcia Triandria.					
álbum. D.D.	(*gr.*)	white-berried.	Portugal.	1774.	4. 6.	H. ♄.	
Empètrum álbum. w.							
CERAT'IOLA. M.		CERAT'IOLA. Monœcia Diandria. N.					
ericoídes. M.	(*st.*)	Heath-like.	N. America.	1810.	4. 6.	H. ♄.	Pursh. fl. amer. 1. t. 13.

ORDO CLXXXII.

ANTIDESMEÆ.

ANTIDE'SMA. W.		ANTIDE'SMA. Diœcia Pentandria. L.					
1 alexitèria. w.	(*gr.*)	Laurel-leaved.	E. Indies.	1793.	5. 6.	S. ♄.	Rheed. mal. 5. t. 11.
2 pubéscens. w.	(*gr.*)	pubescent.	———	1818.	——	S. ♄.	Roxb. corom. 2. t. 167.
3 paniculàta. w.	(*gr.*)	panicled.	———	1800.	——	S. ♄.	
STIL'AGO. W.		STIL'AGO. Diœcia Triandria. W.					
1 Bùnius. w.	(*gr.*)	Laurel-leaved.	E. Indies.	1757.	8. 10.	S. ♄.	Rheed. mal. 4. t. 56.
2 diándra. w.	(*gr.*)	diandrous.	———	1800.	——	S. ♄.	Roxb. corom. 2. t. 166.

ORDO CLXXXIII.

URTICEÆ. *Juss. gen.* 400.

F'ICUS. L.		FIG-TREE. Polygamia Diœcia. W.					
1 religiòsa. R.S.	(*gr.*)	Poplar-leaved.	E. Indies.	1731.	....	S. ♄.	Rheed. mal. 1. t. 27.
2 superstitiòsa. L.en.	(*gr.*)	superstitious.	———	....	....	S. ♄.	

3	nymphæifòlia. R.S.	(*gr.*)	Water-lily-leaved.	E. Indies.	1759.	....	S. ♄.		
4	cordàta. R.S.	(*gr.*)	heart-leaved.	C. B. S.	1802.	4. 6.	S. ♄.	Thunb. fic. p. 8. c. ic.	
5	oblongàta. L.en.	(*gr.*)	oblong-leaved.	———	1816.	....	S. ♄.		
6	crassinérvia. R.S.	(*gr.*)	thick-nerved.	S. America.	——	....	S. ♄.		
7	rubrinérvia. L.en.	(*gr.*)	red-nerved.	Brazil.	1824.	....	S. ♄.		
8	rubiginòsa. R.S.	(*gr.*)	ferrugineous.	N. S. W.	1789.	3. 6.	G. ♄.	Venten. malm. t. 114.	
	austràlis. W.								
9	costàta. R.S.	(*gr.*)	rib-leaved.	E. Indies.	1763.	....	S. ♄.		
10	lævigàta. R.S.	(*gr.*)	smooth.	W. Indies.	1822.	....	S. ♄.		
11	lentiginòsa. R.S.	(*gr.*)	veined-leaved.	———	1826.	....	S. ♄.		
12	citrifòlia. R.S.	(*gr.*)	Citron-leaved.	———	1824.	....	S. ♄.		
13	coronàta. R.S.	(*gr.re.*)	red-crowned.	N. S. W.	1807.	3. 6.	G. ♄.	Colla hort. rip. t. 8.	
14	lùcida. R.S.	(*gr.*)	shining-leaved.	E. Indies.	1772.	4. 6.	S. ♄.		
15	scándens. R.S.	(*gr.*)	climbing.	S. America.	1816.	——	S. ♄.		
16	stipulàta. R.S.	(*gr.*)	trailing.	China.	1771.	——	G. ♄.		
17	pedunculàta. R.S.	(*gr.*)	Willow-leaved.	S. America.	1776.	5. 7.	S. ♄.	Pluk. alm. t. 178. f. 4.	
18	venòsa. R.S.	(*gr.*)	vein-leaved.	E. Indies.	1763.	6. 8.	S. ♄.	Rheed. mal. 3. t. 64.	
19	leucótoma. R.S.	(*gr.*)	white-spotted.	———	....	——	S. ♄.	W. hort. ber. 1. t. 36.	
	venòsa. W. *non* H.K.								
20	tinctòria. R.S.	(*gr.*)	Otaheite.	Society Isl.	1783.	5. 6.	S. ♄.		
21	séptica. R.S.	(*gr.*)	rough.	E. Indies.	1826.	....	S. ♄.	Rumph. amb. 3. t. 96.	
22	benjamìna. R.S.	(*gr.*)	oval-leaved.	———	1757.	....	S. ♄.	Rheed. mal. 1. t. 26.	
23	padifòlia. K.S.	(*gr.*)	Padus-leaved.	Mexico.	1826.	....	S. ♄.		
24	complicàta. K.S.	(*gr.*)	complicated.	———	——	....	S. ♄.		
25	glabràta. K.S.	(*gr.*)	smooth.	Teneriffe.	1816.	....	G. ♄.		
26	benghalénsis. R.S.	(*gr.*)	Bengal.	E. Indies.	1690.	4. 5.	S. ♄.	Rheed. mal. 1. t. 28.	
27	microcárpa. R.S.	(*gr.*)	small-fruited.	Guinea.	1823.	——	S. ♄.		
28	americàna. R.S.	(*gr.*)	American.	Jamaica.	1822.	....	S. ♄.	Plum. spec. t. 132. f. 2.	
29	racemòsa. R.S.	(*gr.*)	clustered.	E. Indies.	1759.	4. 6.	S. ♄.	Rheed. mal. 1. t. 25.	
30	pertùsa. R.S.	(*gr.*)	Laurel-leaved.	S. America.	1780.	——	S. ♄.		
31	terebràta. R.S.	(*gr.*)	obovate-leaved.	Mauritius.	1822.	....	S. ♄.	Bory it t. 17.	
32	pùmila. R.S.	(*gr.*)	dwarf.	China.	1759.	4. 6.	G. ♄.	Kæmp. amœn. t. 804.	
33	comòsa. R.S.	(*gr.*)	tufted.	E. Indies.	1825.	....	S. ♄.	Roxb. corom. 2. t. 125.	
34	glomeràta. R.S.	(*gr.*)	crowded.	———	1818.	....	S. ♄.	————— 2. t. 123.	
35	Hookèri.	(*gr.*)	Hooker's.	W. Indies.	1816.	1. 4.	S. ♄.	Hook. ex. flor. t. 111.	
	nítida. H.E.F. *nec aliorum.*								
36	nítida. R.S.	(*gr.*)	glossy-leaved.	E. Indies.	1786.	3. 6.	S. ♄.	Rheed. mal. 3. t. 55.	
37	ciliolòsa. R.S.	(*gr.*)	fringe stipuled.	....	1816.	——	S. ♄.		
38	rùbra. R.S.	(*re.*)	vermilion-fruited.	E. Indies.	1826.	....	S. ♄.	Rumph. amb. 3. t. 85 ?	
39	cotinifòlia. K.S.	(*gr.*)	Cotinus-leaved.	Mexico.	1824.	3. 6.	S. ♄.		
40	obtusifòlia. K.S.	(*gr.*)	blunt-leaved.	———	——	....	S. ♄.		
41	coriàcea. R.S.	(*gr.*)	leathery-leaved.	E. Indies.	1772.	....	S. ♄.		
42	popúlnea. R.S.	(*gr.*)	Poplar-leaved.	———	....	....	S. ♄.		
43	macrophy'lla. R.S.	(*gr.*)	large-leaved.	N. Holland.	1816.	4. 6.	G. ♄.		
44	elástica. H.B.	(*gr.*)	Indian rubber.	E. Indies.	1815.	....	G. ♄.		
45	Brássii. H.T.	(*gr.*)	Brass's.	Sierra Leon.	1822.	....	S. ♄.		
46	lasiophy'lla. L.en.	(*gr.*)	downy-leaved.	....	1816.	....	S. ♄.		
47	glaucophy'lla. R.S.	(*gr.*)	glaucous-leaved.	S. America.	1824.	....	S. ♄.		
48	insipída. R.S.	(*gr.*)	insipid.	Caracas.	1829.	....	S. ♄.		
49	retùsa. R.S.	(*gr.*)	retuse-leaved.	E. Indies.	1793.	4. 6.	S. ♄.		
50	laurifòlia. R.S.	(*gr.*)	Laurel-leaved.	W. Indies.	1759.	——	S. ♄.	Sloan. jam. 2. t. 223.	
	martinicénsis. W.								
51	calyculàta. R.S.	(*gr.*)	calyculate.	Vera Cruz.	1760.	4. 6.	S. ♄.		
52	brasiliénsis. L.en.	(*gr.*)	Brazilian.	Brazil.	1820.	....	S. ♄.		
53	cuneàta. R.S.	(*gr.*)	wedge-leaved.	———	1826.	....	S. ♄.		
54	índica. R.S.	(*gr.*)	Banyan-tree.	E. Indies.	1759.	....	S. ♄.	Rheed. mal. 3. t. 63.	
55	salicifòlia. R.S.	(*gr.*)	Willow-leaved.	Arabia.	1818.	....	S. ♄.		
56	aurantìaca. Lod.	(*or.*)	orange-coloured.	....	——	....	S. ♄.		
57	péndula. L.en.	(*gr.*)	pendulous.	....	1824.	5. 7.	S. ♄.		
58	myrtifòlia. L.en.	(*gr.*)	Myrtle-leaved.		——	....	S. ♄.		
59	pubéscens. R.S.	(*gr.*)	pubescent.	E. Indies.	——	....	S. ♄.		
60	nervòsa. R.S.	(*gr.*)	strong-nerved.	———	1826.	....	S. ♄.		
61	ellíptica. K.S.	(*gr.*)	elliptic-leaved.	S. America.	1824.	....	S. ♄.		
62	acuminàta. L.C.	(*gr.*)	acuminate.	....	——	....	S. ♄.		
63	longifòlia. L.C.	(*gr.*)	long-leaved.	....	——	....	S. ♄.		
64	caudàta. L.C.	(*gr.*)	tailed.	....	——	....	S. ♄.		
65	dumòsa. L.C.	(*gr.*)	bushy.	....	——	....	S. ♄.		
66	viscifòlia. L.C.	(*gr.*)	Viscum-leaved	....	——	....	S. ♄.		
67	lanceolàta. H.B.	(*gr.*)	spear-leaved.	E. Indies.	1820.	....	S. ♄.		

68 obtusàta. L.en. (*gr.*) obtuse-toothed. .... 1820. .... S. ♄.
69 múntia. L.en. (*gr.*) rough-leaved. N. Holland. 1818. .... G. ♄.
70 ulmifòlia. R.S. (*gr.*) Elm-leaved. Philippine. 1825. .... S. ♄.
71 capénsis. R.S. (*gr.*) Cape. C. B. S. 1816. .... G. ♄.
72 exasperàta. R.S. (*gr.*) very rough. Guinea. 1822. .... S. ♄. Willden. fic. t. 2.
scábra. W. *non* Forst.
73 híspida. R.S. (*gr.*) opposite-leaved. E. Indies. 1802. .... S. ♄. Roxb. corom. 2. t. 124.
oppositifòlia. Rox. *scábra.* Jacq. schœn. 3. t. 315.
74 Lichtensteìnii. L.en. (*gr.*) Lichtenstein's. C. B. S. 1824. .... G. ♄.
75 aquática. W. (*gr.*) water. E. Indies. 1758. 4. 6. S. ♄.
76 heterophy'lla. R.S. (*gr.*) various-leaved. ——— 1816. .... S. ♄. Rheed. mal. 3. t. 62.
77 rèpens. R.S. (*gr.*) creeping-stemmed. ——— 1805. .... S. ♄.
78 virgàta. H.B. (*gr.*) twiggy. ——— 1817. .... G. ♄.
79 Carica. R.S. (*gr.*) common. S. Europe. 1548. 6. 7. H. ♄. Trew-Ehret. t. 73. 74.

BRO'SIMUM. W. BREAD-NUT. Polygamia Diœcia. W.
1 Alicástrum. W. (*gr.*) Jamaica. Jamaica. 1776. .... S. ♄. Sw. fl. ind. oc. 1. t. 1. f. 1.
2 spùrium. W. (*gr.*) Milkwood. ——— 1789. .... S. ♄.
3 Galactodéndron. D.D. (*gr.*) Cow-tree. Caracas. 1829. .... S. ♄.
Galactodéndron ùtile. K.S. Palo de Vacca. *Humboldt.*

ARTOCA'RPUS. W. BREAD-FRUIT. Monœcia Monandria. W.
1 incísa. W. (*gr.*) true. S. Sea Isles. 1793. .... S. ♄. Rumph. amb. 1. t. 33.
α nucífera. (*gr.*) *nut-bearing.* ——— —— .... S. ♄.
β apyrèna. (*gr.*) *seedless.* ——— —— .... S. ♄. Rumph. amb. 1. t. 32.
2 integrifòlia. W. (*gr.*) Jaca-tree. E. Indies. 1778. 6. S. ♄. Rheed. mal. 3. t. 26-28.
β heterophy'lla. Lam. *various-leaved.* ——— —— —— S. ♄. Rumph. amb. 1. t. 30.

MACL'URA. N. MACL'URA. Diœcia Tetrandria. N.
1 aurantìaca. N. (*gr.*) Osage Apple. Missouri. 1824. .... H. ♄. Lamb. pin. v. 2. ap. t. 12.
2 tinctòria. D.D. (*gr.*) Fustick-wood. W. Indies. 1739. .... S. ♄. Plum. ic. t. 204.
Mòrus tinctòria. W. *Broussonètia tinctòria.* K.S.

BROUSSON'ETIA. W. PAPER MULBERRY. Diœcia Tetrandria. W.
papyrífera. W. (*gr.br.*) common. Japan. 1751. 4. 7. H. ♄. Andr. reposit. t. 488.

M'ORUS. W. MULBERRY. Monœcia Tetrandria. L.
1 álba. W. (*gr.*) white. China. 1596. 6. H. ♄. Schk. hand. 3. t. 290.
2 tatárica. W. (*gr.*) Tartarian. Tartary. 1784. —— H. ♄. Pall. ros. 2. t. 52.
3 itálica. W. (*gr.*) Italian. Italy. .... —— H. ♄.
4 constantinopolitàna. (*gr.*) Constantinople. Constantinople. .. —— H. ♄.
5 nìgra. W. (*gr.*) common. Italy. 1548. —— H. 5. Wats. dend. brit. t. 159.
6 rùbra. W. (*gr.*) red. N. America. 1629. 6. 7. H. 5.
7 scábra. W.en. (*gr.*) rough. ——— 1812. —— H. ♄.
8 pensylvánica. L.C. (*gr.*) Pensylvanian. ——— —— —— H. ♄.
9 pùmila. L.C. (*gr.*) dwarf. .... —— —— H. ♄.
10 índica. W. (*gr.*) Indian. E. Indies. 1820. .... S. ♄. Rheed. mal. 1. t. 49.
11 sinénsis. L.C. (*gr.*) Chinese. China. —— .... G. ♄.
12 mauritiàna. W. (*gr.*) Mauritius. Mauritius. —— .... S. ♄. Jacq. ic. 3. t. 617.

CONOCE'PHALUS. B.J. CONOCE'PHALUS. Monœcia Tetrandria.
naucleiflòrus. B.R. (*ye.*) Nauclea-flower'd. E. Indies. 1816. 10. 4. S. ♄. Bot. reg. t. 1203.

BŒHM'ERIA. W. BŒHM'ERIA. Monœcia Tetrandria. W.
1 caudàta. W. (*gr.*) tailed. W. Indies. 1822. 4. 8. S. ♄.
2 cylíndrica. W. (*gr.*) cylindrical. Virginia. 1759. 6. 8. H. ♃. Sloan. jam. 1. t. 82. f. 2.
3 macrophy'lla. D.P. (*gr.*) large-leaved. Nepaul. 1820. —— G. ♄.
4 rotundifòlia. D.P. (*gr.*) round-leaved. ——— 1819. —— G. ♃.
5 platyphy'lla. D.P. (*gr.*) rough-leaved. ——— 1818. 6. 10. G. ♄.
6 spicàta. S.S. (*gr.*) spiked. Japan. 1824. —— H. ♃. Hout. syst. 10. t. 72. f. 2.
7 ramiflòra. W. (*gr.*) branch-flowering. W. Indies. 1822. —— S. ♄. Lam. ill. t. 763.
Catùrus ramiflòrus. L. *Pròcris ramiflòrus.* Lam.
8 elongàta. L.en. (*gr.*) elongated. Brazil. —— 7. 9. H. ⊙.
9 lateriflòra. W. (*gr.*) side-flowering. N. America. 1816. 6. 8. H. ♃.
10 rubéscens. W. (*gr.*) tree. Canaries. 1779. 2. 5. G. ♄. Jacq. frag. n. 30. t. 5. f. 1.
Urtìca arbòrea. L'Her. stirp. t. 20.
11 salicifòlia. D.P. (*gr.*) Willow-leaved. Nepaul. 1826. 7. 10. G. ♄.
12 frondòsa. D.P. (*gr.*) frondose. ——— 1818. —— G. ♄.
13 ternifòlia. D.P. (*gr.*) three-leaved. ——— 1827. —— G. ♄.
14 frutéscens. W. (*gr.*) frutescent. ——— 1819. —— G. ♄.

URT'ICA. W. NETTLE. Monœcia Tetrandria. L.
1 pilulífera. W. (*gr.*) Roman. England. .... 6. 8. H. ⊙. Eng. bot. v. 3. t. 148.
2 baleàrica. W. (*gr.*) Balearic. Balearic Isl. 1731. 6. 7. H. ⊙. Blackw. t. 321. f. 1.
3 Dodártii. W. (*gr.*) Dodart's. S. Europe. · 1683. 7. 8. H. ⊙.
4 pùmila. W. (*gr.*) dwarf. N. America. 1817. —— H. ⊙.
5 crassifòlia. W. (*gr.*) thick-leaved. S. America. 1822. —— S. ♄.

6 grandifòlia. w.	(*gr.*)	great-leaved.	Jamaica.	1793.	7. 8.	S. ♄.	Sloan. jam. 1. t. 83. f.2.
7 reticulàta. w.	(*gr.*)	net-leaved.	———	———	6. 8.	S. ♄.	
8 involucràta. B.M.	(*gr.*)	imbossed.	W. Indies.	1809.	10. 11.	S. ♃.	Bot. mag. t. 2481.
9 diffùsa. w.	(*gr.*)	spreading.	———	———	8. 10.	S. ħ.	
10 rùfa. w.	(*gr.*)	rusty.	Jamaica.	1793.	6. 9.	S. ħ.	
11 scabrélla. L.en.	(*gr.*)	rough.	E. Indies.	1815.	10. 12.	G. ♄.	
12 scrípta. D.P.	(*gr.*)	black-lined.	Nepaul.	1823.	6. 8.	H. ♃.	
13 rupéstris. S.S.	(*gr.*)	rock.	Sicily.	1828.	——	H. ħ.	
14 convéxa. H.H.	(*gr.*)	convex-leaved.	....	1824.	6. 9.	H. ☉.	
15 ùrens. w.	(*gr.*)	small.	Britain.	....	6. 10.	H. ☉.	Eng. bot. v. 18. t. 1236.
16 dioíca. w.	(*gr.*)	common.	———	....	7. 9.	H. ♃.	——— v. 25. t. 1750.
17 chamædryoídes. Ph.	(*gr.*)	Germander-like.	Georgia.	1813.	——	H. ☉.	
18 membranàcea. w.	(*gr.*)	membranaceous.	S. Europe.	1818.	——	H. ♃.	
19 angustifòlia. F.	(*gr.*)	narrow-leaved.	Russia.	1824.	——	H. ♃.	
20 árdens. L.en.	(*gr.*(	burning.	Nepaul.	1819.	8. 11.	H. ☉.	
21 cannabìna. w.	(*gr.*)	Hemp-leaved.	Siberia.	1749.	7. 9.	H. ♃.	Amm. ruth. 249. t. 25.
22 rugòsa. w.	(*gr.*)	rough-leaved.	Jamaica.	1793.	5. 7.	S. ♂.	
23 grácilis. w.	(*gr.*)	slender-stalked.	Hudson'sBay.	1782.	6. 8.	H. ♃.	
24 Parietària. w.	(*gr.*)	Pellitory-leaved.	Jamaica.	1793.	7. 9.	S. ♄.	Sloan. j. 1. t. 93. f. 1.
25 ciliàta. w.	(*gr.*)	ciliated.	———	1815.	——	S. ♄.	
26 rádicans. S.S.	(*gr.*)	rooting.	———	1809.	——	H. ħ.	
27 nummularifòlia. w.	(*gr.*)	moneywort-leaved.	———	1822.	——	S. ħ.	Sloan. hist. 1. t. 131. f. 4.
28 depréssa. w.	(*gr.*)	depressed.	———	———	——	S. ♃.	
29 microphy'lla. w.	(*gr.*)	small-leaved.	W. Indies.	1793.	4. 5.	S. ♃.	Sloan. j. 1. t. 93. f. 2.
30 trianthemoídes. P.S.	(*gr.*)	branching.	———	1824.	——	S. ♃.	
31 serrulàta. S.S.	(*gr.*)	saw-leaved.	Jamaica.	———	——	S. ♃.	
32 lùcida. S.S.	(*gr.*)	glossy-leaved.	———	1808.	——	S. ħ.	
33 recúrva. H.H.	(*gr.*)	recurved.	....	1826.	6. 9.	H. ☉.	
34 macrostàchya. L.O.	(*gr.*)	long-spiked.	S. America.	1824.	....	S. ♄.	
35 pulchélla. L.en.	(*gr.*)	neat.	E. Indies.	1816.	....	S. ♄.	
36 æ'stuans. w.	(*gr.*)	Surinam.	Surinam.	1803.	6. 7.	S. ♂.	Jacq. schœn. 3. t. 388.
37 capitàta. w.	(*gr.*)	headed.	N. America.	1820.	——	H. ♃.	
38 canadénsis. w.	(*gr.*)	Canadian.	Canada.	1656.	8. 10.	H. ♃.	Pluk. alm. t. 239. f. 2.
39 nívea. w.	(*gr.*)	white-leaved.	China.	1739.	8. 9.	H. ♃.	Jacq. vind. 2. t. 166.
40 caracasàna. w.	(*gr.*)	Caracas.	Caracas.	1820.	....	S. ♄.	Jacq. schœn. 3. t. 386.
41 caravellàna. L.en.	(*gr.*)	pointed-leaved.	S. America.	——	8. 10.	H. ☉.	
42 baccífera. w.	(*gr.*)	berry-bearing.	———	1793.	7. 8.	S. ♄.	Andr. reposit. t. 454.
43 elongàta. w.	(*gr.*)	elongated.	Philippines.	1823.	——	S. ♄.	
44 capitellàta. S.S.	(*gr.*)	small-headed.	Java.	1816.	4. 8.	S. ♄.	
45 ferocíssima.	(*gr.*)	ferocious.	Nepaul.	1805.	8. 10.	H. ☉.	
hórrida. L.en. *non* K.S.							
46 pentándra. H.B.	(*gr.*)	pentandrous.	E. Indies.	———	——	S. ♃.	
47 arboréscens. L.en.	(*gr.*)	arborescent.	Manila.	1822.	....	S. ♄.	
48 diversifòlia. L.en.	(*gr.*)	different-leaved.	E. Indies.	1820.	....	S. ♄.	
49 heterophy'lla. D.P.	(*gr.*)	various-leaved.	Nepaul.	1823.	6. 8.	H. ♃.	
FORSKO'HLEA. L.		FORSKO'HLEA.	Monœcia Monandria. J.				
1 fruticòsa. w.	(*gr.*)	frutescent.	Teneriffe.	1816.	6. 8.	G. ♄.	
2 tenacíssima. w.	(*gr.*)	clammy.	Egypt.	1767.	7. 8.	H. ☉.	Jacq. vind. 1. t. 48.
3 cándida. w.	(*gr.*)	rough.	C. B. S.	1774.	6. 7.	G. ♃.	
4 víridis. E.	(*gr.*)	green.	Arabia.	1828.	——	G. ☉.	
5 angustifòlia. w.	(*gr.*)	narrow-leaved.	Teneriffe.	1779.	7. 8.	H. ☉.	Murr. c. gott. p. 24. t. 2.
PARIET'ARIA. W.		PELLITORY.	Polygamia Monœcia. L.				
1 índica. w.	(*gr.*)	Indian.	E. Indies.	1790.	4. 5.	S. ♃.	
2 officinàlis. w.	(*gr.*)	common.	Britain.	....	6. 9.	H. ♃.	Eng. bot. v. 13. t. 879.
3 judàica. w.	(*gr.*)	Basil-leaved.	Germany.	1728.	——	H. ♃.	Schkuh. hand. 3. t. 34.
4 pensylvánica. w.	(*gr.*)	Pensylvanian.	N. America.	1816.	——	H. ☉.	
5 urticæfòlia. w.	(*gr.*)	Nettle-leaved.	Isl. Bourbon.	1700.	——	S. ☉.	
6 lusitánica. w.	(*gr.*)	chickweed-leaved.	Spain.	1710.	7. 8.	H. ☉.	Bocc. sic. t. 24. f. 13.
7 crética. w.	(*gr.*)	Cretan.	Crete.	1818.	——	H. ☉.	
8 prostràta. L.en.	(*gr.*)	prostrate.	....	1823.	——	H. ♃.	
9 polygonoídes. w.	(*gr.*)	Polygonum-leav'd.	Armenia.	1728.	——	H. ☉.	Mart. cent. pl. rar. t. 8.
PI'LEA. L.col.		PI'LEA.	Monœcia Tetrandria.				
muscòsa. L.col.	(*gr.*)	mossy.	W. Indies.	1793.	4. 5.	S. ♃.	Lindl. col. t. 4.
CA'NNABIS. W.		HEMP.	Diœcia Tetrandria. L.				
satìva. w.	(*gr.*)	common.	India.	....	6. 7.	H. ☉.	Schk. hand. 3. t. 325.
H'UMULUS. W.		HOP.	Diœcia Tetrandria. L.				
Lùpulus. w.	(*gr.*)	common.	Britain.	....	6. 8.	H. ♃.	Eng. bot. v. 6. t. 427.
THELY'GONUM. W.		DOG'S-CABBAGE.	Monœcia Polyandria. L.				
Cynocrámbe. w.	(*gr.*)	common.	S. Europe.	1710.	7.	H. ☉.	Schk. hand. 3. t. 299.

GN'ETUM. W.		Gn'etum.	Monœcia Monadelphia. L.				
Gnèmon. w.	(*gr.*)	Indian.	India.	1815.	....	S. ♄.	Rumph. amb. 1. t. 71.
CECR'OPIA. W.		Snake-wood.	Diœcia Diandria. L.				
1 peltàta. w.	(*gr.*)	peltated.	Jamaica.	1778.	....	S. ♄.	Lam. ill. t. 800.
2 palmàta. w.	(*gr.*)	palmated.	Brazil.	1823.	....	S. ♄.	
3 cóncolor. w.	(*gr.*)	self-coloured.	———	———	....	S. ♄.	
GUNN'ERA. W.		Gunn'era.	Gynandria Diandria. L.				
perpénsa. w.	(*gr.*)	marsh-marygold-ld.	C. B. S.	1688.	4. 6.	G. ♃.	Bot. mag. t. 2376.
DORST'ENIA. W.		Dorst'enia.	Monœcia Diandria.				
1 cordifòlia. w.	(*gr.*)	heart-leaved.	W. Indies.	1822.	5. 8.	S. ♃.	Swart. obs. t. 7. f. 2.
2 brasiliénsis. w.	(*gr.*)	Brazilian.	S. America.	1792.	3. 4.	S. ♃.	
3 Houstòni. w.	(*gr.*)	Houstoun's.	———	1747.	6. 7.	S. ♃.	Bot. mag. t. 2017.
4 arifòlia. w.	(*gr.*)	Arum-leaved.	Brazil.	1823.	———	S. ♃.	Lodd. bot. cab. t. 999.
5 Contrayérva. w.	(*gr.*)	Contrajerva-root.	S. America.	1748.	5. 8.	S. ♃.	Jacq. ic. 3. t. 614.
6 Drakèna. w.	(*gr.*)	round-rooted.	Vera Cruz.	1729.	———	S. ♃.	Lodd. bot. cab. t. 667.
7 tubicìna. B.M.	(*gr.*)	tubed.	Trinidad.	1824.	7. 9.	S. ♃.	Bot. mag. t. 2804.
8 ceratosánthes. B.C.	(*gr.pu.*)	horned.	S. America.	———	4. 6.	S. ♃.	Lodd. bot. cab. t. 1216.
MERTE'NSIA. K.S.		Merte'nsia.	Polygamia Monœcia. R.S.				
1 lævigàta. K.S.	(*ye.*)	Mexican.	Mexico.	1825.	....	S. ♄.	H. et B. n. gen. 2. t. 103.
2 iguánea. R.S.	(*ye.*)	Jamaica.	Jamaica.	1791.	4. 6.	S. ♄.	
Rhámnus iguáneus. L. *Zízyphus iguánea.* P.S. *Céltis aculeàta.* W.							
3 rhamnoídes R.S.	(*ye.*)	Rhamnus-like.	———	1791.	4. 6.	S. ♄.	Cavan. ic. 3. t. 294.
CE'LTIS. K.S.		Nettle-tree.	Polygamia Monœcia. L.				
1 micrántha. w.	(*gr.*)	smooth.	Jamaica.	1739.	8. 9.	S. ♄.	Plum. ic. t. 206. f. 1.
2 orientàlis. w.	(*gr.*)	oriental.	E. Indies.	1820.	———	S. ♄.	Rheed. mal. 4. t. 40.
3 pùmila. Ph.	(*gr.*)	dwarf.	N. America.	1812.	5.	H. ♄.	
4 crassifòlia. w.	(*gr.*)	heart-leaved.	———	———	5. 6.	H. 5.	Duham. arb. 2. t. 9.
cordifòlia. Duh.							
5 occidentàlis. w.	(*gr.*)	American.	———	1656.	4. 5.	H. 5.	Wats. dend. brit. t. 147.
β *cordàta.* R.S.	(*gr.*)	*heart-leaved.*	———	....	———	H. 5.	
γ *scabriúscula.* R.S.	(*gr.*)	*rough-leaved.*	———	....	———	H. 5.	
6 Willdenowiàna. R.S.	(*gr.*)	Willdenow's.	China.	....	....	H. 5.	
7 sinénsis. R.S.	(*gr.*)	Chinese.	———	....	....	H. 5.	
8 lævigàta. R.S.	(*gr.*)	smooth.	Louisiana.	1812.	4. 5.	H. 5.	
9 Tournefórtii. w.	(*gr.*)	Tournefort's.	Levant.	1738.	....	H. 5.	Tourn. it. 2. t. 41.
10 australìs. w.	(*gr.*)	European.	S. Europe.	1796.	5.	H. 5.	Wats. dend. brit. t. 105.
PLAN'ERA. M.		Plan'era.	Tetrandria Digynia. S.S.				
1 Richárdi. M.	(*br.gr.*)	Richard's.	N. America.	1766.	4. 5.	H. 5.	
U'lmus nemoràlis. H.K.							
2 carpinifòlia. W.D.	(*gr.*)	Hornbeam-leaved.	Siberia.	———	———	H. 5.	Wats. dend. brit. t. 106.
3 parvifòlia.	(*gr.*)	small-leaved.	Caucasus.	....	———	H. 5.	Jacq. schœnb. t. 262.
U'lmus parvifòlia. Jacq.							
4 Gmelìni. M.	(*br.gr.*)	Gmelin's.	N. America.	1816.	———	H. 5.	
aquática. N.							
U'LMUS. R.S.		Elm-tree.	Pentandria Digynia. L.				
1 campéstris. E.B.	(*pu.*)	common English.	Britain.	....	4. 5.	H. 5.	Eng. bot. v. 27. t. 1886.
2 suberòsa. E.B.	(*pu.*)	cork-barked.	———	....	3. 4.	H. 5.	——— v. 31. t. 2161.
3 glábra. E.B.	(*pu.*)	smooth.	———	....	———	H. 5.	——— v. 32. t. 2248.
4 màjor. E.B.	(*br.*)	declining-branch'd.	———	....	———	H. 5.	——— v. 36. t. 2542.
5 montàna. E.B.	(*gr.*)	Wych.	———	....	4. 5.	H. 5.	——— v. 27. t. 1887.
6 effùsa. w.	(*gr.*)	spreading-flowered.	———	....	———	H. 5.	Schk. hand. t. 57.
7 álba. w.	(*br.*)	white.	Hungary.	1816.	———	H. 5.	
8 críspa. w.	(*br.*)	curl-leaved.	N. America.	....	———	H. 5.	
9 americàna. w.	(*gr.*)	large-leaved.	———	1752.	———	H. 5.	
10 pendùla. W.en.	(*gr.*)	pendulous.	———	....	———	H. 5.	
11 fulva. M.	(*br.*)	slippery.	———	....	———	H. 5.	Mich. arb. 3. t. 6.
12 scabra. Mill.	(*br.*)	rough.	....	....	———	H. 5.	
13 pumila. w.	(*br.*)	dwarf.	Siberia.	1771.	———	H. ♄.	Pall. ros. 1. t. 48.
14 microphy'lla. P.S.	(*br.*)	jagged-leaved.	———	....	———	H. ♄.	Pall. r. 1. t. 48. f. A. B. C.
15 alàta. M.	(*br.*)	winged.	N. America.	1816.	———	H. 5.	Mich. arb. 3. t. 5.
16 chinénsis. P.S.	......	Chinese.	China.	....	....	G. ♄.	

ORDO CLXXXIV.

JUGLANDEÆ. *DC. theor. ed.* 1. *p.* 215.

JU'GLANS. N.		WALNUT-TREE.	Monœcia Polyandria. L.				
1 règia. W.	(*gr.*)	common.	Persia.	1562.	4. 5.	H.5.	Lam. ill. t. 781.
β máxima. P.S.	(*gr.*)	*double-fruited.*	———	——	——	H.5.	
γ serotìna. P.S.	(*gr.*)	*late-fruited.*	———	——	——	H.5.	
2 nìgra. W.	(*gr.*)	black.	N. America.	1629.	——	H.5.	Wats. dend. brit. t. 158.
3 cinèrea. W.	(*gr.*)	shell bark.	———	1656.	——	H.5.	——— v. 1. t. 192.
cathártica. Mich. arb. 1. t. 2.							
4 fraxinifòlia. S.S.	(*gr.*)	Ash-leaved.	———	——	——	H.5.	
5 pterocárpa. S.S.	(*gr.*)	winged-fruited.	Caucasus.	1828.	....	H.5.	
Rhús obscùrum. M.B.							
C'ARYA. N.		HICKORY NUT.	Monœcia Tetr-Hexandria.				
1 olivæfórmis. N.	(*gr.*)	Pecan nut.	N. America.	....	4. 5.	H.5.	Mich. arb. 1. t. 3.
2 angustifòlia. H.K.	(*gr.*)	narrow-leaved.	———	1766.	——	H.5.	
3 sulcàta. N.	(*gr.*)	shell-bark.	———	1804.	——	H.5.	Mich. arb. 1. t. 8.
4 álba. N.	(*gr.*)	shag-bark.	———	1730.	——	H.5.	——— 1. t. 7.
Júglans compréssa. W.							
5 tomentòsa. N.	(*gr.*)	mocker nut.	———	1629.	——	H.5.	Mich. arb. 1. t. 6.
Júglans álba. W. *non* M.							
β máxima. N.	(*gr.*)	*large-fruited.*	———	1824.	——	H.5.	
6 pubéscens. L.en.	(*gr.*)	pubescent.	———	....	——	H.5.	
7 microcárpa. N.	(*gr.*)	small-fruited.	———	1824.	——	H.5.	
8 amára. N.	(*gr.*)	bitter nut.	———	1800.	5.	H.5.	
9 porcìna. N.	(*gr.*)	Pig nut.	———	1812.	4. 5.	H.5.	Mich. arb. 1. t. 9. f. 3-4.
Júglans obcordàta. W.							
10 glábra. W.	(*gr.*)	smooth-leaved.	———	1799.	....	H.5.	Mich. arb. 1. t. 9. f. 1-2.
Júglans porcìna β ficifórmis. M.							
11 aquática. N.	(*gr.*)	water.	Carolina.	1826.	....	H.5.	Mich. arb. 1. t. 5.
12 integrifòlia. S.S.	(*gr.*)	entire-leaved.	Louisiana.	——	....	H.5.	
Hicòrius integrifòlius. Rafin.							

ORDO CLXXXV.

CUPULIFERÆ. *Kth. synops.* 1. *p.* 353.

QUE'RCUS. W.		OAK.	Monœcia Polyandria. L.				
1 Phéllos. W.	(*y.gr.*)	Willow-leaved.	N. America.	1723.	5. 6.	H.5.	Mich. arb. 1. t. 12.
β hùmilis. Ph.	(*y.gr.*)	*dwarf.*	———	1812.	——	H.♄.	Catesb. car. 1. t. 22.
2 marítima. W.	(*y.gr.*)	sea.	———	1811.	——	H.♄.	Mich. querc. t. 13. f. 1.
3 serícea. W.	(*y.gr.*)	running.	———	1724.	——	H.♄.	Mich. arb. 2. t. 15.
4 vìrens. W.	(*y.gr.*)	live.	———	1739.	5.	H.5.	——— 2. t. 11.
5 cinèrea. W.	(*y.gr.*)	ash-coloured.	———	1789.	5. 6.	H.5.	——— 2. t. 14.
6 imbricària. W.	(*y.gr.*)	tile-cupped.	———	1786.	——	H.5.	——— 2. t. 13.
7 laurifòlia. W.	(*y.gr.*)	Laurel-leaved.	———	——	5.	H.5.	Mich. querc. t. 17.
β obtùsa. Ph.	(*y.gr.*)	*blunt-leaved.*	———	——	——	H.5.	——— t. 18.
8 lùtea. W.	(*y.gr.*)	yellow-leaved.	Mexico.	1825.	....	H.5.	
9 glaùca. W.	(*y.gr.*)	glaucous.	Japan.	1822.	....	H.5.	Banks ic. Kæmp. t. 17.
10 gramúntia. W.	(*y.gr.*)	Holly-leaved.	S. France.	1730.	6.	H.♄.	
11 Ballòta. W.	(*y.gr.*)	Barbary.	Barbary.	1818.	——	H.♄.	
12 rotundifòlia. W.	(*y.gr.*)	round-leaved.	Spain.	——	——	H.♄.	
13 'Ilex. W.	(*y.gr.*)	evergreen.	S. France.	1581.	5. 6.	H.♄.	Wats. dend. brit. t. 90.
α integrifòlia. H.K.	(*y.gr.*)	*entire-leaved.*	———	——	——	H.♄.	
β serràta. H.K.	(*y.gr.*)	*saw-leaved.*	———	——	——	H.♄.	Duham. arb. 1. t. 123.
γ oblónga. H.K.	(*y.gr.*)	*long-leaved.*	———	——	——	H.♄.	——— 1. t. 124.
δ fagifòlia. L.C.	(*y.gr.*)	*Beech-leaved.*	———	....	——	H.♄.	
ε críspa. L.C.	(*y.gr.*)	*curl-leaved.*	———	....	——	H.♄.	
14 variifòlia.	(*y.gr.*)	different-leaved.	S. Europe.	....	——	H.♄.	
heterophy'lla. Lam. *non* Ph.							

15	crenàta. Lam.	(*y.gr.*)	crenated.	S. Europe.	....	5. 6.	H.♄.	
16	lusitànica. Lam.	(*y.gr.*)	Portugal.	Portugal.	1824.	——	H.♄.	Cavan. icon. 2. t. 129.
17	calycìna. Poir.	(*y.gr.*)	calycine.	S. Europe.	....	——	H.♄.	
18	expánsa. Poir.	(*y.gr.*)	spreading.	————	....	——	H.♄.	
19	Lezermiàna. Poir.	(*y.gr.*)	glossy-leaved.	————	....	——	H.♄.	
20	castellàna. Bosc.	(*y.gr.*)	subdeciduous.	————	....	——	H.♄.	
21	Sùber. W.	(*y.gr.*)	Cork-tree.	S. France.	1699.	6.	H.♄.	Wats. dend. brit. t. 89.
22	coccífera. W.	(*y.gr.*)	Kermes.	————	1683.	5.	H.♄.	———— t. 91.
23	prasìna. Bosc.	(*y.gr.*)	glaucous-leaved.	Spain.	1824.	——	H.♄.	
24	hùmilis. P.S.	(*y.gr.*)	low.	S. Europe.	....	——	H.♄.	Lobel. ic. 2. p. 157.
25	infectòria. W.	(*y.gr.*)	Levant.	Levant.	1812.	....	H.♄.	
26	Turnèri. W.en.	(*y.gr.*)	Turner's.	————	——	....	H.5.	
27	Prìnus. W.	(*y.gr.*)	Chestnut-leaved.	N. America.	1730.	5. 6.	H.5.	Mich. arb. 2. t. 7.
28	bícolor. W.	(*y.gr.*)	two-coloured.	————	1811.	5.	H.5.	
29	montàna. W.	(*y.gr.*)	rock Chestnut.	————	1800.	——	H.5.	Mich. arb. 2. t. 8.
30	Micháuxii. N.	(*y.gr.*)	swamp white.	————	1812.	....	H.5.	———— 2. t. 6.
	Prìnos díscolor. M.							
31	Castànea. W.	(*y.gr.*)	yellow Chestnut.	————	1816.	5.	H.5.	Mich. arb. 2. t. 9.
32	prinoídes. W.	(*y.gr.*)	dwarf Chestnut.	————	——	——	H.ħ.	———— 2. t. 10.
	Chínquapin. Ph.							
33	agrifòlia. W.	(*y.gr.*)	spiny-leaved.	————	1827.	....	H.5.	Pluk. phyt. t. 196. f. 3.
34	heterophy'lla. Ph.	(*y.gr.*)	various-leaved.	————	....	——	H.5.	Mich. arb. 2. t. 16.
35	aquática. Ph.	(*y.gr.*)	water.	————	1723.	——	H.5.	———— 2. t. 17.
36	hemisph'ærica. W.	(*y.gr.*)	hemispherical.	————	1816.	——	H.5.	Mich. querc. t. 20. f. 2.
37	nàna. W.	(*y.gr.*)	dwarf.	————	1738.	——	H.ħ.	Abb. ins. 2. t. 59.
38	trilòba. W.	(*y.gr.*)	three-lobed.	————	1800.	——	H.5.	Mich. querc. t. 26.
39	nìgra. W.	(*y.gr.*)	black.	————	1739.	——	H.ħ.	Mich. arb. 2. t. 18.
40	tinctòria. W.	(*y.gr.*)	dyer's.	————	1800.	——	H.5.	Mich. querc. t. 24.
41	díscolor. W.	(*y.gr.*)	two-coloured.	————	1763.	——	H.5.	———— t. 25.
42	lanuginòsa. D.P.	(*y.gr.*)	Nepaul.	Nepaul.	1818.	....	G.5.	
43	Phullàta. D.P.	(*y.gr.*)	slender-pointed.	————	1824.	....	G.5.	
44	coccìnea. W.	(*y.gr.*)	scarlet.	N. America.	1691.	5.	H.5.	Mich. arb. 2. t. 23.
45	ambígua. M.	(*y.gr.*)	gray.	————	1800.	——	H.5.	———— 2. t. 24.
46	rùbra. W.	(*y.gr.*)	champion.	————	1739.	——	H.5.	———— 2. t. 26.
47	Catesb'æi. W.	(*y.gr.*)	barren scrub.	————	1820.	——	H.ħ.	———— 2. t. 20.
48	falcàta. M.	(*y.gr.*)	downy-leaved.	————	1763.	——	H.5.	———— 2. t. 21.
	elongàta. W.							
49	palústris. W.	(*y.gr.*)	marsh.	————	1800.	——	H.5.	Mich. arb. 2. t. 25.
50	Banistèri. M.	(*y.gr.*)	Banister's.	————	——	——	H.ħ.	———— 2. t. 19.
	ilicifòlia. W.							
51	stellàta. W.	(*y.gr.*)	starred.	————	——	——	H.5.	Mich. arb. 2. t. 4.
	obtusilòba. M.							
52	macrocárpa. W.	(*y.gr.*)	overcup white.	————	1800.	——	H.5.	Mich. arb. 2. t. 3.
53	pseùdo-sùber. W.	(*y.gr.*)	cork-like.	Spain.	1818.	——	H.♄.	Sant. itin. 156. t. 4.
54	Æ'gilops. W.	(*y.gr.*)	Velanida.	S. Europe.	1731.	....	H.5.	Mill. dict. n. 7. t. 215.
55	álba. W.	(*y.gr.*)	white.	N. America.	1724.	5.	H.5.	Mich. arb. 2. t. 1.
56	repánda. L.en.	(*y.gr.*)	repand.	————	....	——	H.5.	Mich. querc. t. 5. f. 2.
57	E'sculus. W.	(*y.gr.*)	Italian.	S. Europe.	1739.	——	H.5.	
58	Ròbur. W.	(*y.gr.*)	sessile-fruited.	Britain.	....	4. 5.	H.5.	Eng. bot. v. 26. t. 1845.
	sessiliflòra. E.B.							
59	pedunculàta. W.	(*y.gr.*)	common.	————	....	——	H.5.	Eng. bot. v. 19. t. 1342.
	Ròbur. E.B. *non* W.							
60	pubéscens. W.	(*y.gr.*)	Durmast.	England.	....	——	H.5.	Second. d. chen. t. 5.
61	fastigiàta. Lam.	(*y.gr.*)	Cypress.	Pyrenees.	1822.	....	H.5.	
62	conglomeràta. P.S.	(*y.gr.*)	clustered.	Europe.	....	4. 5.	H.5.	
63	olivæfórmis. M.	(*y.gr.*)	mossy-cup.	N. America.	1811.	5.	H.5.	Mich. arb. 2. t. 2.
64	lyràta. W.	(*y.gr.*)	lyre-leaved.	————	1786.	——	H.5.	———— 2. t. 5.
65	fagínea. P.S.	(*y.gr.*)	Beech-like.	S. Europe.	1824.	....	H.5.	
66	haliphlèos. P.S.	(*y.gr.*)	hispid-cupped.	France.	....	4. 5.	H.5.	
67	Taùzin. P.S.	(*y.gr.*)	soft jagged-leav'd.	S. Europe.	....	——	H.5.	
	β *laciniàta.* Desv.	(*y.gr.*)	*cut-leaved.*	————	....	——	H.5.	
	γ *digitàta.* Desv.	(*y.gr.*)	*fingered.*	————	....	——	H.5.	
68	Cérris. W.	(*y.gr.*)	Turkey.	Turkey.	1735.	——	H.5.	Wats. dend. brit. t. 92.
	α *frondòsa.* H.K.	(*y.gr.*)	*common.*	————	——	——	H.5.	
	β *bullàta.* H.K.	(*y.gr.*)	*rough-leaved.*	————	——	——	H.5.	
	γ *sinuàta.* H.K.	(*y.gr.*)	*narrow-leaved.*	————	——	——	H.5.	
	? δ *dentàta.* W.D.	(*y.gr.*)	*Fulham.*	————	....	6.	H.5.	Wats. dend. brit. t. 93.
69	Lucombeàna.	(*y.gr.*)	Lucombe's.	Levant.	....	5. 6.	H.5.	
	exoniénsis. L.C.							

70 austrìaca. W. en.	(*y.gr.*)	Austrian.	Austria.	....	4. 5.	H.5.	Clus. hist. 1. p. 20. c. ic.
71 mexicàna. K.S.	(*y.gr.*)	Mexican.	Mexico.	1824.	....	G. ♄.	H. et B. pl. æq. 2. t. 82.
72 spicàta. K.S.	(*y.gr.*)	spiked.	————	——	....	G. ♄.	———————— 2. t. 89.
73 Bonplandiàna.	(*y.gr.*)	Bonpland's.	————	——	....	G. ♄.	———————— 2. t. 93.
ambígua. K.S. *non* M.							
FAGUS. W.		BEECH.	Monœcia Polyandria. L.				
1 americàna.	(*y.gr.*)	white.	N. America.	....	5.	H.5.	Mich. arb. 2. t. 8.
sylvática β americàna. N.							
2 sylvática. W.	(*y.gr.*)	common.	Britain.	....	4. 5.	H.5.	Eng. bot. v. 26. t. 1846.
β purpùrea. H.K.	(*pu.*)	*dark purple.*	Germany.	....	——	H.5.	
3 ferrugìnea. W.	(*y.gr.*)	ferruginous.	N. America.	1796.	5. 6.	H.5.	Mich. arb. 2. t. 9.
4 comptoniæfòlia. Df.	(*y.gr.*)	Fern-leaved.	....	....	——	H. ♄.	
sylvática γ aspleniifòlia. H.S.L.							
CASTANEA. G.		CHESTNUT.	Monœcia Polyandria. W.				
1 americàna. P.S.	(*y.gr.*)	American.	N. America.	....	5. 6.	H.5.	Mich. arb. 2. t. 6.
vésca β americàna. N.							
2 vésca. W.	(*y.gr.*)	common.	England.	....	——	H.5.	Eng. bot. v. 13. t. 886.
β variegàta.	(*y.gr.*)	*striped-leaved.*	————	....	——	H.5.	
γ aspleniifòlia.	(*y.gr.*)	*fern-leaved.*	....	....	——	H. ♄.	
3 pùmila. W.	(*y.gr.*)	dwarf.	N. America.	1699.	7.	H. ♄.	Mich. arb. 2. t. 7.
CO'RYLUS. W.		NUT-TREE.	Monœcia Polyandria. L.				
1 Avellàna. W.	(*y.re.*)	common Hazel.	Britain.	....	2. 4.	H.5.	Eng. bot. v. 11. t. 723.
β álba. H.K.	(*y.re.*)	*white Filbert.*	....	....	——	H.5.	
γ rùbra. H.K.	(*y.re.*)	*red Filbert.*	....	....	——	H.5.	
δ ovàta. Lam.	(*y.re.*)	*oval-fruited.*	....	....	——	H.5.	
ε barcelonénsis. L.C.	(*y.re.*)	*Barcelona.*	Spain.	....	——	H.5.	
ζ grándis. H.K.	(*y.re.*)	*Cob-nut.*	....	....	——	H.5.	
η glomeràta. H.K.	(*y.re.*)	*clustered.*	....	....	——	H.5.	
2 tubulòsa. W.	(*y.re.*)	Lambert's.	S. Europe.	1759.	3. 4.	H.5.	Lam. ill. t. 780.
3 heterophy'lla. F.	(*y.re.*)	various-leaved.	Danube.	1829.	....	H. ♄.	
4 americàna. W.	(*st.re.*)	Cuckold.	N. America.	1798.	3. 4.	H.5.	Wang. amer. t. 29. f. 63.
5 hùmilis. W.	(*st.re.*)	dwarf cuckold.	————	....	——	H. ♄.	
6 rostràta. W.	(*st.re.*)	common cuckold.	————	1745.	——	H.5.	Willd. arb. t. 1. f. 2.
7 Colúrna. W.	(*st.re.*)	Constantinople.	Constantinople.	1665.	——	H.5.	Wats. dend. brit. t. 99.

ORDO CLXXXVI.

BETULINEÆ. *Kth. synops.* 1. *p.* 363.

O'STRYA. W.		HOP-HORNBEAM.	Monœcia Polyandria. W.				
1 vulgàris. W.	(*gr.*)	common.	Italy.	1724.	5.	H.5.	Wats. dend. brit. t. 143.
2 virgínica. W.	(*gr.*)	American.	N. America.	1692.	5. 6.	H.5.	Abb. ins. 2. t. 75.
CARP'INUS. W.		HORNBEAM.	Monœcia Polyandria. L.				
1 Bétulus. W.	(*gr.*)	common.	Britain.	....	3. 5.	H.5.	Eng. bot. v. 29. t. 2032.
β variegàta.	(*gr.*)	*striped-leaved.*	————	....	——	H.5.	
γ incìsa. H.K.	(*gr.*)	*cut-leaved.*	....	....	——	H.5.	
δ quercifòlia. Desf.	(*gr.*)	*Oak-leaved.*	....	....	——	H.5.	
2 americàna. W.	(*gr.*)	American.	N. America.	1812.	——	H.5.	Wats. dend. brit. t. 157.
3 orientàlis. W.	(*gr.*)	eastern.	Levant.	1739.	5. 6.	H.5.	———————— t. 98.
BE'TULA. W.		BIRCH.	Monœcia Polyandria. L.				
1 álba. W.	(*gr.*)	common.	Britain.	....	4. 6.	H.5.	Eng. bot. v. 31. t. 2198.
β verrucòsa. Ehr.	(*gr.*)	*warted.*	————	....	——	H.5.	
γ péndula. Roth.	(*gr.*)	*weeping.*	————	....	——	H.5.	
δ dalecárlica. L.	(*gr.*)	*palmate-leaved.*	....	....	——	H.5.	
ε macrocárpa.	(*gr.*)	*large-fruited.*	....	....	——	H.5.	
2 póntica. Df.	(*gr.*)	eastern.	Asia.	....	——	H.5.	Wats. dend. brit. t. 94.
3 pubéscens. P.S.	(*gr.*)	pubescent.	Europe.	....	4. 6.	H.5.	
4 populifòlia. W.	(*gr.*)	Poplar-leaved.	N. America.	1750.	7.	H.5.	Wats. dend. brit. t. 151.
5 excélsa. W.	(*gr.*)	tall.	————	1767.	5.	H.5.	———————— t. 95.
6 davùrica. W.	(*gr.*)	Daurian.	Dauria.	1786.	7.	H.5.	Pall. ros. 1. t. 39.
7 nìgra. W.	(*gr.*)	black.	N. America.	1736.	7. 8.	H.5.	Wats. dend. brit. t. 153.
8 lùtea. M.	(*gr.*)	yellow.	————	1816.	5. 6.	H.5.	
9 lanulòsa. M.	(*gr.*)	woolly.	————	1812.	——	H.5.	Willd. arb. t. 2. f. 1.
rùbra. Mich. arb. 2. t. 3.							

10	papyràcea. w.	(*gr.*) paper.	N. America.	1750.	5. 6.	H.5.	Wats. dend. brit. t. 151.
11	lénta. w.	(*gr.*) soft.	————	1759.	7.	H.5.	———————— t. 144.
12	carpinifòlia. W.en.	(*gr.*) Hornbeam-leaved.	————	1825.	——	H.5.	
13	carpáthica. Kit.	(*gr.*) Carpathian.	Carpathian mt.	1825.	——	H.5.	
14	ovàta. w.	(*gr.*) oval-leaved.	Europe.	....	6. 7.	H. ♄.	Wats. dend. brit. t. 96.
	víridis. Vill. *A'lnus víridis.* DC.						
15	fruticòsa. w.	(*gr.*) shrubby.	Siberia.	1816.	——	H. ♄.	Wats. dend. brit. t. 154.
16	glandulòsa. M.	(*gr.*) glandular.	N. America.	——	5.	H. ♄.	
17	pùmila. w.	(*gr.*) hairy dwarf.	————	1762.	4. 5.	H. ♄.	Wats. dend. brit. t. 97.
18	nàna. w.	(*gr.*) smooth dwarf.	Scotland.	....	5.	H. ♄.	Eng. bot. v. 33. t. 2326.
	β *macrophy'lla.* Sch.	(*gr.*) *large-leaved.*	Switzerland.	1823.	——	H. ♄.	
19	trístis. L.en.	(*gr.*) dark.	Kamtschatka.	1824.	....	H. ♄.	

A'LNUS. W. Alder. Monœcia Tetrandria. W.

1	glutinòsa. w.	(*gr.*) common.	Britain.	....	3. 4.	H.5.	Eng. bot. v. 21. t. 1508.
	β *emarginàta.*	(*gr.*) *emarginate.*	————	....	——	H.5.	
	γ *incìsa.* w.	(*gr.*) *cut-leaved.*	————	....	——	H.5.	
	δ *laciniàta.* w.	(*gr.*) *jagged-leaved.*	————	....	——	H.5.	
	ε *quercifòlia.*	(*gr.*) *Oak-leaved.*	————	....	——	H.5.	
2	incàna. w.	(*gr.*) hoary-leaved.	Europe.	1780.	6.	H.5.	
	β *angulàta.* H.K.	(*gr.*) *angular-leaved.*	————	——	——	H.5.	
	γ *pinnàta.*	(*gr.*) *pinnated.*	————	....	——	H.5.	
3	oblongàta. w.	(*gr.*) oblong-leaved.	S. Europe.	1730.	7.	H.5.	
	β *ellíptica.*	(*gr.*) *elliptic-leaved.*	————	——	——	H.5.	
4	undulàta. w.	(*gr.*) undulate.	N. America.	1782.	5. 6.	H. ♄.	
	críspa. Ph. *Bétula críspa.* H.K.						
5	glaùca. s.s.	(*gr.*) glaucous.	————	1812.	3. 4.	H.5.	
6	serrulàta. w.	(*gr.*) notch-leaved.	————	1769.	3.	H. ♄.	Abb. ins. 2. t. 92.
7	rugòsa. s.s.	(*gr.*) rugged-leaved.	————	....	——	H. ♄.	
8	sibírica. F.	(*gr.*) Siberian.	Siberia.	1820.	....	H. ♄.	
9	cordifolia. T.P.	(*gr.*) heart-leaved.	Naples.	——	3. 4.	H. ♄.	Ten. flor. nap. t. 99.
10	macrophy'lla. Df.	(*gr.*) large-leaved.	....	——	....	H. ♄.	
11	rùbra. Df.	(*gr.*) red.	....	——	....	H. ♄.	
12	pùmila. L.C.	(*gr.*) dwarf.	....	——	....	H. ♄.	
13	oxyacanthifòlia. L.C.	(*gr.*) Hawthorn-leaved	....	1812.	....	H. ♄.	
14	subrotúnda. Df.	(*gr.*) roundish-leaved	....	1822.	....	H. ♄.	

ORDO CLXXXVII.

SALICEÆ. *Kth. synops.* 1. *p.* 364.

SA'LIX. W. Willow. Diœcia 1-2-3-5-andra-Monadelpha.

1	pr'æcox. s.s.	(*g.ye.*) early.	Europe.	....	2. 3.	H.5.	Salic. woburn. t. 26.
2	pomeránica. W.en.	(*g.ye.*) Pomeranian.	Pomerania.	1816.	3. 5.	H.5.	
3	cinèrea. E.B.	(*g.ye.*) gray.	Britain.	....	5.	H.5.	Eng. bot. v. 27. t. 1897.
4	angustàta. Ph.	(*g.ye.*) narrow-leaved.	Pensylvania.	1811.	3. 4.	H.5.	
5	Wulfeniàna. s.s.	(*br.st.*) Wulfen's.	Scotland.	——	4. 5.	H.5.	Salic. woburn. t. 48.
6	silesìaca. s.s.	(*g.ye.*) Silesian.	Silesia.	1820.	——	H.5.	
7	Ammanniàna. s.s.	(*g.ye.*) Amman's.	Carinthia.	——	——	H.5.	Hoffm. salic 1. t. 17. f. 2.
8	acutifòlia. s.s.	(*g.ye.*) acute-leaved.	Caucasus.	1824.	2. 3.	H.5.	
9	hastàta. w.	(*g.ye.*) halberd-leaved.	Lapland.	1780.	5.	H.5.	Salic. woburn. t. 35.
10	malifòlia. E.B.	(*g.ye.*) Apple-leaved.	England.	....	4.	H.5.	Eng. bot. v. 23. t. 1617.
11	serrulàta. w.	(*g.ye.*) saw-leaved.	Lapland.	1820.	5.	H.5.	Flor. dan. t. 1238.
	hastàta. F.D.						
12	Pontederàna. w.	(*g.ye.*) Pontedera's.	Germany.	1818.	——	H.5.	Salic. woburn. t. 43.
13	phylicifòlia. E.B.	(*g.ye.*) Tea-leaved.	Scotland.	....	——	H. ♄.	Eng. bot. v. 28. t. 1958.
14	Andersoniàna. E.B.	(*g.ye.*) Anderson's.	————	....	——	H. ♄.	———— v. 33. t. 2343.
15	tenuifòlia. E.B.	(*g.ye.*) slender-leaved.	Britain.	....	5. 6.	H. ♄.	———— v. 31. t. 2186.
16	corúscans. w.	(*g.ye.*) glittering.	Switzerland.	1824.	——	H.5.	Jacq. aust. t. 408.
	Arbúscula. Jacq. *non* E.B.						
17	frágilis. E.B.	(*g.ye.*) crack.	Britain.	....	4. 5.	H.5.	Eng. bot. v. 26. t. 1807.
18	Russelliàna. E.B.	(*g.ye.*) Bedford.	England.	....	——	H.5.	———— v. 26. t. 1808.
19	Treviràni. L.en.	(*g.ye.*) Treviranus's.	Germany.	1825.	——	H.5.	
20	hippophæfòlia. L.en.	(*g.ye.*) Sea-buckthorn-ld.	———	1823.	——	H.5.	
21	virèscens. s.w.	(*g.ye.*) greenish.	Switzerland.	——	——	H.5.	Salic. woburn. t. 7.

22	triándra. E.B.	(*g.ye.*)	long-leaved.	Britain.	....	5. 8.	H.5.	Eng. bot. v. 20. t. 1435.
23	amygdalìna. E.B.	(*g.ye.*)	Almond-leaved.	———	....	4. 5.	H.5.	——— v. 27. t. 1936.
24	Hoppeàna. W.	(*g.ye.*)	Hoppe's.	Switzerland.	1817.	——	H.5.	
25	ambígua. Ph.	(*gr.ye.*)	ambiguous.	N. America.	1813.	3. 4.	H.5.	
26	decípiens. E.B.	(*gr.ye.*)	varnished.	England.	....	5.	H.5.	Eng. bot. v. 27. t. 1937.
27	vitellìna. E.B.	(*gr.ye.*)	yellow twigged.	———	....	3. 5.	H.5.	——— v. 20. t. 1389.
28	álba. E.B.	(*gr.ye.*)	common white.	Britain.	....	4. 5.	H.5.	——— v. 34. t. 2430.
29	cœrùlea. E.B.	(*gr.ye.*)	blue-leaved.	England.	....	5. 8.	H.5.	——— v. 34. t. 2431.
30	pedicellàris. S.S.	(*gr.ye.*)	pedicellated.	N. America.	1811.	4. 5.	H.5.	
	pensylvánica. Salic. Woburn. t. 95. ?							
31	babylònica. W.	(*g.st.*)	weeping.	Levant.	1692.	5. 6.	H.5.	Salic. woburn. t. 22.
32	annulàris. S.W.	(*g.st.*)	annular-leaved.	....	1823.	——	H.5.	——— t. 21.
33	myricoídes. S.S.	(*g.ye.*)	Myrica-like.	N. America.	1811.	4.	H.5.	Ann. bot. 2. t. 5. f. 2.
34	lùcida. S.S.	(*g.ye.*)	shining.	———	——	5.	H.5.	——— 2. t. 5. f. 7.
35	rígida. S.S.	(*g.ye.*)	rigid.	———	——	4. 5.	H.5.	——— 2. t. 5. f. 4.
36	cordàta. S.S.	(*gr.ye.*)	heart-leaved.	———	——	——	H. ♄.	——— 2. t. 5. f. 8.
37	nìgra. S.S.	(*g.ye.*)	black.	———	——	5.	H.5.	——— 2. t. 5. f. 5.
38	ripària. W.	(*g.ye.*)	rivulet.	Europe.	1820.	——	H.5.	
39	incàna. DC.	(*g.ye.*)	hoary.	———	——	4.	H.5.	Salic. woburn. t. 90.
40	incanéscens. Sch.	(*g.ye.*)	incanescent.	Switzerland.	1823.	3. 4.	H.5.	——— t. 120.
41	lanàta. E.F.	(*gr.ye.*)	woolly.	Scotland.	———	——	H. ♄.	——— t. 71.
42	pentándra. E.B.	(*gr.ye.*)	Bay-leaved.	Britain.	....	5. 6.	H. ♄.	Eng. bot. v. 26. t. 1805.
43	Meyeriàna. S.S.	(*g.st.*)	Meyer's.	Germany.	1823.	4. 5.	H. ♄.	Salic. woburn. t. 33.
44	Fórbesii.	(*g.st.*)	Forbes's.	Switzerland.	——	——	H.5.	——— t. 32.
	lùcida. S.W. *nec aliorum*.							
45	tetraspérma. W.	(*g.ye.*)	four-seeded.	E. Indies.	1796.	——	S. ♄.	Roxb. corom. 1. t. 97.
46	gariepìna. B.T.	(*g.ye.*)	Cape.	C. B. S.	1816.	——	G. ♄.	
47	Humboldtiàna. W.	(*bl.*)	Humboldt's.	Peru.	1821.	....	G. ♄.	H. et B. 2. t. 99, 100.
48	Bonplandiàna. K.S.	(*br.*)	Bonpland's.	Mexico.	——	....	G. ♄.	——— 2. t. 101, 102.
49	Houstoniàna. Ph.	(*gr.ye.*)	Houstoun's.	America.	1812.	4. 5.	H. ♄.	Salic. woburn. t. 11.
50	Lyòni. S.W.	(*g.ye.*)	Lyon's.	Switzerland.	1816.	——	H. ♄.	——— t. 10.
51	finmárchica. S.S.	(*g.ye.*)	Finmark.	Sweden.	1821.	——	H. ♄.	Flor. dan. t. 1051.
52	Villarsiàna. W.	(*g.ye.*)	Villars's.	S. France.	1818.	5. 6.	H. ♄.	Salic. woburn. t. 17.
53	myrtilloídes. S.S.	(*br.ye.*)	Myrtle-leaved.	Sweden.	1772.	4. 8.	H. ♄.	——— t. 66.
54	retùsa. W.	(*g.ye.*)	blunt-leaved.	Italy.	1763.	——	H. ♄.	——— f. 139.
55	serpyllifòlia. W.	(*g.ye.*)	Thyme-leaved.	Switzerland.	1818.	——	H. ♄.	——— t. 65.
56	Kitaibeliàna. W.	(*g.ye.*)	Kitaibel's.	Carpathian m.	1816.	——	H. ♄.	——— t. 64.
57	'Uva-úrsi. S.S.	(*g.ye.*)	Bearberry-like.	Labrador.	1811.	4. 5.	H. ♄.	——— f. 151.
58	berberifòlia. W.	(*g.ye.*)	Barberry-leaved.	Davuria.	1824.	——	H. ♄.	Pall. ros. 2. t. 82.
59	herbàcea. E.B.	(*g.ye.*)	least.	Britain.	....	6.	H. ♄.	Eng. bot. v. 27. t. 1907.
60	monándra. Ard.	(*g.ye.*)	monandrous.	Europe.	....	3. 4.	H. ♄.	Salic. woburn. t. 4.
61	purpùrea. E.B.	(*g.ye.*)	bitter purple.	England.	....	3.	H. ♄.	Eng. bot. v. 20. t. 1388.
62	Hèlix. E.B.	(*g.ye.*)	Rose.	Britain.	....	3. 4.	H. ♄.	——— v. 19. t. 1343.
63	Lambertiàna. E.B.	(*g.ye.*)	Boyton.	England.	....	——	H. ♄.	——— v. 19. t. 1359.
64	rùbra. E.B.	(*g.ye.*)	red-stemmed.	———	....	4. 5.	H. ♄.	——— v. 16. t. 1145.
65	Forbiàna. E.B.	(*g.ye.*)	basket osier.	———	....	4.	H. ♄.	——— v. 19. t. 1344.
66	viminàlis. E.B.	(*g.ye.*)	common osier.	Britain.	....	4. 5.	H. ♄.	——— v. 27. t. 1898.
67	mollíssima. S.S.	(*g.ye.*)	softest.	Germany.	....	3. 4.	H. ♄.	
68	Smithiàna. S.S.	(*g.ye.*)	Smith's.	England.	....	——	H. ♄.	Salic. woburn. t. 134.
	mollíssima. Eng. bot. v. 21. t. 1509. *nec aliorum*.							
69	acuminàta. E.B.	(*g.ye.*)	acuminate.	Britain.	....	4.	H.5.	Eng. bot. v. 20. t. 1434.
70	holoserícea. W.	(*g.ye.*)	holosericeous.	Germany.	1822.	——	H.5.	
71	aurìta. W.	(*g.ye.*)	eared.	Britain.	....	4. 5.	H.5.	Hoff. sal. 1. t. 22. f. 1.
72	uliginòsa. W. en.	(*g.ye.*)	marsh.	———	....	4. 6.	H.5.	Eng. bot. v. 21. t. 1487.
	aurìta. E.B. *non* W.							
73	paludòsa. L. en.	(*g.ye.*)	fen.	Germany.	....	——	H.5.	
74	aquática. E.B.	(*g.ye.*)	water.	Britain.	....	4.	H.5.	Eng. bot. v. 20. t. 1437.
75	spatulàta. W. en.	(*g.ye.*)	spatulate.	Germany.	....	——	H.5.	
76	cáprea. E.B.	(*g.ye.*)	great round-leaved.	Britain.	....	4. 5.	H.5.	Eng. bot. v. 21. t. 1488.
77	sphacelàta. E.B.	(*g.ye.*)	withered-pointed.	Scotland.	....	——	H. ♄.	——— v. 33. t. 2333.
78	laurìna. W.	(*g.ye.*)	Laurel-leaved.	England.	....	——	H. ♄.	——— v. 26. t. 1806.
	bícolor. E.B. *non* Ehr.							
79	bícolor. W.	(*g.ye.*)	two-coloured.	Germany.	1818.	——	H. ♄.	
80	Forsteriàna. E.B.	(*g.ye.*)	Forster's.	Scotland.	....	——	H.5.	Eng. bot. v. 33. t. 1403.
81	nígricans. E.B.	(*g.ye.*)	dark broad-leaved.	England.	....	4.	H.5.	——— v. 17. t. 1213.
82	violàcea. S.S.	(*g.ye.*)	violet.	Russia.	....	3. 4.	H.5.	Andr. reposit. t. 581.
83	confórmis. S.W.	(*g.ye.*)	uniform-leaved.	....	....	2. 3.	H.5.	Salic. woburn. t. 24.
84	Weigeliàna. S.S.	(*g.ye.*)	long-styled.	Silesia.	1820.	——	H.5.	——— t. 51.
85	cotinifòlia. E.B.	(*g.ye.*)	Quince-leaved.	Britain.	....	4.	H.5.	Eng. bot. v. 20. t. 1403.

86 conífera. s.s.	(*g.ye.*) cone-bearing.	N. America.	1800.	5.	H.5.	Wang. amer. t. 31. f. 72.
87 prinoídes. s.s.	(*g.ye.*) Prinos-like.	———	1811.	3. 4.	H.5.	Salic. woburn. t. 40.
88 fuscàta. s.s.	(*g.ye.*) brown-stemmed.	———	——	——	H.5.	
89 recurvàta. s.s.	(*g.ye.*) recurve-flowered.	———	——	4.	H.♄.	
90 trístis. s.s.	(*g.ye.*) linear-leaved.	———	1765.	3. 4.	H.♄.	Salic. woburn. f 150.
91 Muhlenbergiàna.	(*re.ye.*) Muhlenberg's.	———	1811.	4.	H.♄.	Ann. bot. 2. t. 5. f. 9.
92 grìsea. W.en.	(*br.ye.*) brown.	———	1816.	——	H.♄.	
93 lanceolàta. E.B.	(*g.ye.*) sharp-leaved.	England.	....	4. 5.	H.♄.	Eng. bot. v. 20. t. 1436.
94 stipulàris. E.B.	(*g.ye.*) auricled.	Britain.	....	3.	H.♄.	——— v. 17. t. 1214.
95 undulàta. s.s.	(*g.ye.*) wave-leaved.	Germany.	1819.	4. 5.	H.♄.	Salic. woburn. t. 13.
96 virgàta. s.w.	(*g.ye.*) twiggy.	....	....	5. 6.	H.♄.	——— t. 12.
97 Hoffmanniàna.	(*g.ye.*) Hoffman's.	Switzerland.	....	5.	H.♄.	——— t. 16.
98 montàna. w.	(*g.ye.*) mountain.	———	....	5. 6.	H.5.	——— t. 19.
99 monspeliénsis. s.w.	(*g.ye.*) Montpelier.	Montpelier.	....	4. 5.	H.♄.	——— t. 30.
100 pátens. s.w.	(*g.ye.*) spreading.	....	....	5. 8.	H.♄.	——— t. 39.
101 Willdenowiàna. s.w.	(*g.ye.*) Willdenow's.	....	....	4. 8.	H.♄.	——— t. 41.
102 críspa. s.w.	(*g.st.*) curled.	....	....	3.	H.♄.	——— t. 42.
103 nìtens. s.w.	(*g.ye.*) glossy.	Scotland.	....	4. 5.	H.♄.	——— t. 44.
104 Borreriàna. s.w.	(*g.st.*) Borrer's.	———	....	4.	H.♄.	——— t. 45.
105 Davalliàna. s.w.	(*br.st.*) Davall's.	———	....	——	H.♄.	——— t. 47.
106 petiolàris. E.B.	(*br.st.*) dark long-leaved.	N. America.	....	4.	H.♄.	Eng. bot. v. 16. t. 1147.
107 pátula. s.s.	(*g.ye.*) spreading.	Italy.	1818.	——	H.♄.	
108 oleifòlia. E.B.	(*g.ye.*) Olive-leaved.	Britain.	....	3.	H.5.	Eng. bot. v. 20. t. 1402.
109 cándida. s.s.	(*r.st.*) hoary.	N. America.	1811.	2. 3.	H.♄.	Salic. woburn. t. 91.
110 díscolor. s.s.	(*g.ye.*) brown-branched.	———	——	4.	H.♄.	Ann. bot. 2. t. 5. f. 1.
111 arenària. E.B.	(*g.ye.*) sand.	Scotland.	....	5.	H.ħ.	Eng. bot. v. 26. t. 1809.
112 rèpens. E.B.	(*br.ye.*) creeping.	Britain.	....	5.	H.ħ.	——— v. 3. t. 183.
113 incubàcea. w.	(*bh.ye.*) trailing.	———	1775.	——	H.ħ.	Salic. woburn. t. 79.
114 rosmarinifòlia. E.B.	(*g.ye.*) Rosemary-leaved.	———	....	4. 5.	H.ħ.	Eng. bot. v. 19. t. 1365.
115 Arbúscula. E.B.	(*g.ye.*) little tree.	Scotland.	....	4.	H.♄.	——— v. 19. t. 1366.
116 fúsca. E.B.	(*g.st.*) brown.	Britain.	....	5.	H.ħ.	——— v. 28. t. 1960.
117 parvifòlia. E.B.	(*g.ye.*) small-leaved.	England.	....	4. 5.	H.ħ.	——— v. 28. t. 1961.
118 adscéndens. E.B.	(*g.ye.*) ascending.	———	....	5.	H.ħ.	——— v. 28. t. 1962.
119 prostràta. E.B.	(*g.ye.*) prostrate.	Britain.	....	5.	H.ħ.	——— v. 28. t. 1959.
120 canéscens. W.en.	(*g.ye.*) canescent.	Germany.	1815.	4. 5.	H.ħ.	
121 leucophy'lla. W.en.	(*g.ye.*) white-leaved.	———	——	——	H.ħ.	
122 elæagnoídes. L.en.	(*g.ye.*) Elæagnus-like.	Switzerland.	1823.	4. 8.	H.ħ.	Salic. woburn. t. 69.
123 argéntea. E.B.	(*g.ye.*) silky sand.	Britain.	....	5.	H.ħ.	Eng. bot. v. 19. t. 1364.
124 procúmbens. s.w.	(*g.ye.*) procumbent.	Scotland.	....	——	H.ħ.	Salic. woburn. t. 61.
125 ramifúsca. s.w.	(*g.ye.*) brown branched.	Britain.	....	4. 7.	H.ħ.	——— t. 53.
126 floribúnda. s.w.	(*g.ye.*) many flowered.	———	....	——	H.ħ.	——— t. 54.
127 versifòlia. s.s.	(*g.ye.*) various-leaved.	N. Europe.	1824.	5.	H.ħ.	
128 prunifòlia. E.B.	(*g.ye.*) Plum-leaved.	Scotland.	....	4. 5.	H.ħ.	Eng. bot. v. 19. t. 1361.
129 Dicksoniàna. E.B.	(*g.ye.*) Dickson's.	———	....	5.	H.ħ.	——— v. 20. t. 1390.
130 hùmilis. W.en.	(*g.ye.*) humble.	....	1824.	——	H.ħ.	
131 formòsa. w.	(*g.ye.*) handsome.	Switzerland.	1820.	——	H.ħ.	
132 myrsinìtes. E.B.	(*g.ye.*) Whortle-leaved.	Scotland.	....	4. 6.	H.♄.	Eng. bot. v. 19. t. 1360.
133 venulòsa. E.B.	(*g.ye.*) veiny-leaved.	———	....	4. 5.	H.ħ.	——— v. 19. t. 1362.
134 vacciniifòlia. E.B.	(*g.ye.*) Bilberry-leaved.	———	....	——	H.ħ.	——— v. 33. t. 2341.
135 carinàta. E.B.	(*g.ye.*) folded-leaved.	———	....	——	H.ħ.	——— v. 19. t. 1363.
136 arbutifòlia. w.	(*g.ye.*) Arbutus-leaved.	Switzerland.	1818.	——	H.ħ.	
137 rupéstris. E.B.	(*g.ye.*) rock.	Scotland.	....	4.	H.ħ.	Eng. bot. v. 33. t. 2342.
138 lappònum. w.	(*g.ye.*) Lapland.	Lapland.	....	——	H.♄.	Salic. woburn. t. 73.
139 helvética. Vil.	(*g.ye.*) Swiss.	Switzerland.	1824.	4. 8.	H.♄.	
140 leucophy'lla. w.	(*g.ye.*) white-leaved.	———	——	——	H.♄.	
141 lívida. s.s.	(*g.ye.*) livid.	Lapland.	——	4.	H.ħ.	
142 polàris. s.s.	(*g.ye.*) polar.	———	——	4. 7.	H.ħ.	Salic. woburn. t. 63.
143 longifòlia. s.s.	(*g.ye.*) long-leaved.	N. America.	1819.	4. 5.	H.♄.	Ann. bot. 2. t. 5. f. 6.
144 glàuca. E.B.	(*g.ye.*) glaucous.	Scotland.	....	5.	H.♄.	Eng. bot. v. 26. t. 1810.
145 serícea. w.	(*wh.ye.*) silky.	Alps Europe.	1816.	——	H.♄.	Salic. woburn. t. 74.
146 Stuartiàna. E.B.	(*w.ye.*) Stuart's.	Scotland.	....	7. 8.	H.♄.	Eng. bot. v. 36. t. 2586.
147 reticulàta. E.B.	(*re.*) wrinkled.	Britain.	....	6. 7.	H.ħ.	——— v. 27. t. 1908.
148 hírta. E.B.	(*g.ye.*) hairy-branched.	England.	....	4. 5.	H.5.	——— v. 20. t. 1404.
149 Croweàna. E.B.	(*g.ye.*) Crowe's.	———	....	——	H.♄.	——— v. 16. t. 1146.
150 fagifòlia. w.	(*g.ye.*) Beech-leaved.	Hungary.	1819.	——	H.♄.	
151 cordifòlia. Ph.	(*g.ye.*) heart-shaped.	N. America.	1811.	——	H.♄.	Salic. woburn. f. 143.
152 planifòlia. Ph.	(*g.ye.*) flat-leaved.	Labrador.	——	——	H.♄.	
153 falcàta. Ph.	(*g.ye.*) sickle-leaved.	N. America.	——	——	H.♄.	Salic. woburn. f. 148.
154 tetrápla. L.en.	(*g.ye.*) elegant.	Scotland.	1824.	4.	H.♄.	——— t. 49.

155	obtùsa. L.en.	(g.ye.)	blunt-leaved.	Switzerland.	1820.	4.	H. ♄.	
156	eriántha.L.en.	(g.ye.)	woolly-flowered.	———	——	——	H. ♄.	
157	c'æsia. Vill.	(g.ye.)	cæsious.	N. Europe.	——	4. 5.	H. ♄.	Vill.delph.3. t.50. f.11.
158	ulmifòlia.W.en.	(g.ye.)	Elm-leaved.	Switzerland.	——	4.	H. ♄.	
159	foliòsa. w.	(g.ye.)	leafy.	Lapland.	1818.	——	H. ♄.	
160	obtusifòlia. w.	(g.ye.)	obtuse-leaved.	———	——	——	H. ♄.	
161	atrovìrens.s.w.	(g.ye.)	dark-green.	Switzerland.	1824.	——	H. ♄.	Salic. woburn. t. 108.
162	australis. s.w.	(g.ye.)	southern.	———	——	5.	H. ♄.	——— t. 103.
163	carpinifòlia.s.w.	(g.ye.)	Hornbeam-leaved.	Germany.	——	3. 4.	H. ♄.	
164	cydoniifòlia.Sch.	(g.ye.)	Cydonia-leaved.	Switzerland.	——	——	H. ♄.	
165	subalpìna.s.w.	(g.ye.)	subalpine.	———	——	4. 5.	H. ♄.	Salic. woburn. t. 93.
166	proteæfòlia.s.w.	(g.ye.)	Protea-leaved.	———	1820.	——	H. ♄.	——— t. 75.
167	strépida. s.w.	(g.ye.)	creaking.	———	——	3. 4.	H. ♄.	——— t. 100.
168	pyrenàica. s.s.	(g.ye.)	Pyrenean.	Pyrenees.	1823.	4.	H. ♄.	
169	Jacquiniàna.w.	(g.ye.)	Jacquin's.	Austria.	1818.	——	H. ♄.	Jacq. aust. t. 409.
	fúsca. Jacquin. *nec aliorum*.							
170	Schraderiànum.w.	(g.ye.)	Schrader's.	Germany.	1820.	——	H. ♄.	
171	finmárchica.w.	(g.ye.)	Swedish.	Sweden.	1825.	4. 5.	H. ♄.	
172	serrulàta. w.	(g.ye.)	serrulate.	Lapland.	1822.	——	H. ♄.	
173	versícolor.s.w.	(g.ye.)	various-coloured.	Switzerland.	——	5.	H. ♄.	Salic. woburn. t. 77.
174	Doniàna.E.F.	(br.ye.)	Mr. Don's.	Scotland.	....	4. 5.	H. ♄.	——— t. 88.
175	lineàris.s.w.	(br.st.)	linear-leaved.	Switzerland.	1820.	——	H. ♄.	——— t. 89.
176	villòsa. s.w.	(g.ye.)	villous.	....	....	——	H. ♄.	——— t. 92.
177	refléxa. s.w.	(g.ye.)	reflexed.	....	....	3.	H. ♄.	——— t. 94.
178	petr'æa. s.w.	(g.ye.)	rock-side.	Britain.	....	4.	H. ♄.	——— t. 97.
179	grisonénsis.s.w.	(g.ye.)	Grisons.	Grisons.	1820.	3. 4.	H. ♄.	——— t. 99.
180	sórdida. s.w.	(g.st.)	sordid.	Switzerland.	——	4.	H. ♄.	——— t. 101.
181	glabràta. Sch.	(g.ye.)	smooth.	———	1823.	4. 5.	H. ♄.	
182	decúmbens.s.w.	(g.ye.)	decumbent.	———	——	5.	H. ♄.	Salic. woburn. t. 88.
183	heterophy'lla.Deb.	(g.ye.)	different-leaved.	......	——	——	H. ♄.	
184	murìna. Sch.	(g.ye.)	wall.	Switzerland.	1824.	3. 4.	H. ♄.	
185	nervòsa. Sch.	(g.ye.)	nerved.	———	——	——	H. ♄.	
186	pannòsa. s.w.	(g.ye.)	cloth-leaved.	———	——	4. 5.	H. ♄.	Salic. woburn. t. 123.
187	palléscens.Sch.	(g.st.)	pallid.	———	——	——	H. ♄.	
188	pállida. s.w.	(g.ye.)	pale.	———	——	5.	H. ♄.	Salic. woburn. t. 96.
189	alaternoídes.s.w.	(ye.)	Alaternus-like.	———	1823.	4. 5.	H. ♄.	——— t. 76.
190	albéscens.Sch.	(g.st.)	whitish.	———	——	——	H. ♄.	
191	rivulàris.s.w.	(g.ye.)	rivulet.	———	1824.	5.	H. ♄.	Salic. woburn. t. 102.
192	rotundàta.s.w.	(g.ye.)	rounded-leaved.	———	——	4. 5.	H. ♄.	——— t. 104.
193	dùra. s.w.	(g.ye.)	hardy.	———	——	——	H.5.	——— t. 105.
194	fírma. s.w.	(g.ye.)	firm.	———	——	3. 4.	H. ♄.	——— t. 106.
195	Ansoniàna.s.w.	(g.ye.)	Anson's.	———	——	——	H. ♄.	——— t. 107.
196	coriàcea. s.w.	(g.ye.)	leathery.	———	1822.	3.	H. ♄.	——— t. 112.
197	crassifòlia.s.w.	(g.ye.)	thick-leaved.	....	....	4. 5.	H. ♄.	——— t. 115.
198	vaudénsis.s.w.	(g.ye.)	Vaudois.	Switzerland.	1824.	3. 4.	H. ♄.	——— t. 117.
199	latifòlia. s.w.	(g.ye.)	broad-leaved.	———	——	3.	H. ♄.	——— t. 118.
200	ferrugínea.s.w.	(br.st.)	ferruginous.	Britain.	....	4.	H. ♄.	——— t. 128.
201	geminàta. s.w.	(g.ye.)	twin-flowered.	———	....	3.	H. ♄.	——— t. 129.
202	macrostipulàcea.s.w.	(g.y.)	large-stipul'd.	Switzerland.	1822.	4. 5.	H.5.	——— t. 130.
203	Micheliàna.s.w.	(g.ye.)	Michel's.	———	——	4.	H.5.	——— t. 135.
204	atropurpùrea.s.w.	(g.ye.)	dark-purple.	———	1824.	——	H. ♄.	
205	damascèna.s.w.	(g.ye.)	Damask.	———	——	3. 4.	H. ♄.	
206	mutábilis. s.w.	(g.ye.)	changeable.	———	——	——	H. ♄.	
207	cerasifòlia.Sch.	(g.ye.)	Cherry-leaved.	———	——	4.	H. ♄.	
208	cinnamòmea.Sch.	(g.ye.)	Cinnamon-scented.	———	——	4. 5.	H. ♄.	
209	clethræfòlia.Sch.	(g.ye.)	Clethra-leaved.	———	——	——	H. ♄.	
210	grisophy'lla.s.w.	(g.ye.)	gray-leaved.	———	——	4.	H. ♄.	Salic. woburn. t. 119.
211	lacústris. s.w.	(g.ye.)	lake.	———	——	3.	H. ♄.	——— t. 116.
212	mespilifòlia.Sch.	(g.ye.)	Mespilus-leaved.	———	——	4. 5.	H. ♄.	
213	obtuseserràta.Sch.	(g.ye.)	bluntly-sawed.	———	——	——	H. ♄.	
214	pyrifòlia. Sch.	(g.ye.)	Pear-leaved.	———	——	——	H. ♄.	
215	Schleicheriàna.s.w.	(g.ye.)	Schleicher's.	———	——	——	H. ♄.	Salic. woburn. t. 98.
216	velutìna.W.en.	(g.ye.)	velvetty.	....	1826.	3. 4.	H. ♄.	
217	Starkeàna. w.	(g.ye.)	Starke's.	Silesia.	1820.	4. 5.	H. ♄.	

PO'PULUS. L. POPLAR. Diœcia Octandria. L.

1	álba. E.B.	(fl.)	Abele-tree.	Britain.	....	3. 4.	H.5.	Eng. bot. v. 23. t. 1618.
2	canéscens. E.B.	(fl.)	gray.	England.	....	——	H.5.	——— v. 23. t. 1619.
3	nívea. w.	(fl.)	snowy.	Europe.	....	——	H.5.	
4	trépida. w.	(fl.)	trembling.	N. America.	1812.	——	H.5.	Mich. arb. 3. t. 8. f. 1.

5 trémula. w.	(*fl.*)	Aspen.	Britain.	....	3. 4.	H.5.	Eng. bot. v. 27. t. 1909.
6 lævigàta. w.	(*fl.*)	smooth.	N. America.	1769.	——	H.5.	Mich. arb. 3. t. 11.
7 tremuloídes. M.	(*fl.*)	trembling-like.	————	1811.	——	H.5.	Duham. arb. 2. t. 53.
8 gr'æca. w.	(*fl.*)	Athenian.	Archipelago.	1779.	——	H.5.	———— 184. t. 54.
9 nìgra. w.	(*fl.*)	black.	Britain.	....	——	H.5.	Eng. bot. v. 27. t. 1910.
10 acladésca. Lind.	(*re.*)	black Italian.	N. America.	....	5.	H.5.	
11 hudsònica. M.	(*fl.*)	American black.	————	....	——	H.5.	Mich. arb. 3. t. 10. f. 1.
betulifòlia. Ph.							
12 dilatàta. w.	(*sc.*)	Lombardy.	Italy.	1758.	——	H.5.	
13 vimínea. Df.	(*fl.*)	slender-twigged.	N. America.	1824.	4. 5.	H.5.	
14 monilífera. w.	(*fl.*)	Canadian.	Canada.	1772.	3. 5.	H.5.	Wats. dend. brit. t. 102.
15 angulàta. w.	(*fl.*)	Carolina.	Carolina.	1738.	3.	H.5.	Mich. arb. 3. t. 12.
16 grandidentàta. M.	(*fl.*)	large-toothed.	N. America.	1812.	4.	H.5.	
β *péndula.* N.	(*fl.*)	*weeping.*	————	1820.	——	H.5.	
17 balsamífera. w.	(*gr.*)	Tacamahac.	————	1692.	——	H.5.	Mich. arb. 3. t. 13. f. 1.
18 cándicans. w.	(*gr.*)	heart-leaved.	————	1772.	3.	H.5.	Catesb. car. 1. t. 34.
19 heterophy'lla. w.	(*gr.*)	various-leaved.	————	1765.	4. 5.	H.5.	Mich. arb. 3. t. 9.
20 laurifòlia. Led.	(*fl.*)	Laurel-leaved.	Altay.	1827.	....	H.5.	

ORDO CLXXXVIII.

PLATANEÆ.

PLA'TANUS. W.		PLANE-TREE.	Monœcia Polyandria. L.				
1 orientàlis. w.	(*gr.*)	oriental.	Levant.	1548.	4. 5.	H.5.	Wats. dend. brit. t. 101.
2 cuneàta. w.	(*gr.*)	wave-leaved.	————	1739.	——	H.5.	
3 acerifòlia. w.	(*gr.*)	Maple-leaved.	————	1724.	....	H.5.	
4 occidentàlis. w.	(*gr.*)	American.	N. America.	1640.	——	H.5.	Wats. dend. brit. t. 100.
5 mexicàna.	(*gr.*)	Mexican.	Mexico.	1826.	....	F.5.	
LIQUIDA'MBAR. W.		SWEET-GUM-TREE.	Monœcia Polyandria. L.				
1 Styraciflùa. w.	(*gr.wh.*)	Maple-leaved.	N. America.	1683.	3. 4.	H.5.	Catesb. car. 2. t. 65.
2 imbérbe. w.	(*gr.wh.*)	oriental.	Levant.	1759.	....	H.5.	

ORDO CLXXXIX.

MYRICEÆ. *Kth. synops.* 1. *p.* 361.

COMPT'ONIA. W.		COMPT'ONIA.	Monœcia Triandria. W.				
aspleniifòlia. w.	(*br.*)	Fern-leaved.	N. America.	1714.	3. 5.	H. ħ.	Wats. dend. brit. t. 166.
MYR'ICA. W.		CANDLEBERRY-MYRTLE.	Diœcia Tetrandria. L.				
1 Gàle. w.	(*br.*)	Sweet Gale.	Britain.	....	5.	H. ♄.	Eng. bot. v. 8. t. 562.
2 cerífera. w.	(*br.*)	common.	N. America.	1699.	5. 6.	H. ♄.	Catesb. car. 1. t. 69.
3 carolinénsis. Ph.	(*br.*)	broad-leaved.	————	1730.	5.	H. ħ.	———— 1. t. 13.
4 pensylvánica. Ph.	(*br.*)	Pensylvanian.	————	....	——	H. ħ.	Duham. arb. ed. n. 2. t. 55.
5 Fáya. w.	(*br.*)	Azorian.	Azores.	1777.	6. 7.	F. ♄.	———— 2. t. 56.
6 mexicàna. W. en.	(*br.*)	Mexican.	Mexico.	1821.	6. 8.	G. ħ.	
7 esculénta. D.P.	(*br.*)	eatable.	Nepaul.	——	....	G. ♄.	
8 segregàta. w.	(*br.*)	netted-leaved.	S. America.	——	....	G. ♄.	Jacq. ic. 3. t. 625.
9 æthiòpica. w.	(*br.*)	African.	C. B. S.	1795.	6. 7.	G. ♄.	Pluk. alm. t. 48. f. 8.
10 serràta. w.	(*br.*)	saw-leaved.	————	1793.	8.	G. ħ.	Burm. afr. t. 98. f. 1.
11 laciniàta. W. en.	(*br.*)	smooth Oak-ld.	————	1752.	6. 7.	G. ♄.	Jacq. frag. 2. t. 1. f. 4.
12 quercifòlia. W. en.	(*br.*)	hairy Oak-leav'd.	————	——	——	G. ħ.	
13 cordifòlia. w.	(*br.*)	heart-leaved.	————	1759.	5. 7.	G. ♄.	Pluk. alm. t. 319. f. 7.
14 integrifòlia. H.B.	(*br.*)	entire-leaved.	E. Indies.	1824.	....	S. ♄.	
NAGE'IA. W.		NAGE'IA.	Diœcia Tetrandria.				
1 japónica. w.	(*gr.*)	Japan.	Japan.	1812.	2. 4.	G. ♄.	
Búxus dioíca. Hort. *Myrìca Nàgi.* Thunb.							
2 Putranjìva. H.B.	(*gr.*)	grey-barked.	E. Indies.	1822.	——	S. ♄.	
CASUAR'INA. W.		SHRUBBY HORSETAIL.	Monœcia Monandria. W.				
1 equisetifòlia. w.	(*br.re.*)	Horse-tail-leav'd.	S. Sea Island.	1793.	....	S. ♄.	Rumph. amb. 3. t. 57.
2 nodiflòra. w.	(*br.re.*)	knotted-flower'd.	————	1818.	....	S. ♄.	

3 glaùca. s.s.	(*br.re.*) glaucous.	N. S. W.	1818.	11. 4.	G. ♄.	
4 paludòsa. s.s.	(*br.re.*) marsh.	———	1822.	——	G. ♄.	
5 strícta. w.	(*br.re.*) upright.	———	1775.	11. 2.	G. ♄.	Andr. reposit. t. 346.
6 dist'yla. w.	(*br.re.*) two-styled.	———	1812.	——	G. ♄.	Vent. dec. pl. nov. t. 62.
7 quadriválvis. P.S.	(*br.re.*) four-valved.	N. Holland.	——	——	G. ♄.	Lab. nov. hol. 2. t. 218.
8 torulòsa. w.	(*br.re.*) cork-barked.	N. S. W.	1772.	4. 6.	G. ♄.	
9 muricàta. L.en.	(*br.re.*) muricate.	E. Indies.	1816.	....	S. ♄.	
10 tenuíssima. s.s.	(*br.re.*) slender.	N. S. W.	1824.	4. 6.	G. ♄.	
11 nàna. s.s.	(*br.re.*) dwarf.	———	——	——	G. ♄.	

ORDO CXC.

HAMAMELIDEÆ. *Brown.*

HAMAM'ELIS. W.	WITCH-HAZEL.	Tetrandria Digynia. L.				
1 virgínica. w.	(*ye.*) Virginian.	N. America.	1736.	11. 5.	H. ♄.	Lodd. bot. cab. t. 598.
2 macrophy'lla. Ph.	(*ye.*) large-leaved.	———	1812.	——	H. ♄.	
FOTHERGI'LLA. W.	FOTHERGI'LLA.	Polyandria Digynia. L.				
1 alnifòlia. w.	(*wh.*) obtuse-leaved.	N. America.	1765.	4. 6.	H. ♄.	Bot. mag. t. 1341.
2 màjor. B.M.	(*wh.*) large-leaved.	———	——	5. 6.	H. ♄.	——— t. 1342.
3 Gardèni. Jacq.	(*wh.*) acute-leaved.	———	——	——	H. ♄.	Jacq. ic. 1. t. 100.
4 serotìna. B.M.	(*wh.*) green-leaved.	———	——	——	H. ♄.	

ORDO CXCI.

CONIFERÆ. *Juss. gen.* 411.

SUBORDO I. *TAXINÆ.* Richard.

SALISB'URIA. R.	MAIDENHAIR-TREE.	Monœcia Polyandria. H.K.				
adiantifòlia. L.T.	(*g.st.*) Japan.	Japan.	1754.	4. 5.	H. 5.	Wats. dend. brit. t. 168.
Gíngko bilòba. L.						
PHYLLOCL'ADUS. R.	PHYLLOCL'ADUS.	Monœcia Monadelphia.				
rhomboidàlis. R.	(*ye.*) Fern-leaved.	V. Diem. Isl.	1825.	.....	G. ♄.	Rich. conif. pl. 3.
Podocárpus aspleniifòlia. Labill. nov. holl. 2. t. 221.						
PODOCA'RPUS. L'H.	PODOCA'RPUS.	Monœcia Monadelphia.				
1 Chilìnus. R.	(*ye.*) sharp-leaved.	C. B. S.	1774.	7.	G. ♄.	Rich. conif. pl. 1. f. 1.
2 elongàtus. R.	(*ye.*) Chinese.	China.	——	5. 7.	H. ♄.	——— pl. 1. f. 2.
Táxus chinénsis. H.B.						
3 coriàceus. R.	(*ye.*) leathery-leaved.	———	1824.	——	G. ♄.	Rich. conif. pl. 1. f. 3.
4 verticillàtus.	(*ye.*) whorl-leaved.	Japan.	1812.	....	G. ♄.	
5 macrophy'llus. L.en.	(*ye.*) large-leaved.	China.	1804.	7. 8.	G. ♄.	Kæmpf. ic. t. 24.
6 nereifòlius. L.P.	(*ye.*) Wax-Dammar.	P. Wales Isl.	1809.	....	S. ♄.	Rumph. amb. 3. t. 26.
TA'XUS. W.	YEW-TREE.	Diœcia Monadelphia. L.				
1 baccàta. w.	(*ye.*) common.	Britain.	....	2. 4.	H. 5.	Eng. bot. v. 11. t. 746.
2 hibérnica.	(*ye.*) Irish.	Ireland.	....	——	H. ♄.	
3 canadénsis. w.	(*ye.*) Canadian.	Canada.	1818.	——	H. 5.	
4 nucífera. w.	(*ye.*) Acorn-bearing.	China.	1764.	....	G. ♄.	Kæmpf. amœn. t. 815.
DACRY'DIUM. L.P.	DACRY'DIUM.	Monœcia Monadelphia.				
cupressìnum. L.P.	(*ye.*) N. Zealand spruce.	N. Zealand.	1825.	....	G. ♄.	Lamb. pin. v. 1. t. 41.
Thalamia cupressína. S.S.						

SUBORDO II. *CUPRESSINÆ.* Richard.

CALLI'TRIS. R.	CALLI'TRIS.	Monœcia Monadelphia.				
1 quadriválvis. R.	(*ye.*) jointed.	Barbary.	1815.	2. 5.	F. ♄.	Rich. conif. pl. 8. f. 1.
Thùja articulàta. Vahl. symb. 2. t. 48.						
2 arenòsa.	(*ye.*) sand.	N. Holland.	1824.	....	F. ♄.	
3 pyramidàlis.	(*ye.*) pyramidal.	———	——	....	F. ♄.	

4 austràlis.	(*ye.*)	southern.	N. S. W.	....	4. 5.	F.♄.	
Cupréssus austràlis. P.S.							
5 oblónga. R.	(*ye.*)	oblong-scaled.	———	1822.	——	G.ħ.	Rich. conif. pl. 18. f. 2.
6 rhomboídea. R.	(*ye.*)	rhomboid-scal'd.	N. Holland.	——	——	G.ħ.	———— pl. 18. f. 1.
TH'UJA. W.		ARBOR-VITÆ.	Monœcia Monadelphia. L.				
1 occidentàlis. W.	(*ye.*)	American.	N. America.	1596.	5.	H.♄.	Wats. dend. brit. t. 150.
2 pyramidàlis. T.P.	(*ye.*)	pyramidal.	....	1824.	——	H.♄.	
3 orientàlis. W.	(*ye.*)	Chinese.	China.	1752.	2. 3.	H.♄.	Wats. dend. brit. t. 149.
4 plicàta. L.P.	(*ye.*)	Nee's.	Nootka sound.	1796.	——	H.♄.	Lamb. pin. v. 2. p. 19. t.
5 cupressoídes. W.	(*ye.*)	African.	C. B. S.	1799.	....	G.♄.	
6 dolabràta. W.	(*ye.*)	ovate-leaved.	China.	1822.	——	H.♄.	Lamb. pin. app. t. 1.
7 sphæroídea. S.S.	(*ye.*)	White Cedar.	N. America.	1736.	4. 5.	H.♄.	Wats. dend. brit. t. 156.
Cupréssus thyoídes. W.							
8 péndula. L.P.	(*ye.*)	weeping.	Tartary.	1810.	2. 5.	H.♄.	Lamb. pin. v. 2. t. 51.
CUPRE'SSUS. W.		CYPRESS.	Monœcia Monadelphia. L.				
1 sempervìrens. W.	(*ye.*)	common.	Candia.	1548.	5.	H.♄.	Pall. ros. 2. t. 53.
α stricta.	(*ye.*)	*upright.*	———	——	——	H.♄.	
β horizontàlis.	(*ye.*)	*horizontal.*	———	——	——	H.♄.	
2 torulòsa. D.P.	(*ye.*)	twisting.	Nepaul.	1826.	....	H.♄.	
3 lusitánica. W.	(*ye.*)	Cedar of Goa.	Goa.	1683.	4. 5.	F.♄.	Lamb. pin. 95. t. 49.
4 péndula. Th.	(*ye.*)	pendulous.	Japan.	1818.	....	F.♄.	———— t. 50.
5 baccifórmis. W.en.	(*ye.*)	berry-like.	....	——	4. 5.	H.♄.	
6 juniperoídes. W.	(*ye.*)	African.	C. B. S.	1756.	——	G.ħ.	
TAX'ODIUM. Rich.		TAX'ODIUM.	Monœcia Monadelphia.				
dístichum. Rich.	(*ye.*)	deciduous.	America.	1640.	5.	H.5.	Mich. arb. 3. t. 1.
Cupréssus dísticha. W. *Schubértia.* Mirb.							
α pátens. H.K.	(*ye.*)	*spreading.*	———	——	——	H.5.	
β nùtans. H.K.	(*ye.*)	*long-leaved.*	———	——	——	H.5.	
JUNI'PERUS. W.		JUNIPER.	Diœcia Monadelphia. L.				
1 thurífera. W.	(*ye.*)	Spanish.	S. Europe.	1752.	5. 6.	H.♄.	
2 barbadénsis. W.	(*ye.*)	Barbadoes.	Barbadoes.	1811.	——	G.♄.	Pluk. alm. t. 197. f. 4.
3 bermudiàna. W.	(*ye.*)	Bermudas Cedar.	Bermudas.	1683.	——	F.♄.	Herm. lugd. t. 347.
4 chinénsis. W.	(*ye.*)	Chinese.	China.	1804.	——	H.♄.	
5 excélsa. W.	(*ye.*)	tall.	Siberia.	1806.	——	H.♄.	
6 Hermánni. S.S.	(*ye.*)	Hermann's.	....	....	——	H.♄.	
7 Sabìna. W.	(*ye.*)	Savin.	S. Europe.	1548.	——	H.♄.	
β variegàta.	(*ye.*)	*variegated.*	———	——	——	H.♄.	
8 tamariscifòlia.	(*ye.*)	Tamarisk-leav'd.	———	——	——	H.♄.	
9 prostràta. P.S.	(*ye.*)	prostrate.	N. America.	....	——	H.♄.	
10 daùrica. A.R.	(*ye.*)	Daurian.	Dauria.	1791.	6. 8.	H.♄.	Andr. reposit. t. 534.
11 virginiàna. W.	(*ye.*)	red Cedar.	N. America.	1664.	5. 6.	H.♄.	Mich. arb. 3. t. 5.
12 glaùca. W.en.	(*ye.*)	glaucous.	China.	1814.	——	H.♄.	
13 squamàta. D.P.	(*ye.*)	scaly-branched.	Nepaul.	1824.	....	H.♄.	
14 recúrva. D.P.	(*ye.*)	recurve-branch'd.	———	1822.	5. 6.	H.♄.	
15 commùnis. W.	(*ye.*)	common.	Britain.	....	——	H.♄.	Eng. bot. v. 16. t. 1100.
16 suècica. Mill.	(*ye.*)	Swedish.	N. Europe.	....	——	H.♄.	
17 nàna. W.	(*ye.*)	dwarf.	Siberia.	1815.	——	H.♄.	Pall. ros. 2. t. 54. f. A. B.
sibírica. Burgsd. *commùnis γ.* H.K.							
18 hemisph'ærica. S.S.	(*ye.*)	hemispherical.	Sicily.	1828.	....	H.♄.	
19 oblónga. M.B.	(*ye.*)	oblong.	Armenia.	1829.	....	H.♄.	
20 Oxycèdrus. W.	(*ye.*)	brown-berried.	Spain.	1739.	——	H.♄.	Duham. arb. 2. t. 128.
21 phœnícea. W.	(*ye.*)	Phœnician.	S. Europe.	1683.	——	H.♄.	Pall. ros. 2. t. 57.
22 ly'cia. W.	(*ye.*)	Lycian.	———	1693.	——	H.♄.	———— 2. t. 56.
23 capénsis. Lam.	(*ye.*)	Cape.	C. B. S.	1816.	....	G.♄.	
24 drupàcea. Lab.	(*ye.*)	drupaceous.	....	——	....	H.♄.	Lab. pl. syr. 2. t. 8.
EPH'EDRA.		EPH'EDRA.	Diœcia Monadelphia. L.				
1 distàchya. W.	(*st.*)	great.	France.	1570.	6. 7.	H.♄.	Schk. hand. 3. t. 339.
2 monostàchya. W.	(*st.*)	small.	Siberia.	1772.	9. 11.	H.♄.	Wats. dend. brit. t. 142.
3 altíssima. W.	(*st.*)	tall.	Barbary.	1823.	....	H.♄.	Desf. atl. 2. t. 253.

SUBORDO III. *ABIETINÆ.* Richard.

DA'MMARA. L.P.		DAMMAR PINE.	Monœcia Monadelphia. L.en.				
1 austràlis. L.P.	(*ye.*)	Cowrie-tree.	N. Zealand.	1823.	....	G.ħ.	Lamb. pin. 2. p. 14. t. 6.
2 orientàlis. L.P.	(*ye.*)	Amboyna pitch-tree.	Amboyna.	1804.	....	S.ħ.	—— v. 1. t. 38. -v. 2. p. 16.
loranthifòlia. L.en. *Pìnus Dámmara.* W. *A'gathis loranthifòlia.* Sal.							
ARAUC'ARIA. J.		ARAUC'ARIA.	Diœcia Monadelphia. W.				
1 imbricàta. W.	(*ye.*)	Chile Pine.	Chile.	1796.	....	H.5.	Lamb. pin. 2. p. 9. t. 4. B.

2 brasiliàna. L.P.	(*ye.*)	Brazil Pine.	Brazil.	1816.	....	G.5.	Lamb. pin. 2. p.12. t. 5.
3 excélsa. H.K.	(*ye.*)	Norfolk Island.	Norfolk Isl.	1793.	....	G.5.	——— 1. t. 39. 40.
Dombèya excélsa. L.P.							
4 Cunninghámii.	(*ye.*)	Cunningham's.	N. Holland.	1827.	....	G.5.	
B'ELIS. L.T.		B'ELIS.	Monœcia Monadelphia.				
lanceolàta.	(*ye.*)	spear-leaved.	China.	1804.	12. 1.	G. ♄.	Lamb. pin. 1. t. 34.
jaculifòlia. L.T. *Pìnus lanceolàta.* L.P. *Cunninghámia lanceolàta.* Bot. mag. t. 2743.							
P'INUS. W.		PINE or FIR.	Monœcia Monadelphia. W.				
P'INUS.		PINE-TREE.					
1 sylvéstris. W.	(*ye.*)	Scotch.	Scotland.	....	5.	H.5.	Lamb. pin. 1. t. 1.
2 Larício. P.S.	(*ye.*)	Corsican.	Corsica.	1814.	4. 5.	H.5.	——— v. 2. p. 28. t. 9.
3 uncinàta. DC.	(*ye.*)	hooked.	Pyrenees.	1820.	——	H.5.	
4 Pumìlio. W.	(*ye.*)	upright-coned.	Carniola.	1779.	——	H.5.	Lamb. pin. p. 5. t. 2.
5 Múghus. W.	(*ye.*)	nodding-coned.	Austria.	....	——	H.5.	Jacq. ic. 1. t. 193.
6 púngens. M.	(*ye.*)	prickly-coned.	N. America.	1804.	....	H.5.	Mich. arb. 1. t. 5.
7 Banksiàna. Ph.	(*ye.*)	Hudson's Bay.	Hudson's bay.	1785.	5. 6.	H.5.	Lamb. pin. p. 7. t. 3.
8 Pallasiàna. L.P.	(*ye.*)	Pallas's.	Crimea.	1804.	——	H.5.	——— v. 2. p. 1. t. 1.
9 Pináster. W.	(*ye.*)	cluster.	S. Europe.	1596.	4. 5.	H.5.	——— v. 1. p. 9. t. 4. 5.
10 Pìnea. W.	(*ye.*)	stone.	———	1548.	5.	H.5.	——— v. 1. p. 11. t. 6. 7. 8.
11 marítima. W.	(*ye.*)	maritime.	———	1759.	5. 6.	H.5.	——— v. 2. p. 30. t. 10.
12 halepénsis. W.	(*ye.*)	Aleppo.	Levant.	1683.	5.	H.5.	——— v. 1. p. 15. t. 11.
13 ínops. Ph.	(*ye.*)	Jersey.	N. America.	1739.	——	H.5.	——— v. 1. p. 18. t. 13.
14 resinòsa. Ph.	(*ye.*)	pitch.	———	1756.	——	H.5.	——— v. 1. p. 20. t. 14.
15 variábilis. Ph.	(*ye.*)	2 and 3-leaved.	———	1739.	5. 6.	H.5.	——— v. 1. p. 22. t. 15.
16 canariénsis. DC.	(*ye.*)	Canary.	Canaries.	1815.	3. 4.	G.5.	DC. pl. rar. gen. t. 1-2.
17 T'æda. Ph.	(*ye.*)	frankincense.	N. America.	1713.	5. 6.	H. ♄.	Lam. p. v. 1. p. 23. t. 16. 17.
18 lùtea. Walt.	(*ye.*)	yellow.	———	....	....	H.5.	
19 serotìna. Ph.	(*ye.*)	Fox-tail.	———	1713.	5. 6.	H.5.	Mich. arb. 1. t. 7.
20 rígida. Ph.	(*ye.*)	three-leaved.	———	1759.	——	H.5.	Lamb. pin. p. 25. t. 18. 19.
21 ponderòsa. D.	(*ye.*)	ponderous.	———	1827.	....	H.5.	
22 palústris. Ph.	(*ye.*)	swamp.	———	1730.	....	H.5.	Lamb. p. 27. t. 20.
23 longifòlia. W.	(*ye.*)	long-leaved.	E. Indies.	1801.	....	S.5.	——— p. 29. t. 21.
24 Gerárdi. Wal.	(*ye.*)	Gerard's.	Nepaul.	1824.	....	H.5.	
25 occidentàlis. W.	(*ye.*)	West Indian.	W. Indies.	1820.	....	S. ♄.	Plum. sp. 17. t. 161.
26 adúnca. Bosc.	(*ye.*)	crooked.	....	1822.	....	H.5.	
27 romàna. H. Bel.	(*ye.*)	Roman.	Italy.	——	....	H.5.	
28 sibírica. Dut.	(*ye.*)	Siberian.	Siberia.	——	....	H.5.	
STR'OBUS.		WEYMOUTH PINE.					
29 Cémbra. W.	(*ye.*)	Siberian stone.	Siberia.	1746.	5.	H.5.	Lamb. pin. p. 34. t. 23. 24.
30 pygm'æa. F.	(*ye.*)	pigmy.	———	1816.	....	H. ♄.	
31 Stròbus. W.	(*ye.*)	Weymouth.	N. America.	1705.	4.	H.5.	Lamb. pin. p. 31. t. 22.
32 Lambertiàna. L.P.	(*ye.*)	Lambert's.	California.	1827.	....	H.5.	——— pin. c. icon.
33 excélsa. D.P.	(*ye.*)	Bhotan.	Nepaul.	1822.	....	H.5.	——— v. 2. p. 5. t. 3.
C'EDRUS.		CEDAR.					
34 Cèdrus. W.	(*ye.*)	Cedar of Lebanon.	Levant.	1683.	5.	H.5.	Lamb. pin. v. 1. p. 59. t. 37
35 Deodàra. L.P.	(*ye.*)	Indian Cedar.	Nepaul.	1822.	....	H.5.	——— v. 2. p. 8. t. 4.
L'ARIX.		LARCH.					
36 péndula. W.	(*ye.re.*)	black Larch.	N. America.	1739.	5.	H.5.	Lamb. pin. v. 1. p. 56. t. 36
37 microcárpa. W.	(*ye.re.*)	red Larch.	———	1760.	——	H.5.	——— p. 58. t. 37.
38 Làrix. W.	(*ye.re.*)	common Larch.	Germany.	1629.	3. 4.	H.5.	——— p. 53. t. 35.
39 dahùrica. F.	(*ye.re.*)	Dahurian.	Dahuria.	1827.	....	H.5.	
40 intermèdia. F.	(*ye.re.*)	intermediate.	Altay.	1828.	....	H.5.	
PE'UCE.		SILVER FIR.					
41 canadénsis. W.	(*ye.*)	hemlock Spruce.	N. America.	1736.	5.	H.5.	Lamb. pin. p. 50. t. 32.
42 dumòsa. D.P.	(*ye.*)	bushy.	Nepaul.	1820.	....	H.5.	
43 taxifòlia. Ph.	(*ye.*)	Yew-leaved.	Columbia.	1822.	....	H.5.	Lamb. pin. v. 1. p. 51. t. 33
44 spectábilis. D.P.	(*ye.*)	purple-coned.	Nepaul.	1825.	....	H.5.	——— v. 2. p. 3. t. 2.
45 Pìcea. W.	(*ye.*)	Silver-Fir.	Germany.	1603.	5.	H.5.	——— 1. p. 36. t. 40.
46 Píchta. F.	(*ye.*)	Fischer's.	Altay.	1824.	....	H. ♄.	
47 Douglásii. D.	(*ye.*)	Douglas's.	N. America.	1827.	....	H.5.	
48 Balsamea. W.	(*ye.*)	Balm of Gilead.	———	1696.	5.	H.5.	Lamb. pin. v. 1. p. 48. t. 31
49 Fraséri. Ph.	(*ye.*)	double Balsam.	Pensylvania.	1811.	....	H.5.	
ABIES.		SPRUCE FIR.					
50 nígra. Ph.	(*ye.*)	black Spruce.	N. America.	1700.	5.	H.5.	Lamb. pin. v. 1. p. 41. t. 27
51 rùbra. Ph.	(*ye.*)	red Spruce.	———	1755.	——	H.5.	——— v. 1. p. 43. t. 28.
52 Clanbrassiliàna. L.en.	(*ye.*)	dwarf.	———	1810.	....	H. ♄.	
53 álba. Ph.	(*ye.*)	white Spruce.	———	1700.	5. 6.	H.5.	Lamb. pin. v. 1. p. 39. t. 26
54 'Abies. W.	(*ye.*)	Norway Spruce.	N. Europe.	1548.	4.	H.5.	——— v. 1. p. 37. t. 25.
55 orientàlis. W.	(*ye.*)	oriental.	Levant.	1824.	....	H.5.	——— v. 1. p. 45. t. 29.

Classis II.

Monocotyledoneæ seu Endogenæ. DC.

* Subclassis 1. *PHANEROGANEÆ.* DC.

ORDO CXCII.

CYCADEÆ. *Brown prodr.* 346.

ZAMIA. W.		ZAMIA. Diœcia Polyandria. W.					
1 púngens. w.	(*br.*)	needle.	C. B. S.	1775.	6. 8.	G. ♄.	Till. pis. 129. t. 45.
2 cycadifòlia. w.	(*br.*)	Cycas-leaved.	——	——	——	G. ♄.	Jacq. frag. 1. t. 25. 26.
3 tridentàta. w.	(*br.*)	three-toothed.	——	1820.	——	G. ♄.	
4 angustifòlia. Jac.	(*br.*)	narrow-leaved.	Bahama Isl.	....	——	S. ♄.	Jacq. ic. 3. t. 636.
5 ténuis. w.	(*br.*)	slender-leaved.	——	1823.	——	S. ♄.	
6 mèdia. w.	(*br.*)	intermediate.	W. Indies.	....	——	S. ♄.	Bot. mag. t. 1838.
7 débilis. w.	(*br.*)	weak.	——	1777.	——	S. ♄.	Comm. hort. 1. t. 58.
8 integrifòlia. w.	(*br.*)	dwarf.	——	1768.	——	S. ♄.	Bot. mag. t. 1851.
9 pygm'æa. B.M.	(*g.br.*)	least.	——	....	5.	S. ♃.	—— t. 1741.
10 furfuràcea. w.	(*br.*)	chaffy.	——	1691.	7. 8.	S. ♄.	Trew. ehret. t. 61.
11 spiràlis. B.P.	(*br.*)	spiral.	N. S. W.	1796.	——	G. ♃.	
12 longifòlia. w.	(*br.*)	long-leaved.	C. B. S.	1820.	6. 7.	G. ♄.	Jacq. frag. 1. t. 29.
13 lanuginòsa. w.	(*br.*)	woolly.	——	....	——	G. ♄.	—— 1. t. 30. 31.
14 Cycàdis. w.	(*br.*)	Cycas-like.	——	1775.	5. 6.	G. ♄.	Thunb. act. ups. 2. t. 6.
15 hórrida. w.	(*br.*)	glaucous spiny.	——	1800.	6. 8.	G. ♄.	Jacq. frag. t. 27. 28.
16 latifòlia. L.C.	(*br.*)	broad-leaved.	....	....	——	G. ♄.	
17 prunífera. L.C.	(*br.*)	Plum-like.	....	....	——	S. ♄.	
18 pùmila. L.C.	(*ye.*)	small.	....	1812.	——	S. ♃.	Bot. mag. t. 2006.
19 repánda. L.C.	(*br.*)	repand.	....	....	——	S. ♄.	
C'YCAS. W.		C'YCAS. Diœcia Polyandria. W.					
1 circinàlis. w.	(*br.*)	broad-leaved.	E. Indies.	1700.	5. 6.	S. ♄.	Bot. mag. t. 2826. 2827.
2 glaùca. L.en.	(*br.*)	glaucous-leaved.	——	1814.	....	S. ♄.	
3 revolùta. w.	(*br.*)	narrow-leaved.	China.	1737.	7. 8.	S. ♄.	Linn. trans. 6. t. 29.
4 squamòsa. L.C.	(*br.*)	scaly.	....	....	....	S. ♄.	

ORDO CXCIII.

HYDROCHARIDEÆ. *Brown prodr.* 344.

VALLISN'ERIA. L.		VALLISN'ERIA. Diœcia Diandria. L.					
spiràlis. L.	(*wh.*)	spiral.	S. Europe.	1818.	....	G. *w.* ♃.	Mich. n. g. 12. t. 10. f. 1. 2.
HYDRO'CHARIS. L.		FROG-BIT. Diœcia Enneandria. L.					
Mórsus-rànæ. L.	(*wh.*)	common.	Britain.	....	6. 7.	H. *w.* ♃.	Eng. bot. v. 12. t. 808.
DAMAS'ONIUM. W.		DAMAS'ONIUM. Hexandria Polygynia. W.					
1 índicum. w.	(*wh.*)	Indian.	E. Indies.	1800.	7. 9.	S. *w.* ♃.	Bot. mag. t. 1201.
2 ovalifòlium. B.P.	(*wh.*)	oval-leaved.	N. S. W.	1820.	——	G. *w.* ♃.	
STRATI'OTES. W.		WATER-SOLDIER. Diœcia Dodecandria. L.					
aloìdes. w.	(*wh.*)	Aloe-like.	England.	....	6. 7.	H. *w.* ♃.	Eng. bot. v. 6. t. 379.

ORDO CXCIV.

ALISMACEÆ. *Kth. synops.* 1. *p.* 261.

SAGITT'ARIA. W.		ARROW-HEAD.	Monœcia Polyandria. L.				
1 sagittifòlia. W.	(*wh.*)	common.	England.	....	6. 8.	H.*w*.♃.	Eng. bot. v. 2. t. 84.
2 latifòlia. Ph.	(*wh.*)	broad-leaved.	N. America.	1816.	——	H.*w*.♃.	
β flòre plèno.	(*wh.*)	*double-flowered.*	———	——	——	H.*w*.♃.	
3 sinénsis. B.M.	(*wh.*)	Chinese.	China.	1812.	9. 11.	G.*w*.♃.	Bot. mag. t. 1731.
4 obtusifòlia. S.S.	(*wh.*)	blunt-leaved.	———	1804.	7. 8.	G.*w*.♃.	Rheed. mal. 11. t. 45.
5 Doniàna.	(*wh.*)	Don's.	Nepaul.	1820.	——	G.*w*.♃.	
hastàta. D.P. *non* Ph.							
6 hastàta. Ph.	(*wh.*)	hastate-leaved.	N. America.	1816.	——	H.*w*.♃.	
7 obtùsa. Ph.	(*wh.*)	obtuse-leaved.	———	1822.	6. 9.	H.*w*.♃.	
8 heterophy'lla. Ph.	(*wh.*)	various-leaved.	———	——	6. 8.	H.*w*.♃.	
9 lancifòlia. W.	(*wh.*)	lance-leaved.	W. Indies.	1787.	6. 7.	G.*w*.♃.	Bot. mag. t. 1792.
10 angustifòlia. B.R.	(*wh.*)	narrow-leaved.	———	1827.	4. 8.	S.*w*.♃.	Bot. reg. t. 1141.
11 falcàta. Ph.	(*wh.*)	falcate-leaved.	Carolina.	1811.	6. 7.	H.*w*.♃.	
12 rígida. B.M.	(*wh.*)	brittle-leaved.	N. America.	1806.	——	H.*w*.♃.	Bot. mag. t. 1632.
13 gramínea. Ph.	(*wh.*)	grass-leaved.	Carolina.	1812.	7. 8.	H.*w*.♃.	
14 nàtans. M.	(*wh.*)	floating.	———	——	——	H.*w*.♃.	
ACTINOCA'RPUS. B.P.		ACTINOCA'RPUS.	Hexandria Hex-Octagynia.				
1 mìnor. B.P.	(*wh.*)	small.	N. S. W.	....	5. 8.	G.*w*.♃.	
2 Damasònium.	(*wh.*)	common.	England.	....	6. 8.	H.*w*.♃.	Eng. bot. t. 1615.
Alísma Damasònium. E.B. *Damasònium stellàtum.* P.S.							
ALI'SMA. B.P.		WATER-PLANTAIN.	Hexandria Polygynia. L.				
1 Plantàgo. W.	(*wh.*)	greater.	Britain.	....	6. 7.	H.*w*.♃.	Eng. bot. t. 837.
2 lanceolàta. With.	(*wh.*)	spear-leaved.	....	....	——	H.*w*.♃.	
3 triviàlis. Ph.	(*wh.*)	blunt-leaved.	N. America.	1816.	——	H.*w*.♃.	
4 parnassifòlia. W.	(*wh.*)	Parnassus-leav'd.	Italy.	1824.	——	H.*w*.♃.	
5 nàtans. E.B.	(*wh.*)	floating.	Wales.	....	7. 8.	H.*w*.♃.	Eng. bot. t. 775.
6 ranunculoídes. E.B.	(*w.*)	lesser.	Britain.	....	8.	H.*w*.♃.	——— t. 326.
7 répens. E.F.	(*wh.*)	creeping.	Wales.	....	9. 10.	H.*w*.♃.	Cavan. ic. 1. t. 55.

ORDO CXCV.

BUTOMEÆ. *Kth. synops.* 1. *p.* 260.

B'UTOMUS. L.		FLOWERING RUSH.	Enneandria Hexagynia. L.				
1 umbellàtus. L.	(*ro.*)	umbelled.	Britain.	....	6. 7.	H.*w*.♃.	Eng. bot. v. 10. t. 651.
2 latifòlius. D.P.	(*wh.*)	broad-leaved.	Nepaul.	1823.	——	H.*w*.♃.	
LIMNOCH'ARIS. K.S.		LIMNOCH'ARIS.	Polyandria Polygynia. B.M.				
Plumièri. K.S.	(*ye.*)	Plumier's.	S. America.	1822.	6. 10.	S.♁.	Bot. mag. t. 2525.
emarginàta. H. et B. pl. æq. 1. t. 34. *Alísma flàva.* L.							

ORDO CXCVI.

JUNCAGINEÆ. *Kth. synops.* 1. *p.* 258.

SCHEUCHZ'ERIA. L.		SCHEUCHZ'ERIA.	Hexandria Trigynia. L.				
palústris. E.B.	(*br.gr.*)	marsh.	England.	....	5. 6.	H.*w*.♃.	Eng. bot. v. 26. t. 1801.
TRIGLO'CHIN. L.		ARROW-GRASS.	Hexandria Trigynia. L.				
1 elàtum. N.	(*gr.*)	long-spiked.	N. America.	1820.	5. 8.	H.*w*.♃.	
2 marítimum. E.B.	(*gr.*)	sea-side.	Britain.	....	——	H.*w*.♃.	Eng. bot. v. 4. t. 255.
3 palústre. E.B.	(*gr.*)	marsh.	———	....	7. 8.	H.*w*.♃.	——— v. 6. t. 366.
4 decípiens. B.P.	(*gr.*)	New Holland.	N. S. W.	1820.	——	G.♃.	
5 Barrelièri. S.S.	(*gr.*)	Barrelier's.	Italy.	1825.	——	H.♃.	
6 bulbòsum. B.M.	(*pu.ye.*)	bulbous-rooted.	C. B. S.	1806.	10.	G.♃.	Bot. mag. t. 1445.

ORDO CXCVII.

FLUVIALES. *Kth. synops.* 1. *p.* 131.

POTAMOG'ETON. L.		Pond-weed.	Tetrandria Tetragynia. L.				
1 nàtans. E.B.	(*br.*)	broad-leaved.	Britain.	....	7. 8.	H.*w*.♃.	Eng. bot. v. 26. t. 1822.
2 heterophy'llum. E.B.	(*br.*)	various-leaved.	———	....	7. 9.	H.*w*.♃.	——— v. 18. t. 1285.
3 perfoliàtum. E.B.	(*br.*)	perfoliate.	———	....	7. 8.	H.*w*.♃.	——— v. 3. t. 168.
4 dénsum. E.B.	(*br.*)	close-leaved.	———	....	5. 7.	H.*w*.♃.	——— v. 6. t. 397.
5 flùitans. E.B.	(*br.*)	long-leaved.	———	....	7. 9.	H.*w*.♃.	——— v. 18. t. 1286.
6 lùcens. E.B.	(*br.*)	shining.	———	....	6. 8.	H.*w*.♃.	——— v. 6. t. 376.
7 lanceolàtum. E.B.	(*br.*)	spear-leaved.	England.	....	7. 8.	H.*w*.♃.	——— v. 28. t. 1985.
8 críspum. E.B.	(*br.*)	curled.	Britain.	....	6. 7.	H.*w*.♃.	——— v. 15. t. 1012.
9 compréssum. E.B.	(*br.*)	flat-stalked.	———	....	——	H.*w*.♃.	——— v. 6. t. 418.
10 cuspidàtum. E.F.	(*br.*)	sharp-pointed.	———	....	7.	H.*w*.♃.	Loes. pruss. t. 66.
11 gramíneum. E.B.	(*br.*)	Grass-leaved.	———	....	7. 8.	H.*w*.♃.	Eng. bot. v. 32. t. 2253.
12 pusíllum. E.B.	(*br.*)	small.	———	....	——	H.*w*.♃.	——— v. 3. t. 215.
13 pectinàtum. E.B.	(*br.*)	Fennel-leaved.	———	....	6. 7.	H.*w*.♃.	——— v. 5. t. 323.
RU'PPIA. L.		Ru'ppia.	Tetrandria Tetragynia. L.				
marítima. E.B.	(*br.*)	sea.	Britain.	....	8. 9.	H.*w*.♃.	Eng. bot. v. 2. t. 136.
APONOG'ETON. L.		Aponog'eton.	Hex-Dodecandria Tri-Tetragynia.				
1 monostàchyon. W.	(*wh.*)	simple-spiked.	E. Indies.	1803.	8. 10.	S.*w*.♃.	Andr. reposit. t. 406.
2 distàchyon. W.	(*wh.*)	broad-leaved.	C. B. S.	1788.	5. 7.	G.*w*.b.	Bot. mag. t. 1293.
3 angustifòlium. W.	(*wh.*)	narrow-leaved.	———	——	4. 9.	G.*w*.b.	——— t. 1268.
N'AJAS. L.		N'ajas.	Monœcia Tetrandria. L.				
monospérma. W.	(*gr.*)	single-seeded.	Europe.	1816.	7. 9.	H.*w*.⊙.	Mich. N. gen. 11. t. 8. f. 2.
ZANNICHE'LLIA. L.		Water-weed.	Monœcia Monandria. L.				
palústris. E.B.	(*gr.*)	horned.	Britain.	....	7. 8.	H.*w*.⊙.	Eng. bot. v. 26. t. 1844.
ZOST'ERA. W.		Sea Matweed.	Monandria Monogynia. L.				
marìna. W.	(*gr.*)	common.	Britain.	....	8. 9.	H.*w*.♃.	Eng. bot. v. 7. t. 467.
LE'MNA. L.		Duck-weed.	Monœcia Diandria.				
1 trisúlca. E.B.	(*wh.*)	Ivy-leaved.	Britain.	....	5. 6.	H.*w*.⊙.	Eng. bot. v. 13. t. 926.
2 mìnor. E.B.	(*wh.*)	lesser.	———	....	6. 7.	H.*w*.⊙.	——— v. 16. t. 1095.
3 gíbba. E.B.	(*wh.*)	thick-leaved.	———	....	——	H.*w*.⊙.	——— v. 18. t. 1233.
4 polyrrhìza. E.B.	(*wh.*)	greater.	———	....	7. 8.	H.*w*.⊙.	——— v. 35. t. 2458.

ORDO CXCVIII.

PANDANEÆ. *Brown prodr.* 340.

PAND'ANUS. L.		Screw-Pine.	Diœcia Monandria. L.				
1 l'ævis. Lour.	(*wh.*)	entire-leaved.	China.	1823.	....	S. ħ.	
2 odoratíssimus. W.	(*wh.*)	green-spined.	———	1771.	....	S. ♄.	Roxb. corom. 1. t. 94--6.
3 ùtilis. W. en.	(*wh.*)	red-spined.	———	....	....	S. ♄.	
4 hùmilis. W.	(*wh.*)	dwarf.	Mauritius.	....	....	S. ħ.	Jacq. fragm. t. 14. f. 2.
5 latifòlius. L.C.	(*wh.*)	broad-leaved.	E. Indies.	1820.	....	S. ħ.	
6 f'œtidus. H.B.	(*wh.*)	strong scented.	———	1822.	....	S. ♄.	
7 marginàtus. H.B.	(*wh.*)	red-margined.	Mauritius.	1823.	....	S. ♄.	
8 furcàtus. H.B.	(*wh.*)	forked.	E. Indies.	——	....	S. ♄.	Rheed. mal. 2. t. 8.
9 pedunculàtus. B.P.	(*wh.*)	peduncled.	N. Holland.	1825.	....	S. ♄.	
10 spiràlis. B.P.	(*wh.*)	spiral.	———	1805.	....	S. ♄.	
11 candelàbrum. P.B.	(*wh.*)	chandelier.	Guinea.	1822.	....	S. ħ.	Beauv. fl. owar. 1. t. 21.
12 séssilis. Boj.	(*wh.*)	sessile-fruited.	Pembu.	1825.	....	S. ♄.	
13 élegans. S.S.	(*wh.*)	elegant.	Isle France.	1826.	....	S. ♄.	
14 muriàtus. S.S.	(*wh.*)	oblong-leaved.	Madagascar.	——	....	S. ♄.	
15 èdulis. S.S.	(*wh.*)	eatable.	———	1825.	....	S. ħ.	
16 amaryllifòlius. H.B.	(*wh.*)	Amaryllis-leav'd.	E. Indies.	1820.	....	S. ♄.	
17 inérmis. H.B.	(*wh.*)	spineless.	———	——	....	S. ħ.	
18 álbus. L.C.	(*wh.*)	white-spined.	———	1823.	....	S. ♄.	
19 refléxus. L.C.	(*wh.*)	reflexed-leaved.	———	1820.	....	S. ♄.	
20 longifòlius. L.C.	(*wh.*)	long-leaved.	———	1823.	....	S. ħ.	
21 turbinàtus. L.C.	(*wh.*)	top-shaped.	———	——	....	S. ♄.	
PHYTE'LEPHAS. R.P.		Phyte'lephas.	Diœcia Polyandria. P.S.				
macrocárpa. R.P.	(*wh.*)	large-fruited.	Peru.	1822.	....	S. ħ.	

ORDO CXCIX.

TYPHINÆ. *Kth. synops.* 1. *p.* 132.

TYPHA. L.		CAT'S-TAIL.	Monœcia Triandria. L.				
1 latifòlia. E.B.	(*br.*)	great.	Britain.	....	7.	H.*w.*♃.	Eng. bot. v. 21. t. 1455.
2 mèdia. DC.	(*br.*)	dwarf.	England.	....	——	H.*w.*♃.	——— v. 21. t. 1457.
mìnor. E.B.							
3 angustifòlia. E.B.	(*br.*)	lesser.	Britain.	....	——	H.*w.*♃.	Eng. bot. v. 21. t. 1456.
4 mínima. W.	(*br.*)	least.	Europe.	1819.	——	H.*w.*♃.	
5 dænática. E.	(*br.*)	Egyptian.	Egypt.	1825.	——	H.*w.*♃.	
SPARG'ANIUM. L.		BUR-REED.	Monœcia Triandria. L.				
1 ramòsum. E.B.	(*wh.*)	branched.	Britain.	....	7. 8.	H.*w.*♃.	Eng. bot. v. 11. t. 744.
2 símplex. E.B.	(*wh.*)	unbranched.	———	....	——	H.*w.*♃.	——— v. 11. t. 745.
3 americànum. N.	(*wh.*)	American.	N. America.	1823.	——	H.*w.*♃.	
4 nàtans. E.B.	(*wh.*)	floating.	England.	....	7.	H.*w.*♃.	Eng. bot. v. 4. t. 273.

ORDO CC.

AROIDEÆ. *Juss. gen.* 23.

Sect. I. *ORONTIACEÆ.* Brown prodr. 1. p. 337.

P'OTHOS. L.		P'OTHOS.	Tetrandria Monogynia. L.				
1 scándens. R.S.	(*gr.*)	climbing.	E. Indies.	1818.	4. 6.	S.♄.◡.	Rheed. mal. 7. t. 40.
2 officinàlis. F.I.	(*st.*)	officinal.	———	1820.	4. 7.	S.♄.◡.	
3 Peèpla. F.I.	(*st.*)	Peeplee.	———	——	——	S.♄.◡.	
4 gigántea. F.I.	(*pu.ye.*)	gigantic.	———	1824.	....	S.♄.◡.	
5 pertùsa. F.I.	(*gr.*)	perforated.	———	——	....	S.♄.◡.	Rheed. mal. 12. t. 20. 21.
6 lanceolàta. R.S.	(*vi.*)	lance-leaved.	Barbadoes.	1790.	4. 7.	S.♃.	Plum. amer. 47. t. 62.
7 acaùlis. R.S.	(*gr.*)	stemless.	W. Indies.	——	——	S.♃.	Lodd. bot. cab. t. 483.
8 grácilis. R.S.	(*gr.*)	slender.	Trinidad.	1826.	——	S.♃.	Rudg. pl. gui. 1. t. 32.
9 violàcea. R.S.	(*g.wh.*)	blue-fruited.	Jamaica.	1793.	4. 6.	S.♃.	Hook. exot. flor. t. 55.
10 cannæfòlia. R.S.	(*wh.*)	sweet-scented.	W. Indies.	1785.	4. 5.	S.♃.	Bot. mag. t. 603.
11 crassinérvia. R.S.	(*gr.*)	thick-nerved.	S. America.	1796.	4. 6.	S.♃.	Jacq. ic. 3. t. 609.
12 cordàta. R.S.	(*gr.*)	heart-leaved.	———	1770.	4.	S.♃.	Plum. ic. 26. t. 38.
13 sagittàta. B.M.	(*g.br.*)	arrow-leaved.	W. Indies.	1800.	8.	S.♃.	Bot mag. t. 1584.
14 macrophy'lla. R.S.	(*gr.*)	large-leaved.	———	1794.	5. 6.	S.♃.	Jacq. ic. 3. t. 610.
15 obtusifòlia. R.S.	(*g.br.*)	blunt-leaved.	Barbadoes.	1790.	——	S.♃.	
16 coriàcea. H.E.F.	(*g.br.*)	leathery.	Brazil.	1824.	6.	S.♃.	Hook. ex. flor. t. 210.
17 Harrísii. H.E.F.	(*g.br.*)	Harris's.	———	——	——	S.♃.	—— ——— t. 211.
18 caudàta. F.I.	(*gr.*)	tailed.	E. Indies.	1824.	....	S.♄.◡.	
19 pinnàta. W.	(*gr.*)	wing-leaved.	———	1820.	....	S.♄.◡.	Rumph. am. 5. t. 183. f. 2.
20 decúrsiva. F.I.	(*gr.*)	pinnatifid-leaved.	———	1824.	....	S.♄.◡.	
21 pinnatífida. F.I.	(*gr.*)	pinnatifid.	———	1825.	....	S.♄.◡.	
22 heterophy'lla. F.I.	(*pu.*)	various-leaved.	———	1822.	....	S.♄.◡.	
23 Làsia. F.I.	(*re.*)	prickly.	———	1819.	....	S.♃.	
24 digitàta. R.S.	(*pu.*)	fingered.	Caracas.	1823.	....	S.♄.	Jacq. ic. 3. t. 611.
25 palmàta. R.S.	(*g.pu.*)	palmated.	S. America.	1803.	6. 7.	S.♃.	Plum. amer. t. 64. 65.
26 pentaphy'lla. R.S.	(*bl.*)	five-leaved.	Cayenne.	——	10. 11.	S.♄.	Bot. mag. t. 1375.
27 rubrinérvia. L.en.	(*pu.*)	red-nerved.	S. America.	1820.	....	S.♃.	
28 microphy'lla. B.M.	(*br.*)	small-leaved.	Brazil.	——	8. 10.	S.♃.	Bot. mag. t. 2953.
SYMPLOCA'RPUS. N.		SCUNKWEED.	Tetrandria Monogynia. N.				
f'œtidus. N.	(*d.pu.*)	fetid.	N. America.	1735.	3. 5.	H.♃.	Swt. br. fl. gar. t. 57.
Pòthos f'œtida B.M. 836. *Dracóntium f'œtidum.* W.							
DRACO'NTIUM. W.		DRAGON.	Heptandria Monogynia. L.				
1 polyphy'llum. W.	(*d.pu.*)	purple-stalked.	India.	1759.	3. 6.	S.♃.	Bot. reg. t. 700.
2 spinòsum. W.	(*d.pu.*)	prickly.	Ceylon.	——	4. 5.	S.♃.	
GYMNOST'ACHYS. B.P.		GYMNOST'ACHYS.	Tetrandria Monogynia.				
ancéps. B.P.	(*gr.*)	flat-stemmed.	N. S. W.	1820.	6. 8.	G.♃.	
HOUTTU'YNIA. J.		HOUTTU'YNIA.	Polyandria Polygynia. P.S.				
cordàta. P.S.	(*wh.*)	heart-leaved.	China.	1820.	6. 8.	G.♃.	Bot. mag. t. 2731.

CARLUDO'VICA. R.P. CARLUDO'VICA. Monœcia Polyandria. P.S.
1 palmàta. R.S. (*br.*) palmate. S. America. 1818. .... S.♄.
Sálmia palmàta. R.S. *Ludòvia palmàta.* P.S.
2 latifòlia. R.P. (*br.*) broad-leaved. Peru. 1819. 4. 6. S.♄. Bot.mag.t.2950.et2951.
3 angustifòlia. R.P. (*br.*) narrow-leaved. ——— —— —— S.♄.
4 palmæfòlia. R.S. (*br.*) Palm-leaved. St. Domingo. 1826. .... S.♃. Plum. ic. t. 39.
Sálmia palmæfòlia. W.
5 jamaicénsis. L.C. (*br.*) Jamaica. Jamaica. 1823. .... S.♄.
CYCLA'NTHUS. S.S. CYCLA'NTHUS. Monœcia Monadelphia.
1 Plumíeri. S.S. (*gr.*) Plumier's. W. Indies. 1820. .... S.♃.
2 bipartìtus. S.S. (*gr.*) cleft-leaved. Trinidad. —— .... S.♃.
A'CORUS. L. SWEET-FLAG. Hexandria Monogynia. L.
1 Cálamus. S.S. (*gr.*) common. Britain. .... 6. 7. H.*w.*♃. Eng. bot. v. 5. t. 356.
2 terréstris. S.S. (*gr.*) Chinese. China. 1822. —— H.*w.*♃.
3 gramíneus. S.S. (*gr.*) grass-leaved. ——— 1786. 2. H.*w.*♃. Smith spic. 15. t. 17.
ORO'NTIUM. L. ORO'NTIUM. Hexandria Monogynia. L.
aquáticum. S.S. (*st.*) aquatic. N. America. 1775. 6. H.*w.*♃. Hook. ex. flor. t. 19.
RO'HDEA. S.S. RO'HDEA. Hexandria Monogynia. S.S.
japónica. S.S. (*ye.*) Japan. Japan. 1783. 1. 4. H.♃. Bot. mag. t. 898.
Oróntium japónicum. B.M.
TUPI'STRA. B.M. TUPI'STRA. Hexandria Monogynia. S.S.
1 squálida. B.M. (*br.*) brown-flowered. E. Indies. 1810. 4. 6. S.♃. Bot. reg. t. 704.
2 nùtans. B.R. (*br.*) nodding-flowered. ——— 1818. 11. 12. S.♃. ——— t. 1223.
ASPIDI'STRA. B.R. ASPIDI'STRA. Octandria Monogynia. S.S.
lùrida. B.R. (*br.pu.*) brown-flowered. China. 1820. 2. 4. S.♃. Bot. reg. t. 628.
punctàta. B.R. 977.

Sect. II. *AROIDEÆ VERÆ.* Brown prodr. 335.

CA'LLA. K.S. CA'LLA. Heptandria Monogynia. L.
1 palústris. W. (*wh.*) marsh. N. Europe. 1768. 7. 8. H.*w.*♃. Bot. mag. t. 1831.
2 pertùsa. K.S. (*st.*) perforated. W. Indies. 1752. 4. 6. S.♃. Jacq. schœn. 2. t. 184. 5.
Dracóntium pertùsum. Jacq.
3 aromática. B.M. (*wh.*) aromatic. E. Indies. 1817. —— S.♃. Bot. mag. t. 2279.
4 occúlta. B.C. (*wh.*) hidden-spiked. China. —— 5. 7. G.♃. Lodd. bot. cab. t. 12.
RICHA'RDIA. Kth. RICHA'RDIA. Heptandria Monogynia.
æthiòpica. Kth. (*wh.*) Ethiopian. C. B. S. 1731. 1. 5. G.♃. Bot. mag. t. 832.
Cálla æthiòpica. B.M.
'ARUM. W. ARUM. Monœcia Polyandria. L.
1 crinìtum. W. (*d.pu.*) hairy-sheathed. Minorca. 1777. 3. 4. F.♃. Bot. reg. t. 831.
2 Dracúnculus. W. (*pu.*) common Dragon. S. Europe. 1548. 6. 7. H.♃. Moris. s. 13. t. 5. f. 46.
3 Dracóntium. W. (*gr.*) green Dragon. N. America. 1759. 6. H.♃. Bot. reg. t. 668.
4 bulbíferum. B.M. (*st.br.*) bulb-bearing. E. Indies. 1813. 3. 6. S.♃. Bot. mag. t. 2508.
5 venòsum. W. (*ye.pu.*) purple-flowered .... 1774. 3. S.♃. Bot. reg. t. 1017.
6 campanulàtum. H.B. (*pu.*) bell-flowered. E. Indies. 1816. 4. 6. S.♃.
7 triphy'llum. W. (*std.*) three-leaved. N. America. 1664. 5. 6. H.♃. Lodd. bot. cab. t. 820.
β *Zebrìnum.* B.M. (*std.*) *striped.* ——— —— —— H.♃. Bot. mag. t. 950.
8 pedàtum. W. en. (*bl.ye.*) pedate-leaved. S. America. 1815. —— S.♃. Pl. sel. hort. ber. t. 8.
9 polyphy'llum. L. en. (*pu.*) many-leaved. .... 1818. —— S.♃.
10 ríngens. W. (*pu.*) gaping. Japan. 1822. —— G.♃.
11 atrorùbens. W. (*d.re.*) purple-stalked. N. America. 1758. 6. 7. H.♃. Pluk. alm. t. 77. f. 5.
12 ternàtum. W. (*pu.*) Japan. Japan. 1774. 5. 7. F.♃.
13 Colocàsia. W. (*br.*) Egyptian. Levant. 1551. 6. 8. G.♃. Rumph. amb. 5. t. 109.
14 orixénse. B.P. (*re.*) Orixian. E. Indies. 1802. 8. 10. S.♃. Bot. reg. t. 450.
15 macrorhìzon. W. (*wh.*) long-rooted. ——— 1803. .... S.♃. Herm. parad. t. 73.
16 píctum. W. (*d.pu.*) white-veined. S. Europe. 1820. 5. 6. H.♃.
17 divaricàtum. W. (*gr.*) divaricated. E. Indies. 1759. 6. 7. S.♃. Rheed. mal. 11. t. 20.
18 trilobàtum. W. (*pu.*) three-lobed. Ceylon. 1714. 5. 6. S.♃. Bot. mag. t. 339.
β *auriculàtum.* B.M. (*pu.*) *eared.* ——— .... —— S.♃. ——— t. 2324.
19 maculàtum. W. (*gr.*) common. Britain. .... 5. 7. H.♃. Eng. bot. v. 19. t. 1298.
20 orientàle. M.B. (*br.*) oriental. Caucasus. 1818. —— H.♃.
21 itálicum. W. (*gr.*) Italian. Italy. 1683. 5. 6. H.♃. Bot. mag. t. 2432.
22 vivìparum. L.C. (*gr.*) viviparous. .... 1817. —— S.♃. Lodd. bot. cab. t. 281.
23 minùtum. W. (*re.*) small. E. Indies. 1812. 5. 7. S.♃. Rheed. mal. 11. t. 17.
24 proboscídeum. W. (*w.pu.* proboscis. Apennines. 1825. —— H.♃. Bocc. mus. 2. t. 50.
25 Arisàrum. W. (*gr.*) Friar's Cowl. S. Europe. 1596. 4. 6. F.♃. Jacq. schœn. 2. t. 192.
26 tenuifòlium. W. (*d.pu.*) grass-leaved. ——— 1570. —— H.♃. Bot. reg. t. 512.

27 spiràle. w. (*pu.*) spiral. E. Indies. 1816. 4. 6. S.♃. Bot. mag. t. 2220.
28 gramíneum. s.s. (*pu.*) grass-leaved. S. Europe. 1824. —— H.♃.
29 flagellifórme. B.C. (*pu.*) whip-cord. E. Indies. —— —— S.♃. Lodd. bot. cab. t. 396.
30 sagittifòlium. L.en. .. arrow-leaved. .... —— .... S.♃.
31 hederàceum. w. (*pu.*) Ivy-leaved. W. Indies. 1793. 5. 6. S.♄. Jacq. amer. t. 152.
32 obtusilòbum. L.en. (*pu.*) blunt-lobed. .... 1816. —— S.♄.
33 índicum. Lour. (*pu.*) Indian. China. —— .... S.♄.
34 ramòsum. L.en. (*pu.*) branching. .... —— .... S.♄.
35 lingulàtum. w. (*ye.*) tongue-leaved. W. Indies. 1793. —— S.♄. Plum. ic. 26. t. 37.
36 integrifòlium. L.en. (*gr.*) entire-leaved. .... 1820. .... S.♄.

CAL'ADIUM. V. CAL'ADIUM. Monœcia Polyandria. W.

1 helleborifòlium. w. (*wh.*) Hellebore-leaved. Caracas. 1796. 6. 7. S.♃. Jacq. ic. 3. t. 613.
2 pinnatífidum. w. (*pu.*) pinnatifid. ———— 1817. —— S.♃. Jacq. schœn. 2. t. 187.
3 bícolor. w. (*wh.*) two-coloured. Brazil. 1773. —— S.♃. Bot. mag. t. 2543.
4 nymphæifòlium. w. (*sn.*) Water-lily-leaved. E. Indies. 1800. .... S.♃. Rheed. mal. 11. t. 22.
5 esculéntum. w. (*g.pu.*) esculent. America. 1739. 6. 8. S.♃. Sloan. jam. 1. t. 106. f. 1.
6 odòrum. B.R. (*gr.*) odoriferous. E. Indies. 1818. 5. 7. S.♃. Bot. reg. t. 641.
7 vivìparum. H.P. (*wh.*) viviparous. ———— 1816. —— S.♃. Lodd. bot. cab. t. 281.
8 sagittifòlium. w. (*wh.*) arrow-leaved. W. Indies. 1710. —— S.♃. Jacq. vind. 2. t. 157.
9 virgínicum. H.E.F. (*st.*) Virginian. N. America. 1759. 6. 7. H.♃. Hook. ex. flor. t. 182.
Arum virgínicum. w.
10 seguìnum. w. (*gr.*) Dumb Cane. America. —— 5. S.♄. Hook. ex. flor. t. 1.
11 maculàtum. B.C. (*gr.*) spotted. Trinidad. 1820. 4. 12. S.♄. Lodd. bot. cab. t. 608.
seguìnum β maculàtum. Bot. mag. t. 2606.
12 scándens. s.s. (*wh.*) climbing. Guinea. 1822. .... S.♄. Beauv. fl. owar. t. 3.
13 cucullàtum. P.S. (*gr.*) hooded. China. 1816. .... S.♄.
14 xanthorhìzum. w. (*st.*) yellow-rooted. .... —— 4. 8. S.♄. Jacq. schœn. 2. t. 188.
15 grandifòlium. w. (*wh.*) large-leaved. Caracas. 1803. 3. 7. S.♄. Bot. mag. t. 2643.
16 arboréscens. w. (*wh.*) tree. W. Indies. 1759. 6. 7. S.♄. Plum. amer. 44. t. 60.
17 lácerum. w. (*gr.*) torn-leaved. Caracas. 1822. .... S.♄.
18 tripartìtum. w. (*wh.*) ternate-leaved. ———— 1816. .... S.♄. Jacq. schœn. 2. t. 190.
19 aurìtum. (*wh.*) ear-leaved. S. America. 1739. 6. 8. S.♄. ———— 2. t. 191.
20 pedàtum. H.E.F. (*ye.*) pedate-leaved. Brazil. 1824. —— S.♄. Hook. ex. flor. t. 206.
21 aculeàtum. s.s. (*wh.*) prickly. Surinam. 1822. .... S.♄.◡.

AMBROSI'NIA. W. AMBROSI'NIA. Monœcia Monandria. W.

1 Bássii. w. (*gr.pu.*) flat oval-leaved. S. Europe. 1823. 6. 8. H.♃. Lam. ill. t. 737.
2 ciliàris. s.s. (*gr.pu.*) fringed-spathed. E. Indies. 1824. .... S.♃.

Sect. III. *TACCACEÆ.*

TA'CCA. W. TA'CCA. Hexandria Monogynia. L.

1 pinnatífida. w. (*pu.*) Salep. E. Indies. 1793. 6. 8. S.♃. Lodd. bot. cab. t. 692.
2 áspera. H.B. (*pu.*) rough-stemmed. ———— 1820. .... S.♃.
3 l'ævis. H.B. (*pu.*) smooth. ———— —— .... S.♃.
4 integrifòlia. B.M. (*pu.*) entire-leaved. ———— 1810. 5. 7. S.♃. Bot. mag. t. 1488.

ORDO CCI.

PIPERACEÆ. *Kth. synops.* 1. *p.* 103.

SAUR'URUS. W. LIZARD'S-TAIL. Heptandria Tetragynia. L.

1 cérnuus. w. (*wh.*) drooping. Virginia. 1759. 9. H.*w.*♃. Lam. ill. t. 1.
2 lùcidus. s.s. (*wh.*) shining. N. America. 1791. 8. 10. H.*w.*♃. Jacq. ecl. t. 18.
3 chinénsis. Hort. (*wh.*) Chinese. China. 1819. —— H.*w.*♃.

PIPER. R.P. PEPPER. Diandria Trigynia. L.

1 coriàceum. R.S. (*gr.*) coriaceous. E. Indies. 1815. 7. 8. S.♄. Lodd. bot. cab. t. 128.
2 nítidum. R.S. (*wh.*) glossy. Jamaica. 1793. 5. 6. S.♄.
3 peltàtum. R.S. (*gr.*) peltated. W. Indies. 1748. .... S.♃. Plum. amer. t. 74.
4 umbellàtum. R.S. (*gr.*) umbelled. ———— —— 5. 7. S.♄. Flor. peruv. 1. t. 59. f. a.
5 macrophy'llum. R.S. (*gr.*) broad-leaved. ———— 1800. .... S.♄. Sloan. hist. 1. t. 88. f. 1.
6 geniculàtum. R.S. (*gr.*) jointed. ———— 1826. .... S.♄.
7 adúncum. R.S. (*gr.*) hooked. Jamaica. 1748. 5. S.♄. Jacq. ic. 2. t. 210.

8 hirsùtum. R.S.	(*br.*) hairy-leaved.	Jamaica.	1793.	7.	S. ♄.	
9 híspidum. R.S.	(*gr.*) hispid.	S. America.	——	——	S. ♄.	
10 tuberculàtum. R.S.	(*gr.*) tuberculate.	——	1816.	....	S. ♄.	Jacq. ic. 2. t. 211.
11 ovàtum. R.S.	(*gr.*) oval-leaved.	Trinidad.	1824.	....	S. ♄.	
12 Amalàgo. R.S.	(*gr.*) rough-leaved.	Jamaica.	1759.	....	S. ♄.	Sloan. hist. 1. t. 87. f. 1.
13 Bètle. R.S.	(*gr.*) Betle.	E. Indies.	1804.	....	S. ♄.	Rheed. mal. 7. t. 15.
14 nìgrum. R.S.	(*gr.*) black.	——	1790.	....	S. ♄.	—— 7. t. 12.
15 trioícum. F.I.	(*gr.*) glaucous-leaved.	——	1818.	....	S. ♄.	
16 reticulàtum. R.S.	(*gr.*) netted.	W. Indies.	1748.	8.	S. ♄.	Plum. amer. t. 75.
17 decumànum. R.S.	(*gr.*) great.	Carthagena.	1768.	....	S. ♄.	Jacq. ic. 2. t. 215.
18 Siribòa. R.S.	(*gr.*) Siriboa.	E. Indies.	——	....	S. ♄.	Rumph. am. 5. t. 117. f. 2.
19 lóngum. R.S.	(*gr.*) long.	——	1788.	....	S. ♄.	Rheed. mal. 7. t. 14.
20 glábrum. H.S.S.	(*gr.*) smooth.	Campeachy.	1768.	....	S. ♄.	
21 laurifòlium. H.S.S.	(*gr.*) Laurel-leaved.	Jamaica.	——	....	S. ♄.	
22 racemòsum. H.S.S.	(*gr.*) racemed.	Campeachy.	——	....	S. ♄.	
Stáphyle. R.S.						
23 acutifòlium. R.S.	(*gr.*) acute-leaved.	Peru.	1823.	....	S. ♄.	Flor. per. 1. t. 64. f. a.
24 colubrìnum. L.en.	(*gr.*) spotted-stalked.	Brazil.	1820.	....	S. ♄.	
25 plantagíneum. R.S.	(*gr.*) Plantain-leaved.	W. Indies.	——	....	S. ♄.	Jacq. ic. 1. t. 8.
mèdium. Jacq.						
26 marginàtum. L.en.	(*gr.*) purple-edged.	S. America.	1811.	....	S. ♄.	Jacq. ic. 2. t. 215.
27 unguiculàtum. R.S.	(*gr.*) claw-pointed.	Peru.	1822.	....	S. ♄.	Flor. per. 1. t. 57. f. b.
glaucéscens. Jacq. ecl. 1. t. 76.						
28 hùmile. Mill.	(*gr.*) humble.	Jamaica.	1768.	....	S. ♄.	
29 tomentòsum. H.S.S.	(*gr.*) oval-leaved downy.	Vera Cruz.	——	....	S. ♄.	
30 sidæfòlium. L.en.	(*gr.*) Sida-leaved.	S. America.	1821.	5. 7.	S. ♄.	Lk. et Ot. ic. sel. 1. t. 6.
PEPER'OMIA. R.P.	PEPER'OMIA.	Diandria Trigynia.				
1 brachyphy'lla.	(*gr.*) short-leaved.	....	1818.	....	S. ♄.	
Pìper brachyphy'llum. R.S.						
2 amplexicaùlis. R.S.	(*gr.*) stem-clasping.	W. Indies.	1793.	6. 9.	S. ♃.	
3 magnoliæfòlia. R.S.	(*gr.*) Magnolia-leaved.	——	——	1. 3.	S. ♃.	Jacq. ic. 2. t. 213.
4 obtusifòlia. R.S.	(*gr.*) blunt-leaved.	S. America.	1739.	4. 7.	S. ♃.	Trew ehret. 54. t. 96.
5 clusiæfòlia. B.M.	(*gr.*) Clusia-leaved.	——	1739.	4. 7.	S. ħ.	Bot. mag. t. 2943.
6 cuneifòlia. R.S.	(*gr.*) wedge-leaved.	Caracas.	1809.	——	S. ♃.	Jacq. ic. 2. t. 214.
7 alàta. R.P.	(*gr.*) winged.	S. America.	1812.	....	S. ħ.	Flor. per. 1. t. 48. f. b.
8 acuminàta. R.S.	(*gr.*) acuminate.	W. Indies.	——	4. 9.	S. ♃.	Bot. mag. t. 1882.
9 maculòsa. R.S.	(*gr.*) spot-stalked.	St. Domingo.	1790.	9.	S. ħ.	Hook. ex. flor. t. 92.
10 pellùcida. K.S.	(*gr.*) pellucid.	S. America.	1748.	4. 9.	S. ⊙.	Plum. amer. t. 72.
11 concínna. K.R.	(*gr.*) small heart-leaved.	Jamaica.	1821.	——	S. ħ.	
12 rotundifòlia. K.S.	(*gr.*) round-leaved.	S. America.	1823.	....	S. ♃.	Plum. amer. t. 69.
13 hispídula. R.S.	(*gr.*) annual hispid.	Jamaica.	1818.	8. 10.	S. ⊙.	Swartz icon. t. 4.
14 nummularifòlia. K.S.	(*gr.*) money-wort-leaved.	——	——	4. 9.	S. ♃.	
15 pubéscens. R.S.	(*gr.*) pubescent.	....	——	——	S. ♃.	
16 subrotúnda. H.S.S.	(*gr.*) small Clusia-leav'd.	......	1811.	——	S. ♃.	
17 rèpens. K.S.	(*gr.*) creeping.	S. America.	1823.	——	S. ♃.	
18 sérpens. R.S.	(*gr.*) trailing.	Jamaica.	——	——	S. ♃.	
19 incàna. H.E.F.	(*gr.*) great downy.	Brazil.	1814.	2. 6.	S. ♄.	Hook. ex. flor. t. 66.
20 amplexifòlia. L.en.	(*gr.*) clasping-leaved.	Jamaica.	1822.	——	S. ħ.	
21 talinifòlia. K.S.	(*gr.*) Talinum-leaved.	S. America.	——	4. 8.	S. ħ.	H. et B. n. gen. 1. t. 8.
22 pulchélla. R.S.	(*gr.*) small-leaved.	Jamaica.	1778.	7. 10.	S. ♃.	Lodd. bot. cab. t. 574.
23 pereskiæfòlia. K.S.	(*gr.*) Pereskia-leaved.	S. America.	1820.	5. 6.	S. ♃.	Hook. ex. flor. t. 67.
24 blánda. K.S.	(*gr.*) villous.	Caracas.	1802.	5. 11.	S. ♃.	—— t. 21.
25 verticillàta. R.S.	(*gr.*) whorl-leaved.	Jamaica.	1816.	6. 10.	S. ⊙.	
26 polystàchia. H.E.F.	(*gr.*) many-spiked.	W. Indies.	1775.	6. 7.	S. ♃.	Hook. ex. flor. t. 23.
27 quadrifòlia. H.E.F.	(*gr.*) four-leaved.	Santa Cruz.	1818.	5. 10.	S. ♃.	—— t. 22.
28 Hawórthii.	(*gr.*) Haworth's.	India.	1809.	9. 10.	S. ♃.	
Pìper pubéscens. H.S. *non* Vahl.						
29 inæqualifòlia. R.P.	(*gr.*) unequal-leaved.	Peru.	1800.	8. 10.	S. ♃.	Flor. peruv. 1. t. 46. f. a.
30 stellàta. R.S.	(*gr.*) starry.	Jamaica.	1802.	5. 7.	S. ♃.	Jacq. vind. 2. t. 217.
31 rubricaùle. L.en.	(*gr.*) red-stalked.	W. Indies.	1815.	——	S. ♃.	Nees. hort. ber. t. 8.
32 rubélla. H.E.F.	(*gr.*) red-leaved.	Jamaica.	1820.	——	S. ♃.	Hook. ex. flor. t. 58.
33 trifòlia. R.S.	(*gr.*) three-leaved.	S. America.	1802.	6. 8.	S. ♃.	Plum. amer. t. 68.
34 distàchya. R.S.	(*gr.*) two-rowed.	W. Indies.	1793.	——	S. ♃.	—— t. 67.
35 díscolor. R.S.	(*gr.*) two-coloured.	Jamaica.	1821.	7. 8.	S. ♄.	Lodd. bot. cab. t. 610.
36 capénsis. R.S.	(*gr.*) Cape.	C. B. S.	1816.	——	G. ħ.	
37 tricarinàta. H.S.S.	(*gr.*) triple-keeled.	....	1817.	——	S. ħ.	

ORDO CCII.

ORCHIDEÆ. *Juss. gen.* 64.

O'RCHIS. B.P. O'RCHIS. Gynandria Monandria. H.K.
1 Mòrio. W. (*pu.*) meadow. Britain. .... 4. 6. H.b.♃. Eng. bot. v. 29. t. 2059.
2 longicórnu. W. (*pu.*) flat-spurred. Barbary. 1815. —— H.b.♃. Swt. br. fl. gar. t. 249.
3 pállens. W. (*st.*) pale. Switzerland. 1825. 5. 6. H.b.♃. Jacq. aust. t. 45.
4 máscula. W. (*pu.*) early purple. Britain. .... 4. 5. H.b.♃. Eng. bot. v. 9. t. 631.
5 palústris. W. (*li.*) marsh. Europe. 1824. —— H.b.♃. Jacq. ic. t. 181.
6 pauciflòra. T.N. (*ye.*) few-flowered. Italy. 1825. 6. 7. H.b.♃. Ten. fl. nap. t. 88.
7 Nicodèmi. T.N. (*pu.*) Nicodemus's. —— —— —— H.b.♃. —— t. 90.
8 Cyrílli. T.N. (*ye.*) Cyrillo's. Naples. —— —— H.b.♃. —— t. 87.
9 parviflòra. T.N. (*pu.*) small-flowered. Italy. —— —— H.b.♃.
10 ustulàta. W. (*pu.*) dwarf. England .... 5. 6. H.b.♃. Eng. bot. v. 1. t. 18.
11 militàris. W. (*pu.*) military. Switzerland. 1825. 4. 6. H.b.♃. Swt. br. fl. gar. t. 163.
12 Rivìni. B.F.G. (*li.pu.*) Rivinus's. —— —— 5. 6. H.b.♃. —— t. 162.
13 Smíthii. B.F.G. (*ro.pa.*) Smith's. England. .... —— H.b.♃. Eng. bot. v. 27. t. 1873.
militàris. E.B. *non* W.
14 fúsca. W. (*br.re.*) brown. —— .... —— H.b.♃. Jacq. aust. 4. t. 307.
15 Sìmia. P.S. (*pa.bh.*) Ape. —— .... —— H.b.♃.
16 undulatifòlia. B.B. (*li.*) wave-leaved. Italy. 1818. —— H.b.♃. Bot. reg. t. 375.
17 tephrosánthos. W. (*pa.*) ash-coloured. England. .... —— H.b.♃. Vaill. bot. t. 31. f. 25-26.
18 acuminàta. W. (*li.wh.*) pointed-flower'd. S. Europe. 1815. 4. 6. H.b.♃. Bot. mag. t. 1932.
19 variegàta. W. (*li.wh.*) variegated. Switzerland. 1825. —— H.b.♃. Hall. helv. t. 30.
20 globòsa. W. (*pu.*) round-spiked. —— 1792. 6. 7. H.b.♃. Jacq. aust. 3. t. 265.
21 corióphora. W. (*pu.*) strong-scented. S. Europe. 1825. 5. 6. H.b.♃. Swt. br. fl. gar. t. 219.
22 hircìna. W. (*gr.pu.*) Lizard. England. .... 6. 7. H.b.♃. Eng. bot. v. 1. t. 34.
Saty'rium hircìnum. E.B.
23 Robertiàna. P.S. (*pu.*) long-bracted. S. Europe. 1818. 5. 6. H.b.♃. Ten. fl. nap. t. 91.
24 longibracteàta. S.S. (*pu.*) long-bracted. —— —— —— H.b.♃. Bot. reg. t. 357.
25 quadripunctàta. T.N. (*pu.*) four-spotted. Italy. 1828. —— H.b.♃. Ten. flor. nap. t. 89.
26 saccàta. T.N. (*pu.*) saccate. Sicily. —— —— H.b.♃.
27 papilionàcea. W. (*pu.*) papilionaceous. S. Europe. 1788. 6. 7. H.b.♃. Bot. reg. t. 1155.
28 pseudosambucìna. T.N. (*pu.*) Elder-smelling. Italy. 1828. —— H.b.♃. Ten. flor. nap. t. 86.
β lutéscens. T.N. (*ye.*) *pale yellow.* —— —— —— H.b.♃. —— t. 86. f. b.
29 sulphùrea. B.M. (*st.*) pale-yellow. S. Europe. 1823. 5. 6. H.b.♃. Bot. mag. t. 2569.
30 sambucìna. W. (*ye.*) Elder-scented. Europe. 1825. 4. 5. H.b.♃. Swt. br. fl. gar. t. 299.
31 Schleichèri. B.F.G. (*re.*) Schleicher's. Switzerland. —— —— H.b.♃. —— t. 199.
sambucìna β rùbra. Sch. *non* Jacq.
32 latifòlia. W. (*pu.*) broad-leaved. Britain. .... 5. 6. H.b.♃. Eng. bot. v. 33. t. 2308.
33 maculàta. W. (*li.pu.*) spotted-palmate. —— .... 6. 7. H.b.♃. —— v. 9. t. 632.
34 spectábilis. W. (*wh.pu.*) showy American. N. America. 1801. —— H.b.♃. Swt. br. fl. gar. t. 65.

ANACA'MPTIS. R.O. ANACA'MPTIS. Gynandria Monandria.
pyramidàlis. R.O. (*cr.*) pyramidal. Britain. .... 6. 7. H.b.♃. Eng. bot. v. 2. t. 110.
O'rchis pyramidàlis. E.B.

GYMNAD'ENIA. H.K. GYMNAD'ENIA. Gynandria Monandria. H.K.
1 conópsea. H.K. (*ro.*) fragrant. Britain. .... 6. 7. H.b.♃. Eng. bot. v. 1. t. 10.
O'rchis conópsea. E.B.
2 odoratíssima. R.O. (*li.*) sweet-scented. Switzerland. 1824. —— H.b.♃. Jacq. aust. 3. t. 264.
3 víridis. R.O. (*gr.*) Frog. Britain. .... —— H.b.♃. Eng. bot. v. 2. t. 94.
Saty'rium víride. E.B. *O'rchis víridis.* H.K.
4 álbida. R.O. (*wh.*) small white. —— .... —— H.b.♃. Eng. bot. v. 8. t. 505.

A'CERAS. H.K. A'CERAS. Gynandria Monandria. H.K.
anthropóphora. H.K. (*gr.*) green-man. England. .... 6. H.b.♃. Swt. br. fl. gar. t. 168.

HERMI'NIUM. H.K. HERMI'NIUM. Gynandria Monandria. H.K.
Monòrchis. H.K. (*st.*) musk. England. .... 6. 7. H.b.♃. Eng. bot. v. 1. t. 71.
'Ophrys Monórchis. E.B.

CHAMO'RCHIS. R.O. CHAMO'RCHIS. Gynandria Monandria.
alpìna. R.O. (*wh.*) Alpine. Switzerland. 1825. 6. 7. H.b.♃. Jacq. vind. 1. t. 9.
Chamærèpes alpìna. S.S. *'Ophrys alpìna.* Jacq.

NIGRITE'LLA. R.O. NIGRITE'LLA. Gynandria Monandria.
angustifòlia. R.O. (*d.pu.*) dark-flowered. N. Europe. 1759. 6. 7. H.b.♃. Flor. dan. t. 998.
Saty'rium nìgrum. L. *O'rchis nìgra.* Sw. *Habenària nìgra.* H.K.

PLATANTH'ERA. R.O. PLATANTH'ERA. Gynandria Monandria.
bifòlia. R.O. (*wh.*) Butterfly. Britain. .... 5. 6. H.b.♃. Eng. bot. v. 1. t. 22.
O'rchis bifòlia. E.B. *Habenària bifòlia.* H.K.

	HABEN'ARIA. H.K.		HABEN'ARIA.	Gynandria Monandria. H.K.				
1	leptocèras. B.M.	(*gr.ye.*)	slender-spurred.	S. America.	1825.	10.	S.b.♃.	Bot. mag. t. 2726.
2	macrocèras. W.	(*g.wh.*)	long-horned.	Jamaica.	1828.	9. 10.	S.b.♃.	——— t. 2947.
3	bracteàta. H.K.	(*gr.*)	long-bracted.	N. America.	1805.	5. 6.	H.b.♃.	Swt. br. fl. gar. t. 62.
4	hyberbòrea. H.K.	(*gr.*)	northern.	N. Europe.	———	6. 7.	H.b.♃.	Retz. obs. 4. t. 3.
5	herbìola. H.K.	(*gr.*)	American.	N. America.	1789.	———	H.b.♃.	
6	dilatàta. H.E.F.	(*wh.*)	dilated.	———	1822.	7. 8.	H.b.♃.	Hook. ex. flor. t. 95.
7	orbiculàta. H.E.F.	(*st.*)	round-leaved.	———	1823.	———	H.b.♃.	——— t. 145.
8	tridentàta. H.E.F.	(*wh.*)	three-toothed.	———	1819.	5. 6.	H.b.♃.	——— t. 81.
9	ciliàris. H.K.	(*ye.*)	yellow-fringed.	———	1796.	6. 7.	H.b.♃.	Bot. mag. t. 1668.
10	blephariglóttis. S.S.	(*w.*)	white fringed.	———	1822.	———	H.b.♃.	Hook. ex. flor. t. 87.
11	cristàta. H.K.	(*ye.*)	yellow crested.	———	1806.	9.	H.b.♃.	
12	psycòdes. S.S.	(*ye.*)	yellow jagged.	———	1823.	6. 8.	H.b.♃.	
	O'rchis psycòdes. W.							
13	lácera. B.C.	(*st.*)	torn-flowered.	———	1818.	7.	H.b.♃.	Lodd. bot. cab. t. 229.
14	fimbriàta. H.K.	(*pu.*)	purple-fringed.	———	1777.	———	H.b.♃.	Bot. reg. t. 405.
	BARTHOL'INA. H.K.		BARTHOL'INA.	Gynandria Monandria. H.K.				
	pectinàta. H.K.	(*wh.*)	pectinated.	C. B. S.	1787.	10.	G.b.♃.	Jour. Sc. v. 4. t. 8. f. 2.
	O'rchis Burmanniàna. Sw. *pectinàta.* W.							
	BON'ATEA. W.		BON'ATEA.	Gynandria Monandria.				
	speciòsa. W.	(*gr.wh.*)	handsome.	C. B. S.	1820.	7. 9.	G.b.♃.	Lodd. bot. cab. t. 284.
	SER'APIAS. H.K.		SER'APIAS.	Gynandria Monandria. H.K.				
1	Língua. W.	(*re.*)	tongue-lipped.	S. Europe.	1786.	5. 6.	F.b.♃.	Hook. ex. flor. t. 11.
2	longipètala.	(*st.*)	long-flowered.	Italy.	1825.	———	H.b.♃.	Bot. reg. t. 1189.
	Helleborìne longipètala. Ten. fl. nap. t. 98.							
3	cordígera. W.	(*pu.*)	heart-lipped.	S. Europe.	1806.	7. 8.	F.b.♃.	Andr. reposit. t. 475.
	OPHRYS. H.K.		'OPHRYS.	Gynandria Monandria. H.K.				
1	apífera. W.	(*bk.*)	Bee.	England.	....	6. 7.	H.b.♃.	Eng. bot. v. 6. t. 383.
2	tenthredinífera. W.	(*re.y.*)	Saw-fly.	S. Europe.	1815.	4. 5.	H.b.♃.	Bot. reg. t. 205.
	β *mìnor.* B.R.	(*br.ye.*)	*lesser.*	———	1826.	———	H.b.♃.	——— t. 1093.
3	grandiflòra. T.N.	(*re.y.br.*)	large-flowered.	Italy.	1828.	———	H.b.♃.	Ten. flor. nap. t. 94.
4	hiúlca. S.S.	(*br.ye.*)	Roman.	Rome.	1826.	———	H.b.♃.	
5	arachnìtes. W.	(*bh.br.*)	black Spider.	Switzerland.	1805.	———	H.b.♃.	Bot. mag. t. 2516.
6	aranífera. W.	(*d.br.*)	common Spider.	England.	....	———	H.b.♃.	Eng. bot. v. 1. t. 65.
	limbàta. B.R.	(*da.*)	*dark-limbed.*	S. Europe.	1826.	———	H.b.♃.	Bot. reg. t. 1197.
7	arachnoídes. A.R.	(*br.bh.*)	Spider-like.	Italy.	1805.	———	H.b.♃.	Andr. reposit. t. 470.
8	ciliàta. T.N.	(*br.wh.*)	fringe-flowered.	———	1826.	———	H.b.♃.	Ten. flor. nap. t. 95.
9	Spéculum. B.R.	(*br.bl.*)	looking-glass.	S. Europe.	1818.	———	H.b.♃.	Bot. reg. t. 370.
10	lùtea. H.E.F.	(*ye.*)	yellow.	———	1821.	———	H.b.♃.	Swt. br. fl. gar. t. 206.
11	exaltàta. T.N.	(*bh.br.*)	tall.	Italy.	1826.	———	H.b.♃.	Ten. flor. nap. t. 96.
12	muscífera. E.B.	(*pu.bl.*)	Fly.	England.	....	5. 6.	H.b.♃.	Eng. bot. v. 1. t. 64.
	myòdes. W.							
13	fúsca. B.R.	(*br.ye.*)	dull purple.	S. Europe.	1826.	3. 4.	H.b.♃.	Bot. reg. t. 1071.
14	atráta. B.R.	(*gr.pu.*)	dark-flowered.	Rome.	———	———	H.b.♃.	——— t. 1087.
	SATY'RIUM. H.K.		SATY'RIUM.	Gynandria Monandria. H.K.				
1	cucullàtum. H.K.	(*g.ye.*)	cucullate.	C. B. S.	1787.	6. 9.	F.b.♃.	Bot. reg. t. 416.
2	parviflòrum. Sw.	(*g.bh.*)	small-flowered.	———	1787.	5. 6.	F.b.♃.	Jacq. schœnb. 2. t. 179.
3	membranàceum. W.	(*g.y.*)	membranaceous.	———	1822.	———	F.b.♃.	
4	coriifòlium. W.	(*ye.or.*)	leather-leaved.	———	1819.	5. 6.	F.b.♃.	Swt. br. fl. gar. s. 2. t. 3.
5	foliòsum. S.S.	(*pu.*)	leafy.	———	1828.	7. 8.	F.b.♃.	
6	cárneum. H.K.	(*bh.*)	great-flowered.	———	———	7. 9.	F.b.♃.	Bot. mag. t. 1512.
	PTERYG'ODIUM. W.		PTERYG'ODIUM.	Gynandria Monandria. H.K.				
1	alàtum. S.S.	(*ye.*)	winged.	C. B. S.	1821.	6. 8.	F.b.♃.	
2	cathólicum. S.S.	(*ye.*)	large-flowered.	———	1826.	———	F.b.♃.	
3	volùcris. S.S.	(*ye.*)	arrow-lipped.	———	1797.	6. 7.	F.b.♃.	
	D'ISA. W.		D'ISA.	Gynandria Monandria. H.K.				
1	grandiflòra. W.	(*sc.*)	large-flowered.	C. B. S.	1823.	7. 8.	F.b.♃.	Bot. reg. t. 926.
2	cornùta. W.	(*bh.*)	horned.	———	1805.	6. 7.	F.b.♃.	
3	longicórnis. W.	(*bl.*)	long-spurred.	———	1824.	5. 7.	F.b.♃.	
4	Dracònis. W.	(*wh.pu.*)	Dragon.	———	1823.	———	F.b.♃.	
5	ferrugínea. W.	(*br.*)	ferruginous.	———	1816.	6. 7.	F.b.♃.	
6	chrysostàchya. W.	(*ye.*)	golden-spiked.	———	1826.	———	F.b.♃.	
7	bracteàta. W.	(*gr.*)	small-flowered.	———	1818.	———	F.b.♃.	Bot. reg. t. 324.
8	flexuòsa. W.	(*st.*)	flexuose.	———	1822.	———	F.b.♃.	
9	prasinàta. B.R.	(*gr.pu.*)	green-flowered.	———	1815.	———	F.b.♃.	Bot. reg. t. 209.
10	spathulàta. W.	(*li.*)	spathulate.	———	1805.	———	F.b.♃.	Journ. sc. v. 4. t. 8. f. 3.
11	physòdes. W.	(*pu.*)	nodding.	———	1824.	———	F.b.♃.	
12	graminifòlia. S.S.	(*bl.*)	Grass-leaved.	———	1822.	———	F.b.♃.	Journ. sc. v. 6. t. 1. f. 2.
13	maculàta. W.	(*bl.*)	spotted-stalked.	———	1816.	7. 8.	F.b.♃.	

DI'SPERIS. W. Di'speris. Gynandria Monandria. H.K.
1 capénsis. w. (*cr.*) Cape. C. B. S. 1816. 5. 6. F.b.♃.
2 cucullàta. w. (*li.*) cucullate. ——— 1822. 6. 7. F.b.♃.
3 secúnda. w. (*st.*) side-flowering. ——— 1797. —— F.b.♃.
CORY'CIUM. W. Cory'cium. Gynandria Monandria.
1 orobanchoídes. w. (*ye.*) Orobanche-like. C. B. S. 1825. 6. 8. F.b.♃.
2 críspum. w. (*ye.*) curled-leaved. ——— —— 5. 6. F.b.♃. Buxb. cent. 3. t. 11.
GLO'SSULA. B.R. Glo'ssula. Gynandria Monandria.
tentaculàta. b.r. (*gr.*) horned. China. 1824. 5. 6. F.b.♃. Bot. reg. t. 862.
GOODY'ERA. H.K. Goody'era. Gynandria Monandria. H.K.
1 rèpens. h.k. (*wh.*) creeping. Scotland. .... 7. 8. H.♃. Eng. bot. v. 5. t. 289.
Saty'rium rèpens. e.b. *Neóttia rèpens.* w.
2 tessellàta. b.c. (*wh.*) tessellated. N. America. 1823. 6. 7. H.♃. Lodd. bot. cab. t. 952.
Neóttia rèpens. Ph. *Goodyèra pubéscens. β mìnor.* b.m. 2540.
3 pubéscens. h.k. (*wh.*) downy. N. America. 1802. —— H.♃. Swt. br. fl. gar. s. 2. t. 47.
4 pròcera. h.e.f. (*wh.*) Nepaul. Nepaul. 1821. 7. 12. G.♃. Hook. ex. flor. t. 39.
Neóttia pròcera. b.r. 639.
5 díscolor. b.r. (*wh.*) red-leaved. S. America. 1815. 10. 12. S.♃. Bot. reg. t. 271.
PELE'XIA. B.R. Pele'xia. Gynandria Monandria. B.R.
spiranthoídes. b.r. (*gr.*) decurrent-lipped. W. Indies. 1823. 3. S.♃. Bot. reg. t. 985.
Neóttia adnàta. Sw.
STENORHY'NCHUS. R.O. Stenorhy'nchus. Gynandria Monandria.
1 speciòsus. r.o. (*sc.*) showy. W. Indies. 1790. 3. 6. S.♃. Bot. mag. t. 1374.
Neóttia speciòsa. Hook. ex. flor. t. 39.
2 orchioídes. r.o. (*pu.*) frosted-flowered. Jamaica. 1806. —— S.♃. Bot. reg. t. 701.
3 plantagìneus. h. (*pu.*) Plantain-leaved. W. Indies. 1825. —— S.♃. Hook. ex. flor. t. 226.
Neóttia plantagìnea. Hooker.
4 aphy'lla. (*br.li.*) leafless. Trinidad. —— —— S.♃. Bot. mag. t. 2797.
Neóttia aphy'lla. b.m.
NEO'TTIA. H.K. Neo'ttia. Gynandria Monandria. H.K.
1 macrántha. (*wh.*) great-flowered. W. Indies. 1827. 3. 6. S.♃. Bot. mag. t. 2956.
Neóttia? grandiflòra. b.m. t. 2956. *non* t. 2730.
2 grandiflòra. b.m. (*gr.*) large-flowered. Brazil. 1825. 2. 5. S.♃. Bot. mag. t. 2730.
Spiránthes grandiflòra. Bot. reg. t. 1043.
3 bícolor. b.r. (*wh.*) two-coloured. W. Indies. 1821. 7. 8. S.♃. Bot. reg. t. 794.
4 elàta. w. (*wh.*) tall. ——— 1790. 4. 6. S.♃. Bot. mag. t. 2026.
5 pícta. h.k. (*wh.*) painted-leaved. Trinidad. 1805. 2. 4. S.♃. ——— t. 1562.
6 cérnua. h.k. (*wh.*) nodding-flowered. N. America. 1796. 8. 10. H.♃. Swt. br. fl. gar. t. 42.
7 tórtilis. w. (*wh.*) twisting-spiked. W. Indies. 1822. 6. 7. S.♃.
8 æstivàlis. m. (*wh.*) summer. N. America. —— —— H.♃.
tórtilis. Ph. *non* Sw.
9 spiràlis. h.k. (*wh.*) Ladies-traces. Britain. .... 8. 9. H.b.♃. Eng. bot. v. 8. t. 541.
Ophrys spiràlis. e.b.
10 pùdica. (*bh.*) modest. China. 1819. 9. 12. H.b.♃. Lindl. coll. t. 30.
austràlis. β. b.r. t. 602. *Spiránthes pùdica.* Lindl. col.
11 austràlis. b.p. (*wh.*) southern. N. S. W. 1826. —— F.b.♃.
CALOCH'ILUS. B.P. Caloch'ilus. Gynandria Monandria. B.P.
1 campéstris. b.p. (*wh.*) field. N. S. W. 1826. 4. 6. F.b.♃.
2 paludòsus. b.p. (*wh.*) marsh. ——— —— —— F.b.♃.
PONTHI'EVA. H.K. Ponthi'eva. Gynandria Monandria. H.K.
1 glandulòsa. h.k. (*gr.*) glandular. W. Indies. 1800. 1. 3. S.♃. Bot. mag. t. 842.
2 petiolàta. b.r. (*br.*) stalked. ——— 1823. 7. 8. S.♃. Bot. reg. t. 760.
PRASOPHY'LLUM. B.P. Prasophy'llum. Gynandria Monandria. B.P.
1 elàtum. b.p. .... tall. N. S. W. 1826. .... F.b.♃.
2 flàvum. b.p. (*ye.*) yellow. ——— —— 4. 6. F.b.♃.
3 striàtum. b.p. (*std.*) streaked. ——— —— .... F.b.♃.
4 fúscum. b.p. (*br.*) brown. ——— —— 5. 6. F.b.♃.
5 pátens. b.p. .... spreading. ——— —— .... F.b.♃.
6 rùfum. b.p. (*br.*) rufous. ——— —— 5. 6. F.b.♃.
7 fimbriàtum. b.p. .... fringed. ——— 1826. .... F.b.♃.
CRYPTO'STYLIS. B.P. Crypto'stylis. Gynandria Monandria. B.P.
1 erécta. b.p. (*br.*) upright. N. S. W. 1826. 4. 6. F.b.♃.
2 longifòlia. b.p. (*br.*) long-leaved. ——— 1822. 5. 6. F.b.♃. Lab. n. holl. 2. t. 212.
Maláxis subulàta. Lab.
ORTHO'CERAS. B.P. Ortho'ceras. Gynandria Monandria. B.P.
stríctum. b.p. .... upright. N. S. W. 1826. .... F.b.♃.
DI'URUS. B.P. Di'urus. Gynandria Monandria. B.P.
1 maculàta. b.p. (*ye.sp.*) spotted. N. S. W. 1823. 3. 5. F.b.♃. Sm. exot. bot. 1. t. 30.
2 aùrea. b.p. (*go.*) golden-flowered. ——— 1810. 4. 5. F.b.♃. ——— 1. t. 9.

3 pedunculàta. B.P. (*ye.*) peduncled. N. S. W. 1822. 2. 4. F.b.♃.
4 sulphùrea. B.P. (*st.*) sulphur-coloured. ——— 1826. —— F.b.♃.
5 elongàta. B.P. (*pu.*) elongated. ——— ——— —— F.b.♃.
THELY'MITRA. B.P. Thely'mitra. Gynandria Monandria. B.P.
1 ixioídes. B.P. (*bl.sp.*) spotted blue. N. S. W. 1810. 4. 6. F.b.♃. Sm. ex. bot. 1. t. 29.
2 mèdia. B.P. (*bl.*) middle. ——— 1820. —— F.b.♃.
3 pauciflòra. B.P. (*bl.*) few-flowered. ——— 1826. —— F.b.♃.
4 angustifòlia. B.P. (*bl.*) narrow-leaved. ——— —— 5. 6. F.b.♃.
5 cárnea. B.P. (*fl.*) flesh-coloured. ——— 1823. —— F.b.♃.
6 venòsa. B.P. (*wh.*) veined. ——— 1826. 3. 5. F.b.♃.
LIST'ERA. H.K. List'era. Gynandria Monandria. H.K.
1 ovàta. H.K. (*gr.*) Tway-blade. Britain. .... 5. 6. H.♃. Eng. bot. v. 22. t. 1548.
'*Ophrys ovàta.* E.B. *Epipáctis ovàta.* W.
2 cordàta. H.K. (*br.gr.*) heart-leaved. Britain. .... 6. 7. H.♃. Eng. bot. v. 5. t. 358.
3 convallarioídes. N. (*br.gr.*) small green. N. America. 1826. 5. 6. H.♃. Bart. fl. amer. t. 39. f. 1.
4 Nìdus àvis. (*br.*) Bird's-nest. Britain. .... 5. H.♃. Eng. bot. v. 1. t. 48.
'*Ophrys Nìdus àvis.* E.B. *Epipáctis Nìdus àvis* E.F. *Neóttia Nìdus àvis.* R.O.
EPIPA'CTIS. H.K. Epipa'ctis. Gynandria Monandria. H.K.
1 latifòlia. W. (*pu.*) broad-leaved. Britain. .... 7. 8. H.♃. Eng. bot. v. 4. t. 269.
Seràpias latifòlia. E.B.
2 palústris. W. (*pu.wh.*) marsh. ——— .... —— H.♃. Eng. bot. v. 4. t. 270.
3 pállens. W. (*wh.*) white. ——— .... 6. H.♃. ——— v. 4. t. 271.
Seràpias grandiflòra. E.B.
4 ensifòlia. W. (*wh.*) narrow-leaved. ——— .... —— H.♃. Eng. bot. v. 7. t. 494.
5 rùbra. W. (*pu.wh.*) purple. ——— .... 6. 7. H.♃. ——— v. 7. t. 437.
ARETH'USA. H.K. Areth'usa. Gynandria Monandria. H.K.
1 bulbòsa. H.K. (*pu.*) bulbous. N. America. .... 5. 6. H.b.♃. Bot. mag. t. 2204.
2? plicàta. A.R. (*pu.*) plaited. E. Indies. 1806. 6. 8. S.b.♃. Andr. reposit. t. 321.
POG'ONIA. H.K. Pog'onia. Gynandria Monandria. H.K.
1 divaricàta. H.K. (*pi.*) Lily-leaved. N. America. 1787. 6. 7. H.♃. Catesb. car. 1. t. 58.
2 ophioglossoídes. B.R. (*pi.*) Adder's-tongue. ——— 1816. —— H.♃. Bot. reg. t. 148.
3 péndula. B.R. (*pi.*) drooping. ——— 1824. —— H.♃. ——— t. 906.
Arethùsa péndula. W. *parviflòra.* M. *Tríphora péndula.* N.
MICR'OTIS. B.P. Micr'otis. Gynandria Monandria. B.P.
parviflòra. B.P. (*gr.*) small-flowered. N. S. W. 1826. 4. 6. F.b.♃.
ACIA'NTHUS. B.P. Acia'nthus. Gynandria Monandria. B.P.
1 fornicàtus. B.P. (*br.*) enclosed. N. S. W. 1822. 4. 6. F.b.♃.
2 exsértus. B.P. (*br.*) exserted. ——— —— —— F.b.♃.
3 caudàtus. B.P. (*br.*) long-awned. ——— 1826. —— F.b.♃.
CYRTO'STYLIS. B.P. Cyrto'stylis. Gynandria Monandria. B.P.
renifórmis. B.P. (*br.*) kidney-leaved. N. S. W. 1824. 4. 5. F.♃.
CHILOGLO'TTIS. B.P. Chiloglo'ttis. Gynandria Monandria. B.P.
diphy'lla. B.P. (*br.*) two-leaved. N. S. W. 1826. 4. 6. F.b.♃.
PRESC'OTIA. H.E.F. Presc'otia. Gynandria Monandria.
plantaginifòlia. H.E.F. (*gr.*) Plantain-ld. RioJaneiro. 1822. 9. 4. S.♃. Hook. ex. flor. t. 115.
LI'PARIS. Rich. Li'paris. Gynandria Monandria. H.K.
1 foliòsa. B.R. (*y.gr.*) leafy. Mauritius. 1823. 9. S.♃. Bot. reg. t. 882.
2 bituberculàta. B.R. (*gr.br.*) two-warted. Nepaul. 1822. 2. 3. S.♃. Hook. ex. flor. t. 116.
3 Loesèlii. B.R. (*ye.*) Loesel's. Britain. .... 7. H.b.♃. Eng. bot. v. 1. t. 47.
Maláxis Loesèlii. E.B.
4 longifòlia. B.A. (*gr.*) long-leaved. N. America. 1827. —— H.b.♃. Bart. flor. amer. t. 75.
5 Correàna. N. (*ye.gr.*) two-leaved. ——— 1823. 5. 6. H.b.♃.
6 liliifòlia. B.R. (*pa.bl.*) Lily-leaved. ——— 1758. 6. 7. H.b.♃. Bot. mag. t. 2004.
Maláxis liliifòlia. B.M. '*Ophrys liliifòlia.* A.R. 65.
7 refléxa. B.R. (*gr.*) reflexed. N. Holland. 1820. —— G.b.♃.
8 elàta. B.R. (*br.pu.*) tall. Brazil. —— —— S.♃. Bot. reg. t. 1175.
ERIOCH'ILUS. B.P. Erioch'ilus. Gynandria Monandria. B.P.
autumnàlis. B.P. (*bh.*) autumnal. N. S. W. 1823. 9. 10. G.b.♃. Lab. n. holl. 2. t. 211. f. 2.
Epipáctis cucullàta. Lab.
CALAD'ENIA. B.P. Calad'enia. Gynandria Monandria. H.K.
1 álba. B.P. (*wh.*) white. N. S. W. 1810. 7. 8. G.b.♃.
2 cárnea. B.P. (*fl.*) flesh-coloured. ——— 1826. 6. 7. G.b.♃.
3 cœrùlea. B.P. (*bl.*) blue-flowered. ——— —— —— G.b.♃.
4 alàta. B.P. (*li.*) winged. ——— 1823. 6. 8. G.b.♃.
5 testàcea. B.P. .... fringe-flowered. ——— 1826. .... G.b.♃.
LYPERA'NTHUS. B.P. Lypera'nthus. Gynandria Monandria. B.P.
1 suavèolens. B.P. (*d.br.*) sweet-scented. N. S. W. 1822. 5. 7. F.b.♃.
2 ellípticus. B.P. (*d.br.*) elliptic-leaved. ——— 1826. —— F.b.♃.
3 nígricans. B.P. (*bk.*) heart-leaved. ——— —— —— F.b.♃.

GLOSS'ODIA. B.P. GLOSS'ODIA. Gynandria Monandria. B.P.
1 màjor. B.P. (*bl.*) larger. N. S. W. 1810. 6. 8. F.b.♃.
2 mìnor. B.P. (*bl.*) lesser. ——— 1826. —— F.b.♃.
PTERO'STYLIS. B.P. PTERO'STYLIS. Gynandria Monandria. B.P.
1 concínna. B.P. (*st.*) neat. N. S. W. 1826. 4. 6. F.b.♃.
2 ophioglóssa. B.P. (*st.*) adder's-tongue. ——— —— —— F.b.♃.
3 cúrta. B.P. (*st.*) short-lipped. ——— 1822. 4. 6. F.b.♃.
4 acuminàta. B.P. (*st.*) taper-pointed. ——— 1826. —— F.b.♃.
5 cucullàta. B.P. (*st.*) hooded. V. Diem. Isl. 1823. 5. 7. F.b.♃.
6 nùtans. B.P. (*st.*) nodding. N. S. W. 1826. —— F.b.♃.
7 obtùsa. B.P. (*st.*) blunt-lipped. ——— 1810. 7. 8. F.b.♃.
8 refléxa. B.P. (*st.*) reflexed. ——— 1826. 4. 6. F.b.♃.
9 grandiflòra. B.P. (*st.*) large-flowered. ——— —— —— F.b.♃.
10 longifòlia. B.P. (*st.*) long-leaved. ——— 1823. 6. 8. F.b.♃.
11 gibbòsa. B.P. (*st.*) gibbous. ——— 1826. 5. 6. F.b.♃.
CORYSA'NTHES. B.P. CORYSA'NTHES. Gynandria Monandria. B.P.
1 fimbriàta. B.P. (*br.*) fringed. N. S. W. 1822. 5. 7. F.b.♃.
2 unguiculàta. B.P. (*br.*) pendulous-flowered. ——— —— —— F.b.♃.
3 bicalcaràta. B.P. (*br.*) two-spurred. ——— .... —— F.b.♃. Sal. par. lond. t. 83. ?
Córybas aconitiflòrus. P.L. ?
C'ALEYA. H.K. C'ALEYA. Gynandria Monandria. H.K.
1 màjor. H.K. (*br.gr.*) smooth-lipped. N. S. W. 1810. 6. 7. F.b.♃.
2 mìnor. (*br.gr.*) tubercle-lipped. ——— 1822. —— F.b.♃.
Caleána mìnor. B.P.
CALOP'OGON. H.K. CALOP'OGON. Gynandria Monandria. H.K.
pulchéllus. H.K. tuberous-rooted. N. America. 1771. 7. 8. H.b.♃. Swt. br. fl. gar. t. 115.
Limodòrum tuberòsum. B.M. 116. *Cymbídium pulchéllum.* W.
BL'ETIA. H.K. BL'ETIA. Gynandria Monandria. H.K.
1 Tankervílliæ. H.K. (*wh.pu.*) Ly Tankerville's. China. 1778. 3. 4. S.♃. Bot. mag. t. 1924.
Limodòrum Tankervíllicæ. W.
2 pállida. B.C. (*li.*) pale-flowered. China. 1820. 4. 8. S.♃. Lodd. bot. cab. t. 629.
3 flòrida. H.K. (*pu.*) purple. W. Indies. 1786. 7. 8. S.♃. Redoute. lil. t. 83.
Limodòrum purpùreum. R.L.
4 guineénsis. G.D. (*pu.*) Guinea. Guinea. 1822. .... S.♃.
5 verecúnda. H.K. (*pu.*) tall. W. Indies. 1733. 1. 5. S.♃. Bot. mag. t. 930.
Limodòrum áltum. B.M. *Cymbídium verecúndum.* W.
6 hyacínthina. H.K. (*pu.*) hyacinthine. China. 1802. 3. 9. G.♃. Bot. mag. t. 1492.
Cymbídium hyacínthinum. B.M.
7 capitàta. H.K. (*pu.*) capitate. W. Indies. 1795. 5. 7. S.♃.
8 Woodfórdii. B.M. (*ye.or.*) Woodford's. Trinidad. 1820. 4. 8. S.♃. Bot. mag. t. 2719.
GEOD'ORUM. H.K. GEOD'ORUM. Gynandria Monandria. H.K.
1 purpùreum. H.K. (*pu.*) purple. E. Indies. 1800. 6. 8. S.♃. Roxb. cor. 1. t. 40.
Limodòrum nùtans. R.C.
2 citrìnum. H.K. (*st.*) Lemon-coloured. E. Indies. —— 10. 12. S.♃. Andr. reposit. t. 626.
3 dilatàtum. H.K. (*pi.*) shovel-lipped. ——— —— 5. 8. S.♃. Bot. reg. t. 675.
Limodòrum recúrvum. R.C. 1. t. 39.
CALY'PSO. H.K. CALY'PSO. Gynandria Monandria. H.K.
americàna. H.K. (*pu.*) American. N. America. 1805. 5. 6. H.b.♃. Sal. par. lond. t. 89.
boreàlis. H.E.F. t. 12.
MALA'XIS. Rich. MALA'XIS. Gynandria Monandria. H.K.
paludòsa. E.B. (*ye.*) marsh. England. .... 7. H.b.♃. Eng. bot. v. 1. t. 47.
MICRO'STYLIS. N. MICRO'STYLIS. Gynandria Monandria. N.
ophioglossoídes. N. (*gr.*) Adder's-tongue. N. America. 1820. 4. 5. H.b.♃. Lodd. bot. cab. t. 1146.
β *mexicàna* B.R. (*gr.*) *Mexican.* Mexico. 1828. —— H.♃. Bot. reg. t. 1290.
CORALLORRH'IZA. H.K. CORALLORRH'IZA. Gynandria Monandria. H.K.
1 innàta. H.K. (*st.*) spurless. Scotland. .... 6. 7. H.♃. Eng. bot. v. 22. t. 1547.
2 multiflòra. H.E.F. (*st.*) many-flowered. N. America. 1824. 5. 6. H.♃. Hook. ex. flor. t. 174.
3 odontorhìza. N. (*w.pu.*) tooth-rooted. ——— —— 6. 7. H.♃.
ST'ELIS. H.K. ST'ELIS. Gynandria Monandria. H.K.
1 ophioglossoídes. H.K. (*st.*) Adder's-tongue. W. Indies. 1791. 5. 6. S.♃. Bot. reg. t. 935.
2 micrántha. H.K. (*st.*) small-flowered. Jamaica. 1805. 11. 12. S.♃. Hook. ex. flor. t. 158.
RODRIGU'EZIA. R.P. RODRIGU'EZIA. Gynandria Monandria. B.R.
secúnda. K.S. (*re.*) side-flowering. S. America. 1819. 6. 12. S.♃. Bot. reg. t. 930.
GOM'EZA. B.M. GOM'EZA. Gynandria Monandria. B.M.
recúrva. B.M. (*ye.*) recurved. Brazil. 1814. 5. 6. S.♃. Bot. mag. t. 1748.
NOTY'LIA. B.R. NOTY'LIA. Gynandria Monandria.
punctàta. B.R. (*wh.*) dotted. Trinidad. 1823. 7. 8. S.♃. Bot. reg. t. 759.
Pleurothállis punctàta. B.R. *Gomèza tenuiflòra.* Lod. bot. cab. t. 806.

CIRRH'ÆA. B.R. CIRRH'ÆA. Gynandria Monandria.
dependens. B.R.(ye.gr.)pendent. China. 1824. 6. 8. S.♃. Lodd. bot. cab. t. 936.
Cymbídium depéndens. B.C.
CYMBI'DIUM. H.K. CYMBI'DIUM. Gynandria Monandria. H.K.
1 aloifòlium. H.K. (*d.pu.*) Aloe-leaved. China. 1789. 5. 6. S.♃. Bot. mag. t. 387.
2 sinénse. H.K. (*d.pu.*) Chinese. ——— 1793. 9. 10. S.♃. ——— t. 888.
3 lancifòlium. H.E.F. (*y.re.*) lance-leaved. E. Indies. 1823. 5. 8. S.♃. Hook. ex. flor. t. 51.
4 ensifòlium. H.K. (*st.br.*) sword-leaved. China. 1780. 6. 10. S.♃. Bot. mag. t. 1751.
5 xiphiifòlium. B.R. (*gr.*) Xiphium-leaved ——— 1814. 5. 8. S.♃. Bot. reg. t. 529.
6 iridioídes. D.P. (*wh.*) Iris-leaved. Nepaul. 1828. —— S.♃.
7 suáve. B.P. (*gr.br.*) sweet-scented. N. S. W. 1826. —— G.♃.
8 trípterum. W. (*wh.*) triangular-fruited. Jamaica. 1790. 6. 7. S.♃. Sm. ic. pict. t. 14.
ANISOPE'TALUM. H.E.F. ANISOPE'TALUM. Gynandria Monandria.
Careyànum. H.E.F. (*br.pu.*) Carey's. Nepaul. 1823. 8. 10. G.♃. Hook. ex. flor. t. 149.
BRA'SSIA. H.K. BRA'SSIA. Gynandria Monandria. H.K.
1 maculàta. H.K. (*w.st.pu.*) spotted. Jamaica. 1806. 6. 7. S.♃. Bot. mag. t. 1691.
2 caudàta. B.R. (*g.y.re.*) long-tailed. W. Indies. 1823. —— S.♃. Bot. reg. t. 832.
ANGUL'OA. K.S. ANGUL'OA. Gynandria Monandria.
grandiflòra. K.S. (*ye.pu.*) large-flowered. S. America. 1823. 6. 8. S.♃. H. et B. pl. æq. 1. t. 27.
CATAS'ETUM. K.S. CATAS'ETUM. Gynandria Monandria. B.M.
1 tridentàtum. H.E.F. (*ye.*) three-toothed. Trinidad. 1823. 8. 12. S.♃. Bot. mag. t. 2559.
2 Claveríngi. B.R. (*y.pu.*) Clavering's. Brazil. —— —— S.♃. Bot. reg. t. 840.
3 floribúndum. H.E.F. (*y.pu.*) many-flowered. ——— —— —— S.♃. Hook. ex. flor. t. 151.
4 semiapértum. H.E.F. (*gr.*) greenish-flowered. —— 1825. 5. S.♃. ——— t. 213.
5 Hookèri. L.col. (*y.pu.*) Hooker's. ——— 1816. —— S.♃. Lindl. coll. t. 40.
6 cristàtum. B.R. (*g.wh.*) crested. ——— 1823. —— S.♃. Bot. reg. t. 966.
CYRTOP'ODIUM. B.R. CYRTOP'ODIUM. Gynandria Monandria. H.K.
1 Andersònii. H.K. (*ye.*) Anderson's. W. Indies. 1804. 4. 8. S.♃. Bot. mag. t. 1800.
2 glutinòsum. Mey. (*ye.*) gummy. S. America. 1825. .... S.♃.
3 Woodfórdii. B.M. (*pu.*) Woodford's. ——— 1814. 10. S.♃. Bot. mag. t. 1814.
LISSOCH'ILUS. B.R. LISSOCH'ILUS. Gynandria Monandria. B.R.
speciòsus. B.R. (*ye.*) showy. C. B. S. 1818. 5. 6. S.♃. Bot. reg. t. 573.
ONCI'DIUM. H.K. ONCI'DIUM. Gynandria Monandria. H.K.
1 altíssimum. W. (*ye.*) sharp-petaled. W. Indies. 1793. 8. 9. S.♃. Jacq. amer. t. 141.
2 carthagenénse. W. (*g.br.*) spread-eagle. ——— 1791. 5. 6. S.♃. Bot. mag. t. 777.
Epidéndrum undulàtum. B.M.
3 lùridum. B.R. (*g.br.*) lurid. Trinidad. 1818. 2. 6. S.♃. Bot. reg. t. 727.
4 variegàtum. W. (*wh.pu.*) variegated. Jamaica. 1825. 7. 8. S.♃. Sloan. hist. 1. t. 148. f. 2.
5 divaricàtum. B.R. (*br.y.re.*) spreading. Brazil. 1826. 9. 11. S.♃. Bot. reg. t. 1050.
6 pùbes. B.R. (*br.re.*) pubescent. ——— 1824. 3. 6. S.♃. ——— t. 1007.
7 ornithorhy'nchum. K.S. (*ye.*) long-beaked. Mexico. 1826. 8. 10. S.♃. H. et B. n. gen. 1. t. 80.
8 Papílio. B.R. (*y.or.pu.*) Butterfly. Trinidad. 1823. 4. 6. S.♃. Bot. reg. t. 910.
9 flexuòsum. B.M. (*ye.*) Zigzag. Brazil. 1818. 6. 7. S.♃. Bot. mag. t. 2203.
10 barbàtum. L.col. (*ye.*) bearded. S. America. —— 4. 5. S.♃. Lindl. coll. t. 27.
11 pùmilum. B.R. (*ye.*) dwarf. Brazil. 1824. 6. 7. S.♃. Bot. reg. t. 920.
12 bifòlium. H.K. (*ye.*) two-leaved. S. America. 1811. 7. 8. S.♃. Bot. mag. t. 1491.
13 pulchéllum. B.M. (*bh.or.*) elegant. Demerara. 1825. 6. 7. S.♃. ——— t. 2773.
14 triquètrum. H.K. (*ye.*) triangular-leaved. Jamaica. 1793. 7. 8. S.♃.
15 tetrapétalum. W. (*pu.ye.*) four-petaled. ——— 1824. —— S.♃. Jacq. amer. t. 142.
16 Cebollèta. W. (*br.*) awl-leaved. Carthagena. —— —— S.♃. ——— t. 131. f. 2.
MACRAD'ENIA. B.R. MACRAD'ENIA. Gynandria Monandria. B.R.
lutéscens. B.R. (*ye.*) yellowish. Trinidad. 1821. 12. 2. S.♃. Bot. reg. t. 612.
CŒLO'GYNE. B.R. CŒLO'GYNE. Gynandria Monandria.
1 fimbriàta. B.R. (*st.*) fringed. China. 1824. 9. S.♃. Bot. reg. t. 868.
2 nítida. L.O. (*ye.*) shining-leaved. E. Indies. 1822. .... S.♃.
3 punctulàta. L.O. (*ye.*) dot-flowered. ——— —— .... S.♃.
MEGACLI'NIUM. B.R. MEGACLI'NIUM. Gynandria Monandria. B.R.
falcàtum. B.R. (*ye.re.*) falcate. Sierra Leone. 1824. 6. 9. S.♃. Bot. reg. t. 989.
TRIZE'UXIS. L.col. TRIZE'UXIS. Gynandria Monandria.
falcàta. L.col. (*gr.*) falcate. Trinidad. 1820. 2. 6. S.♃. Lindl. coll. t. 2.
PLEUROTHA'LLIS. H.K. PLEUROTHA'LLIS. Gynandria Monandria. H.K.
1 ruscifòlia. H.K. (*st.*) Ruscus-leaved. W. Indies. 1791. 5. 6. S.♃. Hook. ex. flor. t. 197.
2 racemiflòra. H.E.F. (*st.*) racemed. ——— 1822. 7. 8. S.♃. ——— t. 123.
3 prolífera. B.R. (*ro.*) proliferous. Brazil. 1825. 1. 12. S.♃. Bot. reg. t. 1298.
4 foliòsa. B.M. (*st.ye.*) leafy. ——— —— 2. 6. S.♃. Bot. mag. t. 2746.
ORNITHOCE'PHALUS. BIRD'S-HEAD. Gynandria Monandria.
gladiàtus. H.E.F. (*gr.*) sword-leaved. Trinidad. 1822. 5. 8. S.♃. Hook. ex. flor. t. 127.
OCTOM'ERIA. H.K. OCTOM'ERIA. Gynandria Monandria. H.K.
1 graminifòlia. H.K. (*st.*) Grass-leaved. W. Indies. 1793. 4. 7. S.♃. Bot. mag. t. 2764.

2 serratifòlia. B.M. (*st.br.*) saw-leaved. Brazil. 1826. 10. 12. S. ♃. Bot. mag. t. 2823.
3 spicàta. D.P. (*wh.*) spiked. Nepaul. 1823. 5. 8. S. ♃.

TRIBRA'CHIA. B.R. Tribra'chia. Gynandria Monandria.
péndula. B.R. (*gr.*) pendulous-spiked. SierraLeon. 1822. 6. 11. S. ♃. Bot. reg. t. 963.

SOPHRON'ITIS. B.R. Sophron'itis. Gynandria Monandria.
cérnua. B.R. (*ro.*) cernuous. Brazil. 1826. 11. 12. S. ♃. Bot. reg. t. 1129.

GONG'ORA. R.P. Gong'ora. Gynandria Monandria.
1 atropurpùrea. H.E.F. (*d.pu.*) dark-flowered. Trinidad. 1824. 6. 7. S. ♃. Hook. ex. flor. t. 178.
2 speciòsa. B.M. (*ye.or.*) beautiful. Brazil. 1826. 4. 6. S. ♃. Bot. mag. t. 2755.
3 viridipurpùrea. B.M. (*g.pu.*) green & purple. ——— 1827. 6. 8. S. ♃. ——— t. 2978.

HETEROTA'XIS. B.R. Heterota'xis. Gynandria Monandria.
crassifòlia. B.R. (*ye.*) thick-leaved. Jamaica. 1823. 6. 8. S. ♃. Bot. reg. t. 1028.

DIP'ODIUM. B.P. Dip'odium. Gynandria Monandria. B.P.
punctàtum. B.P. (*pu.*) spotted. N. S. W. 1822. 6. 8. G. ♃. Sm. exot. bot. 1. t. 12.
Dendròbium punctàtum. Sm.

CAMARI'DIUM. B.R. Camari'dium. Gynandria Monandria. B.R.
ochroleùcum. B.R. (*st.*) pale-flowered. Trinidad. 1823. 7. 10. S. ♃. Bot. reg. t. 844.
Dendròbium? álbum. H.E.F. t. 142.

ERIA. B.R. 'Eria. Gynandria Monandria. B.R.
1 stellàta. B.R. (*st.*) starry. Nepaul. 1818. 5. 6. S. ♃. Bot. reg. t. 904.
2 pubéscens. B.R. (*ye.*) pubescent. E. Indies. 1820. 3. 4. S. ♃. Hook. ex. flor. t. 124.
Dendròbium pubéscens. H.E.F.
3 ròsea. B.R. (*bh.*) rose-coloured. China. 1824. 10. S. ♃. Bot. reg. t. 978.

XYL'OBIUM. B.R. Xyl'obium. Gynandria Monandria.
1 squálens. B.R. (*ye.br.*) squalid. Brazil. 1822. 7. 8. S. ♃. Bot. reg. t. 732.
Dendròbium squálens. B.R. *Maxillària squálens.* Bot. mag. t. 2955.
2 racemòsum. (*ye.or.*) racemed. Brazil. 1825. 6. 8. S. ♃. Bot. mag. t. 2789.
Maxillària racemòsa. B.M.

MAXILL'ARIA. R.P. Maxill'aria. Gynandria Monandria. B.R.
1 pallidiflòra. B.M. (*st.*) pale-flowered. St. Vincent's. 1826. 8. 11. S. ♃. Bot. mag. t. 2806.
2 ciliàta. B.R. (*gr.*) fringe-lipped. S. America. 1827. 5. 8. S. ♃. Bot. reg. t. 1206.
3 Barringtòniæ. B.R. (*gr.*) Barrington's. W. Indies. 1790. 6. 8. S. ♃. Sm. ic. pict. t. 25.
Dendròbium Barringtòniæ. Hook. ex. flor. t. 119.
4 Harrisòniæ. B.R. (*st.pu.*) Mrs. Harrison's. Brazil. 1820. 4. 8. S. ♃. Bot. mag. t. 2927.
5 Parkèri. B.M. (*wh.pu.*) Parker's. W. Indies. 1826. —— S. ♃. ——— t. 2729.
6 párvula. H.E.F. (*st.pu.*) small. Brazil. 1824. 3. 5. S. ♃. Hook. ex. flor. t. 217.
7 aromática. H.E.F. (*ye.*) aromatic. ——— 1825. 5. S. ♃. ——— t. 219.

SARCOCH'ILUS. B.P. Sarcoch'ilus. Gynandria Monandria. B.P.
falcàtus. B.P. (*pu.*) falcate. N. S. W. 1821. 5. 8. G. ♃.

DENDR'OBIUM. H.K. Dendr'obium. Gynandria Monandria. H.K.
1 speciòsum. B.P. (*bh.re.*) showy. N. S. W. 1801. 6. 8. G. ♃. Sm. exot. bot. 1. t. 10.
2 'æmulum. B.P. (*wh.*) rival. ——— 1822. 4. 2. G. ♃. Bot. mag. t. 2906.
3 rígidum. B.P. ...... rigid. N. Holland. 1824. .... S. ♃.
4 linguifòrme. B.P. (*w.st.*) tongue-leaved. N. S. W. 1810. 4. 8. G. ♃. Sm. exot. bot. 1. t. 11.
5 teretifòlium. B.P. (*pu.*) filiform-leaved. ——— 1823. —— G. ♃.
6 longicórnu. B.R. (*bh.*) long-spurred. E. Indies. 1828. 4. 6. S. ♄. Bot. reg. t. 1315.
7 secúndum. B.R. (*ro.*) secund. Sumatra. —— —— S. ♄. ——— t. 1291.
8 cucullàtum. B.R. (*bh.*) cucullate. E. Indies. 1815. 3. 5. S. ♄. Bot. reg. t. 548.
9 Pierárdi. H.E.F. (*bh.*) Pierard's. ——— —— 12. 6. S. ♄. Hook. ex. flor. t. 9.
10 Calceolària. H.E.F. (*re.y.*) hollow-lipped. ——— 1820. —— S. ♄. ——— t. 184.
11 fimbriàtum. H.E.F. (*ye.*) fringed. ——— 1818. 4. 6. S. ♄. ——— t. 71.
12 chrysánthum. B.R. (*ye.*) yellow-flower'd. Nepaul. 1828. 2. 4. S. ♄. Bot. reg. t. 1299.
13 crumenàtum. W. (*pu.*) pouched. Java. 1822. 4. 6. S. ♄. Rumph. amb. 6. t. 47. f. 2.
14 monilifórme. W. (*pu.*) glassy. China. 1822. 6. 12. S. ♄. Bot. reg. t. 1314.
15 ancéps. B.R. (*st.*) two-edged. E. Indies. 1826. 2. 9. S. ♄. ——— t. 1239.

ORNITHI'DIUM. H.K. Ornithi'dium. Gynandria Monandria. H.K.
1 coccíneum. H.K. (*sc.*) scarlet-flower'd. W. Indies. 1790. 1. 12. S. ♃. Hook. ex. flor. t. 38.
Cymbídium coccíneum. Bot. mag. t. 1437.
2 refléxum. Gr. (*sc.*) reflexed. W. Indies. 1825. 9. 12. S. ♃.

ISOCH'ILUS. H.K. Isoch'ilus. Gynandria Monandria. H.K.
1 lineàris. H.K. (*re.*) linear-leaved. W. Indies. 1791. 5. 7. S. ♃. Bot. reg. t. 745.
Epidéndrum lineàre. Jacq. amer. t. 131. f. 1. *Cymbídium lineàre.* W.
2 graminoídes. H.E.F. (*st.*) grass-like. W. Indies. 1823. —— S. ♃. Hook. ex. flor. t. 196.
3 prolífer. H.K. (*wh.*) proliferous. ——— 1793. 6. 11. S. ♃. Bot. reg. t. 825.

PHOLID'OTA. B.R. Pholid'ota. Gynandria Monandria.
imbricàta. B.R. (*br.wh.*) imbricated. Nepaul. 1824. 2. 8. S. ♃. Bot. reg. t. 1213.

POLYST'ACHIA. H.E.F. Polyst'achia. Gynandria Monandria.
1 lutèola. H.E.F. (*y.gr.*) smooth. W. Indies. 1818. 7. 8. S. ♃. Hook. ex. flor. t. 103.
Dendròbium polystàchyum. W.

2 pubérula. B.R. (*y.gr.*) downy. Sierra Leone. 1821. 8. S.♃. Bot. reg. t. 851.
BRASSAV'OLA. H.K. BRASSAV'OLA. Gynandria Monandria. H.K.
1 cucullàta. H.K. (*wh.*) single-flowered. W. Indies. 1793. 6. 7. S.♃. Bot. mag. t. 543.
2 tuberculàta. B.M. (*wh.st.*) tuberculated. Brazil. 1826. 6. 9. S.♃. ——— t. 2878.
3 nodòsa. Ly. (*wh.*) many-flowered. W. Indies. 1824. 10. S.♃.
BROUGHT'ONIA. H.K. BROUGHT'ONIA. Gynandria Monandria. H.K.
1 sanguínea. H.K. (*cr.*) blood-coloured. Jamaica. 1793. 6. 7. S.♃. Lodd. bot. cab. t. 793.
2 nítida. W.H. (*re.*) glossy. E. Indies. 1824. —— S.♃.
ZYGOPE'TALON. B.M. ZYGOPE'TALON. Gynandria Monandria.
1 Mackàii. B.M. (*gr.bl.*) Mackay's. Brazil. 1825. 2. 5. S.♃. Bot. mag. t. 2748.
2 rostràtum. B.M. (*wh.pu.*) beaked. Demarara. —— 9. 11. S.♃. ——— t. 2819.
CA'TTLEYA. L.col. CA'TTLEYA. Gynandria Monandria. H.E.F.
1 labiàta. L.col. (*vi.*) great-lipped. S. America. 1818. 4. 10. S.♃. Lindl. coll. t. 33.
2 Loddigèsii. H.E.F. (*vi.*) Loddiges'. ——— 1810. —— S.♃. Hook. ex. flor. t. 186.
Epidéndrum violàceum. Lodd. bot. cab. t. 337.
3 Fórbesii. B.R. (*ye.*) yellow. Brazil. 1823. 9. S.♃. Bot. reg. t. 953.
4 críspa. B.R. (*wh.pu.*) curled-lipped. Trinidad. 1820. —— S.♃. ——— t. 1172.
5 intermèdia. B.M. (*ro.*) intermediate. Brazil. 1824. 3. 5. S.♃. Bot. mag. t. 2851.
CERATOCH'ILUS. B.C. CERATOCH'ILUS. Gynandria Monandria.
1 grandiflòrus. B.C. (*wh.*) large-flowered. Trinidad. 1827. 3. 8. S.♃. Lodd. bot. cab. t. 1414.
2 insígnis. (*st.pu.*) splendid. S. America. 1818. 9. 4. S.♃. Bot. mag. t. 2948-2949.
Stanhòpea insígnis. B.M.
EPIDE'NDRUM. H.K. EPIDE'NDRUM. Gynandria Monandria. H.K.
1 cochleàtum. W. (*d.pu.*) shell-flowered. W. Indies. 1786. 1. 12. S.♃. Bot. mag. t. 572.
2 fràgrans. W. (*st.*) sweet-scented. ——— 1778. 10. S.♃. ——— t. 1669.
3 secúndum. W. (*re.*) pale-flowered. ——— 1793. 6. 7. S.♃. Jacq. amer. t. 137.
4 polybúlbon. W. (*wh.*) many-bulbed. ——— 1822. —— S.♃. Hook. ex. flor. t. 112.
5 verrucòsum. W. (*st.*) warted. ——— 1824. —— S.♃. Lodd. bot. cab. t. 1084.
6 nùtans. W. (*gr.*) nodding. ——— 1794. —— S.♃. Hook. ex. flor. t. 50.
7 umbellàtum. W. (*st.*) umbelled. Jamaica. 1793. —— S.♃. Bot. reg. t. 80.
8 diffùsum. W. (*gr.*) diffuse. W. Indies. 1818. 4. 5. S.♃. Lodd. bot. cab. t. 846.
9 ancéps. B.C. (*gr.*) two-edged. ——— 1820. 8. 10. S.♃. ——— t. 887.
10 noctúrnum. W. (*gr.*) night-smelling. ——— —— 4. 6. S.♃. ——— t. 713.
11 pallidiflòrum. B.M. (*st.*) pale-flowered. ——— 1828. 5. 6. S.♃. Bot. mag. t. 2980.
12 fuscàtum. W. (*br.*) brown. ——— 1790. 6. 7. S.♃. Bot. reg. t. 67.
13 elongàtum. W. (*re.*) long-stalked. Caracas. 1798. 5. 8. S.♄. Bot. mag. t. 611.
14 ellípticum. H.E.F. (*re.*) elliptic-leaved. Brazil. 1824. 3. 8. S.♄. Hook. ex. flor. t. 207.
15 conópseum. H.K. (*st.*) Florida. Florida. 1775. 8. S.♃.
16 monophy'llum. H.E.F. (*gr.*) one-leaved. W. Indies. 1822. 7. S.♃. Hook. ex. flor. t. 109.
17 ciliàre. B.R. (*wh.*) fringed. ——— 1790. 6. 10. S.♃. Bot. reg. t. 784.
18 cuspidàtum. B.R. (*wh.*) pointed. ——— 1808. —— S.♃. ——— t. 783.
ENC'YCLIA. B.M. ENC'YCLIA. Gynandria Monandria.
viridiflòra. B.M. (*gr.*) green-flowered. Brazil. 1826. 2. 6. S.♃. Bot. mag. t. 2831.
FERNAND'ESIA. R.P. FERNAND'ESIA. Gynandria Monandria.
élegans. B.C. (*st.pu.*) elegant. Trinidad. 1822. 4. 10. S.♃. Lodd. bot. cab. t. 1212.
Lockhártia élegans. Bot. mag. t. 2715.
CRYPTARRH'ENA. B.R. CRYPTARRH'ENA. Gynandria Monandria. B.R.
lunàta. B.R. (*ye.*) crescent-lipped. W. Indies. 1815. 5. 8. S.♃. Bot. reg. t. 153.
EUL'OPHIA. B.R. EUL'OPHIA. Gynandria Monandria. B.R.
1 vìrens. B.R. (*gr.*) green. E. Indies. 1823. 6. 8. S.♃. Roxb. corom. 1. t. 38.
Limodòrum vìrens. R.C.
2 guineénsis. B.R. (*pi.*) Guinea. Sierra Leone. 1822. 11. 2. S.♃. Bot. reg. t. 686.
3 grácilis. B.R. (*gr.*) slender ——— —— 6. 9. S.♃. ——— t. 742.
4 streptopétala. B.R. (*ye.*) twisted-petaled. C. B. S. —— 12. 6. S.♃. ——— t. 1002.
5 ensàta. B.R. (*st.*) sword-leaved. ——— 1826. 7. 8. S.♃. ——— t. 1147.
6 trístis. B.R. (*da.*) dark-flowered. ——— 1825. 5. 6. G.♃.
7 longicórnis. B.R. (*ye.*) long-spurred. ——— —— 6. 8. G.♃.
8 barbàta. B.R. (*ye.*) bearded. ——— 1822. —— G.♃.
CALA'NTHE. B.R. CALA'NTHE. Gynandria Monandria. B.R.
1 veratrifòlia. B.R. (*wh.*) Veratrum-leav'd. E. Indies. 1819. 1. 9. S.♃. Bot. reg. t. 720.
2 sylvéstris. Ly. (*wh.*) loose-spiked. Madagascar. 1823. 6. 7. S.♃.
ANGR'ÆCUM. L.col. ANGR'ÆCUM. Gynandria Monandria. B.R.
1 maculàtum. B.R. (*fl.*) spotted. S. America. 1819. 10. 12. S.♃. Bot. reg. t. 618.
2 falcàtum. L.col. (*wh.*) falcate. China. 1815. —— S.♃. ——— t. 283.
Limodòrum falcàtum. B.R.
3 lùridum. Ly. (*br.*) lurid. Sierra Leone. 1822. 7. S.♃.
IONO'PSIS. K.S. IONO'PSIS. Gynandria Monandria.
utricularioídes. L.col. (*w.*) pale-flowered. Trinidad. 1822. 6. 9. S.♃. Hook. ex. flor. t. 113.
Iántha pállida. H.E.F.

AERA'NTHES. B.R. AERA'NTHES. Gynandria Monandria. B.R.
1 grandiflòra. B.R. (*st.*) large-flowered. Madagascar. 1823. 7. 8. S.♃. Bot. reg. t. 817.
2 sesquipedàlis.Ly.(*wh.*) long-flowered. ——— ——— .... S.♃.
A'ERIDES. H.K. AIR-PLANT. Gynandria Monandria. H.K.
1 odoràtum. H.K. (*pi.*) fragrant. China. 1800. .... S.♃.
2 aráchnites. W. (*br.pu.*) common. Japan. 1793. .... S.♃. Kæmpf. jap. t. 869. f. 1.
3 guttàtum. W.H. (*sp.*) spotted-flowered. E. Indies. 1824. ——— S.♃.
4 costàtum. W.H....... ribbed. ——— ——— ——— S.♃.
SARCA'NTHUS. L.col. SARC'ANTHUS. Gynandria Monandria. B.R.
1 paniculàtus.L.col.(*ye.*) panicled. China. 1812. 5. 8. S.♄. Bot. reg. t. 220.
2 succìsus. B.R. (*ye.br.*) bitten. ——— 1824. ——— S.♄. ——— t. 1014.
3 rostràtus.B.R.(*y.or.li.*) rostrate. ——— 1819. 6. 12. S.♄. ——— t. 981.
4 teretifòlius.B.R.(*pu.ye.*)round-leaved. ——— ——— ——— S.♄. ——— t. 676.
Vánda teretifòlia. B.R.
TRICHORH'IZA. Ly. TRICHORH'IZA. Gynandria Monandria.
teretifòlia. L.C.(*br.pu.*) round-leaved. E. Indies. 1819. 10. S.♄. Hook. ex. flor. t. 72.
Vánda? trichorhìza. H.E.F.
VA'NDA. B.R. VA'NDA. Gynandria Monandria. B.R.
1 cruénta. L.C. (*cr.*) red-flowered. China. 1819. .... S.♄.
2 multiflòra. L.col. (*ye.*) many-flowered. ——— 1800. 6. 10. S.♄. Lindl. coll. t. 38.
3 Roxbúrghii.B.R.(*w.pu.*) Roxburgh's. E. Indies. 1816. 10. 12. S.♄. Bot. reg. t. 506.
4 tessellàta. L.C. (*pu.ye.*) tessellate. ——— ——— .... S.♄. Roxb. cor. 1. t. 42.
Cymbídium tessellàtum. R.C.
RENANTH'ERA. B.R. RENANTH'ERA. Gynandria Monandria.
coccínea. B.R. (*sc.*) scarlet-flowered. China. 1816. 8. 4. S.♄. Bot. reg. t. 1131.
VANI'LLA. W. VANI'LLA. Gynandria Monandria. H.K.
1 aromática. H.K. (*wh.*) aromatic. S. America. 1739. 6. 8. S.♄. Plum. ic. 183. t. 188.
2 planifòlia. H.K. (*wh.*) flat-leaved. W. Indies. 1800. 4. 6. S.♄. Andr. reposit. t. 538.
CYPRIP'EDIUM. W. LADIES'-SLIPPER. Gynandria Diandria. H.K.
1 Calcèolus. W. (*ye.*) common. England. .... 5. 7. H.♃. Eng. bot. v. 1. t. 1.
2 parviflòrum. W. (*ye.*) small-flowered. N. America. 1759. ——— H.♃. Swt. br. fl. gar. t. 80.
3 pubéscens. W. (*ye.*) yellow-downy. ——— 1790. ——— H.♃. ——— t. 71.
4 guttàtum. W. (*ye.*) spotted. Siberia. 1829. 4. 5. H.♃. Amman.ruth.p.133.t.22
5 macránthos.W. (*d.pu.*) great-flowered. ——— ——— ——— H.♃. Bot. mag. t. 2938.
6 ventricòsum.W.(*d.pu.*) bellied. ——— ——— ——— H.♃. Swt. br. fl. gar. s. 2. t. 1.
7 spectábile.W.(*pu.wh.*) showy. N. America. 1731. 6. 7. H.♃. Lodd. bot. cab. t. 697.
álbum. Bot. mag. t. 216.
*α incarnàtum.*B.F.G.(*w.pu.*) *flesh-coloured.* ——— ——— ——— H.♃. Swt. br. fl. gar. t. 240. B.
β álbum. B.F.G. (*wh.*) *white-flowered.* ——— 1827. ——— H.♃. ——— t. 240. A.
8 hùmile. W. (*pu.wh.*) two-leaved. ——— 1786. 5. 6. H.♃. Bot. mag. t. 192.
9 arietìnum. W. (*re.*) ram's-head. ——— 1808. ——— H.♃. Swt. br. fl. gar. t. 213.
10 venústum.B.R.(*pu.gr.*) handsome. Nepaul. 1816. 7. 10. S.♃. Bot. reg. t. 788.
11 insígne. L.col. (*gr.pu.*) noble. ——— 1819. ——— S.♃. Lindl. coll. t. 32.

ORDO CCIII.

SCITAMINEÆ. *Brown prodr.* 1. *p.* 305.

GLO'BBA. L. GLO'BBA. Monandria Monogynia. L.
1 marantìna. W. (*ye.*) marantine. E. Indies. 1800. 7. 8. S.♃. Exot. bot. 2. t. 103.
2 bulbífera. F.I. (*ye.*) bulb-bearing. ——— 1818. ——— S.♃.
3 orixénsis. F.I. (*or.*) Orixian. ——— 1819. ——— S.♃.
4 versícolor.Ex.B.(*y.vi.*) various-colour'd. ——— 1820. ——— S.♃. Sm. exot. bot. 2. t. 117.
Hùra. F.I.
5 péndula. F.I. (*ye.*) pendulous-flower'd. ——— 1822. 6. 8. S.♃.
6 Careyàna. F.I. (*ye.*) Carey's. ——— 1821. ——— S.♃. Rosc. scit. t. 110.
7 sessiliflòra. B.M. (*ye.*) sessile-flower'd. ——— 1807. ——— S.♃. Bot. mag. t. 1428.
MANTI'SIA. B.M. MANTI'SIA. Monandria Monogynia. B.M.
1 saltatòria. B.M. (*bl.ye.*) Opera-girl's. E. Indies. 1808. 6. S.♃. Bot. mag. t. 1320.
Glóbba subulàta. F.I. *purpùrea.* A.R. 615. *saltatòria.* R. scit. t. 111.
2 spathulàta.R.S. (*bl.ye.*) spathulate. E. Indies. 1822. 6. 8. S.♃.
*Glóbba spathulàta.*F.I.
CURC'UMA. W. TURMERIC. Monandria Monogynia. L.
1 aromática. P.L.(*ye.pi.*) aromatic. E. Indies. 1797. 4. 8. S.♃. Sal. par. lond. t. 96.
Zerúmbet. F.I. Rheed. mal. v. 11. t. 7.

2 Zedoària.	B.M.	(*ye.pi.*)	silky-leaved.	E. Indies.	1797.	4. 8.	S.♃.	Bot. mag. t. 1546.
3 zanthorrhìza.	F.I.	(*re.cr.*)	yellow-rooted.	Amboyna.	1819.	4. 5.	S.♃.	
4 latifòlia.	R.Sc.	(*ye.pi.*)	broad-leaved.	E. Indies.	——	——	S.♃.	Rosc. scitam. t. 108.
5 elàta.	F.I.	(*ye.pu.*)	tall.	——	——	——	S.♃.	—— t. 104.
6 c'æsia.	F.I.	(*ye.re.*)	clouded-leaved.	——	——	5. 6.	S.♃.	
7 æruginòsa.	F.I.	(*ye.re.*)	verdigrease.	——	1807.	4. 8.	S.♃.	Rosc. scitam. t. 106.
8 ferrugìnea.	F.I.	(*ye.re.*)	rusty.	——	1819.		S.♃.	—— t. 105.
9 rubéscens.	F.I.	(*re.ye.*)	red.	——	——	4. 6.	S.♃.	—— t. 107.
10 comòsa.	F.I.	(*ye.pi.*)	close-flowered.	——	——	——	S.♃.	
11 leucorrhìza.	F.I.	(*y.ro.*)	white-rooted.	——	——	——	S.♃.	Rosc. scitam. t. 102.
12 angustifòlia.	F.I.	(*y.pu.*)	narrow-leaved.	——	1822.	——	S.♃.	Bot. reg. t. 886.
13 lónga.	F.I.	(*st.bh.*)	long-rooted	——	1759.	6. 8.	S.♃.	Rosc. scitam. t. 99.
14 Amàda.	F.I.	(*st.li.*)	Mango-ginger.	——	1819.	——	S.♃.	
15 viridiflòra.	F.I.	(*gr.st.*)	green-flowered.	——	1820.	——	S.♃.	
16 montàna.	F.I.	(*ye.ro.*)	mountain.	——	1823.	4. 6.	S.♃.	Roxb. corom. 2. t. 151.
17 reclinàta.	F.I.	(*pi.*)	reclining.	——	1824.	——	S.♃.	
18 petiolàta.	F.I.	(*ye.pu.*)	petiolate.	——	1822.	——	S.♃.	Rosc. scitam. t. 100.
19 rubricaùlis.	L.T.	(*re.ye.*)	red-stalked.	——	1818.	——	S.♃.	
20 amaríssima.	R.Sc.	(*re.ye.*)	bitter.	——	1822.	——	S.♃.	Rosc. scitam. t. 101.
KÆMPF'ERIA. L.			KÆMPF'ERIA.	Monandria Monogynia. L.				
1 Galánga.	W.	(*pu.wh.*)	officinal.	E. Indies.	1724.	6. 9.	S.♃.	Bot. mag. t. 850.
2 rotúnda.	W.	(*pu.bh.*)	round-rooted.	——	1768.	7. 8.	S.♃.	—— t. 920.
3 latifòlia.	S.S.	(*wh.*)	broad-leaved.	——	1803.	4. 7.	S.♃.	
4 Roscoeàna.	B.R.	(*wh.*)	Roscoe's.	Ava.	1828.	——	S.♃.	Bot. reg. t. 1212.
5 angustifòlia.	F.I.	(*pu.w.*)	narrow-leaved.	E. Indies.	1797.	——	S.♃.	Rosc. scitam. t. 94.
6 panduràta.	F.I.	(*ro.*)	fiddle-lipped.	——	1812.	——	S.♃.	Bot. reg. t. 173.
7 ovalifòlia.	F.I.	(*pu.wh.*)	oval-leaved.	——	1819.	——	S.♃.	Rosc. scitam. t. 95.
8 marginàta.	R.S.	(*pu.*)	red-margined.	——	1820.	——	S.♃.	—— t. 93.
ROSC'OEA. L.T.			ROSC'OEA.	Monandria Monogynia.				
1 purpùrea.	L.T.	(*pu.*)	purple-flower'd.	Nepaul.	1819.	7. 8.	S.♃.	Rosc. scitam. t. 64.
2 grácilis.	L.T.	(*pu.*)	slender.	——	1820.	——	S.♃.	
3 elàtior.	L.T.	(*pu.*)	tall.	——	——	——	S.♃.	
4 spicàta.	L.T.	(*pu.*)	spiked.	——	1821.	——	S.♃.	
5 capitàta.	L.T.	(*pu.*)	headed.	——	1819.	——	S.♃.	
ZI'NGIBER. L.T.			GINGER.	Monandria Monogynia. H.K.				
1 officinàle.	F.I.	(*pu.st.*)	narrow-leaved.	E. Indies.	1605.	6. 8.	S.♃.	Rosc. scitam. t. 83.
2 Zerúmbet.	F.I.	(*st.*)	broad-leaved.	——	1690.	9. 11.	S.♃.	Sm. exot. bot. 2. t. 112.
3 Cassumùnar.	B.M.	(*wh.*)	downy-leaved.	——	1796.	8.	S.♃.	Bot. mag. t. 1426.
purpùreum. L.T. *Amòmum Cliffòrdiæ.* And. repos. t. 555.								
4 ròseum.	F.I.	(*ro.st.*)	rose-coloured.	E. Indies.	1822.	6. 8.	S.♃.	Roxb. corom. 2. t. 126.
5 ligulàtum.	F.I.	(*pi.ye.*)	ligulate.	——	——	——	S.♃.	
6 rùbens.	F.I.	(*re.*)	red-flowered.	——	1820.	5. 11.	S.♃.	Rosc. scitam. t. 88.
7 squarròsum.	F.I.	(*wh.*)	squarrose.	——	——	——	S.♃.	—— t. 89.
8 panduràtum.	F.I.	(*st.re.*)	fiddle-leaved.	——	1819.	5. 8.	S.♃.	
9 Miòga.	L.T.	(*pi.*)	Japanese.	Japan.	1796.	——	S.♃.	Kæmpf. ic. Banks. t. 1.
10 chrysánthum.	R.Sc.	(*ye.*)	yellow-flower'd.	Nepaul.	1821.	——	S.♃.	Rosc. scitam. t. 86.
11 capitàtum.	F.I.	(*ye.*)	headed.	E. Indies.	1820.	6. 9.	S.♃.	—— t. 90.
12 elàtum.	F.I.	(*ye.*)	tall.	——	——	——	S.♃.	—— t. 91.
AM'OMUM. F.I.			AM'OMUM.	Monandria Monogynia. L.				
1 Afzèlii.	L.T.	(*pi.*)	sweet-scented.	Sierra Leone.	1795.	5. 6.	S.♃.	Ann. bot. 1. t. 13.
2 grandiflòrum.	R.S.	(*pi.w.*)	large-flower'd.	——	——	6. 7.	S.♃.	Sm. exot. bot. t. 3.
3 sylvéstre.	R.S.	(*ye.*)	wood.	Jamaica.	1804.	5. 7.	S.♃.	Sloan. hist. 1. t. 105. f. 2.
4 Cardamòmum.	F.I.	(*ye.*)	bastard Cardamon.	E. Indies.	1820.	4. 6.	S.♃.	Rumph. amb. 5. t. 65. f. 1.
5 angustifòlium.	F.I.	(*re.y.*)	narrow-leaved.	Madagascar.	——	——	S.♃.	Sonnerat. it. 2. f. 137.
6 aculeàtum.	F.I.	(*or.cr.*)	bristly-capsuled.	E. Indies.	1823.	6. 8.	S.♃.	
7 máximum.	F.I.	(*re.ye.*)	great.	——	——	——	S.♃.	Bot. reg. t. 929.
8 dealbàtum.	F.I.	(*wh.*)	insipid-seeded.	——	1819.	——	S.♃.	
9 subulàtum.	F.I.	(*ye.*)	subulate-bracted.	——	——	5. 7.	S.♃.	
10 aromáticum.	F.I.	(*st.*)	aromatic.	——	1823.	——	S.♃.	
11 seríceum.	F.I.	(*wh.pi.*)	silky-leaved.	——	1819.	6. 8.	S.♃.	
12 grána-paradìsi.	W.	(*re.*)	Grains of Paradise.	Madagascar.....		2. 4.	S.♃.	Blackw. t. 385.
CO'STUS. F.I.			CO'STUS.	Monandria Monogynia. L.				
1 speciòsus.	F.I.	(*wh.*)	large-flowered.	E. Indies.	1794.	8. 9.	S.♃.	Jacq. ic. 1. t. 1.
2 nepalénsis.	R.Sc.	(*wh.*)	Nepaul.	Nepaul.	1818.	——	S.♃.	Rosc. scitam. t. 80.
speciòsus β. Bot. reg. 665.								
3 cylíndricus.	R.S.	(*wh.re.*)	cylindrical-spiked.	Brazil.	1822.	5. 9.	S.♃.	Rosc. scitam. t. 78.
Pisònis. Bot. reg. t. 899.								
4 áfer.	B.R.	(*wh.*)	African.	Sierra Leone.	——	——	S.♃.	Bot. reg. t. 683.
5 díscolor.	R.Sc.	(*wh.*)	two-coloured.	Maranham.	1823.	——	S.♃.	Rosc. scitam. t. 81.

6 maculàtus. R.Sc. (*wh.*) spotted-stalked. SierraLeone. 1822. 5. 9. S.♃. Rosc. scitam. t. 82.
7 lanàtus. L.C. (*wh.*) woolly. .... —— —— S.♃.
8 spicàtus. L.T. (*ye.*) spiked. S. America. 1793. 6. S.♃. Rosc. scitam. t. 77.
9 comòsus. L.T. (*re.ye.*) tufted. ——— 1815. —— S.♃. Jacq. ic. 3. t. 202.
Alpínia comòsa. Jacq.
10 spiràlis. L.T. (*re.*) spiral. ——— —— 5. 7. S.♃. Jacq. schœn. 1. t. 1.
11 villosíssima. R.S. (*ye.*) villous. W. Indies. 1823. —— S.♃. Jacq. fragm. t. 80.
12 arábicus. L.T. (*wh.*) Arabian. Arabia. 1752. 8. S.♃.

HELL'ENIA. B.P. Hell'enia. Monandria Monogynia. W.
cœrùlea. B.P. (*bh.*) blue-berried. N.Holland. 1820. 5. 6. G.♃. Rosc. scitam. t. 76.

ALPI'NIA. L.T. Alpi'nia. Monandria Monogynia. L.
1 Allúghas. F.I. (*wh.re.*) glossy-leaved. E. Indies. 1796. 2. 3. S.♃. Rosc. scitam. t. 67.
Hellènia Allúghas. Andr. reposit. t. 501.
2 racemòsa. L.T. (*wh.*) clustered. W. Indies. 1752. 7. 9. S.♃. Plum. ic. 11. t. 20.
3 Galánga. F.I. (*g.wh.*) Galanga-major. E. Indies. 1818. 6. 10. S.♃. Rumph. amb. 5. t. 63.
4 Roxbúrghii. (*w.cr.*) Roxburgh's. ——— 1820. —— S.♃.
bracteàta. F.I. *non* L.T.
5 malaccénsis. F.I. (*w.y.cr.*) woolly-leaved. ——— 1799. 4. 6. S.♃. Bot. reg. t. 328.
6 nùtans. F.I. (*re.ye.*) nodding-flower'd. ——— 1792. —— S.♃. Bot. mag. t. 1903.
Reneàlmia nùtans. A.K. 360.
7 cérnua. B.M. (*y.wh.pu.*) drooping. ——— 1790. —— S.♃. Bot. mag. t. 1900.
8 auriculàta. R.Sc. (*y.re.*) eared. ——— 1815. —— S.♃. Rosc. scitam. t. 74.
9 diffíssa. L.T. (*ye.*) cloven. China. —— 4. 6. S.♃. ——— t. 71.
10 mùtica. F.I. (*ye.re.*) spurless. E. Indies. 1811. 4. 9. S.♃. ——— t. 69.
11 calcaràta. F.I. (*re.or.*) upright spiked. China. 1804. 9. S.♃. Andr. reposit. t. 421.
12 spicàta. F.I. (*wh.pu.*) spiked. E. Indies. 1823. 6. 9. S.♃.
13 striàta. L.en. (*std.*) striated. ——— 1818. —— S.♃.
14 bracteàta. L.T. (*ye.re.*) bracted. China. 1823. 5. 8. S.♃. Rosc. scitam. t. 70.
Roscoeàna. R.S.
15 magnífica. R.Sc. (*sc.*) magnificent. Madagascar. 1826. .... S.♃. Rosc. scitam. t. 75.

RENE'ALMIA. R.Sc. Rene'almia. Monandria Monogynia.
1 occidentàlis. (*pa.*) occidental. Jamaica. 1793. 5. 7. S.♃. Gært. fruct. 1. t. 12.
Alpínia occidentàlis. L.T.
2 exaltàta. R.Sc. (*re.*) tubular-flowered. Demarara. 1820. 7. 8. S.♃. Rosc. scitam. t. 65.
Alpínia tubulàta. Bot. reg. t. 777.
3 punícea. (*cr.*) bright red. E. Indies. —— —— S.♃.
Alpínia punìcea. F.I.
4 Cardamòmum mèdium. (*re.*) 9-winged. ——— —— —— S.♃.
5 linguifórmis. F.I. (*re.*) tongue-lipped. ——— 1825. .... S.♃.

ELETT'ARIA. L.T. Cardamon. Monandria Monogynia.
Cardamòmum. L.T. (*w.pu.*) true. E. Indies. 1815. .... S.♃. Linn. trans. 10. t. 4. 5.
Alpínia Cardamòmum. F.I. *Amòmum Cardamòmum.* L.

HEDY'CHIUM. F.I. Garland-flower. Monandria Monogynia. L.
1 angustifòlium. R.Sc. (*sc.*) narrow-leaved. E. Indies. 1815. 5. 12. S.♃. Rosc. scitam. t. 60.
2 aurantìacum. R.Sc. (*or.*) orange-coloured. ——— —— —— S.♃. ——— t. 61.
angustifòlium. B.R. 157. *non* Roxb.
3 coccíneum. R.Sc. (*sc.*) scarlet. Nepaul. 1818. 5. 12. S.♃. Rosc. scitam. t. 58.
angustifòlium. B.M. *non* Roxb.
4 longifòlium. R.Sc. (*sc.*) long-leaved. E. Indies. 1819. 6. 11. S.♃. Rosc. scitam. t. 59.
5 cárneum. R.Sc. (*re.*) flesh-coloured. ——— —— —— S.♃. Bot. mag. t. 2637.
6 flàvum. R.Sc. (*ye.*) yellow. ——— 1818. —— S.♃. Rosc. scitam. t. 49.
7 flavéscens. R.Sc. (*st.*) yellowish. ——— 1821. —— S.♃. ——— t. 50.
8 thyrsifórme. S.S. (*wh.*) thyrse-flowered. ——— 1819. 7. 11. S.♃. ——— t. 56.
heteromállum. Bot. reg. t. 767.
9 villòsum. F.I. (*st.*) villous. ——— 1815. 6. 11. S.♃. Rosc. scitam. t. 54.
10 glaùcum. R.Sc. (*wh.*) glaucous. ——— 1820. —— S.♃. ——— t. 53.
11 ellípticum. R.Sc. (*wh.*) elliptic-leaved. ——— —— 5. 8. S.♃. ——— t. 55.
12 acuminàtum. R.Sc. (*w.or.*) taper-pointed. ——— 1822. —— S.♃. Bot. mag. t. 2969.
13 elàtum. B.R. (*bh.*) tall. Nepaul. 1818. 7. 12. S.♃. Bot. reg. t. 526.
14 speciòsum. F.I. (*ye.*) beautiful. E. Indies. 1823. 8. 10. S.♃.
15 máximum. R.Sc. (*wh.*) large-flowered. ——— 1819. 5. 9. S.♃. Rosc. scitam. t. 52.
16 coronàrium. F.I. (*wh.*) sweet-scented. ——— 1793. 7. 9. S.♃. Bot. mag. t. 708.
17 Gardneriànum. R.Sc. (*ye.*) Gardner's. Nepaul. 1818. —— S.♃. Bot. reg. t. 771.
18 spicàtum. B.M. (*bh.*) spiked. ——— 1820. 8. 11. S.♃. Bot. mag. t. 2300.
19 grácile. F.I. (*wh.*) slender. E. Indies. 1821. —— S.♃.

ORDO CCIV.

CANNEÆ. *Brown prodr.* 1. *p.* 307.

PHRY'NIUM. R.Sc.		PHRY'NIUM.	Monandria Monogynia. L.				
1 dichótomum. F.I.	(*wh.*)	forked.	E. Indies.	1807.	6. 9.	S.♃.	Rumph. amb. 4. t. 7.
2 virgàtum. F.I.	(*wh.*)	slender-stem'd.	——	1818.	——	S.♃.	
3 spicàtum. F.I.	(*wh.*)	spiked.	——	——	——	S.♃.	
4 imbricàtum. F.I.	(*li.*)	imbricated.	——	1824.	5. 8.	S.♃.	
5 grandiflòrum. R.Sc.	(*st.*)	great-flowered.	Brazil.	1819.	6. 8.	S.♃.	Rosc. scitam. t. 33.
6 parviflòrum. F.I.	(*ye.*)	small-flowered.	E. Indies.	——	5. 9.	S.♃.	—— t. 34.
7 capitàtum. F.I.	(*ro.*)	capitate.	——	1820.	——	S.♃.	Rheed. mal. 11. t. 34.
8 violàceum. R.Sc.	(*vi.*)	violet-coloured.	Brazil.	1823.	6. 8.	S.♃.	Rosc. scitam. t. 37.
Calathèa violàcea. Bot. reg. t. 961.							
9 comòsum. R.Sc.	(*ye.*)	tufted.	Trinidad.	1812.	——	S.♃.	Rosc. scitam. t. 35.
10 zebrìnum. R.Sc.	(*pu.*)	Zebra plant.	Brazil.	1815.	4. 7.	S.♃.	—— t. 36.
Marànta zebrìna. Bot. reg. t. 385. *Calathèa zebrìna.* B.R.							
11 flavéscens.	(*ye.*)	yellowish.	Brazil.	1823.	6. 8.	S.♃.	Bot. reg. t. 932.
Calathèa flavéscens. B.R.							
12 Alloùya. R.Sc.	(*wh.*)	variegated.	Guiana.	1824.	——	S.♃.	Rosc. scitam. t. 38.
13 Myrósma. R.Sc.	(*wh.*)	Canna-like.	Surinam.	1825.	——	S.♃.	—— t. 39.
14 cylíndricum. R.Sc.	(*st.*)	cylindrical.	Brazil.	——	2. 3.	S.♃.	—— t. 40.
15 setòsum. R.Sc.		bristly.	——	1826.	——	S.♃.	—— t. 41.
Marànta secúnda. Gr.							
16 Parkèri. R.Sc.	(*ye.*)	Parker's.	Grenada.	——	——	S.♃.	Rosc. scitam. t. 42.
17 Casùpo. R.Sc.	(*ye.*)	two-coloured.	Trinidad.	1823.	6. 8.	S.♃.	—— t. 43.
Calathèa díscolor. S. *Marànta Casùpo.* Jacq. frag. t. 63. f. 4.							
18 ellípticum. R.Sc.	(*st.*)	elliptic-leaved.	Guiana.	1823.	——	S.♃.	Rosc. scitam. t. 44.
19 Róssii. L.C.	(*ye.*)	Ross's.	Brazil.	1825.	4. 7.	S.♃.	
20 villòsum. L.C.	(*ye.*)	villous.	——	——	——	S.♃.	
21 lùteum.	(*wh.*)	yellow-bracted.	Caracas.	1809.	6. 7.	S.♃.	Jacq. ic. 2. t. 20.
Marànta lùtea. Jacq. *Jacquìni.* R.S.							
22 oblíquum. L.C.	(*ye.*)	oblique-leaved.	E. Indies.	1824.	5. 9.	S.♃.	
23 angustifòlium. L.C.	(*ye.*)	narrow-leaved.	——	——	——	S.♃.	
24 littoràle. Led.	(*ye.*)	sea-side.	Brazil.	1827.	——	S.♃.	
25 longibracteàtum.	(*li.*)	long-bracted.	——	1824.	6. 7.	S.♃.	Bot. reg. t. 1020.
26 grandifòlium.	(*ye.*)	large-leaved.	Brazil.	1825.	——	S.♃.	—— t. 1210.
27 maciléntum. B.R.	(*wh.*)	meagre.	——	——	——	S.♃.	
TH'ALIA. W.		TH'ALIA.	Monandria Monogynia. W.				
1 dealbàta. L.T.	(*da.pu.*)	mealy.	Carolina.	1791.	7. 9.	H.*w.*♃.	Bot. mag. t. 1690.
2 geniculàta. R.Sc.	(*bl.*)	jointed.	Demerara.	1824.	6. 8.	S.♃.	Rosc. scitam. t. 45.
CA'NNA. W.		INDIAN SHOT.	Monandria Monogynia. L.				
1 índica. R.Sc.	(*sc.*)	common.	W. Indies.	1570.	1. 12.	S.♃.	Rosc. scitam. t. 1.
2 coccínea. L.T.	(*sc.*)	scarlet.	——	1731.	——	S.♃.	—— t. 11.
3 lùtea. L.T.	(*ye.*)	yellow.	S. America.	1629.	——	S.♃.	—— t. 18.
4 orientàlis. R.Sc.	(*sc.*)	oriental.	E. Indies.	1570.	——	S.♃.	—— t. 12.
β *flàva.*	(*ye.*)	*yellow Indian.*	——	——	——	S.♃.	—— t. 13.
γ *maculàta.*	(*sc.ye.*)	*spotted-flowered.*	——	——	——	S.♃.	Bot. mag. t. 2085.
5 pàtens. B.R.	(*sc.ye.*)	spreading.	China.	1771.	7. 10.	H.♃.	Bot. reg. t. 576.
6 speciòsa. B.M.	(*sc.ye.*)	showy.	Nepaul.	1820.	——	H.♃.	Bot. mag. t. 2317.
7 gigántea. B.R.	(*or.*)	gigantic.	S. America.	1788.	1. 12.	S.♃.	Bot. reg. t. 206.
8 latifòlia. R.Sc.	(*sc.*)	broad-leaved.	——	——	——	S.♃.	Rosc. scitam. t. 4.
9 èdulis. B.R.	(*sc.*)	eatable.	Peru.	1820.	6. 10.	S.♃.	Bot. reg. t. 775.
10 díscolor. B.R.	(*sc.*)	crimson-leaved.	Trinidad.	1827.	8. 12.	S.♃.	—— t. 1231.
11 Lambérti. B.R.	(*cr.*)	Mr. Lambert's.	——	1818.	5. 8.	S.♃.	—— t. 470.
12 excélsa. B.C.	(*sc.*)	tall.	Brazil.	1820.	1. 12.	S.♃.	Lodd. bot. cab. t. 743.
13 limbàta. R.Sc.	(*sc.ye.*)	bordered.	——	1818.	——	S.♃.	Rosc. scitam. t. 9.
aùro-vittàta. Lodd. bot. cab. t. 449.							
14 pállida. R.Sc.	(*st.*)	pale.	W. Indies.	1823.	——	S.♃.	Rosc. scitam. t. 19.
Lagunénsis. Bot. reg. t. 1311.							
β *punctàta.*	(*ye.*)	*spotted-flowered.*	——	——	——	S.♃.	Rosc. scitam. t. 20.
15 aurantìaca. R.Sc.	(*or.*)	orange-coloured.	Brazil.	——	——	S.♃.	—— t. 21.
16 compácta. R.Sc.	(*cr.*)	compact.	S. America.	——	——	S.♃.	—— t. 22.
17 pedunculàta. R.Sc.	(*ye.*)	stalked.	Brazil.	1820.	——	S.♃.	—— t. 8.
18 denudàta. R.Sc.	(*sc.ye.*)	naked-stalked.	——	——	——	S.♃.	—— t. 23.
19 sylvéstris. R.Sc.	(*cr.*)	wood.	S. America.	1813.	——	S.♃.	—— t. 10.
20 lanuginòsa. R.Sc.	(*sc.*)	woolly-stem'd.	Maranham.	1823.	——	S.♃.	—— t. 16.

21 cárnea. R.Sc.	(*fl.*)	flesh-coloured.	Brazil.	1822.	1. 12.	S.♃.	Rosc. scitam. t. 15.
22 lanceolàta. L.C.	(*sc.*)	spear-leaved.	———	1820.	——	S.♃.	
23 esculénta. L.C.	(*sc.*)	esculent.	S. America.	——	——	S.♃.	
24 polymórpha. L.C.	(*sc.*)	variable.	———	1824.	——	S.♃.	
25 iridiflòra. B.M.	(*cr.*)	nodding-flower'd.	Peru.	1816.	——	S.♃.	Bot. reg. 609.
26 fláccida. R.Sc.	(*ye.*)	flaccid.	S. Carolina.	1788.	——	S.♃.	Rosc. scitam. t. 6.
27 glaùca. L.T.	(*ye.*)	glaucous.	S. America.	1732.	——	S.♃.	Sm. exot. bot. t. 102.
β rùfa. B.M.	(*ye.*)	*brown.*	———	....	——	S.♃.	Bot. mag. t. 2302.
MARA'NTA. R.S.		ARROW-ROOT	Monandria Monogynia. L.				
1 arundinàcea. R.S.	(*wh.*)	Indian.	S. America.	1732.	7. 8.	S.♃.	Bot. mag. t. 2307.
2 índica. R.S.	(*wh.*)	smooth-leaved.	India.	....	——	S.♃.	Sloan. jam. t. 149.
3 divaricàta. R.Sc.	(*wh.*)	spreading.	Brazil.	1825.	— —	S.♃.	Rosc. scitam. t. 27.
β purpùrea. R.Sc.	(*pu.*)	*purple-flowered.*	———	——	——	S.♃.	——— t. 28.
4 Tónchat. R.S.	(*pa.bl.*)	ovate.	E. Indies.	1819.	....	S.♃.	Rumph. amb. 4. t. 7.
5 angustifòlia. B.M.	(*li.*)	narrow-leaved.	W. Indies.	1820.	7. 8.	S.♃.	Bot. mag. t. 2398.
6 cuspidàta. R.Sc.	(*ye.*)	cuspidate.	Sierra Leone.	1822.	——	S.♃.	Rosc. scitam. t. 31.
7 bícolor. B.R.	(*wh.*)	two-coloured.	S. America.	1823.	1. 12.	S.♃.	Bot. reg. t. 786.
8 gíbba. R.S.	(*wh.*)	gibbous.	———	——	——	S.♃.	Rosc. scitam. t. 29.
9 oblìqua. R.S.	(*re.*)	oblique.	Guiana.	1803.	10. 1.	S.♃.	Rudg. gui. t. 11.
10 sylvática. L.T.	(*wh.*)	wood.	W. Indies.	1799.	8. 9.	S.♃.	

ORDO CCV.

MUSACEÆ. *Juss. gen.* 61.

M'USA. L.		PLANTAIN-TREE.	Polygamia Monœcia. L.				
1 paradisìaca. W.	(*br.cr.*)	common.	India.	1690.	10. 12.	S.♄.	Trew. ehret. 3. t. 18-20.
2 sapiéntum. W.	(*br.cr.*)	Banana-tree.	W. Indies.	1729.	10. 3.	S.♄.	——— 4. t. 21-23.
3 maculàta. W.	(*br.pi.*)	spotted-fruited.	Mauritius.	1818.	——	S.♄.	
4 rosàcea. W.	(*ro.*)	rose-coloured.	———	1805.	3. 6.	S.♄.	Bot. reg. t. 706.
5 coccínea. W.	(*sc.*)	scarlet-flower'd.	China.	1792.	12. 3.	S.♄.	Andr. reposit. t. 47.
6 ornàta. F.I.	(*li.or.*)	handsome.	E. Indies.	1823.	....	S.♄.	
7 supérba. F.I.	(*br.pa.*)	superb.	———	——	....	S.♃.	
8 gláuca. F.I.	(*ye.*)	glaucous.	———	1824.	....	S.♃.	
UR'ANIA. W.		UR'ANIA.	Hexandria Monogynia. W.				
speciòsa. W.	(*wh.*)	Plantain-leaved.	Madagascar.	1813.	....	S.♄.	Jacq. schœn. 1. t. 93.
Ravenàlia madagascariénsis. Jacq.							
STRELI'TZIA. H.K.		STRELI'TZIA.	Pentandria Monogynia. H.K.				
1 augústa. H.K.	(*wh.*)	Plantain-like.	C. B. S.	1791.	2. 5.	S.♄.	Ker's Strelitz. ic.
2 regìnæ. H.K.	(*ye.bl.*)	Canna-leaved.	———	1773.	4. 5.	I.♃.	Redout. lil. t. 77. 88.
3 ovàta. H.K.	(*ye.bl.*)	oval-leaved.	———	1777.	2. 4.	I.♃.	Bot. mag. t. 119. 120.
4 hùmilis. L.en.	(*ye.bl.*)	dwarf.	———	....	——	I.♃.	
5 farinòsa. H.K.	(*ye.bl.*)	mealy-stalked.	———	1795.	——	I.♃.	Ker's Strelitz. ic.
6 angustifòlia. H.K.	(*ye.bl.*)	narrow-leaved.	———	1778.	5. 6.	I.♃.	——— ic.
7 parvifòlia. H.K.	(*ye.bl.*)	small-leaved.	———	1796.	5. 7.	I.♃.	——— ic.
β júncea. B.R.	(*ye.bl.*)	*rush-like.*	———	——	——	I.♃.	Bot. reg. t. 516.
HELIC'ONIA. R.S.		HELIC'ONIA.	Pentandria Monogynia. L.				
1 Bihài. W.	(*re.sn.*)	Plantain-leaved.	W. Indies.	1786.	4. 5.	S.♃.	Bot. reg. t. 374.
2 hùmilis. R.S.	(*sc.*)	dwarf.	Caracas.	1798.	5. 6.	S.♃.	Jacq. schœn. 1. t. 48. 49.
3 psittacòrum. R.S.	(*ye.re.*)	Parrot-beaked.	W. Indies.	1797.	8. 9.	S.♃.	Andr. reposit. t. 124.
4 Swartziàna. R.S.	(*or.*)	Swartz's.	Jamaica.	——	——	S.♃.	Bot. mag. t. 502.
psittacòrum β. B.M.							
5 dealbàta. L.C.	(*ye.*)	mealy.	S. America.	1823.	——	S.♃.	
6 hirsùta. R.S.	(*ye.*)	hairy-bracted.	———	——	——	S.♃.	
7 buccinàta. F.I.	(*st.*)	Indian.	E. Indies.	1820.	6. 8.	S.♃.	Rumph. amb. 5. t. 62. f. 2.

ORDO CCVI.

IRIDEÆ. *Brown prodr.* 302.

'IRIS. B.M.		FLOWER-DE-LUCE.	Triandria Monogynia. L.				
1 susiàna. R.S.	(*br.bl.*)	Chalcedonian.	Levant.	1596.	3. 4.	H.♃.	Bot. mag. t. 91.
2 lívida. Trat.	(*pu.bl.*)	livid.	———	——	——	H.♃.	Redout. lil. 1. t. 18.
susiána. Redoute *non* B.M.							

3	florentìna. R.S.	(*wh.*)	Florentine.	S. Europe.	1596.	5. 6.	H.♃.	Bot. mag. t. 671.
4	germánica. R.S.	(*bl.*)	German.	Germany.	1573.	——	H.♃.	———— t. 670.
5	subbiflòra. B.M.	(*bl.*)	double-bearing.	Portugal.	1596.	4. 5.	H.♃.	———— t. 1130.
6	pállida. R.S.	(*pa.bl.*)	pale Turkey.	Turkey.	——	5. 6.	H.♃.	———— t. 685.
	nepalénsis. Bot. reg. 818. *non* Don. prodr.							
7	longiflòra. R.S.	(*pa.bl.*)	long-flowered.	....	1824.	——	H.♃.	
8	flavéscens. R.S.	(*st.*)	pale yellow.	Switzerland.	1825.	——	H.♃.	Swt. br. fl. gar. s.2. t.56.
9	sórdida. W.en.	(*pa.ye.*)	dull-flowered.	————	1818.	——	H.♃.	Bot. mag. t. 2861.
	lutéscens. Bot. mag. t. 2861. *nec aliorum.*							
10	sambucìna. R.S.	(*pu.wh.*)	Elder-scented.	S. Europe.	1658.	6.	H.♃.	Bot. mag. t. 187.
11	squálens. R.S.	(*pu.ye.*)	brown-flowered.	————	1768.	——	H.♃.	———— t. 787.
12	lùrida. R.S.	(*da.pu.*)	dark brown.	————	1758.	4. 5.	H.♃.	———— t. 669. 986.
13	variegàta. R.S.	(*pa.ye.*)	variegated.	Hungary.	1597.	5. 6.	H.♃.	———— t. 16.
	β *limbàta.* R.S.	(*pu.ye.*)	*yellow-limbed.*	N. America.	....	——	H.♃.	
14	neglécta. R.S.	(*bl.wh.*)	Horneman's.	....	1820.	6.	H.♃.	Bot. mag. t. 2435.
15	viréscens. R.L.	(*st.*)	greenish-flowered.	......	1824.	5. 6.	H.♃.	Redout. lil. 5. t. 295.
16	Swértii. R.S.	(*bl.wh.*)	reflex-flowered.	....	1748.	——	H.♃.	———— 6. t. 360.
17	plicàta. R.S.	(*bl.*)	plaited-leaved.	Hungary.	1823.	——	H.♃.	Bot. mag. t. 870.
18	am'œna. R.L.	(*pu.wh.*)	pretty.	....	——	——	H.♃.	Redout. lil. 8. t. 336.
19	biflòra. R.S.	(*vi.pu.*)	two-flowered.	S. Europe.	1596.	4. 9.	H.♃.	Besl. h. eyst. 8. f. 1.
20	furcàta. R.S.	(*bl.*)	forked.	Tauria.	1822.	3. 4.	H.♃.	Bot. reg. t. 801.
21	cristàta. R.S.	(*bl.ye.*)	crested.	N. America.	1756.	5.	H.♃.	Bot. mag. t. 412.
22	fimbriàta. V.	(*bl.*)	fringed.	China.	1792.	5. 6.	F.♃.	Vent. cels. t. 9.
	chinénsis. Bot. mag. t. 373.							
23	dichótoma. R.S.	(*vi.pu.*)	forked.	Dauria.	1784.	6. 8.	H.♃.	Swt. br. fl. gar. t. 96.
24	nudicaùlis. R.S.	(*pu.wh.*)	naked-stalked.	....	1748.	5. 6.	H.♃.	
	aphy'lla. W.							
25	arenària. R.S.	(*ye.*)	sand.	Hungary.	1802.	——	H.♃.	Bot. reg. t. 549.
26	lutéscens. R.S.	(*ye.*)	dwarf yellow.	Germany.	1748.	4. 5.	H.♃.	Redout. lil. 5. t. 263?
	pùmila. var. *lutea.* Bot. mag. t. 1209.							
27	flavíssima. R.S.	(*ye.*)	bright yellow.	Siberia.	1814.	5. 6.	H.♃.	Jacq. ic. 3. t. 220.
28	pùmila. R.S.	(*bl.*)	dwarf.	Austria.	1596.	4. 5.	H.♃.	Bot. mag. t. 9.
29	violàcea.	(*vi.*)	violet-coloured.	S. Europe.	....	——	H.♃.	———— t. 1261.
	pùmila. var. *violàcea.* B.M.							
30	hungàrica. R.S.	(*bl.*)	Hungarian.	Hungary.	1815.	5.	H.♃.	Swt. br. fl. gar. t. 74.
31	ibérica. R.S.	(*vi.*)	Iberian.	Iberia.	1821.	5. 6.	H.♃.	
32	Pseùd-A'corus. R.S.	(*ye.*)	yellow-water.	Britain.	....	6.	H.♃.	Eng. bot. v. 9. t. 578.
	β *pállido-flàva.* B.M.	(*ye.*)	*pale-yellow.*	N. America.	1812.	——	H.♃.	Bot. mag. t. 2239.
33	Monnièrii. R.S.	(*ye.*)	Le Monnier's.	Levant.	1820.	6. 7.	H.♃.	Redout. lil. 4. t. 236.
34	taùrica. B.C.	(*ye.*)	Taurian.	Tauria.	1826.	5. 6.	H.♃.	Lodd. bot. cab. t. 1506.
35	fœtidíssima. R.S.	(*br.bl.*)	Gladwyn.	Britain.	....	6.	H.♃.	Eng. bot. v. 9. t. 596.
	β *variegàta.*	(*br.bl.*)	*striped-leaved.*	————	....	——	H.♃.	
36	versícolor. R.S.	(*pu.ye.*)	various-coloured.	N. America.	1732.	5. 6.	H.♃.	Bot. mag. t. 21.
37	fúlva. B.M.	(*co.*)	copper-coloured.	————	1812.	6. 7.	H.♃.	———— t. 1496.
	cùprea. Ph. published 1814. *fúlva.* B.M. published 1812.							
38	tridentàta. Ph.	(*bl.*)	three-toothed.	N. America.	1824.	7.	H.♃.	Swt. br. fl. gar. t. 274.
	tripétala. Bot. mag. t. 2886. *scarcely the same species.*							
39	virgínica. R.S.	(*bl.pu.*)	Virginian.	Virginia.	1758.	6. 7.	H.♃.	Bot. mag. t. 703.
40	grácilis. Bolt.	(*pu.ye.*)	Bolton's.	N. America.	1812.	——	H.♃.	
	Boltoniàna. R.S.							
41	spùria. B.M.	(*bl.*)	spurious.	Siberia.	1759.	7.	H.♃.	Bot. mag. t. 58.
42	nótha. M.B.	(*bl.*)	hybrid.	————	1780.	——	H.♃.	Bieb. cent. pl. ros. 2. t. 77.
	halóphila. Bot. mag. t. 875.							
43	ochroleùca. L.	(*st.*)	sulphur-coloured.	Levant.	1757.	——	H.♃.	Bot. mag. t. 61.
44	halóphila. Pal.	(*st.*)	long-leaved.	Siberia.	1780.	7. 9.	H.♃.	———— t. 1131.
45	stenógyna. R.L.	(*st.*)	cream-coloured.	Caucasus.	1804.	7.	H.♃.	———— t. 1515.
46	desertòrum. Guld.	(*bl.st.*)	sweet-scented.	Russia.	1811.	——	H.♃.	———— t. 1514.
	spatulàta. W.en.							
47	heterophy'lla. S.S.	(*st.*)	various-leaved.	....	——	——	H.♃.	
48	lævigàta. F.	....	smooth.	Dahuria.	1828.	....	H.♃.	
49	vérna. M.	(*bl.*)	vernal.	N. America.	1748.	4. 5.	H.♃.	Swt. br. fl. gar. t. 68.
50	ventricòsa. R.S.	(*pa.bl.*)	bellied.	Dahuria.	1800.	6.	H.♃.	Pall. it. 3. t. B. f. 1.
51	sibírica. R.S.	(*bl.st.*)	flaccid-leaved.	Siberia.	1596.	5. 6.	H.♃.	Bot. mag. t. 50.
52	flexuòsa. R.S.	(*wh.*)	flexuose.	Germany.	....	——	H.♃.	Murr. c. gott. 7. t. 4.
	sibírica. β *alba.* B.M. t. 1163. *Redoute liliac.* t. 438. *acùta* β *álba.* W.enum.							
53	acùta. W.en.	(*bl.*)	straight-leaved.	Siberia.	....	——	H.♃.	
54	triflòra. W.en.	(*bl.*)	three-flowered.	Italy.	1816.	——	H.♃.	
55	hæmatophy'lla. F.	(*bl.*)	red-leaved.	Siberia.	1790.	4. 5.	H.♃.	Swt. br. fl. gar. t. 118.
	sanguínea. R.S. *sibírica* γ *sanguínea.* Bot. mag. t. 1604.							

58 lacústris. N. (*pa.bl.*) marsh. N. America. 1825. 5. 6. H. ♃.
59 prismática. R.S. (*bl.*) New Jersey. ——— 1812. —— H. ♃. Bot. mag. t. 1504.
60 gramínea. R.S. (*bl.pu.*) grass-leaved. Austria. 1597. 6. H. ♃. ——— t. 681.
61 hùmilis. M.B. (*bl.*) pigmy. Siberia. 1804. 5. 6. H. ♃. ——— t. 1123.
62 ruthénica. R.S. (*bl.*) Russian. Russia. 1810. 4. 5. H. ♃. ——— t. 1393.
63 biglùmis. R.S. (*bh.wh.*) two-glumed. Siberia. 1829. —— H. ♃. Pall. it. 3. app. t. 6. f. 1.
64 tenuifòlia. W. (*pa.bl.*) slender-leaved. Dauria. 1796. 5. H. ♃. ——— 3. t. C. f. 2.
65 nepalénsis. D.P. (*bl.*) Nepaul. Nepaul. 1824. 6. 7. H. ♃. Swt. br. fl. gar. s. 2. t. 11.
66 Pallàsii. B.M. (*bl.*) Pallas's. Siberia. 1820. —— H. ♃.
β *chinénsis*. B.M. (*bl.*) *Chinese*. China. —— —— H. ♃. Bot. mag. t. 2331.
67 longispátha. B.M. (*bl.*) long-spathed. Russia. 1822. —— H. ♃. ——— t. 2528.
68 brachycúspis. B.M. (*bl.*) short-pointed. Siberia. 1819. 6. H. ♃. ——— t. 2326.
69 cœlestìna. N. (*bl.*) sky-blue. N. America. 1824. —— F. ♃.
70 tènax. B.R. (*pu.*) tough-leaved. California. 1827. 4. 5. H. ♃. Bot. reg. t. 1218.
71 tuberòsa. R.S. (*vi.*) snake's-head. Levant. 1597. 3. 4. H. ♃. Bot. mag. t. 531.
72 reticulàta. R.S. (*bl.*) netted-rooted. Iberia. 1821. 4. 5. H.b. ♃.
73 Xíphium. R.S. (*va.*) small bulbous. Spain. 1596. 6. H.b. ♃. Bot. mag. t. 686.
74 lusitánica. R.S. (*ye.*) Portuguese. Portugal. —— 4. 5. H.b. ♃. ——— t. 679.
75 Xiphioídes. R.S. (*va.*) great bulbous. Spain. 1571. 6. H.b. ♃. ——— t. 687.
76 caucásica. R.S. (*st.*) Caucasean. Caucasus. 1821. 2. 3. H.b. ♃. Swt. br. fl. gar. t. 255.
77 alàta. Lam. (*bl.*) winged. S. Europe. 1801. .... H.b. ♃. Desf. atl. 1. t. 6?
scorpioídes. Redout. liliac. t. 211.
78 pérsica. R.S. (*bl.ye.*) Persian. Persia. 1629. 3. H.b. ♃. Bot. mag. t. 1.

DI'ETES. Sal. DI'ETES. Triandria Monogynia.
1 iridioídes. (*wh.bl.*) Iris-flowered. C. B. S. 1758. 4. 8. F. ♃. Bot. mag. t. 693.
Mor'æa iridioídes. B.M. '*Iris moræoídes*. Ker.
2 catenulàta. (*wh.bl.*) chain-dotted. Madagascar. 1826. —— F. ♃. Bot. reg. t. 1074.
Mor'æa catenulàta. B.R.

MOR'ÆA. B.M. MOR'ÆA. Triandria Monogynia. L.
1 Sisyrínchium. B.M. (*bl.*) Spanish nut. S. Europe. 1597. 5. 6. F.b. Bot. mag. t. 1407.
2 Tenoreàna. B.F.G. (*bl.*) Tenore's. Naples. 1824. 6. 7. F.b. Swt. br. fl. gar. t. 110.
'*Iris fùgax*. Ten. flor. nap. t. 4. *nec aliorum*.
3 minùta. B.R. (*ye.*) small. C. B. S. 1825. —— F.b.
4 papilionàcea. B.R. (*ye.*) Butterfly. ——— 1795. 5. 6. F.b. Bot. mag. t. 750.
5 ciliàta. B.R. (*ye.*) fringed-leaved. ——— 1787. 4. 6. F.b. ——— t. 1061.
6 barbígera. Sal. (*ro.*) bearded. ——— —— 4. 5. F.b. ——— t. 1012.
7 críspa. B.R. (*std.*) short-spathed. ——— 1803. —— F.b. ——— t. 1284.
8 trístis. B.R. (*bl.ye.*) dull-coloured. ——— 1768. —— F.b. ——— t. 577.
9 trícolor. A.R. (*br.re.ye.*) three-coloured. ——— 1809. —— F.b. Andr. reposit. t. 83.
10 lùrida. B.R. (*d.pu.*) Mr. Griffin's. ——— 1817. 6. 7. F.b. Bot. reg. t. 312.
11 ramòsa. B.R. (*ye.*) branching. ——— 1789. 5. 6. F.b. Bot. mag. t. 771.
12 plumària. B.R. (*bl.*) feathered. ——— 1825. —— F.b.
13 viscària. B.R. (*br.*) bird-limed. ——— 1800. 6. F.b. Bot. mag. t. 587.
14 bituminòsa. B.R. (*ye.*) clammy. ——— 1787. 4. 5. F.b. ——— t. 1045.
15 polystàchya. B.R. (*bl.*) many-spiked. ——— 1825. 6. 7. F.b.
16 longiflòra. B.R. (*ye.*) long-flowered. ——— 1801. 6. 7. F.b. Bot. mag. t. 712.
17 setàcea. B.R. (*bl.*) bristle-leaved. ——— 1825. —— F.b. Thunb. Ir. t. 1. f. 1.
18 èdulis. B.M. (*li.bl.*) long-leaved. ——— 1792. 5. 6. F.b. Bot. mag. t. 613.
19 odòra. P.L. (*wh.*) sweet-scented. ——— —— —— F.b. Sal. par. lond. t. 10.
20 longifòlia. A.R. (*st.*) pale-yellow. ——— 1808. —— F.b. Andr. reposit. t. 45.
èdulis γ lutéscens. Bot. mag. t. 1238.
21 angústa. B.R. (*st.*) rolled-leaved. ——— 1790. —— F.b. Bot. mag. t. 1276.

MA'RICA. H.K. MA'RICA. Triandria Monogynia. H.K.
1 cœrùlea. B.C. (*bl.*) blue. Brazil. 1810. 4. 10. S. ♃. Bot. reg. t. 713.
2 Sabìni. H.T. (*bl.*) Sabine's. St. Thomas. 1822. 9. 11. S. ♃. Hort. trans. v. 6. pl. 1.
3 Northiána. H.K. (*wh.br.*) broad-stemmed. Brazil. 1789. 4. 8. S. ♃. Bot. mag. t. 654.
Mor'æa Northiána. Andr. rep. t. 255.
4 paludòsa. B.M. (*wh.*) marsh. Guiana. 1792. 7. 8. S. ♃. Bot. mag. t. 646.
5 semiapérta. B.C. (*ye.*) half-opened. Brazil. 1820. 5. 7. S. ♃. Lodd. bot. cab. t. 685.
6 hùmilis. B.C. (*wh.ye.*) dwarf. ——— 1823. —— S. ♃. ——— t. 1081.
7 longifòlia. L.O. (*st.*) long-leaved. Brazil. 1830. 6. 10. S. ♃. Lk. et Ot. abb. t. 58.
8 martinicénsis. H.K. (*ye.*) Martinico. W. Indies. 1782. 5. 7. S. ♃. Bot. mag. t. 416.
'*Iris martinicénsis*. B.M. *Cipùra martinicénsis*. K.S. *Trimèzia lùrida*. Sal.

HERBE'RTIA. B.F.G. HERBE'RTIA. Monadelphia Triandria.
1 pulchélla. B.F.G. (*bl.*) plaited-leaved. Maldonado. 1827. 6. 8. F.b. Swt. br. fl. gar. t. 222.
2 pusílla. (*st.*) small. Brazil. 1830. —— F.b. Lk. et Ot. abb. t. 59.
Ferrària pusílla. Lk. et Otto.

CYPE'LLA. B.M. CYPE'LLA. Monadelphia Triandria. B.M.
Herbérti. B.M. (*or.*) Mr. Herbert's. BuenosAyres. 1823. 6. 9. F.b. Swt. br. fl. gar. s. 2. t. 33.
Tigrídia Herbérti. B.M. t. 2599. *Mor'æa Herbérti*. B.R. t. 949.

VIEUSSE'UXIA. DC. VIEUSSE'UXIA. Triandria-Monadelphia Monogynia.
1 tripetaloídes. DC. (*bh.*) three petal-like. C. B. S. 1802. 4. 5. F.b. Bot. mag. t. 702.
Mor'æa tripétala. B.M. '*Iris tripetàla.* W.
2 pavònia. DC. (*or.gr.*) Peacock. ——— 1790. 5. 6. F.b. Bot. mag. t. 1247.
Mor'æa pavònia. B.M. '*Iris pavònia.* W.
3 villòsa. S.S. (*pu.bl.*) villous-leaved. C. B. S. 1789. 4. 5. F.b. Bot. mag. t. 571.
4 tricùspis. S.S. (*pu.*) trident-petaled. ——— 1776. 6. F.b. ——— t. 696.
5 glaùcopis. DC. (*wh.bl.*) white-flowered. ——— —— 5. 7. F.b. Redout. lil. 1. t. 42.
'*Iris pavònia.* Bot. mag. t. 168. *nec aliorum.*
6 Bellendèni. (*ye.*) Mr. Ker's. ——— 1803. 6. 7. F.b. Bot. mag. t. 772.
Mor'æa tricúspis γ lùtea. B.M. certainly a very distinct species.
7 ténuis. R.S. (*or.*) slender. C. B. S. 1807. 5. 6. F.b. Bot. mag. t. 1047.
8 spiràlis. R.S. (*ye.*) spiral-flowered. ——— 1825. —— F.b. De la Roch. dis. t. 5.
9 unguiculàris. R.S. (*bh.*) long-clawed. ——— 1802. —— F.b. Bot. mag. t. 593.

HOM'ERIA. V. HOM'ERIA. Monadelphia Triandria.
1 spicàta. (*ye.or.pu.*) spiked. C. B. S. 1785. 5. 6. F.b. Bot. mag. t. 1283.
Mor'æa spicàta. B.M.
2 collìna. (*pi.*) hill. ——— 1768. —— F.b. Bot. mag. t. 1033.
Mor'æa collìna. B.M.
3 ochroleùca. Sal. (*ye.*) straw-coloured. ——— —— —— F.b. Bot. mag. t. 1103.
Mor'æa collìna. γ. B.M. 1103.
4 lineàta. B.F.G. (*pi.*) lined-leaved. ——— 1825. —— F.b. Swt. br. fl. gar. t. 178.
5 porrifòlia. B.F.G. (*sc.*) leek-leaved. ——— —— —— F.b. ——— ic.
6 elégans. (*or.gr.*) elegant. ——— —— —— F.b. Jacq. schœn. 1. t. 2.
7 aurantìaca. R.S. (*or.*) flaccid-petaled. ——— 1810. —— F.b. Bot. mag. t. 1612.
Mor'æa collìna. (a) miniàta mìnor. B.M. 1612.
8 miniàta. B.F.G. (*pi.*) spot-flowered. ——— 1799. —— F.b. Swt. br. fl. gar. t. 152.

HEXAGLO'TTIS. Vt. HEXAGLO'TTIS. Monadelphia Triandria.
1 flexuòsa. (*ye.*) flexuose. C. B. S. 1803. 5. 6. F.b. Bot. mag. t. 695.
longifòlia. Vent. *Mor'æa flexuòsa.* B.M.
2 virgàta. (*ye.*) slender. ——— 1825. —— F.b. Jacq. icon. 2. t. 228.

PARDA'NTHUS. S.S. PARDA'NTHUS. Triandria Monogynia. S.S.
1 chinénsis. S.S. (*or.*) Chinese. China. 1759. 6. 7. H.♃. Bot. mag. t. 171.
'*Ixia chinénsis.* B.M. *Belamcánda chinénsis.* DC.
2 nepalénsis. (*or.*) Nepaul. Nepaul. 1823. —— H.♃.

BOBA'RTIA. L. BOBA'RTIA. Triandria Monogynia. L.
1 gladiàta. K.I. (*ye.*) hairy-stalked. C. B. S. 1815. 6. 8. G.♃. Bot. reg. t. 229.
Márica gladiàta. B.R. *Mor'æa gladiàta.* W.
2 spathàcea. L. (*ye.*) sheathed. ——— 1798. 7. G.♃. Thunb. dis. p. 9. t. 1.

SISYRI'NCHIUM. L. SISYRI'NCHIUM. Monadelphia Triandria.
1 latifòlium. H.K. (*wh.*) broad-leaved. W. Indies. 1737. 6. 8. S.♃. Bot. mag. t. 655.
Márica plicàta. B.M. *Mor'æa plicàta.* W.
2 bermudiànum. W. (*bl.*) Iris-leaved. Bermudas. 1732. 5. 7. G.♃. Bot. mag. t. 94.
iridioídes. B.M.
3 Nuttálli. (*bl.*) Nuttall's. N. America. 1823. —— H.♃.
bermudiànum. N. *nec aliorum.*
4 chilénse. B.M. (*bl.*) blue Chilian. Chile. 1826. 6. 7. F.♃. Bot. mag. t. 2786.
5 graminifòlium. B.R. (*ye.*) grass-leaved. ——— 1825. 4. 8. F.♃. Bot. reg. t. 1067.
6 flexuòsum. B.R. (*ye.*) flexuose. ——— —— —— F.♃.
7 ancéps. W. (*bl.*) narrow-leaved. N. America. 1693. 6. 7. H.♃. Bot. mag. t. 464.
8 mucronàtum. M. (*bl.*) pointed-flowered. ——— 1812. —— H.♃.
9 convolùtum. W. (*ye.*) convolute. S. America. 1816. 5. 7. F.♃. Wild. h. ber. t. 91.
10 tenuifòlium. W. (*ye.*) slender-leaved. Mexico. —— —— F.♃. ——— t. 92.
11 lùteum. L.en. (*ye.*) yellow-flowered. ——— 1823. —— H.♃.
12 califórnicum. H.K. (*ye.*) Californian. California. 1796. 5. 2. F.♃. Bot. mag. t. 983.
Márica califórnica. B.M.
13 pedunculàtum. B.M. (*ye.*) long-peduncled. Chile. 1827. 6. 8. F.♃. Bot. mag. t. 2965.
14 micránthum. W. (*st.*) small-flowered. Peru. 1822. 5. 9. F.♃. ——— t. 2116.
15 iridifòlium. K.S. (*st.vi.*) Iris-leaved. S. America. —— —— F.♃. Bot. reg. t. 646.
láxum. Bot. mag. t. 2312. *Márica iridifòlia* B.R.
16 odoratíssimum. B.R. (*std.*) sweet-scented. S. America. 1828. 6. 10. F.♃. Bot. reg. t. 1283.
17 striàtum. H.K. (*ye.std.*) streaked-flowered. Mexico. 1788. —— F.♃. Sm. ic. pict. t. 9.
Márica striàta. B.M.
18 palmifòlium. W. (*wh.*) Palm-leaved. Brazil. 1799. 1. 5. S.♃.

ORTHROSA'NTHUS. MORNING-FLOWER, Monadelphia Triandria.
multiflòrus. S.F.A. (*bl.*) many-flowered. N. Holland. 1825. 5. 8. G.♃. Swt. flor. aust. t. 11.

LIBE'RTIA. S.S. LIBE'RTIA. Triandria-Monadelphia Monogynia.
1 paniculàta. S.S. (*wh.*) panicled. N. S. W. 1823. 5. 8. F.♃.
2 grandiflòra. (*wh.*) large-flowered. N. Zealand. 1822. 4. 9. H.♃. Swt. br. fl. gar. t. 64.
Renealmia grandiflòra. S.F.A. the old genus Renealmia is restored by Roscoe.
3 pulchélla. S.S. (*wh.*) neat. N. S. W. 1823. 4. 9. F.♃.

PATERS'ONIA. B.P. PATERS'ONIA. Monadelphia Triandria. H.K.
1 serícea. B.P. (pu.) silky. N. S. W. 1803. 5. 7. F.♃. Bot. mag. t. 1041.
2 lanàta. B.P. (pu.) woolly-spathed. N. Holland. 1825. 5. 8. F.♃. Swt. flor. aust. t. 15.
3 longifòlia. B.P. (bl.) long-leaved. N. S. W. 1822. 5. 7. F.♃.
4 mèdia. B.P. (bl.) middle. ——— 1824. —— F.♃.
5 glabràta. B.P. (bl.) smooth. ——— 1814. —— F.♃. Bot. reg. t. 50.
6 longiscàpa. S.F.A. (bl.) long-stalked. N. Holland. 1825. 5. 8. F.♃. Swt. flor. aust. t. 39.
7 glaùca. B.P. (bl.) glaucous. N. S. W. 1826. —— F.♃. Lab. nov. hol. 1. t. 9.
WITS'ENIA. H.K. WITS'ENIA. Triandria Monogynia.
1 maùra. H.K. (or.) downy-flowered. C. B. S. 1790. 11. 1. G.♄. Bot. reg. t. 5.
2 corymbòsa. H.K. (bl.) corymbose. ——— 1803. 4. 9. G.♄. Bot. mag. t. 895.
3 ramòsa. S.S. (ye.bl.) branching. ——— 1809. —— G.♄.
4 partìta. K.I. (bl.) parted-flowered. ——— 1822. —— G.♄.
ARIST'EA. Ker. ARIST'EA. Triandria Monogynia. L.
1 cyánea. H.K. (bl.) woolly-headed. C. B. S. 1759. 4. 6. G.♃. Bot. mag. t. 458.
2 pusílla. B.M. (bl.) flat-stemmed. ——— 1806. 6. 7. G.♃. ——— t. 1231.
3 spiràlis. H.K. (wh.br.) spiral-flowered. ——— 1795. 4. 5. G.♃. ——— t. 520.
4 capitàta. H.K. (bl.) tallest. ——— 1790. 7. 8. G.♃. ——— t. 605.
màjor. A.R. 160. *Mor'æa cærùlea*. Th. mor. t. 2. f. 2.
5 melaleùca. H.K. (bl.bk.) three-coloured. C. B. S. 1786. 5. 6. G.♃. Bot. mag. t. 1277.
FERR'ARIA. Ker. FERR'ARIA. Monadelphia Triandria. H.K.
1 angustifòlia. (st.) narrow-leaved. C. B. S. 1825. 4. 6. F.b.
2 antheròsa. B.M. (gr.) green variegated. ——— 1800. 3. 7. F.b. Bot. mag. t. 751.
Ferrariòla H.K. *viridiflòra*. A.R. t. 285.
3 uncinàta. B.F.G. (br.bl.) hook-leaved. ——— 1825. —— F.b. Swt. br. fl. gar. t. 161.
4 undulàta. W. (br.wh.) curled. ——— 1755. —— F.b. Bot. mag. t. 144.
5 atràta. L.C. (da.) dark-flowered. ——— 1825. —— F.b. Lodd. bot. cab. t. 1356.
6 divaricàta. B.F.G. (da.) spreading-anther'd. ——— —— —— F.b. Swt. br. fl. gar. t. 192.
7 obtusifòlia. B.F.G. (pu.br.) blunt-leaved. ——— —— —— F.b. ——— ——— t. 148.
TIGRI'DIA. J. TIGER-FLOWER. Monadelphia Triandria. H.K.
1 pavònia. H.K. (sc.) red-flowered. Mexico. 1796. 5. 9. H.b. Redoute lil. t. 6.
Ferrària pavònia. A.R. 178. *Tigrídia*. B.M. 532.
2 conchiiflòra. B.F.G. (ye.) yellow-flowered. Mexico. 1824. —— H.b. Swt. br. fl. gar. t. 128.
GALA'XIA. Ker. GALA'XIA. Monadelphia Triandria. H.K.
1 ovàta. W. (ye.) oval-leaved. C. B. S. 1799. 5. 9. F.b. Andr. reposit. t. 94.
2 grandiflòra. A.R. (ye.) large-flowered. ——— —— —— F.b. ——— t. 164.
3 mucronulàris. Sal. (ro.) mucronated. ——— —— —— F.b. Jac. ic. t. 291. f. inf. sin.
4 versícolor. Sal. (ro.ye.) various-coloured. ——— —— —— F.b. —— f. inf. ad. dextr.
5 gramínea. W. (ye.) narrow-leaved. ——— 1795. 7. 8. F.b. Bot. mag. t. 1292.
PEYRO'USIA. PEYRO'USIA. Triandria Monogynia. H.K.
1 corymbòsa. B.M. (bl.) level-topped. C. B. S. 1791. 5. 6. F.b. Bot. mag. t. 595.
2 falcàta. B.M. (bl.) sickle-leaved. ——— 1825. —— F.b. Thunb. dis. 4. t. 1. f. 3.
3 fasciculàta. B.M. (wh.) fascicled. ——— —— —— F.b. Jacq. ic. 2. t. 291.
Galáxia plicàta. Jacq. *I'xia heterophy'lla*. Vahl. *Ovièda fasciculàta*. S.S.
4 fissifòlia. B.M. (ro.) leafy-spiked. C. B. S. 1809. 8. 9. F.b. Bot. mag. t. 1246.
5 bracteàta. B.M. (wh.) bracteate. ——— 1825. —— F.b. Thunb. act. haf. 6. c. ic.
6 aculeàta. B.F.G. (wh.) prickly-stemmed. ——— —— 5. 7. F.b. Swt. br. fl. gar. s. 2. t. 39.
7 ancéps. B.F.G. (bl.) flat-stemmed. ——— —— —— F.b. ——— ser. 1. t. 143.
8 silenoídes. B.M. (pu.) Silene-like. ——— 1822. —— F.b. Jacq. ic. 2. t. 270.
9 Fabrícii. B.M. (wh.) slender-branched. ——— 1825. —— F.b.
ANOMATH'ECA. Ker. ANOMATH'ECA. Triandria Monogynia. H.K.
júncea. H.K. (ro.) cut-leaved. C. B. S. 1791. 4. 5. F.b. Bot. mag. t. 606.
Lapeyroùsia júncea. B.M. *Gladìolus polystàchyus*. A.R. 66.
BABI'ANA. Ker. BABI'ANA. Triandria Monogynia. S.S.
1 nàna. S.S. (bl.re.) dwarf. C. B. S. 1807. 4. 6. F.b. Andr. reposit. t. 137.
Gladìolus nànus. A.R.
2 rùbro-cyánea. B.M. (bl.re.) red and blue. ——— 1794. —— F.b. Bot. mag. t. 410.
3 villòsa. B.M. (cr.) dark red. ——— 1778. 8. F.b. ——— t. 583.
I'xia punícea. Jacq. ic. 2. t. 287.
4 obtusifòlia. B.M. (vi.) blunt-leaved. ——— 1825. 5. 6. F.b. Jacq. ic. 2. t. 284.
I'xia villòsa. Jacq. *non* H.K.
5 purpùrea. B.M. (d.pu.) dark purple. ——— 1806. —— F.b. Bot. mag. t. 1052.
6 strícta. B.M. (bl.wh.) upright. ——— 1757. —— F.b. ——— t. 621.
7 angustifòlia. (bl.) narrow-leaved. ——— —— —— F.b. ——— t. 637.
strícta (a) B.M. *I'xia villòsæ var*. Jacq. frag. t. 14. f. 3.
8 tenuiflòra. B.F.G. (bl.) slender-flowered. C. B. S. 1825. —— F.b. Swt. br. fl. gar. ic.
9 sulphùrea. B.M. (st.) pale-flowered. ——— 1803. —— F.b. Bot. mag. t. 1053.
10 mucronàta. B.M. (pu.ye.) mucronate. ——— 1825. —— F.b. Jacq. ic. 2. t. 253.
11 plicàta. B.M. (pa.bl.) sweet-scented. ——— 1774. —— F.b. Bot. mag. t. 576.
12 dísticha. B.M. (pa.bl.) two-ranked. ——— —— 6. 7. F.b. ——— t. 626.
Gladìolus plicàtus. Jacq. ic. t. 237. *nec a'iorum*.

13 pállida. W.H. (*pa.*) pale-flowered. Hybrid. .... 6. 7. F.b.
14 ochroleùca. (*st.*) cream-colour'd. C. B. S. 1825. 5. 7. F.b.
15 spathàcea. B.M.(*bl.wh.*)stiff-leaved. ——— 1801. —— F.b. Bot. mag. t. 638.
16 sambucìna. B.M.(*da.bl.*)Elder-scented. ——— 1799. 4. 5. F.b. ——— t. 1019.
17 tubàta. (*wh.re.*) long-flowered. ——— 1774. —— F.b. ——— t. 680.
Gladìolus tubàtus. Jacq. ic. 2. t. 264.
18 tubiflòra. B.M.(*wh.re.*) long-tubed. ——— —— —— F.b. Bot. mag. t. 847.
19 ríngens. B.M. (*pu.*) gaping-flowered. ——— 1752. 5. 6. F.b. Lodd. bot. cab. t. 1006.
20 Thunbérgii. B.M.(*pu.*) many-spiked. ——— 1774. —— F.b.
Anthol'yza plicàta. W.

ANTHOL'YZA. Ker. ANTHOL'YZA. Triandria Monogynia. L.
1 æthiòpica. B.M.(*sc.or.*) flag-leaved. C. B. S. 1759. 1. 4. F.b. Bot. mag. t. 561.
2 præálta. DC. (*sc.or.*) tall. ——— .... —— F.b. Redout. lil. 7. t. 387.
æthiópica. var. β. B.M. 1172. *vittígera.* Sal.
3 montàna. B.C. (*pu.*) mountain. ——— 1824. 5. 7. F.b. Lodd. bot. cab. t. 1022.
4 lùcidor. B.M. (*pu.*) shining. ——— 1825. —— F.b.

ANISA'NTHUS. B.F.G. ANISA'NTHUS. Triandria Monogynia.
1 spléndens. B.F.G. (*sc.*) splendid. C. B. S. 1825. 4. 6. F.b. Swt. br. fl. gar. ic. ined.
2 Cunònia. B.F.G. (*sc.*) scarlet. ——— 1756. 5. 6. F.b. Bot. mag. t. 343.
Anthol'yza Cunònia. B.M. *Gladìolus Cunònia.* H.K.

PETA'MENES. Sal. PETA'MENES. Triandria Monogynia.
quadrangulàris.(*ye.sc.*) four-sided. C. B. S. 1790. 4. 5. F.b. Bot. mag. t. 567.
Gladìolus quadrangulàris. B.M. *abbreviatus.* Andr. repos. t. 166.

WATS'ONIA. Ker. WATS'ONIA. Triandria Monogynia. H.K.
1 aletroídes. B.M. (*ro.*) Aletris-flower'd. C. B. S. 1774. 5. 7. F.b. Bot. mag. t. 441.
β variegàta. B.M. (*ro.*) *variegated.* ——— —— —— F.b. ——— t. 533.
2 angústa. K.I. (*sc.*) narrow-flowered. ——— 1825. —— F.b. Jacq. ic. 2. t. 231.
3 hùmilis. B.M. (*re.*) lake-coloured. ——— 1754. —— F.b. Bot. mag. t. 631.
4 Merìana. B.M. (*ro.*) red-flowered. ——— 1750. —— F.b. ——— t. 418.
5 iridifòlia. Jac. (*fl.*) Iris-leaved. ——— 1795. —— F.b. Jacq. ic. 2. t. 234.
6 fúlgida. Sal. (*sc.*) scarlet. ——— —— —— F.b. Bot. mag. t. 600.
Anthol'yza fúlgens. A.R. 192.
7 ròseo-álba. B.M.(*w.bh.*) two-coloured. ——— .... 7. 8. F.b. Bot. mag. t. 537.
β variegàta. (*wh.ro.*) *variegated-flower'd.* ——— .... —— F.b. ——— t. 1193.
8 brevifòlia. B.M. (*sc.*) short-leaved. ——— 1794. 5. 7. F.b. ——— t. 601.
9 ròsea. B.M. (*ro.*) pyramidal. ——— 1803. 7. 8. F.b. ——— t. 1072.
10 strictiflòra. B.M. (*ro.*) straight-flower'd. ——— 1810. 6. 7. F.b. ——— t. 1406.
11 marginàta. B.M. (*ro.*) broad-leaved. ——— 1774. —— F.b. ——— t. 608.
β minor. B.M. (*ro.*) *shining-leaved.* ——— 1812. —— F.b. ——— t. 1530.
12 rùbens. K.I. (*re.*) red-spathed. ——— 1825. —— F.b.
13 punctàta. H.K. (*bl.*) dotted-flowered. ——— 1800. 4. 5. F.b. Andr. reposit. t. 177.
I'xia punctàta. A.R.
14 plantagínea. B.M. (*bl.*) Fox-tail. ——— 1774. 6. 7. F.b. Bot. mag. t. 553.
15 spicàta. B.M. (*bh.*) hollow-leaved. ——— 1791. 5. F.b. ——— t. 523.
I'xia fistulòsa. B.M.

GLAD'IOLUS. Ker. CORN-FLAG. Triandria Monogynia. L.
1 viperàtus. B.M.(*gr.pu.*) perfumed. C. B. S. 1787. 4. 5. F.b. Swt. br. fl. gar. t. 156.
2 alàtus. B.M. (*sc.ye.*) wing-flowered. ——— 1795. 5. 6. F.b. ——— t. 187.
3 algoénsis. (*co.pu.y.*) Algoa bay. ——— 1824. 7. F.b. Bot. mag. t. 2608.
alàtus β algoënsis. B.M.
4 namaquénsis. B.M. (*sc.y.*) thick-leaved. ——— 1800. 5. 6. F.b. Bot. mag. t. 592.
5 perneábilis. R.S.(*re.y.*) tall. ——— 1825. —— F.b. De la Roch. dis. t. 2.
6 Watsònius. B.M. (*sc.*) Watson's. ——— 1791. 3. 5. F.b. Bot. mag. t. 450.
β variegàta. (*sc.ye.*) *variegated-flower'd.* ——— 1801. —— F.b. ——— t. 569.
7 brevifòlius. H.K. (*fl.*) short-leaved. ——— 1802. —— F.b. ——— t. 727.
8 hirsùtus. H.K. (*ro.*) rose-coloured. ——— 1795. 4. 6. F.b. ——— t. 574.
ròseus. Andr. reposit. t. 11.
9 merianéllus. K.I. (*pu.ye.*) slender-flower'd. ——— —— —— F.b.
10 puníceus. Lam. (*pu.*) dark purple. ——— —— —— F.b. Breyn. cent. 24. t. 12. f. 1.
villòsus. K.I. *Lamárckii.* R.S.
11 aphy'llus. K.I. (*fl.*) leafless. ——— —— —— F.b. Bot. mag. t. 992.
12 versícolor. B.M.(*re.ye.*) various-colour'd. ——— 1794. 5. 6. F.b. Andr. reposit. t. 19.
13 lævis. K.I. (*br.re.st.*) two-nerved. ——— 1806. —— F.b. Bot. mag. t. 1042. ε.
versícolor ε. binérvis. B.M.
14 Breyniànus. K.I.(*br.y.*) Breynius's. ——— —— —— F.b. Breyn. afr. 2. t. 7. f. 1.
15 suavèolens. K.I. (*br.y.*) sweet-scented. ——— 1799. —— F.b. Bot. mag. t. 556.
16 elongàtus. K.I. (*bl.ye.*) long-leaved. ——— —— —— F.b. Jacq. ic. 2. t. 244.
17 recúrvus. B.M. (*pu.ye.*) Violet-scented. ——— 1758. 4. 5. F.b. Bot. mag. t. 578.
18 inflàtus. K.I. (*pa.pu.*) inflated. ——— 1825. —— F.b.
19 grácilis. B.M. (*pa.bl.*) slender. ——— 1800. —— F.b. Bot. mag. t. 562.

20 débilis. B.M. (*wh.pu.*)	weak.	C. B. S.	1822.	4. 5.	F.b.	Bot. mag. t. 2585.
21 trichonemifòlius. B.M. (*ye.*)	slender-leav'd.	———	1810.	5. 6.	F.b.	——— t. 1483.
22 cóncolor. P.L. (*st.*)	pale-yellow.	———	1790.	——	F.b.	Sal. par. lond. t. 8.
23 trístis. B.M. (*st.br.*)	square-leaved.	———	1745.	——	F.b.	Bot. mag. t. 272.
24 hyalìnus. K.I. (*wh.st.pu.*)	transparent.	———	1825.	——	F.b.	Jacq. ic. 2. t. 242.
25 tenéllus. K.I. (*st.br.*)	feeble.	———	——	——	F.b.	——— t. 248.
26 Colvíllii. B.F.G. (*re.ye.*)	Colvill's.	Hybrid.	1823.	——	F.b.	Swt. br. fl. gar. t. 155.
27 cárneus. B.M. (*pa.pu.*)	flesh-coloured.	C. B. S.	1796.	——	F.b.	Bot. mag. t. 591.
28 cuspidàtus. B.M. (*w.pu.*)	sharp-pointed.	———	1795.	4. 5.	F.b.	——— t. 582.
29 angústus. B.M. (*st.re.*)	narrow-leaved.	———	1757.	——	F.b.	——— t. 602.
hastàtus. Thunb. *non* B.M.						
30 vomérculus. K.I. (*bh.vi.*)	spade.	———	1816.	——	F.b.	Bot. mag. t. 1564.
31 involùtus. K.I. (*ro.pu.*)	involute.	———	1757.	5. 6.	F.b.	Mill. ic. t. 286. f. 1.
32 èdulis. B.R. (*wh.pu.ye.*)	eatable-rooted.	———	——	——	F.b.	Bot. reg. t. 169.
33 undulàtus. B.M. (*st.pu.*)	waved-flowered.	———	1760.	4. 5.	F.b.	Bot. mag. t. 647.
34 fasciàtus. R.S. (*pi.pu.*)	fasciate.	———	——	——	F.b.	——— t. 538.
undulàtus β. Bot. mag.						
35 blándus. B.M. (*bh.ro.*)	fairest.	———	1774.	6.	F.b.	Bot. mag. t. 625.
36 álbidus. W. (*wh.*)	white-flowered.	———	——	——	F.b.	——— t. 648.
37 campanulàtus. A.R. (*pu.*)	bell-flowered.	———	1794.	5. 6.	F.b.	Andr. reposit. t. 188.
blándus γ *purpùreo-albescens*. B.M.						
38 excélsus. K.I. (*pi.*)	painted-flower'd.	———	——	——	F.b.	Bot. mag. t. 1665.
blándus δ. *excélsus*. B.M.						
39 trimaculàtus. K.I. (*bh.pu.*)	three-spotted.	———	——	——	F.b.	Lam. ill. 1. t. 32. f. 3.
40 flexuòsus. B.M. (*bh.*)	flexuose.	———	1825.	——	F.b.	Thunb. diss. t. 1. f. 1.
41 floribúndus. B.M. (*pi.w.*)	large-flowered.	———	1788.	5. 7.	F.b.	Bot. mag. t. 610.
42 Millèri. B.M. (*wh.pu.*)	Miller's.	———	1751.	4. 5.	F.b.	——— t. 632.
43 cardinàlis. B.M. (*sc.*)	superb.	———	1789.	7. 8.	F.b.	——— t. 135.
44 bockveldiénsis. W.H. (*co.*)	Bockveld.	———	1826.	——	F.b.	
45 spofforthiànus. W.H. (*va.*)	Blando-cardinalis.	Hybrid.	....	——	F.b.	
46 mitchamiénsis. W.H. (*va.*)	Tristi-hirsutus.	———	....	——	F.b.	
47 rígidus. W.H. (*va.*)	Tristi-blandus.	———	....	——	F.b.	
48 propínquus. W.H. (*var.*)	Floribundo-blandus.	——	....	——	F.b.	
49 fràgrans. W.H. (*va.*)	Recurvo-tristis.	———	....	——	F.b.	
50 Haylockiànus. W.H. (*va.*)	Versicolori-blandus.	——	....	——	F.b.	
51 Herbertiànus. (*va.*)	Tristi-Spofforthianus.	——	....	——	F.b.	
52 odoràtus. W.H. (*va.*)	Hirsuto-Spofforthianus.	—	....	——	F.b.	
53 delicàtus. W.H. (*va.*)	Recurvo-blandus.	———	....	——	F.b.	
54 byzantìnus. B.M. (*pu.*)	Turkish.	Turkey.	1629.	6. 7.	H.b.	Bot. mag. t. 874.
55 commùnis. B.M. (*re.*)	common.	S. Europe.	1596.	——	H.b.	——— t. 86.
β *cárneus*. B.M. (*bh.*)	*pale-flowered.*	———	——	——	H.b.	——— t. 1575.

SPHÆROSP'ORA. SPHÆROSP'ORA. Triandria Monogynia.

1 imbricàta. (*re.wh.*)	European.	S. Europe.	1596.	6. 7.	H.b.	Bot. mag. t. 719.
Gladìolus ségetum. B.M. *Gladìolus imbricàtus*. K.I.						
2 triphy'lla. (*re.*)	three-leaved.	Greece.	1825.	——	H.b.	Flor. græc. 1. t. 38.

SYNNO'TIA. B.F.G. SYNNO'TIA. Triandria Monogynia.

1 variegàta. B.F.G. (*bl.ye.*)	variegated-flower'd.	C. B. S.	1825.	4. 6.	F.b.	Swt. br. fl. gar. t. 150.
2 bícolor. B.F.G. (*ye.bl.*)	two-coloured.	———	1786.	3. 4.	F.b.	Bot. mag. t. 548.
I'xia bícolor. B.M. *Sparáxis bícolor*. Ker. *Gladìolus bícolor*. W.						
3 galeàta. B.F.G. (*vi.ye.*)	helmet-flower'd.	———	1825.	——	F.b.	Jacq. ic. 2. t. 258.
Gladìolus galeàtus. Jacq.						

STREPTANTH'ERA. B.F.G. STREPTANTH'ERA. Triandria Monogynia.

1 élegans. B.F.G. (*w.ve.y.*)	elegant.	C. B. S.	1825.	6. 7.	F.b.	Swt. br. fl. gar. t. 209.
2 cùprea. (*co.ve.*)	copper-coloured.	———	——	——	F.b.	

SPARA'XIS. Ker. SPARA'XIS. Triandria Monogynia. H.K.

1 trícolor. H.K. (*cr.ve.y.*)	three-coloured.	C. B. S.	1789.	5.	F.b.	Bot. mag. t. 381.
2 versícolor. (*cr.bk.y.*)	various-coloured.	———	1811.	——	F.b.	Swt. br. fl. gar. t. 160.
3 Griff'ìni. (*pu.ve.y.*)	Mr. Griffin's.	———	——	——	F.b.	Bot. mag. t. 1482. f. inf.
trícolor γ *violàceo-purpùrea*. B.M.						
4 blánda. (*bh.*)	blush-flowered.	———	——	——	F.b.	Bot. mag. t. 1482. f. med.
trícolor. δ. *subròseo-álbida*. B.M.						
5 grandiflòra. H.K. (*pu.*)	large-flowered.	———	1758.	4. 5.	F.b.	Bot. mag. t. 541.
α *purpùrea*. (*pu.*)	*velvet-flowered.*	———	——	——	F.b.	Andr. reposit. t. 87.
β *striàta*. (*pu.st.*)	*streaked-flower'd.*	———	——	——	F.b.	Bot. mag. t. 779.
6 Liliàgo. (*wh.bh.*)	Lily-flowered.	———	....	——	F.b.	Redout. lil. 2. t. 109.
I'xia Liliàgo. R.L. *Sparáxis grandiflòra* γ *Liliàgo*. Bot. reg. t. 258.						
7 bulbífera. H.K. (*ye.*)	bulb-bearing.	C. B. S.	1758.	5. 6.	F.b.	Bot. mag. t. 545.
8 fràgrans. B.M. (*ye.*)	fragrant.	———	1825.	——	F.b.	Jacq. ic. 2. t. 274.
9 anemonæflòra. B.M. (*st.*)	Anemone-flower'd.	———	——	——	F.b.	——— 2. t. 273.
I'xia anemonæflòra. Jacq.						

2 K 3

TRIT'ONIA. Ker. Trit'onia. Triandria Monogynia. H.K.
1 críspa. H.K. (ro.wh.) curled-leaved. C. B. S. 1787. 4. 5. F.b. Bot. mag. t. 678.
2 anigozanthiflòra. (gr.) Anigozanthus-fld. ——— 1825. 6. 8. F.b.
3 víridis. B.M. (gr.) green-flowered. ——— 1788. 7. F.b. Bot. mag. t. 1275.
4 ròsea. H.K. (wh.ro.) rosy. ——— 1793. 6. 7. F.b. ——— t. 618.
5 capénsis. B.M. (st.pu.) Cape. ——— 1811. 8. 10. F.b. ——— t. 1531.
6 pállida. B.M. (st.) pale-flowered. ——— 1806. —— F.b. Jacq. ic. 2. t. 262.
7 longiflòra. B.M. (st.re.) long-flowered. ——— 1774. 6. 8. F.b. Bot. mag. t. 256.
Hexaglóttis longiflòra. V. *Gladìolus longiflòrus.* Jacq. *I'xia longiflòra.* B.M.
8 tenuiflòra. R.S. (st.) slender-flower'd. C. B. S. 1811. 4. 6. F.b. Bot. mag. t. 1502.
9 cóncolor. (st.) self-coloured. ——— —— —— F.b. ——— t.1502.f.minor
10 Rocheàna. (st.ro.) bending-flowered. ——— —— 8. F.b. ——— t. 1503.
11 pectinàta. B.M. (bh.) pectinated. ——— 1825. 5. 6. F.b.
12 striàta. B.M. (st.vi.) streaked. ——— —— —— F.b. Jacq. ic. 2. t. 260.
13 lineàta. B.M. (st.br.) pencilled. ——— 1774. —— F.b. Bot. mag. t. 487.
14 securígera. B.M. (co.ye.) copper-coloured. ——— —— 5. F.b. ——— t. 383.
15 flàva. B.M. (ye.) yellow. ——— 1780. 4. 6. F.b. Bot. reg. t. 747.
16 refrácta. B.R. (ye.) reflexed. ——— 1815. 5. 6. F.b. ——— t. 135.
17 squálida. B.M. (pi.bh.) sweet-scented. ——— 1774. —— F.b. Bot. mag. t. 581.
18 fenestràta. B.M. (sn.) open-flowered. ——— 1801. —— F.b. ——— t. 704.
19 purpùrea. K.I. (pu.) purple. ——— 1825. —— F.b.
20 crocàta. B.M. (sn.) Crocus-flowered. ——— 1758. —— F.b. Bot. mag. t. 184.
I'xia crocàta. B.M.
21 deústa. B.M. (sn.ve.) spotted-flower'd. ——— 1774. —— F.b. Bot. mag. t. 622.
22 miniàta. B.M. (ye.) late-flowered. ——— 1795. 8. F.b. ——— t. 609.
23 xanthospìla. S.S. (wh.) yellow-spotted. ——— 1825. 5. 6. F.b. Redout. lil. 3. t. 124.

MORPHI'XIA. K.I. Morphi'xia. Triandria Monogynia.
1 lineàris. K.I. (bh.) slender. C. B. S. 1796. 4. 5. F.b. Bot. mag. t. 570.
I'xia lineàris. R.S. *capillàris α gracíllima.* B.M. *Hyàlis grácilis.* Salisb.
2 capillàris. K.I. (bh.) wire-stemmed. C. B. S. 1774. —— F.b. Bot. mag. t. 617.
Hyàlis latifòlia. Salisb. *I'xia láncea.* Jacq. ic. 2. t. 281. *capillàris β stricta.* B.M.
3 aùlica. K.I. (ro.) rose-coloured. C. B. S. 1774. —— F.b. Bot. mag. t. 1013.
I'xia aùlica. H.K. *capillàris γ aùlica.* B.M. *Hyàlis aùlica.* Salisb.
4 incarnàta. K.I. (fl.) flesh-coloured. C. B. S. 1774. —— F.b. Jacq. ic. 2. t. 282.

I'XIA. Ker. I'xia. Triandria Monogynia. L.
1 fucàta. B.M. (bh.) painted. C. B. S. .... 6. 7. F.b. Bot. mag. t. 1379.
2 aristàta. B.M. (pi.) salver-flowered. ——— 1800. —— F.b. ——— t. 589.
3 flexuòsa. B.M. (va.) flexuose. ——— 1757. 4. 5. F.b. ——— t. 624.
4 hy'brida. B.M. (wh.ro.) spurious. ——— —— —— F.b. ——— t. 128.
5 leucántha. P.S. (wh.) white-flowered. ——— 1779. 5. 8. F.b. Jacq. ic. 2. t. 278.
6 pátens. B.M. (cr.) spreading-flower'd. ——— —— 4. 6. F.b. Bot. mag. t. 522.
7 crateroídes. B.M. (cr.) crimson. ——— 1778. —— F.b. ——— t. 594.
8 monadélpha. B.M. (bl.gr.) monadelphous. ——— 1792. —— F.b. ——— t. 607.
9 cúrta. A.R. (or.) shortened. ——— —— —— F.b. Andr. reposit. t. 554.
monadélpha β. Bot. mag. t. 1378.
10 columellàris. B.M. (pu.b.ve.) variegated. ——— 1790. 6. 8. F.b. Bot. mag. t. 630.
11 cònica. B.M. (or.ve.) orange-color'd. ——— 1757. 4. 5. F.b. ——— t. 539.
12 fúsco-citrìna. R.S. (ye.br.) yellow and brown. ——— —— —— F.b. Redoute liliac. t. 86.
13 dùbia. R.S. (ye.pu.) doubtful. ——— .... —— F.b. ——— 2. t. 64.
14 viridiflòra. R.S. (gr.pu.) green-flowered. ——— 1780. 5. 6. F.b. ——— t. 476.
15 maculàta. W. (va.) spot-flowered. ——— —— —— F.b. Jacq.schœn.t.20.21.&c.
16 ovàta. A.R. (wh.pu.) ovate-spiked. ——— —— —— F.b. Andr. reposit. t. 23.
17 capitàta. A.R. (wh.bk.) headed. ——— —— —— F.b. ——— t. 159.
18 ochroleùca. (st.br.) cream-coloured. ——— 1809. —— F.b. Bot. mag. t. 1285.
19 erécta. H.K. (wh.) upright. ——— 1757. —— F.b. Jacq. schœn. 1. t. 18.
α albiflòra. (wh.) *white-flowered.* ——— —— —— F.b. Bot. mag. t. 623.
β incarnàta. (fl.) *flesh-coloured.* ——— —— —— F.b. Andr. reposit. t. 155.
γ lùtea. (ye.) *yellow-flowered.* ——— —— —— F.b. Bot. mag. t. 846.
20 odoràta. (ye.) sweet-scented. ——— —— —— F.b. ——— t. 1173.
21 retùsa. H.K. (ro.) retuse. ——— 1793. 2. 4. F.b. Jacq. ic. 2. t. 275.
polystàchia. Bot. mag. t. 629.
22 scillàris. B.M. (ro.) Squill-flowered. ——— 1787. —— F.b. Bot. mag. t. 542.
23 críspa. B.M. (ro.) curled-leaved. ——— —— 4. 5. F.b. ——— t. 599.

MELASPH'ÆRULA. Ker. Melasph'ærula. Triandria Monogynia. H.K.
1 iridifòlia. B.F.G. (wh.br.) Iris-leaved. C. B. S. 1825. 2. 4. F.b. Swt. br. fl. gar. ic.
2 intermèdia. (pu.st.) intermediate. ——— 1787. 4. 8. F.b. Bot. mag. t. 615.
gramínea. B.M. *excl. synonym.*
3 gramínea. (wh.br.) grass-leaved. ——— 1825. 2. 5. F.b. Jacq. ic. 2. t. 236.
Gladìolus gramíneus. Jacq. *Diàsia graminifòlia.* Red. lil. t. 163.
4 parviflòra. L.C. (wh.) small-flowered. C. B. S. 1825. —— F.b. Lodd. bot. cab. t. 1444.

HESPERA'NTHA. Ker. EVENING-FLOWER. Triandria Monogynia. H.K.					
1 cinnamòmea. H.K. (*w.br.*) curled-leaved.	C. B. S.	1787.	4. 5.	F.b.	Bot. mag. t. 1054.
2 falcàta. H.K. (*wh.br.*) sickle-leaved.	———	——	——	F.b.	——— t. 566.
3 graminifòlia. (*wh.br.*) Grass-leaved.	———	1808.	8. 9.	F.b.	——— t. 1254.
pilòsa β foliis nudis, floribus minoribus. B.M.					
4 pilòsa. H.K. (*wh.br.*) hairy.	C. B. S.	1811.	4. 5.	F.b.	Bot. mag. t. 1475.
5 angústa. B.M. (*wh.*) narrow-leaved.	———	1825.	——	F.b.	Jacq. ic. 2. t. 279.
I'xia angústa. Jacq.					
6 radiàta. H.K. (*wh.br.*) nodding-flowered.	———	1794.	4. 6.	F.b.	Bot. mag. t. 573.
GEISSORH'IZA. Ker. TILE-ROOT. Triandria Monogynia. H.K.					
1 Rocheàna. (*bl.pu.*) plaid.	C. B. S.	1790.	5.	F.b.	Bot. mag. t. 598.
2 monántha. (*bl.*) one-flowered.	———	——	——	F.b.	Houtt.syst.11.t.78.f.1.
3 setàcea. B.M. (*st.re.*) bristle-leaved.	———	1809.	6. 7.	F.b.	Bot. mag. t. 1255.
4 obtusàta. H.K. (*st.*) pale-yellow.	———	1801.	5.	F.b.	——— t. 672.
5 vaginàta. B.F.G. (*ye.*) sheathed.	———	1825.	6. 9.	F.b.	Swt. br. fl. gar. t. 138.
6 secúnda. B.M. (*bl.*) one-ranked.	———	1795.	5. 6.	F.b.	Jacq. ic. 2. t. 277.
α cærùlea. (*bl.*) *blue-flowered.*	———	——	——	F.b.	Bot. mag. t. 597.
β albéscens. (*wh.bl.*) *whitish-flowered.*	———	——	——	F.b.	——— t. 1105.
7 sublùtea. B.M. (*st.*) yellowish.	———	1825.	——	F.b.	
I'xia sublùtea. Lam.					
8 húmilis. K.I. (*st.bh.*) dwarf.	———	1822.	——	F.b.	
9 erécta. W.H. (*st.bl.*) erect.	———	1824.	——	F.b.	
10 imbricàta. B.M. (*st.re.*) imbricated.	———	——	——	F.b.	
11 hírta. B.M. (*pa.bl.*) hairy.	———	——	——	F.b.	
12 ciliàris. Sal. (*st.*) fringed.	———	....	——	F.b.	
infléxa. K.I. *I'xia infléxa.* De la Roche.					
13 excìsa. B.M. (*wh.pi.*) dwarf cut-leaved.	———	1789.	4. 5.	F.b.	Bot. mag. t. 584.
I'xia excìsa. B.M.					
SPATALA'NTHUS. B.F.G. RIBBON-FLOWER. Monadelphia Triandria.					
speciòsus. B.F.G. (*sc.y.ve.*) beautiful.	C. B. S.	1825.	8. 10.	F.b.	Swt. br. fl. gar. t. 300.
TRICHON'EMA. Ker. TRICHON'EMA. Triandria Monogynia. H.K.					
1 Bulbocòdium. H.K. (*pu.*) channel-leaved.	S. Europe.	1739.	3. 4.	H.b.	Bot. mag. t. 265.
2 purpuráscens. (*pu.*) purple.	Naples.	1825.	——	H.b.	Ten. fl. nap. v. 1. t. 2.
I'xia purpuráscens. T.S.					
3 Colúmnæ. T.P. (*st.li.*) Columna's.	———	——	——	H.b.	Redout. lil. t. 88. f. A.
4 cœlestìnum. (*bl.*) blue.	N. America.	1820.	——	H.b.	Bartr. it. t. 3.
I'xia cœlestìna. Ph.					
5 ròseum. B.M. (*ro.*) rose-coloured.	C. B. S.	1808.	6. 7.	F.b.	Bot. mag. t. 1225.
6 speciòsum. B.M. (*ro.*) crimson.	———	——	3. 4.	F.b.	——— t. 1476.
7 pùdicum. B.M. (*ro.*) blush.	———	——	6. 8.	F.b.	——— t. 1244.
8 cruciàtum. S.S. (*ro.*) square-leaved.	———	1758.	5. 6.	F.b.	Jacq. ic. 2. t. 290.
9 longifòlium. Sal. (*ro.*) small-flowered.	———	——	——	F.b.	Bot. mag. t. 575.
cruciàtum. B.M. *non* Jacq.					
10 filifòlium. K.I. (*ye.*) slender-leaved.	———	1812.	——	F.b.	Redout. lil. t. 251. f. 2.
11 recurvifòlium. K.I. (*st.*) recurve-flowered.	———	——	——	F.b.	——— t. 251. f. 1.
12 tortuòsum. K.I. (*ye.*) twisted-leaved.	———	1822.	——	F.b.	
13 cauléscens. B.M. (*ye.*) caulescent.	———	1810.	6. 7.	F.b.	Bot. mag. t. 1392.
14 chloroleùcum. B.M. (*st.*) milk-coloured.	———	1825.	5. 7.	F.b.	Jacq. ic. 2. t. 270.
CR'OCUS. L. CR'OCUS. Triandria Monogynia. L.					
1 vérnus. E.B. (*va.*) spring.	England.	....	2. 4.	H.b.	Eng. bot. v. 5. t. 344.
β plúmbeus. H.T. (*bl.*) *lead-coloured.*	....	....	——	H.b.	Hort.t.v.7.t.11.12.f.10.
γ álbus màjor. H.T. (*wh.*) *large-white.*	....	....	——	H.b.	——— v.7.t.11.12.f.11.
δ leucorhy'nchus. H.T. (*w.*) *white.*	....	....	——	H.b.	——— v.7.t.11.12.f.12.
ε inflàtus. H.T. (*bl.*) *bellied.*	....	....	——	H.b.	——— v.7.t.11.12.f.13.
ζ fucàtus. H.T. (*bh.pu.*) *painted.*	....	....	——	H.b.	——— v.7.t.11.12.f.14.
η píctus. H.T. (*li.pu.*) *pencilled.*	....	....	——	H.b.	——— v.7.t.11.12.f.15.
θ Andersònii. H.T. (*wh.*) *Anderson's.*	....	....	——	H.b.	——— v.7.t.11.12.f.16.
ι Sabìni. H.T. (*bl.*) *Sabine's.*	....	....	——	H.b.	——— v.7.t.11.12.f.17.
κ Gloriànus. H.T. (*wh.bl.*) *Queen of Portugal's.*	....	....	——	H.b.	——— v.7.t.11.12.f.18.
λ pulchéllus. H.T. (*li.bl.*) *pretty.*	....	....	——	H.b.	——— v.7.t.11.12.f.19.
μ neapolitànus. R.S. (*std.*) *Neapolitan.*	Naples.	....	——	H.b.	Bot. mag. t. 860.
ν obovàtus. R.S. (*wh.*) *obovate-flowered.*	S. Europe.	....	——	H.b.	——— t. 2240.
2 striàtus. L.en. (*wh.vi.*) striated.	....	....	——	H.b.	Bot. mag. t. 1110.
3 versícolor. H.K. (*wh.bl.*) party-coloured.	———	1629.	——	H.b.	Hort.tr.v.7.t.11.12.f.6.
α purpùreus. H.T. (*pu.*) *purple-flowered.*	———	——	——	H.b.	——— v.7.t.11.12.f.7.
β plumòsus. H.T. (*pu.w.*) *feathered.*	....	....	——	H.b.	——— v.7.t.11.12.f.8.
γ élegans. H.T. (*pu.wh.*) *elegant.*	....	....	——	H.b.	——— v.7.t.11.12.f.9.
δ urbànus. H.T. (*wh.li.*) *homebred.*	....	....	——	H.b.	

4 biflòrus. H.K.	(*wh.pu.*)	Scotch.	Crimea.	1629.	2. 3.	H.b.	Bot. mag. t. 845.
β *Parkinsònii.* H.T.	(*wh.bl.*)	Parkinson's.	———	———	———	H.b.	Hort. tr. v. 7. t. 11. f. 4.
5 pusíllus. R.S.	(*wh.pu.*)	small.	Italy.	1824.	———	H.b.	Swt. br. fl. gar. t. 106.
6 mínimus. R.L.	(*pu.wh.*)	small.	Corsica.	1828.	———	H.b.	Redout. lil. t. 81.
7 pr'æcox. E.B.	(*pu.wh.*)	early.	Britain.	....	———	H.b.	Eng. bot. t. 2645.
8 argénteus. H.T.	(*pu.wh.*)	silvery.	....	....	———	H.b.	Hort.tr.v.7.t.11.12.f.8.
9 albiflòrus. R.S.	(*wh.*)	white-flowered.	Hungary.	———	2. 4.	H.b.	
10 variegàtus. R.S.	(*ye.pu.*)	variegated.	Levant.	1828.	———	H.b.	Hop.etHor.a.Meer.c.ic
11 reticulàtus. M.B.	(*ye.vi.*)	netted-bulbed.	Caucasus.	1825.	———	H.b.	M. B. cent. pl. rar. t.1.
12 susiánus. H.K.	(*ye.pu.*)	Cloth of Gold.	Turkey.	1605.	———	H.b.	Bot. mag. t. 652.
13 stellàris. H.T.	(*ye.pu.*)	starry cloth of gold.	S.Europe.	1629.	———	H.b.	Hort. trans. 1. p.136. ic.
14 sulphùreus. B.M.	(*st.*)	pale yellow.	———	———	———	H.b.	Bot. mag. t. 1384.
15 lácteus. H.T.	(*st.vi.*)	milk-coloured.	....	....	———	H.b.	
β *penicillàtus.* H.T.	(*st.vi.*)	*pencilled.*	....	....	———	H.b.	Hort. t. v.7. t.11.12. f.3.
16 lagenæflòrus. P.L.	(*st.*)	Gourd-flowered.	Greece.	....	———	H.b.	Bot. mag. t. 2655.
17 aùreus. F.G.	(*go.*)	golden-flowered.	———	....	———	H.b.	Eng. bot. t. 2646.
18 lùteus. R.S.	(*ye.*)	common yellow.	Levant.	1620.	———	H.b.	Bot. mag. t. 45.
vérnus. B.M. *nec aliorum.*							
19 mæsìacus. B.M.	(*ye.*)	great yellow.	Greece.	1629.	———	H.b.	Bot. mag. t. 1111.
20 satìvus. R.S.	(*vi.*)	saffron.	England.	....	9. 10.	H.b.	Eng. bot. v. 5. t. 343.
21 serotìnus. R.S.	(*vi.*)	late-flowered.	S. Europe.	1629.	9. 11.	H.b.	Bot. mag. t. 1267.
22 Pallàsii. M.B.	(*vi.*)	autumnal.	Tauria.	....	———	H.b.	
autumnàlis. R.S.							
23 nudiflòrus. E.B.	(*vi.*)	naked-flowered.	England.	....	10. 11.	H.b.	Eng. bot. v. 7. t. 491.

ORDO CCVII.

HÆMODORACEÆ. *Brown prodr.* 299.

WACHENDO'RFIA. L.		WACHENDO'RFIA.	Triandria Monogynia. L.				
1 thyrsiflòra. R.S.	(*ye.*)	tall-flowered.	C. B. S.	1759.	5. 6.	G. ♃.	Bot. mag. t. 1060.
2 paniculàta. R.S.	(*ye.*)	panicled.	———	1700.	2. 5.	F.b.	——— t. 616.
3 Herbérti.	(*ye.*)	Herbert's.	———	1823.	———	F.b.	——— t. 2610.
paniculàta. β. H. *in* B.M.							
4 hirsùta. R.S.	(*ye.*)	hairy.	———	1687.	4. 6.	F.b.	Bot. mag. t. 614.
5 brevifòlia. B.M.	(*ye.*)	short-leaved.	———	1795.	3. 4.	F.b.	——— t. 1166.
6 Breyniàna. R.S.	(*pu.ye.*)	Breynius's.	———	1825.	———	F.b.	Breyn. cent. f. 37.
7 gramínea. S.S.	(*ye.*)	Grass-leaved.	———	1810.	6.	G. ♃.	
8 tenélla. R.S.	(*pu.*)	slender-leaved.	———	1816.	———	G. ♃.	
XIPHI'DIUM. J.		XIPHI'DIUM.	Triandria Monogynia. W.				
1 álbidum. S.S.	(*wh.*)	white.	W. Indies.	1787.	6. 8.	S. ♃.	Sw. fl. ind. oc. 1. t. 2.
floribúndum. Swartz.							
2 cœrùleum. S.S.	(*bl.*)	blue.	———	1793.	———	S. ♃.	Aub. gui. 1. t. 11.
HÆMOD'ORUM. B.P.		HÆMOD'ORUM.	Triandria Monogynia. R.S.				
1 planifòlium. B.P.	(*st.*)	flat-leaved.	N. S. W.	1810.	7. 11.	F. ♃.	Bot. mag. t. 1610.
2 teretifòlium. B.P.	(*st.*)	round-leaved.	———	1822.	———	F. ♃.	
DIL'ATRIS. Ker.		DIL'ATRIS.	Triandria Monogynia. R.S.				
1 corymbòsa. W.	(*pu.*)	broad-petaled.	C. B. S.	1790.	5. 6.	F. ♃.	Sm. exot. bot. 1. t. 16.
umbellàta. L.							
2 viscòsa. W.	(*pu.*)	clammy.	———	1795.	———	F. ♃.	
3 paniculàta. W.	(*pu.ye.*)	panicled.	———	1825.	———	F. ♃.	
GYROTH'ECA. H.T.		GYROTH'ECA.	Triandria Monogynia.				
tinctòria.	(*re.*)	red-rooted.	N. America.	1812.	6. 7.	H. ♃.	Mich. amer. 1. t. 4.
Lachnánthes tinctòria. Ell. *Dilàtris tinctòria.* Ph. *Heritièra Gmèlini.* M.							
LOPH'IOLA. B.M.		LOPH'IOLA.	Hexandria Monogynia. B.M.				
aùrea. B.M.	(*ye.*)	golden-flowered.	N. America.	1811.	5. 7.	H. ♃.	Bot. mag. t. 1596.
Conóstylis americàna. Ph.							
LAN'ARIA. W.		LAN'ARIA.	Hexandria Monogynia. W.				
plumòsa. W.	(*pu.*)	woolly.	C. B. S.	1787.	6. 7.	G. ♃.	
CONO'STYLIS. B.P.		CONO'STYLIS.	Hexandria Monogynia. S.S.				
1 aculeàta. B.P.	(*st.*)	prickly-leaved.	N. Holland.	1824.	9. 2.	F. ♃.	Bot. mag. t. 2989.
2 serrulàta. B.P.	(*st.*)	serrulate.	———	———	....	F. ♃.	
ANIGOZA'NTHOS. B.P.		ANIGOZA'NTHOS.	Hexandria Monogynia.				
1 flávida. B.P.	(*br.gr.*)	russet-green.	N. Holland.	1803.	5. 9.	F. ♃.	Bot. mag. t. 1151.
2 rùfa. B.P.	(*br.gr.*)	rufous.	———	1824.	———	F. ♃.	Labill. voy. 1. t. 22.

ORDO CCVIII.

HYPOXIDEÆ. *Brown.*

HYPO'XIS. L.		Hypo'xis.	Hexandria Monogynia. L.				
1 júncea. w.	(*ye.*)	rushy.	Carolina.	1787.	6. 7.	H.b.	Sm. spic. 15. t. 16.
2 hygrométrica. B.P.	(*ye.*)	hairy-leaved.	N. Holland.	1820.	——	F.b.	Lab. nov. holl. 1. t. 8.
3 álba. w.	(*wh.*)	white.	C. B. S.	1806.	——	F.b.	Lodd. bot. cab. t. 1074.
4 lineàris. A.R.	(*ye.*)	linear-leaved.	———	1792.	4. 5.	F.b.	Andr. reposit. t. 171.
5 stellàta. w.	(*ye.*)	yellow starry.	———	1752.	4. 6.	F.b.	Bot. mag. t. 662.
6 élegans. P.S.	(*wh.*)	white starry.	———	——	——	F.b.	——— t. 1223.
stellàta β. B.M.	*certainly a distinct species.*						
7 serràta. w.	(*ye.*)	saw-leaved.	———	1788.	6. 7.	F.b.	Bot. mag. t. 709.
8 ovàta. w.	(*ye.*)	smooth-leaved.	———	1806.	2. 5.	F.b.	——— t. 1010.
9 veratrifòlia. w.	(*ye.*)	plaited-leaved.	———	1788.	6. 7.	F.♃.	Jacq. ic. 2. t. 367.
plicàta. Jacq.							
10 carolinénsis. M.	(*ye.*)	Carolina.	Carolina.	1822.	——	F.b.	
11 decúmbens. w.	(*ye.*)	decumbent.	Jamaica.	1755.	6. 9.	S.♃.	Mill. ic. 1. t. 39. f. 2.
12 stellipìlis. B.R.	(*ye.*)	starry-haired.	C. B. S.	1821.	6. 12.	F.♃.	Bot. reg. t. 663.
13 oblíqua. w.	(*ye.*)	oblique-leaved.	———	1795.	6. 7.	F.♃.	Andr. rep. t. 195.
14 praténsis. B.P.	(*ye.*)	field.	N. S. W.	1822.	——	F.b.	
15 eréсta. S.S.	(*ye.*)	upright.	N. America.	1752.	——	H.b.	Bot. mag. t. 710.
16 grácilis. Ot.	(*ye.*)	slender.	S. America.	1828.	——	G.b.	
17 Sellòii. Ot.	(*ye.*)	Sello's.	Buenos Ayres.	1827.	——	F.b.	
18 sobolífera. w.	(*ye.*)	creeping.	C. B. S.	1774.	6. 9.	F.♃.	Bot. mag. t. 711.
19 villòsa. w.	(*ye.*)	villous.	———	——	——	F.♃.	Jacq. ic. 2. t. 307.
20 obtùsa. B.R.	(*ye.*)	obtuse.	———	1816.	——	F.♃.	Bot. reg. t. 169.
21 aquática. w.	(*ye.*)	water.	———	1787.	6. 7.	F.♃.	
CURC'ULIGO. B.P.		Curc'uligo.	Hexandria Monogynia. S.S.				
1 plicàta. B.R.	(*ye.*)	plaited.	C. B. S.	1788.	6. 7.	F.b.	Jacq. schœn. 1. t. 80.
β glábra.	(*ye.*)	*smooth-leaved.*	———	1816.	——	F.b.	Bot. reg. t. 345.
2 orchioídes. S.S.	(*ye.*)	narrow-leaved.	E. Indies.	1800.	——	S.♃.	Roxb. corom. 1. t. 13.
3 brevifòlia. S.S.	(*ye.*)	short-leaved.	———	1804.	5. 7.	S.♃.	Bot. mag. t. 1076.
4 latifòlia. S.S.	(*ye.*)	broad-leaved.	———	——	5. 8.	S.♃.	Bot. reg. t. 754.
5 sumatràna. B.C.	(*ye.*)	Sumatra.	Sumatra.	1818.	——	S.♃.	Lodd. bot. cab. t. 413.
6 recurvàta. S.S.	(*ye.*)	recurved.	Bengal.	1805.	——	S.♃.	Bot. reg. t. 770.
BARBAC'ENIA. Vand.		Barbac'enia.	Hexandria Monogynia.				
purpùrea. B.M.	(*pu.*)	purple-flower'd.	Brazil.	1825.	8. 9.	S.♄.	Bot. mag. t. 2777.
VELL'OSIA. Vand.		Vell'osia.	Polyandria Monogynia.				
squamàta. P.B.	(*bl.*)	scaly.	Brazil.	1829.	....	S.♄.	Pohl. bras. ic. v. 1. t. 9.

ORDO CCIX.

AMARYLLIDEÆ. *Brown prodr. p.* 296.

STERNBE'RGIA. W.K.		Sternbe'rgia.	Hexandria Monogynia.				
1 colchiciflòra. B.R.	(*ye.*)	Colchicum-flower'd.	Hungary.	1816.	9. 10.	H.b.	W. et K. hung. 2. t. 157.
2 Clusiàna. B.R.	(*ye.*)	Clusius's.	Constantinople...		——	H.b.	Clus. hist. 1. t. 163.
3 lùtea. B.R.	(*ye.*)	common yellow.	S. Europe.	1596.	——	H.b.	Bot. mag. t. 290.
Amary'llis lùtea. B.M.							
β angustifòlia.	(*ye.*)	*narrow-leaved.*	———	——	——	H.b.	
ZEPHYRA'NTHES. H.A.		Zephyra'nthes.	Hexandria Monogynia.				
1 chloroleùca. H.A.	(*st.*)	one-leaved.	....	....	5. 8.	G.b.	Ker's rev. pl. 8. f. 1.
2 Atamásco. H.A.	(*wh.*)	Atamasco Lily.	N. America.	1629.	5. 6.	H.b.	Bot. mag. t. 239.
Amary'llis Atamásco. B.M.							
3 tubispátha. H.A.	(*wh.*)	tube-sheathed.	S. America.	....	5. 7.	S.b.	Bot. mag. t. 1586.
Amary'llis tubispátha. B.M.							
4 cándida. B.M.	(*wh.*)	white-flowered.	Peru.	1822.	7. 10.	H.b.	Bot. mag. t. 2607.
Amary'llis cándida. Bot. reg. t. 724.							
5 ròsea. B.R.	(*ro.*)	rose-coloured.	Havannah.	——	9. 10.	G.b.	Bot. reg. t. 821.

6 striàta. B.M.	(*bh.wh.*)	striped.	Mexico.	1824.	6. 8.	H.b.	Bot. mag. t. 2593.
7 verecúnda. B.M.	(*bh.*)	modest.	———	——	——	H.b.	——— t. 2583.
8 carinàta. B.M.	(*ro.*)	keeled-leaved.	———	——	——	H.b.	Swt. br. fl. gar. s. 2. t. 4.
9 grandiflòra. B.R.	(*ro.*)	large-flowered.	———	1825.	——	H.b.	Bot. reg. excl. fol. t. 902.
HABRA'NTHUS. B.M.		HABRA'NTHUS.	Hexandria Monogynia.				
1 gracilifòlius. B.M.	(*ro.*)	slender-leaved.	S. America.	1823.	9. 11.	G.b.	Bot. mag. t. 2464.
2 angústus. B.M.	(*pu.*)	narrow.	BuenosAyres.	1825.	——	G.b.	——— t. 2639.
3 versícolor. B.M.	(*bh.*)	changeable.	S. America.	1823.	1. 4.	G.b.	——— t. 2485.
4 lorifòlius. B.M.	(*ro.*)	lorate-leaved.	———	1824.	....	G.b.	
5 bífidus. B.M.	(*ro.*)	two-cleft.	BuenosAyres.	1825.	6. 9.	G.b.	Bot. mag. t. 2597.
6 spathàceus. B.M.	(*pu.*)	broad-spathed.	———	——	....	G.b.	
7 robústa. W.H.	(*li.*)	robust.	———	1827.	7. 9.	F.b.	Swt. br. fl. gar. s. 2. t. 14.
8 Andersòni. W.H.	(*ye.*)	Anderson's.	S. America.	1829.	5. 6.	F.b.	——— s. 2.
α aùrea.	(*ye.*)	*golden.*	———	——	——	F.b.	
β cùprea.	(*co.*)	*copper-coloured.*	———	——	——	F.b.	
9 advèna. W.H.	(*std.*)	streaked.	———	1807.	——	F.b.	Bot. mag. t. 1125.
Amary'llis advèna. Bot. reg. t. 849.							
10 intermèdia. W.H.	(*re.*)	intermediate.	Brazil.	1827.	1. 3.	S.b.	Bot. reg. t. 1148.
Amary'llis intermèdia. B.R.							
BELLADO'NNA.		BELLADO'NNA LILY.	Hexandria Monogynia.				
1 purpuráscens.	(*pu.*)	light purple.	C. B. S.	1712.	7. 9.	H.b.	Bot. mag. t. 783.
Amary'llis Belladónna. L.							
2 pállida.	(*pa.li.*)	pale-flowered.	———	....	——	F.b.	Bot. reg. t. 714.
3 pùdica.	(*ro.*)	modest.	———	1795.	5. 7.	F.b.	Ker's rev. t. 8. f. 2.
Amary'llis pùdica. K.R.							
4 blánda.	(*bh.*)	charming.	———	1754.	——	F.b.	Bot. mag. t. 1450.
AMARY'LLIS. L.		AMARY'LLIS.	Hexandria Monogynia. L.				
1 reticulàta. W.	(*pu.wh.*)	netted-veined.	Brazil.	1777.	4. 5.	S.b.	Bot. mag. t. 657.
2 striatifòlia.	(*pu.wh.*)	striped-leaved.	———	1815.	——	S.b.	Bot. reg. t. 352.
3 Sweètii. W.H.	(*va.*)	Striatifòlio-Johnsoni.	Colvill's.	1821.	1. 12.	S.b.	
α brevifòlia. C.C.	(*pu.w.*)	*short-leaved.*	Colvill's.	1820.	——	S.b.	
β supérba. C.C.	(*cr.wh.*)	*superb-flowered.*	———	——	——	S.b.	
γ obscùra. C.C.	(*re.wh.*)	*clouded.*	———	——	——	S.b.	
δ recurvàta. C.C.	(*re.bh.*)	*recurved-flowered.*	———	——	——	S.b.	
ε Wellsiàna. C.C.	(*re.w.*)	*Mr. Wells's.*	———	——	——	S.b.	
ζ Vallèti.	(*pu.wh.*)	*Vallets'.*	———	——	——	S.b.	
η rugòsa. C.C.	(*pu.wh.*)	*rugged-flowered.*	———	——	——	S.b.	
θ versícolor.	(*re.wh.*)	*various-coloured.*	———	——	——	S.b.	
ι am`œna.	(*re.wh.*)	*delightful.*	———	——	——	S.b.	
κ affínis.	(*re.wh.*)	*likened.*	———	——	——	S.b.	
λ undulæflòra.	(*pu.wh.*)	*wave-flowered.*	———	——	——	S.b.	
μ palliiflòra.	(*li.wh.*)	*pale-flowered.*	———	——	——	S.b.	
ν díscolor.	(*re.wh.*)	*two-coloured.*	———	——	——	S.b.	
ξ expánsa.	(*sc.wh.*)	*expanded.*	———	——	——	S.b.	
ο variegàta. C.C.	(*re.wh.*)	*variegated-flowered.*	———	——	——	S.b.	
π nótha.	(*pu.wh.*)	*bastard.*	———	——	——	S.b.	
ρ lineàta.	(*re.wh.*)	*line-petaled.*	———	——	——	S.b.	
σ modésta.	(*wh.re.*)	*modest.*	———	——	——	S.b.	
τ patentíssima.	(*sc.*)	*spreading-flower'd.*	———	——	——	S.b.	
υ salmònea.	(*re.*)	*salmon-coloured.*	———	——	——	S.b.	
φ bistriàta.	(*cr.da.*)	*two-lined.*	———	——	——	S.b.	
ψ ligulæflòra.	(*re.wh.*)	*strap-flowered.*	———	——	——	S.b.	
χ Annesleyàna.	(*ve.wh.*)	*Lord Mountnorris's.*	———	——	——	S.b.	Miss Chapman's ic. ined.
4 Colvíllii.	(*va.*)	Reticulato-Johnsoni.	———	——	——	S.b.	
α atrorùbens.	(*da.pu.w.*)	*dark-red.*	———	——	——	S.b.	
β purpuráscens.	(*pu.re.*)	*purplish.*	———	——	——	S.b.	
γ costàta.	(*re.wh.*)	*ribbed-leaved.*	———	——	——	S.b.	
δ dioíca.	(*re.pu.*)	*barren-anthered.*	———	——	——	S.b.	
ε `æmula.	(*re.wh.*)	*red and white.*	———	——	——	S.b.	
ζ nervifòlia.	(*pu.re.*)	*strong-nerved.*	———	——	——	S.b.	
η phœnícea.	(*da.pu.*)	*dark purple red.*	———	——	——	S.b.	
θ stenántha.	(*pa.pu.w.*)	*narrow-petaled.*	———	——	——	S.b.	
ι fláccida.	(*re.pu.w.*)	*flaccid-leaved.*	———	——	——	S.b.	
κ bracteàta.	(*re.pu.*)	*large-bracted.*	———	——	——	S.b.	
λ párvula.	(*pu.wh.*)	*small-flowered.*	———	——	——	S.b.	
5 præclàra. W.H.	(*va.*)	Johnsoni-striatifòlia.	Colvill's.	——	——	S.b.	
α pátens.	(*pu.wh.*)	*spreading-flowered.*	———	——	——	S.b.	
β Colvilliàna.	(*re.p.w.*)	*Colvill's.*	———	——	——	S.b.	
γ decòra.	(*re.pu.*)	*neat.*	———	——	——	S.b.	

6	Johnsòni.	(*sc. wh.*)	Regìnæ-vittàta.	Johnson.	1810.	1. 12.	G.b.	Lodd. bot. cab. t. 159.
	spectábilis. B.C. *brasiliénsis.* Redout. liliac. c. ic.							
	α *Scaramòuch.* W.H.	(*sc.wh.*)	*Mr. Herbert's.*	Mitcham.	1811.	——	G.b.	
	β *Màter.* W.H.	(*sc.wh.*)	*dame.*	———	——	——	G.b.	
	γ *Sorélla.* W.H.	(*sc.wh.*)	*sister.*	———	——	——	G.b.	
	δ *quárta.* W.H.	(*sc.wh.*)	*fourth.*	———	——	——	G.b.	
	ε *péndula.* W.H.	(*sc.wh.*)	*pendulous.*	———	——	——	G.b.	
	ζ *Ponceaunélta.* W.H.	(*sc.wh.*)	*related.*	———	——	——	G.b.	
	η *bárbe blánche.* W.H.	(*sc.wh.*)	*white-bearded.*	Highclere.	—	——	G.b.	
7	regìnæ. W.	(*sc.*)	Mexican Lily.	S. America.	1725.	5. 7.	S.b.	Bot. mag. t. 453.
8	formòsa. W.H.	(*va.*)	Striatifolio-reginæ.	Colvill's.	1820.	——	S.b.	
	α *grandiflòra.*	(*pu.wh.*)	*large-flowered.*	———	——	——	S.b.	
	β *palléscens.*	(*pa.re.w.*)	*pale red and white.*	———	——	——	S.b.	
	γ *rubicúnda.*	(*ve.re.*)	*velvetty red.*	———	——	——	S.b.	
	δ *coccínea.*	(*sc.*)	*dark-veined scarlet.*	———	——	——	S.b.	
	ε *ríngens.*	(*re.wh.*)	*gaping-flowered.*	———	——	——	S.b.	
	ζ *carnéscens.*	(*fl.wh.*)	*flesh-coloured.*	———	——	——	S.b.	
	η *dilùta.*	(*pa.re.wh.*)	*stained-flowered.*	———	——	——	S.b.	
	θ *imbùta.*	(*re.*)	*stained red.*	———	——	——	S.b.	
	ι *compácta.*	(*pa.re.*)	*compact.*	———	——	——	S.b.	
9	gloriòsa.	(*va.*)	Reticulato-reginæ.	———	——	——	S.b.	
	α *sulcàta.*	(*re.wh.*)	*channelled-leaved.*	———	——	——	S.b.	
	β *rubéscens.*	(*re.wh.*)	*pale red.*	———	——	——	S.b.	
	γ *ròseo-álba.*	(*ro.wh.*)	*rosy and white.*	———	——	——	S.b.	
	δ *accèdens.*	(*ve.re.*)	*velvetty.*	———	——	——	S.b.	
10	consanguínea.	(*sc.wh.*)	Valleti-reginæ.	———	1824.	——	S.b.	
	β *serícea.*	(*pu.*)	*silky-flowered.*	———	——	——	S.b.	
11	intermèdia.	(*re.wh.*)	Reginæ-equestris.	———	1820.	——	S.b.	
12	spathàcea. B.M.	(*sc.*)	large-spathed.	Hybrid.	——	——	S.b.	Bot. mag. t. 2315.
13	vittàta. W.	(*sc.wh.*)	striped.	S. America.	1769.	4. 5.	S.b.	——— t. 129.
	β *latifòlia.* L.col.	(*sc.wh.*)	*broad-leaved.*	———	1816.	——	G.b.	Lindl. coll. bot. ic.
14	lineàta. Colla.	(*sc.wh.*)	lined-flowered.	———	1820.	——	S.b.	Coll. hort. rip. ic.
	vittàta γ. *Harrisòniæ.* Bot. reg. t. 988.							
15	magnífica.	(*sc.wh.*)	Striatifolio-vittata.	Colvill's.	1823.	——	S.b.	
16	púlchra. C.C.	(*sc.wh.*)	Reticulato-vittata.	———	——	——	S.b.	
17	rùtila. K.R.	(*sc.*)	fiery.	Brazil.	1815.	1. 12.	S.b.	Bot. reg. t. 23.
18	spléndens. H.A.	(*sc.wh.*)	Rutilo-equestri-vittata.	Spofforth.	1819.	——	F.b.	Herb. append. p. 52. ic.
	α *Mars.* W.H.	(*sc.wh.*)	*Mars.*	———	1822.	——	F.b.	
	β *Vésta.* W.H.	(*sc.wh.*)	*Vesta.*	———	——	——	F.b.	
	γ *Bellòna.* W.H.	(*sc.wh.*)	*Bellona.*	———	——	——	F.b.	
	δ *álta.* W.H.	(*sc.wh.*)	*tallest.*	———	——	——	F.b.	
	ε *ignéscens.*	(*sc.*)	*flame-coloured.*	Colvill's.	——	——	F.b.	
19	sanguínea. W.H.	(*cr.wh.*)	Rutilo-vittata.	Highclere.	1819.	——	G.b.	
20	mìcans. W.H.	(*re.co.*)	Rutilo-crocata.	———	1820.	——	S.b.	
21	angústa.	(*sc.pu.*)	Johnsoni-rutila.	Colvill's.	1822.	——	S.b.	
22	rígida.	(*re.co.g.*)	Crocato-rutila.	———	——	——	S.b.	
23	incl'yta. W.H.	(*sc.wh.*)	Fulgido-vittata.	Highclere.	1819.	——	S.b.	
24	Carnarvòni. W.H.	(*sc.w.*)	Vittato-reginæ.	———	——	——	G.b.	
25	árdens. W.H.	(*sc.co.*)	Crocato-reginæ.	Spofforth.	1820.	——	S.b.	
	α *salmònea.* W.H.	(*pa.re.*)	*salmon-coloured.*	———	——	——	S.b.	
	β *charbon brulant.*	(*pa.r.*)	*shining-flowered.*	———	——	——	S.b.	
26	fúlgida. K.R.	(*sc.*)	fulgid.	Brazil.	1810.	——	S.b.	Bot. reg. t. 226.
27	subbarbàta.	(*or.sc.*)	slightly bearded.	———	1823.	——	S.b.	Bot. mag. t. 2475.
28	flòrida. W.H.	(*sc.wh.*)	Fulgido-rutila.	Spofforth.	1820.	——	S.b.	
29	màjus. W.H.	(*sc.co.*)	Acuminato-rutila.	———	——	——	S.b.	
	β *fratérna.*	(*sc.*)	*brother.*	Colvill's.	1821.	——	S.b.	
30	venòsa. W.H.	(*sc.wh.*)	Vittato-Johnsoni.	Highclere.	——	——	G.b.	
31	invérsa. W.H.	(*sc.wh.*)	Johnsoni-vittata.	Liverpool.	1812.	——	G.b.	
	β *Sáturn.* W.H.	(*sc.wh.*)	*Saturn.*	———	——	——	G.b.	
	γ *Plùto.* W.H.	(*sc.wh.*)	*Pluto.*	Spofforth.	1814.	——	G.b.	
	δ *Andromache.* W.H.	(*sc.*)	*Andromache.*	———	——	——	G.b.	
32	sidérea. W.H.	(*sc.*)	Sanguineo-subbarbata.	Spofforth.	1820.	——	S.b.	
33	stellàta. W.H.	(*sc.*)	Subbarbato-sanguinea.	———	——	——	S.b.	
34	renovàta. W.H.	(*sc.wh.*)	Splendente-fulgida.	———	——	——	S.b.	
35	expánsa. W.H.	(*sc.wh.*)	Equestri-florida.	———	——	——	S.b.	
36	Bèatum. W.H.	(*sc.co.*)	Johnsoni-acuminata.	Wentworth.	—	——	S.b.	
	β *Beatrìce.* W.H.	(*sc.co.*)	*Beatrice.*	———	——	——	S.b.	
	γ *trícolor.*	(*re.pu.gr.*)	three-coloured.	Colvill's.	1821.	——	S.b.	
	δ *macrántha.*	(*re.co.*)	*large-flowered.*	———	——	——	S.b.	

	ε crispàta.	(*re.co.*)	*curled-edged.*	Colvill's.	1821.	1. 12.	S.b.	
	ζ tricúspis.	(*sc.or.*)	*three-pointed.*	——	——	——	S.b.	
36	lugùbris. W.H.	(*sc.or.*)	Styloso-Johnsoni.	Spofforth.	1814.	——	S.b.	
	α sombrélla. W.H.	(*co.*)	*sombre.*	——	——	——	S.b.	
	β griselìna. W.H.	(*pa.*)	*greyish.*	——	——	——	S.b.	
	γ porporìna. W.H.	(*br.*)	*brownish.*	——	——	——	S.b.	
37	trístis. W.H.	(*sc.co.*)	Johnsoni-stylosa.	——	——	——	S.b.	
38	Broòkesii. W.H.	(*sc.co.*)	Crocato-Johnsoni.	Brookes's.	1820.	——	S.b.	
	β Creusa. W.H.	(*sc.co.*)	*Creusa.*	——	——	——	S.b.	
39	aurantìaca. W.H.	(*or.*)	Reginæ-acuminata.	Spofforth.	——	——	S.b.	
	α Camillus. W.H.	(*or.sc.*)	*Camillus.*	——	——	——	S.b.	
	β Gustavus. W.H.	(*or.sc.*)	*Gustavus.*	——	——	——	S.b.	
	γ amábilis.	(*sc.or.*)	*brilliant.*	Colvill's.	1821.	——	S.b.	
	δ fúlva.	(*co.*)	*coppery.*	——	——	——	S.b.	
40	platypétala.	(*sc.or.*)	Acuminato-reginæ.	——	——	——	S.b.	
41	Hòodii.	(*sc.wh.*)	Reginæ-fulgida.	——	1820.	——	S.b.	
42	intermíxta. C.C.	(*re.wh.*)	Equestri-reginæ.	——	——	——	S.b.	
43	Coopèri. W.H.	(*co.*)	Crocato-acuminata.	Wentworth.	——	——	S.b.	
	β transpàrens.	(*co.gr.*)	*transparent.*	Colvill's.	1821.	——	S.b.	
44	élegans. W.H.	(*sc.wh.*)	Formoso-vittata.	Highclere.	——	——	G.b.	
45	Burghcleriàna. W.H.	(*sc.wh.*)	Splendente-vittata.	Burghclere.		——	G.b.	
46	ancéps. W.H.	(*sc.wh.*)	Florido-Johnsoni.	Spofforth.	——	——	G.b.	
47	concínna. W.H.	(*sc.wh.*)	Formoso-Johnsoni.	——	——	——	G.b.	
	α Roundhead. W.H.	(*sc.wh.*)	*Roundhead.*	——	——	——	G.b.	
	β Peter. W.H.	(*sc.wh.*)	*Peter.*	——	——	——	G.b.	
	γ John. W.H.	(*sc.wh.*)	*John.*	——	——	——	G.b.	
	δ Jeffrey. W.H.	(*sc.wh.*)	*Jeffrey.*	——	——	——	G.b.	
48	incónstans.	(*sc.or.*)	Acuminato-Johnsoni.	Colvill's.	——	——	G.b.	
	α venòsa.	(*re.co.*)	*purple-veined.*	——	——	——	G.b.	
	β laterítia.	(*re.co.*)	*brick-coloured.*	——	——	——	G.b.	
	γ glaucéscens.	(*re.co.*)	*glaucescent.*	——	——	——	G.b.	
	δ cóncolor.	(*co.*)	*self-coloured.*	——	——	——	G.b.	
	ε vària.	(*re.co.*)	*variable-coloured.*	——	——	——	G.b.	
	ζ crispiflòra.	(*sc.*)	*scarlet curl-flowered.*	——	——	——	G.b.	
	η pustulòsa.	(*sc.co.*)	*pustulose-flowered.*	——	——	——	G.b.	
	θ breviflòra.	(*re.co.*)	*short-tubed.*	——	——	——	G.b.	
	ι fúlvo-cíncta.	(*re.co.*)	*brown-edged.*	——	——	——	G.b.	
	κ dilécta.	(*sc.*)	*bright-scarlet.*	——	——	——	G.b.	
	λ tortuliflòra.	(*sc.or.*)	*twisted-flowered.*	——	——	——	G.b.	
	μ veniflòra.	(*ye.re.*)	*veined-flowered.*	——	——	——	G.b.	
	ν lineolàta.	(*re.*)	*dark-lined.*	——	——	——	G.b.	
	ξ basiflòra.	(*re.pu.*)	*purple-bottomed.*	——	——	——	G.b.	
	ο imperiàlis.	(*sc.*)	*imperial.*	——	——	——	G.b.	
	π punctulòsa.	(*pa.sc.*)	*purple spotted.*	——	——	——	G.b.	
	ρ unipárvula.	(*re.pu.*)	*lower small petaled.*	——	——	——	G.b.	
	σ purpùreo-rùbens.	(*pu.re.*)	*purplish red.*	——	——	——	G.b.	
	τ flexuòsa.	(*pu.re.*)	*flexuose-petaled.*	——	——	——	G.b.	
	υ delicàta.	(*sc.pu.*)	*purple-striped scarlet.*	——	——	——	G.b.	
49	crocàta. K.R.	(*sn.*)	saffron-coloured.	Brazil.	1810.	4. 6.	S.b.	Bot. reg. t. 38.
50	stylòsa.	(*co.re.*)	long-styled.	Maranham.	1821.	2. 6.	S.b.	Bot. mag. t. 2278.
	maranénsis. B.R. 719. *Hippeástrum stylòsum.* B.M.							
51	acuminàta. B.R.	(*co.*)	acute-flowered.	Maranham.	——	4. 8.	S.b.	Bot. reg. t. 534.
	Hippeástrum pulveruléntum. Bot. mag. t. 227.							
52	spùria.	(*sc.co.*)	Acuminato-reginæ.	Colvill's.	1822.	——	S.b.	
	α sórdida.	(*co.*)	*dull-copper.*	——	——	——	S.b.	
	β carinàta.	(*re.co.*)	*keeled-flowered.*	——	——	——	S.b.	
53	striatiflòra.	(*sc.wh.*)	Acuminato-Sweetii. β.	——	1823.	——	S.b.	
	β consobrìna.	(*pi.pu.*)	*cousin.*	——	——	——	S.b.	
	γ diversícolor.	(*re.w.pu.*)	*different-coloured.*	——	——	——	S.b.	
54	nòbile. W.H.	(*st.pu.*)	Acuminato-striatifolia.	Spofforth.	1820.	——	S.b.	
55	venústa. W.H.	(*sc.pu.w.*)	Fulgido-striatifolia.	——	——	——	S.b.	
56	dilécta. W.H.	(*sc.pu.*)	Equestri-striatifolia.	Colvill's.	——	——	S.b.	
57	Go'weni. W.H.	(*pu.co.*)	Striatifolio-acuminata.	Highclere.	——	——	S.b.	
	α Cavalier. W.H.	(*pu.co.*)	*Cavalier.*	——	——	——	S.b.	
	β Biondelta. W.H.	(*pu.c.*)	*Biondelta.*	——	——	——	S.b.	
	γ Agatha. W.H.	(*pu.co.*)	*Agatha.*	——	——	——	S.b.	
58	Harrisòni. W.H.	(*sc.pu.*)	Striatifolio-stylosa.	Harrison's.	1824.	——	S.b.	
59	trìplex. W.H.	(*sc.pu.w.*)	Goweni-vittatum.	Highclere.	1820.	——	S.b.	

60 pulchérrima.W.H.(*p.co.*)Sweetii-pulverulenta.Spofforth. 1824. 4. 8. S.b.
α Vezzosa. (*pu.re.co.*) *Vezzosa.* —— —— —— S.b.
β Gratian. (*pu.re.co.*) *Gratian.* —— —— —— S.b.
γ Amorosetta. (*pu.re.co.*)*Amorosetta.* —— —— —— S.b.
δ Salamander.(*pu.re.co.*)*changing-coloured.* —— —— —— S.b.
ε Calisto. (*pu.re.co.*) *Calisto.* —— —— —— S.b.
61 bélla. W.H. (*pu.re.w.*) Sweetii-Carnarvoni. —— —— —— S.b.
62 díscolor. W.H. (*sc.br.*) Johnsoni-tristis. —— —— —— S.b.
63 varìata. W.H. (*sc.pu.*) Formoso-inversum. —— —— —— S.b.
64 quíntuplex.W.H.(*pu.re.*)Harrisoni-tristis. —— 1826. —— S.b.
65 exímia. W.H. (*va.*) Goweni-Highcleriàna. Highclere. .. —— S.b.
66 equéstris. K.R. (*sc.*) Barbadoes Lily. W. Indies. 1710. 7. 10. S.b. Bot. mag. t. 305.
β màjor. B.R. (*sc.*) *larger.* —— .... —— S.b. Bot. reg. t. 234.
γ plèna. (*sc.*) *double-flowered.* —— 1809. —— S.b.
67 miniàta. R.P. (*se.*) vermilion. S. America. .... S.b.
68 solandriflòra.L.col.(*st.*)Solandra-flowered. —— 1820. 4. 6. S.b. Lindl. coll. bot. t. 11.
α chloroleùca. (*st.*) *dull-coloured.* —— —— —— S.b.
β rùbro-striàta. (*st.*) *red-striped.* —— —— —— S.b. Bot. mag. t. 2573.
γ purpuráscens.(*st.pu.*)*purple-tubed.* —— —— —— S.b.
δ striàta. (*pu.st.*) *purple-striped.* —— —— —— S.b. Bot. reg. t. 876.
69 Haylócki.W.H.(*st.pu.*) Solandrifloro-acuminata. —— —— S.b.
70 Herbérti. (*va.*) Solandrifloro-stylosa.Spofforth..... —— S.b.
α lónga. (*va.*) *long-flowered.* —— .... —— S.b.
β carnélla. (*fl.*) *flesh-coloured.* —— .... —— S.b.
γ magnífica. (*va.*) *magnificent.* —— .... —— S.b.
71 Highcleriàna.W.H.(*va.*)Styloso-Solandriflora.Highclere... —— S.b.
β pícta. (*va.*) *painted lady.* Spofforth. .... —— S.b.
γ Dóminie. (*va.*) *Dominie.* —— .... —— S.b.
72 calyptràta. K.R. (*gr.*) green-flowered. Brazil. 1816. 5. 8. S.b. Bot. reg. t. 164.
73 aùlica. K.R. (*sc.gr.*) crowned. —— —— 1. 6. S.b. —— t. 444.
74 psittácina.K.R.(*sc.gr.*) parrot-like. —— —— 1. 12. S.b. —— t. 199.
75 Griffìni. W.H. (*sc.gr.*) Psittacino-Johnsoni.Griffin's. 1820. —— S.b.
*α Napoleòna.*W.H.(*sc.gr.*)*Napoleon's.* —— —— —— S.b.
β Victor. W.H.(*sc.gr.*) *Victor.* —— —— —— S.b.
*γ Alompra.*W.H.(*sc.gr.*)*Alompra.* —— —— —— S.b.
δ retinérvia. (*sc.gr.*) *netted-nerved.* Colvill's. 1822. —— S.b.
ε recurviflòra. (*sc.gr.*) *recurve-pointed.* —— —— —— S.b.
ζ campanulàta.(*re.gr.*) *bell-flowered.* —— —— —— S.b.
η quadrilineàta. (*sc.pu.g.*)*four-lined.* —— —— —— S.b.
θ albocíncta. (*wh.sc.*) *white-edged.* —— —— —— S.b.
ι quadrícolor. (*sc.pu.g.w.*)*four-coloured.* —— —— —— S.b.
76 affíne. (*sc.gr.*) Psittacino-reginæ. —— —— —— S.b.
α canaliculàta. (*sc.gr.*) *channel-leaved.* —— —— —— S.b.
β membranàcea. (*re.pu.g.*)*membranous-edged.* —— —— —— S.b.
γ purpurostriàta.(*re.pu.*)*purple-striped.* —— —— —— S.b.
δ latífida. (*re.pu.g.*) *broad-segmented.* —— —— —— S.b.
77 tortuòsa. (*sc.gr.*) Psittacino-equestris. —— —— —— S.b.
78 multistriàta. (*sc.wh.*) Psittacino-Valleti. —— —— —— S.b. Miss Chapman ic.
β retiflòra. (*sc.w.g.*) *netted-flowered.* —— —— —— S.b.
79 pretiòsum.W.H.(*sc.gr.*) Psittacino-Griffini.Highclere. 1824. —— S.b.
80 sínistrum.W.H. (*sc.w.g.*) Griffini-vittatum. —— —— —— S.b.

SPREIK'ELIA. H.A. JACOBEA LILY. Hexandria Monogynia.
formosíssima.H.A.(*cr.*) crimson. S. America. 1658. 5. 6. F.b. Bot. mag. t. 47.
Amary'llis formosíssima. B.M.

LYC'ORIS. H.A. LYC'ORIS. Hexandria Monogynia.
1 aùrea. H.A. (*ye.*) golden. China. 1777. 8. 12. G.b. Bot. reg. t. 611.
Amary'llis aùrea. B.R.
2 radiàta. H.A. (*pi.*) Snowdrop-leaved. —— 1758. 9. 10. G.b. Bot. reg. t. 596.
Amary'llis radiàta. B.R.

NER'INE. H.A. NER'INE. Hexandria Monogynia.
1 curvifòlia. H.A. (*sc.*) curve-leaved. C. B. S. 1794. 7. 9. F.b. Bot. mag. t. 725.
Amary'llis curvifòlia. B.M. *Fothergíllia.* Andr. reposit. t. 163.
2 corúsca. H.A. (*sc.*) brilliant. C. B. S. 1809. —— F.b. Bot. mag. t. 1089.
3 ròsea. B.M. (*ro.*) rose-coloured. —— .... —— F.b. —— t. 2124.
4 venústa. H.A. (*ro.*) poppy-coloured. —— 1806. 6. 8. F.b. —— t. 1090.
5 sarniénsis. H.A. (*re.*) Guernsey Lily. —— 1659. 9. 10. F.b. —— t. 294.
6 pulchélla. H.A. (*re.*) pretty. —— 1820. 7. 9. F.b. —— t. 2407.
7 flexuòsa. H.A. (*pa.re.*) zigzag. —— 1795. 9. 10. F.b. Bot. reg. t. 172.
8 hùmilis. H.A. (*re.*) small. —— —— 7. 10. F.b. Bot. mag. t. 726.

9 undulàta. H.A. (*pa.re.*) waved-flower'd. C. B. S. 1767. 7. 10. F.b. Bot. mag. t. 369.
10 versícolor. H.A. (*li.*) changeable-col'd. Hybrid. 1815. —— F.b. Herb. append. ic.
IXIOLI'RION. H.A. Ixioli'rion. Hexandria Monogynia.
1 montànum. R.S. (*bl.*) mountain. Persia. 1829. .... H.b. Redout. liliac. t. 241.
Amary'llis montàna. Labill. syr. dec. 2. p. 5. t. 1. *Alstrœmèria montàna.* Ker.
2 tatáricum. R.S. (*bl.*) Tartarian. Tartary. 1822. 4. 6. H.b. Pall.it.3.ap.85. t. D.f.1.
Amary'llis tatárica. M.B.
STRUM'ARIA. Ker. Strum'aria. Hexandria Monogynia. L.
1 truncàta. W. (*wh.bh.*) truncated. C. B. S. 1795. 4. 5. F.b. Jacq. ic. 2. t. 357.
2 rubélla. W. (*li.*) pale red. ——— —— 5. 6. F.b. ——— 2. t. 358.
3 angustifòlia. W. (*wh.li.*) narrow-leaved. ——— —— 4. 5. F.b. ——— 2. t. 359.
4 linguæfòlia. W. (*wh.*) tongue-leaved. ——— 1812. —— F.b. ——— 2. t. 356.
5 filifòlia. B.R. (*wh.*) fine-leaved. ——— 1774. 10. 12. F.b. Bot. reg. t. 440.
6 spiràlis. B.M. (*ro.wh.*) spiral-stalked. ——— —— 4. 8. F.b. Bot. mag. t. 1383.
7 críspa. B.M. (*ro.li.*) curled-flowered. ——— 1790. —— F.b. ——— t. 1363.
Amary'llis críspa. Jacq. schœn. t. 72.
8 undulàta. W. (*wh.ro.*) waved. ——— 1825. —— F.b. Jacq. ic. t. 360.
9 stellàris. B.M. (*pi.*) starry. ——— 1794. 10. 11. F.b. Jacq. schœn. 1. t. 71.
10 gemmàta. B.M. (*wh.*) gynandrous. ——— 1810. 8. 10. F.b. Bot. mag. t. 1620.
IMHO'FIA. H.A. Imho'fia. Hexandria Monogynia.
marginàta. H.A. (*pu.*) red margined. C. B. S. 1795. 9. 10. F.b. Jacq. schœn. 1. t. 65.
Brunsvígia marginàta. H.K. *Amary'llis marginàta.* Jacq.
BUPH'ONE. H.A. Buph'one. Hexandria Monogynia.
1 dísticha. H.A. (*pu.*) distichous. C. B. S. 1823. 9. 10. F.b. Patters. iter. t. 1.
2 toxicària. H.A. (*li.*) Poison-bulb. ——— 1774. —— F.b. Bot. reg. t. 567.
β obtusifòlia. H.A. (*li.*) *wave-leaved.* ——— 1816. —— F.b.
3 ciliàris. H.A. (*pa.pu.*) fringe-leaved. ——— 1752. 6. 8. F.b. Bot. reg. t. 1153.
Brunsvígia ciliàris. B.R. *Amary'llis ciliàris.* L. *Hæmánthus ciliàris.* W.
AMMO'CHARIS. H.A. Ammo'charis. Hexandria Monogynia.
1 coránica. H.A. (*pu.*) Mr. Burchell's. C. B. S. 1816. 7. 8. F.b. Bot. reg. t. 139.
Amary'llis coránica. B.R.
β pállida. B.R. (*li.*) *pale-flowered.* ——— —— —— F.b. Bot. reg. t. 1219.
2 falcàta. H.A. (*pu.*) sickle-leaved. ——— 1774. 5. 6. F.b. Bot. mag. t. 1443.
Brunsvígia falcàta. B.M. *Amary'llis falcàta.* H.K. *Crìnum falcàtum.* Jacq. Jacq. vind. 3. t. 60.
BRUNSVI'GIA. Ker. Brunsvi'gia. Hexandria Monogynia. H.K.
1 laticòma. (*pa.pu.*) broad-headed. C. B. S. 1818. 3. 6. F.b. Bot. reg. t. 497.
lùcida. H.A. *Amary'llis laticòma.* B.R.
2 mìnor. B.R. (*ro.*) lesser. ——— 1820. —— F.b. Bot. reg. t. 954.
3 multiflòra. B.M. (*sc.gr.*) many-flowered. ——— 1752. 6. 8. F.b. Bot. mag. t. 1619.
Amary'llis orientàlis. W.
β rubricaùlis. H.A. (*cr.*) *red-stemmed.* ——— 1812. —— F.b. Jacq. schœn. 1. t. 39.
4 Josephínæ. B.R. (*sc.*) Josephine's. ——— 1814. —— F.b. Redout. lil. t. 370-372.
β mìnor. B.R. (*ro.*) *smaller.* ——— —— —— F.b. Bot. reg. t. 192, 193.
γ striàta. B.M. (*pu.std.*) *striped-flowered.* ——— 1823. – F.b. Bot. mag. t. 2578.
5 striàta. H.K. (*pu.*) striated. ——— 1795. 9. 10. F.b. Jacq. schœn. 1. t. 70.
6 rádula. H.K. (*pu.*) rasp-leaved. ——— 1790. 4. 8. F.b. ——— 1. t. 68.
PHYCE'LLA. B.R. Phyce'lla. Hexandria Monogynia.
1 ígnea. B.R. (*sc.ye.*) fiery. Chili. 1824. 8. 10. G.b. Bot. reg. t. 809.
β glauca. (*sc.ye.*) *glaucous-leaved.* ——— — —— G.b. Bot. mag. t. 2687.
2 cyrtanthoídes. B.R. (*sc.ye.*) Cyrtanthus-like. —— 1821. 6. 8. G.b. ——— t. 2399.
Amary'llis cyrtanthoídes. B.M.
GRIFFI'NIA. B.R. Griffi'nia. Hexandria Monogynia. B.R.
1 hyacínthina. B.R. (*bl.*) largest. Brazil. 1815. 6. 9. S.b. Bot. reg. t. 163.
Amary'llis hyacínthina. B.R.
2 intermèdia. B.R. (*bl.*) intermediate. ——— 1825. 5. 7. S.b. Bot. reg. t. 990.
3 parviflòra. B.R. (*bl.*) small flowered. ——— 1820. —— S.b. ——— t. 511.
HÆMA'NTHUS. W. Blood-flower. Hexandria Monogynia. L.
1 coccíneus. S.S. (*sc.*) salmon-colour'd. C. B. S. 1629. 8. 10. F.b. Bot. mag. t. 1075.
2 coarctàtus. S.S. (*sc.*) compressed. ——— 1795. 2. 3. F.b. Bot. reg. t. 181.
3 tigrìnus. S.S. (*sc.*) Tiger-spotted. ——— 1790. 12. 2. F.b. Bot. mag. t. 1705.
4 rotundifòlius. S.S. (*sc.*) round-leaved. ——— —— 6. 10. F.b. ——— t. 1618.
5 crassìpes. S.S. (*sc.*) thick-leaved. ——— 1816. —— F.b.
6 hyalocárpus. S.S. (*re.*) clear-berried. ——— —— —— F.b. Jacq. schœn. 4. t. 409.
7 quadriválvis. S.S. (*re.*) four-valved. ——— 1790. 9. 10. G.b. Bot. mag. t. 1523.
8 húmilis. S.S. (*re.*) small. ——— 1825. —— F.b.
9 Pumílio. S.S. (*re.*) dwarf. ——— 1789. 8. 9. F.b. Jacq. schœn. 1. t. 61.
10 pubéscens. S.S. (*wh.*) villous-leaved. ——— 1774. —— F.b. Bot. reg. t. 382.
11 albiflós. W. (*wh.*) smooth-broad-ld. ——— 1791. 4. 8. F.b. ——— t. 984.
12 cárneus. S.S. (*ro.*) flesh-coloured. ——— 1819. 6. 8. F.b. ——— t. 509.

13	lanceæfòlius. s.s. (*re.*) spear-leaved.	C. B. S.	1794.	9. 10.	F.b.	Jacq. schœn. 1. t. 60.
14	carinàtus. p.s. (*re.*) keel-leaved.	———	1759.	8. 9.	F.b.	
15	amarylloídes. s.s. (*re.*) Amaryllis-like.	———	1825.	——	F.b.	
16	moschàtus. s.s. (*re.*) musk-scented.	———	1816.	——	F.b.	
17	sanguíneus. s.s. (*sc.*) crimson.	———	1820.	6. 8.	F.b.	
18	puníceus. s.s. (*sc.*) waved-leaved.	———	1722.	5. 9.	G.b.	Bot. mag. t. 1315.
19	multiflòrus. s.s. (*sc.*) many-flowered.	Sierra Leon.	1783.	——	S.b.	——— t. 1995 & 961.
	CR'INUM. W.H. CR'INUM. Hexandria Monogynia.					
1	pedunculàtum. b.p. (*wh.*) New Holland.	N. S. W.	1790.	4. 8.	G.b.	Bot. reg. t. 52.
2	exaltàtum. h.a. (*wh.*) high.	E. Indies.	1820.	——	S.b.	
3	canaliculàtum. k.r. (*wh.*) channelled-leaved.	———	1810.	——	S.b.	
4	brachyándrum. k.r. (*wh.*) short-anthered.	N.Holland.	1819.	——	G.b.	
5	brevilímbum. h.a. (*wh.*) short-limbed.	Pacific Isl.	1820.	6. 8.	S.b.	
6	pròcerum. h.a. (*bh.*) tall.	Rangoon.	1822.	——	S.b.	Bot. mag. t. 2684.
7	sínicum. h.a. (*wh.*) Chinese.	China.	1819.	——	S.b.	
8	anómalum. h.a. (*wh.*) plaited-leaved.	———	1822.	——	S.b.	
9	asiáticum. k.r. (*wh.*) Poison bulb.	———	1732.	——	S.b.	Bot. mag. t. 1073.
10	plicàtum. b.m. (*wh.*) plicate-leaved.	———	1823.	4. 6.	S.b.	——— t. 2908.
11	declinàtum. h.a. (*wh.*) declined.	E. Indies.	1819.	4. 8.	S.b.	——— t. 2231.
12	bracteàtum. w. (*wh.*) short-leaved.	Mauritius.	1810.	6. 8.	S.b.	Bot. reg. t. 179.
	brevifòlium. h.b. published 1814. *bracteàtum.* w. 1799. Jacq. schœn. 4. t. 495 in 1804.					
	β *angustifòlium.* h.a. (*wh.*) *narrow-leaved.*	———	——	——	S.b.	
13	erubéscens. k.r. (*re.wh.*) blush-coloured.	S. America.	1789.	——	S.b.	Bot. mag. t. 1232.
	α *rubrilímbum.* (*re.*) *red-limbed.*	———	——	——	S.b.	
	β *major.* h.a. (*re.wh.*) *larger.*	———	——	——	S.b.	
	γ *minor.* h.a. (*re.wh.*) *lesser.*	———	——	——	S.b.	
	δ *corant'ynum.* w.h. (*bh.*) *Corantyne.*	Berbice.	1820.	——	S.b.	
	ε *aquáticum.* w.h. (*wh.*) *water.*	Brazil.	1823.	——	S.b.	
	ζ *glábrum álbum.* w.h. (*wh.*) *shining-leaved.*	Maranham.	1820.	——	S.b.	
	η *glábrum rùbrum.* w.h. (*re.wh.*) *smooth red.*	———	1823.	——	S.b.	
	θ *undulàtum.* w.h. (*wh.*) *wave-leaved.*	Brazil.	——	——	S.b.	Hook. ex. flor. t. 200.
	Crìnum undulàtum. *Hooker.*					
14	cruéntum. b.r. (*re.*) red-flowered.	S. America.	1810.	——	S.b.	Bot. reg. t. 171.
15	sumatrànum. k.r. (*wh.*) Sumatra.	Sumatra.	——	——	S.b.	——— t. 1049.
16	macrocárpon. h.a. (*wh.*) large-fruited.	Pegu.	1820.	——	S.b.	
17	élegans. h.a. (*wh.*) elegant.	———	——	9. 11.	S.b.	Bot. mag. t. 2592.
18	venústum. h.a. .... pretty.	E. Indies.	——	....	S.b.	
19	canalifòlium. h.a. (*wh.*) channelled.	———	——	6. 9.	S.b.	
	β *major.* w.h. (*wh.*) *larger.*	———	1825.	——	S.b.	
20	longifòlium. k.r. (*wh.*) long-leaved.	———	1810.	——	S.b.	
21	lorifòlium. k.r. (*wh.*) lorate-leaved.	———	1819.	——	S.b.	
22	americànum. w. (*wh.*) American.	S. America.	1752.	7. 9.	S.b.	Bot. mag. t. 1034.
23	Commelìni. k.r. (*wh.*) Commelin's.	———	1798.	6. 9.	S.b.	Jacq. schœn. 2. t. 202.
24	stríctum. b.m. (*wh.*) upright.	S. America?	1820.	8. 10.	S.b.	Bot. mag. t. 2635.
25	defíxum. k.r. (*wh.*) marsh.	E. Indies.	1810.	4. 8.	S.b.	——— t. 2208.
26	ensifòlium. k.r. (*bh.*) sword-leaved.	———	1820.	——	S.b.	——— t. 2301.
27	am'œnum. k.r. (*wh.*) delightful.	———	1810.	——	S.b.	
28	multiflòrum. Df. (*wh.*) many-flowered.	....	1822.	——	S.b.	
29	verecúndum. h.a. (*bh.*) blush.	E. Indies.	1820.	6. 9.	S.b.	
30	hùmile. b.m. (*wh.*) humble.	———	1816.	4. 8.	S.b.	Bot. mag. t. 2636.
31	confértum. b.m. (*wh.*) crowded-flowered.	N.Holland.	1822.	6. 7.	G.b.	——— t. 2522.
32	angustifòlium. b.p. (*wh.*) rough-edged.	———	1823.	——	G.b.	
33	arenàrium. b.m. (*wh.*) sand.	———	1818.	5. 8.	G.b.	Bot. mag. t. 2355.
	β *blándum.* b.m. (*bh.*) *blush-flowered.*	———	1821.	——	G.b.	——— t. 2531.
34	purpuráscens. w.h. (*pu.*) purplish.	Fernando Po.	1826.	——	S.b.	
35	erythrophy'llum. w.h. (*pu.*) red-leaved.	Rangoon.	1824.	——	S.b.	
36	amábile. k.r. (*pu.wh.*) beautiful.	Sumatra.	1810.	1. 12.	S.b.	Bot. mag. t. 1605.
37	augústum. k.r. (*pu.wh.*) august.	Mauritius.	1819.	4. 8.	S.b.	——— t. 2397.
38	mauritiànum. b.c. (*wh.*) Mauritius.	———	1816.	6. 9.	S.b.	Lodd. bot. cab. t. 650.
39	submérsum. b.m. (*pu.wh.*) lake.	Brazil.	1823.	8. 10.	S.b.	Bot. mag. t. 2463.
40	scábrum. b.m. (*cr.wh.*) rough-leaved.	S. America.	1810.	1. 12.	S.b.	——— t. 2180.
41	zeylánicum. h.a. (*cr.wh.*) Ceylon.	Ceylon.	1771.	7. 8.	S.b.	——— t. 1171.
	Amary'llis zeylánica. b.m.					
42	latifòlium. h.a. (*bh.*) broad-leaved.	E. Indies.	1806.	7. 9.	S.b.	Bot. reg. t. 1297.
43	longist'ylum. w.h. (*bh.*) long-styled.	———	1819.	6. 10.	S.b.	
44	insígne. w.h. (*cr.wh.*) noble.	———	——	4. 8.	S.b.	Bot. reg. t. 579.
45	Careyànum. b.m. (*bh.*) Carey's.	Mauritius.	1820.	1. 12.	S.b.	Bot. mag. t. 2466.
46	speciòsum. b.m. (*w.ro.*) specious.	E. Indies.	1819.	6. 9.	S.b.	——— t. 2217.
47	moluccànum. b.m. (*w.bh.*) Molucca.	Moluccas.	——	——	S.b.	——— t. 2292.

48 gigánteum. A.R. (*wh.*) gigantic. SierraLeone. 1792. 7. 10. S.b. Andr. rep. t. 169.
petiolàtum. H.A. *Amary'llis gigántea.* Bot. mag. t. 92.
49 Broussonèti. H.A. (*wh.pu.*)Broussonet's. ——— —— —— S.b. Bot. mag. t. 2121.
50 yuccæìdes. H.A. (*re.wh.*)Yucca-flowered. ——— 1740. 6. 10. S.b.
51 dístichum. H.A. (*re.wh.*)distichous. ——— ---- S.b. Bot. mag. t. 1253.
Amary'llis ornàta a. B.M.
52 Láncei. W.H. (*re.wh.*) Lance's. Surinam. 1825. —— S.b.
53 fláccidum. B.M. (*wh.*) flaccid. N. S. W. 1820. 7. 9. G.b. Bot. mag. t. 2133.
Amary'llis australásica. Bot. reg. t. 426.
54 Fórbesi. W.H. (*re.wh.*) Forbes's. Delagoa Bay. 1824? —— S.b.
55 longiflòrum. H.A. (*re.wh.*)long-flowered. W. Indies. 1816. 5. 8. S.b. Bot. reg. t. 303.
Amary'llis longifò'ia β longiflòra. B.R.
56 capénse. H.A. (*wh.*) common Cape. C. B. S. 1752. 7. 11. H.b. Bot. mag. t. 661.
Amary'llis longifòlia. B.M.
β minor. W.H. (*re.wh.*) *lesser.* Highclere. 1819. —— G.b.
57 ripàrium. H.A. (*cr.wh.*) bank. ——— 1816. 6. 9. H.b. Bot. mag. t. 2688.
58 revolùtum. H.A. (*wh.re.*)revolute. —— —— 1774. 8. 10. G.b. Bot. reg. t. 623.
Amary'llis revolùta. A.—B.R.
59 crassifòlium. H.A. (*wh.ro.*)thick-leaved. ——— —— —— G.b. Bot. reg. t. 615.
Amary'llis revolùta. B.—B.R.
60 algoénse. W.H. (*re.wh.*) Algoa-bay. C. B. S. 1826. —— G.b.
61 cáffrum. W.H. (*re.wh.*) Caffre. —— —— 1825. —— G.b.
62 aquáticum. H.A. (*ro.bh.*)water. ——— 1816. 5. 8. G.b. Bot. mag. t. 2352.
63 Goweniànum. W.H. (*bh.*)zeylanico-capense. Highclere. 1813. 6. 9. G.b. Hort. trans. ic.
64 Herbertiànum. (*re.wh.*)Scabro-capense. Spofforth. 1818. —— G.b.
Spofforthiànum. Herbert Mss.
65 fértile. W.H. (*wh.*) Canaliculato-capense. ——— —— —— G.b.
β grandiflòrum. W.H. (*wh.*) *larger-flowered.* ——— —— —— G.b.
66 íngens. W.H. (*wh.*) Pedunculato-capense. Highclere. —— —— G.b.
67 eréctius. W.H. (*bh.*) Corantyno-capense. Spofforth. —— —— G.b.
68 Highcleriànum. W.H. (*wh.*)Asiatico-capense. Highclere —— —— G.b.
69 elongàtum. W.H. (*wh.*) Brevifolio-capense. Spofforth. 1819. —— G.b.
70 modéstum. W.H. (*bh.*) Specioso-capense. ——— —— —— G.b.
71 pùdicum. W.H. (*bh.*) Careyano-capense. ——— —— —— G.b.
72 cùpitum. W.H. (*re.wh.*) Cruento-capense. ——— —— —— G.b.
73 tortuòsum. W.H. (*wh.*) Defixo-capense. ——— —— —— G.b.
74 Fèlix. W.H. (*bh.*) Latifolio-capense. ———1828. —— G.b.
75 microspérmum. W.H. (*bh.*) Revoluto-capense. ——— —— —— G.b.
76 egrègium. W.H. (*re.wh.*)Pedunculato-riparium. ———1826. —— G.b.
77 præ'stans. W.H. (*re.wh.*)Canaliculato-riparium. ——— —— —— G.b.
78 vittàtum. W.H. (*re.wh.*) Scabro-exaltatum. ———1818. —— S.b.
79 Digweèdi. W.H. (*re.wh.*)Scabro-americanum. Highclere. 1826. —— S.b.
80 Coopèri. W.H. (*re.wh.*) Specioso-longifolium. Wentworth— —— S.b.
81 amàtum. W.H. (*re.wh.*) Zeylanico-pedunculatum. Highclere. —— S.b.
82 Bacòni. W.H. (*re.wh.*) Zeylanico-erubescens. Mr. Bacon's. —— S.b.
83 cándidum. W.H. (*wh.*) Americano-bracteatum. Spofforth. —— S.b.
84 rubricaùle. W.H. (*bh.*) Erubescenti-bracteatum. ——— —— —— S.b.
85 púlchrum. W.H. (*re.wh.*)Scabro-erubescens. Highclere. —— —— S.b.
86 bulbulòsum. W.H. (*bh.*) Capensi-erubescens. ——— —— —— G.b.
87 dùplex. W.H. (*bh.*) Americano-erubescens. Spofforth.— —— S.b.
β minor W.H. (*bh.*) *lesser.* Highclere. —— —— S.b.
88 flaccídulum. W.H. (*bh.*) Specioso-defixum. Spofforth. —— —— S.b.
89 promíssum. W.H. (*re.wh.*)Forbesi-Careyanum. ——— —— —— S.b.
90 mirábile. W.H. (*wh.*) Flaccido-canaliculatum. ——— —— —— S.b.
91 ùnicum. W.H. (*wh.*) Flaccido-bracteatum. ——— —— —— S.b.
92 trìplex. W.H. (*wh.*) Careyano-fertile. ——— —— —— S.b.
93 divérsum. W.H. (*wh.*) Careyano-tortuosum. ——— —— —— S.b.
94 quádruplex. W.H. (*wh.*) Candido-fertile. ——— —— —— S.b.

VALL'OTA. H.A. VALL'OTA. Hexandria Monogynia.
purpùrea. H.A. (*re.*) purple-stalked. C. B. S. 1774. 5. 8. F.b. Jacq. schœn. 1. t. 62.
Amary'llis purpùrea. W. *elàta.* Jacq.
a màjor. (*re.*) *larger.* ——— —— —— F.b. Bot. mag. t. 1430.
β mìnor. (*re.*) *lesser.* ——— —— —— F.b. Bot. reg. t. 552.

GASTRON'EMA. H.A. GASTRON'EMA. Hexandria Monogynia.
clavàtum. H.A. (*re.wh.*) clavate. C. B. S. 1816. 5. 8. F.b. Bot. mag. t. 2291.
Cyrtánthus uniflòrus. Bot. reg. t. 168. *Amary'llis Pumílio.* H.K.

CYRTA'NTHUS. W. CYRTA'NTHUS. Hexandria Monogynia. W.
1 oblíquus. W. (*or.gr.*) oblique-leaved. C. B. S. 1774. 5. 8. G.b. Bot. mag. t. 1133.
2 striàtus. B.M. (*ye.sc.*) striped. ——— 1823. —— G.b. ——— t. 2534.

3 vittàtus. S.S.	(*std.*)	ribbon.	C. B. S.	1823.	5. 8.	G.b.	
4 spiràlis. B.R.	(*sc.*)	spiral-leaved.	———	1790.		G.b.	Bot. reg. t. 167.
5 collìnus. B.R.	(*sc.*)	hill.	———	1816.	——	G.b.	——— t. 162.
6 pállidus. B.M.	(*fl.*)	pale-flowered.	———	1822.	——	G.b.	Bot. mag. t. 2471.
7 ventricòsus. W.	(*sc.*)	bellied.	———	1770.		G.b.	Jacq. schœn. 1. t. 76.
8 angustifòlius. W.	(*sc.*)	narrow-leaved.	———	1774.	5. 6.	G.b.	Bot. mag. t. 271.
9 odòrus. B.R.	(*sc.*)	sweet-scented.	———	1818.	5. 7.	G.b.	Bot. reg. t. 503.
COBU'RGHIA. B.F.G.		COBU'RGHIA.	Hexandria Monogynia.				
incarnàta. B.F.G.	(*sc.*)	splendid.	Quito.	1820.	8. 9.	F.b.	Swt. br. fl. gar. s.2. t.17.
STENOME'SSON. H.A.		STENOME'SSON.	Hexandria Monogynia.				
1 flàvum. B.M.	(*ye.*)	slender-toothed.	Peru.	1823.	5. 11.	G.b.	Bot. mag. t. 2641.
Chrysiphìala flàva. B.R. 778.							
2 curvidentàtum. B.M.	(*ye.*)	curve-toothed.	———	1824.	——	G.b.	Bot. mag. t. 2640.
3 pauciflòrum.	(*ye.*)	few-flowered.	———	1822.	——	G.b.	Hook. ex. flor. t. 132.
Chrysiphìala pauciflòra. H.E.F.							
CHLIDA'NTHUS. H.A.		CHLIDA'NTHUS.	Hexandria Monogynia.				
fràgrans. H.A.	(*ye.*)	yellow fragrant.	S. America.	1821.	6. 8.	G.b.	Bot. reg. t. 640.
EUCR'OSIA. B.R.		EUCR'OSIA.	Hexandria Monogynia.				
bícolor. B.R.	(*or.re.*)	two-coloured.	Cape Horn.	1816.	4. 5.	S.b.	Bot. reg. t. 207.
CALOSTE'MMA. B.P.		CALOSTE'MMA.	Hexandria Monogynia.				
1 lùteum. B.R.	(*ye.*)	yellow.	N. Holland.	1819.	8. 10.	F.b.	Bot. reg. t. 421.
2 purpùreum. B.P.	(*pu.*)	purple.	———	——	——	F.b.	——— t. 422.
3 álbum. B.P.	(*wh.*)	white-flowered.	———	1824.	——	G.b.	
E'URYCLES. Sal.		E'URYCLES.	Hexandria Monogynia.				
1 nùda.	(*wh.*)	naked stamen'd.	E. Indies.	1822.	5. 7.	S.b.	
Crìnum nervòsum. H.B.							
2 alàta.	(*wh.*)	wing-stamen'd.	N. Holland.	1821.	——	S.b.	Bot. reg. t. 715.
Pancràtium australásicum. B.R.							
3 coronàta.	(*wh.*)	crowned.	E. Indies.	1759.	——	S.b.	Bot. mag. t. 1419.
Pancràtium amboinénse. B.M. *Pròiphys amboinénsis.* H.A.							
Three very distinct species, all flowered at Mr. Colvill's.							
HYMENOCA'LLIS. H.A.		HYMENOCA'LLIS.	Hexandria Monogynia.				
1 speciòsa. K.R.	(*wh.*)	large-flowered.	W. Indies.	1759.	5. 9.	S.b.	Bot. mag. t. 1453.
Pancràtium speciòsum. B.M.							
2 ovàta. K.R.	(*wh.*)	broad-leaved.	———	....	——	S.b.	Bot. reg. t. 43.
3 am'œna. K.R.	(*wh.*)	handsome.	Guiana.	1790.	——	S.b.	Bot. mag. t. 1467.
4 fràgrans. L.T.	(*wh.*)	fragrant.	W. Indies.	....	——	S.b.	Lodd. bot. cab. t. 834.
5 guianénsis. B.R.	(*wh.*)	Guiana.	Guiana.	1818.	——	S.b.	Bot. reg. t. 265.
6 carib'æa. K.R.	(*wh.*)	Caribean.	W. Indies.	1730.	——	S.b.	Bot. mag. t. 826.
Pancràtium carib'æum. B.M.							
7 declinàta. B.C.	(*wh.*)	declining.	Brazil.	....	——	S.b.	Lodd. bot. cab. t. 558.
8 undulàta. K.R.	(*wh.*)	wave-leaved.	S. America.	....	——	S.b.	
9 expánsa. B.M.	(*wh.*)	expanded.	W. Indies.	1818.	——	S.b.	Bot. mag. t. 1941.
10 pedàlis. B.C.	(*wh.*)	foot-flowered.	Brazil.	1815.	——	S.b.	Lodd. bot. cab. t. 809.
11 pátens. K.R.	(*wh.*)	spreading.	W. Indies.	....	——	S.b.	Redout. lil. 6. t. 380.
12 littoràlis. K.R.	(*wh.*)	sea side.	S. America.	1758.	——	S.b.	Jacq. vind. 3. t. 75.
13 Dryándri. K.R.	(*wh.*)	Dryander's.	———	....	——	S.b.	Bot. mag. t. 825.
14 dísticha. H.A.	(*wh.*)	two-ranked.	———	....	——	S.b.	
15 angústa. K.R.	(*wh.*)	narrow-leaved.	———	....	——	S.b.	Bot. reg. t. 221.
16 tenuiflòra. H.A.	(*wh.*)	slender-flowered.	....	....	——	S.b.	
17 crassifòlia. H.A.	(*wh.*)	thick-leaved.	S. America.	....	——	S.b.	
18 acutifòlia.	(*wh.*)	acute-leaved.	Mexico.	1824.	——	G.b.	Bot. reg. t. 940.
Pancràtium mexicànum. B.R. *nec aliorum. Hymenocállis litoràlis.* δ *acutifòlia.* B.M. 2621.							
19 rotàta. H.A.	(*wh.*)	large-crowned.	Carolina.	1803.	7. 9.	G.b.	Bot. mag. t. 827.
20 mexicàna. K.R.	(*wh.*)	Mexican.	Mexico.	1732.	8.	G.b.	Ker's review. pl.3. f.2.
21 caroliniàna. H.A.	(*wh.*)	Carolina.	Carolina.	1759.	6. 7.	G.b.	——— pl. 3. f. 1.
22 Stàplesi. W.H.	(*wh.*)	Mr. Staples's.	Mexico.	1826.	——	G.b.	
23 ovalifòlium. W.H.	(*wh.*)	oval-leaved.	S. America.	1820.	——	S.b.	
PANCR'ATIUM. H.A.		PANCR'ATIUM.	Hexandria Monogynia.				
1 marítimum. K.R.		sea Daffodil.	S. Europe.	1597.	5. 7.	G.b.	Bot. reg. t. 161.
caroliniánum. B.R. 927. *nec aliorum.*							
2 canariénse. B.R.		Canary.	Canaries.	1815.	6. 7.	G.b.	Bot. reg. t. 174.
3 illy'ricum. K.R.		Illyrian.	S. Europe.	1615.	5. 6.	G.b.	Bot. mag. t. 718.
4 verecúndum. K.R.		Narcissus-leav'd.	E. Indies.	1776.	6. 8.	G.b.	Bot. reg. t. 413.
5 zeylánicum. K.R.		one-flowered.	Ceylon.	1752.	——	S.b.	Bot. mag. t. 2538.
ISM'ENE. Sal.		ISM'ENE.	Hexandria Monogynia.				
1 nùtans. H.A.	(*wh.*)	nodding.	Brazil.	1810.	6. 8.	S.b.	Bot. mag. t. 1561.
Pancràtium calathìnum. B.M. *non* R.L.							
2 calathìnum. H.A.	(*wh.*)	cup-flowered.	———	——	6. 7.	S.b.	Bot. reg. t. 215.

3 Amáncaes. H.A. (*ye.*) Narcissus-flower'd. Peru. 1804. 6. 7. S.b. Bot. mag. t. 1224.
Pancràtium Amáncaes. B.M. *Narcissus Amáncaes.* R.P. 3. t. 283. f. a.

NARCI'SSUS. L. NARCI'SSUS. Hexandria Monogynia. L.

'*AJAX.* H.S.S.

1 obvallàris. H.S.S. (*ye.*) Sibthorp's. S. Europe. .... 3. 4. H.b. Bot. mag. t. 1301. f. inf.
2 bícolor. L. (*wh.ye.*) two-coloured. Spain. 1629. 4. 5. H.b. —— t. 1187.
'*Ajax lorifòlius.* H.S.S.
3 tubæflòrus. S.P. (*wh.ye.*) tube-flowered. S. Europe. .... 3. 5. H.b.
'*Ajax bícolor.* H.S.S.
4 moschàtus. B.M. (*wh.st.*) greater white. Spain. 1629. 4. H.b. Bot. mag. t. 924.
5 álbus. L.T. (*wh.*) white flowered. —— 1759. 3. 4. H.b. —— t. 1300.
moschàtus. δ. B.M. '*Ajax álbus.* H.S.S.
6 cérnuus. R.B. (*wh.*) drooping-flowered........ 1828. —— H.b.
β *plènus.* (*wh.*) *double-flowered.* .... —— —— H.b.
7 abscíssus. H.S.S. (*ye.*) clipt trunk. .... —— H.b. Park. parad. t. 107. f. 1.
8 màjor. B.M. (*ye.*) large. Spain. 1629. —— H.b. Bot. mag. t. 51.
9 propínquus. S.P. (*ye.*) great jagged. S. Europe. .... —— H.b. —— t. 301. f. sup.
màjor β. B.M. '*Ájax propínquus.* H.S.S.
10 Telamònius. H.S.S. (*ye.*) Telamonian. —— .... —— H.b.
β *plènus.* H.S.S. (*ye.*) *double Daffodil.* —— .... —— H.b.
γ *grandiplènus.* H.S.S. (*ye.*) *great double.* —— .... —— H.b. Park. parad. t. 101. f. 7.
11 nòbilis. H.S.S. (*ye.*) noble. —— .... —— H.b.
12 spùrius. H.S.S. (*ye.*) spurious. England. .... —— H.b.
13 serràtus. H.S.S. (*st.ye.*) serrate. —— .... —— H.b. Eng. bot. ic. ined.
β *suàvis.* H.S.S. (*st.ye.*) *sweet-scented.* —— .... —— H.b.
14 Pseùdo-Narcíssus. (*st.ye.*) Daffodil. —— .... 3. 4. H.b. Eng. bot. v. 1. t. 17.
β *scóticus.* H.S.S. (*st.ye.*) *Scotch Daffodil.* Scotland. .... —— H.b.
γ *plènus.* H.S.S. (*ye.*) *double French.* .... —— H.b. Park. parad. t. 101. f. 9.
δ *pleníssimus.* H.S.S. (*ye.*) *rose double.* .... —— H.b. —— t. 101. f. 6.
15 lobulàris. H.P. (*ye.*) lobe-crowned. England. .... —— H.b. Eng. bot. c. ic. ined.
16 pùmilus. S.P. (*ye.*) small wedge-fld. S. Europe. .... 3. H.b. Besl. eyst. 3. fol. 5. f. 4.
'*Ajax cuneiflòrus.* H.S.S.
17 mìnor. B.M. (*ye.*) small. Spain. 1629. 3. 5. H.b. Bot. mag. t. 6.
α *mínimus.* H.S.S. (*ye.*) *smallest.* —— —— —— H.b.
β *mìnor.* H.S.S. (*ye.*) *small.* —— —— —— H.b. Park. parad. t. 107. f. 2.
γ *mèdius.* H.S.S. (*ye.*) *middle-sized.* —— —— —— H.b.
δ *conspícuus.* H.S.S. (*ye.*) *conspicuous.* —— .... —— H.b.
ε *angústus.* H.S.S. (*ye.*) narrow-leaved. —— .... —— H.b.

DIOM'EDES. H.S.S.

18 Maclèaii. B.M. (*st.ye.*) Mac Leay's. Levant. .... 4. 5. H.b. Bot. mag. t. 2588.
Diomèdes màjor. H.P.
19 Sabìni. B.R. (*st.ye.*) Sabine's. —— .... —— H.b. Bot. reg. t. 762.
Diomèdes mìnor. H.P.

CORBUL'ARIA. H.S.S.

20 tenuifòlius. L.T. (*ye.*) slender-leaved. Biscay. 1760. 3. 4. H.b. Swt. br. fl. gar. t. 114.
21 lobulàtus. H.S.S. (*ye.*) lobed. .... —— H.b. Park. parad. t. 107. f. 7.
22 obèsus. S.P. (*ye.*) inflated. Portugal. 1796. —— H.b.
β *mìnor.* H.S.S. (*ye.*) *smaller.* —— .... —— H.b. Park. parad. t. 107. f. 8?
23 álbicans. H.S.S. (*wh.*) Trompet Marin. .... 1629. —— H.b. Lobel. advers. 462. ic.
24 Bulbocòdium. W. (*ye.*) hoop petticoat. S. Europe. —— —— H.b. Bot. mag. t. 88.

QUE'LTIA. H.S.S.

25 aurántius. H.S.S. (*ye.*) orange rimmed. —— —— —— H.b. Park. parad. t. 71. f. 2.
β *lùteo-aurántius.* (*wh.ye.*) *Butter and eggs.* —— —— —— H.b.
26 incomparábilis. B.M. (*st.ye.*) peerless. Portugal. —— 4. 5. H.b. Bot. mag. t. 121.
27 sulphùreus. (*st.*) self coloured peerless..... 1629. —— H.b.
Quéltia cóncolor. H.S.S.
β *plènus.* (*st.*) *double orange crown.* .... —— —— H.b.
28 semipartìtus. H.S.S. (*st.ye.*) cloven-cupped........ 1818. 3. 4. H.b.
29 Quéltia. (*wh.st.*) peerless white. .... —— 4. H.b.
Quéltia álba. H.S.S.
β *níveo-aurántius.* (*wh.st.*) *Orange Phœnix*....... —— —— H.b.
30 montànus. B.R. (*wh.*) mountain. Pyrenees. 1620. —— H.b. Bot. reg. t. 123.
31 galanthifòlius. H.S. (*wh.*) Snowdrop-leav'd. - —— 1720. 5. H.b. Park. parad. t. 73.
32 cápax. H.S.S. (*wh.*) capacious. Levant. .... —— H.b. Redout. lil. t. 177.
calathìnus. R.L. *Quéltia cápax.* H.S.S.
33 Redoùtei. (*wh.*) long-tubed. —— .... —— H.b. Redout. lil. t. 410.
calathìnus. R.L. *non* L. *nec aliorum.*

	SCHISA'NTHES. H.S.S.						
34	orientàlis. L.	(*st.ye.*) oriental.	Levant.	1778.	4. 5.	H.b.	Bot. mag. t. 948.
	incomparábilis. β. B.M. *Schisánthes orientàlis.* H.S.S.						
	GANYM'EDES. H.S.S.						
35	ochroleùcus. H.R.	(*st.*) drooping.	Portugal.	1620.	4.	H.b.	Bot. mag. t. 48.
	triándrus. B.M. *non* L. *Ganymèdes cérnuus.* H.S.S. *ochroleùcus.* H.R.						
36	triándrus. L.	(*wh.*) white drooping.	S. Europe.	1629.	——	H.b.	Park. parad. t. 93. f. 2.
	Ganymèdes álbus. H.R.						
37	nùtans. H.S.S.	(*ye.*) yellow cupped.	———	1789.	3. 5.	H.b.	Bot. mag. t. 945.
	trilòbus. B.M. *Ganymèdes nùtans.* H.S.S.						
38	cóncolor. H.S.S.	(*st.*) self-coloured.	———	1626.	4.	H.b.	
39	striátulus. H.S.S.	(*ye.st.*) faint-striped.	———	....	——	H.b.	
40	pulchéllus. H.S.S.	(*ye.wh.*) white-cupped.	———	1620.	——	H.b.	
	PHYLO'GYNE. H.S.S.						
41	rugulòsus. H.S.S.	(*ye.*) wrinkle-cupped.	———	....	——	H.b.	
42	odòrus. L.	(*ye.*) starry-flowered.	Corsica.	1720.	3.	H.b.	Bot. mag. t. 934.
	calathìnus. α. B.M. *Phylógyne odòra.* H.S.S.						
43	interjéctus. H.S.S.	(*ye.*) great curled cup.	S. Europe.	....	4.	H.b.	
44	triiòbus. L.	(*ye.*) three-lobed.	Spain.	1629.	3.	H.b.	Bot. mag. t. 78.
	odòrus. B.M. *Phylógyne trilòba.* H.S.S. *calathìna.* Sal.						
45	heminàlis. H.S.S.	(*ye.*) lesser curled cup.........		....	4.	H.b.	Park. parad. t. 93. f. 4.
46	párvulus.	(*ye.*) small.	....	1629.	——	H.b.	
	Phylógyne mìnor. H.S.S.						
	β *regìnæ.*	(*ye.*) *Queen Ann's Jonquil.*	———	——	——	H.b.	
	HERM'IONE. H.S.S.						
47	Jonquílla. B.M.	(*ye.*) Jonquil.	Spain.	1596.	4. 5.	H.b.	Bot. mag. t. 15.
	β *plènus.*	(*ye.*) *double Jonquil.*	———	....	——	H.b.	
48	símilis. H.S.S.	(*ye.*) lesser Jonquil.	S. Europe.	1626.	——	H.b.	
49	bìfrons. B.M.	(*ye.*) double-faced.	———	1807.	4.	H.b.	Bot. mag. t. 1186.
50	compréssus. L.T.	(*ye.*) Jasmine Jonquil.	———	1790.	3. 4.	H.b.	
51	primulìnus. H.S.	(*ye.*) Cowslip-cupped.	———	——	——	H.b.	Bot. mag. t. 1299.
	bìfrons. β. B.M. *Hermìone primulìna.* H.S.S.						
52	calathìnus. L.	(*ye.*) many-flowered.	Levant.	....	——	H.b.	Rudb. elys. 2. p. 60. f. 5.
53	tereticàulis. L.T.	(*st.*) cream-coloured.	———	....	4.	H.b.	Bot. mag. t. 1298 ?
54	citrìnus. H.S.S.	(*wh.ye.*) Citroniere.	———	....	——	H.b.	——— t. 946.
	orientàlis. γ. B.M.						
55	floribúndus. H.S.S.	(*wh.ye.*) grand monarque.	———	....	——	H.b.	
56	Trewiànus. B.M.	(*wh.ye.*) Bazelman major.	———	....	——	H.b.	Bot. mag. t. 940.
	Hermìone grandiflòra. H.S.S.						
57	crenulàtus. H.S.	(*wh.ye.*) Bazelman minor.	———	....	——	H.b.	Park. parad. t. 81. f. 5.
58	Tazétta. L.	(*wh.ye.*) French Daffodil.	———	1626.	——	H.b.	——— t. 81. f. 3.
59	dùbius. P.S.	(*wh.*) doubtful.	S. France.	....	——	H.b.	Redout. lil. t. 429.
60	pátulus. DC.	(*wh.ye.*) spreading.	Isl. Hyeres.	....	——	H.b.	
61	fistulòsus. H.S.	(*wh.ye.*) pipe-stemmed.	Levant.	....	——	H.b.	
62	cerìnus. H.S.	(*wh.ye.*) waxen cupped.	———	....	5.	H.b.	
63	Lùna. H.S.S.	(*wh.*) broad white.	———	....	4.	F.b.	
64	papyràtius. B.M.	(*wh.*) paper white.	Italy.	....	2. 3.	F.b.	Bot. mag. t. 947.
	Hermìone papyràtia. H.S.S.						
65	unícolor. T.N.	(*wh.*) one-coloured.	Naples.	....	——	H.b.	Ten. fl. nap. 1. t. 26.
66	præcox. T.N.	(*st.ye.*) early-flowered.	———	....	——	H.b.	——— 1. t. 27.
67	itálicus. B.M.	(*st.ye.*) Italian.	Italy.	....	——	F.b.	Bot. mag. t. 1188.
	Hermìone itálica. H.S.S.						
	α *semiplènus.*	(*wh.*) *semidouble.*	———	....	3.	F.b.	
	β *plènus.*	(*wh.*) *double Roman.*	———	....	——	F.b.	
68	grácilis. B.R.	(*pa.ye.*) slender.	....	....	4.	H.b.	Bot. reg. t. 816.
69	tenùior. B.M.	(*pa.ye.*) slender 2-flowered.	Italy.	1626.	——	H.b.	Bot. mag. t. 379.
70	latifòlius. H.S.S.	(*ye.*) broad-ld. orange.	Levant.	....	4. 5.	H.b.	
71	multiflòrus. H.S.S.	(*ye.*) common yellow.	———	....	——	H.b.	Bot. mag. t. 925.
	Tazétta. B.M. *non* L.						
72	intermèdius. R.L.	(*pa.ye.*) intermediate.	Pyrenees.	....	——	H.b.	Redout. lil. t. 426.
	orientàlis δ. Bot. mag. t. 1026. *Hermìone multiflòra.* β *aùrea.* H.S.S.?						
	NARCI'SSUS. H.S.S.						
73	poéticus. L.	(*wh.*) Poet's.	Greece.	1620.	4.	F.b.	Park. parad. t. 75. f. 3.
74	angustifòlius. B.M.	(*wh.*) narrow-leaved.	Switzerland.	1626.	——	H.b.	Bot. mag. t. 193.
75	majàlis. B.M.	(*wh.*) May-flowering.	England ?	....	5.	H.b.	Eng. bot. v. 4. t. 275.
	poéticus. E.B. *non* L. *patellàris.* S.P.						
	β *exsértus.* H.S.S.	(*wh.*) *long-stamened.*	———	....	——	H.b.	
	γ *plènus.* H.S.S.	(*wh.*) *double-flowered.*	———	....	——	H.b.	Park. parad. t. 75. f. 1.
76	biflòrus. B.M.	(*wh.*) two-flowered.	———	....	4. 5.	H.b.	Eng. bot. v. 4. t. 276.

77 níveus. W.en. (*wh.*) snowy white. S. Europe. .... 4. 5. H.b.
CHLORA'STER. H.P.
78 viridiflòrus. B.M. (*gr.*) green-flowered. Barbary. 1629. 9. 10. F.b. Bot. mag. t. 1687.
Chloráster físsus. H.P.
79 intèger. H.P. (*gr.*) entire-cupped. ——— ——— —— F.b. Park. parad. t. 93. f. 6.
80? obsolètus. H. S.S. (*w.ye.*) leafy autumnal. Spain. ——— —— F.b. ———— 90. t. 89. f. 4.
81? serotìnus. W. (*wh.*) late-flowered. Barbary. ——— —— F.b. Clus. hist. t. 252.
GALA'NTHUS. L. SNOWDROP. Hexandria Monogynia. L.
1 nivàlis. W. (*wh.*) common. Britain. .... 1. 3. H.b. Eng. bot. v. 1. t. 19.
2 plicàtus. B.R. (*wh.*) plaited. S. Europe. 1818. 2. 4. H.b. Bot. reg. t. 545.
LEUC'OJUM. L. SNOWFLAKE. Hexandria Monogynia. L.
1 vérnum. W. (*wh.*) spring. Germany. 1596. 2. 4. H.b. Bot. mag. t. 46.
2 carpáthicum. (*wh.*) Carpathian. Carpathian mt. 1816. —— H.b. ———— t. 1993.
3 æstìvum. W. (*wh.*) summer. England. .... 4. 5. H.b. ———— t. 1210.
4 pulchéllum. P.L. (*wh.*) neat. ——— .... —— H.b. Sal. par. lond. t. 74.
'ACIS. P.L. 'ACIS. Hexandria Monogynia.
1 ròsea. S.F.G. (*ro.*) rose-coloured. S. Europe. 1826. 8. 9. H.b. Swt. br. fl. gar. t. 297.
2 grandiflòra. S.F.G. (*bh.*) large-flowered. ——— 1828. —— H.b. Redout. liliac. t. 217.
3 trichophy'lla. S.F.G. (*wh.*) many-flowered. Barbary. 1812. 1. 2. F.b. Bot. reg. t. 544.
Leucòjum trichophy'llum. B.R.
4 autumnàlis. P.L. (*bh.*) autumnal. Portugal. 1629. 9. F.b. Bot. mag. t. 960.
Leucòjum autumnàle. B.M.
GETHY'LLIS. L. GETHY'LLIS. Hexandria Monogynia. L.
1 spiràlis. W. (*wh.*) spiral-leaved. C. B. S. 1780. 6. 7. G.b. Bot. mag. t. 1088.
2 ciliàris. W. (*wh.*) fringed. ——— 1788. - G.b. Jacq. schœn. 1. t. 79.
3 villòsa. W. (*wh.*) hairy. ——— 1787. —— G.b.
4 lanceolàta. W. (*wh.*) spear-leaved. ——— 1790. —— G.b.
ALSTRŒM'ERIA. W. ALSTRŒM'ERIA. Hexandria Monogynia. L.
1 Pelegrìna. W. (*re.bh.*) spotted flower'd. Peru. 1753. 6. 9. H.♃. Bot. mag. t. 139.
β álbida. (*wh.gr.*) *white flowered.* ——— 1820. —— H.♃.
2 Símsii. S.S. (*sc.ye.*) red-flowered. Chili. 1822. —— H.♃. Swt. br. fl. gar. t. 267.
pulchélla. B.M. 2353. Hook. ex. flor. t. 64.
3 Hookèri. (*ro.*) rose-coloured. Chili. 1822. —— H.♃. Hook. ex. flor. t. 181.
ròsea. H.E.F. *non* R.P.
4 Flós Martìni. B.R. (*w.p.y.*) St. Martin's. ——— ——— —— H.♃. Bot. reg. t. 731.
púlchra. Bot. mag. t. 2421. *trícolor.* Hook. ex. flor. t. 65.
5 bícolor. B.C. (*ye.wh.*) two-coloured. Chili. 1826. —— H.♃. Lodd. bot. cab. t. 1497.
6 psittácina. S.S. (*sc.gr.*) Parrot-like. Mexico. 1829. 7. 9. H.♃. Swt. br. fl. gar. s. 2. t. 15.
7 acutifòlia. Ot. (*or.*) sharp-leaved. S. America. —— 9. 10. F.♃.
8 pállida. Ot. (*pa.*) pale-flowered. Chili. —— H.♃.
9 Curtisiàna. R.S. (*wh.sc.*) striped-flowered. Brazil. 1776. 2. 4. S.♃. Bot. mag. t. 125.
Lígtu. B.M. *nec aliorum.*
10 ovàta. S.S. (*sc.*) oval-leaved. Peru. 1824. .... H.♃.‿. Cavan. ic. 1. t. 76.
11 Salsílla. W. (*pu.gr.*) climbing. ——— 1806. 6. 7. H.♃.‿. Feuill. per. 2. p. 713. t. 6.
12 èdulis. Tus. (*ro.gr.*) eatable-rooted. W. Indies. 1801. 6. 9. S.♃.‿. Andr. reposit. t. 649.
Salsílla. Bot. mag. t. 1613. *non* L.
13 hirtélla. K.S. (*pi.gr.*) hairy. Mexico. 1824. .... H.♃.‿. Swt. br. fl. gar. t. 228.
CL'IVIA. B.R. CL'IVIA. Hexandria Monogynia.
nòbilis. B.R. (*sc.gr.*) scarlet. C. B. S. 1820. 7. 10. G.♃. Bot. reg. t. 1182.
Imatophy'llum Aitòni. Bot. mag. t. 2856.
DORYA'NTHES. B.P. DORYA'NTHES. Hexandria Monogynia. S.S.
excélsa. B.P. (*sc.*) gigantic. N. S. W. 1800. 7. 10. G.♄. Bot. mag. t. 1685.

ORDO CCX.

HEMEROCALLIDEÆ. *Brown prodr.* 1. *p.* 295.

HEMEROCA'LLIS. S.S. DAY LILY. Hexandria Monogynia. L.
1 gramínea. B.M. (*ye.*) narrow-leaved. Siberia. 1759. 6. 7. H.♃. Bot. mag. t. 873.
2 flàva. B.M. (*ye.*) yellow. ——— 1596. —— H.♃. ———— t. 19.
3 speciòsa. W.H. (*co.*) specious. Jamaica. 1816. —— G.♃.
4 dísticha. B.F.G. (*co.*) fan-like. China. 1798. 5. 7. H.♃. Swt. br. fl. gar. t. 28.
5 fúlva. B.M. (*co.*) copper-coloured. Levant. 1596. 6. 8. H.♃. Bot. mag. t. 64.

FU'NKIA. S.S.		FU'NKIA. Hexandria Monogynia. S.S.					
1 álba.	(*wh.*)	white-flowered.	Japan.	1790.	8. 9.	H. ♃.	Bot. mag. t. 1433.
subcordàta. s.s. *Hemerocállis japónica.* B.M. *álba.* Andr. reposit. t. 75.							
2 cœrùlea.	(*wh.*)	blue-flowered.	———	——	6. 8.	H. ♃.	Bot. mag. t. 894.
ovàta. s.s. *Hemerocállis cœrùlea.* B.M.							
AGAPA'NTHUS. W.		AFRICAN LILY. Hexandria Monogynia. L.					
1 umbellàtus. w.	(*bl.*)	large-flowered.	C. B. S.	1692.	1. 12.	G. ♃.	Bot. mag. t. 500.
β variegàtus.	(*bl.*)	*striped-leaved.*	———	....	——	G. ♃.	
2 pr'æcox. W.en.	(*bl.*)	early-flowering.	———	....	1. 4.	G. ♃.	
3 mìnor. B.C.	(*bl.*)	small.	———	....	——	G. ♃.	Lodd. bot. cab. t. 42.
POLIA'NTHES. L.		POLIA'NTHES. Hexandria Monogynia. L.					
1 grácilis. L.en.	(*wh.*)	slender.	Brazil.	1822.	8. 10.	S. ♃.	
2 tuberòsa. w.	(*wh.*)	common.	E. Indies.	1629.	——	S. ♃.	Bot. reg. t. 63.
β flòre plèno.	(*wh.*)	*double-flowered.*	———	——	——	S. ♃.	
BLANDFO'RDIA. B.P.		BLANDFO'RDIA. Hexandria Monogynia. S.S.					
1 nòbilis. B.P.	(*sc.ye.*)	noble.	N. S. W.	1803.	7. 8.	G. ♃.	Bot. reg. t. 286.
2 grandiflòra. B.P.	(*sc.ye.*)	large-flowered.	———	1824.	——	G. ♃.	——— t. 911.
3 punícea.	(*cr.*)	large-leaved.	N. Holland.	1812.	——	G. ♃.	Lab. nov. hol. 1. t. 111.
Alétris punícea. Labillardiere.							
TRIT'OMA. B.M.		TRIT'OMA. Hexandria Monogynia. S.S.					
1 Uvària. H.K.	(*or.*)	great.	C. B. S.	1707.	8. 9.	H. ♃.	Bot. mag. t. 758.
2 Burchéllii. W.H.	(*or.*)	Mr. Burchell's.	———	1816.	——	H. ♃.	
3 mèdia. H.K.	(*sc.or.*)	lesser.	———	1789.	12. 6.	H. ♃.	Bot. mag. t. 744.
4 pùmila. H.K.	(*sc.or.*)	least.	———	1774.	9. 11.	F. ♃.	——— t. 764.
VELTH'EIMIA. B.M.		VELTH'EIMIA. Hexandria Monogynia. W.					
1 viridifòlia. w.	(*sc.or.*)	green-leaved.	C. B. S.	1768.	11. 4.	G.b.	Bot. mag. t. 501.
Alétris capénsis. B.M.							
2 intermèdia.	(*ro.ye.*)	subglaucous.	———	1819.	——	G.b.	
3 glaùca. w.	(*ro.ye.*)	glaucous.	———	1781.	1. 4.	G.b.	Bot. mag. t. 1091.
ALE'TRIS. W.		COLIC-ROOT. Hexandria Monogynia. L.					
1 farinòsa. w.	(*wh.*)	common.	N. America.	1768.	6.	H. ♃.	Bot. mag. t. 1418.
2 aùrea. Ph.	(*wh.*)	golden-tipped.	———	1811.	7. 8.	H. ♃.	Willd. hort. ber. t. 8.
SANSEVI'ERA. W.		SANSEVI'ERA. Hexandria Monogynia. W.					
1 glaùca. H.S.	(*wh.*)	glaucous.	Sierra Leon.	....	6. 10.	S. ♃.	
2 polyphy'lla. H.S.	(*wh.*)	upright-glaucous.	———	....	——	S. ♃.	
3 guineénsis. H.S.	(*wh.*)	Guinea.	Guinea.	1690.	6. 11.	S. ♃.	Bot. mag. t. 1180.
4 lætevìrens. H.S.	(*wh.*)	light green.	....	....	——	S. ♃.	
5 longiflòra. B.M.	(*wh.*)	long-flowered.	....	....	7. 8.	S. ♃.	Bot. mag. t. 2634.
6 spicàta. H.S.	(*wh.*)	spiked.	....	....	——	S. ♃.	Cavan. ic. 3. t. 246.
7 fulvocíncta. H.S.S.	(*wh.*)	brown-edged.	Brazil.	1818.	6. 9.	S. ♃.	
8 zeylánica. H.S.	(*wh.*)	Ceylon.	Ceylon.	1731.	6. 11.	S. ♃.	Bot. reg. t. 160.
9 lanuginòsa. w.	(*wh.*)	woolly.	E. Indies.	....	——	S. ♃.	Rheed. mal. 11. t. 42.
10 ensifòlia. H.S.	(*wh.*)	sword-leaved.	———	....	——	S. ♃.	
11 grandicúspis. H.S.	(*wh.*)	large-pointed.	....	....	....	S. ♃.	
12 pùmila. H.S.	(*wh.*)	dwarf.	C. B. S.	1796.	....	S. ♃.	
13 stenophy'lla. L.en.	(*wh.*)	slender-leaved.	....	1818.	6. 8.	S. ♃.	
14 cárnea. A.R.	(*ro.*)	flesh-coloured.	China.	1792.	4. 10.	H. ♃.	Andr. reposit. t. 361.
TULB'AGIA. W.		TULB'AGIA. Hexandria Monogynia. L.					
1 alliàcea. w.	(*gr.*)	Narcissus-leaved.	C. B. S.	1774.	5. 7.	G. ♃.	Bot. mag. t. 806.
2 affínis. L.en.	(*gr.*)	linear-leaved.	———	1820.	——	G. ♃.	
3 cepàcea. w.	(*gr.*)	Onion-scented.	———	1795.	4.	G. ♃.	
BRODI'ÆA. L.T.		BRODI'ÆA. Triandria Monogynia. L.T.					
1 grandiflòra. L.T.	(*bl.*)	large-flowered.	Georgia.	1806.	6.	H.b.	Bot. reg. t. 1183.
Hoòkera coronària. Sal. par. lond. t. 98.							
2 congésta. L.T.	(*bl.*)	close-headed.	———	——	5.	H.b.	Linn. trans. v. 10. t. 1.
PHYLL'OMA. B.M.		PHYLL'OMA. Hexandria Monogynia.					
1 aloiflòrum. B.M.	(*ye.*)	Aloe-like.	Bourbon.	1766.	4. 6.	D.S. ♄.	Bot. mag. t. 1585.
Lomatophy'llum borbònicum. w. *Drac'æna marginàta.* H.K. *dentàta.* P.S.							
2 mácrum. L.en.	(*ye.*)	lean.	Mauritius.	1817.	6. 7.	D.S. ♄.	
3 rufocínctum.	(*ye.*)	rosy-edged.	E. Indies.	1818.	....	D.S. ♄.	
A'loe rufocíncta. H.S.S.							
PACHIDE'NDRON. H.R.		PACHIDE'NDRON. Hexandria Monogynia. H.R.					
1 africànum. H.R.	(*ye.*)	African.	C. B. S.	1731.	7.	D.G. ♄.	
α làtum. H.R.	(*ye.*)	*broad-leaved.*	———	——	——	D.G. ♄.	
β angústum. H.R.	(*ye.*)	*narrow-leaved.*	———	——	——	D.G. ♄.	Bot. mag. t. 2517.
A'loe pseùdo-africàna. L.en. *A'loe africàna. β. angústior.* B.M.							
2 príncipis. H.R.	(*ye.*)	revolving red-spined.	———	....	....	D.G. ♄.	
3 angustifòlium. H.R.	(*ye.*)	narrow-leaved.	———	1812.	....	D.G. ♄.	
4 fèrox. H.R.	(*st.re.*)	great hedgehog.	———	1759.	4. 5.	D.G. ♄.	Bot. mag. t. 1975.
A'loe fèrox. B.M.							

5 pseùdo-fèrox. H.R. (st.) false hedgehog. C. B. S. 1759. 4. 5. D.G. ♄. Comm. prælud. t. 20.
6 supral'æve. H.R. (ye.) upright hedgehog. ——— 1731. .... D.G. ♄.

A'LOE. H.R. Aloe. Hexandria Monogynia. L.

1 dichótoma. H.S. smooth-stemmed. C. B. S. 1780. .... D.G. ♄.
2 purpuráscens. H.S. (re.gr.) purple. ——— 1789. 7. 10. D.G. ♄. Bot. mag. t. 1474.
3 soccotrìna. H.S. (sc.g.) soccotrine. ——— 1731. 4. 6. D.G. ♄. ——— t. 472.
4 spicàta. H.R. (wh.gr.) white spotted. ——— 1795. —— D.G. ♄.
5 plùridens. H.P. .... many-toothed. ——— 1822. .... D.G. ♄.
6 arboréscens. H.S. (sc.) tree. ——— 1731. 11. 3. D.G. ♄. Bot. mag. t. 1306.
7 grácilis. H.P. .... slender-stalked. ——— 1822. .... D.G. ♄.
8 flavispìna. H.S. (sc.) yellow-spined. ——— 1790. 8. D.G. ♄.
9 mitræfórmis. H.S. (sc.) mitre. ——— 1732. —— D.G. ♄. Bot. mag. t. 1270.
10 nóbilis. H.S. (sc.) great mitre. ——— 1800. 7. 8. D.G. ♄.
11 Commelìni. H.R. (sc.) Commeline's. ——— 1813. —— D.G. ♄.
β mitræfórmis. H.R. *yellowish-toothed.* ——— .... —— D.G. ♄.
12 dístans. H.S. (sc.gr.) small mitre. ——— 1732. 8. D.G. ♄. Bot. mag. 1362.
γ depréssa. H.R. (sc.gr.) *depressed.* ——— .... —— D.G. ♄.
13 xanthacántha. H.S.S. (sc.) yellow mitre. ——— 1817. —— D.G. ♄.
14 vulgàris. DC. (ye.or.) yellow-flowered. W. Indies. 1596. 8. D.S. ♄. DC. plant. gras. t. 27.
barbadénsis. H.S.
15 chinénsis. H.S.S. (ye.) Chinese. China. 1817. —— D.G. ♄.
16 c'æsia. H.S.S. .... cæsious tree. C. B. S. 1818. .... D.G. ♄.
β elàtior. H.S.S. .... *tall.* ——— ——— .... D.G. ♄.
17 frutéscens. H.S.S. (sc.) lesser tree. ——— .... D.G. ♄.
18 albispìna. H.S. (sc.) white-spined. ——— 1796. .... D.G. ♄.
19 glaùca. H.S. (sc.gr.) glaucous red-spined. ——— 1731. 1. 9. D.G. ♄.
α màjor. H.S. (sc.gr.) *larger.* ——— ——— —— D.G. ♄.
β rhodacántha. B.M. (sc.gr.) *lesser red-spined.* ——— ——— —— D.G. ♄. Bot. mag. t. 1278.
γ spinòsior. H.R. (sc.g.) *long-spined.* ——— 1820. .... D.G. ♄.
20 lineàta. H.S. (sc.gr.) striped red-spined. ——— 1789. .... D.G. ♄.
α víridis. H.R. (sc.gr.) *greenest-leaved.* ——— ——— .... D.G. ♄.
β glaucéscens. H.R. (sc.) *glaucescent.* ——— ——— .... D.G. ♄.
21 dorsàlis. R.S. (ye.) spiny-keeled. ——— 1826. .... D.G. ♄.
22 depréssa. H.S. (sc.ye.) depressed. ——— 1731. 8. D.G. ♄. Bot. mag. t. 1332.
23 brevifòlia. H.S. (sc.gr.) short-leaved. ——— ——— 6. 8. D.G. ♄. Bot. reg. t. 996.
24 prolífera. H.R. (or.y.) proliferous. ——— 1818. —— D.G. ♄.
25 sérra. DC. (re.gr.) saw-leaved. ——— ——— —— D.G. ♄. Jacq. eclog. t. 61.
26 striàta. H.S. (re.gr.) streaked. ——— 1795. —— D.G. ♄. Jacq. fragm. t. 68.
paniculàta. Jacquin.
27 striátula. H.P. .... stripe-sheathed. ——— 1820. .... D.G. ♄.
28 ciliàris. H.P. (sc.gr.) fringe-leaved. ——— ——— .... D.G. ♄.
29 ténuior. H.P. (ye.) green-sheathed. ——— 1819. .... D.G. ♄.
30 albocíncta. H.S.S. (sc.g.) pearl edged. ——— 1812. .... D.G. ♄.
31 serrulàta. H.S. (pi.) serrulated. ——— 1789. 7. 8. D.G. ♄. Bot. mag. t. 1415.
32 obscùra. H.S. (sc.or.) great soap. ——— 1727. 6. 10. D.G. ♄. ——— t. 1323.
pícta. α. B.M.
β glaùcior. H.R. (sc.or.) *glaucous-leaved.* ——— 1819. —— D.G. ♄.
γ mágnidens. H.R. (sc.or.) *large-toothed.* ——— .... —— D.G. ♄.
33 palléscens. H.R. (pi.gr.) pale serrulate. ——— 1820. —— D.G. ♄.
34 latifòlia. H.S. (sc.or.) broad-leaved. ——— 1795. 7. 8. D.G. ♄. Bot. mag. t. 1346.
35 grandidentàta. R.S. (sc.g.) great-toothed. ——— .... —— D.G. ♄.
36 saponària. H.S. (sc.or.) common soap. ——— 1727. —— D.G. ♄. Bot. mag. t. 1460.
β lùteo-striàta. H.R. (sc.) *yellow-striped.* ——— 1821. —— D.G. ♄.
37 microcántha. H.S.S. (re.) small-spined. ——— 1819. —— D.G. ♄. Bot. mag. t. 2272.
38 vìrens. H.S. (sc.) apple green. ——— 1790. 8. 9. D.G. ♃. ——— t. 1355.
39 suberécta. H.S. (sc.ye.) lesser hedgehog. ——— 1789. 3. 6. D.G. ♃.
β semiguttàta. H.R. (sc.g.) *half-warted.* ——— 1819. —— D.G. ♃.
40 acuminàta. H.S. (sc.ye.) middle hedgehog. ——— 1795. —— D.G. ♃. Bot. mag. t. 757.
hùmilis. B.M. *non* H.S.
41 echinàta. H.R. (sc.gr.) great tubercled. ——— 1821. 3. 6. D.G. ♃.
42 tuberculàta. H.S. (sc.g.) tubercled. ——— 1796. —— D.G. ♃.
43 aristata. H.P. (sc.gr.) bearded many-leaved. ——— 1819. —— D.G. ♃.
44 subtuberculàta. H.P. (sc.) rough & smooth. ——— ——— —— D.G. ♃.
45 hùmilis. H.S. (sc.gr.) dwarf hedgehog. ——— 1731. —— D.G. ♃. DC. plant. gras. t. 39.
46 incúrva. H.S. (sc.gr.) incurved hedgehog. ——— 1796. —— D.G. ♃. Bot. mag. t. 828.
hùmilis β incúrva. B.M.
47 tenuifòlia. H.R. (sc.gr.) thin-leaved. ——— 1821. —— D.G. ♃.
48 variegàta. H.S. (sc.gr.) partridge-breast. ——— 1720. 3. 12. D.G. ♃. Bot. mag. t. 513.

RHIPIDODE'NDRON. W. RHIPIDODE'NDRON. Hexandria Monogynia.

1	plicátile. H.R.	(re.gr.)	fan.	Africa.	1723.	6. 7.	D.G. ♄.	Bot. mag. t. 457.
	distichum. W. *A'loe plicátilis.* B.M.							
	β *màjor.* H.R.	(re.)	*larger fan.*	——	1820.	——	D.G. ♄.	
2	dichótomum. W.	(re.gr.)	dichotomous.	C. B. S.	1780.	——	D.G. ♄.	Patt. journ. t. 3. 4. 5.

GAST'ERIA. H.S. TONGUE-ALOE. Hexandria Monogynia. H.S.

1	oblíqua. H.S.	(ro.w.g.)	broad-marbled.	C. B. S.	1759.	6. 8.	D.G. ♄.	Bot. mag. t. 979.
	maculàta. H.P. *A'loe lingua. a.* B.M.							
2	púlchra. H.S.	(ro.gr.)	narrow-marbled.	——	1759.	7. 8.	D.G. ♄.	Bot. mag. t. 765.
	A'loe maculàta. B.M.							
3	pícta. H.P.	(ro.gr.)	painted.	——	1824.	——	D.G. ♃.	
	A'loe Bowieàna. R.S.							
4	formòsa. H.P.	(sc.gr.)	beautiful.	——	1826.	——	D.G. ♃.	
5	fasciàta. H.P.	(ro.gr.)	marbled.	——	1820.	——	D.G. ♃.	
	β *láxa.* H.P.	(ro.gr.)	*loose-leaved.*	——	——	——	D.G. ♃.	
6	retàta. H.P.	(ro.gr.)	chequered.	——	1826.	——	D.G. ♃.	
	A'loe dict'yodes. R.S.							
7	nìgricans. H.S.	(re.gr.)	dark-leaved.	——	1790.	6. 7.	D.G. ♃.	Bot. mag. t. 838.
8	marmoràta. H.P.	(re.g.)	marble-leaved.	——	——	——	D.G. ♃.	
9	crassifòlia. H.P.	(re.gr.)	thick-leaved.	——	1820.	——	D.G. ♃.	
10	brevifòlia. H.S.	(re.gr.)	short-leaved.	——	1809.	7 8.	D.G. ♃.	
	α *lætevìrens.* H.P.	(re.gr.)	*pale-green.*	——	——	——	D.G. ♃.	
	β *pervíridis.* H.P.	(re.gr.)	*darker-green.*	——	——	——	D.G. ♃.	
11	obtusifòlia. H.R.	(re.g.)	blunt-leaved.	——	——	——	D.G. ♃.	
12	móllis. H.R.	(re.gr.)	soft muddy-leaved.	——	1819.	——	D.G. ♃.	
13	subnígricans. H.P.	(re.g.)	dark-leaved.	——	1820.	——	D.G. ♃.	
	β *glábrior.* H.P.	(re.gr.)	*smoother.*	——	——	——	D.G. ♃.	
14	dísticha. H.P.	(re.gr.)	distichous.	——	1819.	——	D.G. ♃.	Comm. hort. ams. 2. t. 8.
	α *mìnor.* H.P.	(re.gr.)	*lesser.*	——	——	——	D.G. ♃.	
	β *màjor.* H.P.	(re.gr.)	*larger.*	——	——	——	D.G. ♃.	
15	conspurcàta. H.R.	(re.g.)	dirtied.	——	1796.	3. 11.	D.G. ♃.	
16	angulàta. H.P.	(re.gr.)	angular.	——	——	——	D.G. ♃.	Bot. mag. t. 1322. f. 3.
17	sulcàta. H.P.	(re.w.g.)	furrowed.	——	1791.	——	D.G. ♃.	—— t. 1322. f. 5.
18	excavàta. H.P.	(re.gr.)	excavated.	——	1824.	——	D.G. ♃.	—— t. 1322. f. 4.
19	angustifòlia. H.S.	(re.g.)	narrow.	——	1731.	——	D.G. ♃.	—— t. 1322. f. 2.
20	l'ævis. H.P.	(re.gr.)	smooth.	——	1820.	——	D.G. ♃.	
21	subverrucòsa. H.P.	(re.g.)	large-warted.	——	——	——	D.G. ♃.	
	α *grandipunctàta.* H.R.		*large-dotted.*	——	——	——	D.G. ♃.	
	β *parvipunctàta.* H.B.		*small-dotted.*	——	——	——	D.G. ♃.	
22	verrucòsa. H.S.	(re.g.)	pearl.	——	1731.	3. 11.	D.G. ♃.	Bot. mag. t. 837.
	β *latifòlia.* H.R.	(re.g.)	*broad-leaved.*	——	1820.	——	D.G. ♃.	
23	intermèdia. H.S.	(re.g.)	intermediate.	——	1790.	——	D.G. ♃.	Bot. mag. t. 1322. f. 1.
	β *aspérrima.* H.R.	(re.g.)	*rough-leaved.*	——	1820.	——	D.G. ♃.	
	γ *l'ævior.* H.P.	(re.gr.)	*smoother.*	——	——	——	D.G. ♃.	
	δ *lóngior.* H.P.	(re.gr.)	*longer-leaved.*	——	——	——	D.G. ♃.	
24	répens. H.R.	(re.gr.)	creeping-rooted.	——	——	....	D.G. ♃.	
25	párva. H.P.	(re.gr.)	smallest.	——	1824.	——	D.G. ♃.	
26	decípiens. H.P.	(re.gr.)	deceiving.	——	1823.	7. 10.	D.G. ♃.	
27	carinàta. H.S.	(sc.gr.)	rough-keeled.	——	1731.	6. 7.	D.G. ♃.	Bot. mag. t. 1331.
28	strigàta. H.P.	(re.gr.)	rough strigate.	——	1824.	7. 10.	D.G. ♃.	
29	lætipunctàta. H.P.	(re.gr.)	pale-spotted.	——	1821.	——	D.G. ♃.	
30	subcarinàta. H.S.S.	(re.)	slightly-keeled.	——	1820.	6. 7.	D.G. ♃.	
	β *striàta.* H.R.	(re.gr.)	*striated.*	——	——	——	D.G. ♃.	
31	undàta. H.P.	(re.gr.)	waved.	——	1825.	——	D.G. ♃.	
32	glábra H.S.	(re.gr.)	smooth-keeled.	——	1796.	6. 7.	D.G. ♃.	
	β *mìnor.* H.R.	(re.gr.)	*smaller.*	——	1820.	——	D.G. ♃.	
33	nítida. B.M.	(sc.gr.)	shining.	——	——	7. 8.	D.G. ♃.	Bot. mag. t. 2304.
	α *parvipunctàta.* H.R.		*small dotted.*	——	——	——	D.G. ♃.	
	β *grandipunctàta.* H.R.		*large-dotted.*	——	——	——	D.G. ♃.	
34	trigòna. H.R.	(li.gr.)	trigonal.	——	——	——	D.G. ♃.	
35	obtùsa. H.P.	(re.gr.)	smooth obtuse.	——	——	——	D.G. ♃.	
36	acinacifòlia. H.R.	(ro.g.)	long sword-leaved.	——	1818.	——	D.G. ♃.	Bot. mag. t. 2369.
37	nìtens. H.P.	(re.gr.)	sparse-spotted.	——	1819.	——	D.G. ♃.	
38	venústa. H.P.	(re.gr.)	charming.	——	1822.	——	D.G. ♃.	
39	pluripunctàta. H.P.	(re.gr.)	many-spotted.	——	——	——	D.G. ♃.	
40	ensifòlia. H.P.	(re.gr.)	long sword-leaved.	——	1820.	....	D.G. ♃.	
41	cándicans. H.R.	(re.)	rough marbled white.	——	1818.	7. 8.	D.G. ♃.	
42	linìta. H.P.	(re.)	smeared.	——	1822.	——	D.G. ♃.	
43	bícolor. H.P.	(re.)	half-marbled.	——	——	——	D.G. ♃.	

BOWI'EA. H.P.			Bowi'ea. Hexandria Monogynia.					
1	africàna. H.P.	(ye.)	African.	C. B. S.	1820.	4. 10.	D.G. ♃.	
2	myriacántha. H.P.	(ro.g.)	many-spined.	———	——	——	D.G. ♃.	
HAWO'RTHIA. H.S.S.			Hawo'rthia. Hexandria Monogynia. H.S.					
1	viscòsa. H.S.	(wh.re.)	clammy.	C. B. S.	1732.	5. 7.	D.G. ♃.	Bot. mag. t. 814.
	A'loe víscòsa. B.M.							
	α màjor.	(wh.re.)	*largest.*	———	——	——	D.G. ♃.	
	β mìnor.	(wh.re.)	*lesser.*	———	——	——	D.G. ♃.	
	γ parvifòlia.	(wh.re.)	*smaller-leaved.*	———	——	——	D.G. ♃.	
2	áspera. H.S.	(wh.re.)	rough-triangular.	———	1795.	——	D.G. ♃.	
3	induràta. H.R.	(wh.re.)	) hard branchy.	———	1819.	——	D.G. ♃.	
4	subtortuòsa. R.S.	(wh.re.)	twisted triangular.	———	1818.	——	D.G. ♃.	
5	torquàta. H.P.	(wh.re.)	long-twisted.	———	1824.	——	D.G. ♃.	
6	concínna. H.S.S.	(wh.re.)	neat triangular.	———	——	——	D.G. ♃.	
7	cordifòlia. H.S.S.	(wh.re.)	heart-leaved.	———	——	——	D.G. ♃.	
8	asperiúscula. H.S.S.	(w.re.)	roughish.	———	1818.	——	D.G. ♃.	
9	tortuòsa. H.S.	(wh.re.)	twisted.	———	1794.	5. 9.	D.G. ♃.	Bot. mag. t. 1337.
	A'loe rígida. B.M. *non* DC.							
	β cúrta. H.R.	(wh.re.)	*short twisted.*	———	1817.	——	D.G. ♃.	
	γ tórtella. H.R.	(wh.re.)	*lesser twisted.*	———	——	——	D.G. ♃.	
10	expánsa. H.S.	(wh.re.)	spreading-leaved.	———	1795.	7. 10.	D.G. ♃.	
11	rígida. H.R.	(wh.gr.)	great rigid.	———	1824.	——	D.G. ♃.	DC. plant. gras. t. 62.
12	hy'brida. H.R.	(wh.re.)	bastard pearl.	———	1821.	5. 9.	D.G. ♃.	
13	scábra. H.R.	(wh.re.)	rugged.	———	1818.	——	D.G. ♃.	
14	sórdida. H.R.	(wh.re.)	sordid.	———	——	——	D.G. ♃.	
15	recúrva. H.R.	(wh.re.)	recurved.	———	1795.	8.	D.G. ♃.	Bot. mag. t. 1353.
16	nígricans. H.P.	(wh.re.)	granulated black.	———	1820.	5. 9.	D.G. ♃.	
17	párva. H.P.	(wh.re.)	dwarf.	———	——	——	D.G. ♃.	
18	tessellàta. H.P.	(wh.re.)	dark chequered.	———	1822.	——	D.G. ♃.	
19	aspérula. H.P.	(wh.re.)	pale rough.	———	——	——	D.G. ♃.	
20	multifària. H.P.	(wh.re.)	many-leaved.	———	——	——	D.G. ♃.	
21	retùsa. H.S.	(wh.gr.)	smooth cushion.	———	1720.	——	D.G. ♃.	Bot. mag. t. 455.
	A'loe retùsa. B.M.							
22	mùtica. H.R.	(wh.gr.)	blunt cushion.	———	1819.	——	D.G. ♃.	
23	mirábilis. H.S.S.	(wh.ro.)	rough cushion.	———	1795.	——	D.G. ♃.	Bot. mag. t. 1354.
24	túrgida. H.S.S.	(wh.ro.)	turgid cushion.	———	1818.	——	D.G. ♃.	
25	l'æte-vìrens. H.S.S.	(w.re.)	light green.	———	——	——	D.G. ♃.	
26	papillòsa. H.S.S.	(w.re.)	papillose.	———	——	——	D.G. ♄.	
	β semipapillòsa. H.R.		*partly warted.*	———	....	——	D.G. ♄.	
27	erécta. H.R.	(wh.re.)	upright pearl.	———	1818.	——	D.G. ♃.	DC. plant. gras. t. 57.
	margaritífera. DC.							
28	fasciàta. H.S.S.	(wh.re.)	barred pearl.	———	——	——	D.G. ♃.	
29	subulàta. R.S.	(wh.re.)	awl-pointed.	———	——	——	D.G. ♃.	
30	claripérla. H.P.	(wh.re.)	bright-pearled.	———	1824.	——	D.G. ♃.	
31	granàta. H.S.S.	(wh.re.)	small pearl.	———	1732.	——	D.G. ♃.	Dill. elth. t. 16. f. 18.
	A'loe margaritífera. γ. mínima. Dill.							
32	brèvis. H.S.S.	(wh.ro.)	short-leaved.	———	1810.	——	D.G. ♃.	Bot. mag. t. 1360.
	A'loe margaritífera. γ. mínima. B.M.							
33	mìnor. H.S.	(pu.gr.)	lesser pearl.	———	....	5. 8.	D.G. ♃.	Bot. mag. t. 815.
34	margaritífera. H.S.S.	(wh.)	pearl.	———	1725.	——	D.G. ♃.	Bradl. succ. 3. t. 21.
35	semiglabràta. H.S.S.	(w.gr.)	half-smoothed.	———	1810.	——	D.G. ♃.	
36	rádula. H.S.	(ro.wh.)	raspy pearl.	———	1805.	6. 9.	D.G. ♃.	Jacq. schœn. t. 35.
37	attenuàta. H.S.	(wh.re.)	chalky pearl.	———	1790.	5. 8.	D.G. ♃.	Bot. mag. t. 1345.
	A'loe rádula. B.M. *non* Jacq.							
38	Reinwárdti. H.R.	(wh.re.)	chequered-pearl.	———	1821.	——	D.G. ♃.	
39	semimargaritífera.	(wh.)	semi-pearly.	———	1801.	——	D.G. ♃.	
	α máxima. H.S.S.	(wh.re.)	*largest.*	———	——	——	D.G. ♃.	Comm. hort. 2. t. 10.
	β màjor. H.S.S.	(wh.re.)	*large.*	———	1817.	——	D.G. ♃.	
	γ mìnor. H.S.S.	(wh.re.)	*smaller.*	———	1818.	——	D.G. ♃.	
	δ multiperlàta.	(wh.re.)	*many pearled.*	———	——	——	D.G. ♃.	
40	coarctàta. H.P.	(w.re.)	upright dull-spotted.	———	1822.	——	D.G. ♃.	
41	álbicans. H.S.	(wh.re.)	white-edged.	———	1795.	9. 10.	D.G. ♃.	Bot. mag. t. 1452.
42	l'ævis H.R.	(wh.re.)	narrow-bordered.	———	1820.	——	D.G. ♃.	Comm. hort. 2. t. 7.
43	ramífera. H.R.	(wh.re.)	branching white.	———	——	——	D.G. ♃.	
44	viréscens. H.R.	(wh.re.)	bordered green.	———	——	——	D.G. ♃.	
	β mìnor. H.R.	(wh.re.)	*lesser.*	———	——	——	D.G. ♃.	
45	arachnoídes. H.S.	(li.gr.)	cobweb.	———	1727.	8. 9.	D.G. ♃.	Bot. mag. t. 756.
46	translùcens. H.S.S.	(bh.)	transparent.	———	1795.	5. 8.	D.G. ♃.	——— t. 1417.
	A'loe arachnoídes β translùcens. B.M.							

47 setàta. H.S.S. (*wh.re.*) bristle-edged. C. B. S. 1818. 5. 8. D.G.♃.
β *nígricans*. H.R. (*wh.re.*) *blackish*. ——— —— —— D.G.♃.
γ *mèdia*. H.R. (*wh.re.*) *middle*. ——— —— —— D.G.♃.
δ *màjor*. H.R. (*wh.re.*) *larger*. ——— —— —— D.G.♃.
48 pállida. H.R. (*wh.re.*) pale green. ——— —— —— D.G.♃.
49 atrovìrens. H.R. (*wh.bh.*) dark green. ——— 1752. 5. D.G.♃. Bot. mag. t. 1361.
A'loe atrovìrens. DC. *arachnoìdes pùmila*. B.M.
50 chlorocántha. H.R. (*w.re.*) green-toothed. ——— 1820. 5. 8. D.G.♃.
51 angustifòlia. H.P. (*w.re.*) slender-leaved. ——— —— 6. 7. D.G.♃.
52 reticulàta. H.S. (*wh.*) netted. ——— 1794. 5. 8. D.G.♃. Bot. mag. t. 1314.
A'loe arachnoìdes γ reticulàta. B.M.
53 cymbifórmis. H.S. (*w.pu.*) boat-leaved. ——— 1795. 6. 8. D.G.♃. Bot. mag. t. 802.
concàva. H.R.
54 altilìnea. H.P. (*wh.re.*) ridge-lined. ——— 1821. —— D.G.♃.
55 obtùsa. H.P. (*wh.re.*) small blunt-leaved. ——— —— —— D.G.♃.
56 planifòlia. H.P. (*wh.re.*) flat oval-leaved. ——— —— —— D.G.♃.
57 mucronàta. H.R. (*w.pu.*) membrane-pointed. C. B. S. 1818. —— D.G.♃.
58 límpida. H.R. (*w.pu.*) limpid. ——— —— —— D.G.♃.
59 aristàta. H.R. (*w.pu.*) pin-pointed. ——— —— —— D.G.♃.
60 cuspidàta. H.R. (*w.pu.*) sharp-pointed. ——— —— —— D.G.♃.
61 denticulàta. H.R. (*w.re.*) pale pin-pointed. ——— —— —— D.G.♃.

A'PICRA. H.S.S. A'PICRA. Hexandria Monogynia. H.S.S.
1 spiràlis. H.S.S. (*wh.g.*) great spiral. C. B. S. 1790. 8. 9. D.G.♄.
2 spirélla. H.S.S. (*wh.g.*) small spiral. ——— 1808. —— D.G.♄.
3 bullulàta. H.S.S. (*st.gr.*) blistered. ——— 1818. —— D.G.♄. Jacq. frag. t. 109.
4 pentagòna. H.S.S. (*wh.*) five sided. ——— 1731. 6. 8. D.G.♄. Bot. mag. t. 1338.
5 quinquangulàris. R.S. (*w.st.*) five-angled. ——— .... —— D.G.♄.
6 rígida. W. (*wh.re.*) gunpowdered. ——— 1818. —— D.G.♄. Jacq. frag. t. 108.
pseùdo-rígida. H.S.S.
7 áspera. H.S.S. (*wh.st.*) rough triangled. ——— —— —— D.G.♄.
α *mìnor*. H.S.S. (*wh.st.*) *lesser*. ——— —— —— D.G.♄.
β *màjor*. H.S.S. (*wh.st.*) *larger*. ——— 1795. —— D.G.♄.
8 bicarinàta. H.S.S. (*w.st.*) double-keeled. ——— 1819. —— D.G.♄.
9 nìgra. H.P. (*wh.*) rough black. ——— 1820. 5. 8. D.G.♄.
10 imbricàta. H.R. (*st.*) rough-flowered. ——— 1731. 6. 7. D.G.♄. Bot. mag. t. 1455.
A'loe spiràlis. B.M. *non* H.S.
11 foliolòsa. H.P. (*wh.*) many small-leaved. ——— 1795. 6. 8. D.S.♄. Bot. mag. t. 1352.

ORDO CCXI.

DIOSCOREÆ. *Brown prodr.* 1. *p.* 294.

RAJ'ANIA. W. RAJ'ANIA. Diœcia Hexandria. L.
1 hastàta. W. (*gr.*) halberd-leaved. W. Indies. 1822. 7. 8. S.♃.⌣. Plum. amer. 84. t. 98.
2 cordàta. W. (*gr.*) heart-leaved. ——— 1786. —— S.♃.⌣. Plum. ic. 155. f. 1.
3 quinquefòlia. W. (*gr.*) five-leaved. ——— 1820. —— S.♃.⌣. ——— 155. f. 2.

DIOSC'OREA. W. YAM. Diœcia Hexandria. L.
1 pentaphy'lla. W. (*gr.*) five-leaved. E. Indies. 1768. .... S.♃.⌣. Rheed. mal. 7. t. 35.
2 triphy'lla. W. (*gr.*) three-leaved. ——— 1818. .... S.♃.⌣. Jacq. ic. t. 627.
3 quinquelòba. W. (*gr.*) five-lobed. Japan. 1822. 7. 9. G.♃.⌣. Kæmpf. ic. t. 15.
4 brasiliénsis. W. (*gr.*) Brazilian. Brazil. 1818. —— S.♃.⌣.
5 heterophy'lla. H.B. (*gr.*) various-leaved. E. Indies. —— .... S.♃.⌣.
6 aculeàta. W. (*gr.*) prickly-stemmed. ——— 1803. 7. 9. S.♃.⌣. Rheed. mal. 7. t. 37.
7 rubélla. H.B. (*re.g.*) reddish. ——— —— —— S.♃.⌣.
8 purpùrea. H.B. (*pu.g.*) purple. ——— 1816. —— S.♃.⌣.
9 angustifòlia. S.S. (*gr.*) narrow-leaved. Peru. 1821. —— S.♃.⌣.
10 nummulària. W. (*gr.*) Money-wort. E. Indies. 1820. .... S.♃.⌣. Rumph. amb. 5. t. 162.
11 alàta. W. (*gr.*) wing-stalked. ——— 1739. .... S.♃.⌣. Rheed. mal. 7. t. 38.
12 bulbífera. W. (*gr.*) bulb-bearing. ——— 1692. 7. 9. S.♃.⌣. Sal. par. lond. t. 17.
13 crispàta. H.B. (*gr.*) curl-stalked. ——— 1818. —— S.♃.⌣.
14 altíssima. S.S. (*gr.*) tallest. Martinica. 1821. —— S.♃.⌣. Plum. ic. 117. f. 2.
15 coriàcea. S.S. (*gr.*) leathery. S. America. 1818. —— S.♃.⌣.
16 anguìna. H.B. (*gr.*) snake. E. Indies. 1803. —— S.♃.⌣.
17 pulchélla. H.B. (*gr.*) pretty. ——— 1819. .... S.♃.⌣.

18 atropurpùrea. H.B. (*pu.g.*) dark purple. C. B. S. 1820. .... S. ♃. ⌣.
19 globòsa. H.B. (*gr.*) globular. ——— 1818. .... S. ♃. ⌣.
20 fasciculàta. H.B. (*gr.*) bundled. ——— 1817. .... S. ♃. ⌣.
21 glábra. H.B. (*gr.*) smooth. ——— 1803. 7. 9. S. ♃. ⌣.
22 satìva. W. (*gr.*) common. W. Indies. 1733. —— S. ♃. ⌣. Rheed. mal. 8. t. 51.
23 piperifòlia. W. (*gr.*) Pepper-leaved. S. America. 1817. .... S. ♃. ⌣. Plum. ic. 117. f. 1.
24 nepalénsis. (*gr.*) Nepaul. Nepaul. 1816. 7. 8. H. ♃. ⌣.
25 cinnamomifòlia. B.M. (*g.*) Cinnamon-leav'd. Brazil. 1826. 11. 12. S. ♃. ⌣. Bot. mag. t. 2825.
26 quaternàta. Ph. (*re.g.*) four-leaved. N. America. 1811. 7. H. ♃. ⌣.
27 villòsa. Ph. (*re.g.*) pubescent. ——— 1752. 8. H. ♃. ⌣. Jacq. ic. 3. t. 626.
28 oppositifòlia. L. (*gr.*) opposite-leav'd. E. Indies. 1803. .... S. ♃. ⌣. Petiv. gaz. t. 31. f. 6.

TESTUDIN'ARIA. B.T. TESTUDIN'ARIA. Diœcia Hexandria.
1 Elephántipes. B.R. (*gr.*) Elephant's-foot. C. B. S. 1774. 7. 8. G. ♃. ⌣. Bot. reg. t. 92.
Tàmus Elephántipes. Bot. mag. t. 1347.
2 montàna. B.T. (*gr.*) glaucous-leaved. ——— 1816. —— G. ♃. ⌣.

ORDO CCXII.

TAMEÆ. *Brown prodr.* 294 *in obs.*

T'AMUS. L. BLACK BRYONY. Diœcia Hexandria. L.
1 commùnis. W. (*gr.*) common. England. .... 5. 8. H. ♃. ⌣. Eng. bot. t. 91.
2 crética. W. (*gr.*) Cretan. Candia. 1739. 7. 8. H. ♃. ⌣.

ORDO CCXIII.

SMILACEÆ. *Brown prodr.* 292.

SM'ILAX. L. SM'ILAX. Diœcia Hexandria. L.
1 áspera. S.S. (*w.gr.*) rough bindweed. S. Europe. 1648. 9. H. ♄. ⌣. Schk. handb. 3. t. 328.
β *auriculàta.* (*w.gr.*) *ear-leaved.* ——— —— —— H. ♄. ⌣. Pluk. alm. t. 110. f. 3.
2 nìgra. S.S. (*w.gr.*) black. ——— 1817. —— H. ♄. ⌣.
3 catalònica. S.S. (*w.gr.*) Catalonian. ——— —— —— H. ♄. ⌣.
4 mauritánica. S.S. (*w.gr.*) red-berried. ——— 1820. 8. 10. F. ♄. ⌣.
5 excélsa. S.S. (*w.gr.*) tall. Syria. 1739. 8. 9. F. ♄. ⌣. Buxb. cent. 1. t. 32.
6 hórrida. S.S. (*w.gr.*) straight-spined. N. America. 1820. —— H. ♄. ⌣.
7 zeylánica. S.S. (*w.gr.*) Ceylon. E. Indies. 1778. —— S. ♄. ⌣. Rumph. amb. 5. t. 161.
8 quadrangulàris. S.S. (*w.g.*) square-stalked. N. America. 1812. 6. 8. H. ♄. ⌣. Wats. dend. brit. t. 109.
9 longifòlia. S.S. (*w.gr.*) long-leaved. Cayenne. 1823. .... S. ♄. ⌣.
10 Watsòni. (*w.gr.*) Watson's. N. America. 1811. 7. 8. H. ♄. ⌣. Wats. dend. brit. t. 110.
longifòlia. W.D. *non* Rich.
11 Sarsaparílla. S.S. (*w.gr.*) medicinal. ——— 1664. —— H. ♄. ⌣. Wats. dend. brit. t. 111.
12 brasiliénsis. S.S. (*w.gr.*) Brazilian. Brazil. 1820. —— S. ♄. ⌣.
13 hastàta. S.S. (*w.gr.*) halberd-leaved. Carolina. 1822. —— F. ♄. ⌣. Pluk. alm. t. 111. f. 3.
14 Chìna. S.S. (*w.gr.*) Chinese. China. 1759. .... G. ♄. ⌣. Kæmpf. amœn. t. 782.
15 acuminàta. S.S. (*w.gr.*) sharp-pointed. W. Indies. 1822. 6. 7. S. ♄. ⌣. Plum. ic. 83.
16 glaùca. B.M. (*wh.*) glaucous-leaved. N. America. 1811. 5. 7. H. ♄. ⌣. Bot. mag. t. 1846.
17 austràlis. B.P. (*w.gr.*) oblong-leaved. N. S. W. 1815. .... G. ♄. ⌣.
18 rotundifòlia. S.S. (*w.g.*) round-leaved. N. America. 1760. 7. 8. H. ♄. ⌣.
19 laurifòlia. S.S. (*w.gr.*) Laurel-leaved. ——— 1739. —— H. ♄. ⌣. Catesb. car. 1. t. 15.
20 tamnoídes. S.S. (*w.gr.*) Tamus-leaved. ——— —— 6. 7. H. ♄. ⌣. ——— 1. t. 52.
21 cadùca. S.S. (*w.gr.*) deciduous. ——— 1759. —— H. ♄. ⌣.
22 havanénsis. S.S. (*w.gr.*) Havannah. S. America. 1823. .... S. ♄. ⌣. Jacq. amer. t. 179. f. 102.
23 ovalifòlia. D.P. (*w.gr.*) oval-leaved. Nepaul. 1822. .... H. ♄. ⌣.
24 Bòna-nóx. S.S. (*w.gr.*) fringed. N. America. 1739. 6. 7. H. ♄. ⌣. Pluk. alm. t. 111. f. 1.
25 maculàta. D.P. (*w.gr.*) spotted. Nepaul. 1819. .... H. ♄. ⌣.
26 prolífera. H.B. (*w.gr.*) proliferous. E. Indies. —— .... S. ♄. ⌣.
27 glábra. H.B. (*w.gr.*) smooth. ——— 1820. .... S. ♄. ⌣.
28 Alpìni. S.S. (*w.gr.*) Grecian. Greece. —— 6. 8. F. ♄. ⌣.
29 latifòlia. B.P. (*w.gr.*) broad-leaved. N. Holland. 1791. .... F. ♄. ⌣.
30 lanceolàta. Ph. (*w.gr.*) spear-leaved. N. America. 1785. 5. 6. H. ♄. ⌣. Catesb. car. 2. t. 84.

31 herbàcea. s.s. (*gr.*) herbaceous. N. America. 1669. 7. H. ♃.◡. Bot. mag. t. 1920.
32 pedunculàris. s.s. (*gr.*) long-peduncled. ——— 1812. 5. 7. H. ♃.◡.
33 globífer. s.s. (*w.gr.*) Surinam. Surinam. 1819. .... S. ♄.◡.
34 macrophy'lla. s.s. (*w.g.*) large-leaved. W. Indies. 1822. .... S. ♄.◡.
35 Pseùdo-chìna. s.s. (*w.g.*) Bastard China. S. America. 1739. 5. 6. S. ♄.◡. Sloan. jam. 1. t. 143. f. 1.
36 glyciphy'lla. B.P. (*w.g.*) Botany-bay Tea. N. S. W. 1815. —— S. ♄.◡.
37 canariénsis. s.s. (*w.gr.*) Canary. Canaries. 1816. —— F. ♄.◡.
38 pùbera. s.s. (*w.gr.*) downy. N. America. 1806. —— H. ♄.◡.
39 rùbens. W.D. (*pi.*) pink-flowered. ——— 1811. 7. 8. H. ♄.◡. Wats. dend. brit. t. 108.

RIPOG'ONUM. B.P. Ripog'onum. Hexandria Monogynia.
álbum. B.P. (*wh.*) white-flowered. N. S. W. 1820. 4. 6. G. ♄.◡.

RU'SCUS. L. Butcher's-broom. Diœcia Monadelphia. L.
1 aculeàtus. W. (*w.gr.*) prickly. England. .... 12. 6. H. ħ. Eng. bot. v. 8. t. 560.
2 láxus. L.T. (*w.gr.*) loose. Europe. .... 1. 6. H. ħ.
3 Hypophy'llum. W. (*g.y.*) thick-leaved. Italy. 1640. 5. 6. H. ħ. Bot. mag. t. 2049.
β latifòlius. W. (*g.y.*) *broad-leaved.* S. Europe. —— —— H. ħ. Dill. elt. t. 251. f. 303.
4 Hypoglóssum. W. (*g.y.*) double-leaved. Italy. 1596. 4. 5. H. ħ. Schk. handb. 3. t. 340.
5 andrógynus. W. (*g.y.*) climbing. Canaries. 1713. —— G. ħ.◡. Bot. mag. t. 1898.
6 reticulàtus. W. (*g.y.*) netted-leaved. C. B. S. 1816. —— G. ħ.◡.
7 volùbilis. W. (*g.y.*) many-nerved. ——— —— —— G. ħ.◡.
8 racemòsus. W. (*g.wh.*) Alexandrian Laurel. Portugal. 1713. 6. H. ħ. Wats. dend. brit. t. 145.

POLYGON'ATUM. D. Solomon's-seal. Hexandria Monogynia.
1 verticillàtum. R.L. (*w.g.*) whorl-leaved. Scotland. .... 6. 7. H. ♃. Redout. liliac. t. 244.
Convallària verticillàta. Eng. bot. v. 2. t. 128.
2 bracteàtum. Th. (*wh.g.*) bracted. Switzerland. 1827. —— H. ♃.
3 leptophy'llum. D.P. slender-leaved. Nepaul. 1816. 6. 7. H. ♃.
4 oppositifòlium. D.P. opposite-leaved. ——— —— —— H. ♃. Lodd. bot. cab. t. 640.
Convallària oppositifòlia. B.C.
5 vulgàre. R.L. (*w.gr.*) common. England. .... 5. 6. H. ♃. Redout. liliac. t. 258.
Convallària Polygonàtum. Eng. bot. v. 4. t. 280.
6 pubéscens. Ph. (*w.g.*) pubescent. N. America. 1812. —— H. ♃. Willd. hort. ber. t. 45.
7 canaliculàtum. Ph. (*w.g.*) channel-leaved. ——— —— 6. H. ♃.
8 angustifòlium. Ph. (*st.g.*) narrow-leaved. ——— 1824. 5. 6. H. ♃.
9 multiflòrum. R.L. (*w.g.*) many-flowered. Britain. .... —— H. ♃. Eng. bot. v. 4. t. 279.
10 latifòlium. R.L. (*st.*) broad-leaved. Europe. 1802. —— H. ♃. Redout. liliac. t. 243.
11 macrophy'llum. (*w.g.*) large-leaved. N. America. 1823. —— H. ♃.
latifòlium. Ph. *nec aliorum.*
12 polyánthemum. M.B. (*w.*) many-flowered. Caucasus. —— —— H. ♃.

SMILAC'INA. Df. Smilac'ina. Hexandria Monogynia. Ph.
1 racemòsa. Df. (*wh.*) cluster-flowered. N. America. 1640. 5. 6. H. ♃. Redout. liliac. t. 230.
Convallària racemòsa. Bot. mag. t. 899.
2 stellàta. Df. (*wh.*) star-flowered. ——— 1633. —— H. ♃. Bot. mag. t. 185.
3 trifòlia. Df. (*wh.*) three-leaved. ——— 1812. 6. 7. H. ♃. Gmel. sib. 1. t. 6.
4 canadénsis. Ph. (*wh.*) Canadian. ——— —— —— H. ♃. Redout. lil. t. 216. f. 1.
5 bifòlia. Df. (*wh.*) least. N. Europe. 1596. 5. 6. H. ♃. Bot. mag. t. 510.
Convallària bifòlia. B.M. *Majánthemum bifòlium.* Redout. liliac. t. 216. f. 2.
6 boreàlis. B.M. (*g.ye.*) oval-leaved. N. America. 1778. 5. 6. H. ♃. Bot. mag. t. 1403.
7 umbellàta. Df. (*wh.*) umbel-flowered. ——— —— —— H. ♃. ——— t. 1155.

OPHIOP'OGON. B.M. Snake's-beard. Hexandria Monogynia. B.M.
1 japónicus. B.M. (*wh.*) Japan. Japan. 1784. 6. F. ♃. Bot. mag. t. 1063.
2 spicàtus. B.R. (*bl.*) spiked. China. 1821. 8. 10. F. ħ. Bot. reg. t. 593.
3 intermèdius. D.P. (*wh.*) intermediate. Nepaul. 1824. —— F. ♃.

CONVALL'ARIA. D. Lily of the Valley. Hexandria Monogynia. L.
majàlis. E.B. (*wh.*) common. Britain. .... 5. 6. H. ♃. Eng. bot. v. 15. t. 1035.
a álba. (*wh.*) *white-flowered.* ——— .... —— H. ♃.
β ròsea. (*ro.*) *rose-coloured.* ——— .... —— H. ♃.
γ plèna. (*wh.*) *double-flowered.* ——— .... —— H. ♃.

STRE'PTOPUS. M. Stre'ptopus. Hexandria Monogynia. Ph.
1 amplexifòlius. R.L. (*g.y.*) heart-leaved. Europe. 1752. 5. H. ♃. W. et K. hung. t. 167.
2 distórtus. M. (*g.y.*) distorted. N. America. —— —— H. ♃.
3 ròseus. M. (*ro.*) rose-coloured. ——— 1806. 6. 7. H. ♃. Bot. mag. t. 1489.
Uvulària ròsea. B.M.
4 lanuginòsus. M. (*g.y.*) woolly. ——— 1812. —— H. ♃. Bot. mag. t. 1490.
5 símplex. D.P. (*br.gr.*) simple-stalked. Nepaul. 1822. —— H. ♃.

MED'EOLA. S.S. Indian Cucumber. Hexandria Trigynia. L.
virgínica. S.S. (*gr.*) Virginian. Virginia. 1759. 6. H. ♃. Bot. mag. t. 1316.
Gyròmia virgínica. N.

TRI'LLIUM. L. Tri'llium. Hexandria Trigynia. L.
1 séssile. S.S. (*da.pu.*) sessile-flowered. N. America. 1759. 4. 5. H. ♃. Bot. mag. t. 40.

2 petiolàtum. s.s. (*d.pu.*) Plantain-leaved. N.America. 1811. 4. 5. H. ♃.
3 erythrocárpum.M.(*w.re.*)painted-flowered. —— —— —— 5. 6. H. ♃. Swt. br. fl. gar. t. 212.
píctum. Ph.
4 undulàtum. s.s. (*wh.*) wave-flowered. —— —— —— H. ♃.
5 pusíllum. s.s. (*pu.*) dwarf. Carolina. 1812. —— F. ♃.
6 ovàtum. s.s. (*pa.pu.*) purple-flowered.N. America. —— —— F. ♃.
7 cérnuum. s.s. (*wh.*) drooping-flowered. —— 1758. 4. 5. H. ♃. Bot. mag. t. 954.
8 eréctum. w. (*d.pu.*) stinking. —— 1759. —— H. ♃. —— t. 470.
9 péndulum. s.s. (*wh.*) white-flowered. —— 1805. —— H. ♃. W. hort. ber. 1. t. 35.
eréctum. β. Bot. mag. t. 1027.
10 obovàtum. s.s. (*d.ro.*) obovate-petaled. —— 1824. —— H. ♃.
11 grandiflòrum. s.s.(*wh.*) large-flowered. —— 1799. 4. 6. H. ♃. Sal. par. lond. t. 1.
erythrocárpum. B.M. *non* M.
12 stylòsum. N. (*ro.*) long-styled. Carolina. 1823. —— F. ♃.

P'ARIS. L. Herb-Paris. Octandria Tetragynia. L.
1 quadrifòlia. E.B. (*gr.*) True-love or 1-berry. England. .... 5. 6. H. ♃. Eng. bot. v. 1. t. 7.
2 verticillàta. M.B. (*gr.*) whorl-leaved. Siberia. 1825. —— H. ♃.
3 polyphy'lla. D.P. (*gr.*) many-leaved. Nepaul. 1816. —— H. ♃.

ORDO CCXIV.

ASPHODELEÆ. *Brown prodr.* 1. *p*. 274.

ASPHO'DELUS. L. Asphodel. Hexandria Monogynia. L.
1 clavàtus. H.B. (*wh.*) club-seeded. E. Indies. 1808. 7. 10. S. ⊙.
2 fistulòsus. s.s. (*wh.*) Onion-leaved. S. Europe. 1596. 6. 9. H. ♃. Bot. mag. t. 984.
3 intermèdius.L.en.(*wh.*)intermediate. Canaries. 1824. —— F. ♃.
4 álbus. s.s. (*wh.*) upright. S. Europe. .... 5. 7. H. ♃. Blackw. t. 238.
5 æstìvus. s.s. (*wh.*) summer. Portugal. 1819. —— H. ♃.
6 asiáticus. R.S. (*wh.*) Asiatic. Levant. .... —— H. ♃.
7 ramòsus. s.s. (*wh.*) branched. S. Europe. 1551. —— H. ♃. Bot. mag. t. 799.
8 prolífer. M.B. (*wh.*) proliferous. Armenia. 1824. 6. 10. H. ⊙. Hoff.com. m.1.p.1.t.43.
9 tenùior. s.s. (*ye.gr.*) slender-leaved. Caucasus. 1823. 7. H. ♃. Bot. mag. t. 2626.
10 créticus. s.s. (*ye.gr.*) Cretan. Candia. 1821. 6. 7. H. ♃. Lodd. bot. cab. t. 915.
11 libúrnicus. R.S. (*ye.gr.*) oriental. Levant. —— —— H. ♃. Jacq. f. eclog. t. 77.
12 taùricus. s.s. (*ye.gr.*) Taurian. Tauria. 1812. 5. 7. H. ♃. Redout. liliac. t. 470.
13 capillàris. R.S. (*ye.gr.*) awl-leaved. —— —— —— H. ♃. —— t. 380.
14 lùteus. s.s. (*ye.*) common yellow. Sicily. 1596. 5. 6. H. ♃. Bot. mag. t. 773.

EREM'URUS. S.S. Erem'urus. Hexandria Monogynia. S.S.
spectábilis. s.s. (*ye.*) channel-leaved. Caucasus. 1800. 5. 6. H. ♃. Swt. br. fl. gar. t. 188.
Asphódelus altàicus. w.

CZA'CKIA. S.S. St. Bruno's Lily. Hexandria Monogynia. S.S.
Liliástrum. s.s. (*wh.*) SavoySpiderwort.Switzerland. 1629. 5. 6. H. ♃. Bot. mag. t. 318.
Anthéricum Liliástrum. B.M. *Hemerocállis Liliástrum*. W.en.

PHALA'NGIUM. J. Phala'ngium. Hexandria Monogynia.
1 Liliàgo. B.M. (*wh.*) grass-leaved. S. Europe. 1596. 5. 6. H. ♃. Redout. lil. t. 269.
α màjor. B.M. (*wh.*) *larger*. —— —— —— H. ♃. Bot. mag. t. 1635.
β mìnor. B.M. (*wh.*) *lesser* —— —— —— H. ♃. —— t. 914.
2 ramòsum. B.M. (*wh.*) branching. —— 1570. —— H. ♃. —— t. 1055.
Anthéricum ramòsum. w.
3 gr'æcum. Lam. (*wh.*) Grecian. Greece. 1828. —— F. ♃. Flor. græc. t. 336.
4 nepalénse. B.R. (*wh.*) Nepaul. Nepaul. 1822. 6. 10. F. ♃. Bot. reg. t. 998.
5 glaùcum. P.S. (*wh.*) glaucous. S. America. 1816. —— G. ♃.
6 pomeridiànum. (*wh.*) afternoon-flower'd. —— 1818. —— G. ♃. Bot. reg. t. 564.
Anthéricum pomeridiànum. B.R. *Scílla pomeridiàna*. DC.

ANTHE'RICUM. Anthe'ricum. Hexandria Monogynia. L.
1 hirsùtum. R.S. (*wh.*) hairy-leaved. —— 1822. 6. 10. G. ♃.
2 blepharóphoron. R.S.(*w.pu.*)fringed. —— —— —— G. ♃.
3 bipedunculàtum.(*wh.*) two-flowered. —— 1825. 4. 5. G. ♃. Jacq. ic. 2. t. 410.
4 triflòrum. w. (*wh.*) three-flowered. —— 1782. 8. 10. G. ♃.
5 undulàtum. w. (*wh.*) wave-leaved. —— 1825. 4. 6. G. ♃.
6 pilòsum. w. (*wh.*) pilose. —— —— —— G. ♃. Jacq. ic. 2. t. 416.
7 longifòlium. w. (*wh.*) long-leaved. —— —— —— G. ♃. —— 2. t. 413.
8 graminifòlium.w.(*wh.*) grass-leaved. —— 1794. 6. 7. G. ♃. —— 2. t. 411.
9 vespertìnum. w.(*w.re.*) afternoon-flower'd. —— 1803. 5. 9. G. ♃. Bot. mag. t. 1040.

10 Jacquiniànum. R.S.(*w.*)Jacquin's. C. B. S. 1795. 5. 9. G.♃. Jacq. ic. 2. t. 412.
11 divaricàtum. R.S.(*wh.*) divaricate. ——— ——— ——— G.♃. Jacq. schœn. 4. t. 414.
12 revolùtum. W. (*wh.*) curled-flowered. ——— 1731. 9. 12. G.♃. Bot. mag. t. 1044.
13 floribúndum. W. (*st.*) thick-spiked. ——— 1774. 3. 4. G.♃.
14 squámeum. W. (*wh.*) scaly. ——— 1820. ——— G.♃.
15 filifórme. W. (*wh.gr.*) thread-leaved. ——— 1774. 4. G.♃.
16 elongàtum. R.S.(*wh.g.*) long-racemed. ——— 1822. ——— G.♃.
17 flexifòlium. W. (*wh.*) flexuose-leaved. ——— 1795. 5. 6. G.♃.

BULB'INE. L. BULB'INE. Hexandria Monogynia. L.
1 híspida. L. (*bh.gr.*) hairy-leaved. C. B. S. 1774. 5. 6. D.G.♃. Jacq. ic. 2. t. 409.
2 canaliculàta. S.S. (*wh.*) channel-leaved. ——— ——— 4. 5. D.G.♃. Bot. mag. t. 1124.
3 semibarbàta. S.S. (*ye.*) half bearded. N. S. W. 1820. 4. 8. D.G.♃. Lodd. bot. cab. t. 320.
Anthéricum semibarbàtum. B.P.
4 australis. S.S. (*ye.*) bulbous. ——— 1819. ——— D.G.b.
Anthéricum bulbòsum. B.P.
5 ánnua. S.S. (*ye.*) annual. C. B. S. 1731. 5. 6. H.⊙. Bot. mag. t. 1451.
Anthéricum ánnuum. B.M.
6 incúrva. R.S. (*ye.*) incurved-leaved. ——— 1822. ——— D.G.♃.
7 longiscàpa. S.S. (*ye.*) glaucous-leaved. ——— 1759. 4. 8. D.G.b. Bot. mag. t. 1339.
8 favòsa. R.S. (*ye.*) conical-rooted. ——— 1824. ——— D.G.♃.
9 asphodeloídes. S.S.(*ye.*) upright-leaved. ——— 1759. 6. 8. D.G.♂. Jacq. vind. t. 181.
10 rostràta. S.S. (*ye.*) glaucous shrubby. ——— 1812. 3. 8. D.G.♄. Jacq. ic. 2. t. 403.
11 frutéscens. W. (*ye.*) shrubby. ——— 1702. ——— D.G.♄. Bot. mag. t. 816.
12 pugionifórmis. S.S.(*ye.*) dagger-leaved. ——— 1793. 4. 6. D.G.b. ——— t. 1454.
13 bisulcàta. H.P. (*ye.*) two-channelled. ——— 1825. ——— D.G.b.
14 triquètra. S.S. (*wh.*) triquetrous. ——— ——— ——— D.G.♃.
15 scábra. S.S. (*wh.*) rough. ——— ——— ——— D.G.♃.
16 dùbia. R.S. (*wh.*) doubtful. ——— 1822. ——— D.G.♃.
17 ciliàta. S.S. (*wh.re.*) fringed. ——— ——— ——— D.G.♃.
18 caûda-fèlis. R.S. (*wh.*) Cats-tail. ——— 1825. ——— D.G.♃.
19 planifòlia. R.S. (*ro.w.*) two-coloured. S. Europe. 1824. ——— H.♃. Desf. atl. 1. t. 90.
Phalángium bícolor. Redout. liliac. t. 215.
20 mesembryanthoídes. H.P.(*ye.*)Figwort-like. C. B. S. 1824. ——— D.G.♃.
21 críspa. R.S. (*ye.*) curled-leaved. ——— 1825. ——— D.G.♃.
22 præmórsa. S.S. (*ye.*) bitten-rooted. ——— 1816. ——— D.G.b. Jacq. ic. 2. t. 406.
23 graminea. H.R. (*ye.*) Grass-leaved. ——— 1819. 6. 8. D.G.b.
24 alooìdes. R.S. (*ye.*) Aloe-leaved. ——— 1732. 4. 8. D.G.b. Bot. mag. t. 1317.
Anthéricum aloìdes. B.M.
25 nùtans. S.S. (*st.gr.*) nodding. ——— 1812. ——— D.G.♃. Jacq. ic. 2. t. 407.
26 latifòlia. S.S. (*ye.gr.*) broad-leaved. ——— ——— ——— D.G.♃. ——— 2. t. 408.

ARTHROP'ODIUM. B.P. ARTHROP'ODIUM. Hexandria Monogynia. S.S.
1 cirrhàtum. B.R. (*wh.*) large-flowered. N. Zealand. 1819. 5. 8. F.♃. Bot. reg. t. 709.
2 paniculàtum. B.P.(*w.pu.*)panicled. N. S. W. 1800. 6. 9. F.♃. Bot. mag. t. 1421.
minùs. Bot. reg. 866. *non* B.P. *Anthéricum milleflòrum.* Redout. liliac. t. 58.
3 minùs. B.P. (*wh.pu.*) simple-racemed. ——— 1819. ——— F.♃.
4 péndulum. S.S. (*ri.*) pendulous-flowr'd. N. Holland. 1817. ——— F.♃.
Anthéricum péndulum. W.en.
5 stríctum. B.P. (*wh.pu.*) upright. V. Diem. Isl. 1826. ——— G.♃.
6 fimbriàtum. B.P. (*wh.*) fringe-flowered. N. S. W. 1825. ——— F.♃.

STYPA'NDRA. B.P. STYPA'NDRA. Hexandria Monogynia.
1 glaùca. B.P. (*bl.*) glaucous. N. S. W. 1819. 4. 6. F.♃.
2 cæspitòsa. B.P. (*bl.*) turfy. ——— 1824. ——— F.♃.
3 umbellàta. B.P. (*bl.*) umbel-flowered. ——— 1820. ——— F.♃.

C'ÆSIA. B.P. C'ÆSIA. Hexandria Monogynia. S.S.
vittàta. B.P. (*wh.*) nodding-flowered. N.S.W. 1816. 7. 8. F.♃.

TRICO'RYNE. B.P. TRICO'RYNE. Hexandria Monogynia. S.S.
1 símplex. B.P. (*ye.*) simple-stalked. N. S. W. 1819. 7. 9. H.⊙.
2 elàtior. B.P. (*ye.*) tall. ——— 1825. ——— G.♃.

CHLORO'PHYTUM. B.P. CHLORO'PHYTUM. Hexandria Monogynia. S.S.
1 inornàtum. B.M. (*gr.*) green-flowered. Sierra Leone. .... 6. 8. S.♃. Bot. mag. t. 1071.
2 orchidástrum. B.R. (*wh.*)Orchis-like. ——— 1822. 1. 12. S.♃. Bot. reg. t. 813.
3 elàtum. B.P. (*wh.*) tall. C. B. S. 1751. 8. 9. G.♃. Redout. lil. t. 191.
Anthéricum elàtum. H.K. *Phalángium elàtum.* R.L.

PUSCHK'INIA. M.B. PUSCHK'INIA. Hexandria Monogynia. M.B.
scilloídes. M.B. (*w.bl.*) Squill-like. Iberia. 1819. 5. 6. H.b. Bot. mag. t. 2244.
Adámsia scilloídes. Lindl. coll. bot. t. 21.

ALB'UCA. Ker. ALB'UCA. Hexandria Monogynia. L.
1 altíssima. W. (*wh.gr.*) tall. C. B. S. 1780. 4. 5. F.b. Jacq. ic. 1. t. 63.
2 cornùta. R.L. (*wh.gr.*) three-horned. ——— 1812. ——— F.b. Redout. liliac. t. 70.
3 màjor. W. (*ye.gr.wh.*) great. ——— 1759. ——— F.b. Bot. mag. t. 804.

4 mìnor. W.	(*ye.gr.*)	small.	C. B. S.	1768.	5. 6.	F.b.	Bot. mag. t. 720.
5 fláccida. W.	(*ye.gr.*)	flaccid.	———	1791.	——	F.b.	Jacq. ic. 2. t. 444.
6 viridiflòra. W.	(*ye.gr.*)	green-flowered.	———	1794.	6. 7.	F.b.	Bot. mag. t. 1656.
7 coarctàta. W.	(*ye.*)	channel-leaved.	———	1774.	5. 6.	F.b.	
8 fastigiàta. W.	(*wh.*)	level-topped.	———	——	——	F.b.	Andr. reposit. t. 450.
9 caudàta. W.	(*wh.gr.*)	upright-flowered.	———	1791.	5. 7.	F.b.	Jacq. ic. 2. t. 442.
10 setòsa. W.	(*gr.ye.*)	bristly-rooted.	———	1795.	——	F.b.	Bot. mag. t. 1481.
11 aùrea. W.	(*ye.gr.*)	golden-flowered.	———	1825.	——	F.b.	Jacq. ic. 2. t. 441.
12 abyssínica. W.	(*w.y.g.*)	African.	———	——	——	F.b.	Redout. liliac. t. 195.
13 fràgrans. W.	(*ye.gr.*)	sweet-scented.	———	1791.	6. 7.	F.b.	Jacq. schœn. 1. t. 81.
14 viscòsa. W.	(*w.y.g.*)	clammy-leaved.	———	1779.	5. 6.	F.b.	Jacq. ic. 2. t. 445.
15 spiràlis. W.	(*st.gr.*)	spiral-leaved.	———	1795.	6.	F.b.	——— 2. t. 439.
16 vittàta. B.M.	(*ye.gr.*)	ribbon.	———	1802.	7. 8.	F.b.	Bot. mag. t. 1329.
17 parviflòra. R.S.	(*ye.gr.*)	small-flowered.	———	1812.	——	F.b.	
18 physòdes. B.M.	(*w.pu.*)	dingy-flowered.	———	1804.	6. 7.	F.b.	Bot. mag. t. 1046.
19 nematòdes.	(*w.pu.*)	slender-leaved.	———	1812.	——	F.b.	
Anthéricum nematòdes. R.S.							
20 exuviàta. B.M.	(*w.pu.*)	Adder skin.	———	1795.	5. 7.	F.b.	Bot. mag. t. 871.
Anthéricum exuviatum. Jacq. ic. 2. t. 415.							
21 anthericoídes.	(*ye.gr.*)	Anthericum-like.	———	1788.	8.	F.b.	
Anthéricum albucoides. W.							
22 filifòlia. Ker.	(*wh.gr.*)	filiform-leaved.	———	1825.	7. 8.	F.b.	Jacq. ic. 2. t. 414.
23 fùgax. B.R.	(*wh.pu.*)	fugacious.	———	1795.	4. 6.	F.b.	Bot. reg. t. 311.
Anthéricum fràgrans. Jacq. schœn. 1. t. 86.							
ORNITHO'GALUM. L. STAR OF BETHLEHEM.			Hexandria Monogynia. L.				
1? uniflòrum. W.	(*ye.*)	one-flowered.	Siberia.	1781.	5. 6.	H.b.	Nov. act. pet. 18. t. 6. f. 3.
2 ixioídes. H.K.	(*wh.*)	Ixia-like.	California.	1796.	——	H.b.	
3 montànum. T.N.	(*w.g.*)	mountain.	Naples.	1825.	——	H.b.	Swt. br. fl. gar. s. 2. t. 42.
4 collìnum. R.S.	(*wh.gr.*)	hill.	Sicily.	1829.	——	H.b.	
5 umbellatum. E.B.	(*w.g.*)	common.	England.	....	4. 6.	H.b.	Eng. bot. v. 2. t. 130.
6 refractum. W. en.	(*w.g.*)	reflexed.	S. Europe.	1825.	——	H.b.	Swt. br. fl. gar. s. 2. t. 58.
7 tenuifòlium. R.S.	(*w.g.*)	slender-leaved.	Sicily.	1830.	——	H.b.	
8 exscàpum. T.N.	(*w.g.*)	short-stalked.	Naples.	1823.	——	H.b.	Ten. fl. nap. v. 1. t. 34.
9 fimbriàtum. M.B.	(*w.g.*)	fringed.	Tauria.	1820.	2. 3.	H.b.	Swt. br. fl. gar. s. 2.
10 comòsum. W.	(*wh.gr.*)	short-spiked.	Austria.	1596.	6. 8.	H.b.	Jacq. ic. 2. t. 426.
11 cònicum. W.	(*wh.*)	conical.	C. B. S.	1818.	——	F.b.	——— 2. t. 428.
12 níveum. W.	(*wh.*)	snowy.	———	1774.	——	F.b.	Bot. reg. t. 235.
13 ovàtum. W.	(*wh.gr.*)	oval-leaved.	———	1816.	4. 8.	F.b.	
14 notàtum. R.S.	(*ye.br.*)	brown-marked.	———	1825.	6. 8.	F.b.	Jacq. coll. 2. t. 18. f. 13.
maculàtum. Jacq. *non* Thunb.							
15 ciliàtum. W.	(*wh.*)	ciliated.	———	1818.	4. 6.	F.b.	
16 crenulàtum. W.	(*st.*)	crenulate.	———	1825.	6. 8.	F.b.	
17 maculàtum. R.S.	(*sn.br.*)	spotted-flowered.	———	——	——	F.b.	
18 secúndum. R.S.	(*st.br.*)	one-ranked.	———	1816.	——	F.b.	Jacq. ic. 2. t. 433.
19 graminifolium. R.S.	(*w.g.*)	grass-leaved.	———	——	——	F.b.	
20 tenéllum. W.	(*st.gr.*)	slender.	———	1816.	5. 8.	F.b.	Jacq. ic. 2. t. 427.
21 pilòsum. W.	(*wh.*)	hairy-leaved.	———	——	——	F.b.	
22 revolùtum. W.	(*wh.gr.*)	revolute-flowered.	———	1795.	4. 8.	F.b.	Bot. reg. t. 315.
23 corymbòsum. B.R.	(*wh.*)	corymbose.	Peru.	1822.	——	H.b.	——— t. 906.
24 Rudólphi. R.S.	(*wh.*)	Rudolph's.	C. B. S.	1812.	——	F.b.	Jacq. ecl. t. 20.
tenuifòlium. Redout. liliac. t. 312. *non* Guss.							
25 prasìnum. B.R.	(*gr.*)	green-flowered.	———	1816.	6. 7.	F.b.	Bot. reg. t. 158.
26 vìrens. B.R.	(*gr.*)	greenish.	———	1823.	——	F.b.	——— t. 814.
27 suavèolens. W.	(*gr.ye.*)	fragrant.	———	1816.	——	F.b.	Jacq. ic. 2. t. 431.
28 barbàtum. W.	(*ye.gr.*)	bearded.	———	1795.	5. 7.	F.b.	Jacq. schœn. 1. t. 91.
29 polyphy'llum. W.	(*g.y.*)	many-leaved.	———	1825.	——	F.b.	Jacq. ic. 2. t. 430.
30 rupéstre. W.	(*wh.gr.*)	rock.	———	1795.	5. 8.	F.b.	
31 juncifòlium. W.	(*w.pu.*)	Rush-leaved.	———	1794.	7. 8.	F.b.	Bot. mag. t. 792.
32 fuscàtum. W.	(*br.gr.*)	brown-flowered.	———	1825.	4. 6.	F.b.	Jacq. ic. 2. t. 429.
33 odoràtum. W.	(*gr.ye.*)	sweet-scented.	———	1795.	5. 6.	F.b.	Andr. reposit. t. 260.
34 elàtum. A.R.	(*wh.*)	Egyptian.	Egypt.	1804.	6. 7.	H.b.	——— t. 528.
35 pyramidàle. W.	(*wh.*)	pyramidal.	Spain.	1752.	——	H.b.	Jacq. ic. 2. t. 425.
36 latifòlium. W.	(*wh.*)	broad-leaved.	Egypt.	1629.	——	H.b.	Bot. mag. t. 876.
37 longebracteàtum. W.	(*w.*)	long-bracted.	C. B. S.	1812.	7. 12.	F.b.	Jacq. vind. 1. t. 29.
38 caudàtum. W.	(*wh.gr.*)	long-spiked.	———	1774.	2. 8.	F.b.	Bot. mag. t. 805.
39 scilloídes. W.	(*wh.gr.*)	Squill-like.	———	1795.	6. 7.	F.b.	Jacq. schœn. 1. t. 888.
40 trig'ynum. R.L.	(*wh.gr.*)	three styled.	....	1825.	——	F.b.	Redout. liliac. t. 417.
41 narbonénse. W.	(*w.g.*)	Narbonne.	S. Europe.	1810.	7. 8.	H.b.	Bot. mag. t. 2510.
42 pyrenàicum. W.	(*st.*)	spiked.	England.	....	6. 7.	H.b.	Eng. bot. v. 7. t. 499.

43 sulfùreum. R.S. (*st.*) straw-coloured. Hungary. 1818. 6. 10. H.b. Bot. mag. t. 2623.
Anthéricum sulphùreum. Walds. et Kit. pl. hung. 1. t. 95. *Phalángium sulfùreum.* Poiret.
44 stachyoídes. W. (*w.g.*) close-spiked. S. Europe. 1771. 4. 7. H.b. Renealm. spec. t. 90.
45 lácteum. W. (*st.gr.*) milk white. C. B. S. 1796. 6. 7. F.b. Bot. mag. t. 1134.
46 nùtans. W. (*wh.gr.*) nodding. Britain. .... 4. 5. H.b. Eng. bot. v. 28. t. 1997.
47 coarctàtum. W. (*wh.*) close-flowered. C. B. S. 1804. 6. 7. F.b. Jacq. ic. 2. t. 435.
48 aùreum. W. (*go.*) golden. —— 1790. —— F.b. Bot. mag. t. 190.
49 flavíssimum. Jacq. (*ye.*) great yellow. —— —— —— F.b. Jacq. ic. 2. t. 436.
50 miniàtum. Jacq. (*ye.*) stained. —— —— —— F.b. —— t. 438.
51 thyrsoídes. W. (*wh.br.*) thyrse-flowered. —— 1757. —— F.b. Bot. mag. t. 1164.
α álbum. (*wh.br.*) *white-flowered.* —— —— —— F.b. Bot. reg. t. 316.
β flavéscens. (*st.br.*) *pale-yellow.* —— —— —— F.b. —— t. 305.
52 bícolor. R.S. (*wh.br.*) two-coloured. —— 1816. —— F.b.
53 tuberòsum. R.S. (*ye.*) large-rooted. —— 1787. —— F.b.
54 arábicum. W. (*wh.*) great-flowered. Egypt. 1629. 3. 4. F.b. Bot. mag. t. 728.
55 brachystàchys. F. (*wh.*) short-spiked. Dahuria. 1821. —— H.b.
56 Bérgii. R.S. (*wh.gr.*) Bergius's. —— 1816. —— F.b.
57 marítimum. Lam. (*wh.*) officinal Squill. S. Europe. 1629. 4. 5. F.b. Bot. mag. t. 918.
Squílla. B.M. *Scílla marítima.* W.
58 concínnum. S.P. (*wh.*) neat. Portugal. 1797. —— H.b. Bot. mag. t. 953.
59 unifòlium. B.M. (*wh.*) one-leaved. Gibraltar. 1805. 6. 7. F.b. —— t. 935.

G'AGEA. S.A.B. G'AGEA. Hexandria Monogynia.
1 fasciculàris. S.A.B. (*ye.*) bunch-rooted. Britain. .... 3. 4. H.b. Bot. mag. t. 1200.
lùtea. B.M. 1200. *Ornithógalum lùteum.* Eng. bot. v. 1. t. 21. *non* L.
2 glaùca. B.F.G. (*ye.*) glaucous-leaved. Switzerland. 1825. —— H.b. Swt. br. fl. gar. t. 177.
3 bracteolàris. S.A.B. (*ye.*) sheathing-bracted. —— .... —— H.b. —— t. 158.
Ornithógalum lùteum. L. *non* E.B. V. S. in herb. linn.
4 mínima. R.S. (*ye.*) corn-field. Europe. 1825. —— H.b. Swt. br. fl. gar. s. 2. t. 29.
stellàris. S.A.B. *Ornithógalum arvénse.* S.S *Sternbérgii.* Hop. V. S. in herb. linn.
5 spathàcea. R.S. (*ye.*) sheathed. Germany. 1759. 5. H.b. Flor. græc. t. 331.
6 reticulàta. R.S. (*ye.*) netted-sheathed. Tauria. 1829. —— H.b. Pall. it. 3. ap. t. D. f. 2.
7 pusílla. R.S. (*ye.*) small. Bohemia. 1823. —— H.b. Reichenb. ic. t. 117.
8 Clusiàna. R.S. (*ye.*) Clusius's. S. Europe. .... —— H.b.
9 podòlica. R.S. (*ye.*) Polish. Poland. 1827. —— H.b.
10 erubéscens. R.S. (*y.bh.*) blush. —— —— —— H.b.
11 Liotárdi. R.S. (*ye.*) fistulous-leaved. S. Europe. 1825. —— H.b. Sternb. denks. 2. t. 3.
Ornithógalum fistulòsum. Redout. liliac. t. 221.
12 pygm'æa. R.S. (*ye.*) pigmy. Spain. 1826. —— H.b.
13 arvénse. R.S. (*ye.*) villous-stalked. Europe. .... —— H.b. Flor. græc. t. 332.
Ornithógalum villòsum. M.B.
14 bohèmicum. R.S. (*ye.g.*) Bohemian. Bohemia. 1828. —— H.b. Sturm. flor. t. 23.
15 saxátilis. R.S. (*gr.ye.*) rock. Germany. —— —— H.b.
16 bulbífera. R.S. (*ye.g.*) bulbiferous. Tauria. 1829. —— H.b. Reichenb. ic. t. 117.

BARNA'RDIA. B.R. BARNA'RDIA. Hexandria Monogynia.
1 scilloídes. B.R. (*ro.*) Squill-like. China. 1824. 9. 10. H.b. Bot. reg. t. 1029.
2 japónica. R.S. (*pu.*) Japan. Japan. 1821. 6. 8. H.b.
Onithógalum japónicum. W.

LLO'YDIA. Sal. LLO'YDIA. Hexandria Monogynia.
1 serotìna. (*wh.std.*) mountain. Wales. .... 6. 7. H.b. Eng. bot. v. 12. t. 793.
Anthéricum serotìnum. E.B. *Phalángium serotìnum.* Redout. liliac. t. 270. *Gàgea serotìna.* Ker.
2 striàta. (*wh.std.*) stripe-flowered. Siberia. 1789. 5. 7. H.b.
Ornithógalum striàtum. S.S.

SCI'LLA. W. SQUILL. Hexandria Monogynia. L.
1 non scrípta. L.en. (*bl.*) Harebells. Britain. .... 3. 6. H.b. Eng. bot. v. 6. t. 377.
nùtans. E.B. *Hyacínthus non scríptus.* L.
2 cérnua. R.S. (*ro.*) flesh-coloured. Spain. .... —— H.b. Bot. mag. t. 1461.
3 pátula. R.S. (*bl.*) spreading. S. Europe. 1633. 4. 6. H.b. Redout. liliac. t. 225.
campanulàta. B.M. t. 1102. *non* t. 128.
4 campanulàta. W. (*bl.*) bell-flowered. Spain. —— 5. 6. H.b. Bot. mag. t. 128.
5 corymbòsa. S.S. (*ro.*) Cape. C. B. S. 1793. 8. 12. F.b. —— t. 991.
Massònia corymbòsa. B.M. *Hyacínthus corymbòsus.* L.
6 brevifòlia. R.S. (*wh.*) short-leaved. —— 1822. 1. 3. F.b.
7 brachyphy'lla. R.S. (*bh.*) blush. —— 1811. —— F.b. Bot. mag. t. 1468.
Scílla brevifòlia. B.M. *non* Thunb.
8 sibírica. H.K. (*bl.*) Siberian. Siberia. 1796. 2. 3. H.b. Andr. rep. t. 365.
9 am'œnula. R.S. (*bl.*) pretty. Russia. 1822. 3. 4. H.b. Bot. mag. t. 2408.
10 am'œna. R.S. (*bl.*) nodding. Levant. 1596. —— H.b. —— t. 341.
11 pùmila. R.S. (*pa.bl.*) small. Portugal. 1819. 4. 5. H.b. Brot. phyt. lus. t. 46. f. 2.
monophyllos. Link *ex* R.S.

12 bifòlia. W.	(*va.*)	two-leaved.	England.	....	2. 4.	H.b.	Eng. bot. v. 1. t. 24.
α cœrùlea.	(*bl.*)	*blue-flowered.*	———	....	——	H.b.	
β álba.	(*wh.*)	*white-flowered.*	———	....	——	H.b.	
13 cárnea.	(*ro.*)	rosy.	....	——	——	H.b.	Bot. mag. t. 746. f.sinis.
14 pr'æcox. W.	(*bl.*)	early-flowering.	....	1790.	——	H.b.	
15 Lílio-hyacínthus. W.	(*bl.*)	Lily-rooted.	S. Europe.	1597.	5. 7.	H.b.	Redout. lil. t. 205.
16 vérna. W.	(*bl.*)	vernal.	Britain.	....	4. 5.	H.b.	Eng. bot. v. 1. t. 23.
17 umbellàta. S.S.	(*bl.*)	umbel-flowered.	Pyrenees.	1817.	——	H.b.	Redout. liliac. t. 166.
18 itálica. S.S.	(*bl.*)	Italian.	Italy.	1605.	4. 7.	H.b.	Bot. mag. t. 663.
19 autumnàlis. W.	(*pu.*)	autumnal.	England.	....	8.	H.b.	Eng. bot. v. 2. t. 78.
20 intermèdia. R.S.	(*bl.*)	intermediate.	Sicily.	1829.	——	H.b.	
21 obtusifòlia. R.S.	(*vi.bl.*)	blunt-leaved.	Barbary.	1828.	——	H.b.	Redout liliac. 4. t. 190.
22 undulàta. R.S.	(*li.*)	wave-leaved.	S. Europe.	——	——	H.b.	Desf. atl. 1. t. 47.
23 lingulàta. R.S.	(*bl.*)	tongue-leaved.	Barbary.	——	——	H.b.	Redout. liliac. t. 321.
24 vincentìna. R.S.	(*bl.*)	spreading-flowered.	Portugal.	1827.	——	H.b.	
25 peruviàna. W.	(*va.*)	corymbose.	Spain.	1607.	5.	H.b.	Bot. mag. t. 749.
α cœrùlea.	(*bl.*)	*blue-flowered.*	———	——	——	H.b.	——— t. 749.
β álba.	(*wh.*)	*white-flowered.*	———	——	——	H.b.	
26 Cupaniána. R.S.	(*bl.*)	Cupani's.	Sicily.	1826.	——	H.b.	Cup.panph.sic.5.t.20.1.
27 lusitánica. B.M.	(*pa.bl.*)	Portugal.	Portugal.	1777.	——	H.b.	Bot. mag. t. 1999.
28 hyacinthoídes. W.	(*bl.*)	Hyacinth-like.	Madeira.	1585.	8.	H.b.	——— t. 1140.
29 esculénta. B.M.	(*pa.bl.*)	Quamash.	N. America.	1811.	5. 7.	H.b.	——— t. 1574.
Phalángium Quamash. Ph.							
HYACI'NTHUS. S.S.		HYACINTH.	Hexandria Monogynia. L.				
1 orientàlis. W.	(*bl.*)	common.	Levant.	1596.	3. 5.	H.b.	Bot. reg. t. 995.
α cœrùleus.	(*bl.*)	*blue-flowered.*	———	——	——	H.b.	
β l'yricus.	(*bl.*)	*Lyra grandis.*	———	....	——	H.b.	Swt. flor. guid. t. 96.
γ rùber.	(*re.*)	*red-flowered.*	———	1596.	——	H.b.	
δ flàvus.	(*ye.*)	*yellow-flowered.*	———	——	——	H.b.	
ε álbus.	(*wh.*)	*white-flowered.*	———	——	——	H.b.	
ζ semiplènus.	(*va.*)	*semidouble.*	———	——	——	H.b.	
η plènus.	(*va.*)	*double-flowered.*	———	——	——	H.b.	
θ múltiplex.	(*va.*)	*full-flowered.*	———	——	——	H.b.	
ι atrocœrùleus.	(*bl.*)	*Pourpre foncé.*	———	....	——	H.b.	Swt. flor. guid. t. 99.
κ sceptriformis.	(*pa.bl.*)	*Porcelaine sceptre.*	———	....	——	H.b.	——— t. 51.
λ sanguinoléntus.	(*re.*)	*Waterloo.*	———	....	——	H.b.	——— t. 1.
μ ornàtus.	(*pi.*)	*L'honneur d'Amsterdam.*		....	——	H.b.	——— t. 108.
ν ranunculiflòrus.	(*ye.*)	*yellow Ophir.*	———	....	——	H.b.	——— t. 123.
2 amethy'stinus. W.	(*bl.*)	Amethyst-coloured.	S. Europe.	1759.	4. 5.	H.b.	Swt. br. fl. gar. t. 135.
UROPE'TALON. B.R.		UROPE'TALON.	Hexandria Monogynia. B.R.				
1 viríde. B.R.	(*gr.*)	green-flowered.	C. B. S.	1774.	8.	F.b.	Redout. lil. t. 203.
Hyacínthus víridis. L. *Zuccágnia víridis.* Th. *Lachenàlia víridis.* W.							
2 glaùcum. B.R.	(*gr.br.*)	glaucous-leaved.	C. B. S.	1816.	7. 8.	F.b.	Bot. reg. t. 156.
3 críspum. B.R.	(*br.gr.*)	curled-leaved.	———	——	——	F.b.	
4 longifòlium. B.R.	(*br.gr.*)	long-leaved.	———	1822.	——	F.b.	Bot. reg. t. 974.
5 hyacinthoìdes. R.S.	(*re.*)	Hyacinth-like.	———	——	——	F.b.	
6 serotìnum. B.R.	(*bl.*)	late-flowering.	Spain.	1629.	6. 8.	F.b.	Bot. mag. t. 859.
7 fúlvum. H.S.L.	(*re.br.*)	tile-red.	Mogador.	1808.	6. 9.	F.b.	——— t. 1185.
Scílla serotìna β. B.M. *Hyacínthus fúlvus.* Cav.							
BELLEV'ALIA. Lap.		BELLEV'ALIA.	Hexandria Monogynia.				
romàna.	(*wh.*)	Roman.	Italy.	1596.	5.	H.b.	Bot. mag. t. 939.
operculàta. Lap. *Scílla romàna.* B.M. *Hyacínthus romànus.* L.							
MUSC'ARI. S.S.		GRAPE HYACINTH.	Hexandria Monogynia. S.S.				
1 moschàtum. S.S.	(*br.*)	musk.	Levant.	1596.	4. 5.	H.b.	Bot. mag. t. 734.
2 ambrosìacum. R.L	(*wh.*)	strong-scented.		....	——	H.b.	Redout. liliac. t. 132.
3 macrocárpum. B.F.G.	(*ye.*)	large-capsuled.	Turkey.	1812.	——	H.b.	Swt. br. fl. gar. t. 210.
moschàtum β flàvum. Bot. mag. t. 1565.							
4 comòsum. S.S.	(*br.pu.*)	purple.	S. Europe.	1596.	——	H.b.	Bot. mag. t. 133.
β monstròsum.	(*pu.*)	*feathered.*	———	——	——	H.b.	Moris. s. 4. t. 11. f. 2.
5 ciliàtum. S.S.	(*w.pu.*)	ciliated.	Tauria.	1818.	——	H.b.	Bot. reg. t. 394.
6 pállens. S.S.	(*pa.bl.*)	pale-flowered.		——	——	H.b.	Swt. br. fl. gar. t. 259.
7 glaùcum. B.R.	(*pa.*)	glaucous-leaved.	Persia.	1825.	——	H.b.	Bot. reg. t. 1085.
8 racemòsum. S.S.	(*bl.*)	starch.	Britain.	....	——	H.b.	Eng. bot. v. 27. t. 1391.
9 commutàtum. R.S.	(*d.bl.*)	confused.	Sicily.	....	——	H.b.	
10 parviflòrum. R.S.	(*bl.*)	small-flowered.	———	1827.	——	H.b.	
11 botryoídes. S.S.	(*va.*)	bunch-flower'd.	Italy.	1596.	——	H.b.	Bot. mag. t. 157.
α azùreum. B.F.G.	(*bl.*)	*bright-blue.*	———	——	——	H.b.	Swt. br. fl. gar. 15. f. a.
β pállidum. B.F.G.	(*pa.*)	*pale blue.*	———	——	——	H.b.	——— 15. f. c.
γ álbum. B.F.G.	(*wh.*)	*white-flowered.*	———	——	——	H.b.	——— 15. f. b.

	LACHEN'ALIA. W.		LACHEN'ALIA.	Hexandria Monogynia. W.				
1	péndula. w.	(*y.re.g.*)	pendulous.	C. B. S.	1774.	3. 4.	F.b.	Jacq. ic. 2. t. 400.
	α viridifòlia.	(*y.re.g.*)	*green-leaved.*	———	——	——	F.b.	Bot. mag. t. 590.
	β maculàta.	(*y.re.g.*)	*spotted-leaved.*	———	——	——	F.b.	
2	quadrícolor. B.M.	(*va.*)	four-coloured.	———	1789.	3. 5.	F.b.	Jacq. ic. 2. t. 388.
	α viridifòlia.	(*y.sc.g.*)	*green-leaved.*	———	——	——	F.b.	Bot. mag. t. 588.
	β colorata.	(*y.sc.g.*)	*coloured-leaved.*	———	——	——	F.b.	——— t. 1097.
3	lutèola. Jacq.	(*ye.*)	yellow-flowered.	———	1774.	4. 5.	F.b.	Jacq. ic. 2. t. 395.
	α viridifòlia.	(*ye.*)	*green-leaved.*	———	——	——	F.b.	Bot. mag. t. 1020.
	β maculàta.	(*ye.*)	*spotted-leaved.*	———	——	——	F.b.	——— t. 1704.
4	trícolor. B.M.	(*y.re.g.*)	three-coloured.	———	——	——	F.b.	——— t. 82.
5	rùbida. W.	(*re.*)	dotted-flowered.	———	1803.	9. 10.	F.b.	——— t. 993.
6	punctàta. S.S.	(*w.re.*)	spotted-flowered.	———	1825.	6. 8.	F.b.	Jacq. ic. 2. t. 397.
7	flàva. A.R.	(*ye.*)	bright-yellow.	———	1805.	5. 6.	F.b.	Andr. reposit. t. 77.
8	isopétala. W.	(*pa.br.*)	equal-flowered.	———	1804.	——	F.b.	Jacq. ic. 2. t. 401.
9	hírta. R.S.	(*w.bl.*)	hairy-leaved.	———	1825.	——	F.b.	
10	sessiliflòra. A.R.	(*bh.*)	sessile-flowered.	———	1804.	——	F.b.	Andr. rep. t. 460.
11	unifòlia. W.	(*bl.re.*)	one-leaved.	———	1795.	3. 4.	F.b.	Bot. mag. t. 766.
12	ròsea. A.R.	(*ro.*)	rose-coloured.	———	1800.	4. 5.	F.b.	Andr. rep. t. 296.
13	pusílla. W.	(*wh.*)	dwarf.	———	1825.	——	F.b.	Jacq. ic. 2. t. 385.
14	refléxa. R.S.	(*ye.*)	reflex-leaved.	———	——	——	F.b.	
15	bifòlia. B.M.	(*ro.wh.*)	cowl-leaved.	———	1813.	3. 4.	F.b.	Bot. mag. t. 1611.
16	contáminata. W.	(*w.br.*)	contaminated.	———	1774.	2. 4.	F.b.	——— t. 1401.
17	purpùrea. W.	(*pu.w.*)	purple.	———	1819.	4. 5.	F.b.	Jacq. ic. 2. t. 393.
18	violàcea. W.	(*g.vi.*)	violet.	———	1795.	3. 4.	F.b.	——— 2. t. 394.
19	lùcida. B.M.	(*w.br.*)	glossy-leaved.	———	1798.	3. 5.	F.b.	Bot. mag. t. 1372.
20	racemòsa. B.M.	(*wh.*)	racemed.	———	1811.	5.	F.b.	——— t. 1517.
21	fràgrans. W.	(*wh.*)	sweet-scented.	———	1798.	3. 5.	F.b.	Jacq. schœn. 1. t. 82.
22	unícolor. B.M.	(*li.*)	self-coloured.	———	1806.	5. 6.	F.b.	Bot. mag. t. 1373.
23	anguínea. S.F.G.	(*w.g.*)	snake.	———	1825.	4. 5.	F.b.	Swt. br. fl. gar. t. 179.
24	purpureo-cœrulea.	(*bl.*)	purple-blue.	———	1798.	——	F.b.	Bot. mag. t. 745.
25	pustulàta. W.	(*w.st.*)	blistered.	———	1790.	1. 4.	F.b.	——— t. 817.
26	liliiflòra. W.	(*wh.*)	Lily-flowered.	———	1820.	4. 6.	F.b.	Jacq. ic. 2. t. 387.
27	orthopétala. R.S.	(*w.gr.*)	wedge-flowered.	———	1825.	——	F.b.	——— 2. t. 383.
28	pátula. W.	(*wh.ro.*)	spreading-flower'd.	———	1795.	4. 5.	F.b.	——— 2. t. 384.
29	nervòsa. B.M.	(*br.pu.*)	nerved-leaved.	———	1810.	6.	F.b.	Bot. mag. t. 1497.
30	angustifòlia. W.	(*wh.*)	narrow-leaved.	———	1793.	4. 5.	F.b.	——— t. 735.
31	hyacinthoídes. W.	(*wh.*)	Hyacinth-flower'd.	———	1812.	3. 4.	F.b.	Jacq. ic. 2. t. 382.
32	pállida. W.	(*pa.bl.*)	pale-flowered.	———	1782.	——	F.b.	Bot. reg. t. 314.
33	mediána. R.S.	(*st.gr.*)	intermediate.	———	——	——	F.b.	Jacq. ic. t. 392.
34	mutabilis. S.F.G.	(*ye.*)	changeable.	———	1825.	2. 5.	F.b.	Swt. br. fl. gar. ser. 2.
35	orchioídes. W.	(*ye.*)	Orchis-like.	———	1752.	2. 4.	F.b.	Bot. mag. t. 1269.
36	glaucìna. W.	(*wh.bl.*)	sea-green.	———	1795.	3. 4.	F.b.	Jacq. ic. 2. t. 391.
	DRI'MIA. S.S.		DRI'MIA.	Hexandria Monogynia. S.S.				
1	lanceolàta. R.S.	(*y.gr.*)	spear-leaved.	C. B. S.	1774.	8. 10.	F.b.	Andr. reposit. t. 299.
	Lachenàlia refléxa. Andr. rep. t. 299.							
2	lanceæfòlia. B.M.	(*g.pu.*)	lance-leaved.	C. B. S.	1800.	9. 10.	F.b.	Bot. mag. t. 643.
3	longipedunculàta.	(*g.pu.*)	long-stalked.	———	——	——	F.b.	Schr. pl. rar. mon. t. 100.
4	acuminàta. B.C.	(*g.br.*)	taper-leaved.	———	——	——	F.b.	Lodd. bot. cab. t. 1041.
5	Gawlèri. R.S.	(*gr.pu.*)	revolute-flowered.	———	1774.	——	F.b.	Bot. mag. t. 1380.
	Hyacínthus revolùtus. H.K.							
6	ovalifòlia. R.S.	(*gr.pu.*)	oval-leaved.	———	1812.	——	F.b.	Lodd. bot. cab. t. 278.
7	undulàta. S.S.	(*g.pu.*)	wave-leaved.	———	1825.	8. 10.	F.b.	Jacq. ic. 2. t. 376.
8	purpurascens. S.S.	(*pu.*)	purplish.	———	——	——	F.b.	
9	ciliàris. S.S.	(*g.pu.*)	ciliated.	———	....	——	F.b.	Bot. mag. t. 1444.
10	pusílla. S.S.	(*gr.*)	dwarf.	———	1793.	5. 6.	F.b.	Jacq. ic. 2. t. 374.
11	mèdia. S.S.	(*w.pu.*)	semicylindrical-ld.	———	1818.	7. 8.	F.b.	——— 2. t. 375.
12	elàta. S.S.	(*g.w.*)	tall.	———	1799.	10. 11.	F.b.	Bot. mag. t. 822.
13	altíssima. S.S	(*wh.*)	tallest.	———	1791.	8. 9.	F.b.	——— t. 1074.
	ERIOSPE'RMUM. W.		ERIOSPE'RMUM.	Hexandria Monogynia. W.				
1	parvifòlium. S.S.	(*wh.*)	small-leaved.	C. B. S.	1795.	6. 8.	F.b.	Jacq. ic. 2. t. 422.
2	latifòlium. S.S.	(*wh.*)	broad-leaved.	———	1800.	——	F.b.	——— 2. t. 420.
3	Bellendèni.	(*wh.*)	Mr. Ker's.	———	——	——	F.b.	Bot. mag. t. 1382.
	latifòlium α. B.M. *non* Jacq.							
4	lanceæfòlium. S.S.	(*st.*)	lance-leaved.	———	1795.	——	F.b.	Jacq. ic. 2. t. 421.
5	pubéscens. S.S.	(*wh.*)	pubescent.	———	1820.	——	F.b.	Bot. reg. t. 578.
6	lanuginòsum. S.S.	(*st.*)	woolly-leaved.	———	——	——	F.b.	Jacq. schœn. 3. t. 265.
7	foliolíferum. B.M.	(*w.g.*)	leaflet-bearing.	———	1806.	——	F.b.	Andr. reposit. t. 521.
8	paradóxicum. B.M.	(*w.g.*)	paradoxical.	———	1820.	——	F.b.	Jacq. col. sup. 81. t. 1.

MASS'ONIA. S.S. Mass'onia. Hexandria Monogynia. W.
1 grandiflòra. B.R. (*wh.*) honey-bearing. C. B. S. 1825. 1. 4. F.b. Bot. reg. t. 958.
2 cándida. B.T. (*wh.*) white-flowered. ——— 1816. 2. 5. F.b. ——— t. 694.
longifòlia β. cándida. B.R.
3 longifòlia. s.s. (*wh.*) long-leaved. ——— —— —— F.b.
4 latifòlia. s.s. (*g.pu.*) broad-leaved. ——— 1775. 3. 4. F.b. Bot. mag. t. 848.
5 cordàta. s.s. (*wh.*) heart-leaved. ——— 1820. —— F.b.
6 muricàta. s.s. (*wh.*) prickly-leaved. ——— 1790. 4. 5. F.b. Bot. mag. t. 559.
7 scábra. H.K. (*wh.*) shagreen-leaved. ——— —— 1. 4. F.b. Andr. reposit. t. 220.
pustulàta. Bot. mag. t. 642.
8 echinàta. s.s. (*wh.*) rough-leaved. ——— —— 5. F.b.
9 pauciflòra. s.s. (*wh.*) few-flowered. ——— —— —— F.b.
10 angustifòlia. s.s. (*wh.*) narrow-leaved. ——— 1775. 3. 4. F.b. Bot. mag. t. 736.
11 undulàta. s.s. (*wh.*) wave-leaved. ——— 1791. 4. F.b.
12 ensifòlia. s.s. (*pi.*) trumpet-flowered. ——— 1790. 9. 2. F.b. Bot. mag. t. 554.
violàcea. A.R. 46. *Agapánthus ensifòlius.* W. *Hyacinthus bifòlius.* Cav.

E'UCOMIS. S.S. E'ucomis. Hexandria Monogynia. W.
1 nàna. s.s. (*gr.*) dwarf. C. B. S. 1774. 5. F.b. Bot. mag. t. 1495.
2 bifòlia. s.s. (*gr.st.*) two-leaved. ——— 1792. 4. 5. F.b. ——— t. 840.
3 purpureocaùlis. s.s. (*gr.*) purple-stalked. ——— 1794. 3. 4. F.b. Andr. reposit. t. 369.
4 règia. s.s. (*g.wh.*) tongue-leaved. ——— 1702. 3. 5. F.b. Dill. elth. t. 92. f. 109.
5 undulàta. s.s. (*g.wh.*) wave-leaved. ——— 1760. —— F.b. Bot. mag. t. 1083.
6 punctàta. s.s. (*g.wh.*) spotted. ——— 1783. 7. F.b. ——— t. 913.
7 striàta. H.K. (*g.wh.*) streaked. ——— 1790. 6. 12. F.b. ——— t. 1539.

A'LLIUM. L. Garlic. Hexandria Monogynia. L.
1 satìvum. G.D. (*re.*) common. Sicily. 1548. 6. 9. H.b. Lobel. ic. t. 158.
2 Ophioscórodon. G.D. (*br.*) Rocambole. Crete. 1596. 8. H.b. Clus. his. p. 191. t. 210.
3 contravérsum. G.D. (*pu.*) simple-rooted. Egypt. 1823. —— H.b. Cord. hist. p. 144. f. 1.
Ophioscórodon. L.en.
4 arenàrium. G.D. (*re.*) sand. Britain. .... 6. 8. H.b. Eng. bot. v. 19. t. 1358.
5 Scorodopràsum. G.D. (*re.*) wild Rocambole. Denmark. 1596. 7. H.b. Flor. dan. t. 219. et 1455.
6 littòreum. G.D. (*re.*) Italian sand. Italy. 1824. 8. H.b.
monspessulànum. W.
7 vineàle. G.D. (*ro.*) Crow. Britain. .... 6. 8. H.b. Eng. bot. v. 28. t. 1974.
8 compáctum. G.D. (*ro.*) two-headed crow. France. 1819. —— H.b. Moris. sect. 4. t. 15. f. 3.
9 Púrshii. G.D. (*da.pu.*) Pursh's crow. N. America. 1824. —— H.b.
vineàle. Ph. *nec aliorum.*
10 prolíferum. G.D. (*wh.*) tree Onion. Egypt. .... 7. H.b. Bot. mag. t. 1469.
Cèpa β bulbífera. B.M.
11 Pórrum. G.D. (*gr.pu.*) Leek. Switzerland. 1562. 5. 7. H.b. Plenck. ic. t. 421.
12 Ampelopràsum. G.D. (*pa.*) blue Leek. England. .... 7. 8. H.b. Eng. bot. v. 24. t. 1657.
13 rotúndum. G.D. (*pu.*) round-headed. S. Europe. 1819. 6. 7. H.b. Clus. hist. p. 195. t. 220.
14 Hallèrii. G.D. (*pa.pu.*) Haller's. ——— 1818. —— H.b. Flor. græc. t. 312.
15 rubicúndum. G.D. (*re.*) marbled leek. C. B. S. .... 7. 8. F.b. Bot. mag. t. 1560.
ampelopràsum β. B.M.
16 Waldstèini. G.D. (*d.pu.*) Waldstein's. Smyrna. 1824. 6. 7. H.b. W. et K. hung. t. 82.
17 exsértum. G.D. (*ro.*) long-stamen'd. Russia. 1819. 7. 8. H.b.
18 eréctum. G.D. (*wh.*) upright leek. C. B. S. 1824. —— F.b.
19 acutiflòrum. G.D. (*re.*) sharp-flowered. France. —— —— H.b. Moris. sect. 4. t. 15. f. 4.
20 Synnótii. G.D. (*li.*) Synnot's. C. B. S. 1825. 6. 7. F.b.
21 verrucòsum. G.D. (*li.*) warted-flower'd. ——— —— F.b.
22 multiflòrum. G.D. (*vi.*) many-flowered. Barbary. 1826. 7. 8. F.b.
23 descéndens. G.D. (*d.pu.*) early flowering. Switzerland. 1766. 6. H.b. Flor. græc. t. 316.
24 sphærocéphalon. G.D. (*pu.*) late-flowered. Germany. 1759. 7. H.b. Bot. mag. t. 1764.
25 pruinàtum. G.D. (*li.*) frosted. Portugal. 1819. 6. 8. H.b.
26 cárneum. G.D. (*li.*) blush-coloured. France. 1816. 6. 7. H.b.
27 ascalonicum. G.D. (*li.*) Shallot. Palestine. 1548. —— H.b. Moris. sect. 4. t. 14. f. 4.
β chinénse. G.D. (*li.*) *Chinese Shallot.* China. .... —— H.b.
28 físsile. (*gr.*) Scallion. Asia minor. .... —— H.b. Matt. opus. p. 420. ic.
Cèpa físsilis. Matt.
29 Cèpa. G.D. (*wh.gr.*) Onion. .... 6. H. ♂. b. Flor. græc. t. 326.
β aggregàtum. G.D. (*w.g.*) *Potatoe Onion.* .... —— H.b.
30 Schœnopràsum. G.D. (*pu.*) Chives. Britain. .... 5. 6. H.b. Eng. bot. v. 34. t. 2441.
31 foliòsum. G.D. (*pu.*) leafy Chives. France. 1820. —— H.b. Redout. lil. 4. t. 214.
32 sibíricum. G.D. (*li.*) Siberian Chives. Siberia. 1778. 6. 8. H.b. Bot. mag. t. 1141.
33 fistulòsum. G.D. (*gr.*) Welsh Onion. ——— 1629. 6. 7. H. ♂. b. ——— t. 1230.
β altàicum. Pallas. (*gr.*) *Altaic.* Altay. 1823. —— H.b. Pall. it. 2. n. 108. t. R.
34 cepæfórme. G.D. (*w.g.*) Onion-like. .... 1824. —— H.b.
35 oblíquum. G.D. (*st.*) oblique-leaved. Siberia. 1759. —— H.b. Bot. mag. t. 1408.

36	ramòsum. G.D.	(*st.*)	branchingChives.	Siberia.	1824.	6.	H.b.	
37	angústum. G.D.	(*pu.*)	narrow-leaved.	——	——	——	H.b.	
38	Pallàsii. G.D.	(*pu.*)	Pallas's Chives.	Russia.	....	——	H.b.	Murr.c.got.6.1775.t.3.
39	pusíllum. G.D.	(*li.*)	small.	Italy.	1824.	6. 7.	H.b.	Redout. liliac. 2. t. 118.
40	rubéllum. G.D.	(*ro.*)	reddish.	Siberia.	1825.	7. 8.	H.b.	
41	reticulàtum. G.D.	(*wh.*)	netted.	N. America.	1811.	——	H.b.	Bot. mag. t. 1840.
42	oxypétalum. G.D.	(*bh.*)	acute-petal'd.	Switzerland.	1822.	7.	H.b.	
43	oleràceum. G.D.	(*st.pu.*)	purple-striped.	England.	....	——	H.b.	Eng. bot. v. 7. t. 488.
44	carinàtum. G.D.	(*st.br.*)	keeled.	——	....	5. 6.	H.b.	—— v. 24. t. 1658.
45	fœ̀tidum. W.	(*da.pu.*)	twisted.	Siberia.	1823.	——	H.b.	
46	violàceum. G.D.	(*vi.*)	violet-flowered.	Hungary.	1819.	7.	H.b.	W. et K. hung. 3. t. 278.
	fléxum. W.K.							
47	ásperum. G.D.	(*vi.*)	rough-leaved.	S. Europe.	1820.	——	H.b.	Red. lil. 7. t. 368.
	carinàtum. R.L. *nec aliorum.*							
48	longispàthum. G.D.	(*g.pu.*)	long-spathed.	——	1779.	6. 7.	H.b.	Red. lil. 6. t. 316.
	pállens. Bot. mag. t. 1420.							
49	paniculàtum. G.D.	(*bh.*)	panicled.	——	1780.	——	H.b.	Bot. mag. t. 1432.
50	pállens. G.D.	(*pa.wh.*)	pale-flowered.	——	1779.	——	H.b.	Redout. lil. 5. t. 272.
51	tenuiflòrum. G.D.	(*ro.*)	slender-flowered.	Italy.	1824.	——	H.b.	Ten. fl. nap. 1. t. 30.
52	pulchéllum. G.D.	(*vi.*)	beautiful.	Russia.	1819.	8.	H.b.	Redout. lil. 5. t. 252.
	paniculàtum. R.L. *coloràtum.* S.S.							
53	flàvum. G.D.	(*ye.*)	yellow-leafy.	S. Europe.	1759.	6. 7.	H.b.	Bot. mag. t. 1330.
54	caucásicum. G.D.	(*ro.*)	Caucasian.	Caucasus.	1810.	——	H.b.	—— t. 973.
55	rupéstre. G.D.	(*ro.*)	small Caucasian.	——	1824.	——	H.b.	
56	globòsum. G.D.	(*li.*)	globular-headed.	Siberia.	1820.	7.	H.b.	Gmel. sib. v. 1. t. 10.
57	montànum. G.D.	(*ro.*)	mountain.	Greece.	——	6. 8.	H.b.	Flor. græc. t. 319.
58	parviflòrum. G.D.	(*br.*)	small-flowered.	S. Europe.	1781.	6. 7.	H.b.	
59	moschàtum. G.D.	(*w.pu.*)	small-leaved.	Hungary.	1805.	7. 8.	H.b.	W. et K. hung. 1. t. 68.
	setàceum. W.K.							
60	Stelleriànum. G.D.	(*g.w.*)	Steller's.	Siberia.	1829.	——	H.b.	Gmel.sib.1.t.16.f.1 et 2.
61	saxátile. G.D.	(*wh.*)	rock.	Caucasus.	1825.	6. 7.	H.b.	
62	stríctum. G.D.	(*ro.*)	upright.	Siberia.	1820.	7.	H.b.	Schrad. h. gott. 1. t. 1.
63	lineàre. G.D.	(*re.*)	linear-leaved.	——	1819.	7. 8.	H.b.	Gmel.sib.1.t.13.14.f.1.
64	nùtans. G.D.	(*fl.*)	great nodding.	——	1785.	——	H.b.	Bot. mag. t. 1143.
65	senéscens. G.D.	(*li.*)	Narcissus-leaved.	——	1596.	7.	H.b.	Gmel. sib. 1. t. 11. f. 2.
66	Andersònii. G.D.	(*li.*)	Anderson's.	——	1819.	——	H.b.	
67	spùrium. G.D.	(*li.*)	spurious.	——	....	——	H.b.	
68	glaùcum. G.D.	(*pu.*)	glaucous.	——	1596.	6. 7.	H.b.	Bot. mag. t. 1150.
	senéscens. B.M. *non* L. *baicalénse.* W.en.							
69	angulòsum. G.D.	(*pu.*)	angular-stalked.	——	1739.	——	H.b.	Bot. mag. t. 1149.
70	acutángulum. G.D.	(*pu.*)	acute-angled.	——	1816.	——	H.b.	Ger. emac. p. 186 ic.
71	danubiàle. G.D.	(*pu.*)	short-leaved.	——	——	——	H.b.	
72	láxum. G.D.	(*pu.*)	loose umbelled.	Russia.	1824.	——	H.b.	
73	serotìnum. G.D.	(*pu.*)	late-flowering.	Switzerland.	——	8.	H.b.	
74	álbidum. G.D.	(*wh.*)	whitish-flower'd.	Caucasus.	1823.	6. 8.	H.b.	
75	flavéscens. G.D.	(*st.*)	half naked.	Siberia.	1819.	8.	H.b.	Redout. lil. 6. t. 357.
	denudàtum. R.L.							
76	rubens. G.D.	(*pu.*)	red-rooted.	S. Europe.	1805.	5.	H.b.	Bot. mag. t. 1318.
	bisúlcum. Redout. liliac. t. 286.							
77	prostràtum. G.D.	(*li.*)	small trailing.	Siberia.	....	6.	H.b.	
78	congéstum. G.D.	(*pu.*)	crowded-flower'd.	——	....	——	H.b.	
79	pedemontànum. G.D.	(*ro.*)	Piedmont.	Piedmont.	1820.	7. 8.	H.b.	Allion. ped. t. 254.
80	narcissiflòrum. G.D.	(*ro.*)	large-flowered.	France.	1818.	——	H.b.	Vill. dauph. 2. t. 6.
81	suavèolens. G.D.	(*w.pu.*)	sweet-scented.	——	1801.	6. 8.	H.b.	Jacq. ic. 2. t. 364.
82	ochroleùcum. G.D.	(*st.*)	cream-colour'd.	Hungary.	1816.	7.	H.b.	W. et K. hung. 2. t. 186.
83	cérnuum. G.D.	(*bh.*)	drooping.	N. America.	1806.	6. 7.	H.b.	Bot. mag. t. 1324.
84	stellàtum. G.D.	(*li.*)	starry-flowered.	——	1811.	——	H.b.	—— t. 1576.
85	caroliniànum. G.D.	(*li.*)	Carolina.	——	1819.	——	H.b.	Redout. lil. 2. t. 101.
86	paradóxum. G.D.	(*wh.*)	paradoxical.	Siberia.	1823.	4.	H.b.	
87	scorzonerifòlium. G.D.	(*ye.*)	Scorzonera-leaved.	S. Europe.	—	6. 7.	H.b.	Redout. liliac. 2. t. 99.
88	canadénse. G.D.	(*ro.*)	Canadian.	N. America.	1739.	——	H.b.	Kalm. it. v. 3. p. 79.
89	mutábile. G.D.	(*bh.*)	changeable.	——	....	7.	H.b.	Redout. lil. 4. t. 240.
90	mágicum. G.D.	(*st.*)	broad-leaved.	Spain.	1596.	6. 7.	H.b.	Park. parad. p. 141.
91	ambíguum. G.D.	(*ro.*)	ambiguous.	Levant.	....	5. 6.	H.b.	Flor. græc. t. 327.
	incarnàtum. S.S. *ròseum* β. Bot. mag. t. 978.							
92	amœ̀num. G.D.	(*ro.*)	fine red-flowered.	Italy.	1820.	——	H.b.	Ten. fl. nap. t. 28.
	cárneum. T.N.							
93	ròseum. G.D.	(*ro.*)	rose-coloured.	Spain.	1752.	6.	H.b.	Redout. lil. 4. t. 213.
94	illy'ricum. G.D.	(*pi.*)	Illyrian.	Austria.	....	6. 7.	H.b.	Jacq. ic. 2. t. 365.

95 subhirsùtum. G.D. (*wh.*)	hairy.	S. Europe.	1596.	6.	H.b.	Cyr. pl. r. nap. 2. t. 3.
β *obtùsum.* (*wh.*)	*blunt-flowered.*	———	——	——	H.b.	Bot. mag. t. 774.
96 Clusiànum. G.D. (*wh.*)	Clusius's.	———	1803.	6. 8.	H.b.	Clus. hist. 1. p. 192.
97 Chamæ-Mòly. G.D. (*w.*)	dwarf-Moly.	———	1774.	2. 4.	F.b.	Bot. mag. t. 1203.
98 majàle. G.D. (*wh.*)	large Moly.	Italy.	1825.	6.	F.b.	Ten. fl. nap. 1. t. 28.
99 brachystèmon. R.L. (*w.*)	short-stamened.	S. France.	1826.	——	H.b.	Redout. liliac. 7. t. 374
100 longifòlium. G.D. (*pu.*)	long-leaved.	Mexico.	——	8. 10.	F.b.	Bot. reg. t. 1034.
101 Cowáni. B.R. (*wh.*)	Cowan's.	Chili.	1824.	5.	F.b.	——— t. 758.
102 neapolitànum. G.D. (*w.*)	Neapolitan.	Italy.	——	4. 5.	H.b.	Swt. br. fl. gar. t. 201.
álbum. Bivon.						
103 triquètrum. G.D. (*wh.*)	triquetrous.	S. Europe.	1759.	5. 6.	H.b.	Bot. mag. t. 869.
104 mèdium. G.D. (*wh.*)	intermediate.	Hungary.	1824.	5.	H.b.	
105 pendulìnum. G.D. (*w.*)	pendulous Italian.	Italy.	——	4. 5.	H.b.	Ten. fl. nap. 1. t. 31.
106 nìgrum. G.D. (*wh.*)	black.	S. Europe.	1596.	6.	H.b.	Bot. mag. t. 1148.
mágicum. B.M.						
β *rubéscens.* (*bh.*)	*pale red-flowered.*	Greece.	....	——	H.b.	Flor. græc. t. 323.
107 atropurpùreum. G.D. (*d.pu.*)	dark purple.	Hungary.	1825.	——	H.b.	W. et K. hung. 1. t. 17.
108 Mòly. G.D. (*ye.*)	large yellow.	S. Europe.	1604.	——	H.b.	Bot. mag. t. 499.
109 tricóccum. G.D. (*wh.*)	three-seeded.	N. America.	1770.	7.	H.b.	
110 ursìnum. G.D. (*wh.*)	Ramson.	Britain.	....	4. 5.	H.b.	Eng. bot. v. 2. t. 122.
111 sulcàtum. G.D. (*wh.*)	furrowed.	....	1823.	——	H.b.	Redout. liliac. 8. t. 482.
112 Victoriàlis. G.D. (*st.*)	long-rooted.	Austria.	1739.	5.	H.b.	Bot. mag. t. 1222.
113 odòrum. G.D. (*wh.pu.*)	sweet-flowered.	Siberia.	1787.	6. 7.	H.b.	Redout. lil. 2. t. 98.
tatáricum. Bot. mag. t. 1142.						
114 inodòrum. B.M. (*bh.*)	fragrant.	Carolina.	1776.	3. 4.	H.b.	Bot. mag. t. 1129.
fràgrans. Vent. cels. t. 26. *grácile.* W.						
115 striàtum. G.D. (*bh.w.*)	striated.	America.	1800.	5. 6.	H.b.	Bot. mag. t. 1524 & 1035.
Ornithógalum gramíneum. B.M. 2419.						
116 cáspium. G.D. (*bh.*)	Caspian.	Astracan.	1820.	4. 5.	H.b.	Pall. it. 2. ap. t. 2.
Amary'llis cáspica. W. *Crìnum cáspicum.* Pal.						
117 Akàka. F.	Persian.	Persia.	1829.	....	H.b.	
SOWERB'ÆA. B.P.	SOWERB'ÆA.	Hexandria Monogynia. S.S.				
júncea. B.P. (*ro.*)	Rush-leaved.	N. S. W.	1792.	5. 7.	G.♃.	Bot. mag. t. 1104.
LAXMA'NNIA. B.P.	LAXMA'NNIA.	Hexandria Monogynia. S.S.				
grácilis. B.P. (*wh.*)	slender.	N. S. W.	1822.	5. 7.	G.♃.	
APHYLLA'NTHES. L.	LILY PINK.	Hexandria Monogynia.				
monspeliénsis. L. (*bl.*)	Rush-like.	France.	1791.	6. 7.	H.♃.	Bot. mag. t. 1132.
THYSAN'OTUS. B.P.	THYSAN'OTUS.	Hexandria Monogynia. S.S.				
1 tuberòsus. B.P. (*bl.*)	tuberous-rooted.	N. S. W.	1822.	7. 9.	G.♃.	
2 elàtior. B.P. (*bl.*)	tall.	———	——	——	G.♃.	
3 isanthèrus. B.P. (*bl.*)	equal-anthered.	V. Diem. Isl.	——	——	G.♃.	Bot. reg. t. 655.
4 júnceus. B.P. (*bl.*)	rush-like.	N. S. W.	——	——	G.♃.	——— t. 656.
Chlamyspòrum juncifòlium. Par. lond. t. 103.						
CYANE'LLA. L.	CYANE'LLA.	Hexandria Monogynia. L.				
1 álba. S.S. (*wh.*)	1-flowered white.	C. B. S.	1825.	4. 6.	F.b.	Th. in a. hol. 1794. t. 7. f. 2
2 lùtea. S.S. (*ye.*)	yellow-flowered.	———	1788.	5. 12.	F.b.	Bot. mag. t. 1252.
3 orchidifórmis. S.S. (*bl.*)	Orchis-like.	———	1825.	4. 8.	F.b.	Jacq. ic. 2. t. 447.
4 lineàta. B.T. (*std.*)	lined.	———	1816.	——	F.b.	Bot. reg. t. 1111.
odoratíssima. B.R.						
5 capénsis. S.S. (*pu.*)	purple-flowered.	———	1768.	——	F.b.	Bot. mag. t. 568.
CONANTH'ERA. R.P.	CONANTH'ERA.	Hexandria Monogynia. P.S.				
bifòlia. R.P. (*bl.*)	two-leaved.	Chile.	1823.	3. 5.	F.b.	Lodd. bot. cab. t. 904.
CUMMI'NGIA. D.D.	CUMMI'NGIA.	Hexandria Monogynia.				
campanulàta. D.D. (*bl.*)	Hyacinth-flowered.	Chile.	1823.	4. 6.	F.b.	Swt. br. fl. gar. t. 257.
Conanthèra campanulàta. Hook. ex. flor. t. 214. *bifòlia.* B.M. 2496. *nec aliorum.*						
LEUCOCO'RYNE. B.R.	LEUCOCO'RYNE.	Hexandria Monogynia.				
1 odoràta. B.R. (*wh.*)	sweet-scented.	Chile.	1826.	8. 9.	F.♃.	Bot. reg. t. 1293.
2 ixioídes. B.R. (*bl.*)	Ixia-like.	———	1821.	7. 9.	F.b.	Bot. mag. t. 2382.
Brodi'æa ixioídes. B.M.						
TRITEL'EIA. B.R.	TRITEL'EIA.	Hexandria Monogynia.				
grandiflòra. B.R. (*bl.*)	large-flowered.	N. America.	1827.	7. 8.	H.♃.	
HESPEROSCO'RDUM. B.R.	MISSOURI HYACINTH.	Hexandria Monogynia.				
hyacínthinum. B.R. (*bl.*)	largest-flowered.	N. America.	1827.	7.	H.b.	
Brodi'æa grandiflòra. Pursh *non* Smith.						
DIANE'LLA. B.P.	DIANE'LLA.	Hexandria Monogynia. W.				
1 ensifòlia. B.M. (*wh.*)	whitish-flowered.	E. Indies.	1731.	8.	S.♃.	Bot. mag. t. 1404.
Drac'æna ensifòlia. H.K.						
2 nemoròsa. Lam. (*bl.*)	blue Indian.	———	....	6. 8.	S.♃.	Rumph. amb. 5. t. 73.
3 cœrùlea. B.P. (*bl.*)	rough-edged.	N. S. W.	1783.	5. 8.	F.♃.	Bot. mag. t. 505.

4 longifòlia. B.P.	(*bl.*)	long-leaved.	N. S. W.	1820.	5. 8.	F. ♃.	Bot. reg. t. 734.
5 l'ævis. B.P.	(*bl.*)	smooth-leaved.	———	1823.	——	F. ♃.	
6 revolùta. B.P.	(*bl.*)	revolute-margined.	———	1822.	——	F. ♃.	
7 divaricàta. B.P.	(*bl.*)	divaricated.	N. Holland.	1805.	7. 8.	G. ♃.	
8 strumòsa. B.R.	(*bl.*)	strumous.	———	1822.	5. 8.	G. ♃.	Bot. reg. t. 751.
LUZUR'IAGA. B.P.		LUZUR'IAGA.	Hexandria Monogynia. S.S.				
1 cymòsa. B.P.	(*pu.*)	cymose.	N. S. W.	1820.	4. 6.	G. ħ.	
2 montàna. B.P.	(*pu.*)	mountain.	———	1823.	——	G. ħ'.	
EUSTR'EPHUS. B.P.		EUSTR'EPHUS.	Hexandria Monogynia. S.S.				
1 latifòlius. B.P.	(*pu.*)	broadest-leaved.	N. S. W.	1800.	6. 9.	G. ħ. ◡.	Bot. mag. t. 1245.
2 angustifòlius. B.P.	(*pu.*)	narrow-leaved.	N. Holland.	1816.	——	G. ħ. ◡.	
MYRSIPHY'LLUM. S.S.		MYRSIPHY'LLUM.	Hexandria Trigynia. S.S.				
1 asparagoídes. S.S.	(*wh.*)	broadest-leaved.	C. B. S.	1702.	10. 3.	G. ħ. ◡.	Redout. liliac. t. 442.
Medèola asparagoídes. L.							
2 angustifòlium. S.S.	(*bh.*)	narrow-leaved.	———	1752.	12. 3.	G. ħ. ◡.	Redout. liliac. t. 393.
HERR'ERIA. R.P.		HERR'ERIA.	Hexandria Monogynia.				
1 stellàta. R.P.	(*ye.*)	prickly-stalked.	Chile.	1825.	6. 9.	G. ♄. ◡.	Flor. peruv. 3. t. 303. f. a.
2 parviflòra. B.R.	(*gr.*)	small-flowered.	Brazil.	——	——	S. ♄. ◡.	Bot. reg. t. 1042.
ASPA'RAGUS. L.		ASPA'RAGUS.	Hexandria Monogynia. L.				
1 amárus. S.S.	(*ye.*)	bitter.	S. France.	1820.	6. 8.	H. ♃.	Redout. liliac. t. 446.
2 officinàlis. S.S.	(*gr.*)	common.	England.	....	——	H. ♃.	Eng. bot. v. 5. t. 339.
3 tenuifòlius. DC.	(*ye.*)	slender-leaved.	Hungary.	....	——	H. ♃.	Redout. liliac. t. 434.
sylváticus. W. et. K. hung. 3. t. 201.							
4 declinàtus. S.S.	(*st.*)	long-leaved.	C. B. S.	1759.	——	F. ♃.	
5 decúmbens. S.S.	(*wh.*)	decumbent.	———	1792.	5.	F. ħ.	Jacq. schœn. 1. t. 97.
6 flexuòsus. S.S.	(*wh.*)	flexuose.	———	....	7. 8.	F. ħ. ◡.	
7 davuricus. S.S.	(*st.*)	Daurian.	Dauria.	1819.	6. 8.	H. ♃.	
8 cáspius. R.S.	(*gr.wh.*)	Caspian.	Caucasus.	1821.	——	H. ♃.	
9 marítimus. S.S.	(*gr.wh.*)	sea-side.	Siberia.	1820.	——	H. ♃.	
10 tricarinàtus. DC.	(*g.y.*)	three-keeled.	....	....	——	G. ♄. ◡.	Redout. liliac. t. 451.
11 verticillàtus. S.S.	(*g.y.*)	whorl-leaved.	Caucasus.	1752.	——	H. ♃. ◡.	Buxb. cent. 5. ap. t. 37.
12 umbellàtus. R.S.	(*g.w.*)	umbel-flowered.	Canaries.	1816.	——	G. ♃.	
13 grandiflòrus. R.S.	(*wh.*)	large-flowered.	Teneriffe.	1828.	——	G. ♄. ◡.	
14 longifòlius. F.	(*gr.ye.*)	long-leaved.	Russia.	1827.	....	H. ♃.	
15 acutifòlius. S.S.	(*ye.*)	needle-leaved.	Spain.	1640.	——	H. ħ.	Flor. græc. t. 337.
16 racemòsus. S.S.	(*wh.*)	branching.	E. Indies.	1808.	....	S. ♄.	
17 sarmentòsus. S.S.	(*wh.*)	linear-leaved.	Ceylon.	1710.	8.	S. ♄.	Redout. liliac. t. 460.
18 falcàtus. S.S.	(*wh.*)	sickle-leaved.	E. Indies.	1792.	....	S. ♄.	Burm. zeyl. t. 13 f. 2.
19 retrofráctus. S.S.	(*wh.*)	Larch-leaved.	Africa.	1759.	8. 9.	G. ♃.	Pluk. alm. t. 375. f. 3.
20 capénsis. S.S.	(*wh.re.*)	Cape.	C. B. S.	1691.	4. 5.	G. ♄.	Jacq. schœn. 3. t. 266.
21 æthiòpicus. W.	(*wh.*)	angular-stalked.	———	1816.	....	G. ♄.	
22 scándens. W.	(*wh.vi.*)	climbing.	———	1795.	5. 6.	G. ♄. ◡.	
23 asiáticus. S.S.	(*wh.*)	Asiatic.	Asia.	1759.	——	G. ♄.	Pluk. alm. t. 14. f. 4.
24 subulàtus. S.S.	(*wh.*)	awl-leaved.	C. B. S.	1811.	——	G. ♄.	
25 Niveniànus. R.S.	(*w.pu.*)	Niven's.	———	——	——	G. ħ.	
26 lánceus. R.S.	(*wh.*)	spear-leaved.	———	——	——	G. ♄. ◡.	
27 laricìnus. R.S.	(*wh.*)	Larch-leaved.	———	1816.	——	G. ♄.	
28 tetragonus. R.S.	(*wh.*)	square-leaved.	———	1822.	——	G. ♄.	
29 álbus. S.S.	(*wh.*)	white.	Spain.	1540.	5. 7.	F. ♄.	Moris. s. 1. t. 1. f. 3.
30 aphy'llus. S.S.	(*ye.*)	prickly.	S. Europe.	1640.	——	F. ♄.	Flor. græc. t. 338.
31 hórridus. S.S.	(*ye.*)	long-spined.	———	1819.	——	F. ♄.	——— t. 239.
32 Broussonèti. L.en.	(*st.*)	Broussonet's.	Canaries.	1815.	——	F. ♃.	Jacq. eclog. 2. t. 40.
CHARLWO'ODIA. S.F.A.		CHARLWO'ODIA.	Hexandria Monogynia.				
1 congésta. S.F.A.	(*pu.*)	crowded-panicled.	N. Holland.	1821.	5. 7.	S. ♄.	Swt. flor. aust. t. 18.
2 stricta. S.F.A.	(*bl.*)	loose-panicled.	———	——	——	S. ♄.	Bot. reg. t. 956.
Drac'æna stricta. Bot. mag. t. 2575.							
DRAC'ÆNA. W.		DRAGON-TREE.	Hexandria Monogynia. L.				
1 Dràco. S.S.	(*wh.*)	common.	E. Indies.	1640.	....	S. ♄.	Blackw. t. 358.
2 ensifòlia. H.S.	(*wh.*)	sword-leaved.	....	1800.	....	S. ♄.	
3? indivìsa. S.S.	(*wh.*)	undivided.	N. Zealand.	1819.	6. 8.	G. ♄.	
4? austràlis. S.S.	(*wh.*)	Forster's.	———	——	——	G. ħ.	Bot. mag. t. 2835
5 marginàta. S.S.	(*wh.*)	red-margined.	Madagascar.	1816.	....	S. ♄.	
tessellàta. W. enum.							
6 arbòrea. S.S.	(*wh.*)	tree.	Africa.	1810.	3. 5.	S. ♄.	
7 frágrans. S.S.	(*wh.*)	fragrant.	———	1768.	2. 5.	S. ♄.	Bot. mag. t. 1081.
8 férrea. L.	(*wh.*)	dark-purple.	China.	1771.	3. 4.	S. ♄.	——— t. 2053.
9 terminàlis. W.	(*wh.*)	red-leaved.	India.	1812.	4. 8.	S. ♄.	Lodd. bot. cab. t. 1224.
10 umbraculífera. S.S.	(*wh.*)	umbrella.	Mauritius.	1788.	4. 6.	S. ♄.	——— t. 289.
11 cérnua. S.S.	(*st.li.*)	drooping.	———	....	——	S. ♄.	Jacq. schœn. 1. t. 96.

12 mauritiàna. s.s.	(*wh.*)	Mauritius.	Mauritius.	1816.	....	S. ♄.	
13 undulàta. s.s.	(*wh.*)	wave-leaved.	C. B. S.	——	....	G. ♄.	
14 ovàta. B.M.	(*li.*)	oval-leaved.	Sierra Leon.	1790.	8. 9.	S. ♄.	Bot. mag. t. 1180.
15 nodòsa. L.C.	(*wh.*)	knotted.	....	1822.	....	S. ♄.	
16 refléxa. s.s.	(*li.*)	reflex-leaved.	Madagascar.	1816.	5. 8.	S. ♄.	Redout. liliac. t. 92.
17 surculòsa. B.R.	(*st.*)	trailing.	Sierra Leone.	1822.	——	S. ♄.	Bot. reg. t. 1169.
CORDYL'INE. B.P.		CORDYL'INE.	Hexandria Monogynia.				
1 hemichr'ysa. P.S.	(*wh.*)	golden-leaved.	Bourbon.	1823.	....	S. ♄.	
2 parviflòra. s.s.	(*wh.*)	small-flowered.	Mexico.	1828.	....	S. ♄.	H. et B.n. gen. 3. t. 674.
PHO'RMIUM. W.		FLAX-LILY.	Hexandria Monogynia. W.				
tènax. W.	(*ye.*)	Iris-leaved.	N. Zealand.	1788.	8.	F. ♃.	Redout. liliac. t. 448-449.
XANTHORRH'ŒA. B.P.		GUM-TREE.	Hexandria Monogynia. S.S.				
1 arbòrea. B.P.	(*wh.*)	tree.	N. S. W.	1828.	....	G. ♄.	
2 austràlis. B.P.	(*wh.*)	southern.	V. Diem. Isl.	1820.	....	F. ♄.	
3 hástile. B.P.	(*wh.*)	yellow gum.	N. S. W.	1803.	4. 5.	F. ♄.	
4 mèdia. B.P.	(*wh.*)	intermediate.	———	1819.	——	F. ♄.	
5 mìnor. B.P.	(*wh.br.*)	small.	———	1804.	4. 6.	F. ♃.	
6 bracteàta. B.P.	(*wh.*)	long-bracted.	———	1810.	4. 5.	F. ♃.	
ROXBU'RGHIA. W.		ROXBU'RGHIA.	Octandria Monogynia.				
1 gloriosoídes. R.C.	(*st.pu.*)	cordate-leaved.	E. Indies.	1803.	4. 6.	S. ♃. ◡.	Roxb. corom. 1. t. 32.
2 viridiflòra. Ex.B.	(*st.cr.*)	elliptic-leaved.	———	——	——	S. ♃. ◡.	Sm. exot. bot. 1. t. 57.
gloriòsa. Bot. mag. t. 1500.							

ORDO CXXV.

BROMELIACEÆ. *Juss. gen.* 49.

AG'AVE. L.		AG'AVE.	Hexandria Monogynia. L.				
1 americàna. H.S.	(*gr.*)	large American.	S. America.	1640.	8. 10.	D.G. ♃.	Andr. reposit. t. 438.
2 Millèri. H.S.	(*gr.*)	Miller's.	———	1768.	——	D.G. ♃.	
3 fláccida. H.S.	(*gr.*)	flaccid.	———	1790.	——	D.S. ♃.	
4 lùrida. B.M.	(*gr.*)	Vera Cruz.	Vera Cruz.	1731.	6. 7.	D.S. ♃.	Bot. mag. t. 1522.
5 brachystàchys. s.	(*gr.*)	short-spiked.	Mexico.	1826.	....	D.G. ♃.	Redout. liliac. t. 485.
spicàta. R.L. *non* Cavan.							
6 spicàta. s.	(*gr.ye.*)	spiked.	Havannah.	1825.	....	D.S. ♃.	Cav.h.reg.mad.ic.ined.
7 angustifòlia. H.S.	(*gr.*)	narrow-leaved.	St. Helena.	1770.	....	D.S. ♃.	
8 yuccæfòlia. H.S.S.	(*y.g.*)	Yucca-leaved.	S. America.	1817.	....	D.S. ♃.	Redout. liliac. t. 328, 329.
9 mexicàna. H.S.S.	(*gr.*)	Mexican.	Mexico.	1818.	....	D.G. ♃.	
10 Karátto. H.S.	(*gr.*)	Karatto.	S. America.	1768.	....	D.S. ♃.	
11 vivìpara. H.S.	(*gr.ye.*)	viviparous.	———	1731.	8. 10.	D.S. ♃.	Comm. præl. t. 15.
12 polyacántha. H.R.	(*g.y.*)	black-spined.	....	1800.	....	D.S. ♃.	
13 virgínica. H.S.	(*gr.ye.*)	Virginian.	N. America.	1765.	9.	F. ♃.	Bot. mag. t. 1157.
FURCR'ŒA. V.		FURCR'ŒA.	Hexandria Monogynia. H.S.				
1 gigántea.	(*gr.wh.*)	gigantic.	S. America.	1690.	9. 1.	D.S. ♄.	DC. plant. gras. t. 126.
Agàve fœ'tida. W.							
2 tuberòsa. H.K.	(*gr.wh.*)	tuberous.	———	1739.	8. 9.	D.S. ♄.	Comm. hort. 2. f. 19.
3 cubénsis. H.S.	(*wh.gr.*)	Cuba.	———	——	....	D.S. ♄.	Jacq. am. pic. t. 260. f. 25.
4 rígida. H.S.	(*gr.ye.*)	Vera Cruz.	———	1768.	....	D.S. ♄.	Spin. cat. 1812. c. ic.
5 austràlis. H.S.	....	entire-leaved.	N. Holland.	1811.	....	D.S. ♃.	
6 madagascariénsis. H.		Madagascar.	Madagascar.	1818.	....	D.S. ♃.	
7 cantàla. H.S.S.	....	Canton.	China.	——	....	D.G. ♃.	
LITT'ÆA. H.S.S.		LITT'ÆA.	Hexandria Monogynia. H.S.S.				
geminiflòra. H.S.S.	(*gr.*)	twin-flowered.	S. America.	1810.	....	S. ♄.	Bot. reg. t. 1145.
Agàve geminiflòra. Ker. *Buonapártea júncea.* H.S. *nec aliorum.*							
ANANA'SSA. B.H.		PINE-APPLE.	Hexandria Monogynia. L.				
1 satìva. B.R.	(*bl.*)	common.	S. America.	1690.	1. 12.	S. ♃.	Bot. mag. t. 1554.
2 semiserràta. s.s.	(*bl.*)	semiserrate.	———	....	——	S. ♃.	
3 lùcida. W.en.	(*bl.*)	shining-leaved.	———	....	——	S. ♃.	
4 débilis. B.R.	(*bl.*)	waved-leaved.	———	1826.	——	S. ♃.	
5 bracteàta. B.R.	(*bl.sc.*)	red-bracted.	Jamaica.	1785.	9. 10.	S. ♄.	Bot. reg. t. 1081.
BILLBE'RGIA. Th.		BILLBE'RGIA.	Hexandria Monogynia.				
1 zebrìna. B.M.	(*ye.*)	white-barred.	Brazil.	1825.	6. 7.	S. ♃.	Bot. mag. t. 2686.
2 fasciàta. B.R.	(*bl.*)	banded.	———	——	8. 10.	S. ♃.	Bot. reg. t. 1130.

3 iridifòlia. B.R.	(ye.bl.)	drooping.	Brazil.	1826.	3. 5.	S.♃.	Bot. reg. t. 1068.
4 am'œna. B.R.	(ye.bl.)	pale-flowered.	S. America.	1810.	3. 11.	S.♃.	Lodd. bot. cab. t. 76.
Bromèlia pállida. Bot. reg. t. 344.							
5 pyramidàlis. B.R.	(cr.)	pyramidal.	Brazil.	....	2. 4.	S.♃.	Bot. reg. t. 203.
Bromèlia nudicaùlis. B.R. *pyramidàlis*. Bot. mag. t. 1732.							
β *bícolor*. B.R.	(cr.bl.)	*two-coloured*.	S. America.	....	——	S.♃.	Bot. reg. t. 1181.
6 nudicaùlis. B.R.	(st.)	naked-stalked.	Trinidad.	1822.	4. 6.	S.♃.	Hook. ex. flor. t. 143.
7 cruénta. B.M.	(bl.)	blood-stained.	Brazil.	1824.	——	S.♃.	Bot. mag. t. 2892.
BROM'ELIA. B.R.		BROM'ELIA.	Hexandria Monogynia.				
1 melanántha. B.R.	(bk.)	black-flowered.	Trinidad.	1823.	5. 7.	S.♃.	Bot. reg. t. 766.
2 fastuòsa. L.col.	(pu.)	noble.	S. America.	1815.	8. 10.	S.♄.	Lindl. col. bot. t. 1.
3 lingulàta. S.S.	(ye.)	tongue-leaved.	————	1759.	5. 6.	S.♄.	Plum. ic. t. 64. f. 1.
4 Karátas. S.S.	(ro.)	upright-leaved.	W. Indies.	1739.	....	S.♄.	Jacq. vind. 1. t. 31. 32.
5 hùmilis. S.S.	(bl.)	dwarf.	————	1789.	3.	S.♃.	Jacq. ic. 1. t. 60.
6 Pínguin. S.S.	(re.)	broad-leaved.	————	1699.	3. 4.	S.♄.	Jacq. am. ed. pict. t. 91.
7 chrysántha. S.S.	(ye.)	yellow-flowered.	S. America.	1820.	——	S.♄.	Jacq. schœn. 1. t. 55.
8 exùdans. B.C.	(ye.)	clammy.	————	1816.	——	S.♃.	Lodd. bot. cab. t. 801.
9 paniculígera. S.S.	(sc.)	panicled.	W. Indies.	1822.	——	S.♃.	
10 sylvéstris. S.S.	(cr.)	wild.	S. America	1820.	7. 8.	S.♃.	Bot. mag. t. 2392.
11 Acánga. S.S.	(ye.)	recurved.	————	1824.	....	S.♃.	Pis. bras. t. 91.
12 sessiliflòra. L.C.	....	sessile-flowered.	————	———	....	S.♃.	
GUZMA'NNIA. R.P.		GUZMA'NNIA.	Hexandria Monogynia. S.S.				
trícolor. R.P.	(g.st.ro.)	three-coloured.	S. America.	1818.	5. 6.	S.♃.	Lind. coll. bot. t. 8.
PITCA'IRNIA. S.S.		PITCA'IRNIA.	Hexandria Monogynia. L.				
1 stamínea. S.S.	(sc.)	long-stamened.	S. America.	1823.	1. 3.	S.♃.	Bot. mag. t. 2411.
2 integrifòlia. S.S.	(sc.)	entire-leaved.	W. Indies.	1800.	8.	S.♃.	———— t. 1462.
3 latifòlia. S.S.	(sc.)	broad-leaved.	————	1785.	8. 9.	S.♃.	———— t. 856.
4 bracteàta. R.L.	(sc.)	large bracted-red.	————	1799.	4. 5.	S.♃.	———— t. 2813.
5 sulphùrea. A.R.	(st.)	yellow-flowered.	————	1797.	6. 8.	S.♃.	Andr. reposit. t. 249.
bracteàta β sulphùrea. Bot. mag. t. 1416.							
6 suavèolens. B.R.	(st.)	sweet-scented.	Brazil.	1826.	6. 7.	S.♃.	Bot. reg. t. 1069.
7 álbiflos. B.M.	(wh.)	white-flowered.	————	1824.	9. 12.	S.♃.	Bot. mag. t. 2642.
8 bromeliæfòlia. S.S.	(sc.)	Pine-apple-leaved.	Jamaica.	1781.	6. 8.	S.♃.	———— t. 824.
9 angustifòlia. S.S.	(sc.)	narrow-leaved.	Santa Cruz.	1777.	12. 1.	S.♃.	———— t. 1547.
10 furfuràcea. B.M.	(ro.)	mealy.	S. America.	1818.	7. 9.	S.♃.	———— t. 2657.
11 flámmea. B.R.	(sc.)	flame-coloured.	Brazil.	1823.	11. 12.	S.♃.	———— t. 1092.
12 mèdia. L.C.	(sc.)	intermediate.	S. America.	——	7. 9.	S.♃.	
13 iridiflòra. L.C.	(sc.)	Iris-flowered.	————	——	——	S.♃.	
14 hùmilis. L.C.	(sc.)	dwarf.	————	——	——	S.♃.	
POURR'ETIA. R.P.		POURR'ETIA.	Hexandria Monogynia. S.S.				
1 lanuginòsa. R.P.	(gr.)	woolly.	S. America.	1826.	4. 8.	S.♃.	Flor. peruv. 3. t. 256.
2 pyramidàta. K.S.	(st.)	pyramidal.	————	1822.	4. 6.	S.♃.	———— 3. t. 257.
3 coarctàta. R.P.	(ye.)	crowded-spiked.	Chili.	——	....	S.♃.	Feuil. per. 3. t. 39.
TILLA'NDSIA. L.		TILLA'NDSIA.	Hexandria Monogynia. L.				
1 usneoídes. S.S.	(ye.gr.)	moss-like.	W. Indies.	1823.	4. 8.	S.♃.	Sloan. hist. t. 122. f. 2-3.
2 Bartràmii. S.S.	......	Bartram's.	Carolina.	1825.	....	S.♃.	
3 recurvàta. S.S.	(pu.)	recurve-leaved.	Jamaica.	1793.	7.	S.♃.	Sloan. jam. 1. t. 121. f. 1.
4 xiphioídes. B.R.	(wh.)	Xiphium-like.	Buenos Ayres.	1815.	——	S.♃.	Bot. reg. t. 105.
5 serràta. S.S.	(ye.)	saw-leaved.	Jamaica.	1793.	6. 7.	S.♃.	Plum. ic. 63. t. 75. f. 1.
6 lingulàta. S.S.	(ye.)	tongue-leaved.	————	1776.	——	S.♃.	Jacq. amer. t. 62.
7 acaùlis. B.R.	(wh.)	stemless.	Brazil.	1826.	7. 9.	S.♃.	Bot. reg. t. 1157.
8 psittácina. B.M.	(or.y.)	gaudy-flowered.	————	1827.	4. 8.	S.♃.	Bot. mag. t. 2841.
9 setàcea. S.S.	(st.)	bristle-leaved.	Jamaica.	1822.	——	S.♃.	
10 strícta. S.S.	(bl.)	stiff-leaved.	Brazil.	1810.	11.	S.♃.	Bot. mag. t. 1529.
11 bulbòsa. H.E.F.	(bl.)	bulbous.	Trinidad.	1823.	8. 9.	S.♃.	Hook. ex. flor. t. 173.
12 aloifòlia. H.E.F.	(st.)	Aloe-leaved.	————	1824.	11. 12.	S.♃.	———— t. 205.
13 púlchra. H.E.F.	(wh.)	pretty.	————	1822.	10. 11.	S.♃.	———— t. 154.
14 ánceps. B.C.	(bl.)	two-edged.	W. Indies.	1819.	7.	S.♃.	Lodd. bot. cab. t. 771.
15 utriculàta. S.S.	(st.)	bottle.	S. America.	1793.	7. 8.	S.♃.	
16 tenuifòlia. S.S.	(sc.)	slender-leaved.	W. Indies.	1822.	——	S.♃.	Plum. ic. 232. f. 2.
17 acutifòlia. L.en.	(st.)	acute-leaved.	Brazil.	1815.	——	S.♃.	Bot. reg. t. 749.
flexuòsa β pállida. B.R.							
18 flexuòsa. S.S.	(sc.)	flexuose.	W. Indies.	1790.	——	S.♃.	Jacq. amer. t. 63.
19 nùtans. S.S.	(bl.)	nodding-spiked.	————	1793.	——	S.♃.	
20 fasciculàta. S.S.	(st.)	bundled.	————	1822.	——	S.♃.	
21 nítida. H.E.F.	(wh.)	shining-leaved.	Jamaica.	1823.	11. 12.	S.♃.	Hook. ex. flor. t. 218.
22 polystàchya. S.S.	(st.)	many-spiked.	W. Indies.	1820.	7. 9.	S.♃.	
23 angustifòlia. S.S.	(st.)	narrow-leaved.	————	1822.	——	S.♃.	
24 paniculàta. S.S.	(st.)	panicled.	————	1817.	——	S.♃.	Plum. ic. t. 237.

25 coarctàta. G.C. ...... crowded. W. Indies. 1822. .... S.♃.
26 compréssa. G.C. .... compressed. ——— —— .... S.♃.
27 grácilis. G.C. .... slender. ——— —— .... S.♃.
28 ramòsa. G.C. .... branching. ——— —— .... S.♃.
29 rígida. G.C. .... rigid. ——— —— .... S.♃.
BUONAPA'RTEA. R.P. BUONAPA'RTEA. Hexandria Monogynia.
júncea. R.P. (bl.) rush-leaved. Peru. 1800. .... S.♃. Flor. peruv. 3. t. 242.

ORDO CCXVI.

TULIPACEÆ. *Kth. synops.* 1. *p.* 292.

YU'CCA. L. ADAM'S-NEEDLE. Hexandria Monogynia. L.
1 conspícua. H.S.S. (*wh.*) broad rough-edged....... 1810. 8. 10. F.♄.
2 aloifòlia. H.S.S. (*wh.*) Aloe-leaved. W. Indies. 1696. —— G.♄. Bot. mag. t. 1700.
3 serrulàta. H.S.S. (*wh.*) narrow saw-leaved. Carolina. 1808. .... F.♄.
4 Dracònis. H.S.S. (*wh.*) drooping-leaved. ——— 1732. 8. 9. F.♄. Dill. elt. t. 324. f. 417.
5 crenulàta. H.S.S. (*wh.*) rough-edged oblique. —— 1818. .... F.♄.
6 arcuàta. H.S.S. (*wh.*) arcuated. ——— 1817. .... F.♄.
7 tenuifòlia. H.S.S. (*wh.*) slender-leaved. Malta? —— .... F.♄.
8 concàva. H.S.S. (*wh.*) hollow-thready. N. America. 1816. 7. 8. H.♃.
9 filamentòsa. H.S.S. (*wh.*) common-thready. Virginia. 1675. 7. 10. H.♃. Bot. mag. t. 900.
10 fláccida. H.S.S. (*wh.*) flaccid-thready. N. America. 1816. —— H.♃.
11 pubérula. H.P. (*wh.*) pubescent-stemmed. —— 1822. —— H.♃. Swt. br. fl. gar. t. 251.
12 glaucéscens. H.S.S. (*wh.*) glaucescent. ——— 1815. 7. 8. H.♃. ——— t. 53.
13 recúrva. P.L. (*wh.*) recurved-leaved. Georgia. 1795. 8. 9. H.♄. Sal. par. lond. t. 31.
14 glaùca. B.M. (*st.*) glaucous-leaved. Carolina. 1812. —— H.♃. Bot. mag. t. 2662.
15 stricta. B.M. (*gr.pu.*) straight-leaved. ——— 1811. —— H.♄. ——— t. 2222.
16 angustifòlia. H.S.S. (*gr.*) narrow-leaved. Missouri. —— —— H.♄. ——— t. 2236.
17 supérba. H.S.S. (*wh.*) superb. .... 8. 10. G.♄. Andr. reposit. t. 473.
gloriòsa. A.R. *nec aliorum.*
18 acuminàta. S.F.G. (*br.w.*) tapering-flowered. America. .... 7. 9. H.♄. Swt. br. fl. gar. t. 195.
19 gloriòsa. H.S.S. (*wh.*) handsome. ——— 1596. 7. 8. H.♄. Bot. mag. t. 1260.
20 rufocíncta. H.S.S. (*wh.*) rufous-edged. ——— 1816. —— H.♃.
21 oblíqua. H.S.S. (*wh.*) oblique-leaved. ——— 1808. —— H.♃.
β màjor. H.S.S. (*wh.*) *larger.* ——— —— —— H.♃.
22 spinòsa. K.S. (*pu.or.*) spiny-leaved. Mexico. 1829. .... G.♄.
T'ULIPA. L. TULIP. Hexandria Monogynia. L.
1 altàica. R.S. (*ye.*) Altaic. Altay mount. 1822. 4. 5. H.b.
2 saxátilis. R.S. (*ye.*) rocky. Crete. 1827. —— H.b. Reich. ic. rar. t. 396.
3 pàtens. R.S. (*wh.g.y.*) spreading-flowered. Siberia. 1829. —— H.b.
4 stellàta. B.M. (*bh.wh.*) starry. E. Indies. 1827. 3. 4. G.b. Bot. mag. t. 2762.
5 montàna. B.R. (*cr.*) mountain. Persia. 1826. —— H.b. Bot. reg. t. 1106.
6 biflòra. S.S. (*wh.vi.ye.*) two-flowered. Russia. 1806. 4. H.b. ——— t. 535.
7 trícolor. F. .... three-coloured. ——— 1823. .... H.b.
8 Celsiàna. S.S. (*ye.re.*) Cels's. Levant. .... 6. 7. H.b. Bot. mag. t. 717.
Breyniàna. B.M. *nec aliorum.*
9 Biebersteiniàna. R.S. (*y.pu.*) Bieberstein's. Siberia. 1820. —— H.b.
répens. F.?
10 sylvéstris. S.S. (*ye.*) wild. England. .... 4. 5. H.b. Eng. bot. v. 1. t. 63.
11 túrcica. Roth. (*ye.*) Florentine. S. Europe. .... —— H.b. Swt. br. fl. gar. t. 186.
12 Clusiàna. R.S. (*sc.wh.*) Clusius's. Sicily. 1636. 5. 6. H.b. Bot. mag. t. 1390.
13 maculàta. R.S. (*ye.gr.*) spotted-flowered. Spain. .... —— H.b.
14 mèdia. R.S. (*sc.wh.*) intermediate. .... 1828. —— H.b.
15 cornùta. R.S. (*va.*) horn-petaled. Levant. 1816. 5. H.b. Bot. reg. t. 127.
16 undulàta. (*sc.ve.*) wave-leaved. Whitley's. 1820. 5. 6. H.b.
17 óculus sòlis. S.S. (*sc.ve.*) Sun's-eye. S. Europe. 1816. —— H.b. Redout. liliac. t. 219.
18 pr'æcox. S.S. (*sc.gr.*) early-flowering. Naples. 1825. 3. 4. H.b. Swt. br. fl. gar. t. 157.
19 pérsica. (*sc.ve.*) Persian. Persia. 1826. —— H.b. Bot. reg. t. 1143.
20 suavèolens. S.S. (*sc.ye.*) Van-Thol. S. Europe. 1603. —— H.b. Bot. mag. t. 839.
21 pubéscens. S.S. (*va.*) woolly-stalked. ——— .... —— H.b. Swt. br. fl. gar. t. 78.
22 Gesneriàna. S.S. (*va.*) common. Levant. 1577. 4. 5. H.b.
α coccínea. (sc.ve.) *scarlet-flowered.* Constantinople. 1816. —— H.b. Bot. reg. t. 380.
β versícolor. (va.) *various-coloured.* Levant. —— —— H.b. Bot. mag. t. 1135.
γ simploniénsis. (st.pu.) *Semplon.* .... —— H.b. Swt. flor. guid. t. 53.

δ *Alexandrìna*.	(st.pu.ve.)	*Alexandrina*.	....	....	——	H.b.	Swt. flor. guid. t. 138.
ε *Burnardiàna*.	(st.pu.ve.)	*Burnard's agitator*	....	....	——	H.b.	—————— t. 132.
ζ *Collingwòodii*.	(y.pu.ve.)	*Lord Collingwood*.	....	....	——	H.b.	—————— t. 85.
η *Leopóldi*.	(ye.ve.)	*Prince Leopold*.	....	....	——	H.b.	—————— t. 128.
θ *Geórgii*.	(ye.ve.)	*George the Fourth*	....	....	——	H.b.	—————— t. 89.
ι *platyántha*.	(ye.ve.)	*Platonia*.	....	....	——	H.b.	—————— t. 77.
κ *imperiàlis*.	(ye.ve.)	*Emperor of Austria*.	....	....	——	H.b.	—————— t. 103.
λ *Migùeli*.	(ye.ve.)	*Don Miguel*.	....	....	——	H.b.	—————— t. 19.
μ *Polyphèmi*.	(ye.ve.)	*Lawrence's Polyphemus*	..	....	——	H.b.	—————— t. 7.
ν *sanguífera*.	(ye.cr.ve.)	*Bataille d' Eyleau*	....	....	——	H.b.	—————— t. 4.
ξ *Duckettiàna*.	(y.sc.ve.)	*Sir George Duckett*.	....	....	——	H.b.	—————— t. 66.
ο *Canningiàna*.	(y.sc.ve.)	*Strong's Canning*.	....	....	——	H.b.	—————— t. 135.
π *règia*.	(ye.sc.ve.)	*Strong's King*.	....	....	——	H.b.	—————— t. 42.
ρ *Titiána*.	(ye.sc.ve.)	*Titian*.	....	....	——	H.b.	—————— t. 109.
σ *Jùno*.	(wh.sc.pu.)	*Rose Juno*.	....	....	——	H.b.	—————— t. 56.
τ *Daphneána*.	(wh.cr.)	*Rose Daphne*.	....	....	——	H.b.	—————— t. 83.
υ *retùsa*.	(wh cr.)	*Walworth*.	....	....	——	H.b.	—————— t. 26.
φ *fulmínea*.	(wh.sc.)	*Bartlett's thunderbolt*.	..	....	——	H.b.	—————— t. 95.
χ *Hilliána*.	(wh.sc.)	*Lord Hill*.	....	....	——	H.b.	—————— t. 82.
ψ *fúlgens*.	(w.ro.)	*Rose brillante*.	....	....	——	H.b.	—————— t. 5.
ω *Creweæ*.	(wh.sc.)	*Sherwood's Lady Crewe*	...	....	——	H.b.	—————— t. 153.
αα *Victoriána*.	(w.ro.cr.)	*Princess Victoria*	....	....	——	H.b.	—————— t. 38.
ββ *Stróngii*.	(w.cr.ve.)	*Strong's high Admiral*.	..	....	——	H.b.	—————— t. 11.
γγ *puélla*.	(wh.ro.)	*Pucelle d'Orleans*.	....	....	——	H.b.	—————— t. 50.
δδ *principíssæ*.	(w.ro.cr.)	*Strong's Duchess of Kent*.		....	——	H.b.	—————— t. 75.
εε *purpùreo-ròsea*.	(w.ro.pu.)	*Rose camusa de craix*.		....	——	H.b.	—————— t. 116.
ζζ *iridiflòra*.	(wh.pu.)	*Strong's rainbow*.	....	....	——	H.b.	—————— t. 124.
ηη *Hollandiána*.	(w.pu.ve.)	*Lord Holland*.	....	....	——	H.b.	—————— t. 106.
θθ *Ludoviàna*.	(wh.ve.)	*Louis XVI*.	....	....	——	H.b.	—————— t. 150.
ιι *lancastriénsis*.	(w.pu.ve.)	*Lancashire hero*.	....	....	——	H.b.	—————— t. 119.
κκ *marginàta*.	(wh.ve.)	*Davey's Trafalgar*.	....	....	——	H.b.	—————— t. 145.
λλ *Penelopeàna*.	(w.ve.)	*Penelope*.	....	....	——	H.b.	—————— t. 57.
μμ *Jenkinsòni*.	(w.ve.)	*Goldham's Earl of Liverpool*.		....	——	H.b.	—————— t. 16.
νν *Fredericiàna*.	(w.ve.)	*Willmer's Duke of York*.		....	——	H.b.	—————— t. 121.
ξξ *Buonaparteàna*.	(w.ve.)	*Buonaparte*.	....	....	——	H.b.	—————— t. 72.
οο *augústa*.	(wh.ve.)	*Grande Monarque*	....	....	——	H.b.	—————— t. 92.
ππ *Lampsoniàna*.	(w.ve.)	*Lampson*.	....	....	——	H.b.	—————— t. 100.
ρρ *Aglaiàna*.	(wh.pu.)	*Aglaia*.	....	....	——	H.b.	—————— t. 87.
σσ *Alexándri*.	(wh.pu.)	*Violet Alexander*.	....	....	——	H.b.	—————— t. 30.
ττ *virgínea*.	(wh.ve.)	*Lady of the Lake*.	....	....	——	H.b.	—————— t. 62.
υυ *Winefrediàna*.	(w.v.)	*Winefred*.	....	....	——	H.b.	—————— t. 141.
φφ *Daveyána*.	(wh.ve.)	*Daveyana*.	....	....	——	H.b.	—————— t. 35.
χχ *Marìa*.	(wh.ve.pu.)	*Goldham's Maria*.	....	....	——	H.b.	—————— t. 48.
ψψ *majésta*.	(wh.ve.)	*Majestueuse*.	....	....	——	H.b.	—————— t. 39.
ωω *álbida*.	(wh.ve.)	*Gloria alborum*.	....	....	——	H.b.	—————— t. 74.
ααα *Charlotteàna*.	(w.pu.)	*Princess Charlotte's cenotaph*.		..	....	H.b.	—————— t. 23.
βββ *plèna*.	(va.)	*double-flowered*.	Levant.	1577.	4. 5.	H.b.	
γγγ *laciniàta*.	(ye.re.pu.)	*Parrot*.	——	——	——	H.b.	

FRITILL'ARIA. L. FRITILLARY. Hexandria Monogynia. L.

1 præ'cox.	(*wh.*)	early white.	Europe.	....	3. 4.	H.b.	
Meleàgris β pr'æcox. P.S.							
2 Meleàgris. E.B.	(*pu.wh.*)	chequered.	Britain.	....	3. 5.	H.b.	Eng. bot. v. 9. t. 622.
3 meleagroídes. R.S.	(*d.vi.*)	Altaic.	Altay.	1827.	——	H.b.	
4 tenélla. M.B.	(*br.pu.*)	slender.	Caucasus.	1824.	——	H.b.	Bot. mag. t. 1216.
racemòsa. α mìnor. B.M. *orientàlis.* Adam.							
5 nervòsa. S.S.	(*d.pu.*)	nerved-leaved.	——	....	4. 5.	H.b.	Bot. mag. t. 853.
latifòlia. β mìnor. B.M.							
6 lùtea. M.B.	(*ye.pu.*)	large yellow.	——	1812.	——	H.b.	Bot. mag. t. 1538.
7 latifòlia. W.	(*pu.br.*)	broad-leaved.	——	1604.	——	H.b.	——— t. 1207.
8 ruthénica. S.S.	(*pu.*)	Russian.	——	1820.	——	H.b.	
9 messanénsis. S.S.	(*pu.ye.*)	Italian.	Italy.	1825.	——	H.b.	
10 lusitánica. S.S.	(*vi.ye.*)	Portugal.	Portugal.	——	——	H.b.	
11 nìgra. B.M.	(*d.pu.ye.*)	dark-flowered.	Pyrenees.	1596.	5.	H.b.	Bot. mag. t. 664.
pyrenàica. B.M. *nec aliorum. aquitánica.* Clusius.							
12 racemòsa. R.S.	(*br.pu.*)	cluster-flowered.	Russia.	1605.	5. 6.	H.b.	Bot. mag. t. 952.
13 pyrenàica. L.	(*pu.*)	Pyrenean.	S. Europe.	....	4. 5.	H.b.	Flor. græc. t. 328.
14 tulipifòlia. M.B.	(*pu.vi.*)	Tulip-leaved.	Caucasus.	1816.	——	H.b.	
15 oblíqua. B.M.	(*pu.*)	oblique-leaved.	——	....	4.	H.b.	Bot. mag. t. 857.
16 pérsica. B.M.	(*pu.*)	Persian.	Persia.	1596.	4. 5.	H.b.	——— t. 1537.
β mìnor. B.M.	(*pu.*)	*lesser.*	——	——	——	H.b.	——— t. 962.

17 verticillàta. F.	(*pu.*)	acute-leaved.	Siberia.	1820.	4. 5.	H.b.	
18 scándens. F.	(*st.*)	blunt-leaved.	———	1827.	——	H.b.	
19 leucántha. F.	(*wh.*)	white-flowered.	Altay.	——	——	H.b.	
20 Imperiàlis. W.	(*y.re.*)	Crown Imperial.	Persia.	1596.	3. 4.	H.b.	
α rùbra.	(*re.*)	*red-flowered.*	———	——	——	H.b.	Bot. mag. t. 194.
β flàva.	(*ye.*)	*yellow-flowered.*	———	——	——	H.b.	——— t. 1215.
CYCLOBO'THRA. S.F.G.		CYCLOBO'THRA.	Hexandria Monogynia.				
1 barbàta. S.F.G.	(*ye.*)	bearded.	Mexico.	1827.	7. 8.	H.b.	Swt. br. fl. gar. t. 273.
Fritillària barbàta. K.S.							
2 purpùrea. S.F.G.	(*pu.y.*)	purple-flowered.	———	——	——	H.b.	Swt. br. fl. gar. ser. 2. t. 20.
CALOCHO'RTUS. Ph.		CALOCHO'RTUS.	Hexandria Monogynia.				
1 macrocárpus. B.R.	(*pu.*)	long-fruited.	Columbia.	1826.	7. 8.	H.b.	Bot. reg. t. 1152.
2 nítidus. D.	(*pu.ye.*)	glossy.	———	——	——	H.b.	
AMBLI'RION. Raf.		AMBLI'RION.	Hexandria Monogynia.				
1 pùdicum.	(*ye.*)	dwarf.	N. America.	1825.	....	H.b.	Pursh. fl. amer. t. 8.
Lílium pùdicum. Ph.							
2 lanceolàtum.	(*pu.*)	spear-leaved.	———	1823.	....	H.b.	
Fritillària lanceolàta. Ph. *Lílium affíne.* R.S.							
3 camtschatcénse.	(*pu.*)	Camtschatka.	Camtschatka.	1759.	5.	H.b.	Linn. trans. 10. t. 11.
Lílium camtschatcénse. L.T.							
4 álbum.	(*wh.*)	white-flowered.	N. America.	1825.	4.	H.b.	
Fritillària álba. N.							
LI'LIUM. L.		LILY.	Hexandria Monogynia. L.				
1 cándidum. W.	(*wh.*)	common white.	Levant.	1596.	6. 7.	H.b.	Bot. mag. t. 278.
β striàtum.	(*wh.re.*)	*striped-flowered.*	....	....	——	H.b.	
γ spicàtum.	(*wh.*)	*monstrous-flowered.*	......	....	——	H.b.	
δ variegàtum.	(*wh.*)	*striped-leaved.*	....	....	——	H.b.	
2 peregrìnum. R.S.	(*wh.*)	drooping-flowered.	———	——	——	H.b.	Hayn. getr. dars. 8. t. 27.
3 lancifòlium. R.S.	(*wh.*)	spear-leaved.	Japan.	1824.	....	F.b.	
4 longiflòrum. R.S.	(*wh.*)	long-flowered.	———	1819.	7. 9.	F.b.	
β suaveòlens. R.S.	(*wh.*)	*one-flowered.*	———	——	——	F.b.	Bot. reg. t. 560.
5 japónicum. R.S.	(*wh.*)	Japan.	———	1804.	——	F.b.	Bot. mag. t. 1591.
6 bulbíferum. R.S.	(*or.*)	bulb-bearing.	Italy.	1596.	——	H.b.	——— t. 36.
β umbellàtum. B.M.	(*or.*)	*umbel-flowering.*	———	——	——	H.b.	——— t. 1018.
7 cròceum. R.S.	(*sn.*)	saffron-color'd.	S. Europe.	....	——	H.b.	Redout. liliac. t. 210.
8 latifòlium. L.en.	(*or.*)	broad-leaved.	———	....	——	H.b.	
9 pubéscens. R.S.	(*or.*)	pubescent.	———	....	——	H.b.	
10 spectábile. F.	(*or.*)	showy Siberian.	Siberia.	1754.	——	H.b.	Swt. br. fl. gar. t. 75.
pensylvánicum. B.M. t. 872. *daùricum.* B.M. 1210. *in note.*							
11 monadélphum. B.M.	(*y.*)	monadelphous.	Caucasus.	1800.	6. 7.	H.b.	Bot. mag. t. 1405.
12 cóncolor. S.S.	(*sc.or.*)	self-coloured.	———	1806.	7. 8.	F.b.	——— t. 1165.
13 pùmilum. S.S.	(*sc.*)	dwarf.	Dahuria.	1816.	6. 8.	H.b.	Bot. reg. t. 132.
linifòlium. Horn. *tenuifòlium.* R.S.							
14 andìnum. N.	(*sc.*)	umbel-flowered.	Andes.	1811.	7. 8.	H.b.	Bot. reg. t. 594.
umbellàtum. Ph. *philadélphicum β andìnum.* B.R.							
15 philadélphicum. W.	(*or.*)	Philadelphian.	N. America.	1757.	——	H.b.	Bot. mag. t. 519.
16 Catesb'æi. W.	(*sc.y.*)	Catesby's.	Carolina.	1787.	7. 9.	H.b.	——— t. 259.
17 caroliniánum. M.	(*ye.*)	Carolina.	———	1820.	——	H.b.	Bot. reg. t. 580.
18 canadénse. B.M.	(*ye.*)	Canadian.	N. America.	1629.	7. 8.	H.b.	Bot. mag. t. 800.
β penduliflòrum. R.L.		*red-flowered.*	———	——	——	H.b.	Redout. lil. t. 105.
canadénse β. rùbra. Bot. mag. t. 858.							
19 supérbum. W.	(*ye.*)	superb.	———	1727.	6. 8.	H.b.	Bot. mag. t. 936.
20 Mártagon. W.	(*va.*)	Turk's-cap.	Germany.	1596.	6. 7.	H.b.	Redout. lil. t. 146.
α glábrum.	(*pu.*)	*smooth-stalked.*	———	——	——	H.b.	Bot. mag. t. 1634.
β pubéscens.	(*pu.*)	*hairy-stalked.*	———	——	——	H.b.	——— t. 893.
γ pállidum. H.	(*li.*)	*pale-flowered.*	———	——	——	H.b.	
δ elàtum. H.	(*pu.*)	*tall.*	———	——	——	H.b.	
ε dorsipúnctum. H.	(*pu.*)	*spotted backed.*	———	——	——	H.b.	
ζ purpùreum. H.	(*pu.*)	*purple.*	———	——	——	H.b.	
η perpurpùreum. H.	(*pu.*)	*dark purple.*	———	——	——	H.b.	
θ ocellàre. H.	(*pu.*)	*eyed.*	———	——	——	H.b.	
ι petiolàre. H.	(*pu.*)	*petioled.*	———	——	——	H.b.	
κ purpùreo-plènum. H.	(*pu.*)	*double purple.*	———	——	——	H.b.	
λ álbum. H.	(*wh.*)	*white-flowered.*	———	——	——	H.b.	
μ álbo-plènum.	(*wh.*)	*double white.*	———	——	——	H.b.	
21 tigrìnum. B.M.	(*or.bh.*)	tiger-spotted.	China.	1804.	7. 9.	H.b.	Bot. mag. t. 1237.
22 pyrenàicum. S.S.	(*ye.*)	Pyrenean.	Pyrenees.	1596.	6. 7.	H.b.	Redout. liliac. t. 145.
23 chalcedónicum. W.	(*sc.*)	scarlet Martagon.	Levant.	——	——	H.b.	Bot. mag. t. 30.
24 pompònium. W.	(*sc.*)	scarlet pompone.	Siberia.	1629.	5. 6.	H.b.	——— t. 971.

25 pulchéllum. F. ...... pretty. Dahuria. 1829. .... H.b.
GLORI'OSA. L. GLORI'OSA. Hexandria Monogynia. L.
1 viréscens. B.M. (gr.) greenish-flowered. Mozambique. 1822. 7. 8. S.♃.◡. Bot. mag. t. 2539.
2 supérba. L. (or.) superb. E. Indies. 1690. —— S.♃.◡. Bot. reg. t. 77.
3 símplex. L. (bl.) simple-leaved. Senegal. 1756. —— S.♃.
Methónica gloriòsa. Sal.
ERYTHR'ONIUM. L. DOG'S TOOTH VIOLET. Hexandria Monogynia. L.
1 Déns cànis. W. (pu.) common. Europe. 1596. 3. 4. H.b. Redout. liliac. t. 194.
2 americànum. B.M. (ye.) yellow-flower'd. N. America. 1665. 4. 5. H.b. Bot. mag. t. 1113.
lanceolatùm. Ph.
3 álbidum. N. (wh.) white American. ——— 1825. —— H.b.

ORDO CCXVII.

MELANTHACEÆ. *Brown prodr.* 272.

BULBOC'ODIUM. W. BULBOC'ODIUM. Hexandria Monogynia. L.
1 vérnum. S.S. (li.) spring-flowering. Spain. 1629. 2. 3. H.b. Bot. mag. t. 153.
2 versícolor. S.S. (li.ye.) various-colour'd. Russia. 1820. 9. 10. H.b. Bot. reg. t. 571.
Cólchicum versícolor. B.R.
CO'LCHICUM. S.S. MEADOW SAFFRON. Hexandria Trigynia. L.
1 byzantìnum. S.S. (li.) broad-leaved. Levant. 1629. 9. 10. H.b. Bot. mag. t. 1122.
2 autumnàle. E.B. (li.) common. Britain. .... —— H.b. Eng. bot. v. 2. t. 133.
β álbum. (wh.) *white flowered.* ——— .... —— H.b.
γ multipléx. (li.) *double-flowered.* ——— .... —— H.b.
3 crociflòrum. B.M. (li.) Crocus-flowered. ——— .... —— H.b. Bot. mag. t. 2673.
4 variegàtum. W. (std.) chequer-flowered. Greece. 1629. 8. 10. H.b. ——— t. 1028.
5 arenàrium. S.S. (li.) sand. Hungary. 1816. 9. 10. H.b. W. et K. hung. 2. t. 179.
6 umbròsum. F. (li.) wood. Siberia. —— —— H.b. Bot. reg. t. 541.
7 montànum. S.S. (li.) mountain. S. Europe. 1818. —— H.b.
UVUL'ARIA. S.S. UVUL'ARIA. Hexandria Monogynia. L.
1 perfoliàta. S.S. (gr.ye.) perfoliate. N. America. 1710. 5. 6. H.♃. Sm. exot. bot. 1. t. 49.
2 flàva. S.S. (ye.) yellow-flowered. ——— .... —— H.♃. ——— 1. t. 50.
3 lanceolàta. H.K. (ye.) spear-leaved. ——— 1710. 7. 8. H.♃. Cornut. canad. t. 41.
4 granditlòra. H.K. (ye.) large yellow. ——— 1802. 5. 6. H.♃. Sm. exot. bot. 1. t. 51.
5 sessilifòlia. S.S. (st.) sessile-leaved. ——— 1790. 6. H.♃. ——— 1. t. 52.
6 pubérula. S.S. (ye.) pubescent. ——— 1818. —— H.♃. Swt. br. fl. gar. s. 2. t. 21.
DISP'ORUM. D.P. DISP'ORUM. Hexandria Monogynia.
1 parviflòrum. D.P. (ye.) small-flowered. Nepaul. 1823. 6. 8. F.♃.
2 Pitsùtum. D.P. (ye.) umbel-flowered. ——— 1826. —— F.♃.
3 fúlvum. Sal. (br.pu.) brown-flowered. China. 1801. 6. 11. F.♃. Bot. mag. t. 916.
Uvulària chinénsis. B.M.
SCHELHAMM'ERA. B.P. SCHELHAMM'ERA. Hexandria Monogynia. S.S.
1 undulàta. B.P. (ro.) wave-leaved. N. S. W. 1825. 6. 9. F.♃. Bot. mag. t. 2712.
2 multiflòra. B.P. (ro.) many-flowered. ——— 1826. —— F.♃.
BURCHA'RDIA. B.P. BURCHA'RDIA. Hexandria Monogynia. S.S.
umbellàta. B.P. (wh.) umbel-flowered. N. S. W. 1825. 6. 8. F.♃.
ANGUILL'ARIA. B.P. ANGUILL'ARIA. Hexandria Trigynia.
1 dioíca. B.P. (br.) dioecious. N. S. W. 1826. .... F.♃.
2 biglandulòsa. B.P. (br.) two-glanded. ——— 1822. 6. 8. F.♃.
3 índica. B.P. (d.pu.) Indian. E. Indies. —— —— F.♃.
Melánthium índicum. L
ORNITHOGLO'SSUM. S.S. BIRD'S-TONGUE. Hexandria Trigynia. S.S.
1 undulàtum. B.F.G. (g.pu.) wave-leaved. C. B. S. 1825. 8. 10. F.b. Swt. br. fl. gar. t. 131.
2 víride. S.S. (gr.br.) green-flowered. ——— 1788. 10. 11. F.b. Sal. par. lond. t. 54.
Melánthium víride. Bot. mag. t. 994.
ANDROCY'MBIUM. W. ANDROCY'MBIUM. Hexandria Trigynia. S.S.
1 melanthioídes. S.S. (gr.) narrow-leaved. C. B. S. 1825. 8. 10. F.b.
2 eucomoídes. S.S. (gr.) Tulip-leaved. ——— 1794. 3. 5. F.b. Swt. br. fl. gar. t. 165.
Melánthium eucomoídes. B.M. 641.
3 volutàre. B.T. (gr.) rolled-leaved. ——— 1816. —— F.b.
MELA'NTHIUM. W. MELA'NTHIUM. Hexandria Trigynia. L.
1 júnceum. W. (li.) Rush-leaved. C. B. S. 1788. 6. 11. F.b. Bot. mag. t. 558.
2 secúndum. W. (wh.) side-flowering. ——— 1812. —— F.b. Lam. ill. t. 269. f. 2.

3 capénse. w. (*wh.*) spotted-flowered. C. B. S. 1768. 5. 6. F.b.
4 ciliàtum. w. (*wh.*) fringed. ——— 1825. —— F.b.
5 gramíneum. s.s. (*wh.*) grassy. Barbary. 1823. —— F.b. Cavan. icon. t. 587. f. 1.
KO'LBEA. R.S. Ko'lbea. Hexandria Trigynia.
Breyniàna. R.S. (*ye.*) Breynius's. C. B. S. 1787. 6. 7. F.b. Bot. mag. t. 767.
Melánthium uniflòrum. w.
WURMB'EA. W. Wurmb'ea. Hexandria Trigynia. W.
1 campanulàta. w. (*wh.*) bell-flowered. C. B. S. 1788. 5. 6. F.b. Bot. mag. t. 1291.
Melánthium monopétalum. B.M.
2 purpùrea. H.K. (*pu.*) purple. ——— —— —— F.b. Bot. mag. t. 694.
Melánthium spicàtum. B.M. *Wurmbèa capénsis.* Andr. reposit. t. 221.
3 pùmila. w. (*wh.*) dwarf. C. B. S. 1800. 5. 6. F.b.
4 longiflòra. w. (*wh.*) long-flowered. ——— —— —— F.b.
TOF'IELDIA. L.T. Tof'ieldia. Hexandria Trigynia. S.S.
1 palústris. L.T. (*st.*) Scotch. Britain. .... 7. 8. H.♃. Eng. bot. v. 8. t. 536.
2 racemòsa. Hop. (*st.*) racemed. Germany. 1827. —— H.♃.
3 alpìna. L.T. (*st.*) Alpine. Europe. .... 7. 8. H.♃. Redout. lil. t. 256.
palústris. DC.
4 pùbens. L.T. (*st.*) downy. N. America. 1790. 4. 5. H.♃. Pluk. mant. 29. t. 342. f. 3
5 glutinòsa. L.T. (*ye.*) glutinous. ——— 1827. —— H.♃. Linn. trans. v. 12. t. 8. f. 2.
XEROPHY'LLUM. M. Xerophy'llum. Hexandria Trigynia. S.S.
1 gramíneum. s.s. (*wh.*) grass-leaved. N. America. 1812. 5. 6. H.♃. Bot. mag. t. 1599.
Helònias gráminea. B.M.
2 tènax. s.s. (*wh.*) tough-leaved. ——— 1811. —— H.♃. Pursh. amer. 1. t. 9.
3 asphodeloídes. s.s. (*wh.*) Asphodel-like. ——— 1765. —— H.♃. Bot. mag. t. 748.
Helònias asphodeloídes. B.M.
HEL'ONIAS. W. Hel'onias. Hexandria Trigynia. L.
1 bullàta. w. (*pu.*) broad-leaved. N. America. 1758. 4. 5. H.♃. Bot. mag. t. 747.
latifòlia. M.
2 erythrospérma. M. (*wh.*) channel-leaved. ——— 1770. 6. H.♃. Bot. mag. t. 803.
l'æta. B.M. *Melánthium l'ætum.* W.
3 angustifòlia. M. (*wh.*) narrow-leaved. ——— 1812. —— H.♃. Bot. mag. t. 1540.
ZIGAD'ENUS. M. Zigad'enus. Hexandria Monogynia. S.S.
1 glabérrimus. w. (*wh.*) smooth-leaved. N. America. 1811. 5. 6. H.♃. Mich. amer. 1. t. 22.
Helònias glabérrima. Bot. mag. t. 1680.
2 bracteàtus. (*wh.*) bracted. ——— —— —— H.♃. Bot. mag. t. 1703.
Helònias bracteàta. B.M.
3 élegans. Ph. (*wh.*) elegant. ——— 1830. 6. 7. H.♃.
NOL'INA. M. Nol'ina. Hexandria Trigynia. S.S.
georgiàna. M. (*wh.*) Georgian. Georgia. 1812. 7. 8. H.♃. Pluk. mant. t. 342. f. 1.
LEIMA'NTHIUM. W. Leima'nthium. Polygamia Monœcia.
virgínicum. w. (*wh.*) Virginian. N. America. 1768. 6. 7. H.♃. Bot. mag. t. 985.
Helònias virgínica. B.M. *Melánthium virgínicum.* Ph.
2 monoícum. (*wh.*) monoecious. ——— 1817. —— H.♃.
Melánthium monoícum. Ph.
3 hy'bridum. (*wh.*) hybrid. ——— 1822. —— H.♃.
Melánthium hy'bridum. Ph. *racemòsum.* M.
PELIOSA'NTHES. A.R. Peliosa'nthes. Hexandria Monogynia.
1 hùmilis. A.R. (*gr.*) small. E. Indies. 1809. 5. 6. S.♃. Bot. mag. t. 1532.
2 Tèta. A.R. (*gr.*) large. ——— 1807. 4. S.♃. ——— t. 1302.
CYM'ATION. S.S. Cym'ation. Hexandria Trigynia. R.S.
lævigàtum. s.s. (*bl.*) smooth. C. B. S. 1824. 4. 5. G.b.
Lichtensteínia lævigàta. w.
VER'ATRUM. L. White Hellebore. Polygamia Monœcia. L.
1 nìgrum. w. (*da.pu.*) dark-flowered. Siberia. 1596. 6. 7. H.♃. Bot. mag. t. 963.
2 álbum. w. (*wh.*) white-flowered. S. Europe. 1548. 6. 8. H.♃. Flor. dan. t. 1120.
3 Lobeliànum. W. en. (*wh.*) Lobel's. ——— 1818. —— H.♃.
4 víride. w. (*gr.*) green-flowered. N. America. 1742. 7. 8. H.♃. Bot. mag. t. 1096.
5 parviflòrum. M. (*gr.*) small-flowered. ——— 1809. —— H.♃.
6 angustifòlium. Ph. (*gr.*) narrow-leaved. ——— 1823. —— H.♃.
CHAMÆLI'RIUM. W. Chamæli'rium. Diœcia Hexandria.
caroliniànum. w. (*st.*) spiked-flowered. N. America. 1759. 7. 8. H.♃. Bot. mag. t. 1062.
Helònias lùtea. B.M. *dioíca.* Ph. *Ophyostàchys virgínica.* Redout. liliac. t. 464.

ORDO CCXVIII.

PONTEDEREÆ. *Kth. synops.* 1. *p.* 273.

PONTED'ERIA. L.		PONTED'ERIA.	Hexandria Monogynia. L.				
1 cordàta. s.s.	(*bl.*)	heart-leaved.	N. America.	1759.	6. 8.	H.*w.* ♃.	Bot. mag. t. 1156.
2 azùrea. s.s.	(*pu.*)	large-flowered.	Jamaica.	1822.	——	S.*w.* ♃.	——— t. 2932.
3 lanceolàta. s.s.	(*bl.*)	lanceolate.	Carolina.	....	8. 9.	G.*w.* ♃.	Lodd. bot. cab. t. 613.
4 angustifòlia. s.s.	(*bl.*)	narrow-leaved.	N. America.	1806.	6. 8.	H.*w.* ♃.	
5 crassipès. s.s.	(*bl.*)	thick-footstalked.	Brazil.	1822.	——	S.*w.* ♃.	Mart. n. pl. bras. t. 4.
6 dilatàta. s.s.	(*bl.*)	spreading.	E. Indies.	1806.	5.	S.*w.* ♃.	Andr. reposit. t. 490.
HETERANTH'ERA. R.S.		HETERANTH'ERA.	Triandria Monogynia. W.				
1 renifórmis. R.S.	(*wh.*)	kidney-leaved.	S. America.	1812.	6. 7.	G.*w.* ♃.	Flor. per. 1. t. 71. f. a.
2 acùta. W.	(*wh.*)	acute-leaved.	N. America.	——	——	H.*w.* ♃.	
3 limòsa. R.S.	(*bl.*)	oval-leaved.	W. Indies.	1822.	——	S.*w.* ♃.	Mich. amer. 1. t. 5. f. 1.
LEPTA'NTHUS. M.		LEPTA'NTHUS.	Triandria Monogynia. Ph.				
gramíneus. M.	(*ye.*)	grass-like.	N. America.	1823.	7. 8.	H.*w.* ♃.	Hook. ex. flor. t. 94.

Heteranthèra gramínea. Ph. *Schóllera graminifòlia.* W.

ORDO CCXIX.

COMMELINEÆ. *Brown prodr.* 1. *p.* 268.

DICHORISA'NDRA. S.S.		DICHORISA'NDRA.	Hexandria Monogynia. S.S.				
1 thyrsiflòra. S.S.	(*bl.*)	bunch-flowered.	Brazil.	1820.	8. 10.	S. ♄.	Bot. reg. t. 682.
2 grácilis. S.S.	(*bl.*)	slender.	———	1824.	——	S. ♄.	
3 oxypétala. B.M.	(*pu.*)	sharp-flowered.	———	1825.	6. 9.	S. ♃.	Bot. mag. t. 2721.
4 pubérula. S.S.	(*bl.*)	pubescent.	———	——	——	S. ♃.	
CAMP'ELIA. K.S.		CAMP'ELIA.	Hexandria Monogynia. S.S.				
Zanònia. K.S.	(*wh.*)	Gentian-leaved.	S. America.	1759.	7. 12.	S. ♃.	Redout. lil. 4. t. 192.

Commelìna Zanònia. R.L. *Tradescántia Zanònia.* W. *Zanònia bibractèata.* Cram.

TRADESCA'NTIA. L.		SPIDERWORT.	Hexandria Monogynia. L.				
1 fuscàta. S.S.	(*bl.*)	rusty.	S. America.	1820.	8. 12.	S. ♃.	Bot. reg. t. 482.
2 malabárica. S.S.	(*pu.*)	Grass-leaved.	E. Indies.	1776.	7. 8.	S. ♃.	Rheed. mal. 9. t. 63.
3 tuberòsa. S.S.	(*bl.*)	tuberous-rooted.	———	1817.	——	S. ♃.	
4 speciòsa. S.S.	(*pu.*)	handsome.	Mexico.	1825.	——	S. ♃.	
5 díscolor. S.S.	(*wh.*)	purple-leaved.	S. America.	1783.	4. 9.	S. ♄.	Bot. mag. t. 1192.
6 cordifòlia. S.S.	(*pu.*)	heart-leaved.	Jamaica.	1822.	——	S. ⊙.	
7 erécta. S.S.	(*bl.*)	upright.	Mexico.	1794.	7. 8.	H. ⊙.	Bot. mag. t. 1340.
8 undàta. S.S.	(*bl.*)	wave-leaved.	S. America.	1817.	8. 10.	H. ⊙.	
9 pulchélla. S.S.	(*pu.*)	pretty.	Mexico.	1825.	——	G. ♃.	
10 crassifòlia. S.S.	(*pu.*)	thick-leaved.	———	1796.	7. 10.	H. ♃.	Bot. mag. t. 1598.
11 subáspera. S.S.	(*bl.*)	Lyon's.	N. America.	1812.	5. 10.	H. ♃.	——— t. 1597.
12 virgínica. S.S.	(*bl.*)	Virginian.	———	1629.	5. 8.	H. ♃.	Park. parad. t. 151. f. 4.
α cærùlea.	(*bl.*)	*blue-flowered.*	———	——	——	H. ♃.	Bot. mag. t. 105.
β rùbra.	(*re.*)	*red-flowered.*	———	——	——	H. ♃.	
γ álbida.	(*wh.*)	*whitish-flowered.*	———	——	——	H. ♃.	
δ nìvea.	(*wh.*)	*snowy white.*	———	1824.	——	H. ♃.	
ε pilòsa. B.R.	(*bl.*)	*hairy.*	———	1827.	——	H. ♃.	Bot. reg. t. 1055.
13 congésta. H.E.	(*bl.*)	close-flowered.	———	1825.	——	H. ♃.	
14 ròsea. S.S.	(*ro.*)	rose-coloured.	Carolina.	1802.	5. 10.	H. ♃.	Swt. br. fl. gar. t. 181.
15 multiflòra. S.S.	(*wh.*)	many-flowered.	Jamaica.	1824.	——	S. ♃.	
16 crássula. B.M.	(*wh.*)	white-flowered.	Brazil.	1826.	12. 1.	S. ♃.	Bot. mag. t. 2935.
17 procúmbens. S.S.	(*wh.*)	procumbent.	Caracas.	1823.	——	S. ♃.	Jacq. ic. 2. t. 355.
multiflòra. Jacq. *non* W.							
18 parviflòra. S.S.	(*bl.*)	small-flowered.	Peru.	1822.	——	H. ♃.	Flor. per. 3. t. 272. f. 2.
19 latifòlia. S.S.	(*vi.*)	broad-leaved.	Lima.	1820.	9. 11.	H. ⊙.	——— 3. t. 272.
20 divaricàta. S.S.	(*bl.*)	spreading.	Trinidad.	——	5. 8.	S. ♃.	Aub. gui. 1. t. 12.
21 geniculàta. S.S.	(*wh.*)	knotted.	W. Indies.	1783.	7. 8.	S. ♃.	Jacq. amer. t. 64.
22 paniculàta. S.S.	(*bl.*)	panicled.	E. Indies.	1824.	——	S. ⊙.	

CYAN'OTIS. D.P. Cyan'otis. Hexandria Monogynia.
1 axillàris. D.P. (*bl.*) axil-flowered. E. Indies. 1822. 8. 10. S.⊙. Rheed. mal. 10. t. 13.
2 cristàta. D.P. (*bl.*) crest-bunched. Ceylon. 1770. 7. 9. S.⊙. Bot. mag. t. 1435.
Tradescántia cristàta. B.M.
3 barbàta. D.P. (*bl.*) bearded. Nepaul. 1824. —— H.⊙.
CARTON'EMA. B.P. Carton'ema. Hexandria Monogynia.
spicàtum. B.P. (*ye.*) spiked. New Holland. 1822. 7. 8. S.♃.
CALLI'SIA. L. Calli'sia. Triandria Monogynia. L.
rèpens. L. (*gr.*) creeping. W. Indies. 1776. 6. 7. S.♃. Jacq. amer. t. 11.
ANEIL'EMA. B.P. Aneil'ema. Triandria Monogynia.
1 biflòra. B.P. (*bl.*) two-flowered. N. S. W. 1820. 7. 8. F.♃.
2 acuminàta. B.P. (*bl.*) sharp-pointed. ———— 1822. —— F.♃.
3 longifòlia. H.E.F. (*bl.*) long-leaved. Zanzibar. 1823. 8. 9. S.♃. Hook. ex. flor. t. 204.
4 sínica. B.R. (*bl.*) Chinese. China. 1820. 5. 6. G.♃. Bot. reg. t. 659.
5 nudiflòra. B.P. (*bl.pu.*) naked-flowered. E. Indies. —— 8. 10. S.⊙. Rheed. mal. 9. t. 63.
6 spiràta. B.P. (*bl.*) spear-leaved. ———— 1783. 7. 8. S.⊙.
Commelìna spiràta. L.
7 affínis. B.P. (*bl.*) rough-angled. N. S. W. 1825. —— F.♃.
COMMEL'INA. B.P. Commel'ina. Triandria Monogynia. L.
1 angustifòlia. R.S. (*bl.*) narrow-leaved. Carolina. 1812. 7. 8. F.♃.
2 truncàta. R.S. (*pa.bl.*) truncate-bracted. ———— 1822. 7. 9. H.♃.
3 cyánea. B.P. (*bl.*) bright blue. N. S. W. 1820. —— F.♃.
4 dianthifòlia. DC. (*bl.*) Pink-leaved. .... 1816. —— S.♃. Redout. lil. 7. t. 390.
5 longicaùlis. R.S. (*bl.*) long-stalked. Caracas. 1806. 8. S.♃. Jacq. ic. 2. t. 294.
6 hirtélla. R.S. (*bl.*) hairy. N. America. 1820. 7. 8. H.♃.
7 virgínica. R.S. (*bl.*) Virginian. ———— 1779. —— H.♃. Pluk. alm. t. 174. f. 4.
8 africàna. R.S. (*ye.*) yellow-flowered. C. B. S. 1759. 5. 10. G.♃. Bot. mag. t. 1431.
9 fasciculàta. R.S. (*bl.*) woolly tubered. Lima. 1817. —— H.♃. Flor. per. 1. t. 72. f. b.
10 communis. R.S. (*bl.*) common. America. 1732. 6. 7. H.⊙. Redout. lil. 4. t. 206.
11 polygàma. R.S. (*bl.*) polygamous. Japan. .... —— H.♃. Kæmpf. am. p. 888. ic.
12 caroliniàna. R.S. (*bl.*) Carolina. Carolina. .... —— H.♃.
13 dùbia. R.S. (*bl.*) dubious. .... 1819. 7. 9. H.♃. Redout. lil. 6. t. 359.
14 tuberòsa. R.S. (*bl.*) tuberous-rooted. Mexico. 1732. —— H.♃. Andr. reposit. t. 399.
15 cœléstis. R.S. (*bl.*) sky-blue. .... 1813. 7. 10. H.♃. Swt. br. fl. gar. t. 3.
16 pállida. R.S. (*bl.*) pale-flowered. Mexico. 1822. —— H.♃. Willd. hort. ber. 2. t. 87.
rùbens. Redoute liliac. 7. t. 367.
17 turbinàta. R.S. (*bl.*) top-shaped. Santa Cruz. 1824. —— S.♃.
18 oblíqua. R.S. (*bl.*) oblique-leaved. S. America. 1827. —— S.♃.
19 erécta. R.S. (*bl.*) upright. Virginia. 1732. 8. 9. H.♃. Dill. elt. 94. t. 77. f. 88.
20 móllis. R.S. (*bl.*) soft. Caracas. 1804. —— H.♃. Jacq. ic. 2. t. 293.
21 bengalénsis. R.S. (*bl.*) Bengal. Bengal. 1794. 6. S.♃. Burm. ind. t. 7. f. 3.
22 parviflòra. L.en. (*bl.*) small-flowered. .... 1823. —— S.♃.
23 deficiens. B.M. (*bl.*) deficient. Brazil. 1825. 1. 12. S.♃. Bot. mag. t. 2644.

ORDO CCXX.

PALMÆ. *Brown prodr. p.* 266.

CHAMÆD'OREA. W. Chamæd'orea. Diœcia Hexandria. W.
1 grácilis. W. (*st.*) slender. Caracas. 1803. .... S.♄. Jacq. schœn. 2. t. 247-8.
2 fràgrans. M.P. (*st.*) fragrant. Peru. 1823. .... Martius Palm. ic.
S'ABAL. P.S. S'abal. Hexandria Trigynia. P.S.
1 Adansòni. B.M. (*st.*) dwarf. Florida. 1810. 6. 8. S.♃. Bot. mag. t. 1434.
Córypha mìnor. Jacq. vind. 3. t. 8. *Cham'ærops acàulis.* M.
2 Blackburniàna. L.C. (*st.*) Blackburn's. S. America. 1824. .... S.♃.
3 graminifòlia. L.C. (*st.*) grass-leaved. ———— 1822. .... S.♃.
4 umbraculífera. L.C. (*st.*) umbrella. ———— —— .... S.♃.
RHA'PIS. W. Rha'pis. Polygamia Monœcia. W.
1 flabellifórmis. W. (*st.*) creeping-rooted. China. 1774. 8. S.♃. Bot. mag. t. 1371.
2 arundinàcea. W. (*st.*) simple-leaved. Carolina. 1765. 9. G.♄.
CHAM'ÆROPS. W. Cham'ærops. Polygamia Diœcia. W.
1 excélsa. L.C. (*st.*) tall. Nepaul. 1822. .... S.♄.
2 grácilis. L.C. (*st.*) slender. S. America. —— .... S.♄.
3 guianénsis. L.C. (*st.*) Guiana. Guiana. 1824. .... S.♄.
4 hùmilis. S.S. (*st.*) dwarf fan palm. S. Europe. 1731. 2. 3. G.♄. Andr. reposit. t. 599.

5 serrulàta. S.S.	(*st.*)	saw-leaved.	N. America.	1809.	....	G. ♄.	
6 Hy'strix. S.S.	(*st.*)	Porcupine.	Georgia.	1801.	....	G. ♄.	
7 Palmétto. S.S.	(*st.*)	smooth-stemm'd.	Carolina.	1809.	....	G. ♄.	
LIVIST'ONA. B.P.		LIVIST'ONA.	Hexandria Monogynia. S.S.				
1 inérmis. B.P.	(*st.*)	spineless.	N. Holland.	1823.	....	S. ♄.	
2 hùmilis. B.P.	(*st.*)	dwarf spiny.	———	———	....	S. ♄.	
LODOI'CEA. S.S.		DOUBLE COCOA-NUT.	Diœcia Monadelphia.				
Sechellàrum. B.M.	(*st.*)	Seychelle's Island.	Seychelle's Isl.	1827.	....	S. ♄.	Bot. mag. t. 2734-2738.
LAT'ANIA. W.		LAT'ANIA.	Diœcia Monadelphia. W.				
1 rùbra. W.	(*st.*)	red.	Mauritius.	1788.	....	S. ♄.	Jacq. frag. 1. t. 8.
2 glaucophy'lla. L.C.	(*st.*)	glaucous-leaved.	— ———	1821.	....	S. ♄.	
3 borbònica. W.	(*st.*)	Bourbon.	Bourbon.	———	....	S. ♄.	Jacq. frag. 1. t. 11.
chinénsis. Jacq. *Livistònæ species?* B.P.							
CO'RYPHA. L.		FAN PALM.	Hexandria Monogynia. W.				
1 austràlis. B.P.	(*st.*)	New Holland.	N. S. W.	1822.	....	S. ♄.	
2 elàta. H.B.	(*st.*)	tall.	E. Indies.	1820.	....	S. ♄.	
3 glaucéscens. L.C.	(*st.*)	glaucescent.	— ———	———	....	S. ♄.	
4 umbraculífera. W.	(*st.*)	great.	———	1742.	....	S. ♄.	Rheed. mal. 3. t. 1-12.
5 Után. S.S.	(*st.*)	Molucca.	———	1822.	....	S. ♄.	
TAL'IERA. S.S.		TAL'IERA.	Hexandria Monogynia. S.S.				
bengalénsis. S.S.	(*st.*)	Bengal.	E. Indies.	1822.	....	S. ♄.	
Córypha Talìera. H.B.							
PH'ŒNIX. W.		DATE PALM.	Diœcia Hexandria. W.				
1 dactylífera. W.	(*st.*)	common.	Levant.	1597.	....	G. ♄.	Kæmpf. am. 686. t. 1. 2.
2 sylvéstris. H.B.	(*st.*)	wood.	E. Indies.	1820.	....	S. ♄.	
3 paludòsa. H.B.	(*st.*)	marsh.	———	———	....	S. ♄.	
4 reclinàta. W.	(*st.*)	reclining.	C. B. S.	1792.	....	G. ♄.	Jacq. frag. 1. t. 24.
5 farinífera. W.	(*st.*)	small.	E. Indies.	1800.	....	S. ♄.	Roxb. corom. 1. t. 74.
6 acaùlis. S.S.	(*st.*)	stalkless.	——— ———	1820.	....	S. ♃.	
7 leonénsis. L.C.	(*st.*)	African.	Sierra Leon.	———	....	S. ♄.	
8 pygm'æa. L.C.	(*st.*)	pigmy.	....	———	....	S. ♃.	
S'AGUS. W.		SAGO PALM.	Monœcia Hexandria. W.				
1 Rúmphii. W.	(*st.*)	Rumphius's.	E. Indies.	1800.	....	S. ♄.	Rumph. amb. 1. t. 17. 18.
2 vinífera. P.S.	(*st.*)	prickly-leaved.	Africa.	1820.	....	S. ♄.	Beauv. fl. d'ow. t. 45.
3 Rúffia. W.	(*st.*)	Madagascar.	Madagascar.	———	....	S. ♄.	Jacq. frag. p. 7. t. 4. t. 2.
4 pedunculàta. Poir.	(*st.*)	peduncled.	——— ———	———	....	S. ♄.	
HYPH'ÆNE. G.		HYPH'ÆNE.	Diœcia Hexandria.				
coriàcea. G.	(*st.*)	leathery.	Egypt.	1822.	....	S. ♄.	
Cucífera thebàica. Del.							
AR'ECA. W.		CABBAGE-TREE.	Monœcia Monadelphia. W.				
1 Cátechu. W.	(*st.*)	medicinal.	E. Indies.	1690.	....	S. ♄.	Roxb. corom. 1. t. 75.
2 hùmilis. W.	(*st.*)	dwarf.	———	1814.	....	S. ♄.	Rumph. amb. 1. t. 7.
3 rùbra. W.	(*st.*)	red.	Isl. France.	1823.	....	S. ♄.	
4 crinìta. W.	(*st.*)	hairy.	———	1819.	....	S. ♄.	
5 oleràcea. W.	(*st.*)	esculent.	W. Indies.	1656.	....	S. ♄.	Jacq. amer. t. 170.
6 triándra. H.B.	(*st.*)	three-anthered.	E. Indies.	1819.	....	S. ♄.	
7 lutéscens. W.	(*st.*)	yellowish.	Mauritius.	———	....	S. ♄.	
8 exìlis. L.C.	(*st.*)	slender.	....	1822.	....	S. ♄.	
9 Mánicot. L.C.	(*st.*)	Manicot.	....	———	....	S. ♄.	
10 montàna. L.C.	(*st.*)	mountain.	....	1819.	....	S. ♄.	
ŒNOCA'RPUS. M.P.		ŒNOCA'RPUS.	Monœcia Hexandria.				
Batàua. M.P.	(*st.*)	American.	S. America.	1822.	....	S. ♄.	Martius. palm. ic.
EUTE'RPE. G.		EUTE'RPE.	Polygamia Monœcia.				
1 globòsa. G.	(*st.*)	globular.	E. Indies.	1819.	....	S. ♄.	
Arèca spicàta. Lam.							
2 pisífera. G.	(*st.*)	Pea-bearing.	———	———	....	S. ♄.	
WALLI'CHIA. S.S.		WALLI'CHIA.	Polygamia Monœcia.				
caryotoídes. S.S.	(*st.*)	Caryota-like.	E. Indies.	1818.	....	S. ♄.	Roxb. corom. ic.
EL'AIS. W.		OILY PALM.	Diœcia Hexandria. W.				
1 guineénsis. W.	(*st.*)	Guinea.	Guinea.	1730.	....	S. ♄.	Jacq. amer. t. 172.
2 melanocócca. G.	(*st.*)	black-fruited.	S. America.	1820.	....	S. ♄.	
3 pernambucàna. L.C.	(*st.*)	Pernambuca.	——— ———	———	....	S. ♄.	
SYA'GRUS. M.P.		SYA'GRUS.	Polygamia Monœcia.				
cocoídes. M.P.	(*st.*)	Cocoa-like.	Brazil.	1823.	....	S. ♄.	Mart. palm. ic.
EL'ATE. W.		EL'ATE.	Monœcia Hexandria.				
sylvéstris. W.	(*st.*)	prickly-leaved.	E. Indies.	1763.	4. 6.	S. ♄.	Rheed. mal. 3. t. 22-25.
ACROC'OMIA. M.P.		ACROC'OMIA.	Monœcia Hexandria.				
1 sclerocárpa. M.P.	(*st.*)	dry-fruited.	Brazil.	1822.	....	S. ♄.	Mart. palm. ic.
2 aculeàta. L.C.	(*st.*)	prickly.	W. Indies.	1796.	....	S. ♄.	Jacq. amer. 278. t. 169.
Còcos aculeàta. W.							

3 fusifórmis.	(*st.*) great Macaw-tree.	W. Indies.	1731.	....	S. ♄.	
Còcos fusifórmis. W.						
4 mìnor. L.C.	(*st.*) lesser.	S. America.	1823.	....	S. ♄.	
C'OCOS. W.	COCOA-NUT-TREE. Monœcia Hexandria. W.					
1 nucífera. W.	(*st.*) common.	E. Indies.	1690.	....	S. ♄.	Roxb. corom. 1. t. 73.
2 flexuòsa. M.P.	(*st.*) flexuose.	Brazil.	1823.	....	S. ♄.	Mart. palm. ic.
3 plumòsa. L.C.	(*st.*) feathered.	W. Indies.	——	....	S. ♄.	
MAXIMILI'ANA. M.P.	MAXIMILI'ANA. Polygamia Monœcia.					
règia. M.P.	(*st.*) membranous-fld.	Brazil.	1822.	....	S. ♄.	Mart. palm. ic.
THR'INAX. W.	THR'INAX. Hexandria Monogynia. W.					
parviflòra. W.	(*st.*) small.	Jamaica.	1778.	....	S. ♄.	
ASTROC'ARYUM. M.P.	ASTROC'ARYUM. Monœcia Hexandria.					
1 acaùle. M.P.	(*st.*) stalkless.	Brazil.	1823.	....	S. ♃.	Mart. palm. ic.
2 campéstre. M.P.	(*st.*) field.	———	1824.	....	S. ♄.	——— ic.
3 aculeàtum. S.S.	(*st.*) prickly.	Guiana.	——	....	S. ♄.	
4 vulgàre. M.P.	(*st.*) common.	Brazil.	1823.	....	S. ♄.	Mart. palm. ic.
5 Murumùru. M.P.	(*st.*) tall.	———	——	....	S. ♄.	——— ic.
GEON'OMA. W.	GEON'OMA. Diœcia Hexandria.					
1 simplícifrons. W.	(*st.*) simple-fronded.	S. America.	1823.	....	S. ♃.	
2 macrostàchys. M.P.	(*st.*) large-spiked.	———	——	....	S. ♃.	Mart. palm. ic.
3 acaùlis. M.P.	(*st.*) stalkless.	Brazil.	1822.	....	S. ♃.	——— ic.
4 pinnàtifrons. W.	(*st.*) pinnated.	S. America.	——	....	S. ♃.	
5 Schottiàna. L.C.	(*st.*) Schott's.	Brazil.	——	....	S. ♃.	
MAURI'TIA. L.	MAURI'TIA. Diœcia Hexandria. W.					
1 flexuòsa. W.	(*st.*) flexuose.	W. Indies.	1819.	....	S. ♄.	
2 armàta. M.P.	(*st.*) spiny.	Brazil.	1822.	....	S. ♄.	Mart. palm. ic.
3 vinífera. M.P.	(*st.*) straight-stem'd.	———	——	....	S. ♄.	——— ic.
LICU'ALA. W.	LICU'ALA. Hexandria Monogynia. W.					
1 spinòsa. W.	(*st.*) spiny.	E. Indies.	1802.	....	S. ♄.	Rumph. amb. 1. t. 9.
2 peltàta. H.E.	(*st.*) peltate.	———	1819.	....	S. ♄.	
CA'LAMUS. W.	CA'LAMUS. Hexandria Monogynia. S.S.					
1 vèrus. S.S.	(*st.*) true.	E. Indies.	1820.	....	S. ♄.	Rumph. amb. 5. t. 54.
2 nìger. S.S.	(*st.*) black.	———	——	....	S. ♄.	——— 5. t. 52.
3 rudéntum. S.S.	(*st.*) common.	China.	1812.	....	S. ♄.	
4 Zalácca. S.S.	(*st.*) Java.	E. Indies.	——	....	S. ♄.	Gært. fr. 2. t. 139. f. 2.
5 álbus. P.S.	(*st.*) common.	———	——	....	S. ♄.	Rumph. amb. 5. f. 52.
6 Dràco. S.S.	(*st.*) Dragon.	———	1819.	....	S. ♄.	——— 5. t. 58. f. 1
SEAFO'RTHIA. B.P.	SEAFO'RTHIA. Polygamia Monœcia.					
élegans. B.P.	(*st.*) elegant.	N. Holland.	1822.	....	S. ♄.	
CARY'OTA. W.	CARY'OTA. Monœcia Polyandria. W.					
1 ùrens. W.	(*st.*) torn-leaved.	E. Indies.	1788.	....	S. ♄.	Rheed. mal. 1. t. 11.
2 hórrida. W.	(*st.*) horrid.	S. America.	1820.	....	S. ♄.	
DIPLOTH'EMIUM. M.P.	DIPLOTH'EMIUM. Polygamia Monœcia.					
1 marítimum. M.P.	(*st.*) sea-side.	Brazil.	1823.	....	S. ♃.	Mart. palm. ic.
2 campéstre. M.P.	(*st.*) field.	———	——	....	S. ♃.	——— ic.
ATT'ALEA. K.S.	ATT'ALEA. Monœcia Polyandria.					
1 hùmilis. M.P.	(*st.*) dwarf.	Brazil.	1823.	....	S. ♄.	Mart. palm. ic.
2 cómpta. M.P.	(*st.*) decked.	———	——	....	S. ♄.	——— ic.
3 funífera. M.P.	(*st.*) rope-bearing.	———	——	....	S. ♄.	——— ic.
4 excélsa. M.P.	(*st.*) lofty.	———	——	....	S. ♄.	——— ic.
5 speciòsa. M.P.	(*st.*) handsome.	———	1824.	....	S. ♄.	——— ic.
6 Róssii. L.C.	(*st.*) Ross's.	———	——	....	S. ♄.	
7 spectábilis. L.C.	(*st.*) showy.	———	——	....	S. ♄.	
GOM'UTUS. S.S.	GOM'UTUS. Monœcia Polyandria.					
saccharífer. S.S.	(*st.*) sugar-bearing.	E. Indies.	1820.	....	S. ♄.	
MANIC'ARIA. G.	MANIC'ARIA. Polygamia Monœcia.					
saccífera. G.	(*st.*) bag-like.	E. Indies.	1822.	....	S. ♄.	Gært. sem. 2. t. 176.
BORA'SSUS. W.	BORA'SSUS. Diœcia Hexandria. W.					
flabellifórmis. W.	(*st.*) fan-leaved.	E. Indies.	1771.	....	S. ♄.	Roxb. corom. 1. t. 71. 72.
N'IPA. W.	N'IPA. Monœcia Monadelphia. W.					
frùticans. W.	(*st.*) shrubby.	E. Indies.	1822.	....	S. ♄.	Rumph. amb. 1. t. 16.
DESMO'NCUS. M.P.	DESMO'NCUS. Monœcia Monadelphia.					
1 polyacánthus. M.P.	(*st.*) many-spined.	Brazil.	1822.	....	S. ♄.	Mart. palm. ic.
2 orthacánthus. M.P.	(*st.*) long-spined.	———	——	....	S. ♄.	——— ic.
3 americànus. L.C.	(*st.*) American.	S. America.	1823.	....	S. ♄.	
4 dùbius. L.C.	(*st.*) doubtful.	———	——	....	S. ♄.	
BA'CTRIS. W.	BA'CTRIS. Monœcia Hexandria. W.					
1 pectinàta. M.P.	(*st.*) pectinated.	Brazil.	1825.	....	S. ♄.	Mart. palm. ic.
2 cuspidàta. M.P.	(*st.*) sharp-pointed.	———	——	....	S. ♄.	

3 mìnor. w.	(*st.*)	lesser.	S. America.	1691.	....	S. ♄.	Jacq. amer. t. 171. f. 1.
4 macracántha. M.P.	(*st.*)	large-spined.	Brazil.	1822.	....	S. ♄.	Mart. palm. ic.
5 màjor. w.	(*st.*)	greater.	Carthagena.	1800.	....	S. ♄.	Jacq. amer. t. 171. f. 2.
6 caryotæfòlia. M.P.	(*st.*)	Caryota-leaved.	Brazil.	1825.	....	S. ♄.	Mart. palm. ic.

ORDO CCXXI.

JUNCEÆ. *Brown prodr.* 1. *p.* 257.

KI'NGIA. L.T.		GRASS-TREE.	Hexandria Monogynia.				
austràlis. L.T.	(*st.*)	southern.	N. Holland.	1830.	....	G. ♄.	
XER'OTES. B.P.		XER'OTES.	Diœcia Hexandria.				
1 flexifòlia. B.P.	(*st.*)	bending-leav'd.	N. S. W.	1819.	6. 8.	F. ♃.	Thunb. diss. drac. 17. f. 2.
2 mucronàta. B.P.	(*st.*)	mucronated.	——	1823.	——	F. ♃.	
3 filifórmis. B.P.	(*st.*)	filiform.	——	——	——	F. ♃.	Th. dis. drac. n. 1. ic.
Drac'æna filifórmis. Th.							
4 denticulàta. B.P.	(*st.*)	toothed.	——	1825.	——	F. ♃.	
5 láxa. B.P.	(*st.*)	loose-flowered.	——.	1819.	——	F. ♃.	
6 rígida. B.P.	(*st.*)	stiff-leaved.	N. Holland.	1796.	6. 7.	F. ♃.	Lodd. bot. cab. t. 798.
Lomándra rígida. Lab. n. hol. 1. t. 120.							
7 montàna. B.P.	(*st.*)	mountain.	N. S. W.	1824.	——	F. ♃.	
8 longifòlia. B.P.	(*st.*)	long-leaved.	——	1796.	——	F. ♃.	Lab. nov. hol. 1. t. 19.
Lomándra longifòlia. Lab.							
9 Hy'strix. B.P.	(*st.*)	Porcupine.	——	1823.	——	F. ♃.	
10 'æmula. B.P.	(*st.*)	rival.	——	——	——	F. ♃.	
11 echinàta. M.C.	(*st.*)	bristly.	N. Holland.	1824.	——	F. ♃.	
PLE'EA. M.		PLE'EA.	Enneandria Trigynia. Ph.				
tenuifòlia. M.	(*br.*)	slender-leaved.	Carolina.	1812.	5. 8.	F. ♃.	Bot. mag. t. 1956.
NARTH'ECIUM. E.B.		NARTH'ECIUM.	Hexandria Monogynia.				
1 ossifràgum. E.B.	(*ye.*)	Lancashire Asphodel.	Britain.	....	7. 8.	H. ♃.	Eng. bot. v. 8. t. 535.
2 americànum. B.M.	(*ye.*)	American.	N. America.	1811.	——	H. ♃.	Bot. mag. t. 1505.
LUZ'ULA. DC.		WOODRUSH.	Hexandria Monogynia. S.S.				
1 pilòsa. R.S.	(*br.*)	spring.	Britain.	....	3. 5.	H. ♃.	Eng. bot. v. 11. t. 736.
2 flavéscens. S.S.	(*st.*)	yellowish.	S. Europe.	1817.	——	H. ♃.	Host. gram. 3. t. 94.
3 Forstèri. DC.	(*br.*)	Forster's.	England.	....	5. 6.	H. ♃.	Eng. bot. v. 18. t. 1293.
4 glabràta. S.S.	(*br.*)	smooth.	Switzerland.	1825.	——	H. ♃.	Desv. j. bot. 1. t. 5. f. 3.
5 álbida. DC.	(*wh.br.*)	white.	——	1815.	——	H. ♃.	Leers. herb. t. 13. f. 6.
6 lùtea. DC.	(*st.*)	yellow.	——	——	——	H. ♃.	
7 spadícea. DC.	(*da.br.*)	light red.	Europe.	1816.	——	H. ♃.	Vill. delph. 2. t. 6.
8 parviflòra. S.	(*br.*)	small-flowered.	Switzerland.	1824.	——	H. ♃.	Rost. junc. p. 26. t. 1. f. 1.
9 nìvea. DC.	(*wh.*)	snowy.	——	1770.	——	H. ♃.	Scheuch. gram. t. 7. f. 7.
10 sylvática. S.	(*br.wh.*)	wood.	Britain.	....	5.	H. ♃.	Eng. bot. v. 11. t. 737.
Júncus sylváticus. E.B.							
11 palléscens. S.	(*br.*)	palish.	N. Europe.	1828.	——	H. ♃.	Linn. fl. lap. t. 10. f. 2.
12 hyperborea. S.	(*br.*)	northern.	——	——	——	H. ♃.	
13 arcuàta. S.S.	(*wh.br.*)	fringe-bracted.	Scotland.	....	7.	H. ♃.	Wahl. lapp. t. 4.
14 spicàta. DC.	(*br.*)	spiked.	——	....	——	H. ♃.	Eng. bot. v. 17. t. 1176.
15 congésta. S.S.	(*br.*)	close-headed.	Britain.	....	6.	H. ♃.	Purt. suppl. t. 9.
Júncus liníger. Purt.							
16 sudética. DC.	(*br.*)	Swedish.	Sweden.	1824.	——	H. ♃.	Host. gram. 4. t. 99.
17 pedifórmis. DC.	(*br.*)	foot-shaped.	S. Europe.	1822.	5. 6.	H. ♃.	
18 campéstris. DC.	(*br.*)	field.	Britain.	....	4. 5.	H. ♃.	Eng. bot. v. 10. t. 672.
19 multiflòra. S.S.	(*br.*)	many-flowered.	Europe.	1822.	5. 6.	H. ♃.	
JU'NCUS. DC.		RUSH.	Hexandria Monogynia. L.				
1 multicapitàtus. S.	(*br.*)	many-headed.	Scotland.	....	6. 7.	H. ♃.	
polycéphalus. E.F. *non* Mich.							
2 fúsco-áter. S.	(*br.*)	dark brown.	Europe.	1825.	——	H. ♃.	
3 obtusiflòrus. L.T.	(*bh.*)	blunt-flowered.	Britain.	....	7. 8.	H. ♃.	Eng. bot. v. 30. t. 2141.
4 lampocárpus. L.T.	(*br.*)	shining-fruited.	——	....	——	H. ♃.	—— v. 30. t. 2143.
5 acutiflòrus. L.T.	(*br.*)	sharp-flowered.	——	....	6. 7.	H. ♃.	—— v. 4. t. 238.
articulàtus. E.B.							
6 Holosch'œnus. B.P.	(*br.*)	stripe-seeded.	N. S. W.	1824.	——	F. ♃.	
7 subverticillàtus. L.T.	(*br.*)	whorled.	Britain.	....	7. 8.	H. ♃.	Flor. dan. t. 817.
8 ténuis. S.	(*gr.pu.*)	slender.	N. America.	1826.	——	H. ♃.	
9 castàneus. E.B.	(*br.*)	black-spiked.	Scotland.	....	7.	H. ♃.	Eng. bot. v. 13. t. 900.

10 Jacquìni. s.	(*br.pu.*)	Jacquin's.	S. Europe.	1824.	7.	H.♃.	Jacq. fl. aust. t. 221.
biglùmis. Jacq. vind. t. 4. f. 2. *nec aliorum.*							
11 biglùmis. E.B.	(*br.*)	two-flowered.	Scotland.	....	8.	H.♃.	Eng. bot. v. 13. t. 898.
12 triglùmis. E.B.	(*br.*)	three-flowered.	Britain.	....	7.	H.♃.	——— v. 13. t. 899.
13 capitàtus. S.S.	(*br.g.*)	headed.	———	....	5. 7.	H.⊙.	Weig. obs. t. 2. f. 5.
14 supìnus. L.T.	(*br.*)	dwarf.	———	....	7.	H.♃.	Flor. dan. t. 1099.
15 uliginòsus. E.B.	(*br.*)	little bulbous.	———	....	7. 8.	H.♃.	Eng. bot. v. 12. t. 801.
16 trífidus. E.B.	(*br.*)	three-leaved.	Scotland.	....	——	H.♃.	——— v. 21. t. 1482.
17 Gésneri. E.F.	(*g.br.*)	Gesner's.	———	....	7.	H.♃.	——— v. 31. t. 2174.
grácilis. E.B. *non* B.P.							
18 bufònius. E.B.	(*pa.*)	toad.	Britain.	....	6. 8.	H.♃.	Eng. bot. v. 12. t. 802.
19 cœnòsus. L.T.	(*br.*)	mud.	———	....	7. 8.	H.♃.	Flor. dan. t. 431.
20 compréssus. L.T.	(*g.br.*)	round-fruited.	———	....	——	H.♃.	Eng. bot. v. 13. t. 934.
bulbòsus. E.B.							
21 squarròsus. E.B.	(*br.*)	Goose-corn.	———	....	6. 7.	H.♃.	Eng. bot v. 13. t. 933.
22 Tenagèja. S.S.	(*g.br.*)	annual.	Europe.	1816.	——	H.⊙.	Flor. dan. t. 1160.
23 aristulatús. M.	(*br.*)	awned.	N. America.	1820.	——	H.♃.	
24 plebèius. B.P.	(*pa.pu.*)	linear-leaved.	N. S. W.	1816.	——	F.♃.	
25 planifòlius. B.P.	(*br.pu.*)	flat-leaved.	———	1822.	7. 8.	F.♃.	
26 pauciflòrus. B.P.	(*pa.*)	few-flowered.	———	——	——	F.♃.	
27 vaginàtus. B.P.	(*st.*)	sheathed.	———	1823.	——	F.♃.	
28 polycéphalus. M.	(*gr.*)	panicled.	N. America.	1828.	——	H.♃.	
29 macrostèmon. S.	(*gr.*)	large-stamened.	———	1826.	——	H.♃.	
30 filifórmis. E.B.	(*gr.*)	least.	Britain.	....	8.	H.♃.	Eng. bot. v. 17. t. 1175.
31 pállidus. B.P.	(*pa.*)	pale-flowered.	N. Holland.	1829.	——	F.♃.	
32 árcticus. W.	(*da.br.*)	arctic.	Scotland.	....	7. 8.	H.♃.	Flor. dan. t. 1095.
33 bálticus. S.	(*br.*)	Baltic.	Baltic.	1821.	——	H.♃.	
34 effùsus. E.B.	(*gr.*)	soft.	Britain.	....	5. 8.	H.♃.	Eng. bot. v. 12. t. 836.
35 conglomeràtus. E.B.	(*br.*)	common.	———	....	6. 7.	H.♃.	——— v. 12. t. 835.
36 glaùcus. E.B.	(*br.*)	glaucous.	———	....	7. 8.	H.♃.	——— v. 10. t. 665.
37 marítimus. E.B.	(*br.*)	sea-side.	———	....	8.	H.♃.	——— v. 24. t. 1725.
38 acùtus. E.B.	(*g.br.*)	great sharp sea.	———	....	7. 8.	H.♃.	——— v. 23. t. 1614.
39 globulòsus. E.	(*br.*)	globular.	Egypt.	1829.	——	H.♃.	
? FLAGELL'ARIA. L.		FLAGELL'ARIA.	Hexandria Trigynia. L.				
índica. L.	(*st.*)	Indian.	India.	1782.	6. 7.	S.♄.	Redout. lil. t. 257.
? PHIL'YDRUM. B.P.		PHIL'YDRUM.	Monandria Monogynia. B.M.				
lanuginòsum. B.P.	(*ye.*)	woolly.	N. S. W.	1801.	6. 7.	S.*w.*♂.	Bot. mag. t. 783.
? BURMA'NNIA. L.		BURMA'NNIA.	Triandria Monogynia.				
1 distàchya. B.P.	(*bl.*)	two-spiked.	N. S. W.	1822.	6. 8.	G.*w.*♃.	Burm. zeyl. t. 20. f. 1.
2 biflòra. L.	(*bl.*)	two-flowered.	Virginia.	....	9.	G.*w.*♃.	
? GILLI'ESIA. B.R.		GILLI'ESIA.	Hexandria Monogynia.				
gramínea. B.R.	(*st.*)	Grass-like.	S. America.	1824.	5. 7.	G.♃.	Bot. reg. t. 992.

ORDO CCXXII.

RESTIACEÆ. *Brown prodr. p.* 243.

X'YRIS. B.P.		X'YRIS.	Triandria Monogynia. L.				
1 l'ævis. B.P.	(*ye.*)	smooth.	N. S. W.	1819.	6. 8.	G.♃.	
2 bractèata. B.P.	(*ye.*)	bracteate.	———	1825.	——	G.♃.	
3 júncea. B.P.	(*ye.*)	rushy.	———	1822.	——	G.♃.	
4 grácilis. B.P.	(*ye.*)	slender.	V. Diem. Isl.	1821.	——	G.♃.	
5 operculàta. B.P.	(*ye.*)	rush-leaved.	N. S. W.	1804.	6. 7.	G.♃.	Bot. mag. t. 1158.
6 Púrshii.	(*ye.*)	Pursh's.	N. America.	1825.	——	F.♃.	
índica. Ph. *non* Vahl.							
7 índica. W.	(*ye.*)	Indian.	E. Indies.	1822.	5. 7.	S.♃.	Rheed. mal. 9. t. 71.
8 ancéps. R.S.	(*ye.*)	flat-edged.	Malabar.	1824.	——	S.♃.	
9 americàna. R.S.	(*bl.*)	blue-flowered.	Trinidad.	1825.	——	S.♃.	Aubl. guian. 40. t. 15.
10 caroliniána. R.S.	(*ye.*)	Carolina.	N. America.	1812.	——	G.♃.	
11 capénsis. R.S.	(*ye.*)	Cape.	C. B. S.	1822.	——	G.♃.	
12 brevifòlia. M.	(*ye.*)	short-leaved.	Carolina.	1812.	6. 8.	F.♃.	
ERIOCA'ULON. L.		PIPEWORT.	Triandria Trigynia. L.				
1 septangulàre. E.B.	(*wh.*)	jointed.	Scotland.	....	9.	H.*w.*♃.	Eng. bot. v. 11. t. 733.
2 austràle. B.P.	(*wh.*)	New Holland.	N. S. W.	1822.	8. 10.	G.*w.*♃.	

DEVA'UXIA. B.P.		Deva'uxia.	Monandria Monogynia.				
1 Patersòni. B.P.	(*gr.*)	Paterson's.	N. S. W.	1825.	6. 7.	F. ♃.	Linn.trans.v.10.t.12.f.2
Centrolèpis `æmula. L.T.							
2 Billardíeri. B.P.	(*gr.*)	Billardiere's.	——	——	——	F. ♃.	Linn.trans.v.10.t.12.f.1
Centrolèpis cuspidígera. L.T. *fasciculàris.* Lab. nov. holl. 1. p. 7. t. 1.							
HYPOL'ÆNA. B.P.		Hypol'æna.	Diœcia Triandria.				
1 fastigiàta. B.P.	(*br.*)	crowded.	N. S. W.	1821.	6. 8.	F. ♃.	
2 exsúlca. B.P.	(*br.*)	round-stalked.	——	——	——	F. ♃.	
LEPTOCA'RPUS. B.P.		Leptoca'rpus.	Diœcia Triandria.				
1 tènax. B.P.	(*br.*)	tough.	N. S. W.	1824.	6. 8.	F. ♃.	Lab. nov. holl. 2. t. 229.
Schœnòdum tènax fémina. Lab.							
2 imbricàtus. S.S.	(*br.*)	imbricated.	C. B. S.	——	——	G. ♃.	
EL'EGIA. W.		El'egia.	Diœcia Triandria. W.				
1 júncea. W.	(*br.*)	Rush-like.	C. B. S.	1789.	7. 8.	F. ♃.	Rottb. gram. 10. t. 3. f. 2.
2 racemòsa. W.	(*br.*)	racemed.	——	1804.	5. 6.	F. ♃.	Lam. ill. t. 884. f. 4.
LEPYR'ODIA. B.P.		Lepyr'odia.	Diœcia Triandria. W.				
grácilis. B.P.	(*br.*)	slender.	N. S. W.	1823.	5. 6.	F. ♃.	
WILLDEN'OVIA. S.S.		Willden'ovia.	Diœcia Triandria. W.				
1 striàta. S.S.	(*br.*)	striated.	C. B. S.	1816.	6. 8.	F. ♃.	Th.act.hol.1790.t.2.f.1.
2 tères. S.S.	(*br.*)	round-stalked.	——	1790.	——	F. ♃.	—— 1790. t. 2. f. 2.
THAMNOCHO'RTUS. B.P.		Thamnocho'rtus.	Diœcia Triandria.				
1 dichótomus. B.P.	(*br.*)	forked.	C. B. S.	1816.	6. 8.	F. ♃.	Rottb. gram. t. 2. f. 1.
2 scariòsus. S.S.	(*br.*)	scariose.	——	1822.	——	G. ♃.	Berg. cap. t. 5. f. 8.
RE'STIO. B.P.		Rope-grass.	Diœcia Triandria. L.				
1 austràlis. B.P.	(*br.*)	southern.	V. Diem. Isl.	1812.	4. 6.	F. ♃.	
2 grácilis. B.P.	(*br.*)	slender.	N. S. W.	1823.	6. 8.	F. ♃.	
3 complanàtus. B.P.	(*br.*)	flat-stemmed.	——	——	——	F. ♃.	
4 tectòrum. S.S.	(*br.*)	thatch.	C. B. S.	1793.	5. 6.	F. ♃.	Rottb.gram.10.t.3.f.2.
5 vaginàtus. S.S.	(*br.*)	sheathed.	——	1816.	——	F. ♃.	
6 fastigiàtus. B.P.	(*br.*)	crowded.	N. S. W.	1824.	——	F. ♃.	
7 virgàtus. S.S.	(*br.*)	twiggy.	C. B. S.	1816.	——	F. ♃.	Rottb. gram. t. 1. f. 2.
8 paniculàtus. S.S.	(*br.*)	panicled.	——	——	——	F. ♃.	—— t. 2. f. 3.
9 lateriflòrus. B.P.	(*br.*)	side-flowering.	N. S. W.	1819.	——	F. ♃.	Labil. nov. hol. 2. t. 228.
Caloròphus elongàta. Lab.							
10 tetraphy'llus. B.P.	(*br.*)	four-leaved.	——	1825.	——	F. ♃.	Labil. nov. hol. 2. t. 226.

ORDO CCXXIII.

CYPÉRACEÆ. *Brown prodr.* 1. *p.* 212.

C'AREX. L.		Sedge.	Monœcia Triandria L.				
1 dioíca. W.	(*br.*)	diœcious.	Britain.	....	5. 6.	H. ♃.	Eng. bot. v. 8. t. 543.
2 Davalliàna. W.	(*br.*)	Davall's.	——	....	——	H. ♃.	—— v. 30. t. 2123.
3 pauciflòra. E.B.	(*st.*)	few-flowered.	——	....	6.	H. ♃.	—— v. 29. t. 2041.
4 pyrenàica. S.S.	(*br.*)	Pyrenean.	Pyrenees.	1816.	——	H. ♃.	Schk. caric. 5. t. D. f. 15.
5 pulicàris. E.B.	(*br.*)	Flea.	Britain.	....	6. 7.	H. ♃.	Eng. bot. v. 15. t. 1051.
6 Fraseriàna. H.K.	(*wh.*)	Fraser's.	N. America.	1809.	4. 6.	H. ♃.	Bot. mag. t. 1391.
7 cyperoídes. S.S.	(*br.*)	Bohemian.	Bohemia.	1801.	6. 7.	H. ♃.	Schk. caric. t. A. f. 5.
8 f'œtida. S.S.	(*br.*)	stinking.	Switzerland.	1791.	7. 8.	H. ♃.	—— t. Hh. f. 96.
9 incúrva. E.B.	(*br.*)	curved.	Scotland.	....	——	H. ♃.	Eng. bot. v. 13. t. 927.
10 stenophy'lla. S.S.	(*br.*)	slender-leaved.	Europe.	1824.	——	H. ♃.	Schk. caric. t. G. f. 32.
11 chordorrhìza. S.S.	(*br.*)	creeping-rooted.	Germany.	——	——	H. ♃.	—— t. G. f. 31.
12 baldénsis. S.S.	(*br.*)	three-styled.	Mount Baldo.	1829.	5. 7.	H. ♃.	—— t. Y. f. 81.
13 stellulàta. E.B.	(*br.*)	little prickly.	Britain.	....	5. 6.	H. ♃.	Eng. bot. v. 12. t. 806.
14 ròsea. S.S.	(*ro.*)	Rose.	N. America.	1812.	——	H. ♃.	Schk. caric. Zzz. f. 179.
15 retrofléxa. S.S.	(*br.*)	retroflexed.	Pensylvania.	1826.	——	H. ♃.	—— t. Kkk. f. 140.
16 trìceps. S.S.	(*br.*)	three-headed.	N. America.	1825.	——	H. ♃.	—— t. Zzz. f. 180.
17 f'œnea. S.S.	(*br.*)	short-spiked.	——	1819.	——	H. ♃.	
18 muricàta. E.B.	(*br.*)	greater prickly.	Britain.	....	——	H. ♃.	Eng. bot. v. 16. t. 1097.
19 brizoídes. S.S.	(*br.*)	Briza-like.	Germany.	1815.	5. 7.	H. ♃.	Host. gram. 36. t. 47.
20 argyroglóchin. S.S.	(*wh.*)	silver-spiked.	——	1823.	——	H. ♃.	
21 stramínea. S.S.	(*st.*)	slender-stalked.	N. America.	1803.	6. 7.	H. ♃.	Sch. caric. t. Xxx. f. 174.
22 stipàta. S.S.	(*br.*)	sharp-scaled.	——	1820.	——	H. ♃.	—— t. Hhh. f. 132.
23 Muhlenbérgii. S.S.	(*gr.*)	Muhlenberg's.	——	1824.	——	H. ♃.	—— t. Yyy. f. 178.

24	Schrèberi. s.s.	(*br.*)	Schreber's.	Germany.	1800.	6. 7.	H. ♃.	Host. gram. 1. t. 46.
25	ovàlis. E.B.	(*gr.br.*)	oval-spiked.	Britain.	....	——	H. ♃.	Eng. bot. v. 5. t. 306.
26	cúrta. E.B.	(*wh.*)	white.	———	....	6.	H. ♃.	——— v. 20. t. 1386.
27	divìsa. E.B.	(*bk.*)	bracteated.	———	....	5. 7.	H. ♃.	——— v. 16. t. 1096.
28	sparganioídes. s.s.	(*br.*)	Sparganium-like.	N. America.	1822.	——	H. ♃.	Schk. caric. t. Lll. f. 142.
29	arenària. E.B.	(*br.*)	sand.	Britain.	....	6. 7.	H. ♃.	Eng. bot. v. 13. t. 928.
30	intermèdia. E.B.	(*br.*)	soft brown.	———	....	5. 7.	H. ♃.	——— v. 29. t. 2042.
31	schœnoídes. W.	(*br.*)	Schœnus-like.	Austria.	1819.	——	H. ♃.	Host. gram. 1. t. 45.
32	lagopodioídes. W.	(*br.*)	Hare's-foot.	N. America.	1805.	6. 7.	H. ♃.	Schk. cari. t. Yyy. f. 177.
33	elongàta. E.B.	(*br.gr.*)	elongated.	England.	....	5. 6.	H. ♃.	Eng. bot. v. 27. t. 1920.
34	remòta. E.B.	(*wh.*)	remote.	Britain.	....	——	H. ♃.	——— v. 12. t. 832.
35	axillàris. E.B.	(*br.*)	axillary.	England.	....	——	H. ♃.	——— v. 14. t. 993.
36	divúlsa. E.B.	(*wh.*)	gray.	Britain.	....	5.	H. ♃.	——— v. 9. t. 629.
37	nemoròsa. s.s.	(*gr.*)	shady wood.	Germany.	....	5. 7.	H. ♃.	
38	vulpìna. E.B.	(*gr.*)	great spiked.	Britain.	....	5. 8.	H. ♃.	Eng. bot. v. 5. t. 307.
39	appréssa. B.P.	(*gr.*)	close-spiked.	N. S. W.	1802.	——	F. ♃.	
40	invérsa. B.P.	(*gr.*)	inversed.	———	1820.	——	F. ♃.	
41	scopària. W.	(*gr.*)	broom.	N. America.	1812.	6. 7.	H. ♃.	Schk. cari. t. Xxx. f. 175.
42	multiflòra. W.	(*gr.*)	many-flowered.	———	1819.	——	H. ♃.	——— t. Lll. f. 144.
43	paradóxa. W.	(*gr.*)	paradoxical.	Germany.	——	——	H. ♃.	——— t. E. f. 21.
44	teretiúscula. E.B.	(*br.*)	lesser-panicled.	Britain.	....	5. 6.	H. ♃.	Eng. bot. v. 15. t. 1065.
45	paniculàta. E.B.	(*br.*)	great panicled.	England.	....	6. 7.	H. ♃.	——— v. 15. t. 1064.
46	tenélla. E.F.	(*br.*)	slender.	Scotland.	....	6.	H. ♃.	Schk. caric. t. P. f. 104.
47	norvègica. W.	(*gr.*)	Norway.	Norway.	1823.	——	H. ♃.	——— t. S. f. 66.
48	atràta. E.B.	(*bk.*)	black.	Britain.	....	6. 7.	H. ♃.	Eng. bot. v. 29. t. 2044.
49	bícolor. W.	(*br.ye.*)	two-coloured.	Switzerland.	1823.	——	H. ♃.	Sch. car. t. Aaaa. f. 181.
50	pedunculàta. W.	(*gr.*)	peduncled.	N. America.	——	5. 7.	H. ♃.	——— t. Ggg. f. 131.
51	longifòlia. B.P.	(*gr.*)	long-leaved.	N. S. W.	——	——	F. ♃.	
52	thuringìaca. W.	(*gr.*)	Thuringian.	Hungary.	——	——	H. ♃.	Schk. cari. t. Ppp. f. 155.
53	Buxbàumii. W.	(*br.*)	Buxbaum's.	S. Europe.	1819.	5. 6.	H. ♃.	——— t. X. f. 76.
54	glareòsa. W.	(*br.*)	narrow-leaved.	Norway.	——	——	H. ♃.	——— t. Aaa. f. 97.
55	clandestìna. E.B.	(*wh.*)	dwarf silvery.	Britain.	....	5.	H. ♃.	Eng. bot. v. 30. t. 2124.
56	digitàta. E.B.	(*pu.*)	fingered.	England.	....	5. 6.	H. ♃.	——— v. 9. t. 615.
57	plantagínea. W.	(*br.*)	broad-leaved.	N. America.	1805.	——	H. ♃.	Schk. caric. t. U. f. 70.
58	pilulífera. E.B.	(*br.*)	round-headed.	Britain.	....	4. 6.	H. ♃.	Eng. bot. v. 13. t. 885.
59	montàna. s.s. *collìna.* W.	(*br.*)	mountain.	Europe.	1819.	5. 6.	H. ♃.	Schk. caric. t. F. f. 29.
60	ciliàta. W.	(*br.*)	fringed.	Germany.	——	——	H. ♃.	Schk. caric. t. I. f. 42.
61	præcox. E.B.	(*st.*)	vernal.	Britain.	....	4.	H. ♃.	Eng. bot. v. 16. t. 1099.
62	tomentòsa. E.B.	(*st.*)	downy-fruited.	England.	....	6.	H. ♃.	——— v. 29. t. 2046.
63	exténsa. E.B.	(*gr.*)	long-bracted.	Britain.	....	——	H. ♃.	——— v. 12. t. 833.
64	flàva. E.B.	(*ye.*)	yellow.	———	....	5. 6.	H. ♃.	——— v. 18. t. 1294.
65	Œdèri. E.B.	(*ye.*)	Œder's.	England.	....	6. 7.	H. ♃.	——— v. 25. t. 1773.
66	álba. W.	(*wh.*)	bristle-leaved.	Germany.	1819.	——	H. ♃.	Host. gram. 1. t. 59.
67	Mielichóferi. E.B.	(*gr.*)	loose-spiked.	Scotland.	....	7. 8.	H. ♃.	Eng. bot. v. 32. t. 2293.
68	ferrugínea. W.	(*br.*)	ferruginous.	Europe.	1820.	6. 8.	H. ♃.	Schk. caric. t. M. f. 48.
69	saxátilis. W.	(*st.*)	rock.	N. Europe.	——	——	H. ♃.	——— t. I. f. 40.
70	rígida. E.B.	(*bk.*)	rigid.	Britain.	....	6. 7.	H. ♃.	Eng. bot. v. 29. t. 2047.
71	púlla. E.B.	(*bk.br.*)	russet.	Scotland.	....	7.	H. ♃.	——— v. 29. t. 2045.
72	cæspitòsa. E.B.	(*gr.*)	tufted bog.	Britain.	....	5. 6.	H. ♃.	——— v. 21. t. 1507.
73	stricta. E.B.	(*br.*)	straight-leaved.	———	....	4. 5.	H. ♃.	——— v. 13. t. 914.
74	alpéstris. W.	(*br.*)	Alpine.	Europe.	1804.	5. 6.	H. ♃.	Schk. caric. t. G. f. 35.
75	nítida. W.	(*br.*)	glossy.	Austria.	1805.	——	H. ♃.	Host. gram. 1. t. 71.
76	umbròsa. W.	(*gr.*)	shade.	———	1818.	6. 7.	H. ♃.	——— 1. t. 69.
77	granulàris. W.	(*gr.*)	grain-seeded.	N. America.	1807.	——	H. ♃.	Sch. caric. t. Vvv. f. 169.
78	ánceps. W.	(*gr.*)	two-edged.	———	1805.	7. 8.	H. ♃.	——— t. Fff. f. 128.
79	demíssa. s.s.	(*br.*)	low.	Norway.	1825.	——	H. ♃.	
80	binérvis. E.B.	(*bk.*)	green-ribbed.	Britain.	....	6.	H. ♃.	Eng. bot. v. 18. t. 1235.
81	dístans. E.B.	(*br.*)	loose.	———	....	——	H. ♃.	——— v. 18. t. 1234.
82	fúlva. E.B.	(*ye.br.*)	tawny.	———	....	6. 7.	H. ♃.	——— v. 18. t. 1295.
83	loliàcea. W.	(*br.*)	Ray-grass-like.	Sweden.	1810.	——	H. ♃.	Schk. car. t. Pp. f. 104.
84	lucòrum. W.en.	(*gr.*)	grove.	N. America.	1825.	——	H. ♃.	
85	Michèlii. W.	(*br.*)	Micheli's.	Europe.	1818.	——	H. ♃.	Host. gram. 1. t. 72.
86	péndula. E.B.	(*gr.*)	great pendulous.	Britain.	....	5. 6.	H. ♃.	Eng. bot. v. 33. t. 2315.
87	strigòsa. E.B.	(*gr.*)	loose pendulous.	———	....	——	H. ♃.	——— v. 14. t. 994.
88	depauperàta. E.B.	(*gr.*)	starved-wood.	England.	....	6. 7.	H. ♃.	——— v. 16. t. 1098.
89	pilòsa. s.s.	(*gr.*)	hairy.	Germany.	1819.	——	H. ♃.	Schk. caric. t. M. f. 49.
90	panícea. E.B.	(*bk.*)	Pink-leaved.	Britain.	....	5. 7.	H. ♃.	Eng. bot. v. 21. t. 1505.
91	oligocárpa. W.	(*gr.*)	few-seeded.	N. America.	1825.	——	H. ♃.	Schk. caric. t. Vvv. f. 178.

92 rostràta. w.	(*gr.*) beaked.	N. America.	1818.	5. 7.	H.♃.	Schk.caric.t.Hhh.f.134
93 conglobàta. w.	(*gr.*) globular-fruited.	Hungary.	——	——	H.♃.	
94 sylvática. E.B.	(*gr.*) wood.	Britain.	....	5. 6.	H.♃.	Eng. bot. v. 14. t. 995.
Drimèja. w.						
95 Pseudocypèrus. w.	(*gr.*) Bastard Cypress.	———	....	6. 7.	H.♃.	Eng. bot. v. 4. t. 242.
96 júncea. S.S.	(*gr.*) rush-like.	N. America.	1823.	——	H.♃.	
97 palléscens. E.B.	(*g.ye.*) pale.	Britain.	....	4. 6.	H.♃.	Eng. bot. v. 31. t. 2185.
98 limòsa. E.B.	(*ye.gr.*) green and gold.	———	....	6.	H.♃.	——— v. 29. t. 2043.
99 rariflòra. E.B.	(*bk.*) loose-flowered.	Scotland.	....	——	H.♃.	——— v. 35. t. 2516.
100 capillàris. E.B.	(*br.*) capillary.	Britain.	....	7. 8.	H.♃.	——— v. 29. t. 2069.
101 lævigàta. E.B.	(*gr.*) smooth-stalked.	———	....	5. 6.	H.♃.	——— v. 20. t. 1387.
102 ustulàta. E.B.	(*bk.*) scorched alpine.	Scotland.	....	6. 7.	H.♃.	——— v. 34. t. 2404.
103 flexuòsa. w.	(*br.*) bending.	N. America.	1807.	——	H.♃.	Schk. car. t. Ddd. f. 124.
104 recúrva. E.B.	(*bk.br.*) glaucous heath.	England.	....	5. 6.	H.♃.	Eng. bot. v. 21. t. 1506.
β *Micheliàna.* E.B.	(*bk.*) *smooth-fruited.*	———	....	——	H.♃.	——— v. 32. t. 2236.
105 nùtans. w.	(*br.*) nodding.	Austria.	1815.	6. 7.	H.♃.	Host. gram. 1. t. 83.
106 acuminàta. w.	(*br.*) taper-pointed.	Istria.	1819.	——	H.♃.	——— 1. t. 97.
107 hírta. E.B.	(*br.*) hairy.	Britain.	....	5. 6.	H.♃.	Eng. bot. v. 10. t. 685.
108 filifórmis. E.B.	(*bk.*) slender-leaved.	———	....	6. 7.	H.♃.	——— v. 13. t. 904.
109 acùta. E.B.	(*bk.*) slender-spiked.	———	....	5. 6.	H.*w.*♃.	——— v. 9. t. 580.
110 paludòsa. E.B.	(*bk.*) lesser common.	———	....	——	H.*w.*♃.	——— v. 12. t. 807.
111 ripària. E.B.	(*bk.*) great common.	———	....	4. 6.	H.*w.*♃.	——— v. 9. t. 579.
112 vesicària. E.B.	(*bk.*) short-spiked.	———	....	5. 6.	H.♃.	——— v. 11. t. 779.
113 ampullàcea. E.B.	(*st.*) slender-beaked.	———	....	——	H.♃.	——— v. 11. t. 780.
114 bullàta. w.	(*st.*) blistered.	N. America.	1811.	6.	H.♃.	Schk. caric. Uuu. f.166.
115 hordeifórmis. w.	(*gr.*) Barley-like.	France.	1805.	6. 7.	H.♃.	
116 crinìta. w.	(*gr.*) haired.	N. America.	1807.	——	H.♃.	Schk.caric.t.Eee f.125.
117 aquátilis. w.	(*gr.*) water.	Lapland.	1819.	——	H.*w.*♃.	
118 secalìna. w.	(*br.*) rye-like.	Europe.	1820.	——	H.♃.	Schk.caric.t.Ddd.f.121
119 salìna. w.	(*br.*) salt marsh.	Norway.	....	6.	H.♃.	
120 erióphora. F.	(*wh.*) wool-bearing.	Dahuria.	1828.	——	H.♃.	
121 pentastàchys. F.	(*br.*) five-spiked.	Kamtschatka.	1821.	——	H.♃.	
122 córsica. Ot.	(*br.*) Corsican.	Corsica.	1825.	——	H.♃.	
KOBR'ESIA. W.	KOBR'ESIA.	Monœcia Triandria. W.				
caricìna. w.	(*gr.*) Carex-like.	Britain.	....	8.	H.♃.	Eng. bot. v. 20. t.1410.
Sch'œnus monoícus. E.B. *Cárex hy'brida.* S.S.						
SCL'ERIA. W.	SCL'ERIA.	Monœcia Triandria. W.				
1 Flagéllum. S.S.	(*gr.*) climbing.	W. Indies.	1822.	....	S.♃.	Gært. sem. 1. p.13. t.2.
2 latifòlia. S.S.	(*gr.*) broad-leaved.	———	1824.	....	S.♃.	
3 verticillàta. S.S.	(*gr.*) whorled.	N. America.	1825.	6. 7.	H.♃.	
4 hirtélla. S.S.	(*gr.*) hairy.	———	1822.	——	H.♃.	
5 auricòma. Ot.	(*ye.*) golden-locks.	———	1826.	7. 9.	H.⊙.	
6 ciliàta. w.	(*gr.*) fringed.	———	1823.	——	H.♃.	
SCH'ŒNUS. B.P.	BOG-RUSH.	Triandria Monogynia. L.				
1 nígricans. w.	(*bk.*) black-headed.	Britain.	....	7.	H.*w.*♃.	Eng. bot. v. 16. t. 1121.
2 pusíllus. R.S.	(*br.*) small.	W. Indies.	1824.	7. 8.	S.♃.	Swtz. fl. ind. oc. ic. t. 6.
3 mucronàtus. R.S.	(*br.pu.*) clustered.	S. Europe.	1781.	4. 5.	H.*w.*♃.	Flor. græc. 1. t. 43.
4 imbérbis. B.P.	(*bk.*) beardless.	N. S. W.	1818.	——	F.♃.	
5 melanostàchys. B.R.	(*br.*) black-spiked.	———	1822.	6. 8.	F.♃.	
CHÆTO'SPORA. B.P.	CHÆTO'SPORA.	Triandria Monogynia. S.S.				
1 ferrugínea. S.S.	(*br.*) brown.	Switzerland.	1818.	7. 8.	H.*w.*♃.	Host. gram. 4. t. 71.
Sch'œnus ferrugíneus. R.S.						
2 turbinàta. B.P.	(*br.*) turbinate.	N. S. W.	1822.	——	F.♃.	
3 paludòsa. B.P.	(*br.*) marsh.	———	1824.	——	F.*w.*♃.	
LEPIDOSPE'RMA. B.P.	LEPIDOSPE'RMA.	Triandria Monogynia. S.S.				
1 gladiàta. B.P.	(*br.*) flat-stemmed.	N. Holland.	1819.	7. 8.	F.*w.*♃.	Lab. nov. hol. 1. t. 12.
2 concàva. B.P.	(*br.*) concave.	———	1824.	——	F.♃.	
3 exaltàta. B.P.	(*gr.*) tall.	———	1816.	——	H.♃.	
DULI'CHIUM. P.S.	DULI'CHIUM.	Triandria Monogynia. S.S.				
1 spathàceum. P.S.	(*br.*) sheathed.	N. America.	1818.	7. 8.	H.*w.*♃.	Pluk. alm. t. 301. f. 1.
2 canadénse. Ph.	(*gr.*) Canadian.	Canada.	1826.	——	H.♃.	
RHYNCHO'SPORA. R.S.	RHYNCHO'SPORA.	Triandria Monogynia. S.S.				
1 aùrea. R.S.	(*ye.*) golden.	W. Indies.	1825.	7. 8.	S.♃.	Rottb. gram. t. 21. f. 1.
2 láxa. R.S.	(*br.*) loose.	N. America.	1827.	——	H.*w.*♃.	
3 álba. R.S.	(*wh.*) white-headed.	Britain.	....	8.	H.*w.*♃.	Eng. bot. v. 14. t. 985.
Sch'œnus álbus. E.B.						
4 fúsca. R.S.	(*br.*) brown-headed.	———	....	——	H.*w.*♃.	Eng. bot. v. 22. t. 1575.
5 dístans. Ph.	(*br.*) filiform-leaved.	N. America.	1820.	7. 8.	H.*w.*♃.	
6 comàta. R.S.	(*br.*) leafy.	Brazil.	——	——	S.*w.*♃.	

	CL'ADIUM. B.P.			Cl'adium.	Diandria Monogynia. S.S.				
1	Maríscus.	B.P.	(*br.*)	prickly.	England.	....	7. 8.	H.*w*.♃.	Eng. bot. v. 14. t. 950.
	germánicum. R.S. *Sch'œnus Maríscus*. E.B.								
2	occidentàle.	R.S.	(*br.*)	West Indian.	W. Indies.	1824.	——	S.*w*.♃.	
3	glomeràtum.	B.P.	(*br.*)	close-headed.	N. S. W.	1819.	——	F.*w*.♃.	
4	júnceum.	B.P.	(*br.*)	rushy.	——	1823.	——	F.*w*.♃.	
5	schœnoídes.	B.P.	(*br.*)	Bogrush-like.	——	——	——	F.*w*.♃.	Lab. nov. hol. 1. t. 18.
	Sch'œnus acùtus. Lab.								
	FIMBRI'STYLIS. B.P.			Fimbri'stylis.	Triandria Monogynia. R.S.				
1	ánnua.	R.S.	(*br.*)	annual.	S. Europe.	1818.	7. 9.	H.*w*.⊙.	Host. gram. 3. t. 63.
2	pubérula.	R.S.	(*br.*)	pubescent.	N. America.	1820.	7. 8.	H.♃.	
3	dichótoma.	R.S.	(*br.*)	forked.	E. Indies.	1817.	7. 9.	S.*w*.⊙.	Rottb. gram. t. 13. f. 1.
4	autumnàlis.	R.S.	(*br.*)	autumnal.	N. America.	1822.	7. 9.	H.♃.	
5	ferrugínea.	R.S.	(*br.*)	rusty.	Jamaica.	1825.	——	S.*w*.♃.	Sloan. hist. 1. t. 77. f. 1.
	ABILDG'AARDIA. B.P.			Abildg'aardia.	Triandria Monogynia. R.S.				
	monostàchya.	B.P.	(*wh.*)	one-spiked.	W. Indies.	1819.	6. 8.	S.♃.	Sloan. jam. 1. t. 79. f. 2.
	ELEO'CHARIS. B.P.			Eleo'charis.	Triandria Monogynia. R.S.				
1	acùta.	B.P.	(*br.*)	acute-scaled.	N. S. W.	1819.	7. 8.	F.*w*.♃.	
2	geniculàta.	R.S.	(*br.*)	jointed.	America.	——	——	F.*w*.♃.	Sloan. hist. 1. t. 83. f. 3.
3	plantagínea.	R.S.	(*br.*)	Plantain-like.	E. Indies.	1826.	7. 9.	S.*w*.♃.	Rottb. gram. t. 15. f. 2.
4	palústris.	R.S.	(*br.*)	marsh.	Britain.	....	7.	H.*w*.♃.	Eng. bot. v. 2. t. 131.
	Scírpus palústris. E.B.								
5	ovàta.	R.S.	(*br.*)	ovate.	Germany.	1818.	7. 8.	H.*w*.⊙.	Host. gram. 3. t. 56.
6	multicaùlis.	E.F.	(*br.pu.*)	many-stalked.	Britain.	....	7.	H.*w*.♃.	Eng. bot. v. 17. t. 1187.
	Scírpus multicaùlis. E.B.								
7	aciculàris.	R.S.	(*br.*)	upright needle.	——	....	——	H.*w*.♃.	Eng. bot. v. 11. t. 749.
8	obtùsa.	R.S.	(*br.*)	obtuse-scaled.	N. America.	1818.	7. 8.	H.*w*.♃.	
9	glaucéscens.	R.S.	(*br.*)	glaucous.	——	1819.	——	H.*w*.♃.	
10	ténuis.	R.S.	(*br.*)	slender.	——	——	——	H.*w*.♃.	
11	capitàta.	R.S.	(*br.*)	headed.	America.	1824.	——	H.*w*.♃.	Rottb. gram. t. 15. f. 3.
12	intermèdia.	R.S.	(*br.*)	intermediate.	Pensylvania.	1826.	——	H.*w*.♃.	
	ISOL'EPIS. B.P.			Isol'epis.	Triandria Monogynia. R.S.				
1	flùitans.	R.S.	(*br.*)	floating.	Britain.	....	7. 8.	H.*w*.♃.	Eng. bot. v. 3. t. 216.
	Scírpus flùitans. E.B.								
2	grácilis.		(*br.*)	knotted.	N. Holland.	1820.	——	G.*w*.♃.	
	nodòsa. B.P. *Scírpus grácilis*. Linn. trans. 10. t. 15. f. 2.								
3	setàcea.	R.S.	(*br.gr.*)	bristle-like.	Britain.	....	——	H.*w*.⊙.	Eng. bot. v. 24. t. 1693.
4	Váhlii.	R.S.	(*br.gr.*)	Vahl's.	S. America.	1828.	——	S.*w*.⊙.	
5	Micheliàna.	R.S.	(*st.*)	Micheli's.	S. Europe.	——	——	H.*w*.⊙.	Host. gram. 3. t. 69.
6	Holosch'œnus.	R.S.	(*br.*)	round-headed.	Britain.	....	——	H.*w*.♃.	Eng. bot. v. 23. t. 1612.
	β *austràlis*.	R.S.	(*br.*)	*southern*.	S. Europe.	....	——	H.*w*.♃.	Pluk. phyt. t. 40. f. 5.
7	romàna.	W.	(*br.*)	Roman.	England.	....	——	H.*w*.♃.	Jacq. aust. 5. t. 448.
8	complanàta.	R.S.	(*br.*)	flat-stemmed.	E. Indies.	1820.	——	S.*w*.♃.	
9	angulàris.	R.S.	(*br.*)	angular-stemmed.	——	1827.	——	S.*w*.♃.	
10	curvifòlia.	R.S.	(*br.*)	curve-leaved.	——	——	——	S.*w*.♃.	
11	pubígera.	Schr.	(*br.*)	pubescent.	....	——	——	S.*w*.♃.	
12	Ehrenbérgii.	Ot.	(*br.*)	Ehrenberg's.	Egypt.	1825.	——	H.*w*.⊙.	
	SCI'RPUS. B.P.			Club-Rush.	Triandria Monogynia. L.				
1	cæspitòsus.	E.B.	(*br.*)	scaly-stalked.	Britain.	....	7.	H.*w*.♃.	Eng. bot. v. 15. t. 1029.
2	campéstris.	R.S.	(*pu.st.*)	field.	S. Europe.	1824.	7. 9.	H.*w*.♃.	Hayn. term. bot. t. 25. f. 4.
3	pauciflòrus.	E.B.	(*br.*)	chocolate-headed.	——	....	8.	H.*w*.♃.	Eng. bot. v. 16. t. 1122.
	Bæóthryon. R.S.								
4	púngens.	R.S.	(*br.*)	pungent.	Europe.	....	8. 9.	H.*w*.♃.	
5	mucronàtus.	R.S.	(*gr.br.*)	mucronated.	Britain.	....	——	H.*w*.♃.	Scheuch. agr. t. 9. f. 14.
6	Luzùlæ.	R.S.	(*br.*)	clustered.	E. Indies.	1776.	7. 9.	S.*w*.♃.	
7	lacústris.	E.B.	(*br.*)	tall.	Britain.	....	7.	H.*w*.♃.	Eng. bot. v. 10. t. 666.
8	glaùcus.	E.B.	(*br.*)	glaucous.	England.	....	7. 8.	H.*w*.♃.	—— v. 33. t. 2321.
9	caricìnus.	S.S.	(*br.*)	compressed.	Britain.	....	7.	H.*w*.♃.	—— v. 11. t. 791.
	Sch'œnus compréssus. E.B.								
10	rùfus.	S.S.	(*br.*)	rusty.	——	....	6. 7.	H.*w*.♃.	Eng. bot. v. 15. t. 1010.
	Sch'œnus rùfus. E.B.								
11	triquèter.	E.B.	(*re.br.*)	triangular.	England.	....	8.	H.*w*.♃.	Eng. bot. v. 24. t. 1694.
12	carinàtus.	E.B.	(*br.*)	blunt-edged.	——	....	7. 8.	H.*w*.♃.	—— v. 28. t. 1983.
13	marítimus.	E.B.	(*br.*)	salt marsh.	Britain.	....	7. 9.	H.*w*.♃.	—— v. 8. t. 542.
14	sylváticus.	E.B.	(*br.gr.*)	wood.	——	....	——	H.*w*.♃.	—— v. 13. t. 919.
15	rádicans.	R.S.	(*br.gr.*)	rooting.	Germany.	....	——	H.*w*.♃.	Schk. in Ust. an. 4. t. 1.
16	atrovìrens.	R.S.	(*da.g.*)	dark green.	N. America.	1818.	——	H.*w*.♃.	
17	quinquangulàris.	R.S.	(*br.*)	five-angled.	E. Indies.	1823.	....	S.*w*.♃.	
18	thyrsiflòrus.	F.	(*br.*)	thyrse-flowered.	Siberia.	1821.	7. 9.	H.*w*.♃.	

	C'YPERUS. L.			C'YPERUS.	Triandria Monogynia. L.				
1	Aitòni.	R.S.	(*br.*)	Aiton's.	C. B. S.	....	6. 8.	G.*w*.♃.	
2	articulàtus.	R.S.	(*st.*)	jointed.	S. America.	1820.	——	S.*w*.♃.	Sloan. hist. 1. t. 81. f. 1.
3	viscòsus.	R.S.	(*gr.br.*)	clammy.	Jamaica.	1781.	5. 8.	S.*w*.♃.	Jacq. ic. 2. t. 295.
4	tenéllus.	R.S.	(*gr.br.*)	slender.	C. B. S.	1819.	7. 9.	H.⊙.	
5	mucronàtus.	R.S.	(*wh.*)	pointed.	Levant.	——	——	F.*w*.♃.	Flor. græc. 1. t. 49.
6	pannònicus.	R.S.	(*g.w.*)	dwarf.	Hungary.	1781.	7. 8.	H.⊙.	Jacq. aust. ap. t. 6.
7	margaritàceus.	R.S.	(*wh.*)	pearl.	Guinea.	1822.	7. 9.	S.*w*.♃.	
8	monocéphalus.	F.I.	(*da.br.*)	one-headed.	E. Indies.	1816.	——	S.*w*.♃.	
9	pygmˈæus.	R.S.	(*gr.*)	dwarf.	———	1827.	——	S.*w*.♃.	Rottb. gr. 20. t. 14. f. 4-5.
10	filifórmis.	R.S.	(*br.*)	filiform.	Jamaica.	1825.	——	S.*w*.♃.	
11	tènuis.	R.S.	(*gr.*)	slender.	———	——	——	S.*w*.♃.	
12	polystàchyus.	R.S.	(*br.*)	many-spiked.	E. Indies.	1819.	——	S.*w*.♃.	Rottb. gram. t. 11. f. 1.
13	brúnneus.	R.S.	(*br.*)	brown-spiked.	W. Indies.	1826.	——	S.*w*.♃.	
14	brizˈæus.	R.S.	(*br.*)	Briza-like.	———	1824.	7. 8.	S.*w*.♃.	
15	diándrus.	R.S.	(*br.*)	two-stamened.	N. America.	1827.	——	H.*w*.♃.	
16	Purshii.	R.S.	(*br.wh.*)	Pursh's.	———	1816.	——	H.*w*.♃.	
17	Háspan.	R.S.	(*br.*)	digitate-spiked.	Trinidad.	1824.	——	S.*w*.♃.	
18	alternifòlius.	R.S.	(*br.*)	alternate-leaved.	Madagascar.	1781.	——	S.*w*.♃.	Jacq. ic. 2. t. 298.
19	élegans.	R.S.	(*re.*)	elegant.	Jamaica.	1801.	5. 9.	S.♂.	Sloan. jam. 1. t. 75. f. 1.
20	compréssus.	R.S.	(*pu.st.*)	compressed.	S. America.	1819.	——	S.*w*.♃.	Rottb. gram. t. 9. f. 3.
21	vegètus.	R.S.	(*br.*)	smooth.	America.	1790.	——	G.*w*.♃.	Jacq. vind. 3. t. 12.
	compréssus. Jacq. *non* W.								
22	formòsus.	R.S.	(*ye.*)	handsome.	N. America.	1827.	7. 9.	H.*w*.♃.	
23	Luzùlæ.	R.S.	(*gr.br.*)	compact.	W. Indies.	....	——	S.*w*.♃.	Rottb. gram. t. 13. f. 2.
	polycéphalus. Lam. *non* L.en.								
24	confértus.	R.S.	(*ye.*)	crowded-spiked.	W. Indies.	1825.	7. 9.	S.*w*.♃.	
25	squarròsus.	R.S.	(*br.*)	globe-headed.	E. Indies.	1828.	——	S.*w*.♃.	Rottb. gram. t. 6. f. 3.
26	paniculàtus.	R.S.	(*br.*)	panicled.	W. Indies.	1804.	5. 7.	S.*w*.♃.	Vahl. ic. pl. am. t. 12.
27	flavéscens.	R.S.	(*br.ye.*)	yellowish.	Germany.	1776.	6. 9.	F.*w*.♃.	Flor. græc. t. 47.
28	fúscus.	R.S.	(*br.*)	brown.	Fulham.	——	7. 9.	H.⊙.	——— t. 48.
29	conglomeràtus.	R.S.		many-flowered.	Java.	1820.	5. 9.	S.*w*.♃.	Rottb. gram. t. 15. f. 7.
30	erubéscens.	L.en.	(*bh.*)	blush.	....	——	——	S.*w*.♃.	
31	tuberòsus.	R.S.	(*g.pu.*)	tuberous-rooted.	N. America.	1827.	——	H.*w*.♃.	Rottb. gram. t. 7. f. 1.
32	rotúndus.	R.S.	(*g.pu.*)	round-rooted.	Africa.	1828.	——	H.*w*.♃.	Scheuch. agrost. t. 9. f. 3.
33	gláber.	R.S.	(*br.*)	smooth.	S. Europe.	1820.	——	H.*w*.♃.	Reich. ic. C. 3. t. 204.
34	austràlis.	R.S.	(*br.*)	southern.	———	——	——	H.⊙.	Seguier. veron. 3. t. 2. f. 2.
35	Tenoreànus.	R.S.	(*ye.*)	Tenore's.	Naples.	1825.	——	H.*w*.♃.	Reich. ic. C. 3. t. 212.
36	punctàtus.	R.S.	(*br.gr.*)	spotted-seeded.	E. Indies.	1820.	——	S.*w*.♃.	Pluck. t. 192. f. 3.
37	Ornithòpus.	R.S.	(*st.pu.*)	bird's-foot.	St. Domingo.	1824.	——	S.*w*.♃.	
38	esculéntus.	R.S.	(*st.*)	Rush-nut.	S. Europe.	1597.	7.	H.♃.	Host. gram. 3. t. 75.
39	Hˈydra.	M.	(*gr.wh.*)	American.	N. America.	1812.	——	H.♃.	
40	Pangòrei.	R.S.	(*br.*)	Ceylon.	Ceylon.	1823.	7. 9.	S.*w*.♃.	Rottb. gram. t. 7. f. 3.
41	bàdius.	R.S.	(*br.*)	saw-leaved.	Algiers.	1800.	——	F.*w*.♃.	Desf. atl. 1. t. 7. f. 2.
42	pátulus.	R.S.	(*br.st.*)	spreading.	Hungary.	1818.	——	H.*w*.⊙.	Host. gram. 3. t. 74.
43	lóngus.	E.B.	(*br.gr.*)	sweet.	England.	....	7.	H.*w*.♃.	Eng. bot. v. 19. t. 1309.
44	pròcerus.	R.S.	(*br.wh.*)	tall rough.	E. Indies.	1826.	7. 9.	S.*w*.♃.	Rheed. mal. 12. t. 50.
45	tenuiflòrus.	R.S.	(*br.*)	slender-flower'd.	S. Europe.	1818.	7. 8.	H.*w*.♃.	Jacq. ic. 2. t. 296.
46	Mónti.	R.S.	(*pu.gr.*)	spreading-spiked.	———	——	——	H.*w*.♃.	Host. gram. 4. t. 67.
47	Paramátta.	R.S.	(*br.pu.*)	tuberous-rooted.	———	——	——	H.*w*.♃.	
48	spéctabilis.	Schreb.	(*pu.*)	showy.	———	1827.	——	H.*w*.♃.	
49	diándrus.	R.S.	(*st.br.*)	diandrous.	N. America.	1828.	——	H.*w*.♃.	
50	planifòlius.	R.S.	(*br.*)	flat-leaved.	W. Indies.	1824.	——	S.*w*.♃.	Sloan. his. 1. t. 74. f. 2. 3.
51	Linkiànus.	R.S.	(*br.gr.*)	Link's.	S. Europe.	1819.	6. 8.	H.*w*.♃.	
	polycéphalus. L.enum. *non* Lam.								
52	strigòsus.	R.S.	(*br.st.*)	bristle-spiked.	W. Indies.	1786.	7. 9.	S.*w*.♃.	
53	Michauxiànus.	R.S.	(*g.w.*)	Michaux's.	N. America.	1812.	——	H.*w*.♃.	
54	flavicòmus.	R.S.	(*ye.*)	yellow-spiked.	———	1826.	——	H.*w*.♃.	
55	ˈIria.	R.S.	(*st.gr.*)	tall.	E. Indies.	1802.	7.	S.*w*.♃.	Rheed. mal. 12. t. 56.
56	dístans.	R.S.	(*br.st.*)	distant.	W. Indies.	1820.	7. 8.	S.*w*.♃.	Jacq. ic. t. 299.
57	tegètum.	R.S.	(*br.*)	creeping-rooted.	E. Indies.	1822.	——	S.*w*.♃.	
58	gigánteus.	R.S.	(*st.*)	gigantic.	Trinidad.	1824.	——	S.*w*.♃.	Rudg. guian. t. 21.
59	fastigiàtus.	R.S.	(*ye.*)	lofty.	E. Indies.	1800.	5. 8.	S.*w*.♃.	Rottb. gram. t. 7. f. 2.
60	alopecuroídes.	R.S.	(*g.br.*)	fox-tailed.	C. B. S.	1804.	——	G.*w*.♃.	——— 38. t. 8. f. 2.
61	cruéntus.	R.S.	(*pu.*)	purple-spotted.	Arabia.	1820.	——	S.*w*.♃.	——— 21. t. 5. f. 1.
62	glomeràtus.	R.S.	(*br.*)	round-headed.	Italy.	1804.	——	H.⊙.	Segu. ver. 3. p. 68. t. 2. f. 2.
63	ligulàris.	R.S.	(*br.gr.*)	ligulate.	C. B. S.	1820.	——	G.*w*.♃.	Rottb. gram. t. 11. f. 2.
64	diffórmis.	R.S.	(*br.*)	deformed.	Levant.	1819.	6. 8.	S.*w*.♃.	Flor. græc. t. 46.
65	protráctus.	L.en.	(*br.*)	protracted.	Egypt.	1820.	——	H.*w*.⊙.	Delil. æg. 152. t. 5. f. 3

66 dìves. s.s.	(*ye.*)	large-bracted.	Egypt.	1820.	7.	H.*w.*☉.	Delil. æg. 149. t. 4. f. 3.
67 chlorostàchys. Stev.	(*gr.*)	green-spiked.	Siberia.	1823.	——	H.*w.*♃.	
68 múltiplex. Lk.	(*br.*)	many-spiked.	....	1829.	——	H.*w.*♃.	
69 scáber. Lag.	(*br.*)	rough.	Spain.	1823.	——	H.*w.*☉.	
70 xanthócoma. Ot.	(*ye.*)	yellow-haired.	Sicily.	1829.	——	H.*w.*♃.	
PAP'YRUS. S.S.		PAPER-RUSH.	Triandria Monogynia. S.S.				
1 antiquòrum. s.s.	(*br.y.*)	common.	Egypt.	1802.	7. 9.	S.*w.*♃.	Michel. gen. pl. t. 19.
C'yperus Pap'yrus. R.S.							
2 laxiflòrus. s.s.	(*br.*)	loose-flowered.	Madagascar.	1822.	——	S.*w.*♃.	
3 odoràtus. s.s.	(*st.*)	fragrant.	W. Indies.	1819.	——	S.*w.*♃.	Rudg. guian. p. 17. t. 20.
MARI'SCUS. R.S.		MARI'SCUS.	Triandria Monogynia. R.S.				
1 capillàris. R.S.	(*st.br.*)	capillary.	Trinidad.	1824.	7. 10.	S.*w.*♃.	
2 aggregàtus. R.S.	(*g.br.*)	clustered.	....	1820.	7. 9.	S.*w.*♃.	
3 l'ævis. B.P.	(*br.g.*)	smooth.	N. S. W.	1819.	6. 8.	G.*w.*♃.	
4 ovulàris. R.S.	(*br.g.*)	oval-spiked.	Carolina.	1818.	——	F.*w.*♃.	Pluk. alm. t. 91. f. 4.
5 retrofráctus. R.S.	(*st.pu.*)	retrofracted.	N. America.	1827.	——	H.*w.*♃.	———— t. 415. f. 4.
6 umbellàtus. R.S.	(*br.g.*)	umbelled.	E. Indies.	1789.	——	S.*w.*♃.	Rottb. gram. t. 4. f. 2.
7 elàtus. R.S.	(*br.g.*)	tall.	———	1805.	——	S.*w.*♃.	Jacq. ic. 2. t. 300.
8 confléxus. L.en.	(*br.g.*)	contracted.	Brazil.	1819.	7. 8.	S.*w.*♃.	
KYLLI'NGA. W.		KYLLI'NGA.	Triandria Monogynia. L.				
1 monocéphala. R.S.	(*wh.*)	one-headed.	India.	1793.	6. 7.	S.*w.*♃.	Rottb. gram. t. 4. f. 4.
2 crucifórmis. R.S.	(*wh.*)	cross-formed.	E. Indies.	1825.	——	S.*w.*♃.	Swartz. hort. gott. ic.
3 intermèdia. B.P.	(*wh.*)	intermediate.	N. S. W.	1822.	——	G.*w.*♃.	
4 trìceps. R.S.	(*wh.pu.*)	three-headed.	E. Indies.	1776.	9. 11.	S.*w.*♃.	Rottb. gram. t. 4. f. 6.
5 uncinàta. L.en.	(*gr.*)	hook-seeded.	Brazil.	1820.	7. 8.	S.*w.*♃.	
6 odoràta. R.S.	(*gr.*)	fragrant.	S. America.	1822.	——	S.*w.*♃.	
7 polycéphala. L.en.	(*wh.*)	many-headed.	Brazil.	1820.	——	S.*w.*♃.	
FUIR'ENA. B.P.		FUIR'ENA.	Triandria Monogynia. L.				
1 umbellàta. R.S.	(*br.*)	umbelled.	W. Indies.	1822.	6. 8.	S.*w.*♃.	Rottb. gram. t. 19. f. 3.
2 glomeràta. R.S.	(*pu.g.*)	crowded.	E. Indies.	1827.	——	S.*w.*☉.	———— t. 17. f. 1.
EL'YNA. R.S.		EL'YNA.	Triandria Monogynia. R.S.				
spicàta. R.S.	(*br.*)	simple-spiked.	Europe.	1819.	6. 8.	H.*w.*♃.	Vill. delph. 2. t. 6.
Cárex Bellárdi. Host. gram. 4. t. 77.							
TRICHO'PHORUM. P.S.		TRICHO'PHORUM.	Triandria Monogynia. P.S.				
1 cyperìnum. P.S.	(*br.st.*)	cyperine.	N. America.	1802.	5. 9.	H.*w.*♃.	Pluk. mant. t. 419. f. 3.
2 alpìnum. P.S.	(*br.st.*)	Alpine.	Scotland.	....	7.	H.*w.*♃.	Eng. bot. v. 5. t. 311.
Erióphorum alpìnum. E.B.							
ERIO'PHORUM. L.		COTTON-GRASS.	Triandria Monogynia. L.				
1 vaginàtum. E.B.	(*br.*)	Hare's-tail.	Britain.	....	3. 4.	H.*w.*♃.	Eng. bot. v. 13. t. 873.
2 capitàtum. E.B.	(*br.*)	round-headed.	Scotland.	....	8. 9.	H.*w.*♃.	———— v. 34. t. 2387.
3 grácile. E.B.	(*bk.g.*)	slender.	———	....	7. 8.	H.*w.*♃.	———— v. 34. t. 2402.
4 angustifòlium.	(*pu.g.*)	narrow-leaved.	Britain.	....	4. 5.	H.*w.*♃.	———— v. 8. t. 564.
5 virgínicum. R.S.	(*pu.g.*)	Virginian.	N. America.	1802.	5. 8.	H.*w.*♃.	Pluk. alm. t. 299. f. 4.
6 polystàchion. E.B.	(*pu.g.*)	broad-leaved.	Britain.	....	4.	H.*w.*♃.	Eng. bot. v. 8. t. 563.
7 pubéscens. E.F.	(*bk.*)	pubescent.	———	....	4. 5.	H.*w.*♃.	

ORDO CCXXIV.

GRAMINEÆ. *Juss. gen. p.* 28.

REMIR'EA. R.S.		REMIR'EA.	Triandria Monogynia.				
marítima. R.S.	....	sea-side.	America.	1822.	7. 8.	G.*w.*♃.	Aubl. guian. 1. t. 16.
NA'RDUS. R.S.		MAT-GRASS.	Triandria Monogynia. L.				
stricta. E.B.	....	upright.	Britain.	....	6. 7.	H.*w.*♃.	Eng. bot. v. 5. t. 290.
ORYZO'PSIS. M.		ORYZO'PSIS.	Triandria Monogynia. R.S.				
asperifòlia. M.	....	rough-leaved.	N. America.	1822.	7. 8.	H.♃.	Mich. amer. 1. t. 9.
LYG'EUM. R.S.		LYG'EUM.	Triandria Monogynia. L.				
Spártum. R.S.	....	rush-leaved.	Spain.	1776.	5. 6.	H.♃.	Schreb. gram. t. 52. 53.
CORNUC'OPIÆ. R.S.		HORN OF PLENTY.	Triandria Monogynia. L.				
cucullàtum. R.S.	....	hooded.	Levant.	1788.	8.	H.☉.	Flor. græc. 1. t. 51.
CE'NCHRUS. R.S.		CE'NCHRUS.	Triandria Monogynia. L.				
1 tribuloídes. R.S.	....	spinous.	N. America.	1818.	5. 8.	H.☉.	Sloan. hist. 1. t. 65. f. 1.
2 echinàtus. R.S.	....	rough-spiked.	W. Indies.	1691.	8. 12.	S.♂.	Cavan. ic. 5. p. 39. t. 462.
3 austràlis. B.P.	....	New Holland.	N. S. W.	1822.	6. 8.	H.☉.	

PENNIS'ETUM. R.S. PENNIS'ETUM. Triandria Monogynia. R.S.
1 setòsum. R.S. .... setose. W. Indies. 1804. 7. 8. H.⊙.
2 polystàchyum. R.S. .. many-spiked. E. Indies. —— —— H.⊙. Rumph. amb. 6. t. 7. f. 2.
3 cenchroídes. R.S. ciliated. C. B. S. 1777. 7. 8. H.⊙. Giseck. ic. 1. t. 23.
Cénchrus ciliàris. L.
4 orientàle. R.S. .... oriental. Levant. 1819. —— F.♃.
5 violàceum. R.S. .... violet-spiked. Senegal. 1829. —— S.⊙.
6 compréssum. B.P. .... flat-stemmed. N. S. W. 1820. —— H.♃.
7 holcoídes. R.S. .... Holcus-like. E. Indies. 1823. —— S.⊙.
Pánicum holcoídes. F.I.
8 barbàtum. R.S. .... bearded. —— —— —— S.⊙.
9 dichótomum. R.S. .... forked. Egypt. 1824. —— H.♃. Delil. ægyp. t. 8. f. 3.
10 nepalénse. S.S. .... Nepaul. Nepaul. 1822. —— H.⊙.
11 ásperum. R.S. .... rough-leaved. C. B. S. 1823. —— G.♄.
PENICILL'ARIA. R.S. PENICILL'ARIA. Polygamia Monœcia. W.
1 spicàta. R.S. .... spiked. E. Indies. 1592. 6. 7. H.⊙. Pluk. alm. t. 32. f. 4.
2 alopecuroídes. .... Fox-tail. Jamaica. 1739. 7. 10. H.⊙. —— t. 92. f. 5.
ciliàta. W.en. *cylíndrica.* R.S. *Pánicum alopecuroídes.* L. *Alopecùrus índicus.* L.
SET'ARIA. R.S. SET'ARIA. Triandria Digynia. R.S.
1 serícea. R.S. .... silky. W. Indies. 1780. 5. 9. H.⊙.
2 verticillàta. R.S. .... rough. England. .... 7. 8. H.⊙. Eng. bot. v. 13. t. 874.
3 víridis. R.S. .... green. —— .... —— H.⊙. —— v. 13. t. 875.
Pánicum víride. E.B. *Pennisètum víride.* B.P.
4 pùmila. R.S. .... dwarf. .... 1819. —— H.⊙.
5 macroch'æta. R.S. .. long-spiked. .... —— —— H.⊙.
6 purpuráscens. R.S. .... purplish. S. America. 1822. —— H.⊙.
7 intermèdia. R.S. .... intermediate. E. Indies. 1819. 8. 10. H.⊙.
8 Weinmánni. R.S. .... Weinman's. Teneriffe. 1816. —— H.⊙.
9 glaùca. R.S. .... glaucous. S. Europe. 1771. 7. 8. H.⊙. Host. gram. 2. t. 16.
10 helvòla. R.S. .... pale red. E. Indies. 1822. 8. 10. H.⊙.
11 corrugàta. R.S. .... rugged-flowered. Carolina. 1828. —— H.⊙.
12 erubéscens. R.S. .... blush. S. America. 1822. —— H.⊙.
13 geniculàta. R.S. .... jointed. Antilles. 1805. 7. 8. H.⊙.
14 marítima. R.S. .... sea-side. .... 1818. —— H.⊙.
15 germánica. R.S. .... German. Germany. 1548. —— H.⊙. Host. gram. 2. t. 15.
16 itálica. R.S. .... Italian. Italy. 1616. —— H.⊙. Rumph. amb. 3. t. 76. f. 2.
Pánicum itálicum. Host. gram. 4. t. 14. *Pennisètum itálicum.* B.P.
17 compósita. R.S. .... compound. S. America. 1823. —— H.⊙.
18 cylíndrica. R.S. .... cylindrical. —— —— —— H.⊙.
19 setòsa. R.S. .... bristly. W. Indies. 1822. —— H.⊙.
20 polystàchia. R.S. .... many-spiked. Brazil. 1824. —— S.♂.
21 tenacíssima. R.S. .... tough. —— 1819. —— H.⊙.
22 Pennisètum. R.S. .... Indian. E. Indies. 1820. 6. 8. S.⊙.
23 scándens. R.S. .... climbing. .... —— —— H.⊙.
24 caudàta. R.S. .... tailed. Trinidad. 1824. —— H.⊙.
25 muricàta. R.S. .... sea-side. N. America. 1819. —— H.⊙.
26 auricoma. Lk. .... golden-headed. .... 1826. —— H.⊙.
ANTHOXA'NTHUM. R.S. SPRING-GRASS. Diandria Digynia. L.
1 odoràtum. E.B. .... sweet-scented. Britain. .... 5. H.♃. Eng. bot. v. 9. t. 647.
β láxum. .... *loose-spiked.* —— .... —— H.♃. Loes. ic. 22.
γ pubéscens. .... *downy-spiked.* —— .... —— H.♃.
δ ramòsum. .... *branching-spiked.* —— .... —— H.♃.
2 amárum. R.S. .... bitter. S. Europe. 1810. 7. H.♃. Trin. ic. gram. t. 14.
3 grácile. R.S. .... slender. Italy. 1825. 7. 8. H.⊙. —— t. 13.
4 ovàtum. R.S. .... oval-spiked. Spain. 1824. —— H.⊙.
SPART'INA. R.S. SPART'INA. Triandria Monogynia. R.S.
1 cynosuroídes. R.S. .... Dog's-tail. N. America. 1781. 8. 9. H.♃. Linn. fil. fasc. 1. t. 9.
Dáctylis cynosuroídes. L. *Trachynòtia cynosuroídes.* M. *Limnètis cynosuroídes.* P.S.
2 stricta. R.S. .... upright. Britain. .... 8. H.♃. Eng. bot. v. 6. t. 380.
Dáctylis stricta. E.B. *Limnètis púngens.* P.S.
3 glàbra. R.S. .... smooth. N. America. 1827. —— H.♃.
4 alterniflòra. R.S. .... alternate-flower'd. S. Europe. —— —— H.♃.
5 polystàchia. R.S. .... many-spiked. N. America. 1781. 8. 9. H.♃.
6 pùmila. R.S. .... dwarf. —— 1820. —— H.♃.
7 pectinàta. .... pectinated. —— 1827. —— H.♃.
8 júncea. R.S. .... rushy. —— 1781. 7. 8. H.♃.
9 pàtens. R.S. .... spreading. —— —— —— H.♃.
10 geniculàta. R.S. .... jointed. Java. 1822. —— S.♃. Burm. ind. t. 12. f. 3.
11 arundinàcea. R.S. .... Reed-like. Tristan d'Acunha. 1820. —— G.♃.

ZO'YSIA. R.S.	Zo'ysia. Triandria Digynia. W.					
púngens. R.S.	pungent.	E. Indies.	1819.	6. 8.	S.♃.	
LE'ERSIA. B.P.	Le'ersia. Triandria Digynia. W.					
1 oryzoídes. S.S.	Rice-like.	S. Europe.	1793.	7. 8.	H.w.♃.	Host. gram. 1. t. 35.
2 virgínica. S.S.	Virginian.	N. America.	——	——	H.w.♃.	
3 austràlis. B.P.	southern.	N. S. W.	1819.	——	H.w.♃.	
4 lenticulàris. S.S.	round-glumed.	N. America.	1825.	——	H.w.♃.	
5 brasiliénsis. R.S.	Brazilian.	Brazil.	1826.	——	S.w.♃.	
PER'OTIS. R.S.	Per'otis. Triandria Digynia. W.					
latifòlia. R.S.	broad-leaved.	E. Indies.	1777.	8. 9.	S.⊙.	Rheed. mal. 12. t. 62.
ALOPEC'URUS. R.S.	Foxtail-grass. Triandria Digynia. L.					
1 agréstis. E.B.	slender.	Britain.	....	7. 8.	H.⊙.	Eng. bot. v. 12. t. 848.
2 utriculàtus. R.S.	bladdered.	Italy.	1777.	——	H.⊙.	Flor. græc. t. 63.
3 echinàtus. R.S.	rough.	C. B. S.	1822.	——	G.⊙.	
4 praténsis. E.B.	meadow.	Britain.	....	5.	H.♃.	Eng. bot. v. 11. t. 759.
5 nígricans. R.S.	blackish.	Europe.	1815.	6. 7.	H.♃.	Jacq. ecl. gram. t. 13.
Tauntoniénsis. Sinc. gr. wob. p. 232.						
6 triviàlis. S.O.	long-awned.	Bohemia.	1827.	——	H.♃.	
7 scáber. S.O.	scabrous.	———	———	——	H.♃.	
8 Giesekeànus. R.S.	Gieseke's.	N. Europe.	1828.	——	H.♃.	Flor. dan. t. 1565.
ovàtus. F.D. *non* Forst.						
9 geniculàtus. E.B.	floating.	Britain.	....	5. 8.	H.w.♃.	Eng. bot. v. 18. t. 1250.
10 fúlvus. E.B.	orange-spiked.	England.	....	6.	H.w.♃.	——— v. 21. t. 1467.
11 alpìnus. E.B.	Alpine.	Scotland.	....	5. 6.	H.♃.	——— v. 16. t. 1126.
12 ramòsus. R.S.	branching.	N. America.	1823.	——	H.♃.	
13 arundinàceus. R.S. ..	Reed-like.	....	1819.	——	H.♃.	
14 macrostàchyos. S.S. ..	large-spiked.	Barbary.	1824.	6. 8.	H.♃.	
15 bulbòsus. E.B.	bulbous.	England.	....	7.	H.♃.	Eng. bot. v. 18. t. 1249.
16 colobachnoídes. R.S...	Siberian.	Siberia.	1826.	6. 7.	H.♃.	
17 Gerárdi. Trin.	Gerard's.	S. Europe.	1820.	6. 8.	H.♃.	Jacq. ic. t. 301.
Phlèum Gerárdi. R.S.						
PHL'EUM. R.S.	Cat's-tail-grass. Triandria Digynia. L.					
1 praténse. R.S.	common.	Britain.	....	7.	H.♃.	Eng. bot. v. 15. t. 1076.
α màjus. S.G.W.	*greater.*	———	....	——	H.♃.	Sinc. gr. wob. p. 195. ic.
β mìnus. S.G.W.	*lesser.*	———	....	7. 8.	H.♃.	——— p. 197. ic.
2 nodòsum. W.	knotted.	———	....	7. 9.	H.♃.	Flor. dan. t. 380.
3 Bertolònii. R.S.	Bertoloni's.	France.	1818.	——	H.♃.	
4 alpìnum. R.S.	alpine.	Scotland.	....	7.	H.♃.	Eng. bot. v. 8. t. 519.
5 commutàtum. R.S.	confused.	Piedmont.	1819.	——	H.♃.	
6 felìnum. R.S.	Grecian.	Greece.	1822.	——	H.⊙.	
7 echinàtum. R.S.	hedgehog.	Dalmatia.	——	8. 9.	H.⊙.	Host. gram. 3. t. 11.
ACHNODO'NTON. R.S.	Achnodo'nton. Triandria Digynia. R.S.					
1 Bellárdi. R.S.	bulbous.	Spain.	1798.	6. 7.	H.⊙.	
Phálaris Bellárdii. W. *bulbòsa.* L.						
2 ténue. R.S.	slender.	S. Europe.	1826.	——	H.⊙.	Host. gram. t. 36.
3 arenària. L.en.	sand.	Britain.	....	6. 7.	H.⊙.	Eng. bot. v. 4. t. 222.
Phálaris arenària. E.B. *Phlèum arenàrium.* F.D. t. 915. *Chilochlòa arenària.* R.S.						
CHILOCHL'OA. R.S.	Chilochl'oa. Triandria Digynia. R.S.					
1 Bœhmèri. R.S.	Phalaris-like.	England.	....	7. 9.	H.♃.	Eng. bot. v. 7. t. 459.
Phálaris phleoídes. E.B. *Phlèum phalaroídes.* Kœl. *Bœhmèri.* W.						
2 cuspidàta. R.S.	sharp-pointed.	Tyrol.	1824.	7. 8.	H.♃.	Host. gram. 4. t. 20.
3 Michèlii. L.en.	Micheli's.	Scotland.	....	6. 7.	H.♃.	Eng. bot. v. 32. t. 2265.
Phlèum Michèlii. E.B.						
4 áspera. R.S.	rough.	England.	....	7.	H.⊙.	Eng. bot. v. 15. t. 1077.
Phlèum paniculatum. E.B. *Phlèum ásperum.* Jacq. ic. t. 14. *Phálaris áspera.* W.						
5 ánnua. R.S.	annual.	Caucasus.	1825.	7. 8.	H.⊙.	
SCHMI'DTIA. S.S.	Schmi'dtia. Diandria Digynia. S.S.					
súbtilis. S.S.	close umbelled.	Bohemia.	1824.	6. 8.	H.⊙.	Tratt. fl. aus. t. 451.
Coleánthus súbtilis. R.S.						
POLYP'OGON. R.S.	Polyp'ogon. Triandria Digynia. W.					
1 monspeliénsis. R.S.	chaffy.	Britain.	....	7. 8.	H.⊙.	Eng. bot. v. 24. t. 1704.
Agróstis panícea. E.B.						
2 littoràlis. E.F.	sea-side.	———	....	8.	H.♃.	Eng. bot. v. 18. t. 1251.
3 Lagáscæ. R.S.	Lagasca's.	Spain.	1824.	——	H.♃.	
4 marítimus. R.S.	shore.	S. Europe.	1822.	——	H.⊙.	
5 glomeràtus. R.S.	clustered.	N. America.	——	7. 8.	H.♃.	
6 tatáricus. F.	Tartarian.	Tartary.	1825.	——	H.⊙.	
CER'ESIA. P.S.	Cer'esia. Triandria Digynia. P.S.					
élegans. P.S.	elegant.	Peru.	1818.	8. 9.	H.♃.	Jacq. frag. t. 86.
Páspalum membranàceum. S.S.						

PA'SPALUM. W.	PA'SPALUM.	Triandria Digynia. L.				
1 disséctum. S.S.	.... dissected.	S. America.	1822.	8. 10.	H.⊙.	
2 scrobiculàtum.S.S.	.... dimpled.	E. Indies.	1778.	7. 9.	S.♃.	Hout. n. his. t. 89. f. 3.
3 stoloníferum. S.S.	.... purple.	Peru.	1794.	——	G.♃.	Jacq. ic. 2. t. 302.
4 paniculàtum. S.S.	.... panicled.	W. Indies.	1782.	——	S.⊙.	Sloan. hist. 1. t. 72. f. 2.
5 virgàtum. S.S.	.... twiggy.	———	1822.	——	S.♃.	———1. t. 69. f. 2.
6 l'æve. M.	.... smooth.	N. America.	1818.	——	H.♃.	Lecomt. pasp. t. 91.
7 mucronàtum.R.S.	.... mucronate.	———	1825.	——	H.⊙.	
8 Lecomteànum. R.S.	.. Lecomte's.	———	1827.	——	H.♃.	Lecomt. pas. t. 91.
9 latifòlium. R.S.	.... broad-leaved.	———	——	——	H.♃.	
10 platénse. R.S.	.... Monte Video.	Monte Video.	1824.	——	H.♃.	
11 diffórme. R.S.	.... deformed.	N. America.	1818.	7. 9.	H.♃.	Lecomt. pas. t. 91. p. 284.
12 pusíllum. R.S.	.... small.	St. Thomas.	1822.	——	S.♃.	
13 conjugàtum. S.S.	.... conjugate.	W. Indies.	——	——	S.♃.	Gært. fruct. 2. t. 80. f. 4.
14 notàtum. S.S.	.... two-spiked.	———	1776.	7.	S.♂.	Swartz obs. t. 2. f. 1.
15 vaginàtum. R.S.	.... sheathed.	———	1824.	——	S.♃.	
16 platycaùle. R.S.	.... flat-stalked.	———	——	——	S.♃.	
17 mólle. S.S.	.... soft.	St. Thomas.	1822.	——	S.♃.	
18 Kòra. R.S.	.... smooth-flowered.	E. Indies.	1818.	7. 8.	S.♃.	
19 longiflòrum. R.S.	.... long-flowered.	———	1824.	——	S.♃.	
20 setàceum. M.	.... bristly.	N. America.	——	——	H.♃.	
21 pilòsum. R.S.	.... hairy.	S. America.	1826.	——	S.♃.	
22 incértum. R.S.	.... uncertain.	....	1812.	——	H.♃.	
23 dùbium. S.S.	.... doubtful.	....	1818.	——	H.♃.	
24 glábrum. S.S.	.... smooth.	W. Indies.	1823.	——	S.♃.	
25 tenéllum. W. *élegans*. S.S.	.... elegant.	....	1817.	——	S.♃.	
26 láxum. R.S.	.... loose.	S. America.	1819.	——	S.♃.	
27 inæquàle. L.en.	.... unequal-valved.	Manilla.	1822.	——	S.♃.	
AMPHIP'OGON. B.P.	AMPHIP'OGON.	Triandria Digynia. S.S.				
stríctus. B.P.	.... upright.	N. S. W.	1823.	5. 6.	G.♃.	
GASTRI'DIUM. R.S.	GASTRI'DIUM.	Triandria Digynia. R.S.				
1 australe. R.S. *Mílium lendígerum*. E.B.	.... yellow.	England.	....	8.	H.♃.	Eng. bot. v. 16. t. 1107.
2 mùticum. R.S.	.... beardless.	Sicily.	1819.	7. 8.	H.⊙.	
CHÆT'URUS. R.S.	CHÆT'URUS.	Triandria Digynia. R.S.				
fasciculàtus. R.S.	.... fascicled.	S. Europe.	1815.	7. 9.	H.⊙.	
MI'LIUM. R.S.	MILLET-GRASS.	Triandria Digynia. L.				
1 nìgricans. R.S.	.... Guinea maize.	Peru.	1824.	6. 8.	H.⊙.	
2 effùsum. E.B.	.... common.	Britain.	....	6. 7.	H.♃.	Eng. bot. v. 16. t. 1106.
3 vernàle. R.S.	.... vernal.	Caucasus.	1819.	4. 5.	H.⊙.	
4 scábrum. R.S.	.... rough.	S. Europe.	1826.	——	H.⊙.	
5 confértum. R.S.	.... close-flowered.	Italy.	——	——	H.♃.	
6 amphicárpon.R.S.	.... earth-fruited.	N. America.	1827.	——	H.♃.	Pursh. am. sept. 2. t. 2.
7 gallècicum. R.S.	.... capillary.	Gallicia.	1819.	6. 8.	H.♃.	
8 velutìnum. R.S.	.... downy.	Mexico.	1823.	——	H.⊙.	
9 filifórme. Lag.	.... slender.	New Spain.	1824.	——	H.♃.	
10 microspérmum.R.S.	.. small-seeded.	———	1822.	——	H.⊙.	
11 phleoídes.	.... Phleum-like.	E. Indies.	1820.	——	S.♃.	
PIPTATH'ERUM. R.S.	PIPTATH'ERUM.	Triandria Digynia. R.S.				
1 paradóxum. R.S. *Agróstis paradóxa*. L. *Mílium paradóxum*. W. *Urάchne virèscens*. Trin.	.... black-seeded.	France.	1771.	6. 7.	H.⊙.	Host. gram. 3. t. 23.
2 cœruléscens. R.S.	.... blueish.	Barbary.	1819.	——	H.♃.	Desf. atl. 1. t. 12.
3 multiflòrum. R.S. *Mílium arundinàceum*. F.G. t. 66. *Agróstis miliàcea*. L.	.... many-flowered.	S. Europe.	1778.	——	H.♃.	Host. gram. 3. t. 45.
4 punctàtum. R.S.	.... dotted.	W. Indies.	1824.	——	S.♃.	
CALAMAGRO'STIS. S.S.	CALAMAGRO'STIS	Triandria Digynia. S.S.				
1 epigèjos. S.S.	.... wood.	Britain.	....	7.	H.♃.	Eng. bot. v. 6. t. 403.
2 intermèdia. R.S.	.... intermediate.	Carlsruhe.	1827.	——	H.♃.	
3 littòrea. S.S.	.... sea-side.	Switzerland.	1824.	7. 8.	H.♃.	Schrad. fl. ger. 1. t. 4. f. 2.
4 láxa. Host.	.... loose.	———	——	——	H.♃.	Host. gram. 4. t. 23.
5 glaùca. R.S.	.... glaucous.	N. America.	——	——	H.♃.	
6 Halleriàna. S.S. *Arúndo pseudophragmìtes*. R.S.	.... Haller's.	Switzerland.	1819.	——	H.♃.	Schrad. fl. ger. 1. t. 4. f. 3.
7 stricta. S.S.	.... upright.	Scotland.	....	7. 8.	H.♃.	Eng. bot. v. 30. t. 2160.
8 vària. Host.	.... various.	S. Europe.	1813.	——	H.♃.	Host. gram. 4. t. 47.
9 acutiflòra. DC.	.... sharp-flowered.	Germany.	1819.	——	H.♃.	
10 sylvática. DC. *arundinàcea*. Roth. *Agróstis arundinàcea*. L. *pyramidàlis*. Host.	.... wood.	Europe.	1813.	——	H.♃.	Host. gram. 4. t. 49.

11	purpuràscens. Br.	purplish.	N. America.	1827.	7. 8.	H. ♃.	
12	Schwábii. s.s.	Saxon.	Saxony.	1828.	——	H. ♃.	
13	Langsdórffii. s.s.	Langsdorff's.	....	1823.	——	H. ♃.	
14	confìnis. s.s.	hairy-flowered.	N. America.	1820.	——	H. ♃.	
15	montàna. Host.	mountain.	Europe.	——	——	H. ♃.	Host. gram. 4. t. 46.
16	Hóstii. R.S.	Host's.	——	——	——	H. ♃.	—— 4. t. 48.
	sylvática. Host. *nec aliorum.*						
17	canadénsis. N.	Canadian.	N. America.	1826.	——	H. ♃.	
18	lanceolàta. s.s.	common.	England.	....	7.	H. ♃.	Eng. bot. v. 30. t. 2159.
	Arúndo Calamagróstis. E.B. *Agróstis lanceolàta.* R.S.						
19	speciòsa. Host.	beautiful.	Germany.	1813.	7. 8.	H. ♃.	Host. gram. 4. t. 45.
	Agróstis Calamagróstis. R.S.						
STI'PA. R.S.		FEATHER-GRASS. Triandria Digynia. L.					
1	pennàta. E.B.	common.	Britain.	....	7. 8.	H. ♃.	Eng. bot. v. 19. t. 1356.
2	tórtilis. R.S.	twisted-awned.	S. Europe.	1823.	——	H. ⊙.	Desf. atl. 1. t. 31. f. 1.
3	júncea. R.S.	rushy.	France.	1772.	——	H. ♃.	Flor græc. 1. t. 85.
4	gigántea. R.S.	gigantic.	Spain.	1823.	——	H. ♃.	
5	Lagáscæ. R.S.	Lagasca's.	——	1822.	——	H. ♃.	Desf. atl. 1. t. 28.
6	capillàta.	capillary.	S. Europe.	——	——	H. ♃.	Host. gram. 3. t. 5.
7	tenacíssima. R.S.	tough.	——	1822.	——	H. ♃.	Desf. atl. 1. t. 30.
8	Aristélla. R.S.	awned.	Montpelier.	1806.	——	H. ♃.	Host. gram. 4. t. 34.
9	Redówskii. L.A.	Redowski's.	Altay.	1828.	——	H. ♃.	Ledeb. ic. t. 98.
10	móllis. B.P.	soft.	N. S. W.	1824.	——	F. ♃.	
11	conférta. R.S.	close panicled.	....	1819.	——	H. ♃.	
12	sibírica. R.S.	Siberian.	Siberia.	1777.	——	H. ♃.	Ledeb. ic. t. 99.
AGRA'ULUS. P.B.		AGRA'ULUS. Triandria Digynia. R.S.					
1	canìna. P.B.	brown.	Britain.	....	7.	H. ♃.	Eng. bot. v. 26. t. 1856.
	Agróstis canìna. E.B. *Trichòdium canìnum.* R.S.						
2	alpìna. P.B.	Alpine.	S. Europe.	1818.	7. 8.	H. ♃.	Host. gram. 3. t. 49.
3	rupéstris. R.S.	rock.	Pyrenees.	——	——	H. ♃.	—— 3. t. 50.
4	neglécta. R.S.	neglected.	Hungary.	1822.	——	H. ♃.	
5	flavéscens.	yellowish.	S. Europe.	——	——	H. ♃.	Host. gram. 4. t. 52.
6	setàcea.	bristly.	Britain.	....	——	H. ♃.	Eng. bot. v. 17. t. 1188.
	Agróstis setàcea. E.B. *Trichòdium setàceum.* R.S.						
7	rùbra. R.S.	red.	Lapland.	1820.	——	H. ♃.	
8	hy'brida. R.S.	hybrid.	Switzerland.	——	——	H. ♃.	
9	diffùsa. L.en.	spreading.	....	1823.	——	H. ♃.	
TRICHO'DIUM. M.		WINTER-GREEN-GRASS. Triandria Digynia.					
1	decúmbens. R.S.	decumbent.	N. America.	1786.	6. 7.	H. ♃.	Fraser mon. c. ic.
2	laxiflòrum. R.S.	loose-flowered.	——	1816.	——	H. ♂.	Mich. amer. 1. t. 8.
3	élegans. R.S.	elegant.	France.	——	7. 9.	H. ⊙.	Lois journ. bot. 2. t. 8. f. 1.
AGRO'STIS. R.S.		BENT-GRASS. Triandria Digynia. R.S.					
1	stolonífera. R.S.	creeping.	Britain.	....	7. 8.	H. ♃.	Eng. bot. v. 22. t. 1532.
	α latifòlia. S.G.W.	*Fiorin grass.*	——	....	——	H. ♃.	Sinc. gr. wob. p. 225. ic.
	β angustifòlia. S.G.W.	*narrow-leaved.*	——	....	——	H. ♃.	—— p. 346. ic.
	γ aristàta. S.G.W.	*awned.*	——	....	——	H. ♃.	—— p. 345. ic.
	δ nemoràlis. S.G.W.	*wood.*	——	....	——	H. ♃.	
	ε palústris. S.G.W.	*marsh.*	——	....	——	H. ♃.	
2	dúlcis. Df.	sweet.	S. France.	1824.	——	H. ♃.	
3	stolonìzans. R.S.	stoloniferous.	Poland.	1827.	——	H. ♃.	
4	diffùsa. R.S.	spreading.	Europe.	1816.	——	H. ♃.	Host. gram. 4. t. 55.
5	neglécta. R.S.	neglected.	Hungary.	——	——	H. ♃.	
6	capillàris. R.S.	capillary.	S. Europe.	——	——	H. ⊙.	Sm. ic. ined. 3. t. 54.
7	tenélla. R.S.	slender.	Switzerland.	1824.	——	H. ♃.	Schad. fl. ger. 1. t. 5. f. 1.
8	álba. R.S.	white rooted.	Britain.	....	7.	H. ♃.	Sinc. gr. wob. p. 342. ic.
	β purpuráscens.	*purplish.*	——	....	——	H. ♃.	Eng. bot. v. 17. t. 1189.
	γ vivìpara.	*viviparous.*	——	....	——	H. ♃.	Leers fl. herb. t. 4. f. 3.
9	párvula. R.S.	small.	Hungary.	1826.	——	H. ♃.	
10	tenuifòlia. R.S.	slender-leaved.	Caucasus.	1819.	7. 8.	H. ♃.	Trin. icon. gr. t. 35.
11	pátula. R.S.	knotted spreading.	Switzerland.	1824.	——	H. ♃.	
12	valentìna. R.S.	Valentia.	Spain.	——	——	H. ♃.	
13	dùbia. R.S.	doubtful.	Germany.	1818.	——	H. ♃.	Leers fl. herb. t. 4. f. 4.
14	gigántea. R.S.	gigantic.	Caucasus.	——	——	H. ♃.	
15	frondòsa. R.S.	frondose.	Naples.	1824.	——	H. ♃.	
16	vulgàris. R.S.	fine.	Britain.	....	——	H. ♃.	Eng. bot. v. 24. t. 1671.
17	ambígua. R.S.	ambiguous.	Germany.	1820.	——	H. ♃.	Høst. gram. t. 56.
18	Juréssi. R.S.	close panicled.	——	——	——	H. ♃.	Trin. ic. gram. t. 29.
19	virgínica. R.S.	Virginian.	N. America.	——	——	H. ♃.	Lab. nov. holl. 1. t. 23.
20	brasiliénsis. R.S. ...	Brazilian.	Brazil.	1828.	——	S. ♃.	

21 marítima. R.S.	.... sea-side.	S. Europe.	1825.	7. 8.	H.♃.	
22 Forstèri. R.S.	.... Forster's.	N. Zealand.	1822.	——	H.♃.	
Avèna filifórmis. Forst.						
23 Billardíeri. B.P.	.... Labillardiere's.	N. Holland.	1806.	——	H.♃.	Lab. nov. holl. 1. t. 31.
24 plebèia. B.P.	.... plebeian.	N. S. W.	1822.	——	H.♃.	
25 ǽmula. B.P.	.... flat-leaved.	————	1806.	——	H.♃.	
26 retrofrácta. R.S.	.... broad-leaved.	————	——	4. 5.	F.♃.	
27 hirsùta. R.S.	.... hairy.	Teneriffe.	1810.	——	F.♃.	
28 Spìca vénti. E.B.	.... silky.	England.	....	6. 7.	H.⊙.	Eng. bot. v. 14. t. 951.
29 pállida. DC.	.... pale.	France.	1818.	——	H.⊙.	
30 interrúpta. R.S.	.... interrupted.	S. Europe.	——	——	H.⊙.	Host. gram. 3. t. 48.
31 purpùrea. R.S.	.... purple.	————	1820.	——	H.⊙.	
32 pauciflòra. R.S.	.... few-flowered.	Switzerland.	1824.	6. 8.	H.♃.	Schrad.fl.ger.1. t.3. f.2.
33 decúmbens.Host.	.... decumbent.	Hungary.	——	——	H.♃.	Host. gram. 4. t. 54.
SPORO'BOLUS. B.P.	Sporo'bolus.	Triandria Digynia.				
1 diándrus. R.S.	.... diandrous.	E. Indies.	1820.	6. 8.	S.⊙.	
2 pulchéllus. R.S.	.... pretty.	N. Holland.	1828.	——	F.♃.	
3 índicus. B.P.	.... Indian.	India.	1773.	8. 10.	H.⊙.	Sloan. hist. 1. t. 73. f. 1.
4 purpuráscens.R.S.	.... purplish.	Jamaica.	1820.	——	S.⊙.	
5 tenacíssimus.L.en.	....tough.	W. Indies.	1801.	——	S.♃.	Jacq. ic. 1. t. 16.
6 elongàtus. B.P.	.... elongated.	N. S. W.	1822.	——	H.⊙.	
7 micránthus.	.... small-flowered.	Monte Video.	1825.	——	H.⊙.	
COLP'ODIUM. Trin.	Colp'odium.	Triandria Digynia.				
1 Stevèni. Trin.	.... various-coloured.	Caucasus.	1823.	7. 8.	H.♃.	
Agróstis versícolor. Stev.						
2 latifòlium. Br.	.... broad-leaved.	Melville Isl.	1821.	——	H.⊙.	
TRICHOCHL'OA. Tr.	Trichochl'oa.	Triandria Digynia.				
1 tenuiflòra.	.... slender-flowered.	N.America.	1820.	7. 8.	H.♃.	Willd. hort. ber. t. 12.
longisèta. Tr. *Agróstis tenuiflòra.* W. *Cínna tenuiflòra.* L.en.						
2 mexicàna. Tr.	.... Mexican.	America.	1780.	6. 9.	H.♃.	
3 foliòsa. Tr.	.... leafy.	N. America.	1819.	7. 8.	H.♃.	
Agróstis filifórmis. W.en. *Cínna filifórmis.* L.en.						
4 sobolífera. Tr.	.... soboliferous.	N. America.	1819.	7. 8.	H.♃.	
5 microspérma. DC.	.... small-seeded.	Mexico.	1818.	8. 10.	H.⊙.	
6 capillàris. DC.	.... capillary.	Carolina.	——	——	H.♃.	
7 expánsa. DC.	.... expanded.	————	1822.	——	H.♃.	P.B.agros. nov. t.8. f.2.
purpùrea. P.B.						
8 polypògon. DC.	.... many-bearded.	————	1812.	——	H.♃.	P.B.agros. nov.t.8. f.3.
CI'NNA. L.	Ci'nna.	Monandria Digynia. L.				
1 arundinàcea. L.	.... Reed-like.	Canada.	1789.	7.	H.♃.	Schreb. gram. t. 49.
2 crinìta. Tr.	.... rough-keeled.	N. Zealand.	1822.	——	F.♃.	Lab. nov. hol. 2. t. 263.
3 monándra. R.S.	.... monandrous.	N. America.	1826.	——	H.⊙.	
KNA'PPIA. E.B.	Kn'appia.	Triandria Digynia. S.S.				
agrostídea. E.B.	.... small.	Wales.	....	3. 4.	H.⊙.	Eng. bot. v. 16. t. 1127.
Agróstis mínima. W. *Chamagróstis mínima.* R.S. *Stúrmia vérna.* P.S.						
COLOBA'CHNE. R.S.	Coloba'chne.	Triandria Digynia. R.S.				
vaginàta. R.S.	.... sheathed.	Levant.	1818.	4.	H.♃.	Flor. græc. t. 64.
Alopecùrus angustifòlius. F.G. *Polypògon vaginàtus.* W.						
HELEOCHL'OA. Host.	Heleochl'oa.	Triandria Monogynia.				
1 alopecuroídes. R.S.	.. Foxtail-like.	Germany.	1816.	7. 8.	H.⊙.	Host gram. 1. t. 29.
2 phalaroídes. P.B.	.... Phalaris-like.	Caucasus.	1827.	——	H.⊙.	
3 schœnoídes. R.S.	.... Schœnus-like.	S. Europe.	1783.	——	H.⊙.	Jacq. ic. 1. t. 15.
CRY'PSIS. Host.	Cry'psis.	Triandria Monogynia.				
aculeàta. R.S.	.... prickly.	S. Europe.	1783.	——	H.⊙.	Host. gram. 1. t. 31.
LAPP'AGO. W.	Lapp'ago.	Triandria Digynia. W.				
1 racemòsa. W.	.... branching.	S. Europe.	1771.	——	H.⊙.	Flor. græc. t. 101.
Tràgus racemòsus. R.S. *Cénchrus racemòsus.* L.						
2 alìena. S.S.	.... Brazilian.	Brazil.	1826.	6. 8.	S.⊙.	
3 biflòra. F.I.	.... two-flowered.	E. Indies.	1824.	——	S.♃.	
MUHLENBE'RGIA. W.	Muhlenbe'rgia.	Triandria Digynia. W.				
diffùsa. R.S.	.... spreading.	N. America.	1816.	5. 6.	H.♃.	Schreb. gram. t. 51.
CHÆT'ARIA. R.S.	Chæt'aria.	Triandria Digynia. R.S.				
1 adscensiònis. R.S.	.... West Indian.	W. Indies.	1822.	7. 8.	S.♃.	Sloan. hist. 1. t. 2. f. 56.
Arístida adscensiònis. W.						
2 Hy'strix. R.S.	.... Porcupine.	E. Indies.	——	....	S.♃.	Pluk. alm. t. 191. f. 5.
3 oligántha. R.S.	.... long-glumed.	N. America.	1826.	——	H.w.♃.	
4 pállens. R.S.	.... pale.	Chile.	1825.	——	H.♃.	
5 divaricàta. R.S.	.... spreading.	S. America.	1818.	7. 8.	G.♃.	Jacq. eclog. gram. t. 6.

6 gigántea. R.S.	gigantic.	Teneriffe.	1824.	7. 8.	F.♃.	
7 cœruléscens. R.S.	blueish.	Canaries.	1818.	——	H.⊙.	Desf. atl. t. 21. f. 2.
CURTOP'OGON. R.S.	CURTOP'OGON.	Triandria Digynia.				
dichótomus. R.S.	forked.	N. America.	1826.	6. 8.	H.♂.	Spr. act. pet. 2. t. 6.
Arístida dichòtoma. Ph.						
ARI'STIDA. R.S.	ARI'STIDA.	Triandria Digynia. L.				
1 obtùsa. R.S.	blunt-awned.	Egypt.	1828.	6. 8.	H.⊙.	Delil. ægyp. t. 13. f. 2.
2 vàgans. R.S.	spreading-panicled.	N.S.W.	1820.	7. 8.	F.♃.	Cavan. ic. 5. t. 471.
HO'RDEUM. R.S.	BARLEY.	Triandria Digynia. L.				
1 vulgàre. R.S.	spring.	Levant.	....	7.	H.⊙.	Host. gram. 3. t. 34.
β gigánteum.	*giant.*	———	....		H.⊙.	
2 cœléste. R.S.	naked-seeded.	Levant.	....	——	H.⊙.	Viborg cereal. t. 1.
3 nìgrum. R.S.	black-seeded.	———	....	——	H.♂.	
4 hexástichon. R.S.	winter.	———	....	——	H.⊙.	Viborg cereal. t. 2.
5 bulbòsum. R.S.	bulbous.	S. Europe.	1770.	——	H.♃.	Flor. græc. 1. t. 98.
6 jubàtum. R.S.	long-bearded.	N. America.	1782.	7. 8.	H.♂.	
7 Sieberiànum. R.S.	Sieber's.	Egypt.	1828.	——	H.⊙.	
8 dístichum. R.S.	common.	Tartary.	....	7.	H.⊙.	Viborg cereal. t. 3.
β imbérbe. R.S.	*beardless.*	———	....	——	H.⊙.	
9 nùdum. R.S.	naked seeded.	———	....	——	H.⊙.	Arduin. sag. t. II. f. 4.
10 Zeócriton. R.S.	Battledore.	....	....	8.	H.⊙.	Viborg cereal. t. 4.
11 murìnum. R.S.	wall.	Britain.	....	4. 8.	H.⊙.	Eng. bot. v. 28. t. 1971.
12 praténse. R.S.	meadow.	———	....	6.	H.♃.	——— v. 6. t. 409.
13 marítimum. E.B.	sea-side.	———	....	6. 7.	H.⊙.	——— v. 17. t. 1205.
14 nepalénse.	Nepaul.	Nepaul.	1825.	7. 8.	H.⊙.	
15 Hy'strix. R.S.	hedgehog.	Spain.	1817.	——	H.⊙.	
16 bifàrium. R.S.	two-ranked.	....	1822.	——	H.⊙.	
ÆGOP'OGON. R.S.	ÆGOP'OGON.	Triandria Digynia. R.S.				
1 pusíllus. R.S.	small.	S. America.	1822.	7. 8.	H.⊙.	Cav. h. r. mad. 1. t. 5. f. 2.
2 trisètus. R.S.	three awned.	———	——	——	H.⊙.	——— 1. t. 5. f. 3.
MICROCHL'OA. B.P.	MICROCHL'OA.	Triandria Digynia. R.S.				
setàcea. B.P.	bristly.	Tropics.	1806.	7.	H.⊙.	Roxb. corom. 2. t. 132.
MEL'INIS. S.S.	MEL'INIS.	Triandria Digynia. S.S.				
minutiflòra. S.S.	glutinous.	Brazil.	1822.	6. 7.	S.♃.	Nees Hor. ber. 54. t. 7.
Tristègis glutinòsa. Nees. *Suárdia pícta.* Schrank h. mon. 11. t. 58.						
DIGIT'ARIA. R.S.	DIGIT'ARIA.	Triandria Digynia. R.S.				
1 sanguinàlis. R.S.	slender-spiked.	Britain.	....	8.	H.⊙.	Eng. bot. v. 12. t. 849.
Pánicum sanguinàle. E.B.						
2 bicórnis. R.S.	two-horned.	E. Indies.	1825.	8. 9.	H.⊙.	
3 affìnis. R.S.	related.	S. America.	1820.	7. 8.	H.⊙.	
4 ægyptìaca. R.S.	Egyptian.	Egypt.	1794.	——	H.⊙.	Jacq. obs. 3. t. 30.
5 filifórmis. S.S.	filiform.	N. America.	1781.	——	H.⊙.	
6 Roxbúrghii. S.S.	Roxburgh's.	E. Indies.	1823.	——	H.⊙.	
Pánicum filifórme. F.I.						
7 villòsa. S.S.	villous.	N. America.	1781.	7. 9.	H.⊙.	
8 débilis. S.S.	weak.	Barbary.	1824.	——	H.⊙.	
9 glábra. R.S.	smooth.	S. Europe.	1817.	——	H.⊙.	Schrad. germ. 1. t. 3. f. 6.
Syntherísma glábrum. Schrad.						
10 paspalòdes. R.S.	trailing.	N. America.	——	——	H.⊙.	
11 ciliàris. R.S.	fringed.	S. Europe.	1804.	——	H.⊙.	Host. gram. 4. t. 15.
12 barbàta. R.S.	bearded.	E. Indies.	1824.	——	H.⊙.	
13 bifórmis. R.S.	two-formed.	Canaries.	1820.	——	H.⊙.	
14 marginàta. R.S.	margined.	Brazil.	1825.	——	H.⊙.	
15 lineàris. R.S.	linear-spiked.	India.	1804.	——	H.⊙.	Burm. ind. 25. t. 10. f. 3.
16 inæquàlis. S.S.	hairy-sheathed.	Manila.	1824.	7. 8.	S.⊙.	
17 serotìna. M.	late-flowered.	N. America.	——	——	H.⊙.	
Páspalum serotìnum. R.S.						
18 Andersòni. Ot.	Anderson's.	MonteVideo.	1825.	6. 8.	F.♃.	
19 eriogòna. Schr.	woolly-angled.	S. America.	——	——	S.♃.	
20 mollíssima. Schr.	soft.	———	1826.	——	S.♃.	
21 pilígera. Lk.	hairy.	MonteVideo.	——	——	H.⊙.	
22 violáscens. Lk.	violet.	———	——	——	H.⊙.	
CYN'ODON. R.S.	CYN'ODON.	Triandria Digynia. R.S.				
1 Dáctylon. R.S.	creeping.	England.	....	7. 8.	H.♃.	Eng. bot. v. 12. t. 850.
β índicus.	*Doob-grass.*	India.	1820.	——	S.♃.	Sinc. gr. wob. p. 290. ic.
2 lineàris. R.S.	linear-leaved.	E. Indies.	1796.	——	S.♃.	
3 stellàtus. R.S.	starry.	St. Helena.	1823.	——	S.♃.	
4 virgàtus. R.S.	twiggy.	E. Indies.	1826.	——	S.♃.	

5 pr'æcox. R.S.	early-flowered.	N. America.	1823.	6. 7.	H.♃.	
Digitària sanguinàlis. M.						
BRACHYEL'YTRUM. R.S.	BRACHYEL'YTRUM.	Triandria Digynia.				
aristàtum. R.S.	awned.	N. America.	1827.	6. 7.	H.♃.	Schreb. gram. t. 50.
Muhlenbérgia erécta. Ph. *Dilep'yrum aristòsum.* M.						
PHA'LARIS. R.S.	CANARY-GRASS.	Triandria Digynia. R.S.				
1 canariénsis. R.S.	common.	Britain.	....	6. 8.	H.⊙.	Eng. bot. v. 19. t. 1310.
2 microstàchya. DC.	small-spiked.	N. America.	1818.	——	H.⊙.	
3 aquática. R.S.	water.	Egypt.	1778.	6. 7.	H.⊙.	Host. gram. 2. t. 38.
4 nítida. R.S.	glossy.	Sicily.	1824.	——	H.⊙.	
5 commutàta. R.S.	changed.	Italy.	1823.	——	H.⊙.	Barrel. ic. 700. f. 2.
6 capénsis. R.S.	Cape.	C. B. S.	1804.	——	G.♃.	
7 bulbòsa. R.S.	bulb-bearing.	Spain.	1824.	——	H.♃.	
8 variegàta. R.S.	variegated.	Sicily.	1825.	——	H.♃.	
9 cœruléscens. R.S.	blueish.	S. Europe.	1824.	——	H.⊙.	Buxb. cent. 4. t. 53.
10 nodòsa. R.S.	knotted.	————	1825.	——	H.♃.	Flor. græc. t. 56.
11 paradóxa. R.S.	paradoxical.	————	——	——	H.⊙.	———— 1. t. 58.
12 appendiculàta. R.S. ..	appendaged.	Egypt.	1823.	——	H.⊙.	
DIGRA'PHIS. Tr.	DIGRA'PHIS.	Triandria Digynia.				
1 arundinàcea. Tr.	Reed-like.	Britain.	....	7.	H.*w*.♃.	Eng. bot. v. 6. t. 402.
Arúndo coloràta. E.B. *Phálaris arundinàcea.* L. *Baldíngera.* Fl. wet.						
β variegàta.	*striped-leaved.*	————	....	——	H.♃.	
2 americàna. S.S.	American.	N. America.	....	——	H.♃.	
LAG'URUS. R.S.	HARE'S-TAIL-GRASS.	Triandria Digynia. L.				
ovàtus. E.B.	oval-spiked.	Guernsey.	....	6.	H.⊙.	Eng. bot. v. 19. t. 1334.
ECHINOP'OGON. R.S.	ECHINOP'OGON.	Triandria Digynia.				
ovàtus. R.S.	ovate panicled.	N. Holland.	1820.	7. 8.	H.♃.	Lab. nov. holl. 1. t. 21.
Agróstis ovàta. Lab.						
HIEROCHL'OA. P.B.	HIEROCHL'OA.	Triandria Digynia. R.S.				
1 boreàlis. R.S.	northern.	Scotland.	....	5. 6.	H.♃.	Sinc. gr. wob. p. 167. ic.
Hólcus odoràtus. F.D. t. 963. *répens.* Host. gram. 3. t. 3.						
2 austràlis. R.S.	southern.	S. Europe.	1777.	6. 7.	H.♃.	Host. gram. 1. t. 4.
3 fràgrans. R.S.	fragrant.	N. America.	——	——	H.♃.	
4 alpìna. R.S.	Alpine.	Melville Isl.	1827.	——	H.♃.	Flor. dan. t. 1508.
5 pauciflòra. R.S.	few-flowered.	————	——	——	H.♃.	
CATABR'OSA. R.S.	CATABR'OSA.	Triandria Digynia. R.S.				
1 aquática. R.S.	water.	Britain.	....	5. 6.	H.*w*.♃.	Eng. bot. v. 22. t. 1557.
Aíra aquática. E.B.						
2 virídula. R.S.	close-panicled.	....	1718.	——	H.*w*.♃.	
3 hùmilis. Tr.	dwarf.	Caucasus.	1824.	6. 7.	H.♃.	
AI'RA. R.S.	HAIR-GRASS.	Triandria Digynia. L.				
1 flexuòsa. E.B.	waved.	Britain.	....	7. 8.	H.♃.	Eng. bot. v. 22. t. 1519.
2 uliginòsa. R.S.	marsh.	Germany.	1825.	——	H.♃.	Reich. ic. r. C. 2. t. 150.
3 vérsicolor. R.S.	various-color'd.	S. Europe.	1823.	——	H.♃.	
4 caryophy'llea. E.B. ..	silver.	Britain.	....	——	H.⊙.	Eng. bot. v. 12. t. 812.
5 élegans. R.S.	elegant.	Switzerland.	1824.	——	H.⊙.	Host. gram. 4. t. 35.
6 pulchélla. R.S.	pretty.	Spain.	1820.	——	H.⊙.	
7 pállens. R.S.	pale.	N. America.	——	——	H.♃.	
8 arundinàcea. R.S.	Reed-like.	Cumana.	1823.	——	S.♃.	
9 semineùtra. S.S.	doubtful.	Hungary.	1819.	6. 7.	H.♃.	
DESCHA'MPSIA. R.S.	DESCHA'MPSIA.	Triandria Digynia. R.S.				
1 cæspitòsa. R.S.	turfy.	Britain.	....	8.	H.♃.	Eng. bot. v. 21. t. 1453.
Aíra cæspitòsa. E.B.						
2 Biebersteiniàna. R.S.	Bieberstein's.	Caucasus.	1823.	6. 8.	H.♃.	
3 refrácta. R.S.	reflexed.	Spain.	1822.	——	H.♃.	
4 alpìna. R.S.	Alpine.	Scotland.	....	5. 6.	H.♃.	Eng. bot. v. 30. t. 2102.
Aíra lævigàta. E.B. *alpìna.* L. *Avèna alpìna.* Tr.						
5 brevifòlia. R.S.	short-leaved.	Melville Isl.	1827.	——	H.♃.	
6 bóttnica. Tr.	bristle-leaved.	Bottnia.	1824.	——	H.♃.	
Aíra bóttnica. R.S. *montàna.* R.S.						
SCHI'SMUS. R.S.	SCHI'SMUS.	Triandria Digynia. R.S.				
1 Gouàni. Tr.	Gouan's.	Montpelier.	1820.	6. 7.	H.♃.	
Aíra mèdia. Gouan. *Deschámpsia mèdia.* et *díscolor.* R.S.						
2 marginàtus. R.S.	margined.	Spain.	1781.	——	H.⊙.	Cavan. ic. 1. t. 44. f. 2.
Festùca calycìna. W. *Kœlèria calycìna.* DC.						
3 fasciculàtus. P.B.	fascicled.	St. Domingo.	1827.	——	S.⊙.	
Digitària horizontàlis. W.						
DUPO'NTIA. R.S.	DUPO'NTIA.	Triandria Digynia.				
Fischèri. R.S.	Fischer's.	Melville Isl.	1827.	6. 8.	H.♃.	

CORYNE'PHORUS. R.S. CORYNE'PHORUS. Triandria Digynia. R.S.
1 canéscens. R.S. .... canescent. England. .... 7. 8. H.♃. Eng. bot. v. 17. t. 1190.
Aíra canéscens. E.B.
2 articulatus. R.S. .... jointed. S. Europe. 1824. —— H.♃. Desf. atl. 1. t. 13.
PERIBA'LLIA. Tr. PERIBA'LLIA. Triandria Digynia. R.S.
hispánica. Tr. .... Spanish. Spain. 1820. 6. 7. H.⊙. Cavan. ic. 1. t. 44. f. 1.
Aíra involucràta. Cav. *Airópsis involucràta.* R.S.
AIRO'PSIS. R.S. AIRO'PSIS. Triandria Digynia. R.S.
1 globòsa. R.S. .... globe-flowered. S. Europe. 1823. 6. 8. H.⊙. Cavan. ic. 3. t. 274. f. 1.
2 obtusàta. R.S. .... blunt-flowered. N. America. —— 6. 7. H.♃.
3 brevifòlia. R.S. .... short-leaved. Missouri. 1818. —— H.♃.
4 agrostídea. R.S. .... Agrostis-like. S. Europe. 1823. —— H.♃. DC. ic. rar. t. 1.
Pòa agrostídea. DC. *Aíra agrostídea.* Lois.
ME'LICA. R.S. MELIC-GRASS. Triandria Digynia. L.
1 ciliàta. R.S. .... ciliated. Europe. 1771. 7. H.♃. Flor. græc. 1. t. 70.
2 nùtans. R.S. .... mountain. Britain. .... 6. 7. H.♃. Eng. bot. v. 15. t. 1059.
3 uniflòra. R.S. .... wood. —— —— .... 5. 6. H.♃. ———— v. 15. t. 1058.
4 minùta. R.S. .... small. S. Europe. 1825. —— H.♃. Cavan. ic. 2. t. 175. f. 2.
5 Bauhìni. R.S. .... Italian. Italy. 1806. 6. 7. H.♃. Host. gram. 4. t. 23.
6 pyramidàlis. R.S. .... pyramidal. Barbary. 1804. —— H.♃. Barrel. ic. t. 96. f. 1.
7 speciòsa. R.S. .... handsome. N. America. 1812. —— H.♃. Schreb. gram. t. 54.
glàbra. Ph. *rariflòra.* Schreb.
8 altíssima. R.S. .... tallest. Siberia. 1770. 7. 8. H.♃. Host. gram. 2. t. 9.
9 diffùsa. R.S. .... spreading. N. America. 1824. —— H.♃.
10 pállida. R.S. .... pale. Quito. 1825. —— S.♃.
11 exasperàta. R.S. .... exasperate. .... —— —— H.♃.
12 digitàta. R.S. .... fingered. E. Indies. 1824. —— S.♃.
13 refrácta. R.S. .... reflexed. —— —— —— S.♃.
14 latifòlia. R.S. .... broad-leaved. —— 1822. —— S.♃.
MOLI'NIA. P.B. MOLI'NIA. Triandria Digynia. R.S.
1 cœrùlea. P.B. .... purple. Britain. .... 8. H.♃. Eng. bot. v. 11. t. 750.
vària. S.S. *Mélica cœrùlea.* E.B. *Enòdium cœrùleum.* R.S.
2 sylvática. R.S. .... wood. Europe. 1824. —— H.♃.
DIN'EBA. R.S. DIN'EBA. Triandria Digynia. R.S.
1 americàna. R.S. .... American. S. America. 1822. 6. 8. S.♃.
2 arábica. R.S. .... Arabian. Levant. 1816. 8. 9. H.⊙. Jacq. fragm. t. 121. f. 1.
3 divaricata. R.S. .... spreading. Egypt. 1823. —— H.⊙.
WANGENHE'IMIA. Tr. WANGENHE'IMIA. Triandria Digynia. R.S.
Lìma. Tr. .... imbricated. Spain. 1776. 7. 8. H.⊙. Cavan. ic. 1. t. 91.
Cynosùrus Lìma. Cav. *Dinèba Lìma.* R.S.
SA'CCHARUM. R.S. SUGAR-CANE. Triandria Digynia. L.
1 officinàrum. R.S. .... common. India. 1597. .... S.♃. Rumph. amb. 5. t. 74. f. 1.
2 sinénse. F.I. .... Chinese. China. 1822. .... S.♃. Roxb. corom. 3. t. 232.
3 pròcerum. F.I. .... lofty. E. Indies. —— .... S.♃.
4 violàceum. R.S. .... violet-coloured. W. Indies. 1816. .... S.♃.
5 caudàtum. R.S. .... tailed. —— —— .... S.♃. Sloan. hist. 1. t. 70. f. 1.
6 contráctum. R.S. .... contracted. S. America. 1823. .... S.♃.
7 exaltàtum. F.I. .... tall. E. Indies. 1826. .... S.♄. Rheed. mal. 12. t. 46.
8 polystàchyum. R.S. .. many-spiked. W. Indies. —— .... S.♄.
9 Múnja. F.I. .... slender. E. Indies. 1805. .... S.♃.
IMPER'ATA. R.S. IMPER'ATA. Triandria Digynia. R.S.
1 arundinàcea. R.S. .... Reed-like. N. Holland. 1820. 7. 8. F.♃. Cyril. ic. fasc. 2. t. 11.
2 Thunbérgii. R.S. .... Thunberg's. E. Indies. 1822. .... S.♄.
3 Kœnígii. R.S. .... Kœnig's. W. Indies. 1818. .... S.♄.
4 spontànea. R.S. .... spontaneous. E. Indies. —— .... S.♄. Rheed. mal. 12. t. 46.
TRICHOL'ÆNA. R.S. TRICHOL'ÆNA. Triandria Digynia. R.S.
micrántha. R.S. .... small-flowered. S. Europe. 1816. 7. 8. H.♃. Jacq. ecl. gram. t. 34.
Sáccharum Teneríffæ. Jacq.
ERIA'NTHUS. R.S. ERIA'NTHUS. Triandria Digynia. R.S.
1 saccharoídes. R.S. .... Sugar-cane-like. N. America. 1816. 7. 8. H.♃.
Sáccharum gigánteum. Ph.
2 brevibárbis. R.S. .... short-bearded. —— 1822. —— H.♃.
3 stríctus. R.S. .... upright. —— 1828. —— H.♃.
RIPI'DIUM. Tr. RIPI'DIUM. Polygamia Monœcia. Tr.
1 Ravénnæ. Tr. .... rough. S. Europe. 1816. 7. 8. H.♃. Host. gram. 3. t. 1.
Andropògon Ravénnæ. Host. *Sáccharum Ravénnæ.* W.
2 stríctum. Tr. .... upright. Hungary. 1802. —— H.♃. Host. gram. 2. t. 2.
Sáccharum adpréssum. Kit. *Andropògon stríctus.* Host.

ANDROP'OGON. R.S. ANDROP'OGON. Polygamia Monœcia. L.
1 alopecuroídes. R.S. .. Foxtail. America. 1818. .... G. ♃. Sloan. hist. 1. t. 14.
2 milifórmis. R.S. .... Millet-like. E. Indies. —— .... S. ♃.
miliàceus. F.I. *non* Forsk.
3 argénteus. DC. ...... silvery. Mexico. 1822. .... G. ♃. DC. hort. m. ined. t. 68.
4 saccharoídes. R.S..... Sugar-cane-like. Jamaica. 1818. —— S. ♃.
5 laguroídes. R.S. .... hares-tail-like. New Spain. 1820. .... S. ♃. DC. h. mons. ined. t. 69.
6 hírtus. R.S. .... hairy. S. Europe. 1802. 7. 8. S. ♃. Host. gram. 4. t. 1.
7 angustifòlius. R.S..... narrow-leaved. ———— 1818. —— H. ♃. Jacq. aust. 4. t. 384.
8 Isch'æmum. R.S. .... woolly. ———— 1768. —— H. ♃. Gerard. gall. p. 107. t. 4.
9 serràtus. R.S. ...... saw-leaved. Japan. 1816. .... G. ♃.
láxus. W.
10 comòsus. R.S. ...... tufted. Egypt. 1824. .... H. ♃.
11 annulàtus. R.S. ...... annular. ———— —— .... H. ♃. Forsk. egyp. t. 7.
12 furcàtus. R.S. ...... forked. N. America. 1818. 7. 8. H. ♃.
CHRYSOP'OGON. Tr. CHRYSOP'OGON. Polygamia Monœcia.
1 Gry'llus. Tr. .... purple-flowered. S. Europe. 1791. 6. 7. H. ♃. Flor. græc. 1. t. 67.
Andropògon Gry'llus. F.G. *Hólcus Gry'llus.* B.P.
2 aciculàtus. Tr. ...... sharp-pointed. E. Indies. 1825. 6. 8. S. ♃. Rheed. mal. 12. t. 43.
LEPEOCE'RCIS. Tr. LEPEOCE'RCIS. Polygamia Monœcia.
serràta. Tr. .... serrated. E. Indies. 1812. 6. 8. S. ♃.
Andropògon serràtum. W.
POLL'INIA. R.S. POLL'INIA. Polygamia Monœcia.
1 distàchya. R.S. ...... two-spiked. S. Europe. 1805. 7. 8. H. ♃. Flor. græc. 1. t. 69.
2 striàta. R.S. .... nerve-glumed. E. Indies. 1793. 8. S. ♃.
3 undàta. R.S. .... waved. Mauritius. 1823. 7. 9. S. ♃. Jacq. ic. 3. t. 361.
CYMBOP'OGON. R.S. CYMBOP'OGON. Polygamia Monœcia.
1 élegans. R.S. .... elegant. E. Indies. 1816. 6. 8. S. ♃.
2 prostràtus. .... prostrate. ———— —— —— S. ♃.
glandulòsus. R.S. *Anthistíria prostràta.* F.I. *Andropògon prostràtus.* L.
3 Schœnánthus. R.S. .. Lemon-grass. E. Indies. 1786. .... S. ♃. Rumph. amb. 5. t. 72. f. 2.
Andropògon Schœnánthus. W.
4 arundinàceus. R.S. .. Reed-like. E. Indies. 1820. 6. 8. S. ♃.
HETEROP'OGON. R.S. HETEROP'OGON. Polygamia Monœcia.
1 Alliònii. R.S. .... Allioni's. S. Europe. 1816. 7. 8. H. ♃. Allion. fl. ped. t. 91. f. 4.
2 contórtus. R.S. ...... twisted. E. Indies. 1779. 7. 9. S. ♃. Schk. hand. t. 342. f. a.
Andropògon contórtus. W.
APL'UDA. R.S. APL'UDA. Polygamia Monœcia. L.
1 aristàta. R.S. .... awned. E. Indies. 1820. .... S. ♃. Schreb. gram. 2. t. 42.
2 villòsa. R.S. .... villous. ———— —— .... S. ♃.
ANATHE'RUM. R.S. ANATH'ERUM. Polygamia Monœcia.
1 muricàtum. R.S. .... muricate. E. Indies. 1820. .... S. ♃.
Andropògon muricàtus. Retz. *squarròsus.* W.
2 virgínicum. R.S. .... Virginian. N. America. 1819. 7. 8. H. ♃. Sloan. jam. 1. t. 68. f. 2?
3 bicórne. R.S. ...... two-horned. W. Indies. 1825. —— S. ♃. Sloan. hist. 1. t. 15.
4 caudàtum. R.S. ...... tailed. ———— 1826. —— S. ♃.
5 mùticum. R.S. ...... smooth-spiked. C. B. S. 1794. 7. 9. G. ♃.
CALAM'INA. R.S. CALAM'INA. Polygamia Monœcia.
1 gigántea. R.S. ...... gigantic. Isl. Luzon. 1822. 7. 8. G. ♃. Cavan. ic. 5. t. 458.
2 mùtica. R.S. ...... awnless. E. Indies. 1816. —— S. ♃.
Aplùda mùtica. L.
CYMBA'CHNE. R.S. CYMBA'CHNE. Polygamia Monœcia.
ciliàta. R.S. .... fringed. E. Indies. 1824. 7. 8. S. ♃.
ANTHIST'IRIA. R.S. ANTHIST'IRIA. Polygamia Monœcia. L.
1 ciliàta. R.S. .... fringed. Jamaica. 1821. 8. 10. S. ⊙. Cavan. ic. 5. t. 459.
2 barbàta. R.S. ...... bearded. Japan. 1816. —— G. ⊙. Desf. j. ph. 40. p. 294. t. 2.
3 polystàchya. R.S. .... many-spiked. E. Indies. 1824. —— S. ⊙.
4 heteroclìta. R.S. .... jointed. ———— —— —— S. ⊙.
5 austràlis. B.P. ...... New Holland. N. Holland. 1818. —— F. ♃.
SO'RGHUM. R.S. SO'RGHUM. Polygamia Monœcia.
1 vulgàre. R.S. .... Indian Millet. India. 1596. 7. H. ⊙. Host. gram. 4. t. 2.
2 nìgrum. R.S. .... black-glumed. .... —— H. ⊙. Ard. sag. pad. 1. t. 5. f. 1.
3 bícolor. R.S. .... two-coloured. Persia. 1731. —— H. ⊙. Mieg. ac. helv. 8. t. 4. f. 4.
4 cérnuum. R.S. ...... nodding. .... —— H. ⊙. Host. gram. 4. t. 3.
5 rùbens. R.S. ...... red. Africa. 1817. —— H. ⊙.
6 nervòsum. R.S. ...... nerved-leaved. China. 1827. —— H. ⊙.
7 saccharàtum. R.S..... yellow-seeded. India. 1759. 7. 8. S. ♂. Host. gram. 4. t. 4.
8 caffròrum. R.S. .... Cape. C. B. S. 1816. —— H. ⊙. Ard. sag. pad. 1. t. 1. f. 1.
Arduìni. Jacq. ecl. gram. t. 18.

9 avenàceum. R.S.	Oat-like.	C. B. S.	1816.	7. 9.	H.⊙.	
10 halepénse. R.S.	panicled.	Syria.	1691.	7. 8.	H.♃.	Flor. græc. 1. t. 68.
11 elongàtum. R.S.	long-panicled.	N. Holland.	1823.	——	F.♃.	
HO'LCUS. R.S.	Soft-grass.	Polygamia Monœcia. L.				
1 móllis. R.S.	creeping.	Britain.	....	7. 8.	H.♃.	Eng. bot. v. 17. t. 1170.
2 lanàtus. R.S.	meadow.	———	....	6. 7.	H.♃.	——— v. 17. t. 1169.
ARRHENATH'ERUM. R.S.	Arrhenath'erum.	Polygamia Monœcia.				
1 avenàceum. R.S.	Oat-like.	Britain.	....	6. 7.	H.♃.	Eng. bot. v. 12. t. 813.
Hólcus avenàceus. E.B. *Avèna elàtior.* F.D. t. 165.						
β mùticum.	*awnless.*	Scotland.	....	——	H.♃.	
2 bulbòsum. Presl.	bulbous.	Germany.	....	——	H.♃.	Host. gram. 4. t. 30.
Hólcus bulbòsus. Host. *Avèna bulbòsa.* W.						
ATHEROP'OGON. R.S.	Atherop'ogon.	Polygamia Monœcia.				
apludoídes. R.S.	Apluda-like.	N. America.	1808.	6. 7.	H.♃.	Jacq. ecl. gram. t. 7.
Chlòris cartipéndula. M.						
CHL'ORIS. R.S.	Chl'oris.	Polygamia Monœcia. L.				
1 foliòsa. R.S.	leafy.	St. Thomas.	1822.	7. 9.	S.♃.	
2 ciliàta. R.S.	ciliated.	Jamaica.	1779.	——	S.⊙.	
3 radiàta. R.S.	many spiked.	W. Indies.	1739.	8. 9.	S.⊙.	Hayn. term. bot. t. 40. f. 2
4 pállida. R.S.	pale-spiked.	S. France.	1816.	——	H.⊙.	
5 barbàta. R.S.	bearded.	E. Indies.	1777.	6. 7.	S.⊙.	Jacq. ecl. gram. 1. t. 8.
6 Roxburghiàna. R.S. ..	Roxburgh's.	———	1824.	7. 8.	S.⊙.	
7 inflàta. R.S.	inflated.	California.	——	——	H.♃.	
8 fasciculàta. R.S.	crowded.	Brazil.	——	——	H.⊙.	
9 carib'æa. R.S.	Caribean.	W. Indies.	1827.	——	S.⊙.	
10 polydáctyla. R.S.	many-spiked.	———	1818.	——	S.⊙.	Jacq. ecl. gram. 1. t. 9.
11 élegans. R.S.	elegant.	Mexico.	1825.	——	H.⊙.	H. et B. n. gen. 1. t. 49.
12 compréssa. R.S.	compressed.	S. France.	1820.	6. 8.	H.⊙.	
13 megastàchya. R.S. ..	long-spiked.	N. Holland.	1825.	——	F.♃.	
14 ventricòsa. B.P.	swelled.	N. S. W.	1818.	——	G.♃.	
15 truncàta. B.P.	truncated.	———	1822.	——	G.♃.	
16 pubéscens. R.S.	pubescent.	Peru.	1824.	——	H.⊙.	
17 filifórmis. R.S.	filiform.	....	1825.	——	H.⊙.	
18 distichophy'lla. R.S. ..	fan-leaved.	Chile.	1822.	——	H.♃.	
19 retùsa. R.S.	retuse-glumed.	———	——	——	H.⊙.	
20 dolichostàchya. R.S. ..	long-spiked.	Philippine Isl.	1824.	7. 8.	H.⊙.	
21 crinìta. R.S.	crowded-spiked.	———	——	——	H.⊙.	
22 rufféscens. R.S.	rufescent.	———	——	——	H.⊙.	
23 tenélla. R.S.	slender.	E. Indies.	1825.	——	S.⊙.	
24 grácilis. S.S.	slender.	S. America.	1824.	——	S.♃.	
25 Durandiàna. R.S.	Durand's.	....	....	——	H.⊙.	
EUST'ACHYS. R.S.	Eust'achys.	Polygamia Monœcia.				
1 petr'æa. R.S.	flat-stalked.	Jamaica.	1779.	7. 8.	S.♃.	Jacq. ecl. gram. 1. t. 11.
Chlòris petr'æa. W.						
2 submùtica. R.S.	Mexican.	Mexico.	1827.	——	H.⊙.	H. et B. pl. am. 1. t. 50.
CHRYS'URUS. R.S.	Chrys'urus.	Polygamia Monœcia.				
1 aùreus. S.S.	golden-spiked.	Levant.	1770.	7.	H.⊙.	Host. gram. 3. t. 4.
cynosuroídes. R.S. *Cynosurùs aùreus.* F.G. 1. t. 79. *Lamárckia aùrea.* DC.						
2 echinàtus. R.S.	rough.	England.	....	8.	H.⊙.	Eng. bot. v. 19. t. 1333.
Cynosùrus echinàtus. E.B.						
3 effùsus. R.S.	spreading.	S. Europe.	1824.	7. 8.	H.⊙.	
4 élegans. R.S.	elegant.	———	——	——	H.⊙.	Desf. atl. 1. t. 17.
ISCH'ÆMUM. S.S.	Isch'æmum.	Polygamia Monœcia. L.				
1 mùticum. R.S.	awnless.	E. Indies.	1824.	6. 8.	S.♃.	Rheed. mal. 12. t. 49.
2 conjugàtum. R.S.	yoked.	———	1826.	——	S.♃.	
3 austràle. B.P.	southern.	N. S. W.	1822.	7. 8.	F.♃.	
4 rugòsum. S.S.	rough.	E. Indies.	1791.	——	S.⊙.	Sal. stirp. rar. 1. t. 1.
MEO'SCHIUM. R.S.	Meo'schium.	Polygamia Monœcia.				
1 aristàtum. R.S.	awned.	E. Indies.	1808.	6. 7.	S.♃.	
Isch'æmum aristàtum. S.S.						
2 barbàtum. R.S.	bearded.	Java.	——	——	S.♃.	Houtt. syst. 12. t. 93. f. 4.
ROTTB'OELLIA. R.S.	Rottbo'ellia.	Triandria Digynia. L.				
1 exaltàta. R.S.	tall.	E. Indies.	1806.	....	S.♂.	
2 perforàta. R.S.	perforated.	———	1822.	....	S.♃.	Roxb. corom. t. 182.
3 glábra. R.S.	smooth.	———	——	——	S.♃.	
HEMA'RTHRIA. B.P.	Hema'rthria.	Triandria Digynia.				
compréssa. B.P.	flat-spiked.	N. S. W.	1822.	6. 7.	G.♃.	
LODICUL'ARIA. P.B.	Lodicul'aria.	Triandria Digynia.				
fasciculàta. P.B.	fascicled.	S. Europe.	1826.	6. 7.	H.♃.	P. Beauv. agros. t. 11. f. 6.
Rottbòellia fasciculàta. Desf. atl. 1. t. 36.						

STENOTA'PHRUM. Tr.	STENOTA'PHRUM.	Polygamia Monœcia.				
1 glábrum. Tr.	smooth.	Carolina.	1826.	6. 7.	H. ♃.	
2 dimidiàtum.	creeping-rooted.	W. Indies.	1822.	——	S. ♃.	Schrank. h. monac. t. 98.
americànum. Schk.						
3 complanàtum. Sck...	flat-spiked.	E. Indies.	1824.	——	S. ♃.	Burm. ind. t. 8. f. 5.
LEPTURUS. B.P.	HARD-GRASS.	Triandria Digynia.				
1 incurvàtus. Tr.	sea-side.	Britain.	....	7.	H. ⊙.	Eng. Bot. v. 11. t. 760.
Rottbòellia incurvàta. E.B. *Ophiùrus incurvàtus.* R.S.						
2 filifórmis. Tr.	thready.	S. Europe.	1800.	——	H. ♃.	Barrel. ic. t. 6.
3 cylíndricus. Tr.	cylindrical.	Portugal.	1806.	——	H. ⊙.	Giorn. pis. 4. f. 4-8.
Rottbòellia cylíndrica. W. *subulàta.* DC. *Monérma subulàta.* R.S.						
PHOLI'URUS. Tr.....	PHOLI'URUS.	Triandria Digynia.				
pannònicus. Tr.	Hungarian.	Hungary.	1804.	——	H. ⊙.	Host. gram. 1. t. 24.
MONE'RMA. R.S.	MONE'RMA.	Monandria Digynia.				
monándra. R.S.	monandrous.	Spain.	1804.	——	H. ⊙.	Cavan. ic. t. 39. f. 1.
Rottbòellia monándra. Cav. *Nárdus aristàta.* L. *Psilùrus nardoídes.* Tr.						
MANIS'URIS. R.S.	MANIS'URIS.	Polygamia Monœcia. W.				
granulàris. R.S.	round grained.	E. Indies.	1784.	6. 7.	S. ⊙.	Roxb. corom. 2. t. 118.
ANTHE'PHORA. R.S.	ANTHE'PHORA.	Triandria Monogynia.				
1 élegans. R.S.	elegant.	Jamaica.	1776.	8. 9.	S. ⊙.	Schreb. gram. t. 44.
Trípsacum hermaphrodìtum. L. *Cénchrus lævigàtus.* S.S.						
2 villòsa. R.S.	villous.	W. Indies.	1824.	——	S. ⊙.	
TRI'PSACUM. S.S.	TRI'PSACUM.	Monœcia Triandria. L.				
1 dactyloídes. S.S.	rough-seeded.	Virginia.	1640.	8.	H. ♃.	Lam. ill. t. 750.
2 monostàchyon. S.S. ..	single-spiked.	N. America.	1815.	——	H. ♃.	Willd. hort. ber. t. 1.
ECHINOCHL'OA. R.S.	ECHINOCHL'OA.	Triandria Digynia. R.S.				
1 stagnìna. R.S.	pond.	E. Indies.	1802.	7. 8.	H. *w.* ⊙.	
2 commutàta. R.S.	Hungarian.	Hungary.	——	——	H. *w.* ⊙.	Host. gram. 3. t. 51.
Pánicum stagnìnum. Host.						
3 intermèdia. R.S.	intermediate.	....	1816.	——	H. ⊙.	
4 erythrospérma. R.S...	red-seeded.	....	1825.	——	H. ⊙.	
5 híspida. R.S.	hispid.	Egypt.	1829.	——	H. ⊙.	
6 Crus-Córvi. R.S.	Crow's-foot.	E. Indies.	1781.	——	H. ⊙.	
7 Crus-Gálli. R.S.	Cock's-foot.	England.	....	——	H. ⊙.	Eng. bot. v. 13. t. 876.
Pánicum Crus-Gálli. E.B.						
8 echinàta. R.S.	bristly.	N. America.	1825.	——	H. ⊙.	Jacq. ecl. gram. t. 20.
9 setígera. R.S.	setigerous.	E. Indies.	1820.	——	S. ⊙.	
ORTHOP'OGON. B.P.	ORTHOP'OGON.	Triandria Digynia. S.S.				
1 hirtéllus. S.S........	hairy.	W. Indies.	1795.	6. 7.	S. ♂.	
2 setàrius. S.S.	bristly.	————	1823.	——	S. ⊙.	
3 undulatifòlius. S.S. ..	wave-leaved.	S. Europe.	1795.	6. 7.	H. ⊙.	Host. gram. 3. t. 52.
Pánicum hirtéllum. Host. *non* L.						
4 Burmánni. R.S.	Burman's.	E. Indies.	1819.	....	S. ⊙.	Burm. ind. t. 12. f. 1.
5 africànus.	African.	Africa.	1822.	7. 9.	S. ⊙.	Pal. fl. d'ow. 2. t. 67. f. 1.
Oplismènus africànus. R.S.						
6 compòsitus. S.S.	compound spiked.	E. Indies.	1819.	——	S. ⊙.	Linn. pfl. sys. 12. t. 96. f. 1.
7 fláccidus. B.P.	flaccid.	N. S. W.	1824.	——	H. ⊙.	
8 imbecíllis. B.P.	feeble.	————	——	——	H. ⊙.	
PA'NICUM. R.S.	PANIC-GRASS.	Triandria Digynia. R.S.				
1 colònum. R.S.	purple.	E. Indies.	1699.	7. 8.	H. ⊙.	Ehret pict. t. 3. f. 3.
2 flùitans. R.S.	floating.	————	1825.	——	S. *w.* ⊙.	
3 brizoídes. R.S.	Briza-like.	————	1801.	6. 7.	S. ♂.	Jacq. gram. fasc. t. 2.
4 flávidum. R.S.	yellowish.	————	1823.	——	S. ♂.	
5 mólle. R.S.	soft.	W. Indies.	1824.	——	H. ⊙.	
6 Michauxiànum. R.S.	Michaux's.	N. America.	——	——	H. ⊙.	
Pánicum mólle. Mich. *non* Swartz.						
7 fasciculàtum. R.S. ..	fascicled.	Jamaica.	1801.	——	H. ⊙.	
8 plicàtum. R.S.	plaited.	E. Indies.	1824.	6. 8.	S. ♃.	
9 palmifòlium. R.S.	Palm-leaved.	————	1804.	——	S. ♃.	Jacq. ecl. gram. t. 1.
10 neuròdes. R.S.	nerved-leaved.	Nepaul.	1823.	——	G. ♃.	
nervòsum. F.I. *non* Lam.						
11 costàtum. F.I.	ribbed.	Mauritius.	1805.	——	S. ♃.	
12 carthaginénse. R.S. ..	Carthagena.	S. America.	1818.	——	S. ♃.	
13 conglomeràtum. R.S.	conglomerate.	E. Indies.	——	——	S. ♃.	
14 índicum. F.I.	Indian.	————	——	——	S. ♃.	
15 hispídulum. R.S.	hispid.	————	1804.	7. 8.	H. ⊙.	
16 frumentàceum. F.I...	cultivated.	————	1810.	——	H. ⊙.	
17 dichótomum. R.S.	forked.	N. America.	1817.	——	H. ⊙.	Sp. m. act. pet. 1810. t. 4.
18 coloràtum. R.S.	coloured.	Egypt.	1771.	7. 9.	H. ⊙.	Jacq. ic. 1. t. 58?

19 répens. R.S.	slender.	S. Europe.	1777.	7. 9.	H.⊙.	Flor. græc. 1. t. 61.
20 prolíferum. R.S.	proliferous.	N. America.	1816.	——	H.⊙.	
21 diffórme. R.S.	deformed.	E. Indies.	1825.	——	S.♃.	
22 numidiànum. R.S. ..	Numidian.	Barbary.	1816.	——	H.⊙.	Desf. atl. 1. t. 11.
23 miliàceum. R.S.	Millet.	E. Indies.	1596.	——	H.⊙.	Host. gram. 2. t. 20.
24 ténue. R.S.	slender.	————	1824.	——	S.♃.	
25 tenuíssimum. R.S. ..	very slender.	Brazil.	1827.	——	S.♃.	
26 attenuàtum. R.S.	attenuated.	....	1825.	——	H.⊙.	
27 gongylòdes. R.S.	branching-flowered.	....	1823.	——	H.♃.	
28 miliàre. R.S.........	Millet-like.	E. Indies.	1824.	——	H.⊙.	
29 muricàtum. R.S.	prickly.	————	1805.	——	H.⊙.	
30 capillàre. R.S.	hair-panicled.	America.	1758.	6. 8.	H.⊙.	
β mìnor. R.S.	*smaller.*	————	———	——	H.⊙.	Host. gram. 4. t. 16.
31 tenéllum. R.S.	slender-branch'd.	Sierra Leon.	1820.	——	S.⊙.	
32 grossàrium. R.S.	cultivated.	W. Indies.	1810.	——	H.⊙.	
33 máximum. R.S.	Guinea-grass.	Tropics.	1822.	——	S.♃.	Jacq. ic. 1. t. 13.
34 aspérrimum. R.S.....	rough.	S. America.	1824.	7. 9.	H.⊙.	
35 fúscum. R.S.	brown.	Jamaica.	1822.	——	S.♃.	
36 latifòlium. R.S.......	broad-leaved.	N. America.	1765.	8. 9.	H.♃.	Moris. his. 3. t. 5. f. 4.
37 macrocárpon. R.S. ..	large-fruited.	————	1825.	——	H.♃.	
38 microcárpon. R.S. ..	small-fruited.	————	1822.	——	H.♃.	
39 diffùsum. R.S.	diffuse.	W. Indies.	1823.	——	S.⊙.	
40 oryzoídes. R.S.	Rice-like.	————	1824.	——	S.⊙.	
41 clandestìnum. R.S. ..	hidden-flowered.	N. America.	1802.	7.	H.♃.	
42 virgàtum. R.S.	twiggy.	————	1781.	8. 9.	H.♃.	Spr. m. act. pet. 2. t. 5.
43 ánceps. R.S.	flat-edged.	————	1816.	7. 8.	H.♃.	
44 uliginòsum. R.S.	marsh.	E. Indies.	1819.	——	S.*w*.♃.	
45 nítidum. R.S.	glossy.	N. America.	1811.	——	H.⊙.	
46 arundinàceum. R.S...	Reed-like.	Jamaica.	1819.	——	S.♃.	
47 triphèron. R.S.	brown-seeded.	E. Indies.	1820.	6. 8.	S.⊙.	
tenéllum. F.I. *nec aliorum. Roxbúrghii.* S.S.						
48 brevifòlium. R.S.	short-leaved.	————	1800.	7. 8.	S.♃.	Pluk. alm. t. 189. f. 4.
49 divaricàtum. R.S.	divaricate.	Jamaica.	——	——	S.♃.	Jacq. schœn. 1. t. 25.
50 interrúptum. R.S....	interrupted.	E. Indies.	1820.	——	S.♃.	
51 arboréscens. R.S.	arborescent.	————	1776.	3. 4.	S.♄.	
52 pàtens. R.S.	spreading.	————	1804.	7. 8.	S.♂.	Burm. ind. t. 10. f. 3.
53 altíssimum. R.S.	lofty.	....	1816.	——	H.⊙.	
54 eriogònum. R.S.	woolly-angled.	N. Holland.	1824.	——	G.♃.	
55 marginàtum. B.P.	margin-leaved.	N. S. W.	1822.	——	F.♃.	
56 bícolor. B.P.	two-coloured.	————	——	——	F.♃.	
ISA'CHNE. B.P.	ISA'CHNE. Triandria Digynia. R.S.					
austràlis. B.P.	New Holland.	N. S. W.	1820.	6. 8.	H.*w*.♃.	
MONA'CHNE. R.S.	MONA'CHNE. Triandria Digynia. R.S.					
unilateràlis. R.S.	one-sided.	Jamaica.	1819.	6. 8.	S.♃.	Sloan. hist. 1. t. 14. f. 2.
Mílium villòsum. W.						
AXO'NOPUS. R.S.	AXO'NOPUS. Triandria Digynia. R.S.					
1 digitàtus. R.S.	fingered.	Jamaica.	1818.	6. 8.	H.⊙.	
2 cimicìnus. R.S.	spotted.	E. Indies.	1778.	7. 9.	H.⊙.	
Pánicum cimicìnum. W. *Mílium cimicìnum.* L.						
BECKMA'NNIA. R.S.	BECKMA'NNIA. Triandria Digynia. R.S.					
erucæfórmis. R.S....	linear-spiked.	Europe.	1773.	7.	H.⊙.	Host. gram. 3. t. 6.
SEC'ALE. R.S.	RYE. Triandria Digynia. R.S.					
1 cereàle. R.S.	common.	Tauria.	....	6. 7.	H.⊙.	
α hybérnum R.S.	*winter.*	————	....	——	H.♂.	Blackw. t. 424.
β vérnum. R.S.	*spring.*	————	....	——	H.⊙.	Host. gram. 2. t. 48.
γ compósitum. R.S. ..	*compound-spiked.*	————	....	——	H.⊙.	
2 frágile. R.S.	brittle.	————	1816.	——	H.⊙.	
3 villòsum. F.G.	villous.	S. Europe.	1790.	——	H.⊙.	Flor. græc. t. 97.
Tríticum villòsum. R.S.						
4 sylvéstre. Host.	wood.	Hungary.	1816.	——	H.♂.	Host. gram. 4. t. 2.
5 créticum. W.	Cretan.	Candia.	——	——	H.♃.	
TETRAP'OGON. S.S.	TETRAP'OGON. Polygamia Monœcia.					
villòsus. S.S.	villous.	Barbary.	1818.	7. 8.	H.⊙.	Desf. atl. 2. t. 255.
TRI'TICUM. R.S.	WHEAT. Triandria Digynia. L.					
1 æ'stivum. R.S.	summer.	Baschiros.	....	6. 7.	H.⊙.	Host. gram. 3. t. 26.

α ambígens. LG. *β Bàhi.* LG. *γ Báncos.* LG. *δ Cálot.* LG. *ε cándidum.* LG. *ζ compáctum.* LG. *η Lamárckii.* LG. *θ Róyo.* LG. *ι truncátulum.* LG. *κ variegàtum.* LG. *λ confértum.* LG. *μ hùmile.* LG. *ν mèdium.* LG. *ξ màjus.* LG. *ο Pástor.* LG. *π versícolor.* LG.

2 hybérnum. R.S. .... Lammas. .... 6. 7. H.♂. Host. gram. 3. t.26. f.8
α candidíssimum. LG. *β compáctum.* LG. *γ Morisòni.* LG. *δ Pòlo.* LG. *ε Tessìeri.* LG.
3 sibíricum. R.S. .... Siberian. Siberia. .... —— H.⊙.
4 velutìnum. R.S. .... velvet-spiked. .... —— H.♂. Rod. Schub.cer.n.4.f.4.
5 compósitum. R.S. .... Egyptian. Egypt. 1799. —— H.⊙. Host. gram. 3. t. 27.
α aùreum. LG. *β compósitum.* LG. *γ edéntulum.* LG. *δ prismáticum.* LG. *ε rùbrum.* LG.
6 Linnæànum. R.S. .... Linnean. Egypt. .... 6. 7. H.⊙.
α Caballèro. LG. *β dùbium.* LG. *γ Lajustícia.* LG. *δ depilàtum.* LG. *ε pìceo-cœrùleum.* LG.
7 túrgidum. R.S. .... turgid. .... 6. 7. H.⊙. Host. gram. 3. t. 28.
α Collàdo. LG. *β compósitum.* LG. *γ montànum.* LG. *δ Tournefórtii.* LG. *ε Caùmels.* LG.
ζ Lapìedra. LG.
8 fastuòsum. R.S. .... thick-grained. .... 6. 7. H.⊙.
α albéllum. LG. *β A'sso.* LG. *γ Cannabàte.* LG. *δ Cástro.* LG. *ε Echeándia.* LG. *ζ Juràdo.* LG.
η Alvàrez. LG. *θ Lobèlii.* LG.
9 Gærtnerìànum. R.S. .. Gærtner's. .... H.⊙.
α Bérnal. LG. *β cálvum.* LG. *γ ciliòsum.* LG. *δ Gràells.* LG. *ε 'Oliet.* LG. *ζ Plà.* LG.
η speltoìdes. LG. *θ truncàtum.* LG. *ι Vasállo.* LG. *κ yèpes.* LG. *λ Hallèri.* LG. *μ Párck.* LG.
ν Valcárcel. LG. *ξ Xixòna.* LG. *ο Cleménte.* LG. *π Colòna.* LG. *ρ Gerárdi.* LG. *σ Gleíchen.* LG.
τ Moràles. LG. *υ Plùkenet.* LG. *φ Sanmártin.* LG. *χ Theophrásti.* LG. *ψ Thouìni.* LG.
10 platystàchyum. R.S. .. flat-spiked. .... 6. 7. H.⊙.
11 cochleàre. R.S. .... spoon-like. .... —— H.⊙.
α aristàtum. LG. *β mucronàtum.* LG.
12 Cevállos. R.S. .... Trigo Moro. .... —— H.⊙.
13 dùrum. R.S. .... Barbary. .... —— H.⊙.
14 villòsum. Host. .... villous. .... —— H.⊙. Host. gram. 4. t. 6.
15 sículum. R.S. .... Sicilian. .... —— H.⊙.
16 hordeifórme. R.S. .... Barley form. .... —— H.⊙. Host. gram. 4. t. 5.
17 compáctum. R.S. .... compact. .... —— H.⊙. ———— 4. t. 7.
18 polònicum. R.S. .... Polish. .... 1692. —— H.⊙. ———— 3. t. 31.
19 monocóccon. R.S. .... one-grained. .... 1648. —— H.⊙. ———— 3. t. 32.
α glábrum. LG. *β pubéscens.* LG.
20 dicóccum. R.S. .... two-grained. .... —— H.⊙. Rod.Schub.cer.n.15.f.2
α álbum. .... *white-spiked.* .... —— H.⊙.
β rùfum. .... *brown-spiked.* .... —— H.⊙.
21 atràtum. R.S. .... dark-spiked. .... —— H.⊙. Host. gram. 4. t. 8.
22 tricóccum. R.S. .... three-grained. .... —— H.⊙. Rod.Schub.cer.n.17.f.3
23 Cienfuègos. R.S. .... tooth-valved. .... —— H.⊙.
24 Bauhìni. R.S. .... truncate-valved. .... —— H.⊙.
25 Spélta. R.S. .... Spelt. .... —— H.⊙. Host. gram. 3. t. 29.
Zèa. Host.
α Astùrum. LG. *β rufléscens.* LG. *γ submùticum.* LG. *δ álbum.* LG. *ε cyáneum.* LG.
26 venulòsum. R.S. .... veined. Egypt. .... 6. 7. H.⊙.
27 ichyostàchyum. R.S. .. slender-spiked. .... —— H.⊙.
AGRÓP'YRUM. R.S. COUCH-GRASS. Triandria Digynia. R.S.
1 júnceum. R.S. .... rushy. England. .... 6. 7. H.♃. Eng. bot. v. 12. t. 814.
Tríticum júnceum. E.B.
2 acùtum. R.S. .... acute-glumed. France. 1816. —— H.♃.
3 littoràle. Host. .... sea-side. Mediterranean. — —— H.♃. Host. gram. 4. t. 9.
4 rígidum. R.S. .... rigid. S. Europe. 1805. —— H.♃.
5 intermèdium. Host. intermediate. Germany. —— —— H.♃. Host. gram. 2. t. 22.
6 glaùcum. R.S. .... glaucous. Switzerland. 1824. —— H.♃. ———— 4. t. 10 ?
7 rupéstre. R.S. .... rock. Siberia. 1822. —— H.♃.
8 obtusiflòrum. R.S. .... blunt-flowered. .... —— H.♃.
9 púngens. R.S. .... pungent. S. Europe. 1824. —— H.♃.
10 densiflòrum. R.S. .... close-flowered. Siberia. —— —— H.♃.
11 gigánteum. R.S. .... gigantic. S. Europe. 1805. —— H.♃. Host. gram. 2. t. 23.
Tríticum elongàtum. Host.
12 rèpens. R.S. .... common. Britain. .... 7. 8. H.♃. Eng. bot. v. 13. t. 909.
α arvénse. R.S. .... *field.* ———— .... —— H.♃. Trattin. tab. t. 566.
β subulàtum. R.S. .... *subulate-glumed.* ———— .... —— H.♃. Schreb. gram. t. 26. f. 2.
γ dumetòrum. R.S. .... *many-flowered.* ———— .... —— H.♃. ————f.II.IV.f.3.
δ Vaillantiànum. R.S. .. *Vaillant's.* ———— .... —— H.♃. ————f. 6.
ε Leersiànum. R.S. .. *Leers's.* ———— .... —— H.♃. Leers fl. herb. t.12. f. 4.
ζ capillàre. R.S. .... *hair-awned.* ———— .... —— H.♃.
13 sibíricum. R.S. .... Siberian. Siberia. 1825. —— H.♃.
14 dasyánthum. R.S. .... thick-flowered. Russia. —— —— H.♃.
15 canìnum. R.S. .... bearded. Britain. .... —— H.♃. Eng. bot. v. 20. t. 1372
16 dístichum. R.S. .... two-ranked. C. B. S. 1816. —— G.♃.
17 prostràtum. R.S. .... prostrate. Siberia. 1780. 6. H.⊙. Jacq. vind. 3. t. 44.

18 orientàle. R.S.	hairy-spiked.	Levant.	1807.	6. 7.	H. ♂.	Will.n.act.ber.2.t.4.f.3.
Secàle orientàle. W.						
19 pátulum. Tr.	spreading.	Egypt.	1800.	——	H.☉.	
Tríticum squarròsum. R.S. *Buonapártis.* Spr.						
20 imbricàtum. R.S.	imbricated.	Caucasus.	1824.	6. 8.	H. ♃.	Gmel. sib. 1. n. 51. t. 23.
21 pectinifórme. R.S.	comb-like.	Tauria.	——	——	H. ♃.	Buxb. cent. 1. t. 50. f. 3.
22 cristàtum. R.S.	crested.	——	——	——	H. ♃.	Eng. bot. v. 32. t. 2267.
23 desertòrum. R.S.	desert.	Altay.	1822.	——	H. ♃.	Ledeb. ic. alt. t. 246.
24 angustifòlium. R.S. ..	narrow-leaved.	Siberia.	——	——	H. ♃.	
25 muricàtum. R.S.	bristly.	——	1825.	——	H. ♃.	
26 pectinàtum. R.S.	pectinated.	V. Diem. Isl.	1822.	——	H. ♃.	
27 variegàtum. R.S.	variegated.	....	1818.	——	H. ♃.	
L'OLIUM. R.S.	DARNEL.	Triandria Digynia.	L.			
1 perénne. R.S.	Rye-grass.	Britain.	....	5. 6.	H. ♃.	Eng. bot. v. 5. t. 315.
α vulgàre. R.S.	*common.*	——	....	——	H. ♃.	Host. gram. 1. t. 25.
β ténue. R.S.	*slender.*	——	....	——	H. ♃.	Sinc. gra. wob. p. 211. ic.
γ compósitum. R.S. ..	*compound.*	——	....	——	H. ♃.	Scheuch. agr. 1. f. 7. D.
δ ramòsum. R.S.	*branching.*	——	....	——	H. ♃.	Leers gram. t. 12. f. 1.
ε Russelliànum. H.G.W.	*Russell's.*	——	....	——	H. ♃.	Sinc. gra. wob. p. 216. ic.
ζ Whitwórthii. H.G.W.	*Whitworth's.*	——	....	——	H. ♃.	
η Stickneyànum. H.G.W.	*Stickney's.*	——	....	——	H. ♃.	
θ paniculàtum. H.G.W.	*panicled.*	——	....	——	H. ♃.	
ι monstròsum. H.G.W.	*monstrous-flowered.*	——	....	——	H. ♃.	
κ vivìparum. R.S.	*viviparous.*	——	....	——	H. ♃.	
λ multiflòrum. R.S. ..	*many-flowered.*	——	——	——	H. ♃.	
μ hùmile. R.S.	*dwarf.*	——	....	——	H. ♃.	
2 multiflòrum. R.S.	many-flowered.	Europe.	1816.	——	H.☉.	Vaill. par. t. 17. f. 3.
3 arvénse. R.S.	beardless.	England.	....	7.	H.☉.	Eng. bot. v. 16. t. 1125.
4 canadénse. Panz.	Canadian.	Canada.	1822.	——	H.☉.	
5 complanàtum. L.en...	rigid.	Switzerland.	1825.	——	H.☉.	
rígidum. R.S.						
6 temuléntum. R.S.	bearded.	Britain.	....	7. 8.	H.☉.	Eng. bot. v. 16. t. 1124.
7 speciòsum. R.S.	pretty.	Tauria.	1824.	——	H.☉.	
Æ'GILOPS. R.S.	HARD-GRASS.	Polygamia Monœcia.	L.			
1 caudàta. R.S.	tailed.	Candia.	1820.	7. 8.	H.☉.	
2 cylíndrica. R.S.	cylindrical.	Hungary.	1805.	6. 7.	H.☉.	Host. gram. 2. t. 7.
3 ovàta. R.S.	oval-spiked.	S. Europe.	1683.	——	H.☉.	Flor. græc. 1. t. 93.
4 triaristàta. R.S.	three-awned.	——	1739.	7. 8.	H.☉.	Host. gram. 2. t. 6.
triunciàlis. Host. *non* W.						
5 triunciàlis. R.S.	long-spiked.	——	——	——	H.☉.	Schreb. gram. t. 10. f. 1.
6 comòsa. R.S.	tufted.	Greece.	1828.	——	H.☉.	
7 squarròsa. W.	rough-spiked.	Levant.	1794.	6. 7.	H. ♃.	Schreb. gram. 2. t. 27. f. 2.
Tríticum squarròsum. R.S.						
8 Hy'strix. N.	hedgehog.	Missouri.	1819.	7. 8.	H. ♃.	
E'LYMUS. R.S.	LYME-GRASS.	Triandria Digynia.	L.			
1 arenàrius. E.B.	sand.	Britain.	....	4. 6.	H. ♃.	Eng. bot. v. 24. t. 1672.
2 geniculàtus. E.B.	pendulous.	England.	....	7.	H. ♃.	—— v. 23. t. 1586.
3 Delileànus. R.S.	Delile's.	Egypt.	1827.	——	H.☉.	Delil. egypt. t. 13. f. 1.
4 sabulòsus. R.S.	glaucous.	Siberia.	1806.	6. 7.	H. ♃.	
5 gigánteus. R.S.	gigantic.	N. America.	1790.	7. 8.	H. ♃.	
6 mexicànus. R.S.	Mexican.	Mexico.	1823.	6. 8.	H. ♃.	Jacq. ecl. gr. t. 19.
7 sibíricus. R.S.	Siberian.	Siberia.	1758.	6. 7.	H. ♃.	Schreb. gram. 1. t. 21. f. 1.
8 móllis. R.S.	soft.	Kamtschatka.	1829.	——	H. ♃.	
9 racemòsus. R.S.	racemed.	....	1816.	——	H. ♃.	
10 téner. R.S.	tender.	Siberia.	1801.	——	H. ♃.	
11 philadélphicus. R.S. ..	Philadelphian.	N. America.	1790.	7. 8.	H. ♃.	
12 canadénsis. R.S.	Canadian.	——	1699.	——	H. ♃.	Moris. s. 8. t. 2. f. 10.
13 glaucifòlius. R.S.	glaucous.	——	1816.	——	H. ♃.	
14 villòsus. R.S.	villous.	——	1802.	6. 7.	H. ♃.	
15 virgínicus. R.S.	Virginian.	——	1781.	——	H. ♃.	
16 striàtus. R.S.	striated.	——	1790.	——	H. ♃.	
β mìnor. R.S.	*lesser.*		....	——	H. ♃.	
17 europ'æus. R.S.	European.	England.	....	——	H. ♃.	Eng. bot. v. 19. t. 1317.
18 crinìtus. R.S.	long-awned.	Smyrna.	1806.	7. 8.	H.☉.	Host. gram. 1. t. 27.
19 intermèdius. R.S.	intermediate.	Iberia.	1819.	7. 9.	H.☉.	Buxb. cent. 1. t. 52.
20 altàicus. F.	Altaic.	Altay.	1829.	——	H. ♃.	
21 pauciflorus. R.S.	few-flowered.	....	1824.	——	H.☉.	
22 Càput Medùsæ. R.S...	Portuguese.	Portugal.	1784.	——	H.☉.	Schreb. gram. 2. t. 24. f. 2.
23 júnceus. R.S.	rushy.	Siberia.	1823.	7. 8.	H. ♃.	

24 hordeifórmis. R.S. .... Barley-like. .... 1816. 7. 8. H.⊙.
25 Hy'strix. R.S. .... rough. Levant. 1770. —— H.⊙. Jacq. ic. 2. t. 305.
Asprélla Hy'strix. W.en.
26 pseùdo-hy'strix. R.S... rough American. N. America. 1818. —— H.⊙.
ENNEAP'OGON. R.S. ENNEAP'OGON. Triandria Digynia. R.S.
1 nígricans. R.S. .... blackish. N. S. W. 1822. 7. 8. F.♃.
2 pállidus. R.S. .... pale. ———— 1824. —— F.♃.
Pappóphorum pállidum. B.P.
3 phleoídes. R.S. .... Phleum-like. S. America. 1822. —— F.♃.
PAPPO'PHORUM. B.P. PAPPO'PHORUM. Triandria Digynia. R.S.
1 alopecuroídeum. R.S... Foxtail-like. S. America. 1825. 6. 8. S.♃.
2 laguroídeum. R.S. .. Harestail-like. W. Indies. 1826. —— S.♃.
ECHIN'ARIA. R.S. ECHIN'ARIA. Triandria Digynia. R.S.
capitàta. R.S. .... headed. S. Europe. 1771. 5. 8. H.⊙. Flor. græc. 1. t. 100.
Cénchrus capitàtus. F.G. *Sésleria echinàta.* Host. gram. 3. t. 8.
SESL'ERIA. R.S. SESL'ERIA. Triandria Digynia. R.S.
1 elongàta. R.S. .... long-spiked. Austria. 1805. 6. 7. H.♃. Host. gram. 2. t. 97.
2 cylíndrica. R.S. .... cylindrical. S. Europe. 1818. —— H.♃. Sav. Ust. ann. 1800. ic.
3 tenuifòlia. R.S. .... slender-leaved. ———— ———— 4. 6. H.♃. Host. gram. 4. t. 22.
4 cœrùlea. R.S. .... blue. Britain. .... 4. 5. H.♃. Eng. bot. v. 23. t. 1613.
5 álbicans. R.S. .... whitish. CarpathianMt1826. —— H.♃.
6 nítida. R.S. .... glossy. Italy. 1825. 5. 6. H.♃.
7 tenélla. R.S. .... weak. Switzerland. 1810. —— H.♃. Host. gram. 2. t. 100.
8 sphærocéphala. R.S. .. round-headed. ———— ——— —— H.♃. ———— 2. t. 99.
9 dísticha. R.S. .... distichous. ———— 1824. —— H.♃. ———— 2. t. 76.
Pòa dísticha. Host.
TRI'ODIA. B.P. TRI'ODIA. Triandria Digynia. R.S.
decúmbens. R.S. .... decumbent. Britain. .... 7. 8. H.♃. Eng. bot. v. 11. t. 792.
Pòa decúmbens. E.B. *Festùca decúmbens.* W. *Danthònia decúmbens.* DC.
TR'IDENS. R.S. REDTOP-GRASS. Triandria Digynia. R.S.
quinquéfida. R.S. .... five-cleft. N. America. 1816. 7. 8. H.♃. Jacq. ecl. gr. 2. t. 16.
Triòdia cupr'æa. Jacq. *Windsòria poæfórmis.* N. *Pòa quinquéfida.* Ph.
DANTH'ONIA. DC. DANTH'ONIA. Triandria Digynia. R.S.
1 longifòlia. B.P. .... long-leaved. N. S. W. 1823. 7. 8. F.♃.
2 semiannulàris. B.P. .. semiannular. N. Holland. 1822. —— F.♃. Labill. n. holl. 1. t. 33.
Arúndo semiannulàris. Lab.
3 pilòsa. B.P. .... hairy. N. S. W. 1823. —— F.♃.
4 provinciàlis. R.S. .... Provence. S. Europe. 1818. 6. 7. H.♃. Vill. delph. 2. t. 2. f. 9.
5 spicàta. R.S. .... spiked. N. America. 1823. —— H.♃.
6 strigòsa. R.S. .... bristle-pointed. Britain. .... —— H.♃. Eng. bot. v. 18. t. 1266.
Avèna strigòsa. E.B.
7 macrántha. R.S. .... large-flowered. C. B. S. 1816. —— H.⊙.
8 curvifòlia. R.S. .... curve-leaved. ———— ——— —— G.♃.
9 calycìna. R.S. .... glossy-calyxed. ———— 1822. —— G.♃.
KŒL'ERIA. R.S. KŒL'ERIA. Triandria Digynia. R.S.
1 cristàta. R.S. .... crested. Europe. .... 6. 8. H.♃. Host. gram. 2. t. 75.
β glábra. R.S. *smooth.* Britain. .... 7. 8. H.♃. Eng. bot. v. 9. t. 648.
Aìra cristàta. E.B. *Pòa cristàta.* W.
2 grandiflòra. R.S. .... great-flowered. S. Europe. 1824. 6. 8. H.♃.
3 glaùca. DC. .... glaucous. Germany. 1816. —— H.♃.
4 pensylvánica. DC.... Pensylvanian. N. America. 1818. —— H.♃. Spr. mem. p 2. 1810. t. 7.
5 nítida. R.S. .... glossy. ———— 1828. —— H.♃.
6 macrántha. R.S. .... large-flowered. Russia. 1824. —— H.♃.
7 albéscens. R.S. .... whitish. S. France. 1829. —— H.♃.
8 valesìaca. R.S. .... Swiss. Switzerland. 1824. —— H.♃.
9 setàcea. R.S. .... bristly. S. Europe. 1825. —— H.♃.
10 tuberòsa. P.S. .... tuberous. ———— 1802. 7. 8. H.♃. Lam. ill. t. 45. f. 4.
11 hirsùta. R.S. .... hairy. Switzerland. 1824. —— H.♃.
12 villòsa. R.S. .... villous. S. Europe. 1800. 6. 7. H.♃.
Aíra pubéscens. W. *Phálaris pubéscens.* Lam.
13 phleoídes. R.S. .... cat's-tail. Portugal. 1802. 7. 8. H.⊙. Desf. atl. 1. t. 23.
14 híspida. R.S. .... hispid. S. Europe. 1819. 6. 7. H.⊙. Savi fl. pis. t. 1. f. 5.
15 brachystàchya. R.S... short-spiked. .... —— —— H.⊙.
16 parviflòra. L.en. .... small-flowered. S. Europe. —— —— H.⊙. Desf. atl. 1. t. 32.
Avèna parviflòra. Desf. *Trisètum parviflòrum.* R.S.
17 ægyptìaca. Lk. .... Egyptian. Egypt. 1826. —— H.⊙.
18 láxa. Th. .... loose-flowered. ———— 1829. —— H.⊙.
AV'ENA. R.S. OAT-GRASS. Triandria Digynia. L.
1 pr'æcox. R.S. .... early-flowering. Britain. .... 5. 6. H.⊙. Eng. bot. v. 18. t. 1296.
Aíra pr'æcox. E.B.

2 brèvis. R.S.	.... short.	Germany.	1804.	5. 6.	H.⊙.	Host. gram. 3. t. 42.
3 álba. R.S.	.... white.	France.	....	6. 7.	H.⊙.	
4 satìva. R.S.	.... common.	S. America.	....	——	H.⊙.	Host. gram. 2. t. 59.
α melanospérma.	.... *black Oat.*	———	....	——	H.⊙.	
β leucospérma.	.... *white Oat.*	———	....	——	H.⊙.	
5 orientàlis. R.S.	.... Tartarian.	Levant.	1798.	——	H.⊙.	Host. gram. 3. t. 44.
6 nùda. R.S.	.... naked.	....	....	——	H.⊙.	——— 3. t. 43.
7 fátua. R.S.	.... wild.	Britain.	....	8.	H.⊙.	Eng. bot. v. 31. t. 2221.
8 hirsùta. R.S.	.... hairy.	S. Europe.	——	6. 8.	H.⊙.	
9 stérilis. R.S.	.... bearded.	———	1640.	7. 8.	H.⊙.	Jacq. ic. 1. t. 23.
10 sempervìrens. R.S.	.. evergreen.	Pyrenees.	1816.	6. 7.	H.♃.	Host. gram. 3. t. 41.
11 fállax. R.S.	.... fallacious.	S. Europe.	1824.	——	H.♃.	
12 pállens. R.S.	.... pale.	Portugal.	——	——	H.♃.	
13 praténsis. R.S.	.... narrow-leaved.	Britain.	....	——	H.♃.	Eng. bot. v. 17. t. 1204.
14 cantábrica. R.S.	.... evergreen.	....	1825.	——	H.♃.	
15 bromoídes. R.S.	.... Bromus-like.	S. Europe.	....	——	H.♃.	
16 planicúlmis. R.S.	.... flat-stalked.	Hungary.	1822.	——	H.♃.	Schrad.fl.ger.1.t.6.f.2.
latifòlia. Host. gram. 4. t. 32.						
17 versícolor. R.S.	.... various-colored.	Europe.	1816.	6. 8.	H.♃.	Host. gram. 2. t. 52.
GAUD'INIA. R.S.	.... GAUD'INIA.	Triandria Digynia. R.S.				
frágilis. R.S.	.... brittle.	S. Europe.	1770.	7. 8.	H.⊙.	Host. gram. 2. t. 54.
Avèna frágilis. Host.						
TRIS'ETUM. R.S.	TRISE'TUM.	Triandria Digynia. R.S.				
1 ténue. R.S.	.... slender.	Europe.	1804.	7. 8.	H.⊙.	Host. gram. 2. t. 55.
striàtum. P.S. *Avèna ténuis.* Host.						
2 pensylvánicum. R.S.	.. Pensylvanian.	N. America.	1785.	6. 7.	H.⊙.	
3 micránthum. R.S.	.... small-flowered.	S. Europe.	1823.	——	H.⊙.	
4 negléctum. R.S.	.... neglected.	———	——	——	H.⊙.	Savi fl. pis. 1. t. 1. f. 4.
5 alopecùrus. R.S.	.... Foxtail.	....	1816.	——	H.⊙.	
6 Loeflingiànum. R.S.	.. Loefling's.	S. Europe.	1770.	——	H.⊙.	W.berl.mag.2.t.8. f.3.
7 condensàtum. R.S.	.... close-flowered.	Sicily.	1824.	——	H.⊙.	
8 pilòsum. R.S.	.... hairy.	Iberia.	——	——	H.⊙.	
9 pubéscens. R.S.	.... pubescent.	Britain.	....	7. 8.	H.♃.	Eng. bot. v. 23. t. 1640.
Avèna pubéscens. E.B. .						
10 rígidum. R.S.	.... rigid.	Caucasus.	1825.	——	H.♃.	
11 alpìnum. R.S.	.... Alpine.	Scotland.	....	7.	H.♃.	Eng. bot. v. 30. t. 2141.
Avèna planicúlmis. E.B. *non* Schrad.						
12 carpáthicum. R.S.	.... Carpathian.	Carp. mount.	1818.	7. 8.	H.♃.	Host. gram. 4. t. 31.
13 flavéscens. R.S.	.... golden.	Britain.	....	6. 7.	H.♃.	Eng. bot. v. 14. t. 952.
praténse. P.S. *Avèna flavéscens.* E.B.						
14 alpéstre. R.S.	.... mountain.	S. Europe.	1816.	——	H.♃.	Host. gram. 3. t. 39.
15 brevifòlium. R.S.	.... short-leaved.	———	——	——	H.♃.	——— 3. t. 40.
16 argénteum. R.S.	.... silvery.	Switzerland.	1791.	——	H.♃.	——— 2. t. 53.
17 distichophy'llum.	.... fan-leaved.	———	1796.	——	H.♃.	Vill. delph. 2. t. 4. f. 4.
18 airoídes. R.S.	.... Aira-like.	———	1800.	——	H.♃.	Flor. dan. t. 228.
Avèna airoídes. DC. *Aìra subspicàta.* W.						
19 glabràtum. R.S.	.... smooth.	Portugal.	1818.	7. 8.	H.⊙.	
CENTOTH'ECA. R.S.	CENTOTH'ECA.	Triandria Digynia. R.S.				
lappàcea. R.S.	.... Bur.	India.	1773.	7.	S.⊙.	Pal.deBeau.ag.t.14.f.7.
Cénchrus lappàceus. L.						
CABR'ERA. LG.	CABR'ERA.	Triandria Digynia.				
chrysoblépharis. LG.	golden-spiked.	S. America.	1822.	6. 7.	S.♃.	
P'OA. R.S.	MEADOW-GRASS.	Triandria Digynia. L.				
1 praténsis. R.S.	.... smooth-stalked.	Britain.	....	5. 6.	H.♃.	Eng. bot. v. 15. t. 1073.
2 angustifòlia. DC.	.... narrow-leaved.	———	....	5.	H.♃.	Sinc.gra.wob.p.184.ic.
3 hùmilis. R.S.	.... short-blueish.	———	....	5. 6.	H.♃.	Eng. bot. v. 14. t. 1004.
subcœrùlea. E.B.						
4 víridis. R.S.	.... green.	N. America.	1812.	——	H.♃.	
5 hy'brida. R.S.	.... hybrid.	Switzerland.	1824.	——	H.♃.	
6 Torreyàna. R.S.	.... Torrey's.	N. America.	1825.	——	H.♃.	
7 triviàlis. E.B.	.... common.	Britain.	....	6. 8.	H.♃.	Eng. bot. v. 15. t. 1072.
8 Gmelìni. R.S.	.... Gmelin's.	Germany.	1825.	——	H.♃.	
9 ánnua. E.B.	.... annual.	———	....	1. 12.	H.⊙.	Eng. bot. v. 16. t. 1141.
10 cæspitòsa. R.S.	.... tufted.	N. Zealand.	1822.	6. 8.	H.♃.	
11 grácilis. R.S.	.... slender.	Siberia.	1818.	——	H.♃.	
12 bulbòsa. R.S.	.... bulbous.	England.	....	5. 6.	H.♃.	Eng. bot. v. 15. t. 1071.
13 láxa. R.S.	.... loose.	Scotland.	....	7.	H.♃.	Host. gram. 3. t. 15.
flexuòsa. Eng. bot. v. 16. t. 1123.						
14 distichophy'lla. R.S.	.. fan-leaved.	Switzerland.	1829.	——	H.♃.	Host. gram. 4. t. 26.

15 árctica. R.S.	.... arctic.	Melville Isl.	1827.	7.	H.♃.	
16 angustàta. R.S.	.... narrow.	———	——	——	H.♃.	
17 abbreviàta. R.S.	.... shortened.	———	——	——	H.♃.	
18 Hallèri. R.S.	.... Haller's.	Switzerland.	——	——	H.♃.	
pállens. Hall.						
19 cenísia. R.S.	.... soft.	Mt. Cenis.	1791.	7. 8.	H.♃.	Host. gram. 3. t. 16.
20 alpìna. R.S.	.... Alpine.	Scotland.	....	6. 7.	H.♃.	Eng. bot. v. 14. t. 1003.
21 collìna. R.S.	.... hill.	Europe.	1800.	——	H.♃.	Host. gram. 2. t. 66.
22 concínna. R.S.	.... neat.	Switzerland.	1815.	——	H.♃.	
23 Molinèri. R.S.	.... Molineri's.	———	1816.	——	H.♃.	Balb. misc. t. 5. f. 1.
24 badénsis. W.	.... Baden.	Baden.	1826.	——	H.♃.	
25 brevifòlia. DC.	.... short-leaved.	France.	1822.	——	H.♃.	
26 supìna. R.S.	.... trailing.	Switzerland.	——	——	H.♃.	Host. gram. 4. t. 27.
27 miliàcea. R.S.	.... Millet like.	Hungary.	1824.	——	H.♃.	
28 nemoràlis. R.S.	.... wood.	———	....	6.	H.♃.	Eng. bot. v. 18. t. 1265.
β *angustifòlia.* H.G.W.	*narrow-leaved.*	———	....	——	H.♃.	Sinc. gr. wob. p.182. ic.
29 Gaudìni. R.S.	.... Gaudin's.	Switzerland.	1820.	——	H.♃.	
áspera. Gaud. *non* Spreng.						
30 versícolor. R.S.	.... various-color'd.	Tauria.	1825.	6. 7.	H.♃.	
31 coarctàta. L.en.	.... crowded.	Europe.	....	——	H.♃.	
32 glaùca. E.B.	.... glaucous.	Britain.	....	——	H.♃.	Eng. bot. v. 24. t. 1720.
33 c'æsia. E.B.	.... sea-green.	Scotland.	....	——	H.♃.	——— v. 24. t. 1719.
34 brachyphy'lla. R.S.	.. short-leaved.	N. America.	1816.	——	H.♃.	
35 depauperàta. R.S.	.. few-flowered.	Hungary.	——	——	H.♃.	
36 stérilis. R.S.	.. barren-ground.	Tauria.	1824.	——	H.♃.	
37 nervàta. R.S.	.... nerved.	N. America.	1812.	——	H.♃.	Sinc. gr. wob. p. 190. ic.
38 imbecílla. R.S.	.... weak.	N. Zealand.	1822.	7. 8.	H.♃.	
39 ánceps. R.S.	.... flat-edged.	———	——	——	H.♃.	
40 flàva. R.S.	.... yellow.	N. America.	——	——	H.♃.	
41 hirsùta. R.S.	.... hairy.	———	——	——	H.♃.	
42 índica. R.S.	.... Indian.	E. Indies.	——	——	S.♃.	
43 Poirètii. R.S.	.... Poiret's.	....	1816.	6. 8.	H.⊙.	
44 serotìna. R.S.	.... late-flowering.	Germany.	1800.	7. 9.	H.♃.	
45 effùsa. R.S.	.... spreading.	Hungary.	1823.	6. 8.	H.♃.	
46 sudética. R.S.	.... broad-leaved.	Germany.	1802.	7. 8.	H.♃.	Vill. delph. 2. t. 3.
47 áspera. S.S.	.... rough-branch'd.	E. Indies.	1829.	——	S.♃.	
48 víridis. Ph.	.... green.	N. America.	1812.	——	H.♃.	
49 crocàta. M.	.... saffron.	———	1818.	——	H.♃.	
50 commutàta. R.S.	.... changed.	Spain.	1822.	——	H.♃.	
51 contrácta. R.S.	.... contracted.	E. Indies.	1823.	——	S.♃.	
52 amboinénsis. R.S.	.... Amboyna.	———	1800.	6. 7.	S.♃.	Rumph. amb. 6. t. 7. f. 3.
53 chinénsis. R.S.	.... China.	China.	——	——	H.⊙.	
54 austràlis. B.P.	.... southern.	N. S. W.	1822.	——	H.♃.	Lab. nov. hol. 1. t. 35.
Arúndo poæfórmis. Lab.						
55 plebèia. B.P.	.... plebeian.	———	——	——	H.♃.	
56 compréssa. E.B.	.... compressed.	Britain.	....	6. 8.	H.♃.	Eng. bot. v. 6. t. 365.
57 complanàta. R.S.	.... flattened.	Sicily.	1827.	——	H.♃.	
58 festucæfórmis. R.S.	.. Fescue-like.	Dalmatia.	1800.	7. 9.	H.♃.	Host. gram. 3. t. 17.
59 convolùta. R.S.	.... convolute.	....	1819.	——	H.♃.	
60 atrovìrens. R.S.	.... dark green.	Barbary.	1825.	——	H.♃.	Desf. atl. 1. t. 17.
61 scariòsa. R.S.	.... scariose.	Spain.	1826.	——	H.♃.	
62 coromandeliàna. R.S.	.. Coromandel.	E. Indies.	1824.	——	S.♃.	
63 bromoídes. R.S.	.... Bromus-like.	....	——	——	G.♃.	
64 Unìolæ. R.S.	.... Uniola-like.	C. B. S.	1826.	——	G.♃.	
65 peruviàna. R.S.	.... Peruvian.	Peru.	1802.	7. 8.	H.⊙.	Jacq. ic. 1. t. 18.
66 papillòsa. R.S.	.... papillose.	C. B. S.	1816.	——	H.⊙.	
67 digitàta. B.P.	.... fingered.	N. S. W.	1800.	——	H.⊙.	
68 nùtans. R.S.	.... nodding.	E. Indies.	1820.	——	S.♂.	
69 pállens. R.S.	.... pale.	BuenosAyres.	1824.	——	H.♃.	
70 diándra. F.I.	.... diandrous.	E. Indies.	1820.	——	S.⊙.	
71 unioloídes. R.S.	.... Uniola-like.	———	1824.	——	S.♃.	
72 pellùcida. R.S.	.... pellucid.	N. Holland.	1829.	——	G.♃.	
73 semineùtra. Tr.	.... Hairgrass-like.	Hungary.	1818.	——	H.♃.	
SCLEROCHL'OA. R.S.	SCLEROCHL'OA.	Triandria Digynia. R.S.				
1 divaricàta. R.S.	.... spreading.	S. Europe.	1802.	7. 8.	H.⊙.	Gouan. ill. 4. t. 2. f. 1.
2 procúmbens. R.S.	.... procumbent.	Britain.	....	——	H.⊙.	Eng. bot. v. 8. t. 532.
Pòa procúmbens. E.B. *Glycèria procúmbens.* E.F.						
3 dùra. R.S.	.... rigid.	S. Europe.	1776.	7. 8.	H.⊙.	Host. gram. 2. t. 73.
Pòa dùra. Host. *Cynosùrus dùrus.* L.						

4 rígida. L.en.	.... hard.	England.	....	6. 7.	H.⊙.	Eng. bot. v. 20. t. 1371.
Pòa rígida. E.B. *Glycèria rígida.* E.F. *Megastàchya rígida.* R.S.						
5 articulàta. L.en.	.... jointed.	Barbary.	1816.	——	H.⊙.	Desf. atl. 1. t. 22.
6 marítima.	.... sea-side.	Europe.	....	——	H.⊙.	Cyrill.neap.fasc.2.t.2.
dichótoma. L.en. *Tríticum marítimum.* L. *Festùca marítima.* DC.						
HYDROCHL'OA. Har.	HYDROCHL'OA.	Triandria Digynia.				
1 dístans. Har.	.... distant.	Britain.	....	7.	H.*w*.♃.	Eng. bot. v. 14. t. 986.
Pòa dístans. E.B. *Glycèria dístans.* E.F.						
2 aquática. Har.	.... water.	———	....	——	H.*w*.♃.	Eng. bot. v. 19. t. 1315.
Pòa aquática. E.B. *Glycèria aquática.* E.F.						
3 arundinàcea.	.... Reed-like.	Caucasus.	1825.	——	H.*w*.♃.	
Pòa arundinàcea. M.B.						
4 marítima. Har.	.... sea-side.	Britain.	....	——	H.*w*.♃.	Eng. bot. v. 16. t. 1140.
Pòa marítima. E.B. *Glycèria marítima.* E.F.						
GLYC'ERIA. B.P.	SWEET-GRASS.	Triandria Digynia. R.S.				
flùitans. B.P.	.... floating.	Britain.	....	5. 8.	H.♃.	Eng. bot. v. 22. t. 1520.
Pòa flùitans. E.B. *Festùca flùitans.* W.						
ERAGRO'STIS. R.S.	ERAGRO'STIS.	Triandria Digynia. R.S.				
1 poæoídes. R.S.	.... Poa-like.	Italy.	1699.	7.	H.⊙.	Flor. græc. t. 73.
Pòa Eragróstis. F.G.						
2 tephrosánthos. R.S.	.. Martinica.	Martinica.	1824.	7. 9.	S.⊙.	
3 pilòsa. R.S.	.... hairy.	Italy.	1804.	7. 8.	H.⊙.	Host. gram. 2. t. 68.
4 verticillàta. R.S.	.... whorled.	S. Europe.	1816.	——	H.⊙.	Cavan. ic. 1. t. 93.
5 punctàta. R.S.	.... spotted.	E. Indies.	——	——	S.⊙.	
6 ægyptìaca. R.S.	.... Egyptian.	Egypt.	1812.	——	H.⊙.	Descr. de l'Egyp. t. 10.
7 purpuráscens. R.S.	.. purplish.	....	1822.	——	H.⊙.	
8 despíciens. R.S.	.... doubtful.	....	——	——	H.⊙.	
9 tenélla. R.S.	.... slender.	E. Indies.	1781.	——	S.⊙.	Rheed. mal. 12. t. 41.
10 cynosuroídes. R.S.	.... Dogstail.	Egypt.	1824.	——	H.⊙.	Descr. de l'Egyp. t. 10.
Pòa cynosuroídes. W.						
11 capillàris. L.en.	.... hair-panicled.	N. America.	1781.	10. 11.	H.⊙.	Moris. s. 3. t. 6. f. 33.
Pòa capillàris. R.S.						
12 mexicàna. L.en.	.... Mexican.	Mexico.	1816.	9. 10.	H.⊙.	
13 abyssínica. L.en.	.... Abyssinian.	Abyssinia.	1824.	——	H.⊙.	Jacq. icon. 1. t. 17.
14 plumòsa. R.S.	.... feathered.	E. Indies.	1826.	——	S.⊙.	Rheed. mal. 12. t. 41.
MEGAST'ACHYA. R.S.	MEGAST'ACHYA.	Triandria Digynia. R.S.				
1 Eragróstis. R.S.	.... Love-grass.	S. Europe.	1776.	7. 8.	H.⊙.	Schreb. gram. t. 38.
Pòa megastàchya. P.S. *Brìza Eragróstis.* W. *Eragróstis màjor.* Host. gram. 4. t. 24.						
2 nígricans. R.S.	.... blackish.	S. America.	1823.	——	S.♃.	
3 amábilis. R.S.	...... purple.	E. Indies.	1802.	——	S.⊙.	L. pfl. syst. 12. t. 91. f. 2.
Pòa amábilis. L.						
4 spectábilis. R.S.	.... showy.	N. America.	1816.	——	H.⊙.	
5 pulchélla. R.S.	.... pretty.	Tauria.	1825.	——	H.⊙.	
6 elongàta. R.S.	.... long-panicled.	E. Indies.	1812.	——	S.♃.	Jacq. ecl. gr. t. 3.
7 ciliàris. R.S.	.... ciliated.	Jamaica.	1776.	——	S.⊙.	Jacq. ic. 2. t. 304.
8 rupéstris. R.S.	.... rock.	E. Indies.	1823.	——	S.⊙.	
9 Boryàna. R.S.	.... Bory's.	Mauritius.	1822.	——	S.⊙.	
10 ripària. R.S.	.... crowded.	E. Indies.	1824.	——	S.⊙.	
11 mucronàta. R.S.	.... mucronated.	Africa.	1822.	7. 9.	S.♃.	Beauv. fl. ow. 1. t. 4.
12 polymórpha. R.S.	.... variable.	Tropics.	1826.	——	S.♃.	
DA'CTYLIS. R.S.	COCK'S-FOOT-GRASS.	Triandria Digynia. L.				
1 glomeràta. R.S.	.... rough.	Britain.	....	6. 7.	H.♃.	Eng. bot. v. 5. t. 335.
2 glaucéscens. R.S.	.... glaucescent.	Europe.	1816.	——	H.♃.	
3 hispánica. R.S.	.... Spanish.	Spain.	1814.	——	H.♃.	
4 glaùca. R.S.	.... glaucous.	....	1823.	——	H.♃.	
5 capitàta. R.S.	.... headed.	Sardinia.	1824.	——	H.♃.	
6 altàica. R.S.	.... Altaic.	Altay.	1826.	——	H.♃.	
7 marítima. R.S.	.... sea-side.	Caucasus.	1823.	——	H.♃.	Schrad. fl. ger. 1. t. 6. f. 1.
8 littoràlis. R.S.	.... shore.	S. Europe.	1816.	——	H.♃.	Lam. ill. t. 45. f. 5.
9 rèpens. R.S.	.... creeping.	———	1824.	——	H.♃.	Desf. atl. 1. t. 15.
10 adscéndens. R.S.	.... ascending.	C. B. S.	1823.	——	G.♃.	
11 lagopodioídes. R.S.	.. harefoot.	E. Indies.	1826.	——	S.♃.	Burm. ind. t. 12. f. 2.
12 brevifòlia. R.S.	.... short-leaved.	———	——	——	S.♃.	
13 cynosuroídes. R.S.	.... Dog's-tailed.	———	1823.	——	S.♃.	
14 aristàta. E.	.... awned.	Egypt.	1828.	——	H.⊙.	
BR'IZA. R.S.	.... QUAKING-GRASS.	Triandria Digynia. R.S.				
1 mìnor. R.S.	.... small.	England.	....	7.	H.♃.	Eng. bot. v. 19. t. 1316.
2 vìrens. R.S.	.... green.	S. Europe.	1800.	7. 8.	H.⊙.	Hayn. term. bot. t. 25. f. 6
3 mèdia. E.B.	.... common.	Britain.	....	5. 6.	H.♃.	Eng. bot. v. 5. t. 340.

4 Clùsii. R.S. .... Clusius's. S. Europe. 1823. 5. 6. H.♃. Fouc. j. bot. 3. t. 24. f. 2.
5 geniculàta. R.S. .... jointed. C. B. S. 1816. 7. 8. H.⊙.
6 elàtior. R.S. .... tall. Greece. 1788. —— H.♃. Flor. græc. 1. t. 75.
7 máxima. R.S. .... great-spiked. S. Europe. 1633. 6. 7. H.⊙. ———— 1. t. 76.
8 hùmilis. R.S. .... dwarf. Tauria. 1818. —— H.⊙. Scheuch. agr. 4. t. 4. f. 10.
9 rùbra. R.S. .... red. S. Europe. —— —— H.⊙.

CYNOS'URUS. R.S. DOGSTAIL-GRASS. Triandria Digynia. L.
cristàtus. E.B. .... crested. Britain. .... 8. H.♃. Eng. bot. v. 5. t. 316.

UN'IOLA. R.S. .... SEASIDE OAT. Triandria Digynia. L.
1 paniculàta. R.S. .... panicled. N. America. 1793. 6. 7. H.♃. Catesb. car. 1. t. 32.
2 latifòlia. R.S. .... broad-leaved. ———— 1809. —— H.♃.
3 spicàta. R.S. .... spiked. ———— 1790. 7. H.♃.
4 distichophy'lla. R.S. .. two-ranked. ———— 1789. 6. 7. H.♃.
Festùca distichophy'lla. Ph.

CERATOCHL'OA. R.S. CERATOCHL'OA. Triandria Digynia. R.S.
unioloídes. R.S. .... Uniola-like. N. America. 1796. 6. 7. H.♃. Willd. h. ber. t. 3.
Festùca unioloídes. W.

GRAPHE'PHORUM. R.S. GRAPHE'PHORUM. Triandria Digynia.
melicoídes. R.S. .... Melic-like. N. America. 1824. 6. 8. H.♃.
A'ira melicoídes. M.

LEPTOCHL'OA. R.S. LEPTOCHL'OA. Triandria Digynia. R.S.
1 cynosuroídes. R.S. .... filiform. E. Indies. 1818. 8. 9. H.⊙.
2 pròcera. R.S. .... tall. Brazil. 1823. —— S.♃.
3 filifórmis. R.S. .... slender. S. America. 1818. —— H.⊙. Jacq. ecl. gram. t. 4.
4 virgàta. R.S. .... fine-spiked. ———— 1727. 7. 9. H.⊙. Sloan. jam. 1. t. 70. f. 2.
5 tenérrima. R.S. .... slenderest. China. 1820. —— H.⊙.
6 domingénsis. L.en. .. close-spiked. W. Indies. —— 6. 8. S.♃.

ELEUS'INE. R.S. ELEUS'INE. Triandria Digynia. R.S.
1 coracàna. R.S. .... thick-spiked. E. Indies. 1714. 7. 9. H.⊙. Schreb. gram. 2. t. 35.
2 stricta. R.S. .... upright. ———— 1822. —— H.⊙.
3 índica. R.S. .... Indian. India. 1714. —— H.⊙. Rheed. mal. 12. t. 69.
4 rígida. R.S. .... rigid. Monte Video. 1826. —— H.⊙.
5 verticillàta. R.S. .... whorled. E. Indies. 1823. —— H.⊙.
6 calycìna. R.S. .... calycine. ———— —— —— H.⊙.
7 cereàlis. E. .... Rye-like. Egypt. 1827. —— H.⊙.
8 ovàlis. E. .... oval-spiked. ———— —— —— H.⊙.
9 tristàchya. E. .... three-spiked. ———— —— —— H.⊙.

DACTYLOCT'ENIUM. DACTYLOCT'ENIUM. Triandria Digynia. R.S.
ægyptìacum. R.S. .... Egyptian. Egypt. 1770. 7. 9. H.⊙. Moris. s. 8. t. 3. f. 7.
Cynosùrus ægy'ptius. L. *Eleusìne ægyptìaca.* P.S.

RHABDOCHL'OA. R.S. RHABDOCHL'OA. Triandria Digynia. R.S.
1 cruciàta. R.S. .... cross-spiked. W. Indies. 1818. 7. 9. H.⊙. Sloan. his. 1. t. 69. f. 1.
Chlòris cruciàta. W. *Agróstis cruciàta.* L.
2 mucronàta. R.S. .... mucronate. N. America. 1820. —— H.⊙. Lam. ill. t. 48. f. 2.
3 virgàta. R.S. .... slender. W. Indies. —— —— H.⊙.
4 domingénsis. R.S. .... West Indian. ———— 1820. —— S.♃. Jacq. icon. 1. t. 22.

CHONDR'OSIUM. K.S. CHONDR'OSIUM. Triandria Digynia. R.S.
1 procùmbens. Desv. .. procumbent. Philippines. 1816. 7. 9. H.⊙. Jacq. ecl. gr. t. 12.
Antinochlòa procúmbens. R.S. *Atheropògon procúmbens.* Jacq.
2 ténue. K.S. ...... slender. Mexico. 1823. —— H.♃. H. et B. n. gen. 1. t. 57.
3 prostràtum. ...... prostrate. ———— —— H.⊙.
Antinochlòa prostràta. R.S. *Bouteloùa prostràta.* Lag.
4 hirsùtum. .... hairy. ———— —— —— H.⊙.

CORYCA'RPUS. Zea. CORYCA'RPUS. Diandria Digynia. S.S.
arundinàceus. S.S. .... Reed-like. N. America. 1810. 6. 7. H.♃. Mich. amer. 1. t. 10.
Festùca diándra. M. *Diarrhèna americàna.* R.S. *Ræmèria Zèæ.* R.S.

DIPLA'CHNE. R.S. DIPLA'CHNE. Triandria Digynia. R.S.
fasciculàris. R.S. .... fascicled. N. America. 1818. 6. 8. H.⊙.
Festùca polystàchya. M.

SCHEDON'ORUS. R.S. SCHEDON'ORUS. Triandria Digynia. R.S.
1 praténsis. R.S. .... meadow. Britain. .... 6. 7. H.⊙. Eng. bot. v. 23. t. 1592.
Festùca praténsis. E.B.
2 phœnicoídes. R.S. .. creeping-rooted. Switzerland. 1824. —— H.♃.
3 elàtior. R.S. ...... tall. Britain. .... —— H.♃. Eng. bot. v. 23. t. 1593.
Festùca elàtior. E.B. *et* Sinc. gr. wob. p. 361. ic.
4 cœruléscens. R.S. .... blueish. Barbary. 1818. 6. 7. H.♃.
5 spadíceus. R.S. .... Gerard's. Italy. 1775. 4. 6. H.♃. Host. gram. 3. t. 20.
6 inérmis. R.S. ...... naked-spiked. Hungary. 1816. 6. 7. H.♃. ———— 1. t. 9.
7 calamàrius. R.S. .... Reed-like. Scotland. .... 7. 8. H.♃. Eng. bot. v. 14. t. 1005.

8 decíduus.	deciduous.	England.	....	6. 7.	H. ♃.	Eng. bot. v. 32. t. 2266.
Festùca decídua. E.B.						
9 exaltàtus. R.S.	high upright.	Sicily.	1827.	——	H. ♃.	
10 sylváticus. R.S.	wood.	Germany.	1804.	7. 8.	H. ♃.	Host. gram. 2. t. 78.
11 serotìnus. R.S.	late-flowering.	S. Europe.	1818.	7. 9.	H. ♃.	———— 2. t. 92.
12 loliàceus. R.S.	spiked.	England.	....	6. 7.	H. ♃.	Eng. bot. v. 26. t. 1821.
13 Scheuchzèri. R.S....	Scheuchzer's.	S. Europe.	1822.	——	H. ♃.	
14 violàceus. R.S.	violet coloured.	Europe.	....	——	H. ♃.	
15 nùtans. R.S.	nodding.	Hungary.	1823.	——	H. ♃.	Host. gram. 4. t. 61.
16 nigréscens. R.S.	blackish.	Switzerland.	——	——	H. ♃.	
17 poæfórmis. R.S.	Poa-like.	————	1818.	——	H. ♃.	Host. gram. 2. t. 81.
18 pùmilus. R.S.	dwarf.	S. Europe.	——	——	H. ♃.	———— 2. t. 91.
19 nítidus. R.S.	glossy.	Hungary.	1827.	——	H. ♃.	
20 tenéllus. R.S.	slender.	N. America.	1804.	7. 8.	H. ☉.	
Festùca tenélla. Ph. *Brachypòdium festucoídes.* L.en.						
FEST'UCA. R.S.	FESCUE-GRASS.	Triandria Digynia.	L.			
1 cynosuroídes. R.S. ..	dogs-tail.	Barbary.	1825.	6. 8.	H. ☉.	Desf. atl. 1. t. 21.
2 Fènas. R.S.	tough.	Spain.	1822.	——	H. ♃.	
Pòa tènax. L.en.						
3 ovìna. R.S.	Sheep's.	Britain.	....	6.	H. ♃.	Eng. bot. v. 9. t. 585.
4 brachyphy'lla. R.S. ..	short-leaved.	Melville Isl.	1827.	——	H. ♃.	
brevifòlia. Br. *non* Muhlenberg.						
5 tenuifòlia. L.en.	slender-leaved.	Britain.	....	——	H. ♃.	
6 c'æsia. E.B.	gray.	England.	....	6. 7.	H. ♃.	Eng. bot. v. 27. t. 1917.
7 vivìpara. E.B.	viviparous.	Britain.	....	7.	H. ♃.	———— v. 19. t. 1355.
8 glaùca. R.S.	glaucous.	S. Europe.	....	6. 7.	H. ♃.	
9 pállens. R.S.	pale.	Germany.	1816.	——	H. ♃.	Host. gram. 2. t. 88.
10 strícta. Host.	upright.	————	——	——	H. ♃.	———— 2. t. 86.
11 hirsùta. Host.	hairy.	————	——	——	H. ♃.	———— 2. t. 85.
12 Hallèri. R.S.	Haller's.	Europe.	1820.	——	H. ♃.	
13 cúrvula. R.S.	curved.	Switzerland.	——	——	H. ♃.	
14 alpìna. R.S.	Alpine.	————	——	——	H. ♃.	
15 amethy'stina. R.S....	blue.	S. Europe.	1804.	——	H. ♃.	
16 pícta. R.S.	painted.	Hungary.	1826.	——	H. ♃.	
17 vària. R.S.	variable.	S. Europe.	1823.	——	H. ♃.	Host. gram. 2. t. 90.
18 vallèsiaca.	Vallesian.	Vallesia.	——	——	H. ♃.	
19 pannònica. R.S.	Hungarian.	Hungary.	——	——	H. ♃.	Host. gram. 4. t. 62.
20 xanthìna. R.S.	yellow.	Switzerland.	1818.	——	H. ♃.	———— 3. t. 19.
flavéscens. Host. *non* Bellardi.						
21 flavéscens. DC.	yellowish.	————	——	——	H. ♃.	
22 duriúscula. E.B.	hard.	Britain.	....	——	H. ♃.	Eng. bot. v. 7. t. 470.
23 dumetòrum. S.G.W. ..	pubescent.	————	....	——	H. ♃.	
24 glábra. S.G.W.	smooth.	————	....	6.	H. ♃.	Sinc. gr. wob. p. 180. ic.
25 hordeifórmis. S.G.W...	Barley-like.	....	....	——	H. ♃.	———— p. 159. ic.
26 cámbrica. S.G.W.	Welsh.	Wales.	....	——	H. ♃.	———— p. 157. ic.
27 heterophy'lla. R.S. ..	various-leaved.	S. Europe.	1812.	6. 7.	H. ♃.	Host. gram. 3. t. 18.
28 rùbra. E.B.	red creeping.	Britain.	....	7.	H. ♃.	Eng. bot. v. 29. t. 2056.
29 vaginàta. R.S.	sheathed.	Hungary.	1804.	6. 7.	H. ♃.	
30 Kitaibeliàna. R.S....	Kitaibel's.	————	1818.	——	H. ♃.	
pubéscens. L.en. *non* Zea.						
31 láxa. R.S.	loose.	Switzerland.	1824.	——	H. ♃.	Host. gram. 2. t. 80.
32 decólorans. R.S.	discoloured.	————	——	——	H. ♃.	
láxa. Gaud. *nec aliorum.*						
33 nùtans. R.S.	nodding.	N. America.	1805.	——	H. ♃.	
34 rùbens. R.S.	red.	S. Europe.	1776.	——	H. ☉.	Flor. græc. t. 83.
Bròmus rùbens. F.G.						
35 scábra. R.S.	rough.	C. B. S.	1826.	——	G. ♃.	
36 bracteàta. DC.	leafy-bracted.	....	....	——	H. ♃.	
37 l'ævis. F.	smooth.	Russia.	1806.	7.	H. ♃.	
38 grandiflòra. Ph.	large-flowered.	N. America.	1812.	6. 7.	H. ♃.	
39 aspérrima. Ot.	roughest.	....	1825.	——	H. ♃.	
40 rupéstris. Ot.	rock.	....	....	——	H. ♃.	
MYGAL'URUS. L.en.	MOUSETAIL-GRASS.	Triandria Digynia.	L.en.			
1 bromoídes. L.en.	barren.	Britain.	....	6.	H. ☉.	Eng. bot. v. 20. t. 1411.
2 caudàtus. L.en.	wall.	England.	....	——	H. ☉.	———— v. 20. t. 1412.
Festùca Myùrus. E.B.						
3 alopecuroídes. L.en...	Fox-tail.	Barbary.	1799.	6. 8.	H. ☉.	Desf. atl. 1. t. 25.
4 ciliàta. Lk.	fringed.	Portugal.	1827.	——	H. ☉.	Host. gram. 4. t. 65.
5 uniglùmis. L.en.	single-husked.	England.	....	——	H. ☉.	Eng. bot. v. 20. t. 1430.

6 delicátulus. L.en.....	delicate.	Spain.	1817.	7. 8.	H.⊙.	
7 geniculàtus. L.en.....	jointed.	S. Europe.	1793.	6. 7.	H.⊙.	
Bròmus geniculàtus. R.S. *Festùca stipoìdes.* R.S.						
BRACHYP'ODIUM. R.S.	BRACHYP'ODIUM.	Triandria Digynia. R.S.				
1 pinnàtum. R.S.	spiked heath.	Britain.	....	6. 8.	H.♃.	Eng. bot. v. 11. t. 730.
Bròmus pinnàtus. E.B. *Tríticum pinnàtum.* DC. *Festùca pinnàta.* E.F.						
2 rupéstre. R.S.	rock.	S. Europe.	1816.	——	H.♃.	Host. gram. 4. t. 17.
3 strigòsum. R.S.	strigose.	Tauria.	1824.	——	H.♃.	
4 cæspitòsum. R.S.	tufted.	S. Europe.	——	——	H.♃.	Host. gram. 4. t. 18.
5 ramòsum. R.S.	branched.	————	1818.	——	H.♃.	Flor. græc. t. 84.
6 phœnicoídes. R.S....	pointed-leaved.	S. France.	1824.	——	H.♃.	
7 obtusifòlium. R.S....	blunt-leaved.	Spain.	1818.	——	H.♃.	
8 Tenoriànum. R.S....	Tenore's.	Naples.	1824.	——	H.♃.	
9 Barrelíeri. R.S.	Barrelier's.	————	1827.	——	H.⊙.	
10 longifòlium. R.S.	long-leaved.	Barbary.	1818.	——	H.♃.	
11 sylváticum. R.S.	wood.	Britain.	....	——	H.♃.	Eng. bot. v. 11. t. 729.
Bròmus sylváticus. E.B. *Tríticum sylváticum.* DC. *Festùca sylvática.* E.F.						
12 distàchyum. R.S.	two-spiked.	S. Europe.	1772.	——	H.♃.	Host. gram. 1. t. 20.
13 megastàchyum. R.S...	long-spiked.	————	1826.	——	H.⊙.	
14 macrostàchyum. R.S...	large-spiked.	————	——	——	H.⊙.	
15 ásperum. R.S.	stiff-leaved.	Spain.	1818.	——	H.⊙.	
16 unioloídes. L.en.	Uniola-like.	Sicily.	1758.	7. 8.	H.⊙.	Jacq. icon. 2. t. 303.
17 brevisètum. R.S.	short-awned.	....	1825.	——	H.⊙.	
18 frágile. R.S.	brittle.	Europe.	1824.	——	H.♃.	
19 Hallèri. R.S.	Haller's.	S. Europe.	1818.	——	H.⊙.	Vivian.fr.fl.it. t.26. f.1.
20 hispánicum. Viv.....	slender-stem'd.	————	——	——	H.⊙.	———— t. 23. f. 2.
tenuículum. R.S.						
21 tenuiflòrum. R.S....	slender-flowered.	————	1823.	——	H.⊙.	Host. gram. 2. t. 26.
22 Nárdus. L.en.	slender.	————	——	——	H.⊙.	
tenéllum. R.S. *Tríticum Nárdus.* DC.						
23 loliàceum. R.S.	spiked sea.	Britain.	....	6. 7.	H.⊙.	Eng. bot. v. 4. t. 221.
Tríticum loliàceum. E.B.						
24 Pòa. R.S.	meadow-grass.	France.	1824.	——	H.⊙.	Pluk. phyt. t. 32. f. 7.
25 biunciàle. R.S.	small.	S. Europe.	1818.	6. 8.	H.⊙.	Viv. fl. it. fr. t. 24.
26 unilateràle. R.S.	one-sided.	————	1824.	——	H.⊙.	Bocc. mus. t. 57. f. 2.
27 mexicànum. Bes.....	Mexican.	Mexico.	1816.	7. 9.	H.⊙.	
Festùca mexicàna. R.S.						
TRICH'ÆTA. R.S.	TRICH'ÆTA.	Triandria Digynia. R.S.				
ovàta. R.S.	oval-spiked.	Spain.	1824.	6. 8.	H.♃.	Cavan. ic. 6. t. 591. f. 2.
Bròmus ovàtus. Cav. *Trisètum ovàtum.* P.S.						
BR'OMUS. R.S.	BROME-GRASS.	Triandria Digynia. R.S.				
1 secalìnus. R.S.	smooth Rye.	England.	....	6. 8.	H.⊙.	Eng. bot. v. 17. t. 1171.
2 elongàtus. R.S.	elongated.	Switzerland.	1823.	——	H.⊙.	
3 commutàtus. R.S....	downy Rye.	Britain.	....	——	H.⊙.	Host. gram. 1. t. 11.
4 velutìnus. R.S.	hairy-headed.	————	....	——	H.⊙.	Eng. bot. v. 27. t. 1884.
multiflòrus. E.B.						
5 Triniànus. R.S.	Trinius's.	Gilan.	1826.	——	H.♃.	Trin. m.ac.pet.1813.t.9.
6 móllis. R.S.	soft.	Britain.	....	5. 7.	H.⊙.	Eng. bot. v. 15. t. 1078.
7 confértus. R.S.	close-spiked.	Iberia.	1825.	——	H.⊙.	Jacq. ecl. gr. t. 14.
8 racemòsus. R.S.	smooth.	England.	....	6. 8.	H.⊙.	Eng. bot. v. 15. t. 1079.
9 eréctus. E.B.	upright.	————	....	——	H.♃.	———— v. 7. t. 471.
10 Biebersteìnii. R.S. ..	Bieberstein's.	Caucasus.	1825.	——	H.♃.	
11 austràlis. B.P.	southern.	N. S. W.	1823.	——	H.⊙.	
12 lanceolàtus. R.S.	spear-leaved.	Canaries.	1798.	——	H.⊙.	Buxb. cent. ap. t. 19.
13 lanuginòsus. R.S.	woolly.	S. Europe.	1824.	——	H.⊙.	
14 variegàtus. R.S.	variegated.	Caucasus.	——	——	H.♃.	
15 squarròsus. R.S.	corn.	England.	....	——	H.♃.	Eng. bot. v. 27. t. 1885.
16 vestìtus. R.S.	clothed.	C. B. S.	1816.	——	G.♃.	
17 wolgénsis. R.S.	Wolga.	Caucasus.	——	——	H.⊙.	Jacq. ecl. gr. t. 15.
18 láxus. R.S.	loose.	....	——	——	H.♃.	
19 alopecuroídes. R.S. ..	Fox-tail.	S. Europe.	1799.	——	H.♃.	Desf. atl. 1. t. 25.
contórtus. Desf.						
20 púrgans. R.S.	purging.	Canada.	1793.	——	H.♃.	
21 pubéscens. R.S.	pubescent.	N. America.	1816.	——	H.♃.	
22 ásper. E.B.	hairy wood.	England.	....	——	H.♃.	Eng. bot. v. 17. t. 1172.
23 hirsutíssimus. Cyr. ..	very hairy.	S. Europe.	1822.	——	H.♃.	
24 pendulìnus. R.S.	pendulous.	New Spain.	1823.	——	H.⊙.	
25 glaùcus. R.S.	glaucous.	Pyrenees.	——	——	H.♃.	Lap. fl. pyren. t. 53. bis.
26 altíssimus. R.S.	tallest.	N. America.	1812.	——	H.♃.	

27 gigánteus. R.S.	.... gigantic.	Britain.	....	7. 8.	H. ♃.	Eng. bot. v. 26. t. 1820.
28 triflòrus.	.... three-flowered.	———	....	——	H. ♃.	——— v. 27. t. 1918.
Festùca triflòra. E.B.						
29 montànus. R.S.	.... mountain.	Switzerland.	1827.	——	H. ⊙.	
30 longiflòrus. R.S.	.... long-flowered.	....	1818.	——	H. ♃.	
31 stenophy'llus. R.S.	.... slender-leaved.	....	1824.	——	H. ♃.	
32 angustifòlius. M.B.	.... narrow-leaved.	Caucasus.	——	——	H. ♃.	
33 ciliàtus. R.S.	.... ciliated.	N. America.	1802.	6. 8.	H. ♃.	
34 arvénsis. R.S.	.... field.	Britain.	....	7.	H. ♃.	Eng. bot. v. 28. t. 1984.
35 stérilis. R.S.	.... barren.	———	....	6. 7.	H. ⊙.	——— v. 15. t. 1030.
36 tectòrum. R.S.	.... nodding.	Europe.	1776.	6. 8.	H. ♂.	Flor. græc. t. 82.
37 matriténsis. R.S.	.... wall.	Britain.	....	6.	H. ⊙.	Eng. bot. v. 14. t. 1006.
diándrus. E.B.						
38 ligústicus. R.S.	.... contracted.	Italy.	1824.	6. 8.	H. ⊙.	Barrel. ic. 76. n. 2.
39 rígidus. R.S.	.... rigid.	S. France.	1822.	——	H. ⊙.	Host. gram. 1. t. 18.
40 pilòsus. R.S.	.... hairy wall.	S. Europe.	——	——	H. ⊙.	
41 jubàtus. R.S.	.... loose panicled.	Italy.	1824.	——	H. ⊙.	
42 máximus. R.S.	.... great.	Morocca.	1804.	——	H. ⊙.	Desf. atl. 1. t. 26.
43 scopàrius. R.S.	.... broom.	S. Europe.	1824.	——	H. ⊙.	Cavan. ic. 1. t. 45. f. 2.
44 hùmilis. R.S.	.... low.	Spain.	——	——	H. ⊙.	——— ic. 6. t. 589. f. 2.
45 scabérrimus. R.S.	.... roughest.	Italy.	——	——	H. ⊙.	
46 coarctàtus. H.H.	.... close-flowered.	....	1826.	——	H. ♃.	
47 parviflòrus. Df.	.... small-flowered.	....	1824.	——	H. ♃.	
48 tórtilis. Presl.	.... twisted.	Sicily.	1825.	——	H. ♃.	
49 caucásicus. F.	.... Caucasean.	Caucasus.	——	——	H. ♃.	
ROSTR'ARIA. Tr.	ROSTR'ARIA.	Triandria Digynia.				
pubéscens. Tr.	.... pubescent.	....	1824.	6. 8.	H. ⊙.	
Bròmus dactyloídes. Roth. *Dáctylis púngens.* Horn.						
PHRAGM'ITES. Tr.	REED.	Triandria Digynia. R.S.				
commùnis. Tr.	common.	Britain.	...	7. 8.	H. *w.* ♃.	Eng. bot. v. 6. t. 401.
Arúndo phragmìtes. E.B.						
ARU'NDO. S.S.	REED-GRASS.	Triandria Digynia. S.S.				
1 Dònax. S.S.	.... manured.	S. Europe.	1648.	7. 8.	H. ♃.	Host. gram. 4. t. 38.
β *versícolor.*	.... *striped.*	———	....	——	H. ♃.	Moris. s. 8. t. 8. f. 9.
2 tènax. S.S.	.... tough.	———	1823.	——	H. ♃.	Cyril. neap. f. 2. t. 12.
3 festucàcea. W.en.	.... Fescue-like.	Germany.	——	——	H. *w.* ♃.	
PSA'MMA. R.S.	SEA MATWEED.	Triandria Trigynia. R.S.				
arenària. R.S.	.... common.	Britain.	....	6. 7.	H. ♃.	Eng. bot. v. 8. t. 520.
Arúndo arenària. E.B.						
ARUNDIN'ARIA. R.S.	CANE-BRAKE.	Triandria Monogynia. S.S.				
1 glaucéscens. R.S.	.... glaucous.	E. Indies.	1818.	....	S. ♄.	
Ludólfia glaucéscens. W.						
OL'YRA. W.	OL'YRA.	Monœcia Triandria. L.				
paniculàta. W.	.... broad-leaved.	W. Indies.	1783.	7.	S. ♃.	Sloan. jam. 1. t. 64. f. 2.
Z'EA. W.	INDIAN CORN.	Monœcia Triandria. L.				
1 Máys. W.	.... common.	America.	1562.	6. 7.	H. ⊙.	Lam. ill. t. 749.
2 Curágua. S.S.	.... saw-leaved.	Chile.	1820.	——	H. ⊙.	
CO'IX. W.	JOB'S-TEARS.	Monœcia Triandria. L.				
1 Láchryma. S.S.	.... common.	E. Indies.	1596.	6. 7.	S. ♃.	Bot. mag. t. 2479.
2 agréstis. S.S.	.... round-fruited.	———	1812.	——	S. ♃.	Rumph. amb. 6. t. 9. f. 1.
3 exaltàta. S.S.	.... tall.	China.	1816.	——	S. ♃.	Jacq. ic. gram.
4 arundinàcea. S.S.	.... reed-like.	Mexico.	1824.	——	S. ♃.	
5 K'œnigii. S.S.	.... Kœnig's.	E. Indies.	1818.	——	S. ♃.	
6 gigántea. H.B.	.... gigantic.	Bengal.	1820.	——	H. ♃.	
BAMB'USA. W.	BAMBOO-CANE.	Hexandria Monogynia. W.				
1 arundinàcea. W.	.... common.	India.	1730.	....	S. ♄.	Roxb. corom. 1. t. 79.
2 verticillàta. W.	.... whorl-flowered.	———	1802.	....	S. ♄.	——— 1. t. 80.
MELOCA'NNA. Tr.	MELOCA'NNA.	Hexandria Monogynia. S.S.				
bambusoídes. Tr.	.... berry bearing.	E. Indies.	1818.	....	S. ♄.	
Bambùsa baccífera. Roxb.						
PH'ARUS. W.	PH'ARUS.	Monœcia Hexandria. W.				
1 latifòlius. W.	.... broad-leaved.	Jamaica.	1793.	7. 8.	S. ♃.	Brown. jam. t. 38. f. 3.
2 angustifòlius. D.L.	.... narrow-leaved.	Trinidad.	1821.	——	S. ♃.	
EHRHA'RTA. W.	EHRHA'RTA.	Hexandria Monogynia. S.S.				
1 panícea. W.	.... Panic-grass.	C. B. S.	1790.	5. 7.	G. ♃.	Sm. ined. 1. t. 9.
2 melicoídes. S.S.	.... Melica-like.	———	1816.	——	G. ♃.	
MICROL'ÆNA. B.P.	MICROL'ÆNA.	Tetrandria Digynia. B.P.				
stipoídes. B.P.	.... Stipa-like.	N. S. W.	1822.	7. 8.	F. ♃.	Lab. n. hol. 1. t. 118.
Ehrhárta stipoídes. Lab.						

POTAMO'PHILA.B.P.	POTAMO'PHILA.	Polygamia Monœcia. B.P.				
parviflòra. B.P.	.... small-flowered.	N. S. W.	1820.	6. 7.	F.w.♃.	
OR'YZA. W.	RICE.	Hexandria Digynia. L.				
satìva. W.	.... common.	Ethiopia.	1596.	7.	S.w.☉.	
ZIZ'ANIA. W.	CANADA RICE.	Monœcia Hexandria. W.				
1 aquática. H.K.	.... water.	N. America.	1790.	7. 9.	H.w.♃.	Linn. trans. 7. t. 13.
2 miliàcea. M.	.... Millet-like.	Carolina.	1816.	——	H.w.♃.	
3 fluítans. M.	.... floating.	N. America.	1824.	7. 8.	H.w.♃.	
PARI'ANA. W.	PARI'ANA.	Monœcia Polyandria. W.				
campéstris. W.	Cayenne.	Cayenne.	1803.	....	S.♄.	Aubl. gui. 2. t. 337.

** *Subclassis* II. *CRYPTOGAMEÆ.* DC.

ORDO CCXXV.

CHARACEÆ. *Kth. synops. v.* 1. *p.* 102.

CH'ARA. L.	CH'ARA.	Monœcia Monandria. W.				
1 vulgàris. E.B.	.... common.	Britain.	....	7. 8.	H.w.☉.	Eng. bot. v. 5. t. 336.
2 híspida. E.B.	.... prickly.	————	....	6. 9.	H.w.☉.	———— v. 7. t. 463.
NITE'LLA. Ag.	.... NITE'LLA.	Monœcia Monandria.				
1 tomentòsa. W.	.... tomentose.	....	....	6. 7.	H.w.☉.	Moris. s. 15. t. 4. f. 9.
2 fléxilis. Ag.	.... *smooth.	Britain.	....	7. 8.	H.w.☉.	
3 opàca. Ag.	.... opaque.	————	....	——	H.w.☉.	Eng. bot. v. 15. t. 1070.
Chàra fléxilis. E.B.						
4 translùcens. Ag.	.... great transparent.	————	....	——	H.w.☉.	Eng. bot. v. 26. t. 1855.
5 nidífica. Ag.	.... proliferous.	————	....	——	H.w.☉.	———— v. 24. t. 1703.
6 grácilis. Ag.	.... slender.	————	....	——	H.w.☉.	———— v. 30. t. 2140.

ORDO CCXXVI.

EQUISETACEÆ. *DC. fl. fr. ed.* 3. *v.* 2. *p.* 28.

EQUIS'ETUM. W.		HORSE-TAIL.	Cryptogamia	Gonopterides. W.			
1 arvénse. W.	(*br.ye.*)	corn.	Britain.	....	3. 4.	H.♃.	Eng. bot. v. 29. t. 2020.
2 fluviátile. W.	(*br.ye.*)	great water.			4. 5.	H.♃.	———— v. 29. t. 2022.
3 umbròsum. W. en.	(*br.y.*)	shade.	Germany.	1818.	——	H.♃.	
4 sylváticum. W.	(*br.ye.*)	wood.	Britain.	....	——	H.♃.	Eng. bot. v. 27. t. 1874.
5 limòsum. W.	(*br.ye.*)	smooth naked.	————	....	6. 7.	H.♃.	———— v. 13. t. 929.
6 palústre. W.	(*br.ye.*)	marsh.	————	....	——	H.♃.	———— v. 29. t. 2021.
7 praténse. W.	(*br.ye.*)	meadow.	Germany.	1820.	——	H.♃.	
8 scirpoídes. W.	(*br.ye.*)	American.	N. America.	1812.	——	H.♃.	
9 hyèmale. W.	(*br.ye.*)	Shave grass.	Britain.	....	7. 8.	H.♃.	Eng. bot. v. 13. t. 915.
10 variegàtum. W.	(*br.ye.*)	variegated.	Scotland.	....	6. 7.	H.♃.	———— v. 28. t. 1987.
11 filifórme.	(*br.ye.*)	thready.	N. America.	1800.	7. 8.	H.♃.	

ORDO CCXXVII.

MARSILEACEÆ. *Brown prodr. p.* 166.

PILUL'ARIA. W.		PILLWORT.	Cryptogamia Hydropterides. W.				
globulífera. W.	(*br.*)	Pepper-grass.	Britain.	....	6. 9.	H.w.♃.	Eng. bot. v. 8. t. 521.
ISO'ETES. W.		QUILLWORT.	Cryptogamia Hydropterides. W.				
lacústris. W.	....	marsh.	Britain.	....	5. 10.	H.w.♃.	Eng. bot. v. 16. t. 1084.

SALVI'NIA. W.		Salvi'nia.	Cryptogamia Hydropterides. W.				
nàtans. w.	....	floating.	Europe.	1818.	5. 8.	H.*w*.♃.	Schkuhr filic. t. 173.
MARSI'LEA. W.		Marsi'lea.	Cryptogamia Hydropterides. W.				
quadrifòlia. w.		four-leaved.	S. Europe.	1820.	6. 8.	H.*w*.♃.	J.mem.ac.sci.par.1740.

ORDO CCXXVIII.

LYCOPODIACEÆ. *Kth. synops.* 1. *p.* 95.

LYCOP'ODIUM. W.		Club-moss.	Cryptogamia Stachyopterides. W.				
1 caroliniánum. w.	(*br.*)	Carolina.	Carolina.	1812.	7. 8.	H.♃.	Dill. musc. t. 62. f. 5.
2 clavàtum. w.	(*br.*)	common.	Britain.	....	——	H.♃.	Eng. bot. v. 4. t. 224.
3 complanàtum. w.	(*br.*)	ArborVitæ-leav'd.	N. America.	1770.	——	H.♃.	Flor. dan. t. 78.
4 alpìnum. w.	(*br.*)	Savin-leaved.	Britain.	....	8.	H.♃.	Eng. bot. v. 4. t. 234.
5 dendroídeum. w.	(*br.*)	fan.	N. America.	1770.	7.	H.♃.	Hook. ex. flor. t. 7.
6 annotìnum. w.	(*br.*)	interrupted.	Britain.	....	6. 8.	H.♃.	Eng. bot. v. 24. t. 1727.
7 inundàtum. w.	(*br.*)	marsh.	———	....	——	H.*w*.♃.	——— v. 4. t. 239.
8 alopecuroídes. w.	(*br.*)	Fox-tail.	N. America.	1816.	——	H.♃.	Schkuhr filic. t. 160.
9 Selaginoídes. w.	(*br.*)	prickly.	Britain.	....	8.	H.♃.	Eng. bot. v. 16. t. 1148.
10 ornithopodioídes. w.	(*br.*)	Bird's-claw.	....	1812.	6. 8.	H.♃.	Dill. musc. t. 66. f. 1. B.
11 helvéticum. w.	(*br.*)	Swiss.	Switzerland.	1779.	——	H.♃.	——— t. 64. f. 2.
12 rupéstre. w.	(*br.*)	rock.	N. America.	1818.	——	H.♃.	——— t. 63. f. 11.
13 denticulàtum. w.	(*br.*)	toothed.	Switzerland.	1779.	7.	H.♃.	——— t. 66. f. 1. A.
14 depréssum. w.	(*br.*)	depressed.	C. B. S.	1816.	7. 8.	G.♃.	
15 Selàgo. w.	(*br.*)	Fir.	Britain.	....	6. 8.	H.♃.	Eng. bot. v. 4. t. 233.
16 apòdum. w.	(*br.*)	distichous.	N. America.	1815.	7.	H.♃.	Dill. musc. t. 64. f. 3.
17 lucídulum. w.	(*br.*)	glossy.	———	1822.	——	H.♃.	Schkuhr filic. t. 159.
18 dénsum. B.P.	(*br.*)	dense.	N. S. W.	——	7. 8.	G.♃.	Lab.nov.hol.2.t.251.f.1
PSIL'OTUM. B.P.		Psil'otum.	Cryptogamia Stachyopterides. W.				
triquètrum. B.P.	(*ye.*)	triangular.	N. S. W.	1793.	7. 8.	G.♃.	Schkuhr filic. t. 165. b.
Lycopòdium nùdum. L. *Bernhárdia dichótoma.* W.							

ORDO CCXXIX.

FILICES. *Juss. gen.* 14.

Subordo I. *OPHIOGLOSSEÆ.* Brown prodr. p. 163.

OPHIOGLO'SSUM. W.		Adder's-tongue.	Cryptogamia Stachyopterides. W.				
1 vulgàtum. E.B.	(*br.*)	common.	Britain.	....	5. 6.	H.♃.	Eng. bot. v. 2. t. 108.
2 bulbòsum. w.	(*br.*)	bulbous-rooted.	N. America.	....	——	H.♃.	
3 lusitánicum. w.	(*br.*)	Portugal.	Portugal.	....	——	H.♃.	
4 gramíneum. w.	(*br.*)	grass-like.	N. Holland.	1822.	——	G.♃.	Willd.s.erf.1802.t.1.f.1
5 costàtum. B.P.	(*br.*)	ribbed.	N. S. W.	——	——	G.♃.	
6 reticulàtum. w.	(*br.*)	netted.	W. Indies.	1793.	——	S.♃.	Plum. fil. t. 164.
7 petiolàtum. w.	(*br.*)	petiolate.	———	1822.	1. 3.	S.♃.	Hook. ex. flor. t. 56.
BOTRY'CHIUM. W.		Moonwort.	Cryptogamia Stachyopterides. W.				
1 Lunària. w.	(*br.ye.*)	common.	Britain.	....	5. 6.	H.♃.	Eng. bot. v. 5. t. 318.
Osmúnda lunària. E.B.							
2 fumarioídes. w.	(*br.ye.*)	Fumitory-leaved.	Carolina.	1806.	7. 8.	H.♃.	Schkuhr fil. t. 157.
3 oblíquum. w.	(*br.ye.*)	oblique.	N. America.	1821.	6. 7.	H.♃.	Pluk. mant. t. 427. f. 7.
4 disséctum. w.	(*br.ye.*)	cut-leaved.	———	1806.	7.	H.♃.	Schkuhr fil. t. 158.
5 virgínicum. w.	(*br.ye.*)	Virginian.	———	1790.	8.	H.♃.	——— t. 156.
6 austràle. B.P.	(*br.ye.*)	southern.	N. S. W.	1823.	7. 8.	H.♃.	
7 grácile. Ph.	(*br.ye.*)	slender.	N. America.	1824.	——	H.♃.	

SUBORDO II. *OSMUNDACEÆ.* Brown prodr. p. 161.

LYG'ODIUM. B.P. SNAKE'S-TONGUE. Cryptogamia Schismatopterides. W.
1 polymórphum. K.S. (*br.y.*) variable. S. America. 1816. 5. 9. S.♃.⌣. Cav. ic. 6. t. 595. f. 1.
Ugèna polymórpha. Cav. *Hydroglóssum hirsùtum.* W.
2 scándens. Sw. (*br.ye.*) climbing. E. Indies. 1793. 5. 9. S.♃.⌣. Lodd. bot. cab. t. 742.
3 microphy'llum. B.P. (*br.y.*) small-leaved. N. Holland. 1818. —— S.♃.⌣. Cavan. icon. 6. t. 595. f. 2.
4 venústum. Sw. (*br.y.*) elegant. S. America. 1812. —— S.♃.⌣.
5 circinàtum. Sw. (*br.y.*) rounded. W. Indies. 1818. —— S.♃.⌣.
6 palmàtum. Sw. (*br.y.*) palmate. N. America. 1823. —— S.♃.⌣. Schkuhr filic. t. 140.
Hydroglóssum palmàtum. W.

ANEI'MIA. W. ANEI'MIA. Cryptogamia Schismatopterides. W.
1 fraxinifòlia. Rad. (*br.y.*) Ash-leaved. Brazil. 1829. 9. 12. S.♃.
2 hírta. W. (*br.ye.*) hairy-leaved. W. Indies. 1824. —— S.♃.
3 collìna. Rad. (*br.ye.*) hill. Brazil. 1829. —— S.♃.
4 hirsùta. W. (*br.ye.*) hairy. W. Indies. 1794. .... S.♃. Plum. fil. t. 162.
5 laciniàta. (*br.ye.*) jagged. —— —— —— S.♃.
6 adiantifòlia. W. (*br.ye.*) Maiden-hair-leav'd. —— 1793. 8. 9. S.♃. Plum. fil. t. 158.
7 lanceolàta. W. (*br.ye.*) spear-leaved. —— 1820. —— S.♃.

OSMU'NDA. W. OSMU'NDA. Cryptogamia Schismatopterides. W.
1 cinnamòmea. W. (*br.*) woolly. N. America. 1772. 6. H.♃. Schkuhr fil. t. 146.
2 regàlis. W. (*br.ye.*) flowering fern. Britain. .... 7. 8. H.♃. Eng. bot. v. 3. t. 209.
3 Claytoniàna. W. (*br.ye.*) Clayton's. N. America. 1772. 8. H.♃.
4 interrúpta. W. (*br.ye.*) interrupted. —— .... 6. 7. H.♃. Schkuhr fil. t. 144.
5 spectábilis. W. (*br.ye.*) showy. —— 1811. 7. H.♃. Pluk. alm. t. 184. f. 4.
6 hùmilis. W. (*br.ye.*) dwarf. —— 1823. 6. 8. H.♃. Hook. ex. flor. t. 28.

T'ODEA. W. T'ODEA. Cryptogamia Schismatopterides. W.
africàna. W. (*br.ye.*) African. C. B. S. 1805. 5. 8. G.♃. Schkuhr fil. t. 147.
Osmúnda barbàta. B.P.

SCHIZ'ÆA. W. SCHIZ'ÆA. Cryptogamia Schismatopterides. W.
1 penicillàta. W. (*br.ye.*) pencilled. S. America. 1816. 6. 7. S.♃.
2 pusílla. Ph. (*br.ye.*) small. N. America. 1825. —— H.♃. H. et G. ic. fil. t. 47.
3 rupéstris. B.P. (*br.ye.*) rock. N. S. W. 1822. —— G.♃. —— t. 48.
4 bífida. B.P. (*br.ye.*) bifid. —— —— —— G.♃. Willd. mem. farr. t. 3. f. 2.

SUBORDO III. *GLEICHENEÆ.* Brown prodr. p. 160.

GLEICH'ENIA. B.P. GLEICH'ENIA. Cryptogamia Poropterides. W.
1 pubéscens. K.S. (*br.*) pubescent. Vera Cruz. 1822. 6. 8. S.♃.
Merténsia pubéscens. W.
2 Spelúncæ. B.P. (*br.*) various-leaved. N. S. W. —— —— G.♃.
3 Hermánni. B.P. (*br.*) Hermann's. India. 1829. .... S.♃. H. et G. ic. fil. t. 14.
4 microphy'lla. B.P. (*br.*) small-leaved. N. S. W. 1825. 6. 8. G.♃.
5 flabellàta. B.P. (*br.*) fan-leaved. —— 1823. —— G.♃.

MARA'TTIA. W. MARA'TTIA. Cryptogamia Poropterides. W.
alàta. W. (*br.ye.*) winged. Jamaica. 1793. 5. 8. S.♃. Sm. ic. ined. t. 46.

SUBORDO IV. *PARKERIACEÆ.* Hook et Grev. ic. fil. 97.

PARK'ERIA. H.F. PARK'ERIA. Cryptogamia Poropterides.
1 pteridoídes. H.F. (*br.*) Brake-like. W. Indies. 1822. 5. 10. S.♃. H. et G. ic. fil. t. 97.
2 Lockhárti. H.F. (*br.*) Lockhart's. Trinidad. 1824. —— S.♃.

CERATO'PTERIS. S.S. CERATO'PTERIS. Cryptogamia Poropterides.
thalictroídes. S.S. (*br.*) Thalictrum-like. E. Indies. 1818. 6. 10. S.♃. Pluk. alm. t. 215. f. 3.
Ellobocárpus oleràceus. Kaulfuss.

SUBORDO V. *POLYPODIACEÆ.* Brown prodr. p. 145.

ACRO'STICHUM. W. ACRO'STICHUM. Cryptogamia Filices. L.
1 viscòsum. H.F. (*br.ye.*) viscous. W. Indies. 1826. 6. 10. S.♃. H. et G. ic. fil. t. 61.
2 símplex. W. (*br.ye.*) simple. Jamaica. 1793. 5. 7. S.♃. Lodd. bot. cab. t. 709.
3 apòdum. H.F. (*br.ye.*) stemless. W. Indies. 1824. —— S.♃. H. et G. ic. fil. t. 99.

4 crinìtum. w. (*br.ye.*) hairy. W. Indies. 1793. 5. 7. S. ♃. H. et G. ic. fil. t. 1.
5 alcicórne. B.P. (*br.*) Elk's-horn. N. S. W. 1808. 8. 10. G. ♃. Bot. reg. t. 262-3.
6 Stemària. Beau. (*br.*) Stag's-horn. Guinea. 1822. 5. 8. S. ♃. Beau. fl. d'ow. v. 1. pl. 2.
7 appendiculàtum. H. (*br.*) appendaged. W. Indies. 1824. 6. 8. S. ♃. Hook. ex. flor. t. 108.
8 aùreum. w. (*br.ye.*) golden. America. 1820. 7. 8. G. ♃. Schkuhr fil. t. 1.

NOTHOL'ÆNA. B.P. NOTHOL'ÆNA. Cryptogamia Filices.
1 Marántæ. B.P. (*br.ye.*) Swiss. Switzerland. 1824. 6. 8. H. ♃. Spreng. crypt. t. 2. f. 18.
Acróstichum Marántæ. w.
2 dístans. B.P. (*br.ye.*) distant. N. S. W. 1823. —— G. ♃.
3 lanuginòsa. s.s. (*br.ye*) woolly. Madeira. 1778. 8. 9. G. ♃. Desf. atl. 2. t. 256.
Acróstichum vélleum. H.K.

HEMION'ITIS. W. HEMION'ITIS. Cryptogamia Filices. W.
palmàta. w. (*br.ye.*) palmated. W. Indies. 1793. 6. 8. S. ♃. Hook. ex. flor. t. 33.

GYMNOGRA'MMA. GYMNOGRA'MMA. Cryptogamia Filices.
1 pedàta. s.s. (*br.ye.*) pedate. New Spain. 1822. 6. 9. S. ♃. Swtz. syn. fil. t. 1. f. 3.
2 rùfa. s.s. (*br.ye.*) rusty-haired. Jamaica. 1793. 6. 8. S. ♃. Schkuhr fil. t. 17. 21.
Hemionìtis rùfa. w.
3 tartàrea. s.s. (*br.ye.*) whitened. S. America. 1817. —— S. ♃. Willd. hort. ber. 1. t. 40.
Acróstichum tartàreum. Swtz. *Hemonìtis dealbàta.* w.
4 peruviàna. s.s. (*br.ye.*) Peruvian. Peru. 1822. 6. 8. G. ♃.
5 trifoliàta. s.s. (*br.ye.*) three-leaved. Jamaica. 1810. —— S. ♃. Plum. fil. t. 144.
6 calomelànos. s.s. (*br.y.*) mealy. W. Indies. 1790. 7. 8. S. ♃. Willd. hort. ber. 1. t. 41.
7 chrysophy'lla. s.s. (*br.y.*) golden-leaved. ——— 1824. —— S. ♃.
8 sulphùrea. s.s. (*br.ye.*) sulphur-colored. Jamaica. 1808. —— S. ♃. Schkuhr crypt. t. 4.
9 leptophy'lla. s.s. (*br.ye.*) slender-leaved. S. Europe. 1824. 5. 8. H. ♃. H. et G. ic. fil. t. 25.
10 chærophy'lla. s.s. (*br.y.*) Chervil-leaved. Brazil. 1825. 6. 8. S. ♃. ——— ——— t. 45.

MENI'SCIUM. W. MENI'SCIUM. Cryptogamia Filices. W.
1 triphy'llum. s.s. (*br.ye.*) three-leaved. E. Indies. 1820. 6. 8. S. ♃. H. et G. ic. fil. t. 120.
2 sorbifòlium. s.s. (*br.ye.*) Service-leaved. S. America. 1825. —— S. ♃.
3 reticulàtum. w. (*br.ye.*) netted. Martinico. 1793. 6. 8. S. ♃. Plum. fil. t. 110.

XIPHO'PTERIS. K.F. SWORD FERN. Cryptogamia Filices.
1 serrulàta. K.F. (*br.ye.*) saw-leaved. W. Indies. 1823. 6. 8. S. ♃. Hook. ex. flor. t. 78.
Grammìtis serrulàta. w.
2 myosuroídes. s.s. (*br.y.*) Mousetail-like. ——— 1824. —— S. ♃.

GRAMM'ITIS. Sw. GRAMM'ITIS. Cryptogamia Filices. W.
1 austràlis. B.P. (*br.ye.*) southern. N. S. W. 1822. 6. 8. G. ♃.
2 lineàris. s.s. (*br.ye.*) linear-leaved. Jamaica. 1823. —— S. ♃.
3 lanceolàta. s.s. (*br.ye.*) spear-leaved. Mauritius. 1824. —— S. ♃. H. et G. ic. fil. t. 43.
4 elongàta. s.s. (*br.ye.*) long-leaved. W. Indies. —— —— S. ♃. Schkuhr crypt. t. 7.
5 furcàta. H.F. (*br.ye.*) forked-leaved. Trinidad. 1825. —— S. ♃. H. et G. ic. fil. t. 62.

CE'TERACH. W. CE'TERACH. Cryptogamia Filices.
officinàrum. w. (*br.*) common. Britain. .... 5. 10. H. ♃. Lodd. bot. cab. t. 15.
Scolopéndrium Céterach. Eng. bot. v. 18. t. 1244.

POLYB'OTRYA. S.S. POLYB'OTRYA. Cryptogamia Filices.
1 vivìpara. H.E.F. (*br.y.*) viviparous. W. Indies. 1823. 5. 8. S. ♃. Hook. ex. flor. t. 107.
2 cervìna. s.s. (*br.ye.*) hart's-tongue. ——— —— —— S. ♃. H. et G. ic. fil. t. 81.

NIPHO'BOLUS. K.F. NIPHO'BOLUS. Cryptogamia Filices.
1 pertùsus. s.s. (*br.ye.*) bored. China. 1817. 7. 10. G. ♃. Hook. ex. flor. t. 162.
Polypòdium pertùsum. Hooker.
2 rupéstris. s.s. (*br.ye.*) rock. N. S. W. 1824. —— G. ♃. H. et G. ic. fil. t. 93.
3 conflúens. s.s. (*br.ye.*) confluent. ——— 1822. —— G. ♃.
4 Língua. s.s. (*br.ye.*) tongue-like. Japan. 1817. —— G. ♃. Thunb. jap. t. 38.

POLYP'ODIUM. W. POLYPODY. Cryptogamia Filices. L.
1 sérpens. w. (*ye.*) trailing. W. Indies. 1822. 3. 8. S. ♃. Plum. fil. 105. t. 121.
2 angustifòlium. w. (*ye.*) narrow-leaved. S. America. —— —— S. ♃.
3 lanceolàtum. w. (*ye.*) spear-leaved. ——— 1824. —— S. ♃. Plum. fil. t. 137.
4 attenuàtum. B.P. (*ye.*) attenuated. N. S. W. 1823. —— G. ♃.
5 piloselloídes. w. (*ye.*) Mouse-ear. W. Indies. 1793. —— S. ♃. Plum. fil. t. 118.
6 vacciniifòlium. s.s. (*ye.*) Whortle-leaved. Brazil. 1826. —— S. ♃.
7 tæniòsum. s.s. (*ye.*) jointed. S. America. 1815. —— S. ♃.
8 lycopodioídes. w. (*ye.*) Club-moss. W. Indies. 1793. —— S. ♃. Schkuhr fil. t. 8. c. p.
9 phyllítidis. w. (*ye.*) Hart's-tongue. ——— —— —— S. ♃. Plum. fil. 114. t. 130.
10 irioídes. s.s. (*ye.*) Iris-leaved. E. Indies. 1824. —— S. ♃. H. et G. ic. fil. t. 125.
11 crassifòlium. w. (*ye.*) thick-leaved. W. Indies. 1816. —— S. ♃. Plum. fil. 107. t. 127.
12 rèpens. w. (*ye.*) creeping. S. America. —— —— S. ♃. ——— 115. t. 131.
13 lineàre. s.s. (*ye.*) linear-leaved. Japan. 1822. —— G. ♃. Thunb. ic. jap. t. 18.
14 Billardíeri. B.P. (*ye.*) La Billardiere's. V. Diem. Isl. 1823. —— G. ♃. Lab. nov. hol. 2. t. 240.
15 aùreum. w. (*ye.*) golden. W. Indies. 1742. 3. 5. S. ♃. Plum. fil. 59. t. 76.
16 phymatòdes. w. (*ye.*) warted. ——— 1816. —— S. ♃.

17 quercifòlium. s.s.	(*ye.*) Oak-leaved.	E. Indies.	1824.	3. 5.	S.♃.	Rumph. amb. 6. t. 36.
18 curvàtum. s.s.	(*ye.*) curve-leaved.	Jamaica.	1822.	——	S.♃.	
19 vulgàre. E.B.	(*ye.*) common.	Britain.	....	5. 10.	H.♃.	Eng. bot. v. 16. t. 1149.
β *cámbricum.*	(*ye.*) *Welsh.*	Wales.	....	——	H.♃.	Bolt. fil. t. 2. f. 5. a.
20 virginiànum. W.	(*ye.*) Virginian.	N. America.	....	7.	H.♃.	Plum. fil. t. 77.
21 Scoulèri. H.F.	(*ye.*) Scouler's.	————	1827.	——	H.♃.	H. et G. ic. fil. t. 56.
22 pectinàtum. W.	(*ye.*) comb-leaved.	W. Indies.	1793.	6. 9.	S.♃.	Lodd. bot. cab. t. 748.
23 Plùmula. W.	(*ye.*) feather.	S. America.	1822.	5. 8.	S.♃.	
24 scolopendrioídes. H.F.	(*y.*) hart's-tongue-ld.	Jamaica.	1824.	——	S.♃.	H. et G. ic. fil. t. 42.
25 tenuifòlium. W.	(*ye.*) slender-leaved.	S. America.	1822.	——	S.♃.	Plum. fil. 66. t. 85.
26 asplenifòlium. W.	(*ye.*) Spleenwort-leav'd.	Martinico.	1790.	——	S.♃.	———— 85. t. 102. f. A.
27 incànum. W.	(*ye.*) hoary.	America.	1811.	7.	H.♃.	Schkuhr fil. t. 11. b.
28 trichomanoídes. W.	(*ye.*) Trichomanes like.	W. Indies.	1822.	5. 8.	S.♃.	———— p. 11. t. 10.
29 decumànum. s.s.	(*ye.*) robust.	Brazil.	1818.	——	S.♃.	
30 juglandifòlium. W.	(*ye.*) Walnut-leaved.	S. America.	1822.	——	S.♃.	
31 prolíferum. s.s.	(*ye.*) proliferous.	Brazil.	1829.	9. 12.	S.♃.	
32 crenàtum. W.	(*ye.*) crenated.	S. America.	1824.	....	S.♃.	
33 fraxinifòlium. W.	(*ye.*) Ash-leaved.	Caracas.	1822.	5. 8.	S.♃.	
34 tenéllum. B.P.	(*ye.*) slender.	N. S. W.	——	5. 7.	S.♃.	
35 tetragònum. s.s.	(*ye.*) four-sided.	W. Indies.	1826.	6. 9.	S.♃.	
36 Phegópteris. E.B.	(*ye.*) sun-fern.	Britain.	....	6. 7.	H.♃.	Eng. bot. v. 31. t. 2224.
37 hexagonópterum. W.	(*y.*) triangular.	N. America.	1811.	——	H.♃.	Pluk. alm. t. 284. f. 2.
38 pruinàtum. W.	(*ye.*) white-leaved.	Jamaica.	1793.	5. 8.	S.♃.	
39 effùsum. W.	(*ye.*) spreading.	————	1769.	11.	S.♃.	Sloan. jam. 1. t. 57. f. 3.
40 Dryópteris. W.	(*ye.*) tender-branched.	Britain.	....	6. 9.	H.♃.	Eng. bot. v. 9. t. 616.
41 connectìle. W.	(*ye.*) connected.	N. America.	1812.	7.	H.♃.	
42 calcàreum. W.	(*ye.*) rigid-branched.	Britain.	....	——	H.♃.	Eng. bot. v. 22. t. 1525.
WO'ODSIA. L.T.	Wo'odsia.	Cryptogamia Filices. L.				
1 pubéscens. s.s.	(*br.ye.*) pubescent.	Brazil.	1826.	1. 12.	S.♃.	
2 glabélla. s.s.	(*br.ye.*) smooth.	N. America.	1827.	6. 9.	H.♃.	
3 hyperbòrea. L.T.	(*br.y.*) hairy Alpine.	Scotland.	....	7. 9.	H.♃.	Eng. bot. v. 29. t. 2023.
4 ilvénsis. L.T.	(*br.ye.*) rock.	N. America.	1812.	6. 7.	H.♃.	Schkuhr fil. t. 19.
5 vestìta. Spr.	(*br.ye.*) clothed.	————	1816.	——	H.♃.	
6 Perriniàna. H.F.	(*br.y.*) Perrin's.	————	1825.	8. 11.	G.♃.	H. et G. ic. fil. t. 68.
Alsóphila Perriniàna. s.s.						
PLEOPE'LTIS. K.S.	Pleope'ltis.	Cryptogamia Filices.				
1 ensifòlia. H.E.F.	(*br.y.*) sword-leaved.	C. B. S.	1823.	5. 6.	S.♃.	Hook. ex. flor. t. 62.
2 latifòlia. L.C.	(*br.ye.*) broad-leaved.	————	——	——	S.♃.	
3 nùda. D.P.	(*br.ye.*) naked.	Nepaul.	——	——	S.♃.	Hook. ex. flor. t. 63.
4 angústa. s.s.	(*br.ye.*) narrow-leaved.	Mexico.	1826.	——	S.♃.	
ASPI'DIUM. B.P.	Shield-fern.	Cryptogamia Filices. W.				
1 Hookèri.	(*br.ye.*) Hooker's.	W. Indies.	1812.	6. 9.	S.♃.	H. et G. ic. fil. t. 96.
prolífera. H.F. *non* B.P.						
2 rhizophy'llum. H.F.	(*br.y.*) rooting-leaved.	Jamaica.	1816.	——	S.♃.	H. et G. ic. fil. t. 59.
3 trifoliàtum. W.	(*br.ye.*) three-leaved.	W. Indies.	1769.	4. 8.	S.♃.	Jacq. ic. 3. t. 638.
4 cicutàrium. W.	(*br.ye.*) Cicuta-leaved.	N. America.	1823.	——	H.♃.	Pluk. alm. t. 289. f. 4.
5 pátens. W.	(*br.ye.*) pubescent.	W. Indies.	1784.	7. 9.	S.♃.	Schkuhr fil. t. 334.
6 dentàtum. W.	(*br.ye.*) toothed.	Scotland.	....	7.	H.♃.	Eng. bot. v. 23. t. 1588.
Cyáthea dentàta. E.B.						
7 Hallèri. W.	(*br.ye.*) Haller's.	Switzerland.	1824.	4. 8.	H.♃.	
8 bulbíferum. W.	(*br.ye.*) bulbiferous.	N. America.	1638.	7. 8.	H.♃.	
9 rhæticum. W.	(*ye.*) stone.	Switzerland.	....	——	H.♃.	
10 frágile. W.	(*ye.*) brittle.	Britain.	....	6. 8.	H.♃.	Eng. bot. v. 23. t. 1587.
11 règium. W.	(*ye.*) laciniated.	————	....	6.	H.♃.	———— v. 3. t. 163.
Cyáthea incìsa. E.B.						
12 'æmulum. W.	(*ye.*) dwarf.	Madeira.	1779.	....	F.♃.	
13 alpìnum. W.	(*ye.*) Alpine.	S. Europe.	1825.	6. 8.	H.♃.	Jacq. ic. 3. t. 642.
14 montànum. W.	(*ye.*) mountain.	————	——	——	H.♃.	
15 irríguum. E.B.	(*ye.*) brook.	Britain.	....	6.	H.♃.	Eng. bot. v. 31. t. 2199.
16 atomàrium. W.	(*ye.*) small.	N. America.	1822.	6. 8.	H.♃.	
NEPHR'ODIUM. B.P.	Nephr'odium.	Cryptogamia Filices.				
1 Lonchìtis. H.S.L.	(*ye.*) rough Alpine.	Britain.	....	5. 8.	H.♃.	Eng. bot. v. 12. t. 797.
Aspídium Lonchìtis. E.B.						
2 auriculàtum. H.S.L.	(*ye.*) eared.	E. Indies.	1793.	7.	S.♃.	Burm. zeyl. t. 44. f. 2.
3 acrostichoídes. M.	(*ye.*) Acrostichum-like.	N. America.	1824.	5. 8.	H.♃.	Schkuhr fil. t. 30.
4 mucronàtum. s.s.	(*ye.*) mucronate.	Jamaica.	1826.	——	S.♃.	
5 exaltàtum. B.P.	(*ye.*) lofty.	————	1793.	5. 8.	S.♃.	Schkuhr fil. t. 32. b.
6 ùnitum. B.P.	(*ye.*) smooth united.	E. Indies.	——	——	S.♃.	Burm. zeyl. t. 44. f. 1.
7 propínquum. B.P.	(*ye.*) pubescent.	————	——	——	S.♃.	

8	mólle. B.P.	(*ye.*)	soft.	N. S. W.	1822.	5. 8.	G.♃.	Jacq. ic. t. 640.
9	sesquipedàle. S.S.	(*ye.*)	black-spotted.	W. Indies.	1824.	——	S.♃.	
10	hirsùtulum. S.S.	(*ye.*)	hairy.	E. Indies.	1823.	——	S.♃.	
11	èdule. D.P.	(*ye.*)	eatable-rooted.	Nepaul.	1826.	——	G.♃.	
12	semicordàtum. S.S.	(*ye.*)	half cordate.	W. Indies.	1822.	——	S.♃.	
13	noveboracénse. H.S.L.	(*y.*)	river-side.	N. America.	1812.	7.	H.♃.	Schkuhr fil. t. 46.
14	pectinàtum.	(*ye.*)	comb-like.	....	1823.	5. 8.	S.♃.	———— t. 296.
	Aspídium trapezoídes. Schk.							
15	Oreópteris. H.S.L.	(*ye.*)	Heath.	Britain.	....	7.	H.♃.	Eng. bot. v. 15. t. 1019.
16	Thely'pteris. H.S.L.	(*ye.*)	Lady fern.	————	....	7. 8.	H.♃.	———— v. 15. t. 1018.
17	sérra. S.S.	(*ye.*)	saw-like.	Jamaica.	1819.	——	S.♃.	
18	parasìticum. W.	(*ye.*)	parasitical.	E. Indies.	1824.	——	S.♃.	
19	Goldiànum. H.F.	(*ye.*)	Goldie's.	N. America.	1822.	——	H.♃.	H. et G. ic. fil. t. 102.
20	cristàtum. H.S.L.	(*ye.*)	crested heath.	Britain.	....	6. 7.	H.♃.	Eng. bot. v. 30. t. 2125.
21	aculeàtum. H.S.L.	(*ye.*)	common prickly.	————	....	6. 8.	H.♃.	———— v. 22. t. 1562.
22	pubéscens. H.F.	(*ye.*)	pubescent.	Jamaica.	1824.	——	S.♃.	H. et G. ic. fil. t. 162.
23	marginàle. H.S.L.	(*ye.*)	marginal-flower'd.	N. America.	1772.	6. 9.	H.♃.	Schkuhr fil. t. 45. b.
24	Fílix-más. H.S.L.	(*ye.*)	Male fern.	Britain.	....	6. 8.	H.♃.	Eng. bot. v. 21. t. 1458.
	β *Smíthii.*	(*ye.*)	*Smith's.*	————	....	——	H.♃.	———— v. 28. t. 1949.
	Aspídium cristàtum. E.B. 1949. *non* 2125.							
	γ *eròsum.* Ph.	(*ye.*)	*jagged.*	N. America.	1812.	——	H.♃.	Schkuhr fil. t. 45.
25	lancastriénse. S.S.	(*ye.*)	Lancaster.	————	1825.	——	H.♃.	
26	fràgrans. H.F.	(*ye.*)	fragrant.	Europe.	....	——	H.♃.	H. et G. ic. fil. t. 70.
27	platyphy'llum. S.S.	(*ye.*)	flat-leaved.	S. America.	1826.	——	S.♃.	
28	lobàtum. H.S.L.	(*ye.*)	close-leaved.	England.	....	6. 9.	H.♃.	Eng. bot. v. 22. t. 1563.
29	intermèdium. W.	(*ye.*)	intermediate.	N. America.	1816.	——	H.♃.	
30	spinulòsum. H.S.L.	(*ye.*)	crested prickly.	Britain.	....	6. 8.	H.♃.	Eng. bot. v. 21. t. 1460.
31	dilatàtum. H.S.L.	(*ye.*)	great crested.	————	....	——	H.♃.	———— v. 21. t. 1461.
32	obtùsum. W.	(*ye.*)	blunt leaved.	N. America.	1827.	——	H.♃.	Schkuhr fil. t. 21.
33	rígidum. W.	(*ye.*)	rigid.	Germany.	1816.	——	H.♃.	
34	elongàtum. H.S.L.	(*ye.*)	cut-leaved.	Madeira.	1799.	7. 8.	F.♃.	
35	villòsum. H.S.L.	(*ye.*)	villous.	W. Indies.	1793.	7.	S.♃.	Schkuhr fil. t. 46. b.
36	decompósitum. B.P.	(*ye.*)	decompound.	N. S. W.	1820.	6. 8.	F.♃.	
37	? Bàromez.	....	Baromez.	Tartary.	1824.	....	S.♃.	
	ALLANT'ODIA. B.P.		ALLANT'ODIA.	Cryptogamia Filices.				
1	umbròsa. B.P.	(*ye.*)	Madeira.	Madeira.	1779.	6. 9.	F.♃.	Schkuhr fil. t. 61.
2	axillàris. S.S.	(*ye.*)	slender.	————	——	——	F.♃.	
	Polypòdium axillàre. H.K. *Aspídium axillàre.* Sw.							
3	austràlis. B.P.	(*ye.*)	southern.	V. Diem. Isl.	1820.	——	F.♃.	
4	ténera. B.P.	(*ye.*)	membranaceous.	N. S. W.	——	——	F.♃.	
	ASPL'ENIUM. B.P.		SPLEENWORT.	Cryptogamia Filices. L.				
1	acùtum. S.S.	(*br.*)	acute.	Canaries.	1818.	4. 7.	F.♃.	
2	físsum. S.S.	(*br.*)	cleft-leaved.	Hungary.	1825.	——	H.♃.	
3	fontànum. B.P.	(*st.*)	smooth rock.	England.	....	6. 8.	H.♃.	Eng. bot. v. 29. t. 2024.
	Aspídium fontànum. E.B.							
4	Fìlix-fémina. B.P.	(*st.*)	female fern.	Britain.	....	6. 9.	H.♃.	Eng. bot. v. 21. t. 1459.
5	Michaùxii. S.S.	(*st.*)	Michaux's.	N. America.	1812.	——	H.♃.	
6	Athy'rium. S.S.	(*st.*)	American.	————	1820.	——	H.♃.	Schkuhr filic. t. 78.
	Nephròdium asplenioídes. M.							
7	Hallèri. S.S.	(*st.*)	Haller's.	Switzerland.	1825.	——	H.♃.	
8	Adiántum-nìgrum. W.	(*br.*)	black.	Britain.	....	4. 10.	H.♃.	Eng. bot. v. 28. t. 1950.
9	lanceolàtum. E.B.	(*br.*)	lanceolate.	England.	....	6. 10.	H.♃.	———— v. 4. t. 240.
10	rádicans. S.S.	(*br.*)	rooting-branched.	W. Indies.	1822.	——	S.♃.	
11	bisséctum. S.S.	(*br.*)	bifid.	Jamaica.	1821.	——	S.♃.	
12	montànum. W.	(*br.*)	mountain.	N. America.	1812.	7.	H.♃.	
13	fràgrans. W.	(*br.*)	fragrant.	Jamaica.	1793.	——	S.♃.	
14	Rùta murària. W.	(*br.*)	Wall-rue.	Britain.	....	6. 10.	H.♃.	Eng. bot. v. 3. t. 150.
15	cicutàrium. H.S.	(*br.*)	Cicuta-like.	S. America.	1822.	——	S.♃.	Plum. fil. t. 48. a.
16	formòsum. W.	(*br.*)	handsome.	Santa Cruz.	——	——	S.♃.	
17	præmórsum. W.	(*br.*)	snip-leaved.	Jamaica.	1793.	——	S.♃.	Pluk. alm. t. 73. f. 5.
18	striàtum. W.	(*br.*)	striated.	W. Indies.	——	6. 8.	S.♃.	Plum. fil. t. 18. 19.
19	thelypteroídes. W.	(*br.*)	Ladyfern-like.	N. America.	1823.	7.	H.♃.	Schkuhr fil. t. 76. b.
20	sulcàtum. S.S.	(*br.*)	furrowed.	W. Indies.	1827.	——	S.♃.	Plum. fil. t. 46.
21	pùmilum. S.S.	(*br.*)	dwarf.	————	1823.	——	S.♃.	———— t. 66. A.
22	canariénse. S.S.	(*br.*)	Canary Island.	Canaries.	1824.	——	G.♃.	
23	rhizóphorum. W.	(*br.*)	root-bearing.	Jamaica.	1793.	6. 8.	S.♃.	Sloan. j. 1. t. 29. 30. f. 1.
24	víride. W.	(*br.*)	green.	Britain.	....	6. 9.	H.♃.	Eng. bot. v. 32. t. 2257.
25	melanocàulon. W.	(*br.*)	black-stalked.	N. America.	1812.	7.	H.♃.	
26	Trichómanes. W.	(*br.*)	Maiden hair.	Britain.	....	5. 10.	H.♃.	Eng. bot. v. 8. t. 576.

27 germánicum. W.		alternate-leaved.	Scotland.	....	6. 10.	H.♃.	Eng. bot. v. 32. t. 2258.
alternifòlium. E.B. *Brèynii.* Sw.							
28 ebèneum. W.	(*br.*)	ivory striped.	N. America.	1779.	9.	H.♃.	Lodd. bot. cab. t. 5.
trichomanoìdes. M.							
29 bipartítum. S.S.	(*br.*)	two-parted.	W. Indies.	1822.	——	S.♃.	
30 dimidiàtum. S.S.	(*br.*)	narrow-leaved.	———	1827.	——	S.♃.	
31 aùritum. S.S.	(*br.*)	eared.	———	1829.	9. 12.	S.♃.	
32 zamiæfòlium. S.S.	(*br.*)	Zamia-leaved.	S. America.	1818.	6. 10.	S.♃.	Lodd. bot. cab. t. 854.
33 falcàtum. B.P.	(*br.*)	sickle-leaved.	N. S. W.	1825.	6. 7.	F.♃.	
34 monánthemum. W.	(*br.*)	one-flowered.	C. B. S.	1790.	7.	G.♃.	Sm. ined. t. 73.
35 reséctum. S.S.	(*br.*)	cut-leaved.	Mauritius.	1820.	——	S.♃.	H. et G. ic. fil t. 114.
36 marìnum. W.	(*br.*)	sea-side.	Britain.	....	6. 10.	H.♃.	Eng. bot. v. 6. t. 392.
37 dentàtum. H.F.	(*br.*)	tooth-leaved.	Jamaica.	1824.	——	S.♃.	H. et G. ic. fil. t. 72.
38 flabellifòlium. W.	(*br.*)	fan-leaved.	N. S. W.	1825.	6. 7.	G.♃.	Hook. ex. flor. t. 208.
39 déntex. W.	(*br.*)	toothed.	C. B. S.	1790.	6. 10.	G.♃.	
40 diffórme. B.P.	(*br.*)	deformed.	N. S. W.	1822.	6. 7.	F.♃.	
41 Shephérdii. S.S.	(*br.*)	Shepherd's.	Jamaica.	1824.	6. 9.	S.♃.	
42 angustifòlium. S.S.	(*br.*)	narrow-leaved.	N. America.	1812.	——	H.♃.	Schkuhr fil. t. 69.
43 septentrionàle. W.	(*br.*)	forked.	Britain.	....	6. 10.	H.♃.	Eng. bot. v. 15. t. 1017.
44 palmàtum. S.S.	(*br.*)	palmated.	S. Europe.	1816.	7. 10.	F.♃.	Lodd. bot. cab. t. 868.
45 attenuàtum. B.P.	(*br.*)	slender.	N. S. W.	1825.	6. 7.	F.♃.	
46 rhizophy'llum. W.	(*br.*)	rooting-leaved.	N. America.	1680.	——	H.♃.	
47 serràtum. W.	(*br.*)	saw-leaved.	W. Indies.	1793.	——	S.♃.	Schkuhr fil. t. 64.
48 Nìdus. B.P.	(*br.*)	nest.	N. Holland.	1822.	——	F.♃.	
CÆNO'PTERIS. S.S.		CÆNO'PTERIS.	Cryptogamia Filices.				
1 Odontìtes. S.S.	(*br.*)	long-leaved.	N. Holland.	1822.	6. 10.	G.♃.	Thunb. n. ac. pet. t. E. f. 1.
Asplènium Odontìtes. B.P. *Dàrea Odontìtes.* W.							
2 appendiculàta. S.S.	(*br.*)	loose.	V. Diem. Isl.	——	——	G.♃.	Lab. n. holl. 2. t. 243.
3 rhizophy'lla. S.S.	(*br.*)	rooting-leaved.	W. Indies.	1827.	——	S.♃.	
SCOLOPE'NDRIUM. W.		HART'S-TONGUE.	Cryptogamia Filices. W.				
1 officinàrum. W.	(*br.ye.*)	common.	Britain.	....	7. 8.	H.♃.	Eng. bot. v. 16. t. 1150.
β *críspum.*	(*br.ye.*)	*curled-leaved.*	———	....	——	H.♃.	
γ *undulàtum.*	(*br.ye.*)	*waved-leaved.*	———	....	——	H.♃.	Pluk. phyt. 248. f. 1.
δ *multífidum.*	(*br.ye.*)	*clustered.*	———	....	——	H.♃.	
ε *ramòsum.*	(*br.ye.*)	*branching.*	———	....	——	H.♃.	Pluk. phyt. 248. f. 2.
2 Hemionìtis. S.S.	(*br.*)	Mule's fern.	Canaries.	1779.	6. 7.	G.♃.	Schkuhr crypt. t. 66.
Asplènium Hemionìtis. L.							
DO'ODIA. B.P.		DO'ODIA.	Cryptogamia Filices.				
1 áspera. B.P.	(*br.ye.*)	rough-stalked.	N. S. W.	1808.	3. 9.	G.♃.	Hook. ex. flor. t. 8.
2 caudàta. B.P.	(*br.ye.*)	tailed.	———	1822.	——	G.♃.	——— t. 25.
WOODWA'RDIA. W.		WOODWA'RDIA.	Cryptogamia Filices. W.				
1 onocleoídes. W.	(*br.ye.*)	Onoclea-like.	N. America.	1812.	8. 10.	H.♃.	Schkuhr fil. t. 111.
2 virgínica. W.	(*br.ye.*)	Virginian.	———	1774.	8. 9.	H.♃.	Pluk. alm. t. 179. f. 2.
3 rádicans. W.	(*br.ye.*)	rooting-leaved.	Madeira.	1779.	9. 10.	G.♃.	Schkuhr fil. t. 112.
BLE'CHNUM. W.		BLE'CHNUM.	Cryptogamia Filices. W.				
1 brasiliénse. S.S.	(*br.ye.*)	Brazilian.	Brazil.	1828.	9. 11.	S.♃.	
2 cartilagíneum. B.P.	(*br.*)	cartilaginous.	N. S. W.	1822.	6. 8.	F.♃.	
3 denticulàtum. S.S.	(*br.y.*)	toothed.	Teneriffe.	1826.	6. 9.	F.♃.	
4 lævigatum. B.P.	(*br.*)	smooth.	N. S. W.	1820.	3. 9.	G.♃.	
5 angustifòlium. S.S.	(*br.y.*)	narrow-leaved.	Trinidad.	1824.	——	S.♃.	
6 longifòlium. S.S.	(*br.*)	long-leaved.	———	1826.	——	S.♃.	Bot. mag. t. 2818.
7 glandulòsum. S.S.	(*br.*)	glandular.	Brazil.	1823.	——	S.♃.	
8 occidentàle. W.	(*br.*)	American.	S. America.	1777.	——	S.♃.	Jacq. ic. 3. t. 644.
9 australe. W.	(*br.*)	Cape.	C. B. S.	1691.	——	G.♃.	Schkuhr fil. t. 110. b.
10 polypodioídes. S.S.	(*br.*)	Polypody-leav'd.	Brazil.	1829.	8. 12.	S.♃.	
11 Lancèola. S.S.	(*br.*)	spear-leaved.	———	———	——	S.♃.	
LOM'ARIA. W.		LOM'ARIA.	Cryptogamia Filices. W.				
1 nùda. W.	(*br.*)	naked.	V. Diem. Isl.	1822.	4. 8.	F.♃.	Lab. n. hol. 2. t. 246.
Onoclèa nùda. Lab. *Stegània nùda.* B.P.							
2 Spìcant. S.S.	(*br.*)	northern.	Britain.	....	7. 9.	H.♃.	Eng. bot. v. 17. t. 1159.
Osmúnda Spìcant. L. *Bléchnum boreàle.* E.B.							
3 onocleoídes. S.S.	(*br.*)	Onoclea-like.	Jamaica.	1824.	——	S.♃.	
4 sorbifòlia. S.S.	(*br.*)	Service-leaved.	W. Indies.	1793.	——	S.♃.	Plum. fil. t. 117.
Acróstichum sorbifòlium. L.							
5 longifòlia. S.S.	(*br.*)	long-leaved.	———	1810.	6. 9.	S.♃.	Plum. fil. t. 117. dextr.
6 pròcera. S.S.	(*br.*)	lofty.	N. Holland.	1822.	4. 8.	F.♃.	Lab. n. hol. 2. t. 247.
Bléchnum pròcerum. Lab. *Stegània pròcera.* B.P.							
ONOCL'EA. W.		ONOCL'EA.	Cryptogamia Filices. L.				
1 sensíbilis. W.	(*br.*)	sensitive fern.	Virginia.	1799.	8.	H.♃.	Schkuhr fil. t. 102.
2 obtusilobàta. Ph.	(*br.*)	blunt-lobed.	N. America.	1812.	7.	H.♃.	——— t. 103.

STRUTHIO'PTERIS. W.		STRUTHIO'PTERIS.	Cryptogamia Filices.				
1 germánica. s.s.	(*br.*)	German.	Europe.	1760.	7. 8.	H. ♃.	Schkuhr fil. t. 105.
Osmúnda Struthiópteris. L. *Onoclèa Struthiópteris.* W.							
2 pensylvánica. w.	(*br.*)	Pensylvanian.	N. America.	1812.	——	H. ♃.	Schkuhr fil. t. 104.
Onoclèa nodulòsa. Schkuhr.							
VITT'ARIA. W.		VITT'ARIA.	Cryptogamia Filices. W.				
lineàta. w.	(*br.*)	linear-leaved.	America.	1793.	....	S. ♃.	Schkur fil. t. 101. b.
ANTR'OPHYUM. S.S.		ANTR'OPHYUM.	Cryptogamia Filices.				
lanceolàtum. s.s.	(*br.*)	spear-leaved.	W. Indies.	1793.	6. 10.	S. ♃.	Schkuhr fil. t. 6.
Hemionìtis lanceolàta. L.							
DIPL'AZIUM. W.		DIPL'AZIUM.	Cryptogamia Filices. W.				
1 plantagíneum. s.s.	(*br.*)	Plantain-leav'd.	W. Indies.	1822.	....	S. ♃.	
2 grandifòlium. w.	(*br.*)	large-leaved.	Jamaica.	1793.	8. 10.	S. ♃.	
3 juglandifòlium. s.s.	(*br.*)	Walnut-leaved.	————	1825.	....	S. ♃.	
4 seramporénse. s.s.	(*br.*)	East Indian.	E. Indies.	1827.	....	S. ♃.	
5 auriculàtum. s.s.	(*br.*)	eared.	S. America.	1820.	8. 10.	S. ♄.	
6 barbadénse. w.	(*br.*)	Barbadoes.	W. Indies.	1822.	....	S. ♃.	
7 malabáricum. s.s.	(*br.*)	Malabar.	E. Indies.	1818.	6. 8.	S. ħ.	
Asplènium ambíguum. Swartz.							
8 arboréscens. s.s.	(*br.*)	arborescent.	Mauritius.	1824.	....	S. ♄.	
9 esculéntum. w.	(*br.*)	eatable.	Ceylon.	1822.	....	S. ħ.	
CRYPTOGRA'MMA. Br.		CRYPTOGRA'MMA.	Cryptogamia Filices.				
1 acrostichoídes. Br.	(*st.*)	Acrostichum-like.	N. America.	1827.	7. 9.	H. ♃.	H. et G. ic. fil. t. 29.
Allosòrus acrostichoídes. s.s.							
2 críspa. Br.	(*st.*)	curled.	Britain.	....	——	H. ♃.	Eng. bot. v. 17. t. 1160.
Allosòrus críspus. s.s. *Ptèris críspa.* E.B. *Osmúnda críspa.* L.							
PT'ERIS. W.		BRAKE.	Cryptogamia Filices. L.				
1 pedàta. w.	(*st.*)	pedate.	W. Indies.	1816.	5. 8.	S. ♃.	Pet. fil. 176. t. 8. f. 12.
2 palmàta. w.	(*br.st.*)	palmated.	————	1823.	6. 7.	S. ♃.	Lodd. bot. cab. t. 1299.
3 argéntea. w.	(*st.br.*)	silvery.	Siberia.	——	——	H. ♃.	
4 longifòlia. w.	(*st.br.*)	long-leaved.	W. Indies.	1770.	7. 9.	S. ♃.	Jacq. sch. 3. t. 339.400.
5 grandifòlia. w.	(*st.br.*)	large-leaved.	————	1793.	....	S. ♃.	Schkuhr fil. t. 89.
6 serrulàta. w.	(*st.br.*)	various-leaved.	India.	1770.	8. 9.	S. ♃.	———— t. 91.
7 denticulàta. s.s.	(*st.br.*)	toothed.	Brazil.	1824.	5. 9.	S. ♃.	H. et G. ic. fil. t. 28.
8 crética. w.	(*st.br.*)	Cretan.	Candia.	1818.	5. 7.	G. ♃.	
9 atropurpùrea. w.	(*br.*)	dark purple.	N. America.	1770.	8. 9.	H. ♃.	Schkuhr fil. t. 101.
10 nemoràlis. W. en.	(*st.br.*)	wood.	Africa.	1823.	5. 8.	S. ♃.	
11 biaùrita. s.s.	(*st.br.*)	two-eared.	W. Indies.	1824.	——	S. ♃.	H. et G. ic. fil. t. 142.
12 Plumíeri. s.s.	(*st.br.*)	Plumier's.	S. America.	1818.	——	S. ♃.	
13 argùta. w.	(*st.br.*)	sharp-notched.	Madeira.	1778.	8. 9.	F. ♃.	Pluk. alm. t. 290. f. 2.
14 umbròsa. B.P.	(*st.br.*)	shade.	N. S. W.	1823.	5. 8.	F. ♃.	
15 cordàta. w.	(*st.br.*)	heart-leaved.	Mexico.	1825.	——	G. ♃.	
16 sagittàta. w.	(*st.br.*)	arrow-leaved.	S. America.	1820.	——	S. ♃.	
17 vespertiliònis. B.P.	(*st.*)	Bat-winged.	N. S. W.	1823.	——	F. ♃.	Lab. nov. hol. 2. t. 245.
18 trémula. B.P.	(*st.br.*)	trembling.	————	1822.	——	F. ♃.	
19 esculénta. B.P.	(*st.br.*)	esculent.	————	1815.	——	F. ♃.	Lab. nov. hol. 2. t. 244.
20 grácilis. w.	(*st.br.*)	slender.	Canada.	1823.	7.	H. ♃.	
21 hastàta. w.	(*st.br.*)	hastate.	W. Indies.	——	5. 8.	S. ♃.	
22 aculeàta. w.	(*st.br.*)	prickly-stem'd.	————	1793.	....	S. ħ.	Plum. fil. t. 5. et 11.
23 falcàta. B.P.	(*st.br.*)	falcate.	N. S. W.	1820.	5. 7.	F. ♃.	
24 caudàta. w.	(*st.br.*)	American.	N. America.	1777.	9. 12.	H. ♃.	Jacq. ic. 3. t. 645.
25 aquilìna. E.B.	(*st.br.*)	common.	Britain.	....	7. 8.	H. ♃.	Eng. bot. v. 24. t. 1679.
26 lanuginòsa. w.	(*st.br.*)	woolly.	Bourbon.	1818.	——	S. ♃.	
27 intramarginàlis. K.F.	(*br.*)	Mexican.	Mexico.	1828.	7. 12.	S. ♃.	
TÆN'ITIS. W.		TÆN'ITIS.	Cryptogamia Filices. W.				
1 lanceolàta. s.s.	(*br.*)	spear-leaved.	W. Indies.	1818.	6. 9.	S. ♃.	Plum. fil. t. 132.
2 angustifòlia. s.s.	(*br.*)	narrow-leaved.	————	1823.	——	S. ♃.	
3 graminifòlia. H.E.F.	(*br.*)	grass-leaved.	Trinidad.	1820.	6. 8.	S. ♃.	Hook. ex. flor. t. 77.
4 furcàta. s.s.	(*br.*)	forked.	————	1824.	——	S. ♃.	H. et G. ic. fil. t. 7.
5 chinénsis. s.s.	(*br.*)	Chinese.	China.	1828.	——	G. ♃.	
ADIA'NTUM. W.		MAIDENHAIR.	Cryptogamia Filices. L.				
1 renifórme. w.	(*br.*)	kidney-leaved.	Madeira.	1699.	6. 9.	G. ♃.	Lodd. bot. cab. t. 841.
2 oblíquum. s.s.	(*br.*)	oblique-leaved.	W. Indies.	1826.	4. 8.	S. ♃.	
3 macrophy'llum. w.	(*br.*)	large-leaved.	Jamaica.	1793.	——	S. ♃.	H. et G. ic. fil. t. 132.
4 lunulàtum. s.s.	(*br.*)	lunulate.	E. Indies.	1826.	——	S. ♃.	———— t. 104.
5 deltoídeum. s.s.	(*br.*)	deltoid-leaved.	Jamaica.	1824.	——	S. ♃.	
6 denticulàtum. s.s.	(*br.*)	toothed-leaved.	W. Indies.	1825.	——	S. ♃.	
7 serrulàtum. w.	(*br.*)	saw-leaved.	Jamaica.	1823.	6. 8.	S. ♃.	Sloan. j. 17. t. 36. f. 2.
8 radiàtum. w.	(*br.*)	radiated.	W. Indies.	1776.	4. 8.	S. ♃.	Plum. fil. t. 100.
9 pedàtum. w.	(*br.*)	Canadian.	N. America.	1640.	8. 9.	H. ♃.	Schkuhr fil. t. 115.

10 hispídulum. B.P.	(br.)	bristly.	N. Holland.	1822.	5. 8.	G.♃.	
11 striàtum. S.S.	(br.)	streaked.	W. Indies.	——	——	S.♃.	
12 villòsum. W.	(br.)	hairy-stalked.	Jamaica.	1775.	6. 9.	S.♃.	Schkuhr fil. t. 120.
13 pulverulèntum. W.	(br.)	dusty.	W. Indies.	1793.	....	S.♃.	———— t. 119.
14 triangulàtum. S.S.	(br.)	triangular.	Trinidad.	1824.	——	S.♃.	
15 Capíllus-véneris. W.	(br.)	true.	Britain.	....	5. 9.	H.♃.	Eng. bot. v. 22. t. 1564.
16 trapezifórme. W.	(br.)	rhomb-leaved.	————	1793.	....	S.♃.	Schkuhr fil. t. 122.
17 pentadáctylon. S.S.	(br.)	five-fingered.	Brazil.	1828.	——	S.♃.	H. et G. ic. fil. t. 98.
18 cuneàtum. S.S.	(br.)	wedge-leaved.	————	1824.	——	S.♃.	——— ic. fil. t. 30.
19 assimile. B.P.	(br.)	likened.	N. S. W.	1823.	5. 8.	G.♃.	Lab.n.hol.2.t.248.f.2.
trigònum. Lab.							
20 formòsum. B.P.	(br.)	handsome.	————	1822.	——	G.♃.	
21 ténerum. W.	(br.)	tender.	Jamaica.	1793.	7.	S.♃.	Pluk. alm. t. 254. f. 1.
CHEILA'NTHES. W.		CHEILA'NTHES.	Cryptogamia Filices. W.				
1 ferrugínea. S.S.	(br.)	ferruginous.	Peru.	1816.	5. 8.	S.♃.	
2 grácilis. S.S.	(br.)	slender.	Canada.	1827.	——	H.♃.	
3 pteroídes. S.S.	(br.)	Pteris-like.	C. B. S.	1775.	7. 9.	G.♃.	Hout. n. hist. t. 96. f. 3.
4 odòra. S.S.	(br.)	sweet-scented.	S. Europe.	1824.	——	F.♃.	
5 suavèolens. S.S.	(br.)	scented.	Madeira.	1778.	——	G.♃.	
6 fràgrans. S.S.	(br.)	fragrant.	E. Indies.	——	——	S.♃.	Sw. syn. fil. t. 3. f. 6.
7 microphy'lla. S.S.	(br.)	small-leaved.	W. Indies.	1822.	5. 8.	S.♃.	
8 vestìta. S.S.	(br.)	clothed.	N. America.	1812.	7.	H.♃.	Schkuhr fil. t. 124.
9 hírta. S.S.	(br.)	hairy.	C. B. S.	1816.	5. 8.	G.♃.	
10 lendigera. W.	(br.)	chaffy.	S. America.	1822.	——	S.♃.	
11 tenuifòlia. B.P.	(br.)	slender-leaved.	N. S. W.	1824.	——	G.♃.	
12 spectábilis. S.S.	(br.)	showy.	Brazil.	1829.	9. 12.	S.♃.	
13 rèpens. S.S.	(br.)	creeping.	W. Indies.	1824.	6. 10.	S.♃.	
LONCH'ITIS. W.		LONCH'ITIS.	Cryptogamia Filices. W.				
1 pedàta. S.S.	(br.)	pedate.	Jamaica.	1793.	6. 8.	S.♃.	Brown. jam. 89. t. 1.
Ptèris podophy'lla. W.							
2 hirsùta. S.S.	(br.)	hairy.	W. Indies.	——	....	S.♃.	Schkuhr fil. t. 86.
LINDS'ÆA. B.P.		LINDS'ÆA.	Cryptogamia Filices. W.				
1 renifórmis. L.T.	(br.)	kidney-leaved.	Trinidad.	1826.	5. 7.	S.♃.	Linn. trans. v.3. t.7. f.1.
2 lineàris. B.P.	(br.)	linear-leaved.	N. S. W.	1820.	4. 6.	G.♃.	Sw. fil. 118. t. 3. f. ?.
3 falcàta. L.T.	(br.)	sickle-leaved.	————	1824.	5. 8.	S.♃.	Linn. trans. v. 3.t.7.f.1.
4 trapezifórmis. L.T.	(br.)	rhomb-leaved.	W. Indies.	1825.	——	S.♃.	———— v. 3. t. 9.
5 microphy'lla. B.P.	(br.)	small-leaved.	N. S. W.	1820.	4. 6.	G.♃.	
DAVA'LLIA. W.		DAVA'LLIA.	Cryptogamia Filices. W.				
1 pyxidàta. B.P.	(ye.)	shining.	N. S. W.	1808.	4. 9.	G.♄.	
2 canariénsis. W.	(ye.)	Hare's-foot fern.	Canaries.	1699.	——	G.♃.	Jacq. ic. 1. t. 200.
3 gibberòsa. S.S.	(br.ye.)	gibbous-flower'd.	N. Zealand.	1825.	——	G.♃.	Swt. flor. aust. t. 31.
4 fumarioídes. S.S.	(br.ye.)	Fumitory-like.	W. Indies.	1828.	6. 10.	S.♃.	
5 fláccida. B.P.	(br.ye.)	flaccid.	N. Holland.	1826.	——	G.♃.	
BALA'NTIUM. K.F.		BALA'NTIUM.	Cryptogamia Filices.				
Cúlcita. K.S.	(br.ye.)	shining-leaved.	Madeira.	1779.	9. 10.	F.♃.	
Dicksònia Cúlcita. W.							
DICKS'ONIA. W.		DICKS'ONIA.	Cryptogamia Filices. W.				
1 pilosiúscula. W.	(br.ye.)	hairy.	N. America.	1811.	7. 9.	H.♃.	Schkuhr fil. t. 131.
2 arboréscens. W.	(br.ye.)	tree.	S. Helena.	1786.	6. 12.	S.♄.	Jacq. ic. t. 201.
3 dissécta. W.	(br.ye.)	cut-leaved.	Jamaica.	1793.	9. 10.	S.♃.	
4 adiantoídes. S.S.	(br.st.)	Adiantum-like.	W. Indies.	1828.	8. 12.	S.♃.	
CYATH'EA. B.P.		CYATH'EA.	Cryptogamia Filices. W.				
1 arbòrea. W.	(br.ye.)	tree.	W. Indies.	1793.	....	S.♃.	Plum. fil. 1. t. 1. 2.
2 excélsa. S.S.	(br.ye.)	tall.	Mauritius.	1825.	....	S.♄.	
ALSO'PHILA. B.P.		ALSO'PHILA.	Cryptogamia Filices.				
1 austràlis. B.P.	(br.st.)	southern.	N. S. W.	1820.	....	G.♄.	
2 áspera. S.S.	(br.ye.)	rough.	W. Indies.	1824.	....	S.♄.	
HEMIT'ELIA. B.P.		HEMIT'ELIA.	Cryptogamia Filices.				
multiflòra. S.S.	(br.ye.)	many-flowered.	Jamaica.	1824.	....	S.♄.	
TRICHO'MANES. W.		TRICHO'MANES.	Cryptogamia Filices. W.				
1 membranàceum. W.	(br.)	membranaceous.	W. Indies.	1820.	5. 8.	S.♃.	Hook. ex. flor. t. 76.
2 floribúndum. S.S.	(br.ye.)	many-flowered.	————	1825.	——	S.♃.	H. et G. ic. fil. t. 9.
3 brevisètum. H.K.	(br.ye.)	short-styled.	Britain.	....	5. 6.	H.♃.	Eng. bot. v. 20. t. 1417
Hymenophy'llum alàtum. E.B.							
4 alàtum. S.S.	(br.ye.)	wing-leaved.	W. Indies.	1824.	——	S.♃.	H. et G. ic. fil. t. 11.
5 críspum. S.S.	(br.ye.)	curled.	————	1828.	——	S.♃.	———— ic. fil. t. 12.
HYMENOPHY'LLUM.		FILMY-LEAF.	Cryptogamia Filices. W.				
1 polyánthos. S.S.	(br.ye.)	many-flowered.	W. Indies.	1824.	6. 9.	S.♃.	H. et G. ic. fil. t. 128.
2 tunbridgénse. W.	(br.)	Tunbridge.	Britain.	....	5. 6.	H.♃.	Eng. bot. v. 3. t. 162.
3 hirsùtum. S.S.	(br.ye.)	hairy.	Trinidad.	1823.	5. 9.	S.♃.	H. et G. ic. fil. t. 84.

ADDENDA et CORRIGENDA.

RANUNCULACEÆ. *p.* 1.

CLE'MATIS. p. 1.		VIRGIN'S BOWER.					
grandiflòra. p. 1.	(*gr.*)	large-flowered.	Sierra Leone.	1822.	7. 9.	S. ♄. ◡.	Bot. reg. t. 1234.
chlorántha. B. Reg.							
43 viornoídes. Jac.	(*li.*)	Viorna-like.	N. America.	1826.	6. 9.	H. ♄. ◡.	
ANEM'ONÆ. p. 3.		ANEM'ONE.					
26 Richardsòni. H.A.	(*ye.*)	Richardson's.	N. America.	1827.	4. 7.	H. ♃.	Hook. fl. b. am. t. 4. A.
27 deltoídea. H.A.	(*wh.*)	deltoid.	Columbia.	——	——	H. ♃.	—————— t. 3. A.
RANU'NCULUS. p. 4.		CROWFOOT.					
98 rhomboídeus. H.A.	(*ye.*)	rhomboid-leav'd.	N. America.	1825.	5. 6.	H. ♃.	Edin. phi. jou. 6. t. 11. f. 1.
99 glabérrimus. H.A.	(*ye.*)	smoothest.	——	1827.	——	H. ♃.	Hook. fl. b. am. t. 5. A.
100 ovàlis. H.A.	(*ye.*)	oval-leaved.	——	——	——	H. ♃.	—————— t. 6. B.
101 brevicàulis. H.A.	(*ye.*)	short-stalked.	——	——	——	H. ♃.	—————— t. 7. A.
102 cardiophy'llus. H.A.	(*y.*)	heart-leaved.	——	——	——	H. ♃.	Bot. mag. t. 2999.
103 pygm'æus. DC.	(*ye.*)	pygmy.	——	——	——	H. ♃.	Wahl. lapp. t. 8. f. 1.
104 Eschschóltzii. H.A.	(*ye.*)	Eschscholtz's.	——	——	——	H. ♃.	
105 pedatífidus. DC.	(*ye.*)	pedatifid.	——	——	——	H. ♃.	
106 recurvàtus. DC.	(*ye.*)	recurved.	——	——	——	H. ♃.	
107 Schlechtendàlii. H.A.	(*ye.*)	Schlechtendahl's.	——	——	——	H. ♃.	Schlec. an. s. 2. t. 2.
fasciculàris. Schlec. *non* Muhl.							
CA'LTHA. p. 7.		MARSH MARYGOLD.					
9 biflòra. H.A.	(*wh.*)	two-flowered.	N. America.	1827.	5. 7.	H. *w.* ♃.	
10 leptosépala. H.A.	(*wh.*)	slender-sepaled.	——	——	——	H. *w.* ♃.	Hook. fl. b. am. t. 10.
CO'PTIS. p. 7.		CO'PTIS.					
2 aspleniifòlia. H.A.	(*wh.*)	Fern-leaved.	N. America.	1827.	6. 7.	H. ♃.	Hook. fl. b. am. t. 11.
ACON'ITUM. p. 9.		MONKSHOOD.					
129 Nuttállii.	(*pa.*)	Nuttall's.	N. America.	1828.	7. 8.	H. ♃.	
pállidum. N. *non* R.A.							
130 glábrum. DC.	(*pa.*)	smooth.	....	1829.	——	H. ♃.	

DILLENIACEÆ. *p.* 12.

TRACHYTE'LLA. DC.		TRACHYTE'LLA. Polyandria Monogynia.					
Act'æa. DC.	(*wh.*)	spear-leaved.	China.	1826.	6. 8.	H. ♄. ◡.	

MAGNOLIACEÆ. *p.* 13.

TALA'UMA. DC.		TALA'UMA. Polyandria Polygynia.					
Candóllii. Bl.	(*st.*)	De Candolle's.	Java.	1827.	4. 6.	S. ♄.	Blum. pl. jav. c. ic.
Magnòlia odoratíssima. Hort.							

BERBERIDEÆ. *p.* 16.

BE'RBERIS. p. 16.		BARBERRY.					
19 caroliniàna. N.	(*ye.*)	Carolina.	Carolina.	1828.	5. 7.	H. ♄.	
20 cratægina. DC.	(*ye.*)	Hawthorn-like.	Levant.	1829.	....	H. ♄.	
21 glumàcea. DC.	(*ye.*)	glumaceous.	N. America.	1827.	....	H. ♄.	
A'CHLYS. DC.		A'CHLYS. Polyandria Monogynia.					
triphy'lla. DC.	(*wh.*)	three-leaved.	N. America.	1827.	4. 6.	H. ♃.	Hook. fl. b. am. t. 12.
EPIM'EDIUM. p. 16.		BARRENWORT.					
2 hexándrum. H.A.	(*li.*)	hexandrous.	N. America.	1827.	4. 5.	H. ♃.	Hook. fl. b. am. t. 13.

NYMPHÆACEÆ. *p.* 17.

NYMPHÆA. p. 17.		WATER-LILY.					
odoràta. p. 18.		sweet-scented.	N. America.				
β máxima. N.	(*wh.*)	*largest.*	———	1828.	7. 8.	H.*w.*♃.	
20 sanguínea. N.	(*re.*)	bloody.	———	——	——	H.*w.*♃.	
N'UPHAR. p. 18.		N'UPHAR.					
advèna. p. 18.		striped-flowered.	N. America.				
β lùtea. N.	(*ye.*)	*yellow-flowered.*	———	1828.	——	H.*w.*♃.	

SARRACENIÆ. *p.* 18.

SARRAC'ENIA. p. 18.		SIDE-SADDLE-FLOWER.					
1 mìnor. N.	(*pu.*)	small.	Carolina.	1829.	6. 7.	H.♃.	Swt. br. fl. gar. s. 2.

PAPAVERACEÆ. *p.* 18.

ARGEM'ONE. p. 19.		ARGEM'ONE.					
5 intermèdia.	(*wh.*)	intermediate.	Mexico.	1828.	7. 10.	H.♃.	
SANGUIN'ARIA. p. 19.		PUCCOON.					
2 grandiflòra. Ros.	(*wh.*)	large-flowered.	N. America.	1812.	3. 5.	H.♃.	Rosc. flor. ill. sea. t. 8.
RŒM'ERIA. p. 19.		RŒM'ERIA.					
vermiculàta. Lehm.	(*re.*)	red-flowered.	Persia.	1829.	6. 9.	H.♂.	
GLA'UCIUM. p. 19.		HORN POPPY.					
5 rùbrum. DC.	(*re.*)	red-flowered.	Levant.	——	——	H.♂.	Flor. græc. t. 488.
6 pérsicum. DC.	(*re.ve.*)	red and velvet.	Persia.	——	——	H.⊙.	

CRUCIFERÆ. *p.* 21.

NASTU'RTIUM. p. 22.		NASTU'RTIUM.					
1 nàtans. DC.	(*st.*)	floating.	N. America.	1827.	5. 9.	H.*w.*♃.	Deles. ic. sel. 2. t. 15.
BRA'YA. p. 22.		BRA'YA.					
3 pilòsa. H.A.	(*pu.*)	hairy.	N. America.	——	4. 6.	H.♃.	Hook.fl.b.am.t.17.f.A.
ARABIS. p. 22.		WALL-CRESS.					
47 retrofrácta. Gr.	(*bh.*)	bending.	N. America.	——	4. 7.	H.♃.	
Turrìtis retrofrácta. H.A.							
48 pátula. W.K.	(*pu.*)	spreading.	Hungary.	——	——	H.♃.	W. et K. hung. 1. t. 59.
49 lævigàta. DC.	(*wh.*)	smooth.	N. America.	——	——	H.♃.	
MACROP'ODIUM. p. 2.		MACROP'ODIUM.					
2 laciniàtum. H.A.	(*wh.*)	cut-leaved.	N. America.	——	6. 9.	H.⊙.	Hook.bot.mis.p.III.ic
PA'RRYA. p. 24.		PA'RRYA.					
3 macrocárpa. Br.	(*pu.*)	large-podded.	N. America.	——	5. 6.	H.♃.	Hook. fl. b. am. t. 15.
DR'ABA. p. 25.		WHILLOW-GRASS.					
35 crassifòlia. Gr.	(*ye.*)	thick-leaved.	N. America.	——	4. 5.	H.♃.	
ERY'SIMUM. p. 30.		ERY'SIMUM.					
crepidifòlium. S.S.	(*ye.*)	Crepis-leaved.	N. America.	1829.	5. 6.	H.♂.	
ÆTHION'EMA. p. 32.		ÆTHION'EMA.					
5 membranàceum. DC.	(*li.*)	membranous-podded.	Persia.	——	6. 8.	H.♃.	Swt. br. fl. gar. s.2. t.69.
SIN'APIS. p. 34.		MUSTARD.					
30 amplexicàulis. DC.	(*ye.*)	stem-clasping.	Algiers.	——	6. 9.	H.⊙.	Desf. atl. 2. t. 153.

CAPPARIDEÆ. *p.* 37.

CLE'OME. p. 37.		CLE'OME.					
20 speciosíssima. B.R.	(*pu.*)	pretty.	Mexico.	1827.	8. 11.	H.⊙.	Bot. reg. t. 1312.
CA'PPARIS. p. 38.		CAPER-TREE.					
acutifòlia.	(*wh.*)	sharp-pointed.	S. America.	——	6. 8.	S.♄.	Bot. reg. t. 1320.
acuminàta. Bot. reg. *nec aliorum.*							

VIOLARIEÆ. *p.* 43.

V'IOLA. p. 43.		VIOLET.				
89 pedatífida. G.D.	(*bl.*)	pedatifid.	N. America.	1827.	5. 7. H. ♃.	
90 attenuàta. B.F.G.	(*wh.*)	attenuated.	————	1812.	4. 8. H. ♃.	Swt. br. fl. gar. s. 2.
lanceolàta. Pursh et Nuttall *non* Lin.						
91 præmórsa. B.R.	(*ye.*)	bitten-rooted.	————	1827.	4. 6. H. ♃.	Bot. reg. t. 1254.
HYMENANTH'ERA. DC.		HYMENANTH'ERA. Monadelphia Pentandria.				
dentàta. DC.	(*ye.*)	toothed.	N. S. W.	1822.	3. 5. G. ♄.	

POLYGALEÆ. *p.* 46.

POLY'GALA. p. 46.		MILKWORT.				
42 rubélla. Ph.	(*ro.*)	pale red.	N. America.	1828.	6. 7. H. ♃.	

FRANKENIACEÆ. *p.* 48.

FRANK'ENIA. p. 48.		SEA HEATH.				
9 corymbòsa. DC.	(*ro.*)	corymb-flowered.	Barbary.	1829.	6. 8. F. ♄.	Desf. atl. 1. t. 93.

CARYOPHYLLEÆ. *p.* 48.

DIA'NTHUS. p. 49.		PINK.				
97 ciliàtus. Lehm.	(*pi.*)	fringed.	....	1829.	6. 8. H. ♃.	
98 sículus. S.S.	(*pi.*)	Sicilian.	Sicily.	——	—— H. ♃.	
AREN'ARIA. p. 56.		SANDWORT.				
68 glomeràta. DC.	(*wh.*)	close-headed.	Tauria.	1829.	7. 9. H. ⊙.	
SPERG'ULA. p. 55.		SPURREY.				
9 pilífera. DC.	(*wh.*)	piliferous.	Corsica.	1829.	6. 8. H. ♃.	

LINEÆ. *p.* 58.

L'INUM. p. 58.		FLAX.				
41 altaícum. F.	(*bl.*)	Altay.	Altay.	1829.	6. 8. H. ♃.	

MALVACEÆ.

MA'LVA. p. 59.		MALLOW.				
72 Munroàna. B.R.	(*sc.*)	Munro's.	N. America.	1827.	5. 10. H. ♃.	Bot. reg. t. 1306.
HIBI'SCUS. p. 61.		HIBI'SCUS.				
86 Telfáiriæ.	(*va.*)	Mrs. Telfair's.	Hybrid.	1825.	1. 12. S. ♄.	
α ròseus.	(*ro.*)	*rose-coloured.*	————	——	—— S. ♄.	Bot. mag. t. 2891.
β lilacìnus.	(*li.*)	*pink.*	————	——	—— S. ♄.	
γ cárneus.	(*fl.*)	*flesh-coloured.*	————	——	—— S. ♄.	
δ fulvéscens.	(*bf.*)	*buff-coloured.*	————	——	—— S. ♄.	
ε òchroleùcus.	(*st.*)	*straw-coloured.*	————	——	—— S. ♄.	
S'IDA. p. 63.		S'IDA.				
59 repánda. DC.	(*st.*)	repanded.	....	1829.	6. 8. S. ♄.	

BUTTNERIACEÆ. *p.* 66.

RE'EVESIA. B.R.		RE'EVESIA. Polyadelphia Polyandria.				
thyrsoídea. B.R.	(*wh.*)	thyrse-flowered.	China.	1818.	1. 7. G. ♄.	Bot. reg. t. 1236.
STERC'ULIA. p. 66.		STERC'ULIA.				
16 lanceolàta. B.R.	(*br.*)	spear-leaved.	————	1822.	5. 6. G. ♄.	Bot. reg. t. 1256.
DOMB'EYA. p. 69.		DOMB'EYA.				
4 angulàta. DC.	(*ro.*)	angle-leaved.	Mauritius.	1826.	4. 8. S. ♄.	Bot. mag. t. 2905.

HYPERICINEÆ. *p.* 76.

HYPE'RICUM. p. 76. St. John's-wort.
77 Geblèri. Led. (*ye.*) Gebler's. Altay. 1829. 6. 8. H. ♄.

GUTTIFERÆ. *p.* 78.

GARCI'NIA. p. 78. Garci'nia.
5 purpùrea. h.b. (*pu.*) purple. E. Indies. 1828. .... S. ♄.

GERANIACEÆ. *p.* 88.

PHYMATA'NTHUS. p. 89. Wart-flower.
6 intertínctus. s.g. (*re.w.v.*) stained. Hybrid. 1827. 2. 10. G. ♄. Swt. ger. ser. 2. t. 54.
PELARGO'NIUM. p. 92. Stork's-bill.
504 nodòsum. s.g. (*cr.v.*) knotted-stalked. Hybrid. 1824. 4. 11. G. ♄. Swt. ger. ser. 2. t. 68.
505 bipinnatífidum. s.g. (*pu.*) bipinnatifid. —— 1827. —— G. ♄. —— ser. 2. t. 62.
506 Darnleyànum. s.g. (*cr.*) Earl of Darnley's. —— 1826. —— G. ♄. —— ser. 2. t. 63.
507 pullàceum. s.g. (*sc.v.*) dark brown. —— 1827. 5. 10. G. ♄. —— ser. 2. t. 76.
508 contíguum. s.g. (*sc.*) contiguous. —— 1828. —— G. ♄. —— ser. 2. t. 73.
509 icònicum. s.g. (*sc.bk.*) figured. —— —— —— G. ♄. —— ser. 2. t. 88.
510 Bluntiànum. s.g. (*sc.ve.*) Miss Blunt's. —— —— —— G. ♄. —— ser. 2. t. 79.
511 erióphoron. s.g. (*sc.ve.*) wool-bearing. —— —— —— G. ♄. —— ser. 2. t. 90.
512 obtusidentàtum. s.g. (*sc.*) blunt-toothed. —— —— —— G. ♄. —— ser. 2. t. 92.
513 melanchólicum. s.g. (*sc.ve.*) dark-edged. —— 1827. —— G. ♄. —— ser. 2. t. 53.
514 Spéculum. s.g. (*re.ve.*) looking-glass. —— —— —— G. ♄. —— ser. 2. t. 52.
515 anacampton. s.g. (*re.ve.*) recurved-calyx'd. —— —— —— G. ♄. —— ser. 2. t. 64.
516 Peytòniæ. s.g. (*re.ve.*) Lady Peyton's. —— —— —— G. ♄. —— ser. 2. t. 46.
517 ursìnum. s.g. (*re.*) shaggy-calyxed. —— 1828. —— G. ♄. —— ser. 2. t. 94.
518 suffùsum. s.g. (*pi.re.*) suffused. —— 1827. —— G. ♄. —— ser. 2. t. 47.
519 láxulum. s.g. (*ro.ve.*) loose-umbelled. —— 1828. —— G. ♄. —— ser. 2. t. 75.
520 conchyllàtum. s.g. (*vi.ve.*) violet-purple. —— —— —— G. ♄. —— ser. 2. t. 95.
521 porphy'reon. s.g. (*pu.*) bright-purple. —— —— —— G. ♄. —— ser. 2. t. 89.
522 mollifòlium. s.g. (*pi.ve.*) soft-leaved. —— 1827. —— G. ♄. —— ser. 2. t. 77.
523 Colleyànum. s.g. (*pu.*) Colley's. —— 1828. —— G. ♄. —— ser. 2. t. 83.
524 Littleànum. s.g. (*pu.li.*) Little's. —— 1827. —— G. ♄. —— ser. 2. t. 72.
525 dædàleum. s.g. (*pi.re.*) variable-coloured. —— —— —— G. ♄. —— ser. 2. t. 100.
526 Veitchiànum. s.g. (*pu.*) Veitch's. —— —— —— G. ♄. —— ser. 2. t. 81.
527 implicàtum. s.g. (*pu.*) implicated. —— —— —— G. ♄. —— ser. 2. t. 86.
528 inscúlptum. s.g. (*li.pu.*) engraved-petal'd. —— 1828. —— G. ♄. —— ser. 2. t. 65.
529 lanòsum. s.g. (*li.ve.*) wool-bearing. —— —— —— G. ♄. —— ser. 2. t. 84.
530 flabellifòlium. s.g. (*std.*) fan-leaved. —— —— —— G. ♄. —— ser. 2. t. 48.
531 cordifórme. s.g. (*bh.pu.*) heart-shaped. —— 1827. —— G. ♄. —— ser. 2. t. 49.
532 præclàrum. s.g. (*wh.pu.*) clear-coloured. —— —— —— G. ♄. —— ser. 2. t. 67.
533 clathràtum. s.g. (*pi.ve.*) burred-petaled. —— —— —— G. ♄. —— ser. 2. t. 85.
534 instràtum. s.g. (*wh.ve.*) spreading-spotted. —— 1828. —— G. ♄. —— ser. 2. t. 78.
535 Yeatmaniànum. s.g. (*pu.ve.*) Miss Yeatman's. —— 1827. —— G. ♄. —— ser. 2. t. 59.
536 mirábile. s.g. (*li.ve.*) admirable. —— —— —— G. ♄. —— ser. 2. t. 57.
537 laùtum. s.g. (*pi.pu.*) genteel. —— —— —— G. ♄. —— ser. 2. t. 91.
538 commíxtum. s.g. (*re.ve.*) mingled. —— —— —— G. ♄. —— ser. 2. t. 71.
539 compáctum. s.g. (*bh.re.*) compact. —— 1828. —— G. ♄. —— set. 2. t. 70.
540 exquísitum. s.g. (*w.pu.*) dainty-flowered. —— 1824. —— G. ♄. —— ser. 2. t. 55.
541 polìtum. s.g. (*bh.re.*) polished. —— 1827. —— G. ♄. —— ser. 2. t. 87.
542 Atkinsiànum. s.g. (*bh.pu.*) Mr. Atkins's. —— 1828. —— G. ♄. —— ser. 2. t. 99.
543 atrovìrens. s.g. (*bh.pu.*) dark green-leav'd. —— 1827. —— G. ♄. —— ser. 2. t. 74.
544 adventítium. s.g. (*pu.ve.*) adventitious. —— 1826. —— G. ♄. —— ser. 2. t. 80.
545 perámplum. s.g. (*li.*) very large-flower'd. —— —— —— G. ♄. —— ser. 2. t. 61.
546 glabréscens. s.g. (*w.pu.*) smoothish-leav'd. —— 1824. —— G. ♄. —— ser. 2. t. 45.
547 Hilliànum. s.g. (*w.pu.*) Hill's. —— 1828. —— G. ♄. —— ser. 2. t. 51.
548 pallídulum. s.g. (*pi.*) pale pink. —— —— —— G. ♄. —— ser. 2. t. 97.
549 urbànum. s.g. (*pi.*) home bred. —— —— —— G. ♄. —— ser. 2. t. 50.
550 dissímile. s.g. (*re.l.ve.*) dissimilar. —— —— —— G. ♄. —— ser. 2. t. 60.
551 fastuòsum. s.g. (*re.ve.*) fastuous. —— —— —— G. ♄. —— ser. 2. t. 93.
552 succuléntum. s.g. (*sc.ve.*) succulent-leav'd. —— 1827. —— G. ♄. —— ser. 2. t. 69.
553 Dràkeæ. s.g. (*cr.ve.*) Mrs. Drake's. —— 1828. —— G. ♄. —— ser. 2. t. 96.

554 Annesleyànum. S.G. (*re.ve.*) Miss Annesley's. Hybrid. 1828. 5. 10. G. ♄. Swt. ger. ser. 2. t. 56.
555 Gloriànum. S.G. (*cr.ve.*) Queen of Portugal's. —— —— —— G. ♄. ———— ser. 2. t. 82.
556 Kenríckæ. S.G. (*re.ve.*) Mrs. Kenrick's. —— —— —— G. ♄. ———— ser. 2. t. 58.
557 nùtans. S.G. (*sc.ve.*) nodding-flowered. —— 1827. —— G. ♄. ———— ser. 2. t. 66.
558 Sweetiànum. Sm. (*cr.ve.*) Sweet's. —— 1829. —— G. ♄.
559 staphysagroídes. S.G. (*pu.*) Stavesacre-ld. C. B. S. 1827. —— G. ♄. Swt. ger. ser. 2. t. 98.

BALSAMINEÆ. *p.* 101.

IMP'ATIENS. p. 102. TOUCH-ME-NOT.
pállida. p. 102. pale-flowered. N. America. 1812. 6. 10. H. ⊙. Eng. bot. v. 14. t. 937.
Nòli-tángere. E.B. *nec aliorum.*
7 boreàlis. (*ye.*) northern. Denmark. .... —— H. ⊙. Flor. dan. t. 582.
Nòli-tángere. F.D. *non* L.
Nòli-tángere. L. (*y.re.*) English. England. .... —— H. ⊙.

OXALIDEÆ. *p.* 102.

O'XALIS. p. 102. WOOD-SORREL.
113 tortuòsa. B.R. (*ye.*) twisted. Chile. 1825. 6. 8. G. ♄. Bot. reg. t. 1249.
114 floribúnda. Leh. (*ro.*) many-flowered. Brazil. 1829. 4. 9. F. ♃. Swt. br. fl. gar. s. 2.
úrbica. Hil. (*li.*) Martius's. —— 1828. 5. 9. F. ♃.
Martiàna. Zuc. p. 102. ex S.S. Lodd. bot. cab. ic.

ZYGOPHYLLEÆ. *p.* 104.

LA'RREA. p. 104. LA'RREA.
2 divaricàta. DC. (*ye.*) spreading-lobed. Buenos Ayres. 1829. 6. 9. G. ♄. Cavan. icon. 6. t. 560. f. 1.

RUTACEÆ. *p.* 105.

APLOPHY'LLUM. p. 105. APLOPHY'LLUM.
patavìna, tuberculàta, villòsa, linifòlia et dahùrica, lege patavìnum, tuberculàtum, villòsum, linifòlium, et dahùricum.
9 Buxbàumii. J.R. (*ye.*) Buxbaum's. Levant. 1828. 6. 8. H. ♃. Buxb. cent. 2. t. 28. f. 2.
Rùta Buxbàumii. DC.
DICTA'MNUS. p. 105. FRAXINELLA.
3 angustifòlius. S.F.G. (*pu.*) narrow-leaved. Siberia. 1829. 5. 7. H. ♃. Swt. br. fl. gar. ser. 2.

CELASTRINEÆ. *p.* 109.

EUO'NYMUS. p. 109. SPINDLE-TREE.
16 nànus. DC. (*st.*) dwarf. Caucasus. 1829. 5. 7. H. ♄.
17 gróssus. F.I. (*wh.*) large. Nepaul. 1828. 6. 8. F. ♄.

RHAMNEÆ. *p.* 112.

TRAV'OA. H.M. TRAV'OA. Pentandria Monogynia.
quinquenérvia. H.M. (*w.*) five-nerved. Chile. 1827. 5. 7. F. ♄.

HOMALINEÆ. *p.* 116.

BLACKWE'LLIA. p. 11. BLACKWE'LLIA.
7 padifòlia. B.R. (*wh.*) Padus-leaved. China. 1826. 8. 9. G. ♄. Bot. reg. t. 1308.

CASSUVIEÆ. *p.* 117.

MELANORRH'ŒA. W. MELANORRH'ŒA. Polygamia Diœcia.
usitàta. W.I. (*re.*) useful. E. Indies. 1828. .... S. ♄. Wall. ind. pl. t. 11. 12.
HOLIGA'RNA. DC. HOLIGA'RNA. Polygamia Monœcia.
longifòlia. DC. (*wh.*) long-leaved. E. Indies. 1828. .... S. ♄. Roxb. corom. 3. t. 282.

BURSERACEÆ. *p.* 120.

IC'ICA. DC.		Ic'ica. Octo-Decandria Monogynia.					
1 enneándria. DC.	(*wh.*)	nine-stamened.	Guiana.	1825.	....	S. ♄.	Aubl. guian. 1. t. 134.
2 decándra. DC.	(*wh.*)	ten-stamened.	——	——	....	S. ♄.	
3 altíssima. DC.	(*wh.*)	tallest.	——	1822.	....	S. ♄.	Aubl. guian. 1. t. 132.

LEGUMINOSÆ. *p.* 122.

Subordo I. *PAPILIONACEÆ.*

S'OPHORA. p. 122.		S'ophora.					
12 velutìna. B.R.	(*pu.*)	velvetty.	China.	1826.	6. 7.	G. ♄.	Bot. reg. t. 1185.
BAPTI'SIA. p. 123.		Bapti'sia.					
10 mìnor. Lehm.	(*bl.*)	lesser.	N. America.	....	6. 7.	H. ♃.	
ANTHY'LLIS. p. 132.		Kidney-Vetch.					
22 polycéphala. DC.	(*ye.*)	many-headed.	Barbary.	1829.	6. 7.	H. ♃.	Desf. atl. 2. t. 195.
HOSA'CKIA. B.R.		Hosa'ckia. Diadelphia Decandria.					
1 bícolor. B.R.	(*y.w.*)	two-coloured.	N. America.	1823.	7. 9.	H. ♃.	Bot. reg. t. 1257.
Lòtus pinnàtus. Bot. mag. t. 2913.							
2 Purshiàna. B.R.	(*ye.*)	Pursh's.	——	1827.	——	H. ♃.	
3 decúmbens. B.R.	(*ye.*)	decumbent.	——	——	——	H. ♃.	
4 parviflòra. B.R.	(*ye.*)	small-flowered.	——	——	——	H. ♃.	
TEPHR'OSIA. p. 141.		Tephr'osia.					
38 chinénsis. H.C.	(*li.*)	Chinese.	China.	1822.	5. 8.	G. ħ.	
ASTRA'GALUS. p. 145.		Milk Vetch.					
120 succuléntus. B.R.	(*pu.*)	succulent.	N. America.	1827.	6. 7.	H. ♃.	Bot. reg. t. 1324.
KENN'EDYA. p. 156.		Kenn'edya.					
monophy'lla. p. 157.		simple-leaved.	N. S. W.	1790.	3. 8.	G. ♄.◡.	Vent. malm. t. 106.
β *longeracemòsa.*	(*pi.*)	*long-racemed.*	——	1828.	——	G. ♄.◡.	Bot. reg. t. 1336.
LUP'INUS. p. 159.		Lupine.					
34 pulchéllus.	(*bl.*)	handsome.	Mexico.	——	6. 10.	F. ♄.	Swt. br. fl. gar. s. 2. t. 67.

Subordo III. *MIMOSEÆ.*

AC'ACIA. p. 164.		Ac'acia.					
209 gravèolens. L.C.	(*ye.*)	strong-scented.	N. Holland.	1825.	3. 5.	G. ♄.	Lodd. bot. cab. t. 1460.
uncinàta. p. 165.	(*st.*)	hooked.	N. Holland.	1823.	8. 1.	G. ♄.	Bot. reg. t. 1332.
210 álbida. B.R.	(*ye.*)	white spiny.	Peru.	1824.	4. 6.	G. ♄.	—— t. 1317.

Subordo IV. *CÆSALPINEÆ.*

CASTANOCA'RPUS.		Moreton Bay Chestnut. Decandria Monogynia.					
austràlis.	(*wh.*)	southern.	N. Holland.	1829.	....	G. ♄.	
CA'SSIA. p. 169.		Ca'ssia.					
116 brevifòlia. DC.	(*ye.*)	short-leaved.	Madagascar.	1824.	6. 7.	S. ♄.	

ROSACEÆ. *p.* 178.

R'UBUS. p. 187.		Bramble and Raspberry.					
85 flàvus. D.P.	(*wh.*)	yellow-fruited.	Nepaul.	1827.	....	F. ħ.	
86 setòsus. DC.	(*wh.*)	bristly-stalked.	N. America.	——	5. 7.	H. ħ.	
87 nutkànus. DC.	(*wh.*)	Nootka sound.	——	——	——	H. ħ.	Swt. br. fl. gar. s. 2.
DR'YAS. p. 190.		Dr'yas.					
3 Drummóndii. B.M.	(*ye.*)	Drummond's.	——	——	4. 8.	H. ♃.	Bot. mag. t. 2972.
POTENTI'LLA. p. 190.		Cinquefoil.					
nívea. p. 190.	(*ye.*)	snowy-leaved.	N. Europe.	1816.	6. 8.	H. ♃.	
β *macrophy'lla.* B.M.	(*y.*)	*large-leaved.*	N. America.	1827.	——	H. ♃.	Bot. mag. t. 2982.
101 grácilis. B.M.	(*ye.*)	slender.	——	——	——	H. ♃.	—— t. 2984.
102 Hopwoodiàna. S.F.G.	(*ro.s.*)	Hopwood's.	Recto-formosa.	1829.	6. 9.	H. ♃.	Swt. br. fl. gar. s. 2. t. 61.

COMBRETACEÆ. *p.* 195.

POI'VREA. p. 196.		Poi'vrea.					
6 Afzèlii. D.D.	(*sc.*)	Afzelius's.	Sierra Leone.	1824.	7. 10.	S. ♄.⌣.	Bot. mag. t. 2944.

Combrètum grandiflòrum. B.M. not G.D. in Linn. trans., which is a true *Combrètum*, with 8 stamens; we have adopted M. Decandolle's division of the genus, as Mr. Don informed us, he found the same distinctions as given by Decandolle, and was inclined to divide them himself, but afterwards arranged them in distinct sections, by which they might be readily divided, one genus having irregular convolute cotyledons, and the other straight plaited ones; this circumstance he mentioned to us before his paper was published, and before M. Decandolle's volume had arrived.

VOCHYSIEÆ. *p.* 197.

QU'ALEA. M.P.		Qu'alea. Monandria Monogynia.					
violàcea. M.P.	(*vi.*)	violet-coloured.	Brazil.	1824.	....	S. ♄.	Mart. pl. bras. 1. t. 81.
ERI'SMA. DC.		Eri'sma. Monandria Monogynia.					
floribúnda. DC.	(*bl.*)	many-flowered.	Guiana.	1825.	....	S. ♄.	Tratt. tab. t. 105.

ONAGRARIEÆ. *p.* 197.

FU'CHSIA. p. 197.		Fu'chsia.					
11 thymifòlia. DC.	(*re.*)	Thyme-leaved.	Mexico.	1829.	4. 10.	F. ♄.	Swt. br. fl. gar. s. 2. t. 25.

MELASTOMACEÆ. *p.* 203.

PLER'OMA. p. 204.		Pler'oma.					
3? villòsum. DC.	(*pu.*)	villous.	S. America.	1820.	5. 10.	S. ♄.	Bot. mag. t. 2630.
Melástoma villòsa. B.M.							

PASSIFLOREÆ. *p.* 216.

PASSIFL'ORA. p. 216.		Passion-flower.					
65 Cavanillèsii. DC.	(*co.*)	Cavanille's.	W. Indies.	1822.	7. 10.	S. ♄.⌣.	Cavan. dis. 10. t. 273.
66 Andersònii. DC.	(*st.*)	Anderson's.	St. Vincent.	1823.	——	S. ♄.⌣.	
67 Raddiàna. DC.	(*st.pu.*)	Raddi's.	Brazil.	1829.	——	S. ♄.⌣.	
68 cephaleíma. DC.		crowded-leaved.	———	1826.	....	S. ♄.⌣.	Bory. an. gen. 2. t. 22. f. 2.
69 obscùra. H.C.	(*gr.st.*)	obscure.	S. America.	1823.	2. 10.	S. ♄.⌣.	

CACTEÆ. *p.* 235.

MAMMILL'ARIA. p. 235.		Mammill'aria.					
23 fulvispìna. H.P.	(*re.*)	brown-spined.	Brazil.	1822.	8. 10.	D.S. ♄.	
24 púlchra. H.B.R.	(*pu.*)	pretty.	Mexico.	1828.	6. 10.	D.S. ♄.	Bot. reg. t. 1329.
25 depréssa. DC.	(*wh.re.*)	depressed.	S. America.	1800.	6. 8.	D.S. ♄.	DC. diss. t. 2. f. 2.
díscolor. DC. dis. *non* Haw. *Spinii.* Colla hort. rip. ic.							
díscolor. H.S.	(*wh.re.*)	two-coloured.	S. America.	——	——	D.S. ♄.	
26 stellàris. H.S.	(*ro.*)	starry.	———	1816.	6. 9.	D.S. ♄.	Lodd. bot. cab. t. 79.
non pusilla. DC. *ex.* H.P.							
ECHINOCA'CTUS. p. 235.		Echinoca'ctus.					
22 cornígerus. DC.	(*wh.*)	horn-spined.	Mexico.	1830.	6. 9.	D.S. ♄.	DC. diss. cact. t. 7.
23 nóbilis. H.P.	(*wh.*)	noble.	———	1796.	——	D.S. ♄.	
24 Hy'strix. H.P.	(*wh.*)	Porcupine.	S. America.	1818.	——	D.S. ♄.	
C'EREUS. p. 236.		C'ereus.					
74 mágnus. H.P.	(*wh.*)	great 12-angled.	St. Domingo.	1830.	6. 10.	D.S. ♄.	
75 fèrox. H.P.	(*wh.*)	fierce upright.	Brazil.	1828.	——	D.S. ♄.	
76 Æ'thiops. H.P.	(*wh.*)	black-spined.	———	——	——	D.S. ♄.	
77 setíger. H.P.	....	small quadrangular.	———	1826.	....	D.S. ♄.	
78 undàtus. H.P.	....	triangular China.	China.	1824.	....	D.S. ♄.	
79 crispàtus. H.P.	....	curly-edged.	Brazil.	1826.	....	D.S. ♄.	
80 Colvíllii.	(*ro.*)	Colvill's.	Grandiflora-speciosissimus.		——	D.S. ♄.	

OPU'NTIA. p. 237. INDIAN FIG.
38 longispìna. H.P. (*ye.*) long-spined. Brazil. 1826. 6. 10. D.S. ♄.
39 glomeràta. H.P. (*ye.*) long flat-spined. ——— —— —— D.S. ♄.
40 ròsea. DC. (*ro.*) rose-coloured. Mexico. 1830. .... D.S. ♄. DC. diss. cact. t. 15.
41 imbricàta. DC. (*ro.*) imbricated. S. America. 1820. .... D.S. ♄.
Cèreus imbricàtus. p. 237.
42 cylíndrica. DC. (*ro.*) cylindrical. Peru. 1779. .... D.S. ♄.
Cèreus cylíndricus. p. 237.
RHIPS'ALIS. p. 238. RHIPS'ALIS.
8 cereúscula. H.P. .... Cereus-like. Brazil. 1828. .... D.S. ♄.

GROSSULARIEÆ. *p.* 238.

R'IBES. p. 238. CURRANT AND GOOSEBERRY.
45 punctàtum. DC. (*st.*) spotted. Chile. 1825. 3. 5. H. ♄. Bot. reg. t. 1278.

ESCALLONEÆ. *p.* 239.

ESCALL'ONIA. p. 239. ESCALL'ONIA.
5 bífida. Ot. (*pi.*) bifid. Mexico. 1827. .... F. ♄.
6 viscòsa. Ot. (*wh.*) viscous. Mendoza. 1829. .... G. ♄.

SAXIFRAGEÆ. *p.* 239.

SAXI'FRAGA. p. 239. SAXIFRAGE.
102 spicàta. L.T. (*sp.*) spiked. N. America. 1827. 5. 6. H. ♃.
103 argùta. L.T. (*sp.*) deep-notched. ——— —— —— H. ♃.
104 nudicaùlis. L.T. (*wh.*) naked-stalked. ——— —— —— H. ♃.
105 ferrugínea. Gr. (*wh.*) ferruginous. ——— —— —— H. ♃.
106 pyrolifòlia. L.T. (*wh.*) Pyrola-leaved. ——— —— —— H. ♃.
107 tricuspidàta. L.T. (*ye.*) three-pointed. Greenland. 1827. —— H. ♃.
HEUCH'ERA. p. 242. HEUCH'ERA.
1 micrántha. B.R. (*gr.*) small-flowered. Columbia. —— 6. 7. H. ♃. Bot. reg. t. 1302.
2 Richardsòni. D. (*gr.*) Richardson's. N. America. —— —— H. ♃.
TIARE'LLA. p. 242. TIARE'LLA.
5 cólorans. Gr. (*wh.*) coloured. N. America. —— 4. 6. H. ♃.
MITE'LLA. p. 242. MITE'LLA.
5 trifida. Gr. (*wh.*) trifid. N. America. —— 5. 7. H. ♃.
DRUMMO'NDIA. DC. DRUMMO'NDIA. Pentandria Digynia.
Mitélla pentándra. Bot. mag. t. 2933.

UMBELLIFERÆ. *p.* 243.

ERY'NGIUM. p. 244. ERYNGO. Pentandria Digynia.
36 Andersòni. LG. (*bl.*) Anderson's. .... 8. 9. H. ♃.
DISCOPLE'URA. p. 245. DISCOPLE'URA.
2 capillàcea. DC. (*wh.*) capillary. N. America. 1824. 7. 9. H. ☉.
A'mmi capillàceum. M.
FER'ULA. p. 250. GIANT-FENNEL.
21 lineàris. Cerv. (*ye.*) linear-leaved. Mexico. 1828. 8. 10. F. ♃.
ANTHRI'SCUS. p. 254. ROUGH CHERVILL.
9 torquàta. DC. (*wh.*) twisted. France. 1726. 6. 7. H. ♂. Jacq. aust. 1. t. 63.
Chærophy'llum bulbosum. Jacq.
ARPI'TIUM. Neck. ARPI'TIUM. Pentandria Digynia.
símplex. (*wh.*) simple-stalked. Switzerland. 1824. 6. 8. H. ♃. Jacq. misc. 2. t. 2.
Laserpítium símplex. L. *Gáya símplex.* Gaud. fl. helv. 2. t. 6. non *Gaya.* Kth.

ARALIACEÆ. *p.* 256.

ADA'MIA. W.I. ADA'MIA. Didynamia Angiospermia.
cyànea. W.I. (*bl.*) blue-flowered. E. Indies. 1829. .... S. ♄. Wall. ind. t. 26.

CAPRIFOLIACEÆ. *p.* 257.

SYMPH'ORIA. p. 258. St. Peter's-wort.
3 montàna. K.S. (*li.*) mountain. Mexico. 1829. 7. 10. H. ♄. H.et B.nov.gen.3.t.296

LORANTHEÆ. *p.* 259.

A'UCUBA. W. A'ucuba. Diœcia Tetrandria.
japónica. (*pu.*) blotch-leaved. Japan. 1783. 5. 7. H. ♄. Bot. mag. t. 1197.
LORA'NTHUS. p. 259. Lora'nthus.
2 terréstris. .... terrestrial. N. Holland. 1830. .... G. ♄.

RUBIACEÆ. *p.* 260.

ASPE'RULA. p. 266. Woodruff.
25 nítida. F.G. (*br.*) glossy. Greece. 1829. 7. 9. H. ♃. Flor. græc. t. 124.

DIPSACEÆ. *p.* 269.

ASTEROCE'PHALUS. p. 270. Starhead.
paucisètus. Jac. (*st.*) few-bristled. S. Europe. 1827. 7. 9. H. ♃.

COMPOSITÆ. *p.* 271.

CR'EPIS. p. 273. Cr'epis.
33 macrorhìza. B.M. (*ye.*) large-rooted. Madeira. 1829. 7. 9. H. ♃. Bot. mag. t. 2988.
CHUQUIR'AGA. p. 280. Chuquir'aga.
2 spinòsa. L.T. (*ye.*) spiny. Peru. 1825. 4. 5. F. ♄.
Bacàzia spinòsa. p. 286.
JU'NGIA. L. Ju'ngia. *LABIATIFLORÆ.* Syngenesia Æqualis.
paniculàta. D.D. (*ye.*) panicled. Chile. 1825. 7. 9. H. ♃. DC. an. mus. 19. t. 7.
Dumerília paniculàta. p. 260.
CENTA'UREA. p. 285. Centaury.
141 argùta. S.S. (*st.*) sharp notched. Canaries. 1829. —— F. ♄.
142 decípiens. Thu. (*pu.*) deceiving. France. —— —— H. ♃.
VERN'ONIA. p. 289. Vern'onia.
14 tomentòsa. Ot. (*pu.*) tomentose. Brazil. 1827. 9. 11. S. ♄.
DELI'SLIA. S.S. Deli'slia. Syngenesia Necessaria.
Bertèrii. S.S. (*ye.*) Bertero's. Isl.Magdalen.1829. 6. 8. G. ⊙. Mart. dec. 47. f. 1.
Millèria biflòra. p. 313.
CALE'NDULA. p. 313. Marygold.
22 tomentòsa. S.S. (*ye.*) tomentose. C. B. S. 1828. —— H. ⊙.
CLADA'NTHUS. p. 319. Clada'nthus.
2 canéscens. S.F.G. (*ye.*) canescent. Canaries. 1829. 3. 8. F. ♄. Swt. br. fl. gar. ser. 2.

LOBELIACEÆ. *p.* 322.

LOB'ELIA. p. 322. Lob'elia.
84 purpùrea. B.R. (*pu.*) purple. Chile. 1825. 8. 9. G. ♄. Bot. reg. t. 1325.

CAMPANULACEÆ. *p.* 325.

LIGHTFO'OTIA. p. 325. Lightfo'otia.
Loddigésii. A.DC. (*pa.*) Loddiges'. C. B. S. 1824. 6. 8. G. ♄. Lodd. bot. cab. t. 1038.
tenélla. B.C.
PLATYC'ODON. A.DC. Platyc'odon. Pentandria Monogynia.
grandiflòrum. DC. (*bl.*) great-flowered. Siberia. 1782. 6. 8. H. ♃. Bot. mag. t. 252.
Campánula grandiflòra. B.M. *Wahlenbérgia grandiflòra.* p. 325.

WAHLENBE'RGIA. A.DC. WAHLENBE'RGIA.						
3 Kitaibèlii. A.DC.	(*vi.*) Kitaibel's.	Hungary.	1823.	6. 8.	H.♃.	W. et K. hung. 2. t.154.
Campánula graminifòlia. p. 327. W. et K. hung. *non* Flor. græc.						
4 tenuifòlia. A.DC.	(*vi.*) slender-leaved.	Hungary.	1820.	5. 9.	H.♃.	W. et K. hung. 2. t.155.
5 capénsis. A.DC.	(*vi.*) Cape.	C. B. S.	1803.	7. 10.	H.⊙.	Bot. mag. t. 782.
Campánula capénsis. B.M. *Roélla decúrrens.* Andr. rep. t. 238.						
6 cérnua. A.DC.	(*wh.*) nodding-flowered.	C. B. S.	1804.	7. 9.	G.♂.	
7 lineàris. A.DC.	(*wh.*) linear-leaved.	———	1822.	7. 10.	H.⊙.	
8 procúmbens. A.DC.	(*bl.*) procumbent.	———	1824.	——	H.⊙.	Alph. DC. camp. t. 15.
9 hederàcea. A.DC.	(*pa.bl.*) Ivy-leaved.	England.	....	5. 6.	H.♃.	Eng. bot. v. 2. t. 73.
Campánula hederàcea. p. 328. *Aikínia hederàcea.* Salisb.						
10 arvática.	(*bl.*) Spanish.	Spain.	1825.	5. 7.	H.♃.	Pluk. phyt. t. 23. f. 1.
Campánula arvática. LG.						
11 grácilis. A.DC.	(*bl.*) slender.	N. S. W.	1794.	4. 8.	G.♂.	Bot. mag. t. 691.
Campánula grácilis. B.M.						
12 strícta.	(*bl.*) upright.	———	——	——	G.♂.	Sm. exot. bot. t. 45.
Campánula strícta. Exot. bot. *non* Linn. *eréctа.* p. 326.						
13 capillàris. B.C.	(*bl.*) capillary.	N. Holland.	1824.	——	G.♂.	Lodd. bot. cab. t. 1406.
14 littoràlis. Lab.	(*bl.*) sea side.	V. Diem. Isl.	1820.	——	G.♂.	Lab. nov. holl. 1. t. 70.
15 quadrífida. A.DC.	(*bl.*) four-cleft.	N. S. W.	1822.	——	G.♂.	
16 saxícola. A.DC.	(*bl.*) small rock.	V. Diem. Isl.	1825.	5. 10.	G.♂.	
17 Siebèri. A.DC.	(*bl.*) Sieber's.	N. S. W.	1826.	——	G.♂.	
18 undulàta. A.DC.	(*bl.*) waved.	C. B. S.	1822.	5. 7.	G.♃.	
Campánula undulàta. p. 326.						
19 hispídula. A.DC.	(*bl.wh.*) hispid-calyxed.	———	1816.	6. 9.	H.⊙.	Comm. hort. 2. t. 37.
Campánula hispídula. p. 326.						
20 capillàcea. A.DC.	(*wh.*) capillary.	C. B. S.	1822.	5. 7.	H.♃.	
lobelioídes. A.DC.	(*bh.*) Lobelia-like.	Canaries.	1777.	7. 8.	H.⊙.	Alph. DC. camp. t. 17.
péndula. p. 323.						
SYMPHIA'NDRA. A.DC. SYMPHIA'NDRA. Monadelphia Pentandria.						
péndula. A.DC.	(*st.*) pendulous.	Caucasus.	1824.	6. 8.	H.♃.	Swt. br. fl. gar. s.2. t.66.
Campánula péndula. p. 327. n. 102.						
SPECUL'ARIA. A.DC. VENUS-LOOKING-GLASS. Pentandria Monogynia.						
1 Spéculum. A.DC.	(*bl.*) common.	S. Europe.	1596.	5. 8.	H.⊙.	Bot. mag. t. 102.
Campánula Spéculum. B.M. *Primatocárpus Spéculum.* L'H. *Lagòusia arvénsis.* Dur.						
β álba.	(*wh.*) *white-flowered.*	———	——	——	H.⊙.	
2 hirsùta. A.DC.	(*bl.*) hairy.	Italy.	1822.	5. 8.	H.⊙.	Ten. fl. nap. t. 19.
3 hy'brida. A.DC.	(*bl.*) corn.	England.	....	——	H.⊙.	Eng. bot. v. 6. t. 375.
Campánula hy'brida. E.B. *Primatocárpus hy'bridus.* L'H.						
4 falcàta. A.DC.	(*bl.*) falcate.	Italy.	1822.	——	H.⊙.	Ten. fl. nap. t. 20.
5 pentagònia. A.DC.	(*bl.*) large-flowered.	Turkey.	1686.	——	H.⊙.	Bot. reg. t. 56.
Campánula pentagònia. B.R.						
6 perfoliàta. A.DC.	(*bl.*) perfoliate.	N. America.	1680.	——	H.⊙.	Moris. s. 5. t. 2. f. 23.
MU'SSCHIA. A.DC. MU'SSCHIA. Pentandria Monogynia.						
aùrea. A.DC.	(*ye.*) golden-flowered.	Madeira.	1777.	7. 9.	G.♄.	Bot. reg. t. 57.
Campánula aùrea. p. 326. n. 38.						
α latifòlia. A.DC.	(*ye.*) *broad-leaved.*	———	——	——	G.♄.	Duham. ed. nov. t. 41.
β angustifòlia. A.DC.	(*ye.*) *narrow-leaved.*	———	——	——	G.♄.	Jacq. schœn. 4. t. 472.
PETROMAR'ULA. A.DC. PETROMAR'ULA. Pentandria Monogynia.						
pinnàta. A.DC.	(*bl.*) winged-leaved.	Candia.	1640.	6. 8.	F.♃.	Flor. græc. t. 220.
Phyteùma pinnàtum. p. 328.						
JAS'IONE. p. 328. JAS'IONE. Monadelphia Pentandria.						
4 hùmilis. A.DC.	(*bl.*) dwarf.	S. France.	1824.	6. 7.	H.♃.	

ERICEÆ. *p.* 330.

ANDRO'MEDA. p. 331. ANDRO'MEDA.						
23 tetragòna. S.S.	(*bh.*) four-sided.	N. America.	1827.	....	H.♄.	Pall. ross. 2. t. 73. f. 4.

EBENACEÆ. *p.* 349.

DIO'SPYROS. p. 349. DATE PLUM.						
19 Mabòla. B.R.	(*st.*) Mabola-tree.	Philippines.	1822.	4. 5.	S.♄.	Bot. reg. t. 1139.

APOCINEÆ. *p.* 353.

PACHYP'ODIUM. B.R. PACHYP'ODIUM. Pentandria Monogynia.

succuléntum. (*wh.re.*) succulent. C. B. S. 1820. 4. 6. D.G. ♄. Bot. reg. t. 1321.

tuberòsum. B.R. *Echìtes succulénta.* Jacq. fragm. t. 117. see p. 355.

ASCLEPIADEÆ. *p.* 356.

ASCL'EPIAS. p. 360. SWALLOW-WORT.

33 Greeniàna. N. Green's. N. America. 1828. 6. 8. H. ♃.

POLEMONIACEÆ. *p.* 367.

POLEM'ONIUM. p. 367. GREEK VALERIAN.

9 pulchérrimum. B.M. (*bl.*) bright blue. N. America. 1827. 7. 8. H. ♃. Bot. mag. t. 2979.

10 moschàtum. Gr. (*bl.*) musk-scented. ——— ——— ——— H. ♃.

GI'LIA. p. 368. GI'LIA.

3 púngens. B.M. (*bl.*) pungent-leaved. N. America. 1827. 6. 8. H. ⊙. Bot. mag. t. 2977.

BONPLA'NDIA. Cav. BONPLA'NDIA. Pentandria Monogynia.

geminiflòra. Cav. (*bl.*) twin-flowered. New Spain. 1813. 5. 12. S. ⊙. Bot. reg. t. 92.

Caldàsia heterophy'lla. W. non *Caldasia.* DC.

BORAGINEÆ. *p.* 374.

ONO'SMA. p. 375. ONO'SMA.

16 Gmelìni. L.A. (*st.*) Gmelin's. Altay. 1829. 4. 8. H. ♃. Ledeb. ic. alt. t. 280.

17 polyphy'llum. L.A. (*st.*) crowded-leav'd. Tauria. ——— 7. 8. H. ♃. ——— t. 24.

18 rígidum. L.A. (*pa.ye.*) rigid. ——— 1826. ——— H. ♃. ——— t. 238.

HYDROPHYLLEÆ. *p.* 380.

E'UTOCA. p. 380. E'UTOCA.

3 serícea. B.M. (*bl.*) silky. N. America. 1827. 5. 7. H. ♃. Bot. mag. t. 3003.

To Franklínii, p. 380, add Bot. mag. t. 2985.

SOLANEÆ. *p.* 380.

CE'LSIA. p. 380. CE'LSIA.

9 heterophy'lla. P.S. (*ye.*) various-leaved. .... 1829. 7. 9. F. ♂.

SALPIGLO'SSIS. p. 383. SALPIGLO'SSIS.

5 intermèdia. S.F.G. (*br.ye.*) intermediate. Chile. 1829. 6. 9. F. ♃. Swt. br. fl. gar. s. 2.

SCROPHULARINÆ. *p.* 389.

SCROPHUL'ARIA. p. 391. FIGWORT.

51 lanàta. F. (*br.*) woolly. Siberia. 1828. 7. 11. H. ♃.

MI'MULUS. p. 394. MONKEY-FLOWER.

9 propínquus. B.R. (*ye.*) related. N. America. 1827. 4. 10. H. ♃. Bot. reg. t. 1330.

CALCEOL'ARIA. p. 396. SLIPPERWORT.

15 Herbertiàna. B.R. (*ye.*) Mr. Herbert's. Chile. 1828. 5. 10. G. ♃. Bot. reg. t. 1313.

LABIATÆ. *p.* 399.

ST'ACHYS. p. 406. HEDGE-NETTLE.

49 serícea. (*li.*) silky. Nepaul. 1830. 7. 9. H. ♃.

O'CYMUM. p. 413. BASIL.

21 montànum. B.M. (*wh.*) mountain. W. Indies. 1825. 5. 8. S. ⊙. Bot. mag. t. 2996.

HY'PTIS. p. 414. HY'PTIS.

14 madagascariénsis. (*bl.*) Madagascar. Madagascar. 1828. 6. 8. S. ♂.

VERBENACEÆ. *p.* 415.

LANT'ANA. p. 417. LANT'ANA.
25 móllis. Gr. (*ro.*) soft. Mexico. 1826. 4. 8. G. ♄.

ACANTHACEÆ. *p.* 419.

RUE'LLIA. p. 420. RUE'LLIA.
28 pícta. B.R. (*bl.*) painted. St. Domingo. 1826. 4. 8. S. ♄. Lodd. bot. cab. t. 1448.
JUSTI'CIA. p. 421. JUSTI'CIA.
60 guttàta. B.R. (*ye.sp.*) spotted. E. Indies. 1828. 4. 8. S. ♃. Bot. reg. t. 1334.

PRIMULACEÆ. *p.* 423.

DODEC'ATHEON. p. 423. AMERICAN COWSLIP.
1 Mèadia. p. 423. Mead's. Virginia. 1744. 4 6. H. ♃. Bot. mag. t. 12.
α lilacìnum. (*li.*) *common lilac.* ——— ——— ——— H. ♃. Trew Ehret. t. 12.
β albiflòrum. (*wh.*) *white-flowered.* .... 1824. ——— H. ♃. Lodd. bot. cab. t. 1489.
γ élegans. S.F.G. (*ro.*) *rose-coloured.* .... 1827. 4. 7. H. ♃. Swt. br. fl. gar. s. 2. t. 60.
δ gigánteum. (*li.*) *gigantic.* .... ——— ——— H. ♃.
2 integrifòlium. (*vi.*) entire-leaved. N. America. 1829. ——— H. ♃. Pluk. alm. t. 79. f. 6.
ANDROS'ACE. p. 425. ANDROS'ACE.
17 carinàta. Tor. (*st.*) keeled. N. America. 1827. 4. 7. H. ♃.
18 lineàris. Gr. (*bh.*) linear-leaved. ——— ——— ——— H. ♃.

BEGONIACEÆ. *p.* 437.

BEG'ONIA. p. 437. BEG'ONIA.
40 lóngipes. B.M. (*wh.*) long-stalked. Mexico. 1828. 3. 8. S. ♄. Bot. mag. t. 3001.

POLYGONEÆ. *p.* 438.

POLYG'ONUM. p. 438. PERSICARIA.
61 injucúndum. B.R. (*st.*) unattractive. .... 1825. 5. 6. G. ♄. Bot. reg. t. 1250.
RHE'UM. p. 440. RHUBARB.
15 leucorhìzum. F. (*st.*) white-rooted. Siberia. 1827. 5. 6. H. ♃.

AROIDEÆ. *p.* 479.

P'OTHOS. p. 479. P'OTHOS.
29 longifòlia. R.S. (*br.*) long-leaved. Brazil. 1828. 9. 10. S. ♃.
30 refléxa. Ot. (*br.*) reflexed. ——— ——— ——— S. ♃.
31 réptans. Ot. (*br.*) creeping. ——— ——— ——— S. ♄. ◡
32 rubéscens. Ot. (*br.*) reddish. ——— ——— ——— S. ♃.

ORCHIDEÆ. *p.* 483.

APLE'CTRUM. N. APLE'CTRUM. Gynandria Monandria.
hyèmale. N. (*br.*) winter. N. America. 1827. .... H. ♃.
RODRIGU'EZIA. p. 487. RODRIGU'EZIA.
2 planifòlia. H.C. (*re.*) flat-leaved. S. America. 1826. 6. 12. S. ♃.

CANNEÆ. *p.* 494.

PHRY'NIUM. p. 494. PHRY'NIUM.
28 coloràtum. B.M. (*or.*) orange-coloured. Brazil. 1828. 4. 5. S. ♃. Bot. mag. t. 3010.

MUSACEÆ. *p.* 495.

M'USA. p. 495.		PLANTAIN-TREE.					
chinénsis.	(*sc.*)	dwarf Chinese.	China.	1829.	....	S. ♃.	

IRIDEÆ. *p.* 495.

TRICHON'EMA. p. 503.		TRICHON'EMA.					
ramiflòra.	(*pu.*)	branching-flower'd.	Naples.	1830.	5. 6.	H.b.	
I'xia ramiflòra. T.N.							
CR'OCUS. p. 503.		CR'OCUS.					
24 Thomàsii. T.N.	(*bl.*)	Thomas's.	Naples.	1830.	....	H.b.	
25 odòrus. T.N.	....	sweet-scented.	———	——	....	H.b.	
26 suavèolens. T.N.	....	fragrant.	———	——	....	H.b.	
27 Imperàti. S.S.	(*vi.*)	Imperati's.	———	——	2. 4.	H.b.	Ten. mem. croc. t. 3.

AMARYLLIDEÆ. *p.* 505.

ZEPHYRA'NTHES. p. 505.		ZEPHYRA'NTHES.					
10 mesochlòa. W.H.	(*bh.*)	green-centred.	S. America.	1828.	6. 9.	F.b.	
BRUNSVI'GIA. p. 510.		BRUNSWICK-LILY.					
7 grandiflòra. B.R.	(*pi.*)	large-flowered.	C. B. S.	1827.	8. 9.	F.b.	Bot. reg. t. 1335.
PHYCE'LLA. p. 510.		PHYCE'LLA.					
3 Herbertiàna. B.R.	(*re.y.*)	Mr. Herbert's.	Chile.	1825.	5. 7.	F.b.	Bot. reg. t. 1341.
4 corúsca. Gr.	(*sc.or.*)	glossy.	———	——	——	F.b.	
NARCI'SSUS. p. 514.		NARCI'SSUS.					
82 negléctus. Ten.	(*wh.y.*)	neglected.	Naples.	1830.	4. 5.	H.b.	

ASPHODELEÆ. *p.* 524.

SCI'LLA. p. 527.		SQUILL.					
30 odoràta. R.S.	(*bl.*)	sweet-scented.	Portugal.	1818.	5. 6.	H.b.	
31 praténsis. R.S.	(*bl.*)	meadow.	Hungary.	1827.	——	H.b.	W. et K. hung. t. 189.
32 ròsea. Leh.	(*ro.*)	rose-coloured.	Numidia.	——	——	H.b.	
33 mauritànica. R.S.	(*bl.*)	Mauritanian.	Africa.	1819.	——	H.b.	
DRI'MIA. p. 513.		DRI'MIA.					
14 villòsa. B.R.	(*g.st.*)	villous.	C. B. S.	1826.	5. 6.	F.b.	Bot. reg. t. 1346.
A'LLIUM. p. 530.		GARLICK.					
118 euósmon. Ot.	(*br.*)	good-scented.	S. America.	1829.	5. 6.	F.b.	
119 glandulòsum. Ot.	(*br.*)	glandular.	———	——	——	F.b.	
120 nudicaùle. Leh.	....	naked-stalked.	Altay.	——	....	H.b.	
121 guttàtum. G.D.	(*wh.*)	spotted.	Odessa.	1819.	7. 8.	H.b.	
122 Fischèri.	....	Fischer's.	Siberia.	1829.	....	H.b.	
123 acùtum. S.S.	(*li.*)	acute.	....	1819.	7. 8.	H.b.	
124 intermèdium. G.D.	(*w.*)	intermediate.	S. Europe.	1827.	——	H.b.	
125 fúscum. W.K.	(*br.*)	brown-flowered.	Hungary.	1820.	——	H.b.	W. et K. hung. t. 241.
126 juncifòlium. G.D.	(*wh.*)	Rush-leaved.	Chile.	1826.	——	H.b.	
127 striatéllum. G.D.	(*st.*)	striped.	———	1823.	——	H.b.	Bot. mag. t. 2419.
128 cinèreum. Lehm.	(*st.*)	grey.	Siberia.	1829.	——	H.b.	
MI'LLA. C.I.		MI'LLA. Hexandria Monogynia.					
biflòra. C.I.	(*wh.*)	two-flowered.	Mexico.	1826.	5. 6.	F. ♃.	Cavan. icon. 2. t. 196.

BROMELIACEÆ. *p.* 534.

POURR'ETIA. p. 535.		POURR'ETIA.					
4 cœrùlea. Miers.	(*bl.*)	blue.	Chile.	1827.	4. 8.	S. ♃.	
5 rubricaùlis. Mier.	(*bl.re.*)	red-stemmed.	———	——	——	S. ♃.	
6 magnispátha. Col.	(*g.w.*)	large-spathed.	S. America.	1829.	——	S. ♃.	Coll. h. rip. t. 19.
CARAGU'ATA. Plum.		CARAGU'ATA. Hexandria Monogynia.					
lingulàta. Ly.	(*ye.*)	tongue-leaved.	Jamaica.	1776.	6. 7.	S. ♃.	Jacq. amer. t. 62.
***Tillándsia lingulàta.* p. 535.**							

BUONAPA'RTEA. p. 536. BUONAPA'RTEA.
2 grácilis. .... slender. Mexico. 1828. .... G. ♃.

TULIPACEÆ. *p.* 536.

YU'CCA. p. 536.		ADAM'S-NEEDLE.				
23 grácilis. Lk.	(*wh.*)	slender.	Mexico.	1829.	.... G. ♃.	
ERYTHR'ONIUM. p. 539.		DOG'S-TOOTH VIOLET.				
Déns cànis. L.	(*pu.*)	common.	Europe.	1596.	3. 4. H.b.	Swt. br. fl. gar. s. 2.
Déns cànis. Redoute liliac. t. 193. *non* Bot. mag.						
4 bífida. S.F.G.	(*ro.*)	cleft-stigma'd.	Switzerland.	....	—— H.b.	Swt. br. fl. gar. s. 2.
β albiflòra.	(*wh.*)	*white-flowered.*	——	....	—— H.b.	—— s. 2.

MELANTHACEÆ. *p.* 539.

ZIGAD'ENUS. p. 540.		ZIGAD'ENUS.	Hexandria Trigynia.		
3 élegans. Ph.	(*wh.*)	elegant.	N. America.	1828.	5. 6. H. ♃.

COMMELINEÆ. *p.* 541.

TRADESCA'NTIA. p. 541.		SPIDERWORT.			
23 crássipes.	(*bl.*)	thick-stalked.	Mexico.	1827.	5. 10. H. ♃.
24 diurética. M.P.	(*bl.*)	diuretic.	Brazil.	1825.	6. 8. S. ♃.
COMMEL'INA. p. 542.		COMMEL'INA.			
24 cucullàta. H.C.	(*bl.*)	hooded.	Brazil.	1825.	6. 9. G. ⊙.
25 caripénsis. H.S.	(*bl.*)	rough-leaved.	Trinidad.	1826.	6. 8. S. ♃.

RESTIACEÆ. *p.* 546.

ERIOCA'ULON. p. 546.		PIPEWORT.				
3 decangulàre. Ph.	(*wh.*)	ten-angled.	N. America.	1826.	8. 10. H. *w.* ♃.	Lodd. bot. cab. t. 1310.
4 fasciculàtum. R.S.	(*wh.*)	fascicled.	Guiana.	1825.	—— S. *w.* ♃.	Lam. enc. 3. t. 50. f. 3.

CYPERACEÆ. *p.* 547.

CA'REX. p. 547.		SEDGE.				
123 rupéstris. S.S.	(*br.*)	rock.	Switzerland.	1819.	6. 7. H. ♃.	All. ped. t. 92. f. 1.
124 cúrvula. S.S.	(*br.*)	curved.	S. Europe.	1824.	—— H. ♃.	Schk. caric. t. D. f. 17.
125 lobàta. W.	(*br.*)	lobed.	Switzerland.	1819.	5. 7. H. ♃.	—— t. D. f. 18.
126 rèpens. S.S.	(*br.*)	creeping.	——	1823.	6. 7. H. ♃.	—— t. I. f. 11.
127 leporìna. W.	(*br.*)	hare.	——	1819.	—— H. ♃.	—— t. Fff. f. 1. 29.
128 lóngipes. D.P.	(*br.*)	long-stalked.	Nepaul.	1823.	—— H. ♃.	
129 microstàchya. S.S.	(*br.*)	small-spiked.	N. Europe.	1822.	5. 6. H. ♃.	Schk. caric. t. 6. f. 11.
130 mucronàta. S.S.	(*br.*)	bristle-pointed.	S. Europe.	1819.	6. 7. H. ♃.	—— t. K. f. 44.
131 phæostàchys. E.F.	(*br.*)	brown-spiked.	Scotland.	....	6. H. ♃.	
132 stictocárpa. E.F.	(*br.*)	dotted-fruited.	——	....	6. 7. H. ♃.	
133 angustifòlia. E.F.	(*br.*)	narrow-leaved.	——	....	—— H. ♃.	
134 podocárpa. S.S.	(*br.*)	foot-podded.	N. America.	1827.	—— H. ♃.	
135 brachystàchys. S.S.	(*br.*)	short-spiked.	S. Europe.	1819.	—— H. ♃.	Schk. car. t. P. f. 58.
SCH'ŒNUS. p. 549.		BOG-RUSH.				
6 ferrugìneus. R.S.	(*br.*)	rusty.	Europe.	1829.	7. 8. H. *w.* ♃.	Host. gram. 4. t. 71.
DICHROM'ENA. R.S.		DICHROM'ENA.	Triandria Monogynia.			
leucocéphala. R.S.	(*wh.*)	white-spiked.	W. Indies.	1822.	9. 12. S. *w.* ♃.	
Sch'œnus stellàtus. W.						
FIMBRI'STYLIS. p. 550.		FIMBRI'STYLIS.				
6 diphy'lla. R.S.	(*br.*)	two-leaved.	E. Indies.	1822.	6. 8. S. *w.* ♃.	
ABILDGAA'RDIA. R.S.		ABILDGAA'RDIA.				
2 tristàchya. R.S.	(*wh.*)	three-spiked.	E. Indies.	1824.	6. 8. S. ♃.	
ISO'LEPIS. p. 550.		ISO'LEPIS.				
13 supìna. R.S.	(*br.*)	trailing.	Europe.	1826.	7. 8. H. *w.* ♃.	Host. gram. 3. t. 64.

SCI'RPUS. p. 550.		CLUB-RUSH.					
19 elongàtus. D.P.	(*br.*)	elongated.	Nepaul.	1825.	7. 9.	H.*w.*♃.	
20 articulàtus. R.S.	(*br.*)	jointed.	E. Indies.	1826.	8. 10.	S.*w.*♃.	Rheed. mal. 12. t. 71.
CYP'ERUS. p. 551.		CYP'ERUS.					
71 nilóticus. R.S.	(*br.*)	Nile.	Egypt.	1810.	8. 10.	H.♃.	
72 compáctus. R.S.	(*br.pu.*)	compact.	China.	1819.	7. 8.	H.♃.	
73 kyllingæoídes. R.S.	(*br.*)	Kyllinga-like.	N. America.	1828.	——	H.♃.	
74 dùbius. W.	(*br.*)	doubtful.	E. Indies.	1802.	7. 9.	S.♃.	Rottb. gram. t. 4. f. 5.
75 leucocéphalus. R.S.	(*wh.*)	white-headed.	——	1819.	——	S.♃.	
76 trisúlcus. D.P.	(*br.*)	three-furrowed.	Nepaul.	1830.	——	H.♃.	
77 caricìnus. D.P.	(*br.*)	Carex-like.	——	1826.	——	H.♃.	
78 rígidus. R.S.	(*re.wh.*)	rigid.	Madagascar.	——	——	S.♃.	
HYP'OELYTRUM. R.		HYP'OELYTRUM. Triandria Monogynia.					
argénteum. R.S.	(*wh.pu.*)	silvery.	E. Indies.	1824.	7. 8.	S.♃.	
MARI'SCUS. p. 552.		MARI'SCUS.					
9 paníceus. R.S.	(*br.st.*)	Panic-like.	E. Indies.	1820.	6. 7.	S.*w.*♃.	Rottb. gram. t. 4. f. 1.
KYLLI'NGA. p. 552.		KYLLI'NGA.					
8 brevifòlia. R.S.	(*gr.*)	short-leaved.	E. Indies.	1817.	5. 6.	S.*w.*♃.	Rottb. gram. t. 4. f. 3.

FILICES. *p.* 576.

LYG'ODIUM. p. 577.		SNAKE'S-TONGUE.					
7 volùbile. S.S.	(*br.*)	twining.	W. Indies.	1819.	5. 9.	S.♃.⌣.	Sloan. jam. 1. t. 46. f. 1.
8 hastàtum. W.	(*br.*)	halberd-leaved.	Maranham.	1820.	——	S.♃.⌣.	
ANEI'MIA. p. 577.		ANEI'MIA.					
8 Phyllítidis. S.S.	(*br.*)	flat-leaved.	Trinidad.	1824.	5. 8.	S.♃.⌣.	Plum. fil. t. 156.
SCHIZ'ÆA. p. 577.		SCHIZ'ÆA.					
5 élegans. S.S.	(*br.*)	elegant.	Trinidad.	1825.	5. 8.	S.♃.	Vahl. symb. 2. t. 50.
GLEICH'ENIA. p. 577.		GLEICH'ENIA.					
6 pectinàta. S.S.	(*br.*)	pectinated.	S. America.	1824.	6. 8.	S.♃.	
DAN'ÆA. S.S.		DAN'ÆA. Cryptogamia Poropterides.					
alàta. S.S.	(*br.*)	winged.	W. Indies.	1823.	6. 8.	S.♃.	Plum. fil. t. 109.
ACRO'STICHUM. p. 577.		ACRO'STICHUM.					
9 longifòlium. S.S.	(*br.*)	long-leaved.	W. Indies.	1827.	5. 10.	S.♃.	
10 glandulòsum. H.F.	(*br.*)	glandular.	Jamaica.	1825.	——	S.♃.	H. et G. ic. fil. t. 3.
11 fimbriàtum. S.S.	(*br.*)	fringed.	S. America.	1824.	——	S.♃.	
NOTHOL'ÆNA. p. 578.		NOTHOL'ÆNA.					
4 piloselloídes. S.S.	(*br.ye.*)	simple-leaved.	E. Indies.	1822.	6. 8.	S.♃.	Houtt. pflanz. t. 96. f. 2.
MENI'SCIUM. p. 578.		MENI'SCIUM.					
4 prolíferum. S.S.	(*br.*)	proliférous.	E. Indies.	1820.	5. 6.	S.♃.	
NIPHO'BOLUS. p. 578.		NIPHO'BOLUS.					
5 adnáscens. S.S.	(*br.*)	woolly-leaved.	E. Indies.	1824.	4. 8.	S.♃.	
ASPL'ENIUM. p. 580.		SPLEENWORT.					
49 vivìparum. H.K.	(*br.*)	viviparous.	Mauritius.	1820.	6. 9.	S.♃.	
50 bulbíferum. S.S.	(*br.*)	bulb-bearing.	N. Zealand.	——	——	G.♃.	Schk. crypt. t. 79.
51 diffórme. B.P.	(*br.*)	two-formed.	N. Holland.	1823.	——	G.♃.	
52 Petrárchæ. DC.	(*br.*)	Petrarch's.	France.	1819.	——	H.♃.	H. et G. ic. fil. t. 152.
53 obtusàtum. S.S.	(*br.*)	obtuse.	N. Zealand.	1824.	——	F.♃.	Lab. nov. hol. 2. t. 242.
54 cultrifòlium. S.S.	(*br.*)	keen-leaved.	W. Indies.	1820.	——	S.♃.	Plum. fil. t. 3. f. 1.
55 brasiliénse. S.S.	(*br.*)	Brazilian.	Brazil.	1822.	——	S.♃.	
DOO'DIA. p. 581.		DOO'DIA.					
3 mèdia. B.P.	(*br.*)	middle-sized.	N. Holland.	1823.	3. 9.	G.♃.	
BLE'CHNUM. p. 581.		BLE'CHNUM.					
12 pectinàtum. S.S.	(*br.*)	pectinate.	S. America.	1827.	5. 8.	S.♃.	
13 serrulàtum. S.S.	(*br.*)	saw-leaved.	N. America.	1818.	——	H.♃.	
14 striàtum. S.S.	(*br.*)	striated.	N. Holland.	1824.	——	F.♃.	
LOM'ARIA. p. 581.		LOM'ARIA.					
7 falcàta. S.S.	(*br.*)	falcate.	——	1823.	5. 8.	F.♃.	Lab. nov. hol. 2. t. 248.
DIPL'AZIUM. p. 582.		DIPL'AZIUM.					
10 castaneæfòlium. S.S.	(*br.*)	Chestnut-leaved.	Guinea.	1824.	8.	S.♃.	
PT'ERIS. p. 582.		BRAKE.					
28 sagittæfòlia. S.S.	(*br.*)	arrow-leaved.	Brazil.	1825.	6. 8.	S.♃.	
29 rotundifòlia. S.S.	(*br.*)	round-leaved.	N. Zealand.	1824.	7. 8.	F.♃.	
30 élegans. S.S.	(*br.*)	elegant.	E. Indies.	——	8. 10.	S.♃.	
31 díscolor. S.S.	(*br.*)	discoloured.	Brazil.	1825.	——	S.♃.	
32 heterophy'lla. S.S.	(*br.*)	various-leaved.	W. Indies.	——	——	S.♃.	Plum. fil. t. 37.
33 leptophy'lla. S.S.	(*br.*)	slender-leaved.	Brazil.	——	——	S.♃.	
34 chinénsis. L.C.	(*br.*)	Chinese.	China.	——	——	G.♃.	

35 crenulàta. L.C.	(*br.*) crenulated.	....	1827.	7. 8.	S. ♃.	
36 Cervantèsii. H.L.	(*br.*) Cervantes'.	Mexico.	1824.	——	G. ♃.	
ADIA'NTUM. p. 582.	MAIDENHAIR.					
22 vàrium. S.S.	(*br.*) variable.	S. America.	1820.	7. 9.	S. ♃.	
23 ternàtum. S.S.	(*br.*) ternate.	——	——	——	S. ♃.	
24 pàtens. W.	(*br.*) spreading.	Brazil.	1824.	8. 10.	S. ♃.	
25 rhomboídeum. S.S.	(*br.*) rhomboid.	S. America.	1820.	6. 9.	S. ♃.	
CHEILA'NTHES. p. 583.	CHEILA'NTHES.					
14 caudàta. B.P.	(*br.*) tailed.	N. Holland.	1824.	6. 8.	F. ♃.	
LINDS'ÆA. p. 583.	LINDS'ÆA.					
6 mèdia. B.P.	(*br.*) intermediate.	N. Holland.	1823.	4. 6.	F. ♃.	
DAVA'LLIA. p. 583.	DAVA'LLIA.					
6 élegans. S.S.	(*br.*) elegant.	N. Holland.	1824.	4. 9.	G. ♃.	Schk. crypt. t. 127.
7 dùbia. B.P.	(*br.*) doubtful.	——	1826.	——	G. ♃.	
8 concavadénsis. L.C.	(*br.*) Concavado.	Brazil.	1823.	——	S. ♃.	
CIB'OTIUM. S.S.	CIB'OTIUM. Cryptogamia Filices.					
Billardièri. S.S.	(*br.*) Labillardiere's.	N. Holland.	1824.	....	G. ♄.	

VIOLARIEÆ. *p.* 43.

ANCHIE'TIA. Hil.	ANCHIE'TIA. Pentandria Monogynia.					
pyrifòlia. Hil.	(*wh.*) Pear-leaved.	Brazil.	1822.	....	S. ♄. ◡.	Mart. bras. t. 16.

Noisèttia pyrifòlia. Martius. see our p. 43.

TILIACEÆ. *p.* 70

LUH'EA. W.	LUH'EA. Polyadelphia Polyandria.					
paniculàta. M.P.	(*ro.*) panicled.	Brazil.	1828.	....	S. ♄.	Mart. bras. t. 62.

RUBIACEÆ. *p.* 260.

LIPO'STOMA. D.D.	LIPO'STOMA. Tetrandria Monogynia.					
campanuliflòra. D.D.	(*bl*) bell-flowered.	Brazil.	1825.	2. 10.	S. ♃.	Bot. mag. t. 2840.

Hedyòtis campanuliflòra. p. 263. *Æginètia capitàta.* Gr.

PROTEACEÆ. *p.* 443.

SI'MSIA. B.P.	SI'MSIA. Tetrandria Monogynia.					
anethifòlia. B.P.	(*ye.*) Dill-leaved.	N. Holland.	1825.	....	G. ♄.	
ANAD'ENIA. B.P.	ANAD'ENIA. Tetrandria Monogynia.					
pulchélla. B.P.	(*ro.*) pretty.	N. Holland.	1824.	....	G. ♄.	
GUEV'INA. Mol.	GUEV'INA. Tetrandria Monogynia.					
Avellàna. L.T.	(*st.*) variable-leaved.	Chile.	1826.	....	G. ♄.	Feuil. per. 3. t. 33.

Quádria heterophy'lla. Flor. per. 1. t. 99. f. b.

NATURAL ORDER NOT DETERMINED.

BRE'XIA. B.R.	BRE'XIA. Pentandria Monogynia.					
1 spinòsa. B.R.	(*wh.*) spiny-leaved.	Madagascar.	1815.	7. 8.	S. ♄.	Bot. reg. t. 872.
2 madagascariénsis.	(*wh.*) Madagascar.	——	——	——	S. ♄.	—— t. 730.
Venána madagascariénsis. Lamarck.						
3 chrysophylla.	(*wh.*) golden-leaved.	——	1820.	....	S. ♄.	
LAUROPHY'LLUS. Th.	LAUROPHY'LLUS. Polygamia Diœcia.					
capénsis. W.	(*gr.*) Cape.	C. B. S.	1801.	....	G. ♄.	
PORANTH'ERA. L.T.	PORANTH'ERA. Pentandria Trigynia.					
ericifòlia. L.T.	(*st.*) Heath-leaved.	N. S. W.	1825.	5. 8.	G. ♄.	Linn. trans. 10. t. 22. f. 2

PHLO'X. p. 368. LYCHNIDEA.
36 speciòsa. B.R. (*ro.*) showy. N. America. 1827. 5. 7. H.♃. Bot. reg. t. 1351.

STERC'ULIA. p. 66. STERC'ULIA.
Tragacántha. B.R. (*st.*) Tragacanth-tree. SierraLeone. 1822. 5. 6. S.♄. Bot. reg. t. 1353.

SCI'LLA. p. 527. SQUILL.
34 plúmbea. B.R. (*bl.*) lead-coloured. C. B. S.? 1812. 4. 6. F.b. Bot. reg. t. 1355.

LOB'ELIA. p. 322. LOB'ELIA.
85 Kraúsii. B.M. (*sc.*) Krauss's. St. Domingo. 1828. 1. 3. S.♃. Bot. mag. t. 3012.

VANGU'ERIA. p. 260. VANGU'ERIA.
3 velutìna. B.M. (*gr.*) velvetty. Madagascar. 1828. 5. 6. S.♄. Bot. mag. t. 3014.

BRACHYSTE'LMA. p. 359. BRACHYSTE'LMA.
2 spathulàtum. B.R. (*br.*) spatulate-leaved. C. B. S. 1824. 8. 10. D.G.♃. Bot. reg. t. 1113.
3 críspum. B.M. (*br.y.*) wave-leaved. ——— 1829. —— D.G.♃. Bot. mag. t. 3016.

CEROP'EGIA. p. 359. CEROP'EGIA.
9 élegans. B.M. (*pu.*) elegant. E. Indies. 1828. 8. 10. S.♄.◡. Bot. mag. t. 3015.

ENCY'CLIA. p. 490. ENCY'CLIA.
2 pàtens. B.M. (*br.*) spreading-flowered. Brazil. 1829. 4. 8. S.♃. Bot mag. t. 3013.

Omitted at page 378.
COLD'ENIA. L. COLD'ENIA. Tetrandria Tetragynia.
procúmbens. L. (*wh.*) procumbent. E. Indies. 1699. 7. 8. S.☉. Lamarck. illus. t. 89.

Page 75.
G'ELA. Lour. G'ELA. Octandria Monogynia
lanceolàta. B.C. (*st.*) spear-leaved. China. 1820. 4. 8. G.♄. Lodd. bot. cab. t. 938.
Ximènia.? lanceolàta. DC. *Sèlas lanceolàta.* S.S.

Page 364.
OPHIORRH'IZA. L. SNAKE-ROOT.
Múngos. L. (*wh.*) petiolate. E. Indies. 1820. 5. 8. S.♃.

INDEX GENERUM.

N.B. Mr. SWEET having now finished several of his Botanical Works, and many of his Friends having expressed a wish, that he would Cultivate some handsome, rare, and choice Plants for Sale, finding them at present so difficult to attain, has some intention of commencing in that way; and will be obliged to any of his Friends for their Orders, which will be punctually Executed, and the Plants sent with their proper Names.

TILLING, PRINTER, GROSVENOR ROW, CHELSEA.

Printed in the United States
By Bookmasters